(Continued at back)

TECHNICAL MATHEMATICS
WITH CALCULUS

SECOND CANADIAN EDITION

TECHNICAL MATHEMATICS WITH CALCULUS

PAUL A. CALTER

Professor Emeritus
Vermont Technical College

MICHAEL A. CALTER, Ph.D.

Associate Professor
Wesleyan University

PAUL D. WRAIGHT

Professor
Durham College

DONALD G. SPENCER

Instructor
SIAST, Palliser Campus

John Wiley & Sons Canada, Ltd.

Library and Archives Canada Cataloguing in Publication
Technical mathematics with calculus / Paul Calter
… [et al.]. — 2nd Canadian ed.
Includes index.

ISBN 978-0-470-67884-8

1. Mathematics—Textbooks.   2. Calculus—Textbooks.
I. Calter, Paul.
TA330.C3435 2012   510   C2011-903656-8

Production Credits
VICE PRESIDENT AND PUBLISHER: Veronica Visentin
ACQUISITIONS EDITOR: Rodney Burke
MARKETING MANAGER: Anita Osborne
EDITORIAL MANAGER: Karen Staudinger
PRODUCTION MANAGER: Tegan Wallace
DEVELOPMENTAL EDITOR: Andrea Grzybowski
MEDIA EDITOR: Channade Fenandoe
EDITORIAL ASSISTANT/PERMISSIONS COORDINATOR: Laura Hwee
INTERIOR DESIGN: Mike Chan
COVER DESIGN: Christine Rae
COVER IMAGE: Courtesy of MDA
TYPESETTING: Aptara
PRINTING & BINDING: COURIER

Printed and bound in the United States of America
1 2 3 4 5 CP 16 15 14 13 12

John Wiley & Sons Canada, Ltd.
6045 Freemont Blvd.
Mississauga, Ontario L5R 4J3
Visit our website at www.wiley.ca

About the Authors

Paul Wraight is a professor of mathematics at Durham College in Oshawa, Ontario, and has been teaching at the post-secondary level for over 13 years. Currently the President of the Ontario College Math Association and past Chair of the Ontario College Mathematics Council, Paul has also provided academic consultation in the development of several mathematics and technical publications. As a passionate, down-to-earth educator who has been involved directly with several research projects dealing with retention and the unprepared student, students are often found lined up down the hallway outside his office seeking his no-nonsense advice and approach to problems in mathematics, chemistry, physics, and the hurdles of everyday life.

Donald Spencer has been teaching mathematics for ten years in the Associated Studies area at the Saskatchewan Institute of Applied Science and Technology (SIAST). Don's primary focus is first and second year calculus; however, he has also taught such courses as systems analysis and design, computer programming, finance, statistics, and—most recently—project management.

Don has degrees from Dalhousie University and St. Mary's University in Halifax and the University of Regina in Saskatchewan, pursuing interests in math, science, and business programs. He has taught these topics in Nova Scotia, Quebec, Alberta, and Saskatchewan. During a 25-year hiatus from teaching, Don pursued his interests in computer programming and finance, including some actuarial analyses and pension fund management.

With his lifelong spirit of adventure, Don is an avid sailor who has cruised throughout North America, the United Kingdom, as well as Greece and the American Virgin Islands.

Born in Nova Scotia's Cape Breton Island, Don shows his pride in his Celtic heritage with memberships in several Scottish associations in Saskatchewan and passion for wearing his trademark Tartan ties.

Paul Calter is Professor Emeritus of Mathematics at Vermont Technical College and Visiting Scholar at Dartmouth College. A graduate of The Cooper Union, New York, he received his M.S. from Columbia University and an MFA from Norwich University. Professor Calter has taught technical mathematics for over twenty-five years. In 1987, he was the recipient of the Vermont State College Faculty Fellow Award.

He is a member of the American Mathematical Association of Two Year Colleges, the Mathematical Association of America, the National Council of Teachers of Mathematics, the International Society for the Arts, Sciences, and Technology, the College Art Association, and the Author's Guild.

Calter is involved in the Mathematics Across the Curriculum movement, and has developed and taught a course called Geometry in Art and Architecture at Dartmouth College under an NSF grant.

Professor Calter is the author of several other mathematics textbooks, among which are *Schaum's Outline of Technical Mathematics*, *Problem Solving with Computers*, *Practical Math Handbook for the Building Trades*, *Practical Math for Electricity and Electronics*, *Mathematics for Computer Technology*, *Introductory Algebra and Trigonometry*, *Technical Calculus*, and *Squaring the Circle: Geometry in Art and Architecture*.

Michael Calter is Associate Professor at Wesleyan University. He received his B.S. from the University of Vermont. After receiving his Ph.D. from Harvard University, he completed a post-doctoral fellowship at the University of California at Irvine.

Michael has been working on his father's mathematics texts since 1983, when he completed a set of programs to accompany *Technical Mathematics with Calculus*.

Since that time, he has become progressively more involved with his father's writing endeavours, culminating with becoming co-author on the second edition of *Technical Calculus* and the fourth edition of *Technical Mathematics with Calculus*.

Michael also enjoys the applications of mathematical techniques to chemical and physical problems as part of his academic research. Michael is a member of the American Mathematical Association of Two Year Colleges, the American Association for the Advancement of Science, and the American Chemical Society.

Michael and Paul enjoy hiking and camping trips together. These have included an expedition up Mt. Washington in January, a hike across Vermont, a walk across England on Hadrian's Wall, and many sketching trips into the mountains.

To Rachel, Christopher, and Kaitlin

Preface

The second Canadian edition of *Technical Mathematics with Calculus* continues to provide instructors with the tools they need to engage students, motivate them, and watch them succeed. With abundant Canadian, real-world technical applications that demonstrate the relevance and usefulness of math outside the classroom, the text equips students with the necessary tools and skills to be successful in this course and in their chosen field. The second Canadian edition of this well-respected text has been carefully developed to build upon the success of the first Canadian edition and to introduce critical updates for Canadian instructors and students as suggested by reviewers and colleagues.

FEATURES OF THE CANADIAN EDITION:

- The Canadian edition of this text is presented predominantly in **SI units,** to complement the national and global use of these units. Where appropriate, there is a small retention of content based on the use of British Imperial units to allow students the opportunity to solve problems in either unitary system of designation.
- **Canadian real-world applications** have been added throughout the text to illustrate the relevance and usefulness of technical mathematics outside of the classroom.
- The text is adapted to reflect Canadian spelling conventions and terminology, and incorporates a Canadian geographical and cultural point of view.

NEW TO THIS EDITION:

Several new features have been added to the second edition aimed at demonstrating the link between the math concepts presented on the pages of the book with their application in the world outside the classroom.

- **Math Around Us Boxes** demonstrate the connection between the math students will learn in class with its use out in the working world by profiling how individuals in a variety of careers use math in their jobs—even though they are not necessarily working in math-focussed careers.
- **Canadian Biography Boxes:** Canadians have made and continue to make innovative contributions to technology and engineering, and math often plays an integral part in the work that they do on their way to such achievements. These biography boxes highlight the achievements of famous Canadians in technology in order to engage and motivate students, helping them make connections between mathematics and the real world.
- **Case Studies** look beyond the classroom by challenging students to take the mathematical concepts explained in a given chapter and apply them to a real-life situation. This will help students to appreciate early on in the chapter the usefulness of the material they are learning. These studies extend students' thinking beyond the chapter material; discussions of how to approach each case study is included at the end of the chapter before the review questions.
- **Updated Index of Applications:** Students and instructors want to connect applications and problem material to real world situations and, in particular, their chosen fields of study and instruction. This index has been expanded to include a much more robust breakdown of the application questions, providing instructors and students with a tool to quickly identify applications throughout the text that are relevant to their program.
- **Additional practice problems and exercises** have been added to *WileyPLUS* providing additional options and flexibility for assigning homework and assessment.

PEDAGOGICAL FEATURES:

- **Learning objectives** are listed at the beginning of each chapter, stating specifically what the student should be able to achieve upon completion of the chapter.
- **Clear explanations:** The authors have focussed on presenting the material as clearly as possible, preferring an intuitive approach rather than an overly rigorous one. Information is presented in small segments and includes many illustrations.
- The numerous **examples** form the backbone of the textbook. They are fully worked out and are chosen to help the student complete the exercises.
- The examples, and often the text itself, include discussions of many **technical applications.** They are included for classes that wish to cover these topics, as well as for providing motivation, showing that mathematics have real uses. The newly updated **Index of Applications**, which is now printed on the endpapers, will help instructors and students find specific applications.
- **Problems with approximate solutions are also available in the text.** These include not only expressions and equations with approximate constants, but also those that do not yield to many of the exact methods that are taught, and must be tackled with an approximate method.
- Also included are suggestions on how to **estimate an answer** in order for students to check their work. Suggestions for estimation are provided in the chapter for word problems. Following that, many applications examples begin with an estimation step or end with a check, or simply to examine the answer for reasonableness.
- **Common Error Boxes** emphasize some of the pitfalls and traps that "get" students year after year.

PROBLEM MATERIAL FEATURES:

- Thousands of **exercises** are included to give students the essential practice they need to learn mathematics. Exercises are given after each section, graded by difficulty and grouped by type, to allow practice in a particular area.
- **Chapter review problems** cover concepts from the chapter in random order and vary by type and difficulty.
- Answers to all odd-numbered problems are given in **Appendix D**: **Answers to Selected Problems**.
- Complete solutions to every problem are contained in the **Instructor Solutions Manual.**
- Complete solutions to every fourth problem are given in the **Student Solutions Manual.**
- For those instructors who like to assign occasional **writing questions,** most chapters provide these in the chapter review.
- For those interested in collaborative learning (students working in groups), most chapters contain at least one **team project.** These, like the writing questions, can serve as models for other projects that teachers and students can make up themselves. An introductory project in Chapter 1 gives general guidelines that can be followed later.

INSTRUCTOR AND STUDENT RESOURCES:

- **Instructor Solutions Manual** contains worked out solutions to every problem in the text.
- **Test bank Word files** and a **computerized test bank**
- **Student Solutions Manual** with a fully worked out solution for every fourth problem.
- **Appendix E: Graphing Calculator and Computer Applications:** Now available as an online resource, this appendix provides graphing calculator instructions, examples, exercises, and problems. It also lists computer projects, problems, and numerical methods for the computer.
- *WileyPLUS:* An innovative, research-based online environment for effective teaching and learning. *WileyPLUS* builds students' confidence because it takes the guesswork out of studying by providing students with a clear roadmap: what to do, how to do it, if they did it right. Students will take more initiative so you'll have greater impact on their achievement in the classroom and beyond.

www.wileyplus.com

ACKNOWLEDGEMENTS:

A textbook like this requires an entire community to see it through to completion.

I would like to thank the awesome people at Wiley for this tremendous opportunity. To Susie Galway for coming up with the idea, Rodney Burke for trusting in me and seeing it through, and Andrea Grzybowski for keeping me on task and making my writing sound eloquent.

Thanks to all my colleagues at Durham College, especially Lauren Fuentes and Tony Van Schyndel, for inspiring me to become a better teacher in the classroom. To my office partner extraordinaire Sarah White, for listening to all my high-flying ideas, monotonous explanations (ad nauseam) and offering the best feedback to me.

To my parents for all of your gifts including the logic, patience, and creative vision to see beauty in numbers and the ability to problem solve. To my wife Irene and daughter Alyssa – you are my anchors. Thank you for always being there for me and for your incredible strength and understanding. I love you.

I dedicate this book to my dad and my daughter Brianne, who left us far too soon. I think of both of you constantly.

Paul Wraight, Oshawa, ON

"Teamwork gets results". This mantra echoed in my mind many times over the development process for the second Canadian edition of this text. The team at John Wiley & Sons Canada, including Acquisitions Editor Rodney Burke, Marketing Manager Anita Osborne, Developmental Editor Andrea Grzybowski, Review Coordinator Deanna Durnford, Copy Editor Doug Linzey, Proofreader Julie Lobb, Indexer Belle Wong, and the many others who held this project on task deserve accolades for their enduring and enthusiastic support of this project. Their continuing faith in the value of the goal and the ability of the team kept everyone actively on task and on time. To them, I say "thank you".

I also want to acknowledge the support of my family for their help in fine-tuning the text's exercises and explanations. The ongoing assistance from Kate, Kali, and David enlarged the scope of the team and improved the details of the text. To them, I say "thank you".

Don Spencer, Moose Jaw, SK

Many thanks to our contributors for supplements and applications:

Tristan Barran, Seneca College

Alexei Gokhman, Humber College

Ioulia Kim, Humber College

Midori Kobayashi, Humber College

Brian Lim, Humber College

Lisa Mackay, SAIT

Richard Mitchell, Humber College

Wendi Morrison, Sheridan College

Rohini Raina, Seneca College

Dave Schuett, Durham College

Siobhan Stynes, Seneca College

Derek Wellington, Lambton College

Tony Van Schyndel, Durham College

We wish to thank all of the reviewers of the second Canadian edition. They are:

Claudia Calin, NAIT
Lynda Graham, Sheridan College
Stuart Hood, Conestoga College
Marlene Hutscal, NAIT
Najam Khaja, Centennial College
Midori Kobayashi, Humber College
Stephen Krizan, SAIT Polytechnic
Brian Lim, Humber College
Lisa MacKay, SAIT Polytechnic

Harry Matsugu, Humber College
Richard Mitchell, Humber College
Kulsoom Mohammadi, Red River College
Frosina Stojanovska-Pocuca, Mohawk College
Sheila Spencer, Mount St. Vincent University
Alan Warren, Lambton College
Don Vander Klok, Lambton College

Contents

36 Solving Differential Equations by the Laplace Transform and by Numerical Methods   1084

37 Infinite Series   1111

Appendices

Index

Numerical Computation

When you have completed this chapter, you should be able to:
- Perform calculations using absolute values and round numbers using the principles of significant digits and precision.
- Add and subtract integers using signed, unsigned, exact, and approximate numbers.
- Multiply integers using signed, unsigned, exact, and approximate numbers.
- Divide integers using signed, unsigned, exact, and approximate numbers and evaluate reciprocals.
- Perform calculations using powers and roots of integers.
- Perform calculations on combined expressions following the proper order of operations.
- Express integers and make calculations using scientific notation and engineering notation.
- Convert units of measurement and round to significant digits.
- Substitute given values into equations and formulas.
- Solve common percentage problems.

This chapter will help you review some basic math as a warm-up for the technical mathematics to come. Like any new language, math is a case of "use it or lose it." If you have not used math for a while, your skills may be a bit rusty. Use Chapter 1 to refresh the language of math. A few topics are also added that you will need for technical mathematics. These topics include scientific and engineering notation, accuracy and precision of numbers, and measurements to name a few.

You will notice that in this text, we are covering material that you have likely seen before. In Science and Engineering Technology, it's important not just to know about these mathematics topics but to be fluent in them. Your studies will be *much* easier if you speak "math," the language used to describe the science and technology world.

WORKING WITH MONEY—AMBER-LEE VALLIE

MATH AROUND US

When Amber-Lee Vallie graduated from high school six years ago, she thought she'd never again have to use the math she'd learned in class. However, working as a server at Shoeless Joe's Restaurant since graduation has proven her wrong: "Working in a restaurant means dealing

1–1 The Real Numbers

Before we start our calculator practice, we must learn a few definitions. In mathematics, as in many fields, you will have trouble understanding the material if you do not clearly understand the meanings of the words that are used.

Integers

The *integers*

$$\ldots, -4, -3, -2, -1, 0, 1, 2, 3, 4, \ldots$$

An ellipsis (the three dots) indicates that the sequence of numbers continues indefinitely.

are the whole numbers, including zero and negative values.

Rational and Irrational Numbers

The *rational numbers* include the integers and all other numbers that can be expressed as the quotient of two integers; for example,

$$\frac{1}{2}, -\frac{3}{5}, \frac{57}{23}, -\frac{98}{99}, \text{ and } 7$$

Numbers that cannot be expressed as the quotient of two integers are called *irrational*. Some irrational numbers are

$$\sqrt{2}, \sqrt[3]{5}, \sqrt{7}, \pi$$

where π is approximately equal to 3.1416.

Real Numbers

The rational and irrational numbers together make up the *real numbers*.

Numbers such as $\sqrt{-4}$ do not belong to the real number system. They are called *imaginary numbers* and are discussed in Chapter 21. Except when otherwise noted, all of the numbers we will work with are real numbers.

Decimal Numbers

Most of our computations are with numbers written in the familiar *decimal* system. The names of the places relative to the *decimal point* are shown in Fig. 1–1. We say that the decimal system uses a *base of 10* because it takes 10 units in any place to equal 1 unit in the next-higher place. For example, 10 units in the hundreds position equals 1 unit in the thousands position.

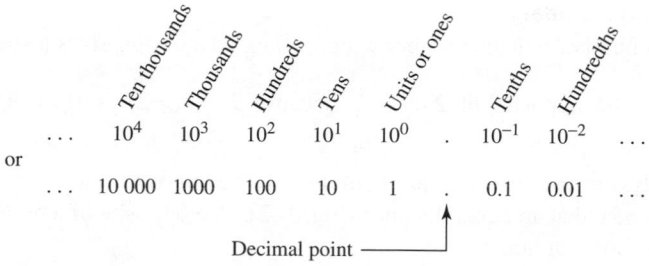

FIGURE 1–1 Values of the positions in a decimal number.

Positional Number Systems

A *positional* number system is one in which the *position* of a digit determines its value. Our decimal system is positional.

◆◆◆ **Example 1:** In the number 351.3, the digit 3 on the right has the value $\frac{3}{10}$ but the digit 3 on the left has the value 300. Thus

$$351.3 = 300 + 50 + 1 + 0.3$$
$$= 3(10^2) + 5(10^1) + 1(10^0) + 3(10^{-1})$$

The position immediately to the left of the decimal point is called position 0. The position numbers then increase to the left and decrease to the right of the 0 position as shown by the values of the exponents of 10. ◆◆◆

> Systems having bases other than 10 are used in computer science: *binary* (base 2), *octal* (base 8), and *hexadecimal* (base 16) (see Chapter 23).

Place Value

Each position in a number has a *place value* equal to the base of the number system raised to the position number. The place values in the decimal number system, as well as the place names, are shown in Fig. 1–2.

> The numbers 10^2, 10^3, etc., are called *powers of 10*. Don't worry if they are unfamiliar to you. We will explain them in Sec. 1–7.

Signed Numbers

A *positive* number is a number that is *greater* than zero, and a *negative* number is *less* than zero. On the *number line* it is customary to show the positive numbers to the right of zero and the negative numbers to the left of zero (Fig. 1–2). Negative numbers may be integers, fractions, rational numbers, or irrational numbers. To distinguish negative numbers from positive numbers, we always place a negative sign (−) in front of a negative number.

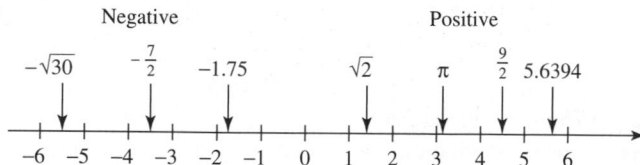

FIGURE 1–2 The number line.

◆◆◆ **Example 2:** Some negative numbers are

$$-5, \ -6.293, \ -\frac{2}{3}, \ -2\frac{7}{8}, \ -\sqrt{5}$$ ◆◆◆

We usually omit writing the positive sign (+) in front of a positive number. Thus a number without a sign is always assumed to be positive. In this chapter we will often write in a + sign for emphasis.

◆◆◆ **Example 3:** Some positive numbers are

$$5, \ +5, \ \frac{2}{3}, \ +\frac{2}{3}, \ \sqrt{5}, \ +\sqrt{5}$$ ◆◆◆

The Opposite of a Number

The *opposite* of a number n is the number which, when added to n, gives a sum of zero.

◆◆◆ **Example 4:** The opposite of 2 is -2, because $2 + (-2) = 0$. The opposite of -6 is $+6$.   ◆◆◆

Geometrically, the opposite of a number n lies on the opposite side of the zero point of the number line from n, and at an equal distance (Fig. 1–2). The opposite of a number is also called the *additive inverse* of that number.

Symbols of Equality and Inequality

Several symbols are used to show the relative positions of two quantities a and b on the number line.

$a = b$ means that a *equals* b and that a and b occupy the same position on the number line.

$a \neq b$ means that a and b are *not equal* and have different locations on the number line.

$a > b$ means that a is *greater than* b and a lies to the right of b on the number line.

$a < b$ means that a is *less than* b and a lies to the left of b on the number line.

$a \cong b$ means that a is *approximately equal to* b and that a and b are near each other on the number line

---

Absolute Value

The *absolute value* of a number n is its *magnitude* regardless of its algebraic sign. It is written $|n|$. It is the distance from n to zero on the number line, without regard to direction.

◆◆◆ **Example 5:**

(a) $|5| = 5$

(b) $|-9| = (+9)$ or 9

(c) $|3 - 7| = |-4| = (+4)$ or 4

(d) $-|-4| = (-4)$ or -4

(e) $-|7 - 21| - |13 - 19| = -(+14) - (+6) = -14 - 6 = -20$ ◆◆◆

Approximate Numbers

Most of the numbers we deal with in technology are *approximate*.

◆◆◆ **Example 6:**

(a) All numbers that represent *measured* quantities are approximate. A certain shaft, for example, is approximately 1.75 inches in diameter.

(b) Many *fractions* can be expressed only approximately in decimal form. Thus $\frac{2}{3}$ is approximately equal to 0.6667.

(c) *Irrational numbers* can be written only approximately in decimal form. The number $\sqrt{3}$ is approximately equal to 1.732.   ◆◆◆

Exact Numbers

Exact numbers are those that *have no uncertainty*.

◆◆◆ **Example 7:**

(a) There are exactly 24 hours in a day; no more, no less.

(b) An automobile has exactly four wheels.

(c) Exact numbers are usually integers, but not always. For example, there are *exactly* 25.4 mm in an inch, by definition.

(d) On the other hand, not all integers are exact. For example, a certain town has a population of *approximately* 3500 people.   ◆◆◆

Significant Digits

In a decimal number, zeros are sometimes used just to locate the decimal point. When zeros are used in that way, we say that they are *not significant*. The remaining digits in the number, including zeros, are called *significant digits*.

◆◆◆ **Example 8:**

(a) The numbers 497.3, 39.05, 8003, and 2.008 each have *four* significant digits.
(b) The numbers 1570, 24 900, 0.0583, and 0.000 583 each have *three* significant digits. The zeros in these numbers serve only to locate the decimal point.
(c) The numbers 18.50, 1.490, and 2.000 each have *four* significant digits. The zeros here are not needed to locate the decimal point. They are placed there to show that those digits are in fact zeros and not something else.   ◆◆◆

An overscore is sometimes placed over the last trailing zero that is significant. Thus the numbers $325\overline{0}$ and $735\,\overline{0}00$ each have four significant digits.

Accuracy and Precision

The *accuracy* of a decimal number is given by the number of *significant digits* in the number; the *precision* of a decimal number is given by the position of the rightmost significant digit.

◆◆◆ **Example 9:**

(a) The number 1.255 is accurate to four significant digits, and precise to three decimal places. We also say it is precise to the nearest thousandth.
(b) The number 23 800 is accurate to three significant digits, and precise to the nearest hundred.   ◆◆◆

Rounding

We will see, in the next few sections, that the numbers we get from a computation often contain *worthless digits* that must be *thrown away*. Whenever we do this, we must *round* our answer.

Round down (do not change the last retained digit) when the first discarded digit is 4 or less. *Round up* (increase the last retained digit by 1) when the first discarded digit is 6 or more, or a 5 followed by a nonzero digit in any of the decimal places to the right.

◆◆◆ **Example 10:**

Number	Rounded to Three Decimal Places
4.3654	4.365
4.3656	4.366
4.365 501	4.366
1.764 999	1.765
1.927 499	1.927

◆◆◆

When the discarded portion is 5 *exactly*, it usually does not matter whether you round up or down. The exception is when you are adding or subtracting a long column of figures, as in statistical computations. If, when discarding a 5, you always rounded up, you could bias the result in that direction. To avoid that, you want to round up about as many times as you round down, and a simple way to do that is to always *round to the nearest even number*.

This is just a convention. We could just as well round to the nearest odd number. When you get out to the workplace, make sure to check with your company what their rounding policy is.

◆◆◆ **Example 11:**

Number	Rounded to Two Decimal Places
4.365	4.36
4.355	4.36
7.765 00	7.76
7.755 00	7.76

◆◆◆

CASE STUDY—ACCURACY AND PRECISION IN CHEMICAL SAMPLES

Use the principles of this chapter to evaluate this case study. You'll find one possible evaluation at the end of the chapter.

A chemical technologist pours 32.5 ml from a collection vessel and separates the liquid samples into seven equal parts in seven test vials. Using a calculator, the technologist calculates the size of each sample by dividing: 32.5 ml ÷ 7 = 4.642 857 143 ml. When the technologist records in the log that each sample is 4.642 857 143 ml, he finds himself in trouble with the lab shift supervisor for recording an inaccurate measurement. Double checking the calculation, he finds it still comes out to 4.642 857 143 ml and decides that he needs to be more accurate by using more digits. Using a calculator with a larger display, he records 4.642 857 142 86 ml. Again an email arrives the next day encouraging a more accurate measurement. Using the calculator application on a PC, the technologist calculates and records 4.642 857 142 857 142 857 142 857 142 8571 ml. Why has the technologist made things even worse? Why would this final result be considered inaccurate when it's reporting the sample size to a precision smaller than a trillionth of a millilitre?

Exercise 1 ◆ The Real Numbers

Equality and Inequality Signs

Insert the proper sign of equality or inequality (= , ≅, >, <) between each pair of numbers.

1. 7 and 10
2. 9 and −2
3. −3 and 4
4. −3 and −5
5. $\frac{3}{4}$ and 0.75
6. $\frac{2}{3}$ and 0.667

Absolute Value

Evaluate the expression.

7. −|9 − 23| − |−7 + 3|
8. −|7 + 45| − |−8 − 34|
9. |12 − 5 + 8| − |−6| + |15|
10. |13 − 6 + 9| − |−8| + |13|
11. −|3 − 9| − |5 − 11| + |21 + 4|
12. −|4 − 8| + |−5 + 11| − |−12 − 6|

Significant Digits

Determine the number of significant digits in each approximate number.

13. 78.3
14. 925.3
15. 9274
16. 29 471
17. 4.008
18. 5.0004
19. 9400
20. 36 000
21. 20 000
22. 800 000
23. 5000.0
24. 60 000.0
25. 0.9972
26. 0.875 32
27. 1.0000
28. 63.0000

Round each number to two decimal places.

29. 38.468
30. 1.996
31. 96.835 001
32. 55.8650
33. 398.372
34. 2.9573
35. 2985.339
36. 278.382

Round each number to one decimal place.

37. 13.98

38. 745.62

39. 5.6501

40. 0.482

41. 398.36

42. 34.927

43. 9839.2857

44. 0.847

Round each number to the nearest hundred.

45. 28 583

46. 7550

47. 3 845 240

48. 274 837

Round each number to three significant digits.

49. 9.284

50. 2857

51. 0.04825

52. 483 982

53. 0.083 75

54. 29.555

55. 29.450 01

56. 8372

1–2 Addition and Subtraction

Adding and Subtracting Integers

◆◆◆ **Example 12:** Evaluate 2845 + 1172 by calculator.

Solution: There are so many types of calculators in use that it would be confusing for us to show keystrokes for any particular one. You must consult your operator's manual. However, the keystrokes for the basic operations are simple and easily learned. For this example you should get the result

$$2845 + 1172 = 4017$$

◆◆◆

Horizontal and Vertical Addition

Numbers to be added may be arranged vertically or horizontally. Of course, when you are using the calculator, it does not make any difference how the numbers are arranged, as they must be keyed in one at a time anyway.

◆◆◆ **Example 13:** The addition of 335, 103, and 224 may be represented horizontally as

$$335 + 103 + 224 = 662$$

or vertically as

$$\begin{array}{r} 335 \\ 103 \\ +224 \\ \hline 662 \end{array}$$

◆◆◆

Adding Signed Numbers

Let us say that we have a shoebox (Fig. 1–3) into which we toss all of our uncashed cheques and unpaid bills until we have time to deal with them. Let us further assume that the total cheques minus the total bills in the shoebox is $500.

We can think of the amount of a cheque as a positive number because it increases our wealth, and the amount of a bill as a negative number because it decreases our wealth. We thus represent a cheque for $100 as (+100), and a bill for $100 as (−100).

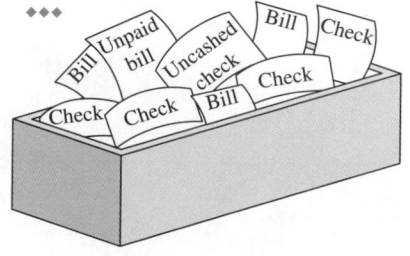

FIGURE 1–3 The shoebox.

Now, let's *add a cheque* for $100 to the box. If we had $500 at first, we must now have $600.

$$500 + (+100) = 600$$

or

$$500 + 100 = 600$$

Here we have added a positive number, and our total has increased by that amount. That is easy to understand. But what does it mean to *add a negative number*?

To find out, let us now *add a bill* for $100 to the box. If we had $500 at first, we must now have $100 less, or $400. Representing the bill by (-100), we have

$$500 + (-100) = 400$$

It seems clear that to add a negative number is no different than subtracting the absolute value of that number.

$$500 + (-100) = 500 - 100$$

This gives us our rule of signs for addition.

If b is a positive quantity, then the operation of adding a negative quantity $(-b)$ to the quantity a is equivalent to the operation of subtracting the positive quantity b from a.

All boxed and numbered formulas are tabulated in numerical order in Appendix A. There they are arranged in logical order, by type, and are numbered consecutively. They may therefore not appear in numerical order here throughout the text.

Rule of Signs for Addition	$a + (-b) = a - b$	6

◆◆◆ **Example 14:**

(a) $7 + (-2) = 7 - 2 = 5$
(b) $-8 + (-3) = -8 - 3 = -11$
(c) $9.92 + (-15.36) = 9.92 - 15.36 = -5.44$ ◆◆◆

Subtracting Signed Numbers

Let us return to our shoebox problem. But now instead of adding cheques or bills to the box, we will *subtract* (remove) cheques or bills from the box.

First we remove (subtract) a cheque for $100 from the box. If we had $500 at first, we must now have $400.

$$500 - (+100) = 400$$

or

$$500 - 100 = 400$$

Here we have subtracted a positive number, and our total has decreased by the amount subtracted, as expected.

Now let us see what it means to *subtract a negative number*. We will remove (subtract) a bill for $100 from the box. If we had $500 at first, we must now have $100 more, or $600, since we have removed a bill. Representing the bill by (-100), we have

$$500 - (-100) = 600$$

It seems clear that to subtract a negative number is the same as to add the absolute value of that number.

$$500 - (-100) = 500 + 100$$

Thus if b is a positive quantity, the operation of subtracting a negative quantity $(-b)$ from a quantity a is equivalent to the operation of adding the positive quantity b to a.

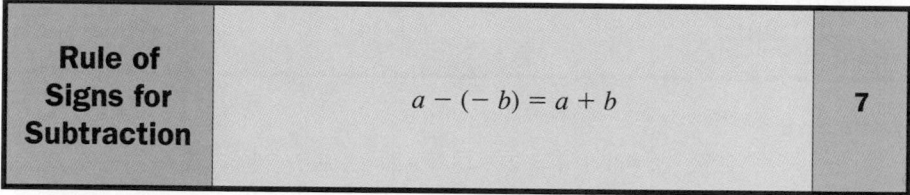

Rule of Signs for Subtraction	$a - (-b) = a + b$	7

◆◆◆ **Example 15:**

(a) $15 - (-3) = 15 + 3 = 18$
(b) $-5 - (-9) = -5 + 9 = 4$
(c) $-25.62 - (-5.15) = -25.62 + 5.15 = -20.47$ ◆◆◆

Subtracting Negative Numbers by Calculator

Note that the $(-)$ sign is used for two different things:

1. For the operation of subtraction.
2. To enter a negative quantity.

This difference is clear on the calculator, which has separate keys for these two functions. The $\boxed{-}$ key is used only for subtraction; the $\boxed{(-)}$ key is used to enter a negative quantity.

Common Error	These two keys look almost alike, so be careful not to confuse them. Note that the key used to enter a negative quantity has parentheses around the negative sign.

Try the following examples on your calculator and see if you get the correct answers.

◆◆◆ **Example 16:**

(a) $15 - (-3) = 15 + 3 = 18$
(b) $-5 - (-9) = -5 + 9 = 4$
(c) $-25 - (-5) = -25 + 5 = -20$ ◆◆◆

Commutative and Associative Laws

The *commutative law* simply says that you can add numbers in *any order*.

These laws are surely familiar to you, even if you do not recognize their names. We will run into them again when studying algebra.

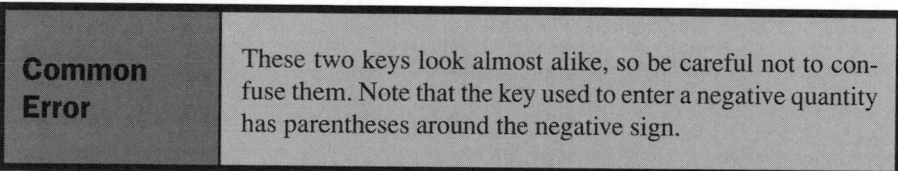

Commutative Law for Addition	$a + b = b + a$	1

◆◆◆ Example 17:

$$2 + 3 = 3 + 2$$
$$= 5$$

◆◆◆

The *associative law* says that you can group numbers to be added in several ways.

Associative Law for Addition	$a + (b + c) = (a + b) + c$ $\qquad\qquad = (a + c) + b$	3

◆◆◆ Example 18:

$$2 + 3 + 4 = 2 + (3 + 4) = 2 + 7 = 9$$
$$= (2 + 3) + 4 = 5 + 4 = 9$$
$$= (2 + 4) + 3 = 6 + 3 = 9$$

◆◆◆

Addition and Subtraction of Approximate Numbers

Addition and subtraction of integers are simple enough. But now let us tackle the problem mentioned earlier: How many digits do we keep in our answer when adding or subtracting *approximate numbers*?

Rule	When adding or subtracting approximate numbers, keep as many decimal places in your answer as contained in the number having the fewest decimal places.

◆◆◆ Example 19:

$$32.4 \text{ cm} + 5.825 \text{ cm} = 38.2 \text{ cm} \ (not \ 38.225 \text{ cm})$$

We do not use the symbol $\cong$ when dealing with approximate numbers. We would not write, for example, $32.4 \text{ cm} + 5.825 \text{ cm} \cong 38.2 \text{ cm}$.

　　Here we can see the reason for our rule for rounding. In one of the original numbers (32.4 cm), we do not know what digit is to the right of the 4, in the hundredths place. We cannot assume that it is zero, for if we *knew* that it was zero, it would have been written in (as 32.40). This uncertainty in the hundredths place in an original number causes uncertainty in the hundredths place in the answer, so we discard that digit and any to the right of it.   ◆◆◆

Students *hate* to throw away those last digits. Remember that by keeping worthless digits, you are telling whoever reads that number that it is more precise than it really is.

◆◆◆ Example 20:

$$
\begin{array}{r}
25.8 \\
18.3\,125 \\
+\ 5.4\,07 \\
\hline
49.5\,195
\end{array}
$$

discard ⟶

◆◆◆

◆◆◆ Example 21: A certain stadium contains about 3500 people. It starts to rain and 372 people leave. How many are left in the stadium?

Solution: Subtracting, we obtain

$$3500 - 372 = 3128$$

which we round to 3100, because 3500 here is known to only two significant digits.   ◆◆◆

It is safer to round the answer *after* adding than to round the original numbers before adding. If you do round before adding, it is prudent to round each original number to *one more* decimal place than you expect in the rounded answer.

Combining Exact and Approximate Numbers

When you are combining an exact number and an approximate one, the accuracy of the result will be limited by the approximate number. Thus, round the result to the number of decimal places found in the approximate number, even though the exact number may *appear* to have fewer decimal places.

♦♦♦ **Example 22:** Express 2 h and 35.8 min in minutes.

Solution: We must add an exact number (120) and an approximate number (35.8).

$$120 + 35.8 = 155.8$$

Since 120 is exact, we do *not* round our answer to the nearest 10 minutes, but we retain as many decimal places as in the approximate number. Our answer is thus 155.8 min.   ♦♦♦

Common Error	Be sure to recognize which numbers in a computation are exact; otherwise, you may perform drastic rounding by mistake.

Exercise 2 ♦ Addition and Subtraction

Combine as indicated.

1. -955	**2.** 8275	**3.** -748
$+212$	-2163	-212
-347	$-\ 874$	-156

Add each column of figures.

4. $99.84	**5.** 96 256	**6.** 98 304
24.96	6 016	6 144
6.24	376	384
1.56	141	24 576
12.48	188	3 072
0.98	1 504	144
3.12	752	49 152

Combine as indicated.

7. $926 + 863$ **8.** $274 + (-412)$

9. $-576 + (-553)$ **10.** $-207 + (-819)$

11. $-575 - 275$ **12.** $-771 - (-976)$

13. $1123 - (-704)$ **14.** $818 - (-207) + 318$

Combine each set of approximate numbers as indicated. Round your answer.

15. $4857 + 73.8$

16. $39.75 + 27.4$

17. $296.44 + 296.997$

18. $385.28 - 692.8$

19. $0.000\ 583 + 0.000\ 8372 - 0.001\ 73$

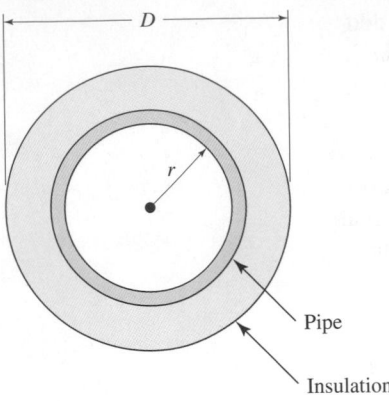

FIGURE 1–4 An insulated pipe.

Equation numbers with the prefix "A" are applications. They are listed toward the end of Appendix A.

20. Mt. Blanc is 15 572 ft. high, and Pike's Peak is about 14 000 ft. high. What is the difference in their heights?

21. Ontario contains 1 068 582 km², and Quebec 1 540 681 km². How much larger is Quebec than Ontario?

22. A man willed $125,000 to his wife and two children. To one child he gave $44,675, to the other $26,380, and to his wife the remainder. What was his wife's share?

23. A circular pipe has an inside radius r of 10.6 cm and a wall thickness of 2.125 cm. It is surrounded by insulation having a thickness of 4.8 cm (Fig. 1–4). What is the outside diameter D of the insulation?

24. A batch of concrete is made by mixing 267 kg of aggregate, 125 kg of sand, 75.5 kg of cement, and 25.25 kg of water. Find the total weight of the mixture.

25. Three resistors, having values of 27.3 ohms (Ω), 4.0155 Ω, and 9.75 Ω, are wired in series. What is the total resistance? (See Eq. A63, which says that the total series resistance is the sum of the individual resistances.)

1–3 Multiplication

Factors and Product
The numbers we multiply to get a *product* are called *factors*. For example,

$$3 \times 5 = 15$$

factors ————————↑ ↑   ↑————————————— product

◆◆◆ **Example 23:** Use your calculator to multiply 183 by 27

Solution: You should get $183 \times 27 = 4941$ ◆◆◆

Commutative, Associative, and Distributive Laws
The *commutative law* states that the *order* of multiplication is not important.

Commutative Law for Multiplication	$ab = ba$	2

◆◆◆ **Example 24:** It is no surprise that

$$2 \times 3 = 3 \times 2$$ ◆◆◆

The *associative law* allows us to group in any way the numbers to be multiplied.

Associative Law for Multiplication	$a(bc) = (ab)c = (ac)b = abc$	4

◆◆◆ **Example 25:**

$$2 \times 3 \times 4 = 2(3 \times 4) = 2(12) = 24$$

$$= (2 \times 3)4 = (6)4 = 24$$
$$= (2 \times 4)3 = (8)3 = 24$$

The *distributive law* shows how a factor may be *distributed* among several terms.

Distributive Law for Multiplication	$a(b + c) = ab + ac$	**5**

◆◆◆ Example 26:

$$2(3 + 4) = 2(7) = 14$$

But the distributive law enables us to do the same computation in a different way.

$$2(3 + 4) = 2(3) + 2(4) = 6 + 8$$
$$= 14 \text{ (as before)}$$

◆◆◆

Multiplying Signed Numbers

To get our rules of signs for multiplication, we use the idea of *multiplication as repeated addition*. For example, to multiply 3 by 4 means to add four 3's (or three 4's)

$$3 \times 4 = 3 + 3 + 3 + 3$$

or

$$3 \times 4 = 4 + 4 + 4$$

Let us return to our shoebox example (Fig. 1–3). Recall that it contains uncashed cheques and unpaid bills. Let's first *add 5 cheques*, each worth $100, to the box. The value of the contents of the box then increases by $500. Multiplying, we have

(number of cheques) × (value of one cheque) = change in value of contents
$$(+5)(+100) = +500$$

Thus a positive number times a positive number gave a positive product. This is nothing new.

Now let's *add 5 bills* to the box, thus decreasing its value by $500. To show this multiplication, we use $(+5)$ for the number of bills, (-100) for the value of each bill, and (-500) for the change in value of the box contents.

$$(+5)(-100) = -500$$

Here, a positive number times a negative number gives a negative product.

Next, we *remove 5 cheques* from the box, thus decreasing its value by $500.

$$(-5)(+100) = -500$$

Here again, the product of a positive number and a negative number is negative. Thus it doesn't matter whether the negative number is the first or the second. This, of course, is what you would expect from the commutative property, which applies to negative numbers as well as to positive numbers.

Finally, we *remove 5 bills* from the box, causing its value to increase by $500.

$$(-5)(-100) = +500$$

Here, the product of two negative numbers is positive.

We summarize these findings to get our *rules of signs for multiplication*.

If a and b are two positive numbers, the product of two quantities of like sign is positive.

Rule of Signs for Multiplication	$(+a)(+b) = (-a)(-b) = +ab$	8

Also, the product of two quantities of unlike sign is negative.

Rule of Signs for Multiplication	$(+a)(-b) = (-a)(+b) = -ab$	9

◆◆◆ **Example 27:**

(a) $2(-3) = -6$

(b) $(-2)3 = -6$

(c) $(-2)(-3) = 6$ ◆◆◆

Multiplying a String of Numbers

When we multiplied two negative numbers, we got a positive product. So when we are multiplying a *string* of numbers, if an even number of them are negative, the answer will be positive, and if an odd number of them are negative, the answer will be negative.

◆◆◆ **Example 28:**

(a) $2(-3)(-1)(-2) = -12$

(b) $2(-3)(-1)(2) = 12$ ◆◆◆

Multiplying Negative Numbers

◆◆◆ **Example 29:** Use your calculator to multiply -96 by -83.

Solution: You should get

$$(-96)(-83) = 7968$$ ◆◆◆

A simpler way to do the last problem would be to multiply $+96$ and $+83$ and determine the sign by inspection.

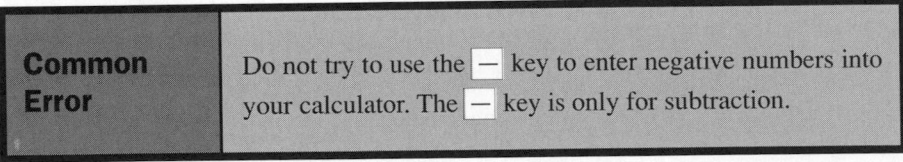

Common Error	Do not try to use the ⎯ key to enter negative numbers into your calculator. The ⎯ key is only for subtraction.

Multiplication of Approximate Numbers

Rule	When multiplying two or more approximate numbers, round the result to as many digits as in the factor having the fewest significant digits.

◆◆◆ Example 30:

$$12.1 \quad \times \quad 15.6 \quad = \quad 189$$

three   three   three
digits   digits   digits   ◆◆◆

When the factors have different numbers of significant digits, keep the same number of digits in your answer as are contained in the factor that has the *fewest* significant digits.

◆◆◆ Example 31:

$$123.56 \quad \times \quad 2.21 \quad = \quad 273$$

five   three   keep
digits   digits   three
                                         digits   ◆◆◆

Common Error	Do not confuse *significant digits* with *decimal places*. The number 274.56 has *five* significant digits and *two* decimal places. Decimal places determine how we round after adding or subtracting. Significant figures determine how we round after multiplying and, as we will soon see, after dividing, raising to a power, or taking roots.

Multiplication with Exact Numbers

When using *exact numbers* in a computation, treat them as if they had *more* significant figures than any of the approximate numbers in that computation.

◆◆◆ Example 32: If a certain car tire weighs 32.2 kg when mounted, how much will four such tires weigh?

Solution: Multiplying, we obtain

$$32.2(4) = 128.8 \text{ kg} \qquad ◆◆◆$$

Since the 4 is an exact number, we retain as many significant figures as contained in 32.2, and round our answer to 129 kg.

Exercise 3 ◆ Multiplication

Multiply each approximate number and retain the proper number of digits in your answer.

1. 3967×1.84 **2.** 4.900×59.3

3. $93.9 \times 0.005\ 5908$ **4.** $4.97 \times 9.27 \times 5.78$

5. $69.0 \times (-258)$ **6.** $-385 \times (-2.2978)$

7. $2.86 \times (4.88 \times 2.97) \times 0.553$ **8.** $(5.93 \times 7.28) \times (8.26 \times 1.38)$

Word Problems

9. What is the cost of 52.5 tonnes (t) of cement at $99.25 per tonne?

10. If 68 t of rail is needed for 1 km of track, how many tonnes will be required for 762 km, and what will be its cost at $1,425 per tonne?

11. Three barges carry 26 t of gravel each, and a fourth carries 35 t. What is the value, to the nearest dollar, of the whole shipment, at $22.75 per tonne?

12. Two cars start from the same place and travel in opposite directions, one at the rate of 45 km/h, the other at 55 km/h. How far apart will they be at the end of 6.0 h?

13. What will be the cost of installing a telephone line 274 km long, at $5,723 per kilometre?

14. The current to a projection lamp is measured at 4.7 A when the line voltage is 115.45 V. Using Eq. A65 (power = voltage × current), find the power dissipated in the lamp.

15. A gear in a certain machine rotates at the speed of 1808 r/min. How many revolutions will it make in 9.500 minutes?

16. How much will 1000 washers weigh if each weighs 2.375 g?

17. One inch equals exactly 2.54 cm. Convert 385.84 in. to centimetres.

18. If there are 360 degrees per revolution, how many degrees are there in 4.863 revolutions?

1–4   Division

Definitions

A quantity *a/b* is also referred to as a *fraction* or the *ratio* of *a* to *b*. Fractions and ratios are discussed in Chapter 8.

The *dividend*, when divided by the *divisor*, gives us the *quotient*.

$$\text{dividend} \div \text{divisor} = \text{quotient}$$

or

$$\frac{\text{dividend}}{\text{divisor}} = \text{quotient}$$

◆◆◆ **Example 33:** Use your calculator to divide 861 by 123.

Solution: You should get

$$861 \div 123 = 7 \qquad\qquad ◆◆◆$$

When we multiplied two integers, we always got an integer for an answer. This is not always the case when dividing.

◆◆◆ **Example 34:** When we divide 2 by 3, we get 0.666 666 666. . . . We must choose how many digits we wish to retain, and we must round our answer. Rounding to, say, three significant digits, we obtain

$$2 \div 3 \cong 0.667 \qquad\qquad ◆◆◆$$

Here it is appropriate to use the $\cong$ symbol.

Division of Approximate Numbers

The rule for rounding with division is almost the same as with multiplication.

Rule	After dividing one approximate number by another, round the quotient to as many digits as there are in the original number having fewer significant digits.

◆◆◆ **Example 35:**

Solution: By calculator,

$$846.2 \div 4.75 = 178.147\ 3684$$

Since 4.75 has three significant digits, we round our quotient to 178.   ◆◆◆

◆◆◆ **Example 36:** Divide 846.2 into three equal parts.

Solution: We divide by the integer 3, and since we consider integers to be exact, we retain in our answer the same number of significant digits as in 846.2.

$$846.2 \div 3 = 282.1$$ ◆◆◆

Dividing Signed Numbers

We will use the rules of signs for multiplication to get the rules of signs for division.

We know that the product of a negative number and a positive number is negative. Thus

$$(-2)(+3) = -6$$

If we divide both sides of this equation by $(+3)$, we get

$$-2 = \frac{-6}{+3}$$

From this we see that *a negative number divided by a positive number gives a negative quotient.*

Again starting with

$$(-2)(+3) = -6$$

we divide both sides by (-2) and get

$$+3 = \frac{-6}{-2}$$

Here we see that *a negative number divided by a negative number gives a positive quotient.*

We also know that the product of two negative numbers is positive. Thus

$$(-2)(-3) = +6$$

Dividing both sides by (-3), we get

$$-2 = \frac{+6}{-3}$$

Thus, *a positive number divided by a negative number gives a negative quotient.* We combine these findings with the fact that the quotient of two positive numbers is positive and get our *rules of signs for division.*

The quotient is positive when dividend and divisor have the same sign.

Rule of Signs for Division	$\dfrac{+a}{+b} = \dfrac{-a}{-b} = \dfrac{a}{b}$	10

The quotient is negative when dividend and divisor have opposite signs.

Rule of Signs for Division	$\dfrac{+a}{-b} = \dfrac{-a}{+b} = -\dfrac{a}{b}$	11

◆◆◆ **Example 37:**

(a) $8 \div (-4) = -2$
(b) $-8 \div 4 = -2$
(c) $-8 \div (-4) = 2$ ◆◆◆

◆◆◆ **Example 38:** Divide 85.4 by -2.5386 on the calculator.

Solution: You should get

$$85.4 \div (-2.5386) = -33.6$$ ◆◆◆

As with multiplication, the sign could also have been found by inspection.

Division by Zero

Zero divided by any quantity (except zero) is zero. Division *by* zero is not defined. It is an illegal operation in mathematics.

◆◆◆ **Example 39:** Using your calculator, divide 5 by zero.

Solution: You should get an error message on your calculator.   ◆◆◆

Reciprocals

The *reciprocal* of any nonzero number n is $1/n$. Thus the product of a quantity and its reciprocal is equal to 1.

Some keys, such as the reciprocal key, might be a *second function* on your calculator, and some other operations might be available only from a menu.

The reciprocal on your calculator can be $1/n$ or x^{-1}. Both will give the same result.

◆◆◆ **Example 40:**

(a) The reciprocal of 10 is 1/10.
(b) The reciprocal of 1/2 is 2.
(c) The reciprocal of $-\frac{3}{4}$ is $-\frac{4}{3}$.   ◆◆◆

◆◆◆ **Example 41:** Use your calculator to verify that

(a) the reciprocal of 6.38 is 0.157.
(b) the reciprocal of -2.754 is -0.3631.   ◆◆◆

Exercise 4 ◆ Division

Divide, and then round your answer to the proper number of digits.

  1. $947 \div 5.82$ **2.** $0.492 \div 0.004\,78$
  3. $-99.4 \div 286.5$ **4.** $-4.8 \div -2.557$
  5. $5836 \div 8264$ **6.** $5.284 \div 3.827$
  7. $94\,840 \div 1.338\,76$ **8.** $3.449 \div (-6.837)$
  9. $2\,497\,000 \div 150\,000$ **10.** $2.97 \div 4.828$

Word Problems Involving Division

11. A stretch of roadway 1858.54 m long is to be divided into 5 equal sections. Find the length of each section.

12. At the rate of 24.5 km in 8.25 h, how many kilometres would a person walk in 12.75 h? (Distance = rate × time.)

13. If 5 masons can build a wall in 8 days, how many masons are needed to build it in 4 days?

14. If 867 shares of pipeline stock are valued at $84,099, what is the value of each share?

Reciprocals

Find the reciprocal of each number, retaining the proper number of digits in your answer.

15. 693

16. 0.006 30

17. −396 000

18. 39.74

19. −0.005 73

20. 938.4

21. 4.992

22. −6.93

23. 11.1

24. −375

25. 1.007

26. 3.98

Word Problems Involving Reciprocals

27. Using the equation; $R_T = \dfrac{R_1 \times R_2}{R_1 + R_2}$, find the equivalent resistance of a 475 Ω resistor (R_1) and a 928 Ω resistor (R_2), when they are connected in a parallel circuit.

28. When an object is placed 126 cm in front of a certain thin lens having a focal length f, the image will be formed 245 cm from the lens. The distances are related by

$$\frac{1}{f} = \frac{1}{126} + \frac{1}{245}$$

Find f.

29. The sine of an angle θ (written sin θ) is equal to the reciprocal of the cosecant of θ (csc θ). Find sin θ if csc θ = 3.58.

30. If two straight lines are perpendicular, the slope of one line is the negative reciprocal of the slope of the other. If the slope of a line is −2.55, find the slope of a line perpendicular to that line.

1–5   Powers and Roots

Definitions

In the expression

$$2^4$$

the number 2 is called the *base*, and the number 4 is called the *exponent*. The expression is read "two to the fourth power." Its value is

$$2^4 = 2\cdot2\cdot2\cdot2 = 16$$

◆◆◆ **Example 42:** Use your calculator to verify that

$$(3.85)^3 = 57.1 \quad \text{(rounded to three digits)} \qquad ◆◆◆$$

Negative Base

A negative base raised to an *even* power gives a *positive* number. A negative base raised to an *odd* power gives a *negative* number.

◆◆◆ **Example 43:**

(a) $(-2)^2 = (-2)(-2) = 4$

(b) $(-2)^3 = (-2)(-2)(-2) = -8$

(c) $(-1)^{24} = 1$

(d) $(-1)^{25} = -1$ ◆◆◆

If you try to do these problems on your calculator, you will probably get an error indication. Some calculators will not work with a negative base, even though this is a valid operation.

Then how do you do it? Simply enter the base as *positive,* find the power, and determine the sign by inspection.

◆◆◆ **Example 44:** Find $(-1.45)^5$.

Solution: From the calculator,

$$(+1.45)^5 = 6.41$$

Since we know that a negative number raised to an odd power is negative, we write

$$(-1.45)^5 = -6.41 \qquad \text{◆◆◆}$$

◆◆◆ **Example 45:** Use your calculator to verify that

$$(3.85)^{-3} = 0.0175 \quad \text{(rounded to three digits)}$$

Notice that raising a positive number to a negative power does *not* result in a negative number. ◆◆◆

Fractional Exponents

We'll see later that fractional exponents are another way of writing radicals. For now, we'll just evaluate fractional exponents on the calculator. The keystrokes are the same as before, but we must be sure that the fractional exponent is enclosed in parentheses, as shown in the following example.

◆◆◆ **Example 46:** Use your calculator to verify that

$$8^{2/3} = 4 \qquad \text{◆◆◆}$$

Roots

If $a^n = b$, then

$$\sqrt[n]{b} = a$$

which is read "the *n*th root of *b* equals *a*." The symbol $\sqrt{}$ is a *radical sign, b* is the *radicand,* and *n* is the *index* of the radical.

◆◆◆ **Example 47:**

(a) $\sqrt{4} = 2$ because $2^2 = 4$
(b) $\sqrt[3]{8} = 2$ because $2^3 = 8$
(c) $\sqrt[4]{81} = 3$ because $3^4 = 81$ ◆◆◆

Principal Root

The *principal root* of a positive number is defined as the *positive* root. Thus $\sqrt{4} = +2$, not ± 2.

The principal root is *negative* when we take an *odd* root of a *negative* number.

◆◆◆ **Example 48:**

$$\sqrt[3]{-8} = -2$$

because $(-2)(-2)(-2) = -8$. ◆◆◆

◆◆◆ Example 49: Use your calculator to verify that

$$\sqrt[5]{28.4} = -1.952\ 826\ 537$$

which we round to 1.95. ◆◆◆

Odd Roots of Negative Numbers by Calculator

An *even* root of a negative number is *imaginary* (such as $\sqrt{-4}$). We will study these in Chapter 18. But an *odd* root of a negative number is *not* imaginary. It is a real, negative number. As with powers, some calculators will not accept a negative radicand. Fortunately, we can outsmart our calculators and take odd roots of negative numbers anyway.

◆◆◆ Example 50: Find $\sqrt[5]{-875}$.

Solution: We know that an odd root of a negative number is real and negative. So we take the fifth root of + 875, by calculator,

$$\sqrt[5]{875} = 3.88\ \text{(rounded)}$$

and we only have to place a minus sign before the number.

$$\sqrt[5]{-875} = -3.88$$ ◆◆◆

Exercise 5 ◆ Powers and Roots

Powers

Evaluate each power without using a calculator. Do not round your answers.

1. 2^3	**2.** 5^3	**3.** $(-2)^3$	**4.** 9^2
5. 1^3	**6.** $(-1)^2$	**7.** $(-1)^{40}$	**8.** 3^2
9. 1^8	**10.** $(-1)^3$	**11.** $(-1)^{41}$	**12.** $(-3)^2$

Evaluate each power of 10.

13. 10^2	**14.** 10^1	**15.** 10^0	**16.** 10^{-4}
17. 10^3	**18.** 10^5	**19.** 10^{-2}	**20.** 10^{-1}
21. 10^4	**22.** 10^{-3}	**23.** 10^{-5}	

Evaluate each expression, retaining the correct number of digits in your answer.

24. $(8.55)^3$	**25.** $(1.007)^5$
26. $(9.55)^3$	**27.** $(-4.82)^3$
28. $(-77.2)^2$	**29.** $(8.28)^{-2}$
30. $(0.0772)^{0.426}$	**31.** $(5.28)^{-2.15}$
32. $(35.2)^{1/2}$	**33.** $(462)^{2/3}$
34. $(88.2)^{-2}$	**35.** $(-37.3)^{-3}$

Powers of 10 will be needed for *scientific notation* later. Arrange these powers of 10 in order, and try to invent a rule that will enable you to write the value of a power of 10 without doing the computation.

Word Problems Involving Powers

36. The distance travelled, in feet, by a falling body, starting from rest, is equal to $16t^2$, where t is the elapsed time. In 5.448 s, the distance fallen is $16(5.448)^2$ ft. Evaluate this quantity. (Treat 16 here as an approximate number.)

37. The power dissipated in a resistance R through which is flowing a current I is equal to I^2R. Therefore the power in a 365 Ω resistor carrying a current of 0.5855 A is $(0.5855)^2(365)$ W. Evaluate this power.

38. The volume of a cube of side 35.8 cm (Fig. 1–5) is $(35.8)^3$ cm³. Evaluate this volume.

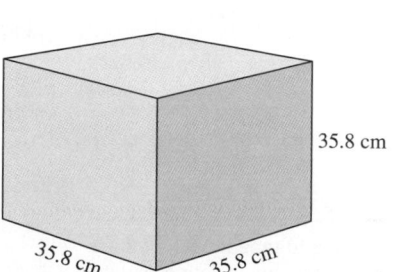

35.8 cm

35.8 cm

35.8 cm

FIGURE 1–5

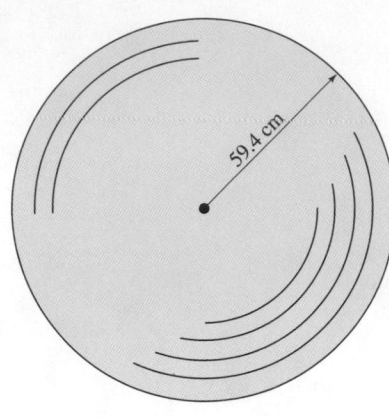

FIGURE 1–6

39. The volume of a 59.4-cm-radius sphere (Fig. 1–6) is $\frac{4}{3}\pi$ (59.4)^3 cm^3. Find this volume.

40. An investment of \$2,000 at a compound interest rate of $6\frac{1}{4}$%, left for $7\frac{1}{2}$ years, will be worth 2,000(1.0625)$^{7.5}$ dollars. Find this amount, to the nearest cent.

Roots

Find each principal root without using your calculator.

41. $\sqrt{25}$ 42. $\sqrt[3]{27}$ 43. $\sqrt{49}$

44. $\sqrt[3]{-27}$ 45. $\sqrt[3]{-8}$ 46. $\sqrt[5]{-32}$

Evaluate each radical by calculator, retaining the proper number of digits in your answer:

47. $\sqrt{49.2}$ 48. $\sqrt{1.863}$ 49. $\sqrt[3]{88.3}$

50. $\sqrt{772}$ 51. $\sqrt{3875}$ 52. $\sqrt[3]{7295}$

53. $\sqrt[3]{-386}$ 54. $\sqrt[5]{-18.4}$ 55. $\sqrt[3]{-2.774}$

Applications of Roots

56. The period T (time for one swing) of a simple pendulum (Fig. 1–7) 2.55 ft. long is

$$T = 2\pi \sqrt{\frac{2.55}{32.0}} \text{ seconds}$$

Evaluate T.

57. The magnitude Z of the impedance in a circuit having a resistance of 3540 Ω and a reactance of 2750 Ω is

$$Z = \sqrt{(3540)^2 + (2750)^2} \text{ ohms}$$

Find Z.

58. The geometric mean B between 3.75 and 9.83 is

$$B = \sqrt{(3.75)(9.83)}$$

Evaluate B.

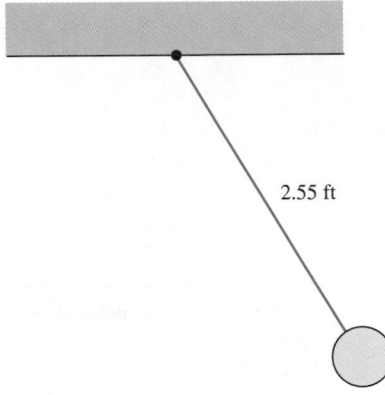

2.55 ft

FIGURE 1–7

1–6   Combined Operations

Order of Operations

If the expression to be evaluated does not contain parentheses, perform the operations in the following order:

1. *Powers and roots, in any order.*
2. *Multiplications and divisions, from left to right.*
3. *Additions and subtractions, from left to right.*

Our first group of calculations will be with integers only, and later we will do some problems that require rounding. We first show a problem containing both addition and multiplication.

◆◆◆**Example 51:** Evaluate $7 + 3 \times 4$.

Solution: The multiplication is done before the addition.

$$7 + 3 \times 4 = 7 + 12 = 19 \qquad\qquad ◆◆◆$$

Next we do a calculation having both a power and multiplication.

◆◆◆**Example 52:** Evaluate 5×3^2.

Be sure to repeat each of these computations on your calculator, consulting your manual where the operations are not clear.

Solution: We raise to the power before multiplying:

$$5 \times 3^2 = 5 \times 9 = 45 \qquad\qquad ◆◆◆$$

Parentheses

When an expression contains parentheses, evaluate first the expression within the parentheses and then the entire expression.

◆◆◆ **Example 53:** Evaluate $(7 + 3) \times 4$.

Solution:

$$(7 + 3) \times 4 = 10 \times 4 = 40$$ ◆◆◆

If the sum or difference of more than one number is to be raised to a power, those numbers must be enclosed in parentheses.

◆◆◆ **Example 54:** Evaluate $(5 + 2)^2$.

Solution: We combine the numbers inside the parentheses before squaring.

$$(5 + 2)^2 = 7^2 = 49$$ ◆◆◆

◆◆◆ **Example 55:** Evaluate $(2 + 6)(7 + 9)$.

Solution: Evaluate the two quantities in parentheses before multiplying.

$$(2 + 6)(7 + 9) = 8 \times 16 = 128$$ ◆◆◆

◆◆◆ **Example 56:** Evaluate $\dfrac{8 + 4}{9 - 3}$.

Solution: Here the fraction line acts like parentheses, grouping the 8 and 4, as well as the 9 and 3. Written on a single line, this problem would be

$$(8 + 4) \div (9 - 3)$$

or

$$12 \div 6 = 2$$ ◆◆◆

Combined Operations with Approximate Numbers

Combined operations with approximate numbers are done the same way. However, we must round our answer properly using the rules given earlier in this chapter.

◆◆◆ **Example 57:** Evaluate the expression

$$\left(\frac{118.8 + 4.23}{\sqrt{136}} \right)^3$$

Solution: By calculator, we get a result of 1174.153 047. But how many digits should we keep? In the numerator, we added a number with one decimal place to another with two decimal places, so we are allowed to keep just one. Thus, the numerator, after addition, is good to one decimal place, or, in this case, four significant digits. The denominator, however, has just three significant digits, so we round our answer to three significant digits, getting 1170. ◆◆◆

◆◆◆ **Example 58:** A rectangular courtyard (Fig. 1–8) having sides of 21.8 ft. and 33.74 ft. has a diagonal measurement x given by the expression

$$x = \sqrt{(21.8)^2 + (33.74)^2}$$

Evaluate the expression to find x.

Solution: By calculator we get a result of 40.169 984. We round our answer to three significant digits because 21.8 has only three, and we get $x = 40.2$ ft. ◆◆◆

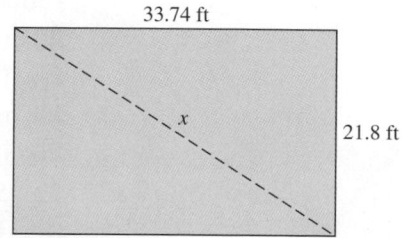

FIGURE 1–8 A rectangular courtyard.

Exercise 6 ◆ Combined Operations

Combined Operations with Exact Numbers

Perform each computation by calculator.

1. $(37)(28) + (36)(64)$
2. $(22)(53) - (586)(4) + (47)(59)$
3. $(63 + 36)(37 - 97)$
4. $(89 - 74 + 95)(87 - 49)$
5. $\dfrac{219}{73} + \dfrac{194}{97}$
6. $\dfrac{228}{38} - \dfrac{78}{26} + \dfrac{364}{91}$
7. $\dfrac{647 + 688}{337 + 108}$
8. $\dfrac{809 - 463 + 1858}{958 - 364 + 508}$
9. $(5 + 6)^2$
10. $(422 + 113 - 533)^4$
11. $(423 - 420)^3$
12. $\left(\dfrac{853 - 229}{874 - 562}\right)^2$
13. $\left(\dfrac{141}{47}\right)^3$
14. $\sqrt{434 + 466}$
15. $\sqrt{(8)(72)}$
16. $\sqrt[3]{657 + 553 - 1085}$
17. $\sqrt[4]{(27)(768)}$
18. $\sqrt{\dfrac{2404}{601}}$
19. $\sqrt[4]{\dfrac{1136}{71}}$
20. $\sqrt{961} + \sqrt{121}$
21. $\sqrt[4]{625} + \sqrt{961} - \sqrt[3]{216}$
22. $\sqrt[4]{256} \times \sqrt{49}$

Combined Operations with Approximate Numbers

Perform the computations, keeping the proper number of digits in your answer.

23. $(7.37)(3.28) + (8.36)(2.64)$
24. $(522)(9.53) - (586)(4.70) + (847)(7.59)$
25. $(63.5 + 83.6)(8.37 - 1.72)$
26. $(8.93 - 3.74 + 9.05)(68.70 - 64.90)$
27. $\dfrac{583}{473} + \dfrac{946}{907}$
28. $\dfrac{6.73}{8.38} - \dfrac{5.97}{8.06} + \dfrac{8.63}{1.91}$

29. $\dfrac{6.47 + 8.604}{3.37 + 90.8}$

30. $\dfrac{809 - 463 + 744}{758 - 964 + 508}$

31. $(5.37 + 2.36)^2$

32. $(4.25 + 4.36 - 5.24)^4$

33. $(6.423 + 1.05)^2$

34. $\left(\dfrac{45.3 - 8.34}{8.74 - 5.62}\right)^{2.5}$

35. $\left(\dfrac{8.90}{4.75}\right)^2$

36. $\sqrt{4.34 + 4.66}$

37. $\sqrt[3]{657 + 553 - 842}$

38. $\sqrt{(28.1)(5.94)}$

39. $\sqrt[5]{(9.06)(4.86)(7.93)}$

40. $\sqrt{\dfrac{653}{601}}$

41. $\sqrt[4]{\dfrac{4.50}{7.81}}$

42. $\sqrt{9.74} + \sqrt{12.5}$

43. $\sqrt[4]{528} + \sqrt{94.2} - \sqrt[3]{284}$

44. $\sqrt[3]{653} \times \sqrt{55.3}$

1-7 Scientific and Engineering Notation

Evaluating Powers of 10

We did some work with powers in Sec. 1–5. We saw, for example, that 2^3 meant

$$2^3 = 2 \cdot 2 \cdot 2 = 8$$

Here, the power 3 tells how many 2's are to be multiplied to give the product. For powers of 10, the power tells how many 10's are to be multiplied to give the product.

◆◆◆ **Example 59:**

(a) $10^2 = 10 \times 10 = 100$

(b) $10^3 = 10 \times 10 \times 10 = 1000$ ◆◆◆

Negative powers, as before, are calculated by means of Eq. 35, $x^{-a} = 1/x^a$.

◆◆◆ **Example 60:**

(a) $10^{-2} = \dfrac{1}{10^2} = \dfrac{1}{100} = 0.01$

(b) $10^{-5} = \dfrac{1}{10^5} = \dfrac{1}{100\,000} = 0.000\,01$ ◆◆◆

Some powers of 10 are summarized in the following table:

Positive Powers	Negative Powers
$1\,000\,000 = 10^6$	$0.1 \qquad = 1/10 = 10^{-1}$
$100\,000 = 10^5$	$0.01 \qquad = 1/10^2 = 10^{-2}$
$10\,000 = 10^4$	$0.001 \qquad = 1/10^3 = 10^{-3}$
$1000 = 10^3$	$0.0001 \qquad = 1/10^4 = 10^{-4}$
$100 = 10^2$	$0.000\,01 \quad = 1/10^5 = 10^{-5}$
$10 = 10^1$	$0.000\,001 = 1/10^6 = 10^{-6}$
$1 = 10^0$	

Table 1–1

Scientific Notation

Let us multiply two large numbers on the calculator: 500 000 and 300 000. We get a display like $\boxed{1.5\,\text{E}+11}$ or $\boxed{1.5\ 11}$, depending on the calculator. What has happened?

Our answer (150 000 000 000) is too large to fit the calculator display, so the machine has automatically switched to *scientific notation*. Our calculator display actually contains *two* numbers: a decimal number (1.5) and an integer (11). Our answer is equal to the decimal number multiplied by 10 raised to the value of the integer.

Calculator display: $\boxed{1.5\quad 11}$

Scientific notation: 1.5×10^{11}

decimal part ————————⌐   ⌐———— power of 10

Ten raised to a power (such as 10^{11}) is called a *power of 10*.

A number is said to be in *scientific notation* when it is written as a number whose absolute value is between 1 and 10, multiplied by a power of 10.

◆◆◆ **Example 61:** The following numbers are written in scientific notation:

(a) 2.74×10^3 (b) 8.84×10^9

(c) 5.4×10^{-6} (d) -1.2×10^{-5} ◆◆◆

Converting Numbers to Scientific Notation

First rewrite the given number with a single digit to the left of the decimal point, discarding any nonsignificant zeros. Then multiply this number by the power of 10 that will make it equal to the original number.

◆◆◆ **Example 62:**

$$346 = 3.46 \times 100$$
$$= 3.46 \times 10^2$$ ◆◆◆

◆◆◆ **Example 63:**

$$2700 = 2.7 \times 1000$$
$$= 2.7 \times 10^3$$

Note that we have discarded the two nonsignificant zeros.   ◆◆◆

When we are converting a number whose absolute value is less than 1, our power of 10 will be negative, as in Example 64.

◆◆◆ **Example 64:**

$$0.000\,009\,50 = 9.50 \times 0.000\,001$$
$$= 9.50 \times 10^{-6}$$ ◆◆◆

The sign of the exponent has nothing to do with the sign of the original number. You can convert a negative number to scientific notation, just as you would a positive number, and put the minus sign on afterward.

◆◆◆ **Example 65:** Convert $-34\,720$ to scientific notation.

Solution: Converting $+34\,720$ scientific notation, we obtain

$$34\,720 = 3.472 \times 10\,000$$
$$= 3.472 \times 10^4$$

Then multiplying both sides by -1 gives

$$-34\ 720 = -3.472 \times 10^4 \qquad \bullet\bullet\bullet$$

Converting Numbers from Scientific Notation
To convert *from* scientific notation, simply reverse the process.

◆◆◆ **Example 66:**

$$4.82 \times 10^5 = 4.82 \times 10\ 000$$
$$= 482\ 000 \qquad \bullet\bullet\bullet$$

◆◆◆ **Example 67:**

$$8.25 \times 10^{-3} = 8.25 \times 0.001$$
$$= 0.008\ 25 \qquad \bullet\bullet\bullet$$

Engineering Notation
Engineering notation is similar to scientific notation. The difference is that

- the exponent is a multiple of three; and
- there can be one, two, or three digits to the left of the decimal point, rather than just one digit.

Having an exponent that is a multiple of 3 makes it easier to use the *SI prefixes* described in Sec. 1–8.

◆◆◆ **Example 68:** Some examples of numbers written in engineering notation are as follows:

(a) 66.3×10^3 (b) 8.14×10^9

(c) 725×10^{-6} (d) 28.72×10^{-12} $\bullet\bullet\bullet$

Converting Numbers to Engineering Notation
Converting to engineering notation is simple if the digits of the decimal number are grouped into sets of three, either by commas or by spaces in the SI convention.

◆◆◆ **Example 69:**

(a) $21\ 840 = 21.84 \times 10^3$

(b) $548\ 000 = 548 \times 10^3$

(c) $72\ 560\ 000 = 72.56 \times 10^6$ $\bullet\bullet\bullet$

For numbers less than 1, it helps to first separate the digits following the decimal point into groups of three. This SI convention is used throughout this text.

◆◆◆ **Example 70:**

(a) $0.87217 = 0.872\ 17 = 872.17 \times 10^{-3}$

(b) $0.000736492 = 0.000\ 736\ 492 = 736.492 \times 10^{-6}$

(c) $0.0000000472 = 0.000\ 000\ 047\ 2 = 47.2 \times 10^{-9}$ $\bullet\bullet\bullet$

Addition and Subtraction
If two or more numbers to be added or subtracted have the *same power of 10*, simply combine the numbers and keep the same power of 10.

This is another application of the distributive law,
$ab + ac = a(b + c)$. Thus

$2 \times 10^5 + 3 \times 10^5$
$= (2 + 3) \times 10^5$
$= 5 \times 10^5$

◆◆◆ **Example 71:**

(a) $(2 \times 10^5) + (3 \times 10^5) = 5 \times 10^5$

(b) $(8 \times 10^3) - (5 \times 10^3) + (3 \times 10^3) = 6 \times 10^3$ $\bullet\bullet\bullet$

If the powers of 10 are different, *they must be made equal* before the numbers can be combined. A shift of the decimal point of one place to the *left* will *increase* the exponent by 1. Conversely, a shift of the decimal point one place to the *right* will *decrease* the exponent by 1.

✦✦✦ **Example 72:**

(a) $(1.5 \times 10^4) + (3 \times 10^3) = (1.5 \times 10^4) + (0.3 \times 10^4)$
$$= 1.8 \times 10^4$$

(b) $(1.25 \times 10^5) - (2 \times 10^4) + (4 \times 10^3)$
$$= (1.25 \times 10^5) - (0.2 \times 10^5) + (0.04 \times 10^5)$$
$$= 1.09 \times 10^5$$ ✦✦✦

Multiplication

We multiply powers of 10 by *adding their exponents*.

✦✦✦ **Example 73:**

> We are really using Eq. 29, $x^a \cdot x^b = x^{a+b}$. This equation is one of the *laws of exponents* that we will study later in more detail.

(a) $10^3 \cdot 10^4 = 10^{3+4} = 10^7$

(b) $10^{-2} \cdot 10^5 = 10^{-2+5} = 10^3$ ✦✦✦

To multiply two numbers in scientific notation, multiply the decimal parts and the powers of 10 *separately*.

✦✦✦ **Example 74:**

$$(2 \times 10^5)(3 \times 10^2) = (2 \times 3)(10^5 \times 10^2)$$
$$= 6 \times 10^{5+2}$$
$$= 6 \times 10^7$$ ✦✦✦

Division

We divide powers of 10 by subtracting the exponent of the denominator from the exponent of the numerator.

✦✦✦ **Example 75:**

(a) $\dfrac{10^5}{10^3} = 10^{5-3} = 10^2$ (b) $\dfrac{10^{-4}}{10^{-2}} = 10^{-4-(-2)} = 10^{-2}$ ✦✦✦

As with multiplication, we divide the decimal parts and the powers of 10 separately.

✦✦✦ **Example 76:**

(a) $\dfrac{8 \times 10^5}{4 \times 10^2} = \dfrac{8}{4} \times \dfrac{10^5}{10^2} = 2 \times 10^{5-2} = 2 \times 10^3$

(b) $\dfrac{12 \times 10^3}{4 \times 10^5} = 3 \times 10^{3-5} = 3 \times 10^{-2}$ ✦✦✦

Exercise 7 ✦ Scientific and Engineering Notation

Powers of 10

Write each number as a power of 10.

  1. 100   **2.** 1 000 000   **3.** 0.0001

  4. 0.001   **5.** 100 000 000

Write each power of 10 as a decimal number.

  6. 10^5 **7.** 10^{-2} **8.** 10^{-5}

  9. 10^{-1} **10.** 10^4

Scientific and Engineering Notation

Write each number in scientific notation and engineering notation.

11. 186 000

12. 24 100

13. 0.0035

14. 0.000 0546

15. 25 742

16. 1.257 42

17. $80\overline{0}0$

18. 16 000

19. 98.3×10^3

20. 87.34×10^3

21. 0.0775×10^{-2}

22. 0.00871×10^{-2}

Convert each number from scientific or engineering notation to decimal notation.

23. 2.85×10^3

24. 1.75×10^{-5}

25. 9×10^4

26. 9.00×10^4

27. 3.667×10^{-3}

28. 76.3×10^3

29. 248.2×10^{-6}

30. 7×10^6

31. 30.00×10^6

32. 942.56×10^{-9}

Multiplication and Division

Multiply the following powers of 10.

33. $10^5 \cdot 10^2$

34. $10^4 \cdot 10^{-3}$

35. $10^{-5} \cdot 10^{-4}$

36. $10^{-2} \cdot 10^5$

37. $10^{-1} \cdot 10^{-4}$

Divide the following powers of 10.

38. $10^8 \div 10^5$

39. $10^4 \div 10^6$

40. $10^5 \div 10^{-2}$

41. $10^{-3} \div 10^5$

42. $10^{-2} \div 10^{-4}$

Multiply without using a calculator.

43. $(3.0 \times 10^3)(5.0 \times 10^2)$

44. $(5 \times 10^4)(8 \times 10^{-3})$

45. $(2 \times 10^{-2})(4 \times 10^{-5})$

46. $(7.0 \times 10^4)(3\overline{0}\ 000)$

Divide without using a calculator.

47. $(8 \times 10^4) \div (2 \times 10^2)$

48. $(6 \times 10^4) \div 0.03$

49. $(3 \times 10^3) \div (6 \times 10^5)$

50. $(8 \times 10^{-4}) \div 400\ 000$

51. $(9 \times 10^4) \div (3 \times 10^{-2})$

52. $49\ 000 \div (7.0 \times 10^{-2})$

Addition and Subtraction

Combine without using a calculator.

53. $(3.0 \times 10^4) + (2.1 \times 10^5)$

54. $(75.0 \times 10^2) + 32\overline{0}0$

55. $(1.557 \times 10^2) + (9.000 \times 10^{-1})$

56. $0.037 - (6.0 \times 10^{-3})$

57. $(7.2 \times 10^4) + (1.1 \times 10^4)$

Scientific Notation and Engineering Notation on the Calculator

Perform the following computations and give your answers in both scientific, and engineering notation. Combine the powers of 10 by hand or with your calculator.

58. $(1.58 \times 10^2)(9.82 \times 10^3)$

59. $(9.83 \times 10^5) \div (2.77 \times 10^3)$

60. $(3.87 \times 10^{-2})(5.44 \times 10^5)$

61. $(2.74 \times 10^3) \div (9.13 \times 10^5)$

62. $(5.6 \times 10^2)(3.1 \times 10^{-1})$

63. $(7.72 \times 10^8) \div (3.75 \times 10^{-9})$

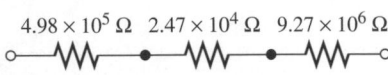

$4.98 \times 10^5 \, \Omega \quad 2.47 \times 10^4 \, \Omega \quad 9.27 \times 10^6 \, \Omega$

FIGURE 1–9   Resistors in series.

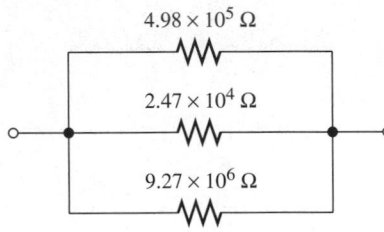

$4.98 \times 10^5 \, \Omega$

$2.47 \times 10^4 \, \Omega$

$9.27 \times 10^6 \, \Omega$

FIGURE 1–10   Resistors in parallel.

Applications

64. Three resistors, having resistances of $4.98 \times 10^5 \, \Omega$, $2.47 \times 10^4 \, \Omega$, and $9.27 \times 10^6 \, \Omega$, are wired in series (Fig. 1–9). Find the total resistance, using the equation for resistors in series, Eq. A63: $R_T = R_1 + R_2 + R_3 + \ldots$

65. Find the equivalent resistance if the three resistors of problem 64 are wired in parallel (Fig. 1–10). Use the equation for "resistors in parallel", Eq. A64: $\left(\frac{1}{R_T}\right) = \left(\frac{1}{R_1}\right) + \left(\frac{1}{R_2}\right) + \left(\frac{1}{R_3}\right) \ldots$ to find the resistance in parallel.

66. Using the equation for finding power $P = V \times I$, find the power dissipated in a resistor if a current of 3.75×10^{-3} *amperes, (I)* produces a voltage drop of 7.24×10^{-4} *volts, (V)* across a resistor.

67. The voltage across an $8.35 \times 10^5 \, \Omega$ resistor is 2.95×10^{-3} V. Find the power dissipated in the resistor, using the equation $P = \frac{V^2}{R}$.

68. Three capacitors, 8.26×10^{-6} farad (F), 1.38×10^{-7} F, and 5.93×10^{-5} F, are wired in parallel. Find the equivalent capacitance by using the following equation $C = C_1 + C_2 + C_3 + \ldots$.

69. A wire 4.75×10^3 cm long is seen to stretch, when loaded, 9.55×10^{-2} cm. Find the strain in the wire, using the equation $\epsilon = \frac{\Delta L}{L}$.

70. How long will it take a rocket travelling at a rate of 3.2×10^6 mph (miles per hour) to travel the 3.8×10^8 miles from the earth to the moon? Use the basic physics equation: $D = R \times t$, *or Distance = Rate × time.* (Remember to manipulate the equation to calculate for time.)

71. The oil sands reserves of Canada are estimated at 3.5×10^{11} tonnes (t). How long (in years) would this supply last at a rate of consumption of 9.1×10^8 tonnes per year?

1–8   Units of Measurement

Systems of Units

A *unit* is a standard of measurement, such as the metre, inch, second, pound, or kilogram. The two main systems of units in use are the *metric system* SI (metres, kilograms, litres, etc.) and the *imperial system* (feet, pounds, gallons, etc.). SI is the *International System of Units,* or *Système international d'unités.* In addition, you may occasionally deal with some special units, such as a *square* of roofing material, and some obsolete units, such as *rods* and *chains.*

The word *denominate* comes from the root *nominare,* to name.

Numbers having units of measure are sometimes called *denominate* numbers. Our main task in this section is to learn how to convert such a number from one unit of measurement to another, say, from feet to metres.

Systems of Units

Each physical quantity can be expressed in any one of a bewildering variety of units. Length, for example, can be measured in metres, inches, nautical miles, angstroms, feet, and so on. In this section we will convert from one metric unit to another, from one imperial unit to another, and from one system to the other.

Symbols and Abbreviations

Most units have symbols (or abbreviated forms, in the case of non-SI units), so that we do not have to write the full word. Thus, the symbol for metres is m, and the symbol for ohms is Ω. In this section we will usually give the full word *and* the abbreviation. The abbreviations for all of the units in this text are given in a table of conversion factors in Appendix B.

Conversion Factors

We can convert from any unit of length to any other unit of length by multiplying by a suitable number, called a *conversion factor*.

◆◆◆ **Example 77:** Convert 654.5 feet (ft.) to metres (m).

Solution: From Appendix B we find the relation between feet and metres:

$$0.3048\text{m} = 1\text{ft.}$$

Dividing both sides by 1 foot, we get the conversion factor.

$$\frac{(0.3048 \text{ m})}{(1 \text{ ft.})} = 1$$

We know that we can multiply any quantity by 1 without changing the value of that quantity. Thus, if we multiply our original quantity (654.5 ft.) by the conversion factor (0.3048 m/ft.), we do not change the value of the original quantity. We will, however, change the units. Multiplying yields

$$654.5 \text{ ft.} = 654.5 \text{ f\!t.} \times \frac{(0.3048 \text{ m})}{(1 \text{ f\!t.})} = 199.5\text{m}$$

Note that we have rounded our answer to four significant digits because all numbers used in the calculation have at least four significant digits (the 0.3048 is exact). ◆◆◆

There are actually six different conversion factors for the foot-metre conversion. This text uses the internationally accepted definition of 1 ft = 0.3048 m. Normally you will use a conversion factor that has at least as many significant digits as your source measurement. Calculators are helpful; but beware of the variations in the imperial and US measurements.

Suppose that in the first step of Example 77, we had divided both sides by 0.3048 m instead of by 1 ft. We could have got another conversion factor:

$$\frac{(1 \text{ ft.})}{(0.3048 \text{ m})} = 1$$

Thus, each relation between two units of measurement gives us *two* conversion factors. Each of these is sometimes called a *unity ratio*—that is, a fraction that is equal to unity, or 1. Since each of these conversion factors is equal to 1, we may multiply any quantity by a conversion factor without changing the value of that quantity. We will, however, cause the units to change. But which of the two conversion factors should we use? It is simple: *Multiply by the conversion factor that will cancel the units you wish to eliminate.*

Significant Digits

You should try to use a conversion factor that is exact, or one that contains at least as many significant digits as (or, preferably, one more than) your original number. Then you should round your answer to as many significant digits as in the original number.

◆◆◆ **Example 78:** Convert 134 acres to hectares (ha).

Solution: From Appendix B we find the equation

$$1 \text{ ha} = 2.471 \text{ acres}$$

We must write our conversion factor so that the unwanted unit (acres) is in the denominator, so the acres will cancel. Our conversion factor is thus

$$\frac{(1\ \text{ha})}{(2.471\ \text{acres})}$$

Multiplying, we obtain

$$134\ \text{acres} = 134\ \cancel{\text{acres}} \times \frac{(1\ \text{ha})}{(2.471\ \cancel{\text{acres}})}$$

$$= 54.2\ \text{ha}$$

The conversion factor used here is not exact, but because it has more significant digits than the original number, we have rounded our answer to three significant digits.   ◆◆◆

Using More Than One Conversion Factor

Sometimes you may not be able to find a *single* conversion factor linking the units you want to convert. You may have to use *more than one*.

◆◆◆ **Example 79:** Convert 7375 yards (yd.) to nautical miles (M).

Solution: In Appendix B we find no conversion factor between nautical miles and yards, but we see that

$$1M = 1852m \quad \text{and} \quad 0.9144\ \text{m} = 1\text{yd.}$$

So

$$7375\ \text{yd.} = 7375\ \cancel{\text{yd.}} \times \frac{0.9144\ \cancel{\text{m}}}{1\ \cancel{\text{yd.}}} \times \frac{1M}{1852\ \cancel{\text{m}}} = 3.641\,M \qquad\qquad ◆◆◆$$

Metric Units

SI is based on the metric system of weights and measures that was developed in France in 1793. It has since been adopted by most countries of the world and is widely used in scientific work in the United States.

The basic SI unit of length is the *metre* (m). The basic unit of mass is the *kilogram* (kg). The basic unit of time is the *second* (s). The remaining basic SI units are the *ampere* (A), *kelvin* (K), *mole* (mol), and *candela* (cd). Among the more common of the derived units are the *litre* (L) and the *millilitre* (mL), units of volume; the *gram* (g) and the *tonne* (t), units of mass; and the *hectare* (ha), a unit of area.

Metric Prefixes

Converting between metric units is made easy because larger and smaller units are related to the basic units by *factors of 10*. These larger or smaller units are indicated by placing a *prefix* before the basic unit. A prefix is a group of letters placed at the beginning of a word to modify the meaning of that word. SI prefixes are given in Table 1-2.

◆◆◆ **Example 80:**

(a) A *kilo*metre (km) is a thousand metres because *kilo* means one thousand (10^3).

$$1\ \text{km} = 1000\ \text{m}$$

(b) A *centi*metre (cm) is one-hundredth of a metre, because *centi* means one hundredth.

$$1\ \text{cm} = 1/100\ \text{m}$$

(c) A *milli*metre (mm) is one-thousandth of a metre, because *milli* means one thousandth.

$$1\ \text{mm} = 1/1000\ \text{m} \qquad\qquad ◆◆◆$$

TABLE 1–2 Metric prefixes

Amount	Multiples and Submultiples	Prefix	Symbol	Pronunciation	Meaning
1 000 000 000 000	10^{12}	tera	T	ter′a	One trillion times
1 000 000 000	10^{9}	giga	G	gi′ga	One billion times
1 000 000	10^{6}	mega	M*	meg′a	One million times
1000	10^{3}	kilo	k*	kil′o	One thousand times
100	10^{2}	hecto	h	hek′to	One hundred times
10	10	deka	da	dek′a	Ten times
0.1	10^{-1}	deci	d	des′i	One tenth of
0.01	10^{-2}	centi	c*	sen′ti	One hundredth of
0.001	10^{-3}	milli	m*	mil′i	One thousandth of
0.000 001	10^{-6}	micro	μ*	mī′kro	One millionth of
0.000 000 001	10^{-9}	nano	n	nan′o	One billionth of
0.000 000 000 001	10^{-12}	pico	p	pē′co	One trillionth of
0.000 000 000 000 001	10^{-15}	femto	f	fem′to	One quadrillionth of
0.000 000 000 000 000 001	10^{-18}	atto	a	at′to	One quintillionth of

*Most commonly used.

Converting Between Metric Units

Converting from one metric unit to another is usually a matter of multiplying or dividing by a power of 10. Most of the time, the names of the units will tell how they are related, so we do not even have to look them up.

◆◆◆ **Example 81:** Convert 72 925 metres (m) to kilometres (km).

Solution: A *kilo*metre is a thousand metres.

$$\frac{1 \text{ km}}{1000 \text{ m}} = 1$$

So, as before,

$$72\ 925 \text{ m} = 72\ 925 \text{ m} \times \frac{1 \text{ km}}{1000 \text{ m}} = 72.925 \text{ km}$$

◆◆◆

For more unusual metric units, simply look up the conversion factor in a table. One of the simplifying features of SI is that each quantity is assigned only one unit that can be expressed in SI units. Although you will still encounter them, the use of older metric units such as the *dyne* (unit of force) and *are* (unit of area) is discouraged. Note also that non-approved units do not have assigned symbols.

◆◆◆ **Example 82:** Convert 2.75×10^{5} dynes to newtons (N).

Solution: We cannot tell from their names how these two units are related to each other. However, from Appendix B we find that

$$1 \text{ N} = 10^{5} \text{ dynes}$$

Converting in the usual way, we obtain

$$2.75 \times 10^{5} \text{ dynes} = 2.75 \times 10^{5} \text{ dynes} \times \frac{(1 \text{ N})}{(10^{5} \text{ dynes})} = 2.75 \text{ N}$$

◆◆◆

Converting from One Imperial Unit to Another

You should be aware that there is a difference in some units between Canadian (imperial) and U.S measures—including the gallon, quart, pint, and fluid ounce.

◆◆◆ **Example 83:** Convert 2.84 cubic feet (cu. ft.) to U.S. gallons (gal.).

Solution: From Appendix B we find

$$1 \text{ cu. ft.} = 7.481 \text{ gal. (U.S.)}$$

Converting gives

$$2.84 \text{ cu. ft.} = 2.84 \text{ cu. ft.} \times \frac{(7.481 \text{ gal.})}{(1 \text{ cu. ft.})} = 21.2 \text{ gal.}$$

rounded to three significant digits.   ◆◆◆

Converting Areas and Volumes

Length may be given in, say, centimetres (cm), *area* in square centimetres (cm²), and *volume* in cubic centimetres (cm³). To get a conversion factor for area or volume, simply square or cube the conversion factor for length.

If we take the equation

$$1 \text{ in.} = 2.54 \text{ cm}$$

and square both sides, we get

$$(1 \text{ in.})^2 = (2.54 \text{ cm})^2$$

or

$$1 \text{ sq. in.} = 6.4516 \text{ cm}^2$$

This gives us a conversion between square centimetres and square inches.

◆◆◆ **Example 84:** Convert 864 square yards to acres.

Solution: Appendix B has no conversion for square yards. However,

$$1 \text{ yd.} = 3 \text{ ft. and } 1 \text{ acre} = 43\,560 \text{ sq. ft.}$$

Squaring 1 yd. yields

$$1 \text{ sq. yd.} = (3 \text{ ft.})^2 = 9 \text{ sq. ft.}$$

So

$$864 \text{ sq. yd.} = 864 \text{ sq. yd.} \times \frac{9 \text{ sq. ft.}}{1 \text{ sq. yd.}} \times \frac{1 \text{ acre}}{43\,560 \text{ sq. ft.}} = 0.179 \text{ acres}$$ ◆◆◆

Converting Rates to Other Units

A *rate* is the amount of one quantity expressed *per unit of some other quantity*. Some rates, with typical units, are

rate of travel (km/h) or (mi./h)	*flow rate (m³/s) or (gal./min)*
application rate (kg/ha)	*unit price ($/kg) or (cents/100 ml)*

Each rate contains *two* units of measure; kilometres per hour, for example, has *kilometres* in the numerator and *hours* in the denominator. It may be necessary to convert *either* or *both* of those units to other units. Sometimes a single conversion factor can be found (such as 1 km/h = 0.2778 m/s), but more often you will have to convert each unit with a *separate* conversion factor.

◆◆◆ **Example 85:** A certain chemical is to be added to a pool at the rate of 1.75 lb. per cubic foot of water. Convert this to grams of chemical per litre of water.

Solution: We write the original quantity as a fraction and multiply by the appropriate factors, themselves written as fractions.

$$1.75 \text{ lb./cu. ft.} = \frac{1.75 \text{ lb.}}{\text{cu. ft.}} \times \frac{453.6 \text{ g}}{\text{lb.}} \times \frac{(1 \text{ cu. ft.})}{(28.32 \text{ L})} = 28.0 \text{ g/L}$$ ◆◆◆

Be sure to write the original quantity as a built-up fraction, $\frac{a}{b}$, rather than on a single line, a/b. This will greatly reduce your chances of making an error.

Exercise 8 ◆ Units of Measurement

Convert the following imperial units.

1. 152 inches to feet
2. 0.153 mile to yards
3. 762.0 feet to inches
4. 627 feet to yards
5. 29 tons to pounds
6. 88.90 pounds to ounces
7. 89 600 pounds to tons
8. 8552 ounces to pounds

Convert the following metric units. Write your answer in scientific notation if the numerical value is greater than 1000 or less than 0.1.

9. 364 000 metres to kilometres
10. 0.000 473 volt to millivolts
11. 735 900 grams to kilograms
12. 7.68×10^{-5} kilowatt to watts
13. 6.2×10^{9} ohms to megohms
14. 825×10^{4} newtons to kilonewtons
15. 9348 picofarads to microfarads
16. 84 398 nanoseconds to milliseconds

Convert between imperial and metric units.

17. 364.0 metres to feet
18. 6.83 inches to millimetres
19. 7.35 pounds to kilograms
20. 2.55 horsepower to kilowatts
21. 4.66 gallons (imp.) to litres
22. 1.28×10^{3} newtons to pounds-force
23. 3.94 yards to metres
24. 834 cubic centimetres to gallons (U.S.)

Convert the following areas and volumes.

25. 2840 square yards to acres
26. 48 300 square metres to hectares
27. 24.8 square feet to square metres
28. 3.72 square metres to square feet
29. 0.982 square kilometres to acres
30. 5.93 acres to square metres
31. 7.360 cubic feet to cubic inches
32. 4.83 cubic metres to cubic yards
33. 73.8 cubic yards to cubic metres
34. 8.220 gallons (U.S.) to cubic feet
35. 267 cubic millimetres to cubic inches
36. 112 litres to gallons (imp.)

Convert units on the following time rates.

37. 4.86 feet per second to miles per hour
38. 777 gallons (U.S) per minute to cubic metres per hour
39. 66.2 miles per hour to kilometres per hour
40. 52.0 knots to miles per minute
41. 953 births per year to births per week

Convert units on the following unit prices.

42. $1.25 per gram to dollars per kilogram
43. $800 per acre to cents per square metre
44. $3.54 per pound to cents per ounce
45. 238 cents per pound to dollars per tonne

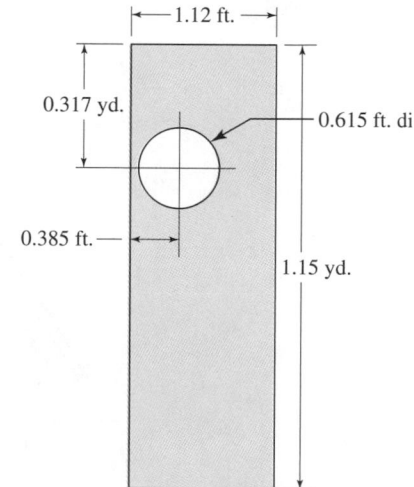

FIGURE 1–11

Applications

46. Convert all of the dimensions for the part in Fig. 1–11 to inches.

47. The jet fuel tank in Fig. 1–12 has a volume of 15.7 cubic feet. How many litres of jet fuel will it hold?

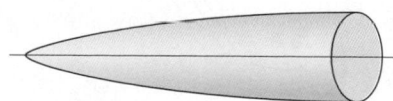

FIGURE 1–12

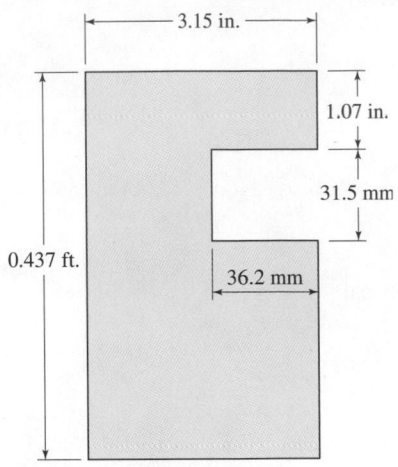

FIGURE 1–13

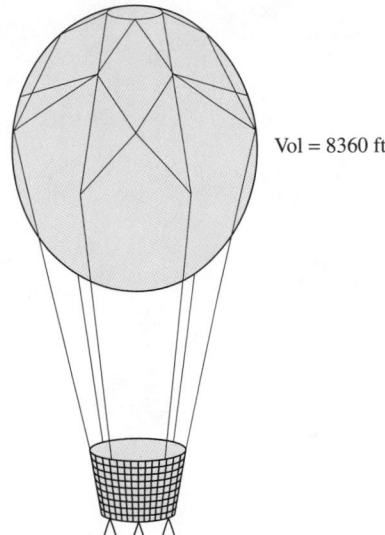

Vol = 8360 ft³

FIGURE 1–16

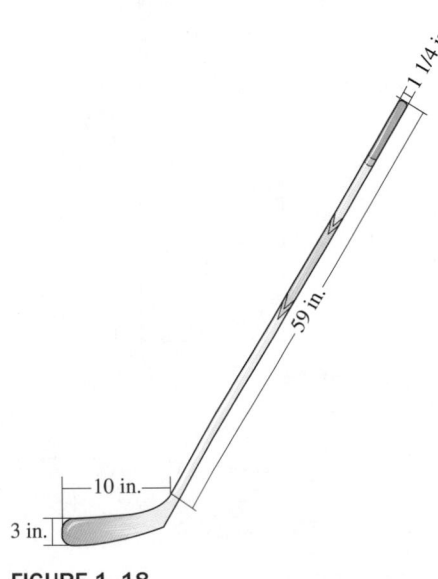

FIGURE 1–18

48. A certain circuit board weighs 0.176 pounds. Find its weight in grams.

49. A certain laptop computer weighs 6.35 kilograms. What is its weight in pounds?

50. A generator has an output of 5.34×10^6 millivolts. What is the output in kilovolts?

51. Convert all of the dimensions in Fig. 1–13 to centimetres.

52. The surface area of a certain lake, shown in Fig. 1–14, is 7361 square yards. Convert this to square metres.

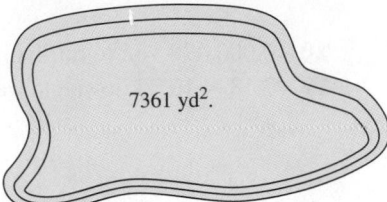

7361 yd².

FIGURE 1–14

53. A solar collector, shown in Fig. 1–15, has an area of 8834 square inches. Convert this to square metres.

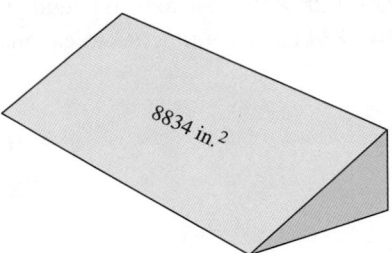

8834 in.²

FIGURE 1–15

54. The volume of a balloon, shown in Fig. 1–16, is 8360 cubic feet. Convert this to cubic metres.

55. The volume of a certain gasoline tank, shown in Fig. 1–17, is 9274 cubic centimetres. Convert this to imperial gallons.

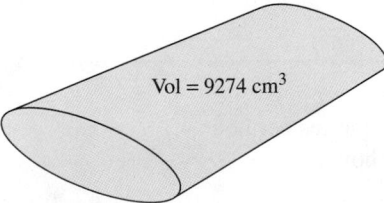

Vol = 9274 cm³

FIGURE 1–17

56. An airplane is cruising at a speed of 785 miles per hour. Convert this speed to kilometres per hour.

57. Algonquin Park in Ontario has an area of 765 345 ha. Convert this area to square kilometres and also to square miles, using the built-in unit conversion features of your calculator.

58. A "tallboy" can of beer holds 900 ml. Express this volume in imperial fluid ounces using the built in unit conversion features of your calculator.

59. A fully grown male black bear can weigh 560 pounds. Convert this weight to kilograms, using the built in unit conversion features of your calculator.

60. Convert all of the dimensions of the hockey stick shown in Figure 1–18 to cm.

1–9   Substituting into Equations and Formulas

Substituting into Equations
We get an *equation* when two expressions are set equal to each other.

◆◆◆ **Example 86:** $x = 5a - 2b + 3c$ is an equation that enables us to find x if we know a, b, and c. ◆◆◆

We will study equations in detail later, but for now we will simply substitute into equations and use our calculators to compute the result. To *substitute into an equation* means to replace the letter quantities in an equation by their given numerical values and to perform the computation indicated.

◆◆◆ **Example 87:** Substitute the values $a = 5$, $b = 3$, and $c = 6$ into the equation

$$x = \frac{3a + b}{c}$$

Solution: Substituting, we obtain

$$x = \frac{3(5) + 3}{6} = \frac{18}{6} = 3$$ ◆◆◆

When substituting approximate numbers, be sure to round your answer to the proper number of digits. Treat any integers in the equation as exact numbers.

Substituting into Formulas
A *formula* is an equation expressing some general mathematical or physical fact, such as the formula for the area of a circle of radius r.

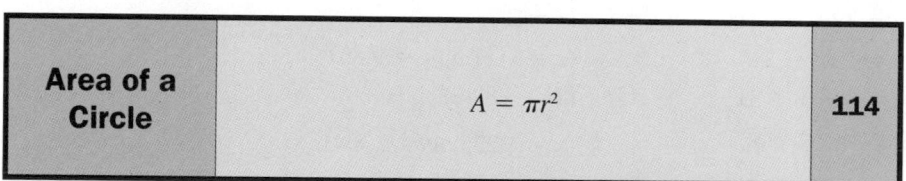

| Area of a Circle | $A = \pi r^2$ | 114 |

We substitute into formulas just as we substituted into equations, except that we now carry *units* along with the numerical values. You will often need conversion factors to make the units cancel properly, so that the answer will be in the desired units.

◆◆◆ **Example 88:** A tensile load of 4500 lb. is applied to a bar that is 5.2 yd. long and has a cross-sectional area of 11.6 cm² (Fig. 1–19). The elongation is 0.38 mm. Using Eq. A54, find the modulus of elasticity E in pounds per square inch.

Solution: Substituting the values *with units* into Eq. A54, we obtain

$$E = \frac{PL}{ae} = \frac{4500 \text{ lb.} \times 5.2 \text{ yd.}}{11.6 \text{ cm}^2 \times 0.38 \text{ mm}}$$

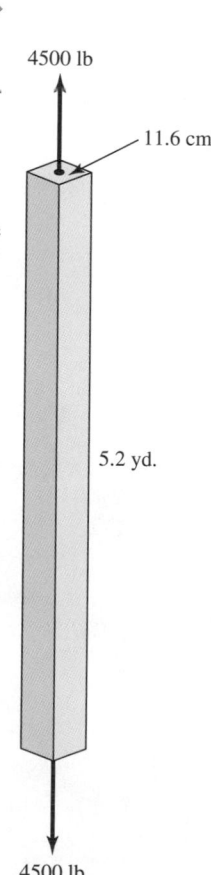

4500 lb

11.6 cm²

5.2 yd.

4500 lb

FIGURE 1–19

Notice that we have a length (5.2 yd.) in the numerator and a length (0.38 mm) in the denominator. To make these units cancel, we use the conversion factors

$$25.4 \text{ mm} = 1 \text{ in.}$$

and

$$36 \text{ in.} = 1 \text{ yd.}$$

Also, our answer is to have square inches in the denominator, not square centimetres. So we use another conversion factor.

$$6.452 \text{ cm}^2 = 1 \text{ sq. in.}$$

| | | | | |

$$E = \frac{4500 \text{ lb.} \times 5.2 \text{ yd.}}{11.6 \text{ cm}^2 \times 0.38 \text{ mm}} \times \frac{25.4 \text{ mm}}{\text{in.}} \times \frac{36 \text{ in.}}{\text{yd.}} \times \frac{6.452 \text{ cm}^2}{\text{sq. in.}}$$

$$= 31\ 000\ 000 \text{ lb./sq. in.} \quad \text{(rounded to two digits)} \qquad \text{◆◆◆}$$

> **Common Error**
>
> Students often neglect to include *units* when substituting into a formula, with the result that the units often do not cancel properly.

If the units to be used in a certain formula are specified, convert all quantities to those specified units before substituting into the formula.	

Exercise 9 ◆ Substituting into Equations and Formulas

Substitute the given integers into each equation. Do not round your answer.

1. $y = 5x + 2$ $(x = 3)$
2. $y = 2m^2 - 3m + 5$ $(m = -2)$
3. $y = 2a - 3x^2$ $(x = 3, a = -5)$
4. $y = 3x^3 - 2x^2 + 4x - 7$ $(x = 2)$
5. $y = 2b + 3w^2 - 5z^3$ $(b = 3, w = -4, z = 2)$
6. $y = \frac{r^2}{x} - \frac{x^3}{r} + \frac{w}{x^2}$ $(x = 5, w = 3, r = -4)$

Substitute the given approximate numbers into each equation. Treat the constants in the equations as exact numbers. Round your answer to the proper number of digits.

7. $y = 7x - 5$ $(x = 2.73)$
8. $y = 2w^2 - 3x^2$ $(x = -11.5, w = 9.83)$
9. $y = 8 - x + 3x^2$ $(x = -8.49)$
10. $y = \sqrt[3]{8x + 7w}$ $(x = 1.225, w = 2.304)$
11. $y = \sqrt{x^3 - 3x}$ $(x = 4.25)$
12. $y = (w - 2x)^{1.6}$ $(x = 1.8, w = 7.2)$

13. Use Eq. A9 to find to the nearest dollar the amount to which $3,000 will accumulate in 5 years at a simple interest rate of 6.5%.
14. Using Eq. A18, find the displacement, after 1.30 s, of a body thrown downward with a speed of 3.60 m/s. Use $g = 9.807$ m/s².
15. Using Eq. A50, convert 128 °F to degrees Celsius.
16. A bar 15.2 m long having a cross-sectional area of 12.7 cm² is subject to a tensile load of 22 500 N (Fig. 1–20). The elongation is 2.75 mm. Use Eq. A54 to find the modulus of elasticity in newtons per square centimetre.
17. Use Eq. A10 to find to the nearest dollar the amount y obtained when $9,570 is allowed to accumulate for 5 years at a compound interest rate of 6.75%.

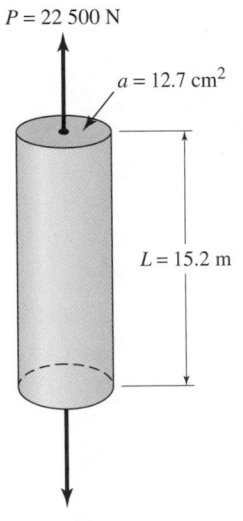

$P = 22\ 500$ N

$a = 12.7$ cm²

$L = 15.2$ m

FIGURE 1–20 A bar in tension.

18. The resistance of a copper coil is 775 Ω at 20.0 °C. The temperature coefficient of resistance is 0.003 93 at 20.0 °C. Use Eq. A70 to find the resistance at 80.0 °C.

1–10 Percentage

Definition of Percent

The word *percent* means *by the hundred*, or *per hundred*. A percent thus gives the number of parts in every hundred.

◆◆◆ **Example 89:** If we say that a certain concrete mix is 12% cement by weight, we mean that 12 kg out of every 100 kg of mix will be cement. ◆◆◆

Rates

The word *rate* is often used to indicate a percent, or percentage rate, as in "rejection rate," "rate of inflation," or "growth rate."

◆◆◆ **Example 90:** A failure rate of 2% means that, on average, 2 parts out of every 100 would be expected to fail. ◆◆◆

Percent as a Fraction

Percent is another way of expressing a *fraction* with 100 as the denominator.

◆◆◆ **Example 91:** If we say that a builder has finished 75% of a house, we mean that he has finished $\frac{75}{100}$ (or $\frac{3}{4}$) of the house. ◆◆◆

Converting Decimals to Percent

Before working some percentage problems, let us first get some practice in converting decimals and fractions to percents, and vice versa. To convert decimals to percent, simply move the decimal point two places to the right and affix the percent symbol (%).

◆◆◆ **Example 92:**

(a) $0.75 = 75\%$

(b) $3.65 = 365\%$

(c) $0.003 = 0.3\%$

(d) $1.05 = 105\%$ ◆◆◆

Converting Fractions or Mixed Numbers to Percent

First write the fraction or mixed number as a decimal, and then proceed as above.

◆◆◆ **Example 93:**

(a) $\frac{1}{4} = 0.25 = 25\%$

(b) $\frac{5}{2} = 2.5 = 250\%$

(c) $1\frac{1}{4} = 1.25 = 125\%$ ◆◆◆

Converting Percent to Decimals

To convert percent to decimals, move the decimal point two places to the left and remove the percent sign.

◆◆◆ **Example 94:**

(a) $13\% = 0.13$

(b) $4.5\% = 0.045$

(c) $155\% = 1.55$

(d) $27\frac{3}{4}\% = 0.2775$

(e) $200\% = 2$ ◆◆◆

Converting Percent to a Fraction

Write a fraction with 100 in the denominator and the percent in the numerator. Remove the percent sign and reduce the fraction to lowest terms.

◆◆◆Example 95:

(a) $75\% = \frac{75}{100} = \frac{3}{4}$

(b) $87.5\% = \frac{87.5}{100} = \frac{875}{1000} = \frac{7}{8}$

(c) $125\% = \frac{125}{100} = \frac{5}{4} = 1\frac{1}{4}$

◆◆◆

Amount, Base, and Rate

Percentage problems always involve three quantities:

1. The *percent rate, P*.
2. The *base, B*: the quantity we are taking the percent of.
3. The *amount, A*, that we get when we take the percent of the base, also called the *percentage*.

In a percentage problem, you will know two of these three quantities (amount, base, and rate), and you will be required to find the third. This is easily done because the rate, base, and amount are related by the following equation:

Percentage	Amount = Rate × Base $A = PB$ where P is expressed as decimal	12

Finding the Amount When the Base and the Rate Are Known

We substitute the given base and rate into Eq. 12 and solve for the amount.

◆◆◆Example 96: What is 35.0 percent of 80.0?

Solution: In this problem the rate is 35.0%, so

$$P = 0.350$$

But is 80.0 the amount or the base?

Tip	In a particular problem, if you have trouble telling which number is the base and which is the amount, look for the key phrase *percent of*. The quantity following this phrase is *always the base*.

Thus we look for the key phrase "percent of."

$$\xrightarrow{\hspace{2cm}} \text{base}$$
What is 35.0 percent of $\boxed{80.0}$?

Since 80.0 immediately follows *percent of*,

$$B = 80.0$$

From Eq. 12,

$$A = PB = (0.350)80.0 = 28.0$$

Common Error	Do not forget to convert the percent rate to a *decimal* when using Eq. 12.

✦✦✦ **Example 97:** Find 3.74% of 5710.

Solution: We substitute into Eq. 12 with

$$P = 0.0374 \text{ and } B = 5710$$

So

$$A = PB = (0.0374)(5710) = 214$$

after rounding to three significant digits. ✦✦✦

Finding the Base When a Percent of It Is Known

We see from Eq. 12 that the base equals the amount divided by the rate (expressed as a decimal), or $B = A/P$.

✦✦✦ **Example 98:** 12% of what number is 78?

Solution: First find the key phrase.

$$12 \text{ percent of } \boxed{\text{what number}} \text{ is } 78?$$
$$\longrightarrow \quad \text{base}$$

It is clear that we are looking for the base. So

$$A = 78 \text{ and } P = 0.12$$

By Eq. 12,

$$B = \frac{A}{P} = \frac{78}{0.12} = 650$$ ✦✦✦

✦✦✦ **Example 99:** 140 is 25% of what number?

Solution: From Eq. 12,

$$B = \frac{A}{P} = \frac{140}{0.25} = 560$$ ✦✦✦

Finding the Percent That One Number Is of Another Number

From Eq. 12, the rate equals the amount divided by the base, or $P = A/B$.

✦✦✦ **Example 100:** 42.0 is what percent of 405?

Solution: By Eq. 12, with $A = 42.0$ and $B = 405$,

$$P = \frac{A}{B} = \frac{42.0}{405} = 0.104 = 10.4\%$$ ✦✦✦

✦✦✦ **Example 101:** What percent of 1.45 is 0.357?

Solution: From Eq. 12,

$$P = \frac{A}{B} = \frac{0.357}{1.45} = 0.246 = 24.6\%$$ ✦✦✦

Percent Change

Percentages are often used to compare two quantities. You often hear statements such as the following:

The price of steel rose 3% over last year's price.
The weights of two cars differed by 20%.
Production dropped 5% from last year.

When the two numbers being compared involve a *change* from one to the other, the *original value* is usually taken as the base.

Percent Change	Percent Change = $\dfrac{\text{New Value} - \text{Original Value}}{\text{Original Value}} \times 100$	13

◆◆◆ Example 102: A certain price rose from $1.55 to $1.75. Find the percentage change in price.

Be sure to show the *direction* of change with a plus or a minus sign, or with words such as *increase* or *decrease*.

Solution: We use the original value, $1.55, as the base. From Eq. 13,

$$\text{percent change} = \frac{1.75 - 1.55}{1.55} \times 100 = 12.9\% \text{ increase} \qquad \text{◆◆◆}$$

A common type of problem is to *find the new value* when the original value is changed by a given percent. We see from Eq. 13 that

$$\text{new value} = \text{original value} + (\text{original value}) \times (\text{percent change})$$

◆◆◆ Example 103: Find the cost of a $156.00 suit after the price increases by $2\frac{1}{2}\%$.

Solution: The original value is 156.00, and the percent change, expressed as a decimal, is 0.025. So

$$\text{new value} = 156.00 + (156.00)(0.025) = \$159.90 \qquad \text{◆◆◆}$$

Percent Efficiency

The power output of any machine or device is always *less* than the power input because of inevitable power losses within the device. The *efficiency* of the device is a measure of those losses.

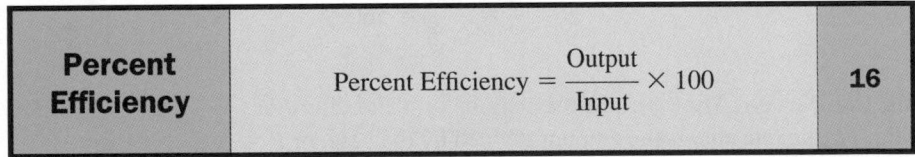

Percent Efficiency	Percent Efficiency = $\dfrac{\text{Output}}{\text{Input}} \times 100$	16

◆◆◆ Example 104: A certain electric motor consumes 865 W and has an output of 1.12 hp (Fig. 1–21). Find the efficiency of the motor. (1 hp = 746 W)

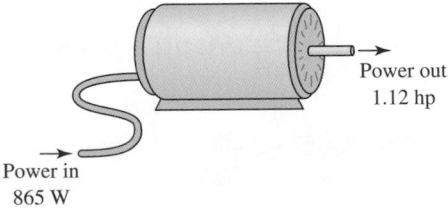

Power out
1.12 hp

Power in
865 W

FIGURE 1–21

Solution: Since output and input must be in the same units, we must convert either to horsepower or to watts. Converting the output to watts, we obtain

$$\text{output} = 1.12\text{hp}\left(\frac{746\text{W}}{\text{hp}}\right) = 836\text{W}$$

By Eq. 20,

$$\text{percent efficiency} = \frac{836}{865} \times 100 = 96.6\% \qquad \text{◆◆◆}$$

Percent Error

The accuracy of measurements is often specified by the *percent error*. The percent error is the difference between the measured value and the known or "true" value, expressed as a percent of the known value.

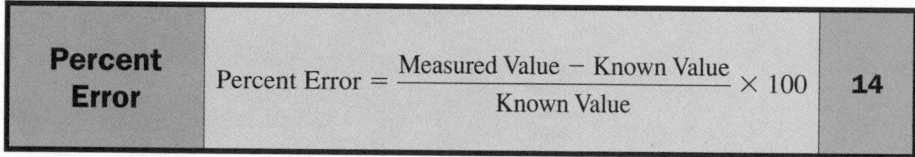

Percent Error	Percent Error $= \dfrac{\text{Measured Value} - \text{Known Value}}{\text{Known Value}} \times 100$	**14**

◆◆◆ **Example 105:** A laboratory weight that is certified to be 500.0 g is placed on a scale (Fig. 1–22). The scale reading is 507.0 g. What is the percent error in the reading?

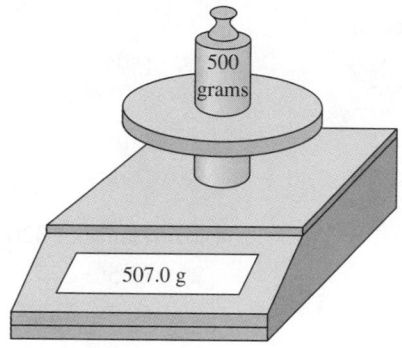

FIGURE 1–22

Solution: From Eq. 14,

$$\text{percent error} = \frac{507.0 - 500.0}{500.0} \times 100 = 1.4\% \text{ high}$$

◆◆◆

Percent Concentration

The following equation applies to a mixture of two or more ingredients:

Percent Concentration	Percent Concentration of Ingredient A $= \dfrac{\text{Amount of } A}{\text{Amount of Mixture}} \times 100$	**15**

As with percent change, be sure to specify the *direction* of the error.

◆◆◆ **Example 106:** A certain fuel mixture contains 18.9 litres of alcohol and 84.7 litres of gasoline. Find the percentage of gasoline in the mixture.

Solution: The total amount of mixture is

$$18.9 \text{ L} + 84.7 \text{ L} = 103.6 \text{ L}$$

So by Eq. 15,

$$\text{percent gasoline} = \frac{84.7}{103.6} \times 100 = 81.8\%$$

◆◆◆

Common Error	The denominator in Eq. 15 must be the *total amount* of mixture, or the sum of *all* of the ingredients. Do not use just one of the ingredients.

Exercise 10 ◆ Percentage

Conversions

Convert each decimal to a percent.

1. 3.72   **2.** 0.877   **3.** 0.0055   **4.** 0.563

Convert each fraction to a percent. Round to three significant digits.

5. $\frac{2}{5}$ **6.** $\frac{3}{4}$ **7.** $\frac{7}{10}$ **8.** $\frac{3}{7}$

Convert each percent to a decimal.

9. 23%   **10.** 2.97%   **11.** $287\frac{1}{2}\%$ **12.** $6\frac{1}{4}\%$

Convert each percent to a fraction.

13. 37.5%   **14.** $12\frac{1}{2}\%$ **15.** 150%   **16.** 3%

Finding the Amount

Find

17. 41.1% of 255 tonnes.   **18.** 15.3% of 326 miles.

19. 33.3% of 662 kilograms.   **20.** 12.5% of 72.0 gallons.

21. 35.0% of 343 litres.   **22.** 50.8% of $245.

23. A resistance, now 7250 Ω, is to be increased by 15.0%. How much resistance should be added?

24. It is estimated that half of one percent of the earth's surface receives more energy than the total projected needs for the year 2015. Assuming the earth's surface area to be 1.97×10^8 sq. mi., find the required area in acres.

25. As an incentive to install solar equipment, a tax credit of 42% of the first $1,100 and 25% of the next $6,400 spent on solar equipment is proposed. How much credit, to the nearest dollar, would a homeowner get when installing $5,500 worth of solar equipment?

26. How much metal will be obtained from 375 tonnes of ore if the metal is 10.5% of the ore?

27. In the province of Ontario, nuclear energy provides 68.7% of the total of 5472 MW of electrical energy consumed during a hot summer day. How many megawatts of electrical energy did the nuclear generating stations produce?

Finding the Base

Find the number of which

28. 86.5 is 16.7%.   **29.** 45.8 is 1.46%.

30. 1.22 is 1.86%.   **31.** 55.7 is 25.2%.

32. 66.6 is 66.6%.   **33.** 58.2 is 75.4%.

34. A federal government report on an experimental electric car gives the range of the car as 161 km and states that this is "49.5% better than on earlier electric vehicles." What was the range of earlier electric vehicles?

35. A man withdrew 25% of his bank deposits and spent 33% of the money withdrawn in the purchase of a radio worth $25. How much money did he have in the bank?

36. Solar panels provide 65% of the heat for a certain building. If $225 per year is now spent for heating oil, what would have been spent if the solar panels were not used?

37. If the United States imports 9.14 million barrels of crude oil per day, and if this is 48.2% of its needs, how much oil is needed per day?

38. The Canadian Legion sells poppies as a fundraiser for Remembrance Day. In 2001, the year of the September 11 World Trade Center attack, the Legion sold 15 million poppies, an increase of 39% over the previous year. How many poppies did the Legion sell in 2000?

Finding the Rate

What percent of

39. 26.8 is 12.3?

40. 36.3 is 12.7?

41. 44.8 is 8.27?

42. 844 is 428?

43. 455 h is 152 h?

44. 483 t is 287 t?

45. A 50 500 L-capacity tank contains 5840 L of water (Fig. 1–23). Express the amount of water in the tank as a percentage of the total capacity.

46. In a journey of 1560 km, a person travelled 195 km by car and the rest of the distance by rail. What percent of the distance was travelled by rail?

47. A power supply has a dc output of 51 V with a ripple of 0.75 V peak to peak. Express the ripple as a percentage of the dc output voltage.

48. The construction of a factory cost $136,000 for materials and $157,000 for labour. What percentage of the total was the labour cost?

49. On a 9346 km campaign trip across Canada, a politician travelled 6178 km by plane and the balance by bus. What percent of the distance did the politician travel in a bus?

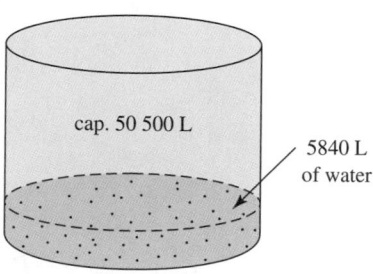

FIGURE 1–23

Percent Change

Find the percent change when a quantity changes

50. from 29.3 to 57.6.

51. from 107 to 23.75.

52. from 227 to 298.

53. from 0.774 to 0.638.

54. The temperature in a building rose from 19.0 °C to 21.0 °C during the day. Find the percent change in temperature.

55. A casting initially weighing 115 lb. has 22.0% of its material machined off. What is its final weight?

56. A certain common stock rose from a value of $35.50 per share to $37.62 per share. Find the percent change in value.

57. A house that costs $635 per year to heat has insulation installed in the attic, causing the fuel bill to drop to $518 per year. Find the percent change in fuel cost.

58. One string of 100 mini incandescent Christmas lights consumes 65 W of electricity. If you were to replace it with a string of 100 LED lights that consumes 5.8 W, what percent change in consumption have you achieved?

Percent Efficiency

59. A certain device (Fig. 1–24) consumes 92.5 W and delivers 62.0 W. Find its efficiency.

60. An electric motor consumes 1250 W. Find the horsepower it can deliver if it is 85.0% efficient. (1 hp = 746 W)

61. A water pump requires an input of 370 W and delivers 5000 kg of water per hour to a house 22 m above the pump. Find its efficiency. (1 W = 0.102 m·kg/s)

62. A certain speed reducer delivers 1.7 kW with a power input of 2.2 kW. Find the percent efficiency of the speed reducer.

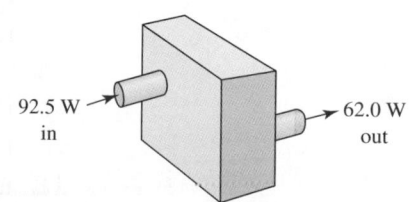

FIGURE 1–24

Percent Error

63. A certain quantity is measured at 125.0 units but is known to be actually 128.0 units. Find the percent error in the measurement.

64. A shaft is known to have a diameter of 35.000 mm. You measure it and get a reading of 34.725 mm. What is the percent error of your reading?

65. A certain capacitor has a working voltage of 125.0 V (dc) − 10%, + 150%. Between what two voltages would the actual working voltage lie?

66. A resistor is labelled as $550\overline{0}$ Ω with a tolerance of $\pm 5\%$. Between what two values is the actual resistance expected to lie?

Percent Concentration

67. A solution is made by mixing 75.0 L of alcohol with 125 L of water. Find the percent concentration of alcohol.

68. 8.0 cubic metres of cement is contained in a concrete mixture that is 12% cement by volume. What is the volume of the total mixture?

69. How many litres of alcohol are contained in 455 L of a gasohol mixture that is 5.5% alcohol by volume?

70. How many litres of gasoline are there in 155 gal. (U.S.) of a methanol-gasoline blend that is 10.0% methanol by volume?

Case Study Discussion—Precision and Accuracy in Chemical Samples

Our technologist assumed that more digits represent a more accurate representation of the amount of each sample. The problem is that when you use many digits, you're telling the reader that you have used instrumentation accurate enough to measure such precise amounts. Now if that is true, fine, you can use the number of digits that correctly represents the precision of your measuring instruments and techniques. In this case, however, the starting amount, 32.5 ml, is precise only to the nearest tenth of a millilitre. Additionally, the lab supervisor seemed to realize that the technologist was very unlikely to be able to divide the sample into so many samples accurate to a million-trillion-trillionth of a millilitre. Reporting such a long number falsely claims a precision that cannot be true.

⬩⬩⬩ CHAPTER 1 REVIEW PROBLEMS ⬩⬩⬩⬩⬩⬩⬩⬩⬩⬩⬩⬩⬩⬩⬩⬩⬩⬩⬩⬩⬩⬩⬩⬩⬩⬩⬩⬩

1. Combine: $1.435 - 7.21 + 93.24 - 4.1116$

2. Give the number of significant digits in:
    (a) 9.886   (b) 1.002   (c) 0.3500   (d) $15.\overline{0}00$

3. Multiply: $21.8(3.775 \times 1.07)$

4. Divide: $88.25 \div 9.15$

5. Find the reciprocal of 2.89.

6. Evaluate: $-|-4 + 2| - |-9 - 7| + 5$

7. Evaluate: $(9.73)^2$

8. Evaluate: $(7.75)^{-2}$

9. Evaluate: $\sqrt{29.8}$

10. Evaluate: $(123)(2.75) - (81.2)(3.24)$

11. Evaluate: $(91.2 - 88.6)^2$

12. Evaluate: $\left(\dfrac{77.2-51.4}{21.6-11.3}\right)^2$

13. Evaluate: $y = 3x^2 - 2x$ when $x = -2.88$

14. Evaluate: $y = 2ab - 3bc + 4ac$ when $a = 5$, $b = 2$, and $c = -6$

15. Evaluate: $y = 2x - 3w + 5z$ when $x = 7.72$, $w = 3.14$, and $z = 2.27$

16. Round to two decimal places.

    (a) 7.977   (b) 4.655   (c) 11.845   (d) 1.004

17. Round to three significant digits.

    (a) 179.2   (b) 1.076   (c) 4.8550   (d) 45 725

18. A news report states that a new hydroelectric generating station in Quebec will produce 140 terajoules of power per year (TJ/a) and that this power, for 20 years of operation, is equivalent to 2.0 million barrels of oil. Using these figures, how many gigajoules is each barrel of oil equivalent to?

19. A certain generator has a power input of 2.50 hp and delivers 1310 W. Find its percent efficiency.

20. Using Eq. A52, find the stress in megapascals for a force of 1.17×10^3 N distributed over an area of 3.14×10^3 mm^2.

21. Combine: $(8.34 \times 10^5) + (2.85 \times 10^6) - (5.29 \times 10^4)$

22. A train running at 25 km/h increases its speed 12.5%. How fast does it then go?

23. The average solar radiation in the continental United States is about 7.4×10^5 joules per square metre per hour (J/m^2·h). How many kilowatts would be collected by 15 acres of solar panels?

24. An item rose in price from $29.35 to $31.59. Find the percent increase.

25. Find the percent concentration of alcohol if 2.0 L of alcohol is added to 57 L of gasoline.

26. A bar, known to be 2.0000 cm in diameter, is measured at 2.0064 cm. Find the percent error in the measurement.

27. The Government of Alberta estimates that there are 170 billion barrels of oil in the oil sands deposits of the province. Express this amount in scientific notation.

28. Multiply: $(7.23 \times 10^5) \times (1.84 \times 10^{-3})$

29. Divide: -39.2 by $-0.003\ 826$

30. Convert 6930 Btu/h to kilowatts.

31. Divide: 8.24×10^{-3} by 1.98×10^7

32. What percent of 40.8 is 11.3?

33. Evaluate: $\sqrt[5]{82.8}$

34. Multiply: $(4.92 \times 10^6) \times (9.13 \times 10^{-3})$

35. Insert the proper sign of equality or inequality between $-\frac{2}{3}$ and -0.660.

36. Convert 0.000 426 mA to microamperes.

37. Find 49.2% of 4827.

38. Combine: $-385 - (227 - 499) - (-102) + (-284)$

39. Find the reciprocal of -0.582.

40. Find the percent change in a voltage that increased from $11\overline{0}$ V to 118 V.

41. A homeowner added insulation, and her yearly fuel consumption dropped from 2210 L to 1425 L. Her present oil consumption is what percent of the former?

42. Write in decimal notation: 5.28×10^4

43. Convert 49.3 pounds-force to newtons.

44. Evaluate: $(45.2)^{-0.45}$

45. Using Eq. A18, find the distance in metres travelled by a falling object in 5.25 s, thrown downward with an initial velocity of 284 m/min.

46. Write in scientific notation: 0.000 374

47. 8460 is what percent of 38 400?

48. The Canadian energy consumption of 4.6 million barrels oil equivalent per day is expected to climb to 6.0 million in 6 years. Find the percent increase in consumption.

49. The population of a certain town is 8118, which is $12\frac{1}{2}\%$ more than it was 3 years ago. What was the population then?

50. The temperature of a room rose from 20.5 °C to 22.0 °C. Find the percent increase.

51. Combine: $4.928 + 2.847 - 2.836$

52. Give the number of significant digits in 2003.0.

53. Multiply: 2.84(38.4)

54. Divide: $48.3 \div 2.841$

55. Find the reciprocal of 4.82.

56. Evaluate: $|-2| - |3 - 5|$

57. Evaluate: $(3.84)^2$

58. Evaluate: $(7.62)^{-2}$

59. Evaluate: $\sqrt{38.4}$

60. Evaluate: $(49.3 - 82.4)(2.84)$

61. Evaluate: $x^2 - 3x + 2$ when $x = 3$

62. Round 45.836 to one decimal place.

63. Round 83.43 to three significant digits.

64. Multiply: $(7.23 \times 10^5) \times (1.84 \times 10^{-3})$

65. Convert 36.82 in. to centimetres.

66. What percent of 847 is 364?

67. Evaluate: $\sqrt[3]{746}$

68. Find 35.8% of 847.

69. 746 is what percent of 992?

70. Write 0.002 74 in scientific notation.

71. Write 73.7×10^{-3} in decimal notation.

72. Evaluate: $(47.3)^{-0.26}$

73. Combine: $6.128 + 8.3470 - 7.236\ 12$

74. Give the number of significant digits in 6013.00.

75. Multiply: 7.184(16.8)

76. Divide: $78.7 \div 8.251$

77. Find the reciprocal of 0.825.

78. Evaluate: $|-5| - |2 - 7| + |-6|$

79. Evaluate: $(0.55)^2$

80. Write 23 800 in scientific notation and in engineering notation.

Writing

81. Suppose you have submitted a report that contains calculations in which you have rounded the answers according to the rules given in this chapter. Jones, your company rival, has sharply attacked your work, calling it "inaccurate" because you did not keep enough digits, and your boss seems to agree.

Write a memo to your boss defending your rounding practices. Point out why it is misleading to retain too many digits. Do not write more than one page. You may use numerical examples to prove your point.

Team Projects

You will find a team project suggested at the end of most chapters. These projects are generally longer and more complicated than the usual exercise, and they are meant to be done by several people working as a team over a few days.

If your class decides to do one or more of these, you should divide into teams of about three to six students. Then each team should

- study the problem
- estimate the answer(s)
- list possible tools and methods for solution; then choose one
- solve the problem
- check the answer
- write a brief report

Students now have more computational tools at their disposal than anyone has had before. Depending on the problem, you might choose from the following:

- manual computation
- trial and error
- scientific calculator
- graphics calculator
- computer, using a programming language such as *BASIC*, *Visual Basic*, *C++*, or *Pascal*; a spreadsheet such as *Excel* or *Quattro Pro*; mathematical computation software such as *MathCad*; or a computer algebra program such as *Derive*, *Maple*, or *Mathematica*.

You might choose *more than one* tool, perhaps doing the computation by calculator and checking the results by computer. For some projects you might want to borrow measuring equipment such as a tape measure or a transit, or you might want to consult reference books. Sometimes it can be useful to make a model. Be sure to consider *any* aids that might help you.

Your first team project is problem 82, the simple one of finding a missing dimension on a template.

82. Starting with the template in Fig. 1–25, you are to make a new template by increasing dimension A by 25.0% and decreasing dimension B by 30.0%. Find the dimension x (in millimetres) for the new template.

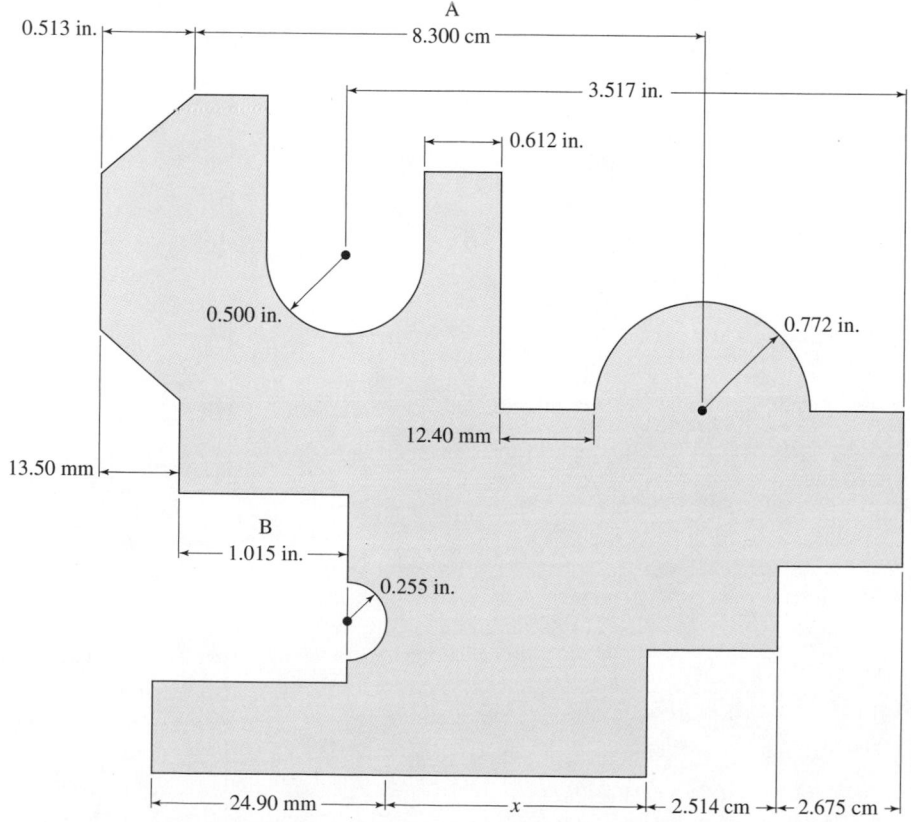

FIGURE 1–25 A template, not drawn to scale.

83. Make a drawing of a cylindrical steel bar, 3 cm in diameter and 8 cm long. Label the diameter as 3.00 cm. Take your drawing to a machine shop and ask for a cost estimate for each of six bars, having lengths of

8 cm	8.000 cm
8.0 cm	8.0000 cm
8.00 cm	8.000 00 cm

Before you go, have each member of your team make cost estimates.

Introduction to Algebra

When you have completed this chapter, you should be able to:

- Determine the degree of an expression, determine the number of terms in the expression, and identify the coefficient of each term.
- Simplify expressions by removing symbols of grouping and by combining like terms by addition and subtraction.
- Use the laws of exponents to simplify and combine expressions containing powers.
- Multiply algebraic expressions.
- Divide algebraic expressions.

Formulas and equations represent the world in the language of science and engineering. A 1000-word description of the path of a meteorite can be summed up in a single equation— and this single equation describes the meteorite path so well that we can tell its distance from the earth thousands of years into the future and thousands and millions of years into the past. The equations and formulas you will be studying in your specialty areas of science and engineering describe the physical world better (more compactly, accurately, and succinctly) than any other language. For this reason, your first step in science and engineering is to learn the language. The language of math is based on representing numbers with letters so that the general characteristics of a formula can be represented. Once you understand an equation or formula, you can then replace the letters with the actual numbers and determine results for a wide variety of situations. The numbers used can make each calculation totally unique from any other set of numbers put into the same formula.

ABORIGINAL WISDOM—ELMER GHOSTKEEPER

Living near Edmonton, Alberta, **Elmer Ghostkeeper** is a proud Metis and civil engineer. Even with his own success, Elmer knows that school can be difficult, especially math and science courses. He says if you "let the numbers dance for you, then math and science classes will become enjoyable." He recalls that in his own student days, "I could not wait for some classes to begin because I enjoyed them so much." He adds, "Learn to follow your heart; it will direct you to your purpose in life. Always keep in mind that engineering and science is just another way to understand Mother Earth and her sacred gifts."

Elmer is the president and CEO of SunStar Resources Inc., an oil and gas company on the Paddle Prairie Metis Settlement. SunStar is wholly owned and operated by the Metis people. Elmer has also been the assistant city engineer of Whitehorse in the Yukon Territory. As SunStar grows it will need engineers and skilled professionals to fill positions. Elmer encourages native youth to stay in school and study math and sciences: "Aboriginal wisdom is extremely important to the economic development and self-sufficiency of Aboriginal communities. With the right education there's every reason to believe our young people can work for companies like SunStar."

Algebra is a generalization of arithmetic. For example, the statement

$$2 \cdot 2 \cdot 2 \cdot 2 = 2^4$$

can be generalized to

$$a \cdot a \cdot a \cdot a = a^4$$

where a can be *any* number, not just 2. We can go further and say that

$$\underbrace{(a)(a)(a)(a) \ldots (a)}_{n \text{ factors}} = a^n$$

where a can be any number, as before, and n can be *any* positive integer, not just 4.

We will learn many new words in this chapter. Every field has its special terms, and algebra is no exception. But since algebra is generalized arithmetic, some of what was said in Chapter 1 (such as rules of signs) will be repeated here.

We will redo the basic operations of addition, subtraction, and so on, but now with symbols rather than numbers. This material is the foundation on which later chapters rest.

2–1 Algebraic Expressions

Mathematical Expressions

A *mathematical expression* is a grouping of mathematical symbols, such as signs of operation, numbers, and letters.

◆◆◆ **Example 1:** The following are mathematical expressions:

(a) $x^2 - 2x + 3$
(b) $4 \sin 3x$
(c) $5 \log x + e^{2x}$

◆◆◆

Algebraic Expressions

An *algebraic expression* is one containing algebraic symbols and operations (addition, subtraction, multiplication, division, roots, and powers), such as in Example 1(a). All other expressions are called *transcendental*, such as in examples 1(b) and 1(c).

Equations

None of the expressions in Example 1 contains an equal sign ($=$). When two expressions are set equal to each other, we get an *equation*.

◆◆◆ **Example 2:** The following are equations:

(a) $2x^2 + 3x - 5 = 0$
(b) $6x - 4 = x + 1$
(c) $y = 3x - 5$
(d) $3 \sin x = 2 \cos x$ ◆◆◆

We mentioned equations briefly in Chapter 1 when we substituted into equations. We treat them in detail in Chapter 3.

Constants and Variables

A *constant* is a quantity that does not change in value in a particular problem. It is usually a number, such as 8, 4.67, or π.

A *variable* is a quantity that may change during a particular problem. A variable is usually represented by a letter from the end of the alphabet (x, y, z, etc.).

◆◆◆ **Example 3:** The constants in the expression

$$3x^2 + 4x + 5$$

are 3, 4, and 5, and the variable is x.   ◆◆◆

A constant can also be represented by a letter. Such a letter is usually chosen from the beginning of the alphabet (a, b, c, etc.). The letter k is often used as a constant. An expression in which the constants are represented by letters is called a *literal expression*.

◆◆◆ **Example 4:** The constants in the literal expression

$$ax^2 + bx + c$$

are a, b, and c, and the variable is x.   ◆◆◆

◆◆◆ **Example 5:** In the expression

$$ax + bx + cz$$

the letters a, b, and c would usually represent constants, and x, y, and z would represent variables.   ◆◆◆

Symbols of Grouping

Mathematical expressions often contain parentheses, ( ), brackets, [ ], and braces, { }. Another grouping symbol is the horizontal line in a fraction; such as $\frac{a+b}{c+d}$. The numerator (top) of the fraction must be evaluated separately from the denominator (bottom). Parentheses are used to group parts of the expression together, and they affect the meaning of the expression.

An easy way to remember this rule is the mnemonic "parents buy braces"— parents (parentheses) buy (brackets) braces (braces).

◆◆◆ **Example 6:** The expression

$$\{[(3 + x^2) - 2] - 2(2x + 3)\} \cdot (x - 3)$$

shows the use of three grouping symbols.   ◆◆◆

We will show how to *remove* symbols of grouping in Sec. 2–2. We will also see in Sec. 2–4 that symbols of grouping can be used to indicate *multiplication*.

◆◆◆ **Example 7:** The value of the expression

$$2 + 3(4 + 5)$$

is *different* from the value of

$$(2 + 3)4 + 5$$

The first expression has a value of 29, and the second expression has a value of 25. As you can see, the placement of symbols of grouping *is* important.   ◆◆◆

CASE STUDY—SURGE PROTECTORS IN HIGH VOLTAGE LINES

When electrical current (represented by the letter I) flows through high voltage transmission lines, one source of power loss in these lines is called "I-squared R loss", a name that comes from the one form of the formula for power in an electrical circuit, $P = I^2R$. In this formula, P is power and R is electrical resistance. The surge protectors safeguarding the lines can handle 10 million times the normal power before melting. During an electrical storm, the transformer near Pickering Nuclear station in Ontario is struck and the current surges 3200 times its normal load. Could a current surge of 3200 times its normal load cause melting of the surge protectors?

Terms
Plus and the minus signs divide an expression into *terms*.

◆◆◆ **Example 8:** The expression

$$4x^2 + 3x + 5$$

$4x^2$	$3x$	5
first term	second term	third term

has *three* terms.   ◆◆◆

An exception is when the plus or minus sign is *within* a symbol of grouping.

◆◆◆ **Example 9:** The expression

$$[(x + 2x^2) + 3] - (2x + 2)$$

$$\underbrace{\hspace{3cm}}_{\text{first term}} \quad \underbrace{\hspace{2cm}}_{\text{second term}}$$

has *two* terms.   ◆◆◆

Factors

The number 1 and the entire expression 3*axy* are also factors. They usually are not stated. See Chapter 8 for a more thorough discussion of factors.

Any divisor of a term is called a *factor* of that term.

◆◆◆ **Example 10:** Some factors of 3*axy* are 3, *a*, *x*, and *y*. The others are combinations such as 3*a* and *axy*.   ◆◆◆

◆◆◆ **Example 11:** The expression

$$2x + 3yz$$

has two terms, 2*x* and 3*yz*. The first term has the *factors* 2 and *x*, and the second term has the *factors* 3, *y*, and *z*.   ◆◆◆

Coefficient
The *coefficient* of a term is the constant part of the term.

◆◆◆ **Example 12:** The coefficient in the term $3axy^2$ is 3*a*, where *a* is a constant. We say that 3*a* is the coefficient of xy^2.   ◆◆◆

When we use the word *coefficient* in this book, it will always refer to the constant part of the term (such as 3*a* in Example 12), also called the *numerical coefficient*. If there is no numerical coefficient written before a term, it is understood to be 1.

◆◆◆ **Example 13:** In the expression: $\dfrac{w}{2} + x - y - 3z$

(a) The coefficient of *w* is $\frac{1}{2}$.
(b) The coefficient of *x* is 1.

(c) The coefficient of y is -1.
(d) The coefficient of z is -3.

◆◆◆ Do not forget to include the minus sign with the coefficient.

Degree

The *degree* of a term refers to the integer power to which the variable is raised.

◆◆◆ **Example 14:**

(a) $2x$ is a first-degree term.
(b) $3x^2$ is a second-degree term.
(c) $5y^9$ is a ninth-degree term.

◆◆◆ The term *degree* is used only when the exponents are *positive integers*. You would not say that $x^{1/2}$ is a "half-degree" term.

If there is more than one variable, we add their powers to obtain the degree of the term.

◆◆◆ **Example 15:**

(a) $2x^2y^3$ is of fifth degree.
(b) $3xyz^2$ is of fourth degree.   ◆◆◆

The degree of an expression is the same as that of the term having the highest degree.

◆◆◆ **Example 16:** $3x^2 - 2x + 4$ is a second-degree expression.   ◆◆◆

A *multinomial* is an algebraic expression having *more than one term*.

◆◆◆ **Example 17:** Some multinomials are

(a) $3x + 5$
(b) $2x^3 - 3x^2 + 7$
(c) $\dfrac{1}{x} + \sqrt{2x}$ ◆◆◆

A *polynomial* is a monomial or multinomial in which the powers to which the unknown is raised are all *nonnegative integers*. The first two expressions in Example 17 are polynomials, but the third is not.

A *binomial* is a polynomial with *two* terms, and a *trinomial* is a polynomial having *three* terms. In Example 17, the first expression is a binomial, and the second is a trinomial.

Exercise 1 ◆ Algebraic Expressions

How many terms are there in each expression?

1. $x^2 - 3x$ **2.** $7y - (y^2 + 5)$
3. $(x + 2)(x - 1)$ **4.** $3(x - 5) + 2(x + 1)$

Write the coefficients of each term. Assume that the letters x and y are variables and that all other letters are constants.

5. $5x^3$ **6.** $2ay^2$ **7.** $\dfrac{bx^3}{4}$

8. $\dfrac{1}{4}(bx)$ **9.** $\dfrac{3y^2}{2a}$ **10.** $\dfrac{2c}{a}(4x^2)$

2-2   Addition and Subtraction of Algebraic Expressions

Constituents

An addition operation consists of a two addends and a resulting sum.

$$\text{addend} + \text{addend} = \text{sum}$$

A subtraction operation consists of a minuend, a subtrahend, and a resulting difference.

$$\text{difference} = \text{minuend} - \text{subtrahend}$$

Like Terms
Like terms are those that differ only in their coefficients.

◆◆◆ **Example 18:**

(a) $2wx$ and $-3wx$ are like terms.
(b) $2wx$ and $-3wx^2$ are *not* like terms.   ◆◆◆

Algebraic expressions are added and subtracted by *combining like terms*. Like terms are added by adding their coefficients.

This process is also referred to as *collecting terms*.

◆◆◆ **Example 19:**

(a) $7x + 5x = 12x$
(b) $8w + 2w - 4w - w = 5w$
(c) $9y - 3y = 6y$
(d) $2x - 3y - 5x + 2y = -3x - y$
(e) $4.21x + 1.23x - 3.11x = 2.33x$ ◆◆◆

In Example 19 we combined the terms on a single line. This method is sometimes called *horizontal* addition and subtraction. For more complicated problems, you may prefer *vertical* addition and subtraction, as shown in the following example.

◆◆◆ **Example 20:** Find the sum of $3x^3 + 2x - 5$ and $x^3 - 3x^2 + 7$. Then subtract $2x^3 + 3x^2 - 4x - 7$.

Solution: Combine the first two expressions.

$$\begin{array}{r} 3x^3 \qquad\; + 2x - 5 \\ + \; x^3 - 3x^2 \qquad + 7 \\ \hline 4x^3 - 3x^2 + 2x + 2 \end{array}$$

Then subtract the third expression from their sum.

$$\begin{array}{r} 4x^3 - 3x^2 + 2x + 2 \\ -(2x^3 + 3x^2 - 4x - 7) \\ \hline 2x^3 - 6x^2 + 6x + 9 \end{array}$$ ◆◆◆

Commutative Law of Addition
We mentioned the commutative law (Eq. 1) when adding numbers in Chapter 1. It simply means that the order of addition does not affect the sum ($2 + 3 = 3 + 2$). The same law applies, of course, when we have letters instead of numbers.

◆◆◆ **Example 21:**

(a) $a + b = b + a$
(b) $2x + 3x = 3x + 2x$
$\qquad\quad = 5x$ ◆◆◆

This law enables us to *rearrange the terms* of an expression for our own convenience. Remember $a - b \neq b - a$. Use caution with negative signs here.

◆◆◆ **Example 22:** Simplify the expression

$$4y - 2x + 3z - 7z + 4x - 6y$$

Solution: Using the commutative law, we rearrange the expression so as to get like terms together.

$$-2x + 4x + 4y - 6y + 3z - 7z$$

Collecting terms, we obtain

$$2x - 2y - 4z$$

   With practice, you will soon be able to omit the first step and collect terms by inspecting the original expression.   ◆◆◆

Removal of Parentheses

To combine quantities within parentheses with other quantities not within those parentheses, we must first remove the parentheses. If the parentheses (or other symbol of grouping) are preceded only by a + sign (or no sign), the parentheses may be removed without any extra effort.

◆◆◆ **Example 23:**

(a) $(a + b - c) = a + b - c$
(b) $(x + y) + (w - z) = x + y + w - z$ ◆◆◆

   When the parentheses are preceded only by a $(-)$ sign, change the sign of every term within the parentheses upon removing the parentheses.

◆◆◆ **Example 24:**

(a) $-(a + b - c) = -a - b + c$
(b) $-(x + y) - (x - y) = -x - y - x + y = -2x$
(c) $(2w - 3x - 2y) - (w - 4x + 5y) - (3w - 2x - y)$
$\quad = 2w - 3x - 2y - w + 4x - 5y - 3w + 2x + y$
$\quad = 2w - w - 3w - 3x + 4x + 2x - 2y - 5y + y$
$\quad = -2w + 3x - 6y$ ◆◆◆

   When there are groups within groups, start simplifying the *innermost* groups and work outward.

◆◆◆ **Example 25:**

$$3x - [2y + 5z - (y + 2)] + 3 - 7z = 3x - [2y + 5z - y - 2] + 3 - 7z$$
$$= 3x - [y + 5z - 2] + 3 - 7z$$
$$= 3x - y - 5z + 2 + 3 - 7z$$
$$= 3x - y - 5z + 5 - 7z$$
$$= 3x - y - 12z + 5$$ ◆◆◆

We'll do more complicated problems of this type later.

Good Working Habits

Start now to develop careful working habits. Take your time! Why make things even harder for yourself by scribbling? Form the symbols with care; work in sequence, from top of page to bottom; do not crowd; use a sharp pencil, not a pen; and erase mistakes instead of crossing them out. What chance do you have if you cannot read your own work?

Common Error	Do not switch between capitals and lowercase letters without reason. Although b and B are the same alphabetic letter, in a math problem they could stand for entirely different quantities.

Exercise 2 ◆ Addition and Subtraction of Algebraic Expressions

Combine as indicated, and simplify.

Remember the "like for like" rule when collecting terms.

1. $7x + 5x$

2. $2x + 5x - 4x + x$

3. $6ab - 7ab - 9ab$

4. $9.4x - 3.7x + 1.4x$

5. Add $7a - 3b + m$ and $3b - 7a - c + m$.

6. What is the sum of $6ab + 12bc - 8cd$, $3cd - 7cd - 9bc$, and $12cd - 2ab - 5bc$?

7. What is the sum of $3a + b - 10$, $c - d - a$, and $-4c + 2a - 3b - 7$?

8. Add $7m + 3n - 11p$, $3a - 9n - 11m$, $8n - 4m + 5p$, and $6n - m + 3p$.

9. Add $8ax + 2(x + a) + 3b$, $9ax + 6(x + a) - 9b$, and $11x + 6b - 7ax - 8(x + a)$.

10. Add $a - 9 - 8a^2 + 16a^3$, $5 + 15a^3 - 12a - 2a^2$, and $6a^2 - 10a^3 + 11a - 13$.

11. Subtract $41x^3 - 2x^2 + 13$ from $15x^3 + x - 18$.

12. Subtract $3b - 6d - 10c + 7a$ from $4d + 12a - 13c - 9b$.

13. Add $x - y - z$ and $y - x + z$.

14. What is the sum of $a + 2b - 3c - 10$, $3b - 4a + 5c + 10$, and $5b - c$?

15. Add $72ax^4 - 8ay^3$, $-38ax^4 - 3ay^4 + 7ay^3$, $8 + 12ay^4$, $-6ay^3 + 12$, and $-34ax^4 + 5ay^3 - 9ay^4$.

16. Add $2a(x - y^2) - 3mz^2$, $4a(x - y^2) - 5mz^2$, and $5a(x - y^2) + 7mz^2$.

17. What is the sum of $9b^2 - 3ac + d$, $4b^2 + 7d - 4ac$, $3d - 4b^2 + 6ac$, $5b^2 - 2ac - 12d$, and $4b^2 - d$?

18. From the sum of $2a + 3b - 4c$ and $3b + 4c - 5d$, subtract the sum of $5c - 6d - 7a$ and $-7d + 8a + 9b$.

19. What is the sum of $7ab - m^2 + q$, $-4ab - 5m^3 - 3q$, $12ab + 14m^2 - z$, and $-6m^2 - 2q$?

20. Add $14(x + y) - 17(y + z)$, $4(y + z) - 19(z + x)$, and $-7(z + x) - 3(x + y)$.

21. Add $3.52(a + b)$, $4.15(a + b)$, and $-1.84(a + b)$.

Symbols of Grouping

Remove symbols of grouping (i.e., braces, brackets, and parentheses), and collect terms. The horizontal line separating a fraction's numerator from its denominator will be discussed later. The next step in questions 22 and 23 has been completed for you.

Remember the mnemonic "parents buy braces" for removing symbols of grouping.

22. $(3x + 5) + (2x - 3) = 3x + 5 + 2x - 3$

23. $3a^2 - (3a - x + b) = 3a^2 - 3a + x - b$

24. $40xy - (30xy - 2b^2 + 3c - 4d)$

25. $(6.4 - 1.8x) - (7.5 + 2.6x)$

26. $7m^2 + 2bc - (3m^2 - bc - x)$

27. $a^2 - a - (4a - y - 3a^2 - 1)$

28. $(2x^4 + 3x^3 - 4) + (3x^4 - 2x^3 - 8)$

29. $a + b - m - (m - a - b)$

30. $(3xy - 2y + 3z) + (-3y + 8z) - (-3xy - z)$

31. $3m - z - y - (2z - y - 3m)$

32. $(2x - 3y) - (5x - 3y - 2z)$

33. $(w + 2z) - (-3w - 5x) + (x + z) - (z - w)$

34. $2a + \{-6b - [3c + (-4b - 6c + a)]\}$

35. $4a - \{a - [-7a - [8a - (5a + 3)] - (-6a - 2a - 9)]\}$

36. $9m - \{3n + [4m - (n - 6m)] - [m + 7n]\}$

Applications

37. The surface area of the box in Fig. 2–1 is

$$2[2w^2 + 3w^2 + 6w^2]$$

Simplify this expression.

38. If a person invests $5,000, x dollars at 12% interest and the rest at 8% interest, the total earnings will be:

$$0.12x + 0.08(5,000 - x)$$

Simplify this expression.

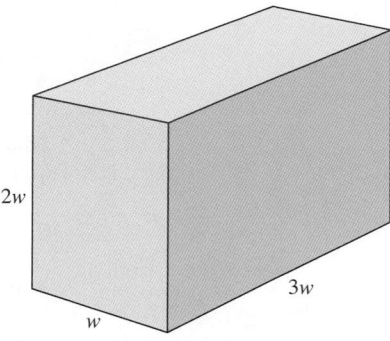

FIGURE 2–1

2–3   Integral Exponents

Definitions

In this section we deal only with expressions that have integers (positive or negative, and zero) as exponents.

◆◆◆ **Example 26:** We study such expressions as

$$x^3 \quad a^{-2} \quad y^0 \quad x^2y^3w^{-1} \quad (m + n)^5 \qquad ◆◆◆$$

A positive *exponent* shows how many times the *base* is to be multiplied by itself.

◆◆◆ **Example 27:** In the expression 3^4, 3 is the base and 4 is the exponent.

$$\text{base} \longrightarrow 3^4 \longleftarrow \text{exponent}$$

Its meaning is

$$3^4 = (3)(3)(3)(3) = 81 \qquad ◆◆◆$$

In general, the following equation applies:

Positive Integral Exponent	a^n means $\underbrace{a \cdot a \cdot a \cdot a \cdot \ldots a}_{n \text{ factors}}$	**28**

Common Error	An exponent applies only to the symbol directly before it. Thus $$2x^2 \neq 2^2x^2 \text{ but it does equal } 2(x)(x)$$ However, $$(2x)^2 = (2x)(2x) = 4x^2$$

	Also, $$-2^2 \neq 4$$
Common Error	However, $$(-2)^2 = (-2)(-2) = 4 \text{ and } -(2)^2 = -(2)(2) = -4$$

Multiplying Powers

Let us multiply the quantity x^2 by x^3. From Eq. 28, we know that

$$x^2 = x \cdot x$$

and that

$$x^3 = x \cdot x \cdot x$$

Multiplying, we obtain

$$x^2 \cdot x^3 = (x \cdot x)\,(x \cdot x \cdot x)$$
$$= x \cdot x \cdot x \cdot x \cdot x$$
$$= x^5$$

Notice that the exponent in the result is the *sum of the two original exponents*.

$$x^2 \cdot x^3 = x^{2+3} = x^5$$

This will always be so. We summarize this rule as our first *law of exponents*.

	$$x^a \cdot x^b = x^{a+b}$$	
Products	When multiplying powers of the same base, we keep the same base and *add the exponents*.	**29**

◆◆◆ **Example 28:**

(a) $x^4(x^3) = x^{4+3} = x^7$
(b) $x^5(x^a) = x^{5+a}$
(c) $y^2(y^n)(y^3) = y^{2+n+3} = y^{5+n}$ ◆◆◆

If nothing else is indicated, every quantity has an exponent of 1.

◆◆◆ **Example 29:**

(a) $x(x^3) = x^{1+3} = x^4$
(b) $a^2(a^4)(a) = a^{2+4+1} = a^7$
(c) $x^a(x)(x^b) = x^{a+1+b}$ ◆◆◆

The "invisible 1" appears again. It is the unwritten coefficient of every term, and now as the unwritten exponent. It is also in the denominator.

$$x = \frac{1x^1}{1}$$

Do not forget about these three invisible 1's. *We use them all the time.*

Quotients

Let us divide x^5 by x^3. By Eq. 29,

$$\frac{x^5}{x^3} = \frac{x \cdot x \cdot x \cdot x \cdot x}{x \cdot x \cdot x} = \frac{x \cdot x \cdot x}{x \cdot x \cdot x} \cdot x \cdot x$$
$$= x \cdot x = x^2$$

We could have obtained the same result by *subtracting exponents:*

$$\frac{x^5}{x^3} = x^{5-3} = x^2$$

This always works, and we state it as the second *law of exponents*.

Quotients	$$\frac{x^a}{x^b} = x^{a-b} \quad (x \neq 0)$$ When dividing powers of the same base, we keep the same base and *subtract the exponents*.	30

◆◆◆ **Example 30:**

(a) $\dfrac{x^5}{x^4} = x^{5-4} = x^1 = x$

(b) $\dfrac{y^4}{y^6} = y^{4-6} = y^{-2}$

(c) $\dfrac{a^4 b^3}{ab^2} = a^{4-1}b^{3-2} = a^3 b$

(d) $\dfrac{y^{n-m}}{y^{n+m}} = y^{n-m-(n+m)} = y^{-2m}$ ◆◆◆

Common Error	Remember when using Rules 29 or 30 that the quantities to be multiplied or divided *must have the same base*.

Power Raised to a Power

Let us take a quantity raised to a power, say, x^2, and raise the entire expression to *another* power, say, 3.

$$(x^2)^3$$

By Eq. 29,

$$(x^2)^3 = (x^2)(x^2)(x^2)$$
$$= (x \cdot x)(x \cdot x)(x \cdot x)$$
$$= x \cdot x \cdot x \cdot x \cdot x \cdot x = x^6$$

a result that could have been obtained by *multiplying the exponents*.

$$(x^2)^3 = x^{2(3)} = x^6$$

In general, the following formula applies for a power raised to a power:

Powers	$$(x^a)^b = x^{ab} = (x^b)^a$$ When raising a power to a power, we keep the same base and *multiply the exponents*.	31

◆◆◆ **Example 31:**

(a) $(2^2)^3 = 2^{2(3)} = 2^6 = 64$

(b) $(w^5)^2 = w^{5(2)} = w^{10}$

(c) $(a^{-3})^2 = a^{(-3)(2)} = a^{-6}$

(d) $(10^4)^3 = 10^{4(3)} = 10^{12}$

(e) $(b^{x+2})^3 = b^{3x+6}$

Product Raised to a Power

We now raise a product, such as xy, to some power, say, 3.

$$(xy)^3$$

By Eq. 28,

$$(xy)^3 = (xy)(xy)(xy)$$
$$= x \cdot y \cdot x \cdot y \cdot x \cdot y$$
$$= x^3 y^3$$

In general, the following equation applies to a product raised to a power:

Product Raised to a Power	$(xy)^n = x^n \cdot y^n$ When a product is raised to a power, each factor may be *separately* raised to the power.	**32**

◆◆◆ **Example 32:**

(a) $(xyz)^5 = x^5 y^5 z^5$
(b) $(2x)^3 = 2^3 x^3 = 8x^3$
(c) $(3.75 \times 10^3)^2 = (3.75)^2 \times (10^3)^2$
$$= 14.1 \times 10^6$$
(d) $(3x^2 y^n)^3 = 3^3 (x^2)^3 (y^n)^3$
$$= 27x^6 y^{3n}$$
(e) $(-x^2 y)^3 = (-1)^3 (x^2)^3 y^3 = -x^6 y^3$ ◆◆◆

A good way to test a "rule" that you are not sure of is to *try it with numbers*. In this case, does $(2 + 3)^2$ equal $2^2 + 3^2$? Evaluating each expression, we obtain

$$(5)^2 \overset{?}{=} 4 + 9$$
$$25 \neq 13$$

Common Error	There is *no* similar rule for the *sum* of two quantities raised to a power. $(x + y)^n \neq x^n + y^n$

Quotient Raised to a Power

Using the same steps as in the preceding section, see if you can show that

$$\left(\frac{x}{y}\right)^3 = \frac{x^3}{y^3}$$

In general, the following equation applies:

Quotient Raised to a Power	$\left(\frac{x}{y}\right)^n = \frac{x^n}{y^n} \quad (y \neq 0)$ When a quotient is raised to a power, the numerator and the denominator may be *separately* raised to the power.	**33**

◆◆◆ **Example 33:**

(a) $\left(\frac{2}{3}\right)^3 = \frac{2^3}{3^3} = \frac{8}{27}$
(b) $\left(\frac{x}{5}\right)^2 = \frac{x^2}{5^2} = \frac{x^2}{25}$

(c) $\left(\dfrac{3a}{2b}\right)^3 = \dfrac{3^3 a^3}{2^3 b^3} = \dfrac{27a^3}{8b^3}$

(d) $\left(\dfrac{2x^2}{5y^3}\right)^3 = \dfrac{2^3 (x^2)^3}{5^3 (y^3)^3} = \dfrac{8x^6}{125y^9}$

(e) $\left(-\dfrac{a}{b^2}\right)^3 = (-1)^3 \dfrac{a^3}{(b^2)^3} = -\dfrac{a^3}{b^6}$

◆◆◆

Zero Exponent

If we divide x^n by itself we get, by Eq. 30,

$$\dfrac{x^n}{x^n} = x^{n-n} = x^0$$

But any expression divided by itself equals 1, so we have the following law:

Zero Exponent	$x^0 = 1 (x \neq 0)$ Any quantity (except 0) raised to the zero power equals 1.	**34**

◆◆◆ **Example 34:**

(a) $(xyz)^0 = 1$
(b) $3862^0 = 1$
(c) $(x^2 - 2x + 3)^0 = 1$
(d) $5x^0 = 5(1) = 5$

◆◆◆

Negative Exponent

We now divide x^0 by x^a. By Eq. 30,

$$\dfrac{x^0}{x^a} = x^{0-a} = x^{-a}$$

Since $x^0 = 1$, we get the following equation:

Negative Exponent	$x^{-a} = \dfrac{1}{x^a} \quad (x \neq 0)$ When taking the reciprocal of a base raised to a power, *change the sign of the exponent.*	**35**

We can, of course, also give Eq. 35 in the form

$$x^a = \dfrac{1}{x^{-a}} \quad (x \neq 0)$$

For practice, let's first rewrite an expression containing negative exponents as one without a negative exponent.

◆◆◆ **Example 35:** Rewrite each expression without negative exponents, using fractions where necessary.

(a) $5^{-1} = \dfrac{1}{5}$

(b) $x^{-4} = \dfrac{1}{x^4}$

(c) $\dfrac{1}{xy^{-2}} = \dfrac{y^2}{x}$

(d) $\dfrac{w^{-3}}{z^{-2}} = \dfrac{z^2}{w^3}$

(e) $\left(\dfrac{3x}{2y^2}\right)^{-b} = \left(\dfrac{2y^2}{3x}\right)^b$

◆◆◆

We can also use negative exponents to rewrite an expression *without* fractions.

⋄⋄⋄ **Example 36:** Rewrite each expression without fractions. Use negative exponents where necessary.

(a) $\dfrac{1}{y} = y^{-1}$

(b) $\dfrac{1}{y^4} = y^{-4}$

(c) $\dfrac{a}{b} = ab^{-1}$

(d) $\dfrac{a^2}{b^3} = a^2 b^{-3}$

(e) $\dfrac{w}{xy^2} = wx^{-1}y^{-2}$

(f) $\dfrac{x^2 y^{-3}}{w^4 z^{-2}} = z^2 y^{-3} w^{-4} z^2$ ⋄⋄⋄

Summary of the Laws of Exponents

Positive Integral Exponent	a^n means $\underbrace{a \cdot a \cdot a \cdot a \cdot \ldots \cdot a}_{n \text{ factors}} = a^n$	28
Products	$x^a \cdot x^b = x^{a+b}$	29
Quotients	$\dfrac{x^a}{x^b} = x^{a-b} \quad (x \neq 0)$	30
Powers	$(x^a)^b = x^{ab} = (x^b)^a$	31
Product Raised to a Power	$(xy)^n = x^n \cdot y^n$	32
Quotient Raised to a Power	$\left(\dfrac{x}{y}\right)^n = \dfrac{x^n}{y^n} \quad (y \neq 0)$	33
Zero Exponent	$x^0 = 1 \quad (x \neq 0)$	34
Negative Exponent	$x^{-a} = \dfrac{1}{x^a} \quad (x \neq 0)$	35

Exercise 3 ⋄ Integral Exponents

Multiply.

1. $10^2 \cdot 10^3$

2. $10^a \cdot 10^b$

3. $x \cdot x^2$

4. $x^3 \cdot x^2$

5. $y^3 \cdot y^4$

6. $w^2 \cdot w^a$

7. $x^2 \cdot x \cdot x^4$

8. $x^2 \cdot x^b \cdot x^3$

9. $w \cdot w^2 \cdot w^3$

10. $y^3 \cdot y^2 \cdot y^4$

11. $x^{n+1} \cdot x^2$

12. $y^{a+2} \cdot y^{2a-1}$

Divide. Write your answers without negative exponents.

13. $\dfrac{y^5}{y^2}$

14. $\dfrac{5^5}{5^3}$

15. $\dfrac{x^{n+2}}{x^{n+1}}$

16. $\dfrac{10^5}{10}$ **17.** $\dfrac{10^{x+5}}{10^{x+3}}$ **18.** $\dfrac{10^2}{10^{-3}}$

19. $\dfrac{x^{-2}}{x^{-3}}$ **20.** $\dfrac{a^{-5}}{a}$

Simplify.

21. $(x^3)^4$ **22.** $(9^2)^3$ **23.** $(a^x)^y$

24. $(x^{-2})^{-2}$ **25.** $(x^{a+1})^2$

Raise to the power indicated.

26. $(xy)^2$ **27.** $(2x)^3$ **28.** $(3x^2y^3)^2$

29. $(3abc)^3$ **30.** $\left(\dfrac{3}{5}\right)^2$ **31.** $\left(-\dfrac{1}{3}\right)^3$

32. $\left(\dfrac{x}{y}\right)^5$ **33.** $\left(\dfrac{2a}{3b^2}\right)^3$ **34.** $\left(\dfrac{3x^2y}{4wz^3}\right)^2$

Write each expression with positive exponents only.

35. a^{-2} **36.** $(-x)^{-3}$ **37.** $\left(\dfrac{3}{y}\right)^{-3}$

38. $a^{-2}bc^{-3}$ **39.** $\left(\dfrac{2a}{3b^3}\right)^{-2}$ **40.** xy^{-4}

41. $2x^{-2} + 3y^{-3}$ **42.** $\left(\dfrac{x}{y}\right)^{-1}$

Express without fractions, using negative exponents where needed.

43. $\dfrac{1}{x}$ **44.** $\dfrac{3}{y^2}$ **45.** $\dfrac{x^2}{y^2}$

46. $\dfrac{x^2y^{-3}}{z^{-2}}$ **47.** $\dfrac{a^{-3}}{b^2}$ **48.** $\dfrac{x^{-2}y^{-3}}{w^{-1}z^{-4}}$

Evaluate.

49. $(a + b + c)^0$ **50.** $8x^0y^2$ **51.** $\dfrac{a^0}{9}$

52. $\dfrac{y}{x^0}$ **53.** $\dfrac{x^{2n} \cdot x^3}{x^{3+2n}}$ **54.** $5\left(\dfrac{x}{y}\right)^0$

Applications

55. We can find the volume of the box in Fig. 2–1 by multiplying length by width by height, getting the following expression

$$(w)(2w)(3w)$$

Simplify this expression.

56. A freely falling body, starting from rest, falls a distance of $4.90t^2$ metres in t seconds. In twice that time it will fall

$$4.90(2t)^2 \text{ m}$$

Simplify this expression.

57. The power in a resistor of resistance R in which flows a current I is I^2R. If the current is reduced to one-third its former value, the power will be

$$\left(\dfrac{I}{3}\right)^2 R$$

Simplify this expression.

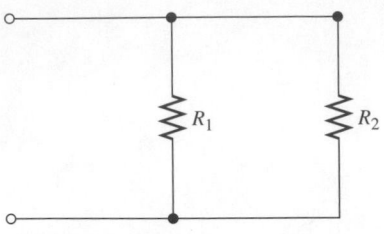

58. The resistance R of two resistors R_1 and R_2 wired in parallel (Fig. 2–2) is found from

$$\frac{1}{R} = \frac{1}{R_1} + \frac{1}{R_2}$$

Use exponents to write this equation without fractions.

FIGURE 2–2

2–4 Multiplication of Algebraic Expressions

Symbols and Definitions

Avoid using the ($\times$) symbol because it could get confused with the letter x.

Multiplication is indicated in several ways: by the usual ($\times$) symbol; by a dot; or by parentheses, brackets, or braces. Thus the product of b and d could be written

$$b \cdot d \qquad b \times d \qquad b(d) \qquad (b)d \qquad (b)(d)$$

The most common convention is to use no symbol at all. The product of b and d would usually be written bd. A calculator or computer usually uses the asterisk or star (*) to indicate multiplication.

The names of the components of a multiplication are

$$\text{multiplicand} \times \text{multiplier} = \text{product}$$

Multiplicands and multipliers are also called factors.
We get a *product* when we multiply two or more *factors*.

$$(\text{factor})(\text{factor})(\text{factor}) = \text{product}$$

Rules of Signs

The product of two factors having the *same* sign is *positive*; the product of two factors having *different* signs is *negative*. The following rules apply:

Rules of Signs		
$(+a)(+b) = (-a)(-b) = +ab$	**8**
$(+a)(-b) = (-a)(+b) = -(+a)(+b) = -ab$	**9**

When we are multiplying more than two factors, the product of every *pair* of negative terms will be positive. Thus if there is an even number of negative factors, the final result will be positive; if there is an odd number of negative factors, the final result will be negative.

♦♦♦ Example 37:

(a) $x(-y)(-z) = xyz$ (Even amount of ($-$) signs)
(b) $(-a)(-b)(-c) = -abc$ (Odd amount of ($-$) signs)
(c) $(-p)(-q)(-r)(-s) = pqrs$ (Even amount of ($-$) signs)   ♦♦♦

Commutative Law

As we saw when multiplying numbers, the *order* of multiplication does not make any difference. In symbols, the law is stated as follows:

Commutative Law of Multiplication	$ab = ba$	2

◆◆◆ **Example 38:** Multiply $3y$ by $-2x$.

Solution:

$$(3y)(-2x) = 3(y)(-2)(x)$$

By Eq. 2,

$$= 3(-2)(y)(x)$$
$$= -6xy$$

It is common practice to write the letters in alphabetical order. ◆◆◆

Multiplying Monomials

A *monomial* is an algebraic expression having *one term*.

◆◆◆ **Example 39:** Some monomials are

(a) $2x^3$ (b) $\dfrac{3wxy}{4}$ (c) $(2b)^3$ ◆◆◆

To multiply monomials, we use the commutative law (Eq. 2), the rules of signs (Eqs. 8 and 9), and the law of exponents for products.

Products	$x^a \cdot x^b = x^{a+b}$	29

◆◆◆ **Example 40:**

$$\begin{aligned}(4a^3b)(-3ab^2) &= (4)(-3)(a^3)(a)(b)(b^2) \\ &= -12a^{3+1}b^{1+2} \\ &= -12a^4b^3\end{aligned}$$

◆◆◆

> The *terms* $4a^3b$ and $-3ab^2$ cannot be *combined* but the factors $4a^3b$ and $(-3ab^2)$ can be *multiplied*.

◆◆◆ **Example 41:** Multiply $5x^2$, $-3xy$, and $-xy^3z$.

Solution: By Eq. 2,

$$5x^2(-3xy)(-xy^3z) = 5(-3)(-1)(x^2)(x)(x)(y)(y^3)(z)$$

By Eq. 29,

$$\begin{aligned}&= 15x^{2+1+1}y^{1+3}z \\ &= 15x^4y^4z\end{aligned}$$

◆◆◆

> With some practice you should be able to omit some steps in many of these solutions. But do not rush it. Use as many steps as you need to understand and implement the process needed to get it right.

◆◆◆ **Example 42:** Multiply $3x^2$ by $2ax^n$.

Solution: By Eq. 2,

$$3x^2(2ax^n) = 3(2a)(x^2)(x^n)$$

By Eq. 29,

$$= 6ax^{2+n}$$

◆◆◆

Multiplying a Multinomial by a Monomial

To multiply a monomial and a multinomial, we use the *distributive law*.

Later we will use this law to remove common factors.

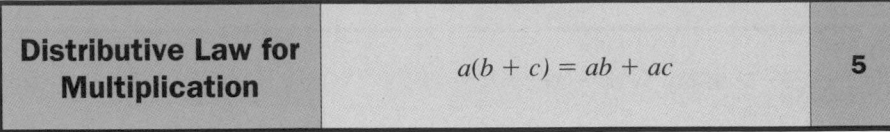

Distributive Law for Multiplication	$a(b + c) = ab + ac$	5

The product of a monomial and a polynomial is equal to the sum of the products of the monomial and each term of the polynomial.

◆◆◆ **Example 43:**

$$2x(x - 3x^2) = 2x(x) + 2x(-3x^2)$$
$$= 2x^2 - 6x^3$$ ◆◆◆

The distributive law, although written for a monomial times a binomial, can be extended for a multinomial having any number of terms. Simply *multiply every term in the multinomial by the monomial*.

◆◆◆ **Example 44:**

(a) $-3x^3(3x^2 - 2x + 4) = -3x^3(3x^2) + (-3x^3)(-2x) + (-3x^3)(4)$
$$= -9x^5 + 6x^4 - 12x^3$$
(b) $4xy(x^2 - 3xy + 2y^2) = 4x^3y - 12x^2y^2 + 8xy^3$
(c) $3.75x(1.83x^2 + 2.27x - 1.49) = 6.86x^3 + 8.51x^2 - 5.59x$ ◆◆◆

Removing Symbols of Grouping

Symbols of grouping, in addition to indicating that the enclosed expression is to be taken as a whole, also indicate multiplication.

◆◆◆ **Example 45:**

(a) $x(y)$ is the product of x and y.
(b) $x(y + 2)$ is the product of x and $(y + 2)$.
(c) $(x - 1)(y + 2)$ is the product of $(x - 1)$ and $(y + 2)$.
(d) $-(y + 2)$ is the product of -1 and $(y + 2)$.   ◆◆◆

The parentheses may be removed after the multiplication has been performed.

◆◆◆ **Example 46:**

(a) $x(y + 2) = xy + 2x$
(b) $5 + 3(x - 2) = 5 + 3x - 6 = 3x - 1$ ◆◆◆

Common Errors	Do not forget to multiply *every* term within the grouping by the preceding factor. $$-(2x + 5) \neq -2x + 5$$
	Multiply the terms within the parentheses *only* by the factor directly preceding it. $$x + 2(a + b) \neq (x + 2)(a + b)$$

If there are groupings within groupings, start simplifying with the *innermost* groupings.

♦♦♦ **Example 47:**

(a) $x - 3[2 - 4(y + 1)] = x - 3[2 - 4y - 4]$
$$= x - 3[-2 - 4y]$$
$$= x + 6 + 12y$$
$$= x + 12y + 6$$

(b) $3\{2[4(w - 4) - (x + 3)] - 2\} - 6x$
$$= 3\{2[4w - 16 - x - 3] - 2\} - 6x$$
$$= 3\{2[4w - x - 19] - 2\} - 6x$$
$$= 3\{8w - 2x - 38 - 2\} - 6x$$
$$= 3\{8w - 2x - 40\} - 6x$$
$$= 24w - 6x - 120 - 6x$$
$$= 24w - 12x - 120$$

♦♦♦

Multiplying a Multinomial by a Multinomial

Using the distributive law (Eq. 5), we multiply one of the multinomials by each term in the other multinomial. We then use the distributive law again to remove the remaining parentheses, and simplify.

♦♦♦ **Example 48:**

$$(x + 4)(x - 3) = x(x - 3) + 4(x - 3)$$
$$= x^2 - 3x + 4x - 12$$
$$= x^2 + x - 12$$

♦♦♦

Some prefer a vertical arrangement for problems like the last one.

♦♦♦ **Example 49:**

$$
\begin{array}{r}
x + 4 \\
x - 3 \\
\hline
- 3x - 12 \\
x^2 + 4x \\
\hline
x^2 + x - 12
\end{array}
$$

as before.

♦♦♦

♦♦♦ **Example 50:**

(a) $(x - 3)(x^2 - 2x + 1)$
$$= x(x^2) + x(-2x) + x(1) - 3(x^2) - 3(-2x) - 3(1)$$
$$= x^3 - 2x^2 + x - 3x^2 + 6x - 3$$
$$= x^3 - 5x^2 + 7x - 3$$

(b) $(2x + 3y - 4z)(x - 2y - 3z)$
$$= 2x^2 - 4xy - 6xz + 3xy - 6y^2 - 9yz - 4xz + 8yz + 12z^2$$
$$= 2x^2 - 6y^2 + 12z^2 - xy - 10xz - yz$$

(c) $(1.24x + 2.03)(3.16x^2 - 2.61x + 5.36)$
$$= 3.92x^3 - 3.24x^2 + 6.65x + 6.41x^2 - 5.30x + 10.9$$
$$= 3.92x^3 + 3.17x^2 + 1.35x + 10.9$$

♦♦♦

When multiplying *more than two* expressions, first multiply any pair, and then multiply that product by the third expression, and so on.

♦♦♦ **Example 51:** Multiply $(x - 2)(x + 1)(2x - 3)$.

Solution: Multiplying the last two binomials yields

$$(x - 2)(x + 1)(2x - 3) = (x - 2)(2x^2 - x - 3)$$

Then multiplying that product by $(x - 2)$ gives

$$(x - 2)(2x^2 - x - 3) = 2x^3 - x^2 - 3x - 4x^2 + 2x + 6$$
$$= 2x^3 - 5x^2 - x + 6 \qquad \blacklozenge\blacklozenge\blacklozenge$$

To raise an expression to a power, simply multiply the expression by itself the proper number of times.

◆◆◆ **Example 52:**

$$(x - 1)^3 = (x - 1)(x - 1)(x - 1)$$
$$= (x - 1)(x^2 - 2x + 1)$$
$$= x^3 - 2x^2 + x - x^2 + 2x - 1$$
$$= x^3 - 3x^2 + 3x - 1 \qquad \blacklozenge\blacklozenge\blacklozenge$$

Product of the Sum and Difference of Two Terms

Let us multiply the binomials $(a - b)$ and $(a + b)$. We get

$$(a - b)(a + b) = a^2 + ab - ab - b^2$$

The two middle terms cancel, leaving us with the following rule:

> When we multiply two binomials, we get expressions that occur frequently enough in our later work for us to take special note of here. These are called *special products*. We will need them especially when we study factoring.

Difference of Two Squares	$(a - b)(a + b) = a^2 - b^2$	41

Thus the product is the *difference of two squares*: the squares of the original numbers.

◆◆◆ **Example 53:**

$$(x + 2)(x - 2) = x^2 - 2^2 = x^2 - 4 \qquad \blacklozenge\blacklozenge\blacklozenge$$

Product of Two Binomials Having the Same First Term

When we multiply $(x + a)$ and $(x + b)$, we get

$$(x + a)(x + b) = x^2 + ax + bx + ab$$

Combining like terms, we obtain the following:

Trinomial, Leading Coefficient = 1	$(x + a)(x + b) = x^2 + (a + b)x + ab$	45

◆◆◆ **Example 54:**

$$(w + 3)(w - 4) = w^2 + (3 - 4)w + 3(-4)$$
$$= w^2 - w - 12 \qquad \blacklozenge\blacklozenge\blacklozenge$$

General Quadratic Trinomial

> Some people like to use the *FOIL* rule. Multiply the *First* terms, then *Outer*, *Inner*, and *Last* terms.

When we multiply two binomials of the form $ax + b$ and $cx + d$, we get

$$(ax + b)(cx + d) = acx^2 + adx + bcx + bd$$

Combining like terms, we obtain the following formula:

General Quadratic Trinomial	$(ax + b)(cx + d) = acx^2 + (ad + bc)x + bd$	46

◆◆◆ **Example 55:**

$$(3x - 2)(2x + 5) = 3(2)x^2 + [(3)(5) + (-2)(2)]x + (-2)(5)$$
$$= 6x^2 + 11x - 10$$

◆◆◆

Powers of Multinomials

As with a monomial raised to a power that is a positive integer, we raise a *multinomial* to a power by multiplying that multinomial by itself the proper number of times.

◆◆◆ **Example 56:**

$$(x - 3)^2 = (x - 3)(x - 3)$$
$$= x^2 - 3x - 3x + 9$$
$$= x^2 - 6x + 9$$

◆◆◆

◆◆◆ **Example 57:**

$$(x - 3)^3 = (x - 3)(x - 3)(x - 3)$$
$$= (x - 3)(x^2 - 6x + 9) \text{ (from Example 56)}$$
$$= x^3 - 6x^2 + 9x - 3x^2 + 18x - 27$$

Combining like terms gives us

$$(x - 3)^3 = x^3 - 9x^2 + 27x - 27$$

◆◆◆

Perfect Square Trinomial

When we *square a binomial* such as $(a + b)$, we get

$$(a + b)^2 = (a + b)(a + b) = a^2 + ab + ab + b^2$$

Collecting terms, we obtain the following formula:

Perfect Square Trinomial	$(a + b)^2 = (a + b)(a + b) = a^2 + 2ab + b^2$	47

This trinomial is called a *perfect square trinomial* because it is the square of a binomial. Its first and last terms are the squares of the two terms of the binomial; its middle term is twice the product of the terms of the binomial. Similarly, when we square $(a - b)$, we get the following:

Perfect Square Trinomial	$(a - b)^2 = a^2 - 2ab + b^2$	48

◆◆◆ **Example 58:**

$$(3x - 2)^2 = (3x)^2 + 3x(-2) - 2(3x) - 2(-2)$$
$$= 9x^2 - 12x + 4$$

◆◆◆

Exercise 4 ♦ Multiplication of Algebraic Expressions

Multiply the following monomials, and simplify. In questions 2 and 3, the first step has been provided for you.

1. $x^3(x^2)$

2. $x^2(3axy) = (1 \cdot 3)(1 \cdot a)(x^2 \cdot x)(1 \cdot y)$

3. $(-5ab)(-2a^2b^3) = (-5 \cdot -2)(a \cdot a^2)(b \cdot b^3)$

4. $(3.52xy)(-4.14xyz)$

5. $(6ab)(2a^2b)(3a^3b^3)$

6. $(-2a)(-3abc)(ac^2)$

7. $2p(-4p^2q)(pq^3)$

8. $(-3xy)(-2w^2x)(-xy^3)(-wy)$

Multiply the polynomial by the monomial, and simplify. The first step has been given to you for questions 10 and 11.

9. $2a(a - 5)$

10. $(x^2 + 2x)x^3 = (x^2 \cdot x^3) + (2x \cdot x^3)$

11. $xy(x - y + xy) = (xy \cdot x) + [xy \cdot (-y)] + (xy \cdot xy)$

12. $3a^2b^3(ab - 2ab^2 + b)$

13. $-4pq(3p^2 + 2pq - 3q^2)$

14. $2.8x(1.5x^3 - 3.2x^2 + 5.7x - 4.6)$

15. $3ab(2a^2b^2 - 4a^2b + 3ab^2 + ab)$

16. $-3xy(2xy^3 - 3x^2y + xy^2 - 4xy)$

Multiply the following binomials, and simplify. The first step has been given to you for questions 17 and 18.

17. $(a + b)(a + c) = (a \cdot a) + (a \cdot c) + (b \cdot a) + (b \cdot c)$

18. $(2x + 2)(5x + 3) = (2x \cdot 5x) + (2x \cdot 3) + (2 \cdot 5x) + (2 \cdot 3)$

19. $(x + 3y)(x^2 - y)$

20. $(a - 5)(a + 5)$

21. $(m + n)(9m - 9n)$

22. $(m^2 + 2c)(m^2 - 5c)$

23. $(7cd^2 + 4y^3z)(7cd^2 - 4y^3z)$

24. $(3a + 4c)(2a - 5c)$

25. $(x^2 + y^2)(x^2 - y^2)$

26. $(3.50a^2 + 1.20b)(3.80a^2 - 2.40b)$

Multiply the following binomials and trinomials, and simplify. In questions 27 and 28, the first step has been started for you.

27. $(a - c - 1)(a + 1) = (a \cdot a) + (a \cdot 1) + [(-c) \cdot (+a)] + [(-c) \cdot (+1)] + [(-1) \bullet (+a)]$
$+ [(-1) \bullet (+1)]$

28. $(x - y)(2x + 3y - 5) = (x \cdot 2x) + (x \cdot 3y) + [(+x) \cdot (-5)] + [(-y) \cdot (+2x)] + [(-y) \cdot$
$(+3y)] + [(-y) \cdot (-5)]$

29. $(m^2 - 3m - 7)(m - 2)$

30. $(-3x^2 + 3x - 9)(x + 3)$

31. $(x^6 + x^4 + x^2)(x^2 - 1)$

32. $(m - 1)(2m - m^2 + 1)$

33. $(1 - z)(z^2 - z + 2)$

34. $(3a^3 - 2ab - b^2)(2a - 4b)$

35. $(a^2 - ay + y^2)(a + y)$

36. $(y^2 - 7.5y + 2.8)(y + 1.6)$

37. $(a^2 + ay + y^2)(a - y)$

38. $(a^2 + ay - y^2)(a - y)$

Multiply the following binomials and polynomials, and simplify. The first step in question 39 has been started for you.

39. $[x^2 - (m - n)x - mn](x - p) = [x^2 - mx + nx - mn](x - p)$

40. $(x^2 + ax - bx - ab)(x + b)$

41. $(m^4 + m^3 + m^2 + m + 1)(m - 1)$

Multiply the following monomials and binomials, and simplify. The first step in question 42 has been started for you.

42. $(2a + 3x)(2a + 3x)9 = [(2a \times 2a) + (2a \cdot 3x) + (3x \cdot 2a) + (3x \cdot 3x)] \times 9$

43. $(3a - b)(3a - b)x$

44. $c(m - n)(m + n)$

45. $(x + 1)(x + 1)(x - 2)$

46. $(m - 2.5)(m - 1.3)(m + 3.2)$

47. $(x + a)(x + b)(x - c)$

48. $(x - 4)(x - 5)(x + 4)(x + 5)$

49. $(1 + c)(1 + c)(1 - c)(1 + c^2)$

Multiply the following trinomials, and simplify. The first step in question 50 has been started for you.

50. $(a + b - c)(a - b + c) = (a \cdot a) + [a \cdot (-b)] + (a \cdot c) + (b \cdot a) + [b \cdot (-b)] + (b \cdot c) + [(-c) \cdot a] + [(-c) \cdot (-b)] + [(-c) \cdot c]$

51. $(2a^3 + 5a)(2a^3 - 5ac^2 + 2a)$

52. $[3(a + b)^2 - (a + b) + 2][4(a + b)^2 - (a + b) - 5]$

53. $(a + b + c)(a - b + c)(a + b - c)$

Multiply the following polynomials, and simplify.

54. $(x^3 - 6x^2 + 12x - 8)(x^2 + 4x + 4)$

55. $(n - 5n^2 + 2 + n^3)(5n + n^2 - 10)$

56. $(mx + my - nx - ny)(mx - my + nx - ny)$

Powers of Algebraic Expressions

Square each binomial. The first step in question 57 has been started for you.

57. $(a + c) = (a + c) \cdot (a + c)$

58. $(p + q)$

59. $(m - n)$

60. $(x - y)$

61. $(A + B)$

62. $(A - C)$

63. $(3.80a - 2.20x)$

64. $(m + c)$

65. $(2c - 3d)$

66. $(x^2 - x)$

67. $(a - 1)$

68. $(a^2x - ax^3)$

69. $(y^2 - 20)$

70. $(x^n - y^2)$

Square each trinomial. The first step in question 71 has been started for you.

71. $(a + b + c) = (a + b + c) \cdot (a + b + c)$

72. $(a - b - c)$

73. $(1 + x - y)$

74. $(x^2 + 2x - 3)$

75. $(x^2 + xy + y^2)$

76. $(x^3 - 2x - 1)$

Cube each binomial. The first step in question 77 has been started for you.

77. $(a + b) = (a + b) \cdot (a + b) \times (a + b)$

78. $(2x + 1)$

79. $(p - 3q)$

80. $(x^2 - 1)$

81. $(a - b)$

82. $(3m - 2n)$

Symbols of Grouping

Remove symbols of grouping, and simplify. The first step has been started for you in questions 83 and 84.

83. $-3[-(a + b) - c] = -3[-a - b - c]$

84. $-[x + (y - 2) - 4] = -[x + y - 2 - 4]$

85. $(p - q) - [p - (p - q) - q]$

86. $-(x - y) - [y - (y - x) - x]$

87. $-5[-2(3y - z) - z] + y$

88. $[(m + 2n) - (2m - n)][(2m + n) - (m - 2n)]$

89. $(x - 1) - \{[x - (x - 3)] - x\}$

90. $(a - b)(a^3 + b^3)[a(a + b) + b^2]$

91. $y - \{4x - [y - (2y - 9) - x] + 2\}$

92. $[2x^2 + (3x - 1)(4x + 5)][5x^2 - (4x + 3)(x - 2)]$

Applications

93. A rectangle has its length L increased by 4.0 units and its width W decreased by 3.0 units. Write an expression for the area of the rectangle, and multiply out.

94. A car travelling at a rate R for a time T will go a distance equal to RT. If the rate is decreased by 3.60 km/h and the time is increased by 4.10 h, write an expression for the new distance traveled, and multiply out.

95. A square of side x has each side increased by 3.50 units. Write an expression for the area of the new square, and multiply out.

96. When a current I flows through a resistance R, the power in the resistance is I^2R. If the current is increased by 3.20 amperes, write an expression for the new power, and multiply out.

97. If the radius of a sphere of radius r is decreased by 3.60 units, write an expression for the new volume of the sphere, and multiply out.

2–5   Division of Algebraic Expressions

Symbols for Division

Division may be indicated by any of the following symbols:

$$x \div y \qquad \frac{x}{y} \qquad x/y$$

The names of the parts are

$$\text{quotient} = \frac{\text{dividend}}{\text{divisor}} = \frac{\text{numerator}}{\text{denominator}}$$

The quantity x/y is called a *fraction*. It is also called the *ratio* of x to y.

Reciprocals

As we saw in Sec. 1–4, the *reciprocal* of a number is 1 divided by that number. The reciprocal of n is $1/n$. We can use the idea of a reciprocal to show how division is related to multiplication. We may write the quotient of $x \div y$ as

$$\frac{x}{y} = \frac{x}{1} \cdot \frac{1}{y} = x \cdot \frac{1}{y}$$

We see that to *divide by a number* is the same thing as to *multiply by its reciprocal*. This fact will be especially useful for dividing by a fraction.

Division by Zero

Division by zero is not a permissible operation.

◆◆◆ **Example 59:** In the fraction

$$\frac{x + 5}{x - 2}$$

x cannot equal 2, or the illegal operation of division by zero will result. ◆◆◆

Rules of Signs

The quotient of two terms of *like* sign is *positive*.

$$\frac{+a}{+b} = \frac{-a}{-b} = \frac{a}{b}$$

The quotient of two terms of *unlike* sign is *negative*.

$$\frac{+a}{-b} = \frac{-a}{+b} = -\frac{a}{b}$$

The fraction itself carries a third sign, which, when negative, reverses the sign of the quotient. These three ideas are summarized in the following rules:

Rules of Signs for Division		
$\dfrac{+a}{+b} = \dfrac{-a}{-b} = -\dfrac{-a}{+b} = -\dfrac{+a}{-b} = \dfrac{a}{b}$		10
$\dfrac{+a}{-b} = \dfrac{-a}{+b} = -\dfrac{-a}{-b} = -\dfrac{a}{b}$		11

These rules show that any *pair* of negative signs may be removed without changing the value of the fraction.

◆◆◆ **Example 60:** Simplify $\dfrac{-ax^2}{-y}$.

Solution: Removing the pair of minus signs, we obtain

$$-\frac{ax^2}{-y} = \frac{ax^2}{y}$$ ◆◆◆

Common Error	Removal of pairs of negative signs pertains only to negative signs that are factors of the numerator, the denominator, or the fraction as a whole. Do not try to remove pairs of signs that apply only to single terms. Thus $$\frac{x - 2}{x - 3} \neq \frac{x + 2}{x + 3}$$

Dividing a Monomial by a Monomial

Any quantity (except zero) divided by itself equals *one*. So if the same factor appears in both the divisor and the dividend, it may be eliminated.

◆◆◆ **Example 61:** Divide $6ax$ by $3a$.

Solution:

$$\frac{6ax}{3a} = \frac{6}{3} \cdot \frac{a}{a} \cdot x = 2x$$ ◆◆◆

To divide quantities having exponents, we use the law of exponents for division.

If division by zero were allowed, we could, for example, divide 2 by zero and get a quotient x:

$$\frac{2}{0} = x$$

or

$$2 = 0 \cdot x$$

but there is no number x which, when multiplied by zero, gives 2, so we cannot allow this operation.

Quotients	$\dfrac{x^a}{x^b} = x^{a-b} \ (x \neq 0)$	30

◆◆◆ **Example 62:** Divide y^5 by y^3.

Solution: By Eq. 30,

$$\frac{y^5}{y^3} = y^{5-3} = y^2 \qquad \text{◆◆◆}$$

When there are numerical coefficients, divide them separately.

◆◆◆ **Example 63:**

(a) $\dfrac{15x^6}{3x^4} = \dfrac{15}{3} \cdot \dfrac{x^6}{x^4} = 5x^{6-4} = 5x^2$

(b) $\dfrac{8.35y^5}{3.72y} = \dfrac{8.35}{3.72} \cdot \dfrac{y^5}{y} = 2.24y^{5-1} = 2.24y^4$ ◆◆◆

If there is more than one unknown, treat each separately.

◆◆◆ **Example 64:** Divide $18x^5y^2z^4$ by $3x^2yz^3$.

Solution:

$$\frac{18x^5y^2z^4}{3x^2yz^3} = \frac{18}{3} \cdot \frac{x^5}{x^2} \cdot \frac{y^2}{y} \cdot \frac{z^4}{z^3} = \left(\frac{18}{3}\right) \cdot x^{(5-2)} \cdot y^{(2-1)} \cdot z^{(4-3)} = 6x^3yz \qquad \text{◆◆◆}$$

Sometimes negative exponents will be obtained.

◆◆◆ **Example 65:**

$$\frac{6x^2}{x^5} = 6x^{2-5} = 6x^{-3}$$

Solution: You should make every effort to not leave your answer with a negative exponent. Thus;

$$6x^{-3} = \frac{6}{1} \times \frac{1}{x^3} = \frac{6}{x^3}$$

$$6x^{-3} = \frac{6}{x^3} \qquad \text{◆◆◆}$$

The process of dividing a monomial by a monomial is also referred to as *simplifying a fraction, or reducing a fraction to lowest terms*. We discuss this in Chapter 9.

◆◆◆ **Example 66:** Simplify the fraction

$$\frac{3x^2yz^5}{9xy^4z^2}$$

Solution: The procedure is no different than if we had been asked to divide $3x^2yz^5$ by $9xy^4z^2$.

$$\frac{3x^2yz^5}{9xy^4z^2} = \frac{3}{9}x^{2-1}y^{1-4}z^{5-2}$$

$$= \frac{1}{3}xy^{-3}z^3$$

or

$$= \frac{xz^3}{3y^3} \qquad \text{◆◆◆}$$

Do not be dismayed if the expressions to be divided have negative exponents. Apply Eq. 30, as before.

◆◆◆ **Example 67:** Divide $21x^2y^{-3}z^{-1}$ by $7x^{-4}y^2z^{-3}$.

Solution: Proceeding as before, we obtain

$$\frac{21x^2y^{-3}z^{-1}}{7x^{-4}y^2z^{-3}} = \frac{21}{7}x^{2-(-4)}y^{-3-2}z^{-1-(-3)}$$

$$= 3x^6y^{-5}z^2$$

$$= \frac{3x^6z^2}{y^5} \qquad \qquad \text{◆◆◆}$$

Dividing a Multinomial by a Monomial
Divide each term of the multinomial by the monomial.

◆◆◆ **Example 68:** Divide $9x^2 + 3x - 2$ by $3x$.

Solution:

$$\frac{9x^2 + 3x - 2}{3x} = \frac{9x^2}{3x} + \frac{3x}{3x} - \frac{2}{3x}$$

This is really a consequence of the distributive law (Eq. 5).

$$\frac{9x^2 + 3x - 2}{3x} = \frac{1}{3x}(9x^2 + 3x - 2) = \left(\frac{1}{3x}\right)(9x^2) + \left(\frac{1}{3x}\right)(3x) + \left(\frac{1}{3x}\right)(-2)$$

Each of these terms is now simplified as in the preceding section.

$$\frac{9x^2}{3x} + \frac{3x}{3x} - \frac{2}{3x} = 3x + 1 - \frac{2}{3x} \qquad \qquad \text{◆◆◆}$$

Common Error	Do not forget to divide *every* term of the multinomial by the monomial. $$\frac{x^2 + 2x + 3}{x} \neq x + 2 + 3$$

◆◆◆ **Example 69:**

(a) $\dfrac{3ab + 2a^2b - 5ab^2 + 3a^2b^2}{15ab} = \dfrac{3ab}{15ab} + \dfrac{2a^2b}{15ab} - \dfrac{5ab^2}{15ab} + \dfrac{3a^2b^2}{15ab}$

$$= \frac{1}{5} + \frac{2a}{15} - \frac{b}{3} + \frac{ab}{5}$$

(b) $\dfrac{6x^3y - 8xy^2 - 2x^4y^4}{-2xy} = -3x^2 + 4y + x^3y^3$

(c) $\dfrac{8.26x^3 - 6.24x^2 + 7.37x}{2.82x} = 2.93x^2 - 2.21x + 2.61 \qquad \qquad$ ◆◆◆

Common Error	There is no similar rule for dividing a monomial by a multinomial. Thus $$\frac{a}{b + c} \neq \frac{a}{b} + \frac{a}{c}$$

Dividing a Polynomial by a Polynomial
Write the divisor and the dividend in the order of descending powers of the variable. Supply any missing terms, using a coefficient of zero. Set up as a long-division problem, as in the following example.

This method is used only for *polynomials* (i.e., expressions in which the powers are all *positive integers*).

◆◆◆ **Example 70:** Divide $x^2 + 4x^4 - 2$ by $1 + x$.

Solution:

1. Write the dividend in descending order of the powers.

$$4x^4 + x^2 - 2$$

2. Supply the missing terms with coefficients of zero.

$$4x^4 + 0x^3 + x^2 + 0x - 2$$

The method works just as well if we write the terms in order of *ascending* powers of the variable, but this is less commonly seen.

3. Write the divisor in descending order of the powers, and set up in long-division format.

$$(x + 1)\overline{)4x^4 + 0x^3 + x^2 + 0x - 2}$$

4. Divide the first term in the dividend $(4x^4)$ by the first term in the divisor (x). Write the result $(4x^3)$ above the dividend, in line with the term having the same power. It is the first term of the quotient.

$$
\begin{array}{r}
4x^3 \\
(x + 1)\overline{)4x^4 + 0x^3 + x^2 + 0x - 2}
\end{array}
$$

5. Multiply the divisor by the first term of the quotient. Write the result below the dividend. Subtract it from the dividend.

$$
\begin{array}{r}
4x^3 \\
(x + 1)\overline{)4x^4 + 0x^3 + x^2 + 0x - 2} \\
-(4x^4 + 4x^3) \\
\hline
-4x^3 + x^2 + 0x - 2
\end{array}
$$

6. Repeat Steps 4 and 5, each time using the new dividend obtained, until the degree of the remainder is less than the degree of the divisor.

$$
\begin{array}{r}
4x^3 - 4x^2 + 5x - 5 \\
(x + 1)\overline{)4x^4 + 0x^3 + x^2 + 0x - 2} \\
\underline{4x^4 + 4x^3} \\
-4x^3 + x^2 + 0x - 2 \\
\underline{-4x^3 - 4x^2} \\
5x^2 + 0x - 2 \\
\underline{5x^2 + 5x} \\
-5x - 2 \\
\underline{-5x - 5} \\
3
\end{array}
$$

The result is written

$$\frac{4x^4 + x^2 - 2}{x + 1} = 4x^3 - 4x^2 + 5x - 5 + \frac{3}{x + 1}$$

or

$$4x^3 - 4x^2 + 5x - 5 \quad \text{(R3)}$$

to indicate a remainder of 3.   ◆◆◆

Common Error

Errors are often made during the subtraction step (Step 5 in Example 70).

$$
\begin{array}{r}
4x^3 \\
(x + 1)\overline{)4x^4 + 0x^3 + x^2 + 0x - 2} \\
\underline{4x^4 + 4x^3} \\
4x^3 + x^2 + 0x - 2
\end{array}
$$

No! Should be $-4x^3$.

◆◆◆ **Example 71:** Divide $x^4 - 5x^3 + 9x^2 - 7x + 2$ by $x^2 - 3x + 2$.

Solution: Using long division, we get

$$
\begin{array}{r}
x^2 - 2x + 1 \\
x^2 - 3x + 2\overline{\smash{\big)}\,x^4 - 5x^3 + 9x^2 - 7x + 2} \\
\underline{x^4 - 3x^3 + 2x^2} \\
-2x^3 + 7x^2 - 7x + 2 \\
\underline{-2x^3 + 6x^2 - 4x} \\
x^2 - 3x + 2 \\
\underline{x^2 - 3x + 2} \\
0
\end{array}
$$

Exercise 5 ◆ Division of Algebraic Expressions

Divide the following monomials.

1. z^5 by z^3

2. $45y^3$ by $15y^2$

3. $-25x^2y^2z^3$ by $5xyz^2$

4. $20a^5b^5c$ by $10abc$

5. $30cd^2f$ by $15cd^2$

6. $36ax^2y$ by $18ay$

7. $-18x^2yz$ by $9xy$

8. $\dfrac{56y^3w^2}{7yw}$

9. $15axy^3$ by $-3ay$

10. $-18a^3x$ by $-3ay$

11. $6acdxy^2$ by $2adxy^2$

12. $12a^2x^2$ by $-3a^2x$

13. $15ay^2$ by $-3ay$

14. $\dfrac{21a^5b}{7a^4b}$

15. $-21vwz^2$ by $7vz^2$

16. $-33r^2s$ by $11rs^2$

17. $35m^2nx$ by $5m^2x$

18. $20x^3y^3z^3$ by $10x^3yz^3$

19. $16ab$ by $4a$

20. $21acd$ by $7c$

21. ab^2c by ac

22. $(a + b)^4$ by $(a + b)$

23. x^m by x^n

24. $-14a^2x^3y^4$ by $7axy^2$

25. $32r^2s^2q$ by $8r^2sq$

26. $-18v^2x^2y$ by v^2xy

27. $24a^{2n}bc^n$ by $(-a^{2n}bc^n)$

28. $36a^{2n}xy^{3n}$ by $(-4a^nxy^{2n})$

29. $25xyz^2$ by $(-5x^2y^2z)$

30. $-28y^2z^{2m}$ by $4y^2z^3x$

31. $-30n^2x^2$ by $6m^2x^3$

32. $28x^2y^2z^2$ by $7xyz$

33. $6abc$ by $2c$

34. ax^3 by ax^2

35. $3mx^6$ by mx

36. $210c^3b$ by $7cd$

37. $42xy$ by xy^2

38. $-21ac$ by $-7a$

39. $-12xy$ by $3y$

40. $72abc$ by $-8c$

Divide the polynomial by the monomial.

41. $2a^3 - a^2$ by a

42. $42a^5 - 6a^2$ by $6a$

43. $21x^4 - 3x^2$ by $3x^2$

44. $35m^4 - 7p^2$ by 7

45. $27x^6 - 45x^4$ by $9x^2$

46. $24x^6 - 8x^3$ by $-8x^3$

47. $34x^3 - 51x^2$ by $17x$

48. $5x^5 - 10x^3$ by $-5x^3$

49. $-3a^2 - 6ac$ by $-3a$

50. $-5x^3 + x^2y$ by $-x^2$

51. $ax^2y - 2xy^2$ by xy

52. $3xy^2 - 3x^2y$ by xy

53. $4x^3y^2 + 2x^2y^3$ by $2x^2y^2$

54. $3a^2b^2 - 6ab^3$ by $3ab$

55. $abc^2 - a^2b^2c$ by $-abc$

56. $9x^2y^2z + 3xyz^2$ by $3xyz$

57. $x^2y^2 - x^3y - xy^3$ by xy

58. $a^3 - a^2b - ab^2$ by $-a$

59. $a^2b - ab + ab^2$ by $-ab$

60. $xy - x^2y^2 + x^3y^3$ by $-xy$

61. $a^2 - 3ab + ac^2$ by a

62. $x^2y - xy^2 + x^2y^3$ by xy

63. $x^2 - 2xy + y^2$ by x

64. $z^2 - 3xz + 3z^3$ by z

65. $m^2n + 2mn - 3m^2$ by mn

66. $c^2d - 3cd^2 + 4d^3$ by cd

Divide the polynomial by the binomial.

67. $x^2 + 15x + 56$ by $x + 7$

68. $x^2 - 15x + 56$ by $x - 7$

69. $2a^2 + 11a + 5$ by $2a + 1$

70. $6a^2 - 7a - 3$ by $2a - 3$

71. $27a^3 - 8b^3$ by $3a - 2b$

72. $4a^2 + 23a + 15$ by $4a + 3$

73. $x^2 - 4x + 3$ by $x + 2$

74. $2 + 4x - x^2$ by $4 - x$

75. $4 + 2x - 5x^2$ by $3 - x$

76. $2x^2 - 5x + 4$ by $x + 1$

Divide each polynomial by the given trinomial.

77. $a^2 - 2ab + b^2 - c^2$ by $a - b - c$

78. $a^2 + 2ab + b^2 - c^2$ by $a + b + c$

79. $x^2 - y^2 + 2yz - z^2$ by $x - y + z$

80. $c^4 + 2c^2 - c + 2$ by $c^2 - c + 1$

81. $x^2 - 4y^2 - 4yz - z^2$ by $x + 2y + z$

82. $x^5 + 37x^2 - 70x + 50$ by $x^2 - 2x + 10$

Case Study Discussion—Surge Protectors in High Voltage Lines

The expression I^2R represents the power in the surge protector. If current increases by 3200 times, we can say the new current I_{new} is equal to 3200 times the old current. So you can write the expression for the new power as $(I_{new})^2R$, or $(3200I)^2R$. Now we can apply the laws of exponents and square both the I and the 3200. The expression for the new surge of power is 10 240 000I^2R because a 3200 times increase in current would mean the power surged by 3200^2, which is 10 240 000 times. This is above the 10 000 000-fold increase in power needed to melt the surge protector. The conclusion is that the surge protector would melt and require replacement after a lighting strike of this magnitude.

✦✦✦ CHAPTER 2 REVIEW PROBLEMS ✦✦✦✦✦✦✦✦✦✦✦✦✦✦✦✦✦✦✦✦✦✦✦✦✦✦✦✦✦

1. Multiply: $(b^4 + b^2x^3 + x^4)(b^2 - x^2)$

2. Square: $(x + y - 2)$

3. Evaluate: $(7.28 \times 10^4)^3$

4. Square: $(xy + 5)$

5. Multiply: $(3x - m)(x^2 + m^2)(3x - m)$

6. Divide: $-x^6 - 2x^5 - x^4$ by $-x^4$

7. Cube: $(2x + 1)$

8. Simplify: $7x - \{-6x - [-5x - (-4x - 3x) - 2]\}$

9. Multiply: $3ax^2$ by $2ax^3$

10. Simplify: $\left(\dfrac{3a^2}{2b^3}\right)^3$

11. Divide: $a^2x - abx - acx$ by ax

12. Divide: $3x^5y^3 - 3x^4y^3 - 3x^2y^4$ by $3x^3y^2$

13. Square: $(4a - 3b)$

14. Multiply: $(x^3 - xy + y^2)(x + y)$

15. Multiply: $(xy - 2)(xy - 4)$

16. Divide: $x^{m+1} + x^{m+2} + x^{m+3} + x^{m+4}$ by x^4

17. Square: $(2a - 3b)$

18. Multiply: $(a^2 - 3a + 8)(a + 3)$

19. Divide: $a^3b^2 - a^2b^5 - a^4b^2$ by a^2b

20. Multiply: $(2x - 5)(x + 2)$

21. Multiply: $(2m - c)(2m + c)(4m^2 + c^2)$

22. Simplify: $\left(\dfrac{8x^5y^{-2}}{4x^3y^{-3}}\right)^3$

23. Divide: $2a^6$ by a^4

24. Multiply: $(a^2 + a^2y + ay^2 + y^3)(a - y)$

25. Simplify: $y - 3[y - 2(4 - y)]$

26. Divide: $-a^7$ by a^5

27. Multiply: $(2x^2 + xy - 2y^2)(3x + 3y)$

28. Divide: $x^4 - \dfrac{1}{2}x^3 - \dfrac{1}{3}x^2 - 2x - 1$ by $2x$

29. Simplify: $-2[w - 3(2w - 1)] + 3w$

30. Multiply: $(a^2 + b)(a + b^2)$

31. Multiply: $(a^4 - 2a^3c + 4a^2c^2 - 8ac^3 + 16c^4)(a + 2c)$

32. Divide: $16x^3$ by $4x$

33. Divide: $(a - c)^m$ by $(a - c)^2$

34. Divide: $7 - 8c^2 + 5c^3 + 8c$ by $5c - 3$

35. Divide: $-x^2y - xy^2$ by $-xy$

36. Simplify: $(-2x^2 - x + 6) - (7x^2 - 2x + 4) + (x^2 - 3)$

37. Cube: $(b - 3)$

38. Simplify: $(2x^2y^3z^{-1})^3$

39. Divide: $2x^{-2}y^3$ by $4x^{-4}y^6$

40. Simplify: $-\{-[-(a - 3) - a]\} + 2a$

41. Multiply: $(x - 2)(x + 4)$

42. Square: $(x^2 + 2)$

43. Divide: $a^2b^2 - 2ab - 3ab^3$ by ab

44. Divide: $3a^3c^3 + 3a^2c - 3ac^2$ by $3ac$

45. Divide: $6a^3x^2 - 15a^4x^2 + 30a^3x^3$ by $-3a^3x^2$

46. Divide: $20x^2y^4 - 14xy^3 + 8x^2y^2$ by $2x^2y^2$

47. Evaluate: $(2.83 \times 10^3)^2$

48. Simplify: $(x - 3) - [x - (2x + 3) + 4]$

49. Multiply: $2xy^3$ by $5x^2y$

50. Divide: $27xy^5z^2$ by $3x^2yz^2$

51. Multiply: $(x - 1)(x^2 + 4x)$

52. Square: $(z^2 - 3)$

53. Evaluate: $(1.33 \times 10^4)^2$

54. Simplify: $(y + 1) - [y(y + 1) + (3y - 1) + 5]$

55. Multiply: $2ab^2$ by $3a^2b$

56. Divide: $64ab^4c^3$ by $8a^2bc^3$

57. Divide: $x^8 + x^4 + 1$ by $x^4 - x$

58. Divide: $1 - a^3b^3$ by $1 - ab$

59. To make 750 kilograms of a new alloy, we take x kg of alloy A, which contains 85% copper, and for the remainder use alloy B, which contains 72% copper. The kilograms of copper in the final batch will be

$$0.85x + 0.72(750 - x)$$

Simplify this expression.

60. The area of a circle of radius r is πr^2. If the radius is tripled, the area will be

$$\pi(3r)^2$$

Simplify this expression.

61. A freely falling body, starting from rest, falls a distance of $16.1t^2$ feet in t seconds. In half that time it will fall

$$16.1\left(\frac{t}{2}\right)^2 \text{ ft.}$$

Simplify this expression.

62. The power in a resistor of resistance R which has a voltage V across it is V^2/R. If the voltage is doubled, the power will be

$$\frac{(2V)^2}{R}$$

Simplify this expression.

Writing

63. Suppose your friend is sick and missed the introduction to algebra and is still out of class. Write a note to your friend explaining in your own words what you think algebra is and how it is related to the arithmetic that you both just finished studying.

64. In simplifying a fraction, we are careful not to change its value. If that's the case, and the new fraction is "the same" as the old, then why bother? Write a few paragraphs explaining why you think it's valuable to simplify fractions (and other expressions as well) or whether there is some advantage to just leaving them alone.

65. Your friend refuses to learn the FOIL rule (Sec. 2-4). "*I want to learn math, not memorize a bunch of tricks!*" he declares. What do you think? Write a paragraph or so giving your opinion on the value or harm in learning devices such as the FOIL rule.

66. In the following chapter we will solve simple equations, something you have probably done before. Without peeking ahead, write down in your own words whatever you remember about solving equations. You may give a list of steps or a description in paragraph form.

Team Projects

67. Write a one-hour exam that will test the most important skills covered in this chapter. Be sure that it is not too long and hard, or not too short and easy. Then work the test and produce an answer key, to be handed in with the test.

68. Fig. 2–3 shows a geometric representation of $(a + b)^2$, the square of the binomial $(a + b)$. Use the figure to evaluate $(a + b)^2$. Then make a similar sketch to evaluate each of the following:
(a) $(2a + b)^2$
(b) $(a + 2)(a + 3)$
(c) $(a + b)(c + d)$
(d) $(a + b + 1)^2$

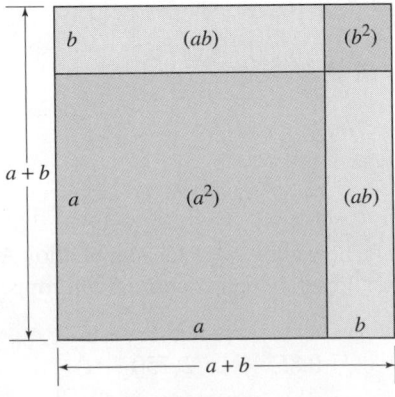

FIGURE 2–3

Simple Equations and Word Problems

When you have completed this chapter, you should be able to:
- Solve simple linear equations, including fractional equations, and check your solution.
- Set up and solve word problems that lead to linear equations.
- Solve financial-type word problems.
- Solve mixture-type word problems.
- Solve statics-type word problems.

Many students have mixed feelings about the material in Chapter 3. In this chapter we create equations by setting two expressions equal to each other, and we can finally start solving equations. We can even check our answers so that we know we have the correct solution. But with solving equations we also must look at the heart of technical mathematics—solving problems.

Very few problems in our careers are presented to us as complete equations. Instead, we have data we can measure or collect and a challenge to be met. To practise these very real problems, we describe the situation and present the dreaded "word problem." Although we look at logical steps to take in solving a problem, there is often room for interpretation, with every problem being unique. To introduce you to word problems we employ several different areas: number problems, mixture problems, financial problems, and statics problems. We have chosen these areas in the hope that you will be able to visualize and understand the situations without special training.

Problem solving is a skill that employers mention over and over again as being absolutely necessary for graduates. You will need to become an expert problem solver. Fortunately, it's a transferrable skill. Whatever program you are in, your career will involve problem solving within a specific area of science and technology. But according to the latest statistics, you will be changing your career (not just your job) several times in your life. Just ask some of the mature students in your class about changing careers!

The best way to become skillful is through practice. The better you understand the process of solving problems, the better your problem-solving skills will be for both your technology and science courses as well as for your career.

When two mathematical expressions are set equal to each other, we get an *equation*. Much of our work in technical mathematics is devoted to *solving equations*. We start in this chapter with the simplest types, and in later chapters we cover more difficult ones (quadratic equations, exponential equations, trigonometric equations, etc.).

Why is it so important to solve equations? The main reason is that equations are used to describe or *model* the way certain things happen in the world. Solving an equation often tells us something important that we would not otherwise know. For example, the note produced by a guitar string is not a whim of nature but can be predicted (solved for) when you know the length, the mass, and the tension in the string, and you know the *equation* relating the pitch, length, mass, and tension.

Thousands of equations exist that link together the various quantities in the physical world, chemistry, finance, factories, and so on—and their number is still increasing. To be able to solve and to manipulate such equations is essential for anyone who has to deal with these quantities on the job.

CASE STUDY—HIRING EMPLOYEES ON A BUDGET

A mechanical engineering technologist, new on the job, finds herself facing clients of her company, Portage la Prairie Engineering Consultants, Ltd. The clients want her opinion on what type of employees wlll be needed for a project her organization will be supervising. The client is providing the employees, and her organization is providing expertise and project management supervision. They want supervisors to gain experience from the project and will also need some staff workers on the team.

The project has a budget of $60,000, with about 25% for salaries, and will take one week to complete. Recommend a mix of supervisors and staff that will closely meet the project budget while still having one supervisor for no more than three staffers. A supervisor earns $1,500 per week, a staff member $1,100 per week. As with many cases, you will not get a perfect answer from the math (for example, if you calculate that you need 3.5 supervisors, we cannot cut a supervisor in half!); however, the math solution will allow you to provide a solution that meets the criteria. Note that the salary category does not have to be exactly 25% of $60,000 but should be as close as possible without going over 25%.

CANADIANS LOVE TO DOWNLOAD—BRENT TOWNSHEND

Brent Townshend is a Canadian artist, inventor, and scientific researcher. Born in Toronto in 1960, he earned an MS in computer science (1983) and a PhD in electrical and biomedical engineering (1987) at Stanford University. A subsequent research career encompassed working at AT&T Bell Laboratories and teaching at McGill and Stanford.

It was AT&T that created the first commercial modem in 1962, the Bell 103. That first modem could transfer data between computers at 300 bits per second. Almost 20 years later, in 1980, the download modem speed had increased to 14.1 kilobits per second (Kbps) and in 1994 doubled to 28.8 Kbps. All this time, the available conduit for transmission was the standard analog telephone line, which was the real stumbling block for transmission of digital information.

Following his stint at Bell Labs, Dr. Townshend began to create a succession of high-tech companies that focused on signal processing. His interest in manipulating and processing digital data led him in 1993 to the ideas that would eventually in 1996 become the 56 Kbps dial-up modem technology known as 56K. This enabled data downloading at a rate two-thirds higher than existing modems at the time (uploading was not as fast). As an independent inventor he was able to apply for and eventually secure patents, and the technology was licensed by modem manufacturers. The first was 3Com, known at the time as U.S. Robotics, still a leader in modem and computer signal-processing technology.

In 1998, the V.90 standard for 56K computer-to-computer communication (based on Townshend's algorithms) was adopted by the International Telecommunications Union.

In that same year, the vice president of advanced development at 3Com, Dale Walsh, stated, "Townshend's groundbreaking work allowed us to create a new product that drastically improved analog modem connection speeds. His invention revolutionized the way people all over the world experience the Web."

3–1 Solving First-Degree Equations

Equations

An equation has two *sides*, or members, and an *equal sign*.

$$3x^2 - 4x = 2x + 5$$

left side | right side
equal sign

Conditional Equations

A *conditional equation* is one whose sides are equal only for certain values of the unknown.

◆◆◆ Example 1: The equation

$$x - 5 = 0$$

is a conditional equation because the sides are equal only when $x = 5$. When we say "equation," we will mean "conditional equation." Other equations that are true for any value of the unknown, such as $x(x + 2) = x^2 + 2x$, are called *identities*. ◆◆◆

The symbol $\equiv$ is used for identities. We would write $x(x + 2) \equiv x^2 + 2x$.

The Solution

The *solution* to an equation is that value (or those values) of the unknown that makes the sides equal.

◆◆◆ Example 2: The solution to the equation

$$x - 5 = 0$$

is

$$x = 5$$

The value $x = 5$ is also called the *root* of the equation. We also say that it *satisfies* the equation. ◆◆◆

First-Degree Equations

A *first-degree* equation (also called a *linear* equation) is one in which the terms containing the unknown are all of first degree (exponents all equal 1). The graph of a linear equation is a straight line.

◆◆◆ Example 3: The equations

$$2x + 3 = 9 - 4x,$$
$$\frac{3x}{2} = 6x + 3,$$

and

$$3x + 5y - 6z = 0$$

are all of first degree. The first and second equations have one variable, but the third equation has more than one variable. ◆◆◆

Solving an Equation

To solve an equation, we *perform the same mathematical operation on each side of the equation.* The object is to get the unknown standing alone on one side of the equation.

◆◆◆ Example 4: Solve $3x = 8 + 2x$.

Solution: Subtracting $2x$ from both sides, we obtain

$$3x - 2x = 8 + 2x - 2x$$

Combining like terms yields

$$x = 8$$

◆◆◆

Checking

To check an apparent solution, substitute it back into the original equation.

Get into the habit of checking your work, for errors creep in everywhere.

◆◆◆ **Example 5:** Checking the apparent solution to Example 4, we substitute 8 for x in the original equation.

$$3x = 8 + 2x$$
$$3(8) \stackrel{?}{=} 8 + 2(8)$$
$$24 = 8 + 16 \quad \text{(checks)}$$

◆◆◆

Common Error	Check your solution only in the *original* equation. Later versions may already contain errors.

◆◆◆ **Example 6:** Solve the equation $3x - 5 = x + 1$.

Solution: We first subtract x from both sides and then add 5 to both sides.

$$\begin{array}{rcr}
3x - 5 &=& x + 1 \\
-x + 5 & & -x + 5 \\
\hline
2x &=& 6
\end{array}$$

Dividing both sides by 2, we obtain

$$x = 3$$

Check: Substituting into the original equation yields

$$3(3) - 5 \stackrel{?}{=} 3 + 1$$
$$9 - 5 = 4 \quad \text{(checks)}$$

◆◆◆

Symbols of Grouping

When the equation contains symbols of grouping, remove them early in the solution.

◆◆◆ **Example 7:** Solve the equation $3(3x + 1) - 6 = 5(x - 2) + 15$.

Solution: First, remove the parentheses:

$$\begin{array}{lrcr}
& 9x + 3 - 6 &=& 5x - 10 + 15 \\
\text{Combine like terms:} & 9x - 3 &=& 5x + 5 \\
\text{Add } -5x + 3 \text{ to both sides:} & -5x + 3 &=& -5x + 3 \\
\hline
& 4x &=& 8 \\
\text{Divide by 4:} & x = 2 & &
\end{array}$$

Check:

$$3(6 + 1) - 6 \stackrel{?}{=} 5(2 - 2) + 15$$
$$21 - 6 \stackrel{?}{=} 0 + 15$$
$$15 = 15 \quad \text{(checks)}$$

◆◆◆

> The mathematical operations that you perform must be done *to both sides* of the equation in order to preserve the equality.

> The mathematical operations that you perform on both sides of an equation must be done to each side as a whole—*not term by term.*

Fractional Equations

A *fractional equation* is one that contains one or more fractions. When an equation contains a single fraction, the fraction can be eliminated by *multiplying both sides by the denominator of the fraction.*

◆◆◆ **Example 8:** Solve:

$$\frac{x}{3} - 2 = 5$$

Solution: Multiplying both sides by 3 gives us

$$3\left(\frac{x}{3} - 2\right) = 3(5)$$

$$3\left(\frac{x}{3}\right) - 3(2) = 3(5)$$

$$x - 6 = 15$$

Add 6 to both sides:
$$\frac{+6 \quad +6}{x = 21}$$

Check:

$$\frac{21}{3} - 2 \overset{?}{=} 5$$

$$7 - 2 = 5 \quad \text{(checks)}$$ ◆◆◆

When there are two or more fractions, multiplying both sides by the product of the denominators (called a *common denominator*) will clear the fractions. You can think of this process as applying the same logic (multiplying by the denominator) for each fraction.

In Chapter 9 we will treat fractional equations in detail. There we will see how to find the *least common denominator.*

◆◆◆ **Example 9:** Solve:

$$\frac{x}{2} + 3 = \frac{x}{3}$$

Solution: First, we multiply both sides by the common factors of 2 and 3. To get the common factors of 2 and 3, we need to multiply them together to get the lowest common factor of 6 (because there are no others lower). Now multiply both sides by 6.

$$6\left(\frac{x}{2} + 3\right) = 6\left(\frac{x}{3}\right)$$

$$6\left(\frac{x}{2}\right) + 6(3) = 6\left(\frac{x}{3}\right)$$

$$3x + 18 = 2x$$

We now subtract $2x$ from both sides, and also subtract 18 from both sides.

$$3x - 2x = -18$$

$$x = -18$$

Check:

$$\frac{-18}{2} + 3 \overset{?}{=} \frac{-18}{3}$$
$$-9 + 3 \overset{?}{=} -6$$
$$-6 = -6 \quad \text{(checks)}$$

◆◆◆

Strategy

While solving an equation, you should keep in mind the objective of *getting x by itself on one side of the equal sign, with no x on the other side*. Use any valid operation toward this end.

Equations can be of so many different forms that we cannot give a step-by-step procedure for their solution, but the following tips should help:

1. Eliminate fractions by multiplying both sides by the common denominator of both sides.
2. Remove any parentheses by performing the indicated multiplication.
3. All terms containing x should be moved to one side of the equation, and all other terms moved to the other side.
4. Like terms on the same side of the equation should be combined at any stage of the solution.
5. Any coefficient of x may be removed by dividing both sides by that coefficient.
6. Check the answer in the original equation.

The following example shows the use of these steps.

◆◆◆ **Example 10:** Solve:

$$\frac{3x - 5}{2} = \frac{2(x - 1)}{3} + 4$$

Solution:

1. We eliminate the fractions by multiplying by 6.

$$6\left(\frac{3x - 5}{2}\right) = 6\left[\frac{2(x - 1)}{3}\right] + 6(4)$$
$$3(3x - 5) = 4(x - 1) + 6(4)$$

2. Then we remove parentheses and combine like terms.

$$9x - 15 = 4x - 4 + 24$$
$$9x - 15 = 4x + 20$$

3. We get all x terms on one side by adding 15 to and subtracting $4x$ from both sides.

$$9x - 4x = 20 + 15$$

4. Combining like terms gives us

$$5x = 35$$

5. Dividing by the coefficient of x yields

$$x = 7$$

Check:

$$\frac{3(7) - 5}{2} \overset{?}{=} \frac{2(7 - 1)}{3} + 4$$
$$\frac{21 - 5}{2} \overset{?}{=} \frac{12}{3} + 4$$
$$\frac{16}{2} \overset{?}{=} 4 + 4$$
$$8 = 8 \quad \text{(checks)}$$

◆◆◆

◆◆◆ **Example 11:** Solve:

$$3(x + 2)(2 - x) = 3(x - 2)(x - 3) + 2x(4 - 3x)$$

Solution: We remove parentheses in two steps. First, using FOIL (Sec. 2-4)

$$3(2x - x^2 + 4 - 2x) = 3(x^2 - 5x + 6) + 8x - 6x^2$$

Then we remove the remaining parentheses.

$$-3x^2 + 12 = 3x^2 - 15x + 18 + 8x - 6x^2$$

Next we combine like terms.

$$-3x^2 + 12 = -3x^2 - 7x + 18$$

Adding $3x^2$ to and subtracting 18 from both sides gives

$$12 - 18 = -7x$$
$$-6 = -7x$$

Dividing both sides by -7 and switching sides, we get

$$x = \frac{6}{7}$$

◆◆◆

Other Variables

Up to now we have used only the familiar x as our variable. Of course, any letter can represent the unknown quantity. When we say to "solve" an equation, we mean to isolate the letter quantity, if only one is present. If there is more than one letter, we must be told which to solve for.

◆◆◆ **Example 12:** Solve the equation

$$4(p - 3) = 3(p + 4) - 5$$

Solution: Removing parentheses and combining like terms, we have

$$4p - 12 = 3p + 12 - 5$$
$$4p - 12 = 3p + 7$$

Adding 12 to and subtracting $3p$ from both sides and combining like terms gives

$$4p - 3p = 7 + 12$$
$$p = 19$$

◆◆◆

Equations with Approximate Numbers

All of our equations so far have had coefficients and constants that were integers, and we assumed them to be exact numbers. But in technology we must often solve equations containing approximate numbers. The method of solution is no different than with exact numbers. Just be sure to round following the rules given in Chapter 1.

◆◆◆ **Example 13:** Solve for x:

$$5.72(x - 1.15) = 3.22x - 7.73$$

Solution: Removing parentheses gives

$$5.72x - 6.58 = 3.22x - 7.73$$

Next, collect terms.

$$5.72x - 3.22x = -7.73 + 6.58$$
$$2.50x = -1.15$$

Dividing both sides by 2.50, we get

$$x = \frac{-1.15}{2.50} = -0.460$$ ◆◆◆

Literal Equations

A *literal equation* is one in which some or all of the constants are represented by letters.

◆◆◆ **Example 14:** The following is a literal equation:

$$3x + b = 5$$ ◆◆◆

To solve a literal equation for a given quantity means to isolate that quantity on one side of the equal sign. The other side of the equation will, of course, contain the other letter quantities. We do this by following the same procedures as for numerical equations.

◆◆◆ **Example 15:** Solve the equation $3x + b = 5$ for x.

Solution: Subtracting b from both sides gives

$$3x = 5 - b$$

Dividing by 3, we obtain

$$x = \frac{5 - b}{3}$$ ◆◆◆

In this chapter we will solve only the simplest kinds of literal equations. Our main treatment will be in Sec. 9–7.

Exercise 1 ◆ Solving First-Degree Equations

Treat any integers in these equations as exact numbers. Leave your answers in fractional (rather than decimal) form. When the equation has decimal coefficients, keep as many significant digits in your answer as contained in those coefficients.

Solve and check each equation.

1. $x - 5 = 28$ **2.** $4x + 2 = 18$

3. $8 - 3x = 7x$ **4.** $9x + 7 = 25$

5. $9x - 2x = 3x + 2$ **6.** $x + 8 = 12 - 3x$

7. $x + 4 = 5x + 2$ **8.** $3y - 2 = 2y$

9. $3x = x + 8$ **10.** $3x = 2x + 5$

11. $3x + 4 = x + 10$ **12.** $y - 4 = 7 + 23y$

13. $2.80 - 1.30y = 4.60$ **14.** $3.75x + 7.24 = 5.82x$

15. $4x + 6 = x + 9$ **16.** $7x - 19 = 5x + 7$

17. $p - 7 = -3 + 9p$ **18.** $-3x + 7 = -15 + 2x$

19. $4x - 11 = 2x - 7$ **20.** $3x - 8 = 12x - 7$

21. $8x + 7 = 4x + 27$ **22.** $3x + 10 = x + 20$

23. $\frac{1}{2}x = 14$ **24.** $\frac{x}{5} = 3x - 25$

25. $5x + 22 - 2x = 31$ **26.** $4x + 20 - 6 = 34$

27. $\frac{n}{3} + 5 = 2n$ **28.** $5x - \frac{2x}{5} = 3$

29. $2x + 5(x - 4) = 6 + 3(2x + 3)$

30. $6x - 5(x - 2) = 12 - 3(x - 4)$

31. $4(3x + 2) - 4x = 5(2 - 3x) - 7$

32. $6 - 3(2x + 4) - 2x = 7x + 4(5 - 2x) - 8$

33. $3(6 - x) + 2(x - 3) = 5 + 2(3x + 1) - x$

34. $3x - 2 + x(3 - x) = (x - 3)(x + 2) - x(2x + 1)$

35. $(2x - 5)(x + 3) - 3x = 8 - x + (3 - x)(5 - 2x)$

36. $x + (1 + x)(3x + 4) = (2x + 3)(2x - 1) - x(x - 2)$

37. $3 - 6(x - 1) = 9 - 2(1 + 3x) + 2x$

38. $7x + 6(2 - x) + 3 = 4 + 3(6 + x)$

39. $3 - (4 + 3x)(2x + 1) = 6x - (3x - 2)(x + 1) - (6 - 3x)(1 - x)$

40. $6.10 - (x + 2.30)(3.20x - 4.30) - 5.20x = 3.30x - x(3.10 + 3.20x) - 4.70$

41. $7r - 2r(2r - 3) - 2 = 2r^2 - (r - 2)(3 + 6r) - 8$

42. $2w - 3(w - 6)(2w - 2) + 6 = 3w - (2w - 1)(3w + 2) + 6$

43. $8.27x - 2.22(2.57 + x)(3.35x - 6.64)$
$= 3.24 - 3.35(2.22x + 5.64)(x - 1.24) - 2.64(2.11x - 3.85)$

44. $1.30x - (x - 2.30)(3.10x - 5.40) - 2.50 = 4.60 - (3.10x - 1.20)(x - 5.30) - 4.30x$

Simple Literal Equations

Solve for x.

45. $ax + 4 = 7$ **46.** $6 + bx = b - 3$

47. $c(x - 1) = 5$ **48.** $ax + 4 = b - 3$

49. $c - bx = a - 3b$

3–2   Solving Word Problems

Problems in Verbal Form

We have already done some simple word problems in earlier chapters, and we will have more of them with each new topic, but their main treatment will be here. After looking at these word problems you may ask, "When will I ever have to solve problems like that on the job?"

Maybe never. But you will most likely have to deal with technical material in written form: instruction manuals, specifications, contracts, building codes, handbooks, and so on. This chapter will aid you in coping with precise technical documents, where a mistaken meaning of a word could cost dollars, your job, or even lives.

Some tips on how to approach any word problem are given in the following paragraphs.

Picture the Problem

Try to *visualize* the situation described in the problem. Form a picture in your mind. *Draw a diagram* showing as much of the given information as possible, including the unknown.

Understand the Words

Look up the meanings of unfamiliar words in a dictionary, handbook, or textbook.

◆◆◆ **Example 16:** A certain problem contains the following statement:

> A simply supported beam carries a concentrated load midway between its centre of gravity and one end.   ◆◆◆

To understand this statement, you must know what a "simply supported beam" is, and what "concentrated load" and "centre of gravity" mean.

Locate the words and expressions standing for mathematical operations, such as

. . . the difference of. . .

. . . is equal to. . .

. . . the quotient of. . .

Math skills are important, but so are reading skills. Most students find word problems difficult, but if you have an *unusual* amount of trouble with them, you should seek out a reading teacher and get help.

♦♦♦ **Example 17:**

(a) "The difference of a and b" means $a - b$ (or $b - a$)
(b) "Four less than x" means $x - 4$
(c) "Three greater than c" means $c + 3$
(d) "The sum of x and 4 is equal to three times x," in symbols, is

$$x + 4 = 3x$$

(e) "Three consecutive integers, starting with x," are

$$x, \quad x + 1, \quad x + 2 \qquad \text{♦♦♦}$$

Many problems will contain a *rate*, either as the unknown or as one of the given quantities. The word *per* is your tipoff. Look for "kilometres *per* hour," "dollars *per* kilogram," "grams *per* litre," and the like. The word *per* also suggests *division*. Miles per gallon, for example, means miles travelled *divided* by gallons of fuel consumed.

Identify the Unknown

Be sure that you know exactly *what is to be found* in a particular problem. Find the sentence that ends with a question mark, or the sentence that starts with "Find . . ." or "Calculate . . ." or "What is . . ." or "How much . . ." or similar phrases.

Label the unknown with a statement such as "Let $x =$ rate of the train, km/h." Be sure to include *units of measure* when defining the unknown.

♦♦♦ **Example 18:** Fifty-five litres of a mixture that is 24% alcohol by volume is mixed with 74 litres of a mixture that is 76% alcohol. Find the amount of alcohol in the final mixture.

Solution:

This step of labelling the unknown annoys many students; they see it as a nitpicking thing that teachers make them do just for show, when they would rather start solving the problem. But realize that this step *is* the start of the problem. It is the crucial first step and *should never be omitted*.

You may be tempted to label the unknown in the following ways:

Let $x =$ alcohol	This is too vague. Are we speaking of volume of alcohol, or weight, or some other property?
Let $x =$ volume of alcohol	The units are missing.
Let $x =$ litres of alcohol	Do we mean litres of alcohol in one of the *initial* mixtures or in the *final* mixture?
Let $x =$ number of litres of alcohol in final mixture	Good.

We will do complete solutions of this type of problem in Sec. 3–4.   ♦♦♦

Define Other Unknowns

If there is more than one unknown, you will often be able to define the additional unknowns *in terms of the original unknown*, rather than introducing another symbol.

♦♦♦ **Example 19:** A word problem contains the following statement: "Find two numbers whose sum is 80 and . . ."

Solution: Here we have two unknowns. We can label them with separate symbols.

$$\text{Let } x = \text{first number}$$
$$\text{Let } y = \text{second number} \qquad \text{♦♦♦}$$

But another way is to label the second unknown in terms of the first, thus avoiding the use of a second variable. Since the sum of the two numbers x and y is 80,

$$x + y = 80$$

from which

$$y = 80 - x$$

This enables us to label both unknowns in terms of the single variable x.

Let x = first number
Let $(80 - x)$ = second number

Sometimes a careful reading of the word problem is needed to *identify* the unknowns before you can define them in terms of one of the variables.

◆◆◆ **Example 20:** A word problem contains the following statement: "A person invests part of his \$10,000 savings in a bank at 6%, and the remainder in a certificate of deposit at 8% . . ." We are then given more information and are asked to find the amount invested at each rate.

Solution: Since the investment is in two parts, both of which are unknown, we can say

Let x = amount invested at 6%

and

$(10,000 - x)$ = amount invested at 8%

But isn't it equally correct to say

Let x = amount invested at 8%

and

$(10,000 - x)$ = amount invested at 6%?

Yes. You can define the variables either way, as long as you are consistent and *do not change the definition part way through the problem.* ◆◆◆

We'll finish this problem in Sec. 3–3.

Estimate the Answer

It is a good idea to *estimate* or *guess* the answer before solving the problem, so that you will have some idea whether the answer you finally get is reasonable.

The kinds of problems you may encounter are so diverse that it is not easy to give general rules for estimating. However, we usually make *simplifying assumptions*, or try to *bracket the answer*.

To make a simplifying assumption means to solve a problem simpler than the given one and then use its answer as an estimate for the given problem.

◆◆◆ **Example 21:** To estimate the surface area of the aircraft nose cone in Fig. 3–1, we could assume that its shape is a right circular cone (shown by dashed lines) rather than the complex shape given. This would enable us to estimate the area using the following simple formula (Eq. 131):

$$\text{area} = \tfrac{1}{2}(\text{perimeter of base}) \times (\text{slant height})$$

For the nose cone,

$$\text{area} \approx \tfrac{1}{2}(3.8)(4.3) = 8.2 \text{ m}^2$$

We should be suspicious if our "exact" calculation gives an answer much different from this. ◆◆◆

To "bracket" an answer means to find two numbers between which your answer must lie. Don't worry that your two values are far apart. This method will catch more errors than you may suppose.

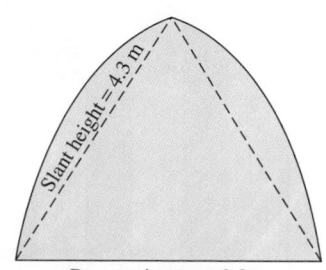

Base perimeter = 3.8 m

FIGURE 3–1 Aircraft nose cone.

◆◆◆ Example 22: It is clear that the length of the bridge cable of Fig. 3–2 must be greater than the straight line distances $AB + BC$, but less than the straight line distances $AD + DE + EC$. So the answer must lie between 286 m and 350 m.

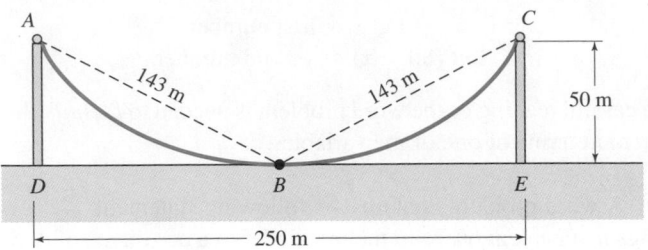

FIGURE 3–2   Suspension bridge.   ◆◆◆

For good problem-solving techniques, read *How to Solve It*, by G. Polya.

Write and Solve the Equation

The unknown quantity must now be related to the given quantities by means of an *equation*. Where does one get an equation? Either the equation is given verbally, right in the problem statement, or the relationship between the quantities in the problem is one you are expected to know (or to be able to find). These could be mathematical relationships, or they could be from any branch of technology. For example:

> *How is the area of a circle related to its radius?*
>
> *How is the current in a circuit related to the circuit resistance?*
>
> *How is the deflection of a beam related to its depth?*
>
> *How is the distance travelled by a car related to its speed?*

The relationships you will need for the problems in this book can be found in Appendix A, Summary of Facts and Formulas. Of course, each branch of technology has many more formulas than can be shown there. Problems that need some of these formulas are treated later in this chapter. Sometimes, as in the number problems we will solve later in this section, the equation is given verbally in the statement of the problem.

Check Your Answer

There are several things we can do to check an answer.

1. Does the answer look *reasonable?* If the problem is a complicated technical one, you might not yet have the experience to know whether an answer is reasonable, but in many cases you can spot a strange answer.

◆◆◆ Example 23: Does your result show an automobile going 600 kilometres per hour? Did a person's age calculate to be 150 years? Did the depth of the bridge girder come out to be 2 millimetres? Don't just write down the answer and run. Look at it hard to see if it makes sense.   ◆◆◆

2. Compare your answer with your initial estimate or guess. Do they agree within reason? If you have found two bracketing numbers within which the answer *must* fall, then you must recheck any answers outside that range.

◆◆◆ Example 24: If a calculation showed the length of the bridge cable in Fig. 3–2 to be 360 m, you could say with certainty that there is an error, because we saw that the length cannot exceed 350 m.   ◆◆◆

3. Check your answer(s) in the original problem statement. Be sure that any answer meets all of the conditions in the problem statement.

◆◆◆ **Example 25:** A problem asked for three consecutive odd integers whose sum is 39. Four students gave the following answers: (a) 12, 13, 14; (b) 13, 15, 17; (c) 9, 11, 19; and (d) 11, 13, 15. Determine which, if any, are correct, without actually solving the problem.

Solution:

(a) 12, 13, 14 are consecutive integers whose sum is 39, but they are not all odd.

(b) 13, 15, 17 are consecutive odd integers, but their sum is not 39.

(c) 9, 11, 19 are odd integers whose sum is 39, but they are not consecutive.

(d) 11, 13, 15 are consecutive odd integers whose sum is 39, and thus (d) is the correct answer.

◆◆◆

Number Problems

Number problems, though of little practical value, will give you practice at picking out the relationship between the quantities in a problem.

◆◆◆ **Example 26:** If twice a certain number is increased by 9, the result is equal to 2 less than triple the number. Find the number.

Solution: Let x = the number. From the problem statement,

$$2x + 9 = 3x - 2$$

Solving for x, we obtain

$$9 + 2 = 3x - 2x$$
$$x = 11$$

Check: We check our answer in the original problem statement. If the number is 11, then twice the number is 22. Thus twice the number increased by 9 is 31. Further, triple the number is 33. Two less than triple the number is 31. So our answer checks.   ◆◆◆

Common Error	Checking an answer by substituting in your *equation* is not good enough. The equation itself may be wrong. Check your answer with the problem statement.

Often the solution of a problem involves finding *more than one quantity*.

◆◆◆ **Example 27:** Find three consecutive odd integers whose sum is 39.

Solution: We first define our unknowns. Since any odd integer is 2 greater than the preceding odd integer, we can say

Let x = first odd integer

Then

$x + 2$ = second consecutive odd integer

and

$x + 4$ = third consecutive odd integer

Since the problem states that the sum of the three integers is 39, we write

$$x + (x + 2) + (x + 4) = 39$$

Solving for x gives

$$3x + 6 = 39$$
$$3x = 33$$
$$x = 11$$

But don't stop here—we are not finished. We're looking for *three* numbers. So

$$x + 2 = 13 \quad \text{and} \quad x + 4 = 15$$

Our three consecutive odd integers are then 11, 13, and 15.

Check:

$$11 + 13 + 15 = 39 \quad \text{(as required)} \qquad \qquad \blacklozenge\blacklozenge\blacklozenge$$

Sometimes there may seem to be too many unknowns to solve a problem. If so, carefully choose one variable to call x, so that the other unknowns can be expressed in terms of x. Then use the remaining information to write your equation.

◆◆◆ **Example 28:** In a group of 102 employees, there are three times as many employees on the day shift as on the night shift, and two more on the swing shift than on the night shift. How many are on each shift?

Estimate: Let's assume that the night and swing shifts have the *same* number N of employees. Then the day shift would have $3N$ employees, making $5N$ in all. So N would be $102 \div 5$, or 20, rounded. So we might expect about 20 on the night shift, a few more than 20 on the swing shift, and about 60 on the day shift.

Solution: Since the numbers of employees on the day and swing shifts are given in terms of the number on the night shift, we write

Let x = number of employees on the night shift

Then

$$3x = \text{number of employees on the day shift}$$

and

$$(x + 2) = \text{number of employees on the swing shift}$$

Since we are told that there are 102 employees in all,

$$x + 3x + (x + 2) = 102$$

from which

$$
\begin{aligned}
5x &= 100 \\
x &= 20 \quad \text{on the night shift} \\
3x &= 60 \quad \text{on the day shift} \\
(x + 2) &= \underline{22} \quad \text{on the swing shift} \\
&= 102 \quad \text{total}
\end{aligned}
$$

These numbers agree well with our estimate. $\qquad \qquad \blacklozenge\blacklozenge\blacklozenge$

The usual steps followed in solving a word problem are summarized here.

> 1. **Study the problem.** Look up unfamiliar words. Make a sketch. Try to visualize the situation in your mind.
> 2. **Estimate the answer.** Make simplifying assumptions, or try to bracket the answer.
> 3. **Identify the unknown(s).** Give it a symbol, such as "Let $x = \ldots$" If there is more than one unknown, try to label the others in terms of the first. Include units.
> 4. **Write and solve an equation.** Look for a relationship between the unknown and the known quantities that will lead to an equation. Write and solve that equation for the unknown. Include units in your answer. If there is a second unknown, be sure to find it also.
> 5. **Check your answer.** See if the answer looks reasonable. See if it agrees with your estimate. Be sure to do a numerical check in the verbal statement itself.

Exercise 2 ◆ Solving Word Problems

Verbal Statements

Rewrite each expression as one or more algebraic expressions.

1. Eight more than 5 times a certain number.
2. Four consecutive integers.
3. Two numbers whose sum is 83.
4. Two numbers whose difference is 17.
5. The amounts of antifreeze and water in 6 L of an antifreeze-water solution.
6. A fraction whose denominator is 7 more than 5 times its numerator.
7. The angles in a triangle, if one angle is half another.
8. The number of litres of antifreeze in a radiator containing x litres of a mixture that is 9% antifreeze.
9. The distance traveled in x hours by a car going 82 km/h.
10. If x equals the length of a rectangle, write an expression for the width of the rectangle if (a) the perimeter is 58 and (b) the area is 200.

Number Problems

11. Nine more than 4 times a certain number is 29. Find the number.
12. Find a number such that the sum of 15 and that number is 4 times that number.
13. Find three consecutive integers whose sum is 174.
14. The sum of 56 and some number equals 9 times that number. Find it.
15. Twelve less than 3 times a certain number is 30. Find the number.
16. When 9 is added to 7 times a certain number, the result is equal to subtracting 3 from 10 times that number. Find the number.
17. The denominator of a fraction is 2 greater than 3 times its numerator, and the reduced value of the fraction is $\frac{3}{10}$. Find the numerator.
18. Find three consecutive odd integers whose sum is 63.
19. Seven less than 4 times a certain number is 9. Find the number.
20. Find the number whose double exceeds 70 by as much as the number itself is less than 80.

21. If $(2x - 3)$ equals 29, what does $(4 + x)$ equal?

22. Separate 197 into two parts, such that the smaller part divides into the greater 5 times with a remainder of 23.

23. After a certain number is diminished by 8, the amount remaining is multiplied by 8; the result is the same as if the number were diminished by 6 and the amount remaining multiplied by 6. What is the number?

24. Thirty-one times a certain number is as much above 40 as 9 times the number is below 40. Find the number.

25. Three times the amount by which a certain number exceeds 6 is equal to the number plus 144. Find the number.

26. In a group of 90 persons, composed of men, women, and children, there are 3 times as many children as men, and twice as many women as men. How many are there of each?

3–3   Financial Problems

We turn now from number problems to ones that have more application to technology: financial, mixture, and statics problems. Uniform motion problems, which often require the solution of a fractional equation, are left for the chapter on fractions (Chapter 9).

◆◆◆ **Example 29:** A consultant had to pay income taxes of $4,867, plus 28% of the amount by which her taxable income exceeded $32,450. Her tax bill was $7,285. What was her taxable income? Work to the nearest dollar.

Solution: Let x = taxable income (dollars). The amount by which her income exceeded $32,450 is then

$$x - 32,450$$

Her tax is 28% of that amount, plus $4,867, so

$$\text{tax} = 4,867 + 0.28(x - 32,450) = 7,285$$

Solving for x, we get

We usually work financial problems to the nearest penny or nearest dollar, regardless of the number of significant digits in the problem.

$$0.28(x - 32,450) = 7,285 - 4,867 = 2,418$$
$$x - 32,450 = \frac{2,418}{0.28} = 8,636$$
$$x = 8,636 + 32,450 = \$41,086$$

Check: Her income exceeds $32,450 by ($41,086 − $32,450) or $8,636. This amount is taxed at 28%, or 0.28($8,636) = $2,418. To this we add the $4,867 tax, getting

$$\$2,418 + \$4,867 = \$7,285$$

as required.   ◆◆◆

◆◆◆ **Example 30:** A person invests part of his $10,000 savings in a bank at 6%, and part in a certificate of deposit at 8%, both simple interest. He gets a total of $750 per year in interest from the two investments. How much is invested at each rate?

Solution: Does this look familiar? It is the problem we started in Example 20. There we defined our variables as

$$x = \text{amount invested at 6\%}$$

and

$$10,000 - x = \text{amount invested at } 8\%$$

If x dollars are invested at 6%, the interest on that investment is $0.06x$ dollars. Similarly, $0.08(10,000 - x)$ dollars are earned on the other investment. Since the total earnings are $750, we write

$$0.06x + 0.08(10,000 - x) = 750$$
$$0.06x + 800 - 0.08x = 750$$
$$-0.02x = -50$$
$$x = \$2,500 \quad \text{at } 6\%$$

and

$$10,000 - x = \$7,500 \quad \text{at } 8\% \qquad \blacklozenge\blacklozenge\blacklozenge$$

Check: Does this look reasonable? Suppose that half of his savings ($5,000) were invested at each rate. Then the 6% deposit would earn $300 and the 8% deposit would earn $400, for a total of $700. But he got more than that ($750) in interest, so we would expect more than $5,000 to be invested at 8% and less than $5,000 at 6%, as we have found. ◆◆◆

Exercise 3 ◆ Financial Problems

1. The labour costs for a certain project were $3,345 per day for 17 technicians and helpers. If the technicians earned $210/day and the helpers $185/day, how many technicians were employed on the project?

2. A company has $86,500 invested in bonds and earns $6,751 in interest annually. Part of the money is invested at 7.4%, and the remainder at 8.1%, both simple interest. How much is invested at each rate?

Financial problems make heavy use of percentage, so you may want to review Sec. 1–10.

3. How much, to the nearest dollar, must a company earn in order to have $95,000 left after paying 27% in taxes?

4. A contractor buys three equal batches of fibreglass insulation for $1,632: the first for $68 per tonne; the second for $64 per tonne; and the third for $72 per tonne. How many tonnes of each does the contractor buy?

5. A water company changed its rates from $1.95 per 1000 L to $1.16 per 1000 L plus a service charge of $45 per month. How much water (to the nearest 1000 L) can you purchase before your new monthly bill will equal the bill under the former rate structure?

6. What salary should a person receive in order to take home $40,000 after deducting 23% for taxes? Work to the nearest dollar.

7. A person was $450.00 in debt, owing a certain sum to a brother, twice that amount to an uncle, and twice as much to a bank as to the uncle. How much was owed to each?

8. A student sold used skis and boots for $210, getting 4 times as much for the boots as for the skis. What was the price of each?

9. A person used part of a $100,000 inheritance to build a house and invested the remainder for 1 year, 1/3 of it at 6% and 2/3 at 5%, both simple interest. The income from both investments was $320. Find the cost of the house.

10. A used sport-utility vehicle (SUV) and a snowplow attachment are worth $7,200, the SUV being worth 7 times as much as the plow. Find the value of each.

11. A person spends 1/4 of her annual income for board, 1/12 for clothes, and 1/2 for other expenses, and saves $10,000. What is her income?

3–4   Mixture Problems

Basic Relationships

The total amount of mixture is equal to the sum of the amounts of the ingredients.

<table>
<tr><td>Total Amount of Mixture = Amount of A + Amount of B + . . .</td><td>A1</td></tr>
</table>

These two ideas are so obvious that it may seem unnecessary to write them down. But because they are obvious, they are often overlooked. They state that the whole is equal to the sum of its parts.

<table>
<tr><td>Final Amount of Each Ingredient
    = Initial Amount + Amount Added − Amount Removed</td><td>A2</td></tr>
</table>

◆◆◆ **Example 31:** From 100.0 kg of solder, half lead and half tin, 20.0 kg are removed. Then 30.0 kg of lead are added. How much lead is contained in the final mixture?

Solution:

$$\text{initial amount of lead} = 0.5(100.0) = 50.0 \text{ kg}$$
$$\text{amount of lead removed} = 0.5(20.0) = 10.0 \text{ kg}$$
$$\text{amount of lead added} = 30.0 \text{ kg}$$

By Eq. A2,

$$\text{final amount of lead} = 50.0 - 10.0 + 30.0 = 70.0 \text{ kg} \qquad ◆◆◆$$

Percent Concentration

The percent concentration of each ingredient is given by the following equation:

<table>
<tr><td>Percent Concentration of Ingredient $A = \dfrac{\text{Amount of } A}{\text{Amount of Mixture}} \times 100$</td><td>15</td></tr>
</table>

◆◆◆ **Example 32:** The total weight of the solder in Example 31 is

$$100.0 - 20.0 + 30.0 = 110.0 \text{ kg}$$

so the percent concentration of lead (of which there are 70.0 kg) is

$$\text{percent lead} = \frac{70.0}{110.0} \times 100 = 63.6\%$$

where 110.0 is the total weight (kg) of the final 63.6% lead mixture. ◆◆◆

Two Mixtures

When *two mixtures* are combined to make a *third* mixture, the amount of any ingredient A in the final mixture is given by the following equation:

<table>
<tr><td>Final Amount of A = Amount of A in First Mixture
    + Amount of A in Second Mixture</td><td>A3</td></tr>
</table>

◆◆◆ **Example 33:** One hundred litres of gasohol containing 12% alcohol are mixed with 200 litres of gasohol containing 8% alcohol. The volume of alcohol in the final mixture is

$$\text{Final amount of alcohol} = 0.12(100) + 0.08(200)$$
$$= 12 + 16 = 28 \text{ L} \qquad ◆◆◆$$

A Typical Mixture Problem

We now use these ideas about mixtures to solve a typical mixture problem.

✦✦✦ **Example 34:** How much steel containing 5.25% nickel must be combined with another steel containing 2.84% nickel to make 3.25 tonnes (t) of steel containing 4.15% nickel?

Estimate: The final steel needs 4.15% of 3.25 t of nickel, about 0.135 t. If we assume that *equal amounts* of each steel were used, the amount of nickel would be 5.25% of 1.625 t or 0.085 t from the first alloy, and 2.84% of 1.625 t or 0.046 t from the second alloy. This gives a total of 0.131 t of nickel in the final steel. This is not enough (we need 0.135 t). Thus *more than half* of the final alloy must come from the higher-nickel alloy, or between 1.625 t and 3.25 t.

Solution: Let x = tonnes of 5.25% steel needed. Fig. 3–3 shows the three alloys and the amount of nickel in each, with the nickel drawn as if it were separated from the rest of the steel. The amount of 2.84% steel is

$$(3.25 - x)$$

The amount of nickel that it contains is

$$0.0284(3.25 - x)$$

The amount of nickel in x tonnes of 5.25% steel is

$$0.0525x$$

The sum of these must give the amount of nickel in the final mixture.

$$0.0525x + 0.0284(3.25 - x) = 0.0415(3.25)$$

Clearing parentheses, we have

$$0.0525x + 0.0923 - 0.0284x = 0.1349$$
$$0.0241x = 0.0426$$
$$x = 1.77 \text{ t of } 5.25\% \text{ steel}$$
$$3.25 - x = 1.48 \text{ t of } 2.84\% \text{ steel}$$

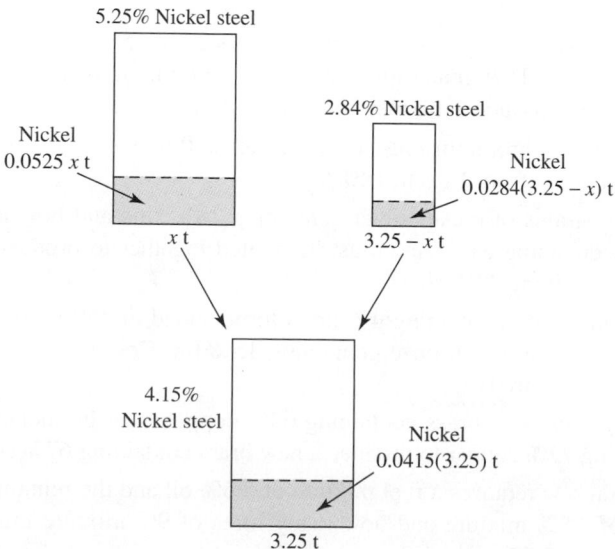

FIGURE 3–3   Three Alloys

Check: First we see that more than half of the final alloy comes from the higher-nickel steel, as predicted in our estimate. Now let us see if the final mixture has the proper percentage of nickel.

$$\text{final amount of nickel} = 0.0525(1.77) + 0.0284(1.48)$$
$$= 0.135 \text{ t}$$
$$\text{percent nickel} = \frac{0.135}{3.25} \times 100 = 0.0415$$

or 4.15%, as required.

Alternate Solution: This problem can also be set up in tabular form, as follows:

	Percent Nickel	×	Amount (t)	=	Amount of Nickel (t)
Steel with 5.25% nickel	5.25%		x		$0.0525x$
Steel with 2.84% nickel	2.84%		$(3.25 - x)$		$0.0284(3.25 - x)$
Final steel	4.15%		3.25		$0.0415(3.25)$

We then equate the sum of the amounts of nickel in the original steels with the amount in the final steel, getting

$$0.0525x + 0.0284(3.25 - x) = 0.0415(3.25)$$

as before.   ◆◆◆

Common Error	If you wind up with an equation that looks like this: ( ) kg nickel + ( ) kg iron = ( ) kg nickel you know that something is wrong. When you are using Eq. A3, *all* of the terms must be for the *same ingredient*.

Exercise 4 ◆ Mixture Problems

Treat the percent amounts given in this exercise as exact numbers, and work to three significant digits.

1. Two different mixtures of gasohol are available, one with 5% alcohol and the other containing 12% alcohol. How many litres of the 12% mixture must be added to $25\overline{0}$ litres of the 5% mixture to produce a mixture containing 9% alcohol?

2. How many tons of chromium must be added to 2.50 tons of stainless steel to raise the percent of chromium from 11% to 18%?

3. How many kilograms of nickel silver containing 18% zinc and how many kilograms of nickel silver containing 31% zinc must be melted together to produce $70\overline{0}$ kg of a new nickel silver containing 22% zinc?

4. A certain bronze alloy containing 4% tin is to be added to $35\overline{0}$ kg of bronze containing 18% tin to produce a new bronze containing 15% tin. How many kilograms of the 4% bronze alloy are required?

5. How many kilograms of brass containing 63% copper must be melted with $11\overline{0}0$ kg of brass containing 72% copper to produce a new brass containing 67% copper?

6. A certain chain saw requires a fuel mixture of 5.5% oil and the remainder gasoline. How many litres of 2.5% mixture and how many litres of 9% mixture must be combined to produce 40.0 L of 5.5% mixture?

7. A certain automobile cooling system contains 11.0 litres of coolant that is 15% antifreeze. How many litres of mixture must be removed so that, when it is replaced with pure anti-freeze, a mixture of 25% antifreeze will result?

8. A vat contains 40$\overline{0}$0 litres of wine with an alcohol content of 10%. How much of this wine must be removed so that, when it is replaced with wine with a 17% alcohol content, the alcohol content in the final mixture will be 12%?

9. A certain paint mixture weighing 30$\overline{0}$ kg contains 20% solids suspended in water. How many kilograms of water must be allowed to evaporate to raise the concentration of solids to 25%?

10. Fifteen litres of fuel containing 3.2% oil is available for a certain two-cycle engine. This fuel is to be used for another engine requiring a 5.5% oil mixture. How many litres of oil must be added?

11. A concrete mixture is to be made which contains 35% sand by weight, and 64$\overline{0}$ kg of mixture containing 29% sand is already on hand. How many kilograms of sand must be added to this mixture to arrive at the required 35%?

12. How many litres of a solution containing 18% sulfuric acid and how many litres of another solution containing 25% sulfuric acid must be mixed together to make 55$\overline{0}$ litres of solution containing 23% sulfuric acid? (All percentages are by volume.)

13. At a blaze, one fire crew has been pumping water at a rate of 225 L/min for 45 minutes when the chief assesses their progress and determines that they will have to put 34 000 L of water on the fire in order to extinguish it. The chief then orders the standby crew to start pumping water, too. The standby crew can pump water at the rate of 200 L/min. How much longer will it take to put out the fire?

14. After a fire has been extinguished, the smoky air needs to be cleared to a level of 335 μg/m^3 of smoke particulates before the fire marshal's investigators can safely enter. The quantity of smoke particulates immediately after this fire is 21 575 μg/m^3. The natural draft has been clearing the air at a rate of 17 μg/m^3/min for 23 minutes when a forced air filter that can clean at a rate of 110 μg/m^3/min is turned on. How much longer will the investigators have to wait?

15. The first fire crew to arrive at a fire has been pumping water at a rate of 180 L/min for 10 minutes when the second crew starts to pump. The second crew pumps at a rate of 205 L/min for 6 minutes, when the third crew starts to pump. The third crew pumps at a rate of 211 L/min. Once all three crews are pumping, it takes 27 minutes to put out the fire. How much water was used?

3–5   Statics Problems

Moments

The *moment of a force* about some point a (M_a) is the product of the force F and the perpendicular distance d from the force to the point.

Moment of a Force about Point a	$M_a = Fd$	A12

◆◆◆ **Example 35:** The moment of the force in Fig. 3–4 about point a is

$$M_a = 875 \text{ N}(1.13 \text{ m}) = 989 \text{ N·m}$$ ◆◆◆

875 N

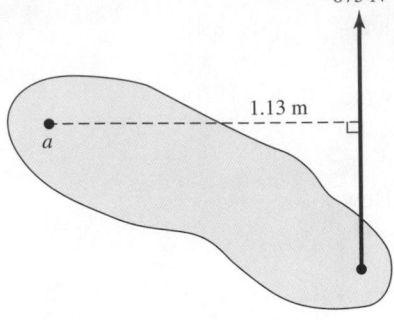

1.13 m

a

FIGURE 3-4 Moment of a force.

Equations A13, A14, and A15 describe Newton's first law of motion.

Equations of Equilibrium

The following equations apply to a body that is in *equilibrium* (i.e., not moving, or moves with a constant velocity):

Equations of Equilibrium	The sum of all horizontal forces acting on the body = 0	**A13**
	The sum of all vertical forces acting on the body = 0	**A14**
	The sum of the moments about any point on the body = 0	**A15**

◆◆◆ **Example 36:** A horizontal uniform beam of negligible weight is 6.35 m long and is supported by columns at either end. A concentrated load of 525 N is applied to the beam. At what distance from one end must this load be located so that the vertical reaction at that same end is 315 N, and what will be the reaction at the other end?

Estimate: If the 525 N load were at the midspan, the two reactions would have equal values of ½ (525) or 262.5 N. Since the left reaction (315 N) is greater than that, we deduce that the load is to the left of the midspan and that the reaction at the right will be less than 262.5 N.

Solution: We draw a diagram (Fig. 3–5) and label the required distance as x. By Eq. A14,

$$R + 315 = 525$$
$$R = 210 \text{ N}$$

Taking moments about p, we set the moments that tend to turn the bar in a clockwise (CW) direction equal to the moments that tend to turn the bar in the counterclockwise (CCW) direction. By Eq. A15,

$$525x = 210(6.35)$$
$$x = \frac{210(6.35)}{525} = 2.54 \text{ m}$$

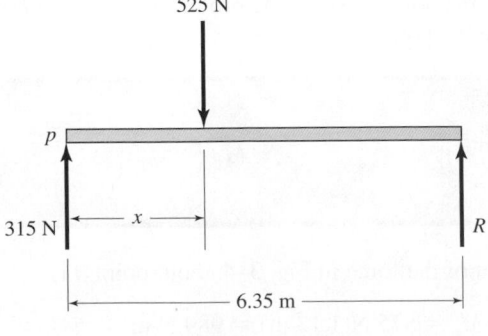

525 N

p

315 N   *x*   *R*

6.35 m

FIGURE 3–5

◆◆◆

◆◆◆ **Example 37:** Find the reactions R_1 and R_2 in Fig. 3–6.

Estimate: Let's assume that the 125 lb. weight is centred on the bar. Then each reaction would equal half the total weight (150 lb. ÷ 2 = 75 lb.). But since the weight is to the left of centre, we expect R_1 to be a bit larger than 75 lb. and R_2 to be a bit smaller than 75 lb.

Solution: In a statics problem, we may consider all of the weight of an object to be concentrated at a single point (called the *centre of gravity*) on that object. For a uniform bar, the centre of gravity is just where you would expect it to be, at the midpoint. Replacing the weights by forces gives the simplified diagram, Fig. 3–7.

The moment of the 125 lb. force about p is, by Eq. A12,

$$125(1.72) \text{ ft.·lb.clockwise}$$

Similarly, the other moments about p are

$$25.0(2.07) \text{ ft.·lb.clockwise}$$

and

$$4.14R_2 \text{ ft.·lb.counterclockwise}$$

But Eq. A15 says that the sum of the clockwise moments must equal the sum of the counterclockwise moments, so

$$4.14R_2 = 125(1.72) + 25.0(2.07)$$
$$= 266.8$$
$$R_2 = 64.4 \text{ lb.}$$

Also, Eq. A14 says that the sum of the upward forces (R_1 and R_2) must equal the sum of the downward forces (125 lb. and 25.0 lb.), so

$$R_1 + R_2 = 125 + 25.0$$
$$R_1 = 125 + 25.0 - 64.4$$
$$= 85.6 \text{ lb.}$$

◆◆◆

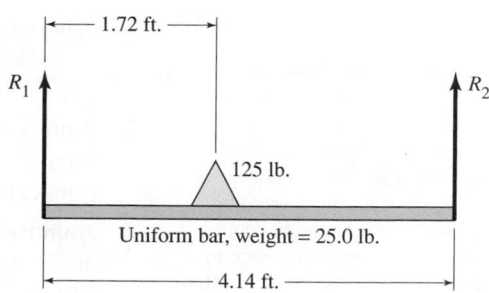

FIGURE 3–6

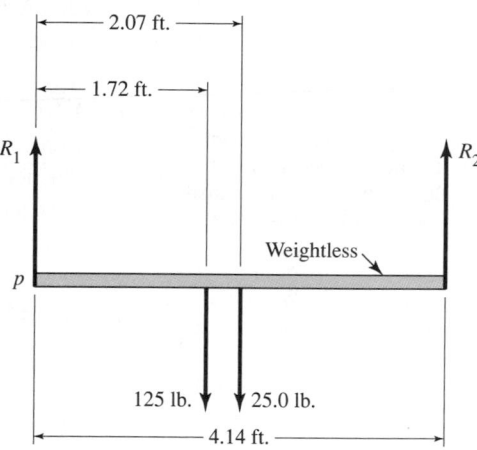

FIGURE 3–7

Exercise 5 ◆ Statics Problems

1. A horizontal beam of negligible weight is 18.0 m long and is supported by columns at either end. A vertical load of 14 500 N is applied to the beam at a distance x from the left end. Find x so that the reaction at the left column is 10 500 N.

2. For the beam of Problem 1, find the reaction at the right column.

3. A certain beam of negligible weight has an additional support 4.52 m from the left end, as shown in Fig. 3–8. The beam is 5.62 m long and has a concentrated load of 9250 N at the free end. Find the vertical reactions R_1 and R_2.

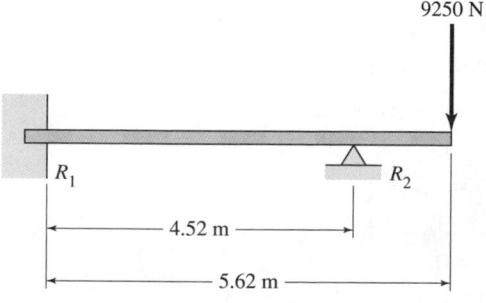

FIGURE 3–8

With technical problems involving approximate numbers, be sure to retain the proper number of significant digits in your answer.

Mass and *weight* are not the same thing: Mass is a measurement of a specific amount of material, which doesn't vary from one place to another; weight is a measure of the force of gravity exerted on a specific mass (and of course, gravity does vary from place to place, even on the surface of the earth). However, in non-scientific, everyday usage we often equate the magnitude of a measure of mass with its weight. Therefore, when we refer to "weight" in terms of the SI mass units of kilograms, grams, or tonnes, we are also assuming a standard earth gravity (acceleration) of 9.80665 m/s², usually rounded to 9.807 or 9.8 (the resulting gravitational force acting on a mass of 1 kg—or weight of 1 kg—is 9.807 newtons). The common U.S. measure of weight is the pound, and the equivalent measure of gravity is 32.2 ft./s². In this text, we mostly differentiate between mass and force using the appropriate SI units. On occasion, though, because of popular or traditional usage, some of our examples and problems will involve pounds or units of mass as measurements of weight. In such cases, the term "weight" or "weighing" will clearly signal that a unit is being used in its role as a force, rather than as a unit of mass.

4. A horizontal bar of negligible weight has a 55.1 kg weight hanging from the left end and a 72.0 kg weight hanging from the right end. The bar is seen to balance 97.5 cm from the left end. Find the length of the bar.

5. A horizontal bar of negligible weight hangs from two cables, one at each end. When a 624 N force is applied vertically downward from a point 185 cm from the left end of the bar, the right cable is seen to have a tension of 341 N. Find the length of the bar.

6. A uniform horizontal beam is 9.74 ft. long and weighs 386 lb. It is supported by columns at either end. A vertical load of 3814 lb. is applied to the beam at a distance x from the left end. Find x so that the reaction at the right column is $20\overline{0}0$ lb.

7. A uniform horizontal beam is 10.40 m long and weighs $136\overline{0}$ kg. It is supported at either end. A vertical load of 13 510 N is applied to the beam 4.250 m from the left end. Find the reaction in newtons at each end of the beam. *Note:* 1 kg (force or weight) = 9.807 N.

8. A bar of uniform cross section is 82.3 cm long and weighs 5.25 kg. A weight of 14.5 kg is suspended from one end. The bar and weight combination is to be suspended from a cable attached at the balance point. How far from the weight should the cable be attached, and what is the tension, in newtons, in the cable? *Note:* 1 kg (force or weight) = 9.807 N.

Case Study Discussion—Hiring Employees on a Budget

The labour budget is 25% of $60,000, which is $0.25 \times \$60,000 = \$15,000$.

Using a three to one ratio, if S is the number of supervisors and F is the number of staffers, $F = 3S$ (three staffers for every supervisor).

Calculating the salaries, $1,500 × number of supervisors + $1,100 × number of staffers = $15,000

or $\$1,500\,S + \$1,100\,F = \$15,000$

But since $F = 3S$, we can substitute $3F$ for S.

$$\$1,500S + \$1,100(3S) = \$15,000$$
$$\$1,500S + \$3,300S = \$15,000$$
$$\$4,800S = \$15,000$$
$$S = \$15,000/\$4,800$$
$$S = 3.125$$

So, we can round down to three supervisors and nine staffers.

Checking our calculations, $1,500 \times 3 + \$1,100 \times 9 = \$14,400$, which is below the 25% budget while complying with the supervisor/staff ratio.

◆◆◆ CHAPTER 3 REVIEW PROBLEMS ◆◆◆◆◆◆◆◆◆◆◆◆◆◆◆◆◆◆◆◆◆◆◆◆◆◆◆◆◆◆

Solve each equation.

1. $2x - (3 + 4x - 3x + 5) = 4$

2. $5(2 - x) + 7x - 21 = x + 3$

3. $3(x - 2) + 2(x - 3) + (x - 4) = 3x - 1$

4. $x + 1 + x + 2 + x + 4 = 2x + 12$

5. $(2x - 5) - (x - 4) + (x - 3) = x - 4$

6. $4 - 5w - (1 - 8w) = 63 - w$

7. $3z - (z + 10) - (z - 3) = 14 - z$

8. $(2x - 9) - (x - 3) = 0$

9. $3x + 4(3x - 5) = 12 - x$

10. $6(x - 5) = 15 + 5(7 - 2x)$

11. $x^2 - 2x - 3 = x^2 - 3x + 1$

12. $(x^2 - 9) - (x^2 - 16) + x = 10$

13. $x^2 + 8x - (x^2 - x - 2) = 5(x + 3) + 3$

14. $x^2 + x - 2 + x^2 + 2x - 3 = 2x^2 - 7x - 1$

15. $10x - (x - 5) = 2x + 47$

16. $7x - 5 - (6 - 8x) + 2 = 3x - 7 + 106$

17. $3p + 2 = \frac{p}{5}$

18. $\frac{4}{n} = 3$

19. $\frac{2x}{3} - 5 = \frac{3x}{2}$

20. $5.90x - 2.80 = 2.40x + 3.40$

21. $4.50(x - 1.20) = 2.80(x + 3.70)$

22. $\frac{x - 4.80}{1.50} = 6.20x$

23. $6x + 3 - (3x + 2) = (2x - 1) + 9$

24. $3(x + 10) + 4(x + 20) + 5x - 170 = 15$

25. $20 - x + 4(x - 1) - (x - 2) = 30$

26. $5x + 3 - (2x - 2) + (1 - x) = 6(9 - x)$

27. Subdivide a metre of tape into two parts so that one part will be 6 cm longer than the other part.

28. A certain mine yields low-grade oil shale containing 18.0 U.S. gallons of oil per ton of rock, and another mine has shale yielding 30.0 gal./ton. How many tons of each must be sent each day to a processing plant that processes 25 $\overline{0}$00 tons of rock per day, so that the overall yield will be 23.0 gal./ton?

29. A certain automatic soldering machine requires a solder containing half tin and half lead. How much pure tin must be added to 55 kg of a solder containing 61% lead and 39% tin to raise the tin content to 50%?

30. A person owed to A a certain sum, to B four times as much, to C eight times as much, and to D six times as much. A total of $570 would pay all of the debts. What was the debt to A?

31. A technician spends $\frac{2}{3}$ of his salary for board and $\frac{2}{3}$ of the remainder for clothing, and saves $2,500 per year. What is his salary?

32. Find four consecutive odd integers such that the product of the first and third will be 64 less than the product of the second and fourth.

33. The front and rear wheels of a tractor are 3 m and 4 m, respectively, in circumference. How many metres will the tractor have travelled when the front wheel has made 250 revolutions more than the rear wheel?

Solve each equation.

34. $3x - (x - 4) + (x + 1) = x - 7$

35. $4(x - 3) = 21 + 7(2x - 1)$

36. $8.20(x - 2.20) = 1.30(x + 3.30)$

37. $5(x - 1) - 3(x + 3) = 12$

38. $(x - 2)(x + 3) = x(x + 7) - 8$

39. $5x + (2x - 3) + (x + 9) = x + 1$

40. $1.20(x - 5.10) = 7.30(x - 1.30)$

41. $2(x - 7) = 11 + 2(4x - 5)$

42. $4.40(x - 1.90) = 8.30(x + 1.10)$

43. $8x - (x - 5) + (3x + 2) = x - 14$

44. $7(x + 1) = 13 + 4(3x - 2)$

Solve for x.

45. $bx + 8 = 2$

46. $c + ax = b - 7$

47. $2(x - 3) = 4 + a$

48. $3x + a = b - 5$

49. $8 - ax = c - 5a$

50. Find the reactions R_1 and R_2 in Fig. 3–9.

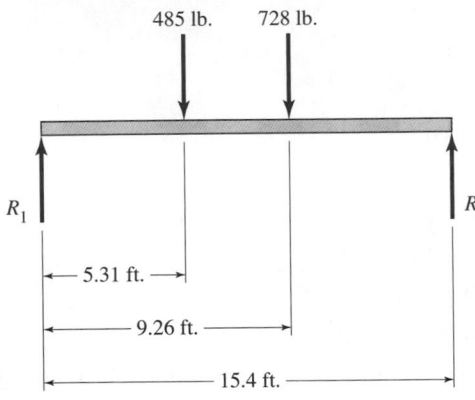

485 lb.   728 lb.

R_1 R_2

5.31 ft.

9.26 ft.

15.4 ft.

FIGURE 3–9

51. A carpenter estimates that a certain deck needs $4,285 worth of lumber, if there is no waste. How much should he buy if he estimates the waste on the amount bought is 7%?

52. How many litres of olive oil costing $4.86 per litre must be mixed with 136 litres of corn oil costing $2.75 per litre to make a blend costing $3.00 per litre?

53. The labour costs for a certain brick wall were $1,118 per day for 10 masons and helpers. If a mason earns $125/day and a helper $92/day, how many masons were on the job?

54. According to a tax table, for your filing status, if your taxable income is over $38,000 but not over $91,850, your tax is $5,700 plus 28% of the amount over $38,100. If your tax bill was $8,126, what was your taxable income? Work to the nearest dollar.

55. A casting weighs 875 kg and is made of a brass that contains 87.5% copper. How many kilograms of copper are in the casting?

56. How much must a person earn to have $45,824 after paying 28% in taxes?

57. 235 L of fuel for a two-cycle engine contain 3.15% oil. To this is added 186 L of fuel containing 5.05% oil. How many litres of oil are in the final mixture?

58. How much steel containing 1.15% chromium must be combined with another steel containing 1.50% chromium to make 8.00 tonnes of steel containing 1.25% chromium?

59. Two gasohol mixtures are available, one with 6.35% alcohol and the other with 11.28% alcohol. How many litres of the 6.35% mixture must be added to 25.0 L of the other mixture to make a final mixture containing 8.00% alcohol?

60. How many tonnes of tin must be added to 2.75 tonnes of bronze to raise the percentage of tin from 10.5% to 16.0%?

61. A student sold a computer and a printer for a total of $995, getting $1\frac{1}{2}$ times as much for the printer as for the computer. What was the price of each?

62. How many kilograms of nickel silver containing 12.5% zinc must be melted with 248 kg of nickel silver containing 16.4% zinc to make a new alloy containing 15.0% zinc?

63. A company has $223,821 invested in two separate accounts and earns $14,817 in simple interest annually from these investments. Part of the money is invested at 5.94%, and the remainder at 8.56%. How much is invested at each rate?

64. From 12.5 L of coolant containing 13.4% antifreeze, 5.50 L are removed. If 5.50 L of antifreeze are then added, how much antifreeze will be contained in the final mixture?

65. A person had $125,815 invested, part in a mutual fund that earned 10.25% per year and the remainder in a bank account that earned 4.25% per year, both simple interest. The amount earned from both investments combined was $11,782. How much was in each investment?

66. The labour costs for a certain project were $3,971 per day for 15 technicians and helpers. If a technician earns $325/day and a helper $212/day, how many technicians were on the job?

67. 15.6 L of cleaner containing 17.8% acid are added to 25.5 L of cleaner containing 10.3% acid. How many litres of acid are in the final mixture?

68. How much brass containing 75.5% copper must be combined with another brass containing 86.3% copper to make 5250 kg of brass containing 80.0% copper?

Writing

69. Make up a word problem. It can be very similar to one given in this chapter or, better, very different. Try to solve it yourself. Then swap with a classmate and solve each other's problem. Note down where the problem may be unclear, unrealistic, or ambiguous. Finally, each of you should rewrite your problem if needed.

70. You probably have some idea what field you plan to enter later: business, engineering, and so on. Look at a book from that field, and find at least one formula commonly used. Write it out; describe what it is used for; and explain the quantities it contains, including their units.

Team Project

71. Fig. 3–10 shows three beams, each carrying a box of a given weight. The weight of each beam is given, which can be considered to be at the midpoint of each beam. The right end of each of the two upper beams rests on the box on the beam below it. A force of 3150 N is needed to support the right end of the lowest beam.

    Find the reactions R_1, R_2, and R_3, and the distance x.

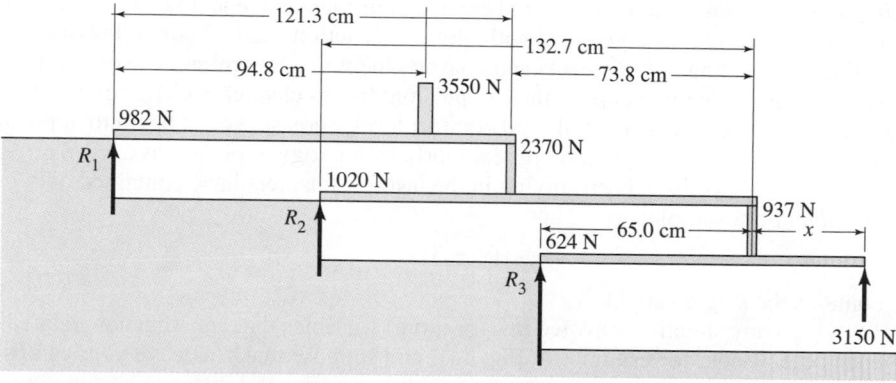

FIGURE 3–10

4

Functions

◆◆◆ OBJECTIVES ◆◆◆

When you have completed this chapter, you should be able to:

- Distinguish between relations and functions and define the range and domain of a function.
- Identify implicit functions, explicit functions, dependent variables, and independent variables.
- Identify and solve composite and inverse functions.

◆◆◆

A function is a math machine that connects one group of variables to another, just like a car transmission connects a motor to the wheels. You put a number into the function and you get another number out. The function "machine" performs one or more mathematical operations on the input to produce the output. If you're thinking this sounds like an equation, you're right. A function has a few unique properties, and the way we write it (the function *notation*) is very powerful. It's a way to label a formula and can, in a very short space, show the reader what formula is being used, which variable is changing, and what was used to make a substitution.

Chapter 4 will provide definitions, and expand your knowledge and skills surrounding functions. In a different arena, you deal with physical functions daily. Your cell phone is a function that accepts finger movements and converts them to a text message. Your bicycle is a function that converts leg movements into propulsion. In this chapter, we'll study much simpler functions. However, as your skills and comfort level increase, you'll begin to appreciate the necessity of function manipulation in real-world technology applications.

The equations we have been solving in the last few chapters have contained only *one variable*. For example,

$$x(x - 3) = x^2 + 7$$

contains the single variable x.

But many situations involve *two* (or more) variables that are somehow related to each other. Look, for example, at Fig. 4–1, and suppose that you leave your campsite and walk a path up the hill. As you walk, both the horizontal distance x from your camp, and the vertical distance y above your camp, will change. You cannot change x without changing y, and vice versa (unless you jump into the air or dig a hole). The variables x and y are *related*.

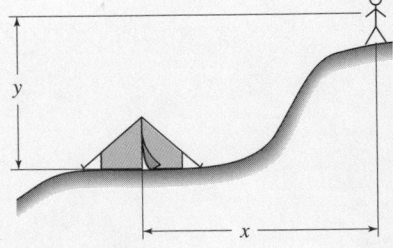

FIGURE 4–1

In this chapter, we study such a relationship between two variables by introducing the concept of a *function*. The idea of a function provides us with a different way of speaking about mathematical relationships. We could say, for example, that the formula for the area of a circle *as a function of* its radius is $A = \pi r^2$.

It may become apparent as you study this chapter that these same problems could be solved without ever introducing the idea of a function. Does the function concept, then, merely give us new jargon for the same old ideas? Not really.

A new way of *speaking* about something can lead to a new way of *thinking* about that thing, and so it is with functions. It will also lead to the powerful and convenient *functional notation,* which is especially useful when studying calculus or computer programming. In addition, this introduction to functions will prepare us for the later study of *functional variation.*

CASE STUDY—SYNCHRONIZING SPECIAL EFFECTS

As an engineering technologist, you are working at Pinewood Toronto Studios, Canada's largest film and television production complex. In more than 23 000 m² (250 000 sq. ft.) of production space, it houses eight stages, including North America's largest purpose-built soundstage. As part of your job of synchronizing special effects for a science fiction movie, a special lighting effect is to be switched on as soon as a test tube (containing an alien virus, according to the script) hits a workbench when dropped from an overhead walkway. You will be doing some dry runs and rehearsals; but it would help if you could do a quick calculation of how long the object will take to fall. Knowing how the director works, you had better do a few calculations because the scene may move and the tube could be dropped from any number of heights (especially if the actor falls and tosses the test tube into the air). Can you come up with a function involving gravity, time, and distance that would be usable in this instance?

4–1 Functions and Relations

Definitions

Simply put, equations can also be expressed as *functions* or *relations*. First, a couple of rules to remember:

- Each *function* is a relation, but not all *relations* are functions.
- Watch for any duplication of x values in a relation: A relation is *not* a function if there are two or more solutions for any x value. For instance, if an x value of 3 results in a y value of 4, but also a y value of 6, then the relation is *not* a function.

A *relation* is simply an informational set of data. For example:

- $\{(x_1, y_1)(x_2, y_2)(x_3, y_3)(x_4, y_4)(x_5, y_5)\}$
- $\{(2, 3) (4, 3) (6, 3) (8, 3) (10, 3)\}$, or even
- {(Kathy, red shoes) (Sarah, white shoes) (Jim, white shoes) (Tony, black shoes) (Kathy, black shoes)}.

Note that the last example is not a function because there's more than one solution for each *x* value: There are two Kathy's in this data set; one wearing red shoes and one wearing black shoes. So this relation, or data set, *is not a function at all*, because of the duplication in the *x* values.

Let's look at the following examples so we can discuss some terminology further:

◆◆◆ **Example 1:** $y = 3x^2 + 4$

This is a function written in equation notation. If we insert any number in for *x*, we get one, and *only* one, answer for *y*.   ◆◆◆

◆◆◆ **Example 2:** $\{(-3, 31), (-2, 16), (0, 4), (2, 16), (3, 31)\}$

This is also a function, but written in *coordinate*, or *relational* notation. For each *x* component, there is only one *y* component. (This is also a relation and is actually the set of coordinates you achieve when plugging numbers into the equation of Example 1).

Domain and Range

Let's look at another example to understand the *x* and *y* components of a relation and how domain and range fit into this picture.   ◆◆◆

◆◆◆ **Example 3:** In the following relation: $\{(2, -3), (4, 6), (3, -1), (6, 6), (2, 3)\}$, we use the first co-ordinate to identify the two components, or values, (*x*, *y*), as shown here:

$$(\ 2, -3 \ )$$

x component   *y* component

Using the same coordinate, let's identify the *domain* and *range*.

$$(\ 2, -3 \ )$$

x component,   *y* component,
<domain>   <range>

From this same example, we can list the domain and range values, and where the domain is *all the x values* and the range is *all the y values*. It is customary to give the domain and range values without duplication and in numerical order.
Let's identify the domain values

All the *x* values, or the *domain*

$$\{(\ 2, -3 \ ), (\ 4, 6 \ ), (\ 3, -1 \ ), (\ 6, 6 \ ), (\ 2, 3 \ )\}$$

and the range values

All the *y* values, or the *range*

$$\{(\ 2, -3 \ ), (\ 4, 6 \ ), (\ 3, -1 \ ), (\ 6, 6 \ ), (\ 2, 3 \ )\}$$

Listing the domain and range values we've identified (without duplication), we get

Domain: $\{2, 3, 4, \text{ and } 6\}$

Range: $\{-3, -1, 3, \text{ and } 6\}$

So now the question remains: Is this relation a function?

The given set represents a relation (in that the x's and y's are related), but it has two y points with the same x value: $(2, -3)$ and $(2, 3)$, so the relation is *not* a function.

In the future, all you have to do to check whether the relation is a function or not is to look for duplicate x values in the given data set. ◆◆◆

Let's look at a visual, such as a plotted graph, for our next example.

◆◆◆ Example 4: State the domain and the range of the following relation. After finding the domain and range, state whether or not it is a function.

$$\{(-4, -3), (-3, -3), (-2, -3), (-1, -3), (0, -3), (1, -3)\}$$

The first step is to list all the x-values for the domain and all the y-values for the range:

Domain: $\{-4, -3, -2, -1, 0, 1\}$

Range: $\{-3\}$

If we plot these, we can see that all the x values are different, and all of the y values are the same (Fig. 4–2).

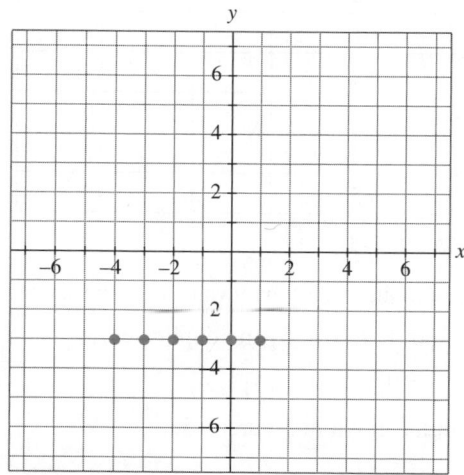

FIGURE 4–2

We are able to draw a straight line through the plotted points (Fig. 4–3). All points are in-line, and we check to see if there are any duplicates of x-values in the data set; there are not, so *the relation is a function.*

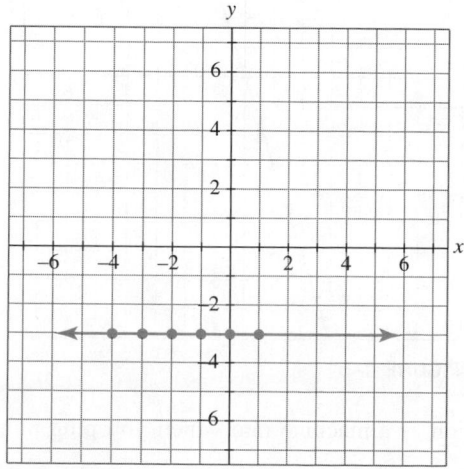

FIGURE 4–3 ◆◆◆

To help visualize how to solve functions, look at the following diagrams. It helps to think of functions as a machine: when you assign values for x, i.e., plug numbers in, you get corresponding values for y.

The first one is an easy function, $y = 3x$, or y is equal to 3 times the value we give to x.

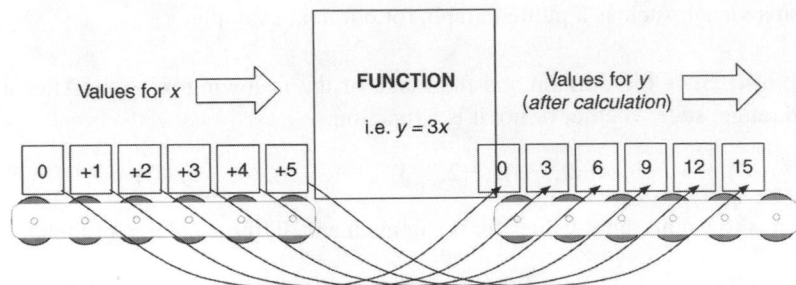

FIGURE 4-4

From the illustration, when we replace x with a number, a corresponding y value results:

$3x$	$=$	y
0	$=$	0
+1	$=$	3
+2	$=$	6
+3	$=$	9
+4	$=$	12
+5	$=$	15

TABLE 4-1

List all the x values for the domain and all the y values for the range:

Domain: {0, 1, 2, 3, 4, 5} (no duplicates)

Range: {0, 3, 6, 9, 12, 15}

Plotting the function, we get the graph in Figure 4–5.

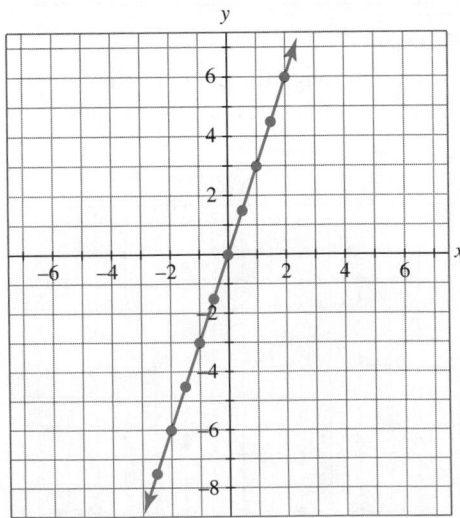

FIGURE 4-5

Again, imagine the function as a machine into which you plug numbers (left) and the result comes out (right).

Using an even easier function such as $y = 0x - 6$, we could imagine solving for x looking like this:

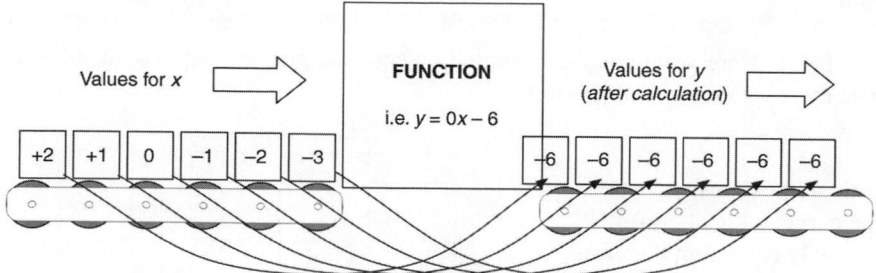

FIGURE 4–6

Using a Graph to Determine if Something is a Function

If we are given a graph, we can use the *vertical line test* to determine whether or not a given data set is a function.

Consider a data set with only 2 points: $\{(-1, 3), (-1, -4)\}$. We know that this is not a function because there are duplicate x values: $x = -1$ is associated to $y = 3$ and $y = -4$.

Plotting the points, we get:

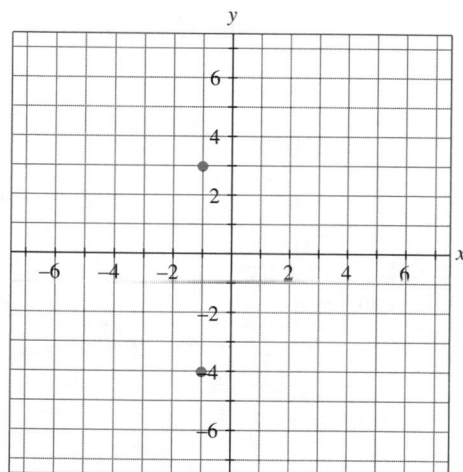

FIGURE 4–7

and with a line it looks like:

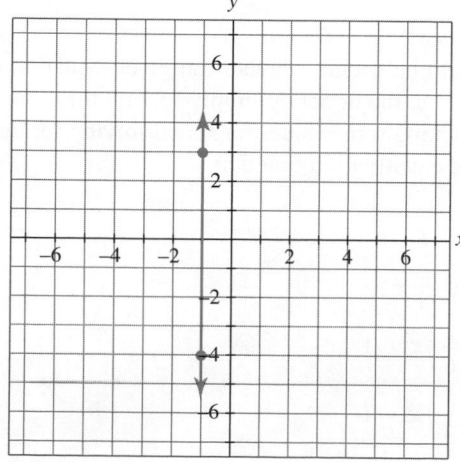

FIGURE 4–8

If a vertical line comes in contact with *two points* of the plotted object, then *the relation is not a function*.

◆◆◆ Example 5 Here are a couple of plotted object examples we can test to see if they are functions, by using the vertical line test.

A vertical line drawn anywhere on the curve in Fig. 4–9(a) will intersect only one point; therefore, it *is* a function. (there is 1 point of intersection).

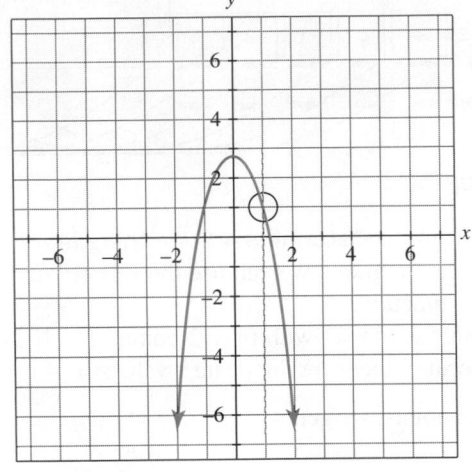

FIGURE 4–9(a)

The curve in Fig. 4–9(b) is *not* a function because a vertical line can intersect two points on this ellipse.

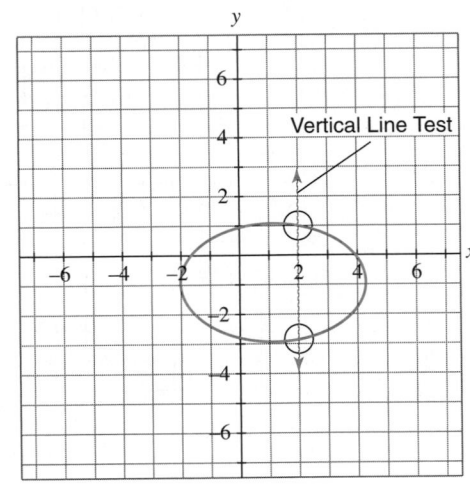

FIGURE 4–9(b)   ◆◆◆

Determining a Function Without a Graph

We can use graphing to help us decide whether the given equation or data set is a function. We can also decide if it's a function or not by simply solving for y from the equation. We can do this by making a T-chart, picking some values for x, and solving for the corresponding values of y. If you can solve for y the equation is a function.

◆◆◆ Example 6: $6y + 2x = 6$

$$6y = -2x + 6$$

$$y = \left(-\frac{2}{6}\right)x + 1$$

$$y = \left(-\frac{1}{3}\right)x + 1$$

Therefore, $6y + 2x = 6$ is a function, because it can be solved for a single y value.   ◆◆◆

◆◆◆ **Example 7:** $y^2 + 5x = -7$

$$y^2 = -5x - 7$$

$$y = \pm\sqrt{-5x - 7}$$

The equation $y^2 + 5x = -7$ is *not* a function because it cannot be solved for only one y value. ◆◆◆

◆◆◆ **Example 8:** In the experiment shown in Fig. 4–10, we change the load on the spring, and for each load we measure and record the distance that the spring has stretched from its unloaded position. If a load of 6 kg causes a stretch of 3 cm, then 6 kg and 3 cm are called *corresponding values*. The value 3 cm has no meaning *by itself*, but is meaningful only when it is *paired* with the load that produced it (6 kg). If we always write the pair of numbers in the *same order* (the load first and the distance second, in this example), it is then called an *ordered pair* of numbers. It is written (6, 3). Can you guess the missing value in the ordered pair (12, ?)?

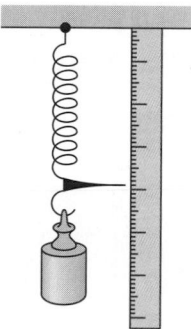

FIGURE 4–10

Solution: The ordered pair (12, 6) is correct. This linear function is called Hooke's law.

A set or table of ordered pairs is a function if for each x value there is only one y value given. The domain of x is the set of all x values in the table, and the range is the set of all y values. ◆◆◆

◆◆◆ **Example 9:** Suppose that the spring experiment of Example 8 gave the following results:

Load (kg)	0	1	2	3	4	5	6	7	8
Stretch (cm)	0	0.5	0.9	1.4	2.1	2.4	3.0	3.6	4.0

This table of ordered pairs is a function. The domain is the set of numbers (0, 1, 2, 3, 4, 5, 6, 7, 8), and the range is the set of numbers (0, 0.5, 0.9, 1.4, 2.1, 2.4, 3.0, 3.6, 4.0). The table associates exactly one number in the range with each number in the domain. Measurement error is apparent here. ◆◆◆

A set of ordered pairs is not always given in table form.

◆◆◆ **Example 10:** The function of Example 9 can also be written

(0, 0), (1, 0.5), (2, 0.9), (3, 1.4), (4, 2.1), (5, 2.4), (6, 3.0), (7, 3.6), (8, 4.0) ◆◆◆

Functions may also be given in verbal form.

◆◆◆ **Example 11:** The shipping charges are $2.25 per kilogram for the first 20 kilograms and $1.95 per kilogram thereafter. This verbal statement is a function relating the shipping costs to the weight of the item. ◆◆◆

We sometimes want to switch from verbal form to an equation, or vice versa.

◆◆◆ **Example 12:** Write y as a function of x if y equals twice the cube of x diminished by half of x.

Solution: We replace the verbal statement by the equation

$$y = 2x^3 - \frac{x}{2}$$

◆◆◆

The function notation is much shorter and clearer than the verbal statement.

◆◆◆ **Example 13:** The equation $y = 5x^2 + 9$ can be stated verbally as, y equals the sum of 9 and 5 times the square of x.

There are, of course, many different ways to express this same relationship.

◆◆◆ **Example 14:** Express the volume (V) of a cone having a base area of 75 units *as a function of* its altitude (H).

Solution: This is another way of saying, " Write an equation for the volume of a cone *in terms of* its base area and altitude."

The formula for the volume of a cone is given by Eq. 130 in Appendix A, Summary of Facts and Formulas.

$$V = \tfrac{1}{3}(\text{base area})(\text{altitude}) = \tfrac{1}{3}(75)H$$

So

$$V = 25H$$

is the required expression.   ◆◆◆

We see, then, that a function or relation can be expressed in several different ways (Fig. 4–11): as an equation, as a table (or set of ordered pairs) or as a verbal statement. A function or relation can also be expressed as a *graph*, as we'll see in Sec. 5–2.

Different forms of a relation or function	
Equation: $y = x^2 - 3$	Table of values: x | 0 1 2 3 y | −3 −2 1 6
Verbal statement: "y is equal to the square of x diminished by 3."	
Set of ordered pairs: $(1, -2)$, $(3, 6)$, $(0, -3)$, $(2, 1)$	Graph:

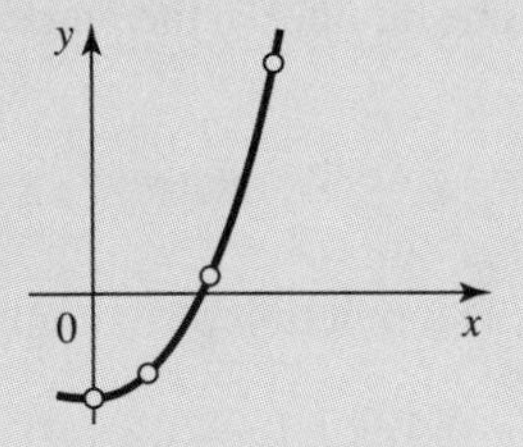

FIGURE 4–11

Other Ways of Finding and Expressing Domain and Range

To be strictly correct, the domain should be stated whenever an equation is written. However, this is often not done, so we follow this convention:

The domain of the function is assumed to be the largest set of real x values that yield real values of y.

◆◆◆ **Example 15:** The function

$$y = x^2$$

gives a real y value for every real x value. Thus the domain is all of the real numbers, which we can write

$$\text{Domain:} \quad -\infty < x < \infty$$

But notice that there is no x that will make y negative. Thus the range of y includes all of the positive numbers and zero. This can be written

$$\text{Range:} \quad y \geq 0, \, or \, 0 \leq y < \infty \qquad \blacklozenge\blacklozenge\blacklozenge$$

Our next example is one in which certain values of x result in a *negative number under a radical sign.*

◆◆◆ **Example 16:** Find the domain and the range of the function

$$y = \sqrt{x - 2}$$

We will deal more with *complex numbers*, including their real and *imaginary* components, in Sec. 21–1.

Solution: Our method is to see what values of x and y do not work (i.e., give a nonreal result or an illegal operation); then the domain and the range will be those values that do work.

Any value of x less than 2 will make the quantity under the radical sign negative, resulting in an *imaginary y.* Thus the domain of x is all of the positive numbers equal to or greater than 2.

$$\text{Domain:} \quad x \geq 2$$

An x equal to 2 gives a y of zero. Any x larger than 2 gives a real y greater than zero. Thus the range of y is

$$\text{Range} \quad y \geq 0 \qquad \blacklozenge\blacklozenge\blacklozenge$$

Our next example shows some values of x that result in *division by zero.*

◆◆◆ **Example 17:** Find the domain and the range of the function

$$y = \frac{1}{x}$$

Solution: Here any value of x but zero will give a real y. Thus the domain is

$$\text{Domain:} \quad x \neq 0$$

Notice here that it is more convenient to state which values of x *do not* work, rather than those that do.

Now what happens to y as x varies over its domain? We see that large x values give small y values, and conversely that small x values give large y values. Also, negative values of x give negatives values of y. However, there is no x that will make y zero, so

$$\text{Range:} \quad y \neq 0 \qquad \blacklozenge\blacklozenge\blacklozenge$$

Our final example shows both division by zero and negative numbers under a radical sign, for certain x values.

◆◆◆ **Example 18:** Find the domain and the range of the function

$$y = \frac{9}{\sqrt{4 - x}}$$

Solution: Since the denominator cannot be zero, x cannot be 4. Also, any x greater than 4 will result in a negative quantity under the radical sign. The domain of x is then

$$\text{Domain:} \quad x < 4$$

Restricted to these values, the quantity $4 - x$ is positive and ranges from very small (when x is nearly 4) to very large (when x is large and negative). Thus the denominator is positive, since we allow only the principal (positive) root and we can vary from near zero to infinity. The range of y, then, includes all of the values greater than zero.

$$\text{Range:} \quad y > 0$$

Exercise 1 ◆ Functions and Relations

Which of the following relations are also functions?

1. $y = 3x^2 - 5$
2. $y = \sqrt{2x}$
3. $y = \pm\sqrt{2x}$
4. $y^2 = 3x - 5$
5. $2x^2 = 3y^2 - 4$
6. $x^2 - 2y^2 - 3 = 0$

7. Is the set of ordered pairs $(1, 3), (2, 5), (3, 8), (4, 12)$ a function?
8. Is the set of ordered pairs $(0, 0), (1, 2), (1, -2), (2, 3), (2, -3)$ a function?

Functions in Verbal Form

For each of the following, write y as a function of x, where the value of y is equal to the given expression.

9. The cube of x
10. The square root of x, diminished by 5
11. x increased by twice the square of x
12. The reciprocal of the cube of x
13. Two-thirds of the amount by which x exceeds 4

Replace each function by a verbal statement of the type given in problems 9 through 13.

14. $y = 3x^3$
15. $y = 5 - x$
16. $y = \dfrac{1}{2} + x$
17. $y = 2\sqrt{x}$
18. $y = 5(4 - x)$

Write the equation called for in each following statement. Refer to Appendix A, Summary of Facts and Formulas, if necessary.

19. Express the area A of a triangle as a function of its base, b, and altitude, h.
20. Express the hypotenuse, c, of a right triangle as a function of its legs, a and b.
21. Express the volume, V, of a sphere as a function of the radius, r.
22. Express the power, P, dissipated in a resistor as a function of its resistance, R, and the current, I, through the resistor.

Write the equations called for in each statement.

23. A car is traveling at a speed of 55 km/h. Write the distance d traveled by the car as a function of time t.
24. To ship its merchandise, a mail-order company charges \$0.65/lb plus \$2.25 for handling and insurance. Express the total shipping charge s as a function of the item weight w.

25. A projectile is shot upward with an initial velocity of 125 m/s. Express the height H of the projectile as a function of time t. (See Eq. A18.)

Domain and Range

State the domain and the range of each function.

26. $(0, 2), (1, 4), (2, 8), (3, 16), (4, 32)$
27. $(-10, 20), (5, 7), (-7, 10), (10, 20), (0, 3)$

28.

x	2	4	6	8	10
y	0	-2	-5	-9	-15

Find the domain and the range for each function.

29. $y = \sqrt{x - 7}$

30. $y = \dfrac{3}{\sqrt{x - 2}}$

31. $y = x - \dfrac{1}{x}$

32. $y = \sqrt{x^2 - 25}$

33. $y = \dfrac{8 - x}{9 - x}$

34. $y = \dfrac{11}{(x + 4)(x + 2)}$

35. $y = \dfrac{4}{\sqrt{1 - x}}$

36. $y = \dfrac{x + 1}{x - 1}$

37. $y = \sqrt{x - 1}$

38. $y = \dfrac{5}{(x + 2)(x - 4)}$

39. Find the range of the function $y = 3x^2 - 5$ whose domain is $0 \le x \le 5$.

4–2 Functional Notation

Implicit and Explicit Forms

When one variable in an equation is isolated on one side of the equal sign, the equation is said to be in *explicit form*.

✦✦✦ **Example 19:** The following equations are all in *explicit form:*

$$y = 2x^3 + 5$$
$$z = ay + b$$
$$x = 3z^2 + 2z - z$$

✦✦✦

When a variable is *not* isolated, the equation is said to be in *implicit form*.

✦✦✦ **Example 20:** The following equations are all in *implicit form:*

$$y = x^2 + 4y$$
$$x^2 + y^2 = 25$$
$$w + x = y + z$$

✦✦✦

Dependent and Independent Variables

In the equation

$$y = x + 5$$

y is called the *dependent variable* because its value depends on the value of x, and x is called the *independent variable*. Of course, the same equation can be written $x = y - 5$, so that x becomes the dependent variable and y the independent variable.

The terms *dependent* and *independent* are used only for an equation in explicit form.

✦✦✦ **Example 21:** In the implicit equation $y - x = 5$, neither x nor y is called dependent or independent.

✦✦✦

Functional Notation

Just as we use the symbol x to represent a *number,* without saying which number we are specifying, we use the notation

$$f(x)$$

to represent a *function* without having to specify which particular function we are talking about.
We can also use functional notation to designate a *particular* function, such as

$$f(x) = x^3 - 2x^2 - 3x + 4$$

A functional relation between two variables x and y, in explicit form, such as

$$y = 5x^2 - 6$$

could be written

$$y = f(x)$$

or

$$f(x) = 5x^2 - 6$$

Where the expression $y = f(x)$ is read "y is a function of x."

Common Error	The expression $y = f(x)$ **does not** mean "y equals f times x."

The independent variable x is sometimes referred to as the *argument* of the function.

◆◆◆ **Example 22:** We may know that the power, P, of an engine depends (somehow) on the engine displacement d. We can express this fact by

$$P = f(d)$$

We are saying that P is a function of d, even though we do not know what the relationship is. ◆◆◆

The letter f is usually used to represent a function, but *other letters* can, of course, be used (g and h being common). Subscripts are also used to distinguish one function from another.

◆◆◆ **Example 23:** You may see functions written as

$$y = f_1(x) \qquad y = g(x)$$
$$y = f_2(x) \qquad y = h(x)$$
$$y = f_3(x)$$

◆◆◆

The letter y itself is often used to represent a function.

◆◆◆ **Example 24:** The function

$$y = 2x^2 - 3$$

will often be written

$$y(x) = 2x^2 - 3$$

to emphasize the fact that y is a function of x. ◆◆◆

Implicit functions can also be represented in functional notation.

◆◆◆ **Example 25:** The equation $x - 3xy + 2y = 0$ can be represented in functional notation by $f(x, y) = 0$. ◆◆◆

Functions relating *more than two variables* can be represented in functional notation, as in the following example.

◆◆◆ **Example 26:**

$y = 2x + 3z$	can be written	$y = f(x, z)$
$z = x^2 - 2y + w^2$	can be written	$z = f(w, x, y)$
$x^2 + y^2 + z^2 = 0$	can be written	$f(x, y, z) = 0$ ◆◆◆

Manipulating Functions

Functional notation provides us with a convenient way of indicating what is to be done with a function or functions. This includes: solving an equation for a different variable; changing from implicit to explicit form (or vice versa); combining two or more functions to make another function; or substituting numerical or literal values into an equation.

◆◆◆ **Example 27:** Write the equation $y = 2x - 3$ in the form $x = f(y)$.

Solution: We are being asked to write the given equation with x, instead of y, as the dependent variable. Solving for x, we obtain

$$2x = y + 3$$

$$x = \frac{y + 3}{2}$$ ◆◆◆

◆◆◆ **Example 28:** Write the equation $x = \dfrac{3}{2y - 7}$ in the form $y = f(x)$.

Solution: Here we are asked to rewrite the equation with y as the dependent variable, so we solve for y. Multiplying both sides by $2y - 7$ and dividing by x gives

$$2y - 7 = \frac{3}{x}$$

or

$$2y = \frac{3}{x} + 7$$

Dividing by 2 gives

$$y = \frac{3}{2x} + \frac{7}{2}$$ ◆◆◆

◆◆◆ **Example 29:** Write the equation $y = 3x^2 - 2x$ in the form $f(x, y) = 0$.

Solution: We are asked here to go from explicit to implicit form. Rearranging gives us

$$3x^2 - 2x - y = 0$$ ◆◆◆

Substituting into Functions

If we have a function, say, $f(x)$, then the notation $f(a)$ means to replace x by a in the same function.

◆◆◆ **Example 30:** Given $f(x) = x^3 - 5x$, find $f(2)$.

Remember that not every function $y = f(x)$ can be solved for x to get a function

$$x = f(y)$$

For example, the function $y = x^2$ when solved for x gives

$$x = \pm\sqrt{y}$$

which is a relation but not a function.

Solution: The notation $f(2)$ means that 2 is to be *substituted for* x in the given function. Wherever an x appears, we replace it with 2.

$$f(x) = x^3 - 5x$$

$$f(2) = (2)^3 - 5(2)$$
$$= 8 - 10 = -2 \qquad \qquad \text{◆◆◆}$$

◆◆◆ **Example 31:** Given $y(x) = 3x^2 - 2x$, find $y(5)$.

Solution: The given notation means to substitute 5 for x, so

$$y(5) = 3(5)^2 - 2(5)$$
$$= 75 - 10 = 65 \qquad \qquad \text{◆◆◆}$$

Often, we must substitute *several values* into a function and combine them as indicated.

◆◆◆ **Example 32:** If $f(x) = x^2 - 3x + 4$, find

$$\frac{f(5) - 3f(2)}{2f(3)}$$

Solution:

$$f(2) = 2^2 - 3(2) + 4 = \ \ 2$$
$$f(3) = 3^2 - 3(3) + 4 = \ \ 4$$
$$f(5) = 5^2 - 3(5) + 4 = 14$$

so

$$\frac{f(5) - 3f(2)}{2f(3)} = \frac{14 - 3(2)}{2(4)}$$
$$= \frac{8}{8} = 1 \qquad \qquad \text{◆◆◆}$$

The substitution might involve *literal* quantities instead of numerical values.

◆◆◆ **Example 33:** Given $f(x) = 3x^2 - 2x + 3$, find $f(5a)$.

Solution: We substitute $5a$ for x.

$$f(5a) = 3(5a)^2 - 2(5a) + 3$$
$$= 3(25a^2) - 10a + 3$$
$$= 75a^2 - 10a + 3 \qquad \qquad \text{◆◆◆}$$

◆◆◆ **Example 34:** If $f(x) = 5x - 2$, find $f(2w)$.

Solution:

$$f(2w) = 5(2w) - 2 = 10w - 2 \qquad \qquad \text{◆◆◆}$$

◆◆◆ **Example 35:** If $f(x) = x^2 - 2x$, find $f(w^2)$.

Solution:

$$f(w^2) = (w^2)^2 - 2(w^2) = w^4 - 2w^2 \qquad \qquad \text{◆◆◆}$$

It can be confusing when the expression to be substituted contains the same variable as in the original function.

◆◆◆ Example 36: If $f(x) = 5x - 2$, then

$$f(x + a) = 5(x + a) - 2$$
$$= 5x + 5a - 2$$

◆◆◆

◆◆◆ Example 37: If $f(x) = x^2 - 2x$, then

$$f(x - 1) = (x - 1)^2 - 2(x - 1)$$
$$= x^2 - 2x + 1 - 2x + 2$$
$$= x^2 - 4x + 3$$

◆◆◆

There may be *more than one function* in a single problem.

◆◆◆ Example 38: Given three different functions,

$$f(x) = 3x \qquad g(x) = x^2 \qquad h(x) = \sqrt{x}$$

evaluate

$$\frac{2g(3) + 4h(9)}{f(5)}$$

Solution: First substitute into each function.

$$f(5) = 3(5) = 15 \qquad g(3) = 3^2 = 9 \qquad h(9) = \sqrt{9} = 3$$

Then combine these as indicated.

$$\frac{2g(3) + 4h(9)}{f(5)} = \frac{2(9) + 4(3)}{15} = \frac{18 + 12}{15} = \frac{30}{15} = 2$$

◆◆◆

Sometimes we must substitute into a function containing *more than one variable*.

◆◆◆ Example 39: Given $f(x, y, z) = 2y - 3z + x$, find $f(3, 1, 2)$.

Solution: We substitute the given numerical values for the variables. Be sure that the numerical values are taken in the *same order* as the variable names in the functional notation.

$$f(x, y, z)$$
$$\updownarrow \; \updownarrow \; \updownarrow$$
$$f(3, 1, 2)$$

Substituting, we obtain

$$f(3, 1, 2) = 2(1) - 3(2) + 3 = 2 - 6 + 3$$
$$= -1$$

◆◆◆

Exercise 2 ◆ Functional Notation

Implicit and Explicit Forms

Which equations are in explicit form and which in implicit form?

1. $y = 5x - 8$ **2.** $x = 2xy + y^2$

3. $3x^2 + 2y^2 = 0$ **4.** $y = wx + wz + xz$

Dependent and Independent Variables

Label the variables in each equation as dependent or independent.

5. $y = 3x^2 + 2x$

6. $x = 3y - 8$

7. $w = 3x + 2y$

8. $xy = x + y$

9. $x^2 + y^2 = z$

Manipulating Functions

10. If $y = 5x + 3$, write $x = f(y)$.

11. If $x = \dfrac{2}{y - 3}$, write $y = f(x)$.

12. If $y = \dfrac{1}{x} - \dfrac{1}{5}$, write $x = f(y)$.

13. If $x^2 + y = x - 2y + 3x^2$, write $y = f(x)$.

14. If $2xw - 3w^2 = 6 + 3x$, write $x = f(w)$.

15. If $5p - q = q - p^2$, write $q = f(p)$.

16. The power, P, dissipated in a resistor is given by $P = I^2R$. Write $R = f(P, I)$.

17. Young's modulus, E, is given by

$$E = \frac{PL}{ae}$$

Write $e = f(P, L, a, E)$.

Substituting into Functions

Substitute the given numerical value(s) into each function.

18. If $f(x) = 2x^2 + 4$, find $f(3)$.   **19.** If $f(x) = 5x + 1$, find $f(1)$.

20. If $f(x) = 15x + 9$, find $f(3)$.   **21.** If $f(x) = 5 - 13x$, find $f(2)$.

22. If $g(x) = 9 - 3x^2$, find $g(-2)$.   **23.** If $h(x) = x^3 - 2x + 1$, find $h(2.55)$.

24. If $f(x) = 7 + 2x$, find $f(3)$.

25. If $f(x) = x^2 - 9$, find $f(-2)$.

26. If $f(x) = 2x + 7$, find $f(-1)$.

27. If $f_1(z) = z^2 - 6.43$, find $f_1(3.02)$.

28. If $g_3(w) = 6.83 + 2.74w$, find $g_3(1.74)$.

Substitute and combine as indicated.

29. Given $f(x) = 2x - x^2$, find $f(5) + 3f(2)$.

30. Given $f(x) = x^2 - 5x + 2$, find $f(2) + f(3) - f(1)$.

31. Given $f(x) = x^2 - 4$, find $\dfrac{f(2) + f(3)}{4}$.

32. Given (a) $f(x) = 2x$ and (b) $f(x) = x^2$, evaluate

$$\frac{f(x + d) - f(x)}{d}$$

for each given function.

We deal with expressions of this form in calculus, when studying the derivative.

Substitute the literal values into each function.

33. If $f(x) = 2x^2 + 4$, find $f(a)$.

34. If $f(x) = 2x - \dfrac{1}{x} + 4$, find $f(2a)$.

35. If $f(x) = 5x + 1$, find $f(a + b)$.

36. If $f(x) = 5 - 13x$, find $f(-2c)$.

Substitute in each function and combine.

37. If $f(x) = x^2$ and $g(x) = \dfrac{1}{x}$, find $f(3) + g(2)$.

38. If $f(x) = \dfrac{2}{x}$ and $g(x) = 3x$, find $f(4) - g\left(\dfrac{1}{3}\right)$.

39. If $f(x) = 8x^2 - 2x + 1$ and $g(x) = x - 5$, find $2f(3) - 4g(3)$.

40. If $h(z) = 2x$ and $g(w) = w^2$, find the following:

(a) $\dfrac{h(4) + g(1)}{5}$ (b) $h[g(3)]$ (c) $g[h(2)]$

Functions of More Than One Variable

Substitute the values into each function.

41. If $f(x, y) = 3x + 2y^2 - 4$, find $f(2, 3)$.

42. If $f(x, y, z) = 3z - 2x + y^2$, find $f(3, 1, 5)$.

43. If $g(a, b) = 2b - 3a^2$, find $g(4, -2)$.

44. If $f(x, y) = y - 3x$, find $3f(2, 1) + 2f(3, 2)$.

45. If $g(a, b, c) = 2b - a^2 + c$, find $\dfrac{3g(1, 1, 1) + 2g(1, 2, 3)}{g(2, 1, 3)}$.

46. If $h(p, q, r) = 6r - 2p + q$, find $h(3, 2, 1) + \dfrac{2h(1, 1, 2) - 3h(3, 4, 5)}{4h(2, 2, 2)}$.

47. If $g(k, l, m) = \dfrac{k + m}{l}$, find $g(2.38, 1.25, 5.38) + 3g(4.27, 8.26, 2.48)$.

Applications

48. The distance travelled in metres by a freely falling body is a function of the elapsed time, t:

$$f(t) = V_0 t + \frac{1}{2}gt^2$$

where V_0 is the initial velocity and g is the acceleration due to gravity (9.807 m/s²). If V_0 is 5.50 m/s, find $f(10.0)$, $f(15.0)$, and $f(20.0)$.

49. The resistance R of a conductor is a function of temperature:

$$f(t) = R_0(1 + \alpha t)$$

where R_0 is the resistance at 0 °C and α is the temperature coefficient of resistance (0.004 27 for copper). If the resistance of a copper coil is $98\overline{0}0\ \Omega$ at 0 °C, find $f(20.0)$, $f(25.0)$, and $f(30.0)$.

50. The maximum deflection in millimetres of a certain cantilever beam, with a concentrated load applied r metres from the fixed end, is a function of r.

$$f(r) = 0.030r^2(24.0 - r)$$

Find the deflections $f(2.\overline{0})$ and $f(3.5)$.

51. The length of a certain steel bar in metres at temperature t is given by

$$L = f(t) = 0.240(1 + 0.0635t)$$

Find $f(115), f(155)$, and $f(215)$.

52. The power, P, in watts, in a resistance, R, in which flows a current, I, is given by

$$P = f(I) = I^2 R$$

If $R = 25.6 \ \Omega$, find $f(1.59), f(2.37)$, and $f(3.17)$.

53. The sound intensity of the siren mounted on a fire truck can be expressed as a function of the distance (in kilometres) from the siren: $f(x) = 12 - 0.2x^2$. Find: $f(3.7)$, $f(6.5)$

54. The temperature in degrees Celsius of a residential fire as a function of distance in metres is $f(x) = 800 - 0.73x^2$. Find the temperatures $f(19)$ and $f(29)$.

55. The distance d (in metres) at which smoke of height h (in metres) from a fire can be seen is specified in the function $d(h) = 2.8h^2 + 25$. Evaluate $d(16)$.

4–3   Composite Functions and Inverse Functions

Composite Functions

Just as we can substitute a constant or a variable into a given function, we can substitute a *function* into a function. This is known as a *composite function*.

◆◆◆ **Example 40:** If $g(x) = x + 1$, find

(a) $g(2)$ (b) $g(z^2)$ (c) $g[f(x)]$

Solution:

(a) $g(2) = 2 + 1 = 3$ (b) $g(z^2) = z^2 + 1$ (c) $g[f(x)] = f(x) + 1$ ◆◆◆

In calculus, we will take derivatives of composite functions by means of the chain rule.

The function $g[f(x)]$ (which we read "g of f of x"), being made up of the two functions $g(x)$ and $f(x)$, is called a *composite function*. If we think of a function as a machine, it is as if we are using the output $f(x)$ of the function machine f as the input of a second function machine g.

$$x \to \boxed{f} \to f(x) \to \boxed{g} \to g[f(x)]$$

We thus obtain $g[f(x)]$ by replacing x in $g(x)$ by the function $f(x)$.

◆◆◆ **Example 41:** Given the functions $g(x) = x + 1$ and $f(x) = x^3$, write the composite function $g[f(x)]$.

Solution: In the function $g(x)$, we replace x by $f(x)$.

$$\begin{array}{ccc} g(x) & = & x + 1 \\ \downarrow & & \downarrow \\ g[f(x)] & = & f(x) + 1 \\ & & \downarrow \\ & = & x^3 + 1 \end{array}$$

since $f(x) = x^3$.

As we have said, the notation $g[f(x)]$ means to substitute $f(x)$ into the function $g(x)$. On the other hand, the notation $f[g(x)]$ means to substitute $g(x)$ into $f(x)$.

$$x \to \boxed{g} \to g(x) \to \boxed{f} \to f[g(x)]$$

In general, $f[g(x)]$ *will not be the same as* $g[f(x)]$.

◆◆◆ **Example 42:** Given $g(x) = x^2$ and $f(x) = x + 1$, find the following:

(a) $f[g(x)]$

(b) $g[f(x)]$

(c) $f[g(2)]$

(d) $g[f(2)]$

Solution

(a) $f[g(x)] = g(x) + 1 = x^2 + 1$

(b) $g[f(x)] = [f(x)]^2 = (x + 1)^2$

(c) $f[g(2)] = 2^2 + 1 = 5$

(d) $g[f(2)] = (2 + 1)^2 = 9$

Notice that here $f[g(x)]$ is not equal to $g[f(x)]$.

Inverse of a Function

Consider a function f that, given a value of x, returns some value of y.

$$x \rightarrow \boxed{f} \rightarrow y$$

If that y is now put into a function g that *reverses the operations* in f so that its output is the original x, then g is called the *inverse* of f.

$$x \rightarrow \boxed{f} \rightarrow y \rightarrow \boxed{g} \rightarrow x$$

The inverse of a function $f(x)$ is often designated by $f^{-1}(x)$.

$$x \rightarrow \boxed{f} \rightarrow y \rightarrow \boxed{f^{-1}} \rightarrow x$$

Common Error	Do not confuse f^{-1} with $\dfrac{1}{f}$.

◆◆◆ **Example 43:** Two such inverse operations are "cube" and "cube root."

$$x \rightarrow \boxed{\text{cube } x} \rightarrow x^3 \rightarrow \boxed{\text{take cube root}} \rightarrow x$$

Thus if a function $f(x)$ has an inverse $f^{-1}(x)$ that reverses the operations in $f(x)$, then the *composite* of $f(x)$ and $f^{-1}(x)$ should have no overall effect. If the input is x, then the output must also be x. In symbols, if $f(x)$ and $f^{-1}(x)$ are inverse functions, then

$$f^{-1}[f(x)] = x$$

and

$$f[f^{-1}(x)] = x$$

Similarly, if $g[f(x)] = x$ and $f[g(x)] = x$, then $f(x)$ and $g(x)$ are inverse functions.

◆◆◆ **Example 44:** Using the example of the cube and cube root, if

$$f(x) = x^3 \quad \text{and} \quad g(x) = \sqrt[3]{x}$$

then

$$g[f(x)] = \sqrt[3]{f(x)} = \sqrt[3]{x^3} = x$$

and

$$f[g(x)] = [g(x)]^3 = (\sqrt[3]{x})^3 = x$$

This shows that $f(x)$ and $g(x)$ are indeed inverse functions. ◆◆◆

To find the inverse of a function $y = f(x)$:

1. Solve the given equation for x.
2. Interchange x and y.

◆◆◆ **Example 45:** We use the cube and cube root example one more time. Find the inverse $g(x)$ of the function

$$y = f(x) = x^3$$

Solution: We solve for x and get

$$x = \sqrt[3]{y}$$

It is then customary to interchange variables so that the dependent variable is y. This gives

$$y = f^{-1}(x) = \sqrt[3]{x}$$

Thus $f^{-1}(x) = \sqrt[3]{x}$, is the inverse of $f(x) = x^3$, as verified earlier.   ◆◆◆

◆◆◆ **Example 46:** Find the inverse $f^{-1}(x)$ of the function

$$y = f(x) = 2x + 5$$

Solution: Solving for x gives

$$x = \frac{y - 5}{2}$$

Interchanging x and y, we obtain

$$y = f^{-1}(x) = \frac{x - 5}{2}$$ ◆◆◆

Sometimes the inverse of a function will *not* be a function itself, but it may be a relation.

◆◆◆ **Example 47:** Find the inverse of the function $y = x^2$.

Solution: Take the square root of both sides.

$$x = \pm \sqrt{y}$$

Interchanging x and y, we get

$$y = \pm \sqrt{x}$$

In this example, a single value of x (say, 4) gives *two* values of y ($+2$ and -2), so the inverse does not meet the definition of a function. It is, however, a relation.   ◆◆◆

Sometimes the inverse of a function gets a special name. The inverse of the sine function, for example, is called the *arcsin*, and the inverse of an exponential function is a *logarithmic* function.

We cover the inverse trigonometric functions in Secs. 7–3, 15–1, and 16–1, and graph them in Sec. 18–6. The exponential function and its inverse, the logarithmic function, are covered in Chapter 20.

Exercise 3 ◆ Composite Functions and Inverse Functions

Composite Functions

1. Given the functions $g(x) = 2x + 3$ and $f(x) = x^2$, write the composite function $g[f(x)]$.

2. Given the functions $g(x) = x^2 - 1$ and $f(x) = 3 + x$, write the composite function $f[g(x)]$.

3. Given the functions $g(x) = 1 - 3x$ and $f(x) = 2x$, write the composite function $g[f(x)]$.

4. Given the functions $g(x) = x - 4$ and $f(x) = x^2$, write the composite function $f[g(x)]$.

Given $g(x) = x^3$ and $f(x) = 4 - 3x$, find:

5. $f[g(x)]$ **6.** $g[f(x)]$ **7.** $f[g(3)]$ **8.** $g[f(3)]$

Inverse Functions

Find the inverse of:

9. $y = 8 - 3x$

10. $y = 5(2x - 3) + 4x$

11. $y = 7x + 2(3 - x)$

12. $y = (1 + 2x) + 2(3x - 1)$

13. $y = 3x - (4x + 3)$

14. $y = 2(4x - 3) - 3x$

15. $y = 4x + 2(5 - x)$

16. $y = 3(x - 2) - 4(x + 3)$

Case Study Discussion—Synchronizing Special Effects

Using the formula $t(d) = \sqrt{2d/g}$ (a function), where gravity $g = 9.8$ m/s², you are able to calculate the time (t) of the fall as a function of the distance (d) it drops. If you calculate for 4 m, 5 m, and 6 m drops you could use function notation to show that you are calculating time for specific distances, like this:

$$\text{Since } t(d) = \sqrt{2d/g}$$
$$t(4) = \sqrt{2(4)/9.8} = 0.904 \text{ s}$$
$$t(5) = \sqrt{2(5)/9.8} = 1.010 \text{ s}$$
$$t(6) = \sqrt{2(6)/9.8} = 1.107 \text{ s}$$

So, in this case, you have about a one second drop with only about a tenth of a second difference for a one metre change in distance.

⟡ CHAPTER 4 REVIEW PROBLEMS ⟡⟡⟡⟡⟡⟡⟡⟡⟡⟡⟡⟡⟡⟡⟡⟡⟡⟡⟡⟡

1. Which of the following relations are also functions?
  (a) $y = 5x^3 - 2x^2$ (b) $x^2 + y^2 = 25$
  (c) $y = \pm\sqrt{2x}$

2. Write y as a function of x if y is equal to half the cube of x, diminished by twice x.

3. Write an equation to express the surface area S of a sphere as a function of its radius r.

4. Find the domain and range for each function.
  (a) $y = \dfrac{5}{\sqrt{3 - x}}$ (b) $y = \dfrac{1 + x}{1 - x}$

5. Label each function as implicit or explicit. If it is explicit, name the dependent and independent variables.
   (a) $w = 3y - 7$
   (b) $x - 2y = 8$
6. Given $y = 3x - 5$, write $x = f(y)$.

7. Given $x^2 + y^2 + 2w = 3$, write $w = f(x, y)$.

8. Write the inverse of the function $y = 9x - 5$.

9. If $y = 3x^2 + 2z$ and $z = 2x^2$, write $y = f(x)$.
10. If $f(x) = 5x^2 - 7x + 2$, find $f(3)$.
11. If $f(x) = 9 - 3x$, find $2f(3) + 3f(1) - 4f(2)$.
12. If $f(x) = 7x + 5$ and $g(x) = x^2$, find $3f(2) - 5g(3)$.
13. If $f(x, y, z) = x^2 + 3xy + z^3$, find $f(3, 2, 1)$.
14. If $f(x) = 8x + 3$ and $g(u) = u^2 - 4$, write $f[g(u)]$.
15. If $f(x) = 5x$, $g(x) = 1/x$, and $h(x) = x^3$, find

$$\frac{5h(1) - 2g(3)}{3f(2)}$$

16. Is the relation $x^2 + 3xy + y^2 = 1$ a function? Why?

17. Replace the function $y = 5x^3 - 7$ by a verbal statement.
18. What are the domain and the range of the function $(10, -8)$, $(20, -5)$, $(30, 0)$, $(40, 3)$, and $(50, 7)$?

19. If $y = 6 - 3x^2$ and $x \geq 0$, write $x = f(y)$.

20. If $x = 6w - 5y$ and $y = 3z + 2w$, write $x = f(w, z)$.
21. Write the inverse of the function $y = 3x + 4(2 - x)$.
22. If $xw - 4w^2 = 5 + 4x$, write $x = f(w)$.

23. If $7p - 2q = 3q - p^2$, write $q = f(p)$.

24. Given the functions $g(x) = x + x^2$ and $f(x) = 5x$, write the composite function $g[f(x)]$.

25. If $h(p, q, r) = 5r + p + 7q$, find $h(1, 2, 3) + \dfrac{2h(1, 1, 1) - 3h(2, 2, 2)}{h(3, 3, 3)}$

26. If $g_3(w) = 1.74 + 9.25w$, find $g_3(1.44)$.
27. Write the inverse of the function $y = 5(7 - 4x) - x$.

28. If $g(k, l, m) = \dfrac{k + 3m}{2l}$, find $g(4.62, 1.39, 7.26)$.
29. Write the inverse of the function $y = 9x + (x + 5)$.

30. The length of a certain steel bar at temperature t is given by

$$L = f(t) = 0.252(1 + 0.0655t) \text{ m}$$

Find $f(112)$, $f(176)$, and $f(195)$.

31. If $f_1(z) = 3.82z^2 - 2.46$, find $f_1(5.27)$.

32. The power, P, in a resistance, R, in which flows a current, I, is given by

$$P = f(I) = I^2 R$$

If $R = 325\ \Omega$, find $f(11.2)$, $f(15.3)$, and $f(21.8)$.

33. Given the functions $g(x) = 3x - 5$ and $f(x) = 7x^2$, write the composite function $f[g(x)]$.

34. Given $g(x) = 5x^2$ and $f(x) = 7 - 2x$ find:
   (a) $f[g(x)]$ (b) $g[f(x)]$
   (c) $f[g(5)]$ (d) $g[f(5)]$

Writing

35. State, in your own words, what is meant by "function" and "relation," and describe how a function differs from a relation. You may use examples to help describe these terms.

5

Graphs

◆◆◆ OBJECTIVES ◆◆

When you have completed this chapter, you should be able to:
- Graph points and empirical data in rectangular coordinates.
- Graph a linear function.
- Make a complete graph of a relation or a function.
- Graph parametric functions in rectangular coordinates.
- Solve equations graphically.

A line not only connects two points; it also contains an infinite number of points within itself. Holding all this information is the role of graphs. Graphs present equations so we can see how two variables interact. When we see a graph we can look for patterns and shapes. We can see similarities between our equation and other equations. Graphs can also show us whether or not there is a connection between two variables. With all the power of computers and calculators, the fact that we still build into them the ability to produce graphs and charts demonstrates how important it is to try to visualize equations. Graphs help connect humans to mathematics.

CASE STUDY—SOLAR CELL EFFICIENCY

The demand for solar and green energy has never been greater. This demand has pushed engineers to develop higher efficiency solar panels. A client of your solar energy company decides to wait until the solar cell efficiency improves to 50% conversion. This chart shows the expected rise in efficiency between 2004 and 2050. Determine the equation for this graph (use a straight line approximation) and calculate when your client might expect to see a 50% conversion rate.

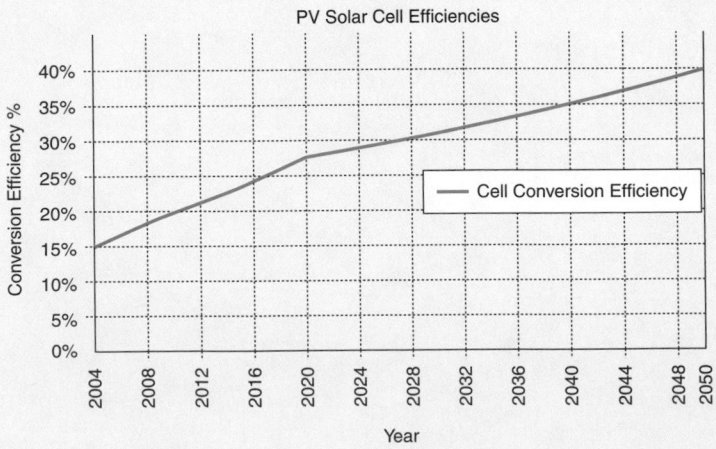

The terms *chart* and *graph* can be used interchangeably.

5–1 Mapping Rectangular Coordinates

The Rectangular Coordinate System

In Chapter 1 we plotted numbers on the number line. Suppose, now, that we take a second number line and place it at right angles to the first one, so that each intersects the other at the zero mark, as in Fig. 5–1. We call this a *rectangular coordinate system*.

Rectangular coordinates are also called *Cartesian* coordinates, after the French mathematician René Descartes (1596–1650). Another type of coordinate system we will use is called the *polar* coordinate system.

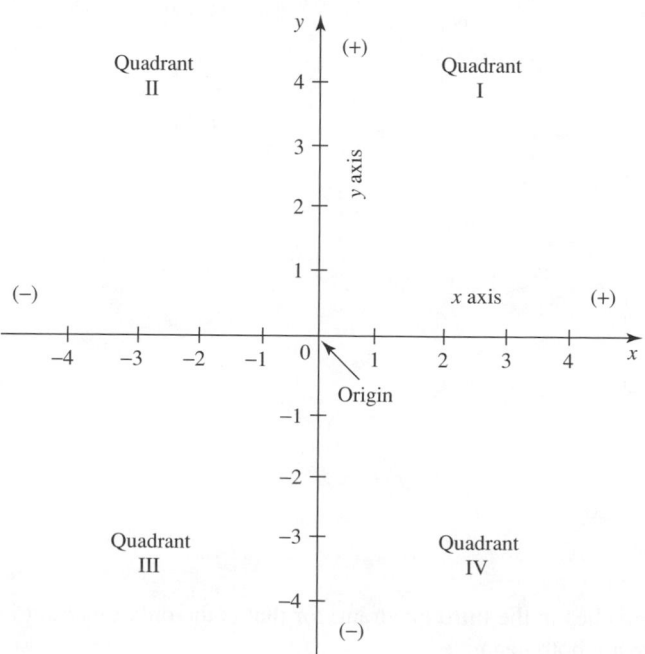

FIGURE 5–1 The rectangular coordinate system.

The horizontal number line is called the *x axis*, and the vertical line is called the *y axis*. They intersect at the *origin*. These two axes divide the plane into four *quadrants*, numbered counterclockwise, as in Fig. 5–1.

Graphing Ordered Pairs

Figure 5–2 shows a point P in the first quadrant. Its horizontal distance from the origin, called the *x coordinate* or *abscissa* of the point, is 3 units. Its vertical distance from the origin, called the *y coordinate* or *ordinate* of the point, is 2 units. The numbers in the *ordered pair* (3, 2) are called the *rectangular coordinates* (or simply *coordinates*) of the point. They are always written in the same order, with the *x* coordinate first (i.e., in alphabetical order). The letter identifying the point is sometimes written before the coordinates, as in $P(3, 2)$.

To plot any ordered pair (h, k), simply place a point at a distance h units from the *y* axis and k units from the *x* axis. Remember that negative values of *x* are located to the left of the origin and that negative *y* values are below the origin.

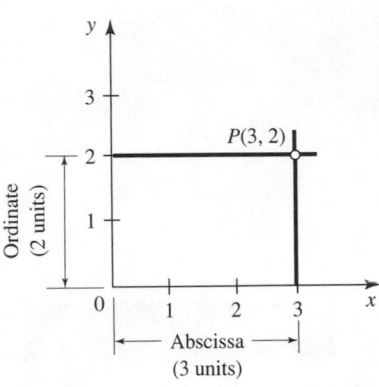

FIGURE 5–2   Is it clear from this figure why we call these *rectangular coordinates?*

◆◆◆ **Example 1:** The points

$$P(4, 1) \quad Q(-2, 3) \quad R(-1, -2) \quad S(2, -3) \quad T(1.3, 2.7)$$

are shown plotted in Fig. 5–3.

Notice that the abscissa is negative in the second and third quadrants and that the ordinate is negative in the third and fourth quadrants. Thus the signs of the coordinates of a point tell us the quadrant in which the point lies.   ◆◆◆

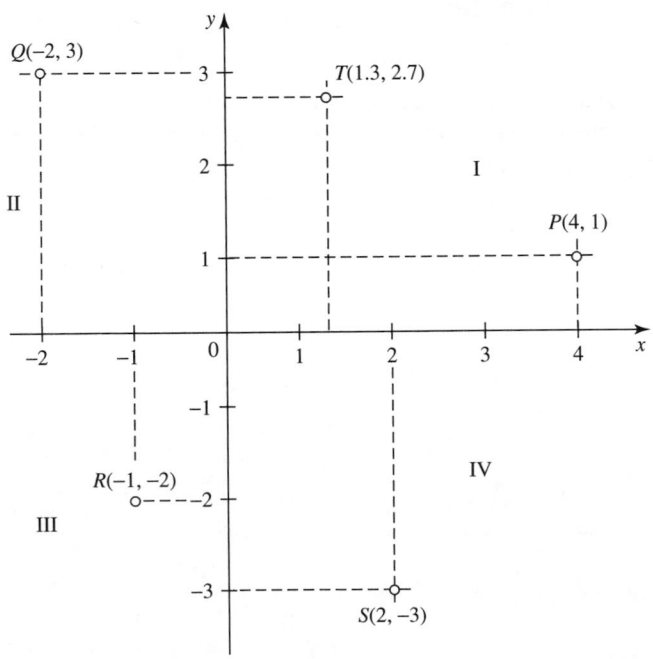

FIGURE 5–3

◆◆◆ **Example 2:** The point $(-3, -5)$ lies in the third quadrant, for that is the only quadrant in which the abscissa and the ordinate are both negative.   ◆◆◆

Exercise 1 ◆ Rectangular Coordinates

Rectangular Coordinates

If h and k are positive quantities, in which quadrants would the following points lie?

1. $(h, -k)$
2. (h, k)
3. $(-h, k)$
4. $(-h, -k)$

5. Which quadrant contains points having a positive abscissa and a negative ordinate?
6. In which quadrants is the ordinate negative?
7. In which quadrants is the abscissa positive?
8. The ordinate of any point on a certain straight line is -5. Give the coordinates of the point of intersection of that line and the y axis.
9. Find the abscissa of any point on a vertical straight line that passes through the point $(7, 5)$.

Graphing Ordered Pairs

10. Write the coordinates of points $A, B, C,$ and D in Fig. 5–4.

11. Write the coordinates of points $E, F, G,$ and H in Fig. 5–4.

12. Write the coordinates of points $A, B, C,$ and D in Fig. 5–5.

13. Write the coordinates of points $E, F, G,$ and H in Fig. 5–5.

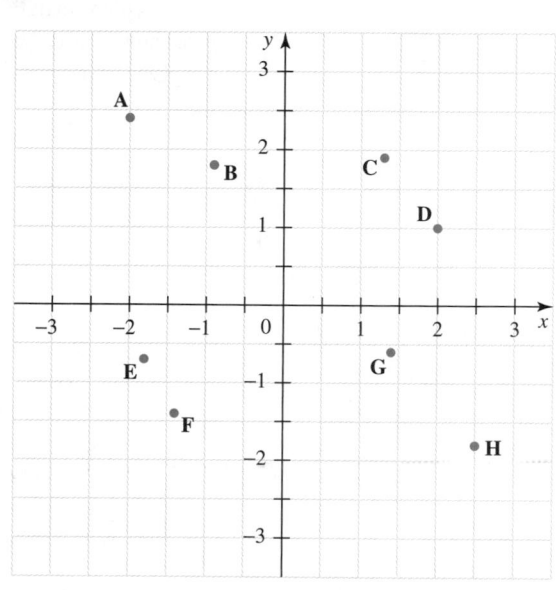

FIGURE 5–4

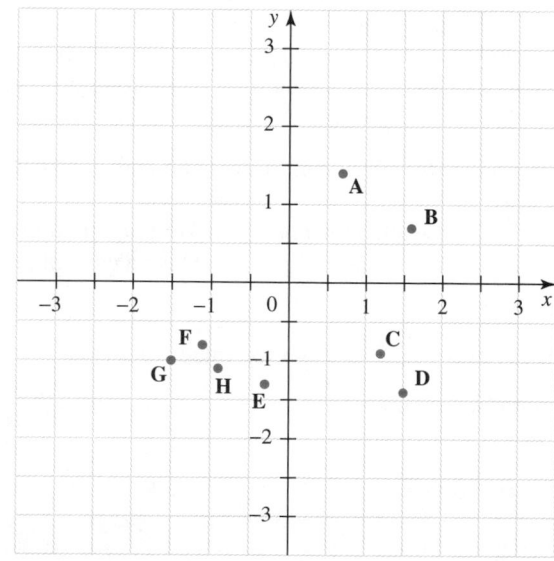

FIGURE 5–5

14. Graph each point.
    (a) $(3, 5)$
    (b) $(4, -2)$
    (c) $(-2.4, -3.8)$
    (d) $(-3.75, 1.42)$
    (e) $(-4, 3)$
    (f) $(-1, -3)$

Graph each set of points, connect them, and identify the geometric figure formed.

15. $(0.7, 2.1), (2.3, 2.1), (2.3, 0.5),$ and $(0.7, 0.5)$
16. $(2, -\frac{1}{2}), (3, -1\frac{1}{2}), (1\frac{1}{2}, -3),$ and $(\frac{1}{2}, -2)$
17. $(-1\frac{1}{2}, 3), (-2\frac{1}{2}, \frac{1}{2}),$ and $(-\frac{1}{2}, \frac{1}{2})$

18. $(-3, -1)$, $(-1, -\frac{1}{2})$, $(-2, -3)$, and $(-4, -3\frac{1}{2})$

19. Three corners of a rectangle have the coordinates $(-4, 9)$, $(8, 3)$, and $(-8, 1)$. Graphically find the coordinate of the fourth corner.

5–2   Graphing a Function Using Ordered Pairs

If the function is given as a set of ordered pairs, simply plot each ordered pair. We usually connect the points with a smooth curve unless we have reason to believe that there are sharp corners, breaks, or gaps in the graph.

If the function is in the form of an *equation*, we obtain a table of ordered pairs by first selecting values of x over the required domain and then computing corresponding values of y. Since we are usually free to select any x values we like, we pick "easy" integer values. We then plot the set of ordered pairs.

Our first graph will be of the straight line.

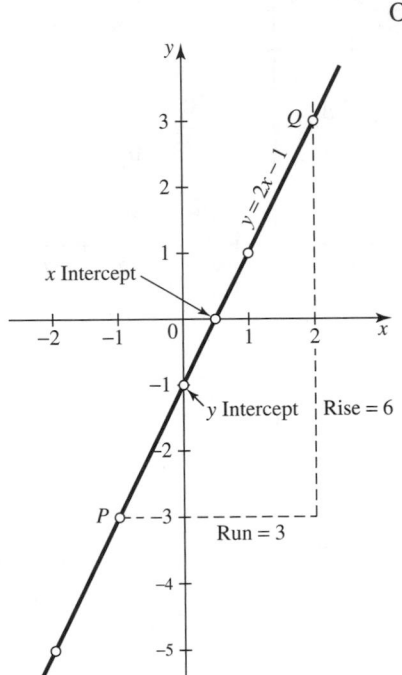

FIGURE 5–6 A first-degree equation will always plot as a straight line—hence the name *linear* equation.

◆◆◆ **Example 3:** Graph the function $y = f(x) = 2x - 1$ for values of x from -2 to 2.

Solution: Substituting into the equation, we obtain

$$f(-2) = 2(-2) - 1 = -5$$
$$f(-1) = 2(-1) - 1 = -3$$
$$f(0) \ = 2(0) \ \ -1 = -1$$
$$f(1) \ = 2(1) \ \ -1 = \ \ 1$$
$$f(2) \ = 2(2) \ \ -1 = \ \ 3$$

Thus our points are $(-2, -5)$, $(-1, -3)$, $(0, -1)$, $(1, 1)$, and $(2, 3)$. Note that each of these pairs of numbers satisfies the given equation.

These points plot as a straight line (Fig. 5–6). Had we known in advance that the graph would be a straight line, we could have saved time by plotting just two points, with perhaps a third as a check. ◆◆◆

In Sec. 5–3 we will discuss the straight line in more detail and show another way to graph it.

Our next graph will be of a curve called the *parabola*. We graph it, and any other function, in the same way that we graphed the straight line: Make a table of point pairs, plot the points, and connect them.

◆◆◆ **Example 4:** Graph the function $y = f(x) = x^2 - 4x - 3$ for values of x from -1 to 5.

Solution: Substituting into the equation, we obtain

$$f(-1) = (-1)^2 - 4(-1) - 3 = 1 + 4 - 3 = 2$$
$$f(0) \ = 0^2 - 0 - 3 = -3$$
$$f(1) \ = 1^2 - 4(1) - 3 = 1 - 4 - 3 = -6$$
$$f(2) \ = 2^2 - 4(2) - 3 = 4 - 8 - 3 = -7$$
$$f(3) \ = 3^2 - 4(3) - 3 = 9 - 12 - 3 = -6$$
$$f(4) \ = 4^2 - 4(4) - 3 = 16 - 16 - 3 = -3$$
$$f(5) \ = 5^2 - 4(5) - 3 = 25 - 20 - 3 = 2$$

The points obtained are plotted in Fig. 5–7.

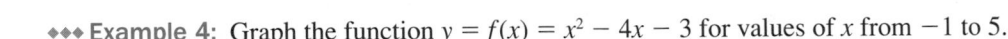

Common Error	Be especially careful when substituting negative values into an equation. It is easy to make an error.

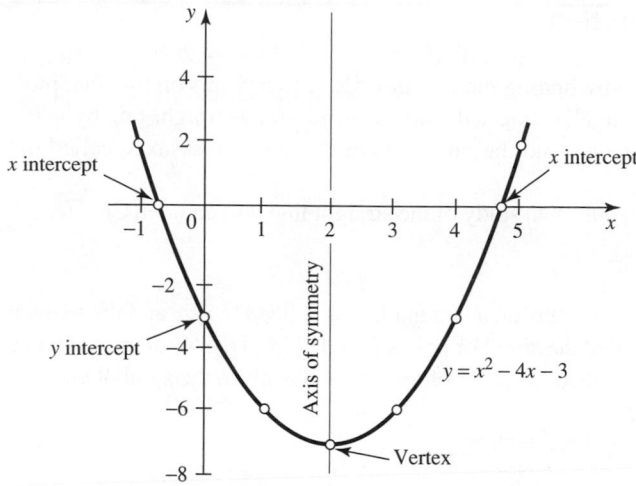

FIGURE 5-7 This curve is called a *parabola,* which we will meet again later. Note the *vertex* and the *axis of symmetry.*

If we had plotted many more points than these—say, billions of them—they would be crowded so close together that they would seem to form a continuous line. The *curve* can be thought of as a *collection* of all points that satisfy the equation. Such a curve (or the set of points) is called a *locus* of the equation.

The straight line and the parabola are covered in much greater detail in analytic geometry in Chapter 22.

Exercise 2 ◆ The Graph of a Function

Graphing Sets of Ordered Pairs

Graph each set of ordered pairs. Connect them with a curve that best fits the data.

1. $(-3, -2), (9, 6), (3, 2), (-6, -4)$
2. $(-7, 3), (0, 3), (4, 10), (-6, 1), (2, 6), (-4, 0)$
3. $(-10, 9), (-8, 7), (-6, 5), (-4, 3), (-2, 4), (0, 5), (2, 6), (4, 7)$
4. $(0, 4), (3, 3.2), (5, 2), (6, 0), (5, -2), (3, -3.2), (0, -4)$

Graph of a Straight Line

Graph each straight line by plotting two points on the line plus a third as a check.

5. $y = 3x + 1$
6. $y = 2x - 2$
7. $y = 3 - 2x$
8. $y = -x + 2$

Graph of Any Function

For each function, make a table of ordered pairs, taking integer values of x from -3 to 3. Plot the point pairs and connect them with a smooth curve.

9. $y = x^2$
10. $y = 4 - 2x^2$
11. $y = \dfrac{x^2}{x + 3}$
12. $y = x^2 - 7x + 10$
13. $y = x^2 - 1$
14. $y = 5x - x^2$
15. $y = x^3$
16. $y = x^3 - 2$

5–3   Graphing a Linear Function

We graphed a straight line in Sec. 5–2 by finding the coordinates of two points on the line, plotting those points, and connecting them. Here, we will show how to plot a straight line by using the steepness of the line, called the *slope*, and the point where it crosses the *y* axis, called the *y intercept*.

   This is still introductory material; our main study of the straight line will come in Chapter 22.

Slope

The horizontal distance between any two points on a straight line is called the *run*, and the vertical distance between the same points is called the *rise*. For points *P* and *Q* in Fig. 5–6, the rise is 6 in a run of 3. The rise divided by the run is called the *slope* of the line and is given the symbol *m*.

$$\text{slope} = m = \frac{\text{rise}}{\text{run}}$$

◆◆◆ **Example 5:** The straight line in Fig. 5–6 has a rise of 6 in a run of 3, so

$$\text{slope } m = \frac{\text{rise}}{\text{run}} = \frac{6}{3} = 2$$ ◆◆◆

◆◆◆ **Example 6:** Find the slope of a straight line that drops half a unit as we move 1 unit in the positive *x* direction.

Solution: The rise is $-1/2$ for a run of 1, so the slope is

$$\text{slope } m = \frac{-1/2}{1} = -1/2$$ ◆◆◆

◆◆◆ **Example 7:** Find the slope of the line in Fig. 5–8. (Do not be thrown off by the different scales on each axis.)

Solution: The rise is $30 - 10 = 20$ and the run is $2 - 0 = 2$, so

$$m = \frac{20}{2} = 10$$

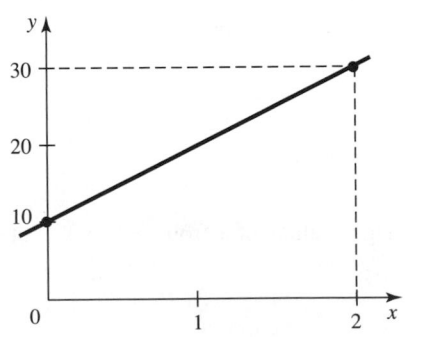

FIGURE 5-8   ◆◆◆

If (x_1, y_1) and (x_2, y_2) are two points on the line, the slope is given by the following formula:

Slope		$m = \dfrac{\text{rise}}{\text{run}} = \dfrac{y_2 - y_1}{x_2 - x_1}$	284

The slope is equal to the rise divided by the run.

◆◆◆ Example 8: Find the slope of the line connecting the points $(-3, 5)$ and $(4, -6)$ in Fig. 5–9.

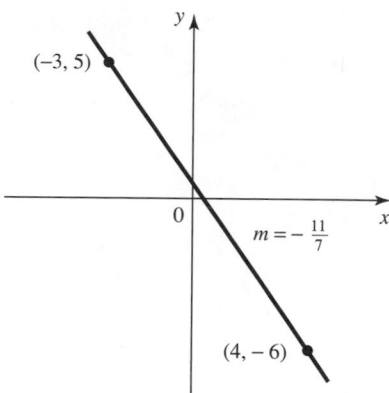

(-3, 5)

$m = -\frac{11}{7}$

(4, -6)

FIGURE 5–9

Solution: We will see that the slope does not depend on what we label as point 1 and point 2. Let us choose

$$x_1 = -3, \qquad y_1 = 5, \qquad x_2 = 4, \qquad y_2 = -6$$

Then by Eq. 284,

$$\text{slope, } m = \frac{-6 - 5}{4 - (-3)} = \frac{-11}{7} = -\frac{11}{7}$$

If we had chosen $(x_1 = 4, y_1 = -6)$ and $(x_2 = -3, y_2 = 5)$, the computation for slope,

$$m = \frac{5 - (-6)}{-3 - 4} = \frac{11}{-7} = -\frac{11}{7}$$

would have given the same result. ◆◆◆

Common Error	Be careful not to mix up the subscripts. $$m \neq \frac{y_2 - y_1}{x_1 - x_2}$$

Horizontal and Vertical Lines

For any two points on a horizontal line, the values of y_1 and y_2 in Eq. 284 are equal, making the slope equal to zero. For a vertical line, the values of x_1 and x_2 are equal, giving division by zero. Hence the slope is undefined for a vertical line. The slopes of various lines are shown in Fig. 5–10.

Intercepts

The *intercepts* of a curve are the places where the curve crosses the x and y axes. The term intercept is used for any curve, not just the straight line.

◆◆◆ Example 9: The x intercept for the straight line in Fig. 5–6 is $(1/2, 0)$, and the y intercept is $(0, -1)$. ◆◆◆

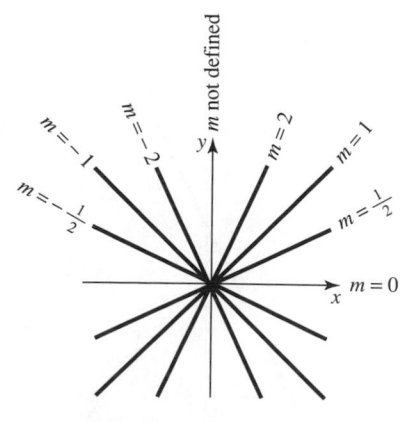

FIGURE 5–10

Equation of a Straight Line

Suppose that $P(x, y)$ is any point on a straight line. We seek an equation that links x and y in a functional relationship so that, for any x, a value of y can be found. We can get such an equation by applying the definition of slope to our point P and some *known* point on the line. Let us use the y intercept $(0, b)$ as the coordinates for the known point. For P, a general point on the line, we use coordinates (x, y). For a line that intersects the y axis b units from the origin (Fig. 5–11), the rise is $y - b$ and the run is $x - 0$. So, by Eq. 284,

$$m = \frac{y - b}{x - 0}$$

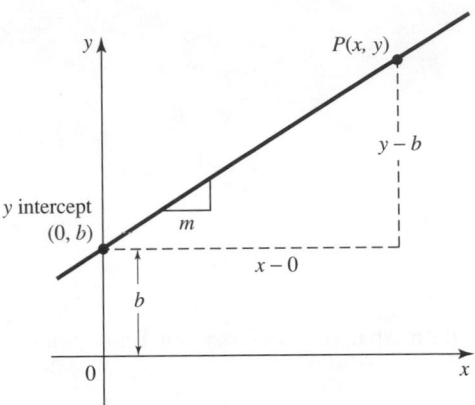

FIGURE 5–11

Simplifying, we have $mx = y - b$, also written in the following form:

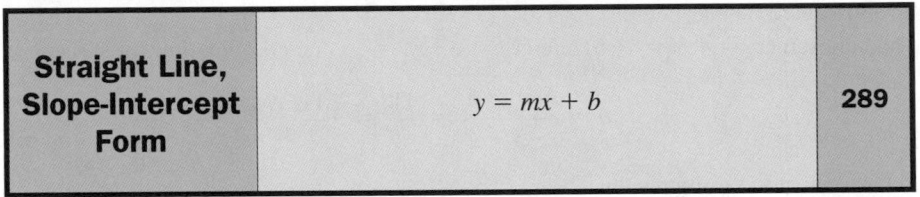

Straight Line, Slope-Intercept Form	$y = mx + b$	289

This is called the slope-intercept form of the equation of a straight line because the slope m and the y intercept b are easily identified once the equation is in this form. For example, in the equation $y = 2x + 1$,

$$y = 2x + 1$$

slope ⟶ ⌐ ⌐ ⟵ y intercept

Here m, the slope, is 2, and b, the y intercept, is 1.

Here, we introduce only the *slope-intercept* form of the equation of a straight line. In Sec. 22–1, we will discuss another presentation of the equation known as the *point-slope* form

◆◆◆ **Example 10:** Write the equation of the straight line, in slope-intercept form, that has a slope of 2 and a y intercept of 3. Make a graph.

Solution: Substituting $m = 2$ and $b = 3$ into Eq. 289 gives

$$y = 2x + 3$$

To graph the line, we first locate the y intercept, 3 units up from the origin. The slope of the line is 2, so we get another point on the line by moving 1 unit to the right and 2 units up from the y intercept. This brings us to $(1, 5)$. Connecting this point to the y intercept gives us our line as shown in Fig. 5–12.   ◆◆◆

◆◆◆ **Example 11:** Find the slope and the y intercept of the line $y = 3x - 4$. Make a graph.

Solution: By inspection, we see that the slope is the coefficient of x and that the y intercept is the constant term. So

$$m = 3 \quad \text{and} \quad b = -4$$

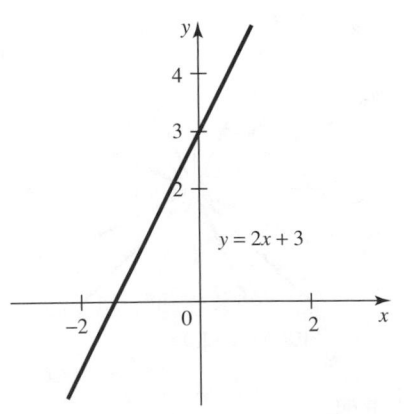

FIGURE 5–12

We plot the y intercept 4 units below the origin, and through it we draw a line with a rise of 3 units in a run of 1 unit, as shown in Fig. 5–13.

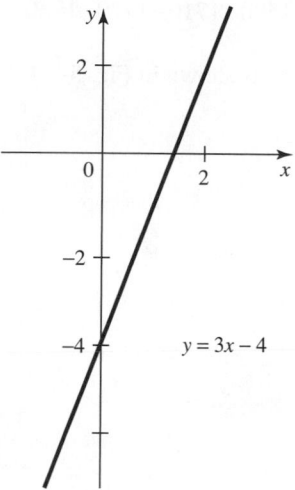

FIGURE 5–13

◆◆◆

Exercise 3 ◆ Graphing the Straight Line

Find the slope of each straight line.

1. Rise = 4; run = 2
2. Rise = 6; run = 4
3. Rise = −4; run = 4
4. Rise = −9; run = −3
5. Connecting (2, 4) and (5, 7)
6. Connecting (5, 2) and (3, 6)
7. Connecting (−2, 5) and (5, −6)
8. Connecting (3, −3) and (−6, 2)

Write the equation of each straight line in slope-intercept form, and make a graph.

9. Slope = 4; y intercept = −3
10. Slope = −1; y intercept = 2
11. Slope = 3; y intercept = −1
12. Slope = −2; y intercept = 3
13. Slope = 2.3; y intercept = −1.5
14. Slope = −1.5; y intercept = 3.7

Find the slope and the y intercept for each equation, and make a graph.

15. $y = 3x - 5$
16. $y = 7x + 2$
17. $y = \dfrac{1}{2}x - \dfrac{1}{4}$
18. $y = -3x + 2$

5–4 Graphing Empirical Data, Formulas, and Parametric Equations

Graphing Empirical Data

Data obtained from an experiment or by observation are called *empirical* data. Empirical data are graphed just as any other set of point pairs. Be sure to label the graph completely, including units on the axes.

◆◆◆ **Example 12:** Graph the following data for the temperature rise in a certain oven:

Time (h)	0	1	2	3	4	5	6	7	8	9	10
Temperature (°F)	102	463	748	1010	1210	1370	1510	1590	1710	1770	1830

Solution: We plot each point, connect the points, and label the graph, as shown in Fig. 5–14.

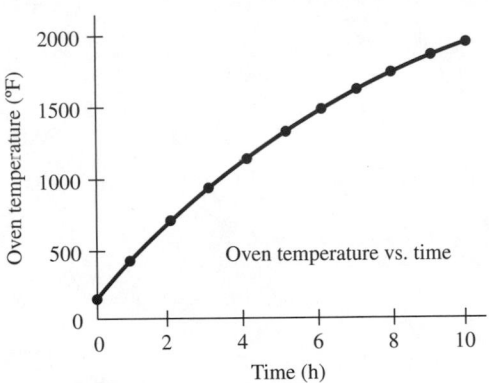

FIGURE 5–14   Note the use of different scales on each axis.   ◆◆◆

Graphing Formulas

A formula is graphed in the same way that we graph any other function: We substitute suitable values of the independent variable into the formula and compute the corresponding values of the dependent variable. We then plot the ordered pairs and connect the points obtained with a smooth curve. However, with formulas we must be careful to handle the units properly and to label the graph more fully.

◆◆◆ **Example 13:** The formula for the power, P, dissipated in a resistor carrying a current of I amperes (A) is

$$P = I^2R \quad \text{watts (W)}$$

where R is the resistance in ohms. Graph P versus I for a resistance of 10 000 Ω. Take I from 0 to 10 A.

Solution: Let us choose values of I of 0, 1, 2, 3, . . . , 10 and make a table of ordered pairs by substituting these into the formula.

I(A)	0	1	2	3	$\cdots$
P(W)	0	10 000	40 000	90 000	$\cdots$

At this point we notice that the figures for wattage are so high that it will be more convenient to work in kilowatts (kW), where 1 kW = 1000 W.

I(A)	0	1	2	3	4	5	6	7	8	9	10
P(kW)	0	10	40	90	160	250	360	490	640	810	1000

These points are plotted in Fig. 5–15. Note the labelling of the graph and of the axes.

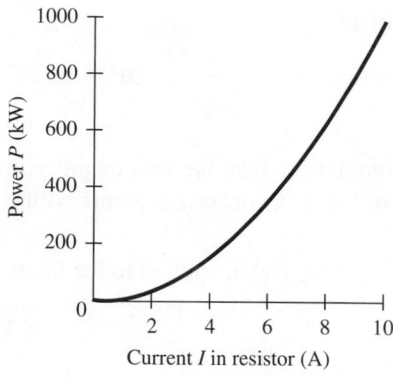

FIGURE 5–15 Power dissipated in a 10 000 Ω resistor calculated using $P = I^2R$ with $R = 10\,000\ \Omega$. ◆◆◆

Graphing Parametric Equations

For the equations we have graphed so far, y has been expressed as a function of x.

$$y = f(x)$$

But x and y can also be related to each other by means of a third variable, say, t, meaning time, if both x and y are given as functions of t.

$$x = g(t)$$

and

$$y = h(t)$$

Such equations are called *parametric equations*. The third variable t is called the *parameter*.

To graph parametric equations, we assign values to the parameter t and compute x and y for each t. We then plot the table of (x, y) pairs.

◆◆◆ **Example 14:** Graph the parametric equations

$$x = 2t \quad \text{and} \quad y = t^2 - 2$$

for $t = -3$ to 3.

Solution: We make a table with rows for t, x, and y. We take values of t from -3 to 3, and for each we compute x and y.

t	-3	-2	-1	0	1	2	3
x	-6	-4	-2	0	2	4	6
y	7	2	-1	-2	-1	2	7

We now plot the (x, y) pairs, $(-6, 7)$, $(-4, 2)$, . . . , $(6, 7)$ and connect them with a smooth curve (Fig. 5–16). The curve obtained is a parabola, as in Example 13, but obtained here with parametric equations.

> In a later chapter, we'll plot parametric equations in *polar* coordinates. We'll also use polar coordinates to describe the trajectory of a projectile.

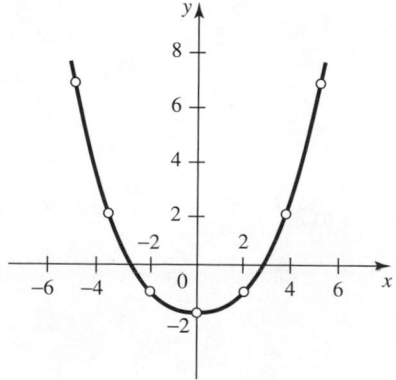

FIGURE 5–16 ◆◆◆

Exercise 4 ◆ Graphing Empirical Data, Formulas, and Parametric Equations

Graphing Empirical Data

Graph the following experimental data. Label the graph completely. Take the first quantity in each table as the abscissa, and the second quantity as the ordinate. Connect the points with a smooth curve.

1. The current, I (mA), through a tungsten lamp and the voltage, V (V), applied to the lamp, are related as follows:

V	10	20	30	40	50	60	70	80	90	100	110	120
I	158	243	306	367	420	470	517	559	598	639	676	710

2. The melting point, T (°C), of a certain alloy and the percent of lead, $P(\%)$, in the alloy, related as follows:

P	40	50	60	70	80	90
T	186	205	226	250	276	304

3. A steel wire in tension, with stress σ (lb./sq. in.) and strain ε (in./in.), related as follows:

ε	0	0.000 19	0.000 57	0.000 94	0.001 34	0.001 73	0.002 16	0.002 56
σ	5000	10 000	20 000	30 000	40 000	50 000	60 000	70 000

Graphing Formulas

4. A milling machine having a purchase price, P, of $15,600 has an annual depreciation, A, of $1,600. Graph the book value y at the end of each year, for $t = 10$ years, using the equation $y = P - At$.

5. The force, f, required to pull a block along a rough surface is given by $f = \mu N$, where N is the normal force and μ is the coefficient of friction. Plot f for values of N from 0 to 1000 N, taking μ as 0.45.

6. A 2580 Ω resistor is wired in parallel with a resistor, R_1. Graph the equivalent resistance, R, for values of R_1 from 0 to 5000 Ω, in steps of 500 Ω. Use Eq. A64.

7. Use Eq. A67 to graph the power, P, dissipated in a 2500 Ω resistor for values of the current, I, from 0 to 10 A.

8. A resistance of 5280 Ω is placed in series with a device that has a reactance of X ohms (Fig. 5–17). Using Eq. A99, plot the absolute value of the impedance, Z, for values of X from 0 to 10 000 Ω.

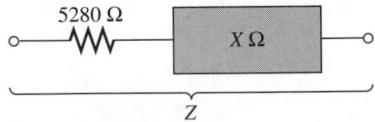

FIGURE 5–17

Graphing Parametric Equations

Plot the following parametric equations for values of t from -3 to 3.

9. $x = t, y = t$

10. $x = 3t, y = t^2$

11. $x = -t, y = 2t^2$

12. $x = -2t, y = t^2 + 1$

5–5 Solving Equations with Graphs

We can use our knowledge of graphing functions to solve equations of the form $f(x) = 0$.

We mentioned earlier that a point at which a graph of a function $y = f(x)$ crosses or touches the x axis is called an x intercept. Such an x intercept is also called a *zero* of that function.

In Fig. 5–18 there are two zeros, since there are two x values for which $y = 0$, and hence $f(x) = 0$. Those x values for which $f(x) = 0$ are called *roots* or *solutions* to the equation $f(x) = 0$.

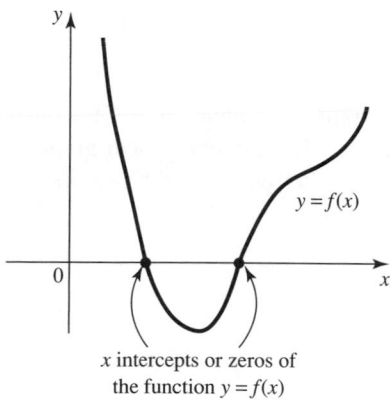

x intercepts or zeros of
the function $y = f(x)$

FIGURE 5–18

Thus if we were to graph the function $y = f(x)$, any value of x at which y is equal to zero would be a solution to $f(x) = 0$. So to solve an equation graphically, we simply put it into the form $f(x) = 0$ and then graph the function $y = f(x)$. Each x intercept is then an approximate solution to the equation.

◆◆◆ **Example 15:** Graphically find the approximate root(s) of the equation

$$4.1x^3 - 5.9x^2 - 3.8x + 7.5 = 0$$

Solution: Let us represent the left side of the given equation by $f(x)$.

$$f(x) = 4.1x^3 - 5.9x^2 - 3.8x + 7.5 \qquad (1)$$

Any value of x for which $f(x) = 0$ will clearly be a solution to Equation (1), so we simply plot $f(x)$ and look for the x intercepts. Not knowing the shape of the curve, we compute $f(x)$ at various values of x until we are satisfied that we have located each region in which an x intercept is located. We then make a table of point pairs for this region.

x	-2	-1	0	1	2	3
$f(x)$	-41.3	1.3	7.5	1.9	9.1	53.7

We graph the function (Fig. 5–19) and read the approximate value of the x intercept.

$$x \cong -1.1$$

This, then, is an approximate solution to the given equation. Have we found *every* root? If we computed y for values of x outside the region that we have graphed, we would see that the curve moves even farther from the x axis, so we are reasonably sure that we have found the only root. With other functions it may not be so easy to tell if the curve will reverse direction somewhere and cross the x axis again.

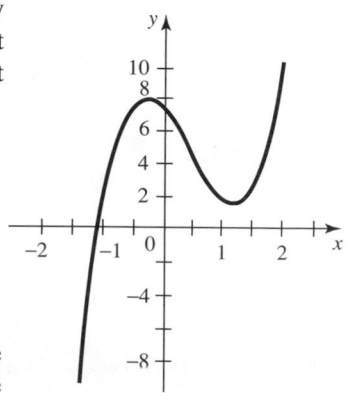

FIGURE 5–19 The shape of this curve is typical of a third-degree (cubic) function.

◆◆◆

Exercise 5 ◆ Graphical Solution of Equations

The following equations all have at least one root between $x = -10$ and $x = 10$. Put each equation into the form $y = f(x)$, and plot just enough of each to find the approximate value of the root(s). (Your graph may appear *inverted* when compared to others, depending on which side of the equals sign the zero term has been placed. The roots, however, will be the same.)

1. $2.4x^3 - 7.2x^2 - 3.3 = 0$
2. $9.4x = 4.8x^3 - 7.2$
3. $25x^2 - 19 = 48x + x^3$
4. $1.2x + 3.4x^3 = 2.8$
5. $6.4x^4 - 3.8x = 5.5$
6. $621x^4 - 284x^3 - 25 = 0$

Can you guess the significance of the point of intersection? Peek ahead at Sec. 10–1.

In each of problems 7 through 10, two equations are given. Graph each equation as if the other were not there, but as if they were on the same coordinate axes. Each equation will graph as a straight line. Give the approximate coordinates of the point of intersection of the two lines.

7. $y = 2x + 1$
   $y = -x + 2$

8. $x - y = 2$
   $x + y = 6$

9. $y = -3x + 5$
   $y = x - 4$

10. $2x + 3y = 9$
    $3x - y = -4$

11. A pump truck is working on a fire. It pumps water at a rate defined by the equation $Y = 225x$ L/min. Graph this equation to determine

    (a) how much water is pumped in 12 min

    (b) how long it takes to pump 8200 L

12. The fire marshal's office has analyzed many residential fires and has found that the equation $Y = 0.848x$ represents the average time (in minutes) that firefighters take to arrive on the scene of residential fires (where x is the distance in kilometres). Graph this equation to determine how long it would take a fire crew to reach a residential fire that is 4.7 km away.

13. The manufacturer of wall-mounted fire extinguishers uses the equation $Y = 0.57^{(x-6)}$ to predict the quantity of its chemicals in the air (in $\mu g/m^3$), where x is the distance (in metres) from which the extinguisher was used. Graph this equation to determine how far away you must be from the spot where a fire extinguisher was deployed in order to not smell its chemicals, (i.e., $Y < 0.5\ \mu g/m^3$).

Case Study Discussion–Solar Cell Efficiency

Your answer may vary slightly depending on how you approximate the straight line. The y intercept is 15%. The average slope is total rise over total run. Total rise is $40\% - 15\% = 25\%$. The total run is $2050 - 2004 = 46$ years. That makes the slope 25%/46 years = 0.543% per year. The general form for a straight line is $y = mx + b$. In our graph the y axis is the conversion efficiency and the x axis is the year. To be accurate, we start at 2004, so the equation should use the number of years since 2004. Let's identify the y variable as E (conversion efficiency percentage) and the x variable as *years* (since 2004). So our equation becomes $E = m$ (*years*) $+ b$. Since m is the slope of the line and b is the y intercept, we can write our formula as

$$E = 0.543\,(\text{year} - 2004) + 15$$

Notice that we have used "year − 2004" as the x variable. Let's test our equation on a known year before we try to predict the unknown. Using our equation to calculate the efficiency in 2028,

$$E = 0.543(\text{year} - 2004) + 15$$
$$E = 0.543(2028 - 2004) + 15$$
$$E = 0.543(24) + 15$$
$$E = 13.03 + 15$$
$$E = 28\% \text{ (rounded)}$$

Looking at the graph, 2028 is 30%, so that's pretty close for a straight-line approximation of a graph.

Now, to get back to the client, finding the year for 50% conversion efficiency,

$$E = 0.543(\text{year} - 2004) + 15$$
$$50 = 0.543(\text{year} - 2004) + 15$$
$$50-15 = 0.543(\text{year} - 2004)$$
$$35 = 0.543(\text{year} - 2004)$$
$$35/0.543 = (\text{year} - 2004)$$
$$64.5 = \text{year} - 2004$$
$$64.5 + 2004 = \text{year}$$
$$2068 = \text{year}$$

So, with a little rounding the client can expect to wait until 2068 for a 50% conversion efficiency. Of course extending data into the future is tricky. We could hit the physical maximum for this type of solar cell and never reach 50%. Or there could be a scientific breakthrough next week that produces a panel with 80% conversion efficiency. However, basing a decision on data would suggest a bit of a wait for 50% conversion.

••• CHAPTER 5 REVIEW PROBLEMS •••••••••••••••••••••••••••••••••

1. Graph the following points and connect them with a smooth curve:

x	0	1	2	3	4	5	6
y	2	$2\frac{1}{4}$	3	4	6	9	13

2. Plot the function $y = x^3 - 2x$ for $x = -3$ to $+3$. Label any roots or intercepts.
3. Graph the function $y = 3x - 2x^2$ from $x = -3$ to $x = 3$. Label any roots or intercepts.
4. Graphically find the approximate value of the roots of the equation

$$(x + 3)^2 = x - 2x^2 + 4$$

5. Graphically find the point of intersection of the following equations:

$$2.5x - 3.7y = 5.2$$
$$6.3x + 4.2y = 2.6$$

6. Make a graph of the pressure p (kPa) in an engine cylinder, where v is the volume (cm³) above the piston.

p	273.4	308.4	371.2	507.1	592.0	781.1	937.0	1229.6
v	174.0	159.6	140.2	114.8	102.5	85.0	75.3	63.5

For problems 7 and 8, plot the parametric equations
for $t = -2$ to 2.

7. $x = 3t$
$y = t^3$

8. $x = 2t^2$
$y = 4t$

For problems 9 through 12, make a complete graph of each function, choosing a domain and a range that include all of the features of interest.

9. $y = 5x^2 + 24x - 12$

10. $y = 9x^2 + 22x - 17$

11. $y = 6x^3 + 3x^2 - 14x - 21$

12. $y = 3x^4 + 21x^2 + 21$

For problems 13 and 14, write the equation of each straight line, and make a graph.

13. Slope $= -2$; y intercept $= 4$

14. Slope $= 4$; y intercept $= -6$

For problems 15 and 16, find the slope and the y intercept of each straight line, and make a graph.

15. $y = 12x - 4$

16. $y = -2x + 11$

Writing

17. The sales of skis by a ski manufacturer are shown for a two-year period in Fig. 5–20. Study the graph and write a verbal summary of the sales pattern for the period shown.

Team Project

18. Graph the function

$$y = 1.24x^3 + 0.482x^2 + 2.85x + 0.394$$

Find, to the nearest hundredth, any roots or maximum and minimum points.

For this type of exercise a spreadsheet is very helpful, in two ways. First, using the spreadsheet, graph the equation to get a rough idea of a detailed solution. Then use the spreadsheet to give a y value for a specific x value. This allows you to quickly determine a very accurate root for the equation.

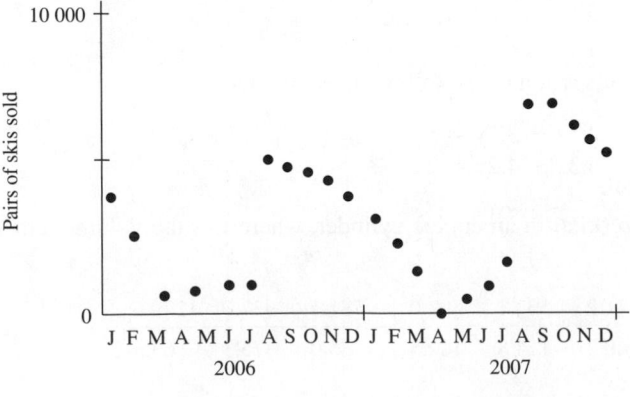

FIGURE 5–20 Ski sales for 2006–2007.

Geometry

OBJECTIVES

When you have completed this chapter, you should be able to:
- Understand the characteristics of lines and angles.
- Solve problems involving triangles.
- Solve problems involving quadrilaterals.
- Solve problems involving circles.
- Solve for volumes and surface areas of solids.

The power of shape! When you see a bridge made from a series of interlocking triangles, a stone arch following a smooth curve, or a droplet of water formed into a perfect sphere, you can feel how even basic shapes can hold solutions to many problems facing technology. To harness this power you need to understand the characteristics of the shapes. You have no doubt seen and worked with these shapes before, but in this chapter we will look over the equations for the properties of the shapes, and you will be challenged to solve problems using these shapes.

THE "GENE MACHINE"—KELVIN OGILVIE

CANADIAN BIOGRAPHY

Canadian inventor, scientist, and teacher, Dr. Kelvin Ogilvie is an internationally recognized and award-winning expert in the areas of biotechnology, bio-organic chemistry, and genetic engineering. In the 1980s, he developed an automated process for sequencing DNA, called the "gene machine," which sped up the procedure by orders of magnitude. His work in developing methods of chemically synthesizing RNA at McGill University resulted in our being able to make drugs that interfere with the ability of specific viruses to replicate themselves. His methods for creating certain types of RNA molecules are standard globally today. Among the 12 patents that Dr. Ogilvie has authored is a drug, used worldwide, that combats infections resulting from weakened immune systems. Dr. Ogilvie was made a member of the Order of Canada in 1991 and was appointed a Canadian senator in 2009. On careers in the future, he says, "I think the opportunities for exciting carriers in science grow every year. The more we understand the world around us at the molecular level, the greater the role of chemistry in helping us further unravel life's secrets and bring opportunities including better health and nutrition to all societies."

6–1   Straight Lines and Angles

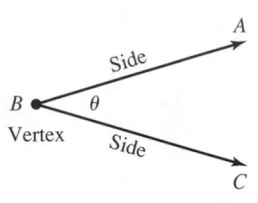

Line segment
(a)

Ray
(b)

FIGURE 6–1

Straight Line

A *line segment* is that portion of a straight line lying between two endpoints [Fig. 6–1(a)]. A *ray*, or *half-line*, is the portion of a line lying to one side of a point (endpoint) on the line [Fig. 6–1(b)].

Angles

An *angle* is formed when two rays or line segments intersect at their endpoints (Fig. 6–2). The point of intersection is called the *vertex* of the angle, and the two rays are called the *sides* of the angle.

The angle shown in Fig. 6–2 can be *designated* in any of the following ways:

angle *ABC*   angle *CBA*   angle *B*   angle *θ*

The symbol ∠ means *angle*, so ∠*B* means angle *B*.

An angle can also be thought of as having been *generated* by a ray turning from some *initial position* to a *terminal position* (Fig. 6–3).

One *revolution* is the amount a ray would turn to return to its original position.

The *units of angular measure* in common use are the *degree* and the *radian* (Sec. 6–4). The *measure* of an angle is the number of units of angular measure it contains. Two angles are *equal* if they have the same measure.

The degree (°) is a unit of angular measure equal to 1/360 of a revolution. Thus there are 360° in one complete revolution.

Figure 6–4 shows a *right* angle ($\frac{1}{4}$ revolution, or 90°), usually marked with a small square at the vertex; an *acute* angle (less than $\frac{1}{4}$ revolution); an *obtuse* angle (greater than $\frac{1}{4}$ revolution but less than $\frac{1}{2}$ revolution); and a *straight* angle ($\frac{1}{2}$ revolution, or 180°).

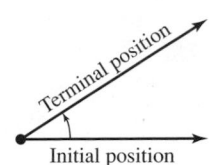

FIGURE 6–2   An angle.

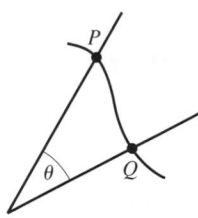

FIGURE 6–3   An angle formed by rotation.

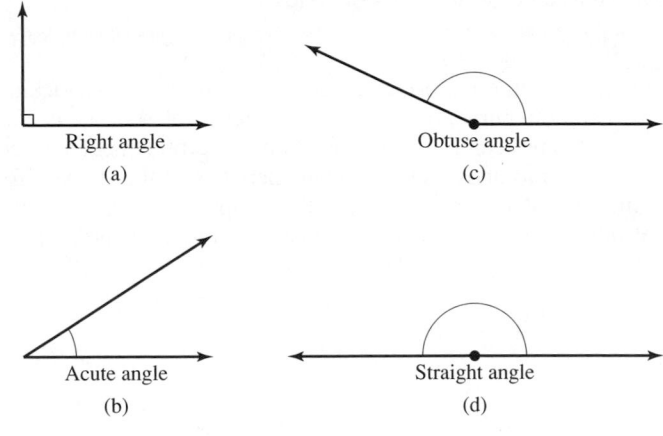

Right angle
(a)

Obtuse angle
(c)

Acute angle
(b)

Straight angle
(d)

FIGURE 6–4   Types of angles.

FIGURE 6–5

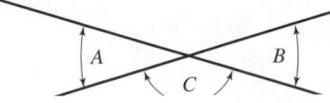

For brevity, we say "angle *A*" instead of the cumbersome but more correct "measure of angle *A*." We will learn how to convert angles from one system of measurement to another (such as degrees to radians) in Chapter 7.

Two lines or rays at right angles to each other are said to be *perpendicular*. Two angles are called *complementary* if their sum is a right angle (i.e., 90°), and *supplementary* if their sum is a straight angle (i.e., 180°).

Two more words we use in reference to angles are *intercept* and *subtend*. In Fig. 6–5 we say that the angle *θ intercepts* the section *PQ* of the curve. Conversely, we say that angle *θ* is *subtended by PQ*.

Angles between Intersecting Lines

Angles *A* and *B* of Fig. 6–6 are called *opposite* or *vertical* angles. It should be apparent that *A* and *B* are equal.

FIGURE 6–6

| Opposite angles of two intersecting straight lines are equal. | **104** |

Two angles are called *adjacent* when they have a common side and a common vertex, such as angles *A* and *C* of Fig. 6–6. When two lines intersect, the adjacent angles are *supplementary*.

◆◆◆ **Example 1:** Find the measure (in degrees) of angles A and B in the structure shown in Fig. 6–7.

Solution: We see from Fig. 6–7 that RQ and PS are straight intersecting lines (we will often assume relationships from the diagram rather than proving them). Since angle A is opposite the 34° angle, angle $A = 34°$. Angle B and the 34° angle are supplementary, so

$$B = 180° - 34° = 146°$$

◆◆◆

◆◆◆ **Example 2:** The top girder PQ in the structure of Fig. 6–7 is parallel to the ground, and angle C is 73°. Find angle D.

Solution: We have two parallel lines, PQ and RS, cut by transversal PS. By statement 105,

$$\angle D = \angle C = 73°$$

◆◆◆

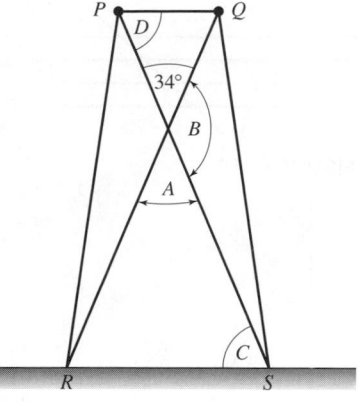

FIGURE 6–7

Families of Lines

A *family* of lines is a group of lines that are related to each other in some way. Figure 6–8(a) shows a family of lines, none of which intersect. They are called *parallel*. Figure 6–8(b) shows a family of lines that all pass through the same point but have different slopes.

In a later chapter we will talk about *families of curves*.

Transversals

A *transversal* is a line that intersects a family or system of lines. In Fig. 6–9, the lines labeled T are transversals. In Fig. 6–9(a), two parallel lines, L_1 and L_2, are cut by transversal T. Angles $A, B, G,$ and H are called *exterior* angles, and $C, D, E,$ and F are called *interior* angles. Angles A and E are called *corresponding* angles. Other corresponding angles in Fig. 6–9(a) are C and G, B and F, and D and H. Angles C and F are called *alternate interior angles*, as are D and E. We have a theorem:

Where two parallel straight lines are cut by a transversal, corresponding angles are equal, and alternate interior angles are equal.	**105**

Thus in Fig. 6–9 (a), $\angle A = \angle E$, $\angle C = \angle G$, $\angle B = \angle F$, and $\angle D = \angle H$.
Also, $\angle D = \angle E$ and $\angle C = \angle F$.

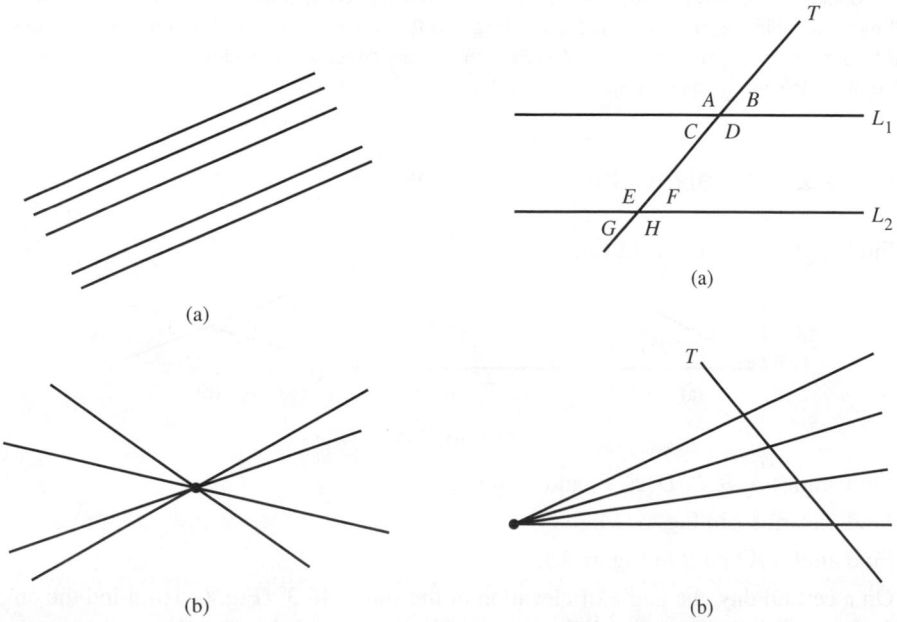

(a)

(b)

FIGURE 6–8 Families of lines.

(a)

(b)

FIGURE 6–9 Transversals.

Another useful theorem applies when a number of parallel lines are cut by *two* transversals, such as in Fig. 6–10. The portions of the transversals lying between the same parallels are called

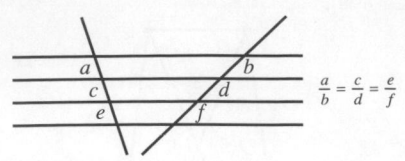

FIGURE 6-10

corresponding segments. (In Fig. 6–10, *a* and *b* are corresponding segments, *c* and *d* are corresponding segments, and *e* and *f* are corresponding segments.)

Where a number of parallel lines are cut by two transversals, corresponding segments of the transversals are proportional.	**106**

In Fig. 6–10,

$$\frac{a}{b} = \frac{c}{d} = \frac{e}{f}$$

◆◆◆ **Example 3:** A portion of a street map is shown in Fig. 6–11. Find the distances *PQ* and *QR*.

Solution: From statement 106,

$$\frac{PQ}{72.0} = \frac{172}{155}$$

$$PQ = \frac{172}{155}(72.0) = 80.0 \text{ m}$$

Similarly,

$$\frac{QR}{198} = \frac{172}{155}$$

$$QR = \frac{172}{155}(198) = 22\overline{0} \text{ m} \qquad ◆◆◆$$

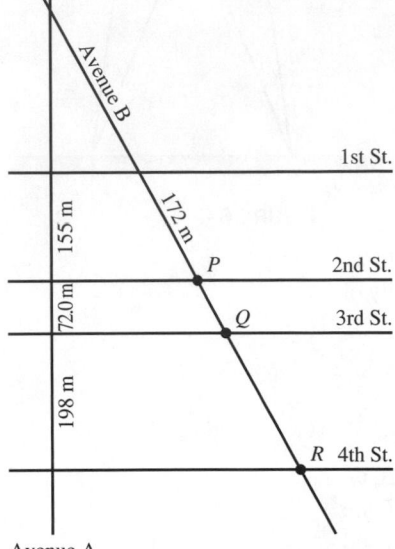

FIGURE 6-11

CASE STUDY—ENGINEERS WITHOUT BORDERS

At your company you have volunteered to work with several co-workers on a project through Engineers Without Borders Canada (www.ewb.ca). Your team is designing a solar powered irrigation system. This will dramatically increase the yield of quality crops, allowing local women farmers to produce enough extra crops to sell and use the money to buy food during the peak dry period. An organization called SELF, Solar Electric Light Fund (www.self.org), has done similar projects with great success. Your task today is to calculate how many solar panels you can ship in a standard 20-foot shipping container. The panels, from Clean Development Group (www.cdgenergy.com), measure 1659 mm × 1000 mm × 50 mm. The shipping containers are 20 ft. long × 8 ft. wide × 8.5 ft. tall. The inside dimensions are 19 ft. 3 in. × 7 ft. 7 in. × 7 ft. 9 in. Add 30 mm in every direction for packaging of the panels and determine how many can be shipped per container.

Exercise 1 ◆ Straight Lines and Angles

1. Find angle *θ* in Figs. 6–12 (a), (b), and (c).

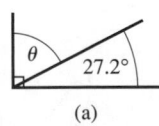

(a)

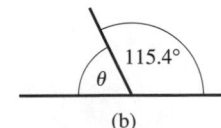

(b)

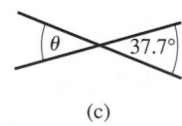

(c)

FIGURE 6-12

2. Find angles *A*, *B*, *C*, *D*, *E*, *F*, and *G* in Fig. 6–13.

3. Find distance *x* in Fig. 6–14.

4. Find angles *A* and *B* in Fig. 6–15.

5. On a certain day, the angle of elevation of the sun is 46.3° (Fig. 6–16). Find the angles *A*, *B*, *C*, and *D* that a ray of sunlight makes with a horizontal sheet of glass. For an explanation of angles of elevation and depression, see Figure 6–18.

6. Three parallel steel cables hang from a girder to the deck of a bridge (Fig. 6–17). Find distance *x*.

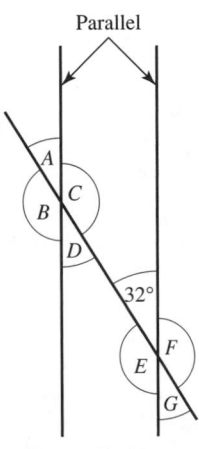

FIGURE 6-13

The *angle of elevation* of some object above the observer is the angle between the line of sight to that object and the horizontal, as in Fig. 6–18.

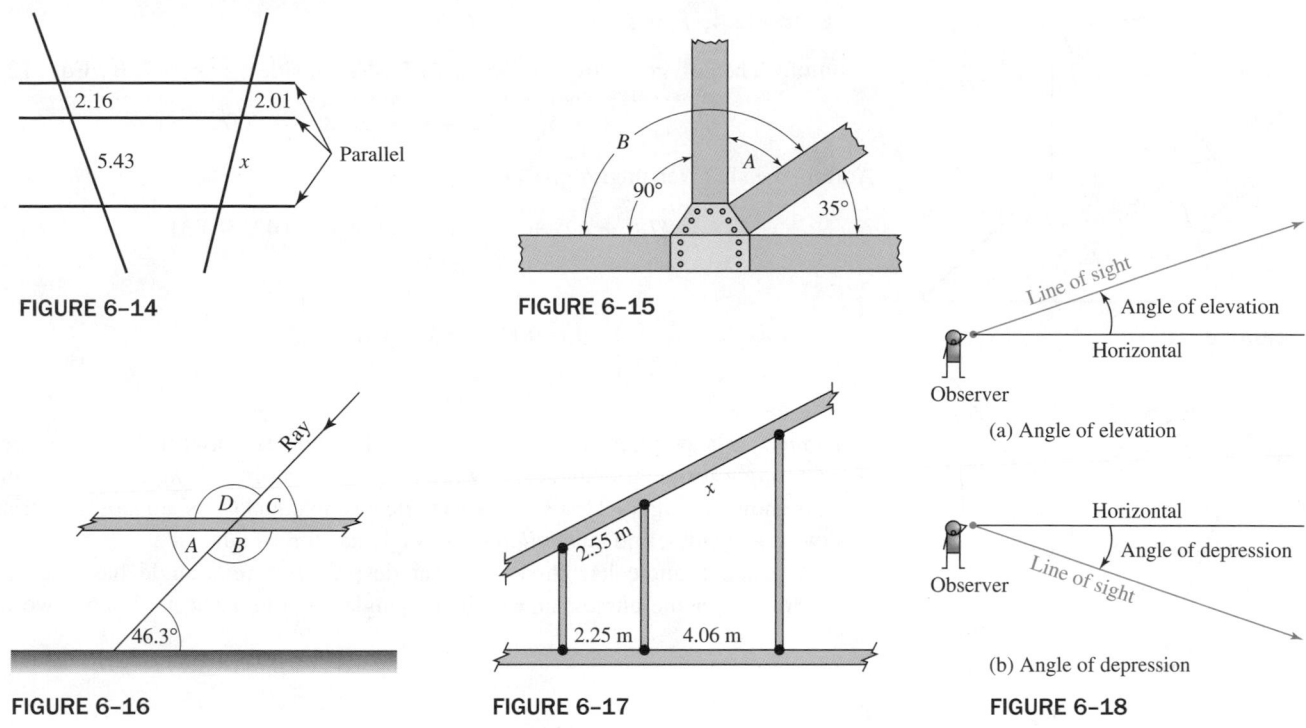

FIGURE 6–14

FIGURE 6–15

(a) Angle of elevation

FIGURE 6–16

FIGURE 6–17

(b) Angle of depression

FIGURE 6–18

6–2 Triangles

Polygons

A *polygon* is a plane figure formed by three or more line segments, called the *sides* of the polygon, joined at their endpoints, as in Fig. 6–19. The points where the sides meet are called *vertices*. We say that the sides of a polygon are equal if their measures (lengths) are equal. If all of the sides and angles of a polygon are equal, it is called a *regular* polygon, as in Fig. 6–20. The *perimeter* of a polygon is simply the sum of its sides.

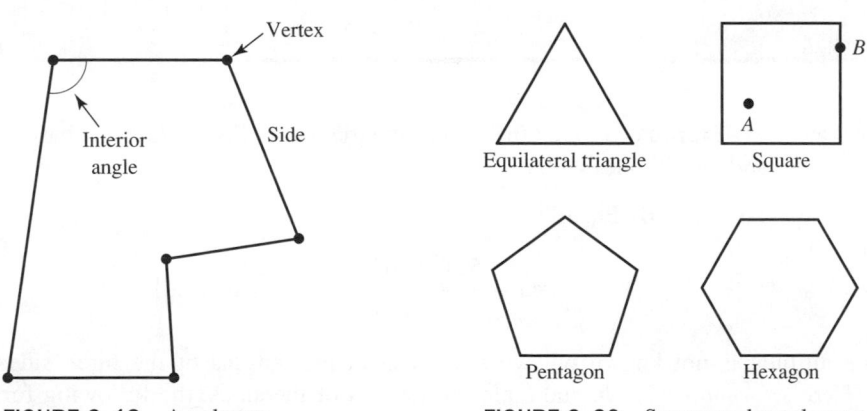

FIGURE 6–19   A polygon

Equilateral triangle

Square

Pentagon

Hexagon

FIGURE 6–20   Some regular polygons.

Modern definitions of plane figures exclude the *interior* as part of the figure. Thus in Fig. 6–20, point A is *not* on the square, while point B *is* on the square. The interior is referred to as a *region*.

Sum of Interior Angles

A polygon of *n* sides has *n* interior angles, such as those shown in Fig. 6–21. Their sum is equal to the following:

Interior Angles of a Polygon	Sum of angles $= (n - 2)180°$	112

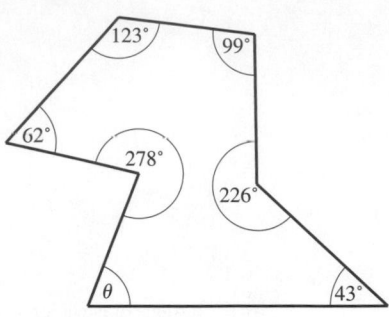

FIGURE 6–21

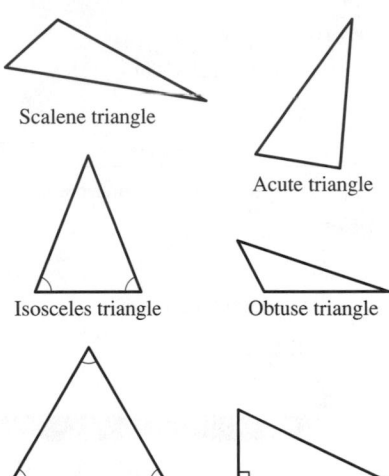

Scalene triangle

Acute triangle

Isosceles triangle   Obtuse triangle

Equilateral triangle   Right triangle

FIGURE 6–22.   Types of triangles.

◆◆◆ Example 4: Find angle θ in Fig. 6–21.

Solution: The polygon shown in Fig. 6–21 has seven sides, so $n = 7$. By Eq. 112,

$$\text{Sum of angles} = (7 - 2)180° = 900°$$

Adding the six given angles gives us

$$278° + 62° + 123° + 99° + 226° + 43° = 831°$$

So

$$\theta = 900° - 831° = 69° \qquad ◆◆◆$$

Triangles

A *triangle* is a polygon having three *sides*. The angles between the sides are the *interior angles* of the triangle, usually referred to as simply the *angles* of the triangle.

As shown in Fig. 6–22, a *scalene* triangle has no equal sides; an *isosceles* triangle has two equal sides; and an *equilateral* triangle has three equal sides.

An *acute* triangle has three acute angles; an *obtuse* triangle has one obtuse angle and two acute angles; and a *right* triangle has one right angle and two acute angles.

Altitude and Base

The *altitude* of a triangle is the perpendicular distance from a vertex to the opposite side called the *base*, or an extension of that side (Fig. 6–23).

Area of a Triangle

We use the following familiar formula:

The measuring of areas, volumes, and lengths is sometimes referred to as *mensuration*.

Area of a Triangle	Area equals one-half the product of the base and the altitude to that base $$A = \frac{bh}{2}$$	**137**

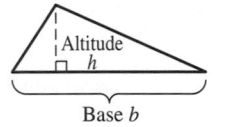

Base *b*

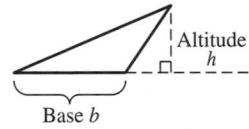

Base *b*

FIGURE 6–23

◆◆◆ Example 5: Find the area of the triangle in Fig. 6–23 if the base is 52.0 and the altitude is 48.0.

Solution: By Eq. 137,

$$\text{area} = \frac{52.0(48.0)}{2} = 1250 \quad \text{(rounded)} \qquad ◆◆◆$$

If the altitude is not known but we have instead the lengths of the three sides, we may use *Hero's formula*. If *a*, *b*, and *c* are the lengths of the sides, the following formula applies:

This formula is named for Hero (or Heron) of Alexandria, a Greek mathematician and physicist of the 1st century A.D.

Hero's Formula	Area of triangle $= \sqrt{s(s - a)(s - b)(s - c)}$ where s is half the perimeter, or $$s = \frac{a + b + c}{2}$$	**138**

◆◆◆ **Example 6:** Find the area of a triangle having sides of lengths 3.25, 2.16, and 5.09.

Solution: We first find s, which is half the perimeter.

$$s = \frac{3.25 + 2.16 + 5.09}{2} = 5.25$$

Thus the area is

$$\text{area} = \sqrt{5.25(5.25 - 3.25)(5.25 - 2.16)(5.25 - 5.09)} = 2.28 \qquad ◆◆◆$$

Sum of the Angles

An extremely useful relationship exists between the interior angles of any triangle.

Sum of the Interior Angles	The sum of the three interior angles A, B, and C of any triangle is 180 degrees. $A + B + C = 180°$	139

◆◆◆ **Example 7:** Find angle A in a triangle if the other two interior angles are 38° and 121°.

Solution: By Eq. 139,

$$A = 180 - 121 - 38 = 21° \qquad ◆◆◆$$

Exterior Angles

An *exterior* angle is the angle between the side of a triangle and an extension of the adjacent side, such as angle θ in Fig. 6–24. The following theorem applies to exterior angles:

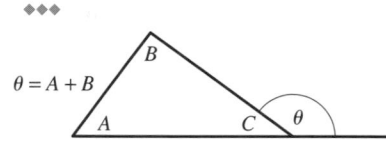

FIGURE 6–24 An exterior angle.

Exterior Angle of a Triangle	The exterior angle of a triangle is the sum of the two angles in the triangle that are not the angle along the extension of the adjacent side. $\theta = A + B$	142

Congruent Triangles

Two triangles (or any other polygons, for that matter) are said to be *congruent* if the angles and sides of one are equal to the angles and sides of the other, as in Fig. 6–25.

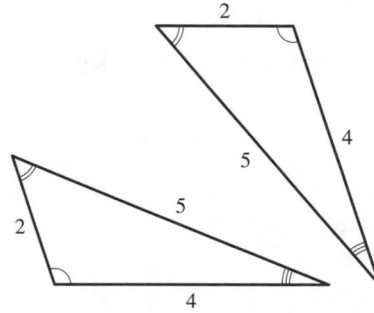

FIGURE 6–25 Congruent triangles.

Similar Triangles

Two triangles are said to be *similar* if they have the *same shape*, even if one triangle is larger than the other. This means that the angles of one of the triangles must equal the angles of the other triangle, as in Fig. 6–26. Sides that lie between the same pair of equal angles are called *corresponding sides*, such as sides a and d. Sides b and e, as well as sides c and f, are also corresponding sides. Thus we have the following two theorems:

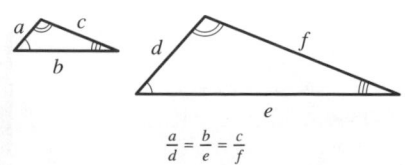

FIGURE 6–26 Similar triangles.

We will see in a later chapter that these relationships also hold for similar figures *other than* triangles, and for similar *solids* as well.

| **Similar Triangles** | If two angles of a triangle equal two angles of another triangle, the triangles are similar. | **143** |
| | Corresponding sides of similar triangles are in proportion. | **144** |

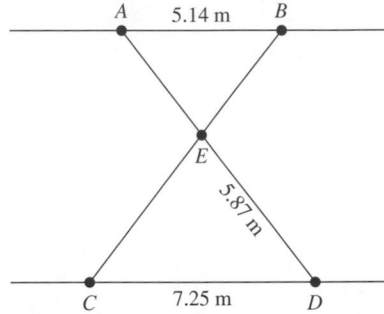

FIGURE 6–27

◆◆◆ **Example 8:** Two beams, *AB* and *CD*, in the framework of Fig. 6–27 are parallel. Find distance *AE*.

Solution: By statement 104, we know that angle *AEB* equals angle *DEC*. Also, by statement 105, angle *ABE* equals angle *ECD*. Thus triangle *ABE* is similar to triangle *CDE*. Since *AE* and *ED* are corresponding sides,

$$\frac{AE}{5.87} = \frac{5.14}{7.25}$$

$$AE = 5.87\left(\frac{5.14}{7.25}\right) = 4.16 \text{ m}$$ ◆◆◆

Right Triangles

In a right triangle (Fig. 6–28), the side opposite the right angle is called the *hypotenuse*, and the other two sides are called *legs*. The legs and the hypotenuse are related by the well-known *Pythagorean theorem*.

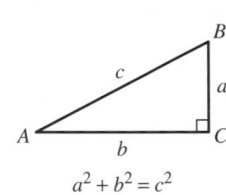

FIGURE 6–28. The Pythagorean theorem, named for the Greek mathematician Pythagoras (ca. 580–500 B.C.).

| **Pythagorean Theorem** | The square of the hypotenuse of a right triangle is equal to the sum of the squares of the two legs. $$a^2 + b^2 = c^2$$ | **145** |

◆◆◆ **Example 9:** A right triangle has legs of length 6 units and 11 units. Find the length of the hypotenuse.

Solution: Letting *c* = the length of the hypotenuse, we have

$$c^2 = 6^2 + 11^2 = 36 + 121 = 157$$
$$c = \sqrt{157} = 12.5 \quad \text{(rounded)}$$ ◆◆◆

◆◆◆ **Example 10:** A right triangle has a hypotenuse of 5 units in length and a leg of 3 units. Find the length of the other leg.

Solution:

$$a^2 = c^2 - b^2$$
$$= 5^2 - 3^2 = 16$$
$$a = \sqrt{16} = 4$$ ◆◆◆

| **Common Error** | Remember that the Pythagorean theorem applies *only to right triangles*. Later we will use trigonometry to find the sides and angles of *oblique* triangles (triangles with no right angles). |

Some Special Triangles

In a *30–60–90 right triangle* [Fig. 6–29(a)], the side opposite the 30° angle is half the length of the hypotenuse.

A *45° right triangle* [Fig. 6–29(b)] is also *isosceles*, and the hypotenuse is $\sqrt{2}$ times the length of either side.

A *3–4–5 triangle* [Fig. 6–29(c)], is a right triangle in which the sides are in the ratio of 3 to 4 to 5.

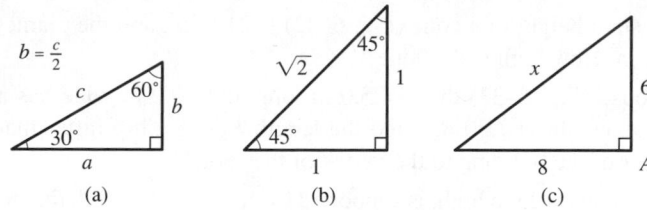

FIGURE 6–29 Some special right triangles.

◆◆◆ **Example 11:** In the 3–4–5 triangle of Fig. 6–29(c), side x is 10 cm. ◆◆◆

Exercise 2 ◆ Triangles

1. What is the cost, to the nearest dollar, of a triangular piece of land whose base is 828 m and altitude 412 m, at $2,925 per acre? Assume all the figures are exact.

2. The house shown in Fig. 6–30 is 12.0 m wide. The ridge is 4.50 m higher than the side walls, and the rafters project 0.750 m beyond the sides of the house. How long are the rafters?

3. What is the length of a handrail for a flight of stairs (Fig. 6–31) where each step is 30.5 cm wide and 22.3 cm high?

For practice, do these problems using only geometry, even if you know some trigonometry.

To help you solve each problem, draw a diagram and label it completely. Look for rectangles, special triangles, or right triangles contained in the diagram. Be sure to look up any word that may be unfamiliar, such as *diagonal, radius,* or *diameter*.

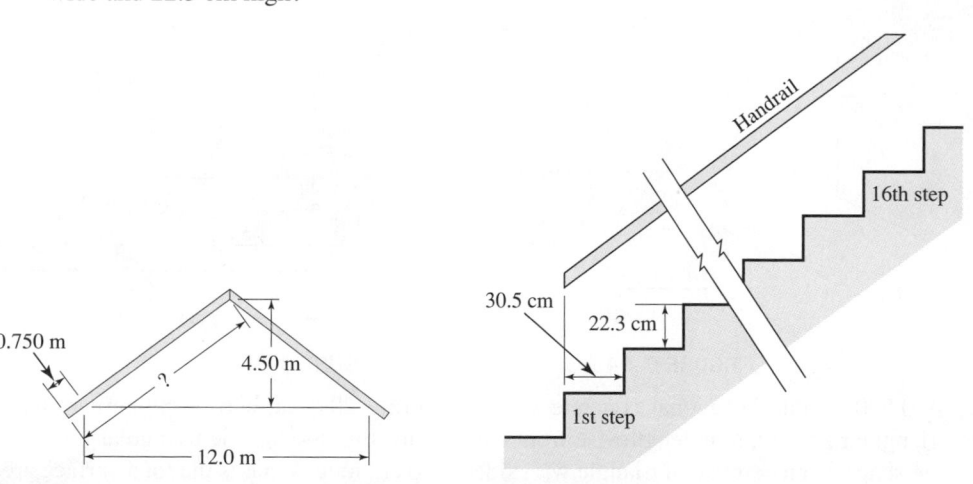

FIGURE 6–30   FIGURE 6–31

4. At $13.50 per square metre, find to the nearest dollar the cost of paving a triangular court, its base being 32.0 m and its altitude 21.0 m.

5. A vertical pole 15.0 m high is supported by three guy wires attached to the top and reaching the ground at distances of 20.0 m, 36.0 m, and 66.7 m from the foot of the pole. What are the lengths of the wires?

6. A ladder 13.0 m long reaches to the top of a building when its foot stands 5.00 m from the building. How high is the building?

7. Two streets, one 16.2 m and the other 31.5 m wide, cross at right angles. What is the diagonal distance between the opposite corners?

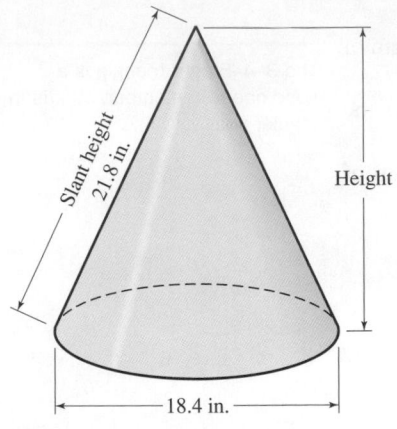

FIGURE 6–32

We have not studied the hexagon as such, but you can solve problems 15 and 16 by dividing the hexagon into right triangles or a rectangle and right triangles. This also works for other regular polygons.

8. A room is 5.00 m long, 4.00 m wide, and 3.00 m high. What is the diagonal distance from one of the lower corners to the opposite upper corner?

9. What is the side of a square whose diagonal is 50.0 m?

10. A rectangular park 125 m long and 233 m wide has a straight walk running through it from opposite corners. What is the length of the walk?

11. A ladder 8.00 m long stands flat against the side of a building. How far must it be drawn out at the bottom to lower the top by 1.00 m?

12. The slant height of a cone (Fig. 6–32) is 21.8 in., and the diameter of the base is 18.4 in. How high is the cone?

13. A house (Fig. 6–33) that is 25.0 m long and 20.0 m wide has a pyramidal roof whose height is 7.50 m. Find the length PQ of a hip rafter that reaches from a corner of the building to the vertex of the roof.

14. An antenna, 125 m high, is supported by cables of length L that reach from the top of the tower to the ground at a distance D from the base of the antenna. If D is $\frac{3}{4}$ the length of a cable, find L and D.

15. A hex head bolt (Fig. 6–34) measures 2.00 cm across the flats. Find the distance x across the corners.

16. A 1.00-m-long piece of steel hexagonal stock measuring 30.0 mm across the flats has a 10.0-mm-diameter hole running lengthwise from end to end. Find the mass of the bar (density = 7.85 g/cm³).

17. Find the distance AB between the centres of the two rollers in Fig. 6–35.

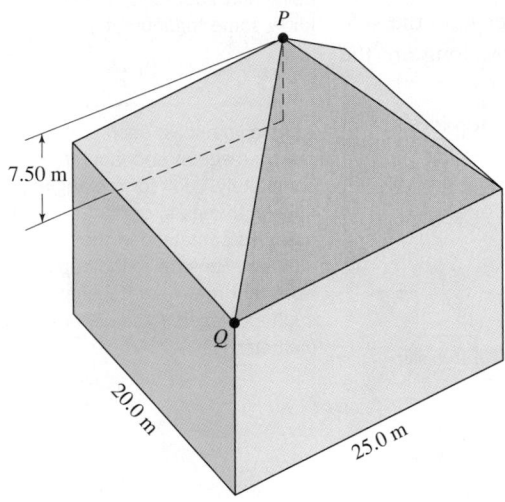

FIGURE 6–33

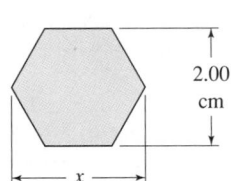

FIGURE 6–34

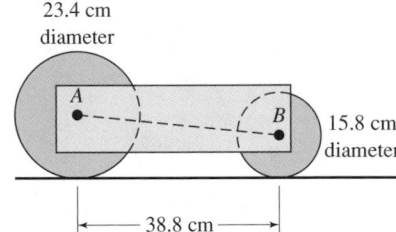

FIGURE 6–35

18. A 0.500-m-long hexagonal concrete shaft measuring 20.0 cm across the corners has a triangular hole running lengthwise from end to end (Fig. 6–36). The triangular hole is in the shape of an equilateral triangle with sides 10.0 cm long. What is the total surface area of the shaft?

Side view

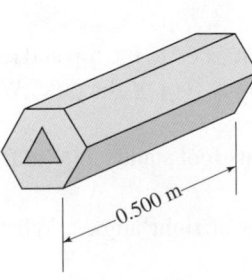

End view

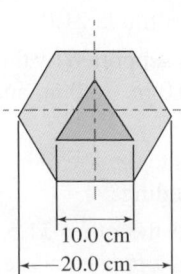

FIGURE 6–36

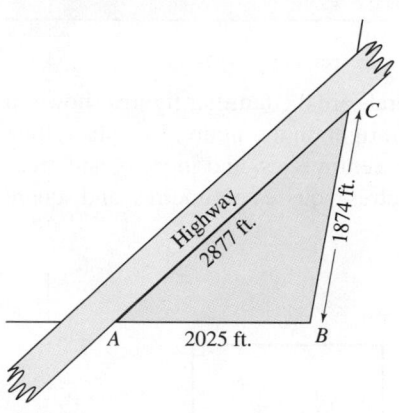

FIGURE 6–37 Note: Angle *B* is not a right angle.

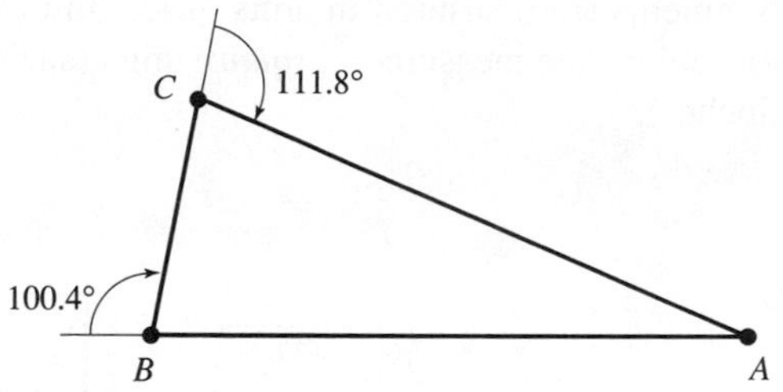

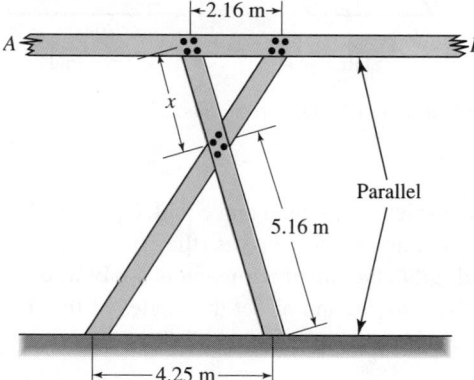

FIGURE 6–38

19. A highway (Fig. 6–37) cuts a corner from a parcel of land. Find the number of acres in the triangular lot *ABC*. (1 acre = 43 560 sq. ft.)

20. A surveyor starts at *A* in Fig. 6–38 and lays out lines *AB*, *BC*, and *CA*. Find the three interior angles of the triangle.

21. A beam *AB* is supported by two crossed beams (Fig. 6–39). Find distance *x*.

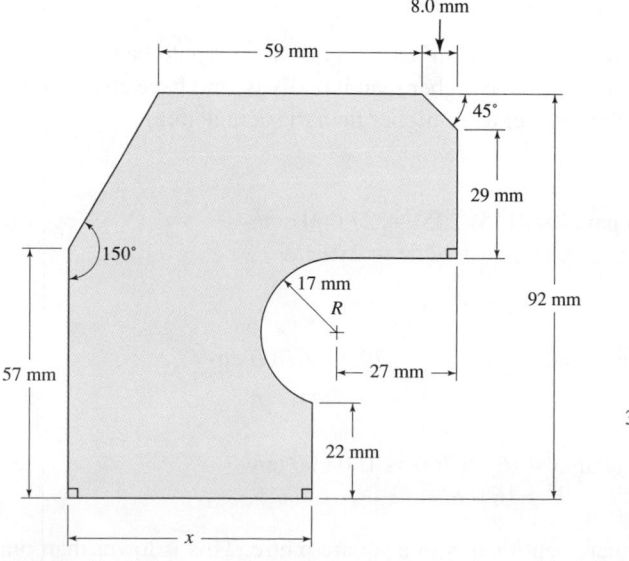

FIGURE 6–39

22. Find dimension *x* in Fig. 6–40.

23. Find dimension *x* on the beveled end of the shaft in Fig. 6–41.

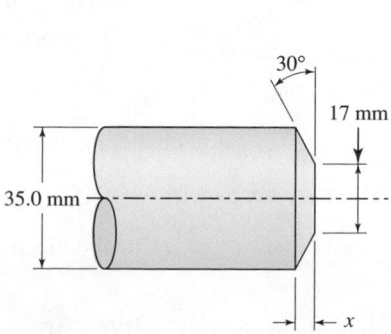

FIGURE 6–40

FIGURE 6–41

6–3   Quadrilaterals

A *quadrilateral* is a polygon having four sides. Quadrilaterals are the familiar figures shown in Fig. 6–42. The formula for the area of each region is given right on the figure. Lengths of line segments are measured in units of length (such as metres, centimetres, and inches), and areas of regions are measured in *square* units (such as square metres, square centimetres, and square inches).

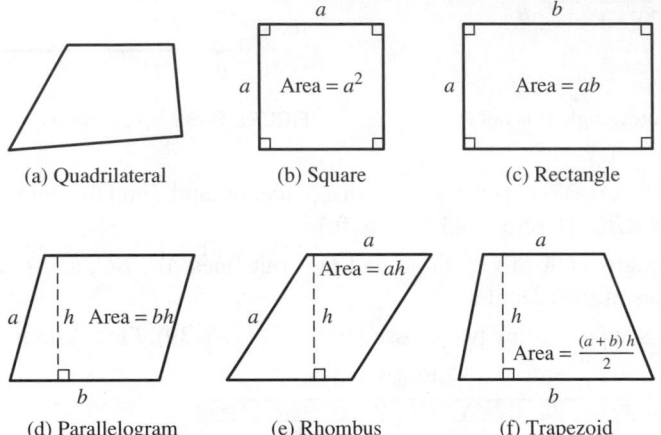

FIGURE 6–42   Quadrilaterals.

For the *parallelogram*, opposite sides are parallel and equal. Opposite angles are equal, and each diagonal cuts the other diagonal into two equal parts (they *bisect* each other).

The *rhombus* is also a parallelogram, so the previous facts apply to it as well. In addition, its diagonals bisect each other at *right angles* and bisect the angles of the rhombus.

The *trapezoid* has two parallel sides, which are called the *bases*, and the altitude is the distance between the bases.

◆◆◆ **Example 12:** A solar collector array consists of six rectangular panels, each 115 cm × 235 cm, each with a blocked rectangular area (needed for connections) measuring 12.0 cm × 22.5 cm. Find the total collecting area in square metres.

Estimate: If the panels were 1.2 m by 2.4 m, or 2.9 m² each, six of them would be 6 × 2.9, or 17.4 m². Since we have estimated each dimension higher than it really is, and have also ignored the blocked areas, we would expect this answer to be higher than the actual area.

Solution:

$$\text{area of each panel} = (115)(235) = 27\,\overline{0}00 \text{ cm}^2$$
$$\text{bocked area} = (12.0)(22.5) = 27\overline{0} \text{ cm}^2$$

Subtracting yields

$$\text{collecting area per panel} = 27\,\overline{0}00 - 27\overline{0} = 26\,700 \text{ cm}^2$$

There are six panels, so

$$\text{total collecting area} = (6)26\,700 = 16\overline{0}\,000 \text{ cm}^2$$
$$= 16.0 \text{ m}^2$$

since there are 100 × 100 = 10⁴ square centimetres in a square metre. This is lower than our estimate, as expected.   ◆◆◆

Exercise 3 ◆ Quadrilaterals

1. Find the area of the parallelogram in Fig. 6–43.
2. Find the area of the rhombus in Fig. 6–44.
3. Find the area of the trapezoid in Fig. 6–45.

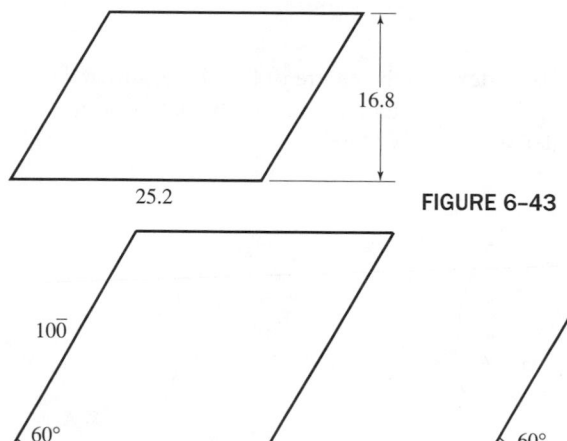

16.8

25.2

FIGURE 6–43

$10\bar{0}$

60°

$10\bar{0}$

FIGURE 6–44

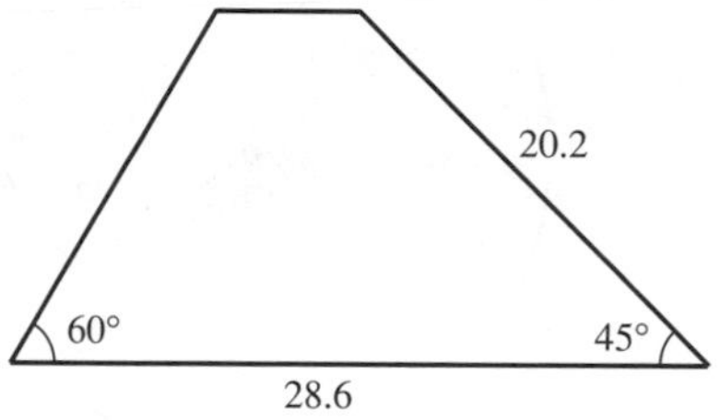

20.2

60°   45°

28.6

FIGURE 6–45

4. Find the cost, to the nearest dollar, of applying roofing membrane to a flat roof, 21.5 m long and 15.2 m wide, at $12.40/m².

5. What will be the cost of flagging a sidewalk 104 m long and 2.2 m wide, at $13.50 per square metre?

6. How many 25-cm-square tiles will cover a floor 16 m by 4 m?

7. What will it cost to carpet a floor, 6.25 m by 7.18 m, at $7.75 per square metre?

8. How many rolls of paper, each 8.00 yd. long and 18.0 in. wide, will paper the sides of a room 16.0 ft. by 14.0 ft. and 10.0 ft. high, deducting 124 sq. ft. for doors and windows?

9. What is the cost of plastering the walls and ceiling of a room 13 m long, 12 m wide, and 7.5 m high, at $12.50 per square metre, allowing 135 m² for doors, windows, and baseboard?

10. What will it cost, to the nearest cent, to cement the floor of a cellar 12.6 m long and 9.2 m wide, at $6.25 per square metre?

11. Find the cost of lining a topless rectangular tank 68 in. long, 54 in. wide, and 48 in. deep, with zinc, weighing 5.2 lb. per square foot, at $1.55 per pound installed.

Work to the nearest dollar in this exercise, unless otherwise asked.

6–4   The Circle

Circumference and Area

For a circle of diameter d and radius r (Fig. 6–46), the following formulas apply:

Circle of Radius r and Diameter d $\pi \cong 3.1416$	Circumference $= 2\pi r = \pi d$	113
	Area $= \pi r^2 = \dfrac{\pi d^2}{4}$	114

◆◆◆ **Example 13:** The circumference and the area of a 25.70-cm-radius circle are

$$C = 2\pi(25.70) = 161.5 \text{ cm}$$

and

$$A = \pi(25.70)^2 = 2075 \text{ cm}^2 \qquad \text{◆◆◆}$$

Radian Measure

A *central angle* (Fig. 6–46) is one whose vertex is at the centre of the circle. An *arc* is a portion of the circle. If an arc is laid off along a circle with a length equal to the radius of the circle, the central angle subtended by this arc is defined as one *radian*.

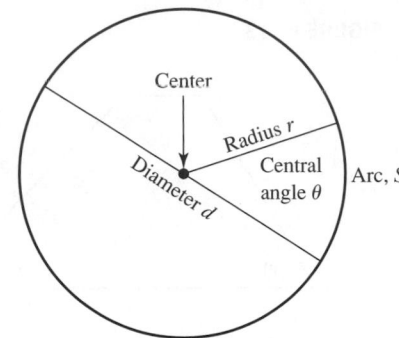

FIGURE 6–46   A circle.

We cover radian measure, and compute areas of sectors and segments, in Chapter 16.

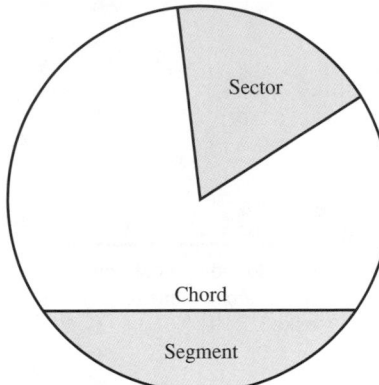

FIGURE 6–47

Other Parts of a Circle

A *sector* (Fig. 6–47) is the region bounded by two radii and one of the arcs that they intercept. A *chord* is a line segment connecting two points on the circle. A *segment* is the region bounded by a chord and one of the arcs that it intercepts. When a circle has two intersecting chords, the following theorem applies:

Intersecting Chords	If two chords in a circle intersect, the product of the parts of one chord is equal to the product of the parts of the other chord. $ab = cd$	**121**

In Fig. 6–48, you can see that $a = 16.3$, $b = x$, $c = 10.1$, and $d = 12.5$

◆◆◆ **Example 14:** Find the length x in Fig. 6–48.

Solution: In Fig. 6–48, we can see that $a = 16.3$, $b = x$, $c = 10.1$, and $d = 12.5$. By statement 121,

$$16.3x = 10.1(12.5)$$
$$x = \frac{10.1(12.5)}{16.3} = 7.75 \qquad \text{◆◆◆}$$

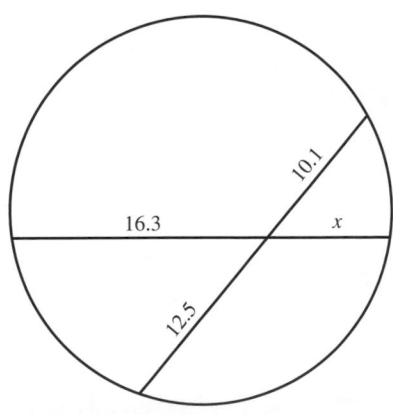

FIGURE 6–48

Tangents and Secants

A *tangent* (Fig. 6–49) is a straight line that touches a circle in just one point. A *secant* cuts the circle in *two* points. A chord is that portion of a secant lying within the circle.

We have two very useful theorems about tangents. First:

Tangents to a Circle	A tangent is perpendicular to the radius drawn through the point of contact.	**119**

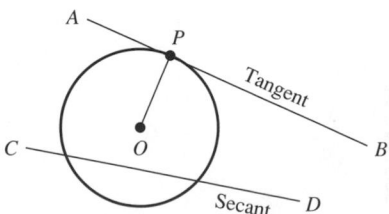

FIGURE 6–49   Tangent and secant.

Thus in Fig. 6–49, angle *OPB* is a right angle.

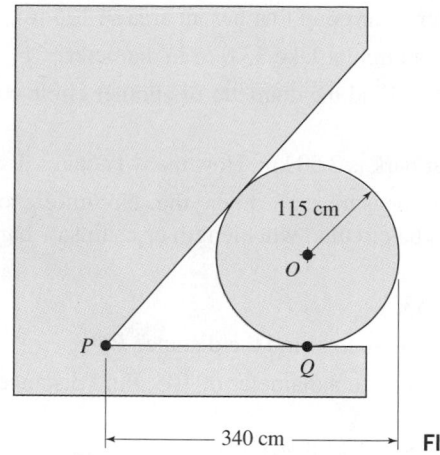

FIGURE 6–50

◆◆◆ **Example 15:** Find the distance *OP* in Fig. 6–50.

Solution: By statement 119, we know that angle *PQO* is a right angle. Also,

$$PQ = 340 - 115 = 225 \text{ cm}$$

So, by the Pythagorean theorem,

$$OP = \sqrt{(115)^2 + (225)^2} = 253 \text{ cm}$$ ◆◆◆

The second theorem follows:

Tangents to a Circle	Two tangents drawn to a circle from a point outside the circle are equal, and they make equal angles with a line from the point to the centre of the circle.	**120**

Thus, in Fig. 6–51,

$$AB = AD \quad \text{and} \quad \angle BAC = \angle CAD$$

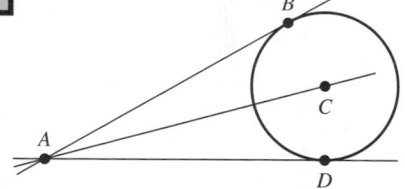

FIGURE 6–51

Semicircle

Any angle inscribed in a semicircle is a right angle.	**118**

◆◆◆ **Example 16:** Find the distance *x* in Fig. 6–52.

Solution: By statement 118, we know the largest angle, is 90°. Then, by Eq. 145,

$$x = \sqrt{(250)^2 - (148)^2} = 201$$ ◆◆◆

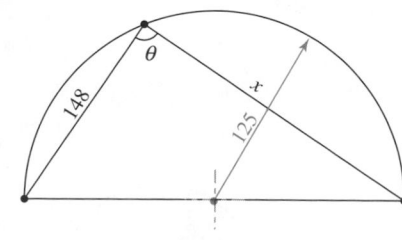

FIGURE 6–52

Exercise 4   ◆   The Circle

1. In a park is a circular fountain whose basin is 22.5 m in circumference. What is the diameter of the basin?

2. How much larger is the side of a square circumscribing a circle 155 cm in diameter than a square inscribed in the same circle?

3. The area of the bottom of a circular pan is 707 cm². What is its diameter?

4. Find the diameter of a circular solar pond that has an area of 125 m².

5. What is the circumference of a circular lake 33.0 m in diameter?

6. The radius of a circle is 5.00 m. Find the diameter of another circle containing 4 times the area of the first.

7. The distance around a circular park is 2.50 km. How many hectares does it contain?

8. A woodcutter uses a tape measure and finds the circumference of a tree to be 248 cm. Assuming the tree to be circular, what length of chainsaw bar is needed to fell the tree?

9. Find the distance x in Fig. 6–53.

10. In Fig. 6–54, distance OP is 8.65 units. Find the distance PQ.

11. Figure 6–55 shows a semicircle with a diameter of 105. Find distance PQ.

By cutting from both sides, one can fell a tree whose diameter is twice the length of the chainsaw bar.

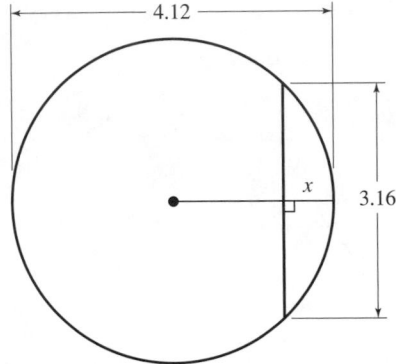

FIGURE 6–53

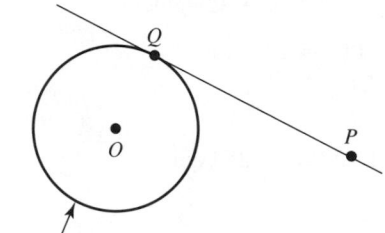

FIGURE 6–54

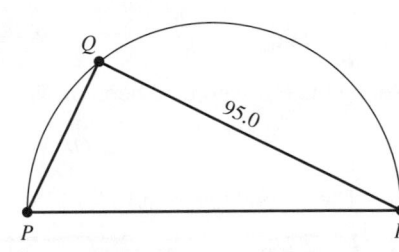

FIGURE 6–55

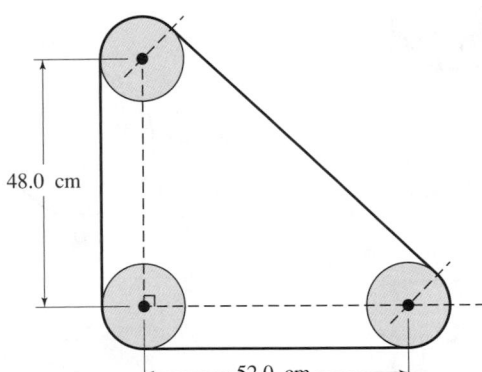

FIGURE 6–56

12. What must be the diameter d of a cylindrical piston so that a pressure of 865 kPa on its circular end will result in a total force of 16 200 N?

13. Find the length of belt needed to connect the three 15.0-cm-diameter pulleys in Fig. 6–56. *Hint:* The total *curved* portion of the belt is equal to the circumference of one pulley.

14. Seven cables of equal diameter are contained within a circular conduit, as in Fig. 6–57. If the inside diameter of the conduit is 25.6 cm, find the cross-sectional area *not* occupied by the cables.

15. A 60° screw thread is measured by placing three wires on the thread and measuring the distance T, as in Fig. 6–58. Find the distance D if the wire diameters are 0.1000 in. and the distance T is 1.500 in., assuming that the root of the thread is a sharp V shape.

16. Figure 6–59 shows two of the supports for a hemispherical dome. Find the length of girder AB.

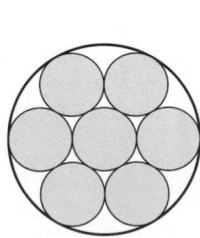

FIGURE 6–57

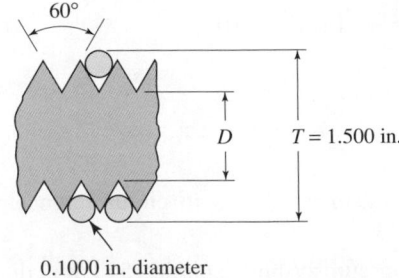

FIGURE 6–58　Measuring a screw thread "over wires."

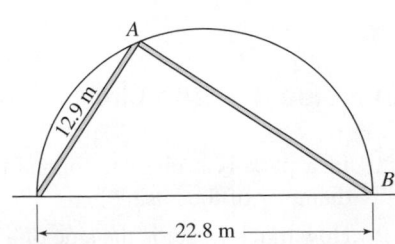

FIGURE 6–59

17. A certain car tire is 78.5 cm in diameter. How far will the car move forward with one revolution of the wheel?

18. Archimedes claimed that the area of a circle is equal to the area of a triangle that has an altitude equal to the radius of the circle and a base equal to the circumference of the circle. Use the formulas of this chapter to show that this is true.

6–5 Volumes and Areas of Solids

Volume

The volume of a solid is a measure of the space it occupies or encloses. It is measured in cubic units (m^3, cm^3, cu. ft., etc.) or, usually for liquids, in litres or gallons. One of our main tasks in this chapter is to compute the volumes of various solids.

Area

We speak about three different kinds of areas. *Surface area* will be the *total* area of the surface of a solid, including the ends, or bases. The *lateral area*, which will be defined later for each solid, does *not* include the areas of the bases. The *cross-sectional area* is the area of the plane figure obtained when we slice the solid in a specified way.

The formulas for the areas and volumes of some common solids are given in Fig. 6–60.

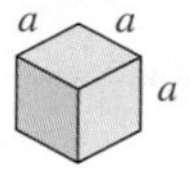 Cube		Volume $= a^3$	122
		Surface area $= 6a^2$	123
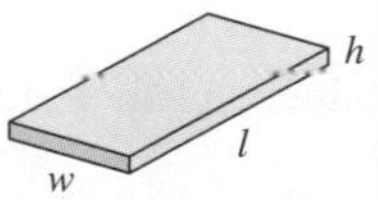 Rectangular parallelepiped		Volume $= lwh$	124
		Surface area $= 2(lw + hw + lh)$	125
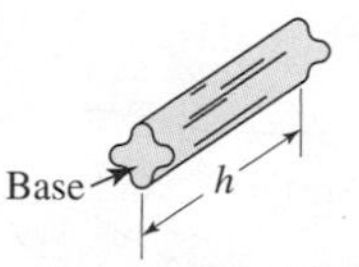	Any cylinder or prism	Volume $=$ (area of base)(altitude)	126
	Right cylinder or prism	Lateral area $=$ (perimeter of base)(altitude) (not incl. bases)	127
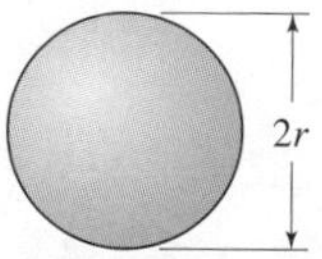 Sphere		Volume $= \frac{4}{3}\pi r^3$	128
		Surface area $= 4\pi r^2$	129
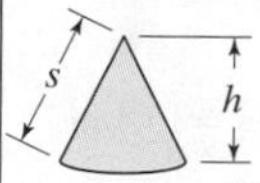	Any cone or pyramid	Volume $= \frac{h}{3}$(area of base)	130
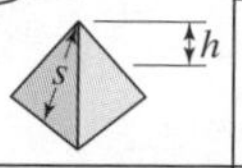	Right circular cone or regular pyramid	Lateral area $= \frac{s}{2}$ (perimeter of base)	131
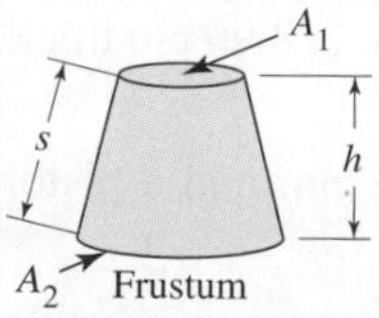	Any cone or pyramid	Volume $= \frac{h}{3}(A_1 + A_2 + \sqrt{A_1 A_2})$	132
Frustum	Right circular cone or regular pyramid	Lateral area $= \frac{s}{2}$(sum of base perimeters) $= \frac{s}{2}(P_1 + P_2)$	133

FIGURE 6–60 Some solids.

◆◆◆ **Example 17:** Find the volume of a cone having a base area of 125 cm² and an altitude of 11.2 cm.

Solution: By Eq. 130,

$$\text{volume} = \frac{125(11.2)}{3} = 467 \text{ cm}^3$$

◆◆◆

Dimension

A geometric figure that has length but no area or volume (such as a line or a curve) is said to be of *one dimension*. A geometric figure having area but not volume (such as a circle) is said to have *two* dimensions, or be *two dimensional*. A figure having volume (such as a sphere) is said to have *three* dimensions, or be *three dimensional*.

Exercise 5 ◆ Volumes and Areas of Solids

Prism and Rectangular Parallelepiped

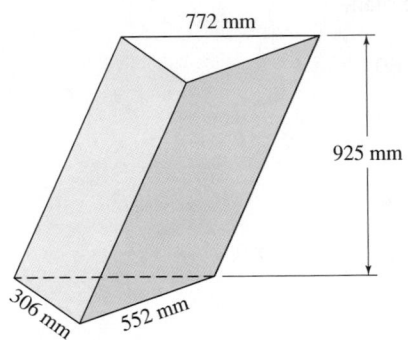

FIGURE 6–61

772 mm

925 mm

306 mm   552 mm

1. Find the volume of the triangular prism in Fig. 6–61.

2. A 2.00-cm cube of steel is placed in a surface grinding machine, and the vertical feed is set so that 0.1 mm of metal is removed from the top of the cube at each cut. How many cuts are needed to reduce the weight of the cube by 8.3 g?

3. A rectangular tank is being filled with liquid, with each cubic metre of liquid increasing the depth by 2.0 cm. The length of the tank is 12.0 m.

   (a) What is the width of the tank?

   (b) How many cubic metres will be required to fill the tank to a depth of 3.0 m?

4. How many loads of gravel will be needed to cover 3.0 km of roadbed, 11 m wide, to a depth of 7.5 cm if one truckload contains 8.0 m³ of gravel?

Cylinder

5. Find the volume of the cylinder in Fig. 6–62.

6. Find the volume and the lateral area of a right circular cylinder having a base radius of 128 and a height of 285.

7. A 3.00-m-long piece of iron pipe has an outside diameter of 7.50 cm and weighs 56.0 kg. Find the wall thickness.

8. A certain bushing is in the shape of a hollow cylinder 18.0 mm in diameter and 25.0 mm long, with an axial hole 12.0 mm in diameter. If the density of the material from which they are made is 2.70 g/cm³, find the mass of 1000 bushings.

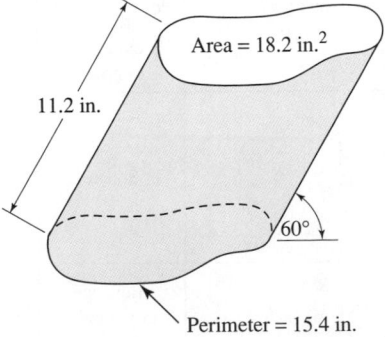

Area = 18.2 in.²

11.2 in.

60°

Perimeter = 15.4 in.

FIGURE 6–62

9. A steel gear is to be lightened by drilling holes through the gear. The gear is 8.90 cm thick. Find the diameter *d* of the holes if each is to remove 340 g.

10. A certain gasoline engine has four cylinders, each with a bore of 82.0 mm and a piston stroke of 95.0 mm. Find the engine displacement in litres.

Cone and Pyramid

11. The circumference of the base of a right circular cone is 40.0 cm, and the slant height is 38.0 cm. What is the area of the lateral surface?

12. Find the volume of a circular cone whose altitude is 24.0 cm and whose base diameter is 30.0 cm.

13. Find the weight of the mast of a ship, its height being $2\overline{0}$ m, the circumference at one end 1.5 m and at the other 1.0 m, if the density of the wood is 935 kg/m³.

14. The slant height of a right pyramid is 11.0 m, and the base is a 4.00-m square. Find the area of the entire surface.

15. How many cubic decimetres are in a piece of timber 10.0 m long, one end being a 38.0-cm square and the other a 30.0-cm square?

16. Find the total surface area and volume of a tapered steel roller 12.0 ft. long and having end diameters of 12.0 in. and 15.0 in.

Sphere

17. Find the volume and the surface area of a sphere having a radius of 744.

18. Find the volume and the radius of a sphere having a surface area of 462.

19. Find the surface area and the radius of a sphere that has a volume of 5.88.

20. How many great-circle areas of a sphere would have the same as that of the surface of the sphere? Note: A *great-circle area* of a sphere is one that has the same diameter as the sphere.

21. Find the weight in kilograms of 100 steel balls each 6.35 cm in diameter.

22. A spherical radome encloses a volume of $900\overline{0}$ m³. Assume that the sphere is complete. (A radome, or radar dome, is a spherical protective enclosure for a microwave or radar antenna.)

   (a) Find the radome radius, r

   (b) If the radome is constructed of a material weighing 2.00 kg/m², find its weight.

Structures

23. A wheat silo complex in Saskatchewan consists of seven silos (Fig. 6–63). Each silo is a cylinder with a cone on top and bottom. Each cylinder is 12 m tall, 4.0 m wide, and the distance from the tip of the bottom cone to the tip of the top cone is 18 m.

   (a) Find the volume of one silo.

   (b) One railway grain car can hold 114 m³. How many cars would be needed to transport the entire contents of seven full silos?

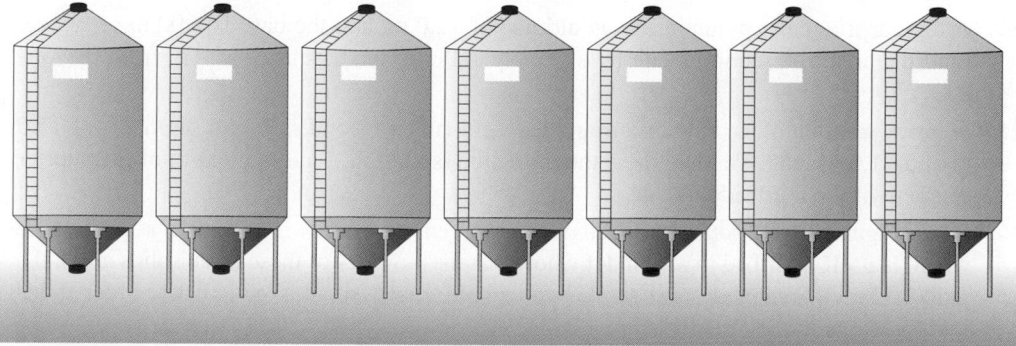

FIGURE 6–63

Case Study Discussion—Engineers Without Borders

First, some unit conversion is needed:

The inside height of the container is 7 ft. 9 in., or 93 in., which, when converted to millimetres, is 93 in. × 25.4 mm/in. = 2362.2, or 2360 mm (rounded down). Because each panel is 80 mm thick, we will be able to stack 2360 mm/80 mm = 29.5 panels. We need to round this down to 29 panels high.

Next, the width of the container interior is 7 ft. 7 in., or 91 in, or 2310 mm rounded (91 in. × 25.4 mm/in. = 2311.4 mm). Because the panels are 1030 mm wide with packaging, we can fit in 2310/1030 = 2.24, or 2 stacks width-wise.

Finally, the interior depth of the container is 19 ft. 3 in., or 231 in., or 5870 mm rounded (231 in. × 25.4 mm/in. = 5867.4). Because each panel is 1689 mm long with packaging, we can fit in 5870/1689 = 3.48, or 3 rows deep. So the container can hold 6 vertical stacks of 29 panels, fitted 2 stacks across and 3 stacks deep, for a total of 174 panels (3 × 2 × 29 = 174).

◆◆◆ CHAPTER 6 REVIEW PROBLEMS ◆◆◆◆◆◆◆◆◆◆◆◆◆◆◆◆◆◆◆◆◆◆◆◆◆◆◆◆◆◆◆

1. A rocket ascends at an angle of 60.0° with the horizontal. After 1.00 min it is directly over a point that is a horizontal distance of 12.0 km from the launch point. Find the speed of the rocket.

2. A rectangular beam 32 cm thick is cut from a log 40 cm in diameter. Find the greatest depth beam that can be obtained.

3. A cylindrical tank 4.00 m in diameter is placed with its axis vertical and is partially filled with water. A spherical diving bell is completely immersed in the tank, causing the water level to rise 1.00 m. Find the diameter of the diving bell.

4. Two vertical piers are $24\overline{0}$ ft. apart and support a circular bridge arch. The highest point of the arch is 30.0 ft. higher than the piers. Find the radius of the arch.

5. Two antenna masts are 10 m and 15 m high and are 12 m apart. How long a wire is needed to connect the tops of the two masts?

6. Find the area and the side of a rhombus whose diagonals are $1\overline{0}0$ cm and 140 cm.

7. Find the area of a triangle that has sides of length 573, 638, and 972.

8. Two concentric circles have radii of 5.00 and 12.0. Find the length of a chord of the larger circle, which is tangent to the smaller circle.

9. A belt that does not cross goes around two pulleys, each with a radius of 10.0 cm and whose centres are 22.0 cm apart. Find the length of the belt.

10. A regular triangular pyramid has an altitude of 12.0 m, and the base is 4.00 m on a side. Find the area of a section made by a plane parallel to the base and 4.00 m from the vertex.

11. A fence parallel to one side of a triangular field cuts a second side into segments of 15.0 m and 21.0 m long. The length of the third side is 42.0 m. Find the length of the shorter segment of the third side.

12. When Figure 6–64 is rotated about axis *AB*, we generate a cone inscribed in a hemisphere which is itself inscribed in a cylinder. Show that the volumes of these three solids are in the ratio 1 : 2 : 3. (Archimedes was so pleased with this discovery, it is said, that he ordered this figure to be engraved on his tomb.)

13. Four interior angles of a certain irregular pentagon are 38°, 96°, 112°, and 133°. Find the fifth interior angle.

14. Find the area of a trapezoid whose bases have lengths of 837 m and 583 m and are separated by a distance of 746 m.

15. A plastic drinking cup has a base diameter of 49.0 mm, is 63.0 mm wide at the top, and is 86.0 mm tall. Find the volume of the cup.

16. A spherical balloon is 5.27 m in diameter and is made of a material that weighs 13.9 kg per $10\overline{0}$ m^2 of surface area. Find the volume and the weight of the balloon.

17. Find the area of a triangle whose base is 38.4 cm and whose altitude is 53.8 cm.

18. Find the volume of a sphere of radius 33.8 cm.

19. Find the area of a parallelogram of base 39.2 m and height 29.3 m.

20. In *The Musgrave Ritual*, Sherlock Holmes calculates the length of the shadow of an elm tree that is no longer standing. He does know that the elm was 64 ft. high and that the shadow was cast at the instant that the sun was grazing the top of a certain oak tree.

   Holmes held a 6-ft.-long fishing rod vertical and measured the length of its shadow at the proper instant. It was 9 ft. long. He then said, "Of course the calculation now was a simple one. If a rod of six feet threw a shadow of nine, a tree of sixty-four feet would throw one of ___ ." How long was the shadow of the elm?

21. Find the volume of a cylinder with base radius 22.3 cm and height 56.2 cm.

22. Find the surface area of a sphere having a diameter of 39.2 cm.

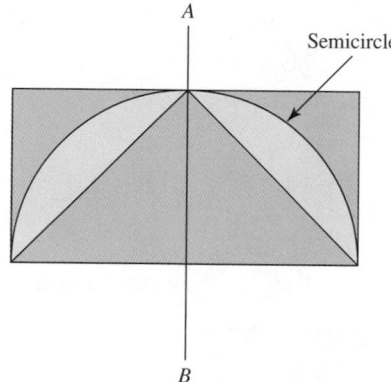

FIGURE 6–64

23. Find the volume of a right circular cone with base diameter 2.84 m and height 5.22 m.

24. Find the volume, in cubic decimetres, of a box measuring 35.8 cm × 37.4 cm × 73.4 cm.

25. Find the volume of a triangular prism having an altitude of 4.65 cm if the area of one end is 24.6 cm^2.

26. To find a circle diameter when the centre is inaccessible, you can place a scale as in Fig. 6–65 and measure the chord c and the perpendicular distance h. Find the radius of the curve if the chord length (c) is 8.25 cm and h is 1.16 cm.

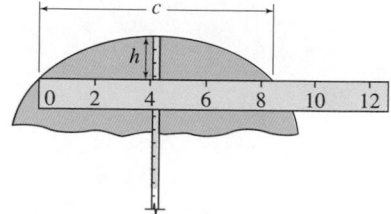

FIGURE 6–65

Writing

27. Write a section in an instruction manual for machinists on how to find the root diameter of a screw thread by measuring "over wires" (Fig. 6–58). Your entry should have two parts: First, a "how-to" section giving step-by-step instructions, and then a "theory" section explaining why this method works. Keep the entry to one page or less.

Team Projects

28. Actually use the method that Sherlock Holmes used in *The Musgrave Ritual*, problem 20, to find the height of a flagpole or building on your campus. Find some way of checking your result.

29. Writing about trees, Leonardo da Vinci said, "every division of the branches, when joined together, adds up to the thickness of the branch at their meeting point" (from the *Codex Urbino*). In other words, the sum of the diameters at *B* in Fig. 6–66 should be approximately equal to the diameter at *A*, and should also be equal to the sum of the diameters at *C*. Try to verify this by actual measurement on some tree branches, assuming that the branch cross sections are circular. Also compare the sum of the *areas* at *A*, *B*, and *C*.

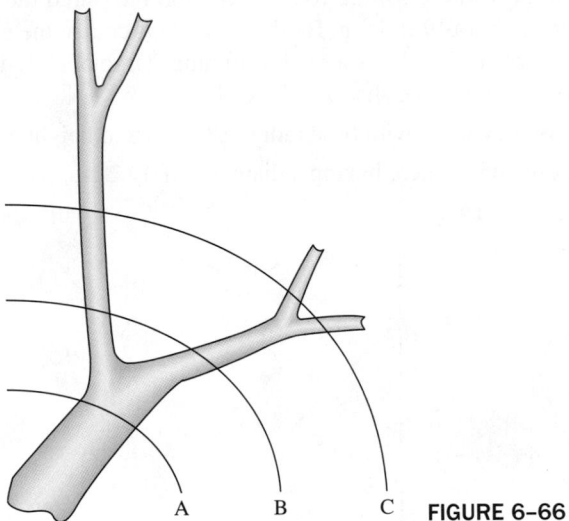

A   B   C   **FIGURE 6–66**

7

Right Triangles and Vectors

••• OBJECTIVES •••

When you have completed this chapter, you should be able to:

- Convert angles between: degrees, minutes, and seconds; decimal degrees; and radians.
- Find the trigonometric functions of an angle.
- Find the acute angle that has a given trigonometric function.
- Find the missing sides and angles of a right triangle.
- Solve practical problems involving the right triangle.
- Resolve a vector into components and, conversely, combine components into a resultant vector.
- Solve practical problems using vectors.

Vectors and right-angled triangles are closely related. Why a chapter on this one shape? Where do we find right-angled triangles? Everywhere! Longitude and latitude, up, down, left and right, we often use a coordinate or grid system to identify position. When you use a grid system the horizontal and vertical lines of the grid are, of course, at right angles to each other. Connect any two points on our grid and we can draw a right-angled triangle. Any square or rectangle can be split into two right-angled triangles with a corner-to-corner diagonal cut. Any solid with corners, like a room or building, can be broken into right-angled triangles. Right-angled triangle ratios (sine, cosine, tangent, and their reciprocals) allow us to solve so many problems that breaking a problem into right-angled triangles is often the first step in solving it. If that isn't enough, vectors, which can be used to describe motion, velocity, acceleration, and more, are also solved using right-angled triangles. A whole chapter on right triangles? Easily!

The terms *right-angled triangle* and *right triangle* mean the same thing, the former being more common in the United Kingdom, the latter here in North America. In this text we will generally use the shorter *right triangle*.

THE FIRST ANTI-GRAVITY SUIT—WILBUR FRANKS

CANADIAN BIOGRAPHY

For pilots engaged in aerial combat, a major problem first noticed as early as 1917 was "fainting in the air." During a sharp turn, the blood is pulled down to the legs and buttocks and stays there during the turn. The faster the turn the greater the centrifugal force pushing everything (including blood in the body) toward the outside of the turn. To the pilots this felt like they weighed more, as if gravity had been increased. The amount of force is rated in *g*'s. One *g* is the normal force of gravity. In a tight turn, you might feel three to five times the force of gravity—3 to 5 *g*'s. In modern fighter aircraft, it can be as high as 9 *g*'s. Pilots were dying because when the blood was pulled down to the lower parts of the body, the brain was starved of blood and oxygen, and the pilot blacked out. A pilot who could not regain consciousness and take back control of the aircraft would crash.

CASE STUDY – BRIDGE BUILDING

A Pratt truss bridge is shown in Figure 7-1. There are many styles of truss bridges. The Pratt design, as you can see, uses right triangles for support. The advantage of this design is that the diagonal members can be smaller and lighter than a truss design that uses equilateral triangles. The calculations for the force distribution can be done with vectors. However, your task in this case study is to calculate the total length of the two types of linear steel parts you would need. The vertical and horizontal members are one type, the diagonal members another.

The bridge is 4 m tall and 60 m between the outermost vertical members. Each of the outermost horizontal base members is 8 m long.

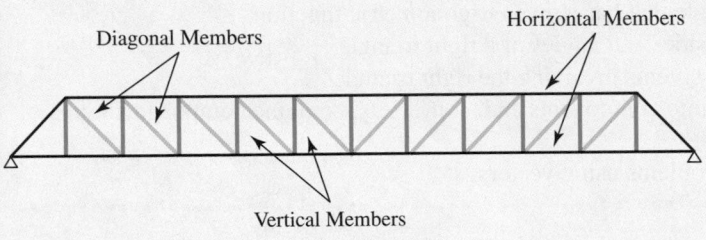

FIGURE 7-1 A Pratt Truss Bridge.

With this chapter we begin our study of *trigonometry*, the branch of mathematics that enables us to solve triangles. The *trigonometric functions* are introduced here and are used to solve right triangles. Other kinds of triangles (i.e., *oblique* triangles) are discussed in Chapter 15, and other applications of trigonometry are given in Chapters 16, 17, and 18.

We build on what we learned about triangles in Chapter 5, mainly the Pythagorean theorem and the fact that the sum of the angles of a triangle is 180°. Here we make use of *coordinate axes*, described in Sec. 5–1.

Also introduced in this chapter are *vectors*, the study of which will be continued in Chapter 15.

7–1 Angles and Their Measures

Angles Formed by Rotation

We can think of an angle as the figure generated by a ray rotating from some initial position to some terminal position (Fig. 7–2). We usually consider angles formed by rotation in the *counterclockwise* direction as *positive*, and angles formed by *clockwise* rotation as *negative*. A rotation of one *revolution* brings the ray back to its initial position.

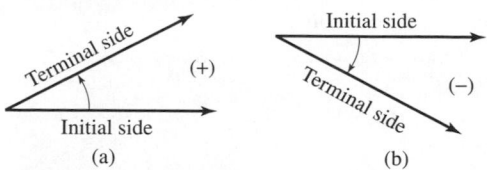

FIGURE 7–2 Angles formed by rotation.

Degrees, Minutes, and Seconds

The degree (°) is a unit of angular measure equal to 1/360 of a revolution; thus $360° =$ one revolution. A fractional part of a degree may be expressed as a common fraction (such as $28\frac{1}{2}°$), as a decimal ($28.50°$), or as *minutes* and *seconds*.

A *minute* (′) is equal to 1/60 of a degree; a *second* (″) is equal to 1/60 of a minute, or 1/3600 of a degree.

◆◆◆ **Example 1:** Some examples of angles written in degrees, minutes, and seconds are

$$85°18'42'' \qquad 62°12' \qquad 69°55'25.4'' \qquad 75°06'03''$$

We will sometimes abbreviate "degrees, minutes, and seconds" as DMS.

◆◆◆ Note that minutes or seconds less than 10 are written with an initial zero. Thus 6′ is written 06′.

Radian Measure

A *central angle* in a circle is one whose vertex is at the center of the circle. An *arc* is a portion of the circle. If an arc is laid off along the circumference of a circle with a length equal to the radius of the circle, the central angle subtended by this arc is defined as one *radian* (Fig. 7–3). A radian (rad) is a unit of angular measure commonly used in trigonometry. We will study radian measure in detail in Chapter 16.

An arc having a length of twice the radius will subtend a central angle of 2 radians. Similarly, an arc with a length of 2π times the radius (the circumference) will subtend a central angle of 2π radians. Thus our conversion factor is

$$2\pi \text{ radians} = 1 \text{ revolution} = 360°$$

Also

$$1 \text{ radian} = 57.295\ 779\ 51\ldots° = \left(\frac{360°}{2\pi}\right) \approx 57.\overset{.}{3}''$$

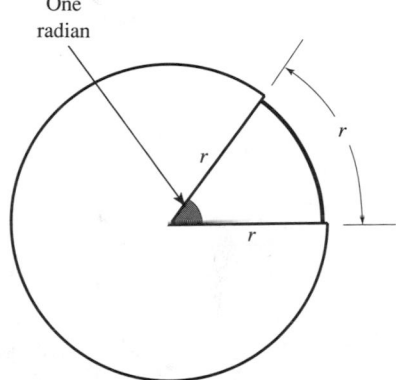

FIGURE 7–3

Conversion of Angles

We convert angles in the same way as we convert any other units. The following is a summary of the needed conversions:

Angle Conversions	$1 \text{ rev} = 360° = 2\pi \text{ rad}$ $1° = 60'$ $1' = 60''$	**117**

◆◆◆ **Example 2:** Convert $47.6°$ to radians and revolutions.

Solution: By Eq. 117,

$$47.6° \times \left(\frac{2\pi \text{ rad}}{360°}\right) = 0.831 \text{ rad}$$

Note that we keep the *same number of significant digits* as we had in the original angle. We convert to revolutions as follows:

$$47.6° \times \left(\frac{1 \text{ rev}}{360°}\right) = 0.132 \text{ rev}$$

◆◆◆

You may prefer to use π rad = 180°. Some calculators have special keys for angle conversions. Check your manual.

◆◆◆Example 3: Convert 1.8473 radians to degrees and revolutions.

Solution: By Eq. 117,

$$1.8473 \text{ rad} \times \left(\frac{360°}{2\pi \text{ rad}}\right) = 105.84°$$

and

$$1.8473 \text{ rad} \left(\frac{1 \text{ rev}}{2\pi \text{ rad}}\right) = 0.294\ 01 \text{ rev}$$ ◆◆◆

◆◆◆Example 4: Convert 1.0000 radian to degrees.

Solution:

$$1.0000 \text{ rad} \left(\frac{360°}{2\pi \text{ rad}}\right) = 57.296°$$

This is a good conversion factor to memorize. ◆◆◆

Conversions involving degrees, minutes, and seconds require several steps. Further, to know how many digits to retain, we note the following:

$$1' = \frac{1}{60} \text{ degree} \quad \text{or } 0.02°$$

and

$$1'' = \frac{1}{3600} \text{ degree} \quad \text{or } 0.0003°$$

Thus an angle known to the nearest minute is about as accurate as a decimal angle known to two decimal places. Also, an angle known to the nearest second is about as accurate as a decimal angle known to the fourth decimal place. Therefore we would treat an angle such as 28°17′37″ as if it had four decimal places and six significant digits.

◆◆◆Example 5: Convert 28°17′37″ to decimal degrees and radians.

Solution: We separately convert the minutes and the seconds to degrees, and add them. Since the given angle is known to the nearest second, we will work to four decimal places.

$$37'' \left(\frac{1°}{3600''}\right) = 0.0103°$$

$$17' \left(\frac{1°}{60'}\right) = 0.2833°$$

$$28° = 28.0000°$$

Add: 28.2936°

Now, by Eq. 117,

$$28.2936° \times \left(\frac{2\pi \text{ rad}}{360°}\right) = 0.493\ 816 \text{ rad}$$ ◆◆◆

◆◆◆Example 6: Convert 1.837 520 radians to degrees, minutes, and seconds.

Solution: We first convert to decimal degrees.

$$1.837\ 520 \text{ rad} \left(\frac{360°}{2\pi \text{ rad}}\right) = 105.2821°$$

Then we convert the decimal part (0.2821°) to minutes,

$$0.2821° \left(\frac{60'}{1°}\right) = 16.93'$$

Note that 1 second of an arc is an extremely small angle. Angle measurements accurate to the nearest second are rare and require special equipment and techniques.

An analog clock can help us understand revolutions. In Fig. 7-4(a), we see that one revolution of the minute hand is 60 minutes. If the minute hand only moves 5 minutes [Fig. 7-4(b)], what would the degree angle be in this instance? This section of time on the clock can be calculated different ways including:

$$\frac{5}{60} = \frac{x}{360},$$
$$\frac{(5 \times 360)}{60} = x,$$
$$x = 30°$$

Therefore, each 5 minute section of a clock is 30 degrees [Fig. 7-4(c)].

(a)

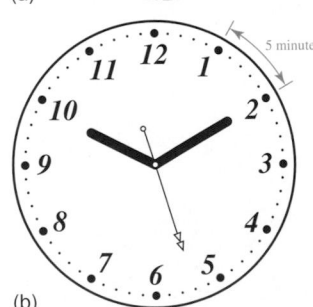
(b)

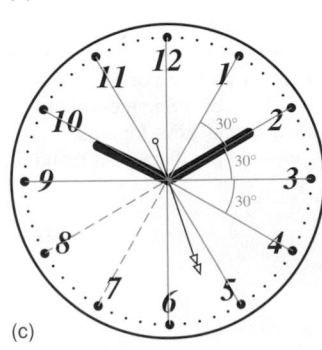
(c)

and convert the decimal part of 16.93' to seconds,

$$0.93'\left(\frac{60''}{1'}\right) = 56''$$

So

$$1.837\ 520\ \text{rad} = 105°16'56''$$

♦♦♦ Another unit of angular measure is the *grad*. There are 400 grads in 1 revolution.

Common Error	When you are reading an angle quickly, it is easy to mistake decimal degrees for degrees and minutes. Don't mistake $28°50'$ for $28.50°$

Exercise 1 ♦ Angles and Their Measures

Angle Conversions

Convert to radians.

1. $27.8°$

2. $38.7°$

3. $35°15'$

4. $270°27'25''$

5. 0.55 rev

6. 0.276 rev

Convert to revolutions.

7. 4.772 rad

8. 2.38 rad

9. $68.8°$

10. $3.72°$

11. $77°18'$

12. $135°27'42''$

Convert to degrees (decimal).

13. 2.83 rad

14. 4.275 rad

15. 0.475 rev

16. 0.236 rev

17. $29°27'$

18. $275°18'35''$

Convert to degrees, minutes, and seconds.

19. 4.2754 rad

20. 1.773 rad

21. 0.449 75 rev

22. 0.784 26 rev

23. $185.972°$

24. $128.259°$

7–2 The Trigonometric Functions

Definitions

We start with coordinate axes upon which we draw an angle, as shown in Fig. 7–5. We always draw such angles the same way, with the vertex at the origin (O) and with one side along the x axis. Such an angle is said to be in *standard position*. We can think of the angle as being generated by a ray rotating counterclockwise, whose initial position is on the x axis. We now select any point P on the terminal side of the angle θ, and we let the coordinates of P be x and y. We form a *right triangle OPQ* by dropping a perpendicular from P to the x axis. The side OP is the *hypotenuse* of the right triangle; we label its length r. Note that by the Pythagorean theorem, $r^2 = x^2 + y^2$. Also, side OQ is *adjacent* to angle θ and has a length x, and side PQ is *opposite* to angle θ and has a length y.

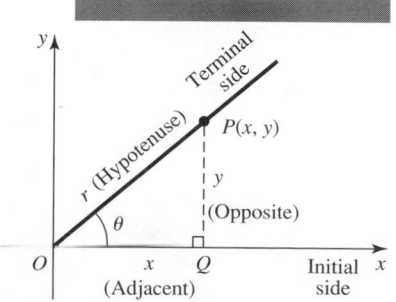

FIGURE 7–5 An angle in standard position.

If we had chosen to place point P farther out on the terminal side of the angle, the lengths x, y, and r would all be greater than they are now. But the *ratio* of any two sides will be the same regardless of where we place P, because all of the triangles thus formed are similar to each other. Such a ratio will change only if we change the angle θ. A ratio between any two sides of a right triangle is called a *trigonometric ratio* or a *trigonometric function*.

The six trigonometric functions are defined as follows:

We discuss only *acute* angles here and discuss *larger* (i.e., *oblique*) angles in Chapter 15.

The cotangent is sometimes abbreviated ctn.

Trigonometric Functions		
$\text{sine } \theta = \sin \theta = \dfrac{y}{r} = \dfrac{\text{opposite side}}{\text{hypotenuse}}$	**146**
$\text{cosine } \theta = \cos \theta = \dfrac{x}{r} = \dfrac{\text{adjacent side}}{\text{hypotenuse}}$	**147**
$\text{tangent } \theta = \tan \theta = \dfrac{y}{x} = \dfrac{\text{opposite side}}{\text{adjacent side}}$	**148**
$\text{cotangent } \theta = \cot \theta = \dfrac{x}{y} = \dfrac{\text{adjacent side}}{\text{opposite side}}$	**149**
$\text{secant } \theta = \sec \theta = \dfrac{r}{x} = \dfrac{\text{hypotenuse}}{\text{adjacent side}}$	**150**
$\text{cosecant } \theta = \csc \theta = \dfrac{r}{y} = \dfrac{\text{hypotenuse}}{\text{opposite side}}$	**151**

Note that each of the trigonometric ratios is *related to* the angle θ. Take the ratio y/r, for example. The values of y/r depend only on the measure of θ, not on the lengths of y or r. To convince yourself of this, draw an angle of, say, 30°, as in Fig. 7–6(a). Then form a right triangle by drawing a perpendicular from any point on side L or from any point on side M [Fig. 7–6(b) and (c)]. No matter how you draw the triangle, the side opposite the 30° angle *will always be half the hypotenuse* ($\sin 30° = \frac{1}{2}$). You would get a similar result for any other angle and for any other trigonometric ratio. *The value of the trigonometric ratio depends only on the measure of the angle.* We have a *relation,* such as the relations we studied in Chapter 4.

Furthermore, for any angle θ, there exists *only one* value for each trigonometric ratio. Thus *each ratio is a function of θ,* which is why we call them the *trigonometric functions* of θ. In functional notation, we could write

$$\sin \theta = f(\theta)$$
$$\cos \theta = g(\theta)$$
$$\tan \theta = h(\theta)$$

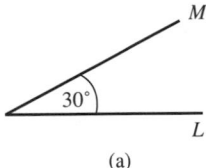

(a)

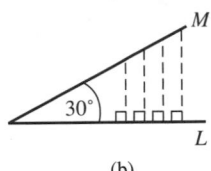

(b)

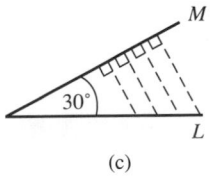

(c)

FIGURE 7–6

and so forth. The angle θ is called the *argument* of the function.

Note that while there is one and only one value of each trigonometric function for a given angle θ, we will see that for any given value of a trigonometric function (say, $\sin \phi = \frac{1}{2}$), there are *infinitely many* values of θ. We treat this idea more fully in Sec. 15–1.

The trigonometric functions are sometimes called *circular* functions.

Reciprocal Relationships

From the trigonometric functions we see that

$$\sin \theta = \frac{y}{r}$$

and that

$$\csc \theta = \frac{r}{y}$$

Hence, sin θ and csc θ are reciprocals.

$$\sin \theta = \frac{1}{\csc \theta}$$

Inspection of the trigonometric functions shows two more sets of reciprocals (see also Fig. 7–7).

Reciprocal Relationships	$\sin \theta = \dfrac{1}{\csc \theta}$	**152a**
	$\cos \theta = \dfrac{1}{\sec \theta}$	**152b**
	$\tan \theta = \dfrac{1}{\cot \theta}$	**152c**

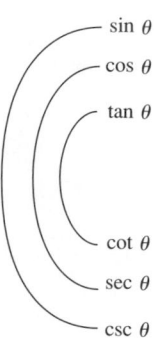

FIGURE 7–7 This diagram shows which functions are reciprocals of each other.

◆◆◆ **Example 7:** Find tan θ if cot $\theta = 0.638$.

Solution: From the reciprocal relationships,

$$\tan \theta = \frac{1}{\cot \theta} = \frac{1}{0.638} = 1.57$$

◆◆◆

Finding the Trigonometric Functions of an Angle

The six trigonometric functions of an angle can be written if we know the angle or if we know two of the sides of a right triangle containing the angle.

◆◆◆ **Example 8:** A right triangle has sides of 4.37 and 2.83 (Fig. 7–8). Write the six trigonometric functions of angle θ.

Solution: We first find the hypotenuse by the Pythagorean theorem.

$$c = \sqrt{(2.83)^2 + (4.37)^2} = \sqrt{27.1}$$
$$= 5.21$$

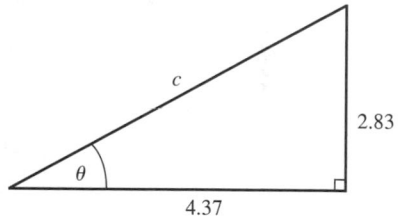

FIGURE 7–8

We also note that the opposite side is 2.83 and the adjacent side is 4.37. Then, by Eqs. 146 through 151,

$$\sin \theta = \frac{\text{opposite side}}{\text{hypotenuse}} = \frac{2.83}{5.21} = 0.543$$

$$\cos \theta = \frac{\text{adjacent side}}{\text{hypotenuse}} = \frac{4.37}{5.21} = 0.839$$

$$\tan \theta = \frac{\text{opposite side}}{\text{adjacent side}} = \frac{2.83}{4.37} = 0.648$$

$$\cot \theta = \frac{\text{adjacent side}}{\text{opposite side}} = \frac{4.37}{2.83} = 1.54$$

$$\sec \theta = \frac{\text{hypotenuse}}{\text{adjacent side}} = \frac{5.21}{4.37} = 1.19$$

$$\csc \theta = \frac{\text{hypotenuse}}{\text{opposite side}} = \frac{5.21}{2.83} = 1.84$$

◆◆◆

Common Error	Do not omit the angle when writing a trigonometric function. To write "sin = 0.543," for example, has no meaning.

◆◆◆ **Example 9:** A point P on the terminal side of an angle B (in standard position) has the coordinates (8.84, 3.15). Write the sine, cosine, and tangent of angle B.

Solution: We are given $x = 8.84$ and $y = 3.15$. We find r from the Pythagorean theorem.

$$r = \sqrt{(8.84)^2 + (3.15)^2} = 9.38$$

Then, by Eqs. 146 through 151,

$$\sin B = \frac{y}{r} = \frac{3.15}{9.38} = 0.336$$

$$\cos B = \frac{x}{r} = \frac{8.84}{9.38} = 0.942$$

$$\tan B = \frac{y}{x} = \frac{3.15}{8.84} = 0.356 \qquad\qquad ◆◆◆$$

Evaluating Trigonometric Expressions

We can combine trigonometric functions with numbers or variables to get more complex expressions, such as

$$2 \sin 35° - 5 \cos 12°$$

In this section we will evaluate some very simple expressions of this kind. Later, we will cover more complex ones.

Simply perform the indicated operations. Just remember that angles are assumed to be in *radians* unless marked otherwise.

◆◆◆ **Example 10:** Evaluate the following to three significant digits:

$$7.82 - 3.15 \cos 67.8°$$

Solution: By calculator, we find that $\cos 67.8° = 0.3778$, so our expression becomes

$$7.82 - 3.15(0.3778) = 6.63$$

When doing such calculations, we keep all of the figures in the calculator whenever possible. If you must write down an intermediate result, as we did here just to show the steps, retain 1 or 2 digits more than required in the final answer. ◆◆◆

◆◆◆ **Example 11:** Evaluate the following to four significant digits:

$$1.836 \sin 2 + 2.624 \tan 3$$

Solution: Here no angular units are indicated, so we assume them to be radians. By calculator, we get

$$1.836 \sin 2 + 2.624 \tan 3 = 1.836(0.909\,30) + 2.624(-0.142\,55)$$
$$= 1.295 \qquad\qquad ◆◆◆$$

Exercise 2 ◆ The Trigonometric Functions

Using a protractor, lay off the given angle on coordinate axes. Drop a perpendicular from any point on the terminal side, and measure the distances x, y, and r, as in Fig. 7–5. Use these measured distances to write the six trigonometric ratios of the angle to two decimal places. Check your answer by calculator.

1. 30°   **2.** 55°
3. 14°   **4.** 37°

Plot the given point on coordinate axes, and connect it to the origin. Measure the angle formed with a protractor. Check by taking the tangent of your measured angle. It should equal the ratio y/x.

5. (3, 4)   **6.** (5, 3)   **7.** (1, 2)   **8.** (5, 4)
9. Write the six trigonometric ratios for angle θ in Fig. 7–9.

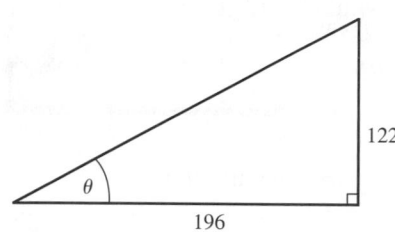

122

θ

196

FIGURE 7–9

Each of the given points is on the terminal side of an angle. Compute the distance r from the origin to the point, and write the sin, cos, and tan of the angle. Work to three significant digits.

10. (3, 5)   **11.** (4, 2)
12. (5, 3)   **13.** (2, 3)
14. (4, 5)   **15.** (4.75, 2.68)

Evaluate. Work to four decimal places

16. sin 47.3°   **17.** tan 44.4°   **18.** sec 29.5°
19. cot 18.2°   **20.** csc 37.5°   **21.** cos 21.27°
22. cos 86.75°   **23.** sec 12.3°   **24.** csc 12.67°

Write the sin, cos, and tan for each angle. Keep three decimal places.

25. 72.8°   **26.** 19.2°
27. 33.1°   **28.** 41.4°
29. 0.744 rad   **30.** 0.385 rad

Evaluate each trigonometric expression to three significant digits.

31. 5.27 sin 45.8° − 1.73   **32.** 2.84 cos 73.4° − 3.83 tan 36.2°
33. 3.72(sin 28.3° + cos 72.3°)   **34.** 11.2 tan 5 + 15.3 cos 3
35. 2.84(5.28 cos 2 − 2.82) + 3.35   **36.** 2.63 sin 2.4 + 1.36 cos 3.5 + 3.13 tan 2.5

7–3   Finding the Angle When the Trigonometric Function Is Given

The operation of finding the angle when the trigonometric function is given is the *inverse* of finding the function when the angle is given. There is special notation to indicate the inverse trigonometric function. If

$$\sin \theta = A$$

we write

$$\theta = \arcsin A$$

or

$$\theta = \sin^{-1} A$$

which is read "θ is the angle whose sine is A." Similarly, we use the symbols arccos A, $\cos^{-1} A$, arctan A, and so on. Some calculators use INV SIN A, for example.

Common Error	Do not confuse the *inverse* with the *reciprocal*. $$\frac{1}{\sin \theta} = (\sin \theta)^{-1} \neq \sin^{-1} \theta$$

◆◆◆ **Example 12:** If $\sin \theta = 0.7337$, find θ in degrees to three significant digits.

Solution:

$$\sin^{-1} 0.7337 = 47.2°$$

Note that a calculator gives us acute angles only.   ◆◆◆

◆◆◆ **Example 13:** If $\tan x = 2.846$, find x in radians to four significant digits.

Solution:

$$\tan^{-1} 2.846 = 1.233$$ ◆◆◆

In this chapter we limit our work with inverse functions to *first-quadrant* angles only. In Chapter 15, we consider any angle.

If the cotangent, secant, or cosecant is given, we first take the reciprocal of that value and then use the appropriate reciprocal relationship.

$$\sin \theta = \frac{1}{\csc \theta} \qquad \cos \theta = \frac{1}{\sec \theta} \qquad \tan \theta = \frac{1}{\cot \theta}$$

◆◆◆ **Example 14:** If $\sec \theta = 1.573$, find θ in degrees to four significant digits.

Solution: Since

$$\cos \theta = \frac{1}{\sec \theta} = \frac{1}{1.573}$$

we first find the reciprocal of 1.573 and then take the inverse cosine.

$$\frac{1}{1.573} = 0.635\ 73$$

Then

$$\cos^{-1} 0.635\ 73 = 50.53$$ ◆◆◆

Exercise 3 ◆ Finding the Angle When the Trigonometric Function Is Given

Find the acute angle (in decimal degrees) whose trigonometric function is given. Keep three significant digits.

1. $\sin A = 0.500$ **2.** $\tan D = 1.53$ **3.** $\sin G = 0.528$

4. $\cot K = 1.77$ **5.** $\sin B = 0.483$ **6.** $\cot E = 0.847$

Without using tables or a calculator, write the sin, cos, and tan of angle A. Leave your answer in fractional form.

7. $\sin A = \dfrac{3}{5}$ **8.** $\cot A = \dfrac{12}{5}$ **9.** $\cos A = \dfrac{12}{13}$

Evaluate the following, giving your answer in decimal degrees to three significant digits.

10. arcsin 0.635 **11.** arcsec 3.86 **12.** $\tan^{-1} 2.85$

13. $\cot^{-1} 1.17$ **14.** $\cos^{-1} 0.229$ **15.** arccsc 4.26

7–4 Solution of Right Triangles

Right Triangles

The right triangle (Fig. 7–10) was introduced in Sec. 6–2, where we also introduced the Pythagorean theorem (Eq. 145) and the equation for the sum of the interior angles (Eq. 139).

Since the angle C is always 90° for right triangles, the equation for the sum of the interior angles can be re-written as

$$A + B + 90° = 180°$$

or

$$A + B = 90°$$

This plus our trigonometric relations for sin, cos, and tan become our tools for solving any right triangle.

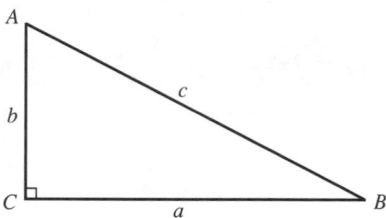

FIGURE 7–10 A right triangle. We will usually *label* a right triangle as shown here. We label the angles with capital letters A, B, and C, with C always the right angle. We label the sides with lowercase letters a, b, and c, with side a opposite angle A, side b opposite angle B, and side c (the hypotenuse) opposite angle C (the right angle).

Pythagorean Theorem	$c^2 = a^2 + b^2$	145
Sum of the Angles	$A + B = 90°$	139
Trigonometric Functions	$\sin \theta = \dfrac{\text{side opposite to } \theta}{\text{hypotenuse}}$	146
	$\cos \theta = \dfrac{\text{side adjacent to } \theta}{\text{hypotenuse}}$	147
	$\tan \theta = \dfrac{\text{side opposite to } \theta}{\text{side adjacent to } \theta}$	148

Solving Right Triangles When One Side and One Angle Are Known

To *solve* a triangle means to find all missing sides and angles (although in most practical problems we need find only *one* missing side or angle). We can solve any right triangle if we know one side and either another side or one angle.

To solve a right triangle when one side and one angle are known:

1. Make a sketch.

2. Find the missing angle by using Eq. 139.

3. Relate the known side to one of the missing sides by one of the trigonometric ratios. Solve for the missing side.
4. Repeat Step 3 to find the second missing side.
5. Check your work with the Pythagorean theorem.

♦♦♦ **Example 15:** Solve right triangle ABC if $A = 18.6°$ and $c = 135$.

Solution:

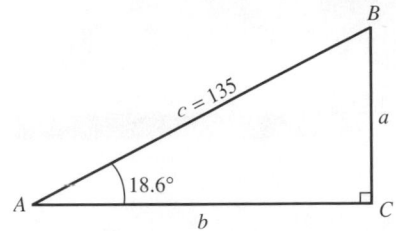

FIGURE 7–11

1. We make a sketch as shown in Figure 7–11.
2. Then, by Eq. 139,

$$B = 90° - A = 90° - 18.6° = 71.4°$$

3. Let us now find side a. We must use one of the trigonometric ratios. *But how do we know which one to use?* And further, *for which of the two angles,* A *or* B, *should we write the trig ratio?*

It is simple. First, always work with the *given angle*, because if you made a mistake in finding angle B and then used it to find the sides, they would be wrong also. Then, to decide which trigonometric ratio to use, we note that side a is *opposite* to angle A and that the given side is the *hypotenuse*. Thus, our trig function must be one that relates the *opposite side* to the *hypotenuse*. Our obvious choice is Eq. 146.

> Realize that either of the two legs can be called *opposite* or *adjacent*, depending on which angle we are referring them to. Here b is adjacent to angle A but opposite to angle B.

$$\sin A = \frac{\text{opposite side}}{\text{hypotenuse}}$$

Substituting the given values, we obtain

$$\sin 18.6 = \frac{a}{135}$$

Solving for a yields

$$a = 135 \sin 18.6°$$
$$= 135(0.3190) = 43.1$$

4. We now find side b. Note that side b is *adjacent* to angle A. We therefore use Eq. 147.

$$\cos 18.6° = \frac{b}{135}$$

So

$$b = 135 \cos 18.6°$$
$$= 135(0.9478) = 128$$

We have thus found all missing parts of the triangle.

5. For a check, we see if the three sides will satisfy the Pythagorean theorem.

> As a rough check of any triangle, see if the longest side is opposite the largest angle and if the shortest side is opposite the smallest angle. Also check that the hypotenuse is greater than each leg but less than their sum.

Check:

$$(43.1)^2 + (128)^2 \stackrel{?}{=} (135)^2$$
$$18\,242 \stackrel{?}{=} 18\,225$$

Since we are working to three significant digits, this is close enough for a check.   ♦♦♦

♦♦♦ **Example 16:** In right triangle ABC, angle $B = 55.2°$ and $a = 207$. Solve the triangle.

Solution:

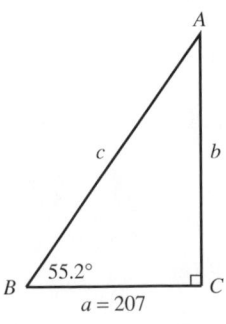

FIGURE 7–12

1. We make a sketch as shown in Fig. 7–12.
2. By Eq. 139,

$$A = 90° - 55.2 = 34.8$$

3. By Eq. 147,

$$\cos 55.2° = \frac{207}{c}$$

$$c = \frac{207}{\cos 55.2°} = \frac{207}{0.5707} = 363$$

4. Then, by Eq. 148,

$$\tan 55.2° = \frac{b}{207}$$

$$b = 207 \tan 55.2 = 207(1.439) = 298$$

5. Checking with the Pythagorean theorem, we have

$$(363)^2 \stackrel{?}{=} (207)^2 + (298)^2$$
$$131\ 769 \stackrel{?}{=} 131\ 653 \text{ (checks to within three significant digits)} \qquad \blacklozenge\blacklozenge\blacklozenge$$

Tip	Whenever possible, use the *given information* for each computation, rather than some quantity previously calculated. This way, any errors in the early computation will not be carried along.

This is good advice for performing *any* computation, not just for solving right triangles.

Solving Right Triangles When Two Sides Are Known

1. Draw a diagram of the triangle.
2. Write the trigonometric ratio that relates one of the angles to the two given sides. Solve for the angle.
3. Subtract the angle just found from 90° to get the second angle.
4. Find the missing side by the Pythagorean theorem.
5. Check the computed side and angles with trigonometric ratios, as shown in the following example.

◆◆◆ **Example 17:** Solve right triangle *ABC* if $a = 1.48$ and $b = 2.25$.

Solution:

1. We sketch the triangle as shown in Fig. 7–13.
2. To find angle *A*, we note that the 1.48 side is *opposite* angle *A* and that the 2.25 side is *adjacent* to angle *A*. The trig ratio relating *opposite* and *adjacent* is the *tangent* (Eq. 148).

$$\tan A = \frac{1.48}{2.25} = 0.6578$$

from which

$$A = 33.3°$$

3. Solving for angle *B*, we have

$$B = 90° - A = 90 - 33.3 = 56.7°$$

4. We find side *c* by the Pythagorean theorem.

$$c^2 = (1.48)^2 + (2.25)^2 = 7.253$$
$$c = 2.69$$

Check:

$$\sin 33.3° \stackrel{?}{=} \frac{1.48}{2.69}$$
$$0.549 \approx 0.550 \quad \text{(checks)}$$

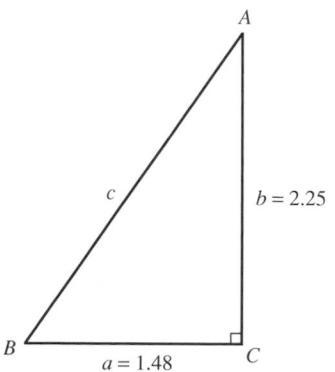

FIGURE 7–13

and

$$\sin 56.7° \overset{?}{=} \frac{2.25}{2.69}$$

$$0.836 = 0.836 \quad \text{(checks)}$$

(Note that we could just as well have used another trigonometric function, such as the cosine, for our check.)   ◆◆◆

Cofunctions

The sine of angle A in Fig. 7–10 is

$$\sin A = \frac{a}{c}$$

But a/c is *also* the cosine of the complementary angle B. Thus we can write

$$\sin A = \cos B$$

Here $\sin A$ and $\cos B$ are called *cofunctions*. Similarly, $\cos A = \sin B$. These cofunctions and others in the same right triangle are given in the following boxes:

Cofunctions Where $A + B = 90°$ or $A + B = \dfrac{\pi}{2}$ rad		
	$\sin A = \cos B$	**154a**
	$\cos A = \sin B$	**154b**
	$\tan A = \cot B$	**154c**
	$\cot A = \tan B$	**154d**
	$\sec A = \csc B$	**154e**
	$\csc A = \sec B$	**154f**

In general, a trigonometric function of an acute angle is equal to the corresponding cofunction of the complementary angle.

◆◆◆ **Example 18:** If the cosine of the acute angle A in a right triangle ABC is equal to 0.725, find the sine of the other acute angle, B.

Solution: By Eq. 154b,

$$\sin B = \cos A = 0.725 \qquad\qquad\qquad ◆◆◆$$

Exercise 4 ◆ Solution of Right Triangles

Right Triangles with One Side and One Angle Given

Sketch each right triangle and find all of the missing parts. Assume the triangles to be labelled as in Fig. 7–10. Work to three significant digits.

1. $a = 155 \quad A = 42.9°$
2. $b = 82.6 \quad B = 61.4°$
3. $a = 1.74 \quad B = 31.9°$
4. $b = 7.74 \quad A = 22.5°$

5. $a = 284$ $A = 1.13$ rad
6. $b = 73.2$ $B = 0.655$ rad
7. $b = 9.26$ $B = 55.2°$
8. $a = 1.73$ $A = 39.3°$
9. $b = 82.4$ $A = 31.4°$
10. $a = 18.3$ $B = 44.1°$

Right Triangles with Two Sides Given

Sketch each right triangle and find all missing parts. Work to three significant digits and express the angles in decimal degrees.

11. $a = 382$ $b = 274$
12. $a = 3.88$ $c = 5.37$
13. $b = 3.97$ $c = 4.86$
14. $a = 63.9$ $b = 84.3$
15. $a = 27.4$ $c = 37.5$
16. $b = 746$ $c = 957$
17. $a = 41.3$ $c = 63.7$
18. $a = 4.82$ $b = 3.28$
19. $b = 228$ $c = 473$
20. $a = 274$ $c = 429$

Cofunctions

Express as a function of the complementary angle.

21. $\sin 38°$
22. $\cos 73°$
23. $\tan 19°$
24. $\sec 85.6°$
25. $\cot 63.2°$
26. $\csc 82.7°$
27. $\tan 35°14'$
28. $\cos 0.153$ rad
29. $\sin 0.475$ rad

7–5   Applications of the Right Triangle

There is, of course, a huge number of applications for the right triangle, a few of which are given in the following examples and exercises. A typical application is finding a distance that cannot be measured directly, as shown in the following example.

♦♦♦ **Example 19:** To find the height of a flagpole (Fig. 7–14), a person measures 12.0 m from the base of the pole and then measures an angle of 40.8° from a point 2.00 m above the ground to the top of the pole. Find the height of the flagpole.

Estimate: If angle A were 45°, then BC would be the same length as AC, or 12 m. But our angle is a bit less than 45°, so we expect BC to be less than 12 m, say, 10 m. Thus our guess for the entire height is about 12 m.

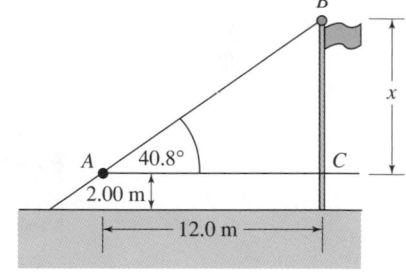

FIGURE 7–14

Solution: In right triangle ABC, BC is opposite the known angle, and AC is adjacent. Using the tangent, we get

$$\tan 40.8° = \frac{x}{12.0}$$

where x is the height of the pole above the observer. Then

$$x = 12.0 \tan 40.8° = 10.4 \text{ m}$$

Adding 2.0 m, we find that the total pole height is 12.4 m, measured from the ground.   ♦♦♦

♦♦♦ **Example 20:** From a plane at an altitude of 1050 m, the pilot observes the angle of depression of a lake to be 18.6°. How far is the lake from a point on the ground directly beneath the plane?

Solution: We first note that an *angle of depression*, like an *angle of elevation*, is defined as the angle between a line of sight and the *horizontal*, as shown in Fig. 7–15. We then make a sketch for our problem (Fig. 7–16). Since the ground and the horizontal line drawn through the plane

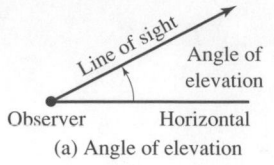

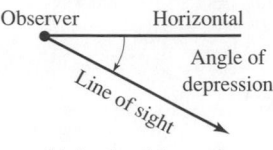

(a) Angle of elevation

(b) Angle of depression

FIGURE 7–15 Angles of elevation and depression.

are parallel, angle A and the angle of depression (18.6°) are alternate interior angles. Therefore $A = 18.6°$. Then

$$\tan 18.6 = \frac{1050}{x}$$

$$x = \frac{1050}{\tan 18.6°} = 3120 \text{ m}$$

◆◆◆

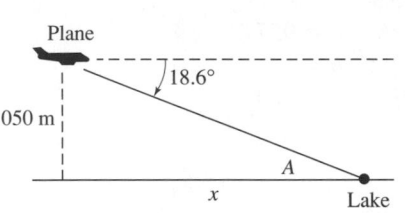

FIGURE 7–16

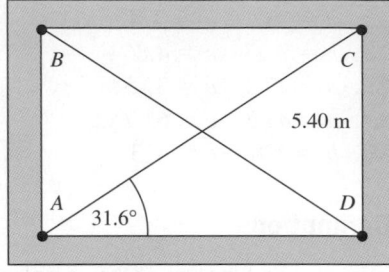

FIGURE 7–17

◆◆◆ **Example 21:** A frame is braced by wires AC and BD, as shown in Fig. 7–17. Find the length of each wire.

Estimate: We can make a quick estimate if we remember that the sine of 30° is 0.5. Thus we expect wire AC to be a little shorter than twice side CD, or around 10 m.

Solution: Noting that in right triangle ACD, CD is opposite the given angle and AC is the hypotenuse, we choose the sine.

$$\sin 31.6° = \frac{5.40}{AC}$$

$$AC = BD = \frac{5.40}{\sin 31.6°} = 10.3 \text{ m}$$

◆◆◆

Exercise 5 ◆ Applications of the Right Triangle

Measuring Inaccessible Distances

1. From a point on the ground 255 m from the base of a tower, the angle of elevation to the top of the tower is 57.6°. Find the height of the tower.

2. A pilot 4220 m directly above the front of a straight train observes that the angle of depression of the end of the train is 68.2°. Find the length of the train.

3. From the top of a lighthouse 49.0 m above the surface of the water, the angle of depression of a boat is observed to be 28.7°. Find the horizontal distance from the boat to the lighthouse.

4. An observer in an airplane 1520 ft. above the surface of the ocean observes that the angle of depression of a ship is 28.8°. Find the straight-line distance from the plane to the ship.

5. The distance PQ across a swamp is desired. A line PR of length 59.3 m is laid off at right angles to PQ. Angle PRQ is measured at 47.6°. Find the distance PQ.

6. The angle of elevation of the top of a building from a point on the ground 105 m from its base is 51.3°. Find the height of the building.

7. The angle of elevation of the top of a building from a point on the ground 75.0 m from its base is 28.0°. How high is the building?

8. From the top of a hill 42.5 m above a stream, the angles of depression of a point on the near shore and of a point on the opposite shore are 42.3° and 40.6°. Find the width of the stream between these two points.

9. From the top of a tree 15.0 m high on the shore of a pond, the angle of depression of a point on the other shore is 6.70°. What is the width of the pond?

10. From the observation deck of Toronto's CN tower, $40\overline{0}$ m above the surface of Lake Ontario, the angle of depression of sighting a sailboat out on the lake is observed to be 21.6°. Determine how far away the sailboat is from the base of the tower.

11. Your survey boat travels along the base of very high cliffs at the Western Brook Pond fjord in Newfoundland. A spectacular waterfall cascades down the side of the rock formation (Fig. 7–18). You can get a very accurate measure of the distance from the boat to the rock face (strong echoes return to your range finder), but you cannot get a good reflection off the top of the falls. You use a sighting device to measure the angle of elevation as 78.0°. If the distance to the rock face is 127 m, how tall are the falls?

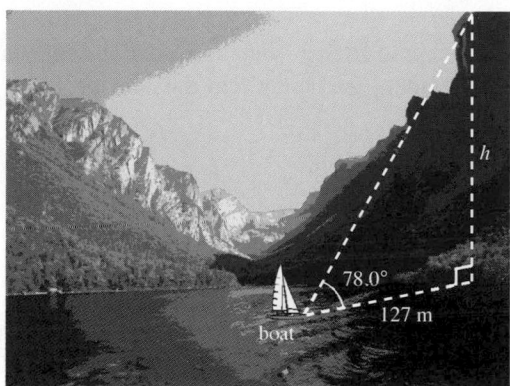

FIGURE 7–18

12. The base-station antenna of a two-way radio is on the roof of a fire hall, 18.0 m above ground level. The signal from a fire truck arrives at the antenna at an angle of 2.10° relative to the road. How far away is the fire truck?

Navigation

13. A passenger on a ship sailing due north at 8.20 km/h noticed that at 1 P.M., a buoy was due east of the ship. At 1:45 P.M., the bearing of the buoy from the ship was S 27°12′ E (Fig. 7–19). How far was the ship from the buoy at 1:00 P.M.?

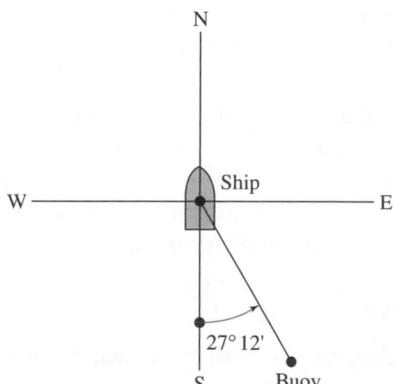

FIGURE 7–19 A compass direction of S 27°12′ E means an angle of 27°12′ measured from due south to the east. Also, remember that 1 minute of arc (1′) is equal to 1/60 degree, or

$$60' = 1°$$

14. An observer at a point P on a coast sights a ship in a direction N 43°15′ E. The ship is at the same time directly east of a point Q, 15.6 km due north of P. Find the distance of the ship from point P and from Q.

15. A ship sailing parallel to a straight coast is directly opposite one of two lights on the shore. The angle between the lines of sight from the ship to these lights is 27°50′, and it is known that the lights are 355 m apart. Find the perpendicular distance of the ship from the shore.

16. An airplane left an airport and flew 315 km in the direction N 15°18′ E, then turned and flew 296 km in the direction S 74°42′ E. How far, and in what direction, should it fly to return to the airport?

17. A ship is sailing due south at a speed of 7.50 km/h when a light is sighted with a bearing S 35°28′ E. One hour later the ship is due west of the light. Find the distance from the ship to the light at that instant.

18. After leaving port, a ship holds a course N 46°12′ E for 225 nautical miles (M). Find how far north and how far east of the port the ship is now located.

Structures

19. A guy wire from the top of an antenna is anchored 16.5 m from the base of the antenna and makes an angle of 85.2° with the ground. Find (a) the height h of the antenna and (b) the length L of the wire.

20. Find the angle θ and length AB in the truss shown in Fig. 7–20.

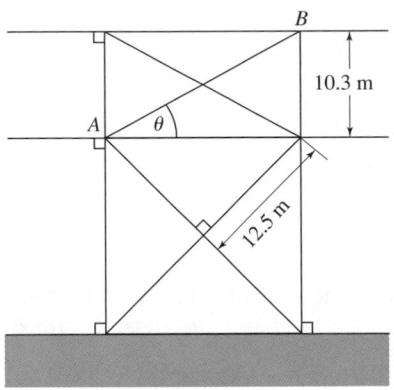

FIGURE 7–20

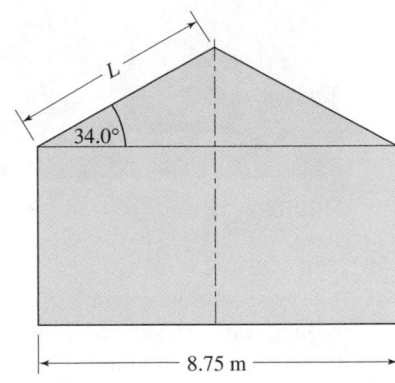

FIGURE 7–21

21. A house is 8.75 m wide, and its roof has an angle of inclination of 34.0° (Fig. 7–21). Find the length L of the rafters.

22. A guy wire 25.0 m long is stretched from the ground to the top of a telephone pole 20.0 m high. Find the angle between the wire and the pole.

23. A fire truck has a ladder that can extend to 16 m. If the truck can get as close as 3 m to the side of a building, could the ladder reach the fourth floor, given that each floor is 3 m high?

24. You are installing a new 50-m-tall radio antenna mast onto Cabot Tower on Signal Hill, Newfoundland. Because of the height of the mast, you need to guy the mast for safety. The safety code for a mast of this height specifies 4 guy wires from the top of the mast. The distance from the base of the mast to the anchor for the guy wire is 15 m. How long must the spool of guy wire be to complete all 4 guys?

Geometry

25. What is the angle, to the nearest tenth of a degree, between a diagonal AB of a cube and a diagonal AC of a face of that cube (Fig. 7–22)?

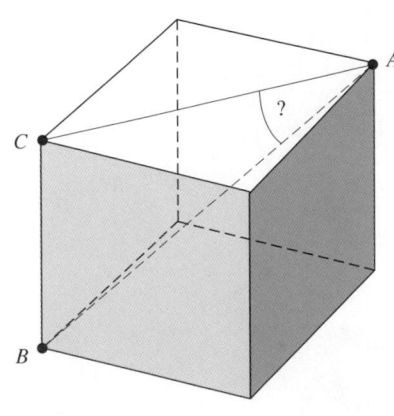

FIGURE 7–22

26. Two of the sides of an isosceles triangle have a length of $15\overline{0}$ units, and each of the base angles is 68.0°. Find the altitude and the base of the triangle.

27. The diagonal of a rectangle is 3 times the length of the shorter side. Find the angle, to the nearest tenth of a degree, between the diagonal and the longer side.

28. Find the angles between the diagonals of a rectangle whose dimensions are $58\overline{0}$ units $\times$ $94\overline{0}$ units.

29. Find the area of a parallelogram if the lengths of the sides are 255 cm and 482 cm and if one angle is 83.2°.

30. Find the length of a side of a regular hexagon inscribed in a 125-cm-radius circle. *Hint:* Draw lines from the center of the hexagon to each vertex and to the midpoint of each side, to form right triangles.

31. Find the length of the side of a regular pentagon circumscribed about a circle of radius 244 cm.

32. The stream of water leaving the nozzle of a fire hose can extend straight out for a significant distance. From a position of 10.0 m from a burning two-storey house, at what angle must the firefighter direct the stream of water to reach a second-storey window located 5.5 m above the ground?

Shop Trigonometry

33. Find the dimension x in Fig. 7–23.

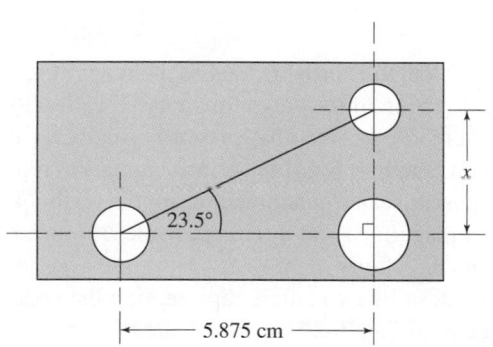

FIGURE 7–23

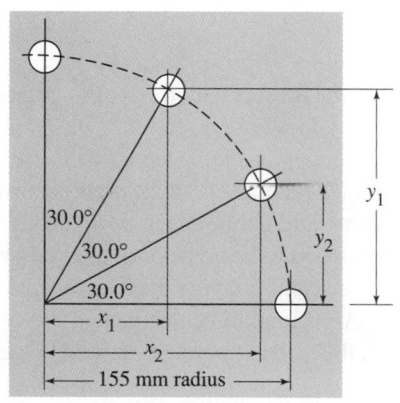

FIGURE 7–24 A bolt circle.

Trigonometry finds many uses in the machine shop. Given in problems 33 through 39 are some standard shop calculations.

34. A bolt circle (Fig. 7–24) is to be made on a jig borer. Find the dimensions x_1, y_1, x_2, and y_2.

35. A bolt circle with a radius of 36.000 cm contains 24 equally spaced holes. Find the straight-line distance between the holes so that they may be stepped off with dividers.

36. A 9.000-cm-long shaft is to taper 3.500° and be 1.000 cm in diameter at the narrow end (Fig. 7–25). Find the diameter d of the larger end. *Hint:* Draw a line (shown dashed in the figure) from the small end to the large, to form a right triangle. Solve that triangle.

37. A measuring square having an included angle of 60.0° is placed over the pulley fragment of Fig. 7–26. The distance d from the corner of the square to the pulley rim is measured at 12.6 cm. Find the pulley radius r.

38. A common way of measuring dovetails is with the aid of round plugs, as in Fig. 7–27. Find the distance x over the 0.5000-in.-diameter plugs. *Hint:* Find the length of side AB in the 30–60–90 triangle ABC, and then use AB to find x.

39. A bolt head (Fig. 7–28) measures 0.750 cm across the flats. Find the distance p across the corners and the width r of each flat.

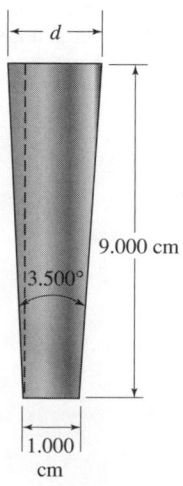

FIGURE 7–25 Tapered shaft.

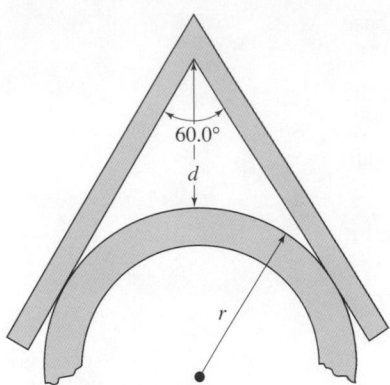

FIGURE 7–26   Measuring the radius of a broken pulley.

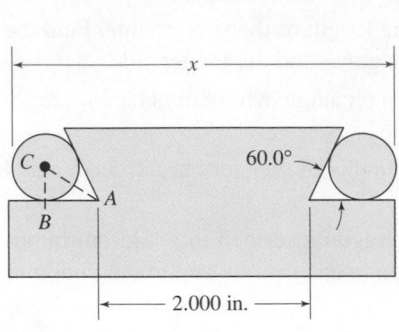

FIGURE 7–27   Measuring a dovetail.

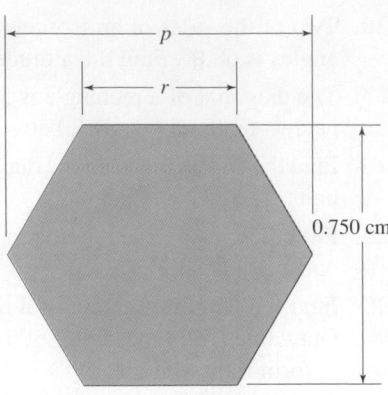

FIGURE 7–28

7–6   Vectors

Definition

A *vector quantity* is one that has *direction* as well as *magnitude*. For example, a velocity cannot be completely described without giving the direction of the motion, as well as the magnitude of the motion (the speed). Other vector quantities are force and acceleration.

Quantities that have magnitude but no direction are called *scalar* quantities. These include time, volume, mass, and so on.

Representation of Vectors

A vector is represented by a line segment whose length is proportional to the magnitude of the vector and whose direction is the same as the direction of the vector quantity (Fig. 7– 29).

Vectors are represented differently in different textbooks, but they are usually written in boldface type. Here, we'll use the most common notation: **boldface** roman capitals to represent vectors, and nonboldface *italic* capitals to represent scalar quantities. So in this textbook, **B** is understood to be a vector quantity, having both magnitude and direction, while *B* is understood to be a scalar quantity, having magnitude but no direction.

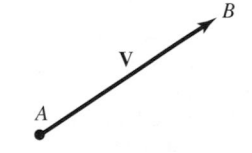

FIGURE 7–29   Representation of a vector.

A vector can be designated by a single letter, or by two letters representing the endpoints, with the starting point given first. Thus the vector in Fig. 7–29 can be labelled

$$\mathbf{V} \quad \text{or} \quad \mathbf{AB}$$

or, when handwritten,

$$\vec{V} \quad \text{or} \quad \vec{AB}$$

The *magnitude* of a vector can be designated with absolute value symbols or by ordinary (nonboldface-italic) type. Thus the magnitude of vector **V** in Fig. 7–29 is

$$|\mathbf{V}| \quad \text{or} \quad V \quad \text{or} \quad |\mathbf{AB}| \quad \text{or} \quad AB$$

Boldface letters are not practical in handwritten work. Instead, it is customary to place an *arrow* over vector quantities.

Polar Form and Rectangular Form

A vector is usually specified by giving its magnitude and direction. It is then said to be in *polar* form.

◆◆◆ **Example 22:**   A certain vector has a magnitude of 5 units and makes an angle of 38° with the positive *x* axis.

We often write such a vector in the form

$$5 \,\underline{/38°}$$

which is read "5 at an angle of 38°."   ◆◆◆

A vector drawn from the origin can also be designated by the *coordinates of its endpoint.* A vector written this way is said to be in *rectangular* form.

◆◆◆ **Example 23:**

(a) The *ordered pair* (3, 5) is a vector. It designates a line segment drawn from the origin to the point (3, 5).
(b) The *ordered triple* of numbers (4, 6, 3) is a vector in three dimensions. It designates a point drawn from the origin to the point (4, 6, 3).
(c) The *set of numbers* (7, 3, 6, 1, 5) is a vector in five dimensions. We cannot draw it, but it is handled mathematically in the same way as the others. ◆◆◆

We cover vectors such as these in Chapter 15.

Drawing a Vector on Coordinate Axes

When a vector is specified by its magnitude and direction, simply draw an arrow whose length is equal to its magnitude (using a suitable scale for the axes) in the given direction. The tail of the vector is placed at the origin, and the angle is measured, as usual, counterclockwise from the positive *x* direction.

◆◆◆ **Example 24:** Draw a vector whose magnitude is 5 and whose direction is 38°.

Solution: We measure 38° counterclockwise from the positive *x* axis and draw the vector from the origin, with length 5 units (Fig. 7–30). ◆◆◆

When a vector is given by the coordinates of its endpoint, simply plot the point and connect it to the origin.

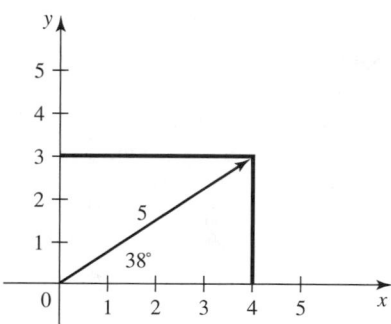

FIGURE 7-30

◆◆◆ **Example 25:** Draw the vector (3.2, 4.8).

Solution: We plot the point (3.2, 4.8) and connect it to the origin, as shown in Fig. 7–31. We draw the arrow at the given point, not at the origin. ◆◆◆

Rectangular Components of a Vector

Any vector can be replaced by two (or more) vectors which, acting together, exactly duplicate the effect of the original vector. They are called the *components* of the vector. The components are usually chosen perpendicular to each other and are then called *rectangular components.* To *resolve* a vector means to replace it by its components.

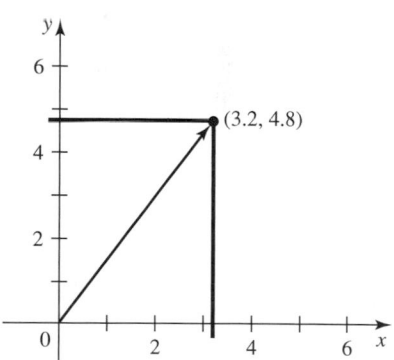

FIGURE 7-31

◆◆◆ **Example 26:** A ball thrown from one person to another moves simultaneously in both the horizontal direction and the vertical direction. Thus the ball has a horizontal and a vertical component of velocity. ◆◆◆

◆◆◆ **Example 27:** Figure 7–32(a) shows a vector **V** resolved into its *x* component $\mathbf{V}_x$ and its *y* component $\mathbf{V}_y$. ◆◆◆

◆◆◆ **Example 28:** Figure 7–32(b) shows a block on an inclined plane with its weight **W** resolved into a component **N** normal (perpendicular) to the plane and a component **T** tangential (parallel) to the plane. ◆◆◆

◆◆◆ **Example 29:** Figure 7–33 shows the vector **V**(2, 3) resolved into its *x* and *y* components, $\mathbf{V}_x$ and $\mathbf{V}_y$. It is clear that the *x* component is 2 units, and the *y* component is 3 units. Note that the coordinates of the endpoint are also the magnitudes of the components. ◆◆◆

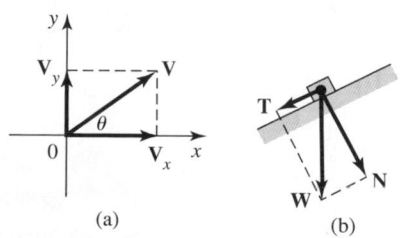

FIGURE 7-32 Rectangular components of a vector.

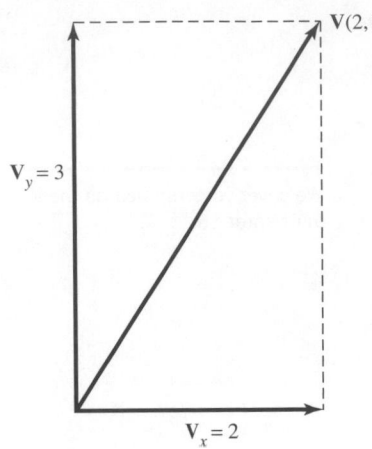

FIGURE 7–33

The component of a vector along an axis is also referred to as the *projection* of the vector onto that axis.

Thus, finding the rectangular components of a vector on coordinate axes is no different than converting the vector from polar to rectangular form.

Resolution of Vectors

We see in Fig. 7–32 that a vector and its rectangular components form a rectangle that has two *right triangles*. Thus we can resolve a vector into its rectangular components with the right-triangle trigonometry of this chapter. In Fig. 7–32(a),

$$\sin \theta = \frac{V_y}{V}$$

Note that we have used nonboldface type to represent the magnitudes V and V_y of vectors $\mathbf{V}$ and $\mathbf{V}_y$. From this we get the magnitude V_y of the vector $\mathbf{V}_y$.

$$V_y = V \sin \theta$$

Similarly,

$$V_x = V \cos \theta$$

So

$$\mathbf{V}_y = V_y \underline{/90°} \quad \text{and} \quad \mathbf{V}_x = V_x \underline{/0°}$$

◆◆◆ **Example 30:** The vector $\mathbf{V}$ in Fig. 7–32(a) has a magnitude of 248 units and makes an angle θ of 38.2° with the x axis. Find the x and y components.

Solution:

$$\sin 38.2° = \frac{V_y}{248}$$

$$V_y = 248 \sin 38.2° = 153$$

Similarly,

$$V_x = 248 \cos 38.2° = 195$$

Thus we can write

$$\mathbf{V} = 248 \underline{/38.3°} = (195, 153)$$

in rectangular form.   ◆◆◆

◆◆◆ **Example 31:** A cable exerts a force of 558 N at an angle of 47.2° with the horizontal [Fig. 7–34(a)]. Resolve this force into vertical and horizontal components.

Solution: We draw a vector diagram as shown in Fig. 7–34(b). Then

$$\sin 47.2° = \frac{V_y}{558}$$

$$V_y = 558 \sin 47.2° = 409 \text{ N}$$

Similarly,

$$V_x = 558 \cos 47.2° = 379 \text{ N}$$

So

$$\mathbf{V}_y = 409 \underline{/90°} \text{ and } \mathbf{V}_x = 379 \underline{/0°}$$ ◆◆◆

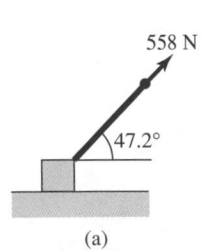
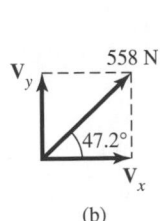

(a)   (b)

FIGURE 7–34

Resultant of Perpendicular Vectors

Just as any vector can be *resolved* into components, so can several vectors be *combined* into a single vector called the *resultant,* or *vector sum.* The process of combining vectors into a resultant is called *vector addition.*

When combining or adding two perpendicular vectors, we can use right-triangle trigonometry, just as we did for resolving a vector into rectangular components.

◆◆◆ **Example 32:** Find the resultant of two perpendicular vectors whose magnitudes are 485 and 627. Also find the angle that it makes with the 627-magnitude vector.

Solution: We draw a vector diagram as shown in Fig. 7–35. Then by the Pythagorean theorem,

$$\mathbf{R} = \sqrt{(485)^2 + (627)^2} = 793$$

and by Eq. 148,

$$\tan \theta = \frac{485}{627} = 0.774$$
$$\theta = 37.7°$$

◆◆◆

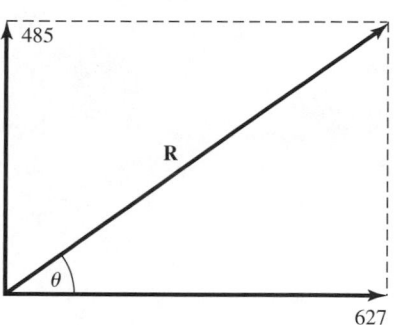

FIGURE 7–35 Resultant of two vectors.

◆◆◆ **Example 33:** If the components of vector **A** are

$$\mathbf{A}_x = 735 \quad \text{and} \quad \mathbf{A}_y = 593$$

find the magnitude of **A** and the angle θ that it makes with the x axis.

Solution: By the Pythagorean theorem,

$$A^2 = A_x^2 + A_y^2$$
$$= (735)^2 + (593)^2 = 891\ 900$$
$$A = 944$$

and

$$\tan \theta = \frac{A_y}{A_x}$$
$$= \frac{593}{735}$$
$$\theta = 38.9°$$

In other words,

$$\mathbf{A} = (735, 593) = 944\ \underline{/38.9°}$$

◆◆◆

Resultant of Non-perpendicular Vectors

We can use right-triangle trigonometry to find the resultant of any number of non-perpendicular vectors, such as those shown in Fig. 7–36(a) as follows:

1. Resolve each vector into its x and y components.
2. Combine the x components into a single vector $\mathbf{R}_x$ in the x direction, and combine the y components into a single vector $\mathbf{R}_y$ in the y direction.
3. Find the resultant $\mathbf{R}$ of $\mathbf{R}_x$ and $\mathbf{R}_y$.

◆◆◆ **Example 34:** Find the resultant of the three vectors in Fig. 7–36(a):

$$6.34\ \underline{/29.5°}, \quad 4.82\ \underline{/47.2°}, \quad \text{and} \quad 5.52\ \underline{/73.0°}$$

Solution: We first resolve each given vector into x and y components. We will arrange the computation in table form for convenience.

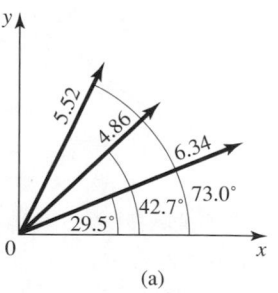

x components	y components
$6.34 \cos 29.5° = 5.52$	$6.34 \sin 29.5° = 3.12$
$4.82 \cos 47.2° = 3.27$	$4.82 \sin 47.2° = 3.54$
$5.52 \cos 73.0° = 1.61$	$5.52 \sin 73.0° = 5.28$
Sums: $\quad R_x = 10.4$	$R_y = 11.9$

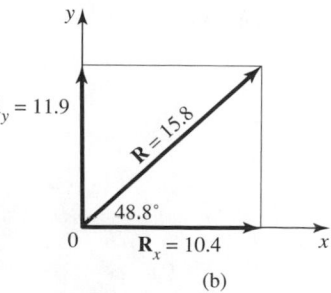

FIGURE 7–36

So

$$\mathbf{R}_x = 10.4 \ \underline{/0°} \qquad \text{and} \qquad \mathbf{R}_y = 11.9 \ \underline{/90°}$$

We now find the resultant of $\mathbf{R}_x$ and $\mathbf{R}_y$ [Fig. 7–36(b)]. The magnitude of the resultant is

$$R = \sqrt{R_x^2 + R_y^2} = \sqrt{(10.4)^2 + (11.9)^2} = 15.8$$

Now we find the angle.

$$\tan \theta = \frac{R_y}{R_x} = \frac{11.9}{10.4} = 1.14$$

Thus $\theta = 48.7°$

So the resultant $\mathbf{R}$ is 15.8 $\underline{/48.7°}$.   ◆◆◆

Exercise 6 ◆ Vectors

Drawing Vectors

Draw each vector on coordinate axes.

1. Magnitude = 3.84;
   angle = 72.4°
2. Magnitude = 9.46;
   angle = 28.4°
3. 36.2 $\underline{/44.8°}$
4. 1.63 $\underline{/29.4°}$
5. (2.84, 5.37)
6. (72.4, 58.3)

Resolution of Vectors

Given the magnitude of each vector and the angle θ that it makes with the x axis, find the x and y components.

7. Magnitude = 4.93 $\theta = 48.3°$
8. Magnitude = 835 $\theta = 25.8°$
9. Magnitude = 1.884 $\theta = 58.24°$
10. Magnitude = 362 $\theta = 13.8°$

11. At the Industrial Manufacturing Centre (IMC) of Durham College, a robotic arm can move in three directions, or axes (Fig. 7–37). One common way of labelling the three axes is x, y, and z. Show how the net distance moved can be expressed as

$$d = (x^2 + y^2 + z^2)^{1/2}$$

Resultant of Perpendicular Vectors

In the following problems, the magnitudes of perpendicular vectors A and B are given. Find the resultant and the angle that it makes with vector B.

12. $A = 483 \ B = 382$
13. $A = 2.85 \ B = 4.82$
14. $A = 7364 \ B = 4837$
15. $A = 46.8 \ B = 38.6$
16. $A = 1.25 \ B = 2.07$
17. $A = 274 \ B = 529$

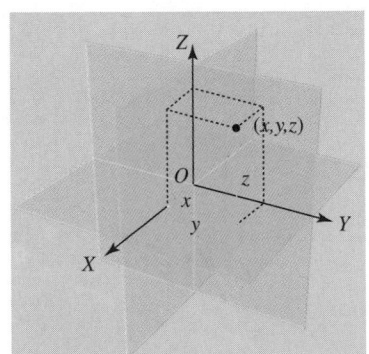

FIGURE 7–37 Three Dimensional Cartesian coordinate system

In the following problems, vector A has x and y components A_x and A_y. Find the magnitude of $\mathbf{A}$ and the angle that it makes with the x axis.

18. $A_x = 483 \ A_y = 382$
19. $A_x = 6.82 \ A_y = 4.83$
20. $A_x = 58.3 \ A_y = 37.2$
21. $A_x = 2.27 \ A_y = 3.97$

Resultant of Non-perpendicular Vectors

Find the resultant of each set of vectors.

22. 4.83 $\underline{/18.3°}$ and 5.99 $\underline{/83.5°}$
23. 13.5 $\underline{/29.3°}$ and 27.8 $\underline{/77.2°}$
24. 635 $\underline{/22.7°}$ and 485 $\underline{/48.8°}$
25. 83.2 $\underline{/49.7°}$ and 52.5 $\underline{/66.3°}$
26. 1.38 $\underline{/22.4°}$, 2.74 $\underline{/49.5°}$, and 3.32 $\underline{/77.3°}$
27. 736 $\underline{/15.8°}$, 586 $\underline{/69.2°}$, and 826 $\underline{/81.5°}$

7-7 Applications of Vectors

Any vector quantity such as force, velocity, or impedance can be combined or resolved by the methods in the preceding section.

Force Vectors

◆◆◆ **Example 35:** A cable running from the top of a telephone pole creates a horizontal pull of 875 N, as shown in Fig. 7–38. A support cable running to the ground is inclined 71.5° from the horizontal. Find the tension in the support cable.

Solution: We draw the forces acting at the top of the pole as shown in Fig. 7–39. We see that the horizontal component of the tension T in the support cable must equal the horizontal pull of 875 N. So

$$\cos 71.5° = \frac{875}{T}$$

$$T = \frac{875}{\cos 71.5°} = 2760 \text{ N}$$

◆◆◆

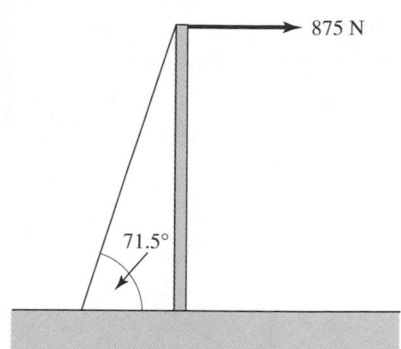

FIGURE 7–38

Velocity Vectors

◆◆◆ **Example 36:** A river flows at the rate of 4.70 km/h. A rower, who can travel 7.51 km/h in still water, heads directly across the current. Find the rate and the direction of travel of the boat.

Solution: The boat is crossing the current and at the same time is being carried downstream (Fig. 7–40). Thus the velocity V has one component of 7.51 km/h across the current and another component of 4.70 km/h downstream. We find the resultant of these two components by the Pythagorean theorem.

$$V^2 = (7.51)^2 + (4.70)^2$$

from which

$$V = 8.86 \text{ km/h}$$

Now finding the angle θ yields

$$\tan \theta = \frac{4.70}{7.51}$$

from which

$$\theta = 32.0°$$

◆◆◆

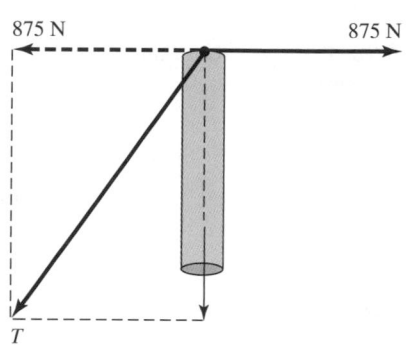

FIGURE 7–39

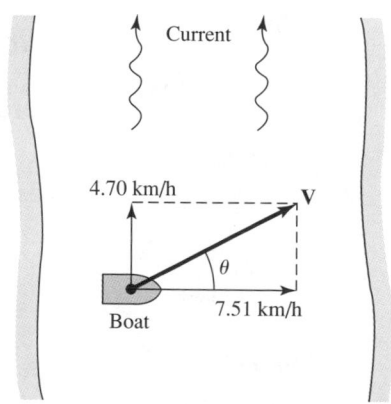

FIGURE 7–40

Impedance Vectors

Vectors find extensive application in electrical technology, and one of the most common applications is in the calculation of impedances. Figure 7–41 shows a resistor, an inductor, and a capacitor, connected in series with an ac source. The *reactance X* is a measure of how much the capacitance and inductance retard the flow of current in such a circuit. It is the difference between the *capacitive reactance X_C* and the *inductive reactance X_L*.

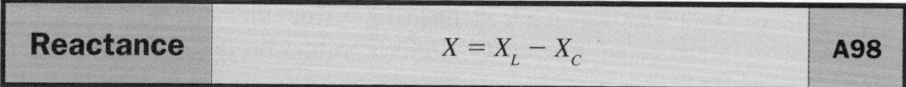

| Reactance | $X = X_L - X_C$ | A98 |

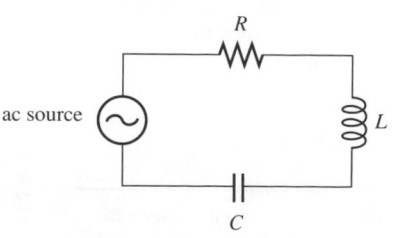

FIGURE 7–41

We certainly are not attempting to teach ac circuits in a few paragraphs, but hope only to reinforce concepts learned in your other courses.

The *impedance Z* is a measure of how much the flow of current in an ac circuit is retarded by all circuit elements, including the resistance. The magnitude of the impedance is related to the total resistance R and reactance X by the following formula:

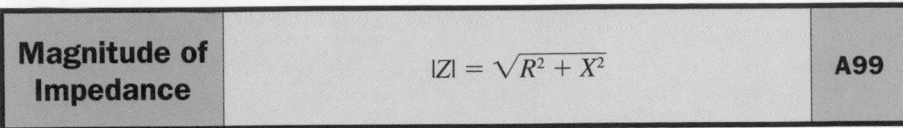

Magnitude of Impedance	$\|Z\| = \sqrt{R^2 + X^2}$	A99

The impedance, resistance, and reactance form the three sides of a right triangle, the *vector impedance diagram* (Fig. 7–42). The angle ϕ between Z and R is called the *phase angle*.

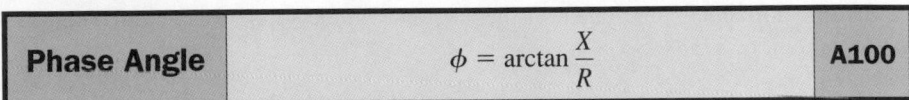

Phase Angle	$\phi = \arctan \dfrac{X}{R}$	A100

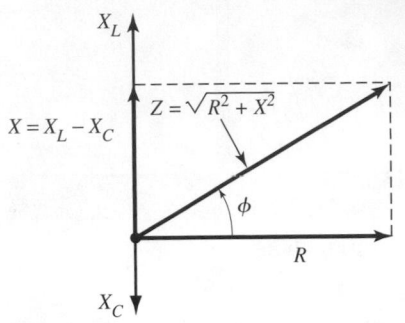

FIGURE 7–42   Vector impedance diagram.

◆◆◆ **Example 37:** The capacitive reactance of a certain circuit is 2720 Ω, the inductive reactance is 3260 Ω, and the resistance is 1150 Ω. Find the reactance and the magnitude of the impedance of the circuit; also find the phase angle.

Solution: By Eq. A98,

$$X = 3260 - 2720 = 540 \ \Omega$$

By Eq. A99,

$$|Z| = \sqrt{(1150)^2 + (540)^2} = 1270 \ \Omega$$

and by Eq. A100,

$$\phi = \arctan \frac{540}{1150} = 25.2°$$

◆◆◆

Exercise 7 ◆ Applications of Vectors

Force Vectors

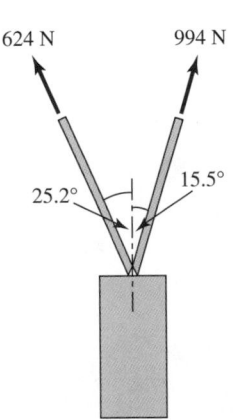

FIGURE 7–43

1. What force, neglecting friction, must be exerted to drag a 56.5 N weight up a slope inclined 12.6° from the horizontal?

2. What is the largest weight, in kilograms (1 kg = 9.807 N), that a tractor can drag up a slope inclined 21.2° from the horizontal if it is able to pull along the incline with a force of 27.5 kN? Neglect the force due to friction.

3. Two ropes hold a crate as shown in Fig. 7–43. One is pulling with a force of 994 N in a direction 15.5° with the vertical, and the other is pulling with a force of 624 N in a direction 25.2° with the vertical. How much does the crate weigh, in newtons?

4. A truck weighing 7280 kg moves up a bridge inclined 4.80° from the horizontal. Find the force of the truck normal (perpendicular) to the bridge.

Remember that when kg is used as a force or weight, 1 kg = 9.807 N.

5. A person has just enough strength to pull a 1270-N weight up a certain slope. Neglecting friction, find the angle at which the slope is inclined to the horizontal if the person is able to exert a pull of 551 N.

6. A person wishes to pull a 255-lb. weight up an incline to the top of a wall 14.5 ft. high. Neglecting friction, what is the length of the shortest incline (measured along the incline) that can be used if the person's pulling strength is 145 lb.?

You will need the equations of equilibrium, Eqs. A13 and A14, for some of these problems.

7. A truck weighing 18.6 t stands on a hill inclined 15.4° from the horizontal. How large a force (in tonnes) must be counteracted by brakes to prevent the truck from rolling downhill?

8. Two opposing hockey players are battling for the puck and both slap their sticks on the puck at the same time, but from slightly different angles. One player strikes the puck with a force of 8.5 N at an angle of 23° while the other hits the puck with a force of 11.3 N at an angle of 77°. Neglecting the friction of the ice surface, what direction does the puck travel in?

Velocity Vectors

9. A plane is headed due west with an air speed of 212 km/h (Fig. 7–44). It is driven from its course by a wind from due north blowing at 23.6 km/h. Find the ground speed of the plane and the actual direction of travel. Refer to Fig. 7–45 for definitions of flight terminology.

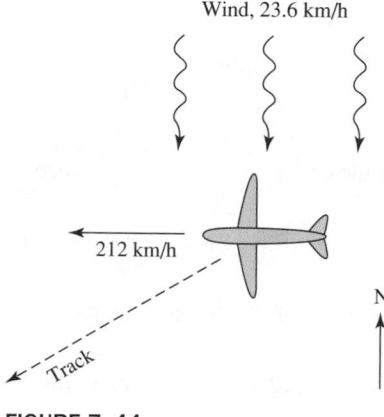

Wind, 23.6 km/h

212 km/h

N

Track

FIGURE 7–44

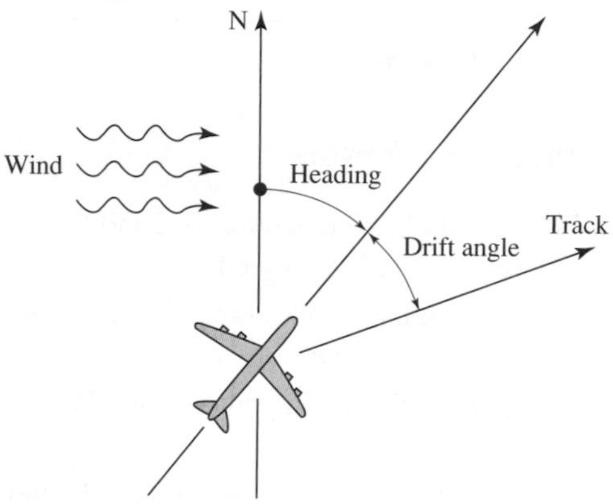

N

Wind

Heading

Drift angle

Track

N

FIGURE 7–45 Flight terminology. The *heading* of an aircraft is the direction in which the craft is *pointed*. Due to air current, it usually will not travel in that direction but in an actual path called the *track*. The angle between the heading and the track is the *drift angle*. The *air speed* is the speed relative to the surrounding air, and the *ground speed* is the craft's speed relative to the earth.

10. At what air speed should an airplane head due west in order to follow a course S 80°15′ W if a 20.5-km/h wind is blowing from due north?

11. A pilot heading his plane due north finds the actual direction of travel to be N 5°12′ E. The plane's air speed is 315 km/h, and the wind is from due west. Find the plane's ground speed and the velocity of the wind.

12. A certain escalator travels at a rate of 10.6 m/min, and its angle of inclination is 32.5°. What is the vertical component of the velocity? How long will it take a passenger to travel 10.0 m vertically?

13. A projectile is launched at an angle of 55.6° to the horizontal with a speed of 7550 m/min. Find the vertical and horizontal components of this velocity.

14. At what speed with respect to the water would a ship have to head due north in order to follow a course N 5°15′ E if a current is flowing due east at the rate of 10.6 km/h?

See Fig. 7–19 if you've forgotten how compass directions are written.

15. In Ontario, Peterborough is located 64 km and 47° north of east from Oshawa. Muskoka is 105 km and 42° north of west from Peterborough (Fig. 7–46). Determine the distance and direction required to travel from Oshawa to Muskoka.

16. On Okanagan Lake, a motorboat leaves a cottage dock and travels 1580 m due west, then turns 35.0° to the south and travels another 1640 m to another cottage dock. Find the distance and direction required to travel from the second dock to the first.

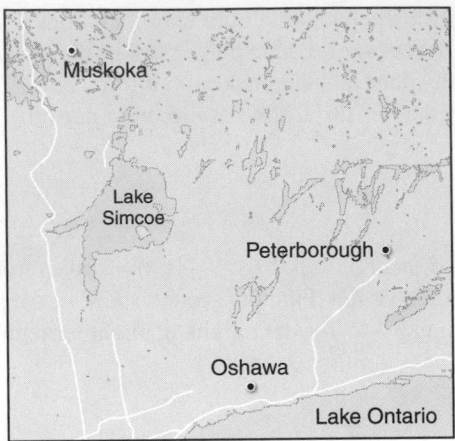

FIGURE 7–46

Impedance Vectors

17. A circuit has a reactance of 2650 Ω and a phase angle of 44.6°. Find the resistance and the magnitude of the impedance.

18. A circuit has a resistance of 115 Ω and a phase angle of 72.0°. Find the reactance and the magnitude of the impedance.

19. A circuit has an impedance of 975 Ω and a phase angle of 28.0°. Find the resistance and the reactance.

20. A circuit has a reactance of 5.75 Ω and a resistance of 4.22 Ω. Find the magnitude of the impedance and the phase angle.

21. A circuit has a capacitive reactance of 1776 Ω, an inductive reactance of 5140 Ω, and a total impedance of 5560 Ω. Find the resistance and the phase angle.

Case Study Discussion – Bridge Building

There are 10 right-angled triangles. The vertical side of each triangle has a height of 4 m. The base of each triangle is 60 m/10 = 6 m. The diagonal member length becomes the hypotenuse, so using the Pythagorean theorem, we take the square root of $(4^2 + 6^2)$ $=\sqrt{52} = 7.21$ m.

Bridge span vertical members: 11 × 4 m = 44 m
Bridge span horizontal members: 20 × 6 m = 120 m
Bridge span diagonal members: 10 × 7.21 m = 72.1 m

Base members at both ends = 2 × 8 m = 16 m
Diagonal members at both ends = 2 × $\sqrt{4^2 + 8^2}$ = 2 × 8.94 m = 17.9 m

Total diagonal = 72.1 + 17.9 = 90 m
Total horizontal/vertical = 44 + 120 + 16 = 180 m

◆◆◆ **CHAPTER 7 REVIEW PROBLEMS** ◆◆◆◆◆◆◆◆◆◆◆◆◆◆◆◆◆◆◆◆◆◆◆◆◆◆◆◆◆◆◆◆◆

Perform the angle conversions, and fill in the missing quantities.

	Decimal Degrees	Degrees–Minutes–Seconds	Radians	Revolutions
1.	38.2			
2.		25°28′45″		
3.			2.745	
4.				0.275

The following points are on the terminal side of an angle θ. Write the six trigonometric functions of the angle, and find the angle in decimal degrees, to three significant digits.

5. (3, 7)

6. (2, 5)

7. (4, 3)

8. (2.3, 3.1)

Write the six trigonometric functions of each angle. Keep four decimal places.

9. 72.9°

10. 35°13′33″

11. 1.05 rad

Find acute angle θ in decimal degrees.

12. $\sin \theta = 0.574$ **13.** $\cos \theta = 0.824$ **14.** $\tan \theta = 1.345$

Evaluate each expression. Give your answer as an acute angle in degrees.

15. arctan 2.86 **16.** $\cos^{-1} 0.385$ **17.** arcsec 2.447

Solve right triangle ABC.

18. $a = 746$ and $A = 37.2°$ **19.** $b = 3.72$ and $A = 28.5°$ **20.** $c = 45.9$ and $A = 61.4°$

21. Find the x and y components of a vector that has a magnitude of 885 and makes an angle of 66.3° with the x axis.

22. Find the magnitude of the resultant of two perpendicular vectors that have magnitudes of 54.8 and 39.4, and find the angle the resultant makes with the 54.8 vector.

23. A vector has x and y components of 385 and 275. Find the magnitude and the direction of that vector.

24. A telephone pole casts a shadow 13.5 m long when the angle of elevation of the sun is 15.4°. Find the height of the pole.

25. From a point 46.5 m in front of a church, the angles of elevation of the top and base of its steeple are 22.5° and 19.6°, respectively. Find the height of the steeple.

26. A circuit has a resistance of 125 Ω, an impedance of 256 Ω, an inductive reactance of 312 Ω, and a positive phase angle. Find the capacitive reactance and the phase angle.

Evaluate to four decimal places.

27. $\cos 59.2°$ **28.** $\csc 19.3°$ **29.** $\tan 52.8°$

30. $\sec 37.2°$ **31.** $\cot 24.7°$ **32.** $\sin 42.9°$

Find, to the nearest tenth of a degree, the acute angle whose trigonometric function is given.

33. $\tan \theta = 1.7362$ **34.** $\sec \theta = 2.9914$ **35.** $\sin \theta = 0.7253$

36. $\csc \theta = 3.2746$ **37.** $\cot \theta = 0.9475$ **38.** $\cos \theta = 0.3645$

Evaluate each trigonometric expression to three significant digits.

39. 8.23 cos 15.8° + 8.73

40. 5.82 − 5.89 sin 16.2°

41. 9.26(cos 88.3° + sin 22.3°)

42. 52.2 sin 3 + 45.3 cos 2

Draw each vector on coordinate axes.

43. Magnitude = 726; angle = 32.4°

44. 12.6 $\angle$ 74.8°

45. (7.84, 8.37)

Find the resultant of each set of vectors.

46. 633 $\angle$ 68.3° and 483 $\angle$ 23.5°

47. 13.2 $\angle$ 22.7° and 22.6 $\angle$ 76.8°

48. 8.28 $\angle$ 82.3°, 9.14 $\angle$ 15.5°, and 7.72 $\angle$ 47.9°

49. Point *C* is due east of point *A* across a lake, as shown in Fig. 7–47. To measure the distance *AC*, a person walks 139.3 m due south to point *B*, then walks 158.2 m to *C*. Find the distance *AC* and the direction the person was walking when going from *B* to *C*.

Writing

50. Write a short paragraph, with examples, on how you might possibly use an idea from this chapter in a real-life situation: on the job, in a hobby, or around the house.

Team Project

51. Make a device for measuring angles in the horizontal plane, using a sighting tube of some sort and a protractor. Use this "transit" and a rope of known length to find some distance on your campus, such as the distance across a ball field. Compare your results with the actual taped distance. Give a prize to the team that gets closest to the taped value.

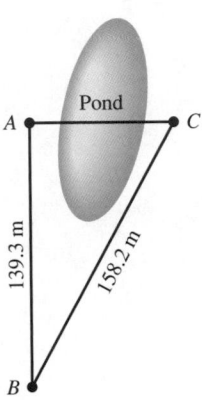

FIGURE 7–47

Factors and Factoring

When you have completed this chapter, you should be able to:
- Factor expressions by removing common factors.
- Factor binomials that are the difference of two squares.
- Factor the sum or difference of two cubes.
- Factor trinomials.
- Factor the perfect square trinomial.
- Factor expressions by grouping.
- Use factoring techniques in applications.

Multiplication combines two or more factors into a product. *Factoring* is simply the reverse process: breaking up a product into its constituent factors. When we take an expression (or number) and break it up into its factors, we are changing the form of the expression without changing its value. This change of form opens up possibilities: we see the expression in a different light, and can often find new ways of working with it or solving it. Technical mathematics is all about problem solving, and as a problem solver or troubleshooter, you need to be able to look at things from many directions. Factoring gives us this ability.

The idea of a factor is not entirely new to us. You might glance back at Secs. 1-3 and 2-1 before starting this chapter.

COUNTING ON A CALCULATOR—JOHN HORTON

MATH AROUND US

Certified General Accountant John Horton has been owner and president of Personal Tax Services since graduating from Fleming College in 1981. His company does both personal and business accounting, auditing, payroll services, and bookkeeping.

As an accountant, math and numbers are a natural part of his day-to-day work. Although the majority of John's work can be done with computer software, he has found it beneficial to keep up with his math skills so that he can double check calculations. He advises, "Keep your math skills sharp and use a calculator to keep any mistakes to a minimum. It will save you time and money in the end. In the days that I went to school, there weren't any calculators and I sure do appreciate them now. I still use one that prints out a tape to recheck figures."

8–1   Common Factors

Factors of an Expression

The *factors of an expression* are those quantities whose product is the original expression.

◆◆◆ **Example 1:** The factors of $x^2 - 9$ are $x + 3$ and $x - 3$, because

$$(x + 3)(x - 3) = x^2 + 3x - 3x - 9$$
$$= x^2 - 9$$

◆◆◆

Prime Factors

Many expressions have no factors other than 1 and themselves. They are called *prime expressions*.

◆◆◆ **Example 2:** Two factors of 30 are 5 and 6. The 5 is prime because it cannot be factored. However, the 6 is not prime because it can be factored into 2 and 3. Thus the prime factors of 30 are 2, 3, and 5.

◆◆◆

◆◆◆ **Example 3:** The expressions x, $x - 3$, and $x^2 - 3x + 7$ are all prime. They cannot be factored further.

◆◆◆

◆◆◆ **Example 4:** The expressions $x^2 - 9$ and $x^2 + 5x + 6$ are not prime because they can be factored. They are said to be *factorable*.

◆◆◆

Factoring

Factoring is the process of finding the factors of an expression. It is *the reverse of finding the product* of two or more quantities.

Multiplication (Finding the Product)	$x(x + 4) = x^2 + 4x$
Factoring (Finding the Factors)	$x^2 + 4x = x(x + 4)$

We factor an expression by *recognizing the form* of that expression and then applying standard rules for factoring. In the first type of factoring we will cover, we look for *common factors*.

Common Factors

If each term of an expression contains the same quantity (called the *common factor*), that quantity may be *factored out*.

This is simply the *distributive law*, which we studied earlier.

Common Factor	$ab + ac = a(b + c)$	**5**

◆◆◆ **Example 5:** In the expression

$$2x + x^2$$

each term contains an x as a common factor. So we write

$$2x + x^2 = x(2 + x)$$

Most of the factoring we will do will be of this type.

◆◆◆

◆◆◆ Example 6:

(a) $3x^3 - 2x + 5x^4 = x(3x^2 - 2 + 5x^3)$
(b) $3xy^2 - 9x^3y + 6x^2y^2 = 3xy(y - 3x^2 + 2xy)$
(c) $3x^3 - 6x^2y + 9x^4y^2 = 3x^2(x - 2y + 3x^2y^2)$
(d) $2x^2 + x = x(2x + 1)$ ◆◆◆

When factoring, don't confuse superscripts (powers) with subscripts.

◆◆◆ Example 7: The expression $x^3 + x^2$ is factorable.

$$x^3 + x^2 = x^2(x + 1)$$

But the expression

$$x_3 + x_2$$

is not factorable. ◆◆◆

Common Error	Students are sometimes puzzled over the "1" in Example 7. Why should it be there? After all, when you remove a chair from a room, it is *gone;* there is nothing (zero) remaining where the chair used to be. If you remove an x by factoring, you might assume that nothing (zero) remains where the x used to be. $$2x^2 + x = x(2x + 0)?$$ Prove to yourself that this is not correct by multiplying the factors to see if you get back the original expression.

Factors in the Denominator

Common factors may appear in the denominators of the terms as well as in the numerators.

◆◆◆ Example 8:

(a) $\dfrac{1}{x} + \dfrac{2}{x^2} = \dfrac{1}{x}\left(1 + \dfrac{2}{x}\right)$

(b) $\dfrac{x}{y^2} + \dfrac{x^2}{y} + \dfrac{2x}{3y} = \dfrac{x}{y}\left(\dfrac{1}{y} + x + \dfrac{2}{3}\right)$ ◆◆◆

Checking

To check if factoring has been done correctly, simply multiply your factors together and see if you get back the original expression.

◆◆◆ Example 9: Are $2xy$ and $x + 3 - y^2$ the factors of

$$2x^2y + 6xy - 2xy^3?$$

Solution: Multiplying the factors, we obtain

$$2xy(x + 3 - y^2) = 2x^2y + 6xy - 2xy^3 \quad \text{(checks)}$$

◆◆◆ This check will tell us whether we have factored *correctly* but, of course, not whether we have factored *completely* (that is, found the prime factors).

CASE STUDY – CHECKING YOUR FACTORING

Here is an interesting look at mathematics: Although it may seem a little strange to have $x = y$, there is actually nothing wrong with it. Sometimes you will see an x changed to a y or a y changed to an x. Normally, you can't do that unless $x = y$, as is the case here. Look over the steps below and notice the factoring. Use the knowledge you gain in this chapter to check whether the factoring is done correctly. If everything is correct, then two is the same as one and all of our mathematics developed over the last several thousand years is wrong. Can you determine what is wrong in this chain of math?

$$x^2 - y^2 = x^2 - xy$$
$$(x + y)(x - y) = x(x - y)$$
$$\frac{(x + y)\cancel{(x - y)}}{\cancel{(x - y)}} = \frac{x\cancel{(x - y)}}{\cancel{(x - y)}}$$
$$x + y = x$$
$$\text{given } x = y$$
$$x + x = x$$
$$2x = x$$
$$2 = 1$$

Exercise 1 ◆ Common Factors

Factor each expression and check your results.

1. $3y^2 + y^3$
2. $6x - 3y$
3. $x^5 - 2x^4 + 3x^3$
4. $9y - 27xy$
5. $3a + a^2 - 3a^3$
6. $8xy^3 - 6x^2y^2 + 2x^3y$
7. $5(x + y) + 15(x + y)^2$
8. $\dfrac{a}{3} - \dfrac{a^2}{4} + \dfrac{a^3}{5}$
9. $\dfrac{3}{x} + \dfrac{2}{x^2} - \dfrac{5}{x^3}$
10. $\dfrac{3ab^2}{y^3} - \dfrac{6a^2b}{y^2} + \dfrac{12ab}{y}$
11. $\dfrac{5m}{2n} + \dfrac{15m^2}{4n^2} - \dfrac{25m^3}{8n}$
12. $\dfrac{16y^2}{9x^2} - \dfrac{8y^3}{3x^3} + \dfrac{8y^4}{3x}$
13. $5a^2b + 6a^2c$
14. $a^2c + b^2c + c^2d$
15. $4x^2y + cxy^2 + 3xy^3$
16. $4abx + 6a^2x^2 + 8ax$
17. $3a^3y - 6a^2y^2 + 9ay^3$
18. $2a^2c - 2a^2c^2 + 3ac$
19. $5acd - 2c^2d^2 + bcd$
20. $4b^2c^2 - 12abc - 9c^2$
21. $8x^2y^2 + 12x^2z^2$

22. $6xyz + 12x^2y^2z$

23. $3a^2b + abc - abd$

24. $5a^3x^2 - 5a^2x^3 + 10a^2x^2z$

Applications

25. When a bar of length L_0 is changed in temperature by an amount t, its new length L will be $L = L_0 + L_0\alpha t$, where α is the coefficient of thermal expansion. Factor the right side of this equation.

26. A sum of money a when invested for t years at an interest rate n will accumulate to an amount y, where $y = a + ant$. Factor the right side of this equation.

27. When a resistance R_1 is heated from a temperature t_1 to a new temperature t, it will increase in resistance by an amount $\alpha(t - t_1)R_1$, where α is the temperature coefficient of resistance. The final resistance will then be $R = R_1 + \alpha(t - t_1)R_1$. Factor the right side of this equation.

Compare your result for problem 27 with Eq. A70.

28. An item costing P dollars is reduced in price by 15%. The resulting price C is then $C = P - 0.15P$. Factor the right side of this equation.

29. The displacement of a uniformly accelerated body is given by

$$s = v_0 t + \frac{a}{2}t^2$$

Factor the right side of this equation.

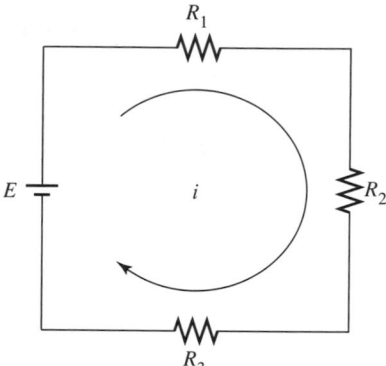

30. The sum of the voltage drops across the resistors in Fig. 8–1 must equal the battery voltage E.

$$E = iR_1 + iR_2 + iR_3$$

Factor the right side of this equation.

31. The mass of a spherical shell having an outside radius of r_2 and an inside radius r_1 is   **FIGURE 8–1**

$$\text{mass} = \frac{4}{3}\pi r_2^3 D - \frac{4}{3}\pi r_1^3 D$$

where D is the mass density of the material. Factor the right side of this equation.

8–2 Difference of Two Squares

Form

As we saw in Sec. 2–4, an expression of the form

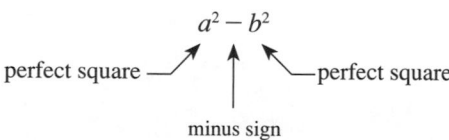

where one perfect square is subtracted from another, is called a *difference of two squares*. It arises when $(a - b)$ and $(a + b)$ are multiplied together.

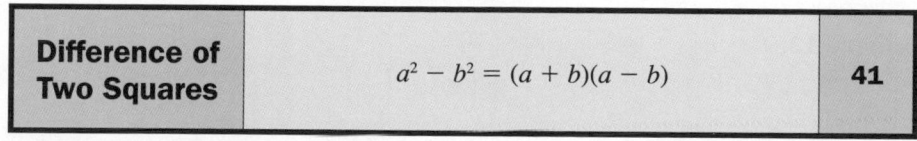

Difference of Two Squares	$a^2 - b^2 = (a + b)(a - b)$	41

Factoring the Difference of Two Squares

Once we recognize its form, the difference of two squares is easily factored.

◆◆◆ **Example 10:**

$$4x^2 - 9 = (2x)^2 - (3)^2$$

$$(2x)^2 - (3)^2 = (2x\ )(2x\ )$$

square root of the first term

$$(2x)^2 - (3)^2 = (2x\ 3)(2x\ 3)$$

square root of the
last term

$$4x^2 - 9 = (2x + 3)(2x - 3)$$

opposite
signs

◆◆◆

◆◆◆ **Example 11:**

(a) $y^2 - 1 = (y + 1)(y - 1)$
(b) $9a^2 - 16b^2 = (3a + 4b)(3a - 4b)$
(c) $49x^2 - 9a^2y^2 = (7x - 3ay)(7x + 3ay)$
(d) $1 - a^2b^2c^2 = (1 - abc)(1 + abc)$

◆◆◆

Common Error	There is no similar rule for factoring the *sum* of two squares, such as $$a^2 + b^2$$

The difference of any two quantities that have *even powers* can be factored as the difference of two squares. *Express each quantity as a square* by means of the following equation:

Powers	$$(x^a)^b = x^{ab} = (x^b)^a$$	**31**

◆◆◆ **Example 12:**

(a) $a^4 - b^6 = (a^2)^2 - (b^3)^2$
     $\quad\quad = (a^2 + b^3)(a^2 - b^3)$
(b) $x^{2a} - y^{6b} = (x^a)^2 - (y^{3b})^2$
     $\quad\quad\ = (x^a + y^{3b})(x^a - y^{3b})$
(c) $4^p - 9^p = 2^{2p} - 3^{2p}$
     $\quad\quad = (2^p)^2 - (3^p)^2$
     $\quad\quad = (2^p + 9^p)(2^p - 9^p)$
(d) $4^2 - 9^2 = (4 + 9)(4 - 9)$
     $\quad\quad = (4 + 9)(2 + 3)(2 - 3)$

◆◆◆

Sometimes one or both terms in the given expression may be a fraction.

◆◆◆ **Example 13:**

$$\frac{1}{x^2} - \frac{1}{y^2} = \left(\frac{1}{x}\right)^2 - \left(\frac{1}{y}\right)^2$$

$$= \left(\frac{1}{x} + \frac{1}{y}\right)\left(\frac{1}{x} - \frac{1}{y}\right)$$

◆◆◆

Sometimes one (or both) of the terms in an expression is itself a binomial.

◆◆◆ Example 14:

$$(a + b)^2 - x^2 = [(a + b) + x][(a + b) - x]$$ ◆◆◆

Factoring Completely

After factoring an expression, see if any of the factors themselves can be factored *again*.

◆◆◆ Example 15: Factor $a - ab^2$.

Solution: Always remove any common factors first. Factoring, we obtain

$$a - ab^2 = a(1 - b^2)$$

Now factoring the difference of two squares, we get

$$a - ab^2 = a(1 + b)(1 - b)$$ ◆◆◆

◆◆◆ Example 16:

$$x^4 - y^4 = (x^2 + y^2)(x^2 - y^2)$$

Now factoring the difference of two squares again, we obtain

$$x^4 - y^4 = (x^2 + y^2)(x + y)(x - y)$$ ◆◆◆

Exercise 2 ◆ Difference of Two Squares

Factor completely.

1. $4 - x^2$
2. $x^2 - 9$
3. $9a^2 - x^2$
4. $25 - x^2$
5. $4x^2 - 4y^2$
6. $9x^2 - y^2$
7. $x^2 - 9y^2$
8. $16x^2 - 16y^2$
9. $9c^2 - 16d^2$
10. $25a^2 - 9b^2$
11. $9y^2 - 1$
12. $4x^2 - 9y^2$

Expressions with Higher Powers or Literal Powers

Factor completely.

13. $m^4 - n^4$
14. $a^8 - b^8$
15. $m^{2n} - n^{2m}$
16. $9^{2n} - 4b^{4n}$
17. $a^{16} - b^8$
18. $9a^2b^2 - 4c^4$

19. $25x^4 - 16y^6$

20. $36y^2 - 49z^6$

21. $16a^4 - 121$

22. $121a^4 - 16$

23. $25a^4b^4 - 9$

24. $121a^2 - 36b^4$

Applications

25. The thrust washer in Fig. 8–2 has a surface area of

$$\text{area} = \pi r_2^{\ 2} - \pi r_1^{\ 2}$$

Factor this expression.

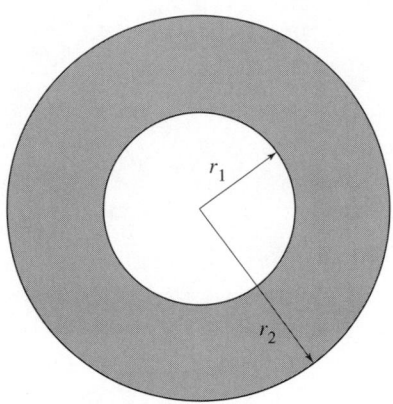

26. A flywheel of diameter d_1 (Fig. 8–3) has a balancing hole of diameter d_2 drilled through it. The mass M of the flywheel is then

$$M = \frac{\pi d_1^{\ 2}}{4}Dt - \frac{\pi d_2^{\ 2}}{4}Dt$$

Factor this expression.

27. A spherical balloon shrinks from radius r_1 to radius r_2. The change in surface area is, from Eq. 129, $4\pi r_1^2 - 4\pi r_2^2$. Factor this expression.

28. When an object is released from rest, the distance fallen between time t_1 and time t_2 is $\frac{1}{2}gt_2^2 - \frac{1}{2}gt_1^2$ where g is the acceleration due to gravity. Factor this expression.

FIGURE 8–2 Thrust washer.

29. When a body of mass m slows down from velocity v_1 to velocity v_2, the decrease in kinetic energy is $\frac{1}{2}mv_1^2 - \frac{1}{2}mv_2^2$. Factor this expression.

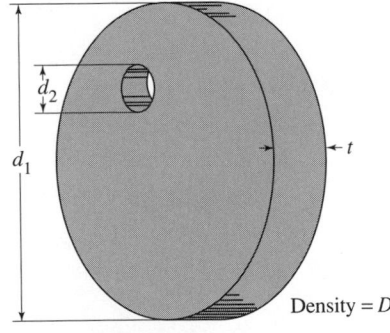

30. The work required to stretch a spring from a length x_1 (measured from the unstretched position) to a new length x_2 is $\frac{1}{2}kx_2^2 - \frac{1}{2}kx_1^2$. Factor this expression.

31. The formula for the volume of a cylinder of radius r and height h is $V = \pi r^2 h$. Write an expression for the volume of a hollow cylinder that has an inside radius r, an outside radius R, and a height h. Then factor the expression completely.

FIGURE 8–3 Flywheel.

32. A body at temperature T will radiate an amount of heat kT^4 to its surroundings and will absorb from the surroundings an amount of heat kT_s^4, where T_s is the temperature of the surroundings. Write an expression for the net heat transfer by radiation (amount radiated minus amount absorbed), and factor this expression completely.

████   **8–3   Factoring Trinomials**

Trinomials

A *trinomial*, you recall, is a polynomial having *three terms*.

◆◆◆ **Example 17:** $2x^3 - 3x + 4$ is a trinomial.   ◆◆◆

A *quadratic trinomial* in x has an x^2 term, an x term, and a constant term.

◆◆◆ **Example 18:** $4x^2 + 3x - 5$ is a quadratic trinomial. ◆◆◆

In this section we factor only quadratic trinomials.

Test for Factorability

Not all quadratic trinomials can be factored. We test for factorability as follows:

Test for Factorability	The trinomial $ax^2 + bx + c$ (where a, b, and c are constants) is factorable if $b^2 - 4ac$ is a perfect square	44

$b^2 - 4ac$ is called the *discriminant* (Eq. 101). We will use it in Chapter 14 to predict the nature of the roots of a quadratic equation.

◆◆◆ **Example 19:** Can the trinomial $6x^2 + x - 12$ be factored?

Solution: Using the test for factorability, Eq. 44, with $a = 6$, $b = 1$, and $c = -12$, we have

$$b^2 - 4ac = 1^2 - 4(6)(-12) = 289$$

By taking the square root of 289 on the calculator, we see that it is a perfect square ($289 = 17^2$), so the given trinomial is factorable. We will see later that its factors are $(2x + 3)$ and $(3x - 4)$. ◆◆◆

The coefficient of the x^2 term in a quadratic trinomial is called the *leading coefficient*. (It is the constant a in Eq. 44). We will first factor the easier type of trinomial, in which the leading coefficient is 1.

Trinomials with a Leading Coefficient of 1

We get such a trinomial when we multiply two binomials if the coefficients of the x terms in the binomials are also 1.

$$(x + a)(x + b) = x^2 + (a + b)x + ab$$

Note that the coefficient of the middle term of the trinomial equals the *sum* of a and b, and that the last term equals the *product* of a and b. Knowing this, we can reverse the process and find the factors when the trinomial is given.

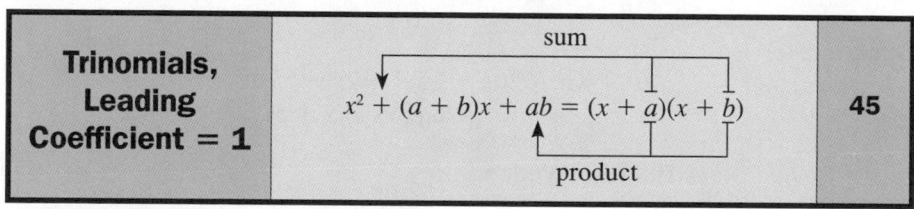

Trinomials, Leading Coefficient = 1	sum $x^2 + (a + b)x + ab = (x + a)(x + b)$ product	45

◆◆◆ **Example 20:** Factor the trinomial $x^2 + 8x + 15$.

Solution: From Eq. 45, this trinomial, if factorable, will factor as $(x + a)(x + b)$, where a and b have a sum of 8 and a product of 15. The integers 5 and 3 have a sum of 8 and a product of 15. Thus

$$x^2 + 8x + 15 = (x + 5)(x + 3)$$ ◆◆◆

Using the Signs to Aid Factoring

The signs of the terms of the trinomial can tell you the signs of the factors.

1. If the sign of the last term is *positive,* both signs in the factors will be the same (both positive or both negative). The sign of the middle term of the trinomial will tell whether both signs in the factors are positive or negative.

2. If the sign of the last term of the trinomial is *negative,* the signs of *a* and *b* in the factors will differ (one positive, one negative). The sign of the middle term of the trinomial will tell which quantity, *a* or *b*, is larger (in absolute value), the positive one or the negative one.

◆◆◆ **Example 21:**

positive, so signs in the factors are the same

(a) $x^2 + 6x + 8 = (x + 4)(x + 2)$

positive, so signs in the factors are positive

positive, so signs in the factors are the same

(b) $x^2 - 6x + 8 = (x - 4)(x - 2)$

negative, so signs in the factors are negative   ◆◆◆

◆◆◆ **Example 22:**

negative, so signs in the factors differ; one positive and one negative

$x^2 - 2x - 8 = (x - 4)(x + 2)$

negative, so the larger number (4) has a negative sign   ◆◆◆

◆◆◆ **Example 23:** Factor $x^2 + 2x - 15$.

Solution: We first look at the sign of the last term (−15). The negative sign tells us that the signs of the factors will differ.

$$x^2 + 2x - 15 = (x-\quad)(x+\quad)$$

We then find two numbers whose product is -15 and whose sum is 2. These two numbers must have opposite signs in order to have a negative product. Also, since the middle coefficient is $+2$, the positive number must be 2 greater than the negative number. Two numbers that meet these conditions are $+5$ and -3. So

$$x^2 + 2x - 15 = (x + 5)(x - 3)$$ ◆◆◆

Remember to remove any *common factors* before factoring a trinomial.

◆◆◆ **Example 24:**

$$2x^3 + 4x^2 - 30x = 2x(x^2 + 2x - 15)$$
$$= 2x(x - 3)(x + 5)$$ ◆◆◆

Trinomials Having More Than One Variable

As with the difference of two squares, we can factor many trinomials that at first glance do not seem to be in the proper form. Examples 25 and 26 are some common types.

◆◆◆ **Example 25:** Factor $x^2 + 5xy + 6y^2$.

Solution: One way to factor such a trinomial is to first temporarily drop the second variable (*y* in this example);

$$x^2 + 5x + 6$$

then factor the resulting trinomial.

$$(x + 2)(x + 3)$$

Finally, note that if we tack *y* onto the second term of each factor, this will cause *y* to appear in the middle term of the trinomial and y^2 to appear in the last term of the trinomial. The required factors are then

$$(x + 2y)(x + 3y)$$ ◆◆◆

◆◆◆ **Example 26:** Factor $w^2x^2 + 3wxy - 4y^2$.

Solution: Temporarily dropping w and y yields

$$x^2 + 3x - 4$$

which factors into

$$(x - 1)(x + 4)$$

We then note that if a w were appended to each x term and a y to the second term in each binomial,

$$(wx - y)(wx + 4y)$$

the product of these factors will give the original expression. ◆◆◆

Trinomials with Higher Powers

Trinomials with powers *greater than 2* may be factored if one exponent can be expressed as a *multiple of 2*, and if that exponent is *twice the other*.

◆◆◆ **Example 27:**

$$x^6 - x^3 - 6 = (x^3)^2 - (x^3) - 6$$
$$= (x^3 - 3)(x^3 + 2) \qquad ◆◆◆$$

Trinomials with Variables in the Exponents

If one exponent is *twice the other*, you may be able to factor such a trinomial.

◆◆◆ **Example 28:**

$$x^{2n} + 5x^n + 6 = (x^n)^2 + 5(x^n) + 6$$
$$= (x^n + 3)(x^n + 2) \qquad ◆◆◆$$

Substitution.

You may prefer to handle some expressions by making a *substitution,* as in the following examples.

This technique will be valuable when we solve *equations of quadratic type* in Chapter 14.

◆◆◆ **Example 29:** Factor $x^{6n} + 2x^{3n} - 3$.

Solution: We see that one exponent is twice the other. Let us substitute a new letter for x^{3n}. Let

$$w = x^{3n}$$

Then

$$x^{6n} = (x^{3n})^2 = w^2$$

and our expression becomes

$$w^2 + 2w - 3$$

which factors into

$$(w + 3)(w - 1)$$

or, substituting back, we have

$$(x^{3n} + 3)(x^{3n} - 1) \qquad ◆◆◆$$

◆◆◆ **Example 30:** Factor $(x + y)^2 + 7(x + y) - 30$.

Solution: We might be tempted to expand this expression, but notice that if we make the substitution

$$w = x + y$$

we get

$$w^2 + 7w - 30$$

which factors into $(w + 10)(w - 3)$. Substituting back, we get the factors

$$(x + y + 10)(x + y - 3)$$ ◆◆◆

Exercise 3 ◆ Factoring Trinomials

Test for Factorability

Test each trinomial for factorability.

1. $3x^2 - 5x - 9$
2. $3a^2 + 7a + 2$
3. $2x^2 - 12x + 18$
4. $5z^2 - 2z - 1$
5. $x^2 - 30x - 64$
6. $y^2 + 2y - 7$

Trinomials with a Leading Coefficient of 1

Factor completely.

7. $x^2 - 10x + 21$
8. $x^2 - 15x + 56$
9. $x^2 - 10x + 9$
10. $x^2 + 13x + 30$
11. $x^2 + 7x - 30$
12. $x^2 + 3x + 2$
13. $x^2 + 7x + 12$
14. $c^2 + 9c + 18$
15. $x^2 - 4x - 21$
16. $x^2 - x - 56$
17. $x^2 + 6x + 8$
18. $x^2 + 12x + 32$
19. $b^2 - 8b + 15$
20. $b^2 + b - 12$
21. $b^2 - b - 12$
22. $3b^2 - 30b + 63$
23. $2y^2 - 26y + 60$
24. $4z^2 - 16z - 84$
25. $3w^2 + 36w + 96$

Trinomials Having More Than One Variable

Factor completely.

26. $x^2 - 13xy + 36y^2$
27. $x^2 + 19xy + 84y^2$
28. $a^2 - 9ab + 20b^2$
29. $a^2 + ab - 6b^2$
30. $a^2 + 3ab - 4b^2$
31. $a^2 + 2ax - 63x^2$
32. $x^2y^2 - 19xyz + 48z^2$
33. $a^2 - 20abc - 96b^2c^2$
34. $x^2 - 10xyz - 96y^2z^2$
35. $a^2 + 49abc + 48b^2c^2$

Trinomials with Higher Powers or with Variables in the Exponents

Factor completely.

36. $a^{6p} + 13a^{3p} + 12$
37. $(a + b)^2 - 7(a + b) - 8$

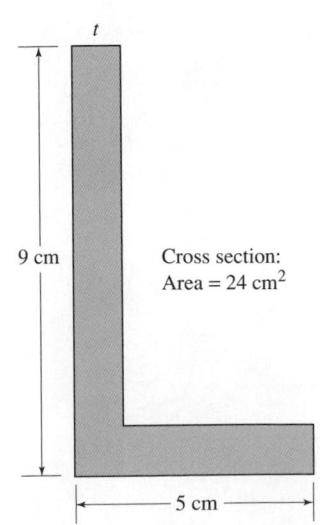

9 cm

Cross section:
Area = 24 cm²

5 cm

FIGURE 8–4

38. $w(x + y)^2 + 5w(x + y) + 6w$

39. $x^4 - 9x^2y^2 + 20y^4$

40. $a^6b^6 - 23a^3b^3c^2 + 132c^4$

Applications

41. To find the thickness t of the angle iron in Fig. 8–4, it is necessary to solve the quadratic equation

$$t^2 - 14t + 24 = 0$$

Factor the left side of this equation. (*Extra:* Can you see where this equation comes from?)

42. To find the dimensions of a rectangular field having a perimeter of 70 m and an area of 300 m², we must solve the equation

$$x^2 - 35x + 300 = 0$$

Factor the left side of this equation.

43. To find the width of the frame in Fig. 8–5, we must solve the equation

$$x^2 - 22x + 40 = 0$$

Factor the left side of this equation.

44. To find two resistors that will give an equivalent resistance of 400 Ω when wired in series and 75 Ω when wired in parallel, we must solve the equation

$$R^2 - 400R + 30\ 000 = 0$$

Factor the left side of this equation.

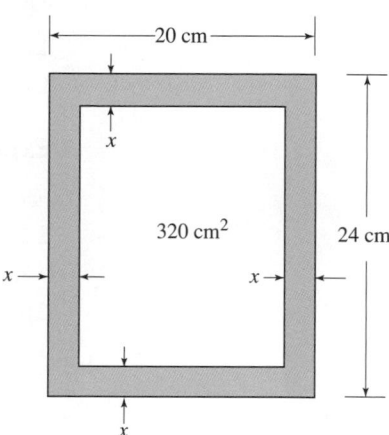

FIGURE 8–5

8–4 Factoring by Grouping

When the expression to be factored has four or more terms, these terms can sometimes be arranged in smaller groups that can be factored separately.

◆◆◆ **Example 31:** Factor $ab + 4a + 3b + 12$.

Solution: Group the two terms containing the factor a, and factor the two containing the factor 3. Remove the common factor from each pair of terms.

We'll use this same grouping idea later to factor certain trinomials.

$$(ab + 4a) + (3b + 12) = a(b + 4) + 3(b + 4)$$

Both terms now have the common factor $(b + 4)$, which we factor out.

$$(b + 4)(a + 3) \qquad ◆◆◆$$

◆◆◆ **Example 32:** Factor $x^2 - y^2 + 2x + 1$.

Solution: Taking our cue from Example 31, we try grouping the x terms together.

$$(x^2 + 2x) + (-y^2 + 1)$$

But we are no better off than before. We then might notice that if the $+1$ term were grouped with the x terms, we would get a trinomial that could be factored. Thus

$$(x^2 + 2x + 1) - y^2$$
$$(x + 1)(x + 1) - y^2$$

or

$$(x + 1)^2 - y^2$$

We now have the difference of two squares. Factoring again gives

$$(x + 1 + y)(x + 1 - y) \qquad ◆◆◆$$

Exercise 4 ◆ Factoring by Grouping

Factor completely.

1. $a^3 + 3a^2 + 4a + 12$
2. $x^3 + x^2 + x + 1$
3. $x^3 - x^2 + x - 1$
4. $2x^3 - x^2 + 4x - 2$
5. $x^2 - bx + 3x - 3b$
6. $ab + a - b - 1$
7. $3x - 2y - 6 + xy$
8. $x^2y^2 - 3x^2 - 4y^2 + 12$
9. $x^2 + y^2 + 2xy - 4$
10. $x^2 - 6xy + 9y^2 - a^2$
11. $m^2 - n^2 - 4 + 4n$
12. $p^2 - r^2 - 6pq + 9q^2$

Expressions with Three or More Letters

Factor completely.

13. $ax + bx + 3a + 3b$
14. $x^2 + xy + xz + yz$
15. $a^2 - ac + ab - bc$
16. $ay - by + ab - y^2$
17. $2a + bx^2 + 2b + ax^2$
18. $2xy + wy - wz - 2xz$
19. $6a^2 + 2ab - 3ac - bc$
20. $x^2 + bx + cx + bc$
21. $b^2 - bc + ab - ac$
22. $bx - cx + bc - x^2$
23. $x^2 - a^2 - 2ab - b^2$
24. $p^2 - y^2 - x^2 - 2xy$

8–5   The General Quadratic Trinomial

When we multiply the two binomials $(ax + b)$ and $(cx + d)$, we get a trinomial with a leading coefficient of ac, a middle coefficient of $(ad + bc)$, and a constant term of bd.

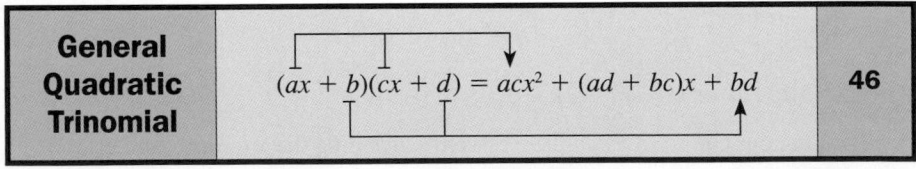

| General Quadratic Trinomial | $(ax + b)(cx + d) = acx^2 + (ad + bc)x + bd$ | 46 |

The general quadratic trinomial may be factored by trial and error or by the grouping method.

Trial and Error

To factor the general quadratic trinomial by trial and error, we look for four numbers, a, b, c, and d, such that

$$ac = \text{the leading coefficient}$$
$$ad + bc = \text{the middle coefficient}$$
$$bd = \text{the constant term}$$

Also, the *signs* in the factors are found in the same way as for trinomials with a leading coefficient of 1.

◆◆◆ **Example 33:** Factor $2x^2 + 5x + 3$.

Solution: The leading coefficient ac is 2, and the constant term bd is 3. Try

$$a = 1, \quad c = 2, \quad b = 3, \quad \text{and} \quad d = 1$$

Then $ad + bc = 1(1) + 3(2) = 7$. No good. It is supposed to be 5. We next try

$$a = 1, \quad c = 2, \quad b = 1, \quad \text{and} \quad d = 3$$

Then $ad + bc = 1(3) + 1(2) = 5$. This works. So

$$2x^2 + 5x + 3 = (x + 1)(2x + 3)$$ ◆◆◆

Grouping Method

The grouping method eliminates the need for trial and error.

◆◆◆ **Example 34:** Factor $3x^2 - 16x - 12$.

Solution:

1. Multiply the leading coefficient and the constant term.

$$3(-12) = -36$$

2. Find two numbers whose product equals -36 and whose sum equals the middle coefficient, -16. Two such numbers are 2 and -18.
3. Rewrite the trinomial, splitting the middle term according to the selected factors ($-16x = 2x - 18x$).

$$3x^2 + 2x - 18x - 12$$

   Group the first two terms together and the last two terms together.

$$(3x^2 + 2x) + (-18x - 12)$$

4. Remove common factors from each grouping.

$$x(3x + 2) - 6(3x + 2)$$

5. Remove the common factor $(3x + 2)$ from the entire expression:

$$(3x + 2)(x - 6)$$

   which are the required factors. ◆◆◆

> **Common Error**
>
> It is easy to make a mistake when factoring out a negative quantity. Thus when we factored out a -6 in going from Step 3 to Step 4 in Example 34, we got
>
> $$(-18x - 12) = -6(3x + 2)$$
>
> but not
>
> $$= -6(3x - 2)$$
> $$\uparrow \text{\_\_\_\_\_ incorrect!}$$

Notice that if we had grouped the terms differently in Step 3 of Example 34, we would have arrived at the same factors as before.

$$3x^2 - 18x + 2x - 12$$
$$3x(x - 6) + 2(x - 6)$$
$$(3x + 2)(x - 6)$$

Sometimes an expression may be simplified before factoring, as we did in Sec. 8–3.

Some people have a knack for factoring and can quickly factor a trinomial by trial and error. Other rely on the longer but surer grouping method.

◆◆◆ Example 35: Factor $12(x + y)^{6n} - (x + y)^{3n}z - 6z^2$.

Solution: If we substitute $a = (x + y)^{3n}$, our expression becomes

$$12a^2 - az - 6z^2$$

Temporarily dropping the z's gives

$$12a^2 - a - 6$$

which factors into

$$(4a - 3)(3a + 2)$$

Replacing the z's, we get

$$12a^2 - az - 6z^2 = (4a - 3z)(3a + 2z)$$

Finally, substituting back $(x + y)^{3n}$ for a, we have

$$12(x + y)^{6n} - (x + y)^{3n}z - 6z^2 = [4(x + y)^{3n} - 3z][4(x + y)^{3n} + 2z] \qquad ◆◆◆$$

Exercise 5 ◆ The General Quadratic Trinomial

Factor completely.

1. $4x^2 - 13x + 3$
2. $5a^2 - 8a + 3$
3. $5x^2 + 11x + 2$
4. $7x^2 + 23x + 6$
5. $12b^2 - b - 6$
6. $6x^2 - 7x + 2$
7. $2a^2 + a - 6$
8. $2x^2 + 3x - 2$
9. $5x^2 - 38x + 21$
10. $4x^2 + 7x - 15$
11. $3x^2 + 6x + 3$
12. $2x^2 + 11x + 12$
13. $3x^2 - x - 2$
14. $7x^2 + 123x - 54$
15. $4x^2 - 10x + 6$
16. $3x^2 + 11x - 20$
17. $4a^2 + 4a - 3$
18. $9x^2 - 27x + 18$
19. $9a^2 - 15a - 14$
20. $16c^2 - 48c + 35$

Expressions Reducible to Quadratic Trinomials

Factor completely.

21. $49x^6 + 14x^3y - 15y^2$
22. $-18y^2 + 42x^2 - 24xy$
23. $4x^6 + 13x^3 + 3$
24. $5a^{2n} - 8a^n + 3$
25. $5x^{4n} + 11x^{2n} + 2$
26. $12(a + b)^2 - (a + b) - 6$
27. $3(a + x)^{2n} - 3(a + x)^n - 6$

Applications

28. An object is thrown upward with an initial velocity of 32 ft./s from a building 128 ft. above the ground. The height s of the object above the ground at any time t is given by

$$s = 128 + 32t - 16t^2$$

Factor the right side of this equation.

29. An object is thrown into the air with an initial velocity of 23 m/s. To find the approximate time it takes for the object to reach a height of 12 m, we must solve the quadratic equation

$$5t^2 - 23t + 12 = 0$$

Factor the left side of this equation.

30. To find the depth of cut h needed to produce a flat of a certain width on a 1-cm-radius bar (Fig. 8–6), we must solve the equation

$$4h^2 - 8h + 3 = 0$$

Factor the left side of this equation.

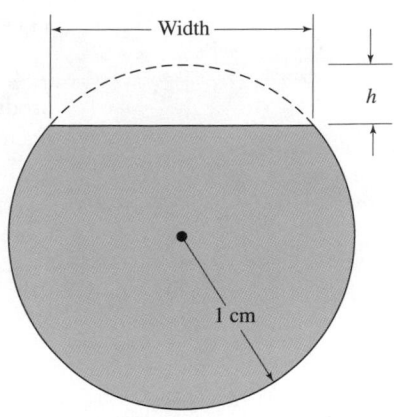

FIGURE 8–6

8–6   The Perfect Square Trinomial

The Square of a Binomial

In Chapter 2, we saw that the expression obtained when a binomial is squared is called a *perfect square trinomial*.

◆◆◆ **Example 36:** Square the binomial $(2x + 3)$.

Solution:

$$(2x + 3)^2 = 4x^2 + 6x + 6x + 9$$
$$= 4x^2 + 12x + 9$$

◆◆◆

Note that in the perfect square trinomial obtained in Example 36, the first and last terms are the squares of the first and last terms of the binomial.

$$(2x + 3)^2$$

square ⟋   ⟍ square

$$4x^2 + 12x + 9$$

The middle term is twice the product of the terms of the binomial.

$$(2x + 3)^2$$

product

$$6x$$

twice the product

$$4x^2 + 12x + 9$$

Also, the constant term is always positive. In general, the following equations apply:

Perfect Square Trinomials	$(a + b)^2 = a^2 + 2ab + b^2$	47
	$(a - b)^2 = a^2 - 2ab + b^2$	48

Any quadratic trinomial can be manipulated into the form of a perfect square trinomial by a procedure called *completing the square*. We will use that method in Sec. 14–2 to derive the quadratic formula.

Factoring a Perfect Square Trinomial

We can factor a perfect square trinomial in the same way we factored the general quadratic trinomial in Sec. 8–5. However, the work is faster if we recognize that a trinomial is a perfect square. If it is, its factors will be the square of a binomial. The terms of that binomial are the square roots of the first and last terms of the trinomial. The sign in the binomial will be the same as the sign of the middle term of the trinomial.

◆◆◆ Example 37: Factor $a^2 - 4a + 4$.

Solution: The first and last terms are both perfect squares, and the middle term is twice the product of the square roots of the first and last terms. Thus the trinomial is a perfect square. Factoring, we obtain

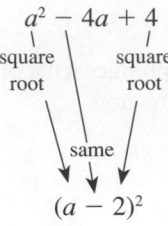

◆◆◆

◆◆◆ Example 38:

$$9x^2y^2 - 6xy + 1 = (3xy - 1)^2$$

◆◆◆

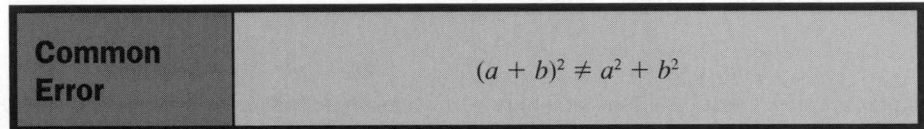

Common Error	$(a + b)^2 \neq a^2 + b^2$

Exercise 6 ◆ The Perfect Square Trinomial

Factor completely.

1. $x^2 + 4x + 4$
2. $x^2 - 30x + 225$
3. $y^2 - 2y + 1$
4. $x^2 + 2x + 1$
5. $2y^2 - 12y + 18$
6. $9 - 12a + 4a^2$
7. $9 + 6x + x^2$
8. $4y^2 - 4y + 1$
9. $9x^2 + 6x + 1$
10. $16x^2 + 16x + 4$
11. $9y^2 - 18y + 9$
12. $16n^2 - 8n + 1$
13. $16 + 16a + 4a^2$
14. $1 + 20a + 100a^2$
15. $49a^2 - 28a + 4$

Perfect Square Trinomials with Two Variables

Factor completely.

16. $a^2 - 2ab + b^2$
17. $x^2 + 2xy + y^2$
18. $a^2 - 14ab + 49b^2$
19. $a^2w^2 + 2abw + b^2$
20. $a^2 - 10ab + 25b^2$
21. $x^2 + 10ax + 25a^2$
22. $c^2 - 6cd + 9d^2$
23. $x^2 + 8xy + 16y^2$

Expressions That Can Be Reduced to Perfect Square Trinomials

Factor completely.

24. $1 + 2x^2 + x^4$
25. $z^6 + 16z^3 + 64$
26. $36 + 12a^2 + a^4$
27. $49 - 14x^3 + x^6$
28. $a^2b^2 - 8ab^3 + 16b^4$
29. $a^4 - 2a^2b^2 + b^4$
30. $16a^2b^2 - 8ab^2c^2 + b^2c^4$
31. $4a^{2n} + 12a^nb^n + 9b^{2n}$

8–7 Sum or Difference of Two Cubes

Definition

An expression such as

$$x^3 + 27$$

is called the *sum of two cubes (x^3 and 3^3)*. In general, when we multiply the binomial $(a + b)$ and the trinomial $(a^2 - ab + b^2)$, we obtain

$$(a + b)(a^2 - ab + b^2) = a^3 - a^2b + ab^2 + a^2b - ab^2 + b^3$$
$$= a^3 + b^3$$

All but the cubed terms drop out, leaving the sum of two cubes.

Sum of Two Cubes	$a^3 + b^3 = (a + b)(a^2 - ab + b^2)$	42
Difference of Two Cubes	$a^3 - b^3 = (a - b)(a^2 + ab + b^2)$	43

When we recognize that an expression is the sum (or difference) of two cubes, we can write the factors immediately.

◆◆◆ **Example 39:** Factor $x^3 + 27$.

Solution: This expression is the sum of two cubes, $x^3 + 3^3$. Substituting into Eq. 42, with $a = x$ and $b = 3$, yields

$$x^3 + 27 = x^3 + 3^3 = (x + 3)(x^2 - 3x + 9)$$

same sign

opposite sign

always +

◆◆◆

◆◆◆ **Example 40:** Factor $27x^3 - 8y^3$.

Solution: This expression is the difference of two cubes, $(3x)^3 - (2y)^3$. Factoring gives us

$$27x^3 - 8y^3 = (3x)^3 - (2y)^3 = (3x - 2y)(9x^2 + 6xy + 4y^2)$$

same sign

opposite sign

always +

◆◆◆

Common Error	The middle term of the trinomials in Eqs. 42 and 43 is often mistaken as $2ab$. $$a^3 + b^3 \neq (a + b)(a^2 - 2ab + b^2)$$ no!

When the powers are multiples of 3, we may be able to factor the expression as the sum or difference of two cubes.

◆◆◆ **Example 41:**

$$a^6 - b^9 = (a^2)^3 - (b^3)^3$$

Factoring, we obtain

$$a^6 - b^9 = (a^2 - b^3)(a^4 + a^2b^3 + b^6)$$ ◆◆◆

Exercise 7 ◆ Sum or Difference of Two Cubes

Factor completely.

1. $64 + x^3$
2. $1 - 64y^3$
3. $2a^3 - 16$
4. $a^3 - 27$
5. $x^3 - 1$
6. $x^3 - 64$
7. $x^3 + 1$
8. $a^3 - 343$
9. $a^3 + 64$
10. $x^3 + 343$
11. $x^3 + 125$
12. $64a^3 + 27$
13. $216 - 8a^3$
14. $343 - 27y^3$
15. $343 + 64x^3$

More Difficult Types

Factor completely.

16. $27x^9 + 512$
17. $y^9 + 64x^3$
18. $64a^{12} + x^{15}$
19. $27x^{15} + 8a^6$
20. $8x^{6p} + 8a^6$
21. $8a^{6x} - 125b^{3x}$
22. $64x^{12a} - 27y^{15a}$
23. $64x^{3n} - y^{9n}$
24. $x^3y^3 - z^3$

Applications

25. The volume of a hollow spherical shell having an inside radius r_1 and an outside radius r_2 is $\frac{4}{3}\pi r_2^3 - \frac{4}{3}\pi r_1^3$. Factor this expression completely.
26. A cistern is in the shape of a hollow cube whose inside dimension is s and whose outside dimension is S. If it is made of concrete of density d, its weight is $dS^3 - ds^3$. Factor completely.

Case Study Discussion – Checking Your Factoring

In this case, all the factoring is correct, yet we still know that something is very wrong. In the fourth line, we have a simple cross multiplication to simplify the expression; however, the denominator is $x - y$. Since $x = y$, $x - y$ has to be zero, and in our mathematical system, dividing by zero is *undefined*. What undefined means is that we cannot write a mathematical definition for dividing by zero.

Sometimes we are so busy factoring we forget a very basic rule. By using the step:

$$\frac{(x + y)\cancel{(x - y)}}{\cancel{(x - y)}} = \frac{x\cancel{(x - y)}}{\cancel{(x - y)}}$$ we are really dividing BOTH sides by "$(x - y)$"."

This leads us to the problem with the next step: if $x = y$, then the denominator here must be zero.

It would really look something like this:

$$\frac{(x + y)(x - y)}{(\text{"0"})} = \frac{x(x - y)}{(\text{"0"})},$$ and we know that anything divided by 0 (zero) is *undefined*.

So, now this means that the next line, $x + y = x$ can't be true.

This is why you have to be careful when factoring mathematical expressions and equations.

This particular equation is *unfactorable*.

In this particular equation where the result is $2 = 1$, we knew there was a problem; in a different equation, the result might not have been that obvious. This is why in complex mathematics there is a lot of checking, to ensure that this type of mistake does not happen.

❖❖❖ CHAPTER 8 REVIEW PROBLEMS ❖❖❖❖❖❖❖❖❖❖❖❖❖❖❖❖❖❖❖❖❖❖❖❖

Factor completely.

1. $x^2 - 2x - 15$
2. $2a^2 + 3a - 2$
3. $x^6 - y^4$
4. $2ax^2 + 8ax + 8a$
5. $2x^2 + 3x - 2$
6. $a^2 + ab - 6b^2$
7. $(a - b)^3 + (c + d)^3$
8. $8x^3 - \dfrac{y^3}{27}$
9. $\dfrac{x^2}{y} - \dfrac{x}{y}$
10. $2ax^2y^2 - 18a$
11. $xy - 2y + 5x - 10$
12. $3a^2 - 2a - 8$
13. $x - bx - y + by$
14. $\dfrac{2a^2}{12} - \dfrac{8b^2}{27}$
15. $(y + 2)^2 - z^2$
16. $2x^2 - 20ax + 50a^2$
17. $x^2 - 7x + 12$
18. $4a^2 - (3a - 1)^2$

19. $16x^{4n} - 81y^{8n}$

20. $xy - y^2 + xz - yz$

21. $9a^2 + 12az^2 + 4z^4$

22. $4a^6 - 4b^6$

23. $x^{2m} + 2x^m + 1$

24. $4x^{6m} + 4x^{3m}y^m + y^{2m}$

25. $(x - y)^2 - z^2$

26. $x^2 - 2x - 3$

27. $1 - 16x^2$

28. $a^2 - 2a - 8$

29. $9x^4 - x^2$

30. $x^2 - 21x + 110$

31. $(p - q)^3 - 27$

32. $5a^2 - 20x^2$

33. $27a^3 - 8w^3$

34. $3x^2 - 6x - 45$

35. $16x^2 - 16xy + 4y^2$

36. $64m^3 - 27n^3$

37. $6ab + 2ay + 3bx + xy$

38. $15a^2 - 11a - 12$

39. $2y^4 - 18$

40. $ax - bx + ay - by$

41. The reduction in power in a resistance R by lowering the voltage across the resistor from V_2 to V_1 is

$$\frac{V_2^2}{R} - \frac{V_1^2}{R}$$

Factor this expression.

42. A conical tank of height h is filled to a depth d. The volume of liquid that can still be put into the tank is

$$V = \frac{\pi}{3}r_1^2h - \frac{\pi}{3}r_2^2d$$

where r_1 is the base radius of the tank and r_2 is the base radius of the liquid. Factor the right side of this equation.

Writing

43. We have studied the factoring of seven different types of expressions in this chapter. List them and give an example of each. State in words how to recognize each and how to tell one from the other. Also list at least four other expressions that are *not* one of the given seven, and state why each is different from those we have studied.

Team Project

44. The area of a rectangle is the *product* of its sides, so we can think of the product $5x$ as the area of a rectangle of sides 5 and x (Fig. 8–7).

Similarly, the trinomial $2x^2 + 5x + 2$ can be thought of as the area of a rectangle of sides $(2x + 1)$ and $(x + 2)$, because

$$2x^2 + 5x + 2 = (2x + 1)(x + 2)$$

as shown in Fig. 8–8.

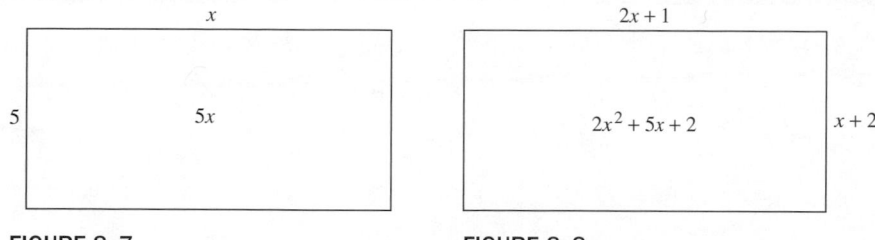

FIGURE 8–7   **FIGURE 8–8**

To factor a trinomial using areas, cut rectangles of paper with areas equal to the terms in the trinomial. Thus the trinomial $2x^2 + 5x + 2$ can be represented by two squares of side x, five rectangles with sides 1 and x, and two squares of side 1 (Fig. 8–9). Choose any size for x, but avoid making it an integer value.

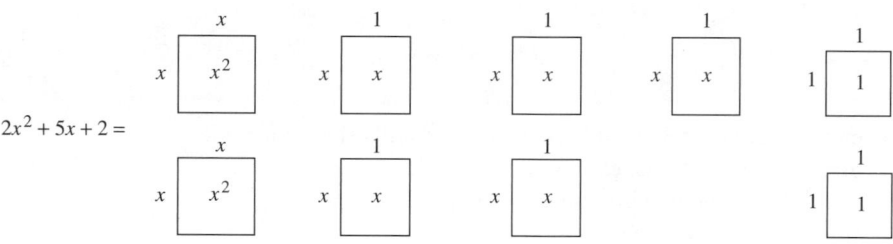

FIGURE 8–9

Arrange the individual squares to form a rectangle. There will be only one way to do this. The sides $(2x + 1)$ and $(x + 2)$ of this rectangle will be the factors of the trinomial, as shown in Fig. 8–10.

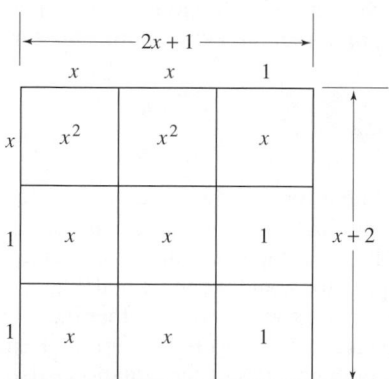

FIGURE 8–10

Use this method to factor the following trinomials:

$$x^2 + 3x + 2$$
$$2x^2 + 7x + 6$$
$$x^2 + 13x + 30$$

Expand the method to handle trinomials with *negative* terms. Use it to factor

$$x^2 + 2x - 3$$
$$x^2 + 7x - 30$$

Fractions and Fractional Equations

When you have completed this chapter, your skills should include:

- Simplifying fractional expressions.
- Multiplying and dividing fractional expressions.
- Adding and subtracting fractional expressions
- Simplifying complex fractional expressions.
- Solving fractional equations.
- Solving word problems using fractional equations.
- Working with literal equations and formulas.

You already know about fractions with numbers. In algebra, however, the numbers are replaced with letters, coefficients, and even entire expressions. Many equations and formulas in science and technology are in the form of a fraction. For example, the force of gravity between two masses m_1 and m_2 is

$$F = k \, \frac{m_1 m_2}{r^2}$$

In fact, any equation for a field force—gravity, electrostatic force, magnetic force—will be in a fractional form following this or a very similar *inverse square* law.

So now the rules of working with the numerators and denominators of fractions must be applied to entire algebraic expressions, and we must make heavy use of factoring techniques in order to simplify them. As we work with formulas that include fractions, we must be careful: it's very easy to make mistakes when we cross multiply. Remember that anything you do must be done to each term on both sides of the equation. Also, don't be intimidated by complex fractions where a numerator or denominator might contain a fraction; use your skills and take it one step at a time.

Not all of this material is new to us. Some was covered in Chapter 2, and simple fractional equations were solved in Chapter 3.

CASE STUDY – SCUBA DIVING AIR TANKS

No matter what area of science or engineering technology you wind up in, you are likely to find yourself working in areas you never thought you would. Your career will follow a path defined by opportunities and your ability to apply your knowledge in new areas.

In this case, we are looking at an air tank for a scuba diver. You could be a mechanical engineering technician or technologist working for the compressor manufacturer. You could be trained in electronics technology and be designing a computer-controlled filling system or even a chemical technologist with the company that monitors gas impurities. You want to confirm that a tank rated for approximately 30 minutes would reasonably last a diver 30 minutes at a depth of 10 metres. The formula you can use is

$$t = \frac{\dfrac{P_i V_i}{P_1 + \rho g h} - V_i}{C}$$

where

t = time in minutes
V_i = Volume of air in the tank in m^3 (use 0.0150 m^3)
P_i = Pressure of air in the tank in Pa (use 2.02 × 10^7 Pa)
C = rate of air consumption in m^3/min
g = 9.8 m/s^2 (gravitational constant)
ρ = mass density of sea water in kg/m^3
P_1 = air pressure at the surface of the water in Pa
h = depth of the diver (given as 10 m)

Based on this formula, would a diver with a 30-minute tank be correct in expecting 30 minutes of air during a 10 m dive?

Some variables are given; the rest you can determine through a little research.

STRUCTURAL ENGINEER—KAREN DECONTIE

Karen Decontie holds a master of engineering degree in civil engineering from the University of Calgary. Her work with Public Works and Government Services Canada keeps the public safe while enjoying the beauty of Canada's national parks. Karen states that what she likes best about her job is that it is "challenging, feeling that you can build something that is useful to many people." As a structural engineer, she is responsible for ensuring the safety of bridges and structures. As a member of the Algonquin Nation, she believes that the more First Nations engineers there are, "the more decisions we can make on our own." In her own case, she says, her engineering work enables her to contribute to society both for the present and for future generations of visitors to the national parks.

Karen grew up in Kitigan Zibi in Quebec and went to McGill for her undergraduate degree before moving to Calgary, where she has lived since. Leaving home was not easy, but the experience gave her an opportunity to "learn different thinking," and the process of becoming a professional engineer "helped me to gain self-esteem and believe in myself as a person."

Source: Native Access, *Role Model: Karen Decontie,*
http://www.nativeaccess.com/allabout/pop_rm/rm_decontie_karen.html

9–1   Simplification of Fractions

Parts of a Fraction

A fraction has a *numerator,* a *denominator,* and a *fraction line.*

$$\text{fraction line} \longrightarrow \frac{a}{b} \begin{array}{l} \leftarrow \text{numerator} \\ \leftarrow \text{denominator} \end{array}$$

Quotient

A fraction is a way of indicating a *quotient* of two quantities. The fraction *a/b* can be read "*a* divided by *b*."

Ratio

The quotient of two numbers or quantities is also spoken of as the *ratio* of those quantities. Thus the ratio of x to y is x/y. The two forms $\frac{a}{b}$ and a/b are equally valid.

Division by Zero

Since division by zero is not permitted, it should be understood in our work with fractions that *the denominator cannot be zero.*

◆◆◆ **Example 1:** What values of x are not permitted in the following fraction:

$$\frac{3x}{x^2 + x - 6}$$

Solution: Factoring the denominator, we get

$$\frac{3x}{x^2 + x - 6} = \frac{3x}{(x - 2)(x + 3)}$$

We see that an x equal to 2 or -3 will make $(x - 2)$ or $(x + 3)$ equal to zero. This will result in division by zero, so these values are not permitted.   ◆◆◆

Common Fractions

A fraction whose numerator and denominator are both integers is called a *common fraction.*

◆◆◆ **Example 2:** The following are common fractions:

$$\frac{2}{3}, \quad \frac{-9}{5}, \quad \frac{-124}{125} \quad \text{and} \quad \frac{18}{-11} \qquad \text{◆◆◆}$$

Algebraic Fractions

An *algebraic fraction* is one whose numerator and/or denominator contain *literal* quantities.

◆◆◆ **Example 3:** The following are algebraic fractions:

$$\frac{x}{y}, \quad \frac{\sqrt{x + 2}}{x}, \quad \frac{3}{y}, \quad \text{and} \quad \frac{x^2}{x - 3} \qquad \text{◆◆◆}$$

Recall that a polynomial is an expression in which the exponents are nonnegative integers.

Rational Algebraic Fractions

An algebraic fraction is called *rational* if the numerator and the denominator are both *polynomials.*

◆◆◆ **Example 4:** The following are rational fractions:

$$\frac{x}{y}, \quad \frac{3}{w^3}, \quad \text{and} \quad \frac{x^2}{x - 3}$$

But

$$\frac{\sqrt{x + 2}}{x}$$

is not a rational fraction, since there is no perfect root for the numerator.   ◆◆◆

Proper and Improper Fractions

A *proper* common fraction is one whose numerator is smaller than its denominator.

◆◆◆ **Example 5:** $\frac{3}{5}, \frac{1}{3}$, and $\frac{9}{11}$ are *proper* fractions, whereas $\frac{8}{5}, \frac{3}{2}$, and $\frac{7}{4}$ are *improper* fractions.   ◆◆◆

A proper *algebraic* fraction is a rational fraction whose numerator is of *lower degree* than the denominator.

◆◆◆ **Example 6:** The following are proper fractions:

$$\frac{x}{x^2 + 2} \quad \text{and} \quad \frac{x^2 + 2x - 3}{x^3 + 9}$$

However,

$$\frac{x^3 - 2}{x^2 + x - 3} \quad \text{and} \quad \frac{x^2}{y}$$

are improper fractions. ◆◆◆

Mixed Form
A *mixed number* is the sum of an integer and a fraction.

◆◆◆ **Example 7:** The following are mixed numbers:

$$2\frac{1}{2}, \quad 5\frac{3}{4}, \quad \text{and} \quad 3\frac{1}{3} \qquad \text{◆◆◆}$$

A *mixed expression* is the sum or difference of a polynomial and a rational algebraic fraction.

◆◆◆ **Example 8:** The following are mixed expressions:

$$3x - 2 + \frac{1}{x} \quad \text{and} \quad y - \frac{y}{y^2 + 1} \qquad \text{◆◆◆}$$

Decimals and Fractions
To change a fraction to an equivalent decimal, simply divide the numerator by the denominator.

◆◆◆ **Example 9:** To write $\frac{9}{11}$ as a decimal, we divide 9 by 11.

$$\frac{9}{11} = 0.818\ 181\ \overline{81} \ldots \text{ or } 0.\overline{81}$$

We get a *repeating decimal;* the dots following the number indicate that the digits continue indefinitely. Repeating decimals are also commonly written with a bar over the repeating part: $0.\overline{81}$ ◆◆◆

To change a decimal number to a fraction, write a fraction with the decimal number in the numerator and 1 in the denominator. Multiply numerator and denominator by a multiple of 10 that will make the numerator a whole number. Finally, reduce to lowest terms.

We'll cover reducing fractions to their lowest terms in more detail later in this section.

◆◆◆ **Example 10:** Express 0.875 as a fraction.

Solution:

$$0.875 = \frac{0.875}{1} = \frac{875}{1000} = \frac{7}{8} \qquad \text{◆◆◆}$$

To express a *repeating* decimal as a fraction, follow the steps in the next example.

◆◆◆ **Example 11:** Change the repeating decimal $0.81\overline{81}$ to a fraction.

Solution: Let $x = 0.81\ \overline{81}$. Multiplying by 100, we have

$$100x = 81.\overline{81}$$

Subtracting the first equation, $x = 0.81\ \overline{81}$ from the second gives us

$$99x = 81 \quad \text{(exactly)}$$

Dividing by 99 yields

$$x = \frac{81}{99} = \frac{9}{11} \quad \text{(reduced)} \qquad \text{◆◆◆}$$

Simplifying a Fraction by Reducing to Lowest Terms

We reduce a fraction to lowest terms by dividing both numerator and denominator by any factor that is contained in both.

Simplifying Fractions	$\dfrac{ad}{bd} = \dfrac{a}{b}$	50

◆◆◆ **Example 12:** Reduce the following to lowest terms. Write the answer without negative exponents.

(a) $\dfrac{9}{12} = \dfrac{3(3)}{4(3)} = \dfrac{3}{4}$

(b) $\dfrac{3x^2yz}{9xy^2z^3} = \dfrac{3}{9} \cdot \dfrac{x^2}{x} \cdot \dfrac{y}{y^2} \cdot \dfrac{z}{z^3} = \dfrac{x}{3yz^2}$ ◆◆◆

When possible, factor the numerator and the denominator. Then divide both numerator and denominator by any factors common to both.

◆◆◆ **Example 13:**

(a) $\dfrac{2x^2 + x}{3x} = \dfrac{(2x + 1)x}{3(x)} = \dfrac{2x + 1}{3}$

(b) $\dfrac{ab + bc}{bc + bd} = \dfrac{b(a + c)}{b(c + d)} = \dfrac{a + c}{c + d}$

(c) $\dfrac{2x^2 - 5x - 3}{4x^2 - 1} = \dfrac{(2x + 1)(x - 3)}{(2x + 1)(2x - 1)} = \dfrac{x - 3}{2x - 1}$

(d) $\dfrac{x^2 - ax + 2bx - 2ab}{2x^2 + ax - 3a^2} = \dfrac{x(x - a) + 2b(x - a)}{(x - a)(2x + 3a)}$

$= \dfrac{(x - a)(x + 2b)}{(x - a)(2x + 3a)} = \dfrac{x + 2b}{2x + 3a}$ ◆◆◆

The process of striking out the same factors from numerator and denominator is called *cancelling*.

Common Errors	Use caution when cancelling: If a factor is missing from *even one term* in the numerator or denominator, that factor *cannot* be cancelled. $$\dfrac{xy - z}{wx} \neq \dfrac{y - z}{w}$$
	We may divide (or multiply) the numerator and denominator by the same quantity (Eq. 50), but we *may not add or subtract* the same quantity in the numerator and denominator, as this will change the value of the fraction. For example, $$\dfrac{3}{5} \neq \dfrac{3 + 1}{5 + 1} = \dfrac{4}{6} = \dfrac{2}{3}$$

Simplifying Fractions by Changing Signs

Recall from Chapter 2 that any two of the three signs of a fraction may be changed without changing the value of a fraction.

		10
Rules of Signs	$\dfrac{+a}{+b} = \dfrac{-a}{-b} = -\dfrac{-a}{+b} = -\dfrac{+a}{-b} = \dfrac{a}{b}$	10
	$\dfrac{+a}{-b} = \dfrac{-a}{+b} = -\dfrac{-a}{-b} = -\dfrac{a}{b}$	11

We can sometimes use this idea to simplify a fraction; that is, to reduce it to lowest terms.

◆◆◆ **Example 14:** Simplify the fraction

$$-\frac{3x - 2}{2 - 3x}$$

Solution: We can change the sign of the denominator *and* the sign of the entire fraction.

$$-\frac{3x - 2}{2 - 3x} = +\frac{3x - 2}{-(2 - 3x)} = \frac{3x - 2}{-2 + 3x} = \frac{3x - 2}{3x - 2} = 1$$

changed

◆◆◆

Exercise 1 ◆ Simplification of Fractions

Hint: Factor the denominators in problems 4, 5, and 6.

In each fraction, what values of x, if any, are not permitted?

1. $\dfrac{12}{x}$

2. $\dfrac{x}{12}$

3. $\dfrac{18}{x - 5}$

4. $\dfrac{5x}{x^2 - 49}$

5. $\dfrac{7}{x^2 - 3x + 2}$

6. $\dfrac{3x}{8x^2 - 14x + 2}$

Change each fraction to a decimal. Work to four decimal places.

7. $\dfrac{7}{12}$

8. $\dfrac{5}{9}$

9. $\dfrac{15}{16}$

10. $\dfrac{125}{155}$

11. $\dfrac{11}{3}$

12. $\dfrac{25}{9}$

Change each decimal to a fraction.

13. 0.4375

14. 0.390 625

15. 0.6875

16. 0.281 25

17. 0.7777 . . .

18. 0.636 363 . . .

Simplify each fraction by manipulating the algebraic signs.

19. $\dfrac{a - b}{b - a}$

20. $-\dfrac{2x - y}{y - 2x}$

21. $\dfrac{(a - b)(c - d)}{b - a}$

22. $\dfrac{w(x - y - z)}{y - x + z}$

Reduce to lowest terms. Write your answers without negative exponents.

23. $\dfrac{14}{21}$

24. $\dfrac{81}{18}$

25. $\dfrac{75}{35}$

26. $\dfrac{36}{44}$

27. $\dfrac{2ab}{6b}$

28. $\dfrac{12m^2n}{15mn^2}$

29. $\dfrac{21m^2p^2}{28mp^4}$

30. $\dfrac{abx - bx^2}{acx - cx^2}$

31. $\dfrac{4a^2 - 9b^2}{4a^2 + 6ab}$

32. $\dfrac{3a^2 + 6a}{a^2 + 4a + 4}$

33. $\dfrac{x^2 + 5x}{x^2 + 4x - 5}$

34. $\dfrac{xy - 3y^2}{x^3 - 27y^3}$

35. $\dfrac{x^2 - 4}{x^3 - 8}$

36. $\dfrac{2a^3 + 6a^2 - 8a}{2a^3 + 2a^2 - 4a}$

37. $\dfrac{2m^3n - 2m^2n - 24mn}{6m^3 + 6m^2 - 36m}$

38. $\dfrac{9x^3 - 30x^2 + 25x}{3x^4 - 11x^3 + 10x^2}$

39. $\dfrac{2a^2 - 2}{a^2 - 2a + 1}$

40. $\dfrac{3a^2 - 4ab + b^2}{a^2 - ab}$

41. $\dfrac{x^2 - z^2}{x^3 - z^3}$

42. $\dfrac{2x^2}{6x - 4x^2}$

43. $\dfrac{2a^2 - 8}{2a^2 - 2a - 12}$

44. $\dfrac{2a^2 + ab - 3b^2}{a^2 - ab}$

45. $\dfrac{x^2 - 1}{2xy + 2y}$

46. $\dfrac{x^3 - a^2x}{x^2 - 2ax + a^2}$

47. $\dfrac{2x^4y^4 + 2}{3x^8y^8 - 3}$

48. $\dfrac{18a^2c - 6bc}{42a^2d - 14bd}$

49. $\dfrac{(x + y)^2}{x^2 - y^2}$

50. $\dfrac{mw + 3w - mz - 3z}{m^2 - m - 12}$

51. $\dfrac{b^2 - a^2 - 6b + 9}{5b - 15 - 5a}$

52. $\dfrac{3w - 3y - 3}{w^2 + y^2 - 2wy - 1}$

9–2   Multiplication and Division of Fractions

Multiplication

We multiply a fraction a/b by another fraction c/d as follows:

Multiplying Fractions	$\dfrac{a}{b} \cdot \dfrac{c}{d} = \dfrac{ac}{bd}$	**51**

The product of two or more fractions is a fraction whose numerator is the product of the numerators of the original fractions and whose denominator is the product of the denominators of the original fractions.

◆◆◆ **Example 15:**

(a) $\dfrac{2}{3} \cdot \dfrac{5}{7} = \dfrac{2(5)}{3(7)} = \dfrac{10}{21}$

(b) $\dfrac{1}{2} \cdot \dfrac{2}{3} \cdot \dfrac{3}{5} = \dfrac{1(2)(3)}{2(3)(5)} = \dfrac{1}{5}$

(c) $5\dfrac{2}{3} \cdot 3\dfrac{1}{2} = \dfrac{17}{3} \cdot \dfrac{7}{2} = \dfrac{119}{6}$

◆◆◆

<table>
<tr><td rowspan="3">**Common Error**</td><td>When multiplying mixed numbers, students sometimes try to multiply the whole parts and the fractional parts separately.</td></tr>
<tr><td>$$2\frac{2}{3} \times 4\frac{3}{5} \neq 8\frac{6}{15}$$</td></tr>
<tr><td>The correct way is to write each mixed number as an improper fraction and to multiply as shown.

$$2\frac{2}{3} \times 4\frac{3}{5} = \frac{8}{3} \times \frac{23}{5} = \frac{184}{15} = 12\frac{4}{15}$$</td></tr>
</table>

◆◆◆ **Example 16:**

(a) $\dfrac{2a}{3b} \cdot \dfrac{5c}{4a} = \dfrac{10ac}{12ab} = \dfrac{5c}{6b}$

(b) $\dfrac{x}{x+2} \cdot \dfrac{x^2-4}{x^3} = \dfrac{x(x^2-4)}{(x+2)x^3}$

> Leave your product in factored form until *after* you simplify.

$\qquad = \dfrac{x(x+2)(x-2)}{(x+2)x^3} = \dfrac{x-2}{x^2}$

(c) $\dfrac{x^2+x-2}{x^2-4x+3} \cdot \dfrac{2x^2-3x-9}{2x^2+7x+6}$

$\qquad = \dfrac{(x-1)(x+2)(x-3)(2x+3)}{(x-1)(x-3)(x+2)(2x+3)} = 1$

◆◆◆

Division

To divide one fraction, a/b, by another fraction, c/d,

$$\frac{\dfrac{a}{b}}{\dfrac{c}{d}}$$

we multiply numerator and denominator by d/c, as follows:

$$\frac{\dfrac{a}{b}\cdot\dfrac{d}{c}}{\dfrac{c}{d}\cdot\dfrac{d}{c}} = \frac{\dfrac{a}{b}\cdot\dfrac{d}{c}}{\dfrac{cd}{cd}} = \frac{\dfrac{a}{b}\cdot\dfrac{d}{c}}{1} = \frac{a}{b}\cdot\frac{d}{c} = \frac{ad}{bc}$$

> Can you say *why* it is permissible to multiply numerator and denominator by d/c?

We see that dividing by a fraction is the same as *multiplying by the reciprocal* of that fraction.

<table>
<tr><td>**Division of Fractions**</td><td>$$\frac{a}{c} \div \frac{c}{d} = \frac{a}{b} \cdot \frac{d}{c} = \frac{ad}{bc}$$</td><td>**52**</td></tr>
</table>

When dividing fractions, invert the divisor and multiply.

◆◆◆ **Example 17:**

(a) $\dfrac{2}{3} \div \dfrac{5}{7} = \dfrac{2}{3} \cdot \dfrac{7}{5} = \dfrac{14}{15}$

(b) $3\dfrac{2}{5} \div 2\dfrac{4}{15} = \dfrac{17}{5} \div \dfrac{34}{15}$

$$= \dfrac{17}{5} \times \dfrac{15}{34} = \dfrac{3}{2}$$

(c) $\dfrac{x}{y} \div \dfrac{x+2}{y-1} = \dfrac{x}{y} \cdot \dfrac{y-1}{x+2}$

$$= \dfrac{x(y-1)}{y(x+2)}$$

(d) $\dfrac{x^2+x-2}{x} \div \dfrac{x+2}{x^2} = \dfrac{x^2+x-2}{x} \cdot \dfrac{x^2}{x+2}$

$$= \dfrac{(x+2)(x-1)x^2}{x(x+2)} = x(x-1)$$

(e) $x \div \dfrac{\pi r^2 x}{4} = \dfrac{x}{1} \div \dfrac{\pi r^2 x}{4} = \dfrac{x}{1} \cdot \dfrac{4}{\pi r^2 x} = \dfrac{4}{\pi r^2}$ ◆◆◆

Common Error	We invert the *divisor* and multiply. Be sure not to invert the *dividend*.

Changing Improper Fractions to Mixed Form

To write an improper fraction in mixed form, divide the numerator by the denominator, and express the remainder as the numerator of a fraction whose denominator is the original denominator.

◆◆◆ **Example 18:** Write $\dfrac{45}{7}$ as a mixed number.

Solution: Dividing 45 by 7, we get 6 with a remainder of 3, so

$$\dfrac{45}{7} = 6\dfrac{3}{7}$$ ◆◆◆

The procedure is the same for changing an *algebraic* fraction to mixed form. Divide the numerator by the denominator.

◆◆◆ **Example 19:**

$$\dfrac{x^2+3}{x} = \dfrac{x^2}{x} + \dfrac{3}{x}$$

$$= x + \dfrac{3}{x}$$ ◆◆◆

Exercise 2 ◆ Multiplication and Division of Fractions

Multiply and reduce.

1. $\dfrac{1}{3} \times \dfrac{2}{5}$

2. $\dfrac{3}{7} \times \dfrac{21}{24}$

3. $\dfrac{2}{3} \times \dfrac{9}{7}$

4. $\dfrac{11}{3} \times 7$

5. $\dfrac{2}{3} \times 3\dfrac{1}{5}$

6. $3 \times 7\dfrac{2}{5}$

7. $3\dfrac{3}{4} \times 2\dfrac{1}{2}$

8. $\dfrac{3}{5} \times \dfrac{2}{7} \times \dfrac{5}{9}$

9. $3 \times \dfrac{5}{8} \times \dfrac{4}{5}$

10. $\dfrac{15a^2}{7b^2} \cdot \dfrac{28ab}{9a^3c}$

11. $\dfrac{a^4b^4}{2a^2y^n} \cdot \dfrac{a^2x}{xy^n}$

12. $\dfrac{3x^2y^2z^3}{4a^2b^2c^2} \cdot \dfrac{8a^3b^2c^2}{9x^2yz^3}$

13. $\dfrac{5m^2n^2p^4}{3x^2yz^3} \cdot \dfrac{21xyz^2}{20m^2n^2p^2}$

14. $\dfrac{x+y}{x-y} \cdot \dfrac{x^2-y^2}{(x+y)^2}$

15. $\dfrac{x^2-a^2}{xy} \cdot \dfrac{xy}{x+a}$

16. $\dfrac{x^4-y^4}{a^2x^2} \cdot \dfrac{ax^3}{x^2+y^2}$

17. $\dfrac{a}{x-y} \cdot \dfrac{b}{x+y}$

18. $\dfrac{x+y}{10} \cdot \dfrac{ax}{3(x+y)}$

19. $\dfrac{c}{x^2-y^2} \cdot \dfrac{d}{x^2-y^2}$

20. $\dfrac{x-2}{2x-3} \cdot \dfrac{2x^2-5x+3}{x^2-4}$

21. $\dfrac{x^2-1}{x^2+x-6} \cdot \dfrac{x^2+2x-8}{x^2-4x-5}$

22. $\dfrac{5x-1}{x^2-x-2} \cdot \dfrac{x-2}{10x^2+13x-3}$

23. $\dfrac{2x^2+x-1}{3x^2-11x-4} \cdot \dfrac{3x^2+7x+2}{2x^2-7x+3}$

24. $\dfrac{x^3+1}{x+1} \cdot \dfrac{x+3}{x^2-x+1}$

25. $\dfrac{2x^2+x-6}{4x^2-3x-1} \cdot \dfrac{4x^2+5x+1}{x^2-4x-12}$

Divide and reduce.

26. $\dfrac{7}{9} \div \dfrac{5}{3}$

27. $3\dfrac{7}{8} \div 2$

28. $\dfrac{7}{8} \div 4$

29. $\dfrac{9}{16} \div 8$

30. $24 \div \dfrac{5}{8}$

31. $\dfrac{5}{8} \div 2\dfrac{1}{4}$

32. $2\dfrac{7}{8} \div 1\dfrac{1}{2}$

33. $2\dfrac{5}{8} \div \dfrac{1}{2}$

34. $50 \div 2\dfrac{2}{3}$

35. $\dfrac{5abc^3}{3x^2} \div \dfrac{10ac^3}{6bx^2}$

36. $\dfrac{7x^2y}{3ad} \div \dfrac{2xy^2}{3a^2d}$

37. $\dfrac{4a^3x}{8dy^2} \div \dfrac{2a^2x^2}{8a^2y}$

38. $\dfrac{5x^2y^3z}{6a^2b^2c} \div \dfrac{10xy^3z^2}{8ab^2c^2}$

39. $\dfrac{3an+cm}{x^2-y^2} \div (x^2+y^2)$

40. $\dfrac{a^2+4a+4}{d+c} \div (a+2)$

41. $\dfrac{ac+ad+bc+bd}{c^2-d^2} \div (a+b)$

42. $\dfrac{5(x+y)^2}{x-y} \div (x+y)$

43. $\dfrac{1}{x^2+17x+30} \div \dfrac{1}{x+15}$

44. $\dfrac{5xy}{a-x} \div \dfrac{10xy}{a^2-x^2}$

45. $\dfrac{3x-6}{2x+3} \div \dfrac{x^2-4}{4x^2+2x-6}$

46. $\dfrac{a^2 - a - 2}{3a - 1} \div \dfrac{a - 2}{6a^2 + 7a - 3}$

47. $\dfrac{2p^2 + p - 6}{4p^2 - 3p - 1} \div \dfrac{2p^2 - 5p + 3}{4p^2 + 9p + 2}$

48. $\dfrac{x^3 + 1}{x + 1} \div \dfrac{x^2 - x + 1}{3x + 9}$

49. $\dfrac{z^2 - 1}{z^2 + z - 6} \div \dfrac{z^2 - 4z - 5}{z^2 + 2z - 8}$

50. $\dfrac{x^2 + 5x + 6}{2x^2 - 3x + 1} \div \dfrac{x^2 - 9}{2x^2 - 7x + 3}$

Improper Fractions and Mixed Form

Change each improper fraction to a mixed number.

51. $\dfrac{5}{3}$ **52.** $\dfrac{11}{5}$ **53.** $\dfrac{29}{12}$

54. $\dfrac{17}{3}$ **55.** $\dfrac{47}{5}$ **56.** $\dfrac{125}{12}$

Change each improper algebraic fraction to a mixed expression.

57. $\dfrac{x^2 + 1}{x}$ **58.** $\dfrac{1 - 4x}{2x}$ **59.** $\dfrac{4x - x^2}{2x^2}$

Applications

60. For a thin lens, the relationship between the focal length f, the object distance p, and the image distance q is

$$f = \frac{pq}{p + q}$$

A second lens of focal length f_1 has the same object distance p but a different image distance q_1.

$$f_1 = \frac{pq_1}{p + q_1}$$

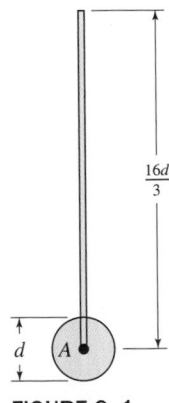

$\frac{16d}{3}$

Find the ratio f/f_1 and simplify.

61. The mass density of an object is its mass divided by its volume. Write and simplify an expression for the density of a sphere having a mass m and a volume equal to $4\pi r^3/3$.

62. The pressure on a surface is equal to the total force divided by the area. Write and simplify an expression for the pressure on a circular surface of area $\pi d^2/4$ subjected to a distributed load F.

d (A)

FIGURE 9–1

63. The stress on a bar in tension is equal to the load divided by the cross-sectional area. Write and simplify an expression for the stress in a bar having a trapezoidal cross section of area $(a + b)h/2$, subject to a load P.

64. The acceleration on a body is equal to the force on the body divided by its mass, and the mass equals the volume of the object times the density. Write and simplify an expression for the acceleration of a sphere having a volume $4\pi r^3/3$ and a density D, subjected to a force F.

65. To find the moment of the area A in Fig. 9–1, we must multiply the area $\pi d^2/4$ by the distance to the pivot, $16d/3$. Multiply and simplify.

66. The circle in Fig. 9–2 has an area $\pi d^2/4$, and the sector has an area equal to $ds/4$. Find the ratio of the area of the circle to the area of the sector.

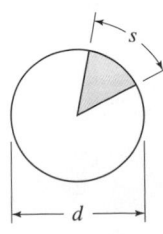

s

d

FIGURE 9–2

9–3   Addition and Subtraction of Fractions

Similar Fractions

Similar fractions (also called *like* fractions) are those having the same (common) denominator.

To add or subtract similar fractions, *combine the numerators and place them over the common denominator.*

Addition and Subtraction of Fractions	$$\frac{a}{b} \pm \frac{c}{b} = \frac{a \pm c}{b}$$	53

◆◆◆ **Example 20:**

(a) $\dfrac{2}{3} + \dfrac{5}{3} = \dfrac{2+5}{3} = \dfrac{7}{3}$

(b) $\dfrac{1}{x} + \dfrac{3}{x} = \dfrac{1+3}{x} = \dfrac{4}{x}$

(c) $\dfrac{3x}{x+1} - \dfrac{5}{x+1} + \dfrac{x^2}{x+1} = \dfrac{3x - 5 + x^2}{x+1}$ ◆◆◆

Least Common Denominator

The *least common denominator,* or LCD (also called the *lowest* common denominator), is the smallest expression that is exactly divisible by each of the denominators. Thus the LCD must contain all of the prime factors of each of the denominators. The common denominator of two or more fractions is simply the product of the denominators of those fractions. To find the *least* common denominator, before multiplying, drop any prime factor from one denominator that also appears in another denominator.

> The least common denominator is also called the *least common multiple* (LCM) of the denominators.

◆◆◆ **Example 21:** Find the LCD for the two fractions $\frac{3}{8}$ and $\frac{1}{18}$.

Solution: Factoring each denominator, we obtain

$$
\begin{array}{cc}
8 & 18 \\
(2)(2)(2) & (2)(3)(3)
\end{array}
$$

↘ duplicates; ↗
include only once
in LCD

Dropping one of the 2's that appear in both sets of factors, we find that our LCD is then the product of these factors.

$$\text{LCD} = (2)(2)(2)(3)(3) = 8(9) = 72 \qquad \text{◆◆◆}$$

For this simple problem, you probably found the LCD by inspection in less time than it took to read this example, and you are probably wondering what all the fuss is about. What we are really doing is developing a *method* that we can use on algebraic fractions when the LCD is *not* obvious.

◆◆◆ **Example 22:** Find the LCD for the fractions

$$\frac{5}{x^2 + x}, \quad \frac{x}{x^2 - 1}, \quad \text{and} \quad \frac{9}{x^3 - x^2}$$

Solution: The denominator $x^2 + x$ has the prime factors x and $x + 1$; the denominator $x^2 - 1$ has the prime factors $x + 1$ and $x - 1$; and the denominator $x^3 - x^2$ has the prime factors x, x, and $x - 1$. Our LCD is thus

$$(x)(x)(x + 1)(x - 1) \quad \text{or} \quad x^2(x^2 - 1)$$ ◆◆◆

Combining Fractions with Different Denominators

To combine fractions with different denominators, first find the LCD. Then multiply numerator and denominator of each fraction by that quantity that will make the denominator equal to the LCD. Finally, combine as shown above, and simplify.

◆◆◆ **Example 23:** Add $\frac{1}{2}$ and $\frac{2}{3}$.

Solution: The LCD is 6, so

$$\frac{1}{2} + \frac{2}{3} = \frac{1}{2}\left(\frac{3}{3}\right) + \frac{2}{3}\left(\frac{2}{2}\right)$$
$$= \frac{3}{6} + \frac{4}{6}$$
$$= \frac{7}{6}$$ ◆◆◆

◆◆◆ **Example 24:** Add $\frac{3}{8}$ and $\frac{1}{18}$.

It is not necessary to write the fractions with the *least* common denominator; any common denominator will work as well. But your final result will then have to be reduced to lowest terms.

Solution: The LCD, from Example 21, is 72. So

$$\frac{3}{8} + \frac{1}{18} = \frac{3}{8}\cdot\frac{9}{9} + \frac{1}{18}\cdot\frac{4}{4}$$
$$= \frac{27}{72} + \frac{4}{72}$$
$$= \frac{31}{72}$$ ◆◆◆

The method for adding and subtracting unlike fractions can be summarized by the formula that follows.

Combining Unlike Fractions	$\dfrac{a}{b} \pm \dfrac{c}{d} = \dfrac{ad}{bd} \pm \dfrac{bc}{bd} = \dfrac{ad \pm bc}{bd}$	**54**

◆◆◆ **Example 25:** Combine $\dfrac{x}{2y} - \dfrac{5}{x}$.

Solution: The LCD will be the product of the two denominators, or $2xy$. So

$$\frac{x}{2y} - \frac{5}{x} = \frac{x}{2y}\left(\frac{x}{x}\right) - \frac{5}{x}\left(\frac{2y}{2y}\right)$$
$$= \frac{x^2}{2xy} - \frac{10y}{2xy}$$
$$= \frac{x^2 - 10y}{2xy}$$ ◆◆◆

The procedure is the same, of course, even when the denominators are more complicated.

◆◆◆ **Example 26:** Combine the fractions

$$\frac{x + 2}{x - 3} + \frac{2x + 1}{3x - 2}$$

Solution: Our LCD is $(x - 3)(3x - 2)$, so

$$\frac{x + 2}{x - 3} + \frac{2x + 1}{3x - 2} = \frac{(x + 2)(3x - 2)}{(x - 3)(3x - 2)} + \frac{(2x + 1)(x - 3)}{(3x - 2)(x - 3)}$$

$$= \frac{(3x^2 + 4x - 4) + (2x^2 - 5x - 3)}{(x - 3)(3x - 2)}$$

$$= \frac{5x^2 - x - 7}{3x^2 - 11x + 6} \qquad \text{◆◆◆}$$

Always factor the denominators completely. This will show whether the same factor appears in more than one denominator.

◆◆◆ **Example 27:** Combine and simplify:

$$\frac{1}{x - 3} + \frac{x - 1}{x^2 + 3x + 9} + \frac{x^2 + x - 3}{x^3 - 27}$$

Solution: Factoring gives

$$\frac{1}{x - 3} + \frac{x - 1}{x^2 + 3x + 9} + \frac{x^2 + x - 3}{(x - 3)(x^2 + 3x + 9)}$$

Our LCD is then $x^3 - 27$ or $(x - 3)(x^2 + 3x + 9)$. Multiplying and adding, we have

$$\frac{(x^2 + 3x + 9) + (x - 1)(x - 3) + (x^2 + x - 3)}{(x - 3)(x^2 + 3x + 9)}$$

$$= \frac{x^2 + 3x + 9 + x^2 - 4x + 3 + x^2 + x - 3}{(x - 3)(x^2 + 3x + 9)}$$

$$= \frac{3x^2 + 9}{x^3 - 27} \qquad \text{◆◆◆}$$

Combining Integers and Fractions

To combine integers and fractions, treat the integer as a fraction having 1 as a denominator, and combine as shown above.

◆◆◆ **Example 28:**

$$3 + \frac{2}{9} = \frac{3}{1} + \frac{2}{9} = \frac{3}{1} \cdot \frac{9}{9} + \frac{2}{9}$$

$$= \frac{27}{9} + \frac{2}{9} = \frac{29}{9} \qquad \text{◆◆◆}$$

The same procedure may be used to change a *mixed number* to an *improper fraction*.

◆◆◆ **Example 29:**

$$2\frac{1}{3} = 2 + \frac{1}{3} = \frac{6}{3} + \frac{1}{3}$$

$$= \frac{7}{3} \qquad \text{◆◆◆}$$

Changing a Mixed Algebraic Expression to an Improper Fraction

This procedure is no different from that in the preceding section. Write the nonfractional expression as a fraction with 1 as the denominator, find the LCD, and combine, as shown in the following example.

◆◆◆ Example 30:

$$x^2 + \frac{5}{2x} = \frac{x^2}{1}\left(\frac{2x}{2x}\right) + \frac{5}{2x}$$

Adding, we obtain

$$x^2 + \frac{5}{2x} = \frac{2x^3 + 5}{2x}$$ ◆◆◆

Partial Fractions

Those fractions whose sum is equal to a given fraction are called *partial fractions*.

◆◆◆ Example 31: If we add the two fractions $2/x$ and $3/x^2$, we get

$$\frac{2}{x} + \frac{3}{x^2} = \frac{2x}{x^2} + \frac{3}{x^2} = \frac{2x + 3}{x^2}$$ ◆◆◆

Turned around, we say that the fraction $(2x + 3)/x^2$ has the partial fractions $2/x$ and $3/x^2$.

To be able to separate a fraction into its partial fractions is important in calculus. We mention them here but will not calculate any until later.

Don't use your calculator for these numerical problems. The practice you get working with common fractions will help you when doing algebraic fractions.

Exercise 3 ◆ Addition and Subtraction of Fractions

Common Fractions and Mixed Numbers

Combine and simplify.

1. $\frac{3}{5} + \frac{2}{5}$

2. $\frac{1}{8} - \frac{3}{8}$

3. $\frac{2}{7} + \frac{5}{7} - \frac{6}{7}$

4. $\frac{5}{9} + \frac{7}{9} - \frac{1}{9}$

5. $\frac{1}{3} - \frac{7}{3} + \frac{11}{3}$

6. $\frac{1}{5} - \frac{9}{5} + \frac{12}{5} - \frac{2}{5}$

7. $\frac{1}{2} + \frac{2}{3}$

8. $\frac{3}{5} - \frac{1}{3}$

9. $\frac{3}{4} + \frac{7}{16}$

10. $\frac{2}{3} + \frac{3}{7}$

11. $\frac{5}{9} - \frac{1}{3} + \frac{3}{18}$

12. $\frac{1}{2} + \frac{1}{3} + \frac{1}{5}$

13. $2 + \frac{3}{5}$

14. $3 - \frac{2}{3} + \frac{1}{6}$

15. $\frac{1}{5} - 7 + \frac{2}{3}$

16. $\frac{1}{5} + 2 - \frac{1}{3} + 5$

17. $\frac{1}{7} + 2 - \frac{3}{7} + \frac{1}{5}$

18. $5 - \frac{3}{5} + \frac{2}{15} - 2$

Rewrite each mixed number as an improper fraction.

19. $2\frac{2}{3}$

20. $1\frac{5}{8}$

21. $3\frac{1}{4}$

22. $2\frac{7}{8}$

23. $5\frac{11}{16}$

24. $9\frac{3}{4}$

Algebraic Fractions

Combine and simplify.

25. $\dfrac{1}{a} + \dfrac{5}{a}$

26. $\dfrac{3}{x} + \dfrac{2}{x} - \dfrac{1}{x}$

27. $\dfrac{2a}{y} + \dfrac{3}{y} - \dfrac{a}{y}$

28. $\dfrac{x}{3a} - \dfrac{y}{3a} + \dfrac{z}{3a}$

29. $\dfrac{5x}{2} - \dfrac{3x}{2}$

30. $\dfrac{7}{x+2} - \dfrac{5}{x+2}$

31. $\dfrac{3x}{a-b} + \dfrac{2x}{b-a}$

32. $\dfrac{a}{x^2} - \dfrac{b}{x^2} + \dfrac{c}{x^2}$

33. $\dfrac{7}{a+1} + \dfrac{9}{1+a} - \dfrac{3}{-a-1}$

34. $\dfrac{2a}{x(x-1)} - \dfrac{3a}{x^2-x} + \dfrac{a}{x-x^2}$

35. $\dfrac{3a}{2x} + \dfrac{2a}{5x}$

36. $\dfrac{1}{x} + \dfrac{1}{y} + \dfrac{1}{z}$

37. $\dfrac{a}{3} + \dfrac{2}{x} - \dfrac{1}{2}$

38. $\dfrac{1}{x+3} + \dfrac{1}{x-2}$

39. $\dfrac{a+b}{3} - \dfrac{a-b}{2}$

40. $\dfrac{3}{a+b} + \dfrac{2}{a-b}$

41. $\dfrac{4}{x-1} - \dfrac{5}{x+1}$

42. $\dfrac{x+1}{x-1} - \dfrac{x-1}{x+1}$

43. $\dfrac{b}{a+b} - \dfrac{a}{b-a} - \dfrac{a-b}{a+b}$

44. $\dfrac{y^2-2xy+x^2}{x^2-xy} + \dfrac{x}{x-y}$

45. $\dfrac{1}{2(x-1)} - \dfrac{1}{2(x+1)} + \dfrac{1}{x^2}$

46. $\dfrac{1+x}{1+x+x^2} + \dfrac{1-x}{1-x+x^2}$

47. $\dfrac{1}{x+1} + \dfrac{2}{x+2} + \dfrac{3}{x-3}$

48. $\dfrac{a-b}{ab} + \dfrac{b-c}{bc} + \dfrac{c-a}{ac}$

49. $\dfrac{x}{x^2-9} + \dfrac{x}{x+3}$

50. $\dfrac{x+3}{x^2-2x-8} - \dfrac{2x}{x-4}$

51. $\dfrac{5}{x-2} - \dfrac{3x+4}{x^3-8}$

52. $\dfrac{3x}{x-1} + \dfrac{5}{x^3+x^2-2x} + \dfrac{2}{x} + \dfrac{x-1}{x+2}$

53. $\dfrac{1}{x+d+1} - \dfrac{1}{x+1}$

54. $\dfrac{1}{(x+d)^2} - \dfrac{1}{x^2}$

55. $x + d + \dfrac{1}{x+d} - \left(x + \dfrac{1}{x}\right)$

56. $\dfrac{x+d}{(x+d-1)^2} - \dfrac{x}{(x-1)^2}$

Expressions of this form arise in calculus, when you are taking the derivative of a fraction by the delta method.

Mixed Form

Rewrite each mixed expression as an improper fraction.

57. $x + \dfrac{1}{x}$

58. $\dfrac{1}{2} - x$

59. $\dfrac{x}{x-1} - 1 - \dfrac{1}{x+1}$

60. $a + \dfrac{1}{a^2 - b^2}$

61. $3 - a - \dfrac{2}{a}$

62. $\dfrac{5}{x} - 2x + \dfrac{3}{x^2}$

63. $\left(3b + \dfrac{1}{x}\right) - \left(2b - \dfrac{b}{ax}\right)$

64. $\left(5x + \dfrac{x+2}{3}\right) - \left(2x - \dfrac{2x-3}{4}\right)$

Applications

Treat the given numbers in these problems as exact, and leave your answers in fractional form.

65. A certain work crew can grade 7 km of roadbed in 3 days, and another crew can do 9 km in 4 days. How much can both crews together grade in 1 day?

Hint: In problem 65, first find the amount that each crew can do in one day (e.g., the first crew can grade 7/3 km per day). Then add the separate amounts to get the daily total. You can use a similar approach to the other work problems in this group.

66. Liquid is running into a tank from a pipe that can fill four tanks in 3 days. Meanwhile, liquid is running out from a drain that can empty two tanks in 4 days. What will be the net change in volume in 1 day?

67. A planer makes a 1 m cutting stroke at a rate of 15 m/min and returns at 75 m/min. How long does it take for the cutting stroke and return? (See Eq. A17.)

68. Two resistors, 5Ω and 15Ω, are wired in parallel. What is the resistance of the parallel combination? (See Eq. A64.)

Do this group of problems without using your calculator, and leave your answers in fractional form.

69. One crew can put together five machines in 8 days. Another crew can assemble three of these machines in 4 days. How much can both crews together assemble in 1 day?

70. One crew can assemble M machines in p days. Another crew can assemble N of these machines in q days. Write an expression for the number of machines that both crews together can assemble in 1 day, combine into a single term, and simplify.

71. A steel plate in the shape of a trapezoid (Fig. 9–3) has a hole of diameter d. The area of the plate, less the hole, is

$$\frac{(a+b)h}{2} - \frac{\pi d^2}{4}$$

Combine these two terms and simplify.

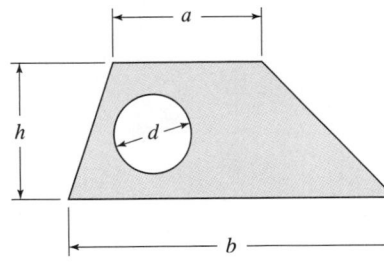

FIGURE 9–3

72. The total resistance R of Fig. 9–4 is

$$\frac{R_1 R_2}{R_1 + R_2} + R_3$$

Combine into a single term and simplify.

73. If a car travels a distance d at a rate V, the time required will be d/V. The car then continues for a distance d_1 at a rate V_1, and a third distance d_2 at rate V_2. Write an expression for the total travel time; then combine the three terms into a single term and simplify.

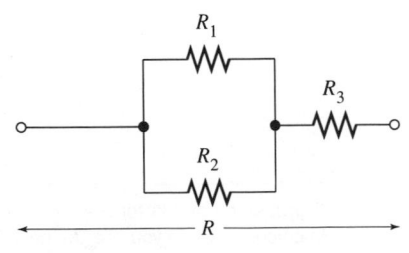

FIGURE 9–4

9–4   Complex Fractions

Fractions that have only *one* fraction line are called *simple* fractions. Fractions with *more than one* fraction line are called *complex* fractions.

◆◆◆ **Example 32:** The following are complex fractions:

$$\frac{\dfrac{a}{b}}{c} \quad \text{and} \quad \frac{x - \dfrac{1}{x}}{\dfrac{x}{y} - x} \qquad\qquad ◆◆◆$$

We show how to simplify complex fractions in the following examples.

◆◆◆ **Example 33:** Simplify the complex fraction

$$\frac{\dfrac{1}{2} + \dfrac{2}{3}}{3 + \dfrac{1}{4}}$$

Solution: We can simplify this fraction by multiplying numerator and denominator by the least common denominator for all of the individual fractions. The denominators are 2, 3, and 4, so the LCD is 12. Multiplying, we obtain

$$\frac{\left(\dfrac{1}{2} + \dfrac{2}{3}\right)12}{\left(3 + \dfrac{1}{4}\right)12} = \frac{6 + 8}{36 + 3} = \frac{14}{39} \qquad\qquad ◆◆◆$$

◆◆◆ **Example 34:** Simplify the complex fraction

$$\frac{1 + \dfrac{a}{b}}{1 - \dfrac{b}{a}}$$

Solution: The LCD for the two small fractions a/b and b/a is ab. Multiplying, we obtain

$$\frac{1 + \dfrac{a}{b}}{1 - \dfrac{b}{a}} \cdot \frac{ab}{ab} = \frac{ab + a^2}{ab - b^2}$$

or, in factored form,

$$\frac{a(b + a)}{b(a - b)} \qquad\qquad ◆◆◆$$

In the following example, we simplify a small section at a time and work outward. Try to follow the steps.

◆◆◆ **Example 35:**

$$\frac{a + \dfrac{1}{b}}{a + \dfrac{1}{b + \dfrac{1}{a}}} \cdot \frac{ab}{ab + 1} = \frac{\dfrac{ab + 1}{b}}{a + \dfrac{1}{\dfrac{ab + 1}{a}}} \cdot \frac{ab}{ab + 1}$$

$$= \frac{\dfrac{ab + 1}{b}}{a + \dfrac{a}{ab + 1}} \cdot \frac{ab}{ab + 1}$$

$$= \frac{\dfrac{ab + 1}{b}}{\dfrac{a(ab + 1) + a}{ab + 1}} \cdot \frac{ab}{ab + 1}$$

$$= \frac{ab+1}{b} \cdot \frac{ab+1}{a(ab+1)+a} \cdot \frac{ab}{ab+1}$$

$$= \frac{a(ab+1)}{a(ab+1)+a} = \frac{ab+1}{ab+2} \qquad \text{⬥⬥⬥}$$

Exercise 4 ⬥ Complex Fractions

Simplify. Leave your answers as improper fractions.

1. $\dfrac{\frac{2}{3}+\frac{3}{4}}{\frac{1}{5}}$

2. $\dfrac{\frac{3}{4}-\frac{1}{3}}{\frac{1}{2}+\frac{1}{6}}$

3. $\dfrac{\frac{1}{2}+\frac{1}{3}+\frac{1}{4}}{3-\frac{4}{5}}$

4. $\dfrac{\frac{4}{5}}{\frac{1}{5}+\frac{2}{3}}$

5. $\dfrac{5-\frac{2}{5}}{6+\frac{1}{3}}$

6. $\dfrac{1}{2}+\dfrac{3}{\frac{2}{5}+\frac{1}{3}}$

7. $\dfrac{x+\frac{y}{4}}{x-\frac{y}{3}}$

8. $\dfrac{\frac{a}{b}+\frac{x}{y}}{\frac{a}{z}-\frac{x}{c}}$

9. $\dfrac{1+\frac{x}{y}}{1-\frac{x^2}{y^2}}$

10. $\dfrac{x+\frac{a}{c}}{x+\frac{b}{d}}$

11. $\dfrac{a^2+\frac{x}{3}}{4+\frac{x}{5}}$

12. $\dfrac{3a^2-3y^2}{\frac{a+y}{3}}$

13. $\dfrac{x+\frac{2d}{3ac}}{x+\frac{3d}{2ac}}$

14. $\dfrac{4a^2-4x^2}{\frac{a+x}{a-x}}$

15. $\dfrac{x^2-\frac{y^2}{2}}{\frac{x-3y}{2}}$

16. $\dfrac{\frac{ab}{7}-3d}{3c-\frac{ab}{d}}$

17. $\dfrac{1+\frac{1}{x+1}}{1-\frac{1}{x-1}}$

18. $\dfrac{xy-\frac{3x}{ac}}{\frac{ac}{x}+2c}$

19. $\dfrac{\frac{2m+n}{m+n}-1}{1-\frac{n}{m+n}}$

20. $\dfrac{\frac{x^3+y^3}{x^2-y^2}}{\frac{x^2-xy+y^2}{x-y}}$

21. $$\dfrac{1}{a + \dfrac{1}{1 + \dfrac{a+1}{3-a}}}$$

22. $$\dfrac{x+y}{x+y+\dfrac{1}{x-y+\dfrac{1}{x+y}}}$$

23. $$\dfrac{3abc}{bc+ac+ab} - \dfrac{\dfrac{a-1}{a}+\dfrac{b-1}{b}+\dfrac{c-1}{c}}{\dfrac{1}{a}+\dfrac{1}{b}+\dfrac{1}{c}}$$

24. $$\dfrac{\dfrac{1}{a}+\dfrac{1}{b+c}}{\dfrac{1}{a}-\dfrac{1}{b+c}}\left(1 + \dfrac{b^2+c^2-a^2}{2bc}\right)$$

25. $$\dfrac{\dfrac{x^2-y^2-z^2-2yz}{x^2-y^2-z^2+2yz}}{\dfrac{x-y-z}{x+y-z}}$$

Applications

26. A car travels a distance d_1 at a rate V_1, then another distance d_2 at a rate V_2. The average speed for the entire trip is

$$\text{average speed} = \dfrac{d_1 + d_2}{\dfrac{d_1}{V_1} + \dfrac{d_2}{V_2}}$$

Simplify this complex fraction.

27. The equivalent resistance of two resistors in parallel is

$$\dfrac{R_1 R_2}{R_1 + R_2}$$

If each resistor is made of wire of resistivity ρ, with R_1 using a wire of length L_1 and cross-sectional area A_1, and R_2 having a length L_2 and area A_2, our expression becomes

$$\dfrac{\dfrac{\rho L_1}{A_1} \cdot \dfrac{\rho L_2}{A_2}}{\dfrac{\rho L_1}{A_1} + \dfrac{\rho L_2}{A_2}}$$

See Eq. A71, $R = \rho L/A$.

Simplify this complex fraction.

9–5 Fractional Equations

Solving Fractional Equations

An equation in which one or more terms is a fraction is called a *fractional equation*. To solve a fractional equation, first eliminate the fractions by multiplying both sides of the equation by the least common denominator (LCD) of every term. We can do this because multiplying both sides of an equation by the same quantity (the LCD in this case) does not unbalance the equation. With the fractions thus eliminated, the equation is then solved as any nonfractional equation is.

◆◆◆ Example 36: Solve for x:

$$\frac{3x}{5} - \frac{x}{3} = \frac{2}{15}$$

Solution: Multiplying both sides of the equation by the LCD (15), we obtain

$$15\left(\frac{3x}{5} - \frac{x}{3}\right) = 15\left(\frac{2}{15}\right)$$
$$9x - 5x = 2$$
$$x = \frac{1}{2}$$

◆◆◆

Equations with the Unknown in the Denominator

The procedure is the same when the unknown appears in the denominator of one or more terms. However, the LCD will now contain the unknown. Here it is understood that x cannot have a value that will make any of the denominators in the problem equal to zero. Such forbidden values are sometimes stated with the problem, but often they are not.

◆◆◆ Example 37: Solve for x:

$$\frac{2}{3x} = \frac{5}{x} + \frac{1}{2}(x \neq 0)$$

Solution: The LCD is $6x$. Multiplying both sides of the equation yields

> We assume the integers in these equations to be exact numbers. Thus we leave our answers in fractional form rather than converting to (approximate) decimals.

$$6x\left(\frac{2}{3x}\right) = 6x\left(\frac{5}{x} + \frac{1}{2}\right)$$
$$6x\left(\frac{2}{3x}\right) = 6x\left(\frac{5}{x}\right) + 6x\left(\frac{1}{2}\right)$$
$$4 = 30 + 3x$$
$$3x = -26$$
$$x = -\frac{26}{3}$$

Check:

$$\frac{2}{3\left(-\frac{26}{3}\right)} \overset{?}{=} \frac{5}{-\frac{26}{3}} + \frac{1}{2}$$
$$-\frac{2}{26} \overset{?}{=} -\frac{15}{26} + \frac{13}{26}$$
$$-\frac{2}{26} = -\frac{2}{26} \quad \text{(checks)}$$

◆◆◆

◆◆◆ Example 38: Solve for x:

$$\frac{8x + 7}{5x + 4} = 2 - \frac{2x}{5x + 1}$$

Solution: Multiplying by the LCD $(5x + 4)(5x + 1)$ yields

$$(8x + 7)(5x + 1) = 2(5x + 4)(5x + 1) - 2x(5x + 4)$$
$$40x^2 + 35x + 8x + 7 = 2(25x^2 + 5x + 20x + 4) - 10x^2 - 8x$$
$$40x^2 + 43x + 7 = 50x^2 + 50x + 8 - 10x^2 - 8x$$
$$43x + 7 = 42x + 8$$
$$x = 1$$

Check:

$$\frac{8(1) + 7}{5(1) + 4} \overset{?}{=} 2 - \frac{2(1)}{5(1) + 1}$$
$$\frac{15}{9} \overset{?}{=} 2 - \frac{2}{6}$$
$$\frac{5}{3} = \frac{6}{3} - \frac{1}{3} \qquad \text{(checks)} \qquad \blacklozenge\blacklozenge\blacklozenge$$

Common Error	The technique of multiplying by the LCD in order to eliminate the denominators is valid *only when we have an equation*. Do not multiply through by the LCD when there is no equation!

◆◆◆ Example 39: Solve for x:

$$\frac{3}{x - 3} = \frac{2}{x^2 - 2x - 3} + \frac{4}{x + 1}$$

Solution: The denominator of the second fraction factors into

$$x^2 - 2x - 3 = (x - 3)(x + 1)$$

The LCD is therefore $(x - 3)(x + 1)$. Multiplying both sides of the given equation by the LCD gives

$$3(x + 1) = 2 + 4(x - 3)$$
$$3x + 3 = 4x - 10$$
$$x = 13 \qquad \blacklozenge\blacklozenge\blacklozenge$$

Exercise 5 ◆ Fractional Equations

Solve for x.

1. $2x + \dfrac{x}{3} = 28$ **2.** $4x + \dfrac{x}{5} = 42$

3. $x + \dfrac{x}{5} = 24$ **4.** $\dfrac{x}{6} + x = 21$

5. $3x - \dfrac{x}{7} = 40$ **6.** $x - \dfrac{x}{6} = 25$

7. $\dfrac{2x}{3} - x = -24$ **8.** $\dfrac{3x}{5} + 7x = 38$

9. $\dfrac{x}{2} + \dfrac{x}{3} + \dfrac{x}{4} = 26$ **10.** $\dfrac{x - 1}{2} = \dfrac{x + 1}{3}$

11. $\dfrac{3x - 1}{4} = \dfrac{2x + 1}{3}$

12. $x + \dfrac{2x}{3} + \dfrac{3x}{4} = 29$

13. $2x + \dfrac{x}{3} - \dfrac{x}{4} = 50$

14. $3x - \dfrac{2x}{3} - \dfrac{5x}{6} = 18$

15. $3x - \dfrac{x}{6} + \dfrac{x}{12} = 70$

16. $\dfrac{x}{4} + \dfrac{x}{6} + \dfrac{x}{8} = 26$

17. $\dfrac{6x - 19}{2} = \dfrac{2x - 11}{3}$

18. $\dfrac{7x - 40}{8} = \dfrac{9x - 80}{10}$

19. $\dfrac{3x - 116}{4} + \dfrac{180 - 5x}{6} = 0$

20. $\dfrac{3x - 4}{2} - \dfrac{3x - 1}{16} = \dfrac{6x - 5}{8}$

21. $\dfrac{x - 1}{8} - \dfrac{x + 1}{18} = 1$

22. $\dfrac{3x}{4} + \dfrac{180 - 5x}{6} = 29$

23. $\dfrac{15x}{4} = \dfrac{9}{4} - \dfrac{3 - x}{2}$

24. $\dfrac{x}{4} + \dfrac{x}{10} + \dfrac{x}{8} = 19$

Equations with Unknown in Denominator

25. $\dfrac{2}{3x} + 6 = 5$

26. $9 - \dfrac{4}{5x} = 7$

27. $4 + \dfrac{1}{x + 3} = 8$

28. $\dfrac{x + 5}{x - 2} = 5$

29. $\dfrac{x - 3}{x + 2} = \dfrac{x + 4}{x - 5}$

30. $\dfrac{3}{x + 2} = \dfrac{5}{x} + \dfrac{4}{x^2 + 2x}$

31. $\dfrac{9}{x^2 + x - 2} = \dfrac{7}{x - 1} - \dfrac{3}{x + 2}$

32. $\dfrac{3x + 5}{2x - 3} = \dfrac{3x - 3}{2x - 1}$

33. $\dfrac{4}{x^2 - 1} + \dfrac{1}{x - 1} + \dfrac{1}{x + 1} = 0$

34. $\dfrac{x}{3} - \dfrac{x^2 - 5x}{3x - 7} = \dfrac{2}{3}$

35. $\dfrac{3}{x + 1} - \dfrac{x + 1}{x - 1} = \dfrac{x^2}{1 - x^2}$

36. $\dfrac{x + 1}{2(x - 1)} - \dfrac{x - 1}{x + 1} = \dfrac{17 - x^2}{2(x^2 - 1)}$

37. $\dfrac{9x + 20}{36} - \dfrac{x}{4} = \dfrac{4x - 12}{5x - 4}$

38. $\dfrac{10x + 13}{18} - \dfrac{x + 2}{x - 3} = \dfrac{5x - 4}{9}$

39. $\dfrac{6x + 7}{10} - \dfrac{3x + 1}{5} = \dfrac{x - 1}{3x - 4}$

40. $\dfrac{5x}{1 - x} - \dfrac{7x}{3 + x} = \dfrac{12(1 - x^2)}{x^2 + 2x - 3}$

9–6   Word Problems Leading to Fractional Equations

Solving Word Problems

The method for solving the word problems in this chapter (and in any other chapter) is no different from that given in Chapter 3. The resulting equations, however, will now contain fractions.

◆◆◆ **Example 40:** One-fifth of a certain number is 3 greater than one-sixth of that same number. Find the number.

Solution: Let x = the number. Then from the problem statement,

$$\frac{1}{5}x = \frac{1}{6}x + 3$$

Multiplying by the LCD (30), we obtain

$$6x = 5x + 90$$
$$x = 90$$

Check: Is one-fifth of 90 (which is 18) 3 greater than one-sixth of 90 (which is 15)? Yes. ◆◆◆

Rate Problems

Fractional equations often arise when we solve problems involving *rates*. Problems of this type include:

Uniform motion:	*amount traveled = rate of travel × time traveled*
Work:	*amount of work done = rate of work × time worked*
Fluid flow:	*amount of fluid = flow rate × duration of flow*
Energy flow:	*amount of energy = rate of energy flow × time*

Notice that these equations are mathematically identical. Word problems involving motion, flow of fluids or solids, flow of heat, electricity, solar energy, mechanical energy, and so on, can be handled in the same way.

To convince yourself of the similarity of all of these types of problems, consider the following work problem:

- If worker M can do a certain job in 5 h, and N can do the same amount in 8 h, how long will it take M and N together to do that same amount?

Are the following problems really any different?

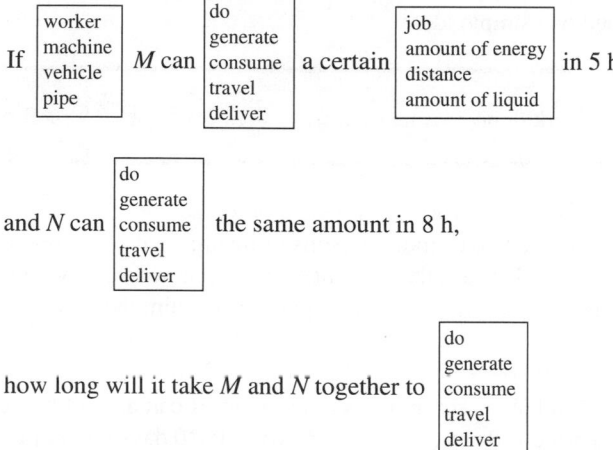

that same amount?

We will first do a uniform motion problem and then proceed to work and flow problems.

Uniform Motion

Motion is called *uniform* when the speed does not change. The distance traveled at constant speed is related to the rate of travel and the elapsed time by the following equation:

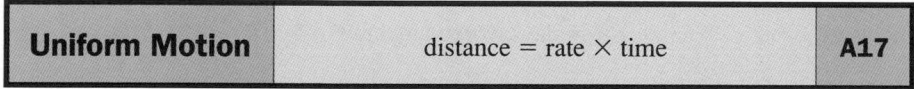

Uniform Motion	distance = rate × time	**A17**

Be careful not to use this formula for anything but *uniform* motion.

◆◆◆ **Example 41:** A train departs at noon travelling at a speed of 64 km/h. A car leaves the same station $\frac{1}{2}$h later to overtake the train, travelling on a road parallel to the track. If the car's speed is 96 km/h, at what time and at what distance from the station will it overtake the train?

Estimate: The train gets a head start of (1/2)(64), or 32 km. But the car goes 32 km/h faster than the train, so every hour, the distance between car and train is reduced by 32 km. Thus it should take 1 hour for the car to overtake the train. The distance will be 1 hour's travel distance for the car, or 96 km.

Solution: Let x = time traveled by car (hours). Then $x + \frac{1}{2}$ = time travelled by train (hours). The distance traveled by the car is

$$\text{distance} = \text{rate} \times \text{time} = 96x$$

and for the train,

$$\text{distance} = 64\left(x + \frac{1}{2}\right)$$

But the distance is the same for car and train, so we can write the equation

$$96x = 64\left(x + \frac{1}{2}\right)$$

Removing parentheses yields

$$96x = 64x + 32$$
$$96x - 64x = 32$$
$$32x = 32$$
$$x = 1\,\text{h} = \text{time for car}$$
$$x + \frac{1}{2} = 1\tfrac{1}{2}\text{h} = \text{time for train}$$

So the car overtakes the train at 1:30 P.M. at a distance of 96 km from the station.　◆◆◆

Work Problems

To tackle work problems, we need one simple idea:

Work	work done = (rate of work) × (time)	A5

We often use this equation to find the rate of work for a person or a machine.

In a typical work problem, there are two or more persons or machines doing work, each at a different rate. For each worker the work rate is the amount done by that worker divided by the time taken to do the work. That is, if a person can stamp 9 parts in 13 min, that person's work rate is $\frac{9}{13}$ part per minute.

◆◆◆ **Example 42:** Crew A can assemble 2 cars in 5 days, and crew B can assemble 3 cars in 7 days. If both crews together assemble 100 cars, with crew B working 10 days longer than crew A, how many days (rounded to the nearest day) must each crew work?

Estimate: Working alone, crew A does 2 cars in 5 days, or 100 cars in 250 days. Similarly, crew B does 100 cars in about 231 days. Together they do 200 cars in 250 + 231 or 481 days, or about 100 cars in 240 days. Thus we would expect each crew to work about half that, or about 120 days each. But crew B works 10 days longer than A, so we guess that they work about 5 days longer than 120 days, whereas crew A works about 5 days less than 120 days.

Solution: Let

$$x = \text{days worked by crew } A$$

and

$$x + 10 = \text{days worked by crew } B$$

The work rate of each crew is

$$\text{crew } A: \quad \text{rate} = \frac{2}{5} \text{ car per day}$$

$$\text{crew } B: \quad \text{rate} = \frac{3}{7} \text{ car per day}$$

The amount done by each crew equals their work rate times the number of days worked. The sum of the amounts done by the two crews must equal 100 cars.

$$\frac{2}{5}x + \frac{3}{7}(x + 10) = 100$$

We clear fractions by multiplying by the LCD, 35.

$$14x + 15(x + 10) = 3500$$
$$14x + 15x + 150 = 3500$$
$$29x = 3350$$
$$x = 116 \quad \text{days for crew } A$$
$$x + 10 = 126 \quad \text{days for crew } B \qquad \text{◆◆◆}$$

Fluid Flow and Energy Flow

For flow problems, we use the simple equation

Fluid or Energy Flow	amount of flow = flow rate × time	**A4**

Here we assume, of course, that the flow rate is constant.

◆◆◆ **Example 43:** A certain small hydroelectric generating station can produce 61 gigajoules (GJ) of energy per year. After 4.0 months of operation, another generator is added which, by itself, can produce 39 GJ in 5.0 months. How many additional months are needed for a total of 95 GJ to be produced?

Solution: Let x = additional months. The original generating station can produce 61/12 GJ per month for 4.0 + x months. The new generator can produce 39/5.0 GJ per month for x months. The problem states that the total amount produced (in GJ) is 95, so

$$\frac{61}{12}(4.0 + x) + \frac{39}{5.0}x = 95$$

Multiplying by the LCD, 60, we obtain

$$5.0(61)(4.0 + x) + 12(39)x = 95(60)$$
$$1220 + 305x + 468x = 5700$$
$$773x = 4480$$
$$x = \frac{4480}{773} = 5.8 \text{ months} \qquad \text{◆◆◆}$$

Exercise 6 ◆ Word Problems

Number Problems

Treat the numbers in Problems 1–5 as exact.

1. There is a number such that if $\frac{1}{5}$ of it is subtracted from 50 and this difference is multiplied by 4, the result will be 70 less than the number. What is the number?

2. The number 100 is separated into two parts such that if $\frac{1}{3}$ of one part is subtracted from $\frac{1}{4}$ of the other, the difference is 11. Find the two parts.

3. The second of two numbers is 1 greater than the first, and $\frac{1}{2}$ of the first plus $\frac{1}{5}$ of the first is equal to the sum of $\frac{1}{3}$ of the second and $\frac{1}{4}$ of the second. What are the numbers?

4. Find two consecutive integers such that one-half of the larger exceeds one-third of the smaller by 9.

5. The difference between two positive numbers is 20, and $\frac{1}{7}$ of one is equal to $\frac{1}{3}$ of the other. What are the numbers?

Uniform Motion Problems

Depending on how you set them up, some of these problems may not result in fractional equations.

6. A person sets out from Boston and walks toward Portland at the rate of 3.0 km/h. Three hours afterward, a second person sets out from the same place and walks in the same direction at the rate of 4.0 km/h. How far from Boston will the second person overtake the first?

7. A courier who goes at the rate of 6.5 mi./h is followed, after 4.0 h, by another who goes at the rate of 7.5 mi./h. In how many hours will the second overtake the first?

8. A person walks to the top of a mountain at the rate of 3.0 km/h and down the same way at the rate of 6.0 km/h. If he walks for a total of 6.0 h, how far is it to the top?

9. In going a certain distance, a train traveling at the rate of $4\overline{0}$ km/h takes 2.0 h less than a train traveling $3\overline{0}$ km/h. Find the distance.

10. Spacecraft A is over Houston at noon on a certain day and travelling at a rate of $3\overline{00}$ km/h. Spacecraft B, attempting to overtake and dock with A, is over Houston at 1:15 P.M. and is traveling in the same direction as A, at 410 km/h. At what time will B overtake A? At what distance from Houston?

11. On board the space shuttle is the Shuttle Remote Manipulator System (SRMS) or "Canadarm." The specifications for the Canadarm velocity state that the manipulator can be moved at 6 cm/s when loaded and 60 cm/s when unloaded. If the Canadarm delivers a load and returns unloaded in the shortest possible time of 3 min 40 s, what is the one-way distance in metres?

12. From downtown Toronto, two people head for Montreal. One person takes the Via express train from Union Station, while the other catches a direct flight from Toronto Island Airport. The airport and the train station are less than 0.5 km apart, so the starting point is essentially the same for both travellers. Because of security checks at the airport, the flying passenger leaves 1.5 hours after the train passenger. The train travels at an average speed of 110 km/h, but the plane arrives in Montreal 45 minutes earlier than the train. If the distance from Toronto to Montreal is 505 km, find the speed of the plane.

13. The track length of the Grand Prix of Toronto race is 2.824 km. One race car travels at a speed of 72 m/s while another travels at 76 m/s. How much longer does it take for the slower car to complete one lap?

14. During the Calgary Stampede, there are pigeon races. The birds fly a 30 km race every day during Stampede, at an average speed of 12.8 km/h. On one race day, after the bird cage was opened to start the race, a particular bird waited for 25 minutes before leaving the cage to start the race. How fast does this pigeon have to fly in order to arrive at the finish line at the same time as the rest of the flock?

Work Problems

15. One labourer can do a certain job in 5 days, a second in 6 days, and a third in 8 days. In what time can the three together do the job?

16. Three masons build 318 m of wall. Mason A builds 7.0 m/day, B builds 6.0 m/day, and C builds 5.0 m/day. Mason B works twice as many days as A, and C works half as many days as A and B combined. How many days did each work?

17. If a carpenter can roof a house in 10 days and another can do the same in 14 days, how many days will it take if they work together?

18. A technician can assemble an instrument in 9.5 h. After working for 2.0 h, she is joined by another technician who, alone, could do the job in 7.5 h. How many additional hours are needed to finish the job?

19. A certain screw machine can produce a box of parts in 3.3 h. A new machine is to be ordered having a speed such that both machines working together would produce a box of parts in 1.4 h. How long would it take the new machine alone to produce a box of parts?

Fluid Flow Problems

20. A tank can be filled by a pipe in 3.0 h and emptied by another pipe in 4.0 h. How much time will be required to fill an empty tank if both are running?

The humourist Stephen Leacock, in his essay "A, B, and C—The Human Element in Mathematics," tells of those word problem heroes: A, full-blooded and having great strength and endurance; B, quiet and easygoing; and C, the weakling. Here's a sample:

"The first time I ever saw these men was one evening after a regatta. They had all been rowing in it, and it had transpired that A could row as much in one hour as B in two or C in four. B and C had come in dead fagged [tired] and C was coughing badly. . . . Just then A came blustering in and shouted, "I say, you fellows, Smith has shown me three cisterns in his garden and he says we can pump them until tomorrow night. I bet I can beat you both . . . I heard B growl . . . but they went, and presently I could tell from the sound of the water that A was pumping four times as fast as C."

In this group of problems, as in the previous ones, we assume that the rates are constant.

21. Two pipes empty into a tank. One pipe can fill the tank in 8.0 h, and the other in 9.0 h. How long will it take both pipes together to fill the tank?

22. A tank has three pipes connected. The first, by itself, could fill the tank in $2\frac{1}{2}$ h, the second in 2 h, and the third in 1 h 40 min. In how many minutes will the tank be filled?

23. A tank can be filled by a certain pipe in 18.0 h. Five hours after this pipe is opened, it is supplemented by a smaller pipe which, by itself, could fill the tank in 24.0 h. Find the total time, measured from the opening of the larger pipe, to fill the tank.

24. At what rate must liquid be drained from a tank in order to empty it in 1.50 h if the tank takes 4.70 h to fill at the rate of 3.50 m³/min?

Energy Flow Problems

25. A certain power plant consumes 1500 tonnes (t) of coal in 4.0 weeks. There is a stockpile of $1\overline{0}\,000$ t of coal available when the plant starts operating. After 3.0 weeks in operation, an additional boiler, capable of using 2300 t in 3.0 weeks, is put on line with the first boiler. In how many more weeks will the stockpile of coal be consumed?

26. A certain array of solar cells can generate 6.0 gigajoules (GJ) in 5.0 months (under standard conditions). After this array has been operating for 3.0 months, another array of cells is added which alone can generate 15.0 GJ in 7.0 months. How many additional months, after the new array has been added, is needed for the total energy generated from both arrays to be $3\overline{0}$ GJ?

27. A landlord owns a house that consumes 6300 L of heating oil in three winters. He buys another (insulated) house, and the two houses together use 5550 L of oil in two winters. How many winters would it take the insulated house alone to use 3750 L of oil?

28. A wind generator can charge 20 storage batteries in 24 h. After the generator has been charging for 6.0 h, another generator, which can charge the batteries in 36 h, is also connected to the batteries. How many additional hours are needed to charge the 20 batteries?

29. A certain solar collector can absorb 9.0 MJ in 7.0 h. Another panel is added, and together they collect 35 MJ in 5.0 h. How long would it take the new panel alone to collect 35 MJ?

Financial Problems

30. After spending $\frac{1}{4}$ of my money, then $\frac{1}{5}$ of the remainder, I had $66 left. How many dollars did I start with?

31. A person who had inherited some money spent $\frac{3}{8}$ of it the first year and $\frac{4}{5}$ of the remainder the next year and had $1420 left. What was the inheritance?

32. Three children were left an inheritance of which the eldest received $\frac{2}{3}$, the second $\frac{1}{5}$, and the third the rest, which was $20,000. How much did each receive?

33. A's capital was $\frac{3}{4}$ of B's. If A's had been $500 less, it would have been only $\frac{1}{2}$ of B's. What was the capital of each?

Here are a few financial problems, similar to those we had in Sec. 3–3, except that these contain fractions. Give your answers to the nearest dollar. Glance back to that section if you need help in setting up these problems.

9–7   Literal Equations and Formulas

Literal Equations

A *literal equation* is one in which some or all of the constants are represented by letters.

♦♦♦ Example 44: The following is a literal equation:

$$a(x + b) = b(x + c)$$

However,

$$2(x + 5) = 3(x + 1)$$

is a numerical equation.

Formulas

A *formula* is a literal equation that relates two or more mathematical or physical quantities. These are the equations, mentioned in the introduction of this chapter, that describe the workings of the physical world. A listing of some of the common formulas used in technology is given in "Appendix A, Summary of Facts and Formulas."

◆◆◆ **Example 45:** Newton's second law is a formula relating the force acting on a body with its mass and acceleration. The equation is given in the following box.

| Newton's Second Law | $F = ma$ | A20 |

◆◆◆

Solving Literal Equations and Formulas

When we solve a literal equation or formula, we cannot, of course, get a *numerical* answer, as we could with equations that had only one unknown. Our object here is to *isolate* one of the letters on one side of the equal sign. We "solve for" one of the literal quantities.

◆◆◆ **Example 46:** Solve for x:

$$a(x + b) = b(x + c)$$

Solution: Our goal is to isolate x on one side of the equation. Removing parentheses, we obtain

$$ax + ab = bx + bc$$

Subtracting bx and then ab will place all of the x terms on one side of the equation.

$$ax - bx = bc - ab$$

Factoring to isolate x yields

$$x(a - b) = b(c - a)$$

Dividing by $(a - b)$, where $a \cdot b$, gives us

$$x = \frac{b(c - a)}{a - b}$$

◆◆◆

◆◆◆ **Example 47:** Solve for x:

$$b\left(b + \frac{x}{a}\right) = d$$

Solution: Dividing both sides by b, we have

$$b + \frac{x}{a} = \frac{d}{b}$$

Subtracting b yields

$$\frac{x}{a} = \frac{d}{b} - b$$

Multiplying both sides by a gives us

$$x = a\left(\frac{d}{b} - b\right)$$

◆◆◆

◆◆◆ **Example 48:** The formula for the amount of heat q flowing by conduction through a wall of thickness L, conductivity k, and cross-sectional area A is

$$q = \frac{kA(t_1 - t_2)}{L}$$

where t_1 and t_2 are the temperatures of the warmer and cooler sides, respectively. Solve this equation for t_1.

Solution: Multiplying both sides by L/kA gives

$$\frac{qL}{kA} = t_1 - t_2$$

then adding t_2 to both sides, we get

$$t_1 = \frac{qL}{kA} + t_2$$

◆◆◆

Checking Literal Equations

As with numerical equations, we can substitute our expression for x back into the original equation and see if it checks.

◆◆◆ **Example 49:** We check the solution of Example 47 by substituting $a(d/b - b)$ for x in the original equation.

$$b\left(b + \frac{x}{a}\right) \overset{?}{=} d$$

$$b\left(b + \frac{a(d/b - b)}{a}\right) \overset{?}{=} d$$

$$b\left(b + \frac{d}{b} - b\right) \overset{?}{=} d$$

$$b\left(\frac{d}{b}\right) \overset{?}{=} d \quad \text{(checks)}$$

◆◆◆

Another simple "check" is to substitute numerical values. But avoid simple values such as 0 and 1 that may make an incorrect solution appear to check, and avoid any values that will make a denominator zero.

◆◆◆ **Example 50:** Check the results from Example 47.

Solution: Let us *choose* values for a, b, and d, say,

$$a = 2, \quad b = 3, \quad \text{and} \quad d = 6$$

With these values,

$$x = a\left(\frac{d}{b} - b\right) = 2\left(\frac{6}{3} - 3\right) = -2$$

We now see if these values check in the original equation.

$$b\left(b + \frac{x}{a}\right) = d$$

$$3\left(3 + \frac{-2}{2}\right) \overset{?}{=} 6$$

$$3(3 - 1) \overset{?}{=} 6$$

$$3(2) = 6 \quad \text{(checks)}$$

Realize that this check is not a rigorous one, because we are testing the solution only for the specific values chosen. However, if it checks for those values, we can be *reasonably* sure that our solution is correct. ◆◆◆

Literal Fractional Equations

To solve literal fractional equations, the procedure is the same as for other fractional equations: Multiply by the LCD to eliminate the fractions.

◆◆◆ **Example 51:** Solve the following for x in terms of a, b, and c:

$$\frac{x}{b} - \frac{a}{c} = \frac{x}{a}$$

Solution: Multiplying by the LCD (abc) yields

$$acx - a^2b = bcx$$

Rearranging so that all x terms are together on one side of the equation, we obtain

$$acx - bcx = a^2b$$

Factoring gives us

$$x(ac - bc) = a^2b$$
$$x = \frac{a^2b}{ac - bc} \qquad\qquad ◆◆◆$$

◆◆◆ **Example 52:** Solve for x:

$$\frac{x}{x - a} - \frac{x + 2b}{x + a} = \frac{a^2 + b^2}{x^2 - a^2}$$

Solution: The LCD is $x^2 - a^2$; that is, $(x - a)(x + a)$. Multiplying each term by the LCD gives

$$x(x + a) - (x + 2b)(x - a) = a^2 + b^2$$

Removing parentheses and rearranging so that all of the x terms are together, we obtain

$$x^2 + ax - (x^2 + 2bx - ax - 2ab) = a^2 + b^2$$
$$x^2 + ax - x^2 - 2bx + ax + 2ab = a^2 + b^2$$
$$2ax - 2bx = a^2 - 2ab + b^2$$

Then we factor:

$$2x(a - b) = (a - b)^2$$

and divide by $2(a - b)$:

$$x = \frac{(a - b)^2}{2(a - b)} = \frac{a - b}{2} \qquad\qquad ◆◆◆$$

Sometimes a fractional equation will have some terms that are not fractions themselves, but the method of solution is the same.

◆◆◆ **Example 53:** Solve for x:

$$\frac{x}{b} + c = \frac{2x}{a + b} + b$$

Solution: We multiply both sides by the LCD, $b(a + b)$.

$$x(a + b) + c(b)(a + b) = 2x(b) + b(b)(a + b)$$

Removing parentheses and moving all of the x terms to one side gives

$$ax + bx + abc + b^2c = 2bx + ab^2 + b^3$$
$$ax + bx - 2bx = ab^2 + b^3 - abc - b^2c$$

Factoring yields

$$x(a - b) = b(ab + b^2 - ac - bc)$$
$$x = \frac{b(ab + b^2 - ac - bc)}{a - b}$$

◆◆◆

Exercise 7 ◆ Literal Equations and Formulas

Solve for x.

1. $2ax = bc$

2. $ax + dx = a - c$

3. $a(x + y) = b(x + z)$

4. $4x = 2x + ab$

5. $4acx - 3d^2 = a^2d - d^2x$

6. $a(2x - c) = a + c$

7. $a^2x - cd = b - ax + dx$

8. $3(x - r) = 2(x + p)$

9. $\frac{a}{2}(x - 3w) = z$

10. $cx - x = bc - b$

11. $3x + m = b$

12. $ax + m = cx + n$

13. $ax - bx = c + dx - m$

14. $3m + 2x - c = x + d$

15. $ax - ab = cx - bc$

16. $p(x - b) = qx + d$

17. $3(x - b) = 2(bx + c) - c(x - b)$

18. $a(bx + d) - cdx = c(dx + d)$

Literal Fractional Equations

19. $\frac{w + x}{x} = w(w + y)$

20. $\frac{ax + b}{c} = bx - a$

21. $\frac{p - q}{x} = 3q$

22. $\frac{bx - c}{ax - c} = 5$

23. $\frac{a - x}{5} = \frac{b - x}{2}$

24. $\frac{x}{a} - b = \frac{c}{d} - x$

25. $\frac{x}{a} - a = \frac{a}{c} - \frac{x}{c - a}$

26. $\frac{x}{a - 1} - \frac{x}{a + 1} = b$

27. $\frac{x - a}{x - b} = \left(\frac{2x - a}{2x - b}\right)^2$

28. $\frac{a - b}{bx + c} + \frac{a + b}{ax - c} = 0$

29. $\frac{x + a}{x - a} - \frac{x - b}{x + b} = \frac{2(a + b)}{x}$

30. $\frac{ax}{b} - \frac{b - x}{2c} + \frac{a(b - x)}{3d} = a$

Word Problems with Literal Solutions

31. Two persons are d kilometres apart. They set out at the same time and travel toward each other. One travels at the rate of m km/h, and the other at the rate of n km/h. How far will each have ◆ traveled when they meet?

32. If A can do a piece of work in p hours, B in q hours, C in r hours, and D in s hours, how many hours will it take to do the work if all work together?

33. After spending $1/n$ and $1/m$ of my money, I had a dollars left. How many dollars did I have at first?

Formulas

34. The correction C for the sag in a surveyor's tape weighing w N/m and pulled with a force of P N is

$$C = \frac{w^2 L^3}{24 P^2} \text{ metres}$$

Solve this equation for the distance measured, L.

35. When a bar of length L_0 having a coefficient of linear thermal expansion α is increased in temperature by an amount Δt, it will expand to a new length L, where

$$L = L_0(1 + \alpha \Delta t)$$

Solve this equation for Δt.

36. Solve the equation given in problem 35 for the initial length L_0.

37. A rod of cross-sectional area a and length L will stretch by an amount e when subject to a tensile load of P. The modulus of elasticity is given by

$$E = \frac{PL}{ae}$$

Solve this equation for a.

38. The formula for the displacement s of a freely falling body having an initial velocity v_0 and acceleration a is

$$s = v_0 t + \frac{1}{2} a t^2$$

Solve this equation for a.

39. The formulas for the amount of heat flowing through a wall by conduction is

$$q = \frac{kA(t_1 - t_2)}{L}$$

Solve this equation for t_2.

40. An amount a invested at a simple interest rate n for t years will accumulate to an amount y, where $y = a + ant$. Solve for a.

41. The formula for the equivalent resistance R for the parallel combination of two resistors, R_1 and R_2, is

$$\frac{1}{R} = \frac{1}{R_1} + \frac{1}{R_2}$$

Solve this formula for R_2.

42. Applying Kirchhoff's law (Eq. A68) to loop 1 in Fig. 9–5, we get

$$E = I_1 R_1 + I_1 R_2 - I_2 R_2$$

Solve for I_1.

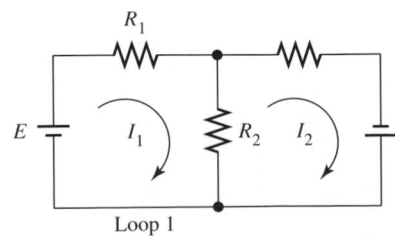

FIGURE 9–5

43. If the resistance of a conductor is R_1 at temperature t_1, the resistance will change to a value R when the temperature changes to t, where

$$R = R_1[1 + \alpha(t - t_1)]$$

and α is the temperature coefficient of resistance at temperature t_1.

Solve this equation for t_1.

44. Solve the equation given in problem 43 for the initial resistance R_1.

45. Taking the moment M about point p in Fig. 9–6, we get

$$M = R_1L - F(L - x)$$

Solve for L.

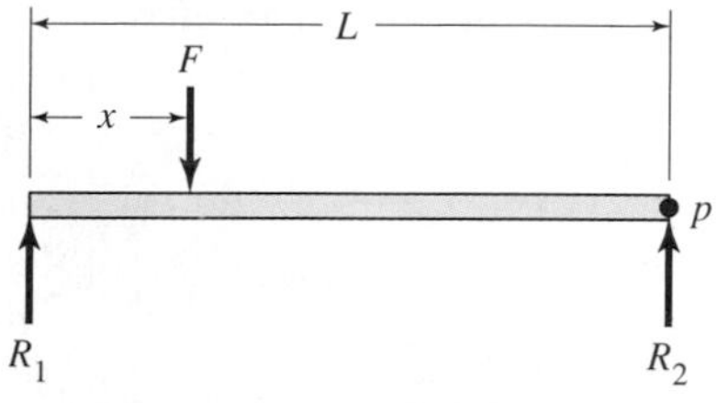

46. Three masses, m_1, m_2, and m_3, are attached together and accelerated by means of a force F, where

$$F = m_1a + m_2a + m_3a$$

Solve for the acceleration a.

47. A bar of mass m_1 is attached to a sphere of mass m_2 (Fig. 9–7). The distance x to the center of gravity (C.G.) is

$$x = \frac{10m_1 + 25m_2}{m_1 + m_2}$$

Solve for m_1.

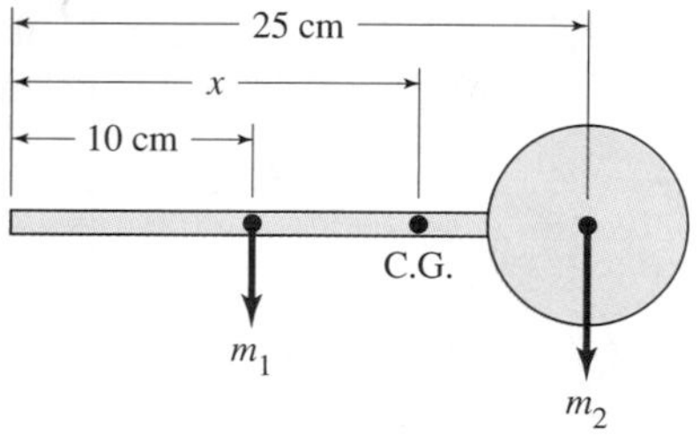

48. A ball of mass m is swung in a vertical circle (Fig. 9–8). At the top of its swing, the tension T in the cord plus the ball's weight mg is just balanced by the centrifugal force mv^2/R.

$$T + mg = \frac{mv^2}{R}$$

Solve for m.

49. The total energy of a body of mass m, moving with velocity v and located at a height y above some datum, is the sum of the potential energy mgy and the kinetic energy $\frac{1}{2}mv^2$.

$$E = mgy + \frac{1}{2}mv^2$$

Solve for m.

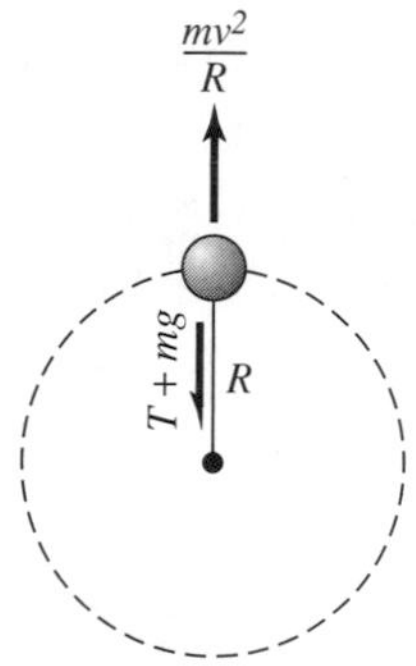

50. Toronto's CN Tower is not just a tourist attraction; it's also a microwave transmitter tower. The tower's microwave dishes are located just under the main pod, at 338 m above the ground. The parabolic shape of a microwave antenna "dish" focuses the signal and amplifies it mechanically, without electronics. The amount of amplification achieved by the dish shape is

$$Ap = 6\left(\frac{D}{\lambda}\right)^2$$

where Ap is the gain or amplification, D is the diameter of the dish and λ is the carrier wavelength.

(a) Solve the formula for D.

(b) What dish diameter would be required to provide a gain of $5\bar{0}$ for a signal with a $2\bar{0}$-cm wavelength?

FIGURE 9–6

FIGURE 9–7

FIGURE 9–8

Case Study Discussion – Scuba Diving Air Tanks

Depending on your research, you may have a slightly different answer. Using the following parameters (from *Physics,* 8th edition, by Cutnell and Johnson) we obtain:

$C = 0.0300 \text{ m}^3/\text{min}$
$\rho = 1025 \text{ kg/m}^3$
$P_1 = 1.01 \times 10^5 \text{ Pa}$

$$t = \frac{\dfrac{(2.02 \times 10^7)(0.0150)}{1.01 \times 10^5 + (1025)(9.80)(10)} - (0.0150)}{0.0300}$$

$t = 49.6 \text{ min}$

Based on this calculation, rating the tank at 30 minutes seems reasonable. (*Note: This does not take into account any applicable government standards; this is purely a math question and is not intended to imply that this calculation alone would legally be enough to rate a tank.*)

Based on this formula, as the value of h rises, the term ρgh gets larger, so the denominator $P_1 + \rho gh$ also gets larger, making the fraction $\frac{P_i V_i}{P_1 + \rho gh}$ smaller. If this fraction in the numerator of the equation gets smaller, the larger fraction (the entire expression on the right hand side) gets smaller. Therefore, the time available to the diver decreases as he or she dives deeper. So a 30 minute rating at 10 m is not a reasonable time for a deeper dive.

◆◆◆ CHAPTER 9 REVIEW PROBLEMS ◆◆◆◆◆◆◆◆◆◆◆◆◆◆◆◆◆◆◆◆◆◆◆◆◆◆

Perform the indicated operations and simplify.

1. $\dfrac{a^2 + b^2}{a - b} - a + b$

2. $\dfrac{3}{5x^2} - \dfrac{2}{15xy} + \dfrac{1}{6y^2}$

3. $\dfrac{3m + 1}{3m - 1} + \dfrac{m - 4}{5 - 2m} - \dfrac{3m^2 - 2m - 4}{6m^2 - 17m + 5}$

4. $\dfrac{\dfrac{1}{1 - x} - \dfrac{1}{1 + x}}{\dfrac{1}{1 - x^2} - \dfrac{1}{1 + x^2}}$

5. $(a^2 + 1 + a)\left(1 - \dfrac{1}{a} + \dfrac{1}{a^2}\right)$

6. $\dfrac{\dfrac{a - 1}{6} - \dfrac{2a - 7}{2}}{\dfrac{3a}{4} - 3}$

7. $\dfrac{1 + \dfrac{a - c}{a + c}}{1 - \dfrac{a - c}{a + c}}$

8. $\left(1 + \dfrac{x + y}{x - y}\right)\left(1 - \dfrac{x - y}{x + y}\right)$

9. $\dfrac{x^2 + 8x + 15}{x^2 + 7x + 10} - \dfrac{x - 1}{x - 2}$

10. $\dfrac{x^2 - 5ax + 6a^2}{x^2 - 8ax + 15a^2} - \dfrac{x - 7a}{x - 5a}$

11. $\dfrac{1}{a^2 - 7a + 12} + \dfrac{2}{a^2 - 4a + 3} - \dfrac{3}{a^2 - 5a + 4}$

12. $\dfrac{x^4 - y^4}{(x - y)^2} \times \dfrac{x - y}{x^2 + xy}$

13. $\dfrac{a^3 - b^3}{a^3 + b^3} \times \dfrac{a^2 - ab + b^2}{a - b}$

14. $\dfrac{3wx^2y^3}{7axyz} \times \dfrac{4a^3xz}{6aw^2y}$

15. $\dfrac{4a^2 - 9c^2}{4a^2 + 6ac}$

16. $\dfrac{b^2 - 5b}{b^2 - 4b - 5}$

17. $\dfrac{3a^2 + 6a}{a^2 + 4a + 4}$

18. $\dfrac{20(a^3 - c^3)}{4(a^2 + ac + c^2)}$

19. $\dfrac{x^2 - y^2 - 2yz - z^2}{x^2 + 2xy + y^2 - z^2}$

20. $\dfrac{2a^2 + 17a + 21}{3a^2 + 26a + 35}$

21. $\dfrac{(a + b)^2 - (c + d)^2}{(a + c)^2 - (b + d)^2}$

22. $\dfrac{1}{x - 2a} + \dfrac{a^2}{x^3 - 8a^3} - \dfrac{x + a}{x^2 + 2ax + 4a^2}$

Solve for x.

23. $cx - 5 = ax + b$

24. $a(x - 3) - b(x + 2) = c$

25. $\dfrac{5 - 3x}{4} + \dfrac{3 - 5x}{3} = \dfrac{3}{2} - \dfrac{5x}{3}$

26. $\dfrac{3x - 1}{11} - \dfrac{2 - x}{10} = \dfrac{6}{5}$

27. $\dfrac{x + 3}{2} + \dfrac{x + 4}{3} + \dfrac{x + 5}{4} = 16$

28. $\dfrac{2x + 1}{4} - \dfrac{4x - 1}{10} + \dfrac{5}{4} = 0$

29. $x^2 - (x - p)(x + q) = r$

30. $mx - n = \dfrac{nx - m}{p}$

31. $p = \dfrac{q - rx}{px - q}$

32. $m(x - a) + n(x - b) + p(x - c) = 0$

33. $\dfrac{x - 3}{4} - \dfrac{x - 1}{9} = \dfrac{x - 5}{6}$

34. $\dfrac{1}{x-5} - \dfrac{1}{4} = \dfrac{1}{3}$

35. $\dfrac{2}{x-4} + \dfrac{5}{2(x-4)} + \dfrac{9}{2(x-4)} = \dfrac{1}{2}$

36. $\dfrac{2}{x-2} = \dfrac{5}{2(x-1)}$

37. $\dfrac{6}{(2x+1)} = \dfrac{4}{(x-1)}$

38. $\dfrac{x+2}{x-2} - \dfrac{x-2}{x+2} = \dfrac{x+7}{x^2-4}$

39. $\dfrac{10-7x}{6-7x} = \dfrac{5x-4}{5x}$

40. $\dfrac{9x+5}{6(x-1)} + \dfrac{3x^2-51x-71}{18(x^2-1)} = \dfrac{15x-7}{9(x+1)}$

41. It is estimated that a bulldozer can prepare a certain site in 15.0 days and that a larger bulldozer can prepare the same site in 11.0 days. If we assume that they do not get in each other's way, how long will it take the two machines, working together, to prepare the site?

42. The acceleration a_1 of a body of mass m_1 subjected to a force F_1 is, by Eq. A20, $a_1 = F_1/m_1$. Write an expression for the difference in acceleration between that body and another having a mass m_2 and force F_2. Combine the terms of that expression and simplify.

43. Multiply: $2\dfrac{3}{5} \times 7\dfrac{5}{9}$

44. Divide: $\dfrac{8x^3y^2z}{5a^4b^3c}$ by $\dfrac{4xy^5z^3}{15abc^2}$

45. Multiply: $\dfrac{3wx^2y^3}{7axyz}$ by $\dfrac{4a^3xz}{6aw^2y}$

46. Divide: $9\dfrac{5}{7} \div 3\dfrac{4}{9}$

47. Multiply: $5\dfrac{2}{7}$ by $2\dfrac{3}{4}$

48. Divide: $3\dfrac{3}{10}$ by $2\dfrac{2}{11}$

Writing

49. Suppose that a vocal member of your local school board says that the study of fractions is no longer important now that we have computers and insists that it be cut from the curriculum to save money.

    Write a short letter to the editor of your local paper in which you agree or disagree. Give your reasons for retaining or eliminating the study of fractions.

Team Project

50. The ancient Egyptians wrote each fraction (except for $\frac{2}{3}$) as the sum of unit fractions (a fraction with a numerator of 1). Thus

$$\frac{3}{5} \text{ would be written as } \frac{1}{2} + \frac{1}{10}$$

and

$$\frac{5}{7} \text{ would be written as } \frac{1}{2} + \frac{1}{7} + \frac{1}{14}$$

This is called the *decomposition* of a fraction. Repetitions were not allowed. Thus $\frac{3}{5}$ would be written as above not as

$$\frac{1}{5} + \frac{1}{5} + \frac{1}{5}$$

Decompose each following proper fraction into unit fractions:

$$\frac{3}{4} \quad \frac{2}{5} \quad \frac{4}{5} \quad \frac{5}{6} \quad \frac{2}{7} \quad \frac{3}{7} \quad \frac{4}{7} \quad \frac{6}{7}$$

$$\frac{3}{8} \quad \frac{5}{8} \quad \frac{7}{8} \quad \frac{2}{9} \quad \frac{4}{9} \quad \frac{5}{9} \quad \frac{7}{9} \quad \frac{8}{9}$$

10

Systems of Linear Equations

◆◆◆ OBJECTIVES ◆◆

When you have completed this chapter, you should be able to:
- Find an approximate graphical solution to a system of two equations.
- Solve a system of two equations in two unknowns by the addition-subtraction method.
- Solve a system of two equations by substitution or by the Gauss-Seidel method.
- Solve a system of two equations having fractional coefficients or unknowns in the denominators.
- Solve a system of three or more equations by addition-subtraction or by substitution.
- Write a system of equations to describe a given word problem and to solve the system.

◆◆

Usually, a physical situation can be described by a single equation. For example, the displacement of a freely falling body is described by Eq. A18, $s = v_o t + \frac{1}{2} at^2$. Often, however, a situation can be described only by *more than one* equation.

For example, to find the two currents I_1 and I_2 in the circuit of Fig. 10–1, we must solve a *set* of two equations:

$$98I_1 - 43I_2 = 5$$
$$43I_1 - 115I_2 = 5$$

To do this, we need to find values for I_1 and I_2 that satisfy both equations at the same time.

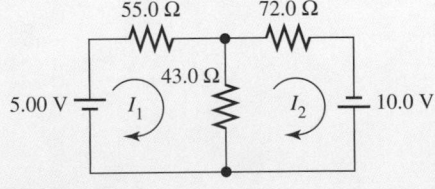

FIGURE 10–1

In this chapter we will learn how to solve such sets of two or three linear equations, and we'll also practise writing *systems of equations* to describe a variety of technical problems. Later, in Chapter 11, we'll learn how to solve systems of equations by means of determinants; and in Chapter 12 we'll solve sets of equations using matrices.

ERGOGRIP INC—SARAH AND ALEXANDRA LEVY

Success happens when you set out to help others. As the Canadian population ages, products and services for elderly people become very important, allowing them to live in dignity with independence. Just ask **Sarah Levy** and her daughter **Alexandra**. These two Montreal entrepreneurs were associated with a company that provided institutional meals to nursing homes. Working with seniors, they saw a need: They learned that normal utensils and dishware used by people with symptoms characteristic of diseases such as arthritis and Parkinson's were not appropriate. They came to believe that the fundamental needs for a pleasant dining experience, independence, and dignity were often denied to the residents.

Together, Sarah and Alexandra obtained financing and designed, manufactured, and marketed a line of products used in homes, hospitals, and nursing homes. The products include mugs and utensils with easy-grip handles and special thermal dishware to keep food warm longer. The Levys' company, Ergogrip Inc., has successfully penetrated the North American nursing home market well beyond the borders of Quebec.

10–1 Systems of Two Linear Equations

Linear Equations

We have previously defined a *linear equation* as one of *first degree.*

♦♦♦ **Example 1:** The equation

$$3x + 5 = 20$$

is a linear equation in *one unknown.* We learned how to solve this kind of equation in Chapter 3. ♦♦♦

♦♦♦ **Example 2:** The equation

$$2x - y = 3$$

is a linear equation in *two unknowns.* A linear equation with the terms written in this order, $Ax + By + C = 0$, is said to be in *general form.* If we graph this equation, we get a *straight line,* as shown in Fig. 10–2. Glance back at Sec. 5–3 if you've forgotten how to make such a graph.

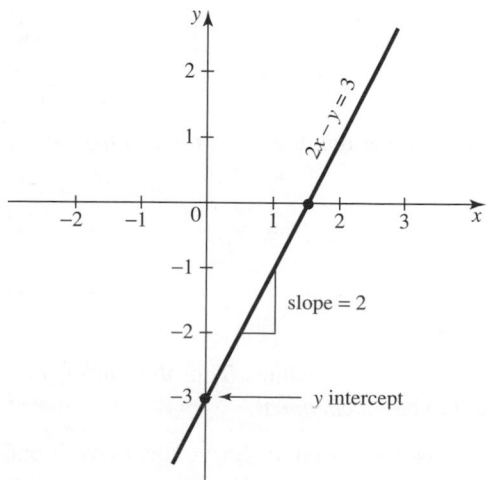

FIGURE 10–2

♦♦♦

If a linear equation in two unknowns is solved for y, we get an equation in *slope-intercept* form, as explained in Sec. 5–3.

◆◆◆ **Example 3:** If we solve the equation from Example 2 for y, we get the same equation in slope-intercept form.

$$y = 2x - 3$$

Recall from Sec. 5–3 that the coefficient of the x term is the *slope* and the constant term is the *y intercept*. Thus in our example,

$$\text{slope} = 2 \qquad y \text{ intercept} = -3$$

The graph is, of course, the same as in Example 2.   ◆◆◆

A linear equation can have *any number* of unknowns.

◆◆◆ **Example 4:**

(a) $x - 3y + 2z = 5$ is a linear equation in three unknowns.
(b) $3x + 2y - 5z - w = 6$ is a linear equation in four unknowns.   ◆◆◆

Systems of Equations

Systems of equations are also called simultaneous equations.

A set of two or more equations that simultaneously impose conditions on all of the variables is called a *system of equations*.

◆◆◆ **Example 5:**

(a) $3x - 2y = 5$
$\quad x + 4y = 1$
is a system of two linear equations in two unknowns.

(b) $\quad x - 2y + 3z = 4$
$\quad 3x + \ y - 2z = 1$
$\quad 2x + 3y - \ z = 3$
is a system of three linear equations in three unknowns.

Note that some variables may have zero coefficients and may not appear in every equation.

(c) $2x - \ y = 5$
$\quad x - 2z = 3$
$\quad 3y - \ z = 1$
is also a system of three linear equations in three unknowns.

(d) $y = 2x^2 - 3$
$\quad y^2 = \ x + 5$
is a system of two quadratic equations in two unknowns.   ◆◆◆

Solution to a System of Equations

The *solution to a system of equations* is a set of values of the unknowns that will *satisfy every equation* in the system.

◆◆◆ **Example 6:** The system of equations

$$x + y = 5 \tag{1}$$
$$x - y = 3 \tag{2}$$

Remember: It is a good idea to number your equations, as in this example, to help keep track of your work.

is satisfied *only* by the values $x = 4$, $y = 1$, and by *no other* set of values. Thus the pair (4, 1) is the solution to the system, and the equations are said to be *independent*.   ◆◆◆

To get a numerical solution for all of the unknowns in a system of linear equations, if one exists, *there must be as many independent equations as there are unknowns*. We first solve two equations in two unknowns; then later, three equations in three unknowns; and then larger systems. But for a solution to be possible, the number of equations must always equal, or exceed, the number of unknowns.

Approximate Graphical Solution to a Pair of Equations

Since any point on a curve satisfies the equation of that curve, the coordinates of the *points of intersection* of two curves will satisfy the equations of *both* curves. Thus we merely have to plot the two curves and find their point or points of intersection; these will be the solution to the pair of equations.

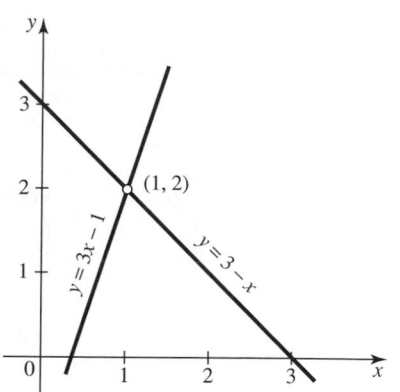

◆◆◆ **Example 7:** Graphically find the approximate solution to the pair of linear equations

$$3x - y = 1 \qquad (1)$$
$$x + y = 3 \qquad (2)$$

Solution: We plot the lines as in Chapter 5 and as shown in Fig. 10–3. Note that the lines intersect at the point (1, 2). Our solution is then $x = 1, y = 2$. ◆◆◆

FIGURE 10–3. Graphical solution to a pair of equations. This is also a good way to get an approximate solution to a pair of *nonlinear* equations.

A *linear* equation in two unknowns plots as a *straight line.* Thus two linear equations plot as two straight lines. If the lines intersect in a single point, the coordinates of that point will satisfy both equations, and hence that point is a solution to the set of equations. If the lines are *parallel,* there is no point whose coordinates satisfy both equations, and the set of equations is said to be *inconsistent.* If the two lines *coincide* at every point, the coordinates of any point on one line will satisfy both equations. Such a set of equations is called *dependent.*

Solving a Pair of Linear Equations by the Addition-Subtraction Method

The method of *addition-subtraction,* and the method of *substitution* that follows, both have the object of *eliminating* one of the unknowns.

In the addition-subtraction method, we eliminate one of the unknowns by first (if necessary) multiplying each equation by such numbers that will make the coefficients of one unknown in both equations equal in absolute value. The two equations are then added or subtracted so as to eliminate that variable.

Let us first use this method for the same system that we solved graphically in Example 7.

◆◆◆ **Example 8:** Solve by the addition-subtraction method:

$$3x - y = 1 \qquad (1)$$
$$x + y = 3 \qquad (2)$$

Solution: Simply adding the two equations causes y to drop out.

$$3x - y = 1$$
$$\underline{x + y = 3}$$
$$\text{Add}: \quad 4x \quad\ = 4$$
$$x = 1$$

Substituting into the second given equation yields

$$1 + y = 3$$
$$y = 2$$

Our solution is then $x = 1, y = 2$, as found graphically in Example 7. ◆◆◆

In the next example we must multiply one equation by a constant before adding.

◆◆◆ **Example 9:** Solve by the addition-subtraction method:

$$2x - 3y = -4 \qquad (1)$$
$$x + \ y = \ \ 3 \qquad (2)$$

Solution: Multiply the second equation by 3.

$$\begin{array}{r} 2x - 3y = -4 \\ 3x + 3y = 9 \\ \hline \text{Add:} \quad 5x = 5 \end{array}$$

We have thus reduced our two original equations to a single equation in one unknown. Solving for x gives

$$x = 1$$

Substituting into the second original equation, we have

$$\begin{array}{r} 1 + y = 3 \\ y = 2 \end{array}$$

So the solution is $x = 1$, $y = 2$.

Check: Substitute into the first original equation.

$$\begin{array}{r} 2(1) - 3(2) \overset{?}{=} -4 \\ 2 - 6 = -4 \quad \text{(checks)} \end{array}$$

Also substitute into the second original equation.

$$1 + 2 = 3 \quad \text{(checks)} \qquad\qquad \blacklozenge\blacklozenge\blacklozenge$$

Often it is necessary to multiply *both* given equations by suitable factors, as shown in the following example.

◆◆◆ **Example 10:** Solve by addition or subtraction:

$$\begin{array}{lr} 5x - 3y = 19 & (1) \\ 7x + 4y = 2 & (2) \end{array}$$

Solution: Multiply the first equation by 4 and the second by 3.

$$\begin{array}{r} 20x - 12y = 76 \\ 21x + 12y = 6 \\ \hline \text{Add:} \quad 41x = 82 \\ x = 2 \end{array}$$

Substituting $x = 2$ into the first equation gives

$$\begin{array}{r} 5(2) - 3y = 19 \\ 3y = -9 \\ y = -3 \end{array}$$

So the solution is $x = 2$, $y = -3$.

These values check when substituted into each of the original equations (work not shown). Notice that we could have eliminated the x terms by multiplying the first equation by 7 and the second by -5, and adding. The results, of course, would have been the same: $x = 2$, $y = -3$. ◆◆◆

The coefficients in the preceding examples were integers, but in applications they will usually be approximate numbers. If so, we must retain the proper number of significant digits, as in the following example.

◆◆◆ **Example 11:** Solve for x and y:

$$\begin{array}{lr} 2.64x + 8.47y = 3.72 & (1) \\ 1.93x + 2.61y = 8.25 & (2) \end{array}$$

Solution: Let us eliminate y. We multiply the first equation by 2.61 and the second equation by -8.47. We should carry at least two decimal places in our calculation, and round our answer to three digits at the end.

$$
\begin{array}{rl}
6.89x + 22.11y = & 9.71 \\
-16.35x - 22.11y = & -69.88 \\
\hline
\text{Add:} \quad -9.46x \qquad\quad = & -60.17 \\
x = & 6.36
\end{array}
$$

> We are less likely to make a mistake adding rather than subtracting, so it is safer to multiply by a negative number and add the resulting equations, as we do here.

Substituting into the first equation yields

$$
\begin{array}{rl}
2.64(6.36) + 8.47y = & 3.72 \\
8.47y = 3.72 - 16.79 = & -13.07 \\
y = & -1.54
\end{array}
$$

> Another approach is to divide each equation by the coefficient of its x term, thus making each x coefficient equal to 1. Then subtract one equation from the other.

◆◆◆

Substitution Method

To use the *substitution method* to solve a pair of linear equations, first solve either original equation for one unknown in terms of the other unknown. Then substitute this expression into the other equation, thereby eliminating one unknown.

◆◆◆ **Example 12:** Solve by substitution:

$$
\begin{align}
7x - 9y &= 1 \tag{1}\\
y &= 5x - 17 \tag{2}
\end{align}
$$

> If, as in this example, one or both of the given equations are already in explicit form, then the substitution method is probably the easiest. Otherwise, many students find the addition-subtraction method easier.

Solution: We substitute $(5x - 17)$ for y in the first equation and get

$$
\begin{array}{rl}
7x - 9(5x - 17) &= 1 \\
7x - 45x + 153 &= 1 \\
-38x &= -152 \\
x &= 4
\end{array}
$$

Substituting $x = 4$ into Equation (2) gives

$$
\begin{array}{l}
y = 5(4) - 17 \\
y = 3
\end{array}
$$

So our solution is $x = 4$, $y = 3$.

◆◆◆

Systems Having No Solution

Certain systems of equations have no unique solution. If you try to solve either of these types, *both variables will vanish.*

◆◆◆ **Example 13:** Solve the system

$$
\begin{align}
2x + 3y &= 5 \tag{1}\\
6x + 9y &= 2 \tag{2}
\end{align}
$$

Solution: Multiply the first equation by 3.

$$
\begin{array}{rl}
6x + 9y = & 15 \\
6x + 9y = & 2 \\
\hline
\text{Subtract:} \qquad 0 = & 13 \quad \text{(no solution)}
\end{array}
$$

◆◆◆

If both variables vanish and an *inequality* results, as in Example 13, the system is called *inconsistent.* The equations would plot as two *parallel lines.* There is no point of intersection and hence no solution.

If both variables vanish and an *equality* results (such as $4 = 4$), the system is called *dependent.* The two equations would plot as a *single line,* indicating that there are infinitely many solutions.

> It does not matter much whether a system is inconsistent or dependent; in either case we get no useful solution. But the practical problems that we solve here will always have numerical solutions, so if your variables vanish, go back and check your work.

A Computer Technique: The Gauss-Seidel Method

The *Gauss-Seidel method* is a simple computer technique for solving systems of equations. Suppose that we wish to solve the set of equations

$$x - 2y + 2 = 0 \tag{1}$$
$$3x - 2y - 6 = 0 \tag{2}$$

We first solve the first equation for x in terms of y, and the second equation for y in terms of x.

$$x = 2y - 2 \tag{3}$$
$$y = \frac{3x - 6}{2} \tag{4}$$

We then *guess* at the value of y, substitute this value into Equation (3), and obtain a value for x. This x is substituted into (4), and a value of y is obtained. These values for x and y will not be the correct values; they are only our *first approximation* to the true values.

The latest value for y is then put into (3), producing a new x; which is then put into (4), producing a new y; which is then put into (3); and so on. We repeat the computation until the values no longer change (we say that they *converge* on the true values). If the values of x and y get very large (*diverge* instead of converge), we solve the first equation for y and the second equation for x, and the computation will converge.

This type of repetitive process is called *iteration*, or the *method of successive approximations*.

◆◆◆ **Example 14:** Calculate x and y for the equations above using the Gauss-Seidel method. Take $y = 0$ for the first guess.

Solution: From Equation (3) we get

$$x = 2(0) - 2 = -2$$

Substituting $x = -2$ into (4) yields

$$y = \frac{3(-2) - 6}{2} = -6$$

With y equal to -6, we then use (3) to find a new x, and so on. We get the following values:

x	y
-2	-6
-14	-24
-50	-78
-158	-240
-482	-726
$-1\,454$	$-2\,184$
$-4\,370$	$-6\,558$
$-13\,118$	$-19\,680$
$-39\,362$	$-59\,046$
.	.
.	.
.	.

Note that the values are *diverging*. We try again, this time solving Equation (1) for y,

$$y = \frac{x + 2}{2}$$

and solving (2) for x,

$$x = \frac{2y + 6}{3}$$

Repeating the computation, starting with $x = 0$, gives the following values:

x	y
2.666 667	1
3.555 556	2.333 334
3.851 852	2.777 778
3.950 617	2.925 926
3.983 539	2.975 309
3.994 513	2.991 769
3.998 171	2.997 256
3.999 39	2.999 086
3.999 797	2.999 695
3.999 932	2.999 899
3.999 977	2.999 966
3.999 992	2.999 989
3.999 998	2.999 996
3.999 999	2.999 999
4	3
4	3
⋮	⋮

which converge on $x = 4$ and $y = 3$. ◆◆◆

How do we know what value to take for our first guess? Choose whatever values seem "reasonable." If you have no idea, then choose any values, such as the zeros in this example. Actually, it does not matter much what values we start with. If the computation is going to converge, it will converge regardless of the initial values chosen. However, the closer your guess, the *faster* the convergence.

CASE STUDY—ELECTRICAL CURRENTS

As an electronics technologist, you are analyzing a robotic servo system. After some initial calculations and analysis, you have reduced the system to the equivalent circuit shown below.

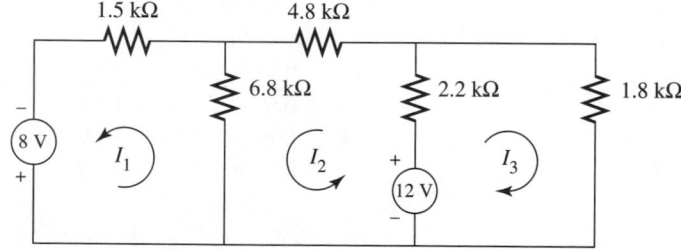

We can write an equation for each loop based on the voltage drops and sources. Electronics technology students may be familiar with these equations:

Loop 1: $6.8(I_1 - I_2) + 1.5I_1 = 8$
Loop 2: $2.2(I_2 + I_3) + 4.8I_2 + 6.8(I_2 - I_1) = 12$
Loop 3: $2.2(I_3 + I_2) + 1.8I_3 = 12$

Now, the "k" in the values near the resistors (the squiggly lines) stands for kilo (or thousand), so 2.2k would be 2200. However, to keep this straightforward, the resistance is divided by 1000, so the answers will be the currents in milliamperes. This is not a concern, but it helps this to makes sense for both electronics and non-electronics students. To determine the currents, remove the parentheses and collect like terms. Now you will have three equations with three unknowns. Solve for I_1, I_2, and I_3. This chapter will show you how.

Exercise 1 ◆ Systems of Two Linear Equations

Graphical Solution

Graphically find the approximate solution to each system of equations.

1. $2x - y = 5$
$x - 3y = 5$

2. $x + 2y = -7$
$5x - y = 9$

3. $x - 2y = -3$
$3x + y = 5$

4. $4x + y = 8$
$2x - y = 7$

5. $2x + 5y = 4$
$5x - 2y = -3$

6. $x - 2y + 2 = 0$
$3x - 6y + 2 = 0$

Algebraic Solution

Solve each system of equations by addition-subtraction, or by substitution.

No applications are given here; they are located in Exercise 3.

7. $2x + y = 11$
$3x - y = 4$

8. $x - y = 7$
$3x + y = 5$

9. $4x + 2y = 3$
$-4x + y = 6$

10. $2x - 3y = 5$
$3x + 3y = 10$

11. $3x - 2y = -15$
$5x + 6y = 3$

12. $7x + 6y = 20$
$2x + 5y = 9$

13. $x + 5y = 11$
$3x + 2y = 7$

14. $4x - 5y = -34$
$2x - 3y = -22$

15. $x = 11 - 4y$
$5x - 2y = 11$

16. $2x - 3y = 3$
$4x + 5y = 39$

17. $7x - 4y = 81$
$5x - 3y = 57$

18. $3x + 4y = 85$
$5x + 4y = 107$

19. $3x - 2y = 1$
$2x + y = 10$

20. $5x - 2y = 3$
$2x + 3y = 5$

21. $y = 9 - 3x$
$x = 8 - 2y$

22. $y = 2x - 3$
$x = 19 - 3y$

23. $29.1x - 47.6y = 42.8$
$11.5x + 72.7y = 25.8$

24. $4.92x - 8.27y = 2.58$
$6.93x + 2.84y = 8.36$

25. $4n = 18 - 3m$
$m = 8 - 2n$

26. $5p + 4q - 14 = 0$
$17p = 31 + 3q$

27. $3w = 13 + 5z$
$4w - 7z - 17 = 0$

28. $3u = 5 + 2v$
$5v + 2u = 16$

29. $3.62x = 11.7 + 4.73y$
$4.95x - 7.15y - 12.8 = 0$

30. $3.03a = 5.16 + 2.11b$
$5.63b + 2.26a = 18.8$

31. $4.17w = 14.7 - 3.72v$
$v = 8.11 - 2.73w$

32. $5.66p + 4.17q - 16.9 = 0$
$13.7p = 32.2 + 3.61q$

Computer

33. Use a spreadsheet for the Gauss-Seidel method described in Sec. 10–1. Use it to solve any of the sets of equations in this exercise set.

10–2   Other Systems of Equations

Systems with Fractional Coefficients

When one or more of the equations in our system have fractional coefficients, simply multiply the entire equation by the LCD, and proceed as before.

◆◆◆ **Example 15:** Solve for x and y:

$$\frac{x}{2} + \frac{y}{3} = \frac{5}{6} \quad (1)$$

$$\frac{x}{4} - \frac{y}{2} = \frac{7}{4} \quad (2)$$

The techniques in this section apply not only to two equations in two unknowns, but to larger systems as well.

Multiply Equation (1) by 6 [because it's the least common multiple of 2, 3, and 6] and Equation (2) by 4 [because it's the LCM of 2 and 4].

$$6\left(\frac{x}{2}\right) + 6\left(\frac{y}{3}\right) = 6\left(\frac{5}{6}\right) \qquad {}^{3}\cancel{6}\left(\frac{x}{\cancel{2}}\right) + {}^{2}\cancel{6}\left(\frac{y}{\cancel{3}}\right) = {}^{1}\cancel{6}\left(\frac{5}{\cancel{6}}\right) \quad (1)$$

$$4\left(\frac{x}{4}\right) - 4\left(\frac{y}{2}\right) = 4\left(\frac{7}{4}\right) \qquad {}^{1}\cancel{4}\left(\frac{x}{\cancel{4}}\right) - {}^{2}\cancel{4}\left(\frac{y}{\cancel{2}}\right) = {}^{1}\cancel{4}\left(\frac{7}{\cancel{4}}\right) \quad (2)$$

$$3x + 2y = 5 \quad (3)$$

$$x - 2y = 7 \qquad \text{—— Keep the minus sign}$$

$$\text{Add: } \overline{4x \qquad = 12} \quad (4)$$

Solution: $x = \dfrac{12}{4} = 3$

Having x, we can now substitute back to get y. It is not necessary to substitute back into one of the original equations. We choose instead the easiest place to substitute, such as Equation (4). (However, when *checking*, be sure to substitute your answers into both of the *original* equations.) Substituting $x = 3$ into (4) gives

$$3 - 2y = 7$$
$$2y = 3 - 7 = -4$$
$$y = -2 \qquad ◆◆◆$$

Fractional Equations with Unknowns in the Denominator

The same method (multiplying by the LCD) can be used to clear fractions when the unknowns appear in the denominators. Note that such equations are not linear (that is, not of first degree) as were the equations that we have solved so far, but we are able to solve them by the same methods.

To save space, we will not show the check in many of these examples.

◆◆◆ **Example 16:** Solve the system

$$\frac{10}{x} - \frac{9}{2y} = 6 \quad (1)$$

$$\frac{20}{3x} + \frac{15}{y} = 1 \quad (2)$$

Of course, neither x nor y can equal zero in this system.

Solution: We multiply Equation (1) by $2xy$ and (2) by $3xy$.

$$20y - 9x = 12xy \quad (3)$$
$$20y + 45x = 3xy \quad (4)$$

Subtracting (4) from (3) gives

$$-54x = 9xy$$
$$y = -6$$

Substituting $y = -6$ back into (1), we have

$$\frac{10}{x} - \frac{9}{2(-6)} = 6$$

$$\frac{10}{x} = 6 - \frac{3}{4} = \frac{21}{4}$$

$$\frac{x}{10} = \frac{4}{21}$$

$$x = \frac{40}{21}$$

So our solution is $x = 40/21$, $y = -6$. ◆◆◆

A convenient way to solve nonlinear systems such as these is to *substitute new variables* that will make the equations linear. Solve in the usual way and then substitute back.

◆◆◆ **Example 17:** Solve the nonlinear system of equations

$$\frac{2}{3x} + \frac{3}{5y} = 17 \tag{1}$$

$$\frac{3}{4x} + \frac{2}{3y} = 19 \tag{2}$$

Solution: We substitute $m = 1/x$ and $n = 1/y$ and get the linear system

$$\frac{2m}{3} + \frac{3n}{5} = 17 \tag{3}$$

$$\frac{3m}{4} + \frac{2n}{3} = 19 \tag{4}$$

Again, we clear fractions by multiplying each equation by its LCD. Multiplying Equation (3) by 15 gives

$$10m + 9n = 255 \tag{5}$$

and multiplying (4) by 12 gives

$$9m + 8n = 228 \tag{6}$$

Using the addition-subtraction method, we multiply (5) by 8 and multiply (6) by -9.

$$
\begin{array}{r}
80m + 72n = 2040 \\
-81m - 72n = -2052 \\
\hline
\end{array}
$$
Add: $-m = -12$
$$m = 12$$

Substituting $m = 12$ into (5) yields

$$120 + 9n = 255$$
$$n = 15$$

Finally, we substitute back to get x and y.

$$x = \frac{1}{m} = \frac{1}{12} \quad \text{and} \quad y = \frac{1}{n} = \frac{1}{15}$$ ◆◆◆

Common Error	Students often forget that last step. We are not solving for m and n, but for x and y.

Literal Equations

We use the method of addition-subtraction or the method of substitution for solving systems of equations with literal coefficients, treating the literals as if they were numbers.

◆◆◆ **Example 18:** Solve for x and y in terms of m and n:

$$2mx + ny = 3 \tag{1}$$
$$mx + 3ny = 2 \tag{2}$$

Solution: We will use the addition-subtraction method. Multiply the second equation by -2.

$$
\begin{array}{r}
2mx + ny = 3 \\
-2mx - 6ny = -4 \\
\hline
\end{array}
$$
Add: $-5ny = -1$
$$y = \frac{1}{5n}$$

This technique of substitution is very useful, and we will use it again later to reduce certain equations to quadratic form. Do not confuse this kind of substitution with the method of substitution that we studied in Sec. 10–1.

Substituting back into the second original equation, we obtain

$$mx + 3n\left(\frac{1}{5n}\right) = 2$$

$$mx + \frac{3}{5} = 2$$

$$mx = 2 - \frac{3}{5} = \frac{7}{5}$$

$$x = \frac{7}{5m}$$

◆◆◆

◆◆◆ **Example 19:** Solve for x and y by the addition-subtraction method:

$$a_1x + b_1y = c_1 \tag{1}$$
$$a_2x + b_2y = c_2 \tag{2}$$

Solution: Multiplying the first equation by b_2 and the second equation by b_1, we obtain

$$a_1b_2x + b_1b_2y = b_2c_1$$
$$a_2b_1x + b_1b_2y = b_1c_2$$

Subtract: $\quad (a_1b_2 - a_2b_1)x = b_2c_1 - b_1c_2$

Dividing by $a_1b_2 - a_2b_1$ gives

$$x = \frac{b_2c_1 - b_1c_2}{a_1b_2 - a_2b_1}$$

Now solving for y, we multiply the first equation by a_2 and the second equation by a_1. Writing the second equation above the first, we get

$$a_1a_2x + a_1b_2y = a_1c_2$$
$$a_1a_2x + a_2b_1y = a_2c_1$$

Subtract: $\quad (a_1b_2 - a_2b_1)y = a_1c_2 - a_2c_1$

Dividing by $a_1b_2 - a_2b_1$ yields

$$y = \frac{a_1c_2 - a_2c_1}{a_1b_2 - a_2b_1}$$

◆◆◆

This result may be summarized as follows:

Two Linear Equations in Two Unknowns	is	The solution to the set of equations $$a_1x + b_1y = c_1$$ $$a_2x + b_2y = c_2$$ $$x = \frac{b_2c_1 - b_1c_2}{a_1b_2 - a_2b_1} \qquad y = \frac{a_1c_2 - a_2c_1}{a_1b_2 - a_2b_1}$$ $$(a_1b_2 - a_2b_1 \neq 0)$$	**63**

Thus Eq. 63 is a formula for solving a pair of linear equations. We simply have to identify the six numbers a_1, b_1, c_1, a_2, b_2, and c_2 and substitute them into the equations. But be sure to put the equations into standard form first.

◆◆◆ **Example 20:** Solve the system:

$$9y = 7x - 15 \tag{1}$$
$$-17 + 5x = 8y \tag{2}$$

Solution: Rewrite the equations in the form given in Eq. 63.

$$7x - 9y = 15 \tag{3}$$
$$5x - 8y = 17 \tag{4}$$

Then substitute, with

$$a_1 = 7 \qquad b_1 = -9 \qquad c_1 = 15$$
$$a_2 = 5 \qquad b_2 = -8 \qquad c_2 = 17$$

$$x = \frac{b_2c_1 - b_1c_2}{a_1b_2 - a_2b_1} = \frac{-8(15) - (-9)(17)}{7(-8) - 5(-9)}$$
$$= -3$$

$$y = \frac{a_1c_2 - a_2c_1}{a_1b_2 - a_2b_1} = \frac{7(17) - (5)(15)}{-11}$$
$$= -4$$

◆◆◆

Exercise 2 ◆ Other Systems of Equations

Fractional Coefficients

Solve simultaneously.

1. $\dfrac{x}{5} + \dfrac{y}{6} = 18$

$\dfrac{x}{2} - \dfrac{y}{4} = 21$

2. $\dfrac{x}{2} + \dfrac{y}{3} = 7$

$\dfrac{x}{3} + \dfrac{y}{4} = 5$

3. $\dfrac{x}{3} + \dfrac{y}{4} = 8$

$x - y = -3$

4. $\dfrac{x}{2} + \dfrac{y}{3} = 5$

$\dfrac{x}{3} + \dfrac{y}{2} = 5$

5. $\dfrac{3x}{5} + \dfrac{2y}{3} = 17$

$\dfrac{2x}{3} + \dfrac{3y}{4} = 19$

6. $\dfrac{x}{7} + 7y = 251$

$\dfrac{y}{7} + 7x = 299$

7. $\dfrac{m}{2} + \dfrac{n}{3} - 3 = 0$

$\dfrac{n}{2} + \dfrac{m}{5} = \dfrac{23}{10}$

8. $\dfrac{p}{6} - \dfrac{q}{3} + \dfrac{1}{3} = 0$

$\dfrac{2p}{3} - \dfrac{3q}{4} - 1 = 0$

9. $\dfrac{r}{6.20} - \dfrac{s}{4.30} = \dfrac{1}{3.10}$

$\dfrac{r}{4.60} - \dfrac{s}{2.30} = \dfrac{1}{3.50}$

Unknowns in the Denominator

Solve simultaneously.

10. $\dfrac{8}{x} + \dfrac{6}{y} = 3$

$\dfrac{6}{x} + \dfrac{15}{y} = 4$

11. $\dfrac{1}{x} + \dfrac{3}{y} = 11$

$\dfrac{5}{x} + \dfrac{4}{y} = 22$

12. $\dfrac{5}{x} + \dfrac{6}{y} = 7$

$\dfrac{7}{x} + \dfrac{9}{y} = 10$

13. $\dfrac{2}{x} + \dfrac{4}{y} = 14$

$\dfrac{6}{x} - \dfrac{2}{y} = 14$ (

14. $\dfrac{6}{x} + \dfrac{8}{y} = 1$

$\dfrac{7}{x} - \dfrac{11}{y} = -9$

15. $\dfrac{2}{5x} + \dfrac{5}{6y} = 14$

$\dfrac{2}{5x} - \dfrac{3}{4y} = -5$

16. $\dfrac{5}{3a} + \dfrac{2}{5b} = 7$

$\dfrac{7}{6a} - 3 = \dfrac{1}{10b}$

17. $\dfrac{1}{5z} + \dfrac{1}{6w} = 18$

$\dfrac{1}{4w} - \dfrac{1}{2x} + 21 = 0$

18. $\dfrac{8.10}{5.10t} + \dfrac{1.40}{3.60s} = 1.80$

$\dfrac{2.10}{1.40s} - \dfrac{1.30}{5.20t} + 3.10 = 0$

Literal Equations

Solve for x and y in terms of the other literal quantities.

19. $3x - 2y = a$
$2x + y = b$

20. $ax + by = r$
$ax + cy = s$

21. $ax - dy = c$
$mx - ny = c$

22. $px - qy + pq = 0$
$2px - 3qy = 0$

10–3 Word Problems with Two Unknowns

Many problems contain two or more unknowns that must be found. To solve such problems, we must write *as many independent equations as there are unknowns*. Otherwise, it is not possible to obtain numerical answers, although one unknown could be expressed in terms of another.

Set up these problems as we did in Chapters 3 and 9, and solve the resulting system of equations by any of the methods of this chapter.

We will start with a **financial** problem. We did financial problems in one unknown in Sec. 3–3. You may want to flip through that section before starting here.

◆◆◆ **Example 21:** A certain investment in bonds had a value of $248,000 after 4 years, and of $260,000 after 5 years, at simple interest (Fig. 10–4). Find the amount invested and the interest rate. [The formula $y = P(1 + nt)$ gives the amount y obtained by investing an amount P for t years at an interest rate n.]

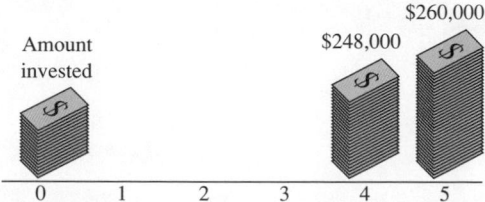

FIGURE 10–4

Estimate: We know that the original investment has to be less than $248,000. We might also reason that the investment cannot be too small, if it is to grow to $248,000 in only 4 years. Also some familiarity with bonds (or a quick inquiry of someone who is familiar with them) reveals that any kind of bond historically has yields somewhere between a few percent and about 15%.

Solution: For the 4-year investment, $t = 4$ years and $y = \$248,000$. Substituting into the given formula yields

$$\$248,000 = P(1 + 4n)$$

Then substitute again, with $t = 5$ years and $y = \$260,000$.

$$\$260,000 = P(1 + 5n)$$

Thus we get two equations in two unknowns: P (the amount invested) and n (the interest rate). Next we remove parentheses.

$$248,000 = P + 4Pn \tag{1}$$
$$260,000 = P + 5Pn \tag{2}$$

At this point we may be tempted to subtract one of these equations from the other to eliminate P. But this will not work because P remains in the term containing Pn. Instead, we multiply Equation (1) by 5 and (2) by -4 to eliminate the Pn term.

$$
\begin{array}{rl}
1,240,000 = & 5P + 20Pn \\
-1,040,000 = & -4P - 20Pn \\
\hline
\text{Add:} \quad \$200,000 = & P
\end{array}
$$

Substituting back into (1) gives us

$$248,000 = 200,000 + 4(200,000)n$$

from which

$$n = 0.06$$

Thus a sum of \$200,000 was invested at 6%.

Check: For the 4-year period,

$$y = 200,000[1 + 4(0.06)] = \$248,000 \quad \text{(checks)}$$

and for the 5-year period,

$$y = 200,000[1 + 5(0.06)] = \$260,000 \quad \text{(checks)} \qquad \blacklozenge\blacklozenge\blacklozenge$$

We'll now do a **mixture** problem. We did mixture problems with one unknown in Sec. 3–4.

◆◆◆ Example 22: Two different gasohol mixtures are available, one containing 5.00% alcohol, and the other 12.0% alcohol. How much of each, to the nearest litre, should be mixed to obtain $10\overline{0}0$ L of gasohol containing 10.0% alcohol (Fig. 10–5)?

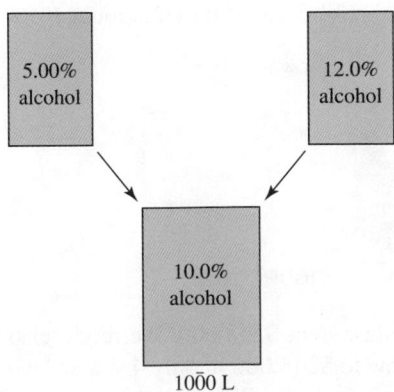

FIGURE 10–5

Estimate: Note that using 500 L of each original mixture would give the 1000 L needed, but with a percent alcohol midway between 5% and 12%, or around 8.5%. This is lower than needed, so we reason that we need more than 500 L of the 12% mixture and less than 500 L of the 5% mixture.

Solution: We let

$$x = \text{litres of } 5.00\% \text{ gasohol needed}$$
$$y = \text{litres of } 12.0\% \text{ gasohol needed}$$

So

$$x + y = 1000 \tag{1}$$

and

$$0.05x + 0.12y = 0.10(1000) \tag{2}$$

are our two equations in two unknowns. Multiplying the first equation by 5 and the second equation by 100, we have

$$5x + 5y = 5000$$
$$5.00x + 12.0y = 10\,000$$

Subtracting the first equation from the second yields

$$7.0y = 5000$$
$$y = 714 \text{ L of the } 5.00\% \text{ gasohol solution}$$

So

$$x = 1000 - y = 286 \text{ L of the } 12.0\% \text{ gasohol solution} \qquad \blacklozenge\blacklozenge\blacklozenge$$

Our next application is a **work** problem. Glance back at Sec. 9–6 to see how we set up work problems in one unknown.

◆◆◆ **Example 23:** During a certain day, two computer printers are observed to process 1705 form letters, with the slower printer in use for 5.5 h and the faster for 4.0 h (Fig. 10–6). On another day, the slower printer works for 6.0 h and the faster for 6.5 h, and together they print 2330 form letters. How many letters can each print in an hour, working alone?

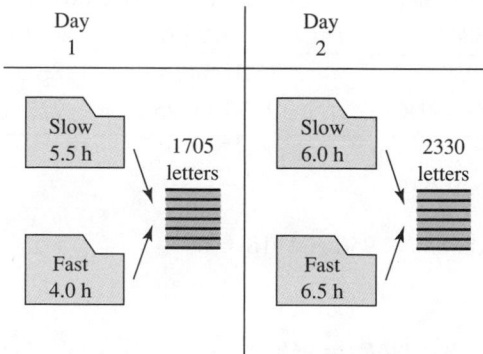

FIGURE 10–6

Estimate: Assume for now that both printers work at the same rate. On the first day they work a total of 9.5 h and print 1705 letters, or $1705 \div 9.5 \approx 180$ letters per hour, and on the second day, $2330 \div 12.5 \approx 186$ letters per hour. But we expect more from the fast printer, say, around 200 letters/h, and less from the slow printer, say, around 150 letters/h.

Solution: We let

$$x = \text{rate of slow printer, letters/h}$$
$$y = \text{rate of fast printer, letters/h}$$

We write two equations to express, for each day, the total amount of work produced, remembering from Eq. A5 that

$$\text{amount of work done} = \text{work rate} \times \text{time elapsed}$$

On the first day, the slow printer produces $5.5x$ letters while the fast printer produces $4.0y$ letters. Together they produce

$$5.5x + 4.0y = 1705 \tag{1}$$

Similarly, for the second day,

$$6.0x + 6.5y = 2330 \tag{2}$$

Using the addition-subtraction method, we multiply the first equation by 6.5 and the second by -4.0.

$$
\begin{aligned}
35.75x + 26y &=\ 11\,082 \\
-24x - 26y &= -9\,320 \\
\hline
\text{Add:}\quad 11.75x\qquad\ &=\ \ 1\,762 \\
x &=\ 150 \text{ letters/h}
\end{aligned}
$$

Substituting back yields

$$
\begin{aligned}
5.5(150) + 4y &= 1705 \\
y &=\ \ 220 \text{ letters/h}
\end{aligned}
$$

◆◆◆

Our next example will be from **statics**. Refer to Sec. 3–5 for the basic formulas and to Sec. 7–6 for the resolution of vectors.

◆◆◆ **Example 24:** Find the forces F_1 and F_2 in Fig. 10–7.

Solution: We resolve each vector into its x and y components. We will take components to the right and upward as positive. As in Sec. 7–6, we will arrange the values in a table, for convenience.

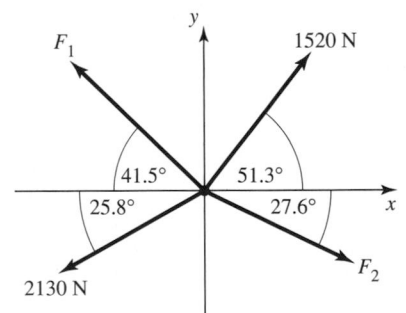

Force	x component	y component
F_1	$-F_1 \cos 41.5° = -0.749F_1$	$F_1 \sin 41.5° = 0.663F_1$
F_2	$F_2 \cos 27.6° = 0.886F_2$	$-F_2 \sin 27.6° = -0.463F_2$
1520 N	$1520 \cos 51.3° = 950$	$1520 \sin 51.3° = 1186$
2130 N	$-2130 \cos 25.8° = -1918$	$-2130 \sin 25.8° = -927$

FIGURE 10-7

We then set the sum of the x components to zero,

$$\Sigma F_x = -0.749F_1 + 0.886F_2 + 950 - 1918 = 0$$

or

$$-0.749F_1 + 0.886F_2 = 968 \tag{1}$$

and set the sum of the y components to zero,

$$\Sigma F_y = 0.663F_1 - 0.463F_2 + 1186 - 927 = 0$$

or

$$0.663F_1 - 0.463F_2 = -259 \tag{2}$$

Now we divide Equation (1) by 0.749 and divide (2) by 0.663 and get

$$
\begin{array}{r}
-F_1 + 1.183F_2 = 1292 \\
F_1 - 0.698F_2 = -391 \\
\hline
\text{Add:} \quad 0.485F_2 = 901 \\
F_2 = 1860\text{N}
\end{array}
$$

(3)

Substituting back into (3) gives

$$
\begin{aligned}
F_1 &= 1.183F_2 - 1292 \\
&= 1.183(1860) - 1292 \\
&= 908\text{N}
\end{aligned}
$$

◆◆◆

Exercise 3 ◆ Word Problems with Two Unknowns

Number Problems

1. The sum of two numbers is 24 and their difference is 8. What are the numbers?

2. The sum of two numbers is 29 and their difference is 5. What are the numbers?

3. There is a fraction such that if 3 is added to the numerator, its value will be $\frac{1}{3}$, and if 1 is subtracted from the denominator, its value will be $\frac{1}{5}$. What is the fraction?

4. The sum of the two digits of a certain number is 9. If 9 is added to the original number, the new number will have the original digits reversed. Find the number. (*Hint:* Let $x = $ the tens digit and $y = $ the ones digit, so that the value of the number is $10x + y$.)

Some of these problems can be set up using just one unknown, but for now use two unknowns for the practice.

5. A number consists of two digits. The number is 2 more than 8 times the sum of the digits, and if 54 is subtracted from the number, the digits will be reversed. Find the number.

6. If the larger of two numbers is divided by the smaller, the quotient is 7 and the remainder is 4. But if 3 times the greater is divided by twice the smaller, the quotient is 11 and the remainder is 4. What are the numbers?

7. The sum of two numbers divided by 2 gives a quotient of 24, and their difference divided by 2 gives a quotient of 17. What are the numbers?

Geometry Problems

8. If the width of a certain rectangle is increased by 3 and the length decreased by 3, the area is seen to increase by 6 [Fig. 10–8(a)]. But if the width is reduced by 5 and the length increased by 3, the area decreases by 90 [Fig. 10–8(b)]. Find the original dimensions.

9. If the width of a certain rectangle is increased by 3 and the length reduced by 4, we get a square with the same area as the original rectangle (Fig. 10–9). Find the length and the width of the original rectangle.

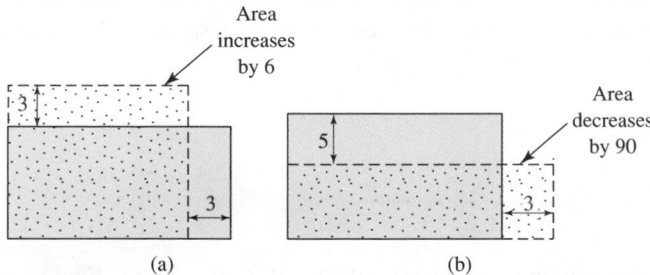

FIGURE 10–8

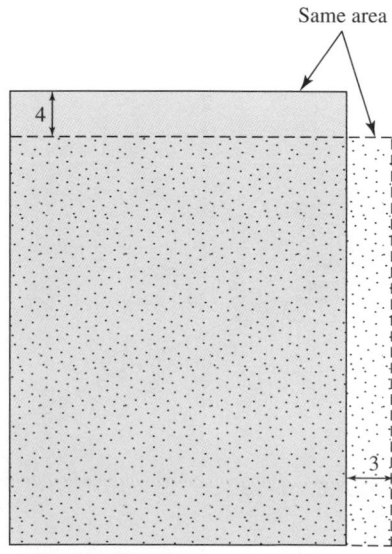

FIGURE 10–9

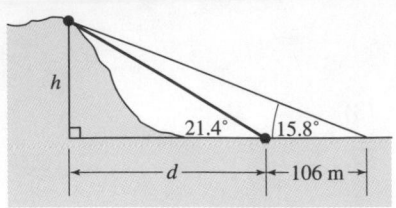

FIGURE 10-10

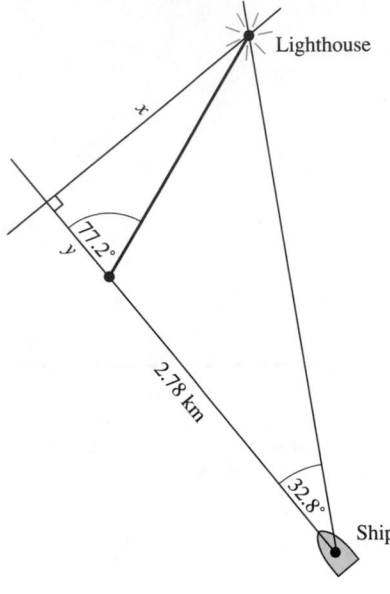

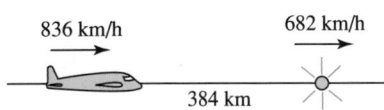

FIGURE 10-11

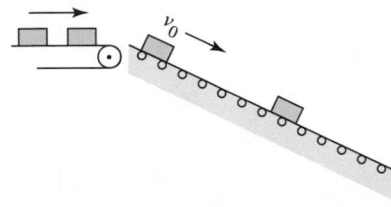

FIGURE 10-12

Trigonometry Problems

10. A surveyor measures the angle of elevation of a hill at 15.8° (Fig. 10–10). She then moves 106 m closer, on level ground, and measures the angle of elevation at 21.4°. Find the height h of the hill and the distance d. (*Hint:* Write the expression for the tangent of the angle of elevation, at each location, and solve the two resulting equations simultaneously.)

11. A ship travelling a straight course sights a lighthouse at an angle of 32.8° (Fig. 10–11). After the ship sails another 2.78 km, the lighthouse is at an angle of 77.2°. Find the distances x and y.

Uniform Motion Problems

12. In a first race, A gives B a headstart of 30.0 m and overtakes him in 4.00 min. In a second race he gains $25\overline{0}$ m on B when B runs $30\overline{0}0$ m. Find the rate at which each runs.

13. A and B run two races of $12\overline{0}$ m. In the first race, A gives B a start of 30.0 m, and neither wins the race. In the second race, A gives B a start of 15.0 m and beats him by $6\frac{2}{3}$ s. How many metres can each run in a second?

14. A certain river has a speed of 2.50 km/h. A rower travels downstream for 1.50 h and returns in 4.50 h. Find his rate in still water, and find the one-way distance travelled.

15. A canoeist can paddle 20.0 km down a certain river and back in 8.0 h (40.0 km round trip). She can also paddle 5.0 km down river in the same time as she paddles 3.0 km up river. Find her rate in still water, and find the rate of the current.

16. A space shuttle and a damaged satellite are 384 km apart (Fig. 10–12). The shuttle travels at 836 km/h, and the satellite at 682 km/h, in the same direction. How long will it take the shuttle to overtake the satellite, and in what distance?

Non-uniform Motion Problems

17. The arm of an industrial robot starts at a speed v_0 and drops 34.8 cm in 4.28 s, at constant acceleration a. Its motion is described by

$$s = v_0 t + at^2/2$$

or

$$34.8 = 4.28 v_0 + a(4.28)^2/2$$

In another trial, the arm is found to drop 58.3 cm in 5.57 s, with the same initial speed and acceleration. Find v_0 and a.

18. Crates start at the top of an inclined roller ramp with a speed of v_0 (Fig. 10–13). They roll down with constant acceleration, reaching a speed of 16.3 ft./s after 5.58 s. The motion is described by Eq. A19,

$$v = v_0 + at$$

or

$$16.3 = v_0 + 5.58a$$

The crates are also seen to reach a speed of 18.5 ft./s after 7.03 s. Find v_0 and a.

FIGURE 10-13

Financial Problems

19. A certain investment, at simple interest, amounted in 5 years to $3,000 and in 6 years to $3,100. Find the amount invested, to the nearest dollar, and the rate of interest (see Eq. A9).

20. A shipment of 21 computer keyboards and 33 monitors cost $35,564.25. Another shipment of 41 keyboards and 36 monitors cost $49,172.50. Find the cost of each keyboard and each monitor.

21. A person invested $4,400, part of it in bank bonds bearing 3.10% interest and the remainder in Canada Savings Bonds bearing 4.85% interest (Fig. 10–14), and she received the same income from each. How much, to the nearest dollar, was invested in each?

22. A farmer bought $1\overline{0}0$ ha of land, part at $1,850 per hectare and part at $2,250 (Fig. 10–15), paying for the whole $211,000. How much land was there in each part?

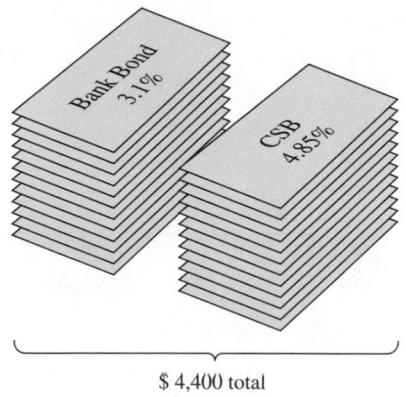

$ 4,400 total

FIGURE 10–14

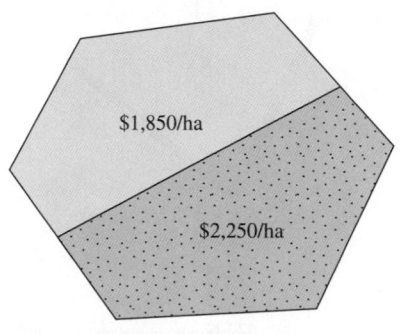

FIGURE 10–15   One hundred hectares for $211,000.

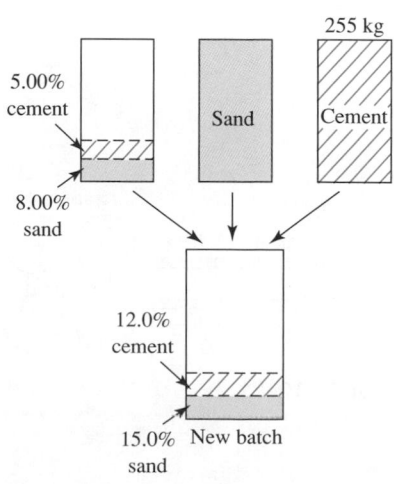

FIGURE 10–16

23. If I lend my money at 6% simple interest for a given time, I shall receive $720 interest; but if I lend it for 3 years longer, I shall receive $1,800. Find the amount of money and the time.

Mixture Problems

24. A certain brass alloy contains 35% zinc and 3.0% lead (Fig. 10–16). Then x kg of zinc and y kg of lead are added to $2\overline{0}0$ kg of the original alloy to make a new alloy that is $4\overline{0}\%$ zinc and 4.0% lead.
    (a) Show that the amount of zinc is given by

    $$0.35(200) + x = 0.40(200 + x + y)$$

    and that the amount of lead is given by

    $$0.03(200) + y = 0.04(200 + x + y)$$

    (b) Solve for x and y.

25. A certain concrete mixture contains 5.00% cement and 8.00% sand (Fig. 10–17). How many kilograms of this mixture and how many kilograms of sand should be combined with 255 kg of cement to make a batch that is 12.0% cement and 15.0% sand?

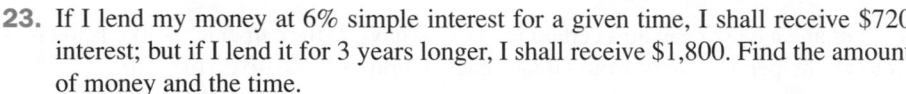

FIGURE 10–17

26. A distributor has two gasohol blends: one that contains 5.00% alcohol and another with 11.0% alcohol. How many litres of each must be mixed to make $50\overline{0}$ L of gasohol containing 9.50% alcohol?

27. A potting mixture contains 12.0% peat moss and 6.00% vermiculite. How much peat and how much vermiculite must be added to $10\overline{0}$ kg of this mixture to produce a new mixture having 15.0% peat and 15.0% vermiculite?

Statics Problems

28. Find the forces F_1 and F_2 in Fig. 10–18.

29. Find the tensions in the ropes in Fig. 10–19.

30. When the 45.3 kg mass in Fig. 10–20 is increased to $10\overline{0}$ kg, the balance point shifts 15.4 cm. Find the length of the bar and the original distance from the balance point to the 45.3-kg mass.

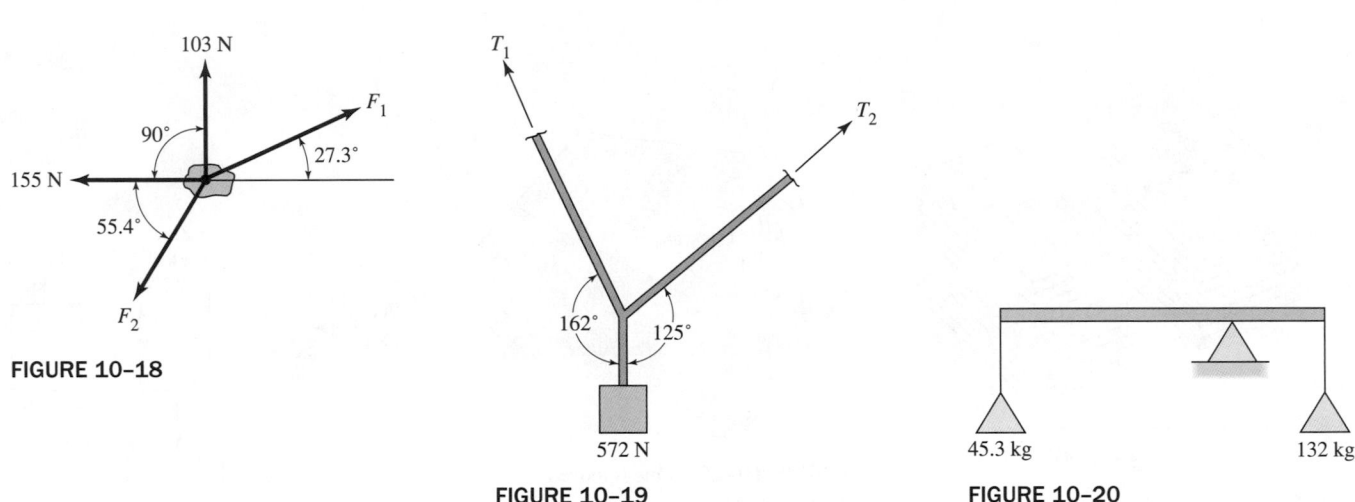

FIGURE 10–18

FIGURE 10–19

FIGURE 10–20

Work Problems

31. A carpenter and a helper can do a certain job in 15.0 days. If the carpenter works 1.50 times as fast as the helper, how long would it take each, working alone, to do the job?

32. During one week, two machines produce a total of 27 210 parts, with the faster machine working for 37.5 h and the slower for 28.2 h. During another week, they produce 59 830 parts, with the faster machine working 66.5 h and the slower machine working 88.6 h. How many parts can each, working alone, produce in an hour?

Flow Problems

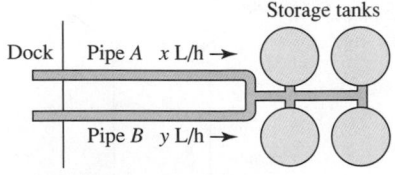

FIGURE 10–21

33. Two different-sized pipes lead from a dockside to a group of oil storage tanks (Fig. 10–21). On one day the two pipes are seen to deliver 117 000 L, with pipe A in use for 3.5 h and pipe B for 4.5 h. On another day the two pipes deliver 151 200 L, with pipe A operating for 5.2 h and pipe B for 4.8 h. Assume that pipe A can deliver x L/h and pipe B can deliver y L/h.

(a) Show that the total amounts of oil delivered for the two days are given by the equations

$$3.5x + 4.5y = 117\ 000$$
$$5.2x + 4.8y = 151\ 200$$

(b) Solve for x and y.

34. Working together, two conveyors can fill a certain bin in 6.00 h. If one conveyor works 1.80 times as fast as the other, how long would it take each to fill the bin working alone?

Energy Flow Problems

35. A hydroelectric generating plant and a coal-fired generating plant together supply a city of 255 000 people, with the hydro plant producing 1.75 times the power of the coal plant. How many people could each service alone?

36. During a certain week, a small wind generator and a small hydro unit together produce 5880 kWh, with the wind generator operating only 85.0% of the time. During another week, the two units produce 6240 kWh, with the wind generator working 95.0% of the time and the hydro unit down 7.50 h for repairs. Assuming that each unit has a constant output when operating, find the number of kilowatts produced by each in 1.00 h.

Electrical Problems

37. To find the currents I_1 and I_2 in Fig. 10–22, we use Eq. A68 in each loop and get the following pair of equations:

$$6.00 - R_1I_1 - R_2I_1 + R_2I_2 = 0$$
$$12.0 - R_3I_2 - R_2I_2 + R_2I_1 = 0$$

Solve for I_1 and I_2 if $R_1 = 736\ \Omega$, $R_2 = 386\ \Omega$, and $R_3 = 375\ \Omega$.

38. Use Eq. A68 to write a pair of equations for the circuit of Fig. 10–23 as in problem 37. Solve these equations for I_1 and I_2.

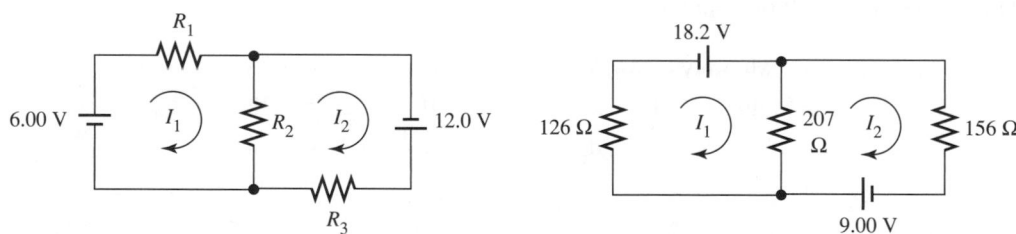

FIGURE 10–22   **FIGURE 10–23**

39. The resistance R of a conductor at temperature t is given by Eq. A70.

$$R = R_1(1 + \alpha t)$$

where R_1 is the temperature at 0°C and α is the temperature coefficient of resistance. A coil of this conductor is found to have a resistance of 31.2 Ω at 25.4 °C and a resistance of 35.7 Ω at 57.3 °C. Find R_1 and α.

40. The resistance R of two resistors in parallel (Fig. 10–24) is given by Eq. A64.

$$1/R = 1/R_1 + 1/R_2$$

R is found to be 283.0 Ω, but if R_1 is doubled and R_2 halved, then R is found to be 291.0 Ω. Find R_1 and R_2.

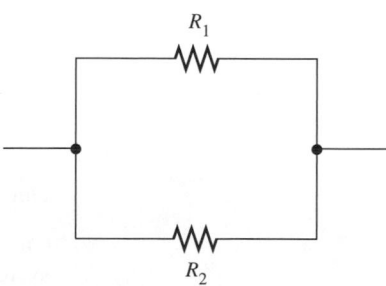

FIGURE 10–24

10–4   Systems of Three Linear Equations

Our strategy here is to reduce a given system of three equations in three unknowns to a system of two equations in two unknowns, which we already know how to solve.

In Chapter 11 we will use *determinants* to solve sets of three or more equations.

Addition-Subtraction Method

We take any two of the given equations and, by addition-subtraction or substitution, eliminate one variable, obtaining a single equation in two unknowns. We then take another pair of equations (which must include the one not yet used, as well as one of those already used) and similarly obtain a second equation in the *same two* unknowns. This pair of equations can then be solved simultaneously, and the values obtained are substituted back to obtain the third variable.

◆◆◆ **Example 25:** Solve the following:

$$
\begin{array}{lr}
\text{Grouping 1:} \quad \boxed{6x - 4y - 7z = 17} & (1) \\
\boxed{9x - 7y - 16z = 29} \quad \text{Grouping 2:} & (2) \\
\boxed{10x - 5y - 3z = 23} & (3)
\end{array}
$$

Solution: Let us start by eliminating x from Equations (1) and (2).

$$
\begin{array}{lrr}
\text{Grouping 1: Multiply Equation (1) by 3:} & 18x - 12y - 21z = 51 & (4) \\
\text{Multiply (2) by } -2: & -18x + 14y + 32z = -58 & (5) \\
\text{Add:} & 2y + 11z = -7 & (6)
\end{array}
$$

We now eliminate the same variable, x, from Equations (1) and (3).

$$
\begin{array}{lrr}
\text{Grouping 2: Multiply (1) by } -5: & -30x + 20y + 35z = -85 & (7) \\
\text{Multiply (3) by 3:} & 30x - 15y - 9z = 69 & (8) \\
\text{Add:} & 5y + 26z = -16 & (9)
\end{array}
$$

Now we solve (6) and (9) simultaneously.

$$
\begin{array}{lr}
\text{Multiply (6) by 5:} & 10y + 55z = -35 \\
\text{Multiply (9) by } -2: & -10y - 52z = 32 \\
\text{Add:} & 3z = -3 \\
& z = -1
\end{array}
$$

Substituting $z = -1$ into (6) gives us

$$
2y + 11(-1) = -7
$$
$$
y = 2
$$

Substituting $y = 2$ and $z = -1$ into (1) yields

$$
6x - 4(2) - 7(-1) = 17
$$
$$
x = 3
$$

Our solution is then $x = 3$, $y = 2$, and $z = -1$.

Check: We check a system of three (or more) equations in the same way that we checked a system of two equations: by substituting back into the original equations.

Substituting into (1) gives

$$
6(3) - 4(2) - 7(-1) = 17
$$
$$
18 - 8 + 7 = 17 \quad \text{(checks)}
$$

Substituting into (2), we get

$$
9(3) - 7(2) - 16(-1) = 29
$$
$$
27 - 14 + 16 = 29 \quad \text{(checks)}
$$

Finally we substitute into (3).

$$10(3) - 5(2) - 3(-1) = 23$$
$$30 - 10 + 3 = 23 \quad \text{(checks)}$$ ◆◆◆

Substitution Method

A *sparse* system (one in which many terms are missing) is often best solved by substitution, as in the following example.

◆◆◆ **Example 26:** Solve by substituting:

$$x - y = 4 \qquad (1)$$
$$x + z = 8 \qquad (2)$$
$$x - y + z = 10 \qquad (3)$$

Solution: From the first two equations, we can write both y and z in terms of x.

$$y = x - 4$$

and

$$z = 8 - x$$

Substituting these back into the third equation yields

$$x - (x - 4) + (8 - x) = 10$$

from which

$$x = 2$$

Substituting back gives

$$y = x - 4 = 2 - 4 = -2$$

and

$$z = 8 - x = 8 - 2 = 6$$

Our solution is then $x = 2$, $y = -2$, and $z = 6$. ◆◆◆

Fractional Equations

We use the same techniques for solving a set of three fractional equations that we did for two equations:

1. Multiply each equation by its LCD to eliminate fractions.
2. If the unknowns are in the denominators, substitute new variables that are the reciprocals of the originals.

◆◆◆ **Example 27:** Solve for x, y, and z:

$$\frac{4}{x} + \frac{9}{y} - \frac{8}{z} = 3 \qquad (1)$$

$$\frac{8}{x} - \frac{6}{y} + \frac{4}{z} = 3 \qquad (2)$$

$$\frac{5}{3x} + \frac{7}{2y} - \frac{2}{z} = \frac{3}{2} \qquad (3)$$

Solution: We make the substitution

$$p = \frac{1}{x} \quad q = \frac{1}{y} \quad \text{and} \quad r = \frac{1}{z}$$

and also multiply Equation (3) by its LCD, 6, to clear fractions.

$$4p + 9q - 8r = 3 \qquad (4)$$
$$8p - 6q + 4r = 3 \qquad (5)$$
$$10p + 21q + 12r = 9 \qquad (6)$$

We multiply (5) by 2 and add it to (4), eliminating the r terms.

$$20p - 3q = 9 \qquad (7)$$

Then we multiply (5) by 3 and add it to (6).

$$34p + 3q = 18 \qquad (8)$$

Adding (7) and (8) gives

$$54p = 27$$
$$p = \frac{1}{2}$$

Then from (8),

$$3q = 18 - 17 = 1$$
$$q = \frac{1}{3}$$

and from (5),

$$4r = 3 - 4 + 2 = 1$$
$$r = \frac{1}{4}$$

Returning to our original variables, we have $x = 2$, $y = 3$, and $z = 4$.   ◆◆◆

Literal Equations

The same techniques apply to literal equations, as shown in the following example.

◆◆◆ **Example 28:** Solve for x, y, and z:

$$x + 3y = 2c \qquad (1)$$
$$y + 2z = a \qquad (2)$$
$$2z - x = 3 \qquad (3)$$

Solution: We first notice that adding Equations (1) and (3) will eliminate x.

$$\begin{array}{lr} x + 3y \qquad\quad = 2c & (1) \\ \underline{-x \qquad\quad + 2z = +3} & (3) \\ 3y + 2z = 2c + 3 & (1) + (3) \end{array}$$

Add (1) + (3):

From this equation we subtract (2), eliminating z:

$$\begin{array}{lr} 3y + 2z = 2c + 3 & (1) + (3) \\ \underline{y + 2z = \qquad\quad + a} & (2) \\ 2y = 2c + 3 - a & \end{array}$$

from which we obtain

$$y = \frac{2c + 3 - a}{2}$$

Substituting back into (1), we have

$$x = 2c - 3\left(\frac{2c + 3 - a}{2}\right)$$

which simplifies to

$$x = \frac{3a - 2c - 9}{2}$$

Substituting this expression for x back into (3) gives

$$2z = 3 + \left(\frac{3a - 2c - 9}{2} \right)$$

which simplifies to

$$z = \frac{3a - 2c - 3}{4}$$ ◆◆◆

◆◆◆ **Example 29:** Find the forces F_1, F_2, and F_3 in Fig. 10–25.

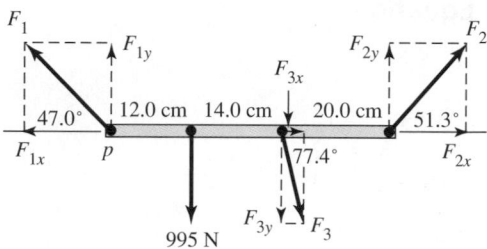

FIGURE 10–25

Solution: We resolve each force into its x and y components (see Sec. 7–6 to refresh your memory). By Eq. A13, the sum of the vertical forces must be zero, so

$$F_{1y} + F_{2y} - F_{3y} - 995 = 0 \qquad (1)$$

By Eq. A14, the sum of the horizontal forces must be zero, giving

$$F_{1x} - F_{2x} - F_{3x} = 0 \qquad (2)$$

By Eq. A15, the sum of the moments about any point, p in this example, must be zero, so we get

$$12.0(995) + 26.0F_{3y} - 46.0F_{2y} = 0 \qquad (3)$$

Using the trigonometric functions, we change these equations as follows:

$$F_1 \sin 47.0° + F_2 \sin 51.3° - F_3 \sin 77.4° - 995 = 0 \qquad (4)$$
$$F_1 \cos 47.0° - F_2 \cos 51.3° - F_3 \cos 77.4° = 0 \qquad (5)$$
$$12.0(995) + 26.0F_3 \sin 77.4° - 46.0F_2 \sin 51.3° = 0 \qquad (6)$$

After evaluating the trigonometric functions and simplifying, we have

$$0.731\,354F_1 + 0.780\,430F_2 - 0.976\,917F_3 = 995 \qquad (7)$$
$$0.682\,000F_1 - 0.625\,243F_2 - 0.218\,143F_3 = 0 \qquad (8)$$
$$-35.899\,799F_2 + 25.373\,866F_3 = -11940 \qquad (9)$$

Solving Equation (9) for F_3 gives

$$F_3 = 1.41F_2 - 470 \qquad (10)$$

Substituting this into (7) and (8) and simplifying, we get

$$F_1 - 0.816F_2 = \quad 733.6$$
$$F_1 - 1.367F_2 = -150.2 \qquad (11)$$

Subtracting, we get $0. 0.551F_2 = 883.8$ from which

$$F_2 = 1604 = 16\overline{0}0 \text{ N}$$

Substituting back into (11) gives

$$F_1 = 732.6 + 0.816(1604) = 2040 \text{ N}$$

and substituting into (10) we get

$$F_3 = 1.41(1604) - 470 = 1790 \text{ N}$$

◆◆◆

Exercise 4 ◆ Systems of Three Linear Equations

Solve each system of equations.

1. $x + y = 35$
$x + z = 40$
$y + z = 45$

2. $x + y + z = 12$
$x - y = 2$
$x - z = 4$

3. $3x + y = 5$
$2y - 3z = -5$
$x + 2z = 7$

4. $x - y = 5$
$y - z = -6$
$2x - z = 2$

5. $x + y + z = 18$
$x - y + z = 6$
$x + y - z = 4$

6. $x + y + z = 90$
$2x - 3y = -20$
$2x + 3z = 145$

7. $x + 2y + 3z = 14$
$2x + y + 2z = 10$
$3x + 4y - 3z = 2$

8. $x + y + z = 35$
$x - 2y + 3z = 15$
$y - x + z = -5$

9. $x - 2y + 2z = 5$
$5x + 3y + 6z = 57$
$x + 2y + 2z = 21$

10. $1.21x + 1.48y + 1.63z = 6.83$
$4.94x + 4.27y + 3.63z = 21.7$
$2.88x + 4.15y - 2.79z = 2.76$

11. $2.51x - 4.48y + 3.13z = 10.8$
$2.84x + 1.37y - 1.66z = 6.27$
$1.58x + 2.85y - 1.19z = 27.3$

12. $5a + b - 4c = -5$
$3a - 5b - 6c = -20$
$a - 3b + 8c = -27$

13. $p + 3q - r = 10$
$5p - 2q + 2r = 6$
$3p + 2q + r = 13$

Fractional Equations

14. $x + \dfrac{y}{3} = 5$

$x + \dfrac{z}{3} = 6$

$y + \dfrac{z}{3} = 9$

15. $\dfrac{1}{x} + \dfrac{1}{y} = 5$

$\dfrac{1}{y} + \dfrac{1}{z} = 7$

$\dfrac{1}{x} + \dfrac{1}{z} = 6$

16. $\dfrac{x}{10} + \dfrac{y}{5} + \dfrac{z}{20} = \dfrac{1}{4}$

$x + y + z = 6$

$\dfrac{x}{3} + \dfrac{y}{2} + \dfrac{z}{6} = 1$

17. $\dfrac{1}{x} + \dfrac{2}{y} - \dfrac{1}{z} = -3$

$\dfrac{3}{x} + \dfrac{1}{y} + \dfrac{1}{z} = 4$

$\dfrac{1}{x} - \dfrac{1}{y} + \dfrac{2}{z} = 6$

Literal Equations

Solve for x, y, and z.

18. $x - y = a$
$y + z = 3a$
$5z - x = 2a$

19. $x + a = y + z$
$y + a = 2x + 2z$
$z + a = 3x + 3y$

20. $ax + by = (a + b)c$
$by + cz = (c + a)b$
$ax + cz = (b + c)a$

21. $x + y + 2z = 2(b + c)$
$x + 2y + z = 2(a + c)$
$2x + y + z = 2(a + b)$

22. $a_1x + b_1y + c_1z = k_1$
$a_2x + b_2y + c_2z = k_2$
$a_3x + b_3y + c_3z = k_3$

> Solving problem 22 gives us formulas for solving a set of three equations. We'll find them useful in the next chapter.

Number Problems

23. The sum of the digits in a certain three-digit number is 21. The sum of the first and third digits is twice the middle digit. If the hundreds and tens digits are interchanged, the number is reduced by 90. Find the number.

24. Two fractions have the same denominator. If 1 is subtracted from the numerator of the smaller fraction, its value will be $\frac{1}{3}$ of the larger fraction; but if 1 is subtracted from the numerator of the larger, its value will be twice that of the smaller fraction. The difference between the fractions is $\frac{1}{3}$. What are the fractions?

25. A certain number is expressed by three digits whose sum is 10. The sum of the first and last digits is $\frac{2}{3}$ of the second digit; and if 198 is subtracted from the number, the digits will be reversed. What is the number?

Electrical Problems

26. When writing Kirchhoff's law (Eq. A68) for a certain three-loop network, we get the set of equations

$$3I_1 + 2I_2 - 4I_3 = 4$$
$$I_1 - 3I_2 + 2I_3 = -5$$
$$2I_1 + I_2 - I_3 = 3$$

where I_1, I_2, and I_3 are the loop currents in amperes. Solve for these currents.

27. For the three-loop network of Fig. 10–26:
(a) Use Kirchhoff's law to show that the currents may be found from

$$159I_1 - 50.9I_2 = 1$$
$$407I_1 - 2400I_2 + 370I_3 = 1$$
$$142I_2 - 380I_3 = 1$$

(b) Solve this set of equations for the three currents.

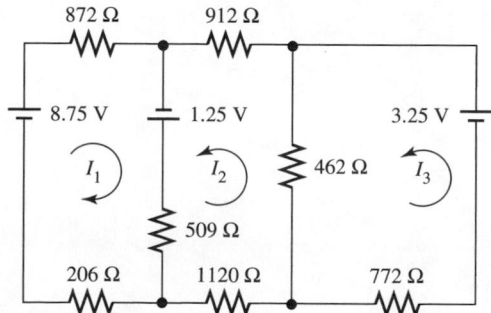

FIGURE 10–26

Statics Problems

28. Find the forces F_1, F_2, and F_3 in Fig. 10–27.

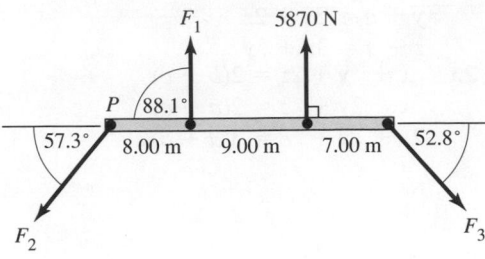

FIGURE 10–27

29. Find the forces F_1, F_2, and F_3 in Fig. 10–28.

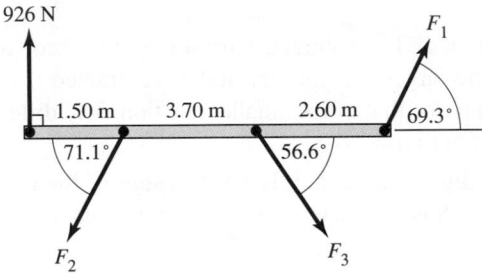

FIGURE 10–28

Computer

30. The Gauss-Seidel method described in Sec. 10–1 works for any number of equations. For three equations in three unknowns, for example, write the equations in the form

$$x = f(y, z)$$
$$y = g(x, z)$$
$$z = h(x, y)$$

and proceed as before using a spreadsheet. Try it on one of the systems given in this exercise set. If the computation diverges, then solve each equation for a different variable than before, and try again. It may take two or three tries to find a combination that works.

Case Study Discussion – Electrical Currents

Let's solve for the currents:

Loop 1: $6.8(I_1 - I_2) + 1.5I_1 = 8$
Loop 2: $2.2(I_2 + I_3) + 4.8I_2 + 6.8(I_2 - I_1) = 12$
Loop 3: $2.2(I_3 + I_2) + 1.8I_3 = 12$

Removing the parentheses,

$6.8I_1 - 6.8I_2 + 1.5I_1 = 8$
$2.2I_2 + 2.2I_3 + 4.8I_2 + 6.8I_2 - 6.8I_1 = 12$
$2.2I_3 + 2.2I_2 + 1.8I_3 = 12$

Combining like terms,

$$8.3I_1 - 6.8I_2 = 8 \tag{1}$$
$$-6.8I_1 + 13.8I_2 + 2.2I_3 = 12 \tag{2}$$
$$2.2I_2 + 4I_3 = 12 \tag{3}$$

Solving the above system produces the following results:

$$I_1 = 2.36, I_2 = 1.70, I_3 = 2.06$$

••• CHAPTER 10 REVIEW PROBLEMS ••••••••••••••••••••••••••••••••••

Solve each system of equations by any method.

1. $4x + 3y = 27$
   $2x - 5y = -19$

2. $\dfrac{x}{3} + \dfrac{y}{2} = \dfrac{4}{3}$
   $\dfrac{x}{2} + \dfrac{y}{3} = \dfrac{7}{6}$

3. $\dfrac{15}{x} + \dfrac{4}{y} = 1$
   $\dfrac{5}{x} - \dfrac{12}{y} = 7$

4. $2x + 4y - 3z = 22$
   $4x - 2y + 5z = 18$
   $6x + 7y - z = 63$

5. $5x + 3y - 2z = 5$
   $3x - 4y + 3z = 13$
   $x + 6y - 4z = -8$

6. $3x - 5y - 2z = 14$
   $5x - 8y - z = 12$
   $x - 3y - 3z = 1$

7. $\dfrac{3}{x+y} + \dfrac{4}{x-z} = 2$
   $\dfrac{6}{x+y} + \dfrac{5}{y-z} = 1$
   $\dfrac{4}{x-z} + \dfrac{5}{y-z} = 2$

8. $4x + 2y - 26 = 0$
   $3x + 4y = 39$

9. $2x - 3y + 14 = 0$
   $3x + 2y = 44$

10. $\dfrac{2}{x} + \dfrac{1}{y} = \dfrac{4}{3}$
   $\dfrac{3}{x} + \dfrac{5}{y} = \dfrac{19}{6}$

11. $\dfrac{9}{x} + \dfrac{8}{y} = \dfrac{43}{6}$
   $\dfrac{3}{x} + \dfrac{10}{y} = \dfrac{29}{6}$

12. $\dfrac{x}{a} + \dfrac{y}{b} = p$
   $\dfrac{x}{b} + \dfrac{y}{a} = q$

13. $x + y - a$
$x + z = b$
$y + z = c$

14. $\dfrac{5x}{6} + \dfrac{2y}{5} = 14$

$\dfrac{3x}{4} - \dfrac{2y}{5} = 5$

15. $\dfrac{2x}{7} + \dfrac{2y}{3} = \dfrac{16}{3}$

$x + y = 12$

16. $x + y + z = 35$
$5x + 4y + 3z = 22$
$3x + 4y - 3z = 2$

17. $2x - 4y + 3z = 10$
$3x + y - 2z = 6$
$x + 3y - z = 20$

18. $5x + y - 4z = -5$
$3x - 5y - 6z = -20$
$x - 3y + 8z = -27$

19. $x + 3y - z = 10$
$5x - 2y + 2z = 6$
$3x + 2y + z = 13$

20. $x + 21y = 2$
$2x + 27y = 19$

21. $2x - y = 9$
$5x - 3y = 14$

22. $6x - 2y + 5z = 53$
$5x + 3y + 7z = 33$
$x + y + z = 5$

23. $9x + 4y = 54$
$4y + 9y = 89$

24. A sum of money was divided between A and B so that A's share was to B's share as 5 is to 3. Also, A's share exceeded $\frac{5}{9}$ of the whole sum by \$50. What was each share?

25. A and B together can do a job in 12 days, and B and C together can do the same job in 16 days. How long would it take them all working together to do the job if A does $1\frac{1}{2}$ times as much as C?

26. If the numerator of a certain fraction is increased by 2 and its denominator diminished by 2, its value will be 1. If the numerator is increased by the denominator and the denominator is diminished by 5, its value will be 5. Find the fraction.

Writing

27. Suppose that you have a number of paired equations to solve in order to find the loop currents in a circuit that you are designing. Each pair of equations has different values for the variables. You must leave for a week but want your assistant to solve the equations in your absence. Write step-by-step instructions to be followed, using the addition-subtraction method, including instructions as to the number of digits to be retained.

Team Project

28. A certain nickel silver alloy contains the following:

Copper	55.9%	Lead	0.10%
Zinc	31.25%	Nickel	12.00%
Tin	0.50%	Manganese	0.25%

How many kilograms of zinc, tin, lead, nickel, and manganese must be added to 400 kg of this alloy to make a new "leaded nickel silver" with the following composition:

Copper	44.50%	Lead	1.00%
Zinc	42.00%	Nickel	10.00%
Tin	0.50%	Manganese	2.00%

11

Determinants

••• OBJECTIVES ••

When you have completed this chapter, you should be able to:
- Evaluate second-order determinants and apply Cramer's Rule.
- Evaluate third-order determinants by minors and apply Cramer's Rule.
- Evaluate higher order determinants up to 5 equations in 5 unknowns.

The use of determinants is an efficient technique for solving large systems of equations. We saw in Chapter 10 that you could solve a system of five equations by reducing it to a system of four, then to a system of three, then to a system of two, and finally solving for the last unknown. There is a way to solve the larger systems with less work, and that is by using determinants. This technique is based on a very long and complicated formula. The techniques, however, allow you to set up all the variables in an array (called a determinant) and then, following some simple rules, calculate the value of the determinant. This value is then used with other matrix values to solve the system.

This might seem to be just a method used in an era before calculators and computers; in fact, your calculator may well have a built-in system-of-equations solver. However, the determinant technique can be used not just for numerical coefficients, but also for literal coefficients or even coefficients that are expressions themselves. You could even use this technique to set up a spreadsheet to solve large systems of equations by having the user fill in the coefficients and letting the formulas you entered in the spreadsheet cells do the rest of the work—if you understand how a matrix works!

CASE STUDY—SETTING UP SYSTEMS OF EQUATIONS

You are working for a company that is installing an industrial automation process system for a manufacturer. One of your assignments is to program a spreadsheet to solve for a system of five equations and five unknowns. Your spreadsheet should include instructions for where to enter the variables. You should let the user know that when an equation does not have one of the variables in it, he or she should enter zero as its coefficient. The final solutions to the variables should be formatted to stand out to the user. Of course, correct and accurate functioning is the top priority, but presentation and being easy to use and understand is also very important. The five variables are L, P, r, h, and n. Here's an example of what a user will want to solve:

$$5L + 2P + r + 6h = 5n + 83.6$$
$$-2L + 6P + 2.5r - 20 + 7h + 2n = 44.2$$
$$3P + 0.5r + 4h = 23.9 - 3n$$
$$3L - 2r + 3h = 12.8$$
$$75 + 8L + P + 5h = 3n + 180.5$$

11–1 Second-Order Determinants

Definitions

In Chapter 10, we studied the set of equations

$$a_1x + b_1y = c_1$$
$$a_2x + b_2y = c_2$$

Now let us refer to the values

$$\begin{vmatrix} a_1 \\ a_2 \end{vmatrix}$$ as the column of x coefficients,

$$\begin{vmatrix} b_1 \\ b_2 \end{vmatrix}$$ as the column of y coefficients, and

$$\begin{vmatrix} c_1 \\ c_2 \end{vmatrix}$$ as the column of constants.

When we solved this set of equations, we got the solution

$$x = \frac{b_2c_1 - b_1c_2}{a_1b_2 - a_2b_1} \text{ and } y = \frac{a_1c_2 - a_2c_1}{a_1b_2 - a_2b_1}$$

The denominator $(a_1b_2 - a_2b_1)$ of each of these fractions is called the *determinant* of the coefficients a_1, b_1, a_2, and b_2. It is commonly expressed by the symbol

$$\begin{vmatrix} a_1 & b_1 \\ a_2 & b_2 \end{vmatrix}$$

which is understood to signify the product of the upper-left and lower-right numbers, minus the product of the upper-right and lower-left numbers. Thus,

$$\begin{vmatrix} a_1 & b_1 \\ a_2 & b_2 \end{vmatrix} = a_1b_2 - a_2b_1$$

It is common practice to use the same word *determinant* to refer to the *symbol*

$$\begin{vmatrix} a_1 & b_1 \\ a_2 & b_2 \end{vmatrix}$$

as well as to the value $(a_1b_2 - a_2b_1)$. We'll follow this practice.

◆◆◆ Example 1: In the determinant

$$\begin{vmatrix} 2 & 5 \\ 6 & 1 \end{vmatrix}$$

each of the numbers 2, 5, 6, and 1 is called an *element*. There are two *rows*, the first row containing the elements 2 and 5, and the second row having the elements 6 and 1. There are also two *columns*, the first with elements 2 and 6, and the second with elements 5 and 1. The *value* of this determinant is

$$2(1) - 5(6) = 2 - 30 = -28$$

◆◆◆

Thus a determinant is written as a *square array*, enclosed between vertical bars. The number of rows equals the number of columns. The *order* of a determinant is equal to the number of rows or columns that it contains. Thus, the determinants above are of second order.

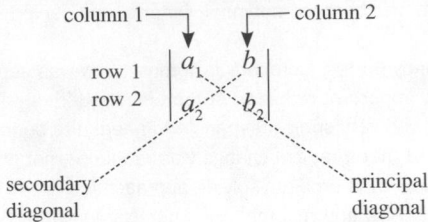

The diagonal running from upper left to lower right is called the *principal diagonal*. The other is called the *secondary diagonal*.

Second-Order Determinants	$\begin{vmatrix} a_1 & b_1 \\ a_2 & b_2 \end{vmatrix} = a_1 b_2 - a_2 b_1$	65

The value of a second-order determinant is equal to the product of the elements on the principal diagonal minus the product of the elements on the secondary diagonal.

◆◆◆ **Example 2:**

$$\begin{vmatrix} 2 & -1 \\ 3 & 5 \end{vmatrix} = 2(5) - 3(-1) = 13 \qquad \text{◆◆◆}$$

Common Errors	Students sometimes add the numbers on a diagonal instead of *multiplying*. Be careful not to do that.
	Don't just ignore a zero element. It causes the product along its diagonal to be zero.

◆◆◆ **Example 3:**

$$\begin{vmatrix} 4 & a \\ x & 3b \end{vmatrix} = 4(3b) - x(a) = 12b - ax \qquad \text{◆◆◆}$$

◆◆◆ **Example 4:**

$$\begin{vmatrix} 0 & 3 \\ 5 & 2 \end{vmatrix} = 0(2) - 3(5) = -15 \qquad \text{◆◆◆}$$

Solving a System of Equations by Determinants
We now return to the set of equations

$$a_1 x + b_1 y = c_1$$
$$a_2 x + b_2 y = c_2 \qquad (1)$$

If, in the solution for x,

$$x = \frac{b_2 c_1 - b_1 c_2}{a_1 b_2 - a_2 b_1} \qquad (2)$$

we replace the denominator by the square array

$$\begin{vmatrix} a_1 & b_1 \\ a_2 & b_2 \end{vmatrix} = a_1 b_2 - a_2 b_1$$

we get

$$x = \frac{b_2 c_1 - b_1 c_2}{\begin{vmatrix} a_1 & b_1 \\ a_2 & b_2 \end{vmatrix}} \tag{3}$$

This denominator is called the *determinant of the coefficients*, or sometimes the *determinant of the system*. It is usually given the Greek capital letter delta (Δ). If this determinant equals zero, there is no solution to the set of equations.

Now look at the numerator of Equation (3), which is $b_2 c_1 - b_1 c_2$. If this expression were the value of some determinant, it is clear that the elements on the principal diagonal must be b_2 and c_1 and that the elements on the secondary diagonal must be b_1 and c_2. One such determinant is

— column of constants

$$\begin{vmatrix} c_1 & b_1 \\ c_2 & b_2 \end{vmatrix} = b_2 c_1 - b_1 c_2$$

Our solution for x can then be expressed as

$$x = \frac{\begin{vmatrix} c_1 & b_1 \\ c_2 & b_2 \end{vmatrix}}{\begin{vmatrix} a_1 & b_1 \\ a_2 & b_2 \end{vmatrix}}$$

An expression for y can be developed in a similar way. Thus the solution to the system of Equations (1), in terms of determinants, is given by the following rule:

| **Cramer's Rule** | $$x = \frac{\begin{vmatrix} c_1 & b_1 \\ c_2 & b_2 \end{vmatrix}}{\begin{vmatrix} a_1 & b_1 \\ a_2 & b_2 \end{vmatrix}} \quad \text{and} \quad y = \frac{\begin{vmatrix} a_1 & c_1 \\ a_2 & c_2 \end{vmatrix}}{\begin{vmatrix} a_1 & b_1 \\ a_2 & b_2 \end{vmatrix}}$$ | **70** |

Named after the Swiss mathematician Gabriel Cramer (1704–52), Cramer's rule, as we will see, is valid for systems with any number of linear equations.

or

| **Cramer's Rule** | The solution for any variable is a fraction whose denominator is the determinant of the coefficients and whose numerator is the same determinant, except that the column of coefficients for the variable for which we are solving is replaced by the column of constants. | **69** |

Of course, the denominator Δ cannot be zero, or we get division by zero, indicating that the set of equations has no unique solution.

◆◆◆ **Example 5:** Solve by determinants:

$$2x - 3y = 1$$
$$x + 4y = -5$$

It's easier to find y by substituting x back into one of the previous equations, but we use determinants instead, to show how it is done.

We would, of course, check our answers by substituting back into the original equations, as we did in Chapter 10.

Solution: We first evaluate the determinant of the coefficients, because if $\Delta = 0$, there is no unique solution, and we have no need to proceed further.

$$\Delta = \begin{vmatrix} 2 & -3 \\ 1 & 4 \end{vmatrix} = 2(4) - 1(-3) = 8 + 3 = 11$$

Solving for x, we have

column of constants

$$x = \frac{\begin{vmatrix} 1 & -3 \\ -5 & 4 \end{vmatrix}}{\Delta} = \frac{1(4) - (-5)(-3)}{11} = \frac{4 - 15}{11} = -1$$

Solving for y yields

$$y = \frac{\begin{vmatrix} 2 & 1 \\ 1 & -5 \end{vmatrix}}{\Delta} = \frac{2(-5) - 1(1)}{11} = \frac{-10 - 1}{11} = -1$$

◆◆◆

Common Error

Don't forget to first arrange the given equations into the form

$$a_1 x + b_1 y = c_1$$
$$a_2 x + b_2 y = c_2$$

before writing the determinants.

◆◆◆ **Example 6:** Solve for x and y by determinants:

$$2ax - by + 3a = 0$$
$$4y = 5x - a$$

Solution: We first rearrange our equations.

$$2ax - by = -3a$$
$$5x - 4y = a$$

The determinant of the coefficients is

$$\Delta = \begin{vmatrix} 2a & -b \\ 5 & -4 \end{vmatrix} = 2a(-4) - 5(-b) = 5b - 8a$$

So

$$x = \frac{\begin{vmatrix} -3a & -b \\ a & -4 \end{vmatrix}}{\Delta} = \frac{(-3a)(-4) - a(-b)}{5b - 8a}$$

$$= \frac{12a + ab}{5b - 8a}$$

and

$$y = \frac{\begin{vmatrix} 2a & -3a \\ 5 & a \end{vmatrix}}{\Delta} = \frac{2a(a) - 5(-3a)}{5b - 8a}$$

$$= \frac{2a^2 + 15a}{5b - 8a}$$

◆◆◆

Exercise 1 ◆ Second-Order Determinants

Find the value of each second-order determinant.

1. $\begin{vmatrix} 3 & 2 \\ 1 & -4 \end{vmatrix}$

2. $\begin{vmatrix} -2 & 5 \\ 3 & -3 \end{vmatrix}$

3. $\begin{vmatrix} 0 & 5 \\ -3 & -4 \end{vmatrix}$

4. $\begin{vmatrix} 8 & 3 \\ -1 & 2 \end{vmatrix}$

5. $\begin{vmatrix} -4 & 5 \\ 7 & -2 \end{vmatrix}$

6. $\begin{vmatrix} 7 & -6 \\ -5 & 4 \end{vmatrix}$

7. $\begin{vmatrix} 4.82 & 2.73 \\ 2.97 & 5.28 \end{vmatrix}$

8. $\begin{vmatrix} 48.7 & -53.6 \\ 4.93 & 9.27 \end{vmatrix}$

9. $\begin{vmatrix} -2/3 & 2/5 \\ -1/3 & 4/5 \end{vmatrix}$

10. $\begin{vmatrix} 1/2 & 1/4 \\ -1/2 & -1/4 \end{vmatrix}$

11. $\begin{vmatrix} a & b \\ c & d \end{vmatrix}$

12. $\begin{vmatrix} 2m & 3n \\ -m & 4n \end{vmatrix}$

13. $\begin{vmatrix} \sin\theta & 3 \\ \tan\theta & \sin\theta \end{vmatrix}$

14. $\begin{vmatrix} 2i_1 & i_2 \\ -3i_1 & 4i_2 \end{vmatrix}$

15. $\begin{vmatrix} 3i_1 & i_2 \\ \sin\theta & \cos\theta \end{vmatrix}$

16. $\begin{vmatrix} 3x & 2 \\ x & 5 \end{vmatrix}$

Solve each pair of equations by determinants.

17. $2x + y = 11$
$3x - y = 4$

18. $5x + 7y = 101$
$7x - y = 55$

19. $3x - 2y = -15$
$5x + 6y = 3$

20. $7x + 6y = 20$
$2x + 5y = 9$

21. $x + 5y = 11$
$3x + 2y = 7$

22. $4x - 5y = -34$
$2x - 3y = -22$

23. $x + 4y = 11$
$5x - 2y = 11$

24. $2x - 3y = 3$
$4x + 5y = 39$

25. $7x - 4y = 81$
$5x - 3y = 57$

26. $3x + 4y = 85$
$5x + 4y = 107$

27. $3x - 2y = 1$
$2x + y = 10$

28. $5x - 2y = 3$
$2x + 3y = 5$

29. $y = 9 - 3x$
$x = 8 - 2y$

30. $y = 2x - 3$
$x = 19 - 3y$

31. $29.1x - 47.6y = 42.8$
$11.5x + 72.7y = 25.8$

32. $4.92x - 8.27y = 2.58$
$6.93x + 2.84y = 8.36$

33. $4n = 18 - 3m$
$m = 8 - 2n$

34. $5p + 4q - 14 = 0$
$17p = 31 + 3q$

35. $3w = 13 + 5z$
$4w - 7z - 17 = 0$

36. $3u = 5 + 2v$
$5v + 2u = 16$

Fractional and Literal Equations

Solve for x and y by determinants.

37. $\dfrac{x}{3} + \dfrac{y}{2} = 1$

$\dfrac{x}{2} - \dfrac{y}{3} = -1$

38. $px - qy + pq = 0$
$2px - 3qy = 0$

39. $ax + by = p$
$cx + dy = q$

40. $\dfrac{x}{2} - \dfrac{3y}{4} = 0$

$\dfrac{x}{3} + \dfrac{y}{4} = 0$

41. $\dfrac{3}{x} + \dfrac{8}{y} = 3$

$\dfrac{15}{x} - \dfrac{4}{y} = 4$

42. $4mx - 3ny = 6$
$3mx + 2ny = -7$

Applications

43. A shipment of 4 cars and 2 trucks cost \$172,172. Another shipment of 3 cars and 5 trucks cost \$209,580. Find the cost of each car and each truck.

44. An airplane and a helicopter are 125 km apart. The airplane is traveling at 226 km/h, and the helicopter at 85.0 km/h, both in the same direction. How long will it take the airplane to overtake the helicopter, and in what distance from the initial position of the airplane?

45. A certain alloy contains 31.0% zinc and 3.50% lead. Then x kg of zinc and y kg of lead are added to 325 kg of the original alloy to make a new alloy that is 45.0% zinc and 4.80% lead. The amount of zinc is given by

$$0.310(325) + x = 0.450(325 + x + y)$$

and the amount of lead is given by

$$0.0350(325) + y = 0.0480(325 + x + y)$$

Solve for x and y.

11–2   Third-Order Determinants

The technique we used for finding the value of a second-order determinant *cannot* be used for higher-order determinants. We will now show two other methods: first, what we'll call the *diagonal* method, which works *only for third-order determinants*; and then *development by minors,* which works for *any* determinant.

The Diagonal Method

The value of a third-order determinant can be found as follows. To the right of the determinant, rewrite the first two columns.

$$\begin{vmatrix} a_1 & b_1 & c_1 \\ a_2 & b_2 & c_2 \\ a_3 & b_3 & c_3 \end{vmatrix} \begin{matrix} a_1 & b_1 \\ a_2 & b_2 \\ a_3 & b_3 \end{matrix}$$

Then multiply the elements along each of the six diagonals. The value of the determinant is the sum of the products along those diagonals running downward to the right, *minus* the products along those diagonals running downward to the left.

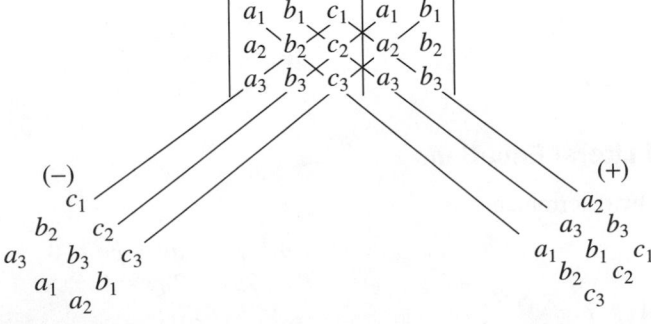

We will not emphasize this method because it works only for third-order determinants.

The value of the determinant is then given by the following equation:

Third-Order Determinant	$\begin{vmatrix} a_1 & b_1 & c_1 \\ a_2 & b_2 & c_2 \\ a_3 & b_3 & c_3 \end{vmatrix} = \begin{aligned} & a_1b_2c_3 + a_3b_1c_2 + a_2b_3c_1 \\ & - a_3b_2c_1 - a_1b_3c_2 - a_2b_1c_3 \end{aligned}$	**66**

◆◆◆ **Example 7:** Evaluate the following third-order determinant:

$$\begin{vmatrix} 2 & 1 & 0 \\ 3 & -2 & 1 \\ 5 & 1 & 0 \end{vmatrix}$$

Solution: Rewriting the first two columns to the right of the determinant, we have

$$\begin{vmatrix} 2 & 1 & 0 \\ 3 & -2 & 1 \\ 5 & 1 & 0 \end{vmatrix} \begin{matrix} 2 & 1 \\ 3 & -2 \\ 5 & 1 \end{matrix}$$

Multiplying along the diagonals, we obtain

$$2(-2)0 + 1(1)5 + 0(3)(1) - [0(-2)(5) + 2(1)(1) + 1(3)(0)]$$
$$= \quad 0 \quad + \quad 5 \quad + \quad 0 \quad - [\quad 0 \quad + \quad 2 \quad + \quad 0]$$
$$= 5 - 2 = 3$$

◆◆◆

Minors

The *minor of an element in a determinant* is a determinant of next lower order, obtained by deleting the row and the column in which that element lies.

◆◆◆ **Example 8:** We find the minor of element c in the determinant

$$\begin{vmatrix} a & b & c \\ d & e & f \\ g & h & i \end{vmatrix}$$

by striking out the first row and the third column.

$$\begin{vmatrix} a & b & c \\ d & e & f \\ g & h & i \end{vmatrix} \quad \text{or} \quad \begin{vmatrix} d & e \\ g & h \end{vmatrix}$$

◆◆◆

Sign Factor

The *sign factor* for an element depends upon the position of the element in the determinant. To find the sign factor, add the row number and the column number of the element. If the sum is even, the sign factor is $+1$; if the sum is odd, the sign factor is -1.

A minor with the sign factor attached is called a *signed minor*. It is also called a *cofactor* of the given element.

◆◆◆ **Example 9:**

(a) For the determinant in Example 8, element c is in the first row and the third column. The sum of the row and column numbers

$$1 + 3 = 4$$

is even, so the sign factor for the element c is $+1$.

(b) The sign factor for the element b in Example 8 is -1.

◆◆◆

The sign factor for the element in the upper left-hand corner of the determinant is $+1$, and the signs alternate according to the pattern

$$\begin{vmatrix} + & - & + & - & + & - & \cdot \\ - & + & - & + & - & \cdot & \cdot \\ + & - & + & - & \cdot & \cdot & \cdot \\ - & + & - & \cdot & \cdot & \cdot & \cdot \\ + & - & \cdot & \cdot & \cdot & \cdot & \cdot \\ - & \cdot & \cdot & \cdot & \cdot & \cdot & \cdot \\ \cdot & \cdot & \cdot & \cdot & \cdot & \cdot & \cdot \end{vmatrix}$$

so that the sign factor may be found simply by *counting off* from the upper left corner.

Thus for any element in a determinant, we can write three quantities:

1. The element itself.
2. The sign factor of the element.
3. The minor of the element.

◆◆◆ **Example 10:** For the element in the second row and the first column in the following determinant:

$$\begin{vmatrix} -2 & 5 & 1 \\ 3 & -2 & 4 \\ 1 & -4 & 2 \end{vmatrix}$$

1. the element itself is 3;
2. the sign factor of that element is −1; and
3. the minor of that element is

$$\begin{vmatrix} 5 & 1 \\ -4 & 2 \end{vmatrix}$$ ◆◆◆

Evaluating a Determinant by Minors

We now define the *value of a determinant* as follows:

Value of a Determinant	To find the value of a determinant of order n: **1.** Multiply each of the n elements in any row (or column) by the sign factor of that element and the minor of that element. **2.** Take the algebraic sum of those n products.	**68**

◆◆◆ **Example 11:** Evaluate by minors:

$$\begin{vmatrix} 3 & 2 & 4 \\ 1 & 6 & -2 \\ -1 & 5 & -3 \end{vmatrix}$$

Solution: We first choose a row or a column for development. The work of expansion is greatly reduced if that row or column contains zeros. Our given determinant has no zeros, so let us choose column 1, which at least contains some 1's.

the elements of column 1

$$\begin{vmatrix} 3 & 2 & 4 \\ 1 & 6 & -2 \\ -1 & 5 & -3 \end{vmatrix} = 3(1)\begin{vmatrix} 6 & -2 \\ 5 & -3 \end{vmatrix} + 1(-1)\begin{vmatrix} 2 & 4 \\ 5 & -3 \end{vmatrix} + (-1)(1)\begin{vmatrix} 2 & 4 \\ 6 & -2 \end{vmatrix}$$

the sign factor of each element

the minor of each element

$$= 3[(6)(-3) - (-2)(5)] - 1[(2)(-3) - (4)(5)] - 1[(2)(-2) - (4)(6)]$$
$$= -24 + 26 + 28$$
$$= 30$$ ◆◆◆

◆◆◆ **Example 12:** Evaluate by minors:

$$\begin{vmatrix} a_1 & b_1 & c_1 \\ a_2 & b_2 & c_2 \\ a_3 & b_3 & c_3 \end{vmatrix}$$

We would get the same result if we chose another row or column for development.

Solution: Choosing, say, the first row for development, we get

$$a_1\begin{vmatrix} b_2 & c_2 \\ b_3 & c_3 \end{vmatrix} - b_1\begin{vmatrix} a_2 & c_2 \\ a_3 & c_3 \end{vmatrix} + c_1\begin{vmatrix} a_2 & b_2 \\ a_3 & b_3 \end{vmatrix}$$

$$= a_1(b_2c_3 - b_3c_2) - b_1(a_2c_3 - a_3c_2) + c_1(a_2b_3 - a_3b_2)$$

$$= a_1b_2c_3 - a_1b_3c_2 - a_2b_1c_3 + a_3b_1c_2 + a_2b_3c_1 - a_3b_2c_1$$

$$= a_1b_2c_3 + a_3b_1c_2 + a_2b_3c_1 - a_3b_2c_1 - a_1b_3c_2 - a_2b_1c_3 \qquad \text{◆◆◆}$$

The value of the determinant is thus given by the following formula:

Third-Order Determinants	$\begin{vmatrix} a_1 & b_1 & c_1 \\ a_2 & b_2 & c_2 \\ a_3 & b_3 & c_3 \end{vmatrix} = \begin{matrix} a_1b_2c_3 + a_3b_1c_2 + a_2b_3c_1 \\ - a_3b_2c_1 - a_1b_3c_2 - a_2b_1c_3 \end{matrix}$	**66**

Solving a System of Three Equations by Determinants

If we were to *algebraically* solve the system of equations

$$a_1x + b_1y + c_1z = k_1$$
$$a_2x + b_2y + c_2z = k_2$$
$$a_3x + b_3y + c_3z = k_3$$

we would get the following solution:

The solution of this system was given as a problem in Exercise 4 of Chapter 10.

Three Equations in Three Unknowns	$x = \dfrac{b_2c_3k_1 + b_1c_2k_3 + b_3c_1k_2 - b_2c_1k_3 - b_3c_2k_1 - b_1c_3k_2}{a_1b_2c_3 + a_3b_1c_2 + a_2b_3c_1 - a_3b_2c_1 - a_1b_3c_2 - a_2b_1c_3}$ $y = \dfrac{a_1c_3k_2 + a_3c_2k_1 + a_2c_1k_3 - a_3c_1k_2 - a_1c_2k_3 - a_2c_3k_1}{a_1b_2c_3 + a_3b_1c_2 + a_2b_3c_1 - a_3b_2c_1 - a_1b_3c_2 - a_2b_1c_3}$ $z = \dfrac{a_1b_2k_3 + a_3b_1k_2 + a_2b_3k_1 - a_3b_2k_1 - a_1b_3k_2 - a_2b_1k_3}{a_1b_2c_3 + a_3b_1c_2 + a_2b_3c_1 - a_3b_2c_1 - a_1b_3c_2 - a_2b_1c_3}$	**64**

Notice that the three denominators are identical and are equal to the value of the determinant formed from the coefficients of the unknowns (Eq. 66). As before, we call it the *determinant of the coefficients*, Δ.

Furthermore, we can obtain the numerator of each from the determinant of the coefficients by replacing the coefficients of the variable in question with the constants k_1, k_2, and k_3. The solution to our set of equations

$$a_1x + b_1y + c_1z = k_1$$
$$a_2x + b_2y + c_2z = k_2$$
$$a_3x + b_3y + c_3z = k_3$$

is then given by Cramer's rule:

Cramer's Rule	$x = \dfrac{\begin{vmatrix} k_1 & b_1 & c_1 \\ k_2 & b_2 & c_2 \\ k_3 & b_3 & c_3 \end{vmatrix}}{\Delta} \quad y = \dfrac{\begin{vmatrix} a_1 & k_1 & c_1 \\ a_2 & k_2 & c_2 \\ a_3 & k_3 & c_3 \end{vmatrix}}{\Delta} \quad z = \dfrac{\begin{vmatrix} a_1 & b_1 & k_1 \\ a_2 & b_2 & k_2 \\ a_3 & b_3 & k_3 \end{vmatrix}}{\Delta}$ where $\Delta = \begin{vmatrix} a_1 & b_1 & c_1 \\ a_2 & b_2 & c_2 \\ a_3 & b_3 & c_3 \end{vmatrix} \neq 0$	**71**

Although we do not prove it, Cramer's rule works for higher-order systems as well. We restate it now in words.

| **Cramer's Rule** | The solution for any variable is a fraction whose denominator is the determinant of the coefficients and whose numerator is the same determinant, except that the column of coefficients for the variable for which we are solving is replaced by the column of constants. | **69** |

◆◆◆ **Example 13:** Solve by determinants:

$$x + 2y + 3z = 14$$
$$2x + y + 2z = 10$$
$$3x + 4y - 3z = 2$$

Solution: We first write the determinant of the coefficients

$$\Delta = \begin{vmatrix} 1 & 2 & 3 \\ 2 & 1 & 2 \\ 3 & 4 & -3 \end{vmatrix}$$

and expand it by minors. Choosing the first row for expansion gives

$$\Delta = \begin{vmatrix} 1 & 2 & 3 \\ 2 & 1 & 2 \\ 3 & 4 & -3 \end{vmatrix} = 1\begin{vmatrix} 1 & 2 \\ 4 & -3 \end{vmatrix} - 2\begin{vmatrix} 2 & 2 \\ 3 & -3 \end{vmatrix} + 3\begin{vmatrix} 2 & 1 \\ 3 & 4 \end{vmatrix}$$

$$= 1(-3 - 8) - 2(-6 - 6) + 3(8 - 3)$$

$$= -11 + 24 + 15 = 28$$

As when we are solving a set of two equations, a Δ of zero would mean that the system had no unique solution, and we would stop here.

We now make a new determinant by replacing the coefficients of x, $\begin{vmatrix} 1 \\ 2 \\ 3 \end{vmatrix}$, by the column of constants $\begin{vmatrix} 14 \\ 10 \\ 2 \end{vmatrix}$, and expand it by, say, the second column.

$$\begin{vmatrix} 14 & 2 & 3 \\ 10 & 1 & 2 \\ 2 & 4 & -3 \end{vmatrix} = -2\begin{vmatrix} 10 & 2 \\ 2 & -3 \end{vmatrix} + 1\begin{vmatrix} 14 & 3 \\ 2 & -3 \end{vmatrix} - 4\begin{vmatrix} 14 & 3 \\ 10 & 2 \end{vmatrix}$$

$$= -2(-30 - 4) + 1(-42 - 6) - 4(28 - 30)$$

$$= 68 - 48 + 8 = 28$$

Dividing this value by Δ gives x.

$$x = \frac{28}{\Delta} = \frac{28}{28} = 1$$

Next we replace the coefficients of y with the column of constants and expand the first column by minors.

$$\begin{vmatrix} 1 & 14 & 3 \\ 2 & 10 & 2 \\ 3 & 2 & -3 \end{vmatrix} = 1\begin{vmatrix} 10 & 2 \\ 2 & -3 \end{vmatrix} - 2\begin{vmatrix} 14 & 3 \\ 2 & -3 \end{vmatrix} + 3\begin{vmatrix} 14 & 3 \\ 10 & 2 \end{vmatrix}$$

$$= 1(-30 - 4) - 2(-42 - 6) + 3(28 - 30)$$

$$= -34 + 96 - 6 = 56$$

Dividing by Δ gives the value of y.

$$y = \frac{56}{\Delta} = \frac{56}{28} = 2$$

We can get z also by determinants or, more easily, by substituting back. Substituting $x = 1$ and $y = 2$ into the first equation, we obtain

$$1 + 2(2) + 3z = 14$$
$$3z = 9$$
$$z = 3$$

The solution is thus $(1, 2, 3)$. ◆◆◆

Often, some terms will have zero coefficients or decimal coefficients. Further, the terms may be out of order, as in the following example.

◆◆◆ **Example 14:** Solve for x, y, and z:

$$23.7y + 72.4x = 82.4 - 11.3x$$
$$25.5x - 28.4z + 19.3 = 48.2y$$
$$13.4 + 66.3z = 39.2x - 10.5$$

Solution: We rewrite each equation in the form $ax + by + cz = k$, combining like terms as we go and putting in the missing terms with zero coefficients. We do this *before* writing the determinant.

$$83.7x + 23.7y + 0z = 82.4 \tag{1}$$
$$25.5x - 48.2y - 28.4z = -19.3 \tag{2}$$
$$-39.2x + 0y + 66.3z = -23.9 \tag{3}$$

Note: You can multiply the second and third equations by –1 to reduce the minus signs. That process should reduce the number of arithmetic errors in the rest of your solution process.

The determinant of the system is then

$$\Delta = \begin{vmatrix} 83.7 & 23.7 & 0 \\ 25.5 & -48.2 & -28.4 \\ -39.2 & 0 & 66.3 \end{vmatrix}$$

Let us develop the first row by minors.

$$\Delta = 83.7 \begin{vmatrix} -48.2 & -28.4 \\ 0 & 66.3 \end{vmatrix} - 23.7 \begin{vmatrix} 25.5 & -28.4 \\ -39.2 & 66.3 \end{vmatrix} + 0$$
$$= 83.7(-48.2)(66.3) - 23.7[25.5(66.3) - (-28.4)(-39.2)]$$
$$= -281\,000 \text{ (to three significant digits)}$$

Note: Reducing the values to three significant figures should be done as the last step in the solution process. It is done earlier in this example to simplify the calculations.

Now solving for x gives us

$$x = \frac{\begin{vmatrix} 82.4 & 23.7 & 0 \\ -19.3 & -48.2 & -28.4 \\ -23.9 & 0 & 66.3 \end{vmatrix}}{\Delta}$$

Let us develop the first row of the determinant by minors. We get

$$82.4 \begin{vmatrix} -48.2 & -28.4 \\ 0 & 66.3 \end{vmatrix} - 23.7 \begin{vmatrix} -19.3 & -28.4 \\ -23.9 & 66.3 \end{vmatrix} + 0$$
$$= 82.4(-48.2)(66.3) - 23.7[(-19.3)(66.3) - (-28.4)(-23.9)]$$
$$= -217\,000$$

Dividing by Δ yields

$$x = \frac{-217\,000}{-281\,000} = 0.772$$

We solve for y and z by substituting back. From Equation (1), we get

$$23.7y = 82.4 - 83.7(0.772) = 17.8$$
$$y = 0.751$$

and from (3)

$$66.3z = -23.9 + 39.2(0.772) = 6.36$$
$$z = 0.0959$$

The solution is then $x = 0.772$, $y = 0.751$, and $z = 0.0959$.   ◆◆◆

Exercise 2 ◆ Third-Order Determinants

Evaluate each determinant.

1. $\begin{vmatrix} 1 & 0 & 2 \\ 3 & 1 & 0 \\ 1 & 2 & 1 \end{vmatrix}$
  2. $\begin{vmatrix} 2 & -1 & 3 \\ 0 & 2 & 1 \\ 3 & -2 & 4 \end{vmatrix}$

3. $\begin{vmatrix} -3 & 1 & 2 \\ 0 & -1 & 5 \\ 6 & 0 & 1 \end{vmatrix}$
  4. $\begin{vmatrix} -1 & 0 & 3 \\ 2 & 0 & -2 \\ 1 & -3 & 4 \end{vmatrix}$

5. $\begin{vmatrix} 5 & 1 & 2 \\ -3 & 2 & -1 \\ 4 & -3 & 5 \end{vmatrix}$
  6. $\begin{vmatrix} 1.0 & 2.4 & -1.5 \\ -2.6 & 0 & 3.2 \\ -2.9 & 1.0 & 4.1 \end{vmatrix}$

7. $\begin{vmatrix} 2 & 1 & 3 \\ 0 & -2 & 4 \\ 0 & 1 & 5 \end{vmatrix}$
  8. $\begin{vmatrix} 1 & 5 & 4 \\ -3 & 6 & -2 \\ -1 & 5 & 3 \end{vmatrix}$

Solve by determinants.

Some of these problems are identical to those given in Exercise 4 of Chapter 10.

9. $x + y + z = 18$
$x - y + z = 6$
$x + y - z = 4$

10. $x + y + z = 12$
$x - y = 2$
$x - z = 4$

11. $x + y = 35$
$x + z = 40$
$y + z = 45$

12. $x + y + z = 35$
$x - 2y + 3z = 15$
$y - x + z = -5$

13. $x + 2y + 3z = 14$
$2x + y + 2z = 10$
$3x + 4y - 3z = 2$

14. $x + y + z = 90$
$2x - 3y = -20$
$2x + 3z = 145$

15. $x - 2y + 2z = 5$
$5x + 3y + 6z = 57$
$x + 2y + 2z = 21$

16. $3x + y = 5$
$2y - 3z = -5$
$x + 2z = 7$

17. $2x - 4y + 3z = 10$
$3x + y - 2z = 6$
$x + 3y - z = 20$

18. $x - y = 5$
$y - z = -6$
$2x - z = 2$

19. $1.15x + 1.95y + 2.78z = 15.3$
$2.41x + 1.16y + 3.12z = 9.66$
$3.11x + 3.83y - 2.93z = 2.15$

20. $1.52x - 2.26y + 1.83z = 4.75$
$4.72x + 3.52y + 5.83z = 45.2$
$1.33x + 2.61y + 3.02z = 18.5$

Fractional Equations

Solve by determinants.

21. $x + \dfrac{y}{3} = 5$

$x + \dfrac{z}{3} = 6$

$y + \dfrac{z}{3} = 9$

22. $\dfrac{x}{10} + \dfrac{y}{5} + \dfrac{z}{20} = \dfrac{1}{4}$

$x + y + z = 6$

$\dfrac{x}{3} + \dfrac{y}{2} + \dfrac{z}{6} = 1$

23. $\dfrac{1}{x} + \dfrac{1}{y} = 5$

$\dfrac{1}{y} + \dfrac{1}{z} = 7$

$\dfrac{1}{x} + \dfrac{1}{z} = 6$

24. $\dfrac{1}{x} + \dfrac{2}{y} - \dfrac{1}{z} = -3$

$\dfrac{3}{x} + \dfrac{1}{y} + \dfrac{1}{z} = 4$

$\dfrac{1}{x} - \dfrac{1}{y} + \dfrac{2}{z} = 6$

Applications

25. Applying Kirchhoff's law to a certain three-loop network gives

$$283I_1 - 274I_2 + 163I_3 = 352$$
$$428I_1 + 163I_2 + 373I_3 = 169$$
$$338I_1 - 112I_2 - 227I_3 = 825$$

Solve this set of equations for the three currents.

26. Find the forces F_1, F_2, and F_3 in Fig. 11–1. Neglect the weight of the beam.

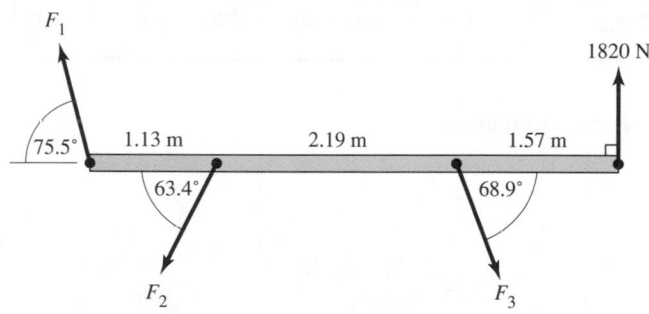

FIGURE 11–1

11–3   Higher-Order Determinants

We can evaluate a determinant of any order by repeated use of the method of minors. Thus any row or column of a fifth-order determinant can be developed, thereby reducing the determinant to five fourth-order determinants. Each of these can then be developed into four third-order determinants, and so on, until only second-order determinants remain.

Obviously, this method involves lots of work. To avoid it, we first *simplify* the determinant, using the *properties of determinants* given in the following section, and then apply a systematic method for reducing the determinants by minors shown after that.

Properties of Determinants

A determinant larger than third order can usually be evaluated faster if we first reduce its order before expanding by minors. We do this by applying the various *properties of determinants* as shown in the following examples.

Zero Row or Column	If all elements in a row (or column) are zero, then the value of the determinant is zero.	**72**

♦♦♦ Example 15:

$$\begin{vmatrix} 4 & 8 & 0 \\ 5 & 2 & 0 \\ 3 & 9 & 0 \end{vmatrix} = 0$$

♦♦♦

Identical Rows or Columns	If two rows (or columns) are identical, then the value of the determinant is zero.	**73**

♦♦♦ Example 16: We find the value of the determinant

$$\begin{vmatrix} 2 & 1 & 3 \\ 4 & 2 & 5 \\ 4 & 2 & 5 \end{vmatrix}$$

by expanding by minors using, say, the first row.

$$2\begin{vmatrix} 2 & 5 \\ 2 & 5 \end{vmatrix} - 1\begin{vmatrix} 4 & 5 \\ 4 & 5 \end{vmatrix} + 3\begin{vmatrix} 4 & 2 \\ 4 & 2 \end{vmatrix}$$
$$= 2(10 - 10) - 1(20 - 20) + 3(8 - 8)$$
$$= 0$$

♦♦♦

Zeros Below the Principal Diagonal	If all elements below the principal diagonal are zeros, then the value of the determinant is the product of the elements along the principal diagonal.	**74**

Expand this determinant by minors and see for yourself that its value is indeed 6.

♦♦♦ Example 17: The determinant

$$\begin{vmatrix} 2 & 9 & 7 & 8 \\ 0 & 1 & 5 & 9 \\ 0 & 0 & 3 & 4 \\ 0 & 0 & 0 & 1 \end{vmatrix}$$

has the value

$$(2)(1)(3)(1) = 6$$

♦♦♦

Interchanging Rows with Columns	If we change the rows to columns and the columns to rows, the value of a determinant will be unchanged.	**75**

♦♦♦ Example 18: The value of the determinant

$$\begin{vmatrix} 5 & 2 \\ 4 & 3 \end{vmatrix}$$

is

$$(5)(3) - (2)(4) = 7$$

If the first column now becomes the first row, and the second column becomes the second row, we get the determinant

$$\begin{vmatrix} 5 & 4 \\ 2 & 3 \end{vmatrix}$$

which has the value

$$(5)(3) - (4)(2) = 7$$

as before.   ◆◆◆

| Interchange of Two Rows (or Columns) | A determinant will change sign when we interchange two rows (or columns) | 76 |

◆◆◆ **Example 19:** We find the value of the determinant

$$\begin{vmatrix} 3 & 0 & 2 \\ 1 & 4 & 3 \\ 2 & 1 & 2 \end{vmatrix}$$

by expanding by minors using the first row.

$$3 \begin{vmatrix} 4 & 3 \\ 1 & 2 \end{vmatrix} - 0 + 2 \begin{vmatrix} 1 & 4 \\ 2 & 1 \end{vmatrix} = 3(8 - 3) + 2(1 - 8) = 1$$

Let us now interchange, say, the first and second columns.

$$\begin{vmatrix} 0 & 3 & 2 \\ 4 & 1 & 3 \\ 1 & 2 & 2 \end{vmatrix}$$

We find its value again by expanding using the first row.

$$0 - 3 \begin{vmatrix} 4 & 3 \\ 1 & 2 \end{vmatrix} + 2 \begin{vmatrix} 4 & 1 \\ 1 & 2 \end{vmatrix} = -3(8 - 3) + 2(8 - 1) = -1$$

So interchanging two columns has reversed the sign of the determinant.   ◆◆◆

| Multiplying by a Constant | If each element in a row (or column) is multiplied by some constant, then the value of the determinant is multiplied by that constant. | 77 |

◆◆◆ **Example 20:** We saw that the value of the determinant in Example 19 was equal to 1.

$$\begin{vmatrix} 3 & 0 & 2 \\ 1 & 4 & 3 \\ 2 & 1 & 2 \end{vmatrix} = 1$$

Let us now multiply the elements in the third row by some constant, say, 3.

$$\begin{vmatrix} 3 & 0 & 2 \\ 1 & 4 & 3 \\ 6 & 3 & 6 \end{vmatrix}$$

This new determinant has the value

$$3(24 - 9) + 2(3 - 24) = 45 - 42 = 3$$

or three times its previous value.   ◆◆◆

We can also use this rule *in reverse,* to *remove a factor* from a row or a column.

◆◆◆ **Example 21:** The determinant

$$\begin{vmatrix} 3 & 1 & 0 \\ 30 & 10 & 15 \\ 3 & 2 & 1 \end{vmatrix}$$

could be evaluated as it is, but let us first factor a 3 from the first column.

$$\begin{vmatrix} 3 & 1 & 0 \\ 30 & 10 & 15 \\ 3 & 2 & 1 \end{vmatrix} = 3\begin{vmatrix} 1 & 1 & 0 \\ 10 & 10 & 15 \\ 1 & 2 & 1 \end{vmatrix}$$

Then we factor a 5 from the second row.

$$3\begin{vmatrix} 1 & 1 & 0 \\ 10 & 10 & 15 \\ 1 & 2 & 1 \end{vmatrix} = 3(5)\begin{vmatrix} 1 & 1 & 0 \\ 2 & 2 & 3 \\ 1 & 2 & 1 \end{vmatrix}$$

$$= 15[(2-6)-(2-3)] = -45 \qquad \text{◆◆◆}$$

Multiples of One Row (or Column) Added to Another	If the elements of a row (or column) are multiplied by some factor and are then added to the corresponding elements of another row or column, the value of a determinant is unchanged.	**78**

◆◆◆ **Example 22:** Again using the determinant from Example 19,

$$\begin{vmatrix} 3 & 0 & 2 \\ 1 & 4 & 3 \\ 2 & 1 & 2 \end{vmatrix} = 1$$

let us get a new second row by multiplying the third row by -4 and adding those elements to the second row. The new elements in the second row are then

$$1 + 2(-4) = -7, \quad 4 + 1(-4) = 0, \quad \text{and} \quad 3 + 2(-4) = -5$$

giving a new determinant

$$\begin{vmatrix} 3 & 0 & 2 \\ -7 & 0 & -5 \\ 2 & 1 & 2 \end{vmatrix}$$

whose value we find by developing the second column by minors.

$$-1\begin{vmatrix} 3 & 2 \\ -7 & -5 \end{vmatrix} = -1[3(-5) - 2(-7)] = -[-15 + 14] = 1 \text{ (as before)}$$

Notice, however, that we have *introduced another zero* into the determinant, making it easier to evaluate. This, of course, is the point of the whole operation. ◆◆◆

Exercise 3 ◆ Higher-Order Determinants

Evaluate each determinant.

1. $\begin{vmatrix} 4 & 3 & 1 & 0 \\ -1 & 2 & -3 & 5 \\ 0 & 1 & -1 & 2 \\ 0 & 2 & -3 & 5 \end{vmatrix}$

2. $\begin{vmatrix} -1 & 3 & 0 & 2 \\ 2 & -1 & 1 & 0 \\ 5 & 2 & -2 & 0 \\ 1 & -1 & 3 & 1 \end{vmatrix}$

3. $\begin{vmatrix} 2 & 0 & -1 & 0 \\ 0 & 0 & 2 & -1 \\ 1 & 3 & 2 & 1 \\ 3 & 1 & 1 & -2 \end{vmatrix}$

4. $\begin{vmatrix} 1 & 2 & -1 & 1 \\ -1 & 1 & 2 & 3 \\ 3 & -1 & 1 & 2 \\ 1 & 2 & -1 & 1 \end{vmatrix}$

5. $\begin{vmatrix} 3 & 1 & 0 & 2 & 4 \\ 1 & 2 & 4 & 0 & 1 \\ 2 & 3 & 1 & 4 & 2 \\ 1 & 2 & 0 & 2 & 1 \\ 3 & 4 & 1 & 3 & 1 \end{vmatrix}$

6. $\begin{vmatrix} 2 & 1 & 5 & 3 & 6 \\ 1 & 4 & 2 & 4 & 3 \\ 3 & 1 & 2 & 4 & 1 \\ 5 & 2 & 3 & 1 & 4 \\ 4 & 5 & 2 & 3 & 1 \end{vmatrix}$

Four Equations in Four Unknowns

Solve by determinants.

7. $x + y + 2z + w = 18$
$x + 2y + z + w = 17$
$x + y + z + 2w = 19$
$2x + y + z + w = 16$

8. $2x - y - z - w = 0$
$x - 3y + z + w = 0$
$x + y - 4z + w = 0$
$x + y + w = 36$

9. $3x - 2y - z + w = -3$
$-x - y + 3z + 2w = 23$
$x + 3y - 2z + w = -12$
$2x - y - z - 3w = -22$

10. $x + 2y = 5$
$y + 2z = 8$
$z + 2u = 11$
$u + 2x = 6$

11. $x + y = a + b$
$y + z = b + c$
$z + w = a - b$
$w - x = c - b$

12. $2x - 3y + z - w = -6$
$x + 2y - z = 8$
$3y + z + 3w = 0$
$3x - y + w = 0$

Five Equations in Five Unknowns

Solve by determinants.

13. $x + y = 9$
$y + z = 11$
$z + w = 13$
$w + u = 15$
$u + x = 12$

14. $3x + 4y + z = 35$
$3z + 2y - 3w = 4$
$2x - y + 2w = 17$
$3z - 2w + v = 9$
$w + y = 13$

15. $w + v + x + y = 14$
$w + v + x + z = 15$
$w + v + y + z = 16$
$w + x + y + z = 17$
$v + x + y + z = 18$

16. Applying Kirchhoff's law to a certain four-loop network gives the following equations:

$$57.2I_1 + 92.5I_2 - 23.0I_3 - 11.4I_4 = 38.2$$
$$95.3I_1 - 14.9I_2 + 39.0I_3 + 59.9I_4 = 29.3$$
$$66.3I_1 + 81.4I_2 - 91.5I_3 + 33.4I_4 = -73.6$$
$$38.2I_1 - 46.6I_2 + 30.1I_3 + 93.2I_4 = 55.7$$

Solve for the four loop currents.

17. The following equations result when Kirchhoff's law is applied to a certain four-loop network:

$$4.27I_1 - 5.27I_2 + 4.27I_3 + 9.63I_4 = 6.82$$
$$7.92I_1 + 9.36I_2 - 9.72I_3 + 4.14I_4 = -8.83$$
$$8.36I_1 - 2.27I_2 + 4.77I_3 + 7.33I_4 = 3.93$$
$$7.37I_1 + 9.36I_2 - 3.82I_3 - 2.73I_4 = 5.04$$

Find the four currents.

18. Four types of computers are made by a company, each requiring the following numbers of hours for four steps:

	Assembly	Burn-In	Inspection	Testing
Model A	4.50 h	12 h	1.25 h	2.75 h
Model B	5.25 h	12 h	1.75 h	3.00 h
Model C	6.55 h	24 h	2.25 h	3.75 h
Model D	7.75 h	36 h	3.75 h	4.25 h
Available	15 157 h	43 680 h	4824 h	8928 h

Also shown in the table are the monthly production hours available for each step. How many of each type of computer, rounded to the nearest unit, can be made each month?

19. We have available four bronze alloys containing the following percentages of copper, zinc, and lead:

	Alloy 1	Alloy 2	Alloy 3	Alloy 4
Copper	52.0	53.0	54.0	55.0
Zinc	30.0	38.0	20.0	38.0
Lead	3.00	2.00	4.00	3.00

How many kilograms of each alloy should be taken to produce $60\overline{0}$ kg of a new alloy that is 53.8% copper, 30.1% zinc, and 3.20% lead?

Case Study Discussion—Setting Up Systems of Equations

Your spreadsheet will need to give some data entry instructions to the user. The equations will need to be rearranged so that all the variables are on one side and the constants are on the other. It would also help avoid mistakes if the variables appeared in the same order (we'll use L, P, r, h, and n, but you could use a different variable order), for example:

$$5L + 2P + 1r + 6h - 5n = 83.6$$
$$-2L + 6P + 2.5r + 7h + 2n = 64.2$$
$$0L + 3P + 0.5r + 4h + 3n = 23.9$$
$$3L + 0P - 2r + 3h + 0n = 12.8$$
$$8L + 1P + 0r + 5h - 3n = 105.5$$

Try this on your spreadsheet program. Remember to test your results to ensure that you have a correct solution. For a full walkthrough of how to complete this case study, visit www.wiley.com/canada/calter

---

◆◆◆ CHAPTER 11 REVIEW PROBLEMS ◆◆◆◆◆◆◆◆◆◆◆◆◆◆◆◆◆◆◆◆◆◆◆◆◆◆◆◆◆◆

Evaluate.

1. $\begin{vmatrix} 6 & \frac{1}{2} & -2 \\ 3 & \frac{1}{4} & 4 \\ 2 & -\frac{1}{2} & 3 \end{vmatrix}$

2. $\begin{vmatrix} 2 & 7 & -2 & 8 \\ 4 & 1 & 1 & -3 \\ 0 & 3 & -1 & 4 \\ 6 & 4 & 2 & -8 \end{vmatrix}$

3. $\begin{vmatrix} 0 & n & m \\ -n & 0 & l \\ -m & -l & 0 \end{vmatrix}$

4. $\begin{vmatrix} 8 & 2 & 0 & 1 & 4 \\ 0 & 1 & 4 & 2 & 7 \\ 2 & 6 & 3 & 8 & 0 \\ 1 & 4 & 2 & 6 & 5 \\ 4 & 6 & 8 & 3 & 5 \end{vmatrix}$

5. $\begin{vmatrix} 25 & 23 & 19 \\ 14 & 11 & 9 \\ 21 & 17 & 14 \end{vmatrix}$

6. $\begin{vmatrix} x + y & x - y \\ x - y & x + y \end{vmatrix}$

7. $\begin{vmatrix} 9 & 13 & 17 \\ 11 & 15 & 19 \\ 17 & 21 & 25 \end{vmatrix}$

8. $\begin{vmatrix} 5 & -3 & -2 & 0 \\ 4 & 1 & -6 & 2 \\ -1 & 4 & 3 & -5 \\ 0 & 6 & -4 & 2 \end{vmatrix}$

9. $\begin{vmatrix} 1 & 2 & 3 \\ 3 & 1 & 2 \\ 2 & 3 & 1 \end{vmatrix}$

10. $\begin{vmatrix} 2 & -3 & 1 \\ -2 & 4 & 5 \\ 3 & -1 & -4 \end{vmatrix}$

11. $\begin{vmatrix} 1 & 2 & 2 & 4 \\ 1 & 4 & 4 & 1 \\ 1 & 1 & 2 & 2 \\ 4 & 8 & 11 & 13 \end{vmatrix}$

12. $\begin{vmatrix} a - b & -2a \\ 2b & a - b \end{vmatrix}$

13. $\begin{vmatrix} 3 & 1 & 5 & 2 \\ 4 & 10 & 14 & 6 \\ 8 & 9 & 1 & 4 \\ 6 & 15 & 21 & 9 \end{vmatrix}$

14. $\begin{vmatrix} 7 & 8 & 9 \\ 28 & 35 & 40 \\ 21 & 26 & 30 \end{vmatrix}$

15. $\begin{vmatrix} 1 & 5 & 2 \\ 4 & 7 & 3 \\ 9 & 8 & 6 \end{vmatrix}$

16. $\begin{vmatrix} 3 & -6 \\ -5 & 4 \end{vmatrix}$

17. $\begin{vmatrix} 6 & 4 & 7 \\ 9 & 0 & 8 \\ 5 & 3 & 2 \end{vmatrix}$

18. $\begin{vmatrix} 3 & 2 & 1 & 3 \\ 4 & 2 & 2 & 4 \\ 2 & 3 & 1 & 6 \\ 10 & 4 & 5 & 8 \end{vmatrix}$

Solve by determinants.

19. $x + y + z + w = -4$
$x + 2y + 3z + 4w = 0$
$x + 3y + 6z + 10w = 9$
$x + 4y + 10z + 20w = 24$

20. $x + 2y + 3z = 4$
$3x + 5y + z = 18$
$4x + y + 2z = 12$

21. $4x + 3y = 27$
$2x - 5y = -19$

22. $8x - 3y - 7z = 85$
$x + 6y - 4z = -12$
$2x - 5y + z = 33$

23. $x + 2y = 10$
$2x - 3y = -1$

24. $x + y + z + u = 1$
$2x + 3y - 4z + 5u = -31$
$3x - 4y + 5z + 6u = -22$
$4x + 5y - 6z - u = -13$

25. $2x - 3y = 7$
$5x + 2y = 27$

26. $6x + y = 60$
$3x + 2y = 39$

27. $2x + 5y = 29$
$2x - 5y = -21$

28. $4x + 3y = 7$
$2x - 3y = -1$

29. $4x - 5y = 3$
$3x + 5y = 11$

30. $x + 5y = 41$
$3x - 2y = 21$

31. $x + 2y = 7$
$x + y = 5$

32. $8x - 3y = 22$
$4x + 5y = 18$

33. $3x + 4y = 25$
$4x + 3y = 21$

34. $37.7x = 59.2 + 24.6y$
$28.3x - 39.8y - 62.5 = 0$

35. $5.03a = 8.16 + 5.11b$
$3.63b + 7.26a = 28.8$

36. $541x + 216y + 412z = 866$
$211x + 483y - 793z = 315$
$215x + 495y + 378z = 253$

37. $72.3x + 54.2y + 83.3z = 52.5$
$52.2x - 26.6y + 83.7z = 75.2$
$33.4x + 61.6y + 30.2z = 58.5$

38. A link in a certain mechanism starts at a speed v_0 and travels 27.5 cm in 8.25 s, at constant acceleration a. Its motion is described by Eq. A18,

$$s = v_0 t + \frac{1}{2} at^2$$

or

$$27.5 = 8.25v_0 + \frac{1}{2} a(8.25)^2$$

In another trial, the link is found to travel 34.6 cm in 11.2 s, with the same initial speed and acceleration. Find v_0 and a.

39. The following equations result when Kirchhoff's law is applied to a certain four-loop network:

$$14.7I_1 - 25.7I_2 + 14.7I_3 + 19.3I_4 = 26.2$$
$$17.2I_1 + 19.6I_2 - 19.2I_3 + 24.4I_4 = -28.3$$
$$18.6I_1 - 22.7I_2 + 24.7I_3 + 17.3I_4 = 23.3$$
$$27.7I_1 + 19.6I_2 - 33.2I_3 - 42.3I_4 = 25.4$$

Find the four currents.

Team Project

40. Solve the following by any method:

$$28.3x + 29.2y - 33.1z + 72.4u + 29.4v = 39.5$$
$$73.2x - 28.4y + 59.3z - 27.4u + 49.2v = 82.3$$
$$33.7x + 10.3y + 72.3z + 29.3u - 21.2v = 28.4$$
$$92.3x - 39.5y + 29.5z - 10.3u + 82.2v = 73.4$$
$$88.3x + 29.3y + 10.3z + 84.2u + 29.3v = 39.4$$

12

Matrices

OBJECTIVES

When you have completed this chapter, you should be able to:
- Describe an array and matrix by its characteristics.
- Perform mathematical operations on matrices.
- Calculate the value of the inverse of a matrix.
- Solve a system of linear equations by matrix inversion.

This chapter takes a look at something similar to the determinants of Chapter 11. This tool for solving systems of equations is called *matrix inversion*. First, let's look at what a matrix is, then we'll learn how to use it.

A matrix is a type of *array*. An array is a set of numbers that can be used to represent many things, including vectors. For example, an array with two elements (representing x and y coordinates) would indicate a vector on a two-dimensional plane, while a three-element array would represent a vector in three dimensions. It's also mathematically possible to have larger arrays, such as a four-element array (representing a vector in four dimensions) or a five-element array (a vector in five dimensions). This is one of the amazing and beautiful properties of mathematics: all the rules which are applied to one, two, three, or four dimensions can be extended to work out the characteristics of any multidimensional space-time combination. In fact, mathematical description may be the only way we can imagine a multidimensional universe (with a possible number of dimensions between 11 to 23, according to string theory and membrane theory). We might not need to solve multidimensional vectors in engineering technology and applied science, but we can use arrays and matrices for problems inside our own four-dimensional space-time universe.

CASE STUDY—3D IMAGING

One use for matrix math is in programming 3-D engines for video games. A single point on the screen can be defined as a vector. The TV or computer screen is two-dimensional only, so to create the look of depth, objects must change size as they come closer to the viewer or go farther back into the scene. An image is made up of hundreds or thousands of vectors. By multiplying this set or matrix, you can change its size. You can also multiply the vector array by another array, which allows you to rotate the image. In rotation, parts of the image become larger and some become smaller as the image rotates about an axis.

The 3-D imaging in games is now crossing into medical imaging technology. Instead of a game programmer designing an initial object, for example, a medical scan inputs data for an organ or blood vessel. The software can then rotate and present the organ or blood vessel from different angles. The same imaging technology can scan a water pump in a nuclear plant and present the operator with a movable 3-D image of a part of the pump. This last application is our case study.

You are a mechanical engineering technologist who has specialized in the area of non-destructive evaluation of engineered parts. (There are courses and complete programs at some colleges in becoming an NDE specialist.) Your company is looking into a software package that will render three-dimensional images of parts scanned using ultrasonic sound waves. Your department manager believes that some of the older mechanical engineer managers and vice-presidents you have for clients will want to know how 3-D imaging works. As a recent college graduate, you are expected to know about computers and programs. You will need to do a brief presentation on how the program can take scanned input and produce a 3-D image that can then be rotated and examined from all sides. For this case study, just prepare the last part of the seminar on the rotation of the image, using matrix mathematics.

12–1 Definitions

Arrays
A set of numbers, called *elements,* arranged in a pattern, is called an *array.* Arrays are named for the shape of the pattern made by the elements.

◆◆◆ Example 1: The array

$$\begin{pmatrix} 3 & 5 & 4 \\ 1 & 3 & 6 \\ 9 & 2 & 4 \end{pmatrix}$$

is called a *square array.*
    The array

$$\begin{pmatrix} 6 & 2 & 8 \\ 3 & 2 & 7 \end{pmatrix}$$

is called a *rectangular array.*
    The array

$$\begin{pmatrix} 7 & 3 & 9 & 1 \\ & 8 & 3 & 2 \\ & & 9 & 1 \\ & & & 5 \end{pmatrix}$$

is called a *triangular array.*   ◆◆◆

Vectors
A *vector* is an array consisting of a single row or a single column.

◆◆◆ **Example 2:** The array $(2, 6, -2, 8)$ is called a *row vector,* and

The array $\begin{pmatrix} 7 \\ 3 \\ 9 \\ 2 \end{pmatrix}$ is called a *column vector.*   ◆◆◆

What do these vectors have to do with the vectors we defined in Chapter 7, which we used to represent directed quantities such as force and velocity?

The two apparently different definitions agree, if we think of a vector as the *coordinates of the endpoint* of a line segment drawn from the origin.

◆◆◆ **Example 3:**

(a) The vector $(4, 7)$ represents a line segment in a plane drawn from the origin to the point $(4, 7)$, as shown in Fig. 12–1(a).

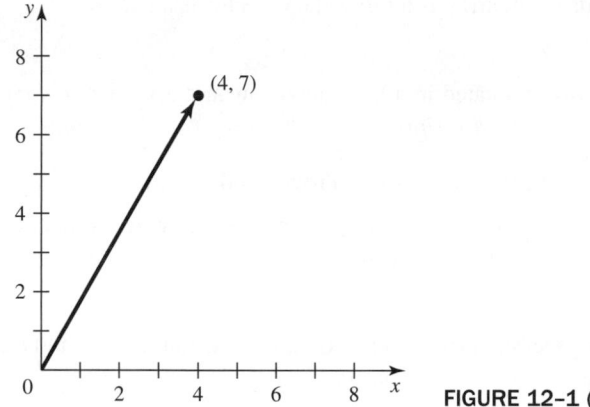

FIGURE 12–1 (a)

(b) The vector $(3, 2, 4)$ represents a line segment in three dimensions, drawn from the origin to the point $(3, 2, 4)$, as shown in Fig. 12–1(b).

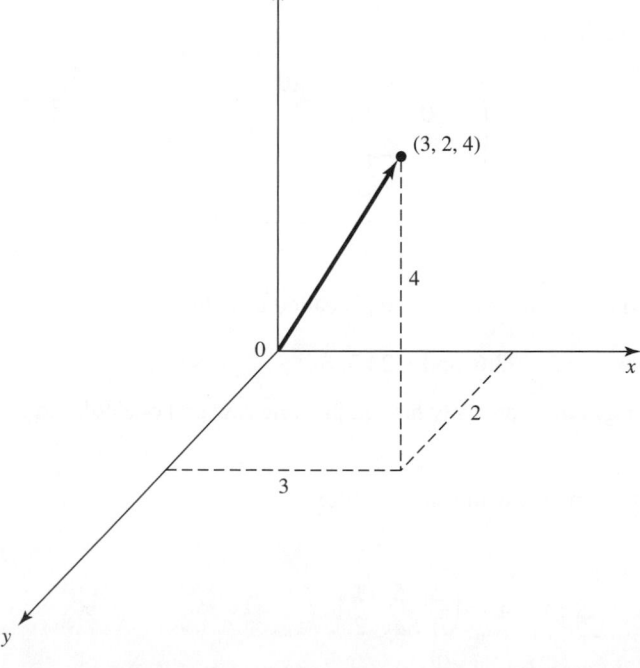

FIGURE 12–1 (b)

(c) The vector $(5, 8, 1, 3, 6)$ can be thought of as representing a line segment in five dimensions, although we cannot draw it. However, we can work mathematically with this vector using the same rules as for the others.   ◆◆◆

Matrices

A *matrix* is a *rectangular* array.

A matrix is sometimes referred
to as a *vector of vectors*.

◆◆◆ Example 4: The following rectangular arrays are matrices:

$$\begin{pmatrix} 2 & 5 & 7 \\ 6 & 3 & 1 \end{pmatrix} \tag{a}$$

$$\begin{pmatrix} 7 & 3 & 5 \\ 1 & 9 & 3 \\ 8 & 3 & 2 \end{pmatrix} \tag{b}$$

Further, (b) is a *square* matrix. For a square matrix, the diagonal running from the upper-left element to the lower-right element is called the *main diagonal*. In (b), the elements on the main diagonal are 7, 9, and 2. The diagonal from upper right to lower left is called the *secondary* diagonal. In our example, the elements 5, 9, and 8 are on the secondary diagonal. ◆◆◆

In everyday language, an array is a *table* and a vector is a *list*.

Subscripts

Each element in an array is located in a horizontal *row* and a vertical *column*. We indicate the row and column by means of *subscripts*.

◆◆◆ Example 5: The element a_{25} is located in row 2 and column 5. ◆◆◆

Thus an element in an array needs *double subscripts* to give its location. An element in a list, such as b_7, needs only a *single subscript*.

Dimensions

A matrix will have, in general, m rows and n columns. The numbers m and n are the *dimensions* of the matrix, as, for example, a 4×5 matrix.

◆◆◆ Example 6: The matrix

$$\begin{pmatrix} 2 & 3 & 5 & 9 \\ 1 & 0 & 8 & 3 \end{pmatrix}$$

has the dimensions 2×4, and the matrix

$$\begin{pmatrix} 5 & 3 & 8 \\ 1 & 0 & 3 \\ 2 & 5 & 2 \\ 1 & 0 & 7 \end{pmatrix}$$

has the dimensions 4×3. ◆◆◆

Scalars

A single number, as opposed to an array of numbers, is called a *scalar*.

◆◆◆ Example 7: Some scalars are 5, 693.6, and -24.3. ◆◆◆

A scalar can also be thought of as an array having just one row and one column.

Naming a Matrix

We will often denote or *name* a matrix with a single letter.

◆◆◆ Example 8: We can let

$$\mathbf{A} = \begin{pmatrix} 2 & 5 & 1 \\ 0 & 4 & 3 \end{pmatrix}$$

Thus we can represent an entire array by a single symbol. ◆◆◆

The Null or Zero Matrix

A matrix in which all elements are zero is called the *null* or *zero* matrix.

◆◆◆ **Example 9:** The matrix

$$\mathbf{O} = \begin{pmatrix} 0 & 0 & 0 \\ 0 & 0 & 0 \end{pmatrix}$$

is a 2×3 zero matrix. We will denote the zero matrix by the boldface letter $\mathbf{O}$. ◆◆◆

The Unit Matrix

If all of the elements *not* on the main diagonal of a square matrix are 0, we have a *diagonal matrix*. If, in addition, the elements on the main diagonal are equal, we have a *scalar matrix*. If the elements on the main diagonal of a scalar matrix are 1's, we have a *unit matrix* or an *identity matrix*.

◆◆◆ **Example 10:** The following matrices are all diagonal matrices:

$$\mathbf{A} = \begin{pmatrix} 3 & 0 & 0 \\ 0 & 5 & 0 \\ 0 & 0 & 7 \end{pmatrix} \qquad \mathbf{B} = \begin{pmatrix} 6 & 0 & 0 \\ 0 & 6 & 0 \\ 0 & 0 & 6 \end{pmatrix} \qquad \mathbf{I} = \begin{pmatrix} 1 & 0 & 0 \\ 0 & 1 & 0 \\ 0 & 0 & 1 \end{pmatrix}$$

$\mathbf{B}$ and $\mathbf{I}$ are, in addition, scalar matrices, and $\mathbf{I}$ is a unit matrix. We usually denote the unit matrix by the letter $\mathbf{I}$. ◆◆◆

Equality of Matrices

When we are comparing two matrices *of the same dimension,* elements in each matrix having the same row and column subscripts are called *corresponding elements*. Two matrices of the same dimension are equal if their corresponding elements are equal.

◆◆◆ **Example 11:** The matrices

$$\begin{pmatrix} x & 2 & 5 \\ 3 & 1 & 7 \end{pmatrix} \text{ and } \begin{pmatrix} 4 & 2 & 5 \\ 3 & 1 & y \end{pmatrix}$$

are equal only if $x = 4$ and $y = 7$. ◆◆◆

Transpose of a Matrix

The *transpose* $\mathbf{A}'$ of a matrix $\mathbf{A}$ is obtained by changing the rows to columns and changing the columns to rows.

◆◆◆ **Example 12:** The transpose of the matrix

$$\begin{pmatrix} 1 & 2 & 3 & 4 \\ 5 & 6 & 7 & 8 \end{pmatrix}$$

is

$$\begin{pmatrix} 1 & 5 \\ 2 & 6 \\ 3 & 7 \\ 4 & 8 \end{pmatrix}$$

◆◆◆

Determinant of a Square Matrix

We form the determinant of a square matrix by writing the same elements enclosed between vertical bars.

◆◆◆ **Example 13:** The determinant of the square matrix

$$\begin{pmatrix} a & b & c & d \\ e & f & g & h \\ i & j & k & l \\ m & n & o & p \end{pmatrix}$$

is

$$\begin{vmatrix} a & b & c & d \\ e & f & g & h \\ i & j & k & l \\ m & n & o & p \end{vmatrix}$$

◆◆◆

We'll see later when we want to find the *inverse* of a square matrix that the inverse exists only for a nonsingular matrix.

In Chapter 11 we learned how to evaluate determinants, so we'll not repeat that material here. We do, however, need a definition that will be useful later in this chapter.

A square matrix is called *singular* if its determinant is zero; it is called *nonsingular* if its determinant is not zero.

Exercise 1 ◆ Definitions

Given the following arrays:

$$A = \begin{pmatrix} 2 & 5 & 1 \\ 6 & 3 & 7 \\ 1 & 6 & 9 \\ 7 & 4 & 2 \end{pmatrix} \qquad B = \begin{pmatrix} 7 \\ 3 \\ 9 \\ 2 \end{pmatrix} \qquad C = \begin{pmatrix} f & i & q & w \\ & g & w & k \\ & & c & z \\ & & & b \end{pmatrix}$$

$$D = \begin{pmatrix} 6 & 2 & 0 & 1 \\ 2 & 8 & 3 & 9 \end{pmatrix} \qquad E = \begin{pmatrix} x & y \\ x & w \end{pmatrix} \qquad F = \begin{pmatrix} 0 & 0 & 0 & 0 \\ 0 & 0 & 0 & 0 \end{pmatrix}$$

$$G = 7 \qquad\qquad H = \begin{pmatrix} 3 & 8 \\ & 5 \end{pmatrix} \qquad I = (3 \quad 8 \quad 4 \quad 6)$$

Which of the nine arrays shown above is:

1. a rectangular array?
2. a square array?
3. a triangular array?
4. a column vector?
5. a row vector?
6. a table?
7. a list?
8. a scalar?
9. a null matrix?
10. a matrix?

For matrix **A** above, find each of the following elements:

11. a_{32}
12. a_{41}

Give the dimensions of:

13. matrix **A**.
14. matrix **B**.
15. matrix **D**.
16. matrix **E**.

Write the transpose of:

17. matrix **F**.
18. matrix **D**.

19. Under what conditions will the matrices

$$\begin{pmatrix} 6 & x \\ w & 8 \end{pmatrix} \quad \text{and} \quad \begin{pmatrix} y & 2 \\ 7 & z \end{pmatrix} \text{ be equal?}$$

12–2 Operations with Matrices

In earlier chapters we learned how to add, subtract, and multiply numbers and algebraic quantities. We now perform these operations on matrices.

Addition and Subtraction of Matrices

To add two matrices of the same dimensions, simply combine corresponding elements. We first show the addition of two vectors.

◆◆◆ **Example 14:**

$$(3 \quad 2 \quad -1 \quad 5) + (2 \quad 4 \quad 5 \quad -2) = (5 \quad 6 \quad 4 \quad 3)$$

◆◆◆

Geometric Interpretation of the Sum of Two Vectors

Recall that we have given a geometric interpretation to a vector as the coordinates of the endpoint of a line segment drawn from the origin (Fig. 12–1). Thus the vector (3 2) can represent the vector **A** shown in Fig. 12–2. Let us plot another vector **B**(1 4) on the same graph as **A** and find the sum **A + B**.

$$\mathbf{A + B} = (3 \quad 2) + (1 \quad 4) = (4 \quad 6)$$

Thus the geometric interpretation of the sum **A + B** is the *resultant* of **A** and **B**.

◆◆◆ **Example 15:** Find the resultant of the vectors (3 2) and (1 4).

Solution: Adding the vectors gives the resultant.

$$(3 \quad 2) + (1 \quad 4) = (4 \quad 6)$$

◆◆◆

We'll now show the addition of matrices.

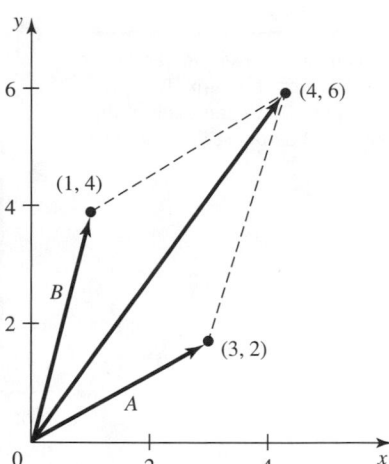

FIGURE 12–2. The resultant of two vectors. In Sec. 15–5 we'll show that this is an illustration of the *parallelogram method* for finding the resultant of two vectors.

◆◆◆ **Example 16:**

$$\begin{pmatrix} 2 & -3 & 4 & 1 \\ 0 & 5 & 6 & -8 \end{pmatrix} + \begin{pmatrix} 5 & 8 & 1 & 2 \\ 9 & -3 & 0 & 1 \end{pmatrix} = \begin{pmatrix} 7 & 5 & 5 & 3 \\ 9 & 2 & 6 & -7 \end{pmatrix}$$

◆◆◆

Subtraction is done in a similar way.

◆◆◆ **Example 17:**

$$\begin{pmatrix} 2 & -3 & 4 & 1 \\ 0 & 5 & 6 & -8 \end{pmatrix} - \begin{pmatrix} 5 & 8 & 1 & 2 \\ 9 & -3 & 0 & 1 \end{pmatrix} = \begin{pmatrix} -3 & -11 & 3 & -1 \\ -9 & 8 & 6 & -9 \end{pmatrix}$$

◆◆◆

As with ordinary addition, the *commutative law* (Eq. 1) and the *associative law* (Eq. 3) apply.

Commutative Law	$\mathbf{A + B = B + A}$	79
Associative Law	$\mathbf{A + (B + C) = (A + B) + C}$ $= \mathbf{(A + C) + B}$	80

Addition and subtraction are not defined for two matrices of different dimensions.

Product of a Scalar and a Matrix

To multiply a matrix by a scalar, multiply each element in the matrix by the scalar. Let's start by multiplying a vector by a scalar.

◆◆◆ **Example 18:**

$$3(2 \quad 4 \quad -3 \quad 1) = (6 \quad 12 \quad -9 \quad 3)$$

◆◆◆

The operation of multiplying a vector by a scalar can be interpreted geometrically. For example, if we multiply vector **A** (see Fig. 12–2) by 2, we get

$$2\mathbf{A} = 2(3 \quad 2) = (6 \quad 4)$$

Notice that this vector is collinear with **A** but of twice the length. So multiplying a vector by a scalar merely changes the length of the vector.

Now let's multiply a matrix by a scalar.

Working in reverse, we can *factor* a scalar from a matrix. Thus in Example 19, we can think of the number 3 as being *factored out*.

◆◆◆ Example 19:

$$3\begin{pmatrix} 2 & 4 & 1 \\ 5 & 7 & 3 \end{pmatrix} = \begin{pmatrix} 6 & 12 & 3 \\ 15 & 21 & 9 \end{pmatrix}$$

◆◆◆

In symbols, the formula is expressed as follows:

Product of a Scalar and Matrix	$k\begin{pmatrix} a & b \\ c & d \end{pmatrix} = \begin{pmatrix} ka & kb \\ kc & kd \end{pmatrix}$	87

Multiplication of Vectors and Matrices

Not all matrices can be multiplied. Further, for matrices **A** and **B**, it may be possible to find the product **AB** but *not* the product **BA**. Two matrices for which multiplication is defined are called *conformable matrices*.

Conformable Matrices	The product **AB** of two matrices **A** and **B** is defined only when the number of columns in **A** equals the number of rows in **B**.	82

◆◆◆ Example 20: If

$$\mathbf{A} = \begin{pmatrix} 3 & 5 \\ 2 & 1 \\ 7 & 4 \\ 9 & 0 \end{pmatrix} \quad \text{and} \quad \mathbf{B} = \begin{pmatrix} 7 & 3 & 8 \\ 4 & 7 & 1 \end{pmatrix}$$

we can find the product **AB**, because the number of columns in **A** (2 columns) equals the number of rows in **B** (2 rows). However, we *cannot* find the product **BA** because the number of columns in **B** (3 columns) does not equal the number of rows in **A** (4 rows).   ◆◆◆

For two or more conformable matrices, the *associative law* and the *distributive law* for multiplication apply.

Associative Law	$\mathbf{A(BC)} = \mathbf{(AB)C} = \mathbf{ABC}$	84

Distributive Law	$\mathbf{A(B + C)} = \mathbf{AB} + \mathbf{AC}$	85

These equations are similar to Eqs. 4 and 5 for ordinary multiplication.

The *dimensions* of the matrix obtained by multiplying an $m \times p$ matrix with a $p \times n$ matrix are equal to $m \times n$. In other words, the product matrix **AB** will have the same number of *rows* as **A** and the same number of *columns* as **B**.

Dimensions of the Product	$(m \times p)(p \times n) = (m \times n)$	86

◆◆◆ Example 21: The product **AB** of a 3×5 matrix **A** and a 5×7 matrix **B** is a 3×7 matrix.

◆◆◆

Scalar Product of Two Vectors

We will show matrix multiplication first with vectors and then extend it to other matrices.

The *scalar product* (also called the *dot product* or *inner product*) of a row vector **A** and a column vector **B** is defined only when the number of columns in **A** is equal to the number of rows in **B**.

If each vector contains m elements, the dimensions of the product will be, by Eq. 86,

$$(1 \times m)(m \times 1) = (1 \times 1)$$

Whatever we say about matrices applies to vectors as well.

Thus the product is a scalar, hence the name *scalar product*.

To multiply a row vector by a column vector, multiply the first elements of each, then the second, and so on, and add the products obtained.

◆◆◆ **Example 22:**

$$(2 \quad 6 \quad 4 \quad 8)\begin{pmatrix} 3 \\ 5 \\ 1 \\ 7 \end{pmatrix} = [(2)(3) + (6)(5) + (4)(1) + (8)(7)]$$

$$= (6 + 30 + 4 + 56) = 96 \qquad \text{◆◆◆}$$

We state this operation as the following formula:

Scalar Product of Two Vectors	$(a \quad b \quad \ldots)\begin{pmatrix} x \\ y \\ \vdots \end{pmatrix} = (ax + by + \ldots)$	88

◆◆◆ **Example 23:** A certain store has four types of radios, priced at \$71, \$62, \$83, and \$49. On a particular day, the store sells quantities of 8, 3, 7, and 6, respectively. If we represent the prices by a row vector and the quantities by a column vector, and multiply, we get

$$(71 \quad 62 \quad 83 \quad 49)\begin{pmatrix} 8 \\ 3 \\ 7 \\ 6 \end{pmatrix} = [71(8) + 62(3) + 83(7) + 49(6)]$$

$$= 1629$$

This number represents the total dollar income from the sale of the four types of radios on that day. ◆◆◆

Product of a Row Vector and a Matrix

By Eq. 86, the product of a row vector and a matrix having n columns will be a row vector having n elements.

$$(1 \times p)(p \times n) = (1 \times n)$$

We saw earlier that the product of a row vector and a column vector is a single number, or scalar. Thus the product of the given row vector and the first column of the matrix will give a scalar. This scalar will be *the first element in the product vector.*

We find the *second* element in the product vector by multiplying the second column in the matrix by the row vector, and so on.

◆◆◆ **Example 24:** Multiply

$$(1 \quad 3 \quad 0)\begin{pmatrix} 7 & 5 \\ 1 & 0 \\ 4 & 9 \end{pmatrix}$$

Solution: The dimensions of the product will be

$$(1 \times 3)(3 \times 2) = (1 \times 2)$$

Multiplying the vector and the first column of the matrix gives

Here we use the asterisk (*) as a placeholder to denote the other element still to be computed.

$$(1 \quad 3 \quad 0)\begin{pmatrix} \mathbf{7} & 5 \\ \mathbf{1} & 0 \\ \mathbf{4} & 9 \end{pmatrix} = [\mathbf{1(7) + 3(1) + 0(4)} \quad * \quad]$$

$$= (\mathbf{10} \quad *)$$

Multiplying the vector and the second column of the matrix yields

$$(1 \quad 3 \quad 0)\begin{pmatrix} 7 & \mathbf{5} \\ 1 & \mathbf{0} \\ 4 & \mathbf{9} \end{pmatrix} = [10 \qquad \mathbf{1(5) + 3(0) + 0(9)}]$$

$$= (10 \quad 5)$$

◆◆◆

◆◆◆ **Example 25:** Continuing Example 23, let us suppose that on the second day, the store sells quantities of 5, 9, 1, and 4, respectively, of the four radios. If we again represent the prices by a row vector, but now represent the quantities by a matrix and multiply, we get

$$(71 \quad 62 \quad 83 \quad 49)\begin{pmatrix} 8 & 5 \\ 3 & 9 \\ 7 & 1 \\ 6 & 4 \end{pmatrix} = [71(8) + 62(3) + 83(7) + 49(6)$$
$$71(5) + 62(9) + 83(1) + 49(4)]$$
$$= (1629 \quad 1192)$$

We interpret these numbers as the total dollar income from the sale of the four radios on each of the two days. ◆◆◆

Product of Two Matrices

We now multiply two matrices **A** and **B**. We already know how to multiply a row vector and a matrix. So let us think of the matrix **A** as several row vectors. Each of these row vectors, multiplied by the matrix **B**, will produce a row vector in the product **AB**.

◆◆◆ **Example 26:** Find the product **AB** of the matrices

$$\mathbf{A} = \begin{pmatrix} 1 & 0 \\ 3 & 2 \end{pmatrix} \qquad \mathbf{B} = \begin{pmatrix} 5 & 4 & 8 \\ 6 & 7 & 9 \end{pmatrix}$$

Solution: The dimensions of the product will be

$$(2 \times 2)(2 \times 3) = (2 \times 3)$$

We think of the matrix **A** as being two row vectors, each of which, when multiplied by matrix **B**, will produce a row vector in the product matrix.

$$\underset{\mathbf{A}}{\begin{pmatrix} 1 & 0 \\ 3 & 2 \end{pmatrix}} \underset{\mathbf{B}}{\begin{pmatrix} 5 & 4 & 8 \\ 6 & 7 & 9 \end{pmatrix}} = \underset{\mathbf{AB}}{\begin{pmatrix} * & * & * \\ * & * & * \end{pmatrix}}$$

Multiplying by the first row vector in **A**, we have

$$\underset{\mathbf{A}}{\begin{pmatrix} \mathbf{1} & \mathbf{0} \\ 3 & 2 \end{pmatrix}} \underset{\mathbf{B}}{\begin{pmatrix} 5 & 4 & 8 \\ 6 & 7 & 9 \end{pmatrix}} = \underset{\mathbf{AB}}{\begin{bmatrix} [1(5) + 0(6) & 1(4) + 0(7) & 1(8) + 0(9)] \\ [\quad * & * & * \quad] \end{bmatrix}}$$

$$= \begin{bmatrix} [5 & 4 & 8] \\ [* & * & *] \end{bmatrix}$$

Then multiplying by the second row vector in **A** yields

$$\begin{pmatrix} 1 & 0 \\ 3 & 2 \end{pmatrix} \begin{pmatrix} 5 & 4 & 8 \\ 6 & 7 & 9 \end{pmatrix} = \begin{bmatrix} 5 & 4 & 8 \\ 3(5) + 2(6) & 3(4) + 2(7) & 3(8) + 2(9) \end{bmatrix}$$

$$\quad\;\text{**A**}\qquad\quad\text{**B**}\qquad\qquad\qquad\quad\text{**AB**}$$

giving the final product

$$\begin{pmatrix} 1 & 0 \\ 3 & 2 \end{pmatrix} \begin{pmatrix} 5 & 4 & 8 \\ 6 & 7 & 9 \end{pmatrix} = \begin{pmatrix} 5 & 4 & 8 \\ 27 & 26 & 42 \end{pmatrix}$$

$$\quad\;\text{**A**}\qquad\quad\text{**B**}\qquad\quad\text{**AB**}$$

◆◆◆

◆◆◆ **Example 27:** Find the product **AB** of the matrices

$$\begin{pmatrix} 1 & 0 & 2 \\ 0 & 1 & 1 \\ 3 & 0 & 2 \\ 1 & 1 & 0 \\ 2 & 1 & 1 \end{pmatrix} \begin{pmatrix} 3 & 0 & 1 & 1 \\ 1 & 2 & 3 & 0 \\ 0 & 1 & 2 & 1 \end{pmatrix}$$

$$\qquad\;\text{**A**}\qquad\qquad\text{**B**}$$

Solution: We think of the matrix **A** as five row vectors. Each of these five row vectors, multiplied by the matrix **B**, will produce a row vector in the product **AB**.

$$\begin{matrix}(1 & 0 & 2)\\(0 & 1 & 1)\\(3 & 0 & 2)\\(1 & 1 & 0)\\(2 & 1 & 1)\end{matrix} \begin{pmatrix} 3 & 0 & 1 & 1 \\ 1 & 2 & 3 & 0 \\ 0 & 1 & 2 & 1 \end{pmatrix} = \begin{matrix}(\quad\quad\quad\;)\\(\quad\quad\quad\;)\\(\quad\quad\quad\;)\\(\quad\quad\quad\;)\\(\quad\quad\quad\;)\end{matrix}$$

$$\qquad\text{**A**}\qquad\qquad\quad\text{**B**}\qquad\qquad\text{**AB**}$$
$$\quad(5 \times 3)\qquad\;(3 \times 4)\qquad\;(5 \times 4)$$

Multiplying the *first* row vector (1 0 2) in **A** by the matrix **B** gives the first row vector in **AB**.

$$\begin{matrix}\mathbf{(1 \;\; 0 \;\; 2)}\\(0 \;\; 1 \;\; 1)\\(3 \;\; 0 \;\; 2)\\(1 \;\; 1 \;\; 0)\\(2 \;\; 1 \;\; 1)\end{matrix} \begin{pmatrix} 3 & 0 & 1 & 1 \\ 1 & 2 & 3 & 0 \\ 0 & 1 & 2 & 1 \end{pmatrix} = \begin{matrix}\mathbf{(3 \;\; 2 \;\; 5 \;\; 3)}\\(\quad\quad\quad\;)\\(\quad\quad\quad\;)\\(\quad\quad\quad\;)\\(\quad\quad\quad\;)\end{matrix}$$

$$\qquad\text{**A**}\qquad\qquad\quad\text{**B**}\qquad\qquad\text{**AB**}$$

Multiplying the *second* row vector in **A** and the matrix **B** gives a second row vector in **AB**.

$$\begin{matrix}(1 \;\; 0 \;\; 2)\\\mathbf{(0 \;\; 1 \;\; 1)}\\(3 \;\; 0 \;\; 2)\\(1 \;\; 1 \;\; 0)\\(2 \;\; 1 \;\; 1)\end{matrix} \begin{pmatrix} 3 & 0 & 1 & 1 \\ 1 & 2 & 3 & 0 \\ 0 & 1 & 2 & 1 \end{pmatrix} = \begin{matrix}(3 \;\; 2 \;\; 5 \;\; 3)\\\mathbf{(1 \;\; 3 \;\; 5 \;\; 1)}\\(\quad\quad\quad\;)\\(\quad\quad\quad\;)\\(\quad\quad\quad\;)\end{matrix}$$

$$\qquad\text{**A**}\qquad\qquad\quad\text{**B**}\qquad\qquad\text{**AB**}$$

The *third* row vector in **A** produces a third row vector in **AB**.

$$\begin{matrix}(1 \;\; 0 \;\; 2)\\(0 \;\; 1 \;\; 1)\\\mathbf{(3 \;\; 0 \;\; 2)}\\(1 \;\; 1 \;\; 0)\\(2 \;\; 1 \;\; 1)\end{matrix} \begin{pmatrix} 3 & 0 & 1 & 1 \\ 1 & 2 & 3 & 0 \\ 0 & 1 & 2 & 1 \end{pmatrix} = \begin{matrix}(3 \;\; 2 \;\; 5 \;\; 3)\\(1 \;\; 3 \;\; 5 \;\; 1)\\\mathbf{(9 \;\; 2 \;\; 7 \;\; 5)}\\(\quad\quad\quad\;)\\(\quad\quad\quad\;)\end{matrix}$$

$$\qquad\text{**A**}\qquad\qquad\quad\text{**B**}\qquad\qquad\text{**AB**}$$

And so on, until we get the final product.

$$\begin{pmatrix} 1 & 0 & 2 \\ 0 & 1 & 1 \\ 3 & 0 & 2 \\ 1 & 1 & 0 \\ 2 & 1 & 1 \end{pmatrix} \begin{pmatrix} 3 & 0 & 1 & 1 \\ 1 & 2 & 3 & 0 \\ 0 & 1 & 2 & 1 \end{pmatrix} = \begin{pmatrix} 3 & 2 & 5 & 3 \\ 1 & 3 & 5 & 1 \\ 9 & 2 & 7 & 5 \\ 4 & 2 & 4 & 1 \\ 7 & 3 & 7 & 3 \end{pmatrix}$$

$$\qquad \mathbf{A} \qquad\qquad \mathbf{B} \qquad\qquad \mathbf{AB}$$

We can state the product of two matrices as a formula as follows:

This formula works, of course, for conformable matrices with any number of elements.

Product of Two Matrices	$\begin{pmatrix} a & b & c \\ d & e & f \end{pmatrix} \begin{pmatrix} u & x \\ v & y \\ w & z \end{pmatrix} = \begin{pmatrix} au + bv + cw & ax + by + cz \\ du + ev + fw & dx + ey + fz \end{pmatrix}$	92

◆◆◆ **Example 28:** Continuing Examples 23 and 25, how much would the income from the four radios be for each of the two days if the prices were discounted to

$$\$69 \quad \$57 \quad \$75 \quad \text{and} \quad \$36$$

and if they were marked up to

$$\$75 \quad \$65 \quad \$90 \quad \text{and} \quad \$57$$

Solution: Our prices can be represented by the matrix

$$\begin{pmatrix} 71 & 62 & 83 & 49 \\ 69 & 57 & 75 & 36 \\ 75 & 65 & 90 & 57 \end{pmatrix}$$

where the four columns represent the four different types of radios, as before, and the three rows represent the regular price, the discounted price, and the marked-up price. The quantities of each radio sold on each of the two days are represented by the same matrix as in Example 25. To find the income from the four radios for each of the three prices for each on each of the two days, we multiply the two matrices.

$$\begin{pmatrix} 71 & 62 & 83 & 49 \\ 69 & 57 & 75 & 36 \\ 75 & 65 & 90 & 57 \end{pmatrix} \begin{pmatrix} 8 & 5 \\ 3 & 9 \\ 7 & 1 \\ 6 & 4 \end{pmatrix} = \begin{pmatrix} 1629 & 1192 \\ 1464 & 1077 \\ 1767 & 1278 \end{pmatrix}$$

The figures in the first row, $1,629 and $1,192, are the amounts from the sale of radios on the first and second days, as in Example 25. The other two rows give the incomes with the discounted, and then the marked-up, prices. ◆◆◆

Product of Square Matrices

We multiply square matrices **A** and **B** just as we do other matrices. But unlike for rectangular matrices, both the products **AB** and **BA** will be defined, although they usually will not be equal.

◆◆◆ **Example 29:** Find the products **AB** and **BA** for the square matrices

$$\mathbf{A} = \begin{pmatrix} 0 & 1 & 2 \\ 1 & 1 & 0 \\ 2 & 3 & 1 \end{pmatrix} \quad \mathbf{B} = \begin{pmatrix} 2 & 1 & 0 \\ 1 & 0 & 3 \\ 3 & 1 & 1 \end{pmatrix}$$

and show that the products are not equal.

Solution: Multiplying as shown earlier, we have

$$\mathbf{AB} = \begin{pmatrix} 0 & 1 & 2 \\ 1 & 1 & 0 \\ 2 & 3 & 1 \end{pmatrix} \begin{pmatrix} 2 & 1 & 0 \\ 1 & 0 & 3 \\ 3 & 1 & 1 \end{pmatrix} = \begin{pmatrix} 7 & 2 & 5 \\ 3 & 1 & 3 \\ 10 & 3 & 10 \end{pmatrix}$$

and

$$\mathbf{BA} = \begin{pmatrix} 2 & 1 & 0 \\ 1 & 0 & 3 \\ 3 & 1 & 1 \end{pmatrix} \begin{pmatrix} 0 & 1 & 2 \\ 1 & 1 & 0 \\ 2 & 3 & 1 \end{pmatrix} = \begin{pmatrix} 1 & 3 & 4 \\ 6 & 10 & 5 \\ 3 & 7 & 7 \end{pmatrix}$$

◆◆◆

We see that multiplying the matrices in the reverse order does *not* give the same result.

Commutative Law	AB ≠ BA Matrix multiplication *is not* commutative	83

Product of a Matrix and a Column Vector

We have already multiplied a row vector **A** by a matrix **B**. Now we multiply a matrix **A** by a column vector **B**.

◆◆◆ **Example 30:** Multiply

$$\begin{pmatrix} 3 & 1 & 2 & 0 \\ 1 & 3 & 5 & 2 \end{pmatrix} \begin{pmatrix} 2 \\ 1 \\ 3 \\ 4 \end{pmatrix}$$

Solution: The dimensions of the product will be

$$(2 \times 4)(4 \times 1) = (2 \times 1)$$

Multiplying the column vector by the first row in the matrix yields

$$\begin{pmatrix} \mathbf{3} & \mathbf{1} & \mathbf{2} & \mathbf{0} \\ 1 & 3 & 5 & 2 \end{pmatrix} \begin{pmatrix} 2 \\ 1 \\ 3 \\ 4 \end{pmatrix} = \begin{pmatrix} \mathbf{3(2) + 1(1) + 2(3) + 0(4)} \end{pmatrix} = \begin{pmatrix} 13 \end{pmatrix}$$

Here, we use the same rules as for finding the product of any two matrices.

Then multiplying by the second row, we have

$$\begin{pmatrix} 3 & 1 & 2 & 0 \\ \mathbf{1} & \mathbf{3} & \mathbf{5} & \mathbf{2} \end{pmatrix} \begin{pmatrix} 2 \\ 1 \\ 3 \\ 4 \end{pmatrix} = \begin{pmatrix} 13 \\ \mathbf{1(2) + 3(1) + 5(3) + 2(4)} \end{pmatrix} = \begin{pmatrix} 13 \\ 28 \end{pmatrix}$$

◆◆◆

Most of the matrix multiplication that we do later will be of a matrix and a column vector.

Note that the product of a matrix and a column vector is a column vector.

◆◆◆ **Example 31:** Five students decided to raise money by selling T-shirts for three weeks. The following table shows the number of shirts sold by each student during each week:

		Week 1	2	3
	1	8	2	6
	2	3	5	2
Student	3	6	5	4
	4	7	3	1
	5	2	3	4

They charged $7.50 per shirt the first week and then lowered the price to $6.00 the second week and to $5.50 the third week. Find the total amount raised by each student for the three weeks.

Solution: We know that to get the total amounts, we must multiply the matrix B representing the number of T-shirts sold,

$$\mathbf{B} = \begin{pmatrix} 8 & 2 & 6 \\ 3 & 5 & 2 \\ 6 & 5 & 4 \\ 7 & 3 & 1 \\ 2 & 3 & 4 \end{pmatrix}$$

by a vector **C** representing the price per shirt. But should **C** be a row vector or a column vector? And should we find the product **BC** or the product **CB**?

If we let **C** be a 1×3 row vector and try to multiply it by the 5×3 matrix **B**, we see that we cannot, because these matrices are not comfortable for either product **BC** or **CB**.

On the other hand, if we let **C** be a 3×1 column vector,

$$\mathbf{C} = \begin{pmatrix} 7.50 \\ 6.00 \\ 5.50 \end{pmatrix}$$

the product **CB** is not defined, but the product **BC** gives a column vector

$$\mathbf{BC} = \begin{pmatrix} 8 & 2 & 6 \\ 3 & 5 & 2 \\ 6 & 5 & 4 \\ 7 & 3 & 1 \\ 2 & 3 & 4 \end{pmatrix} \begin{pmatrix} 7.50 \\ 6.00 \\ 5.50 \end{pmatrix} = \begin{pmatrix} 105.00 \\ 63.50 \\ 97.00 \\ 76.00 \\ 55.00 \end{pmatrix}$$

which represents the total earned by each student.   ◆◆◆

Exercise 2 ◆ Operations with Matrices

Addition and Subtraction of Matrices

Combine the following vectors and matrices.

1. $(3 \quad 8 \quad 2 \quad 1) + (9 \quad 5 \quad 3 \quad 7)$

2. $\begin{pmatrix} 6 & 3 & 9 \\ 2 & 1 & 3 \end{pmatrix} + \begin{pmatrix} 8 & 2 & 7 \\ 1 & 6 & 3 \end{pmatrix}$

3. $\begin{pmatrix} 5 \\ -3 \\ 5 \\ 8 \end{pmatrix} + \begin{pmatrix} 7 \\ 0 \\ -4 \\ 8 \end{pmatrix} - \begin{pmatrix} -2 \\ 7 \\ -5 \\ 2 \end{pmatrix}$

4. $\begin{pmatrix} 6 & 2 & -7 \\ -3 & 5 & 9 \\ 2 & 5 & 3 \end{pmatrix} - \begin{pmatrix} 5 & 2 & 9 \\ 1 & 4 & 8 \\ 7 & -3 & 4 \end{pmatrix} + \begin{pmatrix} 1 & -6 & 7 \\ 2 & 4 & -9 \\ -6 & 2 & 7 \end{pmatrix}$

5. $\begin{pmatrix} 1 & 5 & -2 & 1 \\ 0 & -2 & 3 & 5 \\ -2 & 9 & 3 & -2 \\ 5 & -1 & 2 & 6 \end{pmatrix} - \begin{pmatrix} -6 & 4 & 2 & 8 \\ 9 & 2 & -6 & 3 \\ -5 & 2 & 6 & 7 \\ 6 & 3 & 7 & 0 \end{pmatrix}$

Product of a Scalar and a Matrix

Multiply each scalar and vector.

6. $6 \begin{pmatrix} 9 \\ -2 \\ 6 \\ -3 \end{pmatrix}$

7. $-3(5 \quad 9 \quad -2 \quad 7 \quad 3)$

Multiply each scalar and matrix.

8. $7\begin{pmatrix} 7 & -3 & 7 \\ -2 & 9 & 5 \\ 9 & 1 & 0 \end{pmatrix}$

9. $-3\begin{pmatrix} 4 & 2 & -3 & 0 \\ -1 & 6 & 3 & -4 \end{pmatrix}$

Remove any factors from the vector or matrix.

10. $\begin{pmatrix} 25 \\ -15 \\ 30 \\ -5 \end{pmatrix}$

11. $\begin{pmatrix} 21 & 7 & -14 \\ 49 & 63 & 28 \\ 14 & -21 & 56 \\ -42 & 70 & 84 \end{pmatrix}$

If

$$\mathbf{A} = \begin{pmatrix} 2 & 5 & -1 \\ -3 & 2 & 6 \\ 9 & -4 & 5 \end{pmatrix} \quad \text{and} \quad \mathbf{B} = \begin{pmatrix} 0 & 1 & 6 \\ -7 & 2 & 1 \\ 4 & -2 & 0 \end{pmatrix}$$

find the following.

12. $\mathbf{A} - 4\mathbf{B}$

13. $3\mathbf{A} + \mathbf{B}$

14. $2\mathbf{A} + 2\mathbf{B}$

15. Show that $2(\mathbf{A} + \mathbf{B}) = 2\mathbf{A} + 2\mathbf{B}$, thus demonstrating the distributive law (Eq. 85).

Scalar Product of Two Vectors

Multiply each pair of vectors.

16. $(6 \quad -2)\begin{pmatrix} -1 \\ 7 \end{pmatrix}$

17. $(5 \quad 3 \quad 7)\begin{pmatrix} 2 \\ 6 \\ 1 \end{pmatrix}$

18. $(-3 \quad 6 \quad -9)\begin{pmatrix} 8 \\ -3 \\ 4 \end{pmatrix}$

19. $(-4 \quad 3 \quad 8 \quad -5)\begin{pmatrix} 7 \\ -6 \\ 0 \\ 3 \end{pmatrix}$

Product of a Row Vector and a Matrix

Multiply each row vector and matrix.

20. $(4 \quad -2)\begin{pmatrix} 3 & 6 \\ -1 & 5 \end{pmatrix}$

21. $(3 \quad -1)\begin{pmatrix} 4 & 7 & -2 & 4 & 0 \\ 2 & -6 & 8 & -3 & 7 \end{pmatrix}$

22. $(2 \quad 8 \quad 1)\begin{pmatrix} 5 & 6 \\ 3 & 1 \\ 1 & 0 \end{pmatrix}$

23. $(-3 \quad 6 \quad -1)\begin{pmatrix} 5 & -2 & 7 \\ -6 & 2 & 5 \\ 9 & 0 & -1 \end{pmatrix}$

24. $(9 \quad 1 \quad 4)\begin{pmatrix} 6 & 2 & 1 & 5 & -8 \\ 7 & 1 & -3 & 0 & 3 \\ 1 & -5 & 0 & 3 & 2 \end{pmatrix}$

25. $(2 \quad 4 \quad 3 \quad -1 \quad 0)\begin{pmatrix} 4 & 3 & 5 & -1 & 0 \\ 1 & 3 & 5 & -2 & 6 \\ 2 & -4 & 3 & 6 & 5 \\ 3 & 4 & -1 & 6 & 5 \\ 2 & 3 & 1 & 4 & -6 \end{pmatrix}$

Product of Two Matrices

Multiply each pair of matrices.

26. $\begin{pmatrix} 7 & 2 \\ 1 & 3 \end{pmatrix}\begin{pmatrix} 4 & 1 & 2 \\ 3 & 2 & 5 \end{pmatrix}$

27. $\begin{pmatrix} 3 & 1 & 4 \\ 1 & -3 & 5 \end{pmatrix}\begin{pmatrix} 4 & 2 \\ -1 & 5 \\ 2 & -6 \end{pmatrix}$

28. $\begin{pmatrix} 4 & 2 & -1 \\ -2 & 1 & 6 \end{pmatrix}\begin{pmatrix} 5 & -1 & 2 & 3 \\ 1 & 3 & 5 & 2 \\ 0 & 1 & -2 & 4 \end{pmatrix}$

29. $\begin{pmatrix} 5 & 1 \\ -2 & 0 \\ 1 & 3 \end{pmatrix}\begin{pmatrix} -4 & 1 \\ 0 & 5 \end{pmatrix}$

30. $\begin{pmatrix} 2 & 4 \\ 0 & -1 \\ 2 & 5 \end{pmatrix}\begin{pmatrix} 0 & -2 & 3 & -2 \\ 1 & 3 & -4 & 1 \end{pmatrix}$

31. $\begin{pmatrix} 4 & 1 & -2 & 3 \\ 0 & 2 & 1 & 5 \end{pmatrix}\begin{pmatrix} 3 & -1 \\ 0 & 2 \\ 3 & 1 \\ 0 & -2 \end{pmatrix}$

32. $\begin{pmatrix} -1 & 0 & 3 & 1 \\ 2 & 1 & 0 & -3 \end{pmatrix}\begin{pmatrix} 1 & 2 & 3 & 0 \\ 3 & 0 & -1 & 2 \\ -1 & 4 & 3 & 0 \\ 2 & 0 & -2 & 1 \end{pmatrix}$

33. $\begin{pmatrix} 3 & 1 \\ -2 & 0 \\ 1 & -3 \\ 5 & 0 \end{pmatrix}\begin{pmatrix} -1 & 0 \\ 3 & 2 \end{pmatrix}$

34. $\begin{pmatrix} 1 & 0 & -2 \\ 2 & 4 & 1 \\ 0 & -1 & 3 \end{pmatrix}\begin{pmatrix} 4 & 0 & 1 & -2 \\ 0 & 2 & 1 & 3 \\ 1 & 0 & -3 & 2 \end{pmatrix}$

35. $\begin{pmatrix} 2 & 1 & 0 & -1 \\ 1 & 2 & -1 & 1 \\ 0 & -2 & 4 & 0 \end{pmatrix}\begin{pmatrix} 1 & 2 & 0 & -1 \\ 3 & 1 & -1 & 3 \\ 2 & 1 & 0 & 3 \\ 3 & 5 & 1 & 0 \end{pmatrix}$

Product of Square Matrices

Find the product **AB** and the product **BA** for each pair of square matrices.

36. $\mathbf{A} = \begin{pmatrix} 2 & 3 \\ 5 & 1 \end{pmatrix}$ $\mathbf{B} = \begin{pmatrix} 4 & 6 \\ 7 & 9 \end{pmatrix}$

37. $\mathbf{A} = \begin{pmatrix} 2 & -1 & 0 \\ 1 & 0 & 2 \\ -2 & 1 & 0 \end{pmatrix}$ $\mathbf{B} = \begin{pmatrix} 2 & 1 & 0 \\ 3 & 0 & 1 \\ -2 & 4 & 1 \end{pmatrix}$

Using the matrices in problem 36, find the following.

38. $\mathbf{A}^2$ 39. $\mathbf{B}^3$

Product of a Matrix and a Column Vector

Multiply each matrix and column vector.

40. $\begin{pmatrix} 2 & 1 \\ 0 & 3 \end{pmatrix}\begin{pmatrix} 3 \\ 1 \end{pmatrix}$ 41. $\begin{pmatrix} 2 & 0 & 1 \\ 1 & 5 & 2 \end{pmatrix}\begin{pmatrix} 3 \\ 0 \\ -1 \end{pmatrix}$

42. $\begin{pmatrix} 2 & 1 & 0 \\ 1 & 0 & -2 \\ 3 & 5 & 0 \\ 0 & -1 & 3 \end{pmatrix} \begin{pmatrix} 3 \\ 0 \\ -1 \end{pmatrix}$ **43.** $\begin{pmatrix} 0 & 2 & -1 & 4 \\ 1 & 5 & 3 & 0 \\ -2 & 0 & 1 & 5 \end{pmatrix} \begin{pmatrix} 2 \\ 0 \\ -1 \\ 3 \end{pmatrix}$

44. The unit matrix **I** is a square matrix that has 1's on the main diagonal and 0's elsewhere. Show that the product of **I** and another square matrix **A** is simply **A**.

Multiplying by the Unit Matrix	$AI = IA = A$	94

Applications

45. A certain store has four brands of athletic shoes, priced at \$81, \$72, \$93, and \$69. On a particular day, the store sells quantities of 2, 3, 7, and 4, respectively. Represent the prices and the quantities by vectors, and multiply them to get the total dollar income from the sale of the four brands of shoes on that day.

46. A certain store has computers, monitors, and keyboards priced at \$2,171, \$625, and \$149. On a particular day, the store sells quantities of 2, 3, and 5, respectively, and on the next day 4, 4, and 3. Represent the prices by a row vector and the quantities by a matrix. Multiply to get the receipts per day.

47. A building contractor has orders for 2 one-storey houses, 4 two-storey houses, and 3 three-storey houses. He estimates that the amounts needed (in units) for each are as given in the following table:

	Lumber	Labour	Plumbing	Electrical
One-storey	2.0	6.2	1.6	1.4
Two-storey	3.1	7.5	2.5	2.1
Three-storey	4.2	9.3	3.1	2.9

Represent the number of houses ordered by a row vector and the costs by a matrix. Multiply the two to get a list of needed amounts of the four items for the three kinds of houses combined.

48. In problem 47, suppose that the costs per unit for the four items are \$10,400; \$3,620; \$11,800; and \$9,850, respectively. Represent these costs by a column vector, and multiply it by the matrix to get a vector giving the total costs for each kind of house.

49. Two pizza shops sell the following numbers of pizzas:

On Monday	Cheese	Ham	Veggie
Shop A	32	26	19
Shop B	41	33	43

On Tuesday	Cheese	Ham	Veggie
Shop A	35	18	22
Shop B	33	26	32

The prices of the pizzas are \$9.93, \$10.45, and \$11.20, respectively. Represent the sales for Monday by a matrix **M**, the sales for Tuesday by a matrix **T**, and the prices by a vector **P**.

(a) Add **M** and **T** to get the total sales per pizza per shop for Monday and Tuesday.

(b) Subtract **M** from **T** to get the change in sales from Monday to Tuesday.

(c) Multiply **M** and **P** to get the total income per shop for Monday, and multiply **T** and **P** to get the total income per shop for Tuesday.

50. A computer spreadsheet is ideal for matrix operations because the data are already in the form of lists or tables. Each cell in the spreadsheet corresponds to an element in the matrix.

To multiply two matrices that have already been entered in the spreadsheet (in whatever application you are used to using), locate the command to do matrix multiplication and follow the instructions.

Try this procedure for any of the multiplications in this exercise set.

12–3   The Inverse of a Matrix

Inverse of a Square Matrix

Recall from Sec. 12–1 that the unit matrix **I** is a square matrix that has 1's along its main diagonal and 0's elsewhere, such as

$$\mathbf{I} = \begin{pmatrix} 1 & 0 & 0 & 0 \\ 0 & 1 & 0 & 0 \\ 0 & 0 & 1 & 0 \\ 0 & 0 & 0 & 1 \end{pmatrix}$$

We now define the *inverse* of any given matrix **A** as another matrix $\mathbf{A}^{-1}$ such that the product of the matrix and its inverse is equal to the unit matrix.

Product of a Matrix and Its Inverse	$\mathbf{A}\mathbf{A}^{-1} = \mathbf{A}^{-1}\mathbf{A} = 1$	93

The product of a matrix and its inverse is the unit matrix.

In *ordinary* algebra, if b is some nonzero quantity, then

$$b\left(\frac{1}{b}\right) = bb^{-1} = 1$$

In *matrix* algebra, the unit matrix **I** has properties similar to those of the number 1, and the inverse of a matrix has properties similar to the *reciprocal* in ordinary algebra.

◆◆◆ **Example 32:** The inverse of the matrix

$$\mathbf{A} = \begin{pmatrix} 1 & 0 \\ 2 & 0.5 \end{pmatrix} \quad \text{is} \quad \mathbf{A}^{-1} = \begin{pmatrix} 1 & 0 \\ -4 & 2 \end{pmatrix}$$

because

$$\mathbf{A}\mathbf{A}^{-1} = \begin{pmatrix} 1 & 0 \\ 2 & 0.5 \end{pmatrix}\begin{pmatrix} 1 & 0 \\ -4 & 2 \end{pmatrix} = \begin{pmatrix} 1(1) + 0(-4) & 1(0) + 0(2) \\ 2(1) + 0.5(-4) & 2(0) + 0.5(2) \end{pmatrix}$$

$$= \begin{pmatrix} 1 & 0 \\ 0 & 1 \end{pmatrix}$$

and

$$\mathbf{A}^{-1}\mathbf{A} = \begin{pmatrix} 1 & 0 \\ -4 & 2 \end{pmatrix}\begin{pmatrix} 1 & 0 \\ 2 & 0.5 \end{pmatrix} = \begin{pmatrix} 1(1) + 0(2) & 1(0) + 0(0.5) \\ -4(1) + 2(2) & -4(0) + 2(0.5) \end{pmatrix}$$

$$= \begin{pmatrix} 1 & 0 \\ 0 & 1 \end{pmatrix}$$

◆◆◆

> Not every matrix has an inverse. The inverse exists only for a *nonsingular* matrix, that is, one whose determinant is not zero. Such a matrix is said to be *invertible*.

Finding the Inverse of a Matrix

When we solved sets of linear equations in Chapter 10, we were able to

1. interchange two equations
2. multiply an equation by a nonzero constant
3. add a constant multiple of one equation to another equation

Thus, for a matrix that represents such a system of equations, we may perform the following transformations without altering the meaning of a matrix:

Elementary Transformations of a Matrix	1. Interchange any rows. 2. Multiply a row by a nonzero constant. 3. Add a constant multiple of one row to another row.	96

One method to find the inverse of a matrix is to apply the rules of matrix transformation to transform the given matrix into a unit matrix, while at the same time performing the same operations on a unit matrix. The unit matrix, after the transformations, becomes the inverse of the original matrix.

$$
\begin{array}{cc}
\mathbf{A} & \mathbf{I} \\
\downarrow & \downarrow \\
\mathbf{I} & \mathbf{A}^{-1}
\end{array}
$$

We will not try to prove the method here, but only show its use.

◆◆◆ **Example 33:** Find the inverse of the matrix of Example 32.

$$
\mathbf{A} = \begin{pmatrix} 1 & 0 \\ 2 & 0.5 \end{pmatrix}
$$

Solution: To the given matrix $\mathbf{A}$, we append a unit matrix,

$$
\begin{pmatrix} 1 & 0 & | & 1 & 0 \\ 2 & 0.5 & | & 0 & 1 \end{pmatrix}
$$

obtaining a *double-square* matrix. We then double the first row and subtract it from the second.

$$
\begin{pmatrix} 1 & 0 & | & 1 & 0 \\ 0 & 0.5 & | & -2 & 1 \end{pmatrix}
$$

We now double the second row.

$$
\begin{pmatrix} 1 & 0 & | & 1 & 0 \\ 0 & 1 & | & -4 & 2 \end{pmatrix}
$$

Our original matrix $\mathbf{A}$ is now the unit matrix, and the original unit matrix is now the inverse of $\mathbf{A}$.

$$
\mathbf{A}^{-1} = \begin{pmatrix} 1 & 0 \\ -4 & 2 \end{pmatrix}
$$

◆◆◆

◆◆◆ **Example 34:** Find the inverse of the matrix

$$
\mathbf{A} = \begin{pmatrix} a_1 & b_1 \\ a_2 & b_2 \end{pmatrix}
$$

Solution: We append the unit matrix as follows:

$$
\begin{pmatrix} a_1 & b_1 & | & 1 & 0 \\ a_2 & b_2 & | & 0 & 1 \end{pmatrix}
$$

Then we multiply the first row by b_2 and the second by b_1, and subtract the second from the first.

$$
\begin{pmatrix} a_1 b_2 - a_2 b_1 & 0 & | & b_2 & -b_1 \\ a_2 b_1 & b_1 b_2 & | & 0 & b_1 \end{pmatrix}
$$

We now divide the first row by $a_1b_2 - a_2b_1$.

$$\begin{pmatrix} 1 & 0 & \bigg| & \dfrac{b_2}{a_1b_2 - a_2b_1} & \dfrac{-b_1}{a_1b_2 - a_2b_1} \\ a_2b_1 & b_1b_2 & \bigg| & 0 & b_1 \end{pmatrix}$$

Then we multiply the first row by a_2b_1 and subtract it from the second. After combining fractions, we get

$$\begin{pmatrix} 1 & 0 & \bigg| & \dfrac{b_2}{a_1b_2 - a_2b_1} & \dfrac{-b_1}{a_1b_2 - a_2b_1} \\ 0 & b_1b_2 & \bigg| & \dfrac{-a_2b_1b_2}{a_1b_2 - a_2b_1} & \dfrac{a_1b_1b_2}{a_1b_2 - a_2b_1} \end{pmatrix}$$

We then divide the second row by b_1b_2.

$$\begin{pmatrix} 1 & 0 & \bigg| & \dfrac{b_2}{a_1b_2 - a_2b_1} & \dfrac{-b_1}{a_1b_2 - a_2b_1} \\ 0 & 1 & \bigg| & \dfrac{-a_2}{a_1b_2 - a_2b_1} & \dfrac{a_1}{a_1b_2 - a_2b_1} \end{pmatrix}$$

The inverse of our matrix is then

$$\mathbf{A}^{-1} = \frac{1}{a_1b_2 - a_2b_1} \begin{pmatrix} b_2 & -b_1 \\ -a_2 & a_1 \end{pmatrix}$$

after $1/(a_1b_2 - a_2b_1)$ has been factored out.

Comparing this result with the original matrix, we see that

1. the elements on the main diagonal are interchanged
2. the elements on the secondary diagonal have reversed sign
3. it is divided by the determinant of the original matrix

◆◆◆

Shortcut for Finding the Inverse of a 2 × 2 Matrix

The results of Example 34 provide us with a way to quickly write the inverse of a 2×2 matrix. We obtain the inverse of a 2×2 matrix by

1. interchanging the elements on the main diagonal;
2. reversing the signs of the elements on the secondary diagonal; and
3. dividing by the determinant of the original matrix.

◆◆◆ **Example 35:** Find the inverse of

$$\mathbf{A} = \begin{pmatrix} 5 & -2 \\ 1 & 3 \end{pmatrix}$$

Solution: We interchange the 5 and the 3 and reverse the signs of the -2 and the 1.

$$\begin{pmatrix} 3 & 2 \\ -1 & 5 \end{pmatrix}$$

Then we divide by the determinant of the original matrix.

$$5(3) - (-2)(1) = 17$$

So

$$\mathbf{A}^{-1} = \frac{1}{17} \begin{pmatrix} 3 & 2 \\ -1 & 5 \end{pmatrix}$$

◆◆◆

Exercise 3 ◆ The Inverse of a Matrix

Find the inverse of each matrix. Work to at least three decimal places.

1. $\begin{pmatrix} 4 & 8 \\ -5 & 0 \end{pmatrix}$
                                 2. $\begin{pmatrix} -2 & 4 \\ 3 & 1 \end{pmatrix}$

3. $\begin{pmatrix} 8 & -3 \\ 1 & 6 \end{pmatrix}$
                                 4. $\begin{pmatrix} 1 & 5 & 3 \\ 2 & 3 & 0 \\ 6 & 8 & 1 \end{pmatrix}$

5. $\begin{pmatrix} -1 & 3 & 7 \\ 4 & -2 & 0 \\ 7 & 2 & -9 \end{pmatrix}$
                   6. $\begin{pmatrix} 18 & 23 & 71 \\ 82 & 49 & 28 \\ 36 & 19 & 84 \end{pmatrix}$

7. $\begin{pmatrix} 9 & 1 & 0 & 3 \\ 4 & 3 & 5 & 2 \\ 7 & 1 & 0 & 3 \\ 1 & 4 & 2 & 7 \end{pmatrix}$
          8. $\begin{pmatrix} -2 & 1 & 5 & -6 \\ 7 & -3 & 2 & -6 \\ -3 & 2 & 6 & 1 \\ 5 & 2 & 7 & 5 \end{pmatrix}$

9. As previously mentioned, a spreadsheet is ideal for matrix operations because the data are already in the form of lists or tables. To invert a matrix that had already been entered in a spreadsheet (in whatever application you are used to using), locate the command for matrix inversion and follow the instructions. Try this procedure for any of the matrices in this exercise set.

12–4 Solving a System of Linear Equations by Matrix Inversion

Representing a System of Equations by Matrices

Suppose that we have the set of equations

$$2x + y + 2z = 10$$
$$x + 2y + 3z = 14$$
$$3x + 4y - 3z = 2$$

The left sides of these equations can be represented by the product of a square matrix **A** made up of the coefficients, called the *coefficient matrix,* and a column vector **X** made up of the unknowns. If

$$\mathbf{A} = \begin{pmatrix} 2 & 1 & 2 \\ 1 & 2 & 3 \\ 3 & 4 & -3 \end{pmatrix} \quad \text{and} \quad \mathbf{X} = \begin{pmatrix} x \\ y \\ z \end{pmatrix}$$

then

$$\mathbf{AX} = \begin{pmatrix} 2 & 1 & 2 \\ 1 & 2 & 3 \\ 3 & 4 & -3 \end{pmatrix}\begin{pmatrix} x \\ y \\ z \end{pmatrix} = \begin{pmatrix} 2x + y + 2z \\ x + 2y + 3z \\ 3x + 4y - 3z \end{pmatrix}$$

Note that our product is a column vector having *three* elements. The expression $(2x + y + 2z)$ is a *single element* in that vector.

If we now represent the constants in our set of equations by the column vector

$$\mathbf{B} = \begin{pmatrix} 10 \\ 14 \\ 2 \end{pmatrix}$$

our system of equations can be written, using matrices, as

$$\begin{pmatrix} 2 & 1 & 2 \\ 1 & 2 & 3 \\ 3 & 4 & -3 \end{pmatrix}\begin{pmatrix} x \\ y \\ z \end{pmatrix} = \begin{pmatrix} 10 \\ 14 \\ 2 \end{pmatrix}$$

or in the following more compact form:

Matrix Form for a System of Equations	$AX = B$	95

Solving Systems of Equations by Matrix Inversion

Suppose that we have a system of equations, which we represent in matrix form as

$$AX = B$$

where A is the coefficient matrix, X is the column vector of the unknowns, and B is the column of constants. Multiplying both sides by A^{-1}, the inverse of A, we have

$$A^{-1}AX = A^{-1}B$$

or

$$IX = A^{-1}B$$

since $A^{-1}A = I$. But, by Eq. 94, $IX = X$, so we have the following formula:

Solving a Set of Equations Using the Inverse	$X = A^{-1}B$	98

There are several other methods for using matrices to solve a system of equations, including the *Gauss elimination* and the *unit matrix* method.

    The matrix inversion method is built into many calculators and computer programs.

The solutions to a set of equations can be obtained by multiplying the column of constants by the inverse of the coefficient matrix.

This is called the *matrix inversion* method. To use this method, set up the equations in standard form. Write A, the matrix of the coefficients. Then find A^{-1} by matrix inversion or by the shortcut method (for a 2×2 matrix). Finally, multiply A^{-1} by the column of constants B to get the solution.

◆◆◆ **Example 36:** Solve the system of equations

$$\begin{array}{rcl} x - 2y &=& -3 \\ 3x + y &=& 5 \end{array}$$

by the matrix inversion method.

Solution: The coefficient matrix A is

$$\begin{pmatrix} 1 & -2 \\ 3 & 1 \end{pmatrix}$$

We could have used the shortcut method here, since this is a 2×2 matrix. We have, however, used the longer general method to give another example of its use.

We now find the inverse, A^{-1}, of A. We append the unit matrix,

$$\left(\begin{array}{cc|cc} 1 & -2 & 1 & 0 \\ 3 & 1 & 0 & 1 \end{array} \right)$$

and then multiply the first row by -3 and add it to the second row.

$$\left(\begin{array}{cc|cc} 1 & -2 & 1 & 0 \\ 0 & 7 & -3 & 1 \end{array} \right)$$

Then we multiply the second row by $2/7$ and add it to the first row.

$$\left(\begin{array}{cc|cc} 1 & 0 & 1/7 & 2/7 \\ 0 & 7 & -3 & 1 \end{array} \right)$$

Finally, we divide the second row by 7.

$$\begin{pmatrix} 1 & 0 & | & 1/7 & 2/7 \\ 0 & 1 & | & -3/7 & 1/7 \end{pmatrix}$$

Thus the inverse of our coefficient matrix **A** is

$$\mathbf{A}^{-1} = \begin{pmatrix} 1/7 & 2/7 \\ -3/7 & 1/7 \end{pmatrix}$$

We then get our solution by multiplying the inverse $\mathbf{A}^{-1}$ by the column of constants, **B**.

$$\mathbf{A}^{-1} \qquad \mathbf{B} \quad = \quad \mathbf{X}$$

$$\begin{pmatrix} 1/7 & 2/7 \\ -3/7 & 1/7 \end{pmatrix} \begin{pmatrix} -3 \\ 5 \end{pmatrix} = \begin{pmatrix} -3/7 + 10/7 \\ 9/7 + 5/7 \end{pmatrix} = \begin{pmatrix} 1 \\ 2 \end{pmatrix}$$

so $x = 1$ and $y = 2$.   ◆◆◆

We see that even for a system of only two equations, manual solution by matrix inversion is a lot of work. Our main reason for covering it is for its wide use in computer solutions of sets of equations.

◆◆◆ **Example 37:** Solve the general system of two equations

$$a_1 x + b_1 y = c_1$$
$$a_2 x + b_2 y = c_2$$

by matrix inversion.

Solution: We write the coefficient matrix

$$\mathbf{A} = \begin{pmatrix} a_1 & b_1 \\ a_2 & b_2 \end{pmatrix}$$

The inverse of our coefficient matrix has already been found in Example 34. It is

$$\mathbf{A}^{-1} = \frac{1}{a_1 b_2 - a_2 b_1} \begin{pmatrix} b_2 & -b_1 \\ -a_2 & a_1 \end{pmatrix}$$

By Eq. 98 we now get the column of unknowns by multiplying the inverse by the column of coefficients.

$$\mathbf{X} = \mathbf{A}^{-1}\mathbf{B}$$

$$= \frac{1}{a_1 b_2 - a_2 b_1} \begin{pmatrix} b_2 & -b_1 \\ -a_2 & a_1 \end{pmatrix} \begin{pmatrix} c_1 \\ c_2 \end{pmatrix}$$

$$= \begin{pmatrix} \dfrac{b_2 c_1 - b_1 c_2}{a_1 b_2 - a_2 b_1} \\[2ex] \dfrac{a_1 c_2 - a_2 c_1}{a_1 b_2 - a_2 b_1} \end{pmatrix}$$

Thus we again get the following, now-familiar, equations:

Algebraic Solution for a System of Linear Equations	$x = \dfrac{b_2 c_1 - b_1 c_2}{a_1 b_2 - a_2 b_1} \quad y = \dfrac{a_1 c_2 - a_2 c_1}{a_1 b_2 - a_2 b_1}$	63

◆◆◆

Exercise 4 ◆ Solving a System of Linear Equations by Matrix Inversion

1. Solve any of the sets of equations in Chapter 10 or 11 by matrix inversion.
2. You learned how to invert a square matrix with a spreadsheet in Exercise 3 and how to multiply matrices in Exercise 2. Now combine the two operations to solve a system of equations by entering the coefficient matrix, inverting it, and multiplying it by the column of constants. Try this method on any system of equations.

Case Study Discussion—3D Imaging

When doing a presentation to clients, senior managers or professionals, you have to remember that they have been involved with specialty areas or administration for a while, so you want to keep it straightforward without being too basic and causing insult. It can be a little tricky. Be diplomatic and don't take offence if they ask you to skip ahead (it's a good idea to let the audience know that you are ready to skip ahead if they would like).

The basic concept here is that a matrix can use four elements to represent a three-dimensional point in space by expressing the three angles from the x, y, and z axes and the lengths of the vectors. Then, by multiplying by the sine, cosine, or tangent of the angles, you can rotate the matrix about one of the axes. Here are examples of matrices that can be used.

To rotate along the x axis:

$$\begin{pmatrix} 1 & 0 & 0 & 0 \\ 0 & \cos(x) & -\sin(x) & 0 \\ 0 & \sin(x) & \cos(x) & 0 \\ 0 & 0 & 0 & 1 \end{pmatrix}$$

To rotate along the y axis:

$$\begin{pmatrix} \cos(y) & 0 & \sin(y) & 0 \\ 0 & 1 & 0 & 0 \\ -\sin(y) & 0 & \cos(y) & 0 \\ 0 & 0 & 0 & 1 \end{pmatrix}$$

To rotate along the z axis:

$$\begin{pmatrix} \cos(z) & -\sin(z) & 0 & 0 \\ \sin(z) & \cos(z) & 0 & 0 \\ 0 & 0 & 1 & 0 \\ 0 & 0 & 0 & 1 \end{pmatrix}$$

You can also find matrices for translation (movement along an axis) or for scaling an image.

◆◆◆ CHAPTER 12 REVIEW PROBLEMS ◆◆◆◆◆◆◆◆◆◆◆◆◆◆◆◆◆◆◆◆◆◆◆◆◆◆◆◆◆◆◆◆

1. Combine:
$$(4 \quad 9 \quad -2 \quad 6) + (2 \quad 4 \quad 9 \quad -1) - (4 \quad 1 \quad -3 \quad 7)$$

2. Multiply:
$$\begin{pmatrix} 8 & 4 \\ 9 & -5 \end{pmatrix} \begin{pmatrix} 3 & 7 \\ -3 & 6 \end{pmatrix}$$

3. Multiply:
$$(7 \quad -2 \quad 5) \begin{pmatrix} 6 \\ -2 \\ 8 \end{pmatrix}$$

4. Find the inverse:

$$\begin{pmatrix} 5 & -2 & 5 & 1 \\ 2 & 8 & -3 & -1 \\ -4 & 2 & 7 & 5 \\ 9 & 2 & 5 & 3 \end{pmatrix}$$

5. Find the inverse:

$$\begin{pmatrix} 4 & -4 \\ 3 & 1 \end{pmatrix}$$

6. Combine:

$$\begin{pmatrix} 5 & 9 \\ -4 & 7 \end{pmatrix} + \begin{pmatrix} -2 & 6 \\ 3 & 8 \end{pmatrix}$$

7. Multiply:

$$\begin{pmatrix} 1 & 2 \\ 0 & 3 \end{pmatrix}\begin{pmatrix} 1 & 2 & 4 \\ 3 & 0 & 1 \end{pmatrix}$$

8. Multiply

$$\begin{pmatrix} 8 & 2 & 6 & 1 \\ 9 & 0 & 1 & 4 \\ 1 & 6 & 9 & 3 \\ 6 & 9 & 3 & 5 \end{pmatrix}\begin{pmatrix} 5 \\ 3 \\ 2 \\ 8 \end{pmatrix}$$

9. Solve:

$$\begin{aligned} x + y + z + w &= -4 \\ x + 2y + 3z + 4w &= 0 \\ x + 3y + 6z + 10w &= 9 \\ x + 4y + 10z + 20w &= 24 \end{aligned}$$

10. Multiply:

$$7\begin{pmatrix} 5 & -3 \\ -4 & 8 \end{pmatrix}$$

11. Solve:

$$\begin{aligned} 5x + 3y - 2z &= 5 \\ 3x - 4y + 3z &= 13 \\ x + 6y - 4z &= -8 \end{aligned}$$

12. Find the inverse:

$$\begin{pmatrix} 5 & 1 & 4 \\ -2 & 6 & 9 \\ 4 & -1 & 0 \end{pmatrix}$$

13. Solve:

$$\begin{aligned} 4x + 3y &= 27 \\ 2x - 5y &= -19 \end{aligned}$$

14. Multiply:

$$9(6 \quad 3 \quad -2 \quad 8)$$

15. The four loop currents (mA) in a certain circuit satisfy the equations

$$\begin{aligned} 376I_1 + 374I_2 - 364I_3 - 255I_4 &= 284 \\ 898I_1 - 263I_2 + 229I_3 + 274I_4 &= -847 \\ 364I_1 + 285I_2 - 857I_3 - 226I_4 &= 746 \\ 937I_1 + 645I_2 - 387I_3 + 874I_4 &= -254 \end{aligned}$$

Solve for the four currents.

16. A factory makes TVs, stereos, CD players, and DVD players, which it sells for $181, $62, $33, and $49, respectively. On a particular day, the factory makes quantities of 232, 373, 737, and 244, respectively. Represent the prices and the quantities by vectors, and multiply them to get the total value of the four items made on that day.

17. On the following day, the factory in problem 16 makes quantities of the given items of 273, 836, 361, and 227. Represent the prices by a row vector and the quantities made for both days by a matrix. Multiply to get a table giving the total values per day.

18. A company has two plants that make the following numbers of items:

In April			
	Skis	Snowboards	Surfboards
Plant *A*	132	126	129
Plant *B*	241	133	143

In May			
	Skis	Snowboards	Surfboards
Plant *A*	135	218	122
Plant *B*	133	126	232

The prices of the items are $129.53, $110.65, and $411.60, respectively. Represent the sales for April by a matrix **A**, the sales for May by a matrix **M**, and the prices by a vector **P**.

(a) Add **A** and **M** to get the total production per item per plant for April and May.

(b) Subtract **A** from **M** to get the change in production from April to May.

(c) Multiply **A** and **P** to get the total income for April, and multiply **M** and **P** to get the total income for May.

Writing

19. Suppose that you have submitted a lab report in which you stated that you have used matrix inversion to solve some sets of equations. The report was rejected because you failed to say anything about that method.

Write one paragraph explaining matrix inversion to be inserted into the next draft of your report. Start by telling what matrix inversion is, and then go on to say how it works.

Team Project

20. In Chapter 10 we learned how to solve a system of equations by classical methods. In Chapter 11 we used determinants, and in this chapter we learned matrix inversion. For each method we could do the solution manually, or with a spreadsheet.

Given a system of equations by your instructor, solve that system as many ways as you can. See which team can get the correct solution by the greatest number of means.

Exponents and Radicals

♦♦♦ OBJECTIVES ♦♦

When you have completed this chapter, you should be able to:
- Raise a variable or single term to an integral exponent.
- Simplify a radical.
- Perform basic math operations with radicals.
- Solve equations containing radical terms.

♦♦♦

You have already seen in Chapter 2 how to raise a variable to a power and how to take a root. Now, however, we want to be able to do the same sort of operation on entire expressions or the entire side of an equation. You will have to be able to bring expressions out from under a root sign, or remove the root signs all together. There are so many formulas and equations that use powers or roots that you will need to be able to work with these types of equations. Some of you will express concern that you are learning to do by hand what calculators are now able to complete for you. The problem is that the calculators are very sensitive to syntax; the way an equation or expression is entered into the calculator can make a big difference. Sometimes it's not just enough to enter it mathematically correctly, as some calculators have their own quirks. If you have no idea what the final answer should be, you are at the mercy of the calculator—and when you think about it, that's not what's expected of an expert in engineering technology or applied sciences.

CASE STUDY—RESONANT FREQUENCY IN A CIRCUIT

In electronics when you combine a capacitor (a device that will hold energy in an electrostatic field) and an inductor (a device that will hold energy in an electromagnetic field), you create what's called a resonant circuit. The capacitor will charge and then release into the circuit its energy, which is then stored in the inductor's electromagnetic field and then discharged back to the capacitor. Like a see-saw, the energy transfers back and forth between the two devices. The external power source only has to make up for power loss in the wires, and the circuit will push current back and forth between the capacitor and inductor. Hook this up to a speaker and you get a tone like a ringing bell. The tone, or frequency, of the ringing is determined by the formula

$$f_r = \frac{1}{2\pi}\sqrt{\frac{1}{LC} - \frac{R^2}{L^2}}$$

In a report to a client on the specifications for a circuit, you, as the electronics engineering technologist, have tested the circuit and compared the test results to the theoretical calculated values. The client has a concern; he says that he uses another formula for calculating the resonant frequency:

$$f_r = \frac{1}{2\pi\sqrt{LC}} \sqrt{1 - \frac{CR^2}{L}}$$

Although it works out to the same value in this case, the client is not convinced that your formula could be used in all tests. To prove that they are in fact mathematically the same, you are going to have to show the math converting the first formula to the second.

13–1   Integral Exponents

In this section, we continue the study of exponents that we started in Sec. 2–3. We repeat the laws of exponents that were explained there and use them to simplify harder expressions than before. You should refresh your memory on the laws of exponents covered in that section before going too far into this chapter.

Negative Exponents

The law that we derived in Sec. 2–3 for negative exponents is repeated here.

Negative Exponent	$x^{-a} = \dfrac{1}{x^a} \quad (x \neq 0)$	**35**

When taking the reciprocal of a base raised to a power, *change the sign of the exponent.*

We'll use the laws of exponents mainly to simplify expressions, to make them easier to work with in later computations, such as solving equations containing exponents. For example, we use the law for negative exponents to rewrite an expression so that it does not contain a negative exponent.

◆◆◆ **Example 1:**

(a) $7^{-1} = \dfrac{1}{7}$

(b) $x^{-1} = \dfrac{1}{x}$

(c) $z^{-3} = \dfrac{1}{z^3}$

(d) $xy^{-1} = \dfrac{x}{y}$

(e) $\dfrac{ab^{-2}}{c^{-3}d} = \dfrac{ac^3}{b^2 d}$

◆◆◆

Another use for negative exponents is to rewrite an expression so that it does not contain fractions.

◆◆◆ **Example 2:**

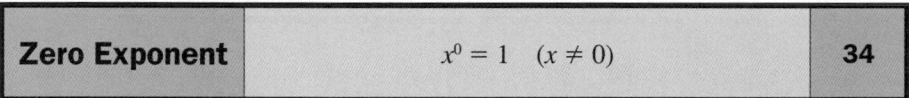

(a) $\dfrac{1}{w} = w^{-1}$

(b) $\dfrac{x}{y} = xy^{-1}$

(c) $\dfrac{p^3}{q^2} = p^3 q^{-2}$

(d) $\dfrac{ab^2 c}{w^3 x^2 y} = ab^2 cw^{-3} x^{-2} y^{-1}$ ◆◆◆

Zero Exponents

The law that we derived in Sec. 2–3 for zero exponents is repeated here.

Zero Exponent	$x^0 = 1 \quad (x \neq 0)$	34

Any quantity (except 0) raised to the zero power equals 1.

We use this law to rewrite an expression so that it contains no zero exponents.

◆◆◆ **Example 3:**

(a) $367^0 = 1$

(b) $(\sin 37.2°)^0 = 1$

(c) $x^2 y^0 z = x^2 z$

(d) $(abc)^0 = 1$ ◆◆◆

We'll use these laws frequently in the examples to come.

Common Error	$x^{-a} \neq x^{1/a}$	

Power Raised to a Power

Another law that we derived in Chapter 1 is for raising a power to a power.

Powers	$(x^a)^b = x^{ab} = (x^b)^a$	31

When raising a power to a power, keep the same base and *multiply the exponents.*

◆◆◆ **Example 4:**

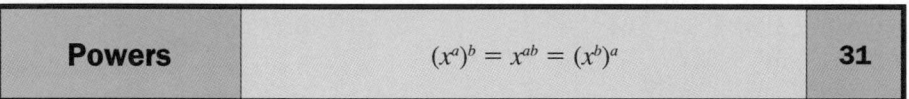

(a) $(2^2)^2 = 2^4 = 16$

(b) $(3^{-1})^4 = 3^{-4} = \dfrac{1}{3^4} = \dfrac{1}{81}$

(c) $(a^2)^3 = a^6$

(d) $(z^{-3})^{-2} = z^6$

(e) $(x^5)^{-2} = x^{-10} = \dfrac{1}{x^{10}}$ ◆◆◆

Products

The two laws of exponents that apply to products are given here again.

Products	$x^a \cdot x^b = x^{a+b}$	29

When multiplying powers of the same base, keep the same base and *add the exponents.*

Product Raised to a Power	$(xy)^n = x^n \cdot y^n$	29

When a product is raised to a power, each factor may be *separately* raised to the power.

◆◆◆ **Example 5:**

(a) $2axy^3z(3a^2xyz^3) = 6a^3x^2y^4z^4$　　　　　　(b) $(3bx^3y^2)^3 = 27b^3x^9y^6$

(c) $\left(\dfrac{xy^2}{3}\right)^2 = \dfrac{x^2y^4}{9}$ 　　　　　　　　　　　　　　　　　　◆◆◆

	Parentheses are important when we are raising products to a power.
Common Error	$(2x)^0 = 1$ 　　but　　 $2x^0 = 2(1) = 2$ $(4x)^{-1} = \dfrac{1}{4x}$ 　　but　　 $4x^{-1} = \dfrac{4}{x}$

We try to leave our expressions without zero or negative exponents.

◆◆◆ **Example 6:**

(a) $3p^3q^2r^{-4}(4p^{-5}qr^4) = 12p^{-2}q^3r^0 = \dfrac{12q^3}{p^2}$

(b) $(2m^2n^3x^{-2})^{-3} = 2^{-3}m^{-6}n^{-9}x^6 = \dfrac{x^6}{8m^6n^9}$ 　　　　　　◆◆◆

Of course, our old rules for multiplying multinomials still apply.

◆◆◆ **Example 7:**

(a) $3x^{-2}y^3(2ax^3 - x^2y + ax^4y^{-3})$
$= 6axy^3 - 3x^0y^4 + 3ax^2y^0$
$= 6axy^3 - 3y^4 + 3ax^2$

(b) $(p^2x - q^2y^{-2})(p^{-2} - q^{-2}y^2)$
$= p^0x - p^2q^{-2}xy^2 - p^{-2}q^2y^{-2} + q^0y^0$
$= x - \dfrac{p^2xy^2}{q^2} - \dfrac{q^2}{p^2y^2} + 1$ 　　　　　　　　　　　　◆◆◆

Letters in the exponents are handled just as if they were numbers.

◆◆◆ **Example 8:**

$$x^{n+1}y^{1-n}(x^{2n-3}y^{n-1}) = x^{n+1+2n-3}y^{1-n+n-1}$$
$$= x^{3n-2}y^0 = x^{3n-2}$$ 　　　　◆◆◆

Quotients

Our two laws of exponents that apply to quotients are repeated here.

Quotients	$\dfrac{x^a}{x^b} = x^{a-b}$ 　　$(x \neq 0)$	30

When dividing powers of the same base, keep the same base and **subtract the exponents**.

| Quotient Raised to a power | $\left(\dfrac{x}{y}\right)^n = \dfrac{x^n}{y^n}$ $(y \neq 0)$ | 33 |

When a quotient is raised to a power, the numerator and the denominator may be **separately** raised to the power.

◆◆◆ **Example 9:**

(a) $\dfrac{x^5}{x^3} = x^2$

(b) $\dfrac{16p^5q^{-3}}{8p^2q^{-5}} = 2p^3q^2$

(c) $\dfrac{x^{2p-1}y^2}{x^{p+4}y^{1-p}} = x^{2p-1-p-4}y^{2-1+p}$
$= x^{p-5}y^{1+p}$

(d) $\left(\dfrac{x^3y^2}{p^2q}\right)^3 = \dfrac{x^9y^6}{p^6q^3}$ ◆◆◆

As before, we try to leave our expressions without zero or negative exponents.

◆◆◆ **Example 10:**

(a) $\dfrac{8a^3b^{-5}c^3}{4a^5b^{-1}c^3} = 2a^{-2}b^{-4}c^0 = \dfrac{2}{a^2b^4}$

(b) $\left(\dfrac{a}{b}\right)^{-2} = \dfrac{a^{-2}}{b^{-2}} = \dfrac{b^2}{a^2} = \left(\dfrac{b}{a}\right)^2$ ◆◆◆

Note in Example 10(b) that the negative exponent had the effect of inverting the fraction.

Don't forget our hard-won skills of factoring, of combining fractions over a common denominator, and of simplifying complex fractions. Use them where needed.

◆◆◆ **Example 11:**

(a) $\dfrac{1}{(5a)^2} + \dfrac{1}{(2a^2b^{-4})^2} = \dfrac{1}{25a^2} + \dfrac{1}{\dfrac{4a^4}{b^8}}$

We simplify the compound fraction and combine the two fractions over a common denominator, getting

$$\dfrac{1}{25a^2} + \dfrac{b^8}{4a^4} = \dfrac{4a^2}{100a^4} + \dfrac{25b^8}{100a^4}$$
$$= \dfrac{4a^2 + 25b^8}{100a^4}$$

(b) $(x^{-2} - 3y)^{-2} = \dfrac{1}{(x^{-2} - 3y)^2} = \dfrac{1}{\left(\dfrac{1}{x^2} - 3y\right)^2}$

$= \dfrac{1}{\left(\dfrac{1 - 3x^2y}{x^2}\right)^2}$

$= \dfrac{x^4}{(1 - 3x^2y)^2}$

(c) $\dfrac{p^{4m} - q^{2n}}{p^{2m} + q^n} = \dfrac{(p^{2m})^2 - (q^n)^2}{p^{2m} + q^n}$

Factoring the difference of two squares in the numerator and cancelling gives us

$$\dfrac{(p^{2m} + q^n)(p^{2m} - q^n)}{p^{2m} + q^n} = p^{2m} - q^n$$ ◆◆◆

Exercise 1 ◆ Integral Exponents

Simplify, and write without negative exponents.

1. $3x^{-1}$

2. $4a^{-2}$

3. $2p^0$

4. $(3m)^{-1}$

5. $a(2b)^{-2}$

6. x^2y^{-1}

7. a^3b^{-2}

8. m^2n^{-3}

9. p^3q^{-1}

10. $(2x^2y^3z^4)^{-1}$

11. $(3m^3n^2p)^0$

12. $(5p^2q^2r^2)^3$

13. $(4a^3b^2c^6)^{-2}$

14. $(a^p - b^q)^2$

15. $(x + y)^{-1}$

16. $(2a + 3b)^{-1}$

17. $(m^{-2} - 6n)^{-2}$

18. $(4p^2 + 5q^{-4})^{-2}$

19. $2x^{-1} + y^{-2}$

20. $p^{-1} - 3q^{-2}$

21. $(3m)^{-3} - 2n^{-2}$

22. $(5a)^{-2} + (2a^2b^{-4})^{-2}$

23. $(x^n + y^m)^2$

24. $(x^{a-1} + y^{a-2})(x^a + y^{a-1})$

25. $(16x^6y^0 \div 8x^4y) \div 4xy^6$

26. $(a^{n-1} + b^{n-2})(a^n + b^{n-1})$

27. $(72p^6q^7 \div 9p^4q) \div 8pq^6$

28. $\dfrac{(p^2 - pq)^8}{(p - q)^4}$

29. $\left(\dfrac{x^{m+n}}{x^n}\right)^m$

30. $\dfrac{27x^{3n} + 8y^{3n}}{3x^n + 2y^n}$

31. $\left(\dfrac{3}{b}\right)^{-1}$

32. $\left(\dfrac{x}{4}\right)^{-1}$

33. $\left(\dfrac{2x}{3y}\right)^{-2}$

34. $\left(\dfrac{3p}{4y}\right)^{-1}$

35. $\left(\dfrac{2px}{3qy}\right)^{-3}$

36. $\left(\dfrac{p^2}{q^0}\right)^{-1}$

37. $\left(\dfrac{3a^4b^3}{5x^2y}\right)^2$

38. $\left(\dfrac{x^2}{y}\right)^{-3}$

39. $\left(\dfrac{-2a^3x^3}{3b^2y}\right)^{2n}$

40. $\left(\dfrac{3p^2y^3}{4qx^4}\right)^{-2}$

41. $\left(\dfrac{2p^{-2}z^3}{3q^{-2}x^{-4}}\right)^{-1}$

42. $\left(\dfrac{9m^4n^3}{3p^3q}\right)^3$

43. $\left(\dfrac{5w^2}{2z}\right)^p$

44. $\left(\dfrac{5a^5b^3}{2x^2y}\right)^{3n}$

45. $\left(\dfrac{3n^{-2}y^3}{5m^{-3}x^4}\right)^{-2}$

46. $\left(\dfrac{z^0z^{2n-2}}{z^n}\right)^3$

47. $\left(\dfrac{1 + a}{1 + b}\right)^3 \div \dfrac{1 + a^3}{b^2 - 1}$

48. $\left(\dfrac{p^2 - q^2}{p + q}\right)^2 \cdot \dfrac{(p^3 + q^3)^2}{(p + q)}$

49. $\dfrac{(x^2 - xy)^7}{(x - y)^5}$

50. $\dfrac{(a^n)^{2n} - (b^{2n})^n}{(a^n)^n + (b^n)^n}$

51. $\dfrac{27a^{3n} - 8b^{3n}}{3a^n - 2b^n}$

52. $\left(\dfrac{a^2 a^{2n-2}}{a^n}\right)^2$

53. $\left(\dfrac{x + 1}{y + 1}\right)^3 \div \dfrac{x^3 + 1}{y^2 - 1}$

54. $\left(\dfrac{a^2 - b^2}{x + y}\right)^2 \cdot \dfrac{(x^3 + y^3)^2}{(a + b)^2}$

55. $\left(\dfrac{7x^3}{15y^3 a^{m-1}} \div \dfrac{x^{n+1}}{a^{m+n}}\right) \dfrac{y^n}{a^{m-1}}$ (

56. $\dfrac{a^7 b^8 c^9 + a^6 b^7 c^8 - 3a^5 b^6 c^7}{a^5 b^6 c^7}$

57. $\dfrac{128a^4 b^3 - 48a^5 b^2 - 40a^6 b + 15a^7}{3a^2 - 8ab}$

58. $\left(\dfrac{x^{m+n}}{x^n}\right)^m \div \left(\dfrac{x^n}{x^{m+n}}\right)^{n+m}$

59. $\left(\dfrac{5a^3}{25b^3 a^{m-1}} \div \dfrac{a^{n+1}}{a^{m+n}}\right) \dfrac{b^n}{a^{m-1}}$

60. $\dfrac{p^4 q^8 r^9 - p^6 q^4 r^8 + 3p^5 q^6 r^4}{p^3 q^2 r^4}$

61. $\dfrac{3x^4 y^3 + 4x^5 y^2 + 3x^6 y + 4x^7}{4x^2 + 3xy}$

62. $\left(\dfrac{p^n}{p^{m+n}}\right)^{n-m} \div \left(\dfrac{p^{m+n}}{p^n}\right)^m$

63. $\dfrac{a^{-1} + b^{-1}}{a^{-1} - b^{-1}} \cdot \dfrac{a^{-2} - b^{-2}}{a^{-2} + b^{-2}} \cdot \dfrac{1}{\left(1 + \dfrac{b}{a}\right)^{-2} + \left(1 + \dfrac{a}{b}\right)^{-2}}$

Applications

64. The resistance R of two resistors R_1 and R_2 wired in parallel, as shown in Fig. 13–1, is given by Eq. A64:

$$\frac{1}{R} = \frac{1}{R_1} + \frac{1}{R_2}$$

Write this equation without fractions.

65. When a current I causes a power P in a resistance R, that resistance is given by Eq. A67.

$$R = \frac{P}{I^2}$$

Write this equation without fractions.

66. The volume of a cube of side a is a^3. If we double the length of the side, the volume becomes

$$(2a)^3$$

Simplify this expression.

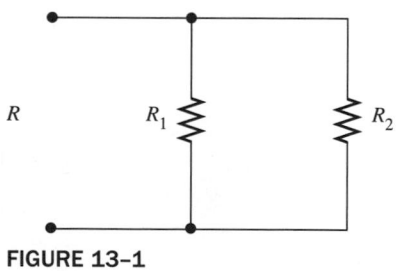

FIGURE 13–1

67. The power to a resistor is given by Eq. A67. If the current is then doubled and the resistance is halved, we have

$$(2I)^2\left(\frac{R}{2}\right)$$

Simplify this expression.

13–2   Simplification of Radicals

Relation between Fractional Exponents and Radicals

A quantity raised to a *fractional exponent* can also be written as a *radical*. A radical consists of a *radical sign*, a quantity under the radical sign called the *radicand*, and the *index* of the radical. A radical is written as follows:

Unless otherwise stated, we'll assume that when the index is even, the radicand is *positive*. We'll consider square roots and fourth roots of negative numbers when we discuss imaginary numbers in Chapter 21.

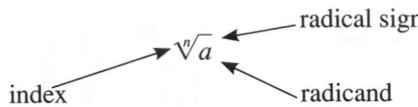

where *n* is an integer. An index of 2, for a square root, is usually not written.

In general, the following equation applies:

Relation between Exponents and Radicals	$a^{1/n} = \sqrt[n]{a}$	36

◆◆◆ **Example 12:**

(a) $4^{1/2} = \sqrt{4} = 2$ (b) $8^{1/3} = \sqrt[3]{8} = 2$

(c) $x^{1/2} = \sqrt{x}$ (d) $y^{1/4} = \sqrt[4]{y}$

(e) $w^{-1/2} = \dfrac{1}{\sqrt{w}}$ ◆◆◆

◆◆◆ **Example 13:** Simplify $\sqrt[5]{382}$.

Solution:

$$\sqrt[5]{382} = 382^{1/5} = 3.28$$

by calculator, rounded to three significant digits.   ◆◆◆

If a quantity is raised to a fractional exponent *m/n* (where *m* and *n* are integers), we have

$$a^{m/n} = (a^m)^{1/n} = \sqrt[n]{a^m}$$

Further,

$$a^{m/n} = (a^{1/n})^m = (\sqrt[n]{a})^m$$

So we have the following formula:

$a^{m/n} = \sqrt[n]{a^m} = (\sqrt[n]{a})^m$	37

◆◆◆ **Example 14:**

(a) $8^{2/3} = (\sqrt[3]{8})^2 = (2)^2 = 4$

(b) $\sqrt[5]{(68.3)^3} = (68.3)^{3/5} = 12.6$, by calculator

(c) $\sqrt[3]{x^4} = x^{4/3}$

(d) $(\sqrt[5]{y^2})^3 = (y^{2/5})^3 = y^{6/5}$ ◆◆◆

We use these definitions to switch between *exponential form* and *radical form*.

◆◆◆ **Example 15:** Express $x^{1/2}y^{-1/3}$ in radical form.

Solution:

$$x^{1/2}y^{-1/3} = \frac{x^{1/2}}{y^{1/3}} = \frac{\sqrt{x}}{\sqrt[3]{y}}$$ ◆◆◆

◆◆◆ **Example 16:** Express $\sqrt[3]{y^2}$ in exponential form.

Solution:

$$\sqrt[3]{y^2} = (y^2)^{1/3} = y^{2/3}$$ ◆◆◆

◆◆◆ **Example 17:** Find the value of $\sqrt[4]{(81)^3}$ without using a calculator.

Solution: Instead of cubing 81 and then taking the fourth root, let us take the fourth root first.

$$\sqrt[4]{(81)^3} = (\sqrt[4]{81})^3 = (3)^3 = 27$$ ◆◆◆

Common Error	Don't confuse the **coefficient** of a radical with the **index** of a radical. $$3\sqrt{x} \neq \sqrt[3]{x}$$

Root of a Product

We have several *rules of radicals*, which are similar to the laws of exponents and, in fact, are derived from them. The first rule is for products. By our definition of a radical,

$$\sqrt[n]{ab} = (ab)^{1/n}$$

Using the law of exponents for a product and then returning to radical form,

$$(ab)^{1/n} = a^{1/n}b^{1/n} = \sqrt[n]{a}\,\sqrt[n]{b}$$

So our first rule of radicals is as follows:

Root of a Product	$\sqrt[n]{ab} = \sqrt[n]{a}\,\sqrt[n]{b}$	**38**

The root of a product equals the product of the roots of the factors.

◆◆◆ **Example 18:** We may split the radical $\sqrt{9x}$ into two radicals, as follows:

$$\sqrt{9x} = \sqrt{9}\,\sqrt{x} = 3\sqrt{x}$$

since $\sqrt{9} = 3$.   ◆◆◆

◆◆◆ **Example 19:** Write as a single radical $\sqrt{7}\,\sqrt{2}\,\sqrt{x}$.

Solution: By Eq. 38,

$$\sqrt{7}\,\sqrt{2}\,\sqrt{x} = \sqrt{7(2)x} = \sqrt{14x}$$ ◆◆◆

Common Error	There is no similar rule for the root of a sum. $$\sqrt[n]{a+b} \neq \sqrt[n]{a} + \sqrt[n]{b}$$

◆◆◆ **Example 20:** $\sqrt{9} + \sqrt{16}$ does *not* equal $\sqrt{25}$.   ◆◆◆

Common Error	Equation 31 does not hold when a and b are both *negative* and the index is even.

◆◆◆ **Example 21:**

$$(\sqrt{-4})^2 = \sqrt{-4}\sqrt{-4} \neq \sqrt{(-4)(-4)}$$
$$\neq \sqrt{16} = +4$$

Instead, we convert to *imaginary numbers*, as we will show in Sec. 21–1.   ◆◆◆

Root of a Quotient

Just as the root of a product can be split up into the roots of the factors, the root of a quotient can be expressed as the root of the numerator divided by the root of the denominator. We first write the quotient in exponential form,

$$\sqrt[n]{\frac{a}{b}} = \left(\frac{a}{b}\right)^{1/n} = \frac{a^{1/n}}{b^{1/n}}$$

by the law of exponents for quotients. Returning to radical form, we have the following:

Root of a Quotient	$$\sqrt[n]{\frac{a}{b}} = \frac{\sqrt[n]{a}}{\sqrt[n]{b}}$$	**39**

The root of a quotient equals the quotient of the roots of numerator and denominator.

◆◆◆ **Example 22:** The radical $\sqrt{\dfrac{w}{25}}$ can be written $\dfrac{\sqrt{w}}{\sqrt{25}}$ or $\dfrac{\sqrt{w}}{5}$.   ◆◆◆

Simplest Form for a Radical

A radical is said to be in *simplest form* when:

Rather than "simplifying," we are actually putting radicals into a *standard form*, so that they may be combined or compared.

1. The radicand has been reduced as much as possible.
2. There are no radicals in the denominator and no fractional radicands.
3. The index has been made as small as possible.

Removing Factors from the Radicand

Using the fact that

Remember that we are doing numerical problems as a way of learning the rules. If you simply want the decimal value of a radical expression containing only numbers, use your calculator.

$$\sqrt{x^2} = x, \qquad \sqrt[3]{x^3} = x, \ldots, \qquad \sqrt[n]{x^n} = x$$

try to factor the radicand so that one or more of the factors is a perfect nth power (where n is the index of the radical). Then use Eq. 38 to split the radical into two or more radicals, some of which can then be reduced.

◆◆◆ Example 23:

$$\sqrt{50} = \sqrt{(25)(2)} = \sqrt{25}\sqrt{2} = 5\sqrt{2} \qquad ◆◆◆$$

◆◆◆ Example 24:

$$\sqrt{x^3} = \sqrt{x^2 x} = \sqrt{x^2}\sqrt{x} = x\sqrt{x} \qquad ◆◆◆$$

◆◆◆ Example 25: Simplify $\sqrt{50x^3}$.

Solution: We factor the radicand so that some factors are perfect squares.

$$\sqrt{50x^3} = \sqrt{(25)(2)x^2 x}$$

Then, by Eq. 38,

$$= \sqrt{25}\sqrt{x^2}\sqrt{2x} = 5x\sqrt{2x} \qquad ◆◆◆$$

◆◆◆ Example 26: Simplify $\sqrt{24y^5}$.

Solution: We look for factors of the radicand that are perfect squares.

$$\sqrt{24y^5} = \sqrt{4(6)y^4 y} = 2y^2\sqrt{6y} \qquad ◆◆◆$$

◆◆◆ Example 27:

$$\sqrt[4]{24y^5} = \sqrt[4]{24y^4 y} = y\sqrt[4]{24y} \qquad ◆◆◆$$

◆◆◆ Example 28: Simplify $\sqrt[3]{24y^5}$.

Solution: We look for factors of the radicand that are perfect cubes.

$$\sqrt[3]{24y^5} = \sqrt[3]{8(3)y^3 y^2}$$
$$= \sqrt[3]{8y^3}\sqrt[3]{3y^2} = 2y\sqrt[3]{3y^2} \qquad ◆◆◆$$

When the radicand contains more than one term, try to *factor out* a perfect n^{th} power (where n is the index).

◆◆◆ Example 29: Simplify $\sqrt{4x^2 y + 12x^4 z}$.

Solution: We factor $4x^2$ from the radicand and then remove it from under the radical sign.

$$\sqrt{4x^2 y + 12x^4 z} = \sqrt{4x^2(y + 3x^2 z)}$$
$$= 2x\sqrt{y + 3x^2 z} \qquad ◆◆◆$$

Rationalizing the Denominator

An expression is considered in simpler form when its denominators contain no radicals. To put it into this form is called *rationalizing the denominator*. We will show how to rationalize the denominator when it is a square root, a cube root, or a root with any index and when it has more than one term.

If the denominator is a square root, multiply numerator and denominator of the fraction by a quantity that will make the radicand in the denominator a perfect square. Note that we are eliminating radicals from the *denominator* and that the numerator may still contain radicals. Further, even though we call this process *simplifying*, the resulting radical may look more complicated than the original.

◆◆◆ **Example 30:**

$$\frac{5}{\sqrt{2}} = \frac{5}{\sqrt{2}} \cdot \frac{\sqrt{2}}{\sqrt{2}} = \frac{5\sqrt{2}}{\sqrt{4}} = \frac{5\sqrt{2}}{2} \qquad \text{◆◆◆}$$

When the *entire* fraction is under the radical sign, we make the denominator of that fraction a perfect square and remove it from the radical sign.

◆◆◆ **Example 31:**

$$\sqrt{\frac{3x}{2y}} = \sqrt{\frac{3x(2y)}{2y(2y)}} = \sqrt{\frac{6xy}{4y^2}} = \frac{\sqrt{6xy}}{2y} \qquad \text{◆◆◆}$$

If the denominator is a *cube* root, we must multiply numerator and denominator by a quantity that will make the quantity under the radical sign a perfect cube.

◆◆◆ **Example 32:** Simplify

$$\frac{7}{\sqrt[3]{4}}$$

Solution: To rationalize the denominator $\sqrt[3]{4}$, we multiply numerator and denominator of the fraction by $\sqrt[3]{2}$. We chose this as the factor because it results in a perfect cube (8) under the radical sign.

$$\frac{7}{\sqrt[3]{4}} = \frac{7}{\sqrt[3]{4}} \cdot \frac{\sqrt[3]{2}}{\sqrt[3]{2}} = \frac{7\sqrt[3]{2}}{\sqrt[3]{8}} = \frac{7\sqrt[3]{2}}{2} \qquad \text{◆◆◆}$$

The same principle applies regardless of the index. In general, if the index is n, we must make the quantity under the radical sign (in the denominator) a perfect nth power.

◆◆◆ **Example 33:**

$$\frac{2y}{3\sqrt[5]{x}} = \frac{2y}{3\sqrt[5]{x}} \cdot \frac{\sqrt[5]{x^4}}{\sqrt[5]{x^4}} = \frac{2y\sqrt[5]{x^4}}{3\sqrt[5]{x^5}} = \frac{2y\sqrt[5]{x^4}}{3x} \qquad \text{◆◆◆}$$

Sometimes the denominator will have more than one term.

◆◆◆ **Example 34:**

$$\sqrt{\frac{a}{a^2 + b^2}} = \sqrt{\frac{a}{a^2 + b^2} \cdot \frac{a^2 + b^2}{a^2 + b^2}} = \frac{\sqrt{a(a^2 + b^2)}}{a^2 + b^2} \qquad \text{◆◆◆}$$

Reducing the Index

We can sometimes reduce the index by writing the radical in exponential form and then reducing the fractional exponent, as in the next example.

◆◆◆ **Example 35:**

(a) $\sqrt[6]{x^3} = x^{3/6}$
    $= x^{1/2} = \sqrt{x}$

(b) $\sqrt[4]{4x^2y^2} = \sqrt[4]{(2xy)^2}$
    $= (2xy)^{2/4} = (2xy)^{1/2}$
    $= \sqrt{2xy}$ ◆◆◆

Exercise 2 ◆ Simplification of Radicals

Exponential and Radical Forms

Express in radical form.

1. $a^{1/4}$
2. $x^{1/2}$
3. $z^{3/4}$
4. $a^{1/2}b^{1/4}$
5. $(m - n)^{1/2}$
6. $(x^2y)^{-1/2}$
7. $\left(\dfrac{x}{y}\right)^{-1/3}$
8. $a^0b^{-3/4}$

Express in exponential form.

9. $\sqrt{b}$

10. $\sqrt[3]{x}$

11. $\sqrt{y^2}$

12. $4\sqrt[3]{xy}$

13. $\sqrt[n]{a + b}$

14. $\sqrt[n]{x^m}$

15. $\sqrt{x^2y^2}$

16. $\sqrt[n]{a^n b^{3n}}$

Simplifying Radicals

Write in simplest form.

17. $\sqrt{18}$

18. $\sqrt{75}$

19. $\sqrt{63}$

20. $\sqrt[3]{16}$

21. $\sqrt[3]{-56}$

22. $\sqrt[4]{48}$

23. $\sqrt{a^3}$

24. $3\sqrt{50x^5}$

25. $\sqrt{36x^2y}$

26. $\sqrt[3]{x^2y^5}$

27. $x\sqrt[3]{16x^3y}$

28. $\sqrt[4]{64m^2n^4}$

29. $3\sqrt[5]{32xy^{11}}$

30. $6\sqrt[3]{16x^4}$

31. $\sqrt{a^3 - a^2b}$

32. $x\sqrt{x^4 - x^3y^2}$

33. $\sqrt{9m^3 + 18n}$

34. $\sqrt{2x^3 + x^4y}$

35. $\sqrt[3]{x^4 - a^2x^3}$

36. $x\sqrt{a^3 + 2a^2b + ab^2}$

37. $(a + b)\sqrt{a^3 - 2a^2b + ab^2}$

38. $\sqrt{\dfrac{2}{5}}$

39. $\sqrt{\dfrac{3}{7}}$

40. $\sqrt{\dfrac{2}{3}}$

41. $\sqrt[3]{\dfrac{1}{4}}$

42. $\sqrt{\dfrac{5}{8}}$

43. $\sqrt[3]{\dfrac{2}{9}}$

44. $\sqrt[4]{\dfrac{7}{8}}$

45. $\sqrt{\dfrac{1}{2x}}$

46. $\sqrt{\dfrac{5m}{7n}}$

47. $\sqrt{\dfrac{3a^3}{5b}}$

48. $\sqrt{\dfrac{5ab}{6xy}}$

49. $\sqrt[3]{\dfrac{1}{x^2}}$

50. $\sqrt[3]{\dfrac{81x^4}{16yz^2}}$

51. $\sqrt[6]{\dfrac{4x^6}{9}}$

52. $\sqrt{x^2 - \left(\dfrac{x}{2}\right)^2}$

53. $(x^2 - y^2)\sqrt{\dfrac{x}{x + y}}$

54. $\sqrt{a^3 + \left(\dfrac{a^3}{2}\right)^3}$

55. $(m + n)\sqrt{\dfrac{m}{m - n}}$

56. $\sqrt{\left(\dfrac{x + 1}{2}\right)^2 - x}$

57. $\sqrt{\dfrac{8a^2 - 48a + 72}{3a}}$

If you are using a calculator for any of these problems, be sure to leave the answers in radical form.

For these more complicated types, you should start by simplifying the expression under the radical sign.

Applications

58. The period ω_n for simple harmonic motion is given by Eq. A37.

$$\omega_n = \sqrt{\dfrac{kg}{W}}$$

Write this equation in exponential form.

59. The magnitude, Z, of the impedance of a series RLC circuit is given by Eq. A99.

$$Z = \sqrt{R^2 + X^2}$$

Write this equation in exponential form.

60. Given Eq. A99 from problem 59, write the expression for Z when $X = 2R$, and simplify.

61. The hypotenuse in right triangle ABC, shown in Fig. 13–2, is given by the Pythagorean theorem.

$$c = \sqrt{a^2 + b^2}$$

Write an expression for c when $b = 3a$, and simplify.

62. In problem 58, rationalize the denominator in Eq. A37.

63. A stone is thrown upward with a horizontal velocity of 40 ft./s and an upward velocity of 60 ft./s. At t seconds it will have a horizontal displacement H equal to $40t$ and a vertical displacement V equal to $60t - 16t^2$. The straight-line distance S from the stone to the launch point is found by the Pythagorean theorem. Write an equation for S in terms of t, and simplify.

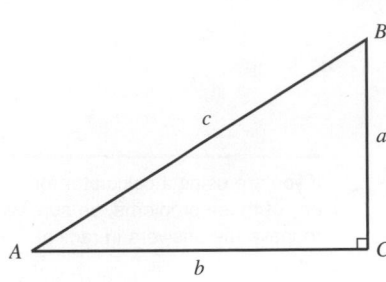

FIGURE 13–2

13–3   Operations with Radicals

Adding and Subtracting Radicals

Radicals are called *similar* if they have the same index and the same radicand, such as $5\sqrt[3]{2x}$ and $3\sqrt[3]{2x}$. We add and subtract radicals by *combining similar radicals*.

One reason we learned to put radicals into a standard form is to combine them.

◆◆◆ **Example 36:**

$$5\sqrt{y} + 2\sqrt{y} - 4\sqrt{y} = 3\sqrt{y}$$ ◆◆◆

Radicals that may look similar at first glance may not actually be similar.

◆◆◆ **Example 37:** The radicals

$$\sqrt{2x} \text{ and } \sqrt{3x}$$

are *not* similar.   ◆◆◆

Common Error	Do not try to combine radicals that are not similar. $$\sqrt{2x} + \sqrt{3x} \neq \sqrt{5x}$$

Radicals that do not appear to be similar at first may turn out to be so after simplification.

◆◆◆ **Example 38:**

(a) $\sqrt{18x} - \sqrt{8x} = 3\sqrt{2x} - 2\sqrt{2x} = \sqrt{2x}$

(b) $\sqrt[3]{24y^4} + \sqrt[3]{81x^3y} = 2y\sqrt[3]{3y} + 3x\sqrt[3]{3y}$
$$= (2y + 3x)\sqrt[3]{3y}$$

(c) $5y\sqrt{\dfrac{x}{y}} - 2\sqrt{xy} + x\sqrt{\dfrac{y}{x}}$

$$= 5y\sqrt{\dfrac{xy}{y^2}} - 2\sqrt{xy} + x\sqrt{\dfrac{yx}{x^2}}$$

$$= 5\sqrt{xy} - 2\sqrt{xy} + \sqrt{xy} = 4\sqrt{xy}$$ ◆◆◆

Multiplying Radicals

Radicals having the *same index* can be multiplied by using Eq. 38.

$$\sqrt[n]{a}\,\sqrt[n]{b} = \sqrt[n]{ab}$$

◆◆◆ **Example 39:**

(a) $\sqrt{x}\,\sqrt{2y} = \sqrt{2xy}$

(b) $(5\sqrt[3]{3a})(2\sqrt[3]{4b}) = 10\sqrt[3]{12ab}$

(c) $(2\sqrt{m})(5\sqrt{n})(3\sqrt{mn}) = 30\sqrt{m^2n^2} = 30mn$

(d) $\sqrt{2x}\sqrt{3x} = \sqrt{6x^2} = x\sqrt{6}$

(e) $\sqrt{3x}\sqrt{3x} = 3x$

(f) $\sqrt[3]{2x}\sqrt[3]{4x^2} = \sqrt[3]{8x^3} = 2x$

◆◆◆

We multiply radicals having *different indices* by first going to exponential form, then multiplying using Eq. 29 ($x^a x^b = x^{a+b}$), and finally returning to radical form.

◆◆◆ **Example 40:**

$$\sqrt{a}\,\sqrt[4]{b} = a^{1/2}b^{1/4}$$
$$= a^{2/4}b^{1/4} = (a^2b)^{1/4}$$

Or, in radical form,

$$= \sqrt[4]{a^2b}$$

◆◆◆

When the radicands are the same, the work is even easier.

◆◆◆ **Example 41:** Multiply $\sqrt[3]{x}$ by $\sqrt[6]{x}$.

Solution:

$$\sqrt[3]{x}\,\sqrt[6]{x} = x^{1/3}x^{1/6} = x^{2/6}x^{1/6}$$

By Eq. 29,

$$= x^{3/6} = x^{1/2}$$
$$= \sqrt{x}$$

◆◆◆

Multinomials containing radicals are multiplied in the way we learned in Sec. 2–4.

◆◆◆ **Example 42:**

(a) $(3 + \sqrt{x})(2\sqrt{x} - 4\sqrt{y}) = 3(2\sqrt{x}) - 3(4\sqrt{y}) + 2\sqrt{x}\,\sqrt{x} - 4\sqrt{x}\,\sqrt{y}$
$$= 6\sqrt{x} - 12\sqrt{y} + 2x - 4\sqrt{xy}$$

(b) $\sqrt{50x}(\sqrt{2x} - \sqrt{xy}) = \sqrt{100x^2} - \sqrt{50x^2y}$
$$= 10x - 5x\sqrt{2y}$$

(c) $(\sqrt{x} - \sqrt{y})(\sqrt{x} + \sqrt{y}) = \sqrt{x^2} + \sqrt{xy} - \sqrt{xy} - \sqrt{y^2}$
$$= \sqrt{x^2} - \sqrt{y^2}$$
$$= x - y$$

◆◆◆

To raise a radical to a *power*, we simply multiply the radical by itself the proper number of times.

◆◆◆ **Example 43:** Cube the expression

$$2x\sqrt{y}$$

Solution:

$$(2x\sqrt{y})^3 = (2x\sqrt{y})(2x\sqrt{y})(2x\sqrt{y})$$
$$= 2^3 x^3 (\sqrt{y})^3$$
$$= 8x^3\sqrt{y^3} = 8x^3 y\sqrt{y}$$

    ◆◆◆

Dividing Radicals

Radicals having the same indices can be divided using Eq. 39.

$$\frac{\sqrt[n]{a}}{\sqrt[n]{b}} = \sqrt[n]{\frac{a}{b}}$$

◆◆◆ **Example 44:**

(a) $\dfrac{\sqrt{4x^5}}{\sqrt{2x}} = \sqrt{\dfrac{4x^5}{2x}} = \sqrt{2x^4} = x^2\sqrt{2}$

(b) $\dfrac{\sqrt{3a^7} + \sqrt{12a^5} - \sqrt{6a^3}}{\sqrt{3a}} = \sqrt{\dfrac{3a^7}{3a}} + \sqrt{\dfrac{12a^5}{3a}} - \sqrt{\dfrac{6a^3}{3a}}$

$$= \sqrt{a^6} + \sqrt{4a^4} - \sqrt{2a^2}$$
$$= a^3 + 2a^2 - a\sqrt{2}$$

    ◆◆◆

It is a common practice to rationalize the denominator after division.

◆◆◆ **Example 45:**

$$\frac{\sqrt{10x}}{\sqrt{5y}} = \sqrt{\frac{10x}{5y}} = \sqrt{\frac{2x}{y}}$$

Rationalizing the denominator, we obtain

$$\sqrt{\frac{2x}{y}} = \sqrt{\frac{2x}{y} \cdot \frac{y}{y}} = \frac{\sqrt{2xy}}{y}$$

    ◆◆◆

If the indices are different, we go to exponential form, as we did for multiplication. Divide using Eq. 30 and then return to radical form.

◆◆◆ **Example 46:**

$$\frac{\sqrt[3]{a}}{\sqrt[4]{b}} = \frac{a^{1/3}}{b^{1/4}} = \frac{a^{4/12}}{b^{3/12}} = \left(\frac{a^4}{b^3}\right)^{1/12}$$

Returning to radical form, we obtain

$$\left(\frac{a^4}{b^3}\right)^{1/12} = \sqrt[12]{\frac{a^4}{b^3}}$$
$$= \sqrt[12]{\frac{a^4}{b^3} \cdot \frac{b^9}{b^9}}$$
$$= \sqrt[12]{\frac{a^4 b^9}{b^{12}}}$$
$$= \frac{1}{b}\sqrt[12]{a^4 b^9}$$

When dividing by a binomial containing *square roots*, multiply the divisor and the dividend by the *conjugate* of that binomial (a binomial that differs from the original dividend only in the sign of one term).

Remember that when we multiply a binomial, for example, $(a + b)$, by its conjugate $(a - b)$,

$$(a + b)(a - b) = a^2 - ab + ab - b^2$$
$$= a^2 - b^2$$

Thus, the conjugate of the binomial $a + b$ is $a - b$. The conjugate of $x - y$ is $x + y$.

we get an expression where the cross-product terms ab and $-ab$ drop out, and the remaining terms are both squares. Thus if a or b were square roots, our final expression would have no square roots.

We use this operation to remove square roots from the denominator, as shown in the following example.

◆◆◆ **Example 47:** Divide $(3 + \sqrt{x})$ by $(2 - \sqrt{x})$.

Solution: The conjugate of the divisor $2 - \sqrt{x}$ is $2 + \sqrt{x}$. Multiplying divisor and dividend by $2 + \sqrt{x}$, we get

$$\frac{3 + \sqrt{x}}{2 - \sqrt{x}} = \frac{3 + \sqrt{x}}{2 - \sqrt{x}} \cdot \frac{2 + \sqrt{x}}{2 + \sqrt{x}}$$

$$= \frac{6 + 3\sqrt{x} + 2\sqrt{x} + \sqrt{x}\sqrt{x}}{4 + 2\sqrt{x} - 2\sqrt{x} - \sqrt{x}\sqrt{x}}$$

$$= \frac{6 + 5\sqrt{x} + x}{4 - x} \qquad ◆◆◆$$

Exercise 3 ◆ Operations with Radicals

Addition and Subtraction of Radicals

Combine as indicated.

1. $2\sqrt{24} - \sqrt{54}$

2. $\sqrt{300} + \sqrt{108} - \sqrt{243}$

3. $\sqrt{24} - \sqrt{96} + \sqrt{54}$

4. $\sqrt{128} - \sqrt{18} + \sqrt{32}$

5. $2\sqrt{50} + \sqrt{72} + 3\sqrt{18}$

6. $\sqrt[3]{384} - \sqrt[3]{162} + \sqrt[3]{750}$

7. $2\sqrt[3]{2} - 3\sqrt[3]{16} + \sqrt[3]{-54}$

8. $3\sqrt[3]{108} + 2\sqrt[3]{32} - \sqrt[3]{256}$

9. $\sqrt[3]{625} - 2\sqrt[3]{135} - \sqrt[3]{320}$

10. $5\sqrt[3]{320} + 2\sqrt[3]{40} \quad 4\sqrt[3]{135}$

11. $\sqrt[4]{768} - \sqrt[4]{48} - \sqrt[4]{243}$

12. $3\sqrt{\dfrac{5}{4}} + 2\sqrt{45}$

13. $\sqrt{128x^2y} - \sqrt{98x^2y} + \sqrt{162x^2y}$

14. $3\sqrt{\dfrac{1}{3}} - 2\sqrt{\dfrac{3}{4}} + 4\sqrt{3}$

15. $7\sqrt{\dfrac{27}{50}} - 3\sqrt{\dfrac{2}{3}}$

16. $4\sqrt{50} - 2\sqrt{72} + \dfrac{3}{\sqrt{2}}$

17. $\sqrt{a^2x} + \sqrt{b^2x}$

18. $\sqrt{2b^2xy} - \sqrt{2a^2xy}$

19. $\sqrt{x^2y} - \sqrt{4a^2y}$

20. $\sqrt{80a^3} - 3\sqrt{20a^3} - 2\sqrt{45a^3}$

21. $4\sqrt{3a^2x} - 2a\sqrt{48x}$

22. $\sqrt[3]{125x^2} - 2\sqrt[3]{8x^2}$

23. $\sqrt[3]{1250a^3b} + \sqrt[3]{270c^3b}$

24. $2\sqrt[3]{ab^7} + 3\sqrt[3]{a^7b} + 2\sqrt[3]{8a^4b^4}$

25. $\sqrt[5]{a^{13}b^{11}c^{12}} - 2\sqrt[5]{a^8bc^2} + \sqrt[5]{a^3b^6c^7}$

26. $\sqrt{\dfrac{9x}{16}} - \dfrac{3\sqrt{x}}{4}$

27. $\sqrt[6]{8x^3} - 2\sqrt[4]{\dfrac{x^2}{4}} + 3\sqrt{8x}$

28. $\sqrt{\dfrac{x}{a^2}} + 2\sqrt{\dfrac{x}{b^2}} - 3\sqrt{\dfrac{x}{c^2}}$

29. $3\sqrt{\dfrac{3x}{4y^2}} + 2\sqrt{\dfrac{x}{27y^2}} - \sqrt{\dfrac{x}{3y^2}}$

30. $\sqrt{(a + b)^2x} + \sqrt{(a - b)^2x}$

As in Exercise 2, if you are using a calculator, remember to leave the answers in radical form.

Multiplication of Radicals

Multiply.

31. $3\sqrt{3}$ by $5\sqrt{3}$

32. $2\sqrt{3}$ by $3\sqrt{8}$

33. $\sqrt{8}$ by $\sqrt{160}$

34. $\sqrt{\dfrac{5}{8}}$ by $\sqrt{\dfrac{3}{4}}$

35. $4\sqrt[3]{45}$ by $2\sqrt[3]{3}$

36. $3\sqrt{3}$ by $2\sqrt[3]{2}$

37. $3\sqrt{2}$ by $2\sqrt[3]{3}$

38. $2\sqrt{3}$ by $\sqrt[4]{5}$

39. $2\sqrt[3]{3}$ by $5\sqrt[4]{4}$

40. $2\sqrt[3]{24}$ by $\sqrt[9]{\dfrac{8}{27}}$

41. $2x\sqrt{3a}$ by $3\sqrt{y}$

42. $3\sqrt[3]{9a^2}$ by $\sqrt[3]{3abc}$

43. $\sqrt[5]{4xy^2}$ by $\sqrt[5]{8x^2y}$

44. $\sqrt{\dfrac{a}{b}}$ by $\sqrt{\dfrac{c}{d}}$

45. $\sqrt{a}$ by $\sqrt[4]{b}$

46. $\sqrt[3]{x}$ by $\sqrt{y}$

47. $\sqrt{xy}$ by $2\sqrt{xz}$ and $\sqrt[3]{x^2y^2}$

48. $\sqrt[3]{a^2b}$ by $\sqrt[3]{2a^2b^2}$ and $\sqrt{3a^3b^2}$

49. $(\sqrt{5} - \sqrt{3})$ by $2\sqrt{3}$

50. $(x + \sqrt{y})$ by $\sqrt{y}$

51. $\sqrt{x^3 - x^4y}$ by $\sqrt{x}$

52. $(\sqrt{x} - \sqrt{2})$ by $(2\sqrt{x} + \sqrt{2})$

53. $(a + \sqrt{b})$ by $(a - \sqrt{b})$

54. $(\sqrt{x} + \sqrt{y})$ by $(\sqrt{x} + \sqrt{y})$

55. $(4\sqrt{x} + 2\sqrt{y})$ by $(4\sqrt{x} - 5\sqrt{y})$

56. $(x + \sqrt{xy})$ by $(\sqrt{x} - \sqrt{y})$

Powers

Square the following expressions.

57. $3\sqrt{y}$

58. $4\sqrt[3]{4x^2}$

59. $3x\sqrt[3]{2x^2}$

60. $5 + 4\sqrt{x}$

61. $3 - 5\sqrt{a}$

62. $\sqrt{a} + 5a\sqrt{b}$

Cube the following radicals.

63. $5\sqrt{2x}$

64. $2x\sqrt{3x}$

65. $5\sqrt[3]{2ax}$

Division of Radicals

Divide.

66. $8 \div 3\sqrt{2}$

67. $6\sqrt{72} \div 12\sqrt{32}$

68. $(4\sqrt[3]{3} - 3\sqrt[3]{2}) \div \sqrt[3]{2}$

69. $5\sqrt[3]{12} \div 10\sqrt{8}$

70. $9\sqrt[3]{18} \div 3\sqrt{6}$

71. $12\sqrt[3]{4a^3} \div 4\sqrt[3]{2a^5}$

72. $12\sqrt{256} \div 18\sqrt{2}$

73. $(4\sqrt[3]{4} + 2\sqrt[3]{3} + 3\sqrt[3]{6}) \div \sqrt[3]{6}$

74. $2 \div \sqrt[3]{6}$

75. $\sqrt{72} \div 2\sqrt[4]{64}$

76. $8 \div 2\sqrt[3]{4}$

77. $8\sqrt[3]{ab} \div 4\sqrt{ac}$

78. $\sqrt[3]{4ab} \div \sqrt[4]{2ab}$

79. $2\sqrt[4]{a^2b^3c^2} \div 4\sqrt[3]{ab^2c^2}$

80. $6\sqrt[3]{x^4y^7z} \div \sqrt[4]{xy}$

81. $(3 + \sqrt{2}) \div (2 - \sqrt{2})$

82. $5 \div \sqrt{3x}$

83. $4\sqrt{x} \div \sqrt{a}$

84. $10 \div \sqrt[3]{9x^2}$

85. $12 \div \sqrt[3]{4x^2}$

86. $5\sqrt{a^2b} \div 3\sqrt[3]{a^2b^2}$

87. $(10\sqrt{6} - 3\sqrt{5}) \div (2\sqrt{6} - 4\sqrt{5})$

88. $\sqrt{x} \div (\sqrt{x} + \sqrt{y})$

89. $a \div (a + \sqrt{b})$

90. $(a + \sqrt{b}) \div (a - \sqrt{b})$

91. $(\sqrt{x} - \sqrt{y}) \div (\sqrt{x} + \sqrt{y})$

92. $(3\sqrt{m} - \sqrt{2n}) \div (\sqrt{3n} + \sqrt{m})$

93. $(5\sqrt{x} + \sqrt{2y}) \div (2\sqrt{3x} + 2\sqrt{y})$

94. $(5\sqrt{2p} - 3\sqrt{3q}) \div (3\sqrt{3p} + 2\sqrt{5q})$

95. $(2\sqrt{3a} + 4\sqrt{5b}) \div (3\sqrt{2a} + 4\sqrt{b})$

13–4 Radical Equations

An equation in which the unknown is under a radical sign is called a *radical equation*. To solve a radical equation, it is necessary to isolate the radical term on one side of the equal sign and then raise both sides to whatever power will eliminate the radical.

◆◆◆ **Example 48:** Solve for x:

$$\sqrt{x - 5} - 4 = 0$$

Solution: Rearranging yields

$$\sqrt{x - 5} = 4$$

Squaring both sides: $\quad x - 5 = 16$

$$x = 21$$

We will solve more difficult radical equations in Chapter 14.

Check:

$$\sqrt{21 - 5} - 4 \overset{?}{=} 0$$
$$\sqrt{16} - 4 \overset{?}{=} 0$$
$$4 - 4 = 0 \quad \text{(checks)}$$

◆◆◆

◆◆◆ **Example 49:** Solve for x:

$$\sqrt{x^2 - 6x} = x - 9$$

Solution: Squaring, we obtain

$$x^2 - 6x = x^2 - 18x + 81$$
$$12x = 81$$
$$x = \frac{81}{12} = \frac{27}{4}$$

Check:

Remember that we take only the principal (positive) root.

$$\sqrt{\left(\frac{27}{4}\right)^2 - 6\left(\frac{27}{4}\right)} \overset{?}{=} \frac{27}{4} - 9$$

$$\sqrt{\frac{729}{16} - \frac{162}{4}} \overset{?}{=} \frac{27}{4} - \frac{36}{4}$$

$$\sqrt{\frac{81}{16}} = \frac{9}{4} \neq -\frac{9}{4} \quad \text{(does not check)}$$

Thus the given equation has no solution. ◆◆◆

Common Error	The squaring process often introduces *extraneous roots*. These are discarded because they do not satisfy the original equation. Check your answer in the original equation.

If the equation has more than one radical, isolate one at a time and square both sides, as in the following example.

◆◆◆ **Example 50:** Solve for x:

$$\sqrt{x-32} + \sqrt{x} = 16$$

Tip: It is usually better to isolate and square the *most complicated* radical first.

Solution: Rearranging gives

$$\sqrt{x-32} = 16 - \sqrt{x}$$

Squaring yields

$$x - 32 = (16)^2 - 32\sqrt{x} + x$$

We rearrange again to isolate the radical.

$$32\sqrt{x} = 256 + 32 = 288$$
$$\sqrt{x} = 9$$

Squaring again, we obtain

$$x = 81$$

Check:

$$\sqrt{81 - 32} + \sqrt{81} \overset{?}{=} 16$$
$$\sqrt{49} + 9 \overset{?}{=} 16$$
$$7 + 9 = 16 \quad \text{(checks)} \qquad\qquad ◆◆◆$$

If a radical equation contains a fraction, we proceed as we did with other fractional equations and multiply both sides by the least common denominator.

◆◆◆ **Example 51:** Solve the following:

$$\sqrt{x-3} + \frac{1}{\sqrt{x-3}} = \sqrt{x}$$

Solution: Multiplying both sides by $\sqrt{x-3}$ gives

$$x - 3 + 1 = \sqrt{x}\,\sqrt{x-3}$$
$$x - 2 = \sqrt{x^2 - 3x}$$

Squaring both sides yields

$$x^2 - 4x + 4 = x^2 - 3x$$
$$x = 4$$

Check:

$$\sqrt{4 - 3} + \frac{1}{\sqrt{4-3}} \overset{?}{=} \sqrt{4}$$
$$1 + \frac{1}{1} = 2 \quad \text{(checks)} \qquad\qquad ◆◆◆$$

Radical equations having indices other than 2 are solved in a similar way.

◆◆◆ **Example 52:** Solve for x:

$$\sqrt[3]{x-5} = 2$$

Solution: Cubing both sides, we obtain

$$x - 5 = 8$$
$$x = 13$$

Check:

$$\sqrt[3]{13 - 5} \stackrel{?}{=} 2$$
$$\sqrt[3]{8} = 2 \quad \text{(checks)}$$

◆◆◆

Exercise 4 ◆ Radical Equations

Radical Equations

Solve for x and check.

1. $\sqrt{x} = 6$
2. $\sqrt{x} + 5 = 9$
3. $\sqrt{7x + 8} = 6$
4. $\sqrt{3x - 2} = 5$
5. $\sqrt{2.95x - 1.84} = 6.23$
6. $\sqrt{5.88x + 4.92} = 7.72$
7. $\sqrt{x + 1} = \sqrt{2x - 7}$
8. $2 = \sqrt[4]{1 + 3x}$
9. $\sqrt{3x + 1} = 5$
10. $\sqrt[3]{2x} = 4$
11. $\sqrt{x - 3} = \dfrac{4}{\sqrt{x - 3}}$
12. $x + 2 = \sqrt{x^2 + 6}$
13. $\sqrt{x^2 - 7.25} = 8.75 - x$
14. $\sqrt{x - 15.5} = 5.85 - \sqrt{x}$
15. $\dfrac{6}{\sqrt{3 + x}} = \sqrt{x + 3}$
16. $\sqrt{12 + x} = 2 + \sqrt{x}$
17. $\sqrt[5]{x - 7} = 1$
18. $\sqrt{x^2 - 2x} - \sqrt{1 - x} = 1 - x$
19. $\sqrt{1.25x} - \sqrt{1.25x + 14.3} + 3.18 = 0$
20. $2.10x + \sqrt{4.41x^2 - 8.36} = 9.37$
21. $\sqrt{x - 2} - \sqrt{x} = \dfrac{1}{\sqrt{x - 2}}$
22. $\sqrt{\dfrac{1}{x}} - \sqrt{x} = \sqrt{1 + x}$
23. $\sqrt{5x - 19} - \sqrt{5x + 14} = -3$
24. $\sqrt[3]{x - 3} = 2$

Applications

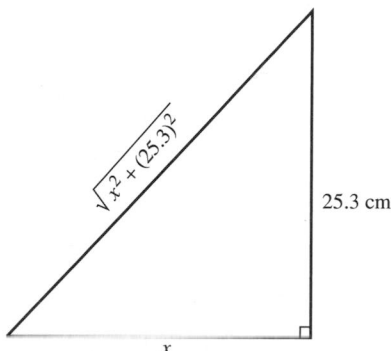

FIGURE 13–3

25. The right triangle in Fig. 13–3 has one side equal to 25.3 cm and a perimeter of 68.4 cm, so

$$x + 25.3 + \sqrt{x^2 + (25.3)^2} = 68.4$$

Solve this equation for x and for the hypotenuse.

26. Find the missing side and the hypotenuse of a right triangle that has one side equal to 293 cm and a perimeter of 994 cm.

27. Find the side and the hypotenuse of a right triangle that has a side of 2.73 m and a perimeter of 11.4 m.

28. The natural frequency f_n of a body under simple harmonic motion (Fig. 13–4) is found from Eqs. A37 and A38.

$$f_n = \dfrac{1}{2\pi} \sqrt{\dfrac{kg}{W}}$$

where g is the gravitational constant, k is a constant, and W is the weight.
Solve for W.

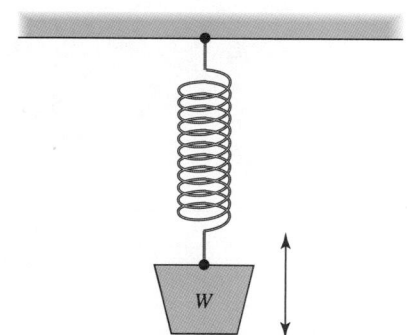

FIGURE 13–4

29. The magnitude Z of the impedance in an RLC circuit (Fig. 13–5) having a resistance R, an inductance L, and a capacitance C, to which is applied a voltage of frequency ω, is given by Eq. A99.

$$Z = \sqrt{R^2 + \left(\omega L - \dfrac{1}{\omega C}\right)^2}$$

Solve for C.

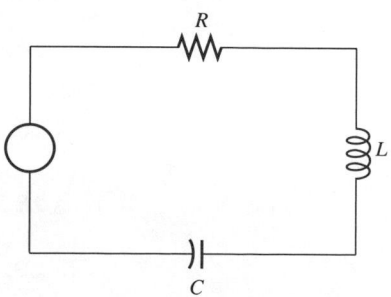

FIGURE 13–5

30. The resonant frequency ω_n for the circuit shown in Fig. 13–5 is given by Eq. A91.

$$\omega_n = \frac{1}{\sqrt{LC}}$$

Solve for L.

Case Study Discussion—Resonant Frequency in a Circuit

Two common formulas are used to determine the resonant frequency in a circuit. The first and the last ones shown are often used. The six intermediate formulas (follow them down the left column, then down the right) show the two formats to be identical. There are many variations to formulas.

You will find that concerns over correct formulas do occur from time to time. The best solution is not to argue over which formula is best but to show mathematically that they are the same. Justifying to a client what you are doing and how you are doing it is very common.

$$f_r = \frac{1}{2\pi}\sqrt{\frac{1}{LC} - \frac{R^2}{L^2}}$$

$$= \frac{1}{2\pi}\sqrt{\frac{L}{L^2C} - \frac{CR^2}{L^2C}}$$

$$= \frac{1}{2\pi}\sqrt{\frac{L - CR^2}{L^2C}}$$

$$= \frac{1}{2\pi}\sqrt{\frac{\dfrac{L}{L} - \dfrac{CR^2}{L}}{LC}}$$

$$= \frac{1}{2\pi}\sqrt{\frac{1 - \dfrac{CR^2}{L}}{LC}}$$

$$= \frac{1}{2\pi}\sqrt{\frac{1}{LC}\left(1 - \frac{CR^2}{L}\right)}$$

$$= \frac{1}{2\pi}\sqrt{\frac{1}{LC}}\sqrt{1 - \frac{CR^2}{L}}$$

$$= \frac{1}{2\pi\sqrt{LC}}\sqrt{1 - \frac{CR^2}{L}}$$

◆◆◆ **CHAPTER 13 REVIEW PROBLEMS** ◆◆◆◆◆◆◆◆◆◆◆◆◆◆◆◆◆◆◆◆◆◆◆◆◆◆◆◆◆◆◆◆

Simplify.

1. $\sqrt{52}$ **2.** $\sqrt{108}$ **3.** $\sqrt[3]{162}$

4. $\sqrt[4]{9}$ **5.** $\sqrt[6]{4}$ **6.** $\sqrt{81a^2x^3y}$

7. $3\sqrt[4]{81x^5}$ **8.** $\sqrt{ab^2 - b^3}$ **9.** $\sqrt[3]{(a - b)^5 x^4}$

10. $5\sqrt{\dfrac{7x}{12y}}$ **11.** $\sqrt{\dfrac{a - 2}{a + 2}}$

Perform the indicated operations and simplify. Don't use a calculator.

12. $4b\sqrt{3y} \cdot 5\sqrt{x}$ **13.** $\sqrt{x} \cdot \sqrt{x^3 - x^4y}$

14. $(\sqrt{x} + \sqrt{y})(\sqrt{x} - \sqrt{y})$ **15.** $\dfrac{3 - 2\sqrt{3}}{2 - 5\sqrt{2}}$

16. $\dfrac{\sqrt{x} - \sqrt{y}}{\sqrt{x} + \sqrt{y}}$ **17.** $\dfrac{x + \sqrt{x^2 - y^2}}{x - \sqrt{x^2 - y^2}}$

18. $\sqrt{98x^2y^2} - \sqrt{128x^2y^2}$ **19.** $4\sqrt[3]{125x^2} + 3\sqrt[3]{8x^2}$

20. $3\sqrt{x^2 - y^2} - 2\sqrt{\dfrac{x + y}{x - y}}$

21. $\sqrt{a}\,\sqrt[3]{b}$ **22.** $\sqrt[3]{2abc^2} \cdot \sqrt{abc}$

23. $(3 + 2\sqrt{x})^2$

24. $(4x\sqrt{2x})^3$

25. $3\sqrt{50} - 2\sqrt{32}$

26. $3\sqrt{72} - 4\sqrt{8} + \sqrt{128}$

27. $5\sqrt{\dfrac{9}{8}} - 2\sqrt{\dfrac{25}{18}}$

28. $2\sqrt{2} \div \sqrt[3]{2}$

29. $\sqrt{2ab} \div \sqrt{4ab^2}$

30. $9 \div \sqrt[3]{7x^2}$

31. $3\sqrt{9} \cdot 4\sqrt{8}$

Solve for x and check.

32. $\dfrac{\sqrt{x} - 8}{\sqrt{x} - 6} = \dfrac{\sqrt{x} - 4}{\sqrt{x} + 2}$

33. $\sqrt{x + 6} = 4$

34. $\sqrt{2x - 7} = \sqrt{x - 3}$

35. $\sqrt[5]{2x + 4} = 2$

36. $\sqrt{4x^2 - 3} = \sqrt{2x - 1}$

37. $\sqrt{x} + \sqrt{x - 9.75} = 6.23$

38. $\sqrt[3]{21.5x} = 2.33$

Simplify, and write without negative exponents.

39. $(x^{n-1} + y^{n-2})(x^n + y^{n-1})$

40. $(72a^6b^7 \div 9a^4b) \div 8ab^6$

41. $\left(\dfrac{9x^4y^3}{6x^3y}\right)^3$

42. $\left(\dfrac{5w^2}{2x}\right)^n$

43. $\left(\dfrac{2x^5y^3}{x^2y}\right)^3$

44. $5a^{-1}$

45. $3w^{-2}$

46. $2r^0$

47. $(3x)^{-1}$

48. $2a^{-1} + b^{-2}$

49. $x^{-1} - 2y^{-2}$

50. $(5a)^{-3} - 3b^{-2}$

51. $(3x)^{-2} + (2x^2y^{-4})^{-2}$

52. $(a^n + b^m)^2$

53. $(p^{a-1} + q^{a-2})(p^a + q^{a-1})$

54. $(16a^6b^0 \div 8a^4b) \div 4ab^6$

55. p^2q^{-1}

56. x^3y^{-2}

57. r^2s^{-3}

58. a^3b^{-1}

59. $(3x^3y^2z)^0$

60. $(5a^2b^2c^2)^3$

61. The volume V of a sphere of radius r is given by Eq. 128.

$$V = \frac{4}{3}\pi r^3$$

If the radius is tripled, the volume is then

$$V = \frac{4}{3}\pi(3r)^3$$

Simplify this expression.

62. Find the missing side and the hypotenuse of a right triangle that has one side equal to 154 cm and a perimeter of 558 cm.

63. The geometric mean B between two numbers A and C is

$$B = \sqrt{AC}$$

Use the rules of radicals to write this equation in a different form.

64. Equation A40 gives the damped angular velocity ω_d of a suspended weight as

$$\omega_d = \sqrt{\omega_n^2 - \frac{c^2 g^2}{W^2}}$$

Rationalize the denominator and simplify.

Writing

65. Explain how exponents and radicals are really two different ways of writing the same expression. Also explain why, if they are the same, we need both.

Team Project

66. Use Hero's formula, Eq. 138, to write an expression for the area of an equilateral triangle of side a, and simplify. Compare your result with what you get when

multiplying half the base times the altitude (Eq. 137).

Quadratic Equations

◆◆◆ **OBJECTIVES** ◆◆

When you have completed this chapter, you should be able to:
- Solve quadratic equations by factoring.
- Solve quadratic equations by completing the square.
- Solve quadratic equations using the quadratic formula.
- Solve applications using quadratic equations.
- Graph a quadratic function.
- Work with quadratic equations
- Solve third-degree and higher equations.
- Solve systems of quadratic equations.

What happens when we have to solve a problem where a variable is squared? In physics, acceleration is measured in how many metres per second the velocity changes per second—or metres per second per second, which is metres per second *squared*. Even some of the most basic formulas in physics will include a t^2 expression for time squared. So far, you have looked at problem solving using equations that usually don't have a variable raised to a power. Now that we've reviewed the properties of powers and radicals, it's time to start solving problems that include variables raised to a power. An equation or expression where a variable is raised to the second power is called a *quadratic*. There are a few methods we can use to solve quadratic equations; each one best for a certain type of quadratic. We will also look at the *quadratic formula*, which has the ability to solve almost any quadratic.

CASE STUDY—GEOMETRIC REPRESENTATION OF A QUADRATIC

This case study is a little different; let's look at an interesting way of proving some mathematical principles. We know that to calculate the area of a square, we take the square of one side. So to demonstrate the solution to $(a + b)^2$, we can draw a square with each side having a length of $a + b$. In this diagram, we have drawn *a* larger than *b* just to make the diagram a little easier to understand.

The area of the square represents multiplying the length by the width, which in this case is $(a + b)^2$. We can represent the area by four quadrilaterals.

① → $a \times a = a^2$
② → $a \times b = ab$
③ → $a \times b = ab$
④ → $b \times b = b^2$

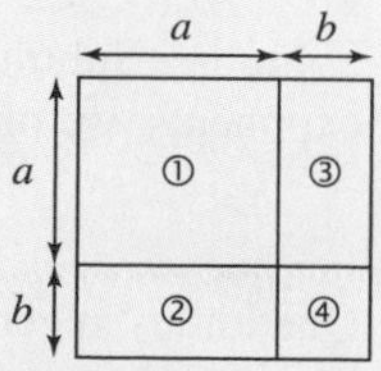

Adding the four areas together we get $a^2 + 2ab + b^2$ for the total area, so we can say that:

$$(a + b)^2 = a^2 + 2ab + b^2$$

FIGURE 14–1

Can you use a similar geometric approach to develop an equation for $(a + b)^3$?

BEING RESOURCEFUL IN HUMAN RESOURCES—ELYSE PAYNE

Elyse Payne works as a human resources generalist, supporting employees at a Crown corporation. Though she didn't expect it, using math is a daily part of her job.

Each day Elyse finds herself doing basic math like adding, subtracting, and calculating percentages. However, for her, the most surprising use of her math skills has come from her job's heavy use of Microsoft Excel® . She says, "I use the Formula bar in Excel daily. It calls upon my math skills because I have to think of how the calculation will come out before I make entries or create functions."

While Elyse did well in her math courses throughout high school, university, and college, it wasn't clear how the subject could eventually apply to her future career aspects. However, now that she is out of school and in the workforce, she believes that education helps contribute to her success. "It's because of my math skills that I can use an application like Excel® easily and successfully, making me a more productive and valuable employee to my employer."

14–1 Solving Quadratics by Factoring

Terminology

A polynomial equation of second degree is called a *quadratic equation*. It is common practice to refer to it simply as a *quadratic*.

Recall that in a polynomial, all powers of x are positive integers.

♦♦♦ **Example 1:** The following equations are quadratic equations:

(a) $4x^2 - 5x + 2 = 0$ (b) $x^2 = 58$

(c) $9x^2 - 5x = 0$ (d) $2x^2 - 7 = 0$

Equation (a) is called a *complete* quadratic; (b) and (d), which have no x terms, are called *pure* quadratics; and (c), which has no constant term, is called an *incomplete* quadratic. ♦♦♦

A quadratic is in *general form* when it is written in the following form, where a, b, and c are constants:

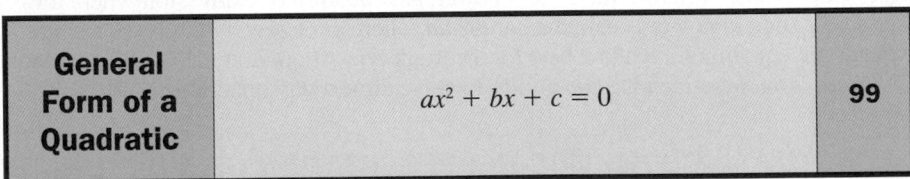

General Form of a Quadratic	$ax^2 + bx + c = 0$	99

♦♦♦ **Example 2:** Write the quadratic equation

$$7 - 4x = \frac{5x^2}{3}$$

in general form, and identify a, b, and c.

Solution: Subtracting $5x^2/3$ from both sides and writing the terms in descending order of the exponents, we obtain

$$-\frac{5x^2}{3} - 4x + 7 = 0$$

Quadratics in general form are usually written without fractions and with the first term positive. Multiplying by -3, we get

$$5x^2 + 12x - 21 = 0$$

The equation is now in general form, with $a = 5$, $b = 12$, and $c = -21$. ◆◆◆

Number of Roots

We'll see later in this chapter that the maximum number of solutions that certain types of equations may have is equal to the degree of the equation. Thus a quadratic equation, being of degree 2, has *two solutions* or *roots*. The two roots are sometimes equal, or they may be imaginary or complex numbers. However, in applications, we'll see that one of the two roots must sometimes be discarded.

◆◆◆ **Example 3:** The quadratic equation

$$x^2 = 4$$

has two roots, $x = 2$ and $x = -2$. ◆◆◆

Solving Pure Quadratics

To solve a *pure* quadratic, we simply isolate the x^2 term and then take the square root of both sides, as in the following example.

◆◆◆ **Example 4:** Solve $3x^2 - 75 = 0$.

Solution: Adding 75 to both sides and dividing by 3, we obtain

$$3x^2 = 75$$
$$x^2 = 25$$

Taking the square root yields

$$x = \pm \sqrt{25} = \pm 5$$

Check: We check our solution the same way as for other equations, by substituting back into the original equation. Now, however, we must check two solutions.

$$\text{Substitute } +5: \quad 3(5)^2 - 75 = 0$$
$$75 - 75 = 0 \quad \text{(checks)}$$
$$\text{Substitute } -5: \quad 3(-5)^2 - 75 = 0$$
$$75 - 75 = 0 \quad \text{(checks)} \qquad ◆◆◆$$

> When taking the square root, be sure to keep both the plus *and* the minus values. *Both* will satisfy the equation.

At this point students usually grumble: "First we're told that $\sqrt{4} = +2$ only, not ± 2 (Sec. 1–5). *Now* we're told to keep both plus and minus. What's going on here?" Here's the difference: When we solve a quadratic, we know that it must have two roots, so we keep both values. When we evaluate a square root, such as $\sqrt{4}$, which is not the solution of a quadratic, we agree to take only the positive value to avoid ambiguity.

The roots might sometimes be irrational, as in the following example.

◆◆◆ **Example 5:** Solve $4x^2 - 15 = 0$.

Solution: Following the same steps as before, we get

$$4x^2 = 15$$
$$x^2 = \frac{15}{4}$$
$$x = \pm \sqrt{\frac{15}{4}} = \pm \frac{\sqrt{15}}{2} \qquad ◆◆◆$$

Solving Incomplete Quadratics

To solve an *incomplete* quadratic, remove the common factor x from each term, and set each factor equal to zero.

◆◆◆ **Example 6:** Solve

$$x^2 + 5x = 0$$

Solution: Factoring yields

$$x(x + 5) = 0$$

Note that this expression will be true if either or both of the two factors equal zero. We therefore set each factor in turn equal to zero.

$$x = 0 \quad x + 5 = 0$$
$$x = -5$$

The two solutions are thus $x = 0$ and $x = -5$.   ◆◆◆

We use this idea often in this chapter. If we have the product of two quantities a and b set equal to zero, $ab = 0$, this equation will be true if $a = 0$ ($0 \cdot b = 0$) or if $b = 0$ ($a \cdot 0 = 0$), or if both are zero ($0 \cdot 0 = 0$).

Common Error	Do not cancel an x from the terms of an incomplete quadratic. That will cause a root to be lost. In the last example, if we had said $$x^2 = -5x$$ and had divided by x, $$x = -5$$ we would have obtained the correct root $x = -5$ but would have lost the root $x = 0$.

Solving Complete Quadratics

We now consider a quadratic that has all of its terms in place: the *complete* quadratic.

General Form	$ax^2 + bx + c = 0$	99

First write the quadratic in general form, as given in Eq. 99. Factor the trinomial (if possible) by the methods of Chapter 8, and set each factor equal to zero.

◆◆◆ **Example 7:** Solve by factoring:

$$x^2 - x - 6 = 0$$

Solution: Factoring gives

$$(x - 3)(x + 2) = 0$$

This equation will be satisfied if either or both of the two factors $(x - 3)$ and $(x + 2)$ are zero. We therefore set each factor in turn to equal zero.

$$x - 3 = 0 \quad x + 2 = 0$$

so the roots are

$$x = 3 \quad \text{and} \quad x = -2$$ ◆◆◆

Common Error	When the product of two quantities is zero, as in $$(x - 3)(x - 2) = 0$$ we can set each factor equal to zero, getting $x - 3 = 0$ and $x + 2 = 0$. But *this is valid only when the product is zero*. Thus if $$(x - 3)(x + 2) = 5$$ we *cannot* say that $x - 3 = 5$ and $x + 2 = 5$.

Often an equation must first be simplified before factoring.

♦♦♦ **Example 8:** Solve for x:

$$x(x - 8) = 2x(x - 1) + 9$$

Solution: Removing parentheses gives

$$x^2 - 8x = 2x^2 - 2x + 9$$

Collecting terms, we get

$$x^2 + 6x + 9 = 0$$

Factoring yields

$$(x + 3)(x + 3) = 0$$

which gives the double root,

$$x = -3 \qquad\qquad ♦♦♦$$

Sometimes at first glance an equation will not look like a quadratic. The following example shows a fractional equation which, after simplification, turns out to be a quadratic.

♦♦♦ **Example 9:** Solve for x:

$$\frac{3x - 1}{4x + 7} = \frac{x + 1}{x + 7}$$

Solution: We start by multiplying both sides by the LCD, $(4x + 7)(x + 7)$. We get

$$(3x - 1)(x + 7) = (x + 1)(4x + 7)$$

or

$$3x^2 + 20x - 7 = 4x^2 + 11x + 7$$

Collecting terms gives

$$x^2 - 9x + 14 = 0$$

Factoring yields

$$(x - 7)(x - 2) = 0$$

so $x = 7$ and $x = 2$. ♦♦♦

Writing the Equation When the Roots Are Known
Given the roots, we simply reverse the process to find the equation.

◆◆◆ **Example 10:** Write the quadratic equation that has the roots $x = 2$ and $x = -5$.

Solution: If the roots are 2 and -5, we know that the factors of the equation must be $(x - 2)$ and $(x + 5)$. So

$$(x - 2)(x + 5) = 0$$

Multiplying gives us

$$x^2 + 5x - 2x - 10 = 0$$

So

$$x^2 + 3x - 10 = 0$$

is the original equation.   ◆◆◆

Solving Radical Equations

In Chapter 13 we solved *simple* radical equations. We isolated a radical on one side of the equation and then squared both sides. Here we solve equations in which this squaring operation results in a quadratic equation.

◆◆◆ **Example 11:** Solve for x:

$$3\sqrt{x - 1} - \frac{4}{\sqrt{x - 1}} = 4$$

Solution: We clear fractions by multiplying through by $\sqrt{x - 1}$.

$$3(x - 1) - 4 = 4\sqrt{x - 1}$$
$$3x - 7 = 4\sqrt{x - 1}$$

Squaring both sides yields

$$9x^2 - 42x + 49 = 16(x - 1)$$

Removing parentheses and collecting terms gives

$$9x^2 - 58x + 65 = 0$$

Factoring gives

$$(x - 5)(9x - 13) = 0$$
$$x = 5 \quad \text{and} \quad x = \frac{13}{9}$$

Check: When $x = 5$,

$$3\sqrt{5 - 1} - \frac{4}{\sqrt{5 - 1}} \stackrel{?}{=} 4$$
$$3(2) - \frac{4}{2} = 4 \quad \text{(checks)}$$

When $x = \frac{13}{9}$,

$$\sqrt{\frac{13}{9} - 1} = \sqrt{\frac{4}{9}} = \frac{2}{3}$$

So

$$3 \cdot \frac{2}{3} - \frac{4}{\frac{2}{3}} \stackrel{?}{=} 4$$
$$2 - 6 \neq 4 \quad \text{(does not check)}$$

Our solution is then $x = 5$.   ◆◆◆

Remember that the squaring operation sometimes gives an extraneous root that will not check.

Exercise 1 • Solving Quadratics by Factoring

Pure and Incomplete Quadratics

Solve for x.

1. $2x = 5x^2$
2. $2x - 40x^2 = 0$
3. $3x(x - 2) = x(x + 3)$
4. $2x^2 - 6 = 66$
5. $5x^2 - 3 = 2x^2 + 24$
6. $7x^2 + 4 = 3x^2 + 40$
7. $(x + 2)^2 = 4x + 5$
8. $5x^2 - 2 = 3x^2 + 6$
9. $8.25x^2 - 2.93x = 0$
10. $284x = 827x^2$

Complete Quadratics

Solve for x.

11. $x^2 + 2x - 15 = 0$
12. $x^2 + 6x - 16 = 0$
13. $x^2 - x - 20 = 0$
14. $x^2 + 13x + 42 = 0$
15. $x^2 - x - 2 = 0$
16. $x^2 + 7x + 12 = 0$
17. $2x^2 - 3x - 5 = 0$
18. $4x^2 - 10x + 6 = 0$
19. $2x^2 + 5x - 12 = 0$
20. $3x^2 - x - 2 = 0$
21. $5x^2 + 14x - 3 = 0$
22. $5x^2 + 3x - 2 = 0$
23. $\dfrac{2 - x}{3} = 2x^2$
24. $(x - 6)(x + 6) = 5x$
25. $x(x - 5) = 36$
26. $(2x - 3)^2 = 2x - x^2$
27. $x^2 - \dfrac{5}{6}x = \dfrac{1}{6}$
28. $\dfrac{7}{x + 4} - \dfrac{1}{4 - x} = \dfrac{2}{3}$

There are no applications in the first few exercise sets. These will come a little later.

Literal Equations

Solve for x.

29. $4x^2 + 16ax + 12a^2 = 0$
30. $14x^2 - 23ax + 3a^2 = 0$
31. $9x^2 + 30bx + 24b^2 = 0$
32. $24x^2 - 17xy + 3y^2 = 0$

Writing the Equation from the Roots

Write the quadratic equations that have the following roots.

33. $x = 4$ and $x = 7$
34. $x = -3$ and $x = -5$
35. $x = \dfrac{2}{3}$ and $x = \dfrac{3}{5}$
36. $x = p$ and $x = q$

Radical Equations

Solve each radical equation for x, and check.

37. $\sqrt{5x^2 - 3x - 41} = 3x - 7$
38. $3\sqrt{x - 1} - \dfrac{4}{\sqrt{x - 1}} = 4$
39. $\sqrt{x + 5} = \dfrac{12}{\sqrt{x + 12}}$
40. $\sqrt{7x + 8} - \sqrt{5x - 4} = 2$

14–2   Solving Quadratics by Completing the Square

If a quadratic equation is not factorable, it is possible to manipulate it into factorable form by a procedure called *completing the square*. The form into which we shall put our expression is the *perfect square trinomial*, Eqs. 47 and 48, which we studied in Sec. 8–6.

In the perfect square trinomial,

1. The first and last terms are perfect squares.
2. The middle term is twice the product of the square roots of the outer terms.

To complete the square, we manipulate our given expression so that these two conditions are met. This is best shown by an example.

The method of *completing the square* is really too cumbersome to be a practical tool for solving quadratics. The main reason we learn it is to derive the *quadratic formula*. Furthermore, the method of completing the square is a useful technique that we'll use again in later chapters.

♦♦♦ **Example 12:** Solve the quadratic $x^2 - 8x + 6 = 0$ by completing the square.

Solution: Subtracting 6 from both sides, we obtain

$$x^2 - 8x = -6$$

We complete the square by *adding the square of half the coefficient of the x term* to both sides. The coefficient of x is -8. We take half of -8 and square it, getting $(-4)^2$ or 16. Adding 16 to both sides yields

$$x^2 - 8x + 16 = -6 + 16 = 10$$

Factoring, we have

$$(x - 4)^2 = 10$$

Taking the square root of both sides, we obtain

$$x - 4 = \pm \sqrt{10}$$

Finally, we add 4 to both sides.

$$x = 4 \pm \sqrt{10}$$
$$\cong 7.16 \quad \text{or} \quad 0.838 \qquad\qquad ♦♦♦$$

> **Common Error**
>
> When you are adding the quantity needed to complete the square to the left-hand side, it's easy to forget to add the same quantity to the right-hand side.
>
> $$x^2 - 8x \boxed{+ 16} = -6 \boxed{+ 16}$$
>
> don't forget

If the x^2 term has a coefficient other than 1, divide through by this coefficient *before* completing the square.

♦♦♦ **Example 13:** Solve:

$$2x^2 + 4x - 3 = 0$$

Solution: Rearranging and dividing by 2 gives

$$x^2 + 2x = \frac{3}{2}$$

Completing the square by adding 1 (half the coefficient of the x term, squared) to both sides of the equation, we have

$$x^2 + 2x + 1 = \frac{3}{2} + 1$$

$$(x + 1)^2 = \frac{5}{2}$$

$$x + 1 = \pm\sqrt{\frac{5}{2}} = \frac{\pm\sqrt{10}}{2}$$

$$x = -1 \pm \frac{1}{2}\sqrt{10}$$

◆◆◆

Common Error	Don't forget to divide through by the coefficient of the x^2 term *before* completing the square.

Exercise 2 ◆ Solving Quadratics by Completing the Square

Solve each quadratic by completing the square.

1. $x^2 - 8x + 2 = 0$ **2.** $x^2 + 4x - 9 = 0$

3. $x^2 + 7x - 3 = 0$ **4.** $x^2 - 3x - 5 = 0$

5. $4x^2 - 3x - 5 = 0$ **6.** $2x^2 + 3x - 3 = 0$

7. $3x^2 + 6x - 4 = 0$ **8.** $5x^2 - 2x - 1 = 0$

9. $4x^2 + 7x - 5 = 0$ **10.** $8x^2 - 4x - 3 = 0$

11. $x^2 + 6.25x - 3.27 = 0$ **12.** $2.45x^2 + 6.88x - 4.17 = 0$

Quadratics having noninteger coefficients can also be solved by completing the square. The method is exactly the same, although the work is harder.

14–3 Solving Quadratics by Formula

Of the several methods we have for solving quadratics, the most useful is the quadratic formula. It will work for any quadratic, regardless of the type of roots, and it can easily be programmed for the computer.

Derivation of the Quadratic Formula

We wish to find the roots of the equation

$$ax^2 + bx + c = 0$$

by completing the square. We start by subtracting c from both sides and dividing by a.

$$x^2 + \frac{b}{a}x = -\frac{c}{a}$$

Completing the square gives

$$x^2 + \frac{b}{a}x + \left(\frac{b}{2a}\right)^2 = \frac{b^2}{4a^2} - \frac{c}{a}$$

We're assuming here that the coefficient a is not zero.

Factoring, we obtain

$$\left(x + \frac{b}{2a}\right)^2 = \frac{b^2 - 4ac}{4a^2}$$

Taking the square root yields

$$x + \frac{b}{2a} = \pm\frac{\sqrt{b^2 - 4ac}}{2a}$$

Rearranging, we get the following well-known formula for finding the roots of the quadratic equation, $ax^2 + bx + c = 0$:

Quadratic Formula	$x = \dfrac{-b \pm \sqrt{b^2 - 4ac}}{2a}$	100

Using the Quadratic Formula

Simply put the given equation into general form (Eq. 99); list a, b, and c; and substitute them into the formula.

◆◆◆ **Example 14:** Solve $2x^2 - 5x - 3 = 0$ by the quadratic formula.

Solution: The equation is already in general form, with

$$a = 2 \qquad b = -5 \qquad c = -3$$

Substituting into Eq. 100, we obtain

$$x = \frac{-(-5) \pm \sqrt{(-5)^2 - 4(2)(-3)}}{2(2)}$$

$$= \frac{5 \pm \sqrt{25 + 24}}{4}$$

$$= \frac{5 \pm \sqrt{49}}{4} = \frac{5 \pm 7}{4}$$

$$= 3 \quad \text{and} \quad -\frac{1}{2}$$

◆◆◆

Common Error	Always rewrite a quadratic in general form before trying to use the quadratic formula.

◆◆◆ **Example 15:** Solve the equation $5.25x - 2.94x^2 + 6.13 = 0$ by the quadratic formula.

Solution: The constants a, b, and c are *not* 5.25, −2.94, and 6.13, as you might think at first glance. We must rearrange the terms.

$$-2.94x^2 + 5.25x + 6.13 = 0$$

Although not a necessary step, dividing through by the coefficient of x^2 simplifies the work a bit.

$$x^2 - 1.79x - 2.09 = 0$$

Substituting into the quadratic formula with $a = 1$, $b = -1.79$, and $c = -2.09$, we have

$$x = \frac{1.79 \pm \sqrt{(-1.79)^2 - 4(1)(-2.09)}}{2(1)}$$

$$= 2.60 \quad \text{and} \quad -0.805$$

◆◆◆

We now have the tools to finish the falling-body problem started in the introduction to this chapter.

◆◆◆ **Example 16:** Solve the equation

$$30 = 5t + 4.9t^2$$

(where t is in seconds) to three significant digits.

Solution: We first go to general form.

$$4.9t^2 + 5t - 30 = 0$$

Substituting into the quadratic formula gives us

$$t = \frac{-5 \pm \sqrt{25 - 4(4.9)(-30)}}{2(4.9)}$$

$$= \frac{-5 \pm \sqrt{613}}{9.8} = \frac{-5 \pm 24.8}{9.8}$$

$$\simeq 2.02 \text{ s} \quad \text{and} \quad -3.04 \text{ s}$$

In this problem, a negative elapsed time makes no sense, so we discard the negative root. Thus it will take 2.02 s for an object to fall 30 m when thrown downward with a speed of 5 m/s. ◆◆◆

Nonreal Roots

Quadratics resulting from practical problems will usually yield real roots if the equation is properly set up. Other quadratics, however, could give *nonreal roots,* as shown in the next example.

◆◆◆ **Example 17:** Solve:

$$x^2 + 2x + 5 = 0$$

Solution: From Eq. 100,

$$x = \frac{-2 \pm \sqrt{4 - 4(1)(5)}}{2} = \frac{-2 \pm \sqrt{-16}}{2}$$

The expression $\sqrt{-16}$ can be written $\sqrt{-1}\sqrt{16}$ or $j4$, where $j = \sqrt{-1}$. So

$$x = \frac{-2 \pm j4}{2} = -1 \pm j2 \qquad ◆◆◆$$

> See Chapter 21 for a complete discussion of imaginary and complex numbers.

Predicting the Nature of the Roots

As shown by Example 17, when the quantity in Eq. 100 under the radical sign ($b^2 - 4ac$) is negative, the roots are not real. It should also be evident that if $b^2 - 4ac$ is zero, the roots will be equal. Thus we can use the quantity $b^2 - 4ac$, called the *discriminant,* to predict what kind of roots an equation will give, without having to actually find the roots.

The Discriminant	If a, b, and c are real, and if $b^2 - 4ac > 0$, the roots are real and unequal. if $b^2 - 4ac = 0$, the roots are real and equal. if $b^2 - 4ac < 0$, the roots are not real.	**101**

◆◆◆ **Example 18:** What sort of roots can you expect from the equation

$$3x^2 - 5x + 7 = 0?$$

Solution: Computing the discriminant, we obtain

$$b^2 - 4ac = (-5)^2 - 4(3)(7) = 25 - 84 = -59$$

A negative discriminant tells us that the roots are not real. ◆◆◆

> Do not lose too much sleep over the discriminant. If you need to know the nature of the roots of a quadratic and cannot remember Eq. 101, all you have to do is calculate the roots themselves. This is not much more work than finding the discriminant.

Exercise 3 ◆ Solving Quadratics by Formula

Solve by quadratic formula to three significant digits.

1. $x^2 - 12x + 28 = 0$
2. $x^2 - 6x + 7 = 0$
3. $x^2 + x - 19 = 0$
4. $x^2 - x - 13 = 0$
5. $3x^2 + 12x - 35 = 0$
6. $29.4x^2 - 48.2x - 17.4 = 0$
7. $36x^2 + 3x - 7 = 0$
8. $28x^2 + 29x + 7 = 0$
9. $49x^2 + 21x - 5 = 0$
10. $16x^2 - 16x + 1 = 0$
11. $3x^2 - 10x + 4 = 0$
12. $x^2 - 34x + 22 = 0$
13. $3x^2 + 5x = 7$
14. $4x + 5 = x^2 + 2x$
15. $x^2 - 4 = 4x + 7$
16. $x^2 - 6x - 14 = 3$
17. $6x - 300 = 205 - 3x^2$
18. $3x^2 - 25x = 5x - 73$
19. $2x^2 + 100 = 32x - 11$
20. $33 - 3x^2 - 10x = 0$
21. $4.26x + 5.74 = 1.27x^2 + 2.73x$
22. $1.83x^2 - 4.26 = 4.82x + 7.28$
23. $x^2 - 6.27x - 14.4 = 3.17$
24. $6.47x - 338 = 205 - 3.73x^2$
25. $x(2x - 3) = 3x(x + 4) - 2$
26. $2.95(x^2 + 8.27x) = 7.24x(4.82x - 2.47) + 8.73$
27. $(4.20x - 5.80)(7.20x - 9.20) = 8.20x + 9.90$
28. $(2x - 1)^2 + 6 = 6(2x - 1)$

Nature of the Roots

Determine whether the roots of each quadratic are real or nonreal and whether they are equal or unequal.

29. $x^2 - 5x - 11 = 0$ 30. $x^2 - 6x + 15 = 0$
31. $2x^2 - 4x + 7 = 0$ 32. $2x^2 + 3x - 2 = 0$
33. $3x^2 - 3x + 5 = 0$ 34. $4x^2 + 4x + 1 = 0$

14–4   Applications and Word Problems

Now that we have the tools to solve any quadratic, let us go on to problems from technology that require us to solve these equations. At this point you might want to take a quick look at Chapter 3 and review some of the suggestions for setting up and solving word problems. You should set up these problems just as you did then.

When you solve the resulting quadratic, you will get two roots, of course. If one of the roots does not make sense in the physical problem (such as a beam having a length of −2000 ft), throw it away. But do not be too hasty. Often a second root will give an unexpected but good answer.

◆◆◆ **Example 19:** The angle iron in Fig. 14–2 has a cross-sectional area of 53.4 cm². Find the thickness x.

Estimate: Let's assume that we have two rectangles of width x, with lengths of 15.6 cm and 10.4 cm. Setting their combined area equal to 53.4 cm² gives

$$x(15.6 + 10.4) = 53.4$$

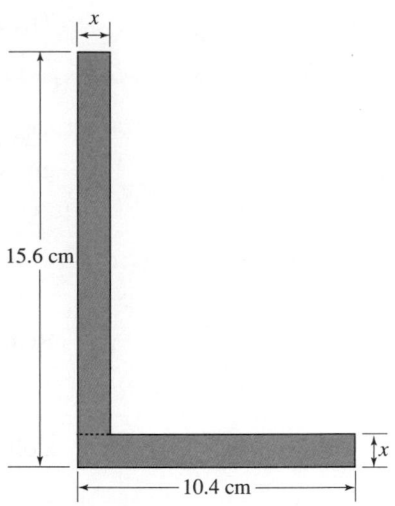

FIGURE 14–2

Thus we get the approximation $x = 2.05$ cm. But since our assumed lengths were too great, because we have counted the small square in the corner twice, our estimated value of x must be too small. We thus conclude that $x > 2.05$ cm.

Solution: We divide the area into two rectangles, as shown by the dashed line at the bottom of Fig. 14–2. One rectangle has an area of $10.4x$, and the other has an area of $(15.6 − x)x$. Since the sum of these areas must be 53.4,

$$10.4x + (15.6 − x)x = 53.4$$

Putting this equation into standard form, we get

$$10.4x + 15.6x − x^2 = 53.4$$
$$x^2 − 26.0x + 53.4 = 0$$

By the quadratic formula,

$$x = \frac{26.0 \pm \sqrt{676 − 4(53.4)}}{2} = \frac{26.0 \pm 21.5}{2}$$

So

$$x = \frac{26.0 − 21.5}{2} = 2.25 \text{ cm}$$

and

$$x = \frac{26.0 + 21.5}{2} = 23.8 \text{ cm}$$

We discard 23.8 cm because it is an impossible solution in this problem.

Check: Does our answer meet the requirements of the original statement? Let us compute the area of the angle iron using our value of 2.25 cm for the thickness. We get

$$\text{area} = 2.25(13.35) + 2.25(10.4)$$
$$= 30.0 + 23.4 = 53.4 \text{ cm}^2$$

which is the required area. Our answer is also a little bigger than our 2.05 cm estimate, as expected. ◆◆◆

◆◆◆ **Example 20:** A certain train is to be replaced with a "bullet" train that goes 40.0 km/h faster than the old train and that will make its regular $85\overline{0}$-km run in 3.00 h less time. Find the speed of each train.

Solution: Let

$$x = \text{rate of old train} \quad \text{(km/h)}$$

Then

$$x + 40.0 = \text{rate of bullet train} \quad \text{(km/h)}$$

The time it takes the old train to travel 850 km at x km/h is, by Eq. A17,

$$\text{time} = \frac{\text{distance}}{\text{rate}} = \frac{850}{x} \quad \text{(h)}$$

The time for the bullet train is then $(850/x - 3.00)$ h. Applying Eq. A17 for the bullet train gives us

$$\text{rate} \times \text{time} = \text{distance}$$

$$(x + 40.0)\left(\frac{850}{x} - 3.00\right) = 850$$

Removing parentheses, we have

$$850 - 3.00x + \frac{34\,000}{x} - 120 = 850$$

Collecting terms and multiplying through by x gives

$$-3.00x^2 + 34\,000 - 120x = 0$$

or

$$x^2 + 40.0x - 11\,333 = 0$$

Solving for x by the quadratic formula yields

$$x = \frac{-40.0 \pm \sqrt{1600 - 4(-11\,333)}}{2}$$

If we drop the negative root, we get

$$x = \frac{-40.0 + 217}{2} = 88.3 \text{ km/h} = \text{speed of old train}$$

and

$$x + 40.0 = 128.3 \text{ km/h} = \text{speed of bullet train} \qquad \blacklozenge\blacklozenge\blacklozenge$$

Exercise 4 ◆ Applications and Word Problems

Number Problems

1. What fraction added to its reciprocal gives $2\frac{1}{6}$?
2. Find three consecutive numbers such that the sum of their squares will be 434.
3. Find two numbers whose difference is 7 and the difference of whose cubes is 1267.
4. Find two numbers whose sum is 11 and whose product is 30.
5. Find two numbers whose difference is 10 and the sum of whose squares is 250.
6. A number increased by its square is equal to 9 times the next higher number. Find the number.

The "numbers" in problems 2 through 6 are all positive integers.

Geometry Problems

3 cm

3 cm

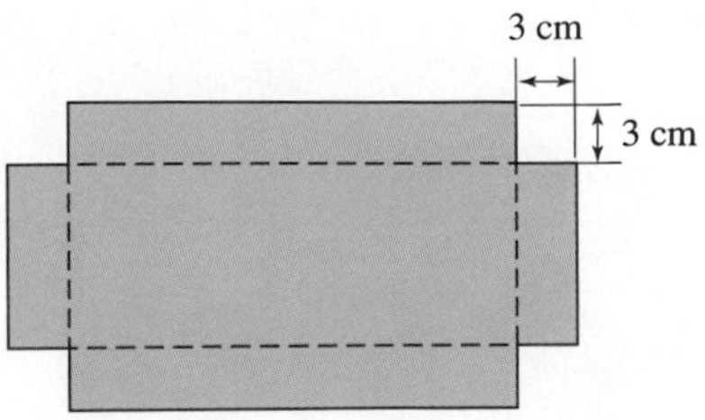

FIGURE 14–3

7. A rectangle is to be 2 m longer than it is wide and have an area of 24 m². Find its dimensions.
8. One leg of a right triangle is 3 cm greater than the other leg, and the hypotenuse is 15 cm. Find the legs of the triangle.
9. A rectangular sheet of brass is twice as long as it is wide. Squares, 3 cm × 3 cm, are cut from each corner (Fig. 14–3), and the ends are turned up to form an open box having a volume of 648 cm³. What are the dimensions of the original sheet of brass?

10. The length, width, and height of a cubical shipping container are all decreased by 1.0 ft.; thereby decreasing the volume of the cube by 37 ft.³. What was the volume of the original container?

11. Find the dimensions of a rectangular field that has a perimeter of 724 m and an area of 32,400 m².

12. A flat of width w is to be cut on a bar of radius r (Fig. 14–4). Show that the required depth of cut x is given by the formula

$$x = r \pm \sqrt{r^2 - \frac{w^2}{4}}$$

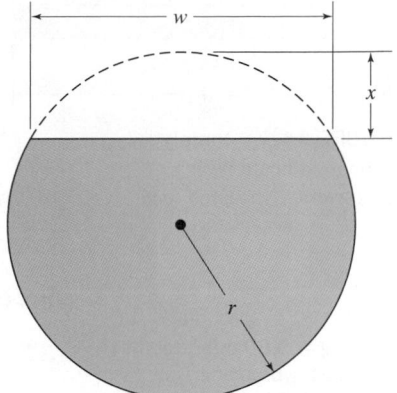

FIGURE 14–4

13. A casting in the shape of a cube is seen to shrink 0.175 cm on a side, with a reduction in volume of 2.48 cm³. Find the original dimensions of the cube.

14. The cylinder in Fig. 14–5 has a surface area of 846 cm², including the ends. Find its radius.

15. A cylindrical tank having a diameter of 75.5 cm is placed so that it touches a wall, as in Fig. 14–6. Find the radius x of the largest pipe that can fit into the space between the tank, the wall, and the floor. *Hint:* First use the Pythagorean theorem to show that $OC = 53.4$ cm and that $OP = 15.6$ cm. Then in triangle OQR, $(OP - x)^2 = x^2 + x^2$.

Uniform Motion

16. A truck travels $35\overline{0}$ km to a delivery point, unloads, and, now empty, returns to the starting point at a speed 8.00 km/h greater than on the outward trip. What was the speed of the outward trip if the total round-trip driving time was 14.4 h?

17. An airplane flies 355 miles to city A. Then, with better winds, it continues on to city B, 448 miles from A, at a speed 15.8 mi./h greater than on the first leg of the trip. The total flying time was 5.20 h. Find the speed at which the plane travelled to city A.

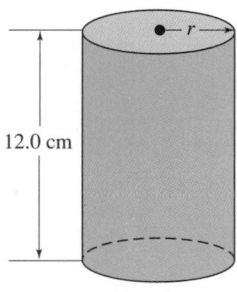

FIGURE 14–5

18. An express bus travels a certain 320-km route in 1.0 h less time than it takes a local bus to travel a $30\overline{0}$-km route. Find the speed of each bus if the speed of the local is $2\overline{0}$ km/h less than that of the express.

19. A trucker calculates that if he increased his average speed by $2\overline{0}$ km/h, he could travel his $8\overline{0}0$-km route in 2.0 h less time than usual. Find his usual speed.

20. A boat sails 30 km at a uniform rate. If the rate had been 1 km/h more, the time of the sailing would have been 1 h less. Find the rate of travel.

Work Problems

21. A certain punch press requires 3 h longer to stamp a box of parts than does a newer-model punch press. After the older press has been punching a box of parts for 5 h, it is joined by the newer machine. Together, they finish the box of parts in 3 additional hours. How long does it take each machine, working alone, to punch a box of parts?

22. Two water pipes together can fill a certain tank in 8.40 h. The smaller pipe alone takes 2.50 h longer than the larger pipe to fill that same tank. How long would it take the larger pipe alone to fill the tank?

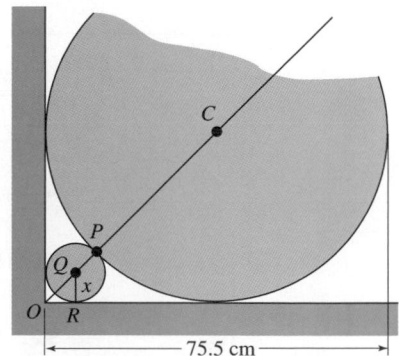

FIGURE 14–6

23. A labourer built 35 m of stone wall. If she had built 2 m less each day, it would have taken her 2 days longer. How many metres did she build each day, working at her usual rate?

24. A woman worked part-time a certain number of days, receiving for her pay $1,800. If she had received $10 per day less than she did, she would have had to work 3 days longer to earn the same sum. How many days did she work?

Simply Supported Beam

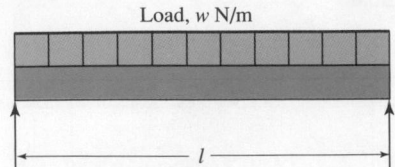

Load, w N/m

FIGURE 14–7 Simply supported beam with a uniformly distributed load.

25. For a simply supported beam of length l having a distributed load of w N/m (Fig. 14–7), the bending moment M at any distance x from one end is given by

$$M = \frac{1}{2}wlx - \frac{1}{2}wx^2$$

Find the locations on the beam where the bending moment is zero.

26. A simply supported beam, 12.5 m long, carries a distributed load of 15.5 kN/m. At what distances from an end of the beam will the bending moment be 112 kN? (Use the equation from problem 25.)

Freely Falling Body

Use $s = v_0t + \frac{1}{2}gt^2$, Eq. A18, where $g = 9.80$ m/s², for these falling-body problems, but be careful of the signs. If you take the upward direction as positive, g will be *negative*.

27. An object is thrown upward with a velocity of 45.0 m/s. When will it be 25.0 m above its initial position?

28. An object is thrown upward with an initial speed of $12\overline{0}$ m/s. Find the time for it to return to its starting point.

Electrical Problems

29. Referring to Fig. 14–8, determine what two resistances will give a total resistance of $78\overline{0}$ Ω when wired in series and 105 Ω when wired in parallel. (See Eqs. A63 and A64.)

Eq. A63: $R_T = R_1 + R_2 + R_3\ldots$

Eq. A64: $\dfrac{1}{R_T} = \dfrac{1}{R_1} + \dfrac{1}{R_2} + \dfrac{1}{R_3} + \ldots$

30. Find two resistances that will give an equivalent resistance of 9070 Ω in series and 1070 Ω in parallel.

31. In the circuit of Fig. 14–9, the power P dissipated in the load resistor R_1 is

$$P = EI - I^2R$$

If the voltage E is 115 V, and if $R = 10\overline{0}$ Ω, find the current I needed to produce a power of 29.3 W in the load.

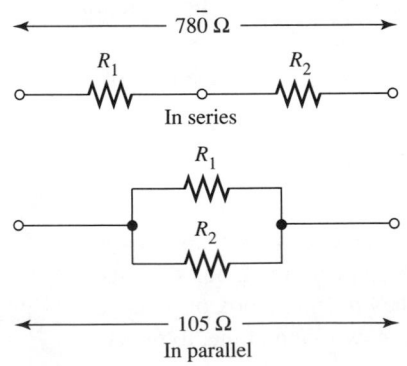

FIGURE 14–8

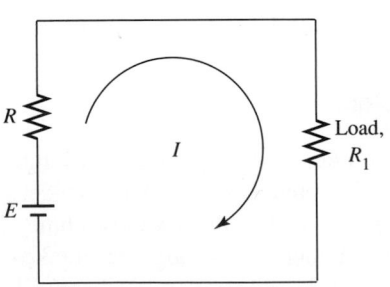

FIGURE 14–9

32. The reactance X of a capacitance C and an inductance L in series (Fig. 14–10) is

$$X = \omega L - \frac{1}{\omega C}$$

FIGURE 14–10

where ω is the angular frequency in rad/s, L is in henries (H), and C is in farads (F). Find the angular frequency needed to make the reactance equal to 1500 Ω, if $L = 0.5$ H and $C = 0.2 \times 10^{-6}$ F (0.2 μF).

33. Figure 14–11 shows two currents flowing in a single resistor R. The total current in the resistor will be $I_1 + I_2$, so the power dissipated is

$$P = (I_1 + I_2)^2 R$$

If $R = 100$ Ω and $I_2 = 0.2$ A, find the current I_1 needed to produce a power of 9.0 W.

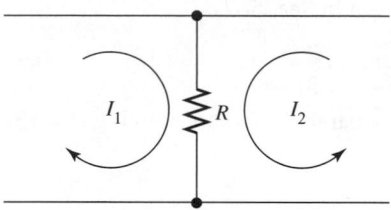

FIGURE 14–11

34. A *square-law device* is one whose output is proportional to the square of the input. A junction field-effect transistor (JFET) (Fig. 14–12) is such a device. The current I that will flow through an n-channel JFET when a voltage V is applied is

$$I = A\left(1 - \frac{V}{B}\right)^2$$

where A is the drain saturation current and B is the gate source pinch-off voltage. Solve this equation for V.

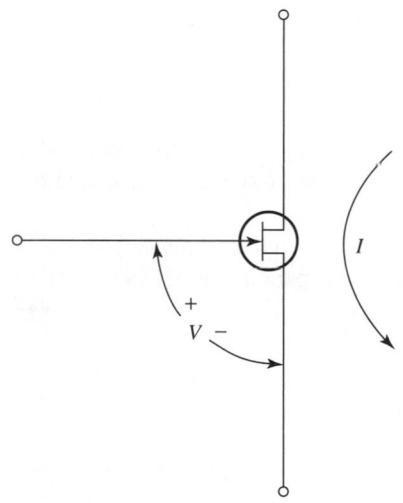

FIGURE 14–12

35. A certain JFET has a drain saturation current of 4.8 mA and a gate source pinch-off voltage of -2.5 V. What input voltage is needed to produce a current of 1.5 mA?

14–5 Graphing the Quadratic Function

Quadratic Functions
A *quadratic function* is one whose highest-degree term is of second degree.

◆◆◆**Example 21:** The following functions are quadratic functions:

(a) $f(x) = 5x^2 - 3x + 2$ (b) $f(x) = 9 - 3x^2$

(c) $f(x) = x(x + 7)$ (d) $f(x) = x - 4 - 3x^2$ ◆◆◆

The Parabola

When we plot a quadratic function, we get the well-known and extremely useful curve called the *parabola*.

◆◆◆ **Example 22:** Plot the quadratic function $y = x^2 + x - 3$ for $x = -3$ to $x = 3$.

Solution: We can graph the function using a graphics calculator or graphing software on a computer, or by computing a table of point pairs as was shown in Sec. 5–2.

x	-3	-2	-1	0	1	2
y	3	-1	-3	-3	-1	3

We plot these points and connect them with a smooth curve, a parabola, as shown in Fig. 14–13.

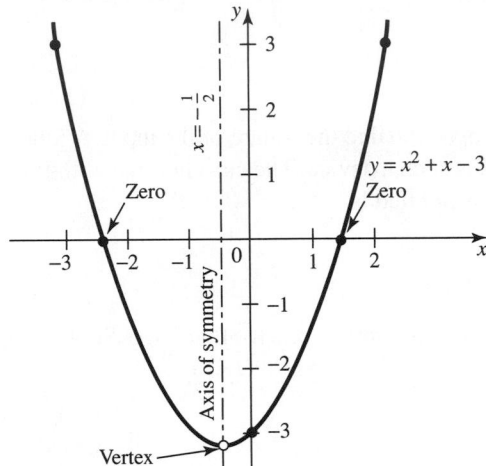

FIGURE 14–13

Note that the curve in this example is symmetrical about the line $x = -\frac{1}{2}$. This line is called the *axis of symmetry*. The point where the parabola crosses the axis of symmetry is called the *vertex*.

Recall that in Chapter 5 we said that a low point in a curve is called a *minimum point,* and that a high point is called a *maximum point.* Thus the vertex of this parabola is also a minimum point.   ◆◆◆

Solving Quadratics Graphically

In Sec. 4–5 we saw that a graphical solution to the equation

$$f(x) = 0$$

could be obtained by plotting the function

$$y = f(x)$$

and locating the points where the curve crosses the x axis. Those points are called the *zeros* of the function. Those points are then the *solution, or roots,* of the equation. Notice that in Fig. 14–13 there are two zeros, corresponding to the two roots of the equation.

◆◆◆ **Example 23:** Graphically find the approximate roots of the equation

$$x^2 + x - 3 = 0$$

Solution: The function

$$y = x^2 + x - 3$$

is already plotted in Fig. 14–13. Reading the x intercepts as accurately as possible, we get for the roots

$$x \simeq -2.3 \quad \text{and} \quad x \simeq 1.3$$

◆◆◆

Exercise 5 ◆ Graphing the Quadratic Function

Plot the following quadratic functions. Label the vertex, any zeros, and the axis of symmetry.

1. $y = x^2 + 3x - 15$ **2.** $y = 5x^2 + 14x - 3$
3. $y = 8x^2 - 6x + 1$ **4.** $y = 2.74x^2 - 3.12x + 5.38$

Parabolic Arch

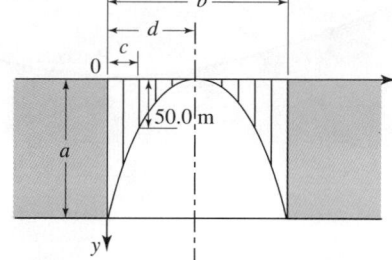

5. The parabolic arch is often used in construction because of its great strength. The equation of the bridge arch in Fig. 14–14 is $y = 0.0625x^2 - 5.00x + 100$. Find the distances a, b, c, and d.

6. Plot the parabolic arch of problem 5, taking values of x every 10 m.

FIGURE 14–14 A parabolic bridge arch. Note that the y axis is positive downward.

7. When firefighters fight a grass fire, they prefer to stand back from the edge of the fire and "lob" the water from the hose onto the fire. The stream of water is under high pressure, and the water can be airborne for some period of time. If the spray of the water leaving the hose is modelled by the equation $d = -1.5\,t^2 + 30$, where d represents the distance (in metres) between the hose and the fire and t represents the time (in seconds) elapsed since the water was sprayed (at a constant high pressure), determine how long the water remains in the air after leaving the nozzle.

8. When firefighters work on a residential fire in an older, established neighbourhood, they sometimes have to contend with mature trees as obstacles to spraying water onto the fire. The equation $Y = -x^2 + 9.7x$ models the height (Y) of the water spray leaving the hose, where x is the distance to the obstacle in metres. How tall a tree can the firefighters spray over from a distance of 4.85 m?

9. A firefighting hose and nozzle manufacturer is evaluating three new nozzle designs. The spray trajectory for each nozzle has been modelled and their associated equations are listed below:
Nozzle 1 : $Y = -0.6x^2 + 10x$
Nozzle 2 : $Y = -0.7x^2 + 12x$
Nozzle 3 : $Y = -0.5x^2 + 8x$
where Y represents the height of the water spray, and x represents the spray distance in metres. Graphically determine the maximum spray height and distance for each nozzle.

Parabolic Reflector

10. The parabolic solar collector in Fig. 14–15 has the equation $x^2 = 125y$. Plot the curve, taking values of x every 10 cm.

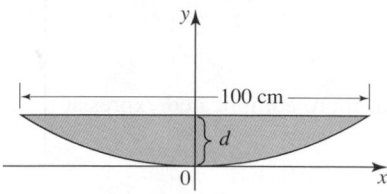

FIGURE 14–15

11. Find the depth d of the collector of problem 10 at its center.

Vertical Highway Curves

Where there is a change in the slope of a highway surface, such as at the top of a hill or the bottom of a dip, the roadway is often made parabolic in shape to provide a smooth transition between the two different grades.

A *parabolic surface* has the property that all of the incoming rays that are parallel to the axis of symmetry, after being reflected from the surface of the parabola, will converge to a single point called the *focus*. This property is used in optical and radio telescopes, solar collectors, searchlights, microwave and radar antennas, and other devices.

To construct a vertical highway curve (Fig. 14–16), we use the following property of the parabola:

The offset C from a tangent to a parabola is proportional to the square of the distance x from the point of tangency, or $C = kx^2$, where k is a constant of proportionality, found from a known offset.

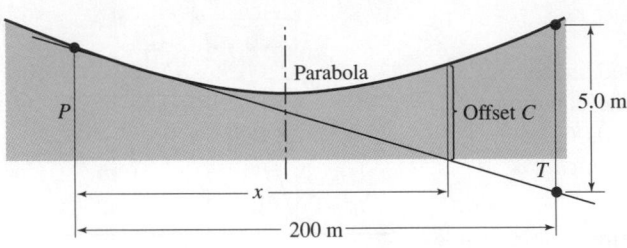

FIGURE 14–16 A dip in a highway.

12. If the offset is 5.00 m at a point $20\overline{0}$ m from the point of tangency, find the constant of proportionality k.
13. For the highway curve of problem 12, find the offset at a point 125 m from the point of tangency.

Square-Law Devices

14. For the *n*-channel JFET of problem 35 in Exercise 4, plot the curve of current I versus applied voltage V, using the vertical axis for current and the horizontal axis for voltage. Compute I for values of V from -2.5 V to 0 V. (The curve obtained is a parabola and is called the *transconductance curve*.) Graphically determine the voltage needed to produce a current of 1.00 mA.

14–6   Equations of Quadratic Type

Recognition of Form

The general quadratic formula, Eq. 99, can be written

$$a(x)^2 + b(x) + c = 0$$

The form of this equation obviously does not change if we put a different symbol in the parentheses, as long as we put the same one in both places. The equations

$$a(Q)^2 + b(Q) + c = 0$$

and

$$a(\$)^2 + b(\$) + c = 0$$

are still quadratics. The form still does not change if we put more complicated expressions in the parentheses, such as

$$a(x^2)^2 + b(x^2) + c = 0$$

or

$$a(\sqrt{x})^2 + b(\sqrt{x}) + c = 0$$

or

$$a(x^{-1/3})^2 + b(x^{-1/3}) + c = 0$$

The preceding three equations can no longer be called quadratics, however, because they are no longer of second degree. They are called *equations of quadratic type*. They can be easily recognized because the power of x in one term is always twice the power of x in another term (and there are no other terms containing powers of x).

Solving Equations of Quadratic Type

The solution of equations of quadratic type is best shown by examples.

◆◆◆ **Example 24:** Solve:

$$x^4 - 5.12x^2 + 4.08 = 0$$

Solution: We see that the power of x in the first term is twice the power of x in the second term. We make the substitution and let

$$w = x^2$$

Our original equation becomes the quadratic

$$w^2 - 5.12w + 4.08 = 0$$

Its roots are

$$w = \frac{-(-5.12) \pm \sqrt{(-5.12)^2 - 4(1)(4.08)}}{2(1)} = 4.133 \quad \text{and} \quad 0.9872$$

Substituting back, we have

$$w = x^2 = 4.133 \quad \text{and} \quad w = x^2 = 0.9872$$

from which we obtain

$$x = \pm 2.03 \quad \text{and} \quad x = \pm 0.994$$

Check: These four values of x each satisfy the original equation (work is not shown). ◆◆◆

◆◆◆ **Example 25:** Solve the equation

$$2x^{-1/3} + x^{-2/3} + 1 = 0$$

Solution: Inspecting the powers of x, we see that one of them, $-\frac{2}{3}$, is twice the other, $-\frac{1}{3}$. We make the substitution and let

$$w = x^{-1/3}$$

Our equation becomes

$$2w + w^2 + 1 = 0$$

or

$$w^2 + 2w + 1 = 0$$

which factors into

$$(w + 1)(w + 1) = 0$$

Setting each factor equal to zero yields

$$w + 1 = 0 \quad \text{and} \quad w + 1 = 0$$

So

$$w = -1 = x^{-1/3}$$

Cubing both sides, we get

$$x^{-1} = (-1)^3 = -1$$
$$= \frac{1}{x} = -1$$

So

$$x = \frac{1}{-1} = -1$$

Check: $2(-1)^{-1/3} + (-1)^{-2/3} + 1 \overset{?}{=} 0$

$$2(-1) \quad + (1) \quad + 1 = 0$$

$$0 = 0 \quad \text{(checks)} \qquad\qquad \blacklozenge\blacklozenge\blacklozenge$$

Exercise 6 ◆ Equations of Quadratic Type

Solve for all real values of x.

1. $x^6 - 6x^3 + 8 = 0$ **2.** $x^4 - 10x^2 + 9 = 0$

3. $x^{-1} - 2x^{-1/2} + 1 = 0$ **4.** $9x^4 - 6x^2 + 1 = 0$

5. $x^{-2/3} - x^{-1/3} = 0$ **6.** $x^{-6} + 19x^{-3} = 216$

7. $x^{-10/3} + 244x^{-5/3} + 243 = 0$ **8.** $x^6 - 7x^3 = 8$

9. $x^{2/3} + 3x^{1/3} = 4$ **10.** $64 + 63x^{-3/2} - x^{-3} = 0$

11. $81x^{-3/4} - 308 - 64x^{3/4} = 0$ **12.** $27x^6 + 46x^3 = 16$

13. $x^{1/2} - x^{1/4} = 20$ **14.** $3x^{-2} + 14x^{-1} = 5$

15. $2.35x^{1/2} - 1.82x^{1/4} = 43.8$ **16.** $1.72x^{-2} + 4.75x^{-1} = 2.24$

14–7   Simple Equations of Higher Degree

Polynomial of Degree n

In Sec. 14–1, we used factoring to find the roots of a quadratic equation. We now expand this idea to include polynomials of higher degree. A polynomial, remember, has exponents that are all positive integers.

Polynomial of Degree n	$a_0x^n + a_1x^{n-1} + \ldots + a_{n-1}x + a_n$	102

The Factor Theorem

In Sec. 14–1, the expression

$$x^2 - x - 6 = 0$$

factored into

$$(x - 3)(x + 2) = 0$$

and hence had the roots

$$x = 3 \quad \text{and} \quad x = -2$$

In a similar way, the cubic equation

$$x^3 - 7x - 6 = 0$$

has the factors

$$(x + 1)(x - 3)(x + 2) = 0$$

and hence has the roots

$$x = -1 \qquad x = 3 \qquad x = -2$$

In general, the following equation applies:

| Factor Theorem | If a polynomial equation $f(x) = 0$ has a root r, then $(x - r)$ is a factor of the polynomial $f(x)$. Conversely, if $(x - r)$ is a factor of a polynomial $f(x)$, then r is a root of $f(x) = 0$. | 103 |

Number of Roots

A polynomial equation of degree n has n and only n roots.

Two or more of the roots may be equal, however, and some may be nonreal, giving the appearance of fewer than n roots; but there can *never* be *more* than n roots.

Solving Polynomial Equations

If an expression cannot be factored directly, we proceed largely by trial and error to obtain one root. Once a root r is obtained (and verified by substituting in the equation), we obtain a *reduced* or *depressed* equation by dividing the original equation by $(x - r)$. The process is then repeated.

A graph of the function is very helpful in finding the roots.

◆◆◆ **Example 26:** The graph of Fig. 14–17 shows that one root of the cubic equation $16x^3 - 13x + 3 = 0$ is $x = -1$. Find the other roots.

Solution: For a cubic, we expect a maximum of three roots. By the factor theorem, since $x = -1$ is a root, then $(x + 1)$ must be a factor of the given equation.

$$(x + 1)(\text{other factor}) = 0$$

We obtain the "other factor" by dividing the given equation by $(x + 1)$, getting $16x^2 - 16x + 3$ (work is not shown). The "other factor," or reduced equation, is then the quadratic,

$$16x^2 - 16x + 3 = 0$$

By the quadratic formula,

$$x = \frac{16 \pm \sqrt{16^2 - 4(16)(3)}}{2(16)} = \frac{1}{4} \text{ and } \frac{3}{4}$$

Our three roots are therefore $x = -1$, $x = \dfrac{1}{4}$, and $x = \dfrac{3}{4}$.   ◆◆◆

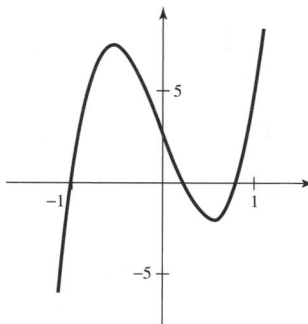

FIGURE 14–17 Graph of $y = 16x^3 - 13x + 3$.

Exercise 7 ◆ Simple Equations of Higher Degree

Find the remaining roots of each polynomial, given one root.

1. $x^3 + x^2 - 17x + 15 = 0$, $x = 1$

2. $x^3 + 9x^2 + 2x - 48 = 0$, $x = 2$

3. $x^3 - 2x^2 - x + 2 = 0$, $\qquad\qquad x = 1$

4. $x^3 + 9x^2 + 26x + 24 = 0$, $\qquad\quad x = -2$

5. $2x^3 + 7x^2 - 7x - 12 = 0$, $\qquad\quad x = -1$

6. $3x^3 - 7x^2 + 4 = 0$, $\qquad\qquad\quad x = 1$

7. $9x^3 + 21x^2 - 20x - 32 = 0$, $\qquad x = -1$

8. $16x^4 - 32x^3 - 13x^2 + 29x - 6 = 0$, $\quad x = -1$

14–8   Systems of Quadratic Equations

In Chapter 10 we solved systems of *linear* equations. We now use the same techniques, substitution or addition-subtraction, to solve sets of two equations in two unknowns where one or both equations are of second degree.

One Equation Linear and One Quadratic
These systems are best solved by substitution.

◆◆◆ **Example 27:** Solve for x and y to two decimal places.

$$2x + y = 5 \qquad\qquad (1)$$
$$2x^2 + y^2 = 17 \qquad\quad (2)$$

Solution: From the first equation,

$$x = \frac{5 - y}{2}$$

Squaring gives

$$x^2 = \frac{25 - 10y + y^2}{4}$$

Substituting into the second equation yields

$$\frac{2(25 - 10y + y^2)}{4} + y^2 = 17$$

$$25 - 10y + y^2 + 2y^2 = 34$$

We get the quadratic,

$$3y^2 - 10y - 9 = 0$$

which, by quadratic formula, gives

$$y = 4.07 \quad \text{and} \quad y = -0.74$$

Substituting back for x gives

$$x = \frac{5 - 4.07}{2} = 0.47 \quad \text{and} \quad x = \frac{5 - (-0.74)}{2} = 2.87$$

So our two solutions are $(0.47, 4.07)$ and $(2.87, -0.74)$. The plots of the two given equations, shown in Fig. 14–18, show points of intersection at those coordinates. ◆◆◆

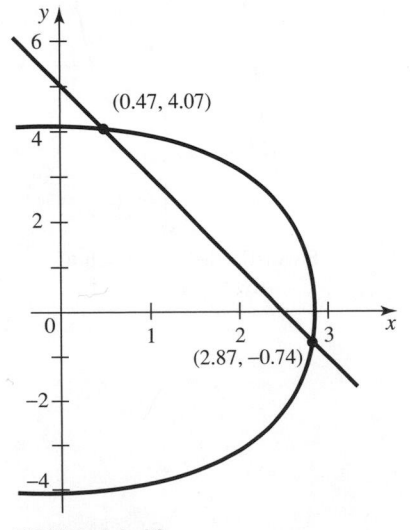

(0.47, 4.07)

(2.87, −0.74)

FIGURE 14–18

◆◆◆ **Example 28:** Solve for x and y:

$$x + y = 2 \qquad\qquad (1)$$
$$xy + 15 = 0 \qquad\quad (2)$$

Solution: From the first equation,

$$y = 2 - x$$

Substituting into the second equation, we have

$$x(2 - x) + 15 = 0$$

from which

$$x^2 - 2x - 15 = 0$$

Factoring gives

$$(x - 5)(x + 3) = 0$$

$$x = 5 \quad \text{and} \quad x = -3$$

Substituting back yields

$$y = 2 - 5 = -3$$

and

$$y = 2 - (-3) = 5$$

So the solutions are

$$x = 5, y = -3 \quad \text{and} \quad x = -3, y = 5$$

as shown in Fig. 14–19.

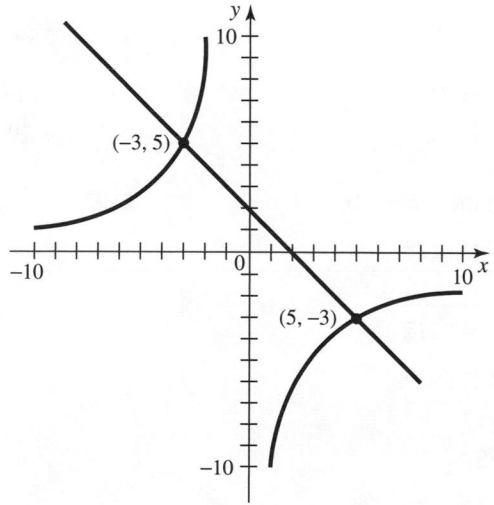

FIGURE 14–19 ◆◆◆

Both Equations Quadratic

These systems may be solved either by substitution or by addition-subtraction.

◆◆◆ **Example 29:** Solve for x and y:

$$3x^2 - 4y = 47 \tag{1}$$
$$7x^2 + 6y = 33 \tag{2}$$

Solution: We multiply the first equation by 3 and the second by 2.

$$9x^2 - 12y = 141 \tag{3}$$
$$14x^2 + 12y = 66 \tag{4}$$

Add: $23x^2 = 207$

$$x^2 = 9$$
$$x = \pm 3$$

Substituting back, we get $y = -5$ for both $x = 3$ and $x = -3$. The plot of the given curves, given in Fig. 14–20, clearly shows the points of intersections $(3, -5)$ and $(-3, -5)$.

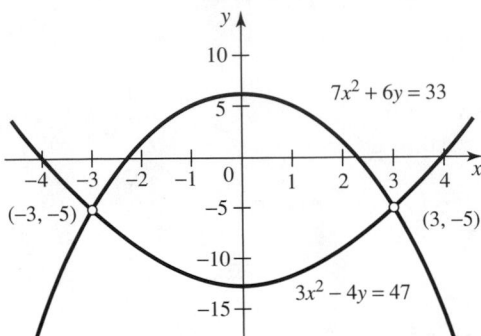

FIGURE 14-20   Intersection of two parabolas. This is a good way to get an approximate solution to any pair of equations that can readily be graphed.   ◆◆◆

The graphs of the equations in the preceding examples intersected in two points. However, the graphs of two second-degree equations can sometimes intersect in as many as four points.

◆◆◆ **Example 30:** Solve for x and y:

$$3x^2 + 4y^2 = 76 \tag{1}$$
$$-11x^2 + 3y^2 = 4 \tag{2}$$

Solution: We multiply the first equation by 3 and the second by -4.

$$9x^2 + 12y^2 = 228 \tag{3}$$
$$44x^2 - 12y^2 = -16 \tag{4}$$

Add: $53x^2 = 212$

$$x^2 = 4$$
$$x = \pm 2$$

From the first equation, when x is either $+2$ or -2,

$$4y^2 = 76 - 12 = 64$$
$$y = \pm 4$$

So our solutions are

$x = 2$	$x = 2$	$x = -2$	$x = -2$
$y = 4$	$y = -4$	$y = 4$	$y = -4$

◆◆◆

Sometimes we can eliminate the squared terms from a pair of equations and obtain a linear equation. This linear equation can then be solved simultaneously with one of the original equations.

◆◆◆ **Example 31:** Solve for x and y:

$$x^2 + y^2 - 3y = 4 \tag{1}$$
$$x^2 + y^2 + 2x - 5y = 2 \tag{2}$$

Solution: Subtracting the second equation from the first gives

$$-2x + 2y = 2$$

or

$$y = x + 1$$

Substituting $x + 1$ for y in the first equation gives

$$x^2 + (x + 1)^2 - 3(x + 1) = 4$$
$$2x^2 - x - 6 = 0$$

Factoring yields

$$(x - 2)(2x + 3) = 0$$

from which

$$x = 2 \quad \text{and} \quad x = -3/2$$

Substituting $x = 2$ into both original equations gives

$$y = 0 \qquad y = 2 \qquad y = 3$$

Substituting $x = -3/2$ into both original equations gives

$$y = 7/2 \qquad y = 11/2 \qquad y = -1/2$$

giving solutions of $(2, 3)$, $(-3/2, -1/2)$, $(2, 0)$, $(2, 2)$, $(-3/2, 7/2)$, and $(-3/2, 11/2)$. However, only two of these,

$$(2, 3) \qquad \text{and} \qquad (-3/2, -1/2)$$

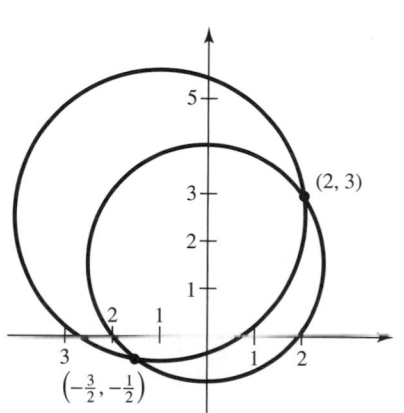

FIGURE 14–21

check in *both* original equations. Figure 14–21 shows plots of the two original equations, which graph as circles. The points found are the coordinates of the points of intersection of the two circles.   ◆◆◆

Exercise 8 • Systems of Quadratic Equations

Solve for x and y.

1. $x - y = 4$
  $x^2 - y^2 = 32$

2. $x + y = 5$
  $x^2 + y^2 = 13$

3. $x + 2y = 7$
  $2x^2 - y^2 = 14$

4. $x - y = 15$
  $x = 2y^2$

5. $x + 4y = 14$
  $y^2 - 2y + 4x = 11$

6. $x^2 + y = 3$
  $x^2 - y = 4$

7. $2x^2 + 3y = 10$
  $x^2 - y = 8$

8. $2x^2 - 3y = 15$
  $5x^2 + 2y = 12$

9. $x + 3y^2 = 7$
  $2x - y^2 = 9$

10. $x^2 - y^2 = 3$
  $x^2 + 3y^2 = 7$

11. $x - y = 11$
  $xy + 28 = 0$

12. $x^2 + y^2 - 8x - 4y + 16 = 0$
  $x^2 + y^2 - 4x - 8y + 16 = 0$

Case Study Discussion—Geometric Representation of a Quadratic

To visually represent $(a + b)^3$, we will construct a cube where each side has a length of $(a + b)$. Again, to make it easier to follow, let's make a longer than b (Fig. 14–22).

We now have eight solids: cubes and rectangular solids. We can see at least one surface for seven of the solids, and there is one tucked in the lower back left side we can't see.

① → $a \times a \times b = a^2b$

② → $a \times b \times b = ab^2$

③ → $a \times b \times b = ab^2$

④ → $b \times b \times b = b^3$

⑤ → $a \times a \times a = a^3$

⑥ → $a \times a \times b = a^2b$

⑦ → $a \times b \times b = ab^2$

⑧ → $a \times a \times b = a^2b$ (not shown)

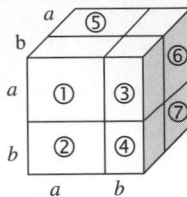

FIGURE 14–22

The total volume is found by multiplying the three sides together or $(a + b)^3$. Adding up the volume of all eight shapes (combining like terms as we go), we get $(a + b)^3 = a^3 + 3a^2b + 3ab^2 + b^3$.

◆◆◆ CHAPTER 14 REVIEW PROBLEMS ◆◆◆◆◆◆◆◆◆◆◆◆◆◆◆◆◆◆◆◆◆◆◆◆◆◆◆◆◆

Solve each equation for any real roots. Give any approximate answers to three significant digits.

1. $y^2 - 5y - 6 = 0$

2. $2x^2 - 5x = 2$

3. $x^4 - 13x^2 + 36 = 0$

4. $(a + b)x^2 + 2(a + b)x = -a$

5. $w^2 - 5w = 0$

6. $x(x - 2) = 2(-2 - x + x^2)$

7. $\dfrac{r}{3} = \dfrac{r}{r + 5}$

8. $6y^2 + y - 2 = 0$

9. $3t^2 - 10 = 13t$

10. $2.73x^2 + 1.47x - 5.72 = 0$

11. $\dfrac{1}{t^2} + 2 = \dfrac{3}{t}$

12. $3w^2 + 2w - 11 = 0$

13. $9 - x^2 = 0$

14. $\dfrac{2y}{3} = \dfrac{3}{5} + \dfrac{2}{y}$

15. $2x(x + 2) = x(x + 3) + 5$

16. $\dfrac{ax^2}{b} + \dfrac{bx}{c} + \dfrac{c}{a} = 0$

17. $y + 6 = 5y^{1/2}$

18. $18 + w^2 + 11w = 0$

19. $9y^2 + y = 5$

20. $\dfrac{5}{w} - \dfrac{4}{w^2} = \dfrac{3}{5}$

21. $\dfrac{z}{2} = \dfrac{5}{z}$

22. $3x - 6 = \dfrac{5x + 2}{4x}$

23. $2x^2 + 3x = 2$

24. $x^2 + Rx - R^2 = 0$

25. $27x = 3x^2$

26. $\dfrac{n}{n + 1} = \dfrac{2}{n + 3} - 1$

27. $z^{2/3} + 8 = 9z^{1/3}$

28. $6w^2 + 13w + 6 = 0$

29. $5y^2 = 125$

30. $\dfrac{t + 2}{t} = \dfrac{4t}{3}$

31. $1.26x^2 - 11.8 = 1.13x$

32. $2.62z^2 + 4.73z - 5.82 = 0$

33. $5.12y^2 + 8.76y - 9.89 = 0$

34. $3w^2 + 2w^4 - 2 = 0$

35. $27.2w^2 + 43.6w = 45.2$

36. $2.35x^{-2} + 4.26x^{-1} - 5.57 = 0$

37. $3.21z^6 - 21.3 = 4.23z^3$

38. If one root is $x = 1$, find the other roots of the following equation:
$$x^3 + x^2 - 5x + 3$$

39. Solve for x and y.
$$x^2 + y^2 = 55$$
$$x + y = 7$$

40. Solve for x and y.
$$x - y^2 = 1$$
$$x^2 + 3y^2 = 7$$

41. If one root is $x = -1$, find the other roots of the following equation:
$$x^4 - 2x^3 - 13x^2 + 14x + 24 = 0$$

42. Plot the parabola $y = 3x^2 + 2x - 6$. Label the vertex, the axis of symmetry, and any zeros.

43. Solve for x and y.
$$x + 2y = 4$$
$$2x^2 - 4y = 1$$

44. Solve for x and y.
$$2x + y = 5$$
$$2x^2 + y = 8$$

45. A person purchased some bags of insulation for $1,000. If she had purchased 5 more bags for the same sum, they would have cost 12 cents less per bag. How many did she buy?

46. The perimeter of a rectangular field is 184 m, and its area 1920 m². Find its dimensions.

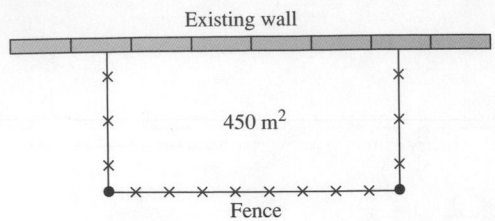

Existing wall

450 m²

Fence

FIGURE 14–23

47. The rectangular yard of Fig. 14–23 is to be enclosed by fence on three sides, and an existing wall is to form the fourth side. The area of the yard is to be 450 m², and its length is to be twice its width. Find the dimensions of the yard.

48. A fast train runs 16 km/h faster than a slow train and takes 3.0 h less to travel 576 km. Find the rates of the trains.

49. A man started to walk 3 km, intending to arrive at a certain time. After walking 1 km, he was detained 10 min and had to walk the rest of the way 1 km/h faster in order to arrive at the intended time. What was his original speed?

50. A rectangular field is 12 m longer than it is wide and contains 448 m². What are the lengths of its sides?

51. A tractor wheel, 5 m in circumference, revolves in a certain number of seconds. If it revolved in a time longer by 1.0 s, the tractor would travel 4800 m less in 1.0 h. In how many seconds does it revolve?

52. The iron counterweight in Fig. 14–24 is to have its weight increased by 50% by plates of iron bolted along the top and side (but not at the ends). The top plate and the side plate have the same thickness. Find their thickness.

53. A 26-cm-wide strip of steel is to have its edges bent up at right angles to form an open trough, as in Fig. 14–25. The cross-sectional area is to be 80 cm². Disregarding the thickness of the steel sheet, find the depth and the width of the trough.

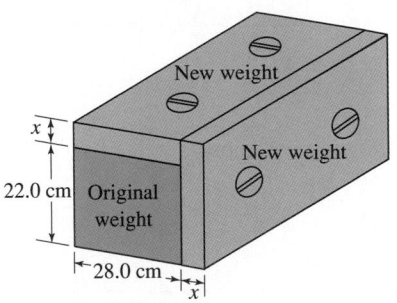

New weight

New weight

x

22.0 cm   Original weight

28.0 cm x

FIGURE 14–24

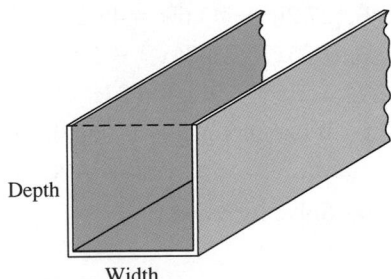

Depth

Width

FIGURE 14–25

54. A boat sails 30 mi. at a uniform rate. If the rate had been 1 mi./h less, the time of the sailing would have been 1 h more. Find the rate of travel.

55. In a certain number of hours a woman travelled 36.0 km. If she had travelled 1.50 km more per hour, it would have taken her 3.00 h less to make the journey. How many kilometres did she travel per hour?

56. The length of a rectangle court exceeds its width by 2 m. If the length and the width were each increased by 3 m, the area of the court would be 80 m². Find the dimensions of the court.

57. The area of a certain square will double if its dimensions are increased by 6.0 m and 4.0 m, respectively. Find its dimensions.

58. A mirror 36 cm by 24 cm is to be set in a frame of uniform width, and the area of the frame is to be equal to that of the glass. Find the width of the frame.

Writing

59. As in Chapter 3, our writing assignment is to make up a word problem. But now we want a word problem that leads to a quadratic equation. As before, swap with a classmate; solve each other's problem; note anything unclear, unrealistic, or ambiguous; and then rewrite your problem if needed.

Team Project

60. The bending moment for a simply supported beam carrying a distributed load w (Fig. 14–7) is $M = \frac{1}{2}wlx - \frac{1}{2}wx^2$. The bending moment curve is therefore a parabola. Plot the curve of M vs. x for a 5-m-long beam carrying a load of 10 kN/m. Take 0.5 m intervals along the beam. Graphically locate a) the points of zero bending moment and b) the point of maximum bending moment.

15

Oblique Triangles and Vectors

❖❖❖ OBJECTIVES ❖❖❖

When you have completed this chapter, you should be able to:
- Solve trigonometric functions for any angle.
- Solve problems using the law of sines.
- Solve problems using the law of cosines.
- Solve applications using oblique triangles.
- Find the resultant of two or more non-perpendicular vectors.

Back to triangles! You have seen that the right triangle is extremely useful in solving many problems. Although almost any geometric problem can be broken down into right-angled triangles to solve, mathematicians have worked out the properties of other types of triangles as well. We can use those properties to solve many problems without converting to right triangles. This gives you the ability (as science and engineering technology professionals) to use the best, most-efficient tools for doing the job. This chapter takes a look at *oblique triangles*—triangles without right angles.

Now, without right angles we should not be able to use sines or cosines, but some very clever geometry shows that an oblique triangle can be split up into right-angled triangles. Then, with a little algebra, we can derive two very powerful laws: the *law of sines* and the *law of cosines*. These two laws free us from having to do the conversion to right-angled triangles. They allow you to solve a problem using oblique triangles. A big time saver!

CASE STUDY—BRACKET FOR SOLAR PANEL FRAME

As a mechanical engineering technologist, you are part of a team that designs custom frames to hold solar panels. The panels are designed to have some motion and sun-tracking capabilities. A triangular brace has one side of 35 cm and an adjacent side of 20 cm. The angle between these two sides can vary from 35° to 55°. The side opposite to this angle can vary in length as the bracket moves in a slider. What is the range of lengths of this opposite side as the angle varies from 35° to 55°?

CANADIAN BIOGRAPHY

THE NITROGEN-FILLED LIGHT BULB—HENRY WOODWARD AND MATHEW EVANS

The invention of the light bulb has a strong Canadian connection. The first light bulb to be produced commercially was invented by Thomas Edison; however, Edison had bought the American patent and Canadian rights to a nitrogen-filled light bulb invented by a Toronto medical student, Henry Woodward, with help of an innkeeper, Mathew Evans. At the time, many inventors were using a carbon filament and different gases to prevent the filament from burning up. Woodward and Evans formed a company but could not attract investors to start production. Their invention was so promising that a year after patenting their light bulb, Edison bought the patent and modified their design to what became the first commercially viable light bulb.

15–1 Trigonometric Functions of Any Angle

Definition of the Trigonometric Functions

We defined the trigonometric functions of any angle in Sec. 7–2 but have so far done problems only with acute angles. We turn now to larger angles.

Figures 15–1(a)–(c) show angles in the second, third, and fourth quadrants, and Fig. 15–1(d) shows an angle greater than 360°. The trigonometric functions of any of these angles are defined exactly as for an acute angle in quadrant I. From any point P on the terminal side of the angle we drop a perpendicular to the x axis, forming a right triangle with legs x and y and with a hypotenuse r. The six trigonometric ratios are then given by Eqs. 146 through 151, just as before, except that some of the ratios might now be negative.

Trigonometric Functions		
$\text{sine } \theta = \sin \theta = \dfrac{y}{r} = \dfrac{\text{opposite side}}{\text{hypotenuse}}$	**146**
$\text{cosine } \theta = \cos \theta = \dfrac{x}{r} = \dfrac{\text{adjacent side}}{\text{hypotenuse}}$	**147**
$\text{tangent } \theta = \tan \theta = \dfrac{y}{x} = \dfrac{\text{opposite side}}{\text{adjacent side}}$	**148**
$\text{cotangent } \theta = \cot \theta = \dfrac{x}{y} = \dfrac{\text{adjacent side}}{\text{opposite side}}$	**149**
$\text{secant } \theta = \sec \theta = \dfrac{r}{x} = \dfrac{\text{hypotenuse}}{\text{adjacent side}}$	**150**
$\text{cosecant } \theta = \csc \theta = \dfrac{r}{y} = \dfrac{\text{hypotenuse}}{\text{opposite side}}$	**151**

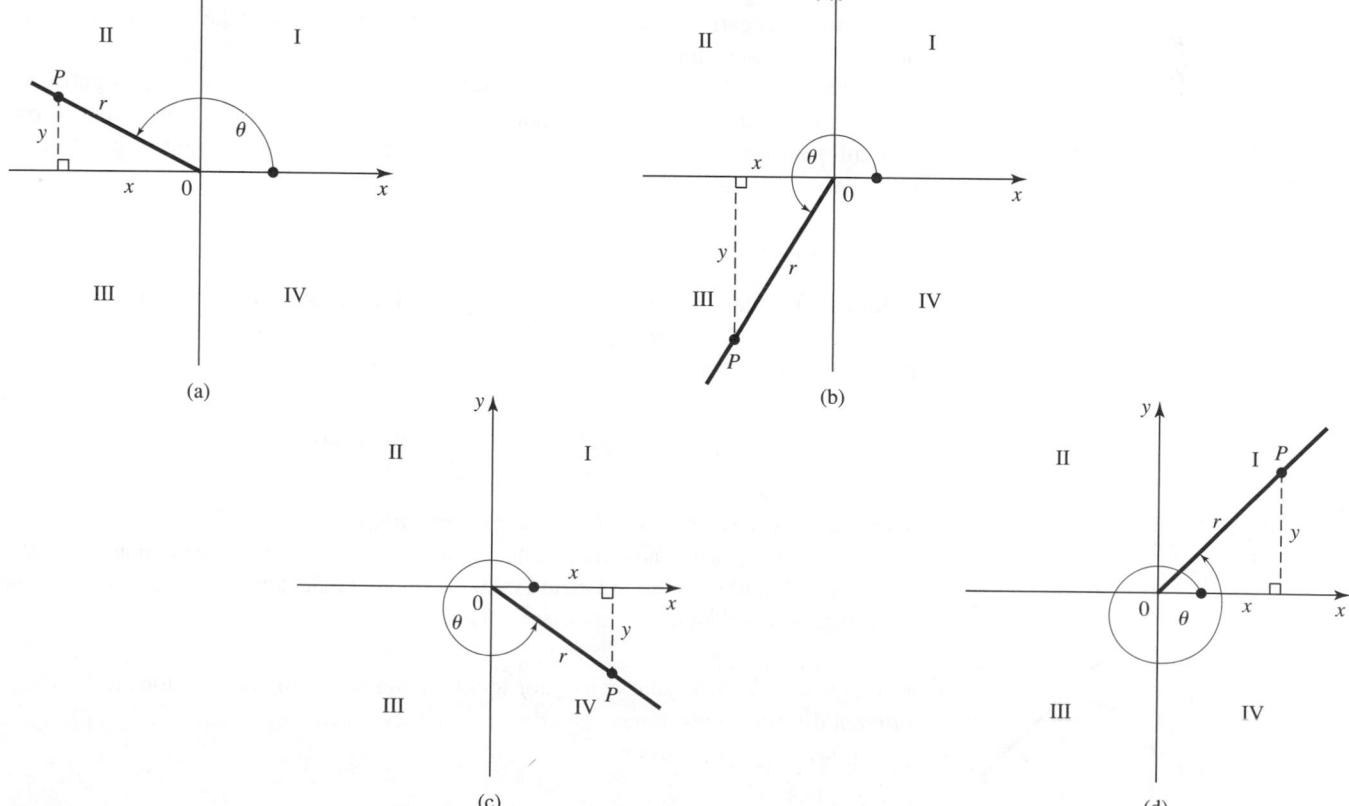

(a)   (b)   (c)   (d)

FIGURE 15–1

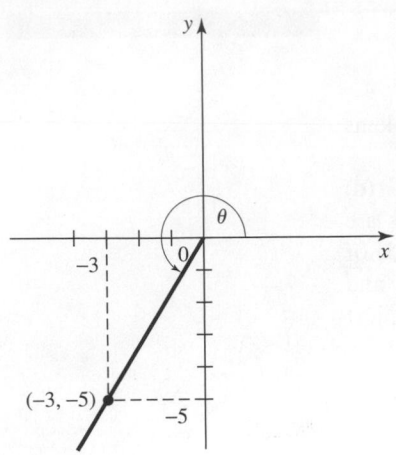

FIGURE 15–2

Why bother learning the signs when a calculator gives them to us automatically? One reason is that you will need them when using a calculator to find the *inverse* of a function, as discussed later.

◆◆◆ **Example 1:** A point on the terminal side of angle θ has the coordinates $(-3, -5)$. Write the six trigonometric functions of θ to three significant digits.

Solution: We sketch the angle as shown in Fig. 15–2 and see that it lies in the third quadrant. We find distance r by the Pythagorean theorem.

$$r^2 = (-3)^2 + (-5)^2 = 9 + 25 = 34$$
$$r = 5.83$$

Then, by Eqs. 146 through 151, with $x = -3$, $y = -5$, and $r = 5.83$,

$$\sin \theta = \frac{y}{r} = \frac{-5}{5.83} = -0.858$$

$$\cos \theta = \frac{x}{r} = \frac{-3}{5.83} = -0.515$$

$$\tan \theta = \frac{y}{x} = \frac{-5}{-3} = 1.67$$

$$\cot \theta = \frac{x}{y} = \frac{-3}{-5} = 0.600$$

$$\sec \theta = \frac{r}{x} = \frac{5.83}{-3} = -1.94$$

$$\csc \theta = \frac{r}{y} = \frac{5.83}{-5} = -1.17 \qquad \text{◆◆◆}$$

Algebraic Signs of the Trigonometric Functions

We saw in Chapter 7 that the trigonometric functions of first-quadrant angles were always positive. From the preceding example, it is clear that some of the trigonometric functions of angles in the second, third, and fourth quadrants are negative, because x or y can be negative (r is always positive). Figure 15–3 shows the signs of the trigonometric functions in each quadrant.

Instead of trying to remember which trigonometric functions are negative in which quadrants, just sketch the angle and note whether x or y is negative. From this information you can figure out whether the function you want is positive or negative.

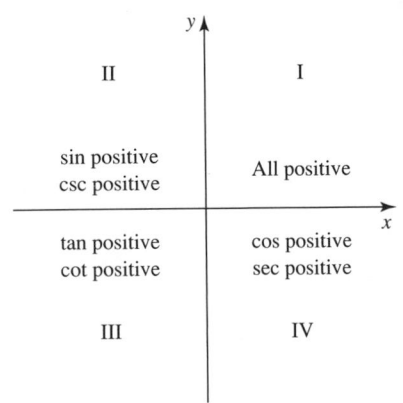

FIGURE 15–3

◆◆◆ **Example 2:** What is the algebraic sign of csc 315°?

Solution: We make a sketch, such as in Fig. 15–4. It is not necessary to draw the angle accurately, but it must be shown in the proper quadrant, quadrant IV in this case. We note that y is negative and that r is (always) positive. So

$$\csc 315° = \frac{r}{y} = \frac{(+)}{(-)} = \text{negative} \qquad \text{◆◆◆}$$

Trigonometric Functions of Any Angle by Calculator

In Sec. 7–2 we learned how to find the six trigonometric functions of an acute angle. Now we will find the trigonometric functions of any angle, acute or obtuse, positive or negative, or less than or greater than 360°.

◆◆◆ **Example 3:** Use your calculator to verify the following calculations to four significant digits:

(a) $\sin 212° = -0.5299$
(b) $\cos 163° = -0.9563$
(c) $\tan 314° = -1.036$ \qquad ◆◆◆

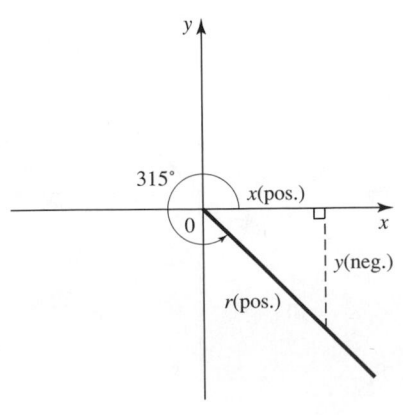

FIGURE 15–4

As before, to find the cosecant, secant, and cotangent, we use the reciprocal relations.

Reciprocal Relationships	$\sin \theta = \dfrac{1}{\csc \theta}$ $\cos \theta = \dfrac{1}{\sec \theta}$ $\tan \theta = \dfrac{1}{\cot \theta}$	152

◆◆◆ **Example 4:** Verify the following calculations to four significant digits:

(a) $\cot 101° = -0.1944$
(b) $\sec 213° = -1.192$
(c) $\csc 294° = -1.095$ ◆◆◆

◆◆◆ **Example 5:** The following calculations are worked to four significant digits:

(a) $\tan(-35.0°) = -0.7002$
(b) $\sec(-125°) = -1.743$ ◆◆◆

Angles greater than 360° are handled in the same manner as any other angle.

◆◆◆ **Example 6:**

(a) $\cos 412° = 0.6157$
(b) $\csc 555° = -3.864$

Evaluating Trigonometric Expressions

In Sec. 7–2. we evaluated some simple trigonometric expressions. Here we will evaluate some that are slightly more complicated and some that contain obtuse angles.

◆◆◆ **Example 7:** Evaluate the expression

$$(\sin^2 48° + \cos 62°)^3$$

to four significant digits.

Solution: The notation $\sin^2 48°$ is the same as $(\sin 48°)^2$. Let us carry five digits and round to four in the last step.

$$(\sin^2 48° + \cos 62°)^3 = [(0.743\ 14)^2 + 0.469\ 47]^3$$
$$= (1.0217)^3 = 1.067$$ ◆◆◆

◆◆◆ **Example 8:** Evaluate the expression

$$\cos 123.5° - \sin^2 242.7°$$

to four significant digits.

Solution: Since

$$\cos 123.5° = -0.551\ 94$$

and

$$\sin 242.7° = -0.888\ 62$$

we get

$$\cos 123.5° - \sin^2 242.7° = -0.551\ 94 - (-0.888\ 62)^2$$
$$= -1.342$$

rounded to four digits. ◆◆◆

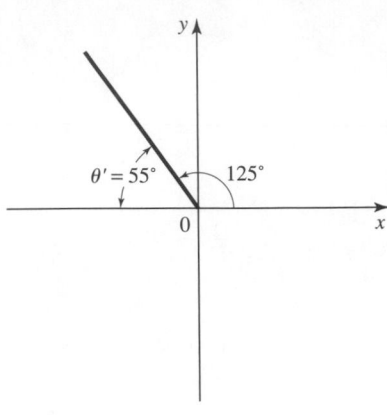

FIGURE 15–5   Reference angle θ'. It is also called the *working* angle.

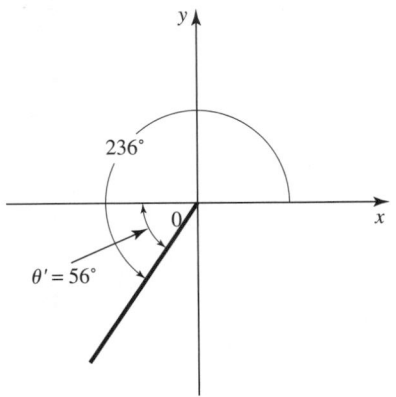

FIGURE 15–6

Note how the prime sign is used *both* as a modifier to the angle and as a unit symbol for the angular minute in Example 12.

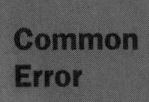

Common Error	Do not confuse an exponent that is on the angle with one that is on the entire function.

$$(\sin\theta)^2 = \sin^2\theta \neq \sin\theta^2$$

Reference Angle

Finding the trigonometric function of any angle by calculator is no problem—the calculator does all of the thinking. This is not the case, however, when we are given the function and are asked to find the angle. For this operation we will use the *reference angle*.

For an angle in standard position on coordinate axes, the *acute* angle that its terminal side makes with the x axis is called the *reference angle*, θ'.

◆◆◆ **Example 9:** The reference angle θ' for an angle of 125° is

$$\theta' = 180° - 125° = 55°$$

as shown in Fig. 15–5.   ◆◆◆

◆◆◆ **Example 10:** The reference angle θ' for an angle of 236° is

$$\theta' = 236° - 180° = 56°$$

as in Fig. 15–6.   ◆◆◆

We treat the quadrantal angles, 180° and 360°, as *exact* numbers. Thus the rounding of our answer is determined by the significant digits in the given angle.

◆◆◆ **Example 11:** The reference angle θ' for an angle of 331.6° (Fig. 15–7) is

$$\theta' = 360° - 331.6° = 28.4°$$ ◆◆◆

◆◆◆ **Example 12:** The reference angle θ' for an angle of 375°15′ is

$$\theta' = 375°15' - 360° = 15°15'$$ ◆◆◆

as in Fig. 15–8.

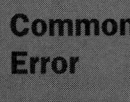

Common Error	The reference angle is measured always from the x axis, never from the y axis.

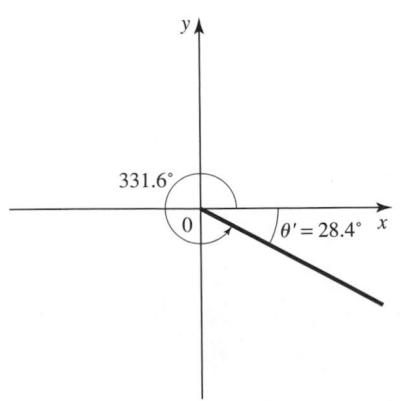

FIGURE 15–7

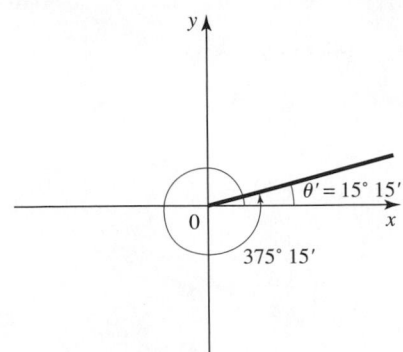

FIGURE 15–8

It might be helpful to remember that for an acute angle θ':

$180° - \theta'$ is in quadrant II.

$180° + \theta'$ is in quadrant III.

$360° - \theta'$ is in quadrant IV.

Finding the Angle When the Function Is Given

We use the same procedure as was given for acute angles. However, a calculator gives us just *one* angle when we enter the value of a trigonometric function. But we know that there are infinitely many angles that have the same value of that trigonometric function. We must use the angle given by calculator to determine the reference angle, which we then use to compute as many larger angles as we wish having the same trigonometric ratio. Usually, we want only the two positive angles less than 360° that have the required trigonometric function.

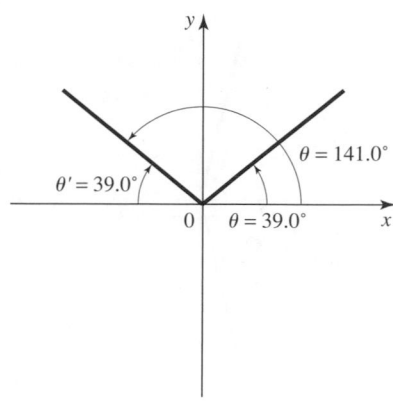

FIGURE 15–9

◆◆◆ **Example 13:** If $\sin \theta = 0.6293$, find two positive values of θ less than 360°. Work to the nearest tenth of a degree.

Solution: From the calculator,

$$\sin^{-1} 0.6293 = 39.0°$$
$$\theta' = 39.0$$

The sine is positive in the first and second quadrants. As shown in Fig. 15–9, our first-quadrant angle is 39.0°, and our second-quadrant angle is, taking 39.0° as the reference angle,

$$\theta = 180° - 39.0° = 141.0° \qquad ◆◆◆$$

> Check your work by taking the sine. In this case, $\sin 141.0° = 0.6293$, which checks.

◆◆◆ **Example 14:** Find, to the nearest tenth of a degree, the two positive angles less than 360° that have a tangent of -2.25.

Solution: From the calculator, $\tan^{-1}(-2.25) = -66.0°$, so our reference angle is 66.0°. The tangent is negative in the second and fourth quadrants. As shown in Fig. 15–10, our second-quadrant angle is

$$180° - 66.0° = 114.0°$$

and our fourth-quadrant angle is

$$360° - 66.0° = 294.0° \qquad ◆◆◆$$

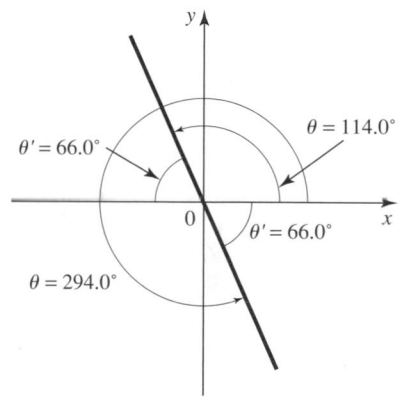

FIGURE 15–10

◆◆◆ **Example 15:** Find $\cos^{-1} 0.575$ to the nearest tenth of a degree.

Solution: From the calculator,

$$\cos^{-1} 0.575 = 54.9°$$

The cosine is positive in the first and fourth quadrants. Our fourth-quadrant angle is

$$360° - 54.9° = 305.1° \qquad ◆◆◆$$

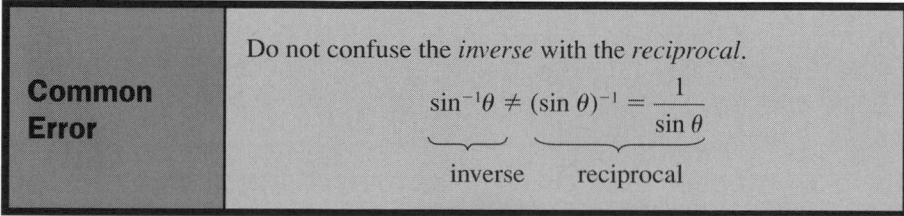

Common Error	Do not confuse the *inverse* with the *reciprocal*. $$\underbrace{\sin^{-1}\theta}_{\text{inverse}} \neq \underbrace{(\sin \theta)^{-1}}_{\text{reciprocal}} = \frac{1}{\sin \theta}$$

When the cotangent, secant, or cosecant is given, we use the reciprocal relationships (Eqs. 152) as in the following example.

◆◆◆ **Example 16:** Find two positive angles less than 360° that have a secant of -4.22. Work to the nearest tenth of a degree.

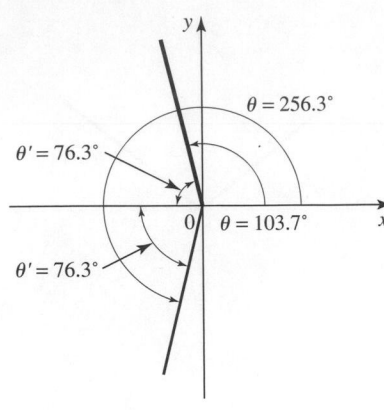

FIGURE 15–11

Solution: If we let the angle be θ, then, by Eq. 152b,

$$\cos \theta = \frac{1}{\sec \theta} = \frac{1}{-4.22} = -0.237$$

By calculator,

$$\theta = 103.7° \qquad \text{◆◆◆}$$

The secant is also negative in the third quadrant. Our reference angle (Fig. 15–11) is

$$\theta' = 180° - 103.7° = 76.3°$$

so the third-quadrant angle is

$$\theta = 180° + 76.3° = 256.3° \qquad \text{◆◆◆}$$

◆◆◆ **Example 17:** Evaluate arcsin (-0.528) to the nearest tenth of a degree.

Solution: As before, we seek only two positive angles less than 360°. By calculator,

$$\arcsin (-0.528) = -31.9°$$

which is a fourth-quadrant angle. As a positive angle, it is

$$\theta = 360° - 31.9° = 328.1°$$

The sine is also negative in the third quadrant. Using 31.9° as our reference angle, we have

$$\theta = 180° + 31.9° = 211.9° \qquad \text{◆◆◆}$$

Special Angles

The angles in Table 15–1 appear so frequently in problems that it is convenient to be able to write their trigonometric functions from memory.

TABLE 15–1

Angle	Sin	Cos	Tan
0°	0	1	0
30°	$\frac{1}{2}$	$\sqrt{3}/2$ (or 0.8660)	$\sqrt{3}/3$ (or 0.5774)
45°	$\sqrt{2}/2$ (or 0.7071)	$\sqrt{2}/2$ (or 0.7071)	1
60°	$\sqrt{3}/2$ (or 0.8660)	$\frac{1}{2}$	$\sqrt{3}$ (or 1.732)
90°	1	0	undefined
180°	0	-1	0
270°	-1	0	undefined
360°	0	1	0

The angles 0°, 90°, 180°, and 360° are called *quadrantal* angles because the terminal side of each of them lies along one of the coordinate axes. Notice that the tangent is undefined for 90° and 270°, angles whose terminal side is on the y axis. The reason is that for any point (x, y) on the y axis, the value of x is zero. Since the tangent is equal to y/x, we have division by zero, which is not defined.

Exercise 1 ◆ Trigonometric Functions of Any Angle

Signs of the Trigonometric Functions

Assume all angles in this exercise set to be in standard position.

State in what quadrant or quadrants the terminal side of θ can lie if:

1. $\theta = 123°$.

2. $\theta = 272°$.

3. $\theta = -47°$.

4. $\theta = -216°$.

5. $\theta = 415°$.   **6.** $\theta = -415°$.

7. $\theta = 845°$.   **8.** $\sin \theta$ is positive.

9. $\cos \theta$ is negative.   **10.** $\sec \theta$ is positive.

11. $\cos \theta$ is positive and $\sin \theta$ is negative.

12. $\sin \theta$ and $\cos \theta$ are both negative.

13. $\tan \theta$ is positive and $\csc \theta$ is negative.

State whether the following expressions are positive or negative. Do not use your calculator, and try not to refer to your book.

14. $\sin 174°$ **15.** $\cos 329°$ **16.** $\tan 227°$

17. $\sec 332°$ **18.** $\cot 206°$ **19.** $\csc 125°$

20. $\sin(-47°)$ **21.** $\tan(-200°)$ **22.** $\cos 400°$

Give the algebraic signs of the sine, cosine, and tangent of the following.

23. $110°$ **24.** $206°$ **25.** $335°$

26. $-48°$ **27.** $500°$

Trigonometric Functions

Sketch the angle and write the six trigonometric functions if the terminal side of the angle passes through the given point.

28. $(3, 5)$ **29.** $(-4, 12)$ **30.** $(24, -7)$

31. $(-15, -8)$ **32.** $(0, -3)$ **33.** $(5, 0)$

Write the sine, cosine, and tangent of θ if:

34. $\sin \theta = -\frac{3}{5}$ and $\tan \theta$ is positive.

35. $\cos \theta = -\frac{4}{5}$ and θ is in the third quadrant.

36. $\csc \theta = -\frac{25}{7}$ and $\cos \theta$ is negative.

37. $\tan \theta = 2$ and θ is not in the first quadrant.

38. $\cot \theta = -\frac{4}{3}$ and $\sin \theta$ is positive.

39. $\sin \theta = \frac{2}{3}$ and θ is not in the first quadrant.

40. $\cos \theta = -\frac{7}{9}$ and $\tan \theta$ is negative.

Leave your answers for problems 34 through 40 in fractional or radical form.

Write, to four significant digits, the sine, cosine, and tangent of each angle.

41. $101°$ **42.** $216°$ **43.** $331°$

44. $125.8°$ **45.** $-62.85°$ **46.** $-227.4°$

47. $486°$ **48.** $-527°$ **49.** $114°23'$

50. $264°15'45''$ **51.** $-166°55'$ **52.** $1.15°$

53. $-11°18'$ **54.** $412°$ **55.** $238°$

Evaluating Trigonometric Expressions

Find the numerical value of each expression to four significant digits.

56. $\sin 35° + \cos 35°$

57. $\sin 125° \tan 225°$

58. $\cos 270° \cos 150° + \sin 270° \sin 150°$

59. $\dfrac{\sin^2 155°}{1 + \cos 155°}$

60. $\sin^2 75°$

61. $\tan^2 125° - \cos^2 125°$

62. $(\cos^2 206° + \sin 206°)^2$

63. $\sqrt{\sin^2 112° - \cos 112°}$

Inverse Trigonometric Functions

Find all positive angles less than 360° whose trigonometric function is given. Give any approximate answers to the nearest tenth of a degree.

64. $\sin \theta = \frac{1}{2}$.

65. $\tan \theta = -1$

66. $\cot \theta = -\sqrt{3}$

67. $\cos \theta = 0.8372$

68. $\csc \theta = -3.85$

69. $\tan \theta = 6.372$

70. $\cos \theta = -\frac{1}{2}$

71. $\cot \theta = 0$

72. $\cos \theta = -1$

73. $\sin \theta = -0.6358$

74. $\cot \theta = -2.8458$

75. $\tan \theta = 1.7361$

76. $\cos \theta = 0.3759$

Evaluate to the nearest tenth of a degree. Find only positive angles less than 360°.

77. $\arcsin(-0.736)$

78. $\operatorname{arcsec} 2.85$

79. $\cos^{-1} 0.827$

80. $\operatorname{arccot} 5.22$

81. $\arctan(-4.48)$

82. $\csc^{-1} 5.02$

Evaluate from memory. Do not use a calculator.

83. $\sin 30°$

84. $\cos 45°$

85. $\tan 180°$

86. $\sin 360°$

87. $\tan 45°$

88. $\cos 30°$

89. $\sin 180°$

90. $\cos 360°$

91. $\tan 0°$

92. $\sin 90°$

93. $\cos 180°$

94. $\sin 45°$

95. $\tan 360°$

96. $\cos 270°$

97. $\cos 60°$

98. $\sin 60°$

99. $\cos 0°$

100. $\tan 90°$

101. $\cos 90°$

15–2   Law of Sines

Derivation

We first derive the *law of sines* for an oblique triangle in which all three angles are acute, such as in Fig. 15–12. We start by breaking the given triangle into two right triangles by drawing altitude h to side AB.

In right triangle ACD,

$$\sin A = \frac{h}{b} \quad \text{or} \quad h = b \sin A$$

In right triangle *BCD*,

$$\sin B = \frac{h}{a} \quad \text{or} \quad h = a \sin B$$

So

$$b \sin A = a \sin B$$

Dividing by sin *A* sin *B*, we have

$$\frac{a}{\sin A} = \frac{b}{\sin B}$$

Similarly, drawing altitude *j* to side *AC*, and using triangles *BEC* and *AEB*, we get

$$j = a \sin C = c \sin A$$

or

$$\frac{a}{\sin A} = \frac{c}{\sin C}$$

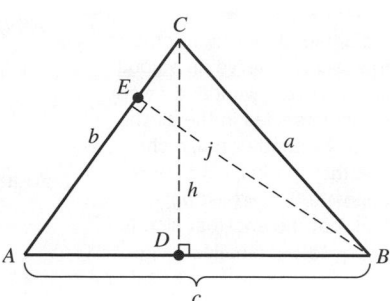

FIGURE 15–12   Derivation of the law of sines.

Combining this with the previous result, we obtain the following equation:

Law of Sines	$\dfrac{a}{\sin A} = \dfrac{b}{\sin B} = \dfrac{c}{\sin C}$	**140**

The sides of a triangle are proportional to the sines of the opposite angles.

When one of the angles of the triangle is obtuse, as in Fig. 15–13, the derivation is nearly the same. We draw an altitude *h* to the extension of side *AB*. Then:

In right triangle *ACD:* $\sin A = \dfrac{h}{b}$ or $h = b \sin A$

In right triangle *BCD:* $\sin(180 - B) = \dfrac{h}{a}$

But sin(180 − *B*) = sin *B*, so

$$h = a \sin B$$

The derivation then proceeds exactly as before.

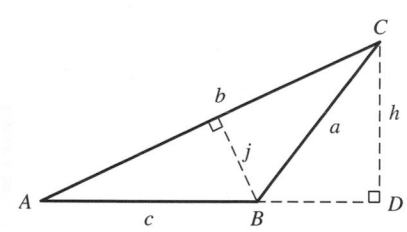

FIGURE 15–13

Solving a Triangle When Two Angles and One Side Are Given (AAS or ASA)

Recall that "solving a triangle" means to find all missing sides and angles. Here we use the law of sines to solve an oblique triangle. To do this, we must have *a known side opposite to a known angle*, as well as another angle. We abbreviate these given conditions as AAS (angle–angle–side) or ASA (angle–side–angle).

◆◆◆ **Example 18:** Solve triangle *ABC* where *A* = 32.5°, *B* = 49.7°, and *a* = 226.

Solution: We want to find the unknown angle *C* and sides *b* and *c*. We sketch the triangle as shown in Fig. 15–14. The missing angle is found by Eq. 139.

$$C = 180° - 32.5° - 49.7° = 97.8°$$

Then, by the law of sines,

$$\frac{a}{\sin A} = \frac{b}{\sin B}$$

$$\frac{226}{\sin 32.5°} = \frac{b}{\sin 49.7°}$$

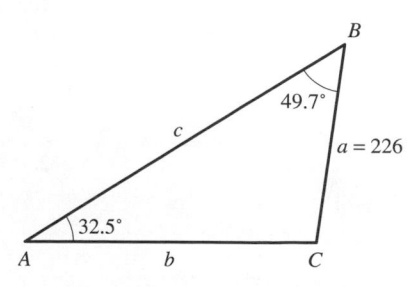

FIGURE 15–14

A diagram drawn more or less to scale can serve as a good check of your work and reveal inconsistencies in the given data. As another rough check, see that the longest side is opposite the largest angle and that the shortest side is opposite the smallest angle.

Solving for b, we get

$$b = \frac{226 \sin 49.7°}{\sin 32.5°} = 321$$

Again using the law of sines, we have

$$\frac{a}{\sin A} = \frac{c}{\sin C}$$

$$\frac{226}{\sin 32.5°} = \frac{c}{\sin 97.8°}$$

So

$$c = \frac{226 \sin 97.8°}{\sin 32.5°} = 417$$ ◆◆◆

Notice that if two angles are given, the third can easily be found, since all angles of a triangle must have a sum of 180°. This means that if two angles and an included side are given (ASA), the problem can be solved as shown above for AAS.

Solving a Triangle When Two Sides and One Angle Are Given (SSA): The Ambiguous Case

In Example 18, *two angles and one side* were given. We can also use the law of sines when *one angle and two sides* are given, provided that the given angle is opposite one of the given sides. But we must be careful here. Sometimes we will get no solution, one solution, or two solutions, depending on the given information. The possibilities are given in Table 15–2.

We see that we will sometimes get *two* correct solutions, both of which may be reasonable in a given application.

Common Error	If it appears that you will get no solution when doing an application of oblique triangles, it probably means that the data are incorrect or that the problem is not properly set up. Don't give up at that point, but go back and check your work.

TABLE 15–2

When angle A is obtuse, and:	Then:	Example:
1. $a \leq c$	Side a is too short to intersect side b, so there is *no solution*.	
2. $a > c$	Side a can intersect side b in only one place, so we get *one solution*.	

TABLE 15–2 (Continued)

When angle A is acute, and:	Then:	Example:
1. $a \geq c$	Side a is too long to intersect side b in more than one place, so there is *one solution.*	
2. $a < h < c$ where the altitude h is $h = c \sin A$	Side a is too short to touch side b, so we get *no solution.*	
3. $a = h < c$ where $h = c \sin A$	Side a just reaches side b, so we get a right triangle having *one solution.*	
4. $a > h < c$ where $h = c \sin A$	Side a can intersect side b in two places, giving *two solutions.* This is the **ambiguous case.**	

Another simple way to check for the number of solutions is to make a sketch. But the sketch must be fairly accurate, as in the following example.

◆◆◆ **Example 19:** Solve triangle ABC where $A = 27.6°$, $a = 112$, and $c = 165$.

Solution: Let's first calculate the altitude h to find out how many solutions we may have.

$$h = c \sin A$$
$$= 165 \sin 27.6° = 76.4$$

We see that side a (112) is greater than h (76.4) but less than c (165), so we have the ambiguous case with two solutions, as verified by Fig. 15–15. We will solve for both possible triangles. By the law of sines,

$$\frac{\sin C}{165} = \frac{\sin 27.6°}{112}$$
$$\sin C = \frac{165 \sin 27.6°}{112} = 0.6825$$
$$C = 43.0°$$

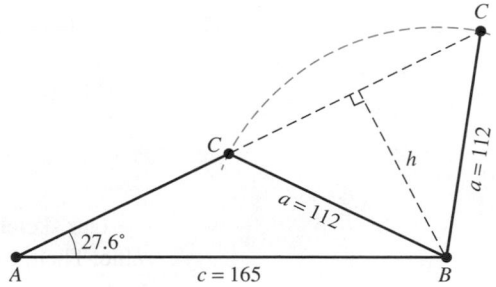

FIGURE 15–15 The ambiguous case.

This is *one* of the possible values for C. The other is

$$C = 180° - 43.0° = 137.0°$$

We now find the two corresponding values for side b and angle B.
    When $C = 43.0°$,

$$B = 180° - 27.6° - 43.0° = 109.4°$$

Recall from Sec. 15–1 that there are two angles less than 180° for which the sine is positive. One of them, θ, is in the first quadrant, and the other, $180° - \theta$, is in the second quadrant.

So

$$\frac{b}{\sin 109.4°} = \frac{112}{\sin 27.6°}$$

from which $b = 228$.

When $C = 137.0°$,

$$B = 180° - 27.6° - 137.0° = 15.4°$$

So

$$\frac{b}{\sin 15.4°} = \frac{112}{\sin 27.6°}$$

from which $b = 64.2$. So our two solutions are given in the following table:

	A	B	C	a	b	c
1.	27.6°	109.4°	43.0°	112	228	165
2.	27.6°	15.4°	137.0°	112	64.2	165

◆◆◆

Common Error	In a problem such as the preceding one, it is easy to forget the second possible solution ($C = 137.0°$), especially since a calculator will give only the acute angle when computing the arc sine.

Not every SSA problem has two solutions, as we'll see in the following example.

◆◆◆ **Example 20:** Solve triangle ABC if $A = 35.2°$, $a = 525$, and $c = 412$.

Solution: When we sketch the triangle (Fig. 15–16), we see that the given information allows us to draw the triangle in only one way. We will get a unique solution. By the law of sines,

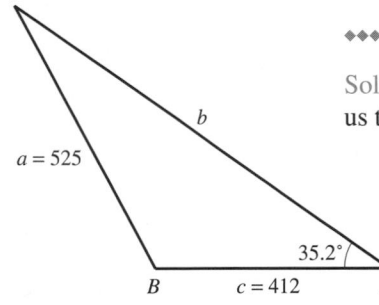

FIGURE 15–16

$$\frac{\sin C}{412} = \frac{\sin 35.2°}{525}$$

$$\sin C = \frac{412 \sin 35.2°}{525} = 0.4524$$

Thus the two possible values for C are

$$C = 26.9° \quad \text{and} \quad C = 180° - 26.9° = 153.1°$$

Our sketch, even if crudely drawn, shows that C cannot be obtuse, so we discard the 153.1° value. Then, by Eq. 139,

$$B = 180° - 35.2° - 26.9° = 117.9°$$

It is sometimes convenient, as in this example, to use the *reciprocals* of the expressions in the law of sines.

Using the law of sines once again, we get

$$\frac{b}{\sin 117.9°} = \frac{525}{\sin 35.2°}$$

$$b = \frac{525 \sin 117.9°}{\sin 35.2°} = 805$$

◆◆◆

Exercise 2 ◆ Law of Sines

Data for triangle *ABC* are given in the following table. Solve for the missing parts.

	Angles			Sides		
	A	**B**	**C**	**a**	**b**	**c**
1.		46.3°		228	304	
2.		65.9°			1.59	1.46
3.			43.0°		15.0	21.0
4.	21.45°	35.17°		276.0		
5.	61.9°		47.0°	7.65		
6.	126°	27.0°			119	
7.		31.6°	44.8°	11.7		
8.	15.0°		72.0°			375
9.		125°	32.0°			58.0
10.	24.14°	38.27°		5562		
11.		55.38°	18.20°		77.85	
12.	44.47°		63.88°			1.065
13.	18.0°	12.0°		50.7		
14.			45.55°		1137	1010

15–3   Law of Cosines

Derivation

Consider an oblique triangle *ABC* as shown in Fig. 15 17. As we did for the law of sines, we start by dividing the triangle into two right triangles by drawing an altitude *h* to side *AC*.

In right triangle *ABD*,

$$c^2 = h^2 + (AD)^2$$

But $AD = b - CD$. Substituting, we get

$$c^2 = h^2 + (b - CD)^2 \qquad (1)$$

Now, in right triangle *BCD*, by Eq. 147,

$$\frac{CD}{a} = \cos C$$

or

$$CD = a \cos C$$

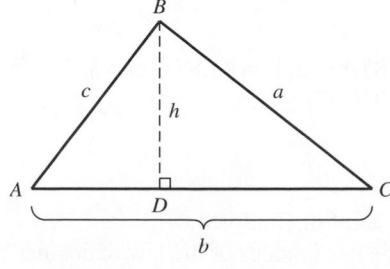

FIGURE 15–17   Derivation of the law of cosines.

Substituting *a* cos *C* for *CD* in Eq. (1) yields

$$c^2 = h^2 + (b - a \cos C)^2$$

Squaring, we have

$$c^2 = h^2 + b^2 - 2ab \cos C + a^2 \cos^2 C \qquad (2)$$

Let us leave this expression for the moment and write the Pythagorean theorem for the same triangle BCD.

$$h^2 = a^2 - (CD)^2$$

Again substituting $a \cos C$ for CD, we obtain

$$h^2 = a^2 - (a \cos C)^2$$
$$= a^2 - a^2 \cos^2 C$$

Substituting this expression for h^2 back into (2), we get

$$c^2 = a^2 - a^2 \cos^2 C + b^2 - 2ab \cos C + a^2 \cos^2 C$$

Collecting terms, we get the law of cosines.

$$c^2 = a^2 + b^2 - 2ab \cos C$$

We can repeat the derivation, with perpendiculars drawn to side AB and to side BC, and get two more forms of the law of cosines.

Most students find it easier to think of the law of cosines in terms of the parts of the triangle, that is, *the given angle, the side opposite to the given angle, and the sides adjacent to the given angle*, rather than by letters of the alphabet.

Law of Cosines	$a^2 = b^2 + c^2 - 2bc \cos A$ $b^2 = a^2 + c^2 - 2ac \cos B$ $c^2 = a^2 + b^2 - 2ab \cos C$	141

Notice that when the angle in these equations is 90°, the law of cosines reduces to the Pythagorean theorem.

The square of any side equals the sum of the squares of the other two sides minus twice the product of the other sides and the cosine of the opposite angle.

◆◆◆ Example 21: Find side x in Fig. 15–18.

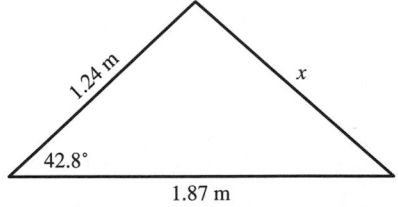

FIGURE 15–18

Solution: By the law of cosines,

$$x^2 = (1.24)^2 + (1.87)^2 - 2(1.24)(1.87) \cos 42.8°$$
$$= 5.03 - 4.64(0.7337) = 1.63$$
$$x = 1.28 \text{ m} \qquad\qquad\qquad ◆◆◆$$

When to Use the Law of Sines or the Law of Cosines

It is sometimes not clear whether to use the law of sines or the law of cosines to solve a triangle. We use the law of sines when we have a *known side opposite a known angle*. We use the law of cosines only when the law of sines does not work, that is, for all other cases. In Fig. 15–19, the heavy lines indicate the known information and might help in choosing the proper law.

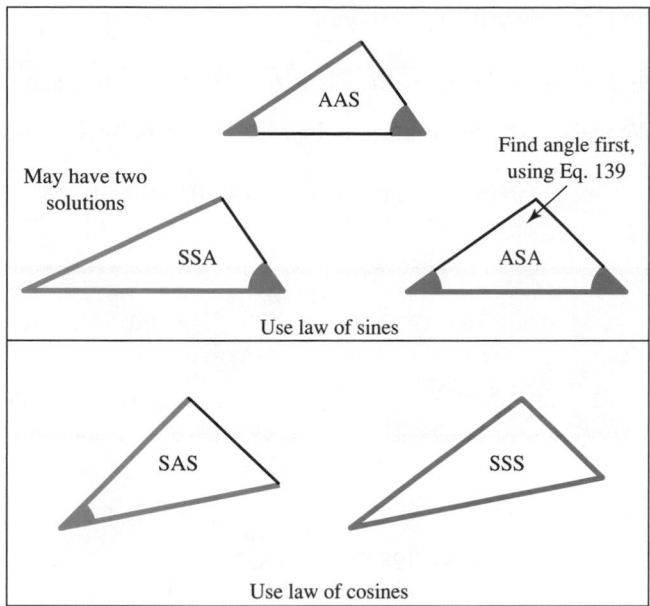

FIGURE 15–19   When to use the law of sines or the law of cosines.

Using the Cosine Law When Two Sides and the Included Angle Are Known

We can solve triangles by the law of cosines if we know *two sides and the angle between them*, or if we know *three sides*. We consider the first of these cases in the following example.

◆◆◆ **Example 22:** Solve triangle *ABC* where $a = 184$, $b = 125$, and $C = 27.2°$.

Solution: We make a sketch as shown in Fig. 15 20.
Then, by the law of cosines,

$$c^2 = a^2 + b^2 - 2ab \cos C$$
$$= (184)^2 + (125)^2 - 2(184)(125) \cos 27.2°$$
$$c = \sqrt{(184)^2 + (125)^2 - 2(184)(125) \cos 27.2°}$$
$$= 92.6$$

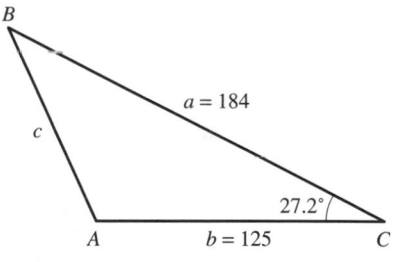

FIGURE 15–20

Now that we have a known side opposite a known angle, we can use the law of sines to find angle *A* or angle *B*. Which shall we find first?

Use the law of sines to *find the acute angle first* (angle *B* in this example). If, instead, you solve for the obtuse angle first, you might forget to subtract the angle obtained from the calculator from 180°. Further, if one of the angles is so close to 90° that you cannot tell from your sketch if it is acute or obtuse, find the other angle first, and then subtract the two known angles from 180° to obtain the third angle.

By the law of sines,

$$\frac{\sin B}{125} = \frac{\sin 27.2°}{92.6}$$

$$\sin B = \frac{125 \sin 27.2°}{92.6} = 0.617$$

$$B = 38.1° \quad \text{and} \quad B = 180 - 38.1 = 141.9°$$

Notice that we cannot initially use the law of sines because we do not have a known side opposite a known angle.

We drop the larger value because our sketch shows us that *B* must be acute. Then, by Eq. 139,

$$A = 180° - 27.2° - 38.1° = 114.7°$$

◆◆◆

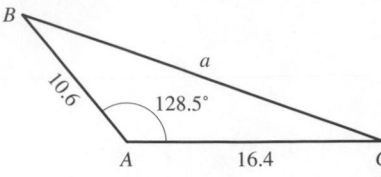

FIGURE 15–21

In our next example, the given angle is *obtuse*.

◆◆◆ **Example 23:** Solve triangle ABC where $b = 16.4$, $c = 10.6$, and $A = 128.5°$.

Solution: We make a sketch as shown in Fig. 15–21. Then, by the law of cosines,

$$a^2 = (16.4)^2 + (10.6)^2 - 2(16.4)(10.6) \cos 128.5°$$

Common Error	The cosine of an obtuse angle is *negative*. Be sure to use the proper algebraic sign when applying the law of cosines to an obtuse angle.

In our example,

$$\cos 128.5° = -0.6225$$

So

$$a = \sqrt{(16.4)^2 + (10.6)^2 - 2(16.4)(10.6)(-0.6225)} = 24.4$$

By the law of sines,

$$\frac{\sin B}{16.4} = \frac{\sin 128.5°}{24.4}$$

$$\sin B = \frac{16.4 \sin 128.5°}{24.4} = 0.526$$

$$B = 31.7° \text{ and } B = 180 - 31.7 = 148.3°$$

We drop the larger value because our sketch shows us that B must be acute. Then, by Eq. 139,

$$C = 180° - 31.7° - 128.5° = 19.8° \qquad ◆◆◆$$

Using the Cosine Law When Three Sides Are Known

When three sides of an oblique triangle are known, we can use the law of cosines to solve for one of the angles. A second angle is found using the law of sines, and the third angle is found by subtracting the other two from 180°.

◆◆◆ **Example 24:** Solve triangle ABC in Fig. 15–22, where $a = 128$, $b = 146$, and $c = 222$.

Solution: We start by writing the law of cosines for any of the three angles. A good way to avoid ambiguity is to find the largest angle first (the law of cosines will tell us if it is acute or obtuse), and then we are sure that the other two angles are acute.

Writing the law of cosines for angle C gives

$$(222)^2 = (128)^2 + (146)^2 - 2(128)(146) \cos C$$

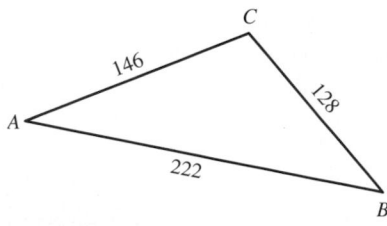

FIGURE 15–22

Solving for $\cos C$ gives

$$\cos C = -0.3099$$

Since the cosine is negative, C must be obtuse, so

$$C = 108.1°$$

Then by the law of sines

$$\frac{\sin A}{128} = \frac{\sin 108.1°}{222}$$

from which $\sin A = 0.5480$. Since we know that A is acute, we get

$$A = 33.2°$$

Finally,

$$B = 180° - 108.1° - 33.2° = 38.7°$$ ◆◆◆

Exercise 3 ◆ Law of Cosines

Data for triangle ABC are given in the following table. Solve for the missing parts.

	Angles			Sides		
	A	**B**	**C**	**a**	**b**	**c**
1.			106.0°	15.7	11.2	
2.		51.4°		1.95		1.46
3.	68.3°				18.3	21.7
4.				728	906	663
5.			27.3°	128	152	
6.		128°		1.16		1.95
7.	135°				275	214
8.			35.2°	77.3	81.4	
9.				11.3	15.6	12.8
10.		41.7°		199		202
11.	115°				46.8	51.3
12.				1.475	1.836	2.017
13.			67.0°	9.08	6.75	
14.		129°		186		179
15.	158°				1.77	1.99
16.				41.8	57.2	36.7
17.			41.77°	1445	1502	
18.		108.8°		7.286		6.187
19.				97.3	81.4	88.5
20.	36.29°				47.28	51.36

15–4   Applications

As with right triangles, oblique triangles have many applications in technology, as you will see in the exercises for this section. Follow the same procedures for setting up these problems as we used for other word problems, and solve the resulting triangles by the law of sines or the law of cosines, or both.

   If an *area* of an oblique triangle is needed, either compute all the sides and use Hero's formula (Eq. 138), or find an altitude with right-triangle trigonometry and use Eq. 137.

◆◆◆ **Example 25:** Find the area of the gusset in Fig. 15–23(a).

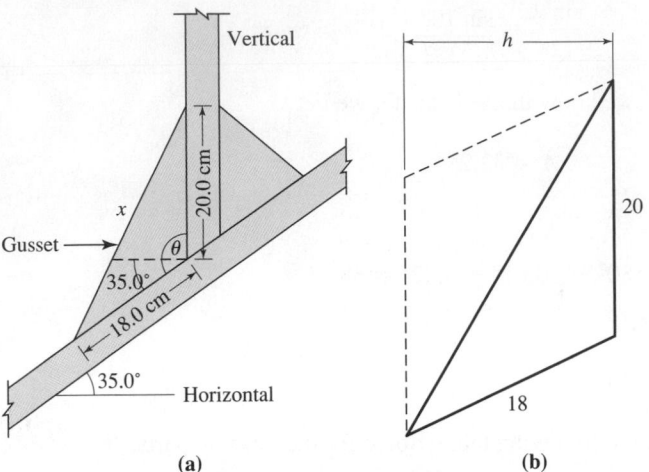

(a)   (b)   **FIGURE 15–23**

Solution: We first find θ.

$$\theta = 90° + 35.0° = 125.0°$$

We now know two sides and the angle between them, so can find side x by the law of cosines.

$$x^2 = (18.0)^2 + (20.0)^2 - 2(18.0)(20.0) \cos 125° = 1137$$
$$x = 33.7 \text{ cm}$$

We find the area of the gusset by Hero's formula (Eq. 138).

$$s = \frac{1}{2}(18.0 + 20.0 + 33.7) = 35.9$$

$$\text{area} = \sqrt{35.9(35.9 - 18.0)(35.9 - 20.0)(35.9 - 33.7)}$$
$$= 150 \text{ cm}^2$$

Check: Does the answer look reasonable? Let's place the gusset inside the parallelogram, as shown in Fig. 15–23(b), and estimate the height h by eye at about 15 cm. Thus the area of the parallelogram would be 15×20, or 300 cm², just double that found for the triangle.   ◆◆◆

◆◆◆ **Example 26:** A ship takes a sighting on two buoys. At a certain instant, the bearing of buoy A is N 44.23° W, and that of buoy B is N 62.17° E. The distance between the buoys is 3.60 km, and the bearing of B from A is N 87.87° E. Find the distance of the ship from each buoy.

Estimate: Let us draw the figure with a ruler and protractor as shown in Fig. 15–24. Notice how the compass directions are laid out, starting from the north and turning in the indicated direction. Measuring, we get $SA = 1.7$ units and $SB = 2.8$ units. If you try it, you will probably get slightly different values.

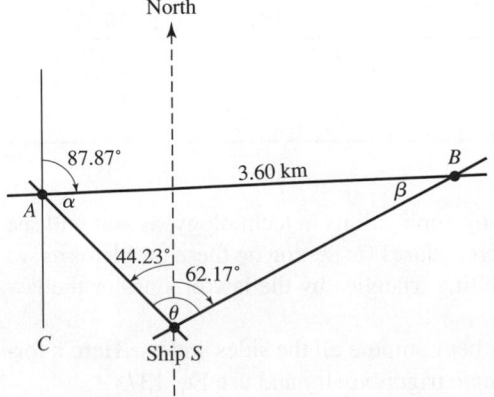

FIGURE 15–24

Solution: Calculating the angles of triangle *ABS* gives

$$\theta = 44.23° + 62.17° = 106.40°$$

Since angle *SAC* is 44.23°,

$$\alpha = 180° - 87.87° - 44.23° = 47.90°$$

and

$$\beta = 180° - 106.40° - 47.90° = 25.70°$$

From the law of sines,

$$\frac{SA}{\sin 25.70°} = \frac{SB}{\sin 47.90°} = \frac{3.60}{\sin 106.40°}$$

from which

$$SA = \frac{3.60 \sin 25.70°}{\sin 106.40°} = 1.63 \text{ km}$$

and

$$SB = \frac{3.60 \sin 47.90°}{\sin 106.40°} = 2.78 \text{ km}$$

which agree with our estimated values.   ◆◆◆

Exercise 4 ◆ Applications

Determining Inaccessible Distances

1. Two stakes, *A* and *B*, are 88.6 m apart. From a third stake *C*, the angle *ACB* is 85.4°, and from *A*, the angle *BAC* is 74.3°. Find the distance from *C* to each of the other stakes.

This set of problems contains no vectors, which are covered in the next section.

2. From a point on level ground, the angles of elevation of the top and the bottom of an antenna standing on top of a building are 32.6° and 27.8°, respectively. If the building is 125 ft. high, how tall is the antenna?

Remember that angles of elevation or depression are always measured from the horizontal.

3. A triangular lot measures 115 m, 187 m, and 215 m along its sides. Find the angles between the sides.

4. Two boats are 45.5 km apart. Both are travelling toward the same point, which is 87.6 km from one of them and 77.8 km from the other. Find the angle at which their paths intersect.

Navigation

5. A ship is moving at 15.0 km/h in the direction N 15.0° W. A helicopter with a speed of 22.0 km/h is due east of the ship. In what direction should the helicopter travel if it is to meet the ship?

Hint: Draw your diagram after t hours have elapsed.

6. City *A* is 215 km N 12.0° E from city *B*. The bearing of city *C* from *B* is S 55.0° E. The bearing of *C* from *A* is S 15.0° E. How far is C from *A*? From *B*?

7. A ship is moving in a direction S 24.25° W at a rate of 8.60 km/h. If a launch that travels at 15.4 km/h is due west of the ship, in what direction should it travel in order to meet the ship?

8. A ship is 9.50 km directly east of a port. If the ship sails southeast for 2.50 km, how far will it be from the port?

9. From a plane flying due east, the bearing of a radio station is S 31.0° E at 1 P.M. and S 11.0° E at 1:20 P.M. The ground speed of the plane is 625 km/h. Find the distance of the plane from the station at 1 P.M.

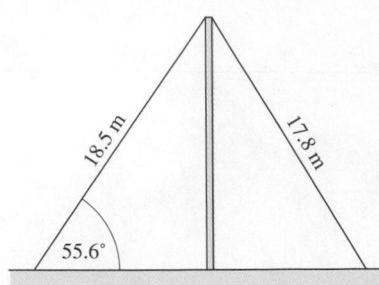

FIGURE 15-25

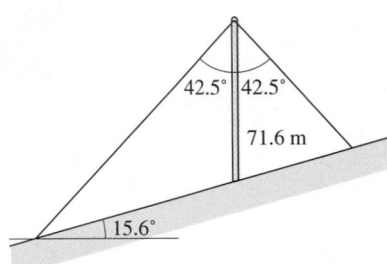

FIGURE 15-26

Structures

10. A tower stands vertically on sloping ground whose inclination with the horizontal is 11.6°. From a point 42.0 m downhill from the tower (measured along the slope), the angle of elevation of the top of the tower is 18.8°. How tall is the tower?

11. A vertical antenna stands on a slope that makes an angle of 8.70° with the horizontal. From a point directly uphill from the antenna, the angle of elevation of its top is 61.0°. From a point 16.0 m farther up the slope (measured along the slope), the angle of elevation of its top is 38.0°. How tall is the antenna?

12. A power pole on level ground is supported by two wires that run from the top of the pole to the ground (Fig. 15–25). One wire is 18.5 m long and makes an angle of 55.6° with the ground, and the other wire is 17.8 m long. Find the angle that the second wire makes with the ground.

13. A 71.6-m-high antenna mast is to be placed on sloping ground, with the cables making an angle of 42.5° with the top of the mast (Fig. 15–26). Find the lengths of the two cables.

14. In the roof truss in Fig. 15–27, find the lengths of members *AB*, *BD*, *AC*, and *AD*.

15. From a point on level ground between two power poles of the same height, cables are stretched to the top of each pole. One cable is 52.6 m long, the other is 67.5 m long, and the angle of intersection between the two cables is 125°. Find the distance between the poles.

16. A pole standing on level ground makes an angle of 85.8° with the horizontal. The pole is supported by a 22.0-ft. prop whose base is 12.5 ft. from the base of the pole. Find the angle made by the prop with the horizontal.

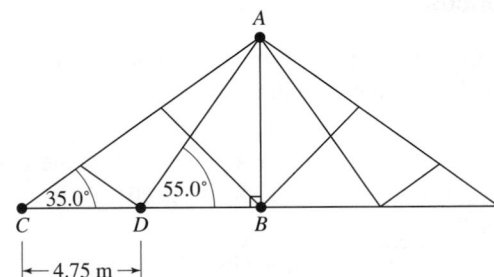

FIGURE 15-27   Roof truss.

Mechanisms

17. In the slider crank mechanism of Fig. 15–28, find the distance *x* between the wrist pin *W* and the crank centre *C* when *θ* = 35.7°.

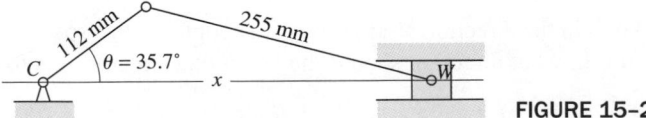

FIGURE 15-28

18. In the four-bar linkage of Fig. 15–29, find angle *θ* when angle *BAD* is 41.5°.

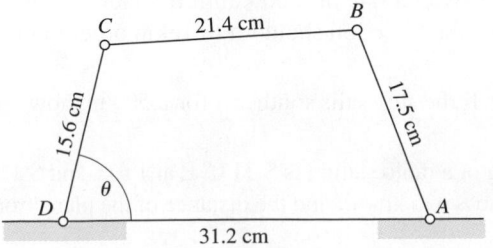

FIGURE 15-29

19 Two links, *AC* and *BC*, are pivoted at *C*, as shown in Fig. 15–30. How far apart are *A* and *B* when angle *ACB* is 66.3°?

20. A robot is used to lift a metal grill from a bin and place it precisely in a housing for welding. When the robot has placed the grill, a welding robot will make the spot weld. A concern is raised over the amount of heat reaching the unshielded shoulder joint of the robot. As shown in Fig. 15–31, the robotic arm has a 65.0 cm arm extended from a base (shoulder) and a second arm 20.0 cm long connected to the first at the "elbow." If the angle between the two arms is 115°, what is the distance between the manipulator holding the grill and the shoulder joint?

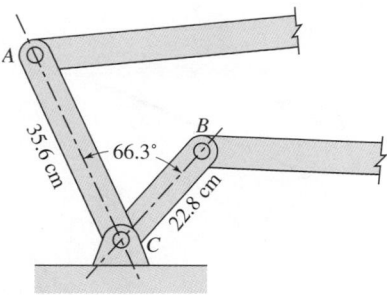

FIGURE 15–30

Geometry

21. Find angles *A*, *B*, and *C* in the quadrilateral in Fig. 15–32.

22. Find side *AB* in the quadrilateral in Fig. 15–33.

23. Two sides of a parallelogram are 22.8 and 37.8 m, and one of the diagonals is 42.7 m. Find the angles of the parallelogram.

24. Find the lengths of the sides of a parallelogram if its diagonal, which is 125 mm long, makes angles with the sides of 22.7° and 15.4°.

25. Find the lengths of the diagonals of a parallelogram, two of whose sides are 3.75 m and 1.26 m; their included angle is 68.4°.

26. A *median* of a triangle is a line joining a vertex to the midpoint of the opposite side. In triangle *ABC*, *A* = 62.3°, *b* = 112, and the median from *C* to the midpoint of *c* is 186. Find *c*.

27. The sides of a triangle are 124, 175, and 208. Find the length of the median drawn to the longest side.

28. The angles of a triangle are in the ratio 3:4:5, and the shortest side is 994. Solve the triangle.

29. The sides of a triangle are in the ratio 2:3:4. Find the cosine of the largest angle.

30. Two solar panels are to be placed as shown in Fig. 15–34. Find the minimum distance *x* so that the first panel will not cast a shadow on the second when the angle of elevation of the sun is 18.5°.

31. Find the overhang *x* so that the window in Fig. 15–35 will be in complete shade when the sun is 60° above the horizontal.

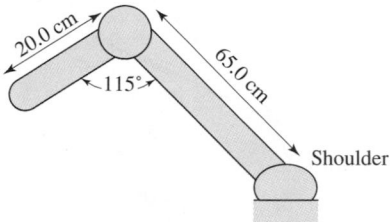

FIGURE 15–31

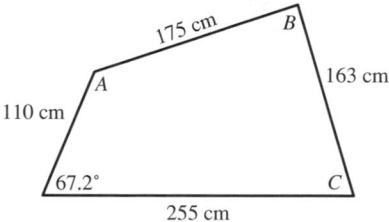

FIGURE 15–32

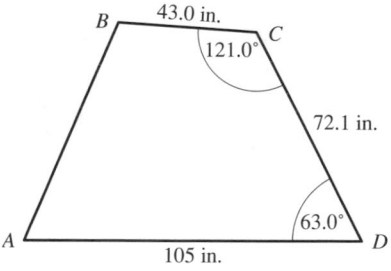

FIGURE 15–33

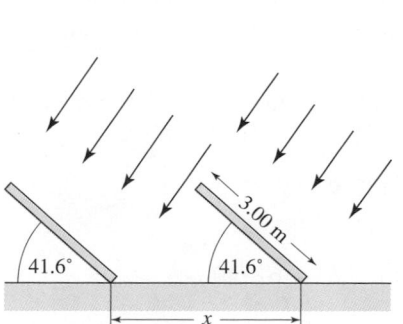

FIGURE 15–34 Solar panels.

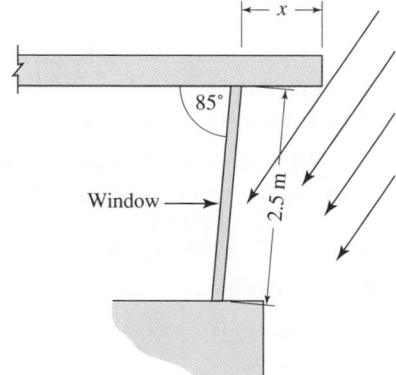

FIGURE 15–35

15–5 Addition of Vectors

In Sec. 7–6 we added two or more nonperpendicular vectors by first resolving each into components, then adding the components, and finally resolving the components into a single resultant.

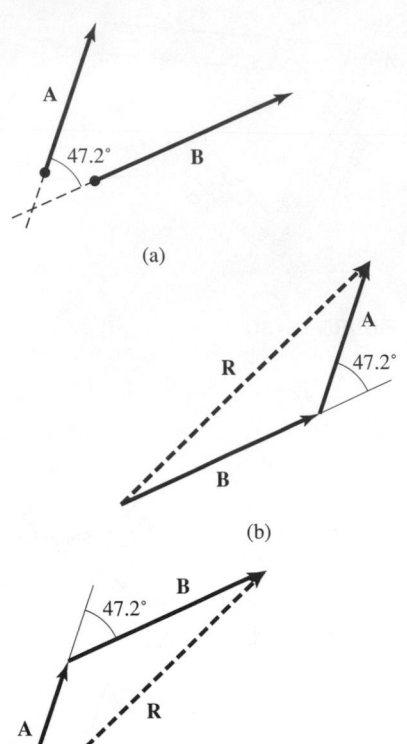

(a)

(b)

(c)

FIGURE 15–36   Addition of vectors.

Now, by using the law of sines and the law of cosines, we can combine two nonperpendicular vectors directly, with much less time and effort. However, when more than two vectors must be added, it is faster to resolve each into its x and y components, combine the x components and the y components, and then find the resultant of those two perpendicular vectors.

In this section we show both methods.

Vector Diagram

We can illustrate the resultant, or vector sum, of two vectors by means of a diagram. Suppose that we wish to add vectors **A** and **B** in Fig. 15–36(a). If we draw the two vectors *tip to tail*, as in Fig. 15–36(b), the resultant **R** will be the vector that will complete the triangle when drawn from the tail of the first vector to the tip of the second vector. It does not matter whether vector **A** or vector **B** is drawn first; the same resultant will be obtained either way, as shown in Fig. 15–36(c).

The *parallelogram method* will give the same result. To add the same two vectors **A** and **B** as before, we first draw the given vectors *tail to tail* (Fig. 15–37) and complete a parallelogram by drawing lines from the tips of the given vectors, parallel to the given vectors. The resultant **R** is then the diagonal of the parallelogram drawn from the intersections of the tails of the original vectors.

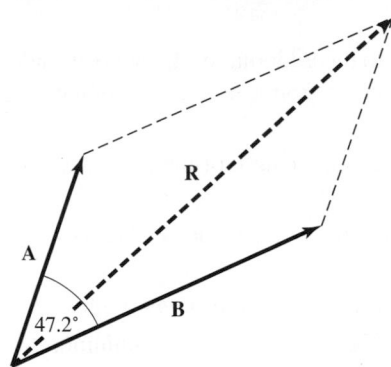

FIGURE 15–37   Parallelogram method.

Finding the Resultant of Two Nonperpendicular Vectors

Whichever method we use for drawing two vectors, *the resultant is one side of an oblique triangle*. To find the length of the resultant and the angle that it makes with one of the original vectors, we simply solve the oblique triangle by the methods we learned earlier in this chapter.

◆◆◆ **Example 27:** Two vectors, **A** and **B**, make an angle of 47.2° with each other as shown in Fig. 15–37. If their magnitudes are $A = 125$ and $B = 146$, find the magnitude of the resultant **R** and the angle that **R** makes with vector **B**.

Solution: We make a vector diagram, either tip to tail or by the parallelogram method. Either way, we must solve the oblique triangle in Fig. 15–38 for R and ϕ. Finding θ yields

$$\theta = 180° - 47.2° = 132.8°$$

By the law of cosines,

$$R^2 = (125)^2 + (146)^2 - 2(125)(146) \cos 132.8° = 61\,740$$
$$R = 248$$

Then, by the law of sines,

$$\frac{\sin \phi}{125} = \frac{\sin 132.8}{248}$$

$$\sin \phi = \frac{125 \sin 132.8}{248} = 0.3698$$

$$\phi = 21.7°$$ ◆◆◆

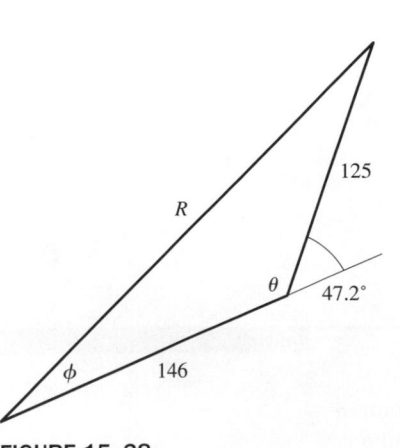

FIGURE 15–38

Addition of Several Vectors

The law of sines and the law of cosines are good for adding *two* nonperpendicular vectors. However, when *several* vectors are to be added, we usually break each into its x and y components and combine them, as in the following example.

◆◆◆ **Example 28:** Find the resultant of the vectors shown in Fig. 15–39(a).

Solution: The x component of a vector of magnitude V at any angle θ is

$$V \cos \theta$$

and the y component is

$$V \sin \theta$$

These equations apply for an angle in any quadrant. We compute and tabulate the x and y components of each original vector and find the sums of each as shown in the following table.

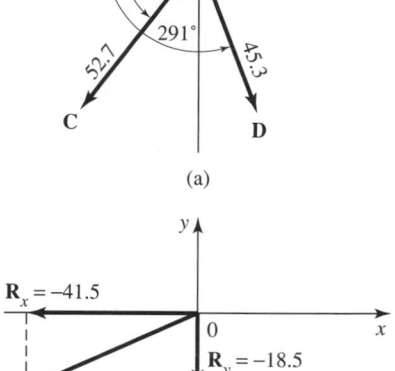

(a)

Vector	x Component	y Component
A	$42.0 \cos 58.0° = 22.3$	$42.0 \sin 58.0° = 35.6$
B	$56.1 \cos 148° = -47.6$	$56.1 \sin 148° = 29.7$
C	$52.7 \cos 232° = -32.4$	$52.7 \sin 232° = -41.5$
D	$45.3 \cos 291° = 16.2$	$45.3 \sin 291° = -42.3$
R	$R_x = -41.5$	$R_y = -18.5$

The two vectors R_x and R_y are shown in Fig. 15–39(b). We find their resultant R by the Pythagorean theorem.

$$R^2 = (-41.5)^2 + (-18.5)^2 = 2065$$
$$R = 45.4$$

We find the angle θ by

$$\theta = \arctan \frac{R_y}{R_x}$$
$$= \arctan \frac{-18.5}{-41.5}$$
$$= 24.0° \text{ or } 204°$$

Since our resultant is in the third quadrant, we drop the $24.0°$ value. Thus the resultant has a magnitude of 45.4 and a direction of 204°. This is often written in the form

$$R = 45.4 \, \underline{/204°}$$

(b)

FIGURE 15–39

◆◆◆   This is called *polar form*, which we'll cover in Chapter 17.

Exercise 5   ◆   Addition of Vectors

The magnitudes of vectors **A** and **B** are given in the following table, as well as the angle between the vectors. For each, find the magnitude R of the resultant and the angle that the resultant makes with vector **B**.

In Sec. 7–7 we gave examples of applications of force vectors, velocity vectors, and impedance vectors. You might want to review that section before trying the applications here. These problems are set up in exactly the same way, only now they will require the solution of an oblique triangle rather than a right triangle.

	Magnitudes		
	A	**B**	**Angle**
1.	244	287	21.8°
2.	1.85	2.06	136°
3.	55.9	42.3	55.5°
4.	1.006	1.745	148.4°
5.	4483	5829	100.0°
6.	35.2	23.8	146°

Force Vectors

7. Two forces of 18.6 N and 21.7 N are applied to a point on a body. The angle between the forces is 44.6°. Find the magnitude of the resultant and the angle that it makes with the larger force.

8. Two forces whose magnitudes are 187 N and 206 N act on an object. The angle between the forces is 88.4°. Find the magnitude of the resultant force.

9. A force of 125 N pulls due west on a body, and a second force pulls N 28.7° W. The resultant force is 212 N. Find the second force and the direction of the resultant.

10. Forces of 675 N and 828 N act on a body. The smaller force acts due north; the larger force acts N 52.3° E. Find the direction and the magnitude of the resultant.

11. Two forces of 925 N and 1130 N act on an object. Their lines of action make an angle of 67.2° with each other. Find the magnitude and the direction of their resultant.

12. Two forces of 136 lb. and 251 lb. act on an object with an angle of 53.9° between their lines of action. Find the magnitude of their resultant and its direction.

13. The resultant of two forces of 1120 N and 2210 N is 2870 N. What angle does the resultant make with each of the two forces?

14. Three forces are in equilibrium: 212 N, 325 N, and 408 N. Find the angles between their lines of action.

Velocity Vectors

See Fig. 7–45 for definitions of the flight terms used in these problems.

15. As an airplane heads west with an air speed of 325 km/h; a wind with a speed of 35.0 km/h causes the plane to travel slightly south of west with a ground speed of 305 km/h. In what direction is the wind blowing? In what direction does the plane travel?

16. A boat heads S 15.0° E on a river that flows due west. The boat travels S 11.0° W with a speed of 25.0 km/h. Find the speed of the current and the speed of the boat in still water.

17. A pilot wishes to fly in the direction N 45.0° E. The wind is from the west at 36.0 km/h, and the plane's speed in still air is 388 km/h. Find the heading and the ground speed.

18. The heading of a plane is N 27.7° E, and its air speed is 255 mi./h. If the wind is blowing from the south with a velocity of 42.0 mi./h, find the actual direction of travel of the plane and its ground speed.

19. A plane flies with a heading of N 48.0° W and an air speed of 584 km/h. It is driven from its course by a wind of 58.0 km/h from S 12.0° E. Find the ground speed and the drift angle of the plane.

Current and Voltage Vectors

20. We will see later that it is possible to represent an alternating current or voltage by a vector whose length is equal to the maximum amplitude of the current or voltage, placed at an angle that we later define as the *phase angle*. Then to add two alternating currents or voltages, we *add the vectors* representing those voltages or currents in the same way that we add force or velocity vectors.

    A current I_1 is represented by a vector of magnitude 12.5 A at an angle of 15.6°, and a second current I_2 is represented by a vector of magnitude 7.38 A at an angle of 132°, as shown in Fig. 15–40. Find the magnitude and the direction of the sum of these currents, represented by the vector I.

21. Figure 15–41 shows two impedances in parallel, with the currents in each represented by

$$I_1 = 18.4 \text{ A} \quad \text{at } 51.5°$$

and

$$I_2 = 11.3 \text{ A} \quad \text{at } 0°$$

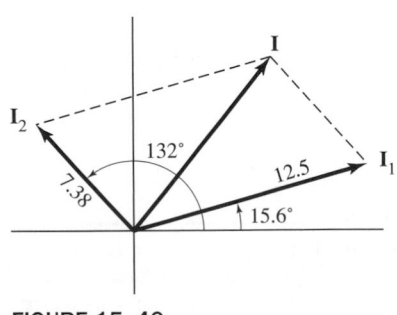

FIGURE 15–40

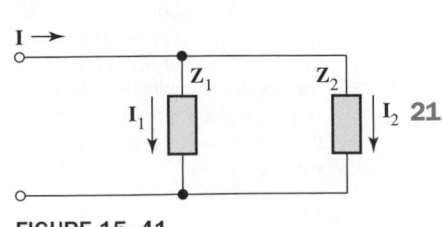

FIGURE 15–41

The current $\mathbf{I}$ will be the vector sum of $\mathbf{I}_1$ and $\mathbf{I}_2$. Find the magnitude and the direction of the vector representing $\mathbf{I}$.

22. Figure 15–42 shows two impedances in series, with the voltage drop $\mathbf{V}_1$ equal to 92.4 V at 71.5° and $\mathbf{V}_2$ equal to 44.2 V at −53.8°. Find the magnitude and the direction of the vector representing the total drop $\mathbf{V}$.

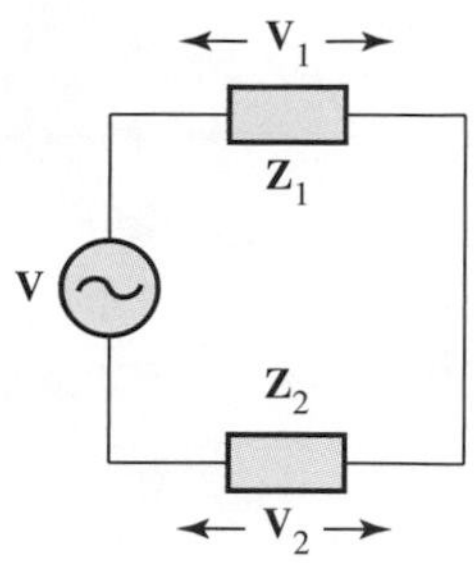

Addition of Several Vectors

Find the resultant of each of the following sets of vectors.

23. $273\underline{/34.0°}$, $179\underline{/143°}$, $203\underline{/225°}$, $138\underline{/314°}$

24. $72.5\underline{/284°}$, $28.5\underline{/331°}$, $88.2\underline{/104°}$, $38.9\underline{/146°}$

FIGURE 15–42

Case Study Discussion—Bracket for Solar Panel Frame

The bracket represents a SAS triangle. The minimum length of the opposite side can be calculated using the lower angle and the formula $c^2 = a^2 + b^2 - 2ab \cos C$, where a and b are the known sides, 35 cm and 20 cm, and c is the side opposite the angle formed at the junction of sides a and b, angle C. The lower value of c would be

$$c = \sqrt{a^2 + b^2 - 2ab \cos C}$$
$$= \sqrt{35^2 + 20^2 - 2(35)(20) \cos 35°} = 21.9 \text{ cm}$$

The longest side can be calculated using the wider angle of 55°

$$c = \sqrt{a^2 + b^2 - 2ab \cos C}$$
$$= \sqrt{35^2 + 20^2 - 2(35)(20) \cos 55°} = 28.7 \text{ cm}$$

So the side opposite this changing angle can range from 21.9 cm to 28.7 cm.

⬥⬥⬥ CHAPTER 15 REVIEW PROBLEMS ⬥⬥⬥⬥⬥⬥⬥⬥⬥⬥⬥⬥⬥⬥⬥⬥⬥⬥⬥⬥⬥⬥⬥⬥⬥⬥⬥⬥

Solve oblique triangle ABC if:

1.	$C = 135°$	$a = 44.9$	$b = 39.1$
2.	$A = 92.4°$	$a = 129$	$c = 83.6$
3.	$B = 38.4°$	$a = 1.84$	$c = 2.06$
4.	$B = 22.6°$	$a = 2840$	$b = 1170$
5.	$A = 132°$	$b = 38.2$	$c = 51.8$

In what quadrant(s) will the terminal side of θ lie if:

6. $\theta = 227°$　　**7.** $\theta = -45°$　　**8.** $\theta = 126°$　　**9.** $\theta = 170°$

10. $\tan \theta$ is negative

Without using book or calculator, state the algebraic sign of:

11. $\tan 275°$　　　　　**12.** $\sec(-58°)$

13. $\cos 183°$　　　　　**14.** $\cos 45°$

15. $\sin 300°$

Write the sin, cos, and tan, to three significant digits, for the angle whose terminal side passes through the given point.

16. $(-2, 5)$　　　　**17.** $(-3, -4)$　　　　**18.** $(5, -1)$

Two vectors of magnitudes A and B are separated by an angle θ. Find the resultant and the angle that the resultant makes with vector $\mathbf{B}$.

19. $A = 837$　　$B = 527$　　$\theta = 58.2°$

20. $A = 2.58$　　$B = 4.82$　　$\theta = 82.7°$

21. $A = 44.9$　　$B = 29.4$　　$\theta = 155°$

22. $A = 8374$ $B = 6926$ $\theta = 115.4°$

23. From a ship sailing north at the rate of 18.0 km/h, the bearing of a lighthouse is N 18°15′ E. Ten minutes later the bearing is N 75°46′ E. How far is the ship from the lighthouse at the time of the second observation?

Write the sin, cos, and tan, to four decimal places, of:

24. 273°   **25.** 175°   **26.** 334°36′

27. 127°22′   **28.** 114°

Evaluate each expression to four significant digits.

29. $\sin 35° \cos 35°$

30. $\tan^2 68°$

31. $(\cos 14° + \sin 14°)^2$

32. What angle does the slope of a hill make with the horizontal if a vertical tower 18.5 m tall, located on the slope of the hill, is found to subtend an angle of 25.5° from a point 35.0 m directly downhill from the foot of the tower, measured along the slope?

33. Three forces are in equilibrium. One force of 457 N acts in the direction N 28.0° W. The second force acts N 37.0° E. Find the direction of the third force of magnitude 638 N.

Find to the nearest tenth of a degree all nonnegative values of θ less than 360°.

34. $\cos \theta = 0.736$

35. $\tan \theta = -1.16$

36. $\sin \theta = 0.774$

Evaluate to the nearest tenth of a degree.

37. $\arcsin 0.737$

38. $\tan^{-1} 4.37$

39. $\cos^{-1} 0.174$

40. A ship wishes to travel in the direction N 38.0° W. The current is from due east at 4.20 mi./h, and the speed of the ship in still water is 18.5 mi./h. Find the direction in which the ship should head and the speed of the ship in the actual direction of travel.

41. Two forces of 483 N and 273 N act on a body. The angle between the lines of action of these forces is 48.2°. Find the magnitude of the resultant and the angle that it makes with the 483-N force.

42. Find the vector sum of two voltages, $\mathbf{V}_1 = 134\underline{/24.5°}$ and $\mathbf{V}_2 = 204\underline{/85.7°}$

43. A point on a rotating wheel has a tangential velocity of 523 cm/s. Find the x and y components of the velocity when the point is in the position shown in Fig. 15–43.

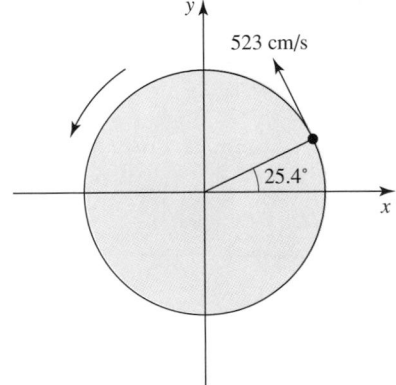

FIGURE 15–43

Writing

44. Suppose that you have analyzed a complex bridge truss made up of many triangles, sometimes using the law of sines and sometimes the more time-consuming law of cosines. Your client's accountant, angry over the size of your bill, has attacked your report for often using the longer law when it is clear to him that the shorter law of sines is also "good for solving triangles." Write a letter to your client explaining why you sometimes had to use one law and sometimes the other.

Team Project

45. Four mutually tangent circles are shown in Fig. 15–44. Find the radius of the shaded circle.

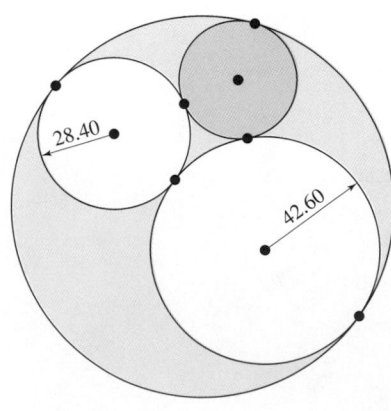

FIGURE 15–44

Radian Measure, Arc Length, and Circular Motion

••• OBJECTIVES •••

When you have completed this chapter, you should be able to:
- Measure angles in radians.
- Calculate the arc length.
- Solve problems involving circular motion.
••

We have looked at a lot of geometry, trigonometry, and angles. In this chapter we look at what happens in rotation—when we start putting angles in motion. In velocity, we measure the change in distance over a period of time, so we have metres per second. In rotation, we measure a change in angle over a period of time, so we have degrees per second to tell us how fast something is spinning. This is called *angular velocity*, and we often see it expressed in revolutions per second or revolutions per minute. We will also look at another way of measuring the size of an angle—in *radians*. Because the radian links the size of an angle with the radius of a circle, it simplifies formulas involving distance travelled along a circular path. So in this chapter we will talk about spinning wheels and pulleys, generators, pumps, and even the earth rotating on its own axis or revolving around the sun.

CASE STUDY—WATER TREATMENT PLANT

You are part of an Engineers without Borders team installing a water treatment plant in Haiti. You have rebuilt and adapted a pump to lift water, but the pump starts to vibrate. You suspect cavitation on the impeller blades. After an exchange of emails, you are advised that the speed at the tip of the blades must not exceed 2.1 m/s. If the impeller has a 12-inch diameter, what is the maximum number of revolutions per minute for smooth operation?

ALOUETTE TOPSIDE SOUNDER—COLIN FRANKLIN   CANADIAN BIOGRAPHY

In 1987, the Centennial Engineering Board of Canada described the Alouette topside sounder as one of the ten most outstanding achievements of Canadian engineering of the past 100 years. This Canadian designed and built satellite was the first step in a multinational effort to explore the upper ionosphere. September 29, 2012, marks the 50th anniversary of the launch of Alouette, the first satellite completely built by a country other than the U.S. or USSR.

The chief electrical engineer for Alouette was Dr. **Colin Franklin**. Born in New Zealand in 1927, he earned degrees in physics (NZ) and electrical engineering (UK) and came to Canada in 1957, joining the Defence Research Telecommunications Establishment (DRTE) in Ottawa. Following the first Alouette, he was involved in engineering subsequent scientific research

16–1 Radian Measure

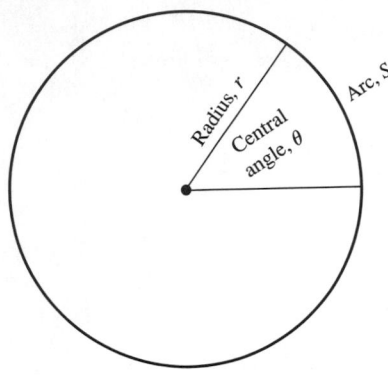

FIGURE 16–1

A *central angle* is one whose vertex is at the centre of the circle (Fig. 16–1). An *arc* is a portion of the circle. In Chapter 6 we introduced the *radian*: If an arc is laid off along a circle with a length equal to the radius of the circle, the central angle subtended by this arc is defined as *one radian*, as shown in Fig. 16–2.

Angle Conversion

An arc having a length equal to the radius of the circle subtends a central angle of one radian; an arc having a length of twice the radius subtends a central angle of two radians; and so on. Thus an arc with a length of 2π times the radius (the entire circumference) subtends a central angle of 2π radians. Therefore 2π radians is equal to 1 revolution, or 360°. This gives us the following conversions for angular measure:

Angle Conversions	$1 \text{ rev} = 360° = 2\pi \text{ rad}$	117

We convert angular units in the same way that we converted other units in Chapter 1.

◆◆◆ **Example 1:** Convert 47.6° to radians and revolutions.

Solution: By Eq. 117,

$$47.6° \left(\frac{2\pi \text{ rad}}{360°} \right) = 0.831 \text{ rad}$$

Note that 2π and 360° are exact numbers, so we keep the same number of digits in our answer as in the given angle. Now converting to revolutions, we obtain

$$47.6° \left(\frac{1 \text{ rev}}{360°} \right) = 0.132 \text{ rev}$$ ◆◆◆

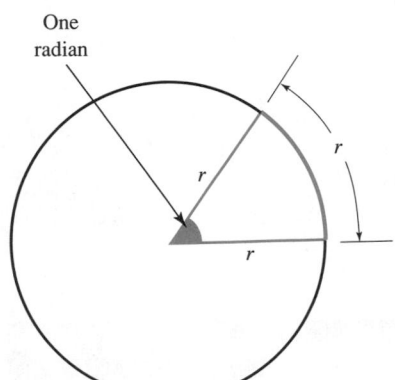

FIGURE 16–2

If we divide 360° by 2π we get

$$1 \text{ rad} \approx 57.3°$$

Some prefer to use Eq. 117 in this approximate form instead of the one given.

◆◆◆ **Example 2:** Convert 1.8473 rad to degrees and revolutions.

Solution: By Eq. 117,

$$1.8473 \text{ rad} \left(\frac{360°}{2\pi \text{ rad}} \right) = 105.84°$$

and

$$1.8473 \text{ rad} \left(\frac{1 \text{ rev}}{2\pi \text{ rad}} \right) = 0.294\ 01 \text{ rev}$$ ◆◆◆

426

◆◆◆ **Example 3:** Convert 1.837 520 rad to degrees, minutes, and seconds.

Solution: We first convert to decimal degrees.

$$1.837\ 520\ \text{rad}\left(\frac{360°}{2\pi\ \text{rad}}\right) = 105.2821°$$

Converting the decimal part (0.2821°) to minutes, we obtain

$$0.2821°\left(\frac{60'}{1°}\right) = 16.93'$$

Converting the decimal part of 16.93′ to seconds yields

$$0.93'\left(\frac{60''}{1'}\right) = 56''$$

So 1.837 520 rad = 105°16′56″.   ◆◆◆

Radian Measure in Terms of π

Not only can we express radians in decimal form, but it is also very common to express radian measure in terms of π. We know that 180° equals π radians, so

$$90° = \frac{\pi}{2}\ \text{rad}$$

$$45° = \frac{\pi}{4}\ \text{rad}$$

$$15° = \frac{\pi}{12}\ \text{rad}$$

and so on. Thus to convert an angle from degrees to radians in terms of π, multiply the angle by (π rad/180°), and reduce to lowest terms.

◆◆◆ **Example 4:** Express 135° in radian measure in terms of π.

Solution:

$$135°\left(\frac{\pi\ \text{rad}}{180°}\right) = \frac{135\pi}{180}\ \text{rad} = \frac{3\pi}{4}\ \text{rad}$$ ◆◆◆

Common Error	Students sometimes confuse the decimal value of π with the degree equivalent of π radians.
	
	    What is the value of π? 180 or 3.1416 . . .?
	    Remember that the approximate decimal value of π is always
	
	$$\pi \cong 3.1416$$
	
	but that π radians converted to degrees equals
	
	$$\pi\ \text{radians} = 180°$$
	
	    Note that we write "radians" or "rad" after π when referring to the angle, but not when referring to the decimal value.

To convert an angle from radians to degrees, multiply the angle by (180°/π rad). Cancel the π in numerator and denominator, and reduce.

◆◆◆ **Example 5:** Convert $7\pi/9$ rad to degrees.

Solution:

$$\frac{7\pi}{9}\ \text{rad}\left(\frac{180°}{\pi\ \text{rad}}\right) = \frac{7(180)}{9}\ \text{deg} = 140°$$ ◆◆◆

Trigonometric Functions of Angles in Radians

We use a calculator to find the trigonometric functions of angles in radians just as we did for angles in degrees. However, we first *switch the calculator into radian mode*. Consult your calculator manual for how to switch to this mode.

◆◆◆ **Example 6:** Use your calculator to verify the following to four decimal places:

(a) sin 2.83 rad = 0.3066
(b) cos 1.52 rad = 0.0508
(c) tan 0.463 rad = 0.4992   ◆◆◆

To find the cotangent, secant, or cosecant, we use the reciprocal relations, just as when working in degrees.

◆◆◆ **Example 7:** Find sec 0.733 rad to three decimal places.

Solution: We put the calculator into radian mode. The reciprocal of the secant is the cosine, so we take the cosine of 0.733 rad and get

$$\cos 0.733 \text{ rad} = 0.7432$$

Then taking the reciprocal of 0.7432 gives

$$\sec 0.733 = \frac{1}{\cos 0.733} = \frac{1}{0.7432} = 1.346$$ ◆◆◆

◆◆◆ **Example 8:** Use your calculator to verify the following to four decimal places:

(a) csc 1.33 rad = 1.0297
(b) cot 1.22 rad = 0.3659
(c) sec 0.726 rad = 1.3372   ◆◆◆

The Inverse Trigonometric Functions

The inverse trigonometric functions are found the same way as when working in degrees. Just be sure that your calculator is in radian mode.

As with the trigonometric functions, some calculators require that you press the function key before entering the number, and some require that you press it after entering the number.

◆◆◆ **Example 9:** Use your calculator to verify the following, in radians, to four decimal places:

(a) arcsin 0.2373 = 0.2396 rad
(b) $\cos^{-1} 0.5152 = 1.0296$ rad
(c) arctan 3.246 = 1.2720 rad   ◆◆◆

To find the arccot, arcsec, and arccsc, we first take the reciprocal of the given function and then find the inverse function, as shown in the following example.

◆◆◆ **Example 10:** If $\theta = \cot^{-1} 2.745$, find θ in radians to four decimal places.

Solution: If the cotangent of θ is 2.745, then

$$\cot \theta = 2.745$$

so

$$\frac{1}{\tan \theta} = 2.745$$

Taking reciprocals of both sides gives

$$\tan \theta = \frac{1}{2.745} = 0.3643$$

Remember that the inverse trigonometric function can be written in two different ways. Thus the inverse sine can be written

$$\text{arcsin } \theta \quad \text{or} \quad \sin^{-1} \theta$$

Also recall that there are infinitely many angles that have a particular value of a trigonometric function. Of these, we are finding just the smallest positive angle.

So

$$\theta = \tan^{-1} 0.3643 = 0.3494 \qquad \text{\tiny ◆◆◆}$$

To find the trigonometric function of an angle in radians expressed in terms of π using the calculator, it is necessary first to convert the angle to decimal form.

◆◆◆ **Example 11:** Find $\cos(5\pi/12)$ to four significant digits.

Solution: Converting $5\pi/12$ to decimal form, we have

$$5(\pi)/12 = 1.3090$$

With our calculator in radian mode, we then take the cosine.

$$\cos 1.3090 \text{ rad} = 0.2588 \qquad \text{\tiny ◆◆◆}$$

◆◆◆ **Example 12:** Evaluate to four significant digits:

$$5\cos(2\pi/5) + 4\sin^2(3\pi/7)$$

Solution: From the calculator,

$$\cos \frac{2\pi}{5} = 0.309\,02 \quad \text{and} \quad \sin \frac{3\pi}{7} = 0.974\,93$$

so

$$5\cos \frac{2\pi}{5} + 4\sin^2 \frac{3\pi}{7} = 5(0.309\,02) + 4(0.974\,93)^2$$
$$= 5.347 \qquad \text{\tiny ◆◆◆}$$

Areas of Sectors and Segments

Sectors and segments of a circle (Fig. 16–3) were defined in Sec. 6–4. Now we will compute their areas.

The area of a circle of radius r is given by πr^2, so the area of a semicircle, of course, is $\pi r^2/2$; the area of a quarter circle is $\pi r^2/4$; and so on. The sector area is the same fractional part of the whole area as the central angle is of the whole.

Thus if the central angle is 1/4 revolution, the sector area is also 1/4 of the total circle area.

If the central angle (in radians) is $\theta/2\pi$ revolution, the sector area is also $\theta/2\pi$ the total circle area. So

$$\text{area of sector} = \pi r^2 \left(\frac{\theta}{2\pi} \right)$$

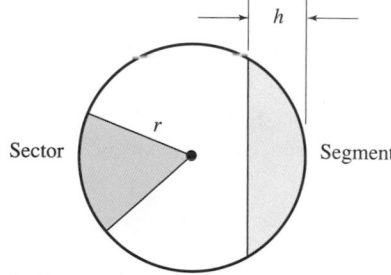

FIGURE 16–3 Segment and sector of a circle.

Area of a Sector	$\text{Area} = \dfrac{r^2\theta}{2}$ where θ is in radians	**116**

◆◆◆ **Example 13:** Find the area of a sector having a radius of 8.25 m and a central angle of 46.8°.

Solution: We first convert the central angle to radians.

$$46.8°(\pi \text{ rad}/180°) = 0.8168 \text{ rad}$$

Then, by Eq. 116,

$$\text{area} = \frac{(8.25)^2(0.8168)}{2} = 27.8 \text{ m}^2 \qquad \text{\tiny ◆◆◆}$$

The area of a *segment* of a circle (Fig. 16–3) is

Area of a Segment	$\text{Area} = r^2 \arccos \dfrac{r-h}{r} - (r-h)\sqrt{2rh - h^2}$
	where $\arccos \dfrac{r-h}{r}$ is in radians

An alternate equation for calculating the area of a segment is:

Area of a Segment (alternate equation)	$\text{Area} = \dfrac{r^2(\theta - \sin\theta)}{2}$
	where $\theta = 2 \arccos \left[\dfrac{(r-h)}{r} \right]$ (in radians)

◆◆◆ **Example 14:** Compute the area of a segment having a height of 10.0 cm in a circle of radius 25.0 cm.

Solution: Substituting into the given equation gives

$$\text{Area} = (25.0)^2 \arccos \frac{25.0 - 10.0}{25.0} - (25.0 - 10.0)\sqrt{2(25.0)(10.0) - (10.0)^2}$$
$$= 625 \arccos 0.600 - 15.0\sqrt{500 - 100}$$
$$= 625(0.9273) - 15.0(20.0) = 28\overline{0} \text{ cm}^2 \qquad\qquad ◆◆◆$$

The same solution can be found using the alternative equation for the area of a segment:
First, find θ

$$\theta = 2 \arccos [(25.0 - 10.0)/25.0]$$
$$= 106.3°$$
$$= 106.3° \, (\pi/180°) \text{ rad}$$
$$= 1.855 \text{ rad}$$

Then, substituting into the equation $r^2(\theta - \sin\theta)/2$, we obtain:

$$\text{Area} = 25.0^2(1.855 - \sin 106.5°)/2 = 279.7 \text{ cm}^2$$
$$= 28\overline{0} \text{ cm}^2$$

Exercise 1 ◆ Radian Measure

Convert to radians.

1. 47.8°	**2.** 18.7°	**3.** 35°15′
4. 0.370 rev	**5.** 1.55 rev	**6.** 1.27 rev

Convert to revolutions.

7. 1.75 rad	**8.** 2.30 rad	**9.** 3.12 rad
10. 0.0633 rad	**11.** 1.12 rad	**12.** 0.766 rad

Convert to degrees (decimal).

13. 2.83 rad	**14.** 4.275 rad	**15.** 0.372 rad
16. 0.236 rad	**17.** 1.14 rad	**18.** 0.116 rad

Convert each angle given in degrees to radian measure in terms of π.

19. 60°   **20.** 130°   **21.** 66°

22. 240°   **23.** 126°   **24.** 105°

25. 78°   **26.** 305°   **27.** 400°

28. 150°   **29.** 81°   **30.** 189°

Convert each angle given in radian measure to degrees.

31. $\dfrac{\pi}{8}$ **32.** $\dfrac{2\pi}{3}$ **33.** $\dfrac{9\pi}{11}$

34. $\dfrac{3\pi}{5}$ **35.** $\dfrac{\pi}{9}$ **36.** $\dfrac{4\pi}{5}$

37. $\dfrac{7\pi}{8}$ **38.** $\dfrac{5\pi}{9}$ **39.** $\dfrac{2\pi}{15}$

40. $\dfrac{6\pi}{7}$ **41.** $\dfrac{\pi}{12}$ **42.** $\dfrac{8\pi}{9}$

Calculator

Use a calculator to evaluate to four significant digits.

43. $\sin \dfrac{\pi}{3}$ **44.** $\tan 0.442$ **45.** $\cos 1.063$

46. $\tan\left(-\dfrac{2\pi}{3}\right)$ **47.** $\cos \dfrac{3\pi}{5}$ **48.** $\sin\left(-\dfrac{7\pi}{8}\right)$

49. $\sec 0.355$ **50.** $\csc \dfrac{4\pi}{3}$ **51.** $\cot \dfrac{8\pi}{9}$

52. $\tan \dfrac{9\pi}{11}$ **53.** $\cos\left(\dfrac{6\pi}{5}\right)$ **54.** $\sin 1.075$

55. $\cos 1.832$ **56.** $\cot 2.846$ **57.** $\sin 0.6254$

58. $\csc 0.8163$ **59.** $\arcsin 0.7263$ **60.** $\arccos 0.6243$

61. $\cos^{-1} 0.2320$ **62.** $\operatorname{arccot} 1.546$ **63.** $\sin^{-1} 0.2649$

64. $\csc^{-1} 2.6263$ **65.** $\arctan 3.7253$ **66.** $\operatorname{arcsec} 2.8463$

67. $\sin^2 \dfrac{\pi}{6} + \cos \dfrac{\pi}{6}$ **68.** $7 \tan^2 \dfrac{\pi}{9}$ **69.** $\cos^2 \dfrac{3\pi}{4}$

70. $\dfrac{\pi}{6} \sin \dfrac{\pi}{6}$ **71.** $\sin \dfrac{\pi}{8} \tan \dfrac{\pi}{8}$ **72.** $3 \sin \dfrac{\pi}{9} \cos^2 \dfrac{\pi}{9}$

73. Find the area of a sector having a radius of 5.92 cm and a central angle of 62.5°.

74. Find the area of a sector having a radius of 3.15 m and a central angle of 28.3°.

75. Find the area of a segment of height 12.4 cm in a circle of radius 38.4 cm.

76. Find the area of a segment of height 55.4 cm in a circle of radius 122.6 cm.

77. A weight bouncing on the end of a spring moves with *simple harmonic motion* according to the equation $y = 4 \cos 25t$, where y is in inches. Find the displacement y when $t = 2.00$ s. (In this equation, the angle $25t$ must be in radians.)

78. The angle D (measured at the earth's centre) between two points on the earth's surface is found by

$$\cos D = \sin L_1 \sin L_2 + \cos L_1 \cos L_2 \cos(M_1 - M_2)$$

where L_1 and M_1 are the latitude and longitude, respectively, of one point, and L_2 and M_2 are the latitude and longitude of the second point. Find the angle between Yarmouth, NS (latitude 44° N, longitude 66° W), and Whitehorse, YT (61° N, 135° W), by substituting into this equation.

79. A grinding machine for granite uses a grinding disk that has four abrasive pads with the dimensions shown in Fig. 16–4. Find the area of each pad.

80. A partial pulley is in the form of a sector with a cylindrical hub, as shown in Fig. 16–5. Using the given dimensions, find the volume of the pulley.

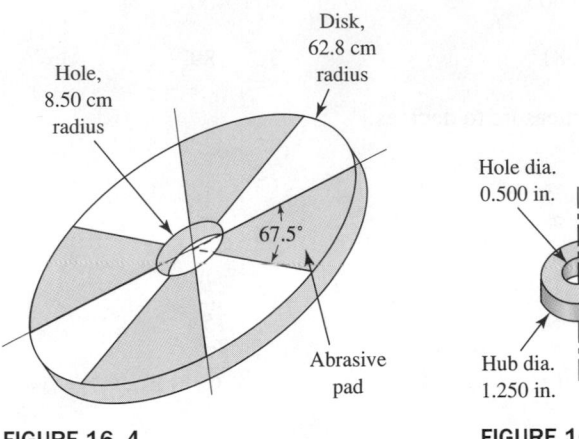

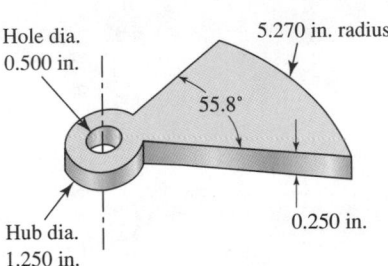

FIGURE 16–4   **FIGURE 16–5**

16–2   Arc Length

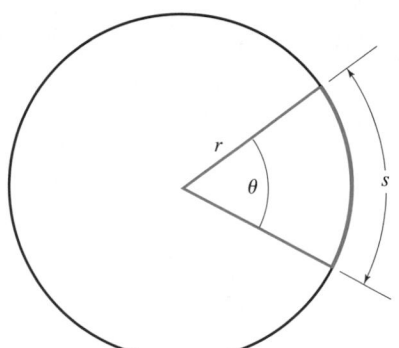

FIGURE 16–6   Relationship between arc length, radius, and central angle.

We have seen that arc length, radius, and central angle are related to each other. In Fig. 16–6, if the arc length s is equal to the radius r, we have θ equal to 1 radian. For other lengths s, the angle θ is equal to the *ratio* of s to r.

Central Angle	$\theta = \dfrac{s}{r}$ where θ is in radians	**115**

The central angle in a circle (in radians) is equal to the ratio of the intercepted arc and the radius of the circle.

When you are dividing s by r to obtain θ, s and r must have the same units. The units cancel, *leaving θ as a dimensionless ratio* of two lengths. Thus the radian is not a unit of measure like the degree or inch, although we usually carry it along as if it were.

We can use Eq. 115 to find any of the quantities θ, r, or s when the other two are known.

◆◆◆ **Example 15:** Find the angle that would intercept an arc of 27.0 m in a circle of radius 21.0 m.

Solution: From Eq. 115,

$$\theta = \frac{s}{r} = \frac{27.0 \text{ m}}{21.0 \text{ m}} = 1.29 \text{ rad} \qquad \text{◆◆◆}$$

◆◆◆ **Example 16:** Find the arc length intercepted by a central angle of 62.5° in a 10.4-cm-radius circle.

Solution: Converting the angle to radians, we get

$$62.5° \left(\frac{\pi \text{ rad}}{180°} \right) = 1.09 \text{ rad}$$

By Eq. 115,

$$s = r\theta = 10.4 \text{ cm}(1.09) = 11.3 \text{ cm} \qquad \text{◆◆◆}$$

◆◆◆ **Example 17:** Find the radius of a circle in which an angle of 2.06 rad intercepts an arc of 115 m.

Solution: By Eq. 115,

$$r = \frac{s}{\theta} = \frac{115 \text{ m}}{2.06} = 55.8 \text{ m}$$ ◆◆◆

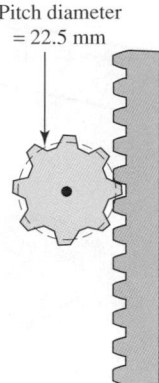

Pitch diameter = 22.5 mm

Common Error	Be sure that r and s have the same units. Convert if necessary.

FIGURE 16–7 Rack and pinion.

◆◆◆ **Example 18:** Find the angle that intercepts a 35.8-in. arc in a circle of radius 49.2 cm.

Solution: We divide s by r and convert units at the same time.

$$\theta = \frac{s}{r} = \frac{35.8 \text{ in.}}{49.2 \text{ cm}} \cdot \frac{2.54 \text{ cm}}{1 \text{ in.}} = 1.85 \text{ rad}$$ ◆◆◆

◆◆◆ **Example 19:** How far will the rack in Fig. 16–7 move when the pinion rotates 300.0°?

Solution: Converting to radians gives us

$$\theta = 300.0° \left(\frac{\pi \text{ rad}}{180°} \right) = 5.236 \text{ rad}$$

Then, by Eq. 115,

$$s = r\theta = 11.25 \text{ mm}(5.236) = 58.9 \text{ mm}$$

As a rough check, we note that the pinion rotates less than 1 revolution, so the rack will travel a distance less than the circumference of the pinion. This circumference is 22.5π or about 71 mm. So our answer of 58.91 mm seems reasonable.

◆◆◆ **Example 20:** How many kilometres north of the equator is a town of latitude 43.6° N. Assume that the earth is a sphere of radius 6370 km, and refer to the definitions of latitude and longitude shown in Fig. 16–8.

Estimate: Our given latitude angle is about 1/8 of a circle, so the required distance must be about 1/8 of the earth's circumference (13 000π or 40 000 km), or about 5000 km.

Solution: The latitude angle, in radians, is

$$\theta = 43.6° \left(\frac{\pi \text{ rad}}{180°} \right) = 0.761 \text{ rad}$$

Then, by Eq. 115,

$$s = r\theta = 6370 \text{ km}(0.761) = 4850 \text{ km}$$ ◆◆◆

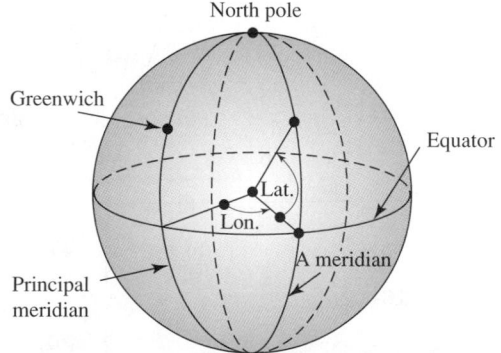

FIGURE 16–8 *Latitude* is the angle (measured at the earth's centre) between a point on the earth and the equator. *Longitude* is the angle between the meridian passing through a point on the earth and the principal (or prime) meridian passing through Greenwich, England.

Assume the earth to be a sphere with a radius of 6370 km. Actually, the distance from pole to pole is about 43 km less than the diameter at the equator.

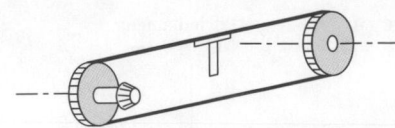

FIGURE 16–9

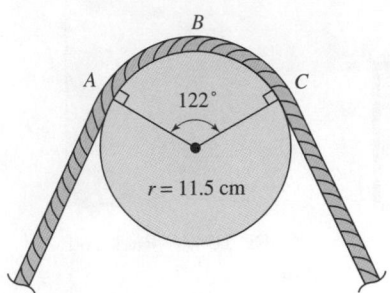

FIGURE 16–10   Belt and pulley.

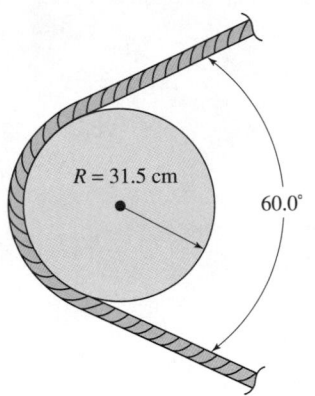

FIGURE 16–11   Brake drum.

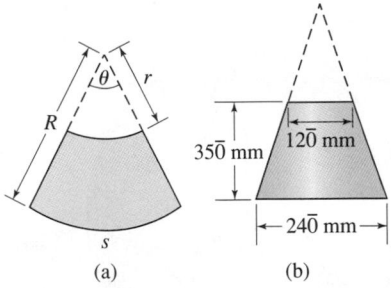

FIGURE 16–12

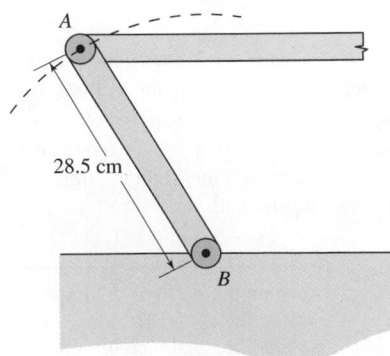

FIGURE 16–13

Exercise 2 ◆ Arc Length

In the following table, s is the length of arc subtended by a central angle θ in a circle of radius r. Fill in the missing values.

	r	θ	s
1.	483 mm	$2\pi/5$ rad	
2.	11.5 cm	1.36 rad	
3.	284 ft.	46°24′	
4.	2.87 m	1.55 rad	
5.	64.8 in.	38.5°	
6.	28.3 m		32.5 m
7.	263 mm		582 mm
8.	21.5 cm		18.2 cm
9.	3.87 m		15.8 ft.
10.		$\pi/12$ rad	88.1 in.
11.		77.2°	1.11 cm
12.		2.08 rad	3.84 m
13.		12°55′	28.2 ft.
14.		$5\pi/6$ rad	125 mm

Applications

15. A certain town is at a latitude of 35.2 °N. Find the distance in kilometres from the town to the north pole.

16. The hour hand of a clock is 85.5 mm long. How far does the tip of the hand travel between 1:00 A.M. and 11:00 A.M.?

17. Find the radius of a circular railroad track that will cause a train to change direction by 17.5° in a distance of $18\overline{0}$ m.

18. The pulley attached to the tuning knob of a radio (Fig. 16–9) has a radius of 35 mm. How far will the needle move if the knob is turned a quarter of a revolution?

19. Find the length of contact ABC between the belt and pulley in Fig. 16–10.

20. One circular "track" on a magnetic disk used for computer data storage is located at a radius of 155 mm from the centre of the disk. If 10 "bits" of data can be stored in 1 mm of track, how many bits can be stored in the length of track subtending an angle of $\pi/12$ rad?

21. If we assume the earth's orbit around the sun to be circular, with a radius of 150 million km, how many kilometres does the earth travel (around the sun) in 125 days?

22. Find the latitude of a city that is 2035 km from the equator.

23. A 1.25-m-long pendulum swings 5.75° on each side of the vertical. Find the length of arc travelled by the end of the pendulum.

24. A brake band is wrapped around a drum (Fig. 16–11). If the band has a width of 92.0 mm, find the area of contact between the band and the drum.

25. Sheet metal is to be cut from the pattern of Fig. 16–12(a) and bent to form the frustum of a cone [Fig. 16–12(b)], with top and bottom open. Find the dimensions r and R and the angle θ in degrees.

26. An isosceles triangle is to be inscribed in a circle of radius 1.000. Find the angles of the triangle if its base subtends an arc of length 1.437.

27. A satellite is in a circular orbit 225 km above the equator of the earth. How many kilometres must it travel for its longitude to change by 85.0°?

28. City B is due north of city A. City A has a latitude of 14°37′ N, and city B has a latitude of 47°12′ N. Find the distance in kilometres between the cities.

29. The link AB in the mechanism of Fig. 16–13 rotates through an angle of 28.3°. Find the distance travelled by point A.

30. Find the radius R of the sector gear of Fig. 16–14.

31. A circular highway curve has a radius of 325.500 ft. and a central angle of 15°25′15″ measured to the centreline of the road. Find the length of the curve.

32. When metal is in the process of being bent, the amount s that must be allowed for the bend is called the *bending allowance*, as shown in Fig. 16–15. Assume that the neutral axis (the line at which there is no stretching or compression of the metal) is at a distance from the inside of the bend equal to 0.4 of the metal thickness t.

    (a) Show that the bending allowance is

    $$s = \frac{A(r + 0.4t)\pi}{180°}$$

    where A is the angle of bend in degrees.

    (b) Find the bending allowance for a 60.0° bend in 1.00-cm-thick steel with a radius of 3.80 cm.

33. In Sec. 16–1 we gave the formula for the area of a circular sector of radius r and central angle θ: area $= r^2\theta/2$ (Eq. 116). Using Eq. 115, $\theta = s/r$, show that the area of a sector is also equal to $rs/2$, where s is the length of the arc intercepted by the central angle.

34. Find the area of a sector having a radius of 34.8 cm and an arc length of 14.7 cm.

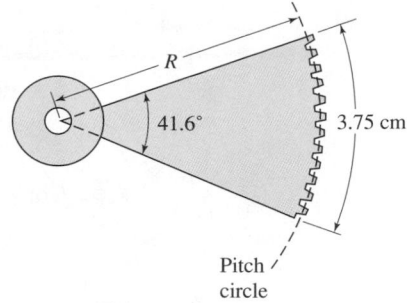

FIGURE 16–14 Sector gear.

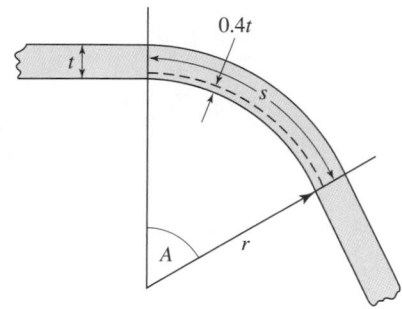

FIGURE 16–15 Bending allowance.

16–3 Uniform Circular Motion

Angular Velocity

Let us consider a rigid body that is rotating about a point O, as shown in Fig. 16–16. The *angular velocity* ω is a measure of the *rate* at which the object rotates. The motion is called *uniform* when the angular velocity is constant. The units of angular velocity are degrees, radians, or revolutions, per unit time.

Angular Displacement

The angle θ through which a body rotates in time t is called the *angular displacement*. It is related to ω and t by the following equation:

Angular Displacement	$\theta = \omega t$	**A26**

The angular displacement is the product of the angular velocity and the elapsed time.

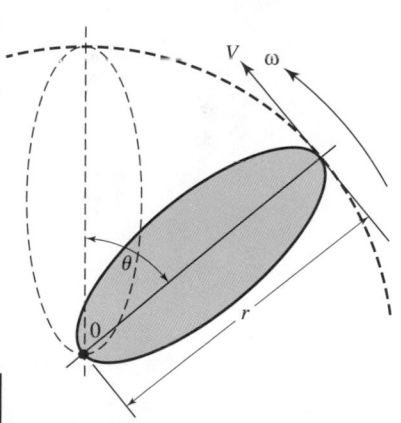

FIGURE 16–16 Rotating body. The symbol ω is lowercase Greek omega, not the letter w.

Equation A26 is similar to our old formula (distance = rate × time) for linear motion.

◆◆◆ **Example 21:** A wheel is rotating with an angular velocity of 1800 rev/min. How many revolutions does the wheel make in 1.5 s?

Solution: We first make the units of time consistent. Converting yields

$$\omega = \frac{1800 \text{ rev}}{\text{min}} \cdot \frac{1 \text{ min}}{60 \text{ s}} = 30 \text{ rev/s}$$

Then, by Eq. A26,

$$\theta = \omega t = \frac{30 \text{ rev}}{\text{s}}(1.5 \text{ s}) = 45 \text{ rev}$$

◆◆◆

◆◆◆ **Example 22:** Find the angular velocity in revolutions per minute of a pulley that rotates 275° in 0.750 s.

Solution: By Eq. A26,

$$\omega = \frac{\theta}{t} = \frac{275°}{0.750 \text{ s}} = 367 \text{ deg/s}$$

Converting to rev/min, we obtain

$$\omega = \frac{367°}{\text{s}} \cdot \frac{60 \text{ s}}{\text{min}} \cdot \frac{1 \text{ rev}}{360°} = 61.2 \text{ rev/min} \qquad ◆◆◆$$

◆◆◆ **Example 23:** How long will it take a spindle rotating at 3.55 rad/s to make 1000 revolutions?

Solution: Converting revolutions to radians, we get

$$\theta = 1000 \text{ rev} \cdot \frac{2\pi \text{ rad}}{1 \text{ rev}} = 6280 \text{ rad}$$

Then, by Eq. A26,

$$t = \frac{\theta}{\omega} = \frac{6280 \text{ rad}}{3.55 \text{ rad/s}} = 1770 \text{ s} = 29.5 \text{ min} \qquad ◆◆◆$$

Linear Speed

For any point on a rotating body, the linear displacement per unit time along the circular path is called the *linear speed*. The linear speed is zero for a point at the centre of rotation and is directly proportional to the distance r from the point to the centre of rotation. If ω is expressed in radians per unit time, the linear speed v is given by the following equation:

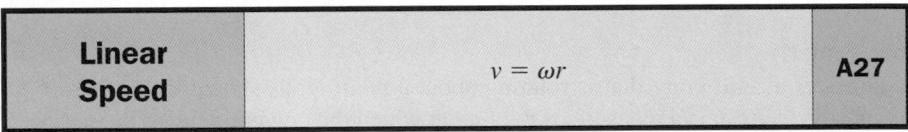

Linear Speed	$v = \omega r$	A27

The linear speed of a point on a rotating body is proportional to the distance from the centre of rotation.

◆◆◆ **Example 24:** A wheel is rotating at 2450 rev/min. Find the linear speed of a point 35.0 cm from the centre.

Solution: We first express the angular velocity in terms of radians.

$$\omega = \frac{2450 \text{ rev}}{\text{min}} \cdot \frac{2\pi \text{ rad}}{\text{rev}} = 15\,400 \text{ rad/min}$$

Then, by Eq. A27,

$$v = \omega r = \frac{15\,400 \text{ rad}}{\text{min}}(35.0 \text{ cm}) = 539\,000 \text{ cm/min}$$

$$= 89.8 \text{ m/s} \qquad ◆◆◆$$

What became of "radians" in our answer? Shouldn't the final units be rad cm/min? No. Remember that the radian is a dimensionless ratio; it is the ratio of two lengths (arc length and radius) whose units cancel.

Common Error	Remember when using Eq. A27 that the angular velocity must be expressed in *radians* per unit time.

◆◆◆ **Example 25:** A belt having a speed of 885 cm/min turns a 12.5-cm-radius pulley. Find the angular velocity of the pulley in rev/min.

Solution: By Eq. A27,

$$\omega = \frac{v}{r} = \frac{885 \text{ cm}}{\text{min}} \div 12.5 \text{ cm} = 70.8 \text{ rad/min}$$

Converting yields

$$\omega = \frac{70.8 \text{ rad}}{\text{min}} \cdot \frac{1 \text{ rev}}{2\pi \text{ rad}} = 11.3 \text{ rev/min} \qquad \text{◆◆◆}$$

Exercise 3 ◆ Uniform Circular Motion

Angular Velocity

Fill in the missing values.

	rev/min	rad/s	deg/s
1.	1850		
2.		5.85	
3.			77.2
4.		$3\pi/5$	
5.			48.1
6.	22 600		

7. A flywheel makes 725 revolutions in a minute. How many degrees does it rotate in 1.00 s?

8. A propeller on a wind generator rotates 60.0° in 1.00 s. Find the angular velocity of the propeller in revolutions per minute.

9. A gear is rotating at 2550 rev/min. How many seconds will it take to rotate through an angle of 2.00 rad?

Linear Speed

10. A milling machine cutter has a diameter of 75.0 mm and is rotating at 56.5 rev/min. What is the linear speed at the edge of the cutter?

11. A sprocket 3.00 cm in diameter is driven by a chain that moves at a speed of 55.5 cm/s. Find the angular velocity of the sprocket in revolutions per minute.

12. A capstan on a magnetic tape drive rotates at $36\overline{0}0$ rad/min and drives the tape at a speed of 45.0 m/min. Find the diameter of the capstan in millimetres.

13. A blade on a water turbine turns 155° in 1.25 s. Find the linear speed of a point on the tip of the blade 0.750 m from the axis of rotation.

14. A steel bar 6.50 in. in diameter is being ground in a lathe. The surface speed of the bar is 55.0 ft./min. How many revolutions will the bar make in 10.0 s?

15. Assuming the earth to be a sphere 12 700 km in diameter, calculate the linear speed in kilometres per hour of a point on the equator due to the rotation of the earth about its axis.

16. Assuming the earth's orbit about the sun to be a circle with a radius of 1.50×10^8 km, calculate the linear speed of the earth around the sun.

17. A car is travelling at a rate of 65.5 km/h. Its tires have a radius of 31.6 cm. Find the angular velocity of the wheels of the car.

18. A wind generator has a propeller 7.03 m in diameter, and the gearbox between the propeller and the generator has a gear ratio of 1:44 (with the generator shaft rotating faster than the propeller). Find the tip speed of the propeller, in metres per second, when the generator is rotating at $18\overline{0}0$ rev/min.

Case Study Discussion—Water Treatment Plant

Here we have many mixed units, which is common in the real world. If we convert the impeller diameter to metres, 12 inches (or 1 foot) becomes 0.3048 m. So the impeller radius is 0.1524 m, and we can use the formula $\omega = v/r$.

$$\omega = \frac{v}{r} = \frac{2.1}{0.1524} = 13.7 \text{ rads/sec}$$

Now we can convert, ω, the angular velocity of 13.7 rad/s into revolutions per minute:

$$\omega = 13.7 \text{ rad/sec} \times \frac{1 \text{ rev}}{2\pi \text{ rad}} \times \frac{60 \text{ sec}}{1 \text{ min}} = 1291 \text{ rev/min}$$

◆◆◆ CHAPTER 16 REVIEW PROBLEMS ◆◆◆◆◆◆◆◆◆◆◆◆◆◆◆◆◆◆◆◆◆◆◆◆◆◆◆◆

Convert to degrees.

1. $\dfrac{3\pi}{7}$ **2.** $\dfrac{9\pi}{4}$ **3.** $\dfrac{\pi}{9}$

4. $\dfrac{2\pi}{5}$ **5.** $\dfrac{11\pi}{12}$

6. Find the central angle in radians that would intercept an arc of 5.83 m in a circle of radius 7.29 m.

7. Find the angular velocity in rad/s of a wheel that rotates 33.5 revolutions in 1.45 min.

8. A wind generator has blades 3.50 m long. Find the tip speed when the blades are rotating at 35 rev/min.

Convert to radians in terms of π.

9. 300°   **10.** 150°   **11.** 230°   **12.** 145°

Evaluate to four significant digits.

13. $\sin \dfrac{\pi}{9}$ **14.** $\cos^2 \dfrac{\pi}{8}$ **15.** $\sin \left(\dfrac{2\pi}{5}\right)^2$

16. $3 \tan \dfrac{3\pi}{7} - 2 \sin \dfrac{3\pi}{7}$ **17.** $4 \sin^2 \dfrac{\pi}{9} - 4 \sin \left(\dfrac{\pi}{9}\right)^2$

18. What arc length is subtended by a central angle of 63.4° in a circle of radius 4.85 cm?

19. A winch has a drum 30.0 cm in diameter. The steel cable wrapped around the drum is to be pulled in at a rate of 3.00 m/s. Ignoring the thickness of the cable, find the angular speed of the drum in revolutions per minute.

20. Find the linear velocity of the tip of a 13.5-cm-long minute hand of a clock.

21. A satellite has a circular orbit around the earth with a radius of 4250 mi. The satellite makes a complete orbit of the earth in 3 days, 7 h, and 35 min. Find the linear speed of the satellite.

Evaluate to four decimal places. (Angles are in radians.)

22. $\cos 1.83$ **23.** $\tan 0.837$

24. $\csc 2.94$ **25.** $\sin 4.22$

26. $\cot 0.371$ **27.** $\sec 3.38$

28. $\cos^2 1.74$ **29.** $\sin(2.84)^2$

30. $(\tan 0.475)^2$ **31.** $\sin^2(2.24)^2$

32. Find the area of a sector with a central angle of 29.3° and a radius of 37.2 cm.

33. Find the radius of a circle in which an arc of length 384 mm intercepts a central angle of 1.73 rad.

34. Find the arc length intercepted in a circle of radius 3.85 m by a central angle of 1.84 rad.

35. A parts-storage tray, shown in Fig. 16–17, is in the form of a partial sector of a circle. Using the dimensions shown, find the inside volume of the tray in cubic centimetres.

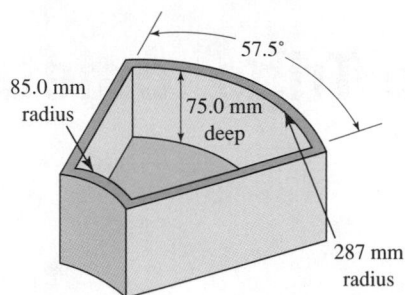

FIGURE 16–17

Writing

36. When we multiply an angular velocity in radians per second by a length in metres, we get a linear speed in metres per second. Explain why radians do not appear in the units for linear speed.

Team Projects

37. The kinetic energy of a rotating body is given by

$$E = \tfrac{1}{2}Mk^2\omega^2 \text{ joules (J)}$$

where M is the mass in kilograms, ω is the angular velocity in rad/s, and k is the radius of gyration in metres, which for a sphere of radius r is $\sqrt{2}/5r$.

Your team should weigh and measure a golf ball, a billiard ball, a baseball, a softball, and a basketball. Then compute the kinetic energy of rotation for each, assuming that it is spinning at a rate of 3.0 revolutions per second.

38. Would you like to tell your friends that you measured the circumference of the earth with just a metre stick? In the third century B.C., the Greek scientific writer, astronomer, mathematician, and poet Eratosthenes did just that. At Syene (now Aswan), he noted that the sun was directly overhead at the solstice by observing that the sun's rays shone directly into a deep well. He also learned that at that instant, the sun cast a shadow of 7.5° at Alexandria 800 km to the north. He used these data to calculate the circumference of the earth.

(a) Calculate the circumference of the earth, using Eratosthenes' data.

(b) Repeat Eratosthenes' experiment. First devise a way to measure the angle of elevation of the sun when it is on your meridian of longitude. (*Caution:* Use the *shadow* cast by a vertical stick for this. Never look directly at the sun, especially with a telescope!) Then enlist an assistant who is located at least 160 km north or south of your location (the farther the better), and determine the north-south distance between the two locations from a map. Have this person determine the angle of the sun, when it is on the meridian, using your method. That same day, do the same at your location. Use the data from both locations to calculate the circumference of the earth.

17

Graphs of the Trigonometric Functions

✦✦✦ OBJECTIVES ✦✦

When you have completed this chapter, you should be able to:
- Identify and calculate the characteristics of periodic waves and the sine curve.
- Calculate the amplitude, period, and phase angle from the general equation
  $y = a \sin(bx + c)$.
- Draw a quick sketch of the sine curve.
- Draw a quick sketch of other trigonometric functions.
- Calculate the amplitude, time period, and phase shift for the sine wave as a function of time.
- Graph on a polar coordinate system.
- Graph parametric equations.

✦✦

If a picture is worth a thousand words, a graph is worth a thousand calculations. For example, a graph can instantaneously show the thousands of points that make up a solid line. The graph can also visually demonstrate many of the characteristics of the equation that defines the graph.

In this chapter, we look at the graphs of trigonometric functions. You already have an idea of how these functions work; the graphs allow us to see how the functions vary through all possible angles. We also introduce periodic waves (think of the heartbeat monitor showing basically the same waveform repeated over and over).

These functions will describe many physical and mathematical properties that repeat regularly, in devices such as pendulums, electrical generators, or mechanical levers and gears.

17–1   The Sine Curve

Periodic Functions

A curve that *repeats* its shape over and over is called a *periodic* curve or *periodic waveform*, as shown in Fig. 17–1. The function that has this graph is thus called a *periodic function*. Each repeated portion of the curve is called a *cycle*.

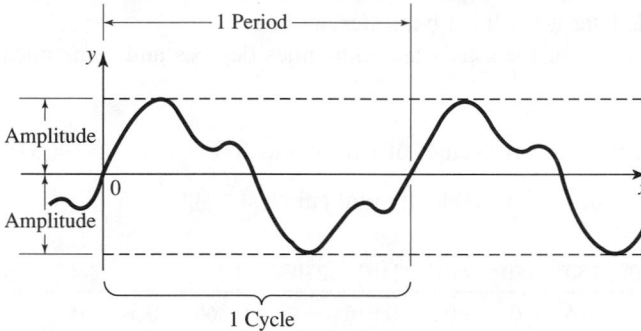

FIGURE 17–1   A periodic waveform.

The *x* axis represents either *an angle* (in degrees or radians) or *time* (usually seconds or milliseconds).

The *period* is the *x* distance taken for one cycle. Depending on the units on the *x* axis, the period is expressed in degrees per cycle, radians per cycle, or seconds per cycle.

The *frequency* of a periodic waveform is the number of cycles that will fit into 1 unit (degree, radian, or second) along the *x* axis. It is the reciprocal of the period.

$$\text{frequency} = \frac{1}{\text{period}}$$

> Recall that a *complete graph* of a function is one that contains all of the features of interest: intercepts, maximum and minimum points, and so forth. Thus a complete graph of a periodic waveform must contain at least one cycle.

Units of frequency are cycles/degree, cycles/radian, or, most commonly, cycles/second (called hertz), depending on the units of the period.

The *amplitude* is half the difference between the greatest and least value of a waveform (Fig. 17–1). The peak-to-peak value of a curve is equal to twice the amplitude.

The *instantaneous value* of a waveform is its *y* value at any given *x*.

◆◆◆ **Example 1:** For the voltage curve in Fig. 17–2, find the period, frequency, amplitude, peak-to-peak value, and the instantaneous value of *y* when *x* is 50 ms.

Solution: By observation of Fig. 17–2,

$$\text{period} = 200 \text{ ms/cycle}$$

$$\text{frequency} = \frac{1 \text{ cycle}}{200 \text{ ms}}$$

$$= 5 \text{ cycles/s} = 5 \text{ hertz (Hz)}$$

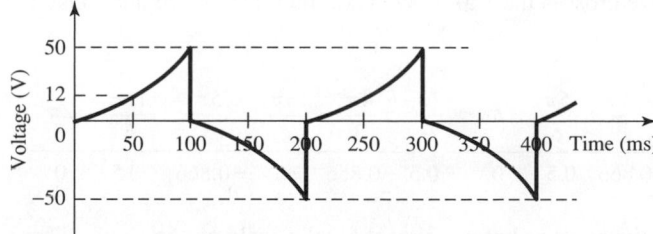

FIGURE 17–2

$$amplitude = 50 \text{ V}$$

$$peak\text{-}to\text{-}peak \text{ value} = 100 \text{ V}$$

$$instantaneous \text{ value when } x \text{ is } 50 \text{ ms} = 12 \text{ V} \qquad \bullet\bullet\bullet$$

Graph of the Sine Function $y = \sin x$

The sine curve or sine wave is also called a *sinusoid*.

We will graph the sine curve just as we graphed other functions in Sec. 5–2. We choose values of x and substitute into the equation to get corresponding values of y, and then we plot the resulting table of point pairs. A little later we will learn a faster method.

The units of angular measure used for the x axis are sometimes degrees and sometimes radians. We will show examples of each.

◆◆◆ **Example 2:** Plot the sine curve $y = \sin x$ for values of x from 0 to 360°.

Solution: Let us choose 30° intervals and make a table of point pairs.

x	0	30°	60°	90°	120°	150°	180°	210°	240°	270°	300°	330°	360°
y	0	0.5	0.866	1	0.866	0.5	0	−0.5	−0.866	−1	−0.866	−0.5	0

Plotting these points gives the periodic curve shown in Fig. 17–3. If we had plotted points from $x = 360°$ to $720°$, we would have gotten another cycle of the sine curve, identical to the one just plotted. We see then that the sine curve is periodic, requiring 360° for each cycle. Thus the period is equal to 360°.

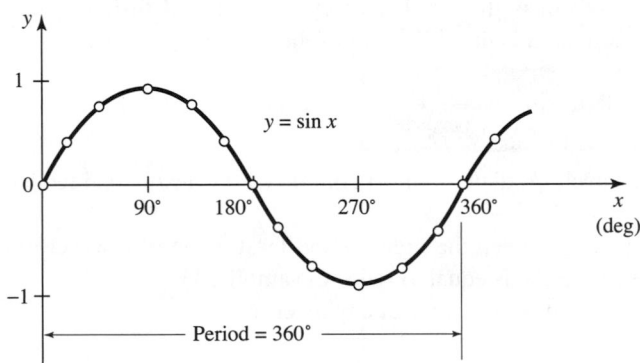

FIGURE 17–3 $y = \sin x$ (x is in degrees).   ◆◆◆

We'll now do an example in radians.

It is not easy to work in radian measure when we are used to thinking of angles in degrees. You might want to glance back at Sec. 16–1.

◆◆◆ **Example 3:** Plot $y = \sin x$ for values of x from 0 to 2π rad.

Solution: We could choose integer values for x, such as 1, 2, 3, . . . , or follow the usual practice of using multiples and fractions of π. This will give us points at the peaks and valleys of the curve and the point where the curve crosses the x axis. We take intervals of $\pi/6$ and make the following table of point pairs:

x (rad)	0	$\dfrac{\pi}{6}$	$\dfrac{\pi}{3}$	$\dfrac{\pi}{2}$	$\dfrac{2\pi}{3}$	$\dfrac{5\pi}{6}$	π	$\dfrac{7\pi}{6}$	$\dfrac{4\pi}{3}$	$\dfrac{3\pi}{2}$	$\dfrac{5\pi}{3}$	$\dfrac{11\pi}{6}$	2π
y	0	0.5	0.866	1	0.866	0.5	0	−0.5	−0.866	−1	−0.866	−0.5	0

Our curve, shown in Fig. 17–4, is the same as before, except for the scale on the x axis. There, we show both integer values of x and multiples of π.

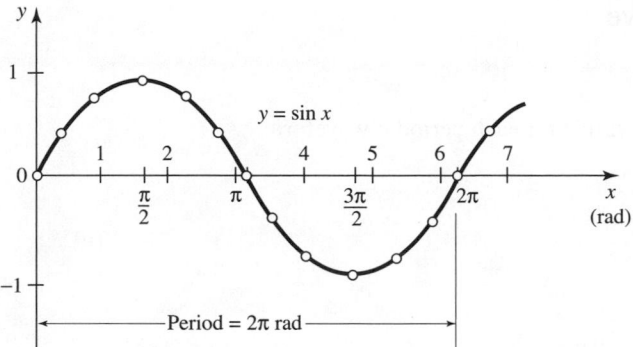

FIGURE 17–4 $y = \sin x$ (x is in radians).

Figure 17–5 shows the same curve with both radian and degree scales on the x axis. Notice that one cycle of the curve fits into a rectangle whose height is twice the amplitude of the sine curve and whose width is equal to the period. We use this later for rapid sketching of periodic curves.

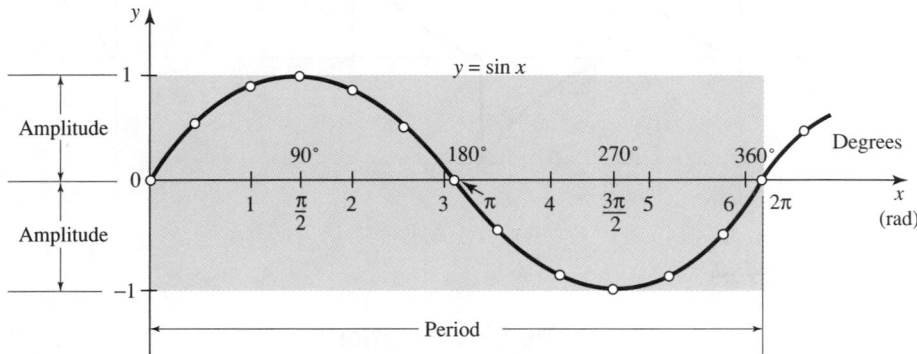

FIGURE 17–5 The sine curve (x is in degrees and radians).

Graph of $y = a \sin x$

When the sine function has a coefficient a, the absolute value of a is the *amplitude* of the sine wave. In the preceding examples, the constant a was equal to 1, and we got sine waves with amplitudes of 1. In examples that follow, a will have other values.

◆◆◆ **Example 4:** Plot the sine curve $y = 3 \sin x$ for $x = 0$ to 2π.

Solution: When we make a table of point pairs, it becomes apparent that all the y values will be three times as large as the ones in the preceding example. These points are plotted in Fig. 17–6.

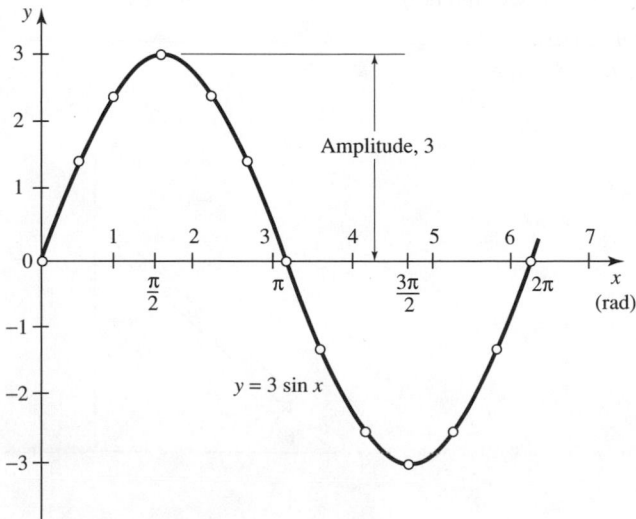

FIGURE 17–6 Graph of $y = 3 \sin x$.

◆◆◆

Exercise 1 • The Sine Curve

Periodic Waveforms

Find the period, frequency, and amplitude for each periodic waveform.

1. Figure 17–7(a).
2. Figure 17–7(b).

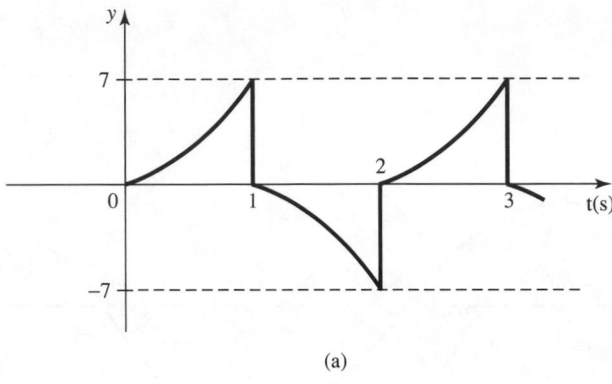

(a)

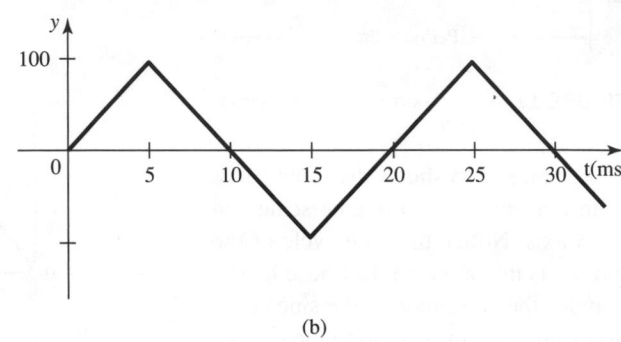

(b)

FIGURE 17-7

The Sine Function

Graph each function. Find the amplitude and period.

3. $y = 2 \sin x$
4. $y = 1.5 \sin x$
5. $y = -3 \sin x$
6. $y = 4.25 \sin x$

Graph each function and find the amplitude.

7. $y = -2.5 \sin x$ 8. $y = -3.75 \sin x$

9. As the ladder in Fig. 17–8 is raised, the height h increases as θ increases.
    (a) Write the function $h = f(\theta)$.
    (b) Graph $h = f(\theta)$ for $\theta = 0$ to $90°$.

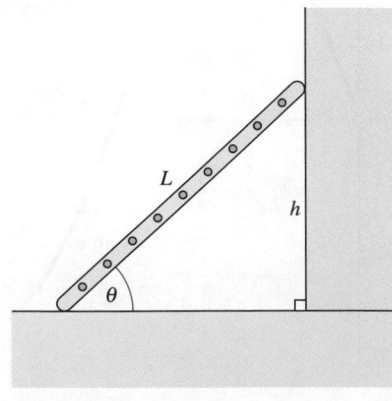

FIGURE 17-8   A ladder.

10. Figure 17–9 shows a Scotch yoke mechanism, or Scotch crosshead.
   (a) Write an expression for the displacement y of the rod PQ as a function of θ and of the length R of the rotating arm.
   (b) Graph $y = f(\theta)$ for $\theta = 0$ to $360°$.

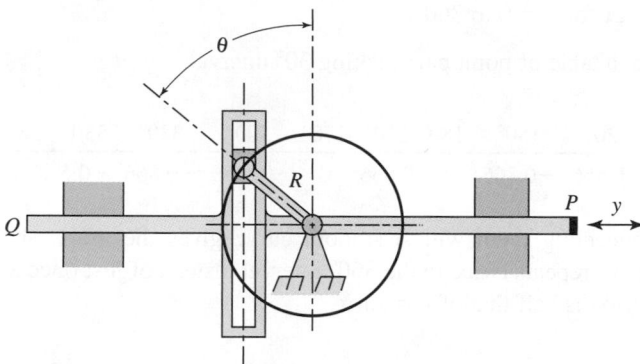

FIGURE 17–9   Scotch crosshead.

11. For the pendulum of Fig. 17–10:
   (a) Write an equation for the horizontal displacement x as a function of θ.
   (b) Graph $x = f(\theta)$ for $\theta = 0$ to $30°$.

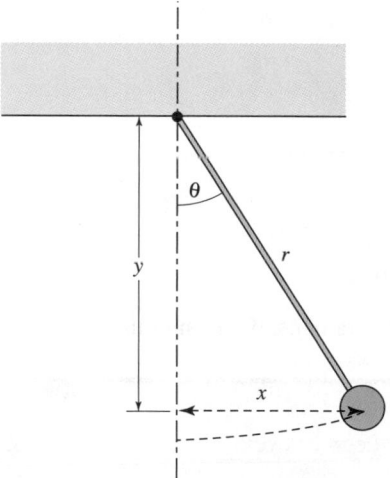

FIGURE 17–10   A pendulum.

CASE STUDY—VOLTAGE OUTPUT OF A GENERATOR

In a three-phase generator, three sine waves are added together. Each sine wave is separated by 120°, so you have one each at 0°, 120°, and 240°. By graphing the sum of the three curves, show how the combination delivers almost continuous output voltage.

17–2   The General Sine Wave, y = a sin(bx + c)

In the preceding section, we considered only sine waves that had a period of 360° and those that passed through the origin. Here we will graph sine waves of any period and those that are *shifted* in the x direction. We'll start with sine waves of any period.

Graph of y = a sin bx

For the sine function $y = a \sin x$, we saw that the curve repeated in an interval of 360° (or 2π radians). We called this interval the *period*. Now we will see how the constant b affects the graph of the equation.

✦✦✦ **Example 5:** Plot $y = \sin 2x$ for $x = 0$ to 360°.

Solution: As before, we make a table of point pairs, taking 30° intervals.

x	0	30°	60°	90°	120°	150°	180°	210°	270°	300°	330°	330°	360°
y	0	0.866	0.866	0	−0.866	−0.866	0	0.866	0	−0.866	−0.866	−0.5	0

Plotting these points and connecting them with a smooth curve gives the graph shown in Fig. 17–11. Notice that the curve repeats twice in the 360° interval instead of just once as it did before. In other words, the period is half that of $y = \sin x$.

The sine wave $y = \sin x$ is sometimes called the *fundamental*, and the sine waves

$y = \sin bx$, where $b = 1, 2, 3,$

. . .

are called the *harmonics*. Thus $y = \sin 2x$ is the second *harmonic*.

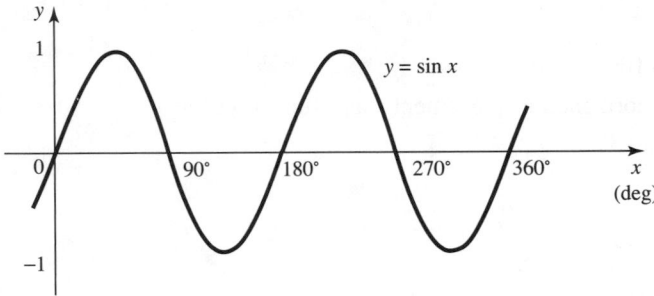

FIGURE 17–11   Graph of $y = \sin 2x$.   ✦✦✦

In general,

$$y = a \sin bx$$

will have a period of 360°/b, if x is in degrees, or $2\pi/b$ radians, if x is in radians.

Period	$P = \dfrac{360}{b}$ degrees/cycle $P = \dfrac{2\pi}{b}$ radians/cycle	211

✦✦✦ **Example 6:** Find the period for the sine curve $y = \sin 4x$ in degrees and also in radians per cycle.

Solution:

$$P = \frac{360°}{4} = 90°/\text{cycle}$$

or

$$P = \frac{2\pi}{4} = \frac{\pi}{2} \text{ rad/cycle} \qquad ✦✦✦$$

The *frequency*, f, is the reciprocal of the period.

Frequency	$f = \dfrac{b}{360}$ cycles/degree $f = \dfrac{b}{2\pi}$ cycles/radian	212

◆◆◆ **Example 7:** Determine the amplitude, period, and frequency of the sine curve $y = 2 \sin(\pi x/2)$, where x is in radians, and graph one cycle of the curve.

Solution: Here, $a = 2$ and $b = \pi/2$, so the amplitude is 2, and the period is

$$P = \frac{2\pi}{\pi/2} = 4 \text{ rad/cycle}$$

or, in degrees,

$$P = \frac{360°}{\pi/2} = 229.2°/\text{cycle}$$

We make a table of point pairs for $x = 0$ to 4 rad.

x (rad)	0	$\frac{1}{2}$	1	$\frac{3}{2}$	2	$\frac{5}{2}$	3	$\frac{7}{2}$	4
y	0	1.41	2	1.41	0	−1.41	−2	−1.41	0

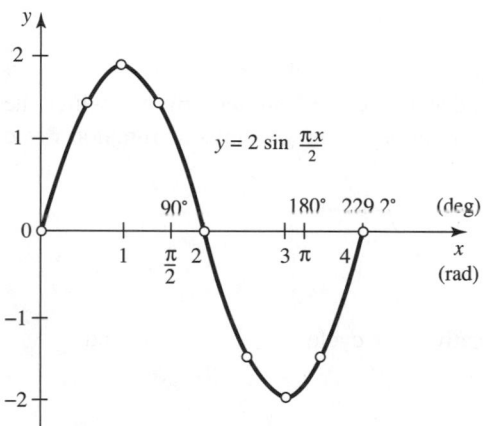

$$y = 2 \sin \frac{\pi x}{2}$$

FIGURE 17-12 Graph of $y = 2 \sin(\pi x/2)$.

These points are plotted in Fig. 17-12. The frequency is

$$f = \frac{1}{P} = \frac{1}{4} \text{ cycle/rad}$$

In degrees,

$$f = \frac{1}{229.2°} = 0.004\ 36 \text{ cycle/deg}$$ ◆◆◆

Graph of the General Sine Function y = a sin(bx + c)

The graph of $y = a \sin bx$, we have seen, passes through the origin. However, the graph of the general sine function, given in the following box, will be shifted in the x direction by an amount called the *phase displacement*, or *phase shift*.

◆◆◆ **Example 8:** Graph $y = \sin(3x - 90°)$.

Solution: Since $a = 1$ and $b = 3$, we expect an amplitude of 1 and a period of $360/b = 120°$. The effect that the constant c has on the curve will be clearer if we put the given expression into the form

$$y = a \sin b(x + c/b)$$

by factoring. Thus our equation becomes

$$y = \sin 3(x - 30°)$$

We make a table of point pairs and plot each point as shown in Fig. 17–13.

x	0	15°	30°	45°	60°	75°	90°	105°	120°	135°	150°
y	−1	−0.707	0	0.707	1	0.707	0	−0.707	−1	−0.707	0

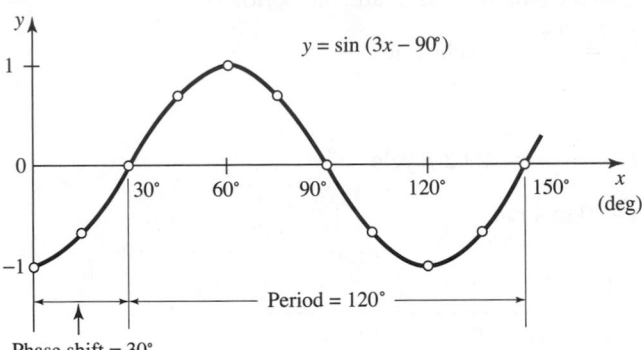

FIGURE 17–13   Sine curve with phase shift.

Our curve, unlike those plotted earlier, does not pass through the origin.   ◆◆◆

We see from Fig. 17–13 that the phase shift is that x distance from the origin at which the positive half-cycle of the sine wave starts. To find this value for the general sine function $y = a \sin(bx + c)$, we set $y = 0$ and solve for x.

$$y = a \sin(bx + c) = 0$$
$$\sin(bx + c) = 0$$
$$bx + c = 0°, 180°, \ldots$$

We discard 180° because that is the start of the negative half-cycle. So $bx + c = 0$, and

$$x = -\frac{c}{b}$$

The formula for phase shift is summarized in the following box:

Phase Shift	Phase shift $= -\dfrac{c}{b}$	**213**
	When the phase shift is *positive*, shift the curve to the *right* of its unshifted position	
	When the phase shift is *negative*, shift the curve to the *left* of its unshifted position	

The phase shift will have the same units, radians or degrees, as the constant c. Where no units are indicated for c, assume it to be in radians. If the phase shift is *negative*, it means that the positive half-cycle starts at a point to the left of the origin, so *the curve is shifted to the left*. Conversely, *a positive* phase shift means that the curve is shifted to the *right*.

◆◆◆ **Example 9:** In the equation $y = \sin(x + 45°)$,

$$a = 1, \qquad b = 1, \qquad \text{and} \qquad c = 45°$$

$$\text{phase shift} = -\frac{c}{b} = -45°$$

so the curve is shifted 45° to the left as shown in Fig. 17–14(b). For comparison, Fig. 17–14 also shows (a) the unshifted sine wave and (c) the sine wave shifted 45° to the right.

> When you study *translation of axes* in analytic geometry, you will see that replacing x in a function by $(x - h)$, where h is a constant, causes the graph of the function to be shifted h units in the x direction.

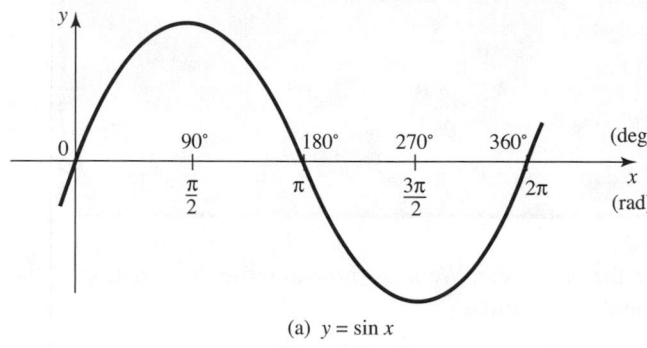

(a) $y = \sin x$
Phase shift = 0

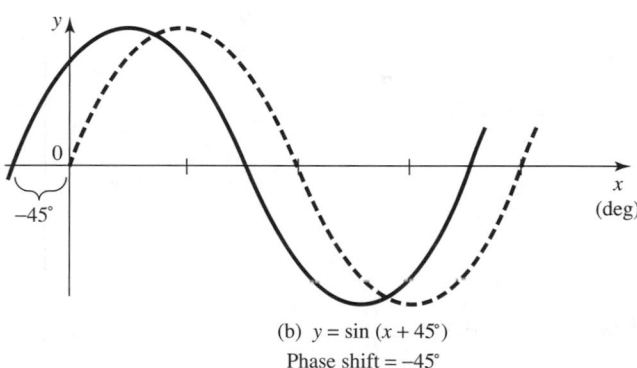

(b) $y = \sin(x + 45°)$
Phase shift = −45°

(c) $y = \sin(x − 45°)$
Phase shift = 45°

FIGURE 17–14 ◆◆◆

◆◆◆ **Example 10:** For the sine function $y = 3 \sin(4x - 2)$, we have

$$a = 3, \qquad b = 4, \quad \text{and} \quad c = -2$$

so

$$\text{phase shift} = \frac{c}{b} = \frac{1}{2} \text{ rad}$$

The curve is thus shifted $\frac{1}{2}$ rad to the right as shown in Fig. 17–15.

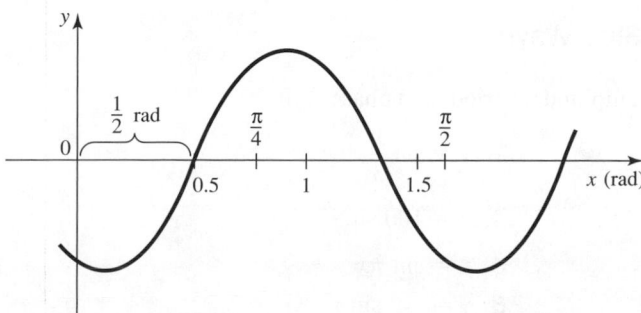

FIGURE 17–15 ◆◆◆

equations 210 through 213 are summarized as follows:

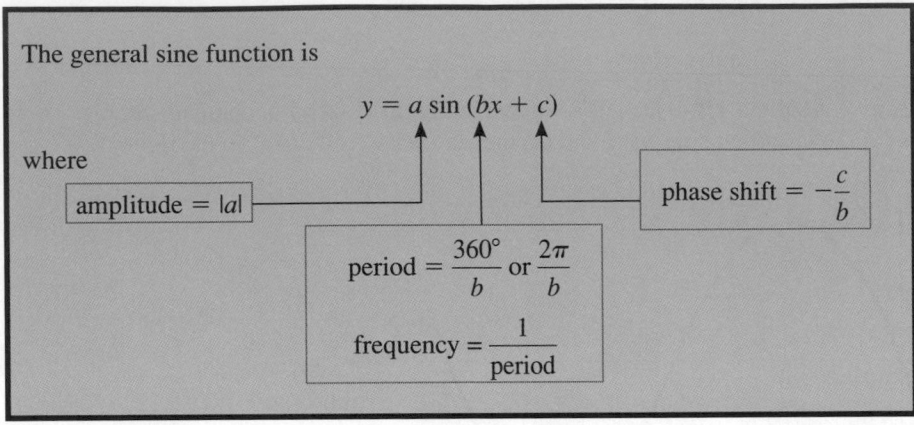

In our next example we reverse the procedure. We show how to write the equation of the curve, given the amplitude, period, and phase shift.

♦♦♦ **Example 11:** Write the equation of a sine curve in which $a = 5$, with a period of π, and a phase shift of -0.5 rad.

Solution: The period is given as π, so

$$P = \pi = \frac{2\pi}{b}$$

so

$$b = \frac{2\pi}{\pi} = 2$$

Then

$$\text{phase shift} = -0.5 = -\frac{c}{b}$$

Since $b = 2$, we get

$$c = -2(-0.5) = 1$$

Substituting into the general equation of the sine function

$$y = a \sin(bx + c)$$

with $a = 5$, $b = 2$, and $c = 1$ gives

$$y = 5 \sin(2x + 1)$$ ♦♦♦

Exercise 2 ♦ The General Sine Wave

Graph each sine function. Find the amplitude, period, and phase shift.

1. $y = \sin 2x$

2. $y = \sin \dfrac{x}{3}$

3. $y = 0.5 \sin 2x$

4. $y = 100 \sin \dfrac{3x}{2}$

5. $y = 2 \sin 3x$

6. $y = \sin \pi x$

7. $y = \sin 3\pi x$

8. $y = -3 \sin \pi x$

9. $y = 4 \sin(x - 180°)$

10. $y = -2 \sin(x + 90°)$

11. $y = \sin(x + 90°)$

12. $y = 20 \sin(2x - 45°)$

13. $y = \sin(x - 1)$

14. $y = 0.5 \sin\left(\dfrac{x}{2} - 1\right)$

15. $y = \sin\left(x + \dfrac{\pi}{2}\right)$

16. $y = \sin\left(x + \dfrac{\pi}{8}\right)$

17. Write the equations of the sine curves in Fig. 17–16.

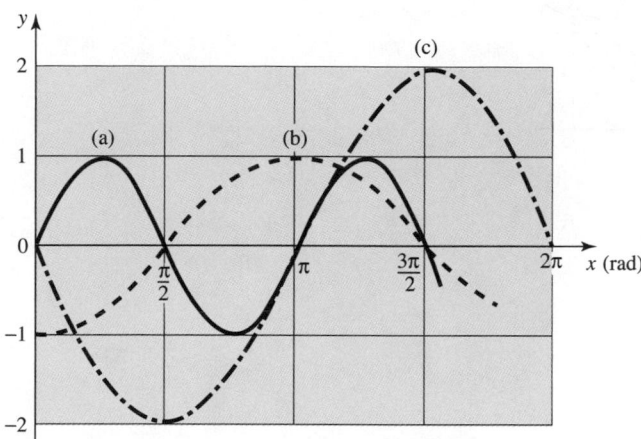

FIGURE 17–16

18. Write the equation of a sine curve with $a = 3$, a period of 4π, and a phase shift of zero.

19. Write the equation of a sine curve with $a = -2$, a period of 6π, and a phase shift of $-\pi/4$.

20. The projectile shown in Fig. 17–17 is fired with an initial velocity v at an angle θ with the horizontal. Its range is given by

$$x = \frac{v^2}{g}\sin 2\theta$$

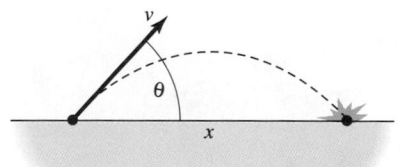

FIGURE 17–17   A projectile.

where g is the acceleration due to gravity.

(a) Graph the range of the projectile for $\theta = 0$ to $90°$. Take $v = 100$ m/s and $g = 9.807$ m/s^2.

(b) What angle gives a maximum range?

21. For the pendulum of Fig. 17–10:

(a) Write an equation for the horizontal displacement x as a function of θ, taking $\theta = 0$ when the pendulum is $15°$ from the vertical.

(b) Graph $x = f(\theta)$ for $\theta = 0$ to $30°$. Take θ as increasing in the counterclockwise direction.

22. The vertical distance x for cog A in Fig. 17–18 is given by $x = r \sin \theta$. Write an expression for the vertical distance to cog B.

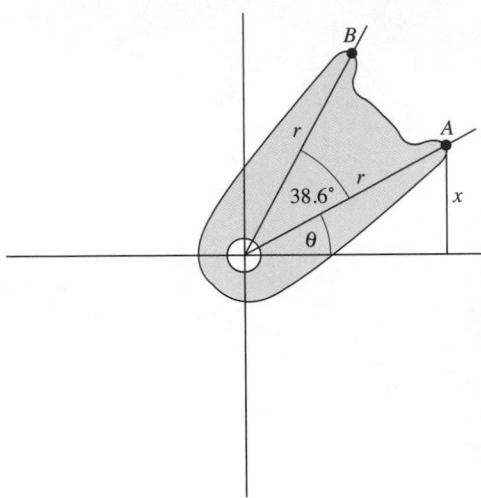

FIGURE 17–18

23. In the mechanism shown in Fig. 17–19, the rotating arm A lifts the follower F vertically.

(a) Write an expression for the vertical displacement y as a function of θ. Take $\theta = 0$ in the extreme clockwise position shown.

(b) Graph $y = f(\theta)$ for $\theta = 0$ to $90°$.

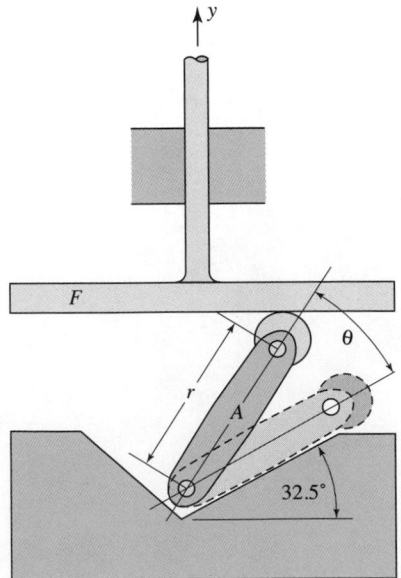

FIGURE 17–19

17–3 Quick Sketching of the Sine Curve

As shown in the preceding examples, we can plot any curve by computing and plotting a set of point pairs. But a faster way to get a sketch is to draw a rectangle of width P and height $2a$, and then sketch the sine curve (whose *shape* does not vary) within that box. First determine the amplitude, period, and phase shift. Then do the following:

1. Draw two horizontal lines each at a distance $|a|$ from the x axis.
2. Draw a vertical line at a distance P from the origin.
3. Subdivide the period P into four equal parts. Label the x axis at these points, and draw vertical lines through them.
4. Lightly sketch in the sine curve.
5. Shift the curve by the amount of the phase shift.

◆◆◆ **Example 12:** Make a quick sketch of $y = 2 \sin(3x + 60°)$.

Solution: We have $a = 2$, $b = 3$, and $c = 60°$. From Eq. 211, the period is $P = 360°/3 = 120°$ and, by Eq. 213, phase shift $= -60°/3 = -20°$. The steps for sketching the curve are shown in Fig. 17–20.

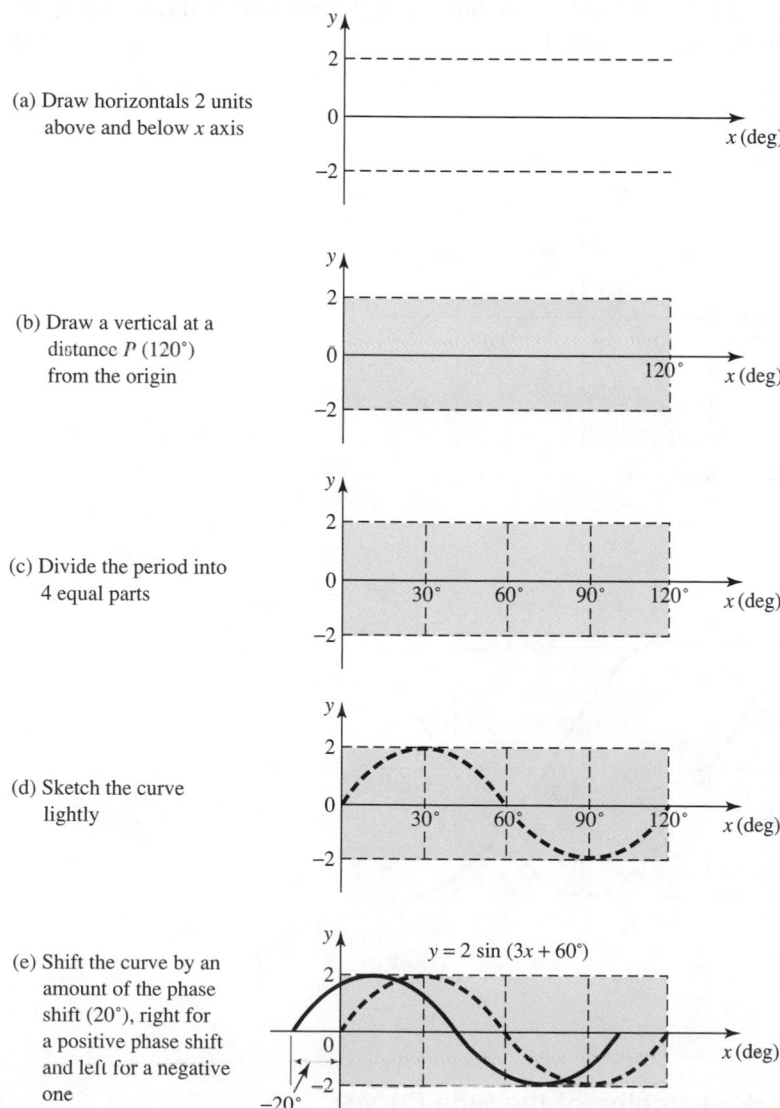

(a) Draw horizontals 2 units above and below x axis

(b) Draw a vertical at a distance P (120°) from the origin

(c) Divide the period into 4 equal parts

(d) Sketch the curve lightly

(e) Shift the curve by an amount of the phase shift (20°), right for a positive phase shift and left for a negative one

FIGURE 17–20   Quick sketching of the sine curve.   ◆◆◆

We now do an example where the units are radians.

◆◆◆ **Example 13:** Make a quick sketch of

$$y = 1.5 \sin(5x - 2)$$

Solution: From the given equation,

$$a = 1.5, \qquad b = 5, \quad \text{and} \quad c = -2 \text{ rad}$$

so

$$P = \frac{2\pi}{5} = 1.26 \text{ rad}$$

and

$$\text{phase shift} = -\left(\frac{-2}{5}\right) = 0.4 \text{ rad}$$

As shown in Fig. 17–21(a), we draw a rectangle whose height is $2 \times 1.5 = 3$ units and whose width is 1.26 rad; then we draw three verticals spaced apart by $1.26/4 = 0.315$ rad. We sketch a sine wave [shown dashed in Fig. 17–21(b)] into this rectangle, and then shift it 0.4 rad to the right to get the final curve as shown in Fig. 17–21(b).

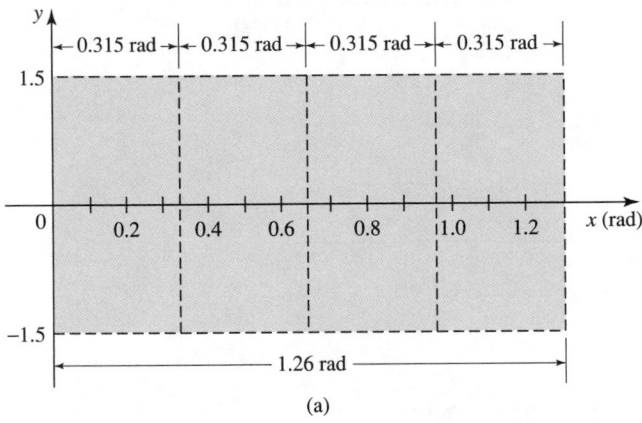

(a)

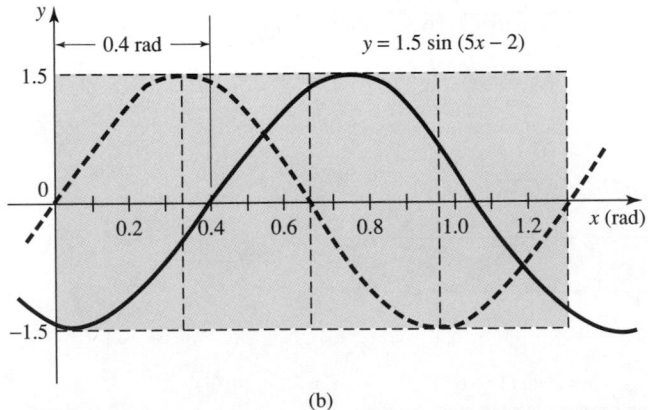

(b)

FIGURE 17–21   ◆◆◆

Exercise 3 ◆ Quick Sketching of the Sine Curve

Make a quick sketch of any of the functions in Exercise 1 or 2, as directed by your instructor.

17–4   Graphs of More Trigonometric Functions

Graph of the Cosine Function $y = \cos x$

Let us graph one cycle of the function $y = \cos x$. We make a table of point pairs and plot each point. The graph of the cosine function is shown in Fig. 17–22.

x	0	$\frac{\pi}{6}$	$\frac{\pi}{3}$	$\frac{\pi}{2}$	$\frac{2\pi}{3}$	$\frac{5\pi}{6}$	π	$\frac{7\pi}{6}$	$\frac{4\pi}{3}$	$\frac{3\pi}{2}$	$\frac{5\pi}{3}$	$\frac{11\pi}{6}$	2π
y	1	0.866	0.5	0	−0.5	−0.866	−1	−0.866	−0.5	0	0.5	0.866	1

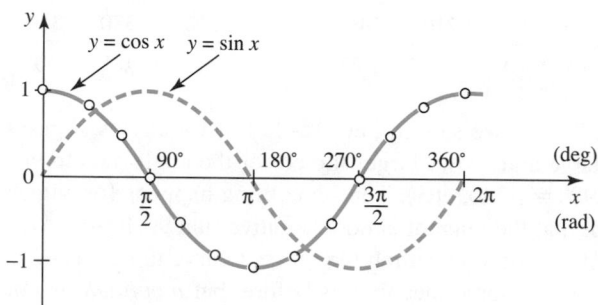

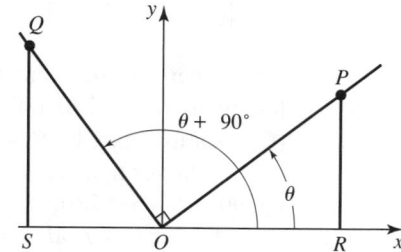

FIGURE 17–22 $y = \cos x$.

FIGURE 17–23

Cosine and Sine Curves Related

Note in Fig. 17–22 that the cosine curve and the sine curve, shown dashed, have the same shape. In fact, the cosine curve appears to be identical to a sine curve shifted 90° to the left, or

$$\cos \theta = \sin(\theta + 90°) \tag{1}$$

We can show that Eq. (1) is true. We lay out the two angles θ and $\theta + 90°$ (Fig. 17–23), choose points P and Q so that $OP = OQ$, and drop perpendiculars PR and QS to the x axis. Since triangles OPR and OQS are congruent, we have $OR = QS$. The cosine of θ is then

$$\cos \theta = \frac{OR}{OP} = \frac{QS}{OQ} = \sin(\theta + 90°)$$

which verifies (1).

Since the sine and cosine curves are identical except for phase shift, we can use either one to describe periodically varying quantities.

Graph of the General Cosine Function $y = a \cos(bx + c)$

The equation for the general cosine function is

$$y = a \cos(bx + c)$$

The amplitude, period, frequency, and phase shift are found the same way as for the sine curve, and, of course, the quick plotting method works here, too.

◆◆◆ **Example 14:** Find the amplitude, period, frequency, and phase shift for the curve $y = 3 \cos(4x − 120°)$, and make a graph.

Solution: From the equation, a = amplitude = 3. Since $b = 4$, the period is

$$\text{period} = \frac{360°}{4} = 90°/\text{cycle}$$

and

$$\text{frequency} = \frac{1}{P} = 0.0111 \text{ cycle/degree}$$

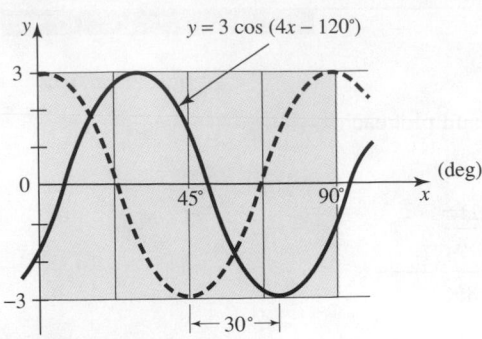

FIGURE 17–24

Since $c = -120°$,

$$\text{phase shift} = -\frac{c}{b} = -\frac{-120°}{4} = 30°$$

So we expect the curve to be shifted 30° to the right. For a quick plot we draw a rectangle with height twice the amplitude and a length of one period as shown in Fig. 17–24. We sketch in the cosine curve, shown dashed, and then shift it 30° to the right to get the final curve, shown solid.   ◆◆◆

Graph of the Tangent Function $y = \tan x$

We now plot the function $y = \tan x$. As usual, we make a table of point pairs.

x	0	30°	60°	90°	120°	150°	180°	210°	240°	270°	300°	330°	360°
y	0	0.577	1.73		−1.73	−0.577	0	0.577	1.73		−1.73	−0.577	0

The tangent is not defined at 90° (or 270°), as we saw in Sec. 15–1. But for angles slightly less than 90° (or 270°), the tangent is large and grows larger the closer the angle gets to 90° (for example, tan 89.9° = 573, tan 89.99° = 5730, etc.). The same thing happens for angles slightly larger than 90° (or 270°) except that the tangent is now negative (tan 90.1° = −573, tan 90.01° = −5730, etc.). With this information, we graph the tangent curve in Fig. 17–25.

The terms *period* and *frequency* have the same meaning as before, but *amplitude* is not used in connection with the tangent function, as the curve extends to infinity in the positive and negative directions.

The vertical lines at $x = 90°$ and 270° that the tangent graph approaches but never touches are called *asymptotes*.

As with the sine and cosine functions, the tangent function may also take the form

$$y = a \tan(bx + c)$$

where the constants, a, b, and c affect the shape, period, and phase angle of the function.

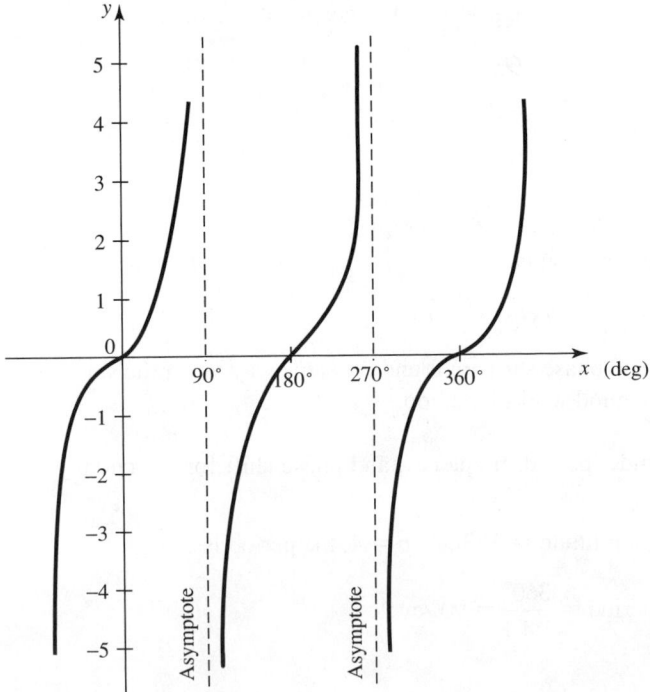

FIGURE 17–25

Graphs of Cotangent, Secant, and Cosecant Functions

For completeness, the graphs of all six trigonometric functions are shown in Fig. 17–26. To make them easier to compare, the same horizontal scale is used for each curve. Only the sine and cosine curves find much use in technology. Notice that of the six, they are the only curves that have no "gaps" or "breaks." They are called *continuous* curves, while the others are called *discontinuous*.

If the need should arise to graph a cotangent, secant, or cosecant function, simply make a table of point pairs and plot them, or use a graphics calculator or graphing utility on a computer.

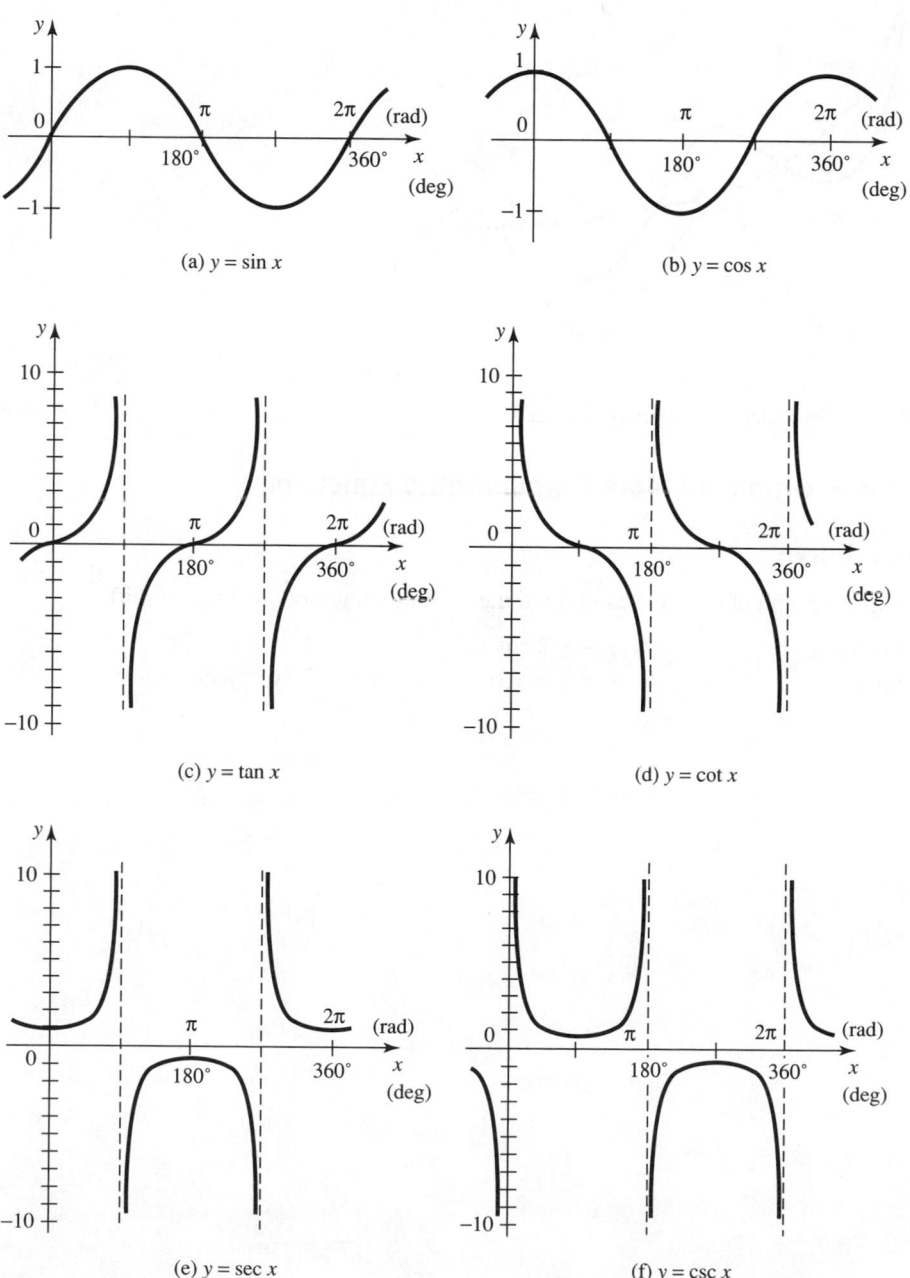

(a) $y = \sin x$

(b) $y = \cos x$

(c) $y = \tan x$

(d) $y = \cot x$

(e) $y = \sec x$

(f) $y = \csc x$

FIGURE 17–26 Graphs of the six trigonometric functions.

Composite Curves

One way to graph a function containing several terms is to graph each term separately and then add (or subtract) them on the graph paper. This method, called the method of *addition of ordinates*, is especially useful when one or more terms of the expression to be graphed are trigonometric functions.

◆◆◆ **Example 15:** Graph the function $y = 8/x + \cos x$, where x is in radians, from $x = 0$ to 12 rad.

Solution: We separately graph $y_1 = 8/x$ and $y_2 = \cos x$, shown dashed in Fig. 17–27. We add the ordinates graphically to get the composite curve, shown as a solid line.

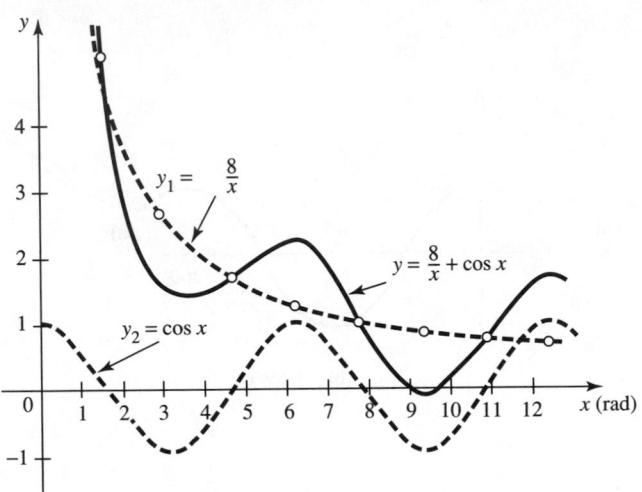

FIGURE 17–27   Graphing by addition of ordinates.   ◆◆◆

Exercise 4 ◆ Graphs of More Trigonometric Functions

The Cosine Curve

Make a complete graph of each function. Find the amplitude, period, and phase shift.

1. $y = 3 \cos x$ **2.** $y = -2 \cos x$ **3.** $y = \cos 3x$

4. $y = \cos 2x$ **5.** $y = 2 \cos 3x$ **6.** $y = 3 \cos 2x$

7. $y = \cos(x - 1)$ **8.** $y = \cos(x + 2)$ **9.** $y = 3\cos\left(x - \dfrac{\pi}{4}\right)$

10. $y = 2\cos(3x + 1)$

The Tangent Curve

Make a complete graph of each function.

11. $y = 2\tan x$ **12.** $y = \tan 4x$ **13.** $y = 3\tan 2x$

14. $y = \tan\left(x - \dfrac{\pi}{2}\right)$ **15.** $y = 2\tan(3x - 2)$ **16.** $y = 4\tan\left(x + \dfrac{\pi}{6}\right)$

Cotangent, Secant, and Cosecant Curves

Make a complete graph of each function.

17. $y = 2\cot 2x$ **18.** $y = 3\sec 4x$ **19.** $y = \csc 3x$

20. $y = \cot(x - 1)$ **21.** $y = 2\sec(3x + 1)$ **22.** $y = \csc(2x + 0.5)$

Composite Curves

Graph by addition of ordinates (x is in radians).

23. $y = x + \sin x$ **24.** $y = \cos x - x$

25. $y = \dfrac{2}{x} - \sin x$ **26.** $y = 2x + 5\cos x$

27. $y = \sin x + \cos 2x$ **28.** $y = \sin x + \sqrt{x}$

29. $y = 2\sin 2x - \cos 3x$ **30.** $y = 3\cos 3x + 2\sin 4x$

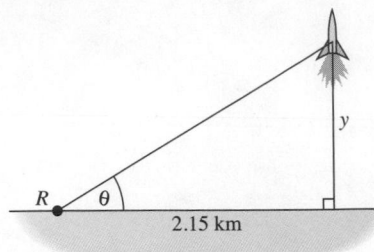

FIGURE 17-28   Tracking a rocket.

31. For the pendulum of Fig. 17–10:
   (a) Write an equation for the vertical distance y as a function of θ.
   (b) Graph $y = f(\theta)$ for $\theta = 0$ to 30°.

32. Repeat problem 31 taking $\theta = 0$ when the pendulum is 20° from the vertical and increasing counterclockwise.

33. A rocket is rising vertically, as shown in Fig. 17–28, and is tracked by a radar dish R.
   (a) Write an expression for the altitude y of the rocket as a function of θ.
   (b) Graph $y = f(\theta)$ for $\theta = 0$ to 60°.

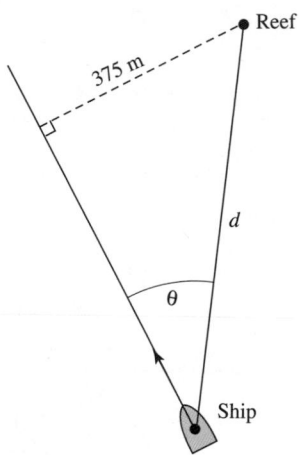

FIGURE 17-29

34. The ship in Fig. 17–29 travels in a straight line while keeping its searchlight trained on the reef.
   (a) Write an expression for the distance d as a function of the angle θ.
   (b) Graph $d = f(\theta)$ for $\theta = 0$ to 90°.

35. If a weight W is dragged along a surface with a coefficient of friction f, as shown in Fig. 17–30, the force needed is

$$F = \frac{fW}{f\sin\theta + \cos\theta}$$

Graph F as a function of θ for $\theta = 0$ to 90°. Take $f = 0.55$ and $W = 5.35$ kg.

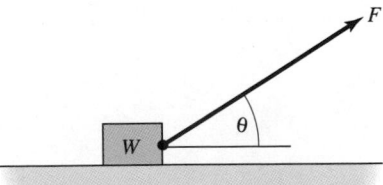

FIGURE 17-30   A weight dragged along a surface.

17–5   The Sine Wave as a Function of Time

The sine curve can be generated in a simple geometric way. Figure 17–31 shows a vector OP rotating with a constant angular velocity ω. A rotating vector is called a *phasor*. Its angular velocity ω is almost always given in radians per second (rad/s).

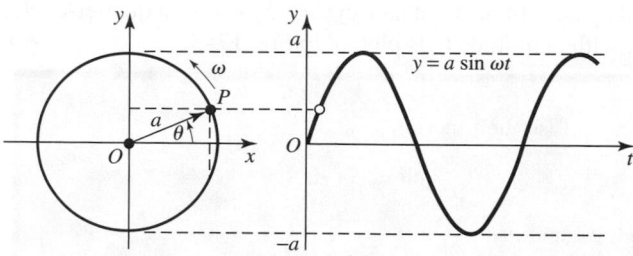

FIGURE 17-31 The sine curve generated by a rotating vector.

If the length of the phasor is a, then its projection on the y axis is

$$y = a \sin \theta$$

But since by Eq. A26 the angle θ at any instant t is equal to ωt,

$$y = a \sin \omega t$$

Further, if the phasor does not start from the x axis but has some *phase angle* ϕ, we get

$$y = a \sin(\omega t + \phi)$$

Notice in this equation that y *is a function of time*, rather than of an angle.

We'll soon see that a sine wave that is a function of time is more appropriate for representing alternating current than a sine wave that is a function of θ.

◆◆◆ **Example 16:** Write the equation for a sine wave generated by a phasor of length 8 rotating with an angular velocity of 300 rad/s and with a phase angle of 0.

Solution: Substituting in the general sine wave equation $y = a \sin(\omega t + \phi)$, with $a = 8$, $\omega = 300$ rad/s, and $\phi = 0$, we have

$$y = 8 \sin 300t$$ ◆◆◆

Period

Recall that the *period* was defined as the distance along the x axis taken by one cycle of the waveform. When the units on the x axis represented an angle, the period was in radians or degrees. Now that the x axis shows time, the period will be in seconds. From Eq. 211,

$$P = \frac{2\pi}{b}$$

But in the equation $y = a \sin \omega t$, $b = \omega$, so we have the following formula:

Period of a Sine Wave	$P = \dfrac{2\pi}{\omega}$ (seconds)	A77

where P is in seconds and ω is in rad/s. ◆◆◆

◆◆◆ **Example 17:** Find the period and the amplitude of the sine wave $y = 6 \sin 10t$, and make a graph with time t shown on the horizontal axis.

Solution: The period, from Eq. A77, is

$$P = \frac{2\pi}{\omega} = \frac{2(3.142)}{10} = 0.628 \text{ s}$$

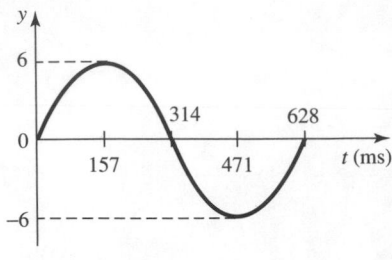

FIGURE 17–32

Thus it takes 628 ms for a full cycle, 314 ms for a half-cycle, 157 ms for a quarter-cycle, and so forth. This sine wave, with amplitude 6, is plotted in Fig. 17–32. ◆◆◆

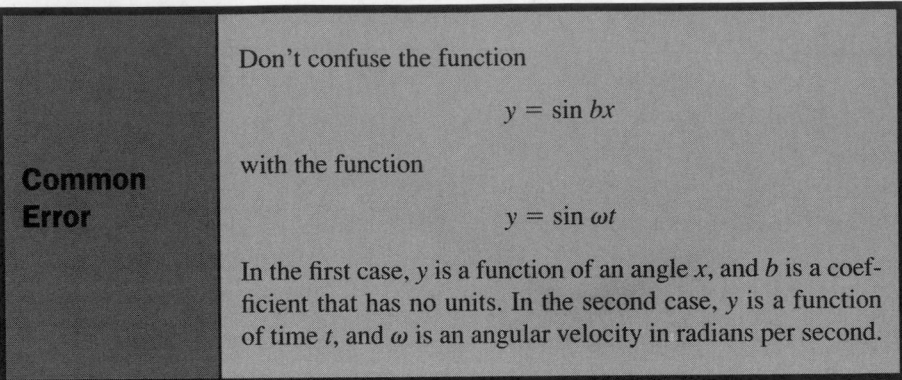

Common Error

Don't confuse the function

$$y = \sin bx$$

with the function

$$y = \sin \omega t$$

In the first case, y is a function of an angle x, and b is a coefficient that has no units. In the second case, y is a function of time t, and ω is an angular velocity in radians per second.

Frequency

From Eq. 212 we see that the frequency f of a periodic waveform is equal to the reciprocal of the period P. So the frequency of a sine wave is given by the following equation:

Frequency of a Sine Wave	$f = \dfrac{1}{P} = \dfrac{\omega}{2\pi}$ (hertz)	A78

where P is in seconds and ω is in rad/s.

The unit of frequency is therefore cycles/s, or *hertz* (Hz).

$$1\ \text{Hz} = 1\ \text{cycle/s}$$

Higher frequencies are often expressed in kilohertz (kHz), where

$$1\ \text{kHz} = 10^3\ \text{Hz}$$

or in megahertz (MHz), where

$$1\ \text{MHz} = 10^6\ \text{Hz}$$

◆◆◆ **Example 18:** The frequency of the sine wave of Example 17 is

$$f = \frac{1}{P} = \frac{1}{0.628} = 1.59\ \text{Hz}$$ ◆◆◆

When the period P is not wanted, the angular velocity can be obtained directly from Eq. A78 by noting that $\omega = 2\pi f$.

◆◆◆ **Example 19:** Find the angular velocity of a 1000-Hz source.

Solution: From Eq. A78,

$$\omega = 2\pi f = 2(3.14)(1000) = 6280\ \text{rad/s}$$ ◆◆◆

A sine wave as a function of time can also have a phase shift, expressed in either degrees or radians, as in the following example.

◆◆◆ **Example 20:** Given the sine wave $y = 5.83 \sin(114t + 15°)$, find (a) the amplitude, (b) the angular velocity, (c) the period, (d) the frequency, and (e) the phase angle; and (f) make a graph.

Solution:

(a) The amplitude is 5.83 units.
(b) The angular velocity is $\omega = 114$ rad/s.
(c) From Eq. A77, the period is

$$P = \frac{2\pi}{\omega} = \frac{2\pi}{114} = 0.0551 \text{ s} = 55.1 \text{ ms}$$

(d) From Eq. A78, the frequency is

$$f = \frac{1}{P} = \frac{1}{0.0551} = 18.1 \text{ Hz}$$

(e) The phase angle is 15°. It is not unusual to see a sine wave given with ωt in radians and the phase angle in degrees. We find the phase shift, in units of time, the same way that we found the phase shift for the general sine function in Sec. 17–1. We find the value of t at which the positive half-cycle starts. As before, we set y equal to zero and solve for t.

$$y = \sin(\omega t + \phi) = 0$$
$$\omega t = -\phi$$

$$t = -\frac{\phi}{\omega} = \text{phase shift}$$

Substituting our values for ω and ϕ (which we first convert to 0.262 radian) gives

$$\text{phase shift} = -\frac{0.262 \text{ rad}}{114 \text{ rad/s}} = -0.002\,30 \text{ s}$$

So our curve is shifted 2.30 milliseconds to the left.

(f) This sine wave is graphed in Fig. 17–33. We draw a rectangle of height 2(5.83) and width 55.1 ms. We subdivide the rectangle into four rectangles of equal width and sketch in the sine wave, shown dashed. We then shift the sine wave 2.30 ms to the left to get the final (solid) curve.

> Even though it is often done, some people object to mixing degrees and radians in the same equation. If you do it, work with extra care.

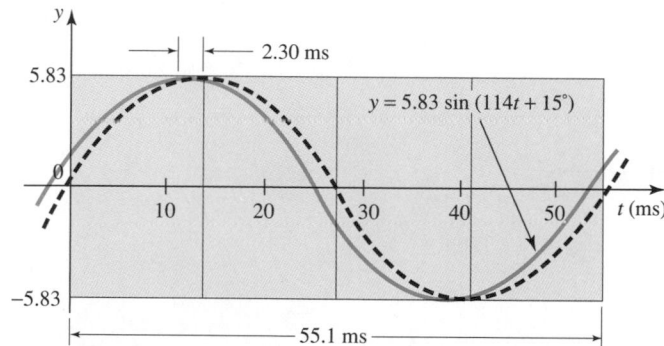

FIGURE 17–33   ◆◆◆

Common Error	It's easy to get confused as to which direction to shift the curve. Check your work by substituting the value of an x intercept (such as $-0.002\,30$ s in Example 20) back into the original equation. You should get a value of zero for y

Adding Sine Waves of the Same Frequency

We have seen that $A \sin \omega t$ is the y component of a phasor of magnitude A rotating at angular velocity ω. Similarly, $B \sin(\omega t + \theta)$ is the y component of a phasor of magnitude B rotating at the same angular velocity ω, but with a phase angle ϕ between A and B. Since each is the y component of a phasor, their sum is equal to the sum of the y components of the two phasors, or, in other words, simply the y component of the resultant of those phasors.

Thus to add two sine waves of the same frequency, we simply *find the resultant of the phasors representing those sine waves.*

◆◆◆ Example 21: Express $y = 2.00 \sin \omega t + 3.00 \sin(\omega t + 60°)$ as a single sine wave.

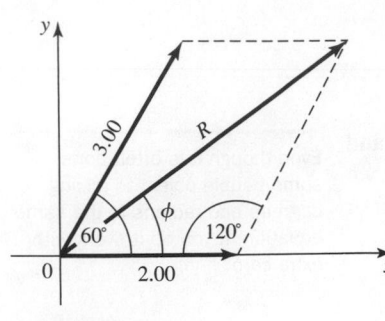

FIGURE 17–34

Solution: We represent $2.00 \sin \omega t$ by a phasor of length 2.00, and we represent 3.00 $\sin(\omega t + 60°)$ as a phasor of length 3.00, at an angle of 60°. These phasors are shown in Fig. 17–34. We wish to add $y = 2.00 \sin \omega t$ and $y = 3.00 \sin(\omega t + 60°)$. By the law of cosines,

$$R = \sqrt{2.00^2 + 3.00^2 - 2(2.00)(3.00) \cos 120°} = 4.36$$

Then, by the law of sines,

$$\frac{\sin \phi}{3.00} = \frac{\sin 120°}{4.36}$$
$$\sin \phi = 0.596$$
$$\phi = 36.6°$$

Thus

$$y = 2.00 \sin \omega t + 3.00 \sin(\omega t + 60°)$$
$$= 4.36 \sin(\omega t + 36.6°)$$

Figure 17–35 shows the two original sine waves as well as a graph of $4.36 \sin(\omega t + 36.6°)$. Note that the sum of the ordinates of the original waves, at any value of ωt, is equal to the ordinate of the final wave.

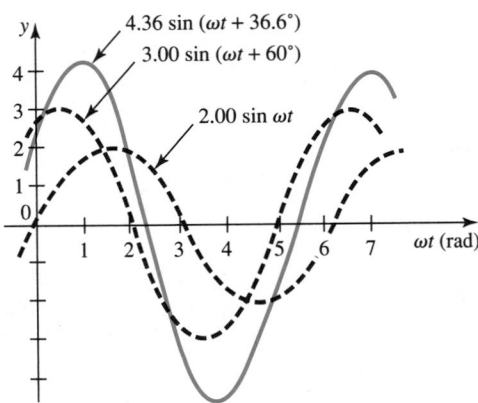

FIGURE 17–35   ◆◆◆

Adding Sine and Cosine Waves of the Same Frequency
In Sec. 17–4 we verified that

$$\cos \omega t = \sin(\omega t + 90°)$$

and saw that a cosine wave $A \cos \omega t$ is identical to the sine wave $A \sin \omega t$, except for a phase difference of 90° between the two curves (see Fig. 17–22). Thus we can find the sum ($A \sin \omega t + B \cos \omega t$) of a sine and a cosine wave the same way that we added two sine waves in the preceding section: by finding the resultant of the phasors representing each sine wave. Since the angle between the two phasors is 90° (Fig. 17–36), the resultant has a magnitude

$$R = \sqrt{A^2 + B^2}$$

and is at an angle

$$\phi = \arctan \frac{B}{A}$$

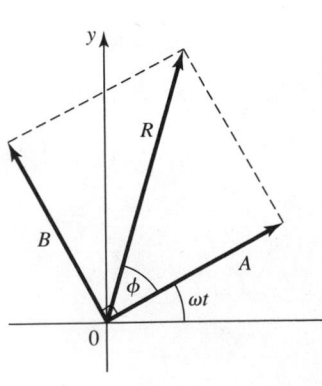

FIGURE 17–36

So we have the following equation:

Addition of a Sine Wave and a Cosine Wave	$A \sin \omega t + B \cos \omega t = R \sin(\omega t + \phi)$ where $R = \sqrt{A^2 + B^2}$ and $\phi = \arctan \dfrac{B}{A}$	**214**

In Sec. 18–2 we will derive this same equation using the trigonometric identities.

Thus we can represent the sum of a sine and a cosine function by a single sine function.

◆◆◆ **Example 22:** Express $y = 274 \sin \omega t + 371 \cos \omega t$ as a single sine function.

Solution: The magnitude of the resultant is

$$R = \sqrt{(274)^2 + (371)^2} = 461$$

and the phase angle is

$$\phi = \arctan \frac{371}{274} = \arctan 1.354 = 53.6°$$

So

Note that you can use the rectangular-to-polar conversion keys on your calculator to do this calculation.

$$y = 274 \sin \omega t + 371 \cos \omega t$$
$$= 461 \sin(\omega t + 53.6°)$$

Figure 17–37 shows the original sine and cosine functions and a graph of $y = 461 \sin(\omega t + 53.6°)$. Note that the sum of the ordinates of the original waves equals the ordinate of the computed wave at any abscissa.

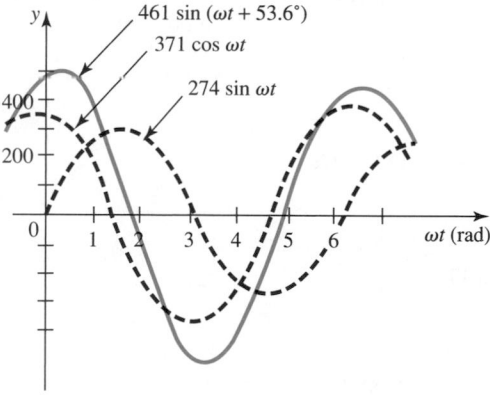

FIGURE 17–37

◆◆◆

Alternating Current

When a loop of wire rotates in a magnetic field, any portion of the wire cuts the field while travelling first in one direction and then in the other direction. Since the polarity of the voltage induced in the wire depends on the direction in which the field is cut, the induced current will travel first in one direction and then in the other: It *alternates*.

The same is true with an armature more complex than a simple loop. We get an *alternating current*. Just as the rotating point P in Fig. 17–31 generated a sine wave, the current induced in the rotating armature will be sinusoidal.

The sinusoidal voltage or current can be described by the equation

$$y = a \sin(\omega t + \phi)$$

or, in electrical terms,

We have given the phase angle ϕ subscripts because, in a given circuit, the current and voltage waves will usually have different phase angles.

$$i = I_m \sin(\omega t + \phi_2) \qquad \text{(Eq. A76)}$$

where i is the current at time t, I_m the maximum current (the amplitude), and ϕ_2 the phase angle. Also, if we let V_m stand for the amplitude of the voltage wave, the instantaneous voltage v becomes

$$v = V_m \sin(\omega t + \phi_1) \qquad \text{(Eq. A75)}$$

Equations A77 and A78 still apply here.

$$\text{period} = \frac{2\pi}{\omega} \quad \text{(s)}$$

$$\text{frequency} = \frac{\omega}{2\pi} \quad \text{(Hz)} \qquad \bullet\bullet\bullet$$

◆◆◆ **Example 23:** Utilities in Canada and the United States supply alternating current at a frequency of 60 Hz. Find the angular velocity ω and the period P.

Solution: By Eq. A78,

$$P = \frac{1}{f} = \frac{1}{60} = 0.0167 \text{ s}$$

and, by Eq. A77,

$$\omega = \frac{2\pi}{P} = \frac{2\pi}{0.0167} = 377 \text{ rad/s}$$

These are good numbers to remember. ◆◆◆

◆◆◆ **Example 24:** A certain alternating current has an amplitude of 1.5 A and a frequency of 60 Hz (cycles/s). Taking the phase angle as zero, write the equation for the current as a function of time, find the period, and find the current at $t = 0.01$ s.

Solution: From Example 23,

$$P = 0.0167 \text{ s} \quad \text{and} \quad \omega = 377 \text{ rad/s}$$

so the equation is

$$i = 1.5 \sin(377t)$$

When $t = 0.01$ s,

$$i = 1.5 \sin(377)(0.01) = -0.882 \text{ A}$$

as shown in Fig. 17–38.

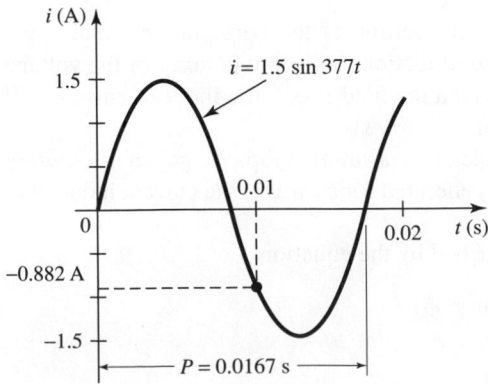

FIGURE 17–38

Phase Shift

When writing the equation of a single alternating voltage or current, we are usually free to choose the origin anywhere along the time axis. So we can place it to make the phase angle equal to zero. However, when there are *two* curves on the same graph that are *out of phase*, we usually locate the origin so that the phase angle of one curve is zero.

If the difference in phase between two sine waves is *t* seconds, then we often say that one curve *leads* the other by *t* seconds. Conversely, we could say that one curve *lags* the other by *t* seconds.

◆◆◆ **Example 25:** Figure 17–39 shows voltage and current waves, with the voltage wave leading a current wave by 5 ms. Write the equations for the two waves.

Solution: For the voltage wave,

$$V_m = 190, \qquad \phi = 0, \qquad \text{and} \qquad P = 0.02 \text{ s}$$

By Eq. A77,

$$\omega = \frac{2\pi}{0.02} = 100\pi = 314 \text{ rad/s}$$

so

$$v = 190 \sin 314t$$

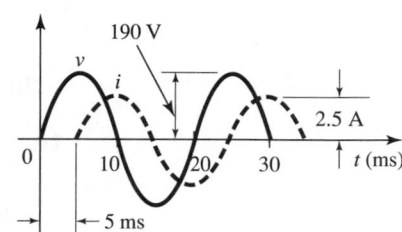

FIGURE 17–39

where *v* is expressed in volts. For the current wave, $I_m = 2.5$ A. Since the curve is shifted to the right by 5 ms,

$$\text{phase shift} = 0.005 \text{ s} = -\frac{\phi}{\omega}$$

so the phase angle ϕ is

$$\phi = -(0.005 \text{ s})(100\pi \text{ rad/s}) = -\frac{\pi}{2} \text{ rad} = -90°$$

The angular frequency is the same as before, so

$$i = 2.5 \sin(314t - 90°)$$

where *i* is expressed in amperes. With ϕ in radians,

$$i = 2.5 \sin\left(314t - \frac{\pi}{2}\right)$$

where *i* is expressed in amperes. ◆◆◆

Exercise 5 ◆ The Sine Wave as a Function of Time

Find the period and the angular velocity of a repeating waveform that has a frequency of:

1. 68 Hz. **2.** 10 Hz. **3.** 5000 Hz.

Find the frequency (in hertz) and the angular velocity of a repeating waveform whose period is:

4. 1 s. **5.** $\frac{1}{8}$ s. **6.** 95 ms.

7. If a periodic waveform has a frequency of 60 Hz, how many seconds will it take to complete 200 cycles?

8. Find the frequency in Hz for a wave that completes 150 cycles in 10 s.

Find the period and the frequency of a sine wave that has an angular velocity of:

9. 455 rad/s.   **10.** 2.58 rad/s.   **11.** 500 rad/s.

Find the period, amplitude, and phase angle for:

12. the sine wave shown in Fig. 17–40(a).

13. the sine wave shown in Fig. 17–40(b).

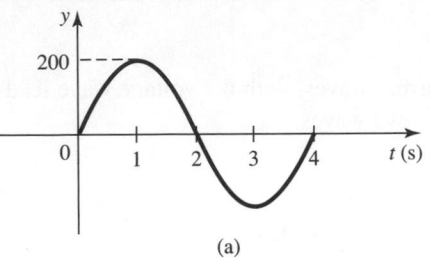   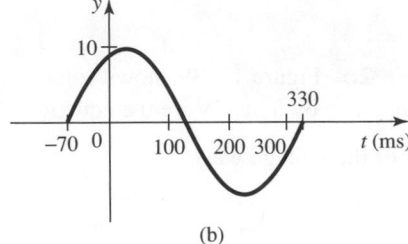

(a)   (b)

FIGURE 17–40

14. Write an equation for a sine wave generated by a phasor of length 5 rotating with an angular velocity of 750 rad/s and with a phase angle of 0°.

15. Repeat problem 14 with a phase angle of 15°.

Plot each sine wave.

16. $y = \sin t$ **17.** $y = 3 \sin 377t$

18. $y = 54 \sin(83t - 20°)$ **19.** $y = 375 \sin\left(55t + \dfrac{\pi}{4}\right)$

Adding Sine Waves and Cosine Waves

Express as a single sine wave.

20. $y = 3.00 \sin \omega t + 5.50 \sin(\omega t + 30°)$

21. $y = 18.3 \sin \omega t + 26.3 \sin(\omega t + 75°)$

22. $y = 2.47 \sin \omega t + 1.83 \sin(\omega t - 24°)$

23. $y = 384 \sin \omega t + 536 \sin(\omega t - 48°)$

24. $y = 8370 \sin \omega t + 7570 \cos \omega t$

25. $y = 7.37 \cos \omega t + 5.83 \sin \omega t$

26. $y = 74.2 \sin \omega t + 69.3 \cos \omega t$

27. $y = 364 \sin \omega t + 550 \cos \omega t$

28. The weight in Fig. 17–41 is pulled down 2.50 cm and released. The distance x is given by

$$x = 2.50 \cos \omega t$$

Graph x as a function of t for two complete cycles. Take $\omega = 42.5$ rad/s.

29. The arm in the Scotch crosshead of Fig. 17–9 is rotating at a rate of 2.55 rev/s.
(a) Write an equation for the displacement y as a function of time.
(b) Graph that equation for two complete cycles.

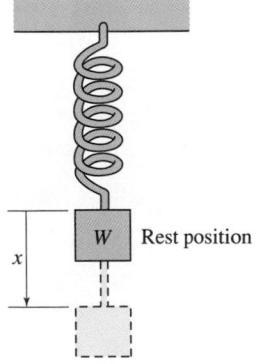

FIGURE 17–41 Weight hanging from a spring.

Alternating Current

30. An alternating current has the equation
$$i = 25 \sin(635t - 18°)$$
where i is given in amperes, A. Find the maximum current, period, frequency, phase angle, and the instantaneous current at $t = 0.01$ s.

31. Given an alternating voltage $v = 4.27 \sin(463t + 27°)$, find the maximum voltage, period, frequency, and phase angle, and the instantaneous voltage at $t = 0.12$ s.

32. Write an equation for an alternating voltage that has a peak value of 155 V, a frequency of 60 Hz, and a phase angle of 22°.

33. Write an equation for an alternating current that has a peak amplitude of 49.2 mA, a frequency of 35 Hz, and a phase angle of 63.2°.

17–6   Polar Coordinates

The Polar Coordinate System

The *polar coordinate system* (Fig. 17–42) consists of a *polar axis*, passing through point O, which is called the *pole*. The location of a point P is given by its distance r from the pole, called the *radius vector*, and by the angle θ, called the *polar angle* (sometimes called the *vectorial* angle or *reference* angle). The polar angle is called *positive* when measured counterclockwise from the polar axis, and *negative* when measured clockwise.

The *polar coordinates* of a point P are thus r and θ, usually written in the form $P(r, \theta)$, or as $r \underline{/\theta}$ (read "r at an angle of θ").

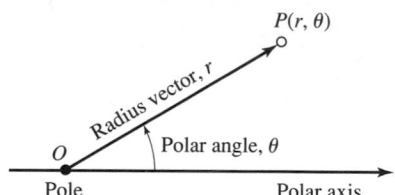

FIGURE 17–42 Polar coordinates.

◆◆◆ **Example 26:** A point at a distance 5 from the origin with a polar angle of 28° can be written

$$P(5, 28°) \qquad \text{or} \qquad 5\underline{/28°}$$

◆◆◆

Most of our graphing will continue to be in rectangular coordinates, but in some cases polar coordinates will be more convenient.

Plotting Points in Polar Coordinates

For plotting in polar coordinates, it is convenient, although not essential, to have *polar coordinate paper* (Fig. 17–43). This paper has concentric circles, equally spaced, and an angular scale in degrees or radians.

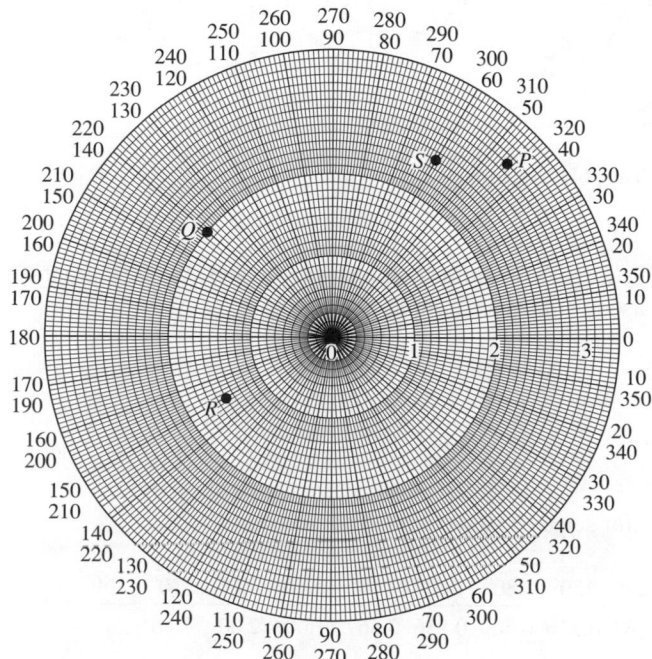

FIGURE 17–43

◆◆◆ Example 27: The points $P(3, 45°)$, $Q(-2, 320°)$, $R(1.5, 7\pi/6)$, and $S(2.5, -5\pi/3)$ are plotted in Fig. 17–43.

Notice that to plot $Q(-2, 320°)$, which has a negative value for r, we first locate $(2, 320°)$ and then find the corresponding point along the diameter and in the opposite quadrant. ◆◆◆

Graphing Equations in Polar Coordinates

To graph a function $r = f(\theta)$, simply assign convenient values to θ and compute the corresponding value for r. Then plot the resulting table of point pairs.

◆◆◆ Example 28: Graph the function $r = \cos \theta$.

Solution: Let us take values for θ every 30° and make a table.

θ	0	30°	60°	90°	120°	150°	180°	210°	240°	270°	300°	330°	360°
r	1	0.87	0.5	0	−0.5	−0.87	−1	−0.87	−0.5	0	0.5	0.87	1

Plotting these points, we get a circle as shown in Fig. 17–44.

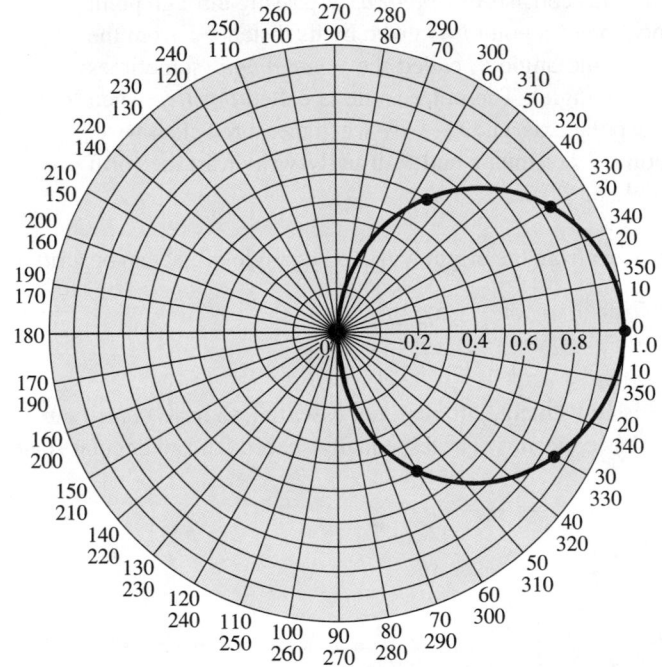

Polar plots like this one are used in telecommunications to plot the radiation patterns of antennas.

FIGURE 17–44   Graph of $r = \cos \theta$. The polar equation of a circle that passes through the pole and has its centre at (c, α) is $r = 2c \cos(\theta - \alpha)$.   ◆◆◆

In Chapters 5 and 14, we graphed a parabola in rectangular coordinates. We now graph an equation in polar coordinates that also results in a parabola.

◆◆◆ Example 29: Graph the parabola

$$r = \frac{p}{1 - \cos \theta}$$

with $p = 1$.

Solution: As before, we compute r for selected values of θ.

θ	0°	30°	60°	90°	120°	150°	180°	210°	240°	270°	300°	330°	360°
r		7.46	2.00	1.00	0.667	0.536	0.500	0.536	0.667	1.00	2.00	7.46	

Note that we get division by zero at $\theta = 0°$ and $360°$, so that the curve does not exist there. Plotting these points (except for 7.46, which is off the graph), we get the parabola shown in Fig. 17–45. ◆◆◆

Some of the more interesting curves in polar coordinates are shown in Fig. 17–46.

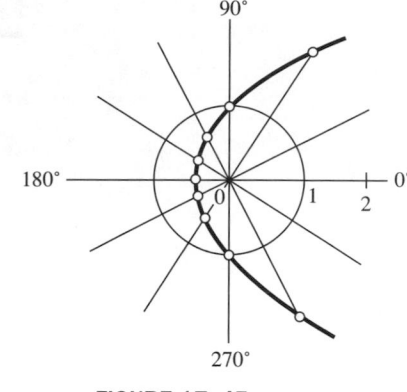

FIGURE 17–45

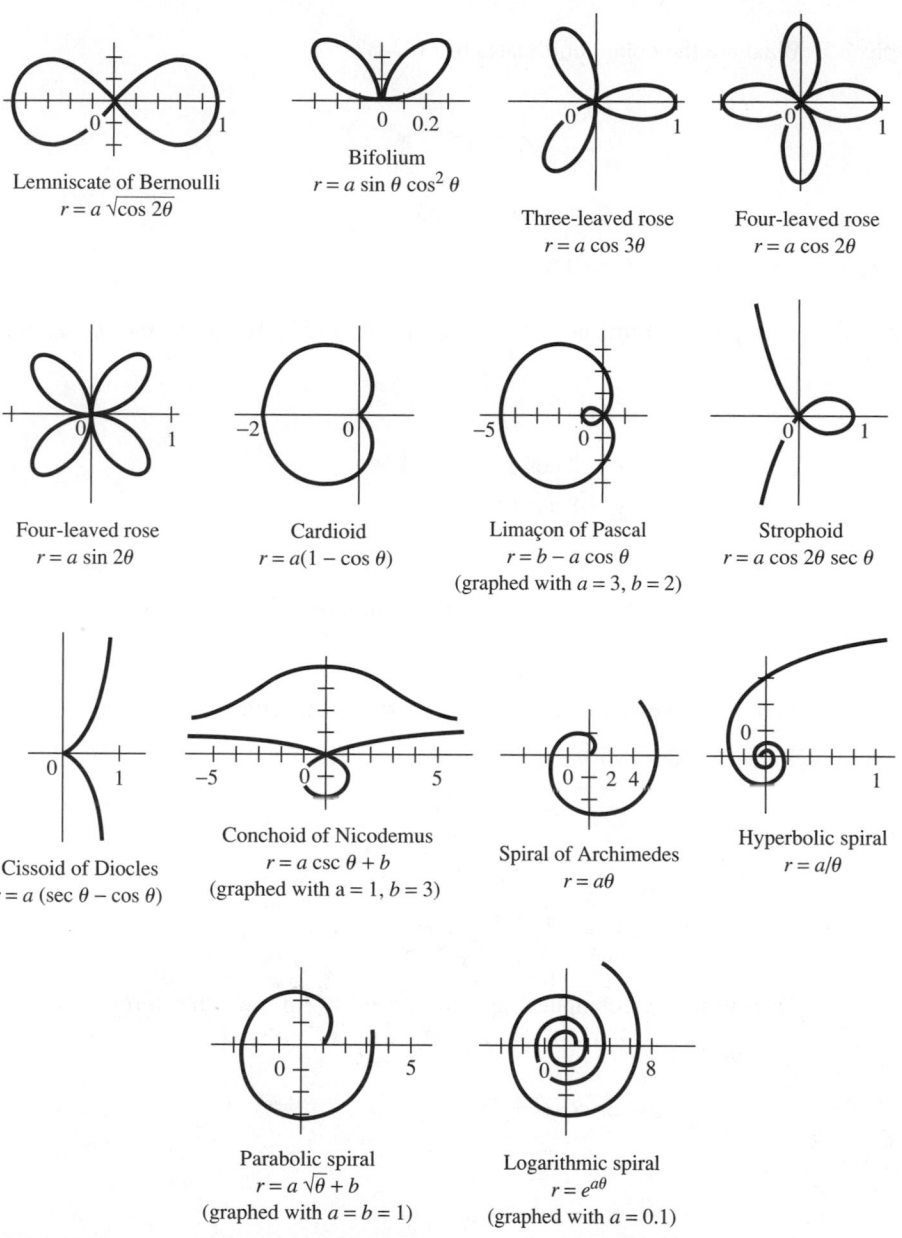

FIGURE 17–46 Some curves in polar coordinates. Unless otherwise noted, the curves were graphed with $a = 1$.

The Lemniscate of Bernoulli and the rose graphs are very similar to what we see when we use directors and reflectors to create directional antennas in telecommunications.

Transforming between Polar and Rectangular Coordinates

We can easily see the relationships between rectangular coordinates and polar coordinates when we put both systems on a single diagram (Fig. 17–47). Using the trigonometric functions and the Pythagorean theorem, we get the following equations:

Rectangular Coordinates		
	$x = r \cos \theta$	158
	$y = r \sin \theta$	159

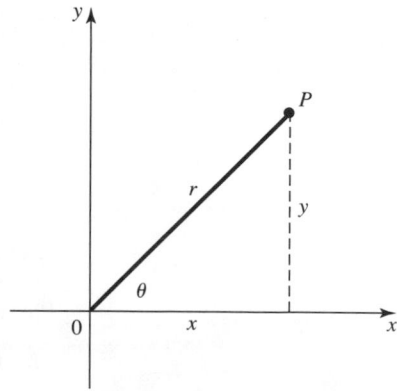

FIGURE 17–47 Rectangular and polar coordinates of a point.

Polar Coordinates	$r = \sqrt{x^2 + y^2}$	160
	$\theta = \arctan \dfrac{y}{x}$	161

◆◆◆ **Example 30:** What are the polar coordinates of $P(3, 4)$?

Solution:

$$r = \sqrt{9 + 16} = 5$$

$$\theta = \arctan \frac{4}{3} = 53.1°$$

So the polar coordinates of P are $(5, 53.1°)$.   ◆◆◆

◆◆◆ **Example 31:** The polar coordinates of a point are $(8, 125°)$. What are the rectangular coordinates?

Solution:

$$x = 8 \cos 125° = -4.59$$
$$y = 8 \sin 125° = 6.55$$

So the rectangular coordinates are $(-4.59, 6.55)$.   ◆◆◆

We may also use Eqs. 158 through 161 to transform *equations* from one system of coordinates to the other.

◆◆◆ **Example 32:** Transform the polar equation $r = \cos \theta$ to rectangular coordinates.

Solution: Multiplying both sides by r yields

$$r^2 = r \cos \theta$$

Then, by Eqs. 158 and 160,

$$x^2 + y^2 = x$$ ◆◆◆

◆◆◆ **Example 33:** Transform the rectangular equation $2x + 3y = 5$ into polar form.

Solution: By Eqs. 158 and 160,

$$2r \cos \theta + 3r \sin \theta = 5$$

or

$$r(2 \cos \theta + 3 \sin \theta) = 5$$

$$r = \frac{5}{2 \cos \theta + 3 \sin \theta}$$ ◆◆◆

Exercise 6 ◆ Polar Coordinates

Plot each point in polar coordinates.

1. $(4, 35°)$ **2.** $(3, 120°)$ **3.** $(2.5, 215°)$

4. $(3.8, 345°)$ **5.** $\left(2.7, \dfrac{\pi}{6}\right)$ **6.** $\left(3.9, \dfrac{7\pi}{8}\right)$

7. $\left(-3, \dfrac{\pi}{2}\right)$ **8.** $\left(4.2, \dfrac{2\pi}{5}\right)$ **9.** $(3.6, -20°)$

10. $(-2.5, -35°)$ **11.** $\left(-1.8, -\dfrac{\pi}{6}\right)$ **12.** $\left(-3.7, -\dfrac{3\pi}{5}\right)$

Graph the following curves from Fig. 17–46.

13. The lemniscate of Bernoulli   **14.** The bifolium

15. The three-leaved rose   **16.** The four-leaved rose, $r = a \cos 2\theta$

17. The cardioid   **18.** The limaçon of Pascal

19. The strophoid   **20.** The cissoid of Diocles

21. The conchoid of Nicodemus   **22.** The spiral of Archimedes

23. The hyperbolic spiral   **24.** The parabolic spiral

25. The logarithmic spiral

Graph each function in polar coordinates.

26. $r = 2 \cos \theta$ **27.** $r = 3 \sin \theta$

28. $r = 3 \sin \theta + 3$ **29.** $r = 2 \cos \theta - 1$

30. $r = 3 \cos 2\theta$ **31.** $r = 2 \sin 3\theta$

32. $r = \sin 2\theta - 1$ **33.** $r = 2 + \cos 3\theta$

Write the polar coordinates of each point.

34. $(2.00, 5.00)$ **35.** $(3.00, 6.00)$

36. $(1.00, 4.00)$ **37.** $(4.00, 3.00)$

38. $(2.70, -1.80)$ **39.** $(-4.80, -5.90)$

40. $(207, 186)$ **41.** $(-312, -509)$

42. $(1.08, -2.15)$

Write the rectangular coordinates of each point.

43. $(5.00, 47.0°)$ **44.** $(6.30, 227°)$

45. $(445, 312°)$ **46.** $\left(3.60, \dfrac{\pi}{5}\right)$

47. $\left(-4.00, \dfrac{3\pi}{4}\right)$ **48.** $\left(18.3, \dfrac{2\pi}{3}\right)$

49. $(15.0, 235.0°)$ **50.** $(-12.0, -48.0°)$

51. $-9.80, -\dfrac{\pi}{5}$

Write each polar equation in rectangular form.

52. $r = 6$

53. $r = 2 \sin \theta$

54. $r = \sec \theta$

55. $r^2 = 1 - \tan \theta$

56. $r(1 - \cos \theta) = 1$

57. $r^2 = 4 - r \cos \theta$

Write each rectangular equation in polar form.

58. $x = 2$

59. $y = -3$

60. $x = 3 - 4y$

61. $x^2 + y^2 = 1$

62. $3x - 2y = 1$

63. $y = x^2$

17–7   Graphing Parametric Equations

Parametric Equations with Trigonometric Functions

In Chapter 5 we discussed parametric equations of the form

$$x = g(t)$$

and

$$y = h(t)$$

where x and y are each expressed in terms of a third variable t, called the *parameter*. There we graphed parametric equations in which the $h(t)$ and $g(t)$ were *algebraic* functions; here we graph some parametric equations in which the functions are *trigonometric*.

The procedure is the same. We assign values to the parameter (which will now be θ) and compute x and y for each θ. The table of (x, y) pairs is then plotted in rectangular coordinates.

◆◆◆ **Example 34:** Plot the parametric equations

$$x = 3 \cos \theta$$
$$y = 2 \sin 2\theta$$

Solution: We select values for θ and compute x and y.

θ	0	$\dfrac{\pi}{4}$	$\dfrac{\pi}{2}$	$\dfrac{3\pi}{4}$	π	$\dfrac{5\pi}{4}$	$\dfrac{3\pi}{2}$	$\dfrac{7\pi}{4}$	2π
x	3	2.12	0	-2.12	-3	-2.12	0	2.12	3
r	0	2	0	-2	0	2	0	-2	0

Each (x, y) pair is plotted in Fig. 17–48.

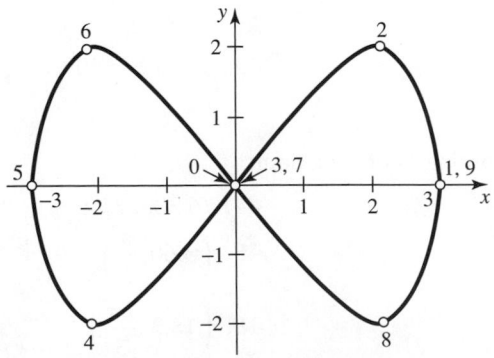

FIGURE 17–48 Patterns of this sort can be obtained by applying ac voltages to both the horizontal and the vertical deflection plates of an oscilloscope. Called *Lissajous figures*, they can indicate the relative amplitudes, frequencies, and phase angles of the two applied voltages.   ◆◆◆

The graphs of some common parametric equations in which the functions are trigonometric are given in Fig. 17–49.

Note that these parametric equations are graphed in *rectangular coordinates*, not in polar coordinates.

Converting a Polar Equation to Parametric Form

Sometimes it is necessary to convert a polar equation to a pair of parametric equations.

To convert the polar equation

$$r = f(\theta)$$

to the parametric equations

$$x = g(\theta) \quad \text{and} \quad y = h(\theta)$$

we use Eqs. 158 and 159. Thus

$$x = r \cos \theta = f(\theta) \cos \theta$$

and

$$y = r \sin \theta = f(\theta) \sin \theta$$

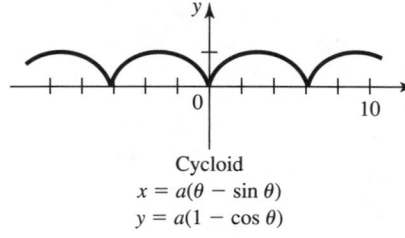

Cycloid
$x = a(\theta - \sin \theta)$
$y = a(1 - \cos \theta)$

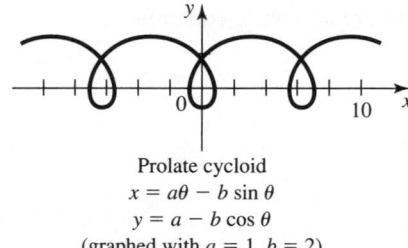

Prolate cycloid
$x = a\theta - b \sin \theta$
$y = a - b \cos \theta$
(graphed with $a = 1$, $b = 2$)

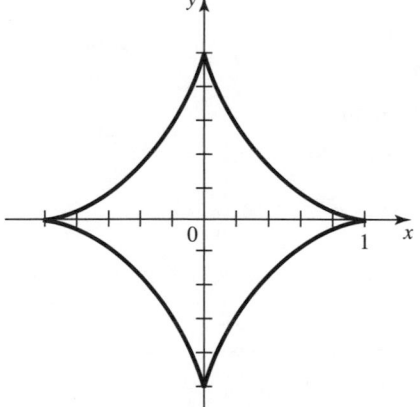

Hypocycloid of four cusps
or astroid
$x = a \cos^3 \theta$
$y = a \sin^3 \theta$

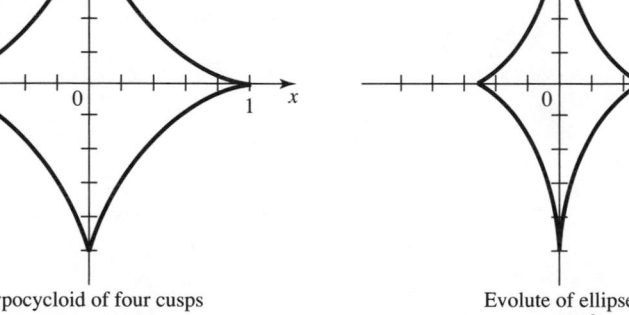

Evolute of ellipse
$x = a \cos^3 \theta$
$y = b \sin^3 \theta$
(graphed with $a = 1$, $b = 2$)

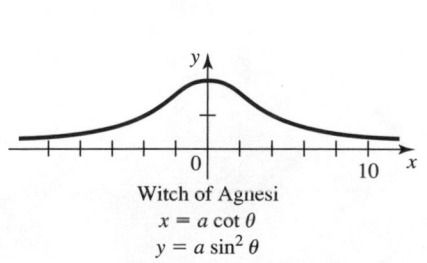

Witch of Agnesi
$x = a \cot \theta$
$y = a \sin^2 \theta$
(graphed with $a = 2$)

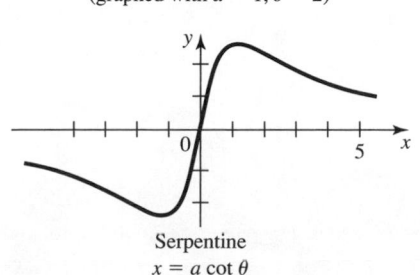

Serpentine
$x = a \cot \theta$
$y = b \sin \theta \cos \theta$
(graphed with $a = 1$, $b = 5$)

FIGURE 17–49 Graphs of some parametric equations. Unless otherwise noted, the curves were graphed with $a = 1$.

◆◆◆ Example 35: Write the polar equation $r = 3 \sin 2\theta$ in parametric form, and plot it.

Solution: Here, $f(\theta) = 3 \sin 2\theta$, so

$$x = f(\theta) \cos \theta = 3 \sin 2\theta \cos \theta$$

and

$$y = f(\theta) \sin \theta = 3 \sin 2\theta \sin \theta$$

Plotting this, we get the four-leaved rose shown in Fig. 17–46.   ◆◆◆

Exercise 7 ◆ Graphing Parametric Equations

Graph each curve from Fig. 17–49.

1. The cycloid
2. The prolate cycloid
3. The hypocycloid of four cusps
4. The evolute of ellipse
5. The witch of Agnesi
6. The serpentine

Graph each pair of parametric equations.

7. $x = \sin \theta$
   $y = \sin \theta$

8. $x = \sin \theta$
   $y = 2 \sin \theta$

9. $x = \sin \theta$
   $y = \sin 2\theta$

10. $x = \sin 2\theta$
    $y = \sin 3\theta$

11. $x = \sin \theta$
    $y = \sin(\theta + \pi/4)$

12. $x = \sin \theta$
    $y = \sin(2\theta - \pi/6)$

Rewrite each polar equation in parametric form, and graph.

13. $r = 3 \cos \theta$
14. $r = 2 \sin \theta$
15. $r = 2 \cos 3\theta$
16. $r = 3 \sin 4\theta$
17. $r = 3 \sin 3\theta - 2$
18. $r = 4 + \cos 2\theta$

Case Study Discussion—Voltage Output of a Generator

With three phases, as one voltage drops away from its peak value, another rises to "take its place". The first diagram shows the three phases. Combined, they produce the output shown at the top of the second diagram.

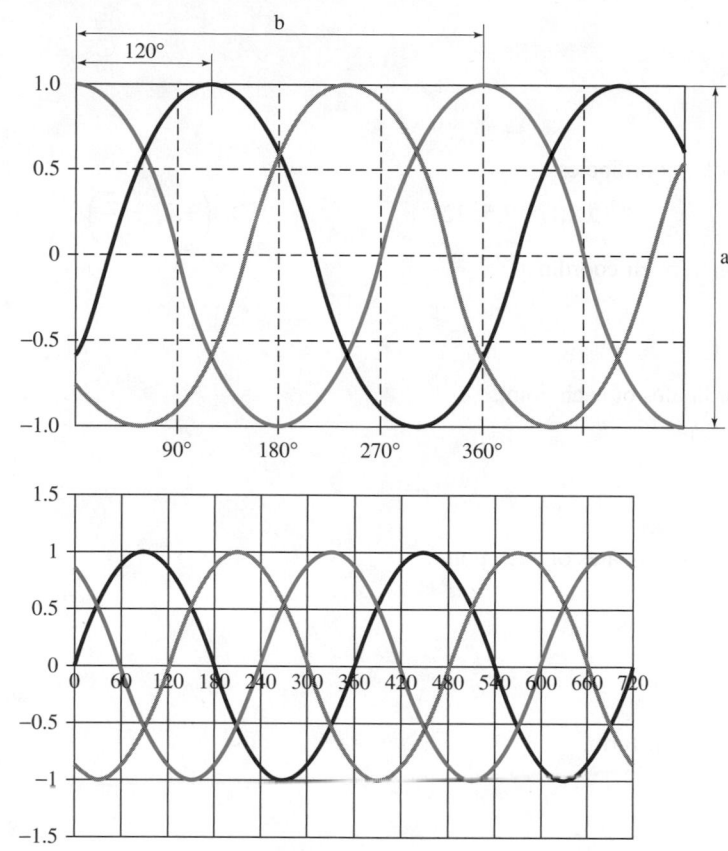

◆◆◆ CHAPTER 17 REVIEW PROBLEMS ◆◆◆◆◆◆◆◆◆◆◆◆◆◆◆◆◆◆◆◆◆◆◆◆◆◆◆◆◆◆◆◆

Graph one cycle of each curve and find the period, amplitude, and phase shift.

1. $y = 3 \sin 2x$

2. $y = 5 \cos 3x$

3. $y = 1.5 \sin\left(3x + \dfrac{\pi}{2}\right)$

4. $y = -5 \cos\left(x - \dfrac{\pi}{6}\right)$

5. $y = 2.5 \sin\left(4x + \dfrac{2\pi}{9}\right)$

6. $y = 4 \tan x$

7. Write the equation of a sine curve with $a = 5$, a period of 3π, and a phase shift of $-\dfrac{\pi}{6}$.

Graph by addition of ordinates.

8. $y = \sin x + \cos 2x$ **9.** $y = x + \cos x$ **10.** $y = 2 \sin x - \dfrac{1}{x}$

Plot each point in polar coordinates.

11. $(3.4, 125°)$ **12.** $(-2.5, 228°)$ **13.** $\left(1.7, -\dfrac{\pi}{6}\right)$

Graph each equation in polar coordinates.

14. $r^2 = \cos 2\theta$
15. $r + 2 \cos \theta = 1$

Write the polar coordinates of each point.

16. $(7, 3)$ $)$
17. $(-5.30, 3.80)$
18. $(-24, -52)$

Write the rectangular coordinates of each point.

19. $(3.80, 48.0°)$
20. $\left(-65, \dfrac{\pi}{9}\right)$
21. $(3.80, -44.0°)$

Transform into rectangular form.

22 $r(1 - \cos \theta) = 2$
23. $r = 2 \cos \theta$

Transform into polar form.

24. $x - 3y = 2$
25. $5x + 2y = 1$
26. Find the frequency and the angular velocity of a sine wave that has a period of 2.5 s.

27. Find the period and the angular velocity of a cosine wave that has a frequency of 120 Hz.

28. Find the period and the frequency of a sine wave that has an angular velocity of 44.8 rad/s.
29. If a sine wave has a frequency of 30 Hz, how many seconds will it take to complete 100 cycles?
30. What frequency must a sine wave have in order to complete 500 cycles in 2 s?
31. Graph the parametric equations $x = 2 \sin \theta$, $y = 3 \sin 4\theta$.
32. Write an equation for an alternating current that has a peak amplitude of 92.6 mA, a frequency of 82 Hz, and a phase angle of 28.3°.
33. Write each of the following as a single sine wave:
    (a) $y = 63.7 \sin \omega t + 42.9 \sin(\omega t - 38°)$
    (b) $y = 1.73 \sin \omega t + 2.64 \cos \omega t$
34. Given an alternating voltage $v = 27.4 \sin(736t + 37°)$, find the maximum voltage, period, frequency, phase angle, and the instantaneous voltage at $t = 0.25$ s.

Convert to parametric form, and plot.

35. $r = 2 \cos 3\theta$

36. $r = 3 \sin 4\theta$

37. Graph: $y = 2.5 \sin(2x + 32°)$

38. Graph: $y = 1.5 \sin(3x - 41°)$

39. Graph: $y = -3.6 \sin\left(x + \dfrac{\pi}{3}\right)$

40. Graph: $y = 12.4 \sin\left(2x - \dfrac{\pi}{4}\right)$

Writing

41. Given the general sine wave $y = a \sin(bx + c)$, explain in your own words how a change in one of the constants (a, b, or c) will affect the appearance of the sine wave.

42. *Doppler effect:* As a train approaches, the sound of its whistle seems higher than when the train recedes. This is called the *Doppler effect*. Read about and write a short paper on the Doppler effect, including an explanation of why it occurs.

Team Projects

43. A force F (Fig. 17–50) pulls the weight along a horizontal surface. If f is the coefficient of friction, then

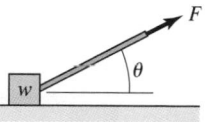

$$F = \frac{fW}{f \sin \theta + \cos \theta}$$

FIGURE 17–50

Graphically find the value of θ, to two significant digits, so that the least force is required. Do this for values of f of 0.50, 0.60, and 0.70.

44. *Beats:* Beats occur when two sine waves of slightly different frequencies are combined.
  (a) Read about beats, and write a short paper on the subject.
  (b) Graph two sine waves, having frequencies of 24 Hz and 30 Hz, for about 20 cycles. Then graph the sum of those two sine waves.
  (c) From the above graphs, what can you deduce about the frequency of the beats?
  (d) Try to produce beats, using a musical instrument, tuning forks, or some other device.
  (e) Name one practical use for beats.

18

Trigonometric Identities and Equations

When you have completed this chapter, you should be able to:
- Simplify trigonometric expressions using fundamental identities.
- Expand or simplify trigonometric expressions containing the sum or difference of two angles.
- Simplify trigonometric expressions containing double angles.
- Simplify trigonometric expressions containing half angles.
- Solve trigonometric equations.
- Evaluate inverse trigonometric functions.

You have used trigonometry in solving vectors and geometry problems, and in working with machinery and navigation. Essentially, anywhere you could model a situation with right triangles you could use trigonometry (and in some cases, using the laws of sines and cosines, you didn't even need to break the situation down into right triangles). As you study more realistic conditions and things get a little more complex, you'll start to see more trigonometric functions in equations. In technical mathematics, we have found some very useful identities that will help you manipulate an equation to find a solution to a problem. An *identity* is something that is true for a wide range of variables.

18–1 Fundamental Identities

Reciprocal Relations

We have already encountered the reciprocal relations in Sec. 7–2, and we repeat them here.

Reciprocal Relations		
$\sin \theta = \dfrac{1}{\csc \theta}$ or $\csc \theta = \dfrac{1}{\sin \theta}$ or $\sin \theta \csc \theta = 1$	**152a**
$\cos \theta = \dfrac{1}{\sec \theta}$ or $\sec \theta = \dfrac{1}{\cos \theta}$ or $\cos \theta \sec \theta = 1$	**152b**
$\tan \theta = \dfrac{1}{\cot \theta}$ or $\cot \theta = \dfrac{1}{\tan \theta}$ or $\tan \theta \cot \theta = 1$	**152c**

◆◆◆ **Example 1:** Simplify $\dfrac{\cos \theta}{\sec^2 \theta}$.

Solution: Using Eq. 152b gives us

$$\frac{\cos \theta}{\sec^2 \theta} = \cos \theta(\cos^2 \theta) = \cos^3 \theta$$

◆◆◆

Quotient Relations

Figure 18–1 shows an angle θ in standard position, as when we first defined the trigonometric functions in Sec. 7–2. We see that

$$\sin \theta = \frac{y}{r} \quad \text{and} \quad \cos \theta = \frac{x}{r}$$

Dividing yields

$$\frac{\sin \theta}{\cos \theta} = \frac{\dfrac{y}{r}}{\dfrac{x}{r}} = \frac{y}{x}$$

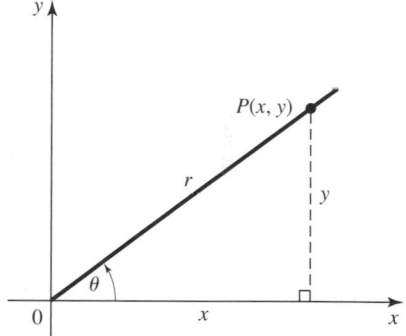

FIGURE 18–1 Angle in standard position.

But $y/x = \tan \theta$, so we have the following equations:

Quotient Relations		
$\tan \theta = \dfrac{\sin \theta}{\cos \theta}$	**162**
$\cot \theta = \dfrac{\cos \theta}{\sin \theta}$	**163**

where $\cot \theta$ is found by taking the reciprocal of $\tan \theta$.

◆◆◆ **Example 2:** Rewrite the expression

$$\frac{\cot \theta}{\csc \theta} - \frac{\tan \theta}{\sec \theta}$$

so that it contains only the sine and cosine, and simplify.

Solution:

$$\frac{\cot \theta}{\csc \theta} - \frac{\tan \theta}{\sec \theta} = \frac{\cos \theta}{\sin \theta} \cdot \frac{\sin \theta}{1} - \frac{\sin \theta}{\cos \theta} \cdot \frac{\cos \theta}{1}$$

$$= \cos \theta - \sin \theta$$

◆◆◆

Pythagorean Relations

We can get three more relations by applying the Pythagorean theorem to the triangle in Figure 18–1.

$$x^2 + y^2 = r^2$$

Dividing through by r^2, we get

$$\frac{x^2}{r^2} + \frac{y^2}{r^2} = 1$$

or

$$\left(\frac{x}{r}\right)^2 + \left(\frac{y}{r}\right)^2 = 1$$

But $x/r = \cos \theta$ and $y/r = \sin \theta$, so the following equation applies:

Pythagorean Relation	$\sin^2 \theta + \cos^2 \theta = 1$	164

Recall that $\sin^2 \theta$ is a short way of writing $(\sin \theta)^2$.

Returning to equation $x^2 + y^2 = r^2$, we now divide through by x^2.

$$1 + \frac{y^2}{x^2} = \frac{r^2}{x^2}$$

But $y/x = \tan \theta$ and $r/x = \sec \theta$, so we have the following:

Pythagorean Relation	$1 + \tan^2 \theta = \sec^2 \theta$	165

Now dividing $x^2 + y^2 = r^2$ by y^2,

$$\frac{x^2}{y^2} + 1 = \frac{r^2}{y^2}$$

But $x/y = \cot \theta$ and $r/y = \csc \theta$. Thus:

Pythagorean Relation	$1 + \cot^2 \theta = \csc^2 \theta$	166

◆◆◆ **Example 3:** Simplify

$$\sin^2 \theta - \csc^2 \theta - \tan^2 \theta + \cot^2 \theta + \cos^2 \theta + \sec^2 \theta$$

Commutative Law	$a + b = b + a$	1

Solution: By the Pythagorean relations,

$$\sin^2 \theta - \csc^2 \theta - \tan^2 \theta + \cot^2 \theta + \cos^2 \theta + \sec^2 \theta$$
$$= (\sin^2 \theta + \cos^2 \theta) - (\tan^2 \theta - \sec^2 \theta) + (\cot^2 \theta - \csc^2 \theta)$$
$$= 1 - (-1) + (-1) = 1$$

◆◆◆

Simplification of Trigonometric Expressions

One use of the trigonometric identities is the simplification of expressions, as in the preceding example. We now give a few more examples.

◆◆◆ **Example 4:** Simplify $(\cot^2 \theta + 1)(\sec^2 \theta - 1)$.

Solution: By Eqs. 165 and 166,

$$(\cot^2 \theta + 1)(\sec^2 \theta - 1) = \csc^2 \theta \tan^2 \theta$$

By Eqs. 152a and 162,

$$= \frac{1}{\sin^2 \theta} \cdot \frac{\sin^2 \theta}{\cos^2 \theta}$$

$$= \frac{1}{\cos^2 \theta}$$

and, by Eq. 152b,

$$= \sec^2 \theta \qquad ◆◆◆$$

◆◆◆ **Example 5:** Simplify $\dfrac{1 - \sin^2 \theta}{\sin \theta + 1}$.

Solution: Factoring the difference of two squares in the numerator gives

$$\frac{1 - \sin^2 \theta}{\sin \theta + 1} = \frac{(1 - \sin \theta)(1 + \sin \theta)}{\sin \theta + 1} = 1 - \sin \theta \qquad ◆◆◆$$

◆◆◆ **Example 6:** Simplify $\dfrac{\cos \theta}{1 - \sin \theta} + \dfrac{\sin \theta - 1}{\cos \theta}$.

Solution: Combining the two fractions over a common denominator, we have

$$\frac{\cos \theta}{1 - \sin \theta} + \frac{\sin \theta - 1}{\cos \theta} = \frac{\cos^2 \theta + (\sin \theta - 1)(1 - \sin \theta)}{(1 - \sin \theta) \cos \theta}$$

$$= \frac{\cos^2 \theta + \sin \theta - \sin^2 \theta - 1 + \sin \theta}{(1 - \sin \theta) \cos \theta}$$

From Eq. 164,

$$= \frac{-\sin^2 \theta - \sin^2 \theta + 2 \sin \theta}{(1 - \sin \theta) \cos \theta}$$

$$= \frac{-2 \sin^2 \theta + 2 \sin \theta}{(1 - \sin \theta) \cos \theta}$$

Factor the numerator:

$$= \frac{2 \sin \theta(-\sin \theta + 1)}{(1 - \sin \theta) \cos \theta}$$

$$= \frac{2 \sin \theta}{\cos \theta}$$

By Eq. 162,

$$= 2 \tan \theta \qquad ◆◆◆$$

> Students tend to forget the rules of *algebraic operations* when working trigonometric expressions. We still need to factor, to combine fractions over a common denominator, and so on.

Trigonometric Identities

We know that an *identity* is an equation that is true for any values of the unknowns, and we have presented eight trigonometric identities so far in this chapter. We now use these eight identities to *verify* or *prove* whether a given identity is true. We do this by trying to transform one side of the identity (usually the more complicated side) until it is identical to the other side.

A good way to start is to rewrite each side so that it contains only sines and cosines.

◆◆◆ Example 7: Prove the identity $\sec \theta \csc \theta = \tan \theta + \cot \theta$.

Solution: Expressing the given identity in terms of sines and cosines gives

$$\sec \theta \csc \theta = \tan \theta + \cot \theta$$

$$\frac{1}{\cos \theta} \cdot \frac{1}{\sin \theta} = \frac{\sin \theta}{\cos \theta} + \frac{\cos \theta}{\sin \theta}$$

If we are to make one side identical to the other, each must finally have the same number of terms. But the left side now has one term and the right has two, so we combine the two fractions on the right over a common denominator.

$$\frac{1}{\sin \theta \cos \theta} = \frac{\sin^2 \theta + \cos^2 \theta}{\sin \theta \cos \theta}$$

By Eq. 164,

$$\frac{1}{\sin \theta \cos \theta} = \frac{1}{\sin \theta \cos \theta}$$ ◆◆◆

Proving identities is *not easy*. It takes a good knowledge of the fundamental identities, lots of practice, and often several false starts. Do not be discouraged if you don't get them right away.

Common Error	Each side of an identity must be worked *separately;* we cannot transpose, or multiply both sides by the same thing, the way we do with equations. The reason is that we do not yet know if the two sides are in fact equal; that is what we are trying to prove.

◆◆◆ Example 8: Prove the identity $\csc x + \tan x = \dfrac{\cot x + \sin x}{\cos x}$.

Solution: As in Example 7, we have a single term on one side of the equals sign and two terms on the other. Now, however, it may be easier to split the single term into two, rather than combine the two terms into one, as we did before.

$$\csc x + \tan x = \frac{\cot x}{\cos x} + \frac{\sin x}{\cos x}$$

Switching to sines and cosines yields

$$\frac{1}{\sin x} + \frac{\sin x}{\cos x} = \frac{\dfrac{\cos x}{\sin x}}{\cos x} + \frac{\sin x}{\cos x}$$

$$= \frac{1}{\sin x} + \frac{\sin x}{\cos x}$$ ◆◆◆

CASE STUDY—AMPLITUDE-MODULATED WAVEFORM

As an electronics engineering technologist, you are analyzing the spectrum of an amplitude-modulated waveform. Besides the original carrier wave, you find that the expression has another term:

$$(mEc \sin 2\pi f_i t) \times (\sin 2\pi f_c t) = mEc (\sin 2\pi f_i) \times (\sin 2\pi f_c t)$$

Use Equation 180 from Appendix A, $\sin \alpha \sin \beta = \frac{1}{2}\cos(\alpha - \beta) - \frac{1}{2}\cos(\alpha + \beta)$, to show that

$$mEc (\sin 2\pi f_i t)(\sin 2pf_c t) = \frac{mEc}{2}\cos 2\pi (f_c - f_i)t \frac{mEc}{2}\cos 2\pi (f_c + f_i)t$$

Exercise 1 ◆ Fundamental Identities

Change to an expression containing only sin and cos.

1. $\tan x - \sec x$

2. $\cot x + \csc x$

3. $\tan \theta \csc \theta$

4. $\sec \theta - \tan \theta \sin \theta$

5. $\dfrac{\tan \theta}{\csc \theta} + \dfrac{\sin \theta}{\tan \theta}$

6. $\cot x + \tan x$

Simplify.

7. $1 - \sec^2 x$

8. $\dfrac{\csc \theta}{\sin \theta}$

9. $\dfrac{\cos \theta}{\cot \theta}$

10. $\sin \theta \csc \theta$

11. $\tan \theta \csc \theta$

12. $\dfrac{\sin \theta}{\csc \theta}$

13. $\sec x \sin x$

14. $\sec x \sin x \cot x$

15. $\csc \theta \tan \theta - \tan \theta \sin \theta$

16. $\dfrac{\cos x}{\cot x \sin x}$

17. $\cot \theta \tan^2 \theta \cos \theta$

18. $\dfrac{\tan x(\csc^2 x - 1)}{\sin x + \cot x \cos x}$

19. $\dfrac{\sin \theta}{\cos \theta \tan \theta}$

20. $\dfrac{\sin^2 x + \cos^2 x}{1 - \cos^2 x}$

21. $\dfrac{1}{\sec^2 x} + \dfrac{1}{\csc^2 x}$

22. $\sin \theta(\csc \theta + \cot \theta)$

23. $\csc x - \cot x \cos x$

24. $1 + \dfrac{\tan^2 \theta}{1 + \sec \theta}$

25. $\dfrac{\sec x - \csc x}{1 - \cot x}$

26. $\dfrac{1}{1 + \sin x} + \dfrac{1}{1 - \sin x}$

27. $\sec^2 x(1 - \cos^2 x)$

28. $\tan x + \dfrac{\cos x}{\sin x + 1}$

29. $\cos \theta \sec \theta - \dfrac{\sec \theta}{\cos \theta}$

30. $\cot^2 x \sin^2 x + \tan^2 x \cos^2 x$

Prove each identity.

31. $\tan x \cos x = \sin x$

32. $\tan x = \dfrac{\sec x}{\csc x}$

33. $\dfrac{\sin x}{\csc x} + \dfrac{\cos x}{\sec x} = 1$

34. $\sin \theta = \dfrac{1}{\cot \theta \sec \theta}$

35. $(\cos^2 \theta + \sin^2 \theta)^2 = 1$

36. $\tan x = \dfrac{\tan x - 1}{1 - \cot x}$

37. $\dfrac{\csc \theta}{\sec \theta} = \cot \theta$

38. $\cot^2 x = \dfrac{\cos x}{\tan x \sin x}$

39. $\cos x + 1 = \dfrac{\sin^2 x}{1 - \cos x}$

40. $\csc x - \sin x = \cot x \cos x$

41. $\cot^2 x - \cos^2 x = \cos^2 x \cot^2 x$

42. $\csc x = \cos x \cot x + \sin x$

43. $1 = (\csc x - \cot x)(\csc x + \cot x)$

44. $\tan x = \dfrac{\tan x + \sin x}{1 + 1 \cos x}$

45. $\dfrac{\tan x + 1}{1 - \tan x} = \dfrac{\sin x + \cos x}{\cos x - \sin x}$

46. $\cot x = \cot x \sec^2 x - \tan x$

47. $\dfrac{\sin \theta + 1}{1 - \sin \theta} = (\tan \theta + \sec \theta)^2$

48. $\dfrac{1 + \sin \theta}{1 - \sin \theta} = \dfrac{1 + \csc \theta}{\csc \theta - 1}$

49. $(\sec \theta - \tan \theta)(\tan \theta + \sec \theta) = 1$

50. $\dfrac{1 + \cot \theta}{\csc \theta} = \dfrac{\tan \theta + 1}{\sec \theta}$

All identities in this chapter can be proven.

18–2   Sum or Difference of Two Angles

Sine of the Sum of Two Angles

We wish now to derive a formula for the sine of the sum of two angles, say, α and β. For example, is it true that

$$\sin 20° + \sin 30° = \sin 50°?$$

Try it on your calculator—you will see that it is not true.

We start by drawing two positive acute angles, α and β (Fig. 18–2), small enough so that their sum ($\alpha + \beta$) is also acute. From any point P on the terminal side of β we draw perpendicular AP to the x axis, and draw perpendicular BP to line OB. Since the angle between two lines equals the angle between the perpendiculars to those two lines, we note that angle APB is equal to α.

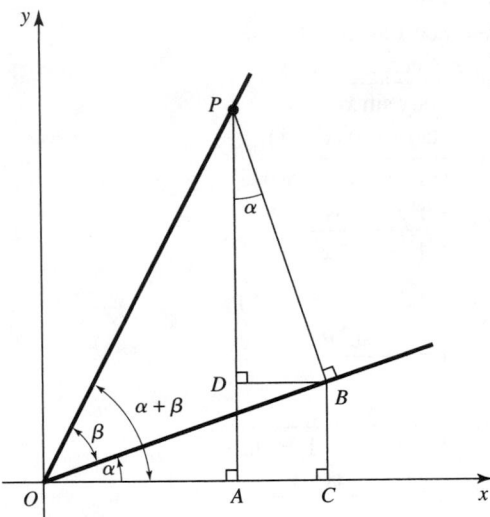

FIGURE 18–2

Then

$$\sin(\alpha + \beta) = \frac{AP}{OP} = \frac{AD + PD}{OP} = \frac{BC + PD}{OP}$$
$$= \frac{BC}{OP} + \frac{PD}{OP}$$

But in triangle OBC,

$$BC = OB \sin \alpha$$

and in triangle PBD,

$$PD = PB \cos \alpha$$

Substituting, we obtain

$$\sin(\alpha + \beta) = \frac{OB \sin \alpha}{OP} + \frac{PB \cos \alpha}{OP}$$

But in triangle OPB,

$$\frac{OB}{OP} = \cos \beta$$

and

$$\frac{PB}{OP} = \sin \beta$$

Thus:

$$\sin(\alpha + \beta) = \sin\alpha\cos\beta + \cos\alpha\sin\beta$$

Common Error	The sine of the sum of two angles *is not* the sum of the sines of each angle. $$\sin(\alpha + \beta) \neq \sin\alpha + \sin\beta$$

We have proven this true for acute angles whose sum is also acute. These identities are, in fact, true for any size angles, positive or negative, although we will not take the space to prove this.

Cosine of the Sum of Two Angles

Again using Fig. 8–2, we can derive an expression for $\cos(\alpha + \beta)$. From Eq. 147,

$$\cos(\alpha + \beta) = \frac{OA}{OP} = \frac{OC - AC}{OP} = \frac{OC - BD}{OP}$$
$$= \frac{OC}{OP} - \frac{BD}{OP}$$

Now, in triangle *OBC*,

$$OC = OB\cos\alpha$$

and in triangle *PDB*,

$$BD = PB\sin\alpha$$

Substituting, we obtain

$$\cos(\alpha + \beta) = \frac{OB\cos\alpha}{OP} - \frac{PB\sin\alpha}{OP}$$

As before,

$$\frac{OB}{OP} = \cos\beta \quad \text{and} \quad \frac{PB}{OP} = \sin\beta$$

Therefore:

$$\cos(\alpha + \beta) = \cos\alpha\cos\beta + \sin\alpha\sin\beta$$

Difference of Two Angles

We can obtain a formula for the sine of the difference of two angles merely by substituting $-\beta$ for β in the equation previously derived for $\sin(\alpha + \beta)$.

$$\sin[\alpha + (-\beta)] = \sin\alpha\cos(-\beta) + \cos\alpha\sin(-\beta)$$

But for β in the first quadrant, $(-\beta)$ is in the fourth, so

$$\cos(-\beta) = \cos\beta \quad \text{and} \quad \sin(-\beta) = -\sin\beta$$

Therefore,

$$\sin(\alpha - \beta) = \sin\alpha\cos\beta + \cos\alpha(-\sin\beta)$$

which is rewritten as follows:

$$\sin(\alpha + \beta) = \sin\alpha\cos\beta + \cos\alpha\sin\beta$$

We see that the result is identical to the formula for $\sin(\alpha + \beta)$ except for a change in sign. This enables us to write the two identities in a single expression using the $\pm$ sign.

When the double signs are used (such as the $\pm$ symbol), it is understood that the upper signs and the lower signs correspond with the elements in the equation.

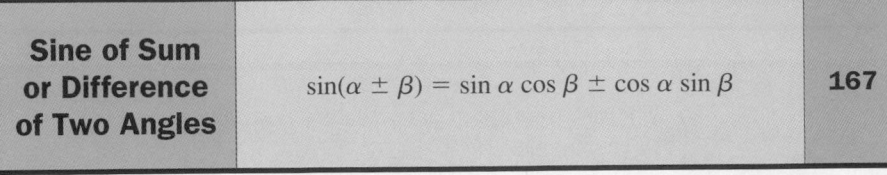

| **Sine of Sum or Difference of Two Angles** | $\sin(\alpha \pm \beta) = \sin \alpha \cos \beta \pm \cos \alpha \sin \beta$ | **167** |

Similarly, finding $\cos(\alpha - \beta)$, we have

$$\cos[\alpha + (-\beta)] = \cos \alpha \cos(-\beta) - \sin \alpha \sin(-\beta)$$

Thus:

$$\cos(\alpha - \beta) = \cos \alpha \cos \beta + \sin \alpha \sin \beta$$

or:

| **Cosine of Sum or Difference of Two Angles** | $\cos(\alpha \pm \beta) = \cos \alpha \cos \beta \mp \sin \alpha \sin \beta$ | **168** |

◆◆◆ **Example 9:** Expand the expression $\sin(x + 3y)$.

Solution: By Eq. 167,

$$\sin(x + 3y) = \sin x \cos 3y + \cos x \sin 3y \qquad \text{◆◆◆}$$

◆◆◆ **Example 10:** Simplify

$$\cos 5x \cos 3x - \sin 5x \sin 3x$$

Solution: We see that this has a similar form to Eq. 168, with $\alpha = 5x$ and $\beta = 3x$, so

$$\cos 5x \cos 3x - \sin 5x \sin 3x = \cos(5x + 3x)$$
$$= \cos 8x \qquad \text{◆◆◆}$$

◆◆◆ **Example 11:** Prove that

$$\cos(180° - \theta) = -\cos \theta$$

Solution: Expanding the left side by means of Eq. 168, we get

$$\cos(180° - \theta) = \cos 180° \cos \theta + \sin 180° \sin \theta$$

But $\cos 180° = -1$ and $\sin 180° = 0$, so

$$\cos(180° - \theta) = (-1) \cos \theta + (0) \sin \theta = -\cos \theta \qquad \text{◆◆◆}$$

◆◆◆ **Example 12:** Prove that

$$\frac{\tan x - \tan y}{\tan x + \tan y} = \frac{\sin(x - y)}{\sin(x + y)}$$

Solution: The right side contains functions of the sum of two angles, but the left side contains functions of single angles. We thus start by expanding the right side by using Eq. 167.

$$\frac{\sin(x - y)}{\sin(x + y)} = \frac{\sin x \cos y - \cos x \sin y}{\sin x \cos y + \cos x \sin y}$$

Our expression now contains $\sin x$, $\sin y$, $\cos x$, and $\cos y$, but we want an expression containing only $\tan x$ and $\tan y$.

To have $\tan x$ instead of $\sin x$, we can divide numerator and denominator by $\cos x$. Similarly, to obtain $\tan y$ instead of $\sin y$, we can divide by $\cos y$. We thus divide numerator and denominator by $\cos x \cos y$.

$$\frac{\sin x \cos y - \cos x \sin y}{\sin x \cos y + \cos x \sin y} = \frac{\dfrac{\sin x \cancel{\cos y}}{\cos x \cancel{\cos y}} - \dfrac{\cancel{\cos x} \sin y}{\cancel{\cos x} \cos y}}{\dfrac{\sin x \cancel{\cos y}}{\cos x \cancel{\cos y}} + \dfrac{\cancel{\cos x} \sin y}{\cancel{\cos x} \cos y}}$$

Then, by Eq. 162,

$$= \frac{\tan x - \tan y}{\tan x + \tan y} \qquad\qquad \blacklozenge\blacklozenge\blacklozenge$$

Tangent of the Sum or Difference of Two Angles
Since, by Eq. 162,

$$\tan \theta = \frac{\sin \theta}{\cos \theta}$$

we simply divide Eq. 167 by Eq. 168.

$$\tan(\alpha + \beta) = \frac{\sin(\alpha + \beta)}{\cos(\alpha + \beta)} = \frac{\sin \alpha \cos \beta + \cos \alpha \sin \beta}{\cos \alpha \cos \beta - \sin \alpha \sin \beta}$$

Dividing numerator and denominator by $\cos \alpha \cos \beta$ yields

$$\tan(\alpha + \beta) = \frac{\dfrac{\sin \alpha}{\cos \alpha} + \dfrac{\sin \beta}{\cos \beta}}{1 - \dfrac{\sin \alpha}{\cos \alpha} \cdot \dfrac{\sin \beta}{\cos \beta}}$$

Applying Eq. 162 again, we get

$$= \frac{\tan \alpha + \tan \beta}{1 - \tan \alpha \tan \beta}$$

A similar derivation (which we will not do) will show that $\tan(\alpha - \beta)$ is identical to the expression just derived, except, as we might expect, for a reversal of signs. We combine the two expressions using double signs as follows:

Tangent of Sum or Difference of Two Angles	$\tan(\alpha \pm \beta) = \dfrac{\tan \alpha \pm \cos \beta}{1 \mp \tan \alpha \tan \beta}$	**169**

◆◆◆ **Example 13:** Simplify

$$\frac{\tan 3x + \tan 2x}{\tan 2x \tan 3x - 1}$$

Solution: This can be put into the form of Eq. 169 if we factor (-1) from the denominator.

$$\frac{\tan 3x + \tan 2x}{\tan 2x \tan 3x - 1} = \frac{\tan 3x + \tan 2x}{-(-\tan 2x \tan 3x + 1)}$$

$$= -\frac{\tan 3x + \tan 2x}{1 - \tan 3x \tan 2x}$$

$$= -\tan(3x + 2x)$$

$$= -\tan 5x \qquad\qquad \blacklozenge\blacklozenge\blacklozenge$$

◆◆◆ **Example 14:** Prove that

$$\tan(45° + x) = \frac{1 + \tan x}{1 - \tan x}$$

Solution: Expanding the left side by Eq. 169, we get

$$\tan(45° + x) = \frac{\tan 45° + \tan x}{1 - \tan 45° \tan x}$$

But tan 45° = 1, so

$$\tan(45° + x) = \frac{1 + \tan x}{1 - \tan x}$$ ◆◆◆

◆◆◆ **Example 15:** Prove that

$$\frac{\cot y - \cot x}{\cot x \cot y + 1} = \tan(x - y)$$

Solution: Using Eq. 152c on the left side yields

$$\frac{\dfrac{1}{\tan y} - \dfrac{1}{\tan x}}{\dfrac{1}{\tan x} \times \dfrac{1}{\tan y} + 1}$$

Multiply numerator and denominator by tan x tan y.

$$\frac{\tan x - \tan y}{1 + \tan x \tan y} = \tan(x - y)$$

Then, by Eq. 169:

$$\tan(x - y) = \tan(x - y)$$ ◆◆◆

Adding a Sine Wave and a Cosine Wave of the Same Frequency

In Sec. 17–3, we used vectors to show that the sum of a sine wave and a cosine wave of the same frequency could be written as a single sine wave, at the original frequency, but with some phase angle. The resulting equation is useful in electrical applications. Here we use the formula for the sum or difference of two angles to derive that equation.

Let $A \sin \omega t$ be a sine wave of amplitude A, and $B \cos \omega t$ be a cosine wave of amplitude B, each of frequency $\omega/2\pi$. If we draw a right triangle (Fig. 18–3) with sides A and B and hypotenuse R, then

$$A = R \cos \phi \quad \text{and} \quad B = R \sin \phi$$

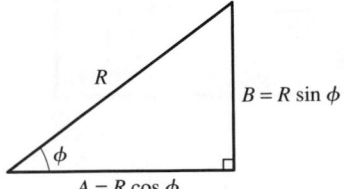

$$B = R \sin \phi$$

$$A = R \cos \phi$$

FIGURE 18–3

The sum of the sine wave and the cosine wave is then

$$A \sin \omega t + B \cos \omega t = R \sin \omega t \cos \phi + R \cos \omega t \sin \phi$$
$$= R(\sin \omega t \cos \phi + \cos \omega t \sin \phi)$$

the expression on the right will be familiar. It is the relationship of the sine of the *sum* of two quantities, ωt and ϕ. Continuing, we have

$$A \sin \omega t + B \cos \omega t = R(\sin \omega t \cos \phi + \cos \omega t \sin \phi)$$
$$= R \sin(\omega t + \phi)$$

by Eq. 167. Thus we have the following equation:

Addition of a Sine Wave and Cosine Wave	$A \sin \omega t + B \cos \omega t = R \sin(\omega t + \phi)$ where $R = \sqrt{A^2 + B^2}$ and $\phi = \arctan \dfrac{B}{A}$	214

This is sometimes referred to as *magnitude* and *phase* form.

◆◆◆ **Example 16:** Express the following as a single sine function:

$$y = 3.46 \sin \omega t + 2.28 \cos \omega t$$

Solution: The magnitude of the resultant is

$$R = \sqrt{(3.46)^2 + (2.28)^2} = 4.14$$

and the phase angle is

$$\phi = \arctan \frac{2.28}{3.46} = \arctan 0.659 = 33.4°$$

So

$$y = 3.46 \sin \omega t + 2.28 \cos \omega t$$
$$= 4.14 \sin(\omega t + 33.4°)$$

◆◆◆

Exercise 2 ◆ Sum or Difference of Two Angles

Expand by means of the addition and subtraction formulas, and simplify.

1. $\sin(\theta + 30°)$
2. $\cos(45° - x)$
3. $\sin(x + 60°)$
4. $\tan(\pi + \theta)$
5. $\cos\left(x + \dfrac{\pi}{2}\right)$
6. $\tan(2x + y)$
7. $\sin(\theta + 2\phi)$
8. $\cos[(\alpha + \beta) + \gamma]$
9. $\tan(2\theta - 3\alpha)$

Simplify.

10. $\cos 2x \cos 9x + \sin 2x \sin 9x$
11. $\cos(\pi + \theta) + \sin(\pi + \theta)$
12. $\sin 3\theta \cos 2\theta - \cos 3\theta \sin 2\theta$
13. $\sin\left(\dfrac{\pi}{3} - x\right) - \cos\left(\dfrac{\pi}{6} - x\right)$

Prove each identity.

14. $\cos x = \sin(x + 90°)$
15. $\sin(30° - x) + \cos(60° - x) = \cos x$
16. $2 \sin \alpha \cos \beta = \sin(\alpha + \beta) + \sin(\alpha - \beta)$
17. $\sin(\theta + 60°) + \cos(\theta + 30°) = \sqrt{3} \cos \theta$
18. $\cos(2\pi - x) = \cos x$

19. $\sin \alpha = \cos(30° - \alpha) - \sin(60° - \alpha)$

20. $\cos(x + y) + \cos(x - y) = 2 \cos x \cos y$

21. $\cos x = \sin\left(x + \dfrac{\pi}{6}\right) - \left(\sin x - \dfrac{\pi}{6}\right)$

22. $\tan(360° - \beta) = -\tan \beta$

23. $\cos(60° + \alpha) + \sin(330° + \alpha) = 0$

24. $\sin 5x \sec x \csc x = \dfrac{\sin 4x}{\sin x} + \dfrac{\cos 4x}{\cos x}$

25. $\dfrac{\sin(x - y)}{\sin(x + y)} = \dfrac{\tan x - \tan y}{\tan x + \tan y}$

26. $\cos(x + 60°) + \cos(60° - x) = \cos x$

27. $\dfrac{\cos(x - y)}{\sin x \cos y} = \tan y + \cot x$

28. $\cot\left(x + \dfrac{\pi}{4}\right) + \tan\left(x - \dfrac{\pi}{4}\right) = 0$

29. $\tan(\alpha + 45°) = \dfrac{\cos \alpha + \sin \alpha}{\cos \alpha - \sin \alpha}$

30. $\dfrac{\cot \alpha \cot \beta - 1}{\cot \alpha + \cot \beta} = \cot(\alpha + \beta)$

31. $\dfrac{1 + \tan x}{1 - \tan x} = \tan\left(\dfrac{\pi}{4} + x\right)$

Express as a single sine function.

32. $y = 47.2 \sin \omega t + 64.9 \cos \omega t$

33. $y = 8470 \sin \omega t + 7360 \cos \omega t$

34. $y = 1.83 \sin \omega t + 2.74 \cos \omega t$

35. $y = 84.2 \sin \omega t + 74.2 \cos \omega t$

18–3 Functions of Double Angles

Sine of 2α

An equation for the sine of 2α is easily derived by setting $\beta = \alpha$ in Eq. 167.

$$\sin(\alpha + \alpha) = \sin \alpha \cos \alpha + \cos \alpha \sin \alpha$$

which is rewritten in the following form:

Sine of Twice an Angle	$\sin 2\alpha = 2 \sin \alpha \cos \alpha$	170

Cosine of 2α

Similarly, setting $\beta = \alpha$ in Eq. 168, we have

$$\cos(\alpha + \alpha) = \cos \alpha \cos \alpha - \sin \alpha \sin \alpha$$

which is also given as follows:

Cosine of Twice an Angle	$\cos 2\alpha = \cos^2 \alpha - \sin^2 \alpha$	171a

There are two alternative forms for Eq. 171a. From Eq. 164,

$$\cos^2 \alpha = 1 - \sin^2 \alpha$$

Substituting into Eq. 171a yields

$$\cos 2\alpha = 1 - \sin^2 \alpha - \sin^2 \alpha$$

Cosine of Twice an Angle	$\cos 2\alpha = 1 - \sin^2 \alpha$	171b

We can similarly use Eq. 164 to eliminate the $\sin^2 \alpha$ term from Eq. 171a.

$$\cos 2\alpha = \cos^2 \alpha - (1 - \cos^2 \alpha)$$

Thus:

Cosine of Twice an Angle	$\cos 2\alpha = 2 \cos^2 \alpha - 1$	171c

◆◆◆ **Example 17:** Express $\sin 3x$ in terms of the single angle x.

Solution: We can consider $3x$ to be $x + 2x$, and, using Eq. 167, we have

$$\sin 3x = \sin(x + 2x) = \sin x \cos 2x + \cos x \sin 2x$$

We now replace $\cos 2x$ by $\cos^2 x - \sin^2 x$, and $\sin 2x$ by $2 \sin x \cos x$.

$$\begin{aligned}
\sin 3x &= \sin x(\cos^2 x - \sin^2 x) + \cos x(2 \sin x \cos x) \\
&= \sin x \cos^2 x - \sin^3 x + 2 \sin x \cos^2 x \\
&= 3 \sin x \cos^2 x - \sin^3 x
\end{aligned}$$

◆◆◆

◆◆◆ **Example 18:** Prove that

$$\cos 2A + \sin(A - B) = 0$$

where A and B are the two acute angles of a right triangle.

Solution: By Eqs. 171a and 167,

$$\cos 2A + \sin(A - B) = \cos^2 A - \sin^2 A + \sin A \cos B - \cos A \sin B$$

But, by the cofunctions of Eq. 154,

$$\cos B = \sin A \qquad \text{and} \qquad \sin B = \cos A$$

So

$$\begin{aligned}
\cos 2A + \sin(A - B) &= \cos^2 A - \sin^2 A + \sin A \sin A - \cos A \cos A \\
&= \cos^2 A - \sin^2 A + \sin^2 A - \cos^2 A = 0
\end{aligned}$$

◆◆◆

◆◆◆ **Example 19:** Simplify the expression

$$\frac{\sin 2x}{1 + \cos 2x}$$

Solution: By Eqs. 170 and 171c,

$$\begin{aligned}
\frac{\sin 2x}{1 + \cos 2x} &= \frac{2 \sin x \cos x}{1 + 2 \cos^2 x - 1} \\
&= \frac{2 \sin x \cos x}{2 \cos^2 x} \\
&= \frac{\sin x}{\cos x} = \tan x
\end{aligned}$$

◆◆◆

Tangent of 2α

Setting $\beta = \alpha$ in Eq. 169 gives

$$\tan(\alpha + \alpha) = \frac{\tan \alpha + \tan \alpha}{1 - \tan \alpha \tan \alpha}$$

Therefore:

Tangent of Twice an Angle	$\tan 2\alpha = \dfrac{2 \tan x}{1 - \tan^2 x}$	**172**

◆◆◆ **Example 20:** Prove that

$$\frac{2 \cot x}{\csc^2 x - 2} = \tan 2x$$

Solution: By Eq. 172,

$$\tan 2x = \frac{2 \tan \alpha}{1 - \tan^2 \alpha}$$

and, by Eq. 152c,

$$\tan 2x = \frac{\dfrac{2}{\cot x}}{1 - \dfrac{1}{\cot^2 x}}$$

Multiply numerator and denominator by $\cot^2 x$.

$$\tan 2x = \frac{2 \cot x}{\cot^2 x - 1}$$

Then, by Eq. 166,

$$\tan 2x = \frac{2 \cot x}{(\csc^2 x - 1) - 1} = \frac{2 \cot x}{\csc^2 x - 2} \qquad\qquad ◆◆◆$$

Common Error	The sine of twice an angle is not twice the sine of the angle. Nor is the cosine (or tangent) of twice an angle equal to twice the cosine (or tangent) of that angle. Remember to use the formulas from this section for all of the trig functions of double angles.

Exercise 3 ◆ Functions of Double Angles

Simplify.

1. $2 \sin^2 x + \cos 2x$

2. $2 \sin 2\theta \cos 2\theta$

3. $\dfrac{2 \tan x}{1 + \tan^2 x}$

4. $\dfrac{2 - \sec^2 x}{\sec^2 x}$

5. $\dfrac{\sin 6x}{\sin 2x} - \dfrac{\cos 6x}{\cos 2x}$

Prove.

6. $\dfrac{2 \tan \theta}{1 - \tan^2 \theta} = \tan 2\theta$

7. $\dfrac{1 - \tan^2 x}{1 + \tan^2 x} = \cos 2x$

8. $2 \csc 2\theta = \tan \theta + \cot \theta$

9. $2 \cot 2x = \cot x - \tan x$

10. $\sec 2x = \dfrac{1 + \cot^2 x}{\cot^2 x - 1}$

11. $\dfrac{1 - \tan x}{\tan x + 1} = \dfrac{\cos 2x}{\sin 2x + 1}$

12. $\dfrac{\sin 2\theta + \sin \theta}{1 + \cos \theta + \cos 2\theta} = \tan \theta$

13. $1 + \cot x = \dfrac{2 \cos 2x}{\sin 2x - 2 \sin^2 x}$

14. $4 \cos^3 x - 3 \cos x = \cos 3x$

15. $\sin \alpha + \cos \alpha = \dfrac{\sin 2\alpha + 1}{\cos \alpha + \sin \alpha}$

16. $\dfrac{\cot^2 x - 1}{2 \cot x} = \cot 2x$

If A and B are the two acute angles in a right triangle, show that:

17. $\sin 2A = \sin 2B$.

18. $\tan(A - B) = -\cot 2A$.

19. $\sin 2A = \cos(A - B)$.

18–4 Functions of Half-Angles

Sine of $\alpha/2$

The double-angle formulas derived in Sec. 18–3 can also be regarded as half-angle formulas, because if one angle is double another, the second angle must be half the first.

Starting with Eq. 171b, we obtain

$$\cos 2\theta = 1 - 2 \sin^2 \theta$$

We solve for $\sin \theta$.

$$2 \sin^2 \theta = 1 - \cos 2\theta$$

$$\sin \theta = \pm \sqrt{\dfrac{1 - \cos 2\theta}{2}}$$

For emphasis, we replace θ by $\alpha/2$.

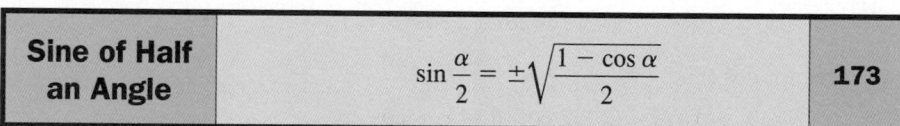

Sine of Half an Angle	$\sin \dfrac{\alpha}{2} = \pm \sqrt{\dfrac{1 - \cos \alpha}{2}}$	173

The $\pm$ sign in Eq. 173 (and in Eqs. 174 and 175c as well) is to be read as plus *or* minus, but *not both*. This sign is different from the $\pm$ sign in the quadratic formula, for example, where we took *both* the positive and the negative values.

The reason for this difference is clear from Fig. 18–4, which shows a graph of $\sin \alpha/2$ and a graph of $+\sqrt{(1 - \cos \alpha)/2}$. Note that the two curves are the same only when $\sin \alpha/2$ is positive. When $\sin \alpha/2$ is negative, it is necessary to use the negative of $\sqrt{(1 - \cos \alpha)/2}$. This occurs when $\alpha/2$ is in the third or fourth quadrant. Thus we choose the plus or the minus according to the quadrant in which $\alpha/2$ is located.

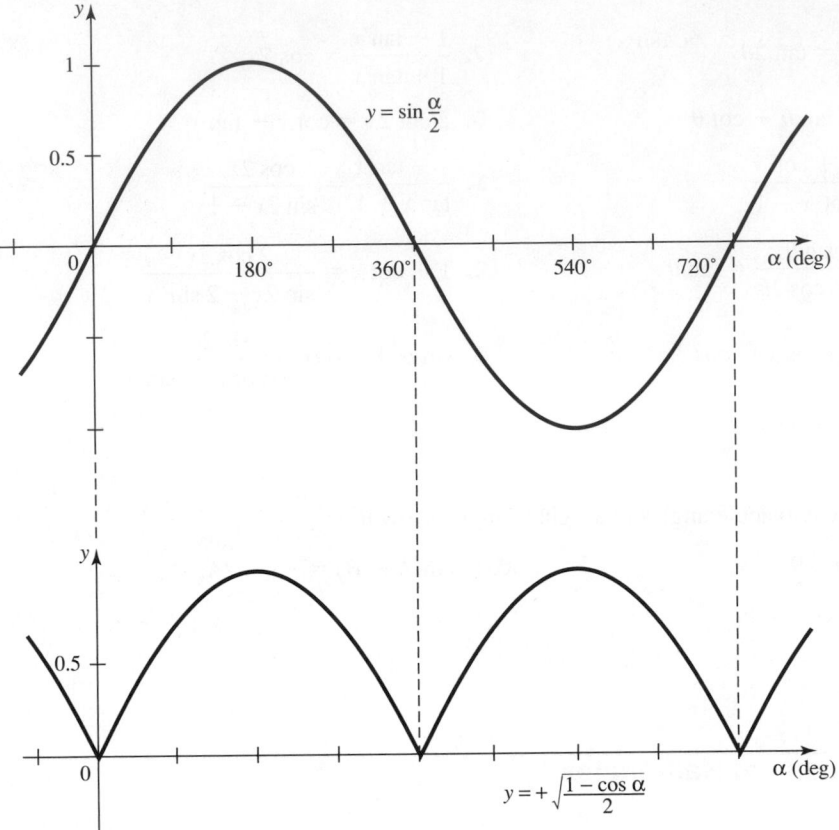

FIGURE 18–4

◆◆◆ **Example 21:** Given that $\cos 200° = -0.9397$, find $\sin 100°$.

Solution: From Eq. 173,

$$\sin \frac{200°}{2} = \pm\sqrt{\frac{1 - \cos 200°}{2}} = \pm\sqrt{\frac{1 - (-0.9397)}{2}} = \pm 0.9848$$

Since 100° is in the second quadrant, the sine is positive, so we drop the minus sign and get

$$\sin 100° = 0.9848$$

◆◆◆

Cosine of $\alpha/2$

Similarly, starting with Eq. 171c, we have

$$\cos 2\theta = 2 \cos^2 \theta - 1$$

Then we solve for $\cos \theta$.

$$2 \cos^2 \theta = 1 + \cos 2\theta$$

$$\cos \theta = \pm\sqrt{\frac{1 + \cos 2\theta}{2}}$$

We next replace θ by $\alpha/2$ and obtain the following:

Here also, we choose the plus *or* the minus according to the quadrant in which $\alpha/2$ is located.

Cosine of Half an Angle	$\cos \dfrac{\alpha}{2} = \pm\sqrt{\dfrac{1 + \cos \alpha}{2}}$	174

Tangent of $\alpha/2$

There are three formulas for the tangent of a half-angle; we will show the derivation of each in turn. First, from Eq. 162,

$$\tan\frac{\alpha}{2} = \frac{\sin\dfrac{\alpha}{2}}{\cos\dfrac{\alpha}{2}}$$

Multiply numerator and denominator by $2\sin(\alpha/2)$.

$$\tan\frac{\alpha}{2} = \frac{2\sin^2\dfrac{\alpha}{2}}{2\sin\dfrac{\alpha}{2}\cos\dfrac{\alpha}{2}}$$

Then, by Eqs. 171b and 170, we have the following:

Tangent of Half an Angle	$\tan\dfrac{\alpha}{2} = \dfrac{1-\cos\alpha}{\sin\alpha}$	**175a**

Another form of this identity is obtained by multiplying numerator and denominator by $1+\cos\alpha$.

$$\tan\frac{\alpha}{2} = \frac{1-\cos\alpha}{\sin\alpha}\cdot\frac{1+\cos\alpha}{1+\cos\alpha}$$

$$= \frac{1-\cos^2\alpha}{\sin\alpha(1+\cos\alpha)} = \frac{\sin^2\alpha}{\sin\alpha(1+\cos\alpha)}$$

also written in the following form:

Tangent of Half an Angle	$\tan\dfrac{\alpha}{2} = \dfrac{\sin\alpha}{1+\cos\alpha}$	**175b**

We can obtain a third formula for the tangent by dividing Eq. 173 by Eq. 174.

$$\tan\frac{\alpha}{2} = \frac{\sin\dfrac{\alpha}{2}}{\cos\dfrac{\alpha}{2}} = \frac{\pm\sqrt{\dfrac{1-\cos\alpha}{2}}}{\pm\sqrt{\dfrac{1+\cos\alpha}{2}}}$$

Tangent of Half an Angle	$\tan\dfrac{\alpha}{2} = \pm\sqrt{\dfrac{1-\cos\alpha}{1+\cos\alpha}}$	**175c**

◆◆◆ **Example 22:** Prove that

$$\frac{1+\sin^2\dfrac{\theta}{2}}{1+\cos^2\dfrac{\theta}{2}} = \frac{3-\cos\theta}{3+\cos\theta}$$

Solution: By Eqs. 173 and 174,

$$\frac{1 + \sin^2 \frac{\theta}{2}}{1 + \cos^2 \frac{\theta}{2}} = \frac{1 + \dfrac{1 - \cos \theta}{2}}{1 + \dfrac{1 + \cos \theta}{2}}$$

Multiply numerator and denominator by 2.

$$= \frac{2 + 1 - \cos \theta}{2 + 1 + \cos \theta}$$

$$= \frac{3 - \cos \theta}{3 + \cos \theta} \qquad \qquad \text{♦♦♦}$$

Exercise 4 ♦ Functions of Half-Angles

Prove each identity.

1. $2 \sin^2 \dfrac{\theta}{2} + \cos \theta = 1$

2. $4 \cos^2 \dfrac{x}{2} \sin^2 \dfrac{x}{2} = 1 - \cos^2 x$

3. $2 \cos^2 \dfrac{x}{2} = \dfrac{\sec x + 1}{\sec x}$

4. $\cot \dfrac{\theta}{2} = \csc \theta + \cot \theta$

5. $\dfrac{\sin \alpha}{\cos \alpha + 1} = \sin \dfrac{\alpha}{2} \sec \dfrac{\alpha}{2}$

6. $\dfrac{\cos^2 \dfrac{\theta}{2} - \cos \theta}{\sin^2 \dfrac{\theta}{2}} = 1$

7. $\sec \theta = \tan \dfrac{\theta}{2} \tan \theta + 1$

8. $\sin x + 1 = \left(\cos \dfrac{x}{2} + \sin \dfrac{x}{2} \right)^2$

9. $2 \sin \alpha + \sin 2\alpha = 4 \sin \alpha \cos^2 \dfrac{\alpha}{2}$

10. $\dfrac{1 - \cos x}{\sin x} = \tan \dfrac{x}{2}$

11. $\dfrac{1 - \tan^2 \dfrac{\theta}{2}}{1 + \tan^2 \dfrac{\theta}{2}} = \cos \theta$

In right triangle ABC, show the following:

12. $\tan \dfrac{A}{2} = \dfrac{a}{b + c}$

13. $\sin \dfrac{A}{2} = \sqrt{\dfrac{c - b}{2c}}$

18–5   Trigonometric Equations

Solving Trigonometric Equations

One use for the trigonometric identities we have just studied is in the solution of trigonometric equations.

A trigonometric equation will usually have an infinite number of roots. However, it is customary to list only *nonnegative values less than 360°* that satisfy the equation.

◆◆◆ **Example 23:** Solve the equation $\sin \theta = \frac{1}{2}$.

Solution: Even though there are infinitely many values of θ,

$$\theta = 30°, 150°, 390°, 510°, \dots$$

whose sine is $\frac{1}{2}$, we will follow the common practice of listing only the nonnegative values less than 360°. Thus our solution is

$$\theta = 30°, 150°$$

The solution could also be expressed in radian measure: $\theta = \pi/6, 5\pi/6$ radians. ◆◆◆

Equations Containing a Single Trigonometric Function and a Single Angle

We start with the simplest type. These equations contain, after simplification, only one trigonometric function (only sine, or only cosine, for example) and have only one angle, as in Example 23. If an equation does not *appear* to be of this type, simplify it first. Then isolate the trigonometric function on one side of the equation, and find the angles.

◆◆◆ **Example 24:** Solve the equation

$$\frac{2.73 \sec \theta}{\csc \theta} + \frac{1.57}{\cos \theta} = 0$$

Solution: As with identities, it is helpful to rewrite the given expression with only sines and cosines. Thus

$$\frac{2.73 \sin \theta}{\cos \theta} + \frac{1.57}{\cos \theta} = 0$$

Multiplying by $\cos \theta$ and rearranging gives an expression with only one function, sine.

$$\sin \theta = -\frac{1.57}{2.73} = -0.575$$

Our reference angle is arcsin(0.575) or 35.1°. The sine is negative in the third and fourth quadrants, so we give the third and fourth quadrant values for θ.

$$\theta = 180° + 35.1° = 215.1°$$

and

$$\theta = 360° - 35.1° = 324.9°$$

Figure 18-5 shows a graph of the function and its zeros.

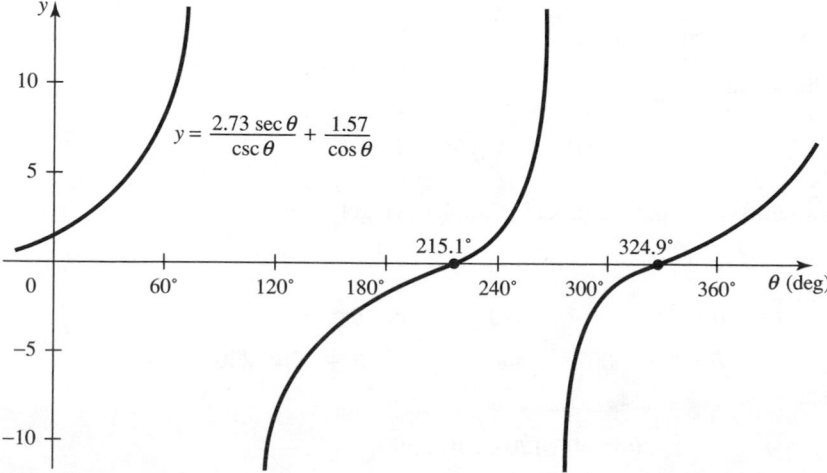

FIGURE 18-5

◆◆◆

Our next example contains a double angle.

◆◆◆ **Example 25:** Solve the equation $2 \cos 2\theta - 1 = 0$.

Solution: Rearranging and dividing, we have

$$\cos 2\theta = \tfrac{1}{2}$$
$$2\theta = 60°, 300°, 420°, 660°, \dots$$
$$\theta = 30°, 150°, 210°, \text{ and } 330°$$

Note that although this problem contained a double angle, *we did not need the double-angle formula*. It would have been needed, however, if the same problem contained both a double angle *and* a single angle.

if we limit our solution to angles less than 360°. Figure 18–6 shows a graph of the function $y = 2 \cos 2\theta - 1$ and its zeros.

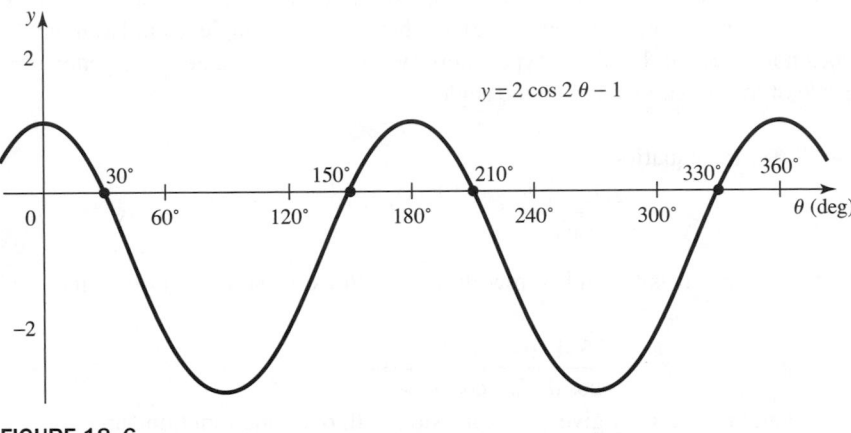

$y = 2 \cos 2\theta - 1$

FIGURE 18–6   ◆◆◆

| **Common Error** | It is easy to forget to find *all of the angles* less than 360°, especially when the given equation contains a double angle, such as in Example 25. |

If one of the functions is *squared*, we may have an equation in *quadratic form*, which can be solved by the methods of Chapter 14.

◆◆◆ **Example 26:** Solve the equation $\sec^2 \theta = 4$.

Solution: Taking the square root of both sides gives us

$$\sec \theta = \pm\sqrt{4} = \pm 2$$

We thus have two solutions. By the reciprocal relations, we get

$$\frac{1}{\cos \theta} = 2 \qquad \text{and} \qquad \frac{1}{\cos \theta} = -2$$
$$\cos \theta = \tfrac{1}{2} \qquad \text{and} \qquad \cos \theta = -\tfrac{1}{2}$$
$$\theta = 60°, 300° \qquad \text{and} \qquad \theta = 120°, 240°$$

Our solution is then

$$\theta = 60°, 120°, 240°, 300°$$

Figure 18–7 shows a graph of the function $y = \sec^2 \theta - 4$ and its zeros.

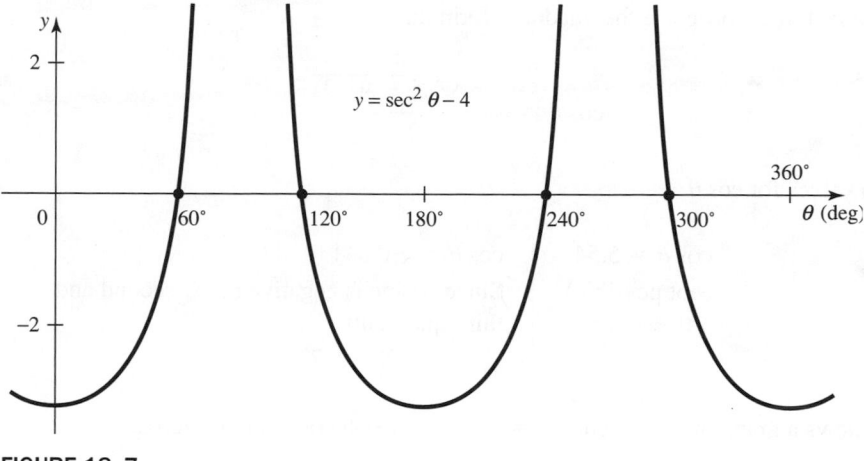

FIGURE 18-7 ◆◆◆

◆◆◆ **Example 27:** Solve the equation $2 \sin^2 \theta - \sin \theta = 0$.

Solution: This is an *incomplete quadratic* in $\sin \theta$. Factoring, we obtain

$$\sin \theta (2 \sin \theta - 1) = 0$$

Setting each factor equal to zero yields

$$
\begin{array}{c|c}
\sin \theta = 0 & 2 \sin \theta - 1 = 0 \\
\theta = 0°, 180° & \sin \theta = \frac{1}{2} \\
& \text{Since sine is positive in the first and second quadrants,} \\
& \theta = 30°, 150°
\end{array}
$$

Figure 18–8 shows a graph of the function $y = 2 \sin^2 \theta - \sin \theta$ and its zeros.

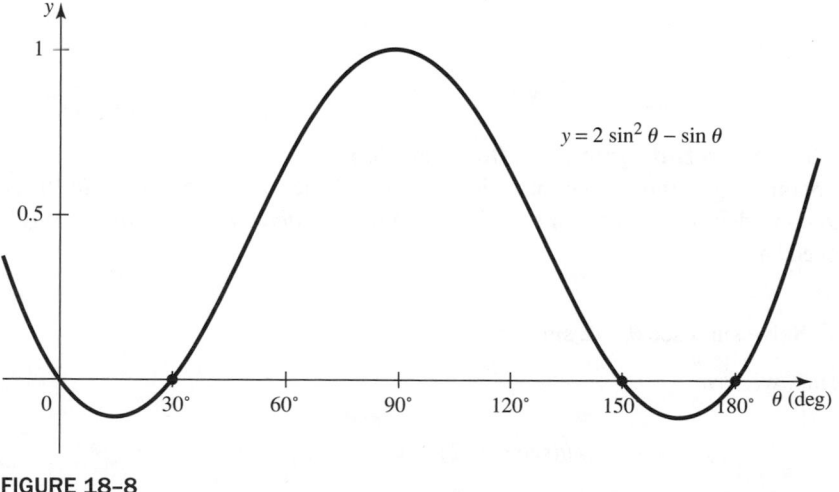

FIGURE 18-8 ◆◆◆

If an equation is in the form of a quadratic that cannot be factored, use the quadratic formula.

◆◆◆ **Example 28:** Solve the equation $\cos^2 \theta = 3 + 5 \cos \theta$.

Solution: Rearranging into standard quadratic form, we have

$$\cos^2 \theta - 5 \cos \theta - 3 = 0$$

This cannot be factored, so we use the quadratic formula.

$$\cos \theta = \frac{5 \pm \sqrt{25 - 4(-3)}}{2}$$

There are two values for $\cos \theta$.

$\cos \theta = 5.54$	$\cos \theta = -0.541$
(not possible)	Since cosine is negative in the second and third quadrants,
	$\theta = 123°, 237°$

Figure 18–9 shows a graph of the function $y = \cos^2 \theta - 3 - 5 \cos \theta$ and its zeros.

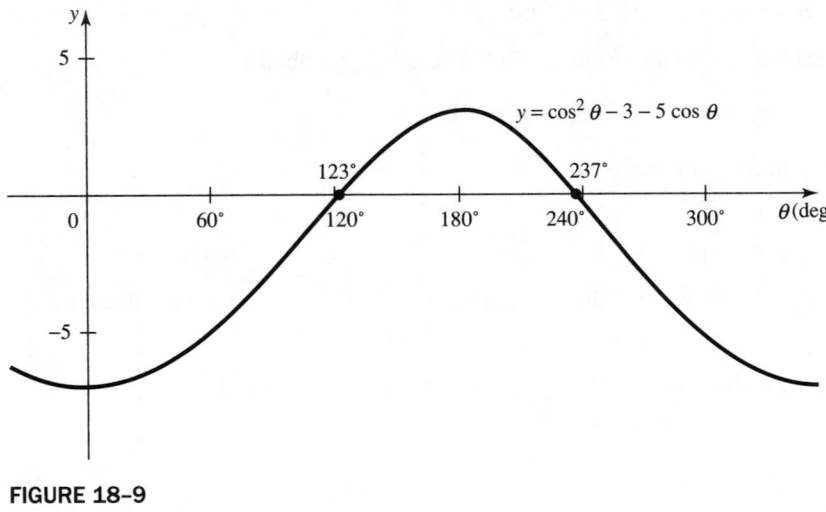

FIGURE 18-9   ◆◆◆

Equations with One Angle But More Than One Function

If an equation contains two or more trigonometric functions of the same angle, first transpose all of the terms to one side and try to factor that side into factors, *each containing only a single function*, and proceed as above.

◆◆◆ **Example 29:** Solve $\sin \theta \sec \theta - 2 \sin \theta = 0$.

Solution: Factoring, we have

$$\sin \theta(\sec \theta - 2) = 0$$

$\sin \theta = 0$	$\sec \theta = 2$
	$\cos \theta = \frac{1}{2}$
	Since cosine is positive in the first and fourth quadrants,
$\theta = 0°, 180°$	$\theta = 60°, 300°$

Figure 18–10 shows a graph of the function $y = \sin \theta \sec \theta - 2 \sin \theta$ and its zeros.

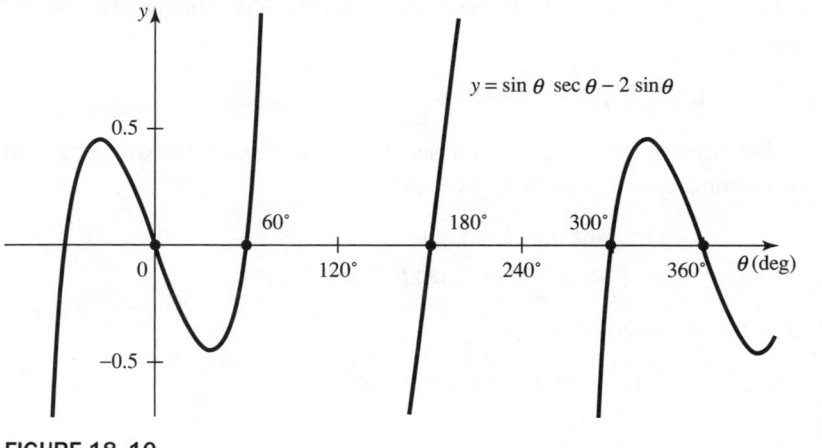

FIGURE 18–10 ◆◆◆

If an expression is *not factorable* at first, use the fundamental identities to *express everything in terms of a single trigonometric function*, and proceed as above.

◆◆◆ **Example 30:** Solve $\sin^2 \theta + \cos \theta = 1$.

Solution: By Eq. 164, $\sin^2 \theta = 1 - \cos^2 \theta$. Substituting gives

$$1 - \cos^2 \theta + \cos \theta = 1$$
$$\cos \theta - \cos^2 \theta = 0$$

Factoring, we obtain

$$\cos \theta (1 - \cos \theta) = 0$$

$$\begin{array}{c|c} \cos \theta = 0 & \cos \theta = 1 \\ \theta = 90°, 270° & \theta = 0° \end{array}$$

Figure 18–11 shows a graph of the function $y = \sin^2 \theta + \cos^2 \theta - 1$ and its zeros.

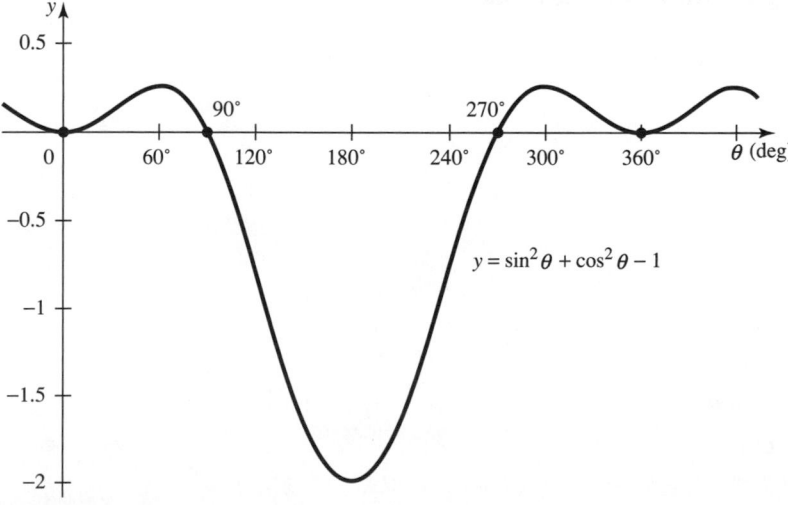

FIGURE 18–11 ◆◆◆

In order to simplify an equation using the Pythagorean relations, it is sometimes necessary to *square both sides*.

◆◆◆ **Example 31:** Solve $\sec \theta + \tan \theta = 1$.

Solution: We have no identity that enables us to write $\sec \theta$ in terms of $\tan \theta$, but we do have an identity for $\sec^2 \theta$. We rearrange and square both sides, getting

$$\sec \theta = 1 - \tan \theta$$
$$\sec^2 \theta = 1 - 2 \tan \theta + \tan^2 \theta$$

Replacing $\sec^2 \theta$ by $1 + \tan^2 \theta$ yields

$$1 + \tan^2 \theta = 1 - 2 \tan \theta + \tan^2 \theta$$
$$\tan \theta = 0$$
$$\theta = 0°, 180°$$

Since squaring can introduce extraneous roots that do not satisfy the original equation, we substitute back to check our answers. We find that the only angle that satisfies the given equation is $\theta = 0°$. Figure 18–12 shows a graph of the function and the zeros.

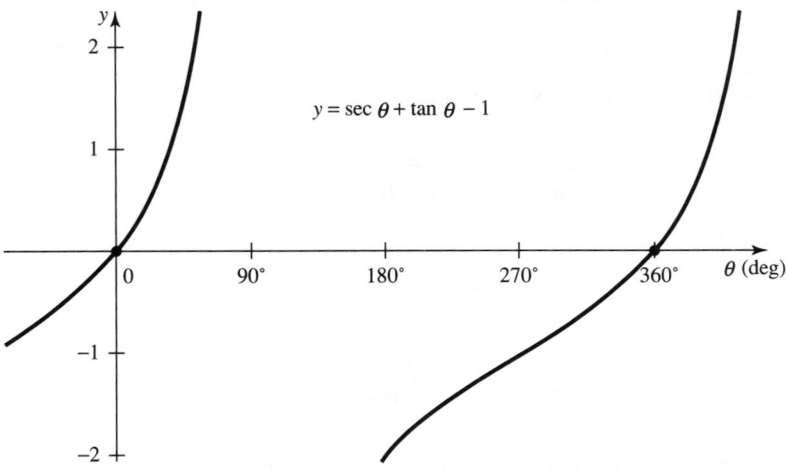

$y = \sec \theta + \tan \theta - 1$

FIGURE 18–12   ◆◆◆

Exercise 5 ◆ Trigonometric Equations

Solve each equation for all nonnegative values of θ less than 360°.

1. $\sin \theta = \frac{1}{2}$
2. $2 \cos \theta - \sqrt{3} = 0$
3. $1 - \tan \theta = 0$
4. $\sin \theta = \sqrt{3} \cos \theta$
5. $4 \sin^2 \theta = 3$
6. $2 \sin 3\theta = \frac{1}{2}$
7. $3 \sin \theta - 1 = 2 \sin \theta$
8. $\csc^2 \theta = 4$
9. $2 \cos^2 \theta = 1 + 2 \sin^2 \theta$
10. $2 \sec \theta = -3 - \cos \theta$
11. $4 \sin^4 \theta = 1$
12. $2 \csc \theta - \cot \theta = \tan \theta$
13. $1 + \tan \theta = \sec^2 \theta$

14. $1 + \cot^2 \theta = \sec^2 \theta$

15. $3 \cot \theta = \tan \theta$

16. $\sin^2 \theta = 1 - 6 \sin \theta$

17. $3 \sin(\theta/2) - 1 = 2 \sin^2(\theta/2)$

18. $\sin \theta = \cos \theta$

19. $4 \cos^2 \theta + 4 \cos \theta = -1$

20. $1 + \sin \theta = \sin \theta \cos \theta + \cos \theta$

21. $3 \tan \theta = 4 \sin^2 \theta \tan \theta$

22. $3 \sin \theta \tan \theta + 2 \tan \theta = 0$

23. $\sec \theta = -\csc \theta$

24. $\sin \theta = 2 \cos(\theta/2)$

25. $\sin \theta = 2 \sin \theta \cos \theta$

26. $\cos \theta \sin 2\theta = 0$

18–6   Inverse Trigonometric Functions

Inverse of the Sine Function: The Arc Sine

Recall from Sec. 4–3 that to find the inverse of a function $y = f(x)$, we perform the following steps:

1. Solve the given equation for x.

2. Interchange x and y.

Thus, starting with the sine function.

$$y = \sin x$$

we solve for x and get

$$x = \arcsin y$$

Now we interchange x and y so that y is, as usual, the dependent variable. The inverse of $y = \sin x$ is thus

$$y = \arcsin x$$

The graph of $y = \arcsin x$ is shown in Fig. 18–13. Note that it is identical to the graph of the sine function but with the x and y axes interchanged.

Glance back at what we have already said about inverse trigonometric functions in Chapters 7 and 15.

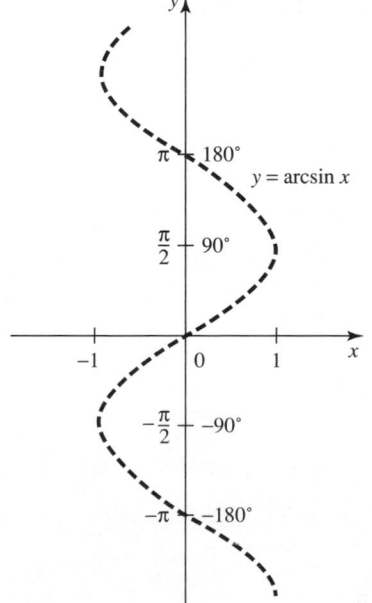

FIGURE 18–13   Graph of arcsin x.

Recalling the distinction between a relation and a function from Sec. 4–1, we see that $y =$ arcsin x is a relation, but not a function, because for a single x there may be more than one y. For this relation to also be a function, we must limit the range so that repeated values of y are not encountered.

The arc sine, for example, is customarily limited to the interval between $-\pi/2$ and $\pi/2$ (or $-90°$ to $90°$), as shown in Fig. 18–14. The numbers within this interval are called the *principal values* of the arc sine. It is also, of course, the *range* of arcsin x.

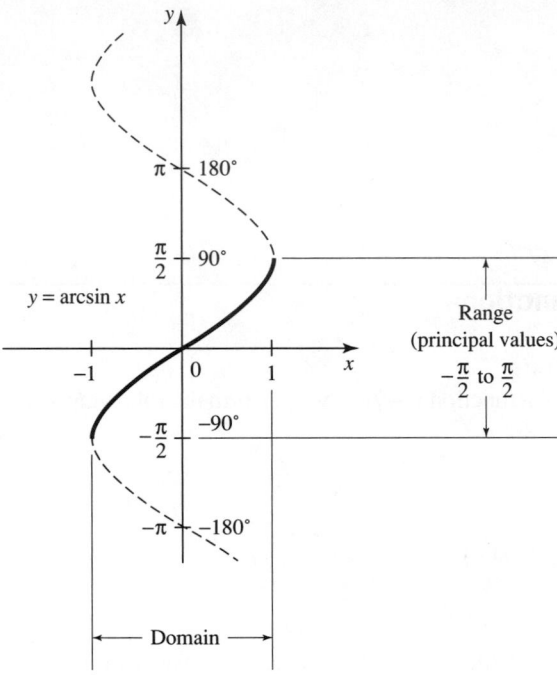

FIGURE 18–14   Principal values.

Given these limits, our relation $y =$ arcsin x now becomes a *function*. The first letter of the function is sometimes capitalized. Thus, $y =$ arcsin x would be the relation, and $y =$ Arcsin x would be the function.

Arc Cosine and Arc Tangent

In a similar way, the inverse of the cosine function and the inverse of the tangent function are limited to principal values, as shown by the solid lines in Figs. 18–15 and 18–16.

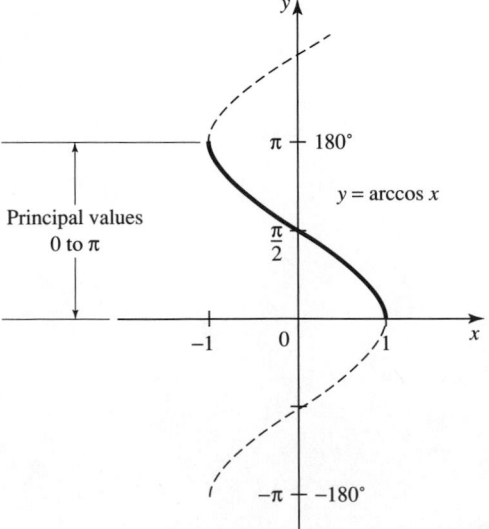

FIGURE 18–15

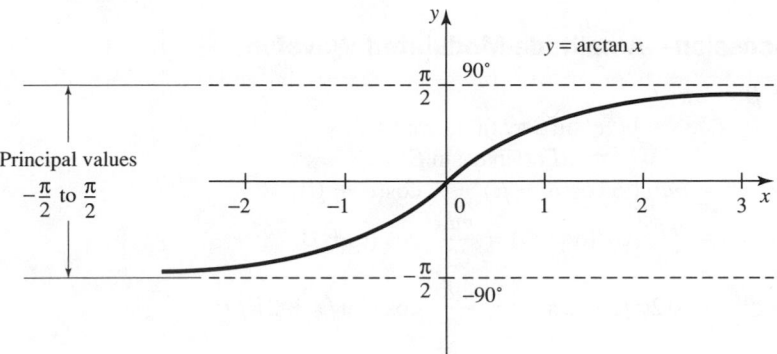

FIGURE 18–16

The numerical values of any of the arc functions are found by calculator, as was shown in Sec. 7–3 and 15–1.

To summarize, if x is *positive*, then arcsin x is an angle between 0 and 90° and hence lies in quadrant I. The same is true of arccos x and arctan x. If x *is negative*, then arcsin x is an angle between $-90°$ and 0° and hence is in quadrant IV. The same is true of arctan x. However, arccos x is in quadrant II when x is negative. These facts are summarized in the following table:

x	Arcsin x	Arccos x	Arctan x
Positive	Quadrant I	Quadrant I	Quadrant I
Negative	Quadrant IV	Quadrant II	Quadrant IV

◆◆◆ **Example 32:** Find (a) arcsin 0.638, (b) tan $^{-1}(-1.63)$, and (c) arcsec(-2.34). Give your answer in degrees to one decimal place.

Solution:

(a) By calculator,

$$\text{arcsin } 0.638 = 39.6°$$

(b) By calculator,

$$\tan^{-1}(-1.63) = -58.5°$$

(c) Using the reciprocal relations, we get

$$\text{arcsec}(-2.34) = \text{arccos } \frac{1}{-2.34} = \text{arccos}(20.427)$$
$$= 115.3°$$ ◆◆◆

Exercise 6 ◆ Inverse Trigonometric Functions

Evaluate each expression. Give your answer in degrees to one decimal place.

1. arcsin 0.374   **2.** arccos 0.826   **3.** arctan(-4.92)

4. $\cos^{-1} 0.449$ **5.** $\tan^{-1} 6.92$ **6.** $\sin^{-1}(-0.822)$

7. arcsec 3.96   **8.** $\cot^{-1} 4.97$ **9.** $\csc^{-1} 1.824$

10. $\sec^{-1} 2.89$

Case Study Discussion—Amplitude-Modulated Waveform

Let $\alpha = 2\pi f_i t$ and $\beta = 2\pi f_c t$

$$mEc(\sin 2\pi f_i t)(\sin 2\pi f_c t)$$
$$= mEc(\sin \alpha \sin \beta)$$
$$= mEc[\tfrac{1}{2} \cos(\alpha - \beta) - \tfrac{1}{2} \cos(\alpha + \beta)]$$
$$= \frac{mEc}{2} \cos(\alpha - \beta) - \frac{mEc}{2} \cos(\alpha + \beta)$$
$$= \frac{mEc}{2} \cos(2\pi f_i t - 2\pi f_c t) - \frac{mEc}{2} \cos(2\pi f_i t + 2\pi f_c t)$$
$$= \frac{mEc}{2} \cos 2\pi (f_c - f_i)t - \frac{mEc}{2} \cos 2\pi (f_c - f_i)t$$

••• CHAPTER 18 REVIEW PROBLEMS •••••••••••••••••••••••••••••••

Prove.

1. $\sec^2 \theta + \tan^2 \theta = \sec^4 \theta - \tan^4 \theta$

2. $\dfrac{1 + \csc \theta}{\cot \theta} = \dfrac{\cot \theta}{\csc \theta - 1}$

3. $\tan^2 \theta \sin^2 \theta = \tan^2 \theta - \sin^2 \theta$

4. $\cot \theta + \csc \theta = \dfrac{1}{\csc \theta - \cot \theta}$

5. $\sin(45° + \theta) - \sin(45° - \theta) = \sqrt{2} \sin \theta$

6. $\dfrac{\cot \alpha \cot \beta + 1}{\cot \beta - \cot \alpha} = \cot(\alpha - \beta)$

7. $\dfrac{\sin^3 \theta + \cos^3 \theta}{\sin \theta + \cos} = 1 - \dfrac{\sin 2\theta}{2}$

8. $\dfrac{1 + \tan \theta}{1 - \tan \theta} = \sec 2\theta + \tan 2\theta$

9. $\sin \theta \cot \dfrac{\theta}{2} = \cos \theta + 1$

10. $\sin^4 \theta + 2 \sin^2 \theta \cos^2 \theta + \cos^4 \theta = 1$

11. $\cot \alpha \cot \beta = \dfrac{\csc \beta + \cot \alpha}{\tan \alpha \sec \beta + \tan \beta}$

12. $\cos \theta \cot \theta = \dfrac{\cos \theta + \cot \theta}{\tan \theta + \sec \theta}$

Simplify.

13. $\dfrac{\sin \theta \sec \theta}{\tan \theta}$

14. $\sec^2 \theta - \sin^2 \theta \sec^2 \theta$

15. $\dfrac{\sec \theta}{\sec \theta - 1} + \dfrac{\sec \theta}{\sec \theta + 1}$

16. $\dfrac{\sin^3 \theta + \cos^3 \theta}{\sin \theta + \cos \theta}$

17. $\cos \theta \tan \theta \csc \theta$

18. $(\sec \theta - 1)(\sec \theta + 1)$

19. $\dfrac{\cot \theta}{1 + \cot^2 \theta} \cdot \dfrac{\tan^2 \theta + 1}{\tan \theta}$

20. $\dfrac{\cos \theta}{1 - \tan \theta} + \dfrac{\sin \theta}{1 - \cot \theta}$

21. $(1 - \sin^2 \theta) \sec^2 \theta$

22. $\tan^2 \theta(1 + \cot^2 \theta)$

23. $\dfrac{\sin \theta - 2 \sin \theta \cos \theta}{2 - 2 \sin^2 \theta - \cos \theta}$

Solve for all positive values of θ less than 360°.

24. $1 + 2 \sin^2 \theta = 3 \sin \theta$

25. $3 + 5 \cos \theta = 2 \cos^2 \theta$

26. $\cos \theta - 2 \cos^3 \theta = 0$

27. $\tan \theta = 2 - \cot \theta$

28. $\sin 3\theta - \sin 6\theta = 0$

29. $\cot 2\theta - 2 \cos 2\theta = 0$

30. $\sin \theta + \cos 2\theta = 1$

31. $\sin \theta = 1 - 3 \cos \theta$

32. $5 \tan \theta = 6 + \tan^2 \theta$

33. $\sin^2 \theta = 1 + 6 \sin \theta$

34. $\tan^2 2\theta = 1 + \sec 2\theta$

35. $16 \cos^4 \dfrac{\theta}{2} = 9$

36. $\tan^2 \theta + 1 = \sec \theta$

Evaluate.

37. $\arcsin 0.825$ **38.** $\cos^{-1} 0.232$ **39.** $\arctan 3.55$

Express as a single sine function.

40. $y = 1.84 \sin \omega t + 2.18 \cos \omega t$

41. $y = 47.6 \sin \omega t + 62.1 \cos \omega t$

Team Project

42. Several sections of rail were welded together to form a continuous straight rail 85 m long. It was installed when the temperature was $-2\overline{0}$ °C, and the crew neglected to allow a gap for thermal expansion. The ends of the rail are fixed so that they cannot move outward, so the rail took the shape shown in Fig. 18–17 when the temperature rose to $5\overline{0}$ °C. Assuming the curve to be circular, find the height x at the midpoint of the track.

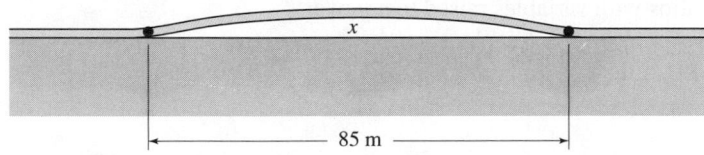

FIGURE 18–17

19

Ratio, Proportion, and Variation

···· OBJECTIVES ····

When you have completed this chapter, you should be able to:
- Represent ratios and proportions as fractions.
- Solve for constants of proportionality.
- Solve ratios with variables raised to a power.
- Solve inversely proportional ratios.
- Work with ratios with more than one variable.

You have seen some ratios in Chapter 9. What you will see in this chapter will not introduce any new algebra, just a new and important way to look at ratios and fractions. Here, we look at how you go from a relationship that you establish between two variables (from experiments or review of data) to an equation. This gives you the ability to create your own equations for specialized areas of science or technology.

We will see that the methods in this chapter also give us another powerful tool for making *estimates*. In particular, we will be able to estimate a new value for one variable when we change the other by a certain amount. Again, we will be interested in such real quantities as the velocity of an object or the resistance of a circuit.

We have, in fact, already studied some aspects of functional variation—first when we substituted numerical values for *x* into a function and computed corresponding values of *y* and later when we graphed functions in Chapter 5. We will do some graphing here, too, and we also learn some better techniques for solving numerical problems.

CANADIAN INNOVATION

CONFEDERATION BRIDGE

The Confederation Bridge is the world's longest fixed link over ice-covered water. It provides a link between mainland Canada and Prince Edward Island via New Brunswick. The bridge traces out a slight curve along its 12.9 km path. Built between 1993 and 1997, the bridge was designed with graceful curves to force drivers to steer and remain attentive, to reduce the potential for accidents that experts believe happen more often on straight highways or bridges.

One of the most important structural elements of the bridge is hidden: a construction technique called post-tensioning was used to tie the pier bases, shafts, and girder together into one solid, continuous structure that is designed to last 100 years. Post-tensioning uses cables running through the structure to actually cause a curvature that will be flattened by loading the structure. As the cables of the sections are connected, the structure becomes one integrated unit as far as stress and load are concerned.

19–1 Ratio and Proportion

Ratio

In our work so far, we have often dealt with expressions of the form

$$\frac{a}{b}$$

There are several different ways of looking at such an expression. We can say that a/b is

- A *fraction*, with a numerator a and a denominator b
- A *quotient*, where a is *divided* by b
- The *ratio* of a to b

Thus a ratio can be thought of as a fraction or as the quotient of two quantities. For a ratio of two physical quantities, it is usual to express the numerator and denominator *in the same units,* so that they cancel and leave the ratio *dimensionless.* The quantities a and b are called the *terms* of the ratio.

◆◆◆ **Example 1:** A corridor is 400 cm wide and 18 m long. Find the ratio of length to width.

Solution: We first express the length and width in the same units, say, metres.

$$400 \text{ cm} = 4 \text{ m}$$

So the ratio of length to width is

$$\frac{18}{4} = \frac{9}{2} \qquad\qquad\qquad ◆◆◆$$

Another way of writing a ratio is with the colon (:). The ratio of a to b is

$$\frac{a}{b} = a:b$$

In Example 1 we could write that the ratio of length to width is 9:2.

Dimensionless ratios are handy because you do not have to worry about units, and they are often used in technology. Some examples are as follows:

Poisson's ratio	fuel-air ratio	turn ratio
endurance ratio	load ratio	gear ratio
radian measure of angles	pi	trigonometric ratio

The word *specific* is often used to denote a ratio when there is a standard unit to which a given quantity is being compared under standard conditions. The following are a few such ratios:

specific heat	specific volume	specific conductivity
specific weight	specific gravity	specific speed

For example, the specific gravity of a solid or liquid is the ratio of the density of the substance to the density of water at a standard temperature.

◆◆◆ **Example 2:** At room temperature, aluminum has a density of 2.70 g/cm³, and water has a density of 0.998 g/cm³. Find the specific gravity (SG) of aluminum.

Solution: Dividing gives

$$\text{SG} = \frac{2.70 \text{ g/cm}^3}{0.998 \text{ g/cm}^3} = 2.71$$

Note that specific gravity is a *dimensionless ratio.* ◆◆◆

Proportion

A *proportion* is an equation obtained when two ratios are set equal to each other. If the ratio $a:b$ is equal to the ratio $c:d$, we have the proportion

This ratio can also be written $a:b::c:d$.

$$a:b = c:d$$

which reads "the ratio of a to b equals the ratio of c to d" or "a is to b as c is to d." We will often write such a proportion in the form

$$\frac{a}{b} = \frac{c}{d}$$

The two inside terms of a proportion are called the *means,* and the two outside terms are the *extremes.*

$$\overset{\displaystyle \overset{\text{extremes}}{\overbrace{}}}{a:b \quad = \quad c:d}$$
$$\underset{\text{means}}{\underbrace{}}$$

Finding a Missing Term

Solve a proportion just as you would any other fractional equation.

◆◆◆ **Example 3:** Find x if $\dfrac{3}{x} = \dfrac{7}{9}$

Solution: Multiplying both sides by the LCD, $9x$, we obtain

$$27 = 7x$$
$$x = \frac{27}{7}$$

◆◆◆

◆◆◆ **Example 4:** Find x if $\dfrac{x+2}{3} = \dfrac{x-1}{5}$.

Solution: Multiplying through by 15 gives

$$5(x + 2) = 3(x - 1)$$
$$5x + 10 = 3x - 3$$
$$2x = -13$$
$$x = -\frac{13}{2}$$

◆◆◆

◆◆◆ **Example 5:** Insert the missing quantity in the following equation:

$$\frac{x-3}{2x} = \frac{?}{4x^2}$$

Solution: Replacing the question mark with a variable, say, z, and solving for z gives us

$$z = \frac{4x^2(x-3)}{2x}$$
$$= 2x(x-3)$$

◆◆◆

Mean Proportional

When the means (the two inside terms) of a proportion are equal, as in

$$a:b = b:c$$

the term b is called the *mean proportional* between a and c. Solving for b, we get

$$b^2 = ac$$

Mean Proportional	$b = \pm\sqrt{ac}$	59

The mean proportional b is also called the *geometric mean* between a and c, because a, b, and c form a *geometric progression* (a series of numbers in which each term is obtained by multiplying the previous term by the same quantity).

Some books define b as positive if both a and c are positive. Here, we'll show both values.

◆◆◆ **Example 6:** Find the mean proportional between 3 and 12.

Solution: From Eq. 59,

$$b = \pm\sqrt{3(12)} = \pm 6 \qquad \text{◆◆◆}$$

Applications

Use the same guidelines for setting up these word problems as were given in Sec. 3–2 for any word problem. Just be careful not to reverse the terms in the proportion. Match the first item in the verbal statement with the first number in the ratio, and the second with the second.

◆◆◆ **Example 7:** A man is 25 years old and his brother is 15. In how many years will their ages be in the ratio $5:4$?

Solution: Let x = required number of years. After x years, the man's age will be

$$25 + x$$

and the brother's age will be

$$15 + x$$

Forming a proportion, we have

$$\frac{25 + x}{15 + x} = \frac{5}{4}$$

Multiplying by $4(15 + x)$, we obtain

$$4(25 + x) = 5(15 + x)$$
$$100 + 4x = 75 + 5x$$
$$x = 25 \text{ years} \qquad \text{◆◆◆}$$

CASE STUDY—WHEATSTONE BRIDGE

Strain gauges are electronic devices that measure the slightest bend or deformation of a surface. If you apply a strain gauge to a beam, it will register these very slight variations, allowing you to determine the stress the beam is under. Besides monitoring strain in actual structures, these devices are used in checking models of structures and prototypes.

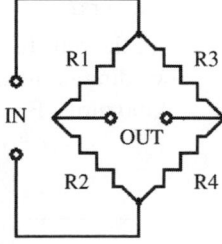

You are an electronics engineering technologist. As part of an engineering research project for the federal Department of Public Works and Government Services, your task is to set up and monitor 50 strain gauges to record the stress on a pylon of the Confederation Bridge.

Let's start with a circuit called a Wheatstone bridge.

The Wheatstone bridge circuit is a diamond shaped circuit where the ratio of the resistances on one side can be compared to the ratio of the resistances on the other side.

$$V_{OUT} = \left(\frac{R_4}{R_3 + R_4} - \frac{R_2}{R_1 + R_2}\right)V_{IN}$$

If R_4 represents our strain gauge, the other three resistors are precision resistors with values known to a very high accuracy. If $R_1 = R_2 = R_3 = 1000\ \Omega$ and $V_{OUT} = 0.012$ V when $V_{IN} = 9$ V, find R_4.

Exercise 1 ◆ Ratio and Proportion

Find the value of x.

1. $3 : x = 4 : 6$ **2.** $x : 5 = 3 : 10$
3. $4 : 6 = x : 4$ **4.** $3 : x = x : 12$
5. $x : (14 - x) = 4 : 3$ **6.** $x : 12 = (x - 12) : 3$
7. $x : 6 = (x + 6) : 10\frac{1}{2}$ **8.** $(x - 7) : (x + 7) = 2 : 9$

Insert the missing quantity.

9. $\dfrac{x}{3} = \dfrac{?}{9}$ **10.** $\dfrac{?}{4x} = \dfrac{7}{16x}$

11. $\dfrac{5a}{7b} = \dfrac{?}{-7b}$ **12.** $\dfrac{a - b}{c - d} = \dfrac{b - a}{?}$

13. $\dfrac{x + 2}{5x} = \dfrac{?}{5}$

Find the mean proportional between the following.

14. 2 and 50   **15.** 3 and 48   **16.** 6 and 150
17. 5 and 45   **18.** 4 and 36

Applications

19. Find two numbers that are to each other as $5 : 7$ and whose sum is 72.

20. Separate $150 into two parts so that the smaller may be to the greater as 7 is to 8.

21. Into what two parts may the number 56 be separated so that one may be to the other as 3 is to 4?

22. The difference between two numbers is 12, and the larger is to the smaller as 11 is to 7. What are the numbers?

23. An estate of $75,000 was divided between two heirs so that the elder's share was to the younger's as 8 is to 7. What was the share of each?

24. Two partners jointly bought some stock. The first put in $400 more than the second, and the stock of the first was to that of the second as 5 is to 4. How much money did each invest?

25. The sum of two numbers is 20, and their sum is to their difference as 10 is to 1. What are the numbers?

26. The sum of two numbers is a, and their sum is to their difference as m is to n. What are the numbers?

27. For the transformer shown in Fig. 19–1, the ratio of the number of turns in the secondary winding to the number of turns in the primary winding is 15. The secondary winding has 4500 turns. Find the number of turns in the primary.

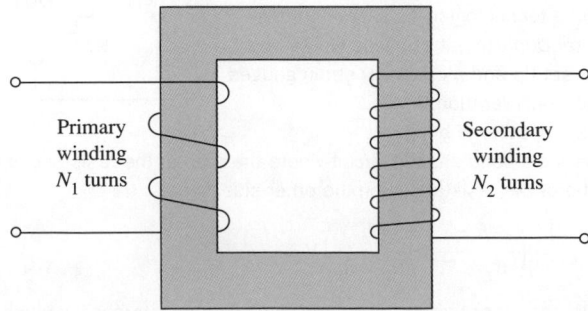

FIGURE 19–1   A transformer.

28. A line, as shown in Fig. 19–2, is subdivided into two segments a and b such that the ratio of the smaller segment a to the larger segment b equals the ratio of the larger b to the whole $(a + b)$. The ratio a/b is called the *golden ratio* or *golden section*. Set up a proportion, based on the above definition, and compute the numerical value of this ratio.

FIGURE 19–2   A line subdivided by the golden ratio.

19–2   Direct Variation

Constant of Proportionality

If two variables are related by an equation of the following form:

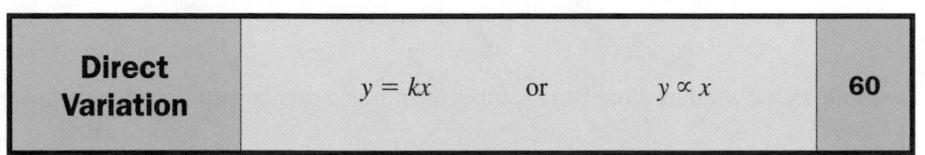

where k is a constant, we say that y *varies directly* as x, or that y is *directly proportional* to x. The constant k is called the *constant* of *proportionality*. As shown above, direct variation may also be written using the special symbol $\propto$:

$$y \propto x$$

It is read "y varies directly as x" or "y is directly proportional to x."

Solving Variation Problems

Variation problems can be solved with or without evaluating the constant of proportionality. We first show a solution in which the constant *is* found by substituting the given values into Eq. 60.

◆◆◆ **Example 8:** If y is directly proportional to x, and y is 27 when x is 3, find y when x is 6.

Solution: Since y varies directly as x, we use Eq. 60.

$$y = kx$$

To find the constant of proportionality, we substitute the given values for x and y, 3 and 27.

$$27 = k(3)$$

So $k = 9$. Our equation is then

$$y = 9x$$

When $x = 6$,

$$y = 9(6) = 54$$ ◆◆◆

We now show how to solve such a problem without finding the constant of proportionality.

◆◆◆ **Example 9:** Solve Example 8 *without* finding the constant of proportionality.

Solution: When quantities vary directly with each other, we can set up a proportion and solve it. Let us represent the initial values of x and y by x_1 and y_1, and the second set of values by x_2 and y_2. Substituting each set of values into Eq. 60 gives us

$$y_1 = kx_1$$

Glance back at Sec. 5–3 where we plotted the equation (Eq. 289) of straight line

$$y = mx + b$$

We see that Eq. 60 is the equation of a straight line with a slope of k and a y intercept of zero.

and

$$y_2 = kx_2$$

We divide the second equation by the first, and k cancels.

$$\frac{y_2}{y_1} = \frac{x_2}{x_1}$$

The same proportion can also be written in the form $y_1/x_1 = y_2/x_2$. Two other forms are also possible:

$$\frac{x_1}{y_1} = \frac{x_2}{y_2}$$

and

$$\frac{x_1}{y_2} = \frac{x_1}{y_2}$$

The proportion says, The new y is to the old y as the new x is to the old x. We now substitute the old x and y (3 and 27), as well as the new x (6).

$$\frac{y_2}{y_1} = \frac{6}{3}$$

Solving yields

$$y_2 = 27\left(\frac{6}{3}\right) = 54 \quad \text{as before} \qquad \bullet\bullet\bullet$$

◆◆◆ **Example 10:** If y varies directly as x, fill in the missing numbers in the table of values.

x	1	2		5	
y			16	20	28

Solution: We find the constant of proportionality from the given pair of values (5, 20). Starting with Eq. 60, we have

$$y = kx$$

and substituting gives

$$20 = k(5)$$

$$k = 4$$

So

$$y = 4x$$

With this equation we find the missing values.

When $x = 1$: $y = 4$

When $x = 2$: $y = 8$

When $y = 16$: $x = \dfrac{16}{4} = 4$

When $y = 28$: $x = \dfrac{28}{4} = 7$

So the completed table is

x	1	2	4	5	7
y	4	8	16	20	28

$\bullet\bullet\bullet$

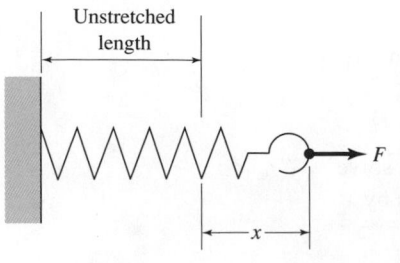

Unstretched length

FIGURE 19–3

Applications

Many practical problems can be solved using the idea of direct variation. Once you know that two quantities are directly proportional, you may assume an equation of the form of Eq. 60. Substitute the two given values to obtain the constant of proportionality, which you then put back into Eq. 60 to obtain the complete equation. From it you may find any other corresponding values.

Alternatively, you may decide not to find k, but to form a proportion in which three values will be known, enabling you to find the fourth.

◆◆◆ **Example 11:** The force F needed to stretch a spring (Fig. 19–3) is directly proportional to the distance x stretched. If it takes 15 N to stretch a certain spring 28 cm, how much force is needed to stretch it 34 cm?

Estimate: We see that 15 N will stretch the spring 28 cm, or about 2 cm per newton. Thus a stretch of 34 cm should take about $34 \div 2$, or 17 N.

Solution: Assuming an equation of the form

$$F = kx$$

and substituting the first set of values, we have

$$15 = k(28)$$

$$k = \frac{15}{28}$$

So the equation is $F = \dfrac{15}{28}x$. When $x = 34$,

$$F = \left(\frac{15}{28}\right)34 = 18\ \text{N} \quad \text{(rounded)}$$

which is close to our estimated value of 17 N.   ◆◆◆

Exercise 2 ◆ Direct Variation

1. If y varies directly as x, and y is 56 when x is 21, find y when x is 74.

2. If w is directly proportional to z, and w has a value of 136 when z is 10.8, find w when z is 37.3.

3. If p varies directly as q, and p is 846 when q is 135, find q when p is 448.

4. If y is directly proportional to x, and y has a value of 88.4 when x is 23.8:
(a) Find the constant of proportionality.
(b) Write the equation $y = f(x)$.
(c) Find y when $x = 68.3$.
(d) Find x when $y = 164$.

Assuming that y varies directly as x, fill in the missing values in each table of ordered pairs.

5.

x	9	11	
y	45		75

6.

x	3.40	7.20		12.3
y		50.4	68.6	

7.

x	115	125		154
y			167	187

8. Graph the linear function $y = 2x$ for values of x from -5 to 5.

Applications

9. The distance between two cities is 828 km, and they are 29.5 cm apart on a map. Find the distance between two points 15.6 cm apart on the same map.

10. If the weight of 2500 steel balls is 3.65 kg, find the number of balls in 10.0 kg.

11. If 80 transformer laminations make a stack 1.75 cm thick, how many laminations are contained in a stack 3.00 cm thick?

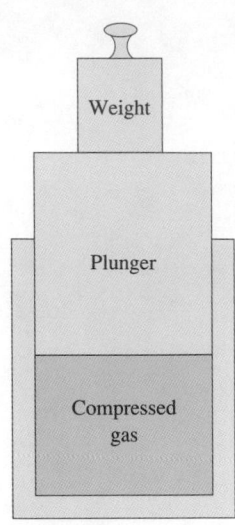

FIGURE 19–4

For problem 17, use Charles' law: *The volume of a gas at constant pressure is directly proportional to its absolute temperature.* K is the abbreviation for kelvin, the SI absolute temperature scale. Add 273.15 to Celsius temperatures to obtain temperatures on the kelvin scale.

12. If your car now gets 9.00 km/L of gasoline, and if you can go 400 km on a tank of gasoline, how far could you drive with the same amount of gasoline in a car that gets 15.0 km/L?

13. A certain automobile engine delivers 53 kW and has a displacement (the total volume swept out by the pistons) of 3.0 L. If the power is directly proportional to the displacement, what power would you expect from a similar engine that has a displacement of 3.8 L?

14. The resistance of a conductor is directly proportional to its length. If the resistance of 2.60 km of a certain transmission line is 155 Ω, find the resistance of 75.0 km of that line.

15. The resistance of a certain spool of wire is 1120 Ω. A piece 10.0 m long is found to have a resistance of 12.3 Ω. Find the length of wire on the spool.

16. If a certain machine can make 1850 parts in 55 min, how many parts can it make in 7.5 h? Work to the nearest part.

17. In Fig. 19–4, the constant force on the plunger keeps the pressure of the gas in the cylinder constant. The piston rises when the gas is heated and falls when the gas is cooled. If the volume of the gas is 1520 cm³ when the temperature is 302 K, find the volume when the temperature is 358 K.

18. The power generated by a hydroelectric plant is directly proportional to the flow rate through the turbines, and a flow rate of 5625 imperial gallons of water per minute produces 41.2 MW. How much power would you expect when a drought reduces the flow to 5000 gal./min?

19–3   The Power Function

Definition

In Sec. 19–2, we saw that we could represent the statement "y varies directly as x" by Eq. 60,

$$y = kx$$

Similarly, if y varies directly as the *square* of x, we have

$$y = kx^2$$

or if y varies directly as the *square root* of x,

$$y = k\sqrt{x} = kx^{1/2}$$

These are all examples of the *power function*.

Power Function	$y = kx^n$	196

The constants can, of course, be represented by any letter. Appendix A shows a instead of k.

For exponents of 4 and 5, we have the *quartic* and *quintic* functions, respectively.

The constants k and n can be any positive or negative number. This simple function gives us a great variety of relations that are useful in technology and whose forms depend on the value of the exponent, n. A few of these are shown in the following table:

When:	We Get the:		Whose Graph Is a:
$n = 1$	Linear function	$y = kx$ (direct variation)	Straight line
$n = 2$	Quadratic function	$y = kx^2$	Parabola
$n = 3$	Cubic function	$y = kx^3$	Cubical parabola
$n = -1$		$y = \dfrac{k}{x}$ (inverse variation)	Hyperbola

The first three of these are special cases of the *general polynomial function of degree n.*

Polynomial of Degree *n*	$y = a_0 x^n + a_1 x^{n-1} + \ldots + a_{n-3} x^3 + a_{n-2} x^2 + a_{n-1} x + a_n$	102

where *n* is a positive integer and the *a*'s are constants. For example, the quadratic function $y = kx^2$ is a polynomial function of second degree, where $a_{n-2} = k$ and all of the other *a*'s are zeros. It is called *incomplete* because it is lacking an *x* term and a constant term.

We also consider power functions that are not polynomials. Those with noninteger positive powers will be covered in this section, and those with negative exponents in the following section on inverse variation.

Graph of the Power Function

The graph of a power function varies greatly, depending on the exponent. We first show some power functions with positive, integral exponents.

◆◆◆ **Example 12:** Graph the power functions $y = x^2$ and $y = x^3$ for a range of *x* from -3 to $+3$.

Solution: Make a table of values.

x	-3	-2	-1	0	1	2	3
$y = x^2$	9	4	1	0	1	4	9
$y = x^3$	-27	-8	-1	0	1	8	27

The curves $y = x^2$ and $y = x^3$ (shown dashed) are plotted in Fig. 19–5. Notice that the plot of $y = x^2$ has no negative values of *y*. This is typical of power functions that have *even* positive integers for exponents. The plot of $y = x^3$ does have negative *y*'s for negative values of *x*. The shape of this curve is typical of a power function that has an *odd* positive integer as exponent.

The graph of $y = kx^3$ is sometimes called a *cubical parabola.*

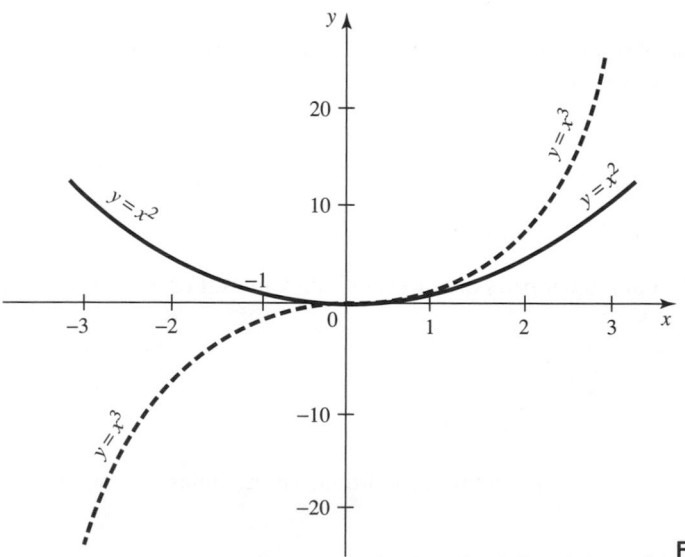

FIGURE 19–5

When we are graphing a power function with a *fractional* exponent, the curve may not exist for negative values of *x*.

◆◆◆ **Example 13:** Graph the function $y = x^{1/2}$ for $x = -9$ to 9.

Solution: We see that for negative values of x, there are no real number values of y. For the nonnegative values, we get

x	0	1	2	3	4	5	6	7	8	9
y	0	1	1.41	1.73	2.00	2.24	2.45	2.65	2.83	3.00

which are plotted in Fig. 19–6.

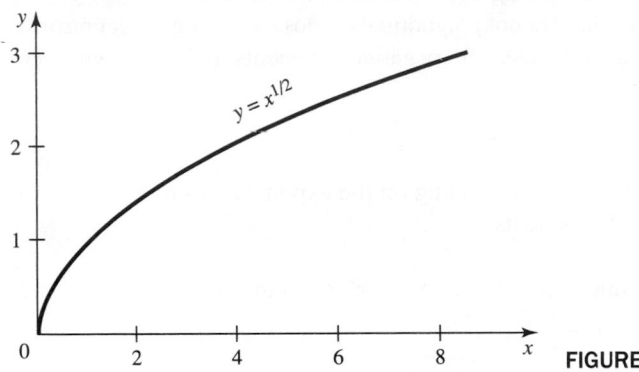

FIGURE 19–6

Solving Power Function Problems

As with direct variation (and with inverse variation and combined variation, treated later), we can solve problems involving the power function with or without finding the constant of proportionality. The following example will illustrate both methods.

◆◆◆ **Example 14:** If y varies directly as the $\frac{3}{2}$ power of x, and y is 54 when x is 9, find y when x is 25.

Solution by Solving for the Constant of Proportionality: Here we set up an equation with k and solve for the constant of proportionality. We let the exponent n in Eq. 196 be equal to $\frac{3}{2}$.

$$y = kx^{3/2}$$

As before, we evaluate the constant of proportionality by substituting the known values into the equation. Using $x = 9$ and $y = 54$, we obtain

$$54 = k(9)^{3/2} = k(27)$$

so $k = 2$. Our power function is then

$$y = 2x^{3/2}$$

This equation can then be used to find other pairs of corresponding values. For example, when $x = 25$,

$$y = 2(25)^{3/2}$$
$$= 2(\sqrt{25})^3$$
$$= 2(5)^3 = 250$$

Solution by Setting Up a Proportion: Here we set up a proportion. Three values are known, and we then solve for the fourth. If

$$y_1 = kx_1^{3/2}$$

and

$$y_2 = kx_2^{3/2}$$

then y_2 is to y_1 as $kx_2^{3/2}$ is to $kx_1^{3/2}$.

$$\frac{y_2}{y_1} = \left(\frac{x_2}{x_1}\right)^{3/2}$$

Substituting $y_1 = 54$, $x_1 = 9$, and $x_2 = 25$ yields

$$\frac{y_2}{54} = \left(\frac{25}{9}\right)^{3/2}$$

from which

$$y_2 = 54\left(\frac{25}{9}\right)^{3/2} = 250 \qquad \text{◆◆◆}$$

◆◆◆ **Example 15:** The horizontal distance S travelled by a projectile is directly proportional to the square of its initial velocity V. If the distance travelled is 1250 m when the initial velocity is 190 m/s, find the distance travelled when the initial velocity is 250 m/s.

Solution: The problem statement implies a power function of the form $S = kV^2$. Substituting the initial set of values gives us

$$1250 = k(190)^2$$

$$k = \frac{1250}{(190)^2} = 0.034\ 63$$

So our relationship is

$$S = 0.034\ 63V^2$$

When $V = 250$ m/s,

$$S = 0.034\ 63(250)^2 = 2160 \text{ m} \qquad \text{◆◆◆}$$

Some problems may contain no numerical values at all, as in the following example.

◆◆◆ **Example 16:** If y varies directly as the cube of x, by what factor will y change if x is increased by 25%?

Solution: The relationship between x and y is

$$y = kx^3$$

If we give subscripts 1 to the initial values and subscripts 2 to the final values, we may write the proportion

$$\frac{y_2}{y_1} = \frac{kx_2^3}{kx_1^3} = \left(\frac{x_2}{x_1}\right)^3$$

If the new x is 25% greater than the old x, then $x_2 = x_1 + 0.25x_1$ or $1.25x_1$. We thus substitute

$$x_2 = 1.25x_1$$

so

$$\frac{y_2}{y_1} = \left(\frac{1.25x_1}{x_1}\right)^3 = \left(\frac{1.25}{1}\right)^3 = 1.95$$

or

$$y_2 = 1.95y_1$$

So y has increased by a factor of 1.95. ◆◆◆

Similar Figures

We considered similar triangles in Chapter 6 and said that corresponding sides were in proportion. We now expand the idea to cover similar plane figures of *any* shape, and also similar *solids*. Similar figures (plane or solid) are those in which the distance between any two points

We are using the term *dimension* to refer only to linear dimensions, such as lengths of sides or perimeters. It does not refer to angles, areas, or volumes.

on one of the figures is a *constant multiple* of the distance between two corresponding points on the other figure. In other words, if two corresponding dimensions are in the ratio of, say, 2 : 1, all other corresponding dimensions must be in the ratio of 2 : 1. In other words:

Dimensions of Similar Figures	Corresponding dimensions of plane or solid similar figures are in proportion.	**134**

◆◆◆ **Example 17:** Two similar solids are shown in Fig. 19–7. Find the hole diameter *D*.

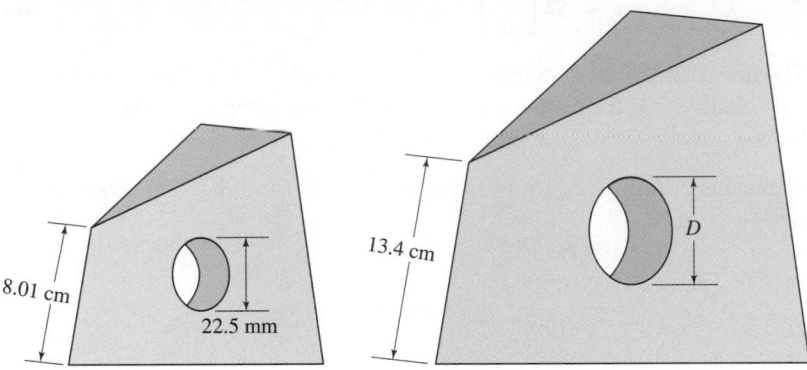

FIGURE 19-7 Similar solids.

Solution: By Statement 134,

$$\frac{D}{22.5} = \frac{13.4}{8.01}$$

$$D = \frac{22.5(13.4)}{8.01} = 37.6 \text{ mm}$$ ◆◆◆

Note that since we are dealing with *ratios* of corresponding dimensions, it was not necessary to convert all dimensions to the same units.

Areas of Similar Figures

The area of a square of side *s* is equal to the square of one side, that is, $s \times s$ or s^2. Thus if a side is multiplied by a factor *k*, the area of the larger square is $(ks) \times (ks)$ or $k^2 s^2$. We see that the area has increased by a factor of k^2.

An area more complicated than a square can be thought of as made up of many small squares, as shown in Fig. 19–8. Then, if a dimension of that area is multiplied by *k*, the area of each small square increases by a factor of k^2, and hence the entire area of the figure increases by a factor of k^2. Thus:

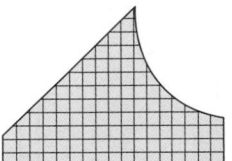

FIGURE 19-8 We can, in our minds, make the squares so small that they completely fill any irregular area. We use a similar idea in calculus when finding areas by integration.

Areas of Similar Figures	Areas of similar plane or solid figures are proportional to the *squares* of any two corresponding dimensions.	**135**

Thus, if a figure has its sides doubled, the new area would be *four* times the original area. This relationship is valid not only for plane areas and for surface areas of solids, but also for cross-sectional areas of solids.

The units of measure for area are square centimetres (cm²), square inches (sq. in.), and so on, depending on the units used for the sides of the figure.

◆◆◆ **Example 18:** The triangular top surface of the smaller solid shown in Fig. 19–7 has an area of 31.2 cm². Find the area *A* of the corresponding surface on the larger solid.

Estimate: To bracket our answer, note that if the dimensions of the larger solid were double those of the smaller solid, the area of the top would be greater by a factor of 4, or about 120 cm². But the larger dimensions are less than double, so the area will be less than 120 cm². Thus we expect the area to be between 35 and 120 cm².

Solution: By Statement 135, the area A is to 31.2 as the *square* of the ratio of 13.4 to 8.01.

$$\frac{A}{31.2} = \left(\frac{13.4}{8.01}\right)^2$$

$$A = 31.2\left(\frac{13.4}{8.01}\right)^2 = 87.3 \text{ cm}^2 \qquad \text{◆◆◆}$$

Volumes of Similar Solids

A cube of side s has volume s^3. We can think of any solid as being made up of many tiny cubes, each of which has a volume equal to the cube of its side. Thus if the dimensions of the solid are multiplied by a factor of k, the volume of each cube (and hence the entire solid) will increase by a factor of k^3.

Volumes of Similar Figures	Volumes of similar solid figures are proportional to the *cubes* of any two corresponding dimensions.	**136**

The units of measure for volume are cubic centimetres (cm³), cubic inches (cu. in.), and so on, depending on the units used for the sides of the solid figure.

◆◆◆ **Example 19:** If the volume of the smaller solid in Fig. 19–7 is 255 cm³, find the volume V of the larger solid.

Solution: By Statement 136, the volume V is to 255 as the *cube* of the ratio of 13.4 to 8.01.

$$\frac{V}{255} = \left(\frac{13.4}{8.01}\right)^3$$

$$V = 255\left(\frac{13.4}{8.01}\right)^3 = 1190 \text{ cm}^3 \qquad \text{◆◆◆}$$

Note that we know very little about the size and shape of these solids, yet we are able to compute the volume of the larger solid. The methods of this chapter give us another powerful tool for making *estimates*.

Common Error	Students often forget to *square* corresponding dimensions when finding areas and to *cube* corresponding dimensions when finding volumes.

Scale Drawings

An important application of similar figures is in the use of *scale drawings,* such as maps, engineering drawings, surveying layouts, and so on. The ratio of distances on the drawing to corresponding distances on the actual object is called the *scale* of the drawing. Use Statement 135 to convert between the areas on the drawing and areas on the actual object.

◆◆◆ **Example 20:** A certain map has a scale of 1 : 5000. How many hectares on the land are represented by 168 cm² on the map?

Solution: If A = the area on the land, then, by Statement 135,

$$\frac{A}{168} = \left(\frac{5000}{1}\right)^2 = 25\,000\,000$$

$$A = 168(25\,000\,000) = 42.0 \times 10^8 \text{ cm}^2$$

Converting to hectares, we have

$$A = 42.0 \times 10^8 \text{ cm}^2 \left(\frac{1 \text{ m}^2}{10^4 \text{ cm}^2} \right) \left(\frac{1 \text{ ha}}{10^4 \text{ m}^2} \right) = 42.0 \text{ ha} \qquad \blacklozenge\blacklozenge\blacklozenge$$

Exercise 3 ◆ The Power Function

1. If y varies directly as the square of x, and y is 726 when x is 163, find y when x is 274.

2. If y is directly proportional to the square of x, and y is 5570 when x is 172, find y when x is 382.

3. If y varies directly as the cube of x, and y is 4.83 when x is 1.33, find y when x is 3.38.

4. If y is directly proportional to the cube of x, and y is 27.2 when x is 11.4, find y when x is 24.9

5. If y varies directly as the square of x, and y is 285.0 when x is 112.0, find y when x is 351.0.

6. If y varies directly as the square root of x, and y is 11.8 when x is 342, find y when x is 288.

7. If y is directly proportional to the cube of x, and y is 638 when x is 145, find y when x is 68.3

8. If y is directly proportional to the five-halves power of x, and y has the value 55.3 when x is 17.3:

   (a) Find the constant of proportionality.
   (b) Write the equation $y = f(x)$.
   (c) Find y when $x = 27.4$.
   (d) Find x when $y = 83.6$.

9. If y varies directly as the fourth power of x, fill in the missing values in the following table of ordered pairs:

x	18.2	75.6	
y	29.7		154

10. If y is directly proportional to the $\frac{3}{2}$ power of x, fill in the missing values in the following table of ordered pairs:

x	1.054	1.135	
y		4.872	6.774

11. If y varies directly as the cube root of x, fill in the missing values in the following table of ordered pairs:

x	315		782
y		148	275

12. Graph the power function $y = 1.04x^2$ for $x = -5$ to 5.

13. Graph the power function $y = 0.553x^5$ for $x = -3$ to 3.

14. Graph the power function $y = 1.25x^{3/2}$ for $x = 0$ to 5.

Freely Falling Body

15. If a body falls 176 m in 6.00 s, how far will it fall in 9.00 s?

16. If a body falls 4.90 m during the first second, how far will it fall during the third second?

17. If a body falls 129 ft. in 2.00 s, how many seconds will it take to fall 525 ft.?

18. If a body falls 738 units in 3.00 s, how far will it fall in 6.00 s?

From Eq. A18, we see that the distance fallen by a body (from rest) varies directly as the square of the elapsed time. Assume an initial velocity of zero for each of these problems.

Power in a Resistor

19. If the power dissipated in a resistor is 486 W when the current is 2.75 A, find the power when the current is 3.45 A.

20. If the current through a resistor is increased by 28%, by what percent will the power increase?

21. By what factor must the current in an electric heating coil be increased to triple the power consumed by the heater?

The power dissipated in a resistor varies directly as the square of the current in the resistor (Eq. A67).

Photographic Exposures

22. A certain photograph will be correctly exposed at a shutter speed of 1/100 s with a lens opening of f5.6. What shutter speed is required if the lens opening is changed to f8?

23. For the photograph in problem 22, what lens opening is needed for a shutter speed of 1/50 s?

24 Most cameras have the following f stops: f2.8, f4, f5.6, f8, f11, and f16. To keep the same correct exposure, by what factor must the shutter speed be increased when the lens is opened one stop?

25. A certain enlargement requires an exposure time of 24 s when the enlarger's lens is set at f22. What lens opening is needed to reduce the exposure time to 8 s?

The exposure time for a photograph is directly proportional to the square of the f stop. (The f stop of a lens is its focal length divided by its diameter.)

Photographer's rule of thumb: *Double the exposure time for each increase in f stop.*

Similar Figures

26. A container for storing propane gas is 0.755 m high and contains 20.0 L. How high must a container of similar shape be to have a volume of 40.0 L?

27. A certain wood stove has a firebox volume of 0.120 m³. What firebox volume would be expected if all dimensions of the stove were increased by a factor of 1.25?

28. If the stove in problem 27 weighed 149 kg, how much would the larger stove be expected to weigh?

29. A certain solar house stores heat in 155 t of stones, which are in a chamber beneath the house. Another solar house is to have a chamber of similar shape but with all dimensions increased by 15%. How many tonnes of stone will it hold?

30. Each side of a square is increased by 15.0 mm, and the area is seen to increase by 2450 mm². What were the dimensions of the original square?

31. The floor plan of a certain building has a scale of 2 cm = 1 m and shows a room having an area of 260 cm². What is the actual room area in square metres?

32. The area of a window of a car is 118 cm² on a drawing having a scale of 1 : 4. Find the actual window area in square metres.

33. A pipe 3.00 cm in diameter discharges $12\overline{0}$ L of water in a certain time. What must be the diameter of a pipe that will discharge $18\overline{0}$ L in the same time?

Assume that the amount of flow through a pipe is proportional to its cross-sectional area.

34. If it cost \$756 to put a fence around a circular pond, what will it cost to enclose another pond having $\frac{1}{5}$ the area?

35. A triangular field whose base is 215 m contains 12 400 m². Find the area of a field of similar form whose base is 328 m.

Fluids

36. At the Darlington nuclear plant in Ontario, operators must know the rate of flow in the pipes cooling the reactors. However, for operational and safety reasons, the pipes cannot be blocked by a measuring instrument. Instead, the pressure is measured before and after a narrowing of the pipe (see Fig. 19–9). The rate of flow is directly proportional to the square root of the difference in pressure ΔP, where $\Delta P = P_2 - P_1$

(a) Write the expression for rate of flow.
(b) The flow in a pipe is 200 L/min. If ΔP increases by 16%, what is the new flow rate?

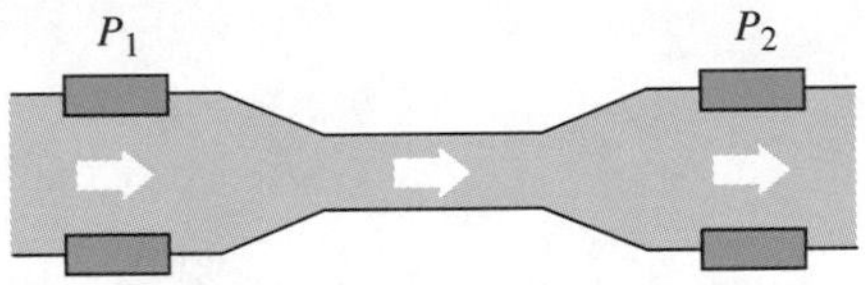

P_1 P_2

FIGURE 19–9 Diaphragms measure pressure without blocking flow.

19–4   Inverse Variation

Definition

When we say that "y varies inversely as x" or that "y is inversely proportional to x," we mean that x and y are related by the following equation, where, as before, k is a constant of proportionality:

We could also say that we have inverse variation when the product of x and y is a constant, that is, when $xy = k$.

Inverse Variation	$y = \dfrac{k}{x}$ or $y \propto \dfrac{1}{x}$	61

The Hyperbola

The graph of the function $y = k/x$ is called a *hyperbola*. It is one of the *conic sections*, which you will study in more detail when you study analytic geometry (Chapter 22).

◆◆◆ **Example 21:** Graph the function $y = 2/x$ from $x = 0$ to $x = 5$.

Solution: Make a table of point pairs.

Do not confuse inverse variation with the inverse of a function.

x	0	$\frac{1}{2}$	1	2	3	4	5
y	Not defined	4	2	1	$\frac{2}{3}$	$\frac{1}{2}$	$\frac{2}{5}$

These points are graphed in Fig. 19–10. Note that as x takes on smaller and smaller values, y gets larger and larger; and as x gets very large, y gets smaller and smaller but never reaches zero.

Thus the curve gets ever closer to the x and y axes but never touches them. The x and y axes are *asymptotes* for this curve.

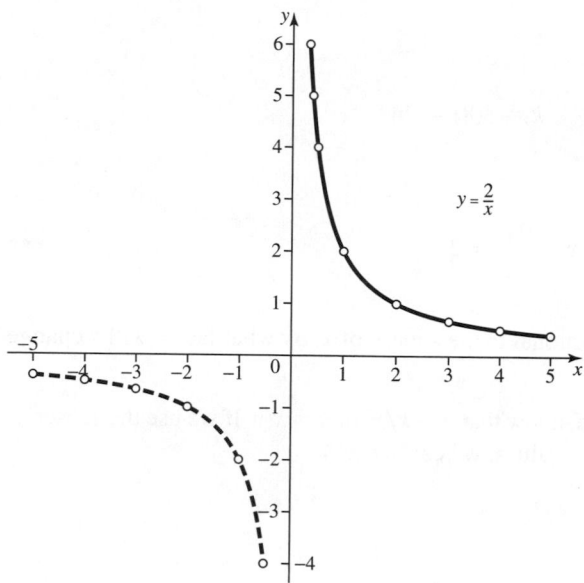

FIGURE 19–10 A hyperbola.

We did not plot the curve for negative values of x, but, if we had, we would have gotten the second branch of the hyperbola, shown dashed in the figure.

Power Function with Negative Exponent

Suppose, for example, that y varies inversely as the cube of x. This would be written

$$y = \frac{k}{x^3}$$

which is the same as

$$y = kx^{-3}$$

Note that this equation is a power function with a negative exponent. Such power functions are sometimes referred to as *hyperbolic*.

◆◆◆ **Example 22:** Rewrite $yx^2 = 10$.

Solution: Dividing by x^2, we see that

$$y = \frac{10}{x^2}$$

which can be rewritten as

$$y = 10x^{-2}$$

Therefore y is 10 times the inverse proportion of the square of x.   ◆◆◆

Solving Inverse Variation Problems

Inverse variation problems are solved by the same methods as for any other power function. As before, we can work these problems with or without finding the constant of proportionality.

◆◆◆ **Example 23:** If y is inversely proportional to x, and y is 8 when x is 3, find y when x is 12.

Solution: x and y are related by $y = k/x$. We find the constant of proportionality by substituting the given values of x and y (3 and 8).

$$8 = \frac{k}{3}$$

$$k = 3(8) = 24$$

So $y = 24/x$. When $x = 12$,

$$y = \frac{24}{12} = 2$$ ◆◆◆

◆◆◆ **Example 24:** If y is inversely proportional to the square of x, by what factor will y change if x is tripled?

Solution: From the problem statement we know that $y = k/x^2$ or $k = x^2 y$. If we use the subscript 1 for the original values and 2 for the new values, we can write

$$k = x^2_1 y_1 = x^2_2 y_2$$

Since the new x is triple the old, we substitute $3x_1$ for x_2.

$$x^2_1 y_1 = (3x_1)^2 y_2$$

$$= 9x^2_1 y_2$$

$$y_1 = 9y_2$$

or

$$y_2 = \frac{y_1}{9}$$

So the new y is one-ninth of its initial value.   ◆◆◆

◆◆◆ **Example 25:** If y varies inversely as x, fill in the missing values in the table:

x	1			9	12
y			12	4	

Solution: Substitute the ordered pair (9, 4) into Eq. 61.

$$4 = \frac{k}{9}$$

$$k = 36$$

So

$$y = \frac{36}{x}$$

Filling in the table, we have

x	1	3	9	12
y	36	12	4	3

◆◆◆ **Example 26:** The number of oscillations N that a pendulum (Fig. 19–11) makes per unit time is inversely proportional to the square root of the length L of the pendulum, and $N = 2.00$ oscillations per second when $L = 85.0$ cm.

(a) Write the equation $N = f(L)$.
(b) Find N when $L = 115$ cm.

Solution:

(a) The equation will be of the form

$$N = \frac{k}{\sqrt{L}}$$

Substituting $N = 2.00$ and $L = 85.0$ gives us

$$2.00 = \frac{k}{\sqrt{85.0}}$$

$$k = 2.00\sqrt{85.0} = 18.4$$

So the equation is

$$N = \frac{18.4}{\sqrt{L}}$$

(b) When $L = 115$,

$$N = \frac{18.4}{\sqrt{115}} = 1.72 \text{ oscillations per second}$$

Check: Does the answer seem reasonable? We see that N has decreased, which we expect with a longer pendulum, but it has not decreased by much. As a check, we note that L has increased by a factor of $115 \div 85$, or 1.35, so we expect N to decrease by a factor of $\sqrt{1.35}$, or 1.16. Dividing, we find that $2.00 \div 1.72 = 1.16$, so the answer checks. ◆◆◆

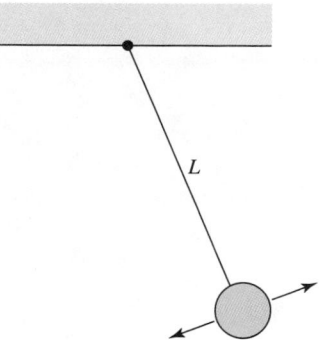

FIGURE 19–11 A simple pendulum.

Exercise 4 ◆ Inverse Variation

1. If y varies inversely as x, and y is 385 when x is 832, find y when x is 226.
2. If y is inversely proportional to the square of x, and y has the value 1.55 when x is 7.38, find y when x is 44.2.
3. If y is inversely proportional to x, how does y change when x is doubled?
4. If y varies inversely as x, and y has the value 104 when x is 532:
   (a) Find the constant of proportionality.
   (b) Write the equation $y = f(x)$.
   (c) Find y when x is 668.
   (d) Find x when y is 226.
5. If y is inversely proportional to x, fill in the missing values.

x	306	622	
y	125		418

6. Fill in the missing values, assuming that y is inversely proportional to the square of x.

x	2.69		7.83
y		1.16	1.52

7. If y varies inversely as the square root of x, fill in the missing values.

x		3567	5725	
y	1136	1828		

8. If y is inversely proportional to the cube root of x, by what factor will y change when x is tripled?

9. If y is inversely proportional to the square root of x, by what percentage will y change when x is decreased by 50.0%?

10. Plot the curve $y = 5/x^2$ for the domain 0 to 5.

11. Plot the function $y = 1/x$ for the domain 0 to 10.

Boyle's Law

Boyle's law states that for a confined gas at a constant temperature, the product of the pressure and the volume is a constant. Another way of stating this law is that the pressure is inversely proportional to the volume or that the volume is inversely proportional to the pressure. Assume a constant temperature in the following problems.

The *pascal* (Pa) is the SI unit of pressure. It equals 1 newton per square metre.

12. A certain quantity of gas, when compressed to a volume of 2.50 m³, has a pressure of 184 Pa. Find the pressure resulting when that gas is further compressed to 1.60 m³.

13. The air in the cylinder in Fig. 19–4 is at a pressure of 101 kPa and occupies a volume of 2870 cm³. Find the pressure when it is compressed to 410 cm³.

14. A balloon contains 320 m³ of gas at a pressure of 140 000 Pa. What would the volume be if the same quantity of gas were at a pressure of 250 000 Pa?

Gravitational Attraction

Newton's law of gravitation states that any two bodies attract each other with a force that is inversely proportional to the square of the distance between them.

15. The force of attraction between two certain steel spheres is 375 pN (1 piconewton $= 10^{-12}$ N) when the spheres are placed 18.0 cm apart. Find the force of attraction when they are 52.0 cm apart.

16. How far apart must the spheres in problem 15 be placed to cause the force of attraction between them to be 3.00 pN?

17. The force of attraction between the earth and some object is called the *weight* of that object. The law of gravitation states, then, that the weight of an object is inversely proportional to the square of its distance from the centre of the earth. If a person weighs 150 lb. on the surface of the earth (assume this to be 3960 mi. from the centre), how much will he weigh 1500 mi. above the surface of the earth?

18. How much will a satellite, whose weight on earth is 675 N, weigh at an altitude of $25\overline{0}$ km above the surface of the earth?

Illumination

The *inverse square law* states that for a surface illuminated by a light source, the intensity of illumination on the surface is inversely proportional to the *square* of the distance between the source and the surface.

19. A certain light source produces an illumination of 800 lx (a *lux* [lx] is 1 lumen per square metre) on a surface. Find the illumination on that surface if the distance to the light source is doubled.

20. A light source located 2.75 m from a surface produces an illumination of 528 lx on that surface. Find the illumination if the distance is changed to 1.55 m.

21. A light source located 7.50 m from a surface produces an illumination of 426 lx on that surface. At what distance must that light source be placed to give an illumination of $85\overline{0}$ lx?

22. When a document is photographed on a certain copy stand, an exposure time of $\frac{1}{25}$ s is needed, with the light source 0.750 m from the document. At what distance must the light be located to reduce the exposure time to $\frac{1}{100}$ s?

Electrical

23. The current in a resistor is inversely proportional to the resistance. By what factor will the current change if a resistor increases 10.0% due to heating?

24. The resistance of a wire is inversely proportional to the square of its diameter. If an AWG size 12 conductor (2.050-mm diameter) has a resistance of 14.8 Ω, what will be the resistance of an AWG size 10 conductor (2.588-mm diameter) of the same length and material?

25. The capacitive reactance X_C of a circuit varies inversely as the capacitance C of the circuit. If the capacitance of a certain circuit is decreased by 25.0%, by what percentage will X_C change?

19–5 Functions of More Than One Variable

Joint Variation

So far in this chapter, we have considered only cases where y was a function of a *single* variable x. In functional notation, this is represented by $y = f(x)$. In this section we cover functions of two or more variables, such as

$$y = f(x, w)$$

$$y = f(x, w, z)$$

and so forth. When y varies directly as x and w, we say that y varies *jointly* as x and w. The three variables are related by the following equation, where, as before, k is a constant of proportionality:

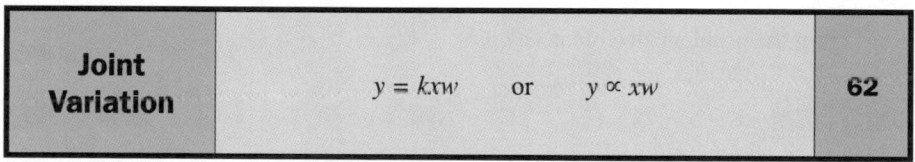

| Joint Variation | $y = kxw$ or $y \propto xw$ | 62 |

◆◆◆ **Example 27:** If y varies jointly as x and w, how will y change when x is doubled and w is one-fourth of its original value?

Solution: Let y' be the new value of y obtained when x is replaced by $2x$ and w is replaced by $w/4$, while the constant of proportionality k, of course, does not change. Substituting in Eq. 62, we obtain

$$y' = k(2x)\left(\frac{w}{4}\right)$$

$$= \frac{kxw}{2}$$

But, since $kxw = y$, then

$$y' = \frac{y}{2}$$

So the new y is half as large as its former value. ◆◆◆

Combined Variation

When one variable varies with two or more variables in ways that are more complex than in Eq. 62, it is referred to as *combined* variation. This term is applied to many different kinds of relationships, so a formula cannot be given that will cover all types. You must carefully read the problem statement in order to write the equation. Once you have the equation, the solution of combined variation problems is no different than for other types of variation.

◆◆◆ **Example 28:** If y varies directly as the cube of x and inversely as the square of z, we write

$$y = \frac{kx^3}{z^2}$$

◆◆◆

◆◆◆ **Example 29:** If y is directly proportional to x and the square root of w, and inversely proportional to the square of z, we write

$$y = \frac{kx\sqrt{w}}{z^2}$$

◆◆◆

◆◆◆ **Example 30:** If y varies directly as the square root of x and inversely as the square of w, by what factor will y change when x is made four times larger and w is tripled?

Solution: First write the equation linking y to x and w, including a constant of proportionality k.

$$y = \frac{k\sqrt{x}}{w^2}$$

We get a new value for y (let's call it y') when x is replaced by $4x$ and w is replaced by $3w$. Thus

$$y' = \frac{k\sqrt{4x}}{(3w)^2} = k\frac{2\sqrt{x}}{9w^2}$$

$$= \frac{2}{9}\left(\frac{k\sqrt{x}}{w^2}\right) = \frac{2}{9}y$$

We see that y' is $\frac{2}{9}$ as large as the original y. ◆◆◆

◆◆◆ **Example 31:** If y varies directly as the square of w and inversely as the cube root of x, and $y = 225$ when $w = 103$ and $x = 157$, find y when $w = 126$ and $x = 212$.

Solution: From the problem statement,

$$y = k\frac{w^2}{\sqrt[3]{x}}$$

Solving for k, we have

$$225 = k\frac{(103)^2}{\sqrt[3]{157}} = 1967k$$

or $k = 0.1144$. So

$$y = \frac{0.1144w^2}{\sqrt[3]{x}}$$

When $w = 126$ and $x = 212$,

$$y = \frac{0.1144(126)^2}{\sqrt[3]{212}} = 305$$

◆◆◆

◆◆◆ **Example 32:** y is directly proportional to the square of x and inversely proportional to the cube of w. By what factor will y change if x is increased by 15% and w is decreased by 20%?

Solution: The relationship between x, y, and w is

$$y = k\frac{x^2}{w^3}$$

so we can write the proportion

$$\frac{y_2}{y_1} = \frac{k\dfrac{x_2^2}{w_2^3}}{k\dfrac{x_1^2}{w_1^3}} = \frac{x_2^2}{w_2^3} \cdot \frac{w_1^2}{x_1^3} = \left(\frac{x_2}{x_1}\right)^2\left(\frac{w_1}{w_2}\right)^3$$

We now replace x_2 with $1.15x_1$ and replace w_2 with $0.8w_1$.

$$\frac{y_2}{y_1} = \left(\frac{1.15x_1}{x_1}\right)^2 \left(\frac{w_1}{0.8w_1}\right)^3 = \frac{(1.15)^2}{(0.8)^3} = 2.58$$

So y will increase by a factor of 2.58. ◆◆◆

Exercise 5 ◆ Functions of More Than One Variable

Joint Variation

1. If y varies jointly as w and x, and y is 483 when x is 742 and w is 383, find y when x is 274 and w is 756.

2. If y varies jointly as x and w, by what factor will y change if x is tripled and w is halved?

3. If y varies jointly as w and x, by what percent will y change if w is increased by 12% and x is decreased by 7.0%?

4. If y varies jointly as w and x, and y is 3.85 when w is 8.36 and x is 11.6, evaluate the constant of proportionality, and write the complete expression for y in terms of w and x.

5. If y varies jointly as w and x, fill in the missing values.

w	x	y
46.2	18.3	127
19.5	41.2	
	8.86	155
12.2		79.8

Combined Variation

6. If y is directly proportional to the square of x and inversely proportional to the cube of w, and y is 11.6 when x is 84.2 and w is 28.4, find y when x is 5.38 and w is 2.28.

7. If y varies directly as the square root of w and inversely as the cube of x, by what factor will y change if w is tripled and x is halved?

8. If y is directly proportional to the cube root of x and to the square root of w, by what percent will y change if x and w are both increased by 7.0%?

9. If y is directly proportional to the $\frac{3}{2}$ power of x and inversely proportional to w, and y is 284 when x is 858 and w is 361, evaluate the constant of proportionality, and write the complete equation for y in terms of x and w.

10. If y varies directly as the cube of x and inversely as the square root of w, fill in the missing values in the table.

w	x	y
1.27		3.05
	5.66	1.93
4.66	2.75	3.87
7.07	1.56	

Geometry

11. The area of a triangle varies jointly as its base and altitude. By what percent will the area change if the base is increased by 15% and the altitude decreased by 25%?

12. If the base and the altitude of a triangle are both halved, by what factor will the area change?

Electrical

13. When an electric current flows through a wire, the resistance to the flow varies directly as the length and inversely as the cross-sectional area of the wire. If the length and the diameter are both tripled, by what factor will the resistance change?

14. If $75\overline{0}$ m of 3.00-mm-diameter wire has a resistance of 27.6 Ω, what length of similar wire 5.00 mm in diameter will have the same resistance?

Gravitation

15. Newton's law of gravitation states that every body in the universe attracts every other body with a force that varies directly as the product of their masses and inversely as the square of the distance between them. By what factor will the force change when the distance is doubled and each mass is tripled?

16. If both masses are increased by 60% and the distance between them is halved, by what percent will the force of attraction increase?

Illumination

17. The intensity of illumination at a given point is directly proportional to the intensity of the light source and inversely proportional to the square of the distance from the light source. If a desk is properly illuminated by a 75.0-W lamp 2.50 m from the desk, what size lamp will be needed to provide the same lighting at a distance of 3.75 m?

18. How far from a 150-candela light source would a picture have to be placed so as to receive the same illumination as when it is placed 12 m from an 85-candela source?

Gas Laws

19. The volume of a given weight of gas varies directly as the absolute temperature t and inversely as the pressure p. If the volume is 4.45 m³ when p = 225 kilopascals (kPa) and t = 305 K, find the volume when p = 325 kPa and t = 354 K.

20. If the volume of a gas is 125 m³, find its volume when the absolute temperature is increased 10% and the pressure is doubled.

Work

21. The amount paid to a work crew varies jointly as the number of persons working and the length of time worked. If 5 workers earn $5,123.73 in 3.0 weeks, in how many weeks will 6 workers earn a total of $6,148.48?

22. If 5 bricklayers take 6.0 days to finish a certain job, how long would it take 7 bricklayers to finish a similar job requiring 4 times the number of bricks?

Strength of Materials

23. The maximum safe load of a rectangular beam (Fig. 19–12) varies jointly as the width and the square of the depth and inversely as the length of the beam. If a beam 8.00 cm wide, 11.5 cm deep, and 2.00 m long can safely support $70\overline{0}0$ kg, find the safe load for a beam 6.50 cm wide, 13.4 cm deep, and 2.60 m long made of the same material.

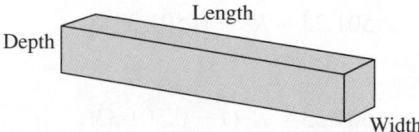

FIGURE 19-12 A rectangular beam.

24. If the width of a rectangular beam is increased by 11%, the depth decreased by 8%, and the length increased 6%, by what percent will the safe load change?

Mechanics

25. The number of vibrations per second made when a stretched wire (Fig. 19–13) is plucked varies directly as the square root of the tension in the wire and inversely as the length. If a 1.00-m-long wire will vibrate 325 times a second when the tension is 115 N, find the frequency of vibration if the wire is shortened to 0.750 m and the tension is decreased to 95.0 N.

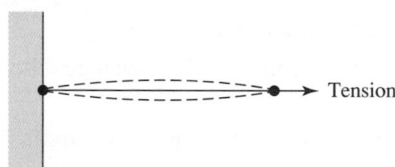

FIGURE 19-13 A stretched wire.

26. The kinetic energy of a moving body is directly proportional to its mass and the square of its speed. If the mass of a bullet is halved, by what factor must its speed be increased to have the same kinetic energy as before?

Fluid Flow

27. The time needed to empty a vertical cylindrical tank (Fig. 19–14) varies directly as the square root of the height of the tank and the square of the radius. By what factor will the emptying time change if the height is doubled and the radius increased by 25%?

28. The power available in a jet of liquid is directly proportional to the cross-sectional area of the jet and to the cube of the velocity. By what factor will the power increase if the area and the velocity are both increased 50%?

FIGURE 19-14 A vertical cylindrical tank.

Case Study Discussion—Wheatstone Bridge

The strain gauge has a nominal resistance of 1000 Ω. When stressed, the resistance will change. Now we could just measure the current flow through the gauge, and divide that into our applied voltage (by Ohm's law) to give us the resistance. The problem is that the accuracy of the measurement depends not only on a very good ammeter (to measure current) but also on a very precise voltage source. The Wheatstone bridge operates on the principle of ratios of the resistances that form the bridge. Regardless of the power supply, we simply measure V_{OUT} and V_{IN} and use ratios. Even if the 9V in our example varied, the V_{OUT} would also vary but the ratio would be the same.

$$V_{OUT} = \left(\frac{R_4}{R_3 + R_4} - \frac{R_2}{R_1 + R_2} \right) V_{IN}$$

$$501.3\dot{3} + 0.501\,33 R_4 = R_4$$

$$0.012 = \left(\frac{R_4}{1000 + R_4} - \frac{1000}{1000 + 1000} \right) 9$$

$$501.3\dot{3} = R_4 - 0.501\,3\dot{3} R_4$$

$$\frac{0.012}{9} = \left(\frac{R_4}{1000 + R_4} - \frac{1000}{2000} \right)$$

$$501.3\dot{3} = R_4 (1 - 0.501\,3\dot{3})$$

$$0.001\,3\dot{3} = \frac{R_4}{1000 + R_4} - 0.5$$

$$501.3\dot{3} = R_4 (0.498\,6\dot{6})$$

$$0.501\,3\dot{3} = \frac{R_4}{1000 + R_4}$$

$$\frac{501.3\dot{3}}{0.498\,6\dot{6}} = R_4$$

$$0.501\,3\dot{3}(1000 + R_4) = R_4$$

$$1005.347\ \Omega = R_4$$

••• CHAPTER 19 REVIEW PROBLEMS •••••••••••••••••••••••••••••••••

1. If y varies inversely as x, and y is 736 when x is 822, find y when x is 583.

2. If y is directly proportional to the $\frac{5}{2}$ power of x, by what factor will y change when x is tripled?

3. If y varies jointly as x and z, by what percent will y change when x is increased by 15% and z is decreased by 4%?

4. The braking distance of an automobile varies directly as the square of the speed. If the braking distance of a certain automobile is 11.0 m at 40.0 km/h, find the braking distance at 90.0 km/h.

5. The rate of flow of liquid from a hole in the bottom of a tank is directly proportional to the square root of the liquid depth. If the flow rate is 225 L/min when the depth is 3.46 m, find the flow rate when the depth is 1.00 m.

A knot is equal to 1 nautical mile (M) per hour (1 M = 1852 m).

6. The power needed to drive a ship varies directly as the cube of the speed of the ship, and a 77.4-kW engine will drive a certain ship at 11.2 knots (kn). Find the power needed to propel that ship at 18.0 kn.

7. If the tensile strength of a cylindrical steel bar varies as the square of its diameter, by what factor must the diameter be increased to triple the strength of the bar?

8. The life of an incandescent lamp varies inversely as the 12th power of the applied voltage, and the light output varies directly as the 3.5th power of the applied voltage. By what factor will the life increase if the voltage is lowered by an amount that will decrease the light output by 10%?

9. One of Kepler's laws states that the time for a planet to orbit the sun varies directly as the $\frac{3}{2}$ power of its distance from the sun. How many years will it take for Saturn, which is about $9\frac{1}{2}$ times as far from the sun as is the earth, to orbit the sun?

10. The volume of a cone varies directly as the altitude and the square of the base radius. By what factor will the volume change if the altitude is doubled and the base radius is halved?

11. The number of oscillations made by a pendulum in a given time is inversely proportional to the length of the pendulum. A certain clock with a 75.00-cm-long pendulum is losing 15.00 min/d. Should the pendulum be lengthened or shortened, and by how much?

12. A trucker usually makes a trip in 18.0 h at an average speed of 90.0 km/h. Find the travelling time if the speed were reduced to 75.0 km/h.

13. The force on the vane of a wind generator varies directly as the area of the vane and the square of the wind velocity. By what factor must the area of a vane be increased so that the wind force on it will be the same in a 12-km/h wind as it was in a 35-km/h wind?

14. The maximum deflection of a rectangular beam varies inversely as the product of the width and the cube of the depth. If the deflection of a beam having a width of 15 cm and a depth of 35 cm is 7.5 mm, find the deflection if the width is made 20 cm and the depth 45 cm.

15. When an object moves at a constant speed, the travel time is inversely proportional to the speed. If a satellite now circles the earth in 26 h 18 min, how long will it take if booster rockets increase the speed of the satellite by 15%?

16. The life of an incandescent lamp is inversely proportional to the 12th power of the applied voltage. If a lamp has a life of 2500 h when run at 115 V, what will its life expectancy be when it is run at 110 V?

17. How far from the earth will a spacecraft be equally attracted by the earth and the moon? The distance from the earth to the moon is approximately 385 000 km, and the mass of the earth is about 82.0 times that of the moon.

18. Ohm's law states that the electric current flowing in a circuit varies directly as the applied voltage and inversely as the resistance. By what percent will the current change when the voltage is increased by 25% and the resistance increased by 15%?

19. The allowable strength F of a column varies directly as its length L and inversely as its radius of gyration r. If $F = 35$ kg when L is 12 m and r is 3.6 cm, find F when L is 18 m and r is 4.5 cm.

20. The time it takes a pendulum to go through one complete oscillation (the *period*) is directly proportional to the square root of its length. If the period of a 1-m pendulum is 1.25 s, how long must the pendulum be to have a period of 2.5 s?

21. If the maximum safe current in a wire is directly proportional to the $\frac{3}{2}$ power of the wire diameter, by what factor will the safe current increase when the wire diameter is doubled?

22. Four workers take ten 6-h days to finish a job. How many workers are needed to finish a similar job that is 3 times as large, in five 8-h days?

23. When a jet of water strikes the vane of a water turbine and is deflected through an angle θ, the force on the vane varies directly as the square of the jet velocity and the sine of $\theta/2$. If θ is decreased from 55° to 40°, and if the jet velocity increases by 40%, by what percentage will the force change?

24. By what factor will the kinetic energy change if the speed of a projectile is doubled and its mass is halved? Use the fact that the kinetic energy of a moving body is directly proportional to the mass and the square of the velocity of the body.

25. The fuel consumption for a ship is directly proportional to the cube of the ship's speed. If a certain tanker uses 1584 U.S. gal. of diesel fuel on a certain run at 15.0 kn, how much fuel would it use for the same run when the speed is reduced to 10.0 kn?

26. The diagonal of a certain square is doubled. By what factor does the area change?

27. A certain parachute is made from 52.0 m² of fabric. If all of its dimensions are increased by a factor of 1.40, how many square metres of fabric will be needed?

28. The rudder of a certain airplane has an area of 2.50 m². By what factor must the dimensions be scaled up to triple the area?

29. A 3.0-cm-diameter pipe fills a cistern in 5.0 h. Find the diameter of a pipe that will fill the same cistern in 9.0 h.

30. If 315 m of fence will enclose a circular field containing 2.0 acres, what length will enclose 8.0 acres?

31. The surface area of a one-quarter model of an automobile measures 1.10 m². Find the surface area of the full-sized car.

32. A certain water tank is 5.00 m high and holds $90\overline{0}0$ L. How much would a 7.00-m-high tank of similar shape hold?

33. A certain tractor weighs 5.80 t. If all of its dimensions were scaled up by a factor of 1.30, what would the larger tractor be expected to weigh?

34. If a trough 2.5 m long holds 12 pailfuls of water, how many pailfuls will a similar trough hold that is 4.0 m long?

35. A ball 4.50 in. in diameter weighs 18.0 oz. What is the weight of another ball of the same density that is 9.00 in. in diameter?

36. A ball 4.00 inches in diameter weighs 9.00 lb. What is the weight of a ball 25.0 cm in diameter made of the same material?

37. A worker's contract has a cost-of-living clause that requires that his salary be proportional to the cost-of-living index. If he earned $450 per week when the index was 9.40, how much should he earn when the index is 12.7?

38. A woodsman pays a landowner $16.00 for each cord of wood he cuts on the landowner's property, and he sells the wood for $150 per cord. When the price of firewood rose to $190, the landowner asked for a proportional share of the price. How much should he get per cord?

39. The series of numbers 3.5, 5.25, 7.875, … form a *geometric progression,* where the ratio of any term to the one preceding it is a constant. This is called the *common ratio.* Find this ratio.

40. The *specific gravity* (SG) of a solid or liquid is the ratio of the density of the substance to the density of water at a standard temperature (Eq. A45). Taking the density of water as 1.00 g/cm³, find the density of copper having a specific gravity of 8.89.

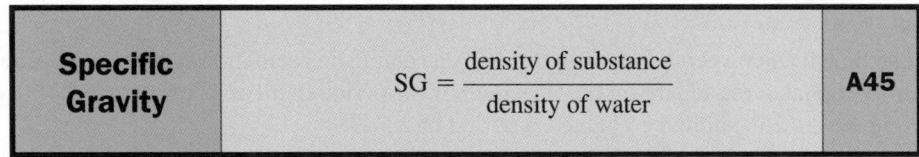

Specific Gravity	$SG = \dfrac{\text{density of substance}}{\text{density of water}}$	A45

41. An iron casting having a surface area of 746 cm² is heated so that all its dimensions increase by 1.00%. Find the new area.

42. A certain boiler pipe will permit a flow of 35.5 L/min. Buildup of scale inside the pipe eventually reduces its diameter to three-fourths of its previous value. Assuming that the flow rate is proportional to the cross-sectional area, find the new flow rate for the pipe.

Writing

43. Suppose that your company makes plastic trays and is planning new ones with dimensions double those now being made. Your company president is convinced that they will need only twice as much plastic as the older version. "Twice the size, twice the plastic," he proclaims, and no one is willing to challenge him. Your job is to make a presentation to the president where you tactfully point out that he is wrong and where you explain that the new trays will require eight times as much plastic. Write your presentation.

Team Project

44. We saw that the volumes (and hence the weights) of solids are proportional to the *cube* of corresponding dimensions. Applying that idea, suppose that a sporting goods company has designed a new line of equipment (ski packages, windsurfers, diving gear, personal watercraft, clothing, etc.) based on the following statement:

The weights of people of similar build are proportional to the cube of their heights.

The goal of this project is to prove or disprove the given statement. Your team will use data gathered from students on your campus in reaching a conclusion.

Exponential and Logarithmic Functions

♦♦♦ OBJECTIVES ♦♦♦

When you have completed this chapter, you should be able to:
- Examine the characteristics and graphing of exponential functions.
- Derive and use equations for exponential growth and decay.
- Use the logarithm form of exponential functions.
- Use the properties of logarithms to change the form of expressions.
- Solve exponential equations.
- Solve logarithmic equations.
- Graph logarithmic and semi-logarithmic scales.

Up until now, you have rearranged equations to solve for unknown variables. The variables could be in a term or expression, in the numerator or denominator of a fraction, or under a radical sign. They cannot hide from you.

To continue your mastery of algebra, we now head for a new frontier: What if the unknown variable is part of an exponent? There is no technique using division, cross-multiplying, or transposition that will pull that variable "down" from the exponent so that we can get it by itself on one side of the equals sign and solve for it. Logarithms allow us to do exactly that! A logarithm rearranges the way we write an exponential expression so that the exponent becomes a coefficient in an equation we can solve.

You might have heard the story about the inventor of chess, who asked that the reward for his invention be a grain of wheat on the first square of the chessboard, two grains on the second square, four on the third, and so on, each square getting double the amount as on the previous square. It turns out that he would receive more wheat than has been grown by humans during all history—an illustration of the remarkable properties of the *exponential function*, which we study in this chapter.

20–1 The Exponential Function

Definition

An *exponential function* is one in which the independent variable appears in the *exponent*. The quantity that is raised to the *power* is called the *base*.

Realize that in the *exponential function* the "x" (or what you're trying to find) is in the exponent (as in $y = 5^x$), whereas in the *power function*, the "x" is in the base (as in $y = x^n$).

◆◆◆ **Example 1:** The following are exponential functions if a, b, and e represent positive constants:

$$y = 5^x \qquad y = b^x \qquad y = e^x \qquad y = 10^x$$
$$y = 3a^x \qquad y = 7^{x-3} \qquad y = 5e^{-2x}$$

◆◆◆

We will see that e, like π, has a specific value, which we will discuss later in this chapter.

Graph of the Exponential Function

We may make a graph of an exponential function in the same way as we graphed other functions in Chapter 5. Choose values of the independent variable x throughout the region of interest, and for each x compute the corresponding value of the dependent variable y. Plot the resulting table of ordered pairs, and connect the points with a smooth curve.

◆◆◆ **Example 2:** Graph the exponential function $y = 2^x$ from $x = -4$ to $x = +3$. For comparison, plot the power function $y = x^2$ on the same axes.

Solution: We make our graphs, shown in Fig. 20–1 by computing and plotting point pairs.

For the given function: $y = x^2$	
−4	−16
−3	−9
−2	−4
−1	−1
0	0
+1	1
+2	4
+3	9

For the given function: $y = 2^x$	
−4	.06
−3	.13
−2	.25
−1	.50
0	1
+1	2
+2	4
+3	8

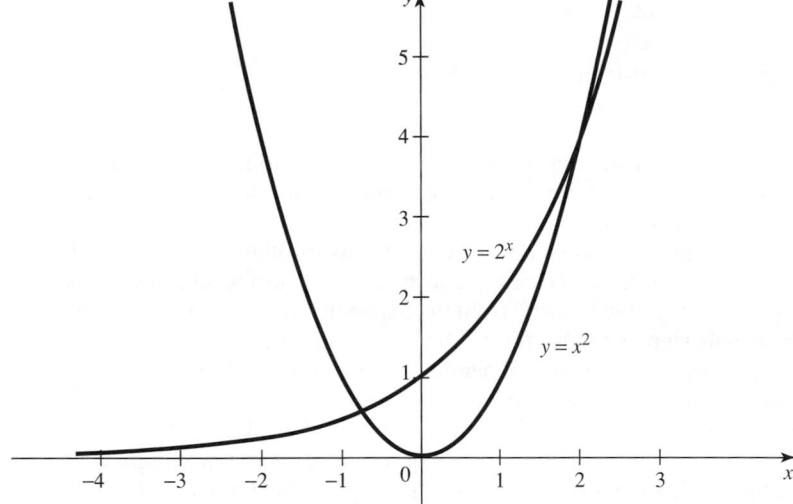

FIGURE 20–1

Note that the graph of the exponential function is completely different from the graph of the power function, whose graph is a parabola. For the exponential function, note that as x increases, y increases even more greatly, and that the curve gets steeper and steeper. This is a characteristic of *exponential growth*, which we cover more fully below. Also note that the y intercept is 1 and that there is no x intercept, the x axis being an asymptote. Finally, note that each successive y value is two times the previous one. We say that y *increases geometrically*.   ◆◆◆

Sometimes the exponent in an exponential function is negative. We will see in the following example what effect a negative exponent has on the graph of the function.

◆◆◆ **Example 3:** Graph the exponential functions $y = 2^x$ and $y = 2^{-x}$, for $x = -4$ to 4, on the same axes.

Solution: We have already graphed $y = 2^x$ in the preceding example, and now we add the graph of $y = 2^{-x}$, shown in Fig. 20–2.

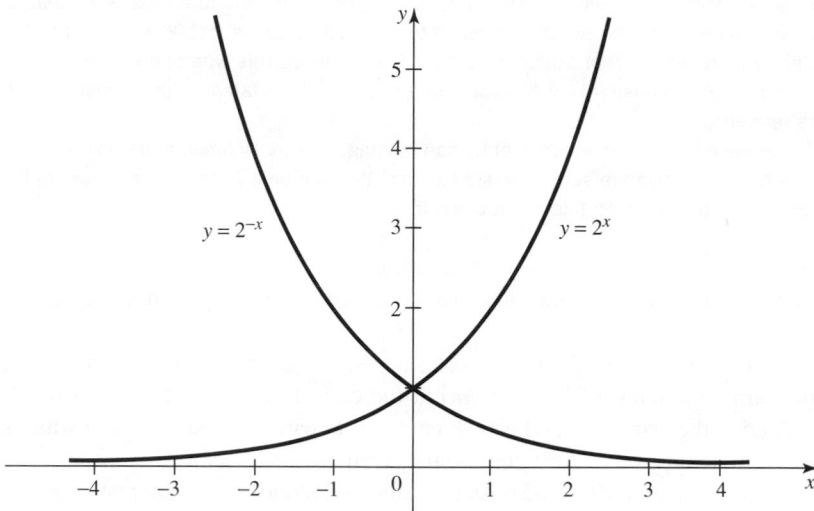

FIGURE 20–2

Note that when the exponent is *positive*, the graph *increases exponentially*, and that when the exponent is *negative*, the graph *decreases exponentially*. Also notice that neither curve reaches the x axis. Thus for the exponential functions $y = 2^x$ and $y = 2^{-x}$, y cannot equal zero.   ◆◆◆

Exponential functions often have the irrational number e as the base. We will see why in the following section. For now, we will treat e like any other number, using $e \approx 2.718$ for its approximate value.

◆◆◆ **Example 4:** Graph the exponential function $y = 1.85e^{-x/2}$, for $x = -4$ to 4, if e equals 2.718.

Solution: We make our graph, shown in Fig. 20–3, by computing and plotting point pairs.

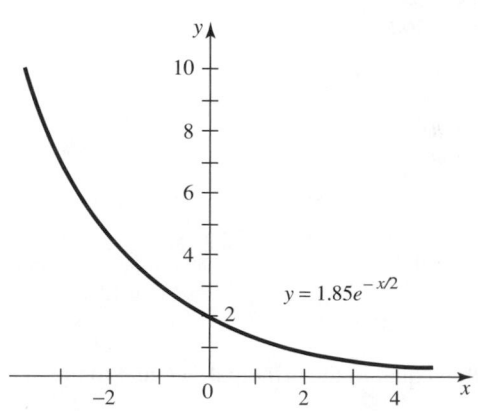

For the given function: $y = 1.85e^{-\frac{x}{2}}$	
-4	14
-3	8.3
-2	5
-1	3.1
0	1.9
$+1$	1.1
$+2$	.68
$+3$	.41
$+4$	.25

FIGURE 20–3

Note that the negative sign in the exponent has made the graph *decrease* exponentially as x increases, and that the curve approaches the x axis as an asymptote. The shape of the curve, however, is that of an exponential function.   ◆◆◆

CASE STUDY

When you go for an audio test of your hearing, the measurement of sound intensity, or volume, is the decibel. The *decibel* (one-tenth of a bel, named in honour of Alexander Graham Bell), is a way of expressing an increase or decrease in power. The scale compares a measurement of electrical power or air pressure to a reference level. If the tested level is 100 times greater than the reference level, that is a "2" on the logarithmic scale, since the exponent needed to raise 10 to 100 is 2 ($100 = 10^2$); a "3" on the logarithmic scale represents 10^3 or 1000 times the reference level. This type of measurement is called a *relative* measurement because it is with respect to a reference level (an *absolute* measurement is independent of a reference).

Your task is to determine the dynamic range of human hearing. How many times more powerful is the loudest sound (before pain) than the softest audible sound? Would it be 100 times more powerful? 1000 times? (Neither of these is correct; this is not a multiple choice question.)

Compound Interest

In this and the next few sections, we will see some problems that require exponential equations for their solution.

When an amount of money *a* (called the *principal*) is invested at an interest rate *n* (expressed as a decimal), it will earn an amount of interest (*an*) at the end of the first interest period. If this interest is then added to the principal, and if both continue to earn interest, we have what is called *compound interest*, the type of interest commonly given on bank accounts.

Suppose that the interest rate is 5% (0.05). As the balance increases, the interest received for the current period (5% of *a* or 0.05*a*) is added to the previous balance (100% of *a* or 1.00*a*) to give a new total of 1.05*a*. So we could say we multiply *a* by (1 + 0.05) or 1.05.

Similarly, for an interest rate of *n*, the balance y_t at the end of any period *t* is obtained by multiplying the preceding balance y_{t-1} by (1 + *n*).

Recursion Relation for Compound Interest	$y_t = y_{t-1}(1 + n)$	A10b

This equation is called a *recursion relation*, from which each term in a series of numbers may be found from the term immediately preceding it. We'll do more with recursion relations in Chapter 25, but now we will use Eq. A10b to make a table showing the balance at the end of a number of interest periods.

Period	Balance *y* at End of Period
1	$a(1 + n)$
2	$a(1 + n)^2$
3	$a(1 + n)^3$
4	$a(1 + n)^4$
.	.
.	.
.	.
t	$a(1 + n)^t$

We thus get the following exponential function, with *a* equal to the principal, *n* equal to the rate of interest, and *t* the number of interest periods:

Compound Interest Computed Once per Period	$y = a(1 + n)^t$	A10a

✦✦✦ **Example 5:** To what amount (to the nearest cent) will an initial deposit of $500 accumulate if invested at a compound interest rate of $6\frac{1}{2}\%$ per year for 8 years?

Solution: We are given

> Principal $a = \$500$
> Interest rate $n = 0.065$, expressed in decimal form
> Number of interest periods $t = 8$

Substituting into Eq. A10a gives

$$y = 500(1 + 0.065)^8$$
$$= \$827.50$$

✦✦✦

In Sec. 20–2, we will use this equation to derive an equation for exponential (continuous) growth.

Compound Interest Computed *m* Times per Period

We can use Eq. A10a to compute, say, interest on a savings account for which the interest is computed once a year. But what if the interest is computed more often, as most banks do, for example, monthly, weekly, or even daily?

Let us modify the compound interest formula. Instead of computing interest *once* per period, let us compute it *m times* per period. The exponent in Eq. A10 thus becomes *mt*. The interest rate *n*, however, has to be reduced to $(1/m)$th of its old value, or to n/m.

Compound Interest Computed *m* Times per Period	$y = a\left(1 + \dfrac{n}{m}\right)^{mt}$	A11

✦✦✦ **Example 6:** Repeat the computation of Example 5 with the interest computed monthly rather than annually.

Solution: As in Example 5, $a = \$500$, annual interest rate $n = 0.065$, and $t = 8$. But n is an annual rate, so we must use Eq. A11, with $m = 12$.

$$y = 500\left(1 + \frac{0.065}{12}\right)^{12(8)} = 500(1.005\ 42)^{96}$$
$$= \$839.83$$

or $12.33 more than in Example 5.

✦✦✦

Exercise 1 ✦ The Exponential Function

Exponential Function

Graph each exponential function over the given domains of x. Take $e = 2.718$.

1. $y = 0.2(3.2)x$ $(x = -4 \text{ to } +4)$
2. $y = 3(1.5)^{-2x}$ $(x = -1 \text{ to } 5)$
3. $y = 5(1 - e^{-x})$ $(x = 0 \text{ to } 10)$
4. $y = 4e^{x/2}$ $(x = 0 \text{ to } 4)$
5. A rope or chain between two points hangs in the shape of a curve called a *catenary* (Fig. 20–4). Its equation is

$$y = \frac{a(e^{x/a} + e^{-x/a})}{2}$$

Using values of $a = 2.5$ and $e = 2.718$, plot y for $x = -5$ to 5.

FIGURE 20–4 A rope or chain hangs in the shape of a catenary. Mount your finished graph for problem 5 vertically, and hang a fine chain from the endpoints. Does its shape (after you have adjusted its length) match that of your graph?

Compound Interest

The amount a is called the *present worth* of the future amount y. This formula, and the three that follow, are standard formulas used in business and finance.

6. Find the amount to which \$500 will accumulate in 6 years at a compound interest rate of 6% per year compounded annually.

7. If our compound interest formula is solved for a, we get

$$a = \frac{y}{(1+n)^t}$$

where a is the amount that must be deposited now at interest rate n to produce an amount y in t years. How much money would have to be invested at $7\frac{1}{2}\%$ per year to accumulate to \$10,000 in 10 years?

8. What interest rate (compounded annually) is needed to enable an investment of \$5,000 to accumulate to \$10,000 in 12 years?

9. Using Eq. A11, find the amount to which \$1 will accumulate in 20 years at a compound interest rate of 10% per year (a) compounded annually, (b) compounded monthly, and (c) compounded daily.

A series of annual payments such as these is called an *annuity*.

10. If an amount of money R is deposited *every year* for t years at a compound interest rate n, it will accumulate to an amount y, where

$$y = R\left[\frac{(1+n)^t - 1}{n}\right]$$

If a worker pays \$2,000 per year into a retirement plan yielding 6% annual interest, what will the value of this annuity be after 25 years?

This is called a *sinking fund*.

11. If the equation for an annuity in problem 10 is solved for R, we get

$$R = \frac{ny}{(1+n)^t - 1}$$

where R is the annual deposit required to produce a total amount y at the end of t years. How much would the worker in problem 10 have to deposit each year (at 6% per year) to have a total of \$100,000 after 25 years?

This is called *capital recovery*.

12. The yearly payment R that can be obtained for t years from a present investment a is

$$R = a\left[\frac{n}{(1+n)^t - 1} + n\right]$$

If a person has \$85,000 in a retirement fund, how much can be withdrawn each year so that the fund will be exhausted in 20 years if the amount remaining in the fund is earning $6\frac{1}{4}\%$ per year?

20–2   Exponential Growth and Decay

Exponential Growth

Equation A11, $y = a(1 + n/m)^{mt}$, allows us to compute the amount obtained, with interest compounded any number of times per period that we choose. We could use this formula to compute the interest if it were compounded every week, or every day, or every second. But what about *continuous* compounding, or continuous growth? What would happen to Eq. A11 if m got very, very large, in fact, infinite? Let us make m increase to large values; but first, to simplify the work, we make the substitution. Let

$$k = \frac{m}{n}$$

Equation A11 then becomes

$$y = a\left(1 + \frac{1}{k}\right)^{knt}$$

which can be written

$$y = a\left[\left(1 + \frac{1}{k}\right)^k\right]^{nt}$$

Then as m grows large, so will k. Let us try some values on the calculator and evaluate only the quantity in the square brackets, to four decimal places.

k	$\left(1 + \dfrac{1}{k}\right)^k$
1	2
10	2.5937 ...
100	2.7048 ...
1000	2.7169 ...
10 000	2.7181 ...
100 000	2.7183 ...
1 000 000	2.7183 ...

We get the surprising result that as m (and k) continue to grow infinitely large, the value of $(1 + 1/k)^k$ does *not* grow without limit, but approaches a specific, well-known value, the value 2.7183. . . . This important number is given the special symbol e. We can express the same idea in *limit notation* by writing the following equation:

Definition of e	$$\lim_{k \to \infty}\left(1 + \frac{1}{k}\right)^k = e$$	**185**

As k grows without bound, the value of $(1 + 1/k)^k$ approaches e (≈ 2.7183 . . .).

Thus when m gets infinitely large, the quantity $(1 + 1/k)^k$ approaches e, and the formula for continuous growth becomes the following:

Exponential Growth	$$y = ae^{nt}$$	**199**

The number e is named after a Swiss mathematician, Leonhard Euler (1707–83). Its value has been computed to thousands of decimal places.

◆◆◆ **Example 7:** Repeat the computation of Example 5 if the interest were compounded *continuously* rather than monthly.

Solution: From Eq. 199,

$$
\begin{aligned}
y &= 500e^{0.065(8)} \\
&= 500e^{0.52} \\
&= \$841.01
\end{aligned}
$$

or $1.18 more than when compounded monthly. ◆◆◆

Banks that offer "continuous compounding" of your account do not have a clerk constantly computing your interest. They use Eq. 199.

Common Error	When you are using Eq. 199, the time units of n and t *must agree*.

◆◆◆ **Example 8:** A quantity grows exponentially at the rate of 2% per minute. By how many times will it have increased after 4 h?

Solution: The units of n and t do *not* agree (yet).

$$n = 2\% \text{ per } minute \quad \text{and} \quad t = 4 \ hours$$

So we convert one of them to agree with the other.

$$t = 4\,h = 240\,min$$

Using Eq. 199,

$$y = ae^{nt} = ae^{0.02(240)}$$

We can find the ratio of y to a by dividing both sides by a.

$$\frac{y}{a} = e^{0.02(240)} = e^{4.8} = 122$$

So the final amount is 122 times as great as the initial amount.   ◆◆◆

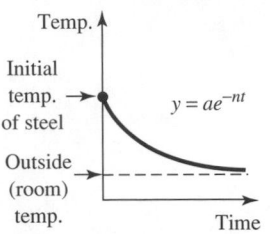

FIGURE 20–5 Exponential decrease in temperature. This is called *Newton's law of cooling*.

Exponential Decay

When a quantity *decreases* so that the amount of decrease is proportional to the present amount, we have what is called *exponential decay*. Take, for example, a piece of steel just removed from a furnace. The amount of heat leaving the steel depends upon its temperature: The hotter it is, the faster heat will leave, and the faster it will cool. So as the steel cools, its lower temperature causes slower heat loss. Slower heat loss causes slower rate of temperature drop, and so on. The result is the typical *exponential decay curve* shown in Fig. 20–5. Finally, the temperature is said to *asymptotically approach* the outside (room) temperature.

The equation for exponential decay is the same as for exponential growth, except that the exponent is negative.

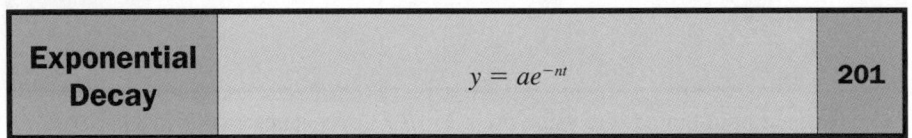

| Exponential Decay | $y = ae^{-nt}$ | 201 |

◆◆◆ **Example 9:** A room initially 80 °F above the outside temperature cools exponentially at the rate of 25% per hour. Find the temperature of the room (above the outside temperature) at the end of 135 min.

Solution: We first make the units consistent: 135 min = 2.25 h. Then, by Eq. 201,

$$y = 80e^{-0.25(2.25)} = 45.6\ °F$$ ◆◆◆

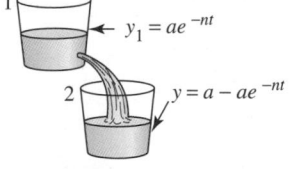

FIGURE 20–6 The amount in bucket 2 grows exponentially to an upper limit a.

Exponential Growth to an Upper Limit

Suppose that the bucket in Fig. 20–6 initially contains a gallons of water. Then assume that y_1, the amount remaining in the bucket, decreases exponentially. By Eq. 201,

$$y_1 = ae^{-nt}$$

The amount of water in bucket 2 thus increases from zero to a final amount a. The amount y in bucket 2 at any instant is equal to the amount that has left bucket 1, or

$$y = a - y_1$$
$$= a - ae^{-nt}$$

which is defined formally in the following equation:

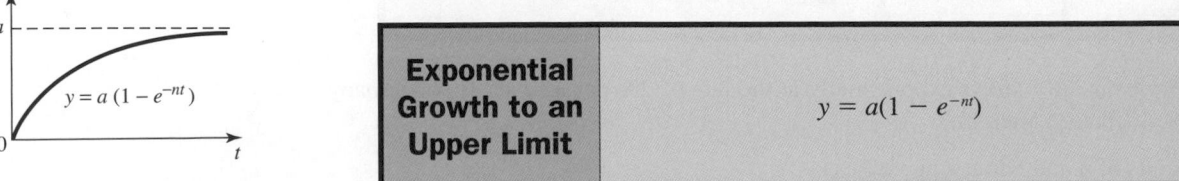

| Exponential Growth to an Upper Limit | $y = a(1 - e^{-nt})$ | 202 |

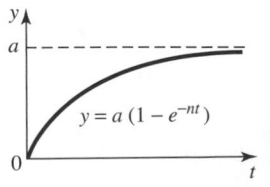

FIGURE 20–7 Exponential growth to an upper limit.

So the equation for this type of growth to an upper limit is exponential, where a equals the upper limit quantity, n is equal to the rate of growth, and t represents time (see Fig. 20–7).

✦✦✦ **Example 10:** The voltage across a certain capacitor grows exponentially from 0 V to a maximum of 75 V, at a rate of 7% per second. Write the equation for the voltage at any instant, and find the voltage at 2.0 s.

Solution: From Eq. 202, with $a = 75$ and $n = 0.07$,

$$v = 75(1 - e^{-0.07t})$$

where v represents the instantaneous voltage. When $t = 2.0$ s,

$$v = 75(1 - e^{-0.14})$$
$$= 9.80 \text{ V}$$

✦✦✦

Time Constant

Our equations for exponential growth and decay all contained a rate of growth n. Often it is more convenient to work with the *reciprocal* of n, rather than n itself. This reciprocal is called the *time constant, T*.

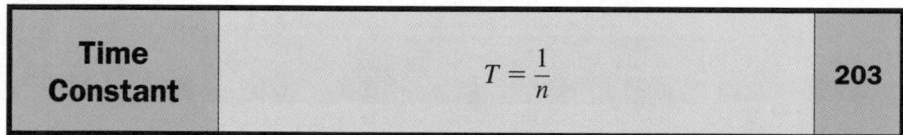

Time Constant	$T = \dfrac{1}{n}$	**203**

✦✦✦ **Example 11:** If a quantity decays exponentially at a rate of 25% per second, what is the time constant?

Solution:

$$T = \frac{1}{0.25} = 4 \text{ s}$$

Notice that the time constant has the units of *time*.

✦✦✦

Universal Growth and Decay Curves

If we replace n by $1/T$ in Eq. 201, we get

$$y = ae^{-t/T}$$

Then we divide both sides by a.

$$\frac{y}{a} = e^{-t/T}$$

We plot this curve with the dimensionless ratio t/T for our horizontal axis and the dimensionless ratio y/a for the vertical axis (Fig. 20–8). Thus the horizontal axis is *the number of time constants* that have elapsed, and the vertical axis is *the ratio of the final amount to the initial amount*. It is called the *universal decay curve*.

We can use the universal decay curve to obtain *graphical solutions* to exponential decay problems.

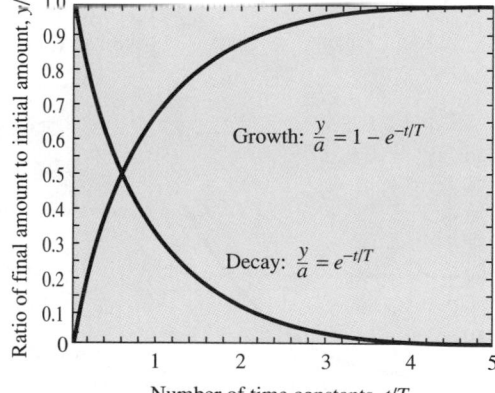

FIGURE 20–8 Universal growth and decay curves.

✦✦✦ **Example 12:** A current decays from an initial value of 300 mA at an exponential rate of 20% per second. Using the universal decay curve, graphically find the current after 7 s.

Solution: We have $n = 0.20$, so the time constant is

$$T = \frac{1}{n} = \frac{1}{0.20} = 5 \text{ s}$$

Our given time $t = 7$ s is then equivalent to

$$\frac{t}{T} = \frac{7}{5} = 1.4 \text{ time constants}$$

From the universal decay curve, at $t/T = 1.4$, we read

$$\frac{y}{a} = 0.25$$

Using the $\frac{y}{a} = e^{\frac{-t}{T}}$ formula gives the same value.

Since $a = 300$ mA,

$$y = 300(0.25) = 75 \text{ mA}$$ ◆◆◆

Figure 20–8 also shows the *universal growth curve*

$$\frac{y}{a} = 1 - e^{-t/T}$$

It is used in the same way as the universal decay curve.

A Recursion Relation for Exponential Growth

We have found that an amount a will grow to an amount y in t periods of time, where

$$y = ae^{nt} \qquad\qquad (199)$$

where n is the rate of growth. Computing y for different values of t gives the following table:

t	y
0	$y_0 = ae^0 = a$
1	$y_1 = ae^n = y_0 e^n$
2	$y_2 = ae^{2n} = (e^n)(ae^n) = e^n y_1$
3	$y_3 = ae^{3n} = (e^n)(ae^{2n}) = e^n y_2$
.	.
.	.
.	.
t	$y_t = e^n y_{t-1}$

Thus each value y_1 may be obtained by multiplying the preceding value y_{t-1} by the constant e^n. If we replace e^n by a constant B (sometimes called the *birthrate*), we have the recursion relation, given in the following equation:

Recursion Relation for Exponential Growth	$y_t = By_{t-1}$	204

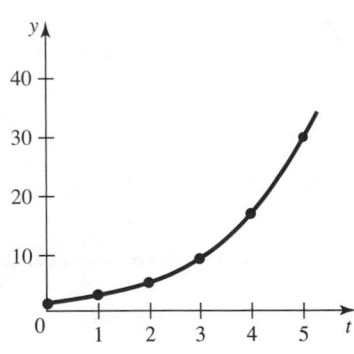

FIGURE 20–9

◆◆◆ **Example 13:** Compute and graph values of y versus t using a starting value of 1 and a birthrate of 2, for five time periods.

Solution: We get each new y by multiplying the preceding y by 2.

t	0	1	2	3	4	5
y	1	2	4	8	16	32

The graph of y versus t, shown in Fig. 20–9, is a typical exponential growth curve.   ◆◆◆

Nonlinear Growth Equation

Equations 199 and 204 represent a population that can grow without being limited by shortage of food or space, by predators, and so forth. This is rarely the case, so biologists have modified the equation to take those limiting factors into account. One model of the growth equation that takes these limits into account is obtained by multiplying the right side of Eq. 204 by the factor $(1 - y_{t-1})$.

Nonlinear Growth Equation	$y_t = By_{t-1}(1-y_{t-1})$	205

This equation is one form of what is called the *Logistic Function,* or the *Verhulst Population Model.* As y_{t-1} gets larger, the factor $(1 - y_{t-1})$ gets smaller. This equation will cause the predicted population y to stabilize at some value, for a range of values of the birthrate B from 1 to 3, regardless of the starting value for y.

The population of 0.8 is a *normalized* population, considered to be 80% of some reference value.

◆◆◆ **Example 14:** Compute the population for 20 time periods using the nonlinear growth equation. Use a starting population of 0.8 and a birthrate of 2.

Solution: Make a table of values.

t	Population y
0	0.8
1	$2(0.8)(1 - 0.8) = 0.32$
2	$2(0.32)(1 - 0.32) = 0.44$
3	$2(0.44)(1 - 0.44) = 0.49$
4	$2(0.49)(1 - 0.49) = 0.50$
5	$2(0.50)(1 - 0.50) = 0.50$
.	.
.	.
.	.
20	$2(0.50)(1 - 0.50) = 0.50$

As shown in Fig. 20–10(b), the population y stabilizes at a value of 0.50. ◆◆◆

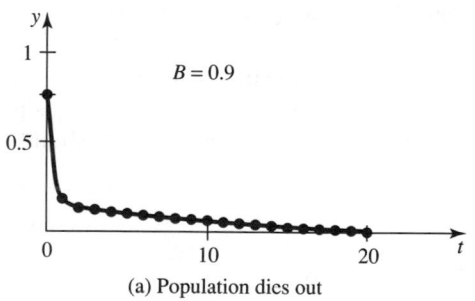

(a) Population dies out

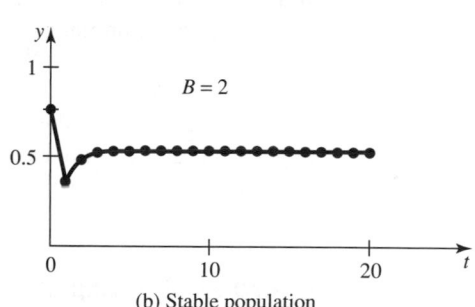

(b) Stable population

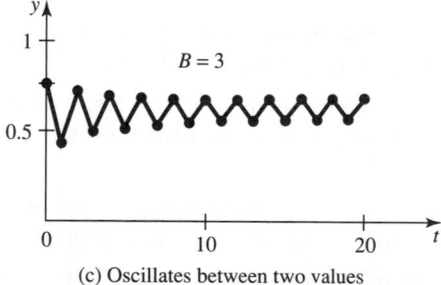

(c) Oscillates between two values

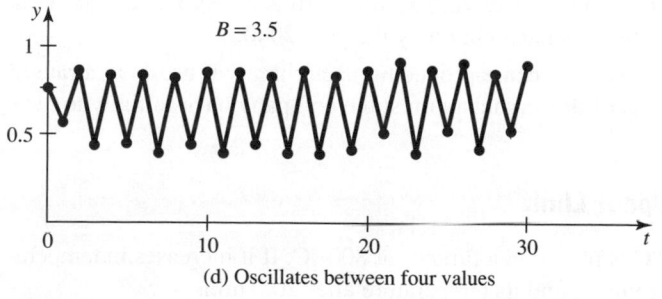

(d) Oscillates between four values

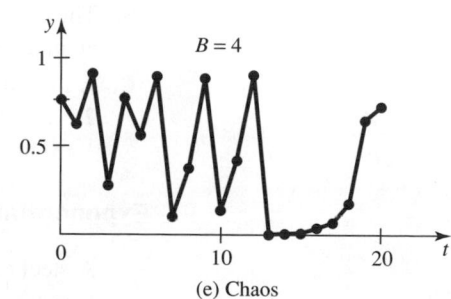

(e) Chaos

FIGURE 20–10

For a birthrate less than 1, the nonlinear growth equation predicts that the population eventually dies out [Fig. 20–10(a)]. For a birthrate between 1 and 3, the population y will stabilize at some value between 0 and 0.667. But for a birthrate of 3 or greater, strange things start to happen.

◆◆◆ **Example 15:** Repeat the computation of Example 14 with a birthrate of 3.

Solution: Again using a starting value of 0.8 and $B = 3$, we get the following table:

t	y	t	y
0	0.8	.	.
1	0.48	.	.
2	0.75	.	.
3	0.56	17	0.62
4	0.74	18	0.71
5	0.58	19	0.62
		20	0.71

We see that the value y *oscillates* between two different values [Fig. 20–10(c)]. We have what is called *period doubling* or *bifurcation*. ◆◆◆

Chaos

For B between 3.4495 and 3.56, the population has been found to oscillate between *four* different values. Figure 20–10(d), for $B = 3.5$, shows oscillation between values of 0.38, 0.5, 0.83, and 0.87. At higher values of B, the population oscillates between 8 values, then 16. Above $B = 3.57$, the population jumps around in a way that is not predictable at all.

◆◆◆ **Example 16:** Figure 20–10(e) shows the first 20 values of y using $B = 4$ and an initial y of 0.8. Note that the values do not repeat. The same simple equation that gave us predictable populations at some values of B now gives us *chaos* at other values of B, just slightly different from before. ◆◆◆

Exercise 2 ◆ Exponential Growth and Decay

Exponential Growth

Use either the formulas or the universal growth and decay curves, as directed by your instructor.

1. A quantity grows exponentially at the rate of 5.00% per year for 7 years. Find the final amount if the initial amount is $20\overline{0}$ units.

2. A yeast manufacturer finds that the yeast will grow exponentially at a rate of 15.0% per hour. How many kilograms of yeast must the manufacturer start with to obtain $50\overline{0}$ kg at the end of 8 h?

3. If the population of a country was 11.4 million in 1995 and grows at an annual rate of 1.63%, find the population by the year 2000.

4. If the rate of inflation is 12.0% per year, how much could a camera that now costs $225 be expected to cost 5 years from now?

5. The oil consumption of a certain country was 12.0 million barrels in 1995, and it grows at an annual rate of 8.30%. Find the oil consumption by the year 2000.

6. A pharmaceutical company makes a vaccine-producing bacterium that grows at a rate of 2.5% per hour. How many units of this organism must the company have initially to have $100\overline{0}$ units after 5 days?

Exponential Growth to an Upper Limit

7. A steel casting initially at 0 °C is placed in a furnace at $80\overline{0}$ °C. If it increases in temperature at the rate of 5.00% per minute, find its temperature after 20.0 min.

8. Equation 202 approximately describes the hardening of concrete, where y is the compressive strength in megapascals (MPa) t days after pouring. Using values of $a = 28.0$ and $n = 0.0696$ in the equation, find the compressive strength after 14 days.

9. When the switch in Fig. 20–11 is closed, the current i will grow exponentially according to the following equation:

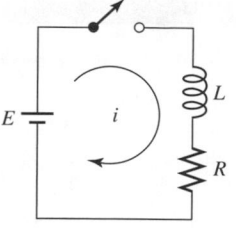

FIGURE 20–11

Inductor Current When Charging	$i = \dfrac{E}{R}(1 - e^{-Rt/L})$ (amperes)	A88

where L is the inductance in henries and R is the resistance in ohms. Find the current at 0.0750 s if $R = 6.25\ \Omega$, $L = 186$ H, and $E = 25\bar{0}$ V.

Exponential Decay

10. A flywheel is rotating at a speed of $18\bar{0}0$ rev/min. When the power is disconnected, the speed decreases exponentially at the rate of 32.0% per minute. Find the speed after 10.0 min.

11. A steel forging is $80\bar{0}$ °C above room temperature. If it cools exponentially at the rate of 2.00% per minute, how much will its temperature drop in 1 h?

12. As light passes through glass or water, its intensity decreases exponentially according to the equation

$$I = I_0 e^{-kx}$$

where I is the intensity at a depth x and I_0 is the intensity before entering the glass or water. If, for a certain filter glass, $k = 0.500$/cm (which means that each centimetre of filter thickness removes half the light reaching it), find the fraction of the original intensity that will pass through a filter glass 2.00 cm thick.

13. The atmospheric pressure p decreases exponentially with the height h (in kilometres) above the earth according to the function $p = 100.8 e^{-h/8}$ kPa. Find the pressure at a height of $984\bar{0}$ m.

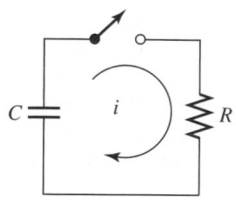

FIGURE 20–12

14. A certain radioactive material loses its radioactivity at the rate of $2\frac{1}{2}$% per year. What fraction of its initial radioactivity will remain after 10.0 years?

15. When a capacitor C, charged to a voltage E, is discharged through a resistor R (Fig. 20–12), the current i will decay exponentially according to the following equation:

Capacitor Current When Charging or Discharging	$i = \dfrac{E}{R} e^{-t/RC}$ (amperes)	A83

Find the current after 45 ms (0.045 s) in a circuit where $E = 220$ V, $C = 130\ \mu$F (130×10^{-6} F), and $R = 2700\ \Omega$.

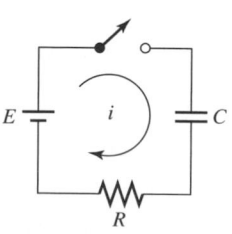

16. When a fully discharged capacitor C (Fig. 20–13) is connected across a battery, the current i flowing into the capacitor will decay exponentially according to Eq. A83. If $E = 115$ V, $R = 350\ \Omega$, and $C = 0.000\ 750$ F, find the current after 75.0 ms.

FIGURE 20–13

17. In a certain fabric mill, cloth is removed from a dye bath and is then observed to dry exponentially at the rate of 24% per hour. What percent of the original moisture will still be present after 5 h?

20–3   Logarithms

In Sec. 20–1, we studied the exponential function

$$y = b^x$$

Given b and x, we know how to find y. Thus if $b = 2.48$ and $x = 3.50$, then, by calculator,

$$y = 2.48^{3.50} = 24.0$$

Given y and x, we are also able to find b. For example, if y were 24.0 and x were 3.50,

$$24.0 = b^{3.50}$$
$$b = 24.0^{1/3.50} = 2.48$$

We will finish this problem in Sec. 20–5.

Suppose, now, that we have y and b and want to find x. Say that $y = 24.0$ when $b = 2.48$. Then

$$24.0 = 2.48^x$$

We have the same relationship between the same quantities as we had before but a different way of expressing it. We have a similar situation in ordinary language. Take "Mary is John's *sister*." If we had only the word *sister* (and not *brother*) to describe the relationship between children, we would have to say something like "John is the person whom Mary is the sister of." That is as awkward as the exponent to which the base must be raised to obtain the number." So we introduce new words, such as *brother* and *logarithm*.

Try it. You will find that none of the math we have learned so far will enable us to find x. To solve this and similar problems, we need *logarithms*, which are explained in this section.

Definition of a Logarithm

The logarithm of some positive number y *is the exponent* to which a base b must be raised to obtain y.

◆◆◆ **Example 17:** Since $10^2 = 100$, we say that "2 is the logarithm of 100, to the base 10" because 2 is the exponent to which the base 10 must be raised to obtain 100. The logarithm is written

$$\log_{10} 100 = 2$$

Notice that the base is written in small numerals below the word *log*. The statement is read "The log of 100 (to the base 10) is 2." This means that 2 is an exponent because *a logarithm is an exponent;* 100 is the result of raising the base to that power. Therefore these two expressions are equivalent:

$$10^2 = 100$$
$$\log_{10} 100 = 2 \qquad\qquad ◆◆◆$$

Given the exponential function $y = b^x$, we say that "x is the logarithm of y, to the base b" because x is the exponent to which the base b must be raised to obtain y.

So, in general, the following equation applies:

Definition of a Logarithm	If $\qquad y = b^x \qquad (y > 0, b > 0, b \neq 1)$ Then $\qquad x = \log_b y$ (x is the logarithm of y to the base b.)	**186**

In practice, the base b is taken as either 10 for common logarithms or e for natural logarithms. If "log" is written without a base, the base is assumed to be 10.

◆◆◆ Example 18:

$$\log 5 \equiv \log_{10} 5$$ ◆◆◆

Converting between Logarithmic and Exponential Forms

In working with logarithmic and exponential expressions, we often have to switch between exponential and logarithmic forms. When converting from one form to the other, remember that (1) the base in exponential form is also the base in logarithmic form and (2) "the *log* is the *exponent*."

◆◆◆ Example 19: Change the exponential expression $3^2 = 9$ to logarithmic form.

Solution: The base (3) in the exponential expression remains the base in the logarithmic expression.

$$3^2 \quad = 9$$
$$\log_3 (\ ) = (\ )$$

The log is the exponent.

$$3^2 \quad = 9$$
$$\log_3 (\ ) = 2$$

So our expression is

$$\log_3 9 = 2$$

> We could also write the same expression in *radical form* as $\sqrt{9} = 3$.

This is read "the log *of* 9 (to the base 3) *is* 2." ◆◆◆

◆◆◆ Example 20: Change $\log_e x = 3$ to exponential form.

Solution:

$$\log_e x = 3$$
$$e^3 \quad = x$$

> Common logarithms are also called *Briggs' logarithms*, after Henry Briggs (1561–1630).

◆◆◆

◆◆◆ Example 21: Change $10^3 = 1000$ to logarithmic form.

Solution:

$$\log_{10} 1000 = 3$$ ◆◆◆

Common Logarithms

Although we may write logarithms to any base, only two bases are in regular use: the base 10 and the base e. Logarithms to the base e, called *natural* logarithms, are discussed later. In this section we study logarithms to the base 10, called *common* logarithms.

> Remember that when a logarithm is written without a base (such as log x), we assume the base to be 10.

◆◆◆ Example 22: Verify the following common logarithms, found to five significant digits:

(a) $\log 74.82 = 1.8740$
(b) $\log 0.037\,55 = -1.4254$ ◆◆◆

> A logarithm such as 1.8740 can be expressed as the sum of an integer 1, called the *characteristic*, and a positive decimal 0.8740, called the *mantissa*. These terms were useful when tables were used to find logarithms.

◆◆◆ Example 23: Try to take the common logarithm of -2. You should get an error indication. We cannot take the logarithm of a negative number. ◆◆◆

Suppose that we have the logarithm of a number and want to find the number itself, say, $\log x = 1.852$. This process is known as *finding the antilogarithm (antilog)* of 1.852.

◆◆◆ **Example 24:** Verify the following antilogarithms to four significant digits:

(a) The antilog of 1.852 is 71.12.

(b) The antilog of -1.738 is 0.018 28.   ◆◆◆

◆◆◆ **Example 25:** Find x to four significant digits if log $x = 2.74$.

Solution: By calculator,

$$x = 10^{2.74} = 549.5$$ ◆◆◆

The Logarithmic Function

We wish to graph the logarithmic function $x = \log_b y$, but first let us follow the usual convention of calling the dependent variable y and the independent variable x. Our function then becomes the following:

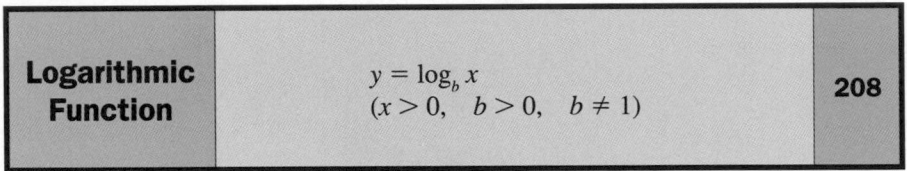

Logarithmic Function	$y = \log_b x$ $(x > 0, \quad b > 0, \quad b \neq 1)$	208

◆◆◆ **Example 26:** Graph the logarithmic function $y = \log x$ for $x = -1$ to 14.

Solution: We make a table of point pairs.

x	-1	0	1	2	3	4	5	6	7	8	9	10	11	12	13	14
y			0	0.30	0.48	0.60	0.70	0.78	0.85	0.90	0.95	1.00	1.04	1.08	1.11	1.15

First we notice that when x is negative or zero, our calculator gives an error indication. In other words, the domain of x is the positive numbers only. *We cannot take the common logarithm of a negative number or zero.* We'll see that this is true for logarithms to other bases as well.

If we try a value of x between 0 and 1, say, $x = 0.5$, we get a negative y (-0.3 in this case). Plotting our point pairs gives the curve in Fig. 20–14.

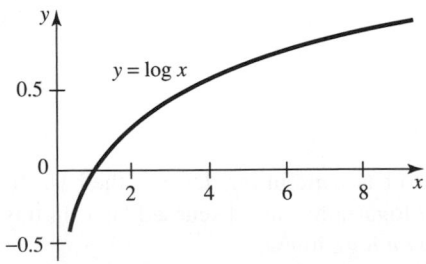

FIGURE 20–14   ◆◆◆

Natural Logarithms

Natural logarithms are also called *Napierian logarithms,* after John Napier (1550–1617), who first wrote logarithms.

You may want to refer to Sec. 20–2 for our earlier discussion of the number e (2.718 approximately). After we have defined e, we can use it as a base; logarithms using e as a base are called *natural* logarithms. They are written ln x. Thus ln x is understood to mean $\log_e x$.

◆◆◆ **Example 27:** Find ln 19.3 to four significant digits.

Solution: ln $19.3 = 2.960$ ◆◆◆

◆◆◆ **Example 28:** Verify the following to four significant digits:

(a) $\ln 25.4 = 3.235$

(b) $\ln 0.826 = -0.1912$

(c) $\ln(-3)$ gives an error indication. ◆◆◆

◆◆◆ **Example 29:** Find x to four digits if $\ln x = 3.825$.

Solution: Rewriting the expression $\ln x = 3.825$ in exponential form, we have

$$x = e^{3.825} = 45.83$$ ◆◆◆

Exercise 3 ◆ Logarithms

Converting between Logarithmic and Exponential Forms

Convert to logarithmic form.

1. $3^4 = 81$
2. $5^3 = 125$
3. $4^6 = 4096$
4. $7^3 = 343$
5. $x^5 = 995$
6. $a^3 = 6.83$

Convert to exponential form.

7. $\log_{10} 100 = 2$
8. $\log_2 16 = 4$
9. $\log_5 125 = 3$
10. $\log_4 1024 = 5$
11. $\log_3 x = 57$
12. $\log_x 54 = 285$
13. $\log_5 x = y$
14. $\log_w 3 = z$

Logarithmic Function

Graph each logarithmic function from $x = \frac{1}{2}$ to 4.

15. $y = \log_e x$
16. $y = \log_{10} x$

Common Logarithms

Find the common logarithm of each number to four decimal places.

17. 27.6
18. 4.83
19. 5.93
20. 9.26
21. 48.3
22. 385
23. 836
24. 1.03
25. 27.4
26. 0.0573
27. 0.007 37
28. 9738
29. 34 970
30. 5.28×10^6

Find x to three significant digits if the common logarithm of x is equal to the given value.

31. 1.584
32. 2.957
33. 5.273
34. 0.886
35. -2.857
36. 0.366
37. -2.227
38. 4.973
39. 3.979
40. 1.295
41. -3.972
42. -0.8835

Natural Logarithms

Find the natural logarithm of each number to four decimal places.

43. 48.3
44. 846
45. 2365
46. 1.005
47. 0.845
48. 4.97
49. 0.008 36
50. 45 900
51. 0.000 0462
52. 82 900
53. 3.84×10^4
54. 8.24×10^{-3}

Find to four significant digits the number whose natural logarithm is the given value.

55. 2.846   **56.** 4.263   **57.** 0.879

58. −2.846   **59.** −0.365   **60.** 5.937

61. 0.936   **62.** −4.97   **63.** 15.84

64. −9.47   **65.** −18.36   **66.** 21.83

20–4   Properties of Logarithms

In this section we use the properties of logarithms along with our ability to convert back and forth between exponential and logarithmic forms. These together will enable us to later solve equations that we could not previously handle, such as exponential equations.

Products

We wish to write an expression for the log of the product of two positive numbers, say, M and N.

$$\log_b MN = ?$$

Let us substitute: $M = b^c$ and $N = b^d$. Then

$$MN = b^c b^d = b^{c+d}$$

Here, $b > 0$ and $b \cdot 1$.

by the laws of exponents. Writing this expression in logarithmic form, we have

$$\log_b MN = c + d$$

But, since $b^c = M$, $c = \log_b M$. Similarly, $d = \log_b N$. Substituting, we have the following equation:

Log of a Product	$\log_b MN = \log_b M + \log_b N$	187

The log of a product equals the sum of the logs of the factors.

◆◆◆ **Example 30:**

(a) $\log 7x = \log 7 + \log x$

(b) $\log 3 + \log x + \log y = \log 3xy$ ◆◆◆

Common Error	The log of a sum is *not* equal to the sum of the logs.
	$\log_b(M + N) \neq \log_b M + \log_b N$
	The product of two logs is *not* equal to the sum of the logs.
	$(\log_b M)(\log_b N) \neq \log_b M + \log_b N$

We don't pretend that this is a practical problem, now that we have calculators. But we recommend doing a few logarithmic computations just to get practice in using the properties of logarithms, which we'll need later to solve logarithmic and exponential equations.

◆◆◆ **Example 31:** Multiply $(2.947)(9.362)$ by logarithms.

Solution: We take the log of the product

$$\log(2.947 \times 9.362) = \log 2.947 + \log 9.362$$
$$= 0.469\,38 + 0.971\,37 = 1.440\,75$$

This gives us the logarithm of the product. To get the product itself, we take the antilog.

$$(2.947)(9.362) = 10^{1.44075} = 27.59$$ ◆◆◆

Quotients

Let us now consider the quotient M divided by N. Using the same substitution as above, we obtain

$$\frac{M}{N} = \frac{b^c}{b^d} = b^{c-d}$$

Going to logarithmic form yields

$$\log_b \frac{M}{N} = c - d$$

Finally, substituting back gives us the following equation:

Log of a Quotient	$\log_b \dfrac{M}{N} = \log_b M - \log_b N$	188

The log of a quotient equals the log of the dividend minus the log of the divisor.

♦♦♦ **Example 32:**

$$\log \frac{3}{4} = \log 3 - \log 4 \qquad\qquad ♦♦♦$$

♦♦♦ **Example 33:** Express the following as the sum or difference of two or more logarithms:

$$\log \frac{ab}{xy}$$

Solution: By Eq. 188,

$$\log \frac{ab}{xy} = \log ab - \log xy$$

and by Eq. 187,

$$\log ab - \log xy = \log a + \log b - (\log x + \log y)$$
$$= \log a + \log b - \log x - \log y \qquad ♦♦♦$$

♦♦♦ **Example 34:** The expression $\log 10x^2 - \log 5x$ can be expressed as a single logarithm as follows:

$$\log \frac{10x^2}{5x} \qquad \text{or} \qquad \log 2x \qquad ♦♦♦$$

Common Error	Similar errors are made with quotients as with products. $\log_b (M - N) \neq \log_b M - \log_b N$ $\dfrac{\log_b M}{\log_b N} \neq \log_b M - \log_b N$

♦♦♦ **Example 35:** Divide using logarithms:

$$28.47 \div 1.883$$

Solution: By Eq. 188,

$$\log \frac{28.47}{1.883} = \log 28.47 - \log 1.883$$
$$= 1.4544 - 0.2749 = 1.1795$$

This gives us the logarithm of the quotient. To get the quotient itself, we take the antilog.

$$28.47 \div 1.883 = 10^{1.1795} = 15.12$$

◆◆◆

Powers

Consider now the quantity M raised to a power p. Substituting b^c for M, as before, we have

$$M^p = (b^c)^p = b^{cp}$$

In logarithmic form,

$$\log_b M^p = cp$$

Substituting back gives us the following equation:

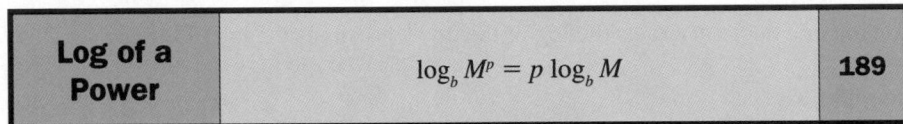

Log of a Power	$\log_b M^p = p \log_b M$	189

The log of a number raised to a power equals the power times the log of the number.

◆◆◆ **Example 36:**

$$\log 3.85^{1.4} = 1.4 \log 3.85$$

◆◆◆

◆◆◆ **Example 37:** The expression $7 \log x$ can be expressed as a single logarithm with a coefficient of 1 as follows:

$$\log x^7$$

◆◆◆

◆◆◆ **Example 38:** Express $2 \log 5 + 3 \log 2 - 4 \log 1$ as a single logarithm.

Solution:

$$2 \log 5 + 3 \log 2 - 4 \log 1 = \log 5^2 + \log 2^3 - \log 1^4$$

$$= \log \frac{25(8)}{1} = \log 200$$

◆◆◆

◆◆◆ **Example 39:** Express as a single logarithm with a coefficient of 1:

$$2 \log \frac{2}{3} + 4 \log \frac{a}{b} - 3 \log \frac{x}{y}$$

Solution:

$$2 \log \frac{2}{3} + 4 \log \frac{a}{b} - 3 \log \frac{x}{y} = \log \frac{2^2}{3^2} + \log \frac{a^4}{b^4} - \log \frac{x^3}{y^3}$$

$$= \log \frac{4a^4 y^3}{9b^4 x^3}$$

◆◆◆

◆◆◆ **Example 40:** Rewrite without logarithms the equation

$$\log(a^2 - b^2) + 2 \log a = \log(a - b) + 3 \log b$$

Solution: We first try to combine all of the logarithms into a single logarithm. Rearranging gives

$$\log(a^2 - b^2) - \log(a - b) + 2 \log a - 3 \log b = 0$$

By the properties of logarithms,

$$\log \frac{a^2 - b^2}{a - b} + \log \frac{a^2}{b^3} = 0$$

or

$$\log \frac{a^2(a - b)(a + b)}{b^3(a - b)} = 0$$

Simplifying yields

$$\log \frac{a^2(a + b)}{b^3} = 0$$

We then rewrite without logarithms by going from logarithmic to exponential form.

$$\frac{a^2(a + b)}{b^3} = 10^0 = 1$$

or

$$a^2(a + b) = b^3 \qquad \qquad \blacklozenge\blacklozenge\blacklozenge$$

◆◆◆ Example 41: Find $35.82^{1.4}$ by logarithms to four significant digits.

Solution: By Eq. 189,

$$\log 35.82^{1.4} = 1.4 \log 35.82$$

$$= 1.4(1.5541) = 2.1758$$

Taking the antilog yields

$$35.82^{1.4} = 10^{2.1758} = 149.9 \qquad \qquad \blacklozenge\blacklozenge\blacklozenge$$

Roots

We can write a given radical expression in exponential form and then use the rule for powers. Thus, by Eq. 36,

$$\sqrt[q]{M} = M^{1/q}$$

Taking the logarithm of both sides, we obtain

$$\log_b \sqrt[q]{M} = \log_b M^{1/q}$$

Then, by Eq. 189, the following equation applies:

Log of a Root	$\log_b \sqrt[q]{M} = \dfrac{1}{q}\log_b M$	190

The log of the root of a number equals the log of the number divided by the index of the root.

◆◆◆ Example 42:

(a) $\log \sqrt[5]{8} = \frac{1}{5}\log 8 \cong \frac{1}{5}(0.9031) \cong 0.1806$

(b) $\frac{\log x}{2} + \frac{\log z}{4} = \log \sqrt{x} + \log \sqrt[4]{z}$ $\qquad \qquad \blacklozenge\blacklozenge\blacklozenge$

> We may take the logarithm of both sides of an equation, just as we took the square root of both sides of an equation or took the sine of both sides of an equation.

◆◆◆ **Example 43:** Find $\sqrt[5]{3824}$ by logarithms.

Solution: By Eq. 190,

$$\log \sqrt[5]{3824} = \frac{1}{5} \log 3824 = \frac{3.5825}{5}$$
$$= 0.7165$$

Taking the antilog gives us

$$\sqrt[5]{3824} = 10^{0.7165} = 5.206$$ ◆◆◆

Log of 1

Let us take the log to the base b of 1 and call it x. Thus

$$x = \log_b 1$$

Switching to exponential form, we have

$$b^x = 1$$

This equation is satisfied, for any positive b, by $x = 0$. Therefore:

Log of 1	$\log_b 1 = 0$	191

The log of 1 is zero.

Log of the Base

We now take the $\log_b$ of its own base b. Let us call this quantity x.

$$\log_b b = x$$

Going to exponential form gives us

$$b^x = b$$

So x must equal 1.

Log of the Base	$\log_b b = 1$	192

The log (to the base b) of b is equal to 1.

◆◆◆ **Example 44:**

(a) $\log 1 = 0$ (b) $\ln 1 = 0$

(c) $\log_5 5 = 1$ (d) $\log 10 = 1$

(e) $\ln e = 1$ ◆◆◆

Log of the Base Raised to a Power

Consider the expression $\log_b b^n$. *We set this expression equal to x and, as before, go to expo-*
nential form.

$$\log_b b^n = x$$
$$b^x = b^n$$
$$x = n$$

Therefore:

Log of the Base Raised to a Power	$\log_b b^n = n$	193

The log (to the base b) of b raised to a power is equal to the power.

◆◆◆ **Example 45:**

(a) $\log_2 2^{4.83} = 4.83$

(b) $\log_e e^{2x} = 2x$

(c) $\log_{10} 0.0001 = \log_{10} 10^{-4} = -4$

To answer Example 45(a) on the Ti-89, press CATALOG, select "log" from the menu and press ENTER. Then type in the numbers. The display will look like this: log(2^.83, 2).

◆◆◆

Base Raised to a Logarithm of the Same Base

We want to evaluate an expression of the form

$$b^{\log_b x}$$

Setting this expression equal to y and taking the logarithms of both sides, gives us

$$y = b^{\log_b x}$$
$$\log_b y = \log_b b^{\log_b x} = \log_b x \log_b b$$

Since $\log_b b = 1$, we have $\log_b y = \log_b x$. Taking the antilog of both sides, we have

$$x = y = b^{\log_b x}$$

Thus:

Base Raised to a Logarithm of the Same Base	$b^{\log_b x} = x$	194

◆◆◆ **Example 46:**

(a) $10^{\log w} = w$ (b) $e^{\log_e 3x} = 3x$ ◆◆◆

Change of Base

We can convert between natural logarithms and common logarithms (or between logarithms to any base) by the following procedure. Suppose that $\log N$ is the common logarithm of some number N. We set it equal to x.

$$\log N = x$$

Going to exponential form, we obtain

$$10^x = N$$

Now we take the natural log of both sides.

$$\ln 10^x = \ln N$$

By Eq. 189,

$$x \ln 10 = \ln N$$
$$x = \frac{\ln N}{\ln 10}$$

But $x = \log N$, so we have the following equation:

Change of Base	$\log N = \dfrac{\ln N}{\ln 10}$ where $\ln 10 \cong 2.3026$	195

The common logarithm of a number is equal to the natural log of that number divided by the natural log of 10.

◆◆◆ **Example 47:** Find $\log N$ if $\ln N = 5.824$.

Solution: By Eq. 195,

$$\log N = \frac{5.824}{2.3026} = 2.529 \qquad \text{◆◆◆}$$

◆◆◆ **Example 48:** What is $\ln N$ if $\log N = 3.825$?

Solution: By Eq. 195,

$$\ln N = \ln 10(\log N)$$
$$= 2.3026(3.825) = 8.807 \qquad \text{◆◆◆}$$

Exercise 4 ◆ Properties of Logarithms

Write as the sum or difference of two or more logarithms.

1. $\log \dfrac{2}{3}$ **2.** $\log 4x$

3. $\log ab$ **4.** $\log \dfrac{x}{2}$

5. $\log xyz$ **6.** $\log 2ax$

7. $\log \dfrac{3x}{4}$ **8.** $\log \dfrac{5}{xy}$

9. $\log \dfrac{1}{2x}$ **10.** $\log \dfrac{2x}{3y}$

11. $\log \dfrac{abc}{d}$ **12.** $\log \dfrac{x}{2ab}$

Express as a single logarithm with a coefficient of 1. Assume that the logarithms in each problem have the same base.

13. $\log 3 + \log 4$ **14.** $\log 7 - \log 5$

15. $\log 2 + \log 3 - \log 4$ **16.** $3 \log 2$

17. $4 \log 2 + 3 \log 3 - 2 \log 4$ **18.** $\log x + \log y + \log z$

19. $3 \log a - 2 \log b + 4 \log c$ **20.** $\dfrac{\log x}{3} + \dfrac{\log y}{2}$

21. $\log \frac{x}{a} + 2 \log \frac{y}{b} + 3 \log \frac{z}{c}$

22. $2 \log x + \log(a + b) - \frac{1}{2} \log(a + bx)$

23. $\dfrac{1}{2} \log(x + 2) + \log(x - 2)$

Rewrite each equation so that it contains no logarithms.

24. $\log x + 3 \log y = 0$

25. $\log_2 x + 2 \log_2 y = x$

26. $2 \log(x - 1) = 5 \log(y + 2)$

27. $\log_2(a^2 - b^2) + 1 = 2 \log_2 a$

28. $\log(x + y) - \log 2 = \log x + \log(x^2 - y^2)$

29. $\log(p^2 - q^2) - \log(p + q) = 2$

30. $\log_2 2$ **31.** $\log_e e$ **32.** $\log_{10} 10$

33. $\log_3 3^2$ **34.** $\log_e e^x$ **35.** $\log_{10} 10^x$

36. $e^{\log_e x}$ **37.** $2^{\log_2 3y}$ **38.** $10^{\log x^2}$

Logarithmic Computation

Evaluate using logarithms.

39. 5.937×92.47

40. $6923 \times 0.003\ 846$

41. $3.97 \times 8.25 \times 9.82$

42. $88.25 \div 42.94$

43. $\sqrt[9]{8563}$

44. $4.836^{3.970}$

45. $83.62^{0.5720}$

46. $\sqrt[3]{587}$

47. $\sqrt[7]{8364}$

48. $\sqrt[4]{62.4}$

Change of Base

Find the common logarithm of the number whose natural logarithm is the given value.

49. 8.36

50. -3.846

51. 3.775

52. 15.36

53. 5.26

54. -0.638

Find the natural logarithm of the number whose common logarithm is the given value.

55. 84.9

56. 2.476

57. -3.82

58. 73.9

59. 2.37

60. -2.63

This may not seem like hot stuff in the age of calculators and computers, but the use of logarithms revolutionized computation in its time. The French mathematician Laplace wrote: "The method of logarithms, by reducing to a few days the labour of many months, doubles, as it were, the life of the astronomer, besides freeing him from the errors and disgust inseparable from long calculation."

20-5   Exponential Equations

Solving Exponential Equations

We return now to the problem we started in Sec. 20–3 but could not finish, that of finding the *exponent* when the other quantities in an exponential equation were known. We tried to solve the equation

$$24.0 = 2.48^x$$

The key to solving exponential equations is to *take the logarithm of both sides*. This enables us to use Eq. 189 to extract the unknown from the exponent. Taking the logarithm of both sides gives

$$\log 24.0 = \log 2.48^x$$

By Eq. 189,

$$\log 24.0 = x \log 2.48$$

$$x = \frac{\log 24.0}{\log 2.48} = \frac{1.380}{0.3945} = 3.50$$

In this example we took *common* logarithms, but *natural* logarithms would have worked just as well.

◆◆◆ **Example 49:** Solve for x to three significant digits:

$$3.25^{x+2} = 1.44^{3x-1}$$

Solution: We take the log of both sides. This time choosing to use natural logs, we get

$$(x + 2) \ln 3.25 = (3x - 1) \ln 1.44$$

$$\frac{\ln 3.25}{\ln 1.44}(x + 2) = 3x - 1$$

By calculator,

$$3.232(x + 2) = 3x - 1$$

Removing parentheses and collecting terms yields

$$0.232x = -7.464$$

$$x = -32.2$$

◆◆◆

Solving Exponential Equations with Base e

If an exponential equation contains the base e, taking *natural* logs rather than common logarithms will simplify the work.

◆◆◆ Example 50: Solve for x:

$$157 = 112e^{3x+2}$$

Solution: Dividing by 112, we have

$$1.402 = e^{3x+2}$$

Taking natural logs gives us

$$\ln 1.402 = \ln e^{3x+2}$$

By Eq. 189,

$$\ln 1.402 = (3x + 2) \ln e$$

But $\ln e = 1$, so

$$\ln 1.402 = 3x + 2$$

$$x = \frac{\ln 1.402 - 2}{3} = -0.554 \qquad\qquad ◆◆◆$$

◆◆◆ Example 51: Solve for x to three significant digits:

$$3e^x + 2e^{-x} = 4(e^x - e^{-x})$$

Solution: Removing parentheses gives

$$3e^x + 2e^{-x} = 4e^x - 4e^{-x}$$

Combining like terms yields

$$e^x - 6e^{-x} = 0$$

Factoring gives

$$e^{-x}(e^{2x} - 6) = 0$$

This equation will be satisfied if either of the two factors is equal to zero. So we set each factor equal to zero (just as when solving a quadratic by factoring) and get

$$e^{-x} = 0 \quad \text{and} \quad e^{2x} = 6$$

There is no value of x that will make e^{-x} equal to zero, so we get no root from that equation. We solve the other equation by taking the natural log of both sides.

$$2x \ln e = \ln 6 = 1.792$$
$$x = 0.896 \qquad\qquad ◆◆◆$$

Our exponential equation might be of *quadratic type,* in which case it can be solved by substitution as we did in Chapter 14.

◆◆◆ Example 52: Solve for x to three significant digits:

$$3e^{2x} - 2e^x - 5 = 0$$

Solution: If we substitute

$$w = e^x$$

our equation becomes the quadratic

$$3w^2 - 2w - 5 = 0$$

By the quadratic formula,

$$w = \frac{2 \pm \sqrt{4 - 4(3)(-5)}}{6} = -1 \quad \text{and} \quad \frac{5}{3}$$

Substituting back, we have

$$e^x = -1 \quad \text{and} \quad e^x = \frac{5}{3}$$

So $x = \ln(-1)$, which we discard, and

$$x = \ln\frac{5}{3} = 0.511 \qquad \qquad \blacklozenge\blacklozenge\blacklozenge$$

Solving Exponential Equations with Base 10

If an equation contains 10 as a base, the work will be simplified by taking *common* logarithms.

◆◆◆ **Example 53:** Solve $10^{5x} = 2(10^{2x})$ to three significant digits.

Solution: Taking common logs of both sides gives us

$$\log 10^{5x} = \log[2(10^{2x})]$$

By Eqs. 187 and 189,

$$5x \log 10 = \log 2 + 2x \log 10$$

By Eq. 192, $\log 10 = 1$, so

$$5x = \log 2 + 2x$$

$$3x = \log 2$$

$$x = \frac{\log 2}{3} = 0.100 \qquad \qquad \blacklozenge\blacklozenge\blacklozenge$$

Doubling Time

Being able to solve an exponential equation for the exponent allows us to return to the formulas for exponential growth and decay (Sec. 20–2) and to derive two interesting quantities: doubling time and half-life.

If a quantity grows exponentially according to the following function:

Exponential Growth	$y = ae^{nt}$	**199**

it will eventually double (y will be twice a). Setting $y = 2a$, we get

$$2a = ae^{nt}$$

or $2 = e^{nt}$. Taking natural logs, we have

$$\ln 2 = \ln e^{nt} = nt \ln e = nt$$

also given in the following form:

Doubling Time	$t = \dfrac{\ln 2}{n}$	**200**

Since $\ln 2 \cong 0.693$, if we let P be the rate of growth expressed as a percent ($P = 100n$), *not* as a decimal, we get a convenient *rule of thumb:*

$$\text{doubling time} \cong \frac{70}{P}$$

◆◆◆ **Example 54:** In how many years will a quantity double if it grows at the rate of 5.0% per year?

Solution:

$$\text{doubling time} = \frac{\ln 2}{0.050}$$

$$= \frac{0.693}{0.050}$$

$$= 14 \text{ years} \qquad \text{◆◆◆}$$

Half-Life

When a material decays exponentially according to the following function:

Exponential Decay	$y = ae^{-nt}$	**201**

the time it takes for the material to be half gone is called the *half-life*. If we let $y = a/2$,

$$\frac{1}{2} = e^{-nt} = \frac{1}{e^{nt}}$$

$$2 = e^{nt}$$

Notice that the equation for half-life is the same as for doubling time. The rule of thumb is, of course, also the same. The rule of thumb gives us another valuable tool for *estimation*.

Taking natural logarithms gives us

$$\ln 2 = \ln e^{nt} = nt$$

Half-Life	$t = \dfrac{\ln 2}{n}$	**200**

◆◆◆ **Example 55:** Find the half-life of a radioactive material that decays exponentially at the rate of 2.0% per year.

Solution:

$$\text{half-life} = \frac{\ln 2}{0.020} = 35 \text{ years} \qquad \text{◆◆◆}$$

Change of Base

If we have an exponential expression, such as 5^x, we can convert it to another exponential expression with base e or any other base.

◆◆◆ **Example 56:** Convert 5^x into an exponential expression with base e.

Solution: We write 5 as e raised to some power, say, n.

$$5 = e^n$$

Taking the natural logarithm, we obtain

$$\ln 5 = \ln e^n = n$$

So $n = \ln 5 \simeq 1.61$, and $5 = e^{1.61}$. Then

$$5^x = (e^{1.61})^x = e^{1.61x} \qquad \text{◆◆◆}$$

♦♦♦ **Example 57:** Convert 2^{4x} to an exponential expression with base e.

Solution: We first express 2 as e raised to some power.

$$2 = e^n$$

$$\ln 2 = \ln e^n = n$$

$$n = \ln 2 = 0.693$$

Then

$$2^{4x} = (e^{0.693})^{4x} = e^{2.773x} \qquad \text{♦♦♦}$$

♦♦♦ **Example 58:** A bacteria sample contains 2850 bacteria and grows at a rate of 5.85% per hour. Assuming that the bacteria grow exponentially:

(a) Write the equation for the number N of bacteria as a function of time.
(b) Find the time for the sample to grow to 10 000 bacteria.

Estimate: Note that the doubling time here is $70 \div 5.85$ or about 12 hours. To grow from 2850 to 10 000 will require nearly two doublings, or nearly 24 hours.

Solution:

(a) We substitute into the equation for exponential growth, $y = ae^{nt}$, with $y = N$, $a = 2850$, and $n = 0.0585$, getting

$$N = 2850e^{0.0585t}$$

(b) Setting $N = 10\,000$ and solving for t, we have

$$10\,000 = 2850e^{0.0585t}$$

$$e^{0.0585t} = 3.51$$

We take the natural log of both sides and set $\ln e = 1$.

$$0.0585t \ln e = \ln 3.51$$

$$t = \frac{\ln 3.51}{0.0585} = 21.5 \text{ h}$$

which agrees with our estimate. ♦♦♦

Exercise 5 ♦ Exponential Equations

Solve for x to three significant digits.

1. $2^x = 7$

2. $(7.26)^x = 86.8$

3. $(1.15)^{x+2} = 12.5$

4. $(2.75)^x = (0.725)^{x^2}$

5. $(15.4)^{\sqrt{x}} = 72.8$

6. $e^{5x} = 125$

7. $5.62e^{3x} = 188$

8. $1.05e^{4x+1} = 5.96$

9. $e^{2x-1} = 3e^{x+3}$

10. $14.8e^{3x^2} = 144$

11. $5^{2x} = 7^{3x-2}$

12. $3^{x^2} = 175^{x-1}$

13. $10^{3x} = 3(10^x)$

14. $e^x + e^{-x} = 2(e^x - e^{-x})$

15. $2^{3x+1} = 3^{2x+1}$

16. $5^{2x} = 3^{3x+1}$

17. $7e^{1.5x} = 2e^{2.4x}$

18. $e^{4x} - 2e^{2x} - 3 = 0$

19. $e^x + e^{-x} = 2$

20. $e^{6x} - e^{3x} - 2 = 0$

Hint: Problems 18 through 20 are *equations of quadratic type.*

Convert to an exponential expression with base e.

21. $3(4^{3x})$

22. $4(2.2^{4x})$

Applications

23. The current i in a certain circuit is given by

$$i = 6.25e^{-125t} \quad \text{(amperes)}$$

where t is the time in seconds. At what time will the current be 1.00 A?

24. The current through a charging capacitor is given by

$$i = \frac{E}{R}e^{-t/RC} \qquad \text{(A83)}$$

If $E = 325$ V, $R = 1.35\ \Omega$, and $C = 3210\ \mu\text{F}$, find the time at which the current through the capacitor is 0.0165 A.

25. The voltage across a charging capacitor is given by

$$v = E(1 - e^{-t/RC}) \qquad \text{(A84)}$$

If $E = 20.3$ V and $R = 4510\ \Omega$, find the time when the voltage across a 545-μF capacitor is equal to 10.1 V.

26. The temperature above its surroundings of an iron casting initially at $11\overline{0}0$ °C will be

$$T = 11\overline{0}0e^{-0.0620t}$$

after t minutes. Find the time for the casting to be at a temperature of 275 °C above its surroundings.

27. A certain long pendulum, released from a height y_0 above its rest position, will be at a height

$$y = y_0e^{-0.75t}$$

at t seconds. If the pendulum is released at a height of 15 cm, at what time will the height be 5.0 cm?

28. A population growing at a rate of 2.0% per year from an initial population of $9\overline{0}00$ will grow in t years to an amount

$$P = 9\overline{0}00e^{0.02t}$$

How many years will it take the population to triple?

29. The barometric pressure in kPa at a height of h metres above sea level is

$$p = 99e^{-kh}$$

where $k = 1.25 \times 10^{-4}$. At what height will the pressure be 33 kPa?

30. The approximate density of seawater at a depth of h km is

$$d = 1024e^{0.004\,23h} \quad \text{(kg/m}^3)$$

Find the depth at which the density will be 1032 kg/m³.

31. A rope passing over a rough cylindrical beam (Fig. 20–15) supports a weight W. The force F needed to hold the weight is

$$F = We^{-\mu\theta}$$

where μ is the coefficient of friction and θ is the angle of wrap in radians. If $\mu = 0.150$, what angle of wrap is needed for a force of $10\overline{0}$ lb. to hold a weight of $20\overline{0}$ lb.?

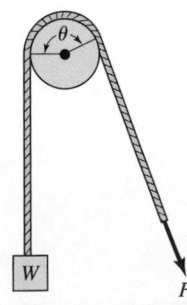

FIGURE 20–15

32. Using the formula for compound interest, Eq. A10, calculate the number of years it will take a sum of money to triple when invested at a rate of 12% per year.

33. Using the formula for present worth, given in problem 7 of Exercise 1, in how many years will $50,000 accumulate to $70,000 at 15% interest?

34. Using the annuity formula, given in problem 10 of Exercise 1, find the number of years it will take a worker to accumulate $100,000 with an annual payment of $3,000 if the interest rate is 8.0%.

35. Using the capital recovery formula from problem 12 of Exercise 1, calculate the number of years a person can withdraw $10,000/y from a retirement fund containing $60,000 if the rate of interest is $6\frac{3}{4}\%$.

36. Find the half-life of a material that decays exponentially at the rate of 3.50% per year.

37. How long will it take the U.S. annual oil consumption to double if it is increasing exponentially at a rate of 7.0% per year?

38. How long will it take the world population to double at an exponential growth rate of 1.64% per year?

39. What is the maximum annual growth in energy consumption permissible if the consumption is not to double in the next 20 years?

20–6 Logarithmic Equations

Often a logarithmic expression can be evaluated, or an equation containing logarithms can be solved, simply by rewriting it in exponential form.

◆◆◆ **Example 59:** Evaluate $x = \log_5 25$.

Solution: Changing to exponential form, we have

$$5^x = 25 = 5^2$$

$$x = 2$$

Since $5^2 = 25$, the answer checks. ◆◆◆

◆◆◆ **Example 60:** Solve for x:

$$\log_x 4 = \frac{1}{2}$$

Solution: Going to exponential form gives us

$$x^{1/2} = 4$$

Squaring both sides yields

$$x = 16$$

Since $16^{1/2} = 4$, the answer checks. ◆◆◆

◆◆◆ **Example 61:** Solve for x:

$$\log_{25} x = -\frac{3}{2}$$

Solution: Writing the equation in exponential form, we obtain

$$25^{-3/2} = x$$

$$x = \frac{1}{25^{3/2}} = \frac{1}{5^3} = \frac{1}{125}$$

Since $25^{-3/2} = \frac{1}{125}$, the answer checks. ◆◆◆

◆◆◆ **Example 62:** Evaluate $x = \log_{81} 27$.

Solution: Changing to exponential form yields

$$81^x = 27$$

But 81 and 27 are both powers of 3.

$$(3^4)^x = 3^3$$
$$3^{4x} = 3^3$$
$$4x = 3$$
$$x = 3/4$$

To check, we verify that $81^{3/4} = 3^3 = 27$. ◆◆◆

If only one term in our equation contains a logarithm, we isolate that term on one side of the equation and then go to exponential form.

◆◆◆ **Example 63:** Solve for x to three significant digits:

$$3 \log(x^2 + 2) - 6 = 0$$

Solution: Rearranging and dividing by 3 gives

$$\log(x^2 + 2) = 2$$

Going to exponential form, we obtain

$$x^2 + 2 = 10^2 = 100$$

$$x^2 = 98$$

$$x = \pm 9.90$$ ◆◆◆

If every term contains "log," we use the properties of logarithms to combine those terms into a single logarithm on each side of the equation and then take the antilog of both sides.

◆◆◆ **Example 64:** Solve for x to three significant digits:

$$3 \log x - 2 \log 2x = 2 \log 5$$

Solution: Using the properties of logarithms gives

$$\log \frac{x^3}{(2x)^2} = \log 5^2$$

$$\log \frac{x}{4} = \log 25$$

Taking the antilog of both sides, we have

$$\frac{x}{4} = 25$$

$$x = 100 \qquad\qquad \blacklozenge\blacklozenge\blacklozenge$$

If one or more terms do not contain a log, combine all of the terms that do contain a log on one side of the equation. Then go to exponential form.

◆◆◆ **Example 65:** Solve for x:

$$\log(5x + 2) - 1 = \log(2x - 1)$$

Solution: Rearranging yields

$$\log(5x + 2) - \log(2x - 1) = 1$$

By Eq. 188,

$$\log\frac{5x + 2}{2x - 1} = 1$$

Expressing in exponential form gives

$$\frac{5x + 2}{2x - 1} = 10^1 = 10$$

Solving for x, we have

$$5x + 2 = 20x - 10$$

$$12 = 15x$$

$$x = \frac{4}{5} \qquad\qquad \blacklozenge\blacklozenge\blacklozenge$$

◆◆◆ **Example 66:** *The Richter Scale.* The Richter magnitude R, used to rate the intensity of an earthquake, is given by

$$R = \log\frac{a}{a_0}$$

where a and a_0 are the vertical amplitudes of the ground movement of the measured earthquake and of an earthquake taken as reference. If two earthquakes measure 4 and 5 on the Richter scale, by what factor is the amplitude of the stronger quake greater than that of the weaker?

Solution: Let us rewrite the equation for the Richter magnitude using the laws of logarithms.

$$R = \log a - \log a_0$$

Then, for the first earthquake,

$$4 = \log a_1 - \log a_0$$

and for the second,

$$5 = \log a_2 - \log a_0$$

Subtracting the first equation from the second gives

$$1 = \log a_2 - \log a_1 = (\log a_2/a_1)$$

Going to exponential form, we have

$$\frac{a_2}{a_1} = 10^1 = 10$$

$$a_2 = 10\, a_1$$

So the second earthquake has 10 times the amplitude of the first.

Check: Does this seem reasonable? From our study of logarithms, we know that as a number increases tenfold, its common logarithm increases only by 1. For example, $\log 45 = 1.65$ and $\log 450 = 2.65$. Since the Richter magnitude is proportional to the common log of the amplitude, it seems reasonable that a tenfold increase in amplitude increases the Richter magnitude by just 1 unit.   ◆◆◆

Exercise 6 • Logarithmic Equations

Find the value of x in each expression.

1. $x = \log_3 9$
2. $x = \log_2 8$
3. $x = \log_8 2$
4. $x = \log_9 27$
5. $x = \log_{27} 9$
6. $x = \log_4 8$
7. $x = \log_8 4$
8. $x = \log_{27} 81$
9. $\log_x 8 = 3$
10. $\log_3 x = 4$
11. $\log_x 27 = 3$
12. $\log_x 16 = 4$
13. $\log_5 x = 2$
14. $\log_x 2 = \dfrac{1}{2}$
15. $\log_{36} x = \frac{1}{2}$
16. $\log_2 x = 3$
17. $x = \log_{25} 125$
18. $x = \log_5 125$

Solve for x. Give any approximate results to three significant digits. Check your answers.

19. $\log(2x + 5) = 2$
20. $2 \log(x + 1) = 3$
21. $\log(2x + x^2) = 2$
22. $\ln x - 2 \ln x = \ln 64$
23. $\ln 6 + \ln(x - 2) = \ln(3x - 2)$
24. $\ln(x + 2) - \ln 36 = \ln x$
25. $\ln(5x + 2) - \ln(x + 6) = \ln 4$
26. $\log x + \log 4x = 2$
27. $\ln x + \ln(x + 2) = 1$
28. $\log 8x^2 - \log 4x = 2.54$
29. $2 \log x - \log(1 - x) = 1$
30. $3 \log x - 1 = 3 \log(x - 1)$
31. $\log(x^2 - 4) - 1 = \log(x + 2)$
32. $2 \log x - 1 = \log(20 - 2x)$
33. $\log(x^2 - 1) - 2 = \log(x + 1)$
34. $\ln 2x - \ln 4 + \ln(x - 2) = 1$
35. $\log(4x - 3) + \log 5 = 6$

Applications

The equation in problem 36 is derived from the compound interest formula. The equations in problems 37 and 38 are obtained from the equations for an annuity and for capital recovery from Exercise 1. Can you see how they were derived?

36. An amount of money a invested at a compound interest rate of n per year will take t years to accumulate to an amount y, where t is

$$t = \frac{\log y - \log a}{\log(1 + n)} \quad \text{(years)}$$

How many years will it take an investment to triple in value when deposited at 8.00% per year?

37. If an amount R is deposited once every year at a compound interest rate n, the number of years it will take to accumulate to an amount y is

$$t = \frac{\log\left(\dfrac{ny}{R} + 1\right)}{\log(1 + n)} \quad \text{(years)}$$

How many years will it take an annual payment of $1500 to accumulate to $13,800 at 9.0% per year?

38. If an amount a is invested at a compound interest rate n, it will be possible to withdraw a sum R at the end of every year for t years until the deposit is exhausted. The number of years is given by

$$t = \frac{\log\left(\dfrac{an}{R - an} + 1\right)}{\log(1 + n)} \quad \text{(years)}$$

If $200,000 is invested at 12% interest, for how many years can an annual withdrawal of $30,000 be made before the money is used up?

39. The *magnitude M* of a star of intensity I is

$$M = 2.5 \log\frac{I_1}{I} + 1$$

where I_1 is the intensity of a first-magnitude star. What is the magnitude of a star whose intensity is one-tenth that of a first-magnitude star?

40. The difference in elevation h (m) between two locations having barometer readings of B_1 and B_2, in kilopascals, is given by the logarithmic equation $h = 18\,430 \log(B_2/B_1)$, where B_1 is the pressure at the *upper* station. Find the difference in elevation between two stations having barometer readings of 96.16 kPa at the lower station and 86.53 kPa at the upper.

41. What will the barometer reading be 248.0 m above a station having a reading of 93.13 kPa?

Use the following information for problems 42 through 45. If the power input to a network or device is P_1 and the power output is P_2, the number of decibels gained or lost in the device is given by the following logarithmic equation:

Decibels Gained or Lost	$G = 10 \log_{10}\dfrac{P_2}{P_1}$ (dB)	**A104**

42. A certain amplifier gives a power output of $1\overline{0}00$ W for an input of $5\overline{0}$ W. Find the dB gain.

43. A transmission line has a loss of 3.25 dB. Find the power transmitted for an input of 2750 kW.

44. What power input is needed to produce a 250-W output with an amplifier having a $5\overline{0}$-dB gain?

45. The output of a certain device is half the input. How many decibels does this loss represent?

46. The heat loss q per metre of cylindrical pipe insulation (Fig. 20–16) having an inside radius r_1 and outside radius r_2 is given by the logarithmic equation

$$q = \frac{2\pi k(t_1 - t_2)}{\ln(r_2/r_1)} \quad \text{(W)}$$

where t_1 and t_2 are the inside and outside temperatures (°C) and k is the conductivity of the insulation. Find q for a 10-cm-thick insulation having a conductivity of 0.065 and wrapped around a 22-cm-diameter pipe at 290 °C if the surroundings are at 32 °C.

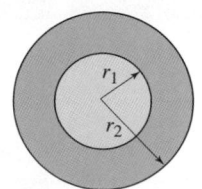

FIGURE 20–16 An insulated pipe.

47. If a resource is being used up at a rate that increases exponentially, the time it takes to exhaust the resource (called the *exponential expiration time,* EET) is

$$\text{EET} = \frac{1}{n}\ln\left(\frac{nR}{r} + 1\right)$$

where n is the rate of increase in consumption, R the total amount of the resource, and r the initial rate of consumption. If we assume that Canada has oil reserves of 185×10^9 barrels and that our present rate of consumption is 1.20×10^9 barrels/y, how long will it take to exhaust these reserves if our consumption increases by 7.00% per year?

Use the following information for problems 48 through 50. The pH value of a solution having a concentration C of hydrogen ions is given by the following equation:

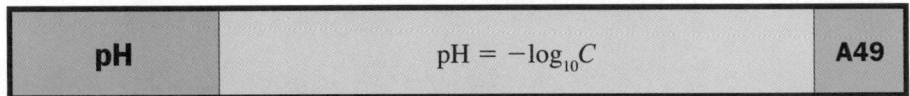

pH	$\text{pH} = -\log_{10} C$	A49

48. Find the concentration of hydrogen ions in a solution having a pH of 4.65.

49. A pH of 7 is considered *neutral,* while a lower pH is *acidic* and a higher pH is *alkaline.* What is the hydrogen ion concentration at a pH of 7.0?

50. The acid rain during a certain storm had a pH of 3.5. Find the hydrogen ion concentration. How does it compare with that for a pH of 7.0? Show that:
(a) when the pH doubles, the hydrogen ion concentration is *squared.*
(b) when the pH increases by a factor of n, the hydrogen ion concentration is raised *to the nth power.*

51. Two earthquakes have Richter magnitudes of 4.6 and 6.2. By what factor is the amplitude of the stronger quake greater than that of the weaker? See Example 66.

52. The bel is a unit named after Alexander Graham Bell that describes an increase in power (gain) or a loss in power (attenuation). It's a relative measurement, which means it describes the gain or attenuation with respect to a reference power level. Because the unit is quite large for telecommunication or sound level applications, the more practical decibel (dB) is commonly used. To calculate the gain or loss between P_1, the input (or reference) power level, and P_2, the output power level, we use the formula

$$G = 10 \log (P_2/P_1) \quad \text{(in dB)}$$

(a) An amplifier in an MP3 player receives a 12.0 μW (12.0×10^{-6} W) signal and amplifies it to 5.00 mW (5.00×10^{-3} W) for the ear buds. What is the power gain to the amplifier in decibels?
(b) If a telephone line causes a signal loss of 3 dB (which is treated as -3dB), what is the power loss as a percentage?

20–7   Graphs on Logarithmic and Semilogarithmic Paper

Logarithmic and Semilogarithmic Paper

Our graphing so far has all been done on ordinary graph paper, on which the lines are equally spaced. For some purposes, though, it is better to use *logarithmic* paper (Fig. 20–17), also called *log-log* paper, or *semilogarithmic* paper (Fig. 20–18), also called *semilog* paper. Looking at the logarithmic scales of these graphs, we note the following:

1. The lines are not equally spaced. The distance in inches from, say, 1 to 2, is equal to the distance from 2 to 4, which, in turn, is equal to the distance from 4 to 8.

2. Each tenfold increase in the scale, say, from 1 to 10 or from 10 to 100, is called a *cycle.* Each cycle requires the same distance in inches along the scale.

3. The log scales do not include zero.

Looking at Fig. 20–18, notice that although the numbers on the vertical scale are in equal increments (1, 2, 3, . . . , 10), the spacing on that scale is proportional to the logarithms of those numbers. So the numeral "2" is placed at a position corresponding to log 2 (which is 0.3010, or about one-third of the distance along the vertical); "4" is placed at log 4 (about 0.6 of the way); and "10" is at log 10 (which equals 1, at the top of the scale).

When to Use Logarithmic or Semilog Paper
We use these special papers when:

1. The range of the variables is too large for ordinary paper.
2. We want to graph a power function or an exponential function. Each of these will plot as a straight line on the appropriate paper, as shown in Fig. 20–19.
3. We want to find an equation that will approximately represent a set of empirical data.

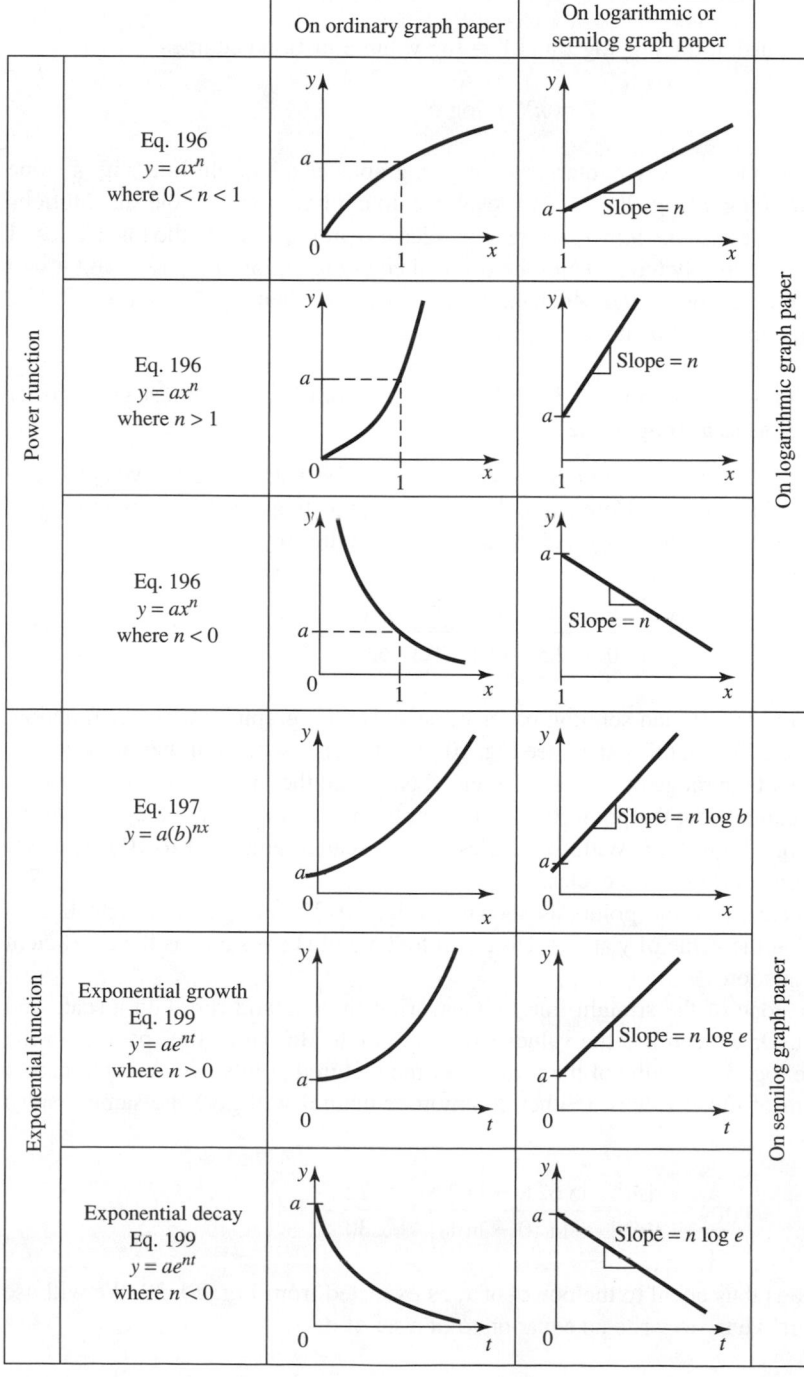

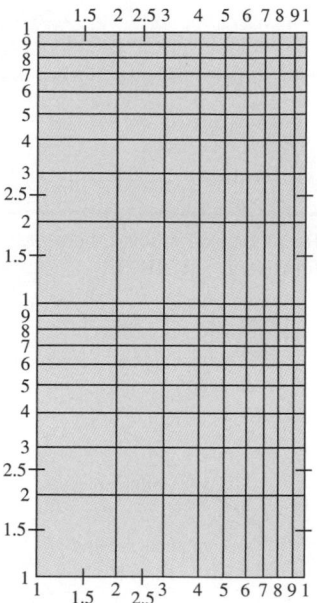

FIGURE 20–17
Logarithmic graph paper.

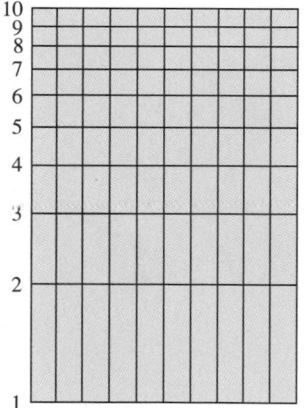

FIGURE 20–18
Semilogarithmic graph paper

FIGURE 20–19 The power function and the exponential function, graphed on ordinary paper and log-log or semilog paper.

Graphing the Power Function

A *power function* is one whose equation is of the following form:

Power Function	$y = ax^n$	196

Recall that we studied the power function in Sec. 19–3.

where a and n are nonzero constants. This equation is nonlinear (except when $n = 1$), and the shape of its graph depends upon whether n is positive or negative and whether n is greater than or less than 1. Figure 20–19 shows the shapes that this curve can have for various ranges of n. If we take the logarithm of both sides of Eq. 196, we get

$$\log y = \log(ax^n)$$
$$= \log a + n \log x$$

If we now make the substitution $X = \log x$ and $Y = \log y$, our equation becomes

$$Y = nX + \log a$$

This equation *is* linear and, on rectangular graph paper, graphs as a straight line with a slope of n and a y intercept of $\log a$ (Fig. 20–20). However, we do not have to make the substitutions shown above *if we use logarithmic paper,* where the scales are proportional to the logarithms of the variables x and y. We simply have to plot the original equation on log-log paper, and it will be a straight line which has slope n and which has a value of $y = a$ when $x = 1$ (see Fig. 20–19). This will be illustrated in the following example.

FIGURE 20–20 Graph of $y = ax^n$ on rectangular graph paper, where $X = \log x$ and $Y = \log y$.

◆◆◆ **Example 67:** Plot the equation $y = 2.5x^{1.4}$ for values of x from 1 to 10. Choose graph paper so that the equation plots as a straight line.

Solution: We make a table of point pairs. Since the graph will be a straight line, we need only two points, with a third as a check. Here, we will plot four points to show that all points do lie on a straight line. We choose values of x and for each compute the value of y.

x	0	1	4	7	10
y	0	2.5	17.4	38.1	62.8

We choose log-log rather than semilog paper because we are graphing a power function, which plots as a straight line on this paper (see Fig. 20–19). We choose the number of cycles for each scale by looking at the range of values for x and y. Note that the logarithmic scales do not contain zero, so we cannot plot the point (0, 0). Thus on the x axis we need one cycle. On the y axis we must go from 2.5 to 62.8. With two cycles we can span a range of 1 to 100. Thus we need log-log paper, one cycle by two cycles.

We mark the scales, plot the points as shown in Fig. 20–21, and get a straight line as expected. We note that the value of y at $x = 1$ is equal to 2.5 and is the same as the coefficient of $x^{1.4}$ in the given equation.

We can get the slope of the straight line by measuring the rise and run with a scale and dividing rise by run. Or we can use the values from the graph. But since the spacing on the graph really tells the logarithm value of the position of the pictured points, we must remember to take the logarithm of those values. (Either common or natural will give the same result.) Thus

$$\text{slope} = \frac{\text{rise}}{\text{run}} = \frac{\ln 62.8 - \ln 2.5}{\ln 10 - \ln 1} = \frac{3.22}{2.30} = 1.40$$

The slope of the line is thus equal to the power of x, as expected from Fig. 20–20. We will use these ideas later when we try to write an equation to fit a set of data.   ◆◆◆

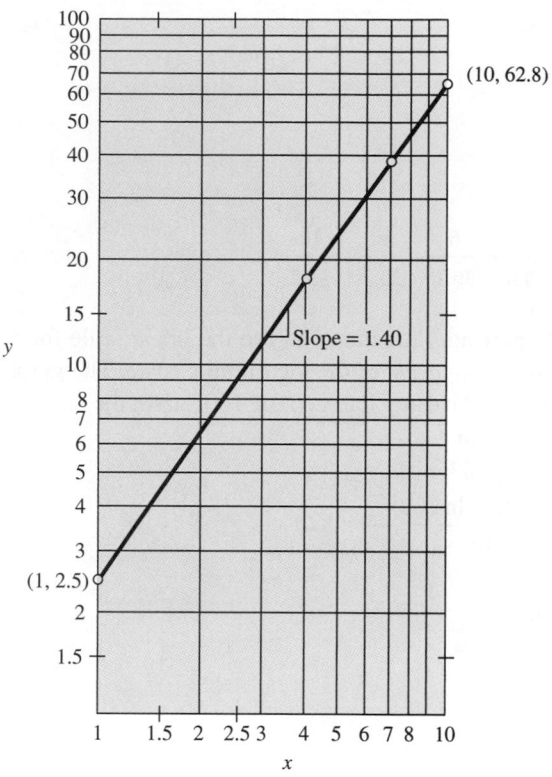

FIGURE 20–21 Graph of $y = 2.5x^{1.4}$. ◆◆◆

Common Error	Be sure to take the *logs* of the values on the x and y axes when computing the slope.

Graphing the Exponential Function

Consider the exponential function given by the following equation:

Exponential Function	$y = a(b)^{nx}$	**197**

If we take the logarithm of both sides, we get

$$\log y = \log b^{nx} + \log a$$
$$= (n \log b)x + \log a$$

If we replace $\log y$ with Y, we get the linear equation

$$Y = (n \log b)\, x + \log a$$

If we graph the given equation on semilog paper with the logarithmic scale along the y axis, we get a straight line with a slope of $n \log b$ which cuts the y axis at a (Fig. 20–19). Also shown are the special cases where the base b is equal to e (2.718 . . .). Here, the independent variable is shown as t, because exponential growth and decay are usually functions of time.

◆◆◆ Example 68: Plot the exponential function

$$y = 100e^{-0.2x}$$

for values of x from 0 to 10.

Solution: We make a table of point pairs.

x	0	2	4	6	8	10
y	100	67.0	44.9	30.1	20.2	13.5

We choose semilog paper for graphing the exponential function and use the linear scale for x. The range of y is from 13.5 to 100; thus we need *one cycle* of the logarithmic scale. The graph is shown in Fig. 20–22. Note that the line obtained has a y intercept of 100. Also, the slope is equal to $n \log e$ or, if we use natural logs, is equal to n.

> When computing the slope on semilog paper, we take the logarithms of the values on the log scale, but not on the linear scale. Computing the slope using common logs, we get
>
> slope $= n \log e$
>
> $= \dfrac{\log 13.5 - \log 100}{10}$
>
> $= -0.0870$
>
> $n = \dfrac{-0.0870}{\log e} = -0.2$
>
> as we got using natural logs.

$$\text{slope} = n = \frac{\ln 13.5 - \ln 100}{10} = -0.2$$

This is the coefficient of x in our given equation.

◆◆◆

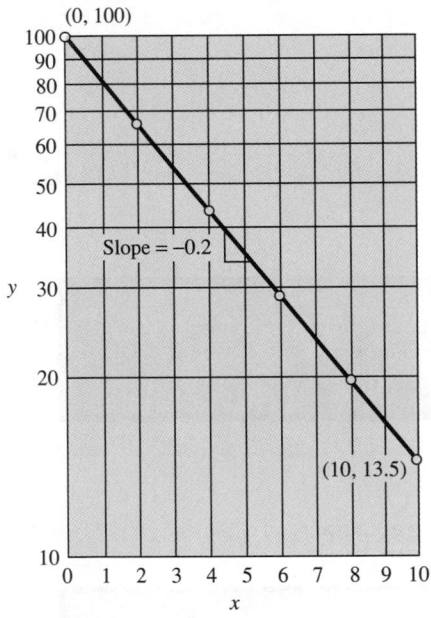

FIGURE 20–22   Graph of $y = 100e^{-0.2x}$.

◆◆◆

Empirical Functions

We choose logarithmic or semilog paper to plot a set of empirical data when:

1. The range of values is too large for ordinary paper.
2. We suspect that the relation between our variables may be a power function or an exponential function, and we want to find that function.

We show the second case by means of an example.

> The process of fitting an approximate equation to fit a set of data points is called *curve fitting*. In statistics it is referred to as *regression*. Here we will do only some very simple cases.

◆◆◆ Example 69: A test of a certain electronic device shows it to have an output current i versus input voltage v as shown in the following table:

v (V)	1	2	3	4	5
i (A)	5.61	15.4	27.7	42.1	58.1

Plot the given empirical data, and try to find an approximate formula for y in terms of x.

Solution: We first make a graph on linear graph paper (Fig. 20–23) and get a curve that is concave upward. Comparing its shape with the curves in Fig. 20–19, we suspect that the equation of the curve (if we can find one at all) may be either a power function or an exponential function.

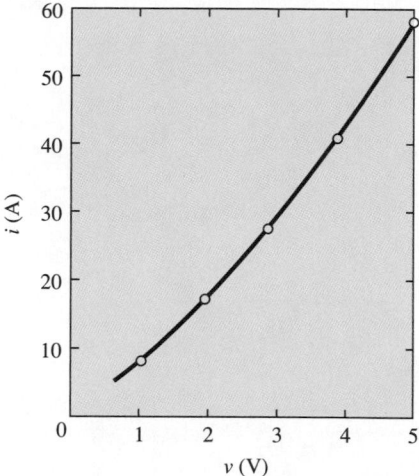

FIGURE 20–23 Plot of table of points on linear graph paper.

Next, we make a plot on semilog paper (Fig. 20–24) and do *not* get a straight line. However, a plot on log-log paper (Fig. 20–25) *is* linear. We thus assume that our equation has the form

$$i = av^n \quad \text{or} \quad \ln i = n \ln v + \ln a$$

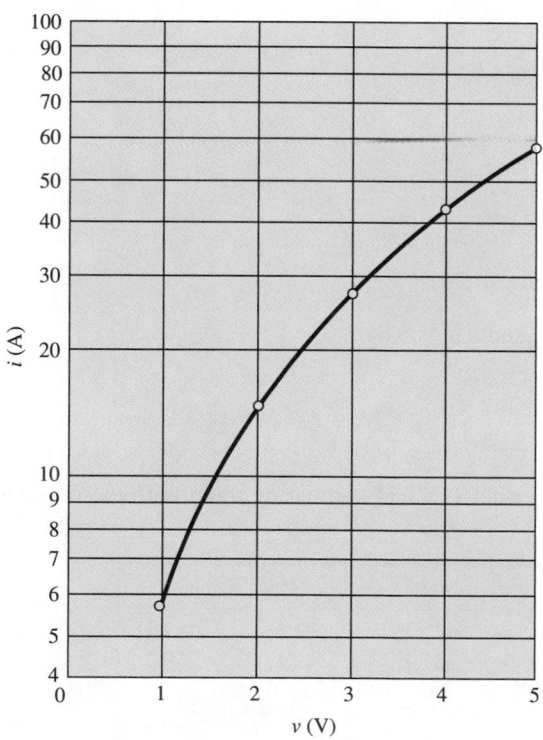

FIGURE 20–24 Plot of table of points on semilog paper.

In Example 67 we showed how to compute the slope of the line to get the exponent n, and we also saw that the coefficient a was the value of the function at $x = 1$. Now we show a different method for finding a and n which can be used even if we do not have the y value at $x = 1$. We choose two points on the curve, say, (5, 58.1) and (2, 15.4), and substitute each into

$$\ln i = n \ln v + \ln a$$

getting

$$\ln 58.1 = n \ln 5 + \ln a$$

and

$$\ln 15.4 = n \ln 2 + \ln a$$

Here our data plotted as a nice straight line on logarithmic paper. But with real data we are often unable to draw a straight line that passes through every point. The *method of least squares*, not shown here, is often used to draw a line that is considered the "best fit" for a scattering of data points.

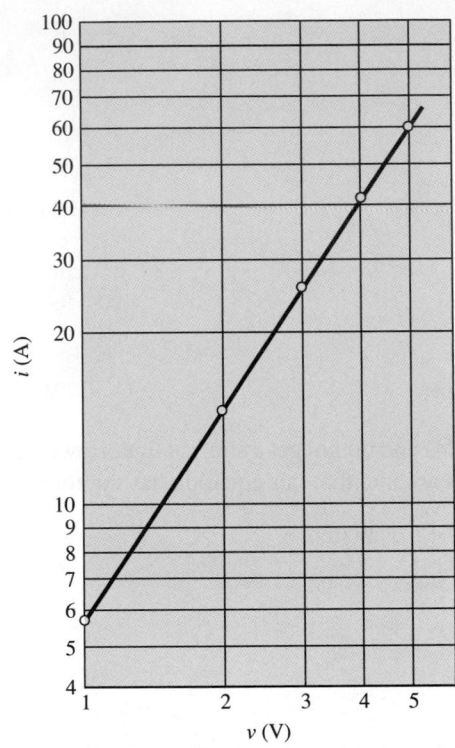

FIGURE 20–25   Plot of table of points on log-log paper.

Here, again, we could have used common logarithms and gotten the same result.

A simultaneous solution, not shown, for n and a yields

$$n = 1.45 \quad \text{and} \quad a = 5.64$$

Our equation is then

$$i = 5.64 v^{1.45}$$

Finally, we test this formula by computing values of i and comparing them with the original data, as shown in the following table:

The fitting of a power function to a set of data is called *power regression* in statistics.

v	1	2	3	4	5
Original i	5.61	15.4	27.7	42.1	58.1
Calculated i	5.64	15.4	27.7	42.1	58.2

We get values very close to the original.   ◆◆◆

Exercise 7 ◆ Graphs on Logarithmic and Semilogarithmic Paper

Graphing the Power Function

Graph each power function on log-log paper for $x = 1$ to 10.

1. $y = 2x^3$

2. $y = 3x^4$

3. $y = x^5$

4. $y = x^{3/2}$

5. $y = 5\sqrt{x}$

6. $y = 2\sqrt[3]{x}$

7. $y = 1/x$

8. $y = 3/x^2$

9. $y = 2x^{2/3}$

Graph each set of data on log-log paper, determine the coefficients graphically, and write an approximate equation to fit the given data.

10.

x	2	4	6	8	10
y	3.48	12.1	25.2	42.2	63.1

11.

x	2	4	6	8	10
y	240	31.3	9.26	3.91	2.00

Graphing the Exponential Function

Graph each exponential function on semilog paper.

12. $y = 3^x$

13. $y = 5^x$

14. $y = e^x$

15. $y = e^{-x}$

16. $y = 4^{x/2}$

17. $y = 3^{x/4}$

18. $y = 3e^{2x/3}$

19. $y = 5e^{-x}$

20. $y = e^{x/2}$

Graphing Empirical Functions

Graph each set of data on log-log or semilog paper, determine the coefficients graphically, and write an approximate equation to fit the given data.

21.

x	0	1	2	3	4	5
y	1.00	2.50	6.25	15.6	39.0	97.7

22.

x	1	2	3	4
y	48.5	29.4	17.6	10.8

23. Current in a tungsten lamp, i, for various voltages, v:

v (V)	2	8	25	50	100	150	200
i (mA)	24.6	56.9	113	172	261	330	387

24. Difference in temperature, T, between a cooling body and its surroundings at various times, t:

t (s)	0	3.51	10.9	19.0	29.0	39.0	54.0	71.0
T (°F)	20.0	19.0	17.0	15.0	13.0	11.0	8.50	6.90

25. Pressure, p, of 1 lb of saturated steam at various volumes, v:

v (ft.³)	26.4	19.1	14.0	10.5	8.00
p (lb./in.²)	14.7	20.8	28.8	39.3	52.5

26. Maximum height y reached by a long pendulum t seconds after being set in motion:

t (s)	0	1	2	3	4	5	6
y (cm)	10	4.97	2.47	1.22	0.61	0.30	0.14

Case Study Discussion

There is room for variations in the answer to this case. Loudness, what is too loud, is often a matter of opinion. Variables such as sound frequency or humidity can be a factor. Below is a table of some common minimum and maximum values quoted from different resources. (Note: although 90 dB is below the pain level, it is the level at which sustained exposure without hearing protection may result in hearing loss, and 140 dB is the maximum *with* hearing protection.)

Power Range $= 10^{(dB/10)}$

Minimum	Maximum	Range in Decibels	Power Ratio
6 dB: lowest audible sound	120 dB: threshold of pain	114 dB	$10^{11.4} = 2.5 \times 10^{11}$
0 dB: threshold of hearing	130 dB: threshold of pain	130 dB	10^{13}
0 dB: weakest sound heard	90 dB: level at which sustained exposure may result in hearing loss	90 dB	10^{9}
0 dB: weakest sound heard	140 dB: even short-term exposure can cause permanent damage—loudest recommended exposure *with* hearing protection	140 dB	10^{14}

◆◆◆ CHAPTER 20 REVIEW PROBLEMS ◆◆◆◆◆◆◆◆◆◆◆◆◆◆◆◆◆◆◆◆◆◆◆◆◆◆◆◆◆◆

Convert to logarithmic form.

1. $x^{5.2} = 352$

2. $4.8^x = 58$

3. $24^{1.4} = x$

Convert to exponential form.

4. $\log_3 56 = x$

5. $\log_x 5.2 = 124$

6. $\log_2 x = 48.2$

Solve for x.

7. $\log_{81} 27 = x$

8. $\log_3 x = 4$

9. $\log_x 32 = -\dfrac{5}{7}$

Write as the sum or difference of two or more logarithms.

10. $\log xy$

11. $\log \dfrac{3x}{z}$

12. $\log \dfrac{ab}{cd}$

Express as a single logarithm.

13. $\log 5 + \log 2$

14. $2 \log x - 3 \log y$

15. $\frac{1}{2} \log p - \frac{1}{4} \log q$

Find the common logarithm of each number to four decimal places.

16. 6.83

17. 364

18. 0.006 38

Find x if $\log x$ is equal to the given value.

19. 2.846

20. 1.884

21. -0.473

Evaluate using logarithms.

22. 5.836×88.24

23. $\dfrac{8375}{2846}$

24. $(4.823)^{1.5}$

Find the natural logarithm of each number to four decimal places.

25. 84.72

26. 2846

27. 0.00 873

Find the number whose natural logarithm is the given value.

28. 5.273

29. 1.473

30. -4.837

Find $\log x$ if

31. $\ln x = 4.837$

32. $\ln x = 8.352$

Find $\ln x$ if

33. $\log x = 5.837$

34. $\log x = 7.264$

Solve for x to three significant digits.

35. $(4.88)^x = 152$

36. $e^{2x+1} = 72.4$

37. $3 \log x - 3 = 2 \log x^2$

38. $\log(3x - 6) = 3$

39. A sum of $1,500 is deposited at a compound interest rate of $6\frac{1}{2}\%$ compounded quarterly. How much will it accumulate to in 5 years?

40. A quantity grows exponentially at a rate of 2.00% per day for 10.0 weeks. Find the final amount if the initial amount is $50\overline{0}$ units.

41. A flywheel decreases in speed exponentially at the rate of 5.00% per minute. Find the speed after 20.0 min if the initial speed is 2250 rev/min.

42. Find the half-life of a radioactive substance that decays exponentially at the rate of 3.50% per year.

43. Find the doubling time for a population growing at the rate of 3.0% per year.

44. Graph the power function $y = 3x^{2/3}$ on log-log paper for $x = 1$ to 6.

45. Graph the exponential function $y = e^{x/3}$ on semilog paper for $x = 0$ to 6.

Writing

46. In this chapter we have introduced two more kinds of equations: the *exponential* equation and the *logarithmic* equation. Earlier we have had five other types. List the seven types, give an example of each, and write one sentence for each telling how it differs from the others.

Team Project

47. The temperatures of a kettle of water, measured every minute during cooling from $1\overline{0}0$ °C at $t = 0$ to 23 °C, are as follows:

Time (min.)	Temperature (°C)
0	100
1	92.6
2	85.7
3	79.4
4	73.5
5	68.0
6	63.0
7	58.3
8	54.0
9	50.0
10	46.3
11	42.9
12	39.7
13	36.8
14	34.0
15	31.5
16	29.2
17	27.0
18	25.0
19	23.2

Write an equation for the kettle temperature as a function of time, with all constants evaluated.

21

Complex Numbers

OBJECTIVES

When you have completed this chapter, you should be able to:
*   Identify the characteristics of imaginary and complex numbers.
*   Graph complex numbers.
*   Express complex numbers in trigonometric and polar form.
*   Express complex numbers in exponential form.
*   Represent vectors using complex numbers.
*   Apply the characteristics of complex numbers and solve alternating-current problems.

Complex numbers—a poor name for this mathematical technique, as they are actually very straightforward. A complex number is used as a coordinate on a two-dimensional plane, almost exactly like an x, y coordinate. The difference is that the x axis is the *real* number line and the y axis is the *imaginary* number line. Complex numbers allow you to work much more easily with vectors.

Mathematicians developed this imaginary number line to solve the mathematical problem of finding the square root of a negative number. The square root of any negative number is not solvable by a real number because there are no two identical roots that when multiplied together will give you a negative number. However, if you multiply a positive number, say 3, by the square root of -1 *twice*, you get

$$3 \times \sqrt{-1} \times \sqrt{-1} = 3 \times -1 = -3$$

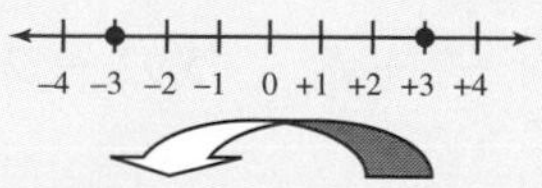

FIGURE 21–1

And if you multiply a negative number, say -2, by the square root of -1 twice, you get

$$-2 \times \sqrt{-1} \times \sqrt{-1} = -2 \times -1 = 2$$

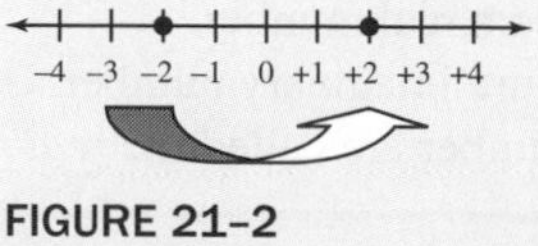

FIGURE 21–2

Even if we can't write an actual number value for $\sqrt{-1}$, we see that the effect of multiplying a number by the square root of -1 twice is like rotating 180° onto the opposite side of the number line. From $+3$ to -3 (Fig. 21–1) or from -2 to $+2$ (Fig. 21–2), it seems that we are not so much multiplying by a number as rotating about the zero of the number line. So, if multiplying twice

585

is a 180° rotation, multiplying *once* must be a 90° rotation. But a rotation to where? This is when mathematicians proposed an imaginary number line 90° to the real number line. In other words, a number line that acts just like the *y* axis if you think of the real number line as a horizontal axis.

So mathematicians developed a way of keeping track of which number was on the real number line and which one was on the imaginary number line. They then worked out the math behind a system of two axes. Well, almost immediately, the engineering world jumped on any math that would help with a two-axis system. For years, technology had been using *x* and *y* coordinate systems, horizontal and vertical directions. The tricky part, though, was keeping track of which number was which once you took one number out of the ordered pair. Without the pair, where the first number is always *x* and the second number always *y*, it was difficult to tell a reader of your equation which was which any more. The complex-number system puts the indicator right next to the number (in engineering technology we use a "*j*") and even sets up the rules that treat the *j*-operator like a variable. The complex-number system solves many problems in anything that uses a two-dimensional coordinate system, including much of our work with vectors.

21–1 Complex Numbers in Rectangular Form

The Imaginary Unit
Recall that in the real number system, the equation

$$x^2 = -1$$

had no solution because there was no real number such that its square was -1. Now we extend our number system to allow the quantity $\sqrt{-1}$ to have a meaning. We define the *imaginary unit* as the square root of -1 and represent it by the symbol *j*.

$$\text{Imaginary unit: } j = \sqrt{-1}$$

Complex Numbers
A *complex number* is any number, real or imaginary, that can be written in the form

$$a + jb$$

where *a* and *b* are real numbers, and $j = \sqrt{-1}$ is the *imaginary unit*.

> The letter *i* is often used for the imaginary unit. In technical work, however, we save *i* for electric current. Further, *j* (or *i*) is sometimes written before the *b*, and sometimes after. So the number *j*5 may also be written 5*i*, 5*j*, or *i*5.

♦♦♦ **Example 1:** The following numbers are complex numbers:

$$4 + j2 \qquad -7 + j8 \qquad 5.92 - j2.93 \qquad 83 \qquad j27$$ ♦♦♦

Real and Imaginary Numbers
When $b = 0$ in a complex number $a + jb$, we have a *real number*. When $a = 0$, the number is called a *pure imaginary number*.

♦♦♦ **Example 2:**
(a) The complex number 48 is also a real number.
(b) The complex number *j*9 is a pure imaginary number. ♦♦♦
   The two parts of a complex number are called the *real part* and the *imaginary part*.

> This is called the *rectangular form* of a complex number.

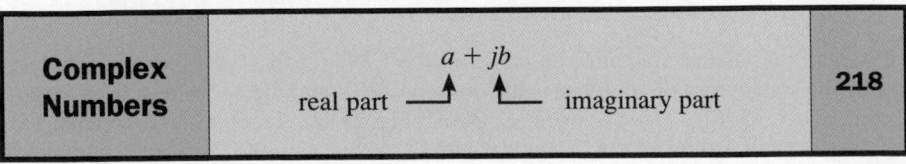

| **Complex Numbers** | $a + jb$
 real part —↑ ↑— imaginary part | **218** |

586

Addition and Subtraction of Complex Numbers

To combine complex numbers, separately combine the real parts, then combine the imaginary parts, and express the result in the form $a + jb$.

Addition of Complex Numbers	$(a + jb) + (c + jd) = (a + c) + j(b + d)$	219
Subtraction of Complex Numbers	$(a + jb) - (c + jd) = (a - c) + j(b - d)$	220

♦♦♦ **Example 3:** Add or subtract, as indicated.

(a) $j3 + j5 = j8$

(b) $j2 + (6 - j5) = 6 - j3$

(c) $(2 - j5) + (-4 + j3) = (2 - 4) + j(-5 + 3) = -2 - j2$

(d) $(-6 + j2) - (4 - j) = (-6 - 4) + j[2 - (-1)] = -10 + j3$ ♦♦♦

Powers of j

We often have to evaluate powers of the imaginary unit, especially the square of j. Since

$$j = \sqrt{-1}$$

then

$$j^2 = \sqrt{-1} \cdot \sqrt{-1} = -1$$

Higher powers are easily found.

$$j^3 = \quad j^2 j = (-1)j = -j$$
$$j^4 = (j^2)^2 = (-1)^2 = \quad 1$$
$$j^5 = \quad j^4 j = (1)j = \quad j$$
$$j^6 = j^4 j^2 = (1)j^2 = -1$$

We see that the values are starting to repeat: $j^5 = j$, $j^6 = -1$, and so on. The first four values keep repeating.

Note that when the exponent n is a multiple of 4, then $j^n = 1$.

Powers of j	$j = \sqrt{-1}$ $j^2 = -1$ $j^3 = -1$ $j^4 = 1$ $j^5 = j$ ...	217

♦♦♦ **Example 4:** Evaluate j^{17}.

Solution: Using the laws of exponents, we express j^{17} in terms of one of the first four powers of j.

$$j^{17} = j^{16} j = (j^4)^4 j = (1)^4 j = j$$ ♦♦♦

Multiplication of Imaginary Numbers

Multiply as with ordinary numbers, but use Eq. 217 to simplify any powers of j.

♦♦♦ **Example 5:**

(a) $5 \times j3 = j15$

(b) $j2 \times j4 = j^2 8 = (-1)8 = -8$

(c) $3 \times j4 \times j5 \times j = j^3 60 = (-j)60 = -j60$

(d) $(j3)^2 = j^2 3^2 = (-1)9 = -9$ ♦♦♦

Multiplication of Complex Numbers

Multiply complex numbers as you would any algebraic expressions: Replace j^2 by -1, and put the expression into the form $a + jb$.

Multiplication of Complex Numbers	$(a + jb)(c + jd) = (ac - bd) + j(ad + bc)$	221

◆◆◆ **Example 6:**

(a) $3(5 + j2) = 15 + j6$

(b) $(j3)(2 - j4) = j6 - j^2 12 = j6 - (-1)12 = 12 + j6$

(c) $(3 - j2)(-4 + j5) = 3(-4) + 3(j5) + (-j2)(-4) + (-j2)(j5)$
$$= -12 + j15 + j8 - j^2 10$$
$$= -12 + j15 + j8 - (-1)10$$
$$= -2 + j23$$

(d) $(3 - j5)^2 = (3 - j5)(3 - j5)$
$$= 9 - j15 - j15 + j^2 25$$
$$= 9 - j30 - 25$$
$$= -16 - j30$$

◆◆◆

To multiply radicals that contain negative quantities in the radicand, first express all quantities in terms of j, then proceed to multiply. Always be sure to convert radicals to imaginary numbers *before* performing other operations, or contradictions can result.

◆◆◆ **Example 7:** Multiply $\sqrt{-4}$ by $\sqrt{-4}$.

Solution: Converting to imaginary numbers, we obtain

We will see in the next section that multiplication and division are easier in *polar* form.

$$\sqrt{-4} \cdot \sqrt{-4} = (j2)(j2) = j^2 4$$

Since $j^2 = -1$,

$$j^2 4 = -4$$

◆◆◆

Common Error	It is *incorrect* to write $\sqrt{-4}\,\sqrt{-4} = \sqrt{(-4)(-4)} = \sqrt{16} = +4$ Our previous rule of $\sqrt{a} \cdot \sqrt{b} = \sqrt{ab}$ applied only to positive a and b.

CASE STUDY—SERIES RESONANT CIRCUIT

This case study is geared a little bit more toward electronics students; however, anyone can work with these equations. A series resonant circuit can be used to tune a circuit to a specific frequency. The impedance (Z) of the circuit is the vector sum of the resistance (R), capacitive reactance ($-jX_C$), and inductive reactance ($+jX_L$). The equation then is $Z = R + jX_L - jX_C$. The unit for all these variables is the ohm (Ω). Let's assume that the resistance is 2000 Ω, the capacitive reactance is 1500 Ω, and the inductive reactance is 1500 Ω. If frequency changes, the two reactance values will no longer be equal. Show how the impedance is a minimum only at the frequency where the two reactance values are equal. Also show how at resonance the circuit (or vector) phase angle is zero.

The Conjugate of a Complex Number

The *conjugate* of a complex number is obtained by changing the sign of its imaginary part.

◆◆◆ **Example 8:**
(a) The conjugate of $2 + j3$ is $2 - j3$.
(b) The conjugate of $-5 - j4$ is $-5 + j4$.
(c) The conjugate of $a + jb$ is $a - jb$.

Multiplying any complex number by its conjugate will eliminate the j term.

A computer algebra system can find the conjugate of a complex number. See problem 68 in ◆◆◆ Exercise 1.

◆◆◆ **Example 9:**

$$(2 + j3)(2 - j3) = 4 - j6 + j6 - j^2 9 = 4 - (-1)(9) = 13$$

◆◆◆ This equation has the same form as the difference of two squares

Division of Complex Numbers

Division involving single terms, real or imaginary, is shown by examples.

◆◆◆ **Example 10:**
(a) $j8 \div 2 = j4$
(b) $j6 \div j3 = 2$
(c) $(4 - j6) \div 2 = 2 - j3$ ◆◆◆

◆◆◆ **Example 11:** Divide 6 by $j3$.

Solution:

$$\frac{6}{j3} = \frac{6}{j3} \cdot \frac{j3}{j3} = \frac{j18}{j^2 9} = \frac{j18}{-9} = -j2$$ ◆◆◆

To divide by a complex number, multiply dividend and divisor by the conjugate of the divisor. This will make the divisor a real number, as in the following example.

◆◆◆ **Example 12:** Divide $3 - j4$ by $2 + j$.

Solution: We multiply numerator and denominator by the conjugate $(2 - j)$ of the denominator.

$$\frac{3 - j4}{2 + j} = \frac{3 - j4}{2 + j} \cdot \frac{2 - j}{2 - j}$$

$$= \frac{6 - j3 - j8 + j^2 4}{4 + j2 - j2 - j^2}$$

$$= \frac{2 - j11}{5}$$

$$= \frac{2}{5} - j\frac{11}{5}$$ ◆◆◆

This is very similar to *rationalizing the denominator* of a fraction containing radicals (Sec. 13–2).

In general, the following equation applies:

Division of Complex Numbers	$\dfrac{a + jb}{c + jd} = \dfrac{ac + bd}{c^2 + d^2} + j\dfrac{bc - ad}{c^2 + d^2}$	222

A CAS will give complex roots of a quadratic equation. See problem 69 in Exercise 1.

Quadratics with Complex Roots

When we solved quadratic equations in Chapter 14, the roots were real numbers. Some quadratics, however, will yield complex roots, as in the following example.

◆◆◆ **Example 13:** Solve for x to three significant digits:
$$2x^2 - 5x + 9 = 0$$

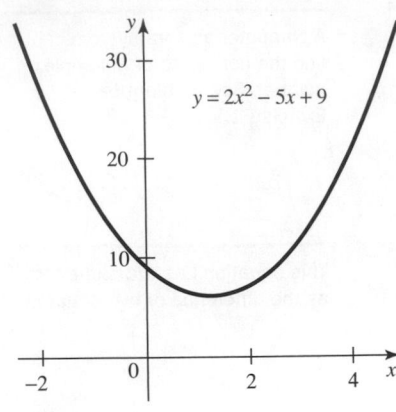

FIGURE 21–3

Solution: By the quadratic formula (Eq. 100),

$$x = \frac{5 \pm \sqrt{25 - 4\,(2)\,(9)}}{4} = \frac{5 \pm \sqrt{-47}}{4}$$

$$= \frac{5 \pm j6.86}{4} = 1.25 \pm j1.72$$

A graph of the function $y = 2x^2 - 5x + 9$ (Fig. 21–3) shows that the curve does not cross the x axis and hence has no real zeros.   ◆◆◆

Complex Factors

It is sometimes necessary to factor an expression into *complex factors*, as shown in the following example.

◆◆◆ **Example 14:** Factor the expression $a^2 + 4$ into complex factors.

Solution:

$$a^2 + 4 = a^2 - j^2 4 = a^2 - (j2)^2$$
$$= (a - j2)(a + j2) \qquad\qquad ◆◆◆$$

Exercise 1 ◆ Complex Numbers in Rectangular Form

Write as imaginary numbers.

1. $\sqrt{-9}$ **2.** $\sqrt{-81}$ **3.** $\sqrt{-\dfrac{4}{16}}$ **4.** $\sqrt{-\dfrac{1}{5}}$

Write as a complex number in rectangular form.

5. $4 + \sqrt{-4}$ **6.** $\sqrt{-25} + 3$

7. $\sqrt{-\dfrac{9}{4}} - 5$ **8.** $\sqrt{-\dfrac{1}{3}} + 2$

Combine and simplify.

9. $\sqrt{-9} + \sqrt{-4}$ **10.** $\sqrt{-4a^2} - a\sqrt{-25}$

11. $(3 - j2) + (-4 + j3)$ **12.** $(-1 - j2) - (j + 4)$

13. $(a - j3) + (a + j5)$ **14.** $(p + jq) + (q + jp)$

15. $\left(\dfrac{1}{2} + \dfrac{j}{3}\right) + \left(\dfrac{1}{4} - \dfrac{j}{6}\right)$ **16.** $(-84 + j91) - (28 + j72)$

17. $(2.28 - j1.46) + (1.75 + j2.66)$

Evaluate each power of j.

18. j^{11} **19.** j^5

20. j^{10} **21.** j^{21}

22. j^{14}

Multiply and simplify.

23. $7 \times j2$ **24.** $9 \times j3$

25. $j3 \times j5$ **26.** $j \times j4$

27. $4 \times j2 \times j3 \times j4$ **28.** $j \times 5 \times j4 \times j3 \times 5$

29. $(j5)^2$ **30.** $(j3)^2$

31. $2(3 - j4)$ **32.** $-3(7 + j5)$

33. $j4(5 - j2)$ **34.** $j5(2 + j3)$

35. $(3 - j5)(2 + j6)$

36. $(5 + j4)(4 + j2)$

37. $(6 + j3)(3 - j8)$

38. $(5 - j3)(8 + j2)$

39. $(5 - j2)^2$

40. $(3 + j6)^2$

Write the conjugate of each complex number.

41. $2 - j3$

42. $-5 - j7$

43. $p + jq$

44. $-j5 + 6$

45. $-jm + n$

46. $5 - j8$

Divide and simplify.

47. $j8 \div 4$

48. $9 \div j3$

49. $j12 \div j6$

50. $j44 \div j2$

51. $(4 + j2) \div 2$

52. $8 \div (4 - j)$

53. $(-2 + j3) \div (1 - j)$

54. $(5 - j6) \div (-3 + j2)$

55. $(j7 + 2) \div (j3 - 5)$

56. $(-9 + j3) \div (2 - j4)$

Quadratics with Complex Roots

Solve for x to three significant digits.

57. $3x^2 - 5x + 7 = 0$

58. $2x^2 + 3x + 5 = 0$

59. $x^2 - 2x + 6 = 0$

60. $4x^2 + x + 8 = 0$

Complex Factors

Factor each expression into complex factors.

61. $x^2 + 9$

62. $b^2 + 25$

63. $4y^2 + z^2$

64. $25a^2 + 9b^2$)

21-2   Graphing Complex Numbers

The Complex Plane

A complex number in rectangular form is made up of two parts: a real part a and an imaginary part jb. To graph a complex number, we can use a rectangular coordinate system (Fig. 21–4) in which the horizontal axis is the *real* axis and the vertical axis is the *imaginary* axis. Such a coordinate system defines what is called the *complex plane*. Real numbers are graphed as points on the horizontal axis; pure imaginary numbers are graphed as points on the vertical axis. Complex numbers, such as $3 + j5$ (Fig. 21–4), are graphed elsewhere within the plane.

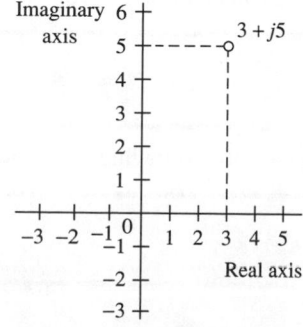

FIGURE 21–4   The complex plane.

Graphing a Complex Number

To plot a complex number $a + jb$ in the complex plane, simply locate a point with a horizontal coordinate of a and a vertical coordinate of b.

◆◆◆ **Example 15:** Plot the complex numbers $(2 + j3)$, $(-1 + j2)$, $(-3 - j2)$, and $(1 - j3)$.

Solution: The points are plotted in Fig. 21–5.

Such a plot is called an *Argand diagram*, named for Jean Robert Argand (1768–1822).

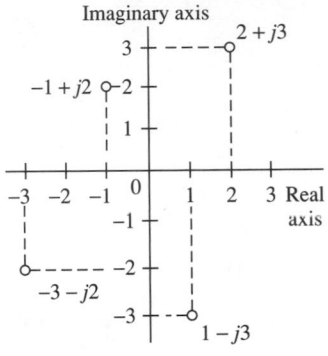

FIGURE 21-5   ◆◆◆

Exercise 2 ◆ Graphing Complex Numbers

Graph each complex number.

1. $2 + j5$
3. $3 - j2$
5. $j5$
7. $-2.7 - j3.4$
9. $1.46 - j2.45$

2. $-1 - j3$
4. $2 - j$
6. $2.25 - j3.62$
8. $5.02 + j$

21–3   Complex Numbers in Trigonometric and Polar Forms

Polar Form

In some books, the terms *polar form* and *trigonometric form* are used interchangeably. Here, we distinguish between them.

In Chapter 17 we saw that a point in a plane could be located by *polar coordinates*, as well as by *rectangular coordinates*, and we learned how to convert from one set of coordinates to the other. We will do the same thing now with complex numbers, converting between *rectangular* and *polar* form.

In Fig. 21–6, we plot a complex number $a + jb$. We connect that point to the origin with a line of length r at an angle of θ with the horizontal axis. Now, in addition to expressing the complex number in terms of a and b, we can express it as follows in terms of r and θ:

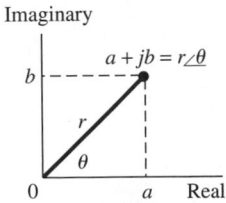

FIGURE 21-6 Polar form of a complex number shown on a complex plane.

Polar Form	$a + jb = r\underline{/\theta}$	228

Since $\cos \theta = a/r$ and $\sin \theta = b/r$, we have the following equations:

$a = r \cos \theta$	224

$b = r \sin \theta$	225

The radius r is called the *absolute value*. It can be found from the Pythagorean theorem.

$$r = \sqrt{a^2 + b^2} \qquad\qquad \textbf{226}$$

The angle θ is called the *argument* of the complex number. Since $\tan\theta = b/a$:

$$\theta = \arctan\frac{b}{a} \qquad\qquad \textbf{227}$$

◆◆◆ **Example 16:** Write the complex number $2 + j3$ in polar form.

Solution: The absolute value is

$$r = \sqrt{2^2 + 3^2} \cong 3.61$$

The argument is found from

$$\tan\theta = \frac{3}{2}$$

$$\theta \cong 56.3°$$

So

$$2 + j3 = 3.61\underline{/56.3°} \qquad\qquad\qquad ◆◆◆$$

Trigonometric Form

If we substitute the values of a and b from Eqs. 224 and 225 into $a + jb$, we get

$$a + jb = (r\cos\theta) + j(r\sin\theta)$$

Factoring gives us the following equation:

Trigonometric Form	$a + jb = r(\cos\theta) + j\sin\theta)$ rectangular trigonometric form form	**223**

The angle θ is sometimes written $\theta = \arg z$, which means "θ is the argument of the complex number z." Also, the expression in parentheses in Eq. 223 is sometimes abbreviated as cis θ, so cis $\theta = \cos\theta + j\sin\theta$.

◆◆◆ **Example 17:** Write the complex number $2 + j3$ in polar and trigonometric forms.

Solution: We already have r and θ from Example 16.

$$r = 3.61 \quad\text{and}\quad \theta = 56.3°$$

That is,

$$2 + j3 = 3.61\underline{/56.3°}$$

So, by Eq. 224,

$$2 + j3 = 3.61(\cos 56.3° + j\sin 56.3°) \qquad\qquad ◆◆◆$$

◆◆◆ **Example 18:** Write the complex number $6(\cos 30° + j\sin 30°)$ in polar and rectangular forms.

Solution: By inspection,

$$r = 6 \quad\text{and}\quad \theta = 30°$$

So
$$a = r \cos \theta = 6 \cos 30° = 5.20$$

and
$$b = r \sin \theta = 6 \sin 30° = 3.00$$

Thus our complex number in rectangular form is $5.20 + j3.00$, and in polar form, $6\underline{/30°}$. ◆◆◆

Arithmetic Operations in Polar Form

It is common practice to switch back and forth between rectangular and polar forms during a computation, using that form in which a particular operation is easier. We saw that addition and subtraction are fast and easy in rectangular form, and we now show that multiplication, division, and raising to a power are best done in polar form.

Products

We will use trigonometric form to work out the formula for multiplication and then express the result in the simpler polar form.

Let us multiply $r(\cos \theta + j \sin \theta)$ by $r'(\cos \theta' + j \sin \theta')$.

$$[r(\cos \theta + j \sin \theta)][r'(\cos \theta' + j \sin \theta')]$$
$$= rr'(\cos \theta \cos \theta' + j \cos \theta \sin \theta' + j \sin \theta \cos \theta' + j^2 \sin \theta \sin \theta')$$
$$= rr'[(\cos \theta \cos \theta' - \sin \theta \sin \theta') + j(\sin \theta \cos \theta' + \cos \theta \sin \theta')]$$

But, by Eq. 168,

$$\cos \theta \cos \theta' - \sin \theta \sin \theta' = \cos(\theta + \theta')$$

and by Eq. 167,

$$\sin \theta \cos \theta' + \cos \theta \sin \theta' = \sin(\theta + \theta')$$

So

$$r(\cos \theta + j \sin \theta) \cdot r'(\cos \theta' + j \sin \theta') = rr'[\cos(\theta + \theta') + j \sin(\theta + \theta')]$$

Switching now to polar form, we get the following equation:

Products	$r\underline{/\theta} \cdot r'\underline{/\theta'} = rr'\underline{/\theta + \theta'}$	229

The absolute value of the product of two complex numbers is the product of their absolute values, and the argument is the sum of the individual arguments.

◆◆◆ **Example 19:** Multiply $5\underline{/30°}$ by $3\underline{/20°}$.

Solution: The absolute value of the product will be $5(3) = 15$, and the argument of the product will be $30° + 20° = 50°$. So

$$5\underline{/30°} \cdot 3\underline{/20°} = 15\underline{/50°}$$ ◆◆◆

The angle θ is not usually written greater than 360°. Subtract multiples of 360° if necessary.

◆◆◆ **Example 20:** Multiply $6.27\underline{/300°}$ by $2.75\underline{/125°}$.

Solution:

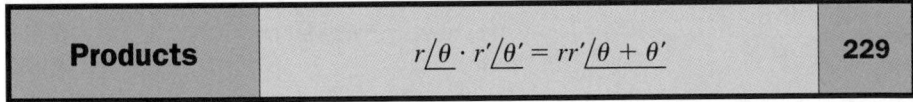

$$6.27\underline{/300°} \times 2.75\underline{/125°} = (6.27)(2.75)\underline{/300° + 125°}$$
$$= 17.2\underline{/425°}$$
$$= 17.2\underline{/65°}$$ ◆◆◆

Quotients

The rule for division of complex numbers in trigonometric form is similar to that for multiplication.

You might try deriving this yourself.

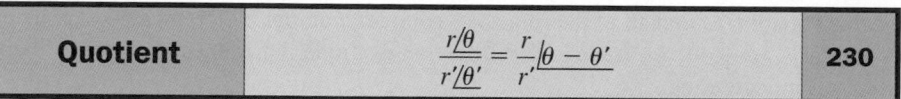

| Quotient | $\dfrac{r\underline{/\theta}}{r'\underline{/\theta'}} = \dfrac{r}{r'}\underline{/\theta - \theta'}$ | 230 |

The absolute value of the quotient of two complex numbers is the quotient of their absolute values, and the argument is the difference (numerator minus denominator) of their arguments.

◆◆◆ **Example 21:**

$$\frac{6\underline{/70°}}{2\underline{/50°}} = 3\underline{/20°}$$

◆◆◆

Powers

To raise a complex number to a power, we merely have to multiply it by itself the proper number of times, using Eq. 229.

$$(r\underline{/\theta})^2 = r\underline{/\theta} \cdot r\underline{/\theta} = r \cdot r\underline{/\theta + \theta} = r^2\underline{/2\theta}$$

$$(r\underline{/\theta})^3 = r\underline{/\theta} \cdot r^2\underline{/2\theta} = r \cdot r^2\underline{/\theta + 2\theta} = r^3\underline{/3\theta}$$

Do you see a pattern developing? In general, we get the following equation:

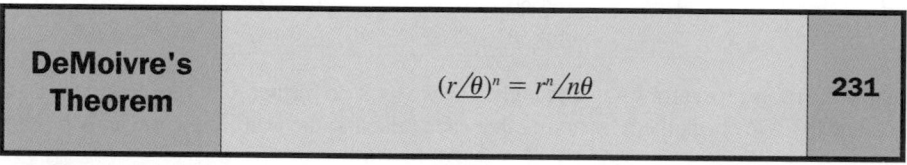

| DeMoivre's Theorem | $(r\underline{/\theta})^n = r^n\underline{/n\theta}$ | 231 |

DeMoivre's theorem is named after Abraham DeMoivre (1667–1754).

When a complex number is raised to the nth power, the new absolute value is equal to the original absolute value raised to the nth power, and the new argument is n times the original argument.

◆◆◆ **Example 22:**

$$(2\underline{/10°})^5 = 2^5\underline{/5(10°)} = 32\underline{/50°}$$

◆◆◆

Roots

We know that $1^4 = 1$, so, conversely, the fourth root of 1 is 1.

$$\sqrt[4]{1} = 1$$

But we have seen that $j^4 = 1$, so shouldn't it also be true that

$$\sqrt[4]{1} = j \quad ?$$

In fact, both 1 and j are fourth roots of 1, and there are two other roots as well. In this section we learn how to use DeMoivre's theorem to find all roots of a number.

First we note that a complex number $r\underline{/\theta}$ in polar form is unchanged if we add multiples of $360°$ to the angle θ (in degrees). Thus

$$r\underline{/\theta} = r\underline{/\theta + 360°} = r\underline{/\theta + 720°} = r\underline{/\theta + 360k}$$

where k is an integer. Rewriting DeMoivre's theorem with an exponent $1/p$ (where p is an integer), we get

$$(r\underline{/\theta})^{1/p} = (r\underline{/\theta + 360k})^{1/p} = r^{1/p}\underline{/(\theta + 360k)/p}$$

so

$$(r\underline{/\theta})^{1/p} = r^{1/p}\underline{/(\theta + 360k)/p} \qquad (1)$$

We use this equation to find all of the roots of a complex number by letting k take on the values $k = 0, 1, 2, \dots, (p - 1)$. We'll see in the following example that values of k greater than $(p - 1)$ will give duplicate roots.

◆◆◆ **Example 23:** Find the fourth roots of 1.

Solution: We write the given number in polar form; thus $1 = 1\underline{/0°}$. Then, using Eq. (1) with $p = 4$,

$$1^{1/4} = (1\underline{/0})^{1/4} = 1^{1/4}\underline{/(0 + 360k)/4} = 1\underline{/90k}$$

We now let $k = 0, 1, 2, \dots$.

	Root	
k	**Polar Form**	**Rectangular Form**
0	$1\underline{/0°}$	1
1	$1\underline{/90°}$	j
2	$1\underline{/180°}$	-1
3	$1\underline{/270°}$	$-j$
4	$1\underline{/360°}$	$1\underline{/0°} = 1$
5	$1\underline{/450°}$	$1\underline{/90°} = j$

Notice that the roots repeat for $k = 4$ and higher, so we look no further. Thus the fourth roots of 1 are $1, j, -1$, and $-j$. We check them by noting that each raised to the fourth power equals 1.   ◆◆◆

Thus the number 1 has four fourth roots. In general, any number has two square roots, three cube roots, and so on.

> There are p pth roots of a complex number.

If a complex number is in rectangular form, we convert to polar form before taking the root.

◆◆◆ **Example 24:** Find $\sqrt[3]{256 + j192}$.

Solution: We convert the given complex number to polar form.

$$r = \sqrt{(256)^2 + (192)^2} = 320 \qquad \theta = \arctan(192/256) = 36.9°$$

Then, using Eq. (1), we have

$$\sqrt[3]{256 + j192} = (320\underline{/36.9°})^{1/3} = (320)^{1/3}\underline{/(36.9° + 360k)/3}$$

$$= 6.84\ \underline{/12.3° + 120k}$$

We let $k = 0, 1, 2$, to obtain the three roots.

	Root	
k	**Polar Form**	**Rectangular Form**
0	$6.84\underline{/12.3° + 0°} \quad = 6.84\underline{/12.3°}$	$6.68 + j1.46$
1	$6.84\underline{/12.3° + 120°} = 6.84\underline{/132.3°}$	$-4.60 + j5.06$
2	$6.84\underline{/12.3° + 240°} = 6.84\underline{/252.3°}$	$-2.08 - j6.52$

◆◆◆

Exercise 3 ◆ Complex Numbers in Trigonometric and Polar Forms

Write each complex number in polar and trigonometric forms.

1. $5 + j4$

2. $-3 - j7$

3. $4 - j3$

4. $8 + j4$

5. $-5 - j2$

6. $-4 + j7$

7. $-9 - j5$

8. $7 - j3$

9. $-4 - j7$

Round to three significant digits, where necessary, in this exercise.

Write in rectangular and in polar forms.

10. $4(\cos 25° + j \sin 25°)$

11. $3(\cos 57° + j \sin 57°)$

12. $2(\cos 110° + j \sin 110°)$

13. $9(\cos 150° + j \sin 150°)$

14. $7(\cos 12° + j \sin 12°)$

15. $5.46(\cos 47.3° + j \sin 47.3°)$

Write in rectangular and trigonometric forms.

16. $5\underline{/28°}$

17. $9\underline{/59°}$

18. $4\underline{/63°}$

19. $7\underline{/-53°}$

20. $6\underline{/-125°}$

Multiplication and Division

Multiply.

21. $3(\cos 12° + j \sin 12°)$ by $5(\cos 28° + j \sin 28°)$

22. $7(\cos 48° + j \sin 48°)$ by $4(\cos 72° + j \sin 72°)$

23. $5.82(\cos 44.8° + j \sin 44.8°)$ by $2.77(\cos 10.1° + j \sin 10.1°)$

24. $5\underline{/30°}$ by $2\underline{/10°}$

25. $8\underline{/45°}$ by $7\underline{/15°}$

26. $2.86\underline{/38.2°}$ by $1.55\underline{/21.1°}$

Divide.

27. $8(\cos 46° + j \sin 46°)$ by $4(\cos 21° + j \sin 21°)$

28. $49(\cos 27° + j \sin 27°)$ by $7(\cos 15° + j \sin 15°)$

29. $58.3(\cos 77.4° + j \sin 77.4°)$ by $12.4(\cos 27.2° + j \sin 27.2°)$

30. $24\underline{/50°}$ by $12\underline{/30°}$

31. $50\underline{/72°}$ by $5\underline{/12°}$

32. $71.4\underline{/56.4°}$ by $27.7\underline{/15.2°}$

Powers and Roots

Evaluate.

33. $[2(\cos 15° + j \sin 15°)]^3$

34. $[9(\cos 10° + j \sin 10°)]^2$

35. $(7\underline{/20°})^2$

36. $(1.55\underline{/15°})^3$

37. $\sqrt{57\underline{/52°}}$

38. $\sqrt{22\underline{/12°}}$

39. $\sqrt[3]{38\underline{/73°}}$

40. $\sqrt[3]{15\underline{/89°}}$

41. $\sqrt{135 + j204}$

42. $\sqrt[3]{56.3 + j28.5}$

21–4   Complex Numbers in Exponential Form

Euler's Formula

We have already expressed a complex number in *rectangular* form

$$a + jb$$

in *polar* form

$$r\underline{/\theta}$$

and in *trigonometric* form

$$r(\cos\theta + j\sin\theta)$$

Our next (and final) form for a complex number is *exponential* form, given by Euler's formula.

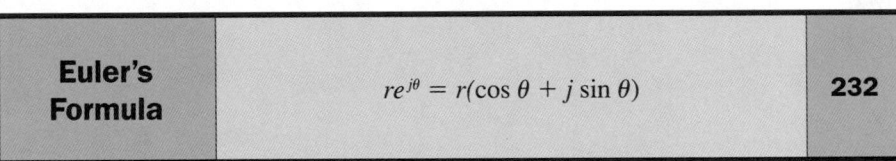

Euler's Formula	$re^{j\theta} = r(\cos\theta + j\sin\theta)$	232

This formula is named after Leonard Euler (1707–83). We give it without proof here.

Here e is the base of natural logarithms (approximately 2.718), discussed in Chapter 20, and θ is the argument expressed in radians.

◆◆◆ **Example 25:** Write the complex number $5(\cos 180° + j\sin 180°)$ in exponential form.

Solution: We first convert the angle to radians: $180° = \pi$ rad. So

$$5(\cos\pi + j\sin\pi) = 5e^{j\pi} \qquad\qquad ◆◆◆$$

Setting $\theta = \pi$ and $r = 1$ in Euler's formula gives

$$e^{j\pi} = \cos\pi + j\sin\pi$$
$$= -1 + j0$$
$$= -1$$

The constant e arises out of natural growth, j is the square root of -1, and π is the ratio of the circumference of a circle to its diameter. Thus we get the astounding result that the irrational number e raised to the product of the imaginary unit j and the irrational number π give the integer -1.

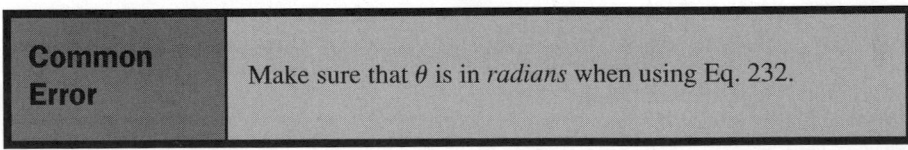

Common Error	Make sure that θ is in *radians* when using Eq. 232.

◆◆◆ **Example 26:** Write the complex number $3e^{j2}$ in rectangular, trigonometric, and polar forms.

Solution: Changing θ to degrees gives us

$$\theta = 2 \text{ rad} = 115°$$

and from Eq. 232,

$$r = 3$$

From Eq. 224,

$$a = r \cos \theta = 3 \cos 115° = -1.27$$

and from Eq. 225,

$$b = r \sin \theta = 3 \sin 115° = 2.72$$

So, in rectangular form,

$$3e^{j2} = -1.27 + j2.72$$

In polar form,

$$3e^{j2} = 3\underline{/115°}$$

In trigonometric form,

$$3e^{j2} = 3(\cos 115° + j \sin 115°) \qquad \blacklozenge\blacklozenge\blacklozenge$$

Products

We now find products, quotients, powers, and roots of complex numbers in exponential form. These operations are quite simple in this form, as we merely have to use the laws of exponents. Thus by Eq. 29:

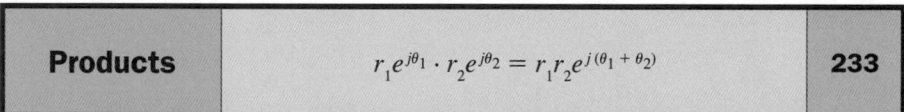

| Products | $r_1 e^{j\theta_1} \cdot r_2 e^{j\theta_2} = r_1 r_2 e^{j(\theta_1 + \theta_2)}$ | 233 |

◆◆◆ **Example 27:**

(a) $2e^{j3} \cdot 5e^{j4} = 10e^{j7}$ (b) $2.83e^{j2} \cdot 3.15e^{j4} = 8.91e^{j6}$

(c) $-13.5e^{-j3} \cdot 2.75e^{j5} - -37.1e^{j2}$ ◆◆◆

Quotients

By Eq. 30:

| Quotients | $\dfrac{r_1 e^{j\theta_1}}{r_2 e^{j\theta_2}} = \dfrac{r_1}{r_2} e^{j(\theta_1 - \theta_2)}$ | 234 |

◆◆◆ **Example 28:**

(a) $\dfrac{8e^{j5}}{4e^{j2}} = 2e^{j3}$ (b) $\dfrac{63.8e^{j2}}{13.7e^{j5}} = 4.66e^{-j3}$

(c) $\dfrac{5.82e^{-j4}}{9.83e^{-j7}} = 0.592e^{j3}$ ◆◆◆

Powers and Roots

By Eq. 31:

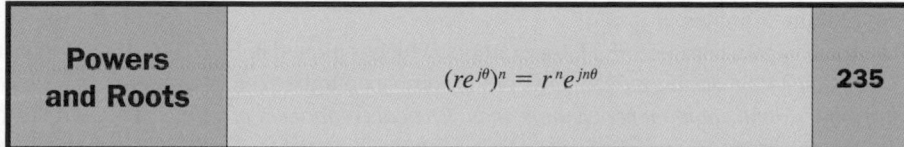

| Powers and Roots | $(re^{j\theta})^n = r^n e^{jn\theta}$ | 235 |

◆◆◆ Example 29:

(a) $(2e^{j3})^4 = 16e^{j12}$ 　　　　　　　　　　　　(b) $(-3.85e^{-j2})^3 = -57.1e^{-j6}$

(c) $(0.223e^{-j3})^{-2} = \dfrac{e^{j6}}{(0.223)^2} = \dfrac{e^{j6}}{0.0497} = 20.1e^{j6}$ 　　　　　　　◆◆◆

Exercise 4 ◆ Complex Numbers in Exponential Form

Express each complex number in exponential form.

1. $2 + j3$ 　　　　　　　　　　　　　　　　　2. $-1 + j2$

3. $3(\cos 50° + j \sin 50°)$ 　　　　　　　　　4. $12\underline{/14°}$

5. $2.5\underline{/\pi/6}$ 　　　　　　　　　　　　　　6. $7\left(\cos \dfrac{\pi}{3} + j \sin \dfrac{\pi}{3}\right)$

7. $5.4\underline{/\pi/12}$ 　　　　　　　　　　　　　8. $5 + j4$

Express in rectangular, polar, and trigonometric forms.

9. $5e^{j3}$
10. $7e^{j5}$
11. $2.2e^{j1.5}$
12. $4e^{j2}$

Operations in Exponential Form

Multiply.

13. $9e^{j2} \cdot 2e^{j4}$ 　　　　　　　　　　　　14. $8e^j \cdot 6e^{j3}$

15. $7e^j \cdot 3e^{j3}$ 　　　　　　　　　　　　　16. $6.2e^{j1.1} \cdot 5.8e^{j2.7}$

17. $1.7e^{j5} \cdot 2.1e^{j2}$ 　　　　　　　　　　18. $4e^{j7} \cdot 3e^{j5}$

Divide.

19. $18e^{j6}$ by $6e^{j3}$ 　　　　　　　　　　　20. $45e^{j4}$ by $9e^{j2}$

21. $55e^{j9}$ by $5e^{j6}$ 　　　　　　　　　　　22. $123e^{j6}$ by $105e^{j2}$

23. $21e^{j2}$ by $7e^j$ 　　　　　　　　　　　　24. $7.7e^{j4}$ by $2.3e^{j2}$

Evaluate.

25. $(3e^{j5})^2$ 　　　　　　　　　　　　　　　26. $(4e^{j2})^3$

27. $(2e^j)^3$

21–5　Vector Operations Using Complex Numbers

Vectors Represented by Complex Numbers

One of the major uses of complex numbers is that they can represent vectors and, as we will soon see, can enable us to manipulate vectors in ways that are easier than we learned when studying oblique triangles.

　　Take the complex number $2 + j3$, for example, which is plotted in Fig. 21–7. If we connect that point with a line to the origin, we can think of the complex number $2 + j3$ *as representing a vector* **R** *having a horizontal component of 2 units and a vertical component of 3 units*. The complex number used to represent a vector can, of course, be expressed in any of the forms of a complex number.

Vector Addition and Subtraction

Let us place a second vector on our diagram, represented by the complex number $3 - j$ (Fig. 21–8). We can add them graphically by the parallelogram method, and we get a resultant $5 + j2$. But let us add the two original complex numbers, $2 + j3$ and $3 - j$.

$$(2 + j3) + (3 - j) = 5 + j2$$

This result is the same as what we obtained graphically. In other words, *the resultant of two vectors is equal to the sum of the complex numbers representing those vectors.*

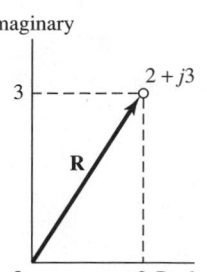

FIGURE 21–7 A vector represented by a complex number.

◆◆◆ Example 30: Subtract the vectors $25 \underline{/48°} - 18 \underline{/175°}$.

Solution: Vector addition and subtraction are best done in rectangular form. Converting the first vector, we have

$$a_1 = 25 \cos 48° = 16.7$$
$$b_1 = 25 \sin 48° = 18.6$$

so

$$25 \underline{/48°} = 16.7 + j18.6$$

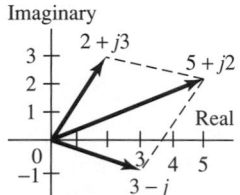

FIGURE 21–8 Vector addition with complex numbers.

Similarly, for the second vector,

$$a_2 = 18 \cos 175° = -17.9$$
$$b_2 = 18 \sin 175° = 1.57$$

so

$$18 \underline{/175°} = -17.9 + j1.57$$

Combining, we have

$$(16.7 + j18.6) - (-17.9 + j1.57) = (16.7 + 17.9) + j(18.6 - 1.57)$$
$$= 34.6 + j17.0$$

These vectors are shown in Fig. 21–9.

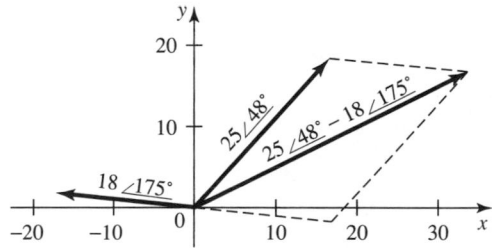

FIGURE 21–9　　　　　　　　　　　　　　　　　　　　◆◆◆

Multiplication and Division of Vectors

Multiplication or division is best done if the vector is expressed as a complex number in polar form.

◆◆◆ Example 31: Multiply the vectors $2 \underline{/25°}$ and $3 \underline{/15°}$.

Solution: By Eq. 229, the product vector will have a magnitude of $3(2) = 6$ and an angle of $25° + 15° = 40°$. The product is thus

$$6 \underline{/40°}$$

◆◆◆

◆◆◆ Example 32: Divide the vector $8 \underline{/44°}$ by $4 \underline{/12°}$.

Solution: By Eq. 230,

$$(8\,\underline{/44°}) \div (4\,\underline{/12°}) = \frac{8}{4}\,\underline{/44° - 12°} = 2\,\underline{/32°}$$ ◆◆◆

◆◆◆ **Example 33:** A certain current **I** is represented by the complex number $1.15\,\underline{/23.5°}$ amperes, and a complex impedance **Z** is represented by $24.6\,\underline{/14.8°}$ ohms. Multiply **I** by **Z** to obtain the voltage **V**.

Solution:

$$\mathbf{V} = \mathbf{IZ} = (1.15\,\underline{/23.5°})(24.6\,\underline{/14.8°}) = 28.3\,\underline{/38.3°}\ (\text{V})$$ ◆◆◆

The *j* Operator

Let us multiply a complex number $r\,\underline{/\theta}$ by j. First we must express j in polar form. By Eq. 224,

$$j = 0 + j1 = 1(\cos 90° + j \sin 90°) = 1\,\underline{/90°}$$

After we multiply by j, the new absolute value will be

$$r \cdot 1 = r$$

and the argument will be

$$\theta + 90°$$

So

$$r\,\underline{/\theta} \cdot j = r\,\underline{/\theta} + 90°$$

Thus the only effect of multiplying our original complex number by j was to increase the angle by 90°. If our complex number represents a vector, *we may think of j as an operator that causes the vector to rotate through one-quarter revolution.*

Exercise 5 ◆ Vector Operations Using Complex Numbers

Express each vector in rectangular and polar forms.

Magnitude	Angle
1. 7.00	49.0°
2. 28.0	136°
3. 193	73.5°
4. 34.2	1.10 rad
5. 39.0	2.50 rad
6. 59.4	58.0°

Combine the vectors.

  7. $(3 + j2) + (5 - j4)$

  8. $(7 - j3) - (4 + j)$

  9. $58\,\underline{/72°} + 21\,\underline{/14°}$

10. $8.60\,\underline{/58°} - 4.20\,\underline{/160°}$

11. $9(\cos 42° + j \sin 42°) + 2(\cos 8° + j \sin 8°)$

12. $8(\cos 15° + j \sin 15°) - 5(\cos 9° - j \sin 9°)$

Multiply the vectors.

Round to three significant digits, where needed, in this exercise set.

13. $(7 + j3)(2 - j5)$

14. $(2.50\,\underline{/18°})(3.70\,\underline{/48°})$

15. $2(\cos 25° - j \sin 25°) \cdot 6(\cos 7° - j \sin 7°)$

Divide the vectors.

16. $(25 - j2) \div (3 + j4)$

17. $(7.70 \;\underline{/47°}) \div (2.50 \;\underline{/15°})$

18. $5(\cos 72° + j \sin 72°) \div 3(\cos 31° + j \sin 31°)$

21–6   Alternating Current Applications

Rotating Vectors in the Complex Plane

We have already shown that a vector can be represented by a complex number, $r\;\underline{/\theta}$. For example, the complex number $5.00\;\underline{/28°}$ represents a vector of magnitude 5.00 at an angle of 28° with the real axis.

Glance back at Sec. 17–5 where we first introduced alternating current.

A phasor (a rotating vector) may also be represented by a complex number $R\;\underline{/\omega t}$ by replacing the angle θ by ωt, where ω is the angular velocity and t is time.

◆◆◆ **Example 34:** The complex number

$$11\;\underline{/5t}$$

represents a phasor of magnitude 11 rotating with an angular velocity of 5 rad/s.   ◆◆◆

In Sec. 17–5 we showed that a phasor of magnitude R has a projection on the y axis of $R \sin \omega t$. Similarly, a phasor $R\;\underline{/\omega t}$ in the complex plane (Fig. 21–10) will have a projection on the imaginary axis of $R \sin \omega t$. Thus

$$R \sin \omega t = \text{imaginary part of } R\omega t$$

Similarly,

$$R \cos \omega t = \text{real part of } R\;\underline{/\omega t}$$

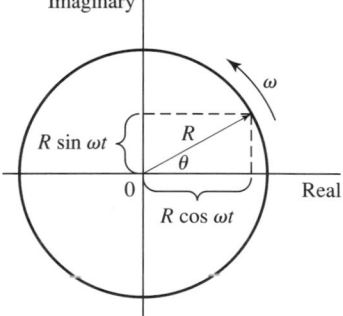

FIGURE 21–10

Thus, either a sine or a cosine wave can be represented by a complex number $R\;\underline{/\omega t}$, depending upon whether we project onto the imaginary or the real axis. It usually does not matter which we choose, because the sine and cosine functions are identical except for a phase difference of 90°. What does matter is that we are consistent. Here we will follow the convention of projecting the phasor onto the imaginary axis, and we will drop the phrase "imaginary part of." Thus we say that

$$R \sin \omega t = R\;\underline{/\omega t}$$

Similarly, if there is a phase angle ϕ,

$$R \sin(\omega t + \phi) = R\;\underline{/\omega t + \phi}$$

One final simplification: It is customary to draw phasors at $t = 0$, so that ωt vanishes from the expression. Thus we write

$$R \sin(\omega t + \phi) = R\;\underline{/\phi}$$

Effective or Root Mean Square (rms) Values

Before we write currents and voltages in complex form, we must define the *effective value* of current and voltage. The power P delivered to a resistor by a direct current I is $P = I^2R$. We now define an effective current I_{eff} that delivers the same power from an alternating current as that delivered by a direct current of the same number of amperes.

$$P = I_{eff}^{2}R$$

Figure 21–11 shows an alternating current $i = I_m \sin \omega t$ and i^2, the square of that current. The mean value of this alternating quantity is found to deliver the same power in a resistor as a steady quantity of the same value.

$$P = I_{eff}^{2}R = (\text{mean } i^2)R$$

We get the effective value by taking the square root of the mean value squared. Hence it is also called a *root mean squared* value. Thus I_{eff} is often written I_{rms} instead.

But the mean value of i^2 is equal to $I_m^2/2$, half the peak value, so

$$I_{eff} = \sqrt{I_m^2/2} = I_m/\sqrt{2}$$

Thus *the effective current* I_{eff} *is equal to the peak current* I_m *divided by* $\sqrt{2}$. Similarly, *the effective voltage* V_{eff} *is equal to the peak voltage* V_m *divided by* $\sqrt{2}$.

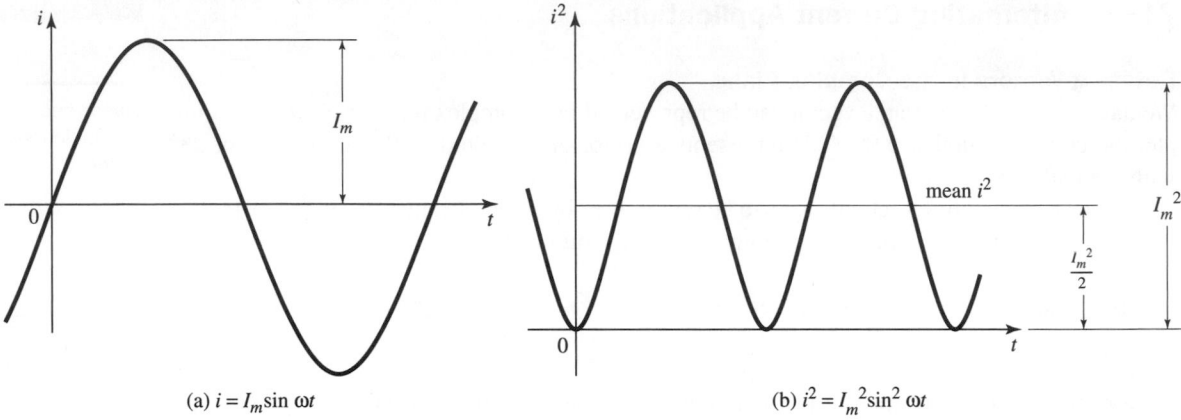

(a) $i = I_m \sin \omega t$ (b) $i^2 = I_m^2 \sin^2 \omega t$

FIGURE 21–11

◆◆◆ **Example 35:** An alternating current has a peak value of 2.84 A. What is the effective current?

Solution:

$$I_{eff} = \frac{I_m}{\sqrt{2}} = \frac{2.84}{\sqrt{2}} = 2.01 \text{ A}$$ ◆◆◆

Alternating Current and Voltage in Complex Form

Next we will write a sinusoidal current or voltage expression in the form of a complex number because computations are easier in that form. At the same time, we will express the magnitude of the current or voltage in terms of effective value. The effective current I_{eff} is the quantity that is always implied when a current is given, rather than the peak current I_m. It is the quantity that is read when using a meter. The same is true of effective voltage.

To write an alternating current in complex form, we give the effective value of the current and the phase angle. The current is written in boldface type: **I**. Thus we have the following equations:

Complex Voltage and Current	The current is represented by $i = I_m \sin(\omega t + \phi_2)$ $\mathbf{I} = I_{eff} \angle \phi_2 = \dfrac{I_m}{\sqrt{2}} \angle \phi_2$	**A76**
	Similarly, the voltage is represented by $v = V_m \sin(\omega t + \phi_1)$ $\mathbf{V} = V_{eff} \angle \phi_1 = \dfrac{V_m}{\sqrt{2}} \angle \phi_1$	**A75**

Note that the complex expressions for voltage and current do not contain t. We say that the voltage and current have been converted from the *time domain* to the *phasor domain*.

♦♦♦ **Example 36:** The complex expression for the alternating current

$$i = 2.84 \sin(\omega t + 33°) \, \text{A}$$

is

$$\mathbf{I} = I_{\text{eff}} \underline{/33°} = \frac{2.84}{\sqrt{2}} \underline{/33°} \, \text{A}$$

$$= 2.01 \underline{/33°} \, \text{A} \qquad ♦♦♦$$

♦♦♦ **Example 37:** The sinusoidal expression for the voltage

$$\mathbf{V} = 84.2 \underline{/\!-49°} \, \text{V}$$

is

$$v = 84.2\sqrt{2} \sin(\omega t - 49°) \, \text{V}$$

$$= 119 \sin(\omega t - 49°) \, \text{V} \qquad ♦♦♦$$

Complex Impedance

In Sec. 7–7 we drew a vector impedance diagram in which the impedance was the resultant of two perpendicular vectors: the resistance R along the horizontal axis and the reactance X along the vertical axis. If we now use the complex plane and draw the resistance along the *real* axis and the reactance along the *imaginary* axis, we can represent impedance by a complex number.

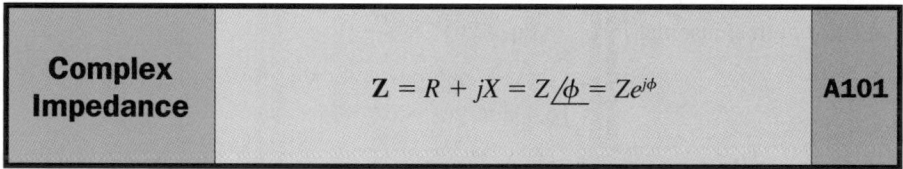

| Complex Impedance | $\mathbf{Z} = R + jX = Z\underline{/\phi} = Ze^{j\phi}$ | A101 |

Note that the "complex" expressions are the simpler. This happy situation is used in the ac computations to follow, which are much more cumbersome without complex numbers.

where

R = circuit resistance
X = circuit reactance = $X_L - X_C$ (A98)
$|Z|$ = magnitude of impedance = $\sqrt{R^2 + X^2}$ (A99)
ϕ = phase angle = $\arctan\dfrac{X}{R}$ (A100)

♦♦♦ **Example 38:** A circuit has a resistance of 5 Ω in series with a reactance of 7 Ω, as shown in Fig. 21–12(a). Represent the impedance by a complex number.

Solution: We draw the vector impedance diagram as shown in Fig. 21–12(b). The impedance in rectangular form is

$$\mathbf{Z} = 5 + j7$$

The magnitude of the impedance is

$$\sqrt{5^2 + 7^2} = 8.60$$

The phase angle is

$$\arctan\frac{7}{5} = 54.5°$$

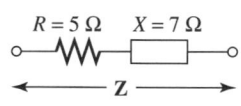

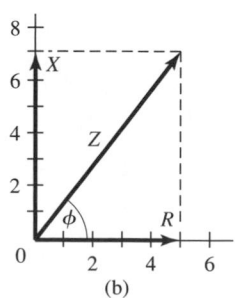

FIGURE 21–12

So we can write the impedance as a complex number in polar form as follows:

$$\mathbf{Z} = 8.60 \underline{/54.5°} \qquad ♦♦♦$$

Ohm's Law for Alternating Current (ac)

We stated at the beginning of this section that the use of complex numbers would make calculations with alternating current almost as easy as for direct current. We do this by means of the following relationship:

| Ohm's Law for ac | V = ZI | A103 |

Note the similarity between this equation and Ohm's law, Eq. A62. It is used in the same way.

◆◆◆ **Example 39:** A voltage of 142 sin 200t is applied to the circuit of Fig. 21–12(a). Write a sinusoidal expression for the current i.

Solution: We first write the voltage in complex form. By Eq. A75,

$$\mathbf{V} = \frac{142}{\sqrt{2}} \underline{/0°} = 100 \underline{/0°}$$

Next we find the complex impedance $\mathbf{Z}$. From Example 38,

$$\mathbf{Z} = 8.60 \underline{/54.5°}$$

The complex current $\mathbf{I}$, by Ohm's law for ac, is

$$\mathbf{I} = \frac{\mathbf{V}}{\mathbf{Z}} = \frac{100 \underline{/0°}}{8.60 \underline{/54.5°}} = 11.6 \underline{/-54.5°}$$

Then the current in sinusoidal form is (Eq. A76)

$$i = 11.6\sqrt{2} \sin(200t - 54.5°)$$
$$= 16.4 \sin(200t - 54.5°)$$

The current and voltage phasors are plotted in Fig. 21–13, and the instantaneous current and voltage are plotted in Fig. 21–14. Note the phase difference between the voltage and current waves. This phase difference, 54.5°, converted to time, is

$$54.5° = 0.951 \text{ rad} = 200t$$
$$t = 0.951/200 = 0.00475 \text{ s} = 4.75 \text{ ms}$$

We say that the current lags the voltage by 54.5° or 4.75 ms.

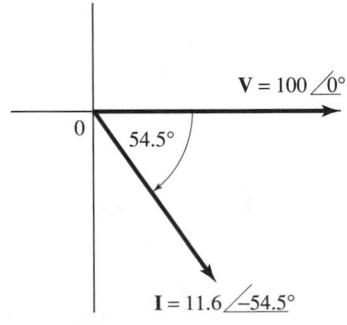

FIGURE 21–13

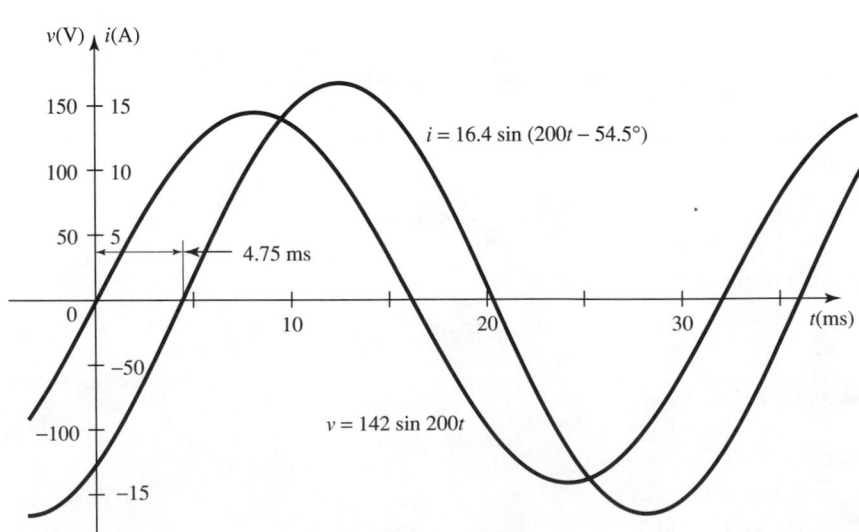

FIGURE 21–14   ◆◆◆

Exercise 6 ◆ Alternating Current Applications

Express each current or voltage in complex form.

1. $i = 250 \sin(\omega t + 25°)$

2. $v = 1.5 \sin(\omega t - 30°)$

3. $v = 57 \sin(\omega t - 90°)$

4. $i = 2.7 \sin \omega t$

5. $v = 144 \sin \omega t$

6. $i = 2.7 \sin(\omega t - 15°)$

Express each current or voltage in sinusoidal form.

7. $\mathbf{V} = 150 \underline{/0°}$

8. $\mathbf{V} = 1.75 \underline{/70°}$

9. $\mathbf{V} = 300 \underline{/-90°}$

10. $\mathbf{I} = 25 \underline{/30°}$

11. $\mathbf{I} = 7.5 \underline{/0°}$

12. $\mathbf{I} = 15 \underline{/-130°}$

Express the impedance of each circuit as a complex number in rectangular form and in polar form.

13. Fig. 21–15(a)

14. Fig. 21–15(b)

15. Fig. 21–15(c)

16. Fig. 21–15(d)

17. Fig. 21–15(e)

18. Fig. 21–15(f)

19. Fig. 21–15(g)

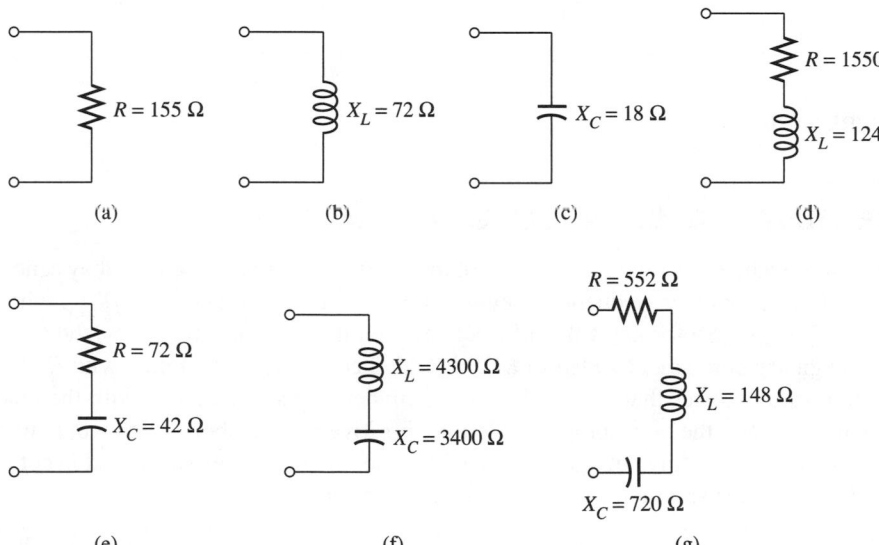

FIGURE 21–15

20. Write a sinusoidal expression for the current i in each part of Fig. 21–16.

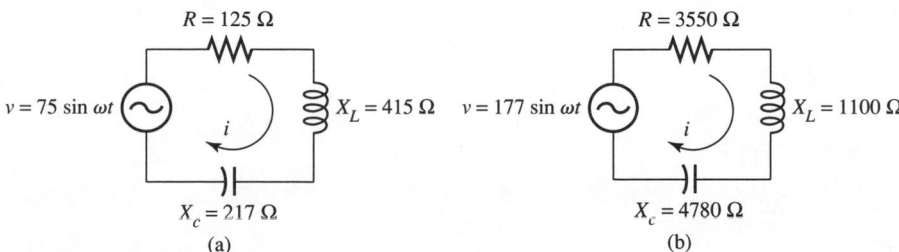

FIGURE 21–16

21. Write a sinusoidal expression for the voltage v in each part of Fig. 21–17.

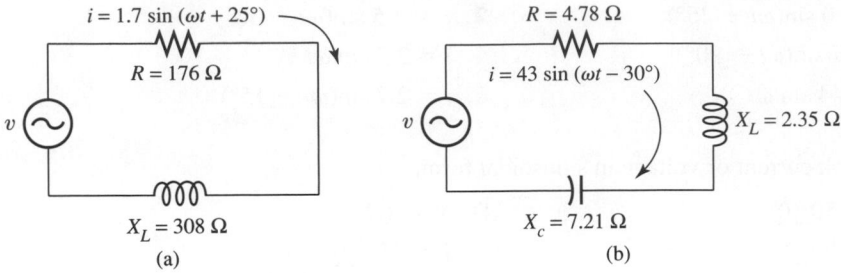

(a)   (b)

FIGURE 21–17

22. Find the complex impedance $\mathbf{Z}$ in each part of Fig. 21–18.

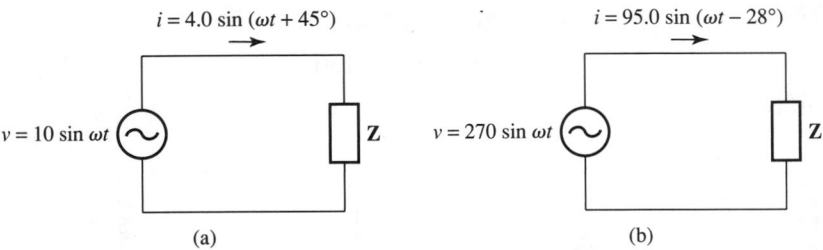

(a)   (b)

FIGURE 21–18

CASE STUDY DISCUSSION—SERIES RESONANT CIRCUIT

The reactance values are on opposite sides of the j axis. When they are equal, they cancel, and $Z = R$. At any other frequency, they will not be equal, and $Z = R \pm jX_{\text{NET}}$, where $jX_{\text{NET}} = jX_{\text{L}} - jX_{\text{C}}$. So for any value of frequency not the resonant frequency, there will be an imaginary component added to R. This added component combined with R must result in a value greater than R and therefore a higher impedance. Also, with the reactance cancelled out, the only component left is a positive real number, which would have a phase angle of zero. Any net reactance would cause a vector resultant (Z) that is not on the real number line and therefore there would be a phase angle.

◆◆◆ CHAPTER 21 REVIEW PROBLEMS ◆◆◆◆◆◆◆◆◆◆◆◆◆◆◆◆◆◆◆◆◆◆◆◆◆◆◆◆◆

Express the following as complex numbers in rectangular, polar, trigonometric, and exponential forms.

1. $3 + \sqrt{-4}$

2. $-\sqrt{-9} - 5$

3. $-2 + \sqrt{-49}$

4. $3.25 + \sqrt{-11.6}$

Evaluate.

5. j^{17}

6. j^{25}

Combine the complex numbers. Leave your answers in rectangular form.

7. $(7 - j3) + (2 + j5)$

8. $4.8\,\underline{/28°} - 2.4\,\underline{/72°}$

9. $52(\cos 50° + j \sin 50°) + 28(\cos 12° + j \sin 12°)$

10. $2.7e^{j7} - 4.3e^{j5}$

Multiply. Leave your answer in the same form as the complex numbers.

11. $(2 - j)(3 + j5)$

12. $(7.3\,\underline{/21°})(2.1\,\underline{/156°})$

13. $2(\cos 20° + j \sin 20°) \cdot 6(\cos 18° + j \sin 18°)$

14. $(93e^{j2})(5e^{j7})$

Divide and leave your answer in the same form as the complex numbers.

15. $(9 - j3) \div (4 + j)$

16. $(18\,\underline{/72°}) \div (6\,\underline{/22°})$

17. $16(\cos 85° + j \sin 85°) \div 8(\cos 40° + j \sin 40°)$

18. $127e^{j8} \div 4.75e^{j5}$

Graph each complex number.

19. $7 + j4$

20. $2.75\,\underline{/44°}$

21. $6(\cos 135° + j \sin 135°)$

22. $4.75e^{j2.2}$

Evaluate each power.

23. $(-4 + j3)^2$

24. $(5\,\underline{/12°})^3$

25. $[5(\cos 10° + j \sin 10°)]^3$

26. $[2e^{j3}]^5$

27. Express in complex form: $i = 45 \sin(\omega t + 32°)$

28. Express in sinusoidal form: $\mathbf{V} = 283\,\underline{/-22°}$

29. Write a complex expression for the current i in Fig. 21–19.

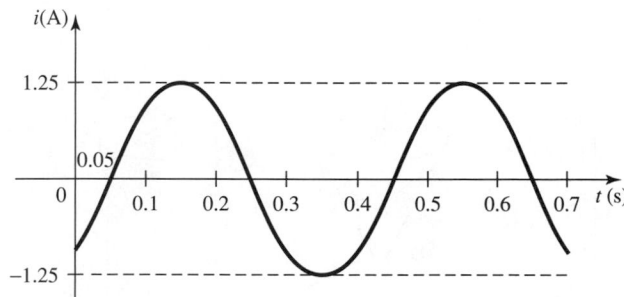

FIGURE 21–19

Writing

30. Suppose that you are completing a job application to an electronics company which asks you: "Explain, in writing, how you would use complex numbers in an electrical calculation, and illustrate your explanation with an example." How would you respond?

22

Analytic Geometry

This chapter begins our study of *analytic geometry*. The main idea in analytic geometry is to place geometric figures (points, lines, circles, and so forth) onto coordinate axes, where they may be studied using the methods of algebra.

We start with the *straight line*, building on what we have already covered in Chapter 5. We study the idea of the *slope* of a line and of a *curve*, ideas that will be crucial to our understanding of the *derivative*. Next come the *conic sections*: the circle, parabola, ellipse, and hyperbola. They are called *conic sections*, or simply *conics*, because each can be obtained by passing a plane through a right circular cone, as shown in Fig. 22–1.

When the plane is perpendicular to the cone's axis, it intercepts a *circle*. When the plane is tilted a bit, but not so much as to be parallel to an element (a line on the cone that passes through the vertex) of the cone, we get an *ellipse*. When the plane is parallel to an element of

You might want to try making these shapes by cutting a cone from a form made of modeling clay or damp sand. Also, what position of the plane would you use to make a straight line? A point?

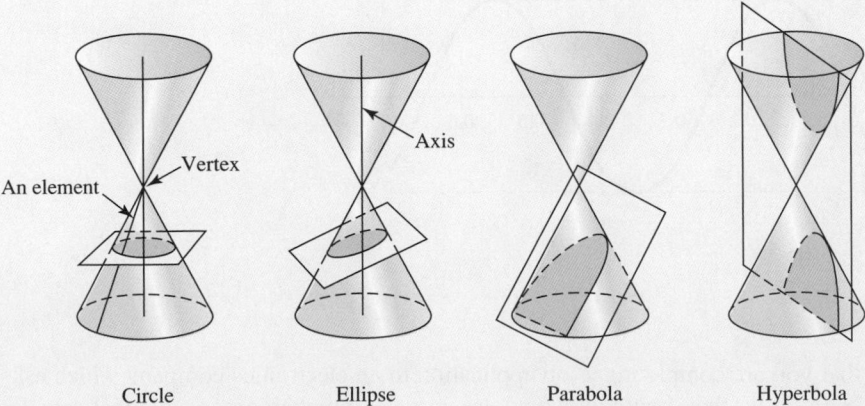

Circle   Ellipse   Parabola   Hyperbola

FIGURE 22–1 Conic sections.

the cone, we get a *parabola*, and when the plane is steep enough to cut the upper nappe of the cone, we get the two-branched *hyperbola*.

In this chapter we learn to write the equations of the straight line and the conics. These equations enable us to understand and use these geometric figures in ways that are not possible without the equations. We also give some interesting applications for each curve.

22–1 The Straight Line

Length of a Line Segment Parallel to a Coordinate Axis

When we speak about the length of a line segment or about the distance between two points, we usually mean the *magnitude* of that length or distance.

To find the magnitude of the length of a line segment lying on or parallel to the x axis, simply subtract the abscissa (the x value) of either endpoint from the abscissa of the other endpoint, and take the absolute value of this result.

The procedure is similar for a line segment lying on or parallel to the y axis: Subtract the ordinate (the y value) of either endpoint from the ordinate of the other endpoint, and take the absolute value of this result.

◆◆◆ **Example 1:** The magnitudes of the lengths of the lines in Fig. 22–2 are as follows:

(a) $AB = |5 - 2| = 3$

(b) $PQ = |5 - (-2)| = 7$

(c) $RS = |-1 - (-5)| = 4$

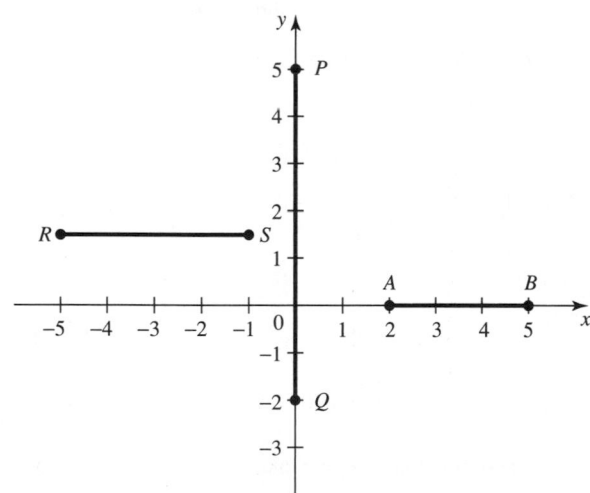

FIGURE 22–2   ◆◆◆

Note that it doesn't matter if we reverse the order of the endpoints. We get the same result.

◆◆◆ **Example 2:** Repeating Example 1(a) with the endpoints reversed gives

$$AB = |2 - 5| = |-3| = 3$$

as before.   ◆◆◆

Directed Distance

Sometimes when speaking about the length of a line segment or the distance between two points, it is necessary to specify *direction* as well as magnitude. When we specify the *directed distance AB*, for example, we mean the distance *from A to B*, sometimes written $\overline{AB}$.

Note that directed distance applies only to a line parallel to a coordinate axis.

◆◆◆ **Example 3:** For the two points $P(5, 2)$ and $Q(1, 2)$, the directed distance PQ is $PQ = 1 - 5 = -4$, and the directed distance QP is $QP = 5 - 1 = 4$.   ◆◆◆

Increments

Let us say that a particle is moving along a curve from point P to point Q, as shown in Fig. 22–3. As it moves, its abscissa changes from x_1 to x_2. We call this change an *increment* in x and label it Δx (read "delta x"). Similarly, the ordinate changes from y_1 to y_2 and is labelled Δy. The increments are found simply by subtracting the coordinates at P from those at Q.

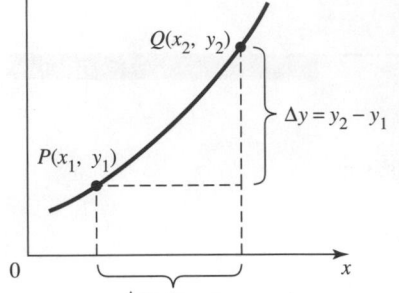

FIGURE 22–3 Increments.

Increments	$\Delta x = x_2 - x_1$ and $\Delta y = y_2 - y_1$	337

◆◆◆ **Example 4:** A particle moves from $P_1(2, 5)$ to $P_2(7, 3)$. The increments in its coordinates are

$$\Delta x = x_2 - x_1 = 7 - 2 = 5$$

and

$$\Delta y = y_2 - y_1 = 3 - 5 = -2$$ ◆◆◆

Distance Formula

We now find the length of a line inclined at any angle, such as PQ in Fig. 22–4. We first draw a horizontal line through P and drop a perpendicular from Q, forming a right triangle PQR. The sides of the triangle are d, Δx, and Δy. By the Pythagorean theorem,

$$d^2 = (\Delta x)^2 + (\Delta y)^2 = (x_2 - x_1)^2 + (y_2 - y_1)^2$$

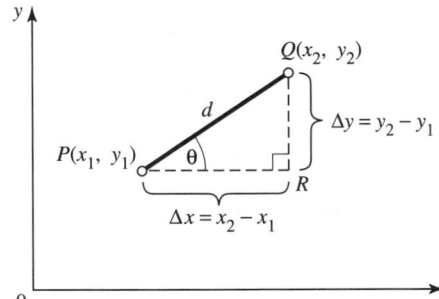

FIGURE 22–4 Length of a line segment.

Since we want the magnitude of the distance, we take only the positive root and get the following:

Distance Formula	$d = \sqrt{(\Delta x)^2 + (\Delta y)^2}$ $= \sqrt{(x_2 - x_1)^2 + (y_2 - y_1)^2}$	283

◆◆◆ **Example 5:** Find the length of the line segment with the endpoints

$$(3, -5) \text{ and } (-1, 6).$$

Solution: Let us give the first point $(3, -5)$ the subscripts 1, and the other point the subscripts 2 (it does not matter which we label 1). So

$$x_1 = 3 \qquad y_1 = -5 \qquad x_2 = -1 \qquad y_2 = 6$$

Substituting into the distance formula, we obtain

$$d = \sqrt{(-1-3)^2 + [6-(-5)]^2} = \sqrt{(-4)^2 + (11)^2} = 11.7 \text{ (rounded)} \qquad \blacklozenge\blacklozenge\blacklozenge$$

Common Error	Do not take the square root of each term separately. $$d \neq \sqrt{(x_2 - x_1)^2} + \sqrt{(y_2 + y_1)^2}$$

CASE STUDY—PARABOLIC MICROPHONE

A parabolic microphone is a microphone mounted at the focal point of a parabolic reflector. The reflector collects the acoustical energy and focuses it onto the microphone. This mechanical amplification of sound allows sounds from a few hundred metres to several kilometres to be picked up and recorded. In this case study, you are challenged to build your own parabolic microphone and provide a mathematical analysis of its operation.

Slope and Angle of Inclination of a Straight Line

Recall from Sec. 15–3 that the slope of a straight line for which

$$\text{rise} = \Delta y = y_2 - y_1$$
$$\text{run} = \Delta x = x_2 - x_1$$

is given by the following equation:

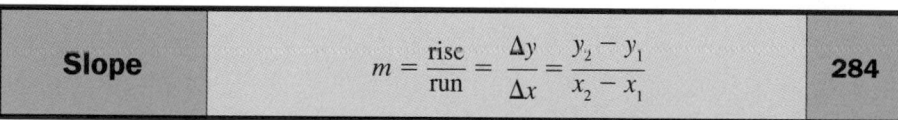

Slope	$$m = \frac{\text{rise}}{\text{run}} = \frac{\Delta y}{\Delta x} = \frac{y_2 - y_1}{x_2 - x_1}$$	**284**

The slope is equal to the rise divided by the run.

We now relate the slope to the angle of inclination. The smallest positive angle that a line makes with the positive x direction is called the angle of inclination θ of the line (Fig. 22–5). A line parallel to the x axis has an angle of inclination of zero. Thus the angle of inclination can have values from 0° up to 180°. From Fig. 22–4,

$$\tan \theta = \frac{\text{opposite side}}{\text{adjacent side}} = \frac{y_2 - y_1}{x_2 - x_1} = \frac{\text{rise}}{\text{run}} = \frac{\Delta y}{\Delta x}$$

But this is the slope m of the line, so we have the following equations:

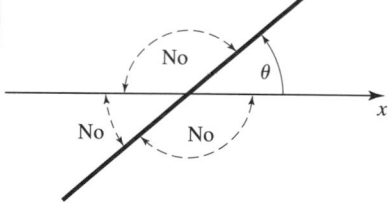

FIGURE 22–5 Angle of inclination, θ.

Slope	$$m = \tan \theta$$ $$0° \leq \theta < 180°$$	**285**

The slope of a line is equal to the tangent of the angle of inclination θ (except when $\theta = 90°$).

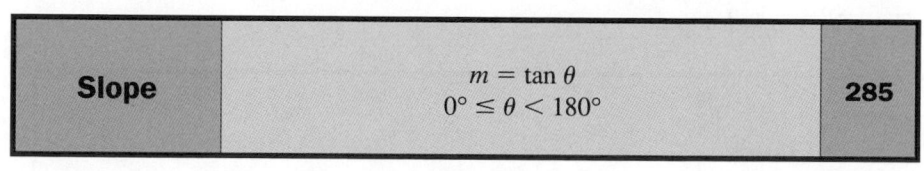

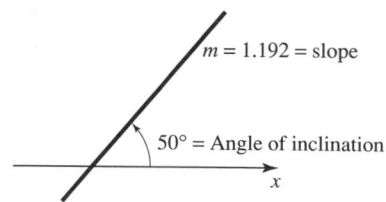

FIGURE 22–6

◆◆◆ **Example 6:** The slope of a line having an angle of inclination of 50° (Fig. 22–6) is, by Eq. 285,

$$m = \tan 50° = 1.192 \text{ (rounded)} \qquad \blacklozenge\blacklozenge\blacklozenge$$

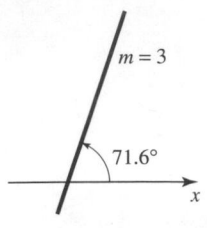

FIGURE 22–7

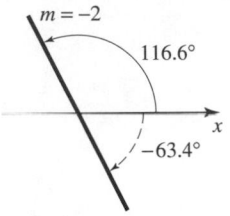

FIGURE 22–8

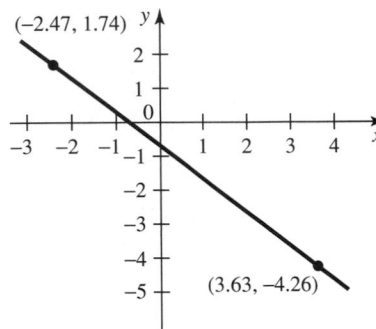

FIGURE 22–9

◆◆◆ Example 7: Find the angle of inclination of a line having a slope of 3 (Fig. 22–7).

Solution: By Eq. 285, $\tan \theta = 3$, so

$$\theta = \arctan 3 = 71.6° \quad \text{(rounded)}$$

When the slope is negative, our calculator will give us a negative angle, which we then use to obtain a positive angle of inclination less than 180°. ◆◆◆

◆◆◆ Example 8: For a line having a slope of 22 (Fig. 22–8), $\arctan (22) = 263.4°$ (rounded), so

$$\theta = 180° - 63.4° = 116.6°$$ ◆◆◆

◆◆◆ Example 9: Find the angle of inclination of the line passing through (22.47, 1.74) and (3.63, 24.26) (Fig. 22–9).

Solution: The slope, from Eq. 284, is

$$m = \frac{-4.26 - 1.74}{3.63 - (-2.47)} = -0.984$$

from which $\theta = 135.5°$ ◆◆◆

When *different scales* are used for the x and y axes, the angle of inclination will appear distorted.

◆◆◆ Example 10: The angle of inclination of the line in Fig. 22–10 is $\theta = \arctan 10 = 84.3°$. Notice that the angle in the graph appears much smaller than 84.3° because of the different scales on each axis.

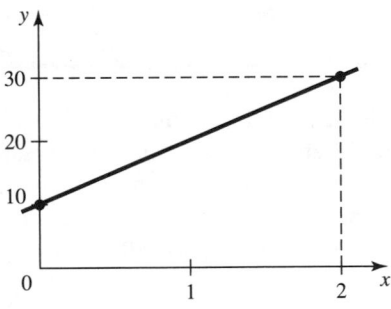

FIGURE 22–10

Slopes of Parallel and Perpendicular Lines
Parallel lines have *equal* slopes.

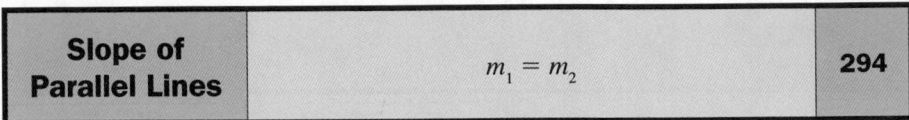

Slope of Parallel Lines	$m_1 = m_2$	294

Now consider a line L_1, with slope m_1 and angle of inclination θ_1 (Fig. 22–11), and a perpendicular line L_2, with slope m_2 and angle of inclination $\theta_2 = \theta_1 + 90°$. Note that $\angle QOS = \theta_2 - 90° = \theta_1 + 90° - 90° = \theta_1$, so right triangles OPR and OQS are congruent. Thus the magnitudes of their corresponding sides a and b are equal. The slope m_1 of L_1 is a/b, and the slope m_2 of L_2 is $-b/a$, so we have the following equation:

Slope of Perpendicular Lines	$m_1 = -\dfrac{1}{m_2}$	**295**

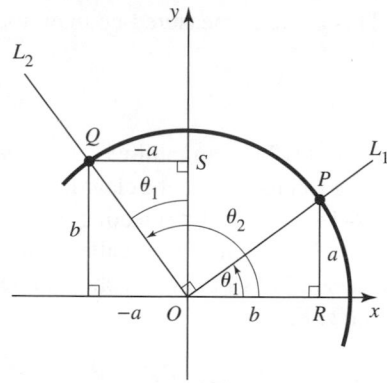

FIGURE 22–11 Slopes of perpendicular lines.

The slope of a line is the negative reciprocal of the slope of a perpendicular to that line.

◆◆◆ **Example 11:** Any line perpendicular to a line whose slope is 5 has a slope of $-\frac{1}{5}$. ◆◆◆

Common Error	The minus sign in Eq. 295 is often forgotten. $$m_1 \neq \frac{1}{m_2}$$

◆◆◆ **Example 12:** Is angle ACB of Fig. 22–12 a right angle?

Solution: The slope of AC is

$$\frac{8-3}{5-(-15)} = \frac{5}{20} = \frac{1}{4}$$

To be perpendicular, BC must have a slope of -4. Its actual slope is

$$\frac{-10-8}{10-5} = \frac{-18}{5} = -3.6$$

Although lines AC and BC appear to be perpendicular, we have shown that they are not. Thus $\underline{/ACB}$ is not a right angle.

◆◆◆

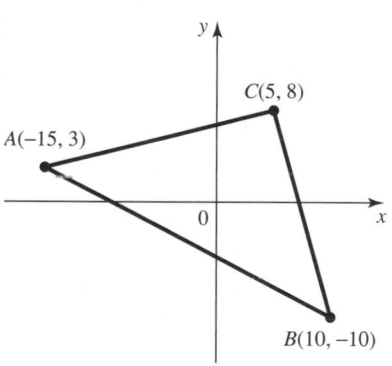

FIGURE 22–12

Angle of Intersection between Two Lines

Figure 22–13 shows two lines L_1 and L_2 intersecting at an angle ϕ, measured counterclockwise from line 1 to line 2. We want a formula for ϕ in terms of the slopes m_1 and m_2 of the lines. By Eq. 142, θ_2 equals the sum of the angle of inclination θ_1 and the angle of intersection ϕ. So $\phi = \theta_2 - \theta_1$. Taking the tangent of both sides gives $\tan\phi = \tan(\theta_2 - \theta_1)$, or

$$\tan\phi = \frac{\tan\theta_2 - \tan\theta_1}{1 + \tan\theta_1 \tan\theta_2}$$

by the trigonometric identity for the tangent of the difference of two angles (Eq. 169). But the tangent of an angle of inclination is the slope, so we have the following equation:

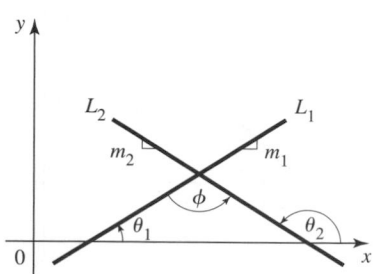

FIGURE 22–13 Angle of intersection between two lines.

Angle of Intersection between Two Lines	$\tan\phi = \dfrac{m_2 - m_1}{1 + m_1 m_2}$	**296**

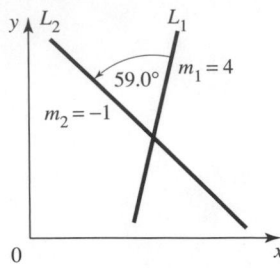

FIGURE 22–14

◆◆◆ **Example 13:** The tangent of the angle of intersection between line L_1, having a slope of 4, and line L_2, having a slope of -1, is

$$\tan \phi = \frac{-1 - 4}{1 + 4(-1)} = \frac{-5}{-3} = \frac{5}{3}$$

from which $\phi = \arctan \frac{5}{3} = 59.0°$ (rounded). This angle is measured counterclockwise from line 1 to line 2, as shown in Fig. 22–14.   ◆◆◆

Tangents and Normals to Curves

We know what the tangent to a circle is, and we probably have an intuitive idea of what the *tangent to a curve* is. We'll speak a lot about tangents, so let us try to get a clear idea of what one is.

To determine the tangent line at P in Fig. 22–15, we start by selecting a second point, Q, on the curve. As shown, Q can be on either side of P. The line PQ is called a *secant line*. We then let Q approach P (from either side). If the secant lines PQ approach a single line T as Q approaches P, then line T is called the *tangent line at P*.

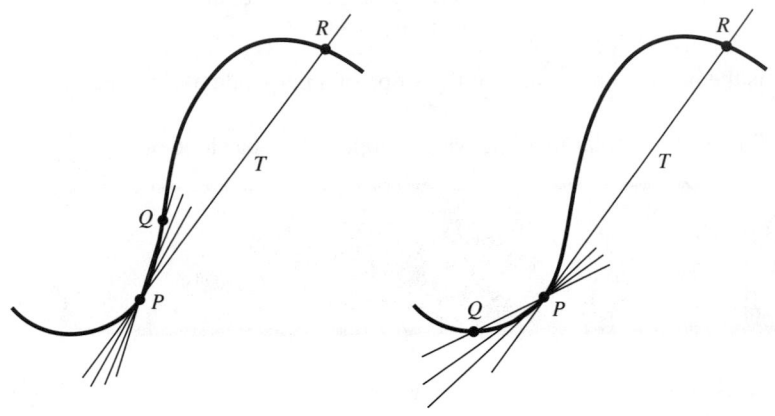

In calculus you will see that the slope gives the *rate of change* of the function, an extremely useful quantity to find. You will see that it is found by taking the *derivative* of the function.

FIGURE 22–15   A tangent line T at point P, here defined as the limiting position of a secant line PQ, as Q approaches P.

Note that the tangent line can intersect the curve at more than one point, such as point R. However, at points "near" P, the tangent line intersects the curve just once while a secant line intersects the curve twice. The *slope of a curve* at some point P is defined as the *slope of the tangent* to the curve drawn through that point.

◆◆◆ **Example 14:** Plot the curve $y = x^2$ for integer values of x from 0 to 4. By eye, draw the tangent to the curve at $x = 2$, and find the approximate value of the slope.

x	0	1	2	3	4
y	0	1	4	9	16

Solution: We make a table of point pairs.
The plot and the tangent line are shown in Fig. 22–16. Using the scales on the x and y axes, we measure a rise of 12 units for the tangent line in a run of 3 units. The slope of the tangent line is then

$$m_t \approx \frac{12}{3} = 4$$ ◆◆◆

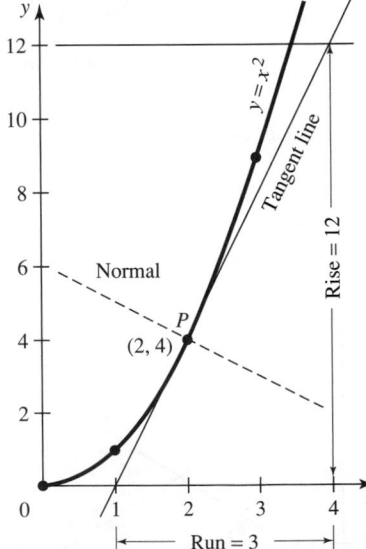

FIGURE 22–16   Slope of a curve.

The normal to a curve at a given point is the line perpendicular to the tangent at that point, as in Fig. 22–16. Its slope is the negative reciprocal of the slope of the tangent. For Example 14, $m_n = -\frac{1}{4}$.

Equation of a Straight Line

This is a good time to glance back at Sec. 5–3 and review some of the exercises there.

In Sec. 5–3 we introduced the *slope-intercept form* for the equation of a straight line. The equation for a line with a slope m and y intercept b is as follows:

Slope-Intercept Form	$y = mx + b$	**289**

Using this equation, we can quickly write the equation of a straight line if we know the slope and y intercept. However, we might have *other* information about the line, such as the coordinates of two points on that line, and want to write the equation using that information. To enable us to do this, we'll now derive several other forms for the straight-line equation.

General Form

The slope-intercept form is the equation of a straight line in *explicit* form, $y = \phi(x)$. We can also write the equation of a straight line in implicit form, $\phi(x, y) = 0$, simply by transposing all terms to one side of the equals sign and simplifying. We usually write the x term first, then the y term, and finally the constant term. This form is referred to as the *general form* of the equation of a straight line.

General Form	$Ax + By + C = 0$	**286**

Here A, B, and C are constants.

You will see as we go along that there are several forms for the equation of a straight line, and we use the one that is most convenient in a particular problem. But to make it easy to compare answers, we usually rearrange the equation into general form.

◆◆◆ **Example 15:** Change the equation $y = \dfrac{6}{7}x - 5$ from slope-intercept form to general form.

Solution: Subtracting y from both sides, we have

$$0 = \frac{6}{7}x - 5 - y$$

Multiplying by 7 and rearranging gives $6x - 7y - 35 = 0$. ◆◆◆

Point-Slope Form

Let us write an equation for a line that has slope m, but that passes through a given point (x_1, y_1) which is not, in general, on either axis, as in Fig. 22–17. Again using the definition of slope (Eq. 284), with a general point (x, y), we get the following form:

Point-Slope Form	$m = \dfrac{y - y_1}{x - x_1}$ or $y - y_1 = m(x - x_1)$	**291**

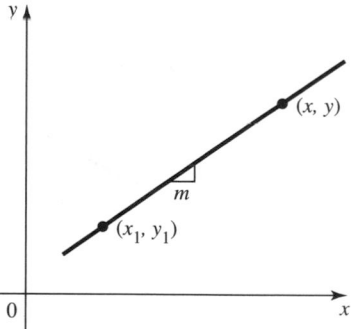

FIGURE 22–17

This form of the equation of a straight line is most useful when we know the slope of a line and one point through which the line passes.

◆◆◆ **Example 16:** Write the equation in general form of the line having a slope of 2 and passing through the point $(1, -3)$ (Fig. 22–18).

Solution: Substituting $m = 2$, $x_1 = 1$, $y_1 = -3$ into Eq. 291 gives us

$$2 = \frac{y - (-3)}{x - 1}$$

Multiplying by $x - 1$ yields

$$2x - 2 = y + 3$$

or, in general form, $2x - y - 5 = 0$.

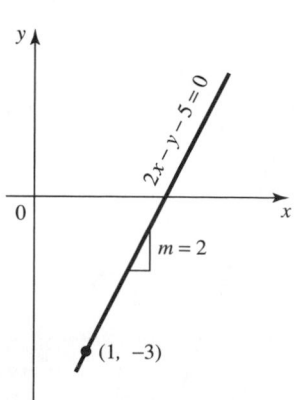

◆◆◆ **FIGURE 22–18**

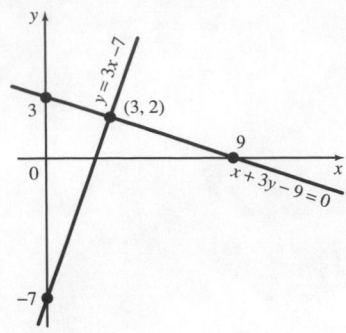

FIGURE 22-19

◆◆◆ **Example 17:** Write the equation in general form of the line passing through the point (3, 2) and perpendicular to the line $y = 3x - 7$ (Fig. 22–19).

Solution: The slope of the given line is 3, so the slope of our perpendicular line is $-\frac{1}{3}$. Using the point-slope form, we obtain

$$-\frac{1}{3} = \frac{y - 2}{x - 3}$$

Going to the general form, $x - 3 = -3y + 6$, or

$$x + 3y - 9 = 0$$

◆◆◆

◆◆◆ **Example 18:** Write the equation in general form of the line passing through the point (3.15, 5.88) and having an angle of inclination of 27.8° (Fig. 22–20).

Solution: From Eq. 285, $m = \tan 27.8° = 0.527$. From Eq. 291,

$$0.527 = \frac{y - 5.88}{x - 3.15}$$

Changing to general form, $0.527x - 1.66 = y - 5.88$, or

$$0.527x - y + 4.22 = 0$$

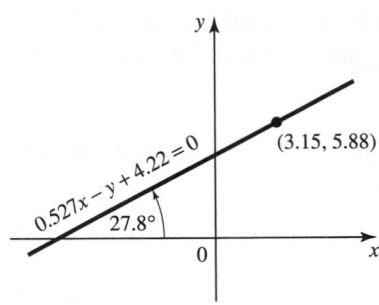

FIGURE 22-20   ◆◆◆

Two-Point Form

If two points on a line are known, the equation of the line is easily written using the two-point form, which we now derive. If we call the points P_1 and P_2 in Fig. 22–21, the slope of the line is

$$m = \frac{y_2 - y_1}{x - x_1}$$

FIGURE 22-21

The slope of the line segment connecting P_1 with any other point P on the same line is

$$m = \frac{y - y_1}{x - x_1}$$

Since these slopes must be equal, we get the following equation:

Two Point Form	$\dfrac{y - y_1}{x - x_1} = \dfrac{y_2 - y_1}{x_2 - x_1}$	290

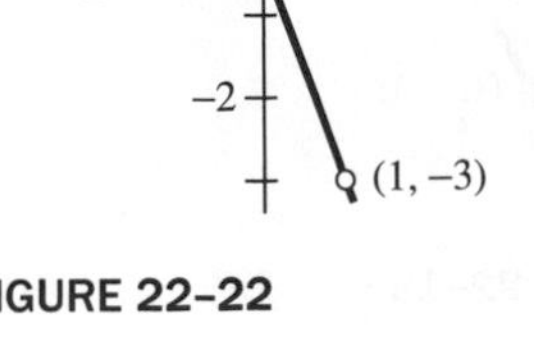

◆◆◆ **Example 19:** Write the equation in general form of the line passing through the points (1, −3) and (−2, 5) (Fig. 22–22).

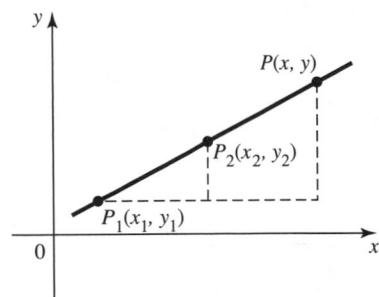

FIGURE 22-22

Solution: Calling the first given point P_1 and the second P_2 and substituting into Eq. 290, we have

$$\frac{y - (-3)}{x - 1} = \frac{5 - (-3)}{-2 - 1} = \frac{8}{-3}$$

$$\frac{y + 3}{x - 1} = -\frac{8}{3}$$

Putting the equation into general form, we have $3y + 9 = -8x + 8$, or

$$8x + 3y + 1 = 0 \qquad \text{◆◆◆}$$

Lines Parallel to the Coordinate Axes

Line 1 in Fig. 22–23 is parallel to the x axis. Its slope is 0 and it cuts the y axis at $(0, b)$. From the point-slope form, $y - y_1 = m(x - x_1)$, we get $y - b = 0(x - 0)$, or the following:

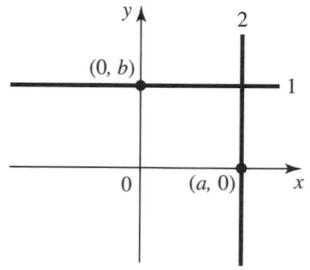

FIGURE 22–23

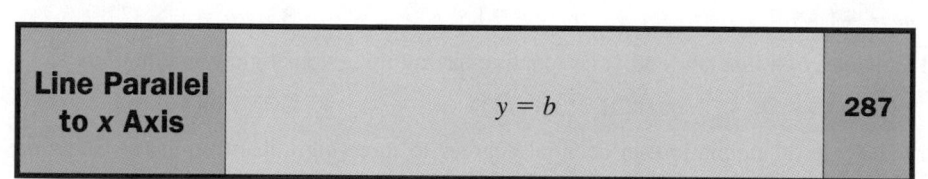

| Line Parallel to x Axis | $y = b$ | 287 |

Line 2 has an undefined slope, but we get its equation by noting that $x = a$ at every point on the line, *regardless of the value of y*. Thus:

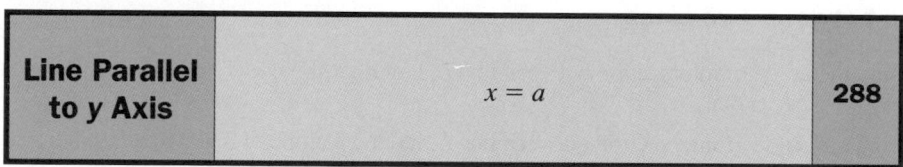

| Line Parallel to y Axis | $x = a$ | 288 |

◆◆◆ **Example 20:**

(a) A line that passes through the point $(5, -2)$ and is parallel to the x axis has the equation $y = -2$.

(b) A line that passes through the point $(5, -2)$ and is parallel to the y axis has the equation $x = 5$. ◆◆◆

Exercise 1 ◆ The Straight Line

Length of a Line Segment

Find the distance between the given points.

1. $(5, 0)$ and $(2, 0)$ **2.** $(0, 3)$ and $(0, -5)$

3. $(-2, 0)$ and $(7, 0)$ **4.** $(-4, 0)$ and $(-6, 0)$

5. $(0, -2.74)$ and $(0, 3.86)$ **6.** $(55.34, 0)$ and $(25.38, 0)$

7. $(5.59, 3.25)$ and $(8.93, 3.25)$

8. $(-2.06, -5.83)$ and $(-2.06, -8.34)$

9. $(8.38, -3.95)$ and $(2.25, -4.99)$

Find the directed distance AB.

10. $A(3,0)$; $B(5, 0)$ **11.** $A(3.95, -2.07)$; $B(-3.95, -2.07)$

12. $B(-8, -2)$; $A(-8, -5)$ **13.** $A(-9, -2)$; $B(17, -2)$

14. $B(-6, -6)$; $A(-6, -7)$ **15.** $B(11.5, 3.68)$; $A(11.5, -5.38)$

Slope and Angle of Inclination

Find the slope of the line having the given angle of inclination.

16. $38.2°$

17. $77.9°$

18. 1.83 rad

19. $58°14'$

20. $156.3°$

21. $132.8°$

Find the angle of inclination, in decimal degrees to three significant digits, of a line having the given slope.

22. $m = 3$

23. $m = 1.84$

24. $m = -4$

25. $m = -2.75$

26. $m = 0$

27. $m = -15$

Find the slope of a line perpendicular to a line having the given slope.

28. $m = 5$

29. $m = 2$

30. $m = 4.8$

31. $m = -1.85$

32. $m = -2.85$

33. $m = -5.372$

Find the slope of a line perpendicular to a line having the given angle of inclination.

34. $58.2°$

35. 1.84 rad

36. $136°44'$

Find the angle of inclination, in decimal degrees to three significant digits, of a line passing through the given points.

37. $(5, 2)$ and $(-3, 4)$

38. $(-2.5, -3.1)$ and $(5.8, 4.2)$

39. $(6, 3)$ and $(-1, 5)$

40. $(x, 3)$ and $(x + 5, 8)$

Angle between Two Lines

41. Find the angle of intersection between line L_1 having a slope of 1 and line L_2 having a slope of 6.

42. Find the angle of intersection between line L_1 having a slope of 3 and line L_2 having a slope of -2.

43. Find the angle of intersection between line L_1 having an angle of inclination of $35°$ and line L_2 having an angle of inclination of $160°$.

44. Find the angle of intersection between line L_1 having an angle of inclination of $22°$ and line L_2 having an angle of inclination of $86°$.

Tangents to Curves

Plot the functions from $x = 0$ to $x = 4$. Graphically find the slope of the curve at $x = 2$.

45. $y = x^3$

46. $y = -x^2 + 3$

47. $y = \dfrac{x^2}{4}$

Equation of a Straight Line

Write the equation, in general form, of each line.

48. slope $= 4$; y intercept $= 23$

49. slope $= 2.25$; y intercept $= -1.48$

50. passes through points $(3, 5)$ and $(-1, 2)$

51. passes through points $(4.24, -1.25)$ and $(3.85, 4.27)$

52. slope $= -4$; passes through $(-2, 5)$

53. slope $= -2$; passes through $(-2, -3)$

54. x intercept $= 5$; y intercept $= -3$

55. x intercept $= -2$; y intercept $= 6$

56. y intercept $= 3$; parallel to $y = 5x - 2$

57. y intercept $= -2.3$; parallel to $2x - 3y + 1 = 0$

58. y intercept $= -5$; perpendicular to $y = 3x - 4$

59. y intercept $= 2$; perpendicular to $4x - 3y = 7$

60. passes through $(-2, 5)$; parallel to $y = 5x - 1$

61. passes through $(4, -1)$; parallel to $4x - y = 23$

62. passes through $(-4, 2)$; perpendicular to $y = 5x - 3$

63. passes through $(6, 1)$; perpendicular to $6y - 2x = 3$

64. y intercept $= 4.0$; angle of inclination $= 48°$

65. y intercept $= -3.52$; angle of inclination $= 154°44'$

66. passes through $(4, -1)$; angle of inclination $= 22.8°$

67. passes through $(-2.24, 5.17)$; angle of inclination $= 68°14'$

68. passes through $(5, 2)$; is parallel to the x axis.

69. passes through $(-3, 6)$; is parallel to the y axis.

70. line A in Fig. 22–24

71. line B in Fig. 22–24

72. line C in Fig. 22–24

73. line D in Fig. 22–24

74. Find the length of girder AB in Fig. 22–25.

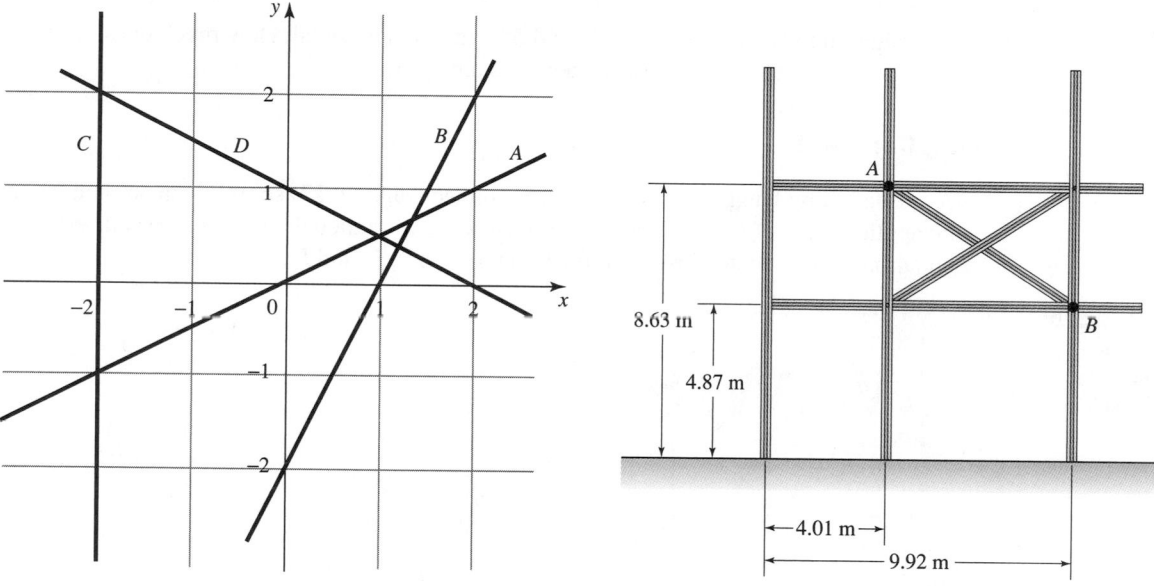

FIGURE 22–24   **FIGURE 22–25**

75. Find the distance between the centres of the holes in Fig. 22–26.

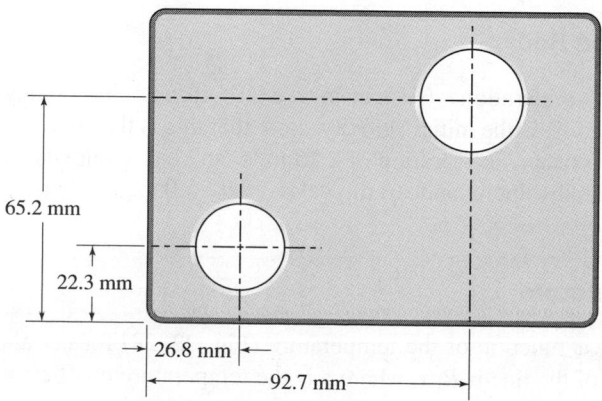

FIGURE 22–26

76. A triangle has vertices at $(3, 5)$, $(-2, 4)$, and $(4, -3)$. Find the length of each side. Then compute the area using Hero's formula (Eq. 138). Work to three significant digits.

77. The distance between two stakes on a slope is taped at 2055 ft., and the angle of the slope with the horizontal is $12.3°$. Find the horizontal distance between the stakes.

78. What is the angle of inclination with the horizontal of a roadbed that rises 15.0 m in each $25\overline{0}$ m, measured horizontally?

79. How far apart must two stakes on a $7°$ slope be placed so that the horizontal distance between them is 1250 m?

80. On a 5% road grade, at what angle is the road inclined to the horizontal? How far does one rise in travelling upward $50\overline{0}$ m, measured along the road?

81. A straight tunnel under a river is 755 m long and descends 12 m in this distance. What angle does the tunnel make with the horizontal?

82. A straight driveway slopes downward from a house to a road and is 28.0 m in length. If the angle of inclination from the road to the house is $3.6°$, find the height of the house above the road.

83. An escalator is built so as to rise 2 m for each 3 m of horizontal travel. Find its angle of inclination.

84. A straight highway makes an angle of $4.5°$ with the horizontal. How much does the highway rise in a distance of 1.20 km, measured along the road?

Spring Constant

85. A spring whose length is L_0 with no force applied (Fig. 22–27) stretches an amount x with an applied force of F, where $F = kx$. The constant k is called the *spring constant*. Write an equation, in slope-intercept form, for F in terms of k, L, and L_0.

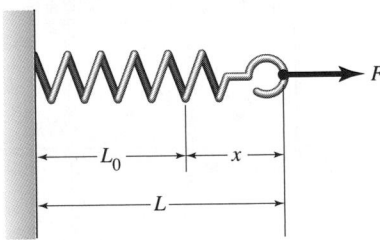

FIGURE 22–27

86. What force would be needed to stretch a spring ($k = 14.5$ N/cm) from an unstretched length of 8.50 cm to a length of 12.50 cm?

Velocity of Uniformly Accelerated Body

87. When a body moves with constant acceleration a (such as in free fall), its velocity v at any time t is given by $v = v_0 + at$, where v_0 is the initial velocity. Note that this is the equation of a straight line. If a body has a constant acceleration of 2.15 m/s^2 and has a velocity of 21.8 m/s at 5.25 s, find (a) the initial velocity and (b) the velocity at 25.0 s.

Resistance Change with Temperature

88. The resistance of metals is a linear function of the temperature (Fig. 22–28) for certain ranges of temperature. The slope of the line is $R_1\alpha$, where a is the temperature coefficient of resistance at temperature t_1. If R is the resistance of any temperature t, write an equation for R as a function of t.

89. Using a value for α of $\alpha = 1/(234.5t_1)$ (for copper), find the resistance of a copper conductor at 75.0 °C if its resistance at 20.0 °C is 148.4 Ω.

90. If the resistance of the copper conductor in problem 89 is 1255 Ω at 20.0 °C, at what temperature will the resistance be 1265 Ω?

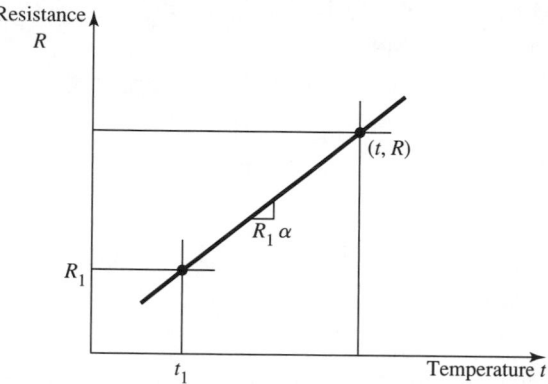

FIGURE 22–28

Thermal Expansion

91. When a bar is heated, its length will increase from an initial length L_0 at temperature t_0 to a new length L at temperature t. The plot of L versus t is a straight line (Fig. 22–29) with a slope of $L_0\alpha$, where α is the coefficient of thermal expansion. Derive the equation $L = L_0(1 + \alpha \, \Delta t)$, where Δt is the change in temperature, $t - t_0$.

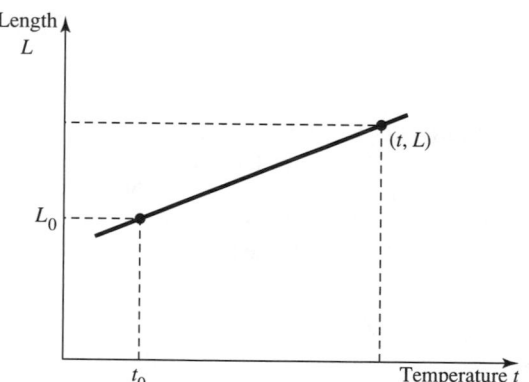

FIGURE 22–29

92. A steel pipe is 21.50 m long at 0 °C. Find its length at 75.0 °C if α for steel is 12.0×10^{-6} per degree Celsius.

Fluid Pressure

93. The pressure at a point located at a depth x ft. from the surface of the liquid varies directly as the depth. If the pressure at the surface is 20.6 lb./sq. in. and increases by 0.432 lb./sq. in. for every foot of depth, write an equation for P as a function of the depth x (in feet). At what depth will the pressure be 30.0 lb./sq. in.?

94. A straight pipe slopes downward from a reservoir to a water turbine (Fig. 22–30). The pressure head at any point in the pipe, expressed in metres, is equal to the vertical distance between the point and the surface of the reservoir. If the reservoir surface is 25 m above the upper end of the pipe, write an expression for the head H as a function of the horizontal distance x. At what distance x will the head be 35 m?

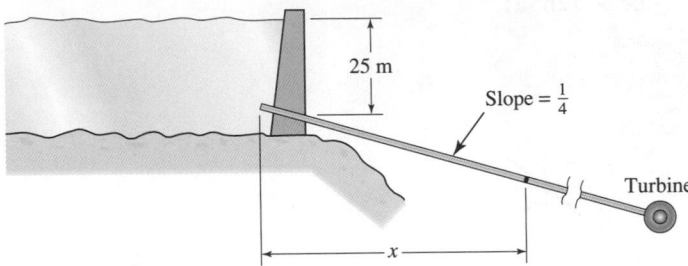

FIGURE 22–30

Temperature Gradient

95. Figure 22–31 shows a uniform wall whose inside face is at temperature t_i and whose outside face is at t_o. The temperatures within the wall plot as a straight line connecting t_i and t_o. Write the equation $t = f(x)$ of that line, taking $x = 0$ at the inside face, if $t_i = 25.0\ °\text{C}$ and $t_o = 30.0\ °\text{C}$. At what x will the temperature be 0 °C? What is the slope of the line?

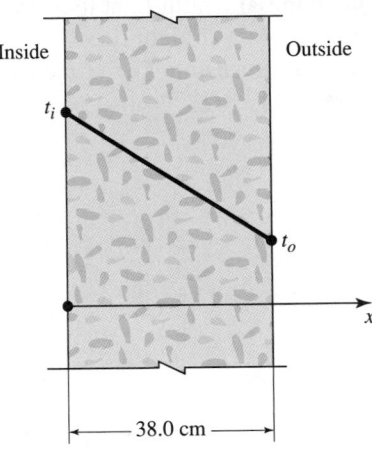

FIGURE 22–31   Temperature drop in a wall. The slope of the line (in °C/cm) is called the *temperature gradient*. The amount of heat flowing through the wall is proportional to the temperature gradient.

Hooke's Law

96. The increase in length of a wire in tension is directly proportional to the applied load P. Write an equation for the length L of a wire that has an initial length of 3.00 m and that stretches 1.00 mm for each 12.5 N. Find the length of the wire with a load of 750 N.

Straight-Line Depreciation

97. The straight-line method is often used to depreciate a piece of equipment for tax purposes. Starting with the purchase price P, the item is assumed to drop in value the same amount each year (the annual depreciation) until the salvage value S is reached (Fig. 22–32). Write an expression for the book value y (the value at any time t) as a function of the number of years t. For a lathe that cost $15,428 and has a salvage value of $2,264, find the book value after 15 years if it is depreciated over a period of 20 years.

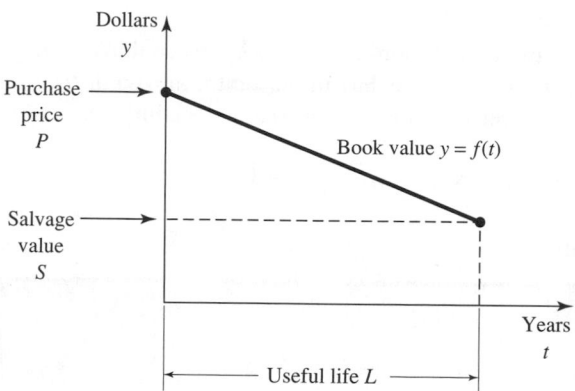

FIGURE 22–32 Straight-line depreciation.

22–2 The Circle

Definition

Anyone who has drawn a circle using a compass will not be surprised by the following definition of the circle:

Definition of a Circle	A *circle* is a plane curve all points of which are at a fixed distance (the *radius*) from a fixed point (the *centre*).	303

Circle with Centre at the Origin

We will now derive an equation for a circle of radius r, starting with the simplest case of a circle whose centre is at the origin (Fig. 22–33). Let x and y be the coordinates of any point P on the circle. The equation we develop will give a relationship between x and y that will have two meanings: geometrically, (x, y) will represent a point on the circle; and algebraically, the numbers corresponding to those points will satisfy that equation.

We observe that, by the definition of a circle, the distance OP must be constant and equal to r. But, by the distance formula (Eq. 283),

$$OP = \sqrt{x^2 + y^2} = r$$

Squaring, we get a *standard equation of a circle* (also called *standard form of the equation*).

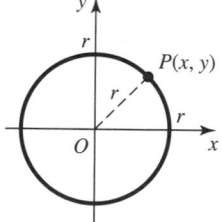

FIGURE 22–33 Circle with centre at origin.

Note that both x^2 and y^2 have the same coefficient. Otherwise, the graph is not a circle.

Standard Equation, Circle of Radius *r*: Centre at Origin (*O*)	$x^2 + y^2 = r^2$	304

◆◆◆ **Example 21:** Write, in standard form, the equation of a circle of radius 3, whose centre is at the origin.

Solution:

$$x^2 + y^2 = 3^2 = 9$$

◆◆◆

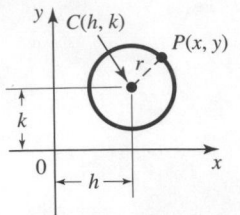

FIGURE 22-34 Circle with centre at (h, k).

Circle with Centre Not at the Origin

Figure 22–34 shows a circle whose centre has the coordinates (h, k). We can think of the difference between this circle and the one in Fig. 22–33 as having its centre moved or *translated h* units to the right and k units upward. Our derivation is similar to the preceding one.

$$CP = r = \sqrt{(x - h)^2 + (y - k)^2}$$

Squaring, we get the following equation:

Standard Equation, Circle of Radius *r*: Centre at (*h*, *k*)	$(x - h)^2 + (y - k)^2 = r^2$	**305**

◆◆◆ **Example 22:** Write, in standard form, the equation of a circle of radius 5 whose centre is at $(3, -2)$.

Solution: We substitute into Eq. 305 with $r = 5$, $h = 3$, and $k = -2$.

$$(x - 3)^2 + [y - (-2)]^2 = 5^2$$
$$(x - 3)^2 + (y + 2)^2 = 25$$

◆◆◆

◆◆◆ **Example 23:** Find the radius and the coordinates of the centre of the circle $(x + 5)^2 + (y - 3)^2 = 16$.

Solution: We see that $r^2 = 16$, so the radius r is 4. Also, since

$$x - h = x + 5$$

then

$$h = -5$$

and since

$$y - k = y - 3$$

then

$$k = 3$$

So the centre is at $(-5, 3)$ as shown in Fig. 22–35.

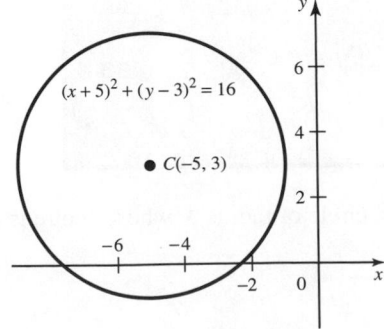

FIGURE 22–35

◆◆◆

Common Error	It is easy to get the signs of h and k wrong. In Example 23, do *not* take $$h = +5 \quad \text{and} \quad k = -5$$

Translation of Axes

We see that Eq. 305 for a circle with its centre at (h, k) is almost identical to Eq. 304 for a circle with centre at the origin, *except that x has been replaced by* $(x - h)$ *and y has been replaced by* $(y - k)$. This same substitution will, of course, work for curves other than the circle, and we will use it to translate or *shift* the axes for the other conic sections.

Translation of Axes	To translate or shift the axes of a curve to the left by a distance h and downward by a distance k, replace x by $(x - h)$ and y by $(y - k)$ in the equation of the curve.	**298**

General Equation of a Circle

A second-degree equation in x and y which has all possible terms would have an x^2 term, a y^2 term, an xy term, and all terms of lesser degree as well. The general second-degree equation is usually written with terms in the following order, where $A, B, C, D, E,$ and F are constants:

General Second-Degree Equation	$Ax^2 + Bxy + Cy^2 + Dx + Ey + F = 0$	**297**

There are six constants in this equation, but only five are independent. We can divide through by any constant and thus make it equal to 1.

If we expand Eq. 305, we get

$$(x - h)^2 + (y - k)^2 = r^2$$
$$x^2 - 2hx + h^2 + y^2 - 2ky + k^2 = r^2$$

Rearranging gives

$$x^2 + y^2 - 2hx - 2ky + (h^2 + k^2 - r^2) = 0$$

or an equation with $D, E,$ and F as constants.

General Equation of a Circle	$x^2 + y^2 + Dx + Ey + F = 0$	**306**

Comparing this with the general second-degree equation, we see that the general second-degree equation

$$Ax^2 + Bxy + Cy^2 + Dx + Ey + F = 0$$

represents a circle if $B = 0$ and $A = C$.

◆◆◆ **Example 24:** Write the equation of Example 23 in general form.

Solution: The equation, in standard form, was

$$(x + 5)^2 + (y - 3)^2 = 16$$

Expanding, we obtain

$$x^2 + 10x + 25 + y^2 - 6y + 9 = 16$$

or

$$x^2 + y^2 + 10x - 6y + 18 = 0 \qquad \text{◆◆◆}$$

Changing from General to Standard Form

When we want to go from general form to standard form, we must complete the square, both for x and for y.

We learned how to complete the square in Sec. 14–2 of Chapter 14, "Quadratic Equations." Glance back there if you need to refresh your memory.

◆◆◆ **Example 25:** Write the equation $2x^2 + 2y^2 - 18x + 16y + 60 = 0$ in standard form. Find the radius and centre, and plot the curve.

Solution: We first divide by 2:

$$x^2 + y^2 - 9x + 8y + 30 = 0$$

and separate the x and y terms:

$$(x^2 - 9x) + (y^2 + 8y) = 230$$

We complete the square for the x terms with $\frac{81}{4}$ because we take $\frac{1}{2}$ of (-9) and square it. Similarly, for completing the square for the y terms, $\left(\frac{1}{2} \times 8\right)^2 = 16$. Completing the square on the left and compensating on the right gives

$$\left(x^2 - 9x + \frac{81}{4}\right) + (y^2 + 8y + 16) = -30 + \frac{81}{4} + 16$$

Factoring each group of terms yields

$$\left(x - \frac{9}{2}\right)^2 + (y + 4)^2 = \left(\frac{5}{2}\right)^2$$

So

$$r = \frac{5}{2} \qquad h = \frac{9}{2} \qquad k = -4$$

The circle is shown in Fig. 22–36. ◆◆◆

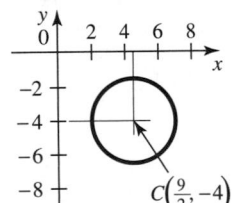

FIGURE 22–36.

Applications

◆◆◆ **Example 26:** Write an equation for the circle shown in Fig. 22–37, and find the dimensions A and B produced by the circular cutting tool.

Estimate: The easiest way to get an estimate here is to make a sketch and measure the distances. We get

$$A \approx 2.6 \text{ cm} \qquad \text{and} \qquad B \approx 2.7 \text{ cm}$$

Note that the orientations of the x and y axes are reversed in this example.

Solution: In order to write an equation, we must have coordinate axes. Otherwise, the quantities x and y in the equation will have no meaning. Since no axes are given, we are free to draw them anywhere we please. Our first impulse might be to place the origin at the corner p, since most of the dimensions are referenced from that point, but our equation will be simpler if we place the origin at the centre of the circle. Let us draw axes as shown, so that not all our numbers will be negative. The equation of the circle is then

$$x^2 + y^2 = (1.500)^2$$
$$= 2.2500$$

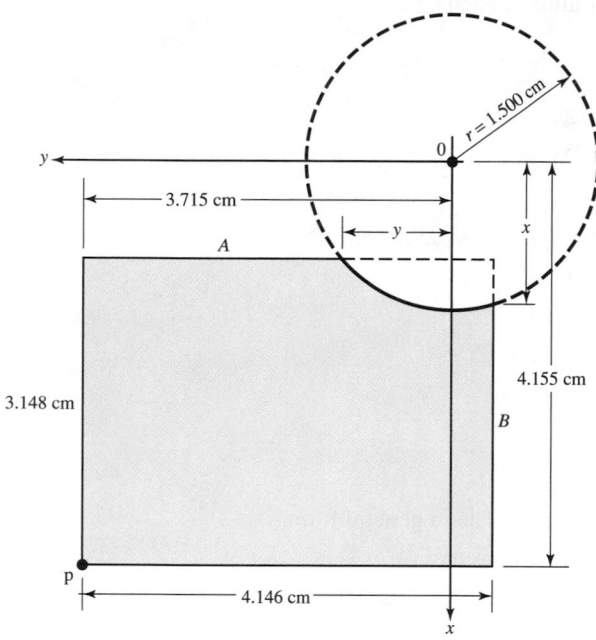

FIGURE 22–37

The distance from the y axis to the top edge of the block is $4.155 - 3.148 - 1.007$ cm. Substituting this value for x into the equation of the circle will give the corresponding value for y. When $x = 1.007$,

$$y^2 = 2.250 - (1.007)^2 = 1.236$$
$$y = 1.112 \text{ cm}$$

Similarly, the distance from the x axis to the right edge of the block is $4.146 - 3.715 = 0.431$ cm. When $y = 20.431$,

$$x^2 = 2.250 - (20.431)^2 = 2.064$$
$$x = 1.437$$

Finally,

$$A = 3.715 - 1.112 = 2.603 \text{ cm}$$

and

$$B = 4.155 - 1.437 = 2.718 \text{ cm} \qquad \qquad \text{◆◆◆}$$

Exercise ✦ The Circle

Equation of a Circle

Write the equation of each circle in standard form.

1. centre at $(0, 0)$; radius $= 7$

2. centre at $(0, 0)$; radius $= 4.82$

3. centre at (2, 3); radius = 5
4. centre at (5, 2); radius = 10
5. centre at (5, −3); radius = 4
6. centre at (−3, −2); radius = 11

Find the centre and radius of each circle.

7. $x^2 + y^2 = 49$
8. $x^2 + y^2 = 64.8$
9. $(x − 2)^2 + (y + 4)^2 = 16$
10. $(x + 5)^2 + (y − 2)^2 = 49$
11. $(y + 5)^2 + (x − 3)^2 = 36$
12. $(x − 2.22)^2 + (y + 7.16)^2 = 5.93$
13. $x^2 + y^2 − 8x = 0$
14. $x^2 + y^2 − 2x − 4y = 0$
15. $x^2 + y^2 − 10x + 12y + 25 = 0$
16. $x^2 + y^2 − 4x + 2y = 36$
17. $x^2 + y^2 + 6x − 2y = 15$
18. $x^2 + y^2 − 2x − 6y = 39$

Write the equation of each circle in general form.

19. centre at (1, 1); passes through (4, −3)
20. diameter joins the points (−2, 5) and (6, − 1)

Tangent to a Circle

Write the equation of the tangent to each circle at the given point. (*Hint:* The slope of the tangent is the negative reciprocal of the slope of the radius to the given point.)

21. $x^2 + y^2 = 25$ at (4, 3)
22. $(x − 5)^2 + (y − 6)^2 = 100$ at (11, 14)
23. $x^2 + y^2 + 10y = 0$ at (−4, −2)

Intercepts

Find the x and y intercepts for each circle. (*Hint:* Set x and y, in turn, equal to zero to find the intercepts.)

24. $x^2 + y^2 − 6x + 4y + 4 = 0$
25. $x^2 + y^2 − 5x − 7y + 6 = 0$

Intersecting Circles

Find the point(s) of intersection. (*Hint:* Solve each pair of equations simultaneously as we did in Exercise 8 in Chapter 14.)

26. $x^2 + y^2 − 10x = 0$ and $x^2 + y^2 + 2x − 6y = 0$
27. $x^2 + y^2 − 3y − 4 = 0$ and $x^2 + y^2 + 2x − 5y − 2 = 0$)
28. $x^2 + y^2 + 2x − 6y + 2 = 0$ and $x^2 + y^2 − 4x + 2 = 0$

Applications

Even though you might be able to solve some of these with only the Pythagorean theorem, we suggest that you use analytic geometry for the practice.

29. Prove that any angle inscribed in a semicircle is a right angle.
30. Write the equation of the circle in Fig. 22–38, taking the axes as shown. Use your equation to find A and B.

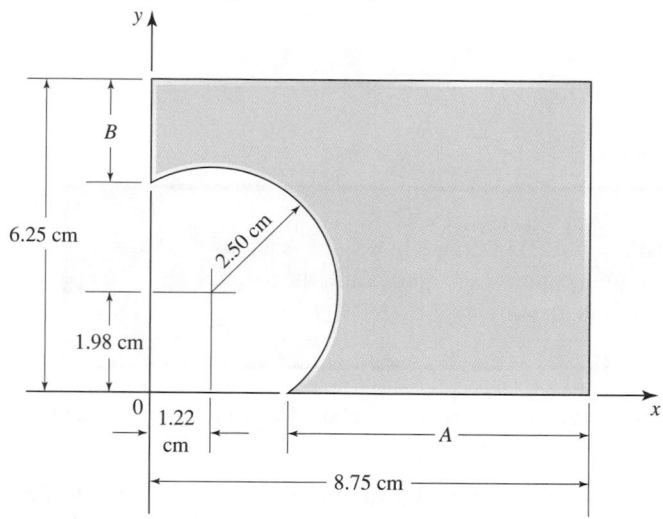

FIGURE 22-38

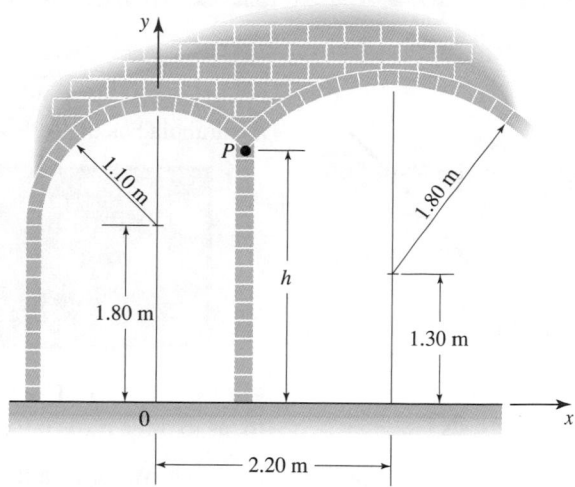

FIGURE 22-39

31. Write the equations for each of the circular arches in Fig. 22–39, taking the axes as shown. Solve simultaneously to get the point of intersection P, and compute the height h of the column.

32. Write the equation of the centreline of the circular street shown in Fig. 22–40, taking the origin at the intersection O. Use your equation to find the distance y.

33. Each side of a Gothic arch is a portion of a circle. Write the equation of one side of the Gothic arch shown in Fig. 22–41. Use the equation to find the width w of the arch at a height of 3.00 m.

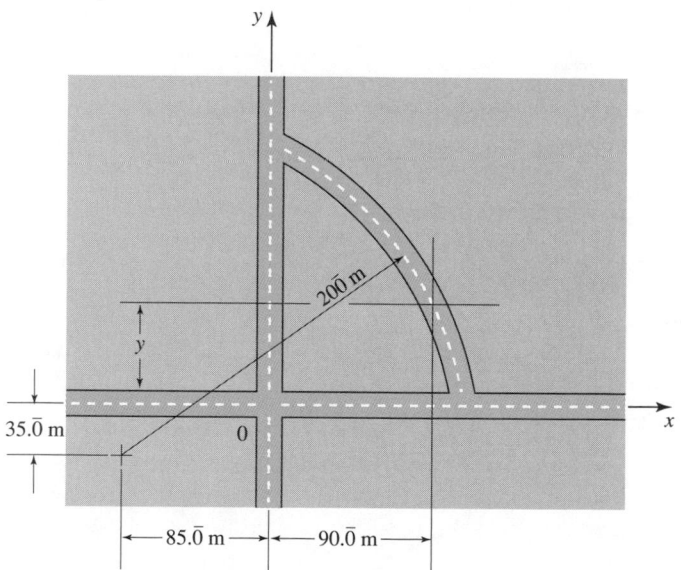

FIGURE 22-40 Circular street.

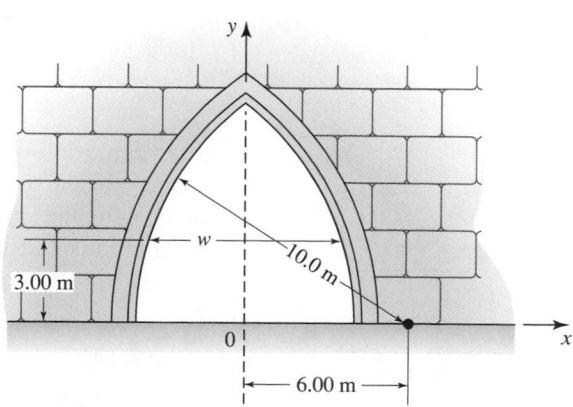

FIGURE 22-41 Gothic arch.

22-3 The Parabola

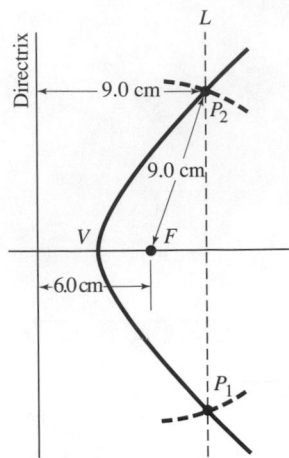

FIGURE 22–42   Construction of a parabola.

Definition

A parabola has the following definition:

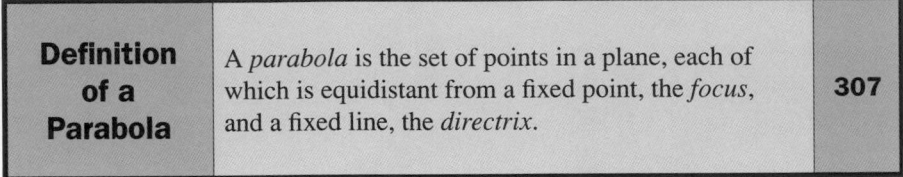

| **Definition of a Parabola** | A *parabola* is the set of points in a plane, each of which is equidistant from a fixed point, the *focus*, and a fixed line, the *directrix*. | **307** |

◆◆◆ **Example 27:** Use Definition 307 above to construct a parabola for which the distance from focus to directrix is 6.0 cm.

Solution: We draw a line to represent the directrix as shown in Fig. 22–42, and we indicate a focus 6.0 cm from that line. Then we draw a line L parallel to the directrix at some arbitrary distance, say, 9.0 cm. With that same 9.0-cm distance as radius and F as centre, we use a compass to draw arcs intersecting L at P_1 and P_2. Each of these points is now at the same distance (9.0 cm) from F and from the directrix, and is hence a point on the parabola. Repeat the construction with distances other than 9.0 cm to get more points on the parabola.   ◆◆◆

Figure 22–43 shows the typical shape of a parabola. The parabola has an *axis of symmetry* which intersects it at the *vertex*. The distance p from directrix to vertex is equal to the directed distance from the vertex to the focus.

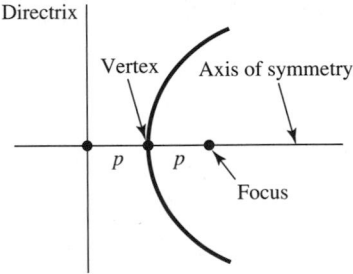

FIGURE 22–43   Parabola.

◆◆◆

Standard Equation of a Parabola with Vertex at the Origin

Let us place the parabola on coordinate axes with the vertex at the origin and with the axis of symmetry along the x axis, as shown in Fig. 22–44. Choose any point P on the parabola. Then, by the definition of a parabola, $FP = AP$. But in right triangle FBP,

$$FP = \sqrt{(x-p)^2 + y^2}$$

and

$$AP = p + x$$

But, since $FP = AP$,

$$\sqrt{(x-p)^2 + y^2} = p + x$$

Squaring both sides yields

$$(x-p)^2 + y^2 = p^2 + 2px + x^2$$

$$x^2 - 2px + p^2 + y^2 = p^2 + 2px + x^2$$

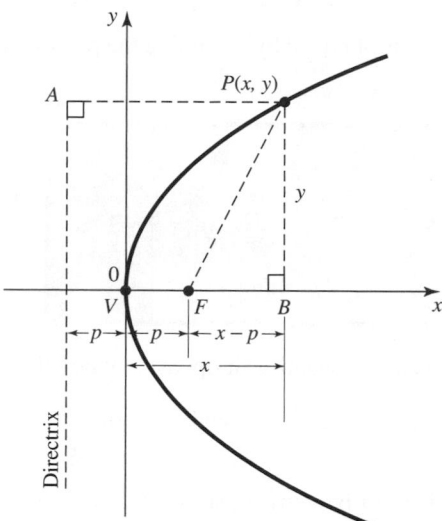

FIGURE 22–44

Collecting terms, we get the standard equation of a parabola with vertex at the origin.

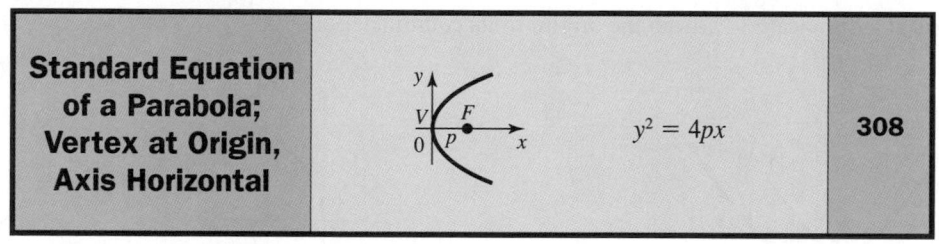

| **Standard Equation of a Parabola; Vertex at Origin, Axis Horizontal** | $y^2 = 4px$ | **308** |

We have defined p as the directed distance VF from the vertex V to the focus F. Thus if p is positive, the focus must lie to the right of the vertex, and hence the parabola opens to the right. Conversely, if p is negative, the parabola opens to the left.

◆◆◆ **Example 28:** Find the coordinates of the focus of the parabola $2y^2 + 7x = 0$.

Solution: Subtracting $7x$ from both sides and dividing by 2 gives us

$$y^2 = -3.5x$$

Thus $4p = -3.5$ and $p = -0.875$. Since p is negative, the parabola opens to the left. The focus is thus 0.875 unit to the left of the vertex (Fig. 22–45) and has the coordinates $(-0.875, 0)$. ◆◆◆

◆◆◆ **Example 29:** Find the coordinates of the focus and write the equation of a parabola that has its vertex at the origin, that has a horizontal axis of symmetry, and that passes through the point $(5, -6)$.

Solution: Our given point must satisfy the equation $y^2 = 4px$. Substituting = for x and 26 for y gives

$$(-6)^2 = 4p(5)$$

So $4p = \dfrac{36}{5}$ and $p = \dfrac{9}{5}$. The equation of the parabola is then $y^2 = 36x/5$, or

$$5y^2 = 36x$$

Since the axis is horizontal and p is positive, the focus must be on the x axis and is a distance $p\left(\dfrac{9}{5}\right)$ to the right of the origin. The coordinates of F are thus $\left(\dfrac{9}{5}, 0\right)$, as shown in Fig. 22–46. ◆◆◆

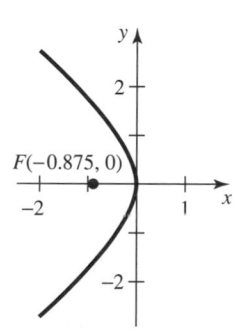

FIGURE 22–45

We have graphed the parabola before. Glance back at Sec. 5–2.

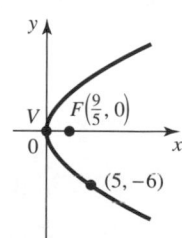

FIGURE 22–46

Standard Equation of a Parabola with Vertical Axis

The standard equation for a parabola having a vertical axis is obtained by switching the positions of x and y in Eq. 308.

| Standard Equation of a Parabola; Vertex at Origin, Axis Vertical | 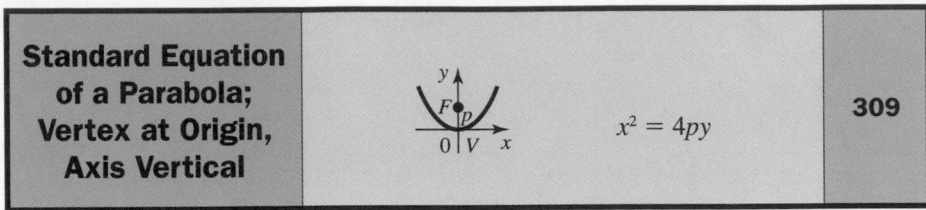 $x^2 = 4py$ | 309 |

When p is positive, the parabola opens upward; when p is negative, it opens downward.

◆◆◆ **Example 30:** A parabola has its vertex at the origin and passes through the points $(3, 2)$ and $(-3, 2)$. Write its equation and find the focus.

Solution: The sketch of Fig. 22–47 shows that the axis must be vertical, so our equation is of the form $x^2 = 4py$. Substituting 3 for x and 2 for y gives

$$3^2 = 4p(2)$$

from which $4p = \dfrac{9}{2}$ and $p = \dfrac{9}{8}$. Our equation is then $x^2 = 9y/2$ or $2x^2 = 9y$. The focus is on the y axis at a distance $p = \dfrac{9}{8}$ from the origin, so its coordinates are $\left(0, \dfrac{9}{8}\right)$ ◆◆◆

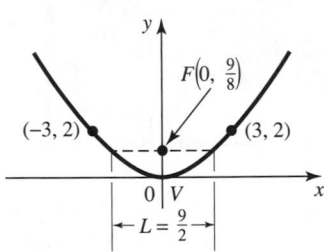

FIGURE 22–47

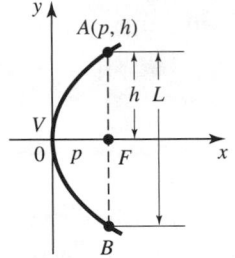

FIGURE 22–48

Focal Width of a Parabola

The *latus rectum* of a parabola is a line through the focus which is perpendicular to the axis of symmetry, such as line AB in Fig. 22–48. The length of the latus rectum is also called the *focal width*. We will find the focal width, or length L of the latus rectum, by substituting the coordinates (p, h) of point A into Eq. 308.

$$h^2 = 4p(p) = 4p^2$$
$$h = \pm 2p$$

The focal width is twice h, so we have the following equation:

| Focal Width of a Parabola | $L = |4p|$ | 110 |

The focal width (length of the latus rectum) of a parabola is four times the distance from vertex to focus.

◆◆◆ **Example 31:** The focal width for the parabola in Fig. 22–47 is $\dfrac{9}{2}$, or 4.5 units.   ◆◆◆

Shift of Axes

As with the circle, when the vertex of the parabola is not at the origin but at (h, k), our equations will be similar to Eqs. 308 and 309, except that x is replaced with $x - h$ and y is replaced with $y - k$.

Standard Equation of a Parabola; Vertex at (*h, k*)	Axis Horizontal	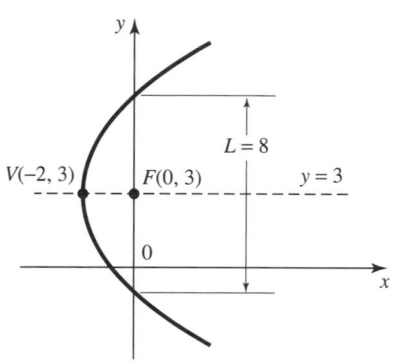	$(y - k)^2 = 4p(x - h)$	**310**
	Axis Vertical		$(x - h)^2 = 4p(y - k)$	**311**

◆◆◆ **Example 32:** Find the vertex, focus, focal width, and equation of the axis for the parabola $(y - 3)^2 = 8(x + 2)$.

Solution: The given equation is of the same form as Eq. 310, so the axis is horizontal. Also, $h = -2$, $k = 3$, and $4p = 8$. So the vertex is at $V(-2, 3)$ (Fig. 22–49). Since $p = \frac{8}{4} = 2$, the focus is 2 units to the right of the vertex, at $F(0, 3)$. The focal width is $4p$, so $L = 8$. The axis is horizontal and 3 units from the x axis, so its equation is $y = 3$.   ◆◆◆

◆◆◆ **Example 33:** Write the equation of a parabola that opens upward, with vertex at $(-1, 2)$, and that passes through the point $(1, 3)$ (Fig. 22–50). Find the focus and the focal width.

Solution: We substitute $h = -1$ and $k = 2$ into Eq. 311.

$$(x + 1)^2 = 4p(y - 2)$$

Now, since $(1, 3)$ is on the parabola, these coordinates must satisfy our equation. Substituting yields

$$(1 + 1)^2 = 4p(3 - 2)$$

Solving for p, we obtain $2^2 = 4p$, or $p = 1$. So the equation is

$$(x + 1)^2 = 4(y - 2)$$

The focus is p units above the vertex, at $(-1, 3)$. The focal width is, by Eq. 313,

$$L = |4p| = 4(1) = 4 \text{ units}$$ ◆◆◆

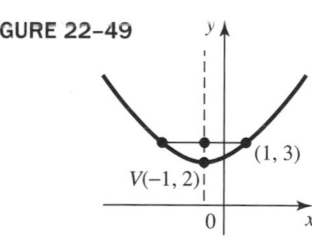

FIGURE 22–49

FIGURE 22–50

General Equation of a Parabola

We get the general equation of the parabola by expanding the standard equation (Eq. 310) as follows:

$$(y - k)^2 = 4p(x - h)$$
$$y^2 - 2ky + k^2 = 4px - 4ph$$

or

$$y^2 - 4px - 2ky + (k^2 + 4ph) = 0$$

which is of the following general form ($C, D, E,$ and F are constants):

General Equation of a Parabola with Horizontal Axis	$Cy^2 + Dx + Ey + F = 0$	**312**

Compare this with the general second-degree equation.

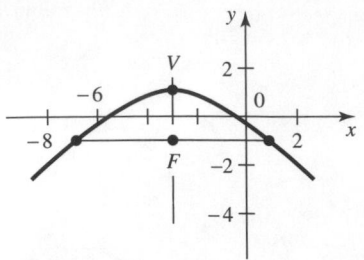

FIGURE 22–51

We see that the equation of a parabola having a horizontal axis of symmetry has a y^2 term but no x^2 term. Conversely, the equation for a parabola with vertical axis has an x^2 term but not a y^2 term. The parabola is the only conic for which there is only one variable squared.

If the coefficient B of the xy term in the general second-degree equation (Eq. 297) were not zero, it would indicate that the axis of symmetry was *rotated* by some amount and was no longer parallel to a coordinate axis. The presence of an xy term indicates rotation of the ellipse and hyperbola as well.

Completing the Square

As with the circle, we go from general to standard form by completing the square.

◆◆◆ **Example 34:** Find the vertex, focus, and focal width for the parabola

$$x^2 + 6x + 8y + 1 = 0$$

Solution: Separating the x and y terms, we have

$$x^2 + 6x = -8y - 1$$

Completing the square by adding 9 to both sides, we obtain

$$x^2 + 6x + 9 = -8y - 1 + 9$$

Factoring yields

$$(x + 3)^2 = -8(y - 1)$$

which is the form of Eq. 311, with $h = -3$, $k = 1$, and $p = 22$.

The vertex is $(-3, 1)$. Since the parabola opens downward (Fig. 22–51), the focus is 2 units below the vertex, at $(-3, -1)$. The focal width is $4|p|$, or 8 units.   ◆◆◆

Exercise 3 ◆ The Parabola

Parabola with Vertex at Origin

Find the focus and focal width of each parabola.

1. $y^2 = 8x$ 　　　　　　　　　　　**2.** $x^2 = 16y$

3. $7x^2 + 12y = 0$ 　　　　　　　　**4.** $3y^2 + 5x = 0$

Write the equation of each parabola.

5. focus at $(0, -2)$

6. passes through $(25, 20)$; axis horizontal

7. passes through $(3, 2)$ and $(3, -2)$

8. passes through $(3, 4)$ and $(-3, 4)$

Parabola with Vertex Not at Origin

Find the vertex, focus, focal width, and equation of the axis for each parabola.

9. $(y - 5)^2 = 12(x - 3)$ 　　　　　**10.** $(x + 2)^2 = 16(y - 6)$

11. $(x - 3)^2 = 24(y + 1)$ 　　　　　**12.** $(y + 3)^2 = 4(x + 5)$

13. $3x + 2y^2 + 4y - 4 = 0$ 　　　　**14.** $y^2 + 8y + 4 - 6x = 0$

15. $y - 3x + x^2 + 1 = 0$ 　　　　　**16.** $x^2 + 4x - y - 6 = 0$

Write the equation, in general form, for each parabola.

17. vertex at $(1, 2)$; $L = 8$; axis is $y = 2$; opens to the right

18. axis is $y = 3$; passes through $(6, -1)$ and $(3, 1)$

19. passes through $(-3, 3)$, $(-6, 5)$, and $(-11, 7)$; axis horizontal

20. vertex $(0, 2)$; axis is $x = 0$; passes through $(24, 22)$

21. axis is $y = 21$; passes through $(24, 22)$ and $(2, 1)$

Trajectories

22. A ball thrown into the air will, neglecting air resistance, follow a parabolic path, as shown in Fig. 22–52. Write the equation of the path, taking axes as shown. Use your equation to find the height of the ball when it is at a horizontal distance of 30.0 m from O.

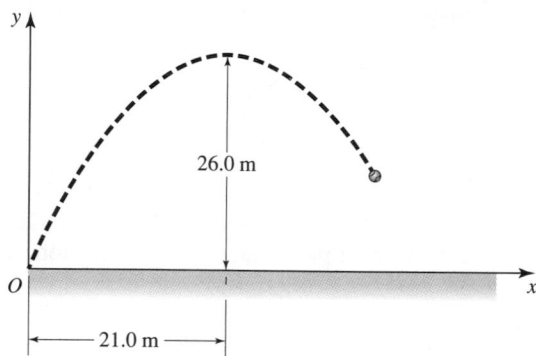

FIGURE 22–52   Ball thrown into the air.

23. Some comets follow a parabolic orbit with the sun at the focal point (Fig. 22–53). Taking axes as shown, write the equation of the path if the distance p is 75 million kilometres.

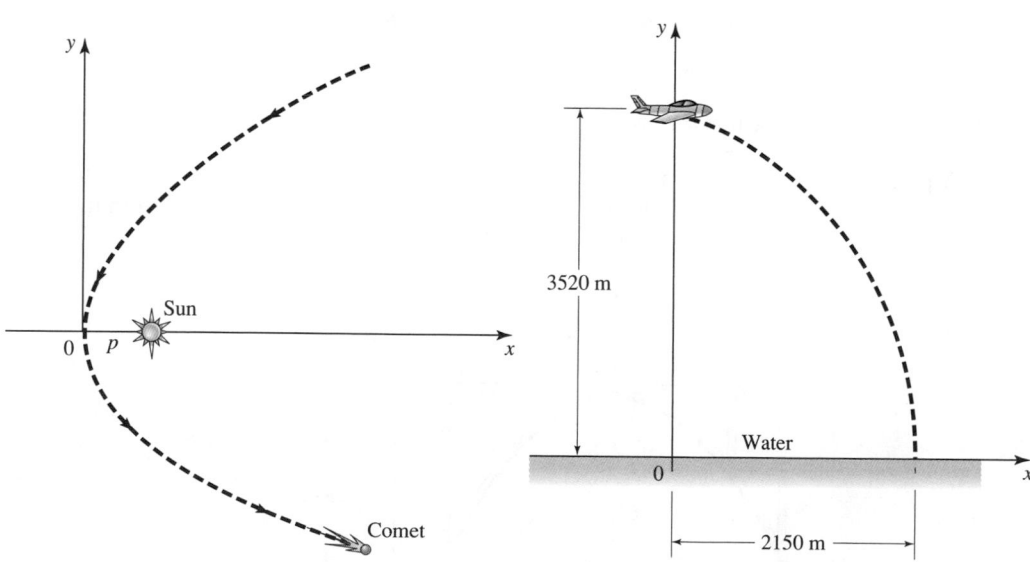

FIGURE 22–53   Path of a comet.   **FIGURE 22–54**   Object dropped from an aircraft.

24. An object dropped from a moving aircraft (Fig. 22–54) will follow a parabolic path if air resistance is negligible. A weather instrument released at a height of 3520 m is observed to strike the water at a distance of 2150 m from the point of release. Write the equation of the path, taking axes as shown. Find the height of the instrument when x is 1000 m.

Parabolic Arch

25. A 3.0-m-high truck passes under a parabolic arch, as shown in Fig. 22–55. Find the maximum distance x that the side of the truck can be from the centre of the road.

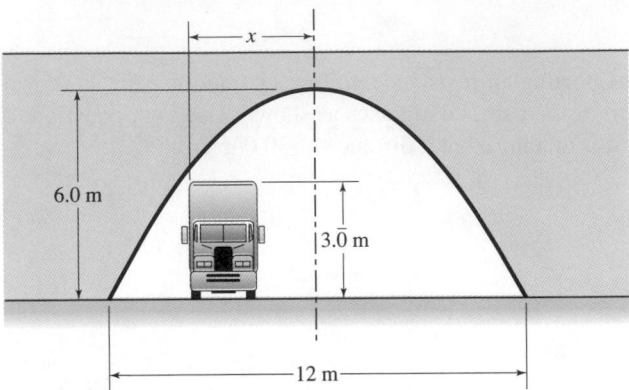

FIGURE 22–55   Parabolic arch.

26. Assuming the bridge cable AB of Fig. 22–56 to be a parabola, write its equation, taking axes as shown.

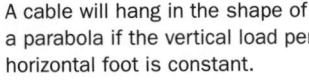

A cable will hang in the shape of a parabola if the vertical load per horizontal foot is constant.

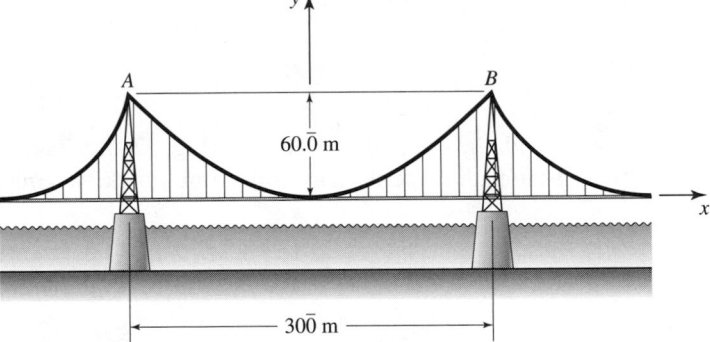

FIGURE 22–56   Parabolic bridge cable.

27. A parabolic arch supports a roadway as shown in Fig. 22–57. Write the equation of the arch, taking axes as shown. Use your equation to find the vertical distance from the roadway to the arch at a horizontal distance of 50.0 m from the centre.

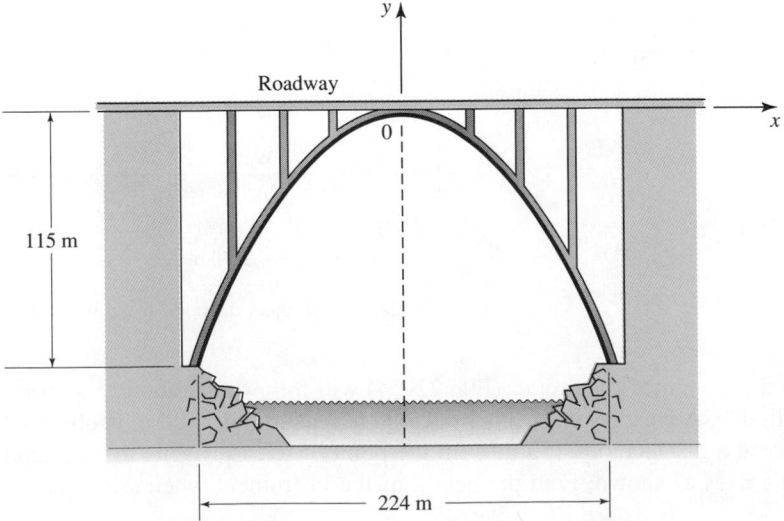

FIGURE 22–57   Parabolic arch.

Parabolic Reflector

28. A certain solar collector consists of a long panel of polished steel bent into a parabolic shape (Fig. 22–58), which focuses sunlight onto a pipe P at the focal point of the parabola. At what distance x should the pipe be placed?

29. A parabolic collector for receiving television signals from a satellite is shown in Fig. 22–59. The receiver R is at the focus, 1.00 m from the vertex. Find the depth d of the collector.

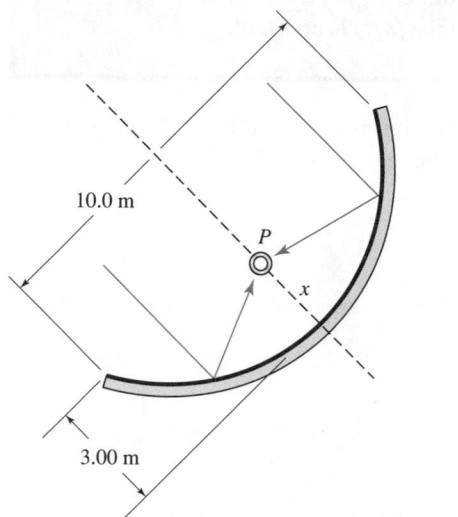

FIGURE 22–58 Parabolic solar collector.

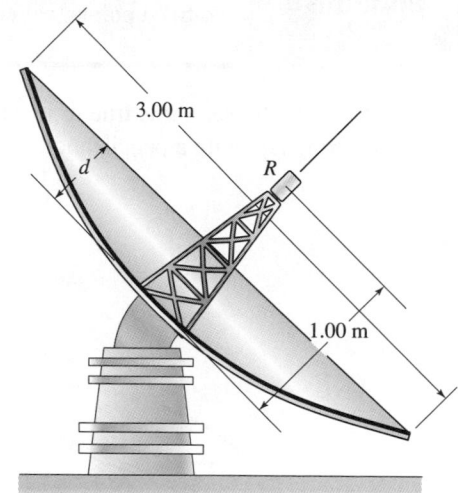

FIGURE 22–59 Parabolic antenna.

Vertical Highway Curves

30. A parabolic curve is to be used at a dip in a highway. The road dips 32.0 m in a horizontal distance of 125 m and then rises to its previous height in another 125 m. Write the equation of the curve of the roadway, taking the origin at the bottom of the dip and the y axis vertical.

31. Write the equation of the vertical highway curve in Fig. 22–60, taking axes as shown.

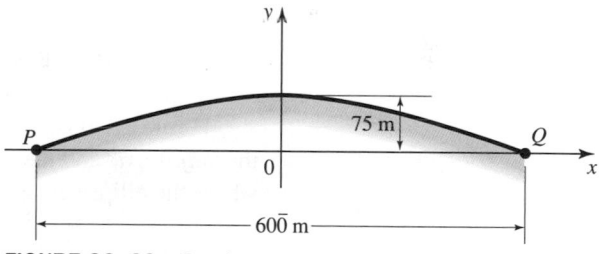

FIGURE 22–60 Road over a hill.

Beams

32. A simply supported beam with a concentrated load at its midspan will deflect in the shape of a parabola, as shown in Fig. 22–61. If the deflection at the midspan is 1.00 in., write the equation of the parabola (called the *elastic curve*), taking axes as shown.

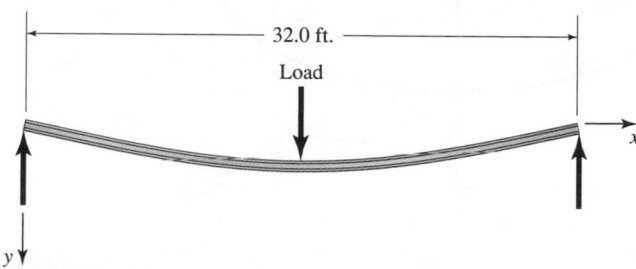

FIGURE 22–61 Deflection of a beam.

33. Using the equation found in problem 32, find the deflection of the beam in Fig. 22–61 at a distance of 10.0 ft. from the left end.

22–4   The Ellipse

Definition

The ellipse is defined in the following way:

Definition of an Ellipse	An *ellipse* is the set of all points in a plane such that the sum of the distances from each point on the ellipse to two fixed points (called the *foci*) is constant.	**315**

One way to draw an ellipse that is true to its definition is to pass a loop of string around two tacks, pull the string taut with a pencil, and trace the curve (Fig. 22–62).

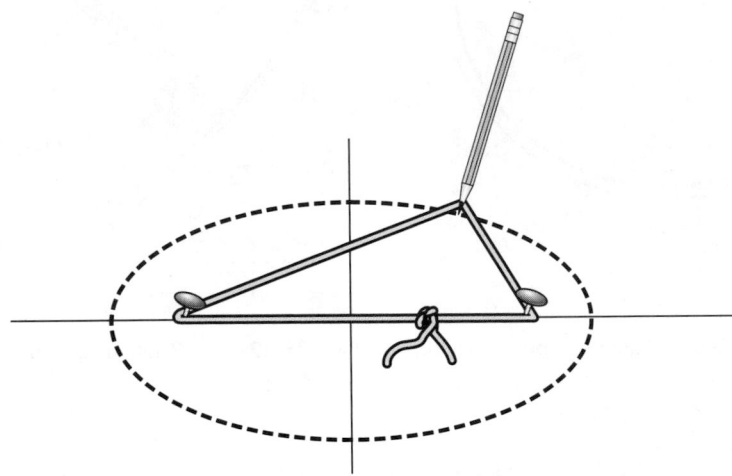

FIGURE 22–62   Construction of an ellipse.

Figure 22–63 shows the typical shape of the ellipse. An ellipse has two axes of symmetry: the *major axis* and the *minor axis,* which intersect at the centre of the ellipse. A *vertex* is a point where the ellipse crosses the major axis.

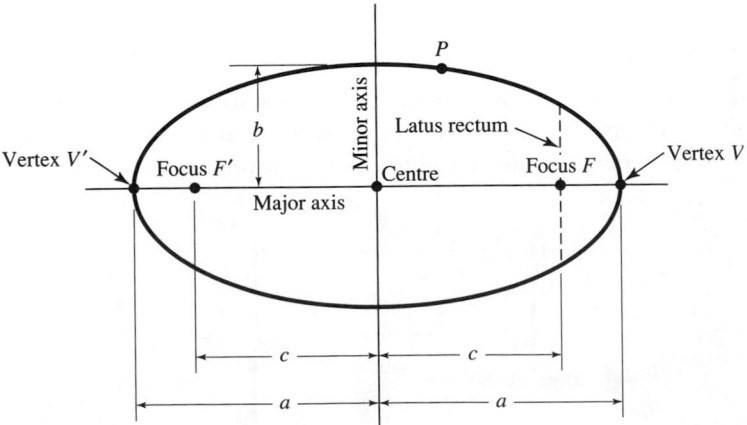

FIGURE 22–65   Ellipse

It is often convenient to speak of *half* the lengths of the major and minor axes, and these are called the *semimajor* and *semiminor* axes, whose lengths we label a and b, respectively. The distance from either focus to the centre is labelled c.

Distance to Focus

Before deriving an equation for the ellipse, let us first write an expression for the distance c from the centre to a focus in terms of the semimajor axis a and the semiminor axis b.

From our definition of an ellipse, if P is any point on the ellipse, then

$$PF + PF' = k \tag{1}$$

where k is a constant. If P is taken at a vertex V, then Eq. (1) becomes

$$VF + VF' = k = 2a \tag{2}$$

because $VF + VF\infty$ is equal to the length of the major axis. Substituting back into (1) gives us

$$PF + PF' = 2a \tag{3}$$

Figure 22–64 shows our point P moved to the intersection of the ellipse and the minor axis. Here PF and PF' are equal. But since their sum is $2a$, PF and PF' must each equal a. By the Pythagorean theorem, $c^2 + b^2 = a^2$, or

$$c^2 = a^2 - b^2$$

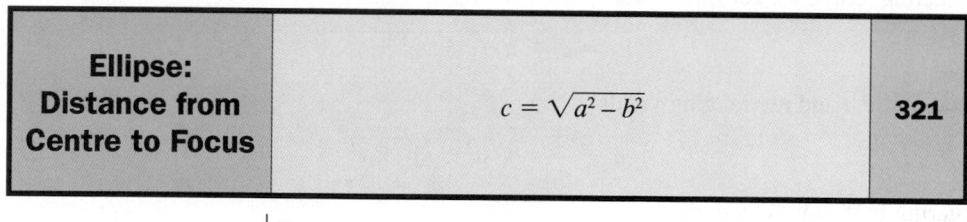

| **Ellipse: Distance from Centre to Focus** | $c = \sqrt{a^2 - b^2}$ | **321** |

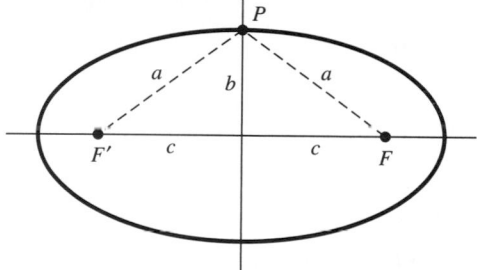

FIGURE 22–64

Ellipse with Centre at the Origin, Major Axis Horizontal

Let us place an ellipse on coordinate axes with its centre at the origin and major axis along the x axis, as shown in Fig. 22–65. If $P(x, y)$ is any point on the ellipse, then by the definition of the ellipse,

$$PF + PF' = 2a$$

A line from a focus to a point on the ellipse is called a *focal radius*.

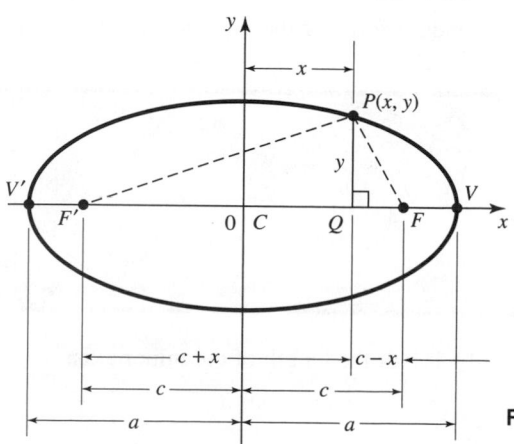

FIGURE 22–65 Ellipse with centre at origin.

To get PF and PF' in terms of x and y, we first drop a perpendicular from P to the x axis. Then in triangle PQF,

$$PF = \sqrt{(c - x)^2 + y^2}$$

and in triangle PQF',

$$PF' = \sqrt{(c + x)^2 + y^2}$$

Substituting yields

$$PF + PF' = \sqrt{(c - x)^2 + y^2} + \sqrt{(c + x)^2 + y^2} = 2a$$

Rearranging, we obtain

$$\sqrt{(c + x)^2 + y^2} = 2a - \sqrt{(c - x)^2 + y^2}$$

Squaring, and then expanding the binomials, we have

$$x^2 + 2cx + c^2 + y^2 = 4a^2 - 4a\sqrt{(c - x)^2 + y^2} + c^2 - 2cx + x^2 + y^2$$

Collecting terms, we get

$$cx = a^2 - a\sqrt{(c - x)^2 + y^2}$$

Dividing by a and rearranging yields

$$a - \frac{cx}{a} = \sqrt{(c - x)^2 + y^2}$$

Squaring both sides again yields

$$a^2 - 2cx + \frac{c^2 x^2}{a^2} = c^2 - 2cx + x^2 + y^2$$

Collecting terms, we have

$$a^2 + \frac{c^2 x^2}{a^2} = c^2 + x^2 + y^2$$

But $c^2 = a^2 + b^2$. Substituting, we have

$$a^2 + \frac{x^2}{a^2}(a^2 - b^2) = a^2 - b^2 + x^2 + y^2$$

or

$$a^2 + x^2 - \frac{b^2 x^2}{a^2} = a^2 - b^2 + x^2 + y^2$$

Collecting terms and rearranging gives us

$$\frac{b^2 x^2}{a^2} + y^2 = b^2$$

Finally, dividing through by b^2, we get the standard form of the equation of an ellipse with centre at origin.

Note that if $a = b$, this equation reduces to the equation of a circle.

| **Standard Equation of an Ellipse: Centre at Origin, Major Axis Horizontal** | | $\dfrac{x^2}{a^2} + \dfrac{y^2}{b^2} = 1$

 $a > b$ | **316** |

Here a is half the length of the major axis, and b is half the length of the minor axis.

✦✦✦ Example 35 Find the vertices and foci for the ellipse

$$16x^2 + 36y^2 = 576$$

Solution: To be in standard form, our equation must have + (unity) on the right side. Dividing by 576 and simplifying gives us

$$\frac{x^2}{36} + \frac{y^2}{16} = 1$$

from which $a = 6$ and $b = 4$. The vertices are then $V(6, 0)$ and $V'(-6, 0)$, as shown in Fig. 22–66. The distance c from the centre to a focus is

$$c = \sqrt{6^2 - 4^2} = \sqrt{20} \cong 4.47$$

So the foci are $F(4.47, 0)$ and $F'(-4.47, 0)$.

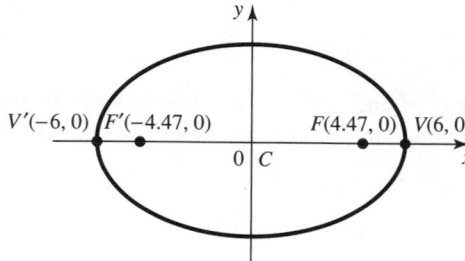

FIGURE 22–66 ✦✦✦

✦✦✦ Example 36 An ellipse whose centre is at the origin and whose major axis is on the x axis has a minor axis of 10 units and passes through the point (6, 4), as shown in Fig. 22–67. Write the equation of the ellipse in standard form.

Solution: Our equation will have the form of Eq. 316. Substituting, with $b = 5$, we have

$$\frac{x^2}{a^2} + \frac{y^2}{25} = 1$$

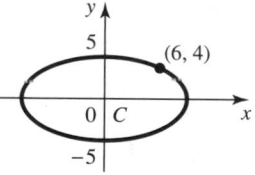

FIGURE 22–67

Since the ellipse passes through (6, 4), these coordinates must satisfy our equation. Substituting yields

$$\frac{36}{a^2} + \frac{16}{25} = 1$$

Solving for a^2, we multiply by the LCD, $25a^2$.

$$36(25) + 16a^2 = 25a^2$$

$$9a^2 = 36(25)$$

$$a^2 = 100$$

So our final equation is

$$\frac{x^2}{100} + \frac{y^2}{25} = 1$$ ✦✦✦

Ellipse with Centre at Origin, Major Axis Vertical
When the major axis is vertical rather than horizontal, the only effect on the standard equation is to interchange the positions of x and y.

<table>
<tr><td>

Standard Equation of an Ellipse: Centre at Origin, Major Axis Vertical

</td><td>

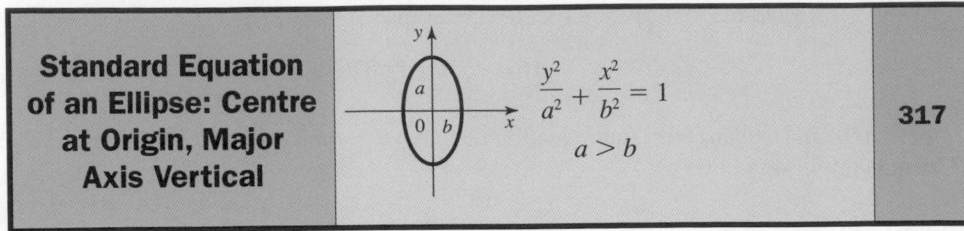

$$\frac{y^2}{a^2} + \frac{x^2}{b^2} = 1$$

$$a > b$$

</td><td>

317

</td></tr>
</table>

Notice that the quantities a and b are dimensions of the semimajor and semiminor axes and remain so as the ellipse is turned or shifted. Therefore for an ellipse in any position, the distance c from centre to focus is found the same way as before.

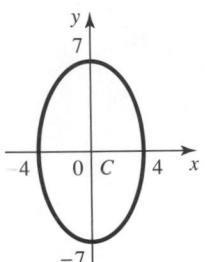

FIGURE 22–68

◆◆◆ **Example 37** Find the lengths of the major and minor axes and the distance from centre to focus for the ellipse

$$\frac{x^2}{16} + \frac{y^2}{49} = 1$$

Solution: How can we tell which denominator is a^2 and which is b^2? It is easy: a is always greater than b. So $a = 7$ and $b = 4$. Thus the major and minor axes are 14 units and 8 units long (Fig. 22–68). From Eq. 321,

$$c = \sqrt{a^2 - b^2} = \sqrt{49 - 16} = \sqrt{33} \cong 5.74$$

◆◆◆ **Example 38** Write the equation of an ellipse with centre at the origin, whose major axis is 12 units on the y axis and whose minor axis is 10 units (Fig. 22–69).

Solution: Substituting into Eq. 317, with $a = 6$ and $b = 5$, we obtain

$$\frac{y^2}{36} + \frac{x^2}{25} = 1$$

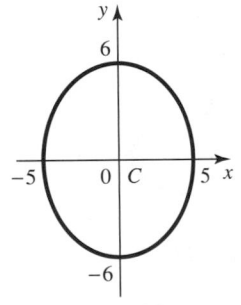

FIGURE 22–69

Common Error	Do not confuse a and b in the ellipse equations. The larger denominator is always a^2. Also, the variable (x or y) in the same term with a^2 tells the direction of the major axis.

Shift of Axes

Now consider an ellipse whose centre is not at the origin, but at (h, k). The equation for such an ellipse will be the same as before, except that x is replaced by $x - h$ and y is replaced by $y - k$.

<table>
<tr><td rowspan="2">

Standard Equation of an Ellipse: Centre at (h, k)

</td><td>Major Axis Horizontal</td><td>

$$\frac{(x-h)^2}{a^2} + \frac{(y-k)^2}{b^2} = 1$$

$$a > b$$

</td><td>

318

</td></tr>
<tr><td>Major Axis Vertical</td><td>

$$\frac{(y-k)^2}{a^2} + \frac{(x-h)^2}{b^2} = 1$$

$$a > b$$

</td><td>

319

</td></tr>
</table>

◆◆◆ **Example 39:** Find the centre, vertices, and foci for the ellipse

$$\frac{(x-5)^2}{9} + \frac{(y+3)^2}{16} = 1$$

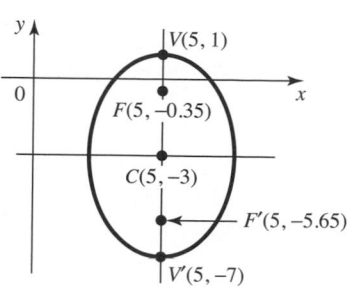

Solution: From the given equation, $h = 5$ and $k = -3$, so the centre is $C(5, -3)$, as shown in Fig. 22–70. Also, $a = 4$ and $b = 3$, and the major axis is vertical because a is with the y term. By setting $x = 5$ in the equation of the ellipse, we find that the vertices are $V(5, 1)$ and $V'(5, -7)$. The distance c to the foci is

$$c = \sqrt{4^2 - 3^2} = \sqrt{7} \cong 2.65$$

so the foci are $F(5, -0.35)$ and $F'(5, -5.65)$.   ◆◆◆

FIGURE 22–70

◆◆◆ **Example 40:** Write the equation in standard form of an ellipse with vertical major axis 10 units long, centre at $(3, 5)$, and whose distance between focal points is 8 units (Fig. 22–71).

Yes, Eq. 321 still applies here.

Solution: From the information given, $h = 3$, $k = 5$, $a = 5$, and $c = 4$. By Eq. 321,

$$b = \sqrt{a^2 - c^2} = \sqrt{25 - 16} = 3 \text{ units}$$

Substituting into Eq. 319, we get

$$\frac{(y-5)^2}{25} + \frac{(x-3)^2}{9} = 1$$ ◆◆◆

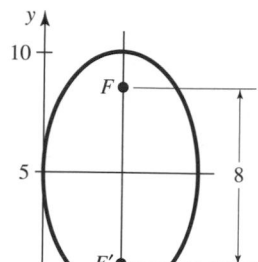

FIGURE 22–71

General Equation of an Ellipse

As we did with the circle, we now expand the standard equation for the ellipse to get the general equation. Starting with Eq. 318,

$$\frac{(x-h)^2}{a^2} + \frac{(y-k)^2}{b^2} = 1$$

we multiply through by a^2b^2 and expand the binomials:

$$b^2(x^2 - 2hx + h^2) + a^2(y^2 - 2ky + k^2) = a^2b^2$$
$$b^2x^2 - 2b^2hx + b^2h^2 + a^2y^2 - 2a^2ky + a^2k^2 - a^2b^2 = 0$$

or

$$b^2x^2 + a^2y^2 - 2b^2hx - 2a^2ky + (b^2h^2 + a^2k^2 - a^2b^2) = 0$$

which is of the following general form:

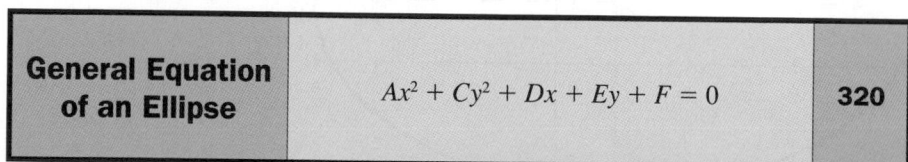

General Equation of an Ellipse	$Ax^2 + Cy^2 + Dx + Ey + F = 0$	320

Comparing this with the general second-degree equation, we see that $B = 0$. As for the parabola (and the hyperbola, as we will see), this indicates that the axes of the curve are parallel to the coordinate axes. In other words, the curve is not rotated.

Also note that neither A nor C is zero. That tells us that the curve is not a parabola. Further, A and C have different values, telling us that the curve cannot be a circle. We'll see that A and C will have the same sign for the ellipse and opposite signs for the hyperbola.

Completing the Square

As before, we go from general to standard form by completing the square.

◆◆◆ **Example 41:** Find the centre, foci, vertices, and major and minor axes for the ellipse $9x^2 + 25y^2 + 18x - 50y - 191 = 0$.

Solution: Grouping the x terms and the y terms gives us

$$(9x^2 + 18x) + (25y^2 - 50y) = 191$$

Factoring yields

$$9(x^2 + 2x) + 25(y^2 - 2y) = 191$$

Completing the square, we obtain

$$9(x^2 + 2x + 1) + 25(y^2 - 2y + 1) = 191 + 9 + 25$$

Factoring gives us

$$9(x + 1)^2 + 25(y - 1)^2 = 225$$

Finally, dividing by 225, we have

$$\frac{(x + 1)^2}{25} + \frac{(y - 1)^2}{9} = 1$$

We see that $h = 21$ and $k = 1$, so the centre is at $(-1, 1)$. Also, $a = 5$, so the major axis is 10 units and is horizontal, and $b = 3$, so the minor axis is 6 units. A vertex is located = units to the right of the centre, at $(4, 1)$, and = units to the left, at $(-6, 1)$. From Eq. 321,

$$c = \sqrt{a^2 - b^2}$$
$$= \sqrt{25 - 9}$$
$$= 4$$

So the foci are at $(3, 1)$ and $(-5, 1)$. This ellipse is shown in Fig. 22–72.

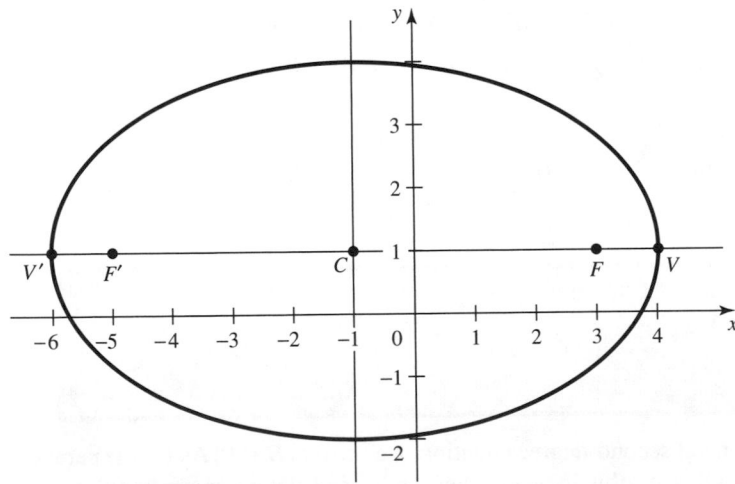

FIGURE 22–72   ◆◆◆

Common Error	In Example 41 we needed + to complete the square on x, but we added 9 to the right side because the expression containing the + was multiplied by a factor of 9. Similarly, for y, we needed + on the left but added 25 to the right. It is very easy to forget to multiply by those factors.

Focal Width of an Ellipse

The focal width L, or length of the latus rectum, is the width of the ellipse through the focus (Fig. 22–73). The sum of the focal radii PF and PF' from a point P at one end of the latus rectum must equal $2a$, by the definition of an ellipse. So $PF' = 2a - PF$, or, since $L/2 = PF$, $PF' = 2a - L/2$. Squaring both sides, we obtain

$$(PF')^2 = 4a^2 - 2aL + \frac{L^2}{4} \tag{1}$$

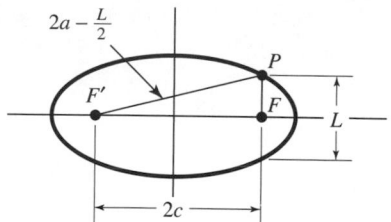

FIGURE 22–73
Derivation of focal width.

But in right triangle PFF',

$$(PF')^2 = \left(\frac{L}{2}\right)^2 + (2c)^2 = \frac{L^2}{4} + 4c^2 = \frac{L^2}{4} + 4a^2 - 4b^2 \tag{2}$$

since $c^2 = a^2 - b^2$. Equating (1) and (2) and collecting terms gives $2aL = 4b^2$, or the following:

Focal Width of an Ellipse	$L = \dfrac{2b^2}{a}$	**322**

As with the parabola, the main use for L is for quick sketching of the ellipse.

The focal width of an ellipse is twice the square of the semiminor axis, divided by the semimajor axis.

◆◆◆ **Example 42:** The focal width of an ellipse that is 25 m long and 10 m wide is

$$L = \frac{2(5^2)}{12.5} = \frac{50}{12.5} = 4 \text{ m}$$

◆◆◆

Exercise 4 • The Ellipse

Ellipse with Centre at Origin

Find the coordinates of the vertices and foci for each ellipse.

1. $\dfrac{x^2}{25} + \dfrac{y^2}{16} = 1$

2. $\dfrac{x^2}{49} + \dfrac{y^2}{36} = 1$

3. $3x^2 + 4y^2 = 12$

4. $64x^2 + 15y^2 = 960$

5. $4x^2 + 3y^2 = 48$

6. $8x^2 + 25y^2 = 200$

Write the equation of each ellipse in standard form.

7. vertices at $(65, 0)$; foci at $(\pm 4, 0)$

8. vertical major axis 8 units long; a focus at $(0, 2)$

9. horizontal major axis 12 units long; passes through $(3, \sqrt{3})$

10. passes through $(4, 6)$ and $(2, 3\sqrt{5})$

11. horizontal major axis 26 units long; distance between foci $= 24$

12. vertical minor axis 10 units long; distance from focus to vertex $= 1$

13. passes through $(1, 4)$ and $(26, 1)$

14. distance between foci $= 18$; sum of axes $= 54$; horizontal major axis

Ellipse with Centre Not at Origin

Find the coordinates of the centre, vertices, and foci for each ellipse. Round to three significant digits where needed.

15. $\dfrac{(x-2)^2}{16} + \dfrac{(y+2)^2}{9} = 1$

16. $\dfrac{(x+5)^2}{25} + \dfrac{(y-3)^2}{49} = 1$

17. $5x^2 + 20x + 9y^2 - 54y + 56 = 0$

18. $16x^2 - 128x + 7y^2 + 42y = 129$

19. $7x^2 - 14x + 16y^2 + 32y = 89$

20. $3x^2 - 6x + 4y^2 + 32y + 55 = 0$

21. $25x^2 + 150x + 9y^2 - 36y + 36 = 0$

Write the equation of each ellipse.

22. minor axis $= 10$; foci at $(13, 2)$ and $(-11, 2)$

23. centre at $(0, 3)$; vertical major axis $= 12$; length of minor axis $= 6$

24. centre at $(2, -1)$; a vertex at $(2, 5)$; length of minor axis $= 3$

25. centre at $(-2, -3)$; a vertex at $(-2, 1)$; a focus halfway between vertex and centre

Intersections of Curves

Solve simultaneously to find the points of intersection of each ellipse with the given curve.

26. $3x^2 + 6y^2 = 11$ with $y = x + 1$

27. $2x^2 + 3y^2 = 14$ with $y^2 = 4x$

28. $x^2 + 7y^2 = 16$ with $x^2 + y^2 = 10$

Construction of an Ellipse

29. Draw an ellipse using tacks and string, as shown in Fig. 22–62, with major axis of 84.0 cm and minor axis of 58.0 cm. How far apart must the tacks be placed?

Applications

For problems 30 through 33, first write the equation for each ellipse.

30. A certain bridge arch is in the shape of half an ellipse 40.0 m wide and 10.0 m high. At what horizontal distance from the centre of the arch is the height equal to 5.00 m?

31. A curved mirror in the shape of an ellipse will reflect all rays of light coming from one focus onto the other focus (Fig. 22–74). A certain spot heater is to be made with a heating element at A and the part to be heated at B, contained within an ellipsoid (a solid obtained by rotating an ellipse about one axis). Find the width x of the chamber if its length is 25 cm and the distance from A to B is 15 cm.

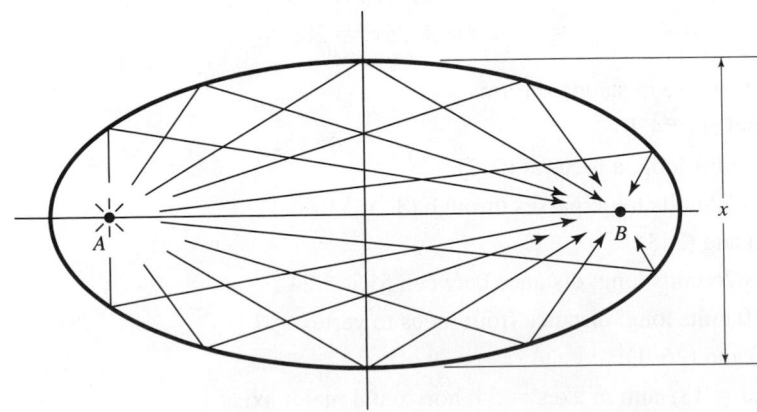

FIGURE 22–74
Focusing property of the ellipse.

32. The paths of the planets and certain comets are ellipses, with the sun at one focal point. The path of Halley's comet is an ellipse with a major axis of 36.18 AU and a minor axis of 9.12 AU. What is the greatest distance that Halley's comet gets from the sun?

> An *astronomical unit*, AU, is the distance between the earth and the sun, about 150×10^6 km.

33. A large elliptical culvert (Fig. 22–75) is filled with water to a depth of 1.0 m. Find the width w of the stream.

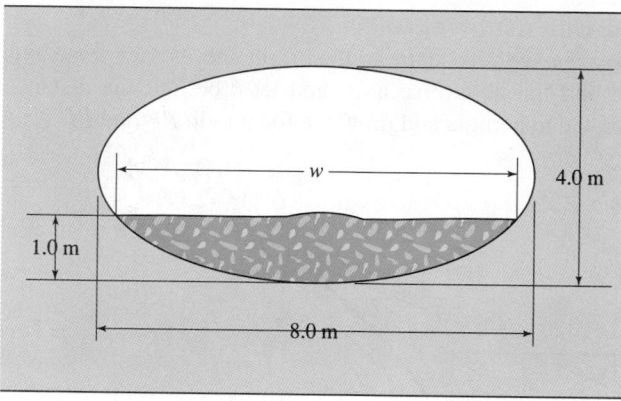

FIGURE 22–75

22–5 The Hyperbola

The hyperbola is defined as follows:

Definition of a Hyperbola	A *hyperbola* is the set of all points in a plane such that the distances from each point to two fixed points, the *foci*, have a constant difference.	325

Figure 22–76 shows the typical shape of the hyperbola. The point midway between the foci is called the *centre* of the hyperbola. The line passing through the foci and the centre is one axis of the hyperbola. The hyperbola crosses that axis at points called the *vertices*. The line

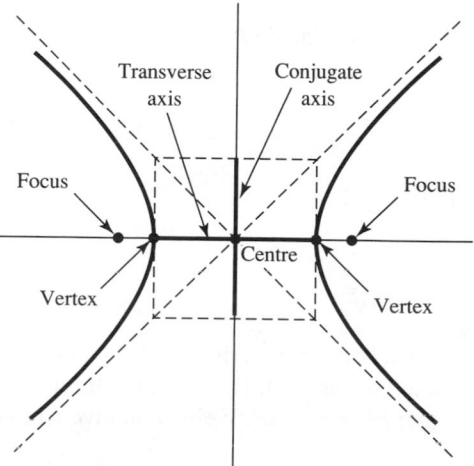

FIGURE 22–76 Hyperbola.

segment connecting the vertices is called the *transverse axis*. A second axis of the hyperbola passes through the centre and is perpendicular to the transverse axis. The segment of this axis, shown in Fig. 22–76, is called the *conjugate axis*. Half the lengths of the transverse and conjugate axes are the *semitransverse* and *semiconjugate* axes, respectively. They are also referred to as *semiaxes*.

Standard Equations of a Hyperbola with Centre at Origin

We place the hyperbola on coordinate axes, with its centre at the origin and its transverse axis on the x axis (Fig. 22–77). Let a be half the transverse axis, and let c be half the distance between foci. Now take any point P on the hyperbola and draw the focal radii PF and PF'.

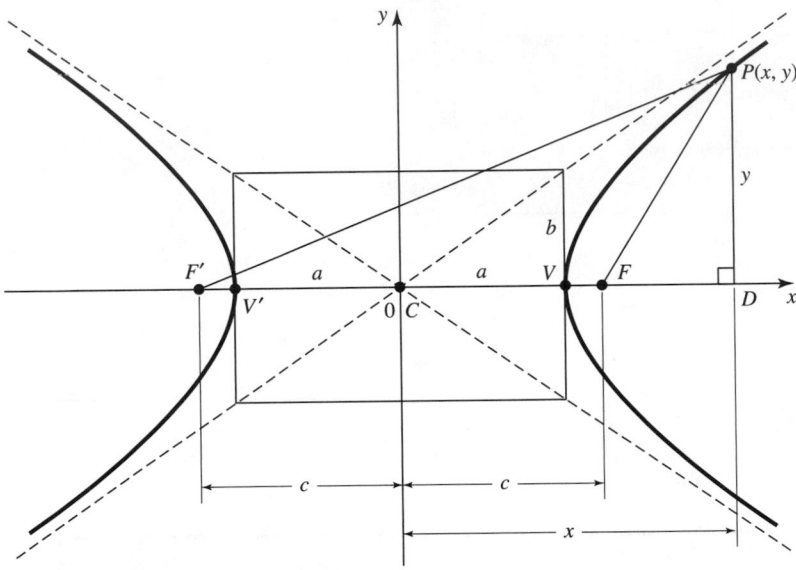

FIGURE 22-77   Hyperbola with centre at origin.

Then, by the definition of the hyperbola, $|PF' - PF| =$ constant. The constant can be found by moving P to the vertex V, where

$$PF' - PF = VF' - VF$$
$$= 2a + V'F' - VF$$
$$= 2a$$

since $V'F'$ and VF are equal. So $|PF' - PF| = 2a$. But in right triangle $PF'D$,

$$PF' = \sqrt{(x + c)^2 + y^2}$$

and in right triangle PFD,

$$PF = \sqrt{(x - c)^2 + y^2}$$

so

$$\sqrt{(x + c)^2 + y^2} - \sqrt{(x - c)^2 + y^2} = 2a$$

Notice that this equation is almost identical to the one we had when deriving the equation for the ellipse in Sec. 22–4, and the derivation is almost identical as well. However, to eliminate c from our equation, we define a new quantity b, such that $b^2 = c^2 - a^2$. We'll soon give a geometric meaning to the quantity b.

The remainder of the derivation is left as an exercise. The resulting equations are very similar to those for the ellipse.

Try to derive these equations. Follow the same steps we used for the ellipse.

Standard Equations of a Hyperbola: Centre at Origin	Transverse Axis Horizontal	$\dfrac{x^2}{a^2} - \dfrac{y^2}{b^2} = 1$	326
	Transverse Axis Vertical	$\dfrac{y^2}{a^2} - \dfrac{x^2}{b^2} = 1$	327

Asymptotes of a Hyperbola

In Fig. 22–77, the dashed diagonal lines are asymptotes of the hyperbola. (The two branches of the hyperbola approach, but do not intersect, the asymptotes.)

We find the slope of these asymptotes from the equation of the hyperbola. Solving Eq. 326 for y gives $y^2 = b^2(x^2/a^2 - 1)$, or

$$y = \pm b\sqrt{\frac{x^2}{a^2} - 1}$$

As x gets large, the $+$ under the radical sign becomes insignificant in comparison with x^2/a^2, and the equation for y becomes

$$y = \pm\frac{b}{a}x$$

This is the equation of a straight line having a slope of b/a. Thus as x increases, the branches of the hyperbola more closely approach straight lines of slopes $\pm b/a$.

In general, if the distance from a point P on a curve to some line L approaches zero as the distance from P to the origin increases without bound, then L is an asymptote. The hyperbola has two such asymptotes, of slopes $\pm b/a$.

For a hyperbola whose transverse axis is *vertical*, the slopes of the asymptotes can be shown to be $\pm a/b$.

Standard Equations of a Hyperbola: Centre at Origin	Transverse Axis Horizontal	slope $= \pm\dfrac{b}{a}$	332
	Transverse Axis Vertical	slope $= \pm\dfrac{a}{b}$	333

Common Error	The asymptotes are not (usually) perpendicular to each other. The slope of one is not the negative reciprocal of the other.

Looking back at Fig. 22–77, we can now give meaning to the quantity b. If an asymptote has a slope b/a, it must have a rise of b in a run equal to a. Thus b is the distance, perpendicular to the transverse axis, from vertex to asymptote. It is half the length of the conjugate axis.

Hyperbola: Distance from Center to Focus	$c = \sqrt{a^2 + b^2}$	331

◆◆◆ **Example 43:** Find a, b, and c, the coordinates of the centre, the vertices, and the foci, and the slope of the asymptotes for the hyperbola

$$\frac{x^2}{25} - \frac{y^2}{36} = 1$$

Solution: This equation is of the same form as Eq. 326, so we know that the centre is at the origin and that the transverse axis is on the x axis. Also, $a^2 = 25$ and $b^2 = 36$, so

$$a = 5 \quad \text{and} \quad b = 6$$

The vertices then have the coordinates

$$V(5, 0) \quad \text{and} \quad V'(-5, 0)$$

Then, from Eq. 331,

$$c^2 = a^2 + b^2 = 25 + 36 = 61$$

so $c = \sqrt{61} \approx 7.81$. The coordinates of the foci are then

$$F(7.81, 0) \quad \text{and} \quad F'(-7.81, 0)$$

The slopes of the asymptotes are

$$\pm \frac{b}{a} = \pm \frac{6}{5}$$

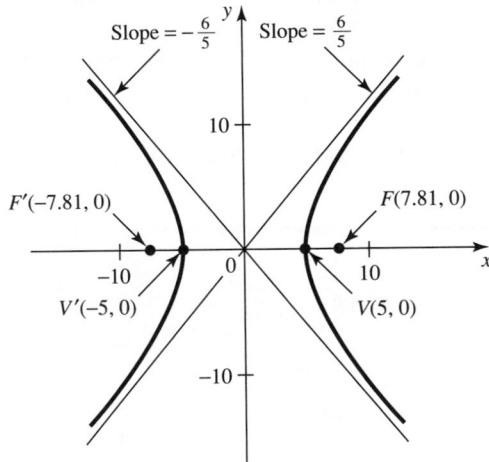

FIGURE 22-78   Graph of $\dfrac{x^2}{25} - \dfrac{y^2}{36} = 1$.

This hyperbola is shown in Fig. 22–78. In the following section we show how to make such a graph.   ◆◆◆

Graphing a Hyperbola

A good way to start a sketch of the hyperbola is to draw a rectangle whose dimension along the transverse axis is $2a$ and whose dimension along the conjugate axis is $2b$. The asymptotes are then drawn along the diagonals of this rectangle. Half the diagonal of the rectangle has a length $\sqrt{a^2 + b^2}$, which is equal to c, the distance to the foci. Thus an arc of radius c will cut extensions of the transverse axis at the focal points.

As with the parabola and ellipse, a perpendicular through a focus connecting two points on the hyperbola is called a *latus rectum*. Its length, called the *focal width*, is $2b^2/a$, the same as for the ellipse.

◆◆◆ **Example 44:** Find the vertices, foci, semiaxes, slope of the asymptotes, and focal width, and graph the hyperbola $64x^2 - 49y^2 = 3136$.

Solution: We write the equation in standard form by dividing by 3136.

$$\frac{x^2}{49} - \frac{y^2}{64} = 1$$

This is the form of Eq. 326, so the transverse axis is horizontal, with $a = 7$ and $b = 8$. We draw a rectangle of width 14 and height 16 (shown shaded in Fig. 22–79), thus locating the vertices at $(\pm 7, 0)$. Diagonals through the rectangle give us the asymptotes of slopes $\pm \frac{8}{7}$. We locate the foci by swinging an arc of radius c, equal to half the diagonal of the rectangle. Thus

$$c = \sqrt{7^2 + 8^2} = \sqrt{113} \cong 10.6$$

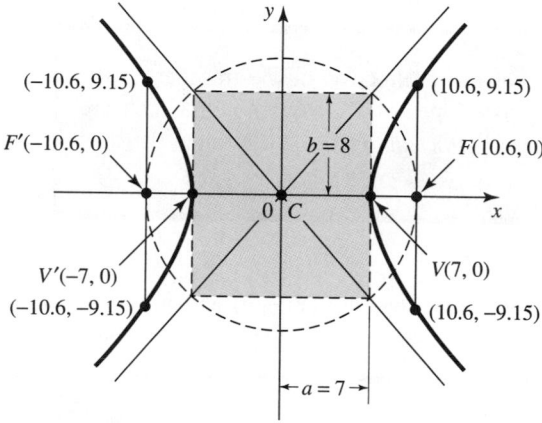

FIGURE 22–79

The foci then are at $(\pm 10.6, 0)$. We obtain a few more points by computing the focal width.

$$L = \frac{2b^2}{a} = \frac{2(64)}{7} \cong 18.3$$

This gives us the additional points $(10.6, 9.15)$, $(10.6, -9.15)$, $(-10.6, 9.15)$, and $(-10.6, -9.15)$. ◆◆◆

◆◆◆ **Example 45:** A hyperbola whose centre is at the origin has a focus at $(0, -5)$ and a transverse axis 8 units long, as shown in Fig. 22–80. Write the standard equation of the hyperbola, and find the slope of the asymptotes.

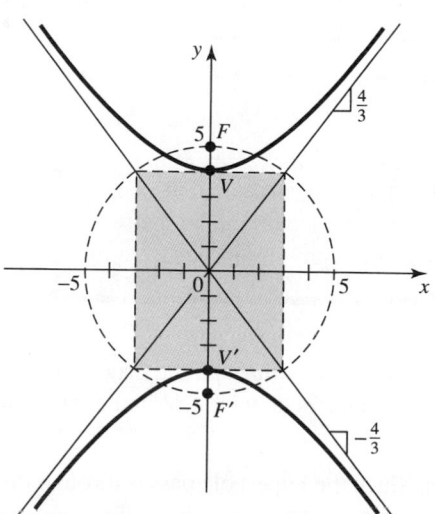

FIGURE 22–80

Solution: The transverse axis is 8, so $a = 4$. A focus is 5 units below the origin, so the transverse axis must be vertical, and $c = 5$. From Eq. 331,

$$b = \sqrt{c^2 - a^2} = \sqrt{25 - 16} = 3$$

Substituting into Eq. 327 gives us

$$\frac{y^2}{16} - \frac{x^2}{9} = 1$$

and from Eq. 333,

$$\text{slope of asymptotes} = \pm\frac{a}{b} = \pm\frac{4}{3}$$

◆◆◆

Common Error	Be sure that you know the direction of the transverse axis before computing the slopes of the asymptotes, which are $\pm b/a$ when the axis is horizontal, but $\pm a/b$ when the axis is vertical.

Shift of Axes

As with the other conics, we shift the axes by replacing x by $(x - h)$ and y by $(y - k)$ in Eqs. 326 and 327.

Standard Equations of a Hyperbola: Centre at (h, k)	Transverse Axis Horizontal		$\dfrac{(x - h)^2}{a^2} - \dfrac{(y - k)^2}{b^2} = 1$	**328**
	Transverse Axis Vertical		$\dfrac{(y - k)^2}{a^2} - \dfrac{(x - h)^2}{b^2} = 1$	**329**

Common Error	Do not confuse these equations with those for the ellipse. Here the terms have **opposite signs**. Also, a^2 is always the denominator of the **positive** term, even though it can be smaller than b^2. As with the ellipse, the variable in the same term as a^2 tells the direction of the transverse axis.

◆◆◆ **Example 46:** A certain hyperbola has a focus at $(1, 0)$, passes through the origin, has its transverse axis on the x axis, and has a distance of 10 between its focal points. Write its standard equation.

Solution: In Fig. 22–81 we plot the given focus $F(1, 0)$. Since the hyperbola passes through the origin, and the transverse axis also passes through the origin, a vertex must be at the origin as well. Then, since $c = 5$, we go 5 units to the left of F and plot the centre, $(-4, 0)$. Thus $a = 4$ units, from centre to vertex. Then, by Eq. 331,

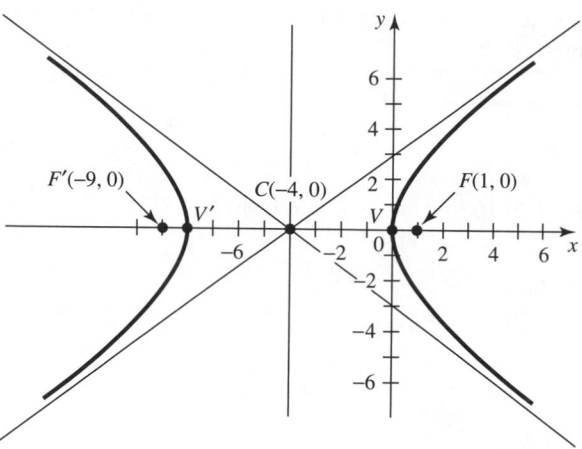

FIGURE 22–81

$$b = \sqrt{c^2 - a^2}$$

$$= \sqrt{25 - 16} = 3$$

Substituting into Eq. 328 with $h = -4$, $k = 0$, $a = 4$, and $b = 3$, we get

$$\frac{(x + 4)^2}{16} - \frac{y^2}{9} = 1$$

as the required equation.

◆◆◆

General Equation of a Hyperbola

Since the standard equations for the ellipse and hyperbola are identical except for the minus sign, it is not surprising that their general equations should be identical except for a minus sign. The general second-degree equation for ellipse or hyperbola is as follows:

General Equation of a Hyperbola	$Ax^2 + Cy^2 + Dx + Ey + F = 0$	330

It's easy to feel overwhelmed by the large number of formulas in this chapter. Study the summary of formulas in Appendix A, where you'll find them side by side and can see where they differ and where they are alike.

Comparing this with the general second-degree equation, we see that $B = 0$, indicating that the axes of the curve are parallel to the coordinate axes.

Also note that neither A nor C is zero, which tells us that the curve is not a parabola. Further, A and C have different values, telling us that the curve cannot be a circle. But here we will see that A and C have opposite signs, while for the ellipse, A and C have the same sign.

As before, we complete the square to go from general form to standard form.

◆◆◆ **Example 47:** Write the equation $x^2 - 4y^2 - 6x + 8y - 11 = 0$ in standard form. Find the centre, vertices, foci, and asymptotes.

Solution: Rearranging gives us

$$(x^2 - 6x) - 4(y^2 - 2y) = 11$$

Completing the square and factoring yields

$$(x^2 - 6x + 9) - 4(y^2 - 2y + 1) = 11 + 9 - 4$$

$$(x - 3)^2 - 4(y - 1)^2 = 16$$

(Take care, when completing the square, to add -4 to the right side to compensate for 1 times -4 added on the left.) Dividing by 16 gives us

$$\frac{(x-3)^2}{16} - \frac{(y-1)^2}{4} = 1$$

from which $a = 4$, $b = 2$, $h = 3$, and $k = 1$.

From Eq. 331, $c = \sqrt{a^2 + b^2} = \sqrt{16 + 4} \approx 4.47$, and from Eq. 332,

$$\text{slope of asymptotes} = \pm\frac{b}{a} = \pm\frac{1}{2}$$

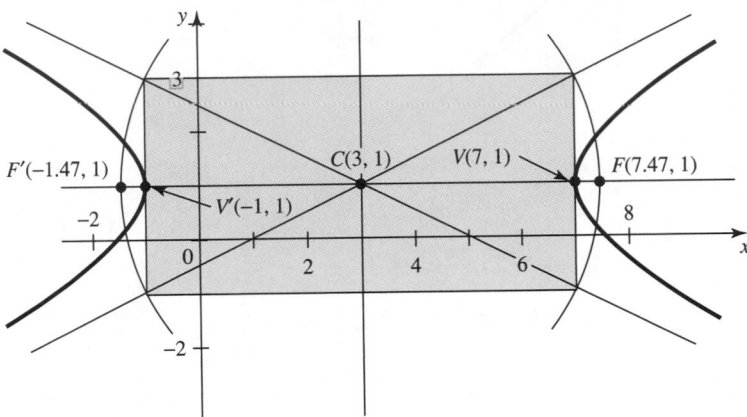

FIGURE 22–82

This hyperbola is plotted in Fig. 22–82.   ◆◆◆

Hyperbola Whose Asymptotes Are the Coordinate Axes
The graph of an equation of the form

$$xy = k$$

where k is a constant, is a hyperbola similar to those we have studied in this section, but rotated so that the coordinate axes are now the asymptotes, and the transverse and conjugate axes are at $45°$.

Hyperbola: Axes Rotated 45°	$xy = k$	**335**

◆◆◆ **Example 48:** Plot the equation $xy = -4$.

Solution: We select values for x and evaluate $y = -4/x$.

x	-4	-3	-2	-1	0	1	2	3	4
y	1	$\frac{4}{3}$	2	4	$-$	-4	-2	$-\frac{4}{3}$	-1

The graph of Fig. 22–83 shows vertices at $(-2, 2)$ and $(2, -2)$. We see from the graph that a and b are equal and that each is the hypotenuse of a right triangle of side 2. Therefore

$$a = b = \sqrt{2^2 + 2^2} = \sqrt{8}$$

and from Eq. 331,

$$c = \sqrt{a^2 + b^2} = \sqrt{16} = 4$$

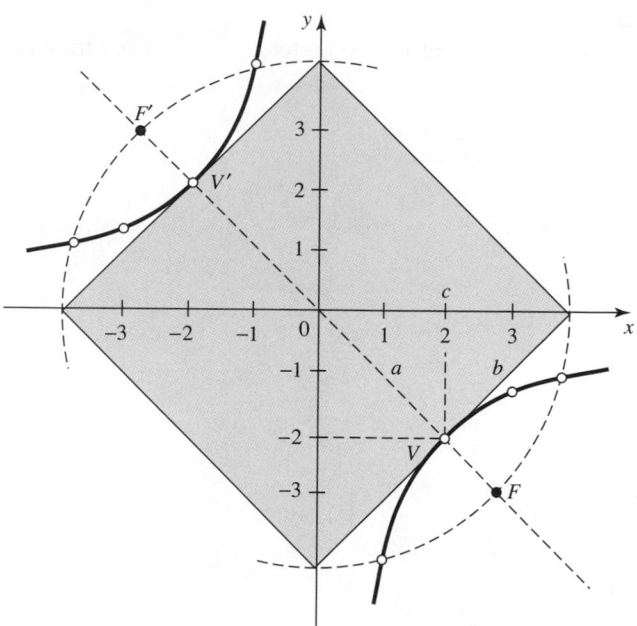

FIGURE 22–83 Hyperbola whose asymptotes are the coordinate axes.

◆◆◆

When the constant k in Eq. 335 is negative, the hyperbola lies in the second and fourth quadrants, as in Example 48. When k is positive, the hyperbola lies in the first and third quadrants.

Exercise 5 ◆ The Hyperbola

Hyperbola with Centre at Origin

Find the vertices, the foci, the lengths a and b of the semiaxes, and the slope of the asymptotes for each hyperbola.

1. $\dfrac{x^2}{16} - \dfrac{y^2}{25} = 1$

2. $\dfrac{y^2}{25} - \dfrac{x^2}{49} = 1$

3. $16x^2 - 9y^2 = 144$

4. $9x^2 - 16y^2 = 144$

5. $x^2 - 4y^2 = 16$

6. $3x^2 - y^2 + 3 = 0$

Write the equation of each hyperbola in standard form.

7. vertices at $(\pm 5, 0)$; foci at $(\pm 13, 0)$

8. vertices at $(0, \pm 7)$; foci at $(0, \pm 10)$

9. distance between foci = 8; transverse axis = 6 and is horizontal

10. conjugate axis = 4; foci at $(\pm 2.5, 0)$

11. transverse and conjugate axes equal; passes through $(3, 5)$; transverse axis vertical

12. transverse axis = 16 and is horizontal; conjugate axis = 14

13. transverse axis = 10 and is vertical; curve passes through $(8, 10)$

14. transverse axis = 8 and is horizontal; curve passes through $(5, 6)$

Hyperbola with Centre Not at Origin
Find the centre, lengths of the semiaxes, foci, slope of the asymptotes, and vertices for each hyperbola.

15. $\dfrac{(x-2)^2}{25} - \dfrac{(y+1)^2}{16} = 1$

16. $\dfrac{(y+3)^2}{25} - \dfrac{(x-2)^2}{49} = 1$

17. $16x^2 - 64x - 9y^2 - 54y = 161$

18. $16x^2 + 32x - 9y^2 + 592 = 0$

19. $4x^2 + 8x - 5y^2 - 10y + 19 = 0$

20. $y^2 - 6y - 3x^2 - 12x = 12$

Write the equation of each hyperbola in standard form.

21. centre at $(3, 2)$; length of the transverse axis $= 8$ and is vertical; length of the conjugate axis $= 4$

22. foci at $(-1, -1)$ and $(-5, -1)$; $a = b$

23. foci at $(5, -2)$ and $(-3, -2)$; vertex halfway between centre and focus

24. length of the conjugate axis $= 8$ and is horizontal; centre at $(-1, -1)$; length of the transverse axis $= 16$

Hyperbola Whose Asymptotes Are Coordinate Axes

25. Write the equation of a hyperbola with centre at the origin and with asymptotes on the coordinate axes, whose vertex is at $(6, 6)$.

26. Write the equation of a hyperbola with centre at the origin and with asymptotes on the coordinate axes, which passes through the point $(-9, 2)$.

Applications

27. A hyperbolic mirror (Fig. 22–84) has the property that a ray of light directed at one focus will be reflected so as to pass through the other focus. Write the equation of the mirror shown, taking the axes as indicated.

28. A ship (Fig. 22–85) receives simultaneous radio signals from stations P and Q, on the shore. The signal from station P is found to arrive 375 microseconds (μs) later than that from Q. Assuming that the signals travel at a rate of 0.298 km/μs, find the distance from the ship to each station (*Hint:* The ship will be on one branch of a hyperbola H whose foci are at P and Q. Write the equation of this hyperbola, taking axes as shown, and then substitute $12\overline{0}$ km for y to obtain the distance x.)

29. Boyle's law states that under certain conditions, the product of the pressure and volume of a gas is constant, or $pv = c$. This equation has the same form as the hyperbola (Eq. 335). If a certain gas has a pressure of 25.0 kPa at a volume of 1000 cm³, write Boyle's law for this gas, and make a graph of pressure versus volume.

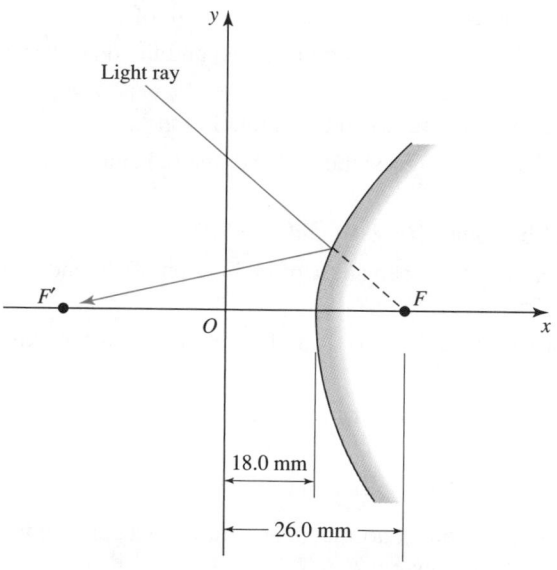

FIGURE 22-84 Hyperbolic mirror. This type of mirror is used in the *Cassegrain* form of reflecting telescope.

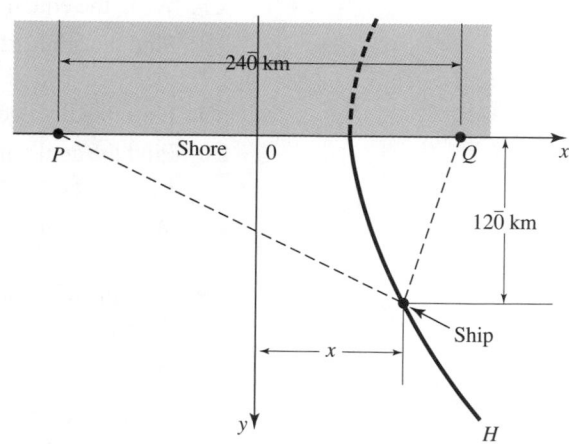

FIGURE 22-85

Case Study Discussion—Parabolic Microphone

The parabolic microphone can be made from inexpensive materials. A few sites on the Internet describe "dollar store" projects that use a small umbrella hat to form the reflector dish. We don't have room for a full set of instructions here, but you will be able to vary the position of the microphone to see just how well the reflector works.

◆◆◆ CHAPTER 22 REVIEW PROBLEMS ◆◆◆◆◆◆◆◆◆◆◆◆◆◆◆◆◆◆◆◆◆◆◆◆◆◆◆

1. Find the distance between the points (3,0) and (7,0).

2. Find the distance between the points (4, −4) and (1, −7).

3. Find the slope of the line perpendicular to a line that has an angle of inclination of 34.8°.

4. Find the angle of inclination in degrees of a line with a slope of −3.

5. Find the angle of inclination of a line perpendicular to a line having a slope of 1.55.

6. Find the angle of inclination of a line passing through (−3, 5) and (5, −6).

7. Find the slope of a line perpendicular to a line having a slope of $a/2b$.

8. Write the equation of a line having a slope of −2 and a y intercept of 5.

9. Find the slope and y intercept of the line $2y − 5 = 3(x − 4)$.

10. Write the equation of a line having a slope of $2p$ and a y intercept of $p − 3q$.

11. Write the equation of the line passing through (−5, −1) and (−2, 6).

12. Write the equation of the line passing through (−r, s) and (2r, −s).

13. Write the equation of the line having a slope of 5 and passing through the point (−4, 7).

14. Write the equation of the line having a slope of $3c$ and passing through the point $(2c, c − 1)$.

15. Write the equation of the line having an x intercept of -3 and a y intercept of 7.

16. Find the acute angle between two lines if one line has a slope of 1.50 and the other has a slope of 3.40.

17. Write the equation of the line that passes through $(2, 5)$ and is parallel to the x axis.

18. Find the angle of intersection between line L_1 having a slope of 2 and line L_2 having a slope of 7.

19. Find the directed distance AB between the points $A(-2, 0)$ and $B(-5, 0)$.

20. Find the angle of intersection between line L_1 having an angle of inclination of $18°$ and line L_2 having an angle of inclination of $75°$.

21. Write the equation of the line that passes through $(-3, 6)$ and is parallel to the y axis.

22. Find the increments in the coordinates of a particle that moves along a curve from $(3, 4)$ to $(5, 5)$.

23. Find the area of a triangle with vertices at $(6, 4)$, $(5, -2)$, and $(-3, -4)$.

In problems 24 through 33, a tangent T, of slope m, and a normal N are drawn to a curve at the point $P(x_1, y_1)$, as shown in Fig. 22–86. Show the following:

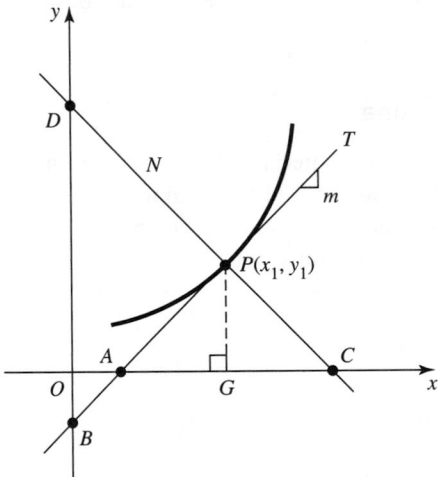

FIGURE 22-86   Tangent and normal to a curve.

24. The equation of the tangent is

$$y - y_1 = m(x - x_1)$$

25. The equation of the normal is

$$x - x_1 + m(y - y_1) = 0$$

26. The x intercept A of the tangent is

$$x_1 - y_1/m$$

27. The y intercept B of the tangent is

$$y_1 - mx_1$$

28. The length of the tangent from P to the x axis is

$$PA = \frac{y_1}{m}\sqrt{1 + m^2}$$

29. The length of the tangent from P to the y axis is

$$PB = x_1\sqrt{1 + m^2}$$

30. The x intercept C of the normal is

$$x_1 + my_1$$

31. The y intercept D of the normal is

$$y_1 + x_1/m$$

31. The length of the normal from P to the x axis is

$$PC = y_1 \sqrt{1 + m^2}$$

33. The length of the normal from P to the y axis is

$$PD = \frac{x_1}{m} \sqrt{1 + m^2}$$

Identify the curve represented by each equation. Find, where applicable, the centre, vertices, foci, radius, semiaxes, and so on.

34. $x^2 - 2x - 4y^2 + 16y = 19$

35. $x^2 + 6x + 4y = 3$

36. $x^2 + y^2 = 8y$

37. $25x^2 - 200x + 9y^2 - 90y = 275$

38. $16x^2 - 9y^2 = 144$

39. $x^2 + y^2 = 9$

40. Write the equation for an ellipse whose centre is at the origin, whose major axis ($2a$) is 20 and is horizontal, and whose minor axis ($2b$) equals the distance ($2c$) between the foci.

41. Write an equation for the circle passing through $(0, 0)$, $(8, 0)$, and $(0, -6)$.

42. Write the equation for a parabola whose vertex is at the origin and whose focus is $(-4.25, 0)$.

43. Write an equation for a hyperbola whose transverse axis is horizontal, with centre at $(1, 1)$ passing through $(6, 2)$ and $(-3, 1)$.

44. Write the equation of a circle whose centre is $(-5, 0)$ and whose radius is 5.

45. Write the equation of a hyperbola whose centre is at the origin, whose transverse axis = 8 and is horizontal, and passing through $(10, 25)$.

46. Write the equation of an ellipse whose foci are $(2, 1)$ and $(-6, 1)$, and the sum of the focal radii is 10.

47. Write the equation of a parabola whose axis is the line $y = -7$, whose vertex is 3 units to the right of the y axis, and passing through $(4, -5)$.

48. Find the intercepts of the curve $y^2 + 4x - 6y = 16$.

49. Find the points of intersection of $x^2 + y^2 + 2x + 2y = 2$ and $3x^2 + 3y^2 + 5x + 5y = 10$.

50. A stone bridge arch is in the shape of half an ellipse and is 15.0 m wide and 5.00 m high. Find the height of the arch at a distance of 6.00 m from its centre.

51. Write the equation of a hyperbola centred at the origin, where the conjugate axis = 12 and is vertical, and the distance between foci is 13.

52. A parabolic arch is 5.00 m high and 6.00 m wide at the base. Find the width of the arch at a height of 2.00 m above the base.

53. A stone thrown in the air follows a parabolic path and reaches a maximum height of 0 ft. in a horizontal distance of 48.0 ft. At what horizontal distances from the launch point will the height be 25.0 ft.?

Writing

54. Write a short paragraph explaining, in your own words, what is meant by the *slope of a curve*.

55. Suppose that the day before your visit to your former high school math class, the teacher unexpectedly asks you to explain how that day's topic, the conic sections, got that name.

Write a paragraph on what you will tell the class about how the circle, ellipse, parabola, and hyperbola (and the point and straight line, too) can be formed by intersecting a plane and a cone. You may plan to illustrate your talk with a clay model or a piece of cardboard rolled into a cone. You will probably be asked what the conic sections are good for, so write out at least one use for each curve.

Team Projects

56. There are an infinite number of number pairs p and q (such as 3 and 1.5) whose product equals their sum. Write an expression for q as a function of p, and make a graph of p versus q for all of the real numbers, not just integer values, from -5 to 5. With a suitable shift of axes, show that the graph is a hyperbola of the form $y = 1/x$.

57. We have already shown how to graph each of the conic sections. Now graph, in the same viewing window, the ellipse

$$\frac{x^2}{a^2} + \frac{y^2}{b^2} = 1$$

the hyperbola

$$\frac{x^2}{a^2} - \frac{y^2}{b^2} = 1$$

and the hyperbola

$$\frac{y^2}{a^2} - \frac{x^2}{b^2} = 1$$

using the same values of a and b for each. Also graph the asymptotes of the hyperbolas.

58. If the area of an ellipse is twice the area of the inscribed circle, find the length of the semi-major axis of the ellipse.

59. If a projectile is launched with a horizontal velocity v_x and vertical velocity v_y, its position after t seconds is given by

$$x = v_x t$$

$$y = v_y t - (g/2)t^2$$

where g is the acceleration due to gravity. Eliminate t from this pair of equations to get $y = f(x)$, and show that this equation represents a parabola.

23

Binary, Hexadecimal, Octal, and BCD Numbers

♦♦♦ OBJECTIVES ♦♦

When you have completed this chapter, you should be able to:
- Convert between binary and decimal numbers.
- Convert between decimal and binary fractions.
- Convert between binary and hexadecimal numbers.
- Convert between decimal and hexadecimal numbers.
- Convert between binary and octal numbers.
- Convert between binary and 8421 BCD numbers.

Email, texting, audio and video streaming, and memory storage of songs on MP3 players and pictures on camera memory cards—all of this is done in binary number format. Binary is also the underlying format of all programs running any computer. A programmer might work in a language such as C++ or Java, but the program compilers convert the program to binary code for the computer to actually read and follow instructions. Binary is a number system with only two numbers, 1 and 0. This system is not as convenient for humans to use as is our base-10 number system, but for a computer it is much easier to work with an "on" or "off" state, high voltage or low voltage.

To store a song without digital encoding, you have to be able to record every slight variation or level of input, which makes it hard to transmit or store, since any slight extraneous variation in power level could distort your signal (like scratchy vinyl records). If you have ever watched TV using an antenna for input, you know how much these analog signals can vary; the result is obvious from the picture and sound quality. A digital signal, on the other hand, can vary considerably in level without causing changes in quality as long as the circuitry in your TV or computer or cell phone can detect the difference between a high and a low voltage.

Hexadecimal, octal, and BCD (binary-coded decimal) are different systems of expressing binary numbers in a way that makes it easier for humans to read and work with them.

23–1   The Binary Number System

Binary Numbers

A *binary number* is a sequence of the digits 0 and 1, such as

$$1101001$$

The number shown has no fractional part and so is called a *binary integer*. A binary number having a fractional part contains a *binary point* (also called a *radix point*), as in the number

$$1001.01$$

Base or Radix

The *base* of a number system (also called the *radix*) is equal to the number of digits used in the system.

◆◆◆ **Example 1:**

(a) The decimal system uses the ten digits

$$0 \quad 1 \quad 2 \quad 3 \quad 4 \quad 5 \quad 6 \quad 7 \quad 8 \quad 9$$

and has a base of 10.

(b) The binary system uses two digits

$$0 \quad 1$$

and has a base of 2.   ◆◆◆

Bits, Bytes, and Words

Each of the digits is called a *bit,* from *binary digit.* A *byte* is a group of 8 bits, and a *word* is the largest string of bits that a computer can handle in one operation. The number of bits in a word is called the *word length.* Different computers have different word lengths. Earlier desktop and laptop computers had 8- and 16-bit processors, but now 32- and 64-bit processors are common. The longer words are often broken down into bytes for easier handling. Half a byte (4 bits) is called a *nibble.*

A *kilobyte* (Kbyte or KB) is 1024 (2^{10}) bytes, and a *megabyte* (Mbyte or MB) is 1 048 576 (2^{20}, or 1024×1024) bytes.

Writing Binary Numbers

A binary number is sometimes written with a subscript 2 when there is a chance that the binary number would otherwise be mistaken for a decimal number.

◆◆◆ **Example 2:** The binary number 110 could easily be mistaken for the decimal number 110, unless we write it

$$110_2$$

Similarly, a decimal number that could be mistaken for binary is often written with a subscript 10, as in

$$101_{10}$$ ◆◆◆

Long binary numbers are sometimes written with their bits in groups of four for easier reading.

◆◆◆ **Example 3:** The number 100100010100.001001 is easier to read when written as

$$1001 \quad 0001 \quad 0100.0010 \quad 01$$ ◆◆◆

The leftmost bit in a binary number is called the *high-order* or *most significant bit* (*MSB*). The bit at the extreme right of the number is the *low-order* or *least significant bit* (*LSB*).

Place Value

As noted in Chapter 1, a positional number system is one in which the position of a digit determines its value, and each position in a number has a *place value* equal to the base of the number system raised to the position number.

The place values in the binary number system are

$$\ldots 2^4 \quad 2^3 \quad 2^2 \quad 2^1 \quad 2^0 \quad . \quad 2^{-1} \quad 2^{-2} \ldots$$

or

$$\ldots 16 \quad 8 \quad 4 \quad 2 \quad 1 \quad . \quad \tfrac{1}{2} \quad \tfrac{1}{4} \ldots$$

binary point

A more complete list of place values for a binary number is given in Table 23–1.

Expanded Notation

Thus the value of any digit in a number is the product of that digit and the place value. The value of the entire number is then the sum of these products.

◆◆◆ **Example 4:**

(a) The decimal number 526 can be expressed as

$$5 \times 10^2 + 2 \times 10^1 + 6 \times 10^0$$

or

$$5 \times 100 + 2 \times 10 + 6 \times 1$$

or

$$500 \quad + \quad 20 \quad + \quad 6$$

(b) The binary number 1011 can be expressed as

$$1 \times 2^3 + 0 \times 2^2 + 1 \times 2^1 + 1 \times 2^0$$

or

$$1 \times 8 + 0 \times 4 + 1 \times 2 + 1 \times 1$$

or

$$8 \quad + \quad 0 \quad + \quad 2 \quad + \quad 1$$

Numbers written in this way are said to be in *expanded notation*. ◆◆◆

Converting Binary Numbers to Decimal

To convert a binary number to decimal, simply write the binary number in expanded notation (omitting those where the bit is 0), and add the resulting values.

◆◆◆ **Example 5:** Convert the binary number

$$1001.011$$

to decimal.

Solution: In expanded notation,

$$1001.011 = 1 \times 8 + 1 \times 1 + 1 \times \tfrac{1}{4} + 1 \times \tfrac{1}{8}$$

$$= 8 \quad + \quad 1 \quad + \quad \tfrac{1}{4} \quad + \quad \tfrac{1}{8}$$

$$= 9\tfrac{3}{8}$$

$$= 9.375 \qquad\qquad ◆◆◆$$

Largest Decimal Number Obtainable with *n* Bits

The largest possible 3-bit binary number is

$$1\,1\,1 = 7$$

or

$$2^3 - 1$$

TABLE 23–1

Position	Place Value
10	$2^{10} = 1024$
9	$2^9 = 512$
8	$2^8 = 256$
7	$2^7 = 128$
6	$2^6 = 64$
5	$2^5 = 32$
4	$2^4 = 16$
3	$2^3 = 8$
2	$2^2 = 4$
1	$2^1 = 2$
0	$2^0 = 1$
−1	$2^{-1} = 0.5$
−2	$2^{-2} = 0.25$
−3	$2^{-3} = 0.125$
−4	$2^{-4} = 0.0625$
−5	$2^{-5} = 0.031\ 25$
−6	$2^{-6} = 0.015\ 625$

Compare these place values to those in the decimal system shown in Fig. 1–1.

The largest possible 4-bit number is

$$1\,1\,1\,1 = 15$$

or

$$2^4 - 1$$

Similarly, the largest n-bit binary number is given by the following:

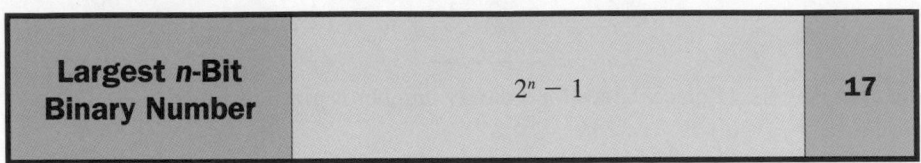

Largest n-Bit Binary Number	$2^n - 1$	17

◆◆◆ **Example 6:** If a computer stores numbers with 15 bits, what is the largest decimal number that can be represented?

Solution: The largest decimal number is

$$2^{15} - 1 = 32\,768 \qquad\qquad\qquad ◆◆◆$$

Significant Digits

In Example 6, we saw that a 15-bit binary number could represent a decimal number no greater than 32 768. Thus it took 15 binary digits to represent 5 decimal digits. As a rule of thumb, *it takes about 3 bits for each decimal digit.*

Stated another way, if we have a computer that stores numbers with 15 bits, we should assume that the decimal numbers that it prints do not contain more than 5 significant digits.

◆◆◆ **Example 7:** If we want a computer to print decimal numbers containing 7 significant digits, how many bits must it use to store those numbers?

Solution: Using our rule of thumb, we need

$$3 \times 7 = 21 \text{ bits} \qquad\qquad\qquad ◆◆◆$$

CASE STUDY—IP ADDRESSES

To connect to the Internet, a computer or device must have an address. There are two addressing systems in use: Internet Protocol version 4 (IP4) and Internet Protocol version 6 (IP6). IP addresses for PCs, printers, etc. are in the IP4 format, but current operating systems are capable of using IP6. IP4 uses four binary octets (8 bits). *Octet* is the term commonly used when addressing; *byte* is the term used for 8 bits in memory, etc.

To make it easier for humans to work with, the IP address is presented as a decimal number, each octet separated by a dot. A typical binary IP address might be, for example, 0100100000 00111011001100100010011—not easy for humans to read. If we break it up into 8-bit units— 01001000.00001110.11001100.10010011—then convert each 8-bit octet to its decimal equivalent, we get 72.14.204.147, which is much easier for human use.

IP6 is a longer address to accommodate the reality that more and more addresses will be needed in the future. The IP6 address is 128 bits long. To display it in the same fashion as IP4, we need 16 octets. Your task is to research the format for IP6 and determine some of the rules for display.

Converting Decimal Integers to Binary

To convert a decimal *integer* to binary, we first divide it by 2, obtaining a quotient and a remainder. We write down the remainder and divide the quotient by 2, getting a new quotient and remainder. We then repeat this process until the quotient is zero.

◆◆◆ **Example 8:** Convert the decimal integer 59 to binary.

Solution: We divide 59 by 2, getting a quotient of 29 and a remainder of 1. Then dividing 29 by 2 gives a quotient of 14 and a remainder of 1. These calculations, and those that follow, can be arranged in a table, as follows:

Remainder

$$
\begin{array}{ll}
2\overline{)59} & \\
2\overline{)29} & 1\,\text{LSB} \\
2\overline{)14} & 1 \\
2\overline{)7} & 0 \\
2\overline{)3} & 1 \\
2\overline{)1} & 1 \\
2\overline{)0} & 1\,\text{MSB}
\end{array}
$$

Read up

Our binary number then consists of the digits in the remainders, with those at the *top* of the column appearing to the *right* in the binary number. Thus

$$59_{10} = 111011_2$$

◆◆◆

The conversion can, of course, be checked by converting back to decimal.

To convert a decimal *fraction* to binary, we first multiply it by 2, remove the integer part of the product, and multiply by 2 again. We then repeat the procedure.

◆◆◆ **Example 9:** Convert the decimal fraction 0.546 875 to binary.

Solution: We multiply the given number by 2, getting 1.093 75. We remove the integer part, 1, leaving 0.093 75, which we again multiply by 2, getting 0.1875. We repeat the computation until we get a product that has a fractional part of zero, as in the following table:

Integer Part

$$
\begin{array}{ll}
0.546\,875 & \\
\times \quad\;\; 2 & \\
\hline
\mathbf{1}.093\,75 & 1\,\text{MSB} \\
\times \quad\;\; 2 & \\
\hline
\mathbf{0}.1875 & 0 \\
\times \quad\;\; 2 & \\
\hline
\mathbf{0}.375 & 0 \\
\times \quad\;\; 2 & \\
\hline
\mathbf{0}.75 & 0 \\
\times \quad\; 2 & \\
\hline
\mathbf{1}.50 & 1 \\
\times \quad\; 2 & \\
\hline
\mathbf{1}.00 & 1\,\text{LSB}
\end{array}
$$

Read down

Notice that we remove the integral part, shown in boldface, before multiplying by 2.

Note that this is the reverse of what we did when converting a decimal integer to binary.

We stop now that the fractional part of the product (1.00) is zero. The column containing the integer parts is now our binary number, with the digits at the top appearing at the *left* of the binary number. So

$$0.546\,875_{10} = 0.1000\,11_2$$

◆◆◆

In Example 9 we were able to find an exact binary equivalent of a decimal fraction. This is not always possible, as shown in the following example.

◆◆◆ **Example 10:** Convert the decimal fraction 0.743 to binary.

Solution: We follow the same procedure as before and get the following values.

Integer Part

$$
\begin{array}{rl}
0.743 & \\
\times \quad 2 & \\
\hline
\mathbf{1}.486 & 1 \\
\times \quad 2 & \\
\hline
\mathbf{0}.972 & 0 \\
\times \quad 2 & \\
\hline
\mathbf{1}.944 & 1 \\
\times \quad 2 & \\
\hline
\mathbf{1}.888 & 1 \\
\times \quad 2 & \\
\hline
\mathbf{1}.776 & 1 \\
\times \quad 2 & \\
\hline
\mathbf{1}.552 & 1 \\
\times \quad 2 & \\
\hline
\mathbf{1}.104 & 1 \\
\times \quad 2 & \\
\hline
\mathbf{0}.208 & 0 \\
\times \quad 2 & \\
\hline
\mathbf{0}.416 & 0 \\
\end{array}
$$

It is becoming clear that this computation can continue indefinitely, demonstrating that *not all decimal fractions can be exactly converted to binary*. This inability to make an exact conversion is an unavoidable source of inaccuracy in some computations. To decide how far to carry out computation, we use the rule of thumb that each decimal digit requires about 3 binary bits to give the same accuracy. Thus our original 3-digit number requires about 9 bits, which we already have. The result of our conversion is then

$$0.743_{10} = 0.1011\ 1110\ 0_2 \qquad \text{◆◆◆}$$

To convert a decimal number having both an integer part and a fractional part to binary, convert each part separately as shown above, and combine.

◆◆◆ **Example 11:** Convert the number 59.546 875 to binary.

Solution: From the preceding examples,

$$59 = 11\ 1011$$

and

$$0.546\ 875 = 0.1000\ 11$$

So

$$59.546\ 875 = 11\ 1011.1000\ 11 \qquad \text{◆◆◆}$$

Converting Binary Fractions to Decimal

To convert a binary fraction to decimal, we use Table 23–1 to find the decimal equivalent of each binary bit located to the right of the binary point, and add.

◆◆◆ **Example 12:** Convert the binary fraction 0.101 to decimal.

Solution: From Table 23–1,

$$
\begin{array}{rcl}
0.1 = 2^{-1} & = & 0.5 \\
0.001 = 2^{-3} & = & \underline{0.125} \\
\text{Add:} \quad 0.101 & = & 0.625
\end{array}
\qquad \text{◆◆◆}
$$

The procedure is no different when converting a binary number that has both a whole and a fractional part.

◆◆◆ Example 13: Convert the binary number 10.01 to decimal.

Solution: From Table 23–1,

$$10 = 2$$
$$0.01 = 0.25$$
$$\text{Add:} \quad 10.01 = 2.25$$

◆◆◆

Exercise 1 ◆ The Binary Number System

Conversion between Binary and Decimal

Convert each binary number to decimal.

1. 10
2. 01
3. 11
4. 1001
5. 0110
6. 1111
7. 1101
8. 0111
9. 1100
10. 1011
11. 0101
12. 1101 0010
13. 0110 0111
14. 1001 0011
15. 0111 0111
16. 1001 1010 1000 0010

Convert each decimal number to binary.

17. 5
18. 9
19. 2
20. 7
21. 72
22. 28
23. 93
24. 17
25. 274
26. 937
27. 118
28. 267
29. 8375
30. 2885
31. 82 740
32. 72 649

Convert each decimal fraction to binary. Retain 8 bits to the right of the binary point when the conversion is not exact.

33. 0.5
34. 0.25
35. 0.75
36. 0.375
37. 0.3
38. 0.8
39. 0.55
40. 0.35
41. 0.875
42. 0.4375
43. 0.3872
44. 0.8462

Convert each binary fraction to decimal.

45. 0.1
46. 0.11
47. 0.01
48. 0.011
49. 0.1001
50. 0.0111

Convert each decimal number to binary. Keep 8 bits to the right of the binary point when the conversion is not exact.

51. 5.5
52. 2.75

53. 4.375   **54.** 29.381

55. 948.472   **56.** 2847.228 53

Convert each binary number to decimal.

57. 1.1   **58.** 10.11

59. 10.01   **60.** 101.011

61. 1 1001.0110 1   **62.** 10 1001.1001 101

23–2   The Hexadecimal Number System

Hexadecimal Numbers

Hexadecimal numbers (or *hex* for short) are obtained by grouping the bits in a binary number into sets of four and representing each such set by a single number or letter. A hex number one-fourth the length of the binary number is thus obtained.

Base 16

Since a 4-bit group of binary digits can have a value between 0 and 15, we need 16 symbols to represent all of these values. The base of hexadecimal numbers is thus 16. We use the digits from 0 to 9 and the capital letters A to F, as shown in Table 23–2.

Converting Binary to Hexadecimal

To convert binary to hexadecimal, group the bits into sets of four starting at the binary point, adding zeros as needed to fill out the groups. Then assign to each group the appropriate letter or number from Table 23–2.

◆◆◆ **Example 14:** Convert 10110100111001 to hexadecimal.

Solution: Grouping, we get

$$10\quad 1101\quad 0011\quad 1001$$

or

$$0010\quad 1101\quad 0011\quad 1001$$

From Table 23–2, we obtain

$$2\qquad D\qquad 3\qquad 9$$

TABLE 23–2   Hexadecimal and octal number systems.

Decimal	Binary	Hexadecimal	Octal
0	0000	0	0
1	0001	1	1
2	0010	2	2
3	0011	3	3
4	0100	4	4
5	0101	5	5
6	0110	6	6
7	0111	7	7
8	1000	8	10
9	1001	9	11
10	1010	A	12
11	1011	B	13
12	1100	C	14
13	1101	D	15
14	1110	E	16
15	1111	F	17

So the hexadecimal equivalent is 2D39. Hexadecimal numbers are sometimes written with the subscript 16, as in

$$2D39_{16}$$ ◆◆◆

The procedure is no different for binary fractions.

◆◆◆ **Example 15:** Convert 101111.0011111 to hexadecimal.

Solution: Grouping from the binary point gives us

$$10 \quad 1111 \, . \, 0011 \quad 111$$

or

$$0010 \quad 1111 \, . \, 0011 \quad 1110$$

From Table 23–2, we have

$$2 \quad F \, . \, 3 \quad E$$

or 2F.3E. ◆◆◆

Converting Hexadecimal to Binary

To convert hexadecimal to binary, we simply reverse the procedure.

◆◆◆ **Example 16:** Convert 3B25.E to binary.

Solution: We write the group of 4 bits corresponding to each hexadecimal symbol.

$$3 \quad B \quad 2 \quad 5 \, . \, E$$
$$0011 \quad 1011 \quad 0010 \quad 0101 \, . \, 1110$$

or 11 1011 0010 0101.111. ◆◆◆

Converting Hexadecimal to Decimal

As with decimal and binary numbers, each hex digit has a *place value,* equal to the base, 16, raised to the position number. Thus the place value of the first position to the left of the decimal point is

$$16^0 = 1$$

and the next is

$$16^1 = 16$$

and so on.

To convert from hex to decimal, first replace each letter in the hex number by its decimal equivalent. Write the number in expanded notation, multiplying each hex digit by its place value. Add the resulting numbers.

Instead of converting directly between decimal and hex, many find it easier to first convert to binary.

◆◆◆ **Example 17:** Convert the hex number 3B.F to decimal.

Solution: We replace the hex B with 11, and the hex F with 15, and write the number in expanded form.

$$3 \qquad B \quad . \quad F$$
$$3 \qquad 11 \quad . \quad 15$$
$$(3 \times 16^1) + (11 \times 16^0) + (15 \times 16^{-1})$$
$$= \quad 48 \quad + \quad 11 \quad + \quad 0.9375$$
$$= 59.9375$$ ◆◆◆

Converting Decimal to Hexadecimal

To convert decimal to hexadecimal, we repeatedly divide the given decimal number by 16 and convert each remainder to hex. The remainders form our hex number, the last remainder obtained being the most significant digit.

◆◆◆ **Example 18:** Convert the decimal number 83759 to hex.

Solution:

$$83759 \div 16 = 5234 \quad \text{with remainder of } 15$$
$$5234 \div 16 = 327 \quad \text{with remainder of } 2$$
$$327 \div 16 = 20 \quad \text{with remainder of } 7 \qquad \text{Read}$$
$$20 \div 16 = 1 \quad \text{with remainder of } 4 \qquad \text{up}$$
$$1 \div 16 = 0 \quad \text{with remainder of } 1$$

Changing the number 15 to hex F and reading the remainders from bottom up, we find that our hex equivalent is 1472F. ◆◆◆

Exercise 2 ◆ The Hexadecimal Number System

Binary-Hex Conversions

Convert each binary number to hexadecimal.

1. 1101

2. 1010

3. 1001

4. 1111

5. 1001 0011

6. 0110 0111

7. 1101 1000

8. 0101 1100

9. 1001 0010 1010 0110

10. 0101 1101 0111 0001

11. 1001.0011

12. 10.0011

13. 1.0011 1

14. 101.101

Convert each hex number to binary.

15. 6F

16. B2

17. 4A

18. CC

19. 2F35

20. D213

21. 47A2

22. ABCD

23. 5.F

24. A4.E

25. 9.AA

26. 6D.7C

Decimal-Hex Conversions

Convert each hex number to decimal.

27. F2

28. 5C

29. 33

30. DF

31. 37A4

32. A3F6

33. F274

34. C721

35. 3.F

36. 22.D

37. ABC.DE

38. C.284

Convert each decimal number to hex.

39. 39   **40.** 13

41. 921   **42.** 554

43. 2741   **44.** 9945

45. 1736   **46.** 2267

23–3 The Octal Number System

The Octal System

The *octal* number system uses eight digits, 0 to 7, and hence has a base of eight. A comparison of the decimal, binary, hex, and octal digits is given in Table 23–2.

Binary-Octal Conversions

To convert from binary to octal, write the bits of the binary number in groups of three, starting at the binary point. Then write the octal equivalent for each group.

◆◆◆ **Example 19:** Convert the binary number 1 1010 0011 0110 to octal.

Solution: Grouping the bits in sets of three from the binary point, we obtain

$$001 \ 101 \ 000 \ 110 \ 110$$

and the octal equivalents,

$$1 \quad 5 \quad 0 \quad 6 \quad 6$$

so the octal equivalent of 1 1010 0011 0110 is 15066.   ◆◆◆

To convert from octal to binary, simply reverse the procedure shown in Example 19.

◆◆◆ **Example 20:** Convert the octal number 7364 to binary.

Solution: We write the group of 3 bits corresponding to each octal digit.

$$7 \quad 3 \quad 6 \quad 4$$
$$111 \ 011 \ 110 \ 100$$

or 1110 1111 0100.   ◆◆◆

Exercise 3 ◆ The Octal Number System

Convert each binary number to octal.

1. 110   **2.** 010

3. 111   **4.** 101

5. 11 0011   **6.** 01 1010

7. 0110 1101   **8.** 1 1011 0100

9. 100 1001   **10.** 1100 1001

Convert each octal number to binary.

11. 26   **12.** 35

13. 623   **14.** 621

15. 5243   **16.** 1153

17. 63150   **18.** 2346

23–4   BCD Codes

In Sec. 23–1, we converted decimal numbers into binary. Recall that each bit had a place value equal to 2 raised to the position number. Thus the binary number 10001 is equal to

$$2^4 + 2^0 = 16 + 1 = 17$$

We will now refer to numbers such as 10001 as *straight binary*.

    With a BCD or *binary-coded-decimal* code, a decimal number is not converted *as a whole* to binary, but rather *digit by digit*. For example, the 1 in the decimal number 17 is 0001 in 4-bit binary, and the 7 is equal to 0111. Thus

$$17 = 10001 \text{ in straight binary}$$

and

$$17 = 0001\ 0111 \text{ in BCD}$$

    When we converted the digits in the decimal number 17, we used the same 4-bit binary equivalents as in Table 23–2. The bits have the place values 8, 4, 2, and 1, and BCD numbers written in this manner are said to be in *8421 code*. The 8421 code and its decimal equivalents are listed in Table 23–3.

Converting Decimal Numbers to BDC

To convert decimal numbers to BCD, simply convert each decimal digit to its equivalent 4-bit code.

◆◆◆ **Example 21:** Convert the decimal number 25.3 to 8421 BCD code.

Solution: We find the BCD equivalent of each decimal digit from Table 23–3.

$$\begin{array}{ccccc} 2 & 5 & . & 3 \\ 0010 & 0101 & . & 0011 \end{array}$$

or   0010 0101.0011 .   ◆◆◆

TABLE 23-3 Binary-coded-decimal numbers.

Decimal	8421	2421	5211
0	0000	0000	0000
1	0001	0001	0001
2	0010	0010	0011
3	0011	0011	0101
4	0100	0100	0111
5	0101	1011	1000
6	0110	1100	1010
7	0111	1101	1100
8	1000	1110	1110
9	1001	1111	1111

Converting from BCD to Decimal

To convert from BCD to decimal, separate the BCD number into 4-bit groups, and write the decimal equivalent of each group.

◆◆◆ **Example 22:** Convert the 8421 BCD number 1 0011.0101 to decimal.

Solution: We have

$$0001 \quad 0011 \quad . \quad 0101$$

From Table 23–3, we have

$$1 \quad 3 \quad . \quad 5$$

or 13.5. ◆◆◆

Other BCD Codes

Other BCD codes are in use which represent each decimal digit by a 6-bit binary number or an 8-bit binary number, and other 4-bit codes in which the place values
are other than 8421. Some of these are shown in Table 23–3. Each is used in the same way as shown for the 8421 code.

◆◆◆ **Example 23:** Convert the decimal number 825 to 2421 BCD code.

Solution: We convert each decimal digit separately,

$$8 \quad 2 \quad 5$$
$$1110 \quad 0010 \quad 1011$$

◆◆◆

Other Computer Codes

In addition to the binary, hexadecimal, octal, and BCD codes we have discussed, many other codes are used in the computer industry. Numbers can be represented by the *excess-3 code* or the *Gray code*. There are error-detecting and parity-checking codes. Other codes are used to represent letters of the alphabet and special symbols, as well as numbers. Some of these are *Morse code*, the American Standard Code for Information Interchange or *ASCII* (pronounced "as-key"), the Extended Binary-Coded-Decimal Interchange Code or *EBCDIC* (pronounced "eb-si-dik"), and the *Hollerith code* used for punched cards.

Space does not permit discussion of each code. However, if you have understood the manipulation of the codes we have described, you should have no trouble when faced with a new one.

Exercise 4 ◆ BCD Codes

Convert each decimal number to 8421 BCD.

1. 62
2. 25
3. 274
4. 284
5. 42.91
6. 5.014

Convert each 8421 BCD number to decimal.

7. 1001
8. 101
9. 110 0001
10. 100 0111
11. 11 0110.1000
12. 11 1000.1000 1

CASE STUDY DISCUSSION—IP ADDRESSES

IP6 is presented to humans as 8 sets of 16 bits, where every 4 bits are represented by a hexadecimal number and each group of 16 bits (4 hex numbers) is separated from the next by a colon. There are several rules for handling zeros and reducing the length of an address presentation (of course the actual binary address is always 128 bits).

Without writing all 128 bits, let's look at just the first 32 bits of a 128 bit address:

00100000000000010000110110111000

Every 4 bits are converted to a hex number, and a colon follows each group of 16 bits (4 hex):

2001:0db8

So now let's look at a full address using this notation:

2001:0db8:1f70:0000:0999:0de8:7648:06e8

To reducc the length, the first rule is that we may represent a group of consecutive zeros by two colons:

2001:0db8:1f70::0999:0de8:7648:06e8

This may encompass more than one group of zeros, but can be used only once in an address.

The second rule is that leading zeros within each group may be omitted:

2001:db8:1f70::999:de8:7648:6e8

◆◆◆ CHAPTER 23 REVIEW PROBLEMS ◆◆◆◆◆◆◆◆◆◆◆◆◆◆◆◆◆◆◆◆◆◆◆◆◆◆◆◆◆◆

1. Convert the binary number 110.001 to decimal.
2. Convert the binary number 1100.0111 to hex.
3. Convert the hex number 2D to BCD.
4. Convert the binary number 1100 1010 to decimal.
5. Convert the hex number 2B4 to octal.
6. Convert the binary number 110 0111.011 to octal.
7. Convert the decimal number 26.875 to binary.
8. Convert the hexadecimal number 5E.A3 to binary.
9. Convert the octal number 534 to decimal.
10. Convert the BCD number 1001 to hex.
11. Convert the octal number 276 to hex.
12. Convert the 8421 BCD number 1001 0100 0011 to decimal.
13. Convert the decimal number 8362 to hex.
14. Convert the octal number 2453 to BCD.
15. Convert the BCD number 1001 0010 to octal.
16. Convert the decimal number 482 to octal.
17. Convert the octal number 47135 to binary.
18. Convert the hexadecimal number 3D2.A4F to decimal.
19. Convert the BCD number 0111 0100 to straight binary.
20. Convert the decimal number 382.4 to 8421 BCD.

Writing

21. A legislator, trying to trim the college budget, has suggested dropping the study of binary numbers. "Why do we need other numbers?" he asks. "Plain numbers were good enough when I went to school."

    Write a letter to this legislator, explaining why, in this computer age, "plain" numbers are not enough. Include specific examples of the use of binary numbers. Be polite but forceful, and limit your letter to one page.

Inequalities and Linear Programming

When you have completed this chapter, you should be able to:
- Identify characteristics of inequalities.
- Solve inequalities.
- Solve problems using linear programming.

For almost 23 chapters, you have been working with and solving equations. In most cases, the solution to an equation is the one value (sometimes two values, as you saw in quadratic equations) that makes the equation true. However, in the real world there is sometimes more to a formula or equation. For example, if a company makes any amount of income that is above the total costs of its products or services, the company makes a profit. "Profit" potentially encompasses a huge number of values (any positive number of dollars or fractions of dollars). Hence an infinite number of values will satisfy a generic profit equation. A mathematical formula representing the statement that the company is in profit if income is greater than cost could therefore use the > sign instead of the = sign. Now we no longer have an equation; we have an inequality.

We can work with inequalities and solve many problems in both technology and business applications.

We then introduce linear programming (a term that should not be confused with computer programming), which plays an important part in certain business decisions. It helps us decide how to allocate resources in order to reduce costs and maximize profit—a concern of technical people as well as management. We then write a computer program to do linear programming.

24–1   Definitions

Inequalities

An *inequality* is a statement that one quantity is greater than or less than another quantity.

◆◆◆ **Example 1:** The statement $a > b$ means "*a* is greater than *b*."   ◆◆◆

Parts of an Inequality

$$a > b$$

left side or member ⌐ ↑ ⌐ right side or member

inequality sign

Inequality Symbols

In Sec. 1–1, we introduced the inequality symbols $>$ and $<$. These symbols can be combined with the equals sign to produce the compound symbols $\geq$ and $\leq$, which mean "greater than or equal to" and "less than or equal to." Further, a slash through any of these symbols indicates the negation of the inequality.

◆◆◆ **Example 2:** The statement $a \not> b$ means "*a* is not greater than *b*."   ◆◆◆

Sense of an Inequality

The *sense* of an inequality refers to the *direction* in which the inequality sign points. We use this term mostly when operating on an inequality, where certain operations will *change the sense* but others will not.

Conditional and Unconditional Inequalities

Just as we have conditional equations and identities, we also have conditional and unconditional inequalities. As with equations, a *conditional inequality* is one that is satisfied only by certain values of the unknown, whereas *unconditional equalities* are true for any values of the unknown.

◆◆◆ **Example 3:** The statement $x - 2 > 5$ is a conditional inequality because it is true only for values of x greater than 7.   ◆◆◆

| In this chapter we limit ourselves to the *real* numbers. |

◆◆◆ **Example 4:** The statement $x^2 + 5 > 0$ is an unconditional inequality (also called an *absolute* inequality) because for any value of x, positive, negative, or zero, x^2 cannot be negative. Thus $x^2 + 5$ is always greater than zero.   ◆◆◆

Linear and Nonlinear Inequalities

A *linear* inequality is defined in the same way as a linear equation, where all variables are of first *degree* or less. Other inequalities are called *nonlinear*.

◆◆◆ **Example 5:** The inequalities

$$x + 2 > 3x - 5$$

and

$$2x - 3y \geq 4$$

are linear, while the inequalities

$$x^2 - 3 < 2x$$

and

$$y \leq x^3 + 2x - 3$$

are nonlinear.   ◆◆◆

Inequalities with Three Members

The inequality $3 < x$ has two members, left and right. However, many inequalities have *three* members.

◆◆◆ **Example 6:** The statement $3 < x < 9$ means "x is greater than 3 and less than 9," or "x is a number between 3 and 9." The given inequality is really a combination of the two inequalities

$$x > 3$$

and

$$x < 9$$

An inequality such as the one in Example 6 is sometimes called a *double inequality*. ◆◆◆

Common Error	It is incorrect to write the two inequalities $$x < 3 \quad \text{and} \quad x > 9$$ as $$3 > x > 9$$ because there is no number that is less than 3 and also greater than 9.

◆◆◆ **Example 7:** Combine the two inequalities

$$x > -6 \quad \text{and} \quad x \le 22$$

into a double inequality.

Solution: We write

$$-6 < x \le 22$$ ◆◆◆

Intervals

The *interval* of an inequality is the set of all numbers between the left member and the right member, called the *endpoints* of the interval. The interval can include one or both endpoints. An interval is called *open* or *closed* depending on whether or not the endpoint is included.

An interval is often denoted by enclosing the endpoints in parentheses or brackets. We use parentheses for an open interval and brackets for a closed interval, or a combination of the two.

◆◆◆ **Example 8:** Some inequalities and their notations are as follows:

Inequality	Notation	Open or Closed
(a) $2 \le x \le 3$	$[2, 3]$	Closed
(b) $4 < x < 5$	$(4, 5)$	Open
(c) $6 < x \le 7$	$(6, 7]$	Open on left, closed on right

◆◆◆

Graphing Intervals

An inequality can be represented on the number line by graphing the interval. An open endpoint is usually shown as an open circle, while a closed endpoint is shown solid.

◆◆◆ **Example 9:** The three inequalities of Example 8 are graphed in Fig. 24–1.

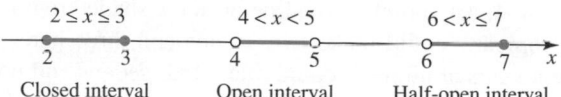

FIGURE 24–1 ◆◆◆

An interval is called unbounded if it has only one endpoint and extends indefinitely in one direction.

◆◆◆ **Example 10:**

(a) The inequality $x < 1$ is shown graphically in Fig. 24–2. It is said to be *open on the right and unbounded on the left*. The interval notation for the inequality $x < 1$ is $(-\infty, 1)$.

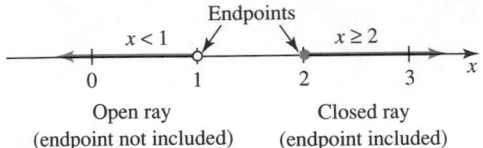

FIGURE 24–2

(b) The inequality $x \geq 2$ is also shown in Fig. 24–2. It is said to be *closed on the left and unbounded on the right*. The interval notation for this inequality is $[2, \infty)$.   ◆◆◆

Graphing Inequalities Having Two Variables

An inequality having *two variables* can be represented by a *region* in the xy plane. To graph an inequality, we first graph the corresponding equality with a dashed curve. Then, if the equality is included in the graph, we replace the dashed curve by a solid one. If the points above (or below) the curve satisfy the inequality, we shade the region above (or below) the curve.

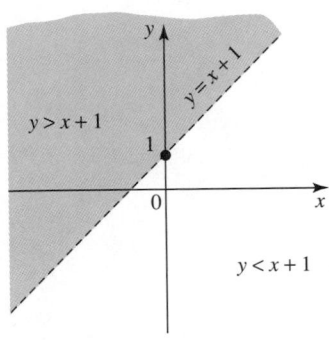

FIGURE 24–3   Points within the shaded region satisfy the inequality $y > x + 1$.

◆◆◆ **Example 11:** Graph the inequality $y > x + 1$.

Solution: We graph the line $y = x + 1$ as shown in Fig. 24–3 using a solid line if points on the line satisfy the given inequality, or a dashed line if they do not. Here, we need a dashed line. All points in the shaded region above the dashed line satisfy the given equality. For example, if we substitute the values for the point $(1, 3)$ into our inequality, we get

$$3 > 1 + 1$$

which is true. So $(1, 3)$ will be in the region that satisfies the inequality $y > x + 1$. This trial point helps us choose the region to shade (here, the region above the line) and serves as a check for our work.   ◆◆◆

◆◆◆ **Example 12:** A certain automobile company can make a sedan in 3 h and a hatchback in 2 h. How many of each type of car can be made if the working time is not to exceed 12 h?

Solution: We let

$$x = \text{number of sedans made in 12 h}$$

and

$$y = \text{number of hatchbacks made in 12 h}$$

We look at the time it takes to make each vehicle. Since each sedan takes 3 h to make, it will take $3x$ hours to make x sedans. Similarly, it takes $2y$ hours for the hatchbacks. So we get the inequality which states that the sum of the times is less than or equal to 12 h.

$$3x + 2y \leq 12$$

We plot the equation $3x + 2y = 12$ in Fig. 24–4. Any point on the line or in the shaded region satisfies the inequality. For example, the company could make two sedans and three hatchbacks in the 12-h period, or it could make no cars at all, or it could make four sedans and no hatchbacks, and so on. It could not, however, make three sedans and three hatchbacks in the allotted 12 h.   ◆◆◆

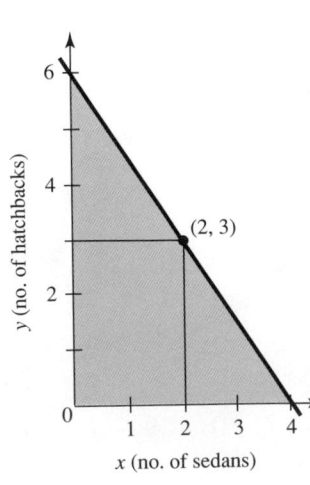

FIGURE 24–4

Sometimes we want to graph two inequalities on the same axes. This is a way to solve two inequalities simultaneously.

◆◆◆ **Example 13:** Figure 24–5 shows a graph of the two inequalities

$$y < x + 1 \quad \text{and} \quad y \geq -2x - 2$$

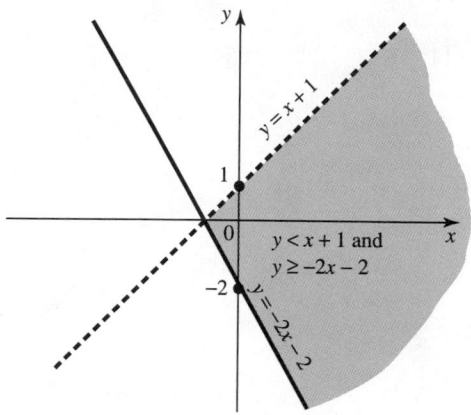

FIGURE 24–5

Points within the shaded region simultaneously satisfy both inequalities. ◆◆◆

Exercise 1 ◆ Definitions

Kinds of Inequalities

Label each inequality as conditional, unconditional, linear, or nonlinear.

1. $x^2 > -4$

2. $x + 2 < 3 - 4x$

3. $3x^2 - 2x \leq 6$

4. $3x^4 + 5 > 0$

5. $3x + 3 \geq x - 7$

6. $x^2 - 3x + 2 > 7$

Inequalities with Three Members

Rewrite each double inequality as two separate inequalities.

7. $5 < x < 9$

8. $-2 \leq x < 7$

9. $-11 < x \leq 1$

10. $22 < x \leq 53$

Rewrite each pair of inequalities as a double inequality.

11. $x > 5, x < 8$

12. $x > 24, x < 1$

13. $x \geq -1, x < 24$

14. $x > 12, x \leq 13$

Intervals

Graph each inequality on the number line.

15. $x > 5$

16. $2 < x < 5$

17. $x < 3$

18. $-1 \leq x \leq 3$

19. $x \geq -2$

20. $4 \leq x < 7$

21. $x \leq 1$

22. $-4 < x \leq 0$

Write the inequality corresponding to each given interval, and graph.

23. $(-\infty, 8)$ **24.** $(-\infty, 5]$

25. $(-3, \infty)$ **26.** $[10, \infty)$

27. $(3, 12)$ **28.** $[-2, 7]$

29. $(4, 22]$ **30.** $[-11, 5)$

Graphing Inequalities Having Two Variables

Graph each inequality.
31. $y < 3x - 2$
32. $x + y \geq 1$
33. $2x - y < 3$
34. $y > -x^2 + 4$

Graph the region in which both inequalities are satisfied.
35. $y > -x + 3$
     $y < 2x - 2$
36. $y < x^2 - 2$
     $y \geq x$
37. $y > x^2 - 4$
     $y < 4 - x^2$

38. The supply voltage V to a certain device is required to be over 100 V, but not over 150 V. Express this as an inequality.

39. A manufacturer makes skis and snowshoes. A pair of skis takes 2.5 h to make, and a pair of snowshoes takes 3.1 h. Write and graph an inequality to represent the number of pairs of skis and snowshoes that can be made in 8 h.

24–2   Solving Inequalities

Unlike the solution to a conditional equation, which is one or more discrete values of the variable, the solution to a conditional inequality will usually be an infinite set of values. We first show graphical methods for finding these values, and then we give an algebraic method. For each method we will first solve linear inequalities and then nonlinear inequalities.

Graphical Solution of Inequalities
We will show two graphical methods. The first method is to graph both sides of the inequality and then identify the intervals for which the inequality is true. We can make our graphs by a point-to-point plot or with a graphics calculator or graphing utility on a computer.

Our first example will be the solution of a linear inequality.

◆◆◆ **Example 14:** Graphically solve the linear inequality

$$3x - 4 > x + 2$$

Solution: On the same axes, we graph

$$y = 3x - 4 \quad \text{and} \quad y = x + 2$$

as shown in Fig. 24–6. The interval for which $3x - 4$ is greater than $x + 2$ is clearly

$$x > 3$$

or

$$(3, \infty)$$

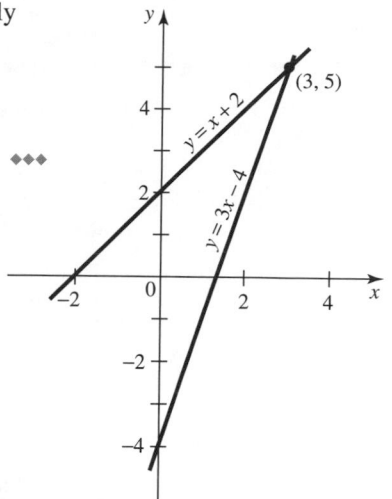

Our second graphical method is to move all terms to one side of the inequality and to graph the resulting expression. We then identify the intervals for which this inequality is true. We will illustrate this method by solving a nonlinear inequality.

◆◆◆ Example 15: Solve the nonlinear inequality

$$x^2 - 3 > 2x$$

Solution: Rearranging gives

$$x^2 - 2x - 3 > 0$$

We now plot the function

$$y = x^2 - 2x - 3$$

as shown in Fig. 24–7.

FIGURE 24–6

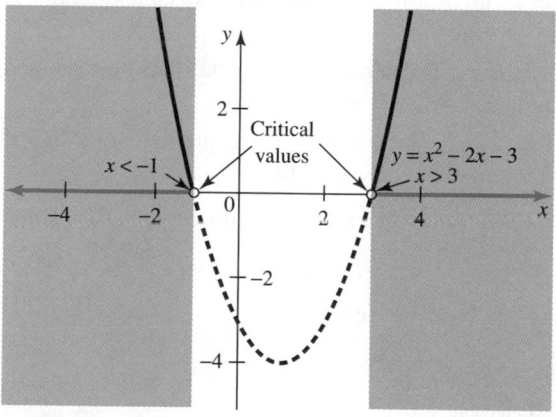

FIGURE 24–7 The points $x = -1$ and $x = 3$ where the curve crosses the x axis are called *critical values*. $x < -1$ and $x > 3$ are solutions because those x values satisfy the inequality.

Note that our solutions to the inequality are *all values of* x that make the inequality true; that is, those values of x that here make y positive. The curve (a parabola) is above the x axis when $x < -1$ and $x > 3$. These values of x are the solutions to our inequality, because, for these values, $x^2 - 2x - 3 > 0$. The intervals containing the solutions are thus

$$(-\infty, -1) \quad \text{and} \quad (3, \infty) \qquad \text{◆◆◆}$$

Checking a Solution

We can gain confidence in our solution by substituting values of x into the inequality. Those values within the range of our solution should check, while all others should not.

◆◆◆ Example 16: For the inequality of Example 15, the values of, say, $x = 4$ and $x = -2$ should check, while $x = 1$ should not. Substituting $x = 4$ into our original inequality gives

$$4^2 - 3 > 2(4)$$

or

$$16 - 3 > 8$$

which is true. Similarly, the value $x = -2$ gives

$$(-2)^2 - 3 > 2(-2)$$

or

$$4 - 3 > -4$$

which is also true. On the other hand, a value between -1 and 3, such as $x = 1$, gives

$$1^2 - 3 > 2(1)$$

or

$$-2 \not> 2$$

which is not true, as expected.   ◆◆◆

Algebraic Solution

The object of an *algebraic solution* to a conditional inequality is to locate the *critical values* (the points such as in Fig. 24–7 where the curve crosses the x axis). To do this, it is usually a good idea to transpose all terms to one side of the inequality and simplify the expression as much as possible. We use the following principles for operations with inequalities:

1. *Addition or subtraction*: The sense of an inequality is not changed if the same quantity is added to or subtracted from both sides.

2. *Multiplication or division by a positive quantity*: The sense of an inequality is not changed if both sides are multiplied or divided by the same *positive* quantity.

3. *Multiplication or division by a negative quantity*: The sense of an inequality *is reversed* if both sides are multiplied or divided by the same *negative* quantity.

As we did with the graphical solution, we will first solve a linear inequality and then some nonlinear inequalities.

◆◆◆ **Example 17:** Solve the linear inequality $21x - 23 < 2x + 15$.

Solution: Rearranging gives us

$$19x < 38$$

Dividing by 19, we get

$$x < 2$$

or $(-\infty, 2)$, as shown in Fig. 24–8.   ◆◆◆

When solving a nonlinear inequality, we usually get more than one critical value, which thus divide the x axis into more than two intervals. To decide which of these intervals contain the correct values, we test each for algebraic sign, as shown in the following example.

◆◆◆ **Example 18:** Solve the nonlinear inequality $2x - \dfrac{x^2}{2} > -\dfrac{21}{2}$.

Solution: Multiplying by 2, we have

$$4x - x^2 > -21$$

Rearranging yields

$$4x - x^2 + 21 > 0$$

or

$$-x^2 + 4x + 21 > 0$$

Multiplying by -1, which reverses the sense of the inequality, we obtain

$$x^2 - 4x - 21 < 0$$

Factoring gives us

$$(x - 7)(x + 3) < 0$$

The critical values are those that make the left side equal zero, or $x = 7$ and $x = 23$. These divide the x axis into three intervals.

$$x < -3 \qquad -3 < x < 7 \qquad x > 7$$

Which of these is correct? We can decide by computing the sign of $(x - 7)(x + 3)$ in each interval. Since this product is supposed to be less than zero, we choose the interval(s) in which the product is negative. Thus the signs are as follows:

Interval	$(x - 7)$	$(x + 3)$	$(x - 7)(x + 3)$
$x < -3$	$-$	$-$	$+$
$-3 < x < 7$	$-$	$+$	$-$
$x > 7$	$+$	$+$	$+$

The *sign pattern* for $(x - 7)(x + 3)$ is shown in Fig. 24–9. The signs are correct in only one interval, so our solution is

$$x > -3 \quad \text{and} \quad x < 7$$

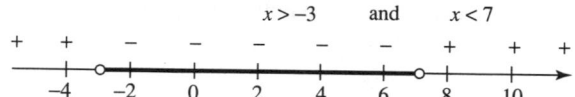

FIGURE 24–9

or $(-3, 7)$ in interval notation. Note that this is the interval for which the graph of $y = x^2 - 4x - 21$ is *below* the x axis (Fig. 24–10).

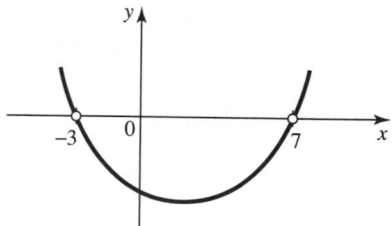

FIGURE 24–10 ◆◆◆

Inequalities containing the $\geq$ or $\leq$ signs are solved the same way as in the preceding examples, except that now the endpoint of an interval may be included in the solution.

◆◆◆ **Example 19:** Solve the nonlinear inequality $x^3 - x^2 \geq 2x$.

Solution: Rearranging and factoring gives

$$x^3 - x^2 - 2x \geq 0$$

$$x(x^2 - x - 2) \geq 0$$

$$x(x + 1)(x - 2) \geq 0$$

The critical values are

$$x = -1 \quad x = 0 \quad x = 2$$

We now analyze the signs, as in Example 18, looking for those intervals in which the product $x(x + 1)(x - 2)$ is positive.

Interval	x	$(x + 1)$	$(x - 2)$	$x(x + 1)(x - 2)$
$x < -1$	−	−	−	−
$-1 < x < 0$	−	+	−	+
$0 < x < 2$	+	+	−	−
$x > 2$	+	+	+	+

Our solution is thus

$$-1 \leq x \leq 0 \quad \text{and} \quad x \geq 2$$

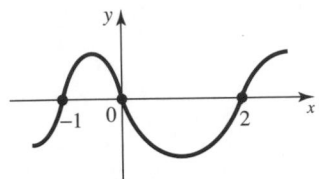

FIGURE 24–11

Note that the endpoints x = −1, 0, and 2 are included in the solution. In interval notation, the solutions are $[-1, 0]$ and $[2, \infty)$.

The inequality is graphed in Fig. 24–11. Note that we selected the two intervals that correspond to values on the curve that are on or above the x axis.   ◆◆◆

Our next example shows the algebraic and graphical solutions of a double inequality.

◆◆◆ **Example 20:** Solve the inequality

$$-5 < 3x + 4 \leq 9$$

Algebraic Solution: We can manipulate both inequalities at the same time.

$$-9 < 3x \leq 5$$

$$-3 < x \leq 5/3$$

or $(-3, 5/3]$ in interval notation.

Graphical Solution: We graph the three functions

$$y = -5 \quad y = 3x + 4 \quad y = 9$$

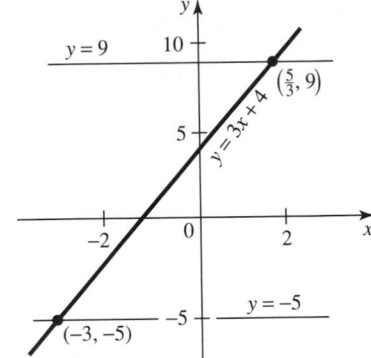

FIGURE 24–12

on the same axes (Fig. 24–12). The interval for which $3x + 4$ lies between −5 and 9 is $(-3, 5/3]$.   ◆◆◆

Our final example shows a graphical solution to a difficult nonlinear inequality.

CASE STUDY—FUNDRAISING REVENUES

As a volunteer for a wildlife rescue team, you are asked to assess your team's ability to optimize efforts for a fundraiser. The team has decided to sell handmade carved wall plaques. Each plaque is made from driftwood. One is a simply carved but large dog. The other is a smaller but more-finely crafted horse. The dog will take about an hour to carve, the horse twice as long. Wood pieces large enough for the dog sell for $9, and for the horse, $4. Volunteer woodcarvers can donate a total of 44 hours, and $180 is available to buy the driftwood. Paint has been donated, but you have only 25 hours of painter time available. The selling price is the profit, since we will consider all material and time as donated.

	Time to carve	Material Costs	Painting Time	Price (Profit)
Dog	1 hour	$9	1 hour	$65
Horse	2 hours	$4	1 hour	$45
Available	**44 hours**	**$180**	**25 hours**	

Calculate the combination of dog and horse statues that would maximize the revenue of the fundraising event.

Inequalities Containing Absolute Values

Recall from Sec. 1–1 that the absolute value of a quantity is its magnitude and is always positive.

◆◆◆ **Example 21:**

(a) $|5| = 5$
(b) $|-5| = 5$ ◆◆◆

It would thus be correct to say that in Example 21, the quantity inside the absolute value sign could be 5 or -5. So if we see the expression $|x| > 2$, we know that either x or $-x$ will be greater than 2.

$$\underbrace{|x| > 2}_{\substack{\text{expression with} \\ \text{absolute value signs}}} \quad \text{is equivalent to} \quad \underbrace{x > 2 \quad \text{and} \quad 2x > 2}_{\substack{\text{two inequalities without} \\ \text{absolute value signs}}}$$

The two inequalities can be written $x > 2$ and $x < -2$ because multiplying the right-hand inequality by -1 reverses the sense of the inequality. (i.e., reverses the direction of the inequality sign)

Similarly, $|x| < 2$ is equivalent to $x < 2$ and $-x < 2$, or (after multiplying on the right by -1) $x < 2$ and $x > -2$. This rule is stated as follows:

$$|x| < a \quad \text{is equivalent to} \quad 2-a < x < a$$
$$|x| > a \quad \text{is equivalent to} \quad x < 2-a \quad \text{or} \quad x > a$$

These rules are, of course, equally true for the $\leq$ or $\geq$ relationships.

To solve algebraically an inequality containing an absolute value sign, replace the inequality by a double inequality or by two inequalities having no absolute value signs. Solve each inequality as shown before.

To graphically solve an inequality containing an absolute value sign, simply graph both sides of the inequality, and note the intervals in which the inequality is true. If you wish, you may move all terms to one side of the inequality before graphing, as in the second graphical method.

We will show both methods in the following examples.

◆◆◆ **Example 22:** Solve the inequality $|3x - 2| \leq 5$.

Algebraic Solution: This can be written as

$$(3x - 2) \leq 5 \quad \text{and} \quad -(3x - 2) \leq 5$$

By leaving the first inequality as is and by adding $(3x - 2)$ and subtracting 5 from each side of the second inequality, we get

$$(3x - 2) \leq 5 \quad \text{and} \quad -5 \leq (3x - 2),$$

which can be rewritten as

$$-5 \leq (3x - 2) \leq 5.$$

Graphical Solution: We graph

$$y = |3x - 2| \quad \text{and} \quad y = 5$$

FIGURE 24–13

on the same axes (Fig. 24–13). The interval for which $|3x - 2|$ is less than or equal to 5 is the same as found algebraically. ◆◆◆

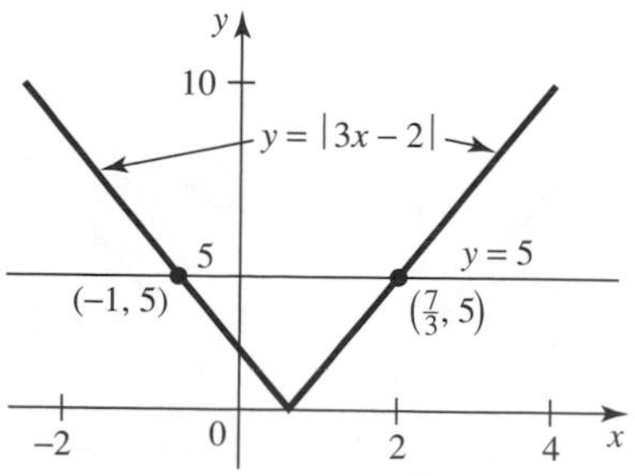

◆◆◆ **Example 23:** Solve the inequality $|7 - 3x| > 4$.

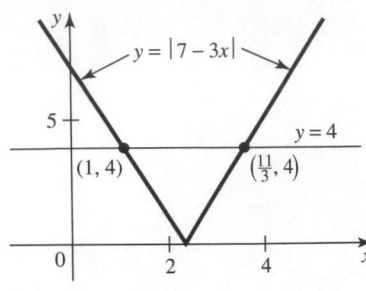

FIGURE 24–14

Algebraic Solution: We replace this inequality with

$$(7 - 3x) > 4 \quad \text{or} \quad (7 - 3x) < -4$$

Subtracting 7, we obtain

$$-3x > -3 \quad \text{or} \quad -3x < -11$$

Dividing by -3 and reversing the sense yields

$$x < 1 \quad \text{or} \quad x > \frac{11}{3}$$

So x may have values less than 1 or greater than $3\frac{2}{3}$.
In interval notation, the solution is

$$(-\infty, 1) \quad \text{or} \quad (11/3, \infty)$$

Graphical Solution: We graph

$$y = |7 - 3x| \quad \text{and} \quad y = 4$$

on the same axes (Fig. 24–14). The interval for which $|7 - 3x|$ is greater than 4 is the same as found algebraically.   ◆◆◆

Finding the Domain of a Function

We can use inequalities to find the *domain* of a function or relation, those values of x for which the function or relation has real values. The domain is useful in graphing to locate, in advance, those regions in which the function exists.

◆◆◆ **Example 24:** Find the values of x for which the relation $x^2 - y^2 - 4x + 3 = 0$ is real.

Solution: We solve for y.

$$y^2 = x^2 - 4x + 3$$

$$y = \pm\sqrt{x^2 - 4x + 3}$$

For y to be real, the quantity under the radical sign must be nonnegative, so we seek the values of x that satisfy the inequality

$$x^2 - 4x + 3 \geq 0$$

Factoring gives

$$(x - 1)(x - 3) \geq 0$$

The critical values are then $x = 1$ and $x = 3$. Analyzing the signs, we obtain the following table:

Interval	$(x - 1)$	$(x - 3)$	$(x - 1)(x - 3)$
$x < -1$	$-$	$-$	$+$
$1 < x < 3$	$+$	$-$	$-$
$x > +3$	$+$	$+$	$+$

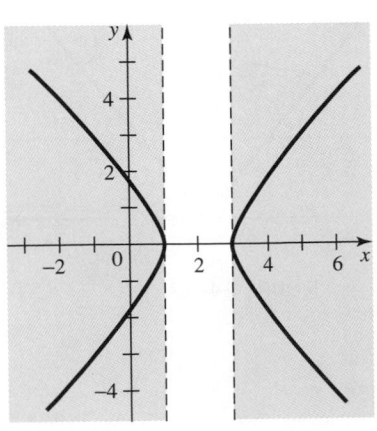

FIGURE 24–15

We conclude that the function exists for the ranges $x \leq 1$ and $x \geq 3$, but not for $1 < x < 3$. This information helps when graphing the curve, for we now know where the curve does and does not exist. (The curve is actually a hyperbola, as shown in Fig. 24–15, which is studied in analytic geometry.)   ◆◆◆

Exercise 2 ◆ Solving Inequalities

Solve each inequality.

1. $2x - 5 > x + 4$

2. $x + 2 < 3 - 2x$

3. $x^2 + 4 \geq 2x^2 - 5$

4. $2x^2 + 15x \leq -25$

5. $2x^2 - 5x + 3 > 0$

6. $6x^2 < 19x - 10$

7. $\dfrac{x}{5} + \dfrac{x}{2} \geq 4$

8. $\dfrac{x+1}{2} - \dfrac{x-2}{3} \leq 4$

9. $1 < x + 3 < 8$

10. $12 > 2x - 1 > -3$

11. $-2 \leq 2x + 4 \leq 6$

12. $10 > 8 + x \geq 2$

Inequalities Containing Absolute Values

Solve each inequality.

13. $|3x| > 9$

14. $2|x + 3| < 8$

15. $5|x| > 8$

16. $|3x + 2| \leq 11$

17. $|2 - 3x| \geq 8$

18. $3|x + 2| > 21$

19. $|x + 5| > 2x$

20. $|x + 7| > 3x$

21. $|3 - 5x| \leq 4 - 3x$

22. $|x + 1| \geq 2x - 1$

Finding the Domain of a Function

For what values of x is each expression real?

23. $\sqrt{x^2 + 4x - 5}$

24. $\sqrt{x^3 - 4x^2 - 12x}$

25. $y^2 = x^2 - x - 6$

26. $x^2 + y^2 + x - 2 = 0$

Applications

27. A certain machine is rented for $155 per day, and it costs $25 per hour to operate. The total cost per day for this machine is not to exceed $350. Write an inequality for this situation, and find the maximum permissible hours of operation.

28. A computer salesperson earns an annual salary of $25,500 and also gets a 37.5% commission on sales. How much must she sell in order to make an annual income of at least $50,000?

29. A company can make a camera lens for $12.75 and sell it for $39.99. It also has overhead costs of $2,884 per week. Find the minimum number of lenses that must be sold for the company to make a profit.

24–3   Linear Programming

We will do problems with only two variables here. To see how to *solve* problems with more than two variables, look up the *simplex method* in a text on linear programming.

Linear programming is a method for finding the maximum value of some function (usually representing *profit*) when the variables that it contains are themselves restricted to within certain limits. The function that is maximized is called the *objective function*, and the limits on the variables are called *constraints*. The constraints will be in the form of inequalities.

We show the method by means of examples.

In applications, *x* and *y* usually represent amounts of a certain resource, or the number of items produced, and thus will be restricted to *positive* values.

◆◆◆ **Example 25:** Find the values of x and y that will make z a maximum, where the objective function is

$$z = 5x + 10y$$

and x and y are positive and subject to the constraints

$$x + y \leq 5$$

and

$$2y - x \leq 4$$

Solution: We plot the two inequalities as shown in Fig. 24–16. The only permissible values for x and y are the coordinates of points on the edges of or within the shaded region. These are called *feasible solutions*.

But which (x, y) pair selected from all those in the shaded region will give the greatest value of z? In other words, which of the infinite number of feasible solutions is the *optimum* solution?

To help us answer this, let us plot the equation $z = 5x + 10y$ for different values of z. For example:

When $z = 0$,

$$5x + 10y = 0$$

$$y = -\frac{x}{2}$$

When $z = 10$,

$$5x + 10y = 10$$

$$y = -\frac{x}{2} + 1$$

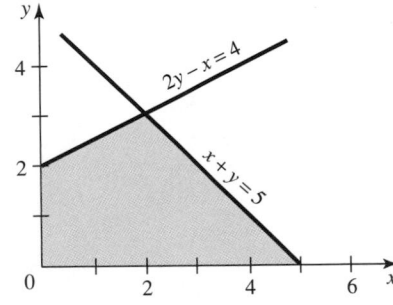

FIGURE 24–16   Feasible solutions.

and so on.

We get a *family* of parallel straight lines, each with a slope of $-\frac{1}{2}$ (Fig. 24–17). The point P at which the line having the greatest z value intersects the region gives us the x and y values

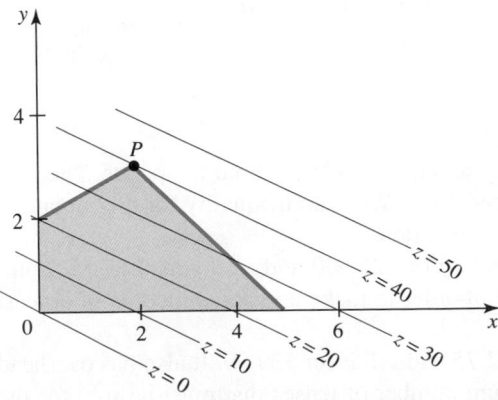

FIGURE 24–17   Optimum solution at P.

we seek. Such a point will always occur at a *corner* or *vertex* of the region. Thus instead of plotting the family of lines, it is necessary only to compute z at the vertices of the region. The coordinates of the vertices are called the *basic feasible solutions*. Of these, the one that gives the greatest z is the *optimum solution*.

In our problem, the vertices are $(0, 0)$, $(0, 2)$, $(5, 0)$, and P, whose coordinates we get by simultaneous solution of the equations of the boundary lines.

$$x + y = 5$$
$$x - 2y = -4$$

Subtract: $\quad 3y = 9$
$$y = 3$$

Substituting back, we have

$$x = 2$$

So P has the coordinates $(2, 3)$. Now compute z at each vertex.

$$\text{At } (0, 0) \quad z = 5(0) + 10(0) = 0$$
$$\text{At } (0, 2) \quad z = 5(0) + 10(2) = 20$$
$$\text{At } (5, 0) \quad z = 5(5) + 10(0) = 25$$
$$\text{At } (2, 3) \quad z = 5(2) + 10(3) = 40$$

So, $x = 2$, $y = 3$, which gives the largest z, is our optimum solution. ♦♦♦

♦♦♦ **Example 26:** Find the nonnegative values of x and y that will maximize the function $z = 2x + y$ with the following *four* constraints:

$$y - x \le 6$$
$$x + 4y \le 40$$
$$x + y \le 16$$
$$2x - y \le 20$$

Solution: We graph the inequalities as shown in Fig. 24–18 and compute the points of intersection. This gives the six basic feasible solutions.

$$(0, 0) \quad (0, 6) \quad (3.2, 9.2) \quad (8, 8) \quad (12, 4) \quad (10, 0)$$

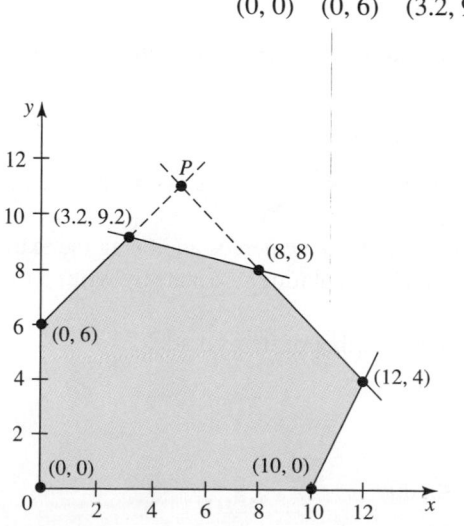

FIGURE 24–18 The coordinates of the corners of the shaded region are the basic feasible solutions. One of these is the optimum solution.

We now compute z at each point of intersection.

Point	$z = 2x + y$
(0,0)	$z = 2(0) + 0 \quad = 0$
(0,6)	$z = 2(0) + 6 \quad = 6$
(3.2,9.2)	$z = 2(3.2) + 9.2 = 15.6$
(8,8)	$z = 2(8) + 8 \quad = 24$
(12,4)	$z = 2(12) + 4 \quad = 28$
(10,0)	$z = 2(10) + 0 \quad = 20$

The greatest z (28) occurs at $x = 12$, $y = 4$, which is thus our optimum solution.   ◆◆◆

Applications

Our applications will usually be business problems, where our objective function z (the one we maximize) represents profit, and our variables x and y represent the quantities of two products to be made. A typical problem is to determine how much of each product should be manufactured so that the profit is a maximum.

◆◆◆ **Example 27:** Each month a company makes x stereos and y television sets. The manufacturing hours, material costs, testing time, and profit for each are as follows:

	Mfg. Time	Material Costs	Test Time	Profit
Stereo	4.00 h	$95.00	1.00 h	$45.00
Television	5.00 h	$48.00	2.00 h	$38.00
Available	180 h	$3,600	60.0 h	

Also shown in the table are the maximum amount of manufacturing and testing time and the cost of material available per month. How many of each item should be made for maximum profit?

Solution: First, why should we make any TV sets at all if we get more profit on stereos? The answer is that we are limited by the money available. In 180 h we could make

$$180 \div 4.00 = 45.0 \text{ stereos}$$

But 45 stereos, at $95 each, would require $4275, but there is only $3600 available. Then why not make

$$\$3600 \div 95.00 = 37.0 \text{ stereos}$$

using all our money on the most profitable item? However, this would leave us with manufacturing time unused but no money left to make any TV sets.

Thus we wonder if we could do better by making fewer than 37 stereos and using the extra time and money to make some TV sets. We seek the *optimum* solution. We start by writing the equations and inequalities to describe our situation.

Our total profit z will be the profit per item times the number of items made.

$$z = 45.00x + 38.00y$$

The manufacturing time must not exceed 180 h, so

$$4.00x + 5.00y \le 180$$

The materials cost must not exceed $3,600, so

$$95.00x + 48.00y \le 3600$$

The test time must not exceed 60 h, so

$$x + 2.00y \leq 60.0$$

This gives us our objective function and three constraints. We plot the constraints in Fig. 24–19.

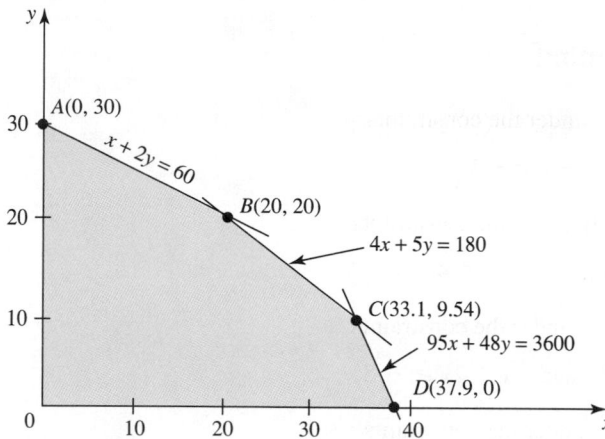

FIGURE 24–19

We find the vertices of the region by simultaneously solving pairs of equations. To find point A, we set $x = 0$ in the equation

$$x + 2.00y = 60.0$$

and get $y = 30.0$, or

$$A(0, 30.0)$$

To find point B, we solve simultaneously

$$x + 2.00y = 60.0$$
$$4.00x + 5.00y = 180$$

and get

$$B(20.0, 20.0)$$

Again, the actual work is omitted here. To find point C, we solve the equations

$$4.00x + 5.00y = 180$$
$$95.00x + 48.00y = 3600$$

and get $x = 33.1$, $y = 9.54$, or

$$C(33.1, 9.54)$$

To find point D, we use the equation

$$95.00x + 48.00y = 3600$$

Letting $y = 0$ gives $x = 37.9$, or

$$D(37.9, 0)$$

We then evaluate the profit z at each vertex (omitting the point 0, 0 at which $z = 0$).

At $A(0, 30.0)$ $z = 45.00(0) \quad + 38.00(30.0) = \$1,140$

At $B(20.0, 20.0)$ $z = 45.00(20.0) + 38.00(20.0) = \$1,660$

$$\text{At } C(33.1, 9.54) \quad z = 45.00(33.1) + 38.00(9.54) = \$1,852$$
$$\text{At } D(37.9, 0) \qquad z = 45.00(37.9) + 38.00(0) \quad = \$1,706$$

Thus our maximum profit, \$1,852, is obtained when we make 33.1 stereos and 9.54 television sets. ◆◆◆

Exercise 3 ◆ Linear Programming

As in the examples, we seek only nonnegative values of x and y.

1. Maximize the function $z = x + 2y$ under the constraints
$$x + y \le 5 \quad \text{and} \quad x - y \ge -2$$

2. Maximize the function $z = 4x + 3y$ under the constraints
$$x - 2y \ge -2 \quad \text{and} \quad x + 2y \le 4$$

3. Maximize the function $z = 5x + 4y$ under the constraints
$$7x + 4y \le 35 \qquad 2x - 3y \le 7 \quad \text{and} \quad x - 2y \ge -5$$

4. Maximize the function $z = x + 2y$ under the constraints
$$x - y \ge -5 \qquad x + 5y \le 45 \quad \text{and} \quad x + y \le 15$$

For problems 5, 6, and 7, round your answer to the nearest integer.

5. A company makes pulleys and sprockets. A pulley takes 3 min to make, requires \$1.25 in materials, and gives a profit of \$2.10. A sprocket takes 4 min to make, requires \$1.30 in materials, and gives a profit of \$2.35. The time to make x pulleys and y sprockets is not to exceed 8 h, and the cost for materials must not exceed \$175. How many of each item should be made for maximum profit?

6. A company makes two computers. Model A requires 225 h of labour and \$223 in material and yields a profit of \$1,975. Model B requires 67 h of labour and \$118 in material and yields a profit of \$824. There are 172000 h and \$240,000 available per month to make the computers. How many of each computer must be made each month for maximum profit?

7. Each week a company makes x disk drives and y monitors. The manufacturing hours, material costs, inspection time, and profit for each are as follows:

	Manufacturing Time (h)	Material Costs	Inspection Time (h)	Profit
Disk drive	5	\$ 21	4.5	\$34
Monitor	3	28	4	22
Available	50	320	50	

Also shown in the table are the maximum amount of manufacturing and inspection time available per week and the cost of materials. How many of each item should be made per week for maximum profit?

Case Study Discussion—Fundraising Revenues

In this case, we have three areas that are constrained: carving time, material costs, and painting time. We can use inequalities to establish the effects of the limits and then determine which combination of the limited resources produces the highest profit.

Let the number of dog statues be y, the number of horse statues x, and the profits z.

The revenue would be: $z = 45y + 65x$

Inequalities:

$$\text{Time to carve: } 2x + y \le 44$$
$$\text{Material costs: } 4x + 9y \le 180$$
$$\text{Painting time: } x + y \le 25$$

Plotting these limits, we produce the following graph. The area under the limits represents all combinations of dog and horse carvings that can be produced with the available resources.

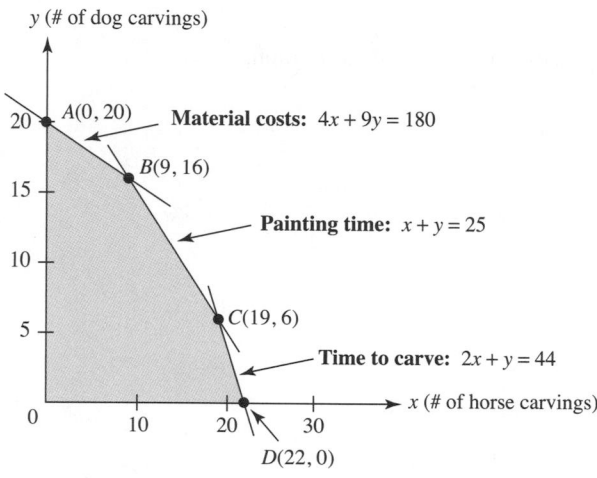

FIGURE 24–20

We graph each limit (Fig. 24-20) and label four points, A to D, found by solving the intersecting equations simultaneously as we did in Chapter 10.

A (0, 20) represents the lowest x intercept of the three limits and describes making no horses and 20 dogs. If we only make dogs, the limiting factor is material cost. After making 20 dogs, we have used up the $180 (even though we still have carving time and painting time left over). Revenue (made by making only dogs) $= 45x + 65y = 45(0) + 65(20) = \$1,300$.

B (9, 16) represents 9 horses and 16 dogs, using all of the paint time, all of the money, and 34 of the 44 carving hours. So, unlike point A, we have completely used up two of the three limiting resources. Revenue $= 45x + 65y = 45(9) + 65(16) = \$1,445$.

C (19, 6) represents 16 horses and 9 dogs, using all of the paint time, $145 of the money, and 41 of the 44 carving hours. Although there is time and money to carve more, we can't paint any more carvings. Revenue $= 45x + 65y = 45(16) + 65(9) = \$1,305$.

D (22, 0) represents the other extreme: making only horses and no dogs. Twenty-two horses completely use up all the carving time (even though we do have money and paint time left over). Revenue (making only horses) $= 45x + 65y = 45(22) + 65(0) = \$990$.

The maximum revenue is realized when 9 horses and 16 dogs are produced. The conditions at point B also allow you to report back to the team that there are extra carving hours available and that if they can get a little more driftwood and painting time, they could raise a little more money for the wildlife rescue work.

❖❖ CHAPTER 24 REVIEW PROBLEMS ❖❖❖❖❖❖❖❖❖❖❖❖❖❖❖❖❖❖❖❖❖❖❖❖❖❖❖❖❖❖❖

1. Rewrite as two separate inequalities:

$$-11 < x \leq 1$$

2. Rewrite as an inequality:

$$x \geq -1 \qquad x < 24$$

Graph on the number line, and give the interval notation for each.

3. $x < 8$ 4. $2 \leq x < 6$

Write the inequality corresponding to each given interval, and graph.

5. $(8, \infty)$ 6. $[-4, 9)$

Solve each inequality algebraically or graphically.

7. $2x - 5 > x + 3$

8. $|2x| < 8$

9. $|x - 4| > 7$

10. $x^2 > x + 6$

11. Maximize the function $z = 7x + y$ under the constraints

$$2x + 3y \leq 5 \quad \text{and} \quad x - 4y \geq -4$$

12. A company makes four-cylinder engine blocks and six-cylinder engine blocks. A "four" requires 4 h of machine time and 5 h of labour and gives a profit of $172. A "six" needs 5 h on the machine and 3 h of labour and yields $195 in profit. If 130 h of labour and 155 h of machine time are available, how many (to the nearest integer) of each engine block should be made for maximum profit?

Find the values of x for which each relation is real.

13. $\sqrt{x^2 + 8x + 15}$

14. $\sqrt{x^2 - 10x + 16}$

15. $y^2 = x^2 + 5x - 14$

16. $x^2 - y^2 - 10x + 9 = 0$

Solve each double inequality.

17. $-4.73 < x - 3 \leq 5.82$

18. $2.84 \leq x + 5 < 7.91$

19. Solve graphically to three significant digits:

$$1.27 - 2.24x + 1.84x^2 < 3.87x - 2.44x^2$$

20. A gear factory can make a bevel gear for $1.32 and sell it for $4.58. It also has overhead costs of $1,845 per week for the bevel gear department. Find the minimum number that must be sold for the company to make a profit on bevel gears.

Writing

21. A job applicant says on her resumé that she knows linear programming, and the personnel director of your company insists that she fits your request for a "programmer." Write a

memo to the personnel director explaining the difference between linear programming and computer programming.

Team Project

22. Search newspapers, magazines, and advertising flyers for inequalities in verbal form. Clip as many as you can find. For each, write an algebraic expression with inequality symbols, as we've done in this chapter, and make a graph.

25

Sequences, Series, and the Binomial Theorem

••• OBJECTIVES ••

When you have completed this chapter, you should be able to:
- Identify characteristics of sequences and series.
- Solve for the general term, specific term, sum, and mean of arithmetic progressions.
- Solve for specific terms, sum of sets of terms, mean, and applications of geometric progressions.
- Calculate limits and sums of infinite geometric progressions.
- Use binomial theorem properties to expand binomials.

Many things in our world follow repeated patterns with slight but predictable variations. A simple example would be a ball bouncing, the height of each bounce getting smaller and smaller. The height of succeeding bounces would follow a mathematical series, as would the vibrations of a spring or an earthquake. A complex wave such as that produced by music or voice or a 3-D TV signal can be analyzed as a series of mathematical terms (this is called Fourier analysis). This chapter looks at these very important math techniques, both arithmetic and geometric series and sequences.

We finish this chapter with an introduction to the *binomial theorem*. The binomial theorem enables us to expand a binomial, such as $(3x^2 - 2y)^5$, without actually having to multiply the terms. It is also useful in deriving formulas in calculus, statistics, and probability.

25–1   Sequences and Series

Sequences

A *sequence* is a set of quantities, called *terms*, arranged in the same order as the positive integers.

$$u_1, u_2, u_3, \ldots, u_n$$

◆◆◆ **Example 1:**

(a) The sequence $1, \frac{1}{2}, \frac{1}{3}, \frac{1}{4}, \frac{1}{5}, \ldots, 1/n$ is called a *finite sequence* because it has a finite number of terms, n.

(b) The sequence $1, \frac{1}{2}, \frac{1}{3}, \frac{1}{4}, \frac{1}{5}, \ldots, 1/n, \ldots$ is called an *infinite sequence*. The three dots at the end indicate that the sequence continues indefinitely.

(c) The sequence $3, 7, 11, 15, \ldots$ is called an *arithmetic sequence*, or *arithmetic progression* (AP), because each term after the first is equal to the sum of the preceding term and a constant. That constant (4, in this example) is called the *common difference*.

(d) The sequence $2, 6, 18, 54, \ldots$ is called a *geometric sequence*, or *geometric progression* (GP), because each term after the first is equal to the *product* of the preceding term and a constant. That constant (3, in this example) is called the *common ratio*.

(e) The sequence $1, 1, 2, 3, 5, 8, \ldots$ is called a *Fibonacci sequence*. Each term after the first is the sum of the two preceding terms.   ◆◆◆

Series

A *series* is the indicated sum of a sequence.

$$u_1 + u_2 + u_3 + \cdots + u_n + \cdots \qquad \mathbf{441}$$

◆◆◆ **Example 2:**

(a) The series $1 + \frac{1}{2} + \frac{1}{3} + \frac{1}{4} + \frac{1}{5}$ is called a *finite series*. It is also called a *positive* series because all of its terms are positive.

(b) The series $x^2 - x^2 + x^3 - \cdots + x^n + \cdots$ is an *infinite series*. It is also called an *alternating* series because the signs of the terms alternate in sign.

(c) The series $6 + 9 + 12 + 15 + \cdots$ is an *infinite arithmetic series*. The terms of this series form an AP.

(d) $1 - 2 + 4 - 8 + 16 - \cdots$ is an *infinite, alternating, geometric series*. The terms of this series form a GP.   ◆◆◆

General Term

The *general term* u_n in a sequence or series is an expression involving n (where $n = 1, 2, 3, \ldots$) by which we can obtain any term. The general term is also called the nth term. If we have an expression for the general term, we can then find any specific term of the sequence or series.

◆◆◆ **Example 3:** The general term of a certain series is

$$u_n = \frac{n}{2n + 1}$$

Write the first three terms of the series.

Solution: Substituting $n = 1$, $n = 2$, and $n = 3$, in turn, we get

$$u_1 = \frac{1}{2(1) + 1} = \frac{1}{3} \qquad u_2 = \frac{2}{2(2) + 1} = \frac{2}{5} \qquad u_3 = \frac{3}{2(3) + 1} = \frac{3}{7}$$

so our series is

$$\frac{1}{3} + \frac{2}{5} + \frac{3}{7} + \cdots + \frac{n}{2n} + 1 + \cdots$$ ◆◆◆

Graphing a Sequence

We can graph a sequence if, as in Example 3, the general term is known.

◆◆◆ **Example 4:** Graph the sequence given in Example 3.

Solution: We let $n = 1, 2, 3, \ldots$ and compute the value of as many terms as we wish. We then plot the points,

$$(1, \tfrac{1}{3}), (2, \tfrac{2}{5}), (3, \tfrac{3}{7}), \ldots$$

as in Fig. 25–1, and join them with a smooth curve (Fig. 25–2).

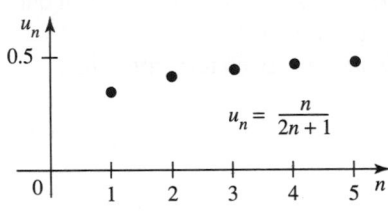

FIGURE 25–1

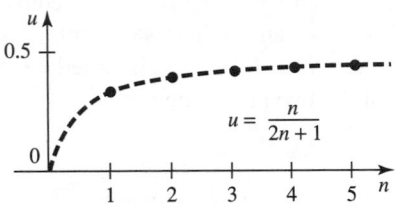

FIGURE 25–2

◆◆◆

The words *recurrence* and *recursive* are sometimes used instead of *recursion*. We have used recursion relations before, for compound interest in Sec. 20–1 and for exponential growth in Sec. 20–2.

Recursion Relations

We have seen that we can find the terms of a series, given an expression for the nth term. Sometimes we can find each term from one or more immediately preceding terms. The relationship between a term and those preceding it is called a *recursion relation* or *recursion formula*.

The *Fibonacci sequence* is named for Leonardo Fibonacci (ca. 1170–ca. 1250), an Italian number theorist and algebraist who studied this sequence. He is also known as Leonardo of Pisa.

◆◆◆ **Example 5:** Each term (after the first) in the series $1 + 4 + 13 + 40 + \cdots$ is found by multiplying the preceding term by 3 and adding 1. The recursion relation is then

$$u_n = 3u_{n-1} + 1$$ ◆◆◆

◆◆◆ **Example 6:** In a *Fibonacci sequence*

$$1, 1, 2, 3, 5, 8, 13, 21, 34, \ldots$$

each term after the first is the sum of the two preceding terms. Its recursion relation is

$$u_n = u_{n-1} + u_{n-2}$$ ◆◆◆

◆◆◆ **Example 7:** Table 25–1 shows a portion of the first 80 terms of the series

$$\frac{1}{e} + \frac{2}{e^2} + \frac{3}{e^3} + \cdots + \frac{n}{e^n} + \cdots$$

Also shown for each term u_n is the sum of that term and all preceding terms, called the *partial sum*, and the ratio of that term to the one preceding. We notice several things about this series.

TABLE 25-1

n	Term u_n	Partial Sum S_n	Ratio
1	0.367 879	0.367 879	
2	0.270 671	0.638 550	0.735 759
3	0.149 361	0.787 911	0.551 819
4	0.073 263	0.861 174	0.490 506
5	0.033 690	0.894 864	0.459 849
6	0.014 873	0.909 736	0.441 455
7	0.006 383	0.916 119	0.429 193
8	0.002 684	0.918 803	0.420 434
9	0.001 111	0.919 914	0.413 864
10	0.000 454	0.920 368	0.408 755
⋮	⋮	⋮	⋮
75	0.000 000	0.920 674	0.372 851
76	0.000 000	0.920 674	0.372 785
77	0.000 000	0.920 674	0.372 720
78	0.000 000	0.920 674	0.372 657
79	0.000 000	0.920 674	0.372 596
80	0.000 000	0.920 674	0.372 536

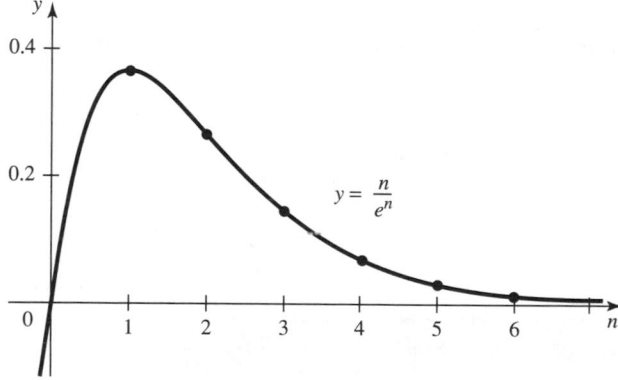

$$y = \frac{n}{e^n}$$

FIGURE 25-3

1. The terms get smaller and smaller and approach a value of zero (Fig. 25-3).
2. The partial sums approach a limit (0.920 674).
3. The ratio of two successive terms approaches a limit that is less than 1.

What do those facts tell us about this series? We'll see in the following section. ◆◆◆

CASE STUDY—FOREVER PAYMENTS

Geometric series can sometimes show us surprising things. For example, what if you where to receive $1,000 every year—forever? What is the value of $1,000 per year for you and then your decedents if the payments never stop? The first thought that might come to mind is that any amount paid regularly forever would represent an infinite amount. But you could have this "forever payment" for the price of a car! A geometric series deals with things like infinite payments. Can you show how an infinite number of payments can be made for a finite amount of money?

Convergence or Divergence of an Infinite Series

When we use a series to do computations, we cannot, of course, work with an infinite number of terms. In fact, for practical computation, we prefer as few terms as possible, as long as we get the needed accuracy. We want a series in which the terms decrease rapidly and approach zero, and for which the sum of the first several terms is not too different from the sum of all of the terms of the series. Such a series is said to *converge* on some limit. A series that does not converge is said to *diverge*.

There are many tests for convergence. A particular test might tell us with certainty if the series converges or diverges. Sometimes, though, a test will fail and we must try another test. Here we consider only three tests.

1. *Magnitude of the terms*: If the terms of an infinite series do not approach zero, the series diverges. However, if the terms do approach zero, we cannot say with certainty if the series converges. Thus having the terms approach zero is a necessary but not a sufficient condition for convergence.
2. *Sum of the terms*: If the sum of the first n terms of a series reaches a limiting value as we take more and more terms, the series converges. Otherwise, the series diverges.
3. *Ratio test*: We compute the ratio of each term to the one preceding it and see how that ratio changes as we go further out in the series. If that ratio approaches a number that is

   (a) less than 1, the series converges
   (b) greater than 1, the series diverges.
   (c) equal to 1, the test is not conclusive.

♦♦♦ Example 8: Does the series of Example 7 converge or diverge?

Solution: We apply the three tests by inspecting the computed values.

1. The terms approach zero, so this test is not conclusive.
2. The sum of the terms appears to approach a limiting value (0.920 674), indicating convergence.
3. The ratio of the terms appears to approach a value of less than 1, indicating convergence.

Thus the series appears to converge.   ♦♦♦

Note that we say *appears* to converge. The terms in Table 25–2, while convincing, is still not a *proof of convergence*.

♦♦♦ Example 9: Table 25–2 shows computed values for the series

$$1 + \frac{1}{2} + \frac{1}{3} + \cdots + \frac{1}{n} + \cdots$$

TABLE 25–2

n	Term u_n	Partial Sum S_n	Ratio
1	1.000 000	1.000 000	
2	0.500 000	1.500 000	0.500 000
3	0.333 333	1.833 333	0.666 667
4	0.250 000	2.083 334	0.750 000
5	0.200 000	2.283 334	0.800 000
6	0.166 667	2.450 000	0.833 333
7	0.142 857	2.592 857	0.857 143
8	0.125 000	2.717 857	0.875 000
9	0.111 111	2.828 969	0.888 889
10	0.100 000	2.928 969	0.900 000
.	.	.	.
.	.	.	.
.	.	.	.
75	0.013 333	4.901 356	0.986 667
76	0.013 158	4.914 514	0.986 842
77	0.012 987	4.927 501	0.987 013
78	0.012 821	4.940 322	0.987 180
79	0.012 658	4.952 980	0.987 342
80	0.012 500	4.965 480	0.987 500
.	.	.	.
.	.	.	.
.	.	.	.

495	0.002 020	6.782 784	0.997 980
496	0.002 016	6.784 800	0.997 984
497	0.002 012	6.786 813	0.997 988
498	0.002 008	6.788 821	0.997 992
499	0.002 004	6.790 825	0.997 996
500	0.002 000	6.792 825	0.998 000

Does the series converge or diverge?

Solution: Again we apply the three tests.

1. We see that the terms are getting smaller, so this test is not conclusive.
2. The ratio of the terms appears to approach 1, so this test also is not conclusive.
3. The partial sum S_n does not seem to approach a limit, but appears to keep growing even after 495 terms.

It thus appears that *this series diverges*. ◆◆◆

Exercise 1 ◆ Sequences and Series

Write the first five terms of each series, given the general term.

1. $u_n = 3n$
2. $u_n = 2n + 3$
3. $u_n = \dfrac{n + 1}{n^2}$
4. $u_n = \dfrac{2^n}{n} =$

Deduce the general term of each series. Use it to predict the next two terms.

5. $2 + 4 + 6 + \cdots$
6. $1 + 8 + 27 + \cdots$
7. $\dfrac{2}{4} + \dfrac{4}{5} + \dfrac{8}{6} + \dfrac{16}{7} + \cdots$
8. $2 + 5 + 10 + \cdots$

Deduce a recursion relation for each series. Use it to predict the next two terms.

9. $1 + 5 + 9 + \cdots$
10. $5 + 15 + 45 + \cdots$
11. $3 + 9 + 81 + \cdots$
12. $5 + 8 + 14 + 26 + \cdots$

25–2   Arithmetic Progressions

Recursion Formula

We stated earlier that an *arithmetic progression* (or AP) is a sequence of terms in which each term after the first equals the sum of the preceding term and a constant, called the *common difference*, d. If a_n is any term of an AP, the recursion formula for an AP is as follows:

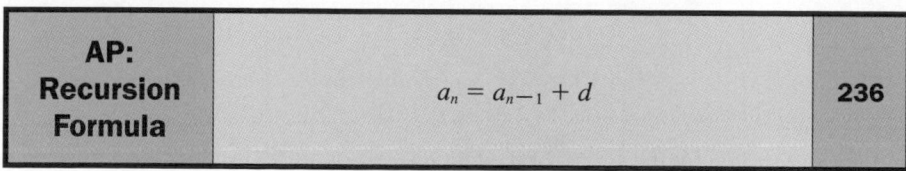

| **AP:** **Recursion Formula** | $a_n = a_{n-1} + d$ | **236** |

Each term of an AP after the first equals the sum of the preceding term and the common difference.

◆◆◆ **Example 10:** The following sequences are arithmetic progressions. The common difference for each is given in parentheses:

(a) $1, 5, 9, 13, \ldots$ $(d = 4)$
(b) $20, 30, 40, 50, \ldots$ $(d = 10)$
(c) $75, 70, 65, 60, \ldots$ $(d = 25)$

We see that each series is increasing when d is positive and decreasing when d is negative.

◆◆◆

General Term

For an AP whose first term is a and whose common difference is d, the terms are

$$a, a + d, a + 2d, a + 3d, a + 4d, \ldots$$

We see that each term is the sum of the first term and a multiple of d, where the coefficient of d is one less than the number n of the term. So the nth term a_n is given by the following equation:

The general term is sometimes called the *last* term, but this is not an accurate name since the AP continues indefinitely.

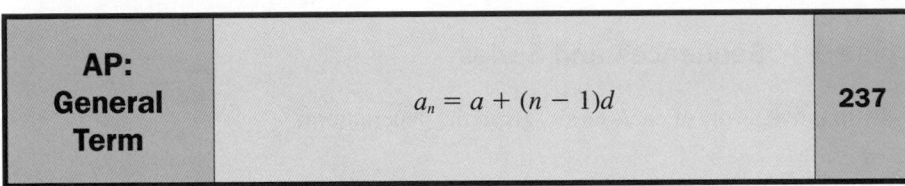

| AP: General Term | $a_n = a + (n - 1)d$ | 237 |

The nth term of an AP is found by adding the first term and $(n - 1)$ times the common difference.

◆◆◆ **Example 11:** Find the twentieth term of an AP that has a first term of 5 and common difference of 4.

Solution: By Eq. 237, with $a = 5$, $n = 20$, and $d = 4$,

$$a_{20} = 5 + 19(4) = 81$$ ◆◆◆

Of course, Eq. 237 can be used to find any of the four quantities $(a, n, d,$ or $a_n)$ given the other three.

◆◆◆ **Example 12:** Write the AP whose eighth term is 19 and whose fifteenth term is 33.

Solution: Applying Eq. 237 twice gives

$$33 = a + 14d$$
$$19 = a + 7d$$

We now have two equations in two unknowns, which we solve simultaneously. Subtracting the second from the first gives $14 = 7d$, so $d = 2$. Also, $a = 33 - 14d = 5$, so the general term of our AP is

$$a_n = 5 + (n - 1)2$$

and the AP is then

$$5, 7, 9, \ldots$$ ◆◆◆

AP: Sum of n Terms

Let us derive a formula for the sum s_n of the first n terms of an AP. Adding term by term gives

$$s_n = a + (a + d) + (a + 2d) + \cdots + (a_n - d) + a_n \tag{1}$$

or, written in reverse order,

$$s_n = a_n + (a_n - d) + (a_n - 2d) + \cdots + (a + d) + a \qquad (2)$$

Adding Eqs. (1) and (2) term by term gives

$$2s_n = (a + a_n) + (a + a_n) + (a + a_n) + \cdots$$
$$= n(a + a_n)$$

Dividing both sides by 2 gives the following formula:

AP: Sum of n Terms	$s_n = \dfrac{n}{2}(a + a_n)$	238

The sum of n terms of an AP is half the product of n and the sum of the first and nth terms.

◆◆◆ **Example 13:** Find the sum of 10 terms of the AP

$$2, 5, 8, 11, \ldots$$

Solution: From Eq. 237, with $a = 2$, $d = 3$, and $n = 10$,

$$a_{10} = 2 + 9(3) = 29$$

Then, from Eq. 238,

$$s_{10} = \frac{10(2 + 29)}{2} = 155 \qquad \text{◆◆◆}$$

We get another form of Eq. 238 by substituting the expression for a_n from Eq. 237, as follows:

AP: Sum of n Terms	$s_n = \dfrac{n}{2}[2a + (n - 1)d]$	239

This form is useful for finding the sum without first computing the nth term.

◆◆◆ **Example 14:** We repeat Example 13 without first having to find the tenth term. From Eq. 239,

$$s_{10} = \frac{10[2(2) + 9(3)]}{2} = 155 \qquad \text{◆◆◆}$$

Sometimes the sum is given and we must find one of the other quantities in the AP.

◆◆◆ **Example 15:** How many terms of the AP 5, 9, 13, . . . give a sum of 275?

Solution: We seek n so that $s_n = 275$. From Eq. 239, with $a = 5$ and $d = 4$,

$$275 = \frac{n}{2}[2(5) + (n - 1)4]$$
$$= 5n + 2n(n - 1)$$

Removing parentheses and collecting terms gives the quadratic equation

$$2n^2 + 3n - 275 = 0$$

From the quadratic formula,

$$n = \frac{-3 \pm \sqrt{9 - 4(2)(-275)}}{2(2)} = 11 \quad \text{or} \quad -12.5$$

We discard the negative root and get 11 terms as our answer.   ◆◆◆

Arithmetic Means

The first term a and the last (nth) term a_n of an AP are sometimes called the *extremes*, while the intermediate terms, a_2, a_3, . . ., a_{n-1}, are called *arithmetic means*. We now show, by example, how to insert any number of arithmetic means between two extremes.

◆◆◆ **Example 16:** Insert five arithmetic means between 3 and −9.

Solution: Our AP will have seven terms, with a first term of 3 and a seventh term of −9. From Eq. 237,

$$-9 = 3 + 6d$$

from which $d = -2$. The progression is then

$$3, 1, -1, -3, -5, -7, -9$$

and the five arithmetic means are 1, −1, −3, −5, and −7.   ◆◆◆

Average Value

Let us insert a single arithmetic mean between two numbers.

◆◆◆ **Example 17:** Find a single arithmetic mean m between the extremes a and b.

Solution: The sequence is a, m, b. The common difference d is $m - a = b - m$. Solving for m gives

$$2m = b + a$$

Dividing by 2 gives

$$m = \frac{a + b}{2}$$

which agrees with the common idea of an *average* of two numbers as the sum of those numbers divided by 2.   ◆◆◆

Harmonic Progressions

A sequence is called a *harmonic progression* if the reciprocals of its terms form an arithmetic progression.

◆◆◆ **Example 18:** The sequence

$$1, \ \frac{1}{3}, \ \frac{1}{5}, \ \frac{1}{7}, \ \frac{1}{9}, \ \frac{1}{11}, \dots$$

is a harmonic progression because the reciprocals of the terms, 1, 3, 5, 7, 9, 11, . . ., form an AP.   ◆◆◆

It is not possible to derive an equation for the nth term or for the sum of a harmonic progression. However, we can solve problems involving harmonic progressions by taking the reciprocals of the terms and using the formulas for the AP.

◆◆◆ **Example 19:** Find the tenth term of the harmonic progression

$$2, \ \frac{2}{3}, \ \frac{2}{5}, \ldots$$

Solution: We write the reciprocals of the terms,

$$\frac{1}{2}, \ \frac{3}{2}, \ \frac{5}{2}, \ldots$$

and note that they form an AP with $a = \frac{1}{2}$ and $d = 1$. The tenth term of the AP is then

$$a_n = a + (n - 1)d$$
$$a_{10} = \frac{1}{2} + (10 - 1)(1)$$
$$= \frac{19}{2}$$

The tenth term of the harmonic progression is the reciprocal of $\frac{19}{2}$, or $\frac{2}{19}$.   ◆◆◆

Harmonic Means

To insert *harmonic means* between two terms of a harmonic progression, we simply take the reciprocals of the given terms, insert arithmetic means between those terms, and take reciprocals again.

◆◆◆ **Example 20:** Insert three harmonic means between $\frac{2}{9}$ and 2.

Solution: Taking reciprocals, our AP is

$$\frac{9}{2}, \ , \ , \ , \ \frac{1}{2}$$

In this AP, $a = \frac{9}{2}$, $n = 5$, and $a_5 = \frac{1}{2}$. We can find the common difference d from the equation

$$a_n = a + (n - 1)d$$
$$\frac{1}{2} = \frac{9}{2} + 4d$$
$$-4 = 4d$$
$$d = -1$$

Having the common difference, we can fill in the missing terms of the AP,

$$\frac{9}{2}, \ \frac{7}{2}, \ \frac{5}{2}, \ \frac{3}{2}, \ \frac{1}{2}$$

Taking reciprocals again, our harmonic progression is

$$\frac{2}{9}, \ \frac{2}{7}, \ \frac{2}{5}, \ \frac{2}{3}, \ 2$$ ◆◆◆

Exercise 2 ◆ Arithmetic Progressions

General Term

1. Find the fifteenth term of an AP with first term 4 and common difference 3.
2. Find the tenth term of an AP with first term 8 and common difference 2.
3. Find the twelfth term of an AP with first term -1 and common difference 4.
4. Find the ninth term of an AP with first term -5 and common difference -2.
5. Find the eleventh term of the AP

$$9, 13, 17, \ldots$$

6. Find the eighth term of the AP

$$-5, -8, -11, \ldots$$

7. Find the ninth term of the AP

$$x, x + 3y, x + 6y, \ldots$$

8. Find the fourteenth term of the AP

$$1, \frac{6}{7}, \frac{5}{7}, \ldots$$

For problems 9 through 12, write the first five terms of each AP.

9. First term is 3 and thirteenth term is 55.
10. First term is 5 and tenth term is 32.
11. Seventh term is 41 and fifteenth term is 89.
12. Fifth term is 7 and twelfth term is 42.
13. Find the first term of an AP whose common difference is 3 and whose seventh term is 11.
14. Find the first term of an AP whose common difference is 6 and whose tenth term is 77.

Sum of an Arithmetic Progression

15. Find the sum of the first 12 terms of the AP

$$3, 6, 9, 12, \ldots$$

16. Find the sum of the first five terms of the AP

$$1, 5, 9, 13, \ldots$$

17. Find the sum of the first nine terms of the AP

$$5, 10, 15, 20, \ldots$$

18. Find the sum of the first 20 terms of the AP

$$1, 3, 5, 7, \ldots$$

19. How many terms of the AP 4, 7, 10, . . . will give a sum of 375?
20. How many terms of the AP 2, 9, 16, . . . will give a sum of 270?

Arithmetic Means

21. Insert two arithmetic means between 5 and 20.
22. Insert five arithmetic means between 7 and 25.
23. Insert four arithmetic means between -6 and -9.
24. Insert three arithmetic means between 20 and 56

Harmonic Progressions

25. Find the fourth term of the harmonic progression

$$\frac{3}{5}, \frac{3}{8}, \frac{3}{11}, \ldots$$

26. Find the fifth term of the harmonic progression

$$\frac{4}{19}, \frac{4}{15}, \frac{4}{11}, \ldots$$

Harmonic Means

27. Insert two harmonic means between $\frac{7}{9}$ and $\frac{7}{15}$.

28. Insert three harmonic means between $\frac{6}{21}$ and $\frac{6}{5}$.

Applications

29. *Loan Repayment:* A person agrees to repay a loan of $10,000 with an annual payment of $1,000 plus 8% of the unpaid balance.
(a) Show that the interest payments alone form the AP: $800, $720, $640, ...
(b) Find the total amount of interest paid.

30. *Simple Interest:* A person deposits $50 in a bank on the first day of each month, at the same time withdrawing all interest earned on the money already in the account.
(a) If the rate is 1% per month, computed monthly, write an AP whose terms are the amounts withdrawn each month.
(b) How much interest will have been earned in the 36 months following the first deposit?

31. *Straight-Line Depreciation:* A certain milling machine has an initial value of $150,000 and a scrap value of $10,000 twenty years later. Assuming that the machine depreciates the same amount each year, find its value after 8 years.

To find the amount of depreciation for each year, divide the total depreciation (initial value – scrap value) by the number of years of depreciation.

32. *Salary or Price Increase:* A person is hired at a salary of $40,000 and receives a raise of $2,500 at the end of each year. Find the total amount earned during 10 years.

33. *Freely Falling Body:* A freely falling body falls $g/2$ metres during the first second, $3g/2$ m during the next second, $5g/2$ m during the third second, and so on, where $g \cong 9.807$ m/s². Find the total distance the body falls during the first 10 s.

34. Using the information of problem 33, show that the total distance s fallen in t seconds is $s = \frac{1}{2}gt^2$.

25–3 Geometric Progressions

Recursion Formula

A geometric sequence or *geometric progression* (GP) is one in which each term after the first is formed by multiplying the preceding term by a factor r, called the *common ratio*. Thus if a_n is any term of a GP, the recursion relation is as follows:

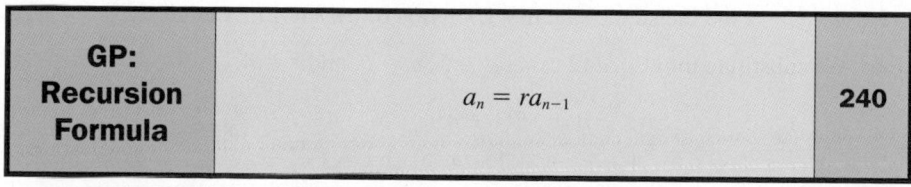

| GP: Recursion Formula | $a_n = ra_{n-1}$ | 240 |

Each term of a GP after the first equals the product of the preceding term and the common ratio.

◆◆◆ **Example 21:** Some geometric progressions, with their common ratios given, are as follows:

(a) $2, 4, 8, 16, \ldots$ $(r = 2)$

(b) $27, 9, 3, 1, \frac{1}{3}, \ldots$ $(r = \frac{1}{3})$

(c) $-1, 3, -9, 27, \ldots$ $(r = -3)$ ◆◆◆

General Term

For a GP whose first term is a and whose common ratio is r, the terms are

$$a, ar, ar^2, ar^3, ar^4, \ldots$$

We see that each term after the first is the product of the first term and a power of r, where the power of r is one less than the number n of the term. So the nth term a_n is given by the following equation:

GP: General Term	$a_n = ar^{n-1}$	241

The nth term of a GP is found by multiplying the first term by the $n-1$ power of the common ratio.

◆◆◆ **Example 22:** Find the sixth term of a GP with first term 5 and common ratio 4.

Solution: We substitute into Eq. 241 for the general term of a GP, with $a = 5$, $n = 6$, and $r = 4$.

$$a_6 = 5(4^5) = 5(1024) = 5120 \qquad\qquad ◆◆◆$$

GP: Sum of n Terms

We find a formula for the sum s_n of the first n terms of a GP (also called the sum of n terms of a geometric series) by adding the terms of the GP.

$$s_n = a + ar + ar^2 + ar^3 + \cdots + ar^{n-2} + ar^{n-1} \qquad (1)$$

Multiplying each term in Eq. (1) by r gives

$$rs_n = ar + ar^2 + ar^3 + \cdots + ar^{n-1} + ar^n \qquad (2)$$

Subtracting (2) from (1) term by term, we get

$$(1 - r)s_n = a - ar^n$$

Dividing both sides by $(1 - r)$ gives us the following formula:

GP: Sum of n Terms	$s_n = \dfrac{a(1 - r^n)}{1 - r}$	242

◆◆◆ **Example 23:** Find the sum of the first six terms of the GP in Example 22.

Solution: We substitute into Eq. 242 using $a = 5$, $n = 6$, and $r = 4$.

$$s_n = \frac{a(1 - r^n)}{1 - r} = \frac{5(1 - 4^6)}{1 - 4} = \frac{5(-4095)}{-3} = 6825 \qquad ◆◆◆$$

We can get another equation for the sum of a GP in terms of the nth term a_n. We substitute into Eq. 242, using $ar^n = r(ar^{n-1}) = ra_n$, as follows:

| **GP: Sum of** | | |
| **n Terms** | $s_n = \dfrac{a - ra_n}{1 - r}$ | **243** |

◆◆◆ **Example 24:** Repeat Example 23, given that the sixth term (found in Example 22) is $a_6 = 5120$.

Solution: Substitution, with $a = 5$, $r = 4$, and $a_6 = 5120$, yields

$$s_n = \frac{a - ra_n}{1 - r} = \frac{5 - 4(5120)}{1 - 4} = 6825$$

as before. ◆◆◆

Geometric Means

As with the AP, the intermediate terms between any two terms are called *means*. A single number inserted between two numbers is called the *geometric mean* between those numbers.

◆◆◆ **Example 25:** Insert a geometric mean b between two numbers a and c.

Solution: Our GP is a, b, c. The common ratio r is then

$$r = \frac{b}{a} = \frac{c}{b}$$

from which $b^2 = ac$, which is rewritten as follows:

| **Geometric** | | |
| **Mean** | $b = \pm\sqrt{ac}$ | **59** |

This is not new. We studied the mean proportional in Chapter 19.

The geometric mean, or mean proportional, between two numbers is equal to the square root of their product. ◆◆◆

◆◆◆ **Example 26:** Find the geometric mean between 3 and 48.

Solution: Letting $a = 3$ and $c = 48$ gives us

$$b = \pm\sqrt{3(48)} = \pm 12$$

Our GP is then

$$3, \mathbf{12}, 48$$

or

$$3, \mathbf{-12}, 48$$

Note that we get *two* solutions. ◆◆◆

To insert *several* geometric means between two numbers, we first find the common ratio.

◆◆◆ **Example 27:** Insert four geometric means between 2 and $15\frac{3}{16}$.

Solution: Here $a = 2$, $a_6 = 15\frac{3}{16}$, and $n = 6$. Then

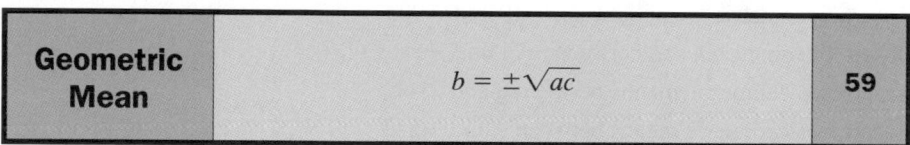

$$a_6 = ar^5$$
$$15\frac{3}{16} = 2r^5$$

$$r^5 = \frac{243}{32}$$

$$r = \frac{3}{2}$$

Having r, we can write the terms of the GP. They are

$$2, \quad 3, \quad 4\frac{1}{2}, \quad 6\frac{3}{4}, \quad 10\frac{1}{8}, \quad 15\frac{3}{16} \qquad\qquad •••$$

Exercise 3 ◆ Geometric Progressions

1. Find the fifth term of a GP with first term 5 and common ratio 2.
2. Find the fourth term of a GP with first term 7 and common ratio −4.
3. Find the sixth term of a GP with first term −3 and common ratio 5.
4. Find the fifth term of a GP with first term −4 and common ratio −2.
5. Find the sum of the first ten terms of the GP in problem 1.
6. Find the sum of the first nine terms of the GP in problem 2.
7. Find the sum of the first eight terms of the GP in problem 3.
8. Find the sum of the first five terms of the GP in problem 4.

Geometric Means

9. Insert a geometric mean between 5 and 45.
10. Insert a geometric mean between 7 and 112.
11. Insert a geometric mean between −10 and −90.
12. Insert a geometric mean between −21 and −84.
13. Insert two geometric means between 8 and 216.
14. Insert two geometric means between 9 and −243.
15. Insert three geometric means between 5 and 1280.
16. Insert three geometric means between 144 and 9.

Applications

17. *Exponential Growth:* Using the equation for exponential growth:

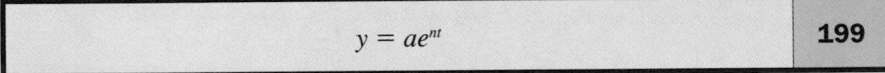

$$y = ae^{nt} \qquad\qquad \mathbf{199}$$

    with $a = 1$ and $n = 0.5$, compute values of y for $t = 0, 1, 2, \ldots, 10$. Show that while the values of t form an AP, the values of y form a GP. Find the common ratio.

18. *Exponential Decay:* Repeat problem 17 with the formula for exponential decay:

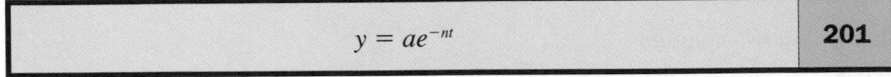

$$y = ae^{-nt} \qquad\qquad \mathbf{201}$$

One of the most famous and controversial references to arithmetic and geometric progressions was made by Thomas Malthus in 1798. He wrote: *"Population, when unchecked, increases in a geometrical ratio, and subsistence for man in an arithmetical ratio."*

19. *Cooling:* A certain iron casting is at 950 °C and cools so that its temperature at each minute is 10% less than its temperature the preceding minute. Find its temperature after 1 h.
20. *Light through an Absorbing Medium:* Sunlight passes through a glass filter. Each millimetre of glass absorbs 20% of the light passing through it. What percentage of the original sunlight will remain after passing through 5.0 mm of the glass?
21. *Radioactive Decay:* A certain radioactive material decays so that after each year the radioactivity is 8% less than at the start of that year. How many years will it take for its radioactivity to be 50% of its original value?

22. *Pendulum:* Each swing of a certain pendulum is 85.0% as long as the one before. If the first swing is 12.0 cm, find the entire distance travelled in eight swings.

23. *Bouncing Ball:* A ball dropped from a height of 3.00 m rebounds to half its height on each bounce. Find the total distance travelled when it hits the ground for the fifth time.

24. *Population Growth:* Each day the size of a certain colony of bacteria is 25% larger than on the preceding day. If the original size of the colony was 10 000 bacteria, find its size after 5 days.

25. *Ancestry:* A person has two parents, and each parent has two parents, and so on. We can write a GP for the number of ancestors as 2, 4, 8, ... Find the total number of ancestors in five generations, starting with the parents' generation.

26. *Musical Scale:* The frequency of the "A" note above middle C is, by international agreement, equal to 440 Hz. A note one *octave* higher is at twice that frequency, or 880 Hz. The octave is subdivided into 12 *half-tone* intervals, where each half-tone is higher than the one preceding by a factor equal to the twelfth root of 2. Write a GP showing the frequency of each half-tone, from 440 to 880 Hz. Work to two decimal places.

This is called the *equally tempered scale* and is usually attributed to Johann Sebastian Bach (1685–1750).

27. *Chemical Reactions:* Increased temperature usually causes chemicals to react faster. If a certain reaction proceeds 15% faster for each 10 °C increase in temperature, by what factor is the reaction speed increased when the temperature rises by 50 °C?

28. *Mixtures:* A radiator contains 30% antifreeze and 70% water. One-fourth of the mixture is removed and replaced by pure water. If this procedure is repeated three more times, find the percent antifreeze in the final mixture.

29. *Energy Consumption:* If energy consumption in Canada is 7.00% higher each year, by what factor will the energy consumption have increased after 10.0 years?

30. *Atmospheric Pressure:* The pressures measured at 1-km intervals above sea level form a GP, with each value smaller than the preceding by a factor of 0.882. If the pressure at sea level is 101.1 kPa, find the pressure at an altitude of 8 km.

31. *Compound Interest:* A person deposits $10,000 in a bank giving 6% interest, compounded annually. Find to the nearest dollar the value of the deposit after 50 years.

32. *Inflation:* The price of a certain house, now $126,000, is expected to increase by 5% each year. Write a GP whose terms are the value of the house at the end of each year, and find the value of the house after 5 years.

33. *Depreciation:* When calculating depreciation by the *declining-balance method*, a taxpayer claims as a deduction a fixed percentage of the book value of an asset each year. The new book value is then the last book value less the amount of the depreciation. Thus for a machine having an initial book value of $100,000 and a depreciation rate of 40%, the first year's depreciation is 40% of $100,000, or $40,000, and the new book value is $100,000 − $40,000 = $60,000. Thus the book values for each year form the following GP:

$$\$100,000, \ \$60,000, \ \$36,000, \ldots$$

Find the book value after 5 years.

25–4 Infinite Geometric Progressions

Sum of an Infinite Geometric Progression

Before we derive a formula for the sum of an infinite geometric progression, let us explore the idea graphically and numerically.

We have already determined that the sum of n terms of any geometric progression with a first term a and a common ratio r is

$$s_n = \frac{a(1 - r^n)}{1 - r} \qquad \text{(Eq. 242)}$$

Thus a graph of s_n versus n should tell us about the sum changes as n increases.

◆◆◆ **Example 28:** Graph the sum s_n versus n for the GP

$$1, 1.2, 1.2^2, 1.2^3, \ldots$$

for $n = 0$ to 20.

Solution: Here, $a = 1$ and $r = 1.2$, so

$$s_n = \frac{1 - 1.2^n}{1 - 1.2} = \frac{1.2^n - 1}{0.8}$$

The sum s_n is graphed in Fig. 25–4. Note that the sum continues to increase. We say that this progression *diverges*.

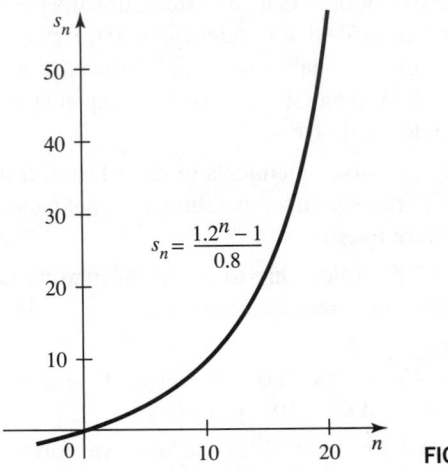

FIGURE 25-4   ◆◆◆

Decreasing GP

If the common ratio r in a GP is less than 1, each term in the progression will be less than those preceding it. Such a progression is called a *decreasing* GP.

◆◆◆ **Example 29:** Graph the sum s_n versus n for the GP

$$1, 0.8, 0.8^2, 0.8^3, \ldots$$

for $n = 0$ to 20.

Solution: Here, $a = 1$ and $r = 0.8$, so

$$s_n = \frac{1 - 0.8^n}{1} - 0.8 = \frac{1 - 0.8^n}{0.2}$$

The sum is graphed in Fig. 25–5. Note that the sum appears to reach a limiting value, so we say that this progression *converges*. The sum appears to be approaching a value of 5.

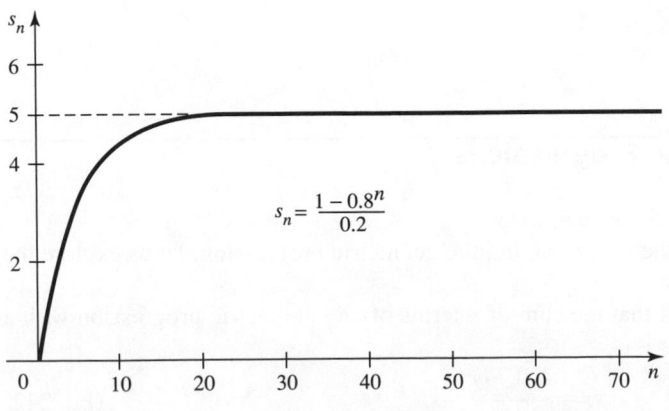

FIGURE 25-5   ◆◆◆

Our graphs have thus indicated that the sum of an infinite number of terms of a decreasing GP appears to approach a limit. Let us now verify that fact numerically.

◆◆◆ **Example 30:** Find the sum of infinitely many terms of the GP

$$9, 3, 1, \frac{1}{3}, \ldots$$

Solution: Knowing that $a = 9$ and $r = \frac{1}{3}$, we can use a computer to compute each term and keep a running sum as shown in Table 25–3. Notice that the terms get smaller and smaller, and the sum appears to approach a value of around 13.5. ◆◆◆

We will confirm the value found in Example 30 after we have derived a formula for the sum of an infinite geometric progression.

Limit Notation

In order to derive a formula for the sum of an infinite geometric progression, we need to introduce limit notation. Such notation will also be of great importance in the study of calculus.

First, we use an arrow ($\rightarrow$) to indicate that a quantity *approaches* a given value.

We used limit notation briefly in Sec. 20–2, and here we show a bit more. Limits are usually covered more completely in a calculus course.

◆◆◆ **Example 31:**

(a) $x \rightarrow 5$ means that x gets closer and closer to the value 5.
(b) $n \rightarrow \infty$ means that n grows larger and larger, without bound.
(c) $y \rightarrow 0$ means that y continuously gets closer and closer to zero. ◆◆◆

If some function $f(x)$ approaches some value L as x approaches a, we could write this as

$$f(x) \rightarrow L \quad \text{as} \quad x \rightarrow a$$

which is written in more compact notation as follows:

Limit Notation	$\lim_{x \to a} f(x) = L$	**336**

TABLE 25–3

	Term	Sum
1	9.0000	9.0000
2	3.0000	12.0000
3	1.0000	13.0000
4	0.3333	13.3333
5	0.1111	13.4444
6	0.0370	13.4815
7	0.0123	13.4938
8	0.0041	13.4979
9	9.0014	13.4993
10	0.0005	13.4998
11	0.0002	13.4999
12	0.0001	13.5000
13	0.0000	13.5000
14	0.0000	13.5000
15	0.0000	13.5000
16	0.0000	13.5000
17	0.0000	13.5000
18	0.0000	13.5000

Limit notation lets us compactly write the value to which a function is tending, as in the following example.

◆◆◆ **Example 32:**

(a) $\lim_{b \to 0} b = 0$

(b) $\lim_{b \to 0} (a + b) = a$

(c) $\lim_{b \to 0} \dfrac{a + b}{c + d} = \dfrac{a}{c + d}$

(d) $\lim_{x \to 0} \dfrac{x + 1}{x + 2} = \dfrac{1}{2}$ ◆◆◆

A Formula for the Sum of an Infinite, Decreasing Geometric Progression

As the number of terms n of an infinite, decreasing geometric progression increases without bound, the term a_n will become smaller and will approach zero. Using our new notation, we may write

$$\lim_{n \to \infty} a_n = 0 \quad \text{when} \quad |r| < 1$$

Thus in the equation for the sum of n terms,

$$s_n = \frac{a - ra_n}{1 - r} \qquad \text{(Eq. 243)}$$

the expression ra_n will approach zero as n approaches infinity, so the sum s_∞ of infinitely many terms (called the *sum to infinity*) is

$$s_\infty = \lim_{n \to \infty} \frac{a - ra_n}{1 - r} = \frac{a}{1 - r}$$

Thus letting $S = s_\infty$, the sum to infinity is given by the following equation:

GP: Sum to Infinity	$S = \dfrac{a}{1-r}$ when $|r| < 1$	**244**

◆◆◆ **Example 33:** Find the sum of infinitely many terms of the GP

$$9, \quad 3, \quad 1, \quad \frac{1}{3}, \ldots$$

Solution: Here $a = 9$ and $r = \dfrac{1}{3}$, so

$$S = \frac{a}{1-r} = \frac{9}{1 - 1/3} = \frac{9}{2/3} = 13\frac{1}{2}$$

This agrees with the value we found numerically in Example 30.   ◆◆◆

Repeating Decimals

A *finite decimal* (or terminating decimal) is one that has a finite number of digits to the right of the decimal point, followed by zeros. An *infinite decimal* (or nonterminating decimal) has an infinite string of digits to the right of the decimal point. A *repeating decimal* (or periodic decimal) is one that has a block of one or more digits that repeats indefinitely. It can be proved that when one integer is divided by another, we get either a finite decimal or a repeating decimal. Irrational numbers, on the other hand, yield *nonrepeating decimals*.

◆◆◆ **Example 34:**

(a) $\frac{3}{4} = 0.750\,000\,000\,00 \ldots$ (finite decimal)

(b) $\frac{3}{7} = 0.428\,571\,428\,571\,428\,571 \ldots$ (infinite repeating decimal)

(c) $\sqrt{2} = 1.414\,213\,562 \ldots$ (infinite nonrepeating decimal)   ◆◆◆

We can write a repeating decimal as a rational fraction by writing the repeating part as an infinite geometric series, as shown in the following example.

◆◆◆ **Example 35:** Write the number $4.851\,51 \ldots$ as a rational fraction.

Solution: We rewrite the given number as

$$4.8 + 0.051 + 0.000\,51 + \cdots$$

The terms of this series, except for the first, form an infinite GP where $a = 0.051$ and $r = 0.01$. The sum S to infinity is then

$$S = \frac{a}{1-r} = \frac{0.051}{1 - 0.01} = \frac{0.051}{0.99} = \frac{51}{990} = \frac{17}{330}$$

Our given number is then equal to $4.8 + S$, so

$$4.851\,51 = 4 + \frac{8}{10} + \frac{17}{330} = 4\frac{281}{330}$$ ◆◆◆

Exercise 4 ◆ Infinite Geometric Progressions

Evaluate each limit.

1. $\lim\limits_{b \to 0} (b - c + 5)$

2. $\lim\limits_{b \to 0} (a + b^2)$

3. $\lim\limits_{b \to 0} \dfrac{3 + b}{c + 4}$

4. $\lim\limits_{b \to 0} \dfrac{a + b + c}{b + c - 5}$

Find the sum of the infinitely many terms of each GP.

5. $144, 72, 36, 18, \ldots$

6. $8, 4, 1, \frac{1}{4}, \ldots$

7. $10, 2, 0.4, 0.08, \ldots$

8. $1, \frac{1}{4}, \frac{1}{16}, \ldots$

Write each decimal number as a rational fraction.

9. $0.57\overline{57}\ldots$ **10.** $0.69\overline{69}\ldots$

11. $7.681\,\overline{81}\ldots$ **12.** $5.861\,11\ldots$

13. Each swing of a certain pendulum is 78% as long as the one before. If the first swing is 20 cm, find the entire distance travelled by the pendulum before it comes to rest.

14. A ball dropped from a height of 10 m rebounds to three-fourths of its height on each bounce. Find the total distance travelled by the ball before it comes to rest.

25–5 The Binomial Theorem

Powers of a Binomial

Recall that a *binomial*, such as $(a + b)$, is a polynomial with two terms. By actual multiplication we can show that

$$(a + b)^1 = a + b$$
$$(a + b)^2 = a^2 + 2ab + b^2$$
$$(a + b)^3 = a^3 + 3a^2b + 3ab^2 + b^3$$
$$(a + b)^4 = a^4 + 4a^3b + 6a^2b^2 + 4ab^3 + b^4$$

We now want a formula for $(a + b)^n$ with which to expand a binomial without actually carrying out the multiplication. In the expansion of $(a + b)^n$, where n is a positive integer, we note the following patterns:

1. There are $n + 1$ terms.

2. The power of a is n in the first term, decreases by 1 in each later term, and reaches 0 in the last term.

3. The power of b is 0 in the first term, increases by 1 in each later term, and reaches n in the last term.

4. Each term has a total degree of n. (That is, the sum of the degrees of the variables is n.)

5. The first coefficient is 1. Further, the product of the coefficient of any term and its power of a, divided by the number of the term, gives the coefficient of the next term. (This property gives a *recursion formula* for the coefficients of the binomial expansion.)

The formula is expressed as follows:

$$(a + b)^n = a^n + na^{n-1}b + \frac{n(n - 1)}{2} a^{n-2} b^2 + \frac{n(n - 1)(n - 2)}{2(3)} a^{n-3} b^3 + \cdots + b^n$$

◆◆◆ **Example 36:** Use the binomial theorem to expand $(a + b)^5$.

Solution: From the binomial theorem,

$$(a + b)^5 = a^5 + 5a^4b + \frac{5(4)}{2} a^3b^2 + \frac{5(4)(3)}{2(3)} a^2b^3 + \frac{5(4)(3)(2)}{2(3)(4)} ab^4 + b^5$$

$$= a^5 + 5a^4b + 10a^3b^2 + 10a^2b^3 + 5ab^4 + b^5$$

which can be verified by actual multiplication.   ◆◆◆

If the expression contains only one variable, we can verify the expansion graphically.

◆◆◆ **Example 37:** Use the result from Example 36 to expand $(x + 1)^5$, and verify graphically.

Solution: Substituting $a = x$ and $b = 1$ into the preceding result gives

$$(x + 1)^5 = x^5 + 5x^4 + 10x^3 + 10x^2 + 5x + 1$$

To verify graphically, we plot the following on the same axes:

$$y_1 = (x + 1)^5$$

> We will use the binomial theorem to obtain the values in a *binomial probability distribution* in the following chapter on statistics.

and

$$y_2 = x^5 + 5x^4 + 10x^3 + 10x^2 + 5x + 1$$

The two graphs, shown in Fig. 25–6, are identical. This is not a *proof* that our expansion is correct, but it certainly gives us more confidence in our result.

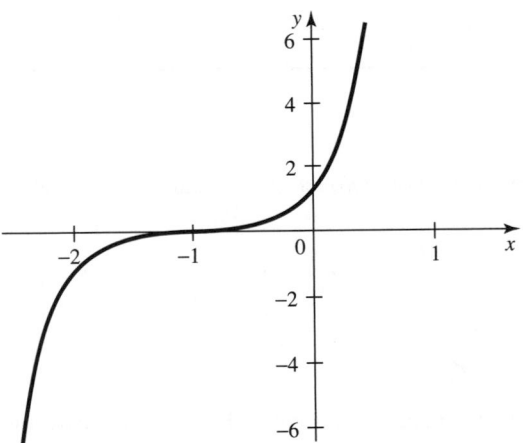

FIGURE 25–6   Graphs of $y = (x + 1)^5$ and $y = x^5 + 5x^4 + 10x^3 + 10x^2 + 5x + 1$.   ◆◆◆

Factorial Notation

We now introduce *factorial notation*. For a positive integer n, factorial n, written $n!$, is the product of all of the positive integers less than or equal to n.

$$n! = 1 \cdot 2 \cdot 3 \cdots n$$

Of course, it does not matter in which order we multiply the numbers. We may also write

$$n! = n(n - 1) \cdots 3 \cdot 2 \cdot 1$$

◆◆◆ **Example 38:**

(a) $3! = 1 \cdot 2 \cdot 3 = 6$ (b) $5! = 1 \cdot 2 \cdot 3 \cdot 4 \cdot 5 = 120$

(c) $\dfrac{6!\,3!}{7!} = \dfrac{6!\,(2)(3)}{7(6!)} = \dfrac{6}{7}$ (d) $0! = 1$ by definition ◆◆◆

Binomial Theorem Written with Factorial Notation

The use of factorial notation allows us to write the binomial theorem more compactly. Thus:

Binomial Theorem	$(a+b)^n = a^n + na^{n-1}b + \dfrac{n(n-1)}{2!}a^{n-2}b^2 + \dfrac{n(n-1)(n-2)}{3!}a^{n-3}b^3 + \cdots + b^n$	245

◆◆◆ **Example 39:** Using the binomial formula, expand $(a+b)^6$.

Solution: There will be $(n+1)$ or seven terms. Let us first find the seven coefficients. The first coefficient is always 1, and the second coefficient is always n, or 6. We calculate the remaining coefficients in Table 25–4.

TABLE 25–4

Term	Coefficient
1	1
2	$n = 6$
3	$\dfrac{n(n-1)}{2!} = \dfrac{6(5)}{1(2)} = 15$
4	$\dfrac{n(n-1)(n-2)}{3!} = \dfrac{6(5)(4)}{1(2)(3)} = 20$
5	$\dfrac{n(n-1)(n-2)(n-3)}{4!} = \dfrac{6(5)(4)(3)}{1(2)(3)(4)} = 15$
6	$\dfrac{n(n-1)(n-2)(n-3)(n-4)}{5!} = \dfrac{6(5)(4)(3)(2)}{1(2)(3)(4)(5)} = 6$
7	$\dfrac{n(n-1)(n-2)(n-3)(n-4)(n-5)}{6!} = \dfrac{6(5)(4)(3)(2)(1)}{1(2)(3)(4)(5)(6)} = 1$

The seven terms thus have the coefficients

$$1 \quad 6 \quad 15 \quad 20 \quad 15 \quad 6 \quad 1$$

and our expansion is

$$(a+b)^6 = a^6 + 6a^5b + 15a^4b^2 + 20a^3b^3 + 15a^2b^4 + 6ab^5 + b^6$$ ◆◆◆

Pascal's Triangle

We now have expansions for $(a+b)^n$, for values of n from 1 to 6, to which we add $(a+b)^0 = 1$. Let us now write the coefficients only of the terms of each expansion.

```
Exponent n
   0                         1
   1                      1     1
   2                   1     2     1
   3                1     3     3     1
   4             1     4     6     4     1
   5          1     5    10    10     5     1
   6       1     6    15    20    15     6     1
```

Pascal's triangle is named for Blaise Pascal (1623–62), the French geometer, probabilist, combinatorist, physicist, and philosopher.

This array is called *Pascal's triangle*. We may use it to predict the coefficients of expansions with powers higher than 6 by noting that each number in the triangle is equal to the sum of the two numbers above it (thus the 15 in the lowest row is the sum of the 5 and the 10 above it). When expanding a binomial, we may take the coefficients directly from Pascal's triangle, as shown in the following example.

◆◆◆ **Example 40:** Expand $(1 - x)^6$, obtaining the binomial coefficients from Pascal's triangle.

Solution: From Pascal's triangle, for $n = 6$, we read the coefficients

$$1 \quad 6 \quad 15 \quad 20 \quad 15 \quad 6 \quad 1$$

We now substitute into the binomial formula with $n = 6$, $a = 1$, and $b = -x$. When one of the terms of the binomial is negative, as here, we must be careful to include the minus sign whenever that term appears in the expansion. Thus

$$
\begin{aligned}
(1 - x)^6 &= [1 + (-x)]^6 \\
&= 1^6 + 6(1^5)(-x) + 15(1^4)(-x)^2 + 20(1^3)(-x)^3 \\
&\quad + 15(1^2)(-x)^4 + 6(1)(-x)^5 + (-x)^6 \\
&= 1 - 6x + 15x^2 - 20x^3 + 15x^4 - 6x^5 + x^6
\end{aligned}
$$

◆◆◆

If either term in the binomial is itself a power, a product, or a fraction, it is a good idea to enclose that entire term in parentheses before substituting.

◆◆◆ **Example 41:** Expand $(3/x^2 + 2y^3)^4$.

Solution: The binomial coefficients are

$$1 \quad 4 \quad 6 \quad 4 \quad 1$$

We substitute $3/x^2$ for a and $2y^3$ for b.

$$
\begin{aligned}
\left(\frac{3}{x^2} + 2y^3\right)^4 &= \left[\left(\frac{3}{x^2}\right) + (2y^3)\right]^4 \\
&= \left(\frac{3}{x^2}\right)^4 + 4\left(\frac{3}{x^2}\right)^3 (2y^3) + 6\left(\frac{3}{x^2}\right)^2 (2y^3)^2 + 4\left(\frac{3}{x^2}\right)(2y^3)^3 + (2y^3)^4 \\
&= \frac{81}{x^8} + \frac{216y^3}{x^6} + \frac{216y^6}{x^4} + \frac{96y^9}{x^2} + 16y^{12}
\end{aligned}
$$

◆◆◆

Sometimes we might not need the entire expansion but only the first several terms.

◆◆◆ **Example 42:** Find the first four terms of $(x^2 - 2y^3)^{11}$.

Solution: From the binomial formula, with $n = 11$, $a = x^2$, and $b = (-2y^3)$,

$$
\begin{aligned}
(x^2 - 2y^3)^{11} &= [(x^2) + (-2y^3)]^{11} \\
&= (x^2)^{11} + 11(x^2)^{10}(-2y^3) + \frac{11(10)}{2}(x^2)^9(-2y^3)^2 \\
&\quad + \frac{11(10)(9)}{2(3)}(x^2)^8(-2y^3)^3 + \cdots \\
&= x^{22} - 22x^{20}y^3 + 220x^{18}y^6 - 1320x^{16}y^9 + \cdots
\end{aligned}
$$

◆◆◆

A *trinomial* can be expanded by the binomial formula by grouping the terms, as in the following example.

◆◆◆ **Example 43:** Expand $(1 + 2x - x^2)^3$ by the binomial formula.

Solution: Let $z = 2x - x^2$. Then

$$(1 + 2x - x^2)^3 = (1 + z)^3 = 1 + 3z + 3z^2 + z^3$$

Substituting back, we have

$$(1 + 2x - x^2)^3 = 1 + 3(2x - x^2) + 3(2x - x^2)^2 + (2x - x^2)^3$$
$$= 1 + 6x + 9x^2 - 4x^3 - 9x^4 + 6x^5 - x^6$$

◆◆◆

General Term

We can write the general or rth term of a binomial expansion $(a + b)^n$ by noting the following pattern:

1. The power of b is 1 less than r. Thus a fifth term would contain b^4, and the rth term would contain b^{r-1}.
2. The power of a is n minus the power of b. Thus a fifth term would contain a^{n-5+1}, and the rth term would contain a^{n-r+1}.
3. The coefficient of the rth term is

$$\frac{n!}{(r-1)!(n-r+1)!}$$

The formula is expressed as follows:

General Term	In the expansion for $(a + b)^n$, $$r \text{ th term} = \frac{n!}{(r-1)!\,(n-r+1)!}\, a^{n-r+1}b^{r-1}$$	**246**

◆◆◆ **Example 44:** Find the eighth term of $(3a + b^5)^{11}$.

Solution: Here, $n = 11$ and $r = 8$. So $n - r + 1 = 4$. Substitution yields

$$\frac{n!}{(r-1)!\,(n-r+1)!} = \frac{11!}{7!\,4!} = 330$$

Then

$$\text{the eighth term} = 330(3a)^{11-8+1}(b^5)^{8-1}$$

$$= 330(81a^4)(b^{35}) = 26{,}730a^4b^{35}$$

◆◆◆

Proof of the Binomial Theorem

We have shown that the binomial theorem is true for $n = 1, 2, 3, 4$, or 5. But what about some other exponent, say, $n = 50$? To show that the theorem works for any positive integral exponent, we give the following proof.

We first show that if the binomial theorem is true for some exponent $n = k$, it is also true for $n = k + 1$. We assume that

$$(a + b)^k = a^k + ka^{k-1}b + \cdots + Pa^{k-r}b^r + Qa^{k-r-1}b^{r+1}$$
$$+ Ra^{k-r-2}b^{r+2} + \cdots + b^k \tag{1}$$

where we show three intermediate terms with coefficients P, Q, and R. Multiplying both sides of Eq. (1) by $(a + b)$ gives

$$(a + b)^{k+1} = a^{k+1} + ka^k b + \cdots + Pa^{k-r+1}b^r + Qa^{k-r}b^{r+1}$$

$$+ Ra^{k-r-1}b^{r+2} + \cdots + ab^k + \cdots + a^k b + \cdots$$

$$+ Pa^{k-r}b^{r+1} + Qa^{k-r-1}b^{r+2} + Ra^{k-r-2}b^{r+3} + \cdots + b^{k+1}$$

$$(a + b)^{k+1} = a^{k+1} + (k + 1)a^k b + \cdots + (P + Q)a^{k-r}b^{r+1}$$

$$+ (Q + R)a^{k-r-1}b^{r+2} + \cdots + ab^k + b^{k+1} \tag{2}$$

after combining like terms.

Does Eq. (2) obey the binomial theorem? First, the power of b is 0 in the first term and increases by 1 in each later term, reaching $k + 1$ in the last term, so property 3 is met. Thus there are $k + 1$ terms containing b, plus the first term that does not, making a total of $k + 2$ terms, so property 1 is met.

Next, the power of a is $k + 1$ in the first term and decreases by 1 in each succeeding term until it equals 0 in the last term, making the total degree of each term equal to $k + 1$, so properties 2 and 4 are met. Further, the coefficient of the first term is 1, and the coefficient of the second term is $k + 1$, which agrees with property 5.

Finally, in a binomial expansion, the product of the coefficient of any term and its power of a, divided by the number of the term, must give the coefficient of the next term. Here, the first of our two intermediate terms has a coefficient $(P + Q)$. The coefficient of its a term is $k - r$, the number of that term is $r + 2$, and the coefficient of the next term is $(Q + R)$. Thus the identity

$$(P + Q)\frac{k - r}{r + 2} = Q + R$$

must be true. But since

$$Q = \frac{P(k - r)}{r + 1} \quad \text{and} \quad R = \frac{Q(k - r - 1)}{r + 2}$$

we must show that

$$\left[P + \frac{P(k - r)}{r + 1} \right] \frac{k - r}{r + 2} = \frac{P(k - r)}{r + 1} + \frac{P(k - r)(k - r - 1)}{(r + 1)(r + 2)}$$

Working each side separately, we combine each over a common denominator.

$$\frac{P(k - r)(r + 1)}{(r + 1)(r + 2)} + \frac{P(k - r)(k - r)}{(r + 1)(r + 2)} = \frac{P(k - r)(r + 2)}{(r + 1)(r + 2)} + \frac{P(k - r)(k - r - 1)}{(r + 1)(r + 2)}$$

Since the denominators are the same on both sides, we simply have to show that the numerators are identical. Factoring gives us

$$P(k - r)(r + 1 + k - r) = P(k - r)(r + 2 + k - r - 1)$$
$$P(k - r)(k + 1) = P(k - r)(k + 1)$$

This completes the proof that if the binomial theorem is true for some exponent $n = k$, it is also true for the exponent $n = k + 1$. But since we have shown, by direct multiplication, that the binomial theorem *is true* for $n = 5$, it must be true for $n = 6$. And since it is true for $n = 6$, it must be true for $n = 7$, and so on, for all of the positive integers.

Fractional and Negative Exponents

We have shown that the binomial formula is valid when the exponent n is a positive integer. We will not prove it here, but the binomial formula also holds when the exponent is negative or fractional. However, we now obtain an infinite series, called a *binomial series,* with no last term. Further, the binomial series is equal to $(a + b)^n$ only if the series converges.

◆◆◆ **Example 45:** Write the first four terms of the infinite binomial expansion for $\sqrt{5 + x}$.

Solution: We replace the radical with a fractional exponent and substitute into the binomial formula in the usual way.

$$(5 + x)^{1/2} = 5^{1/2} + \left(\frac{1}{2}\right) 5^{-1/2} x + \frac{(1/2)(-1/2)}{2} 5^{-3/2} x^2$$
$$+ \frac{(1/2)(-1/2)(-3/2)}{6} 5^{-5/2} x^3 + \cdots$$
$$= \sqrt{5} + \frac{x}{2\sqrt{5}} - \frac{x^2}{8(5)^{3/2}} + \frac{x^3}{16(5)^{5/2}} - \cdots$$

Switching to decimal form and working to five decimal places, we get

$$(5 + x)^{1/2} \approx 2.236\ 07 + 0.223\ 61x - 0.011\ 18x^2 + 0.001\ 12x^3 - \cdots \qquad \text{◆◆◆}$$

We now ask if this series is valid for any x. Let's try a few values. For $x = 0$,

$$(5 + 0)^{1/2} \cong 2.236\ 07$$

which checks within the number of digits retained. For $x = 1$,

$$(5 + 1)^{1/2} = \sqrt{6} \cong 2.236\ 07 + 0.223\ 61 - 0.011\ 18 + 0.001\ 12 - \cdots$$
$$\cong 2.44962$$

which compares with a value for $\sqrt{6}$ of 2.449 49. Not bad, considering that we are using only four terms of an infinite series.

To determine the effect of larger values of x, we have used a computer to generate eight terms of the binomial series as listed in Table 25–5.

TABLE 25–5

x	$\sqrt{5 + x}$ by Series	$\sqrt{5 + x}$ by Calculator	Error
0	2.236 068	2.236 068	0
2	2.645 766	2.645 751	0.000 015
4	3.002 957	3	0.002 957
6	3.379 908	3.316 625	0.063 283
8	4.148 339	3.605 551	0.543 788
11	9.723 964	4	5.723 964

Into it we have substituted various x values, from 0 to 11, and compared the value obtained by series with that from a calculator. The values are plotted in Fig. 25–7.

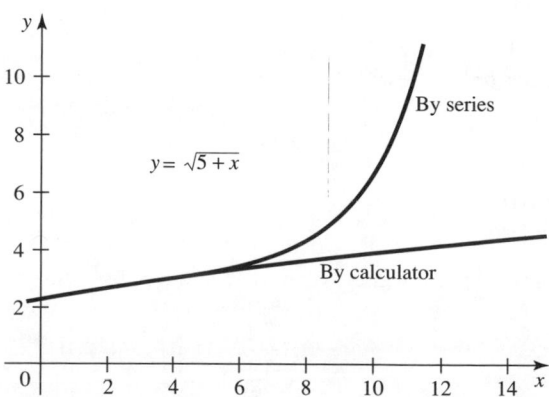

FIGURE 25–7

Notice that the accuracy gets worse as x gets larger, with useless values being obtained when x is 6 or greater. This illustrates the following:

> A binomial series is equal to $(a + b)^n$ only if the series converges, and that occurs when $|a| > |b|$.

Our series in Example 45 thus converges when $x < 5$.

The binomial series, then, is given by the following formula:

Binomial Series	$$(a + b)^n = a^n + na^{n-1}b + \frac{n(n-1)}{2!}a^{n-2}b^2 \\ + \frac{n(n-1)(n-2)}{3!}a^{n-3}b^3 + \cdots \\ \text{where } |a| > |b|$$	**247**

The binomial series is often written with $a = 1$ and $b = x$, so we give that form here also.

Binomial Series	$$(1 + x)^n = 1 + nx + \frac{n(n-1)}{2!}x^2 + \frac{n(n-1)(n-2)}{3!}x^3 + \cdots \\ \text{where } |x| < 1$$	**248**

◆◆◆ **Example 46:** Expand to four terms:

$$\frac{1}{\sqrt[3]{1/a - 3b^{1/3}}}$$

Solution: We rewrite the given expression without radicals or fractions, and then expand.

$$\frac{1}{\sqrt[3]{1/a - 3b^{1/3}}} = [(a^{-1}) + (-3b^{1/3})]^{-1/3}$$

$$= (a^{-1})^{-1/3} - \frac{1}{3}(a^{-1})^{-4/3}(-3b^{1/3}) + \frac{2}{9}(a^{-1})^{-7/3}(-3b^{1/3})^2$$

$$- \frac{14}{81}(a^{-1})^{-10/3}(-3b^{1/3})^3 + \cdots$$

$$= a^{1/3} + a^{4/3}b^{1/3} + 2a^{7/3}b^{2/3} + \frac{14}{3}a^{10/3}b + \cdots \qquad ◆◆◆$$

Exercise 5 ◆ The Binomial Theorem

Factorial Notation

Evaluate each factorial.

1. $6!$

2. $8!$

3. $\dfrac{7!}{5!}$

4. $\dfrac{11!}{9!\,2!}$

5. $\dfrac{7!}{3!\,4!}$

6. $\dfrac{8!}{3!\,5!}$

Binomials Raised to an Integral Power

Verify each expansion. Obtain the binomial coefficients by formula or from Pascal's triangle as directed by your instructor.

7. $(x + y)^7 = x^7 + 7x^6y + 21x^5y^2 + 35x^4y^3 + 35x^3y^4 + 21x^2y^5 + 7xy^6 + y^7$

8. $(4 + 3b)^4 = 256 + 768b + 864b^2 + 432b^3 + 81b^4$

9. $(3a - 2b)^4 = 81a^4 - 216a^3b + 216a^2b^2 - 96ab^3 + 16b^4$

10. $(x^3 + y)^7 = x^{21} + 7x^{18}y + 21x^{15}y^2 + 35x^{12}y^3 + 35x^9y^4 + 21x^6y^5 + 7x^3y^6 + y^7$

11. $(x^{1/2} + y^{2/3})^5 = x^{5/2} + 5x^2y^{2/3} + 10x^{3/2}y^{4/3} + 10xy^2 + 5x^{1/2}y^{8/3} + y^{10/3}$

12. $(1/x + 1/y^2)^3 = 1/x^3 + 3/x^2y^2 + 3/xy^4 + 1/y^6$

13. $(a/b - b/a)^6 = (a/b)^6 - 6(a/b)^4 + 15(a/b)^2 - 20 + 15(b/a)^2 - 6(b/a)^4 + (b/a)^6$

14. $(1/\sqrt{x} + y^2)^4 = x^{-2} + 4x^{-3/2}y^2 + 6x^{-1}y^4 + 4x^{-1/2}y^6 + y^8$

15. $(2a^2 + \sqrt{b})^5 = 32a^{10} + 80a^8b^{1/2} + 80a^6b + 40a^4b^{3/2} + 10a^2b^2 + b^{5/2}$

16. $(\sqrt{x} - c^{2/3})^4 = x^2 - 4x^{3/2}c^{2/3} + 6xc^{4/3} - 4x^{1/2}c^2 + c^{8/3}$

Verify the first four terms of each binomial expansion.

17. $(x^2 + y^3)^8 = x^{16} + 8x^{14}y^3 + 28x^{12}y^6 + 56x^{10}y^9 + \cdots$

18. $(x^2 - 2y^3)^{11} = x^{22} - 22x^{20}y^3 + 220x^{18}y^6 - 1320x^{16}y^9 + \cdots$

19. $(a - b^4)^9 = a^9 - 9a^8b^4 + 36a^7b^8\ 2\ 84a^6b^{12} + \cdots$

20. $(a^3\ 1\ 2b)^{12} = a^{36} + 24a^{33}b + 264a^{30}b^2 + 1760a^{27}b^3 + \cdots$

Trinomials

Verify each expansion.

21. $(2a^2 + a + 3)^3 = 8a^6 + 12a^5 + 42a^4 + 37a^3 + 63a^2 + 27a + 27$

22. $(4x^2 - 3xy - y^2)^3 = 64x^6 - 144x^5y + 60x^4y^2 + 45x^3y^3 - 15x^2y^4 - 9xy^5 - y^6$

23. $(2a^2 + a - 4)^4 = 16a^8 + 32a^7 - 104a^6 - 184a^5 + 289a^4 + 368a^3$
$$- 416a^2 - 256a + 256$$

General Term

Write the requested term of each binomial expansion, and simplify.

24. Seventh term of $(a^2 - 2b^3)^{12}$

25. Eleventh term of $(2 - x)^{16}$

26. Fourth term of $(2a - 3b)^7$

27. Eighth term of $(x + a)^{11}$

28. Fifth term of $(x - 2\sqrt{y})^{25}$

29. Ninth term of $(x^2 + 1)^{15}$

Binomials Raised to a Fractional or Negative Power

Verify the first four terms of each infinite binomial series.

30. $(1 - a)^{2/3} = 1 - 2a/3 - a^2/9 - 4a^3/81 \cdots$

31. $\sqrt{1 + a} = 1 + a/2 - a^2/8 + a^3/16 \cdots$

32. $(1 + 5a)^{-5} = 1 - 25a + 375a^2 - 4375a^3 \cdots$

33. $(1 + a)^{-3} = 1 - 3a + 6a^2 - 10a^3 \cdots$

34. $1/\sqrt[6]{1-a} = 1 + a/6 + 7a^2/72 + 91a^3/1296 \cdots$

Case Study Discussion—Forever Payments

In economics, we find a discussion of "present value of money." The idea is that $1,000 a year from now would have a present value of $952.38, assuming that you can invest this amount at 5% interest. That is, $952.38 invested at 5% would grow to $1,000 in one year. How do we calculate this?

This formula can calculate the value of any amount accruing interest: $A = P(1 + n)$. The final amount A is the principle P times $1 +$ interest rate (expressed as a decimal). So if we know the final amount A and want to calculate the principle, we can use the equation $P = A/(1 + i)$. To earn $1,000 in one interest period at, say, 5% interest, we would need $P = \$1,000/(1 + 0.05) = \$952.38$.

If the payment were two years away it would be $1000/(1 + 0.05)^2 = \$907.03$. The power of two represents the two interest payments. So, if we receive payments of $1,000 in year one and $1,000 in the second year, the total cost of that $2,000 would be $952.38 + $907.03 = $1,859.41 (again assuming an investment of 5% per year, simple interest).

As you can see, we can write the present value of the two $1,000 payments as

$$\frac{1000}{(1 + i)} + \frac{1000}{(1 + i)^2}.$$

But how do you calculate a present value of money if the money keeps going and going forever?

If we extend the formula above to infinite payments, the formula becomes a series:

$$P = \sum_{n=1}^{\infty} \frac{1000}{(1+i)^n}$$

This is a geometric series with the common ratio of $1/(1 + i)$. Remember that the present value of the first payment, a, is $\dfrac{1000}{(1+ i)}$. The sum of the series is

$$\frac{a}{1 - r} = \frac{\dfrac{1000}{(1 + i)}}{1 - \dfrac{1}{(1+ i)}} = \frac{1000}{i}$$

This states that an infinite number of $1,000 payments would have a present value of $1,000/0.05 = $20,000. Now, this actually does make sense because $20,000 invested perpetually at 5% would generate $1,000 worth of interest every year, forever.

♦♦♦ CHAPTER 25 REVIEW PROBLEMS ♦♦♦♦♦♦♦♦♦♦♦♦♦♦♦♦♦♦♦♦♦♦♦♦♦♦♦♦♦♦♦

1. Find the sum of seven terms of the AP: $-4, -1, 2, \ldots$

2. Find the tenth term of $(1 - y)^{14}$.

3. Insert four arithmetic means between 3 and 18.

4. How many terms of the AP $-5, -2, 1, \ldots$ will give a sum of 63?

5. Insert four harmonic means between 2 and 12.

6. Find the twelfth term of an AP with first term 7 and common difference 5.

7. Find the sum of the first 10 terms of the AP $1, 2\frac{2}{3}, 4\frac{1}{3}, \ldots$

Expand each binomial.

8. $(2x^2 - \sqrt{x}/2)^5$;

9. $(a - 2)^5$

10. $(x - y^3)^7$

11. Find the fifth term of a GP with first term 3 and common ratio 2.

12. Find the fourth term of a GP with first term 8 and common ratio -4.

Evaluate each limit.

13. $\lim\limits_{b \to 0} (2b - x + 9)$

14. $\lim\limits_{b \to 0} (a^2 + b^2)$

Find the sum of infinitely many terms of each GP.

15. $4, 2, 1, \ldots$

16. $\frac{1}{4}, -\frac{1}{16}, \frac{1}{64}, \ldots$

17. $1, -\frac{2}{5}, \frac{4}{25}, \ldots$

Write each decimal number as a rational fraction.

18. $0.86\overline{86}\ldots$

19. $0.54\overline{44}\ldots$

20. Find the seventh term of the harmonic progression $3, 3\frac{3}{7}, 4, \ldots$

21. Find the sixth term of a GP with first term 5 and common ratio 3.

22. Find the sum of the first seven terms of the GP in problem 21.

23. Find the fifth term of a GP with first term -4 and common ratio 4.

24. Find the sum of the first seven terms of the GP in problem 23.

25. Insert two geometric means between 8 and 125.

26. Insert three geometric means between 14 and 224.

Evaluate each factorial.

27. $\dfrac{4!\,5!}{2!}$

28. $\dfrac{8!}{4!\,4!}$

29. $\dfrac{7!\,3!}{5!}$

30. $\dfrac{7!}{2!\,5!}$

Find the first four terms of each binomial expansion.

31. $\left(\dfrac{2x}{y^2} - y\sqrt{x}\right)^7$

32. $(1 + a)^{-2}$

33. $\dfrac{1}{\sqrt{1 + a}}$

34. Expand using the binomial theorem:
$$(1 - 3a + 2a^2)^4$$

35. Find the eighth term of $(3a - b)^{11}$.

36. Find the tenth term of an AP with first term 12 if the sum of the first ten terms is 10.

Writing

37. State in your own words the difference between an arithmetic progression and a geometric progression. Give a real-world example of each.

Team Project

Zeno's Racetrack Paradox is named for Zeno of Elea (ca. 490–ca. 435 B.C.), a Greek philosopher and mathematician. He wrote several famous paradoxes, including this one.

38. *Zeno's Racetrack Paradox:* A runner can never reach a finish line 1 kilometre away because first he would have to run half a kilometre, and then he must run half of the remaining distance, or one-quarter kilometre, and then half of that, and so on. Since there are an infinite number of distances that must be run, it will take an infinite length of time, and so the runner will never reach the finish line. Show that the distances to be run form an infinite series,

$$\frac{1}{2}, \frac{1}{4}, \frac{1}{8}, \cdots$$

Then disprove the paradox by actually finding the sum of that series and showing that the sum is *not* infinite.

26

Introduction to Statistics and Probability

Why statistics? Because modern organizations—from businesses to governments, and from university and college researchers to non-government organizations working around the world to improve conditions—all make the same demand: Prove it! Organizations are data driven. If a decision is important, it requires facts.

Statistics is a branch of mathematics that attempts to analyze and present large amounts information in a form that is understandable by humans. It also can be used to extrapolate and infer beyond known data, within certain limits. It can indicate that a process is starting to go out of control even before any harm or waste occurs. Information or recommendations presented to your clients, customers, employers, or colleagues will be subject to analysis, questioning, and challenges. Your knowledge of statistics will enable you to present information accurately and realistically. Your knowledge of statistics will also help you read and understand technical evaluations and claims made by suppliers or the competition. It will help you understand whether or not a data set provides sufficient evidence for a decision.

In this chapter we introduce many new terms. We then show how to display statistical data in graphical and numerical forms. How reliable are those statistical data? To answer that question, we include a brief introduction to *probability*, which enables us to describe the *confidence* we can place on those data.

26–1   Definitions and Terminology

Populations and Samples

In statistics, an entire group of people or things is called a *population* or *universe*. A population can be *infinite* or *finite*. A small part of the entire group chosen for study is called a *sample*.

◆◆◆ **Example 1:** In a study of the heights of students at Tech College, the entire student body is our *population*. This is a *finite* population. Instead of studying every student in the population, we may choose to work with a *sample* of only 100 students.   ◆◆◆

◆◆◆ **Example 2:** Suppose that we want to determine what percentage of tosses of a coin will be heads. All possible tosses of the coin are an example of an *infinite* population. Theoretically, there is no limit to the number of tosses that can be made.   ◆◆◆

Parameters and Statistics

We often use just a few numbers to describe an entire population. Such numbers are called *parameters*. The branch of statistics in which we use parameters to describe a population is referred to as *descriptive statistics*.

◆◆◆ **Example 3:** For the population of all students at Tech College, a computer read the height of each student and computed the average height of all students. It found this *parameter*, average height, to be 173 cm.   ◆◆◆

Similarly, we use just a few numbers to describe a sample. Such numbers are called *statistics*. When we then use these sample statistics to *infer* things about the entire population, we are engaging in what is called *inductive statistics*.

◆◆◆ **Example 4:** From the population of all students in Tech College, a sample of 100 students was selected. The average of their heights was found to be 172 cm. This is a sample *statistic*. From it we infer that for the entire population, the average height is 172 cm plus or minus some *standard error*. We show how to compute standard error in Sec. 26–6.   ◆◆◆

Thus *parameters* describe *populations*, while *statistics* describe *samples*. The first letters of the words help us to keep track of which goes with which.

$$\text{Population} \longrightarrow \text{Parameter}$$

versus

$$\text{Sample} \longrightarrow \text{Statistic}$$

Variables

In statistics we distinguish between data that are *continuous* (obtained by measurement), *discrete* (obtained by counting), or *categorical* (belonging to one of several categories).

◆◆◆ **Example 5:** A survey at Tech College asked each student for (a) height (*continuous data*); (b) number of courses taken (*discrete data*); and (c) whether they were in favor (yes or no) of eliminating final exams (*categorical data*).   ◆◆◆

A *variable*, as in algebra, is a quantity that can take on different values within a problem; *constants* do not change value. We call variables *continuous*, *discrete*, or *categorical*, depending on the type of data they represent.

◆◆◆ **Example 6:** If, in the survey of Example 5, we let

$$H = \text{student's height}$$
$$N = \text{number of courses taken}$$
$$F = \text{opinion on eliminating finals}$$

then H is a continuous variable, N is a discrete variable, and F is a categorical variable.   ◆◆◆

Ways to Display Data

Statistical data are usually presented as a table of values or are displayed in a graph or a chart. Common types of graphs are the *x-y* graph, scatter plot, bar graph, or pie chart. Here we give examples of the display of discrete and categorical data. We cover display of continuous data in Sec. 26–2.

♦♦♦**Example 7:** Table 26–1 shows the number of skis made per month by the Ace Ski Company. We show these discrete data as an *x-y* graph, a scatter plot, a bar chart, and a pie chart in Fig. 26–1.

TABLE 26–1

Month	Pairs Made
Jan.	13 946
Feb.	7 364
Mar.	5 342
Apr.	3 627
May	1 823
June	847
July	354
Aug.	425
Sept.	3 745
Oct.	5 827
Nov.	11 837
Dec.	18 475
Total	73 612

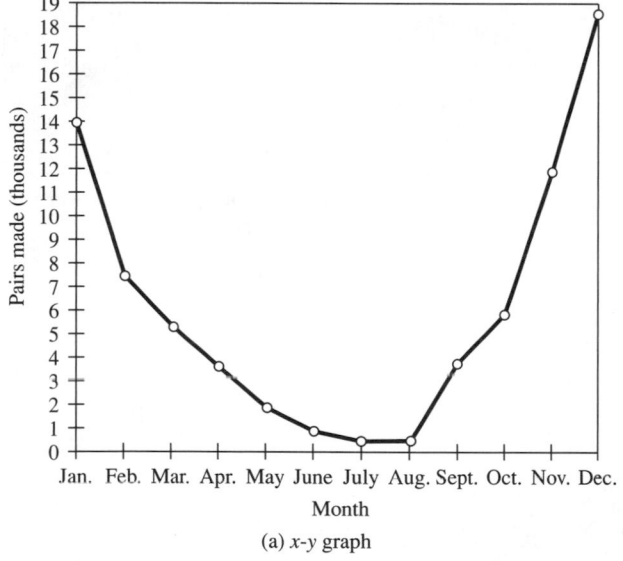

(a) *x-y* graph

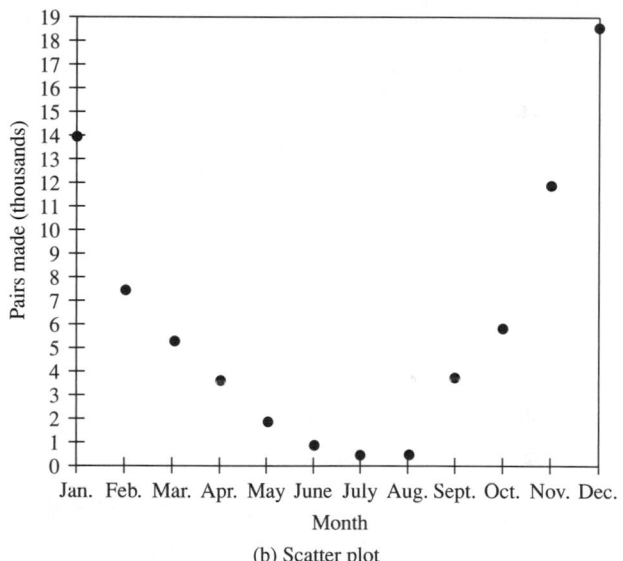

(b) Scatter plot

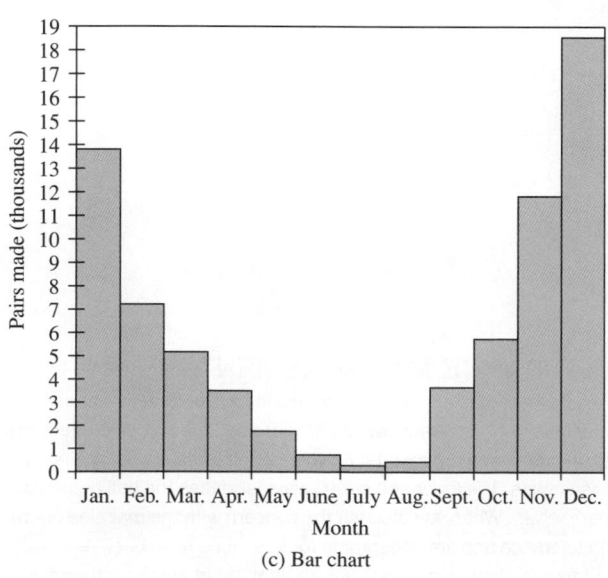

(c) Bar chart

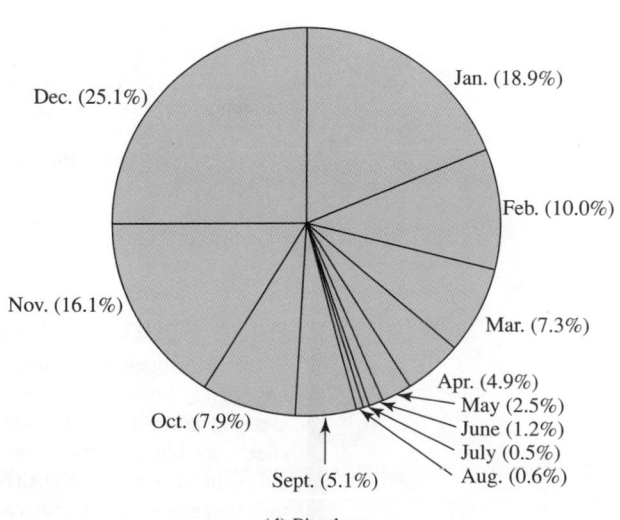

(d) Pie chart

FIGURE 26–1 ♦♦♦

TABLE 26–2

Type	Pairs Made
Downhill	35 725
Racing	7 735
Jumping	2 889
Cross-country	27 263
Total	73 612

◆◆◆ **Example 8:** Table 26–2 shows the number of skis of various types made in a certain year by the Ace Ski Company. We show these discrete data as an *x-y* graph, a bar chart, and a pie chart (Fig. 26–2).

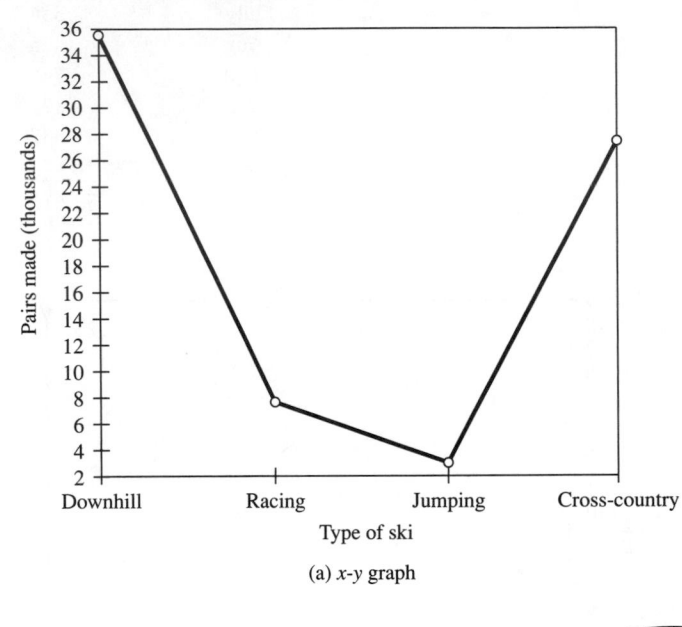

(a) *x-y* graph

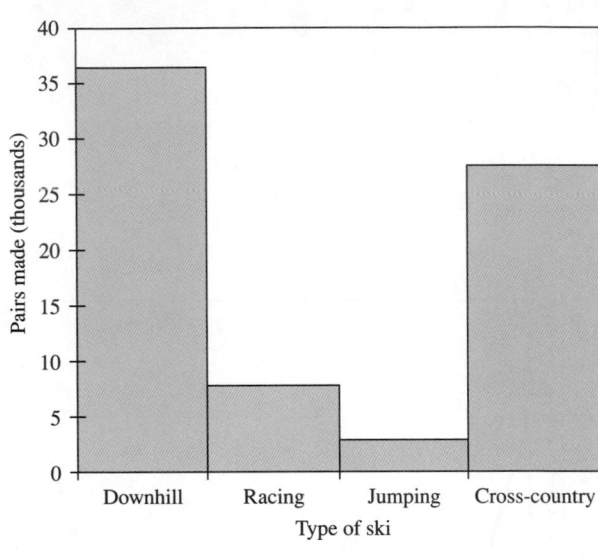

(b) Bar chart

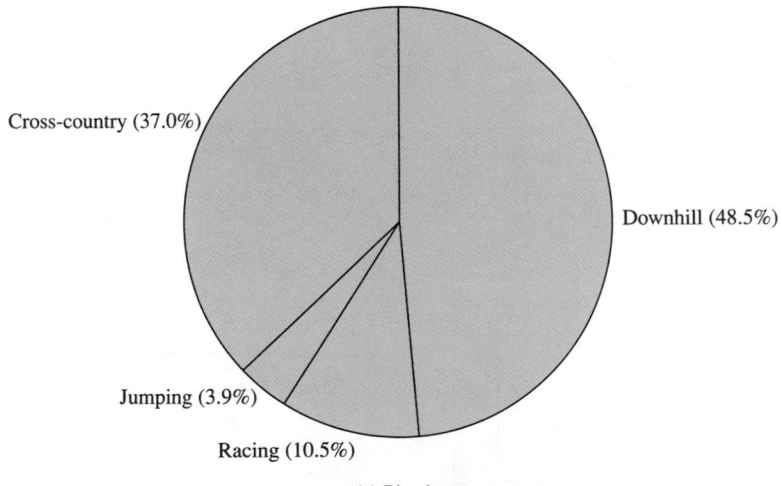

(c) Pie chart

FIGURE 26–2　　　　　　　　　　　　　　　◆◆◆

CASE STUDY—QUALITY ASSURANCE MEASUREMENTS

You are a mechanical engineering technologist working as a quality assurance inspector for a mechanical engineering firm specializing in single- and double-row ring bearings. You have trained the machine operators in taking quality-assurance measurements and recording the results. As you track the results using $\overline{X}$ and R charts, you are concerned with the cage assemblies. The inner and outer races (surfaces the ball bearings roll on) are fine, as are the ball bearings themselves. When you discuss the concern with the machine operators, they state that the cages are all within tolerance and are acceptable for use, none have been rejected for use, and none have been returned by customers. You agree that they are right. What are three things you could have found in the data that would indicate a problem, even though there is no problem?

Exercise 1 ◆ Definitions and Terminology

Types of Data

Label each type of data as continuous, discrete, or categorical.

1. The number of people in each constituency who voted for Jones.
2. The life of certain 100-W light bulbs.
3. The blood types of patients at a certain hospital.
4. The number of Ski-Doos sold each day.
5. The models of snowmobiles sold each day.
6. The weights of steers in a given herd.

Graphical Representation of Data

7. The following table shows the population of a certain town for the years 1920–30:

Year	Population
1920	5364
1921	5739
1922	6254
1923	7958
1924	7193
1925	6837
1926	7245
1927	7734
1928	8148
1929	8545
1930	8623

   Make a scatter plot and an *x-y* graph of the population versus the year.

8. Make a bar graph for the data of problem 7.

9. In a certain election, the tally was as listed in the following table:

Candidate	Number of Votes
Smith	746 000
Jones	623 927
Doe	536 023
Not voting	163 745

   Make a bar chart showing the election results.

10. Make a pie chart showing the election results of problem 9. To make the pie chart, simply draw a circle and subdivide the area of the circle into four sectors. Make the central angle (and hence the area) of each sector proportional to the votes obtained by each candidate as a percentage of the total voting population. Label the percent in each sector.

26–2   Frequency Distributions

Raw Data

Data that have not been organized in any way are called *raw data*. Data that have been sorted into ascending or descending order are called an *array*. The *range* of the data is the difference between the largest and the smallest number in that array.

◆◆◆**Example 9:** Five students were chosen at random and their heights (in metres) were measured. The *raw data* obtained were

$$1.42 \quad 1.83 \quad 1.47 \quad 1.43 \quad 1.70$$

If we sort the numbers into ascending order, we get the *array*

$$1.42 \quad 1.43 \quad 1.47 \quad 1.70 \quad 1.83$$

The *range* of the data is

$$\text{range} = 1.83 - 1.42 = 0.41 \text{ m} \qquad ◆◆◆$$

Grouped Data

For discrete and categorical data, we have seen that *x-y* plots and bar charts are effective. These graphs can also be used for *continuous* data, but first we must collect those data into groups. Such groups are usually called *classes*. We create classes by dividing the range into a convenient number of *intervals*. Each interval has an upper and a lower *limit* or *class boundary*. The *class width* is equal to the upper class boundary minus the lower class limit. The intervals are chosen so that a given data point cannot fall directly on a class boundary. We call the centre of each interval the *class midpoint*.

Frequency Distribution

Once the classes have been defined, we may tally the number of measurements falling within each class. Such a tally is called a *frequency distribution*. The number of entries in a given class is called the *absolute frequency*. That number, divided by the total number of entries, is called the *relative frequency*. The relative frequency is often expressed as a percent.

To Make a Frequency Distribution:

1. Subtract the smallest number in the data from the largest to find the range of the data.
2. Divide the range into class intervals, usually between five and twenty. Avoid class limits that coincide with actual data.
3. Get the class frequencies by tallying the number of observations that fall within each class.
4. Get the relative frequency of each class by dividing its class frequency by the total number of observations.

◆◆◆**Example 10:** The grades of 30 students are as follows:

$$85.0 \quad 62.0 \quad 94.6 \quad 76.3 \quad 77.4 \quad 82.3 \quad 58.4 \quad 86.4 \quad 84.4 \quad 69.8$$
$$91.3 \quad 75.0 \quad 69.6 \quad 85.4 \quad 84.1 \quad 74.7 \quad 65.7 \quad 86.4 \quad 90.6 \quad 69.7$$
$$86.7 \quad 70.5 \quad 78.4 \quad 86.4 \quad 81.5 \quad 60.7 \quad 79.8 \quad 97.2 \quad 85.7 \quad 78.5$$

Group these raw data into convenient classes. Choose the class width, the class midpoints, and the class limits. Find the absolute frequency and the relative frequency of each class.

Solution:

1. The highest value in the data is 97.2 and the lowest is 58.4, so

$$\text{range} = 97.2 - 58.4 = 38.8$$

2. If we divide the range into five classes, we get a class width of 38.8/5 or 7.76. If we divide the range into twenty classes, we get a class width of 1.94. Picking a convenient width between those two values, we let

$$\text{class width} = 6$$

Then we choose class limits of

$$56, 62, 68, \ldots, 98$$

Sometimes a given value, such as 62.0 in the above grade list, will fall right on a class limit. Although it does not matter much into which of the two adjoining classes we put it, we can easily avoid such ambiguity. One way is to make our class limits the following:

$$56\text{–}61.9, \quad 62\text{–}67.9, \quad 68\text{–}73.9, \ldots, 92\text{–}97.9$$

Thus 62.0 would fall into the second class. Another way to say the same thing is to use the interval notation from Sec. 24–1.

$$[56, 62), [62, 68), [68, 74), \ldots, [92, 98)$$

Recall that the interval [56, 62) is *closed* on the left, meaning that it includes 56, and *open* on the right, meaning that it does *not* include 62. Again, a data value of 62.0 would go into the second class.

Our class midpoints are then

$$59, 65, 71, \ldots, 95$$

3. Our next step is to tally the number of grades falling within each class and get a frequency distribution, shown in Table 26–3.

TABLE 26–3   Frequency distribution of student grades.

Class Limits	Class Midpoint	Tally	Absolute Frequency	Relative Frequency
56–61.9	59	//	2	2/30 = 6.7%
62–67.9	65	//	2	2/30 = 6.7%
68–73.9	71	////	4	4/30 = 13.3%
74–79.9	77	ℳℍ //	7	7/30 = 23.3%
80–85.9	83	ℳℍ //	7	7/30 = 23.3%
86–91.9	89	//// /	6	6/30 = 20.0%
92–97.9	95	//	2	2/30 = 6.7%
			Total 30	30/30 = 100%

4. We then divide each frequency by 30 to get the relative frequency.   ◆◆◆

Frequency Histogram

A *frequency histogram* is a bar chart showing the frequency of occurrence of each class. The width of each bar is equal to the class width, and the height of each bar is equal to the frequency. The horizontal axis can show the class limits, the class midpoints, or both. The vertical axis can show the absolute frequency, the relative frequency, or both.

◆◆◆**Example 11:** We show a frequency histogram for the data of Example 10 in Fig. 26–3. Note that the horizontal scale shows both the class limits and the class midpoints, and the vertical scales show both absolute frequency and relative frequency.

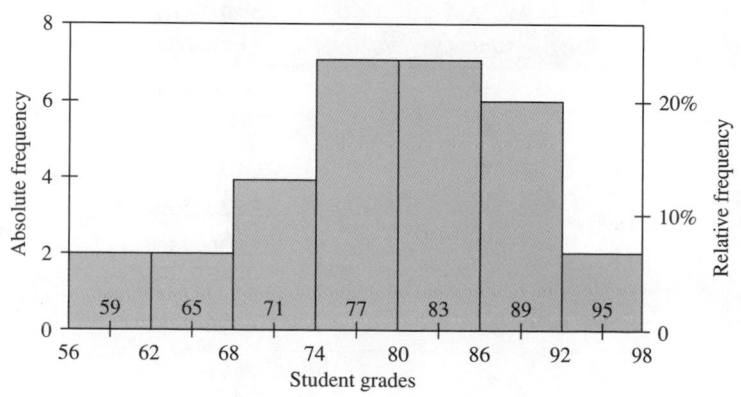

FIGURE 26–3

Since all of the bars in a frequency histogram have the same width, the *areas* of the bars are proportional to the heights of the bars. The areas are thus proportional to the class frequencies. So if one bar in a relative frequency histogram has a height of 20%, that bar will contain 20% of the area of the whole graph. Thus there is a 20% chance that one grade chosen at random will fall within that class. We will make use of this connection between areas and
◆◆◆ probabilities in Sec. 26–4.

Frequency Polygon

A *frequency polygon* is simply an *x-y* graph in which frequency is plotted against class midpoint. As with the histogram, it can show either absolute or relative frequency, or both.

◆◆◆ **Example 12:** We show a frequency polygon for the data of Example 10 in Fig. 26–4.

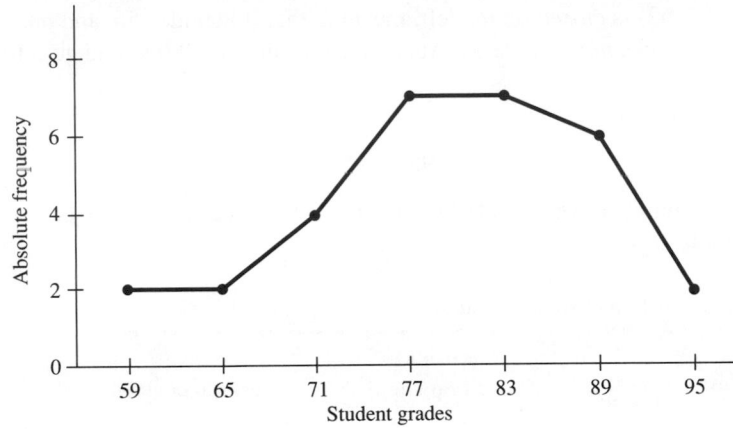

FIGURE 26-4   A frequency polygon for the data of Example 10. Note that this polygon could have been obtained by connecting the midpoints of the tops of the bars in Fig. 26–3.

◆◆◆

Cumulative Frequency Distribution

We obtain a *cumulative* frequency distribution by summing all values *less than* a given class limit. The graph of a cumulative frequency distribution is called a *cumulative frequency polygon*, or *ogive*.

◆◆◆ **Example 13:** Make and graph a cumulative frequency distribution for the data of Example 10.

Solution: We compute the cumulative frequencies by adding the values below the given limits as shown in Table 26–4. Then we plot them in Fig. 26–5.

TABLE 26-4   Cumulative frequency distribution for student grades.

Grade	Cumulative Absolute Frequency	Cumulative Relative Frequency
Under 62.45	2	2/30 (6.7%)
Under 68	4	4/30 (13.3%)
Under 74	8	8/30 (26.7%)
Under 80	15	15/30 (50%)
Under 86	22	22/30 (73.3%)
Under 92	28	28/30 (93.3%)
Under 98	30	30/30 (100%)

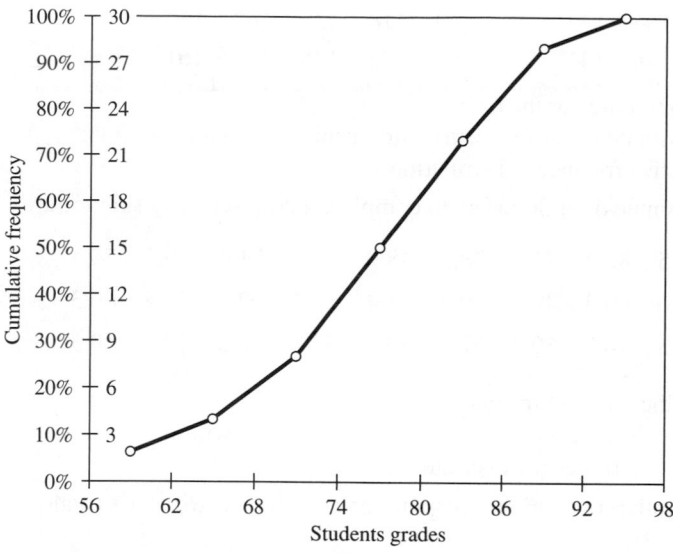

FIGURE 26–5 ◆◆◆

Stem and Leaf Plot

A fast and easy way to get an idea of the distribution of a new set of data is to make a *stem and leaf* plot. We will show how by means of an example.

◆◆◆ **Example 14:** Make a stem and leaf plot for the data in Example 10.

Solution: Let us choose as the *stem* of our plot the first digit of each grade. These range from 5 to 9, representing grades from 50 to 90. Then opposite each number on the stem we write all of the values in the table starting with that number, as shown in Table 26–5.

TABLE 26–5

Stem	Leaves									
5	8.4									
6	2.0	9.8	9.6	5.7	9.7	0.7				
7	6.3	7.4	5.0	4.7	0.5	8.4	9.8	8.5		
8	5.0	2.3	6.4	4.4	5.4	4.1	6.4	6.7	6.4	1.5	5.7
9	4.6	1.3	0.6	7.2						

We sometimes go on to arrange the leaves in numerical order (not shown here). You may also choose to drop the decimal points.

To see the shape of the distribution, simply turn the page on its side. Notice the similarity in shape to that of Fig. 26–3.

Also note that the stem and leaf plot *preserves the original data*. Thus we could exactly reconstruct the data list of Example 10 from this plot, whereas such a reconstruction is not possible from the grouped data of Table 26–3.

Exercise 2 ◆ Frequency Distributions

Frequency Distributions

1. The weights (in pounds) of 40 students at Tech College are

$$127 \quad 136 \quad 114 \quad 147 \quad 158 \quad 149 \quad 155 \quad 162$$
$$154 \quad 139 \quad 144 \quad 114 \quad 163 \quad 147 \quad 155 \quad 165$$
$$172 \quad 168 \quad 146 \quad 154 \quad 111 \quad 149 \quad 117 \quad 152$$

$$
\begin{array}{cccccccc}
166 & 172 & 158 & 149 & 116 & 127 & 162 & 153 \\
118 & 141 & 128 & 153 & 166 & 125 & 117 & 161
\end{array}
$$

(a) Determine the range of the data.
(b) Make an absolute frequency distribution using class widths of 5 lb.
(c) Make a relative frequency distribution.

2. The times (in minutes) for 30 racers to complete a cross-country ski race are

$$
\begin{array}{cccccccccc}
61.4 & 72.5 & 88.2 & 71.2 & 48.5 & 48.9 & 54.8 & 71.4 & 99.2 & 74.5 \\
84.6 & 73.6 & 69.3 & 49.6 & 59.3 & 71.4 & 89.4 & 92.4 & 48.4 & 66.3 \\
85.7 & 59.3 & 74.9 & 59.3 & 72.7 & 49.4 & 83.8 & 50.3 & 72.8 & 69.3
\end{array}
$$

(a) Determine the range of the data.
(b) Make an absolute frequency distribution using class widths of 5 min.
(c) Make a relative frequency distribution.

3. The prices (in dollars) of various computer printers in a distributer's catalog are

$$
\begin{array}{cccccccccc}
850 & 625 & 946 & 763 & 774 & 789 & 584 & 864 & 884 & 698 \\
913 & 750 & 696 & 864 & 795 & 747 & 657 & 886 & 906 & 697 \\
867 & 705 & 784 & 864 & 946 & 607 & 798 & 972 & 857 & 785
\end{array}
$$

(a) Determine the range of the data.
(b) Make an absolute frequency distribution using class widths of $50.
(c) Make a relative frequency distribution.

Frequency Histograms and Frequency Polygons

Draw a histogram, showing both absolute and relative frequency, for the data of

4. problem 1.

5. problem 2.

6. problem 3.

Draw a frequency polygon, showing both absolute and relative frequency, for the data of

7. problem 1.
8. problem 2.
9. problem 3.

Cumulative Frequency Distributions

Make a cumulative frequency distribution showing
(a) absolute frequency and
(b) relative frequency, for the data of

10. problem 1.
11. problem 2.
12. problem 3.

Draw a cumulative frequency polygon, showing both absolute and relative frequency, for the data of

13. problem 10.
14. problem 11.
15. problem 12.

Stem and Leaf Plots

Make a stem and leaf plot for the data of

16. problem 1.

17. problem 2.

18. problem 3.

26–3   Numerical Description of Data

We saw in Sec. 26–2 how we can describe a set of data by frequency distribution, a frequency histogram, a frequency polygon, or a cumulative frequency distribution. We can also describe a set of data with just a few numbers, and this more compact description is more convenient for some purposes. For example, if, in a report, you wanted to describe the heights of a group of students and did not want to give the entire frequency distribution, you might simply say:

The mean height is 58 inches, with a standard deviation of 3.5 inches.

The *mean* is a number that shows the *centre* of the data; the *standard deviation* is a measure of the *spread* of the data. As mentioned earlier, the mean and the standard deviation may be found either for an entire population (and are thus population parameters) or for a sample drawn from that population (and are thus sample statistics).

Thus to describe a population or a sample, we need numbers that give both the centre of the data and the spread. Further, for sample statistics, we need to give the uncertainty of each figure. This will enable us to make inferences about the larger population.

◆◆◆ **Example 15:** A student recorded the running times for a sample of participants in a race. From that sample she inferred (by methods we'll learn later) that for the entire population of racers

$$\text{mean time} = 23.65 \pm 0.84 \text{ minutes}$$
$$\text{standard deviation} = 5.83 \pm 0.34 \text{ minutes} \qquad ◆◆◆$$

The mean time is called a *measure of central tendency*. We show how to calculate the mean, and other measures of central tendency, later in this section. The standard deviation is called a *measure of dispersion*. We also show how to compute it, and other measures of dispersion, in this section. The "plus-or-minus" values show a degree of uncertainty called the *standard error*. The uncertainty ± 0.84 min is called the *standard error of the mean*, and the uncertainty ± 0.34 min is called the *standard error of the standard deviation*. We show how to calculate standard errors in Sec. 26–6.

Measures of Central Tendency: The Mean

Some common measures of the centre of a distribution are the mean, the median, and the mode. The arithmetic mean, or simply the *mean*, of a set of measurements is equal to the sum of the measurements divided by the number of measurements. It is what we commonly call the *average*. It is the most commonly used measure of central tendency. If our n measurements are $x_1, x_2, \ldots, x_n$, then

$$\Sigma x = x_1 + x_2 + \cdots + x_n$$

There is a story, probably untrue, about a statistician who drowned in a lake that had an average depth of 1 ft. (30 cm).

where we use the Greek symbol Σ (sigma) to represent "the sum of." The mean, which we call $\bar{x}$ (read "x bar"), is then given by the following:

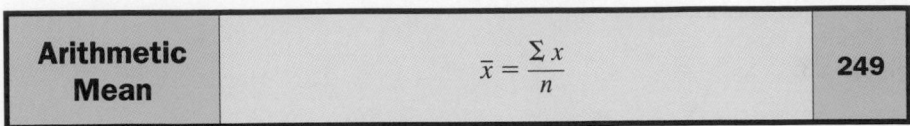

Arithmetic Mean	$\bar{x} = \dfrac{\Sigma x}{n}$	249

The arithmetic mean of n measurements is the *sum* of those measurements divided by n.

We can calculate the mean for a sample or for the entire population. However, we use a *different symbol* in each case.

Σ is the Greek capital letter "sigma".

$\bar{x}$ is the *sample* mean

μ (mu) is the *population* mean

◆◆◆ Example 16: Find the mean of the following sample:

$$746 \quad 574 \quad 645 \quad 894 \quad 736 \quad 695 \quad 635 \quad 794$$

Solution: Adding the values gives

$$\Sigma x = 746 + 574 + 645 + 894 + 736 + 695 + 635 + 794 = 5719$$

Then, with $n = 8$,

$$\bar{x} = \frac{5719}{8} = 715$$

rounded to three significant digits.   ◆◆◆

Weighted Mean

When not all of the data are of equal importance, we may compute a *weighted mean*, where the values are weighted according to their importance.

◆◆◆ Example 17: A student has grades of 83, 59, and 94 on 3 one-hour exams, a grade of 82 on the final exam which is equal in weight to 2 one-hour exams, and a grade of 78 on a laboratory report, which is worth 1.5 one-hour exams. Compute the weighted mean.

Solution: If a one-hour exam is assigned a weight of 1, then we have a total of the weights of

$$\Sigma w = 1 + 1 + 1 + 2 + 1.5 = 6.5$$

To get a weighted mean, we add the products of each grade and its weight, and divide by the total weight.

$$\text{weighted mean} = \frac{83(1) + 59(1) + 94(1) + 82(2) + 78(1.5)}{6.5} = 79.5 \qquad ◆◆◆$$

In general, the weighted mean is given by the following:

$$\text{weighted mean} = \frac{\Sigma\,(wx)}{\Sigma\,w}$$

Midrange

We have already noted that the range of a set of data is the difference between the highest and the lowest numbers in the set. The *midrange* is simply the value midway between the two extreme values.

$$\text{midrange} = \frac{\text{highest value} + \text{lowest value}}{2}$$

◆◆◆ **Example 18:** The midrange for the values

$$3, 5, 6, 6, 7, 11, 11, 15$$

is

$$\text{midrange} = \frac{3 + 15}{2} = 9$$

◆◆◆

Mode

Our next measure of central tendency is the *mode*.

Mode	The mode of a set of numbers is the value(s) that occur most often in the set.	251

A set of numbers can have no mode, one mode, or more than one mode.

◆◆◆ **Example 19:**

(a) The set   1   3   3   5   6   7   7   7   9   has the mode 7.
(b) The set   1   1   3   3   5   5   7   7   has no mode.
(c) The set   1   3   3   3   5   6   7   7   7   9   has two modes, 3 and 7.
                                                              It is called *bimodal*.   ◆◆◆

Median

To find the *median*, we simply arrange the data in order of magnitude and pick the *middle value*. For an even number of measurements, we take the mean of the two middle values.

Median	The median of a set of numbers arranged in order of magnitude is the middle value of an odd number of measurements, or the mean of the two middle values of an even number of measurements.	250

◆◆◆ **Example 20:** Find the median of the data in Example 16.

Solution: We rewrite the data in order of magnitude.

$$574 \quad 635 \quad 645 \quad 695 \quad 736 \quad 746 \quad 794 \quad 894$$

The two middle values are 695 and 736. Taking their mean gives

$$\text{median} = \frac{695 + 736}{2} = 715.5$$

◆◆◆

Five-Number Summary

The median of that half of a set of data from the minimum value up to and including the median is called the *lower hinge*. Similarly, the median of the upper half of the data is called the *upper hinge*.

The lowest value in a set of data, together with the highest value, the median, and the two hinges, is called a *five-number summary* of the data.

◆◆◆ **Example 21:** For the data

12   17   18   20   22   28   32   34   49   52   59   66

the median is $(28 + 32)/2 = 30$, so

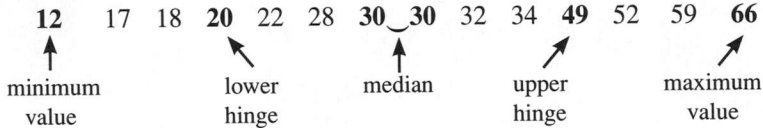

The five-number summary for this set of data may be written

[12, 20, 30, 49, 66]

◆◆◆

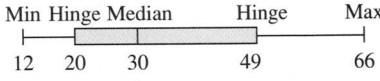

FIGURE 26–6 A boxplot. It is also called a *box and whisker* diagram.

Boxplots

A graph of the values in the five-number summary is called a *boxplot*.

◆◆◆ **Example 22:** A boxplot for the data of Example 21 is given in Fig. 26–6.   ◆◆◆

Measures of Dispersion

We will usually use the mean to describe a set of numbers, but used alone it can be misleading.

◆◆◆ **Example 23:** The sets of numbers

1   1   1   1   1   1   1   1   1   91

and the set

10   10   10   10   10   10   10   10   10   10

each have a mean value of 10 but are otherwise quite different. Each set has a sum of 100. In the first set most of this sum is concentrated in a single value, but in the second set the sum is *dispersed* among all the values.   ◆◆◆

Thus we need some *measure of dispersion* of a set of numbers. Four common ones are the range, the percentile range, the variance, and the standard deviation. We will cover each of these.

Range

We have already introduced the range in Sec. 26–2, and we define it here.

| **Range** | The range of a set of numbers is the difference between the largest and smallest numbers in the set. | **252** |

◆◆◆ **Example 24:** For the set of numbers

$$6 \quad 3 \quad 9 \quad 12 \quad 44 \quad 2 \quad 53 \quad 1 \quad 8$$

the largest value is 53 and the smallest is 1, so the range is

$$\text{range} = 53 - 1 = 52 \qquad \text{◆◆◆}$$

We sometimes give the range by stating the end values themselves. Thus, in Example 24 we might say that the range is from 1 to 53.

Quartiles, Deciles, and Percentiles

We have seen that for a set of data arranged in order of magnitude, the median divides the data into two equal parts. There are as many numbers below the median as there are above it.

Similarly, we can determine two more values that divide each half of the data in half again. We call these values *quartiles*. Thus one-fourth of the values will fall into each quartile. The quartiles are labelled Q_1, Q_2, and Q_3. Thus Q_2 is the median.

Those values that divide the data into 10 equal parts we call *deciles* and label D_1, D_2, ... Those values that divide the data into 100 equal parts are *percentiles*, labelled P_1, P_2, ... Thus the 25th percentile is the same as the first quartile ($P_{25} = Q_1$). The median is the 50th percentile, the fifth decile, and the second quartile ($P_{50} = D_5 = Q_2$). The 70th percentile is the seventh decile ($P_{70} = D_7$).

One measure of dispersion sometimes used is to give the range of values occupied by some given percentiles. Thus we have a *quartile range*, *decile range*, and *percentile range*. For example, the quartile range is the range from the first to the third quartile.

◆◆◆ **Example 25:** Find the quartile range for the set of data

$$2 \quad 4 \quad 5 \quad \quad 7 \quad 8 \quad 11 \quad \quad 15 \quad 18 \quad 19 \quad \quad 21 \quad 24 \quad 25$$

Solution: The quartiles are

$$Q_1 = \frac{5 + 7}{2} = 6$$

$$Q_2 = \text{the median} = \frac{11 + 15}{2} = 13$$

$$Q_3 = \frac{19 + 21}{2} = 20$$

Then

$$\text{quartile range} = Q_3 - Q_1 = 20 - 6 = 14 \qquad \text{◆◆◆}$$

Note that the quartiles are not in the same locations as the upper and lower hinges. The hinges here would be at 7 and 19.

We see that half the values in Example 25 fall within the quartile range. We may similarly compute the range of any percentiles or deciles. The range from the 10th to the 90th percentile is the one commonly used.

Variance

We now give another measure of dispersion called the *variance*, but we must first mention deviation. We define the *deviation* of any number x in a set of data as the difference between that number and the mean $\bar{x}$ of that set of data.

◆◆◆ **Example 26:** A certain set of measurements has a mean of 48.3. What are the deviations of the values 24.2 and 69.3 in that set?

Solution: The deviation of 24.2 is

$$24.2 - 48.3 = -24.1$$

and the deviation of 69.3 is

$$69.3 - 48.3 = 21.0$$ ◆◆◆

To get the *variance* of a population of n numbers, we add up the squares of the deviations of each number in the set and divide by n.

σ is the Greek small letter sigma.

Population Variance	$\sigma^2 = \dfrac{\Sigma(x - \bar{x})^2}{n}$	253

To find the variance of a *sample*, it is more accurate to divide by $n - 1$ rather than n. As with the mean, we use one symbol for the sample variance and a different symbol for the population variance.

s^2 is the *sample* variance
σ^2 is the *population* variance

For large samples (over 30 or so), the variance found by either formula is practically the same.

Sample Variance	$s^2 = \dfrac{\Sigma(x - \bar{x})^2}{n - 1}$	254

◆◆◆ Example 27: Compute the variance for the population

1.74   2.47   3.66   4.73   5.14   6.23   7.29   8.93   9.56

Solution: We first compute the mean, $\bar{x}$.

$$\bar{x} = \frac{1.74 + 2.47 + 3.66 + 4.73 + 5.14 + 6.23 + 7.29 + 8.93 + 9.56}{9}$$

$$= 5.53$$

We then subtract the mean from each of the nine values to obtain deviations. The deviations are then squared and added, as shown in Table 26–6.

TABLE 26–6

Measurement x	Deviation $x - \bar{x}$	Deviation Squared $(x - \bar{x})^2$
1.74	−3.79	14.36
2.47	−3.06	9.36
3.66	−1.87	3.50
4.73	−0.80	0.64
5.14	−0.39	0.15
6.23	0.70	0.49
7.29	1.76	3.10
8.93	3.40	11.56
9.56	4.03	16.24
$\Sigma x = 49.75$	$\Sigma(x - \bar{x}) = -0.02$	$\Sigma(x - \bar{x})^2 = 59.41$

The variance σ^2 is then

$$\sigma^2 = \frac{59.41}{9} = 6.60$$ ◆◆◆

Standard Deviation

Once we have the variance, it is a simple matter to get the *standard deviation*. It is the most common measure of dispersion.

Standard Deviation	The standard deviation of a set of numbers is the positive square root of the variance.	255

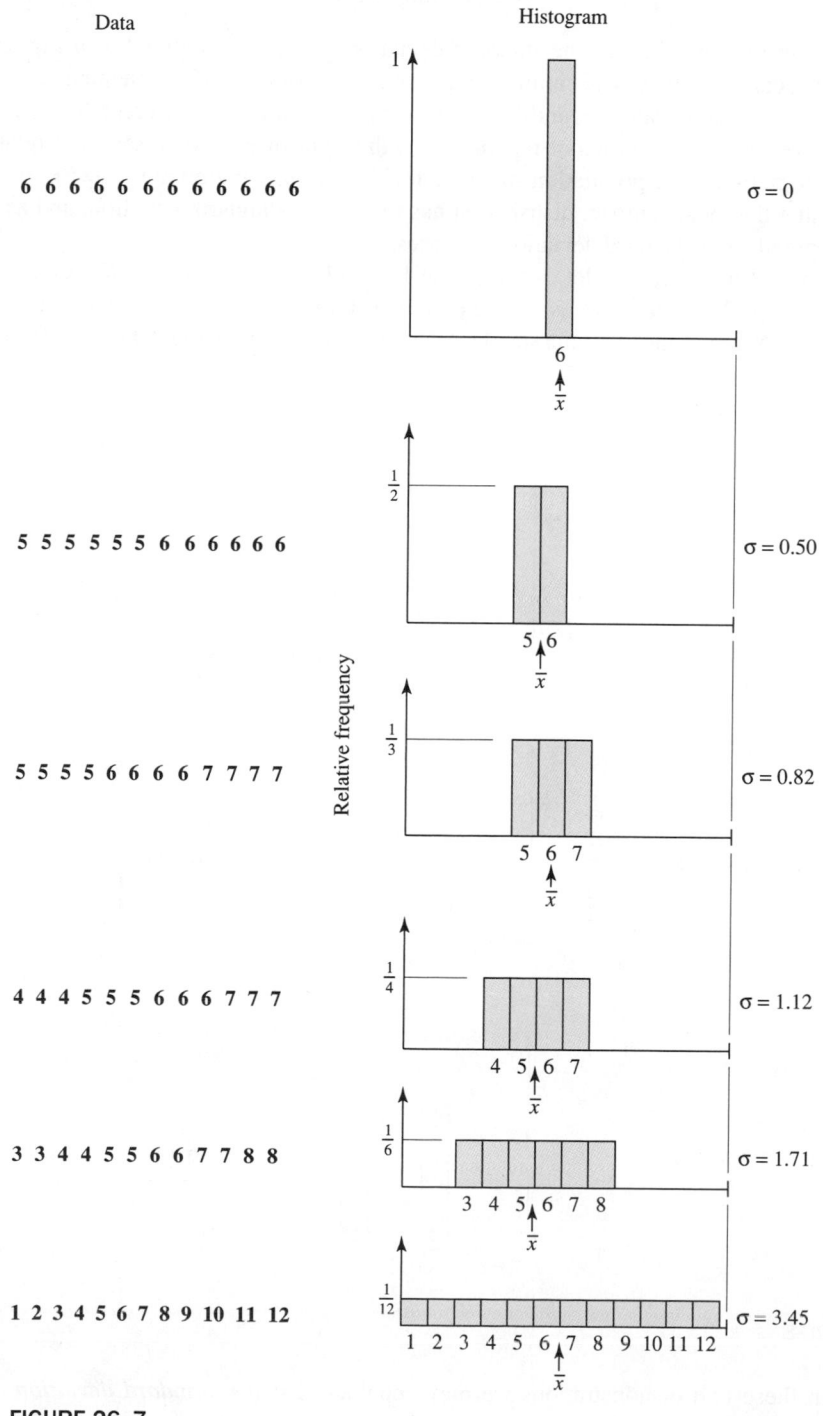

FIGURE 26–7

As before, we use *s* for the *sample* standard deviation and σ for the *population* standard deviation.

◆◆◆ **Example 28:** Find the standard deviation for the data of Example 27.

Solution: We have already found the variance in Example 27.

$$\sigma^2 = 6.60$$

Taking the square root gives the standard deviation.

$$\sigma = \sqrt{6.60} = 2.57 \qquad\qquad ◆◆◆$$

To get an intuitive feel for the standard deviation, we have computed it in Fig. 26–7 for several data sets consisting of 12 numbers that can range from 1 to 12. In the first set, all of the numbers have the same value, 6, and in the last set every number is different. The data sets in between have differing amounts of repetition. To the right of each data set are a relative frequency histogram and the population standard deviation (computation not shown).

Note that the most compact distribution has the lowest standard deviation, and as the distribution spreads, the standard deviation increases.

For a final demonstration, let us again take 12 numbers in two groups of six equal values, as shown in Fig. 26–8. Now let us separate the two groups, first by one interval and then by two intervals. Again, notice that the standard deviation increases as the data move further from the mean.

> The mean shifts slightly from one distribution to the next, but this does not affect our conclusions.

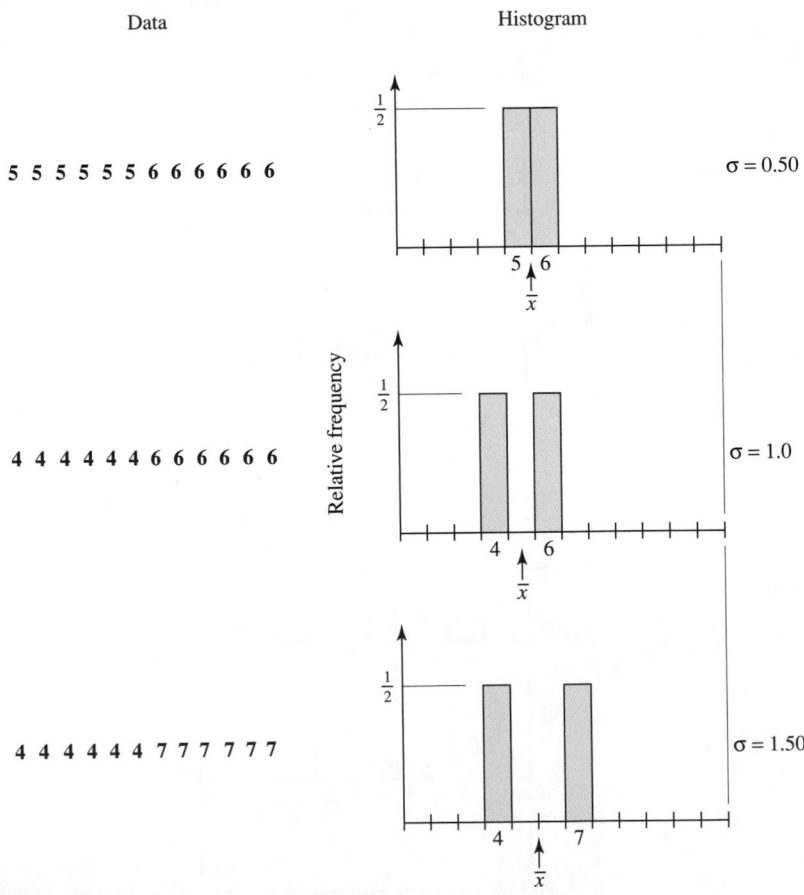

FIGURE 26-8

From these two demonstrations we may conclude that *the standard deviation increases whenever data move away from the mean.*

Exercise 3 ◆ Numerical Description of Data

Mean

1. Find the mean of the following set of grades:

   85 74 69 59 60 96 84 48 89 76 96 68 98 79 76

2. Find the mean of the following set of weights:

   173 127 142 164 163 153 116 199

3. Find the mean of the weights in problem 1 of Exercise 2.
4. Find the mean of the times in problem 2 of Exercise 2
5. Find the mean of the prices in problem 3 of Exercise 2.

Weighted Mean

6. A student's grades and the weight of each grade are given in the following table. Find their weighted mean.

	Grade	Weight
Hour exam	83	5
Hour exam	74	5
Quiz	93	1
Final exam	79	10
Report	88	7

7. A student receives hour-test grades of 86, 92, 68, and 75, a final exam grade of 82, and a project grade of 88. Find the weighted mean if each hour-test counts for 15% of his grade, the final exam counts for 30%, and the project counts for 10%.

Midrange

8. Find the midrange of the grades in problem 1.
9. Find the midrange of the weights in problem 2.

Mode

10. Find the mode of the grades in problem 1.
11. Find the mode of the weights in problem 2.
12. Find the mode of the weights in problem 1 of Exercise 2.
13. Find the mode of the times in problem 2 of Exercise 2.
14. Find the mode of the prices in problem 3 of Exercise 2.

Median

15. Find the median of the grades in problem 1.
16. Find the median of the weights in problem 2.
17. Find the median of the weights in problem 1 of Exercise 2.

18. Find the median of the times in problem 2 of Exercise 2.
19. Find the median of the prices in problem 3 of Exercise 2.

Five-Number Summary

20. Give the five-number summary for the grades in problem 1.
21. Give the five-number summary for the weights in problem 2.

Boxplot

22. Make a boxplot using the results of problem 20.
23. Make a boxplot using the results of problem 21.

Range

24. Find the range of the grades in problem 1.
25. Find the range of the weights in problem 2.

Percentiles

26. Find the quartiles and give the quartile range of the following data:

        28   39   46   53   69   71   83   94   102   117   126

27. Find the quartiles and give the quartile range of the following data:

        1.33   2.28   3.59   4.96   5.23   6.89

        7.91   8.13   9.44   10.6   11.2   12.3

Variance and Standard Deviation

28. Find the variance and standard deviation of the grades in problem 1. Assume that these grades are a sample drawn from a larger population.
29. Find the variance and standard deviation of the weights in problem 2. Assume that these weights are a sample drawn from a larger population.
30. Find the population variance and standard deviation of the weights in problem 1 of Exercise 2.
31. Find the population variance and standard deviation of the times in problem 2 of Exercise 2.
32. Find the population variance and standard deviation of the prices in problem 3 of Exercise 2.

26–4   Introduction to Probability

Why do we need probability to learn statistics? In Sec. 26–3 we learned how to compute certain sample statistics, such as the mean and the standard deviation. Knowing, for example, that the mean height $\bar{x}$ of a sample of students is 69.5 in., we might *infer* that the mean height μ of the entire population is 69.5 in. But how reliable is that number? Is μ equal to 69.5 in. *exactly?* We'll see later that we give population parameters such as μ as a *range* of values, say, 69.5 ± 0.6 in., and we state the *probability* that the true mean lies within that range. We might say, for example, that there is a 68% chance that the true value lies within the stated range.

Further, we may report that the sample standard deviation is, say, 2.6 in. What does that mean? Of 1000 students, for example, how many can be expected to have a height falling within, say, 1 standard deviation of the mean height at that school? We use probability to help us answer questions such as that.

Further, statistics are often used to "prove" many things, and it takes a knowledge of probability to help decide whether the claims are valid or are merely a result of chance.

✦✦✦ **Example 29:** Suppose that a nurse measures the heights of 14-year-old students in a town located near a chemical plant. Of 100 students measured, she finds 75 students to be shorter than 59 in. Parents then claim that this is caused by chemicals in the air, but the company claims that such heights can occur by chance. We cannot resolve this question without a knowledge of probability. ✦✦✦

We distinguish between *statistical* probability and *mathematical* probability.

✦✦✦ **Example 30:** We can toss a die 6000 times and count the number of times that "one" comes up, say, 1006 times. We conclude that the chance of tossing a "one" is 1006 out of 6000, or about 1 in 6. This is an example of what is called *statistical* probability.

On the other hand, we may reason, without even tossing the die, that each face has an equal chance of coming up. Since there are six faces, the chance of tossing a "one" is 1 in 6. This is an example of *mathematical* probability. ✦✦✦

We first give a brief introduction to mathematical probability and then apply it to statistics.

Probability of a Single Event Occurring

Suppose that an event A can happen in m ways out of a total of n equally likely ways. The probability $P(A)$ of A occurring is defined as follows:

Probability of a Single Event Occurring	$P(A) = \dfrac{m}{n}$	256

Note that the probability $P(A)$ is a number between 0 and 1. $P(A) = 0$ means that there is no chance that A will occur. $P(A) = 1$ means that it is certain that A will occur. Also note that if the probability of A occurring is $P(A)$, the probability of A *not* occurring is $1 - P(A)$.

✦✦✦ **Example 31:** A bag contains 15 balls, 6 of which are red. Assuming that any ball is as likely to be drawn as any other, the probability of drawing a red ball from the bag is $\frac{6}{15}$ or $\frac{2}{5}$. The probability of drawing a ball that is not red is $\frac{9}{15}$ or $\frac{3}{5}$. ✦✦✦

✦✦✦ **Example 32:** A loonie and a quarter are each tossed once. Determine (a) the probability $P(A)$ that both coins will show heads and (b) the probability $P(B)$ that not both coins will show heads.

Solution: The possible outcomes of a single toss of each coin are as follows:

Loonie	Quarter
Heads	Heads
Heads	Tails
Tails	Heads
Tails	Tails

In this book, weights and measures are predominantly given in SI. This reflects current practice and education policy in Canada. We recognize, however, that in the realm of personal measurement (like height and weight) the imperial measures of feet, inches, and pounds are still common. Therefore, a number of examples and exercises in this chapter use imperial measures.

Thus there are four equally likely outcomes. For event A (both heads), there is only one favorable outcome (heads, heads). Thus

$$P(A) = \frac{1}{4} = 0.25$$

Event B (that not both coins show heads) can happen in three ways (HT, TH, and TT). Thus $P(B) = \frac{3}{4} = 0.75$. Or, alternatively,

$$P(B) = 1 - P(A) = 0.75 \qquad \text{◆◆◆}$$

Probability of Two Events Both Occurring

If $P(A)$ is the probability that A will occur, and $P(B)$ is the probability that B will occur, the probability $P(A, B)$ that *both* A and B will occur is as follows:

Probability of Two Events Both Occurring	$P(A, B) = P(A)P(B)$	257

Independent events are those in which the outcome of the first in no way affects the outcome of the second.

The probability that two independent events will both occur is the product of their individual probabilities.

The rule above can be extended as follows:

Probability of Several Events All Occurring	$P(A, B, C, \dots) = P(A)P(B)P(C)\dots$	258

The probability that several independent events, $A, B, C, \dots$, will all occur is the product of the probabilities for each.

◆◆◆ **Example 33:** What is the probability that in three tosses of a coin, all will be heads?

Solution: Each toss is independent of the others, and the probability of getting a head on a single toss is $\frac{1}{2}$. Thus the probability of three heads being tossed is

$$\left(\frac{1}{2}\right)\left(\frac{1}{2}\right)\left(\frac{1}{2}\right) = \frac{1}{8}$$

Why subtract $P(A,B)$ in this formula? Since $P(A)$ includes all occurrences of A, it must include $P(A,B)$. Since $P(B)$ includes all occurrences of B, it also includes $P(A,B)$. Thus if we add $P(A)$ and $P(B)$ we will mistakenly be counting $P(A,B)$ twice.

For example, if we count all aces and all spades in a deck of cards, we get 16 cards (13 spades plus 3 additional aces). But if we had added the number of aces (4) to the number of spades (13), we would have gotten the incorrect answer of 17 by counting the ace of spades twice.

or 1 in 8. If we were to make, say, 800 sets of tosses, we would expect to toss three heads about 100 times.

◆◆◆

Probability of Either of Two Events Occurring

If $P(A)$ and $P(B)$ are the probabilities that events A and B will occur, the probability $P(A + B)$ that *either A or B* or both will occur is as follows:

Probability of Either or Both of Two Events Occurring	$P(A + B) = P(A) + P(B) - P(A, B)$	259

The probability that either A or B or both will occur is the sum of the individual probabilities for A and B, less the probability that both will occur.

◆◆◆ **Example 34:** In a certain cup production line, 6% are chipped, 8% are cracked, and 2% are both chipped and cracked. What is the probability that a cup picked at random will be either chipped or cracked, or both chipped and cracked?

Solution: If

$$P(C) = \text{probability of being chipped} = 0.06$$

and

$$P(S) = \text{probability of being cracked} = 0.08$$

and

$$P(C, S) = 0.02$$

then

$$P(C + S) = 0.06 + 0.08 - 0.02 = 0.12$$

We would thus predict that 12 cups out of 100 chosen would be chipped or cracked, or both chipped and cracked. ◆◆◆

Probability of Two Mutually Exclusive Events Occurring

Mutually exclusive events A and B are those that cannot both occur at the same time. Thus the probability $P(A, B)$ of both A and B occurring is zero.

◆◆◆ **Example 35:** A tossed coin lands either heads or tails. Thus the probability $P(H, T)$ of a single toss being both heads and tails is

$$P(H, T) = 0$$

◆◆◆

Setting $P(A, B) = 0$ in Eq. 259 gives the following equation:

Probability of Two Mutually Exclusive Events Occurring	$P(A + B) = P(A) + P(B)$	260

The probability that either of two mutually exclusive events will occur in a certain trial is the sum of the two individual probabilities.

We can generalize this to n mutually exclusive events. The probability that one of n mutually exclusive events will occur in a certain trial is the sum of the n individual probabilities.

◆◆◆ **Example 36:** In a certain school, 10% of the students have red hair, and 20% have black hair. If we select one student at random, what is the probability that this student will have either red hair or black hair?

Solution: The probability $P(A)$ of having red hair is 0.1, and the probability $P(B)$ of having black hair is 0.2. The two events (having red hair or having black hair) are mutually exclusive. Thus the probability $P(A + B)$ of a student having either red or black hair is

$$P(A + B) = 0.1 + 0.2 = 0.3$$

◆◆◆

Random Variables

In Sec. 26–1 we defined continuous, discrete, and categorical variables. Now we define a *random* variable.

Suppose that there is a process or an experiment whose outcome is a real number X. If the process is repeated, then X can assume specific values, each determined by chance and each with a certain probability of occurrence. X is then called a *random variable*.

◆◆◆ **Example 37:** The process of rolling a pair of dice once results in a single numerical value X, whose value is determined by chance but whose value has a definite probability of occurrence. Repeated rollings of the dice may result in other values of X, which can take on the values

$$2, 3, 4, 5, 6, 7, 8, 9, 10, 11, 12$$

Here X is a random variable.   ◆◆◆

Discrete and Continuous Random Variables

A *discrete* random variable can take on a finite number of variables. Thus the value X obtained by rolling a pair of dice is a discrete random variable.

A *continuous* random variable may have infinitely many values on a continuous scale, with no gaps.

◆◆◆ **Example 38:** An experiment measuring the temperatures T in major Canadian cities at any given hour gives values that are determined by chance, depending on the whims of nature, but each value has a definite probability of occurrence. The random variable T is continuous because temperatures are read on a continuous scale.   ◆◆◆

Probability Distributions

A probability distribution is sometimes called a *probability density function*, a *relative frequency function*, or a *frequency function*.

If a process or an experiment has X different outcomes, where X is a discrete random variable, we say that the probability of a particular outcome is $P(X)$. If we calculate $P(X)$ for each X, we have what is called a *probability distribution*. Such a distribution may be presented as a table, a graph, or a formula.

◆◆◆ **Example 39:** Make a probability distribution for a single roll of a single die.

Solution: Let X equal the number on the face that turns up, and $P(X)$ be the probability of that face turning up. That is, $P(5)$ is the probability of rolling a five. Since each number has an equal chance of being rolled, we have the following:

X	$P(X)$
1	$\frac{1}{6}$
2	$\frac{1}{6}$
3	$\frac{1}{6}$
4	$\frac{1}{6}$
5	$\frac{1}{6}$
6	$\frac{1}{6}$

We graph this probability distribution in Fig. 26–9.

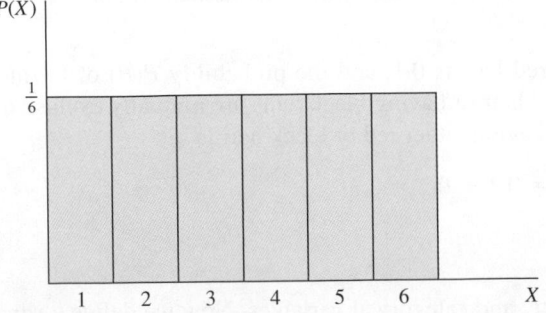

FIGURE 26–9 Probability distribution for the toss of a single die.   ◆◆◆

♦♦♦ **Example 40:** Make a probability distribution for the two-coin toss experiment of Example 32.

Solution: Let X be the number of heads tossed, and $P(X)$ be the probability of X heads being tossed. The experiment has four possible outcomes (HH, HT, TH, TT). The outcome $X = 0$ (no heads) can occur in only one way (TT), so the probability $P(0)$ of not tossing a head is one in four, or $\frac{1}{4}$. The probabilities of the other outcomes are found in a similar way. They are as follows:

X	$P(X)$
0	$\frac{1}{4}$
1	$\frac{1}{2}$
2	$\frac{1}{4}$

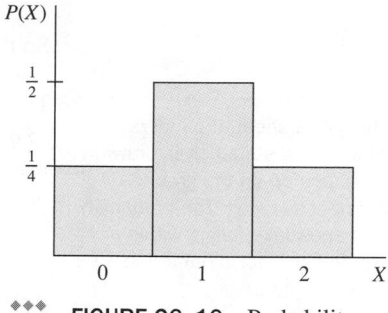

♦♦♦ **FIGURE 26–10** Probability distribution.

They are shown graphically in Fig. 26–10.

Probabilities as Areas on a Probability Distribution

The probability of an event occurring can be associated with a geometric *area* of the probability distribution or relative frequency distribution.

♦♦♦ **Example 41:** For the distribution in Fig. 26–10, each bar has a width of 1 unit, so the total shaded area is

$$1\left(\frac{1}{4}\right) + 1\left(\frac{1}{2}\right) + \left(\frac{1}{4}\right) = 1 \text{ square unit}$$

The bar representing the probability of tossing a "zero" has an area of $\frac{1}{4}$, which is equal to the probability of that event occurring. The areas of the other two bars also give the value of their probabilities.   ♦♦♦

Probabilities from a Relative Frequency Histogram

Suppose we were to repeat the two-coin toss experiment of Example 40 a very large number of times and draw a relative frequency histogram of our results. That relative frequency histogram *would be identical to the probability distribution of Fig. 26–10.* The relative frequency of tossing a "zero" would be $\frac{1}{4}$, of a "one" would be $\frac{1}{2}$, and of a "two" would be $\frac{1}{4}$.

We thus may think of a probability distribution as being the limiting value of a relative frequency distribution as the number of observations increases—in fact, the distribution for a *population*.

Thus we can obtain probabilities from a relative frequency histogram for a population or from one obtained using a large number of observations.

♦♦♦ **Example 42:** For the relative frequency histogram for the population of students' grades (Fig. 26–3), find the probability that a student chosen at random will have received a grade between 92 and 98.

Solution: The given grades are represented by the bar having a class midpoint of 95 and a relative frequency of 6.7%. Since the relative frequency distribution for a population probability is no different from the probability distribution, the grade of a student chosen at random has a 6.7% chance of falling within that class.   ♦♦♦

Probability of a Variable Falling between Two Limits

Further, the probability of a variable falling *between two limits* is given by the area of the probability distribution or relative frequency histogram for that variable, between those limits.

◆◆◆ **Example 43:** Using the relative frequency histogram shown in Fig. 26–3, find the chance of a student chosen at random having a grade that falls between 62 and 80.

Solution: This range spans three classes in Fig. 26–3, having relative frequencies of 6.7%, 13.3%, and 23.3%. The chance of falling within one of the three is

$$6.7 + 13.3 + 23.3 = 43.3\%$$

So the probability is 0.433.   ◆◆◆

Thus we have shown that *the area between two limits on a probability distribution or relative frequency histogram for a population is a measure of the probability of a measurement falling within those limits.* This idea will be used again when we compute probabilities from a continuous probability distribution such as the normal curve.

> Think of a sheet of paper on which the distribution is drawn lying face up on the ground during a snowfall. The probability of a snowflake falling within any region is proportional to the area of that region.

Binomial Experiments

A *binomial experiment* is one that can have only two possible outcomes.

◆◆◆ **Example 44:** The following are examples of binomial experiments:

(a) A tossed coin shows either heads or tails.
(b) A manufactured part is either accepted or rejected.
(c) A voter either chooses Smith for Congress or does not choose Smith for Congress.   ◆◆◆

Binomial Probability Formula

We call the occurrence of an event a *success* and the nonoccurrence of that event a *failure*. Let us denote the probability of success of a single trial of a binomial experiment by p. Then the probability of failure, q, must be $(1 - p)$.

◆◆◆ **Example 45:** If the probability is 80% ($p = 0.8$) that a single part drawn at random from a production line is accepted, then the probability is 20% ($q = 1 - 0.8$) that a part is rejected.   ◆◆◆

If we repeat a binomial experiment n times, we will have x successes, where

$$x = 0, 1, 2, 3, \ldots, n$$

The probability $P(x)$ of having x successes in n trials is given by the following:

> There are tables given in statistics books that can be used instead of this formula.

Binomial Probability Formula	$P(x) = \dfrac{n!}{(n - x)!\, x!}\, p^x\, q^{n-x}$	262

> Does this formula look a little familiar? We will soon connect the binomial distribution with what we have already learned about the binomial theorem.

◆◆◆ **Example 46:** For the production line of Example 45, find the probability that of 10 parts taken at random from the line, exactly 6 will be acceptable.

Solution: Here, $n = 10$, $x = 6$, and $p = 0.8$, $q = 0.2$, from before. Substituting into the formula gives

$$P(6) = \frac{10!}{(10 - 6)!\, 6!}\, (0.8)^6\, (0.2)^{10-6}$$

$$= \frac{10!}{4!\, 6!}\, (0.8)^6\, (0.2)^4 = 0.0881$$

or about 9%.   ◆◆◆

Binomial Distribution

In Example 46, we computed a single probability, that 6 out of 10 parts drawn at random will be acceptable. We can also compute the probabilities of having 0, 1, 2, . . . , 10 acceptable parts in a group of 10. If we compute all 10 possible probabilities, we will then have a *probability distribution*.

◆◆◆ **Example 47:** For the production line of Example 45, find the probability that of 10 parts taken at random from the line, 0, 1, 2, . . . , 10 will be acceptable.

Solution: Computing the probabilities in the same way as in the preceding example (work not shown), we get the following binomial distribution:

x	$P(x)$
0	0
1	0
2	0.0001
3	0.0008
4	0.0055
5	0.0264
6	0.0881
7	0.2013
8	0.3020
9	0.2684
10	0.1074
Total	1.0000

This binomial distribution is shown graphically in Fig. 26–11.

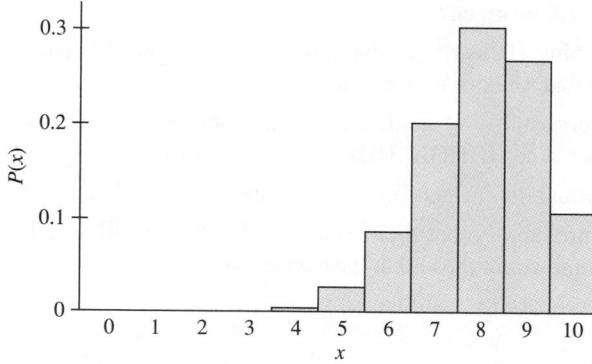

FIGURE 26–11 ◆◆◆

Binomial Distribution Obtained by Binomial Theorem

The n values in a binomial distribution for n trials may be obtained by using the binomial theorem to expand the binomial

$$(q + p)^n$$

where p is the probability of success of a single trial and $q = (1 - p)$ is the probability of failure of a single trial, as before.

Refer to Sec. 25–6 for a refresher on the binomial theorem.

◆◆◆ **Example 48:** Use the binomial theorem, Eq. 245, to obtain the binomial distribution of Example 47.

Solution: We expand $(q + p)^n$ using $q = 0.2$, $p = 0.8$, and $n = 10$.

$$(0.2 + 0.8)^{10}$$

$$= (0.2)^{10} (0.8)^0 + 10(0.2)^9 (0.8)^1 + \frac{10(9)}{2!} (0.2)^8 (0.8)^2$$

$$+ \frac{10(9)(8)}{3!} (0.2)^7 (0.8)^3 + \frac{10(9)(8)(7)}{4!} (0.2)^6 (0.8)^4$$

$$+ \frac{10(9)(8)(7)(6)}{5!} (0.2)^5 (0.8)^5 + \frac{10(9)(8)(7)(6)(5)}{6!} (0.2)^4 (0.8)^6$$

$$+ \frac{10(9)(8)(7)(6)(5)(4)}{7!} (0.2)^3 (0.8)^7 + \frac{10(9)(8)(7)(6)(5)(4)(3)}{8!} (0.2)^2 (0.8)^8$$

$$+ \frac{10(9)(8)(7)(6)(5)(4)(3)(2)}{9!} (0.2)^1 (0.8)^9 + \frac{10!}{10!} (0.2)^0 (0.8)^{10}$$

$$= 0 + 0 + 0.0001 + 0.0008 + 0.0055 + 0.0264 + 0.0881$$

$$+ 0.2013 + 0.3020 + 0.2684 + 0.1074$$

As expected, we get the same values as given by binomial probability formula.   ◆◆◆

Exercise 4 ✦ Introduction to Probability

Probability of a Single Event

1. We draw a ball from a bag that contains 8 green balls and 7 blue balls. What is the probability that a ball drawn at random will be green?

2. A card is drawn from a deck containing 13 hearts, 13 diamonds, 13 clubs, and 13 spades. What is the chance that a card drawn at random will be a heart?

3. If we toss four coins, what is the probability of getting two heads and two tails? [*Hint:* This experiment has 16 possible outcomes (HHHH, HHHT, . . . , TTTT).]

4. If we toss four coins, what is the probability of getting one head and three tails?

5. If we throw two dice, what is the probability that their sum is 9? [Hint: List all possible outcomes, (1, 1), (1, 2), and so on, and count those that have a sum of 9.]

6. If two dice are thrown, what is the probability that their sum is 7?

Probability That Several Events Will All Occur

7. A die is rolled twice. What is the probability that both rolls will give a six?

8. We draw four cards from a deck, replacing each before the next is drawn. What chance is there that all four draws will be a red card?

Probability That Either of Two Events Will Occur

9. At a certain school, 55% of the students have brown hair, 15% have blue eyes, and 7% have both brown hair and blue eyes. What is the probability that a student chosen at random will have either brown hair or blue eyes, or both brown hair and blue eyes?

10. Find the probability that a card drawn from a deck will be either a "picture" card (jack, queen, or king) or a spade picture card.

Probability of Mutually Exclusive Events

11. What chance is there of tossing a sum of 7 or 9 when two dice are tossed? You may use your results from problems 5 and 6 here.

12. On a certain production line, 5% of the parts are underweight and 8% are overweight. What is the probability that one part selected at random is either underweight or overweight?

Probability Distributions

13. For the relative frequency histogram for the population of students' grades (Fig. 26–3), find the probability that a student chosen at random will have gotten a grade between 68 and 74.

14. Using the relative frequency histogram of Fig. 26–3, find the chance of a student chosen at random having a grade between 74 and 86.

Binomial Distribution

A binomial experiment is repeated n times, with a probability p of success on one trial. Find the probability $P(x)$ of x successes, if:

15. $n = 5$, $p = 0.4$, and $x = 3$.

16. $n = 7$, $p = 0.8$, and $x = 4$.

17. $n = 7$, $p = 0.8$, and $x = 6$.

18. $n = 9$, $p = 0.3$, and $x = 6$.

For problems 19 and 20, find all of the terms in each probability distribution, and graph.

19. $n - 5$, $p - 0.8$ **20.** $n = 6$, $p = 0.3$

21. What is the probability of tossing 7 heads in 10 tosses of a fair coin?

22. If male and female births are equally likely, what is the probability of five births being all girls?

23. If the probability of producing a defective disk is 0.15, what is the probability of getting 10 bad disks in a sample of 15?

24. A certain multiple-choice test has 20 questions, each of which has four choices, only one of which is correct. If a student were to guess every answer, what is the probability of getting 10 correct?

26–5   The Normal Curve

In the preceding section, we showed some discrete probability distributions and studied the binomial distribution, one of the most useful of the discrete distributions. Now we introduce *continuous* probabilities and go on to study the most useful of these, the *normal curve*.

Continuous Probability Distributions

Suppose now that we collect data from a "large" population. For example, suppose that we record the weight of every university student in a certain province. Since the population is large, we can use very narrow class widths and still have many observations that fall within each class. If we do so, our relative frequency histogram, Fig. 26–12(a), will have very narrow bars, and our relative frequency polygon, Fig. 26–12(b), will be practically a *smooth curve*.

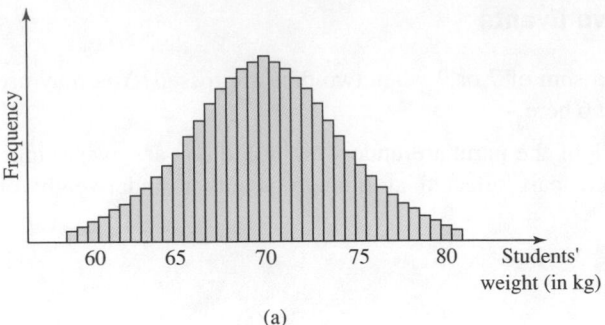

(a)

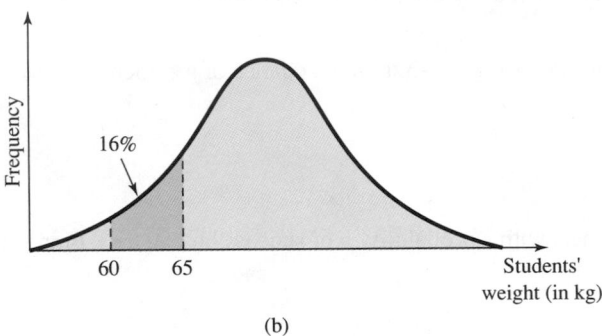

(b)

FIGURE 26–12

We get probabilities from the nearly smooth curve just as we did from the relative frequency histogram. *The probability of an observation falling between two limits* a *and* b *is proportional to the area bounded by the curve and the* x *axis, between those limits.* Let us designate the total area under the curve as 100% (or $P = 1$). Then the probability of an observation falling between two limits is equal to the area between those two limits.

◆◆◆ **Example 49:** Suppose that 16% of the entire area under the curve in Fig. 26–12(b) lies between the limits 60 and 65. That means there is a 16% chance ($P = 0.16$) that one student chosen at random will weigh between 60 and 65 kg.

Normal Distribution

The frequency distribution for many real-life observations, such as a person's height, is found to have a typical "bell" shape. We have seen that the chance of one observation falling between two limits is equal to the area under the relative frequency histogram [such as Fig. 26–12(a)] between those limits. But how are we to get those areas for a curve such as in Fig. 26–12(b)?

One way is to make a *mathematical model* of the probability distribution and from it get the areas of any section of the distribution. One such mathematical model is called the *normal distribution, normal curve,* or *Gaussian distribution.*

The normal distribution is obtained by plotting the following equation:

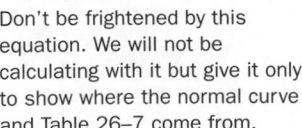

The *Gaussian distribution* is named for Carl Friedrich Gauss (1777–1855), whose *Gauss-Seidel method* we studied in Chapter 10. Some consider him to be one of the greatest mathematicians of all time. Many theorems and proofs are named after him.

Don't be frightened by this equation. We will not be calculating with it but give it only to show where the normal curve and Table 26–7 come from.

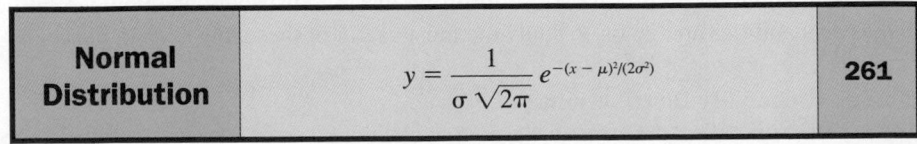

| Normal Distribution | $y = \dfrac{1}{\sigma \sqrt{2\pi}} e^{-(x - \mu)^2/(2\sigma^2)}$ | 261 |

The normal curve, shown in Fig. 26–13, is bell-shaped and is symmetrical about its mean, μ. Here, σ is the standard deviation of the population. This curve closely matches many frequency distributions in the real world.

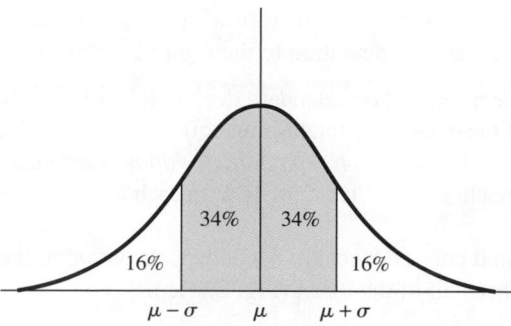

FIGURE 26–13   The normal curve.

To use the normal curve to determine probabilities, we must know the areas between any given limits. Since the equation of the curve is known, it is possible, by methods too advanced to show here, to compute these areas. They are given in Table 26–7. The areas are given "within *z* standard deviations of the mean." This means that each area given is that which lies between the mean and z standard deviations *on one side* of the mean, as shown in Fig. 26–14.

TABLE 26–7   Areas under the normal curve within *z* standard deviations of the mean

z	Area	*z*	Area
0.1	0.0398	2.1	0.4821
0.2	0.0793	2.2	0.4861
0.3	0.1179	2.3	0.4893
0.4	0.1554	2.4	0.4918
0.5	0.1915	2.5	0.4938
0.6	0.2258	2.6	0.4952
0.7	0.2580	2.7	0.4965
0.8	0.2881	2.8	0.4974
0.9	0.3159	2.9	0.4981
1.0	0.3413	3.0	0.4987
1.1	0.3643	3.1	0.4990
1.2	0.3849	3.2	0.4993
1.3	0.4032	3.3	0.4995
1.4	0.4192	3.4	0.4997
1.5	0.4332	3.5	0.4998
1.6	0.4452	3.6	0.4998
1.7	0.4554	3.7	0.4999
1.8	0.4641	3.8	0.4999
1.9	0.4713	3.9	0.5000
2.0	0.4772	4.0	0.5000

More extensive tables can be found in most statistics books.

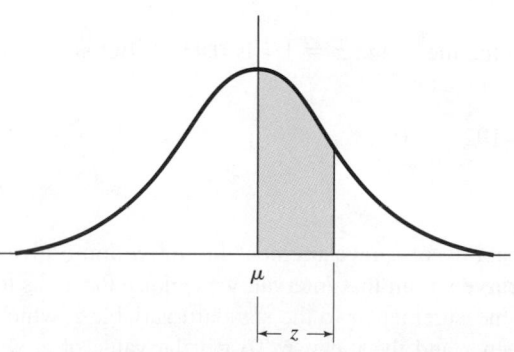

FIGURE 26–14

The number *z* is called the *standardized variable,* or *normalized variable*. When used for educational testing, it is often called the *standard score,* or *z score*. Since the normal curve is symmetrical about the mean, the given values apply to either the left or the right side.   ◆◆◆

◆◆◆ **Example 50:** Find the area under the normal curve between the following limits: from one standard deviation to the left of the mean, to one standard deviation to the right of the mean.

Solution: From Table 26–7, the area between the mean and one standard deviation is 0.3413. By symmetry, the same area lies to the other side of the mean. The total is thus 2(0.3413), or 0.6826. Thus *about 68% of the area under the normal curve lies within one standard deviation of the mean* (Fig. 26–13). Therefore 100-68 or 32% of the area lies in the "tails," or 16% in each tail.   ◆◆◆

◆◆◆ **Example 51:** Find the area under the normal curve between 0.8 standard deviation to the left of the mean, to 1.1 standard deviations to the right of the mean (Fig. 26–15).

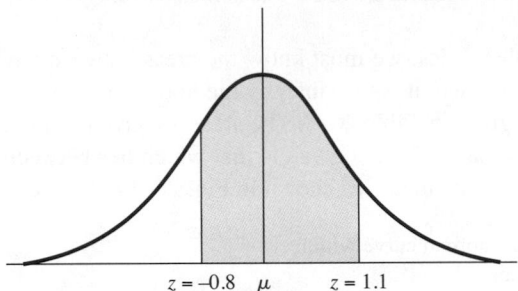

$z = -0.8$ μ $z = 1.1$

FIGURE 26–15

Solution: From Table 26–7, the area between the mean and 0.8 standard deviation ($z = 0.8$) is 0.2881. Also, the area between the mean and 1.1 standard deviations ($z = 1.1$) is 0.3643. The combined area is thus

$$0.2881 + 0.3643 = 0.6524$$

or 65.24% of the total area.   ◆◆◆

◆◆◆ **Example 52:** Find the area under the tail of the normal curve to the right of $z = 1.4$ (Fig. 26–16).

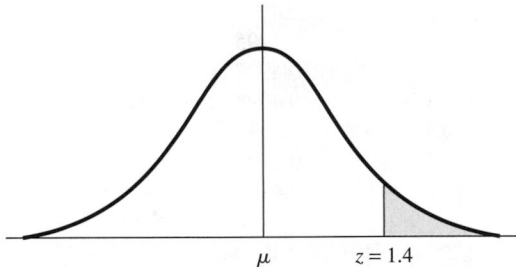

μ $z = 1.4$

FIGURE 26–16

Solution: From Table 26–7, the area between the mean and $z = 1.4$ is 0.4192. But since the total area to the right of the mean is $\frac{1}{2}$, the area in the tail is

$$0.5000 - 0.4192 = 0.0808$$

or 8.08% of the total area.   ◆◆◆

Since, for a normal distribution, the probability of a measurement falling within a given interval is equal to the area under the normal curve within that interval, we can use the areas to assign probabilities. We must first convert the measurement x to the standard variable z, which tells the number of standard deviations between x and the mean $\bar{x}$. To get the value of z, we simply find the difference between x and $\bar{x}$ and divide that difference by the standard deviation.

$$z = \frac{x - \bar{x}}{s}$$

◆◆◆ **Example 53:** Suppose that the heights of 2000 students have a mean of 176 cm with a standard deviation of 8.1 cm. Assuming the heights to have a normal distribution, predict the number of students (a) shorter than 165 cm, (b) taller than 178 cm, and (c) between 165 cm and 178 cm.

Solution:

(a) Let us compute z for $x = 165$ cm. The mean $\bar{x}$ is 176 cm and the standard deviation s is 8.1 cm, so

$$z = \frac{165 - 176}{8.1} = -1.4$$

From Table 26–7, with $z = 1.4$, we read an area of 0.4192. The area in the tail to the left of $z = -1.4$ is thus

$$0.5000 - 0.4192 = 0.0808$$

Thus there is a 8.08% chance that a single student will be shorter than 165 cm. Since there are 2000 students, we predict that

$$8.08\% \text{ of } 2000 \text{ students} = 162 \text{ students}$$

will be within this range.

Since the normal curve is symmetrical about the mean, either a positive or a negative value of z will give the same area. Thus for compactness, Table 26–7 shows only positive values of z.

(b) For a height of 178 cm, the value of z is

$$z = \frac{178 - 176}{8.1} = 0.2$$

The area between the mean and $z = 0.2$ is 0.0793, so the area to the right of $z = 0.2$ is

$$0.5000 - 0.0793 = 0.4207$$

Since there is a 42.07% chance that a student will be taller than 178 cm, we predict that 0.4207(2000) or 841 students will be within that range.

(c) Since $162 + 841$ or 1003 students are either shorter than 165 cm or taller than 178 cm, we predict that 997 students will be between these two heights. ◆◆◆

Exercise 5 ◆ The Normal Curve

Use Table 26–7 for the following problems.

1. Find the area under the normal curve between the mean and 1.5 standard deviations in Fig. 26–17.

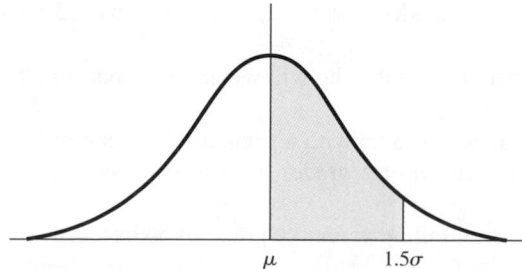

FIGURE 26–17

2. Find the area under the normal curve between 1.6 standard deviations on both sides of the mean in Fig. 26–18.

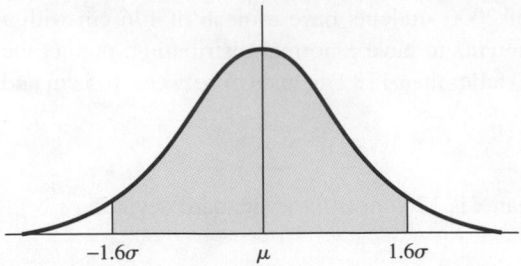

FIGURE 26–18

3. Find the area in the tail of the normal curve to the left of $z = -0.8$ in Fig. 26–19.

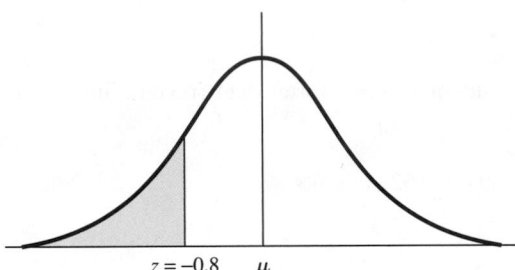

FIGURE 26–19

4. Find the area in the tail of the normal curve to the right of $z = 1.3$ in Fig. 26–20.

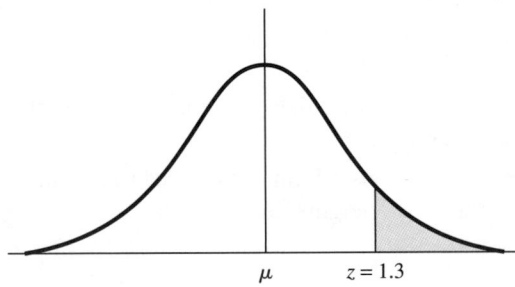

FIGURE 26–20

5. The distribution of the weights of 1000 students at Tech College has a mean of 163 lb. with a standard deviation of 18 lb. Assuming that the weights are normally distributed, predict the number of students who have weights between 130 and 170 lb.

6. For the data of problem 5, predict the number of students who will weigh less than 130 lb.

7. For the data of problem 5, predict the number of students who will weigh more than 195 lb.

8. On a test given to 500 students, the mean grade was 82.6 with a standard deviation of 7.4. Assuming a normal distribution, predict the number of A grades (a grade of 90 or over).

9. For the test of problem 8, predict the number of failing grades (a grade of 60 or less).

10. Computation of the gas consumption of a new car model during a series of trials gives a set of data that is normally distributed with a mean of 7.1 L/100 km and a standard deviation of 0.7 L/100 km. Find the probability that the same car will consume less than 6.5 L/100 km on the next trial.

26–6 Standard Errors

When we draw a random sample from a population, it is usually to *infer* something about that population. Typically, from our sample we compute a *statistic* (such as the sample mean $\bar{x}$) and use it to infer a population *parameter* (such as the population mean μ).

◆◆◆ **Example 54:** A researcher measured the heights of a randomly drawn sample of students at Tech College and calculated the mean height of that sample. It was

$$\text{mean height } \bar{x} \cong 67.50 \text{ in.}$$

From this he inferred that the mean height of the entire population of students at Tech College was

$$\text{mean height } \mu \cong 67.50 \text{ in.} \hspace{2cm} ◆◆◆$$

In general,

$$\text{population parameter} = \text{sample statistic}$$

How *accurate* is the population parameter that we find in this way? We know that all measurements are approximate, and we usually give some indication of the accuracy of a measurement. In fact, a population parameter such as the mean height is often given in the form

$$\text{mean height } \mu = 67.50 \pm 0.24 \text{ in.}$$

where ± 0.24 is called the *standard error* of the mean. In this section we show how to compute the standard error of the mean and the standard error of the standard deviation.

Further, when we give a range of values for a statistic, we know that it is possible that another sample can have a mean that lies outside the given range. In fact, the true mean itself can lie outside the given range. That is why a statistician will *give the probability* that the true mean falls within the given range.

◆◆◆ **Example 55:** Using the same example of heights at Tech College, we might say that

$$\text{mean height } \mu = 67.50 \pm 0.24 \text{ in.}$$

with a 68% chance that the mean student height μ falls within the range

$$67.50 - 0.24 \quad \text{to} \quad 67.50 + 0.24$$

or from 67.26 to 67.74 in. ◆◆◆

We call the interval 67.50 ± 0.24 in Example 55 a *68% confidence interval*. This means that there is a 68% chance that the true population mean falls within the given range. So the complete way to report a population parameter is

$$\boxed{\text{population parameter} = \text{sample statistic} \pm \text{standard error}}$$

within a given confidence interval. We will soon show how to compute this confidence interval, and others as well, but first we must lay some groundwork.

Frequency Distribution of a Statistic

Suppose that you draw a sample of, say, 40 heights from the population of students in your school, and find the mean $\bar{x}$ of those 40 heights, say, 65.6 in. Now measure another 40 heights from the same population and get another $\bar{x}$, say, 69.3 in. Then repeat, drawing one sample of 40 after another, and for each compute the statistic $\bar{x}$. This collection of $\bar{x}$'s now has a frequency distribution of its own (called the $\bar{x}$ distribution) with its own mean and standard deviation.

We now ask, *How is the $\bar{x}$ distribution related to the original distribution?* To answer this, we need a statistical theorem called the *central limit theorem*, which we give here without proof.

Standard Error of the Mean

Suppose that we were to draw *all possible* samples of size n from a given population, and for each sample compute the mean $\bar{x}$, and then make a frequency distribution of the $\bar{x}$'s. The central limit theorem states the following:

1. The mean of the $\bar{x}$ distribution is the same as the mean μ of the population from which the samples were drawn.
2. The standard deviation of the $\bar{x}$ distribution is equal to the standard deviation σ of the population divided by the square root of the sample size n. The standard error or a sampling distribution of a statistic such as $\bar{x}$ is often called its *standard error*. We will denote the standard error of $\bar{x}$ by $SE_{\bar{x}}$. So

$$SE_{\bar{x}} = \frac{\sigma}{\sqrt{n}}$$

3. For a "large" sample (over 30 or so), the $\bar{x}$ distribution is approximately a normal distribution, even if the population from which the samples were drawn is not a normal distribution.

We will illustrate parts of the central limit theorem with an example.

◆◆◆ **Example 56:** Consider the population of 256 integers from 100 to 355 given in Table 26–8. The frequency distribution of these integers [Fig. 26–21(a)] shows that each class has a frequency of 32. We computed the population mean μ by adding all of the integers and dividing by 256 and got 227.5. We also got a population standard deviation σ of 74.

TABLE 26–8

174	230	237	173	214	242	113	177	238	175	303	172	114	179	307	176
190	246	253	189	194	258	129	193	254	191	319	188	130	111	108	103
206	262	269	205	210	274	145	209	270	207	335	204	146	127	124	119
222	278	285	221	226	290	161	225	286	223	351	220	162	143	140	135
300	229	302	228	296	233	298	232	301	239	236	231	304	241	306	240
316	245	318	244	312	249	314	248	317	255	252	247	320	257	322	256
332	261	334	260	328	265	330	264	333	271	268	263	336	273	338	272
348	277	350	276	344	281	346	280	349	287	284	279	352	289	354	288
166	102	109	165	170	234	105	169	110	167	295	164	106	171	299	168
182	118	125	181	186	250	121	185	126	183	311	180	122	187	315	184
198	134	141	197	202	266	137	201	142	199	327	196	138	203	331	200
178	150	157	213	218	282	153	217	158	215	343	212	154	219	347	216
292	101	294	100	297	107	104	235	293	195	323	192	305	115	112	243
308	117	310	116	313	123	120	251	309	211	339	208	321	131	128	259
324	133	326	132	329	139	136	267	325	227	355	224	337	147	144	275
340	149	342	148	345	155	152	283	341	159	156	151	353	163	160	291

Then 32 samples of 16 integers each were drawn from the population. For each sample we got the sample mean $\bar{x}$. The 32 values we got for $\bar{x}$ are as follows:

234 229 236 235 227 232 229 224 266 230 218 212 234 226 237 216
170 222 220 218 232 247 262 277 265 252 238 225 191 204 216 223

We now make a frequency distribution for the sample means, which is called a *sampling distribution of the mean* [Fig. 26–21(b)]. We further compute the mean of the sample means and get 229.6 nearly the same value as for the population mean μ. The standard deviation of the sample means is computed to be 21.1. Thus we have the following results:

1. The mean of the $\bar{x}$ distribution (229.6) is approximately equal to the mean μ of the population (227.5).

2. By the central limit theorem, the standard deviation $SE_{\bar{x}}$ of the $\bar{x}$ distribution is found by dividing σ by the square root of the sample size.

$$SE_{\bar{x}} = \frac{\sigma}{\sqrt{n}} = \frac{74}{\sqrt{16}} = 18.5$$

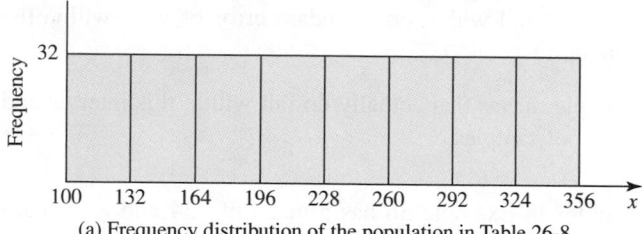

(a) Frequency distribution of the population in Table 26-8

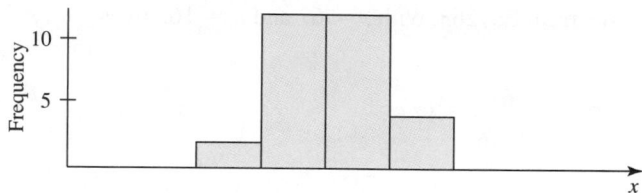

(b) Sampling distribution of the mean (frequency distribution of the means of 32 samples drawn from the population in Table 26-8)

FIGURE 26–21

This compares well with the actual standard deviation of the $\bar{x}$ distribution, which was 21.1. ◆◆◆

We now take the final step and show how to predict the population mean μ from the mean $\bar{x}$ and standard deviation s of a *single sample*. We also give the probability that our prediction is correct.

Predicting μ from a Single Sample

If the standard deviation of the $\bar{x}$ distribution, $SE_{\bar{x}}$, is small, most of the $\bar{x}$'s will be near the centre μ of the population. Thus a particular $\bar{x}$ has a good chance of being close to μ and will hence be a good estimator of μ. Conversely, a large $SE_{\bar{x}}$ means that a given x will be a poor estimator of μ.

However, since the frequency distribution of the $\bar{x}$'s is normal, the chance of a single $\bar{x}$ lying within one standard deviation of the mean μ is 68%. Conversely, the mean μ has a 68% chance of lying within one standard deviation of *a single randomly chosen* $\bar{x}$. Thus there is a 68% chance that the true population mean μ falls within the interval

$$\bar{x} \pm SE_{\bar{x}}$$

In this way we can estimate the population mean μ from a *single sample*. Not only that, but we are able to give a range of values within which μ must lie and to give *the probability* that μ will fall within that interval.

There is just one difficulty: To compute the standard error $SE_{\bar{x}}$ by the central limit theorem, we must divide the population standard deviation σ by the square root of the sample size. But σ *is not usually known*. Thus it is common practice to use the sample standard deviation s instead. Thus the standard error of the mean is approximated by the following formula (valid if the population is large):

Standard Error of the Mean $\bar{x}$	$SE_{\bar{x}} = \dfrac{\sigma}{Rn} \cong \dfrac{s}{\sqrt{n}}$	263

◆◆◆ **Example 57:** For the population of 256 integers (Table 26–8), the central limit theorem says that 68% of the sample means should fall within one standard error of μ, or within the interval 227.5 ± 18.5. Is this statement correct?

Solution: We count the number of sample means that actually do fall within this interval and get 23, which is 72% of the total number of samples.　　◆◆◆

◆◆◆ **Example 58:** One of the 32 samples in Example 56 has a mean of 234 and a standard deviation s of 67. Use these figures to estimate the mean μ of the population.

Solution: Computing the standard error from Eq. 263, with $s = 67$ and $n = 16$, gives

$$SE_{\bar{x}} \cong \frac{67}{\sqrt{16}} = 17$$

We can then predict that there is a 68% chance that μ will fall within the range

$$\mu = 234 \pm 17$$

We see that the true population mean 227.5 does fall within that range.　　◆◆◆

◆◆◆ **Example 59:** The heights of 64 randomly chosen students at Tech College were measured. The mean $\bar{x}$ was found to be 68.25 in., and the standard deviation s was 2.21 in. Estimate the mean μ of the entire population of students at Tech College.

Solution: From Eq. 263, with $s = 2.21$ and $n = 64$,

$$SE_{\bar{x}} \cong \frac{s}{\sqrt{n}} = \frac{2.21}{\sqrt{64}} = 0.28$$

We can then state that there is a 68% chance that the mean height μ of all students at Tech College is

$$\mu = 68.25 \pm 0.28 \text{ in.}$$

In other words, the true mean has a 68% chance of falling within the interval

$$67.97 \text{ to } 68.53 \text{ in.}$$　　◆◆◆

Confidence Intervals

The interval 67.97 to 68.53 in. found in Example 59 is called the *68% confidence interval*. Similarly, there is a 95% chance that the population mean falls within *two* standard errors of the mean.

$$\mu = \bar{x} \pm 2SE_{\bar{x}}$$

So we call it the *95% confidence interval*.

◆◆◆**Example 60:** For the data of Example 59, find the 95% confidence interval.

Solution: We had found that SE_x was 0.28 in. Thus we may state that μ will lie within the interval

$$68.25 \pm 2(0.28)$$

or

$$68.25 \pm 0.56 \text{ in.}$$

This is the 95% confidence interval.   ◆◆◆

We find other confidence intervals in a similar way.

Standard Error of the Standard Deviation

We can compute standard errors for sample statistics other than the mean. The most common is for the standard deviation. The standard error of the standard deviation is called SE_s and is given by the following equation:

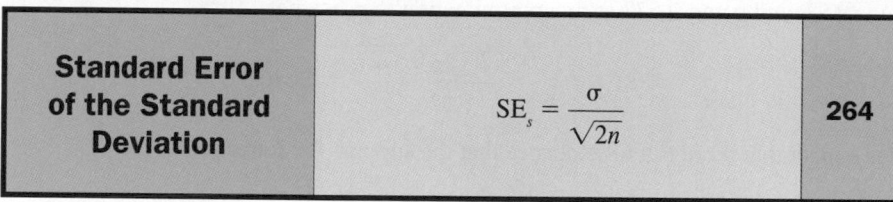

Standard Error of the Standard Deviation	$SE_s = \dfrac{\sigma}{\sqrt{2n}}$	264

where again we use the sample standard deviation s if we don't know σ.

◆◆◆**Example 61:** A single sample of size 32 drawn from a population is found to have a standard deviation s of 21.55. Give the population standard deviation σ with a confidence level of 68%.

Solution: From Eq. 264, with $\sigma \cong s = 21.55$ and $n = 32$,

$$SE_s = \frac{21.55}{\sqrt{2(32)}} = 2.69$$

Thus there is a 68% chance that σ falls within the interval

$$21.55 \pm 2.69$$ ◆◆◆

Standard Error of a Proportion

Consider a binomial experiment in which the probability of occurrence (called a *success*) of an event is p and the probability of nonoccurrence (a *failure*) of that event is q (or $1 - p$). For example, for the toss of a die, the probability of success in rolling a three is $p = \frac{1}{6}$, and the probability of failure in rolling a three is $q = \frac{5}{6}$.

If we threw the die n times and recorded the proportion of successes, it would be close to, but probably not equal to, p. Suppose that we then threw the die another n times and got another proportion, and then another n times, and so on. If we did this enough times, the proportion of successes of our samples would form a normal distribution with a mean equal to p and a standard deviation given by the following:

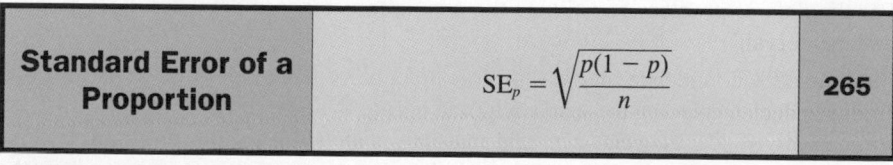

Standard Error of a Proportion	$SE_p = \sqrt{\dfrac{p(1 - p)}{n}}$	265

◆◆◆**Example 62:** Find the 68% confidence interval for rolling a three for a die tossed 150 times.

Solution: Using Eq. 265, with $p = \frac{1}{6}$ and $n = 150$,

$$SE_p = \sqrt{\frac{\left(\frac{1}{6}\right)\left(\frac{5}{6}\right)}{150}} = 0.030$$

Thus there is a 68% probability that in 150 rolls of a die, the proportion of threes will lie between $\frac{1}{6} + 0.03$ and $\frac{1}{6} - 0.03$. ◆◆◆

We can use the proportion of success in a single sample to estimate the proportion of success of an entire population. Since we will not usually know p for the population, we use the proportion of successes found from the sample when computing SE_p.

◆◆◆ **Example 63:** In a poll of 152 students at Tech College, 87 said that they would vote for Jones for president of the student union. Estimate the support for Jones among the entire student body.

Solution: In the sample, $\frac{87}{152} = 0.572$ support Jones. So we estimate that her support among the entire student body is 57.2%. To compute the confidence interval, we need the standard error. With $n = 152$, and using 0.572 as an approximation to p, we get

$$SE_p \approx \sqrt{\frac{0.572(1 - 0.572)}{152}} = 0.040$$

We thus expect that there is a 68% chance that the support for Jones is

$$0.572 \pm 0.040$$

or that she will capture between 53.2% and 61.2% of the vote of the entire student body. ◆◆◆

Exercise 6 ◆ Standard Errors

1. The heights of 49 randomly chosen students at Tech College were measured. Their mean $\bar{x}$ was found to be 69.47 in., and their standard deviation s was 2.35 in. Estimate the mean μ of the entire population of students at Tech College with a confidence level of 68%.

2. For the data of problem 1, estimate the population mean with a 95% confidence interval.

3. For the data of problem 1, estimate the population standard deviation with a 68% confidence interval.

4. For the data of problem 1, estimate the population standard deviation with a 95% confidence interval.

5. A single sample of size 32 drawn from a population is found to have a mean of 164.0 and a standard deviation s of 16.31. Give the population mean with a confidence level of 68%.

6. For the data of problem 5, estimate the population mean with a 95% confidence interval.

7. For the data of problem 5, estimate the population standard deviation with a 68% confidence interval.

8. For the data of problem 5, estimate the population standard deviation with a 95% confidence interval.

9. Find the 68% confidence interval for drawing a heart from a deck of cards for 200 draws from the deck, replacing the card each time before the next draw.

10. In a survey of 950 viewers, 274 said that they watched a certain program. Estimate the proportion of viewers in the entire population that watched that program, given the 68% confidence interval.

26–7   Process Control

Statistical Process Control

Statistical process control, or SPC, is perhaps the most important application of statistics for technology students. Process control involves continuous testing of samples from a production line. Any manufactured item will have chance variations in weight, dimensions, colour, and so forth. As long as the variations are due only to chance, we say that the process is *in control*.

◆◆◆ Example 64: The diameters of steel balls on a certain production line have harmless chance variations between 1.995 mm and 2.005 mm. The production process is said to be in control. **◆◆◆**

However, as soon as there is variation due to causes other than chance, we say that the process is no longer in control.

◆◆◆ Example 65: During a certain week, some of the steel balls from the production line of Example 64 were found to have diameters over 2.005. The process was then *out of control*. **◆◆◆**

One problem of process control is to detect when a process is out of control so that those other causes may be eliminated.

◆◆◆ Example 66: The period during which the process was out of control in Example 65 was found to occur when the factory air conditioning was out of operation during a heat wave. Now the operators are instructed to stop the production line during similar occurrences. **◆◆◆**

Control Charts

The main tool of SPC is the *control chart*. The idea behind a control chart is simple. We pull samples off a production line at regular time intervals, measure them in some way, and graph those measurements over time. We then establish upper and lower limits between which the sample is acceptable, but outside of which it is not acceptable.

◆◆◆Example 67: Figure 26–22 shows a control chart for the diameters of samples of steel balls of the preceding examples. One horizontal line shows the mean value of the diameter, while the other two give the upper and lower *control limits*. Note that the diameters fluctuate randomly between the control limits until the start of the heat wave, and then go out of control.

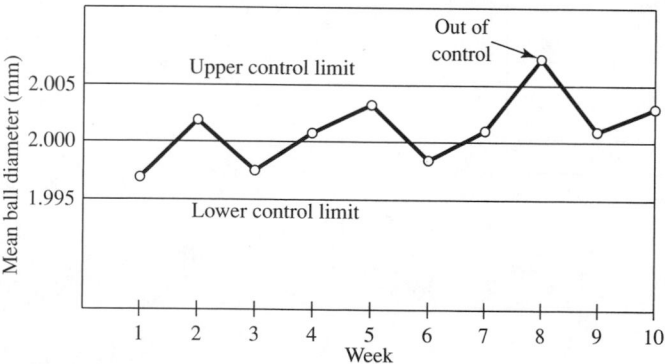

FIGURE 26–22   A control chart.   **◆◆◆**

To draw a control chart, we must decide what variables we want to measure and what statistical quantities we want to compute. Then we must calculate the control limits. A sample can be tested for a categorical variable (such as pass/fail) or for a continuous variable (such as the ball diameter). When testing a categorical variable, we usually compute the *proportion* of those items that pass. When testing a continuous variable, we usually compute the *mean, standard deviation,* or *range.*

To cover the subject of statistics usually requires an entire textbook, and SPC is often an entire chapter in such a book or can take an entire textbook by itself. Obviously we can only scratch the surface here.

We will draw three control charts: one for the proportion of a categorical variable, called a *p chart;* a second for the mean of a continuous variable, called an $\overline{X}$ *chart;* and a third for the range of a continuous variable, called an *R chart.* The formulas for computing the central line and the control limits are summarized in Fig. 26–23. We will illustrate the construction of each chart with an example.

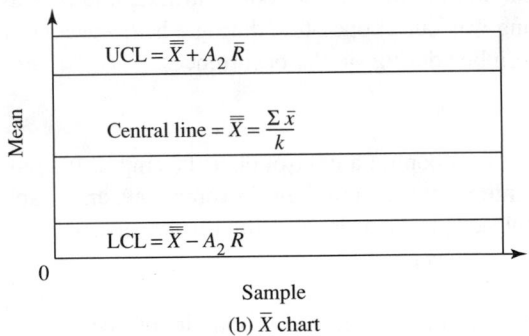

(a) *p* chart

$\overline{X}$ means the average of averages and is pronounced "x double bar".

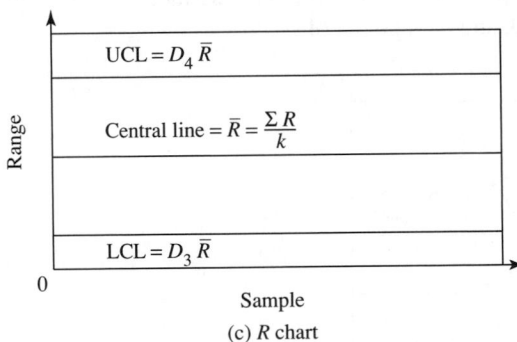

(b) $\overline{X}$ chart

(c) *R* chart

FIGURE 26–23

The *p* Chart

◆◆◆ **Example 68:** Make a control chart for the proportion of defective light bulbs coming off a certain production line.

Solution:

1. *Choose a sample size n and the testing frequency.*

Let us choose to test a sample of 1000 bulbs every day for 21 days.

2. *Collect the data. Count the number d of defectives in each sample of 1000. Obtain the proportion defective p for each sample by dividing the number of defectives d by the sample size n.*

The collected data and the proportion defective are given in Table 26–9.

TABLE 26–9 Daily samples of 1000 light bulbs per day for 21 days.

Day	Number Proportion	Defective Defective
1	63	0.063
2	47	0.047
3	59	0.059
4	82	0.082
5	120	0.120
6	73	0.073
7	58	0.058
8	55	0.055
9	39	0.039
10	99	0.099
11	85	0.085
12	47	0.047
13	46	0.046
14	87	0.087
15	90	0.090
16	67	0.067
17	85	0.085
18	24	0.024
19	77	0.077
20	58	0.058
21	103	0.103
	Total 1464	

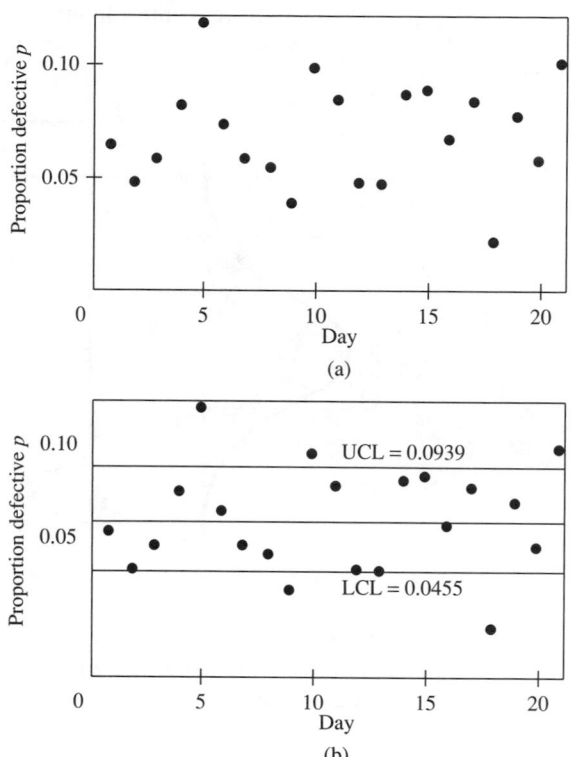

FIGURE 26–24

3. *Start the control chart by plotting p versus time.*

The daily proportion defective p is plotted in Fig. 26–24(a)

> We are assuming here that our sampling distribution is approximately normal.

4. *Compute the average proportion defective $\bar{p}$ by dividing the total number of defectives for all samples by the total number of items measured.*

The total number of defectives is 1464, and the total number of bulbs tested is 1000(21) = 21 000. So

> If the calculation of the lower limit gives a negative number, we have no lower control limit.

$$\bar{p} = \frac{1464}{21\,000} = 0.0697$$

5. *Compute the standard error SE_p. From Sec. 26–6, the standard error for a proportion in which the probability, p, of success of a single event is given by*

$$SE_p = \sqrt{\frac{p(1-p)}{n}} \tag{265}$$

Since we do not know q, we use $\bar{p}$ as an estimator for p.

$$SE_p = \sqrt{\frac{\bar{p}(1-\bar{p})}{n}} = \sqrt{\frac{0.0697(1-0.0697)}{1000}} = 0.008\,05$$

6. *Choose a confidence interval. Recall from Sec. 26–6 that one standard error will give a 68% confidence interval, two standard errors a 95% confidence interval, three a 99.7% level, and so forth (Fig. 26–25). Set the control limits at these values.*

> The next step would be to *analyze* the finished control chart. We would look for points that are out of control and try to determine their cause. We do not have space here for this discussion, which can be found in any SPC book.
>
> We would also *adjust* the *control limits* if in our calculation of the limits we used values that we now see are outside those limits. Since those values are not the result of chance but are caused by some production problem, they should not be used in calculating the permissible variations due to chance. Thus we would normally recompute the control limits without using those values. We omit that step here.

We will use three-sigma limits, which are the most commonly used. Thus

$$\text{upper control limit} = \bar{p} + 3\text{SE}_p = 0.0697 + 3(0.008\,05) = 0.0939$$

$$\text{lower control limit} = \bar{p} - 3\text{SE}_p = 0.0697 - 3(0.008\,05) = 0.0455$$

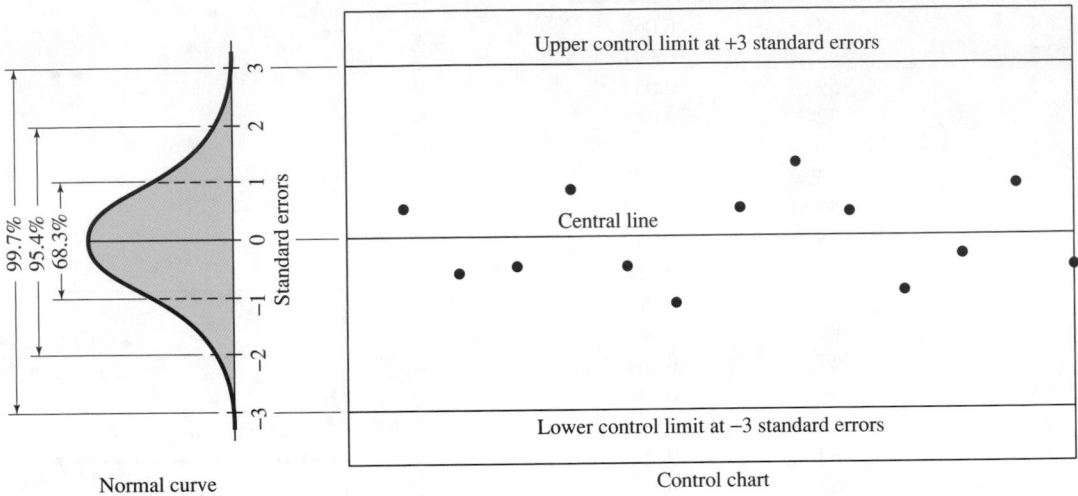

FIGURE 26–25   Relationship between the normal curve and the control chart. Note that the upper and lower limits are placed at three standard errors from the mean so that there is a 99.7% chance that a sample will fall between those limits.

7. *Draw a horizontal line at $\bar{p}$ and at the upper and lower control limits at $\bar{p} \pm 3SE_p$.*

These are shown in the control chart in Fig. 26–25.   ◆◆◆

Control Chart for a Continuous Variable

We construct a control chart for a continuous variable in much the same way as for a categorical variable. The variables commonly charted are the mean and either the range or the standard deviation. The range is usually preferred over the standard deviation because it is simple to calculate and easier to understand by factory personnel who might not be familiar with statistics. We will now make control charts for the mean and range, commonly called the $\bar{X}$ and *R charts*.

The $\bar{X}$ and R Charts

We again illustrate the method with an example. As is usually done, we will construct both charts at the same time.

◆◆◆ **Example 69:** Make $\bar{X}$ and *R* control charts for a sampling of the wattages of the lamps on a production line.

Solution:

1. *Choose a sample size n and the testing frequency.*

Since the measurement of wattage is more time-consuming than just simple counting of defectives, let us choose a smaller sample size than before—say, 5 bulbs every day for 21 days.

2. *Collect the data.*

We measure the wattage in each sample of 5 bulbs. For each sample we compute the sample mean $\bar{X}$ and the sample range *R*. These are given in Table 26–10.

TABLE 26–10 Daily samples of wattages of 5 lamps per day for a period of 21 days.

Day	Wattage of 5 Samples					Mean $\overline{X}$	Range R
1	105.6	92.8	92.6	101.5	102.5	99.0	13.0
2	106.2	100.4	106.6	108.3	109.2	106.1	8.8
3	108.6	103.3	101.8	96.0	98.8	101.7	12.6
4	95.5	106.3	106.1	97.8	101.0	101.3	10.8
5	98.4	91.5	106.3	91.6	93.2	96.2	14.8
6	101.9	102.5	94.5	107.7	108.6	103.0	14.1
7	96.7	94.8	106.9	103.4	96.2	99.6	12.2
8	99.3	107.5	94.8	102.1	108.2	102.4	13.4
9	98.0	108.3	94.8	98.6	102.8	100.5	13.5
10	109.3	98.6	92.2	106.7	96.9	100.8	17.1
11	98.8	95.0	99.4	104.8	96.2	98.8	9.8
12	91.2	103.3	95.3	103.2	108.9	100.4	17.7
13	98.5	102.8	106.4	108.6	94.2	102.1	14.4
14	100.8	108.1	104.7	108.6	97.0	103.8	11.6
15	109.0	102.6	90.5	91.3	90.5	96.8	18.5
16	100.0	104.0	99.5	92.9	93.4	98.0	11.1
17	106.3	106.2	107.6	90.4	102.9	102.7	17.2
18	103.4	91.8	105.7	106.9	106.3	102.8	15.1
19	99.7	106.8	99.6	106.6	90.1	100.6	16.7
20	98.4	104.5	94.5	101.7	96.2	99.0	10.0
21	102.2	103.0	90.3	94.0	103.5	98.6	13.2
					Sums	2114	285.5
					Averages	100.7	13.6

3. *Start the control chart by plotting $\overline{X}$ and R versus time.*

The daily mean and range are plotted as shown in Fig. 26–26.

4. *Compute the average mean $\overline{\overline{X}}$ and the average range $\overline{R}$.*

We take the average of the 21 means.

$$\overline{\overline{X}} = \frac{\Sigma \overline{X}}{21} = \frac{2114}{21} = 100.7$$

The average of the ranges is

$$\overline{R} = \frac{\Sigma R}{21} = \frac{285.5}{21} = 13.60$$

5. *Compute the control limits.*

When making an $\overline{X}$ and R chart, we normally do not have the standard deviation to use in computing the control limits. The average range $\overline{R}$, multiplied by a suitable constant, is usually used to give the control limits for both the mean and the range charts. For the mean, the three-sigma control limits are

$$\overline{\overline{X}} \pm A_2 \overline{R}$$

where A_2 is a constant depending on sample size, given in Table 26–11. The control limits for the range are

$$D_3 \overline{R} \quad \text{and} \quad D_4 \overline{R}$$

where D_3 and D_4 are also from the table.

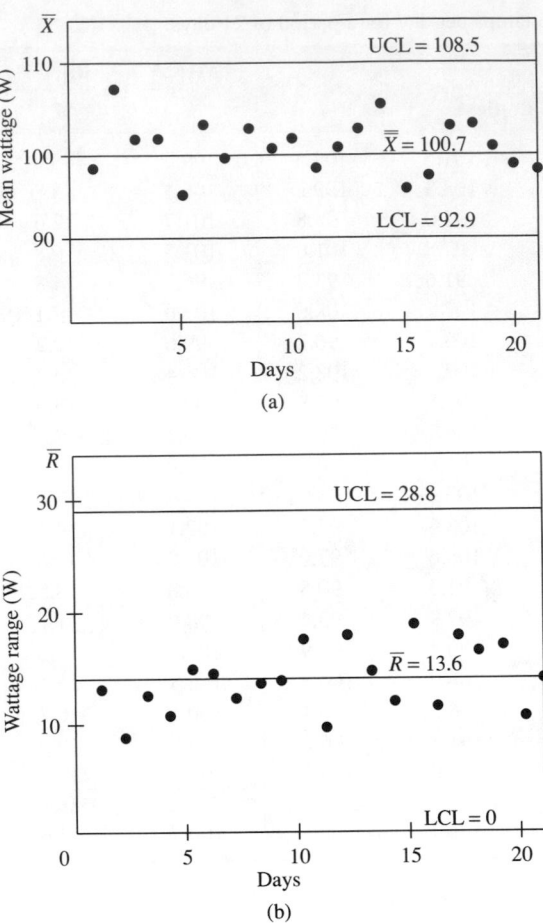

FIGURE 26-26

TABLE 26–11   Factors for computing control chart lines.

n	2	3	4	5	6	7	8	9	10
A_2	1.880	1.023	0.729	0.577	0.483	0.419	0.373	0.337	0.308
D_3	0	0	0	0	0	0.076	0.136	0.184	0.223
D_4	3.268	2.574	2.282	2.114	2.004	1.924	1.864	1.816	1.777

Excerpted from the *ASTM Manual on Presentation of Data and Control Chart Analysis* (1976).
For larger sample sizes consult a book on SPC.

For a sample size of 5, $A_2 = 0.577$, so the control limits for the mean are

$$UCL = 100.7 + 0.577(13.60) = 108.5$$

and

$$LCL = 100.7 - 0.577(13.60) = 92.9$$

Also from the table, $D_3 = 0$ and $D_4 = 2.114$, so the control limits for the range are

$$UCL = 2.114(13.60) = 28.8$$

and

$$LCL = 0(13.58) = 0$$

6. *Draw the central line and the upper and lower control limits.*

These are shown in the control chart [Fig. 26–26(b)].   ◆◆◆

Exercise 7 ◆ Process Control

1. The proportion of defectives for samples of 1000 tennis balls per day for 20 days is as follows:

Day	Number Defective	Proportion Defective
1	31	0.031
2	28	0.028
3	26	0.026
4	27	0.027
5	30	0.030
6	26	0.026
7	26	0.026
8	27	0.027
9	29	0.029
10	27	0.027
11	27	0.027
12	26	0.026
13	31	0.031
14	30	0.030
15	27	0.027
16	27	0.027
17	25	0.025
18	26	0.026
19	30	0.030
20	28	0.028
Total	554	

Find the values for the central line, and determine the upper and lower control limits.

2. Draw a p chart for the data of problem 1.

3. The proportion of defectives for samples of 500 calculators per day for 20 days is as follows:

Day	Number Defective	Proportion Defective
1	153	0.306
2	174	0.348
3	139	0.278
4	143	0.286
5	156	0.312
6	135	0.270
7	141	0.282
8	157	0.314
9	125	0.250
10	126	0.252
11	155	0.310
12	174	0.348
13	138	0.276
14	165	0.330
15	144	0.288
16	166	0.332
17	145	0.290
18	153	0.306
19	132	0.264
20	169	0.338
Total	2990	

Find the values for the central line, and determine the upper and lower control limits.

4. Draw a p chart for the data of problem 3.

5. Five pieces of pipe are taken from a production line each day for 21 days, and their lengths are measured as shown in the following table:

Day	Measurements of 5 Samples (cm)				
1	201.7	184.0	201.3	183.9	192.6
2	184.4	184.2	207.2	194.4	193.1
3	217.2	212.9	198.7	185.1	196.9
4	201.4	201.8	182.5	195.0	215.4
5	214.3	197.2	219.8	219.9	194.5
6	190.9	180.6	181.9	203.6	185.1
7	189.5	209.5	207.6	215.2	200.0
8	182.2	216.4	190.7	189.0	204.2
9	184.9	217.3	185.7	183.0	189.7
10	211.2	194.8	208.6	209.4	183.3
11	197.1	189.5	200.1	197.7	187.9
12	207.1	203.1	218.4	199.3	219.1
13	199.1	207.1	205.8	197.4	189.4
14	198.9	196.1	210.3	210.6	217.0
15	215.2	200.8	205.0	186.6	181.6
16	180.2	216.8	182.5	206.3	191.7
17	219.3	195.1	182.2	207.4	191.9
18	204.0	209.9	196.4	202.3	206.9
19	202.2	216.3	217.8	200.7	215.1
20	185.0	194.1	191.2	186.3	201.6
21	181.6	187.8	188.2	191.7	200.1

Find the values for the central line, and determine the upper and lower control limits for the mean.

6. Find the values for the central line for the data of problem 5, and determine the upper and lower control limits for the range.

7. Draw an $\overline{X}$ chart for the data of problem 5.

8. Draw an R chart for the data of problem 5.

9. Five circuit boards are taken from a production line each day for 21 days and are weighed (in grams) as shown in the following table:

Day	Measurements of 5 Samples (g)				
1	16.7	19.5	11.1	18.8	11.8
2	15.0	14.5	14.5	18.4	11.5
3	14.3	16.5	13.3	14.0	14.4
4	19.3	14.0	14.6	15.0	18.0
5	16.2	13.6	16.8	12.4	16.6
6	14.3	16.7	18.4	18.3	16.5
7	13.5	15.5	13.7	19.8	11.6
8	15.6	12.3	16.8	18.3	11.8
9	13.3	15.4	16.1	17.0	12.2
10	17.6	18.6	12.4	19.9	19.3
11	19.0	12.8	14.7	19.6	15.8
12	12.3	12.2	14.9	15.5	12.9
13	16.3	12.3	18.6	14.0	11.3
14	11.4	15.8	13.9	19.1	10.5
15	12.5	11.4	14.1	12.4	14.9
16	18.1	15.1	19.2	15.2	17.6
17	18.7	13.7	11.0	17.0	12.7
18	11.3	11.5	13.4	13.1	13.9
19	16.6	18.7	12.1	14.7	11.2
20	11.2	10.5	16.2	11.0	16.2
21	12.8	18.7	10.4	16.8	17.1

Find the values for the central line, and determine the upper and lower control limits for the mean.

10. Find the values for the central line for the data of problem 9, and determine the upper and lower control limits for the range.

11. Draw an $\overline{X}$ chart for the data of problem 9.

12. Draw an R chart for the data of problem 9.

26–8   Regression

Curve Fitting

In statistics, *curve fitting*, the fitting of a curve to a set of data points, is called *regression*. The fitting of a straight line to data points is called *linear regression*, while the fitting of some other curve is called *nonlinear regression*. Fitting a curve to *two* sets of data is called *multiple regression*. We will cover linear regression in this section.

> The word *regression* was first used in the nineteenth century when these methods were used to determine the extent by which the heights of children of tall or short parents *regressed* or got closer to the mean height of the population

Scatter Plot

We mentioned *scatter plots* in Sec. 26–1. A scatter plot is simply a plot of all of the data points. It can be made by hand, by graphics calculator, or by computer.

> Glance back at Sec. 20–7 where we did some curve fitting using logarithmic and semilogarithmic graph paper.

◆◆◆ **Example 70:** Scatter plots for three sets of data are shown in Fig. 26–27. In Fig. 26–27(b), the point (6, 7) is called an *outlier*. Such points are usually suspected as being the result of an error and are sometimes discarded.

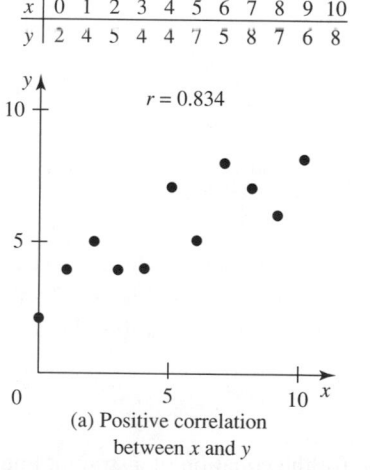

x	0	1	2	3	4	5	6	7	8	9	10
y	2	4	5	4	4	7	5	8	7	6	8

$r = 0.834$

(a) Positive correlation between x and y

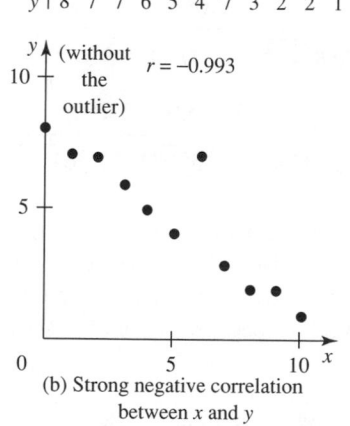

x	0	1	2	3	4	5	6	7	8	9	10
y	8	7	7	6	5	4	7	3	2	2	1

(without the outlier) $r = -0.993$

(b) Strong negative correlation between x and y

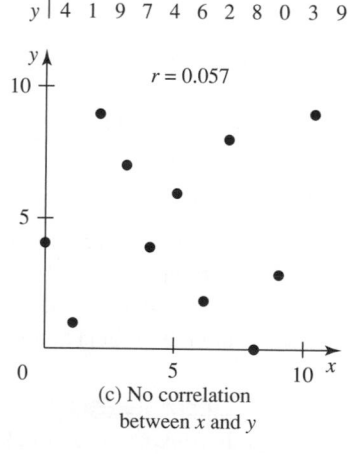

x	0	1	2	3	4	5	6	7	8	9	10
y	4	1	9	7	4	6	2	8	0	3	9

$r = 0.057$

(c) No correlation between x and y

FIGURE 26–27   ◆◆◆

Correlation Coefficient

In Fig. 26–27 we see that some data are more scattered than others. The *correlation coefficient* r gives a numerical measure of this scattering. For a set of n xy pairs, the correlation coefficient is given by the following equation:

Correlation Coefficient	$$r = \frac{n \sum xy - \sum x \sum y}{\sqrt{n \sum x^2 - (\sum x)^2} \sqrt{n \sum y^2 - (\sum y)^2}}$$	266

The formula looks complicated but is actually easy to use when doing the computation in table form.

◆◆◆ **Example 71:** Find the correlation coefficient for the data in Fig. 26–27(a).

Solution: We make a table giving x and y, the product of x and y, and the squares of x and y. We then sum each column as shown in Table 26–12.

TABLE 26–12

x	y	xy	x^2	y^2
0	2	0	0	4
1	4	4	1	16
2	5	10	4	25
3	4	12	9	16
4	4	16	16	16
5	7	35	25	49
6	5	30	36	25
7	8	56	49	64
8	7	56	64	49
9	6	54	81	36
10	8	80	100	64
Sums 55	60	353	385	364
Σx	Σy	Σxy	Σx^2	Σy^2

Then, from Eq. 266, with $n = 11$,

$$r = \frac{n \Sigma xy - \Sigma x \Sigma y}{\sqrt{n \Sigma x^2 - (\Sigma x)^2} \, \sqrt{n \Sigma y^2 - (\Sigma y)^2}}$$

$$= \frac{11(353) - 55(60)}{\sqrt{11(385) - (55)^2} \, \sqrt{11(364) - (60)^2}} = 0.834$$

The correlation coefficient can vary from $+1$ (perfect positive correlation) through 0 (no correlation) to -1 (perfect negative correlation). For comparison with the data in Fig. 26–27(a), the correlation coefficients for the data in (b) and (c) are found to be $r = -0.993$ (after discarding the outlier) and $r = 0.057$, respectively (work not shown).   ◆◆◆

Linear Regression

In *linear regression*, the object is to find the constants m and b for the equation of a straight line that best fits a given set of data points. The simplest way to do this is to fit a straight line by eye on a scatter plot and to measure the slope and the y intercept of that line.

Slope Intercept Form	$y = mx + b$	**289**

◆◆◆ **Example 72:** Find, by eye, the equation of the straight line that best fits the following data:

x	1	2	3	4	5	6	7	8
y	3.249	3.522	4.026	4.332	4.900	5.121	5.601	5.898

Solution: We draw a scatter diagram as shown in Fig. 26–28. Then using a straight-edge, we draw a line through the data, trying to balance those points above the line with those below. We then read the y intercept and the rise of the line in a chosen run.

$$y \text{ intercept } b = 2.8$$

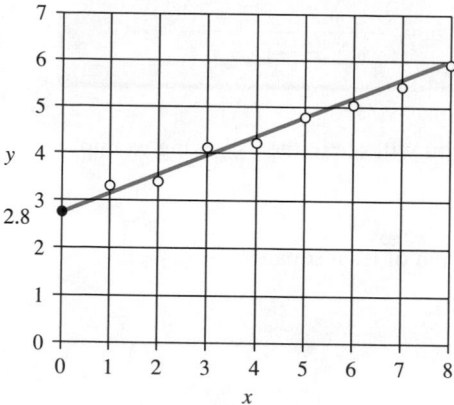

FIGURE 26–28

Our line appears to pass through (8, 6), so in a run of 8 units, the line rises 3.2 units. Its slope is then

$$\text{slope } m = \frac{\text{rise}}{\text{run}} = \frac{3.2}{8} = 0.40$$

The equation of the fitting line is then

$$y = 0.40x + 2.8 \qquad\qquad ◆◆◆$$

Method of Least Squares

Sometimes the points can be too scattered to enable drawing a line with a good fit, or perhaps we desire more accuracy than can be obtained when fitting by eye. Then we might want a method that does not require any manual steps, so that it can be computerized. One such method is the *method of least squares*.

Let us define a *residual* as the vertical distance between a data point and the approximating curve (Fig. 26–29). The method of least squares is a method to fit a straight line to a data set so that *the sum of the squares of the residuals is a minimum*—hence the name *least squares*.

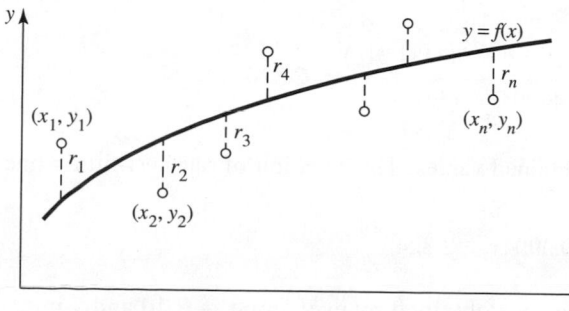

FIGURE 26–29   Definition of a residual.

With this method, the slope and the y intercept of the least squares line are given by the following equations:

We give these equations without proof here. They are derived using calculus.

Least Squares Line	slope $m = \dfrac{n \sum xy - \sum x \sum y}{n \sum x^2 - (\sum x)^2}$ y intercept $b = \dfrac{\sum x^2 \sum y - \sum x \sum xy}{n \sum x^2 - (\sum x)^2}$	**267**

These equations are easier to use than they look, as you will see in the following example.

◆◆◆ **Example 73:** Repeat Example 72 using the method of least squares.

Solution: We tabulate the given values in Table 26–13.

TABLE 26–13

x	y	x^2	xy
1	3.249	1	3.249
2	3.522	4	7.044
3	4.026	9	12.08
4	4.332	16	17.33
5	4.900	25	24.50
6	5.121	36	30.73
7	5.601	49	39.21
8	5.898	64	47.18
$\sum x = 36$	$\sum y = 36.649$	$\sum x^2 = 204$	$\sum xy = 181.316$

In the third column of Table 26–13, we tabulate the squares of the abscissas given in column 1, and in the fourth column we list the products of x and y. The sums are given below each column.

Substituting these sums into Eq. 267 and letting $n = 8$, we have

This is not the only way to fit a line to a data set, but it is widely regarded as the one that gives the "best fit."

$$\text{slope} = \frac{8(181.316) - 36(36.649)}{8(204) - (36)^2} = 0.3904$$

and

$$y \text{ intercept} = \frac{204(36.649) - 36(181.316)}{8(204) - (36)^2} = 2.824$$

which agree well with our graphically obtained values. The equation of our best-fitting line is then

$$y = 0.3904x + 2.824$$

Note the close agreement with the values we obtained by eye: slope = 0.40 and y intercept = 2.8.

◆◆◆

Exercise 8 ◆ Regression

Correlation Coefficient

Find the correlation coefficient for each set of data.

1.

−8	−6.238
−6.66	−3.709
−5.33	−0.712
−4	1.887
−2.66	4.628
−1.33	7.416
0	10.2
1.33	12.93
2.66	15.70
4	18.47
5.33	21.32
6.66	23.94
8	26.70
9.33	29.61
10.6	32.35
12	35.22

2.

−20	82.29
−18.5	73.15
−17.0	68.11
−15.6	59.31
−14.1	53.65
−12.6	45.90
−11.2	38.69
−9.73	32.62
−8.26	24.69
−6.8	18.03
−5.33	11.31
−3.86	3.981
−2.4	−2.968
−0.93	−9.986
0.53	−16.92
2	−23.86

3.

−11.0	−65.30
−9.33	−56.78
−7.66	−47.26
−6.00	−37.21
−4.33	−27.90
−2.66	−18.39
−1.00	−9.277
0.66	0.081
2.33	9.404
4.00	18.93
5.66	27.86
7.33	37.78
9.00	46.64
10.6	56.69
12.3	64.74
14.0	75.84

Method of Least Squares

Find the least squares line for each set of data

4. From problem 1.
5. From problem 2.
6. From problem 3.

Case Study Discussion—Quality Assurance Measurements

The key here is that there is no problem yet. The $\overline{X}$ and R charts are designed to measure normal variations. Whether for manufacturing, customer service, or wildlife birth-defect rates, control charts are designed to show patterns where none should be present. If slight variations in the ring bearing cage measurements are truly random, we have controlled all the manufacturing processes. However, if a pattern is starting to emerge from the data, something is creating that pattern, and you will have to find out what that something is before it pushes the measurements out of specifications. This is the main point of quality assurance: finding and correcting situations that could lead to problems, before they become problems. Patterns that indicate an out-of-control situation include the following:

- A single point lying outside the control limits.
- Two out of three successive points on the same side of the centreline and farther than 2σ from it.
- Four out of five successive points on the same side of the centreline and farther than 1σ from it.
- A run of eight points in a row on the same side of the centreline—or 10 out of 11, 12 out of 14, 16 out of 20.
- Obvious consistent or persistent patterns that suggest something unusual about your data and your process.

••• CHAPTER 26 REVIEW PROBLEMS •••••••••••••••••••••••••••••••••••

Label each type of data as continuous, discrete, or categorical.

1. The life of certain radios.

2. The number of houses sold each day.

3. The colors of the cars at a certain dealership.

4. The sales figures for a certain product for the years 1970–80 are as follows:

Year	Number Sold
1970	1344
1971	1739
1972	2354
1973	2958
1974	3153
1975	3857
1976	3245
1977	4736
1978	4158
1979	5545
1980	6493

Make an x-y graph of the sales versus the year.

5. Make a bar graph for the data of problem 4.

6. If we toss three coins, what is the probability of getting one head and two tails?

7. If we throw two dice, what chance is there that their sum is 8?

8. A die is rolled twice. What is the probability that both rolls will give a two?

9. At a certain factory, 72% of the workers have brown hair, 6% are left-handed, and 3% both have brown hair and are left-handed. What is the probability that a worker chosen at random will either have brown hair or be left-handed, or both?

10. What is the probability of tossing a five or a nine when two dice are tossed?

11. Find the area under the normal curve between the mean and 0.5 standard deviation to one side of the mean.

12. Find the area under the normal curve between 1.1 standard deviations on both sides of the mean.

13. Find the area in the tail of the normal curve to the left of $z = 0.4$

14. Find the area in the tail of the normal curve to the right of $z = 0.3$.

Use the following table of data for problems 15 through 29.

```
146   153   183   148   116   127   162   153
168   161   117   153   116   125   173   131
117   183   193   137   188   159   154   112
174   182   144   144   133   167   192   145
162   138   137   154   141   129   137   152
```

15. Determine the range of the data.

16. Make a frequency distribution using class widths of 5. Show both absolute and relative frequency.

17. Draw a frequency histogram showing both absolute and relative frequency.

18. Draw a frequency polygon showing both absolute and relative frequency.

19. Make a cumulative frequency distribution.

20. Draw a cumulative frequency polygon.

21. Find the mean.
22. Find the median.
23. Find the mode.
24. Find the sample variance.
25. Find the sample standard deviation.
26. Predict the population mean with a 68% confidence interval.
27. Predict the population mean with a 95% confidence interval.
28. Predict the population standard deviation with a 68% confidence interval.
29. Predict the population standard deviation with a 95% confidence interval.
30. On a test given to 300 students, the mean grade was 79.3 with a standard deviation of 11.6. Assuming a normal distribution, estimate the number of A grades (a score of 90 or over).
31. For the data of problem 30 estimate the number of failing grades (a score of 60 or lower).
32. Determine the quartiles and give the quartile range of the following data:

> 118   133   135   143   164   173   179   199   212   216   256

33. Determine the quartiles and give the quartile range of the following data:

> 167   245   327   486   524   639   797   853   974   1136   1162   1183

34. The proportion defective for samples of 700 keyboards per day for 20 days is as follows:

Day	Number Defective	Proportion Defective
1	123	0.176
2	122	0.175
3	125	0.178
4	93	0.132
5	142	0.203
6	110	0.157
7	98	0.140
8	120	0.172
9	139	0.198
10	142	0.203
11	128	0.183
12	132	0.188
13	92	0.132
14	114	0.163
15	92	0.131
16	105	0.149
17	146	0.208
18	113	0.162
19	122	0.175
20	92	0.131
Total	2351	

Find the values of the central line, determine the upper and lower control limits, and make a control chart for the proportion defective.

35. Five castings per day are taken from a production line each day for 21 days and are weighed. Their weights, in grams, are as follows:

Day	Measurements of 5 Samples				
1	1134	1168	995	992	1203
2	718	1160	809	432	650
3	971	638	1195	796	690
4	598	619	942	1009	833
5	374	395	382	318	329
6	737	537	692	562	960
7	763	540	738	969	786
8	777	1021	626	786	472
9	1015	1036	1037	1063	1161
10	797	1028	1102	796	536
11	589	987	765	793	481
12	414	751	1020	524	1100
13	900	613	1187	458	661
14	835	957	680	845	1023
15	832	557	915	934	734
16	353	829	808	626	868
17	472	798	381	723	679
18	916	763	599	338	1026
19	331	372	318	304	354
20	482	649	1133	1022	320
21	752	671	419	715	413

Find the values of the central line, determine the upper and lower control limits, and make a control chart for the mean of the weights.

36. Find the values of the central line, determine the upper and lower control limits, and make a control chart for the range of the weights.

37. Find the correlation coefficient and the least squares fit for the following data:

x	y
5	6.882
11.2	−7.623
17.4	−22.45
23.6	−36.09
29.8	−51.13
36.0	−64.24
42.2	−79.44
48.4	−94.04
54.6	−107.8
60.8	−122.8
67.0	−138.6
73.2	−151.0
79.4	−165.3
85.6	−177.6
91.8	−193.9
98	−208.9

38. For the relative frequency histogram for the population of student grades (Fig. 26–3), find the probability that a student chosen at random will have received a grade between 74 and 86.

39. Find the 68% confidence interval for drawing a king from a deck of cards, for 200 draws from the deck, replacing the card each time before the next draw.

40. In a survey of 120 shoppers, 71 said that they preferred Brand *A* potato chips over Brand *B*. Estimate the proportion of shoppers in the entire population that would prefer Brand *A*, given the 68% confidence interval.

Writing

41. Give an example of one statistical claim (such as an advertisement, a commercial, or a political message) that you have heard or read lately that has made you sceptical. State your reasons for being suspicious, and point out how the claim could otherwise have been presented to make it more plausible.

Team Project

42. A certain diode has the following characteristics:

Voltage across Diode (V)	Current through Diode (mA)
0	0
5	2.063
10	5.612
15	10.17
20	15.39
25	21.30
30	27.71
35	34.51
40	42.10
45	49.95
50	58.11

Linearize the data using the methods of Sec. 20–7. Then apply the method of least squares to find the slope and the y intercept of the straight line obtained. Using those values, write and graph an equation for the current as a function of the voltage. How does your graphed equation compare with the plot of the original points?

27

Derivatives of Algebraic Functions

◆◆◆ OBJECTIVES ◆◆

When you have completed this chapter, you should be able to:
- Evaluate limits for algebraic, trigonometric, and exponential functions.
- Calculate a derivative by the delta method.
- Calculate a derivative using the rules for derivatives.
- Find the derivative of a function raised to a power.
- Use the Product and Quotient rules for finding a derivative.
- Take the derivative of implicit functions.
- Find the derivatives of higher-order derivatives.

◆◆◆

This is the first chapter in the study of calculus. *Calculus* is a branch of mathematics that begins with the concept of calculating *change*. Most of the early work by Newton and Leibniz focused on physics problems having to do with the motion of the planets. Although we covered some change (such as displacement), and even rates of change (such as velocity and acceleration), in algebra, calculus is designed to work with more complicated aspects of change. Once we can deal with change—represented as the slope of a graph—we start to find many applications where calculus simplifies calculations dealing with changes over distance or over time.

27–1 Limits

Limit Notation

Suppose that x and y are related by a function, such as $y = 3x$. Then if we are given a value of x, we can find a corresponding value for y. For example, when x is 2, y is 6.

We now want to extend our mathematical language to be able to say what will happen to y *not* when x *is* a certain value, but when x *approaches* a certain value. For example, when x approaches 2, y approaches 6. The *notation* we use to say the same thing is

$$\lim_{x \to 2} y = \lim_{x \to 2} (3x) = 6$$

We read this as "the limit, as x approaches 2, of $3x$, is 6."

In general, if the function $f(x)$ approaches some value L as x approaches a, we indicate that condition with the following notation:

Limit Notation	$\lim_{x \to a} f(x) = L$	336

This is not our first use of the limit idea. We used it to find the value of e, the base of natural logarithms (Sec. 20–2), and to find the sum of an infinite geometric progression (Sec. 25–4).

Why bother with new notation? Why not just say, in the preceding example, that y is 6 when x is 2? Why is it necessary to creep up on the answer like that?

It is true that limit notation offers no advantage in an example such as the last one. But we really need it when our function *cannot* reach the limit or is *not even defined* at the limit.

♦♦♦ **Example 1:** The function $y = 1/x^2$ is graphed in Fig. 27–1. Even though y never reaches 0, we can still write

$$\lim_{x \to \infty} \frac{1}{x^2} = 0$$

Further, even though our function is not even defined at $x = 0$, because it results in division by zero, we can still write

$$\lim_{x \to 0} \frac{1}{x^2} = \infty \qquad \qquad ♦♦♦$$

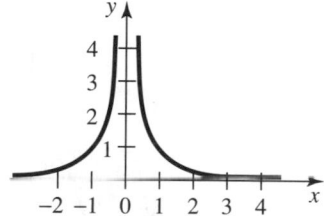

FIGURE 27–1 Graph of $y = 1/x^2$.

Visualizing a Limit

The simplest way to find the approximate limit of a function is to graph that function and observe graphically what it does at the limiting value.

♦♦♦ **Example 2:** Graphically evaluate

$$\lim_{x \to 0} \frac{(x + 1)^2 - 1}{x}$$

Solution: We graph the given expression in Fig. 27–2. We get no value for $x = 0$, as that results in division by zero. But we note that the value of the expression approaches 2 as x approaches zero, so we write

$$\lim_{x \to 0} \frac{(x + 1)^2 - 1}{x} = 2 \qquad \qquad ♦♦♦$$

Limits Involving Zero or Infinity

Limits involving zero or infinity can usually be evaluated using the following facts. If C is a nonzero constant, we have the following equations:

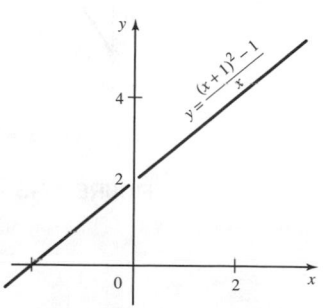

FIGURE 27–2

(1) $\lim_{x \to 0} Cx = 0$	(4) $\lim_{x \to \infty} Cx = \infty$
(2) $\lim_{x \to 0} \dfrac{x}{C} = 0$	(5) $\lim_{x \to \infty} \dfrac{x}{C} = \infty$
(3a) $\lim_{x \to 0^+} \dfrac{C}{x} = +\infty$	(6) $\lim_{x \to \infty} \dfrac{C}{x} = 0$
(3b) $\lim_{x \to 0^-} \dfrac{C}{x} = -\infty$	

◆◆◆ **Example 3:**

(a) $\lim_{x \to 0} 5x^3 = 0$ (b) $\lim_{x \to 0} \left(7 + \dfrac{x}{2}\right) = 7$

(c) $\lim_{x \to 0} \left(3 + \dfrac{25}{x^2}\right) = \infty$ (d) $\lim_{x \to \infty} (3x - 2) = \infty$

(e) $\lim_{x \to \infty} \left(8 + \dfrac{x}{4}\right) = \infty$ (f) $\lim_{x \to \infty} \left(\dfrac{5}{x} - 3\right) = -3$

Each function is graphed in Fig. 27–3.

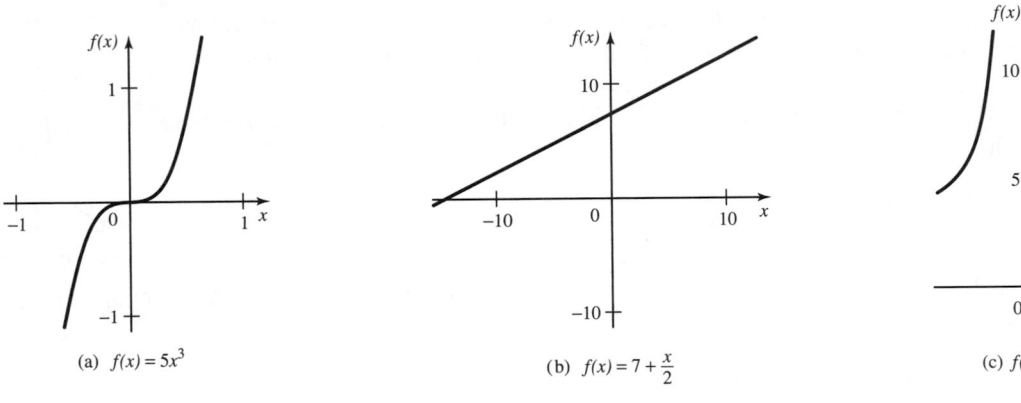

(a) $f(x) = 5x^3$

(b) $f(x) = 7 + \dfrac{x}{2}$

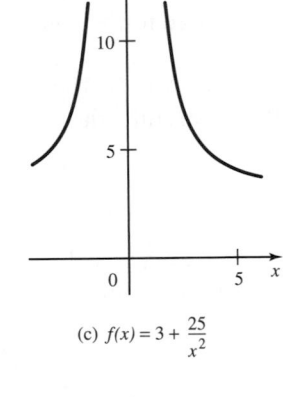

(c) $f(x) = 3 + \dfrac{25}{x^2}$

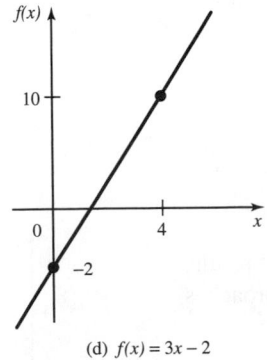

(d) $f(x) = 3x - 2$

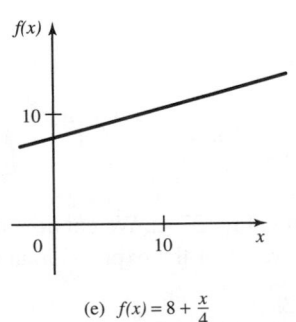

(e) $f(x) = 8 + \dfrac{x}{4}$

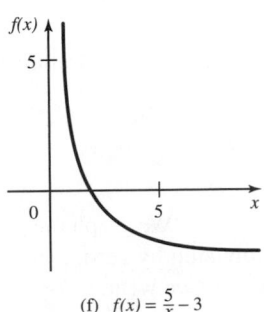

(f) $f(x) = \dfrac{5}{x} - 3$

FIGURE 27–3

Rules 3a and 3b need more explanation. The limit of $y = 3x$, as x approached 2, was 6. It did not matter whether x approached 2 from below or above (from values less than 2 or from values greater than 2). But sometimes it does matter.

◆◆◆ **Example 4:** The function $y = 1/x$ is graphed in Fig. 27–4. When x approaches zero from "above" (that is, from the *right* in Fig. 27–4), which we write $x \to 0^+$, we get

$$\lim_{x \to 0^+} \frac{1}{x} = +\infty$$

which means that y grows without bound in the positive direction. But when x approaches zero from "below" (from the *left* in Fig. 27–4), we write

$$\lim_{x \to 0^-} \frac{1}{x} = -\infty$$

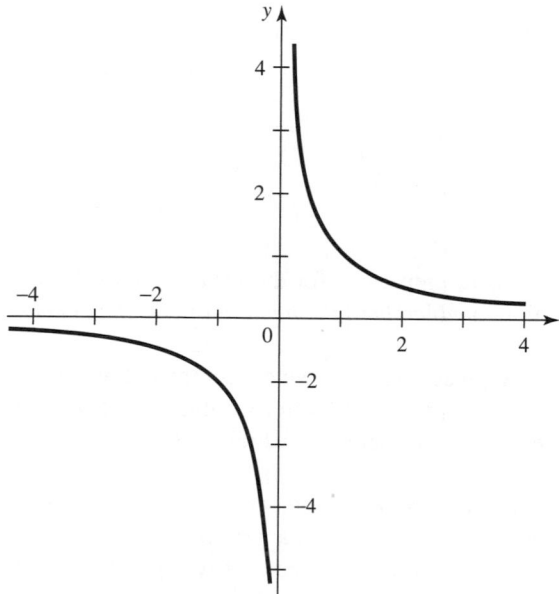

FIGURE 27–4   Graph of $y = 1/x$.

which means that y grows without bound in the negative direction.   ◆◆◆

When we want the limit, as x becomes infinite, of the quotient of two polynomials, such as

$$\lim_{x \to \infty} \frac{4x^3 - 3x^2 + 5}{3x - x^2 - 5x^3}$$

we see that both numerator and denominator become infinite. However, the limit of such an expression can be found if *we divide both numerator and denominator by the highest power of x occurring in either*.

◆◆◆ **Example 5:** Evaluate $\lim_{x \to \infty} \dfrac{4x^3 - 3x^2 + 5}{3x - x^2 - 5x^3}$.

Solution: Dividing both numerator and denominator by x^3 gives us

$$\lim_{x \to \infty} \frac{4x^3 - 3x^2 + 5}{3x - x^2 - 5x^3} = \lim_{x \to \infty} \frac{4 - \dfrac{3}{x} + \dfrac{5}{x^3}}{\dfrac{3}{x^2} - \dfrac{1}{x} - 5}$$

$$= \frac{4 - 0 + 0}{0 - 0 - 5} = -\frac{4}{5}$$

One of the theorems on limits (which we will not prove) is that the limit of the sum of several functions is equal to the sum of their individual limits.

as shown in Fig. 27–5.

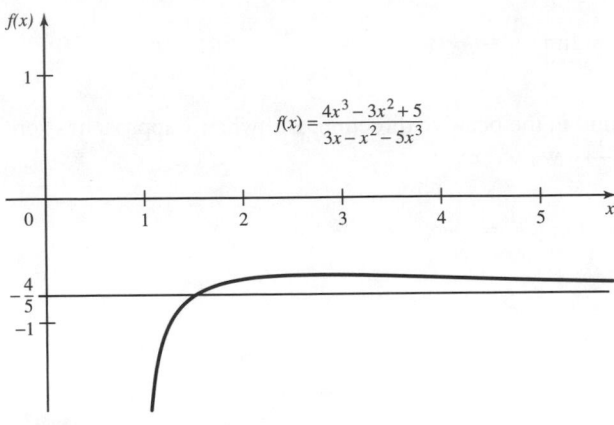

$$f(x) = \frac{4x^3 - 3x^2 + 5}{3x - x^2 - 5x^3}$$

FIGURE 27–5   ◆◆◆

Limits of the Form 0/0

The purpose of this entire discussion of limits is to prepare us for the idea of the *derivative*. There we will have to find the limit of a fraction in which *both the numerator and the denominator approach zero*.

At first glance, when we see a numerator approaching zero, we expect the entire fraction to approach zero. But when we see a denominator approaching zero, we throw up our hands and cry "division by zero." What then do we make of a fraction in which *both* numerator and denominator approach zero?

First, keep in mind that the denominator is not *equal to* zero; it is only *approaching* zero. Second, even though a shrinking numerator would make a fraction approach zero, in this case the denominator is also shrinking. So, in fact, our fraction will not necessarily approach infinity, or zero, but will often approach some useful finite value.

If you can't find the limit of an expression in its given form, try to manipulate the expression into another form in which the limit can be found. This is often done simply by performing the indicated operations. Factoring the given expression will often help us find a limit.

◆◆◆ **Example 6:** Evaluate $\lim\limits_{x \to 3} \dfrac{x^2 - 9}{x - 3}$.

Solution: As x approaches 3, both numerator and denominator approach zero, which tells us nothing about the limit of the entire expression. However, if we factor the numerator, we get

$$\lim_{x \to 3} \frac{x^2 - 9}{x - 3} = \lim_{x \to 3} \frac{(x - 3)(x + 3)}{x - 3}$$

$$= \lim_{x \to 3} \left(\frac{x - 3}{x - 3} \right)(x + 3)$$

Now as x approaches 3, both the numerator and the denominator of the fraction $(x - 3)/(x - 3)$ approach zero. But since the numerator and denominator are equal, the fraction will equal 1 for any nonzero value of x, no matter how small. So

$$\lim_{x \to 3} \frac{x^2 - 9}{x - 3} = \lim_{x \to 3} (x + 3) = 6$$

as shown in Fig. 27–6.

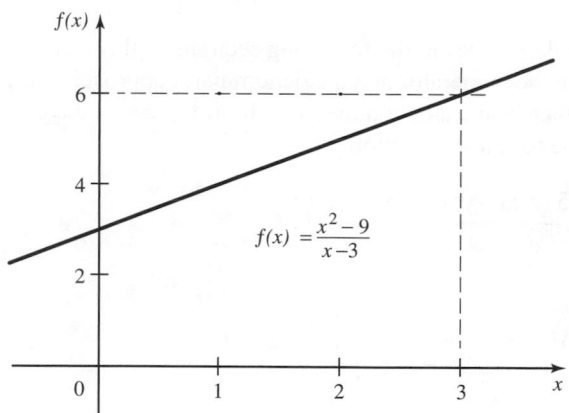

FIGURE 27-6

◆◆◆ Example 7:

$$\lim_{x \to 2} \frac{x^2 - 3x + 2}{x - 2} = \lim_{x \to 2} \frac{(x - 2)(x - 1)}{x - 2}$$

$$= \lim_{x \to 2} \left(\frac{x - 2}{x - 2}\right)(x - 1)$$

$$= \lim_{x \to 2} (x - 1) = 1$$

as shown in Fig. 27–7.

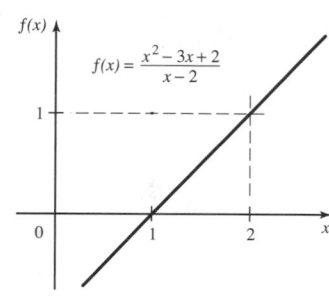

◆◆◆ FIGURE 27-7

Limits Found Numerically

A simple numerical way to evaluate a limit is to substitute values of x closer and closer to the value that it is to approach and see if the expression approaches a limit. We will use such a numerical method in the next example, where we evaluate a limit of the form 0/0.

This, of course, is not a *proof* that the limit found is the correct one or even that a limit exists.

◆◆◆ Example 8: Evaluate $\lim_{x \to 0} \dfrac{\sqrt{9 + x} - 3}{3x}$.

Solution: We see that when x *equals* zero, we get

$$\frac{\sqrt{9 + 0} - 3}{3(0)} = \frac{3 - 3}{0}$$

$$= \frac{0}{0}$$

a result that is indeterminate. But what happens when x *approaches* zero? Let us use the calculator to substitute smaller and smaller values of x. Working to five decimal places, we get the following values:

x	10	1	0.1	0.01	0.001	0.0001
	0.045 30	0.054 09	0.055 40	0.055 54	0.055 55	0.055 56

So as x approaches zero, the given expression appears to approach $0.055\,5\overline{5}$ as a limit (Fig. 27–8). ◆◆◆

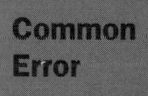

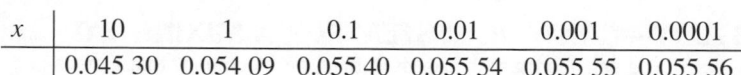

Be sure to distinguish between a denominator that is zero and one that is **approaching** zero. In the first case we have division by zero, but in the second case we might get a useful answer.

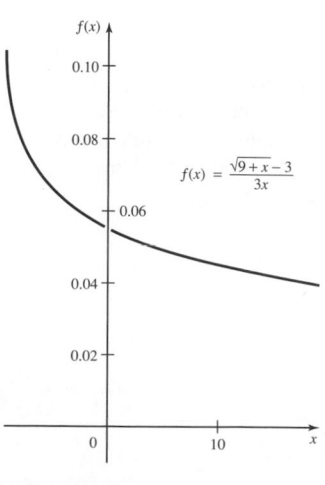

FIGURE 27-8

When the Limit Is an Expression

Our main use for limits will be for finding derivatives in the following sections of this chapter. There we will evaluate limits in which both the numerator and the denominator approach zero, and the resulting limit is an *expression* rather than a single number. A limit typical of the sort we will have to evaluate later is given in the following example.

◆◆◆ **Example 9:** Evaluate $\displaystyle\lim_{d \to 0} \frac{(x+d)^2 + 5(x+d) - x^2 - 5x}{d}$.

Solution: If we try to set d equal to zero,

$$\frac{(x+0)^2 + 5(x+0) - x^2 - 5x}{0} = \frac{x^2 + 5x - x^2 - 5x}{0} = \frac{0}{0}$$

we get the indeterminate expression $0/0$. Instead, let us remove parentheses in the original expression. We get

$$\lim_{d \to 0} \frac{(x+d)^2 + 5(x+d) - x^2 - 5x}{d}$$

$$= \lim_{d \to 0} \frac{x^2 + 2dx + d^2 + 5x + 5d - x^2 - 5x}{d}$$

$$= \lim_{d \to 0} \frac{2dx + d^2 + 5d}{d} = \lim_{d \to 0} (2x + d + 5)$$

after dividing through by d. Now letting d approach zero, we obtain

$$\lim_{d \to 0} \frac{(x+d)^2 + 5(x+d) - x^2 - 5x}{d} = 2x + 5 \qquad\qquad ◆◆◆$$

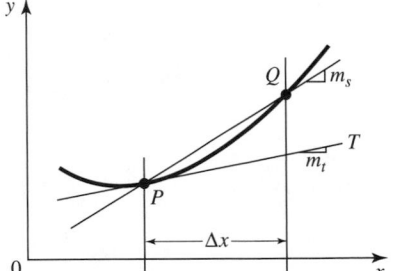

FIGURE 27–9 Tangent to a curve as the limiting position of the secant. This figure will have an important role in our introduction of the derivative in the following section.

Slope of Tangent Line as Limit of Slope of Secant

One of our most important uses of limits will come in the following section, where we find the slope of a tangent line drawn to a curve at some point P. In Sec. 22–1, we defined the tangent line at P as the limiting position of the secant line PQ as Q approached P. If Δx is the horizontal distance between P and Q (Fig. 27–9), we say that Q approaches P as Δx approaches zero. We can thus say that the slope m_t of the tangent line is the limit of the slope m_s of the secant line as Δx approaches zero. We can now restate this idea in compact form using our new limit notation, as follows:

$$m_t = \lim_{\Delta x \to 0} m_s$$

CASE STUDY—CONTROL SYSTEM FOR A MIXING VAT

You are an electronics engineering technologist working for a company that installs and configures instrumentation and control systems for industrial processes. In a vat for mixing a slurry compound, you have installed low- and high-level alarms to warn if the vat gets too close to empty or is in danger of overflowing. The plant operators point out that if the largest of the intake valves gets stuck in the open position, by the time it reaches the high-level alarm, they will not have enough time to react. It is not an option to set the high-level alarm to a lower position because in order to give the operators enough time to react, the alarm would need to be set below normal operating levels and would be going off constantly. Studying the process, you determine that the largest valve would never be open continuously during normal operation. You also notice that a combination of faults in the smaller inlets would also have the level rising too fast for the operators to react. Suggest a solution using principles from this chapter.

Exercise 1 ◆ Limits

Evaluate each limit.

1. $\lim\limits_{x \to 2} (x^2 + 2x - 7)$

2. $\lim\limits_{x \to -1} (x^3 - 3x^2 - 5x - 5)$

3. $\lim\limits_{x \to 2} \dfrac{x^2 - x - 1}{x + 3}$

4. $\lim\limits_{x \to 5} \dfrac{5 + 4x - x^2}{5 - x}$

5. $\lim\limits_{x \to 5} \dfrac{x^2 - 25}{x - 5}$

6. $\lim\limits_{x \to -7} \dfrac{49 - x^2}{x + 7}$

7. $\lim\limits_{x \to 1} \dfrac{x^2 + 2x - 3}{x - 1}$

8. $\lim\limits_{x \to 5} \dfrac{x^2 - 12x + 35}{5 - x}$

Limits Involving Zero or Infinity

9. $\lim\limits_{x \to 0} (4x^2 - 5x - 8)$

10. $\lim\limits_{x \to 0} \dfrac{3 - 2x}{x + 4}$

11. $\lim\limits_{x \to 0} \dfrac{\sqrt{x} - 4}{\sqrt[3]{x} + 5}$

12. $\lim\limits_{x \to 0} \dfrac{3 + x - x^2}{(x + 3)(5 - x)}$

13. $\lim\limits_{x \to 0} \left(\dfrac{1}{2 + x} - \dfrac{1}{2} \right) \cdot \dfrac{1}{x}$

14. $\lim\limits_{x \to 0} x \cos x$

15. $\lim\limits_{x \to \infty} \dfrac{2x + 5}{x - 4}$

16. $\lim\limits_{x \to \infty} \dfrac{5x - x^2}{2x^2 - 3x}$

17. $\lim\limits_{x \to \infty} \dfrac{x^2 + x - 3}{5x^2 + 10}$

18. $\lim\limits_{x \to \infty} \dfrac{4 + 2^x}{3 + 2^x}$

19. $\lim\limits_{x \to -\infty} 10^x$

20. $\lim\limits_{x \to \infty} 10^x$

Limits of the Form 0/0

21. $\lim\limits_{x \to 0} \dfrac{\sin x}{\tan x}$

22. $\lim\limits_{x \to 4} \dfrac{x^2 - 16}{x - 4}$

23. $\lim\limits_{x \to -3} \dfrac{x^2 + 2x - 3}{x + 3}$

24. $\lim\limits_{x \to 1} \dfrac{x^3 - x^2 + 2x - 2}{x - 1}$

Limits Depending on Direction of Approach

25. $\lim\limits_{x \to 0^+} \dfrac{7}{x}$

26. $\lim\limits_{x \to 0^+} \dfrac{e^x}{x}$

27. $\lim\limits_{x \to 0^+} \dfrac{x + 1}{x}$

28. $\lim\limits_{x \to 0^-} \dfrac{x + 1}{x}$

29. $\lim\limits_{x \to 2^-} \dfrac{5 + x}{x - 2}$

30. $\lim\limits_{x \to 1^+} \dfrac{2x + 3}{1 - x}$

When the Limit Is an Expression

31. $\lim\limits_{d \to 0} x^2 + 2d + d^2$

32. $\lim\limits_{d \to 0} 2x + d$

33. $\lim\limits_{d \to 0} \dfrac{(x + d)^2 - x^2}{x^2(x + d)}$

34. $\lim\limits_{d \to 0} \dfrac{(x + d)^2 - x^2}{d}$

35. $\lim\limits_{d \to 0} \dfrac{3(x + d) - 3x}{d}$

36. $\lim\limits_{d \to 0} \dfrac{[2(x + d) + 5] - (2x + 5)}{d}$

37. $\lim\limits_{d \to 0} 3x + d - \dfrac{1}{(x + d + 2)(x - 2)}$

38. $\lim\limits_{d \to 0} \dfrac{[(x + d)^2 + 1] - (x^2 + 1)}{d}$

39. $\lim\limits_{d \to 0} \dfrac{(x + d)^3 - x^3}{d}$

40. $\lim\limits_{d \to 0} \dfrac{\sqrt{x + d} - \sqrt{x}}{d}$

41. $\lim\limits_{d \to 0} \dfrac{(x + d)^2 - 2(x + d) - x^2 + 2x}{d}$

42. $\lim\limits_{d \to 0} \dfrac{\dfrac{7}{x + d} - \dfrac{7}{x}}{d}$

27–2   The Derivative

Rate of Change

We will begin with some familiar ideas about motion to introduce the idea of *rate of change,* which will then lead us to the derivative.

Recall that in *uniform motion,* the rate of change of displacement with respect to time (the velocity) is constant. The graph of displacement versus time, shown in Fig. 27–10(a), is a straight line whose *slope is equal to the rate of change of displacement with respect to time.*

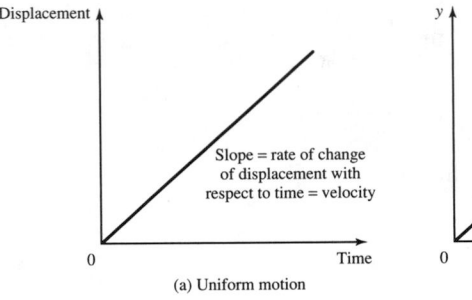
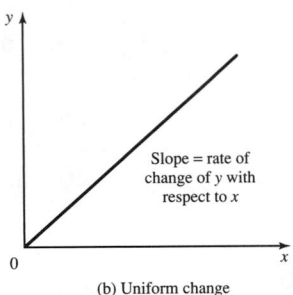

FIGURE 27–10

In general, for a function whose graph is a straight line [Fig. 27–10(b)], *the slope gives the rate of change of y with respect to x.*

Average Rate of Change

For a function $y = f(x)$ whose graph is *not* a straight line (Fig. 27–11), we cannot give the rate of change for the entire function. However, we can choose an interval on the curve and give the *average rate of change over that interval.* For the interval PQ (Fig. 27–12), the average rate of change is equal to the change in y divided by the change in x over that interval. Thus *the slope of the secant line* PQ *gives us the average rate of change from* P *to* Q.

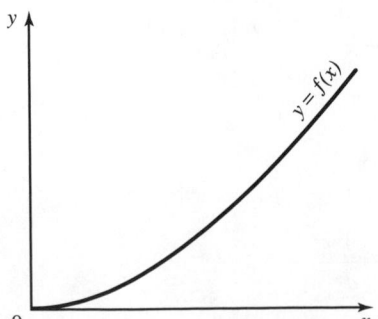

FIGURE 27–11   Nonuniform change.

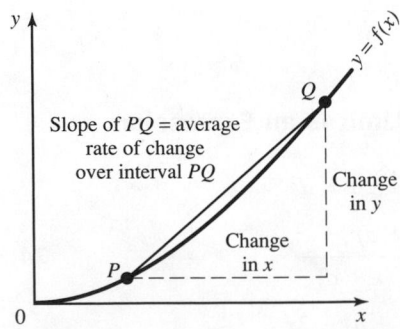

FIGURE 27–12   Average rate of change.

◆◆◆**Example 10:** Find the average rate of change of the function $y = 3x^2 + 1$ over the interval $P(1, 4)$ to $Q(3, 28)$ (Fig. 27–13).

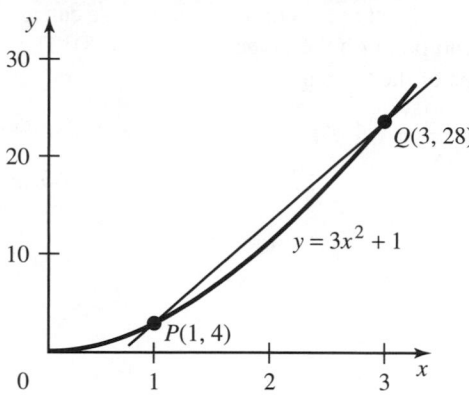

FIGURE 27–13

Solution: The average rate of change is equal to the slope of secant line PQ.

$$\text{average rate of change} = \text{slope of } PQ = \frac{28 - 4}{3 - 1} = 12$$

◆◆◆

The Tangent Gives Instantaneous Rate of Change: A Graphical Method

In Sec. 22–1 we defined the slope of a curve at a point P as equal to the *slope of the tangent* T *to the curve* drawn through P. In this case, it is the slope of the tangent line T that is tangent to the curve at point P. (Fig. 22–14).

Also recall that in our study of limits, we said that the slope m_t of a tangent line to a curve (Fig. 27–9) is the limiting value of the slope m_s of the secant line, as Δx approaches zero.

$$m_t = \lim_{\Delta x \to 0} m_s$$

This is illustrated again in Fig. 27–14. Since slope gives us rate of change, *the slope of the tangent at a point* P *on a curve gives us the instantaneous rate of change at* P.

In the following example we will estimate the slope of the tangent of a given curve.

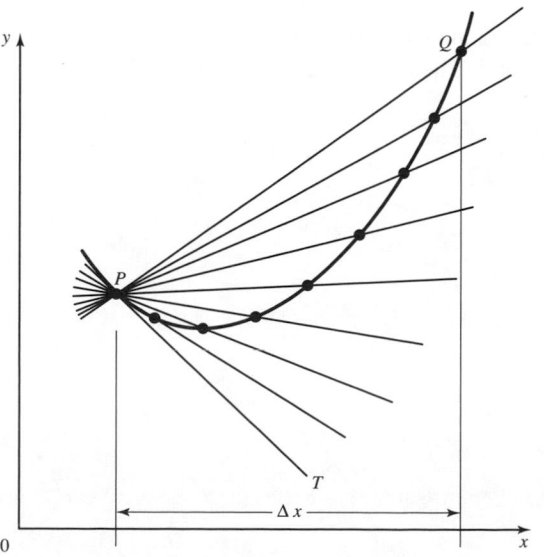

FIGURE 27–14

In some fields, such as mechanism analysis, we might have a graph of displacement versus time for a point on the mechanism, but no equation. To get the velocity of that point, we need the rate of change of displacement, given by the tangent to the curve. In the laboratory, tangents can be drawn more accurately by placing a front-surface mirror, such as a polished piece of metal, perpendicular to the curve at the desired point. The mirror is then adjusted so that the curve appears to merge with its mirror image, with no corner. The normal is then traced, and the tangent is constructed at right angles to the normal.

◆◆◆ **Example 11:** For the function given graphically in Fig. 27–15, graphically estimate the instantaneous rate of change at $P(2, 2)$.

Solution: We place a straightedge at $(2, 2)$ so that it is tangent to the curve, as well as we can tell by eye, and draw the tangent. We then locate a second point on the tangent, say, $(4, 2.8)$. Then the instantaneous rate of change is given by the slope of the tangent.

$$\text{slope} = \frac{\text{rise}}{\text{run}} \cong \frac{2.8 - 2}{4 - 2} = 0.4$$

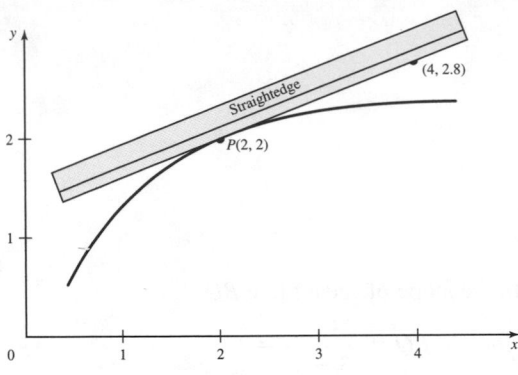

FIGURE 27–15   ◆◆◆

The Tangent as the Limiting Value of the Secant:
A Numerical Method

If we have an equation for the given function, we can compute the approximate slope of the tangent line by the following numerical method.

We place Q at a horizontal distance Δx from P. Then we let Q approach P in steps, at each step computing the slope m_s of the secant line PQ. We then reduce Δx and repeat the computation.

By watching how the slope changes as we proceed, we can get a measure of the accuracy of our answer. We continue until the computed value does not change within the accuracy we want.

◆◆◆ **Example 12:** Use a numerical method to find the instantaneous rate of change for the function $y = x^2$ at $P(1, 1)$, approximate to three significant digits.

Solution: Let us place Q to the right of P at a distance of $\Delta x = 1$ as shown in Fig. 27–16. The abscissa of Q is then 2, and the ordinate is 2^2 or 4. The slope of the secant PQ from $P(1, 1)$ to $Q(2, 4)$ is

$$\text{slope} = \frac{\Delta y}{\Delta x} = \frac{4 - 1}{2 - 1} = 3$$

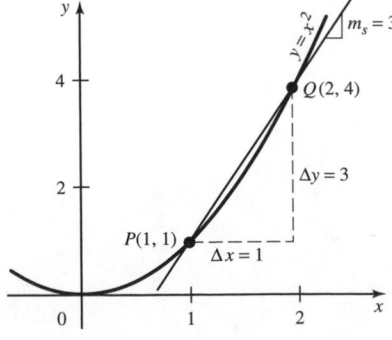

FIGURE 27–16

Let us then repeat the computation by computer, halving Δx each time. We get the values shown in Table 27–1.

TABLE 27–1

Δx	x	y	Δy	Slope
1.0000	2.0000	4.0000	3.0000	3.0000
0.5000	1.5000	2.2500	1.2500	2.5000
0.2500	1.2500	1.5625	0.5625	2.2500
0.1250	1.1250	1.2656	0.2656	2.1248
0.0625	1.0625	1.1289	0.1289	2.0624
0.0313	1.0313	1.0636	0.0636	2.0319
0.0156	1.0156	1.0314	0.0314	2.0128
0.0078	1.0078	1.0157	0.0157	2.0128
0.0039	1.0039	1.0078	0.0078	2.0000
0.0020	1.0020	1.0040	0.0040	2.0000
0.0010	1.0010	1.0020	0.0020	2.0000
0.0005	1.0005	1.0010	0.0010	2.0000
0.0002	1.0002	1.0004	0.0004	2.0000
0.0001	1.0001	1.0002	0.0002	2.0000

Notice that the value of the slope appears to be reaching a limiting value of 2. Also, if we wanted only three significant digits, we could have stopped at $\Delta x = 0.0039$ because the second decimal place in the slope no longer changes beyond that value.   ◆◆◆

The Derivative

We see that we can get the approximate slope of the tangent line, and hence the rate of change of the function, at any point we wish. Our next step is to derive a *formula* for finding the same thing. The advantages of having a formula or equation are that (1) it will give us the rate of change *everywhere* on the curve (we merely have to substitute the point into the equation); (2) it will give us the *exact* value; and (3) it is much less work to use an equation than the numerical or graphical methods.

We will derive the formula in exactly the same way that we found the slope of the tangent in Example 12. The difference is that now, instead of working with specific numbers, we will work with *symbols*.

Starting from $P(x, y)$ on the curve $y = f(x)$ (Fig. 27–17), we locate a second point Q spaced from P by a horizontal increment Δx. In a run of Δx, the graph of $y = f(x)$ rises by an amount that we call Δy. PQ then has a rise Δy in a run of Δx, so

$$\text{slope of } PQ = \frac{\text{rise}}{\text{run}} = \frac{\Delta y}{\Delta x} = \frac{f(x + \Delta x) - f(x)}{\Delta x}$$

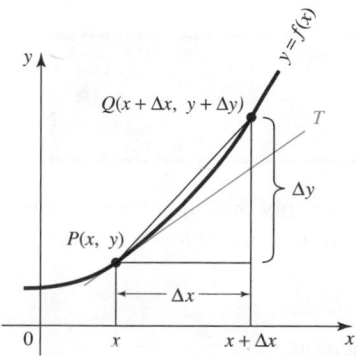

FIGURE 27–17 Derivation of the derivative.

We now let Δx approach zero, and point Q will approach P along the curve. Since the tangent PT is the limiting position of the secant PQ, the slope of PQ will approach the slope of PT.

$$\text{slope of } PT = \lim_{\Delta x \to 0} \frac{\Delta y}{\Delta x} = \lim_{\Delta x \to 0} \frac{f(x + \Delta x) - f(x)}{\Delta x}$$

This important quantity is called the *derivative* of y with respect to x and is given the symbol $\dfrac{dy}{dx}$.

We will see later that *dy* and *dx* in the symbol *dy/dx* have separate meanings of their own. But for now we will treat *dy/dx* as a single symbol.

The Derivative	or	$\dfrac{dy}{dx} = \lim\limits_{\Delta x \to 0} \dfrac{\Delta y}{\Delta x}$ $\dfrac{dy}{dx} = \lim\limits_{\Delta x \to 0} \dfrac{f(x + \Delta x) - f(x)}{\Delta x}$	338

The derivative is the instantaneous rate of change of the function $y = f(x)$ with respect to the variable x.

◆◆◆ **Example 13:** Find the derivative of $f(x) = x^2$.

Solution: By Eq. 338,

$$\begin{aligned}
\frac{dy}{dx} &= \lim_{\Delta x \to 0} \frac{f(x + \Delta x) - f(x)}{\Delta x} \\
&= \lim_{\Delta x \to 0} \frac{(x + \Delta x)^2 - x^2}{\Delta x} \\
&= \lim_{\Delta x \to 0} \frac{x^2 + 2x\,\Delta x + (\Delta x)^2 - x^2}{\Delta x} \\
&= \lim_{\Delta x \to 0} \frac{2x\,\Delta x + (\Delta x)^2}{\Delta x} \\
&= \lim_{\Delta x \to 0} (2x + \Delta x) = 2x
\end{aligned}$$

◆◆◆

Common Error	The symbol Δx is a *single symbol*. It is not Δ times x. Thus $x \cdot \Delta x \neq \Delta x^2$

Derivatives by the Delta Method

Since $y = f(x)$, we can also write Eq. 338 using y instead of $f(x)$.

The Derivative	$\dfrac{dy}{dx} = \lim\limits_{\Delta x \to 0} \dfrac{(y + \Delta y) - y}{\Delta x}$	338

Our main use for the delta method will be to derive rules with which we can quickly find derivatives.

Instead of substituting directly into Eq. 338, some prefer to apply this equation as a series of steps. This is sometimes referred to as the *delta method, delta process,* or *four-step rule.*

◆◆◆ **Example 14:**

(a) Find the derivative of the function $y = 3x^2$ by the delta method.

(b) Evaluate the derivative at $x = 1$.

Solution:

(a) Solve using the following steps:

1. Starting from $P(x, y)$, shown in Fig. 27–18(a), locate a second point Q, spaced from P by a horizontal distance Δx and by a vertical distance Δy. Since the coordinates of $Q(x + \Delta x, y + \Delta y)$ must satisfy the given function, we may substitute $x + \Delta x$ for x and $y + \Delta y$ for y in the original equation.

$$y + \Delta y = 3(x + \Delta x)^2$$
$$= 3[x^2 + 2x\,\Delta x + (\Delta x)^2]$$
$$= 3x^2 + 6x\,\Delta x + 3(\Delta x)^2$$

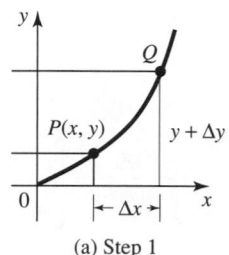

(a) Step 1

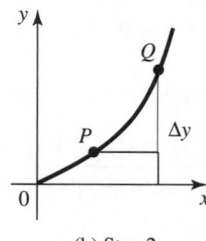

(b) Step 2

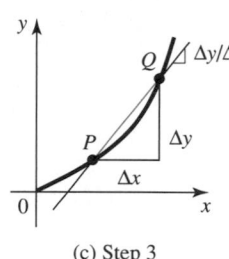

(c) Step 3

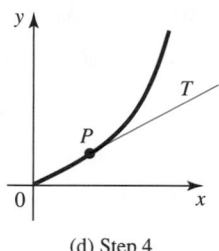

(d) Step 4

FIGURE 27–18 Derivatives by the delta method.

2. Find the rise Δy from P to Q [Fig. 27–18(b)] by subtracting the original function

$$(y + \Delta y) - y = 3x^2 + 6x\,\Delta x + 3(\Delta x)^2 - 3x^2$$
$$\Delta y = 6x\,\Delta x + 3(\Delta x)^2$$

Of course, rise can be either positive or negative. Thus a curve could, in fact, fall in a given run, but we still refer to it as the rise.

3. Find the slope of the secant line PQ [Fig. 27–18(c)] by dividing the rise Δy by the run Δx.

$$\frac{\Delta y}{\Delta x} = \frac{6x\,\Delta x + 3(\Delta x)^2}{\Delta x}$$

4. Let Δx approach zero. This causes Δy also to approach zero and appears to make $\Delta y/\Delta x$ equal to the indeterminate expression $0/0$. But recall from our study of limits in the preceding section that a fraction can often have a limit *even when both numerator and denominator approach zero*. To find it, we divide through by Δx and get $\Delta y/\Delta x = 6x + 3\,\Delta x$. Then the slope of the tangent T [Fig. 27–18(d)] is

$$\frac{dy}{dx} = \lim_{\Delta x \to 0} \frac{\Delta y}{\Delta x} = 6x + 3(0) = 6x$$

(b) When $x = 1$,

$$\left.\frac{dy}{dx}\right|_{x=1} = 6(1) = 6$$

The curve $y = 3x^2$ is graphed in Fig. 27–19 with a tangent line of slope 6 drawn at the point $(1, 3)$.

◆◆◆

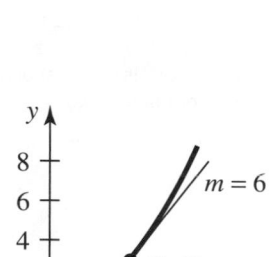

FIGURE 27–19 Graph of $y = 3x^2$.

If our expression is a fraction with x in the denominator, Step 2 will require us to find a common denominator, as in the following example.

◆◆◆ **Example 15:** Find the derivative of $y = \dfrac{3}{x^2 + 1}$.

Solution:

1. Substitute $x + \Delta x$ for x and $y + \Delta y$ for y.

$$y + \Delta y = \frac{3}{(x + \Delta x)^2 + 1}$$

2. Subtracting yields

$$y + \Delta y - y = \frac{3}{(x + \Delta x)^2 + 1} - \frac{3}{x^2 + 1}$$

Combining the fractions over a common denominator gives

$$\Delta y = \frac{3(x^2 + 1) - 3[(x + \Delta x)^2 + 1]}{[(x + \Delta x)^2 + 1](x^2 + 1)}$$

which simplifies to

$$\Delta y = \frac{-6x\,\Delta x - 3(\Delta x)^2}{[(x + \Delta x)^2 + 1](x^2 + 1)}$$

3. Dividing by Δx, we obtain

$$\frac{\Delta y}{\Delta x} = \frac{-6x - 3\,\Delta x}{[(x + \Delta x)^2 + 1](x^2 + 1)}$$

4. Letting Δx approach zero gives

$$\frac{dy}{dx} = -\frac{6x}{(x^2 + 1)^2} \qquad\qquad ◆◆◆$$

The following example shows how to use the delta method to differentiate an expression containing a radical.

◆◆◆ **Example 16:** Find the slope of the tangent to the curve $y = \sqrt{x}$ at the point $(4, 2)$.

Solution: We first find the derivative in four steps.

1. $y + \Delta y = \sqrt{x + \Delta x}$
2. $\Delta y = \sqrt{x + \Delta x} - \sqrt{x}$

> When simplifying radicals, we used to rationalize the denominator. Here we rationalize the numerator instead!

The later steps will be easier if we now write this expression as a fraction with no radicals in the numerator. To do this, we multiply by the conjugate and get

$$\Delta y = (\sqrt{x + \Delta x} - \sqrt{x}) \cdot \frac{\sqrt{x + \Delta x} + \sqrt{x}}{\sqrt{x + \Delta x} + \sqrt{x}}$$

$$= \frac{(x + \Delta x) - x}{\sqrt{x + \Delta x} + \sqrt{x}}$$

$$= \frac{\Delta x}{\sqrt{x + \Delta x} + \sqrt{x}}$$

3. Dividing by Δx yields

$$\frac{\Delta y}{\Delta x} = \frac{1}{Rx + \Delta x + \sqrt{x}}$$

4. Letting Δx approach zero gives

$$\frac{dy}{dx} = \frac{1}{2\sqrt{x}}$$

When $x = 4$,

$$\frac{dy}{dx}\Big|_{x=4} = \frac{1}{4}$$

which is the slope of the tangent to $y = \sqrt{x}$ at the point (4, 2).   ◆◆◆

Other Symbols for the Derivative

Other symbols are used instead of dy/dx to indicate a derivative. A prime (') symbol is often used. For example, if a certain function is equal to y, then the symbol y' can represent the derivative of that function. Further, if $f(x)$ is a certain function, then $f'(x)$ is the derivative of that function. Thus

$$\frac{dy}{dx} = y' = f'(x)$$

The y' or f' notation is handy for specifying the derivative at a particular value of x.

◆◆◆**Example 17:** To specify a derivative evaluated at $x = 2$, we can write $y'(2)$, or $f'(2)$, instead of the clumsier

$$\frac{dy}{dx}\Big|_{x=2}$$

   ◆◆◆

The Derivative as an Operator

We can think of the derivative as an *operator:* one that operates on a function to produce the derivative of that function. The symbol d/dx or D in front of an expression indicates that the expression is to be differentiated. For example, the symbols

$$\frac{d}{dx}(\ ) \quad \text{or} \quad D_x(\ ) \quad \text{or} \quad D(\ )$$

mean to find the derivative of the expression enclosed in parentheses. $D(\ )$ means to differentiate the function with respect to its independent variable.

◆◆◆**Example 18:** We saw in Example 14(a) that if $y = 3x^2$, then $dy/dx = 6x$. This same result can be written

$$\frac{d(3x^2)}{dx} = 6x$$
$$D_x(3x^2) = 6x$$
$$D(3x^2) = 6x$$

   ◆◆◆

Keep in mind that even though the notation is different, we find the derivative in exactly the same way.

Continuity and Discontinuity

A curve is called *continuous* if it contains no breaks or gaps, and it is said to be *discontinuous* at a value of x where there is a break or gap. The derivative does not exist at such points. It also does not exist where the curve has a jump, corner, cusp, or any other feature at which it is not possible to draw a tangent line, as at the points shown in Fig. 27–20. At such points, we say that the function is *not differentiable.*

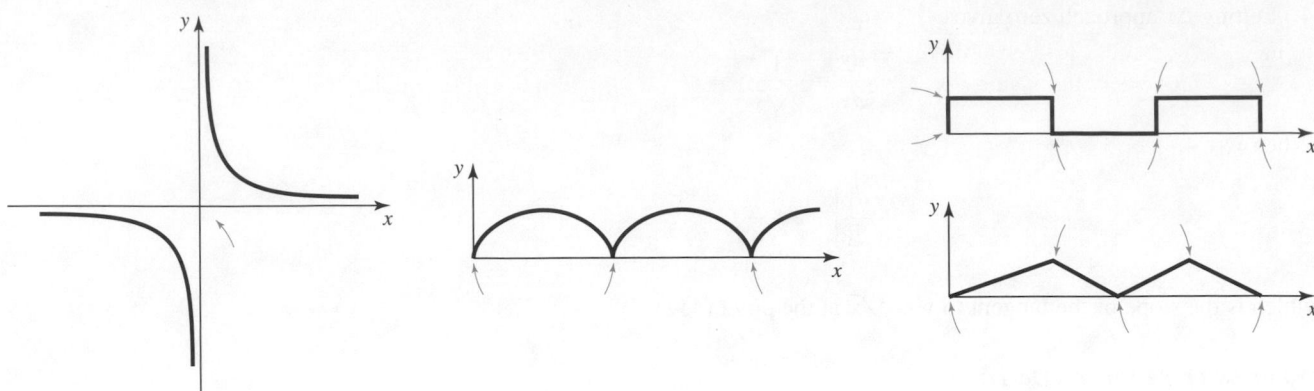

FIGURE 27-20   The arrows show the points at which the derivative does not exist.

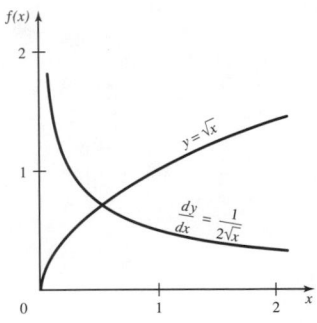

FIGURE 27-21

Graph of a Derivative

A derivative function can be graphed just as any other function. We will sometimes graph the derivative on the same axes as the original function in order to point out various relationships between the two.

◆◆◆ **Example 19:** Graph the function $y = \sqrt{x}$ and its derivative $dy/dx = 1/(2\sqrt{x})$ (which we found in Example 16) on the same axes.

Solution: The graphs are shown in Fig. 27–21.   ◆◆◆

Derivative of a Function That Is Given Numerically

We have taken the derivative of a function given in the form of a graph, and of a function given as a formula. But sometimes we want the derivative of a function given as a table of point pairs. We can get an estimate of such a function by computing the slopes of the lines joining adjacent points in the table.

◆◆◆ **Example 20:** Estimate the derivative of the following function:

x	1.00	1.10	1.20	1.30	1.40	1.50	1.60	1.70
y	2.05	2.20	2.60	3.10	3.70	4.30	5.00	6.10

Solution: In Table 27–2, we show the difference Δy between adjacent y values in column 3 and that quantity divided by the difference $\Delta x = 0.1$ between x values in column 4.

Some computer algebra systems can take derivatives of a set of data such as this. *Derive*, for example, uses the DIF_DATA( ) command.

TABLE 27-2

x	y	Δy	**Slope**
1.00	2.05		
1.10	2.20	0.15	1.50
1.20	2.60	0.40	4.00
1.30	3.10	0.50	5.00
1.40	3.70	0.60	6.00
1.50	4.30	0.60	6.00
1.60	5.00	0.70	7.00
1.70	6.10	1.10	11.00

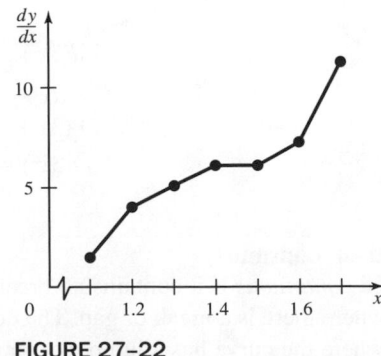

FIGURE 27-22

The approximate derivative, column 4, is graphed in Fig. 27–22.   ◆◆◆

In the above computation, we found the slope of the line connecting any point and the point to its *right*. We could just as well have chosen to use the point to the *left*. A refinement of this method, which we will not show, uses the *average* of the slopes found to the right and the left.

Summary of Methods for Finding Derivatives

Never before have there been so many ways for finding a derivative. Table 27–3 might help to keep track of these methods.

The first column indicates whether the function is given graphically, numerically, or by formula. Only three of these methods give an exact derivative, indicated in the third column. Whether the method gives the derivative at a single point or gives a derivative function over the entire domain of x is indicated by *point* or *function* in the fourth column. The last column shows where in this chapter the method can be found.

TABLE 27–3

Function	Method of Differentiation			Example or Section
Function given graphically	Draw tangent with straightedge; measure slope	Approximate	Point	Example 11
Function given numerically	Numerical, manual	Approximate	Function	Example 20
Function given by formula	Numerical, done manually	Approximate	Point	Example 12
	Delta method	Exact	Function	Examples 13–16
	By rule	Exact	Function	Secs. 27–3 through 27–5

Exercise 2 ◆ The Derivative

Draw a tangent to the curve at the given point in Fig. 27–23 and estimate the slope.

1. At (1, 2) **2.** At (3, 3)

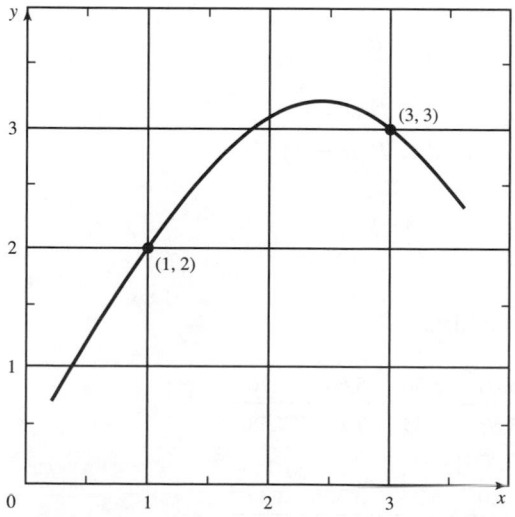

FIGURE 27–23

Delta Method

Find the derivative by the delta method.

3. $y = 3x - 2$ **4.** $y = 4x - 3$

5. $y = 2x + 5$ **6.** $y = 7 - 4x$

7. $y = x^2 + 1$ **8.** $y = x^2 - 3x + 5$

9. $y = x^3$ **10.** $y = x^3 - x^2$

11. $y = \dfrac{3}{x}$ **12.** $y = \dfrac{x}{(x - 1)^2}$

13. $y = \sqrt{3 - x}$ **14.** $y = \dfrac{1}{\sqrt{x}}$

Find the slope of the tangent at the given value of x.

15. $y = \dfrac{1}{x^2}$ at $x = 1$ **16.** $y = \dfrac{1}{x + 1}$ at $x = 2$

17. $y = x + \dfrac{1}{x}$ at $x = 2$ **18.** $y = \dfrac{1}{x}$ at $x = 3$

19. $y = 2x - 3$ at $x = 3$ **20.** $y = 16x^2$ at $x = 1$

21. $y = 2x^2 - 6$ at $x = 3$ **22.** $y = x^2 + 2x$ at $x = 1$

Other Symbols for the Derivative

23. If $y = 2x^2 - 3$, find y'.   **24.** In problem 23, find $y'(3)$.

25. In problem 23, find $y'(-1)$.   **26.** If $f(x) = 7 - 4x^2$, find $f'(x)$.

27. In problem 26, find $f'(2)$.   **28.** In problem 26, find $f'(-3)$.

Operator Notation

Find the derivative.

29. $\dfrac{d}{dx}(3x + 2)$ **30.** $\dfrac{d}{dx}(x^2 - 1)$

31. $D_x(7 - 5x)$ **32.** $D_x(x^2)$

33. $D(3x + 2)$ **34.** $D(x^2 - 1)$

Function Given Numerically

Numerically find the derivative of each set of data.

35.

x	5.00	5.10	5.20	5.30	5.40	5.50	5.60	5.70
y	3.96	4.18	4.66	5.07	5.68	6.35	7.12	7.58

36.

x	3.00	3.05	3.10	3.15	3.20	3.25	3.30	3.35
y	4.15	4.42	5.12	6.21	7.12	8.63	9.10	9.22

27–3 Rules for Derivatives

We now apply the delta method to various kinds of simple functions and derive a rule for each with which we can then find the derivative.

The Derivative of a Constant

To find the derivative of the function $y = c$, we substitute c for $f(x)$ and c for $f(x + \Delta x)$ in Eq. 338, and get

$$\frac{d(c)}{dx} = \lim_{\Delta x \to 0} \frac{c - c}{\Delta x} = 0$$

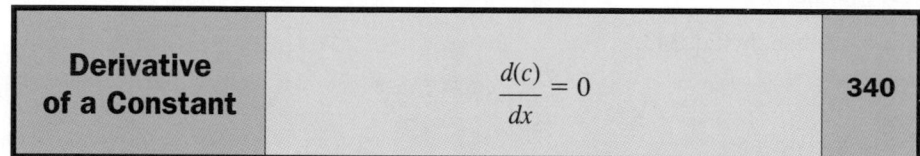

Derivative of a Constant	$\dfrac{d(c)}{dx} = 0$	340

The derivative of a constant is zero.

This is not surprising, because the graph of the function $y = c$ is a straight line parallel to the x axis (Fig. 27–24) whose slope is, of course, zero.

◆◆◆ **Example 21:** If $y = 2\pi^2$, then

$$\frac{dy}{dx} = 0 \qquad \qquad ◆◆◆$$

FIGURE 27–24

Derivative of a Constant Times a Power of x

We let $y = cx^n$, where n is a positive integer and c is any constant. Using the delta method:

1. We substitute $x + \Delta x$ for x and $y + \Delta y$ for y.

$$y + \Delta y = c(x + \Delta x)^n$$

By the binomial formula, Eq. 245,

$$y + \Delta y = c\left[x^n + nx^{n-1}(\Delta x) + \frac{n(n-1)}{2}x^{n-2}(\Delta x)^2 + \cdots + (\Delta x)^n\right]$$

2. Subtracting $y = cx^n$, we get

$$\Delta y = c\left[nx^{n-1}(\Delta x) + \frac{n(n-1)}{2}x^{n-2}(\Delta x)^2 + \cdots + (\Delta x)^n\right]$$

The binomial formula shows the expression we get when a binomial $(a + b)$ is raised to any positive integral power n.

3. Dividing by Δx yields

$$\frac{\Delta y}{\Delta x} = c\left[nx^{n-1} + \frac{n(n-1)}{2}x^{n-2}\Delta x + \cdots + (\Delta x)^{n-1}\right]$$

4. Finally, letting Δx approach zero, all terms but the first vanish. Thus

$$\frac{dy}{dx} = cnx^{n-1}$$

or:

Derivative of a Constant Times a Power of x	$\dfrac{d}{dx}cx^n = cnx^{n-1}$	343

The derivative of a constant times a power of x is equal to the product of the constant, the exponent, and x raised to the exponent reduced by 1.

◆◆◆ **Example 22:**

(a) If $y = x^3$, then, by Eq. 343,

$$\frac{dy}{dx} = 3x^{3-1} = 3x^2$$

(b) If $y = 5x^2$, then

$$\frac{dy}{dx} = 5(2)x^{2-1} = 10x$$ ◆◆◆

Power Function with Negative Exponent

In our derivative of Eq. 343, we had required that the exponent n be a positive integer. We'll now show that the rule is also valid when n is a *negative* integer.

If n is a negative integer, then $m = -n$ is a positive integer, and

$$y = cx^n = cx^{-m} = \frac{c}{x^m}$$

We again use the delta method.

1. We substitute $x + \Delta x$ for x, and $y + \Delta y$ for y.

$$y + \Delta y = \frac{c}{(x + \Delta x)^m}$$

2. Subtracting gives us

$$\Delta y = \frac{c}{(x + \Delta x)^m} - \frac{c}{x^m} = c\,\frac{x^m - (x + \Delta x)^m}{x^m(x + \Delta x)^m}$$

$$= c\,\frac{x^m - \left(x^m + mx^{m-1}\Delta x + \dfrac{m(m-1)}{2}x^{m-2}\Delta x^2 + \cdots + \Delta x^m\right)}{x^m(x + \Delta x)^m}$$

3. Dividing by Δx yields

$$\frac{\Delta y}{\Delta x} = -c\,\frac{mx^{m-1} + \dfrac{m(m-1)}{2}x^{m-2}\Delta x + \cdots + \Delta x^{m-1}}{x^m(x + \Delta x)^m}$$

4. Letting Δx go to zero, we get

$$\frac{dy}{dx} = -c\,\frac{mx^{m-1}}{x^{2m}} = -cmx^{m-1-2m} = -cmx^{-m-1} = cnx^{n-1}$$

This shows that Eq. 343 is valid when the exponent n is negative as well as when it is positive.

◆◆◆ **Example 23:**

(a) If $y = x^{-4}$, then

$$\frac{dy}{dx} = -4x^{-5} = -\frac{4}{x^5}$$

(b) $\dfrac{d(3x^{-2})}{dx} = 3(-2)x^{-3} = -\dfrac{6}{x^3}$

(c) If $y = -3/x^2$, then $y = -3x^{-2}$ and

$$y' = -3(-2)x^{-3} = \dfrac{6}{x^3}$$

◆◆◆

Power Function with Fractional Exponent

We have shown that the exponent n in Eq. 343 can be a positive or a negative integer. The rule is also valid when n is a positive or a negative rational number. We'll prove it in Sec. 27–4.

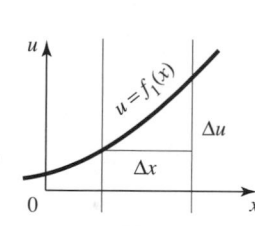

(a)

◆◆◆ **Example 24:** If $y = x^{-5/3}$, then

$$\dfrac{dy}{dx} = -\dfrac{5}{3}x^{-8/3}$$

◆◆◆

To find the derivative of a radical, write it in exponential form, and use Eq. 343.

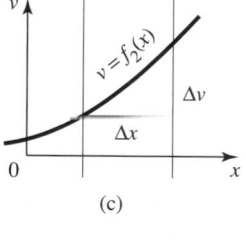

(b)

◆◆◆ **Example 25:** If $y = \sqrt[3]{x^2}$, then

$$\dfrac{dy}{dx} = \dfrac{d}{dx}x^{2/3} = \dfrac{2}{3}x^{-1/3} = \dfrac{2}{3\sqrt[3]{x}}$$

◆◆◆

Derivative of a Sum

We want the derivative of the function

$$y = u + v + w$$

where u, v, and w are all functions of x. These can be visualized as shown in Figs. 27–25(b), (c), and (d). We use the delta method as before. Starting from $P(x, y)$ on the curve $y = f(x)$, we locate a second point Q spaced from P by a horizontal increment Δx. In a run of Δx, the graph of $y = f(x)$ rises by an amount that we call Δy [Fig. 27–25(a)].

But in a run of Δx, the graph of u also has a rise, and we call this rise Δu. Similarly, the graphs of v and w will rise by amounts that we call Δv and Δw.

Thus at $(x + \Delta x)$, the values of u, v, w, and y are $(u + \Delta u)$, $(v + \Delta v)$, $(w + \Delta w)$, and $(y + \Delta y)$. Substituting these values into the original function $y = u + v + w$ gives

$$y + \Delta y = (u + \Delta u) + (v + \Delta v) + (w + \Delta w)$$

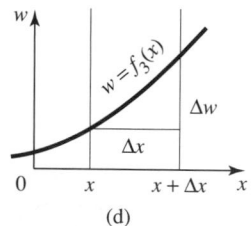

(c)

Subtracting the original function yields

$$y + \Delta y - y = u + \Delta u + v + \Delta v + w + \Delta w - u - v - w$$

$$\Delta y = \Delta u + \Delta v + \Delta w$$

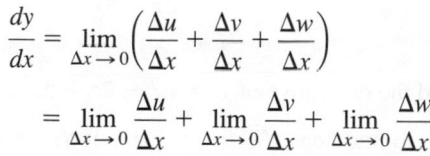

(d)

FIGURE 27–25

Dividing by Δx gives us

$$\dfrac{\Delta y}{\Delta x} = \dfrac{\Delta u + \Delta v + \Delta w}{\Delta x}$$

So

$$\dfrac{dy}{dx} = \lim_{\Delta x \to 0}\left(\dfrac{\Delta u}{\Delta x} + \dfrac{\Delta v}{\Delta x} + \dfrac{\Delta w}{\Delta x}\right)$$

$$= \lim_{\Delta x \to 0}\dfrac{\Delta u}{\Delta x} + \lim_{\Delta x \to 0}\dfrac{\Delta v}{\Delta x} + \lim_{\Delta x \to 0}\dfrac{\Delta w}{\Delta x}$$

Here we use the idea that the limit of a sum of several functions is equal to the sum of the individual limits.

This leads us to the following sum rule:

Derivative of a Sum	$\dfrac{d}{dx}(u + v + w) = \dfrac{du}{dx} + \dfrac{dv}{dx} + \dfrac{dw}{dx}$	**344**

The derivative of the sum of several functions is equal to the sum of the derivatives of those functions.

In addition to using the rule for sums here, we also need the rules for a power function and for a constant. We usually have to apply several rules in one problem.

◆◆◆ **Example 26:**

$$\frac{d}{dx}(2x^3 - 3x^2 + 5x + 4) = 6x^2 - 6x + 5 \qquad \text{◆◆◆}$$

◆◆◆ **Example 27:** Find the derivative of $y = \dfrac{x^2 + 3}{x}$.

Solution: At first glance it looks as if none of the rules learned so far apply here. But if we rewrite the function as

$$y = \frac{x^2 + 3}{x} = x + \frac{3}{x} = x + 3x^{-1}$$

we may write

$$\frac{dy}{dx} = 1 + 3(-1)x^{-2}$$

$$= 1 - \frac{3}{x^2} \qquad \text{◆◆◆}$$

Common Error	Students often forget to simplify an expression as much as possible **before** taking the derivative.

Other Variables

So far we have been using x for the independent variable and y for the dependent variable, but now let us get some practice using other variables.

◆◆◆ **Example 28:**

(a) If $s = 3t^2 - 4t + 5$, then

$$\frac{ds}{dt} = 6t - 4$$

(b) If $y = 7u - 5u^3$, then

$$\frac{dy}{du} = 7 - 15u^2$$

(c) If $u = 9 + x^2 - 2x^4$, then

$$\frac{du}{dx} = 2x - 8x^3 \qquad \text{◆◆◆}$$

◆◆◆ **Example 29:** Find the derivative of $T = 4z^2 - 2z + 5$.

Solution: What are we to find here, dT/dz, or dz/dT, or dT/dx, or something else?

> When we say "derivative," we mean the derivative of the given function **with respect to the independent variable,** unless otherwise noted.

Here we find the derivative of T with respect to the independent variable z.

$$\frac{dT}{dz} = 8z - 2$$

◆◆◆

Exercise 3 ◆ Rules for Derivatives

Find the derivative of each function.

1. $y = a^2$
2. $y = 3b + 7c$
3. $y = x^7$
4. $y = x^4$
5. $y = 3x^2$
6. $y = 5.4x^3$
7. $y = x^{-5}$
8. $y = 2x^{-3}$
9. $y = \dfrac{1}{x}$
10. $y = \dfrac{1}{x^2}$
11. $y = \dfrac{3}{x^3}$
12. $y = \dfrac{3}{2x^2}$
13. $y = 7.5x^{1/3}$
14. $y = 4x^{5/3}$
15. $y = 4\sqrt{x}$
16. $y = 3\sqrt[3]{x}$
17. $y = -17\sqrt{x^3}$
18. $y = -2\sqrt[5]{x^4}$
19. $y = 3 - 2x$
20. $y = 4x^2 + 2x^3$
21. $y = 3x - x^3$
22. $y = x^4 + 3x^2 + 2$
23. $y = 3x^3 + 7x^2 - 2x + 5$
24. $y = x^3 - x^{3/2} + 3x$
25. $y = ax + b$
26. $y = ax^5 - 5bx^3$
27. $y = \dfrac{x^2}{2} - \dfrac{x^7}{7}$
28. $y = \dfrac{x^3}{1.75} + \dfrac{x^2}{2.84}$
29. $y = 2x^{3/4} + 4x^{-1/4}$
30. $y = \dfrac{2}{x} - \dfrac{3}{x^2}$
31. $y = 2x^{4/3} - 3x^{2/3}$
32. $y = x^{2/3} - a^{2/3}$
33. $y = \dfrac{x + 4}{x}$
34. $y = \dfrac{x^3 + 1}{x}$
35. If $y = 2x^3 - 3$, find y'.
36. In problem 35, find $y'(2)$.
37. In problem 35, find $y'(-3)$.
38. If $f(x) = 7 - 4x^2$, find $f'(x)$.
39. In problem 38, find $f'(1)$.
40. In problem 38, find $f'(-3)$.

> Here, as usual, the letters a, b, c, . . . , represent constants.

Find the derivative.

41. $\dfrac{d}{dx}(3x^5 + 2x)$
42. $\dfrac{d}{dx}(2.5x^2 - 1)$
43. $D_x(7.8 - 5.2x^{-2})$
44. $D_x(4x^2 - 1)$
45. $D(3x^2 + 2x)$
46. $D(1.75x^{-2} - 1)$
47. Find the slope of the tangent to the curve $y = x^2 - 2$, where x equals 2.
48. Find the slope of the tangent to the function $y = x - x^2$ at $x = 2$.
49. If $y = x^3 - 5$, find $y'(1)$.
50. If $f(x) = 1/x^2$, find $f'(2)$.
51. If $f(x) = 2.75x^2 - 5.02x$, find $f'(3.36)$.
52. If $y = \sqrt{83.2x^3}$, find $y'(1.74)$.

Other Variables

Find the derivative with respect to the independent variable.

53. $v = 5t^2 - 3t + 4$

54. $z = 9 - 8w + w^2$

55. $s = 58.3t^3 - 63.8t$

56. $x = 3.82y + 6.25y^4$

57. $y = \sqrt{5w^3}$

58. $w = \dfrac{5}{x} - \dfrac{3}{x^2}$

59. $v = \dfrac{85.3}{t^4}$

60. $T = 3.55\sqrt{1.06w^5}$

27–4   Derivative of a Function Raised to a Power

Composite Functions

In Sec. 27–3 we derived Rule 343 for finding the derivative of x raised to a power. But Rule 343 does not apply when we have an *expression* raised to a power, such as

$$y = (2x + 7)^5 \tag{1}$$

We might consider this function as being made up of two parts. One part is the function, which we will call u, that is being raised to the power.

$$u = 2x + 7 \tag{2}$$

Then y, which is a function of u, is seen to be

$$y = u^5 \tag{3}$$

Our original function (1), which can be obtained by combining (2) and (3), is called a *composite function*.

You might want to review composite functions in Chapter 4 before going further.

The Chain Rule

The *chain rule* will enable us to take the derivative of a composite function, such as (1). Consider the situation where y is a function of u,

$$y = g(u)$$

and u, in turn, is a function of x,

$$u = h(x)$$

so that y is therefore a function of x.

$$y = f(x)$$

The graphs of $u = h(x)$ and $y = f(x)$ are shown in Fig. 27–26. We now start our derivation of the chain rule by recalling our definition of the derivative as the limit of the quotient $\Delta y/\Delta x$ as Δx approaches zero.

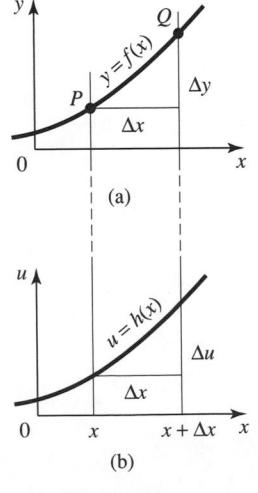

FIGURE 27–26

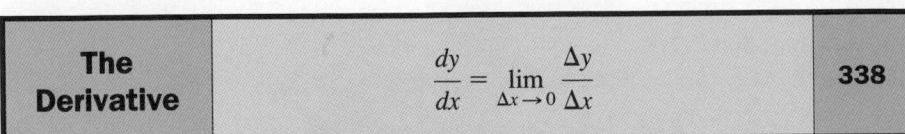

| The Derivative | $\dfrac{dy}{dx} = \lim\limits_{\Delta x \to 0} \dfrac{\Delta y}{\Delta x}$ | 338 |

where Δy is the rise of the graph of $y = f(x)$ in a run of Δx [Fig. 27–26(a)]. But in a run of Δx, the graph of u also has a rise, and we call this rise Δu [Fig. 27–26(b)]. Before taking the limit in Eq. 338, let us multiply the quotient $\Delta y/\Delta x$ by $\Delta u/\Delta u$, assuming here that Δu is not zero.

$$\frac{\Delta y}{\Delta x} = \frac{\Delta y}{\Delta x} \cdot \frac{\Delta u}{\Delta u}$$

Rearranging gives us

$$\frac{\Delta y}{\Delta x} = \frac{\Delta y}{\Delta u} \cdot \frac{\Delta u}{\Delta x}$$

> The chain rule is true even for those rare cases where Δu may be zero, but it takes a more complicated proof to show this.

We now let Δx approach zero and apply a theorem (which we'll use without proof) that the limit of a product of two functions is equal to the product of the limits of the two functions.

$$\lim_{\Delta x \to 0} \frac{\Delta y}{\Delta x} = \lim_{\Delta x \to 0} \frac{\Delta y}{\Delta u} \cdot \frac{\Delta u}{\Delta x} = \left(\lim_{\Delta x \to 0} \frac{\Delta y}{\Delta u} \right) \left(\lim_{\Delta x \to 0} \frac{\Delta u}{\Delta x} \right)$$

But Δu also approaches zero as Δx approaches zero, so we may write

$$\lim_{\Delta x \to 0} \frac{\Delta y}{\Delta x} = \lim_{\Delta u \to 0} \frac{\Delta y}{\Delta u} \cdot \lim_{\Delta x \to 0} \frac{\Delta u}{\Delta x}$$

Then, by our definition of the derivative, Eq. 338,

$$\lim_{\Delta x \to 0} \frac{\Delta y}{\Delta x} = \frac{dy}{dx}$$

Similarly,

$$\lim_{\Delta u \to 0} \frac{\Delta y}{\Delta u} = \frac{dy}{du} \quad \text{and} \quad \lim_{\Delta x \to 0} \frac{\Delta u}{\Delta x} = \frac{du}{dx}$$

So we get the following equation:

Chain Rule	$$\frac{dy}{dx} = \frac{dy}{du} \cdot \frac{du}{dx}$$	**339**

If y is a function of u, and u is a function of x, then the derivative of y with respect to x is the product of the derivative of y with respect to u and the derivative of u with respect to x.

The Power Rule

We now use the chain rule to find the derivative of a function raised to a power, $y = cu^n$. If, in Eq. 343, we replace x by u, we get

$$\frac{dy}{du} = \frac{d}{du}(cu^n) = cnu^{n-1}$$

> We'll have other uses for the chain rule later.

Then, by the chain rule, we get dy/dx by multiplying by du/dx.

$$\frac{dy}{dx} = \frac{dy}{du} \cdot \frac{du}{dx}$$

Derivative of a Function Raised to a Power	$$\frac{d(cu^n)}{dx} = cnu^{n-1}\frac{du}{dx}$$	**345**

> We'll refer to Eq. 345 as the "power rule."

The derivative (with respect to x) of a constant times a function raised to a power is equal to the product of the constant, the power, the function raised to the power less 1, and the derivative of the function (with respect to x).

♦♦♦ **Example 30:** Take the derivative of $y = (x^3 + 1)^5$.

Solution: We use the rule for a function raised to a power, with

$$n = 5 \quad \text{and} \quad u = x^3 + 1$$

Then

$$\frac{du}{dx} = 3x^2$$

so

$$\frac{dy}{dx} = nu^{n-1}\frac{du}{dx}$$

$$= 5(x^3 + 1)^4(3x^2)$$

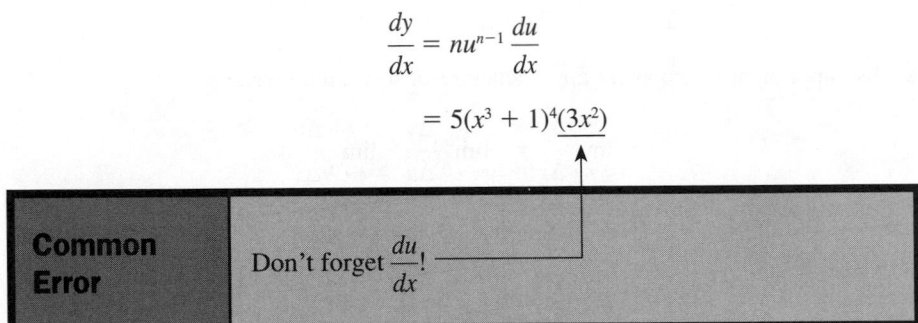

Common Error	Don't forget $\dfrac{du}{dx}$!

Now simplifying our answer, we get

$$\frac{dy}{dx} = 15x^2(x^3 + 1)^4 \qquad \text{♦♦♦}$$

Our next example shows the use of the power rule when *the exponent is negative*.

♦♦♦ **Example 31:** Take the derivative of $y = \dfrac{3}{x^2 + 2}$.

Solution: Rewriting our function as $y = 3(x^2 + 2)^{-1}$ and applying Rule 345 with $u = x^2 + 2$, we get

$$\frac{dy}{dx} = 3(-1)(x^2 + 2)^{-2}(2x)$$

$$= -\frac{6x}{(x^2 + 2)^2} \qquad \text{♦♦♦}$$

We'll now use the power rule for *fractional exponents*, even though we will not prove that the rule is valid for them until later in this chapter.

♦♦♦ **Example 32:** Differentiate $y = \sqrt[3]{1 + x^2}$.

Solution: We rewrite the radical in exponential form.

$$y = (1 + x^2)^{1/3}$$

Then, by Rule 345, with $u = 1 + x^2$, and $du/dx = 2x$,

$$\frac{dy}{dx} = \frac{1}{3}(1 + x^2)^{-2/3}(2x)$$

Or, returning to radical form,

$$\frac{dy}{dx} = \frac{2x}{3\sqrt[3]{(1 + x^2)^2}}$$

◆◆◆

◆◆◆ **Example 33:** If $f(x) = \dfrac{5}{\sqrt{x^2 + 3}}$, find $f'(1)$.

Solution: Rewriting our function without radicals gives us

$$f(x) = 5(x^2 + 3)^{-1/2}$$

Taking the derivative yields

$$f'(x) = 5\left(-\frac{1}{2}\right)(x^2 + 3)^{-3/2}(2x)$$

$$= -\frac{5x}{(x^2 + 3)^{3/2}}$$

Substituting $x = 1$, we obtain

$$f'(1) = -\frac{5}{(1 + 3)^{3/2}} = -\frac{5}{8}$$

◆◆◆

Exercise 4 ◆ Derivative of a Function Raised to a Power

Find the derivative of each function.

As before, the letters $a, b, c, \ldots,$ represent constants.

1. $y = (2x + 1)^5$

2. $y = (2 - 3x^2)^3$

3. $y = (3x^2 + 2)^4 - 2x$

4. $y = (x^3 + 5x^2 + 7)^2$

5. $y = (2 - 5x)^{3/5}$

6. $y = -\dfrac{2}{x + 1}$

7. $y = \dfrac{2.15}{x^2 + a^2}$

8. $y = \dfrac{31.6}{1 - 2x}$

9. $y = \dfrac{3}{x^2 + 2}$

10. $y = \dfrac{5}{(x^2 - 1)^2}$

11. $y = \left(a - \dfrac{b}{x}\right)^2$

12. $y = \left(a + \dfrac{b}{x}\right)^3$

13. $y = \sqrt{1 - 3x^2}$

14. $y = \sqrt{2x^2 - 7x}$

15. $y = \sqrt{1 - 2x}$

16. $y = \dfrac{b}{a}\sqrt{a^2 - x^2}$

17. $y = \sqrt[3]{4 - 9x}$

18. $y = \sqrt[3]{a^3 - x^3}$

19. $y = \dfrac{1}{\sqrt{x + 1}}$

20. $y = \dfrac{1}{\sqrt{x^2 - x}}$

Find the derivative.

21. $\dfrac{d}{dx}(3x^5 + 2x)^2$

22. $\dfrac{d}{dx}(1.5x^2 - 3)^3$

23. $D_x(4.8 - 7.2x^{-2})^2$

24. $D(3x^3 - 5)^3$

Find the derivative with respect to the independent variable.

25. $v = (5t^2 - 3t + 4)^2$ **26.** $z = (9 - 8w)^3$

27. $s = (8.3t^3 - 3.8t)^{-2}$ **28.** $x = \sqrt{3.2y + 6.2y^4}$

29. Find the derivative of the function $y = (4.82x^2 - 8.25x)^3$ when $x = 3.77$.

30. Find the slope of the tangent to the curve $y = 1/(x + 1)$ at $x = 2$.

31. If $y = (x^2 - x)^3$, find $y'(3)$.

32. If $f(x) = \sqrt[3]{2x} + (2x)^{2/3}$, find $f'(4)$.

27–5   Derivatives of Products and Quotients

Derivative of a Product

We often need the derivative of the product of two expressions, such as $y = (x^2 + 2)\sqrt{x - 5}$, where each of the expressions is itself a function of x. Let us label these expressions u and v. So our function is

$$y = uv$$

where u and v are functions of x. These can be visualized as in Figs. 27–25(a), (b), and (c). We use the delta method as before. Starting from $P(x, y)$ on the curve $y = f(x)$, we locate a second point Q spaced from P by a horizontal increment Δx. In a run of Δx, the graph of $y = f(x)$ is seen to rise by an amount that we call Δy [Fig. 27–25(a)].

But in a run of Δx, the graph of u also has a rise, and we call this rise Δu. Similarly, the graph of v will rise by an amount that we call Δv.

Thus, at $(x + \Delta x)$, the values of u, v, and y are $(u + \Delta u)$, $(v + \Delta v)$, and $(y + \Delta y)$. Substituting these values into the original function $y = uv$ gives

$$y + \Delta y = (u + \Delta u)(v + \Delta v)$$
$$= uv + u\,\Delta v + v\,\Delta u + \Delta u\,\Delta v$$

Subtracting $y = uv$ gives us

$$\Delta y = u\,\Delta v + v\,\Delta u + \Delta u\,\Delta v$$

Dividing by Δx yields

$$\frac{\Delta y}{\Delta x} = u\frac{\Delta v}{\Delta x} + v\frac{\Delta u}{\Delta x} + \Delta u\frac{\Delta v}{\Delta x}$$

As Δx now approaches 0, Δy, Δu, and Δv also approach 0, and

$$\lim_{\Delta x \to 0} \frac{\Delta y}{\Delta x} = \frac{dy}{dx}, \quad \lim_{\Delta x \to 0} \frac{\Delta u}{\Delta x} = \frac{du}{dx}, \quad \lim_{\Delta x \to 0} \frac{\Delta v}{\Delta x} = \frac{dv}{dx}$$

so

$$\frac{dy}{dx} = u\frac{dv}{dx} + v\frac{du}{dx} + 0\frac{dv}{dx}$$

which can be rewritten as follows:

Derivative of a Product of Two Factors	$$\frac{d(uv)}{dx} = u\frac{dv}{dx} + v\frac{du}{dx}$$	**346**

The derivative of a product of two factors is equal to the first factor times the derivative of the second factor, plus the second factor times the derivative of the first.

◆◆◆ **Example 34:** Find the derivative of $y = (x^2 + 2)(x - 5)$.

Solution: We let the first factor be u, and the second be v.

$$y = \underbrace{(x^2 + 2)}_{u} \underbrace{(x - 5)}_{v}$$

So

$$\frac{du}{dx} = 2x \text{ and } \frac{dv}{dx} = 1$$

Using the product rule,

$$\frac{dy}{dx} = (x^2 + 2)\frac{d}{dx}(x - 5) + (x - 5)\frac{d}{dx}(x^2 + 2)$$

$$= (x^2 + 2)(1) + (x - 5)(2x)$$

$$= x^2 + 2 + 2x^2 - 10x$$

$$= 3x^2 - 10x + 2$$ ◆◆◆

> We could have multiplied the two factors together and taken the derivative term by term. Try it and see if you get the same result. But sometimes this is not possible, as in Example 39.

◆◆◆ **Example 35:** Differentiate $y = (x + 5)\sqrt{x - 3}$.

Solution: By the product rule,

$$\frac{dy}{dx} = (x + 5) \cdot \frac{1}{2}(x - 3)^{-1/2}(1) + \sqrt{x - 3}(1)$$

$$= \frac{x + 5}{2\sqrt{x - 3}} + \sqrt{x - 3}$$ ◆◆◆

Products with More Than Two Factors

Our rule for the derivative of a product having two factors can be easily extended. Take an expression with three factors, for example.

$$y = uvw = (uv)w$$

By the product rule,

$$\frac{dy}{dx} = uv\frac{dw}{dx} + w\frac{d(uv)}{dx}$$

$$= uv\frac{dw}{dx} + w\left(u\frac{dv}{dx} + v\frac{du}{dx}\right)$$

Thus:

Derivative of a Product of Three Factors	$$\frac{d(uvw)}{dx} = uv\frac{dw}{dx} + uw\frac{dv}{dx} + vw\frac{du}{dx}$$	**347**

The derivative of the product of three factors is an expression of three terms, each term being the product of two of the factors and the derivative of the third factor.

◆◆◆ **Example 36:** Differentiate $y = x^2 (x - 2)^5 \sqrt{x + 3}$.

Solution: By Eq. 347,

$$\frac{dy}{dx} = x^2 (x - 2)^5 \left[\frac{1}{2}(x + 3)^{-1/2} \right] + x^2 (x + 3)^{1/2} [5(x - 2)^4]$$

$$+ (x - 2)^5 (x + 3)^{1/2} (2x)$$

$$= \frac{x^2 (x - 2)^5}{2\sqrt{x + 3}} + 5x^2 (x - 2)^4 \sqrt{x + 3} + 2x (x - 2)^5 \sqrt{x + 3} \qquad ◆◆◆$$

We now generalize this result (without proof) as follows to cover any number of factors:

Derivative of a Product of n Factors	The derivative of the product of n factors is an expression of n terms, each term being the product of $n - 1$ of the factors and the derivative of the other factor.	348

Derivative of a Constant Times a Function

Let us use the product rule for the product cu, where c is a constant and u is a function of x.

$$\frac{d}{dx} cu = c \frac{du}{dx} + u \frac{dc}{dx} = c \frac{du}{dx}$$

since the derivative dc/dx of a constant is zero. Thus:

Derivative of a Constant Times a Function	$\dfrac{d(cu)}{dx} = c \dfrac{du}{dx}$	342

The derivative of the product of a constant and a function is equal to the constant times the derivative of the function.

◆◆◆ **Example 37:** If $y = 3(x^2 - 3x)^5$, then

$$\frac{dy}{dx} = 3 \frac{d}{dx} (x^2 - 3x)^5$$

$$= 3(5)(x^2 - 3x)^4 (2x - 3)$$

$$= 15(x^2 - 3x)^4 (2x - 3) \qquad ◆◆◆$$

Tip	If one of the factors is a constant, it is much easier to use Rule 342 for a constant times a function, rather than the product rule

Derivative of a Quotient

To find the derivative of the function

$$y = \frac{u}{v}$$

where u and v are functions of x, we first rewrite it as a product.

$$y = uv^{-1}$$

Now, using the rule for products and the rule for a power function,

$$\frac{dy}{dx} = u(-1)\, v^{-2}\, \frac{dv}{dx} + v^{-1}\, \frac{du}{dx}$$

$$= -\frac{u}{v^2}\frac{dv}{dx} + \frac{1}{v}\frac{du}{dx}$$

We combine the two fractions over the LCD, v^2, and rearrange.

$$\frac{dy}{dx} = \frac{v}{v^2}\frac{du}{dx} - \frac{u}{v^2}\frac{dv}{dx}$$

which can be rewritten as follows:

Derivative of a Quotient	$\dfrac{d}{dx}\left(\dfrac{u}{v}\right) = \dfrac{v\,\dfrac{du}{dx} - u\,\dfrac{dv}{dx}}{v^2}$	349

The derivative of a quotient equals the denominator times the derivative of the numerator minus the numerator times the derivative of the denominator, all divided by the square of the denominator.

◆◆◆ **Example 38:** Take the derivative of $y = 2x^3/(4x + 1)$.

Solution: The numerator is $u = 2x^3$, so

$$\frac{du}{dx} = 6x^2$$

and the denominator is $v = 4x + 1$, so

$$\frac{dv}{dx} = 4$$

Some prefer to use the product rule to do quotients, treating the quotient u/v as the product uv^{-1}.

Common Error	It is very easy, in Eq. 349, to interchange u and v, by mistake

Applying the quotient rule yields

$$\frac{dy}{dx} = \frac{(4x + 1)(6x^2) - (2x^3)(4)}{(4x + 1)^2}$$

Simplifying, we get

$$\frac{dy}{dx} = \frac{24x^3 + 6x^2 - 8x^3}{(4x + 1)^2} = \frac{16x^3 + 6x^2}{(4x + 1)^2}$$

◆◆◆

Tip	If the numerator or denominator is a constant, it is easier to rewrite the expression as a product and use the power rule.

◆◆◆ **Example 39:** Find the derivative of $y = \dfrac{2x^3}{3} + \dfrac{4}{x^2}$.

Solution: Rather than use the quotient rule, we rewrite the expression as

$$y = \frac{2x^3}{3} + 4x^{-2}$$

Using the power rule and the sum rule,

$$\frac{dy}{dx} = \frac{2}{3}(3)x^2 + 4(-2)x^{-3} = 2x^2 - \frac{8}{x^3}$$

◆◆◆

◆◆◆ **Example 40:** Find $s'(3)$ if $s = \dfrac{(t^3 - 3)^2}{\sqrt{t + 1}}$.

Solution: By the quotient rule,

$$s' = \frac{\sqrt{t + 1}\,(2)(t^3 - 3)(3t^2) - (t^3 - 3)^2\left(\dfrac{1}{2}\right)(t + 1)^{-1/2}}{t + 1}$$

$$= \frac{6t^2\,(t^3 - 3)\,\sqrt{t + 1} - \dfrac{(t^3 - 3)^2}{2\sqrt{t + 1}}}{t + 1}$$

We could simplify now, but it will be easier to just substitute in the unsimplified expression. Letting $t = 3$ gives

$$s'(3) = \frac{6(9)(27 - 3)\,\sqrt{3 + 1} - \dfrac{(27 - 3)^2}{2\sqrt{3 + 1}}}{3 + 1} = 612$$

◆◆◆

Some of these can be multiplied out. For a few of these, take the derivative both before and after multiplying out, and compare the two.

Exercise 5 ◆ Derivatives of Products and Quotients

Products

Find the derivative of each function.

1. $y = x(x^2 - 3)$
2. $y = x^3(5 - 2x)$
3. $y = (5 + 3x)(3 + 7x)$
4. $y = (7 - 2x)(x + 4)$
5. $y = x(x^2 - 2)^2$
6. $y = x^3(8.24x - 6.24x^3)$
7. $y = (x^2 - 5x)^3\,(8x - 7)^2$
8. $y = (1 - x^2)(1 + x^2)^2$
9. $y = x\sqrt{1 + 2x}$
10. $y = 3x\sqrt{5 + x^2}$
11. $y = x^2\sqrt{3 - 4x}$
12. $y = 2x^2 - 2x\sqrt{x^2 - 5}$
13. $y = x\sqrt{a + bx}$
14. $y = \sqrt{x}\,(3x^2 + 2x - 3)$

15. $y = (3x + 1)^3 \sqrt{4x - 2}$

16. $y = (2x^2 - 3)\sqrt[3]{3x + 5}$

17. $v = (5t^2 - 3t)(t + 4)^2$

18. $z = (6 - 5w)^3 (w^2 - 1)$

19. $s = (81.3t^3 - 73.8t)(t - 47.2)^2$

20. $x = (y - 49.3)\sqrt{23.2y + 16.2y^4}$

Find the derivative.

21. $\dfrac{d}{dx}(2x^5 + 5x)^2(x - 3)$

22. $D_x(11.5x + 49.3)(14.8 - 27.2x^{-2})^2$

Products with More Than Two Factors

Find the derivative of each function.

23. $y = x(x + 1)^2(x - 2)^3$

24. $y = x\sqrt{x + 1}\sqrt[3]{x}$

25. $y = (x - 1)^{1/2}(x + 1)^{3/2}(x + 2)$

26. $y = x\sqrt{x + 2}\sqrt[3]{x - 1}$

Quotients

Find the derivative of each function.

27. $y = \dfrac{x}{x + 2}$

28. $y = \dfrac{x}{x^2 + 1}$

29. $y = \dfrac{x^2}{4 - x^2}$

30. $y = \dfrac{x - 1}{x + 1}$

31. $y = \dfrac{x + 2}{x - 3}$

32. $y = \dfrac{2x - 1}{(x - 1)^2}$

33. $y = \dfrac{2x^2 - 1}{(x - 1)^2}$

34. $y = \dfrac{a - x}{n + x}$

35. $y = \dfrac{x^{1/2}}{x^{1/2} + 1}$

36. $s = \sqrt{\dfrac{t - 1}{t + 1}}$

37. $w = \dfrac{z}{\sqrt{z^2 - a^2}}$

38. $v = \sqrt{\dfrac{1 + 2t}{1 - 2t}}$

The answers to some of these problems, especially the quotients, might need a lot of simplification to match the book answer. Don't be discouraged if your answer does not appear to check at first.

39. Find the slope of the tangent to the curve $y = \sqrt{16 + 3x}/x$ at $x = 3$.

40. Find the derivative of the function $y = x/(7.42x^2 - 2.75x)$ when $x = 1.47$.

41. If $y = x\sqrt{8 - x^2}$, find $y'(2)$.

42. If $f(x) = x^2/\sqrt{1 + x^3}$, find $f'(2)$.

27–6   Derivatives of Implicit Relations

Derivatives with Respect to Other Variables

Our rule for the derivative of a function raised to a power is as follows:

$$\frac{d(cu^n)}{dx} = cnu^{n-1}\frac{du}{dx}$$ **345**

Up to now, the variable in the function u has been the *same variable* that we take the derivative with respect to, as in the following example.

◆◆◆ **Example 41:**

(a) $\dfrac{d}{dx}(x^3) = 3x^2\dfrac{dx}{dx} = 3x^2$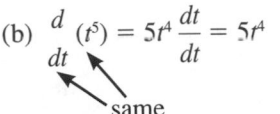

(b) $\dfrac{d}{dt}(t^5) = 5t^4\dfrac{dt}{dt} = 5t^4$

same   same   ◆◆◆

Since $dx/dx = dt/dt = 1$, we have not bothered to write these in. Of course, Rule 345 is just as valid when our independent variable is *different* from the variable that we are taking the derivative with respect to. The following example shows the power rule being applied to such cases.

◆◆◆ **Example 42:**

(a) $\dfrac{d}{dx}(u^5) = 5u^4\dfrac{du}{dx}$

different

(b) $\dfrac{d}{dt}(w^4) = 4w^3\dfrac{dw}{dt}$

different

(c) $\dfrac{d}{dx}(y^6) = 6y^5\dfrac{dy}{dx}$

different

(d) $\dfrac{d}{dx}(y) = 1y^0\dfrac{dy}{dx} = \dfrac{dy}{dx}$

different   ◆◆◆

> Mathematical ideas do not, of course, depend upon which letters of the alphabet we happen to have chosen when doing a derivation. Thus in any of our rules you can replace any letter, say, x, with any other letter, such as z or t.

Common Error

It is very easy to forget to include the dy/dx in problems such as the following:

$$\frac{d}{dx}(y^6) = 6y^5\frac{dy}{dx}$$

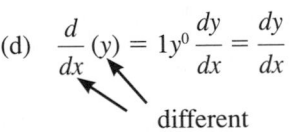 Don't forget!

Our other rules for derivatives (for sums, products, quotients, etc.) also work when the independent variable(s) of the function is different from the variable we are taking the derivative with respect to.

◆◆◆ **Example 43:**

(a) $\dfrac{d}{dx}(2y + z^3) = 2\dfrac{dy}{dx} + 3z^2\dfrac{dz}{dx}$

(b) $\dfrac{d}{dt}(wz) = w\dfrac{dz}{dt} + z\dfrac{dw}{dt}$

> We will have to find dx/dy when solving arc-length problems in Sec. 32–1.

(c) $\dfrac{d}{dx}(x^2y^3) = x^2(3y^2)\dfrac{dy}{dx} + y^3(2x)\dfrac{dx}{dx}$

$$= 3x^2y^2\frac{dy}{dx} + 2xy^3$$

◆◆◆

We usually think of x as being the independent variable and y the dependent variable, and we have been taking the derivative dy/dx of y with respect to x. Sometimes, however, the positions of x and y will be reversed, and we will want to find dx/dy, as in the following example.

◆◆◆ **Example 44:** Find the derivative of x with respect to y, dx/dy, if

$$x = y^2 - 3y + 2$$

Solution: We take the derivative exactly as before, except that x is where y usually is, and vice versa.

$$dx/dy = 2y - 3$$ ◆◆◆

Implicit Relations

Recall that in an implicit relation, neither variable is isolated on one side of the equals sign.

◆◆◆ **Example 45:** $x^2 + y^2 = y^3 - x$ is an implicit relation. ◆◆◆

To find the derivative of an implicit relation, take the derivative of both sides of the equation, and then solve for dy/dx. When taking the derivatives, keep in mind that the derivative of x with respect to x is 1, and that the derivative of y with respect to x is dy/dx.

> We need to be able to differentiate implicit relations because we cannot always solve for one of the variables before differentiating.

◆◆◆ **Example 46:** Given the implicit relation in Example 45, find dy/dx.

Solution: We take the derivative term by term.

$$2x\frac{dx}{dx} + 2y\frac{dy}{dx} = 3y^2\frac{dy}{dx} - \frac{dx}{dx}$$

or

$$2x + 2y\frac{dy}{dx} = 3y^2\frac{dy}{dx} - 1$$

Solving for dy/dx gives us

$$2y\frac{dy}{dx} - 3y^2\frac{dy}{dx} = -2x - 1$$

$$(2y - 3y^2)\frac{dy}{dx} = -2x - 1$$

$$\frac{dy}{dx} = \frac{2x + 1}{3y^2 - 2y}$$ ◆◆◆

> Note that the derivative, unlike those for explicit functions, contains both x and y.

When taking implicit derivatives, it is convenient to use the y' notation instead of dy/dx.

◆◆◆ **Example 47:** Find the derivative dy/dx for the relation

$$x^2y^3 = 5$$

Solution: Using the product rule, we obtain

$$x^2(3y^2)y' + y^3(2x) = 0$$

$$3x^2y^2y' = -2xy^3$$

$$y' = -\frac{2xy^3}{3x^2y^2} = -\frac{2y}{3x}$$ ◆◆◆

◆◆◆ **Example 48:** Find dy/dx, given $x^2 + 3x^3y + y^2 = 4xy$.

Solution: Taking the derivative of each term, we obtain

$$2x + 3x^3y' + y(9x^2) + 2yy' = 4xy' + 4y$$

Moving all terms containing y' to the left side and the other terms to the right side yields

$$3x^3y' + 2yy' - 4xy' = 4y - 2x - 9x^2y$$

Factoring gives us

$$(3x^3 + 2y - 4x)y' = 4y - 2x - 9x^2y$$

So

$$y' = \frac{4y - 2x - 9x^2y}{3x^3 + 2y - 4x} \qquad \text{◆◆◆}$$

Power Function with Fractional Exponent

Now that we are able to take derivatives implicitly, we can show that the power rule—

$$\boxed{\frac{d}{dx}x^n = nx^{n-1} \qquad \textbf{341}}$$

—and hence Eq. 345 are both valid when the exponent n is a fraction.

Let $n = p/q$, where p and q are both integers, positive or negative. Then

$$y = x^n = x^{p/q}$$

Raising both sides to the qth power, we have

$$y^q = x^p$$

Using Eq. 343, we take the derivative of each side.

$$qy^{q-1}\frac{dy}{dx} = px^{p-1}\frac{dx}{dx}$$

Solving for dy/dx yields

$$\frac{dy}{dx} = \frac{p}{q}\frac{x^{p-1}}{y^{q-1}} = n\frac{x^{p-1}}{[x^{(p/q)}]^{(q-1)}}$$

since $p/q = n$ and $y = x^{p/q}$. Applying the laws of exponents gives

$$\frac{dy}{dx} = n\frac{x^{p-1}}{x^{p-p/q}} = nx^{p-1-p+n} = nx^{n-1}$$

We have now shown that the power rule works for any rational exponent, positive or negative. It is also valid for an irrational exponent (such as π), as we will show in Sec. 33–5.

Differential Form

Up to now, we have treated the symbol dy/dx as a *whole*, and not as the quotient of two quantities dy and dx. Here, we give dy and dx separate names and meanings of their own. The quantity dy is called the differential of y, and dx is called the differential of x.

These two differentials, dx and dy, have a simple geometric interpretation. Figure 27−27 shows a tangent drawn to a curve $y = f(x)$ at some point P. The slope of the curve is found by evaluating dy/dx at P. The *differential dy is then the rise of the tangent line, in some arbitrary run dx.* Since the rise of a line is equal to the slope of the line times the run, we have the following equation:

$$\boxed{\textbf{Differential of } \textbf{\textit{y}} \qquad dy = f'(x)dx \qquad \textbf{372}}$$

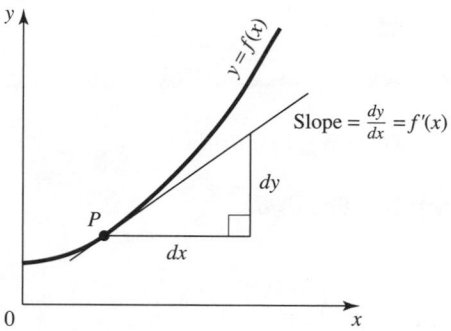

FIGURE 27–27

where we have represented the slope by $f'(x)$ instead of dy/dx, to avoid confusion. If we take the derivative of some function, say, $y = x^3$, we get

$$\frac{dy}{dx} = 3x^2$$

Since we may now think of dy and dx as separate quantities (differentials), we can multiply both sides by dx.

$$dy = 3x^2\, dx$$

This expression is said to be in *differential form*.

Thus to find dy, the differential of y, given some function of x, simply take the derivative of the function and multiply by dx.

◆◆◆ **Example 49:** If $y = 3x^2 - 2x + 5$, find the differential dy.

Solution: Taking the derivative gives us

$$\frac{dy}{dx} = 6x - 2$$

Multiplying by dx, we get the differential of y.

$$dy = (6x - 2)\, dx$$ ◆◆◆

An equation containing a derivative (called a *differential equation*) is usually written in differential form before it is solved.

To find the differential of an *implicit* function, simply find the derivative as in Examples 50 through 52, and then multiply both sides by dx.

◆◆◆ **Example 50:** Find the differential of the implicit function

$$x^3 + 3xy - 2y^2 = 8$$

Solution: Differentiating term by term gives us

$$3x^2 + 3x\frac{dy}{dx} + 3y - 4y\frac{dy}{dx} = 0$$

$$(3x - 4y)\frac{dy}{dx} = -3x^2 - 3y$$

$$\frac{dy}{dx} = \frac{-3x^2 - 3y}{3x - 4y} = \frac{3x^2 + 3y}{4y - 3x}$$

$$dy = \frac{3x^2 + 3y}{4y - 3x}\, dx$$ ◆◆◆

Exercise 6 ✦ Derivatives of Implicit Relations

Derivatives with Respect to Other Variables

1. If $y = 2u^3$, find dy/dw.

2. If $z = (w + 3)^2$, find dz/dy.

3. If $w = y^2 + u^3$, find dw/du.

4. If $y = 3x^2$, find dy/du.

Find the derivative.

5. $\dfrac{d}{dx}(x^3y^2)$

6. $\dfrac{d}{dx}(w^2 - 3w - 1)$

7. $\dfrac{d}{dt}\sqrt{3z^2 + 5}$

8. $\dfrac{d}{dz}(y - 3)\sqrt{y - 2}$

Find dx/dy.

9. $x = y^2 - 7y$

10. $x = (y - 3)^2$

11. $x = (y - 2)(y + 3)^5$

12. $x = \dfrac{y^2}{(4 - y)^3}$

Derivatives of Implicit Relations

Find dy/dx. (Treat a and r as constants.)

13. $5x - 2y = 7$

14. $2x + 3y^2 = 4$

15. $xy = 5$

16. $x^2 + 3xy = 2y$

17. $y^2 = 4ax$

18. $y^2 - 2xy = a^2$

19. $x^3 + y^3 - 3axy = 0$

20. $x^2 + y^2 = r^2$

21. $y + y^3 = x + x^3$

22. $x + 2x^2y = 7$

23. $y^3 - 4x^2y^2 + y^4 = 9$

24. $y^{3/2} + x^{3/2} = 16$

Find the slope of the tangent to each curve at the given point.

25. $x^2 + y^2 = 25$ at $x = 2$ in the first quadrant

26. $x^2 + y^2 = 25$ at $(3, 4)$

27. $2x^2 + 2y^3 - 9xy = 0$ at $(1, 2)$

28. $x^2 + xy + y^2 - 3 = 0$ at $(1, 1)$

Differentials

Write the differential dy for each function.

29. $y = x^3$

30. $y = x^2 + 2x$

31. $y = \dfrac{x - 1}{x + 1}$

32. $y = (2 - 3x^2)^3$

33. $y = (x + 1)^2(2x + 3)^3$

34. $y = \sqrt{1 - 2x}$

35. $y = \dfrac{\sqrt{x - 4}}{3 - 2x}$

Write the differential dy in terms of x, y, and dx for each implicit relation.

36. $3x^2 - 2xy + 2y^2 = 3$

37. $x^3 + 2y^3 = 5$

38. $2x^2 + 3xy + 4y^2 = 20$

39. $2\sqrt{x} + 3\sqrt{y} = 4$

27–7 Higher-Order Derivatives

After taking the derivative of a function, we may then take the derivative of the derivative. That is called the *second derivative*. Our original derivative we now call the *first derivative*. The symbols used for the second derivative are

$$\frac{d^2y}{dx^2} \quad \text{or} \quad y'' \quad \text{or} \quad f''(x) \quad \text{or} \quad D^2y$$

We can then go on to find third, fourth, and higher derivatives.

We will have many uses for the second derivative in the next few chapters, but derivatives higher than second order are rarely needed.

♦♦♦ **Example 51:** Given the function

$$y = x^4 + 2x^3 - 3x^2 + x - 5$$

we get

$$y' = 4x^3 + 6x^2 - 6x + 1$$

and

$$y'' = 12x^2 + 12x - 6$$
$$y''' = 24x + 12$$
$$y^{(iv)} = 24$$
$$y^{(v)} = 0$$

and all higher derivatives will also be zero. ♦♦♦

♦♦♦ **Example 52:** Find the second derivative of $y = (x + 2)\sqrt{x - 3}$.

Solution: Using the product rule, we obtain

$$y' = (x + 2)\left(\frac{1}{2}\right)(x - 3)^{-1/2} + \sqrt{x - 3} \quad (1)$$

and

$$y'' = \frac{1}{2}\left[(x + 2)\left(-\frac{1}{2}\right)(x - 3)^{-3/2} + (x - 3)^{-1/2)}(1)\right] + \frac{1}{2}(x - 3)^{-1/2}$$

$$= -\frac{x + 2}{4(x - 3)^{3/2}} + \frac{1}{\sqrt{x - 3}} \quad ♦♦♦$$

Tip	You can usually save a lot of work if you *simplify the first derivative* before taking the second.

Exercise 7 ◆ Higher-Order Derivatives

Find the second derivative of each function.

1. $y = 3x^4 - x^3 + 5x$　　　　**2.** $y = x^3 - 3x^2 + 6$

3. $y = \dfrac{x^2}{x + 2}$　　　　**4.** $y = \dfrac{3 + x}{3 - x}$

5. $y = \sqrt{5 - 4x^2}$　　　　**6.** $y = \sqrt{x + 2}$

7. $y = (x - 7)(x - 3)^3$　　　　**8.** $y = x^2\sqrt{2.3x - 5.82}$

9. If $y = x(9 + x^2)^{1/2}$, find $y''(4)$.　　　　**10.** If $f(x) = 1/\sqrt{x} + \sqrt{x}$, find $f''(1)$.

Case Study Discussion—Control System for a Mixing Vat

You can provide early indication of problems by monitoring rate of change using either analog or digital systems.

If you have a level indicator that provides a continuous output of the slurry level (like an ultrasonic level or laser optical level indicator mounted inside the roof of the vat), you can monitor the rate of change of the level. If the output of the level measuring device is an analog voltage, you can send that signal through a differentiator op-amp circuit that produces an output voltage equal to the time rate of change of the input voltage:

$$V_{OUT} = d(V_{IN})/dt$$

If the valves are simply completely open or completely closed, you could use an integrator circuit where the output is the integral of the input voltage. This input voltage would simply be present if the valve is open and 0V if the valve is closed.

$$V_{OUT} = \int V_{IN}\, dt$$

If the valve is open longer than would be required for a normal mixing process, the output level would reach an alarm level.

If the level indicator provides a digital output, then you can use a microcontroller which would read the data and perform the differential equation on the input digital values and alarm at a preset level, or program the integration of the input representing the state of the valves.

◆◆◆ CHAPTER 27 REVIEW PROBLEMS ◆◆◆◆◆◆◆◆◆◆◆◆◆◆◆◆◆◆◆◆◆◆◆◆◆◆◆◆◆◆

1. Differentiate $y = \sqrt{\dfrac{x^2 - 1}{x^2 + 1}}$.

2. Evaluate $\lim\limits_{x \to 1} \dfrac{x^2 + 2 - 3x}{x - 1}$.

3. Find dy/dx if $x^{2/3} + y^{2/3} = 9$.

4. Find y'' if $y = \dfrac{x}{\sqrt{2x + 1}}$.

5. If $f(x) = \sqrt{4x^2 + 9}$, find $f'(2)$.

6. Evaluate $\lim\limits_{x \to \infty} \dfrac{5x + 3x^2}{x^2 - 1 - 3x}$.

7. Find dy/dx if $\dfrac{x^2}{a^2} + \dfrac{y^2}{b^2} = 1$.

8. Find dy/dx if $y = \sqrt[3]{\dfrac{3x + 2}{2 - 3x}}$.

9. If $f(x) = \sqrt{25 - 3x}$, find $f''(3)$.

10. Find dy/dx if $y = \dfrac{x^2 + a^2}{a^2 - x^2}$.

11. Find dy/dx by the delta method if $y = 5x - 3x^2$.

12. Evaluate $\lim\limits_{x \to 0} \dfrac{\sin x}{x + 1}$.

13. Find the slope of the tangent to the curve $y = \dfrac{1}{\sqrt{25 - x^2}}$ at $x = 3$.

14. Evaluate $\lim\limits_{x \to 0} \dfrac{e^x}{x}$. ∞

15. If $y = 2.15x^3 - 6.23$, find $y'(5.25)$.

16. Evaluate $\lim\limits_{x \to -7} \dfrac{x^2 + 6x - 7}{x + 7}$.

17. Find the derivative $\dfrac{d}{dx}(3x + 2)$.

18. Find the derivative $D(3x^4 + 2)^2$.

19. Find the derivative $D_x(x^2 - 1)(x + 3)^{-4}$.

20. If $v = 5t^2 - 3t + 4$, find dv/dt.

21. If $z = 9 - 8w + w^2$, find dz/dx.

22. Evaluate $\lim\limits_{x \to 5} \dfrac{25 - x^2}{x - 5}$.

23. If $f(x) = 7x - 4x^3$, find $f''(x)$.

24. Find the following derivative:

$$D_x(21.7x + 19.1)(64.2 - 17.9x^{-2})^2$$

25. If $s = 58.3t^3 - 63.8t$, find ds/dt.

26. Find $\dfrac{d}{dx}(4x^3 - 3x + 2)$.

Find the derivative.

27. $y = 6x^2 - 2x + 7$

28. $y = (3x + 2)(x^2 - 7)$

29. $y = 16x^3 + 4x^2 - x - 4$

30. $y = \dfrac{2x}{x^2 - 9}$

31. $y = 2x^{17} + 4x^{12} - 7x + 1/(2x^2)$

32. $y = (2x^3 - 4)^2$

33. $y = \dfrac{x^2 + 3x}{x - 1}$

34. $y = x^{2/5} + 2x^{1/3}$

Find $f''(x)$.

35. $f(x) = 6x^4 + 4x^3 - 7x^2 + 2x - 17$

36. $f(x) = (2x + 1)(5x^2 - 2)$

Write the differential dy for each function.

37. $y = 3x^4$

38. $y = (2x - 5)^2$

39. $x^2 + 3y^2 = 36$

40. $x^3 - y - 2x = 5$

Writing

41. We find the derivative by the delta method by first finding Δy divided by Δx and then letting Δx approach zero. Explain in your own words why this doesn't give division by zero, causing us to junk the whole calculation.

42. Do you feel it is reasonable to learn rules for derivatives, now that we can use calculators or computers to find derivatives? Can you think of any reasons for continuing to learn them? Write a memo to the head of your mathematics department stating whether or not you think the present method of learning derivatives should be changed, and give your reasons.

Team Project

43. For a function assigned by your instructor, find the derivative in as many ways as you can: analytically, graphically, or numerically.

28

Graphical Applications of the Derivative

••• OBJECTIVES ••

- Write the equation of the tangent or the normal to a curve.
- Use the derivative to find the maximum, minimum, and inflection points of an equation.
- Sketch, verify, and interpret graphs using derivatives.
- Use Newton's method to find roots of an equation.

You saw in the last chapter that the derivative is the instantaneous slope of a curve. In this first chapter of applications, you will see how the instantaneous slope can be used to find characteristics of functions that would otherwise require careful graphing to identify and visualize. By using the derivative and second and even third derivatives, too, you can calculate the locations of a function's maximum points, minimum points, and the equation for any line tangent to, or forming a normal to, the function.

The second derivative is also used here both to locate points on a curve where the curvature changes direction and to tell whether a point at which the tangent is horizontal is a peak or a valley.

The derivative is further used in *Newton's method* for finding roots of equations, a technique that works very well on the computer. Finally, the chapter closes with a summary of steps to aid in curve sketching.

28–1   Tangents and Normals

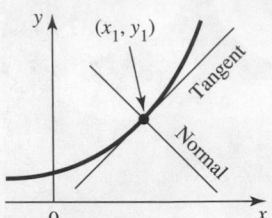

FIGURE 28-1   Tangents and normal to the curve.

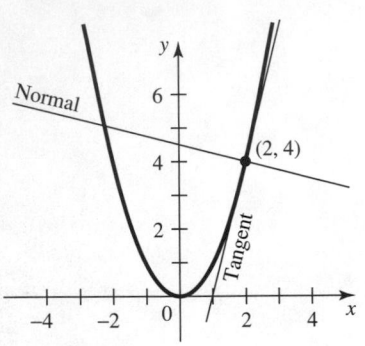

FIGURE 28-2

Tangent to a Curve

We saw in Chapter 27 that the slope m_t of the tangent to a curve at some point [(x_1, y_1) in Fig. 28–1] is given by the derivative of the equation of the curve evaluated at that point.

$$\text{slope of tangent} = m_t = y'(x_1)$$

Knowing the slope and the coordinates of the given point, we then use the point-slope form of the straight-line equation (Eq. 291) to write the equation of the tangent.

◆◆◆ **Example 1:**   Write the equation of the tangent to the curve $y = x^2$ at $x = 2$ (Fig. 28–2).

Solution: The derivative is $y' = 2x$. When $x = 2$, $y(2) = 2^2 = 4$, and

$$y'(2) = 2(2) = 4 = m_t$$

Using the point-slope form of the straight-line equation, we obtain

$$\frac{y - 4}{x - 2} = 4$$

or

$$y - 4 = 4x - 8$$

So $4x - y - 4 = 0$ is the equation of the tangent.   ◆◆◆

Normal to a Curve

The *normal* to a curve at a given point is the line perpendicular to the tangent at that point, as in Fig. 28–1. From Eq. 295, we know that the slope m_n of the normal will be the negative reciprocal of the slope of the tangent.

◆◆◆ **Example 2:**   Find the equation of the normal at the same point as in Example 1.

Solution: The slope of the normal is

$$m_n = -\frac{1}{m_t} = -\frac{1}{4}$$

Again using the point-slope form, we have

$$\frac{y - 4}{x - 2} = -\frac{1}{4}$$
$$4y - 16 = -x + 2$$

so $x + 4y - 18 = 0$ is the equation of the normal.   ◆◆◆

Implicit Relations

When the equation of the curve is an implicit relation, you may choose to solve for y, when possible, before taking the derivative. Often, though, it is easier to take the derivative implicitly, as in the following example.

◆◆◆ **Example 3:**   Find (a) the equation of the tangent to the ellipse $4x^2 + 9y^2 = 40$ at the point $(1, -2)$ and (b) the x intercept of the tangent (Fig. 28–3).

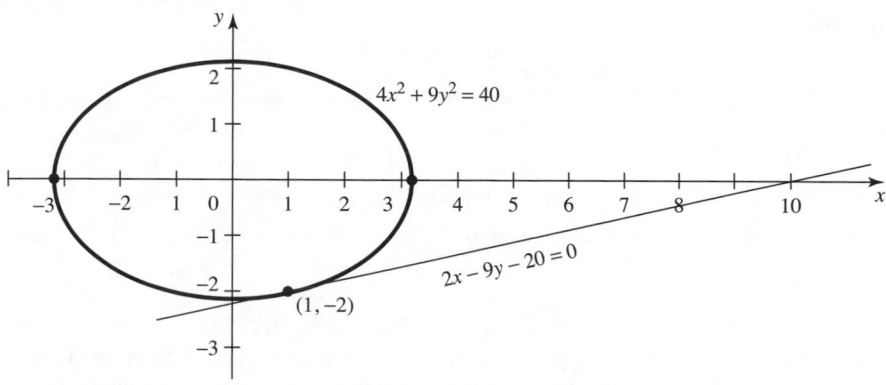

FIGURE 28–3

Solution:

(a) Taking the derivative implicitly, we have $8x + 18yy' = 0$, or

$$18yy' = -8x$$

$$y' = -\frac{8x}{18y} = -\frac{4x}{9y}$$

At $(1, -2)$,

$$y'(1, -2) = -\frac{4(1)}{9(-2)} = \frac{2}{9} = m_t$$

Using the point-slope form gives us

$$\frac{y - (-2)}{x - 1} = \frac{2}{9}$$

$$9y + 18 = 2x - 2$$

so $2x - 9y - 20 = 0$ is the equation of the tangent.

(b) Setting y equal to zero in the equation of the tangent gives $2x - 20 = 0$, or an x intercept of

$$x = 10$$

◆◆◆

Angle of Intersection of Two Curves

The angle between two curves is defined as the angle between their tangents at the point of intersection, which is found from Eq. 296.

> If the points of intersection are not known, solve the two equations simultaneously.

◆◆◆ **Example 4:** Find the angle of intersection between the parabolas (a) $y^2 = x$ and (b) $y = x^2$ at the point of intersection $(1, 1)$ (Fig. 28–4).

Solution:

(a) Taking the derivative, we have $2yy' = 1$, or

$$y' = \frac{1}{2y}$$

At $(1, 1)$,

$$y'(1, 1) = \frac{1}{2} = m_1$$

(b) Taking the derivative gives us $y' = 2x$. At $(1, 1)$,

$$y'(1, 1) = 2 = m_2$$

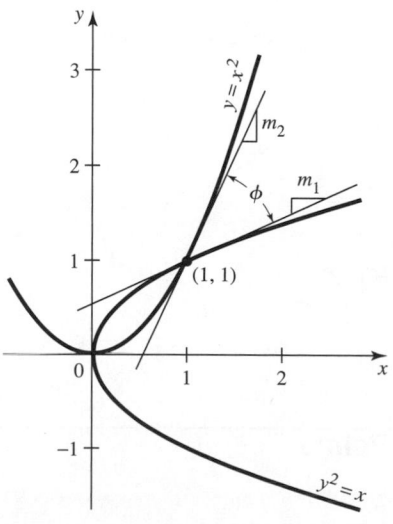

FIGURE 28–4 Angle of intersection of two curves.

From Eq. 296,

$$\tan \phi = \frac{m_2 - m_1}{1 + m_2 m_1}$$

$$= \frac{2 - \frac{1}{2}}{1 + \frac{1}{2}(2)} = 0.75$$

$$\phi = 36.9°$$

◆◆◆

CASE STUDY—POWER OUTPUT OF A SOUND SYSTEM

You are an electronics technologist designing a sound system for an outdoor concert. Because of the shape of the outdoor area, including some trees and rock outcroppings, you would like to add some more speakers. You currently have 40 speakers, each putting out 300 W of power. For every speaker you add, the power of every speaker drops by 5 W. How many speakers can you add to maximize the power output, and how much power is going to be required?

Exercise 1 ◆ Tangents and Normals

For problems 1 through 6, write the equations of the tangent and normal at the given point.

1. $y = x^2 + 2$ at $x = 1$
2. $y = x^3 - 3x$ at $(2, 2)$
3. $y = 3x^2 - 1$ at $x = 2$
4. $y = x^2 - 4x + 5$ at $(1, 2)$
5. $x^2 + y^2 = 25$ at $(3, 4)$
6. $16x^2 + 9y^2 = 144$ at $(2, 2.98)$
7. Find the first quadrant point on the curve $y = x^3 - 3x^2$ at which the tangent to the curve is parallel to $y = 9x + 7$.
8. Write the equation of the tangent to the parabola $y^2 = 4x$ that makes an angle of 45° with the x axis.
9. The curve $y = 2x^3 - 6x^2 - 2x + 1$ has two tangents parallel to the line $2x + y = 12$. Find their equations.
10. Each of two tangents to the circle $x^2 + y^2 = 25$ has a slope $\frac{3}{4}$. Find the points of contact.
11. Find the equation of the line tangent to the parabola $y = x^2 - 3x + 2$ and perpendicular to the line $5x - 2y - 10 = 0$.

Angles between Curves

Find the angle(s) of intersection between the given curves.

12. $y = x^2 + x - 2$ and $y = x^2 - 5x + 4$ at $(1, 0)$
13. $y = -2x$ and $y = x^2(1 - x)$ at $(0, 0)$, $(2, -4)$, and $(-1, 2)$
14. $y = 2x + 2$ and $x^2 - xy + y^2 = 4$ at $(0, 2)$ and $(-2, -2)$

28–2   Maximum, Minimum, and Inflection Points

As usual, we limit this discussion to smooth curves, without cusps or corners at which there is no derivative.

Some Definitions

Figure 28–5 shows a path over the mountains from A to H. It goes over three peaks, B, D, and F. These are called *maximum points* on the curve from A to H. The highest, peak D, is called the *absolute maximum* on the curve from A to H, while peaks B and F are called *relative maximum* points. Similarly, valley G is called an *absolute minimum,* while valleys C and E are *relative minimums.* All of the peaks and valleys are referred to as *extreme values.* Figure 28–6 shows maximum and minimum points on a graph of $y = f(x)$.

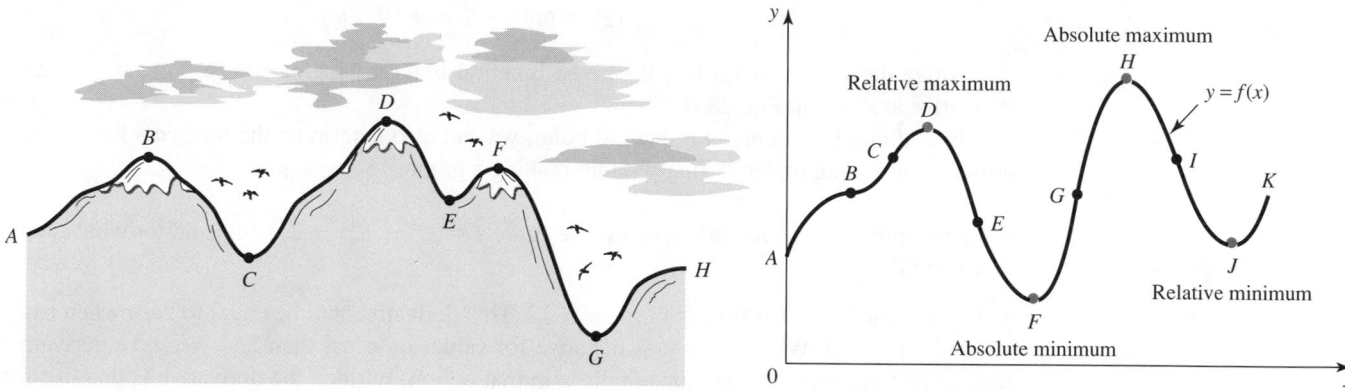

FIGURE 28–5 Path over the mountains.

FIGURE 28–6 Maximum, minimum, and inflection points.

Increasing and Decreasing Functions

Within the interval from A to K in Fig. 28–6, the function is said to be *increasing* from A to D, because the curve rises as we move in the positive x direction. The curve function is also increasing from F to H and from J to K. The function is *decreasing* from D to F and from H to J.

The first derivative tells us whether a function is increasing or decreasing at any point.

◆◆◆ **Example 5:** Graph the function $y = 2x^3 - 5x + 7$ and its first derivative on the same axes.

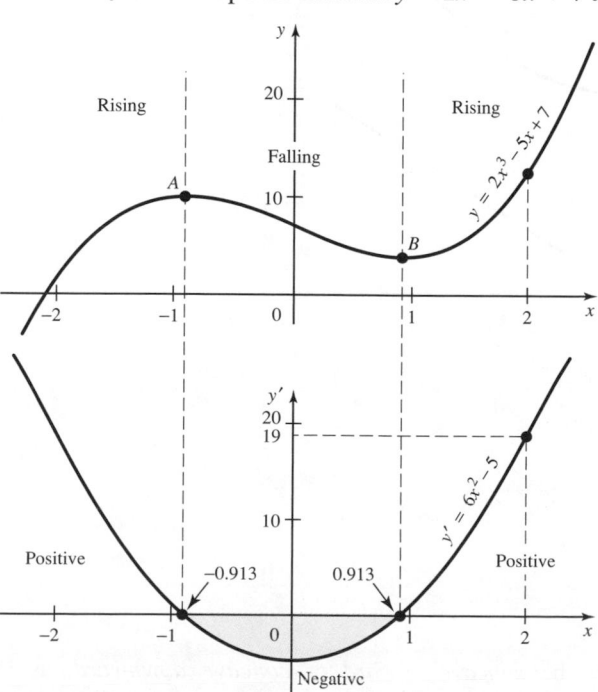

Solution: We graph the given function and its derivative curve as shown in Fig. 28–7. Note that the derivative is negative in the interval $-0.913 < x < 0.913$ and that the given function is decreasing within that interval. Also note that y' is positive for $x > 0.913$ and $x < -0.913$ and that the function is increasing within those regions.

FIGURE 28–7

◆◆◆

Thus the slope of a curve, and hence y', is *positive* for an *increasing* function. Conversely, y' is *negative* for a *decreasing* function. We can use these facts to determine whether a curve is increasing or not at a particular point, without having to graph the function.

⬧⬧⬧ **Example 6:** Is the function $y = 2x^3 - 5x + 7$ increasing or decreasing at $x = 2$?
Solution: The derivative is $y' = 6x^2 - 5$. At $x = 2$,

$$y'(2) = 6(4) - 5 = +19$$

A positive derivative means that the given function is increasing at $x = 2$. Note the positive derivative at $x = 2$ in Fig. 28–7.   ⬧⬧⬧

In addition to looking at a specific point, we can also determine the *intervals* for which a curve is increasing or decreasing without making a graph.

⬧⬧⬧ **Example 7:** For what values of x is the curve $y = 3x^2 - 12x - 2$ rising, and for what values of x is it falling?

Solution: The first derivative is $y' = 6x - 12$. This derivative will be equal to zero when $6x - 12 = 0$, or $x = 2$. We see that y' is negative for values of x less than 2. A *negative* derivative tells us that our original function is falling in that region. Further, the derivative is positive for $x > 2$, so the given function is rising in that region, as shown in Fig. 28–8.

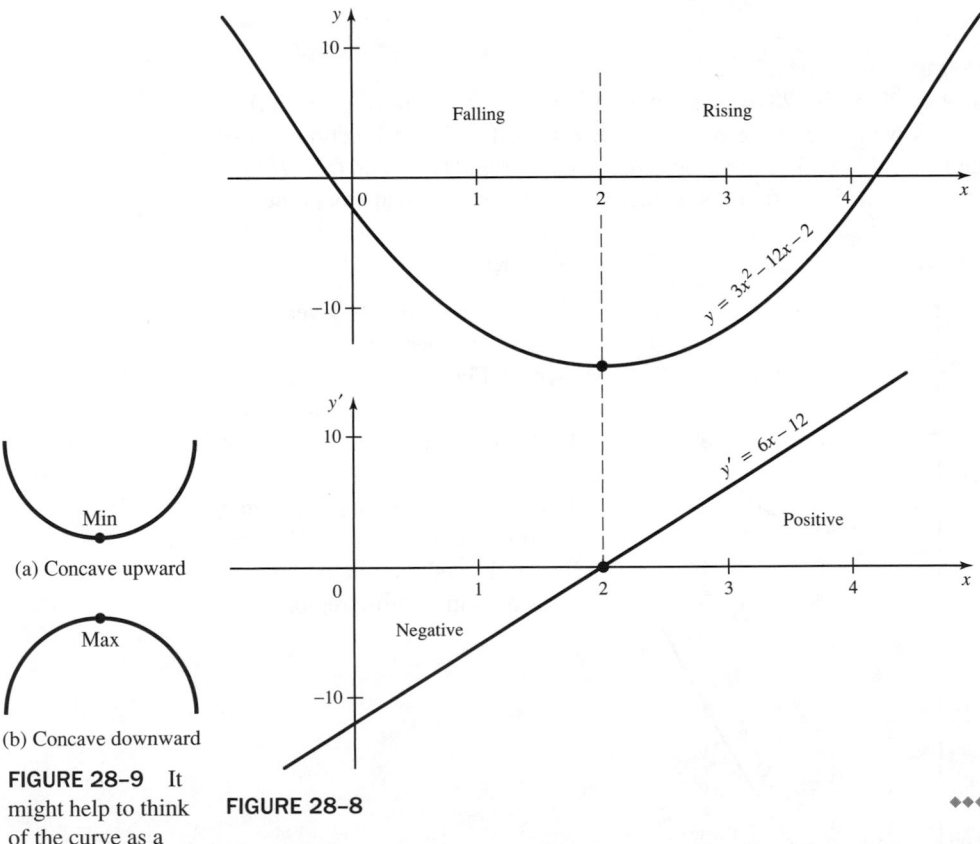

FIGURE 28–9 It might help to think of the curve as a bowl. When it is concave upward, it will hold (+) water, and when concave downward it will spill (−) water.

FIGURE 28–8   ⬧⬧⬧

Concavity

Portions of a curve may be said to be *concave upward* or *concave downward,* as in Fig. 28–9. Thus for the function in Fig. 28–10(a), the curve is concave upward to the right of the origin and concave downward to the left of the origin.

Also graphed is the first derivative of the given function in Fig. 28–10(b). Note that in the region where the function is concave upward, the slope of the tangent is increasing, and hence the first derivative y' (which gives the value of the slope) is also increasing. Since y' is increasing, then the derivative of the derivative, y'', must be positive. In other words, *the second derivative is positive where a curve is concave upward*. This is verified by the graph of y'' in Fig. 28–10(c).

The reverse is also true—that *the second derivative is negative where a curve is concave downward*. We can also infer that the second derivative is zero where the curve is neither concave upward nor concave downward, such as at the origin in Fig. 28–10(a).

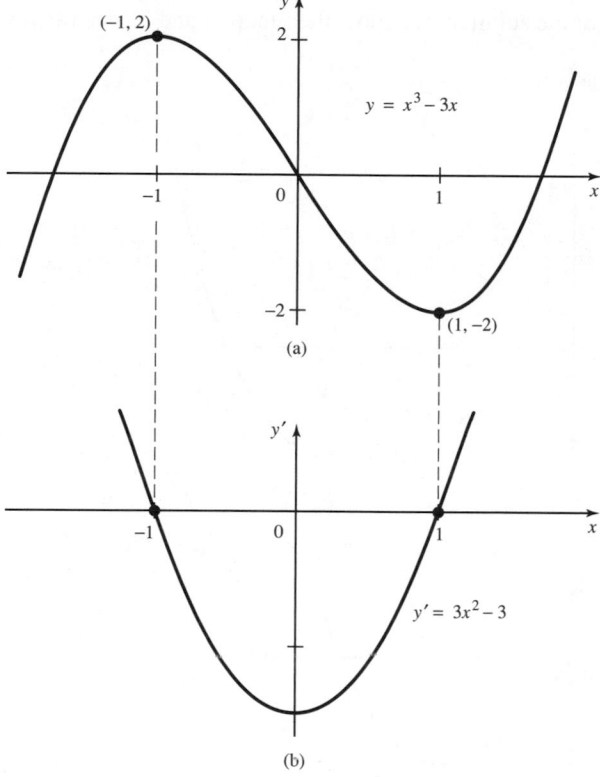

(a)

(b)

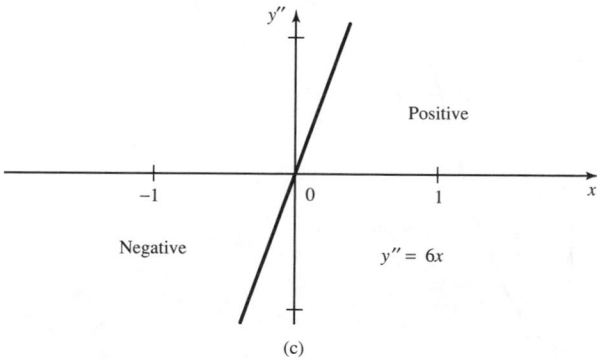

(c)

FIGURE 28–10

We can use these ideas to determine the concavity of curves at particular points or regions without making a graph.

◆◆◆ **Example 8:** Is the curve $y = 3x^4 - 7x^2 - 2$ concave upward or concave downward at (a) $x = 0$ and (b) $x = 1$?

Solution: The first derivative is $y' = 12x^3 - 14x$, and the second derivative is $y'' = 36x^2 - 14$.

(a) At $x = 0$,

$$y''(0) = -14$$

A negative second derivative tells that the given curve is concave downward at that point.

(b) At $x = 1$,

$$y''(1) = 36 - 14 = 22$$

or positive, so the curve is concave upward at that point.

Although a graph was not needed for the solution, we show the function and its derivatives in Fig. 28–11 for verification.

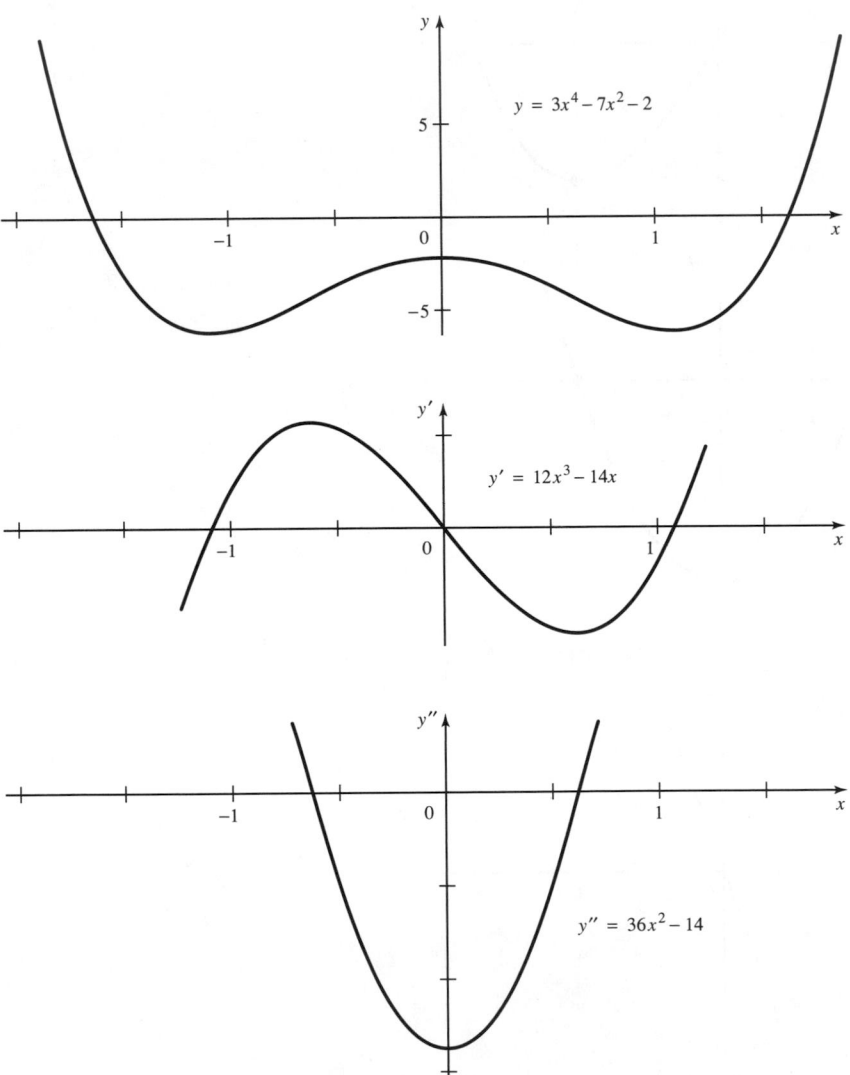

FIGURE 28–11   ◆◆◆

The idea of maximum and minimum values will lead to some interesting applications in Chapter 29.

Finding Maximum and Minimum Points

Points having horizontal tangents are called *stationary points*. They include maximums, minimums, and rare points such as B in Fig. 28–6. Looking back at Fig. 28–7, notice that the maximum point A occurs at a value of x for which y' is zero. The minimum point B also occurs where y' is zero. In Fig. 28–8 as well, the minimum point occurs where y' is zero.

This should not be surprising. Since the slope is positive to one side of a maximum or minimum and negative to the other side, the slope, and hence y', must be zero right at the maximum or minimum.

We use this fact to locate a maximum or minimum point.

| To find maximum and minimum points (and other stationary points), find the values of x for which the first derivative is zero. | **366** |

We can find such points graphically or analytically, as in the following example.

◆◆◆ **Example 9:** Find the maximum and minimum points for the function $y = x^3 - 3x$.
We take the first derivative,

$$y' = 3x^2 - 3$$

We find the value of x that makes the derivative zero by *setting the derivative equal to zero and solving for x.*

$$3x^2 - 3 = 0$$
$$x = \pm 1$$

Solving for y, we have

$$y(1) = 1 - 3 = -2$$

and

$$y(-1) = -1 + 3 = 2$$

So the points of zero slope are $(1, -2)$ and $(-1, 2)$, as in Fig. 28–10.   ◆◆◆

Testing for Maximum or Minimum

The simplest way to tell whether a stationary point is a maximum, a minimum, or neither, is to graph the curve, as in Example 9. There are also a few tests that we can use to identify maximum and minimum points without having to graph the function. They are the first-derivative test, the second-derivative test, and the ordinate test.

In the *first-derivative test,* we look at the slope of the tangent on either side of a stationary point. At the minimum point B, shown in Fig. 28–12, for example, we see that the slope is negative to the left (at A) and positive to the right (at C) of that point. The reverse is true for a maximum point. If the first derivative has the same sign on the left and right side of a stationary point, the point is neither a maximum nor a minimum. Since the slope is given by the first derivative, we have the following:

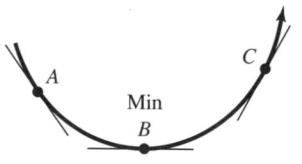

FIGURE 28–12

| **First-Derivative Test** | The first derivative is negative to the left of, and positive to the right of, a minimum point. The reverse is true for a maximum point. | **367** |

We use this test only on points close to the suspected maximum or minimum. If our test points are too far away, the curve might have already changed direction.

The *second-derivative test* uses the fact that a minimum point occurs in a region of a curve that is concave upward (y'' is positive), while a maximum point occurs where a curve is concave downward (y'' is negative).

This test will not work on rare occasions. For example, the function $y = x^4$ has a minimum point at the origin, but the second derivative there is zero, not a positive number as expected.

Second-Derivative Test	If the first derivative at some point is zero, then, if the second derivative is 1. positive, the point is a minimum. 2. negative, the point is a maximum. 3. zero, the test fails.	368

Common Error	It is tempting to group *maximum* with *positive*, and *minimum* with *negative*. Remember that they are just the *reverse* of this.	

To use the *ordinate* test to distinguish between a maximum point and a minimum point, look at the height of the curve to either side of the point.

Ordinate Test	Find y a small distance to either side of the point to be tested. If y is greater there, we have a minimum; if it is less, we have a maximum.	369

◆◆◆ **Example 10:** The function $y = x^3 - 3x$ was found in Example 9 to have stationary points at $(1, -2)$ and $(-1, 2)$. Use all three tests to determine if $(1, -2)$ is a maximum or a minimum.

Solution: *First-derivative test:* We evaluate the first derivative a small amount to either side of the point, say, at $x = 0.9$ and $x = 1.1$.

$$y' = 3x^2 - 3$$
$$y'(0.9) = 3(0.9)^2 - 3 = -0.57$$
$$y'(1.1) = 3(1.1)^2 - 3 = 0.63$$

Slopes that are negative to the left and positive to the right of the point indicate a minimum point.
  Second-derivative test: Evaluating y'' at the point gives

$$y'' = 6x$$
$$y''(1) = 6$$

A positive second derivative also indicates a minimum point.
  Ordinate test: We compute y at $x = 0.9$ and $x = 1.1$.

$$y(0.9) = (0.9)^3 - 3(0.9) = -1.97$$
$$y(1.1) = (1.1)^3 - 3(1.1) = -1.97$$

Thus the curve is higher a small distance to either side of the extreme value $(1, -2)$, again indicating a minimum point.   ◆◆◆

Implicit Relations

The procedure for finding maximum and minimum points is no different for an implicit relation, although it usually takes more work.

◆◆◆ **Example 11:** Find any maximum and minimum points on the curve
$$x^2 + 4y^2 - 6x + 2y + 3 = 0.$$

Solution: Taking the derivative implicitly gives

$$2x + 8yy' - 6 + 2y' = 0$$
$$y'(8y + 2) = 6 - 2x$$
$$y' = \frac{3 - x}{4y + 1}$$

Setting this derivative equal to zero gives $3 - x = 0$, or

$$x = 3$$

Substituting $x = 3$ into the original equation, we get

$$9 + 4y^2 - 18 + 2y + 3 = 0$$

Collecting terms yields

$$2y^2 + y - 3 = 0$$

Factoring gives us

$$(y - 1)(2y + 3) = 0$$
$$y = 1 \qquad y = -\frac{3}{2}$$

So the points of zero slope are $(3, 1)$ and $(3, -\frac{3}{2})$.

We now apply the second-derivative test. Using the quotient rule gives

$$y'' = \frac{(4y + 1)(-1) - (3 - x)4y'}{(4y + 1)^2}$$
$$= -\frac{(4y + 1) + 4(3 - x)y'}{(4y + 1)^2}$$

Replacing y' by $(3 - x)/(4y + 1)$ and simplifying, we have

$$y'' = -\frac{(4y + 1) + 4(3 - x)\dfrac{3 - x}{4y + 1}}{(4y + 1)^2}$$
$$= -\frac{(4y + 1)^2 + 4(3 - x)^2}{(4y + 1)^3}$$

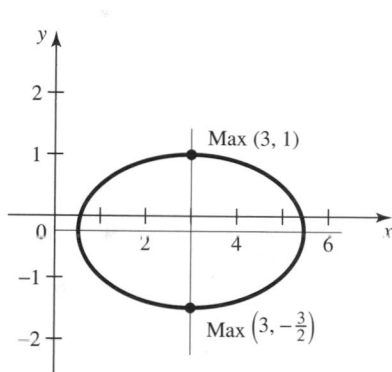

FIGURE 28–13   Graph of $x^2 + 4y^2 - 6x + 2y + 3 = 0$.

At $(3, 1)$, $y'' = -0.2$. The negative second derivative tells us that $(3, 1)$ is a maximum point. At $(3, -\frac{3}{2})$, $y'' = 0.2$. This tells us that we have a minimum point, as shown in Fig. 28–13.   ◆◆◆

Inflection Points

A point where the curvature changes from concave upward to concave downward (or vice versa) is called an *inflection point,* or *point of inflection.* In Fig. 28–6 they are points B, C, E, G, and I.

We saw that the second derivative was positive where a curve was concave upward and negative where concave downward. In going from positive to negative, the second derivative must somewhere be zero, and this is at the point of inflection.

Inflection Points	To find points of inflection, set the second derivative to zero and solve for x. Test by seeing if the second derivative changes sign a small distance to either side of the point.	**370**

Unfortunately, a curve can have a point of inflection where the second derivative does not exist, such as the curve $y = x^{1/3}$ at $x = 0$. Fortunately, these cases are rare.

As with finding maxima and minima, we can use either an analytical or a graphical method to find inflection points.

◆◆◆ **Example 12:** Find any points of inflection on the curve

$$y = x^3 - 3x^2 - 5x + 7$$

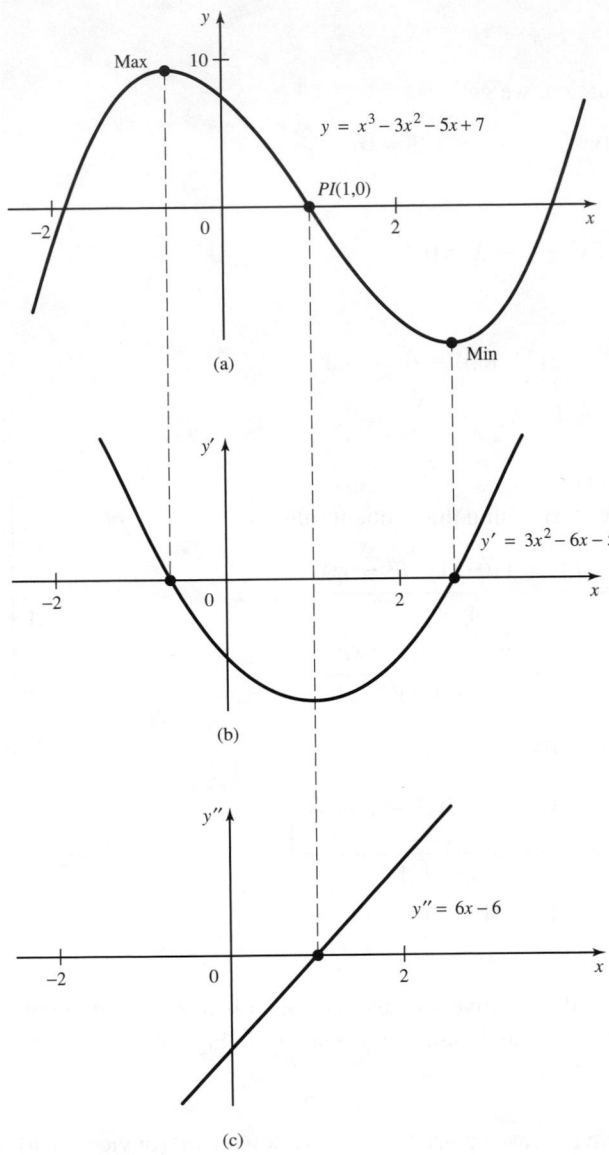

FIGURE 28-14

We take the derivative twice.

$$y' = 3x^2 - 6x - 5$$
$$y'' = 6x - 6$$

We now set y'' to 0 and solve for x.

$$6x - 6 = 0$$
$$x = 1$$

and

$$y(1) = 1 - 3 - 5 + 7 = 0$$

A graph of the given function [Fig. 28–14(a)] shows the point (1, 0) clearly to be a point of inflection. If there were any doubt, we would test it by seeing if the second derivative has opposite signs on either side of the point. ◆◆◆

Summary

Figure 28–15 summarizes the ideas of this section.

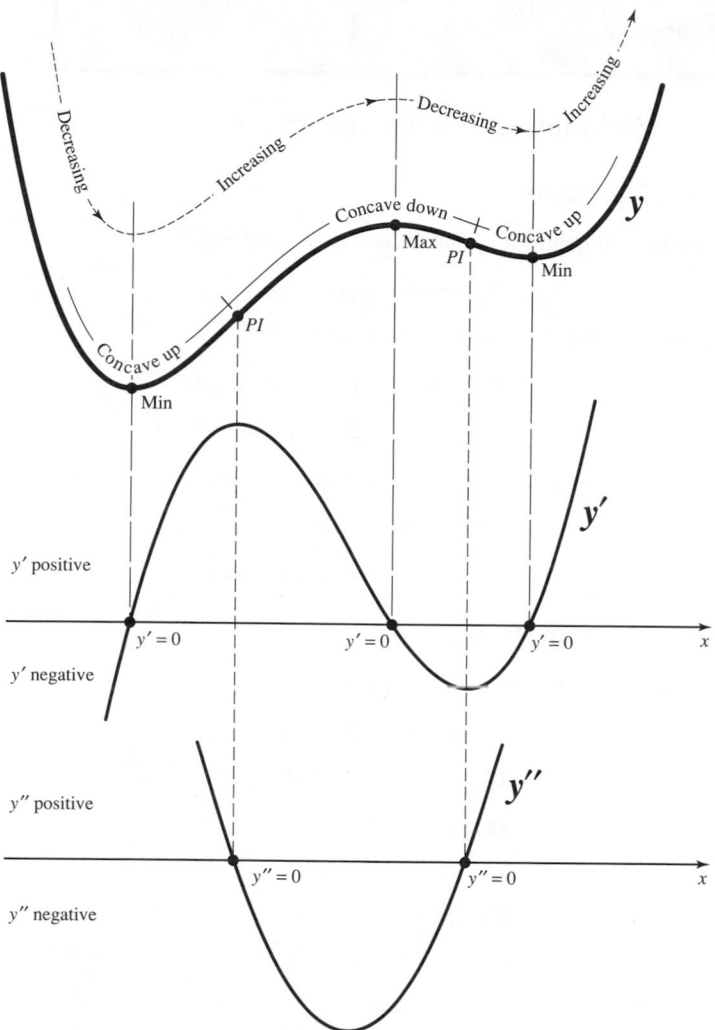

FIGURE 28–15

The function y is:

- higher on both sides near a minimum point.
- lower on both sides near a maximum point.

The first derivative y′ is:

- positive when y is increasing and negative when y is decreasing.
- zero at a maximum or minimum point.

(continues on next page)

(continued)

The second derivative y'' is:

- positive when the y curve is concave upward.
- negative when the y curve is concave downward.
- positive at a minimum point.
- negative at a maximum point.
- zero at an inflection point.

Exercise 2 ◆ Maximum, Minimum, and Inflection Points

Increasing and Decreasing Functions

For what values of x is each curve rising, and for what values is each falling?

1. $y = 3x + 5$ **2.** $y = 4x^2 + 16x - 7$

Is each function increasing or decreasing at the value indicated?

3. $y = 3x^2 - 4$ at $x = 2$ **4.** $y = x^2 - x - 3$ at $x = 0$

5. $y = 4x^2 - x$ at $x = -2$ **6.** $y = x^3 + 2x - 4$ at $x = -1$

State whether each curve is concave upward or concave downward at the given value of x.

7. $y = x^4 + x^2$ at $x = 2$ **8.** $y = 4x^5 - 5x^4$ at $x = 1$

9. $y = -2x^3 - 2\sqrt{x + 2}$ at $x = \frac{1}{4}$ **10.** $y = \sqrt{x^2 + 3x}$ at $x = 2$

Maximum, Minimum, and Inflection Points

Find the maximum, minimum, and inflection points for each function.

11. $y = x^2$ **12.** $y = x^3 + 3x^2 - 2$

13. $y = 6x - x^2 + 4$ **14.** $y = x^3 - 3x + 4$

15. $y = x^3 - 7x^2 + 36$ **16.** $y = x^4 - 4x^3$

17. $y = 2x^2 - x^4$ **18.** $16y = x^2 - 32x$

19. $y = x^4 - 4x$ **20.** $2y = x^2 - 4x + 6$

21. $y = 3x^4 - 4x^3 - 12x^2$ **22.** $y = 2x^3 - 9x^2 + 12x - 3$

23. $y = x^3 + 3x^2 - 9x + 5$ **24.** $y = \dfrac{x^2 - 7x + 6}{x - 10}$

25. $y = (x - 1)^4(x + 2)^3$ **26.** $y = (x - 2)^2(2x + 1)$

Find the maximum and minimum points for each implicit relation.

27. $4x^2 + 9y^2 = 36$ **28.** $x^2 + y^2 - 2x + 4y = 4$

29. $x^2 - x - 2y^2 + 36 = 0$ **30.** $x^2 + y^2 - 8x - 6y = 0$

28–3   Sketching, Verifying, and Interpreting Graphs

It is not enough to be able to make a graph. We must be able to *interpret* a graph—to discover hidden behaviour, to explain it to others, and to confirm that it is correct. We must also be sure that we have a complete graph, with no features of interest undiscovered outside the viewing window.

With our new calculus tools we are now able to do all of these things. We will review them here, along with some that we've had earlier.

Curve-Sketching Tips

A. Identify the **type of equation**. This will often save you time.
B. Find any **intercepts**.
C. Take advantage of **symmetry**.
D. Determine the domain and **range** of the function to locate any places where the curve does not exist.
E. Sketch in any **asymptotes**.
F. Locate the regions in which the function is **increasing** or **decreasing**.
G. Locate any **maximum**, **minimum**, and **inflection points**.
H. Determine the **end behaviour** as x gets very large.
I. Be sure to make a **complete graph**. Plot extra points if needed.

We will now give examples of each.

A. Type of Equation: Does the equation look like any we have studied before—a linear function, power function, quadratic function, trigonometric function, exponential function, logarithmic function, or equation of a conic? For a polynomial, the *degree* will tell the maximum number of extreme points you can expect.

 **Example 13:** The function $y = x^5 - 3x^2 + 4$ is a polynomial of fifth degree, so its graph (Fig. 28–16) can have up to four extreme points and three inflection points. There may, however, be fewer. ◆◆◆

B. Intercepts: Find the x intercept(s), or root(s), by setting y equal to zero and solving the resulting equation for x. Find the y intercept(s) by setting x equal to zero and solving for y.

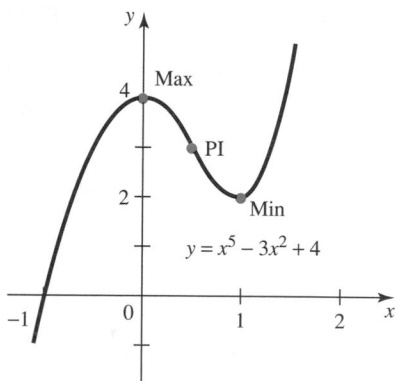

FIGURE 28–16

◆◆◆ **Example 14:**

(a) The function $3x + 2y = 6$ has a y intercept at

$$3(0) + 2y = 6$$

or at $y = 3$, and an x intercept at

$$3x + 2(0) = 6$$

or at $x = 2$ (Fig. 28–17).

(b) The function of Example 13 has a y intercept at

$$y = 0 - 0 + 4$$

or at $y = 4$ (Fig. 28–16), but to find the x intercept involves solving a fifth-degree equation, so we don't bother. ◆◆◆

C. Symmetry: A curve is symmetrical about the

(a) x axis, if the equation does not change when we substitute $-y$ for y.
(b) y axis, if the equation does not change when we substitute $-x$ for x.
(c) origin, if the equation does not change when we substitute $-x$ for x and $-y$ for y.

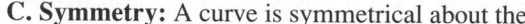 **Example 15:** Given the function $x^2 + y^2 - 3x - 8 = 0$:

(a) Substituting $-y$ for y gives $x^2 + (-y)^2 - 3x - 8 = 0$, or $x^2 + y^2 - 3x - 8 = 0$. This is the same as the original, indicating symmetry about the x axis.

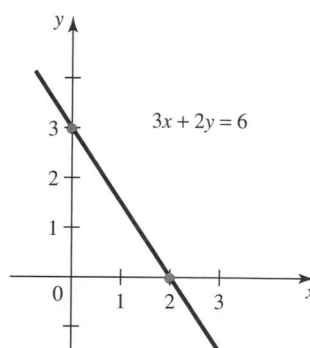

FIGURE 28–17

A function whose graph is symmetrical about the y axis is called an *even function*. One symmetrical about the origin is called an *odd function*.

(b) Substituting $-x$ for x gives $(-x)^2 + y^2 - 3(-x) - 8 = 0$, or $x^2 + y^2 + 3x - 8 = 0$. This equation is different from the original, indicating no symmetry about the y axis.

(c) Substituting $-x$ for x and $-y$ for y gives $(-x)^2 + (-y)^2 - 3(-x) - 8 = 0$ or $x^2 + y^2 + 3x - 8 = 0$. The equation is different from the original, indicating no symmetry about the origin, as shown in Fig. 28–18.

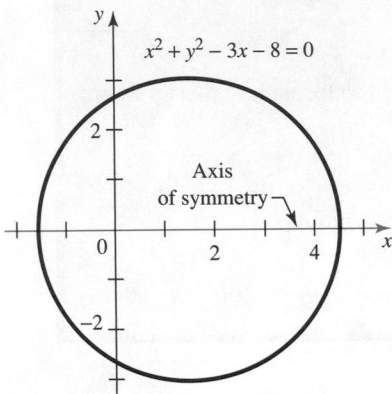

FIGURE 28–18   ◆◆◆

D. Range: Look for values of the variables that give division by zero or that result in negative numbers under a radical sign. The curve will not exist at these values.

◆◆◆ **Example 16:** For the function

$$y = \frac{\sqrt{x + 2}}{x - 5}$$

y is not real for $x < -2$ and does not exist for $x = 5$ (Fig. 28–19).

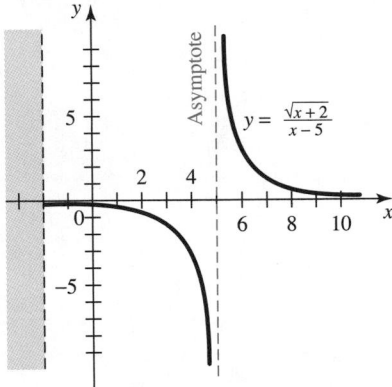

FIGURE 28–19   ◆◆◆

E. Asymptotes: Look for some value of x that when *approached* from either the left or right, will cause y to become infinite. You will then have found a *vertical asymptote*. Then if y approaches some particular value as x becomes infinite, you will have found a *horizontal asymptote*. Similarly, check what happens to y when x becomes infinite in the negative direction.

◆◆◆ **Example 17:** The function in Example 16 becomes infinite as x approaches 5, so we expect a vertical asymptote at $x = 5$ (Fig. 28–19).   ◆◆◆

F. Increasing or Decreasing Function: The first derivative will be positive in regions where the function is increasing, and negative where the function is decreasing. Thus inspection of the first derivative can show where the curve is rising and where it is falling.

◆◆◆ **Example 18:** For the function $y = x^2 - 4x + 2$, the first derivative $y' = 2x - 4$ is positive for $x > 2$ and negative for $x < 2$. Thus we expect the curve to fall in the region to the left of $x = 2$ and rise in the region to the right of $x = 2$ (Fig. 28–20). ◆◆◆

G. Maximum, Minimum, and Inflection Points: Maxima and minima are found where y' is zero. Inflection points are found where y'' is zero. These steps may be done analytically or graphically.

◆◆◆ **Example 19:** To find the maximum and minimum points on the curve $y = x^3 - 2x^2 + x + 1$, we set the first derivative

$$y' = 3x^2 - 4x + 1$$

to zero and solve for x, getting

$$y = \tfrac{1}{3} \quad \text{and} \quad x = 1$$

giving a maximum at $(\tfrac{1}{3}, 1.15)$ and a minimum at $(1, 1)$. To get the inflection point, we set the second derivative

$$y'' = 6x - 4$$

to zero, getting $x = \tfrac{2}{3}$. The inflection point is then $(\tfrac{2}{3}, 1.07)$. For a graphical solution, we graph y, y', and y'' on the same axes, as shown in Fig. 28–21. The zeros on the y' graph locate the maxima and minima, and the zero on the y'' graph locates the inflection point. ◆◆◆

H. End Behavior: What happens to y as x gets very large in both the positive and the negative directions? Does y continue to grow without bound or approach some asymptote, or is it possible that the curve will turn and perhaps cross the x axis again? Similarly, can you say what happens to x as y gets very large?

◆◆◆ **Example 20:** As x grows, for the function of Example 19, the second, third, and fourth terms on the right side become less significant compared to the x^3 term. So, far from the origin, the function will have the appearance of the function $y = x^3$ (Fig. 28–22).

I. Complete Graph: Once we have found all the zeros, maxima, minima, inflection points, and asymptotes, and have investigated the end behavior, we can be quite sure that we have not missed any features of interest.

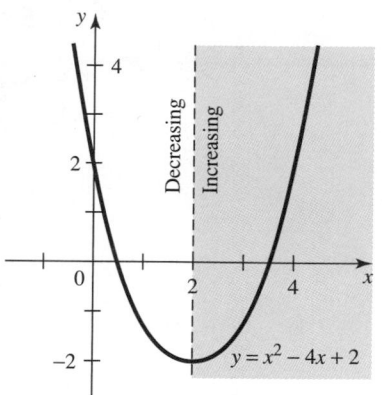

FIGURE 28–20

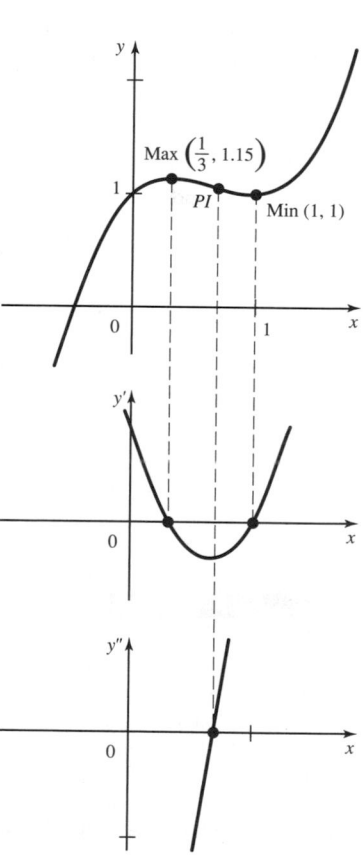

FIGURE 28–21

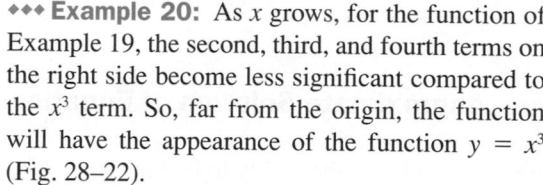

FIGURE 28–22

Graphing Regions

In later applications, we will have to identify *regions* that are bounded by two or more curves. Here we get some practice in doing that.

◆◆◆ **Example 21:** Locate the region bounded by the y axis and the curves $y = x^2$, $y = 1/x$, and $y = 4$.

Solution: The three curves given are, respectively, a parabola, a hyperbola, and a straight line (Fig. 28–23). We see that three different closed areas are formed, but the shaded area is the only one *bounded by each one of the given curves* and the y axis. ◆◆◆

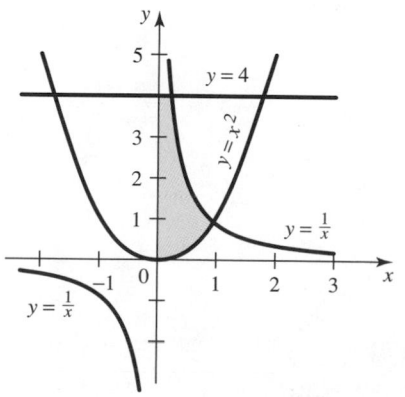

FIGURE 28–23 Graphing a region bounded by several curves.

Exercise 3 ◆ Sketching, Verifying, and Interpreting Graphs

Make a complete graph of each function.

1. $y = 4x^2 - 5$

2. $y = 3x - 2x^2$

3. $y = 5 - \dfrac{1}{x}$

4. $y = \dfrac{3}{x} + x^2$

5. $y = x^4 - 8x^2$

6. $y = \dfrac{1}{x^2 - 1}$

7. $y = x^3 - 9x^2 + 24x - 7$

8. $y = x\sqrt{1 - x}$

9. $y = 5x - x^5$

10. $y = \dfrac{9}{x^2 + 9}$

11. $y = \dfrac{6x}{3 + x^2}$

12. $y = \dfrac{x}{\sqrt{4 - x^2}}$

13. $y = x^3 - 6x^2 + 9x + 3$

14. $y = x^2\sqrt{6 - x^2}$

15. $y = \dfrac{96x - 288}{x^2 + 2x + 1}$

16. $y = \dfrac{\sqrt{x}}{x - 1}$

17. $y = \left(x/\sqrt{x^2 + 1} \right)$

18. $y = \dfrac{x^3}{(1 + x^2)^2}$

19. $y = x^2 + 2x$

20. $y = x^2 - 3x + 2$

21. $y = x^3 + 4x^2 - 5$

22. $y = x^4 - x^2$

Graph the region bounded by the given curves.

23. $y = 3x^2$ and $y = 2x$

24. $y^2 = 4x$, $x = 5$, and the x axis, in the first quadrant

25. $y = 5x^2 - 2x$, the y axis, and $y = 4$, in the second quadrant

26. $y = 4/x$, $y = x$, $x = 6$, and the x axis

28–4　Approximate Solution of Equations by Newton's Method

We have had so many laws and methods named "Newton" that you might think there were twenty Newtons all working in different fields. But no. All are from the same Isaac Newton (1642–1727).

Newton's method uses the ability to find the slope of a tangent to a curve simply by taking the derivative. With this method we can quickly find the root of an equation of the form $y = f(x)$, even though we cannot solve the equation for this root. The result is *approximate*, but this is hardly a drawback since we get as many significant digits as we want.

Figure 28–24 shows a graph of the function $y = f(x)$ and the zero where the curve crosses the x axis. The solution to the equation $f(x) = 0$ is then the value a.

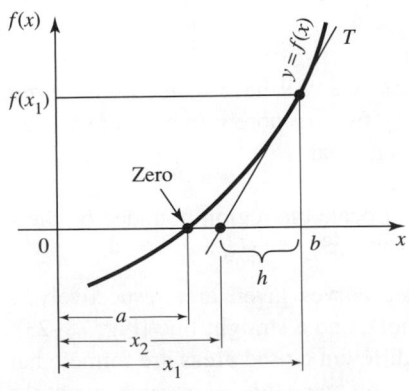

FIGURE 28–24　Newton's method.

Suppose that our first approximation to a is the value x_1 (our first guess is too high). We can *correct* our first guess by subtracting from it the amount h, taking for our second approximation the point where the tangent line T crosses the x axis. The slope of the tangent line at x_1 is

$$m = f'(x_1) = \frac{\text{rise}}{\text{run}} = \frac{f(x_1)}{h}$$

So

$$h = \frac{f(x_1)}{f'(x_1)}$$

and

$$\text{second approximation} = \text{first guess} - h$$

or

$$x_2 = x_1 - \frac{f(x_1)}{f'(x_1)}$$

You can stop at this point or repeat the calculation several times to get the accuracy needed, with each new value x_{n+1} obtained from the old value x_n by the following recursion formula:

Newton's Method	$x_{n+1} = x_n - \dfrac{f(x_n)}{f'(x_n)}$	371

A good way to get a first guess is to graph the function. Note that Equation 371 works whether your first guess is to the left or to the right of the zero. Also, the method is the same regardless of whether the curve is rising or falling where it crosses the x axis.

◆◆◆ **Example 22:** Use Newton's method to find one root of the equation $x^3 - 5x = 5$ to one decimal place. Take $x = 3$ as a first guess.

Solution: We first move all terms to one side of the equation and express them as a function of x.

$$f(x) = x^3 - 5x - 5$$

This function is graphed in Fig. 28–25, which shows a root near $x = 3$. Taking the derivative yields

$$f'(x) = 3x^2 - 5$$

At the value 3,

$$f'(3) = 3(9) - 5 = 22$$

and

$$f(3) = 27 - 15 - 5 = 7$$

The first correction is then

$$h = \frac{7}{22} = 0.32$$

Our second approximation is then $3 - 0.32 = 2.68$. We now repeat the entire calculation, using 2.68 as our guess.

$$f(2.68) = 0.8488$$

$$f'(2.68) = 16.55$$

$$h = \frac{0.8488}{16.55} = 0.0513$$

$$\text{third approximation} = 2.68 - 0.0513 = 2.629$$

Since h is 0.0513 and getting smaller, we know that it will have no further effect on the first decimal place, so we stop here. Our root, then, to one decimal place is $x = 2.6$.   ◆◆◆

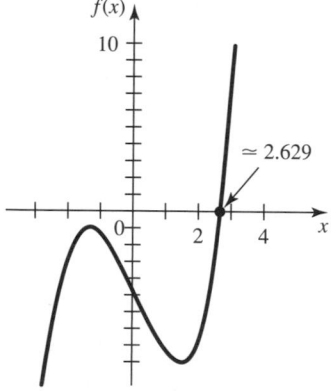

FIGURE 2–28 A root found by Newton's method.

Newton's method will usually fail when the root you seek is near a maximum, minimum, or inflection point, or near a discontinuity. If h does not shrink rapidly as you repeat the computation, this might be the cause.

Exercise 4 ◆ Approximate Solution of Equations by Newton's Method

Each equation has at least one root between $x = -10$ and $x = 10$. Find one such root to two decimal places using Newton's method.

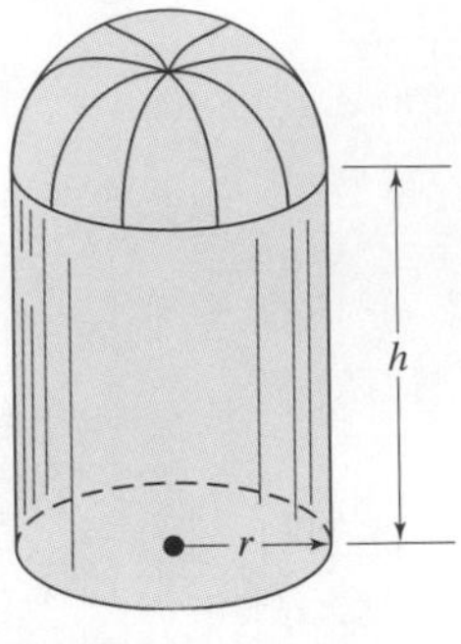

1. $x^3 + 3x^2 - 40 = 0$ 2. $x^3 + 3x^2 - 10 = 0$

3. $x^3 - 3x^2 + 8x - 25 = 0$ 4. $x^3 - 6x + 12 = 0$

5. $x^3 + 2x - 8 = 0$ 6. $3x^3 - 4x = 1$

7. $x^3 + 4x + 12 = 0$ 8. $x^4 + 8x - 12 = 0$

9. A square of side x is cut from each corner of a square sheet of metal 16 cm on a side, and the edges are turned up to form a tray whose volume is 282 cm³. Write an equation for the volume as a function of x, and use Newton's method to find x to two decimal places.

10. A solid sphere of radius r and specific gravity S will sink in water to a depth x, where

$$x^3 - 3rx^2 + 4r^3S = 0$$

Use Newton's method to find x to two decimal places if $r = 4$ m and $S = 0.700$.

11. A certain silo consists of a cylindrical base of radius r and height h, topped by a hemisphere (Fig. 28–26). It contains a volume V equal to

$$V = \pi r^2 h + \frac{2}{3}\pi r^3$$

If $h = 20.0$ m and $V = 800$ m³, use Newton's method to find r to two decimal places.

12. A segment of height x cut from a sphere of radius r (Fig. 28–27) has a volume

$$V = \pi\left(rx^2 - \frac{x^3}{3}\right)$$

Find x to two decimal places if $r = 4.0$ cm and $V = 150$ cm³.

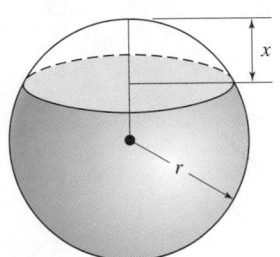

FIGURE 28–26   Silo.

FIGURE 28–27   Segment cut from a sphere.

Case Study Discussion—Power Output of a Sound System

This is an optimization problem. We know that adding one speaker would add 295 W of audio power but drop the original 40 speakers by $5 \times 40 = 200$ W, so we still get a gain of power. To calculate the optimum number of speakers, we first write the formula for the output power, since output power is the characteristic we are trying to maximize. Let s be the number of speakers we add to the original 40:

Total power = Number of speakers × power per speaker

$$P = (40 + s)(300 - 5s)$$
$$P = 12\,000 - 200s + 300s - 5s^2$$
$$P = -5s^2 + 100s + 12\,000$$

We can now take the derivative of total power.

$$P' = -10s + 100$$

Let the derivative equal zero to find the maximum point.

$$0 = -10s + 100$$
$$10s = 100$$
$$s = 10$$

Therefore, we add 10 more speakers for a total of 50 speakers. The total power will be

$$P = (40 + s)(300 - 5s)$$
$$P = (40 + 10)[300 - 5(10)]$$
$$P = (50)(300 - 50)$$
$$P = (50)(250)$$
$$P = 12\ 500$$

So 12 500 W is the maximum power. We can verify this by checking before and after the maximum.

If $s = 8$, the power would be 12 480 W, and if $s = 12$, the power would be 12 480 W.

••• CHAPTER 28 REVIEW PROBLEMS ••••••••••••••••••••••••••••••••••

1. Find the roots of $x^3 - 4x + 2 = 0$ to two decimal places.
2. Find any points of inflection on the curve $y = x^3(1 + x^2)$.
3. Find the maximum and minimum points on the curve $y = 3\sqrt[3]{x} - x$.
4. Find any maximum points, minimum points, and points of inflection for the function $3y = x^3 - 3x^2 - 9x + 11$.
5. Write the equations of the tangent and normal to the curve $y = \dfrac{1 + 2x}{3 - x}$ at the point $(2, 5)$.
6. Find the coordinates of the point on the curve $y = \sqrt{13 - x^2}$ at which the slope of the tangent is $-\frac{2}{3}$.
7. Find the x intercept of the tangent to the curve $y = \sqrt{x^2 + 7}$ at the point $(3, 4)$.
8. Graph the function $y = \dfrac{4 - x^2}{\sqrt{1 - x^2}}$, and locate any features of interest (roots, asymptotes, maximum/minimum points, and points of inflection).
9. For which values of x is the curve $y = \sqrt{4x}$ rising, and for what values is it falling?

10. Is the function $y = \sqrt{5 - 3x}$ increasing or decreasing at $x = 1$?
11. Is the curve $y = x^2 - x^5$ concave upward or concave downward at $x = 1$?
12. Graph the region bounded by the curves $y = 3x^3$, $x = 1$, and the x axis.
13. Write the equations of the tangent and normal to the curve $y = 3x^3 - 2x + 4$ at $x = 2$.

14. Find the x intercept of the tangent and normal in problem 13.

15. Find the angle of intersection between the curves $y = x^2/4$ and $y = 2/x$.
16. Graph the function $y = 3x^3\sqrt{9 - x^2}$, and locate any features of interest.

Writing

17. In this chapter we gave nine things to look for when graphing a function. Can you list at least seven from memory? Write a paragraph explaining how just one of the nine is useful in graphing.

More Applications of the Derivative

What is this stuff good for? This chapter gives the main answer to that question, at least for the derivative.

You saw in Chapter 28 how applications using the derivative were helpful in solving mathematical and geometric problems. In this chapter, we will spread out more and look at other applications. These applications have to do with rates of change. This is where calculus shines at problem solving. You saw that the derivative represents the slope of a line—the amount y changes with respect to a change in x. If the x axis becomes a time axis, anything that changes over time, such as velocity or acceleration, can become a derivative calculation. Calculus is the best way of describing anything in motion.

The inspiration for the development of calculus came from Newton's study of gravity. Gravity follows an inverse square law and plots not as a straight line but as a curved line. The curved line has a slope that is changing. Algebra is not very good at describing rates of change, so the need to describe rates of change led Newton to develop calculus.

For electrical applications, we use the rate of change of charge to find current, the rate of change of current to find the voltage across an inductor, and the rate of change of voltage to find the current in a capacitor.

We will then go on to use the fact that the derivative is zero at a maximum or minimum point to find maximum and minimum values of a varying quantity.

29–1　Rate of Change

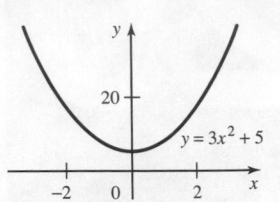

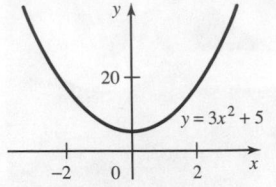

FIGURE 29–1

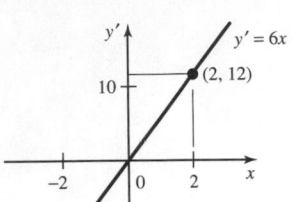

Rate of Change Given by the Derivative

We already established in Chapter 27 that the first derivative of a function gives the *rate of change* of that function.

◆◆◆ **Example 1:** Find the instantaneous rate of change of the function $y = 3x^2 + 5$ when $x = 2$.

Solution: Taking the derivative,

$$y' = 6x$$

When $x = 2$,

$$y'(2) = 6(2) = 12 = \text{instantaneous rate of change when } x = 2$$

Graphical Solution: Graph the derivative of the given function and determine its value at the required point. Figure 29–1 shows graphs of $y = 3x^2 + 5$, and its derivative $y' = 6x$. Notice that the derivative has a value of 12 at $x = 2$. ◆◆◆

Of course, rates of change can involve variables other than x and y. When two or more related quantities are changing, we often speak about the *rate of change* of one quantity with respect to one of the other quantities. For example, if a steel rod is placed in a furnace, its temperature and its length both increase. Since the length varies with the temperature, we can speak about the *rate of change of length with respect to temperature*. But the length of the bar is also varying with time, so we can speak about the *rate of change of length with respect to time*. Time rates are the most common rates of change with which we have to deal.

◆◆◆ **Example 2:** The temperature T (°F) in a certain furnace varies with time t (s) according to the function $T = 4.85t^3 + 2.96t$. Find the rate of change of temperature at $t = 3.75$ s.

Solution: Taking the derivative of T with respect to t,

$$\frac{dT}{dt} = 14.55t^2 + 2.96 \text{ °F/s}$$

At $t = 3.75$ s,

$$\left.\frac{dT}{dt}\right|_{t = 3.75 \text{ s}} = 14.55(3.75)^2 + 2.96 = 208 \text{ °F/s}$$

Graphical Solution: We graph the derivative in Fig. 29–2 and determine the value of dT/dt when $t = 3.75$ s. ◆◆◆

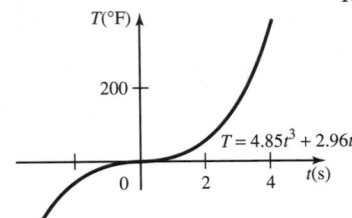

FIGURE 29–2

Students not specializing in electricity might wish to skip this section.

Electric Current

The idea of rate of change finds many applications in electrical technology, a few of which we cover here. The *coulomb* (C) is the unit of electrical *charge*. The *current* in amperes (A) is the number of coulombs passing a point in a circuit in 1 second. If the current varies with time, then the *instantaneous current* is given by the following equation:

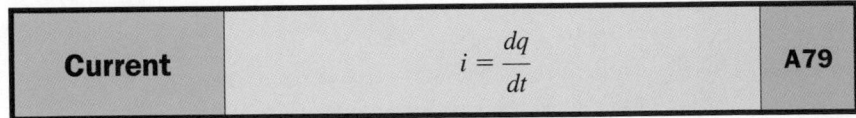

| Current | $i = \dfrac{dq}{dt}$ | A79 |

Current is the rate of change of charge.

◆◆◆ **Example 3:** The charge through a 2.85 Ω resistor is given by

$$q = 1.08t^3 - 3.82t \text{ C}$$

Write an expression for (a) the instantaneous current through the resistor, (b) the instantaneous voltage across the resistor, and (c) the instantaneous power in the resistor. (d) Evaluate each at 2.00 s.

Solution:

(a) $i = dq/dt = 3.24t^2 - 3.82$ A
(b) By Ohm's law,

$$v = Ri = 2.85(3.24t^2 - 3.82)$$
$$= 9.23t^2 - 10.9 \text{ V}$$

(c) Since $P = vi$,

$$P = (9.23t^2 - 10.9)(3.24t^2 - 3.82)$$
$$= 29.9t^4 - 70.6t^2 + 41.6 \text{ W}$$

(d) At $t = 2.00$ s,

$$i = 3.24(4.00) - 3.82 = 9.14 \text{ A}$$
$$v = 9.23(4.00) - 10.9 = 26.0 \text{ V}$$
$$P = 29.9(16.0) - 70.6(4.00) + 41.6 = 238 \text{ W}$$

Graphical Check: Graphs of i, v, and P, given in Fig. 29–3, show these values at $t = 2.00$ s. ◆◆◆

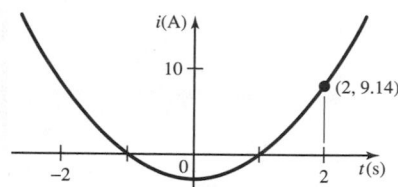

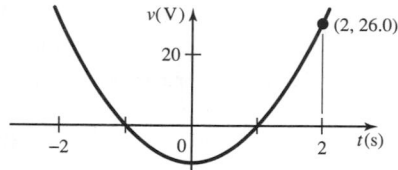

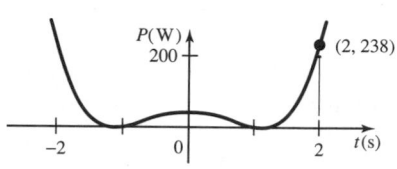

FIGURE 29–3

Current in a Capacitor

If a steady voltage is applied across a capacitor, no current will flow into the capacitor (after the initial transient currents have died down). But if the applied voltage varies with time, the instantaneous current i to the capacitor will be proportional to the rate of change of the voltage. The constant of proportionality is called the *capacitance*, C.

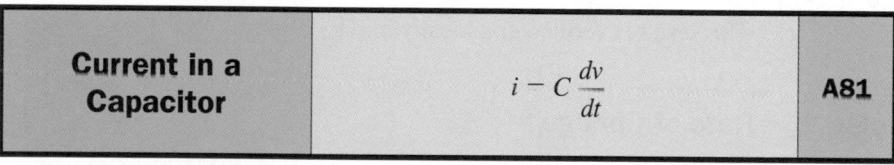

| Current in a Capacitor | $i - C\dfrac{dv}{dt}$ | A81 |

The current to a capacitor equals the capacitance times the rate of change of the voltage.

The sign convention is that the current is assumed to flow in the direction of the voltage drop, as shown in Fig. 29–4. If the current is assumed to be in the direction of the voltage *rise*, then one of the sides in Eq. A81 is taken as negative.

◆◆◆ **Example 4:** The voltage applied to a 2.85-microfarad (μF) capacitor is $v = 1.47t^2 + 48.3t - 38.2$ V. Find the current at $t = 2.50$ s.

Solution: The derivative of the voltage equation is $dv/dt = 2.94t + 48.3$. Then, from Eq. A81,

$$i = C\frac{dv}{dt} = (2.85 \times 10^{-6})(2.94t + 48.3) \text{ A}$$

At $t = 2.50$ s,

$$i = (2.85 \times 10^{-6})[2.94(2.50) + 48.3] = 159 \times 10^{-6} \text{ A}$$
$$= 0.159 \text{ mA}$$

Graphical Check: This result is verified graphically in Fig. 29–5. ◆◆◆

Voltage across an Inductor

If the current through an inductor (such as a coil of wire) is steady, there will be no voltage drop across the inductor. But if the current varies, a voltage will be induced that is proportional to the rate of change of the current. The constant of proportionality L is called the *inductance* and is measured in henrys (H).

The units here are volts for v, seconds for t, farads for C, and amperes for i.

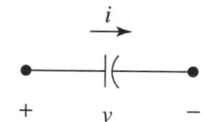

FIGURE 29–4 Current in a capacitor.

A microfarad equals 10^{-6} farad.

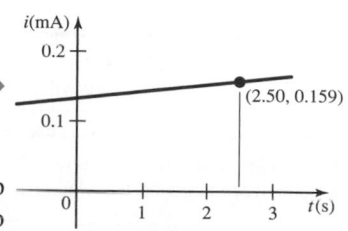

FIGURE 29–5

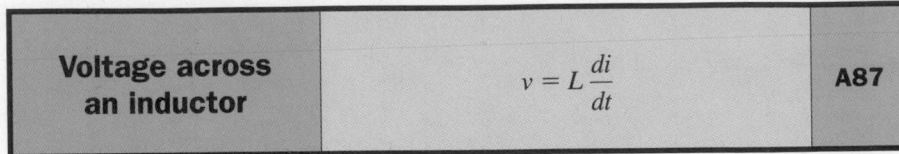

| Voltage across an inductor | $v = L\dfrac{di}{dt}$ | A87 |

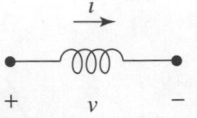

FIGURE 29-6   Voltage across an inductor.

The voltage across an inductor equals the inductance times the rate of change of the current.

The sign convention is similar to that for the capacitor: The current is assumed to flow in the direction of the voltage drop, as shown in Fig. 29–6. Otherwise, one term in Eq. A87 is taken as negative.

◆◆◆ **Example 5:** The current in a 8.75-H inductor is given by

$$i = \sqrt{t^2 + 5.83t}$$

Find the voltage across the inductor at $t = 5.00$ s.

Solution: By Eq. A87,

$$v = 8.75\frac{di}{dt}$$

$$= 8.75\left(\frac{1}{2}\right)(t^2 + 5.83t)^{-1/2}(2t + 5.83)$$

At $t = 5.00$ s,

$$v = 8.75\left(\frac{1}{2}\right)[25.0 + 5.83(5.00)]^{-1/2}(10.0 + 5.83)$$

$$= 9.41\ \text{V}$$

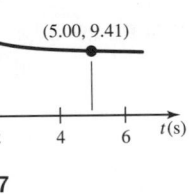

FIGURE 29-7

Graphical Check: This result is verified graphically in Fig. 29–7.   ◆◆◆

Exercise 1 ◆ Rate of Change

Use Boyle's law, $pv = k$.

1. The air in a certain cylinder is at a pressure of 25.5 kPa when its volume is 146 cm³. Find the rate of change of the pressure with respect to volume as the piston descends farther.

Use the inverse square law, $I = k/d^2$.

2. A certain light source produces an illumination of 655 lux on a surface at a distance of 2.75 m. Find the rate of change of illumination with respect to distance, and evaluate it at 2.75 m.

3. A spherical balloon starts to shrink as the gas escapes. Find the rate of change of its volume with respect to its radius when the radius is 1.00 m.

Use Eq. A67, $P = I^2R$.

4. The power dissipated in a certain resistor is 865 W at a current of 2.48 A. What is the rate of change of the power with respect to the current as the current starts to increase?

5. The period (in seconds) for a pendulum of length L cm to complete one oscillation is equal to $P = 0.224\sqrt{L}$. Find the rate of change of the period with respect to length when the length is 23.0 cm.

The rate of change dT/dx is called the *temperature gradient*.

6. The temperature T at a distance x in. from the end of a certain heated bar is given by $T = 2.24x^3 + 1.85x + 95.4$ (°C). Find the rate of change of temperature with respect to distance at a point 3.75 cm from the end.

The shape taken by the axis of a bent beam is called the *elastic curve*.

7. The cantilever beam in Fig. 29–8 has a deflection y at a distance x from the built-in end of

$$y = \frac{wx^2}{24EI}(x^2 + 6L^2 - 4Lx)$$

where E is the modulus of elasticity and I is the moment of inertia. Write an expression for the rate of change of deflection with respect to the distance x. Regard E, I, w, and L as constants.

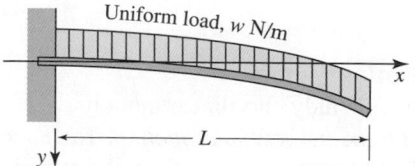

FIGURE 29-8 Cantilever beam with uniform load.

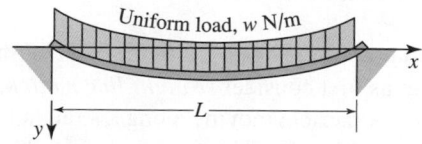

FIGURE 29-9 Simply supported beam with uniform load.

8. The equation of the elastic curve for the beam of Fig. 29–9 is

$$y = \frac{wx}{24EI}(L^3 - 2Lx^2 + x^3)$$

Write an expression for the rate of change of deflection (the slope) of the elastic curve at $x = L/4$. Regard E, I, w, and L as constants.

9. The charge q (in coulombs) through a 4.82-Ω resistor varies with time according to the function $q = 3.48t^2 - 1.64t$. Write an expression for the instantaneous current through the resistor.

10. Evaluate the current in problem 9 at $t = 5.92$ s.

11. Find the voltage across the resistor of problem 9. Evaluate it at $t = 1.75$ s.

12. Find the instantaneous power in the resistor of problem 9. Evaluate it at $t = 4.88$ s.

13. The charge q (in coulombs) at a resistor varies with time according to the function $q = 22.4t + 41.6t^3$. Write an expression for the instantaneous current through the resistor, and evaluate it at 2.50 s.

14. The voltage applied to a 33.5-μF capacitor is $v = 6.27t^2 - 15.3t + 52.2$ V. Find the current at $t = 5.50$ s.

15. The voltage applied to a 1.25-μF capacitor is $v = 3.17 + 28.3t + 29.4t^2$ V. Find the current at $t = 33.2$ s.

16. The current in a 1.44-H inductor is given by $i = 5.22t^2 - 4.02t$. Find the voltage across the inductor at $t = 2.00$ s.

17. The current in a 8.75-H inductor is given by $i = 8.22 + 5.83t^3$. Find the voltage across the inductor at $t = 25.0$ s.

18. A capacitor is an electronic component that can hold a charge for a short time. When we apply electrical pressure (voltage, v), charge flows onto the plates at a rate known as electrical current, i. The magnitude of the current is directly proportional to the time rate change of voltage, dv/dt. The formula is

$$i_c = C\frac{dv}{dt}$$

where i_c is current in amperes and C is capacitance in farads.
Using the formula above, answer the following:
a) What current flow would occur when the charging voltage is changing at a rate of 12 V/s in a 0.000 012 F capacitor?
b) If the capacitor is rated at 450 μF (450 $\times$ 10^{-6} F), what rate of change of voltage would cause a 15 A fuse to blow?
c) What capacitance would be required to limit the current flow in a circuit to 100 mA (100 $\times$ 10^{-3} A) for a 30 V/s charging voltage?

CASE STUDY—ROBOT MOTION

You are an electromechanical engineering technologist on a robotic design team. You team is working on humanoid robots that can assist patients in a health facility. The robots are being designed to gently lift and carry patients from bed to chair and back again, or to do any lifting and transportation needed. To be reassuring and nonthreatening to the patient, the robots will look human and, by using facial recognition software, be able to watch for signs of fear or stress in the patients. The control system must ensure that there are no jerks in the motion of the robot. A jerk would be a change in acceleration. If the robot scans the area between bed and chair and computes a path, how can we be sure there are no sudden jerking motions?

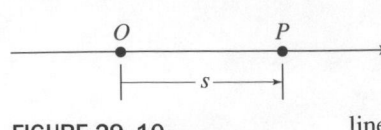

FIGURE 29-10

29–2   Motion of a Point

Displacement and Velocity in Straight-Line Motion

Let us first consider *straight-line motion;* later we will study curvilinear motion.

A particle moving along a straight line is said to be in *rectilinear motion.* To define the position P of the particle in Fig. 29–10, we first choose an origin O on the straight line. Then the distance s from O to P is called the *displacement* of the particle. Further, the displacement of the particle at any instant of time t is determined if we have a function $s = f(t)$.

◆◆◆ **Example 6:** The displacement of a particle is given by

$$s = f(t) = 3t^2 + 4t \text{ m}$$

where t is in seconds. Find the displacement at 2.15 s.

Solution: Substituting 2.15 for t gives

$$s = 3(2.15)^2 + 4(2.15) = 22.5 \text{ m} \qquad\qquad ◆◆◆$$

Next we distinguish between speed and velocity. As an object moves along some path, the distance travelled *along the path* per unit time is called the *speed.* No account is taken of any change in direction; hence speed is a *scalar* quantity.

Velocity, on the other hand, is a *vector* quantity, having both *magnitude* and *direction.* For an object moving along a curved path, the magnitude of the velocity along the path is equal to the speed, and the *direction* of the velocity is the same as that of the *tangent to the curve* at that point. We will also speak of the components of that velocity in directions other than along the path, usually in the x and y directions.

As with average and instantaneous rates of change, we also can have average speed and average velocity, or instantaneous speed and instantaneous velocity. These terms have the same meaning as in Sec. 29–1.

Velocity is the rate of change of displacement and hence *is given by the derivative* of the displacement. If we give displacement the symbol s, then we have the following equation:

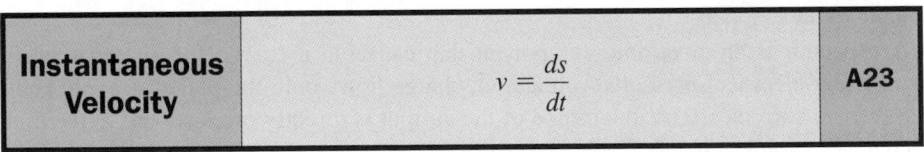

| **Instantaneous Velocity** | $v = \dfrac{ds}{dt}$ | **A23** |

The velocity is the rate of change of the displacement.

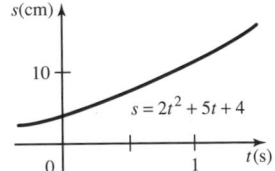

◆◆◆ **Example 7:** The displacement of an object is given by $s = 2t^2 + 5t + 4$ (cm), where t is the time in seconds. Find the velocity at 1 s.

Solution: We take the derivative

$$v = \frac{ds}{dt} = 4t + 5$$

At $t = 1$ s,

$$v(1) = \left.\frac{ds}{dt}\right|_{t=1} = 4(1) + 5 = 9 \text{ cm/s}$$

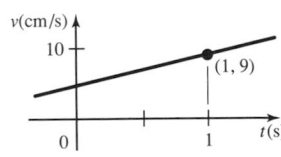

FIGURE 29-11

Graphical Solution: As with other rate of change problems, we can graph the first derivative and determine its value at the required point. Thus the graph for velocity, shown in Fig. 29–11, shows a value of 9 cm/s at $t = 1$ s.   ◆◆◆

Acceleration in Straight-Line Motion

Acceleration is defined as the time rate of change of velocity (i.e., the rate of change of velocity with respect to time). It is also a vector quantity. Since the velocity is itself the derivative of the displacement, the acceleration is the derivative of the derivative of the displacement, or the *second derivative* of displacement, with respect to time. We write the second derivative of *s* with respect to *t* as follows:

$$\frac{d^2s}{dt^2}$$

Istantaneous Acceleration	$a = \dfrac{dv}{dt} = \dfrac{d^2s}{dt^2}$	**A25**

The acceleration is the rate of change of the velocity.

◆◆◆ **Example 8:** One point in a certain mechanism moves according to the equation $s = 3t^3 + 5t - 3$ cm, where *t* is in seconds. Find the instantaneous velocity and acceleration at $t = 2$ s.

Solution: We take the derivative twice with respect to *t*.

$$v = \frac{ds}{dt} = 9t^2 + 5$$

and

$$a = \frac{dv}{dt} = 18t$$

At $t = 2$ s,

$$v(2) = 9(2)^2 + 5 = 41 \text{ cm/s}$$

and

$$a(2) = 18(2) = 36 \text{ cm/s}^2$$

Graphical Check: These values are verified graphically in Fig. 29–12.   ◆◆◆

Later we will do the reverse of differentiation, called *integration*, to find the velocity from the acceleration, and then the displacement from the velocity.

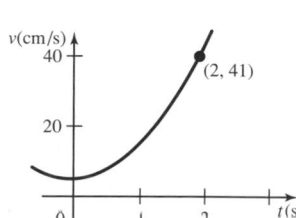

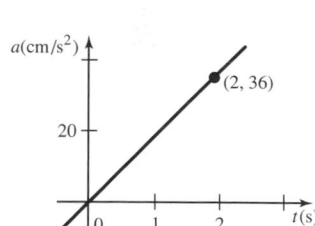

Velocity in Curvilinear Motion

At any instant we may think of a point as moving in a direction *tangent* to the path, as in Fig. 29–13. Thus if the speed is known and the direction of the tangent can be found, the *instantaneous velocity* (a vector having both magnitude and direction) can be found.

A more useful way of giving the instantaneous velocity, however, is by its *x* and *y* components (Fig. 29–14). If the magnitude and direction of the velocity are known, the components can be found by resolving the velocity vector into its *x* and *y* components.

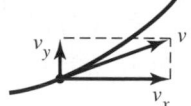

FIGURE 29–13

FIGURE 29–14   *x* and *y* components of velocity.

FIGURE 29–12

◆◆◆ **Example 9:** A point moves along the curve $y = 2x^3 - 5x^2 - 1$ cm.

(a) Find the direction of travel at $x = 2.00$ cm.
(b) If the speed *v* of the point along the curve is 3.00 cm/s, find the *x* and *y* components of the velocity when $x = 2.00$ cm.

Solution:

(a) Taking the derivative of the given function gives

$$\frac{dy}{dx} = 6x^2 - 10x$$

When $x = 2.00$,

$$y'(2.00) = 6(2.00)^2 - 10(2.00) = 4.00$$

The slope of the curve at that point is thus 4.00, and the direction of travel is $\tan^{-1} 4.00 = 76.0°$, as shown in Fig. 29–15.

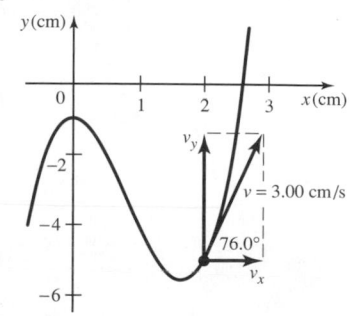

FIGURE 29–15

(b) Resolving the velocity vector v into x and y components gives us

$$v_x = 3.00 \cos 76.0° = 0.726 \text{ cm/s}$$
$$v_y = 3.00 \sin 76.0° = 2.91 \text{ cm/s}$$

◆◆◆

Displacement Given by Parametric Equations

If the x displacement and y displacement are each given by a separate function of time (parametric equations), we may find the x and y components directly by taking the derivative of each equation.

x and y Components of Velocity	$v_x = \dfrac{dx}{dt}$	$v_y = \dfrac{dy}{dt}$	A33

Once we have expressions for the x and y components of velocity, we simply have to take the derivative again to get the x and y components of acceleration.

x and y Components of Acceleration	$a_x = \dfrac{dv_x}{dt} = \dfrac{d^2x}{dt^2}$	$a_y = \dfrac{dv_y}{dt} = \dfrac{d^2y}{dt^2}$	A35

◆◆◆ **Example 10:** A point moves along a curve such that its horizontal displacement is

$$x = 4t + 5 \text{ cm}$$

and its vertical displacement is

$$y = 3 + t^2 \text{ cm}$$

Find the instantaneous velocity at $t = 1.00$ s, its horizontal and vertical components, and its magnitude and direction.

Solution: Using Eqs. A33, we take derivatives.

$$v_x = 4 \quad \text{and} \quad v_y = 2t$$

At $t = 1.00$ s,

$$v_x = 4.00 \text{ cm/s} \quad \text{and} \quad v_y = 2(1.00) = 2.00 \text{ cm/s}$$

We get the resultant of v_x and v_y by vector addition.

$$v = \sqrt{v_x^2 + v_y^2} = \sqrt{16.0 + 4.00} = 4.47 \text{ cm/s}$$

Finding the direction of the resultant, we have

$$\tan \theta = v_y/v_x = 2.00/4.00 = 1/2$$
$$\theta = 26.6°$$

Figure 29–16 shows a parametric plot of the given equations. At $t = 1.00$ s, the moving point is at (9, 4). The figure also shows the velocity tangent to the curve and the x and y components of the velocity.

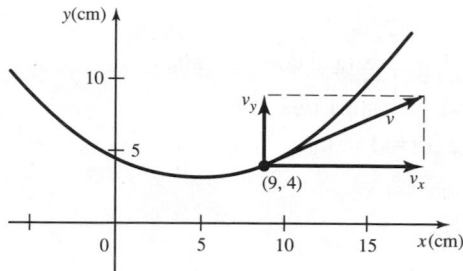

FIGURE 29–16

Trajectories

A *trajectory* is the path followed by a projectile, such as a ball thrown into the air. Trajectories are usually described by parametric equations, with the horizontal and vertical motions considered separately.

If air resistance is neglected, a projectile will move horizontally with constant velocity and will fall with constant acceleration, just like any other freely falling body. Thus if the initial velocities are v_{0x} and v_{0y} in the horizontal and vertical directions, the parametric equations of motion are

$$x = v_{0x}t \quad \text{and} \quad y = v_{0y}t - \frac{gt^2}{2}$$

◆◆◆ **Example 11:** A projectile launched with initial velocity v_0 at an angle θ_0 (Fig. 29–17) has horizontal and vertical displacements of

$$x = (v_0 \cos \theta_0)t \quad \text{and} \quad y = (v_0 \sin \theta_0)t - \frac{gt^2}{2}$$

Find the horizontal and vertical velocities and accelerations.

Solution: We take derivatives, remembering that θ_0 is a constant.

$$v_x = \frac{dx}{dt} = v_0 \cos \theta_0 \quad \text{and} \quad v_y = \frac{dy}{dt} = v_0 \sin \theta_0 - gt$$

We take derivatives again to get the accelerations.

$$a_x = \frac{dv_x}{dt} = 0 \quad \text{and} \quad a_y = \frac{dv_y}{dt} = -g$$

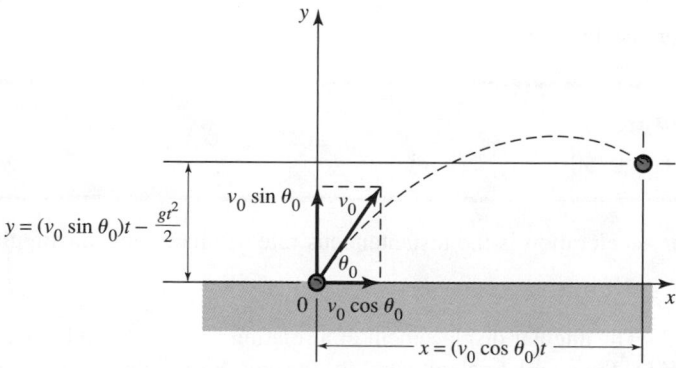

FIGURE 29–17 A trajectory.

As expected, we get a horizontal motion with constant velocity and a vertical motion with constant acceleration. ◆◆◆

◆◆◆ **Example 12:** A projectile is launched at an angle of 62.0° to the horizontal with an initial velocity of 177 m/s. Find (a) the horizontal and vertical positions of the projectile and (b) the horizontal and vertical velocities after 1.55 s.

Solution:

(a) We first resolve the initial velocity into horizontal and vertical components.

$$v_{0x} = v_0 \cos \theta = 177 \cos 62.0° = 83.1 \text{ m/s}$$
$$v_{0y} = v_0 \sin \theta = 177 \sin 62.0° = 156 \text{ m/s}$$

The horizontal and vertical displacements are then

$$x = v_{0x}t = 83.1t \text{ m}$$

and

$$y = v_{0y}t - gt^2/2 = 156t - gt^2/2 \text{ m}$$

At $t = 1.55$ s, and with $g = 9.81$ m/s^2,

$$x = 83.1(1.55) = 129 \text{ m}$$

and

$$y = 156(1.55) - 9.81(1.55)^2/2 = 230 \text{ m}$$

(b) The horizontal velocity is constant, so

$$v_x = v_{0x} = 83.1 \text{ m/s}$$

and the vertical velocity is

$$v_y = v_{0y} - gt$$
$$= 156 - 9.81(1.55) = 141 \text{ m/s}$$ ◆◆◆

Rotation

We saw for straight-line motion that velocity is the instantaneous rate of change of displacement. Similarly, for rotation (Fig. 29–18), the *angular velocity, ω*, is the instantaneous rate of change of *angular displacement, θ.*

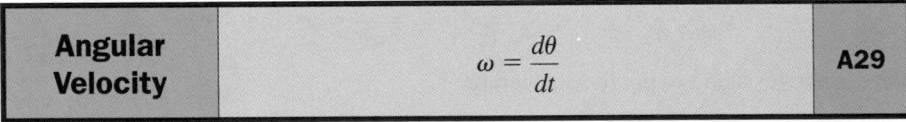

Angular Velocity	$\omega = \dfrac{d\theta}{dt}$	A29

The angular velocity is the instantaneous rate of change of the angular displacement with respect to time.

Similarly for acceleration:

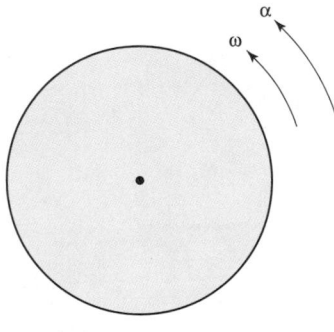

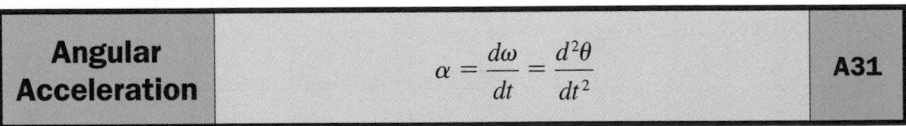

Angular Acceleration	$\alpha = \dfrac{d\omega}{dt} = \dfrac{d^2\theta}{dt^2}$	A31

FIGURE 29–18

The angular acceleration is the instantaneous rate of change of the angular velocity with respect to time.

◆◆◆ **Example 13:** The angular displacement of a rotating body is given by $\theta = 1.75t^3 + 2.88t^2 + 4.88$ rad. Find (a) the angular velocity and (b) the angular acceleration at $t = 2.00$ s.

Solution:

(a) From Eq. A29,

$$\omega = \frac{d\theta}{dt} = 5.25t^2 + 5.76t$$

At 2.00 s, $\omega = 5.25(4.00) + 5.76(2.00) = 32.5$ rad/s.

(b) From Eq. A31,

$$\alpha = \frac{d\omega}{dt} = 10.5t + 5.76$$

At 2.00 s, $\alpha = 10.5(2.00) + 5.76 = 26.8$ rad/s^2.

Graphical Check: These results are verified graphically in Fig. 29–19. ◆◆◆

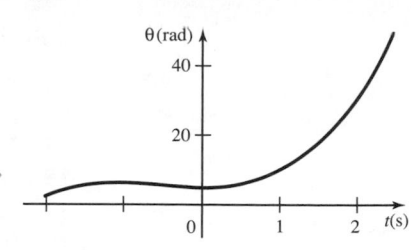

Exercise 2 ◆ Motion of a Point

Straight-Line Motion

Find the instantaneous velocity and acceleration at the given time for the straight-line motion described by each equation, where s is in centimetres and t is in seconds.

1. $s = 32t - 8t^2$ at $t = 2$
2. $s = 6t^2 - 2t^3$ at $t = 1$
3. $s = t^2 + t^{-1} + 3$ at $t = \frac{1}{2}$
4. $s = (t + 1)^4 - 3(t + 1)^3$ at $t = -1$
5. $s = 120t - 16t^2$ at $t = 4$
6. $s = 3t - t^4 - 8$ at $t = 1$
7. The distance in metres travelled in time t seconds by a point moving in a straight line is given by the formula $s = 40t + 4.9t^2$. Find the velocity and the acceleration at the end of 2.0 s.
8. A car moves according to the equation $s = 250t^2 - \frac{5}{4}t^4$, where t is measured in minutes and s in metres.
   (a) How far does the car go in the first 10.0 min?
   (b) What is the maximum speed?
   (c) How far has the car moved when its maximum speed is reached?

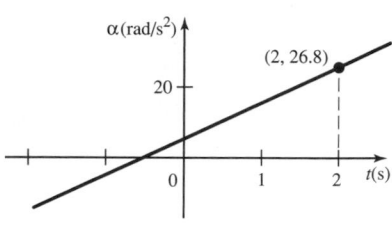

FIGURE 29–19

Solve or verify some of these problems graphically.

9. If the distance travelled by a ball rolling down an incline in t seconds is s metres, where $s = 6t^2$, find its speed when $t = 5$ s.
10. The height s in feet reached by a ball t seconds after being thrown vertically upward at 320 ft./s is given by $s = 320t - 16t^2$. Find (a) the greatest height reached by the ball and (b) the velocity with which it reaches the ground.
11. A bullet was fired straight upward so that its height in metres after t seconds was $s = 700t - 4.9t^2$.
   (a) What was its initial velocity?
   (b) What was its greatest height?
   (c) What was its velocity at the end of 10 s?
12. The height h kilometres to which a balloon will rise in t minutes is given by the formula

$$h = \frac{10t}{\sqrt{4000 + t^2}}$$

   At what rate is the balloon rising at the end of 30.0 min?
13. If the equation of motion of a point is $s = 16t^2 - 64t + 64$, find the position and acceleration at which the point first comes to rest.

Curvilinear Motion

14. A point moves along the curve $y = 2x^3 - 3x^2 - 2$ cm.
   (a) Find the direction of travel at $x = 1.50$ cm.
   (b) If the speed of the point along the curve is 3.75 cm/s, find the x and y components of the velocity when $x = 1.50$ cm.
15. A point moves along the curve $y = x^4 + x^2$ m.
   (a) Find the direction of travel at $x = 2.55$ m
   (b) If the speed of the point along the curve is 1.25 m/s, find the x and y components of the velocity when $x = 2.55$ m

Equations Given in Parametric Form

16. A point has horizontal and vertical displacements (in cm) of $x = 3t^2 + 5t$ and $y = 13 - 3t^2$, respectively. Find the x and y components of the velocity and acceleration at $t = 4.55$ s.

17. Find the magnitude and direction of the resultant velocity in problem 16.

18. A point has horizontal and vertical displacements (in cm) of $x = 4 - 2t^2$ and $y = 5t^2 + 3$, respectively.
(a) Find the x and y components of the velocity and acceleration at $t = 2.75$ s.
(b) Find the magnitude and direction of the resultant velocity.

Trajectories

19. A projectile is launched at an angle of 43.0° to the horizontal with an initial velocity of 6350 ft./s. Find (a) the horizontal and vertical positions of the projectile and (b) the horizontal and vertical velocities after 7.00 s.

20. A projectile is launched at an angle of 27.0° to the horizontal with an initial velocity of 385 m/s. Find (a) the horizontal and vertical positions of the projectile and (b) the horizontal and vertical velocities after 3.25 s.

21. An aircraft approaching Vancouver International Airport is flying at a horizontal speed of 200 km/hr and is descending at a rate of 4 m/s. If the angle of elevation between the control tower and the aircraft is 9°, how quickly is the distance between the control tower and the plane decreasing when the aircraft is 25 km away and at an altitude of 4000 m?

Rotation

22. The angular displacement of a rotating body is given by $\theta = 44.8t^3 + 29.3t^2 + 81.5$ rad. Find the angular velocity at $t = 4.25$ s.

23. Find the angular acceleration in problem 21 at $t = 22.4$ s.

24. The angular displacement of a rotating body is given by $\theta = 184 + 271t^3$ rad. Find (a) the angular velocity and (b) the angular acceleration at $t = 1.25$ s.

25. The angular displacement of a rotating body is given by $\theta = 2.84t^3 - 7.25$ rad. Find (a) the angular velocity and (b) the angular acceleration at $t = 4.82$ s.

29–3   Related Rates

In *related rate* problems, there are *two* quantities changing with time. The rate of change of one of the quantities is given, and the other must be found. A procedure that can be followed is:

1. Locate the *given rate*. Since it is a rate, it can be *expressed as a derivative* with respect to time.

2. Determine the *unknown* rate. Express it also as a derivative with respect to time.

You would, of course, study the problem statement and make a diagram, as you would for other word problems.

3. Find an *equation* linking the variable in the given rate with that in the unknown rate. If there are other variables in the equation, they must be eliminated by means of other relationships.

4. Take the derivative of the equation *with respect to time*.

5. Substitute the given values and solve for the unknown rate.

◆◆◆ **Example 14:** A 20.0-ft. ladder leans against a building (Fig. 29–20). The foot of the ladder is pulled away from the building at a rate of 2.00 ft./s. How fast is the top of the ladder falling when its foot is 10.0 ft. from the building?

Estimate: Note that when the foot of the ladder is 10 ft. from the wall, the ladder's angle is arccos (10/20) or 60°. If the ladder is at 45°, the top and foot should move at the same speed; when the angle is steeper, the top should move more slowly than the foot. Thus we expect the top to move at a speed less than 2 ft./s. Also, since y is decreasing, we expect the rate of change of y to be negative.

Solution:

1. If we let x be the distance from the foot of the ladder to the building, we have

$$\frac{dx}{dt} = 2.00 \text{ ft./s}$$

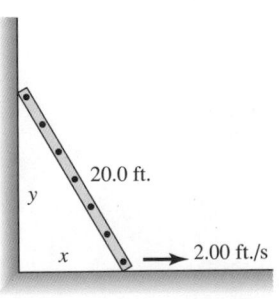

FIGURE 29–20 The ladder problem.

20.0 ft.

y

x ⟶ 2.00 ft./s

2. If y is the distance from the ground to the top of the ladder, we are looking for dy/dt.

3. The equation linking x and y is the Pythagorean theorem.

$$x^2 + y^2 = 20.0^2$$

Solving for y,

$$y = (400 - x^2)^{1/2}$$

4. Taking the derivative with respect to t,

$$\frac{dy}{dt} = \frac{1}{2}(400 - x^2)^{-1/2}(-2x)\frac{dx}{dt}$$

5. Substituting $x = 10.0$ ft. and $dx/dt = 2.00$ ft./s gives

$$\frac{dy}{dt} = \frac{1}{2}(400 - 100)^{-1/2}(-20.0)(2.00) = -1.15 \text{ ft./s}$$

The negative sign indicates that y is decreasing.

Graphical Solution: The derivative dy/dt, after substituting 2.00 for dx/dt, is

$$dy/dt = -2x(400 - x^2)^{-1/2}$$

We plot this derivative as shown in Fig. 29–21, and graphically determine that the value of dy/dt at $x = 10.0$ ft. is -1.15 ft./s.

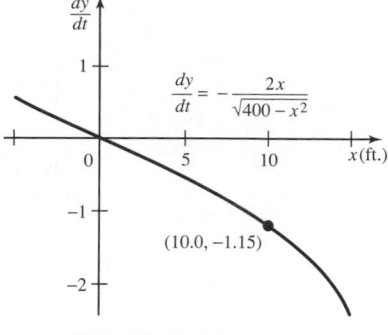

FIGURE 29–21

Common Error	In Step 4 of Example 14, when taking the derivative of x^2, it is tempting to take the derivative with respect to x, rather than t. Also, don't forget the dx/dt in the derivative. $$\frac{d}{dt}x^2 = 2x \boxed{\frac{dx}{dt}}$$ ——— Don't forget!
	Students often substitute the given values too soon. For example, if we had substituted $x = 10.0$ and $y = 17.3$ before taking the derivative, we would have gotten $$(10.0)^2 + (17.3)^2 = (20.0)^2$$ Taking the derivative now gives us $$0 = 0!$$ Do not substitute the given values until *after* you have taken the derivative.

Alternate Solution: Steps 1 through 3 are the same as above, but instead of solving for y as we did in Step 3, we now take the derivative *implicitly*.

$$2x\frac{dx}{dt} + 2y\frac{dy}{dt} = 0$$

When $x = 10.0$ ft., the height of the top of the ladder is

$$y = \sqrt{400 - (10.0)^2} = 17.32 \text{ ft.}$$

We now substitute 2.00 for dx/dt, 10.0 for x, and 17.32 ft. for y.

It will often be easier to take the derivative implicitly rather than first to solve for one of the variables.

$$2(10.0)(2.00) + 2(17.32)\frac{dy}{dt} = 0$$

$$\frac{dy}{dt} = -1.15 \text{ ft./s} \qquad ◆◆◆$$

In the next example when we find an equation linking the variables, it contains *three* variables. In such a case we need a *second equation* with which to eliminate one variable.

◆◆◆ **Example 15:** A conical tank with vertex down has a base radius of 3.00 m and a height of 6.00 m (Fig. 29–22). Water flows in at a rate of 2.00 m³/h. How fast is the water level rising when the depth y is 3.00 m?

Estimate: Suppose that our tank was *cylindrical* with a radius of 1.5 m and cross-sectional area of $\pi(1.5)^2 \cong 7$ m². The water would rise at a constant rate, equal to the incoming flow rate divided by the cross-sectional area of the tank, or

$$2 \text{ m}^3/\text{h} \div 7 \text{ m}^2 \cong 0.3 \text{ m/h}$$

The conical tank should have about this value when half full, as in our example, less where the tank is wider and greater where narrower.

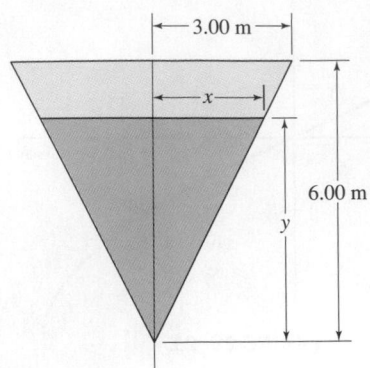

FIGURE 29–22   Conical tank.

Solution: Let x equal the base radius of the water surface and V the volume of water when the tank is partially filled. Then

1. Given: $dV/dt = 2.00$ m³/h.
2. Unknown: dy/dt when $y = 3.00$ m.
3. The equation linking V and y is that for the volume of a cone, $V = (\pi/3)x^2y$. But, in addition to the two variables in our derivatives, V and y, we have the third variable, x. We must eliminate x by means of another equation. By similar triangles,

$$\frac{x}{y} = \frac{3}{6}$$

from which $x = \dfrac{y}{2}$. Substituting yields

$$V = \frac{\pi}{3} \cdot \frac{y^2}{4} \cdot y = \frac{\pi}{12}y^3$$

4. Now we have V as a function of y only. Taking the derivative gives us

$$\frac{dV}{dt} = \frac{\pi}{4}y^2\frac{dy}{dt}$$

5. Substituting 3.00 for y and 2.00 for dV/dt, we obtain

$$2.00 = \frac{\pi}{4}(9.00)\frac{dy}{dt}$$

$$\frac{dy}{dt} = \frac{8.00}{9.00\ \pi} = 0.283 \text{ m/h}$$

This agrees well with our estimate.

Graphical Solution: The derivative dy/dt, after substituting 2.00 for dV/dt, is

$$\frac{dy}{dt} = 2\left(\frac{4}{\pi y^2}\right) = \frac{8}{\pi y^2}$$

We plot dy/dx as shown in Fig. 29–23 and graphically find that dy/dt is 0.283 when y is 3.00 m.

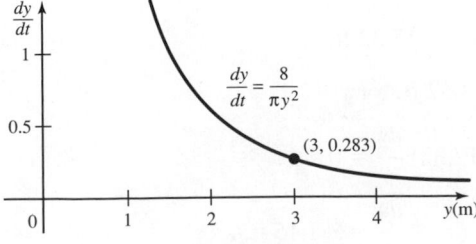

FIGURE 29–23   ◆◆◆

Some problems have two *independently* moving objects, as in the following example.

◆◆◆**Example 16:** Ship *A* leaves a port *P* and travels west at 11.5 km/h. After 2.25 h, ship *B* leaves *P* and travels north at 19.4 km/h. How fast are the ships separating 5.00 h after the departure of *A*?

Solution: Figure 29–24 shows the ships *t* h after *A* has left. Ship *A* has gone 11.5*t* km and *B* has gone 19.4(*t* − 2.25) km. The distance *S* between them is given by

$$S^2 = (11.5t)^2 + [19.4(t - 2.25)]^2$$
$$= 132t^2 + 376(t - 2.25)^2$$

Taking the derivative, we have

$$2S \frac{dS}{dt} = 2(132)t + 2(376)(t - 2.25)$$
$$\frac{dS}{dt} = \frac{132t + 376t - 846}{S} = \frac{508t - 846}{S}$$

At *t* = 5.00 h, *A* has gone 11.5(5.00) = 57.5 km, and *B* has gone

$$19.4(5.00 - 2.25) = 53.4 \text{ km}$$

The distance *S* between them is then

$$S = \sqrt{(57.5)^2 + (53.4)^2} = 78.5 \text{ km}$$

Substituting gives

$$\frac{dS}{dt} = \frac{508(5.00) - 846}{78.5} = 21.6 \text{ km/h}$$

Graphical Check: After substituting *S* = 78.5 km/h, our derivative is

$$dS/dt = 6.47t - 10.8$$

A graph of *dS/dt* (Fig. 29–25) shows a value of 21.6 at *t* = 5.00 s.

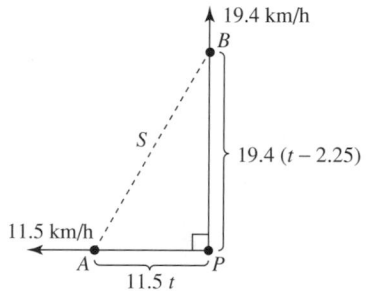

FIGURE 29–24   Note that we give the position of each ship after *t* hours have elapsed.

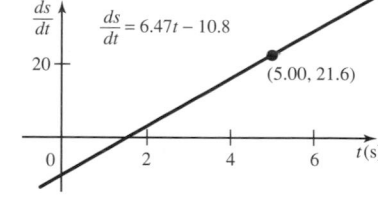

◆◆◆ **FIGURE 29–25**

Exercise 3 ◆ Related Rates

One Moving Object

1. An airplane flying horizontally at a height of 8000 m and at a rate of 100 m/s passes directly over a pond. How fast is its straight-line distance from the pond increasing 1 min later?

2. A ship moving 30.0 km/h is 6.00 km from a straight beach and is moving parallel to the beach. How fast is the ship approaching a lighthouse on the beach when 10.0 km (straight-line distance) from it?

3. A person is running at the rate of 12.0 km/h on a horizontal street directly toward the foot of a tower 30.0 m high. How fast is the person approaching the top of the tower when 15.0 m from the foot?

Ropes and Cables

4. A boat is fastened to a rope that is wound around a winch 10.0 m above the level at which the rope is attached to the boat. The boat is drifting away at the horizontal rate of 4.00 m/s. How fast is the rope increasing in length when 15.0 m of rope is out?

5. A boat with its anchor on the bottom at a depth of 40.0 m is drifting away from the anchor at 4.00 m/s, while the anchor cable slips from the boat at water level. At what rate is the cable leaving the boat when 50.0 m of cable is out? Assume that the cable is straight.

6. A kite is at a constant height of 120 m and moves horizontally, at 4.00 km/h, in a straight line away from the person holding the cord. Assuming that the cord remains straight, how fast is the cord being paid out when its length is 130 m?

7. A rope runs over a pulley at A and is attached at B as shown in Fig. 29–26. The rope is being wound in at the rate of 2.00 m/s. How fast is B rising when AB is horizontal?

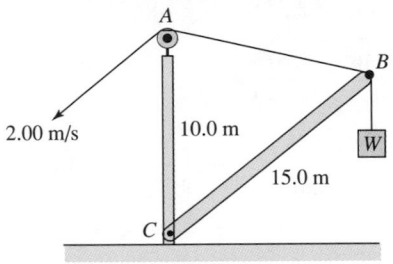

FIGURE 29–26   FIGURE 29–27

8. A weight W is being lifted between two poles as shown in Fig. 29–27. How fast is W being raised when A and B are 10.0 m apart if they are being drawn together, each moving at the rate of 18.0 m/s?

9. A bucket is raised by a person who walks away from the building at 1.00 ft./s (Fig. 29–28). How fast is the bucket rising when $x = 80.0$ in.?

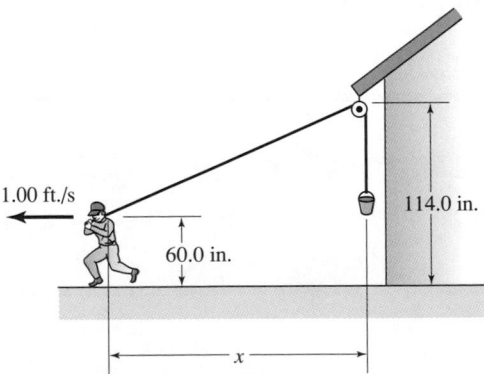

FIGURE 29–28

Two Moving Objects

10. Two trains start from the same point at the same time, one going east at a rate of 40.0 km/h and the other going south at 60.0 km/h. Find the rate at which they are separating after 1.00 h of travel.

11. An airplane leaves a field at noon and flies east at $1\overline{0}0$ km/h. A second airplane leaves the same field at 1 P.M. and flies south at 150 km/h. How fast are the airplanes separating at 2 P.M.?

12. An elevated train on a track 30.0 m above the ground crosses a street (which is at right angles to the track) at the rate of 20.0 m/s. At that instant, an automobile, approaching at the rate of 30.0 m/s, is 40.0 m from a point directly beneath the track. Find how fast the train and the automobile are separating 2.00 s later.

13. As a person is cycling over a bridge at 5.00 m/s, a boat, 30.0 m beneath the cyclist and moving perpendicular to the bridge, passes directly underneath at 10.0 m/s. How fast are the person and the boat separating 3.00 s later?

Moving Shadows

14. A light is $10\overline{0}$ ft. from a wall (Fig. 29–29). A person runs at 13.0 ft./s away from the wall. Find the speed of the shadow on the wall when the person's distance from the wall is 50.0 ft.

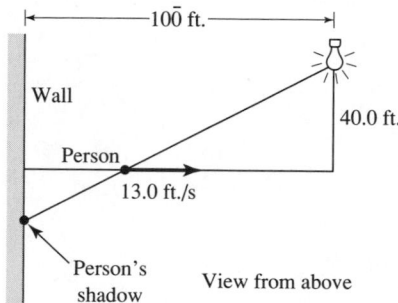

FIGURE 29–29

15. A lamp is located on the ground 10.0 m from a building. A person 2.00 m tall walks from the light toward the building at a rate of 2.00 m/s. Find the rate at which the person's shadow on the wall is shortening when the person is 5.00 m from the building.

16. A ball dropped from a height of $10\overline{0}$ m is at height s at t seconds, where $s = 10\overline{0} - gt^2/2$ m (Fig. 29–30). The sun, at an altitude of 40.0°, casts a shadow of the ball on the ground. Find the rate dx/dt at which the shadow is travelling along the ground when the ball has fallen 50.0 m.

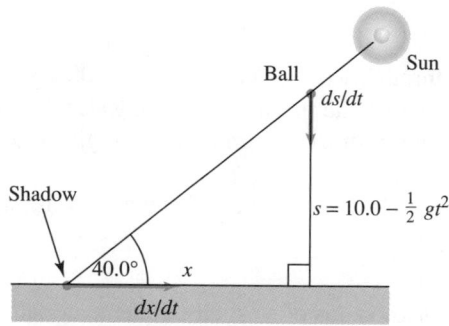

FIGURE 29–30

17. A square sheet of metal 10.0 cm on a side is expanded by increasing its temperature so that each side of the square increases 0.005 00 cm/s. At what rate is the area of the square increasing at 20.0 s?

18. A circular plate in a furnace is expanding so that its radius is changing 0.010 cm/s. How fast is the area of one face changing when the radius is 5.0 cm?

19. The volume of a cube is increasing at 10.0 cm³/min. At the instant when its volume is 125 cm³, what is the rate of change of its edge?

20. The edge of an expanding cube is changing at the rate of 3.0 mm/s. Find the rate of change of its volume when its edge is 5.0 m long.

21. At some instant the diameter x of a cylinder (Fig. 29–31) is 10.0 cm and is increasing at a rate of 1.00 cm/min. At that same instant, the height y is 20.0 cm and is decreasing at a rate (dy/dt) such that the volume is not changing ($dV/dt = 0$). Find dy/dt.

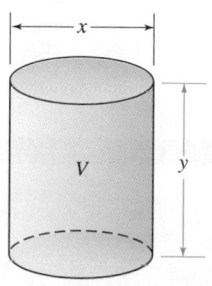

FIGURE 29–31

Fluid Flow

22. Water is running from a vertical cylindrical tank 3.00 m in diameter at the rate of $3\pi\sqrt{h}$ m³/min, where h is the depth of the water in the tank. How fast is the surface of the water falling when $h = 9.00$ m?

23. Water is flowing into a conical reservoir 20.0 m deep and 10.0 m across the top, at a rate of 15.0 m³/min. How fast is the surface rising when the water is 8.00 m deep?

24. Sand poured on the ground at the rate of 3.00 m³/min forms a conical pile whose height is one-third the diameter of its base. How fast is the altitude of the pile increasing when the radius of its base is 2.00 m?

25. A horizontal trough 5.00 m long has ends in the shape of an isosceles right triangle (Fig. 29–32). If water is poured into it at the rate of 0.500 m³/min, at what rate is the surface of the water rising when the water is 1.00 m deep?

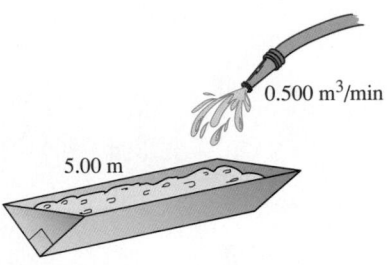

FIGURE 29–32

Gas Laws

Use Boyle's law, pv = constant.

26. A tank contains 30.0 m³ of gas at a pressure of 35.0 kPa. If the pressure is decreasing at the rate of 0.350 kPa/h, find the rate of increase of the volume.

27. The adiabatic law for the expansion of air is $pv^{1.4} = C$. If at a given time the volume is observed to be 2.00 m³, and the pressure is $30\overline{0}$ kPa, at what rate is the pressure changing if the volume is decreasing 0.100 m³/s?

Miscellaneous

28. The speed v m/s of a certain bullet passing through wood is given by $v = 300\sqrt{1 - 3x}$, where x is the depth in metres. Find the rate at which the speed is decreasing after the bullet has penetrated 10.0 cm. (*Hint:* When substituting for dx/dt, simply use the given expression for v.)

29. As a man walks a distance x along a board (Fig. 29–33), he sinks a distance of y in., where

$$y = \frac{Px^3}{3EI}$$

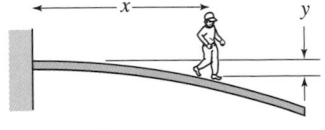

FIGURE 29–33

Here, P is the person's weight, 165 lb.; E is the modulus of elasticity of the material in the board, 1 320 000 lb./sq. in.; and I is the modulus of elasticity of the cross section, 10.9 (in.⁴). If he moves at the rate of 25.0 in./s, how fast is he sinking when $x = 75.0$ in.?

30. A stone dropped into a calm lake causes a series of circular ripples. The radius of the outer one increases at 0.600 m/s. How rapidly is the disturbed area changing at the end of 3.00 s?

29–4   Optimization

In Chapter 28, we found *extreme points,* the peaks and valleys on a curve. We did this by finding the points where the slope (and hence the first derivative) was zero. We now apply the same idea to problems in which we find, for example, the point of minimum cost, or the point of maximum efficiency, or the point of maximum carrying capacity.

Suggested Steps for Maximum-Minimum Problems

1. Locate the quantity (which we will call Q) to be maximized or minimized, and locate the independent variable, say, x, which is varied in order to maximize or minimize Q.
2. Write an equation linking Q and x. If this equation contains another variable as well, that variable must be eliminated by means of a second equation. A graph of $Q = f(x)$ will show any maximum or minimum points.
3. Take the derivative dQ/dx.
4. Locate the values of x for which the derivative is zero. This may be done graphically by finding the zeros for the graph of the derivative or analytically by setting the derivative to zero and solving for x.
5. Check any extreme points found to see if they are maxima or minima. This can be done simply by looking at your graph of $Q = f(x)$, by applying your knowledge of the physical problem, or by using the first- or second-derivative tests. Also check whether the maximum or minimum value you seek is at one of the endpoints. An endpoint can be a maximum or minimum point in the given interval, even though the slope there is not zero.

   A list of general suggestions such as these is usually so vague as to be useless without examples. Our first example is one in which the relation between the variables is given verbally in the problem statement.

◆◆◆ **Example 17:** What two positive numbers whose product is 100 have the least possible sum?

Solution:

1. We want to minimize the sum S of the two numbers by varying one of them, which we call x. Then

$$\frac{100}{x} = \text{other number}$$

2. The sum of the two numbers is

$$S = x + \frac{100}{x}$$

3. Taking the derivative yields

$$\frac{dS}{dx} = 1 - \frac{100}{x^2}$$

4. Setting the derivative to zero and solving for x gives us

$$x^2 = 100$$
$$x = \pm 10$$

Since we are asked for positive numbers, we discard the -10. The other number is $100/10 = 10$.

5. But have we found those numbers that will give a *minimum* sum, as requested, or a *maximum* sum? We can check this by means of the second-derivative test. Taking the second derivative, we have

$$S'' = \frac{200}{x^3}$$

When $x = 10$,

$$S''(10) = \frac{200}{1000} = 0.2$$

which is *positive,* indicating a *minimum* point.

The sum of two positive numbers whose product is a constant is always a minimum when the numbers are equal, as in this example. Can you prove the general case?

Graphical Solution: A graph of S versus x, shown in Fig. 29–34(a), shows a minimum at $x \approx 10$. To locate the minimum more accurately, we graph the derivative as shown in Fig. 29–34(b). Note that it has a zero at $x = 10$.

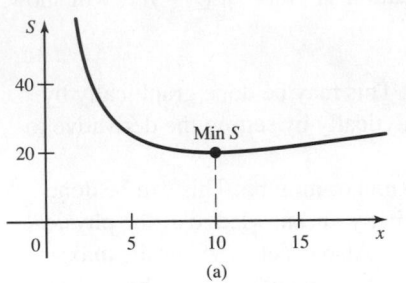 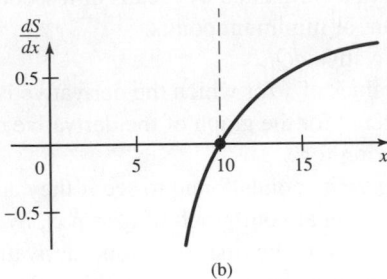

FIGURE 29–34   ◆◆◆

In the next example, the equation linking the variables is easily written from the geometrical relationships in the problem.

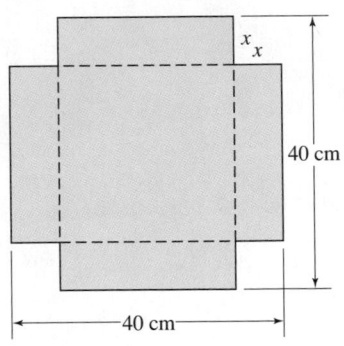

FIGURE 29–35

◆◆◆ **Example 18:** An open-top box is to be made from a square of sheet metal 40 cm on a side by cutting a square from each corner and bending up the sides along the dashed lines (Fig. 29–35). Find the dimension x of the cutout that will result in a box of the greatest volume.

Solution:

1. We want to maximize the volume V by varying x.
2. The equation is

$$V = \text{length} \cdot \text{width} \cdot \text{depth}$$
$$= (40 - 2x)(40 - 2x)x$$
$$= x(40 - 2x)^2$$

3. Taking the derivative, we obtain

$$\frac{dV}{dx} = x(2)(40 - 2x)(-2) + (40 - 2x)^2$$
$$= -4x(40 - 2x) + (40 - 2x)^2$$
$$= (40 - 2x)(-4x + 40 - 2x)$$
$$= (40 - 2x)(40 - 6x)$$

4. Setting the derivative to zero and solving for x, we have

$$
\begin{array}{c|c}
40 - 2x = 0 & 40 - 6x = 0 \\
x = 20 \text{ cm} & x = \dfrac{20}{3} = 6.67 \text{ cm}
\end{array}
$$

We discard $x = 20$ cm, because it is a minimum value and results in the entire sheet of metal being cut away. Thus we keep $x = 6.67$ cm as our answer.

Graphical Check: The graph of the volume equation, shown in Fig. 29–36, shows a maximum at $x \cong 6.67$, so we don't need a further test of that point. The graph of the derivative has a zero at $x = 6.67$, verifying our solution.   ◆◆◆

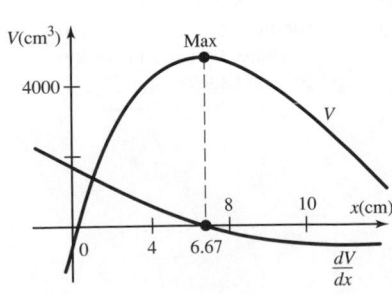

FIGURE 29–36

Our next example is one in which the equation is given.

◆◆◆ **Example 19:** The deflection y of a simply supported beam with a concentrated load P (Fig. 29–37) at a distance x from the left end of the beam is given by

$$y = \frac{Pbx}{6LEI}(L^2 - x^2 - b^2)$$

where E is the modulus of elasticity, and I is the moment of inertia of the beam's cross section. Find the value of x at which the deflection is a maximum for a 8.00-m-long beam with a concentrated load 2.00 m from the right end.

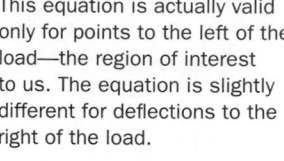

This equation is actually valid only for points to the left of the load—the region of interest to us. The equation is slightly different for deflections to the right of the load.

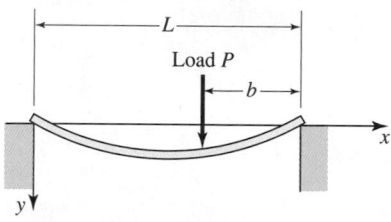

FIGURE 29–37 Simply supported beam with concentrated load.

Estimate: It seems reasonable that the maximum deflection would be at the centre, at $x = 4$ m from the left end of the beam. But it is equally reasonable that the maximum deflection would be at the concentrated load, at $x = 6$. So our best guess might be that the maximum deflection is somewhere between 4 and 6 m.

Solution:

1. We want to find the distance x from the left end at which the deflection is a maximum.
2. The equation is given in the problem statement.
3. We take the derivative using the product rule, noting that every quantity but x or y is a constant.

$$\frac{dy}{dx} = \frac{Pb}{6LEI}[x(-2x) + (L^2 - x^2 - b^2)(1)]$$

4. We set this derivative equal to zero and solve for x.

$$2x^2 = L^2 - x^2 - b^2$$

$$3x^2 = L^2 - b^2$$

$$x = \pm \sqrt{\frac{L^2 - b^2}{3}}$$

We drop the negative value, since x cannot be negative in this problem. Now substituting $L = 8.00$ m and $b = 2.00$ m gives us

$$x = \pm \sqrt{\frac{64.0 - 4.00}{3}} = 4.47 \text{ m}$$

Thus the maximum deflection occurs between the load and the midpoint of the beam, as expected.

5. It is clear from the physical problem that our point is a maximum, so no test is needed. ◆◆◆

The equation linking the variables in Example 19 had only two variables, x and y. In the following example, our equation has *three* variables, one of which must be eliminated before we take the derivative. We eliminate the third variable by means of a second equation.

◆◆◆ **Example 20:** Assume that the strength of a rectangular beam varies directly as its width and the square of its depth. Find the dimensions of the strongest beam that can be cut from a round log 30.0 cm in diameter (Fig. 29–38).

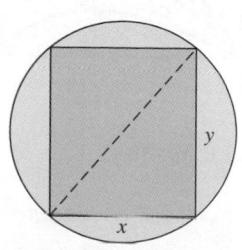

FIGURE 29–38 Beam cut from a log.

Estimate: A square beam cut from a 30-cm log would have a width of about 20 cm. But we are informed, and know from experience, that depth contributes more to the strength than width, so we expect a width less than 20 cm. But from experience we would be suspicious of a very narrow beam, say, 5 cm or less. So we expect a width between 5 and 20 cm. The depth has to be greater than 20 cm but cannot exceed 30 cm. These numbers bracket our answer.

Solution:

1. We want to maximize the strength S by varying the width x.
2. The strength S is $S = kxy^2$, where k is a constant of proportionality. Note that we have hree variables, S, x, and y. We must eliminate x or y. By the Pythagorean theorem,

$$x^2 + y^2 = 30.0^2$$

so

$$y^2 = 900 - x^2$$

Substituting, we obtain

$$S = kx(900 - x^2)$$

3. The derivative is

$$\frac{dS}{dx} = kx(-2x) + k(900 - x^2)$$
$$= -3kx^2 + 900k$$

4. Setting the derivative equal to zero and solving for x gives us $3x^2 = 900$, so

$$x = \pm\ 17.3 \text{ cm}$$

We discard the negative value, of course. The depth is

$$y = \sqrt{900 - 300} = 24.5 \text{ cm}$$

5. But have we found the dimensions of the beam with *maximum* strength, or minimum strength? Let us use the second-derivative test to tell us. The second derivative is

$$\frac{d^2 S}{dx^2} = -6kx$$

When $x = 17.3$,

$$\frac{d^2 S}{dx^2} = -104k$$

Since k is positive, the second derivative is negative, which tells us that we have found a *maximum*.

Graphical Solution or Check: As before, we can solve the problem or check the analytical solution graphically by graphing the function and its derivative. But here the equations contain an unknown constant k!

  Figure 29–39 shows the function graphed with assumed values of $k = 1$, 2, and 3. Notice that *changing k does not change the horizontal location of the maximum point.* Thus we can do a graphical solution or check with any value of k, with $k = 1$ being the simplest choice.   ◆◆◆

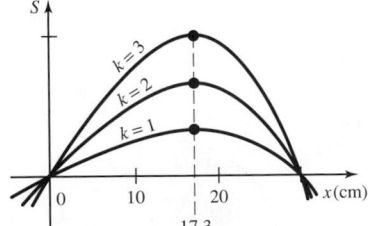

FIGURE 29–39

Sometimes a graphical solution is the only practical one, as in the following example.

◆◆◆ **Example 21:** Find the dimensions x and y and the minimum cost for a 10.0-cu. ft.-capacity, open-top box with square bottom (Fig. 29–40). The sides are aluminum at $1.28/sq. ft., and the bottom is copper at $2.13/sq. ft. An aluminum-to-aluminum joint costs $3.95/ft., and a copper-to-aluminum joint costs $2.08/ft.

Solution: The costs are as follows:

an aluminum side	$1.28xy$
the copper bottom	$2.13x^2$
one alum.-to-alum. joint	$3.95y$
one alum.-to-copper joint	$2.08x$

the total cost C is then

$$C = 4(1.28xy) + 2.13x^2 + 4(3.95y) + 4(2.08x)$$

We can eliminate y from this equation by noting that the volume, 10.0 cu. ft., is equal to x^2y, so

$$y = 10.0/x^2$$

Substituting gives

$$C = 4(1.28x)\frac{10.0}{x^2} + 2.13x^2 + 4(3.95)\frac{10.0}{x^2} + 4(2.08x)$$

This simplifies to

$$C = 2.13x^2 + 8.32x + 51.2x^{-1} + 158x^{-2}$$

Taking the derivative,

$$C' = 4.26x + 8.32 - 51.2x^{-2} - 316x^{-3}$$

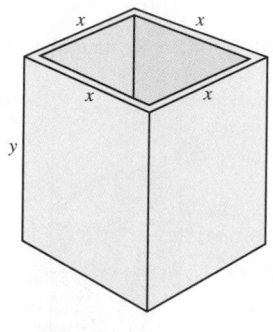

FIGURE 29–40

For *an analytical solution*, we would set C' to zero and attempt to solve for x. For a *numerical solution*, we would use Newton's method, the midpoint method, or a similar technique for finding the root. We will do a *graphical solution*. As in the preceding examples, we graph the function and its derivative as shown in Fig. 29–41 and find a minimum point at $x = 2.83$ ft.

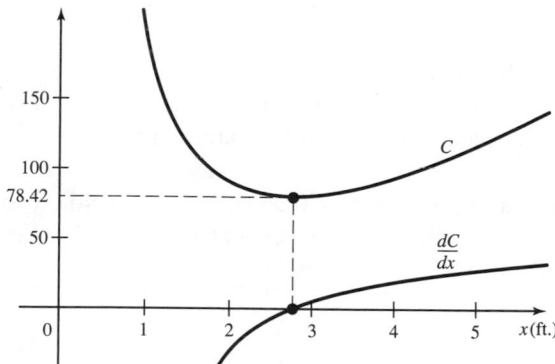

FIGURE 29–41

Substituting back to get y,

$$y = \frac{10.0}{x^2} = \frac{10.0}{(2.83)^2} = 1.25 \text{ ft.}$$

We can get the total cost C by substituting back, or we can read it off the graph. Either way we get $C = \$78.42$.

◆◆◆

Exercise 4 ◆ Optimization

Number Problems

1. What number added to half the square of its reciprocal gives the smallest sum?

2. Separate the number 10 into two parts such that their product will be a maximum.

3. Separate the number 20 into two parts such that the product of one part and the square of the other part is a maximum.

4. Separate the number 5 into two parts such that the square of one part times the cube of the other part will be a maximum.

Minimum Perimeter

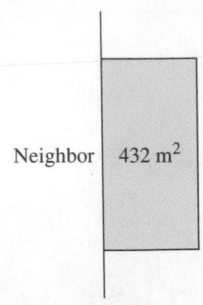

Neighbor 432 m²

FIGURE 29–42   Rectangular garden.

5. A rectangular garden (Fig. 29–42) laid out along your neighbor's lot contains 432 m². It is to be fenced on all sides. If the neighbor pays for half the shared fence, what should be the dimensions of the garden so that your cost is a minimum?

6. It is required to enclose a rectangular field by a fence (Fig. 29–43) and then divide it into two lots by a fence parallel to the short sides. If the area of the field is 2.50 ha, find the lengths of the sides so that the total length of fence will be a minimum.

7. A rectangular pasture 162 m² in area is built so that a long, straight wall serves as one side of it. If the length of the fence along the remaining three sides is the least possible, find the dimensions of the pasture.

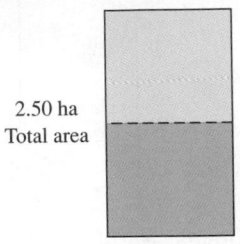

2.50 ha
Total area

FIGURE 29–43   Rectangular field.

Maximum Volume of Containers

8. Find the volume of the largest open-top box that can be made from a rectangular sheet of metal 6.00 cm by 16.0 cm (Fig. 29–44) by cutting a square from each corner and turning up the sides.

9. Find the height and base diameter of a cylindrical, topless tin cup of maximum volume if its area (sides and bottom) is $10\overline{0}$ cm².

10. The slant height of a certain cone is 50.0 cm. What cone height will make the volume a maximum?

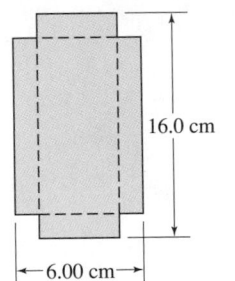

16.0 cm

6.00 cm

FIGURE 29–44   Sheet metal for box.

Maximum Area of Plane Figures

11. Find the maximum area of a rectangle with a perimeter of 20.0 cm.

12. A window composed of a rectangle surmounted by an equilateral triangle is 10.0 m in perimeter (Fig. 29–45). Find the dimensions that will make its total area a maximum.

13. Two corners of a rectangle are on the x axis between $x = 0$ and 10 (Fig. 29–46). The other two corners are on the lines whose equations are $y = 2x$ and $3x + y = 30$. For what value of y will the area of the rectangle be a maximum?

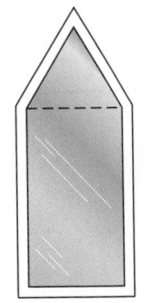

FIGURE 29–45   Window.

Maximum Cross-Sectional Area

14. A trough is to be made of a long rectangular piece of metal by bending up two edges so as to give a rectangular cross section. If the width of the original piece is 14.0 cm, how deep should the trough be made in order that its cross-sectional area be a maximum?

15. A gutter is to be made of a strip of metal 30.0 cm. wide, with the cross section having the form shown in Fig. 29–47. What depth x gives a maximum cross-sectional area?

Minimum Distance

16. Ship A is travelling due south at 40.0 km/h, and ship B is travelling due west at the same speed. Ship A is now 10.0 km from the point at which their paths will eventually cross, and ship B is now 20.0 km from that point. What is the closest that the two ships will get to each other?

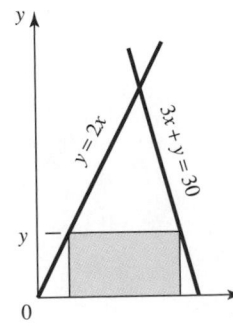

FIGURE 29–46

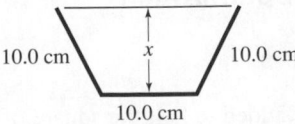

10.0 cm x 10.0 cm

10.0 cm

FIGURE 29–47

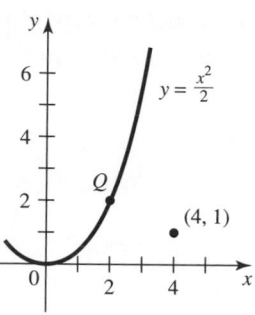

FIGURE 29–48

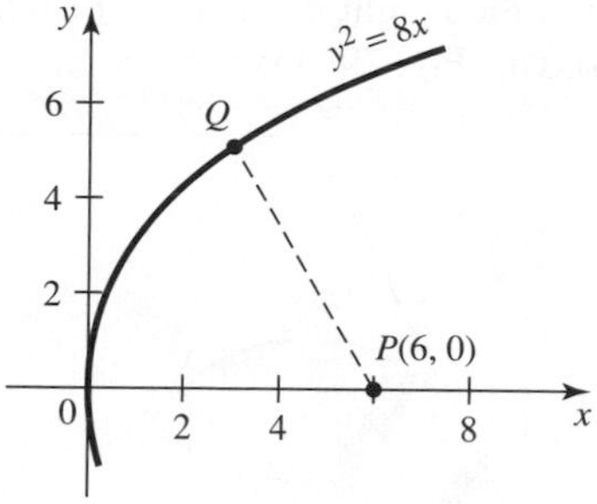

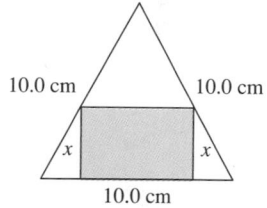

FIGURE 29–49

17. Find the point Q on the curve $y = x^2/2$ that is nearest the point $(4, 1)$ (Fig. 29–48).

18. Given one branch of the parabola $y^2 = 8x$ and the point $P(6, 0)$ on the x axis (Fig. 29–49), find the coordinates of point Q so that PQ is a minimum.

Inscribed Plane Figures

19. Find the dimensions of the rectangle of greatest area that can be inscribed in an equilateral triangle, each of whose sides is 10.0 cm, if one of the sides of the rectangle is on a side of the triangle (Fig. 29–50). (*Hint:* Let the independent variable be the height x of the rectangle.)

20. Find the area of the largest rectangle with sides parallel to the coordinate axes which can be inscribed in the figure bounded by the two parabolas $3y = 12 - x^2$ and $6y = x^2 - 12$ (Fig. 29–51).

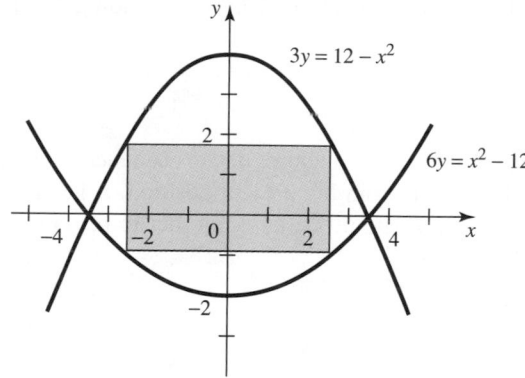

FIGURE 29–51

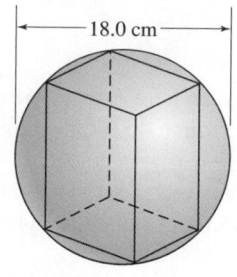

FIGURE 29–50

21. Find the dimensions of the largest rectangle that can be inscribed in an ellipse whose major axis is 20.0 units and whose minor axis is 14.0 units (Fig. 29–52).

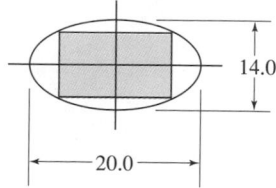

FIGURE 29–52 Rectangle inscribed in ellipse.

Inscribed Volumes

22. Find the dimensions of the rectangular parallelepiped of greatest volume and with a square base that can be cut from a solid sphere 18.0 cm in diameter (Fig. 29–53).

23. Find the dimensions of the right circular cylinder of greatest volume that can be inscribed in a sphere with a diameter of 10.0 cm (Fig. 29–54).

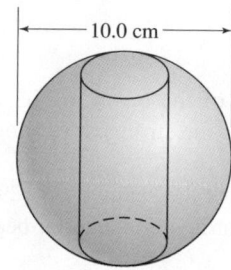

FIGURE 29–53
Parallelepiped
inscribed in sphere.

FIGURE 29–54
Cylinder inscribed
in sphere.

24. Find the height of the cone of minimum volume circumscribed about a sphere of radius 10.0 m (Fig. 29–55).

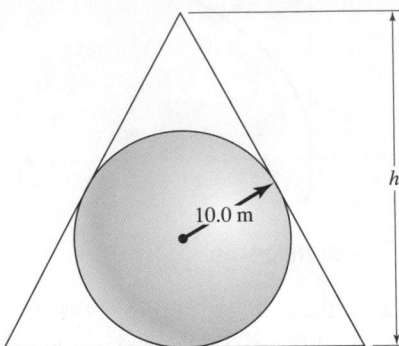

FIGURE 29–55   Cone circumscribed about sphere.

25. Find the altitude of the cone of maximum volume that can be inscribed in a sphere of radius 9.00 m.

Most Economical Dimensions of Containers

26. What should be the diameter of a can holding 1 L and requiring the least amount of metal if the can is open at the top?

27. A silo (Fig. 28–26) has a hemispherical roof, cylindrical sides, and circular floor, all made of steel. Find the dimensions for a silo having a volume of 755 m³ (including the dome) that needs the least steel.

Minimum Travel Time

28. A man in a rowboat at P (Fig. 29–56), 6.00 km from shore, desires to reach point Q on the shore at a straight-line distance of 10.0 km from his present position. If he can walk 4.00 km/h and row 3.00 km/h, at what point L should he land in order to reach Q in the shortest time?

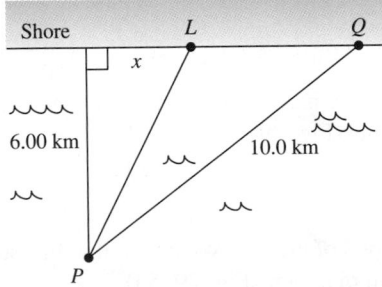

FIGURE 29–56   Find the fastest route from boat P to shore at Q.

Beam Problems

29. The strength S of the beam in Fig. 29–38 is given by $S = kxy^2$, where k is a constant. Find x and y for the strongest rectangular beam that can be cut from a 45.0-cm-diameter cylindrical log.

30. The stiffness Q of the beam in Fig. 29–38 is given by $Q = kxy^3$, where k is a constant. Find x and y for the stiffest rectangular beam that can be cut from a 45.0-cm-diameter cylindrical log.

Light

31. The intensity E of illumination at a point, from a light source at a distance x from the point, is given by $E = kI/x^2$, where k is a constant and I is the intensity of the source. A light M has an intensity three times that of N (Fig. 29–57). At what distance from M is the illumination a minimum?

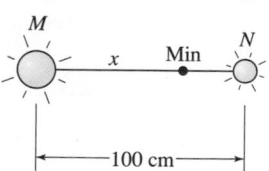

32. The intensity of illumination I from a given light source is given by

$$I = \frac{k \sin \phi}{d^2}$$

FIGURE 29–57

where k is a constant, ϕ is the angle at which the rays strike the surface, and d is the distance between the surface and the light (Fig. 29–58). At what height h should a light be suspended directly over the centre of a circle 3.00 m in diameter so that the illumination at the circumference will be a maximum?

Electrical

33. The power delivered to a load by a $3\overline{0}$-V source of internal resistance 2.0 Ω is $3\overline{0}i - 2.0i^2$ W, where i is the current in amperes. For what current will this source deliver the maximum power?

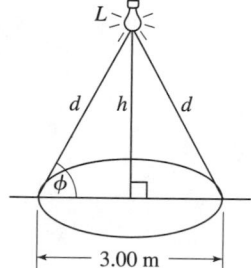

34. When 12 cells, each having an EMF of e and an internal resistance r, are connected to a variable load R, as in Fig. 29–59, the current in R is

FIGURE 29–58

$$i = \frac{3e}{\dfrac{3r}{4} + R}$$

Show that the maximum power (i^2R) delivered to the load is a maximum when the load R is equal to the equivalent internal resistance of the source, $3r/4$.

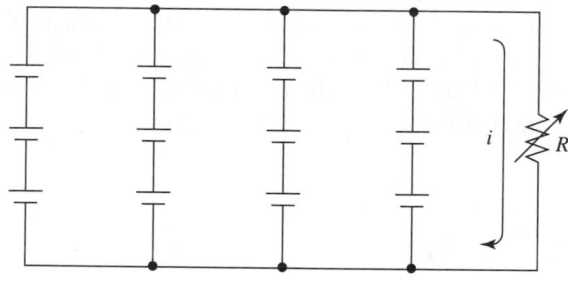

FIGURE 29–59

35. A certain transformer has an efficiency E when delivering a current i, where

$$E = \frac{115i - 25 - i^2}{115i}$$

At what current is the efficiency of the transformer a maximum?

Mechanisms

36. If the lever in Fig. 29–60 weighs 168 N/m, find its length so as to make the lifting force F a minimum.

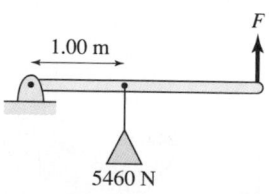

37. The efficiency E of a screw is given by

$$E = \frac{x - \mu x^2}{x + \mu}$$

FIGURE 29–60

where μ is the coefficient of friction and x the tangent of the pitch angle of the screw. Find x for maximum efficiency if $\mu = 0.45$.

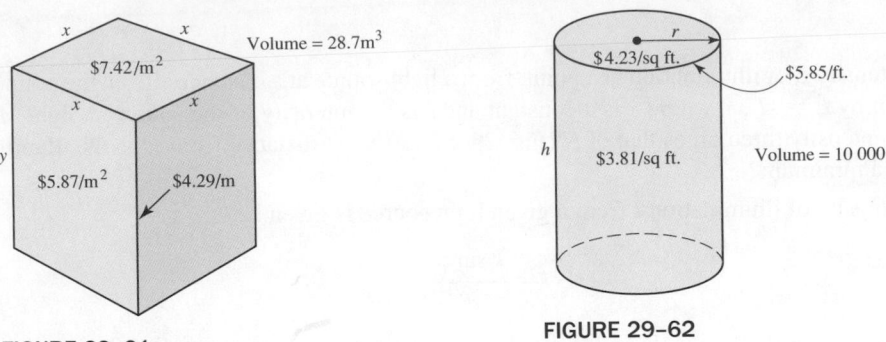

FIGURE 29–61

FIGURE 29–62

Graphical Solution

The following problems require a graphical solution.

38. Find the dimensions x and y and the minimum cost for a 28.7-m³-capacity box with square bottom (Fig. 29–61). The material for the sides costs $5.87/m², and the joints cost $4.29/m. The box has both a top and a bottom, at $7.42/m².

39. A cylindrical tank (Fig. 29–62) with capacity 10 000 cu. ft. has ends costing $4.23/sq. ft. and cylindrical side costing $3.81/sq. ft. The welds cost $5.85/ft. Find the radius, height, and total cost, for a tank of minimum cost.

Case Study Discussion—Robot Motion

Although the implementation of this concept would be done by a team programmer, you can come up with the idea based on your knowledge of calculus and physics found in this chapter. If the robot scans the area of the lift and calculates a path, adding intended displacement and time, the calculation can be checked for "jerk." The second derivative of displacement versus time is acceleration. The third derivative of displacement (the derivative of acceleration) is jerk, the rate of change of acceleration with respect to time. A program line or subroutine can be added to check the third derivative of displacement. If it is not equal to zero, the plan is rejected and a new calculation done.

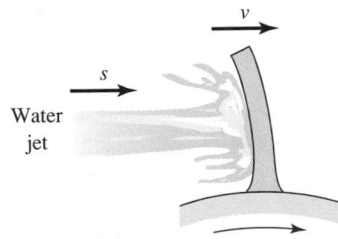

FIGURE 29–63 Turbine blade.

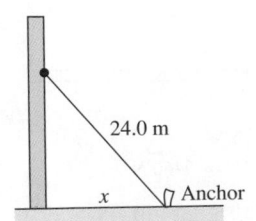

FIGURE 29–64 Pole braced by a cable.

••• CHAPTER 29 REVIEW PROBLEMS •••••••••••••••••••••••••••••••••

1. Airplane A is flying south at a speed of 120 m/s. It passes over a bridge 12 min before another airplane, B, which is flying east at the same height at a speed of 160 m/s. How fast are the airplanes separating 12 min after B passes over the bridge?

2. Find the approximate change in the function $y = 3x^2 - 2x + 5$ when x changes from 5 to 5.01.

3. A person walks toward the base of a 60-m-high tower at the rate of 5.0 km/h. At what rate does the person approach the top of the tower when 80 m from the base?

4. A point moves along the hyperbola $x^2 - y^2 = 144$ with a horizontal velocity $v_x = 15.0$ cm/s. Find the total velocity when the point is at (13.0, 5.00). 41.8 cm/s

5. A conical tank with vertex down has a vertex angle of 60.0°. Water flows from the tank at a rate of 5.00 cm³/min. At what rate is the inner surface of the tank being exposed when the water is 6.00 cm deep?

6. Find the instantaneous velocity and acceleration at $t = 2.0$ s for a point moving in a straight line according to the equation $s = 4t^2 - 6t$.

7. A turbine blade (Fig. 29–63) is driven by a jet of water having a speed s. The power output from the turbine is given by $P = k(sv - v^2)$, where v is the blade speed and k is a constant. Find the blade speed v for maximum power output.

8. A pole (Fig. 29–64) is braced by a cable 24.0 m long. Find the distance x from the foot of the pole to the cable anchor so that the moment produced by the tension in the cable about the foot of the pole is a maximum. Assume that the tension in the cable does not change as the anchor point is changed.

9. Find the height of a right circular cylinder of maximum volume that can be inscribed in a sphere of radius 6.

10. Each of three sides of a trapezoid has a length of 10 units. What length must the fourth side be to make the area a maximum?

11. The air in a certain balloon has a pressure of 40.0 lb./sq. in. and a volume of 5.0 cu. ft. and is expanding at the rate of 0.20 cu. ft./s. If the pressure and volume are related by the equation $pv^{1.41} = $ constant, find the rate at which the pressure is changing.

12. A certain item costs $10 to make, and the number that can be sold is estimated to be inversely proportional to the cube of the selling price. What selling price will give the greatest net profit?

13. The distance s of a point moving in a straight line is given by

$$ s = -t^3 + 3t^2 + 24t + 28 $$

At what times and at what distances is the point at rest?

14. A stone dropped into water produces a circular wave that increases in radius at the rate of 1.50 m/s. How fast is the area within the ripple increasing when its diameter is 6.00 m?

15. What is the area of the largest rectangle that can be drawn with one side on the x axis and with two corners on the curve $y = 8/(x^2 + 4)$?

16. The power P delivered to a load by a 120-V source having an internal resistance of 5 Ω is $P = 120I - 5I^2$, where I is the current to the load. At what current will the power be a maximum?

17. Separate the number 10 into two parts so that the product of the square of one part and the cube of the other part is a maximum.

18. The radius of a circular metal plate is increasing at the rate of 0.01 cm/s. At what rate is the area increasing when the radius is 2.0 m?

19. Find the dimensions of the largest rectangular box with square base and open top that can be made from 300 cm^2 of metal.

20. The voltage applied to a 3.25-μF capacitor is $v = 1.03t^2 + 1.33t + 2.52$ V. Find the current at $t = 15.0$ s.

21. The angular displacement of a rotating body is given by $\theta = 18.5t^2 + 12.8t + 14.8$ rad. Find (a) the angular velocity and (b) the angular acceleration at $t = 3.50$ s.

22. The charge through an 8.24-Ω resistor varies with time according to the function $q = 2.26t^3 - 8.28t$ C. Write an expression for the instantaneous current through the resistor.

23. The charge at a resistor varies with time according to the function $q = 2.84t^2 + 6.25t^3$ C. Write an expression for the instantaneous current through the resistor, and evaluate it at 1.25 s.

Writing

24. Once again, our writing assignment is to make up a word problem, but now one that leads to a max/min or a related rate problem. As before, swap with a classmate; solve each other's problems; note anything unclear, unrealistic, or ambiguous; and then rewrite your problem if needed.

Team Project

25. Lay out a race course (Fig. 29–65) on your school athletic field. The object is to get from *A* to *C* in the shortest time. You may take any path, but you may run only on line *AB* and must walk anywhere else.

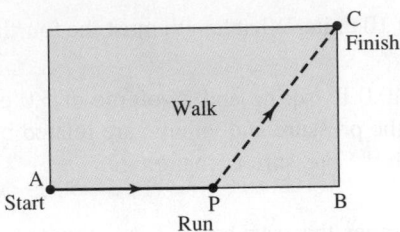

FIGURE 29-65   An athletic field.

Each member of your team should clock his or her rate of running and of walking. Then, using the ideas from problem 28 of Exercise 4 as a guide, compute for each of you the point *P* at which you should leave line *AB* for minimum time. Mark these points.

When ready, challenge another class to a race. Be careful to avoid students who have taken calculus or who are members of the track team.

Integration

When you have completed this chapter, you should be able to:
- Find the indefinite integral.
- Find the integral of an equation.
- Calculate the constant of integration.
- Find the definite integral.
- Approximate the area under a curve.
- Find the exact area under a curve.

Every operation in mathematics has its inverse. For example, we reverse the squaring operation by taking the square root, the arcsin is the inverse of the sine, and so on. In this chapter we learn how to reverse the process of differentiation with the process of *integration*.

In the last three chapters, you saw how being able to calculate the instantaneous slope of a curve had many practical applications. In this and the following two chapters, we will see how being able to reverse differentiation and find the area under a curve has many practical applications.

We then define the *definite integral* and show how to evaluate it. Next we discuss the problem of finding the area bounded by a curve and the *x* axis between two given values of *x*. We find such areas, first approximately by the *midpoint method* and then exactly by means of the definite integral. In the process we develop the *fundamental theorem of calculus,* which ties together the derivative, the integral, and the area under a curve.

30–1   The Indefinite Integral

Reversing the Process of Differentiation

An *antiderivative* of an expression is a new expression which, if differentiated, gives the original expression.

◆◆◆ **Example 1:** The derivative of x^3 is $3x^2$, so an antiderivative of $3x^2$ is x^3.

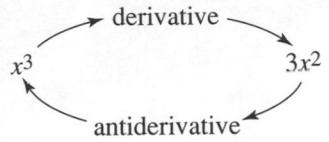

◆◆◆

The derivative of x^3 is $3x^2$, but the derivative of $x^3 + 6$ is also $3x^2$. The derivatives of $x^3 - 99$ and of $x^3 + $ *any constant* are also $3x^2$. This constant, called the *constant of integration,* must be included when we find *the* antiderivative of a function.

We will learn how to evaluate the constant of integration in Sec. 30–3.

◆◆◆ **Example 2:** The derivative of $x^3 + $ any constant is $3x^2$, so the antiderivative of $3x^2$ is $x^3 + $ a constant. We use C to stand for the constant.

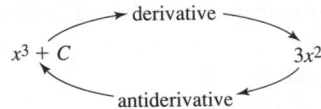

◆◆◆

The antiderivative is also called the *indefinite integral,* and we will use both names. The process of finding the integral or antiderivative is called *integration.* The variable, x in this case, is called the *variable of integration.*

The Integral Sign

We have seen above that the derivative of $x^3 + C$ is $3x^2$, so the antiderivative of $3x^2$ is $x^3 + C$. Let us now state this same idea more formally.

Let there be a function $F(x) + C$. Its derivative is

$$\frac{d}{dx}[F(x) + C] = F'(x) + 0 = F'(x)$$

where we have used the familiar prime (′) notation to indicate a derivative. Going to differential form, we have

$$d[F(x) + C] = F'(x)\,dx$$

It is no accident that the integralsign looks like the letter S. We will see later that it stands for *summation.*

Thus the differential of $F(x) + C$ is $F'(x)\,dx$. Conversely, the antiderivative of $F'(x)\,dx$ is $F(x) + C$.

$$\text{antiderivative of } F'(x)\,dx = F(x) + C$$

Instead of writing "antiderivative of," we use the *integral sign* $\int$ to indicate the antiderivative.

The integral is called indefinite because of the unknown constant. In Sec. 30–4, we study the definite integral, which has no unknown constant.

Indefinite Integral or Antiderivative	$\int F'(x)\,dx = F(x) + C$	374

The integral of the differential of a function is equal to the function itself, plus a constant.

We read Eq. 374 as "the integral of $F'(x)\,dx$ with respect to x is $F(x) + C$." The expression $F'(x)\,dx$ to be integrated is called the *integrand,* and C is the *constant of integration.*

◆◆◆ **Example 3:** What is the indefinite integral of $3x^2\,dx$?

Solution: The function x^3 has $3x^2\,dx$ as a differential, so the integral of $3x^2\,dx$ is $x^3 + C$.

$$\int 3x^2\,dx = x^3 + C$$

◆◆◆

If we let $F'(x)$ be denoted by $f(x)$, Eq. 374 becomes the following:

Indefinite Integral or Antiderivative	$$\int f(x)\,dx = F(x) + C$$ where $f(x) = F'(x)$	375

We use both capital and lowercase F in this section. Be careful not to mix them up.

Some Rules for Finding Integrals

We can obtain rules for integration by reversing the rules we had previously derived for differentiation. Such a list of rules, called a *table of integrals,* is given in Appendix C. This is a very short table of integrals; some fill entire books.

Let there be a function $u = F(x)$ whose derivative is

$$\frac{du}{dx} = F'(x)$$

or

$$du = F'(x)\,dx$$

Just as we can take the derivative of both sides of an equation, we can take the integral of both sides. This gives

$$\int du = \int F'(x)\,dx = F(x) + C$$

by Eq. 374. Substituting $F(x) = u$, we get our first rule for finding integrals.

Rule 1	$$\int du = u + C$$

The integral of the differential of a function is equal to the function itself, plus a constant of integration.

◆◆◆ **Example 4:** We use Rule 1 to integrate the following expressions:

(a) $\displaystyle\int dy = y + C$

(b) $\displaystyle\int dz = z + C$

(c) $\displaystyle\int d(x^3) = x^3 + C$

(d) $\displaystyle\int d(x^2 + 2x) = x^2 + 2x + C$

(e) $\displaystyle\int d\left(\frac{u^{n+1}}{n+1}\right) = \frac{u^{n+1}}{n+1} + C$

◆◆◆

For our second rule, we use the fact that the derivative of a constant times a function equals the constant times the derivative of the function. Reversing the process, we have the following:

Rule 2	$$\int a\,f(x)\,dx = a\int f(x)\,dx = a\,F(x) + C$$

The integral of a constant times a function is equal to the constant times the integral of the function, plus a constant of integration.

Rule 2 says that we may move constants to the left of the integral sign, as in the following example.

◆◆◆ **Example 5:** Find $\int 3kx^2\,dx$.

Solution: From Example 3 we know that $\int 3x^2\,dx = x^3 + C_1$, so if k is some constant,

$$\int 3kx^2\,dx = k\int 3x^2\,dx = k(x^3 + C_1) = kx^3 + kC_1$$
$$= kx^3 + C$$

where we have replaced the constant kC_1 by another constant, C.   ◆◆◆

◆◆◆ **Example 6:** Find $\int 5\,dx$.

Solution: By Rule 2,

$$\int 5\,dx = 5\int dx$$

Then, by Rule 1,

$$5\int dx = 5(x + C_1)$$
$$= 5x + C$$

where we have replaced $5C_1$ by C.   ◆◆◆

We get our third rule by noting that the derivative of a function with several terms is the sum of the derivatives of each term. Reversing gives the following:

The constant C comes from combining the individual constants, $C_1, C_2, \ldots$ into a single term.

Rule 3	$$\int [f(x) + g(x) + h(x) + \cdots]\,dx$$ $$= \int f(x)\,dx + \int g(x)\,dx + \int h(x)\,dx + \cdots + C$$

The integral of a function with several terms equals the sum of the integrals of those terms, plus a constant.

Rule 3 says that when integrating an expression having several terms, we may integrate each term separately.

◆◆◆ **Example 7:** Integrate $\int (3x^2 + 5)\,dx$.

Solution:
By Rule 3:

$$\int (3x^2 + 5)\,dx = \int 3x^2\,dx + \int 5\,dx$$

By Rule 2:

$$= \int 3x^2 \, dx + 5 \int dx$$
$$= x^3 + C_1 + 5x + C_2$$
$$= x^3 + 5x + C$$

From now on, we will not bother writing C_1 and C_2, but will combine them immediately into a single constant C.

where we have combined C_1 and C_2 into a single constant C. ◆◆◆

Our fourth rule is for a power of x. We know that the derivative of $x^{n+1}/(n+1)$ is

$$\frac{d}{dx}\left(\frac{x^{n+1}}{n+1}\right) = (n+1)\frac{x^n}{n+1} = x^n$$

or in differential form

$$d\left(\frac{x^{n+1}}{n+1}\right) = x^n \, dx$$

Taking the integral of both sides gives

$$\int d\left(\frac{x^{n+1}}{n+1}\right) = \int x^n \, dx$$

This gives us Rule 4.

Rule 4	$\displaystyle \int x^n \, dx = \frac{x^{n+1}}{n+1} + C \qquad (n \neq -1)$

The integral of x raised to a power is x raised to that power increased by 1, divided by the new power, plus a constant.

Note that Rule 4 is not valid for $n = -1$, for this would give division by zero. We'll use a later rule when n is -1.

We'll now give some examples of these rules for integration.

◆◆◆ **Example 8:** Integrate $\int x^3 \, dx$.

Solution: We use Rule 4 with $n = 3$.

$$\int x^3 \, dx = \frac{x^{3+1}}{3+1} + C = \frac{x^4}{4} + C$$ ◆◆◆

◆◆◆ **Example 9:** Integrate $\int x^5 \, dx$.

Solution: This is similar to Example 8, except that the exponent is 5. By Rule 4,

$$\int x^5 \, dx = \frac{x^{5+1}}{5+1} + C = \frac{x^6}{6} + C$$ ◆◆◆

◆◆◆ **Example 10:** Integrate $\int 7x^3 \, dx$.

Solution: This is similar to Example 8, except that here our function is multiplied by 7. By Rule 2, we may move the constant factor 7 to the left of the integral sign.

$$\int 7x^3 \, dx = 7 \int x^3 \, dx$$
$$= 7\left(\frac{x^4}{4} + C_1\right)$$
$$= \frac{7x^4}{4} + 7C_1$$
$$= \frac{7x^4}{4} + C$$

where $C = 7C_1$. Since C_1 is an unknown constant, $7C_1$ is also an unknown constant that can be more simply represented by C. From now on we will not even bother with individual constants such as C_1 but will write the final constant C directly.   ◆◆◆

◆◆◆ **Example 11:** Integrate $\int (x^3 + x^5)\, dx$.

Solution: By Rule 3, we can integrate each term separately.

$$\int (x^3 + x^5)\, dx = \int x^3\, dx + \int x^5\, dx$$
$$= \frac{x^4}{4} + \frac{x^6}{6} + C$$

Even though each of the two integrals has produced its own constant of integration, we have combined them immediately into the single constant C.   ◆◆◆

◆◆◆ **Example 12:** Integrate $\int (5x^2 + 2x - 3)\, dx$.

Solution: Integrating term by term yields

$$\int (5x^2 + 2x - 3)\, dx = \frac{5x^3}{3} + x^2 - 3x + C$$ ◆◆◆

Rule 4 is also used when the exponent n is not an integer.

◆◆◆ **Example 13:** Integrate $\int \sqrt[3]{x}\, dx$.

Solution: We write the radical in exponential form with $n = \frac{1}{3}$.

$$\int \sqrt[3]{x}\, dx = \int x^{1/3}\, dx$$

By Rule 4:

$$= \frac{x^{4/3}}{\frac{4}{3}} + C$$
$$= \frac{3x^{4/3}}{4} + C$$ ◆◆◆

The exponent n can also be negative (with the exception of -1, which would result in division by zero).

◆◆◆ **Example 14:** Integrate $\int \frac{1}{t^3}\, dt$.

The variable of integration can, of course, be any letter (not just x), as in this example.

Solution:

$$\int \frac{1}{t^3}\, dt = \int t^{-3}\, dt$$
$$= \frac{t^{-2}}{-2} + C = -\frac{1}{2t^2} + C$$ ◆◆◆

Checking an Integral by Differentiating

Many rules for integration are presented without derivation or proof, so you would be correct in being suspicious of them. However, you can convince yourself that a rule works (and that you have used it correctly) simply by taking the derivative of your result. You should get back the original expression.

◆◆◆ **Example 15:** Taking the derivative of the expression obtained in Example 14, we have

$$\frac{d}{dt}\left(-\frac{t^{-2}}{2} + C \right) = -\frac{1}{2}(-2t^{-3}) + 0 = t^{-3} = \frac{1}{t^3}$$

which is the expression we started with, so our integration was correct.   ◆◆◆

Simplify before Integrating

If an expression does not seem to fit any given rule at first, try changing its form by performing the indicated operations (squaring, removing parentheses, and so on).

◆◆◆ **Example 16:** Integrate $\int (x^2 + 3)^2 \, dx$.

Solution: None of our rules (so far) seem to fit. Rule 3, for example, is for x raised to a power, *not* for $(x^2 + 3)$ raised to a power. However, if we square $x^2 + 3$, we get

$$\int (x^2 + 3)^2 \, dx = \int (x^4 + 6x^2 + 9) \, dx$$

$$= \frac{x^5}{5} + \frac{6x^3}{3} + 9x + C$$

$$= \frac{x^5}{5} + 2x^3 + 9x + C \qquad ◆◆◆$$

◆◆◆ **Example 17:** Integrate $\int \dfrac{x^5 - 2x^3 + 5x}{x} \, dx$.

Solution: This problem looks complicated at first, but let us perform the indicated division first.

$$\int \frac{x^5 - 2x^3 + 5x}{x} \, dx = \int (x^4 - 2x^2 + 5) \, dx$$

Now Rules 3 and 4 can be used.

$$= \frac{x^5}{5} - \frac{2x^3}{3} + 5x + C \qquad ◆◆◆$$

◆◆◆ **Example 18:** Evaluate $\int \dfrac{x^3 - x^2 + 5x - 5}{x - 1} \, dx$.

Solution: Again, no rule seems to fit, so we try to simplify the given expression by long division.

We can also simplify here by factoring the numerator and cancelling.

$$
\begin{array}{r}
x^2 + 5 \\
x - 1 \overline{)\, x^3 - x^2 + 5x - 5} \\
\underline{x^3 - x^2 } \\
0 + 5x - 5 \\
\underline{5x - 5}
\end{array}
$$

So our quotient is $x^2 + 5$, which we now integrate.

$$\int \frac{x^3 - x^2 + 5x - 5}{x - 1} \, dx = \int (x^2 + 5) \, dx = \frac{x^3}{3} + 5x + C \qquad ◆◆◆$$

Simple Differential Equations

Suppose that we have the derivative of a function and we wish to find the original function from its derivative. Let's use our familiar

This equation, as well as any other that contains a derivative, is called a differential equation. In this example we are solving a differential equation.

$$\frac{dy}{dx} = 3x^2$$

We now seek the equation $y = F(x)$, of which $3x^2$ is the derivative. Proceed as follows. First write the given differential equation in differential form by multiplying both sides by dx.

$$dy = 3x^2 \, dx$$

Now take the integral of both sides of the equation.

$$\int dy = \int 3x^2 \, dx$$

so

$$y + C_1 = x^3 + C_2$$

Next combine the two arbitrary constants, C_1 and C_2.

$$y = x^3 + C_2 - C_1 = x^3 + C$$

This is the function whose derivative is $3x^2$. This function is also called the *solution* to the differential equation $dy/dx = 3x^2$.

Exercise 1 ◆ The Indefinite Integral

Find the indefinite integral.

1. $\displaystyle\int dx$

2. $\displaystyle\int dy$

3. $\displaystyle\int x \, dx$

4. $\displaystyle\int x^4 \, dx$

5. $\displaystyle\int \frac{dx}{x^2}$

6. $\displaystyle\int x^{2/3} \, dx$

7. $\displaystyle\int 3x \, dx$

8. $\displaystyle\int 6x \, dx$

9. $\displaystyle\int 3x^3 \, dx$

10. $\displaystyle\int x^5 \, dx$

11. $\displaystyle\int x^n \, dx$

12. $\displaystyle\int x^{1/2} \, dx$

13. $\displaystyle\int \frac{dx}{\sqrt{x}}$

14. $\displaystyle\int 3x^2 \, dx$

15. $\displaystyle\int 4x^3 \, dx$

16. $\displaystyle\int \left(\frac{x^2}{2} - \frac{2}{x^2}\right) dx$

17. $\displaystyle\int \frac{7}{2} x^{5/2} \, dx$

18. $\displaystyle\int 5x^4 \, dx$

19. $\displaystyle\int 2(x + 1) \, dx$

20. $\displaystyle\int x^{4/3} \, dx$

21. $\displaystyle\int \frac{4}{3} x^{1/3} \, dx$

22. $\displaystyle\int (x^{3/2} - 2x^{2/3} + 5\sqrt{x} - 3) \, dx$

23. $\displaystyle\int 2u \, du$

24. $\displaystyle\int \sqrt[3]{t} \, dt$

25. $\displaystyle\int 4s^{1/2} \, ds$

26. $\displaystyle\int 3\sqrt[3]{y} \, dy$

Simplify and integrate.

27. $\displaystyle\int \sqrt{x} \, (3x - 2) \, dx$

28. $\displaystyle\int (x + 1)^2 \, dx$

29. $\displaystyle\int \frac{4x^2 - 2\sqrt{x}}{x} \, dx$

30. $\displaystyle\int (t + 2)(t - 3) \, dt$

31. $\displaystyle\int (1 - s)^3 \, ds$

32. $\displaystyle\int \frac{v^3 + 2v^2 - 3v - 6}{v + 2} \, dv$

Find the function whose derivative is given (solve each differential equation).

33. $\dfrac{dy}{dx} = 4x^2$

34. $\dfrac{dy}{dx} = 2x(x^2 + 6)$

35. $\dfrac{dy}{dx} = x^{-3}$

36. $\dfrac{ds}{dt} = 10t^{-6}$

37. $\dfrac{ds}{dt} = \dfrac{1}{2}t^{-2/3}$

38. $\dfrac{dv}{dt} = 6t^3 - 3t^{-2}$

30–2   Rules for Finding Integrals

Our small but growing table of integrals has four rules so far. To these we will now add rules for integrating a power function and for integrating an expression in the form du/u.

Integral of a Power Function

If u is some function of x, the derivative of $u^{n+1}/(n + 1)$ is

$$\frac{d}{dx}\left(\frac{u^{n+1}}{n + 1}\right) = (n + 1)\frac{u^n}{n + 1}\frac{du}{dx} = u^n\frac{du}{dx}$$

or

$$u^n\, du = d\left(\frac{u^{n+1}}{n+1}\right)$$

Taking the integral of both sides yields

$$\int d\left(\frac{u^{n+1}}{n + 1}\right) = \int u^n\, du$$

Since the integral of the differential of a function is the function itself, we get the following rule:

Rule 5	$\displaystyle\int u^n\, du = \frac{u^{n+1}}{n + 1} + C \qquad (n \neq -1)$

The expression u in Rule = can be any function of x, say, $x^3 + 3$. However, in order for us to use Rule 5, *the quantity u^n must be followed by the derivative of u.* Thus if $u = x^3 + 3$, then $du = 3x^2\, dx$, as in the following example.

◆◆◆ **Example 19:** Find the integral $\int (x^3 + 3)^6(3x^2)\, dx$.

Solution: If we let

$$u = x^3 + 3$$

we see that the derivative of u is

$$\frac{du}{dx} = 3x^2$$

or, in differential form,

$$du = 3x^2\, dx$$

Notice now that our given integral exactly matches Rule 5.

$$\int \underbrace{(x^3 + 3)^6}_{u}\overset{\overset{\displaystyle n}{\frown}}{}\underbrace{(3x^2)dx}_{du}$$

We now apply Rule 5.

$$\int (x^3 + 3)^6 (3x^2)\, dx = \frac{(x^3 + 3)^7}{7} + C$$ ◆◆◆

Students are often puzzled at the "disappearance" of the $3x^2\, dx$ in Example 19.

$$\int (x^3 + 3)^6 \underbrace{(3x^2)dx} = \frac{(x^3 + 3)^7}{7} + C$$

Where did this go?

The $3x^2\, dx$ is the differential of x^3 and does not remain after integration. Do not be alarmed when it vanishes. As before, we can check our integration by taking its derivative. We should get back the original expression, including the part that seemed to "vanish."

◆◆◆ **Example 20:** Check the integration in Example 19 by taking the derivative.

Solution: If

$$y = \frac{1}{7}(x^3 + 3)^7 + C$$

then

$$\frac{dy}{dx} = \frac{1}{7}(7)(x^3 + 3)^6 (3x^2) + 0$$

or

$$dy = (x^3 + 3)^6 (3x^2)\, dx$$

This is the same expression that we started with.   ◆◆◆

Very often an integral will not exactly match the form of Rule 5 or any other rule in the table. However, if all we lack is a constant factor, it can usually be supplied as in the following examples.

◆◆◆ **Example 21:** Integrate $\int (x^3 + 3)^6 x^2\, dx$.

Solution: This is almost identical to Example 19, except that the factor 3 is missing. Realize that we cannot use Rule 5 *yet,* because if

$$u = x^3 + 3$$

then

$$du = 3x^2\, dx$$

Our integral contains $x^2\, dx$ but not $3x^2\, dx$, which we need in order to use Rule 5. But we can insert a factor of 3 into our integrand as long as we compensate for it by multiplying the whole integral by $\frac{1}{3}$.

$$\frac{1}{3} \int (x^3 + 3)^6 (3x^2)\, dx$$

Compensate. ————   ———— Insert.

Integrating gives us

$$\frac{1}{3} \int (x^3 + 3)^6 (3x^2)\, dx = \left(\frac{1}{3}\right)\frac{(x^3 + 3)^7}{7} + C$$

$$= \frac{(x^3 + 3)^7}{21} + C$$ ◆◆◆

◆◆◆ **Example 22:** Evaluate $\displaystyle\int \frac{5z\,dz}{\sqrt{3z^2-7}}$.

Solution: We rewrite the given expression in exponential form and move the 5 outside the integral.

$$\int \frac{5z\,dz}{\sqrt{3z^2-7}} = 5\int (3z^2-7)^{-1/2}z\,dz$$

Here $n = -\frac{1}{2}$. Let $u = 3z^2 - 7$, so du is $6z\,dz$. Comparing with our actual integral, we see that we need to insert a 6 before the z and compensate with $\frac{1}{6}$ outside the integral sign.

$$5\int (3z^2-7)^{-1/2}\,z\,dz = 5\left(\frac{1}{6}\right)\int (3z^2-7)^{-1/2}(6z\,dz)$$

$$= \frac{5}{6}\frac{(3z^2-7)^{1/2}}{\frac{1}{2}} + C = \frac{5}{3}\sqrt{3z^2-7} + C \qquad\qquad ◆◆◆$$

Common Error	Rule 2 allows us to move only *constant factors* to the left of the integral sign. You cannot use this same procedure with variables.

Summary of Integration Rules So Far

Our table of integrals now contains five rules, summarized in the following table. These will be enough for us to do the applications in the following chapter. We will learn more rules in Chapter 34.

1. $\displaystyle\int du = u + C$

2. $\displaystyle\int af(x)\,dx = a\int f(x)\,dx = aF(x) + C$

3. $\displaystyle\int [f(x) + g(x) + h(x) + \cdots]\,dx = \int f(x)\,dx + \int g(x)\,dx + \int h(x)\,dx + \cdots + C$

4. $\displaystyle\int x^n\,dx = \frac{x^{n+1}}{n+1} + C \qquad (n \neq -1)$

5. $\displaystyle\int u^n\,du = \frac{u^{n+1}}{n+1} + C \qquad (n \neq -1)$

Miscellaneous Rules from the Table of Integrals

Now that you can use Rules 1 through 5, you should be able to use many of the rules from the table of integrals in Appendix C. We will restrict ourselves here to finding integrals of algebraic expressions, for which you should be able to identify the du portion in the rule. We will do transcendental functions in Chapter 34.

◆◆◆ **Example 23:** Integrate $\displaystyle\int \frac{5}{x}\,dx$.

Solution: From the table of integrals we find the following:

Rule 7	$\displaystyle\int \frac{du}{u} = \ln|u| + C \qquad (u \neq 0)$

which matches our integral if we let $x = u$. So du then equals dx. Then

$$\int \frac{5}{x} \, dx = 5 \int \frac{dx}{x} = 5 \ln|x| + C$$

by Rule 7. ◆◆◆

When we took derivatives of ln u, we didn't care about negative values because u had to be positive for the logarithm to exist. But we sometimes want the integral of $y = 1/u$ at negative values of u.

◆◆◆ **Example 24:** Integrate $\displaystyle \int \frac{x \, dx}{3 - x^2}$.

Solution: Let $u = 3 - x^2$. Then du is $-2x \, dx$. Our integral does not have the -2 factor, so we insert a factor of -2 and compensate with a factor of $-\frac{1}{2}$.

$$\int \frac{x \, dx}{3 - x^2} = -\frac{1}{2} \int \frac{-2x \, dx}{3 - x^2} = -\frac{1}{2} \ln|3 - x^2| + C$$

by Rule 7. ◆◆◆

The use of absolute value will be made clearer when we substitute values into the integral, as in the following example.

◆◆◆ **Example 25:** Find the equation of the curve whose derivative is $6x/(x^2 - 4)$ and which passes through the point $(0, 5)$.

Solution: Our derivative, in differential form, is

$$dy = \frac{6x \, dx}{x^2 - 4}$$

To integrate, we let $u = x^2 - 4$ and $du = 2x \, dx$. Then, by Rule 7,

$$y = \int \frac{6x}{x^2 - 4} \, dx = 3 \int \frac{2x}{x^2 - 4} \, dx = 3 \ln|x^2 - 4| + C$$

At $(0, 5)$ we get $C = 5 - 3 \ln|0 - 4| = 5 - 3 \ln 4 = 0.841$, so

$$y = 3 \ln|x^2 - 4| + 0.841$$

This function (graphed as a solid line in Fig. 30–1) has a domain from $x = -2$ to 2.

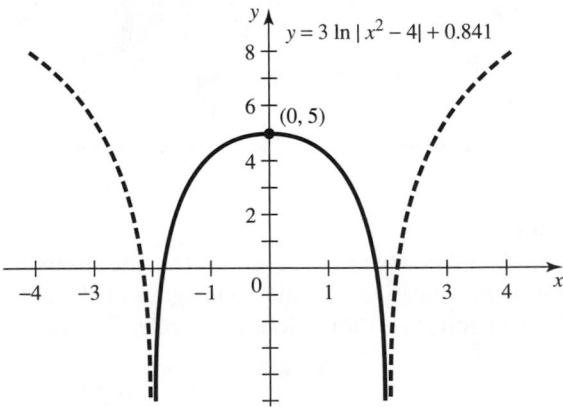

FIGURE 30–1

Do not try to use our function $y = 3 \ln|x^2 - 4| + 0.841$ for $x < -2$ or $x > 2$. Those parts of the curve (shown dashed) *are not continuous* with the section for which we had the known value, so the function we found might not apply there. Our function applies only to $-2 < x < 2$. ◆◆◆

◆◆◆ **Example 26:** Integrate $\displaystyle\int \frac{dx}{4x^2 + 25}$.

Solution: From the table we find the following:

Rule 56	$\displaystyle\int \frac{du}{a^2 + b^2u^2} = \frac{1}{ab}\,\mathrm{Tan}^{-1}\frac{bu}{a} + C$

Letting $a = 5$, $b = 2$, $u = x$, and $du = dx$, our integral matches Rule 56 if we rearrange the denominator.

$$\int \frac{dx}{25 + 4x^2} = \frac{1}{10}\,\mathrm{Tan}^{-1}\frac{2x}{5} + C \qquad ◆◆◆$$

◆◆◆ **Example 27:** Integrate $\displaystyle\int \frac{dx}{(4x^2 + 9)2x}$.

Solution: We match this with the following rule:

Rule 60	$\displaystyle\int \frac{du}{u(u^2 + a^2)} = \frac{1}{2a^2}\,\ln\left|\frac{u^2}{u^2 + a^2}\right| + C$

with $a = 3$, $u = 2x$, and $du = 2dx$. Thus

$$\int \frac{dx}{(4x^2 + 9)2x} = \frac{1}{2}\int \frac{2\,dx}{(2x)[(2x)^2 + 3^2]} = \frac{1}{36}\,\ln\left|\frac{4x^2}{4x^2 + 9}\right| + C \qquad ◆◆◆$$

Common Error	When using the table of integrals, be sure that the integral chosen completely matches the given integral, and be sure especially that *all factors of* du *are present.*

◆◆◆ **Example 28:** Integrate $\displaystyle\int \frac{dx}{x\sqrt{x^2 + 16}}$.

Solution: We use Rule 64.

Rule 64	$\displaystyle\int \frac{du}{u\sqrt{u^2 + a^2}} = \frac{1}{a}\,\ln\left|\frac{u}{a + \sqrt{u^2 + a^2}}\right| + C$

with $u = x$, $a = 4$, and $du = dx$. Thus

$$\int \frac{dx}{x\sqrt{x^2 + 16}} = \frac{1}{4}\,\ln\left|\frac{x}{4 + \sqrt{x^2 + 16}}\right| + C \qquad ◆◆◆$$

Exercise 2 ◆ Rules for Finding Integrals

Integral of a Power Function

Integrate.

1. $\displaystyle\int (x^4 + 1)^3 4x^3 \, dx$

2. $\displaystyle\int (2x^2 - 6)^3 4x \, dx$

3. $\displaystyle\int 2(x^2 + 2x)(2x + 2) \, dx$

4. $\displaystyle\int (1 - 2x)^5 \, dx$

5. $\displaystyle\int \frac{dx}{(1 - x)^2}$

6. $\displaystyle\int 6(x^2 - 1)^2 x \, dx$

7. $\displaystyle\int 9(x^3 + 1)^2 x^2 \, dx$

8. $\displaystyle\int 3x \sqrt{x^2 - 1} \, dx$

9. $\displaystyle\int \frac{y^2 \, dy}{\sqrt{1 - y^3}}$

10. $\displaystyle\int (x + 1)(x^2 + 2x + 6)^2 \, dx$

11. $\displaystyle\int \frac{4x \, dx}{(9 - x^2)^2}$

12. $\displaystyle\int z \sqrt{z^2 - 2} \, dz$

Integral of *du/u*

13. $\displaystyle\int \frac{3}{x} \, dx$

14. $\displaystyle\int \frac{dx}{x - 1}$

15. $\displaystyle\int \frac{5 \, z^2}{z^3 - 3} \, dz$

16. $\displaystyle\int \frac{t \, dt}{6 - t^2}$

17. $\displaystyle\int \frac{x + 1}{x} \, dx$

18. $\displaystyle\int \frac{w^2 + 5}{w} \, dw$

Find the function whose derivative is given.

19. $\dfrac{dy}{dx} = x(1 - x^2)^3$

20. $\dfrac{dy}{dx} = x^2(x^3 - 2)^{1/2}$

21. $\dfrac{dy}{dx} = (3x^2 + x)(2x^3 + x^2)^3$

22. $\dfrac{ds}{dt} = \dfrac{t^2}{\sqrt{5t^3 + 7}}$

23. $\dfrac{dv}{dt} = 5t \sqrt{7 - t^2}$

24. $\dfrac{dy}{dx} = \dfrac{x}{(9 - 4x^2)^2}$

25. $\dfrac{dy}{dx} = \dfrac{7}{x}$

26. $\dfrac{dy}{dx} = \dfrac{4}{x + 3}$

27. $\dfrac{dy}{dx} = \dfrac{x}{x^2 + 5}$

28. $\dfrac{dy}{dx} = \dfrac{7 - x^2}{x}$

Miscellaneous Integrals from the Table

29. $\displaystyle\int \frac{dt}{4 - 9t^2}$

30. $\displaystyle\int \frac{ds}{\sqrt{s^2 - 16}}$

31. $\displaystyle\int \sqrt{25 - 9x^2} \, dx$

32. $\displaystyle\int \sqrt{4x^2 + 9} \, dx$

33. $\displaystyle\int \frac{dx}{x^2 + 9}$

34. $\displaystyle\int \frac{dx}{x^2 + 2x}$

35. $\displaystyle\int \frac{dx}{16x^2 + 9}$

36. $\displaystyle\int x \sqrt{1 + 3x} \, dx$

37. $\displaystyle\int \frac{dx}{9 - 4x^2}$

38. $\displaystyle\int \sqrt{1 + 9x^2}\, dx$

39. $\displaystyle\int \frac{dx}{x^2 - 4}$

40. $\displaystyle\int \frac{dy}{\sqrt{25 - y^2}}$

41. $\displaystyle\int \sqrt{\frac{x^2}{4} - 1}\, dx$

42. $\displaystyle\int \frac{x^2\, dx}{\sqrt{3x + 5}}$

43. $\displaystyle\int \frac{5x\, dx}{\sqrt{1 - x^4}}$

30–3   Constant of Integration

Graphing the Integral

We have seen in the preceding sections that we get a constant of integration whenever we find an indefinite integral. We now give a geometric meaning to the constant of integration by graphing the integral curve.

◆◆◆ **Example 29:** Find the function whose derivative is $dy/dx = 2x$, and make a graph of the function.

Solution: Proceeding as in Sec. 27–6, we write the given differential equation in differential form, $dy = 2x\, dx$.

Integrating, we get

$$y = \int 2x\, dx = 2\left(\frac{x^2}{2}\right) + C = x^2 + C$$

From Sec. 22–3 we recognize this as a parabola opening upward. Further, its vertex is on the y axis at a distance of C units from the origin. When we graph this function as shown in Fig. 30–3, we get not a single curve, but a *family of curves,* each of which has a different value for C.   ◆◆◆

Boundary Conditions

In order for us to evaluate the constant of integration, more information must be given. Such additional information is called a *boundary condition* or *initial condition.*

The term *initial* is more appropriate when our variable is *time.*

◆◆◆ **Example 30:** Find the constant of integration in Example 29 if the curve whose derivative is $2x$ is to pass through the point $(2, 9)$.

Solution: The equation of the curve was found to be $y = x^2 + C$. Letting $x = 2$ and $y = 9$, we have

$$C = 9 - 2^2 = 5$$

as can be verified from Fig. 30–2.   ◆◆◆

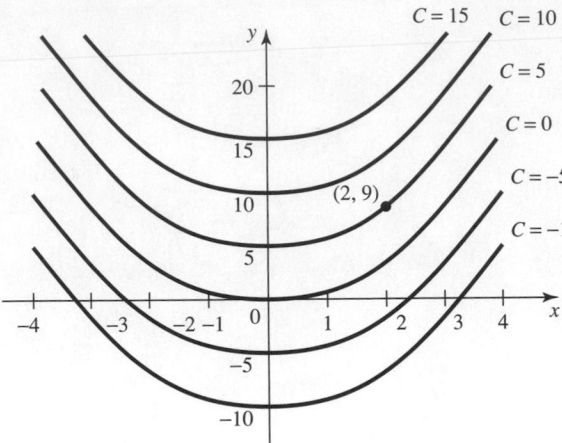

FIGURE 30–2   Family of curves for $y = x^2 + C$. Although we show integer values of C here, C need not be an integer.

In Chapter 28 we graphed a given function, and then *below it* we graphed the derivative of that function. Thus, the lower curve was the derivative of the upper curve. Conversely, the upper curve was the *integral* of the lower curve. Now let us graph a given function $f(x)$, and *above it* graph its integral $F(x)$. As before, the lower curve will be the derivative of the upper curve.

◆◆◆ **Example 31:** Graph the function $f(x) = 3x^2 - 4x - 5$ and its integral $F(x)$, taking the constant of integration equal to zero.

Solution: Integrating gives

$$F(x) = \int (3x^2 - 4x - 5)\, dx = x^3 - 2x^2 - 5x + C$$

We graph $f(x)$, and, aligned above it, we graph $F(x)$ with $C = 0$, as shown in Fig. 30–3. Our original curve $f(x)$ is the derivative of $F(x)$, and it is seen to behave just as we would expect of a derivative curve. Notice that $f(x)$ is positive when $F(x)$ is increasing, negative when $F(x)$ is decreasing, and zero at the maximum and minimum points on $F(x)$.

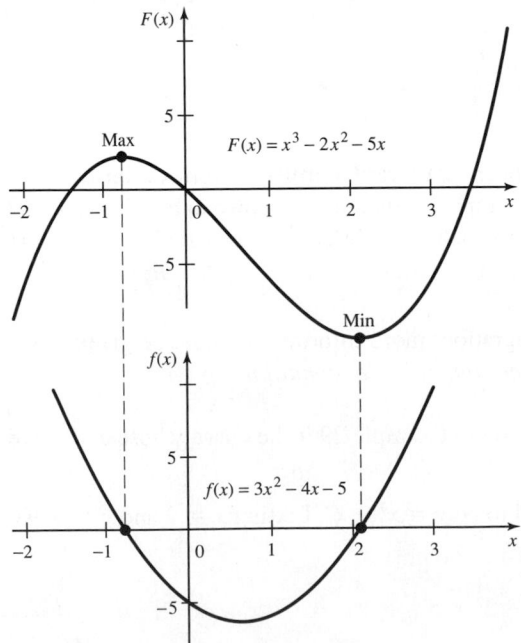

FIGURE 30–3   ◆◆◆

Successive Integration

If we are given the *second* derivative of some function and wish to find the function itself, we simply have to *integrate twice*. However, each time we integrate, we get another constant of integration. These constants can be found if we are given enough additional information, as in the following example.

✦✦✦ **Example 32:** Find the equation that has a second derivative $y'' = 4x - 2$ and that has a slope of 1 at the point (2, 9).

Solution: We can write the second derivative as

$$d(y')/dx = 4x - 2$$

or, in differential form,

$$d(y') = (4x - 2)\, dx$$

Integrating gives

$$y' = \int (4x - 2)\, dx = 2x^2 - 2x + C_1$$

But the slope, and hence y', is 1 when $x = 2$, so

$$C_1 = 1 - 2(2)^2 + 2(2) = -3$$

This gives $y' = 2x^2 - 2x - 3$ or, in differential form,

$$dy = (2x^2 - 2x - 3)\, dx$$

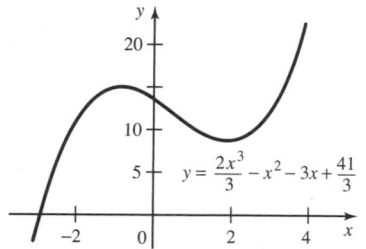

Integrating again gives

$$y - \int (2x^2 - 2x - 3)\, dx = 2x^3/3 - x^2 - 3x + C_2$$

But $y = 9$ when $x = 2$, so

$$C_2 = 9 - 2(2^3)/3 + 2^2 + 3(2) = 41/3$$

Our final equation, with all of the constants evaluated, is thus

$$y = 2x^3/3 - x^2 - 3x + 41/3$$

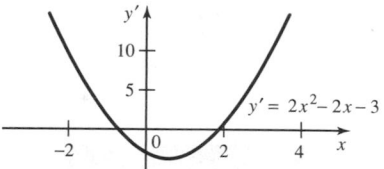

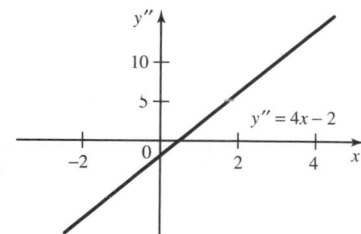

FIGURE 30–4

Figure 30–4 shows a graph of y, the first derivative y', and the second derivative y''. Note how the first derivative curve correctly identifies regions where y is increasing or decreasing and the extreme points. Also note how the second derivative curve locates the inflection point. ✦✦✦

Exercise 3 ✦ Constant of Integration

Families of Curves

Write the function that has the given derivative. Then graph that function for $C = -1$, $C = 0$, and $C = 1$.

1. $\dfrac{dy}{dx} = 3$

2. $\dfrac{dy}{dx} = 5x$

3. $\dfrac{dy}{dx} = 3x^2$

Constant of Integration

For problems 4 through 6, write the function that has the given derivative and passes through the given point.

4. $y' = 3x$, passes through $(2, 6)$

5. $y' = x^2$, passes through $(1, 1)$

6. $y' = \sqrt{x}$, passes through $(2, 4)$

7. If $dy/dx = 2x + 1$, and $y = 7$ when $x = 1$, find the value of y when $x = 3$.

8. If $dy/dx = \sqrt{2x}$, and $y = \frac{1}{3}$ when $x = \frac{1}{2}$, find the value of y when $x = 2$.

Successive Integration

9. Find the equation of a curve that passes through the point $(3, 0)$, has the slope $\frac{7}{2}$ at that point, and has a second derivative $y'' = x$.

10. Find the equation of the curve for which $y'' = 4$ if the curve is tangent to the line $y = 3x$ at $(2, 6)$.

11. Find the equation of a curve that passes through the point $(1, 0)$, is tangent to the line $6x + y = 6$ at that point, and has a second derivative $y'' = 12/x^3$.

12. The second derivative of a curve is $y'' = 12x^2 - 6$. The tangent to this curve at $(2, 4)$ is perpendicular to the line $x + 20y - 40 = 0$. Find the equation of the curve.

30–4   The Definite Integral

In Sec. 30–1 we learned how to find the indefinite integral, or antiderivative, of a function. For example,

$$\int x^2 \, dx = \frac{x^3}{3} + C \qquad (1)$$

We can, of course, evaluate the antiderivative at some particular value, say, $x = 6$. Substituting into Eq. (1) gives us

$$\int x^2 \, dx \Big|_{x=6} = \frac{6^3}{3} + C = 72 + C$$

Similarly, we can evaluate the same integral at, say, $x = 3$. Again substituting into (1) yields

$$\int x^2 \, dx \Big|_{x=3} = \frac{3^3}{3} + C = 9 + C$$

Suppose, now, that we subtract the second antiderivative from the first. We get

$$72 + C - (9 + C) = 63$$

Although we do not know the value of the constant C, we do know that it has the same value in both expressions since both were obtained from (1), so C will drop out when we subtract.

We now introduce new notation. We let

$$\int x^2 \, dx \Big|_{x=6} - \int x^2 \, dx \Big|_{x=3} = \int_3^6 x^2 \, dx$$

The expression on the right is called a *definite integral*. Here 6 is called the *upper limit*, and 3 is the *lower limit*. This notation tells us to evaluate the antiderivative at the upper limit and, from that, subtract the antiderivative evaluated at the lower limit. Notice that a definite integral (unlike the indefinite integral) has a *numerical* value, in this case, 63.

In general, if

$$\int f(x) \, dx = F(x) + C$$

But what is this number, and what is it good for? We'll soon see that the definite integral gives us the area under the curve $y = x^2$, from $x = 3$ to 6, and that it has many applications.

then

$$\int f(x)\, dx \bigg|_{x=a} = F(a) + C$$

$$\int f(x)\, dx \bigg|_{x=b} = F(b) + C$$

and

$$\int_a^b f(x)\, dx = F(b) + C - F(a) - C$$

The constants drop out, leaving the following equation:

Definite Integral	$\displaystyle\int_a^b f(x)\, dx = F(b) - F(a)$	**376**

We require, as usual, that the function $f(x)$ be *continuous* in the interval under consideration.

The definite integral of a function is equal to the antiderivative of that function evaluated at the upper limit b minus the antiderivative evaluated at the lower limit a.

Equation 376 is called the *fundamental theorem of calculus* because it connects the seemingly unrelated processes of differentiation and integration.

Evaluating a Definite Integral

To *evaluate* a definite integral, first integrate the expression (omitting the constant of integration), and write the upper and lower limits on a vertical bar or bracket to the right of the integral. Next substitute the upper limit and then the lower limit, and subtract.

◆◆◆ **Example 33:** Evaluate $\displaystyle\int_2^4 x^2\, dx$

Solution:

$$\int_2^4 x^2\, dx = \frac{x^3}{3}\bigg|_2^4$$

$$= \frac{4^3}{3} - \frac{2^3}{3} = \frac{56}{3} \qquad\qquad ◆◆◆$$

◆◆◆ **Example 34:** Evaluate $\displaystyle\int_0^3 (2x - 5)^3\, dx$.

Solution: We insert a factor of 2 in the integral and compensate with a factor of $\dfrac{1}{2}$ in front.

$$\int_0^3 (2x - 5)^3\, dx = \frac{1}{2}\int_0^3 (2x - 5)^3 (2\, dx)$$

$$= \frac{1}{2}\left(\frac{1}{4}\right)(2x - 5)^4 \big|_0^3$$

$$= (1/8)[(6 - 5)^4 - (0 - 5)^4]$$

$$= (1/8)[1 - 625] = -78 \qquad\qquad ◆◆◆$$

We have seen that integrals will sometimes produce expressions with absolute value signs, as in the following example.

◆◆◆ **Example 35:**

$$\int_{-3}^{-2} \frac{dx}{x} = \ln|x|\bigg|_{-3}^{-2} = \ln|-2| - \ln|-3|$$

The logarithm of a negative number is not defined. But here we are taking the logarithm of the *absolute value* of a negative number. Thus

$$\ln|-2| - \ln|-3| = \ln 2 - \ln 3 \approx -0.405 \qquad \blacklozenge\blacklozenge\blacklozenge$$

Discontinuity

Recall that a function is called *discontinuous* wherever there is a break, gap, corner, cusp, or jump. If a function is discontinuous between two limits a and b, then the definite integral is *not defined* over that interval.

◆◆◆ **Example 36:** The integral $\displaystyle\int_{-3}^{2} \frac{dx}{x}$ is not defined because the function $y = 1/x$ is discontinuous at $x = 0$. ◆◆◆

Common Error	Be sure that your function is continuous and differentiable between the given limits before evaluating a definite integral.

Exercise 4 ◆ The Definite Integral

Evaluate each definite integral to three significant digits.

1. $\displaystyle\int_{1}^{2} x \, dx$

2. $\displaystyle\int_{-2}^{2} x^2 \, dx$

3. $\displaystyle\int_{1}^{3} 7x^2 \, dx$

4. $\displaystyle\int_{-2}^{2} 3x^4 \, dx$

5. $\displaystyle\int_{0}^{4} (x^2 + 2x) \, dx$

6. $\displaystyle\int_{-2}^{2} x^2(x + 2) \, dx$

7. $\displaystyle\int_{2}^{4} (x + 3)^2 \, dx$

8. $\displaystyle\int_{0}^{a} (a^2x - x^3) \, dx$

9. $\displaystyle\int_{1}^{10} \frac{dx}{x}$

10. $\displaystyle\int_{1}^{e} \frac{dx}{x}$

11. $\displaystyle\int_{0}^{1} \frac{x \, dx}{4 + x^2}$

12. $\displaystyle\int_{-2}^{-3} \frac{2t \, dt}{1 + t^2}$

13. $\displaystyle\int_{0}^{1} \frac{dx}{\sqrt{3 - 2x^2}}$

14. $\displaystyle\int_{0}^{1} \frac{x \, dx}{\sqrt{2 - x^2}}$

30–5   Approximate Area Under a Curve

In this section we will find the approximate area under a curve both graphically and numerically. These methods are valuable not only as a lead-in to finding exact areas by integration, but also for finding areas under the many functions that cannot be integrated. Other approximation methods are given in Sec. 34–9.

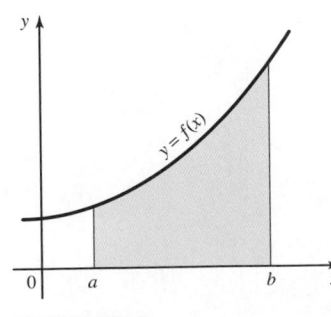

FIGURE 30–5

Estimating Areas

We want to find the approximate value of the area under a curve, such as $y = f(x)$ in Fig. 30–5, between two limits a and b. Our graphical approach is simple. After plotting the curve, we will subdivide the required area into rectangles, find the area of each rectangle, and add them.

◆◆◆ **Example 37:** Find the approximate area under the curve $y = x^2/3$ between the limits $x = 0$ and $x = 3$.

Solution: We draw the curve as shown in Fig. 30–6 between the upper and lower limits. We subdivide the required area into squares $\frac{1}{2}$ unit on a side, and we count them, estimating the fractional part of those that are incomplete. We count 12 squares, each with an area of $\frac{1}{4}$ square units, getting

We have chosen very large squares for this illustration. Of course, smaller squares would give greater accuracy.

$$\text{area} \cong 3 \text{ square units}$$

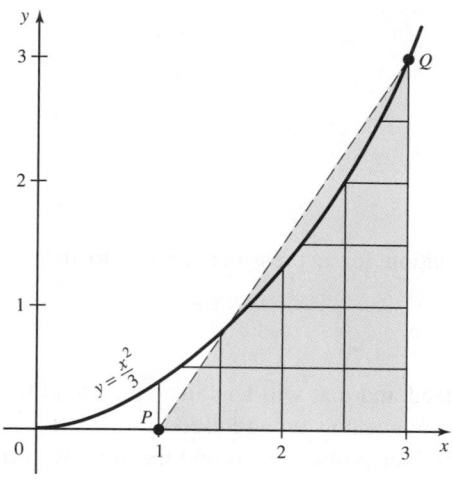

FIGURE 30–6 ◆◆◆

Another way to estimate the area under a curve is to simply sketch a rectangle or triangle of roughly the same area right on the graph of the given curve and compute its area.

◆◆◆ **Example 38:** Make a quick estimate of the area in Example 37.

Solution: We draw line PQ as shown in Fig. 30–6, trying to balance the excluded and included areas, and compute the area of the triangle formed.

$$\text{shaded area} \cong \text{area of } PQR = \frac{1}{2}(2)(3)$$
$$= 3 \text{ square units} \qquad ◆◆◆$$

Summation Notation

Before we derive an expression for the area under a curve, we must learn some new notation to express the sum of a string of terms. We use the Greek capital sigma Σ to stand for summation, or adding up. Thus

$$\sum n$$

means to sum a string of n's. Of course, we must indicate a starting and an ending value for n, and these values are placed on the sigma symbol. Thus

$$\sum_{n=1}^{5} n$$

means to add up the n's starting with $n = 1$ and ending with $n = 5$.

$$\sum_{n=1}^{5} n = 1 + 2 + 3 + 4 + 5 = 15$$

◆◆◆ **Example 39:** Evaluate $\displaystyle\sum_{n=1}^{4} (n^2 - 1)$.

Solution:

$$\sum_{n=1}^{4}(n^2 - 1) = (1^2 - 1) + (2^2 - 1) + (3^2 - 1) + (4^2 - 1)$$
$$= 0 + 3 + 8 + 15 = 26 \qquad \blacklozenge\blacklozenge\blacklozenge$$

◆◆◆ **Example 40:**

(a) $\displaystyle\sum_{k=2}^{5} k^2 = 2^2 + 3^2 + 4^2 + 5^2 = 54$

(b) $\displaystyle\sum_{x=1}^{4} f(x) = f(1) + f(2) + f(3) + f(4)$

(c) $\displaystyle\sum_{i=1}^{n} f(x_i) = f(x_1) + f(x_2) + f(x_3) + \cdots + f(x_n)$ ◆◆◆

In the following section, we use the sigma notation for expressions similar to that of Example 40(c).

A Numerical Technique: The Midpoint Method

We will now show a method that is easily computerized and that will lead us to a method for finding exact areas later. Also, the notation that we introduce now will be used again later.

Figure 30–7 shows a graph of some function $f(x)$. Our problem is to find the area (shown lightly shaded) bounded by that curve, the x axis, and the lines $x = a$ and $x = b$.

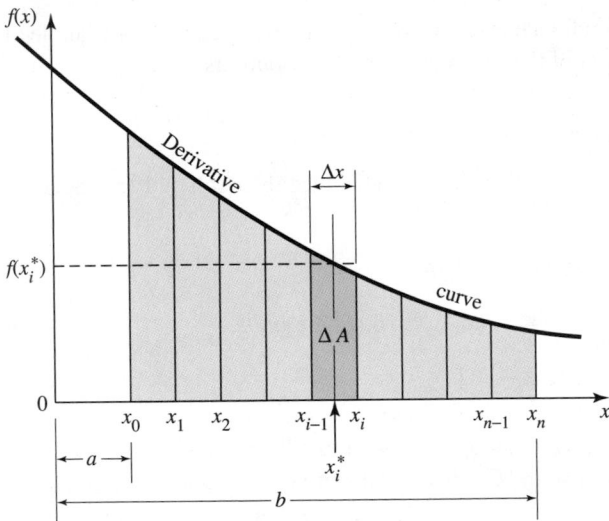

FIGURE 30-7

We start by subdividing that area into n vertical strips, called *panels,* by drawing vertical lines at $x_0, x_1, x_2, \ldots, x_n$. The panels do not have to be of equal width, but we make them equal for simplicity. Let the width of each panel be Δx.

Now look at one particular panel, the one lying between x_{i-1} and x_i (shown shaded darkly). At the midpoint of this panel we choose a point x_i^*. The height of the curve at this value of x is then $f(x_i^*)$. The area ΔA of the dark panel is then *approximately* equal to the area of a rectangle of width Δx and height $f(x_i^*)$.

$$\Delta A \cong f(x_i^*)\, \Delta x$$

Our approximate area might be greater than or less than the actual area ΔA, depending on where we chose x_i^*. (Later we show that x_i^* is chosen in a particular place, but for now, consider it to be anywhere between x_{i-1} and x_i.)

The area of the first panel is, similarly, $f(x_1^*) \Delta x$; of the second panel, $f(x_2^*) \Delta x$; and so on. To get an approximate value for the total area, we add up the areas of each panel.

$$A \cong f(x_1^*) \Delta x + f(x_2^*) \Delta x + f(x_3^*) \Delta x + \cdots + f(x_n^*) \Delta x$$

Rewriting this expression using our sigma notation gives

$$A \cong \sum_{i=1}^{n} f(x_i^*) \Delta x \qquad (381)$$

These are called *Riemann sums*, after Georg Friedrich Bernhard Riemann (1826–66).

Midpoint Method	$A \cong \sum_{i=1}^{n} f(x_i^*) \Delta x$ where $f(x_i^*)$ is the height of the i th panel at its midpoint	381

◆◆◆ **Example 41:** Use the midpoint method to calculate the approximate area under the curve $f(x) = 3x^2$ from $x = 0$ to 10, taking panels of width 2.

Solution: Our graph (Fig. 30–8) shows the panels, with midpoints at 1, 3, 5, 7, and 9. At each midpoint x^*, we compute the height $f(x^*)$ of the curve.

x^*	1	3	5	7	9
$f(x^*)$	3	27	75	147	243

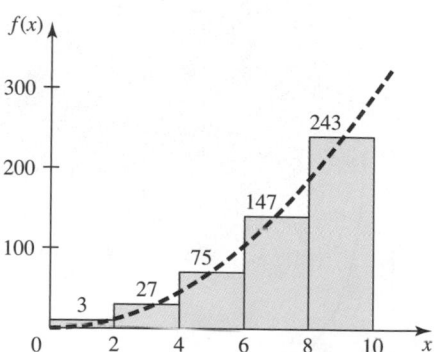

FIGURE 30–8 Area by midpoint method.

The approximate area is then the sum of the areas of each panel.

$$A \cong 3(2) + 27(2) + 75(2) + 147(2) + 243(2)$$

$$= 495(2) = 990 \text{ square units} \qquad ◆◆◆$$

We'll see later that the exact area is 1000 square units.

We can obtain greater accuracy in computing the area under a curve simply by reducing the width of each rectangular panel. Clearly, the panels in Fig. 30–9(b) are a better fit to the curve than those in Fig. 30–9(a).

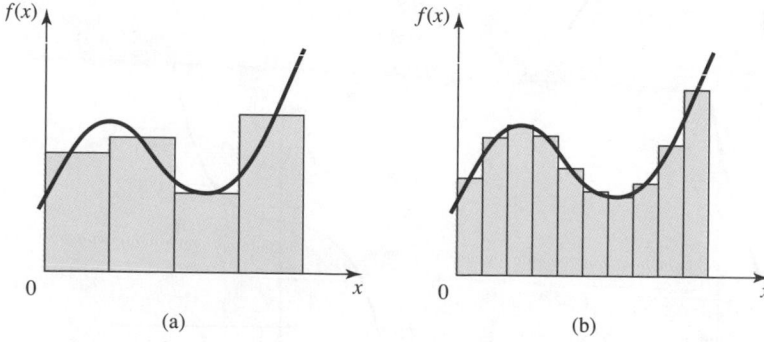

(a)   (b)

FIGURE 30–9 More panels give greater accuracy.

◆◆◆ **Example 42:** Compute the area under the curve in Example 41 by the midpoint method, using panel widths of 2, 1, $\frac{1}{2}$, $\frac{1}{4}$, and so on.

Solution: We compute the approximate area just as in Example 41. We omit the tedious computations (which were done by computer) and show only the results in Table 30–1.

TABLE 30–1

Panel Width	Area
2.0000	990.0000
1.0000	997.5000
0.5000	999.3750
0.2500	999.8438
0.1250	999.9609
0.0625	999.9902
0.0313	999.9968
0.0156	999.9996

Notice that as the panel width decreases, the computed area seems to be approaching a limit of 1000. We'll see in the next section that 1000 is the exact area under the curve. ◆◆◆

Exercise 5 ◆ Approximate Area Under a Curve

Estimation of Areas

Estimate the approximate area of each shaded region.

1.

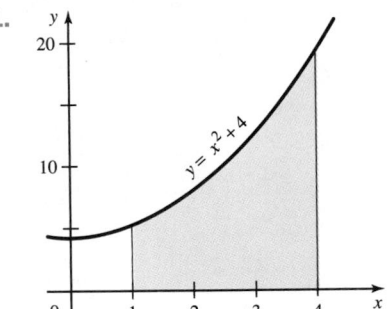

2.

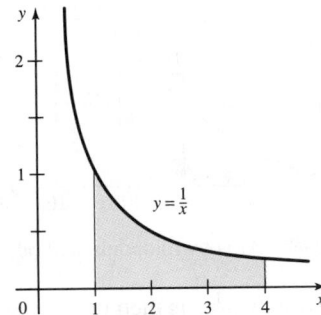

3.

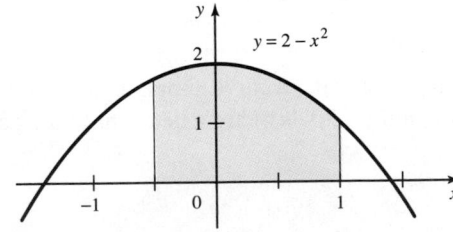

4.

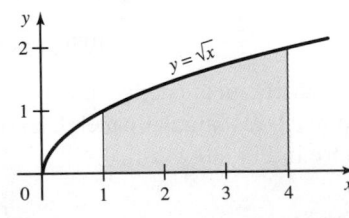

5.

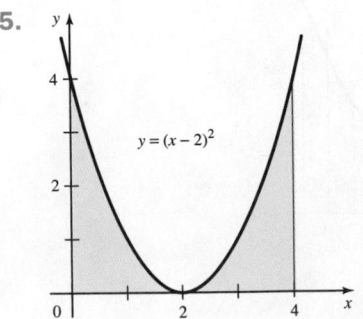

6.

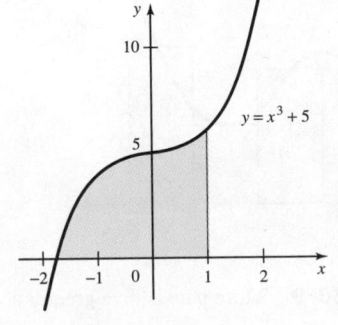

Graph each region. Make a quick estimate of the indicated area, and then use a graphical method to find its approximate value.

7. $y = x^2 + 1$ from $x = 0$ to 8
8. $y = x^2 + 3$ from $x = -4$ to 4
9. $y = 1/x$ from $x = 2$ to 10
10. $y = 2 + x^4$ from $x = -2$ to 0

Sigma Notation

Evaluate each expression.

11. $\displaystyle\sum_{n=1}^{5} n$

12. $\displaystyle\sum_{r=1}^{9} r^2$

13. $\displaystyle\sum_{n=1}^{7} 3n$

14. $\displaystyle\sum_{m=1}^{4} \frac{1}{m}$

15. $\displaystyle\sum_{n=1}^{5} n(n-1)$

16. $\displaystyle\sum_{q=1}^{6} \frac{q}{q+1}$

Approximate Areas by Midpoint Method

Using panels 2 units wide, find the approximate area (in square units) under each curve by the midpoint method.

17. $y = x^2 + 1$ from $x = 0$ to 8
18. $y = x^2 + 3$ from $x = -4$ to 4
19. $y = \dfrac{1}{x}$ from $x = 2$ to 10
20. $y = 2 + x^4$ from $x = -10$ to 0

30–6   Exact Area Under a Curve

We saw with the midpoint method that we could obtain a more accurate area by using a greater number of narrower panels. In the limit, we find the *exact* area by using an infinite number of panels of infinitesimal width.

Starting with Eq. 381,

$$A \cong \sum_{i=1}^{n} f(x_i^*) \, \Delta x$$

we let the panel width Δx approach zero. As it does so, the number of panels approaches infinity, and the sum of the areas of the panels approaches the exact area A under the curve.

Exact Area under a Curve	$A = \displaystyle\lim_{\Delta x \to 0} \sum_{i=1}^{n} f(x_i^*) \, \Delta x$	387

Exact Area by Integration

Equation 387 gives us the exact area under a curve, but not a practical way to find it. We will now derive a formula that will give us the exact area. The derivation will be long, but the formula itself will be very simple. In fact, it is one that you have already used.

Any function $f(x)$ can be regarded as the derivative of some other function $F(x)$. We get $F(x)$ by integrating $f(x)$, if possible. In fact, in Sec. 30–3 we integrated a function $f(x)$ in order to get $F(x)$, and we graphed $f(x)$ and $F(x)$ one above the other as shown in Fig. 30–4.

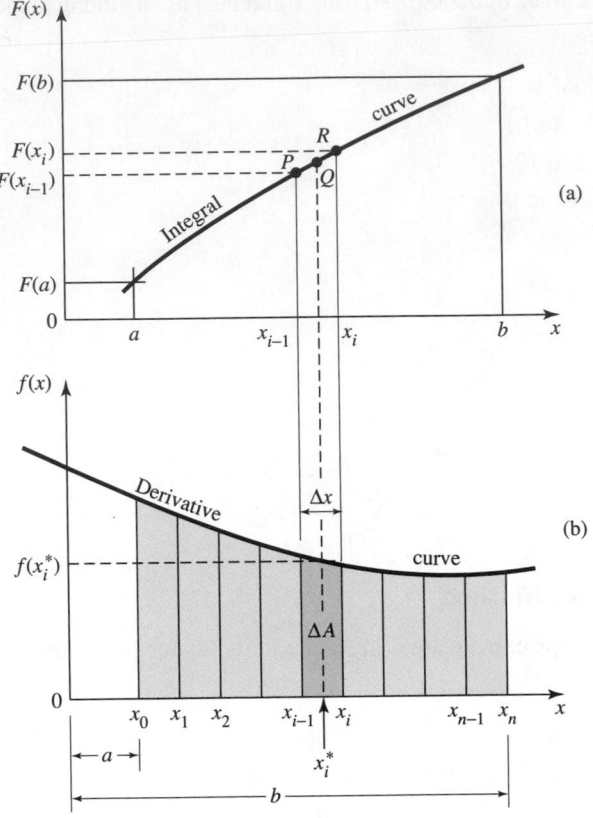

FIGURE 30–10   Area as the limit of a sum.

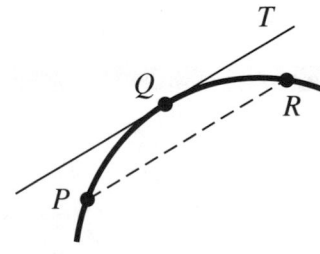

FIGURE 30–11   There is a theorem, called the *mean value theorem*, that says there must be at least one point Q between P and R at which the slope is equal to the slope of PR. We won't prove it, but can you see intuitively that it must be so?

Let us now graph a function $f(x)$ as shown in Fig. 30–10(b) with the area we wish to find between the limits a and b, shown shaded. But now we draw above it the curve $F(x)$ [Fig. 30–10(a)], assuming that such a curve exists. Thus the lower curve $f(x)$ is the derivative of the upper curve $F(x)$, and conversely, the upper curve $F(x)$ is the integral of the lower curve $f(x)$.

For the midpoint method, we arbitrarily selected x_i^* at the midpoint of each panel. We now do it differently. *We select x_i^* so that the slope at point Q on the integral (upper) curve is equal to the slope of the straight line PR,* as shown in Fig. 30–11.

The slope at Q is equal to $f(x_i^*)$, and the slope of PR is equal to

$$f(x_i^*) = \frac{\text{rise}}{\text{run}} = \frac{F(x_i) - F(x_i - 1)}{\Delta x}$$

or

$$f(x_i^*)\,\Delta x = F(x_i) - F(x_{i-1})$$

If we write this expression for each panel, we get

$$f(x_1^*)\,\Delta x = F(x_1) - F(a)$$
$$f(x_2^*)\,\Delta x = F(x_2) - F(x_1)$$
$$f(x_3^*)\,\Delta x = F(x_3) - F(x_2)$$
$$\vdots$$
$$f(x_{n-1}^*)\,\Delta x = F(x_{n-1}) - F(x_{n-2})$$
$$f(x_n^*)\,\Delta x = F(b) - F(x_{n-1})$$

If we add all of these equations, every term on the right drops out except $F(a)$ and $F(b)$.

$$f(x_1^*)\,\Delta x + f(x_2^*)\,\Delta x + f(x_3^*)\,\Delta x + \cdots + f(x_n^*)\,\Delta x = F(b) - F(a)$$

$$\sum_{i=1}^{n} f(x_i^*)\,\Delta x = F(b) - F(a)$$

As before, we let Δx approach zero.

$$\lim_{\Delta x \to 0} \sum_{i=1}^{n} f(x_i^*)\,\Delta x = F(b) - F(a)$$

The left side of this equation is equal to the exact area A under the curve (Eq. 387). The right side is equal to the definite integral from a to b of the function $f(x)$ in Eq. 376. Thus:

Exact Area under a Curve	$A = \displaystyle\int_a^b f(x)\,dx = F(b) - F(a)$	388

We get the amazingly simple result that the area under a curve between two limits is equal to the change in the integral between the same limits, as shown graphically in Fig. 30–12.

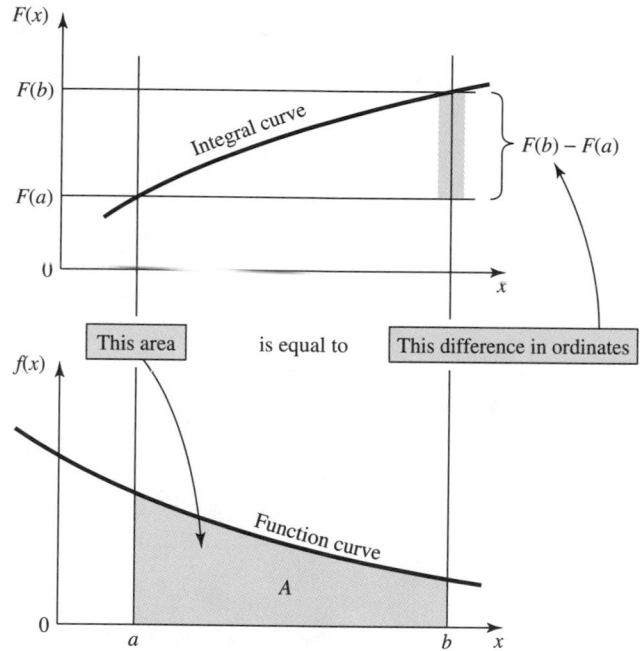

FIGURE 30–12

So to find the exact area under a curve, simply evaluate the definite integral of the given function between the given limits.

◆◆◆ **Example 43:** Find the area bounded by the curve $y = 3x^2 + x + 1$, the x axis, and the lines $x = 2$ and $x = 5$.
Solution: By Eq. 388,

$$A = \int_2^5 (3x^2 + x + 1)\,dx = x^3 + \frac{x^2}{2} + x \Big|_2^5$$

$$= \left(5^3 + \frac{5^2}{2} + 5\right) - \left(2^3 + \frac{2^2}{2} + 2\right)$$

$$= 130.5 \text{ square units}$$

◆◆◆

Exercise 6 ◆ Exact Area under a Curve

Find the area (in square units) bounded by each curve, the given lines, and the x axis.

1. $y = 2x$ from $x = 0$ to 10

2. $y = x^2 + 1$ from $x = 1$ to 20

3. $y = 3 + x^2$ from $x = -5$ to 5

4. $y = x^4 + 4$ from $x = -10$ to -2

5. $y = x^3$ from $x = 0$ to 4

6. $y = 9 - x^2$ from $x = 0$ to 3

7. $y = 1/\sqrt{x}$ from $x = \frac{1}{2}$ to 8

8. $y = x^3 + 3x^2 + 2x + 10$ from $x = -3$ to 3

9. $y = x^2 + x + 1$ from $x = 2$ to 3

10. $y = \sqrt{3x}$ from $x = 2$ to 8

11. $y = 2x + \dfrac{1}{x^2}$ from $x = 1$ to 4

12. $y = \dfrac{10}{\sqrt{x + 4}}$ from $x = 0$ to 5

13. $y = 3x^2$ from $x = 0$ to 10.

Sketch the curve for some of these, and try to make a quick estimate of the area. Also check some graphically.

We are doing only simple area problems here. In the following chapter we will find areas between curves, areas below the x axis, and other more complicated types of problems.

◆◆◆ CHAPTER 30 REVIEW PROBLEMS ◆◆◆◆◆◆◆◆◆◆◆◆◆◆◆◆◆◆◆◆◆◆◆◆◆◆◆◆◆◆◆

Perform each integration.

1. $\displaystyle\int \dfrac{dx}{\sqrt[3]{x}}$

2. $\displaystyle\int \dfrac{x^4 + x^3 + 1}{x^3}\, dx$

3. $\displaystyle\int 3.1\, y^2\, dy$

4. $\displaystyle\int \dfrac{2dt}{t^2}$

5. $\displaystyle\int \sqrt{4x}\, dx$

6. $\displaystyle\int x^2(x^3 - 4)^2\, dx$

7. $\displaystyle\int (x^4 - 2x^3)(2x^3 - 3x^2)\, dx$

8. $\displaystyle\int \dfrac{x\, dx}{x^2 + 3}$

9. $\displaystyle\int \dfrac{dx}{x + 5}$

10. $\displaystyle\int_{-1}^{0} \dfrac{dx}{1 - x}$

11. $\displaystyle\int_{2}^{7} (x^2 - 2x + 3)\, dx$

12. $\displaystyle\int_{0}^{2} \dfrac{x\, dx}{4 + x^2}$

13. $\displaystyle\int_{0}^{a} (\sqrt{a} - \sqrt{x})^2\, dx$

14. $\displaystyle\int_{1}^{2} \sqrt{7 - 3x}\, dx$

15. $\displaystyle\int_{0}^{3} \sqrt{9 + 25x^2}\, dx$

16. $\displaystyle\int_{0}^{2} \sqrt{x^2 + 25}\, dx$

17. $\displaystyle\int_{1}^{4} x\sqrt{1 + 5x}\, dx$

18. $\displaystyle\int_0^2 \sqrt{1 + 9x^2}\, dx$

19. $\displaystyle\int \frac{dx}{9 - 4x^2}$

20. $\displaystyle\int \frac{dx}{\sqrt{x^2 - 25}}$

21. $\displaystyle\int \frac{dx}{9 + 4x^2}$

22. $\displaystyle\int_1^4 \frac{dx}{x^2 + 4x}$

23. $\displaystyle\int_2^6 \frac{dx}{\sqrt{4 + 9x^2}}$

24. Find the equation of a curve that passes through the point (3, 0), has a slope of 9 at that point, and has a second derivative $y'' = x$.

25. The rate of growth of the number N of bacteria in a culture is $dN/dt = 0.5N$. If $N = 100$ when $t = 0$, derive the formula for N at any time.

26. Evaluate $\displaystyle\sum_{n=1}^{5} n^2(n - 1)$.

27. Evaluate $\displaystyle\sum_{d=1}^{4} \frac{d^2}{d + 1}$.

28. Use the midpoint method to find the approximate area under the curve $y = 5 + x^2$ from $x = 1$ to 9. Use panels 2 units wide.

29. Find the positive area bounded by the parabola $y^2 = 8x$, the x axis, and the line $x = 2$.

Writing

30. Integration is the inverse of differentiation. List as many other pairs of inverse mathematical operations as you can. Describe in your own words when the inverse of each operation gives an indefinite result.

Team Project

31. Given a function by your instructor that can be integrated, and an upper and lower limit, (a) make an accurate graph, and estimate the area of the given region by counting boxes on the graph paper, (b) use the midpoint method with different panel widths to calculate the same area, and (c) find the exact area by integrating. Compare the results obtained by the various methods.

31

Applications of the Integral

◆◆◆ **OBJECTIVES** ◆◆

When you have completed this chapter, you should be able to:
- Calculate the motion of a point.
- Use the integral to solve problems in electric circuits.
- Calculate the area of curved objects.
- Calculate the volume of curved objects.

◆◆

We begin this chapter with two applications of the indefinite integral: *motion* and *electric circuits*. We'll see that these problems are the reverse of those we studied for the derivative. For example, we took the derivative of a displacement equation to get velocity; here we take the integral of a velocity equation to get displacement.

In Chapter 30 we defined the definite integral in terms of the area under a curve, and we used the definite integral to compute simple plane areas. In this chapter we will learn a fast way to set up the integral and use it to find more complex areas, such as the area between two curves.

Finally we will compute volumes of solids of revolution. We will set up our integral using the shortcut method of Sec. 31–3; that is, we define a small element of the volume and then sum up all such elements by integration.

This chapter contains just our first batch of applications of the integral. In the following chapter we will compute surface area, length of arc, moments and centroids of areas, force due to fluid pressure, work required for various tasks, and moment of inertia for an area.

31–1   Applications to Motion

Displacement and Velocity

In Sec. 29–2 we saw that the velocity v of a moving point was defined as the rate of change of the displacement s of the point. The velocity was thus equal to the derivative of the displacement, or $v = ds/dt$. We now reverse the process and find the displacement when given the velocity. Since $ds = v\, dt$, integrating gives the following equation:

| Displacement | $s = \int v\, dt$ | A22 |

Similarly, the acceleration a is given by dv/dt, so $dv = a\, dt$. Integrating we get the following:

| Instantaneous Velocity | $v = \int a\, dt$ | A24 |

Thus, if given an equation for acceleration, we can integrate to get an equation for velocity and integrate again to get displacement. We evaluate the constants of integration by substituting boundary conditions, as shown in the following example.

♦♦♦ **Example 1:** A particle moves with a constant acceleration of 4 m/s². It has an initial velocity of 6 m/s and an initial displacement of 2 m. Find the equations for velocity and displacement, and graph the displacement, velocity, and acceleration for $t = 0$ to 10 s.

Solution: We are given $a = dv/dt = 4$, so

$$v = \int 4\, dt = 4t + C_1$$

Since $v = 6$ m/s when $t = 0$, we get $C_1 = 6$, so

$$v = 4t + 6 \text{ m/s}$$

is our equation for the velocity of the particle. Now, since $v = ds/dt$,

$$ds = (4t + 6)\, dt$$

Integrating again gives us

$$s = \frac{4t^2}{2} + 6t + C_2$$

Since the initial displacement is 2 m when $t = 0$, we get $C_2 = 2$. The complete equation for the displacement of the particle is then

$$s = 2t^2 + 6t + 2 \text{ m}$$

The three curves are graphed one above the other in Fig. 31–1 with the same scale used for the horizontal axis of each. Note that each curve is the derivative of the one below it, and conversely, each curve is the integral of the one above it.

Check: We can check our work by taking derivatives. Thus

$$v = ds/dt = 4t + 6$$

and

$$a = dv/dt = 4$$

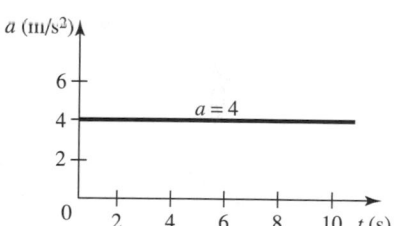

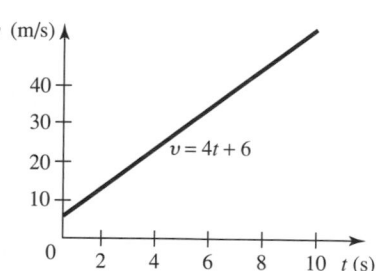

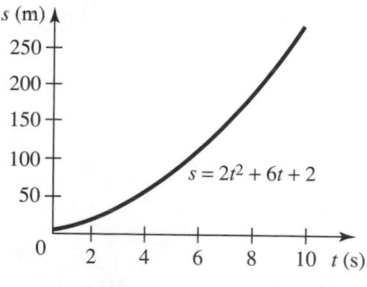

FIGURE 31–1

We could also use a graphing calculator to graph the displacement equation and its first and second derivatives and compare them with Fig. 31–1.   ◆◆◆

Freely Falling Body

Integration provides us with a slick way to derive the equations for the displacement and velocity of a freely falling body.

◆◆◆ **Example 2:** An object falls with constant acceleration, g, due to gravity. Write the equations for the displacement and velocity of the body at any time.

Solution: We are given that $a = dv/dt = g$, so

$$dv = g \, dt$$

Integrating, we find that

$$v = \int g \, dt = gt + C_1$$

When $t = 0$, $v = C_1$, so C_1 is the initial velocity. Let us relabel the constant v_0.

Velocity at Time t	$v = v_0 + gt$	A19

But $v = ds/dt$, so

$$ds = (v_0 + gt) \, dt$$

$$s = \int (v_0 + gt) \, dt = v_0 t + \frac{gt^2}{2} + C_2$$

When $t = 0$, $s = C_2$, so we interpret C_2 as the initial displacement. Let us call it s_0. So the displacement is the following:

Displacement at Time t	$s = s_0 + v_0 t + \frac{gt^2}{2}$	A18

   ◆◆◆

Motion along a Curve

In Sec. 29–2 the motion of a point along a curve was described by parametric equations, with the x and y displacements each given by a separate function of time. We saw that dx/dt gave the velocity v_x in the x direction and that dy/dt gave the velocity v_y in the y direction. Now, given the velocities, we integrate to get the displacements.

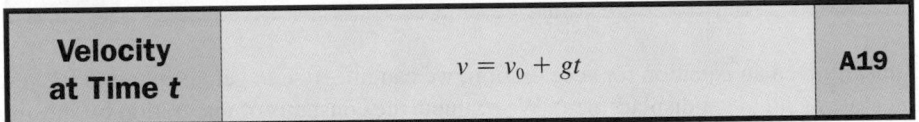

	(a)	(b)	
Displacement in x and y Directions	$x = \int v_x \, dt$	$y = \int v_y \, dt$	A32

Similarly, if we have parametric equations for the accelerations in the x and y directions, we integrate to get the velocities.

	(a)	(b)	
Velocity in x and y Directions	$v_x = \int a_x \, dt$	$v_y = \int a_y \, dt$	A34

◆◆◆ **Example 3:** An object starts from $(2, 4)$ with initial velocities $v_x = 7$ cm/s and $v_y = 5$ cm/s, and it moves along a curved path. It has x and y accelerations of $a_x = 3t$ cm/s² and $a_y = 5$ cm/s². Write expressions for the x and y components of velocity and displacement.

Solution: We integrate to find the velocities.

$$v_x = \int 3t\, dt = 3t^2/2 + C_1 \quad \text{and} \quad v_y = \int 5\, dt = 5t + C_2$$

At $t = 0$, $v_x = 7$ and $v_y = 5$, so

$$v_x = \frac{3t^2}{2} + 7 \text{ cm/s} \quad \text{and} \quad v_y = 5t + 5 \text{ cm/s}$$

Integrating again gives the displacements.

$$x = \int \left(\frac{3t^2}{2} + 7\right) dt \quad \text{and} \quad y = \int (5t + 5)\, dt$$

$$= \frac{t^3}{2} + 7t + C_3 \qquad\qquad = \frac{5t^2}{2} + 5t + C_4$$

At $t = 0$, $x = 2$ and $y = 4$, so our complete equations for the displacements are

$$x = \frac{t^3}{2} + 7t + 2 \text{ cm} \quad \text{and} \quad y = \frac{5t^2}{2} + 5t + 4 \text{ cm} \qquad ◆◆◆$$

Rotation

In Sec. 29–2 we saw that the angular velocity ω of a rotating body was given by the derivative $d\theta/dt$ of the angular displacement θ. Thus, θ is the integral of the angular velocity.

Angular Displacement	$\theta = \int \omega\, dt$	A28

Similarly, the angular velocity is the integral of the angular acceleration α.

Angular Velocity	$\omega = \int \alpha\, dt$	A30

◆◆◆ **Example 4:** A flywheel in a machine starts from rest and accelerates at $3.85t$ rad/s². Find the angular velocity and the total number of revolutions after 10.0 s.

Solution: We integrate to get the angular velocity.

$$\omega = \int 3.85t\, dt = \frac{3.85\, t^2}{2} + C_1 \text{ rad/s}$$

Since the flywheel starts from rest, $\omega = 0$ at $t = 0$, so $C_1 = 0$. Integrating again gives the angular displacement.

$$\theta = \int \frac{3.85\, t^2}{2}\, dt = \frac{3.85\, t^3}{6} + C_2 \text{ rad}$$

Since θ is 0 at $t = 0$, we get $C_2 = 0$. Now evaluating ω and θ at $t = 10.0$ s yields

$$\omega = \frac{3.85(10.0)^2}{2} = 192 \text{ rad/s}$$

and

_____

Recall that 2π radians equals 1 revolution.

$$\theta = \frac{3.85(10.0)^3}{6} = 642 \text{ rad} = 102 \text{ revolutions}$$ ◆◆◆

Exercise 1 ◆ Applications to Motion

Displacement and Velocity

1. The acceleration of a point is given by $a = 4 - t^2$ m/s². Write an equation for the velocity if $v = 2$ m/s when $t = 3$ s.

2. A car starts from rest and continues at a rate of $v = \frac{1}{8} t^2$ m/s. Find the function that relates the distance s the car has travelled to the time t in seconds. How far will the car go in 4 s?

3. A particle starts from the origin. Its x component of velocity is $t^2 - 4$, and its y component is $4t$ cm/s.
   (a) Write equations for x and y.
   (b) Find the distance between the particle and the origin when $t = 2.0$ s.

4. A body is moving at the rate $v = \frac{3}{2}t^2$ m/s. Find the distance that it will move in t seconds if $s = 0$ when $t = 0$.

5. The acceleration of a falling body is $a = -32$ ft./s². Find the relation between s and t if $s = 0$ and $v = 20$ ft./s when $t = 0$.

Motion along a Curve

6. A point starts from rest at the origin and moves along a curved path with x and y accelerations of $a_x = 2.00$ cm/s² and $a_y = 8.00t$ cm/s². Write expressions for the x and y components of velocity.

7. A point starts from rest at the origin and moves along a curved path with x and y accelerations of $a_x = 5.00t^2$ cm/s² and $a_y = 2.00t$ cm/s². Find the x and y components of velocity at $t = 10.0$ s.

8. A point starts from (5, 2) with initial velocities of $v_x = 2.00$ cm/s and $v_y = 4.00$ cm/s and moves along a curved path. It has x and y accelerations of $a_x = 7t$ and $a_y = 2$. Find the x and y displacements at $t = 5.00$ s.

9. A point starts from (9, 1) with initial velocities of $v_x = 6.00$ cm/s and $v_y = 2.00$ cm/s and moves along a curved path. It has x and y accelerations of $a_x = 3t$ and $a_y = 2t^2$. Find the x and y components of velocity at $t = 15.0$ s.

Rotation

10. A wheel starts from rest and accelerates at 3.00 rad/s². Find the angular velocity after 12.0 s.

11. A certain gear starts from rest and accelerates at $8.5t^2$ rad/s². Find the total number of revolutions after 20.0 s.

12. A link in a mechanism rotating with an angular velocity of 3.00 rad/s is given an acceleration of 5.00 rad/s² at $t = 0$. Find the angular velocity after 20.0 s.

13. A pulley in a magnetic tape drive is rotating at 1.25 rad/s when it is given an acceleration of 7.24 rad/s² at $t = 0$. Find the angular velocity at 2.00 s.

31–2 Applications to Electric Circuits

Charge

We stated in Sec. 29–1 that the current i (amperes, A) at some point in a conductor was equal to the time rate of change of the charge q (coulombs, C) passing that point, or

You may want to refer to Sec. 29–1 before starting this section.

$$i = \frac{dq}{dt}$$

We can now solve this equation for q. Multiplying by dt gives

$$dq = i\, dt$$

Integrating, we get the following:

Charge	$q = \int i\, dt$ C	A80

◆◆◆ **Example 5:** The current to a certain capacitor is given by $i = 2t^3 + t^2 + 3$. The initial charge on the capacitor is 6.83 C. Find (a) an expression for the charge on the capacitor and (b) the charge when $t = 5.00$ s.

Solution:

(a) Integrating the expression for current, we obtain

We will use the letter k for the constant of integration in electrical problems to avoid confusion with C for capacitance.

$$q = \int i\, dt = \int (2t^3 + t^2 + 3)\, dt = \frac{t^4}{2} + \frac{t^3}{3} + 3t + k \text{ C}$$

We find the constant of integration by substituting the initial conditions, $q = 6.83$ C at $t = 0$. So $k = 6.83$ C. Our complete equation is then

$$q = \frac{t^4}{2} + \frac{t^3}{3} + 3t + 6.83 \text{ C}$$

(b) When $t = 5.00$ s,

$$q = \frac{(5.00)^4}{2} + \frac{(5.00)^3}{3} + 3(5.00) + 6.83 = 376 \text{ C}$$

◆◆◆

Voltage across a Capacitor

The current in a capacitor has already been given by Eq. A81, $i = C\, dv/dt$, where i is in amperes (A), C in farads (F), v in volts (V), and t in seconds (s). We now integrate to find the voltage across the capacitor.

$$dv = \frac{1}{C}\, i\, dt$$

Voltage across a Capacitor	$v = \frac{1}{C} \int i\, dt$ V	A82

◆◆◆ **Example 6:** A 1.25-F capacitor that has an initial voltage of 25.0 V is charged with a current that varies with time according to the equation $i = t \sqrt{t^2 + 6.83}$. Find the voltage across the capacitor at 1.00 s.

Solution: By Eq. A82,

$$v = \frac{1}{1.25} \int t \sqrt{t^2 + 6.83}\, dt = 0.80 \left(\frac{1}{2}\right) \int (t^2 + 6.38)^{1/2} (2t\, dt)$$

$$= \frac{0.40(t^2 + 6.83)^{3/2}}{3/2} + k = 0.267(t^2 + 6.83)^{3/2} + k$$

where we have again used k for the constant of integration to avoid confusion with the symbol for capacitance. Since $v = 25.0$ V when $t = 0$, we get

$$k = 25.0 - 0.267(6.83)^{3/2} = 20.2 \text{ V}$$

When $t = 1.00$ s,

$$v = 0.267(1.00^2 + 6.83)^{3/2} + 20.2 = 26.0 \text{ V} \qquad \text{◆◆◆}$$

Current in an Inductor

The voltage across an inductor was given by Eq. A87 as $v = L \, di/dt$, where L is the inductance in henrys (H). From this equation we get the following:

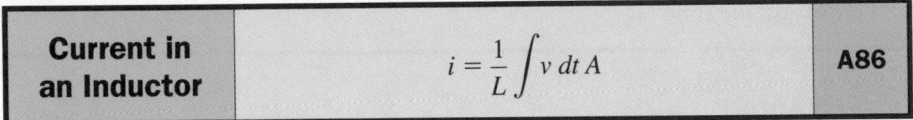

Current in an Inductor	$i = \dfrac{1}{L} \displaystyle\int v \, dt \text{ A}$	**A86**

◆◆◆ **Example 7:** The voltage across a 10.6-H inductor is $v = \sqrt{3t + 25.4}$ V. Find the current in the inductor at 5.25 s if the initial current is 6.15 A.

Solution: From Eq. A86,

$$i = \frac{1}{10.6}\int \sqrt{3t + 25.4} \; dt = 0.09434\left(\frac{1}{3}\right)\int (3t + 25.4)^{1/2}(3dt)$$

$$= \frac{0.03145(3t + 25.4)^{3/2}}{3/2} + k = 0.0210(3t + 25.4)^{3/2} + k$$

When $t = 0$, $i = 6.15$ A, so

$$k = 6.15 - 0.0210(25.4)^{3/2} = 3.46 \text{ A}$$

When $t = 5.25$ s,

$$i = 0.0210[3(5.25) + 25.4]^{3/2} + 3.46 = 9.00 \text{ A} \qquad \text{◆◆◆}$$

Exercise 2 ◆ Applications to Electric Circuits

1. The current to a capacitor is given by $i = 2t + 3$. The initial charge on the capacitor is 8.13 C. Find the charge when $t = 1.00$ s.

2. The current to a certain circuit is given by $i = t^2 + 4$. If the initial charge is zero, find the charge at 2.50 s.

3. The current to a certain capacitor is $i = 3.25 + t^3$. If the initial charge on the capacitor is 16.8 C, find the charge when $t = 3.75$ s.

4. A 21.5-F capacitor with zero initial voltage has a charging current of $i = \sqrt{t}$. Find the voltage across the capacitor at 2.00 s.

5. A 15.2-F capacitor has an initial voltage of 2.00 V. It is charged with a current given by $i = t\sqrt{5 + t^2}$. Find the voltage across the capacitor at 1.75 s.

6. A 75.0-μF capacitor has an initial voltage of 125 V and is charged with a current equal to $i = \sqrt{t + 16.3}$ mA. Find the voltage across the capacitor at 4.00 s.

7. The voltage across a 1.05-H inductor is $v = \sqrt{23t}$ V. Find the current in the inductor at 1.25 s if the initial current is zero.

8. The voltage across a 52.0-H inductor is $v = t^2 - 3t$ V. If the initial current is 2.00 A, find the current in the inductor at 1.00 s.

9. The voltage across a 15.0-H inductor is given by $v = 28.5 + \sqrt{6t}$ V. Find the current in the inductor at 2.50 s if the initial current is 15.0 A.

31–3 Finding Areas by Means of the Definite Integral

A Fast Way to Set Up the Integral

In Sec. 30–6 we found areas bounded by simple curves and the x axis. In this section we go on to find areas bounded by the y, rather than the x, axis, and areas *between* curves. Later, we find volumes, surface areas, and so on, where Eq. 376 cannot be used directly. We will have to set up a *different integral each time*. Looking back at the work we had to do to get Eq. 376 makes us wish for an easier way to set up an integral. We showed in Sec. 30–5 that the definite integral can be thought of as the sum of many small elements of area. This provides us with an intuitive shortcut for setting up an integral without having to go through a long derivation each time.

Think of the integral sign as the letter S, standing for *sum*. It indicates that we are to *add up* the elements that are written after the integral sign. Thus

$$A = \int_a^b f(x)\,dx$$

can be read, "The area A is the sum of all of the elements having a height $f(x)$ and a width dx, between the limits of a and b," as shown in Fig. 31–2.

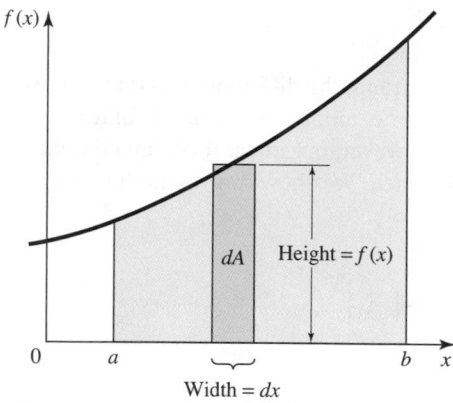

FIGURE 31–2

◆◆◆ **Example 8:** Find the area bounded by the curve $y = x^2 + 3$, the x axis, and the lines $x = 1$ and $x = 4$ (Fig. 31–3).

Estimate: Let us enclose the given area in a rectangle of width 3 and height 19, shown dashed in the figure. We see that the given area occupies more than half the area of the rectangle, say, about 60%. Thus a reasonable estimate of the shaded area would then be 60% of (3)(19), or 34 square units.

Solution: The usual steps are as follows:

1. Make a sketch showing the bounded area, as in Fig. 31–3. Locate a point (x, y) on the curve.
2. Through (x, y) draw a rectangular element of area, which we call dA. Give the rectangle dimensions. We call the width dx because it is measured in the x direction, and we call the height y. The area of the element is thus

$$dA = y\,dx = (x^2 + 3)\,dx$$

3. We think of A as the sum of all the small dA's. We accomplish the summation by integration.

This looks like a long procedure, but it will save us a great deal of time later.

$$A = \int dA = \int (x^2 + 3)\,dx$$

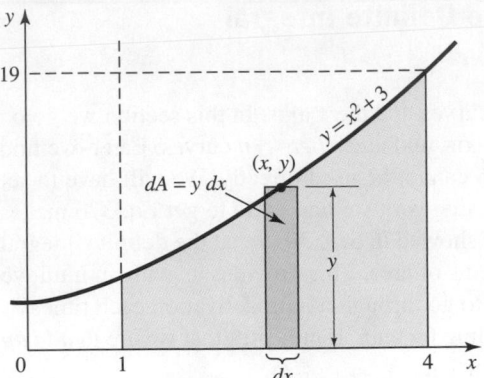

FIGURE 31-3

4. We locate the limits of integration from the figure. Since we are summing the elements in the x direction, our limits must be on x. It is clear that we start the summing at $x = 1$ and end it at $x = 4$. So

$$A = \int_1^4 (x^2 + 3)\, dx$$

5. Check that all parts of the integral, including the integrand, the differential, and the limits of integration, are in terms of the *same variable*. In our example, everything is in terms of x, and the limits are on x, so we can proceed. If, however, our integral contained both x and y, one of the variables would have to be eliminated. We show how to do this in a later example.

6. Our integral is now set up. Integrating gives

$$A = \int_1^4 (x^2 + 3)\, dx = \frac{x^3}{3} + 3x \Big|_1^4$$

$$= \frac{4^3}{3} + 3(4) - \left[\frac{1^3}{3} + 3(1) \right] = 30 \text{ square units} \qquad \blacklozenge\blacklozenge\blacklozenge$$

Area between two Curves

Suppose that we want the area A bounded by an upper curve $y = f(x)$ and a lower curve $y = g(x)$, between the limits a and b (Fig. 31–4). We draw a vertical element whose width is dx, whose height is $f(x) - g(x)$, and whose area dA is

$$dA = [f(x) - g(x)]\, dx$$

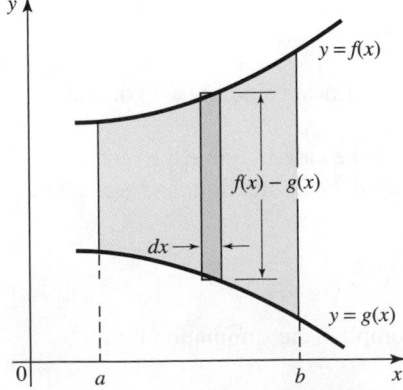

FIGURE 31-4   Area between two curves.

Integrating from a to b gives the total area.

Area between Two Curves	$$A = \int_b^a [f(x) - g(x)]\,dx$$	389

It is important to get *positive* lengths for the elements. To do this, be sure to properly identify the "upper curve" for the region and subtract from it the values on the "lower curve."

◆◆◆ **Example 9:** Find the first-quadrant area bounded by the curves

$$y = x^2 + 3 \qquad y = 3x - x^2 \qquad x = 0 \qquad x = 3$$

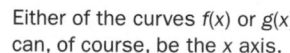
Either of the curves $f(x)$ or $g(x)$ can, of course, be the x axis.

Estimate: We make a sketch as shown in Fig. 31–5, and we "box in" the given area in a 3×12 rectangle, shown dashed, whose area is 36 square units. Let us estimate that the given area is less than half of that, or less than 18 square units.

Solution: Letting $f(x) = x^2 + 3$ and $g(x) = 3x - x^2$, we get

$$f(x) - g(x) = (x^2 + 3) - (3x - x^2) = 2x^2 - 3x + 3$$

The area bounded by the curves is then

$$A = \int_0^3 (2x^2 - 3x + 3)\,dx = \frac{2x^3}{3} - \frac{3x^2}{2} + 3x$$

$$= 18 - \frac{27}{2} + 9 = 13.5 \text{ square units}$$

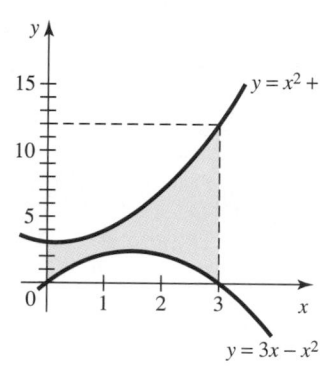
◆◆◆ **FIGURE 31–5**

Don't worry if all or part of the desired area is below the x axis. Just follow the same procedure as when the area is above the axis, and the signs will work out by themselves. Be sure, however, that the lengths of the elements are *positive* by subtracting the lower curve from the upper curve.

◆◆◆ **Example 10:** Find the area bounded by the curves $y = \sqrt{x}$ and $y = x - 3$ between $x = 1$ and 4 (Fig. 31–6).

Estimate: Here our enclosing rectangle, shown dashed, is 3 by 4 units, with an area of 12 square units. The shaded area is about half of that, or about 6 square units.

Solution: Letting $f(x) = \sqrt{x}$ and $g(x) = x - 3$, we get

$$A = \int_a^b [f(x) - g(x)]\,dx$$

$$= \int_1^4 [\sqrt{x} - x + 3]\,dx$$

$$= \frac{2x^{3/2}}{3} - \frac{x^2}{2} + 3x \Big|_1^4$$

$$= 6\frac{1}{6} \text{ square units}$$

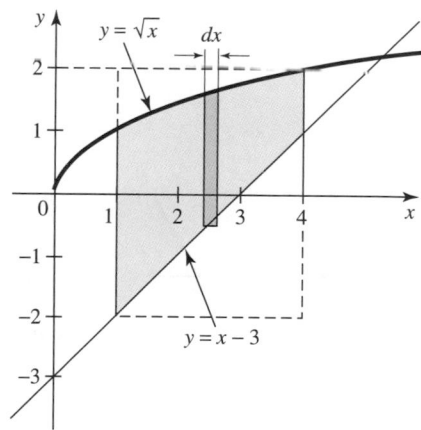
FIGURE 31–6

Limits Not Given

If we must find the area bounded by two curves, and if the limits a and b are not given, we must solve the given equations simultaneously to find their points of intersection. These points will provide the limits we need.

◆◆◆ **Example 11:** Find the area bounded by the parabola $y = 2 - x^2$ and the straight line $y = x$.

Solution: Our sketch, shown in Fig. 31–7, shows two points of intersection. To find them, let us solve the equations simultaneously by setting one equation equal to the other.

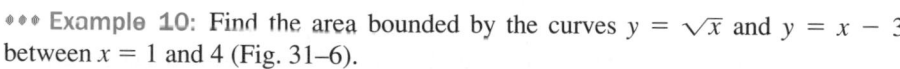

$$2 - x^2 = x$$

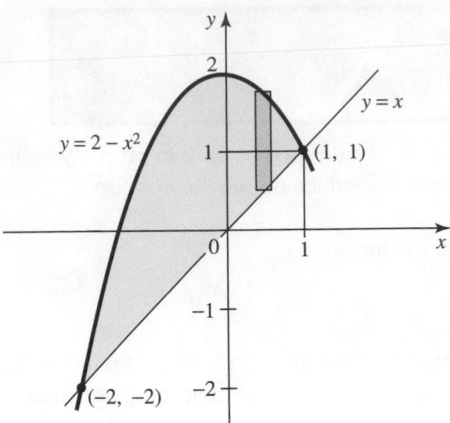

FIGURE 31-7

or $x^2 + x - 2 = 0$. We solve this quadratic by factoring.

$$(x + 2)(x - 1) = 0$$

so $x = -2$ and $x = 1$. Substituting back gives the points of intersection $(1, 1)$ and $(-2, -2)$.

We draw a vertical element whose width is dx and whose height is the upper curve minus the lower, or

$$2 - x^2 - x$$

The area dA of the strip is then

$$dA = (2 - x^2 - x)\, dx$$

We integrate, taking as limits the values of x (-2 and 1) found earlier by simultaneous solution of the given equations.

$$A = \int_{-2}^{1} (2 - x^2 - x)dx$$

$$= 2x - \frac{x^3}{3} - \frac{x^2}{2} \Big|_{-2}^{1} = 4\frac{1}{2} \qquad \text{◆◆◆}$$

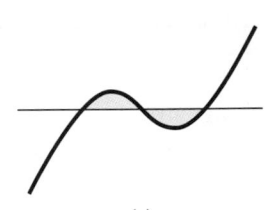

(a)

(b)

FIGURE 31–8   Intersecting curves may bound more than one region.

Several Regions

Sometimes the given curves may cross in several places, or the limits may be specified in such a way as to define *more than one region,* as in Fig. 31–8. In such cases we find the area of each region separately. However, in a region where the curve that we designated as "upper" lies below the other curve, the area of that region will be negative. We then add the absolute values of the areas of each region.

◆◆◆ **Example 12:** Find the area bounded by the curve $y = x^2 - 4$ and the x axis between $x = 1$ and 3.

Solution: Our sketch (Fig. 31–9) shows two regions. If we designate $y = x^2 - 4$ as the upper curve, then A_1 will be negative and A_2 will be positive. We set up two separate integrals.

$$A_1 = \int_{1}^{2} [(x^2 - 4) - 0]\, dx = \frac{x^3}{3} - 4x \Big|_{1}^{2} = -\frac{5}{3}$$

$$A_2 = \int_{2}^{3} [(x^2 - 4) - 0]\, dx = \frac{x^3}{3} - 4x \Big|_{2}^{3} = \frac{7}{3}$$

Adding absolute values gives

$$A = |A_1| + |A_2| = |-5/3| + |7/3| = 4 \text{ square units}$$

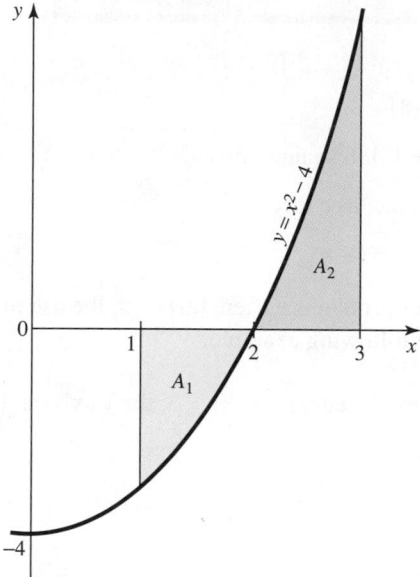

FIGURE 31–9

We will now do an example in which the integral contains absolute value signs.

◆◆◆ **Example 13:** Find the area bounded by the curve $y = 1/x$ and the x axis (a) from $x = 1$ to 4, (b) from $x = -1$ to 4, and (c) from $x = -4$ to -1.

Solution: Integrating, we obtain

$$A = \int_a^b \frac{dx}{x} = \ln |x| \Big|_a^b = \ln|b| - \ln|a|$$

(a) For the limits 1 to 4,

$$A = \ln 4 - \ln 1 = 1.386 \text{ square units}$$

(b) For the limits -1 to 4,

$$A = \ln|4| - \ln|-1| = \ln 4 - \ln 1 = 1.386 \text{ square units (?)}$$

We appear to get the same area between the limits -1 and 4 as we did for the limits 1 and 4. However, a graph (Fig. 31–10) shows that the curve $y = 1/x$ *is discontinuous* at $x = 0$, so we cannot integrate over the interval -1 to 4.

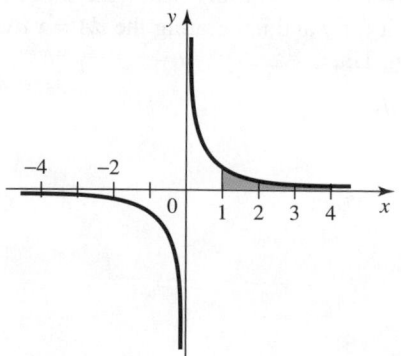

FIGURE 31–10

Common Error	Don't try to set up these area problems without making a sketch. Don't integrate across a discontinuity.

(c) For the limits -4 to -1,

$$A = \ln|-1| - \ln|-4|$$

$$= \ln 1 - \ln 4 = -1.386 \text{ square units}$$

This area lies below the x axis and has a negative value, as expected.   ◆◆◆

Horizontal Elements

So far we have used vertical elements of area to set up our problems. Often, however, the use of horizontal elements will make our work easier, as in the following example.

◆◆◆ **Example 14:** Find the first-quadrant area bounded by the curve $y = x^2 + 3$, the y axis, and the lines $y = 7$ and $y = 12$ (Fig. 31–11).

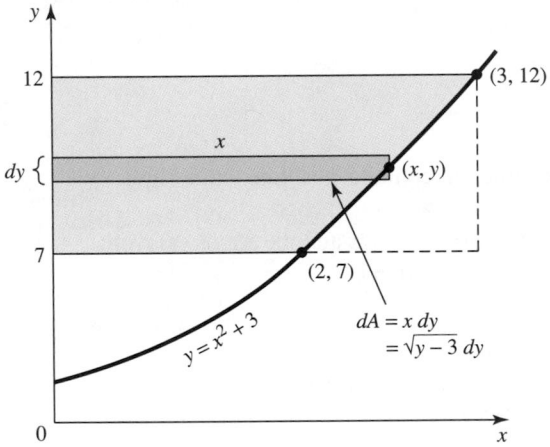

FIGURE 31–11

Estimate: We enclose the given area in a 5×3 rectangle, shown dashed, of area 15. From this we subtract a roughly triangular portion whose area is $\frac{1}{2}(1)(5)$, or 2.5, getting an estimate of 12.5 square units.

Solution: First, establish a point on the curve; here we use (x, y). If we try a vertical element (i.e., $dA = y\,dx$, not shown in the diagram), the rectangle through (x, y) would be partly within the shaded region and partly outside. This is not a good choice because at least two integrals would be needed. If we try a horizontal element of area $dA = x\,dy$, it is entirely within the shaded region. Using 12 as the upper limit and 7 as the lower limit for y and integrating the $dA = x\,dy$ expression, we have completely covered the shaded region. Thus,

$$dA = x\,dy = (y - 3)^{1/2}\,dy$$

Integrating, we have

$$A = \int_{7}^{12} (y - 3)^{1/2}\,dy$$

$$= \left. \frac{2(y - 3)^{3/2}}{3} \right|_{7}^{12}$$

$$= \frac{2}{3}(9)^{3/2} - \frac{2}{3}(4)^{3/2} = \frac{38}{3} \text{ square units}$$ ◆◆◆

◆◆◆ **Example 15:** Find the area bounded by the curves $y^2 = 12x$ and $y^2 = 24x - 36$.

Solution: We first plot the two curves as shown in Fig. 31–12. We recognize from their equations that they are parabolas opening to the right. We find their points of intersection by solving simultaneously.

$$24x - 36 = 12x$$

$$x = 3$$

$$y = \pm \sqrt{12x} = \pm 6$$

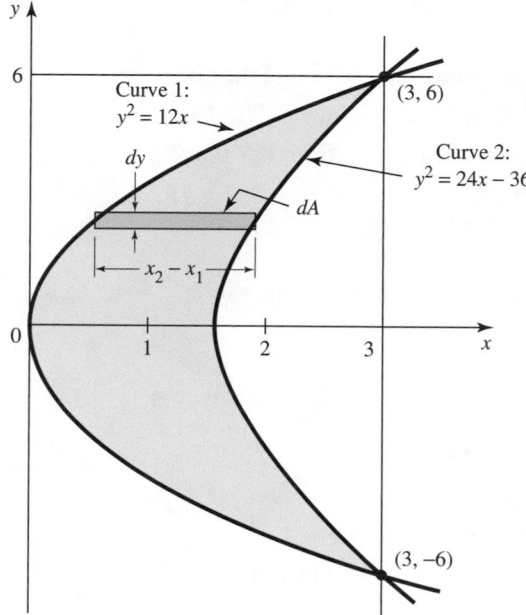

FIGURE 31–12

Since we have symmetry about the x axis, let us solve for the first-quadrant area and later double it. We draw a horizontal strip of width dy and length $x_2 - x_1$. The area dA is then $dA = (x_2 - x_1)\, dy$, where x_1 and x_2 are found by solving each given equation for x.

$$x_1 = \frac{y^2}{12} \quad \text{and} \quad x_2 = \frac{y^2 + 36}{24}$$

Integrating, we get

$$A = \int_0^6 (x_2 - x_1)\, dy$$

$$= \int_0^6 \left(\frac{y^2}{24} + \frac{36}{24} - \frac{y^2}{12} \right) dy$$

$$= \int_0^6 \left(\frac{3}{2} - \frac{y^2}{24} \right) dy$$

$$= \frac{3y}{2} - \frac{y^3}{72} \bigg|_0^6$$

$$= \frac{3(6)}{2} - \frac{(6)^3}{72} = 6$$

By symmetry, the total area between the two curves is twice this, or 12 square units.   ◆◆◆

Common Error	Don't always assume that a vertical element is the best choice. Try setting up the integral in Example 14 or 15 using a vertical element. What problems arise?

Exercise 3 ◆ Finding Areas by Means of the Definite Integral

Areas Bounded by the y Axis

Give any approximate answers to at least three significant digits.

Find the first-quadrant area bounded by the given curve, the y axis, and the given lines.

1. $y = x^2 + 2$ from $y = 3$ to 5 **2.** $8y^2 = x$ from $y = 0$ to 10

3. $y^3 = 4x$ from $y = 0$ to 4 **4.** $y = 4 - x^2$ from $y = 0$ to 3

Areas Bounded by Both Axes

Find the first-quadrant area bounded by each curve and both coordinate axes.

5. $y^2 = 16 - x$ **6.** $y = x^3 - 8x^2 + 15x$

7. $x + y + y^2 = 2$ **8.** $\sqrt{x} + \sqrt{y} = 1$

Areas Bounded by Two Branches of a Curve

Find the area bounded by the given curve and given line.

9. $y^2 = x$ and $x = 4$ **10.** $y^2 = 2x$ and $x = 4$

11. $4y^2 = x^3$ and $x = 8$ **12.** $y^2 = x^2(x^2 - 1)$ and $x = 2$

Areas above and below the x Axis

13. Find the area bounded by the curve $10y = x^2 - 80$, the x axis, and the lines $x = 1$ and $x = 6$

14. Find the area bounded by the curve $y = x^3$, the x axis, and the lines $x = -3$ and $x = 0$.

Find only the portion of the area below the x axis.

15. $y = x^3 - 4x^2 + 3x$ **16.** $y = x^2 - 4x + 3$

Areas between Curves

17. Find the area bounded by the lines $y = 3x$, $y = 15 - 3x$, and the x axis.

18. Find the area bounded by the curves $y^2 = 4x$ and $2x - y = 4$.

19. Find the area bounded by the parabola $y = 6 + 4x - x^2$ and the chord joining $(-2, -6)$ and $(4, 6)$.

20. Find the area bounded by the curve $y^2 = x^3$ and the line $x = 4$.

21. Find the area bounded by the curve $y^3 = x^2$ and the chord joining $(-1, 1)$ and $(8, 4)$.

22. Find the area between the parabolas $y^2 = 4x$ and $x^2 = 4y$.

23. Find the area between the parabolas $y^2 = 2x$ and $x^2 = 2y$.

Areas of Geometric Figures

Hint: For problem 28 and several of those to follow, integrate using Rule 69.

Use integration to verify the formula for the area of each figure.

24. square of side a **25.** rectangle with sides a and b

26. right triangle [Fig. 31–13(a)]

27. triangle [Fig. 31–13(b)]

28. circle of radius r

29. segment of circle [Fig. 31–13(c)]

30. ellipse [Fig. 31–13(d)]

31. parabola [Fig. 31–13(e)]

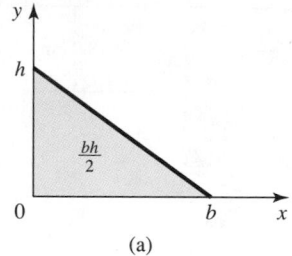
(a)

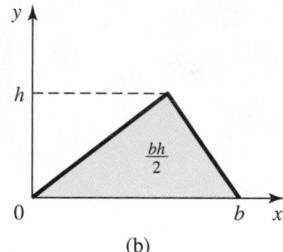
(b)

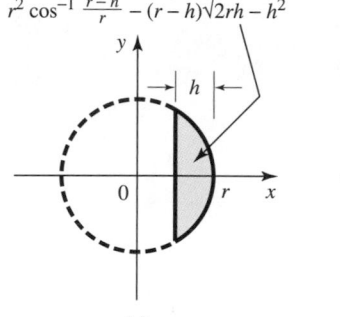
(c)

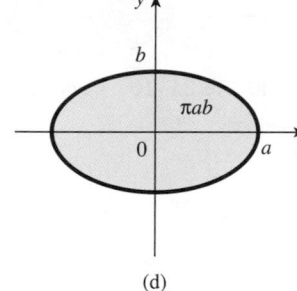
(d)

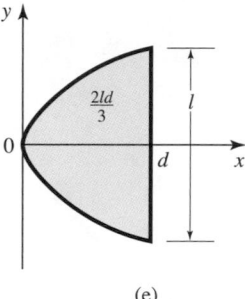
(e)

FIGURE 31–13 Areas of some geometric figures.

Applications

32. An elliptical culvert is partly full of water (Fig. 31–14). Find, by integration, the cross-sectional area of the water.

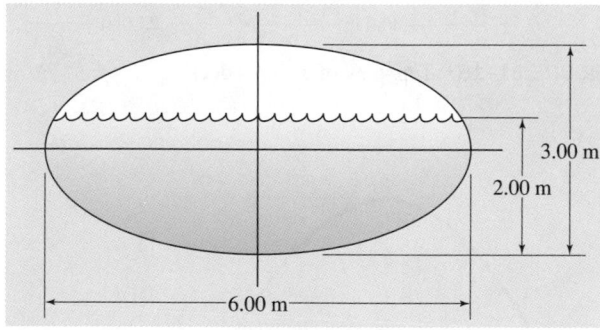

FIGURE 31–14 Elliptical culvert.

33. A mirror (Fig. 31–15) has a parabolic face. Find the volume of glass in the mirror.

34. Figure 31–16 shows a concrete column that has an elliptical cross section. Find the volume of concrete in the column.

35. A concrete roof beam for an auditorium has a straight top edge and a parabolic lower edge (Fig. 31–17). Find the volume of concrete in the beam.

36. The deck of a certain ship has the shape of two intersecting parabolic curves (Fig. 31–18). Find the area of the deck.

37. A lens (Fig. 31–19) has a cross section formed by two intersecting circular arcs. Find by integration the cross-sectional area of the lens.

38. A window (Fig. 31–20) has the shape of a parabola above and a circular arc below. Find the area of the window.

Note that no equations are given for the curves in these applications. Thus before setting up each integral, you must draw axes, put the equation in simplest form, and then write the equation of the curve.

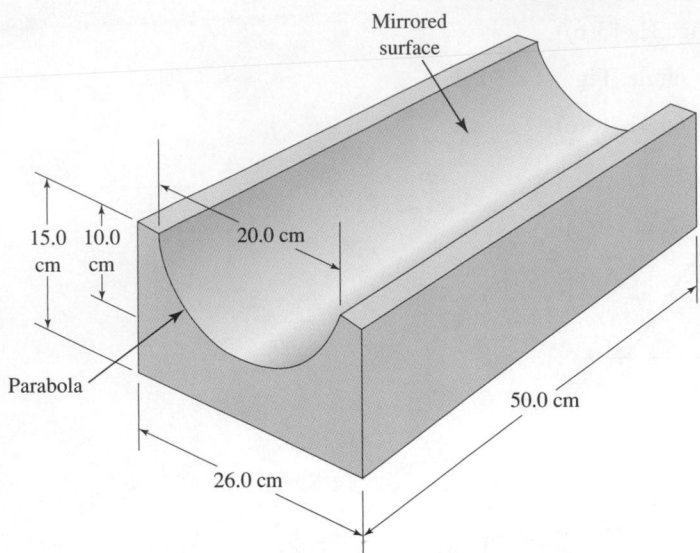

FIGURE 31–15   Parabolic mirror.

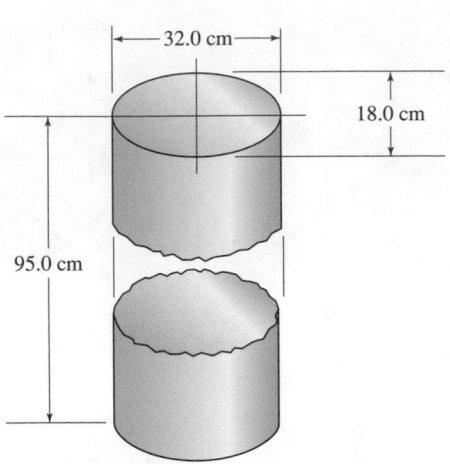

FIGURE 31–16   Concrete column.

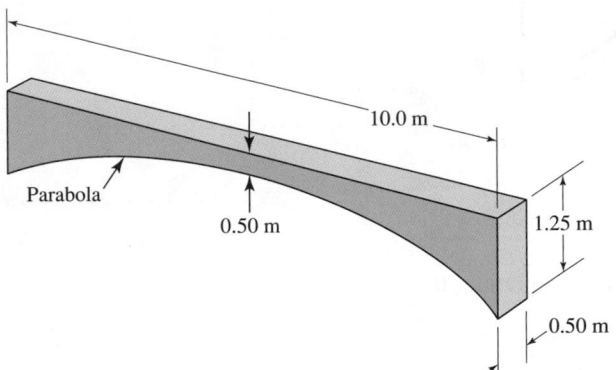

FIGURE 31–17   Roof beam.

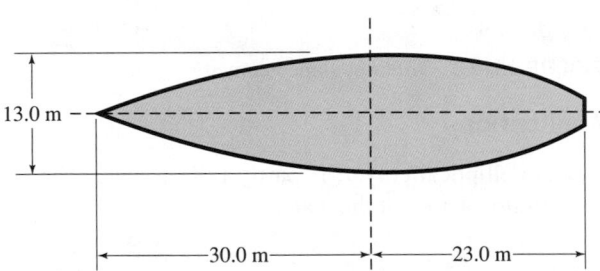

FIGURE 31–18   Top view of a ship's deck.

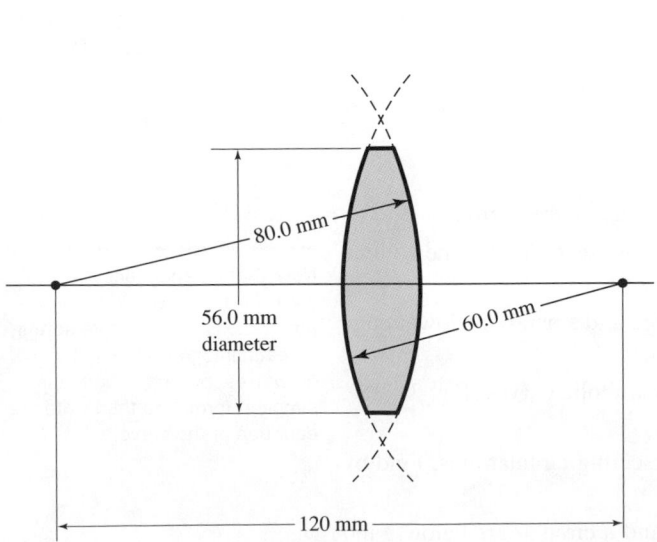

FIGURE 31–19   Cross section of a lens.

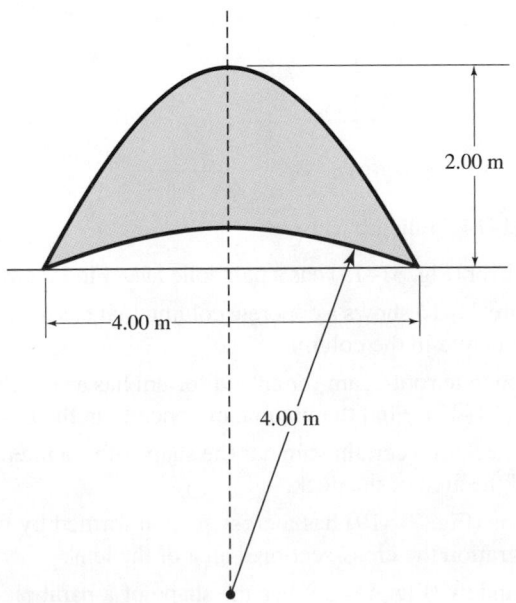

FIGURE 31–20   Window.

31–4 Volumes by Integration

Solids of Revolution

When an area is rotated about some axis *L*, it sweeps out a *solid of revolution*. It is clear from Fig. 31–21 that every cross section of that solid of revolution at right angles to the axis of rotation is a circle.

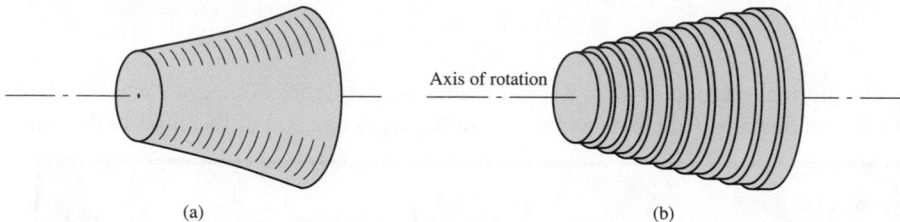

Axis of rotation

(a)   (b)

FIGURE 31–21 (a) Solid of revolution. (b) Solid of revolution approximated by a stack of thin disks.

When the area *A* is rotated about an axis *L* located at some fixed distance from the area, we get a solid of revolution with a cylindrical hole down its centre [Fig. 31–22(b)]. The cross section of this solid, at right angles to the axis of rotation, consists of a ring bounded by an outer circle and an inner concentric circle. When the area *B* is rotated about axis *L* [Fig. 31–22(c)], we get a solid of revolution with a hole of varying diameter down its centre. We first learn how to calculate the volume of a solid with no hole, and then we cover "hollow" solids of revolution.

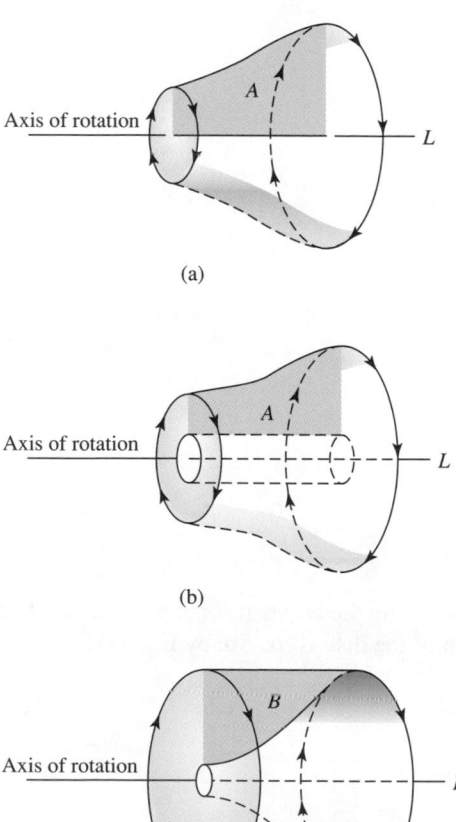

Axis of rotation

A

L

(a)

Axis of rotation

A

L

(b)

Axis of rotation

B

L

(c)

FIGURE 31–22 (a) Solid of revolution. (b) Solid of revolution with an axial hole. (c) Solid of revolution with axial hole of varying diameter.

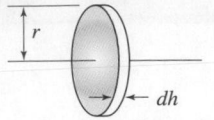

FIGURE 31–23 One disk.

Volumes by the Disk Method: Rotation about the x Axis

We may think of a solid of revolution [Fig. 31–21(a)] as being made up of a stack of thin disks, like a stack of coins of different sizes [Fig. 31–21(b)]. Each disk is called an *element* of the total volume. We let the radius of one such disk be r (which varies with the disk's position in the stack) and let the thickness be equal to dh. Since a disk (Fig. 31–23) is a cylinder, we calculate its volume dV by Eq. 390.

$$dV = \pi r^2 \, dh$$

Now using the shortcut method of Sec. 31–3 for setting up a definite integral, we "sum" the volumes of all such disk-shaped elements by integrating from one end of the solid to the other.

Volumes by the Disk Method	$V = \pi \displaystyle\int_a^b r^2 dh$	**391**

In an actual problem, we must express r and h in terms of x and y, as in the following example.

◆◆◆ **Example 16:** The area bounded by the curve $y = 8/x$, the x axis, and the lines $x = 1$ and $x = 8$ is rotated about the x axis. Find the volume generated.

Estimate: We sketch the solid as shown in Fig. 31–24 and try to visualize a right circular cone having roughly the same volume, say, with height 7 and base radius 5. Such a cone would have a volume of $(\frac{1}{3})\pi(5^2)(7)$, or 58π.

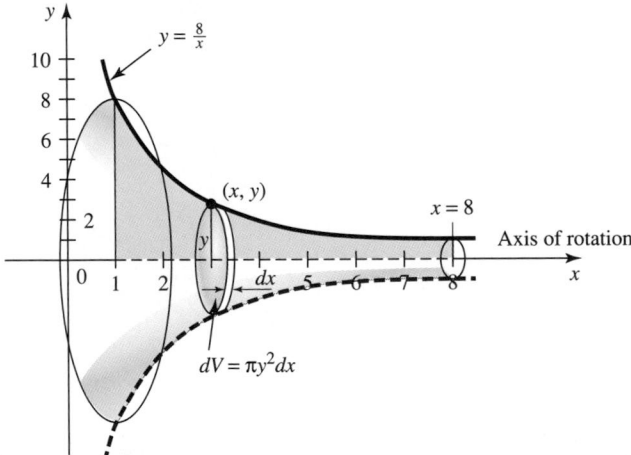

FIGURE 31–24 Volumes by the disk method.

Solution: On our figure we sketch in a typical disk touching the curve at some point (x, y). The radius r of the disk is equal to y, and the thickness dh of the disk is dx. So, by Eq. 391,

$$V = \pi \int_a^b y^2 dx$$

For y we substitute $8/x$, and for a and b, the limits 1 and 8.

$$V = \pi \int_1^8 \left(\frac{8}{x}\right)^2 dx = 64\pi \int_1^8 x^{-2} dx$$

Integrate:

$$= -64\pi \left[x^{-1}\right]_1^8$$

$$= -64\pi \left(\frac{1}{8} - 1\right) = 56\pi \text{ cubic units} \qquad ◆◆◆$$

Common Error	Remember that all parts of the integral, including the limits, must be expressed in terms of the *same variable*.

Volumes by the Disk Method: Rotation about the y Axis

When our area is rotated about the y axis rather than the x axis, we take our element of area as a horizontal disk rather than a vertical one.

◆◆◆ **Example 17:** Find the volume generated when the first-quadrant area bounded by $y = x^2$, the y axis, and the lines $y = 1$ and $y = 4$ is rotated about the y axis.

Solution: Our disk-shaped element of volume, shown in Fig. 31–25, now has a radius x and a thickness dy.

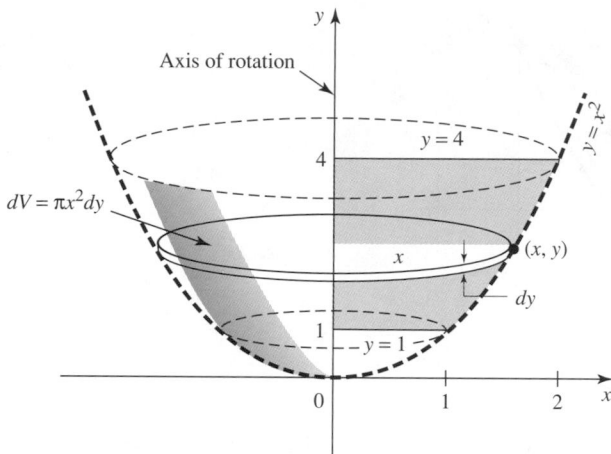

FIGURE 31–25

So, by Eq. 391,

$$V = \pi \int_a^b x^2 dy$$

Substituting y for x^2 and inserting the limits 1 and 4, we have

$$V = \pi \int_1^4 y\,dy = \pi \left[\frac{y^2}{2} \right]_1^4$$

$$= \frac{\pi}{2}(4^2 - 1^2) = \frac{15\pi}{2} \text{ cubic units} \qquad ◆◆◆$$

Volumes by the Shell Method

Instead of using a thin disk for our element of volume, it is sometimes easier to use a *thin-walled shell* (Fig. 31–26). To visualize such shells, imagine a solid of revolution to be turned from a log, with the axis of revolution along the centreline of the log. Each annual growth ring would have the shape of a thin-walled shell, with the solid of revolution being made up of many such shells nested one inside the other.

The volume dV of a thin-walled cylindrical shell of radius r, height h, and wall thickness dr is

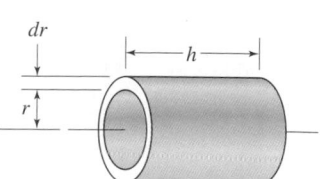

FIGURE 31–26 Thin-walled shell.

$$dV = \text{circumference} \times \text{height} \times \text{wall thickness}$$

$$= 2\pi rh\,dr$$

Integrating gives the following:

Volumes by the Shell Method	$V = 2\pi \int_a^b rh\, dr$	395

Notice that here the integration limits are in terms of r. As with the disk method, r and h must be expressed in terms of x and y in a particular problem.

◆◆◆ **Example 18:** The first-quadrant area bounded by the curve $y = x^2$, the y axis, and the line $y = 4$ (Fig. 31–27) is rotated about the y axis. Use the shell method to find the volume generated.

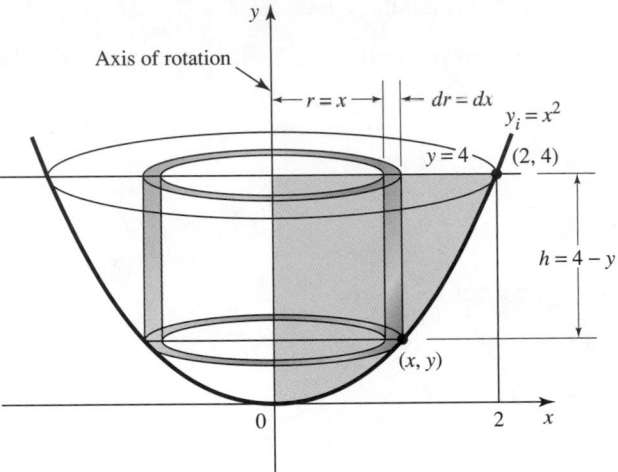

FIGURE 31–27   Volumes by the shell method.

Estimate: The volume of the given solid must be less than that of the circumscribed cylinder, $\pi(2^2)(4)$ or 16π, and greater than the volume of the inscribed cone, $16\pi/3$. Thus we have bracketed our answer between $5\frac{1}{3}\pi$ and 16π.

Solution: Through a point (x, y) on the curve, we sketch an element of area of height $h = 4 - y$ and thickness dx. As the first-quadrant area rotates about the y axis, generating a solid of revolution, our element of area sweeps out a shell of radius $r = x$ and thickness $dr = dx$. The volume dV of the shell is then

$$dV = 2\pi rh\, dr = 2\pi x(4 - y)\, dx$$
$$= 2\pi x(4 - x^2)\, dx$$

since $y = x^2$. Integrating gives the total volume.

$$V = 2\pi \int_0^2 x(4 - x^2)\, dx$$
$$= 2\pi \int_0^2 (4x - x^3)\, dx$$
$$= 2\pi \left[2x^2 - \frac{x^4}{4} \right]_0^2 = 2\pi \left[2(2)^2 - \frac{2^4}{4} \right]$$
$$= 8\pi \text{ cubic units}$$

◆◆◆

Solid of Revolution with Hole

To find the volume of a solid of revolution with an axial hole, such as in Figs. 31–22(b) and (c), we can first find the volume of hole and solid separately, and then subtract. Or we can find the volume of the solid of revolution directly by either the ring or the shell method, as in the following example.

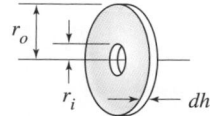

FIGURE 31–28 Ring or washer.

◆◆◆ **Example 19:** The first-quadrant area bounded by the curve $y^2 = 4x$, the x axis, and the line $x = 4$ is rotated about the y axis. Find the volume generated (a) by the ring method and (b) by the shell method.

Solution:

(a) *Ring method:* The volume dV of a thin ring (Fig. 31–28) is given by

$$dV = \pi(r_o^2 - r_i^2)\, dh$$

where r_o is the outer radius and r_i is the inner radius. Integrating gives the following:

The ring method is also called the washer method.

Volumes by the Ring Method	$V = \pi \displaystyle\int_a^b (r_o^2 - r_i^2)\, dh$	393

On our given solid [Fig. 31–29(a)] we show an element of volume in the shape of a ring, centred on the y axis. Its outer and inner radii are x_o and x_i, respectively, and its thickness dh is here dy. Then, by Eq. 393,

$$V = \pi \int_a^b (x_o^2 - x_i^2)\, dy$$

In our problem, $x_o = 4$ and $x_i = y^2/4$. Substituting these values and placing the limits of 0 and 4 on y gives

$$V = \pi \int_0^4 \left(16 - \frac{y^4}{16}\right) dy$$

Integrating, we obtain

$$V = \pi \left[16y - \frac{y^5}{80} \right]_0^4 = \pi \left[16(4) - \frac{4^5}{80} \right] = \frac{256}{5}\pi \text{ cubic units}$$

(b) *Shell method:* On the given solid we indicate an element of volume in the shape of a shell [Fig. 31–29(b)]. Its inner radius r is x, its thickness dr is dx, and its height h is y. Then, by Eq. 394,

$$dV = 2\pi xy\, dx$$

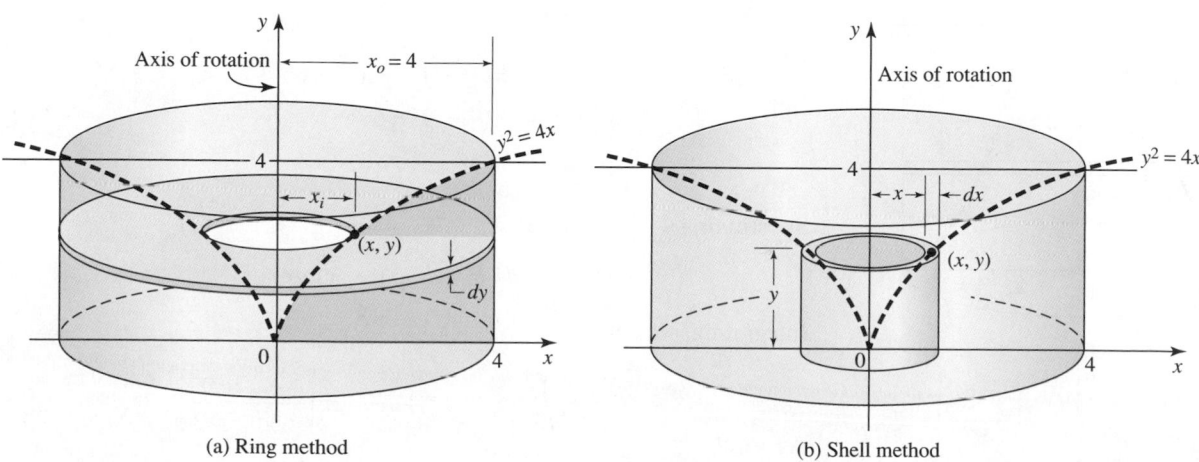

(a) Ring method

(b) Shell method

FIGURE 31–29

So the total volume is

$$V = 2\pi \int xy \, dx$$

Replacing y with $2\sqrt{x}$, we have

$$V = 2\pi \int_0^4 x(2x^{1/2}) \, dx$$

$$= 4\pi \int_0^4 x^{3/2} \, dx$$

$$= 4\pi \left[\frac{x^{5/2}}{5/2}\right]_0^4$$

$$= \frac{8\pi}{5}(4)^{5/2} = \frac{256}{5} \pi \text{ cubic units}$$

as by the ring method.   ◆◆◆

> **Common Error**   In Example 19 the radius x in the shell method varies from 0 to 4. The limits of integration are therefore 0 to 4, not -4 to 4.

Rotation about a Noncoordinate Axis

We can, of course, get a volume of revolution by rotating a given area about some axis other than a coordinate axis. This often results in a solid with a hole in it. As with the preceding hollow figure, these can usually be set up with shells or, as in the following example, with rings.

◆◆◆ **Example 20:** The first-quadrant area bounded by the curve $y = x^2$, the y axis, and the line $y = 4$ is rotated about the line $x = 3$. Use the ring method to find the volume generated.

Estimate: From the cylinder of volume $\pi(3^2)(4)$, let us subtract a cone of volume $(\frac{1}{3})\pi(3^2)(4)$, getting 24π as our estimate.

Solution: Through the point (x, y) on the curve (Fig. 31–30), we draw a ring-shaped element of volume, with outside radius of 3 units, inside radius of $(3 - x)$ units, and thickness dy. The volume dV of the element is then, by Eq. 392,

$$dV = \pi[3^2 - (3 - x)^2] \, dy$$
$$= \pi(9 - 9 + 6x - x^2) \, dy$$
$$= \pi(6x - x^2) \, dy$$

Substituting $\sqrt{y}$ for x gives

$$dV = \pi(6\sqrt{y} - y) \, dy$$

Integrating, we get

$$V = \pi \int_0^4 (6y^{1/2} - y) \, dy = \pi \left[\frac{12y^{3/2}}{3} - \frac{y^2}{2}\right]_0^4$$

$$= \pi \left[4(4)^{3/2} - \frac{16}{2}\right] = 24\pi \text{ cubic units}   ◆◆◆$$

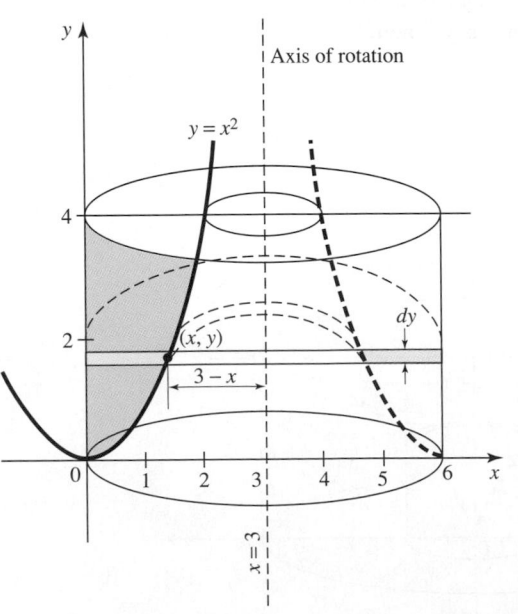

FIGURE 31–29

Exercise 4 ◆ Volumes by Integration

Rotation about the *x* Axis

Find the volume generated by rotating the first-quadrant area bounded by each set of curves and the *x* axis about the *x* axis. Use either the disk or the shell method.

Give any approximate answers to at least three significant digits.

1. $y = x^3$ and $x = 2$

2. $y = \dfrac{x^2}{4}$ and $x = 4$

3. $y = \dfrac{x^{3/2}}{2}$ and $x = 2$

4. $y^2 = x^3 - 3x^2 + 2x$ and $x = 1$

5. $y^2(2 - x) = x^3$ and $x = 1$

6. $\sqrt{x} + \sqrt{y} = 1$

7. $x^{2/3} + y^{2/3} = 1$ from $x = 0$ to 1

8. $y^2 = x\left(\dfrac{x - 3}{x - 4}\right)$

Rotation about the *y* Axis

Find the volume generated by rotating about the *y* axis the first-quadrant area bounded by each set of curves.

9. $y = x^3$, the *y* axis, and $y = 8$

10. $2y^2 = x^3$, $x = 0$, and $y = 2$

11. $9x^2 + 16y^2 = 144$

12. $\left(\dfrac{x}{2}\right)^2 + \left(\dfrac{y}{3}\right)^{2/3} = 1$

13. $y^2 = 4x$ and $y = 4$

Rotation about a Noncoordinate Axis

Find the volume generated by rotating about the indicated axis the first-quadrant area bounded by each set of curves.

14. $x = 4$ and $y^2 = x^3$, about $x = 4$

15. above $y = 3$ and below $y = 4x - x^2$, about $y = 3$

16. $y^2 = x^3$ and $y = 8$, about $y = 9$

17. $y = 4$ and $y = 4 + 6x - 2x^2$, about $y = 4$

Applications

18. The nose cone of a certain rocket is a paraboloid of revolution (Fig. 31–31). Find its volume.

A *paraboloid of revolution* is the figure formed by revolving a parabola about its axis.

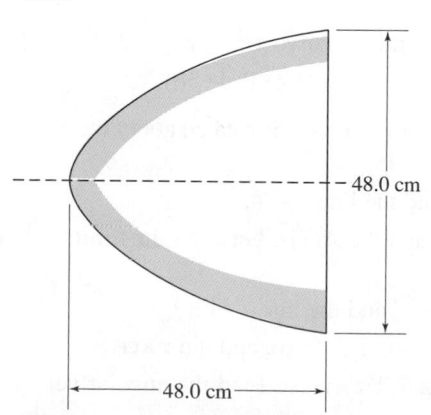

48.0 cm

48.0 cm

FIGURE 31–31 Rocket nose cone.

19. A wing tank for an airplane is a solid of revolution formed by rotating the curve $8y = 4x - x^2$, from $x = 0$ to 3.50, about the *x* axis (Fig. 31–32). Find the volume of the tank.

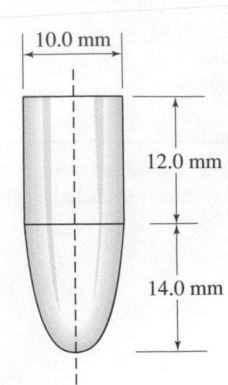

FIGURE 31-33 Bullet.

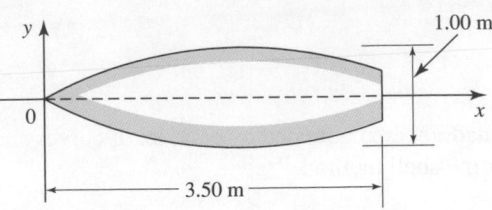

FIGURE 31-32 Airplane wing tank.

20. A bullet (Fig. 31–33) consists of a cylinder and a paraboloid of revolution and is made of lead having a density of 11.3 g/cm³. Find its mass.

21. A telescope mirror, shown in cross section in Fig. 31–34, is formed by rotating the area under the hyperbola $y^2/100 - x^2/1225 = 1$ about the y axis and has a 20.0-cm-diameter hole at its centre. Find the volume of glass in the mirror.

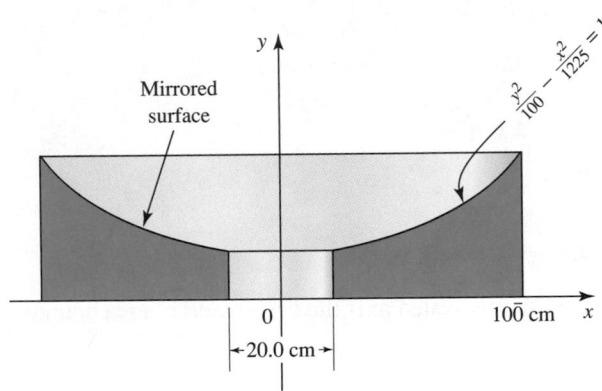

FIGURE 31-34 Telescope mirror.

◆◆◆ CHAPTER 31 REVIEW PROBLEMS ◆◆◆◆◆◆◆◆◆◆◆◆◆◆◆◆◆◆◆◆◆◆◆◆◆◆◆◆◆◆◆◆◆

Give any approximate answers to at least three significant digits.

1. Find the volume generated when the area bounded by the parabolas $y^2 = 4x$ and $y^2 = 5 - x$ is rotated about the x axis.

2. Find the area bounded by the coordinate axes and the curve $x^{1/2} + y^{1/2} = 2$.

3. Find the area between the curve $x^2 = 8y$ and $y = \dfrac{64}{x^2 + 16}$ (Use Rule 56).

4. The area bounded by the parabola $y^2 = 4x$, from $x = 0$ to 8, is rotated about the x axis. Find the volume generated.

5. Find the area bounded by the curve $y^2 = x^3$ and the line $x = 4$.

6. Find the area bounded by the curve $y = 1/x$ and the x axis, between the limits $x = 1$ and $x = 3$.

7. Find the area bounded by the curve $y^2 = x^3 - x^2$ and the line $x = 2$.

8. Find the area bounded by $xy = 6$, the lines $x = 1$ and $x = 6$, and the x axis.

9. A flywheel starts from rest and accelerates at $7.25t^2$ rad/s². Find the angular velocity and the total number of revolutions after 20.0 s.

10. The current to a certain capacitor is $i = t^3 + 18.5$ A. If the initial charge on the capacitor is 6.84 C, find the charge when $t = 5.25$ s.

11. A point starts from (1, 1) with initial velocities of $v_x = 4$ cm/s and $v_y = 15$ cm/s and moves along a curved path. It has x and y components of acceleration of $a_x = t$ and $a_y = 5t$. Write expressions for the x and y components of velocity and displacement.

12. A 15.0-F capacitor has an initial voltage of 25.0 V and is charged with a current equal to $i = \sqrt{4t + 21.6}$ A. Find the voltage across the capacitor at 14.0 s.

13. Find the volume of the solid generated by rotating the ellipse $x^2/16 + y^2/9 = 1$ about the x axis.

14. Find the area bounded by the curves $y^2 = 8x$ and $x^2 = 8y$.

15. Find the volume generated when the area bounded by $y^2 = x^3$, $x = 4$, and the x axis is rotated about the line $y = 8$.

16. Find the volume generated when the area bounded by the curve $y^2 = 16x$, from $x = 0$ to 4, is rotated about the x axis.

17. Find the entire area of the ellipse

$$\frac{x^2}{16} + \frac{y^2}{9} = 1$$

(Use Rule 69).

18. The acceleration of an object that starts from rest is given by $a = 3t$. Write equations for the velocity and displacement of the object.

19. The voltage across a 25.0-H inductor is given by $v = 8.9 + \sqrt{3t}$ V. Find the current in the inductor at 5.00 s if the initial current is 1.00 A.

20. Find the volume generated when the area bounded by $y^2 = x^3$, the y axis, and $y = 8$ is rotated about the line $x = 4$.

Writing

21. List the steps needed to find the area between two curves, and give a short description of each step.

Team Project

22. A projectile is launched with initial velocity v_0 at an angle v_0 with the horizontal, as shown in Fig. 31–35. Derive the equations for the displacement of the projectile.

$$x = (v_0 \cos \theta_0)t + x_0$$

and

$$y = (v_0 \sin \theta_0)t - gt^2/2 + y_0$$

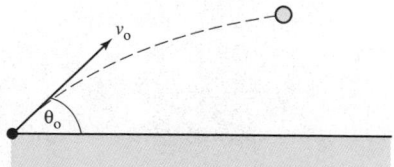

FIGURE 31–35 A projectile

32

More Applications of the Integral

This chapter is a continuation of the preceding one. We continue our applications of the definite integral by computing the length of a curve between two given points, the area of a surface of revolution, and the centroid of a plane area and of a solid of revolution, finding moments and centroids of areas.

In each case, we will set up our integral using the shortcut method of the preceding chapter. We will define a small element of the quantity we want to compute and then sum up all such elements by integration.

The section on centroids is followed by two applications involving centroids: finding the force due to fluid pressure and computing the work required for various tasks. We conclude by computing the moment of inertia for an area and the polar moment of inertia of solids of revolution, both useful quantities when studying strength of materials or the motion of rigid bodies.

32–1   Length of Arc

In this section we develop a way to find the length of a curve between two endpoints, such as the distance PQ in Fig. 32–1. We will find the distance measured *along the curve,* as if the curve were stretched out straight and then measured.

Finding the length of arc is sometimes referred to as *rectifying* a curve.

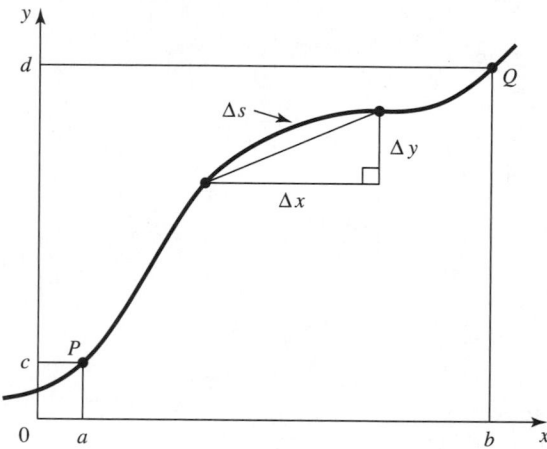

FIGURE 32–1   Length of arc. We are considering here (as usual) only smooth, continuous curves.

Again we use our intuitive method to set up the integral. We think of the curve as being made up of many short sections, each of length Δs. By the Pythagorean theorem,

$$(\Delta s)^2 \cong (\Delta x)^2 + (\Delta y)^2$$

Dividing by $(\Delta x)^2$ gives us

$$\frac{(\Delta s)^2}{(\Delta x)^2} \cong 1 + \frac{(\Delta y)^2}{(\Delta x)^2}$$

Taking the square root yields

$$\frac{\Delta s}{\Delta x} \cong \sqrt{1 + \left(\frac{\Delta y}{\Delta x}\right)^2}$$

We now let the number of these short sections of curve approach infinity as we let Δx approach zero.

$$\frac{ds}{dx} = \lim_{\Delta x \to 0} \frac{\Delta s}{\Delta x} = \sqrt{1 + \left(\frac{dy}{dx}\right)^2}$$

or

$$ds = \sqrt{1 + \left(\frac{dy}{dx}\right)^2}\, dx$$

Now thinking of integration as a summing process, we use it to add up all of the small segments of length ds as follows:

Length of Arc	$$s = \int_a^b \sqrt{1 + \left(\frac{dy}{dx}\right)^2}\, dx$$	396

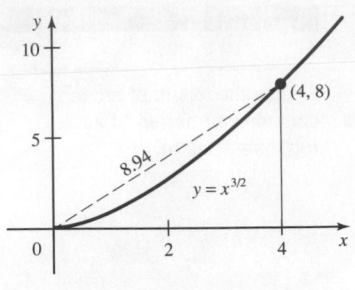

FIGURE 32–2

◆◆◆ **Example 1:** Find the first-quadrant length of the curve $y = x^{3/2}$ from $x = 0$ to 4 (Fig. 32–2).

Estimate: Computing the straight-line distance between the two endpoints $(0, 0)$ and $(4, 8)$ of our arc gives $\sqrt{4^2 + 8^2}$, or about 8.94. Thus, we expect the curve connecting those points to be slightly longer than 8.94.

Solution: Let us first find $(dy/dx)^2$.

$$y = x^{3/2}$$

$$\frac{dy}{dx} = \frac{3}{2} x^{1/2}$$

$$\left(\frac{dy}{dx}\right)^2 = \frac{9x}{4}$$

Then, by Eq. 396,

$$s = \int_0^4 \sqrt{1 + \left(\frac{dy}{dx}\right)^2}\, dx = \int_0^4 \sqrt{1 + \frac{9x}{4}}\, dx$$

$$= \frac{4}{9} \int_0^4 \left(1 + \frac{9x}{4}\right)^{1/2} \left(\frac{9}{4}\, dx\right) = \frac{8}{27} \left[\left(1 + \frac{9x}{4}\right)^{3/2}\right]_0^4$$

$$= \frac{8}{27}(10^{3/2} - 1) \cong 9.07$$

◆◆◆

Another Form of the Arc Length Equation

Another form of the equation for arc length, which can be derived in a similar way to Eq. 396, follows.

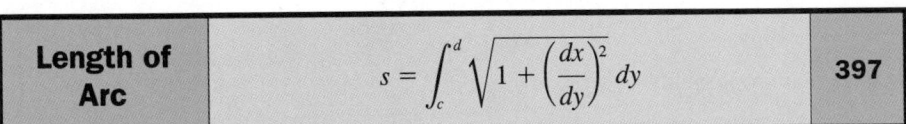

Length of Arc	$s = \int_c^d \sqrt{1 + \left(\dfrac{dx}{dy}\right)^2}\, dy$	397

This equation is more useful when the equation of the curve is given in the form $x = f(y)$, instead of the more usual form $y = f(x)$.

◆◆◆ **Example 2:** Find the length of the curve $x = 4y^2$ between $y = 0$ and 4.

Estimate: The straight-line distance between the endpoints $(0, 0)$ and $(64, 4)$ is $\sqrt{64^2 + 4^2}$, or about 64.1 (Fig. 32–3). As before, we expect the curve to be slightly longer than that.

Solution: Taking the derivative, we have

$$\frac{dx}{dy} = 8y$$

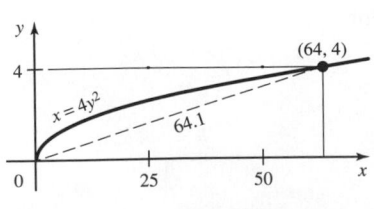

FIGURE 32–3

so $(dx/dy)^2 = 64y^2$. Substituting into Eq. 397, we have

$$s = \int_0^4 \sqrt{1 + 64y^2}\, dy = \frac{1}{8} \int_0^4 \sqrt{1 + (8y)^2}\, (8\, dy)$$

Using Rule 66, with $u = 8y$ and $a = 1$, yields

$$s = \frac{1}{8}\left[\frac{8y}{2}\sqrt{1 + 64y^2} + \frac{1}{2}\ln|8y + \sqrt{1 + 64y^2}|\right]_0^4 =$$

$$= \frac{1}{8}[4(4)\sqrt{1 + 64(16)} + \frac{1}{2}\ln|32 + \sqrt{1 + 64(16)}| - 0] \approx 64.3$$

◆◆◆

> When finding length of arc, we usually wind up with messy integrals. Often they can be evaluated using Rule 66.

Exercise 1 ◆ Length of Arc

Find the length of each curve.

1. $y^2 = x^3$ in the first quadrant from $x = 0$ to $\frac{5}{9}$
2. $6y = x^2$ from the origin to the point $(4, \frac{8}{3})$
3. the circle $x^2 + y^2 = 36$ (use Rule 61)
4. $x^{2/3} + y^{2/3} = 1$ from $x = 0.1$ to 1
5. $4y = x^2$ from $x = 0$ to 4
6. $2y^2 = x^3$ from the origin to $x = 10$
7. the arch of the parabola $y = 4x - x^2$ that lies above the x axis
8. the length in one quadrant of the curve $\left(\frac{x}{2}\right)^{2/3} + \left(\frac{y}{3}\right)^{2/3} = 1$
9. $y^3 = x^2$ between points $(0, 0)$ and $(8, 4)$
10. $y^2 = 8x$ from its vertex to one end of its latus rectum

> Many of these problems will require the use of Rule 66.

> As usual, round all approximate answers in this chapter to at least three significant digits.

> *Hint:* Use Eq. 397 for problems 9 and 10.

Applications

11. Find the length of the cable AB in Fig. 32–4 that is in the shape of a parabola.

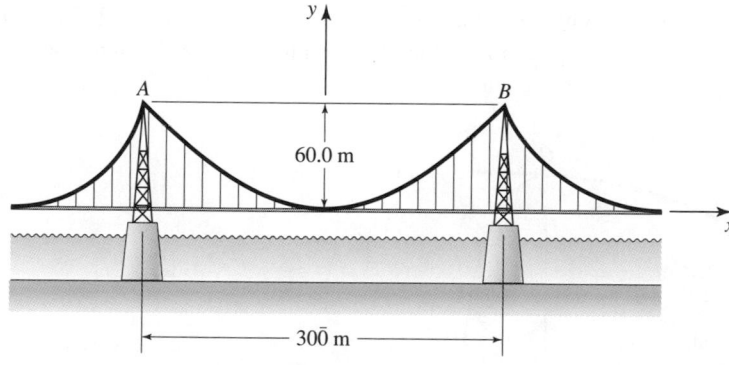

FIGURE 32–4   Suspension bridge.

12. A roadway has a parabolic shape at the top of a hill (Fig. 32–5). If the road is exactly 10 m wide, find the cost of paving from P to Q at the rate of $470/m^2$.

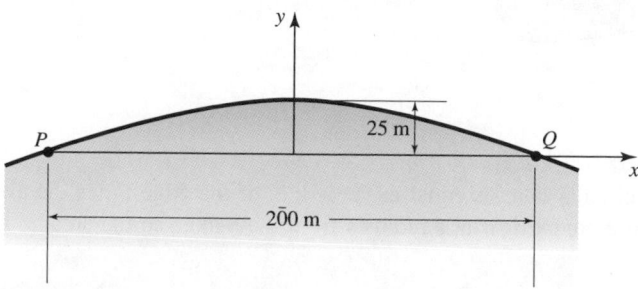

FIGURE 32–5   Road over a hill.

13. The equation of the bridge arch in Fig. 32–6 is $y = 0.0625x^2 - 5x + 100$. Find its length.

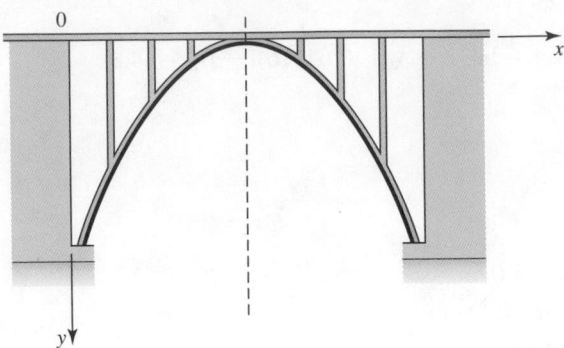

FIGURE 32-6   Parabolic bridge arch. Note that the y axis is positive downward.

14 Find the surface area of the curved portion of the mirror in Fig. 31–15 (in Chapter 31).

15. Find the perimeter of the window in Fig. 31–20 (in Chapter 31).

32–2   Area of Surface of Revolution

Surface of Revolution

We are using the capital letter S for surface area and lowercase s for arc length. Be careful not to confuse the two.

In Sec. 31–4 we rotated an *area* about an axis and got a solid of revolution. We now find the *area* of the surface of such a solid. Alternatively, we can think of a surface of revolution as being generated when a curve rotates about some axis. Let us take the curve of Fig. 32–7 and rotate it about the x axis. The arc PQ sweeps out a surface of revolution while the small section ds sweeps out a hoop-shaped element of that surface.

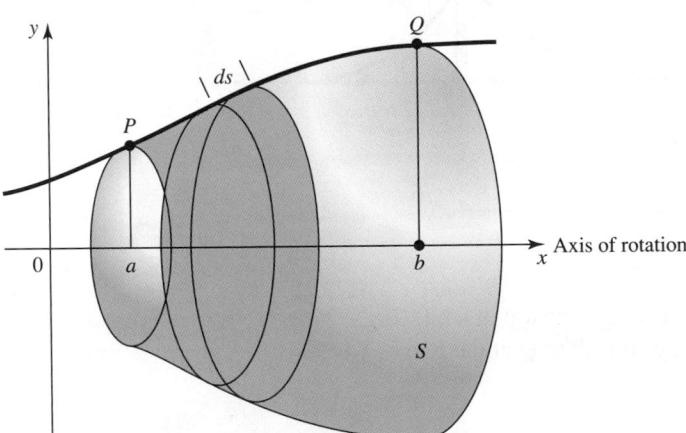

FIGURE 32-7   Surface of revolution.

Surface Area: Rotation about the x Axis

The area of a hoop-shaped or circular band is equal to the length of the edge times the average circumference of the hoop. Our element ds is at a radius y from the x axis, so the area dS of the hoop is

$$dS = 2\pi y \, ds$$

But from Sec. 32–1, we saw that

$$ds = \sqrt{1 + \left(\frac{dy}{dx}\right)^2}\, dx$$

So

$$ds = 2\pi y \sqrt{1 + \left(\frac{dy}{dx}\right)^2}\, dx$$

We then integrate from a to b to sum up the areas of all such elements.

Area of Surface of Revolution about x Axis	$S = 2\pi \displaystyle\int_a^b y \sqrt{1 + \left(\frac{dy}{dx}\right)^2}\, dx$	398

◆◆◆ **Example 3:** The upper half of a parabola, between $x = 0$ and $x = 16$ is rotated about the x axis (Fig. 32–8). Find the area of the surface of revolution generated.

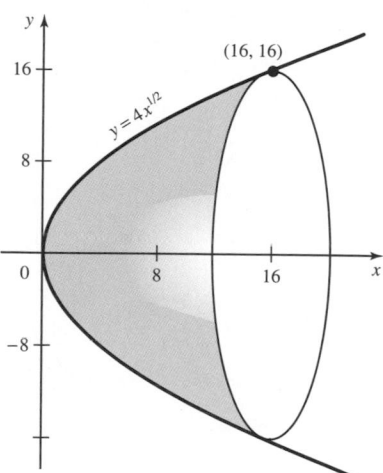

FIGURE 32–8

Estimate: Let us approximate the given surface by a hemisphere of radius 16. Its area is $\frac{1}{2}(4\pi)16^2$, or about 1600 square units.

Solution: We first find the derivative.

$$y = 4x^{1/2}$$
$$\frac{dy}{dx} = 2x^{-1/2}$$
$$\left(\frac{dy}{dx}\right)^2 = 4x^{-1}$$

Surface area problems, like length of arc problems, usually result in difficult integrals.

Substituting into Eq. 398, we obtain

$$S = 2\pi \int_0^{16} 4x^{1/2} \sqrt{1 + 4x^{-1}}\, dx = 8\pi \int_0^{16} (x + 4)^{1/2}\, dx$$

$$= \frac{16\pi}{3}(20^{3/2} - 4^{3/2}) \cong 1365 \text{ square units} \qquad\qquad ◆◆◆$$

Surface Area: Rotation about the y Axis

The equation for the area of a surface of revolution whose axis of revolution is the *y* axis can be derived in a similar way.

Area of Surface of Revolution about y Axis	$$S = 2\pi \int_a^b x \sqrt{1 + \left(\frac{dy}{dx}\right)^2}\, dx$$	399

◆◆◆ **Example 4:** The portion of the curve $y = x^2$ lying between the points $(0, 0)$ and $(2, 4)$ is rotated about the *y* axis (Fig. 32–9). Find the area of the surface generated.

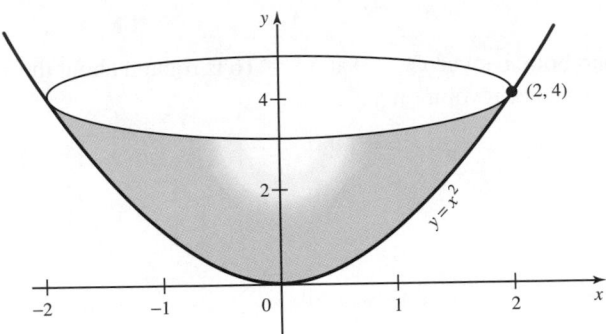

FIGURE 32–9

Solution: Taking the derivative gives $dy/dx = 2x$, so

$$\left(\frac{dy}{dx}\right)^2 = 4x^2$$

Substituting into Eq. 399, we have

$$S = 2\pi \int_0^2 x\sqrt{1 + 4x^2}\, dx = 2\pi/8 \int_0^2 (1 + 4x^2)^{1/2}(8x\, dx)$$

$$= \frac{\pi}{4} \cdot \frac{2(1 + 4x^2)^{3/2}}{3}\Bigg|_0^2 = \frac{\pi}{6}(17^{3/2} - 1^{3/2}) \approx 36.2 \text{ square units}$$ ◆◆◆

Exercise 2: ◆ Area of Surface of Revolution

Find the area of the surface generated by rotating each curve about the *x* axis.

1. $y = \dfrac{x^3}{9}$ from $x = 0$ to 3

2. $y^2 = 2x$ from $x = 0$ to 4

3. $y^2 = 9x$ from $x = 0$ to 4

4. $y^2 = 4x$ from $x = 0$ to 1

5. $y^2 = 4 - x$ in the first quadrant

6. $y^2 = 24 - 4x$ from $x = 3$ to 6
7. $x^{2/3} + y^{2/3} = 1$ in the first quadrant
8. $y = \dfrac{x^3}{6} + \dfrac{1}{2x}$ from $x = 1$ to 3

Find the area of the surface generated by rotating each curve about the y axis.

9. $y = 3x^2$ from $x = 0$ to 5
10. $y = 6x^2$ from $x = 2$ to 4
11. $y = 4 - x^2$ from $x = 0$ to 2
12. $y = 24 - x^2$ from $x = 2$ to 4

Geometric Figures

13. Find the surface area of a sphere by rotating the curve $x^2 + y^2 = r^2$ about a diameter.
14. Find the area of the curved surface of a cone by rotating about the x axis the line connecting the origin and the point (a, b).

Applications

15. Find the surface area of the nose cone shown in Fig. 31–31 (in Chapter 31).
16. Find the cost of copper plating 10 000 bullets (Fig. 31–33 in Chapter 31) at the rate of $45 per square metre.

32–3   Centroids

Centre of Gravity and Centroid

The *centre of gravity* (or *centre of mass*) of a body is the point where all of the mass can be thought to be concentrated, without altering the effect that the earth's gravity has upon it. For simple shapes such as a sphere, cube, or cylinder, the centre of gravity is exactly where you would expect it to be, at the centre of the object. Figure 32–10 shows how the centre of gravity of an irregular figure would be located.

A plane area has no thickness and hence has no weight or mass. Since it makes no sense to speak of centre of mass or centre of gravity for a weightless figure, we use instead the word *centroid*.

First Moment

To calculate the position of the centroid in figures of various shapes, we need the idea of the *first moment*, often referred to as *the moment*. It is not new to us. We know from our previous work that the *moment of a force* about some point a is the product of the force F and the perpendicular distance d from the point to the line of action of the force. In a similar way, we speak of the *moment of an area* about some axis [Fig. 32–11(a)]:

$$\text{moment of an area} = \text{area} \times \text{distance to the centroid}$$

or *moment of a volume* [Fig. 32–11(b)]:

$$\text{moment of a volume} = \text{volume} \times \text{distance to the centroid}$$

or *moment of a mass* [Fig. 32–11(c)]:

$$\text{moment of a mass} = \text{mass} \times \text{distance to the centroid}$$

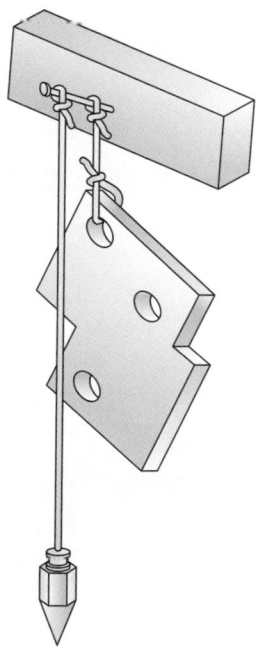

FIGURE 32–10   Any object hung from a point will swing to where its centre of gravity is directly below the point of suspension.

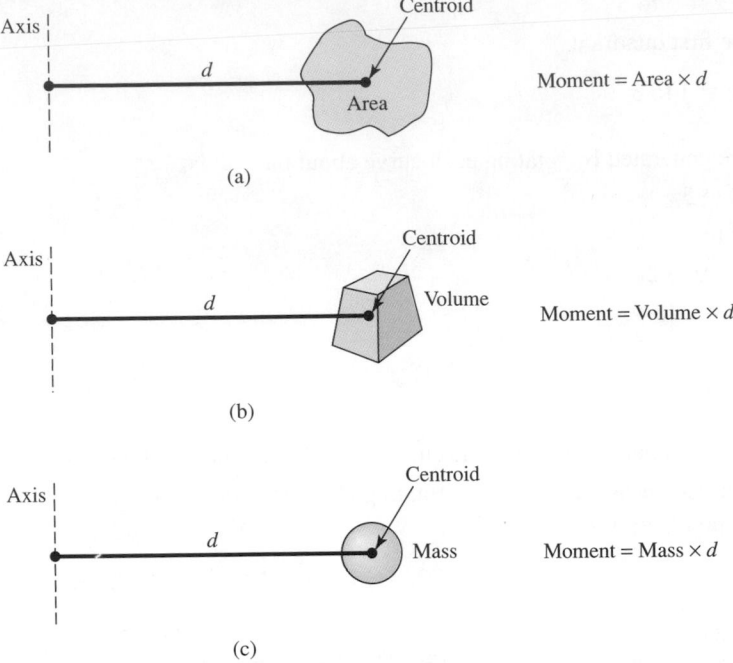

FIGURE 32–11   Moment of (a) an area, (b) a volume, and (c) a mass.

We will learn about second moments (moments of inertia) in Sec. 32–6.

In each case, the distance is that from the axis about which we take the moment, measured to some point on the area, volume, or mass. But to which point on the figure do we measure? To the *centroid or centre of gravity.*

Centroids of Simple Shapes

If a shape has an axis of symmetry, the centroid will be located on that axis. If the centroid is on the x axis, its coordinates will be $(\bar{x}, 0)$; if on the y axis, $(0, \bar{y})$. If there are two or more axes of symmetry, the centroid is found at the intersection of those axes.

We can easily find the centroid of an area that can be subdivided into simple regions, each of whose centroid location is known by symmetry. We first find the moment of each region about some convenient axis by multiplying the area of that region by the distance of its centroid from the axis. We then use the following fact, which we give without proof: For an area subdivided into smaller regions, *the moment of that area about a given axis is equal to the sum of the moments of the individual regions about that same axis.*

◆◆◆ **Example 5:** Find the location of the centroid of the shape in Fig. 32–12(a).

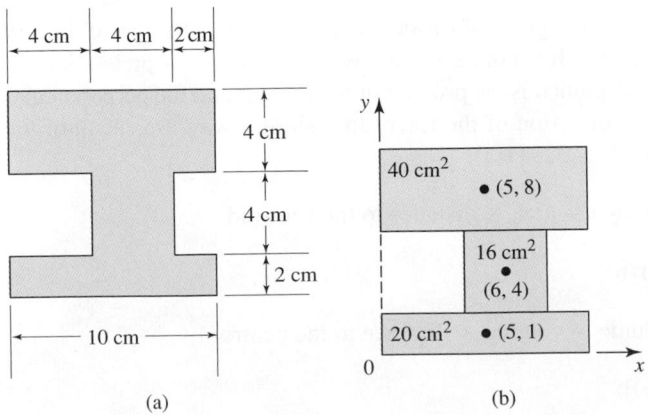

FIGURE 32–12

Solution: We subdivide the area into rectangles as shown in Fig. 32–12(b), locate the centroid, and compute the area of each. We also choose axes from which we will measure the coordinates of the centroid, $\bar{x}$ and $\bar{y}$. The total area is

$$40 + 16 + 20 = 76 \text{ cm}^2$$

The moment M_y about the y axis of the entire region is the sum of the moments of the individual regions.

$$M_y = 40(5) + 16(6) + 20(5) = 396 \text{ cm}^3$$

Since

$$\text{area} \times \text{distance to centroid} = \text{moment}$$

then

$$\bar{x} = \frac{\text{moment about } y \text{ axis}, M_y}{\text{area}}$$

$$= \frac{396 \text{ cm}^3}{76 \text{ cm}^2} = 5.21 \text{ cm}$$

The moment M_x about the x axis is

$$M_x = 40(8) + 16(4) + 20(1) = 404 \text{ cm}^3$$

So

$$\bar{y} = \frac{\text{moment about } x \text{ axis}, M_x}{\text{area}}$$

$$= \frac{404 \text{ cm}^3}{76 \text{ cm}^2} = 5.32 \text{ cm}$$

Centroids by Integration

If an area does not have axes of symmetry whose intersection gives us the location of the centroid, we can often find it by integration. We subdivide the area into thin strips, compute the first moment of each, sum these moments by integration, and then divide by the total area to get the distance to the centroid.

Consider the area bounded by the curves $y_1 = f_1(x)$ and $y_2 = f_2(x)$ and the lines $x = a$ and $x = b$ (Fig. 32–13). We draw a vertical element of area of width dx and height $(y_2 - y_1)$. Since the strip is narrow, all points on it may be considered to be the same distance x from the y axis. The moment dM_y of that strip about the y axis is then

$$dM_y = x \, dA = x(y_2 - y_1) \, dx$$

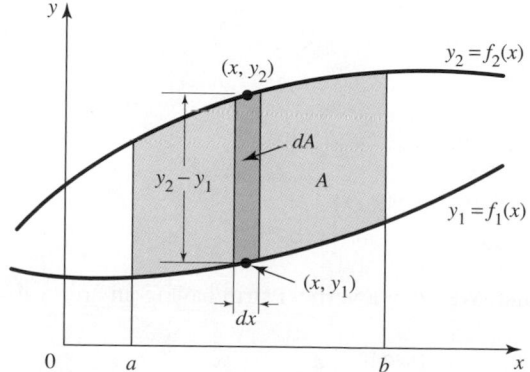

FIGURE 32–13   Centroid of irregular area found by integration.

since $dA = (y_2 - y_1) \, dx$. We get the total moment M_y by integrating.

$$M_y = \int_a^b x(y_2 - y_1) \, dx$$

But since the moment M_y is equal to the area A times the distance $\bar{x}$ to the centroid, we get $\bar{x}$ by dividing the moment by the area. Thus,

Horizontal Distance to Centroid	$\bar{x} = \dfrac{1}{A} \displaystyle\int_a^b x(y_2 - y_1) \, dx$	400

To find $\bar{x}$, we must have the area A. It can be found by integration as in Chapter 31.

We now find the moment about the x axis. The centroid of the vertical element is at its midpoint, which is at a distance of $(y_1 + y_2)/2$ from the x axis. Thus, the moment of the element about the x axis is

$$dM_x = \frac{y_1 + y_2}{2} (y_2 - y_1) \, dx$$

Integrating and dividing by the area gives us $\bar{y}$.

Equations 400 and 401 apply only for *vertical elements*. When using horizontal elements, interchange x and y in these equations.

Vertical Distance to Centroid	$\bar{y} = \dfrac{1}{2A} \displaystyle\int_a^b (y_1 + y_2)(y_2 - y_1) \, dx$	401

Our first example is for an area bounded by one curve and the x axis.

◆◆◆ **Example 6:** Find the centroid of the area bounded by the parabola $y^2 = 4x$, the x axis, and the line $x = 1$ (Fig. 32–14).

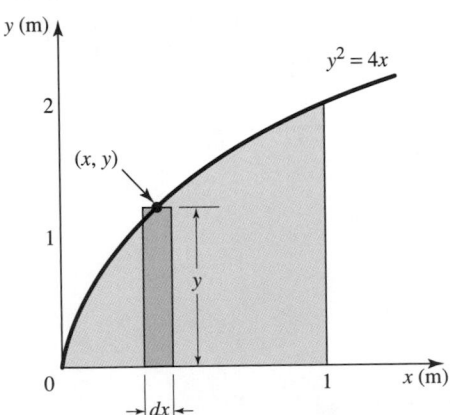

FIGURE 32–14

Solution: We need the area, so we find that first. We draw a vertical strip having an area $y \, dx$ and integrate.

$$A = \int y \, dx = 2 \int_0^1 x^{1/2} \, dx = 2 \left[\frac{x^{3/2}}{3/2} \right]_0^1 = \frac{4}{3} \, \text{m}^2$$

Then, by Eq. 400,

$$x = \frac{1}{A} \int_0^1 x(2x^{1/2} - 0)\, dx$$

$$= \frac{3}{4} \int_0^1 2x^{3/2} dx = \frac{3}{2} \cdot \frac{x^{5/2}}{(5/2)}\Big|_0^1 = \frac{3}{5}\, \text{m}$$

and, by Eq. 401,

$$y = \frac{1}{2A} \int_0^1 (2x^{1/2} + 0)(2x^{1/2} - 0)\, dx$$

$$= \frac{3}{8} \int_0^1 4x\, dx = \frac{3}{8} \cdot \frac{4x^2}{2}\Big|_0^1 = \frac{3}{4}\, \text{m}$$

Check: Does the answer seem reasonable? The given area extends from $x = 0$ to 1, with more area lying to the right of $x = \frac{1}{2}$ than to the left. Thus, we would expect $\bar{x}$ to be between $x = \frac{1}{2}$ and 1, which it is. Similarly, we would expect the centroid to be located between $y = 0$ and 1, which it is.   ◆◆◆

For areas that are not bounded by a curve and a coordinate axis, but are instead bounded by two curves, the work is only slightly more complicated, as shown in the following example.

◆◆◆ **Example 7:** Find the coordinates of the centroid of the area bounded by the curves $6y = x^2 - 4x + 4$ and $3y = 16 - x^2$.

Solution: We plot the curves as shown in Fig. 32–15 and find their points of intersection by solving simultaneously. Multiplying the second equation by -2 and adding the resulting equation to the first gives

$$3x^2 - 4x - 28 = 0$$

Solving by quadratic formula (work not shown), we find that the points of intersection are at $x = -2.46$ and $x = 3.79$. We take a vertical strip whose width is dx and whose height is $y_2 - y_1$, where

$$y_2 - y_1 = \frac{16}{3} - \frac{x^2}{3} - \frac{x^2}{6} + \frac{4x}{6} - \frac{4}{6} = \frac{28 + 4x - 3x^2}{6}$$

The area is then

$$A = \int_{-2.46}^{3.79} (y_2 - y_1)\, dx$$

$$= \frac{1}{6} \int_{-2.46}^{3.79} (28 + 4x - 3x^2)\, dx = 20.4$$

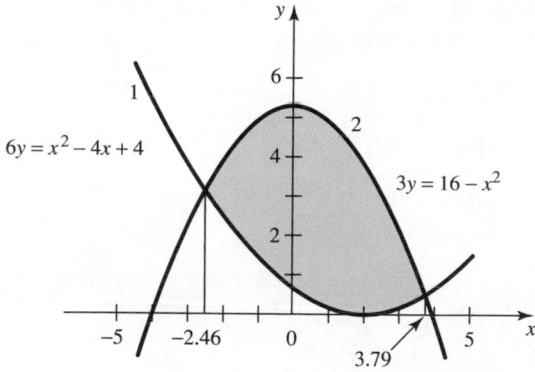

$6y = x^2 - 4x + 4$

$3y = 16 - x^2$

FIGURE 32–15

Then, from Eq. 400,

$$M_y = A\overline{x} = \int_{-2.46}^{3.79} x(y_2 - y_1)\,dx$$

$$= \frac{1}{6}\int_{-2.46}^{3.79} (28x + 4x^2 - 3x^3)\,dx = 13.6$$

$$\overline{x} = \frac{M_y}{A} = \frac{13.6}{20.4} = 0.667$$

We now substitute into Eq. 401, with

$$y_1 + y_2 = \frac{36 - 4x - x^2}{6}$$

Thus,

$$M_x = A\overline{y} = \frac{1}{72}\int_{-2.46}^{3.79} (36 - 4x - x^2)(28 + 4x - 3x^2)\,dx$$

$$= \frac{1}{72}\int_{-2.46}^{3.79} (3x^4 + 8x^3 - 152x^2 + 32x + 1008)\,dx$$

$$= \frac{1}{72}\left[\frac{3x^5}{5} + \frac{8x^4}{4} - \frac{152x^3}{3} + \frac{32x^2}{2} + 1008x \right]_{-2.46}^{3.79} = 52.5$$

$$\overline{y} = \frac{M_x}{A} = \frac{52.5}{20.4} = 2.57$$

Check by Physical Model: Here our check will be to make a model of the given figure. Make a graph as in Fig. 32–15, being sure that you use the same scale for the x and y axes, and paste the graph to cardboard or thin metal. Cut out the shaded area and locate the centroid by suspending it as in Fig. 32–10 or by balancing it on the point of a tack. Compare your experimental result with the calculated one.　　　　　◆◆◆

Centroids of Solids of Revolution by Integration

A solid of revolution is, of course, symmetrical about the axis of revolution, so the centroid must be on that axis. We only have to find the position of the centroid along that axis. The procedure is similar to that for an area. We think of the solid as being subdivided into many small elements of volume, find the sum of the moments for each element by integration, and set this equal to the moment of the entire solid (the product of its total volume and the distance to the centroid). We then divide by the volume to obtain the distance to the centroid.

◆◆◆ **Example 8:** Find the centroid of a hemisphere of radius r.

Solution: We place the hemisphere on coordinate axes as shown in Fig. 32–16 and consider it as the solid obtained by rotating the first-quadrant portion of the curve $x^2 + y^2 = r^2$ about the x axis. Through the point (x, y) we draw an element of volume of radius y and thickness dx, at a distance x from the base of the hemisphere. Its volume is thus

$$dV = \pi y^2\,dx$$

Of course, these methods work only for a solid that is homogeneous, that is, one whose density is the same throughout.

and its moment about the base of the hemisphere (the y axis) is

$$dM_y = \pi x y^2\,dx = \pi x(r^2 - x^2)\,dx$$

Integrating gives us the total moment.

$$M_y = \pi\int_0^r x(r^2 - x^2)\,dx = \pi\int_0^r (r^2 x - x^3)\,dx$$

$$= \pi\left[\frac{r^2 x^2}{2} - \frac{x^4}{4} \right]_0^r = \frac{\pi r^4}{4}$$

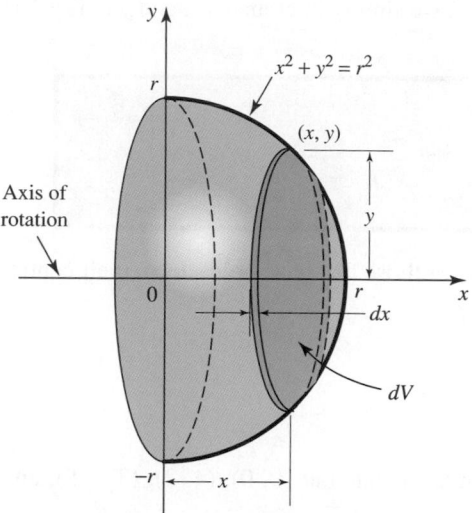

FIGURE 32-16 Finding the centroid of a hemisphere.

The total moment also equals the volume $(\frac{2}{3}\pi r^3$ for a hemisphere) times the distance $\bar{x}$ to the centroid, so

$$\bar{x} = \frac{\pi r^4/4}{2\pi r^3/3} = \frac{3r}{8}$$

Thus, the centroid is located at $(3r/8, 0)$. ◆◆◆

As with centroids of areas, we can write a formula for finding the centroid of a solid of revolution. For the volume V (Fig. 32–17) formed by rotating the curve $y = f(x)$ about the x axis, the distance to the centroid is the following:

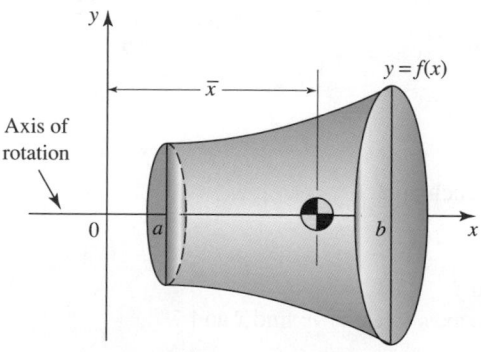

FIGURE 32-17 Centroid of a solid of revolution.

Distance to the Centroid of Volume of Revolution about *x* Axis	$\bar{x} = \dfrac{\pi}{V}\displaystyle\int_a^b xy^2\,dx$	402

For volumes formed by rotation about the y axis, we simply interchange x and y in Eq. 402 and get

Distance to the Centroid of Volume of Revolution about y Axis	$\bar{y} = \dfrac{\pi}{V} \displaystyle\int_c^d yx^2 \, dy$	403

These formulas work for solid figures only, not for those that have holes down their centre.

Exercise 3 ◆ Centroids

Without Integration

1. Find the centroid of four particles of equal mass located at $(0, 0)$, $(4, 2)$, $(3, -5)$, and $(-2, -3)$.
2. Find the centroid in Fig. 32–18(a).
3. Find the centroid in Fig. 32–18(b).
4. Find the centroid in Fig. 32–18(c).

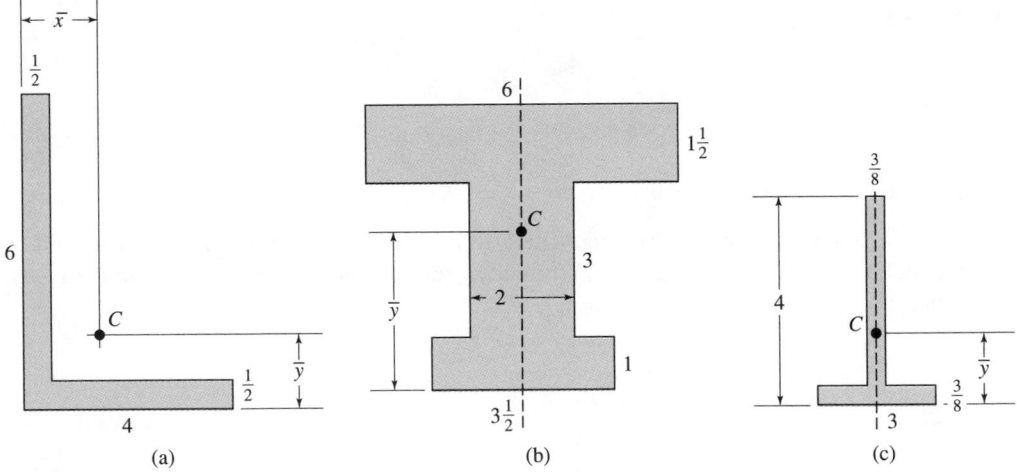

(a)   (b)   (c)

FIGURE 32–18

Centroids of Areas by Integration

Find the specified coordinate(s) of the centroid of each area.
5. bounded by $y^2 = 4x$ and $x = 4$; find $\bar{x}$ and $\bar{y}$.
6. bounded by $y^2 = 2x$ and $x = 5$; find $\bar{x}$ and $\bar{y}$.
7. bounded by $y = x^2$, the x axis, and $x = 3$; find $\bar{y}$.
8. bounded by $\sqrt{x} + \sqrt{y} = 1$ and the coordinate axes (area $= \frac{1}{6}$); find $\bar{x}$ and $\bar{y}$.

Areas Bounded by Two Curves

Find the specified coordinate(s) of the centroid of each area.
9. bounded by $y = x^3$ and $y = 4x$ in the first quadrant; find $\bar{x}$.
10. bounded by $x = 4y - y^2$ and $y = x$; find $\bar{y}$.
11. bounded by $y^2 = x$ and $x^2 = y$; find $\bar{x}$ and $\bar{y}$.
12. bounded by $y^2 = 4x$ and $y = 2x - 4$; find $\bar{y}$.
13. bounded by $2y = x^2$ and $y = x^3$; find $\bar{x}$.
14. bounded by $y = x^2 - 2x - 3$ and $y = 6x - x^2 - 3$ (area $= 21.33$); find $\bar{x}$ and $\bar{y}$.
15. bounded by $y = x^2$ and $y = 2x + 3$ in the first quadrant; find $\bar{x}$.

Centroids of Volumes of Revolution

Find the distance from the origin to the centroid of each volume.

16. formed by rotating the area bounded by $x^2 + y^2 = 4$, $x = 0$, $x = 1$, and the x axis about the x axis.

17. formed by rotating the area bounded by $6y = x^2$, the line $x = 6$, and the x axis about the x axis.

18. formed by rotating the first-quadrant area under the curve $y^2 = 4x$, from $x = 0$ to 1, about the x axis.

19. formed by rotating the area bounded by $y^2 = 4x$, $y = 6$, and the y axis about the y axis.

20. formed by rotating the first-quadrant portion of the ellipse $x^2/64 + y^2/36 = 1$ about the x axis.

21. a right circular cone of height h, measured from its base.

Applications

22. A certain airplane rudder (Fig. 32–19) consists of one quadrant of an ellipse and a quadrant of a circle. Find the coordinates of the centroid.

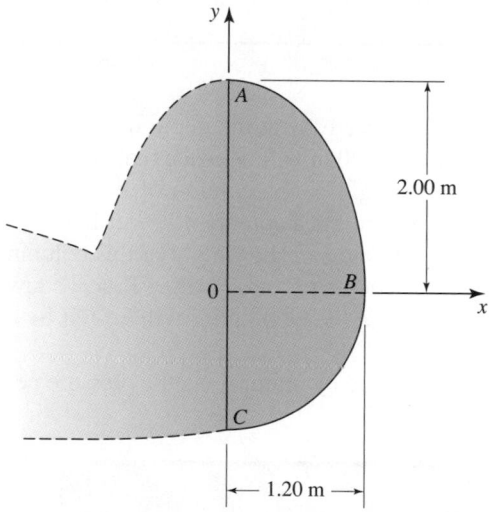

FIGURE 32–19   Airplane rudder.

23. The vane on a certain wind generator has the shape of a semicircle attached to a trapezoid (Fig. 32–20). Find the distance $\bar{x}$ to the centroid.

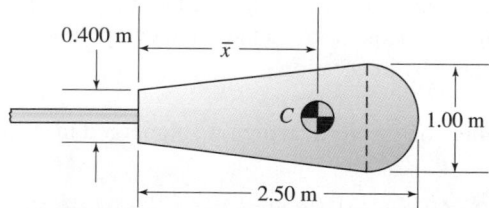

FIGURE 32–20   Wind vane.

24. A certain rocket (Fig. 32–21) consists of a cylinder attached to a paraboloid of revolution. Find the distance from the nose to the centroid of the total volume of the rocket.

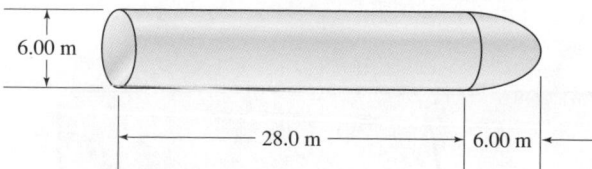

FIGURE 32–21   Rocket.

25. An optical instrument contains a mirror in the shape of a paraboloid of revolution (Fig. 32–22) hollowed out of a cylindrical block of glass. Find the distance from the flat bottom of the mirror to the centroid of the mirror.

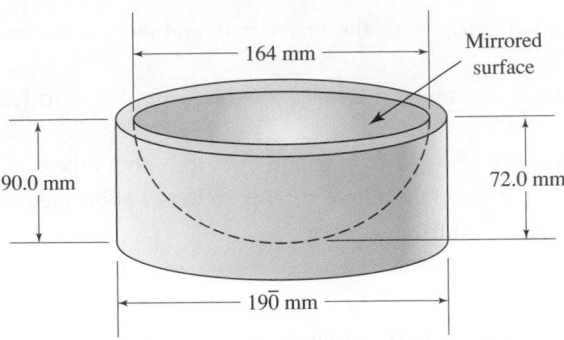

FIGURE 32–22   Paraboloidal mirror.

32–4   Fluid Pressure

The pressure at any point on a submerged surface varies directly with the depth of that point below the surface. Thus, the pressure on a diver at a depth of 20 m will be twice that at 10 m.

The pressure on a submerged area is equal to the weight of the column of fluid above that area. Thus, a square metre of area at a depth of 7 m supports a column of water having a volume of $7 \times 1^2 = 7$ m^3. Since the density of water is 1000 kg/m^3, the weight of this column is $7 \times 1000 = 7000$ kg, so the pressure is 7000 kg/m^2 at a depth of 7 m. Further, Pascal's law says that the pressure is the same in *all directions*, so the same 7000 kg/m^2 will be felt by a surface that is horizontal, vertical, or at any angle.

The *force* exerted by the fluid (*fluid pressure*) can be found by multiplying the pressure per unit area at a given depth by the total area.

Recall that

weight = volume × density

(Eq. A43).

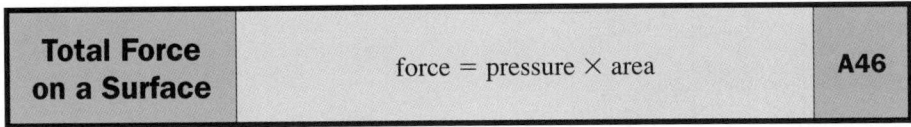

| Total Force on a Surface | force = pressure × area | A46 |

A complication arises from an area that has points at various depths and hence has different pressures over its surface. To compute the force on such a surface, we first compute the force on a narrow horizontal strip of area, assuming that the pressure is the same everywhere on that strip, and then add up the forces on all such strips by integration.

◆◆◆ **Example 9:** Find an expression for the force on the vertical area A submerged in a fluid of density δ (Fig. 32–23).

Solution: Let us take our origin at the surface of the fluid, with the y axis downward. We draw a horizontal strip whose area is dA, located at a depth y below the surface. The pressure at depth y is $y\delta$, so the force dF on the strip is, by Eq. A46,

$$dF = y\delta \, dA = \delta(y \, dA)$$

Integrating, we get the following equation:

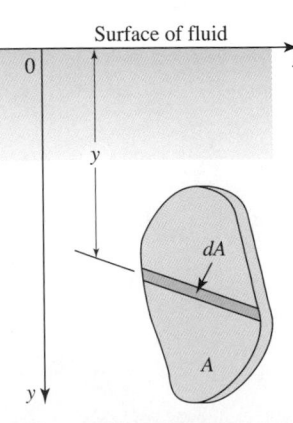

FIGURE 32–23   Force on a submerged vertical surface. Sometimes it is more convenient to take the origin at the surface of the liquid.

| Force of Pressure | $F = \delta \int y \, dA$ | A47 |

But the product $y\, dA$ is nothing but *the first moment of the area dA about the x axis*. Thus, integration will give us the moment M_x of the entire area, about the x axis, multiplied by the density.

$$F = \delta \int y\, dA = \delta M_x$$

But the moment M_x is also equal to the area A times the distance $\bar{y}$ to the centroid, so we have the following:

Force of Pressure	$F = \delta \bar{y} A$	A48

Thus, *the force of pressure on a submerged, vertical surface is equal to the product of its area, the distance to its centroid, and the density of the fluid.* ◆◆◆

We can compute the force on a submerged surface by using either Eq. A47 or A48. We give now an example of each method.

◆◆◆ **Example 10:** A vertical wall in a dam (Fig. 32–24) holds back water whose level is at the top of the wall. Use integration to find the total force of pressure on the wall.

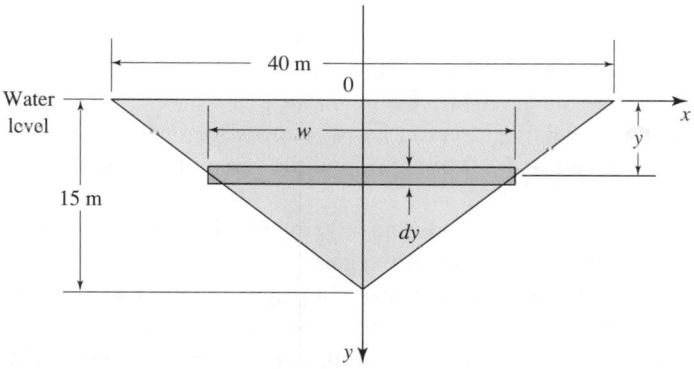

FIGURE 32–24

Estimate: The centroid of a triangle is at the one-third point, here 5 m below the surface. Then force = (density) × (distance to centroid) × (area), or $1000(5)\left(\frac{1}{2}\right)(40)(15) = 1\,500\,000$ kg.

Solution: We sketch an element of area with width w and height dy. So

$$dA = w\, dy \qquad (1)$$

We wish to express w in terms of y. By similar triangles,

$$\frac{w}{40} = \frac{15 - y}{15}$$

$$w = \frac{40}{15}(15 - y)$$

Substituting into Equation (1) yields

$$dA = \frac{40}{15}(15 - y)\, dy$$

Then, by Eq. A47,

$$F = \delta \int y \, dA = \frac{40\delta}{15} \int_0^{15} y(15 - y) \, dy$$

$$= \frac{40\delta}{15} \int_0^{15} (15y - y^2) \, dy$$

$$= \frac{40\delta}{15} \left[\frac{15y^2}{2} - \frac{y^3}{3} \right]_0^{15}$$

$$= \frac{40(1000)}{15} \left[\frac{15(15)^2}{2} - \frac{(15)^3}{3} \right]$$

$$= 1\,500\,000 \text{ kg, or } 1500 \text{ t} \qquad \blacklozenge\blacklozenge\blacklozenge$$

◆◆◆ **Example 11:** The area shown in Fig. 32–14 is submerged in water (density = 1000 kg/m³) so that the origin is 10.0 m below the surface. Find the force on the area using Eq. A48.

Solution: The area and the distance to the centroid have already been found in Example 6 of Sec. 32–3: area = $\frac{4}{3}$ m² and $\bar{y} = \frac{3}{4}$ m up from the origin, as shown in Fig. 32–25. The depth of the centroid below the surface is then

$$10.0 - \frac{3}{4} = 9\frac{1}{4} \text{ m}$$

Thus, by Eq. A48,

$$\text{force of pressure} = \frac{1000 \text{ kg}}{\text{m}^3} \cdot 9\frac{1}{4} \text{ m} \cdot \frac{4}{3} \text{ m}^2 = 12\,300 \text{ kg, or } 12.3\text{t} \qquad \blacklozenge\blacklozenge\blacklozenge$$

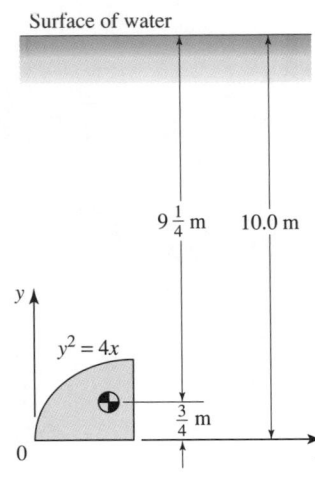

FIGURE 32–25

Exercise 4 ◆ Fluid Pressure

1. A circular plate 2.00 m in diameter is placed vertically in a dam with its centre 15.0 m below the surface of the water. Find the force of pressure on the plate.
2. A vertical rectangular plate in the wall of a reservoir has its upper edge 10.0 m below the surface of the water. It is 3.0 m wide and 2.0 m high. Find the force of pressure on the plate.
3. A vertical rectangular gate in a dam is 5.00 m wide and 3.00 m high. Find the force on the gate when the water level is 4.00 m above the top of the gate.
4. A vertical cylindrical tank has a diameter of 10.0 m and a height of 16.0 m. Find the total force on the curved surface when the tank is full of water.
5. A trough, whose cross section is an equilateral triangle with a vertex down, has sides 2.00 m long. Find the total force on one end when the trough is full of water.
6. A horizontal cylindrical boiler 4.00 m in diameter is half full of water. Find the force on one end.
7. A horizontal tank of oil (density = 60.0 lb/ft³) has ends in the shape of an ellipse with horizontal axis 12.0 ft. long and vertical axis 6.00 ft. long (Fig. 32–26). Find the force on one end when the tank is half full.

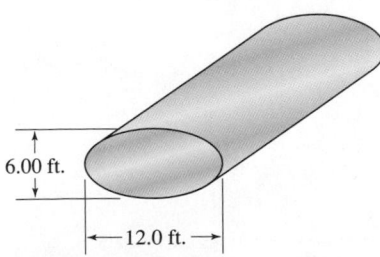

FIGURE 32–26　Oil tank.

8. The cross section of a certain trough is a parabola with vertex down. It is 2.00 m deep and 4.00 m wide at the top. Find the force on one end when the trough is full of water.

32–5 Work

Definition

When a constant force acts on an object that moves in the direction of the force, the *work* done by the force is defined as the product of the force and the distance moved by the object.

Work Done by a Constant Force	work = force × distance	A6

◆◆◆ **Example 12:** The work needed to lift a 100-N weight a distance of 2 m is 200 N·m. ◆◆◆

Variable Force

Equation A6 applies when the force is *constant,* but this is not always the case. For example, the force needed to stretch or compress a spring increases as the spring gets extended (Fig. 32–27). Or, as another example, the expanding gases in an automobile cylinder exert a variable force on the piston.

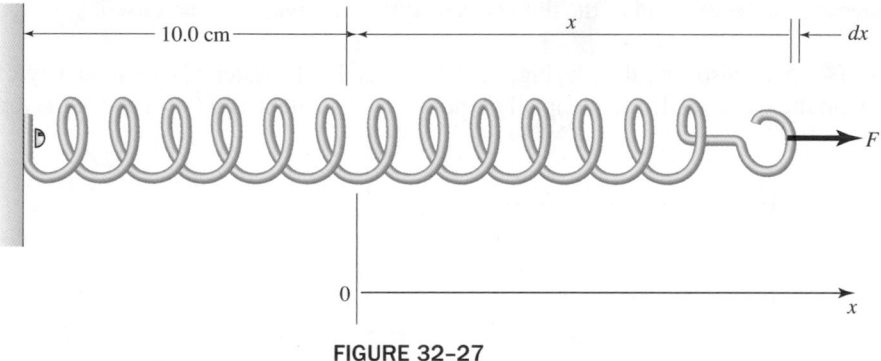

FIGURE 32–27

If we let the variable force be represented by $F(x)$, acting in the x direction from $x = a$ to $x = b$, the work done by this force may be defined as follows:

Work Done by a Variable Force	$$W = \int_{a}^{b} F(x)\,dx$$	A7

We first apply this formula to find the work done in stretching or compressing a spring.

◆◆◆ **Example 13:** A certain spring (Fig. 32–27) has a free length (when no force is applied) of 10.0 cm. The spring constant is 12.0 N/cm. Find the work needed to stretch the spring from a length of 12.0 cm to a length of 14.0 cm.

Estimate: The amount of stretch starts at 2 cm, requiring a force of 2(12), or 24 N, and ends at 4 cm with a force of 48 N. Assuming an average force of 36 N over the 2-cm travel gives an estimate of 72 N·cm.

Solution: We draw the spring partly stretched, as shown, taking our x axis in the direction of movement of the force, with zero at the free position of the end of the spring. The force needed to hold the spring in this position is equal to the spring constant k times the deflection x.

Force Needed to Deform a Spring	$F = kx$	A61

If we assume that the force does not change when stretching the spring an additional small amount dx, the work done is

$$dW = F\,dx = kx\,dx$$

We get the total work by integrating.

$$W = \int F\,dx = k\int_{2}^{4} x\,dx = \frac{kx^2}{2}\Big|_{2}^{4}$$

$$= \frac{12.0}{2}(4^2 - 2^2) = 72.0\ \text{N·cm}$$ ◆◆◆

Common Error	Be sure to measure spring deflections from the free end of the spring, not from the fixed end.

Another typical problem is that of finding the work needed to pump the fluid out of a tank. Such problems may be solved by noting that the work required is equal to the weight of the fluid times the distance that the centroid of the fluid (when still in the tank) must be raised.

◆◆◆ **Example 14:** A hemispherical tank (Fig. 32–28) is filled with water having a density of 1000 kg/m³. Find the work needed to pump all of the water to a height of 3.25 m above the top of the tank.

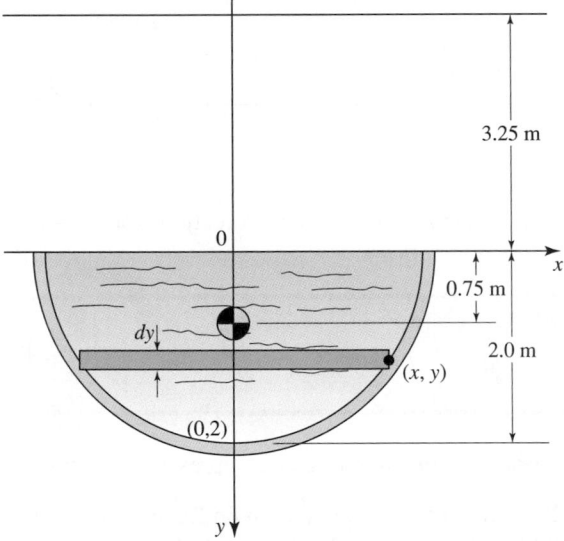

FIGURE 32–28   Hemispherical tank.

Solution: The distance to the centroid of a hemisphere of radius r was found in Example 8 to be $3r/8$, so the centroid of our hemisphere is at a distance

$$\overline{y} = \frac{3}{8}\cdot 2 = 0.75\ \text{m}$$

as shown. It must therefore be raised a distance of 4.00 m. We find the weight of the tankful of water by multiplying its volume times its density.

$$\text{weight} = \text{volume} \times \text{density}$$

$$= \frac{2}{3}\pi(2^3)(1000) = 16\,755\ \text{kg}$$

The work done is then

$$\text{work} = 16\ 755\ \text{kg}(9.81\ \text{N/kg})(4.00\ \text{m}) = 657\ 000\ \text{N·m}$$

◆◆◆

In Example 14, the factor 9.81 N/kg is used to convert kg·m to N·m, the accepted SI unit of work.

If we do not know the location of the centroid, we can find the work directly by integration.

◆◆◆ **Example 15:** Repeat Example 14 by integration, assuming that the location of the centroid is not known.

Solution: We choose coordinate axes as shown in Fig. 32–28. Through some point (x, y) on the curve, we draw an element of volume whose volume dV is

$$dV = \pi x^2\ dy$$

and whose weight is

$$1000(9.81)\ dV = 9810\pi x^2\ dy\ (\text{N})$$

Since this element must be lifted a distance of $(3.25 + y)$ m, the work required is

$$dW = (3.25 + y)(9810\pi x^2\ dy)$$

Integrating gives us

$$W = 9810\pi \int (3.25 + y)x^2\ dy$$

But using the equation for a circle (Eq. 304), we have

$$x^2 = r^2 - y^2 = 4 - y^2$$

We substitute to get the expression in terms of y.

$$W = 9810\pi \int_0^2 (3.25 + y)(4 - y^2)\ dy$$

$$= 9810\pi \int_0^2 (13 + 4y - 3.25y^2 - y^3)\,dy$$

$$= 9810\pi \left[13y + 2y^2 - \frac{3.25y^3}{3} - \frac{y^4}{4} \right]_0^2 = 657\ 000\ \text{N·m}$$

as by the method of Example 14.

◆◆◆

Exercise 5 ◆ Work

Springs

1. A force of 50.0 N will stretch a spring that has a free length of 12.0 cm to a length of 13.0 cm. How much work is needed to stretch the spring from a length of 14.0 cm to 16.0 cm?

2. A spring whose free length is 10.0 cm has a spring constant of 12.0 N·cm. Find the work needed to stretch this spring from 12.0 cm to 15.0 cm.

3. A spring has a spring constant of 8.0 N/cm and a free length of 5.0 cm. Find the work required to stretch it from 6.0 cm to 8.0 cm.

Tanks

4. Find the work required to pump all of the water to the top and out of a vertical cylindrical tank that is 6.00 m in diameter, 8.00 m deep, and completely filled at the start.

5. A hemispherical tank 4.00 m in diameter is filled with water to a depth of 1.50 m. How much work is needed to pump the water to the top of the tank?

6. A conical tank 8.00 m deep and 8.00 m across the top is full of water. Find the work needed to pump the water to a height of 6.00 m above the top of the tank.

7. A tank has the shape of a frustum of a cone, with a bottom diameter of 8.00 ft., a top diameter of 12.0 ft., and a height of 10.0 ft. How much work is needed to pump the contents to a height of 10.0 ft. above the tank, if it is filled with oil of density 50.0 lb./cu. ft.?

Gas Laws

8. Find the work needed to compress air initially at a pressure of 105 kPa from a volume of 10.0 m³ to 3.00 m³. [*Hint:* The work dW done in moving the piston (Fig. 32–29) is $dW = F\,dx$. Express both F and dx in terms of v by means of $F = pA$ and $A\,dx = dv$, and integrate.]

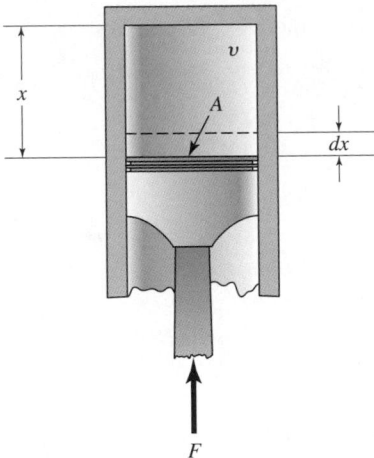

FIGURE 32–29   Piston and cylinder. Assume that the pressure and volume are related by the equation $pv =$ constant.

9. Air is compressed (Fig. 32–29) from an initial pressure of 105 kPa and volume of 10.0 m³ to a pressure of 525 kPa. How much work was needed to compress the air?

10. If the pressure and volume of air are related by the equation $pv^{1.4} = k$, find the work needed to compress air initially at 102 kPa and 0.500 m³ to a pressure of 692 kPa.

Miscellaneous

11. The force of attraction (in newtons) between two masses separated by a distance d is equal to k/d^2, where k is a constant. If two masses are 50 m apart, find the work needed to separate them another 50 m. Express your answer in terms of k.

12. Find the work needed to wind up a vertical cable $10\overline{0}$ m long, weighing 3.0 N/m.

13. A 200-m-long cable weighs 1.00 kg/m and is hanging from a tower with a $10\overline{0}$-kg weight at its end. How much work is needed to raise the weight and the cable a distance of 10.0 m?

32–6   Moment of Inertia

Moment of Inertia of an Area

Figures 32–30(a) and (b) each show a small area at a distance r from a line L in the same plane. In each case, the dimensions of the area are such that we may consider all points on the area as being at the same distance r from the line L.

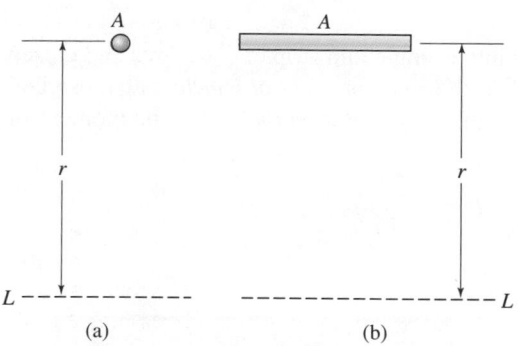

FIGURE 32–30

In Sec. 32–3 we defined the *first moment* of the area about L as being the product of the area times the distance to the line. We now define the *second moment, or moment of inertia I,* as the product of the area times the *square* of the distance to the line.

Moment of Inertia of an Area	$I = Ar^2$	404

The distance r, which is the distance to the line, is called the *radius of gyration.* We write I_x to denote the moment of inertia about the x axis, and I_y for the moment of inertia about the y axis.

◆◆◆**Example 16:** Find the moment of inertia of the thin strip in Fig. 32–30(b) if it has a length of 8.0 cm and a width of 0.20 cm and is 7.0 cm from axis L.

Solution: The area of the strip is $8.0(0.20) = 1.6 \text{ cm}^2$, so the moment of inertia is, by Eq. 404,

$$I = 1.6(7.0)^2 = 78.4 \text{ cm}^4 \qquad \text{◆◆◆}$$

Moment of Inertia of a Rectangle

In Example 16, our area was a thin strip parallel to the axis, with all points on the area at the same distance r from the axis. But what will we use for r when dealing with an extended area, such as the rectangle in Fig. 32–31? Again calculus comes to our aid. Since we can easily compute the moment of inertia of a thin strip, we slice our area into many thin strips and add up their individual moments of inertia by integration.

◆◆◆**Example 17:** Compute the moment of inertia of a rectangle about its base (Fig. 32–31).

Solution: We draw a single strip of area parallel to the axis about which we are taking the moment. This strip has a width dy and a length a. Its area is then

$$dA = a\,dy$$

All points on the strip are at a distance y from the x axis, so the moment of inertia, dI_x, of the single strip is

$$dI_x = y^2(a\,dy)$$

We add up the moments of all such strips by integrating from $y = 0$ to b.

$$I_x = a \int_0^b y^2 dy = \frac{ay^3}{3}\bigg|_0^b = \frac{ab^3}{3}$$

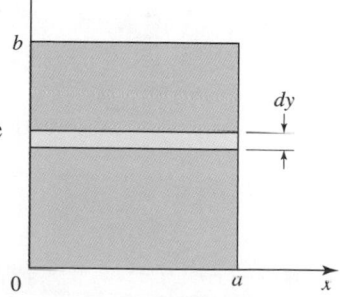

FIGURE 32–31 Moment of inertia of a rectangle.

Radius of Gyration

If all of the area in a plane figure were squeezed into a single thin strip of equal area and placed parallel to the x axis at such a distance that it had the *same moment of inertia* as the original rectangle, it would be at a distance r that we call the *radius of gyration*. If I is the moment of inertia of some area A about some axis, then

$$Ar^2 = I$$

Thus,

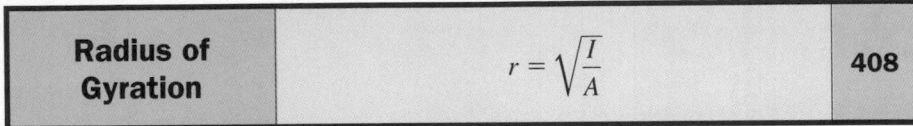

| Radius of Gyration | $r = \sqrt{\dfrac{I}{A}}$ | 408 |

We use r_x and r_y to denote the radius of gyration about the x and y axes, respectively.

◆◆◆ **Example 18:** Find the radius of gyration for the rectangle in Example 17.

Solution: The area of the rectangle is ab, so, by Eq. 408,

$$r_x = \sqrt{\frac{I}{A}} = \sqrt{\frac{ab^3}{3ab}} = \frac{b}{\sqrt{3}} \cong 0.577b$$

Note that the *centroid* is at a distance of $0.5b$ from the edge of the rectangle. This shows that *the radius of gyration is not equal to the distance to the centroid.*   ◆◆◆

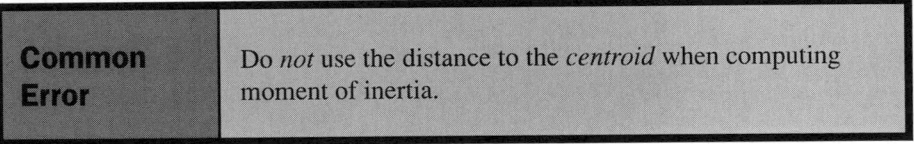

| Common Error | Do *not* use the distance to the *centroid* when computing moment of inertia. |

Moment of Inertia of an Area by Integration

Since we are now able to write the moment of inertia of a rectangular area, we can derive formulas for the moment of inertia of other areas, such as in Fig. 32–32. The area of the vertical strip shown in Fig. 32–32 is dA, and its distance from the y axis is x, so its moment of inertia about the y axis is, by Eq. 404,

$$dI_y = x^2\, dA = x^2 y\, dx$$

since $dA = y\, dx$. The total moment is then found by integrating.

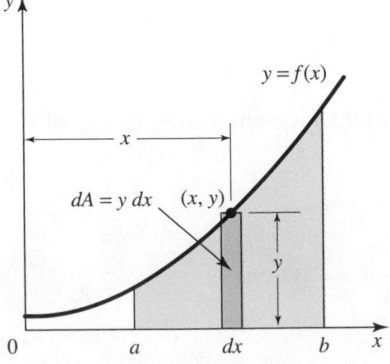

FIGURE 32–32

Moment of Inertia of an Area about y Axis	$I_y = \int x^2 y\, dx$	406

To find the moment of inertia about the x axis, we use the result of Example 17: that the moment of inertia of a rectangle about its base is $ab^3/3$. Thus, the moment of inertia of the rectangle in Fig. 32–32 is

$$dI_x = \frac{1}{3} y^3\, dx$$

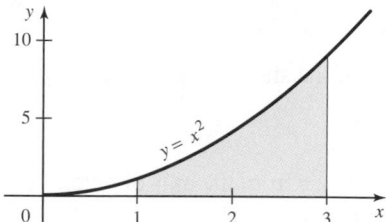

FIGURE 32–33

As before, the total moment of inertia is found by integrating.

Moment of Inertia of an Area about x Axis	$I_x = \frac{1}{3} \int y^3\, dx$	405

◆◆◆ **Example 19:** Find the moment of inertia about the x and y axes for the area under the curve $y = x^2$ from $x = 1$ to 3 (Fig. 32–33).

Solution: From Eq. 406,

$$I_y = \int_1^3 x^2 (x^2)\, dx$$

$$= \int_1^3 x^4\, dx$$

$$= \left.\frac{x^5}{5}\right|_1^3 \cong 48.4$$

Now using Eq. 405, with $y^3 = (x^2)^3 = x^6$,

$$I_x = \frac{1}{3} \int_1^3 x^6\, dx = \frac{1}{3}\left[\frac{x^7}{7}\right]_1^3 = \frac{1}{21}(3^7 - 1^7) = \frac{2186}{21} \cong 104.1 \qquad \text{◆◆◆}$$

◆◆◆ **Example 20:** Find the radius of gyration about the x axis of the area in Example 19.

Solution: We first find the area of the figure. By Eq. 376,

$$A = \int_1^3 x^2\, dx = \left.\frac{x^3}{3}\right|_1^3 = \frac{3^3}{3} - \frac{1^3}{3} \cong 8.667$$

From Example 19, $I_x = 104.1$. So, by Eq. 408,

$$r_x = \sqrt{\frac{I_x}{A}} = \sqrt{\frac{104.1}{8.667}} = 3.466 \qquad \text{◆◆◆}$$

Polar Moment of Inertia

Polar moment of inertia is needed when studying rotation of rigid bodies.

In the preceding sections we found the moment of inertia of an area about some line in the plane of that area. Now we find moment of inertia of a *solid of revolution* about its axis of revolution. We call this the *polar moment of inertia*. We first find the polar moment of inertia for a thin-walled shell and then use that result to find the polar moment of inertia for any solid of revolution.

Polar Moment of Inertia by the Shell Method

A thin-walled cylindrical shell (Fig. 32–34) has a volume equal to the product of its circumference, wall thickness, and height.

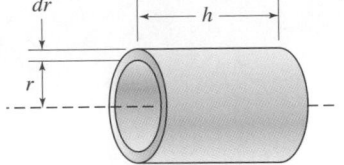

FIGURE 32–34

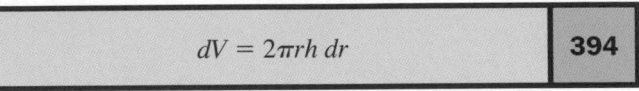

$dV = 2\pi rh\,dr$	**394**

If we let m represent the mass per unit volume, the mass of the shell is then

$$dM = 2\pi mrh\,dr$$

Since we may consider all particles in this shell to be at a distance r from the axis of revolution, we obtain the moment of inertia of the shell by multiplying the mass by r^2.

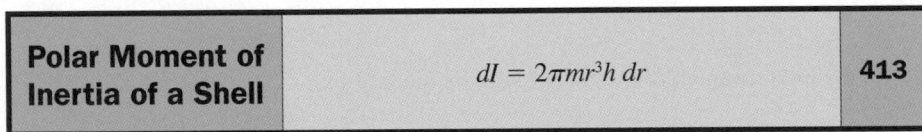

Polar Moment of Inertia of a Shell	$dI = 2\pi mr^3 h\,dr$	**413**

We then think of the entire solid of revolution as being formed by concentric shells. The polar moment of inertia of an entire solid of revolution is then found by integration.

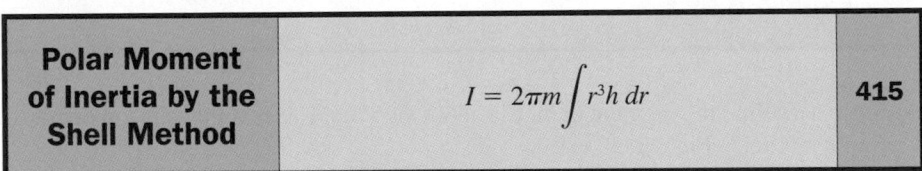

Polar Moment of Inertia by the Shell Method	$I = 2\pi m \displaystyle\int r^3 h\,dr$	**415**

◆◆◆ **Example 21:** Find the polar moment of inertia of a solid cylinder (Fig. 32–35) about its axis.

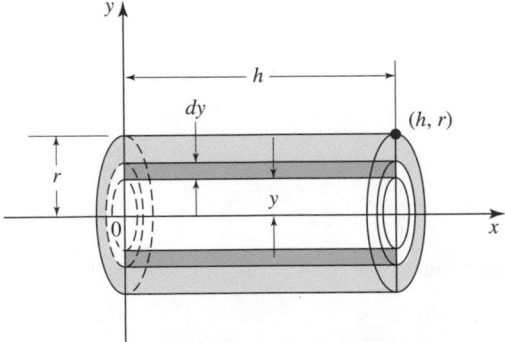

FIGURE 32–35   Polar moment of inertia of a solid cylinder.

Solution: We draw elements of volume in the form of concentric shells. The moment of inertia of each is

$$dI_x = 2\pi my^3 h\,dy$$

Integrating yields

$$I_x = 2\pi m h \int_0^r y^3\, dy = 2\pi m h \left[\frac{y^4}{4}\right]_0^r$$

$$= \frac{\pi m h r^4}{2}$$

But since the volume of the cylinder is $\pi r^2 h$, we get

$$I_x = \frac{1}{2} m V r^2$$

In other words, *the polar moment of inertia of a cylindrical solid is equal to half the product of its density, its volume, and the square of its radius.* ◆◆◆

◆◆◆ **Example 22:** The first-quadrant area under the curve $y^2 = 8x$, from $x = 0$ to 2, is rotated about the x axis. Use the shell method to find the polar moment of inertia of the paraboloid generated (Fig. 32–36).

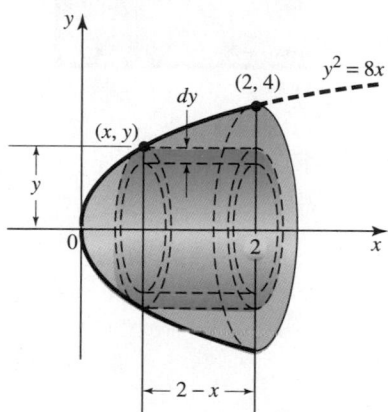

FIGURE 32–36 Polar moment of inertia by shell method.

Solution: Our shell element has a radius y, a length $2 - x$, and a thickness dy. By Eq. 413,

$$dI_x = 2\pi m y^3 (2 - x)\, dy$$

Replacing x by $y^2/8$ gives

$$dI_x = 2\pi m y^3 \left(2 - \frac{y^2}{8}\right) dy$$

Integrating gives us

$$I_x = 2\pi m \int_0^4 \left(2y^3 - \frac{y^5}{8}\right) dy = 2\pi m \left[\frac{2y^4}{4} - \frac{y^6}{48}\right]_0^4 \cong 268m \qquad \text{◆◆◆}$$

Polar Moment of Inertia by the Disk Method

Sometimes the disk method will result in an integral that is easier to evaluate than that obtained by the shell method.

For the solid of revolution in Fig. 32–37, we choose a disk-shaped element of volume of radius r and thickness dh. Since it is a cylinder, we use the moment of inertia of a cylinder found in Example 21.

$$dI = \frac{m\pi r^4\, dh}{2}$$

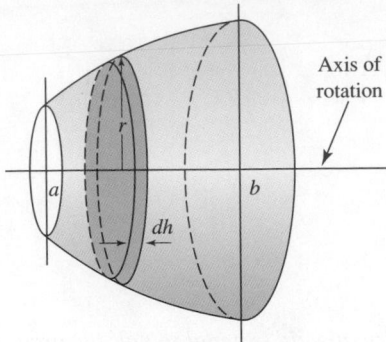

Axis of rotation

FIGURE 32–37   Polar moment of inertia by disk method.

Integrating gives the following:

Polar Moment of Inertia by the Disk Method	$I = \dfrac{m\pi}{2} \displaystyle\int_a^b r^4 \, dh$	414

◆◆◆ **Example 23:** Repeat Example 22 by the disk method.

Solution: We use Eq. 414 with $r = y$ and $dh = dx$.

$$I_x = \frac{m\pi}{2} \int y^4 \, dx$$

But $y^2 = 8x$, so $y^4 = 64x^2$. Substituting yields

$$I_x = 32m\pi \int_0^2 x^2 \, dx$$

$$= 32m\pi \left[\frac{x^3}{3} \right]_0^2$$

$$= 32m\pi \left(\frac{2^3}{3} \right) \cong 268m$$

as before.   ◆◆◆

Exercise 6 ◆ Moment of Inertia

Moment of Inertia of Plane Areas

1. Find the moment of inertia about the x axis of the area bounded by $y = x$, $y = 0$, and $x = 1$.

2. Find the radius of gyration about the x axis of the first-quadrant area bounded by $y^2 = 4x$, $x = 4$, and $y = 0$.

3. Find the radius of gyration about the y axis of the area in problem 2.

4. Find the moment of inertia about the y axis of the area bounded by $x + y = 3$, $x = 0$, and $y = 0$.

5. Find the moment of inertia about the x axis of the first-quadrant area bounded by the curve $y = 4 - x^2$ and the coordinate axes.

6. Find the moment of inertia about the y axis of the area in problem 5.

7. Find the moment of inertia about the x axis of the first-quadrant area bounded by the curve $y^3 = 1 - x^2$.

8. Find the moment of inertia about the y axis of the area in problem 1.

Polar Moment of Inertia

Find the polar moment of inertia of the volume formed when a first-quadrant area with the following boundaries is rotated about the x axis.

9. bounded by $y = x$, $x = 2$, and the x axis

10. bounded by $y = x + 1$, from $x = 1$ to 2, and the x axis

11. bounded by the curve $y = x^2$, the line $x = 2$, and the x axis

12. bounded by $\sqrt{x} + \sqrt{y} = 2$ and the coordinate axes

Find the polar moment of inertia of each solid with respect to its axis in terms of the total mass M of the solid.

13. a right circular cone of height h and base radius r

14. a sphere of radius r

15. a paraboloid of revolution bounded by a plane through the focus perpendicular to the axis of symmetry

••• CHAPTER 32 REVIEW PROBLEMS •••••••••••••••••••••••••••••••

1. Find the distance from centre to the centroid of a semicircle of radius 10.0.

2. A bucket that weighs 3.00 lb. and has a volume of 2.00 cu. ft. is filled with water. It is being raised at a rate of 5.00 ft./s while water leaks from the bucket at a rate of 0.0100 cu. ft./s. Find the work done in raising the bucket $10\overline{0}$ ft.

3. Find the centroid of the area bounded by one quadrant of the ellipse $x^2/16 + y^2/9 = 1$.

4. Find the radius of gyration of the area bounded by the parabola $y^2 = 16x$ and its latus rectum, with respect to its latus rectum.

5. A conical tank is 4.00 m in diameter at the top and is 6.00 m deep. It is filled with a liquid having a density of 1260 kg/m³. How much work is required to pump all of the liquid to the top of the tank?

6. Find the polar moment of inertia of a right circular cone of base radius r, height h, and mass M with respect to its axis.

7. A cylindrical, horizontal tank is 4.00 m in diameter and is half full of water. Find the force on one end of the tank.

8. Find the polar moment of inertia of a sphere of radius r and density m with respect to a diameter.

9. A spring has a free length of 12.0 cm and a spring constant of 5.45 N/cm. Find the work needed to compress it from a length of 11.0 cm to 9.00 cm.

10. Find the coordinates of the centroid of the area bounded by the curve $y^2 = 2x$ and the line $y = x$.

11. A horizontal cylindrical tank 3.00 m in diameter is half full of oil (962 kg/m³). Find the force on one end.

12. Find the length of the curve $y = \frac{4}{5}x^2$ from the origin to $x = 4$.

13. The area bounded by the parabolas $y^2 = 4x$ and $y^2 = x + 3$ is rotated about the x axis. Find the surface area of the solid generated.

14. Find the length of the parabola $y^2 = 8x$ from the vertex to one end of the latus rectum.

15. The area bounded by the curves $x^2 = 4y$ and $x - 2y + 4 = 0$ and by the y axis is rotated about the y axis. Find the surface area of the volume generated.

16. The cables on a certain suspension bridge hang in the shape of a parabola. The towers are $10\overline{0}$ m apart, and the cables dip 10.0 m below the tops of the towers. Find the length of the cables.

17. The first-quadrant area bounded by the curves $y = x^3$ and $y = 4x$ is rotated about the x axis. Find the surface area of the solid generated.

18. Find the centroid of a paraboloid of revolution bounded by a plane through the focus perpendicular to the axis of symmetry.

Writing

19. Suppose that you have designed a tank in the shape of a solid of revolution and have found its centroid by integration. Your manager, a practical person, insists that it is impossible to find the centre of gravity of something that isn't even built yet. Write a memo to your manager explaining how it is done.

Team Project

20. Find the perimeter of an ellipse whose major axis is 10 units and whose minor axis is 6 units. Check your answer in as many ways as you can think of.

Derivatives of Trigonometric, Logarithmic, and Exponential Functions

••• OBJECTIVES ••

When you have completed this chapter, you should be able to:
- Find the derivative of the sine and cosine functions.
- Solve applied problems requiring derivatives of the trigonometric functions.
- Find derivatives of the inverse trigonometric functions.
- Find derivatives of logarithmic functions.
- Find derivatives of exponential functions.

In this chapter we extend our ability to take derivatives to include the trigonometric, logarithmic, and exponential functions. This will enable us to solve a larger range of problems than was possible before. After learning the rules for taking derivatives of these functions, we apply them to problems quite similar to those in Chapters 28 and 29; that is, tangents, related rates, maximum-minimum, and the rest.

If you have not used the trigonometric, logarithmic, or exponential functions for a while, you might want to thumb quickly through that material before starting here.

33–1   Derivatives of the Sine and Cosine Functions

Derivative of sin *u* Approximated Graphically

Before deriving a formula for the derivative of the sine function, let us use a sketch to get an indication of its shape.

◆◆◆ **Example 1:** Graph $y = \sin x$, with x in radians. Use the slopes of the tangents at points on that graph to sketch the graph of the derivative.

Solution: We graph $y = \sin x$ as shown in Fig. 33–1(a) and the slopes as shown in Fig. 33–1(b). Note that the slope of the sine curve is zero at points A, B, C, and D, so the derivative curve must cross the x axis at points A', B', C', and D'. We estimate the slope to be 1 at points E and F, which gives us points E' and F' on the derivative curve. Similarly, the slope is -1 at G, giving us point G' on the derivative curve. We then note that the sine curve is rising from A to B and from C to D, so the derivative curve must be positive in those intervals. Similarly, the sine curve is falling from B to C, so the derivative curve is negative in this interval. Using all of this information, we sketch in the derivative curve.

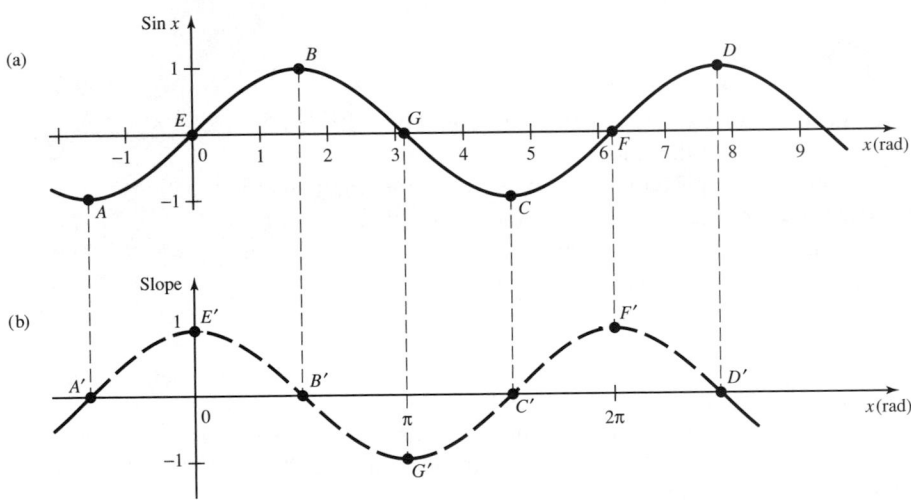

FIGURE 33–1   ◆◆◆

We see that the derivative of the sine curve seems to be another sine curve, but shifted by $\pi/2$ radians. We might recognize it as a cosine curve. So from our graphs it appears that

$$\frac{d(\sin x)}{dx} = \cos x$$

We'll soon see that this is, in fact, correct.

Limit of (sin x)/x

We will soon derive a formula for the derivative of $\sin x$. In that derivation we will need the limit of $(\sin x)/x$ as x approaches zero.

◆◆◆ **Example 2:** Find the limit of $(\sin x)/x$ (where x is in radians) as x approaches zero.

Solution: Let x be a small angle, in radians, in a circle of radius 1 (Fig. 33–2), in which

$$\sin x = \frac{BC}{1} = BC \quad \cos x = \frac{OC}{1} = OC \quad \tan x = \frac{AD}{1} = AD$$

We see that the area of the sector OAB is greater than the area of triangle OBC but less than the area of triangle OAD. But, by Eq. 137,

$$\text{area of triangle } OBC = \frac{1}{2} BC \cdot OC$$

$$= \frac{1}{2} \sin x \cos x$$

and

$$\text{area of triangle } OAD = \frac{1}{2} \cdot OA \cdot AD$$

$$= \frac{1}{2} \tan x$$

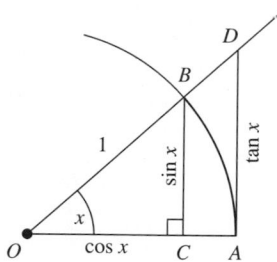

FIGURE 33–2

Also, by Eq. 116,

$$\text{area of sector } OAB = \frac{1}{2} r^2 x$$

$$= \frac{x}{2}$$

So

$$\frac{1}{2} \sin x \cos x < \frac{x}{2} < \frac{1}{2} \tan x$$

Dividing by $\frac{1}{2} \sin x$ gives us

$$\cos x < \frac{x}{\sin x} < \frac{1}{\cos x}$$

Now letting x approach 0, both $\cos x$ and $1/\cos x$ approach 1. Therefore $x/\sin x$, which is "sandwiched" between them, also approaches 1.

$$\lim_{x \to 0} \frac{x}{\sin x} = 1$$

Taking reciprocals yields

$$\lim_{x \to 0} \frac{\sin x}{x} = 1$$

Graphical Check: To check that this result is correct, we plot $(\sin x)/x$ as shown in Fig. 33–3 and see that the y intercept coordinates are (0, 1).

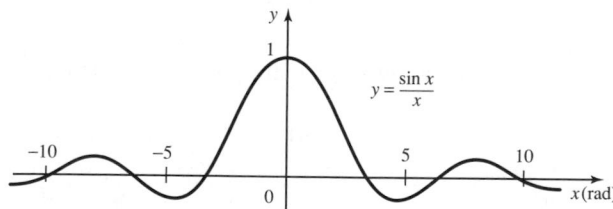

FIGURE 33–3

◆◆◆

Derivative of sin u Found Analytically

We want to be able to find derivatives of expressions such as

$$y = \sin(x^2 + 3x)$$

or expressions of the form $y = \sin u$, where u is a function of x. We will use the delta method to derive a rule for finding this derivative.

Recall that we start the delta method by giving an increment Δu to u, thus causing y to change by an amount Δy.

$$y = \sin u$$

$$y + \Delta y = \sin(u + \Delta u)$$

You might want to glance back at the delta method in Sec. 27–2.

Subtracting the original equation gives

$$\Delta y = \sin(u + \Delta u) - \sin u$$

We now make use of the identity (Eq. 177),

$$\sin \alpha - \sin \beta = 2 \cos \frac{\alpha + \beta}{2} \sin \frac{\alpha - \beta}{2}$$

to transform our equation for Δy into a more useful form. To use this identity, we let

$$\alpha = u + \Delta u \quad \text{and} \quad \beta = u$$

so

$$\Delta y = 2 \cos\left(\frac{u + \Delta u + u}{2}\right) \sin\left(\frac{u + \Delta u - u}{2}\right)$$

$$= 2[\cos(u + \Delta u/2)] \sin(\Delta u/2)$$

Now dividing by Δu, we have

$$\frac{\Delta y}{\Delta u} = \frac{2[\cos(u + \Delta u/2)] \sin(\Delta u/2)}{\Delta u}$$

Dividing numerator and denominator by 2, we get

$$\frac{\Delta y}{\Delta u} = \frac{[\cos(u + \Delta u/2)] \sin(\Delta u/2)}{\Delta u/2}$$

$$= [\cos(u + \Delta u/2)] \frac{\sin(\Delta u/2)}{\Delta u/2}$$

When we evaluated this limit in the preceding section, we required the angle ($\Delta u/2$ in this case) to be in radians. Thus, the formula we derive here also requires the angle to be in radians.

If we now let Δu approach zero, the quantity

$$\frac{\sin(\Delta u/2)}{\Delta u/2}$$

approaches 1, as we saw in Example 2. Also, the quantity $\Delta u/2$ will approach 0, leaving us with

$$\frac{dy}{du} = \cos u$$

Do you remember the chain rule? Glance back at Sec. 27–4.

Now, by the chain rule,

$$\frac{dy}{dx} = \frac{dy}{du} \cdot \frac{du}{dx}$$

Thus,

Derivative of the Sine	$\dfrac{d(\sin u)}{dx} = \cos u \dfrac{du}{dx}$	**350**

The derivative of the sine of some function is the cosine of that function, multiplied by the derivative of that function.

◆◆◆ **Example 3:**

(a) If $y = \sin 3x$, then

$$\frac{dy}{dx} = (\cos 3x) \cdot \frac{d}{dx} 3x$$

$$= 3 \cos 3x$$

(b) If $y = \sin(x^3 + 2x^2)$, then

$$y' = \cos(x^3 + 2x^2) \cdot \frac{d}{dx}(x^3 + 2x^2)$$

$$= \cos(x^3 + 2x^2)(3x^2 + 4x)$$

$$= (3x^2 + 4x) \cos(x^3 + 2x^2)$$

(c) If $y = \sin^3 x$, then by the power rule

$$y' = 3(\sin x)^2 \cdot \frac{d}{dx}(\sin x)$$

$$= 3 \sin^2 x \cos x \qquad ◆◆◆$$

> Recall that $\sin^3 x$ is the same as $(\sin x)^3$.

◆◆◆ **Example 4:** Find the slope of the tangent to the curve $y = x^2 - \sin^2 x$ at $x = 2$ (Fig. 33–4).

Solution: Taking the derivative yields

$$y' = 2x - 2 \sin x \cos x$$

At $x = 2$ rad,

$$y'(2) = 4 - 2 \sin 2 \cos 2$$

$$= 4 - 2(0.909)(-0.416)$$

$$= 4.757 \qquad ◆◆◆$$

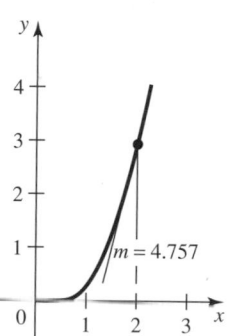

FIGURE 33–4 Graph of $y = x^2 - \sin^2 x$.

Common Error	Remember that x is in *radians* unless otherwise specified. Be sure that your calculator is in radian mode.

Derivative of cos u

We now take the derivative of $y = \cos u$.

We do not need the delta method again because we can relate the cosine to the sine with Eq. 154b, $\cos A = \sin B$, where A and B are complementary angles ($B = \pi/2 - A$; see Fig. 33–5). So

$$y = \cos u = \sin\left(\frac{\pi}{2} - u\right)$$

Then, by Eq. 350,

$$\frac{dy}{dx} = \cos\left(\frac{\pi}{2} - u\right)\left(-\frac{du}{dx}\right)$$

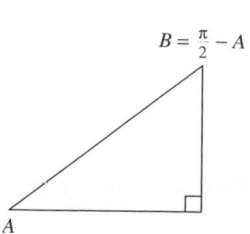

FIGURE 33–5

But $\cos(\pi/2 - u) = \sin u$, so we have the following equation:

Derivative of the Cosine	$\dfrac{d(\cos u)}{dx} = -\sin u \, \dfrac{du}{dx}$	351

The derivative of the cosine of some function is the negative of the sine of that function, multiplied by the derivative of that function.

◆◆◆ **Example 5:** If $y = \cos 3x^2$, then

$$y' = (-\sin 3x^2) \frac{d}{dx}(3x^2)$$

$$= (-\sin 3x^2)6x = -6x \sin 3x^2 \qquad \text{◆◆◆}$$

Common Error	Recall that $\quad \cos 3x^2$ *is not the same* as $(\cos 3x)^2$ and $\quad \cos 3x^2$ *is not the same* as $\cos(3x)^2$ but that $\quad \cos 3x^2$ *is the same* as $\cos(3x^2)$

Of course, our former rules for products, quotients, powers, and so forth, work for trigonometric functions as well.

◆◆◆ **Example 6:** Differentiate $y = \sin 3x \cos 5x$.

Solution: Using the product rule, we have

$$\frac{dy}{dx} = (\sin 3x)(-\sin 5x)(5) + (\cos 5x)(\cos 3x)(3)$$

$$= -5 \sin 3x \sin 5x + 3 \cos 5x \cos 3x \qquad \text{◆◆◆}$$

◆◆◆ **Example 7:** Differentiate $y = \dfrac{2 \cos x}{\sin 3x}$.

Solution: Using the quotient rule gives us

$$y' = \frac{(\sin 3x)(-2 \sin x) - (2 \cos x)(3 \cos 3x)}{(\sin 3x)^2}$$

$$= \frac{-2 \sin 3x \sin x - 6 \cos x \cos 3x}{\sin^2 3x} \qquad \text{◆◆◆}$$

◆◆◆ **Example 8:** Find the maximum and minimum points on the curve $y = 3 \cos x$.

Solution: We take the derivative and set it equal to zero.

$$y' = -3 \sin x = 0$$

$$\sin x = 0$$

$$x = 0, \pm\pi, \pm2\pi, \pm3\pi, \ldots$$

The second derivative is

$$y'' = -3 \cos x$$

which is negative when x equals 0, $\pm 2\pi$, $\pm 4\pi$, ..., so these are the locations of the maximum points. The others, where the second derivative is positive, are minimum points. These agree with the plot of the cosine curve, shown in Fig. 33–6.

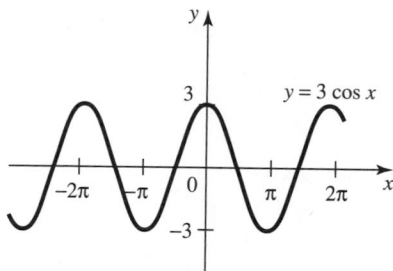

FIGURE 33–6

Implicit derivatives involving trigonometric functions are handled the same way as shown in Chapter 27.

◆◆◆ **Example 9:** Find dy/dx if $\sin xy = x^2 y$.

Solution:

$$(\cos xy)(xy' + y) = x^2 y' + 2xy$$

$$xy' \cos xy + y \cos xy = x^2 y' + 2xy$$

$$xy' \cos xy - x^2 y' = 2xy - y \cos xy$$

$$(x \cos xy - x^2)y' = 2xy - y \cos xy$$

$$y' = \frac{2xy - y \cos xy}{x \cos xy - x^2}$$

◆◆◆

Exercise 1 ◆ Derivatives of the Sine and Cosine Functions

First Derivatives

Find the derivative.

1. $y = \sin x$

2. $y = 3 \cos 2x$

3. $y = \cos^3 x$

4. $y = \sin x^2$

5. $y = \sin 3x$

6. $y = \cos 6x$

7. $y = \sin x \cos x$

8. $y = 15.4 \cos^5 x$

9. $y = 3.75 \, x \cos x$

10. $y = x^2 \cos x^2$

11. $y = \sin^2(\pi - x)$

12. $y = \dfrac{\sin \theta}{\theta}$

13. $y = \sin 2\theta \cos \theta$

14. $y = \sin^5 \theta$

15. $y = \sin^2 x \cos x$

16. $y = \sin 2x \cos 3x$

17. $y = 1.23 \sin^2 x \cos 3x$

18. $y = \dfrac{1}{2} \sin^2 x$

19. $y = \sqrt{\cos 2t}$

20. $y = 42.7 \sin^2 t$

Second Derivatives

For problems 21 through 23, find the second derivative of each function.

21. $y = \sin x$

22. $y = \dfrac{1}{4} \cos 2\theta$

23. $y = x \cos x$

24. If $f(x) = x^2 \cos^3 x$, find $f''(0)$.

25. If $f(x) = x \sin(\pi/2)x$, find $f''(1)$.

Implicit Functions

Find dy/dx for each implicit function.

26. $y \sin x = 1$

27. $xy - y \sin x - x \cos y = 0$

28. $y = \cos(x - y)$

29. $x = \sin(x + y)$

30. $x \sin y - y \sin x = 0$

Tangents

Find the slope of the tangent to four significant digits at the given value of x.

31. $y = \sin x$ at $x = 2$ rad

32. $y = x - \cos x$ at $x = 1$ rad

33. $y = x \sin \dfrac{x}{2}$ at $x = 2$ rad

34. $y = \sin x \cos 2x$ at $x = 1$ rad

Extreme Values and Inflection Points

For each curve, find the maximum, minimum, and inflection points between $x = 0$ and 2π.

35. $y = \sin x$

36. $y = \dfrac{x}{2} - \sin x$

37. $y = 3 \sin x - 4 \cos x$

Newton's Method

Sketch each function from $x = 0$ to 10. Calculate a root to three decimal places by Newton's method. If there is more than one root, find only the smallest positive root.

38. $\cos x - x = 0$

39. $\cos 2x - x = 0$

40. $3 \sin x - x = 0$

41. $2 \sin x - x^2 = 0$

42. $\cos x - 2x^2 = 0$

33–2   Derivatives of the Tangent, Cotangent, Secant, and Cosecant Functions

Derivative of tan *u*

We seek the derivative of $y = \tan u$. By Eq. 162,

$$y = \frac{\sin u}{\cos u}$$

Using the rule for the derivative of a quotient (Eq. 349), we have

$$\frac{dy}{dx} = \frac{\cos^2 u \dfrac{du}{dx} + \sin^2 u \dfrac{du}{dx}}{\cos^2 u}$$

$$= \frac{(\sin^2 u + \cos^2 u) \dfrac{du}{dx}}{\cos^2 u}$$

and, since $\sin^2 u + \cos^2 u = 1$,

$$\frac{dy}{dx} = \frac{1}{\cos^2 u}\frac{du}{dx}$$

and, since $1/\cos^2 u = \sec^2 u$,

$$\frac{dy}{dx} = \sec^2 u\,\frac{du}{dx}$$

Thus,

Derivative of the Tangent	$$\dfrac{d(\tan u)}{dx} = \sec^2 u\,\dfrac{du}{dx}$$	352

The derivative of the tangent of some function is the secant squared of that function, multiplied by the derivative of that function.

◆◆◆ **Example 10:**

(a) If $y = 3\tan x^2$, then

$$y' = 3(\sec^2 x^2)(2x) = 6x\sec^2 x^2$$

(b) If $y = 2\sin 3x\tan 3x$, then, by the product rule,

$$y' = 2[\sin 3x(\sec^2 3x)(3) + \tan 3x(\cos 3x)(3)]$$
$$= 6\sin 3x\sec^2 3x + 6\cos 3x\tan 3x$$

◆◆◆

Derivatives of cot u, sec u, and csc u

Each of these derivatives can be obtained using the rules for the sine, cosine, and tangent already derived and using the following identities:

$$\cot u = \frac{\cos u}{\sin u} \qquad \sec u = \frac{1}{\cos u} \qquad \csc u = \frac{1}{\sin u}$$

We list those derivatives here, together with those already found.

Derivatives of the Trigonometric Functions	$$\dfrac{d(\sin u)}{dx} = \cos u\,\dfrac{du}{dx}$$	350
	$$\dfrac{d(\cos u)}{dx} = -\sin u\,\dfrac{du}{dx}$$	351
	$$\dfrac{d(\tan u)}{dx} = \sec^2 u\,\dfrac{du}{dx}$$	352
	$$\dfrac{d(\cot u)}{dx} = -\csc^2 u\,\dfrac{du}{dx}$$	353
	$$\dfrac{d(\sec u)}{dx} = \sec u\tan u\,\dfrac{du}{dx}$$	354
	$$\dfrac{d(\csc u)}{dx} = -\csc u\cot u\,\dfrac{du}{dx}$$	355

You should memorize at least the first three of these. Notice how the signs alternate. The derivative of each cofunction is negative.

◆◆◆ Example 11:

(a) If $y = \sec(x^3 - 2x)$, then

$$y' = [\sec(x^3 - 2x) \tan(x^3 - 2x)](3x^2 - 2)$$

$$= (3x^2 - 2) \sec(x^3 - 2x) \tan(x^3 - 2x)$$

(b) If $y = \cot^3 5x$, then, by the power rule,

$$y' = 3(\cot 5x)^2(-\csc^2 5x)(5)$$

$$= -15 \cot^2 5x \csc^2 5x$$ ◆◆◆

◆◆◆ **Example 12:** In Fig. 33–7, link L is pivoted at B but slides along the fixed pin P. As slider C moves in the slot at a constant rate of 4.26 cm/s, the angle θ changes. Find the rate at which θ is changing when $S = 6.00$ cm.

Estimate: When $S = 6.00$ cm, θ is equal to $\arctan(8.63/6.00) = 55.2°$. After, say, 0.1 s, S has decreased by 0.426 cm, so then θ is equal to $\arctan(8.63/5.57) = 57.2°$. Thus, θ has increased by about 2° in 0.1 s, or about 20 deg/s.

Solution: S and θ are related by the tangent function, so

$$\tan \theta = \frac{8.63}{S}$$

or

$$S = \frac{8.63}{\tan \theta} = 8.63 \cot \theta$$

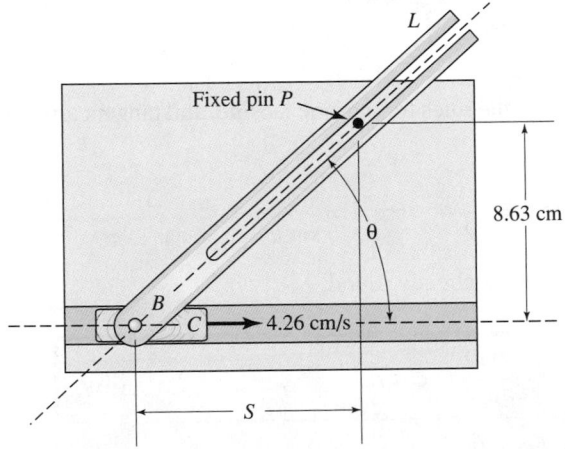

FIGURE 33–7   Pivoted link and slider.

Taking the derivative with respect to *time* gives

$$\frac{dS}{dt} = 8.63 \, (-\csc^2 \theta) \, \frac{d\theta}{dt}$$

Solving for $d\theta/dt$ gives

$$\frac{d\theta}{dt} = -\frac{1}{8.63 \csc^2 \theta} \frac{dS}{dt} = -\frac{\sin^2 \theta}{8.63} \frac{dS}{dt}$$

When $S = 6.00$ cm, $dS/dt = -4.26$ cm/s, and $\theta = 55.2°$ (from our estimate). Substituting, we get

$$\frac{d\theta}{dt} = -\frac{(\sin 55.2°)^2}{8.63}(-4.26) = 0.333 \text{ rad/s}$$

or 19.1 deg/s, which agrees well with our estimate. ◆◆◆

Exercise 2 ◆ Derivatives of the Tangent, Cotangent, Secant, and Cosecant Functions

First Derivative

Find the derivative.

1. $y = \tan 2x$ **2.** $y = \sec 4x$

3. $y = 5 \csc 3x$ **4.** $y = 9 \cot 8x$

5. $y = 3.25 \tan x^2$ **6.** $y = 5.14 \sec 2.11x^2$

7. $y = 7 \csc x^3$ **8.** $y = 9 \cot 3x^3$

9. $y = x \tan x$ **10.** $y = x \sec x^2$

11. $y = 5x \csc 6x$ **12.** $y = 9x^2 \cot 2x$

13. $w = \sin \theta \tan 2\theta$ **14.** $s = \cos t \sec 4t$

15. $v = 5 \tan t \csc 3t$ **16.** $z = 2 \sin 2\theta \cot 8\theta$

17. If $y = 5.83 \tan^2 2x$, find $y'(1)$.   **18.** If $f(x) = \sec^3 x$, find $f'(3)$.

19. If $f(x) = 3 \csc^3 3x$, find $f'(3)$.   **20.** If $y = 9.55x \cot^2 8x$, find $y'(1)$.

Second Derivative

Find the second derivative.

21. $y = 3 \tan x$ **22.** $y = 2 \sec 5\theta$

23. If $y = 3 \csc 2\theta$, find $y''(1)$.   **24.** If $f(x) = 6 \cot 4x$, find $f''(3)$.

Implicit Functions

Find dy/dx for each implicit function.

25. $y \tan x = 2$ **26.** $xy + y \cot x = 0$

27. $\sec(x + y) = 7$ **28.** $x \cot y = y \sec x$

Tangents

29. Find the equation of the tangent to the curve $y = \tan x$ at $x = 1$ rad.

30. Find the equation of the tangent to the curve $y = \sec 2x$ at $x = 2$ rad.

Extreme Values and Inflection Points

For each function, find any maximum, minimum, or inflection points between 0 and π.

31. $y = 2x - \tan x$

32. $y = \tan x - 4x$

Rate of Change

33. An object moves with simple harmonic motion so that its displacement y at time t is $y = 6 \sin 4t$ cm. Find the velocity and acceleration of the object when $t = 0.0500$ s.

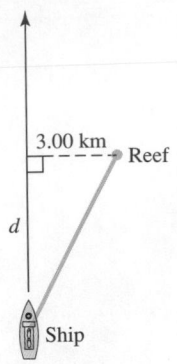

FIGURE 33-8

Related Rates

34. A ship is sailing at 10.0 km/h in a straight line (Fig. 33–8). It keeps its searchlight trained on a reef that is at a perpendicular distance of 3.00 km from the path of the ship. How fast (rad/h) is the light turning when the distance d is 5.00 km?

35. Two cables pass over fixed pulleys A and B (Fig. 33–9), forming an isosceles triangle ABP. Point P is being raised at the rate of 9.00 cm/min. How fast is θ changing when h is 1.50 m?

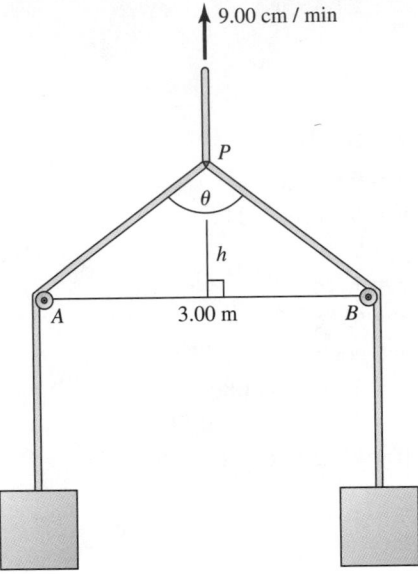

FIGURE 33-9

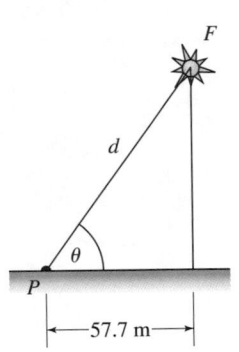

FIGURE 33-10   Falling flare.

36. The illumination at a point P on the ground (Fig. 33–10) from a flare F is $I = k \sin \theta / d^2$ lux, where k is a constant. Find the rate of change of I when the flare is 100 m above the ground and falling at a rate of 1.0 m/s if the illumination at P at that instant is 65 lux.

Maximum-Minimum Problems

37. Find the length of the shortest ladder that will touch the ground, the wall, and the house in Fig. 33–11.

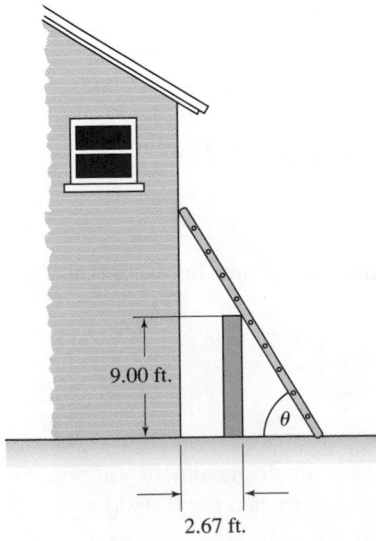

FIGURE 33-11

38. The range x of a projectile fired at an angle θ with the horizontal at a velocity v (Fig. 33–12) is $x = (v^2/g) \sin 2\theta$, where g is the acceleration due to gravity. Find θ for the maximum range.

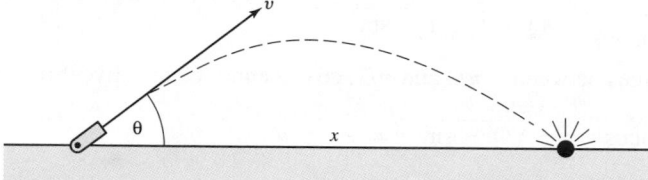

FIGURE 33–12

39. A force F (Fig. 33–13) pulls the weight along a horizontal surface. If f is the coefficient of friction, then

$$F = \frac{fW}{f \sin \theta + \cos \theta}$$

Find θ for a minimum force when $f = 0.60$.

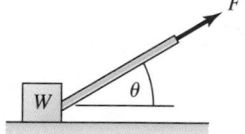

FIGURE 33–13

40. A 6.00-m-long steel girder is dragged along a corridor 3.00 m wide and then around a corner into another corridor at right angles to the first (Fig. 33–14). Neglecting the thickness of the girder, what must the width of the second corridor be to allow the girder to turn the corner?

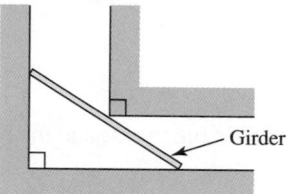

FIGURE 33–14   Top view of a girder dragged along a corridor.

41. If the girder in Fig. 33–14 is to be dragged from a 3.20-m-wide corridor into another 1.35-m-wide corridor, find the length of the longest girder that will fit around the corner. (Neglect the thickness of the girder.)

33–3 Derivatives of the Inverse Trigonometric Functions

Derivative of Arcsin u

We now seek the derivative of $y = \sin^{-1} u$ where y is some angle whose sine is u, as in Fig. 33–15, whose value we restrict to the range $-\pi/2$ to $\pi/2$. We can then write

$$\sin y = u$$

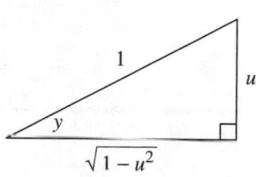

FIGURE 33–15

Taking the derivative yields

$$\cos y \frac{dy}{dx} = \frac{du}{dx}$$

so

$$\frac{dy}{dx} = \frac{1}{\cos y} \frac{du}{dx} \tag{1}$$

However, from Eq. 164,

$$\cos^2 y = 1 - \sin^2 y$$

$$\cos y = \pm \sqrt{1 - \sin^2 y}$$

But since y is restricted to values between $-\pi/2$ and $\pi/2$, $\cos y$ cannot be negative. So

$$\cos y = + \sqrt{1 - \sin^2 y} = \sqrt{1 - u^2}$$

Substituting into Eq. (1), we have the following equation:

Derivative of the Arcsin	$\dfrac{d(\mathrm{Sin}^{-1} u)}{dx} = \dfrac{1}{\sqrt{1 - u^2}} \dfrac{du}{dx}$ $-1 < u < 1$	**356**

◆◆◆**Example 13:** If $y = \mathrm{Sin}^{-1} 3x$, then

$$y' = \frac{1}{\sqrt{1 - (3x)^2}} \quad (3)$$

$$= \frac{3}{\sqrt{1 - 9x^2}}$$ ◆◆◆

Derivatives of Arccos, Arctan, Arccot, Arcsec, and Arccsc

The rules for taking derivatives of the remaining inverse trigonometric functions, and the Arcsin as well, are as follows:

Try to derive one or more of these equations. Follow steps similar to those we used for the derivative of Arcsin.

Derivatives of the Inverse Trigonometric Functions	$\dfrac{d(\mathrm{Sin}^{-1} u)}{dx} = \dfrac{1}{\sqrt{1 - u^2}} \dfrac{du}{dx}$ $-1 < u < 1$	**356**
	$\dfrac{d(\mathrm{Cos}^{-1} u)}{dx} = \dfrac{-1}{\sqrt{1 - u^2}} \dfrac{du}{dx}$ $-1 < u < 1$	**357**
	$\dfrac{d(\mathrm{Tan}^{-1} u)}{dx} = \dfrac{1}{1 + u^2} \dfrac{du}{dx}$	**358**
	$\dfrac{d(\mathrm{Cot}^{-1} u)}{dx} = \dfrac{-1}{1 + u^2} \dfrac{du}{dx}$	**359**
	$\dfrac{d(\mathrm{Sec}^{-1} u)}{dx} = \dfrac{1}{u\sqrt{u^2 - 1}} \dfrac{du}{dx}$ $|u| > 1$	**360**
	$\dfrac{d(\mathrm{Csc}^{-1} u)}{dx} = \dfrac{-1}{u\sqrt{u^2 - 1}} \dfrac{du}{dx}$ $|u| > 1$	**361**

♦♦♦Example 14:

(a) If $y = \text{Cot}^{-1}(x^2 + 1)$, then

$$y' = \frac{-1}{1 + (x^2 + 1)^2}(2x)$$

$$= \frac{-2x}{2 + 2x^2 + x^4}$$

(b) If $y = \text{Cos}^{-1}\sqrt{1-x}$, then

$$y' = \frac{-1}{\sqrt{1-(1-x)}} \cdot \frac{-1}{2\sqrt{1-x}}$$

$$= \frac{1}{2\sqrt{x-x^2}}$$

♦♦♦

Exercise 3 ♦ Derivatives of the Inverse Trigonometric Functions

Find the derivative.

1. $y = x\,\text{Sin}^{-1}x$

2. $y = \text{Sin}^{-1}\dfrac{x}{a}$

3. $y = \text{Cos}^{-1}\dfrac{x}{a}$

4. $y = \text{Tan}^{-1}(\sec x + \tan x)$

5. $y = \text{Sin}^{-1}\dfrac{\sin x - \cos x}{\sqrt{2}}$

6. $y = \sqrt{2ax - x^2} + a\,\text{Cos}^{-1}\dfrac{\sqrt{2ax - x^2}}{a}$

7. $y = t^2\,\text{Cos}^{-1}t$

8. $y = \text{Arcsin}\,2x$

9. $y = \text{Arctan}(1 + 2x)$

10. $y = \text{Arccot}(2x + 5)^2$

11. $y = \text{Arccot}\dfrac{x}{a}$

12. $y = \text{Arcsec}\dfrac{1}{x}$

13. $y = \text{Arccsc}\,2x$

14. $y = \text{Arcsin}\sqrt{x}$

15. $y = t^2\,\text{Arcsin}\dfrac{t}{2}$

16. $y = \text{Sin}^{-1}\dfrac{x}{\sqrt{1+x^2}}$

17. $y = \text{Sec}^{-1}\dfrac{a}{\sqrt{a^2 - x^2}}$

Find the slope of the tangent to each curve.

18. $y = x\,\text{Arcsin}\,x$ at $x = \dfrac{1}{2}$

19. $y = \dfrac{\text{Arctan}\,x}{x}$ at $x = 1$

20. $y = x^2\,\text{Arccsc}\sqrt{x}$ at $x = 2$

21. $y = \sqrt{x}\,\text{Arccot}\dfrac{x}{4}$ at $x = 4$

22. Find the equations of the tangents to the curve $y = \text{Arctan}\,x$ having a slope of $\frac{1}{4}$.

33–4 Derivatives of Logarithmic Functions

Derivative of $\log_b u$

Let us now use the delta method again to find the derivative of the logarithmic function $y = \log_b u$. We first let u take on an increment Δu and y an increment Δy.

$$y = \log_b u$$
$$y + \Delta y = \log_b (u + \Delta u)$$

Subtracting gives us

$$\Delta y = \log_b (u + \Delta u) - \log_b u = \log_b \frac{u + \Delta u}{u}$$

by the law of logarithms for quotients. Now dividing by Δu yields

$$\frac{\Delta y}{\Delta u} = \frac{1}{\Delta u} \log_b \frac{u + \Delta u}{u}$$

We now do some manipulation to get our expression into a form that will be easier to evaluate. We start by multiplying the right side by u/u.

$$\frac{\Delta y}{\Delta u} = \frac{u}{u} \cdot \frac{1}{\Delta u} \log_b \frac{u + \Delta u}{u}$$

$$= \frac{1}{u} \cdot \frac{u}{\Delta u} \log_b \frac{u + \Delta u}{u}$$

Then, using the law of logarithms for powers (Eq. 189), we have

$$\frac{\Delta y}{\Delta u} = \frac{1}{u} \log_b \left(\frac{u + \Delta u}{u} \right)^{u/\Delta u}$$

We now let Δu approach zero.

$$\lim_{\Delta u \to 0} \frac{\Delta y}{\Delta u} = \frac{dy}{du} = \lim_{\Delta u \to 0} \frac{1}{u} \log_b \left(\frac{u + \Delta u}{u} \right)^{u/\Delta u}$$

$$= \frac{1}{u} \log_b \left[\lim_{\Delta u \to 0} \left(1 + \frac{\Delta u}{u} \right)^{u/\Delta u} \right]$$

Let us simplify the expression inside the brackets by making the substitution

$$k = \frac{u}{\Delta u}$$

Then k will approach infinity as Δu approaches zero, and the expression inside the brackets becomes

$$\lim_{\Delta u \to 0} \left(1 + \frac{\Delta u}{u} \right)^{u/\Delta u} = \lim_{k \to \infty} \left(1 + \frac{1}{k} \right)^{k}$$

This limit defines the number e, the familiar base of natural logarithms.

Glance back at Sec. 20–2 where we developed this expression.

Definition of e	$e = \lim\limits_{k \to \infty} \left(1 + \dfrac{1}{k} \right)^{k}$	185

Our derivative thus becomes

$$\frac{dy}{du} = \frac{1}{u} \log_b e$$

We are nearly finished now. Using the chain rule, we get dy/dx by multiplying dy/du by du/dx. Thus,

Derivative of $\log_b u$	$\dfrac{d(\log_b u)}{dx} = \dfrac{1}{u} \log_b e \, \dfrac{du}{dx}$	362a

Or, since $\log_b e = 1/\ln b$,

This form is more useful when the base b is a number other than 10.

Derivative of $\log_b u$	$\dfrac{d(\log_b u)}{dx} = \dfrac{1}{u \ln b} \dfrac{du}{dx}$	362b

♦♦♦**Example 15:** Take the derivative of $y = \log(x^2 - 3x)$.

Solution: The base b, when not otherwise indicated, is taken as 10. By Eq. 362a,

$$\frac{dy}{dx} = \frac{1}{x^2 - 3x}(\log e)(2x - 3)$$

$$= \frac{2x - 3}{x^2 - 3x}\log e$$

Or, since $\log e \cong 0.4343$,

$$\frac{dy}{dx} \cong 0.4343\left(\frac{2x - 3}{x^2 - 3x}\right)$$

♦♦♦

♦♦♦**Example 16:** Find the derivative of $y = x\log_3 x^2$.

Solution: By the product rule,

$$y' = x\left(\frac{1}{x^2\ln 3}\right)(2x) + \log_3 x^2$$

$$= \frac{2}{\ln 3} + \log_3 x^2 \cong 1.82 + \log_3 x^2$$

♦♦♦

Derivative of ln u

Our efforts in deriving Eqs. 362a and 362b will now pay off, because we can use that result to find the derivative of the natural logarithm of a function, as well as the derivatives of exponential functions in the following sections. To find the derivative of $y = \ln u$ we use Eq. 362a.

$$y = \ln u$$

$$\frac{dy}{dx} = \frac{1}{u}\ln e\,\frac{du}{dx}$$

But, by Eq. 192, $\ln e = 1$. Thus,

Derivative of ln u	$\dfrac{d(\ln u)}{dx} = \dfrac{1}{u}\dfrac{du}{dx}$	363

The derivative of the natural logarithm of a function is the reciprocal of that function multiplied by its derivative.

♦♦♦**Example 17:** Differentiate $y = \ln(2x^3 + 5x)$.

Solution: By Eq. 363,

$$\frac{dy}{dx} = \frac{1}{2x^3 + 5x}(6x^2 + 5)$$

$$= \frac{6x^2 + 5}{2x^3 + 5x}$$

♦♦♦

The rule for derivatives of the logarithmic function is often used with other rules for derivatives.

♦♦♦**Example 18:** Take the derivative of $y = x^3\ln(5x + 2)$.

Solution: Using the product rule together with our rule for logarithms gives us

$$\frac{dy}{dx} = x^3\left(\frac{1}{5x + 2}\right)5 + [\ln(5x + 2)]3x^2 = \frac{5x^3}{5x + 2} + 3x^2\ln(5x + 2)$$

♦♦♦

Our work is sometimes made easier if we first use the laws of logarithms to simplify a given expression.

◆◆◆Example 19: Take the derivative of $y = \ln \dfrac{x\sqrt{2x-3}}{\sqrt[3]{4x+1}}$.

Solution: We first rewrite the given expression using the laws of logarithms.

$$y = \ln x + \frac{1}{2}\ln(2x-3) - \frac{1}{3}\ln(4x+1)$$

We now take the derivative term by term.

$$\frac{dy}{dx} = \frac{1}{x} + \frac{1}{2}\left(\frac{1}{2x-3}\right)2 - \frac{1}{3}\left(\frac{1}{4x+1}\right)4$$

$$= \frac{1}{x} + \frac{1}{2x-3} - \frac{4}{3(4x+1)}$$ ◆◆◆

Logarithmic Differentiation

Here we use logarithms to aid in differentiating nonlogarithmic expressions. Derivatives of some complicated expressions can be found more easily if we first take the logarithm of both sides of the given expression and simplify by means of the laws of logarithms. (These operations will not change the meaning of the original expression.) We then take the derivative.

◆◆◆ Example 20: Differentiate $y = \dfrac{\sqrt{x-2}\,\sqrt[3]{x+3}}{\sqrt[4]{x+1}}$.

Solution: We will use logarithmic differentiation here. Instead of proceeding in the usual way, we first take the natural log of both sides of the equation and apply the laws of logarithms.

We could instead take the common log of both sides, but the natural log has a simpler derivative.

$$\ln y = \frac{1}{2}\ln(x-2) + \frac{1}{3}\ln(x+3) - \frac{1}{4}\ln(x+1)$$

Taking the derivative, we have

$$\frac{1}{y}\frac{dy}{dx} = \frac{1}{2}\left(\frac{1}{x-2}\right) + \frac{1}{3}\left(\frac{1}{x+3}\right) - \frac{1}{4}\left(\frac{1}{x+1}\right)$$

We could now use the original expression to replace the y in the answer.

Finally, multiplying by y to solve for dy/dx gives us

$$\frac{dy}{dx} = y\left[\frac{1}{2(x-2)} + \frac{1}{3(x+3)} - \frac{1}{4(x+1)}\right]$$ ◆◆◆

In other cases, this method will allow us to take derivatives not possible with our other rules.

◆◆◆Example 21: Find the derivative of $y = x^{2x}$.

Solution: This is not a power function because the exponent is not a constant. Nor is it an exponential function because the base is not a constant. So neither Rule 345 nor 364 applies. But let us use logarithmic differentiation by first taking the natural logarithm of both sides.

$$\ln y = \ln x^{2x} = 2x \ln x$$

Now taking the derivative by means of the product rule yields

$$\frac{1}{y}\frac{dy}{dx} = 2x\left(\frac{1}{x}\right) + (\ln x)2$$

$$= 2 + 2\ln x$$

Multiplying by y, we get

$$\frac{dy}{dx} = 2(1 + \ln x)y$$

Replacing y by x^{2x} gives us

$$\frac{dy}{dx} = 2(1 + \ln x)x^{2x}$$ ◆◆◆

Exercise 4 ◆ Derivatives of Logarithmic Functions

Derivative of $\log_b u$

Differentiate.

1. $y = \log 7x$

2. $y = \log x^{-2}$

3. $y = \log_b x^3$

4. $y = \log_a(x^2 - 3x)$

5. $y = \log(x\sqrt{5 + 6x})$

6. $y = \log_a\left(\dfrac{1}{2x + 5}\right)$

7. $y = x \log \dfrac{2}{x}$

8. $y = \log \dfrac{(1 + 3x)}{x^2}$

Derivative of ln u

Differentiate.

9. $y = \ln 3x$

10. $y = \ln x^3$

11. $y = \ln(x^2 - 3x)$

12. $y = \ln(4x - x^3)$

13. $y = 2.75x \ln 1.02x^3$

14. $w = z^2 \ln(1 - z^2)$

15. $y = \dfrac{\ln(x + 5)}{x^2}$

16. $y = \dfrac{\ln x^2}{3 \ln(x - 4)}$

17. $s = \ln \sqrt{t - 5}$

18. $y = 5.06 \ln \sqrt{x^2 - 3.25x}$

With Trigonometric Functions

Differentiate.

19. $y = \ln \sin x$

20. $y = \ln \sec x$

21. $y = \sin x \ln \sin x$

22. $y = \ln(\sec x + \tan x)$

Implicit Relations

Find dy/dx.

23. $y \ln y + \cos x = 0$

24. $\ln x^2 - 2x \sin y = 0$

25. $x - y = \ln(x + y)$

26. $xy = a^2 \ln \dfrac{x}{a}$

27. $\ln y + x = 10$

Logarithmic Differentiation

Differentiate.

28. $y = \dfrac{\sqrt{x + 2}}{\sqrt[3]{2 - x}}$

29. $y = \dfrac{\sqrt{a^2 - x^2}}{x}$

30. $y = x^x$

31. $y = x^{\sin x}$

32. $y = (\cot x)^{\sin x}$

33. $y = (\text{Cos}^{-1} x)^x$

Remember to start these by taking the logarithm of both sides.

Tangent to a Curve

Find the slope of the tangent at the point indicated.

34. $y = \log x$ at $x = 1$ **35.** $y = \ln x$ where $y = 0$

36. $y = \ln(x^2 + 2)$ at $x = 4$ **37.** $y = \log(4x - 3)$ at $x = 2$

38. Find the equation of the tangent to the curve $y = \ln x$ at $y = 0$.

Angle of Intersection

Find the angle of intersection of each pair of curves.

39. $y = \ln(x + 1)$ and $y = \ln(7 - 2x)$ at $x = 2$

40. $y = x \ln x$ and $y = x \ln(1 - x)$ at $x = \dfrac{1}{2}$

Extreme Values and Points of Inflection

Find the maximum, minimum, and inflection points for each curve.

41. $y = x \ln x$ **42.** $y = x^3 \ln x$

43. $y = \dfrac{x}{\ln x}$ **44.** $y = \ln(8x - x^2)$

Newton's Method

Find the smallest positive root between $x = 0$ and 10 to two decimal places.

45. $x - 10 \log x = 0$ **46.** $\tan x - \log x = 0$

Applications

47. A certain underwater cable has a core of copper wires covered by insulation. The speed of transmission of a signal along the cable is

$$S = x^2 \ln \frac{1}{x}$$

where x is the ratio of the radius of the core to the thickness of the insulation. What value of x gives the greatest signal speed?

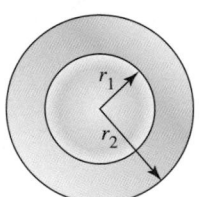

FIGURE 33–16   Insulated pipe.

48. The heat loss q per foot of cylindrical pipe insulation (Fig. 33–16) having an inside radius r_1 and outside radius r_2 is given by the logarithmic equation

$$q = \frac{2\pi k(t_1 - t_2)}{\ln(r_2/r_1)} \quad \text{kJ/h}^2$$

where t_1 and t_2 are the inside and outside temperatures (°C) and k is the conductivity of the insulation. A 10-cm-thick insulation having a conductivity of 0.04 is wrapped around a 30-cm-diameter pipe at 282°C, and the surroundings are at 32°C. Find the rate of change of heat loss q if the insulation thickness is decreasing at the rate of 0.3 cm/h.

49. The pH value of a solution having a concentration C of hydrogen ions is pH $= -\log_{10} C$. Find the rate at which the pH is dropping when the concentration is 2.0×10^{-5} and decreasing at the rate of 5.5×10^{-5} per minute.

Hint: Treat B_2 as a constant in problem 50.

50. The difference in elevation, in metres, between two locations having barometric readings of B_1 and B_2 mm of mercury is given by $h = 18\,413 \log B_2/B_1$, where B_1 is the pressure at the upper location. At what rate is an airplane changing in elevation when the barometric pressure outside the airplane is 546 mm of mercury and decreasing at the rate of 12.7 mm per minute?

51. The power input to a certain amplifier is 2.0 W; the power output is $4\overline{0}0$ W but because of a defective component is dropping at the rate of 0.50 W per day. Use Eq. A104 to find the rate (decibels per day) at which the decibels are decreasing.

33–5 Derivatives of the Exponential Function

Derivative of b^u

We seek the derivative of the exponential function $y = b^u$, where b is a constant and u is a function of x. We can get the derivative without having to use the delta method by using the rule we derived for the logarithmic function (Eq. 363). We first take the natural logarithm of both sides.

$$y = b^u$$

$$\ln y = \ln b^u = u \ln b$$

by Eq. 189. We now take the derivative of both sides, remembering that $\ln b$ is a constant.

$$\frac{1}{y}\frac{dy}{dx} = \frac{du}{dx} \ln b$$

Multiplying by y, we have

$$\frac{dy}{dx} = y\frac{du}{dx} \ln b$$

Finally, replacing y by b^u gives us the following:

Derivative of b^u	$\dfrac{d(b^u)}{dx} = b^u \dfrac{du}{dx} \ln b$	364

The derivative of a base b raised to a function u is the product of b^u, the derivative of the function, and the natural log of the base.

◆◆◆ **Example 22:** Find the derivative of $y = 10^{x^2+2}$.

Solution: By Eq. 364,

$$\frac{dy}{dx} = 10^{x^2+2}(2x) \ln 10 = 2x(\ln 10)10^{x^2+2} \qquad ◆◆◆$$

Common Error	Do not confuse the exponential function $y = b^x$ with the power function $y = x^n$. The derivative of b^x **is not** xb^{x-1}! The derivative of b^x **is** $b^x \ln b$.

Derivative of e^u

We will use Eq. 364 mostly when the base b is the base of natural logarithms, e.

$$y = e^u$$

Taking the derivative by Eq. 364 gives us

$$\frac{dy}{dx} = e^u \frac{du}{dx} \ln e$$

But since $\ln e = 1$, we get the following:

Derivative of e^u	$\dfrac{d}{dx}e^u = e^u \dfrac{du}{dx}$	365

The derivative of an exponential function e^u is the same exponential function, multiplied by the derivative of the exponent.

◆◆◆ **Example 23:** Find the first, second, and third derivatives of $y = 2e^{3x}$.

Solution: By Rule 365,

$$y' = 6e^{3x} \qquad y'' = 18e^{3x} \qquad y''' = 54e^{3x}$$

◆◆◆

> Note that $y = c_1 e^{mx}$ and its derivatives $y' = c_2 e^{mx}$, $y' = c_3 e^{mx}$, etc., are all like terms, since c_1, c_2, and c_3 are constants. We use this fact when solving differential equations in Chapter 35.

◆◆◆ **Example 24:** Find the derivative of $y = e^{x^3 + 5x^2}$.

Solution:

$$\frac{dy}{dx} = e^{x^3 + 5x^2}(3x^2 + 10x)$$

◆◆◆

◆◆◆ **Example 25:** Find the derivative of $y = x^3 e^{x^2}$.

Solution: Using the product rule together with Eq. 365, we have

$$\frac{dy}{dx} = x^3 (e^{x^2})(2x) + e^{x^2}(3x^2)$$
$$= 2x^4 e^{x^2} + 3x^2 e^{x^2} = x^2 e^{x^2}(2x^2 + 3)$$

◆◆◆

Derivative of $y = x^n$, Where n Is Any Real Number

We now return to some unfinished business regarding the power rule. We have already shown that the derivative of the power function x^n is nx^{n-1}, when n is any rational number, positive or negative. Now we show that n can also be an irrational number, such as e or x. Using the fact that $e^{\ln z} = z$, we let

$$x = e^{\ln x}$$

Raising to the nth power gives

$$x^n = (e^{\ln x})^n = e^{n \ln x}$$

Then

$$\frac{d}{dx} x^n = \frac{d}{dx} e^{n \ln x}$$

By Eq. 365,

$$\frac{d}{dx} x^n = \frac{n}{x} e^{n \ln x}$$

Substituting x^n for $e^{n \ln x}$ gives

$$\frac{d}{dx} x^n = \frac{n}{x} x^n = nx^{n-1}$$

Thus, the power rule holds when the exponent n is any real number.

◆◆◆ **Example 26:** The derivative of $3x^\pi$ is

$$\frac{d}{dx} 3x^\pi = 3\pi x^{\pi - 1}$$

◆◆◆

Exercise 5 ◆ Derivatives of the Exponential Function

Derivative of b^u

Differentiate.

1. $y = 3^{2x}$

2. $y = 10^{2x+3}$

3. $y = (x)(10^{2x+3})$

4. $y = 10^{3x}$

5. $y = 2^{x^2}$

6. $y = 7^{2x}$

Derivative of e^u

Differentiate.

7. $y = e^{2x}$

8. $y = e^{x^2}$

9. $y = e^{e^x}$

10. $y = e^{3x^2+4}$

11. $y = e^{\sqrt{1-x^2}}$

12. $y = xe^x$

13. $y = \dfrac{2}{e^x}$

14. $y = (3x + 2)e^{-x^2}$

15. $y = x^2e^{3x}$

16. $y = xe^{-x}$

17. $y = \dfrac{e^x}{x}$

18. $y = \dfrac{x^2 - 2}{e^{3x}}$

19. $y = \dfrac{e^x - e^{-x}}{x^2}$

20. $y = \dfrac{e^x - x}{e^{-x} + x^2}$

21. $y = (x + e^x)^2$

22. $y = (e^x + 2x)^3$

23. $y = \dfrac{(1 + e^x)^2}{x}$

24. $y = \left(\dfrac{e^x + 1}{e^x - 1}\right)^2$

With Trigonometric Functions

Differentiate.

25. $y = \sin^3 e^x$

26. $y = e^x \sin x$

27. $y = e^\theta \cos 2\theta$

28. $y = e^x(\cos bx + \sin bx)$

Implicit Relations

Find dy/dx.

29. $e^x + e^y = 1$

30. $e^x \sin y = 0$

31. $e^y = \sin(x + y)$

Evaluate each expression.

32. $f'(2)$ where $f(x) = e^{\sin(\pi x/2)}$

33. $f''(0)$ where $f(t) = e^{\sin t} \cos t$

34. $f'(1)$ and $f''(1)$ where $f(x) = e^{x-1} \sin \pi x$

With Logarithmic Functions

Differentiate.

35. $y = e^x \ln x$

36. $y = \ln(x^2e^x)$

37. $y = \ln x^{e^x}$

38. $y = \ln e^{2x}$

Find the second derivative.

39. $y = e^t \cos t$

40. $y = e^{-t} \sin 2t$

41. $y = e^x \sin x$

42. $y = \frac{1}{2}(e^x + e^{-x})$

Newton's Method

Find the smallest root that is greater than zero to two decimal places.

43. $e^x + x - 3 = 0$

44. $xe^{-0.02x} = 1$

45. $5e^{-x} + x - 5 = 0$

46. $e^x = \tan x$

Maximum, Minimum, and Inflection Points

47. Find the minimum point of $y = e^{2x} + 5e^{-2x}$.

48. Find the maximum point and the points of inflection of $y = e^{-x^2}$.

49. Find the maximum and minimum points for one cycle of $y = 10e^{-x} \sin x$.

Applications

50. If $10,000 is invested for t years at an annual interest rate of 10% compounded continuously, it will accumulate to an amount y, where $y = 10{,}000e^{0.1t}$. At what rate, in dollars per year, is the balance growing when (a) $t = 0$ years and (b) $t = 10$ years?

51. If we assume that the price of an automobile is increasing or "inflating" exponentially at an annual rate of 8%, at what rate in dollars per year is the price of a car that initially cost $9,000 increasing after 3 years?

52. When a certain object is placed in an oven at $50\overline{0}$ °C, its temperature T rises according to the equation $T = 500(1 - e^{-0.1t})$, where t is the elapsed time (minutes). How fast is the temperature rising (in degrees per minute) when (a) $t = 0$ and (b) $t = 10.0$ min?

53. A catenary has the equation $y = \frac{1}{2}(e^x + e^{-x})$. Find the slope of the catenary when $x = 5$.

54. Verify that the minimum point on the catenary described in problem 53 occurs at $x = 0$.

55. The speed N of a certain flywheel is decaying exponentially according to the equation $N = 1855e^{-0.5t}$ (rev/min), where t is the time (min) after the power is disconnected. Find the angular acceleration (the rate of change of N) when $t = 1$ min.

56. The height y of a certain pendulum released from a height of 50.0 cm is $y = 50.0e^{-0.75t}$ cm, where t is the time after release in seconds. Find the vertical component of the velocity of the pendulum when $t = 1.00$ s.

57. The number of bacteria in a certain culture is growing exponentially. The number N of bacteria at any time t (h) is $N = 10{\,}000e^{0.1t}$. At what rate (number of bacteria per hour) is the population increasing when (a) $t = 0$ and (b) $t = 10\overline{0}$ h?

58. The force F needed to hold a weight W (Fig. 33–17) is $F = We^{-\mu\theta}$ where μ = the coefficient of friction. For a certain beam with $\mu = 0.150$, an angle wrap of 4.62 rad is needed to hold a weight of 800 N with a force of 400 N. Find the rate of change of F if the rope is unwrapping at a rate of 15°/s.

59. The atmospheric pressure at a height of h miles above the earth's surface is given by $p = 101.1e^{-h/6.4}$ kPa. Find the rate of change of the pressure on a rocket that is at 29.0 km and climbing at a rate of $240\overline{0}$ km/h.

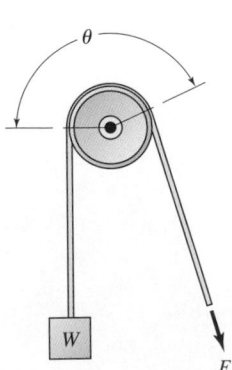

FIGURE 33–17

60. The equation in problem 59 becomes $p = 2121e^{-0.000\ 037h}$ when h is in feet and p is in pounds per square foot. Find the rate of change of pressure on an aircraft at $5\overline{0}00$ ft. climbing at a rate of $1\overline{0}$ ft./s.

61. The approximate density of seawater at a depth of h kilometres is $d = 1025e^{0.004\ 25h}$ kg/m³. Find the rate of change of density, with respect to depth, at a depth of 1.00 km.

❖❖❖ CHAPTER 33 REVIEW PROBLEMS ❖❖❖❖❖❖❖❖❖❖❖❖❖❖❖❖❖❖❖❖❖❖❖❖❖❖❖❖

Find dy/dx.

1. $y = \dfrac{a}{2}(e^{x/a} - e^{-x/a})$

2. $y = 5^{2x+3}$

3. $y = 8 \tan\sqrt{x}$

4. $y = \sec^2 x$

5. $y = x \operatorname{Arctan} 4x$

6. $y = \dfrac{1}{\sqrt{\operatorname{Arcsin} 2x}}$

7. $y^2 = \sin 2x$

8. $y = x\,e^{2x}$

9. $y = x \sin x$

10. $y = x^2 \sin x$

11. $y = x^3 \cos x$

12. $y = \ln \sin(x^2 + 3x)$

13. $y = \dfrac{\sin x}{x}$

14. $y = (\log x)^2$

15. $y = \log x(1 + x^2)$

16. $\cos(x - y) = 2x$

17. $y = \ln(x + \sqrt{x^2 + a^2})$

18. $y = \ln(x + 10)$

19. $y = \csc 3x$

20. $y = \ln(x^2 + 3x)$

21. $y = \dfrac{\sin x}{\cos x}$

22. $y = \dfrac{1}{\cos^2 x}$

23. $y = \ln(2x^3 + x)$

24. $y = x \operatorname{Arcsin} 2x$

25. $y = x^2 \operatorname{Arccos} x$

26. Find the minimum points of the curve $y = \ln(x^2 - 2x + 3)$.

27. Find the points of inflection of the curve $xy = 4 \log(x/2)$.

28. Find a minimum point and a point of inflection on the curve $y \ln x = x$. Write the equation of the tangent at the point of inflection.

29. At what x between $-\pi/2$ and $\pi/2$ is there a maximum on the curve $y = 2 \tan x - \tan^2 x$?

Find the value of dy/dx for the given value of x.

30. $y = x \operatorname{Arccos} x$ at $x = -\frac{1}{2}$

31. $y = \dfrac{\operatorname{Arcsec} 2x}{\sqrt{x}}$ at $x = 1$

32. If $x^2 + y^2 = \ln y + 2$, find y' and y'' at the point $(1, 1)$.

Find the equation of the tangent to each curve.

33. $y = \sin x$ at $x = \pi/6$

34. $y = x \ln x$ parallel to the line $3x - 2y = 5$

35. At what x is the tangent to the curve $y = \tan x$ parallel to the line $y = 2x + 5$?

Find the smallest positive root between $x = 0$ and 10 to three decimal places.

36. $\sin 3x - \cos 2x = 0$

37. $2 \sin \frac{1}{2} x - \cos 2x = 0$

38. Find the angle of intersection between $y = \ln(x^3/8 - 1)$ and $y = \ln(3x - x^2/4 - 1)$ at $x = 4$.

39. A casting is taken from one oven at 900 °C and placed in another oven whose temperature is 0 °C and rising at a linear rate of 60 °C per hour. The temperature T of the casting after t h is then $T = 60t + 900e^{-0.2t}$. Find the minimum temperature reached by the casting and the time at which it occurs.

40. A statue 11 m tall is on a pedestal so that the bottom of the statue is 25 m above eye level. How far from the statue (measured horizontally) should an observer stand so that the statue will subtend the greatest angle at the observer's eye?

Team Project

41. A capacitor (Fig. 33–18) is charged to 300 V. When the switch is closed, the voltage across R is initially 300 V but then drops according to the equation $V_1 = 300e^{-t/RC}$, where t is the time (in seconds), R is the resistance, and C is the capacitance. The voltage V_2 also starts to rise at the instant of switch closure so that $V_2 = 100t$.

(a) Graph V_1, V_2, and V.
(b) Show that the total voltage V is $300e^{-t/RC} + 100t$.
(c) Graphically find t when V is a minimum.
(d) Find t when V is a minimum by setting the derivative dV/dt equal to zero and solving for t.

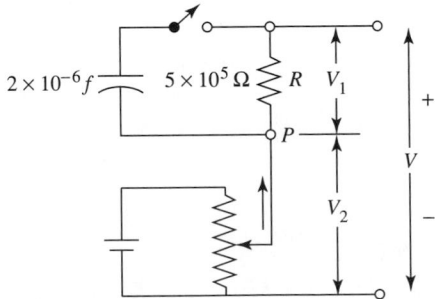

FIGURE 33–18

34

Methods of Integration

♦♦♦ OBJECTIVES ♦♦
When you have completed this chapter, you should be able to:
- Integrate exponential and logarithmic functions.
- Integrate trigonometric functions.
- Compute the average value and root mean square (rms) value of a function.
- Evaluate integrals using integration by parts.
- Integrate rational fractions with denominators having nonrepeated linear factors, repeated linear factors, or quadratic factors.
- Use algebraic substitution to evaluate an integral.
- Integrate using trigonometric substitution.
- Evaluate improper integrals containing infinite limits or discontinuous integrands.
- Integrate approximately using the average ordinate method, the trapezoidal rule, or Simpson's rule.

Unless we use a computer or graphics calculator to do integration, we will usually look up an integral in a table of integrals. We begin this chapter by finding integrals in the table first for the exponential and logarithmic functions and then for the trigonometric functions.

However, a given expression might not match any of the forms given in a table and might even look completely different. Therefore, much of this chapter is concerned with how to manipulate a given expression so that it matches a table entry.

But even with all of our new methods there will always be expressions that we cannot integrate, so we conclude this chapter with a few "approximate" methods of integration; that is, methods that find the value of an integral by numerical approximations. These methods have taken on added importance because they can be used on the computer.

34–1   Integrals of Exponential and Logarithmic Functions

Integral of e^u du

Since the derivative of e^u is

$$\frac{d(e^u)}{dx} = e^u \frac{du}{dx}$$

or $d(e^u) = e^u\, du$, then integrating gives

$$\int e^u\, du = \int d(e^u) = e^u + C$$

or the following:

Rule 8	$\int e^u\, du = e^u + C$

◆◆◆ **Example 1:** Integrate $\int e^{6x}\, dx$.

Solution: To match the form $\int e^u\, du$, let

$$u = 6x$$
$$du = 6\, dx$$

We insert a factor of 6 and compensate with $\frac{1}{6}$; then we use Rule 8.

$$\int e^{6x}\, dx = \frac{1}{6}\int e^{6x}\,(6\, dx) = \frac{1}{6}e^{6x} + C \qquad\qquad ◆◆◆$$

We now do a definite integral. Simply substitute the limits, as before.

◆◆◆ **Example 2:** Integrate $\displaystyle\int_0^3 \frac{6e^{\sqrt{3x}}}{\sqrt{3x}}\, dx$.

Solution: Since the derivative of $\sqrt{3x}$ is $\frac{3}{2}(3x)^{-1/2}$ we insert a factor of $\frac{3}{2}$ and compensate.

$$\int_0^3 \frac{6e^{\sqrt{3x}}}{\sqrt{3x}}\, dx = 6\int_0^3 e^{\sqrt{3x}}\,(3x)^{-1/2}\, dx$$

$$= 6\left(\frac{2}{3}\right)\int_0^3 e^{\sqrt{3x}}\left[\frac{3}{2}(3x)^{-1/2}\, dx\right]$$

$$= 4e^{\sqrt{3x}}\,\Big|_0^3$$

$$= 4e^3 - 4e^0 \cong 76.3 \qquad\qquad ◆◆◆$$

Integral of b^u du

The derivative of $b^u/\ln b$ is

$$\frac{d}{dx}\left(\frac{b^u}{\ln b}\right) = \frac{1}{\ln b}(b^u)\,(\ln b)\frac{du}{dx} = b^u\frac{du}{dx}$$

or, in differential form, $d(b^u/\ln b) = b^u\, du$. Thus the integral of $b^u\, du$ is as follows:

Rule 9	$\int b^u\, du = \dfrac{b^u}{\ln b} + C\;(b > 0,\, b \neq 1)$

◆◆◆ **Example 3:** Integrate $\int 3xa^{2x^2}\,dx$.

Solution:

$$\int 3xa^{2x^2}\,dx = 3\int a^{2x^2}x\,dx$$

$$= 3\left(\frac{1}{4}\right)\int a^{2x^2}(4x\,dx)$$

$$= \frac{3a^{2x^2}}{4\ln a} + C \qquad\qquad ◆◆◆$$

Integral of ln *u*

To integrate the natural logarithm of a function, we use Rule 43, which we give here without proof.

Rule 43	$\displaystyle\int \ln u\,du = u\,(\ln u - 1) + C$

> The argument of the natural logarithm function must be positive. For ease of reading, the absolute value signs have been omitted in this section.

◆◆◆ **Example 4:** Integrate $\int x\ln(3x^2)\,dx$.

Solution: We put our integral into the form of Rule 43 and integrate.

$$\int x\ln(3x^2)\,dx = \left(\tfrac{1}{6}\right)\int \ln(3x^2)(6x\,dx)$$

$$= \left(\tfrac{1}{6}\right)(3x^2)(\ln 3x^2 - 1) + C$$

$$= \tfrac{1}{2}x^2(\ln 3x^2 - 1) + C \qquad\qquad ◆◆◆$$

Integral of log *u*

To integrate the *common* logarithm of a function, we first convert it to a natural logarithm using Eq. 195.

$\displaystyle \log N = \frac{\ln N}{\ln 10} \cong \frac{\ln N}{2.3026}$	**195**

◆◆◆ **Example 5:** Integrate $\displaystyle\int_3^4 \log(3x - 7)\,dx$.

Solution: We convert the common log to a natural log and apply Rule 43.

$$\int_3^4 \log(3x - 7)\,dx = \int_3^4 \frac{\ln(3x - 7)}{\ln 10}\,dx$$

$$= \frac{1}{3\ln 10}\int_3^4 \ln(3x - 7)(3\,dx)$$

$$= \frac{1}{3\ln 10}(3x - 7)\big[\ln(3x - 7) - 1\big]_3^4$$

$$= \frac{1}{3\ln 10}\big[5(\ln 5 - 1) - 2(\ln 2 - 1)\big]$$

$$\cong 0.52997 \qquad\qquad ◆◆◆$$

Exercise 1   ◆   Integrals of Exponential and Logarithmic Functions

Integrate.

Exponential Functions

1. $\displaystyle\int a^{5x}\,dx$

2. $\displaystyle\int a^{9x}\,dx$

Each of these problems has been contrived to match a table entry. But what if you had one that did not match? We'll learn how to deal with those later in this chapter.

---

3. $\displaystyle\int 5^{7x}\,dx$

4. $\displaystyle\int 10^{x}\,dx$

5. $\displaystyle\int a^{3y}\,dy$

6. $\displaystyle\int xa^{3x^2}\,dx$

7. $\displaystyle\int 4e^{x}\,dx$

8. $\displaystyle\int e^{2x}\,dx$

9. $\displaystyle\int xe^{x^2}\,dx$

10. $\displaystyle\int x^2 e^{x^3}\,dx$

11. $\displaystyle\int_{1}^{2} xe^{3x^2}\,dx$

12. $\displaystyle\int_{3}^{4} \sqrt{e^{t}}\,dt$

13. $\displaystyle\int \frac{e^{\sqrt{x}}\,dx}{\sqrt{x}}$

14. $\displaystyle\int (x+3)\,e^{x^2+6x-2}\,dx$

15. $\displaystyle\int_{0}^{1} (e^{x}-1)^2\,dx$

16. $\displaystyle\int_{2}^{3} xe^{-x^2}\,dx$

17. $\displaystyle\int \frac{e^{\sqrt{x-2}}}{\sqrt{x-2}}\,dx$

18. $\displaystyle\int \frac{(e^{x/2}-e^{-x/2})^2}{4}\,dx$

19. $\displaystyle\int (e^{x/a}+e^{-x/a})\,dx$

20. $\displaystyle\int (e^{x/a}-e^{-x/a})^2\,dx$

Logarithmic Functions

21. $\displaystyle\int \ln 3x\,dx$

22. $\displaystyle\int \ln 7x\,dx$

23. $\displaystyle\int_{1}^{2} x \ln x^2\,dx$

As usual, round all approximate answers in this chapter to at least three significant digits.

---

24. $\displaystyle\int \log(5x-3)\,dx$

25. $\displaystyle\int_{2}^{4} x \log(x^2+1)\,dx$

26. $\displaystyle\int_{1}^{3} x^2 \log(2+3x^3)\,dx$

27. Find the area under the curve $y = e^{2x}$ from $x = 1$ to 3.

28. The first-quadrant area bounded by the catenary $y = \frac{1}{2}(e^x + e^{-x})$ from $x = 0$ to 1 is rotated about the x axis. Find the volume generated.

29. The first-quadrant area bounded by $y = e^x$ and $x = 1$ is rotated about the line $x = 1$. Find the volume generated.

30. Find the length of the catenary $y = (a/2)(e^{x/a} + e^{-x/a})$ from $x = 0$ to 6. Use $a = 3$.

31. The curve $y = e^{-x}$ is rotated about the x axis. Find the area of the surface generated, from $x = 0$ to 100.

32. Find the horizontal distance $\bar{x}$ to the centroid of the area formed by the curve $y = \frac{1}{2}(e^x + e^{-x})$, the coordinate axes, and the line $x = 1$.

33. Find the vertical distance $\bar{y}$ to the centroid of the area formed by the curve $y = e^x$ between $x = 0$ and 1.

34. A volume of revolution is formed by rotating the curve $y = e^x$ between $x = 0$ and 1 about the x axis. Find the distance from the origin to the centroid.

35. Find the moment of inertia about the x and y axes for the area bounded by the curve $y = e^x$, the line $x = 1$, and the coordinate axes.

34–2 Integrals of the Trigonometric Functions

To our growing list of rules we add those for the six trigonometric functions. By Eq. 351,

$$\frac{d(-\cos u)}{dx} = \sin u \frac{du}{dx}$$

or $d(-\cos u) = \sin u \, du$. Taking the integral of both sides gives

$$\int \sin u \, du = \cos u + C$$

The integrals of the other trigonometric functions are found in the same way. We thus get the following rules:

Rule 10	$\int \sin u \, du = -\cos u + C$
Rule 11	$\int \cos u \, du = \sin u + C$
Rule 12	$\int \tan u \, du = -\ln|\cos u| + C$
Rule 13	$\int \cot u \, du = \ln|\sin u| + C$
Rule 14	$\int \sec u \, du = \ln|\sec u + \tan u| + C$
Rule 15	$\int \csc u \, du = \ln|\csc u - \cot u| + C$

We use these rules just as we did the preceding ones: Match the given integral *exactly* with one of the rules, inserting a factor and compensating when necessary, and then use the result in the rule. Remember that the value of the variable is in radians.

◆◆◆ **Example 6:** Integrate $\int_0^1 x \sin x^2 \, dx$.

Solution:

$$\int_0^1 x \sin x^2 \, dx = \int_0^1 \sin x^2 (x \, dx)$$

$$= \frac{1}{2} \int_0^1 \sin x^2 (2x \, dx) = -\frac{1}{2} \cos x^2 \Big|_0^1$$

$$= -\frac{1}{2}(\cos 1 - \cos 0) \cong 0.2298 \qquad ◆◆◆$$

Sometimes the trigonometric identities can be used to simplify an expression before integrating.

◆◆◆ **Example 7:** Integrate $\int \dfrac{\cot 5x}{\cos 5x} \, dx$.

Solution: We replace $\cot 5x$ by $\cos 5x / \sin 5x$.

$$\int \frac{\cot 5x}{\cos 5x} \, dx = \int \frac{\cos 5x}{\sin 5x \cos 5x} \, dx = \int \frac{1}{\sin 5x} \, dx$$

$$= \int \csc 5x \, dx$$

$$= \frac{1}{5} \int \csc 5x (5 \, dx) = \frac{1}{5} \ln|\csc 5x - \cot 5x| + C$$

by Rule 15. ◆◆◆

Miscellaneous Rules from the Table

Now that you can use Rules 1 through 15, you should find it no harder to use any rule from the table of integrals.

◆◆◆ **Example 8:** Integrate $\int e^{3x} \cos 2x \, dx$.

Solution: We search the table for a similar form and find the following:

Rule 42	$\displaystyle \int e^{au} \cos bu \, du = \frac{e^{au}}{a^2 + b^2} (a \cos bu + b \sin bu) + C$

This matches our integral if we set

$$a = 3 \qquad b = 2 \qquad u = x \qquad du = dx$$

so

$$\int e^{3x} \cos 2x \, dx = \frac{e^{3x}}{3^2 + 2^2} (3 \cos 2x + 2 \sin 2x) + C$$

$$= \frac{e^{3x}}{13} (3 \cos 2x + 2 \sin 2x) + C \qquad ◆◆◆$$

Exercise 2 ◆ Integrals of the Trigonometric Functions

Integrate.

1. $\displaystyle\int \sin 3x \, dx$

2. $\displaystyle\int \cos 7x \, dx$

3. $\displaystyle\int \tan 5\theta \, d\theta$

4. $\displaystyle\int \sec 2\theta \, d\theta$

5. $\displaystyle\int \sec 4x \, dx$

6. $\displaystyle\int \cot 8x \, dx$

7. $\displaystyle\int 3 \tan 9\theta \, d\theta$

8. $\displaystyle\int 7 \sec 3\theta \, d\theta$

9. $\displaystyle\int x \sin x^2 \, dx$

10. $\displaystyle\int 5x \cos 2x^2 \, dx$

11. $\displaystyle\int \theta^2 \tan \theta^3 \, d\theta$

12. $\displaystyle\int \theta \sec 2\theta^2 \, d\theta$

13. $\displaystyle\int \sin (x + 1) \, dx$

14. $\displaystyle\int \cos (7x - 3) \, dx$

15. $\displaystyle\int \tan (4 - 5\theta) \, d\theta$

16. $\displaystyle\int \sec (2\theta + 3) \, d\theta$

17. $\displaystyle\int x \sec (4x^2 - 3) \, dx$

18. $\displaystyle\int 3x^2 \cot (8x^3 + 3) \, dx$

19. $\displaystyle\int_0^1 xe^{x^2} \, dx$

20. $\displaystyle\int_0^1 \frac{dx}{e^{3x}}$

21. $\displaystyle\int_0^\pi \sin \phi \, d\phi$

22. $\displaystyle\int_0^{\pi/2} \cos \phi \, d\phi$

23. $\displaystyle\int_0^\pi \cos \frac{\theta}{2} \, d\theta$

24. $\displaystyle\int_{\pi/3}^{\pi/2} \sin^2 x \cos x \, dx$

Find the area under each curve in problems 25 through 27.

25. $y = \sin x$ from $x = 0$ to π

26. $y = 2 \cos x$ from $x = -\pi/2$ to $\pi/2$

27. $y = 2 \sin \frac{1}{2}\pi x$ from $x = 0$ to 2 rad

28. Find the area between the curve $y = \sin x$ and the x axis from $x = 1$ rad to 3 rad.

29. Find the area between the curve $y = \cos x$ and the x axis from $x = 0$ to $\frac{3}{2}\pi$.

30. The area bounded by one arch of the sine curve $y = \sin x$ and the x axis is rotated about the x axis. Find the volume of the solid generated.

31. Find the surface area of the volume of revolution of problem 30.

32. The area bounded by one arch of the sine curve $y = \sin x$ and the x axis is rotated about the y axis. Find the volume generated.

33. Find the coordinates of the centroid of the area bounded by the x axis and a half-cycle of the sine curve $y = \sin x$.

34. Find the radius of gyration of the area under one arch of the sine curve $y = \sin x$ with respect to the x axis.

34–3   Average and Root Mean Square Values

We are now able to do two applications that usually require integration of a trigonometric function.

Average Value of a Function

The area A under the curve $y = f(x)$ (Fig. 34–1) between $x = a$ and b is, by Eq. 388,

$$A = \int_a^b f(x)\, dx$$

Within the same interval, the *average ordinate* of that function, y_{avg}, is that value of y that will cause the rectangle $abcd$ to have the same area as that under the curve, or

$$(b - a)\, y_{avg} = A = \int_a^b f(x)\, dx$$

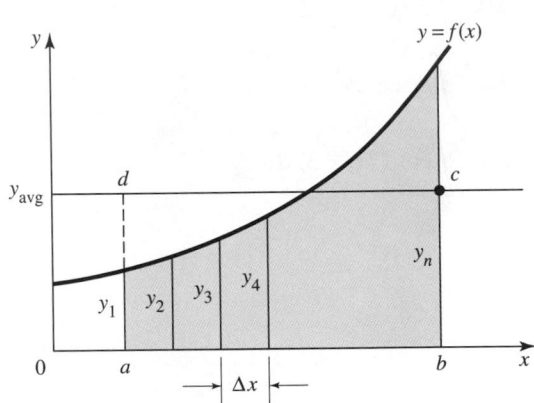

FIGURE 34–1   Average ordinate.

Thus:

Average Ordinate	$y_{avg} = \dfrac{1}{b - a} \int_a^b f(x)\, dx$	417

◆◆◆ **Example 9:** Find the average ordinate of a half-cycle of the sinusoidal voltage

$$v = V \sin \theta \qquad \text{(volts)}$$

Solution: By Eq. 417, with $a = 0$ and $b = \pi$,

$$V_{avg} = \frac{V}{\pi - 0} \int_0^\pi \sin \theta\, d\theta$$

$$= \frac{V}{\pi} \left[-\cos \theta \right]_0^\pi$$

$$= \frac{V}{\pi} (-\cos \pi + \cos 0) = \frac{2}{\pi} V \cong 0.637 V \qquad \text{(volts)}$$

◆◆◆

Root Mean Square Value of a Function

The *root mean square* (rms) value of a function is the square root of the average of the squares of the ordinates. In Fig. 34–1, if we take n values of y spaced apart by a distance Δx, the rms value is approximately

$$\text{rms} \cong \sqrt{\frac{y_1^2 + y_2^2 + y_3^2 + \cdots + y_n^2}{n}}$$

or, using summation notation,

$$\text{rms} \cong \sqrt{\frac{\sum_{i=1}^{n} y_i^2}{n}}$$

Multiplying numerator and denominator of the fraction under the radical by Δx, we obtain

$$\text{rms} \cong \sqrt{\frac{\sum_{i=1}^{n} y_i^2 \Delta x}{n \Delta x}}$$

But $n \Delta x$ is simply the width $(b - a)$ of the interval. If we now let n approach infinity, we get

$$\lim_{n \to \infty} \sum_{i=1}^{n} y_i^2 \Delta x = \int_a^b [f(x)]^2 \, dx$$

Therefore:

Root Mean Square Value	$\text{rms} = \sqrt{\dfrac{1}{b-a} \displaystyle\int_a^b [f(x)]^2 \, dx}$	418

♦♦♦ **Example 10:** Find the rms value for the sinusoidal voltage of Example 9.

Solution: We substitute into Eq. 417 with $a = 0$ and $b = \pi$.

$$\text{rms} = \sqrt{\frac{1}{\pi - 0} \int_0^\pi V^2 \sin^2 \theta \, d\theta} = \sqrt{\frac{V^2}{\pi} \int_0^\pi \sin^2 \theta \, d\theta}$$

But, by Rule 16,

$$\int_0^\pi \sin^2 \theta \, d\theta = \frac{\theta}{2} - \frac{\sin 2\theta}{4} \bigg|_0^\pi$$

$$= \frac{\pi}{2} - \frac{\sin 2\pi}{4} = \frac{\pi}{2}$$

So

$$\text{rms} = \sqrt{\frac{V^2}{\pi} \cdot \frac{\pi}{2}} = \frac{V}{\sqrt{2}} \cong 0.707V \qquad \text{(volts)}$$

♦♦♦

In electrical work, the rms value of an alternating current or voltage is also called the *effective* value. We see then that the effective value is 0.707 of the peak value.

Exercise 3 ♦ Average and Root Mean Square Values

Find the average ordinate for each function in the given interval.

1. $y = x^2$ from 0 to 6

2. $y = x^3$ from -5 to 5

3. $y = \sqrt{1 + 2x}$ from 4 to 12

4. $y = \dfrac{x}{\sqrt{9 + x^2}}$ from 0 to 4

5. $y = \sin^2 x$ from 0 to $\pi/2$

6. $2y = \cos 2x + 1$ from 0 to π

Find the rms value for each function in the given interval.

7. $y = 2x + 1$ from 0 to 6

8. $y = \sin 2x$ from 0 to $\pi/2$

9. $y = x + 2x^2$ from 1 to 4

10. $y = 3 \tan x$ from 0 to $\pi/4$

11. $y = 2 \cos x$ from $\pi/6$ to $\pi/2$

12. $y = 5 \sin 2x$ from 0 to $\pi/6$

34–4   Integration by Parts

Derivation of the Rule

Integration by parts is a useful method that enables us to split products into two parts for easier integration. We start with the rule for the derivative of a product (Eq. 346) in differential form.

$$d(uv) = u \, dv + v \, du$$

Rearranging, we get

$$u \, dv = d(uv) - v \, du$$

Integrating gives us

$$\int u \, dv = \int d(uv) - \int v \, du$$

or the following:

Rule 6	$\int u \, dv = uv - \int v \, du$

This rule is listed as number 6 in our table of integrals.

◆◆◆ **Example 11:** Integrate $\int x \cos 3x \, dx$.

Solution: The integrand is a product to which none of our previous rules apply. We use integration by parts. We separate $x \cos 3x \, dx$ into two parts, one of which we call u and the other dv. But which part shall we call u and which dv?

A good rule of thumb is to take dv as *the more complicated part,* but one that you can still *easily integrate.* So we let

$$u = x \quad \text{and} \quad dv = \cos 3x \, dx$$

Now our Rule 6 requires that we know du and v in addition to u and dv. We obtain du by differentiating u, and we find v by integrating dv.

$$u = x \qquad dv = \cos 3x \, dx$$
$$du = dx \quad \text{and} \quad v = \int \cos 3x \, dx$$
$$= \frac{1}{3} \int \cos 3x (3 \, dx)$$
$$= \frac{1}{3} \sin 3x + C_1$$

Applying Rule 6, we have

$$\int \underbrace{x}_{u} \, \underbrace{\cos 3x \, dx}_{dv} = \underbrace{x}_{u} \underbrace{\left(\frac{1}{3} \sin 3x + C_1 \right)}_{v} - \int \underbrace{\left(\frac{1}{3} \sin 3x + C_1 \right)}_{v} \underbrace{dx}_{du}$$

$$= \frac{x}{3} \sin 3x + C_1 x - \frac{1}{9} \int \sin 3x (3 \, dx) - \int C_1 \, dx$$

$$= \frac{x}{3} \sin 3x + C_1 x + \frac{1}{9} \cos 3x - C_1 x + C$$

$$= \frac{x}{3} \sin 3x + \frac{1}{9} \cos 3x + C$$

◆◆◆

Note that the constant C_1 obtained when integrating *dv does not appear in the final integral,* and, in fact, the final result would be the same as if we had not introduced C_1 in the first place. So in the following examples we will follow the usual practice of dropping the constant C_1 when integrating *dv*.

◆◆◆ **Example 12:** Integrate $\int x \ln x\, dx$.

Solution: Let us choose

$$u = \ln x \quad \text{and} \quad dv = x\, dx$$

Then

$$du = \frac{1}{x}\, dx \quad \text{and} \quad v = \frac{x^2}{2} \quad \text{(plus a constant which we drop)}$$

By Rule 6,

$$\int \underline{\ln x}\,\underline{(x\, dx)} = \underline{(\ln x)}\left(\underline{\frac{x^2}{2}}\right) - \int \underline{\frac{x^2}{2}}\left(\underline{\frac{dx}{x}}\right)$$

$$\int \quad u \quad dv \quad = \quad u \quad v \quad - \int \quad v \quad du$$

$$= \frac{x^2}{2}\ln x - \frac{1}{2}\int x\, dx$$

$$= \frac{x^2}{2}\ln x - \frac{x^2}{4} + C$$

◆◆◆

Sometimes our first choice of *u* and *dv* will result in an integral no easier to evaluate than our original.

◆◆◆ **Example 13:** Integrate $\int xe^x\, dx$.

Solution: *First try:*
Let

$$u = e^x \quad \text{and} \quad dv = x\, dx$$

Then

$$du = e^x\, dx \quad \text{and} \quad v = \frac{x^2}{2}$$

By Rule 6,

$$\int xe^x\, dx = \frac{x^2 e^x}{2} - \frac{1}{2}\int x^2 e^x\, dx$$

we get an integral that is more difficult than the one we started with.

Second try: Let

$$u = x \quad \text{and} \quad dv = e^x\, dx$$

Then

$$du = dx \quad \text{and} \quad v = e^x$$

By Rule 6,

$$\int xe^x\, dx = xe^x - \int e^x\, dx$$

$$= xe^x - e^x + C$$

◆◆◆

Sometimes we may have to integrate by parts *twice* to get the final result.

◆◆◆ **Example 14:** Integrate $\int x^2 \sin x \, dx$.

Solution: Let

$$u = x^2 \quad \text{and} \quad dv = \sin x \, dx$$

Then

$$du = 2x \, dx \quad \text{and} \quad v = -\cos x$$

By Rule 6,

$$\int x^2 \sin x \, dx = -x^2 \cos x + 2 \int x \cos x \, dx \tag{1}$$

Our new integral in Equation (1) is similar to the original, but we note that the power of x has been reduced from 2 to 1. This suggests that integrating by parts *again* may reduce the exponent to zero. Let's try

$$u = x \quad \text{and} \quad dv = \cos x \, dx$$

Then

$$du = dx \quad \text{and} \quad v = \sin x$$

By Rule 6,

$$\int x \cos x \, dx = x \sin x - \int \sin x \, dx$$

$$= x \sin x + \cos x + C_1$$

Substituting back into (1) gives us

$$\int x^2 \sin x \, dx = -x^2 \cos x + 2(x \sin x + \cos x + C_1)$$

$$= -x^2 \cos x + 2x \sin x + 2 \cos x + C \qquad\qquad ◆◆◆$$

Exercise 4 ◆ Integration by Parts

Integrate by parts for the practice, even if you find a rule that fits.

Integrate by parts.

1. $\int x \sin x \, dx$

2. $\int x \sqrt{1-x} \, dx$

3. $\int x \sec^2 x \, dx$

4. $\int_0^2 x \sin \dfrac{x}{2} \, dx$

5. $\int x \cos x \, dx$

6. $\int x^2 \ln x \, dx$

7. $\int x \sin^2 3x \, dx$

8. $\int \cos^3 x \, dx$

9. $\int_1^3 \dfrac{\ln (x+1) \, dx}{\sqrt{x+1}}$

10. $\int \dfrac{x^3 \, dx}{\sqrt{1-x^2}}$

11. $\int \dfrac{\ln (x+1) \, dx}{(x+1)^2}$

12. $\int x^3 \ln x \, dx$

13. $\int_0^4 x e^{2x} \, dx$

14. $\int x \tan^2 x \, dx$

15. $\int x^3 \sqrt{1-x^2} \, dx$

16. $\displaystyle\int \frac{x^2\,dx}{(1+x^2)^2}$

17. $\displaystyle\int \frac{x^2\,dx}{(1-x^2)^{3/2}}$

18. $\displaystyle\int_{-2}^{2} \frac{xe^x\,dx}{(1+x)^2}$

19. $\displaystyle\int \cos x \ln \sin x\,dx$

Integrate by parts twice.

20. $\displaystyle\int x^2 e^{2x}\,dx$

21. $\displaystyle\int x^2 e^{-x}\,dx$

22. $\displaystyle\int x^2 e^x\,dx$

23. $\displaystyle\int e^x \sin x\,dx$

24. $\displaystyle\int x^2 \cos x\,dx$

25. $\displaystyle\int_{1}^{3} e^{-x} \sin 4x\,dx$

26. $\displaystyle\int e^x \cos x\,dx$

27. $\displaystyle\int e^{-x} \cos \pi x\,dx$

34–5 Integrating Rational Fractions

Rational Algebraic Fractions

A rational algebraic fraction is one in which both numerator and denominator are polynomials.

> Recall that in a polynomial, all powers of x are positive integers.

◆◆◆ **Example 15:**

(a) $\dfrac{x^2}{x^3-2}$ is a *proper* rational fraction because the numerator is of lower degree than the denominator.

(b) $\dfrac{x^3-2}{x^2}$ is an *improper* rational fraction.

(c) $\dfrac{\sqrt{x}}{x^3-2}$ is *not* a rational fraction.   ◆◆◆

Integrating Improper Rational Fractions

To integrate an improper rational fraction, perform the indicated division, and integrate term by term.

◆◆◆ **Example 16:** Integrate $\displaystyle\int \frac{x^3-2x^2-5x-2}{x+1}\,dx$.

Solution: When we carry out the long division (work not shown), we get a quotient of $x^2 - 3x - 2$, so

$$\int \frac{x^3-2x^2-5x-2}{x+1}\,dx = \int (x^2 - 3x - 2)\,dx$$

$$= \frac{x^3}{3} - \frac{3x^2}{2} - 2x + C \qquad ◆◆◆$$

Common Error	It is easy to overlook the simple operation of dividing the numerator by the denominator of the given expression. Be sure to consider it whenever you have to integrate an algebraic fraction.

Long division works fine if there is *no remainder,* as in Example 16. When there is a remainder, it will be a *proper* rational fraction. We now learn how to integrate such a proper rational fraction.

Fractions with a Quadratic Denominator

We know that by *completing the square,* we can write a quadratic trinomial $Ax^2 + Bx + C$ in the form $u^2 \pm a^2$. This sometimes allows us to use Rules 56 through 60, as in the following example.

◆◆◆ **Example 17:** Integrate $\displaystyle\int \frac{4\,dx}{x^2 + 3x - 1}$.

Solution: We start by writing the denominator in the form $u^2 \pm a^2$ by completing the square.

$$x^2 + 3x - 1 = \left(x^2 + 3x + \frac{9}{4}\right) - \frac{9}{4} - 1$$

$$= \left(x + \frac{3}{2}\right)^2 - \frac{13}{4}$$

$$= \left(x + \frac{3}{2}\right)^2 - \left(\frac{\sqrt{13}}{2}\right)^2$$

Our integral is then

$$\int \frac{4\,dx}{\left(x + \frac{3}{2}\right)^2 - \left(\frac{\sqrt{13}}{2}\right)^2}$$

Using Rule 57,

$$\int \frac{du}{u^2 - a^2} = \frac{1}{2a} \ln \left|\frac{u - a}{u + a}\right| + C$$

with $u = x + 3/2$ and $a = \sqrt{13}/2$, we get

$$\int \frac{4\,dx}{x^2 + 3x - 1} = 4 \cdot \frac{1}{\frac{2\sqrt{13}}{2}} \ln \left|\frac{x + \frac{3}{2} - \frac{\sqrt{13}}{2}}{x + \frac{3}{2} + \frac{\sqrt{13}}{2}}\right| + C$$

or, in decimal form,

$$\cong 1.109 \ln \left|\frac{x - 0.303}{x + 3.303}\right| + C \qquad\qquad ◆◆◆$$

If the numerator in Example 17 had contained an x term, the integral would not have matched any rule in our table. Sometimes, however, the numerator can be separated into two parts, one of which is the derivative of the denominator.

◆◆◆ **Example 18:** Integrate $\int \dfrac{2x + 7}{x^2 + 3x - 1} \, dx$.

Solution: The derivative of the denominator is $2x + 3$. We can get this in the numerator by splitting the 7 into 3 and 4.

$$\int \frac{(2x + 3) + 4}{x^2 + 3x - 1} \, dx$$

We can now separate this integral into two integrals.

$$\int \frac{(2x + 3)}{x^2 + 3x - 1} \, dx + \int \frac{4}{x^2 + 3x - 1} \, dx$$

The first of these two integrals is, by Rule 7,

$$\int \frac{2x + 3}{x^2 + 3x - 1} \, dx = \ln |x^2 + 3x - 1| + C_1$$

The second integral is the same as that in Example 17. Our complete integral is then

$$\int \frac{2x + 7}{x^2 + 3x - 1} \, dx = \ln |x^2 + 3x - 1| + 1.109 \ln \left| \frac{x - 0.303}{x + 3.303} \right| + C$$ ◆◆◆

Partial Fractions

When studying algebra, we learned how to combine several fractions into a single fraction by writing all fractions with a common denominator; then we added or subtracted the numerators as indicated. Now we do the *reverse*. Given a proper rational fraction, we *separate it* into several simpler fractions whose sum is the original fraction. These simpler fractions are called *partial fractions*.

◆◆◆ **Example 19:** The sum of

$$\frac{2}{x + 1} \quad \text{and} \quad \frac{3}{x - 2}$$

is

$$\frac{2(x - 2) + 3(x + 1)}{(x + 1)(x - 2)} = \frac{5x - 1}{x^2 - x - 2}$$

Therefore

$$\frac{2}{x + 1} \quad \text{and} \quad \frac{3}{x - 2}$$

are the partial fractions of

$$\frac{5x - 1}{x^2 - x - 2}$$ ◆◆◆

We'll need partial fractions again when we study the Laplace transform.

The *first step* in finding partial fractions is to *factor the denominator* of the given fraction to find the set of all possible factors. The remaining steps then depend on the nature of those factors.

1. For each linear factor $ax + b$ in the denominator, there will be a partial fraction $A/(ax + b)$.
2. For the repeated linear factors $(ax + b)^n$, there will be n partial fractions.

$$\frac{A_1}{ax + b} + \frac{A_2}{(ax + b)^2} + \cdots + \frac{A_n}{(ax + b)^n}$$

3. For each quadratic factor $ax^2 + bx + c$, there will be a partial fraction $(Ax + B)/(ax^2 + bx + c)$.
4. For the repeated quadratic factors $(ax^2 + bx + c)^n$, there will be n partial fractions.

$$\frac{A_1 x + B_1}{ax^2 + bx + c} + \frac{A_2 x + B_2}{(ax^2 + bx + c)^2} + \cdots + \frac{A_n x + B_n}{(ax^2 + bx + c)^n}$$

The A's and B's are constants here.

Denominator with Nonrepeated Linear Factors

This is the simplest case, and the method of partial fractions is best shown by an example.

◆◆◆ **Example 20:** Separate $\dfrac{x+2}{x^3-x}$ into partial fractions.

Solution: We first factor the denominator into

$$x \qquad x+1 \qquad x-1$$

each of which is a nonrepeated linear factor. Next we assume that each of these factors will be the denominator of a partial fraction whose numerator is yet to be found. Thus we let

$$\frac{x+2}{x^3-x} = \frac{A}{x} + \frac{B}{x+1} + \frac{C}{x-1} \tag{1}$$

where A, B, and C are the unknown numerators. If our partial fractions are to be *proper,* the numerator of each must be of lower degree than the denominator. Since each denominator is of first degree, A, B, and C *must be constants.*

We now clear fractions by multiplying both sides of Equation (1) by $x^3 - x$.

$$x + 2 = A(x^2 - 1) + Bx(x - 1) + Cx(x + 1)$$

or

$$x + 2 = (A + B + C)x^2 + (-B + C)x - A$$

For this equation to be true, the coefficients of like powers of x on both sides of the equation must be equal. The coefficient of x is 1 on the left side and $(-B + C)$ on the right side. Therefore

$$1 = -B + C \tag{2}$$

> This is called the *method of undetermined* coefficients.

Similarly, for the x^2 term,

$$0 = A + B + C \tag{3}$$

and for the constant term,

$$2 = -A \tag{4}$$

Solving Equations (2), (3), and (4) simultaneously gives

$$A = -2 \qquad B = \frac{1}{2} \quad \text{and} \quad C = \frac{3}{2}$$

> You can check your work by recombining the partial fractions.

Substituting back into (1) gives us

$$\frac{x+2}{x^3-x} = -\frac{2}{x} + \frac{1}{2(x+1)} + \frac{3}{2(x-1)} \qquad\qquad ◆◆◆$$

◆◆◆ **Example 21:** Integrate $\displaystyle\int \frac{x+2}{x^3-x}\,dx.$

Solution: From Example 20,

$$\int \frac{x+2}{x^3-x}\,dx = \int \left[-\frac{2}{x} + \frac{1}{2(x+1)} + \frac{3}{2(x-1)} \right] dx$$

$$= -2\ln|x| + \frac{1}{2}\ln|x+1| + \frac{3}{2}\ln|x-1| + C \qquad\qquad ◆◆◆$$

Denominator with Repeated Linear Factors

We now consider fractions where a linear factor appears more than once in the denominator. Of course, there may also be one or more nonrepeated linear factors, as in the following example.

◆◆◆ **Example 22:** Separate the fraction $\dfrac{x^3 + 1}{x(x - 1)^3}$ into partial fractions.

Solution: We assume that

$$\frac{x^3 + 1}{x(x - 1)^3} = \frac{A}{x} + \frac{B}{(x - 1)} + \frac{C}{(x - 1)^2} + \frac{D}{(x - 1)^3}$$

because $x(x - 1)^3$ has x, $(x - 1)$, $(x - 1)^2$, and $(x - 1)^3$ as its set of possible factors. Multiplying both sides by $x(x - 1)^3$ yields

$$x^3 + 1 = A(x - 1)^3 + Bx(x - 1)^2 + Cx(x - 1) + Dx$$

$$= Ax^3 - 3Ax^2 + 3Ax - A + Bx^3 - 2Bx^2 + Bx + Cx^2 - Cx + Dx$$

$$= (A + B)x^3 + (-3A - 2B + C)x^2 + (3A + B - C + D)x - A$$

Equating coefficients of like powers of x gives

$$A + B = 1$$

$$-3A - 2B + C = 0$$

$$3A + B - C + D = 0$$

$$-A = 1$$

Solving this set of equations simultaneously gives

$$A = -1 \qquad B = 2 \qquad C = 1 \qquad D = 2$$

so

$$\frac{x^3 + 1}{x(x - 1)^3} = -\frac{1}{x} + \frac{2}{(x - 1)} + \frac{1}{(x - 1)^2} + \frac{2}{(x - 1)^3} \qquad\qquad ◆◆◆$$

If a denominator has a factor x^n, treat it as having the linear factor x repeated:

$$x(x)(x)(x) \ldots$$

◆◆◆ **Example 23:** Separate $\dfrac{2}{s^3(3s - 2)}$ into partial fractions.

Solution: We have the linear factor $(3s - 2)$ and the repeated linear factors s, s, and s. So

$$\frac{2}{s^3(3s - 2)} = \frac{A}{3s - 2} + \frac{B}{s} + \frac{C}{s^2} + \frac{D}{s^3}$$

Multiplying through by $s^3(3s - 2)$ gives

$$2 = As^3 + Bs^2(3s - 2) + Cs(3s - 2) + D(3s - 2)$$

Removing parentheses and collecting terms, we obtain

$$2 = (A + 3B)s^3 + (-2B + 3C)s^2 + (3D - 2C)s - 2D$$

Equating coefficients gives

$$A + 3B = 0 \qquad\qquad (1)$$

$$3C - 2B = 0 \qquad\qquad (2)$$

$$3D - 2C = 0 \qquad\qquad (3)$$

$$-2D = 2 \qquad\qquad (4)$$

From Equation (4), we get $D = -1$. Then, from (3),

$$2C = 3D = -3$$

or $C = -\frac{3}{2}$. Then, from (2),

$$2B = 3C = -\frac{9}{2}$$

or $B = -\frac{9}{4}$. Finally, from (1),

$$A = -3B = \frac{27}{4}$$

Our partial fractions are then

$$\frac{2}{s^3(3s - 2)} = \frac{27}{4(3s - 2)} - \frac{9}{4s} - \frac{3}{2s^2} - \frac{1}{s^3}$$

♦♦♦

Denominator with Quadratic Factors

We now deal with fractions whose denominators have one or more quadratic factors in addition to any linear factors.

♦♦♦ **Example 24:** Separate the fraction $\dfrac{x + 2}{x^3 + 3x^2 - x}$ into partial fractions.

Solution: The factors of the denominator are x and $x^2 + 3x - 1$, so we assume that

$$\frac{x + 2}{(x^2 + 3x - 1)x} = \frac{Ax + B}{x^2 + 3x - 1} + \frac{C}{x}$$

Multiplying through by $(x^2 + 3x - 1)x$, we have

$$x + 2 = Ax^2 + Bx + Cx^2 + 3Cx - C$$

$$= (A + C)x^2 + (B + 3C)x - C$$

Equating the coefficients of x^2 gives

$$A + C = 0$$

and of x,

$$B + 3C = 1$$

and the constant terms,

$$-C = 2$$

Solving simultaneously gives

$$A = 2 \qquad B = 7 \qquad C = -2$$

So our partial fractions are

$$\frac{x + 2}{x^3 + 3x^2 - x} = \frac{2x + 7}{x^2 + 3x - 1} - \frac{2}{x}$$

♦♦♦

♦♦♦ **Example 25:** Integrate $\displaystyle\int \frac{x + 2}{x^3 + 3x^2 - x}\, dx$.

Solution: Using the results of Examples 24 and 18, we have

$$\int \frac{x + 2}{x^3 + 3x^2 - x}\, dx = \int \frac{2x + 7}{x^2 + 3x - 1}\, dx - \int \frac{2}{x}\, dx$$

$$= \ln|x^2 + 3x - 1| + 1.109 \ln\left|\frac{x - 0.303}{x + 3.303}\right| - 2\ln|x| + C$$

♦♦♦

Exercise 5 ◆ Integrating Rational Fractions

Nonrepeated Linear Factors

Integrate.

1. $\displaystyle \int \frac{(4x - 2)\, dx}{x^3 - x^2 - 2x}$

2. $\displaystyle \int_2^3 \frac{(3x - 1)\, dx}{x^3 - x}$

3. $\displaystyle \int_1^3 \frac{x + 1}{x^3 + 3x^2 + 2x}\, dx$

4. $\displaystyle \int \frac{x^2\, dx}{x^2 - 4x + 5}$

5. $\displaystyle \int \frac{x + 1}{x^3 + 2x^2 - 3x}\, dx$

Repeated Linear Factors

Integrate.

6. $\displaystyle \int_2^4 \frac{(x^3 - 2)\, dx}{x^3 - x^2}$

7. $\displaystyle \int_0^1 \frac{x - 5}{(x - 3)^2}\, dx$

8. $\displaystyle \int_2^3 \frac{x^2\, dx}{(x - 1)^3}$

9. $\displaystyle \int \frac{(2x - 5)\, dx}{(x - 2)^3}$

10. $\displaystyle \int \frac{dx}{x^3 + x^2}$

11. $\displaystyle \int_2^3 \frac{dx}{(x^2 + x)(x - 1)^2}$

12. $\displaystyle \int_0^2 \frac{x - 2}{(x + 1)^3}\, dx$

Quadratic Factors

Integrate.

13. $\displaystyle \int_1^2 \frac{dx}{x^4 + x^2}$

14. $\displaystyle \int_0^1 \frac{(x^2 - 3)\, dx}{(x + 2)(x^2 + 1)}$

15. $\displaystyle \int_1^2 \frac{(4x^2 + 6)\, dx}{x^3 + 3x}$

16. $\displaystyle \int_1^4 \frac{(5x^2 + 4)\, dx}{x^3 + 4x}$

17. $\displaystyle \int \frac{x^2\, dx}{1 - x^4}$

18. $\displaystyle \int_1^2 \frac{(x - 3)\, dx}{x^3 + x^2}$

19. $\displaystyle \int_0^1 \frac{5x\, dx}{(x + 2)(x^2 + 1)}$

20. $\displaystyle \int_3^4 \frac{(5x^2 - 4)\, dx}{x^4 - 16}$

21. $\displaystyle \int \frac{x^5\, dx}{(x^2 + 4)^2}$

This is also called integration by rationalization.

Expressions containing radicals are usually harder to integrate than those that do not. In this section we learn how to remove all radicals and fractional exponents from an expression by means of a suitable algebraic substitution.

Expressions Containing Fractional Powers of *x*

If the expression to be integrated contains a factor of, say, $x^{1/3}$, we would substitute

$$z = x^{1/3}$$

and the fractional exponents would vanish. If the same problem should also contain $x^{1/2}$, for example, we would substitute

$$z = x^{1/6}$$

because 6 is the lowest common denominator of the fractional exponents.

◆◆◆ **Example 26:** Integrate $\int \dfrac{x^{1/2}\, dx}{1 + x^{3/4}}$.

Solution: Here x is raised to the $\frac{1}{2}$ power in one place and to the $\frac{3}{4}$ power in another. The LCD of $\frac{1}{2}$ and $\frac{3}{4}$ is 4, so we substitute

$$z = x^{1/4}$$

So

$$x^{1/2} = (x^{1/4})^2 = z^2$$

and

$$x^{3/4} = (x^{1/4})^3 = z^3$$

and since

$$x = z^4$$

$$dx = 4z^3\, dz$$

Substituting into our given integral gives us

$$\int \frac{x^{1/2}\, dx}{1 + x^{3/4}} = \int \frac{z^2}{1 + z^3}\,(4z^3\, dz)$$

$$= 4 \int \frac{z^5}{1 + z^3}\, dz$$

The integrand is now a rational fraction, such as those we studied in Sec. 34–5. As before, we divide numerator by denominator.

$$4 \int \frac{z^5}{1 + z^3}\, dz = 4 \int \left(z^2 - \frac{z^2}{1 + z^3} \right) dz$$

$$= \frac{4z^3}{3} - \frac{4}{3} \ln |1 + z^3| + C$$

Finally, we substitute back $z = x^{1/4}$.

$$\int \frac{x^{1/2}\, dx}{1 + x^{3/4}} = \frac{4x^{3/4}}{3} - \frac{4}{3} \ln |1 + x^{3/4}| + C$$

◆◆◆

Expressions Containing Fractional Powers of a Binomial

The procedure here is very similar to that of the preceding case, where only x was raised to a fractional power.

◆◆◆ **Example 27:** Integrate $\int x \sqrt{3 + 2x}\ dx$.

Solution: Let $z = \sqrt{3 + 2x}$. Then

$$x = \frac{z^2 - 3}{2}$$

So

$$dx = \frac{1}{2}(2z)\ dz = z\ dz$$

Substituting, we get

$$\int x\sqrt{3 + 2x}\ dx = \int \frac{z^2 - 3}{2} \cdot z \cdot z\ dz$$

$$= \frac{1}{2}\int (z^4 - 3z^2)\ dz$$

$$= \frac{z^5}{10} - \frac{3z^3}{6} + C$$

Now substituting $\sqrt{3 + 2x}$ for z, we obtain

$$\int x\sqrt{3 + 2x}\ dx = \frac{1}{10}(3 + 2x)^{5/2} - \frac{1}{2}(3 + 2x)^{3/2} + C$$

◆◆◆

Definite Integrals

When using the method of substitution with a definite integral, we may also substitute in the limits of integration. This eliminates the need to substitute back to the original variable.

◆◆◆ **Example 28:** Evaluate $\displaystyle\int_0^3 \frac{x\ dx}{\sqrt{1 + x}}$.

Solution: We let $z = \sqrt{1 + x}$. So

$$x = z^2 - 1$$

and

$$dx = 2z\ dz$$

We now compute the limits on z.

When $x = 0$,

$$z = \sqrt{1 + 0} = 1$$

and when $x = 3$,

$$z = \sqrt{1 + 3} = 2$$

Making the substitution yields

$$\int_0^3 \frac{x\ dx}{\sqrt{1 + x}} = \int_1^2 \frac{z^2 - 1}{z} \cdot 2z\ dz$$

$$= 2\int_1^2 (z^2 - 1)\ dz = \left[\frac{2z^3}{3} - 2z\right]_1^2 = \frac{8}{3}$$

◆◆◆

Exercise 6 ◆ Integrating by Algebraic Substitution

Integrate.

1. $\displaystyle\int \frac{dx}{1 + \sqrt{x}}$

2. $\displaystyle\int \frac{dx}{x - \sqrt{x}}$

3. $\displaystyle\int \frac{dx}{1 + \sqrt[3]{x}}$

4. $\displaystyle\int \frac{dx}{\sqrt{x} + \sqrt[4]{x^3}}$

5. $\displaystyle\int \frac{x\,dx}{\sqrt[3]{1 + x}}$

6. $\displaystyle\int \frac{x\,dx}{\sqrt{x - 1}}$

7. $\displaystyle\int \frac{x\,dx}{\sqrt{2 - 7x}}$

8. $\displaystyle\int \frac{\sqrt{x^2 - 1}\,dx}{x}$

9. $\displaystyle\int \frac{x^2\,dx}{(4x + 1)^{5/2}}$

10. $\displaystyle\int \frac{x\,dx}{(1 + x)^{3/2}}$

11. $\displaystyle\int \frac{(x + 5)\,dx}{(x + 4)\sqrt{x + 2}}$

12. $\displaystyle\int \frac{dx}{x^{5/8} + x^{3/4}}$

13. $\displaystyle\int \frac{dx}{x\sqrt{1 - x^2}}$

14. $\displaystyle\int \frac{(x^{3/2} - x^{1/3})\,dx}{6x^{1/4}}$

15. $\displaystyle\int x\sqrt[3]{1 + x}\,dx$

Evaluate.

16. $\displaystyle\int_0^3 \frac{dx}{(x + 2)\sqrt{x + 1}}$

17. $\displaystyle\int_0^1 \frac{x^{3/2}\,dx}{x + 1}$

18. $\displaystyle\int_0^4 \frac{dx}{1 + \sqrt{x}}$

19. $\displaystyle\int_0^{1/2} \frac{dx}{\sqrt{2x}\,(9 + \sqrt[3]{2x})}$

20. $\displaystyle\int_1^4 \frac{x\,dx}{\sqrt{4x + 2}}$

34–7 Integrating by Trigonometric Substitution

The two legs a and b of a right triangle are related to the hypotenuse c by the familiar Pythagorean theorem,

$$a^2 + b^2 = c^2$$

Thus when we have an expression of the form $\sqrt{u^2 + a^2}$, we can consider it as the hypotenuse of a right triangle whose legs are u and a. Doing so enables us to integrate expressions involving the sum or difference of two squares, especially when they are under a radical sign. A similar substitution will work for other radical expressions, as given in the following table:

Expression	Substitution
$\sqrt{u^2 + a^2}$	Let $u = a \tan \theta$.
$\sqrt{a^2 - u^2}$	Let $u = a \sin \theta$.
$\sqrt{u^2 - a^2}$	Let $u = a \sec \theta$.

These are not the only possible substitutions, but they work well, as you will see in the following examples.

Integrals Involving $u^2 + a^2$

♦♦♦ **Example 29:** Integrate $\displaystyle\int \frac{dx}{\sqrt{x^2 + 9}}$.

Solution: Our substitution, from the table above, is

$$x = 3 \tan \theta \tag{1}$$

We sketch a right triangle as shown in Fig. 34–2 with x as the leg opposite to θ and 3 as the leg adjacent to θ. The hypotenuse is then $\sqrt{x^2 + 9}$.

We now write our integral in terms of θ. Since $\cos \theta = 3/\sqrt{x^2 + 9}$,

$$\sqrt{x^2 + 9} = \frac{3}{\cos \theta} = 3 \sec \theta \tag{2}$$

For the numerator, we find dx by taking the derivative of Equation (1).

$$dx = 3 \sec^2 \theta \, d\theta \tag{3}$$

Substituting (2) and (3) into our original integral, we obtain

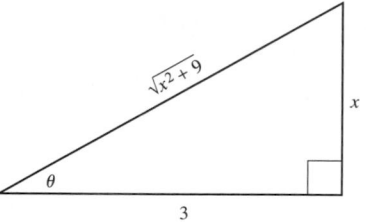

FIGURE 34–2

$$\int \frac{dx}{\sqrt{x^2 + 9}} = \int \frac{3 \sec^2 \theta \, d\theta}{3 \sec \theta}$$

$$= \int \sec \theta \, d\theta$$

$$= \ln |\sec \theta + \tan \theta| + C_1$$

by Rule 14. We now return to our original variable, x. From Fig. 34–2,

$$\sec \theta = \frac{\sqrt{x^2 + 9}}{3} \quad \text{and} \quad \tan \theta = \frac{x}{3}$$

So

$$\int \frac{dx}{\sqrt{x^2 + 9}} = \ln \left| \frac{\sqrt{x^2 + 9}}{3} + \frac{x}{3} \right| + C_1$$

$$= \ln \left| \sqrt{x^2 + 9} + x \right| - \ln 3 + C_1$$

$$= \ln \left| \sqrt{x^2 + 9} + x \right| + C$$

where C_1 and $-\ln 3$ have been combined into a single constant C. ♦♦♦

Integrals Involving $a^2 - u^2$

♦♦♦ **Example 30:** Integrate $\int \sqrt{9 - 4x^2} \, dx$.

Solution: Our radical is of the form $\sqrt{a^2 - u^2}$, with $a = 3$ and $u = 2x$. Our substitution is then

$$2x = 3 \sin \theta$$

or

$$x = \frac{3}{2} \sin \theta$$

Then

$$dx = \frac{3}{2} \cos \theta \, d\theta$$

We sketch a right triangle with hypotenuse 3 and one leg $2x$, as in Fig. 34–3. Then

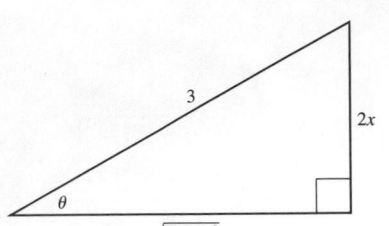

FIGURE 34–3

$$\cos \theta = \frac{\sqrt{9 - 4x^2}}{3}$$

so

$$\sqrt{9 - 4x^2} = 3 \cos \theta$$

Substituting into our given integral gives us

$$\int \sqrt{9 - 4x^2} \, dx = \int (3 \cos \theta)\left(\frac{3}{2} \cos \theta \, d\theta\right)$$

$$= \frac{9}{2} \int \cos^2 \theta \, d\theta$$

By Rule 17,

$$= \frac{9}{2}\left(\frac{\theta}{2} + \frac{\sin 2\theta}{4}\right) + C$$

We now use Eq. 170 to replace $\sin 2\theta$ with $2 \sin \theta \cos \theta$.

$$\int \sqrt{9 - 4x^2} \, dx = \frac{9}{4}(\theta + \sin \theta \cos \theta) + C$$

We now substitute back to return to the variable x. From Fig. 34–3, we see that

$$\sin \theta = \frac{2x}{3}$$

and that

$$\theta = \sin^{-1}\left(\frac{2x}{3}\right)$$

and

$$\cos \theta = \frac{\sqrt{9 - 4x^2}}{3}$$

So

$$\int \sqrt{9 - 4x^2} \, dx = \frac{9}{4}\left(\sin^{-1}\frac{2x}{3} + \frac{2x}{3} \cdot \frac{\sqrt{9 - 4x^2}}{3}\right) + C$$

$$= \frac{9}{4}\left(\sin^{-1}\frac{2x}{3} + \frac{2x\sqrt{9 - 4x^2}}{9}\right) + C \qquad \text{◆◆◆}$$

Definite Integrals

When evaluating a *definite* integral, we may use the limits on x after substituting back into x, or we may *change the limits* when we change variables, as shown in the following example.

◆◆◆ **Example 31:** Integrate $\int_2^4 \sqrt{x^2 - 4}\, dx$.

Solution: We substitute $x = 2 \sec \theta$, so

$$dx = 2 \sec \theta \tan \theta\, d\theta$$

We sketch the triangle as shown in Fig. 34–4, from which

$$\sqrt{x^2 - 4} = 2 \tan \theta$$

Our integral is then

$$\int \sqrt{x^2 - 4}\, dx = \int (2 \tan \theta)(2 \sec \theta \tan \theta\, d\theta)$$

$$= 4 \int \tan^2 \theta \sec \theta\, d\theta$$

$$= 4 \int (\sec^2 \theta - 1) \sec \theta\, d\theta$$

$$= 4 \int (\sec^3 \theta - \sec \theta)\, d\theta$$

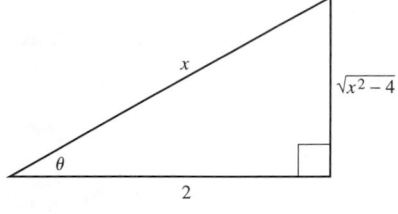

FIGURE 34–4

which we can integrate by Rules 14 and 26. But what about limits? What happens to θ when x goes from 2 to 4? From Fig. 34–4, we see that

$$\cos \theta = \frac{2}{x}$$

When $x = 2$: $\cos \theta = 1$ | When $x = 4$: $\cos \theta = \dfrac{1}{2}$

$\theta = 0°$ | $\theta = 60°$

Note that θ is always acute, so when determining θ, we need consider only angles less than 90°.

So our integral is then

$$\int_{x=2}^{x=4} \sqrt{x^2 - 4}\, dx = 4 \int_{\theta=0°}^{\theta=60°} (\sec^3 \theta - \sec \theta)\, d\theta$$

By Rules 14 and 26,

$$= 4 \left[\frac{1}{2} \sec \theta \tan \theta + \frac{1}{2} \ln|\sec \theta + \tan \theta| - \ln|\sec \theta + \tan \theta| \right]_{0°}^{60°}$$

$$= 2 \left[\sec \theta \tan \theta - \ln|\sec \theta + \tan \theta| \right]_{0°}^{60°}$$

$$= 2[2(1.732) - \ln|2 + 1.732|] - 2[0 - \ln|1 + 0|]$$

$$= 4.294$$

◆◆◆

Exercise 7 ◆ Integrating by Trigonometric Substitution

Integrate.

1. $\displaystyle\int \frac{5\, dx}{(5 - x^2)^{3/2}}$

2. $\displaystyle\int \frac{\sqrt{9 - x^2}\, dx}{x^4}$

3. $\displaystyle\int \frac{dx}{x\sqrt{x^2 + 4}}$

4. $\displaystyle\int_3^4 \frac{dx}{x^3 \sqrt{x^2 - 9}}$

5. $\displaystyle\int \frac{dx}{x^2 \sqrt{4 - x^2}}$

6. $\displaystyle\int \frac{dx}{(x^2 + 2)^{3/2}}$

7. $\displaystyle\int_1^4 \frac{dx}{x\sqrt{25 - x^2}}$

8. $\displaystyle\int \frac{x^2\,dx}{\sqrt{4 - x^2}}$

9. $\displaystyle\int \frac{\sqrt{16 - x^2}\,dx}{x^2}$

10. $\displaystyle\int \frac{dx}{\sqrt{x^2 + 2}}$

11. $\displaystyle\int_3^6 \frac{x^2\,dx}{\sqrt{x^2 - 6}}$

12. $\displaystyle\int_3^5 \frac{x^2\,dx}{(x^2 + 8)^{3/2}}$

13. $\displaystyle\int \frac{dx}{x^2\sqrt{x^2 - 7}}$

34–8   Improper Integrals

Up to now, we have assumed that the limits in the definite integral

$$\int_a^b f(x)\,dx \tag{1}$$

were finite and that the integrand $f(x)$ was continuous for all values of x in the interval $a \le x \le b$. If either or both of these conditions are not met, the integral is called an *improper integral*.

Thus an integral is called improper if

1. one limit or both limits are infinite, or
2. the integrand is discontinuous at some x within the interval $a \le x \le b$.

◆◆◆ **Example 32:**

(a) The integral

$$\int_2^\infty x^2\,dx$$

is improper because one limit is infinite.

(b) The integral

$$\int_0^5 \frac{x^2}{x - 3}\,dx$$

is improper because the integrand, $f(x) = x^2/(x - 3)$, is discontinuous at $x = 3$.   ◆◆◆
We now treat each type separately.

Infinite Limit
The value of a definite integral with one or both limits infinite is defined by the following relationships:

$$\int_a^\infty f(x)\,dx = \lim_{b \to \infty} \int_a^b f(x)\,dx \tag{2}$$

$$\int_{-\infty}^b f(x)\,dx = \lim_{a \to -\infty} \int_a^b f(x)\,dx \tag{3}$$

$$\int_{-\infty}^\infty f(x)\,dx = \lim_{\substack{b \to \infty \\ a \to -\infty}} \int_a^b f(x)\,dx \tag{4}$$

When the limit has a finite value, we say that the integral *converges* or *exists*. When the limit is infinite, we say that the integral *diverges* or *does not exist*.

◆◆◆ **Example 33:** Does this improper integral converge or diverge?

$$\int_1^\infty \frac{dx}{x}$$

Solution: From Equation (2),

$$\int_1^\infty \frac{dx}{x} = \lim_{b \to \infty} \int_1^b \frac{dx}{x}$$

$$= \lim_{b \to \infty} \left[\ln x \right]_1^b$$

$$= \lim_{b \to \infty} \left[\ln b \right] = \infty$$

Since there is no limiting value, we say that the given integral diverges or does not exist.   ◆◆◆

◆◆◆ **Example 34:** Does the following integral converge or diverge?

Solution:

$$\int_{-\infty}^0 e^x \, dx = \lim_{a \to -\infty} \left[e^x \right]_a^0$$

$$= \lim_{a \to -\infty} [1 - e^a]$$

$$= 1$$

Thus the given limit converges and has a value of 1.   ◆◆◆

◆◆◆ **Example 35:** Find the area under the curve $y = 1/x^2$ (Fig. 34–5) from $x = 2$ to ∞.

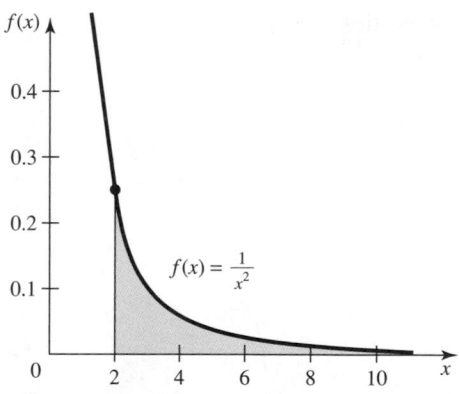

FIGURE 34–5

Solution:

$$A = \int_2^\infty \frac{dx}{x^2} = \lim_{b \to \infty} \int_2^b \frac{dx}{x^2}$$

$$= \lim_{b \to \infty} \left(-\frac{1}{x} \right) \Big|_2^b$$

$$= \lim_{b \to \infty} \left(-\frac{1}{b} + \frac{1}{2} \right) = \frac{1}{2}$$

Thus the area under the curve is not infinite, as might be expected, but has a finite value of $\frac{1}{2}$.

◆◆◆

In many cases we need not carry out the limiting process as in Example 35, but can treat the infinite limit as if it were a very large number. Thus, repeating Example 35, we have

$$\int_2^\infty \frac{dx}{x^2} = \left(-\frac{1}{x}\right)\Big|_2^\infty = -\frac{1}{\infty} + \frac{1}{2} = \frac{1}{2}$$

Examples of where integrals with limits of infinity are used include finding the Laplace transform of a function (Sec. 36–1) and finding areas under the normal curve (Sec. 26–5). (Also see the Team Project in the Review Problems at the end of the chapter.)

Discontinuous Integrand

We now cover the case where the integrand is discontinuous at one of the limits. If the integrand $f(x)$ in Equation (1) is continuous for all values of x in the interval $a \le x \le b$ except at b, then the integral (1) is defined by

$$\int_a^b f(x)\,dx = \lim_{x \to -b} \int_x^b f(x)\,dx \tag{5}$$

where the notation $x \to b^-$ means that x approaches b from values that are less than b. Similarly, if the integrand is discontinuous only at the lower limit a, we have

$$\int_a^b f(x)\,dx = \lim_{x \to a^+} \int_x^b f(x)\,dx \tag{6}$$

where $x \to a^+$ means that x approaches a from values greater than a.

◆◆◆ **Example 36:** Evaluate the improper integral $\displaystyle\int_0^4 \frac{dx}{\sqrt{x}}$.

Solution: Here our function $f(x) = 1/\sqrt{x}$ is discontinuous at the lower limit, 0. Thus from (6),

$$\int_0^4 \frac{dx}{\sqrt{x}} = \lim_{x \to 0^+} \int_x^4 \frac{dx}{\sqrt{x}} = \lim_{x \to 0^+} \left[2\sqrt{x}\right]_0^4$$
$$= 4 - \lim_{x \to 0^+} 2\sqrt{x} = 4 \qquad \text{◆◆◆}$$

Exercise 8 ◆ Improper Integrals

Evaluate each improper integral that converges.

Infinite Limit

1. $\displaystyle\int_1^\infty \frac{dx}{x^3}$

2. $\displaystyle\int_3^\infty \frac{dx}{(x-2)^3}$

3. $\displaystyle\int_{-\infty}^1 e^x\,dx$

4. $\displaystyle\int_5^\infty \frac{dx}{\sqrt{x-1}}$

5. $\displaystyle\int_{-\infty}^\infty \frac{dx}{1+x^2}$

6. $\displaystyle\int_0^\infty \frac{dx}{(x^2+4)^{3/2}}$

Discontinuous Integrand

7. $\displaystyle\int_0^1 \frac{dx}{\sqrt{1-x}}$

8. $\displaystyle\int_0^1 \frac{dx}{x^3}$

9. $\displaystyle\int_0^{\pi/2} \tan x\,dx$

10. $\displaystyle\int_1^2 \frac{dx}{x\sqrt{x^2-1}}$

11. $\displaystyle\int_{-1}^1 \frac{dx}{\sqrt{1-x^2}}$

12. $\displaystyle\int_1^2 \frac{dx}{(x-1)^{2/3}}$

13. The area bounded by the x axis and the curve $y = 1/x$ from $x = 1$ to ∞ is rotated about the x axis. Find the volume of the solid of revolution generated.

Laplace Transforms

The Laplace transform (Sec. 36–1) involves the use of indefinite integrals, such as in problems 14 through 16. Evaluate these, taking t as the variable of integration and treating s and a as constants. (These are Transforms 5, 6, and 17 in Table 36–1.)

14. $\displaystyle\int_0^\infty e^{-st}\, dt$ **15.** $\displaystyle\int_0^\infty te^{-st}\, dt$ **16.** $\displaystyle\int_0^\infty e^{-st}\sin at\, dt$

34–9   Approximate Value of a Definite Integral

The Need for Additional Approximate Methods

So far we have obtained exact values of definite (and indefinite) integrals by using rules from a table of integrals, but only when the function was in the form of an equation that was integrable (Table 34–1). Sometimes, however, we may not have an equation. Our function may be given in the form of a graph or a table of point pairs. We need other methods for these cases.

TABLE 34–1   Methods of integration usable for different types of functions.

	Table of Integrals	Methods of This Section
Indefinite integrals:		
Function integrable	Exact solution	No solution
Function not integrable	No solution	No solution
Definite integrals:		
Function integrable	Exact solution	Approximate solution
Function not integrable	No solution	Approximate solution
Function given graphically	No solution	Approximate solution
Function given by table of point pairs	No solution	Approximate solution

All of the approximate methods presented in this section are based on the fact that the definite integral can be interpreted as the area bounded by the function and the x axis between the given limits. Thus even though the problem we are solving may have nothing to do with area (we could be finding work or a volume, for example), we attack it by finding the area under a curve.

There are many methods for finding the area under a curve, and even mechanical devices such as the planimeter for doing the same. We have already described the *midpoint method* in Sec. 30–5, and here we cover the *average ordinate method*, the *trapezoidal rule*, the *prismoidal formula*, and *Simpson's rule*.

Average Ordinate Method

In Sec. 34–3, we showed how to compute the average ordinate by dividing the area under a curve by the total interval width $b - a$.

$$y_{\text{avg}} = \frac{1}{b-a}\int_a^b f(x)\, dx \qquad \boxed{417}$$

Here we reverse the procedure. We compute the average ordinate (simply by averaging the ordinates of points on the curve) and use it to compute the approximate area. Thus:

Average Ordinate Method	$A \cong y_{avg}(b - a)$	382

Notice that the data points do not have to be equally spaced, although accuracy is usually better if they are.

◆◆◆ **Example 37:** A curve passes through the following points:

x	1	2	4	5	7	7.5	9
y	3.51	4.23	6.29	7.81	7.96	8.91	9.25

Find the area under the curve.

Solution: The average of the seven ordinates is

$$y_{avg} \cong \frac{3.51 + 4.23 + 6.29 + 7.81 + 7.96 + 8.91 + 9.25}{7} = 6.85$$

So $A \cong 6.85(9 - 1) = 54.8$ square units. ◆◆◆

Trapezoidal Rule

FIGURE 34–6

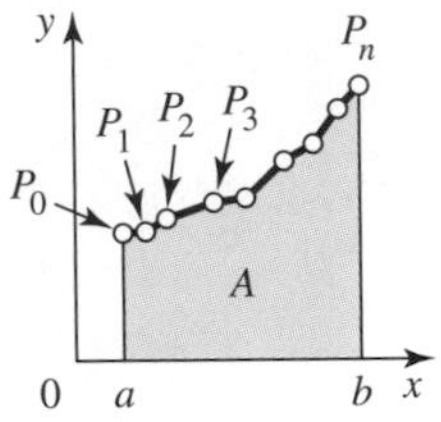

Let us suppose that our function is in the form of a table of point pairs and that we plot each of these points as in Fig. 34–6. Joining these points with a smooth curve would give us a graph of the function, and it is the area under this curve that we seek.

We subdivide this area into a number of vertical *panels* by connecting the points with straight lines and by dropping a perpendicular from each point to the x axis (Fig. 34–7). Each panel then has the shape of a *trapezoid*—hence the name of the rule. If our first data point has the coordinates (x_0, y_0) and the second point is (x_1, y_1), the area of the first panel is, by Eq. 111,

$$\frac{1}{2}(x_1 - x_0)(y_1 + y_0)$$

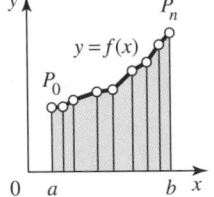

and the sum of all of the areas of all the panels gives us the approximate area under the curve.

FIGURE 34–7
Trapezoidal rule.

Trapezoidal Rule for Unequal Spacing	$A \cong \frac{1}{2}[(x_1 - x_0)(y_1 + y_0) + (x_2 - x_1)(y_2 + y_1) + \cdots + (x_n - x_{n-1})(y_n + y_{n-1})]$	383

If the points are equally spaced in the horizontal direction at a distance of

$$h = x_1 - x_0 = x_2 - x_1 = \cdots = x_n - x_{n-1}$$

Equation 383 reduces to the following:

Trapezoidal Rule for Equal Spacing	$A \cong h\left[\frac{1}{2}(y_0 + y_n) + y_1 + y_2 + \cdots + y_{n-1}\right]$ Where $h = x_1 - x_0$	384

◆◆◆ **Example 38:** Find the area under the graph of the function given by the following table of ordered pairs:

x	0	1	2	3	4	5	6	7	8	9	10
y	4	3	4	7	12	19	28	39	52	67	84

Solution: By Eq. 384 with $h = 1$,

$$A \cong 1\left[\frac{1}{2}(4 + 84) + 3 + 4 + 7 + 12 + 19 + 28 + 39 + 52 + 67\right]$$

$$= 275$$

As a check, we note that the points given are actually points on the curve $y = x^2 - 2x + 4$. Integrating gives us

$$\int_0^{10} (x^2 - 2x + 4)\, dx = \frac{x^3}{3} - x^2 + 4x \bigg|_0^{10} = 273\frac{1}{3}$$

Our approximate area thus agrees with the exact value within about 0.6%.   ◆◆◆

When the function is given in the form of an equation or a graph, we are free to select as many points as we want, for use in Eq. 384. The more points selected, the greater will be the accuracy, and the greater will be the labour.

◆◆◆ **Example 39:** Evaluate the integral

$$\int_0^{10} \left(\frac{x^2}{10} + 2\right) dx$$

taking points spaced 1 unit apart.

Solution: Make a table of point pairs for the function

$$y = \frac{x^2}{10} + 2$$

x	0	1	2	3	4	5	6	7	8	9	10
y	2	2.1	2.4	2.9	3.6	4.5	5.6	6.9	8.4	10.1	12

sum = 46.5

Then, by Eq. 384,

$$\int_0^{10} \left(\frac{x^2}{10} + 2\right) dx \cong 1\left[\frac{1}{2}(2 + 12) + 46.5\right] \cong 53.5$$

          ◆◆◆

Prismoidal Formula

The prismoidal formula gives us the area under a curve that is a parabola if we are given three equally spaced points on the curve. Let us find the area under a parabola [Fig. 34–8(a)]. Since the area does not depend on the location of the origin, let us place the y axis halfway between a and b [Fig. 34–8(b)] so that it divides our area into two panels of width h. Let y_0, y_1, and y_2 be the heights of the curve at $x = -h$, 0, and h, respectively. Then

$$\text{Area} = \int_{-h}^{h} y\, dx = \int_{-h}^{h} (Ax^2 + Bx + C)\, dx = \frac{Ax^3}{3} + \frac{Bx^2}{2} + Cx \bigg|_{-h}^{h}$$

$$= \frac{2}{3}Ah^3 + 2Ch = \frac{h}{3}(2Ah^2 + 6C) \tag{1}$$

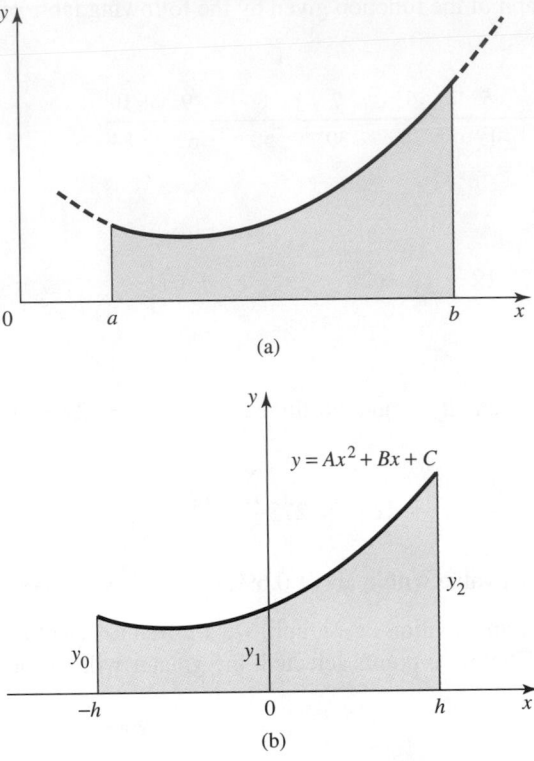

FIGURE 34-8   Areas by prismoidal formula.

We now get A and C in terms of y_0, y_1, and y_2 by substituting coordinates of the three points on the curve, one by one, into the equation of the parabola, $y = Ax^2 + Bx + C$.

$$y(-h) = y_0 = A(-h)^2 + B(-h) + C = Ah^2 - Bh + C$$

$$y(0) = y_1 = A(0) + B(0) + C = C$$

$$y(h) = y_2 = Ah^2 + Bh + C$$

Adding y_0 and y_2 gives

$$y_0 + y_2 = 2Ah^2 + 2C = 2Ah^2 + 2y_1$$

from which $A = (y_0 - 2y_1 + y_2)/2h^2$. Substituting into Equation (1) yields

$$\text{Area} = \frac{h}{3}(2Ah^2 + 6C) = \frac{h}{3}\left(2h^2\frac{y_0 - 2y_1 + y_2}{2h^2} + 6y_1\right)$$

which is also given by the following equation:

Our main use for the prismoidal formula will be in the derivation of Simpson's rule.

Prismoidal Formula	$\text{Area} \cong \dfrac{h}{3}(y_0 + 4y_1 + y_2)$	385

◆◆◆ **Example 40:** A curve passes through the points $(1, 2.05)$, $(3, 5.83)$, and $(5, 17.9)$. Assuming the curve to be a parabola, find the area bounded by the curve and the x axis from $x = 1$ to 5.

Solution: Substituting into the prismoidal formula with $h = 2$, $y_0 = 2.05$, $y_1 = 5.83$, and $y_2 = 17.9$, we have

$$A \cong \frac{2}{3}[2.05 + 4(5.83) + 17.9] = 28.8 \text{ square units}$$

◆◆◆

Simpson's Rule

The prismoidal formula gives us the area under a curve that is a parabola, or *the approximate area under a curve that is assumed to be a parabola,* if we know (or can find) the coordinates of three equally spaced points on the curve. When there are more than three points on the curve, as in Fig. 34–6, we apply the prismoidal formula to each group of three, in turn.

Thus the area of each pair of panels is as follows:

Panels 1 and 2:
$$\frac{h(y_0 + 4y_1 + y_2)}{3}$$

Panels 3 and 4:
$$\frac{h(y_2 + 4y_3 + y_4)}{3}$$

⋮

Panels $n - 1$ and n:
$$\frac{h(y_{n-2} + 4y_{n-1} + y_n)}{3}$$

Adding these areas gives the approximate area under the entire curve, assuming it to be made up of parabolic segments.

Simpson's Rule	$A \cong \dfrac{h}{3}(y_0 + 4y_1 + 2y_2 + 4y_3$ $+ \cdots + 4y_{n-1} + y_n)$	386

Simpson's rule is named for the English mathematician Thomas Simpson (1710–61).

⬩⬩⬩ **Example 41:** A curve passes through the following data points:

x	0	1	2	3	4	5	6	7	8	9	10
y	4.02	5.23	5.66	6.05	5.81	5.62	5.53	5.71	6.32	7.55	8.91

Find the approximate area under the curve using Simpson's rule.

Solution: Substituting, with $h = 1$, we obtain

$$A \cong \frac{4.02 + 4(5.23) + 2(5.66) + 4(6.05) + 2(5.81) + 4(5.62) + 2(5.53) + 4(5.71) + 2(6.32) + 4(7.55) + 8.91}{3}$$

$$= 60.1 \text{ square units}$$

⬩⬩⬩

Exercise 9 ⬩ Approximate Value of a Definite Integral

Evaluate each integral using the average ordinate method, the trapezoidal rule, or Simpson's rule. Choose 10 panels of equal width. Check your answer by integrating.

1. $\displaystyle\int_1^3 \frac{1 + 2x}{x + x^2}\,dx$

2. $\displaystyle\int_1^{10} \frac{dx}{x}$

3. $\displaystyle\int_1^8 (4x^{1/3} + 3)\,dx$

4. $\displaystyle\int_0^\pi \sin x\,dx$

If we had an equation rather than a table of data, we would start by choosing an even number of intervals and then make a table of point pairs.

Using any method, find the area bounded by each curve and the x axis. Use 10 panels.

5. $y = \dfrac{1}{\sqrt{x}}$ from $x = 4$ to 9

6. $y = \dfrac{2x}{1 + x^2}$ from $x = 2$ to 3

7. $y = x\sqrt{1 - x^2}$ from $x = 0$ to 1

8. $y = \dfrac{x + 6}{\sqrt{x + 4}}$ from $x = 0$ to 5

Find the area bounded by the given data points and the x axis. Use any method.

9.

x	0	1	2	3	4	5	6	7	8	9	10
y	415	593	577	615	511	552	559	541	612	745	893

10.

x	0	1	2	3	4	5	6	7	8	9	10
y	3.02	4.63	4.76	5.08	6.31	6.60	6.23	6.48	7.27	8.93	9.11

11.

x	0	1	2	3	4	5	6	7	8	9	10
y	24.0	25.2	25.6	26.0	25.8	25.6	25.5	25.7	26.3	27.5	28.9

12.

x	0	1	2	3	4	5	6	7	8	9	10
y	462	533	576	625	591	522	563	511	602	745	821

Applications

13. A ship's deck, measured every 10.0 m, has the following widths in metres:

$$1.80 \quad 11.3 \quad 17.8 \quad 19.1 \quad 20.3 \quad 19.8 \quad 19.7 \quad 18.9 \quad 15.1 \quad 4.00$$

Find the area of the deck.

14. The dimensions, in centimetres, of a streamlined strut are shown in Fig. 34–9. Find the cross-sectional area.

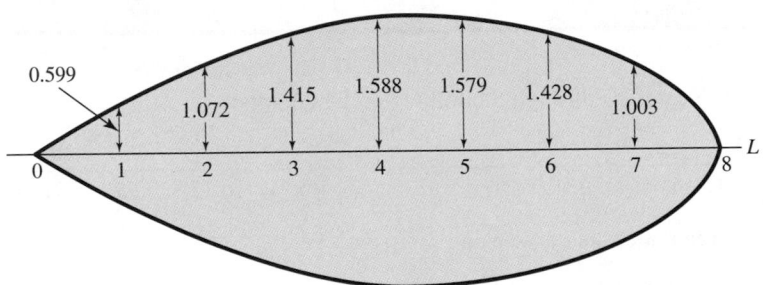

FIGURE 34–9

15. The cross-sectional areas A of a certain ship at various depths D below the waterline are

D (m)	0	2	4	6	8	10	12
A (m²)	2500	2380	2210	1890	1410	810	87

Find the volume of the ship's hull below the waterline.

16. The pressure and volume of an expanding gas in a cylinder are measured and found to be

Volume (cm³)	320	640	960	1280	1600
Pressure (kPa)	474	217	137	98.7	78.0

Find the work done (in joules) by the gas by finding the area under the p-v curve. (Hint: The product of 1 kPa and 1 cm³ is 1×10^{-3} N·m or 1×10^{-3} J.)

17. The light output of a certain lamp (W/nm) as a function of the wavelength of the light (nm) is given as

Wavelength	460	480	500	520	540	560	580	600	620
Output	0	21.5	347	1260	1360	939	338	64.6	0

Find the total output wattage of the lamp by finding the area under the output curve.

One nanometre (nm) equals 10^{-9} m.

18. A certain tank has the following cross-sectional areas at a distance x from one end:

x (m)	0	0.3	0.6	0.9	1.2	1.5	1.8
Area (m^2)	2.83	2.73	2.64	2.42	2.16	1.73	1.39

Find the volume of the tank.

✦✦✦ CHAPTER 34 REVIEW PROBLEMS ✦✦✦✦✦✦✦✦✦✦✦✦✦✦✦✦✦✦✦✦✦✦✦✦✦✦✦✦✦✦✦✦

Integrate. Try some of the definite integrals by an approximate method.

1. $\displaystyle \int \cot^2 x \csc^4 x \, dx$

2. $\displaystyle \int \sqrt{5 - 3x^2} \, dx$

3. $\displaystyle \int \frac{dx}{9x^2 - 4}$

4. $\displaystyle \int \sin x \sec^2 x \, dx$

5. $\displaystyle \int x \ln x^2 \, dx$

6. $\displaystyle \int_1^3 x^2 \log(3 + x^3) \, dx$

7. $\displaystyle \int_0^\pi x \sin(3x^2 - 3) \, dx$

8. $\displaystyle \int e^{3x+4} \, dx$

9. $\displaystyle \int \frac{(3x - 1) \, dx}{x^2 + 9}$

10. $\displaystyle \int \frac{x^2 \, dx}{(9 - x^2)^{3/2}}$

11. $\displaystyle \int \frac{dx}{x \sqrt{1 + x^2}}$

12. $\displaystyle \int \frac{(5x^2 - 3) \, dx}{x^3 - x}$

13. $\displaystyle \int \frac{dx}{2x^2 - 2x + 1}$

14. $\displaystyle \int \frac{dx}{\sqrt{1 + x^2}}$

15. $\displaystyle \int \frac{x \, dx}{\cos^2 x}$

16. $\displaystyle \int \sqrt{1 - 4x^2} \, dx$

17. $\displaystyle \int \frac{(x + 2) \, dx}{x^2 + x + 1}$

18. $\displaystyle \int \frac{\sqrt{x^2 - 1} \, dx}{x^3}$

19. $\int_0^5 \dfrac{(x^2 - 3)\, dx}{(x + 2)(x + 1)^2}$

20. $\int \dfrac{dx}{x^2 \sqrt{5 - x^2}}$

21. $\int_0^4 \dfrac{9x^2\, dx}{(2x + 1)(x + 2)^2}$

22. $\int \dfrac{dx}{\sqrt{4 - (x + 3)^2}}$

23. $\int \dfrac{dx}{9x^2 - 1}$

24. $\int \dfrac{(3x - 7)\, dx}{(x - 2)(x - 3)}$

25. $\int \dfrac{x\, dx}{(1 - x)^5}$

26. $\int \dfrac{dx}{1 - \sqrt{x}}$

27. $\int \cot^3 x \sin x\, dx$

28. $\int \dfrac{3x\, dx}{\sqrt[3]{x + 1}}$

29. $\int x \ln(3x^2 - 2)\, dx$

30. $\int_1^4 5x^2 \log(3x^3 + 7)\, dx$

31. $\int_0^\pi x^2 \cos(x^3 + 5)\, dx$

32. $\int e^{4 - 2x}\, dx$

33. $\int_0^1 \dfrac{(3x^2 + 7x)\, dx}{(x + 1)(x + 2)(x + 3)}$

34. $\int \dfrac{(5x + 9)\, dx}{(x - 9)^{3/2}}$

35. $\int_0^1 \dfrac{(2x^2 + x + 3)\, dx}{(x + 1)(x^2 + 1)}$

36. $\int_2^\infty \dfrac{dx}{x \sqrt{x^2 - 4}}$

37. $\int_0^2 \dfrac{dx}{(x - 2)^{2/3}}$

38. $\int_0^\infty e^{-x} \sin x\, dx$

39. $\int_0^1 \dfrac{dx}{x^2 \sqrt{x^2 + 1}}$

40. Find the length of the curve $y = \ln \sec x$ from the origin to the point $(\pi/3, \ln 2)$.

41. Find the average ordinate for the function $y = \sin^2 x$ for $x = 0$ to 2π.

42. Find the rms value for the function $y = 2x + x^2$ for the interval $x = -1$ to 3.

Writing

43. Assume that there is just one week of class left, time enough to study either exact methods of integration or approximate methods. Your instructor asks your opinion. Write a letter explaining which you think is more important and why.

Team Projects

44. Given an expression by your instructor, find the integral by as many ways as you can: analytically, numerically, or graphically.

45. The equation of the normal curve (Eq. 261) is

$$y = \frac{1}{\sigma\sqrt{2\pi}}\, e^{-(x-\mu)^2/(2\sigma^2)}$$

(a) Simplify this expression by taking the standard deviation σ equal to 1 and the mean μ equal to 0.

(b) Graph this expression, choosing suitable axes. Compare your result with Fig. 26–13.

(c) Write an improper integral expressing the area under this curve between the limits $-\infty$ and $+\infty$.

(d) Evaluate the integral by any method. You should get a value of 1 unit.

(e) Select other limits from Table 26–7. See if you can duplicate by integration some of the entries in that table.

35

Differential Equations

◆◆◆ **OBJECTIVES** ◆◆

When you have completed this chapter, you should be able to:
- Solve simple differential equations graphically or numerically.
- Solve first-order differential equations that have separable variables or that are exact.
- Solve homogeneous first-order differential equations.
- Solve first-order linear differential equations and Bernoulli's equation.
- Solve applications of first-order differential equations.
- Solve second-order differential equations that have separable variables.
- Use the auxiliary equation to determine the general solution to second-order differential equations with right side equal to zero.
- Use the method of undetermined coefficients to solve a second-order differential equation with right side not equal to zero.
- Use second-order differential equations to solve RLC circuits.

◆◆

A *differential equation* is one that contains one or more derivatives. We have already solved some simple differential equations in Chapter 30, and here we go on to solve more difficult types. We begin with first-order differential equations. Our main applications of these equations will be in *exponential growth and decay, motion,* and *electrical circuits*. We then continue our study with second-order differential equations, those that have second derivatives. They can, of course, also have first derivatives, but no third or higher derivatives. Here we solve types of second-order equations that are fairly simple but of great practical importance just the same. Their importance will be borne out by the applications to electrical circuits presented in this chapter.

35–1 Definitions

A *differential equation* is one that contains one or more derivatives. We sometimes refer to differential equations by the abbreviation DE.

◆◆◆ **Example 1:** Some differential equations, using different notation for the derivative, are

(a) $\dfrac{dy}{dx} + 5 = 2xy$

(b) $y'' - 4y' + xy = 0$

(c) $Dy + 5 = 2xy$

(d) $D^2y - 4Dy + xy = 0$

Differential Form

A differential equation containing the derivative $\dfrac{dy}{dx}$ is put into *differential form* simply by multiplying through by dx.

◆◆◆ **Example 2:** Example 1(a) in differential form is

$$dy + 5\,dx = 2xy\,dx \qquad\qquad ◆◆◆$$

Ordinary versus Partial Differential Equations

We get an *ordinary* differential equation when our differential equation contains only two variables and "ordinary" derivatives, as in Examples 1 and 2. When the equation contains *partial* derivatives, because of the presence of three or more variables, we have a *partial differential equation*.

◆◆◆ **Example 3:** $\dfrac{\partial x}{\partial t} = 5\,\dfrac{\partial y}{\partial t}$ is a partial differential equation.

The symbol ∂ is used for partial derivatives. ◆◆◆

We cover only *ordinary* differential equations in this book.

Order

The *order* of a differential equation is the order of the highest-order derivative in the equation.

◆◆◆ **Example 4:**

(a) $\dfrac{dy}{dx} - x = 2y$ is of *first* order.

(b) $\dfrac{d^2y}{dx^2} - \dfrac{dy}{dx} = 3x$ is of *second* order.

(c) $5y''' - 3y'' = xy$ is of *third* order. ◆◆◆

Degree

The *degree of a derivative* is the power to which that derivative is raised.

◆◆◆ **Example 5:** $(y')^2$ is of second degree. ◆◆◆

The *degree of a differential equation* is the degree of the highest-order derivative in the equation. The equation must be rationalized and cleared of fractions before its degree can be determined.

It's important to recognize the type of DE here for the same reason as for other equations. It lets us pick the right method of solution.

◆◆◆ **Example 6:**

(a) $(y'')^3 - 5(y')^4 = 7$ is a second-order equation of *third* degree.

(b) To find the degree of the differential equation

$$\dfrac{x}{\sqrt{y' - 2}} = 1$$

we clear fractions and square both sides, getting

$$\sqrt{y' - 2} = x$$

or

$$y' - 2 = x^2$$

which is of first degree.   ◆◆◆

Common Error	Don't confuse order and degree. The symbols $\dfrac{d^2y}{dx^2}$ and $\left(\dfrac{dy}{dx}\right)^2$ have different meanings.

Solving a Simple Differential Equation

We have already solved some simple differential equations by multiplying both sides of the equation by dx and integrating.

◆◆◆ **Example 7:** Solve the differential equation $dy/dx = x^2 - 3$.

Solution: We first put our equation into differential form, getting $dy = (x^2 - 3)\, dx$. Integrating, we get

$$y = \int (x^2 - 3)dx = \frac{x^3}{3} - 3x + C$$ ◆◆◆

Checking a Solution

Any *function* that satisfies a differential equation is called a *solution* of that equation. Thus, to check a solution, we substitute it and its derivatives into the original equation and see if an identity is obtained.

◆◆◆ **Example 8:** Is the function $y = e^{2x}$ a solution of the differential equation $y'' - 3y' + 2y = 0$?

Solution: Taking the first and second derivatives of the function gives

$$y' = 2e^{2x} \quad \text{and} \quad y'' = 4e^{2x}$$

Substituting the function and its derivatives into the differential equation gives

$$4e^{2x} - 6e^{2x} + 2e^{2x} = 0$$

which is an identity. Thus, $y = e^{2x}$ is a solution to the differential equation. We will see shortly that it is one of many solutions to the given equation.   ◆◆◆

General and Particular Solutions

We have just seen that the function $y = e^{2x}$ is a solution of the differential equation $y'' - 3y' + 2y = 0$. But is it the *only* solution? Try the following functions yourself, and you will see that they are also solutions to the given equation.

$$y = 4e^{2x} \tag{1}$$
$$y = Ce^{2x} \tag{2}$$
$$y = C_1e^{2x} + C_2e^x \tag{3}$$

where C, C_1, and C_2 are *arbitrary constants*.

There is a simple relation, which we state without proof, between the order of a differential equation and the number of constants in the solution.

> The solution of an *n*th-order differential equation can have at most *n* arbitrary constants. A solution having the maximum number of constants is called the *general solution* or *complete solution*.

The differential equation in Example 8 is of second order, so the solution can have up to two arbitrary constants. Thus, Eq. (3) is the general solution, whereas Eqs. (1) and (2) are called *particular solutions*. When we later solve a differential equation, we will first obtain the general solution. Then, by using other given information, we will evaluate the arbitrary constants to obtain a particular solution. The "other given information" is referred to as *boundary conditions* or *initial conditions*.

Exercise 1 ◆ Definitions

Give the order and degree of each equation, and state whether it is an ordinary or partial differential equation.

1. $\dfrac{dy}{dx} + 3xy = 5$

2. $y'' + 3y' = 5x$

3. $D^3y - 4Dy = 2xy$

4. $\dfrac{\partial^2 y}{\partial x^2} + 4y = 7$

5. $3(y'')^4 - 5y' = 3y$

6. $4\dfrac{dy}{dx} - 3\left(\dfrac{d^2y}{dx^2}\right)^3 = x^2 y$

Solve each differential equation.

7. $\dfrac{dy}{dx} = 7x$

8. $2y' = x^2$

9. $4x - 3y' = 5$

10. $3Dy = 5x + 2$

11. $dy = x^2\, dx$

12. $dy - 4x\, dx = 0$

Show that each function is a solution to the given differential equation.

13. $y' = \dfrac{2y}{x}, \; y = Cx^2$

14. $\dfrac{dy}{dx} = \dfrac{x^2}{y^3}, \; 4x^3 - 3y^4 = C$

15. $Dy = \dfrac{2y}{x}, \; y = Cx^2$

16. $y' \cot x + 3 + y = 0, \quad y = C \cos x - 3$

35–2 Graphical and Numerical Solution of Differential Equations

We have said that any function that satisfies a differential equation is a solution to that equation. That function can be given

(a) graphically, as a plotted curve;
(b) numerically, as a table of point pairs; or
(c) analytically, as an equation.

In this section we will find graphical and numerical solutions, and in later sections we will find analytical solutions where possible. The graphical and numerical solutions are especially important because analytical solutions cannot be found for many differential equations.

Graphical Solution by Slope Fields

A differential equation such as

$$dy/dx = x - 2y$$

relates the variables x and y to the derivative dy/dx. The derivative graphically gives us the slope m of the curve. Replacing dy/dx with m gives

$$m = x - 2y$$

Using this equation, we can compute the slope at any point pair. The graph of the slopes is called a *slope field* (also called a *direction field* or *tangent field*). To get the solutions to a differential equation, we simply sketch the curves that have the slopes of the surrounding slope field.

◆◆◆ **Example 9:**

(a) Construct a slope field for the differential equation

$$dy/dx = x - 2y$$

for $x = 0$ to 5 and $y = 0$ to 5.

(b) Sketch the solution that has the boundary conditions $y = 2$ when $x = 0$.

This is obviously a lot of work, which usually makes it an impractical method of solution. It is, however, a good way to visualize a solution to a DE.

Solution:

(a) Computing slopes gives the following:

$$\text{At } (0, 0) \quad m = 0 - 2(0) = 0$$
$$\text{At } (0, 1) \quad m = 0 - 2(1) = -2$$
$$\vdots \qquad \vdots$$
$$\text{At } (5, 5) \quad m = 5 - 2(5) = -5$$

It takes 25 computations to get all the points. We save time by using a computer and get the following:

y						
5	-10	-9	-8	-7	-6	-5
4	-8	-7	-6	-5	-4	-3
3	-6	-5	-4	-3	-2	-1
2	-4	-3	-2	-1	0	1
1	-2	-1	0	1	2	3
0	0	1	2	3	4	5
	0	1	2	3	4	5
				x		

We make our slope field by drawing a short line with the required slope at each point, as shown in Fig. 35–1.

(b) Several solutions are shown dashed in Fig. 35–1. The solution that has the boundary conditions $y = 2$ when $x = 0$ is shown as a solid line.

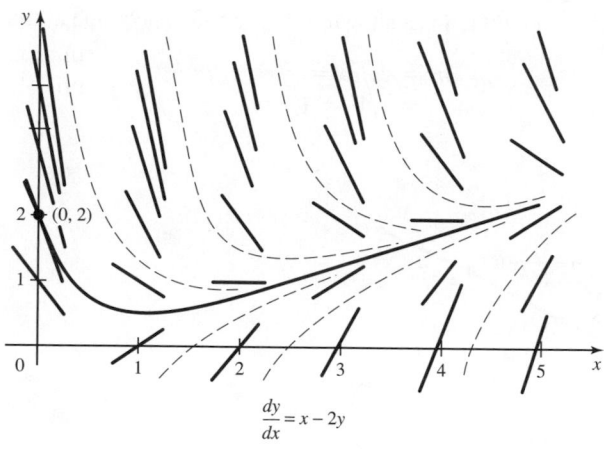

$$\frac{dy}{dx} = x - 2y$$

FIGURE 35–1   ♦♦♦

Euler's Method: Graphical Solution

We will now describe a technique, called *Euler's method*, for solving a differential equation approximately. We use it here for a graphical solution of a DE, and in the next section we use it for a numerical solution.

Suppose that we have a first-order differential equation, which we write in the form

$$y' = f(x, y) \qquad (1)$$

Euler's method is named after the Swiss mathematician Leonhard Euler (1707–83).

and a boundary condition that $x = x_p$ when $y = y_p$. We seek a solution $y = F(x)$ such that the graph of this function (Fig. 35–2) passes through $P(x_p, y_p)$ and has a slope at P given by Eq. (1), $m_p = f(x_p, y_p)$.

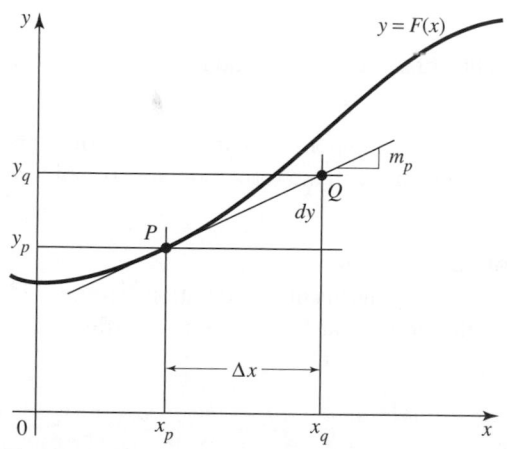

FIGURE 35–2 Euler's method.

Having the slope at P, we then step a distance Δx to the right. The rise dy of the tangent line is then

$$\text{rise} = (\text{slope})(\text{run}) = m_p \, \Delta x$$

Thus, the point Q has the coordinates $(x_p + \Delta x, y_p + m_p \, \Delta x)$. *Point Q is probably not on the curve $y = F(x)$ but might be close enough to use as an approximation to the curve.*

From Q, we repeat the process enough times to reconstruct as much of the curve $y = F(x)$ as needed.

♦♦♦ **Example 10:** Use Euler's method to graphically solve the DE, $dy/dx = x/y^2$, from the boundary value $(1, 1)$ to $x = 5$. Increase x in steps of $\Delta x = 1$ unit.

Solution: We plot the initial point (1, 1) as shown in Fig. 35–3. The slope at that point is then

$$m = dy/dx = x/y^2$$
$$= 1/1^2 = 1$$

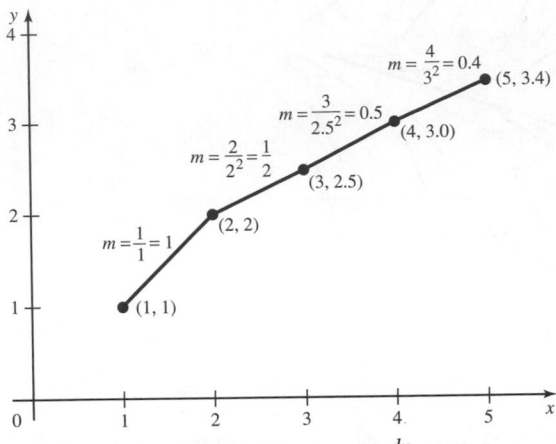

FIGURE 35–3   Graphical solution of $\dfrac{dy}{dx} = \dfrac{x}{y^2}$.

Through (1, 1) we draw a line of slope 1. We assume the slope to be constant over the interval from $x = 1$ to 2. We extend the line to $x = 2$ and get a new point (2, 2). The slope at that point is then

$$m = 2/2^2 = 1/2$$

Through (2, 2) we draw a line of slope $\frac{1}{2}$ and extend it to get $(3, 2\frac{1}{2})$. We continue in this manner to get the final point (5, 3.4).   ◆◆◆

Note that our solution in Example 10 was in the form of a graph. With the numerical method to follow, our solution will be in the form of a table of point pairs.

Euler's Method: Numerical Solution

We can express Euler's method by means of two iteration formulas. If we have the coordinates (x_p, y_p) at any point P, and the slope m_p at P, we find the coordinates (x_q, y_q) of Q by the following iteration formulas:

Euler's Method	$x_q = x_p + \Delta x$ $y_q = y_p + m_p \Delta x$	438

◆◆◆ **Example 11:** Find an approximate solution to $y' = x^2/y$, with the boundary condition that $y = 2$ when $x = 3$. Calculate y for $x = 3$ to 10 in steps of 1.

Solution: The slope at (3, 2) is

$$m = y'(3, 2) = \frac{9}{2} = 4.5$$

If $\Delta x = 1$, the rise is

$$dy = m\,\Delta x = 4.5(1) = 4.5$$

The ordinate of our next point is then $2 + 4.5 = 6.5$. The abscissa of the next point is $3 + 1 = 4$. So the coordinates of our next point are $(4, 6.5)$. Repeating the process using $(4, 6.5)$ as (x_p, y_p), we calculate the next point by first getting m and dy.

$$m = y'(4, 6.5) = \frac{16}{6.5} = 2.462$$
$$dy = 2.462(1) = 2.462$$

So the next point is $(5, 8.962)$. The remaining values are given in Table 35–1, which was computer generated.

The solution to our differential equation then is in the form of a set of (x, y) pairs, the first two columns in Table 35–1. We do not get an equation for our solution.

Table 35–1

x	Approximate y	Exact y	Error
3	2.000 00	2.000 00	0.000 00
4	6.500 00	5.354 13	1.145 87
5	8.961 54	8.326 66	0.634 87
6	11.751 24	11.401 75	0.349 48
7	14.814 75	14.651 51	0.163 24
8	18.122 26	18.092 36	0.029 90
9	21.653 83	21.725 56	−0.071 73
10	25.394 51	25.547 34	−0.152 83

We normally use numerical methods for a differential equation whose exact solution cannot be found. Here we have used an equation whose solution is known so that we can check our answer. The solution is $3y^2 = 2x^3 - 42$, which we use to compute the values shown in the third column of the table, with the difference between exact and approximate values in the third column. Note that the error at $x = 10$ is about 0.6%. The exact and the approximate values are graphed in Fig. 35–4.

Reducing the step size will result in better accuracy (and more work). However, more practical numerical methods for solving DEs are given in the following chapter.

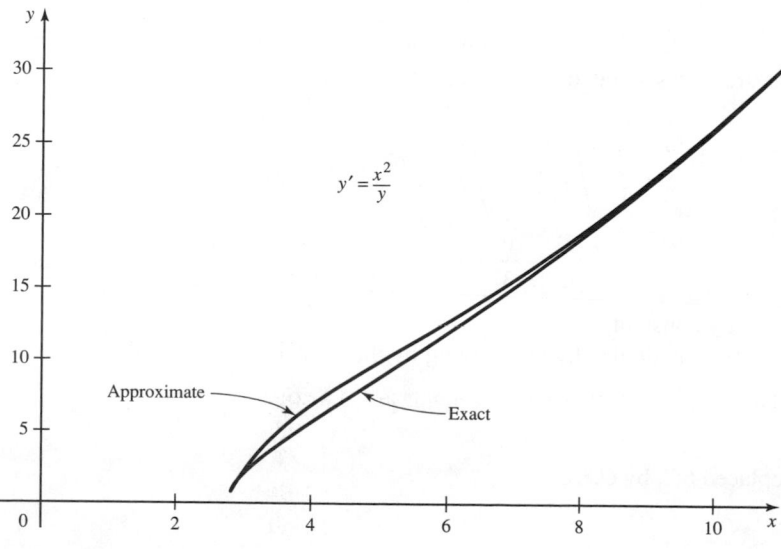

$$y' = \frac{x^2}{y}$$

Approximate

Exact

FIGURE 35–4

Exercise 2 ◆ Graphical and Numerical Solution of Differential Equations

Solve each differential equation and find the approximate value of y requested. Start at the given boundary value and use a slope field or Euler's graphical or numerical method, as directed by your instructor.

1. $y' = x$ Start at $(0, 1)$. Find $y(2)$.
2. $y' = y$ Start at $(0, 1)$. Find $y(3)$.
3. $y' = x - 2y$ Start at $(0, 4)$. Find $y(3)$.
4. $y' = x^2 - y^2 - 1$ Start at $(0, 2)$. Find $y(2)$.

35–3 First-Order Differential Equation, Variables Separable

Given a differential equation of first order, $dy/dx = f(x, y)$, it is sometimes possible to *separate* the variables x and y. That is, when we multiply both sides by dx, the resulting equation can be put into a form that has dy multiplied by a function of y only and dx multiplied by a function of x only.

Form of First-Order DE, Variables Separable	$f(y)\, dy = g(x)\, dx$	419

If this is possible, we can obtain a solution simply by integrating term by term.

◆◆◆ **Example 12:** Solve the differential equation $y' = x^2/y$.

Solution:

1. Rewrite the equation in differential form. We do this by replacing y' with dy/dx and multiplying by dx.

$$dy = \frac{x^2 \, dx}{y}$$

2. Separate the variables. We do this here by multiplying both sides by y, and we get $y \, dy = x^2 \, dx$. The variables are now separated, and our equation is in the form of Eq. 419.
3. Integrate.

$$\int y \, dy = \int x^2 \, dx$$

$$\frac{y^2}{2} = \frac{x^3}{3} + C_1$$

where C_1 is an arbitrary constant.

4. Simplify the answer. We can do this by multiplying by the LCD 6.

$$3y^2 = 2x^3 + C$$

where we have replaced $6C_1$ by C.   ◆◆◆

We'll often leave our answer in implicit form, as we have done here.

Simplifying the Solution

Often the simplification of a solution to a differential equation will involve several steps.

◆◆◆ **Example 13:** Solve the differential equation $dy/dx = 4xy$.

Solution: Multiplying both sides by dx/y and integrating gives us

$$\frac{dy}{y} = 4x \, dx$$

$$\ln |y| = 2x^2 + C$$

We can leave our solution in this implicit form or solve for y.

$$|y| = e^{2x^2+C} = e^{2x^2}e^C$$

Let us replace e^C by another constant k. Since k can be positive or negative, we can remove the absolute value symbols from y, getting

$$y = ke^{2x^2}$$

◆◆◆

The solution to a DE can take on different forms, depending on how you simplify it. Don't be discouraged if your solution does not at first match the one in the answer key.

> We will often interchange one arbitrary constant with another of a different form, such as replacing C_1 by $8C$, or by $\ln C$, or by $\sin C$, or by e^C, or by any other form that will help us to simplify an expression.

The laws of exponents or of logarithms will often be helpful in simplifying an answer, as in the following example.

◆◆◆ **Example 14:** Solve the differential equation $dy/dx = y/(5 - x)$.

Solution:

1. Going to differential form, we have

$$dy = \frac{y \, dx}{5 - x}$$

2. Separating the variables yields

$$\frac{dy}{y} = \frac{dx}{5 - x}$$

3. Integrating gives

$$\ln |y| = -\ln |5 - x| + C_1$$

4. Simplifying gives

$$\ln |y| + \ln |5 - x| = C_1$$

Then using our laws of logarithms we get

$$\ln |y(5 - x)| = C_1$$

Going to exponential form,

$$y(5 - x) = e^{C_1}$$

$$y = \frac{e^{C_1}}{5 - x}$$

where $x \neq 5$. We can leave this expression as it is or simplify it by letting $C = e^{C_1}$, getting

$$y = \frac{C}{5 - x}$$

◆◆◆

Logarithmic, Exponential, or Trigonometric Equations

To separate the variables in certain equations, it might be necessary to use our laws of exponents or logarithms, or the trigonometric identities.

◆◆◆ **Example 15:** Solve the equation $4y'e^{4y} \cos x = e^{2y} \sin x$.

Solution: We replace y' with dy/dx and multiply through by dx, getting $4e^{4y} \cos x \, dy = e^{2y} \sin x \, dx$. We can eliminate x on the left side by dividing through by $\cos x$. Similarly, we can eliminate y from the right by dividing by e^{2y}.

$$\frac{4e^{4y}}{e^{2y}} dy = \frac{\sin x}{\cos x} dx$$

or $4e^{2y} \, dy = \tan x \, dx$. Integrating gives the solution

$$2e^{2y} = -\ln |\cos x| + C$$

◆◆◆

Particular Solution

We can evaluate the constants in our solution to a differential equation when given suitable *boundary conditions*, as in the following example.

◆◆◆ **Example 16:** Solve the equation $2y(1 + x^2)y' + x(1 + y^2) = 0$, subject to the condition that $y = 2$ when $x = 0$.

Solution: Separating variables and integrating, we obtain

$$\frac{2y \, dy}{1 + y^2} + \frac{x \, dx}{1 + x^2} = 0$$

$$\int \frac{2y \, dy}{1 + y^2} + \frac{1}{2} \int \frac{2x \, dx}{1 + x^2} = 0$$

$$\ln |1 + y^2| + \frac{1}{2} \ln |1 + x^2| = C$$

Simplifying, we drop the absolute value signs since $(1 + x^2)$ and $(1 + y^2)$ cannot be negative. Then we multiply by 2 and apply the laws of logarithms.

$$2 \ln (1 + y^2) + \ln (1 + x^2) = 2C$$

$$\ln (1 + x^2)(1 + y^2)^2 = 2C$$

$$(1 + x^2)(1 + y^2)^2 = e^{2C}$$

Applying the boundary conditions that $y = 2$ when $x = 0$ gives

$$e^{2C} = (1 + 0)(1 + 2^2)^2 = 25$$

Our particular solution is then $(1 + x^2)(1 + y^2)^2 = 25$.

◆◆◆

Exercise 3 ◆ First-Order Differential Equation, Variables Separable

General Solution

Find the general solution to each differential equation.

1. $y' = \dfrac{x}{y}$

2. $\dfrac{dy}{dx} = \dfrac{2y}{x}$

We'll have applications later in this chapter.

3. $dy = x^2 y\, dx$

4. $y' = xy^3$

5. $y' = \dfrac{x^2}{y^3}$

6. $y' = \dfrac{x^2 + x}{y - y^2}$

7. $xy\, dx - (x^2 + 1)\, dy = 0$

8. $y' = x^3 y^5$

9. $(1 + x^2)\, dy + (y^2 + 1)\, dx = 0$

10. $y' = x^2 e^{-3y}$

11. $\sqrt{1 + x^2}\, dy + xy\, dx = 0$

12. $y^2\, dx = (1 - x)\, dy$

13. $(y^2 + 1)\, dx = (x^2 + 1)\, dy$

14. $y^3\, dx = x^3\, dy$

15. $(2 + y)\, dx + (x - 2)\, dy = 0$

16. $y' = \dfrac{e^{x-y}}{e^x + 1}$

17. $(x - xy^2)\, dx = -(x^2 y + y)\, dy$

With Exponential Functions

18. $dy = e^{-x}\, dx$

19. $ye^{2x} = (1 + e^{2x})y'$

20. $e^y(y' + 1) = 1$

21. $e^{x-y}\, dx + e^{y-x}\, dy = 0$

With Trigonometric Functions

22. $(3 + y)\, dx + \cot x\, dy = 0$

23. $\tan y\, dx + (1 + x)\, dy = 0$

24. $\tan y\, dx + \tan x\, dy = 0$

25. $\cos x \sin y\, dy + \sin x \cos y\, dx = 0$

26. $\sin x \cos^2 y\, dx + \cos^2 x\, dy = 0$

27. $4 \sin x \sec y\, dx = \sec x\, dy$

Particular Solution

Using the given boundary condition, find the particular solution to each differential equation.

28. $x\, dx = 4y\, dy$, $x = 5$ when $y = 2$

29. $y^2 y' = x^2$, $x = 0$ when $y = 1$

30. $\sqrt{x^2 + 1}\, y' + 3xy^2 = 0$, $x = 1$ when $y = 1$

31. $y' \sin y = \cos x$, $x = \pi/2$ when $y = 0$

32. $x(y + 1)y' = y(1 + x)$, $x = 1$ when $y = 1$

35–4 Exact First-Order Differential Equations

When the left side of a first-order differential equation is the exact differential of some function, we call that equation an *exact differential equation*. Even if we cannot separate the variables in such a differential equation, we might still solve it by integrating a *combination* of terms.

◆◆◆ **Example 17:** Solve $y\,dx + x\,dy = x\,dx$.

Solution: The variables are not now separated, and we see that no amount of manipulation will separate them. However, the combination of terms $y\,dx + x\,dy$ on the left side might ring a bell. In fact, it is the derivative of the product of x and y (Eq. 346).

$$\frac{d(xy)}{dx} = x\frac{dy}{dx} + y\frac{dx}{dx}$$

or

$$d(xy) = x\,dy + y\,dx$$

This, then, is the left side of our given equation. The right side contains only x's, so we integrate.

$$\int (y\,dx + x\,dy) = \int x\,dx$$

$$\int d(xy) = \int x\,dx$$

$$xy = \frac{x^2}{2} + C$$

or

$$y = \frac{x}{2} + \frac{C}{x}$$ ◆◆◆

Integrable Combinations

The expression $y\,dx + x\,dy$ from Example 17 is called an *integrable combination*. Some of the most frequently used combinations are as follows:

Integrable Combinations		
	$x\,dy + y\,dx = d\,(xy)$	**420**
	$\dfrac{x\,dy - y\,dx}{x^2} = d\left(\dfrac{y}{x}\right)$	**421**
	$\dfrac{y\,dx - x\,dy}{y^2} = d\left(\dfrac{x}{y}\right)$	**422**
	$\dfrac{x\,dy - y\,dx}{x^2 + y^2} = d\left(\tan^{-1}\dfrac{y}{x}\right)$	**423**

◆◆◆ **Example 18:** Solve $dy/dx = y(1 - xy)/x$.

Solution: We go to differential form and clear denominators by multiplying through by $x\,dx$, getting $x\,dy = y(1 - xy)\,dx$. Removing parentheses, we obtain

$$x\,dy = y\,dx - xy^2\,dx$$

On the lookout for an integrable combination, we move the $y\,dx$ term to the left side.

$$x\,dy - y\,dx = -xy^2\,dx$$

Any expression that we multiply by (such as $-1/y^2$ here) to make our equation exact is called an integrating factor.

We see now that the left side will be the differential of x/y if we multiply by $-1/y^2$.

$$\frac{y\,dx - x\,dy}{y^2} = x\,dx$$

Integrating gives $x/y = x^2/2 + C_1$, or

$$y = \frac{2x}{x^2 + C} \qquad \blacklozenge\blacklozenge\blacklozenge$$

Particular Solution

As before, we substitute boundary conditions to evaluate the constant of integration.

$\blacklozenge\blacklozenge\blacklozenge$ **Example 19:** Solve $2xy\,dy - 4x\,dx + y^2\,dx = 0$ such that $x = 1$ when $y = 2$.

Solution: It looks as though the first and third terms might be an integrable combination, so we transpose the $-4x\,dx$.

| A certain amount of trial-and-error work might be needed to get the DE into a suitable form. |

$$2xy\,dy + y^2\,dx = 4x\,dx$$

The left side is the derivative of the product xy^2.

$$d(xy^2) = 4x\,dx$$

Integrating gives $xy^2 = 2x^2 + C$. Substituting the boundary conditions, we get $C = xy^2 - 2x^2 = 1(2)^2 - 2(1)^2 = 2$, so our particular solution is

$$xy^2 = 2x^2 + 2 \qquad \blacklozenge\blacklozenge\blacklozenge$$

Exercise 4 ✦ Exact First-Order Differential Equations

Integrable Combinations

Find the general solution of each differential equation.

1. $y\,dx + x\,dy = 7\,dx$

2. $x\,dy = (4 - y)\,dx$

3. $x\dfrac{dy}{dx} = 3 - y$

4. $y + xy' = 9$

5. $2xy' = x - 2y$

6. $x\dfrac{dy}{dx} = 2x - y$

7. $x\,dy = (3x^2 + y)\,dx$

8. $(x + y)\,dx + x\,dy = 0$

9. $3x^2 + 2y + 2xy' = 0$

10. $(1 - 2x^2y)\dfrac{dy}{dx} = 2xy^2$

11. $(2x - y)y' = x - 2y$

12. $\dfrac{y\,dx - x\,dy}{y^2} = x\,dx$

13. $y\,dx - x\,dy = 2y^2\,dx$

14. $(x - 2x^2y)\,dy = y\,dx$

15. $(4y^3 + x)\dfrac{dy}{dx} = y$

16. $(y - x)y' + 2xy^2 + y = 0$

17. $3x - 2y^2 - 4xyy' = 0$

18. $\dfrac{x\,dy - y\,dx}{x^2 + y^2} = 5\,dy$

Using the given boundary condition, find the particular solution to each differential equation.

19. $4x = y + xy'$, $x = 3$ when $y = 1$

20. $y\,dx = (x - 2x^2y)\,dy$, $x = 1$ when $y = 2$

21. $y = (3y^3 + x)\,\dfrac{dy}{dx}$, $x = 1$ when $y = 1$

22. $4x^2 = -2y - 2xy'$, $x = 5$ when $y = 2$

23. $3x - 2y = (2x - 3y)\,\dfrac{dy}{dx}$, $x = 2$ when $y = 2$

24. $5x = 2y^2 + 4xyy'$, $x = 1$ when $y = 4$

35–5   First-Order Homogeneous Differential Equations

Recognizing a Homogeneous Differential Equation

If each variable in a function is replaced by t times the variable, and a power of t can be factored out, we say that the function is *homogeneous*. The power of t that can be factored out of the function is the *degree of homogeneity* of the function.

◆◆◆ **Example 20:** Is $\sqrt{x^4 + xy^3}$ a homogeneous function?

Solution: We replace x with tx and y with ty and get

$$\sqrt{(tx)^4 + (tx)(ty)^3} = \sqrt{t^4x^4 + t^4xy^3} = t^2\sqrt{x^4 + xy^3}$$

Since we were able to factor out a t^2, we say that our function is homogeneous to the second degree.   ◆◆◆

A *homogeneous polynomial* is one in which every term is of the same degree.

◆◆◆ **Example 21:**

(a) $x^2 + xy - y^2$ is a homogeneous polynomial of degree 2.
(b) $x^2 + x - y^2$ is not homogeneous.   ◆◆◆

A first-order *homogeneous differential equation* is one of the following form:

Form of First-Order Homogeneous DE	$M\,dx + N\,dy = 0$	**424**

Here M and N are functions of x and y and are homogeneous functions of the same degree.

◆◆◆ **Example 22:**

(a) $(x^2 + y^2)\,dx + xy\,dy = 0$ is a first-order homogeneous differential equation.
(b) $(x^2 + y^2)\,dx + x\,dy = 0$ is not homogeneous.   ◆◆◆

Solving a First-Order Homogeneous Differential Equation

Sometimes we can transform a homogeneous differential equation whose variables cannot be separated into one whose variables can be separated by making the substitution

$$y = vx$$

as explained in the following example.

⬥⬥⬥ **Example 23:** Solve

$$x\frac{dy}{dx} - y = \sqrt{x^2 + y^2}$$

Solution: We first check if the given equation is homogeneous of first degree. We put the equation into the form of Eq. 424 by multiplying by dx and rearranging.

$$x\,dy - y\,dx = \sqrt{x^2 + y^2}\,dx$$

$$(\sqrt{x^2 + y^2} + y)\,dx - x\,dy = 0$$

This is now in the form of Eq. 424,

$$M\,dx + N\,dy = 0$$

with $M = \sqrt{x^2 + y^2} + y$ and $N = -x$. To test if M is homogeneous, we replace x by tx and y by ty.

$$\sqrt{(tx)^2 + (ty)^2} + ty = \sqrt{t^2x^2 + t^2y^2} + ty$$

$$= t\sqrt{x^2 + y^2} + ty$$

$$= t[\sqrt{x^2 + y^2} + y]$$

We see that t can be factored out and is of first degree. Thus, M is homogeneous of first degree, as is N, so the given differential equation is homogeneous. To solve, we make the substitution $y = vx$ to transform the given equation into one whose variables can be separated. However, when we substitute for y, we must also substitute for dy/dx. Let

$$y = vx$$

Then by the product rule

$$\frac{dy}{dx} = v + x\frac{dv}{dx}$$

Substituting into the given equation yields

$$x\left(v + x\frac{dv}{dx}\right) - vx = \sqrt{x^2 + v^2x^2}$$

which simplifies to

$$x\frac{dv}{dx} = \sqrt{1 + v^2}$$

Separating variables, we have

$$\frac{dv}{\sqrt{1 + v^2}} = \frac{dx}{x}$$

Integrating by Rule 62, we obtain

$$\ln\left|v + \sqrt{1 + v^2}\right| = \ln|x| + C_1$$

$$\ln\left|\frac{v + \sqrt{1 + v^2}}{x}\right| = C_1$$

$$\frac{v + \sqrt{1 + v^2}}{x} = e^{C_1}$$

$$v + \sqrt{1 + v^2} = Cx$$

where $C = e^{C_1}$. Now subtracting v from both sides, squaring, and simplifying gives

$$C^2x^2 - 2Cvx = 1$$

Finally, substituting back, $v = y/x$, we get

$$C^2x^2 - 2Cy = 1 \qquad\qquad ◆◆◆$$

Exercise 5 ◆ First-Order Homogeneous Differential Equations

Find the general solution to each differential equation.

1. $(x - y)\,dx - 2x\,dy = 0$
2. $(3y - x)\,dx = (x + y)\,dy$
3. $(x^2 - xy)y' + y^2 = 0$
4. $(x^2 - xy)\,dy + (x^2 - xy + y^2)\,dx = 0$
5. $xy^2\,dy - (x^3 + y^3)\,dx = 0$
6. $2x^3y' + y^3 - x^2y = 0$

Using the given boundary condition, find the particular solution.

7. $x - y = 2xy', \quad x = 1, y = 1$
8. $3xy^2\,dy = (3y^3 - x^3)\,dx, \quad x = 1, y = 2$
9. $(2x + y)\,dx = y\,dy, \quad x = 2, y = 1$
10. $(x^3 + y^3)\,dx - xy^2\,dy = 0, \quad x = 1, y = 0$

35–6   First-Order Linear Differential Equations

When describing the degree of a term, we usually add the degrees of each variable in the term. Thus, x^2y^3 is of fifth degree.

Sometimes, however, we want to describe the degree of a term with regard to just one of the variables. Thus, we say that x^2y^3 is of second degree in x and of third degree in y.

In determining the degree of a term, we must also consider any derivatives in that term. We consider dy/dx to contribute one "degree" to y in this computation; d^2y/dx^2 to contribute two; and so on. Thus, the term $xy\,dy/dx$ is of first degree in x and of second degree in y.

A first-order differential equation is called *linear* if each term is of first degree or less *in the dependent variable y.*

◆◆◆ **Example 24:**

(a) $y' + x^2y = e^x$ is linear.

(b) $y' + xy^2 = e^x$ is not linear because y is squared in the second term.

(c) $y\,dy/dx - xy = 5$ is not linear because we must add the exponents of y and dy/dx, making the first term of second degree.   ◆◆◆

A first-order linear differential equation can always be written in the following standard form:

Form of First-Order Linear DE	$$\frac{dy}{dx} + Py = Q$$	425

where P and Q are functions of x only.

◆◆◆ **Example 25:** Write the equation $xy' - e^x + y = xy$ in the form of Eq. 425.

Solution: Rearranging gives $xy' + y - xy = e^x$. Factoring, we have

$$xy' + (1 - x)y = e^x$$

Dividing by x gives us the standard form,

$$y' + \frac{1 - x}{x}y = \frac{e^x}{x}$$

where $P = (1 - x)/x$ and $Q = e^x/x$. ◆◆◆

Integrating Factor

The left side of Eq. 425, a first-order linear differential equation, can always be made into an integrable combination by multiplying by an *integrating factor R*. We now find such a factor.

Multiplying Eq. 425 by R, the left side becomes

$$R\frac{dy}{dx} + yRP \tag{1}$$

Let us try to make the left side *the exact derivative of the product Ry of y and the integrating factor R*. The derivative of Ry is

$$R\frac{dy}{dx} + y\frac{dR}{dx} \tag{2}$$

But expressions (1) and (2) will be equal if $dR/dx = RP$. Since P is a function of x only, we can separate variables.

$$\frac{dR}{R} = P\,dx$$

Integrating, $\ln R = \int P\,dx$, which can be rewritten as follows:

We omit the constant of integration because we seek only one integrating factor.

Integrating Factor	$$R = e^{\int P\,dx}$$	426

Thus, *multiplying a given first-order linear equation by the integrating factor $R = e^{\int P\,dx}$ will make the left side of Eq. 425 the exact derivative of Ry*. We can then proceed to integrate to get the solution of the given differential equation.

◆◆◆ **Example 26:** Solve $\dfrac{dy}{dx} + \dfrac{4y}{x} = 3$.

Solution: Our equation is in standard form with $P = 4/x$ and $Q = 3$. Then

$$\int P\,dx = 4\int\frac{dx}{x} = 4\ln|x| = \ln x^4$$

Can you show why $e^{\ln x^4} = x^4$?

Our integrating factor R is then

$$R = e^{\int P\,dx} = e^{\ln x^4} = x^4$$

Multiplying our given equation by x^4 and going to differential form gives

$$x^4\,dy + 4x^3 y\,dx = 3x^4\,dx$$

Notice that the left side is now the derivative of y times the integrating factor, $d(x^4 y)$. Integrating, we obtain

$$x^4 y = \frac{3x^5}{5} + C$$

or

$$y = \frac{3x}{5} + \frac{C}{x^4}$$

In summary, to solve the first-order linear equation

$$y' + Py = Q$$

multiply by an integrating factor $R = e^{\int P\,dx}$, and the solution to the equation becomes

$$yR = \int QR\,dx$$

which can be expressed as follows:

Solution to First-Order Linear DE	$ye^{\int P\,dx} = \int Qe^{\int P\,dx}dx$	427

Next we try an equation having trigonometric functions.

◆◆◆ **Example 27:** Solve $y' + y\cot x = \csc x$.

Solution: This is a first-order differential equation in standard form, with $P = \cot x$ and $Q = \csc x$. Thus, $\int P\,dx = \int \cot x\,dx = \ln|\sin x|$. The integrating factor R is e raised to the $\ln|\sin x|$, or simply $\sin x$. Then, by Eq. 427, the solution to the differential equation is

$$y\sin x = \int \csc x \sin x\,dx$$

Since $\csc x \sin x = 1$, this simplifies to

$$y\sin x = \int dx$$

so

$$y\sin x = x + C$$

Particular Solution

As before, we find the general solution and then substitute the boundary conditions to evaluate C.

◆◆◆ **Example 28:** Solve $y' + \left(\frac{1-2x}{x^2}\right)y = 1$ given that $y = 2$ when $x = 1$.

Solution: We are in standard form with $P = (1 - 2x)/x^2$. We find the integrating factor by first integrating $\int P\,dx$. This gives

$$P = x^{-2} - 2x^{-1}$$

$$\int P\,dx = -\frac{1}{x} - 2\ln|x|$$

$$= -\frac{1}{x} - \ln x^2$$

Our integrating factor is then

$$R = e^{\int P\,dx} = e^{-1/x - \ln x^2} = \frac{e^{-1/x}}{e^{\ln x^2}} = \frac{1}{x^2\,e^{1/x}}$$

Substituting into Eq. 427, the general solution is

$$\frac{y}{x^2\,e^{1/x}} = \int \frac{dx}{x^2\,e^{1/x}}$$

$$= \int e^{-1/x}x^{-2}\,dx = e^{-1/x} + C$$

When $x = 1$ and $y = 2$, $C = 2/e - 1/e = 1/e$, so the particular solution is

$$\frac{y}{x^2 e^{1/x}} = e^{-1/x} + \frac{1}{e}$$

or

$$y = x^2\left(1 + \frac{e^{1/x}}{e}\right)$$ ◆◆◆

Equations Reducible to Linear Form

Sometimes a first-order differential equation that is nonlinear can be expressed in linear form. One type of nonlinear equation easily reducible to linear form is one in which the right side contains a power of y as a factor, an equation known as a *Bernoulli equation*.

Bernoulli's Equation	$\dfrac{dy}{dx} + Py = Qy^n$	**435**

This equation is named after James Bernoulli (1654–1705), the most famous of this family of Swiss mathematicians.

Here P and Q are functions of x, as before. We solve such an equation by making the substitution

$$z = y^{1-n}$$

◆◆◆ **Example 29:** Solve the equation $y' + y/x = x^2y^6$.

Solution: This is a Bernoulli equation with $n = 6$. We first divide through by y^6 and get

$$\frac{1}{y^6}\frac{dy}{dx} + \frac{y^{-5}}{x} = x^2 \qquad (1)$$

We let $z = y^{1-6} = y^{-5}$. The derivative of z with respect to x is

$$\frac{dz}{dx} = -\frac{5}{y^6}\frac{dy}{dx}$$

or

$$\frac{dy}{dx} = -\frac{y^6}{5}\frac{dz}{dx}$$

Substituting for y and dy/dx into Eq. (1) gives

$$-\frac{1}{5}\frac{dz}{dx} + \frac{1}{x}z = x^2$$

or

$$\frac{dz}{dx} - \frac{5}{x}z = -5x^2 \tag{2}$$

This equation is now linear in z. We proceed to solve it as before, with $P = -5/x$ and $Q = -5x^2$. We now find the integrating factor. The integral of P is $\int(-5/x)\,dx = -5\ln x = \ln x^{-5}$, so the integrating factor is

$$R = e^{\int P\,dx} = x^{-5} = \frac{1}{x^5}$$

From Eq. 427,

$$\frac{z}{x^5} = -5\int x^{-3}\,dx = \frac{-5x^{-2}}{-2} + C$$

This is the solution of the differential equation (2). We now solve for z and then substitute back $z = 1/y^5$.

$$z = \frac{5}{2}x^3 + Cx^5 = \frac{1}{y^5}$$

Our solution of (1) is then

$$y^5 = \frac{1}{5x^3/2 + Cx^5}$$

or

$$y^5 = \frac{2}{5x^3 + C_1 x^5}$$

◆◆◆

Summary

To solve a first-order differential equation:

1. If you can *separate the variables*, integrate term by term.
2. If the DE is *exact* (contains an *integrable combination*), isolate that combination on one side of the equation and take the integral of both sides.
3. If the DE is *homogeneous*, start by substituting $y = vx$ and $dy/dx = v + x\,dv/dx$.
4. If the DE is *linear*, $y' + Py = Q$, multiply by the integrating factor $R = e^{\int P\,dx}$, and the solution will be $y = (1/R)\int QR\,dx$.
5. If the DE is a *Bernoulli equation*, $y' + Py = Qy^n$, substitute $z = y^{1-n}$ to make it first-order linear, and proceed as in Step 4.
6. If *boundary conditions* are given, substitute them to evaluate the constant of integration.

Exercise 6 ◆ First-Order Linear Differential Equations

Find the general solution to each differential equation.

1. $y' + \dfrac{y}{x} = 4$

2. $y' + \dfrac{y}{x} = 3x$

3. $xy' = 4x^3 - y$

4. $\dfrac{dy}{dx} + xy = 2x$

5. $y' = x^2 - x^2 y$

6. $y' - \dfrac{y}{x} = \dfrac{-1}{x^2}$

7. $y' = \dfrac{3 - xy}{2x^2}$

8. $y' = x + \dfrac{2y}{x}$

9. $xy' = 2y - x$

10. $(x + 1)y' - 2y = (x + 1)^4$

11. $y' = \dfrac{2 - 4x^2y}{x + x^3}$

12. $(x + 1)y' = 2(x + y + 1)$

13. $xy' + x^2y + y = 0$

14. $(1 + x^3)\,dy = (1 - 3x^2y)\,dx$

15. $y' + y = e^x$

16. $y' = e^{2x} + y$

17. $y' = 2y + 4e^{2x}$

18. $xy' - e^x + y + xy = 0$

19. $y' = \dfrac{4 \ln x - 2x^2y}{x^3}$

With Trigonometric Expressions

20. $y' + y \sin x = 3 \sin x$

21. $y' + y = \sin x$

22. $y' + 2xy = 2x \cos x^2$

23. $y' = 2 \cos x - y$

24. $y' = \sec x - y \cot x$

Bernoulli's Equation

25. $y' + \dfrac{y}{x} = 3x^2y^2$

26. $xy' + x^2y^2 + y = 0$

27. $y' = y - xy^2(x + 2)$

28. $y' + 2xy = xe^{-x^2}y^3$

Particular Solution

Using the given boundary condition, find the particular solution to each differential equation.

29. $xy' + y = 4x, \quad x = 1$ when $y = 5$

30. $\dfrac{dy}{dx} + 5x = x - xy, x = 2$ when $y = 1$

31. $y' + y/x = 5, x = 1$ when $y = 2$

32. $y' = 2 + \dfrac{3y}{x}, x = 2$ when $y = 6$

33. $y' = \tan^2 x + y \cot x, x = \dfrac{\pi}{4}$ when $y = 2$

34. $\dfrac{dy}{dx} + 5y = 3e^x, x = 1$ when $y = 1$

35–7 Geometric Applications of First-Order Differential Equations

Now that we are able to solve some simple differential equations of first order, we turn to applications. Here we not only must solve the equation but also must first *set up* the differential equation. The geometric problems we do first will help to prepare us for the physical applications that follow.

Setting Up a Differential Equation
When reading the problem statement, look for the words "slope" or "rate of change." Each of these can be represented by the first derivative.

♦♦♦ Example 30:

(a) The statement "the slope of a curve at every point is equal to twice the ordinate" is represented by the differential equation

$$\frac{dy}{dx} = 2y$$

(b) The statement "the ratio of abscissa to ordinate at each point on a curve is proportional to the rate of change at that point" can be written

$$\frac{x}{y} = k\frac{dy}{dx}$$

(c) The statement "the slope of a curve at every point is inversely proportional to the square of the ordinate at that point" can be described by the equation

$$\frac{dy}{dx} = \frac{k}{y^2}$$

♦♦♦

Finding an Equation Whose Slope Is Specified

Once the equation is written, it is solved by the methods of the preceding sections.

♦♦♦ Example 31: The slope of a curve at each point is one-tenth the product of the ordinate and the square of the abscissa, and the curve passes through the point (2, 3). Find the equation of the curve.

Solution: The differential equation is

$$\frac{dy}{dx} = \frac{x^2 y}{10}$$

In solving a differential equation, we first see if the variables can be separated. In this case they can be.

$$\frac{10\,dy}{y} = x^2 dx$$

Integrating gives us

$$10 \ln|y| = \frac{x^3}{3} + C$$

At (2, 3),

$$C = 10 \ln 3 - \frac{8}{3} = 8.32$$

Our curve thus has the equation $\ln|y| = x^3/30 + 0.832$.

♦♦♦

Tangents and Normals to Curves

In setting up problems involving tangents and normals to curves, recall that

$$\text{slope of the tangent} = \frac{dy}{dx} = y'$$

$$\text{slope of the normal} = -\frac{1}{y'}$$

♦♦♦ Example 32: A curve passes through the point (4, 2), as shown in Fig. 35–5. If from any point P on the curve, the line OP and the tangent PT are drawn, the triangle OPT is isosceles. Find the equation of the curve.

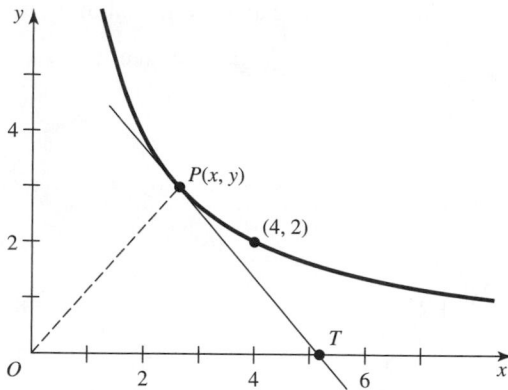

FIGURE 35–5

Solution: The slope of the tangent is dy/dx, and the slope of OP is y/x. Since the triangle is isosceles, these slopes must be equal but of opposite signs.

$$\frac{dy}{dx} = -\frac{y}{x}$$

We can solve this equation by separation of variables or as the integrable combination $x\,dy + y\,dx = 0$. Either way gives the hyperbola

$$xy = C$$

(see Eq. 335). At the point $(4, 2)$, $C = 4(2) = 8$. So our equation is $xy = 8$.   ◆◆◆

Orthogonal Trajectories

In Sec. 30–3 we graphed a relation that had an arbitrary constant, and we got a *family of curves*. For example, the relation $x^2 + y^2 = C^2$ represents a family of circles of radius C, whose centre is at the origin. Another curve that cuts each curve of the family at right angles is called an *orthogonal trajectory* to that family.

To find the orthogonal trajectory to a family of curves:

1. Differentiate the equation of the family to get the slope.
2. Eliminate the constant contained in the original equation.
3. Take the negative reciprocal of that slope to get the slope of the orthogonal trajectory.
4. Solve the resulting differential equation to get the equation of the orthogonal trajectory.

◆◆◆ **Example 33:** Find the equation of the orthogonal trajectories to the parabolas $y^2 = px$.

Solution:

1. The derivative is $2yy' = p$, so

$$y' = \frac{p}{2y} \tag{1}$$

2. The constant p, from the original equation, is y^2/x. Substituting y^2/x for p in Eq. (1) gives $y' = y/2x$.

Common Error	Be sure to eliminate the constant (p in this example) before continuing.

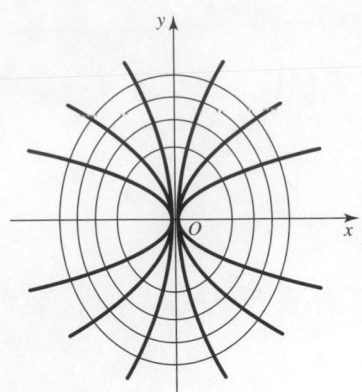

FIGURE 35–6 Orthogonal ellipses and parabolas. Each intersection is at 90°.

3. The slope of the orthogonal trajectory is, by Eq. 295, the negative reciprocal of the slope of the given family.

$$y' = \frac{-2x}{y}$$

4. Separating variables yields

$$y\, dy = -2x\, dx$$

Integrating gives the solution

$$\frac{y^2}{2} = \frac{-2x^2}{2} + C_1$$

which is a family of ellipses, $2x^2 + y^2 = C$ (Fig. 35–6).

◆◆◆

Exercise 7 • Geometric Applications of First-Order Differential Equations

Slope of Curves

1. Find the equation of the curve that passes through the point $(2, 9)$ and whose slope is $y' = x + 1/x + y/x$.

2. The slope of a certain curve at any point is equal to the reciprocal of the ordinate at that point. Write the equation of the curve if it passes through the point $(1, 3)$.

3. Find the equation of a curve whose slope at any point is equal to the abscissa of that point divided by the ordinate and which passes through the point $(3, 4)$.

4. Find the equation of a curve that passes through $(1, 1)$ and whose slope at any point is equal to the product of the ordinate and abscissa.

5. A curve passes through the point $(2, 3)$ and has a slope equal to the sum of the abscissa and ordinate at each point. Find its equation.

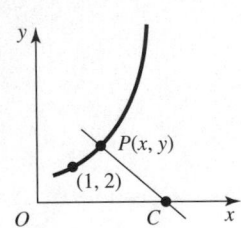

FIGURE 35–7

Tangents and Normals

6. The distance to the x intercept C of the normal (Fig. 35–7) is given by $OC = x + y'y$. Write the equation of the curve passing through $(1, 2)$ for which OC is equal to three times the abscissa of P.

7. A certain first-quadrant curve (Fig. 35–8) passes through the point $(4, 1)$. If a tangent is drawn through any point P, the portion AB of the tangent that lies between the coordinate axes is bisected by P. Find the equation of the curve, given that

$$AP = -\left(\frac{y}{y'}\right)\sqrt{1 + (y')^2} \quad \text{and} \quad BP = x\sqrt{1 + (y')^2}$$

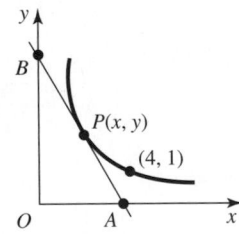

FIGURE 35–8

8. A tangent PT is drawn to a curve at a point P (Fig. 35–9). The distance OT from the origin to a tangent through P is given by

$$OT = \frac{|xy' - y|}{\sqrt{1 + (y')^2}}$$

Find the equation of the curve passing through the point $(2, 4)$ so that OT is equal to the abscissa of P.

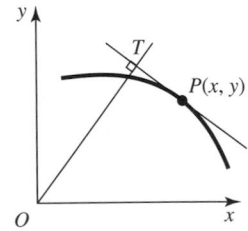

FIGURE 35–9

9. Find the equation of the curve passing through $(0, 1)$ for which the length PC of a normal through P equals the square of the ordinate of P (Fig. 35–10), where

$$PC = y\sqrt{1 + (y')^2}$$

10. Find the equation of the curve passing through $(4, 4)$ such that the distance OT (Fig. 35–9) is equal to the ordinate of P.

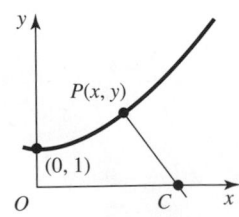

FIGURE 35–10

Orthogonal Trajectories

Write the equation of the orthogonal trajectories to each family of curves.

11. the circles, $x^2 + y^2 = r^2$

12. the parabolas, $x^2 = ay$

13. the hyperbolas, $x^2 - y^2 = Cy$

The equations for the various distances associated with tangents and normals were given in Chapter 22 Review Problems.

35–8 Exponential Growth and Decay

In Chapter 20 we derived Eqs. 199, 201, and 202 for exponential growth and decay and exponential growth to an upper limit by means of compound interest formulas. Here we derive the equation for exponential growth to an upper limit by solving the differential equation that describes such growth. The derivations of equations for exponential growth and decay are left as an exercise.

◆◆◆ **Example 34:** A quantity starts from zero and grows with time such that its rate of growth is proportional to the difference between the final amount a and the present amount y. Find an equation for y as a function of time.

Solution: The amount present at time t is y, and the rate of growth of y we write as dy/dt. Since the rate of growth is proportional to $(a - y)$, we write the differential equation

$$\frac{dy}{dt} = n(a - y)$$

where n is a constant of proportionality. Separating variables, we obtain

$$\frac{dy}{a - y} = n\, dt$$

Integrating gives us

$$-\ln(a - y) = nt + C$$

Going to exponential form and simplifying yields

$$a - y = e^{-nt-C} = e^{-nt}e^{-C} = C_1 e^{-nt}$$

where $C_1 = e^{-C}$. Applying the initial condition that $y = 0$ when $t = 0$ gives $C_1 = a$, so our equation becomes $a - y = ae^{-nt}$, which can be rewritten as follows:

Exponential Growth to an Upper Limit	$y = a\left(1 - e^{-nt}\right)$	202

◆◆◆

Motion in a Resisting Fluid

Here we continue our study of motion. In Chapter 29 we showed that the instantaneous velocity is given by the derivative of the displacement and that the instantaneous acceleration is given by the derivative of the velocity (or by the second derivative of the displacement). In Chapter 31 we solved simple differential equations to find displacement given the velocity or acceleration. Here we do a type of problem that we were not able to solve in Chapter 31.

In this type of problem, an object falls through a fluid (usually air or water) which exerts a *resisting force* that is proportional to the velocity of the object and in the opposite direction. We set up these problems using Newton's second law, $F = ma$, and we'll see that the motion follows the law for exponential growth described by Eq. 202.

◆◆◆ **Example 35:** A crate falls from rest from an airplane. The air resistance is proportional to the crate's velocity, and the crate reaches a limiting speed of 66.4 m/s.

(a) Write an equation for the crate's velocity.
(b) Find the crate's velocity after 0.75 s.

Solution:

(a) By Newton's second law,

$$F = ma = \frac{W}{g}\frac{dv}{dt}$$

where W is the weight of the crate, dv/dt is the acceleration, and $g = 9.806$ m/s^2. Taking the downward direction as positive, the resultant force F is equal to $W - kv$, where k is a constant of proportionality. So

$$W - kv = \frac{W}{9.806}\frac{dv}{dt}$$

We can find k by noting that the acceleration must be zero when the limiting speed (66.4 m/s) is reached. Thus,

$$W - 66.4k = 0$$

So $k = W/66.4$. Our differential equation, after multiplying by 66.4/W, is then $66.4 - v = 6.77 \, dv/dt$. Separating variables and integrating, we have

$$\frac{dv}{66.4 - v} = 0.148 \, dt$$

$$\ln |66.4 - v| = -0.148t + C$$

$$66.4 - v = e^{-0.148t+C} = e^{-0.148t}e^{C}$$

$$= C_1 e^{-0.148t}$$

Since $v = 0$ when $t = 0$, $C_1 = 66.4 - 0 = 66.4$. Then

$$v = 66.4(1 - e^{-0.148t})$$

(b) When $t = 0.75$ s,

$$v = 66.4(1 - e^{-0.111}) = 6.98 \text{ m/s}$$

◆◆◆

Notice that the weight W has dropped out.

Our equation for v is of the same form as Eq. 202 for exponential growth to an upper limit.

Exercise 8 ◆ Exponential Growth and Decay

Exponential Growth

1. A quantity grows with time such that its rate of growth dy/dt is proportional to the present amount y. Use this statement to derive the equation for exponential growth, $y = ae^{nt}$.

2. A biomedical company finds that a certain bacterium used for crop insect control will grow exponentially at the rate of 12.0% per hour. Starting with $10\overline{0}0$ bacteria, how many will the company have after 10.0 h?

3. If the Canadian energy consumption in 1980 was 4.7 million barrels (bbl.) per day oil equivalent and is growing exponentially at a rate of 1.3% per year, estimate the daily oil consumption in the year 2006.

We'll have more problems involving exponential growth and decay in the following section on electric circuits.

Exponential Decay

4. A quantity decreases with time such that its rate of decrease dy/dt is proportional to the present amount y. Use this statement to derive the equation for exponential decay, $y = ae^{-nt}$.

5. An iron ingot is 1030 °C above room temperature. If it cools exponentially at 3.50% per minute, find its temperature (above room temperature) after 2.50 h.

6. A certain pulley in a tape drive is rotating at 2550 rev/min. After the power is shut off, its speed decreases exponentially at a rate of 12.5% per second. Find the pulley's speed after 5.00 s.

Exponential Growth to an Upper Limit

7. A forging, initially at 0 °C, is placed in a furnace at 865 °C, where its temperature rises exponentially at the rate of 6.50% per minute. Find its temperature after 25.0 min.

8. If we assume that the compressive strength of concrete increases exponentially with time to an upper limit of 27.6 MPa, and that the rate of increase is 52.5% per week, find the strength after 2 weeks.

Motion in a Resisting Medium

9. A 202-N carton is initially at rest. It is then pulled horizontally by a 86.3-N force in the direction of motion, and is resisted by a frictional force equal (in newtons) to four times the carton's velocity (in m/s). Show that the differential equation of motion is $dv/dt = 4.19 - 0.194v$.

10. For the carton in problem 9, find the velocity after 1.25 s.

11. An instrument package is dropped from an airplane. It falls from rest through air whose resisting force is proportional to the speed of the package. The terminal speed is 47.2 m/s. Show that the acceleration is given by the differential equation $a = dv/dt = g - gv/47.2$.

12. Find the speed of the instrument package in problem 11 after 0.50 s.

13. Find the displacement of the instrument package in problem 11 after 1.00 s.

14. A 698-N stone falls from rest from a cliff. If the air resistance is proportional to the square of the stone's speed, and the limiting speed of the stone is 38.1 m/s, show that the differential equation of motion is $dv/dt = g - gv^2/38.1^2$.

15. Find the time for the velocity of the stone in problem 14 to be 18.3 m/s.

16. A 66.7-N ball is thrown downward from an airplane with a speed of 6.4 m/s. If we assume the air resistance to be proportional to the ball's speed, and the limiting speed is 41.1 m/s, show that the velocity of the ball is given by $v = 41.1 - 34.7e^{-t/4.19}$ m/s.

17. Find the time at which the ball in problem 16 is going at a speed of 21.3 m/s.

18. A box falls from rest and encounters air resistance proportional to the cube of the speed. The limiting speed is 12.5 ft./s. Show that the acceleration is given by the differential equation $60.7\, dv/dt = 1950 - v^3$.

35–9   Series *RL* and *RC* Circuits

Series *RL* Circuit

Figure 35–11 shows a resistance of R ohms in series with an inductance of L henrys. The switch can connect these elements either to a battery of voltage E (position 1, charge) or to a short circuit (position 2, discharge). In either case, our objective is to find the current i in the circuit. We will see that it is composed of two parts: a *steady-state* current that flows long after the switch has been thrown, and a *transient* current that dies down shortly after the switch is thrown.

⬥⬥⬥ **Example 36:** After being in position 2 for a long time, the switch in Fig. 35–11 is thrown into position 1 at $t = 0$. Write an expression for (a) the current i and (b) the voltage across the inductor.

Solution:

(a) The voltage V_L across an inductance L is given by Eq. A87, $v_L = L\, di/dt$. Using Kirchhoff's voltage law (Eq. A68) gives

$$Ri + L\frac{di}{dt} = E \tag{1}$$

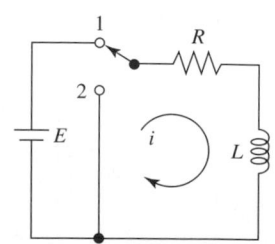

FIGURE 35–11 *RL* circuit.

or

$$\frac{di}{dt} + \frac{R}{L}i = \frac{E}{L} \tag{2}$$

We recognize this as a first-order linear differential equation. Our integrating factor is

$$e^{\int (R/L)\,dt} = e^{Rt/L}$$

In electrical problems, we will use k for the constant of integration, saving C for capacitance. Similarly, R here is resistance, not the integrating constant.

Multiplying Eq. (2) by the integrating factor and integrating, we obtain

$$ie^{Rt/L} = \frac{E}{L}\int e^{Rt/L}\,dt = \frac{E}{R}\,e^{Rt/L} + k$$

Dividing by $e^{Rt/L}$ gives us

$$i = \frac{E}{R} + ke^{-Rt/L} \tag{3}$$

We now evaluate k by noting that $i = 0$ at $t = 0$, so $ke^0 = k = -E/R$. Substituting into Eq. (3) yields the following:

Current in a Charging Inductor	$i = \dfrac{E}{R}(1 - e^{-Rt/L})$	A88

Note that the equation for the current is the same as that for exponential growth to an upper limit (Eq. 202), and that the equation for the voltage across the inductor is of the same form as for exponential decay (Eq. 201).

The first term in this expression (E/R) is the steady-state current, and the second term (E/R) $(e^{-Rt/L})$ is the transient current.

(b) From (1), the voltage across the inductor is

$$v_L = L\frac{di}{dt} = E - Ri$$

Using Eq. A88, we see that $Ri = E - Ee^{-Rt/L}$. Then

$$v_L = E - E + Ee^{-Rt/L}$$

Thus,

Voltage across a Charging Inductor	$v_L = Ee^{-Rt/L}$	A90

Series *RC* Circuit
We now analyze the *RC* circuit as we did the *RL* circuit.

◆◆◆ **Example 37:** A fully charged capacitor (Fig. 35–12) is discharged by throwing the switch from 1 to 2 at $t = 0$. Write an expression for (a) the voltage across the capacitor and (b) the current i.

Solution:

(a) By Kirchhoff's law, the voltage v_R across the resistor at any instant must be equal to the voltage v across the capacitor, but of opposite sign. Further, the current through the resistor, v_R/R, or $-v/R$, must be equal to the current $C\,dv/dt$ in the capacitor.

$$-\frac{v}{R} = C\frac{dv}{dt}$$

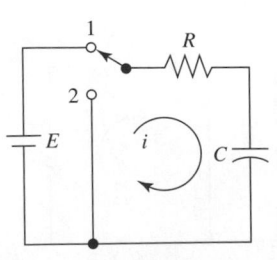

FIGURE 35–12 *RC* circuit.

Separating variables and integrating gives

$$\frac{dv}{v} = -\frac{dt}{RC}$$

$$\ln v = -\frac{t}{RC} + k$$

But at $t = 0$ the voltage across the capacitor is the battery voltage E, so $k = \ln E$. Substituting, we obtain

$$\ln v - \ln E = \ln\frac{v}{E} = -\frac{t}{RC}$$

Or, in exponential form, $v/E = e^{-t/RC}$, which is also expressed as follows:

Voltage across a Discharging Capacitor	$v = Ee^{-t/RC}$	A85

(b) We get the current through the resistor (and the capacitor) by dividing the voltage v by R.

Current in a Discharging Capacitor	$i = \dfrac{E}{R}e^{-t/RC}$	A83

Equations A83 and A85 are both for exponential decay.

◆◆◆

◆◆◆ **Example 38:** For the circuit of Fig. 35–12, $R = 1540\ \Omega$, $C = 125\ \mu\text{F}$, and $E = 115$ V. If the switch is thrown from position 1 to position 2 at $t = 0$, find the current and the voltage across the capacitor at $t = 60$ ms.

Solution: We first compute $1/RC$.

$$\frac{1}{RC} = \frac{1}{1540 \times 125 \times 10^{-6}} = 5.19$$

Then, from Eqs. A83 and A85,

$$i = \frac{E}{R}e^{-t/RC} = \frac{115}{1540}e^{-5.19t}$$

and

$$v = Ee^{-t/RC} = 115e^{-5.19t}$$

At $t = 0.060$ s, $e^{-5.19t} = 0.732$, so

$$i = \frac{115}{1540}(0.732) = 0.0547\ \text{A} = 54.7\ \text{mA}$$

and

$$v = 115(0.732) = 84.2\ \text{V}$$

◆◆◆

Alternating Source

We now consider the case where the *RL* circuit or *RC* circuit is connected to an alternating rather than a direct source of voltage.

◆◆◆ **Example 39:** A switch (Fig. 35–13) is closed at $t = 0$, thus applying an alternating voltage of amplitude E to a resistor and an inductor in series. Write an expression for the current.

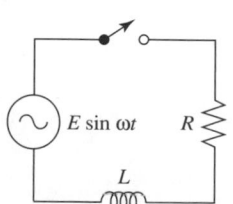

FIGURE 35–13 *RL* circuit with ac source.

Solution: By Kirchhoff's voltage law, $Ri + L\,di/dt = E\sin\omega t$, or

$$\frac{di}{dt} + \frac{R}{L}i = \frac{E}{L}\sin\omega t$$

This is a first-order linear differential equation. Our integrating factor is $e^{\int R/L\,dt} = e^{Rt/L}$. Thus,

$$ie^{Rt/L} = \frac{E}{L}\int e^{Rt/L}\sin\omega t\,dt$$

$$= \frac{E}{L}\frac{e^{Rt/L}}{(R^2/L^2 + \omega^2)}\left(\frac{R}{L}\sin\omega t - \omega\cos\omega t\right) + k$$

by Rule 41. We now divide through by $e^{Rt/L}$ and after some manipulation get

$$i = E\cdot\frac{R\sin\omega t - \omega L\cos\omega t}{R^2 + \omega^2 L^2} + ke^{-Rt/L}$$

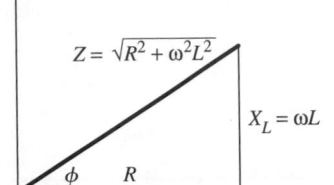

$$Z = \sqrt{R^2 + \omega^2 L^2}$$

$$X_L = \omega L$$

$$\phi \quad R$$

FIGURE 35–14 Impedance triangle.

From the impedance triangle (Fig. 35–14), we see that $R^2 + \omega^2 L^2 = Z^2$, the square of the impedance. Further, by Ohm's law for ac, $E/Z = I$, the amplitude of the current wave. Thus,

$$i = \frac{E}{Z}\left(\frac{R}{Z}\sin\omega t - \frac{\omega L}{Z}\cos\omega t\right) + ke^{-Rt/L}$$

$$= I\left(\frac{R}{Z}\sin\omega t - \frac{\omega L}{Z}\cos\omega t\right) + ke^{-Rt/L}$$

Again from the impedance triangle, $R/Z = \cos\phi$ and $\omega L/Z = \sin\phi$. Substituting yields

$$i = I(\sin\omega t\cos\phi - \cos\omega t\sin\phi) + ke^{-Rt/L}$$

$$= I\sin(\omega t - \phi) + ke^{-Rt/L}$$

which we get by means of the trigonometric identity for the sine of the difference of two angles (Eq. 167). Evaluating k, we note that $i = 0$ when $t = 0$, so

$$k = -I\sin(-\phi) = I\sin\phi = \frac{IX_L}{Z}$$

where, from the impedance triangle, $\sin\phi = X_L/Z$. Substituting, we obtain

$$i = \underbrace{I\sin(\omega t - \phi)}_{\substack{\text{steady-state}\\\text{current}}} + \underbrace{\frac{IX_L}{Z}e^{-Rt/L}}_{\substack{\text{transient}\\\text{current}}}$$

Our current thus has two parts: (1) a steady-state alternating current of magnitude I, out of phase with the applied voltage by an angle ϕ; and (2) a transient current with an initial value of IX_L/Z, which decays exponentially.   ◆◆◆

Exercise 9 ◆ Series *RL* and *RC* Circuits

Series *RL* Circuit

1. If the inductor in Fig. 35–11 is discharged by throwing the switch from position 1 to 2, show that the current decays exponentially according to the function $i = (E/R)e^{-Rt/L}$.

2. The voltage across an inductor is equal to $L\,di/dt$. Show that the magnitude of the voltage across the inductance in problem 1 decays exponentially according to the function $v = Ee^{-Rt/L}$.

3. We showed that when the switch in Fig. 35–11 is thrown from 2 to 1 (charging), the current grows exponentially to an upper limit and is given by $i = (E/R)(1 - e^{-Rt/L})$. Show that the voltage across the inductance ($L\, di/dt$) decays exponentially and is given by $v = Ee^{-Rt/L}$.

4. For the circuit of Fig. 35–11, $R = 382\ \Omega$, $L = 4.75$ H, and $E = 125$ V. If the switch is thrown from 2 to 1 (charging), find the current and the voltage across the inductance at $t = 2.00$ ms.

Series *RC* Circuit

5. The voltage v across the capacitor in Fig. 35–12 during charging is described by the differential equation $(E - v)/R = C\, dv/dt$. Solve this differential equation to show that the voltage is given by $v = E(1 - e^{-t/RC})$. (*Hint:* Write the given equation in the form of a first-order linear differential equation, and solve using Eq. 427.)

6. Show that in problem 5 the current through the resistor (and hence through the capacitor) is given by $i = (E/R)e^{-t/RC}$.

7. For the circuit of Fig. 35–12, $R = 538\ \Omega$, $C = 525\ \mu$F, and $E = 125$ V. If the switch is thrown from 2 to 1 (charging), find the current and the voltage across the capacitor at $t = 2.00$ ms.

Circuits in Which *R*, *L*, and *C* Are Not Constant

8. For the circuit of Fig. 35–11, $L = 2.00$ H, $E = 60.0$ V, and the resistance decreases with time according to the expression

$$R = 4.00/(t + 1)$$

Show that the current i is given by

$$i = 10(t + 1) - 10(t + 1)^{-2}$$

9. For the circuit in problem 8, find the current at $t = 1.55$ ms.

10. For the circuit of Fig. 35–11, $R - 10.0\ \Omega$, $E = 10\overline{0}$ V, and the inductance varies with time according to the expression $L = 5.00t + 2.00$. Show that the current i is given by the expression $i = 10.0 - 40.0/(5.00t + 2.00)^2$.

11. For the circuit in problem 10, find the current at $t = 4.82$ ms.

12. For the circuit of Fig. 35–11, $E = 30\overline{0}$ V, the resistance varies with time according to the expression $R = 4.00t$, and the inductance varies with time according to the expression $L = t^2 + 4.00$. Show that the current i as a function of time is given by $i = 10\overline{0}t\,(t^2 + 12.0)/(t^2 + 4.00)^2$.

13. For the circuit in problem 12, find the current at $t = 1.85$ ms.

14. For the circuit of Fig. 35–12, $C = 2.55\ \mu$F, $E = 625$ V, and the resistance varies with current according to the expression $R = 493 + 372i$. Show that the differential equation for current is $di/dt + 1.51i\, di/dt + 795i = 0$.

Series *RL* or *RC* Circuit with Alternating Current

15. For the circuit of Fig. 35–13, $R = 233\ \Omega$, $L = 5.82$ H, and $e = 58.0 \sin 377t$ V. If the switch is closed when e is zero and increasing, show that the current is given by $i = 26.3 \sin(377t - 83.9°) + 26.1e^{-40t}$ mA.

16. For the circuit in problem 15, find the current at $t = 2.00$ ms.

17. For the circuit in Fig. 35–12, the applied voltage is alternating and is given by $e = E \sin \omega t$. If the switch is thrown from 2 to 1 (charging) when e is zero and increasing, show that the current is given by

$$i = (E/Z)[\sin(\omega t + \phi) - e^{-t/RC} \sin \phi]$$

In your derivation, follow the steps used for the *RL* circuit with an ac source.

18. For the circuit in problem 17, $R = 837\ \Omega$, $C = 2.96\ \mu$F, and $e = 58.0 \sin 377t$. Find the current at $t = 1.00$ ms.

35–10   Second-Order Differential Equations

The General Second-Order Linear Differential Equation

A linear differential equation of second order can be written in the form

$$Py'' + Qy' + Ry = S$$

where P, Q, R, and S are constants or functions of x.

A second-order linear differential equation *with constant coefficients is one where P, Q, and R are constants*, although S can be a function of x, such as in the following equation:

<table>
<tr><td rowspan="1">**Form of Second-Order Linear DE, Right Side Not Zero**</td><td>$ay'' + by' + cy = f(x)$</td><td>**433**</td></tr>
</table>

where a, b, and c are constants. This is the type of equation we will solve in this chapter.

Equation 433 is said to be *homogeneous* if $f(x) = 0$, and it is called *nonhomogeneous* if $f(x)$ is not zero. We will, instead, usually refer to these equations as "right-hand side zero" and "right-hand side not zero."

> We will show a numerical method for approximately solving second-order differential equations in the following chapter.

Operator Notation

> We'll usually use the more familiar y' notation rather than the D operator.

Differential equations are often written using the D operator that we first introduced in Chapter 27, where

$$Dy = y' \qquad D^2y = y'' \qquad D^3y = y''' \qquad \text{etc.}$$

Thus, Eq. 433 can be written

$$aD^2y + bDy + cy = f(x)$$

Second-Order Differential Equations with Variables Separable

We develop methods for solving the general second-order equation in Sec. 35–11. However, simple differential equations of second order that are lacking a first derivative term can be solved by separation of variables, as in the following example.

◆◆◆ **Example 40:** Solve the equation $y'' = 3 \cos x$ (where x is in radians) if $y' = 1$ at the point $(2, 1)$.

Solution: Replacing y'' by $d(y')/dx$ and multiplying both sides by dx to separate variables, we have

$$d(y') = 3 \cos x \, dx$$

Integrating gives us

$$y' = 3 \sin x + C_1$$

> Notice that we have to integrate twice to solve a second-order DE and that two constants of integration have to be evaluated.

Since $y' = 1$ when $x = 2$ rad, $C_1 = 1 - 3 \sin 2 = -1.73$, so

$$y' = 3 \sin x - 1.73$$

or $dy = 3 \sin x \, dx - 1.73 \, dx$. Integrating again, we have

$$y = -3 \cos x - 1.73x + C_2$$

At the point $(2, 1)$, $C_2 = 1 + 3 \cos 2 + 1.73(2) = 3.21$. Our solution is then

$$y = -3 \cos x - 1.73x + 3.21$$

◆◆◆

Exercise 10 ◆ Second-Order Differential Equations

Solve each equation for y.

1. $y'' = 5$

2. $y'' = x$

3. $y'' = 3e^x$

4. $y'' = \sin 2x$

5. $y'' - x^2 = 0$ where $y' = 1$ at the point $(0, 0)$

6. $xy'' = 1$ where $y' = 2$ at the point $(1, 1)$

35–11   Second-Order Differential Equations with Constant Coefficients and Right Side Zero

Solving a Second-Order Equation with Right Side Equal to Zero

If the right side, $f(x)$, in Eq. 433 is zero, and a, b, and c are constants, we have the following (homogeneous) equation:

Form of Second-Order DE, Right Side Zero	$ay'' + by' + cy = 0$	428

To solve this equation, we note that the sum of the three terms on the left side must equal zero. Thus, a solution must be a value of y that will make these terms alike, so they may be combined to give a sum of zero. Also note that each term on the left contains y or its first or second derivative. Thus, *a possible solution is a function such that it and its derivatives are like terms*. Recall from Chapter 33 that one such function was the exponential function

$$y = e^{mx}$$

It has derivatives $y' = me^{mx}$ and $y'' = m^2 e^{mx}$, which are all like terms. We thus try this for our solution. Substituting $y = e^{mx}$ and its derivatives into Eq. 428 gives

$$am^2 e^{mx} + bme^{mx} + ce^{mx} = 0$$

Factoring, we obtain

$$e^{mx}(am^2 + bm + c) = 0$$

Since e^{mx} can never be zero, this equation is satisfied only when $am^2 + bm + c = 0$. This is called the *auxiliary* or *characteristic equation*.

Auxiliary Equation of Second-Order DE, Right Side Zero	$am^2 + bm + c = 0$	429

Note that the coefficients in the auxiliary equation are the same as those in the original DE. We can thus get the auxiliary equation by inspection of the DE.

The auxiliary equation is a quadratic and has two roots which we call m_1 and m_2. Either of the two values of m makes the auxiliary equation equal to zero. Thus, we get *two solutions* to Eq. 428: $y = e^{m_1 x}$ and $y = e^{m_2 x}$.

General Solution

We now show that if y_1 and y_2 are each solutions to $ay'' + by' + cy = 0$, then $y = c_1 y_1 + c_2 y_2$ is also a solution.

We first observe that if $y = c_1 y_1 + c_2 y_2$, then

$$y' = c_1 y'_1 + c_2 y'_2 \quad \text{and} \quad y'' = c_1 y''_1 + c_2 y''_2$$

Substituting into the differential equation (428), we have

$$a(c_1 y''_1 + c_2 y''_2) + b(c_1 y'_1 + c_2 y'_2) + c(c_1 y_1 + c_2 y_2) = 0$$

This simplifies to

$$c_1(ay''_1 + by'_1 + cy_1) + c_2(ay''_2 + by'_2 + cy_2) = 0$$

But $ay''_1 + by'_1 + cy_1 = 0$ and $ay''_2 + by'_2 + cy_2 = 0$ from Eq. 428, so

$$c_1(0) + c_2(0) = 0$$

showing that if y_1 and y_2 are solutions to the differential equation, then $y = c_1 y_1 + c_2 y_2$ is also a solution to the differential equation. Since $y = e^{m_1 x}$ and $y = e^{m_2 x}$ are solutions to Eq. 428, the complete solution is then of the form $y = c_1 e^{m_1 x} + c_2 e^{m_2 x}$.

General Solution of Second-Order DE, Right Side Zero	$y = c_1 e^{m_1 x} + c_2 e^{m_2 x}$	430

We see that the solution to a second-order differential equation depends on the nature of the roots of the auxiliary equation. We thus look at the roots of the auxiliary equation in order to quickly write the solution to the differential equation.

The roots of a quadratic can be real and unequal, real and equal, or nonreal. We now give examples of each case.

Roots Real and Unequal

◆◆◆ **Example 41:** Solve the equation $y'' - 3y' + 2y = 0$.

Solution: We get the auxiliary equation by inspection.

$$m^2 - 3m + 2 = 0$$

It factors into $(m - 1)$ and $(m - 2)$. Setting each factor equal to zero gives $m = 1$ and $m = 2$. These roots are real and unequal; our solution, by Eq. 430, is then

$$y = c_1 e^x + c_2 e^{2x}$$

◆◆◆

Sometimes one root of the auxiliary equation will be zero, as in the next example.

◆◆◆ **Example 42:** Solve the equation $y'' - 5y' = 0$.

Solution: The auxiliary equation is $m^2 - 5m = 0$, which factors into $m(m - 5) = 0$. Setting each factor equal to zero gives

$$m = 0 \quad \text{and} \quad m = 5$$

Our solution is then

$$y = c_1 + c_2 e^{5x}$$

◆◆◆

If the auxiliary equation cannot be factored, we use the quadratic formula to find its roots.

◆◆◆ **Example 43:** Solve $4.82y'' + 5.85y' - 7.26y = 0$.

Solution: The auxiliary equation is $4.82m^2 + 5.85m - 7.26 = 0$. By the quadratic formula,

$$m = \frac{-5.85 \pm \sqrt{(5.85)^2 - 4(4.82)(-7.26)}}{2(4.82)} = 0.762 \quad \text{and} \quad -1.98$$

Our solution is then

$$y = c_1 e^{0.762x} + c_2 e^{-1.98x} \qquad \text{◆◆◆}$$

Roots Real and Equal

If $b^2 - 4ac$ is zero, the auxiliary equation has the double root

$$m = \frac{-b \pm \sqrt{b^2 - 4ac}}{2a} = \frac{-b}{2a}$$

Our solution $y = c_1 e^{mx} + c_2 e^{mx}$ seems to contain two arbitrary constants, but it actually does not. Factoring gives

$$y = (c_1 + c_2)e^{mx} = c_3 e^{mx}$$

where $c_3 = c_1 + c_2$. A solution with one constant cannot be a complete solution for a second-order equation.

Let us *assume* that there is *a second solution* $y = ue^{mx}$, where u is some function of x that we are free to choose. Differentiating, we have

$$y' = mue^{mx} + u'e^{mx}$$

and

$$y'' = m^2 ue^{mx} + mu'e^{mx} + mu'e^{mx} + u''e^{mx}$$

Substituting into $ay'' + by' + cy = 0$ gives

$$a(m^2 ue^{mx} + 2mu'e^{mx} + u''e^{mx}) + b(mue^{mx} + u'e^{mx}) + cue^{mx} = 0$$

which simplifies to

$$e^{mx}[u(am^2 + bm + c) + u'(2am + b) + au''] = 0$$

But $am^2 + bm + c = 0$. Also, $m = -b/2a$, so $2am + b = 0$. Our equation then becomes $e^{mx}(u'') = 0$. Since e^{mx} cannot be zero, we have $u'' = 0$. Thus, any u that has a zero second derivative will make ue^{mx} a solution to the differential equation. The simplest u (not a constant) for which $u'' = 0$ is $u = x$. Thus, xe^{mx} is a solution to the differential equation, and the complete solution to Eq. 428 is as follows:

Equal Roots	$y = c_1 e^{mx} + c_2 x e^{mx}$	431

◆◆◆ **Example 44:** Solve $y'' - 6y' + 9y = 0$.

Solution: The auxiliary equation $m^2 - 6m + 9 = 0$ has the double root $m = 3$. Our solution is then

$$y = c_1 e^{3x} + c_2 x e^{3x} \qquad \text{◆◆◆}$$

Euler's Formula

When the roots of the auxiliary equation are nonreal, our solution will contain expressions of the form e^{jbx}. In the following section we will want to simplify such expressions using *Euler's formula*, which we derive here.

We already used Euler's formula, without proof, in Sec. 21–4. We will derive it again using series in Sec. 37–4.

Let $z = \cos\theta + j\sin\theta$, where $j = \sqrt{-1}$ and θ is in radians. Then

$$\frac{dz}{d\theta} = -\sin\theta + j\cos\theta$$

Multiplying by j (and recalling that $j^2 = -1$) gives

$$j\frac{dz}{d\theta} = -j\sin\theta - \cos\theta = -z$$

Multiplying by $-j$, we get $dz/d\theta = jz$. We now separate variables and integrate.

$$\frac{dz}{z} = j\,d\theta$$

Integrating, we obtain

$$\ln z = j\theta + c$$

When $\theta = 0$, $z = \cos 0 + j\sin 0 = 1$. So $c = \ln z - j\theta = \ln 1 - 0 = 0$. Thus, $\ln z = j\theta$, or, in exponential form, $z = e^{j\theta}$. But $z = \cos\theta + j\sin\theta$, so we arrive at Euler's formula.

Euler's Formula	$e^{j\theta} = \cos\theta + j\sin\theta$	**232**

We can get two more useful forms of Eq. 232 for e^{jbx} and e^{-jbx}. First we set $\theta = bx$. Thus,

$$e^{jbx} = \cos bx + j\sin bx \tag{1}$$

Further,

$$e^{-jbx} = \cos(-bx) + j\sin(-bx) = \cos bx - j\sin bx \tag{2}$$

since $\cos(-A) = \cos A$, and $\sin(-A) = -\sin A$.

Roots Not Real

We return to our second-order differential equation whose solution we are finding by means of the auxiliary equation. We now see that if the auxiliary equation has the nonreal roots $a + jb$ and $a - jb$, our solution becomes

$$y = c_1 e^{(a+jb)x} + c_2 e^{(a-jb)x}$$

$$= c_1 e^{ax} e^{jbx} + c_2 e^{ax} e^{-jbx} = e^{ax}(c_1 e^{jbx} + c_2 e^{-jbx})$$

Using Euler's formula gives

$$y = e^{ax}[c_1(\cos bx + j\sin bx) + c_2(\cos bx - j\sin bx)]$$

$$= e^{ax}[(c_1 + c_2)\cos bx + j(c_1 - c_2)\sin bx]$$

Replacing $c_1 + c_2$ by C_1, and $j(c_1 - c_2)$ by C_2 gives the following:

Nonreal Roots	$y = e^{ax}(C_1\cos bx + C_2\sin bx)$	**432a**

A more compact form of the solution may be obtained by using the equation for the sum of a sine wave and a cosine wave of the same frequency. In Sec. 17–5 we derived the formula (Eq. 214)

$$A\sin\omega t + B\cos\omega t = R\sin(\omega t + \phi)$$

where

$$R = \sqrt{A^2 + B^2} \quad \text{and} \quad \phi = \arctan\frac{B}{A}$$

Thus, the solution to the differential equation can take the following alternative form:

Nonreal Roots	$y = Ce^{ax} \sin(bx + \phi)$	**432b**

We'll find this form handy for applications.

where $C = \sqrt{C_1^2 + C_2^2}$ and $\phi = \arctan C_1/C_2$.

◆◆◆ **Example 45:** Solve $y'' - 4y' + 13y = 0$.

Solution: The auxiliary equation $m^2 - 4m + 13 = 0$ has the roots $m = 2 \pm j3$, two nonreal roots. Substituting into Eq. 432a with $a = 2$ and $b = 3$ gives

$$y = e^{2x}(C_1 \cos 3x + C_2 \sin 3x)$$

or

$$y = Ce^{2x} \sin(3x + \phi)$$

in the alternative form of Eq. 432b.   ◆◆◆

Summary

The types of solutions to a second-order differential equation with right side zero and with constant coefficients are summarized here.

Second-Order DE with Right Side Zero		
$ay'' + by' + cy = 0$		
If the roots of auxiliary equation $am^2 + bm + c = 0$ are:	Then the solution to $ay'' + by' + cy = 0$ is:	
Real and unequal	$y = c_1 e^{m_1 x} + c_2 e^{m_2 x}$	**430**
Real and equal	$y = c_1 e^{mx} + c_2 x e^{mx}$	**431**
Non Real	$y = e^{ax}(C_1 \cos bx + C_2 \sin bx)$	**432a**
	$y = Ce^{ax} \sin(bx + \phi)$	**432b**

Particular Solution

As before, we use the boundary conditions to find the two constants in the solution.

◆◆◆ **Example 46:** Solve $y'' - 4y' + 3y = 0$ if $y' = 5$ at $(1, 2)$.

Solution: The auxiliary equation $m^2 - 4m + 3 = 0$ has roots $m = 1$ and $m = 3$. Our solution is then

$$y = c_1 e^x + c_2 e^{3x}$$

At $(1, 2)$ we get

$$2 = c_1 e + c_2 e^3 \qquad (1)$$

Here we have one equation and two unknowns. We get a second equation by taking the derivative of y.

$$y' = c_1 e^x + 3c_2 e^{3x}$$

Substituting the boundary condition $y' = 5$ when $x = 1$ gives

$$5 = c_1 e + 3c_2 e^3 \tag{2}$$

We solve Eqs. (1) and (2) simultaneously. Subtracting (1) from (2) gives $2c_2 e^3 = 3$, or

$$c_2 = \frac{3}{2e^3}$$
$$= 0.0747$$

Then, from (1),

$$c_1 = \frac{2 - (0.0747)e^3}{e}$$
$$= 0.184$$

Our particular solution is then

$$y = 0.184e^x + 0.0747e^{3x} \qquad \blacklozenge\blacklozenge\blacklozenge$$

Third-Order Differential Equations

We now show (without proof) how the method of the preceding sections can be extended to simple third-order equations that can be easily factored.

◆◆◆ **Example 47:** Solve $y''' - 4y'' - 11y' + 30y = 0$.

Solution: We write the auxiliary equation by inspection.

$$m^3 - 4m^2 - 11m + 30 = 0$$

which factors, by trial and error, into

$$(m - 2)(m - 5)(m + 3) = 0$$

giving roots of 2, 5, and -3. The solution to the given equation is then

$$y = C_1 e^{2x} + C_2 e^{5x} + C_3 e^{-3x} \qquad \blacklozenge\blacklozenge\blacklozenge$$

Exercise 11 ◆ Second-Order Differential Equations with Constant Coefficients and Right Side Zero

Find the general solution to each differential equation.

Second-Order DE, Roots of Auxiliary Equation Real and Unequal

1. $y'' - 6y' + 5y = 0$
2. $2y'' - 5y' - 3y = 0$
3. $y'' - 3y' + 2y = 0$
4. $y'' + 4y' - 5y = 0$
5. $y'' - y' - 6y = 0$
6. $y'' + 5y' + 6y = 0$
7. $5y'' - 2y' = 0$
8. $y'' + 4y' + 3y = 0$
9. $6y'' + 5y' - 6y = 0$
10. $y'' - 4y' + y = 0$

Second-Order DE, Roots of Auxiliary Equation Real and Equal

11. $y'' - 4y' + 4y = 0$
12. $y'' - 6y' + 9y = 0$
13. $y'' - 2y' + y = 0$
14. $9y'' - 6y' + y = 0$
15. $y'' + 4y' + 4y = 0$
16. $4y'' + 4y' + y = 0$
17. $y'' + 2y' + y = 0$
18. $y'' - 10y' + 25y = 0$

Second-Order DE, Roots of Auxiliary Equation Not Real

19. $y'' + 4y + 13y = 0$

20. $y'' - 2y' + 2y = 0$

21. $y'' - 6y' + 25y = 0$

22. $y'' + 2y' + 2y = 0$

23. $y'' + 4y = 0$

24. $y'' + 2y = 0$

25. $y'' - 4y' + 5y = 0$

26. $y'' + 10y' + 425y = 0$

Particular Solution

Solve each differential equation. Use the given boundary conditions to find the constants of integration.

27. $y'' + 6y' + 9y = 0$, $\quad y = 0$ and $y' = 3$ when $x = 0$

28. $y'' + 3y' + 2y = 0$, $\quad y = 0$ and $y' = 1$ when $x = 0$

29. $y'' - 2y' + y = 0$, $\quad y = 5$ and $y' = -9$ when $x = 0$

30. $y'' + 3y' - 4y = 0$, $\quad y = 4$ and $y' = -2$ when $x = 0$

31. $y'' - 2y' = 0$, $\quad y = 1 + e^2$ and $y' = 2e^2$ when $x = 1$

32. $y'' + 2y' + y = 0$, $\quad y = 1$ and $y' = -1$ when $x = 0$

33. $y'' - 4y = 0$, $\quad y = 1$ and $y' = -1$ when $x = 0$

34. $y'' + 9y = 0$, $\quad y = 2$ and $y' = 0$ when $x = \pi/6$

35. $y'' + 2y' + 2y = 0$, $\quad y = 0$ and $y' = 1$ when $x = 0$

36. $y'' + 4y' + 13y = 0$, $\quad y = 0$ and $y' = 12$ when $x = 0$

Third-Order DE

Solve each differential equation.

37. $y''' - 2y'' - y' + 2y = 0$

38. $y''' - y' = 0$

39. $y''' - 6y' + 11y' - 6y = 0$

40. $y''' + y'' - 4y' - 4y = 0$

41. $y''' - 3y'' - y' + 3y = 0$

42. $y''' - 7y' + 6y = 0$

43. $4y''' - 3y' + y = 0$

44. $y''' - y'' = 0$

35–12 Second-Order Differential Equations with Right Side Not Zero

Complementary Function and Particular Integral

We'll now see that the solution to a differential equation is made up of *two parts:* the *complementary function* and the *particular integral*. We show this now for a first-order equation and later for a second-order equation.

The solution to a first-order differential equation, say,

$$y' + \frac{y}{x} = 4 \tag{1}$$

can be found by the methods of Secs. 35–1 through 35–6. The solution to Eq. (1) is

$$y = c/x + 2x$$

Note that the solution has two parts. Let us label one part y_c and the other y_p. Thus, $y = y_c + y_p$, where $y_c = c/x$ and $y_p = 2x$.

Let us substitute for y only $y_c = c/x$ into the left side of (1), and for y' the derivative $-c/x^2$.

$$-\frac{c}{x^2} + \frac{c}{x^2} = 0$$

We get 0 instead of the required 4 on the right-hand side. Thus, y_c *does not satisfy Eq. (1)*. It does, however, satisfy what we call the *reduced equation*, obtained by setting the right side equal to zero. We call y_c the *complementary function*.

We now substitute only $y_p = 2x$ as y in the left side of (1). Similarly, for y', we substitute 2. We now have

$$2 + \frac{2x}{x} = 4$$

We see that y_p does satisfy (1) and is hence a solution. But it cannot be a complete solution because it has no arbitrary constant. We call y_p a *particular integral*. The quantity y_c had the required constant but did not, by itself, satisfy (1).

However, *the sum of y_c and y_p satisfies Eq. (1) and has the required number of constants, and it is hence the complete solution.*

Don't confuse *particular integral* with *particular solution*.

Complete Solution	$y = y_c + y_p$ $= \left(\begin{array}{c}\text{complementary} \\ \text{function}\end{array}\right) + \left(\begin{array}{c}\text{particular} \\ \text{integral}\end{array}\right)$	434

Second-Order Differential Equations

We have seen that the solution to a first-order equation is made up of a complementary function and a particular integral. But is the same true of the second-order equation?

$$ay'' + by' + cy = f(x) \tag{433}$$

If a particular integral y_p is a solution to Eq. 433, we get, on substituting,

$$ay''_p + by'_p + cy_p = f(x) \tag{2}$$

If the complementary function y_c is a complete solution to the reduced equation

$$ay'' + by' + cy = 0 \tag{428}$$

we get

$$ay''_c + by'_c + cy_c = 0 \tag{3}$$

Adding Eqs. (2) and (3) gives us

$$a(y''_p + y''_c) + b(y'_p + y'_c) + c(y_p + y_c) = f(x)$$

Since the sum of the two derivatives is the derivative of the sum, we get

$$a(y_p + y_c)'' + b(y_p + y_c)' + c(y_p + y_c) = f(x)$$

This shows that $y_p + y_c$ is a solution to Eq. 433.

◆◆◆ **Example 48:** Given that the solution to $y'' - 5y' + 6y = 3x$ is

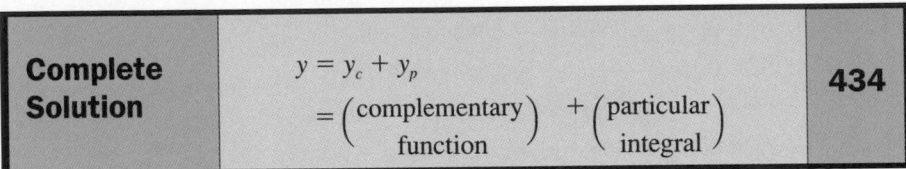

$$y = \underbrace{c_1 e^{3x} + c_2 e^{2x}}_{\substack{\text{complementary} \\ \text{function}}} + \underbrace{\frac{x}{2} + \frac{5}{12}}_{\substack{\text{particular} \\ \text{integral}}}$$

prove by substitution that (a) the complementary function will make the left side of the given equation equal to zero and that (b) the particular integral will make the left side equal to $3x$.

Solution: Given $y'' - 5y' + 6y = 3x$:

(a) Let $\quad y_c = c_1 e^{3x} + c_2 e^{2x}$
$\qquad\quad y'_c = 3c_1 e^{3x} + 2c_2 e^{2x}$
$\qquad\quad y''_c = 9c_1 e^{3x} + 4c_2 e^{2x}$

Substituting on the left gives

$$9c_1e^{3x} + 4c_2e^{2x} - 5(3c_1e^{3x} + 2c_2e^{2x}) + 6(c_1e^{3x} + c_2e^{2x})$$

which equals zero.

(b) Let $\quad y_p = \dfrac{x}{2} + \dfrac{5}{12}$

$$y'_p = \frac{1}{2}$$
$$y''_p = 0$$

Substituting gives

$$0 - 5\left(\frac{1}{2}\right) + 6\left(\frac{x}{2} + \frac{5}{12}\right)$$

or $3x$.   ◆◆◆

Finding the Particular Integral

We already know how to find the complementary function y_c. We set the right side of the given equation to zero and then solve that (reduced) equation just as we did in the preceding section.

To see how to find the particular integral y_p, we start with Eq. 433 and isolate y. We get

$$y = \frac{1}{c}[f(x) - ay'' - by']$$

For this equation to balance, y must contain terms similar to those in $f(x)$. Further, y must contain terms similar to those in its own first and second derivatives. Thus, it seems reasonable to try a solution consisting of the sum of $f(x), f'(x),$ and $f''(x)$, each with an (as yet) undetermined (constant) coefficient.

◆◆◆ **Example 49:** Find y_p for the equation $y'' - 5y' + 6y = 3x$.

Solution: Here $f(x) = 3x, f'(x) = 3,$ and $f''(x) = 0$. Then

$$y_p = Ax + B$$

where A and B are constants yet to be found.   ◆◆◆

> This is sometimes called the *trial function.*

Method of Undetermined Coefficients

To find the constants in y_p, we use the method of undetermined coefficients. We substitute y_p and its first and second derivatives into the differential equation. We then equate coefficients of like terms and solve for the constants.

> We used the method of undetermined coefficients before when finding partial fractions.

◆◆◆ **Example 50:** Determine the constants in y_p in Example 49, and find the complete solution to the given equation given the complementary function $y_c = c_1e^{3x} + c_2e^{2x}$.

Solution: In Example 49 we had $y_p = Ax + B$. Taking derivatives gives

$$y'_p = A \quad \text{and} \quad y''_p = 0$$

Substituting into the original equation, we obtain

$$0 - 5A + 6(Ax + B) = 3x$$
$$6Ax + (6B - 5A) = 3x$$

Equating the coefficients on both sides of this equation, we see that the equation is satisfied when $6A = 3$, or $A = \frac{1}{2}$, and $6B - 5A = 0$. Thus, $B = 5A/6 = \frac{5}{12}$. Our particular integral is then $y_p = x/2 + \frac{5}{12}$, and the complete solution is

$$y = y_c + y_p = c_1e^{3x} + c_1e^{2x} + \frac{x}{2} + \frac{5}{12}$$

◆◆◆

General Procedure

Thus, to solve a second-order linear differential equation with constant coefficients (right side not zero),

$ay'' + by' + cy = f(x)$	**433**

1. Find the *complementary function* y_c by solving the auxiliary equation.
2. Write the *particular integral* y_p. It should contain each term from the right side $f(x)$ (less coefficients) as well as the first and higher derivatives of each term of $f(x)$ (less coefficients). Discard any duplicates.
3. If a term in y_p is a duplicate of one in y_c, multiply that term in y_p by x^n, using the lowest n that *eliminates that duplication* and any new duplication with other terms in y_p.
4. *Write y_p*, each term with an undetermined coefficient.
5. *Substitute y_p* and its first and second derivatives into the differential equation.
6. Evaluate the *coefficients* by the method of undetermined coefficients.
7. Combine y_c and y_p to obtain the complete solution.

When we say "duplicate," we mean terms that are alike, regardless of numerical coefficient. These are discussed in the following section.

We illustrate these steps in the following example.

◆◆◆ **Example 51:** Solve $y'' - y' - 6y = 36x + 50 \sin x$.

Solution:

1. The auxiliary equation $m^2 - m - 6 = 0$ has the roots $m = 3$ and $m = -2$, so the complementary function is $y_c = c_1 e^{3x} + c_2 e^{-2x}$.
2. The terms in $f(x)$ and their derivatives (less coefficients) are

Term in $f(x)$	x	$\sin x$
First derivative	constant	$\cos x$
Second derivative	0	$\sin x$
Third derivative	0	$\cos x$
.	.	.
.	.	.
.	.	.

We see that the sine and cosine terms will keep repeating, so eliminating duplicates, we retain an x term, a constant term, a $\sin x$ term, and a $\cos x$ term.

3. Comparing the terms from y_c and these possible terms that will form y_p, we see that no term in y_p is a duplicate of one in y_c.
4. Our particular integral, y_p, is then

$$y_p = A + Bx + C \sin x + D \cos x$$

5. The derivatives of y_p are

$$y'_p = B + C \cos x - D \sin x$$

and

$$y''_p = -C \sin x - D \cos x$$

Substituting into the differential equation gives

$$y''_p - y'_p - 6y_p = 36x + 50 \sin x$$

$(-C \sin x - D \cos x) - (B + C \cos x - D \sin x)$
$$- 6(A + Bx + C \sin x + D \cos x) = 36x + 50 \sin x$$

6. Collecting terms and equating coefficients of like terms from the left and the right sides of this equation gives the equations

$$-6B = 36 \qquad -B - 6A = 0 \qquad D - 7C = 50 \qquad -7D - C = 0$$

from which $A = 1$, $B = -6$, $C = -7$, and $D = 1$. Our particular integral is then $y_p = 1 - 6x - 7 \sin x + \cos x$.

7. The complete solution is thus $y_c + y_p$, or

$$y = c_1 e^{3x} + c_2 e^{-2x} + 1 - 6x - 7 \sin x + \cos x \qquad \blacklozenge\blacklozenge\blacklozenge$$

Duplicate Terms in the Solution

The terms in the particular integral y_p must be *independent*. If y_p has duplicate terms, only one should be kept. However, if a term in y_p is a duplicate of one in the complementary function y_c (except for the coefficient), *multiply that term in y_p by x^n, using the lowest n that will eliminate that duplication* and any new duplication with other terms in y_p.

◆◆◆ **Example 52:** Solve the equation $y'' - 4y' + 4y = e^{2x}$.

Solution:

1. The complementary function (work not shown) is

$$y_c = c_1 e^{2x} + c_2 x e^{2x}$$

2. Our particular integral should contain e^{2x} and its derivatives, which are also of the form e^{2x}. But since these are duplicates, we need e^{2x} only once.

3. But e^{2x} is a duplicate of the first term, $c_1 e^{2x}$, in y_c. If we multiply by x, we see that xe^{2x} is now a duplicate of the *second* term in y_c. We thus need $x^2 e^{2x}$.

   We now see that in the process of eliminating duplicates with y_c, we might have created a new duplicate within y_p. If so, we again multiply by x^n, using the lowest n that would eliminate that new duplication as well. Here, $x^2 e^{2x}$ does not duplicate any term in y_p, so we proceed.

4. Our particular integral is thus $y_p = Ax^2 e^{2x}$.

5. Taking derivatives

$$y_p' = 2Ax^2 e^{2x} + 2Axe^{2x}$$

   and

$$y_p'' = 4Ax^2 e^{2x} + 8Axe^{2x} + 2Ae^{2x}$$

   Substituting into the differential equation gives

$$4Ax^2 e^{2x} + 8Axe^{2x} + 2Ae^{2x} - 4(2Ax^2 e^{2x} + 2Axe^{2x}) + 4(Ax^2 e^{2x}) = e^{2x}$$

6. Collecting terms and solving for A gives $A = \dfrac{1}{2}$.

7. Our complete solution is then

$$y = c_1 e^{2x} + c_2 x e^{2x} + \frac{1}{2} x^2 e^{2x} \qquad \blacklozenge\blacklozenge\blacklozenge$$

Exercise 12 ◆ Second-Order Differential Equations with Right Side Not Zero

Solve each second-order differential equation.

With Algebraic Expressions

1. $y'' - 4y = 12$

2. $y'' + y' - 2y = 3 - 6x$

3. $y'' - y' - 2y = 4x$

4. $y'' + y' = x + 2$

5. $y'' - 4y = x^3 + x$

With Exponential Expressions

6. $y'' + 2y' - 3y = 42e^{4x}$

7. $y'' - y' - 2y = 6e^x$

8. $y'' + y' = 6e^x + 3$

9. $y'' - 4y = 4x - 3e^x$

10. $y'' - y = e^x + 2e^{2x}$

11. $y'' + 4y' + 4y = 8e^{2x} + x$

12. $y'' - y = xe^x \quad y =$

With Trigonometric Expressions

13. $y'' + 4y = \sin 2x$

14. $y'' + y' = 6 \sin 2x$

15. $y'' + 2y' + y = \cos x$

16. $y'' + 4y' + 4y = \cos x$

17. $y'' + y = 2 \cos x - 3 \cos 2x$

18. $y'' + y = \sin x + 1$

With Exponential and Trigonometric Expressions

19. $y'' + y = e^x \sin x$

20. $y'' + y = 10e^x \cos x$

21. $y'' - 4y' + 5y = e^{2x} \sin x$

22. $y'' + 2y' + 5y = 3e^{-x} \sin x - 10$

Particular Solution

Find the particular solution to each differential equation, using the given boundary conditions.

23. $y'' - 4y' = 8$, $y = y' = 0$ when $x = 0$

24. $y'' + 2y' - 3y = 6$, $y = 0$ and $y' = 2$ when $x = 0$

25. $y'' + 4y = 2$, $y = 0$ when $x = 0$ and $y = \frac{1}{2}$ when $x = \pi/4$

26. $y'' + 4y' + 3y = 4e^{-x}$, $y = 0$ and $y' = 2$ when $x = 0$

27. $y'' - 2y' + y = 2e^x$, $y' = 2e$ at $(1, 0)$

28. $y'' - 9y = 18 \cos 3x + 9$, $y = -1$ and $y' = 3$ when $x = 0$

29. $y'' + y = -2 \sin x$, $y = 0$ at $x = 0$ and $\pi/2$

35–13   *RLC* Circuits

In Sec. 35–8 we studied the *RL* circuit and the *RC* circuit. Each gave rise to a first-order differential equation. We'll now see that the *RLC* circuit will result in a second-order differential equation.

A switch (Fig. 35–15) is closed at $t = 0$. The sum of the voltage drops must equal the applied voltage, so

$$Ri + L\frac{di}{dt} + \frac{q}{C} = E \tag{1}$$

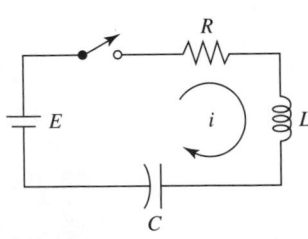

FIGURE 35–15   *RLC* circuit with dc source.

Replacing q by $\int i\, dt$ and differentiating gives

$$R\frac{di}{dt} + L\frac{d^2i}{dt^2} + \frac{i}{C} = 0$$

or

$$Li'' + Ri' + \left(\frac{1}{C}\right)i = 0 \tag{2}$$

This is a second-order linear differential equation, which we now solve as we did before. The auxiliary equation $Lm^2 + Rm + 1/C = 0$ has the roots

$$m = \frac{-R \pm \sqrt{R^2 - 4L/C}}{2L}$$

$$= -\frac{R}{2L} \pm \sqrt{\frac{R^2}{4L^2} - \frac{1}{LC}}$$

We now let

$$a^2 = \frac{R^2}{4L^2} \text{ and } \omega_n^2 = \frac{1}{LC}$$

so that we have the following:

Resonant Frequency of *RLC* Circuit with dc Source	$\omega_n = \sqrt{\dfrac{1}{LC}}$	A91

and our roots become

$$m = -a \pm \sqrt{a^2 - \omega_n^2}$$

or

$$m = -a \pm j\omega_d$$

where $\omega_d^2 = \omega_n^2 - a^2$. We have three possible cases, listed in Table 35–2.

Table 35–2 *RLC* circuit with dc source: second-order differential equation.

Roots of Auxiliary Equation	Type of Solution	
Nonreal	Underdamped	$a < \omega_n$
	No damping (series *LC* circuit)	$a = 0$
Real, equal	Critically damped	$a = \omega_n$
Real, unequal	Overdamped	$a > \omega_n$

We consider the underdamped and overdamped cases with a dc source, and the underdamped case with an ac source.

Underdamped with dc Source, $a < \omega_n$

We first consider the case where R is not zero but is low enough so that $a < \omega_n$. The roots of the auxiliary equation are then nonreal, and the current is

$$i = e^{-at} (k_1 \sin \omega_d t + k_2 \cos \omega_d t)$$

Since i is zero at $t = 0$, we get $k_2 = 0$. The current is then

$$i = k_1 e^{-at} \sin \omega_d t$$

Taking the derivative yields

$$\frac{di}{dt} = k_1 \omega_d e^{-at} \cos \omega_d t - ak_1 e^{-at} \sin \omega_d t$$

At $t = 0$, the capacitor behaves as a short circuit, and there is also no voltage drop across the resistor (since $i = 0$). The entire voltage E then appears across the inductor. Since $E = L \, di/dt$, then $di/dt = E/L$, so

$$k_1 = \frac{E}{\omega_d L}$$

The current is then given by the following:

Underdamped *RLC* Circuit	$i = \dfrac{E}{\omega_d L} e^{-at} \sin \omega_d t$	**A93**

where, from our previous substitution:

Underdamped *RLC* Circuit	$\omega_d = \sqrt{\omega_n^2 - a^2} = \sqrt{\omega_n^2 - \dfrac{R^2}{4L^2}}$	**A94**

We get a damped sine wave whose amplitude decreases exponentially with time.

◆◆◆ **Example 53:** A switch (Fig. 35–15) is closed at $t = 0$. If $R = 225 \, \Omega$, $L = 1.50 \, \text{H}$, $C = 4.75 \, \mu\text{F}$, and $E = 75.4 \, \text{V}$, write an expression for the instantaneous current.

Solution: We first compute LC.

$$LC = 1.50(4.75 \times 10^{-6}) = 7.13 \times 10^{-6}$$

Then, by Eq. A91,

$$\omega_n = \sqrt{\frac{1}{LC}} = \sqrt{\frac{10^6}{7.13}} = 375 \, \text{rad/s}$$

and

$$a = \frac{R}{2L} = \frac{225}{2(1.50)} = 75.0 \, \text{rad/s}$$

Then, by Eq. A94,

$$\omega_d = \sqrt{\omega_n^2 - a^2} = \sqrt{(375)^2 - (75.0)^2} = 367 \, \text{rad/s}$$

The instantaneous current is then

$$i = \frac{E}{\omega_d L} e^{-at} \sin \omega_d t = \frac{75.4}{367(1.50)} e^{-75.0t} \sin 367t \, \text{A}$$

$$= 137 e^{-75.0t} \sin 367t \, \text{mA}$$

This curve is plotted in Fig. 35–16 showing the damped sine wave.

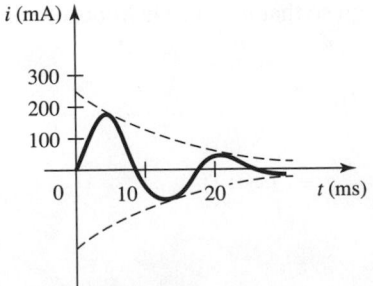

FIGURE 35–16   ◆◆◆

No Damping: The Series *LC* Circuit

When the resistance R is zero, $a = 0$ and $\omega_d = \omega_n$. From Eq. A93 we get the following:

Series *LC* Circuit	$i = \dfrac{E}{\omega_n L} \sin \omega_n t$	**A92**

> This, of course, is a theoretical case, because a real circuit always has some resistance.

which represents a sine wave with amplitude $E/\omega_n L$.

◆◆◆ **Example 54:** Repeat Example 53 with $R = 0$.

Solution: The value of ω_n, from before, is 375 rad/s. The amplitude of the current wave is

$$\frac{E}{\omega_n L} = \frac{75.4}{375(1.50)} = 0.134 \text{ A}$$

so the instantaneous current is

$$i = 134 \sin 375t \text{ mA}$$

◆◆◆

Overdamped with dc Source, $a > \omega_n$

If the resistance is relatively large, so that $a > \omega_n$, the auxiliary equation has the real and unequal roots

$$m_1 = -a + j\omega_d \text{ and } m_2 = -a - j\omega_d$$

The current is then

$$i = k_1 e^{m_1 t} + k_2 e^{m_2 t} \tag{1}$$

Since $i(0) = 0$, we have

$$k_1 + k_2 = 0 \tag{2}$$

Taking the derivative of Eq. (1), we obtain

$$\frac{di}{dt} = m_1 k_1 e^{m_1 t} + m_2 k_2 e^{m_2 t}$$

Since $di/dt = E/L$ at $t = 0$,

$$\frac{E}{L} = m_1 k_1 + m_2 k_2 \tag{3}$$

Solving (2) and (3) simultaneously gives

$$k_1 = \frac{E}{(m_1 - m_2)L} \quad \text{and} \quad k_2 = -\frac{E}{(m_1 - m_2)L}$$

where $m_1 - m_2 = -a + j\omega_d + a + j\omega_d = 2j\omega_d$. The current is then given by the following equation:

Overdamped *RLC* Circuit	$i = \dfrac{E}{2j\omega_d L} [e^{(-a+j\omega_d)t} - e^{(-a-j\omega_d)t}]$	**A95**

◆◆◆ **Example 55:** For the circuit of Examples 53 and 54, let $R = 2550 \ \Omega$, and compute the instantaneous current.

Solution: $a = \dfrac{R}{2L} = \dfrac{2550}{2(1.50)} = 850$ rad/s. Since $\omega_n = 375$ rad/s, we have

$$\omega_d = \sqrt{(375)^2 - (850)^2} = j\,763 \text{ rad/s}$$

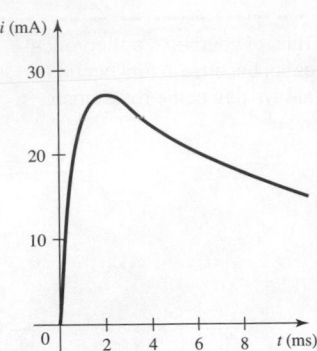

FIGURE 35–17   Current in an overdamped circuit.

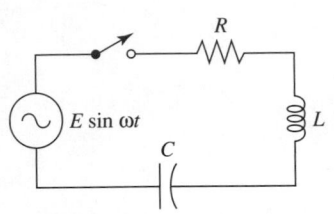

FIGURE 35–18   *RLC* circuit with ac source.

In Chapter 36 we'll do this type of problem by the Laplace transform.

Then $-a - j\omega_d = -87.0$ and $-a + j\omega_d = -1613$. From Eq. A93,

$$i = \frac{75.4}{2(-763)(1.50)} (e^{-1613t} - e^{-87.0t})$$

$$= 32.9(e^{-87.0t} - e^{-1613t}) \text{ mA}$$

This equation is graphed in Fig. 35–17.   ◆◆◆

Underdamped with ac Source

Up to now we have considered only a dc source. We now repeat the underdamped (low-resistance) case with an alternating voltage $E \sin \omega t$.

A switch (Fig. 35–18) is closed at $t = 0$. The sum of the voltage drops must equal the applied voltage, so

$$Ri + L\frac{di}{dt} + \frac{q}{C} = E \sin \omega t \tag{1}$$

Replacing q by $\int i \, dt$ and differentiating gives

$$R\frac{di}{dt} + L\frac{d^2i}{dt^2} + \frac{i}{C} = \omega E \cos \omega t$$

or

$$Li'' + Ri' + \left(\frac{1}{C}\right)i = \omega E \cos \omega t \tag{2}$$

The complementary function is the same as was calculated for the dc case.

$$i_c = e^{-at}(k_1 \sin \omega_d t + k_2 \cos \omega_d t)$$

The particular integral i_p will have a sine term and a cosine term.

$$i_p = A \sin \omega t + B \cos \omega t$$

Taking first and second derivatives yields

$$i_p' = \omega A \cos \omega t - \omega B \sin \omega t$$
$$i_p'' = -\omega^2 A \sin \omega t - \omega^2 B \cos \omega t$$

Substituting into Eq. (2), we get

$$-L\omega^2 A \sin \omega t - L\omega^2 B \cos \omega t - R\omega B \sin \omega t + R\omega A \cos \omega t + \frac{A}{C} \sin \omega t$$
$$+ \frac{B}{C} \cos \omega t = \omega E \cos \omega t$$

Equating the coefficients of the sine terms gives

$$-L\omega^2 A - R\omega B + \frac{A}{C} = 0$$

from which

$$-RB = A\left(\omega L - \frac{1}{\omega C}\right) = AX \tag{3}$$

where X is the reactance of the circuit. Equating the coefficients of the cosine terms gives

$$-L\omega^2 B + R\omega A + \frac{B}{C} = \omega E$$

or

$$RA - E = B\left(\omega L - \frac{1}{\omega C}\right) = BX \qquad (4)$$

Solving (3) and (4) simultaneously gives

$$A = \frac{RE}{R^2 + X^2} = \frac{RE}{Z^2}$$

where Z is the impedance of the circuit, and

$$B = -\frac{EX}{R^2 + X^2} = -\frac{EX}{Z^2}$$

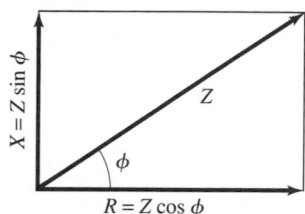

Our particular integral thus becomes

$$i_p = \frac{RE}{Z^2} \sin \omega t - \frac{EX}{Z^2} \cos \omega t$$

$$= \frac{E}{Z^2}(R \sin \omega t - X \cos \omega t)$$

FIGURE 35–19 Impedance triangle.

From the impedance triangle (Fig. 35–19),

$$R = Z \cos \phi \quad \text{and} \quad X = Z \sin \phi$$

where ϕ is the phase angle. Thus,

$$R \sin \omega t - X \cos \omega t = Z \sin \omega t \cos \phi - Z \cos \omega t \sin \phi$$

$$= Z \sin(\omega t - \phi)$$

by the trigonometric identity (Eq. 167). Thus, i_p becomes $E/Z \sin(\omega t - \phi)$, or

$$i_p = I_{max} \sin(\omega t - \phi)$$

since E/Z gives the maximum current I_{max}. The total current is the sum of the complementary function i_c and the particular integral i_p.

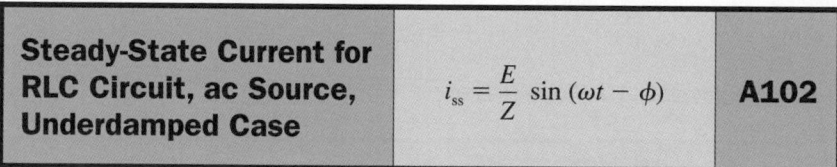

$$i = \underbrace{e^{-at}(k_1 \sin \omega_d t + k_2 \cos \omega_d t)}_{\substack{\text{transient} \\ \text{current}}} + \underbrace{I_{max} \sin(\omega t - \phi)}_{\substack{\text{steady-state} \\ \text{current}}}$$

Thus, the current is made up of two parts: a *transient* part that dies quickly with time and a *steady-state* part that continues as long as the ac source is connected. We are usually interested only in the steady-state current.

Steady-State Current for RLC Circuit, ac Source, Underdamped Case	$i_{ss} = \dfrac{E}{Z} \sin(\omega t - \phi)$	**A102**

♦♦♦ **Example 56:** Find the steady-state current for an *RLC* circuit if $R = 345\ \Omega$, $L = 0.726$ H, $C = 41.4\ \mu$F, and $e = 155 \sin 285t$.

Solution: By Eqs. A96 and A97,

$$X_L = \omega L = 285(0.726) = 207\ \Omega$$

and

$$X_C = \frac{1}{\omega C} = \frac{1}{285(41.4 \times 10^{-6})} = 84.8 \ \Omega$$

The total reactance is, by Eq. A98,

$$X = X_L - X_C = 207 - 84.8 = 122 \ \Omega$$

The impedance Z is found from Eq. A99.

$$Z = \sqrt{R^2 + X^2} = \sqrt{(345)^2 + (122)^2} = 366 \ \Omega$$

The phase angle, from Eq. A100, is

$$\phi = \tan^{-1}\frac{X}{R} = \tan^{-1}\frac{122}{345} = 19.5°$$

The steady-state current is then

$$i_{ss} = \frac{E}{Z} \sin(\omega t - \phi)$$

$$= \frac{155}{366} \sin(285t - 19.5°)$$

$$= 0.423 \sin(285t - 19.5°) \ \text{A}$$

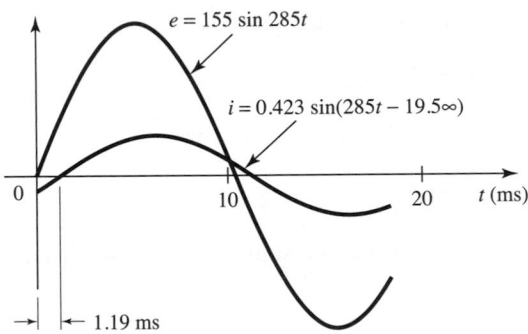

FIGURE 35–20

Figure 35–20 shows the applied voltage and the steady-state current, with a phase angle of 19.5° and the phase difference of 1.19 ms.   ◆◆◆

Resonance

The current in a series RLC circuit will be a maximum when the impedance Z is zero. This will occur when the reactance X is zero, so

$$X = \omega L - \frac{1}{\omega C} = 0$$

Solving for ω, we get $\omega^2 = 1/LC$ or ω_n^2. Thus,

Resonant Frequency	$\omega = \dfrac{1}{\sqrt{LC}} = \omega_n$	A91

◆◆◆ **Example 57:** Find the resonant frequency for the circuit of Example 56, and write an expression for the steady-state current at that frequency.

Solution: The resonant frequency is

$$\omega_n = \frac{1}{\sqrt{LC}} = \frac{1}{\sqrt{0.726\,(41.4 \times 10^{-6})}} = 182 \text{ rad/s}$$

Since $X_L = X_C$, then $X = 0$, $Z = R$, and $\phi = 0$. Thus, I_{max} is $155/345 = 0.449$ A, and the steady-state current is then

$$i_{ss} = 449 \sin 182t \text{ mA} \qquad\qquad \blacklozenge\blacklozenge\blacklozenge$$

Exercise 13 ◆ *RLC* Circuits

Series LC Circuit with dc Source

1. In an *LC* circuit, $C = 1.00\ \mu\text{F}$, $L = 1.00$ H, and $E = 10\overline{0}$ V. At $t = 0$, the charge and the current are both zero. Using Eq. A92, show that $i = 0.1.00 \sin 10\overline{0}0t$ A.
2. For problem 1, take the integral of i to show that

$$q = 10^{-4}(1 - \cos 10\overline{0}0t) \text{ C}$$

3. For the circuit of problem 1, with $E = 0$ and an initial charge of 255×10^{-6} C (i_0 is still 0), the differential equation, in terms of charge, is $Lq'' + q/C = 0$. Solve this DE for q, and show that $q = 255 \cos 10\overline{0}0t\ \mu\text{C}$.
4. By differentiating the expression for charge in problem 3, show that the current is $i = -255 \sin 10\overline{0}0t$ mA.

Series RLC Circuit with dc Source

5. In an *RLC* circuit, $R = 1.55\ \Omega$, $C = 250\ \mu\text{F}$, $L = 0.125$ H, and $E = 100$ V. The current and charge are zero when $t = 0$.
   (a) Show that the circuit is underdamped.
   (b) Using Eq. A93, show that $i = 4.47\ e^{-6.2t} \sin 179t$ A.
6. Integrate i in problem 5 to show that

$$q = -e^{-6.2t}(0.864 \sin 179t + 24.9 \cos 179t) + 24.9 \text{ mC}$$

7. In an *RLC* circuit, $R = 1.75\ \Omega$, $C = 4.25$ F, $L = 1.50$ H, and $E = 100$ V. The current and charge are 0 when $t = 0$.
   (a) Show that the circuit is overdamped.
   (b) Using Eq. A95, show that

$$i = -77.9e^{-1.01t} + 77.9e^{-0.155t} \text{ A}$$

8. Integrate the expression for i for the circuit in problem 7 to show that

$$q = 77.1e^{-1.01t} - 503e^{-0.155t} + 426 \text{ C}$$

Series RLC Circuit with ac Source

9. For an *RLC* circuit, $R = 10.5\ \Omega$, $L = 0.125$ H, $C = 225\ \mu\text{F}$, and $e = 100 \sin 175t$. Using Eq. A102, show that the steady-state current is $i = 9.03 \sin(175t + 18.5°)$.
10. Find the resonant frequency for problem 9.
11. In an *RLC* circuit, $R = 1550\ \Omega$, $L = 0.350$ H, $C = 20.0\ \mu\text{F}$, and $e = 250 \sin 377t$. Using Eq. A102, show that the steady-state current is $i = 161 \sin 377t$ mA.
12. If $R = 10.5\ \Omega$, $L = 0.175$ H, $C = 1.50 \times 10^{-3}$ F, and $e = 175 \sin 55t$, find the amplitude of the steady-state current.
13. Find the resonant frequency for the circuit in problem 12.
14. Find the amplitude of the steady-state current at resonance for the circuit in problem 12.
15. For the circuit of Fig. 35–18, $R = 110\ \Omega$, $L = 5.60 \times 10^{-6}$ H, and $\omega = 6.00 \times 10^5$ rad/s. What value of C will produce resonance?
16. If an *RLC* circuit has the values $R = 10\ \Omega$, $L = 0.200$ H, $C = 500\ \mu\text{F}$, and $e = 100 \sin \omega t$, find the resonant frequency.
17. For the circuit in problem 16, find the amplitude of the steady-state current at resonance.

••• CHAPTER 35 REVIEW PROBLEMS •••••••••••••••••••••••••••••••••••

Find the general solution to each first-order differential equation.

1. $xy + y + xy' = e^x$

2. $y + xy' = 4e^3$

3. $y' + y - 2 \cos x = 0$

4. $y + x^2y^2 + xy' = 0$

5. $2y + 3x^2 + 2xy' = 0$

6. $y' - 3x^2y^2 + \dfrac{y}{x} = 0$

7. $y^2 + (x^2 - xy)y' = 0$

8. $y' = e^{-y} - 1$

9. $(1 - x)\, dy/dx = y^2$

10. $y' \tan x + \tan y = 0$

11. $y + 2xy^2 + (y - x)y' = 0$

Using the given boundary condition, find the particular solution to each differential equation.

12. $x\, dx = 2y\, dy$, $\quad x = 3$ when $y = 1$

13. $y' \sin y = \cos x$, $\quad x = \dfrac{\pi}{4}$ when $y = 0$

14. $xy' + y = 4x$, $x = 2$ when $y = 1$

15. $y\, dx = (x - 2x^2y)\, dy$, $\quad x = 2$ when $y = 1$

16. $3xy^2\, dy = (3y^3 - x^3)\, dx$, $\quad x = 3$ when $y = 1$

17. A gear is rotating at 1550 rev/min. Its speed decreases exponentially at a rate of 9.50% per second after the power is shut off. Find the gear's speed after 6.00 s.

18. For the circuit of Fig. 35–11, $R = 1350\ \Omega$, $L = 7.25$ H, and $E = 225$ V. If the switch is thrown from position 2 to position 1, find the current and the voltage across the inductance at $t = 3.00$ ms.

19. Write the equation of the orthogonal trajectories to the family of parabolas $x^2 = 4y$.

20. For the circuit of Fig. 35–12, $R = 2550\ \Omega$, $C = 145\ \mu$F, and $E = 95.0$ V. If the switch is thrown from position 2 to position 1, find the current and the voltage across the capacitor at $t = 5.00$ ms

21. Find the equation of the curve that passes through the point $(1, 2)$ and whose slope is $y' = 2 + y/x$.

22. An object is dropped and falls from rest through air whose resisting force is proportional to the speed of the package. The terminal speed is 275 ft./s. Show that the acceleration is given by the differential equation $a = dv/dt = g - gv/275$.

23. A certain yeast is found to grow exponentially at the rate of 15.0% per hour. Starting with $50\overline{0}$ g of yeast, how many grams will there be after 15.0 h?

Solve each differential equation.

24. $y'' + 2y' - 3y = 7$, $\quad y = 1$ and $y' = 2$ when $x = 0$

25. $y'' - y' - 2y = 5x$

26. $y'' - 2y = 2x^3 + 3x$

27. $y'' - 4y' - 5y = e^{4x}$,

28. $y''' - 2y' = 0$

29. $y'' - 5y' = 6$, $y = y' = 1$ when $x = 0$

30. $y' + 2y' - 3y = 0$

31. $y'' - 7y' - 18y = 0$

32. $y'' + 6y' + 9y = 0$, $y = 0$ and $y' = 2$ when $x = 0$

33. $y''' - 6y' + 11y' - 6y = 0$

34. $y'' - 2y = 3x^3e^x$

35. $y'' + 2y = 3 \sin 2x$

36. $y'' + y' = 5 \sin 3x$

37. $y'' + 3y' + 2y = 0$, $y = 1$ and $y' = 2$ when $x = 0$

38. $y'' + 9y = 0$

39. $y'' + 3y' - 4y = 0$

40. $y'' - 2y' + y = 0$, $y = 0$ and $y' = 2$ when $x = 0$

41. $y'' + 5y' - y = 0$

42. $y'' - 4y' + 4y = 0$

43. $y'' + 3y' - 4y = 0$, $y = 0$ and $y' = 2$ when $x = 0$

44. $y'' + 4y' + 4y = 0$

45. $y'' + 6y' + 9y = 0$

46. $y'' - 2y' + y = 0$

47. $9y'' - 6y' + y = 0$

48. $y'' - x^2 = 0$, where $y' = 4$ at the point $(1, 2)$

49. For an *RLC* circuit, $R = 3.75 \ \Omega$, $C = 15\overline{0} \ \mu F$, $L = 0.100$ H, and $E = 12\overline{0}$ V. The current and charge are zero when $t = 0$. Find i as a function of time.

50. In problem 49, find an expression for the charge.

51. For an *LC* circuit, $C = 1.75 \ \mu F$, $L = 2.20$ H, and $E = 11\overline{0}$ V. At $t = 0$, the charge and the current are both zero. Find i as a function of time.

52. In problem 51, find the charge as a function of time.

Writing

53. We have given seven steps for the solution of a second-order linear differential equation with constant coefficients. List as many of these steps as you can, and write a one-sentence explanation of each.

Team Project

54. The deflection curve for a beam is given by

$$\frac{d^2 y}{dx^2} = \frac{M}{EI}$$

where E is the modulus of elasticity of the material, I is the moment of inertia of the beam cross section, y is the deflection of the beam, and M is the bending moment at a distance x from one end.

(a) For the cantilever beam of Fig. 35–21, show that the deflection curve is

$$\frac{d^2 y}{dx^2} = \frac{Px - PL}{EI}$$

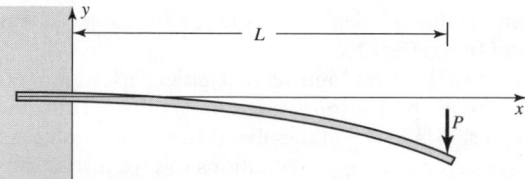

FIGURE 35–21 Cantilever beam.

(b) Solve this DE for the slope dy/dx and for the deflection y. Evaluate any constants of integration.

36

Solving Differential Equations by the Laplace Transform and by Numerical Methods

The methods for solving differential equations that we learned in Chapter 35 are often called *classical methods*. In this chapter we learn other powerful techniques: the *Laplace transform* and *numerical methods*. The Laplace transform enables us to transform a differential equation into an algebraic one. We can then solve the algebraic equation and apply an inverse transform to obtain the solution to the differential equation.

The Laplace transform is good for finding particular solutions of differential equations. For example, it enables us to solve *initial value* problems; that is, when the value of the function is known at $t = 0$. Thus the equations that we deal with here are functions of time, such as electric circuit problems, rather than functions of x.

We first use the Laplace transform to solve first- and second-order differential equations with constant coefficients. We then do some applications and compare our results with those obtained by classical methods in Chapter 35. We go on to solve differential equations using *numerical methods*. These methods use successive approximations to give a solution in the form of *numbers*, rather than equations. One great advantage of numerical methods is that they can be programmed on a computer.

36–1   The Laplace Transform of a Function

Laplace Transform of Some Simple Functions

If y is some function of time, so that $y = f(t)$, the Laplace transform of that function is defined by the following improper integral:

Laplace Transform	$\mathcal{L}[f(t)] = \int_0^\infty f(t)e^{-st}\,dt$	436

The Laplace transform is named after the French analyst, probabilist, astronomer, and physicist Pierre Laplace (1749–1827).

The transformed expression is *a function of s*, which we call $F(s)$. Thus,

$$\mathcal{L}[f(t)] = F(s)$$

We can write the transform of an expression by direct integration; that is, by writing an integral using Eq. 436 and then evaluating it.

Glance back at Sec. 34–8 if you have forgotten how to evaluate an improper integral.

◆◆◆ **Example 1:** If $y = f(t) = 1$, then

$$\mathcal{L}[f(t)] = \mathcal{L}[1] = \int_0^\infty (1)e^{-st}\,dt$$

To evaluate the integral, we multiply by $-s$ and compensate with $-1/s$.

$$\mathcal{L}[1] = -\frac{1}{s}\int_0^\infty e^{-st}(-s\,dt) = -\frac{1}{s}e^{-st}\Big|_0^\infty$$

Figure 36–1 shows a graph of $y = e^{-u}$, which will help us to evaluate the limits. Note that as u approaches infinity, e^{-u} approaches zero, and as u approaches zero, e^{-u} approaches one. So we evaluate e^{-st} equal to zero at the upper limit $t = \infty$ and $e^{-st} = 1$ at the lower limit $t = 0$. Thus

$$\mathcal{L}[1] = -\frac{1}{s}(0 - 1) = \frac{1}{s}$$

◆◆◆

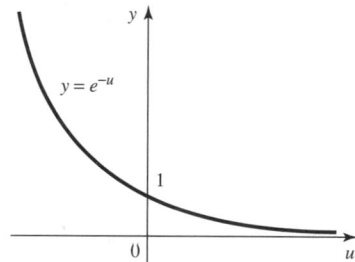

FIGURE 36–1

Rule 2 from our table of integrals in Appendix C says that $\int af(x)\,dx = a\int f(x)\,dx$. It follows then that

$$\mathcal{L}[af(t)] = a\mathcal{L}[f(t)]$$

The Laplace transform of a constant times a function is equal to the constant times the transform of the function.

◆◆◆ **Example 2:** If $\mathcal{L}[1] = 1/s$, as above, then

$$\mathcal{L}[5] = 5\mathcal{L}[1] = \frac{5}{s}$$

◆◆◆

◆◆◆ **Example 3:** If $y = f(t) = t$, then

$$\mathcal{L}[f(t)] = \mathcal{L}[t] = \int_0^\infty te^{-st}\,dt$$

We evaluate the right side using Rule 37 with $u = t$ and $a = -s$.

$$\mathcal{L}[t] = \frac{e^{-st}}{s^2}(-st - 1)\Big|_0^\infty$$

At the upper limit, e^{-st} approaches zero, while st approaches infinity. But e^{-st} is shrinking *faster* than st is growing, so their product approaches zero. This is confirmed by Fig. 36–2, which shows the product $t(e^{-st})$ first increasing but then quickly approaching zero. So we get

$$\mathcal{L}[t] = \frac{e^{-st}}{s^2}(-st - 1)\Big|_0^\infty = 0 - \left(-\frac{1}{s^2}\right) = \frac{1}{s^2}$$

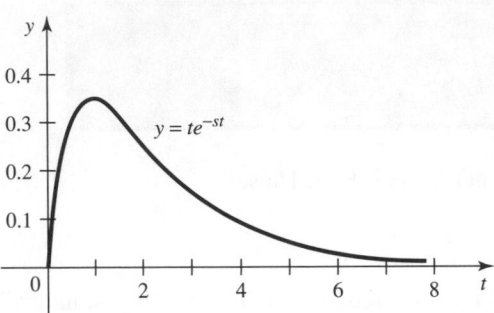

FIGURE 36–2   ◆◆◆

Note that in each case the transformed expression is a function of s only.

◆◆◆ **Example 4:** Find $\mathcal{L}[\sin at]$.

Solution: Using Eq. 436, we find that

$$\mathcal{L}[\sin at] = \int_0^\infty e^{-st} \sin at \, dt$$

Using Rule 41 gives

$$\mathcal{L}[\sin at] = \frac{e^{-st}}{s^2 + a^2}(-s \sin at - a \cos at)\Big|_0^\infty$$

At the upper limit of infinity, e^{-st} approaches zero, causing the entire expression to have a value of zero at that limit. At the lower limit of zero, e^{-st} approaches one, $\sin at$ approaches zero, and $\cos at$ approaches one. Our expression then becomes

$$\mathcal{L}[\sin at] = 0 - \frac{1}{s^2 + a^2}(-a) = \frac{a}{s^2 + a^2} \qquad \text{◆◆◆}$$

Transform of a Sum

If we have the sum of several terms,

$$f(t) = a\, g(t) + b\, h(t) + \cdots$$

then

$$\mathcal{L}[f(t)] = \int_0^\infty [a\, g(t) + b\, h(t) + \cdots] e^{-st} \, dt$$

$$= a \int_0^\infty g(t) e^{-st} \, dt + b \int_0^\infty h(t) e^{-st} \, dt + \cdots$$

Thus,

$$\mathcal{L}[a\, g(t) + b\, h(t) + \cdots] = a\mathcal{L}[g(t)] + b\mathcal{L}[h(t)] + \cdots$$

The Laplace transform of a sum of several terms is the sum of the transforms of each term.

This property allows us to work term by term in writing the transform of an expression made up of sums or differences.

◆◆◆ **Example 5:** If $y = 4 + 5t - 2 \sin 3t$, then

$$\mathcal{L}[y] = \frac{4}{s} + \frac{5}{s^2} - 2\frac{3}{s^2 + 9}$$ ◆◆◆

Transform of a Derivative

To solve differential equations, we must be able to take the transform of a derivative. Let $f'(t)$ be the derivative of some function $f(t)$. Then

$$\mathcal{L}[f'(t)] = \int_0^\infty f'(t)e^{-st}\, dt$$

> When we write the Laplace transform of a function, we often say that we are "taking the transform" of the function.

Integrating by parts, we let $u = e^{-st}$ and $dv = f'(t)\, dt$. So $du = -se^{-st}\, dt$ and $v = \int f'(t)\, dt = f(t)$. Then

$$\mathcal{L}[f'(t)] = f(t)e^{-st}\Big|_0^\infty + s\int_0^\infty f(t)e^{-st}\, dt$$

$$= f(t)e^{-st}\Big|_0^\infty + s\mathcal{L}[f(t)]$$

$$= 0 - f(0) + s\mathcal{L}[f(t)]$$

Thus:

Transform of a Derivative	$\mathcal{L}[f'(t)] = s\mathcal{L}[f(t)] - f(0)$

The transform of a derivative of a function equals s times the transform of the function minus the function evaluated at $t = 0$.

Thus, if we have a function $y = f(t)$, then

$$\mathcal{L}[y'] = s\mathcal{L}[y] - y(0)$$

Note that the transform of a derivative contains $f(0)$ or $y(0)$, the value of the function at $t = 0$. Thus, *to evaluate the Laplace transform of a derivative, we must know the initial conditions.*

◆◆◆ **Example 6:** Take the Laplace transform of both sides of the differential equation $y' = 6t$, with the initial condition that $y(0) = 5$.

Solution:

$$y' = 6t$$

> If two functions are equal, their transforms are equal. Thus, we can take the transform of both sides of an equation without changing the meaning of the equation.

$$\mathcal{L}[y'] = \mathcal{L}[6t]$$

$$s\mathcal{L}[y] - y(0) = \frac{6}{s^2}$$

or, substituting,

$$s\mathcal{L}[y] - 5 = \frac{6}{s^2}$$

since $y(0) = 5$. Later, we will solve such an equation for $\mathcal{L}[y]$ and then find the inverse of the transform to obtain y. ◆◆◆

Similarly, we can find the transform of the second derivative since the second derivative of $f(t)$ is the derivative of $f'(t)$.

$$\mathcal{L}[f''(t)] = s\mathcal{L}[f'(t)] - f'(0)$$

Substituting, we get

$$\mathcal{L}[f''(t)] = s\{s\mathcal{L}[f(t)] - f(0)\} - f'(0)$$

Transform of the Second Derivative	$\mathcal{L}[f''(t)] = s^2\mathcal{L}[f(t)] - sf(0) - f'(0)$

Another way of expressing this equation, if $y = f(t)$, is

$$\mathcal{L}[y''] = s^2\mathcal{L}[y] - sy(0) - y'(0)$$

◆◆◆ **Example 7:** Transform both sides of the second-order differential equation $y'' - 3y' + 4y = 5t$ if $y(0) = 6$ and $y'(0) = 7$.

Solution:

$$y'' - 3y' + 4y = 5t$$

$$\mathcal{L}[y''] - 3\mathcal{L}[y'] + 4\mathcal{L}[y] = \mathcal{L}[5t]$$

$$s^2\mathcal{L}[y] - sy(0) - y'(0) - 3\{s\mathcal{L}[y] - y(0)\} + 4\mathcal{L}[y] = \frac{5}{s^2}$$

Substituting $y(0) = 6$ and $y'(0) = 7$, we obtain

$$s^2\mathcal{L}[y] - 6s - 7 - 3s\mathcal{L}[y] + 3(6) + 4\mathcal{L}[y] = \frac{5}{s^2}$$

or

$$s^2\mathcal{L}[y] - 3s\mathcal{L}[y] + 4\mathcal{L}[y] - 6s + 11 = \frac{5}{s^2} \qquad\qquad ◆◆◆$$

Transform of an Integral

Let us now find the transform of the integral $\int_0^t f(t)\,dt$. By our definition of the Laplace transform, Eq. 436,

$$\mathcal{L}\left[\int_0^t f(t)\,dt\right] = \int_0^\infty \left[\int_0^t f(t)\,dt\right] e^{-st}\,dt$$

Integrating by parts, we let

$$u = \int_0^t f(t)\,dt \quad \text{and} \quad dv = e^{-st}\,dt$$

so that

$$du = f(t)\,dt \quad \text{and} \quad v = -e^{-st}/s$$

Integrating yields

$$\mathcal{L}\left[\int_0^t f(t)\,dt\right] = \left\{(-1/s)e^{-st}\int_0^t f(t)\,dt\right\}\Big|_0^\infty + (1/s)\int_0^\infty f(t)e^{-st}\,dt$$

At the upper limit of infinity, the quantity in the curly brackets vanishes because e^{-st} goes to zero. At the lower limit of zero, the quantity in the curly brackets vanishes because the upper and lower limits of the integral are equal. Further, the integral in the second term is the definition (Eq. 436) of the Laplace transform of $f(t)$. Thus:

Transform of an Integral	$\mathcal{L}\left[\displaystyle\int_0^t f(t)\,dt\right] = (1/s)\mathcal{L}[f(t)]$

◆◆◆ **Example 8:** The voltage across a capacitor of capacitance C is given by

$$V_c = (1/C)\int_0^t i\,dt$$

Find the transform of this voltage.

We will need this equation later for solving electrical problems.

Solution: Using the transform of an integral, we get

$$\mathcal{L}\left[(1/C)\int_0^t i\,dt\right] = (1/Cs)\mathcal{L}[i]$$

◆◆◆

Table of Laplace Transforms

The transforms of some common functions are given in Table 36–1. Instead of transforming a function step by step, we simply look it up in the table.

TABLE 36–1 Short table of Laplace transforms. Here n is a positive integer.

Transform Number	$f(t)$	$\mathcal{L}[f(t)] = F(s)$
1	$f(t)$	$F(s) = \displaystyle\int_0^\infty e^{-st} f(t)\,dt$
2	$f'(t)$	$s\mathcal{L}[f(t)] - f(0)$
3	$f''(t)$	$s^2\mathcal{L}[f(t)] - sf(0) - f'(0)$
4	$a\,g(t) + b\,h(t) + \cdots$	$a\mathcal{L}[g(t)] + b\mathcal{L}[h(t)] + \cdots$
5	1	$\dfrac{1}{s}$
6	t	$\dfrac{1}{s^2}$
7	t^n	$\dfrac{n!}{s^{n+1}}$
8	$\dfrac{t^{n-1}}{(n-1)!}$	$\dfrac{1}{s^n}$
9	e^{at}	$\dfrac{1}{s-a}$
10	$1 - e^{-at}$	$\dfrac{a}{s(s+a)}$
11	te^{at}	$\dfrac{1}{(s-a)^2}$
12	$e^{at}(1+at)$	$\dfrac{s}{(s-a)^2}$
13	$t^n e^{at}$	$\dfrac{n!}{(s-a)^{n+1}}$
14	$t^{n-1}e^{-at}$	$\dfrac{(n-1)!}{(s+a)^n}$
15	$e^{-at} - e^{-bt}$	$\dfrac{b-a}{(s+a)(s+b)}$

TABLE 36–1 *Continued*

Transform Number	$f(t)$	$\mathcal{L}[f(t)] = F(s)$
16	$ae^{-at} - be^{-bt}$	$\dfrac{s(a-b)}{(s+a)(s+b)}$
17	$\sin at$	$\dfrac{a}{s^2 + a^2}$
18	$\cos at$	$\dfrac{s}{s^2 + a^2}$
19	$t \sin at$	$\dfrac{2as}{(s^2 + a^2)^2}$
20	$t \cos at$	$\dfrac{s^2 - a^2}{(s^2 + a^2)^2}$
21	$1 - \cos at$	$\dfrac{a^2}{s(s^2 + a^2)}$
22	$at - \sin at$	$\dfrac{a^3}{s^2(s^2 + a^2)}$
23	$e^{-at} \sin bt$	$\dfrac{b}{(s+a)^2 + b^2}$
24	$e^{-at} \cos bt$	$\dfrac{s+a}{(s+a)^2 + b^2}$
25	$\sin at - at \cos at$	$\dfrac{2a^3}{(s^2 + a^2)^2}$
26	$\sin at + at \cos at$	$\dfrac{2as^2}{(s^2 + a^2)^2}$
27	$\cos at - \frac{1}{2} at \sin at$	$\dfrac{s^3}{(s^2 + a^2)^2}$
28	$\dfrac{b}{a^2}(e^{-at} + at - 1)$	$\dfrac{b}{s^2(s+a)}$
29	$\displaystyle\int_0^t f(t)\, dt$	$\dfrac{\mathcal{L}[f(t)]}{s}$

◆◆◆ **Example 9:** Find the Laplace transform of $t^3 e^{2t}$ by using Table 36–1.

Solution: Our function matches Transform 13, with $n = 3$ and $a = 2$, so

$$\mathcal{L}[t^3 e^{2t}] = \frac{3!}{(s-2)^{3+1}} = \frac{6}{(s-2)^4}$$

   ◆◆◆

Exercise 1 ◆ The Laplace Transform of a Function

Transforms by Direct Integration

Find the Laplace transform of each function by direct integration.

1. $f(t) = 6$ **2.** $f(t) = t$

3. $f(t) = t^2$ **4.** $f(t) = 2t^2$

5. $f(t) = \cos 5t$ **6.** $f(t) = e^t \sin t$

Transforms by Table

Use Table 36–1 to find the Laplace transform of each function.

7. $f(t) = t^2 + 4$

8. $f(t) = t^3 - 2t^2 + 3t - 4$

9. $f(t) = 3e^t + 2e^{-t}$

10. $f(t) = 5te^{3t}$

11. $f(t) = \sin 2t + \cos 3t$

12. $f(t) = 3 + e^{4t}$

13. $f(t) = 5e^{3t} \cos 5t$

14. $f(t) = 2e^{-4t} - t^2$

15. $f(t) = 5e^t - t \sin 3t$

16. $f(t) = t^3 + 4t^2 - 3e^t$

Transforms of Derivatives

Find the Laplace transform of each expression, and substitute the given initial conditions.

17. $y' + 2y, \quad y(0) = 1$

18. $3y' + 2y, \quad y(0) = 3$

19. $y' - 4y, \quad y(0) = 0$

20. $5y' - 3y, \quad y(0) = 2$

21. $y'' + 3y' - y, \quad y(0) = 1$ and $y'(0) = 3$

22. $y'' - y' + 2y, \quad y(0) = 1$ and $y'(0) = 0$

23. $2y'' + 3y' + y, \quad y(0) = 2$ and $y'(0) = 3$

24. $3y'' - y' + 2y, \quad y(0) = 2$ and $y'(0) = 1$

36–2 Inverse Transforms

Before we can use the Laplace transform to solve differential equations, we must be able to transform a function of s back to a function of t. The *inverse Laplace transform* is denoted by $\mathcal{L}^{-1}$. Thus if $\mathcal{L}[f(t)] = F(s)$, then we have the following equation:

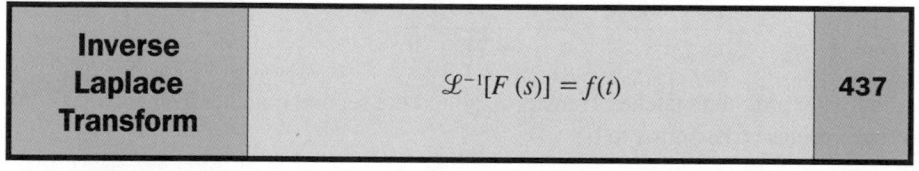

| Inverse Laplace Transform | $\mathcal{L}^{-1}[F(s)] = f(t)$ | 437 |

We'll see that finding the inverse is often harder than finding the transform itself.

We use Table 36–1 to find the inverse of some Laplace transforms. We put our given expression in a form that matches a right-hand entry of Table 36–1 and then read the corresponding entry at the left.

◆◆◆ **Example 10:** If $F(s) = 4/(s^2 + 16)$, find $f(t)$.

Solution: We search Table 36–1 in the right-hand column for an expression of similar form and find Transform 17,

$$\mathcal{L}[\sin at] = \frac{a}{s^2 + a^2}$$

which matches our expression if $a = 4$. Thus

$$f(t) = \sin 4t$$

◆◆◆ **Example 11:** Find y if $\mathcal{L}[y] = 5/(s - 7)^4$.

Solution: From the table we find Transform 13,

$$\mathcal{L}[t^n e^{at}] = \frac{n!}{(s - a)^{n+1}}$$

In order to match our function, we must have $a = 7$ and $n = 3$. The numerator then must be $3! = 3(2) = 6$. So we insert 6 in the numerator and compensate with 6 in the denominator. We then rewrite our function as

$$\mathcal{L}[y] = \frac{5}{6} \frac{6}{(s - 7)^4}$$

which now is the same as Transform 13 with a coefficient of $\frac{5}{6}$. The inverse is then

$$y = \frac{5}{6} t^3 e^{7t} \qquad \bullet\bullet\bullet$$

Completing the Square

To get our given function to match a right-hand table entry, we may have to do some algebra. Sometimes we must complete the square to make our function match one in the table.

◆◆◆ **Example 12:** Find y if $\mathcal{L}[y] = \dfrac{s - 1}{s^2 + 4s + 20}$.

Solution: This does not now match any functions in our table, but it will if we complete the square on the denominator.

> This is no different from the method we used earlier when we completed the square.

$$s^2 + 4s + 20 = (s^2 + 4s \quad\;\;) + 20$$
$$= (s^2 + 4s + 4) + 20 - 4 = (s + 2)^2 + 4^2$$

The denominator is now of the same form as in Transform 24. However, to use Transform 24, the numerator must be $s + 2$, not $s - 1$. So let us add 3 and subtract 3 from the numerator, so that $s - 1 = (s + 2) - 3$. Then

$$\mathcal{L}[y] = \frac{s - 1}{s^2 + 4s + 20} = \frac{(s + 2) - 3}{(s + 2)^2 + 4^2} = \frac{(s + 2)}{(s + 2)^2 + 4^2} - \frac{3}{(s + 2)^2 + 4^2}$$
$$= \frac{(s + 2)}{(s + 2)^2 + 4^2} - \frac{3}{4} \cdot \frac{4}{(s + 2)^2 + 4^2}$$

Now the first expression matches Transform 24 and the second matches Transform 23. We find the inverse of these transforms to be

$$y = e^{-2t} \cos 4t - \frac{3}{4} e^{-2t} \sin 4t = e^{-2t} \left(\cos 4t - \frac{3}{4} \sin 4t \right) \qquad \bullet\bullet\bullet$$

Partial Fractions

We used partial fractions in Sec. 34–5 to make a given expression match one listed in the table of integrals. Now we use partial fractions to make an expression match one listed in our table of Laplace transforms.

> We could also complete the square here, but we would find that the resulting expression would not match a table entry. Some trial-and-error work in algebra might be necessary to get the function to match a table entry.

◆◆◆ **Example 13:** Find y if $\mathcal{L}[y] = 12/(s^2 - 2s - 8)$.

Solution: We separate $\mathcal{L}[y]$ into partial fractions.

$$\mathcal{L}[y] = \frac{12}{s^2 - 2s - 8}$$
$$= \frac{A}{s - 4} + \frac{B}{s + 2}$$

so

$$12 = A(s + 2) + B(s - 4)$$
$$= (A + B)s + (2A - 4B)$$

from which $A + B = 0$ and $2A - 4B = 12$. Solving simultaneously gives $A = 2$ and $B = -2$, so

$$\mathcal{L}[y] = \frac{12}{s^2 - 2s - 8}$$

$$= \frac{2}{s - 4} - \frac{2}{s + 2}$$

Using Transform 9 twice, we get

$$y = 2e^{4t} - 2e^{-2t}$$

◆◆◆

Exercise 2 ◆ Inverse Transforms

Find the inverse of each transform.

1. $\dfrac{1}{s}$

2. $\dfrac{3}{s^2}$

3. $\dfrac{2}{s^3}$

4. $\dfrac{1}{s^2 + 3s}$

5. $\dfrac{4}{4 + s^2}$

6. $\dfrac{s}{(s - 6)^2}$

7. $\dfrac{3s}{s^2 + 2}$

8. $\dfrac{4}{s^3 + 9s}$

9. $\dfrac{s + 4}{(s - 9)^2}$

10. $\dfrac{3s}{(s^2 + 4)^2}$

11. $\dfrac{5}{(s + 2)^2 + 9}$

12. $\dfrac{s + 2}{s^2 - 6s + 8}$

13. $\dfrac{2s^2 + 1}{s(s^2 + 1)}$

14. $\dfrac{s}{s^2 + 2s + 1}$

15. $\dfrac{5s - 2}{s^2(s - 1)(s + 2)}$

16. $\dfrac{1}{(s + 1)(s + 2)^2}$

17. $\dfrac{2}{s^2 + s - 2}$

18. $\dfrac{s + 1}{s^2 + 2s}$

19. $\dfrac{2s}{s^2 + 5s + 6}$

20. $\dfrac{3s}{(s^2 + 4)(s^2 + 1)}$

36–3 Solving Differential Equations by the Laplace Transform

To solve a differential equation with the Laplace transform:
1. Take the transform of each side of the equation.
2. Solve for $\mathcal{L}[y] = F(s)$.
3. Manipulate $F(s)$ until it matches one or more table entries.
4. Take the inverse transform to find $y = f(t)$.

We start with a very simple example.

◆◆◆ **Example 14:** Solve the first-order differential equation $y' + y = 2$ if $y(0) = 0$.

Solution:

1. We write the Laplace transform of both sides and get

$$\mathcal{L}[y'] + \mathcal{L}[y] = \mathcal{L}[2]$$

$$s\mathcal{L}[y] - y(0) + \mathcal{L}[y] = \frac{2}{s}$$

2. We substitute 0 for $y(0)$ and solve for $\mathcal{L}[y]$.

$$(s + 1)\mathcal{L}[y] = \frac{2}{s}$$

$$\mathcal{L}[y] = \frac{2}{s(s + 1)}$$

3. Our function will match Transform 10 if we write it as

$$\mathcal{L}[y] = 2\,\frac{1}{s(s + 1)}$$

4. Taking the inverse gives

$$y = 2(1 - e^{-t}) \qquad \bullet\bullet\bullet$$

In our next example we express $F(s)$ as three partial fractions in order to take the inverse transform.

◆◆◆ **Example 15:** Solve the first-order differential equation

$$y' - 3y + 4 = 9t$$

if $y(0) = 2$.

Solution:

1. We take the Laplace transform of both sides and get

$$\mathcal{L}[y'] - \mathcal{L}[3y] + \mathcal{L}[4] = \mathcal{L}[9t]$$

$$s\mathcal{L}[y] - y(0) - 3\mathcal{L}[y] + \frac{4}{s} = \frac{9}{s^2}$$

2. We substitute 2 for $y(0)$ and solve for $\mathcal{L}[y]$.

$$(s - 3)\mathcal{L}[y] = \frac{9}{s^2} - \frac{4}{s} + 2$$

$$\mathcal{L}[y] = \frac{9}{s^2(s - 3)} - \frac{4}{s(s - 3)} + \frac{2}{(s - 3)}$$

3. We match our expressions with Transforms 28, 10, and 9.

$$\mathcal{L}[y] = \frac{9}{s^2(s - 3)} + \frac{4}{3}\frac{-3}{s(s - 3)} + \frac{21}{(s - 3)}$$

4. Taking the inverse gives

$$y = e^{3t} - 3t - 1 + \frac{4}{3}(1 - e^{3t}) + 2e^{3t} = \frac{5}{3}e^{3t} - 3t + \frac{1}{3} \qquad \bullet\bullet\bullet$$

We now use the Laplace transform to solve a second-order differential equation.

◆◆◆ **Example 16:** Solve the second-order differential equation

$$y'' + 4y - 3 = 0$$

where y is a function of t and y and y' are both zero at $t = 0$.

Solution:

1. Taking the Laplace transform of both sides,

$$\mathcal{L}[y''] + \mathcal{L}[4y] - \mathcal{L}[3] = 0$$

$$s^2\mathcal{L}[y] - sy(0) - y'(0) + 4\mathcal{L}[y] - \frac{3}{s} = 0$$

Substituting $y(0) = 0$ and $y'(0) = 0$ gives

$$s^2\mathcal{L}[y] - 0 - 0 + 4\mathcal{L}[y] - \frac{3}{s} = 0$$

2. Solving for $\mathcal{L}[y]$, we have

$$(s^2 + 4)\mathcal{L}[y] = \frac{3}{s}$$

$$\mathcal{L}[y] = \frac{3}{s(s^2 + 4)}$$

3. We match it to Transform 21.

$$\mathcal{L}[y] = \frac{3}{s(s^2 + 4)} = \frac{3}{4}\frac{4}{s(s^2 + 4)}$$

4. Taking the inverse gives

$$y = \frac{3}{4}(1 - \cos 2t) \qquad \blacklozenge\blacklozenge\blacklozenge$$

We have seen that the hardest part in solving a differential equation by Laplace transform is often in doing the algebra necessary to get the given function to match a table entry. We must often use partial fractions or completing the square, and sometimes both, as in the following example.

◆◆◆ **Example 17:** Solve $y'' + 4y' + 5y = t$, where $y(0) = 1$ and $y'(0) = 2$.

Solution:

1. Taking the Laplace transform of both sides and substituting initial conditions, we find that

$$s^2\mathcal{L}[y] - sy(0) - y'(0) + 4s\mathcal{L}[y] - 4y(0) + 5\mathcal{L}[y] = \frac{1}{s^2}$$

$$\mathcal{L}[y](s^2 + 4s + 5) = \frac{1}{s^2} + s + 6$$

2. Solving for $\mathcal{L}[y]$ gives

$$\mathcal{L}[y] = \frac{1}{s^2(s^2 + 4s + 5)} + \frac{s + 6}{s^2 + 4s + 5} \qquad (1)$$

3. Taking for now just the first fraction in Eq. (1), which we'll call $F_1(s)$, we obtain

$$F_1(s) = \frac{1}{s^2(s^2 + 4s + 5)} = \frac{As + B}{s^2} + \frac{Cs + D}{s^2 + 4s + 5}$$

Multiplying by $s^2(s^2 + 4s + 5)$ gives

$$1 = As^3 + 4As^2 + 5As + Bs^2 + 4Bs + 5B + Cs^3 + Ds^2$$

$$= (A + C)s^3 + (4A + B + D)s^2 + (5A + 4B)s + 5B$$

Equating coefficients gives

$$A = \frac{-4}{25} \qquad B = \frac{1}{5} \qquad C = \frac{4}{25} \qquad D = \frac{11}{25}$$

So

$$F_1(s) = \frac{-4s/25 + 1/5}{s^2} + \frac{4s/25 + 11/25}{s^2 + 4s + 5}$$

$$= \frac{1}{25}\left(\frac{5}{s^2} - \frac{4}{s} + \frac{4s + 11}{s^2 + 4s + 5}\right)$$

We can use Transforms 5 and 6 for $4/s$ and $5/s^2$, but there is no match for the third term in the parentheses. However, by completing the square, we can get the denominator to match those in Transforms 23 and 24. Thus

$$s^2 + 4s + 5 = s^2 + 4s + 4 - 4 + 5$$
$$= (s + 2)^2 + 1^2$$

We now manipulate that third term into the form of Transforms 23 and 24.

$$\frac{4s + 11}{s^2 + 4s + 5} = \frac{4s + 11}{(s + 2)^2 + 1^2} = (4)\frac{s + 11/4 + 2 - 2}{(s + 2)^2 + 1^2}$$

$$= (4)\frac{s + 2 + 3/4}{(s + 2)^2 + 1^2} = (4)\frac{s + 2}{(s + 2)^2 + 1^2} + (3)\frac{1}{(s + 2)^2 + 1^2}$$

Combining the third term with the first and second gives

$$F_1(s) = \frac{1}{25}\left[\frac{5}{s^2} - \frac{4}{s} + (4)\frac{s + 2}{(s + 2)^2 + 1^2} + (3)\frac{1}{(s + 2)^2 + 1^2}\right]$$

Returning to the remaining fraction in Eq. (1), which we call $F_2(s)$, we complete the square in the denominator and use partial fractions to get

$$F_2(s) = \frac{s + 6}{(s + 2)^2 + 1^2} = \frac{s + 2 - 2 + 6}{(s + 2)^2 + 1^2}$$

$$= \frac{s + 2}{(s + 2)^2 + 1^2} + \frac{4}{(s + 2)^2 + 1^2}$$

4. Taking the inverse transform of $F_1(s)$ and $F_2(s)$, we obtain

$$y = \frac{1}{25}[5t - 4 + 4e^{-2t}\cos t + 3e^{-2t}\sin t] + e^{-2t}\cos t + 4e^{-2t}\sin t$$

which simplifies to

$$y = \frac{1}{25}(5t - 4 + 29e^{-2t}\cos t + 103e^{-2t}\sin t) \qquad \blacklozenge\blacklozenge\blacklozenge$$

Exercise 3 ◆ Solving Differential Equations by the Laplace Transform

First-Order Equations

Solve each differential equation by the Laplace transform.

1. $y' - 3y = 0$, $y(0) = 1$
2. $2y' + y = 1$, $y(0) = 3$
3. $4y' - 2y = t$, $y(0) = 0$
4. $y' + 5y = e^{2t}$, $y(0) = 2$
5. $3y' - 2y = t^2$, $y(0) = 3$
6. $y' - 3y = \sin t$, $y(0) = 0$
7. $y' + 2y = \cos 2t$, $y(0) = 0$
8. $4y' - y = 3t^3$, $y(0) = 0$

Second-Order Equations

Solve each differential equation by the Laplace transform.

9. $y'' + 2y' - 3y = 0$, $y(0) = 0$ and $y'(0) = 2$
10. $y'' + y' + y = 1$, $y(0) = 0$ and $y'(0) = 0$

11. $y'' + 3y' = 3$, $y(0) = 1$ and $y'(0) = 2$
12. $y'' + 2y = 2$, $y(0) = 0$ and $y'(0) = 3$
13. $2y'' + y = 4t$, $y(0) = 3$ and $y'(0) = 0$
14. $y'' + y' + 5y = t$, $y(0) = 1$ and $y'(0) = 2$
15. $y'' + 4y' + 3y = t$, $y(0) = 2$ and $y'(0) = 2$
16. $y'' + 4y = 8t^3$, $y(0) = 0$ and $y'(0) = 0$
17. $3y'' + y' = \sin t$, $y(0) = 2$ and $y'(0) = 3$
18. $2y'' + y' = -3$, $y(0) = 2$ and $y'(0) = 1$
19. $y'' - 2y' + y = e^t$, $y(0) = 0$ and $y'(0) = 0$
20. $2y'' + 32y = \cos 2t$, $y(0) = 0$ and $y'(0) = 1$
21. $y'' + 2y' + 3y = te^t$, $y(0) = 0$ and $y'(0) = 0$
22. $3y'' + 2y' - y = \sin 3t$, $y(0) = 0$ and $y'(0) = 0$

Motion in a Resisting Medium

23. A 15.0-kg object is dropped from rest through air whose resisting force is equal to 1.85 times the object's speed (in m/s). Find its speed after 1.00 s.

Exponential Growth and Decay

24. A certain steel ingot is 1040 °C and cools at a rate (in °C/min) equal to 1.25 times its present temperature. Find its temperature after 5.00 min.

25. The rate of growth (bacteria/h) of a colony of bacteria is equal to 2.5 times the number present at any instant. How many bacteria are there after 24 h if there are $5\overline{0}00$ at first?

Mechanical Vibrations

26. A weight that hangs from a spring is pulled down 1.00 cm at $t = 0$ and then is released from rest. The differential equation of motion is $x'' + 6.25x' + 25.8x = 0$. Write the equation for x as a function of time.

27. An alternating force is applied to a weight such that the equation of motion is $x'' + 6.25x' = 45.3 \cos 2.25t$. If v and x are zero at $t = 0$, write an equation for x as a function of time.

36–4   Electrical Applications

Many of the differential equations for motion and electric circuits covered in Chapter 35 are nicely handled by the Laplace transform. However, since the Laplace transform is used mainly for electric circuits, we emphasize that application here. The method is illustrated by examples of several types of circuits with dc and ac sources.

Series *RC* Circuit with dc Source

◆◆◆ Example 18: A capacitor (Fig. 36–3) is discharged through a resistor by throwing the switch from position 1 to position 2 at $t = 0$. Find the current i.

Solution: Summing the voltages around the loop gives

$$Ri + \frac{1}{C}\int_0^t i \, dt - v_c = 0$$

Substituting values and rearranging yields

$$1540i + \frac{1}{125 \times 10^{-6}}\int_0^t i \, dt = 115$$

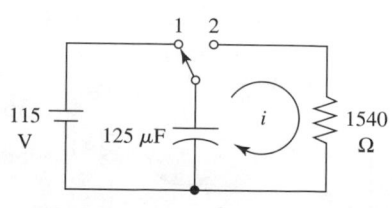

FIGURE 36–3 *RC* circuit.

Taking the transform of each term, we obtain

$$1540\mathscr{L}[i] + \frac{8000\mathscr{L}[i]}{s} = \frac{115}{s}$$

We solve for $\mathscr{L}[i]$ and rewrite our expression in the form of a table entry.

$$\mathscr{L}[i]\left(1540 + \frac{8000}{s}\right) = \frac{115}{s}$$

$$\mathscr{L}[i]\left(\frac{1540s + 8000}{s}\right) = \frac{115}{s}$$

$$\mathscr{L}[i] = \frac{115}{1540s + 8000} = \frac{0.0747}{s + 5.19}$$

Taking the inverse transform using Transform 9 gives us

$$i = 0.0747e^{-5.19t}$$

This is the same result obtained by classical methods in Chapter 35, Example 38.   ◆◆◆

◆◆◆ **Example 19:** A capacitor has an initial charge of 35 V, with the polarity as marked in Fig. 36–4. Find the current if the switch is closed at $t = 0$.

Solution: Summing the voltages around the loop gives

$$Ri + \frac{1}{C}\int_0^t i\,dt - v_c - E = 0$$

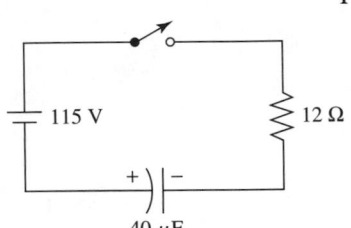

FIGURE 36–4   *RC* circuit.

Substituting values and rearranging, we see that

$$12i + \frac{1}{40 \times 10^{-6}}\int_0^t i\,dt = 150$$

Taking the transform of each term, we have

$$12\mathscr{L}[i] + 25\,000\,\frac{\mathscr{L}[i]}{s} = \frac{150}{s}$$

Solving for $\mathscr{L}[i]$ yields

$$\mathscr{L}[i]\left(12 + \frac{25\,000}{s}\right) = \frac{150}{s}$$

$$\mathscr{L}[i]\left(\frac{12s + 25\,000}{s}\right) = \frac{150}{s}$$

$$\mathscr{L}[i] = \frac{150}{12s + 25\,000} = \frac{12.5}{s + 2080}$$

Taking the inverse transform using Transform 9, we obtain

$$i = 12.5e^{-2080t}$$ ◆◆◆

Series *RL* Circuit with dc Source

◆◆◆ **Example 20:** The initial current in an *RL* circuit is 1 A in the direction shown in Fig. 36–5. The switch is closed at $t = 0$. Find the current.

Solution: Summing the voltages around the loop gives

$$E = Ri + L\frac{di}{dt}$$

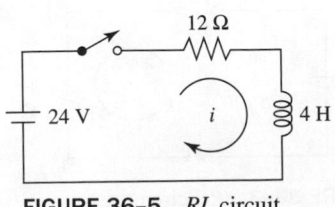

FIGURE 36–5   *RL* circuit.

Substituting values and rearranging, we find that

$$24 = 12i + 4i'$$

or

$$6 = 3i + i'$$

Taking the transform of each term yields

$$\frac{6}{s} = 3\mathcal{L}[i] + s\mathcal{L}[i] - i(0)$$

Substituting 1 for $i(0)$ and solving for $\mathcal{L}[i]$, we have

$$\frac{6}{s} = \mathcal{L}[i](s + 3) - 1$$

$$\mathcal{L}[i] = \frac{6}{s(s + 3)} + \frac{1}{s + 3}$$

Taking the inverse transform using Transforms 9 and 10 gives

$$i = 2(1 - e^{-3t}) + e^{-3t}$$

or

$$i = 2 - e^{-3t}$$

The current is graphed in Fig. 36–6.

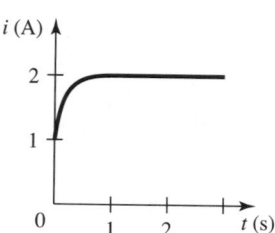

◆◆◆ **FIGURE 36–6**

Series *RL* Circuit with ac Source

◆◆◆ **Example 21:** A switch (Fig. 36–7) is closed at $t = 0$ when the applied voltage is zero and increasing. Find the current.

Solution: Summing the voltages around the loop gives

$$Ri + L\frac{di}{dt} = v$$

Substituting values and rearranging, we find that

$$4i + 0.02\frac{di}{dt} = 80 \sin 400t$$

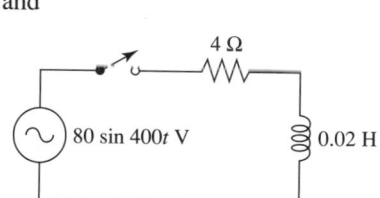

FIGURE 36-7 *RL* circuit with ac source.

Taking the transform of each term yields

$$4\mathcal{L}[i] + 0.02[s\mathcal{L}[i] - i(0)] = \frac{80(400)}{s^2 + (400)^2}$$

Substituting 0 for $i(0)$ and solving for $\mathcal{L}[i]$, we have

$$\mathcal{L}[i](0.02s + 4) = \frac{32\,000}{s^2 + (400)^2}$$

$$\mathcal{L}[i] = \frac{32\,000}{[s^2 + (400)^2](0.02s + 4)} = \frac{1\,600\,000}{[s^2 + (400)^2](s + 200)}$$

We now use the method of Sec. 34–5 for separating into partial fractions a rational fraction that has quadratic factors in the denominator.

$$\frac{1\,600\,000}{[s^2 + (400)^2](s + 200)} = \frac{As + B}{s^2 + (400)^2} + \frac{C}{(s + 200)}$$

Multiplying by $[s^2 + (400)^2](s + 200)$ gives

$$1\,600\,000 = (As + B)(s + 200) + C[s^2 + (400)^2]$$

which simplifies to

$$1\,600\,000 = (A + C)s^2 + (200A + B)s + 200B + 400^2C$$

Equating cocfficients gives

$$A + C = 0$$
$$200A + B = 0$$
$$200B + 400^2C = 1\,600\,000$$

Solving simultaneously for A, B, and C gives

$$A = -8$$
$$B = 1600$$
$$C = 8$$

Substituting back, we get

$$\mathcal{L}[i] = \frac{1600 - 8s}{s^2 + (400)^2} + \frac{8}{s + 200}$$

$$= 4\,\frac{400}{s^2 + (400)^2} - 8\,\frac{s}{s^2 + (400)^2} + 8\,\frac{1}{s + 200}$$

Taking the inverse transform using Transforms 9, 17, and 18, we find that

$$i = 4 \sin 400t - 8 \cos 400t + 8e^{-200t}$$

By Eq. 214, combine the first two terms in the form $I \sin(400t + \theta)$, where

$$I = \sqrt{4^2 + 8^2} = 8.94 \qquad \text{and} \qquad \theta = \tan^{-1}\left(-\frac{8}{4}\right) = -63.4° = -1.107 \text{ rad}$$

so

$$i = 8.94 \sin(400t - 1.107) + 8e^{-200t} \text{ A}$$

A graph of this wave, as well as the applied voltage wave, is shown in Fig. 36–8.

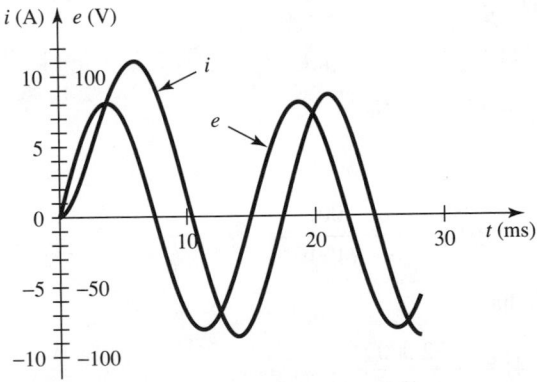

FIGURE 36–8   ◆◆◆

Series *RLC* Circuit with dc Source

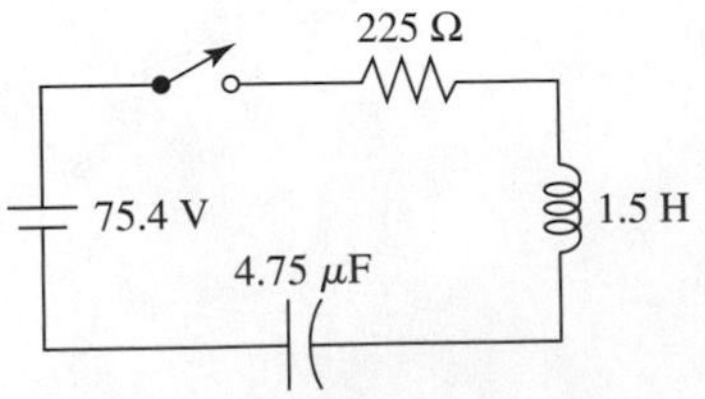

◆◆◆ **Example 22:** A switch (Fig. 36–9) is closed at $t = 0$, and there is no initial charge on the capacitor. Write an expression for the current.

Solution:

1.   The differential equation for this circuit is

$$Ri + L\frac{di}{dt} + \frac{1}{C}\int_0^t i\,dt = E$$

FIGURE 36–9   *RLC* circuit.

225 Ω
75.4 V
4.75 μF
1.5 H

2. Transforming each term and substituting values gives

$$225\mathscr{L}[i] + 1.5\{s\,\mathscr{L}[i] - i(0)\} + \frac{10^6\mathscr{L}[i]}{4.75s} = \frac{75.4}{s}$$

3. Setting $i(0)$ equal to zero and solving for $\mathscr{L}[i]$ yields

$$\mathscr{L}[i] = \frac{50.3}{s^2 + 150s + 140\,000}$$

4. We now try to match this expression with those in the table. Let us complete the square in the denominator.

$$s^2 + 150s + 140\,000 = (s^2 + 150s + 5625) + 140\,000 - 5625$$
$$= (s + 75)^2 + 134\,375$$
$$= (s + 75)^2 + (367)^2$$

Then,

$$\mathscr{L}[i] = \frac{50.3}{s^2 + 150s + 140\,000}$$
$$= \frac{50.3}{(s + 75) + 367^2}$$
$$= 0.137\,\frac{367}{(s + 75)^2 + 367^2}$$

5. We finally have a match, with Transform 23. Taking the inverse transform gives

$$i = 137e^{-75t}\sin 367t \text{ mA}$$

which agrees with the result found in Chapter 35, Example 53.

We see that most of the work when using the Laplace transform is in algebra, rather than in calculus.

◆◆◆

Exercise 4 ◆ Electrical Applications

1. The current in a certain RC circuit satisfies the equation $172i' + 2750i = 115$. If i is zero at $t = 0$, show that

$$i = 41.8(1 - e^{-16.0t}) \text{ mA}$$

2. In an RL circuit, $R = 3750\ \Omega$ and $L = 0.150$ H. It is connected to a dc source of 250 V at $t = 0$. If i is zero at $t = 0$, show that

$$i = 66.7(1 - e^{-25\,000t}) \text{ mA}$$

3. The current in an RLC circuit satisfies the equation

$$0.55i'' + 482i' + 7350i = 120$$

If i and i' are zero at $t = 0$, show that

$$i = 16.1(1 - e^{-16t}) \text{ mA}$$

4. In an RLC circuit, $R = 2950\ \Omega$, $L = 0.120$ H, and $C = 1.55\ \mu$F. It is connected to a dc source of 115 V at $t = 0$. If i is zero at $t = 0$, show that

$$i = 39.6e^{-221t} \text{ mA}$$

5. The current in a certain RLC circuit satisfies the equation

$$3.15i + 0.0223i' + 1680\int_0^t i\,dt = 1$$

If i is zero at $t = 0$, show that

$$i = 169e^{-70.6t}\sin 265t \text{ mA}$$

6. In an RL circuit, $R = 4150\ \Omega$ and $L = 0.127$ H. It is connected to a dc source of 28.4 V at $t = 0$. If i is zero at $t = 0$, show that

$$i = 6.84(1 - e^{-32\,700t})\ \text{mA}$$

7. In an RL circuit, $R = 8.37\ \Omega$ and $L = 0.250$ H. It is connected to a dc source of 50.5 V at $t = 0$. If i is zero at $t = 0$, show that

$$i = 6.03(1 - e^{-33.5t})\ \text{A}$$

36–5   Numerical Solution of First-Order Differential Equations

In Sec. 35–2, we showed how to use Euler's method to graphically and numerically solve a differential equation, but we mentioned there that it is not the most accurate method available. It will, however, serve as a good introduction to other methods. We will now show both the *modified* Euler's method that includes predictor-corrector steps and the Runge-Kutta method. We will apply these methods to both first- and second-order equations.

Modified Euler's Method with Predictor-Corrector Steps

The various *predictor-corrector* methods all have two main steps. First, the *predictor* step tries to predict the next point on the curve from preceding values, and then the *corrector* step tries to improve the predicted values.

Here we modify Euler's method of the preceding section as the predictor, using the coordinates and slope at the present point P to estimate the next point Q. The corrector step then recomputes Q *using the average of the slopes at P and Q.* Our iteration formula is then the following:

Modified Euler's Method	$x_q = x_p + \Delta x$ $$y_q = y_p + \frac{m_p + m_q}{2}\Delta x$$	439

◆◆◆ **Example 23:** Find an approximate solution to $y' = x^2/y$, with the boundary condition that $y = 2$ when $x = 3$. Calculate y for $x = 3$ to 10 in steps of 1. (This is a repeat of Example 11 from Sec. 35–2.)

Solution: Using data from before, the slope m_1 at (3, 2) was 4.5, the second point was predicted to be (4, 6.5), and the slope m_2 at (4, 6.5) was 2.462. The recomputed ordinate at the second point is then, by Eq. 439,

$$y_2 = 2 + \frac{4.5 + 2.462}{2}(1)$$

$$= 5.481$$

so our corrected second point is (4, 5.481). The slope at this point is

$$m_2 = y'(4, 5.481) = 2.919$$

We use this to predict y_3.

$$y_3\ (\text{predicted}) = y_2 + m_2\,\Delta x$$

$$= 5.481 + 2.919(1)$$

$$= 8.400$$

The slope at (5, 8.400) is

$$m_3 = y'(5, 8.400) = 2.976$$

The corrected y_3 is then

$$y_3 \text{ (corrected)} = 5.481 + \frac{2.919 + 2.976}{2}(1) = 8.429$$

The remaining values are given in Table 36–2.

TABLE 36–2

x	Approximate y	Exact y	Error
3	2.000 00	2.000 00	0.000 00
4	5.480 77	5.354 13	0.126 64
5	8.428 50	8.326 66	0.101 84
6	11.491 26	11.401 75	0.089 50
7	14.732 98	14.651 51	0.081 48
8	18.167 90	18.092 36	0.075 55
9	21.796 42	21.725 56	0.070 86
10	25.614 34	25.547 34	0.067 00

Note that the error at $x = 10$ is about 0.26%, compared with 0.6% before. We have gained in accuracy at the cost of added computation. ◆◆◆

Runge-Kutta Method

The Runge-Kutta method, like the modified Euler's, is a predictor-corrector method.

Prediction: Starting at P (Fig. 36–10), we use

the slope m_p to predict R,
the slope m_r to predict S, and
the slope m_s to predict Q.

This method is named after two German mathematicians, Carl Runge (1856–1927) and Wilhelm Kutta (1867–1944).

Correction: We then take a weighted average of the slopes at P, Q, R, and S, giving twice the weight to the slopes at the midpoints R and S as at the endpoints P and Q.

$$m_{avg} = \frac{1}{6}(m_p + 2m_r + 2m_s + m_q)$$

We use the average slope and the coordinates of P to find Q.

$$y_q = y_p + m_{avg}\, \Delta x$$

Runge-Kutta Method	where and	$x_q = x_p + \Delta x$ $y_q = y_p + m_{avg}\, \Delta x$ $m_{avg} = \frac{1}{6}(m_p + 2m_r + 2m_s + m_q)$ $m_p = f'(x_p, y_p)$ $m_r = f'\left(x_p + \frac{\Delta x}{2}, y_p + m_p \frac{\Delta x}{2}\right)$ $m_s = f'\left(x_p + \frac{\Delta x}{2}, y_p + m_r \frac{\Delta x}{2}\right)$ $m_q = f'(x_p + \Delta x, y_p + m_s\, \Delta x)$	440

◆◆◆ **Example 24:** Repeat Example 23 using the Runge-Kutta method and a step of 1.

Solution: Starting at $P(3, 2)$, with $m_p = 4.5$,

$$y_r = 2 + 4.5(0.5) = 4.25$$

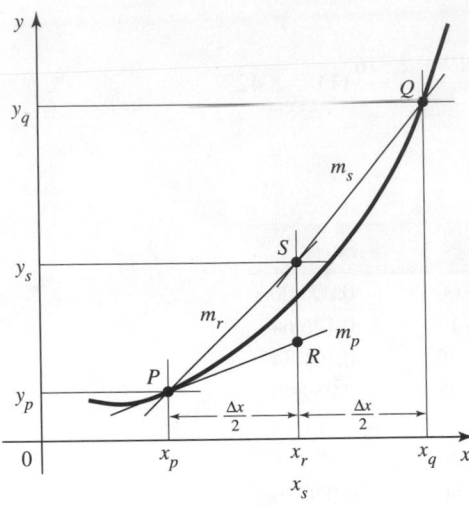

FIGURE 36–10   Runge-Kutta method.

$$m_r = f'(3.5, 4.25) = \frac{(3.5)^2}{4.25} = 2.882$$

$$y_s = 2 + 2.882(0.5) = 3.441$$

$$m_s = f'(3.5, 3.441) = \frac{(3.5)^2}{3.441} = 3.560$$

$$y_q = 2 + 3.560(1) = 5.560$$

$$m_q = f'(4, 5.560) = \frac{4^2}{5.560} = 2.878$$

$$m_{avg} = \frac{4.5 + 2(2.882) + 2(3.560) + 2.878}{6} = 3.377$$

$$y_q \text{ (corrected)} = y_p + m_{avg} \Delta x = 2 + 3.377(1) = 5.377$$

So our second point is (3, 5.377). The rest of the calculation is given in Table 36–3.

TABLE 36–3

x	Approximate y	Exact y	Error
3	2.000 00	2.000 00	0.000 00
4	5.377 03	5.354 13	0.022 90
5	8.341 41	8.326 66	0.014 75
6	11.412 55	11.401 75	0.010 79
7	14.659 92	14.651 51	0.008 41
8	18.099 18	18.092 36	0.006 82
9	21.731 25	21.725 56	0.005 69
10	25.552 19	25.547 34	0.004 84

The error here at $x = 10$ with the Runge-Kutta method is about 0.02%, compared with 0.26% for the modified Euler method.   ◆◆◆

Comparison of the Three Methods

Table 36–4 shows the errors obtained when solving the same differential equation by our three different methods, each with a step size of 1.

TABLE 36–4

x	Euler	Modified Euler	Runge-Kutta
3	0.000 00	0.000 00	0.000 00
4	1.145 87	0.126 64	0.022 90
5	0.634 87	0.101 84	0.014 75
6	0.349 48	0.089 50	0.010 79
7	0.163 24	0.081 48	0.008 41
8	0.029 90	0.075 55	0.006 82
9	0.071 73	0.070 86	0.005 69
10	0.152 83	0.067 00	0.004 84

Divergence

As with most numerical methods, the computation can diverge, with our answers getting farther and farther from the true values. The computations shown here can diverge if either the step size Δx is too large or the range of x over which we compute is too large. Numerical methods can also diverge if the points used in the computation are near a discontinuity, a maximum or minimum point, or a point of inflection.

When doing the computations on the computer (the only practical way), it is easy to change the step size. A good practice is to reduce the step size by a factor of 10 each time (1, 0.1, 0.01, . . .) until the answers do not change in a given decimal place. Further reduction of the step size will eventually result in numbers too small for the computer to handle, and unreliable results will occur.

Exercise 5 ◆ Numerical Solution of First-Order Differential Equations

Solve each differential equation and find the value of y asked for. Use either the modified Euler method or the Runge-Kutta method, with a step size of 0.1.

1. $y' + xy = 3$; $y(4) = 2$. Find $y(5)$.
2. $y' + 5y^2 = 3x$; $y(0) = 3$. Find $y(1)$.
3. $x^4y' + ye^x = xy^3$; $y(2) = 2$. Find $y(3)$.
4. $xy' + y^2 = 3\ln x$; $y(1) = 3$. Find $y(2)$.
5. $1.94x^2y' + 8.23x\sin y - 2.99y = 0$; $y(1) = 3$. Find $y(2)$.
6. $y'\ln x + 3.99(x - y)^2 = 4.28y$; $y(2) = 1.76$. Find $y(3)$.
7. $(x + y)y' + 2.76x^2y^2 = 5.33y$; $y(0) = 1$. Find $y(1)$.
8. $y' - 2.97xe^x - 1.92x\cos y = 0$; $y(2) = 1$. Find $y(3)$.
9. $xy' + xy^2 = 8y\ln|xy|$; $y(1) = 3$. Find $y(2)$.
10. $x^2y' - 8.85x^2\ln y + 3.27x = 0$; $y(1) = 23.9$. Find $y(2)$.

Your answers might differ from those given for this exercise, depending on the method used and roundoff errors.

36–6 Numerical Solution of Second-Order Differential Equations

To solve a second-order differential equation numerically, we first transform the given equation into two first-order equations by substituting another variable, m, for y'.

◆◆◆ **Example 25:** By making the substitution $y' = m$, we can transform the second-order equation

$$y'' + 3y' + 4xy = 0$$

into the two first-order equations

$$y' = m$$

and

$$m' = -3m - 4xy$$

$$= f(x, y, m)$$ ◆◆◆

As before, we start with the given point $P(x_p, y_p)$, at which we must also know the slope m_p. We proceed from P to the next point $Q(x_q, y_q)$ as follows:

1. Compute the average first derivative, m_{avg}, over the interval Δx.
2. Compute the average second derivative m'_{avg} over that interval.
3. Find the coordinates x_q and y_q at Q by using the equations

$$x_q = x_p + \Delta x$$

and

$$y_q = y_p + m_{avg} \Delta x$$

4. Find the first derivative m_q at Q by using

$$m_q = m_p + m'_{avg} \Delta x$$

5. Return to Step 1 and repeat the sequence for the next interval.

We can find m_{avg} and m'_{avg} in Steps 1 and 2 either by the modified Euler's method or by the Runge-Kutta method. The latter method, as before, requires us to find values at the intermediate points R and S. The equations for second-order equations are similar to those we had for first-order equations.

At P:	$x_p, y_p,$ and m_p	are given	$m'_p = f(x_p, y_p, m_p)$
At R:	$x_r = x_p + \dfrac{\Delta x}{2}$		$y_r = y_p + m_p \dfrac{\Delta x}{2}$
	$m_r = m_p + m'_p \dfrac{\Delta x}{2}$		$m'_r = f(x_r, y_r, m_r)$
At S:	$x_s = x_p + \dfrac{\Delta x}{2}$		$y_s = y_p + m_r \dfrac{\Delta x}{2}$
	$m_s = m_p + m'_r \dfrac{\Delta x}{2}$		$m'_s = f(x_s, y_s, m_s)$
At Q:	$x_q = x_p + \Delta x$		$y_q = y_p + m_s \Delta x$
	$m_q = m_p + m'_s \Delta x$		$m'_q = f(x_q, y_q, m_q)$

◆◆◆ **Example 26:** Find an approximate solution to the equation

$$y'' - 2y' + y = 2e^x$$

with boundary conditions $y' = 2e$ at $(1, 0)$. Find points from 1 to 3 in steps of 0.1.

Solution: Replacing y' by m and solving for m' gives

$$m' = f(x, y, m)$$

$$= 2e^x - y + 2m$$

At P: The boundary values are $x_p = 1$, $y_p = 0$, and $m_p = 5.4366$. Then

$$m'_p = f(x_p, y_p, m_p) = 2e^1 - 0 + 2(5.4366) = 16.3098$$

At R: $x_r = x_p + \dfrac{\Delta x}{2} = 1 + 0.05 = 1.05$

$y_r = y_p + m_p \dfrac{\Delta x}{2} = 0 + 5.4366(0.05) = 0.2718$

$m_r = m_p + m_p' \dfrac{\Delta x}{2} = 5.4366 + 16.3098(0.05) = 6.2521$

$m_r' = 2e^{1.05} - 0.2718 + 2(6.2521) = 17.9477$

At S: $x_s = 1.05$

$y_s = y_p + m_r \dfrac{\Delta x}{2} = 0 + 6.2521(0.05) = 0.3126$

$m_s = m_p + m_r' \dfrac{\Delta x}{2} = 5.4366 + 17.9477(0.05) = 6.3340$

$m_s' = 2e^{1.05} - 0.3126 + 2(6.3340) = 18.0707$

At Q: $x_q = x_p + \Delta x = 1.1$

$y_q = y_p + m_s \Delta x = 0 + 6.3360(0.1) = 0.6334$

$m_q = m_p + m_s' \Delta x = 5.4366 + 18.0707(0.1) = 7.2437$

$m_q' = 2e^{1.1} - 0.6334 + 2(7.2437) = 19.8623$

Then

$$m_{\text{avg}} = \frac{m_p + 2m_r + 2m_s + m_q}{6}$$

$$- \frac{5.4366 + 2(6.2521) + 2(6.3340) + 7.2437}{6} = 6.3088$$

and

$$m_{\text{avg}}' = \frac{m_p' + 2m_r' + 2m_s' + m_q'}{6}$$

$$= \frac{16.3098 + 2(17.9477) + 2(18.0707) + 19.8623}{6} = 18.0348$$

So

$$y_q = y_p + m_{\text{avg}} \Delta x = 0 + 6.3088(0.1) = 0.6309$$

and

$$m_q = m_p + m_{\text{avg}}' \Delta x = 5.4366 + 18.0348(0.1) = 7.2401$$

The computation is then repeated. The remaining values are given in Table 36–5, along with the exact y and exact slope, found by analytical solution of the given equation.

TABLE 36–5

x	y	Exact y	Slope	Exact Slope
1.0	0.000	0.000	5.437	5.437
1.1	0.631	0.631	7.240	7.240
1.2	1.461	1.461	9.429	9.429
1.3	2.532	2.532	12.072	12.072
1.4	3.893	3.893	15.248	15.248
1.5	5.602	5.602	19.047	19.047
1.6	7.727	7.727	23.576	23.576
1.7	10.346	10.346	28.957	28.957

x	y	Exact y	Slope	Exact Slope
1.8	13.551	13.551	35.330	35.330
1.9	17.450	17.450	42.856	42.857
2.0	22.167	22.167	51.723	51.723
2.1	27.846	27.847	62.144	62.145
2.2	34.656	34.656	74.366	74.366
2.3	42.789	42.789	88.670	88.670
2.4	52.470	52.470	105.381	105.382
2.5	63.958	63.958	124.870	124.871
2.6	77.550	77.551	147.562	147.563
2.7	93.593	93.594	173.943	173.944
2.8	112.480	112.481	204.570	204.571
2.9	134.669	134.670	240.079	240.080
3.0	160.683	160.684	281.196	281.198

◆◆◆

Exercise 6 ◆ Numerical Solution of Second-Order Differential Equations

Solve each equation by the Runge-Kutta method, and find the value of y at $x = 2$. Take a step size of 0.1 unit.

1. $y'' + y' + xy = 3$, $y'(1, 1) = 2$
2. $y'' - xy' + 3xy^2 = 2y$, $y'(1, 1) = 2$
3. $3y'' + y' + ye^x = xy^3$, $y'(1, 2) = 2$
4. $2y'' - xy' + y^2 = 5 \ln x$, $y'(1, 0) = 1$
5. $xy'' - 2.4x^2y' + 5.3x \sin y - 7.4y = 0$, $y'(1, 1) = 0.1$
6. $y'' + y' \ln x + 2.69(x - y)^2 = 6.26y$, $y'(1, 3.27) = 8.26$
7. $x^2y'' - (x + y)y' + 8.26x^2y^2 = 1.83y$, $y'(1, 1) = 2$
8. $y'' - y' + 0.533xye^x + 0.326x \cos y = 0$, $y'(1, 73.4) = 273$
9. $y'' - xy' + 59.2xy = 74.1y \ln|xy|$, $y'(1, 1) = 2$
10. $xy'' - x^2y' + 11.4x^2 \ln y + 62.2x = 0$, $y'(1, 8) = 52.1$

Because of roundoff errors, your answers might differ from those given for this exercise set.

◆◆◆ **CHAPTER 36 REVIEW PROBLEMS** ◆◆◆◆◆◆◆◆◆◆◆◆◆◆◆◆◆◆◆◆◆◆◆◆◆◆◆◆◆◆◆

Find the Laplace transform by direct integration.

1. $f(t) = 2t$ 2. $f(t) = 3t^2$ 3. $f(t) = \cos 3t$

Find the Laplace transform by using Table 36–1.

4. $f(t) = 2t^3 + 3t$ 5. $f(t) = 3te^{2t}$

6. $f(t) = 2t + 3e^{2t}$ 7. $f(t) = 3e^{-t} - 2t^2$

8. $f(t) = 3t^3 + 5t^2 + 4e^t$ 9. $y' + 3y$, $y(0) = 1$

10. $y' - 6y$, $y(0) = 1$ 11. $3y' + 2y$, $y(0) = 0$

12. $2y'' + y' - 3y$, $y(0) = 1$ and $y'(0) = 3$

13. $y'' + 3y' + 4y$, $y(0) = 1$ and $y'(0) = 3$

Find the inverse transform.

14. $F(s) = \dfrac{6}{4 + s^2}$

15. $F(s) = \dfrac{s}{(s - 3)^2}$

16. $F(s) = \dfrac{s + 6}{(s - 2)^2}$

17. $F(s) = \dfrac{2s}{(s^2 + 5)^2}$

18. $F(s) = \dfrac{3}{(s + 4)^2 + 5}$

19. $F(s) = \dfrac{s + 1}{s^2 - 6s + 8}$

20. $F(s) = \dfrac{4s^2 + 2}{s(s^2 + 3)}$

21. $F(s) = \dfrac{3s}{s^2 + 2s + 1}$

22. $F(s) = \dfrac{3}{s^2 + s + 4}$

23. $F(s) = \dfrac{4s}{s^2 + 4s + 4}$

Solve each differential equation by the Laplace transform.

24. $y' + 2y = 0, \quad y(0) = 1$

25. $y' - 3y = t^2, \quad y(0) = 2$

26. $3y' - y = 2t, \quad y(0) = 1$

27. $y'' + 2y' + y = 2, \quad y(0) = 0$ and $y'(0) = 0$

28. $y'' + 3y' + 2y = t, \quad y(0) = 1$ and $y'(0) = 2$

29. $y'' + 2y = 4t, \quad y(0) = 3$ and $y'(0) = 0$

30. $y'' + 3y' + 2y = e^{2t}, \quad y(0) = 2$ and $y'(0) = 1$

31. A 81.0-N ball is dropped from rest through air whose resisting force is equal to 2.75 times the ball's speed in m/s. Write an expression for the speed of the ball.

32. A weight that hangs from a spring is pulled down 0.500 cm at $t = 0$ and released from rest. The differential equation of motion is $x'' + 3.22x' + 18.5 = 0$. Write the equation for x as a function of time.

33. The current in a certain RC circuit satisfies the equation $8.24i' + 149i = 100$. If i is zero at $t = 0$, write an equation for i.

34. The current in an RLC circuit satisfies the equation $5.45i'' + 2820i' + 9730i = 10$. If i and i' are zero at $t = 0$, write an equation for i.

35. In an RLC circuit, $R = 3750\ \Omega$, $L = 0.150$ H, and $C = 1.25\ \mu$F. The circuit is connected to a dc source of 150 V at $t = 0$. If i is zero at $t = 0$, write an equation for the instantaneous current.

36. In an RL circuit, $R = 3750\ \Omega$ and $L = 0.150$ H. The circuit is connected to an ac source of $28.4 \sin 84t$ V at $t = 0$. If i is zero at $t = 0$, write an equation for the instantaneous current.

Solve each differential equation. Use any numerical method with a step size of 0.1 unit.

37. $y' + xy = 4y, \quad y(2) = 2$. Find $y(3)$.

38. $xy' + 3y^2 = 5xy, \quad y(1) = 5$. Find $y(2)$.

39. $3y'' - y' + xy^2 = 2 \ln y, \quad y'(1, 2) = 8$. Find $y(2)$.

40. $yy'' - 7.2yy' + 2.8x \sin x - 2.2y = 0, \quad y'(1, 5) = 8.2$. Find $y(2)$.

41. $y'' + y' \ln y + 5.27(x - 2y)^2 = 2.84x, \quad y'(1, 1) = 3$. Find $y(2)$.

42. $x^2y'' - (1 + y)y' + 3.67x^2 = 5.28x, \quad y'(1, 2) = 3$. Find $y(2)$.

43. $y'' - xy' + 6e^x + 126 \cos y = 0, \quad y'(1, 2) = 5$. Find $y(2)$.

44. $y' + xy^2 = 8 \ln y, \quad y(1) = 7$. Find $y(2)$.

45. $7.35y' + 2.85x \sin y - 7.34x = 0, \quad y(2) = 4.3$. Find $y(3)$.

46. $yy'' - y' + 77.2y^2 = 28.4x \ln |xy|, \quad y'(3, 5) = 3$. Find $y(4)$.

Writing

47. Suppose that your company gets a new project requiring you to solve many differential equations. You see this as a chance to get that promotion you've always wanted. Write a memo to your boss explaining how those differential equations can be solved using your computer.

Team Project

48. Given

$$y' - y - e^{2t} = 0$$

where $y(0) = 2$, solve this DE by all of the methods at your disposal:

(a) analytically
(b) graphically
(c) numerically
(d) by the Laplace transform

Find y when $t = 2$.

Infinite Series

When you have completed this chapter, you should be able to:
- Use the partial sum test, the limit test, or the ratio test to determine if a series converges.
- Use the ratio test to determine the interval of convergence of a power series.
- Expand a function using a Maclaurin series or a Taylor series.
- Compute numerical values using a Maclaurin series or a Taylor series.
- Add, subtract, multiply, divide, differentiate, and integrate series.
- Describe waveforms using a Fourier series.

To find the sine of an angle or the logarithm of a number, you would probably use a calculator, a computer, or a table. But where did the table come from? And how can the chips in a calculator or computer find sines or logs when all they can do are the four arithmetic operations of addition, subtraction, multiplication, and division?

We can, in fact, find the sine of an angle x (in radians) from a polynomial, such as

$$\sin x \cong x - \frac{x^3}{6} + \frac{x^5}{120}$$

and the natural logarithm from

$$\ln x \cong (x - 1) - \frac{(x - 1)^2}{2} + \frac{(x - 1)^3}{3}$$

But where did these formulas come from? How accurate are they, and can the accuracy be improved? Are they good for all values of x or for just a small range of x? Can similar formulas be derived for other functions, such as e^x? In this chapter we answer such questions.

The technique of representing a function by a polynomial is called *polynomial approximation*, and each of these formulas is really the first several terms of an infinitely long series. We learn how to write Maclaurin and Taylor series for such functions and how to estimate and improve the accuracy of computation.

Why bother with infinite series when we can get these functions so easily using a calculator or computer? First, we might want to understand how a computer finds certain functions, even if

Look back at Chapter 25, where we examined infinite series by computer. Here we treat the same ideas analytically.

we never have to design the logic to do it. Perhaps more important, many numerical methods rely on polynomial approximation of functions. Finally, some limits, derivatives, or integrals are best found by this method.

We first give some tests to see if a series converges or diverges, because a series that diverges is of no use for computation. We then write polynomial approximations for some common functions and use them for computation, and we look at other uses for series. We conclude with infinite series of sine and cosine terms, called *Fourier series*. These series enable us to write series to represent periodic functions, such as waveforms; they are extremely useful in electronics and in mechanical vibrations.

37–1   Convergence and Divergence of Infinite Series

Convergence of a series is an extensive topic. We're only scratching the surface here.

In many applications, we first replace a given function by an infinite series and then do any required operations on the series rather than on the original function. But we cannot work with an infinite number of terms; nor do we want to work with many terms of the series. For practical applications, we want to represent our original function *by only the first few terms* of an infinite series. Thus, we need a series in which the terms decrease in magnitude rapidly and for which the sum of the first several terms is not too different from the sum of all of the terms of the series. Such a series is said to *converge*. A series that does not converge is said to *diverge*. In this section we give three tests for convergence: the partial sum test, the limit test, and the ratio test.

Partial Sum Test

In Chapter 25 we said that an infinite series

$$u_1 + u_2 + u_3 + \cdots + u_n + \cdots \qquad \textbf{441}$$

is the indicated sum of an infinite number of terms.

Now let S_n stand for the sum of the first n terms of an infinite series, so that

$$S_1 = u_1 \qquad S_2 = u_1 + u_2 \qquad S_3 = u_1 + u_2 + u_3 \qquad \text{etc.}$$

These are called *partial sums*. The infinite sequence of partial sums, $S_1, S_2, S_3, \ldots$, might or might not have a definite limit S as n becomes infinite. If a limit S exists, we call it the *sum of the series*. If an infinite series has a sum S, the series is said to *converge*. Otherwise, the series *diverges*.

| **Partial Sum Test** | $\lim\limits_{n \to \infty} S_n = S$ | **443** |

If the partial sums of an infinite series have a limit S, the series converges. Otherwise, the series diverges.

It is not hard to see that a finite series will have a finite sum. But is it possible that the sum of an infinite number of terms can be finite? Yes; however, we must look at the infinite sequence of partial sums $S_1, S_2, S_3, \ldots, S_n$ to determine if this sequence approaches a specific limit S, as in the following example.

◆◆◆ **Example 1:** Does the following geometric series converge?

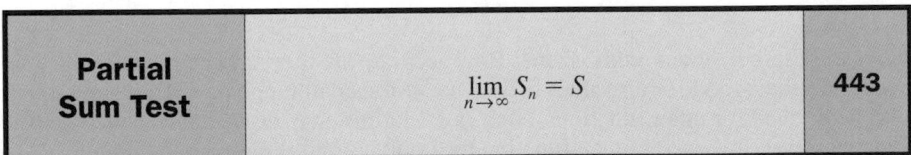

$$1 + \frac{1}{2} + \frac{1}{2^2} + \frac{1}{2^3} + \cdots + \frac{1}{2^{n-1}} + \cdots$$

Solution: The sum of n terms of a geometric series is given by Eq. 242. We substitute, with common ratio $r = \frac{1}{2}$ and $a = 1$.

$$S_n = \frac{a(1 - r^n)}{1 - r}$$

$$= \frac{1 - (1/2)^n}{1 - 1/2} = 2 - \frac{1}{2^{n-1}}$$

Since the limit of this sum is 2 as n approaches infinity, the given series converges. ◆◆◆

Limit Test

You might want to review limits (Sec. 27–1) before going further.

We have defined a convergent infinite series as one for which the partial sums approach zero. For this to happen, it is clear that the terms themselves must get smaller as n increases, and they must approach zero as n becomes infinite. We may use this fact, which we call the *limit test*, as our second test for convergence.

Limit Test	$\lim_{n \to \infty} u_n = 0$	**442**

If the terms of an infinite series do not approach zero as n approaches infinity, the series diverges.

However, if the terms do approach zero, we cannot say for sure that the series converges. That the terms should approach zero is a *necessary* condition for convergence, but not a *sufficient* condition for convergence.

◆◆◆ **Example 2:** Does the infinite series

$$1 + \frac{2}{3} + \frac{3}{5} + \cdots + \frac{n}{2n - 1} + \cdots$$

converge or diverge?

Solution: The series diverges because the general term does not approach zero as n approaches infinity.

$$\lim_{n \to \infty} \frac{n}{2n - 1} = \lim_{n \to \infty} \frac{1}{2 - 1/n} = \frac{1}{2}$$

◆◆◆

◆◆◆ **Example 3:** The harmonic series

$$1 + \frac{1}{2} + \frac{1}{3} + \cdots + \frac{1}{n} + \cdots$$

does have terms that approach zero as n approaches infinity. However, we cannot tell from this information alone whether this series converges. In fact, it does *not* converge, as we found by computer in Chapter 25, Example 9. ◆◆◆

Ratio Test

A third important test for convergence is the *ratio test*. We look at the ratio of two successive terms of the series as n increases. If the limit of that ratio has an absolute value less than 1, the series converges. If the limit is greater than 1, the series diverges. However, a limit of 1 gives insufficient information.

The ratio test is also called the *Cauchy ratio test*. There are many tests for convergence that we cannot give here, but the ratio test will be the most useful for our later work.

Ratio Test	If $\qquad\qquad\qquad \lim\limits_{n \to \infty}\left\|\dfrac{u_{n+1}}{u_n}\right\|$ (a) is less than 1, the series converges. (b) is greater than 1, the series diverges. (c) is equal to 1, the test fails to give conclusive information.	**444**

◆◆◆ Example 4: Use the ratio test on

$$\frac{1}{e} + \frac{2}{e^2} + \frac{3}{e^3} + \cdots + \frac{n}{e^n} + \cdots$$

to determine if this series converges or diverges.

Solution: By the ratio test,

$$\frac{u_{n+1}}{u_n} = \frac{n+1}{e^{n+1}} \div \frac{n}{e^n}$$

$$= \frac{n+1}{e^{n+1}} \cdot \frac{e^n}{n}$$

$$= \frac{n+1}{ne} = \frac{1 + 1/n}{e}$$

or

$$\lim_{n \to \infty}\left|\frac{u_{n+1}}{u_n}\right| = \frac{1}{e}$$

Since $1/e$ is less than 1, the series converges. ◆◆◆

Exercise 1 ◆ Convergence and Divergence of Infinite Series

Partial Sum Test

Use the partial sum test to determine if each series converges or diverges.

  1. The geometric series $9 + 3 + 1 + \cdots$
  2. The geometric series $100 + 10 + 1 + \cdots$
  3. The geometric series $2 + 2.02 + 2.0402 + \cdots$
  4. The geometric series $8.00 + 7.20 + 6.48 + \cdots$

Limit Test

Use the limit test to decide, if possible, whether each series converges or diverges.

  5. The geometric series $16 + 8 + 4 + 2 + \cdots$
  6. The geometric series $1 + 1.05 + 1.1025 + \cdots$
  7. The geometric series $100 + 50 + 25 + \cdots$
  8. The harmonic series $1/6 + 1/8 + 1/10 + \cdots$

Ratio Test

Use the ratio test to determine, if possible, if each series converges or diverges.

  9. $\dfrac{1}{2!} + \dfrac{2}{3!} + \dfrac{3}{4!} + \cdots + \dfrac{n}{(n+1)!} + \cdots$
 10. $1 + \dfrac{3}{2!} + \dfrac{5}{3!} + \dfrac{7}{4!} + \cdots + \dfrac{2n-1}{n!} + \cdots$

11. $\dfrac{3}{2} + \dfrac{9}{8} + \dfrac{9}{8} + \dfrac{81}{64} + \cdots + \dfrac{3^n}{n \cdot 2^n} + \cdots$

12. $3 - \dfrac{3^3}{3 \cdot 1!} + \dfrac{3^5}{5 \cdot 2!} - \dfrac{3^7}{7 \cdot 3!} + \cdots + \dfrac{3^{2n-1}(-1)^{n+1}}{(2n-1)(n-1)!} + \cdots$

13. $1 + \dfrac{1}{2} + \dfrac{1}{3} + \dfrac{1}{4} + \cdots + \dfrac{1}{n} + \cdots$

14. $1 + \dfrac{4}{7} + \dfrac{9}{49} + \dfrac{16}{343} + \cdots + \dfrac{n^2}{7^{n-1}} + \cdots$

37–2   Maclaurin Series

Power Series

An infinite series of the form

$$c_0 + c_1 x + c_2 x^2 + \cdots + c_n x^n + \cdots$$

in which the c's are constants and x is a variable, is called a *power series* in x. Note that this series is a function of x, whereas in our previous examples the terms did not contain x.

♦♦♦ **Example 5:** Is the series $x - \dfrac{x^2}{2} + \dfrac{x^3}{3} - \cdots + (-1)^{n-1} \dfrac{x^n}{n} + \cdots$ a power series?

Solution: Since it matches the form of a power series with $c_0 = 0$, $c_1 = 1$, $c_2 = -\frac{1}{2}$, and so on, the given series is a power series.   ♦♦♦

Another way of writing a power series is

$$c_0 + c_1(x - a) + c_2(x - a)^2 + \cdots + c_n(x - a)^n + \cdots$$

where a and the c's are constants.

Interval of Convergence

A power series, if it converges, does so only for specific values of x. The range of x for which a particular power series converges is called the *interval of convergence*. We find this interval by means of the ratio test.

♦♦♦ **Example 6:** Find the interval of convergence of the series given in Example 5.

Solution: Applying the ratio test,

$$\lim_{n \to \infty} \left| \frac{u_{n+1}}{u_n} \right| = \lim_{n \to \infty} \left| \frac{x^{n+1}}{n+1} \left(\frac{n}{x_n} \right) \right| = \lim_{n \to \infty} \left| \frac{n}{n+1}(x) \right| = |x|$$

It is usual to test for convergence at the endpoints of the interval as well, but this requires tests that we have not covered. So in Example 6 we do not test for $x = 1$ or $x = -1$.

Thus, the series converges when x lies between 1 and -1 and diverges when x is greater than 1 or less than -1. Our interval of convergence is thus $-1 < x < 1$.   ♦♦♦

Representing a Function by a Series

We know that it is possible to represent certain functions by an infinite power series. Take the function

$$f(x) = \frac{1}{1 + x}$$

By ordinary division (try it), we get the infinite power series

$$(1 + x)^{-1} = 1 - x + x^2 - x^3 + \cdots$$

Although we do not show it here, this power series can be shown to converge when x is less than 1. It's possible to represent many (but not all) functions by a power series. We will now demonstrate this idea by using a graphing calculator or some other graphing utility.

◆◆◆ **Example 7:** The exponential function $y = e^x$ is graphed as a solid line in Fig. 37–1(a). Also graphed as a dashed line is the series

$$y = 1 + x + x^2/2!$$

Notice that the graph of the series approximates the graph of the exponential function. If we add another term to the series, getting

$$y = 1 + x + x^2/2! + x^3/3!$$

its graph, shown dotted, is an even better match to the exponential function. If we add two more terms to the series, its graph, shown in Fig. 37–1(b), matches still better. Clearly, the more terms we use, the better is the match.

This series is listed as Number 25 in Table 37–1. It is also listed as Eq. 207.

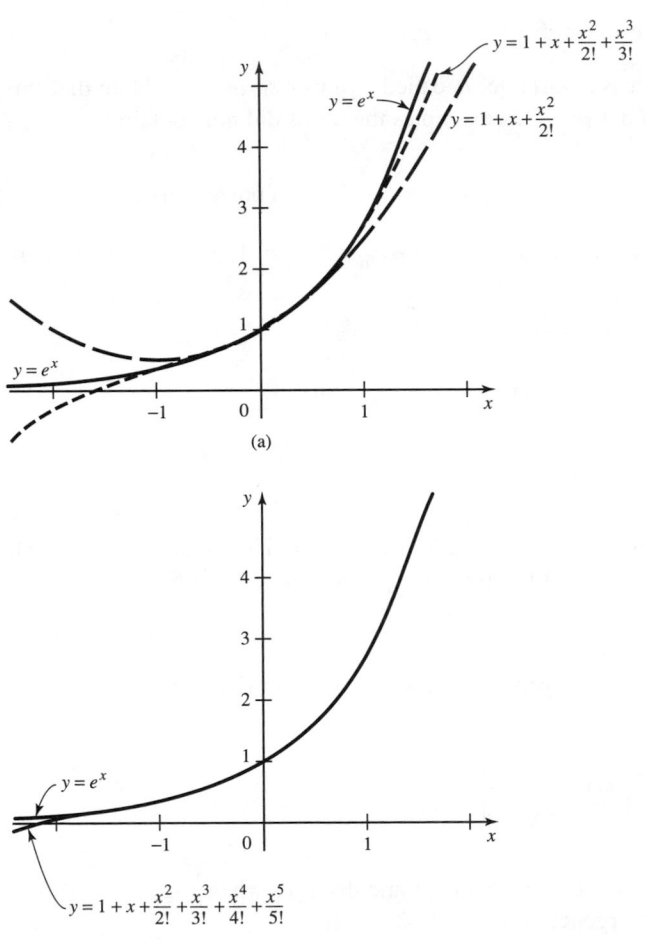

FIGURE 37-1   ◆◆◆

Writing a Maclaurin Series

We require that the function being represented is continuous and differentiable over the interval of interest. Later in this chapter we use Fourier series to represent discontinuous functions.

Let us assume that a certain function $f(x)$ can be represented by an infinite power series, so that

$$f(x) = c_0 + c_1 x + c_2 x^2 + c_3 x^3 + \cdots + c_{n-1} x^{n-1} + \cdots \qquad (1)$$

We now evaluate the constants in this equation. We get c_0 by setting x to zero. Thus,

$$f(0) = c_0$$

Next, taking the derivative of Eq. (1) gives

$$f'(x) = c_1 + 2c_2x + 3c_3x^2 + 4c_4x^3 + \cdots$$

from which we find c_1.

$$f'(0) = c_1$$

We continue taking derivatives and evaluating each at $x = 0$. Thus,

$$f''(x) = 2c_2 + 3(2)c_3x + 4(3)c_4x^2 + \cdots$$
$$f''(0) = 2c_2$$
$$f'''(x) = 3(2)c_3 + 4(3)(2)c_4x + \cdots$$
$$f'''(0) = 3(2)c_3$$

and so on. The constants in (1) are thus

$$c_0 = f(0) \qquad c_1 = f'(0) \qquad c_2 = \frac{f''(0)}{2} \qquad c_3 = \frac{f'''(0)}{3(2)}$$

and, in general,

$$c_n = \frac{f^{(n)}(0)}{n!}$$

Substituting the c's back into (1) gives us what is called a *Maclaurin series*.

This series is named for a Scottish mathematician and physicist, Colin Maclaurin (1698–1746).

Maclaurin Series	$f(x) = f(0) + f'(0)x + \dfrac{f''(0)}{2!}x^2 + \cdots + \dfrac{f^{(n)}(0)}{n!}x^n + \cdots$	**445**

The Maclaurin series expresses a function $f(x)$ in terms of an infinite power series whose nth coefficient is the nth derivative of $f(x)$, evaluated at $x = 0$, divided by $n!$.

Notice that in order for us to write a Maclaurin series for a function, the function and its derivatives must exist at $x = 0$.

We now apply the Maclaurin series expansion to two important functions mentioned previously: the sine and the natural logarithm.

♦♦♦ Example 8: Write a Maclaurin series for the function

$$f(x) = \sin x$$

Solution: We take successive derivatives and evaluate each at $x = 0$.

$$
\begin{aligned}
f(x) &= \sin x & f(0) &= 0 \\
f'(x) &= \cos x & f'(0) &= 1 \\
f''(x) &= -\sin x & f''(0) &= 0 \\
f'''(x) &= -\cos x & f'''(0) &= -1 \\
f^{(iv)}(x) &= \sin x & f^{(iv)}(0) &= 0
\end{aligned}
$$

Substituting into Eq. 445, we have the following:

$\sin x = x - \dfrac{x^3}{3!} + \dfrac{x^5}{5!} - \dfrac{x^7}{7!} + \cdots$	**215**

♦♦♦

This Maclaurin series is also called a *Taylor series* because Maclaurin series are special cases of the Taylor series. This series is the first listed in Table 37–1. Put a bookmark at this table, for we'll be referring to it often in this chapter.

TABLE 37–1   Taylor series expansions for some common functions. The column "Expanded About" gives the value about which the series is expanded. If no value is given, the series has been expanded about $a = 0$, and the series is also called a Maclaurin series.

	Expanded About:
1. $\sin x = x - \dfrac{x^3}{3!} + \dfrac{x^5}{5!} - \dfrac{x^7}{7!} + \cdots$ (Eq. 215)	
2. $\sin x = \dfrac{1}{2} + \dfrac{\sqrt{3}}{2}\left(x - \dfrac{\pi}{6}\right) - \dfrac{1}{4}\left(x - \dfrac{\pi}{6}\right)^2 - \dfrac{\sqrt{3}}{12}\left(x - \dfrac{\pi}{6}\right)^3 + \cdots$	$a = \dfrac{\pi}{6}$
3. $\cos x = 1 - \dfrac{x^2}{2!} + \dfrac{x^4}{4!} - \dfrac{x^6}{6!} + \cdots$ (Eq. 216)	
4. $\cos x = \dfrac{\sqrt{2}}{2}\left[1 - \left(x - \dfrac{\pi}{4}\right) - \dfrac{(x - \pi/4)^2}{2} + \cdots\right]$	$a = \dfrac{\pi}{4}$
5. $\tan x = x + \dfrac{x^3}{3} + \dfrac{2x^5}{15} + \dfrac{17x^7}{315} + \cdots$	
6. $\tan x = 1 + 2\left(x - \dfrac{\pi}{4}\right) + 2\left(x - \dfrac{\pi}{4}\right)^2 + \cdots$	$a = \dfrac{\pi}{4}$
7. $\sec x = 1 + \dfrac{x^2}{2} + \dfrac{5x^4}{24} + \cdots$	
8. $\sec x = 2 + 2\sqrt{3}\left(x - \dfrac{\pi}{3}\right) + 7\left(x - \dfrac{\pi}{3}\right)^2 + \cdots$	$a = \dfrac{\pi}{3}$
9. $\sin^2 x = x^2 - \dfrac{x^4}{3} + \dfrac{2x^6}{45} - \dfrac{x^8}{315} + \cdots$	
10. $\cos^2 x = 1 - x^2 + \dfrac{x^4}{3} - \dfrac{2x^6}{45} + \cdots$	
11. $\sin 2x = 2x - \dfrac{2^3 x^3}{3!} + \dfrac{2^5 x^5}{5!} - \dfrac{2^7 x^7}{7!} + \cdots$	
12. $\cos 2x = 1 - 2x^2 + \dfrac{2^4 x^4}{4!} - \dfrac{2^6 x^6}{6!} + \cdots$	
13. $\sin x^2 = x^2 - \dfrac{x^6}{3!} + \dfrac{x^{10}}{5!} - \cdots$	
14. $\cos x^2 = 1 - \dfrac{x^4}{2!} + \dfrac{x^8}{4!} - \dfrac{x^{12}}{6!} + \cdots$	
15. $\sec^2 x = 1 + x^2 + 2x^4/3 + \cdots$	
16. $\sinh x = \dfrac{e^x - e^{-x}}{2} = x + \dfrac{x^3}{3!} + \dfrac{x^5}{5!} + \dfrac{x^7}{7!} + \cdots$	
17. $\cosh x = \dfrac{e^x + e^{-x}}{2} = 1 + \dfrac{x^2}{2!} + \dfrac{x^4}{4!} + \dfrac{x^6}{6!} + \cdots$	
18. $\dfrac{1}{x} = 1 - (x - 1) + (x - 1)^2 - (x - 1)^3 + \cdots$	$a = 1$
19. $\dfrac{1}{x} = \dfrac{1}{2} - \dfrac{x - 2}{4} + \dfrac{(x - 2)^2}{8} + \cdots$	$a = 2$
20. $\dfrac{1}{1 + x} = 1 - x + x^2 - x^3 + \cdots$	
21. $\sqrt{x} = 1 + \dfrac{1}{2}(x - 1) - \dfrac{1}{8}(x - 1)^2 + \cdots$	$a = 1$

sinh x and cosh x are called *hyperbolic functions*.

TABLE 37–1 *Continued*

	Expanded About:
22. $\sqrt[3]{x} = -1 + \dfrac{x+1}{3} + \dfrac{(x+1)^2}{9} + \dfrac{5(x+1)^3}{81} + \cdots$	$a = -1$
23. $\sqrt{1+x} = 1 + \dfrac{x}{2} - \dfrac{x^2}{8} + \dfrac{x^3}{16} - \cdots$	
24. $\sqrt{x^2 + 1} = \sqrt{2} + \dfrac{(x-1)}{\sqrt{2}} + \dfrac{(x-1)^2}{4\sqrt{2}} + \cdots$	$a = 1$
25. $e^x = 1 + x + \dfrac{x^2}{2!} + \dfrac{x^3}{3!} + \cdots$ (Eq. 207)	
26. $e^x = e\left[1 + (x-1) + \dfrac{(x-1)^2}{2!} + \cdots\right]$	$a = 1$
27. $e^{-x} = 1 - x + \dfrac{x^2}{2!} - \dfrac{x^3}{3!} + \cdots$	
28. $e^{-x} = e\left[1 - (x+1) + \dfrac{(x+1)^2}{2!} + \cdots\right]$	$a = -1$
29. $e^{2x} = 1 + 2x + 2x^2 + \dfrac{4x^3}{3} + \cdots$	
30. $e^{-2x} = 1 - 2x + 2x^2 - \dfrac{4x^3}{3} + \cdots$	
31. $e^{-x^2} = 1 - x^2 + \dfrac{x^4}{2!} - \dfrac{x^6}{3!} + \cdots$	
32. $xe^{2x} = x + 2x^2 + 2x^3 + \cdots$	
33. $xe^{-2x} = x - 2x^2 + 2x^3 - \cdots$	
34. $x^2 e^{2x} = x^2 + 2x^3 + 2x^4 + \cdots$	
35. $e^{jx} = 1 + jx - \dfrac{x^2}{2!} - j\dfrac{x^3}{3!} + \dfrac{x^4}{4!} + \cdots$	
36. $e^x \sin x = x + x^2 + \dfrac{x^3}{3} + \cdots$	
37. $e^x \cos x = 1 + x - \dfrac{x^3}{3} - \dfrac{x^4}{6} + \cdots$	
38. $e^{-x} \cos 2x = 1 - x - \dfrac{3x^2}{2} + \dfrac{11x^3}{6} + \cdots$	
39. $e^{\sin x} = 1 + \sin x + \dfrac{\sin^2 x}{2!} + \dfrac{\sin^3 x}{3!} + \cdots$	
40. $\ln(1+x) = x - \dfrac{x^2}{2} + \dfrac{x^3}{3} - \dfrac{x^4}{4} + \cdots$	
41. $\ln(1-x) = -x - \dfrac{x^2}{2} - \dfrac{x^3}{3} - \dfrac{x^4}{4} - \cdots$	
42. $\ln(1-x^2) = -x^2 - \dfrac{x^4}{2} - \dfrac{x^6}{3} - \dfrac{x^8}{4} - \cdots$	
43. $\ln x = (x-1) - \dfrac{(x-1)^2}{2!} + \dfrac{2(x-1)^3}{3!} - \dfrac{6(x-1)^4}{4!} + \cdots$	$a = 1$
44. $\ln \cos x = -\dfrac{x^2}{2} - \dfrac{x^4}{12} - \dfrac{x^6}{45} - \cdots$	
45. $b^x = 1 + x \ln b + \dfrac{(x \ln b)^2}{2!} + \dfrac{(x \ln b)^3}{3!} + \cdots$	$b > 0$

Now we turn to the natural logarithm.

◆◆◆ **Example 9:** Write a Maclaurin series for $f(x) = \ln x$.

Solution: We proceed as before.

$$f(x) = \ln x \qquad \text{but } f(0) \text{ is undefined!}$$

Therefore, we cannot write a Maclaurin series for $\ln x$. In order to write a Maclaurin series for a function, *the function and all its derivatives must exist at $x = 0$.* In the next section, we'll see how to write an expression for $\ln x$ using a Taylor series.   ◆◆◆

Computing with Maclaurin Series

Maclaurin series are useful in computations because after we have represented a function by a Maclaurin series, we can find the value of the function simply by substituting into the series.

◆◆◆ **Example 10:** Find the sine of 0.5 rad using the first three terms of the Maclaurin series. Work to seven decimal places.

Solution: Substituting into Eq. 215, we find that

$$\sin 0.5 = 0.5 - \frac{(0.5)^3}{6} + \frac{(0.5)^5}{120}$$
$$= 0.5 - 0.020\,8333 + 0.000\,2604 = 0.479\,4271$$

By calculator (which we assume to be correct to seven decimal places), the sine of 0.5 rad is 0.479 4255. Our answer is thus too high by 0.000 0016, or about 0.0003%.   ◆◆◆

The error caused by discarding terms of a series is called the *truncation error*.

Accuracy

In Example 10 we could estimate the truncation error by finding the exact value by calculator. But what if we were designing a computer to calculate sines and must know *in advance* what the accuracy is?

Table 37–2 Sin x calculated by Maclaurin series.

Angle	1 Term	2 Terms	3 Terms	4 Terms	By Computer
0	0.000 000	0.000 000	0.000 000	0.000 000	0.000 000
10	0.174 533	0.173 647	0.173 648	0.173 648	0.173 648
20	0.349 066	0.341 977	0.342 020	0.342 020	0.342 020
30	0.523 599	0.499 674	0.500 002	0.500 000	0.500 000
40	0.698 132	0.641 422	0.642 804	0.642 788	0.642 788
50	0.872 665	0.761 903	0.766 120	0.766 044	0.766 044
60	1.047 198	0.855 801	0.866 295	0.866 021	0.866 025
70	1.221 730	0.917 799	0.940 482	0.939 676	0.939 693
80	1.396 263	0.942 582	0.986 806	0.984 753	0.984 808
90	1.570 796	0.924 832	1.004 525	0.999 843	1.000 000

Because the sine function is periodic, we need to calculate sines of angles only up to 90°. Sines of larger angles can be found from these.

Table 37–2 shows the sines of the angles from zero to 90°, computed using one, two, three, and four terms of the Maclaurin series (Eq. 215). In the last column is the sine, given by the computer, which we assume to be accurate to the number of places given. Notice that the values obtained by series differ from the computer values. The absolute value of the percent error of each is given in Table 37–3 and plotted in Fig. 37–2.

Note that the error decreases rapidly as more terms of the series are used. Also note that the series gives the exact answer at $x = 0$ regardless of the number of terms, and that the error increases as we go farther from zero. With a Maclaurin series, we say that the function is *expanded about $x = 0$.*

Table 37-3 Percent error.

Angle	1 Term	2 Terms	3 Terms	4 Terms
0	0.0000	0.0000	0.0000	0.0000
10	0.5095	0.0008	0.0000	0.0000
20	2.0600	0.0126	0.0000	0.0000
30	4.7198	0.0652	0.0004	0.0000
40	8.6100	0.2125	0.0025	0.0000
50	13.9183	0.5407	0.0099	0.0001
60	20.9200	1.1806	0.0312	0.0005
70	30.0138	2.3298	0.0840	0.0018
80	41.7803	4.2877	0.2029	0.0056
90	57.0796	7.5158	0.4525	0.0157

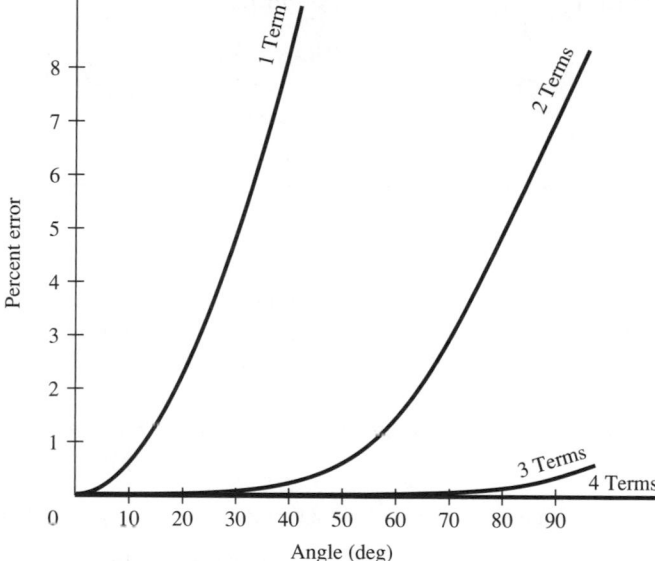

FIGURE 37-2 Percent error in the sine of an angle using one, two, three, and four terms of a Maclaurin series.

In the next section we'll be able to expand a function about any point $x = a$ using a Taylor series, and the best accuracy, as here, will be near that point.

Exercise 2 ◆ Maclaurin Series

Use the ratio test to find the interval of convergence of each power series.

1. $1 + x + x^2 + \cdots + x^{n-1} + \cdots$

2. $1 + \dfrac{2x}{5} + \dfrac{3x^2}{5^2} + \cdots + \dfrac{nx^{n-1}}{5^{n-1}} + \cdots$

3. $x + 2!x^2 + 3!x^3 + \cdots + n!x^n + \cdots$

4. $2x + 4x^2 + 8x^3 + \cdots + 2^n x^n + \cdots$

5. $x - \dfrac{x^3}{3} + \dfrac{x^5}{5} - \dfrac{x^7}{7} + \cdots$

6. $\dfrac{x}{1!} + \dfrac{x^2}{2!} + \cdots + \dfrac{x^n}{n!} + \cdots$

Verify each Maclaurin series expansion from Table 37–1. Starting from $f(x)$, derive the Maclaurin series. Compare your answer with the series in Table 37–1.

7. Series 20	**8.** Series 23	**9.** Series 25
10. Series 27	**11.** Series 31	**12.** Series 32
13. Series 40	**14.** Series 11	**15.** Series 3
16. Series 5	**17.** Series 7	**18.** Series 14
19. Series 9	**20.** Series 44	

Using three terms of the appropriate Maclaurin series, compute each number to four significant digits.

21. $\sqrt{e}$	**22.** $\sin 1$	**23.** $\ln 1.2$
24. $\cos 8°$	**25.** $\sqrt{1.1}$	**26.** $\sec \dfrac{1}{4}$

37–3   Taylor Series

Writing a Taylor Series

The Taylor series is named for Brook Taylor (1685–1731), an English analyst, geometer, and philosopher.

We have seen that certain functions, such as $f(x) = \ln x$, cannot be represented by a Maclaurin series. Further, a Maclaurin series might converge too slowly to be useful. However, if we cannot represent a function by a power series in x (a Maclaurin series), we can often represent the function by a power series in $(x - a)$, called a *Taylor series*.

Let us assume that a function $f(x)$ can be represented by a power series in $(x - a)$, where a is some constant.

$$f(x) = c_0 + c_1(x - a) + c_2(x - a)^2 + \cdots + c_n(x - a)^n + \cdots \tag{1}$$

Since the Maclaurin series is a special case of the Taylor series, they are both referred to as *Taylor series*.

As with the Maclaurin series, we take successive derivatives. We find the c's by evaluating each derivative at $x = a$.

$$f(x) = c_0 + c_1(x - a) + c_2(x - a)^2 + \cdots \qquad\qquad f(a) = c_0$$

$$f'(x) = c_1 + 2c_2(x - a) + \cdots \qquad\qquad f'(a) = c_1$$

$$f''(x) = 2c_2 + 3(2)c_3(x - a) + \cdots \qquad\qquad f''(a) = 2c_2$$

$$f'''(x) = 3(2)c_3 + 4(3)(2)c_4(x - a) + \cdots \qquad\qquad f'''(a) = 3 \cdot 2 \cdot c_3$$

$$\vdots \qquad\qquad\qquad\qquad\qquad \vdots$$

$$f^{(n)}(x) = n!c_n + \cdots + \qquad\qquad f^{(n)}(a) = n!c_n$$

The values of c obtained are substituted back into Eq. (1), and we get the following standard form of the Taylor series:

Taylor Series	$f(x) = f(a) + f'(a)(x - a) + \dfrac{f''(a)}{2!}(x - a)^2 + \cdots + \dfrac{f^{(n)}(a)}{n!}(x - a)^n + \cdots$	**446**

Notice that when $a = 0$, this equation becomes identical to Eq. 445 for a Maclaurin series.

◆◆◆ **Example 11:** Write four terms of a Taylor series for $f(x) = \ln x$ expanded about $a = 1$.

Solution: We take derivatives and evaluate each at $x = 1$.

$$f(x) = \ln x \qquad f(1) = 0$$

$$f'(x) = \frac{1}{x} \qquad f'(1) = 1$$

$$f''(x) = -x^{-2} \qquad f''(1) = -1$$

$$f'''(x) = 2x^{-3} \qquad f'''(1) = 2$$

$$f^{(iv)}(x) = -6x^{-4} \qquad f^{(iv)}(1) = -6$$

Substituting into Eq. 446 gives

$$\ln x = 0 + 1(x - 1) + \frac{-1}{2}(x - 1)^2 + \frac{2}{6}(x - 1)^3 + \frac{-6}{24}(x - 1)^4 + \cdots$$

$$= (x - 1) - \frac{(x - 1)^2}{2} + \frac{(x - 1)^3}{3} - \frac{(x - 1)^4}{4} + \cdots$$

◆◆◆

This is Series 43 in Table 37–1. Recall that we were not able to write a Maclaurin series for ln x.

Computing with Taylor Series

As with the Maclaurin series, we simply substitute the required x into the proper Taylor series. For best accuracy, we expand the Taylor series about a point that is close to the value we wish to compute. Of course, a must be chosen such that $f(a)$ is known.

◆◆◆ **Example 12:** Evaluate ln 0.9 using four terms of a Taylor series. Work to seven decimal places.

Solution: We know from Example 11 that

$$\ln x = (x - 1) - \frac{(x - 1)^2}{2} + \frac{(x - 1)^3}{3} - \frac{(x - 1)^4}{4} + \cdots$$

We choose $a = 1$ since 1 is close to 0.9 and ln 1 is known. Substituting into the series with $x = 0.9$ and with

$$x - 1 = 0.9 - 1 = -0.1$$

gives

$$\ln 0.9 = -0.1 - \frac{(0.1)^2}{2} + \frac{(0.1)^3}{3} - \frac{(0.1)^4}{4} + \cdots$$

$$= -0.1 - \frac{0.01}{2} + \frac{0.001}{3} - \frac{0.0001}{4} + \cdots$$

$$= -0.1 - 0.005 - 0.000\,3333 - 0.000\,0250 = -0.105\,3583$$

So ln $0.9 \cong -0.105\,3583$. By calculator, ln $0.9 = -0.105\,3605$ to seven decimal places, a difference of about 0.0002% from our answer. ◆◆◆

◆◆◆ **Example 13:** Calculate the sines of the angles in 10° intervals from zero to 90°, using two, three, and four terms of a Taylor series expanded about 45°.

Solution: We take derivatives and evaluate each at $x = \pi/4$, remembering to work in radians.

$$f(x) = \sin x \qquad f\left(\frac{\pi}{4}\right) = \frac{\sqrt{2}}{2}$$

$$f'(x) = \cos x \qquad f'\left(\frac{\pi}{4}\right) = \frac{\sqrt{2}}{2}$$

$$f''(x) = -\sin x \qquad f''\left(\frac{\pi}{4}\right) = -\frac{\sqrt{2}}{2}$$

$$f'''(x) = -\cos x \qquad f'''\left(\frac{\pi}{4}\right) = -\frac{\sqrt{2}}{2}$$

Substituting into Eq. 446 gives

$$\sin x = \frac{\sqrt{2}}{2} + \frac{\sqrt{2}}{2}\left(x - \frac{\pi}{4}\right) - \frac{\sqrt{2}}{2(2!)}\left(x - \frac{\pi}{4}\right)^2 - \frac{\sqrt{2}}{2(3!)}\left(x - \frac{\pi}{4}\right)^3 + \cdots$$

A computer calculation of sin x, using two, three, and four terms of this series, is given in Table 37–4. The absolute value of the truncation error, as a percent, is shown in Table 37–5 and plotted in Fig. 37–3.

Table 37–4 Sin x calculated by Taylor series.

Angle	2 Terms	3 Terms	4 Terms	By Computer
0	0.151 746	−0.066 343	−0.009 247	0.000 000
10	0.275 160	0.143 229	0.170 093	0.173 648
20	0.398 573	0.331 262	0.341 052	0.342 020
30	0.521 987	0.497 754	0.499 869	0.500 000
40	0.645 400	0.642 708	0.642 786	0.642 788
50	0.768 813	0.766 121	0.766 043	0.766 044
60	0.892 227	0.867 995	0.865 880	0.866 025
70	1.015 640	0.948 329	0.938 539	0.939 693
80	1.139 054	1.007 123	0.980 259	0.984 808
90	1.262 467	1.044 378	0.987 282	1.000 000

Table 37–5 Percent error.

Angle	2 Terms	3 Terms	4 Terms
10	58.4582	17.5176	2.0473
20	16.5350	3.1456	0.2832
30	4.3973	0.4491	0.0262
40	0.4064	0.0124	0.0003
50	0.3615	0.0100	0.0002
60	3.0255	0.2274	0.0168
70	8.0822	0.9190	0.1228
80	15.6625	2.2659	0.4619
90	26.2467	4.4378	1.2718

Note that we get the best accuracy near the point about which the series is expanded ($x = 45°$). Also note that at other values, the Taylor series gives *less accurate* results than the Maclaurin series with the same number of terms (compare Figs. 37–2 and 37–3). The reason is that the Maclaurin series converges much faster than the Taylor series.

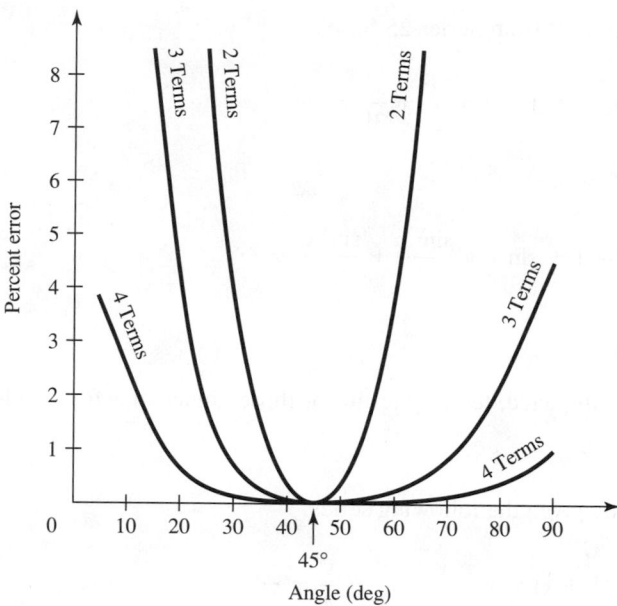

FIGURE 37–3 Percent error in sin x, computed with a Taylor series expanded about 45°.

Exercise 3 ◆ Taylor Series

Verify each Taylor series from Table 37–1. Starting from $f(x)$, derive the Taylor series. Compare your answer with the series in Table 37–1.

1. Series 18 **2.** Series 2 **3.** Series 4 **4.** Series 6

5. Series 8 **6.** Series 21 **7.** Series 22 **8.** Series 26

9. Series 28 **10.** Series 43

Compute each value using three terms of the appropriate Taylor series. Compare your answer with that from a calculator.

11. $\dfrac{1}{1.1}$ **12.** $\sin 32°$

13. $\cos 46.5°$ (use $\cos 45° = 0.70717$) **14.** $\tan 44.2°$

15. $\sec 61.4°$ **16.** $\sqrt{1.24}$

17. $\sqrt[3]{-0.8}$ **18.** $e^{1.1}$

19. $\ln(1.25)$

20. Calculate $\sin 65°$ using **(a)** three terms of a Maclaurin series (Series 1), **(b)** three terms of a Taylor series expanded about $\pi/6$ (Series 2), and **(c)** a calculator. Compare the results of each.

37–4 Operations with Power Series

To expand a function in a power series, we can write a Maclaurin or a Taylor series by the methods of Sec. 37–2 or 37–3. A faster way to get a new series is to modify or combine, when possible, series already derived.

Substitution

Given a series for $f(x)$, a new series can be obtained simply by substituting a different expression for x.

◆◆◆**Example 14:** We get Series 39 for $e^{\sin x}$ from Series 25 for e^x.

$$\text{Series 25: } e^x = 1 + x + \frac{x^2}{2!} + \frac{x^3}{3!} + \cdots$$

Substituting $\sin x$ for x gives

$$\text{Series 39: } e^{\sin x} = 1 + \sin x + \frac{\sin^2 x}{2!} + \frac{\sin^3 x}{3!} + \cdots$$

◆◆◆

Addition and Subtraction

Two power series may be added or subtracted, term by term, for those values of x for which both series converge.

◆◆◆**Example 15:** Suppose that we are given the following series:

$$\text{Series 40: } \ln(1 + x) = x - \frac{x^2}{2} + \frac{x^3}{3} - \frac{x^4}{4} + \cdots$$

$$\text{Series 41: } \ln(1 - x) = -x - \frac{x^2}{2} - \frac{x^3}{3} - \frac{x^4}{4} - \cdots$$

Adding these term by term gives

$$\text{Series 42: } \ln(1 - x^2) = -x^2 - \frac{x^4}{2} - \cdots$$

since $\ln(1 + x) + \ln(1 - x) = \ln[(1 + x)(1 - x)] = \ln(1 - x^2)$.

◆◆◆

Multiplication and Division

One power series may be multiplied or divided by another, term by term, for those values of x for which both converge, provided that division by zero does not occur.

> A series can also be multiplied by a constant or by a polynomial. Thus, a series for $x \ln x$ can be found by multiplying the series for $\ln x$ by x.

◆◆◆**Example 16:** Find the power series for $e^x \sin x$. Discard any terms of fifth degree or higher.

Solution: We will multiply Series 25 for e^x and Series 1 for $\sin x$ to get Series 36 for $e^x \sin x$.

We trim each series to terms of fourth degree or less and multiply term by term, discarding any product of fifth degree or higher.

$$e^x \sin x = \left(1 + x + \frac{x^2}{2!} + \frac{x^3}{3!} + \frac{x^4}{4!} + \cdots\right)\left(x - \frac{x^3}{3!} + \cdots\right) = x - \frac{x^3}{3!} + x^2 + \frac{x^3}{2!} + \cdots$$

$$\text{Series 36: } e^x \sin x = x + x^2 + \frac{x^3}{3} + \cdots$$

◆◆◆

Differentiation and Integration

A power series may be differentiated or integrated term by term for values of x within its interval of convergence (but not, usually, at the endpoints).

◆◆◆**Example 17:** Derive a power series for $\ln(1 + x)$.

Solution: Series 20,

$$\frac{1}{1 + x} = 1 - x + x^2 - x^3 + \cdots$$

converges for $|x| < 1$. Integrating term by term between the limits 0 and x gives

$$\int_0^x \frac{dx}{1+x} = \left[x - \frac{x^2}{2} + \frac{x^3}{3} - \frac{x^4}{4} + \cdots \right]_0^x$$

But, from Rule 7,

$$\int_0^x \frac{dx}{1+x} = \ln(1+x)$$

So we have derived Series 40.

$$\text{Series 40:} \quad \ln(1+x) = x - \frac{x^2}{2} + \frac{x^3}{3} - \frac{x^4}{4} + \cdots \qquad \blacklozenge\blacklozenge\blacklozenge$$

Evaluating Definite Integrals

If an expression cannot be integrated by our former methods, it still might be possible to evaluate its definite integral. We replace the function by a power series and then integrate term by term.

♦♦♦ **Example 18:** Find the approximate area under the curve $y = (\sin x)/x$ from $x = 0$ to 1.

Solution: The area is given by the definite integral

$$A = \int_0^1 \frac{\sin x}{x} dx$$

This expression is not in our table of integrals; nor do we have any earlier method of dealing with it. Instead, let us take the series for $\sin x$ (Eq. 215) and divide it by x, getting

$$\frac{\sin x}{x} = 1 - \frac{x^2}{3!} + \frac{x^4}{5!} - \cdots$$

We integrate the power series term by term, getting

$$A = \int_0^1 \frac{\sin x}{x} dx$$

$$= \left[x - \frac{x^3}{3(3!)} + \frac{x^5}{5(5!)} - \frac{x^7}{7(7!)} + \cdots \right]_0^1$$

$$\simeq 1 - 0.055\ 56 + 0.001\ 67 - 0.000\ 03 + \cdots$$

$$= 0.946\ 08 \qquad \blacklozenge\blacklozenge\blacklozenge$$

Euler's Formula

Here we show how Euler's formula (Eq. 232) can be derived by substituting into the power series for e^x. If we replace x by jx in Series 25, we get

Recall that we used differential equations to derive Euler's formula in Sec. 35–11.

$$e^{jx} = 1 + jx + \frac{(jx)^2}{2!} + \frac{(jx)^3}{3!} + \frac{(jx)^4}{4!} + \cdots$$

$$= 1 + jx + \frac{j^2 x^2}{2!} + \frac{j^3 x^3}{3!} + \frac{j^4 x^4}{4!} + \cdots$$

But by Eq. 217, $j^2 = -1$, $j^3 = -j$, $j^4 = 1$, and so on. So

$$e^{jx} = 1 + jx - \frac{x^2}{2!} - j\frac{x^3}{3!} + \frac{x^4}{4!} + \cdots$$

$$= \left[1 - \frac{x^2}{2!} + \frac{x^4}{4!} - \cdots\right] + j\left[x - \frac{x^3}{3!} + \frac{x^5}{5!} - \cdots\right]$$

However, the series in the brackets are those for $\cos x$ and $\sin x$ (Series 3 and 1). Thus,

Euler's Formula	$e^{jx} = \cos x + j \sin x$	**232**

Exercise 4 ◆ Operations with Power Series

Substitution

By substituting into Series 25 for e^x, verify the following.

1. $e^{-x} = 1 - x + \frac{x^2}{2!} - \frac{x^3}{3!} + \cdots$

2. $e^{2x} = 1 + 2x + 2x^2 + \frac{4x^3}{3} + \cdots$

3. $e^{-x^2} = 1 - x^2 + \frac{x^4}{2!} - \frac{x^6}{3!} + \cdots$

By substituting into Series 1 for $\sin x$, verify the following.

4. $\sin 2x = 2x - \frac{2^3x^3}{3!} + \frac{2^5x^5}{5!} - \frac{2^7x^7}{7!} + \cdots$

5. $\sin x^2 = x^2 - \frac{x^6}{3!} + \frac{x^{10}}{5!} - \cdots$

Addition and Subtraction

Add or subtract two earlier series to verify the following.

6. $\sec x + \tan x = 1 + x + \frac{x^2}{2} + \frac{x^3}{3} + \frac{5x^4}{24} + \cdots$
(Use Series 5 and 7.)

7. $\sin x + \cos x = 1 + x - \frac{x^2}{2!} - \frac{x^3}{3!} + \frac{x^4}{4!} + \cdots$
(Use Series 1 and 3.)

> sinh x is called the *hyperbolic sine* of x, and cosh x is called the *hyperbolic cosine* of x.

8. $\sinh x = \frac{e^x - e^{-x}}{2} = x + \frac{x^3}{3!} + \frac{x^5}{5!} + \frac{x^7}{7!} + \cdots$
(Use Series 25 and 27.)

9. $\cosh x = \frac{e^x + e^{-x}}{2} = 1 + \frac{x^2}{2!} + \frac{x^4}{4!} + \frac{x^6}{6!} + \cdots$
(Use Series 25 and 27.)

10. Using Series 12 for $\cos 2x$ and the identity $\sin^2 x = (1 - \cos 2x)/2$, show that
$$\sin^2 x = x^2 - \frac{x^4}{3} + \frac{2x^6}{45} - \frac{x^8}{315} + \cdots.$$

11. Using Series 12 for $\cos 2x$ and the identity $\cos^2 x = (1 + \cos 2x)/2$, show that
$$\cos^2 x = 1 - x^2 + \frac{x^4}{3} - \frac{2x^6}{45} + \cdots.$$

Multiplication and Division

Multiply or divide earlier series to obtain the following.

12. $xe^{2x} = x + 2x^2 + 2x^3 + \cdots$ (Use Series 29.)

13. $xe^{-2x} = x - 2x^2 + 2x^3 - \cdots$ (Use Series 30.)

14. $e^x \cos x = 1 + x - \dfrac{x^3}{3} - \dfrac{x^4}{6} + \cdots$ (Use Series 3 and 25.)

15. $e^{-x} \cos 2x = 1 - x - \dfrac{3x^2}{2} + \dfrac{11x^3}{6} + \cdots$ (Use Series 12 and 27.)

16. $\dfrac{\sin x}{x} = 1 - \dfrac{x^2}{3!} + \dfrac{x^4}{5!} - \dfrac{x^6}{7!} + \cdots$ (Use Series 1.)

17. $\dfrac{e^x - 1}{x} = 1 + \dfrac{x}{2!} + \dfrac{x^2}{5!} + \dfrac{x^3}{4!} + \cdots$ (Use Series 25.)

18. $\tan x = \dfrac{\sin x}{\cos x} = x + \dfrac{x^3}{3} + \dfrac{2x^5}{15} + \dfrac{17x^7}{315} + \cdots$ (Use Series 1 and 3.)

19. Square the first four terms of Series 1 for $\sin x$ to obtain the series for $\sin^2 x$ given in problem 10.

20. Cube the first three terms of Series 3 for $\cos x$ to obtain the series for $\cos^3 x$.

Differentiation and Integration

21. Find the series for $\cos x$ by differentiating the series for $\sin x$.

22. Differentiate the series for $\tan x$ to show that Series 15 for $\sec^2 x$ is
$\sec^2 x = 1 + x^2 + 2x^4/3 + \cdots$.

23. Find the series for $1/(1 + x)$ by differentiating the series for $\ln(1 + x)$.

24. Find the series for $\ln \cos x$ by integrating the series for $\tan x$.

25. Find the series for $\ln(1 - x)$ by integrating the series for $1/(1 - x)$.

26. Find the series for $\ln(\sec x + \tan x)$ by integrating the series for $\sec x$.

Evaluate each integral by integrating the first three terms of each series.

27. $\displaystyle\int_1^2 \frac{e^x}{\sqrt{x}}\, dx$ **28.** $\displaystyle\int_0^1 \sin x^2\, dx$ **29.** $\displaystyle\int_0^{1/4} e^x \ln(x + 1)\, dx$

37–5 Fourier Series

Thus far, we have used power series to approximate continuous functions. But a power series cannot be written for a *discontinuous function* such as a square wave that is defined by a different expression in different parts of the interval. For these, we use a Fourier series. In fact, most periodic functions can be represented by an infinite series of sine and>or cosine terms if the amplitude and the frequency of each term are properly chosen. To demonstrate this, let us use our graphing calculator or other graphics utility to graph a periodic function as well as a Fourier series intended to represent it in the same viewing window.

◆◆◆ **Example 19:** The full-wave rectifier waveform is shown graphed in Fig. 37–4. To graph this waveform, you must take the absolute value of the cosine.

$$y = |\cos (x)|$$

Portions of the curve that were negative for the cosine function are now positive. On the same graph, we plot the first three terms of the Fourier series

$$y = \frac{2}{\pi}\left(1 + \frac{2}{3}\cos 2x - \frac{2}{15}\cos 4x + \cdots\right)$$

This waveform is listed as Number 12 in Table 37–6.

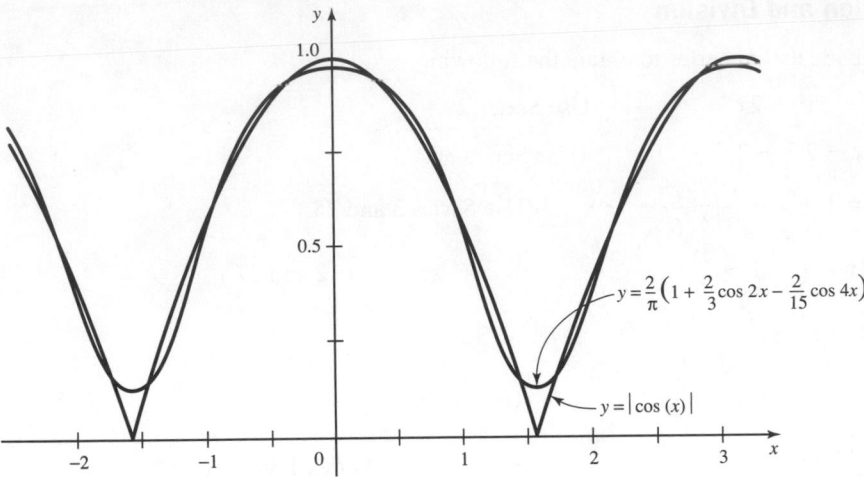

FIGURE 37–4   Full-wave rectifier wave form approximated by three terms of a Fourier series.

Notice how well the series matches the given curve. As with the Taylor series, the match becomes better when we use more terms.   ◆◆◆

In this section we will learn how to write such a series. Then we show how to use any symmetry in the waveform to reduce the work of computation. We conclude with a method for writing a Fourier series for a waveform for which we have no equation, such as from an oscillogram trace.

Definition of Fourier Series

A Fourier series is an infinite series of sine and cosine terms of the following form:

Fourier Series	$$f(x) = \frac{a_0}{2} + a_1 \cos x + a_2 \cos 2x + a_3 \cos 3x + \cdots + a_n \cos nx + \cdots$$ $$+ \, b_1 \sin x + b_2 \sin 2x + b_3 \sin 3x + \cdots + b_n \sin nx + \cdots$$	448

The Fourier series is named for a French analyst and mathematical physicist, Jean Baptiste Joseph Fourier (1768–1830).

Here $a_0/2$ is the *constant term*, $a_1 \cos x$ and $b_1 \sin x$ are called the *fundamental* components, and the remaining terms are called the *harmonics*: second harmonic, third harmonic, and so on. (We have let the constant term be $a_0/2$ instead of just a_0 because it will lead to a simpler expression for a_0 in a later derivation.)

The terms of the Fourier series are made up of sine and cosine functions. The fundamental terms ($a_1 \cos x$ and $b_1 \sin x$) have a period of 2π; that is, the values repeat every 2π radians. The values of the harmonics, having a higher frequency than the fundamental, will repeat more often than 2π. Thus, the fifth harmonic has a period of $2\pi/5$. Therefore, the values of the entire series repeat every 2π radians. Thus, the behavior of the entire series can be found by studying only one cycle, from 0 to 2π, or from $-\pi$ to π, or any other interval containing a full cycle. To write a Fourier series for a periodic function, we observe that the period of the Fourier series is the same as that of the waveform for which it is written.

The Fourier series is an infinite series, but as with Taylor series, we use only a few terms for a given computation.

Finding the Coefficients

To write a Fourier series, we must evaluate the coefficients: the a's and the b's. We first find a_0. To do this, let us integrate both sides of the Fourier series over one cycle. We can choose any two limits of integration that are spaced 2π radians apart, such as $-\pi$ and π.

$$\int_{-\pi}^{\pi} f(x)\, dx = \frac{a_0}{2} \int_{-\pi}^{\pi} dx + a_1 \int_{-\pi}^{\pi} \cos x\, dx$$

$$+ a_2 \int_{-\pi}^{\pi} \cos 2x\, dx + \cdots + a_n \int_{-\pi}^{\pi} \cos nx\, dx + \cdots$$

$$+ b_1 \int_{-\pi}^{\pi} \sin x\, dx + b_2 \int_{-\pi}^{\pi} \sin 2x\, dx + \cdots$$

$$+ b_n \int_{-\pi}^{\pi} \sin nx\, dx + \cdots \tag{1}$$

But when we evaluate a definite integral of a sine wave or a cosine wave over one cycle the result is zero, so Eq. (1) reduces to

$$\int_{-\pi}^{\pi} f(x)\, dx = \frac{a_0}{2} \int_{-\pi}^{\pi} dx = \frac{a_0}{2} x \Big|_{-\pi}^{\pi} = a_0 \pi$$

from which we obtain the following:

$$a_0 = \frac{1}{\pi} \int_{-\pi}^{\pi} f(x)\, dx \qquad \textbf{449}$$

Next we find the general a coefficient, a_n. Multiplying Eq. 448 by $\cos nx$ and integrating gives

$$\int_{-\pi}^{\pi} f(x) \cos nx\, dx = \frac{a_0}{2} \int_{-\pi}^{\pi} \cos nx\, dx + a_1 \int_{-\pi}^{\pi} \cos x \cos nx\, dx$$

$$+ a_2 \int_{-\pi}^{\pi} \cos 2x \cos nx\, dx + \cdots$$

$$+ a_n \int_{-\pi}^{\pi} \cos nx \cos nx\, dx + \cdots$$

$$+ b_1 \int_{-\pi}^{\pi} \sin x \cos nx\, dx + b_2 \int_{-\pi}^{\pi} \sin 2x \cos nx\, dx + \cdots$$

$$+ b_n \int_{-\pi}^{\pi} \sin nx \cos nx\, dx + \cdots \tag{2}$$

As before, when we evaluate the definite integral of the cosine function over one cycle, the result is zero. Further, the trigonometric identity (Eq. 181)

$$\cos A \cos B = 1/2 \cos(A - B) + 1/2 \cos(A + B)$$

shows that the product of two cosine functions can be written as the sum of two cosine functions. Thus, the integral of the product of two cosines is equal to the sum of the integrals of two cosines, each of which is zero, for $A \neq B$, when integrated over one cycle. Thus, Eq. (2) reduces to

$$\int_{-\pi}^{\pi} f(x) \cos nx\, dx = a_n \int_{-\pi}^{\pi} \cos nx \cos nx\, dx$$

$$+ b_1 \int_{-\pi}^{\pi} \sin x \cos nx\, dx$$

$$+ b_2 \int_{-\pi}^{\pi} \sin 2x \cos nx\, dx + \cdots$$

$$+ b_n \int_{-\pi}^{\pi} \sin nx \cos nx\, dx + \cdots \tag{3}$$

In a similar way, the identity (Eq. 182)

$$2 \sin A \cos B = \sin(A + B) + \sin(A - B)$$

shows that those terms containing the integral of the sine times the cosine will vanish. This leaves

$$\int_{-\pi}^{\pi} f(x) \cos nx \, dx = a_n \int_{-\pi}^{\pi} \cos^2 nx \, dx$$

$$= \frac{a_n}{n} \left[\frac{nx}{2} + \frac{\sin 2nx}{4} \right]_{-\pi}^{\pi} = \pi a_n$$

by Rule 17. Thus,

$$a_n = \frac{1}{\pi} \int_{-\pi}^{\pi} f(x) \cos nx \, dx \qquad \mathbf{450}$$

We now use a similar method to find the b coefficients. We multiply Eq. (1) by $\sin nx$ and get

$$\int_{-\pi}^{\pi} f(x) \sin nx \, dx = \frac{a_0}{2} \int_{-\pi}^{\pi} \sin nx \, dx + a_1 \int_{-\pi}^{\pi} \cos x \sin nx \, dx$$

$$+ a_2 \int_{-\pi}^{\pi} \cos 2x \sin nx \, dx + \cdots$$

$$+ a_n \int_{-\pi}^{\pi} \cos nx \sin nx \, dx + \cdots$$

$$+ b_1 \int_{-\pi}^{\pi} \sin x \sin nx \, dx + b_2 \int_{-\pi}^{\pi} \sin 2x \sin nx \, dx + \cdots$$

$$+ b_n \int_{-\pi}^{\pi} \sin nx \sin nx \, dx + \cdots \qquad (4)$$

The trigonometric identity (Eq. 180)

$$2 \sin A \sin B = \cos(A - B) - \cos(A + B)$$

states that the product of two sines is equal to the difference of two cosines, each of which has an integral that is zero over a full cycle, for $A \neq B$. Other integrals vanish for reasons given before, so (4) reduces to

$$\int_{-\pi}^{\pi} f(x) \sin nx \, dx = b_n \int_{-\pi}^{\pi} \sin^2 nx \, dx$$

$$= \frac{b_n}{n} \left[\frac{nx}{2} - \frac{\sin 2nx}{4} \right]_{-\pi}^{\pi}$$

$$= \pi b_n$$

by Rule 16. Thus,

$$b_n = \frac{1}{\pi} \int_{-\pi}^{\pi} f(x) \sin nx \, dx \qquad \mathbf{451}$$

The Fourier series is summarized as follows:

Fourier Series	A periodic function $f(x)$ can be replaced with the Fourier series $$f(x) = \frac{a_0}{2} + a_1 \cos x + a_2 \cos 2x + a_3 \cos 3x + \cdots$$ $$+ b_1 \sin x + b_2 \sin 2x + b_3 \sin 3x + \cdots$$	**448**
	where $$a_0 = \frac{1}{\pi} \int_{-\pi}^{\pi} f(x)\, dx$$	**449**
	$$a_n = \frac{1}{\pi} \int_{-\pi}^{\pi} f(x) \cos nx\, dx$$	**450**
	$$b_n = \frac{1}{\pi} \int_{-\pi}^{\pi} f(x) \sin nx\, dx$$	**451**

When we represent a wave by a Fourier series, we say we synthesize the wave by the series. We get a better synthesis of the original waveform by using more terms of the series.

◆◆◆ **Example 20:** Write a Fourier series to represent a square wave (Waveform 1 in Table 37–6).

Solution: The first step is to write one or more equations that describe the waveform over a full cycle. For Waveform 1, these are

$$f(x) = \begin{cases} -1 \text{ for } -\pi \le x < 0 \\ 1 \text{ for } 0 \le x < \pi \end{cases}$$

Since two equations describe $f(x)$ over a full cycle, we need two separate integrals to find each a or b: one with limits from $-\pi$ to 0, and another from 0 to π. We first find a_0 from Eq. 449.

$$a_0 = \frac{1}{\pi} \int_{-\pi}^{\pi} f(x)\, dx = \frac{1}{\pi} \int_{-\pi}^{0} (-1)\, dx + \frac{1}{\pi} \int_{0}^{\pi} (1)\, dx$$

$$= -\frac{x}{\pi} \Big|_{-\pi}^{0} + \frac{x}{\pi} \Big|_{0}^{\pi} = 0$$

We then get a_n from Eq. 450.

$$a_n = \frac{1}{\pi} \int_{-\pi}^{0} (-1) \cos nx\, dx + \frac{1}{\pi} \int_{0}^{\pi} (1) \cos nx\, dx$$

$$= -\frac{\sin nx}{n\pi} \Big|_{-\pi}^{0} + \frac{\sin nx}{n\pi} \Big|_{0}^{\pi} = 0$$

since the sine is zero for any angle that is a multiple of π. Thus, all of the a coefficients are zero, and our series contains no cosine terms. Then, by Eq. 451,

$$b_1 = \frac{1}{\pi} \int_{-\pi}^{0} (-1) \sin x\, dx + \frac{1}{\pi} \int_{0}^{\pi} (1) \sin x\, dx$$

$$= \frac{\cos x}{\pi} \Big|_{-\pi}^{0} - \frac{\cos x}{\pi} \Big|_{0}^{\pi} = \frac{4}{\pi}$$

$$b_2 = \frac{1}{\pi} \int_{-\pi}^{0} (-1) \sin 2x\, dx + \frac{1}{\pi} \int_{0}^{\pi} (1) \sin 2x\, dx$$

$$= \frac{\cos 2x}{2\pi} \Big|_{-\pi}^{0} - \frac{\cos 2x}{2\pi} \Big|_{0}^{\pi} = \frac{1}{2\pi} [(1-1) - (1-1)] = 0$$

Keep in mind that the limits can be any two values spaced 2π apart. We'll sometimes use limits of 0 and 2π. In Sec. 37–7 we'll rewrite the series for functions having any period.

In Sec. 37–6 we'll learn how to tell in advance if all the sine or cosine terms will vanish.

Table 37–6   Fourier series for common waveforms.

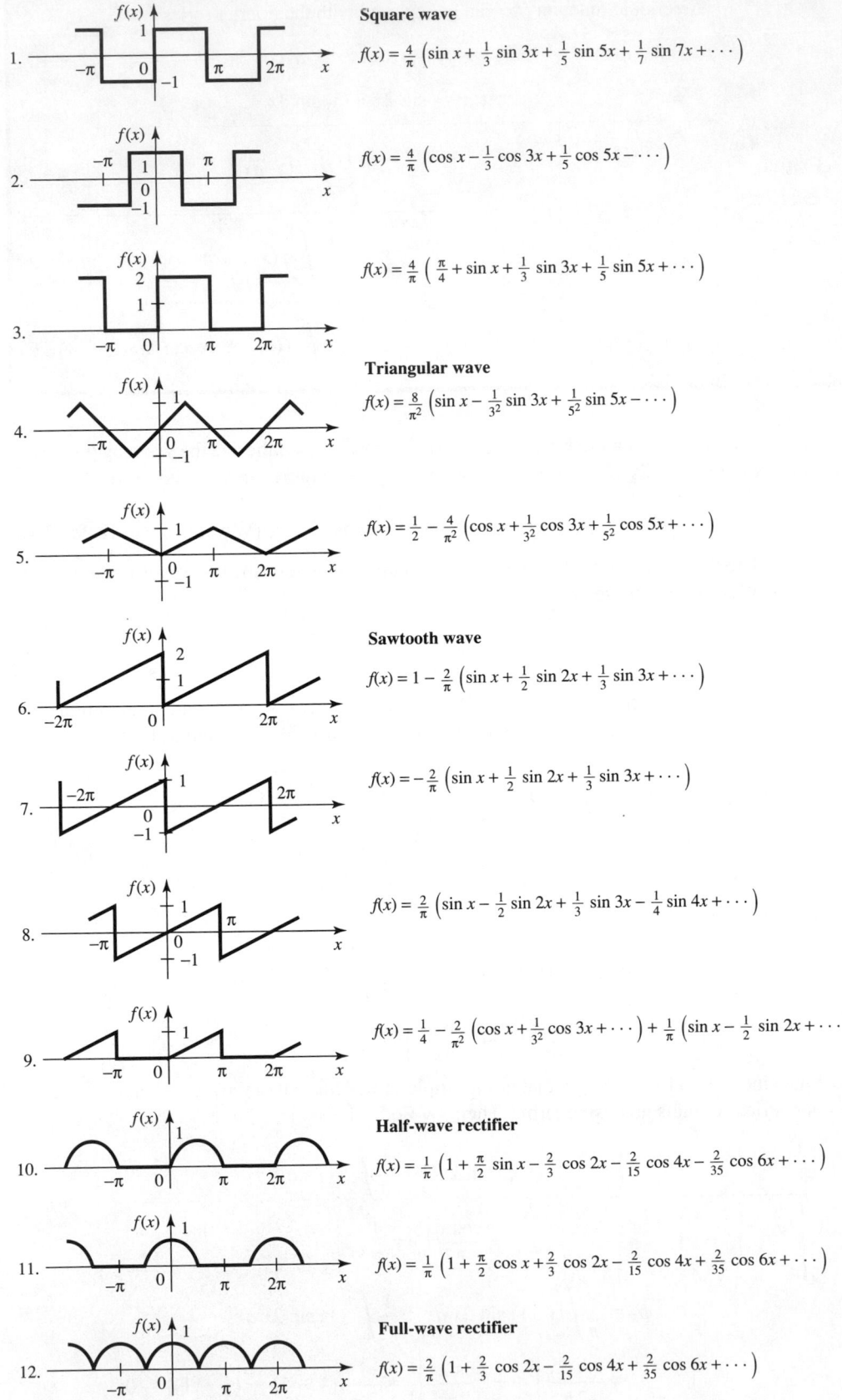

Square wave

1. $f(x) = \frac{4}{\pi}\left(\sin x + \frac{1}{3}\sin 3x + \frac{1}{5}\sin 5x + \frac{1}{7}\sin 7x + \cdots\right)$

2. $f(x) = \frac{4}{\pi}\left(\cos x - \frac{1}{3}\cos 3x + \frac{1}{5}\cos 5x - \cdots\right)$

3. $f(x) = \frac{4}{\pi}\left(\frac{\pi}{4} + \sin x + \frac{1}{3}\sin 3x + \frac{1}{5}\sin 5x + \cdots\right)$

Triangular wave

4. $f(x) = \frac{8}{\pi^2}\left(\sin x - \frac{1}{3^2}\sin 3x + \frac{1}{5^2}\sin 5x - \cdots\right)$

5. $f(x) = \frac{1}{2} - \frac{4}{\pi^2}\left(\cos x + \frac{1}{3^2}\cos 3x + \frac{1}{5^2}\cos 5x + \cdots\right)$

Sawtooth wave

6. $f(x) = 1 - \frac{2}{\pi}\left(\sin x + \frac{1}{2}\sin 2x + \frac{1}{3}\sin 3x + \cdots\right)$

7. $f(x) = -\frac{2}{\pi}\left(\sin x + \frac{1}{2}\sin 2x + \frac{1}{3}\sin 3x + \cdots\right)$

8. $f(x) = \frac{2}{\pi}\left(\sin x - \frac{1}{2}\sin 2x + \frac{1}{3}\sin 3x - \frac{1}{4}\sin 4x + \cdots\right)$

9. $f(x) = \frac{1}{4} - \frac{2}{\pi^2}\left(\cos x + \frac{1}{3^2}\cos 3x + \cdots\right) + \frac{1}{\pi}\left(\sin x - \frac{1}{2}\sin 2x + \cdots\right)$

Half-wave rectifier

10. $f(x) = \frac{1}{\pi}\left(1 + \frac{\pi}{2}\sin x - \frac{2}{3}\cos 2x - \frac{2}{15}\cos 4x - \frac{2}{35}\cos 6x + \cdots\right)$

11. $f(x) = \frac{1}{\pi}\left(1 + \frac{\pi}{2}\cos x + \frac{2}{3}\cos 2x - \frac{2}{15}\cos 4x + \frac{2}{35}\cos 6x + \cdots\right)$

Full-wave rectifier

12. $f(x) = \frac{2}{\pi}\left(1 + \frac{2}{3}\cos 2x - \frac{2}{15}\cos 4x + \frac{2}{35}\cos 6x + \cdots\right)$

Note that b would be zero for $n = 4$ or, in fact, for all even values of n, since $\cos 2x = \cos 4x = \cos 6x$, and so on. Our series, then, contains no even harmonics.

$$b_3 = \frac{1}{\pi}\int_{-\pi}^{0}(-1)\sin 3x\,dx + \frac{1}{\pi}\int_{0}^{\pi}(1)\sin 3x\,dx$$

$$= \left.\frac{\cos 3x}{3\pi}\right|_{-\pi}^{0} - \left.\frac{\cos 3x}{3\pi}\right|_{0}^{\pi}$$

$$= \frac{4}{3\pi}$$

Continuing the same way gives values of b of $4/\pi$, $4/3\pi$, $4/5\pi$, and so on. Our final Fourier series is then

$$f(x) = \frac{4}{\pi}\sin x + \frac{4}{3\pi}\sin 3x + \frac{4}{5\pi}\sin 5x + \frac{4}{7\pi}\sin 7x + \cdots$$

as in Table 37–6. Figure 37–5(b) shows the first two terms (the fundamental and the third harmonic) of our Fourier series. Figure 37–5(c) shows the first four terms (fundamental plus third, fifth, and seventh harmonics), and Fig. 37–5(d) shows the first six terms (fundamental plus third, fifth, seventh, ninth, and eleventh harmonics).

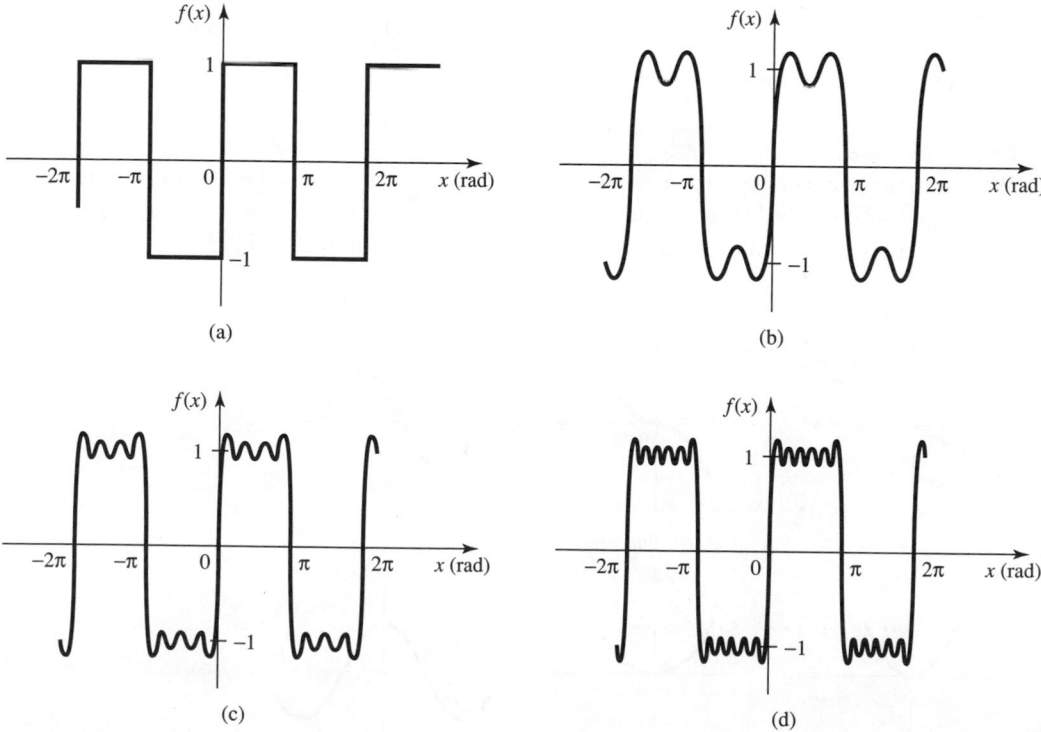

FIGURE 37–5 A square wave synthesized using two, four, and six terms of a Fourier series. ◆◆◆

Exercise 5 ◆ Fourier Series

Write seven terms of the Fourier series given the following coefficients.

1. $a_0 = 4$, $a_1 = 3$, $a_2 = 2$, $a_3 = 1$; $b_1 = 4$, $b_2 = 3$, $b_3 = 2$

2. $a_0 = 1.6$, $a_1 = 5.2$, $a_2 = 3.1$, $a_3 = 1.4$; $b_1 = 7.5$, $b_2 = 5.3$, $b_3 = 2.8$

Write a Fourier series for the indicated waveforms in Table 37–6, first writing the functions that describe the waveform. Check your Fourier series against those given in the table.

3. Waveform 3   **4.** Waveform 4   **5.** Waveform 6

6. Waveform 7   **7.** Waveform 8   **8.** Waveform 9

9. Waveform 10

37–6 Waveform Symmetries

Kinds of Symmetry

We have seen that for the square wave, all of the a_n coefficients of the Fourier series were zero, and thus the final series had no cosine terms. If we could tell in advance that all of the cosine terms would vanish, we could simplify our work by not even concerning ourselves with them. And, in fact, we can tell in advance which terms vanish by noticing the kind of symmetry the given function has. The three kinds of symmetry that are most useful here are as follows:

1. If a function is *symmetric about the origin*, it is called an *odd* function, and its Fourier series has only sine terms.
2. If a function is *symmetric about the y axis*, it is called an *even* function, and its Fourier series has no sine terms.
3. If a function has *half-wave symmetry*, its Fourier series has only odd harmonics.

We now cover each type of symmetry.

Odd Functions

We noted in Chapter 28 that a function $f(x)$ is symmetric about the origin if it remains unchanged when we substitute both $-x$ for x and $-y$ for y. Another way of saying this is that $f(x) = -f(-x)$. Such functions are called *odd functions*, as shown in Fig. 37–6(a).

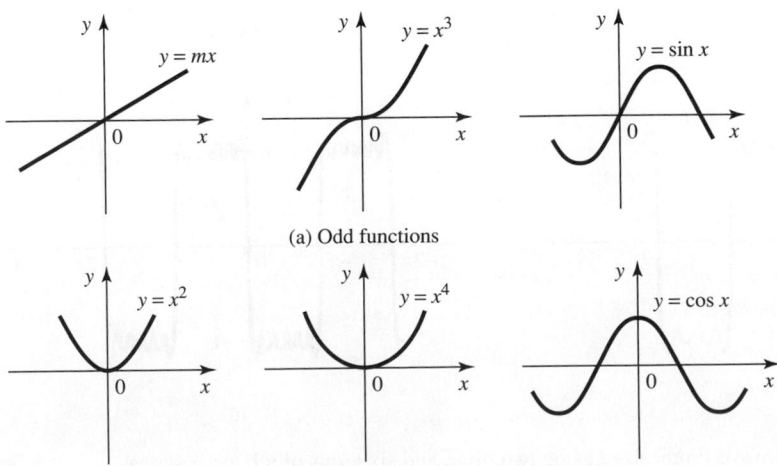

(a) Odd functions

(b) Even functions

FIGURE 37–6

◆◆◆ **Example 21:** The function $y = x^3$ is odd because $(-x)^3 = -x^3$. Similarly, x, x^5, x^7, ... are odd, with the name "odd" coinciding with whether the exponent is odd or even. ◆◆◆

Even Functions
We noted earlier that a function is symmetric about the y axis [Fig. 37–6(b)] if it remains unchanged when we substitute $-x$ for x. In other words, $f(x) = f(-x)$. These are called *even functions*.

◆◆◆ **Example 22:** The function $y = x^2$ is even because $(-x)^2 = x^2$. Similarly, other even powers of x, such as x^4, x^6, ..., are even. ◆◆◆

Fourier Expansion of Odd and Even Functions
In the power series expansion for the sine function

$$\sin x = x - \frac{x^3}{6} + \frac{x^5}{120} - \cdots \qquad \textbf{215}$$

every term is an odd function, and thus $\sin x$ itself is an odd function. In the power series expansion for the cosine function

$$\cos x = 1 - \frac{x^2}{2} + \frac{x^4}{24} - \cdots \qquad \textbf{216}$$

every term is an even function, so $\cos x$, made up of all even functions, is itself an even function. Thus, we conclude the following:

Odd and Even Functions	(a) Odd functions have Fourier series with only sine terms (and no constant term). (b) Even functions have Fourier series with only cosine terms (and may have a constant term).	**456**

◆◆◆ **Example 23:** In Table 37–6, Waveforms 1, 4, 7, and 8 are odd. When deriving their Fourier series from scratch, we would save work by not even looking for cosine terms. Waveforms 2, 11, and 12 are even, while the others are neither odd nor even. ◆◆◆

Shift of Axes
A typical problem is one in which a periodic waveform is given and we are to write a Fourier series for it. When writing a Fourier series for a periodic waveform, we are often free to choose the position of the y axis. Our choice of axis can change the function from even to odd, or vice versa.

◆◆◆ **Example 24:** In Table 37–6, notice that Waveforms 1 and 2 are the same waveform but with the axis shifted by $\pi/2$ radians. Thus, Waveform 1 is an odd function, while Waveform 2 is even. A similar shift is seen with the half-wave rectifier waveforms (10 and 11). ◆◆◆

A *vertical* shift of the x axis will affect the constant term in the series.

◆◆◆ **Example 25:** Waveform 3 is the same as Waveform 1, except for a vertical shift of 1 unit. Thus, the series for Waveform 3 has a constant term of 1 unit. The same vertical shift can be seen with Waveforms 6 and 7. ◆◆◆

Half-Wave Symmetry

When the negative half-cycle of a periodic wave has the same shape as the positive half-cycle (except that it is, of course, inverted), we say that the wave has *half-wave symmetry*. The sine wave, the cosine wave, as well as Waveforms 1, 2, and 4, have half-wave symmetry.

A quick graphical way to test for half-wave symmetry is to measure the ordinates on the wave at two points spaced half a cycle apart. The two ordinates should be equal in magnitude but opposite in sign.

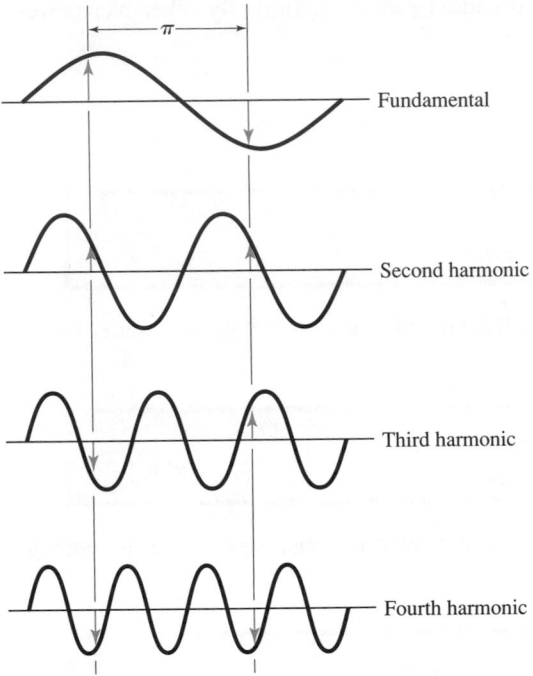

FIGURE 37–7   The fundamental and the third harmonic have half-wave symmetry.

Figure 37–7 shows a sine wave and second, third, and fourth harmonics. For each, one arrow shows the ordinate at an arbitrarily chosen point, and the other arrow shows the ordinate half a cycle away (π radians for the sine wave). For half-wave symmetry, the arrows on a given waveform should be equal in length but opposite in direction. Note that the fundamental and the odd harmonics have half-wave symmetry, while the even harmonics do not. This is also true for harmonics higher than those shown. Thus, for the Fourier series representing a waveform, we conclude the following:

Half-Wave Symmetry	A waveform that has half-wave symmetry has only odd harmonics in its Fourier series.	457

Common Error	Don't confuse odd functions with odd harmonics. Thus, $\sin 4x$ is an even harmonic but an odd function.

By seeing if a waveform is an odd or even function or if it has half-wave symmetry, we can reduce the work of computation. Or, when possible, we can choose our axes to create symmetry, as in the following example.

◆◆◆ **Example 26:** Choose the y axis and write a Fourier series for the triangular wave in Fig. 37–8(a).

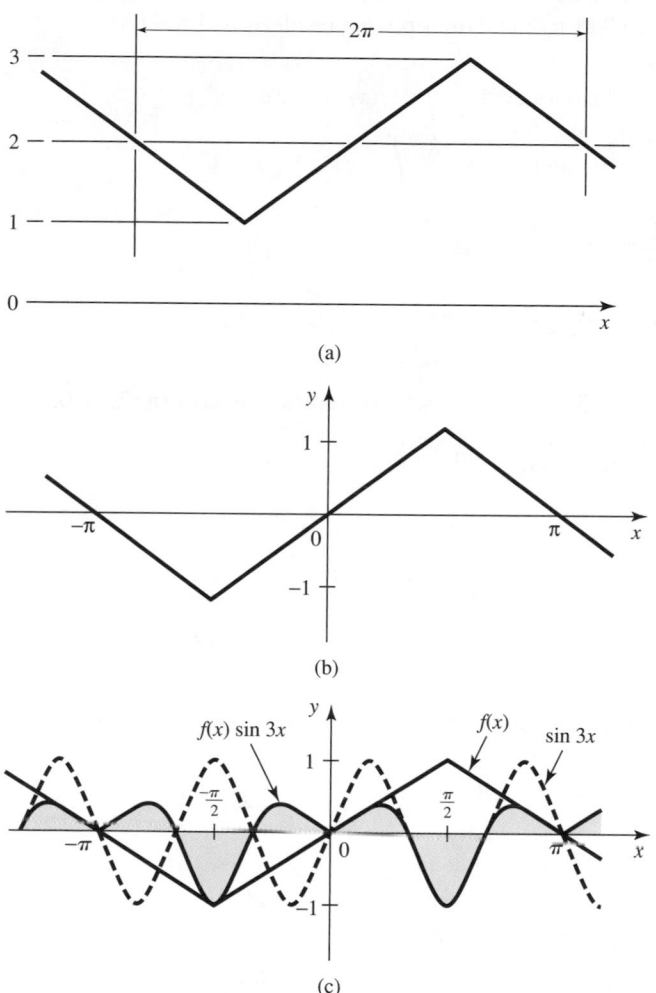

(a)

(b)

(c)

FIGURE 37–8 Synthesis of a triangular wave.

Solution: The given waveform has no symmetry, so let us temporarily shift the x axis up 2 units, which will eliminate the constant term from our series. We will add 2 to our final series to account for this shift. We are free to locate the y axis, so let us place it as in Fig. 37–8(b). Our waveform is now an odd function and has half-wave symmetry, so we expect that the series we write will be composed of odd-harmonic sine terms.

$$y = b_1 \sin x + b_3 \sin 3x + b_5 \sin 5x + \cdots$$

From Eq. 451,

$$b_n = \frac{1}{\pi} \int_{-\pi}^{\pi} f(x) \sin nx \, dx$$

From 0 to $\pi/2$ our waveform is a straight line through the origin with a slope of $2/\pi$, so

$$f(x) = \frac{2}{\pi}x \qquad \left(0 < x < \frac{\pi}{2}\right)$$

The equation of the waveform is different elsewhere in the interval $-\pi$ to π, but we'll see that we need only the portion from 0 to $\pi/2$.

If we graph $f(x) \sin nx$ for any odd value of n, say, 3, we get a curve such as in Fig. 37–8(c). The integral of $f(x) \sin nx$ corresponds to the shaded portion under the curve. Note that this area repeats and that the area under the curve between 0 and $\pi/2$ is one-fourth the area from 2π to π. Thus, we need only integrate from 0 to $\pi/2$ and multiply the result by 4. Therefore

$$
\begin{aligned}
b_n &= \frac{1}{\pi} \int_{-\pi}^{\pi} f(x) \sin nx \, dx = \frac{4}{\pi} \int_0^{\pi/2} f(x) \sin nx \, dx \\
&= \frac{4}{\pi} \int_0^{\pi/2} \frac{2}{\pi} x \sin nx \, dx = \frac{8}{\pi^2} \int_0^{\pi/2} x \sin nx \, dx \\
&= \frac{8}{\pi^2} \left[\frac{1}{n^2} \sin nx - \frac{x}{n} \cos nx \right]_0^{\pi/2} \\
&= \frac{8}{\pi^2} \left[\frac{1}{n^2} \sin \frac{n\pi}{2} - \frac{\pi}{2n} \cos \frac{n\pi}{2} \right]
\end{aligned}
$$

We have only odd harmonics, so $n = 1, 3, 5, 7, \ldots$. For these values of n, $\cos(n\pi/2) = 0$, so

$$
b_n = \frac{8}{\pi^2} \left(\frac{1}{n^2} \sin \frac{n\pi}{2} \right)
$$

from which

$$
b_1 = \frac{8}{\pi^2} \left(\frac{1}{1^2} \right) = \frac{8}{\pi^2} \qquad\qquad b_3 = \frac{8}{\pi^2} \left(\frac{1}{3^2} \right)(-1) = -\frac{8}{9\pi^2}
$$

$$
b_5 = \frac{8}{\pi^2} \left(\frac{1}{5^2} \right)(1) = \frac{8}{25\pi^2} \qquad\qquad b_7 = \frac{8}{\pi^2} \left(\frac{1}{7^2} \right)(-1) = -\frac{8}{49\pi^2}
$$

Our final series, remembering to add a constant term of 2 units for the vertical shift of axis, is then

$$
y = 2 + \frac{8}{\pi^2} \left(\sin x - \frac{1}{9} \sin 3x + \frac{1}{25} \sin 5x - \frac{1}{49} \sin 7x + \cdots \right)
$$

This is the same as Waveform 4, except for the 2-unit shift.   ◆◆◆

Exercise 6 ◆ Waveform Symmetries

Label each function as odd, even, or neither.

1. Fig. 37–9(a)
2. Fig. 37–9(b)
3. Fig. 37–9(c)
4. Fig. 37–9(d)
5. Fig. 37–9(e)
6. Fig. 37–9(f)

Which functions have half-wave symmetry?

  7. Fig. 37–9(a)   8. Fig. 37–9(b)

  9. Fig. 37–9(c)   10. Fig. 37–9(d)

11. Fig. 37–9(e)   12. Fig. 37–9(f)

Using symmetry to simplify your work, write a Fourier series for the indicated waveforms in Table 37–6. As before, start by writing the functions that describe the waveform. Check your series against those given in the table.

13. Waveform 2
14. Waveform 5
15. Waveform 11
16. Waveform 12

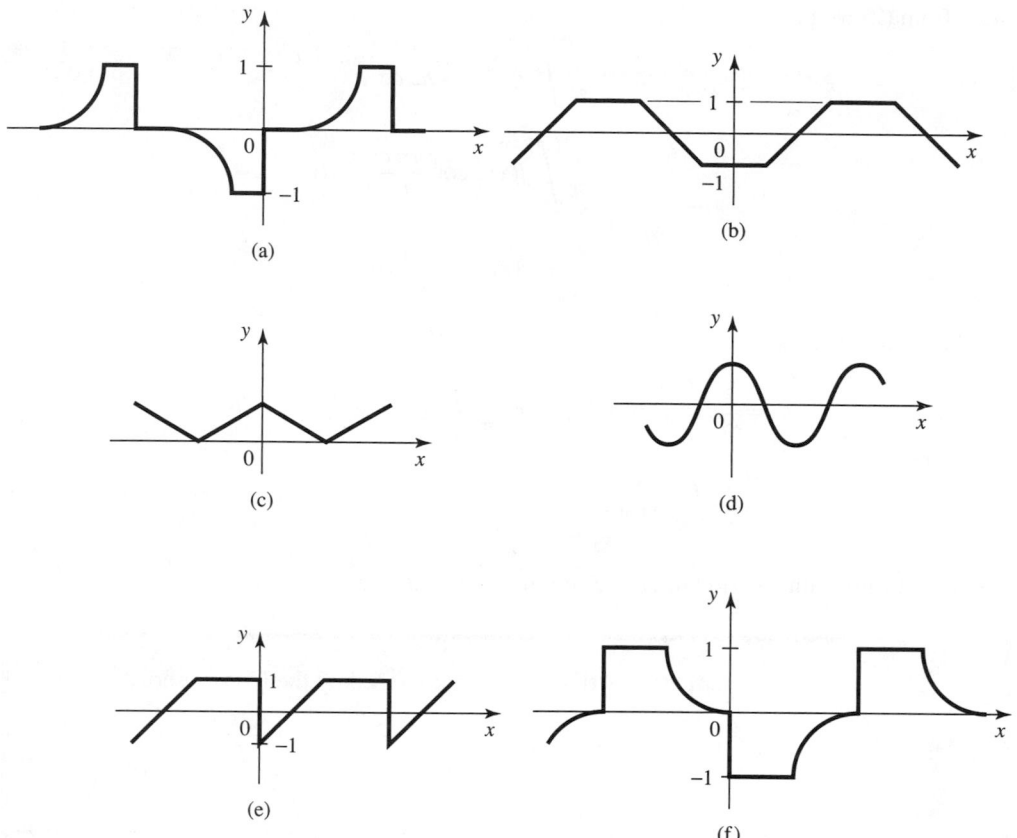

FIGURE 37–9

37–7 Waveforms with Period of 2L

So far we have written Fourier series only for functions with a period of 2π. But a waveform could have some other period, say, 8. Here we modify our formulas so that functions with any period can be represented.

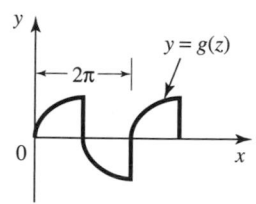

(a)

Figure 37–10(a) shows a function $y = g(z)$ with a period of 2π. The Fourier coefficients for this function are

$$a_0 = \frac{1}{\pi} \int_{-\pi}^{\pi} g(z)\, dz \tag{1}$$

$$a_n = \frac{1}{\pi} \int_{-\pi}^{\pi} g(z) \cos nz\, dz \tag{2}$$

and

$$b_n = \frac{1}{\pi} \int_{-\pi}^{\pi} g(z) \sin nz\, dz \tag{3}$$

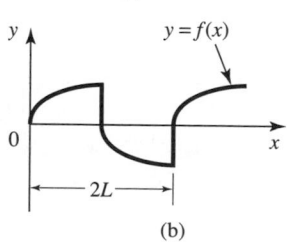

(b)

FIGURE 37–10

Figure 37–10(b) shows a function $y = f(x)$ with a period $2L$, where x and z are related by the proportion

$$\frac{x}{2L} = \frac{z}{2\pi}$$

or $z = \pi x/L$. Substituting into Eq. (1), with $g(z) = y = f(x)$ and $dz = d(\pi x/L) = (\pi/L)\, dx$, we obtain

$$a_0 = \frac{1}{\pi} \int_{-\pi}^{\pi} g(z)\, dz = \frac{1}{\pi} \int_{-L}^{L} f(x) \frac{\pi}{L}\, dx = \frac{1}{L} \int_{-L}^{L} f(x)\, dx$$

Next, from (2) we get

$$a_n = \frac{1}{\pi} \int_{-\pi}^{\pi} g(z) \cos nz \, dz$$

$$= \frac{1}{\pi} \int_{-L}^{L} f(x) \left(\cos \frac{n\pi x}{L} \right) \frac{\pi}{L} \, dx$$

$$= \frac{1}{L} \int_{-L}^{L} f(x) \cos \frac{n\pi x}{L} \, dx$$

Similarly, from (3),

$$b_n = \frac{1}{\pi} \int_{-\pi}^{\pi} g(z) \sin nz \, dz = \frac{1}{\pi} \int_{-L}^{L} f(x) \left(\sin \frac{n\pi x}{L} \right) \frac{\pi}{L} \, dx$$

$$= \frac{1}{L} \int_{-L}^{L} f(x) \sin \frac{n\pi x}{L} \, dx$$

Waveforms with a period of $2L$ are summarized as follows:

Waveforms with Period of 2L	A periodic function can be represented by the Fourier series $$f(x) = \frac{a_0}{2} + a_1 \cos \frac{\pi x}{L} + a_2 \cos \frac{2\pi x}{L} + a_3 \cos \frac{3\pi x}{L} + \cdots \\ + b_1 \sin \frac{\pi x}{L} + b_2 \sin \frac{2\pi x}{L} + b_3 \sin \frac{3\pi x}{L} + \cdots$$	452
	where $\quad a_0 = \frac{1}{L} \int_{-L}^{L} f(x) \, dx$	453
	$a_n = \frac{1}{L} \int_{-L}^{L} f(x) \cos \frac{n\pi x}{L} \, dx$	454
	$b_n = \frac{1}{L} \int_{-L}^{L} f(x) \sin \frac{n\pi x}{L} \, dx$	455

◆◆◆ **Example 27:** Write the Fourier series for the waveform in Fig. 37–11(a).

Solution: Since the function is odd, we expect only sine terms and no constant term. The equation of the waveform from -6 to 6 is $f(x) = x/6$. From Eq. 455, with $L = 6$,

$$b_n = \frac{1}{L} \int_{-L}^{L} f(x) \sin \frac{n\pi x}{L} \, dx$$

$$= \frac{1}{6} \int_{-6}^{6} \frac{x}{6} \sin \frac{n\pi x}{6} \, dx$$

Integrating by Rule 31 and evaluating at the limits gives

$$b_n = \frac{1}{(n\pi)^2} \left[\sin \frac{n\pi x}{6} - \frac{n\pi x}{6} \cos \frac{n\pi x}{6} \right]_{-6}^{6}$$

$$= -\frac{2}{n\pi} \cos n\pi$$

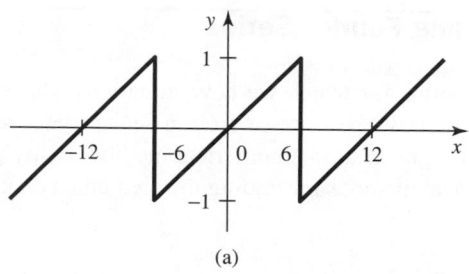

FIGURE 37–11   Synthesis of a sawtooth wave.

from which

$$b_1 = \frac{2}{\pi} \qquad b_2 = \frac{-1}{\pi} \qquad b_3 = \frac{2}{3\pi} \qquad b_4 = \frac{-1}{2\pi} \qquad \text{etc.}$$

Our series is then

$$y = \frac{2}{\pi}\left(\sin\frac{\pi x}{6} - \frac{1}{2}\sin\frac{\pi x}{3} - \frac{1}{3}\sin\frac{\pi x}{2} - \frac{1}{4}\sin\frac{2\pi x}{3} + \cdots \right)$$

Figure 37–11(b) shows a graph of the first five terms of this series.   ◆◆◆

Exercise 7 ◆ Waveforms with Period of 2L

Write a Fourier series for each waveform in Fig. 37–12. Check your series against those given.

1. Fig. 37–12(a)   **2.** Fig. 37–12(b)   **3.** Fig. 37–12(c)

4. Fig. 37–12(d)   **5.** Fig. 37–12(e)   **6.** Fig. 37–12(f)

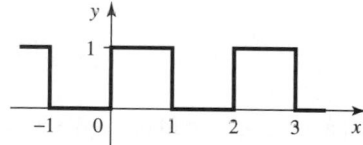

(a) $y = \frac{1}{2} + \frac{2}{\pi}\left(\sin \pi x + \frac{1}{3}\sin 3\pi x + \cdots \right)$

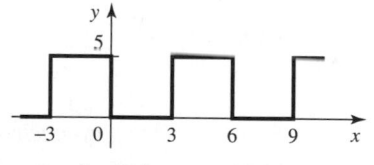

(b) $y = \frac{5}{2} - \frac{10}{\pi}\left(\sin\frac{\pi x}{3} + \frac{1}{3}\sin 3\pi x - \cdots \right)$

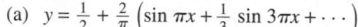
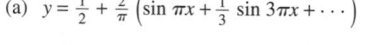

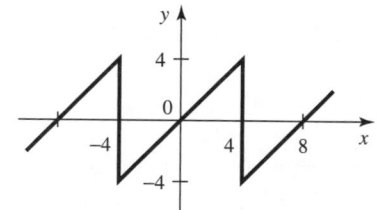

(c) $y = \frac{8}{\pi}\left(\sin\frac{\pi x}{4} - \frac{1}{2}\sin\frac{2\pi x}{4} + \frac{1}{3}\sin\frac{3\pi x}{4} - \cdots \right)$

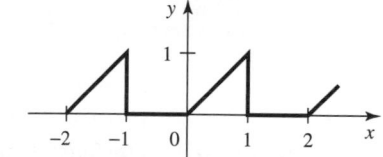

(d) $y = \frac{1}{4} - \frac{2}{\pi^2}\left(\cos \pi x + \frac{1}{9}\cos 3\pi x + \frac{1}{25}\cos 5\pi x + \cdots \right)$

         $+ \frac{1}{\pi}\left(\sin \pi x - \frac{1}{2}\sin 2\pi x + \frac{1}{3}\sin 3\pi x - \cdots \right)$

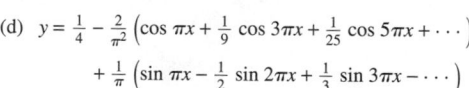

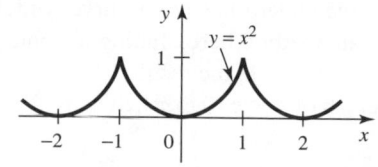

(e) $y = \frac{1}{3} - \frac{4}{\pi^2}\left(\cos \pi x - \frac{1}{4}\cos 2\pi x + \frac{1}{9}\cos 3\pi x - \cdots \right)$

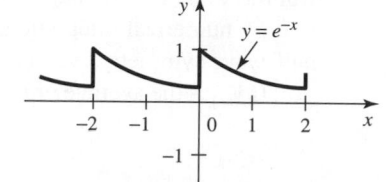

(f) $y = 0.4324 + 0.0795 \cos \pi x + 0.0214 \cos 2\pi x$

         $+ 0.009\,63 \cos 3\pi x + \cdots$

         $+ 0.2499 \sin \pi x + 0.1342 \sin 2\pi x$

         $+ 0.0907 \sin 3\pi x + \cdots$

FIGURE 37–12

37–8   A Numerical Method for Finding Fourier Series

So far we have written Fourier series for waveforms for which we have equations. But what about a wave whose shape we know from an oscilloscope trace or from a set of meter readings but for which we have no equation? Here we give a simple numerical method, easily programmed for the computer. It is based on numerical methods for finding the area under a curve, covered in Sec. 34–9.

◆◆◆ **Example 28:** Find the first eight terms of the Fourier series for the waveform in Fig. 37–13.

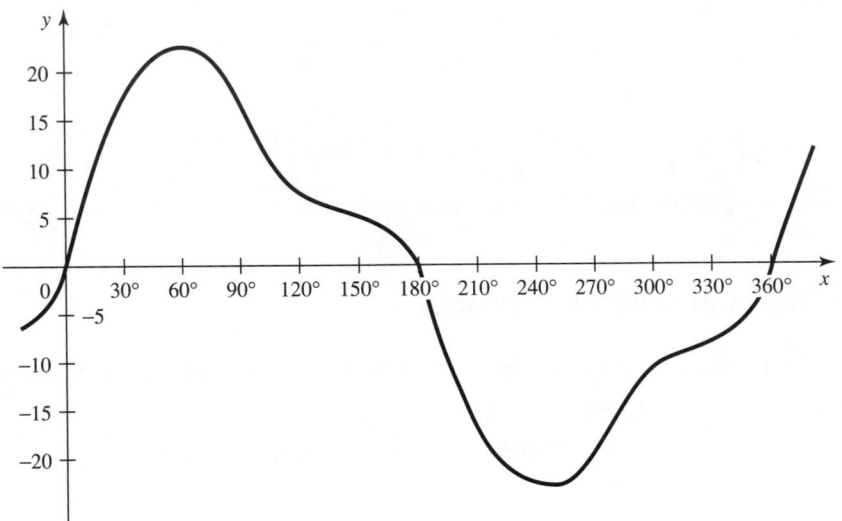

FIGURE 37–13

Solution: Using symmetry: We first note that the wave does not appear to be odd or even, so we expect our series to contain both sine and cosine terms. There is half-wave symmetry, however, so we look only for odd harmonics. The first eight terms, then, will be

$$y = a_1 \cos x + a_3 \cos 3x + a_5 \cos 5x + a_7 \cos 7x$$

$$+ b_1 \sin x + b_3 \sin 3x + b_5 \sin 5x + b_7 \sin 7x$$

Finding the coefficients: From Eq. 450 the first coefficient is

$$a_1 = \frac{1}{\pi} \int_0^{2\pi} y \cos x \, dx$$

But the value of the integral $\int y \cos x \, dx$ is equal to the area under the $y \cos x$ curve, which we find by numerical integration. We use the average ordinate method here. Taking advantage of half-wave symmetry, we integrate over only a half-cycle and double the result.

If y_{avg} is the average ordinate of the $y \cos x$ curve over a half-cycle (0 to π), then, by Eq. 417,

$$\int_0^{2\pi} y \cos x \, dx = 2\pi y_{\text{avg}}$$

so

$$a_1 = \frac{1}{\pi} \int_0^{2\pi} y \cos x \, dx = 2y_{\text{avg}}$$

We divide the half-cycle into a number of intervals, say, at every 10°. At each x we measure the ordinate y and calculate $y \cos x$. For example, at $x = 20°$, y is measured at 14.8, so

$$y \cos x = 14.8 \cos 20° = 14.8(0.93969) = 13.907$$

The computed values of $y \cos x$ for the remaining angles are shown in Table 37–7.

TABLE 37–7

x	y	$\sin x$	$y \sin x$	$\cos x$	$y \cos x$	$\sin 3x$	$y \sin 3x$	$\cos 3x$	$y \cos 3x$
0	0.0	0.0000	0.000	1.0000	0.000	0.0000	0.000	1.0000	0.000
10	10.0	0.1736	1.736	0.9848	9.848	0.5000	5.000	0.8660	8.660
20	14.8	0.3420	5.062	0.9397	13.907	0.8660	12.817	0.5000	7.400
30	16.1	0.5000	8.050	0.8660	13.943	1.0000	16.100	0.0000	0.000
40	20.0	0.6428	12.856	0.7660	15.321	0.8660	17.321	−0.5000	−10.000
50	21.0	0.7660	16.087	0.6428	13.499	0.5000	10.500	−0.8660	−18.187
60	21.2	0.8660	18.360	0.5000	10.600	0.0000	0.000	−1.0000	−21.200
70	20.3	0.9397	19.076	0.3420	6.943	−0.5000	−10.150	−0.8660	−17.580
80	17.7	0.9848	17.431	0.1736	3.074	−0.8660	−15.329	−0.5000	−8.850
90	14.1	1.0000	14.100	0.0000	0.000	−1.0000	−14.100	0.0000	0.000
100	11.2	0.9848	11.030	−0.1736	−1.945	−0.8660	−9.699	0.5000	5.600
110	8.8	0.9397	8.269	−0.3420	−3.010	−0.5000	−4.400	0.8660	7.621
120	7.5	0.8660	6.495	−0.5000	−3.750	0.0000	0.000	1.0000	7.500
130	6.6	0.7660	5.056	−0.6428	−4.242	0.5000	3.300	0.8660	5.716
140	6.1	0.6428	3.921	−0.7660	−4.673	0.8660	5.283	0.5000	3.050
150	5.5	0.5000	2.750	−0.8660	−4.763	1.0000	5.500	0.0000	0.000
160	4.6	0.3420	1.573	−0.9397	−4.323	0.8660	3.984	−0.5000	−2.300
170	3.6	0.1736	0.625	−0.9848	−3.545	0.5000	1.800	−0.8660	−3.118
180	0.0	0.0000	0.000	−1.0000	0.000	0.0000	0.000	−1.0000	0.000
		Sum:	152.477		56.884		27.926		−35.687
		y_{avg}:	8.025		2.994		1.470		−1.878
		$b_1 = 16.050$		$a_1 = 5.988$		$b_3 = 2.940$		$a_3 = -3.757$	

x	$\sin 5x$	$y \sin 5x$	$\cos 5x$	$y \cos 5x$	$\sin 7x$	$y \sin 7x$	$\cos 7x$	$y \cos 7x$
0	0.0000	0.000	1.0000	0.000	0.0000	0.000	1.0000	0.000
10	0.7660	7.660	0.6428	6.428	0.9397	9.397	0.3420	3.420
20	0.9848	14.575	−0.1736	−2.570	0.6428	9.513	−0.7660	−11.337
30	0.5000	8.050	−0.8660	−13.943	−0.5000	−8.050	−0.8660	−13.943
40	−0.3420	−6.840	−0.9397	−18.794	−0.9848	−19.696	0.1736	3.473
50	−0.9397	−19.734	−0.3420	−7.182	−0.1736	−3.647	0.9848	20.681
60	−0.8660	−18.360	0.5000	10.600	0.8660	18.360	0.5000	10.600
70	−0.1736	−3.525	0.9848	19.992	0.7660	15.551	−0.6428	−13.049
80	0.6428	11.377	0.7660	13.559	−0.3420	−6.054	−0.9397	−16.633
90	1.0000	14.100	0.0000	0.000	−1.0000	−14.100	0.0000	0.000
100	0.6428	7.199	−0.7660	−8.580	−0.3420	−3.831	0.9397	10.525
110	−0.1736	−1.528	−0.9848	−8.666	0.7660	6.741	0.6428	5.657
120	−0.8660	−6.495	−0.5000	−3.750	0.8660	6.495	−0.5000	−3.750
130	−0.9397	−6.202	0.3420	2.257	−0.1736	−1.146	−0.9848	−6.500
140	−0.3420	−2.086	0.9397	5.732	−0.9848	−6.007	−0.1736	−1.059
150	0.5000	2.750	0.8660	4.763	−0.5000	−2.750	0.8660	4.763
160	0.9848	4.530	0.1736	0.799	0.6428	2.957	0.7660	3.524
170	0.7660	2.758	−0.6428	−2.314	0.9397	3.383	−0.3421	−1.231
180	0.0000	0.000	−1.0000	0.000	0.0000	0.000	−1.0000	0.000
		8.230		−1.669		7.116		−4.860
		0.433		−0.088		0.375		−0.256
	$b_5 = 0.866$		$a_5 = -0.176$		$b_7 = 0.750$		$a_7 = -0.512$	

We then add all of the values of $y \cos x$ over the half-cycle and get an average value by dividing that sum (56.883) by the total number of data points, 19. Thus,

$$y_{\text{avg}} = \frac{56.884}{19} = 2.994$$

Our first coefficient is then

$$a_1 = 2y_{\text{avg}} = 2(2.994) = 5.988$$

TABLE 37–8

x	Original y	y from Series
0	0.0	1.5
10	10.0	8.0
20	14.8	13.5
30	16.1	16.8
40	20.0	18.4
50	21.0	19.5
60	21.2	20.2
70	20.3	19.5
80	17.7	16.8
90	14.1	13.2
100	11.2	10.3
110	8.8	8.6
120	7.5	7.4
130	6.6	6.2
140	6.1	5.3
150	5.5	5.2
160	4.6	5.2
170	3.6	3.3
180	0.0	−1.5

The remainder of the computation is similar, and the results are shown in Table 37–7. Our final Fourier series is then

$$y = 5.988 \cos x - 3.757 \cos 3x - 0.176 \cos 5x - 0.512 \cos 7x$$
$$+ 16.050 \sin x + 2.940 \sin 3x + 0.866 \sin 5x + 0.750 \sin 7x$$

Check: To test our result, we compute y by series and get the values listed in Table 37–8 and graphed in Fig. 37–14. For better accuracy we could compute more terms of the series or use finer spacing when finding y_{avg}. This would not be much more work when done by computer.

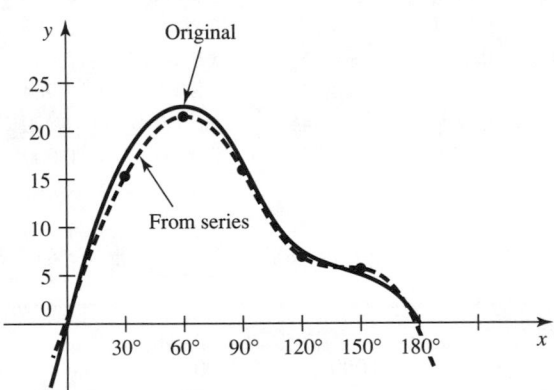

FIGURE 37–14   ◆◆◆

Exercise 8 ◆ A Numerical Method for Finding Fourier Series

Find the first six terms of the Fourier series for each waveform. Assume half-wave symmetry.

1.

x	0°	20°	40°	60°	80°	100°	120°	140°	160°	180°
y	0	2.1	4.2	4.5	7.0	10.1	14.3	14.8	13.9	0

2.

x	0°	20°	40°	60°	80°	100°	120°	140°	160°	180°
y	0	31	52	55	80	111	153	108	49	0

3.

x	0°	20°	40°	60°	80°	100°	120°	140°	160°	180°
y	0	12.1	24.2	34.5	47.0	30.1	24.3	11.8	9.9	0

4.

x	0°	20°	40°	60°	80°	100°	120°	140°	160°	180°
y	0	4.6	9.4	15.4	12.3	10.8	14.1	16.8	10.9	0

◆◆◆ CHAPTER 37 REVIEW PROBLEMS ◆◆◆◆◆◆◆◆◆◆◆◆◆◆◆◆◆◆◆◆◆◆◆◆◆◆◆◆◆

Compute each number using three terms of the appropriate Maclaurin series.

1. $e^{0.3}$

2. $\sin 0.5$

Compute each value using three terms of the appropriate Taylor series from Table 37–1.

3. $\dfrac{1}{2.1}$

4. $\sin 31.5°$

5. Use the ratio test to find the interval of convergence of the power series

$$1 + x + x^2 + x^3 + \cdots + x^n + \cdots$$

6. Use the ratio test to determine if the following series converges or diverges.

$$1 + \frac{3}{3!} + \frac{5}{5!} + \frac{7}{7!} + \cdots$$

Write the first five terms of each series, given the general term.

7. $u_n = 2n^2$

8. $u_n = 3n - 1$

Multiply or divide earlier series to verify the following.

9. $x^2 e^{2x} = x^2 + 2x^3 + 2x^4 + \cdots$ (Use Series 29.)

10. $\dfrac{\sin x}{x^2} = \dfrac{1}{x} - \dfrac{x}{3!} + \dfrac{x^3}{5!} - \cdots$

Deduce the general term of each series. Use it to predict the next two terms.

11. $-1 + 2 + 7 + 14 + 23 + \cdots$

12. $4 + 7 + 10 + 13 + 16 + \cdots$

13. Using series, evaluate $\displaystyle\int_0^1 \frac{e^x - 1}{x}\, dx$

Deduce a recursion relation for each series. Use it to predict the next two terms.

14. $3 + 5 + 9 + 17 + \cdots$

15. $3 + 7 + 47 + \cdots$

16. By substituting into Series 3 for $\cos x$, verify the following series:

$$\text{Series 14:} \quad \cos x^2 = 1 - \frac{x^4}{2!} + \frac{x^8}{4!} - \frac{x^{12}}{6!} + \cdots$$

17. Verify the Maclaurin expansion

$$(1 + x)^3 = 1 + 3x + 3x^2 + x^3$$

18. Verify the following Taylor series, expanded about $a = 1$:

$$\text{Series 24:} \quad \sqrt{x^2 + 1} = \sqrt{2} + \frac{(x - 1)}{\sqrt{2}} + \frac{(x - 1)^2}{4\sqrt{2}} + \cdots$$

Label each function as odd, even, or neither.

19. Fig. 37–15(a)

20. Fig. 37–15(b)

21. Fig. 37–15(c)

22. Fig. 37–15(d)

Which functions have half-wave symmetry?

23. Fig. 37–15(a)

24. Fig. 37–15(b)

25. Fig. 37–15(c)

26. Fig. 37–15(d)

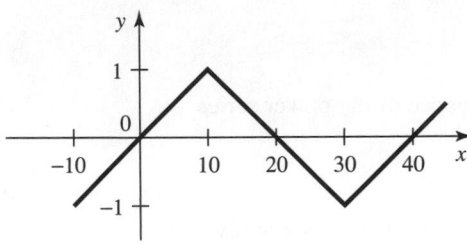

(a) $y = \frac{8}{\pi^2} \left(\sin \frac{\pi x}{20} - \frac{1}{9} \sin \frac{3\pi x}{20} + \frac{1}{25} \sin \frac{5\pi x}{20} - \cdots \right)$

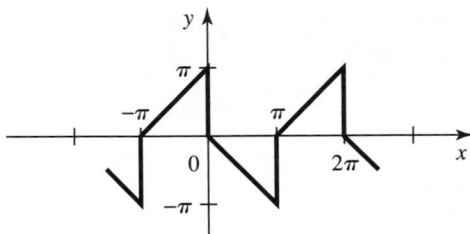

(b) $y = \frac{4}{\pi} \cos x + \frac{4}{9\pi} \cos 3x + \frac{4}{25\pi} \cos 5x + \cdots$

$\quad\quad -2 \sin x - \frac{2}{3} \sin 3x - \frac{2}{5} \sin 5x - \cdots$

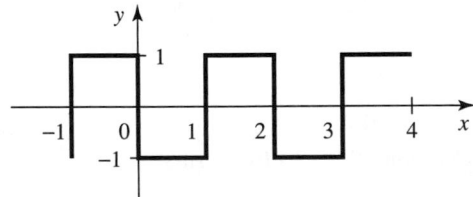

(c) $y = -\frac{4}{\pi} \left(\sin \pi x + \frac{1}{3} \sin 3\pi x + \frac{1}{5} \sin 5\pi x + \cdots \right)$

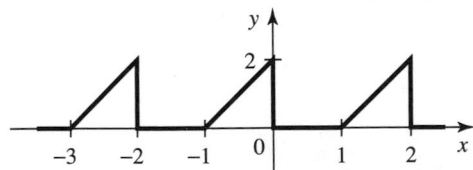

(d) $y = \frac{1}{2} + \frac{4}{\pi^2} \cos \pi x + \frac{4}{9\pi^2} \cos 3\pi x + \frac{4}{25\pi^2} \cos 5\pi x + \cdots$

$\quad\quad - \frac{2}{\pi} \sin \pi x - \frac{1}{\pi} \sin 2\pi x - \frac{2}{3\pi} \sin 3\pi x - \cdots$

FIGURE 37–15

Verify the Fourier series for each waveform.

27. Fig. 37–15(a)

28. Fig. 37–15(b)

29. Fig. 37–15(c)

30. Fig. 37–15(d)

31. Find the first six terms of the Fourier series for the given waveform. Assume half-wave symmetry.

x	0°	20°	40°	60°	80°	100°	120°	140°	160°	180°
y	0	3.2	5.5	6.2	8.3	11.7	15.1	14.3	12.6	0

Writing

32. Your friend writes saying that you are crazy to study infinite series. "They're useless," she says, "because there is no way to compute an infinite number of terms. What can you do with a string of numbers that goes on forever?" Reply in writing to your friend, and either accept or reject her statement. Give specific reasons for your choice.

Team Project

33. Obtain a trace of a periodic waveform from the electronics lab on campus, or from your instructor. Then, (a) write a Fourier series for that waveform, using the methods of this chapter, (b) combine the terms of the series to reconstruct the given waveform, and (c) compare the reconstructed waveform with the original and give some assessment as to the accuracy. Then, (d) add more terms to the Fourier series, and observe their effect on the accuracy of the reconstructed waveform.

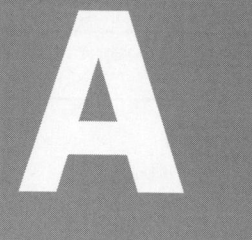

A Summary of Facts and Formulas

Many mathematics courses cover only a fraction of the topics in this text. Further, some of these formulas are included for reference even though they may not appear elsewhere in the text. We hope that this Summary will provide a handy reference, not only for your current course but for others, and for your technical work after graduation.

No.

ALGEBRAIC LAWS	1	Commutative Law — Addition	$a + b = b + a$
	2	Commutative Law — Multiplication	$ab = ba$
	3	Associative Law — Addition	$a + (b + c) = (a + b) + c = (a + c) + b$
	4	Associative Law — Multiplication	$a(bc) = (ab)c = (ac)b = abc$
	5	Distributive Law	$a(b + c) = ab + ac$
RULES OF SIGNS	6	Addition and Subtraction	$a + (-b) = a - (+b) = a - b$
	7		$a + (+b) = a - (-b) = a + b$
	8	Multiplication	$(+a)(+b) = (-a)(-b) = +ab$
	9		$(+a)(-b) = (-a)(+b) = -(+a)(+b) = -ab$
	10	Division	$\dfrac{+a}{+b} = \dfrac{-a}{-b} = -\dfrac{-a}{+b} = -\dfrac{+a}{-b} = \dfrac{a}{b}$
	11		$\dfrac{+a}{-b} = \dfrac{-a}{+b} = -\dfrac{-a}{-b} = -\dfrac{a}{b}$
PERCENTAGE	12		Amount = base × rate $A = BP$
	13		Percent change $= \dfrac{\text{new value} - \text{original value}}{\text{original value}} \times 100$
	14		Percent error $= \dfrac{\text{measured value} - \text{known value}}{\text{known value}} \times 100$
	15		Percent concentration of ingredient $A = \dfrac{\text{amount of } A}{\text{amount of mixture}} \times 100$
	16		Percent efficiency $= \dfrac{\text{output}}{\text{input}} \times 100$

No.

BINARY NUMBERS	17	Largest n-Bit Binary Number	$2^n - 1$
	18	Addition	$0 + 0 = 0$ $0 + 1 = 1$ $1 + 0 = 1$ $1 + 1 = 0$ with carry to next column
	19	Subtraction	$0 - 0 = 0$ $0 - 1 = 1$ with borrow from next column $1 - 0 = 1$ $1 - 1 = 0$ $10 - 1 = 1$
	20		Subtracting B from A is equivalent to adding the two's complement of B to A
	21	Multiplication	$0 \times 0 = 0$ $0 \times 1 = 0$ $1 \times 0 = 0$ $1 \times 1 = 1$
	22	Division	$0 \div 0$ is not defined $1 \div 0$ is not defined $0 \div 1 = 0$ $1 \div 1 = 1$
	23	n-Bit One's Complement of x	$(2^n - 1) - x$
	24		The one's complement of a number can be written by changing the 1s to 0s and the 0s to 1s.
	25	n-Bit Two's Complement of x	$2^n - x$
	26		To find the two's complement of a number first write the one's complement, and add 1.
	27	Negative Binary Numbers	If M is a positive binary number, then $-M$ is the two's complement of M.
EXPONENTS	28	Definition	$a^n = \underbrace{a \cdot a \cdot a \cdot \ldots \cdot a}_{n \text{ factors}}$
	29	Products	$x^a \cdot x^b = x^{a+b}$
	30	Quotients	$\dfrac{x^a}{x^b} = x^{a-b} \quad (x \neq 0)$
	31	Powers	$(x^a)^b = x^{ab} = (x^b)^a$
	32	Product Raised to a Power	$(xy)^n = x^n \cdot y^n$
	33	Quotient Raised to a Power	$\left(\dfrac{x}{y}\right)^n = \dfrac{x^n}{y^n} \quad (y \neq 0)$
	34	Zero Exponent	$x^0 = 1 \quad (x \neq 0)$
	35	Negative Exponent	$x^{-a} = \dfrac{1}{x^a} \quad (x \neq 0)$
	36	Fractional Exponents	$a^{1/n} = \sqrt[n]{a}$
	37		$a^{m/n} = \sqrt[n]{a^m} = (\sqrt[n]{a})^m$

(Complements — rows 23–26; Law of Exponents — rows 29–35)

No.

	No.			
RADICALS	38	Rules of Radicals	Root of a Product	$\sqrt[n]{ab} = \sqrt[n]{a}\,\sqrt[n]{b}$
	39		Root of a Quotient	$\sqrt[n]{\dfrac{a}{b}} = \dfrac{\sqrt[n]{a}}{\sqrt[n]{b}}$
	40		Root of a Power	$\sqrt[n]{a^m} = (\sqrt[n]{a})^m$
SPECIAL PRODUCTS AND FACTORING	41	Binomials	Difference of Two Squares	$a^2 - b^2 = (a - b)(a + b)$
	42		Sum of Two Cubes	$a^3 + b^3 = (a + b)(a^2 - ab + b^2)$
	43		Difference of Two Cubes	$a^3 - b^3 = (a - b)(a^2 + ab + b^2)$
	44	Trinomials	Test for Factorability	$ax^2 + bx + c$ is factorable if $b^2 - 4ac$ is a perfect square
	45		Leading Coefficient = 1	$x^2 + (a + b)x + ab = (x + a)(x + b)$
	46		General Quadratic Trinomial	$acx^2 + (ad + bc)x + bd = (ax + b)(cx + d)$
	47		Perfect Square Trinomials	$a^2 + 2ab + b^2 = (a + b)^2$
	48			$a^2 - 2ab + b^2 = (a - b)^2$
	49	Factoring by Grouping		$ac + ad + bc + bd = (a + b)(c + d)$
FRACTIONS	50	Simplifying		$\dfrac{ad}{bd} = \dfrac{a}{b}$
	51	Multiplication		$\dfrac{a}{b} \cdot \dfrac{c}{d} = \dfrac{ac}{bd}$
	52	Division		$\dfrac{a}{b} \div \dfrac{c}{d} = \dfrac{a}{b} \cdot \dfrac{d}{c} = \dfrac{ad}{bc}$
	53	Addition and Subtraction	Same Denominators	$\dfrac{a}{b} \pm \dfrac{c}{b} = \dfrac{a \pm c}{b}$
	54		Different Denominators	$\dfrac{a}{b} \pm \dfrac{c}{d} = \dfrac{ad}{bd} \pm \dfrac{bc}{bd} = \dfrac{ad \pm bc}{bd}$
PROPORTION	55	In the Proportion $a:b = c:d$	The product of the means equals the product of the extremes.	$ad = bc$
	56		The extremes may be interchanged.	$d:b = c:a$
	57		The means may be interchanged.	$a:c = b:d$
	58		The means may be interchanged with the extremes.	$b:a = d:c$
	59	Mean Proportional	In the Proportion	$a:b = b:c$
			Geometric Mean	$b = \pm\sqrt{ac}$
VARIATION	60	k = Constant of Proportionality	Direct	$y \propto x$ or $y = kx$
	61		Inverse	$y \propto \dfrac{1}{x}$ or $y = \dfrac{k}{x}$
	62		Joint	$y \propto xw$ or $y = kxw$

No.

<table>
<tr><td rowspan="2">SYSTEMS OF LINEAR EQUATIONS</td><td>63</td><td rowspan="2">Algebraic solution</td><td colspan="2">$a_1x + b_1y = c_1$
$a_2x + b_2y = c_2$

$x = \dfrac{b_2c_1 - b_1c_2}{a_1b_2 - a_2b_1}$, and $y = \dfrac{a_1c_2 - a_2c_1}{a_1b_2 - a_2b_1}$　where $a_1b_2 - a_2b_1 \neq 0$</td></tr>
</table>

63 — Algebraic solution

$$a_1x + b_1y = c_1$$
$$a_2x + b_2y = c_2$$

$$x = \frac{b_2c_1 - b_1c_2}{a_1b_2 - a_2b_1}, \quad \text{and} \quad y = \frac{a_1c_2 - a_2c_1}{a_1b_2 - a_2b_1} \quad \text{where } a_1b_2 - a_2b_1 \neq 0$$

64 — Algebraic solution

$$a_1x + b_1y + c_1z = k_1$$
$$a_2x + b_2y + c_2z = k_2$$
$$a_3x + b_3y + c_3z = k_3$$

$$x = \frac{b_2c_3k_1 + b_1c_2k_3 + b_3c_1k_2 - b_2c_1k_3 - b_3c_2k_1 - b_1c_3k_2}{a_1b_2c_3 + a_3b_1c_2 + a_2b_3c_1 - a_3b_2c_1 - a_1b_3c_2 - a_2b_1c_3}$$

$$y = \frac{a_1c_3k_2 + a_3c_2k_1 + a_2c_1k_3 - a_3c_1k_2 - a_1c_2k_3 - a_2c_3k_1}{a_1b_2c_3 + a_3b_1c_2 + a_2b_3c_1 - a_3b_2c_1 - a_1b_3c_2 - a_2b_1c_3}$$

$$z = \frac{a_1b_2k_3 + a_3b_1k_2 + a_2b_3k_1 - a_3b_2k_1 - a_1b_3k_2 - a_2b_1k_3}{a_1b_2c_3 + a_3b_1c_2 + a_2b_3c_1 - a_3b_2c_1 - a_1b_3c_2 - a_2b_1c_3}$$

Determinants — Value of a Determinant

65 — Second Order

$$\begin{vmatrix} a_1 & b_1 \\ a_2 & b_2 \end{vmatrix} = a_1b_2 - a_2b_1$$

66 — Third Order

$$\begin{vmatrix} a_1 & b_1 & c_1 \\ a_2 & b_2 & c_2 \\ a_3 & b_3 & c_3 \end{vmatrix} = a_1b_2c_3 + a_3b_1c_2 + a_2b_3c_1 - (a_3b_2c_1 + a_1b_3c_2 + a_2b_1c_3)$$

67 — Minors

The signed minor of element b in the determinant

$$\begin{vmatrix} a & b & c \\ d & e & f \\ g & h & i \end{vmatrix} \quad \text{is} \quad -\begin{vmatrix} d & f \\ g & i \end{vmatrix}$$

68 — Minors

To find the value of a determinant:

1. Choose any row or any column to develop by minors.
2. Write the product of every element in that row or column and its signed minor.
3. Add these products to get the value of the determinant.

Cramer's Rule

69

The solution for any variable is a fraction whose denominator is the determinant of the coefficients, and whose numerator is also the determinant of the coefficients, except that the column of coefficients of the variable being solved for is replaced by the column of constants.

70 — Two Equations

$$x = \frac{\begin{vmatrix} c_1 & b_1 \\ c_2 & b_2 \end{vmatrix}}{\begin{vmatrix} a_1 & b_1 \\ a_2 & b_2 \end{vmatrix}} \quad \text{and} \quad y = \frac{\begin{vmatrix} a_1 & c_1 \\ a_2 & c_2 \end{vmatrix}}{\begin{vmatrix} a_1 & b_1 \\ a_2 & b_2 \end{vmatrix}}$$

71 — Three Equations

$$x = \frac{\begin{vmatrix} k_1 & b_1 & c_1 \\ k_2 & b_2 & c_2 \\ k_3 & b_3 & c_3 \end{vmatrix}}{\Delta}, \quad y = \frac{\begin{vmatrix} a_1 & k_1 & c_1 \\ a_2 & k_2 & c_2 \\ a_3 & k_3 & c_3 \end{vmatrix}}{\Delta}, \quad z = \frac{\begin{vmatrix} a_1 & b_1 & k_1 \\ a_2 & b_2 & k_2 \\ a_3 & b_3 & k_3 \end{vmatrix}}{\Delta}$$

Where $\Delta = \begin{vmatrix} a_1 & b_1 & c_1 \\ a_2 & b_2 & c_2 \\ a_3 & b_3 & c_3 \end{vmatrix} \neq 0$

Properties of Determinants

No.		
72	Zero Row or Column	If all elements in a row (or column) are zero, the value of the determinant is zero.
73	Identical Rows or Columns	The value of a determinant is zero if two rows (or columns) are identical.
74	Zeros below the Principal Diagonal	If all elements below the principal diagonal are zeros, then the value of the determinant is the product of the elements along the principal diagonal.
75	Interchanging Rows with Columns	The value of a determinant is unchanged if we change the rows to columns and the columns to rows.
76	Interchange of Rows or Columns	A determinant will change sign when we interchange two rows (or columns).
77	Multiplying by a Constant	If each element in a row (or column) is multiplied by some constant, the value of the determinant is multiplied by that constant.
78	Multiples of One Row or Column Added to Another	The value of a determinant is unchanged when the elements of a row (or column) are multiplied by some factor, and then added to the corresponding elements of another row or column.

	No.			
MATRICES	79	Addition	Commutative Law	$\mathbf{A} + \mathbf{B} = \mathbf{B} + \mathbf{A}$
	80		Associative Law	$\mathbf{A} + (\mathbf{B} + \mathbf{C}) = (\mathbf{A} + \mathbf{B}) + \mathbf{C}$ $= (\mathbf{A} + \mathbf{C}) + \mathbf{B}$
	81		Addition and Subtraction	$\begin{pmatrix} a & b \\ c & d \end{pmatrix} + \begin{pmatrix} w & x \\ y & z \end{pmatrix} = \begin{pmatrix} a+w & b+x \\ c+y & d+z \end{pmatrix}$
	82	Multiplication	Conformable Matrices	The product $\mathbf{AB}$ of two matrices $\mathbf{A}$ and $\mathbf{B}$ is defined only when the number of columns in $\mathbf{A}$ equals the number of rows in $\mathbf{B}$.
	83		Commutative Law	$\mathbf{AB} \neq \mathbf{BA}$ Matrix multiplication is *not* commutative.
	84		Associative Law	$\mathbf{A(BC)} = (\mathbf{AB})\mathbf{C} = \mathbf{ABC}$
	85		Distributive Law	$\mathbf{A(B + C)} = \mathbf{AB} + \mathbf{AC}$
	86		Dimensions of the Product	$(m \times p)(p \times n) = (m \times n)$
	87		Product of a Scalar and a Matrix	$k\begin{pmatrix} a & b \\ c & d \end{pmatrix} = \begin{pmatrix} ka & kb \\ kc & kd \end{pmatrix}$
	88		Scalar Product of a Row Vector and a Column Vector	$(a \quad b \quad \cdots)\begin{pmatrix} x \\ y \\ \vdots \end{pmatrix} = (ax + by + \cdots)$
	89		Tensor Product of a Column Vector and a Row Vector	$(2 \times 1)(1 \times 2) \quad (2 \times 2)$ $\begin{pmatrix} a \\ b \end{pmatrix} (x \quad y) = \begin{pmatrix} ax & ay \\ bx & by \end{pmatrix}$
	90		Product of a Row Vector and a Matrix	$(1 \times 2) \quad (2 \times 3) \qquad (1 \times 3)$ $(a \quad b)\begin{pmatrix} u & \boxtimes & w \\ x & y & z \end{pmatrix} = (au + bx \quad d\boxtimes + by \quad aw + bz)$
	91		Product of a Matrix and a Column Vector	$(2 \times 2)(2 \times 1) \quad (2 \times 1)$ $\begin{pmatrix} a & b \\ c & d \end{pmatrix}\begin{pmatrix} x \\ y \end{pmatrix} = \begin{pmatrix} ax + by \\ cx + dy \end{pmatrix}$
	92		Product of Two Matrices	$(2 \times 3) \quad (3 \times 2) \qquad (2 \times 2)$ $\begin{pmatrix} a & b & c \\ d & e & f \end{pmatrix}\begin{pmatrix} u & x \\ \boxtimes & y \\ w & z \end{pmatrix} = \begin{pmatrix} au + b\boxtimes + cw & ax + by + cz \\ du + e\boxtimes + fw & dx + ey + fz \end{pmatrix}$
	93		Product of a Matrix and Its Inverse	$\mathbf{AA^{-1}} = \mathbf{A^{-1}A} = \mathbf{I}$
	94		Multiplying by the Unit Matrix	$\mathbf{AI} = \mathbf{IA} = \mathbf{A}$
	95	Solving Systems of Equations	Matrix Form for a System of Equations	$\mathbf{AX} = \mathbf{B}$
	96		Elementary Transformations of a Matrix	1. Interchange any rows. 2. Multiply a row by a nonzero constant. 3. Add a constant multiple of one row to another row.
	97		Unit Matrix Method	When we transform the coefficient matrix $\mathbf{A}$ into the unit matrix $\mathbf{I}$, the column vector $\mathbf{B}$ gets transformed into the solution set.
	98		Solving a Set of Equations Using the Inverse	$\mathbf{X} = \mathbf{A^{-1}B}$

No.

QUADRATICS	**99**	General Form	$ax^2 + bx + c = 0$
	100	Quadratic Formula	$x = \dfrac{-b \pm \sqrt{b^2 - 4ac}}{2a}$
	101	The Discriminant	If a, b, and c are real, and if $b^2 - 4ac > 0$ the roots are real and unequal if $b^2 - 4ac = 0$ the roots are real and equal if $b^2 - 4ac < 0$ the roots are not real
	102	Polynomial of Degree n	$a_0 x^n + a_1 x^{n-1} + \cdots + a_{n-1}x + a_n$
	103	Factor Theorem	If a polynomial equation $f(x) = 0$ has a root r, then $(x - r)$ is a factor of the polynomial $f(x)$; conversely, if $(x - r)$ is a factor of a polynomial $f(x)$, then r is a root of $f(x) = 0$.
INTERSECTING LINES	**104**		Opposite angles of two intersecting straight lines are equal.
	105		Where two parallel straight lines are cut by a transversal, corresponding angles are equal and alternate interior angles are equal.
	106		Where a number of parallel lines are cut by two transversals, corresponding segments of the transversals are proportional.
QUADRILATERALS	**107**	Square	Area $= a^2$
	108	Rectangle	Area $= ab$
	109	Parallelogram: Diagonals Bisect Each Other	Area $= bh$
	110	Rhombus: Diagonals Intersect at Right Angles	Area $= ah$
	111	Trapezoid	Area $= \dfrac{(a+b)h}{2}$
POLYGON	**112**	n sides	Sum of Angles $= (n-2)180°$
CIRCLES	**113**		Circumference $= 2\pi r = \pi d$
	114		Area $= \pi r^2 = \dfrac{\pi d^2}{4}$
	115		Central angle θ (radians) $= \dfrac{s}{r}$
	116		Area of sector $= \dfrac{rs}{2} = \dfrac{r^2\theta}{2}$ (where θ is in radians)
	117		1 revolution $= 2\pi$ radians $= 360°$ $1° = 60$ minutes 1 minute $= 60$ seconds
	118		Any angle inscribed in a semicircle is a right angle.
	119	Tangents to a Circle	Tangent AP is perpendicular to radius OA.
	120		Tangent AP = tangent BP OP bisects angle APB
	121	Intersecting Chords	$ab = cd$

No.

	No.			
SOLIDS	122		Cube	Volume $= a^3$
	123			Surface area $= 6a^2$
	124		Rectangular Parallelepiped	Volume $= lwh$
	125			Surface area $= 2(lw + hw + lh)$
	126		Any Cylinder or Prism	Volume $=$ (area of base)(altitude)
	127		Right Cylinder or Prism	Lateral area (not incl. bases) $=$ (perimeter of base)(altitude)
	128		Sphere	Volume $= \frac{4}{3}\pi r^3$
	129			Surface area $= 4\pi r^2$
	130		Any Cone or Pyramid	Volume $= \frac{1}{3}$ (area of base)(altitude)
	131		Right Circular Cone or Regular Pyramid	Lateral area $= \frac{1}{2}$ (perimeter of base) $\times$ (slant height)
	132		Any Cone or Pyramid	Volume $= \frac{h}{3}\left(A_1 + A_2 + \sqrt{A_1 A_2}\right)$
	133		Right Circular Cone or Regular Pyramid	Lateral area $= \frac{s}{2}$ (sum of base perimeters) $= \frac{s}{2}(P_1 + P_2)$

	No.		
SIMILAR FIGURES	134		Corresponding dimensions of plane or solid similar figures are in proportion.
	135		Areas of similar plane or solid figures are proportional to the squares of any two corresponding dimensions.
	136		Volumes of similar solid figures are proportional to the cubes of any two corresponding dimensions.

	No.			
ANY TRIANGLE	137	Areas		Area $= \frac{1}{2}bh$
	138		Hero's Formula:	Area $= \sqrt{s(s-a)(s-b)(s-c)}$ where $s = \frac{1}{2}(a+b+c)$
	139		Sum of the Angles	$A + B + C = 180°$
	140		Law of Sines	$\dfrac{a}{\sin A} = \dfrac{b}{\sin B} = \dfrac{c}{\sin C}$
	141		Law of Cosines	$a^2 = b^2 + c^2 - 2bc \cos A$ $b^2 = a^2 + c^2 - 2ac \cos B$ $c^2 = a^2 + b^2 - 2ab \cos C$
	142		Exterior Angle	$\theta = A + B$

	No.	
SIMILAR TRIANGLES	143	If two angles of a triangle equal two angles of another triangle, the triangles are similar.
	144	Corresponding sides of similar triangles are in proportion.

No.

RIGHT TRIANGLES	145		Pythagorean Theorem	$a^2 + b^2 = c^2$
	146	Trigonometric Ratios	Sine	$\sin \theta = \dfrac{y}{r} = \dfrac{\text{opposite side}}{\text{hypotenuse}}$
	147		Cosine	$\cos \theta = \dfrac{x}{r} = \dfrac{\text{adjacent side}}{\text{hypotenuse}}$
	148		Tangent	$\tan \theta = \dfrac{y}{x} = \dfrac{\text{opposite side}}{\text{adjacent side}}$
	149		Cotangent	$\cot \theta = \dfrac{x}{y} = \dfrac{\text{adjacent side}}{\text{opposite side}}$
	150		Secant	$\sec \theta = \dfrac{r}{x} = \dfrac{\text{hypotenuse}}{\text{adjacent side}}$
	151		Cosecant	$\csc \theta = \dfrac{r}{y} = \dfrac{\text{hypotenuse}}{\text{opposite side}}$

	152	Reciprocal Relations	(a) $\sin \theta = \dfrac{1}{\csc \theta}$	(b) $\cos \theta = \dfrac{1}{\sec \theta}$	(c) $\tan \theta = \dfrac{1}{\cot \theta}$

153		In a right triangle, the altitude to the hypotenuse forms two right triangles which are similar to each other and to the original triangle.

154	A and B are Complementary Angles	Cofunctions	(a) $\sin A = \cos B$ (b) $\cos A = \sin B$ (c) $\tan A = \cot B$	(d) $\cot A = \tan B$ (e) $\sec A = \csc B$ (f) $\csc A = \sec B$

	155		Two angles and a side of one are equal to two angles and the corresponding side of the other (ASA), (AAS).
CONGRUENT TRIANGLES	156	Two Triangles Are Congruent If	Two sides and the included angle of one are equal, respectively to two sides and the included angle of the other (SAS).
	157		Three sides of one are equal to the three sides of the other (SSS).

	158		Rectangular	$x = r \cos \theta$
COORDINATE SYSTEMS	159			$y = r \sin \theta$
	160		Polar	$r = \sqrt{x^2 + y^2}$
	161			$\theta = \arctan \dfrac{y}{x}$

	162	Quotient Relations	$\tan \theta = \dfrac{\sin \theta}{\cos \theta}$
TRIGONOMETRIC IDENTITIES	163		$\cot \theta = \dfrac{\cos \theta}{\sin \theta}$
	164	Pythagorean Relations	$\sin^2 \theta + \cos^2 \theta = 1$
	165		$1 + \tan^2 \theta = \sec^2 \theta$
	166		$1 + \cot^2 \theta = \csc^2 \theta$
	167	Sum or Difference of Two Angles	$\sin (\alpha \pm \beta) = \sin \alpha \cos \beta \pm \cos \alpha \sin \beta$
	168		$\cos (\alpha \pm \beta) = \cos \alpha \cos \beta \mp \sin \alpha \sin \beta$
	169		$\tan (\alpha \pm \beta) = \dfrac{\tan \alpha \pm \tan \beta}{1 \mp \tan \alpha \tan \beta}$
	170	Double-Angle Relations	$\sin 2\alpha = 2 \sin \alpha \cos \alpha$

171	Double-Angle Relations	(a) $\cos 2\alpha = \cos^2 \alpha - \sin^2 \alpha$	(b) $\cos 2\alpha = 1 - 2 \sin^2 \alpha$	(c) $\cos 2\alpha = 2 \cos^2 \alpha - 1$
172		$\tan 2\alpha = \dfrac{2 \tan \alpha}{1 - \tan^2 \alpha}$		

No.

TRIGONOMETRIC IDENTITIES *(continued)*				
173	Half-Angle Relations	$\sin \dfrac{\alpha}{2} = \pm \sqrt{\dfrac{1-\cos\alpha}{2}}$		
174		$\cos \dfrac{\alpha}{2} = \mp \sqrt{\dfrac{1+\cos\alpha}{2}}$		
175		(a) $\tan \dfrac{\alpha}{2} = \dfrac{1-\cos\alpha}{\sin\alpha}$	(b) $\tan \dfrac{\alpha}{2} = \dfrac{\sin\alpha}{1+\cos\alpha}$	(c) $\tan \dfrac{\alpha}{2} = \pm\sqrt{\dfrac{1-\cos\alpha}{1+\cos\alpha}}$
176	Sum or Difference of Two Functions	$\sin\alpha + \sin\beta = 2\sin\tfrac{1}{2}(\alpha+\beta)\cos\tfrac{1}{2}(\alpha-\beta)$		
177		$\sin\alpha - \sin\beta = 2\cos\tfrac{1}{2}(\alpha+\beta)\sin\tfrac{1}{2}(\alpha-\beta)$		
178		$\cos\alpha + \cos\beta = 2\cos\tfrac{1}{2}(\alpha+\beta)\cos\tfrac{1}{2}(\alpha-\beta)$		
179		$\cos\alpha - \cos\beta = 2\sin\tfrac{1}{2}(\alpha+\beta)\sin\tfrac{1}{2}(\alpha-\beta)$		
180	Product of Two Functions	$\sin\alpha \sin\beta = \tfrac{1}{2}\cos(\alpha-\beta) - \tfrac{1}{2}\cos(\alpha+\beta)$		
181		$\cos\alpha \cos\beta = \tfrac{1}{2}\cos(\alpha-\beta) + \tfrac{1}{2}\cos(\alpha+\beta)$		
182		$\sin\alpha \cos\beta = \tfrac{1}{2}\sin(\alpha+\beta) + \tfrac{1}{2}\sin(\alpha-\beta)$		
183	Inverse Trigonometric Functions	$\theta = \arcsin C = \arctan \dfrac{C}{\sqrt{1-C^2}}$		
184		$\theta = \arccos D = \arctan \dfrac{\sqrt{1-D^2}}{D}$		
LOGARITHMS				
185	Definition of e	$\displaystyle\lim_{k\to\infty}\left(1+\dfrac{1}{k}\right)^k = e$		
186	Exponential to Logarithmic Form	If $b^x = y$ then $x = \log_b y$ $(y>0,\ b>0,\ b\neq 1)$		
187	Laws of Logarithms — Products	$\log_b MN = \log_b M + \log_b N$		
188	Laws of Logarithms — Quotients	$\log_b \dfrac{M}{N} = \log_b M - \log_b N$		
189	Laws of Logarithms — Powers	$\log_b M^p = p \log_b M$		
190	Laws of Logarithms — Roots	$\log_b \sqrt[q]{M} = \dfrac{1}{q}\log_b M$		
191	Laws of Logarithms — Log of 1	$\log_b 1 = 0$		
192	Laws of Logarithms — Log of the Base	$\log_b b = 1$		
193	Log of the Base Raised to a Power	$\log_b b^n = n$		
194	Base Raised to a Logarithm of the Same Base	$b^{\log_b x} = x$		
195	Change of Base	$\log N = \dfrac{\ln N}{\ln 10} \cong \dfrac{\ln N}{2.3026}$		

No.

SOME USEFUL FUNCTIONS

No.			
196	Power Function		$y = ax^n$
197	Exponential Function		$y = a(b)^{nx}$
198	Series Approximation	$b^x = 1 + x \ln b + \dfrac{(x \ln b)^2}{2!} + \dfrac{(x \ln b)^3}{3!} + \cdots$ $(b > 0)$	
199	Exponential Growth		$y = ae^{nt}$
200		Doubling Time or Half-Life	$t = \dfrac{\ln 2}{n}$
201	Exponential Decay		$y = ae^{-nt}$
202	Exponential Growth to an Upper Limit		$y = a(1 - e^{-nt})$
203	Time Constant	$T = \dfrac{1}{\text{growth rate } n}$	
204	Recursion Relation for Exponential Growth	$y_t = By_{t-1}$	
205	Nonlinear Growth Equation	$y_t = By_{t-1}(1 - y_{t-1})$	
206	Series Approximations	$e = 2 + \dfrac{1}{2!} + \dfrac{1}{3!} + \dfrac{1}{4!} + \cdots$	
207		$e^x = 1 + x + \dfrac{x^2}{2!} + \dfrac{x^3}{3!} + \dfrac{x^4}{4!} + \cdots$	
208	Logarithmic Function		$y = \log_b x$ $(x > 0,\ b > 0,\ b \neq 1)$

No.				
209		Series Approximation		$\ln x = 2a + \dfrac{2a^3}{3} + \dfrac{2a^5}{5} + \dfrac{2a^7}{7} + \cdots$ where $a = \dfrac{x-1}{x+1}$
210				$y = a \sin(bx + c)$
211		Sine Wave of Amplitude a		$\text{Period} = \dfrac{360}{b}$ deg/cycle $= \dfrac{2\pi}{b}$ rad/cycle
212				$\text{Frequency} = \dfrac{b}{360}$ cycle/deg $= \dfrac{b}{2\pi}$ cycle/rad
213				$\text{Phase shift} = -\dfrac{c}{b}$
214		Addition of a Sine Wave and a Cosine Wave		$A \sin \omega t + B \cos \omega t = R \sin(\omega t + \phi)$ where $R = \sqrt{A^2 + B^2}$ and $\phi = \arctan \dfrac{B}{A}$
215		Series Approximation		$\sin x = x - \dfrac{x^3}{3!} + \dfrac{x^5}{5!} - \dfrac{x^7}{7!} + \cdots$
216				$\cos x = 1 - \dfrac{x^2}{2!} + \dfrac{x^4}{4!} - \dfrac{x^6}{6!} + \cdots$

SOME USEFUL FUNCTIONS (*Continued*)

No.				
217		Powers of j		$j = \sqrt{-1}, \quad j^2 = -1, \quad j^3 = -j, \quad j^4 = 1, \quad j^5 = j, \text{ etc.}$
218	Rectangular Form			$a + jb$
219			Sums	$(a + jb) + (c + jd) = (a + c) + j(b + d)$
220			Differences	$(a + jb) - (c + jd) = (a - c) + j(b - d)$
221			Products	$(a + jb)(c + jd) = (ac - bd) + j(ad + bc)$
222			Quotients	$\dfrac{a + jb}{c + jd} = \dfrac{ac + bd}{c^2 + d^2} + j\dfrac{bc - ad}{c^2 + d^2}$
223	Trigonometric Form			$a + jb = r(\cos \theta + j \sin \theta)$
224				$a = r \cos \theta$
225			where	$b = r \sin \theta$
226				$r = \sqrt{a^2 + b^2}$
227				$\theta = \arctan \dfrac{b}{a}$
228	Polar Form			$r \,\underline{/\theta} = a + jb$
229			Products	$r \,\underline{/\theta} \cdot r' \,\underline{/\theta'} = rr' \,\underline{/\theta + \theta'}$
230			Quotients	$r \,\underline{/\theta} \div r' \,\underline{/\theta'} = \dfrac{r}{r'} \,\underline{/\theta - \theta'}$
				DeMoivre's Theorem: $(r \,\underline{/\theta})^n = r^n \,\underline{/n\theta}$
				$re^{j\theta} = r(\cos \theta + j \sin \theta)$
				$r_1 e^{j\theta_1} \cdot r_2 e^{j\theta_2} = r_1 r_2 e^{j(\theta_1 + \theta_2)}$
				$\dfrac{r_1 e^{j\theta_1}}{r_2 e^{j\theta_2}} = \dfrac{r_1}{r_2} e^{j(\theta_1 - \theta_2)}$
				$(re^{j\theta})^n = r^n e^{jn\theta}$

COMPLEX NUMBERS

No.

236	Arithmetic Progression Common Difference $= d$	Recursion Formula	$a_n = a_{n-1} + d$
237		General Term	$a_n = a + (n-1)d$
238		Sum of n Terms	$s_n = \dfrac{n(a + a_n)}{2}$
239			$s_n = \dfrac{n}{2}[2a + (n-1)d]$
240	Geometric Progression Common Ratio $= r$	Recursion Formula	$a_n = ra_{n-1}$
241		General Term	$a_n = ar^{n-1}$
242		Sum of n Terms	$s_n = \dfrac{a(1 - r^n)}{1 - r}$
243			$s_n = \dfrac{a - ra_n}{1 - r}$
244		Sum to Infinity	$S = \dfrac{a}{1 - r} \quad\text{where}\quad |r| < 1$

PROGRESSIONS

245	Binomial Expansion	$(a + b)^n = a^n + na^{n-1}b + \dfrac{n(n-1)}{2!}a^{n-2}b^2 + \dfrac{n(n-1)(n-2)}{3!}a^{n-3}b^3 + \cdots + b^n$
246	General Term	$r\text{th term} = \dfrac{n!}{(r-1)!\,(n-r+1)!}a^{n-r+1}b^{r-1}$
247	Binomial Series	$(a + b)^n = a^n + na^{n-1}b + \dfrac{n(n-1)}{2!}a^{n-2}b^2 + \dfrac{n(n-1)(n-2)}{3!}a^{n-3}b^3 + \cdots$ where $|a| > |b|$
248		$(1 + x)^n = 1 + nx + \dfrac{n(n-1)}{2!}x^2 + \dfrac{n(n-1)(n-2)}{3!}x^3 + \cdots$ where $|x| < 1$

BINOMIAL THEOREM

249	Measures of Central Tendency	Arithmetic Mean	$\bar{x} = \dfrac{\sum x}{n}$
250		Median	The median of a set of numbers arranged in order of magnitude is the middle value, or the mean of the two middle values.
251		Mode	The mode of a set of numbers is the measurement(s) that occurs most often in the set.
252	Measures of Dispersion	Range	The range of a set of numbers is the difference between the largest and the smallest number in the set.
253		Population Variance σ^2	$\sigma^2 = \dfrac{\sum (x - \bar{x})^2}{n}$
254		Sample Variance s^2	$s^2 = \dfrac{\sum (x - \bar{x})^2}{n - 1}$
255		Standard Deviation s	The standard deviation of a set of numbers is the positive square root of the variance.
256	Probability	Of a Single Event	$P(A) = \dfrac{\text{number of ways in which } A \text{ can happen}}{\text{number of equally likely ways}}$
257		Of Two Events Both Occurring	$P(A, B) = P(A)P(B)$
258		Of Several Events All Occurring	$P(A, B, C, \ldots) = P(A)P(B)P(C)\cdots$
259		Of Either of Two Events Occurring	$P(A + B) = P(A) + P(B) - P(A, B)$
260		Of Two Mutually Exclusive Events Occurring	$P(A + B) = P(A) + P(B)$
261	Gaussian Distribution		$y = \dfrac{1}{\sigma\sqrt{2\pi}}e^{-(x-\mu)^2/2\sigma^2}$ where μ = population mean and σ = population standard deviation

STATISTICS AND PROBABILITY

No.

<table>
<tr><td rowspan="6">STATISTICS AND PROBABILITY (Continued)</td><td>262</td><td colspan="2">Binomial Probability Formula</td><td colspan="2">$P(x) = \dfrac{n!}{(n-x)!\,x!}\,p^x q^{n-x}$</td></tr>
<tr><td>263</td><td rowspan="3">Standard Error</td><td>Of the Mean</td><td colspan="2">$SE_{\bar{x}} = \dfrac{\sigma}{\sqrt{n}} = \dfrac{s}{\sqrt{n}}$</td></tr>
<tr><td>264</td><td>Of the Standard Deviation</td><td colspan="2">$SE_s = \dfrac{\sigma}{\sqrt{2n}}$</td></tr>
<tr><td>265</td><td>Of a Proportion</td><td colspan="2">$SE_p = \sqrt{\dfrac{p(1-p)}{n}}$</td></tr>
<tr><td>266</td><td colspan="2">Correlation Coefficient</td><td colspan="2">$r = \dfrac{n\,\Sigma xy - \Sigma x\,\Sigma y}{\sqrt{n\,\Sigma x^2 - (\Sigma x)^2}\ \sqrt{n\,\Sigma y^2 - (\Sigma y)^2}}$</td></tr>
<tr><td>267</td><td colspan="2">Least Squares Line</td><td colspan="2">Slope $m = \dfrac{n\Sigma xy - \Sigma x\,\Sigma y}{n\,\Sigma x^2 - (\Sigma x)^2}$

y intercept $b = \dfrac{\Sigma x^2\,\Sigma y - \Sigma x\,\Sigma xy}{n\,\Sigma x^2 - (\Sigma x)^2}$</td></tr>
</table>

			Truth Table	Venn Diagram	Switch Diagram	Logic Gate
BOOLEAN ALGEBRA AND SETS	Boolean Operators	**268** AND	$A\ \ B\ \ A\cdot B$ $0\ \ 0\ \ \ 0$ $0\ \ 1\ \ \ 0$ $1\ \ 0\ \ \ 0$ $1\ \ 1\ \ \ 1$	Intersection $A \cap B$ $= \{x : x \in A, x \in B\}$	In Series	AB AND gate
		269 OR	$A\ \ B\ \ A+B$ $0\ \ 0\ \ \ 0$ $0\ \ 1\ \ \ 1$ $1\ \ 0\ \ \ 1$ $1\ \ 1\ \ \ 1$	Union $A \cup B$ $= \{x : x \in A, \text{ or } x \in B\}$	In Parallel	$A + B$ OR Gate
		270 NOT	$A\ \ \bar{A}$ $0\ \ \ 1$ $1\ \ \ 0$	A^c Complement A^c $= \{x : x \in U, x \in A\}$	Inverter	$\bar{A}$ Inverter
		271 Exclusive OR	$A\ \ B\ \ A+B$ $0\ \ 0\ \ \ 0$ $0\ \ 1\ \ \ 1$ $1\ \ 0\ \ \ 1$ $1\ \ 1\ \ \ 0$	$A \oplus B = \{x : x \in A$ $\text{or } x \in B,$ $x \in A \cap B\}$		$A \oplus B$ Exclusive OR Gate

No.

BOOLEAN ALGEBRA AND SETS (Continued) — Boolean Algebra (Continued)

No.		(a)	(b)
275	Idempotent Laws	$AA \equiv A$	$A + A \equiv A$
276	Complement Laws	$A \cdot \overline{A} \equiv 0$	$A + \overline{A} \equiv 1$
277	Associative Laws	$A(BC) \equiv (AB)C$	$A + (B + C) \equiv (A + B) + C$
278	Distributive Laws	$A(B + C) \equiv AB + AC$	$A + BC \equiv (A + B)(A + C)$
279	Distributive Laws	$A(\overline{A} + B) \equiv AB$	$A + \overline{A} \cdot B \equiv A + B$
280	Absorption Laws	$A(A + B) \equiv A$	$A + (AB) \equiv A$
281	DeMorgan's Laws	$\overline{A \cdot B} \equiv \overline{A} + \overline{B}$	$\overline{A + B} \equiv \overline{A} \cdot \overline{B}$
282	Involution Law	$\overline{\overline{A}} \equiv A$	

THE STRAIGHT LINE

No.			
283		Distance Formula	$d = \sqrt{(\Delta x)^2 + (\Delta y)^2} = \sqrt{(x_2 - x_1)^2 + (y_2 - y_1)^2}$
284		Slope m	$m = \dfrac{\text{rise}}{\text{run}} = \dfrac{\Delta y}{\Delta x} = \dfrac{y_2 - y_1}{x_2 - x_1}$
285			$m = \tan(\text{angle of inclination}) = \tan \theta$ $0 \le \theta < 180°$
286		General Form	$Ax + By + C = 0$
287		Parallel to x axis	$y = b$
288		Parallel to y axis	$x = a$
289	Equation of Straight Line	Slope-Intercept Form	$y = mx + b$
290		Two-Point Form	$\dfrac{y - y_1}{x - x_1} = \dfrac{y_2 - y_1}{x_2 - x_1}$
291		Point-Slope Form	$m = \dfrac{y - y_1}{x - x_1}$
292		Intercept Form	$\dfrac{x}{a} + \dfrac{y}{b} = 1$
293		Polar Form	$r \cos(\theta - \beta) = p$
294		If L_1 and L_2 are parallel, then	$m_1 = m_2$
295		If L_1 and L_2 are perpendicular, then	$m_1 = -\dfrac{1}{m_2}$
296		Angle of Intersection	$\tan \phi = \dfrac{m_2 - m_1}{1 + m_1 m_2}$

CONIC SECTIONS — Any Conic

No.		
297	General Second-Degree Equation	$Ax^2 + Bxy + Cy^2 + Dx + Ey + F = 0$
298	Translation of Axes	To translate or shift the axes of a curve to the left by a distance h and downward by a distance k, replace x by $(x - h)$ and y by $(y - k)$ in the equation of the curve.

		No.			

		299	Eccentricity		$e = \dfrac{\cos \beta}{\cos \alpha}$
CONIC SECTIONS (Continued)	Any Conic (continued)	300			$e = 0$ for the Circle $0 < e < 1$ for the Ellipse $e = 1$ for the Parabola $e > 1$ for the Hyperbola
		301	Definition of a Conic		$PF = e \cdot PD$
		302	Polar Equation for the Conics		$r = \dfrac{ke}{1 - e \cos \theta}$
	Circle of Radius r	303	The set of points in a plane equidistant from a fixed point		
		304		Standard Form	$x^2 + y^2 = r^2$
		305			$(x - h)^2 + (y - k)^2 = r^2$
		306	General Form		$x^2 + y^2 + Dx + Ey + F = 0$
	Parabola	307	The set of points in a plane such that the distance PF from each point P to a fixed point F (the focus) is equal to the distance PD to a fixed line (the directrix) $PF = PD$		
		308		Standard Form	$y^2 = 4px$
		309			$x^2 = 4py$
		310			$(y - k)^2 = 4p(x - h)$
					$(x - h)^2 = 4p(y - k)$

No.

CONIC SECTIONS (Continued)		No.		
		314	Area	$\dfrac{2}{3}\,ab$
		315	The set of points in a plane such that the sum of the distances PF and PF' from each point P to two fixed points F & F' (the foci) is constant, and equal to the length $2a$ of the major axis $$PF + PF' = 2a$$	
	Ellipse	**316**		Standard Form: $\dfrac{x^2}{a^2} + \dfrac{y^2}{b^2} = 1$ $a > b$
		317		$\dfrac{y^2}{a^2} + \dfrac{x^2}{b^2} = 1$ $a > b$
		318		$\dfrac{(x - h)^2}{a^2} + \dfrac{(y - k)^2}{b^2} = 1$ $a > b$
		319		$\dfrac{(y - k)^2}{a^2} + \dfrac{(x - h)^2}{b^2} = 1$ $a > b$
		320	General Form	$Ax^2 + Cy^2 + Dx + Ey + F = 0$ $A \neq C$, but have same signs
		321	Distance from Center to Focus	$c = \sqrt{a^2 - b^2}$
		322	Focal Width or length of latus rectum	$L = \dfrac{2b^2}{a}$
		323		Eccentricity $e = \dfrac{a}{d} = \dfrac{c}{a}$
		324	Area	πab

No.

<table>
<tr>
<td rowspan="11" style="writing-mode: vertical">CONIC SECTIONS (Continued)</td>
</tr>
</table>

No.			
325		The set of points in a plane such that the difference of the distances PF and PF' from each point P to two fixed points F & F' (the foci) is constant, and equal to the distance $2a$ between the vertices $$PF' - PF = 2a$$	

No.	Hyperbola	Diagram	Standard Form	Equation
326			Standard Form	$$\dfrac{x^2}{a^2} - \dfrac{y^2}{b^2} = 1$$
327				$$\dfrac{y^2}{a^2} - \dfrac{x^2}{b^2} = 1$$
328				$$\dfrac{(x-h)^2}{a^2} - \dfrac{(y-k)^2}{b^2} = 1$$
329				$$\dfrac{(y-k)^2}{a^2} - \dfrac{(x-h)^2}{b^2} = 1$$

No.			
330	General Form		$Ax^2 + Cy^2 + Dx + Ey + F = 0$ $A \neq C$, and have opposite signs
331	Distance from Center to Focus		$$c = \sqrt{a^2 + b^2}$$
332	Slope of Asymptotes	Transverse Axis Horizontal	$$\text{Slope} = \pm\dfrac{b}{a}$$
333		Transverse Axis Vertical	$$\text{Slope} = \pm\dfrac{a}{b}$$
334	Length of Latus Rectum		$$L = \dfrac{2b^2}{a}$$
335			Axes Rotated 45° $xy = k$

No.

No.			
339		The Chain Rule	$$\dfrac{dy}{dx} = \dfrac{dy}{du} \cdot \dfrac{du}{dx}$$
340		Of a Constant	$$\dfrac{d(c)}{dx} = 0$$
341		Of a Power Function	$$\dfrac{d}{dx}\, x^n = nx^{n-1}$$
342		Of a Constant Times a Function	$$\dfrac{d(cu)}{dx} = c\dfrac{du}{dx}$$
343		Of a Constant Times a Power of x	$$\dfrac{d}{dx}\, cx^n = cnx^{n-1}$$
344		Of a Sum	$$\dfrac{d}{dx}\,(u + v + w) = \dfrac{du}{dx} + \dfrac{dv}{dx} + \dfrac{dw}{dx}$$
345		Of a Function Raised to a Power	$$\dfrac{d(cu^n)}{dx} = cnu^{n-1}\dfrac{du}{dx}$$
346		Of a Product	$$\dfrac{d(uv)}{dx} = u\dfrac{dv}{dx} + v\dfrac{du}{dx}$$
347		Of a Product of Three Factors	$$\dfrac{d(uvw)}{dx} = uv\dfrac{dw}{dx} + uw\dfrac{dv}{dx} + vw\dfrac{du}{dx}$$
348		Of a Product of n Factors	The derivative is an expression of n terms, each term being the product of $n-1$ of the factors and the derivative of the other factor.
349		Of a Quotient	$$\dfrac{d}{dx}\left(\dfrac{u}{v}\right) = \dfrac{v\dfrac{du}{dx} - u\dfrac{dv}{dx}}{v^2}$$
350			$$\dfrac{d(\sin u)}{dx} = \cos u\,\dfrac{du}{dx}$$
351			$$\dfrac{d(\cos u)}{dx} = -\sin u\,\dfrac{du}{dx}$$
352		Of the Trigonometric Functions	$$\dfrac{d(\tan u)}{dx} = \sec^2 u\,\dfrac{du}{dx}$$
353			$$\dfrac{d(\cot u)}{dx} = -\csc^2 u\,\dfrac{du}{dx}$$
354			$$\dfrac{d(\sec u)}{dx} = \sec u\,\tan u\,\dfrac{du}{dx}$$
355			$$\dfrac{d(\csc u)}{dx} = -\csc u\,\cot u\,\dfrac{du}{dx}$$
356			$$\dfrac{d(\sin^{-1} u)}{dx} = \dfrac{1}{\sqrt{1-u^2}}\dfrac{du}{dx} \qquad -1 < u < 1$$
357			$$\dfrac{d(\cos^{-1} u)}{dx} = \dfrac{-1}{\sqrt{1-u^2}}\dfrac{du}{dx} \qquad -1 < u < 1$$
358		Of the Inverse Trigonometric Functions	$$\dfrac{d(\tan^{-1} u)}{dx} = \dfrac{1}{1+u^2}\dfrac{du}{dx}$$
359			$$\dfrac{d(\cot^{-1} u)}{dx} = \dfrac{-1}{1+u^2}\dfrac{du}{dx}$$
360			$$\dfrac{d(\sec^{-1} u)}{dx} = \dfrac{1}{u\sqrt{u^2-1}}\dfrac{du}{dx} \qquad |u| > 1$$
361			$$\dfrac{d(\csc^{-1} u)}{dx} = \dfrac{-1}{u\sqrt{u^2-1}}\dfrac{du}{dx} \qquad |u| > 1$$

DIFFERENTIAL CALCULUS (*Continued*)

Rules for Derivatives

No.

362			(a) $\dfrac{d}{dx}(\log_b u) = \dfrac{1}{u}\log_b e\,\dfrac{du}{dx}$ | (b) $\dfrac{d}{dx}(\log_b u) = \dfrac{1}{u\ln b}\dfrac{du}{dx}$
363	Of Logarithmic and Exponential Functions (*Continued*)		$\dfrac{d}{dx}(\ln u) = \dfrac{1}{u}\dfrac{du}{dx}$
364			$\dfrac{d}{dx}b^u = b^u\dfrac{du}{dx}\ln b$
365			$\dfrac{d}{dx}e^u = e^u\dfrac{du}{dx}$
366		Maximum and Minimum Points	To find maximum and minimum points (and other stationary points) set the first derivative equal to zero and solve for x.
367		First-Derivative Test	The first derivative is negative to the left of, and positive to the right of, a minimum point. The reverse is true for a maximum point.
368	Graphical Applications	Second-Derivative Test	If the first derivative at some point is zero, then, if the second derivative is 1. Positive, the point is a minimum. 2. Negative, the point is a maximum. 3. Zero, the test fails.
369		Ordinate Test	Find y a small distance to either side of the point to be tested. If y is greater there, we have a minimum; if less, we have a maximum.
370		Inflection Points	To find points of inflection, set the second derivative to zero and solve for x. Test by seeing if the second derivative changes sign a small distance to either side of the point.
371		Newton's Method	$x_{n+1} = x_n - \dfrac{f(x_n)}{f'(x_n)}$
372	Differentials	Differential of y	$dy = f'(x)\,dx$
373		Approximations Using Differentials	$\Delta y \cong \dfrac{dy}{dx}\Delta x$
374		The Indefinite Integral	$\displaystyle\int F'(x)\,dx = F(x) + C$
375			$\displaystyle\int f(x)\,dx = F(x) + C$ where $f(x) = F'(x)$
376	The Definite Integral	The Fundamental Theorem	$\displaystyle\int_a^b f(x)\,dx = F(b) - F(a)$
377			$\displaystyle\int_a^b c\,f(x)\,dx = c\int_a^b f(x)\,dx$
378		Properties of the Definite Integral	$\displaystyle\int_a^b [f(x) + g(x)]\,dx = \int_a^b f(x)\,dx + \int_a^b g(x)\,dx$
379			$\displaystyle\int_a^b f(x)\,dx = -\int_b^a f(x)\,dx$
380			$\displaystyle\int_a^b f(x)\,dx = \int_a^c f(x)\,dx + \int_c^b f(x)\,dx$

Left vertical labels: **DIFFERENTIAL CALCULUS (Continued)** (rows 362–373), **INTEGRAL CALCULUS** (rows 374–380).

No.

INTEGRAL CALCULUS (*Continued*)	**Approximate Area under a Curve**		Midpoint Method	$A \cong \sum_{i=1}^{n} f(x_i^*) \, \Delta x$ where $f(x_i^*)$ is the height of the *i*th panel at its midpoint
381				
382			Average Ordinate Method	$A \cong y_{\text{avg}} \, (b - a)$
383			Trapezoidal Rule	With unequal spacing $A \cong \frac{1}{2}[(x_1 - x_0)(y_1 + y_0) + (x_2 - x_1)(y_2 - y_1) + \cdots + (x_n - x_{n-1})(y_n + y_{n-1})]$
384				With equal spacing, $h = x_1 - x_0$: $A \cong h[\frac{1}{2}(y_0 + y_n) + y_1 + y_2 + \cdots + y_{n-1}]$
385			Prismoidal Formula	$A = \dfrac{h}{3}\,(y_0 + 4y_1 + y_2)$
386			Simpson's Rule	$A \cong \dfrac{h}{3}\,(y_0 + 4y_1 + 2y_2 + 4y_3 + \cdots + 4y_{n-1} + y_n)$
387	**Exact Area under a Curve**	Defined by Riemann Sums		$A = \lim\limits_{\Delta x \to 0} \sum_{i=1}^{n} f(x_i^*) \, \Delta x = \int_a^b f(x) \, dx$
388		By Integration		$A = \int_a^b f(x) \, dx = F(b) - F(a)$
389		Areas between Two Curves		$A = \int_a^b [f(x) - g(x)] \, dx$
APPLICATIONS OF THE DEFINITE INTEGRAL	**Volumes of Solids of Revolution**	**Disk Method**		Volume $= dV = \pi r^2 \, dh$
390				
391				$V = \pi \int_a^b r^2 \, dh$
392		**Ring Method**		$dV = \pi(r_o^2 - r_i^2) \, dh$
393				$V = \pi \int_a^b (r_o^2 - r_i^2) \, dh$

394	Volumes of Solids of Revolution	Shell Method		$dV = 2\pi rh\, dr$
395				$V = 2\pi \displaystyle\int_a^b rh\, dr$
396	Length of Arc			$s = \displaystyle\int_a^b \sqrt{1 + \left(\dfrac{dy}{dx}\right)^2}\, dx$
397				$s = \displaystyle\int_c^d \sqrt{1 + \left(\dfrac{dx}{dy}\right)^2}\, dy$
398	Surface Area			About x Axis: $S = 2\pi \displaystyle\int_a^b y \sqrt{1 + \left(\dfrac{dy}{dx}\right)^2}\, dx$
399				About y Axis: $S = 2\pi \displaystyle\int_a^b x \sqrt{1 + \left(\dfrac{dy}{dx}\right)^2}\, dx$
400			Of Plane Area:	$\bar{x} = \dfrac{1}{A} \displaystyle\int_a^b x(y_2 - y_1)\, dx$
401				$\bar{y} = \dfrac{1}{2A} \displaystyle\int_a^b (y_1 + y_2)(y_2 - y_1)\, dx$

APPLICATIONS OF THE DEFINITE INTEGRAL (Continued)

troids

Of Solid of Revolution of Volume V:

About x Axis: $\bar{x} = \dfrac{\pi}{V}\displaystyle\int_a^b xy^2\, dx$

No.

APPLICATIONS OF THE DEFINITE INTEGRAL (*Continued*)	**Moment of Inertia**	**Of Areas**	Thin Strip 	$I_p = Ar^2$
			Extended Area 	$I_x = \dfrac{1}{3}\int y^3\,dx$
				$I_y = \int x^2\,y\,dx$
				Polar $I_0 = I_x + I_y$
				Radius of Gyration: $r = \sqrt{\dfrac{I}{A}}$
				Parallel Axis Theorem: $I_B = I_A + As^2$

No.				About Axis of Revolution (Polar Moment of Inertia)	
410					$I_p = Mr^2$
411		**Of Masses**	Disk:		$dI = \dfrac{m\pi}{2}\,r^4\,dh$
412			Ring:		$dI = \dfrac{m\pi}{2}\,(r_0^4 - r_i^4)\,dh$
413			Shell:		$dI = 2\pi m r^3 h\,dr$
414			Solid of Revolution:		Disk Method: $I = \dfrac{m\pi}{2}\displaystyle\int_a^b r^4\,dh$
415					Shell Method: $I = 2\pi m \displaystyle\int r^3 h\,dr$
416					Parallel Axis Theorem: $I_B = I_A + Ms^2$
417	Average and rms Values				Average Ordinate: $y_{avg} = \dfrac{1}{b-a}\displaystyle\int_a^b f(x)\,dx$
418					Root-Mean-Square Value: $rms = \sqrt{\dfrac{1}{b-a}\displaystyle\int_a^b [f(x)]^2\,dx}$

Row numbers (left column): 404, 405, 406, 407, 408, 409, 410, 411, 412, 413, 414, 415, 416, 417, 418

No.

419	First-Order		Variables Separable		$f(y)\,dy = g(x)\,dx$
420			Integrable Combinations		$x\,dy + y\,dx = d(xy)$
421					$\dfrac{x\,dy - y\,dx}{x^2} = d\left(\dfrac{y}{x}\right)$
422					$\dfrac{y\,dx - x\,dy}{y^2} = d\left(\dfrac{x}{y}\right)$
423					$\dfrac{x\,dy - y\,dx}{x^2 + y^2} = d\left(\tan^{-1}\dfrac{y}{x}\right)$
424			Homogeneous		$M\,dx + N\,dy = 0$ (Substitute $y = vx$)
425			First-Order Linear	Form	$y' + Py = Q$
426				Integrating Factor	$R = e^{\int P\,dx}$
427				Solution	$ye^{\int P\,dx} = \int Qe^{\int P\,dx}\,dx$
428	Second-Order	Right Side Zero	Form		$ay'' + by' + cy = 0$
429			Auxiliary Equation		$am^2 + bm + c = 0$

				Roots of Auxiliary Equation	Solution
430	Second-Order	Right Side Zero	Form of Solution	Real and Unequal	$y = c_1 e^{m_1 x} + c_2 e^{m_2 x}$
431				Real and Equal	$y = c_1 e^{mx} + c_2 x e^{mx}$
432				Nonreal	(a) $y = e^{ax}(C_1 \cos bx + C_2 \sin bx)$ or (b) $y = Ce^{ax}\sin(bx + \phi)$

433	Second-Order	Right Side Not Zero	Form		$ay'' + by' + cy = f(x)$
434			Complete Solution		$y = \underset{\substack{\uparrow \\ \text{complementary} \\ \text{function}}}{y_c} + \underset{\substack{\uparrow \\ \text{particular} \\ \text{integral}}}{y_p}$
435			Bernoulli's Equation		$\dfrac{dy}{dx} + Py = Qy^n$ (Substitute $z = y^{1-n}$)
436	Laplace			Definition	$\mathscr{L}[f(t)] = \displaystyle\int_0^\infty f(t)e^{-st}\,dt$

(Left vertical label: **DIFFERENTIAL EQUATIONS**)

No.

		No.			
DIFFERENTIAL EQUATIONS (*Continued*)	Numerical Solution	**438**	Euler's Method		$x_q = x_p + \Delta x$ $y_q = y_p + m_p\,\Delta x$
		439	Modified Euler's Method		$x_q = x_p + \Delta x$ $y_q = y_p + \left(\dfrac{m_p + m_q}{2}\right)\Delta x$
		440	Runge–Kutta Method		$x_q = x_p + \Delta x$ $y_q = y_p + m_{\text{avg}}\,\Delta x$ where $\quad m_{\text{avg}} = \left(\tfrac{1}{6}\right)(m_p + 2m_r + 2m_s + m_q)$ and $\quad m_p = f'(x_p, y_p)$ $m_r = f'\!\left(x_p + \dfrac{\Delta x}{2},\, y_p + m_p\dfrac{\Delta x}{2}\right)$ $m_s = f'\!\left(x_p + \dfrac{\Delta x}{2},\, y_p + m_r\dfrac{\Delta x}{2}\right)$ $m_q = f'(x_p + \Delta x,\, y_p + m_s\,\Delta x)$
INFINITE SERIES	Power Series	**441**	Notation		$u_1 + u_2 + u_3 + \cdots + u_n + \cdots$
		442	Tests for Convergence	Limit Test	$\lim\limits_{n \to \infty} u_n = 0$
		443		Partial Sum Test	$\lim\limits_{n \to \infty} S_n = S$
		444		Ratio Test	If $\lim\limits_{n \to \infty}\left|\dfrac{u_n + 1}{u_n}\right|$ (a) is less than 1, the series converges. (b) is greater than 1, the series diverges. (c) is equal to 1, the test fails.
		445	Maclaurin's Series		$f(x) = f(0) + f'(0)x + \dfrac{f''(0)}{2!}x^2 + \cdots + \dfrac{f^{(n)}(0)}{n!}x^n + \cdots$
		446	Taylor's Series		$f(x) = f(a) + f'(a)(x - a) + \dfrac{f''(a)}{2!}(x - a)^2 + \cdots + \dfrac{f^{(n)}(a)}{n!}(x - a)^n + \cdots$
		447		Remainder after n Terms	$R_n = \dfrac{(x - a)^n}{n!}f^{(n)}(c)$ where c lies between a and x.
	Fourier Series	**448**	Period of 2π		$f(x) = a_0/2 + a_1\cos x + a_2\cos 2x + a_3\cos 3x + \cdots + a_n\cos nx + \cdots$ $\qquad + b_1\sin x + b_2\sin 2x + b_3\sin 3x + \cdots + b_n\sin nx + \cdots$
		449		where	$a_0 = \dfrac{1}{\pi}\displaystyle\int_{-\pi}^{\pi} f(x)\,dx$
		450			$a_n = \dfrac{1}{\pi}\displaystyle\int_{-\pi}^{\pi} f(x)\cos nx\,dx$
		451			$b_n = \dfrac{1}{\pi}\displaystyle\int_{-\pi}^{\pi} f(x)\sin nx\,dx$
		452	Period of $2L$		$f(x) = \dfrac{a_0}{2} + a_1\cos\dfrac{\pi x}{L} + a_2\cos\dfrac{2\pi x}{L} + a_3\cos\dfrac{3\pi x}{L} + \cdots$ $\qquad + b_1\sin\dfrac{\pi x}{L} + b_2\sin\dfrac{2\pi x}{L} + b_3\sin\dfrac{3\pi x}{L} + \cdots$
		453			$a_0 = \dfrac{1}{L}\displaystyle\int_{-L}^{L} f(x)\,dx$
		454		where	$a_n = \dfrac{1}{L}\displaystyle\int_{-L}^{L} f(x)\cos\dfrac{n\pi x}{L}\,dx$
		455			$b_n = \dfrac{1}{L}\displaystyle\int_{-L}^{L} f(x)\sin\dfrac{n\pi x}{L}\,dx$

No.

	No.				
POWER SERIES	456	Fourier Series	Waveform Symmetries	Odd and Even Functions	(a) Odd functions have Fourier Series with only sine terms (and no constant term). (b) Even functions have Fourier Series with only cosine terms (and may have a constant term).
	457			Half-Wave Symmetry	A waveform that has half-wave symmetry has only odd harmonics in its Fourier Series.
MIXTURES	A1		Mixture Containing Ingredients A, B, C, . . .		Total amount of mixture = amount of A + amount of B + $\cdots$
	A2				Final amount of each ingredient = initial amount + amount added − amount removed
	A3		Combination of Two Mixtures		Final amount of A = amount of A from mixture 1 + amount of A from mixture 2
	A4		Fluid Flow		Amount of flow = flow rate × elapsed time $A = QT$
WORK	A5				Amount done = rate of work × time worked
	A6			Constant Force	Work = force × distance = Fd
	A7			Variable Force	Work = $\int_a^b F(x)\, dx$
FINANCIAL	A8		Unit Cost		Unit cost = $\dfrac{\text{total cost}}{\text{number of units}}$
	A9		Interest: Principal a Invested at Rate n for t years Accumulates to Amount y	Simple	$y = a\,(1 + nt)$
	A10			Compounded Annually	(a) $y = a\,(1 + n)^t$ (b) Recursion Relation $y_t = y_{t-1}\,(1 + n)$
	A11			Compounded m times/yr	$y = a\left(1 + \dfrac{n}{m}\right)^{mt}$
STATICS	A12			Moment about Point a	$M_a = Fd$
	A13		Equations of Equilibrium (Newton's First Law)		The sum of all horizontal forces = 0
	A14				The sum of all vertical forces = 0
	A15				The sum of all moments about any point = 0
	A16			Coefficient of Friction	$\mu = \dfrac{f}{N}$

Distance = rate × time

No.					
A21	MOTION (*Continued*)	Linear Motion (*Continued*)	Nonuniform Motion	Average Speed	Average speed = $\dfrac{\text{total distance traveled}}{\text{total time elapsed}}$
A22				Displacement	$s = \int v\, dt$
A23				Instantaneous Velocity	$v = \dfrac{ds}{dt}$
A24					$v = \int a\, dt$
A25				Instantaneous Acceleration	$a = \dfrac{dv}{dt} = \dfrac{d^2 s}{dt^2}$
A26		Rotation	Uniform Motion	Angular Displacement	$\theta = \omega t$
A27				Linear Speed of Point at Radius r	$v = \omega r$
A28			Nonuniform Motion	Angular Displacement	$\theta = \int \omega\, dt$
A29				Angular Velocity	$\omega = \dfrac{d\theta}{dt}$
A30					$\omega = \int \alpha\, dt$
A31				Angular Acceleration	$\alpha = \dfrac{d\omega}{dt} = \dfrac{d^2\theta}{dt^2}$
A32		Curvilinear Motion	x and y Components	Displacement	(a) $x = \int v_x\, dt$ (b) $y = \int v_y\, dt$
A33				Velocity	(a) $v_x = \dfrac{dx}{dt}$ (b) $v_y = \dfrac{dy}{dt}$
A34					(a) $v_x = \int a_x\, dt$ (b) $v_y = \int a_y\, dt$
A35				Acceleration	(a) $a_x = \dfrac{dv_x}{dt}$ (b) $a_y = \dfrac{dv_y}{dt}$ $= \dfrac{d^2 x}{dt^2}$ $= \dfrac{d^2 y}{dt^2}$
A36	MECHANICAL VIBRATIONS		Free Vibrations ($P = 0$)	Simple Harmonic Motion (No Damping)	$x = x_0 \cos \omega_n t$
A37				Undamped Angular Velocity	$\omega_n = \sqrt{\dfrac{kg}{W}}$
A38				Natural Frequency	$f_n = \dfrac{\omega_n}{2\pi}$
A39				Under-damped	$x = x_0 e^{-at} \cos \omega_d t$
A40				Damped Angular Velocity	$\omega_d = \sqrt{\omega_n^2 - \dfrac{c^2 g^2}{W^2}}$
A41				Overdamped	$x = C_1 e^{m_1 t} + C_2 e^{m_2 t}$
A42			Forced Vibrations	Maximum Deflection	$x_0 = \dfrac{Pg}{W \sqrt{4a^2\, \omega^2 + (\omega_n^2 - \omega^2)^2}}$

Spring–mass diagram (A36–A42 column): spring k, mass W, displacement x, forcing $P \sin \omega t$; Coefficient of friction $= c$.

No.

MATERIAL PROPERTIES	A43	Density		$\text{Density} = \dfrac{\text{weight}}{\text{volume}}$ or $\dfrac{\text{mass}}{\text{volume}}$
	A44	Mass		$\text{Mass} = \dfrac{\text{weight}}{\text{acceleration due to gravity}}$
	A45	Specific Gravity or Relative Density		$\dfrac{\text{Specific}}{\text{Gravity}} = \dfrac{\text{density of substance}}{\text{density of water}}$ or $\dfrac{\text{Relative}}{\text{Density}} = \dfrac{\text{density of substance}}{\text{density of water}}$
	A46	Pressure	Total Force on a Surface	$\text{Force} = \text{pressure} \times \text{area}$
	A47		Force on a Submerged Surface	$F = \delta \displaystyle\int y \, dA$
	A48			$F = \delta \bar{y} A$
	A49	pH		$\text{pH} = -10 \log \text{concentration}$
TEMPERATURE	A50	Conversions between Degrees Celsius (C) and Degrees Fahrenheit (F)		$C = \tfrac{5}{9}(F - 32)$
	A51			$F = \tfrac{9}{5}C + 32$
STRENGTH OF MATERIALS	A52	Tension or Compression	Normal Stress	$\sigma = \dfrac{P}{a}$
	A53		Strain	$\epsilon = \dfrac{e}{L}$
	A54		Modulus of Elasticity and Hooke's Law	$E = \dfrac{PL}{ae}$
	A55			$E = \dfrac{\sigma}{\epsilon}$
	A56	Thermal Expansion	Elongation	$e = \alpha L \, \Delta t$
	A57		New Length	$L = L_0 \, (1 + \alpha \Delta t)$
	A58		Strain	$\epsilon = \dfrac{e}{L} = \alpha \Delta t$
	A59		Stress, if Restrained	$\sigma = E\epsilon = E\alpha \, \Delta t$
	A60	Temperature change $= \Delta t$ Coefficient of thermal expansion $= \alpha$	Force, if Restrained	$P = a\sigma = aE\alpha \, \Delta t$
			Force needed	$F = \text{spring constant} \times \text{distance} = kx$

No.

No.				
A62	Ohm's Law		$\text{Current} = \dfrac{\text{voltage}}{\text{resistance}} \qquad I = \dfrac{V}{R}$	
A63	Combinations of Resistors	In Series	$R = R_1 + R_2 + R_3 + \cdots$	
A64		In Parallel	$\dfrac{1}{R} = \dfrac{1}{R_1} + \dfrac{1}{R_2} + \dfrac{1}{R_3} + \cdots$	
A65	Power Dissipated in a Resistor		$\text{Power} = P = VI$	
A66			$P = \dfrac{V^2}{R}$	
A67			$P = I^2 R$	
A68	Kirchhoff's Laws	Loops	The sum of the voltage rises and drops around any closed loop is zero	
A69		Nodes	The sum of the currents entering and leaving any node is zero	
A70	Resistance Change with Temperature		$R = R_1\,[1 + \alpha\,(t - t_1)]$	
A71	Resistance of a wire		$R = \dfrac{\rho l}{A}$	
A72	Combinations of Capacitors	In Series	$\dfrac{1}{C} = \dfrac{1}{C_1} + \dfrac{1}{C_2} + \dfrac{1}{C_3} + \cdots$	
A73		In Parallel	$C = C_1 + C_2 + C_3 + \cdots$	
A74	Charge on a Capacitor at Voltage V		$Q = CV$	
A75	Alternating Voltage		Sinusoidal Form	Complex Form
			$v = V_m \sin(\omega t + \phi_1)$	$\mathbf{V} = V_{eff}\,\underline{/\phi_1}$
A76	Alternating Current		$i = I_m \sin(\omega t + \phi_2)$	$\mathbf{I} = I_{eff}\,\underline{/\phi_2}$
A77	Period of a Sine Wave		$P = \dfrac{2\pi}{\omega}$ seconds	
A78	Frequency of a Sine Wave		$f = \dfrac{1}{P} = \dfrac{\omega}{2\pi}$ hertz	
A79	Current		$i = \dfrac{dq}{dt}$	
A80	Charge		$q = \displaystyle\int i\, dt$ coulombs	
A81	Capacitor — Instantaneous Current		$i = C\dfrac{dv}{dt}$	
A82	Capacitor — Instantaneous Voltage		$v = \dfrac{1}{C}\displaystyle\int i\, dt$ volts	
A83	Capacitor — Current when Charging or Discharging		Series RC Circuit	$i = \dfrac{E}{R}\,e^{-t/RC}$
A84	Capacitor — Voltage when Charging			$v = E(1 - e^{-t/RC})$
A85	Capacitor — Voltage when Discharging			$v = Ee^{-t/RC}$

ELECTRICAL TECHNOLOGY

No.

A86	Inductor	Instantaneous Current		$i = \dfrac{1}{L} \int v \, dt$ amperes
A87		Instantaneous Voltage		$v = L \dfrac{di}{dt}$
A88		Current when Charging	Series *RL* Circuit	$i = \dfrac{E}{R}(1 - e^{-Rt/L})$
A89		Current when Discharging		$i = \dfrac{E}{R} e^{-Rt/L}$
A90		Voltage when Charging or Discharging		$v = Ee^{-Rt/L}$
A91	Series *RLC* Circuit	DC Source	Resonant Frequency	$\omega_n = \sqrt{\dfrac{1}{LC}}$
A92			No Resistance: The Series *LC* Circuit:	$i = \dfrac{E}{\omega_n L} \sin \omega_n t$
A93			Underdamped	$i = \dfrac{E}{\omega_d L} e^{-at} \sin \omega_d t$
A94				where $\omega_d = \sqrt{\omega_n{}^2 - \dfrac{R^2}{4L^2}}$
A95			Overdamped	$i = \dfrac{E}{2j\omega_d L}\left[e^{(-a+j\omega_d)t} - e^{(-a-j\omega_d)t} \right]$
A96		AC Source	Inductive Reactance	$X_L = \omega L$
A97			Capacitive Reactance	$X_C = \dfrac{1}{\omega C}$
A98			Total Reactance	$X = X_L - X_C$
A99			Magnitude of Impedance	$\lvert Z \rvert = \sqrt{R^2 + X^2} = \sqrt{R^2 + \left(\omega L - \dfrac{1}{\omega C}\right)^2}$
A100			Phase Angle	$\phi = \arctan \dfrac{X}{R}$
A101			Complex Impedance	$\mathbf{Z} = R + jX = Z \, \underline{/\phi} = Ze^{j\phi}$
A102			Steady-State Current	$i_{ss} = \dfrac{E}{Z} \sin(\omega t - \phi)$
A103	Ohm's Law for AC			$\mathbf{V} = \mathbf{ZI}$
A104	Decibels Gained or Lost			$G = 10 \log_{10} \dfrac{P_2}{P_1}$ dB

ELECTRICAL TECHNOLOGY (*Continued*)

Conversion Factors

UNIT	EQUALS
LENGTH	
1 angstrom	100 pm
1 centimetre (cm)	0.3937 inch
1 foot	12 inches
	0.3048 m
1 inch	**25.4** mm
	2.54 cm
1 kilometre (km)	3281 feet
	0.6214 mile
	0.5400 M
	1094 yards
1 light-year	9.461×10^{12} km
	5.879×10^{12} miles
1 metre (m)	3.281 feet
	39.37 inches
	1.094 yards
1 micron	1 μm
1 nautical mile (International) (M)	6076 feet
	1852 m
	1.151 miles
1 mile	5280 feet
	1.609 km
1 yard	3 feet
	0.9144 m

Note: Factors in **boldface** are exact conversions

UNIT	EQUALS
ANGLES	
1 degree (°)	60′
	$\pi/180$ rad
	0.017 45 rad
	0.002 778 rev
	3600″
1 minute (′)	0.01667°
	0.000 2909 rad
	60″
1 radian (rad)	0.1592 rev
	180°/π
	1800′/π
	1/2 π rad
	57.296°
	3438′
1 revolution (rev)	360°
	2π rad
1 second (″)	0.000 2778°
AREA	
1 acre	4047 m^2
	0.4047 ha
	43 560 square feet
1 hectare (ha)	2.471 acres
	104 m^2
1 square foot	0.092 90 m^2
1 square inch	**6.4516** cm^2
1 square kilometre (km^2)	247.1 acres
	100 ha
1 square metre (m^2)	10.764 square feet
1 square mile	640 acres
	2.590 km^2
VOLUME	
1 barrel (oil, 42 U.S gallon)	0.159 m^3
	28.32 L

UNIT	EQUALS
MASS	
1 gram (g)	10^{-3} kg
1 slug	14.59 kg
1 tonne (t)	10^3 kg
FORCE	
1 dyne	10 µN
1 newton (N)	0.2248 pound
1 pound	4.448 N
	16 ounces
1 ton	2000 pounds
VELOCITY	
1 foot/minute	**5.08** mm/s
1 foot/second	**304.8** mm/s
	0.6818 mile/hour
1 kilometre/hour (km/h)	0.6214 mile/hour
1 knot (kn)	**1.852** km/h
	1.151 miles/hour
1 metre/second (m/s)	3.281 feet/second
	3.6 km/h
	2.2369 miles/hour
1 mile/hour	**1.609 344** km/h
	0.447 04 m/s
POWER	
1 Btu/hour	0.293 W
1 Btu/second	1.054 kW
1 Btu/pound	**2.326** kJ/kg
1 horsepower	746 W
	550 footpounds/second
1 kilowatt (kW)	3414 Btu/hour
	1.341 horsepower
1 watt (W)	1 J/s
PRESSURE	
1 atmosphere	14.70 pounds/square inch
	101.325 kPa
	760 mm of mercury (Hg)
1 bar	**100** kPa
1 inch of mercury (68 °F, 20 °C)	3.374 kPa
1 pound/square foot	47.880 Pa
1 pound/square inch	6.895 kPa
1 pascal (Pa)	1 N/m^2

UNIT	EQUALS
ENERGY	
1 Btu	1.054 kJ
1 calorie (thermochemical)	**4.184** J
1 foot-pound	1.356 J
1 kilowatt hour (kW·h)	1.341 horsepowerhours
	3.6 MJ
1 watt hour (W·h)	**3.6** kJ
1 watt second (W·s)	**1** J
TORQUE OR MOMENT	
1 foot-pound	1.356 N·m

Source: Adapted with the permission of Canadian Standards Association (CSA) from *CAN/CSA-Z234.1-00 (R2011),* *Metric Practice Guide*, which is copyrighted by CSA, 5060 Spectrum Way, Mississauga, ON, L4W 5N6 Canada. This reprinted material is not the complete and official position of CSA on the reference subject, which is represented solely by the standard in its entirety. For more information on CSA or to purchase standards, please visit its website at www.shopcsa.ca or call 1-800-463-6727.

C

Table of Integrals

Note: Many integrals have alternate forms that are not shown here. Don't be surprised if another table of integrals gives an expression that looks very different than one listed here. Further, a computer algebra system may give integrals that look much different than these, but that will result in the same numerical answers after substituting limits.

Basic forms

1. $\int du = u + C$

2. $\int af(x)\,dx = a\int f(x)\,dx = aF(x) + C$

3. $\int [f(x) + g(x) + h(x) + \cdots]\,dx = \int f(x)\,dx + \int g(x)\,dx + \int h(x)\,dx + \cdots + C$

4. $\int x^n\,dx = \dfrac{x^{n+1}}{n+1} + C \ (n \neq -1)$

5. $\int u^n\,du = \dfrac{u^{n+1}}{n+1} + C \ (n \neq -1)$

6. $\int u\,dv = uv - \int v\,du$

7. $\int \dfrac{du}{u} = \ln |u| + C$

8. $\int e^u\,du = e^u + C$

9. $\int b^u\,du = \dfrac{b^u}{\ln b} + C \ (b > 0, b \neq 1)$

Trigonometric functions

10. $\int \sin u\,du = -\cos u + C$

11. $\int \cos u\,du = \sin u + C$

12. $\int \tan u\,du = -\ln |\cos u| + C$

13. $\int \cot u\,du = \ln |\sin u| + C$

14. $\int \sec u\,du = \ln |\sec u + \tan u| + C$

15. $\int \csc u\,du = \ln |\csc u - \cot u| + C$

Squares of the trigonometric functions	16. $\int \sin^2 u \, du = \dfrac{u}{2} - \dfrac{\sin 2u}{4} + C$
	17. $\int \cos^2 u \, du = \dfrac{u}{2} + \dfrac{\sin 2u}{4} + C$
	18. $\int \tan^2 u \, du = \tan u - u + C$
	19. $\int \cot^2 u \, du = -\cot u - u + C$
	20. $\int \sec^2 u \, du = \tan u + C$
	21. $\int \csc^2 u \, du = -\cot u + C$

Cubes of the trigonometric functions	22. $\int \sin^3 u \, du = \dfrac{\cos^3 u}{3} - \cos u + C$
	23. $\int \cos^3 u \, du = \sin u - \dfrac{\sin^3 u}{3} + C$
	24. $\int \tan^3 u \, dx = \dfrac{1}{2}\tan^2 u + \ln |\cos u| + C$
	25. $\int \cot^3 u \, dx = -\dfrac{1}{2}\cot^2 u - \ln |\sin u| + C$
	26. $\int \sec^3 u \, du = \dfrac{1}{2}\sec u \tan u + \dfrac{1}{2}\ln |\sec u + \tan u| + C$
	27. $\int \csc^3 u \, du = -\dfrac{1}{2}\csc u \cot u + \dfrac{1}{2}\ln |\csc u - \cot u| + C$

Miscellaneous trigonometric forms	28. $\int \sec u \tan u \, du = \sec u + C$
	29. $\int \csc u \cot u \, du = -\csc u + C$
	30. $\int \sin^2 u \cos^2 u \, du = \dfrac{u}{8} - \dfrac{1}{32}\sin 4u + C$
	31. $\int u \sin u \, du = \sin u - u \cos u + C$
	32. $\int u \cos u \, du = \cos u + u \sin u + C$
	33. $\int u^2 \sin u \, du = 2u \sin u + (2 - u^2) \cos u + C$
	34. $\int u^2 \cos u \, du = 2u \cos u + (u^2 - 2) \sin u + C$
	35. $\int \sin^{-1} u \, du = u \sin^{-1} u + \sqrt{1 - u^2} + C$
	36. $\int \tan^{-1} u \, du = u \tan^{-1} u - \ln \sqrt{1 + u^2} + C$
	37. $\int u e^{au} \, du = \dfrac{e^{au}}{?}(au - 1) + C$

Exponential and logarithmic forms

41. $\int e^{au} \sin bu \, du = \dfrac{e^{au}}{a^2 + b^2} (a \sin bu - b \cos bu) + C$

42. $\int e^{au} \cos bu \, du = \dfrac{e^{au}}{a^2 + b^2} (a \cos bu + b \sin bu) + C$

43. $\int \ln u \, du = u(\ln u - 1) + C$

44. $\int u^n \ln |u| \, du = u^{n+1} \left[\dfrac{\ln |u|}{n+1} - \dfrac{1}{(n+1)^2} \right] + C \quad (n \neq -1)$

Forms involving $a + bu$

45. $\int \dfrac{u \, du}{a + bu} = \dfrac{1}{b^2}[a + bu - a \ln |a + bu|] + C$

46. $\int \dfrac{u^2 \, du}{a + bu} = \dfrac{1}{b^3} \left[\dfrac{1}{2}(a + bu)^2 - 2a(a + bu) + a^2 \ln |a + bu| \right] + C$

47. $\int \dfrac{u \, du}{(a + bu)^2} = \dfrac{1}{b^2} \left[\dfrac{a}{a + bu} + \ln |a + bu| \right] + C$

48. $\int \dfrac{u^2 \, du}{(a + bu)^2} = \dfrac{1}{b^3} \left[a + bu - \dfrac{a^2}{a + bu} - 2a \ln |a + bu| \right] + C$

49. $\int \dfrac{du}{u(a + bu)} = \dfrac{-1}{a} \ln \left| \dfrac{a + bu}{u} \right| + C$

50. $\int \dfrac{du}{u^2(a + bu)} = \dfrac{-1}{au} + \dfrac{b}{a^2} \ln \left| \dfrac{a + bu}{u} \right| + C$

51. $\int \dfrac{du}{u(a + bu)^2} = \dfrac{1}{a(a + bu)} - \dfrac{1}{a^2} \ln \left| \dfrac{a + bu}{u} \right| + C$

52. $\int u \sqrt{a + bu} \, du = \dfrac{2(3bu - 2a)}{15b^2} (a + bu)^{3/2} + C$

53. $\int u^2 \sqrt{a + bu} \, du = \dfrac{2(15b^2u^2 - 12abu + 8a^2)}{105b^3} (a + bu)^{3/2} + C$

54. $\int \dfrac{u \, du}{\sqrt{a + bu}} = \dfrac{2(bu - 2a)}{3b^2} \sqrt{a + bu} + C$

55. $\int \dfrac{u^2 \, du}{\sqrt{a + bu}} = \dfrac{2(3b^2u^2 - 4abu + 8a^2)}{15b^3} \sqrt{a + bu} + C$

Forms involving $u^2 \pm a^2$
$(a > 0)$

56. $\int \dfrac{du}{a^2 + b^2u^2} = \dfrac{1}{ab} \tan^{-1} \dfrac{bu}{a} + C$

57. $\int \dfrac{du}{u^2 - a^2} = \dfrac{1}{2a} \ln \left| \dfrac{u - a}{u + a} \right| + C$

58. $\int \dfrac{u^2 \, du}{u^2 - a^2} = u + \dfrac{a}{2} \ln \left| \dfrac{u - a}{u + a} \right| + C$

59. $\int \dfrac{u^2 \, du}{u^2 + a^2} = u - a \tan^{-1} \dfrac{u}{a} + C$

60. $\int \dfrac{du}{u(u^2 \pm a^2)} = \dfrac{\pm 1}{2a^2} \ln \left| \dfrac{u^2}{u^2 \pm a^2} \right| + C$

Forms involving
$\sqrt{a^2 \pm u^2}$
and
$\sqrt{u^2 \pm a^2}$
$(a > 0)$

61. $\int \dfrac{du}{\sqrt{a^2 - u^2}} = \sin^{-1} \dfrac{u}{a} + C$

62. $\int \dfrac{du}{\sqrt{u^2 \pm a^2}} = \ln |u + \sqrt{u^2 \pm a^2}| + C$

63. $\int \dfrac{u^2 \, du}{\sqrt{u^2 \pm a^2}} = \dfrac{u}{2} \sqrt{u^2 \pm a^2} \mp \dfrac{a^2}{2} \ln |u + \sqrt{u^2 \pm a^2}| + C$

64. $\int \dfrac{du}{u\sqrt{u^2 + a^2}} = \dfrac{1}{a} \ln \left| \dfrac{u}{a + \sqrt{u^2 + a^2}} \right| + C$

Forms involving

$\sqrt{a^2 \pm b^2}$

and

$\sqrt{b^2 \pm a^2}$

(a > 0)
(continued)

65. $\displaystyle\int \frac{du}{u\sqrt{u^2 - a^2}} = \frac{1}{a} \sec^{-1} \frac{u}{a} + C$

66. $\displaystyle\int \sqrt{u^2 \pm a^2}\, du = \frac{u}{2} \sqrt{u^2 \pm a^2} \pm \frac{a^2}{2} \ln |u + \sqrt{u^2 \pm a^2}| + C$

67. $\displaystyle\int \frac{\sqrt{u^2 + a^2}\, du}{u} = \sqrt{u^2 + a^2} - a \ln \left| \frac{a + \sqrt{u^2 + a^2}}{u} \right| + C$

68. $\displaystyle\int \frac{\sqrt{u^2 - a^2}\, du}{u} = \sqrt{u^2 - a^2} - a \sec^{-1} \frac{u}{a} + C$

69. $\displaystyle\int \sqrt{a^2 - u^2}\, du = \frac{u}{2}\sqrt{a^2 - u^2} + \frac{a^2}{2} \sin^{-1} \frac{u}{a} + C$

70. $\displaystyle\int u^2 \sqrt{a^2 - u^2}\, du = \frac{-u}{4}(a^2 - u^2)^{3/2} + \frac{a^2 u}{8} \sqrt{a^2 - u^2} + \frac{a^4}{8} \sin^{-1} \frac{u}{a} + C$

71. $\displaystyle\int \frac{\sqrt{a^2 - u^2}\, du}{u} = \sqrt{a^2 - u^2} - a \ln \left| \frac{a + \sqrt{a^2 - u^2}}{u} \right| + C$

72. $\displaystyle\int \frac{\sqrt{a^2 - u^2}\, du}{u^2} = \frac{-\sqrt{a^2 - u^2}}{u} - \sin^{-1} \frac{u}{a} + C$

73. $\displaystyle\int \frac{u^2\, du}{\sqrt{a^2 - u^2}} = \frac{-u}{2} \sqrt{a^2 - u^2} + \frac{a^2}{2} \sin^{-1} \frac{u}{a} + C$

Answers to Selected Problems

••• **CHAPTER 1** ••

Exercise 1

1. $7 < 10$ **3.** $-3 < 4$ **5.** $\frac{3}{4} = 0.75$ **7.** -18 **9.** 24   **11.** 13 **13.** 3   **15.** 4   **17.** 4   **19.** 2.00   **21.** 1   **23.**
5 **25.** 4   **27.** 5   **29.** 38.47   **31.** 96.84   **33.** 398.37   **35.** 2985.34   **37.** 14.0   **39.** 5.7   **41.** 398.4
43. 9839.3   **45.** 28 600   **47.** 3 845 200   **49.** 9.28   **51.** 0.0482   **53.** 0.0838   **55.** 29.5

Exercise 2

1. -1090 **3.** -1116 **5.** 105 233   **7.** 1789   **9.** -1129 **11.** -850 **13.** 1827   **15.** 4931   **17.** 593.44
19. $-0.000\,31$ **21.** 472 099 km^2 **23.** 35.0 cm   **25.** 41.1 Ω

Exercise 3

1. $73\overline{0}0$ **3.** 0.525   **5.** $-17\,800$ **7.** 22.9   **9.** \$5,210   **11.** \$2,571   **13.** \$1,570,000   **15.** 17 180 r/min   **17.** 980.03 cm

Exercise 4

1. 163   **3.** -0.347 **5.** 0.7062   **7.** 70 840   **9.** 17   **11.** 371.708 m   **13.** 10   **15.** 0.001 44   **17.** $-0.000\,002\,53$
19. -175 **21.** 0.2003   **23.** 0.0901   **25.** 0.9930   **27.** 314 Ω **29.** 0.279

Exercise 5

1. 8   **3.** -8 **5.** 1   **7.** 1   **9.** 1   **11.** -1 **13.** 100   **15.** 1   **17.** 1000   **19.** 0.01   **21.** 10 000
23. 0.000 01   **25.** 1.035   **27.** -112 **29.** 0.0146   **31.** 0.0279   **33.** 59.8   **35.** $-0.000\,0193$ **37.** 125 W
39. 878 000 cm^3 **41.** 5   **43.** 7   **45.** -2 **47.** 7.01   **49.** 4.45   **51.** 62.25   **53.** -7.28 **55.** -1.405
57. 4480 Ω

Exercise 6

1. 3340   **3.** -5940 **5.** 5   **7.** 3   **9.** 121   **11.** 27   **13.** 27   **15.** 24   **17.** 12   **19.** 2   **21.** 30   **23.** 46.2
25. 978   **27.** 2.28   **29.** 0.160   **31.** 59.8   **33.** 55.8   **35.** 3.51   **37.** 7.17   **39.** 3.23   **41.** 0.871   **43.** 7.93

Exercise 7

1. 10^2 **3.** 10^{-4} **5.** 10^8 **7.** 0.01   **9.** 0.1   **11.** 1.86×10^5; 186×10^3 **13.** 5.5×10^{-3}; 3.5×10^{-3}
15. 2.5742×10^4; 25.742×10^3 **17.** 8.00×10^3; 8.00×10^3 **19.** 9.83×10^4; 98.3×10^3 **21.** 7.75×10^{-4}; 775×10^{-6}
23. 2850   **25.** 90 000   **27.** 0.003 667 **29.** 0.000 248 2 **31.** 30 000 000 **33.** 10^7 **35.** 10^{-9} **37.** 10^{-5}
39. 10^{-2} **41.** 10^{-8} **43.** 1.5×10^6 **45.** 8×10^{-7} **47.** 4×10^2 **49.** 5×10^3 **51.** 3×10^6 **53.** 2.4×10^5
55. 1.566×10^2 **57.** 8.3×10^4 **59.** $3.55 \times 10^2, 0.355 \times 10^3$ **61.** $3.00 \times 10^{-3}, 3.00 \times 10^{-3}$ **63.** $2.06 \times$
$10^{17}, 0.206 \times 10^{18}$ or 206×10^{15} **65.** $9.79 \times 10^6 \Omega$ **67.** 1.04×10^{-11}W **69.** 2.01×10^{-5} cm/cm **71.** 380 years

Exercise 8

1. 12.7 ft.   **3.** 9144 in.   **5.** 58 000 lb.   **7.** 44.8 tons   **9.** 364 km   **11.** 735.9 kg   **13.** 6.2×10^3 MΩ **15.** 9.348
$\times 10^{-3}$ μF   **17.** 1194 ft.   **19.** 3.33 kg   **21.** 21.2 L   **23.** 3.60 m   **25.** 0.587 acre   **27.** 2.30 m^2 **29.** 243 acres

A-38

31. 12 720 cu. in.   **33.** 56.4 m^3 **35.** 1.63×10^{-2} cu. in.   **37.** 3.31 mi./h   **39.** 107 km/h   **41.** 18.3 births/week
43. 19.8 ¢/m^2 **45.** \$4,760/t   **47.** 445 L   **49.** 14.0 lb.   **51.** 8.00 cm, 2.72 cm, 3.15 cm, 3.62 cm, 13.3 cm
53. 5.699 m^2 **55.** 2.040 gal.   **57.** 7653.45 km^2; 2955.01 sq. mi. **59.** 254.012 kg

Exercise 9

1. 17   **3.** −37   **5.** 14   **7.** 14.1   **9.** 233   **11.** 8.00   **13.** \$3,975   **15.** 53.3 °C   **17.** \$13,266

Exercise 10

1. 372%   **3.** 0.55%   **5.** 40.0%   **7.** 70.0%   **9.** 0.23   **11.** 2.875   **13.** $\frac{3}{8}$ **15.** $1\frac{1}{2}$ **17.** 105 t   **19.** $22\overline{0}$ kg
21. $12\overline{0}$ L   **23.** 1090 Ω   **25.** \$1,562   **27.** 3759 MW   **29.** 3140   **31.** 221   **33.** 77.2   **35.** \$303
37. 19.0 million bbl.   **39.** 45.9%   **41.** 18.5%   **43.** 33.4%   **45.** 11.6%   **47.** 1.5%   **49.** 33.9%   **51.** −77.8%
53. −17.6%   **55.** 89.7 lb   **57.** −18.4%   **59.** 67.0%   **61.** 82%   **63.** −2.344% = 2.344% low **65.** 112.5 V and
312.5 V   **67.** 37.5%   **69.** 25.0 L

Review Problems

1. 83.35   **3.** 88.1   **5.** 0.346   **7.** 94.7   **9.** 5.46   **11.** 6.8   **13.** 30.6   **15.** 17.4   **17.** (a) 179   (b) 1.08   (c)
4.86   (d) 45 700   **19.** 70.2%   **21.** 3.63×10^6 **23.** 12 000 kW   **25.** 3.4%   **27.** 1.7×10^{11} bbl.   **29.** 10 200
31. 4.16×10^{-10} **33.** 2.42   **35.** $-\frac{2}{3} < -0.660$ **37.** 2370   **39.** −1.72   **41.** 64.5%   **43.** 219 N   **45.** $16\overline{0}$ m
47. 22.0%   **49.** 7216   **51.** 4.939   **53.** 109   **55.** 0.207   **57.** 14.7   **59.** 6.20   **61.** 2   **63.** 83.4   **65.** 93.52
cm   **67.** 9.07   **69.** 75.2%   **71.** 0.0737   **73.** 7.239   **75.** 121   **77.** 1.21   **79.** 0.30

◆◆◆ CHAPTER 2 ◆◆

Exercise 1

1. 2   **3.** 1   **5.** 5   **7.** $b/4$ **9.** $3/2a$

Exercise 2

1. $12x$ **3.** $-10ab$ **5.** $2m - c$ **7.** $4a - 2b - 3c - d - 17$ **9.** $11x + 10ax$ **11.** $-26x^3 + 2x^2 + x - 31$
13. 0   **15.** $20 - 2ay^3$ **17.** $18b^2 - 3ac - 2d$ **19.** $7m^2 - 5m^3 + 15ab - 4q - z$ **21.** $5.83(a + b)$ **23.** $3a^2 - 3a$
$+ x - b$ **25.** $-1.1 - 4.4x$ **27.** $4a^2 - 5a + y + 1$ **29.** $2a + 2b - 2m$ **31.** $6m - 3z$ **33.** $5w + 2z + 6x$
35. $a + 12$ **37.** $22w^2$

Exercise 3

1. 10^5 **3.** x^3 **5.** y^7 **7.** x^7 **9.** w^6 **11.** x^{n+3} **13.** y^3 **15.** x **17.** 100   **19.** x **21.** x^{12} **23.** a^{xy}

25. x^{2a+2} **27.** $8x^3$ **29.** $27a^3b^3c^3$ **31.** $-\dfrac{1}{27}$ **33.** $\dfrac{8a^3}{27b^6}$ **35.** $\dfrac{1}{a^2}$ **37.** $\dfrac{y^3}{27}$ **39.** $\dfrac{9b^6}{4a^2}$ **41.** $\dfrac{2}{x^2} + \dfrac{3}{y^3}$ **43.** x^{-1}

45. x^2y^{-2} **47.** $a^{-3}b^{-2}$ **49.** 1   **51.** $\dfrac{1}{9}$ **53.** 1   **55.** $6w^3$ **57.** $\dfrac{I^2R}{9}$

Exercise 4

1. x^5 **3.** $10a^3b^4$ **5.** $36a^6b^5$ **7.** $-8p^4q^4$ **9.** $2a^2 - 10a$ **11.** $x^2y - xy^2 + x^2y^2$ **13.** $-12p^3q - 8p^2q^2 +$
$12pq^3$ **15.** $6a^3b^3 - 12a^3b^2 + 9a^2b^3 + 3a^2b^2$ **17.** $a^2 + ac + ab + bc$ **19.** $x^3 - xy + 3x^2y - 3y^2$ **21.** $9m^2 -$
⋯ **27.** ⋯ ⋯$c - 1$ **29.** $m^3 - 5m^2 - m + 14$ **31.** $x^8 - x^2$

Exercise 5

1. z^2 **3.** $-5xyz$ **5.** $2f$ **7.** $-2xz$ **9.** $-5xy^2$ **11.** $3c$ **13.** $-5y$ **15.** $-3w$ **17.** $7n$ **19.** $4b$ **21.** b^2
23. x^{m-n} **25.** $4s$ **27.** -24 **29.** $-5z/xy$ **31.** $-5n^2/m^2x$ **33.** $3ab$ **35.** $3x^5$ **37.** $42/y$ **39.** $-4x$ **41.** $2a^2 - a$
43. $7x^2 - 1$ **45.** $3x^4 - 5x^2$ **47.** $2x^2 - 3x$ **49.** $a + 2c$ **51.** $ax - 2y$ **53.** $2x + y$ **55.** $ab - c$ **57.** $xy - x^2 -$
y^2 **59.** $1 - a - b$ **61.** $a - 3b + c^2$ **63.** $x - 2y + y^2/x$ **65.** $m + 2 - 3m/n$ **67.** $x + 8$ **69.** $a + 5$
71. $9a^2 + 6ab + 4b^2$ **73.** $x - 6 + 15/(x + 2)$ **75.** $5x + 13 - 35/(3 - x)$ **77.** $a - b + c$ **79.** $x + y - z$
81. $x - 2y - z$

Review Problems

1. $b^6 - b^4x^2 + b^4x^3 - b^2x^5 + b^2x^4 - x^6$ **3.** 3.86×10^{14} **5.** $9x^4 - 6mx^3 - 6m^3x + 10m^2x^2 + m^4$ **7.** $8x^3 +$
$12x^2 + 6x + 1$ **9.** $6a^2x^5$ **11.** $a - b - c$ **13.** $16a^2 - 24ab + 9b^2$ **15.** $x^2y^2 - 6xy + 8$ **17.** $4a^2 - 12ab + 9b^2$
19. $ab - b^4 - a^2b$ **21.** $16m^4 - c^4$ **23.** $2a^2$ **25.** $24 - 8y$ **27.** $6x^3 + 9x^2y - 3xy^2 - 6y^3$ **29.** $13w - 6$
31. $a^5 + 32c^5$ **33.** $(a - c)^{m-2}$ **35.** $x + y$ **37.** $b^3 - 9b^2 + 27b - 27$ **39.** $x^2/2y^3$ **41.** $x^2 + 2x - 8$ **43.** ab
$- 2 - 3b^2$ **45.** $5a - 10x - 2$ **47.** 8.01×10^6 **49.** $10x^3y^4$ **51.** $x^3 + 3x^2 - 4x$ **53.** 1.77×10^8 **55.** $6a^3b^3$
57. $x^4 + x + 1 + R(x^2 + x + 1)$ or $x^4 + x + 1 + (x^2 + x + 1)(x^4 - x)$ **59.** $0.13x + 540$ **61.** $4.02t^2$

◆◆◆ CHAPTER 3 ◆◆◆

Exercise 1

1. 33   **3.** 4/5   **5.** 1/2   **7.** 1/2   **9.** 4   **11.** 3   **13.** -1.38 **15.** 1   **17.** $-1/2$ **19.** 2   **21.** 5   **23.** 28
25. 3   **27.** 3   **29.** 35   **31.** $-5/23$ **33.** 5/6   **35.** 19/5   **37.** 1   **39.** 3/25   **41.** 0   **43.** -0.0543 **45.** $3/a$
47. $\dfrac{5}{c} + 1$ **49.** $\dfrac{3b + c - a}{b}$

Exercise 2

1. $5x + 8$ **3.** x and $83 - x$ **5.** x and $6 - x$ **7.** A, $2A$, and $180 - 3A$ **9.** $82x$ **11.** 5   **13.** 57, 58, and 59
15. 14   **17.** 6   **19.** 4   **21.** 20   **23.** 14   **25.** 81

Exercise 3

1. 8 technicians   **3.** \$130,137   **5.** 57 000 L   **7.** \$64.29 to brother; \$128.57 to uncle; \$257.14 to bank   **9.** \$94,000
11. \$60,000

Exercise 4

1. 333 L   **3.** 485 kg of 18% alloy; 215 kg of 31% alloy   **5.** 1380 kg   **7.** 1.29 L   **9.** 60.0 kg   **11.** 59.1 kg
13. 56.2 min   **15.** 20 202 L

Exercise 5

1. 4.97 m   **3.** $R_1 = 2250$ N downward; $R_2 = 11\ 500$ N upward   **5.** 339 cm   **7.** 12 190 N at right; 14 660 N at left

Review Problems

1. 12   **3.** 5   **5.** 0   **7.** $10\frac{1}{2}$ **9.** 2   **11.** 4   **13.** 4   **15.** 6   **17.** $-5/7$ **19.** -6 **21.** 9.27   **23.** 7   **25.** 6
27. 47 cm; 53 cm   **29.** 12 kg   **31.** \$22,500   **33.** 3000 m   **35.** $-13/5$ **37.** 13   **39.** $-5/7$ **41.** $-5/2$
43. $-7/3$ **45.** $-6/b$ **47.** $\dfrac{a + 10}{2}$ **49.** $\dfrac{5a - c + 8}{a}$ **51.** \$4,608 worth   **53.** 6 masons   **55.** 766 kg   **57.** 16.8 L
59. 49.7 L   **61.** \$398 for computer; \$597 for printer   **63.** \$58,093 at 8.56%; \$165,728 at 5.94%   **65.** \$107,248 at
10.25%; \$18,567 at 4.25%   **67.** 5.40 L   **71.** $R_1 = 1270$ N; $R_2 = 3840$ N; $R_3 = 1580$ N; $x = 29.0$ cm

◆◆◆ CHAPTER 4 ◆◆◆

Exercise 1

1. Is a function   **3.** Not a function   **5.** Not a function   **7.** Yes   **9.** $y = x^3$ **11.** $y = x + 2x^2$ **13.** $y = \dfrac{2}{3}(x - 4)$
15. y is a function of five less x.   **17.** y is a function of twice the square root of x.
19. $A = \dfrac{bh}{2}$ **21.** $V = \dfrac{4\pi r^3}{3}$ **23.** $d = 55t$ km   **25.** $H = 125t - \dfrac{9.807t^2}{2}$ **27.** Domain $= -10, -7, 0, 5, 10$;

Range = 3, 7, 10, 20   **29.** $x \geq 7; y \geq 0$ **31.** $x \neq 0; all\ y$ **33.** $x \neq 9; y \neq 1$ **35.** $x < 1; y > 0$
37. $x \geq 1; y \geq 0$ **39.** $-5 \leq y \leq 70$

Exercise 2

1. Explicit   **3.** Implicit   **5.** x is independent; y is dependent.   **7.** x and y are independent; w is dependent.

9. x and y are independent; z is dependent.   **11.** $y = \dfrac{2}{x} + 3$ **13.** $y = \dfrac{2x^2 + x}{3}$ **15.** $q = \dfrac{p(p+5)}{2}$ **17.** $e = \dfrac{PL}{aE}$

19. 6   **21.** -21 **23.** 12.5   **25.** -5 **27.** 2.69   **29.** -15 **31.** $\dfrac{5}{4}$ **33.** $2a^2 + 4$ **35.** $5a + 5b + 1$ **37.** $9\dfrac{1}{2}$

39. 142   **41.** 20   **43.** -52 **45.** 18   **47.** 8.66   **49.** $1.06 \times 10^4\ \Omega$; $1.08 \times 10^4\ \Omega$; $1.11 \times 10^4\ \Omega$ **51.** 1.99 m;
2.60 m; 3.52 m   **53.** 9.3; 3.6   **55.** 742 m

Exercise 3

1. $2x^2 + 3$ **3.** $1 - 6x$ **5.** $4 - 3x^3$ **7.** -77 **9.** $y = \dfrac{8 - x}{3}$ **11.** $y = \dfrac{x - 6}{5}$ **13.** $y = -x - 3$ **15.** $y = \dfrac{x - 10}{2}$

Review Problems

1. (a) Is a function (b) Not a function (c) Not a function   **3.** $S = 4\pi r^2$ **5.** (a) Explicit; y independent, w dependent

(b) Implicit   **7.** $w = \dfrac{3 - x^2 - y^2}{2}$ **9.** $7x^2$ **11.** 6   **13.** 28   **15.** $\dfrac{13}{90}$ **17.** y is 7 less than 5 times the cube of x.

19. $x = \sqrt{\dfrac{6 - y}{3}}$ **21.** $y = 8 - x$ **23.** $q = \dfrac{p(p+7)}{5}$ **25.** $28\dfrac{2}{3}$ **27.** $y = \dfrac{35 - x}{21}$ **29.** $y = \dfrac{x - 5}{10}$ **31.** 104

33. $f[g(x)] = 63x^2 - 210x + 175$

◆◆◆ **CHAPTER 5** ◆◆

Exercise 1

1. Fourth   **3.** Second   **5.** Fourth   **7.** First and fourth   **9.** $x = 7$ **11.** $E(-1.8, -0.7); F(-1.4, -1.4); G(1.4,\ \ 0.6);$
$H(2.5, -1.8)$ **13.** $E(-0.3, -1.3); F(-1.1, -0.8); G(-1.5, -1); H(-0.9, -1.1)$ **15.**
17.   **19.** $(4, -5)$

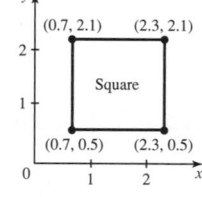

Exercise 2

1.   **3.**   **5.**   **7.**   **9.**

 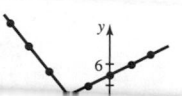 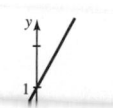 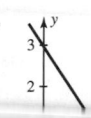 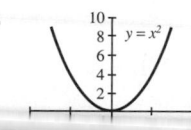

Exercise 3

1. 2   **3.** −1   **5.** 1   **7.** −11/7

9. 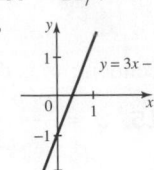   **11.** 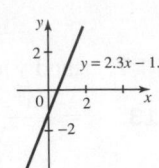   **13.** 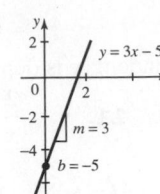   **15.** 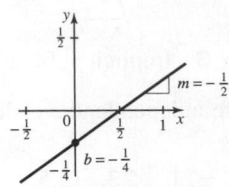   **17.**

Exercise 4

1. 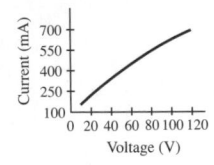   **3.** 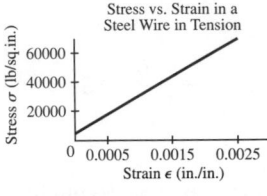   **5.** 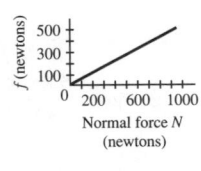   **7.**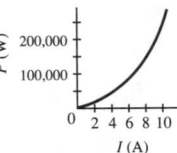

9. 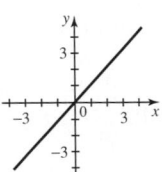   **11.**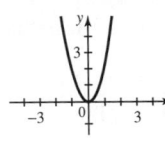

Exercise 5

1. 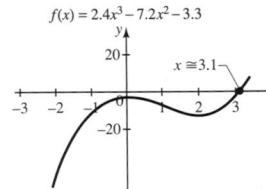   **3.** 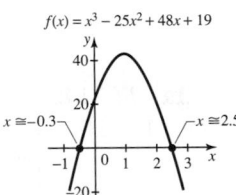   **5.** 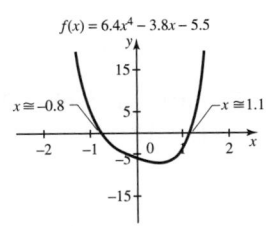   **7.**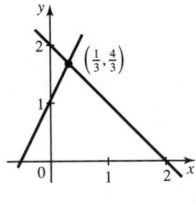

9. 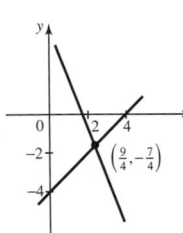   **11.** (a) 2700 (b) 36.4 min   **13.** 7.3 m

Review Problems

1. 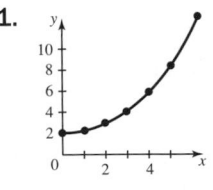   **3.** 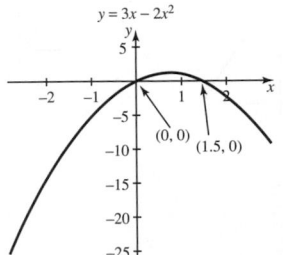   **5.** 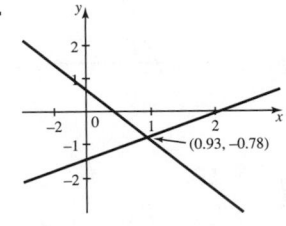   **7.**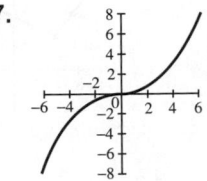

9.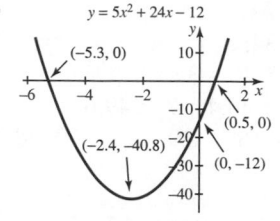

11.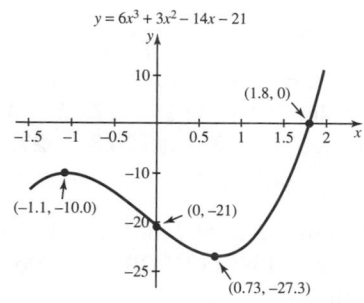

13.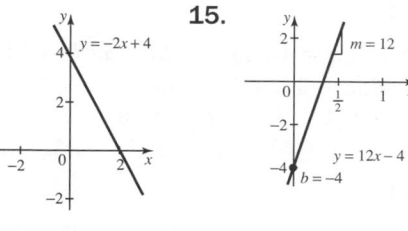

15.

◆◆◆ CHAPTER 6

Exercise 1

1. (a) 62.8° (b) 64.6° (c) 37.7°   **3.** 5.05   **5.** $A = C = 46.3°; B = D = 134°$

Exercise 2

1. $123,280   **3.** 605 cm   **5.** 25.0 m; 39.0 m; 68.4 m   **7.** 35.4 m   **9.** 35.4 m   **11.** 3.87 m   **13.** 17.7 m   **15.** 2.31 cm   **17.** 39.0 cm   **19.** 43.4 acres   **21.** 2.62 m   **23.** 5.2 mm

Exercise 3

1. 423 square units   **3.** 248 square units   **5.** $3,089   **7.** $348   **9.** $4,950   **11.** $861

Exercise 4

1. 7.16 m   **3.** 30.0 cm   **5.** 104 m   **7.** 49.7 ha   **9.** 0.738 units   **11.** 44.7 units   **13.** 218 cm   **15.** 1.200 in.   **17.** 247 cm

Exercise 5

1. 6.34×10^7 mm^3 (0.0634 m^3)   **3.** (a) 4.2 m (b) 150 m^3 **5.** 177 in.^{3} **7.** 1.21 cm   **9.** 2.5 cm   **11.** $76\overline{0}$ cm^2 **13.** 2400 kg   **15.** 1160 dm^3 **17.** 1.72×10^9; 6.96×10^6 **19.** 15.8; 1.12   **21.** 105 kg   **23.** (a) 176 m^3 (b) 11 railway grain cars

Review Problems

1. 1440 km/h   **3.** 2.88 m   **5.** 13 m   **7.** 175 000 square units   **9.** 107 cm   **11.** 17.5 m   **13.** 161°   **15.** 213 cm^3 **17.** 1030 cm^2 **19.** 1150 m^2 **21.** 87 800 cm^3 **23.** 11.0 m^3 **25.** 114 cm^3

◆◆◆ CHAPTER 7

Exercise 1

1. 0.485 rad   **3.** 0.6152 rad   **5.** 3.5 rad   **7.** 0.7595 rev   **9.** 0.191 rev   **11.** 0.2147 rev   **13.** 162°   **15.** 171°   **17.** 29.45°   **19.** 244°57′45″   **21.** 161°54′36″   **23.** 185°58′19″

Exercise 2

1. 1.15, 2.00   **3.** 0.24, 0.97, 0.25, 4.01, 1.03, 4.13   **5.** 53°   **7.** 63°   **9.** 0.528, 0.849, 0.622,

Exercise 4

1. $B = 47.1°$; $b = 167$; $c = 228$ **3.** $b = 1.08$; $A = 58.1°$; $c = 2.05$ **5.** $B = 0.441$ rad; $b = 134$; $c = 314$
7. $a = 6.44$; $c = 11.3$; $A = 34.8°$ **9.** $a = 50.3$; $c = 96.5$; $B = 58.6°$ **11.** $c = 470$; $A = 54.3°$; $B = 35.7°$
13. $a = 2.80$; $A = 35.2°$; $B = 54.8°$ **15.** $b = 25.6$; $A = 46.9°$; $B = 43.1°$ **17.** $b = 48.5$; $A = 40.4°$; $B = 49.6°$
19. $a = 414$; $A = 61.2°$; $B = 28.8°$ **21.** $\cos 52°$ **23.** $\cot 71°$ **25.** $\tan 26.8°$ **27.** $\cot 54°46'$ **29.** $\cos 1.10$ rad

Exercise 5

1. 402 m   **3.** 89.5 m   **5.** 64.9 m   **7.** 39.9 m   **9.** 128 m   **11.** 597 m   **13.** 3.16 km   **15.** 672 m   **17.** 5.34 km
19. 196 m; 197 m   **21.** 5.28 m   **23.** yes   **25.** 35.3   **27.** 139; 112   **29.** 122 000 cm^2 **31.** 355 cm   **33.** 2.55 cm
35. 77.5 mm; 134 mm; 134 mm; 77.5 mm   **37.** 1.550 cm   **39.** 3.366 in.

Exercise 6

1. 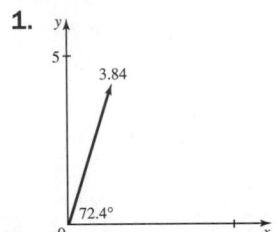   **3.** 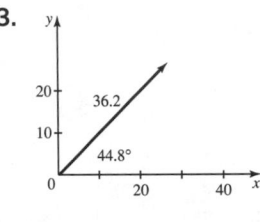   **5.** 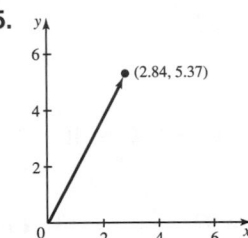   **7.** 3.28; 3.68   **9.** 0.9917; 1.602

13. 5.60; 30.6°   **15.** 60.7; 50.5°   **17.** 596; 27.4°   **19.** 8.36; 35.3°   **21.** 4.57; 60.2°   **23.** 38.2; 62.0°
25. 134; 56.1°   **27.** 1880; 56.4°

Exercise 7

1. 12.3 N   **3.** 1520 N   **5.** 25.7°   **7.** 4.94 t   **9.** 213 km/h; S 83°39′ W   **11.** 316 km/h; 28.7 km/h
13. 4270 m/min; 6230 m/min   **15.** 122 km; 106° north of east   **17.** $R = 2690\ \Omega$; $Z = 3770\ \Omega$ **19.** $R = 861\ \Omega$;
$X = 458\ \Omega$ **21.** 4430 Ω; 37.2°

Review Problems

1. 38°12′; 0.667 rad; 0.106 rev   **3.** 157.3°; 157°18′; 0.4369 rev   **5.** 0.919, 0.394, 2.33, 0.429, 2.54, 1.09, 66.8°   **7.** 0.600,
0.800, 0.750, 1.33, 1.25, 1.67, 36.9°   **9.** 0.9558, 0.2940, 3.2506, 0.3076, 3.4009, 1.0463   **11.** 0.8674, 0.4976, 1.7433, 0.5736,
2.0098, 1.1528   **13.** 34.5°   **15.** 70.7°   **17.** 65.88°   **19.** $B = 61.5°$; $a = 2.02$; $c = 4.23$ **21.** 356; 810   **23.** 473; 35.5°
25. 2.70 m   **27.** 0.5120   **29.** 1.3175   **31.** 2.1742   **33.** 60.1°   **35.** 46.5°   **37.** 46.5°   **39.** 16.6   **41.** 3.79
43. 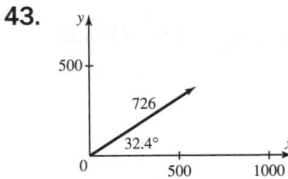   **45.** 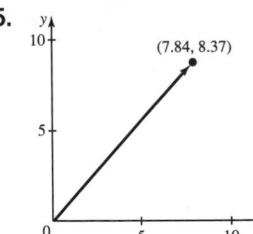   **47.** 32.2; 57.4°   **49.** $AC = 74.98$ m; N 28°18′ E

♦♦♦ **CHAPTER 8** ♦♦♦

Exercise 1

1. $y^2(3 + y)$ **3.** $x^3(x^2 - 2x + 3)$ **5.** $a(3 + a - 3a^2)$ **7.** $5(x + y)[1 + 3(x + y)]$ **9.** $\left(\dfrac{1}{x}\right)\left(3 + \dfrac{2}{x} - \dfrac{5}{x^2}\right)$

11. $\left(\dfrac{5m}{2n}\right)\left(1 + \dfrac{3m}{2n} - \dfrac{5m^2}{4}\right)$ **13.** $a^2(5b + 6c)$ **15.** $xy(4x + cy + 3y^2)$ **17.** $3ay(a^2 - 2ay + 3y^2)$

19. $cd(5a - 2cd + b)$ **21.** $4x^2(2y^2 + 3z^2)$ **23.** $ab(3a + c - d)$ **25.** $L_0(1 + \alpha t)$ **27.** $R_1[1 + \alpha(t - t_1)]$

29. $t\left(v_0 + \dfrac{at}{2}\right)$ **31.** $\dfrac{4\pi D}{3}(r_2^3 - r_1^3)$

Exercise 2

1. $(2 - x)(2 + x)$ **3.** $(3a - x)(3a + x)$ **5.** $4(x - y)(x + y)$ **7.** $(x - 3y)(x + 3y)$ **9.** $(3c - 4d)(3c + 4d)$
11. $(3y - 1)(3y + 1)$ **13.** $(m - n)(m + n)(m^2 + n^2)$ **15.** $(m^n - n^m)(m^n + n^m)$ **17.** $(a^2 - b)(a^2 + b)(a^4 + b^2)(a^8 + b^4)$
19. $(5x^2 - 4y^3)(5x^2 + 4y^3)$ **21.** $(4a^2 - 11)(4a^2 + 11)$ **23.** $(5a^2b^2 - 3)(5a^2b^2 + 3)$ **25.** $\pi(r_2 - r_1)(r_2 + r_1)$

27. $4\pi(r_1 - r_2)(r_1 + r_2)$ **29.** $\dfrac{m(v_1 - v_2)(v_1 + v_2)}{2}$ **31.** $\pi h(R - r)(R + r)$

Exercise 3

1. Not factorable   **3.** Factorable   **5.** Factorable   **7.** $(x - 7)(x - 3)$ **9.** $(x - 9)(x - 1)$ **11.** $(x + 10)(x - 3)$
13. $(x + 4)(x + 3)$ **15.** $(x - 7)(x + 3)$ **17.** $(x + 4)(x + 2)$ **19.** $(b - 5)(b - 3)$ **21.** $(b - 4)(b + 3)$
23. $2(y - 10)(y - 3)$ **25.** $3(w + 4)(w + 8)$ **27.** $(x + 7y)(x + 12y)$ **29.** $(a + 3b)(a - 2b)$ **31.** $(a + 9x)(a - 7x)$
33. $(a - 24bc)(a + 4bc)$ **35.** $(a + 48bc)(a + bc)$ **37.** $(a + b + 1)(a + b - 8)$ **39.** $(x^2 - 5y^2)(x - 2y)(x + 2y)$
41. $(t - 2)(t - 12)$ **43.** $(x - 2)(x - 20)$

Exercise 4

1. $(a^2 + 4)(a + 3)$ **3.** $(x - 1)(x^2 + 1)$ **5.** $(x - b)(x + 3)$ **7.** $(x - 2)(3 + y)$ **9.** $(x + y - 2)(x + y + 2)$
11. $(m - n + 2)(m + n - 2)$ **13.** $(x + 3)(a + b)$ **15.** $(a + b)(a - c)$ **17.** $(2 + x^2)(a + b)$ **19.** $(2a - c)(3a + b)$
21. $(a + b)(b - c)$ **23.** $(x - a - b)(x + a + b)$

Exercise 5

1. $(4x - 1)(x - 3)$ **3.** $(5x + 1)(x + 2)$ **5.** $(3b + 2)(4b - 3)$ **7.** $(2a - 3)(a + 2)$ **9.** $(x - 7)(5x - 3)$
11. $3(x + 1)(x + 1)$ **13.** $(3x + 2)(x - 1)$ **15.** $2(2x - 3)(x - 1)$ **17.** $(2a - 1)(2a + 3)$ **19.** $(3a - 7)(3a + 2)$
21. $(7x^3 - 3y)(7x^3 + 5y)$ **23.** $(4x^3 + 1)(x^3 + 3)$ **25.** $(5x^{2n} + 1)(x^{2n} + 2)$ **27.** $3[(a + x)^n + 1][(a + x)^n - 2]$
29. $(t - 4)(5t - 3)$

Exercise 6

1. $(x + 2)^2$ **3.** $(y - 1)^2$ **5.** $2(y - 3)^2$ **7.** $(3 + x)^2$ **9.** $(3x + 1)^2$ **11.** $9(y - 1)^2$ **13.** $4(2 + a)^2$
15. $(7a - 2)^2$ **17.** $(x + y)^2$ **19.** $(aw + b)^2$ **21.** $(x + 5a)^2$ **23.** $(x + 4y)^2$ **25.** $(z^3 + 8)^2$ **27.** $(7 - x^3)^2$
29. $(u - b)^2(a + b)^2$ **31.** $(2a^n + 3b^n)^2$

Exercise 7

1. $(4 + x)(16 - 4x + x^2)$ **3.** $2(a - 2)(a^2 + 2a + 4)$ **5.** $(x - 1)(x^2 + x + 1)$ **7.** $(x + 1)(x^2 - x + 1)$
9. $(a + 4)(a^2 - 4a + 16)$ **11.** $(x + 5)(x^2 - 5x + 25)$ **13.** $8(3 - a)(9 + 3a + a^2)$ **15.** $(7 + 4x)(49 - 28x + 16x^2)$
17. $(y^3 + 4x)(y^6 - 4xy^3 + 16x^2)$ **19.** $(3x^5 + 2a^2)(9x^{10} - 6x^5a^2 + 4a^4)$ **21.** $(2a^{2x} - 5b^x)(4a^{4x} + 10a^{2x}b^x + 25b^{2x})$

23. $(4x^n - y^{3n})(16x^{2n} + 4x^ny^{3n} + y^{6n})$ **25.** $\left(\dfrac{4\pi}{3}\right)(r_2 - r_1)(r_2^2 + r_2r_1 + r_1^2)$

Review Problems

1. $(x - 5)(x + 3)$ **3.** $(x^3 - y^2)(x^3 + y^2)$ **5.** $(2x - 1)(x + 2)$ **7.** $(a - b + c + d)[(a - b)^2 - (a - b)(c + d) + (c + d)^2]$

9. $\left(\dfrac{x}{y}\right)(x - 1)$ **11.** $(y + 5)(x - 2)$ **13.** $(x - y)(1 - b)$ **15.** $(y + 2 - z)(y + 2 + z)$ **17.** $(x - 3)(x - 4)$

19. $(4x^{2n} + 9y^{4n})(2x^n + 3y^{2n})(2x^n - 3y^{2n})$ **21.** $(3a + 2z^2)^2$ **23.** $(x^m + 1)^2$ **25.** $(x - y - z)(x - y + z)$
27. $(1 + 4x)(1 - 4x)$ **29.** $x^2(3x + 1)(3x - 1)$ **31.** $(p - q - 3)[(p - q)^2 + 3(p - q) + 9]$
33. $(3a - 2w)(9a^2 + 6aw + 4w^2)$ **35.** $4(2x - y)^2$ **37.** $(3b + y)(2a + x)$ **39.** $2(y^2 - 3)(y^2 + 3)$

21. $d - c$ **23.** $\frac{2}{3}$ **25.** $\frac{15}{7}$ **27.** $\frac{a}{3}$ **29.** $\frac{3m}{4p^2}$ **31.** $\frac{2a-3b}{2a}$ **33.** $\frac{x}{x-1}$ **35.** $\frac{x+2}{x^2+2x+4}$ **37.** $\frac{n(m-4)}{3(m-2)}$

39. $\frac{2(a+1)}{a-1}$ **41.** $\frac{x+z}{x^2+xz+z^2}$ **43.** $\frac{a-2}{a-3}$ **45.** $\frac{x-1}{2y}$ **47.** $\frac{2}{3(x^4y^4-1)}$ **49.** $\frac{x+y}{x-y}$ **51.** $\frac{b-3+a}{5}$

Exercise 2

1. $\frac{2}{15}$ **3.** $\frac{6}{7}$ **5.** $2\frac{2}{15}$ **7.** $\frac{75}{8}$ **9.** $1\frac{1}{2}$ **11.** $\frac{a^4b^4}{2y^{2n}}$ **13.** $\frac{7p^2}{4xz}$ **15.** $x-a$ **17.** $\frac{ab}{x^2-y^2}$ **19.** $\frac{cd}{(x^2-y^2)^2}$

21. $\frac{(x-1)(x+4)}{(x+3)(x-5)}$ **23.** $\frac{(x+1)(x+2)}{(x-4)(x-3)}$ **25.** $\frac{(2x-3)(x+1)}{(x-1)(x-6)}$ **27.** $1\frac{15}{16}$ **29.** $\frac{9}{128}$ **31.** $\frac{5}{18}$ **33.** $5\frac{1}{4}$

35. b^2 **37.** $\frac{2a^3}{dxy}$ **39.** $\frac{3an+cm}{x^4-y^4}$ **41.** $\frac{1}{c-d}$ **43.** $\frac{1}{x+2}$ **45.** $\frac{6(x-1)}{x+2}$ **47.** $\frac{(p+2)^2}{(p-1)^2}$ **49.** $\frac{(z-1)(z+4)}{(z+3)(z-5)}$

51. $1\frac{2}{3}$ **53.** $2\frac{5}{12}$ **55.** $9\frac{2}{5}$ **57.** $x+\frac{1}{x}$ **59.** $\frac{2}{x}-\frac{1}{2}$ **61.** $\frac{3m}{4\pi r^3}$ **63.** $\frac{2P}{(a+b)h}$ **65.** $\frac{4\pi d^3}{3}$

Exercise 3

1. 1   **3.** $\frac{1}{7}$ **5.** $\frac{5}{3}$ **7.** $\frac{7}{6}$ **9.** $\frac{19}{16}$ **11.** $\frac{7}{18}$ **13.** $\frac{13}{5}$ **15.** $-\frac{92}{15}$ **17.** $\frac{67}{35}$ **19.** $\frac{8}{3}$ **21.** $\frac{13}{4}$ **23.** $\frac{91}{16}$ **25.** $\frac{6}{a}$

27. $\frac{a+3}{y}$ **29.** x **31.** $\frac{x}{a-b}$ **33.** $\frac{19}{a+1}$ **35.** $\frac{19a}{10x}$ **37.** $\frac{2ax-3x+12}{6x}$ **39.** $\frac{5b-a}{6}$ **41.** $\frac{9-x}{x^2-1}$

43. $\frac{2b(2a-b)}{a^2-b^2}$ **45.** $\frac{2x^2-1}{x^4-x^2}$ **47.** $\frac{2(3x^2+2x-3)}{x^3-7x-6}$ **49.** $\frac{x^2-2x}{(x-3)(x+3)}$ **51.** $\frac{5x^2+7x+16}{x^3-8}$

53. $-\dfrac{d}{(x+1)(x+d+1)}$ **55.** $\frac{d(x^2+dx-1)}{x(x+d)}$ **57.** $\frac{x^2+1}{x}$ **59.** $\frac{2}{x^2-1}$ **61.** $\frac{3a-a^2-2}{a}$

63. $\frac{abx+a+b}{ax}$ **65.** $4\frac{7}{12}$ km   **67.** $\frac{2}{25}$ min   **69.** $\frac{11}{8}$ machines   **71.** $\frac{2h(a+b)-\pi d^2}{4}$

73. $\frac{V_1V_2d+VV_2d_1+VV_1d_2}{VV_1V_2}$

Exercise 4

1. $\frac{85}{12}$ **3.** $\frac{65}{132}$ **5.** $\frac{69}{95}$ **7.** $\frac{3(4x+y)}{4(3x-y)}$ **9.** $\frac{y}{(y-x)}$ **11.** $\frac{5(3a^2+x)}{3(20+x)}$ **13.** $\frac{2(3acx+2d)}{3(2acx+3d)}$ **15.** $\frac{2x^2-y^2}{x-3y}$

17. $\frac{(x+2)(x-1)}{(x+1)(x-2)}$ **19.** 1   **21.** $\frac{4}{3(a+1)}$ **23.** 1   **25.** $\frac{x+y+z}{x-y+z}$ **27.** $\frac{\rho L_1L_2}{A_2L_1+A_1L_2}$

Exercise 5

1. 12   **3.** 20   **5.** 14   **7.** 72   **9.** 24   **11.** 7   **13.** 24   **15.** 24   **17.** $\frac{5}{2}$ **19.** 12   **21.** 17   **23.** $\frac{3}{13}$ **25.** $-\frac{2}{3}$

27. $-\frac{11}{4}$ **29.** $\frac{1}{2}$ **31.** No solution   **33.** -2 **35.** 4   **37.** 8   **39.** 2

Exercise 6

1. 150   **3.** 5 and 6   **5.** 15 and 35   **7.** 26 h   **9.** 240 km   **11.** 12 m   **13.** 2.0 s   **15.** 2 days   **17.** $5\frac{5}{8}$ days
19. 2.4 h   **21.** 4.2 h   **23.** 12.4 h   **25.** 7.8 weeks   **27.** 5.6 winters   **29.** 6.1 h   **31.** \$11,360   **33.** \$1500; \$2000

Exercise 7

1. $\frac{bc}{2a}$ **3.** $\frac{bz-ay}{a-b}$ **5.** $\frac{a^2d+3d^2}{4ac+d^2}$ **7.** $\frac{b+cd}{a^2+a-d}$ **9.** $\frac{2z}{a}+3w$ **11.** $\frac{b-m}{3}$ **13.** $\frac{c-m}{a-b-d}$ **15.** b

17. $\frac{2c+bc+3b}{3-2b+c}$ **19.** $\frac{w}{w(w+y)-1}$ **21.** $\frac{p-q}{3p}$ **23.** $\frac{5b-2a}{3}$ **25.** $\frac{a^2(c+1)(c-a)}{c^2}$ **27.** $\frac{ab}{a+b}$

29. $\dfrac{ab}{b-a}$ **31.** $\dfrac{md}{(m+n)}\,km$ and $\dfrac{nd}{(m+n)}\,km$ **33.** $\dfrac{(anm)}{(nm-m-n)}$ **35.** $\dfrac{L-L_0}{L_0\alpha}$ **37.** $\dfrac{PL}{Ee}$ **39.** $\dfrac{kAt_1-qL}{kA}$

41. $\dfrac{RR_1}{R_1-R}$ **43.** $t_1=\dfrac{R_1(1+\alpha t)-R}{R_1\alpha}$ **45.** $\dfrac{M-Fx}{R_1-F}$ **47.** $\dfrac{m_2(25-x)}{x-10}$ **49.** $\dfrac{E}{gy+\frac12 v^2}$

Review Problems

1. $\dfrac{2ab}{a-b}$ **3.** $\dfrac{1}{(3m-1)}$ **5.** $a^2+1+\dfrac{1}{a^2}$ **7.** $\dfrac{a}{c}$ **9.** $\dfrac{4}{(4-x^2)}$ **11.** 0 **13.** $\dfrac{a^2+ab+b^2}{a+b}$ **15.** $\dfrac{2a-3c}{2a}$

17. $\dfrac{3a}{a+2}$ **19.** $\dfrac{x-y-z}{x+y-z}$ **21.** $\dfrac{a+b-c-d}{a-b+c-d}$ **23.** $\dfrac{b+5}{c-a}$ **25.** 1 **27.** 11 **29.** $\dfrac{r-pq}{p-q}$ **31.** $\dfrac{q(p+1)}{p^2+r}$

33. 7 **35.** 22 **37.** -5 **39.** 3 **41.** 6.35 days   **43.** $19\frac{29}{45}$ **45.** $\dfrac{2ax^2y}{7w}$ **47.** $14\frac{15}{28}$

◆◆◆ CHAPTER 10 ◆◆

Exercise 1

1. $(2,-1)$ **3.** $(1,2)$ **5.** $(-0.24,0.90)$ **7.** $(3,5)$ **9.** $(-\frac34,3)$ **11.** $(-3,3)$ **13.** $(1,2)$ **15.** $(3,2)$ **17.** $(15,6)$
19. $(3,4)$ **21.** $(2,3)$ **23.** $(1.63,0.0971)$ **25.** $m=2,n=3$ **27.** $w=6,z=1$ **29.** $(9.36,4.69)$ **31.** $v=1.06$
$w=2.58$

Exercise 2

1. $(60,36)$ **3.** $\left(\dfrac{87}{7},\dfrac{108}{7}\right)$ **5.** $(15,12)$ **7.** $m=4,n=3$ **9.** $r=3.77,s=1.23$ **11.** $\left(\dfrac12,\dfrac13\right)$ **13.** $\left(\dfrac13,\dfrac12\right)$

15. $\left(\dfrac{1}{10},\dfrac{1}{12}\right)$ **17.** $w=\dfrac{1}{36},z=\dfrac{1}{60}$ **19.** $\left(\dfrac{a+2b}{7},\dfrac{3b-2a}{7}\right)$ **21.** $\left(\dfrac{c(n-d)}{an-dm},\dfrac{c(m-a)}{an-dm}\right)$

Exercise 3

1. 8 and 16   **3.** $4/21$ **5.** 82   **7.** 7 and 41   **9.** 9 by 16   **11.** $x=2.10$ km, $y=0.477$ km   **13.** 3.33 m and 2.50 m
15. 5.3 km/h and 1.3 km/h   **17.** $v_0=0.381$ cm/s; $a=3.62$ cm/s^2 **19.** \$2,500 at 4%   **21.** \$2,684 at 3.10% and \$1,716
at 4.85%   **23.** \$6,000 for 2 years   **25.** 2740 kg mixture; 271 kg sand   **27.** 5.57 kg peat; 11.6 kg vermiculite
29. $T_1=490$ N; $T_2=185$ N   **31.** carpenter: 25.0 days; helper: 37.5 days   **33.** 18 000 L/h and 12 000 L/h
35. 92 700 people and 162 000 people   **37.** $I_1=13.1$ mA; $I_2=22.4$ mA   **39.** $R_1=27.6\ \Omega$; $\alpha=0.00511$

Exercise 4

1. $(15,20,25)$ **3.** $(1,2,3)$ **5.** $(5,6,7)$ **7.** $(1,2,3)$ **9.** $(3,4,5)$ **11.** $(6.18,13.7,18.1)$ **13.** $(2,3,1)$

15. $\left(\dfrac12,\dfrac13,\dfrac14\right)$ **17.** $\left(1,-1,\dfrac12\right)$ **19.** $(a/11,5a/11,7a/11)$ **21.** $[(a+b-c),(a-b+c),(-a+b+c)]$ **23.** 876

25. 361   **27.** 6.38 mA, 0.275 mA, -2.53 mA   **29.** $F_1=426$ N, $F_2=1080$ N, $F_3=362$ N

Review Problems

1. $(3,5)$ **3.** $(5,-2)$ **5.** $(2,-1,1)$ **7.** $\left(-\dfrac95,\dfrac{54}{5},-\dfrac{21}{5}\right)$ **9.** $(8,10)$ **11.** $(2,3)$ **13.** $[(a+b-c)/2,$

33. $m = 2, n = 3$ **35.** $w = 6, z = 1$ **37.** $\left(-\dfrac{6}{13}, \dfrac{30}{13}\right)$ **39.** $\left(\dfrac{dp - bq}{ad - bc}, \dfrac{aq - cp}{ad - bc}\right)$ **41.** $(3, 4)$ **43.** \$31,550; \$22,986   **45.** $x = 90.1$ kg; $y = 8.98$ kg

Exercise 2

1. 11   **3.** 45   **5.** 48   **7.** -28 **9.** $(5, 6, 7)$ **11.** $(15, 20, 25)$ **13.** $(1, 2, 3)$ **15.** $(3, 4, 5)$ **17.** $(6, 8, 10)$ **19.** $(-2.30, 4.80, 3.09)$ **21.** $(3, 6, 9)$ **23.** $\left(\dfrac{1}{2}, \dfrac{1}{3}, \dfrac{1}{4}\right)$ **25.** $1.54, -0.377, -1.15$

Exercise 3

1. 2   **3.** 18   **5.** -66 **7.** $x = 2, y = 3, z = 4, w = 5$ **9.** $x = -4, y = -3, z = 2, w = 5$ **11.** $x = a - c, y = b + c, z = 0, w = a - b$ **13.** $x = 4, y = 5, z = 6, w = 7, u = 8$ **15.** $v = 3, w = 2, x = 4, y = 5, z = 6$ **17.** $I_1 = -6.24$; $I_2 = 10.4$; $I_3 = 8.21$; $I_4 = 5.53$ **19.** 102 kg, 98.0 kg, 218 kg, 182 kg

Review Problems

1. 20   **3.** 0   **5.** -3 **7.** 0   **9.** 18   **11.** 15   **13.** 0   **15.** -29 **17.** 133   **19.** $x = -4, y = -3, z = 2, w = 1$ **21.** $(3, 5)$ **23.** $(4, 3)$ **25.** $(5, 1)$ **27.** $(2, 5)$ **29.** $(2, 1)$ **31.** $(3, 2)$ **33.** $(9/7, 37/7)$ **35.** $(3.19, 1.55)$ **37.** $(-15.1, 3.52, 11.4)$ **39.** $0.558, -1.34, -0.240, -0.667$

◆◆◆ CHAPTER 12 ◆◆◆

Exercise 1

1. A, B, D, E, F, I   **3.** C, H   **5.** I   **7.** B, I   **9.** F   **11.** 6   **13.** 4×3 **15.** 2×4

17. $\begin{pmatrix} 0 & 0 \\ 0 & 0 \\ 0 & 0 \\ 0 & 0 \end{pmatrix}$ **19.** If $y = 6, x = 2, w = 7,$ and $z = 8$

Exercise 2

1. $(12 \quad 13 \quad 5 \quad 8)$ **3.** $\begin{pmatrix} 14 \\ -10 \\ 6 \\ 14 \end{pmatrix}$ **5.** $\begin{pmatrix} 7 & 1 & -4 & -7 \\ -9 & -4 & 9 & 2 \\ 3 & 7 & -3 & -9 \\ -1 & -4 & -5 & 6 \end{pmatrix}$ **7.** $(-15 \quad -27 \quad 6 \quad -21 \quad -9)$

9. $\begin{pmatrix} -12 & -6 & 9 & 0 \\ 3 & -18 & -9 & 12 \end{pmatrix}$ **11.** $7\begin{pmatrix} 3 & 1 & -2 \\ 7 & 9 & 4 \\ 2 & -3 & 8 \\ -6 & 10 & 12 \end{pmatrix}$ **13.** $\begin{pmatrix} 6 & 16 & 3 \\ -16 & 8 & 19 \\ 31 & -14 & 15 \end{pmatrix}$ **17.** 35   **19.** -61

21. $(10 \quad 27 \quad -14 \quad 15 \quad -7)$ **23.** $(-60 \quad 18 \quad 10)$ **25.** $(15 \quad 2 \quad 40 \quad 2 \quad 34)$ **27.** $\begin{pmatrix} 19 & -13 \\ 17 & -43 \end{pmatrix}$

29. $\begin{pmatrix} -20 & 10 \\ 8 & -2 \\ -4 & 16 \end{pmatrix}$ **31.** $\begin{pmatrix} 6 & -10 \\ 3 & -5 \end{pmatrix}$ **33.** $\begin{pmatrix} 0 & 2 \\ 2 & 0 \\ -10 & -6 \\ -5 & 0 \end{pmatrix}$ **35.** $\begin{pmatrix} 2 & 0 & -2 & 1 \\ 8 & 8 & -1 & 2 \\ 2 & 2 & 2 & 6 \end{pmatrix}$

37. $\mathbf{AB} = \begin{pmatrix} 1 & 2 & -1 \\ -2 & 9 & 2 \\ -1 & -2 & 1 \end{pmatrix}$ $\mathbf{BA} = \begin{pmatrix} 5 & -2 & 2 \\ 4 & -2 & 0 \\ -2 & 3 & 8 \end{pmatrix}$ **39.** $\mathbf{B}^3 \begin{pmatrix} 778 & 1050 \\ 1225 & 1653 \end{pmatrix}$ **41.** $\begin{pmatrix} 5 \\ 1 \end{pmatrix}$ **43.** $\begin{pmatrix} 13 \\ -1 \\ 10 \end{pmatrix}$ **45.** \$1,305

47. 29, 70.3, 22.5, 19.9   **49.** (a) $\begin{pmatrix} 67 & 44 & 41 \\ 74 & 59 & 75 \end{pmatrix}$ (b) $\begin{pmatrix} 3 & -8 & 3 \\ -8 & -7 & -11 \end{pmatrix}$ (c) $\mathbf{MP} = \begin{pmatrix} 802.26 \\ 1233.58 \end{pmatrix}$ $\mathbf{TP} = \begin{pmatrix} 782.05 \\ 957.79 \end{pmatrix}$

Exercise 3

1. $\begin{pmatrix} 0 & -0.200 \\ 0.125 & 0.100 \end{pmatrix}$ **3.** $\begin{pmatrix} 0.118 & 0.059 \\ -0.020 & 0.157 \end{pmatrix}$ **5.** $\begin{pmatrix} 0.074 & 0.168 & 0.057 \\ 0.148 & -0.164 & 0.115 \\ 0.090 & 0.094 & -0.041 \end{pmatrix}$

7. $\begin{pmatrix} 0.500 & 0 & -0.500 & 0 \\ 10.273 & -0.545 & -13.091 & 1.364 \\ -4.727 & 0.455 & 5.909 & -0.636 \\ -4.591 & 0.182 & 5.864 & -0.455 \end{pmatrix}$

Review Problems

1. $(2 \quad 12 \quad 10 \quad -2)$ **3.** 86 **5.** $\begin{pmatrix} 0.062 & 0.250 \\ -0.188 & 0.250 \end{pmatrix}$ **7.** $\begin{pmatrix} 7 & 2 & 6 \\ 9 & 0 & 3 \end{pmatrix}$ **9.** $x = -4, y = -3, z = 2, w = 1$

11. $(2, -1, 1)$ **13.** $(3, 5)$ **15.** -0.557 mA, 0.188 mA, -0.973 mA, -0.271 mA **17.** \$101,395; \$124,281

◆◆◆ **CHAPTER 13** ◆◆◆

Exercise 1

1. $\dfrac{3}{x}$ **3.** 2 **5.** $\dfrac{a}{4b^2}$ **7.** $\dfrac{a^3}{b^2}$ **9.** $\dfrac{p^3}{q}$ **11.** 1 **13.** $\dfrac{1}{16a^6b^4c^{12}}$ **15.** $\dfrac{1}{x+y}$ **17.** $\dfrac{m^4}{(1-6m^2n)^2}$ **19.** $\dfrac{2}{x} + \dfrac{1}{y^2}$

21. $\dfrac{1}{27m^3} - \dfrac{2}{n^2}$ **23.** $x^{2n} + 2x^ny^m + y^{2m}$ **25.** $\dfrac{x}{2y^7}$ **27.** p **29.** x^{m2} **31.** $\dfrac{b}{3}$ **33.** $\dfrac{9y^2}{4x^2}$ **35.** $\dfrac{27q^3y^3}{8p^3x^3}$

37. $\dfrac{9a^8b^6}{25x^4y^2}$ **39.** $\dfrac{4^na^{6n}x^{6n}}{9^nb^{4n}y^{2n}}$ **41.** $\dfrac{3p^2}{2q^2x^4z^3}$ **43.** $\dfrac{5^pw^{2p}}{2^pz^p}$ **45.** $\dfrac{25n^4x^8}{9m^6y^6}$ **47.** $\dfrac{(1+a)^2(b-1)}{(1-a+a^2)(b+1)^2}$ **49.** $x^7(x-y)^2$

51. $9a^{2n} + 6a^nb^n + 4b^{2n}$ **53.** $-\dfrac{(x+1)^2(y-1)}{(y+1)^2(x^2-x+1)}$ **55.** $\left(\dfrac{7}{15}\right)a^{n-m+2}x^{2-n}y^{n-3}$ **57.** $a^3(5a^2 - 16b^2)$

59. $\dfrac{(a^{4-m}b^{n-3})}{5}$ **61.** $x^3(x^2 + y^2)$ **63.** $\dfrac{(a+b)^4}{(a^2+b^2)^2}$ **65.** $R = PI^{-2}$ **67.** $2I^2R$

Exercise 2

1. $\sqrt[4]{a}$ **3.** $\sqrt[4]{z^3}$ **5.** $\sqrt{m-n}$ **7.** $\sqrt[3]{\dfrac{y}{x}}$ **9.** $b^{1/2}$ **11.** y **13.** $(a+b)^{1/n}$ **15.** xy **17.** $3\sqrt{2}$ **19.** $3\sqrt{7}$

21. $-2\sqrt[3]{7}$ **23.** $a\sqrt{a}$ **25.** $6x\sqrt{y}$ **27.** $2x^2\sqrt[3]{2y}$ **29.** $6y^2\sqrt[5]{xy}$ **31.** $a\sqrt{a-b}$ **33.** $3\sqrt{m^3 + 2n}$

35. $x\sqrt[3]{x-a^2}$ **37.** $(a^2 - b^2)\sqrt{a}$ **39.** $\dfrac{\sqrt{21}}{7}$ **41.** $\dfrac{\sqrt[3]{2}}{2}$ **43.** $\dfrac{\sqrt[3]{6}}{3}$ **45.** $\dfrac{\sqrt{2x}}{2x}$ **47.** $\dfrac{a\sqrt{15ab}}{5b}$ **49.** $\dfrac{\sqrt[3]{x}}{x}$

51. $\dfrac{x\sqrt[3]{18}}{3}$ **53.** $(x-y)\sqrt{x^2 + xy}$ **55.** $\dfrac{(m+n)\sqrt{m^2 - mn}}{m-n}$ **57.** $\dfrac{(2a-3)\sqrt{6a}}{3a}$ **59.** $Z = (R^2 + X^2)^{1/2}$

61. $a\sqrt{10}$ **63.** $S = 4t\sqrt{16t^2 - 120t + 325}$

73. $\dfrac{4\sqrt[3]{18}}{3} + \sqrt[3]{4} + 3$ **75.** $\dfrac{3}{2}$ **77.** $\dfrac{2\sqrt[6]{a^5b^2c^3}}{ac}$ **79.** $\dfrac{1}{2c}\sqrt[12]{a^2bc^{10}}$ **81.** $\dfrac{8 + 5\sqrt{2}}{2}$ **83.** $\dfrac{4\sqrt{ax}}{a}$ **85.** $\dfrac{6\sqrt[3]{2x}}{x}$

87. $-\dfrac{30 + 17\sqrt{30}}{28}$ **89.** $\dfrac{a^2 - a\sqrt{b} + b}{a^2 - b}$ **91.** $\dfrac{x - 2\sqrt{xy} + y}{x - y}$ **93.** $\dfrac{5\sqrt{3x} - 5\sqrt{xy} + R6xy - \sqrt{2y}}{6x - 2y}$

95. $\dfrac{3\sqrt{6a} - 4\sqrt{3ab} + 6\sqrt{10ab} - 8\sqrt{5b}}{9a - 8b}$

Exercise 4

1. 36   **3.** 4   **5.** 13.8   **7.** 8   **9.** 8   **11.** 7   **13.** 4.79   **15.** 3   **17.** 8   **19.** 0.347   **21.** No solution   **23.** 7

25. 14.1 cm; 29.0 cm   **27.** 3.91 m; 4.77 m   **29.** $C = \dfrac{1}{\omega^2 L \pm \omega\sqrt{Z^2 - R^2}}$

Review Problems

1. $2\sqrt{13}$ **3.** $3\sqrt[3]{6}$ **5.** $\sqrt[3]{2}$ **7.** $9x\sqrt[4]{x}$ **9.** $(a - b)x\sqrt[3]{(a - b)^2 x}$ **11.** $\dfrac{\sqrt{a^2 - 4}}{a + 2}$ **13.** $x^2\sqrt{1 - xy}$

15. $-\dfrac{1}{46}(6 + 15\sqrt{2} - 4\sqrt{3} - 10\sqrt{6})$ **17.** $\dfrac{2x^2 + 2x\sqrt{x^2 - y^2} - y^2}{y^2}$ **19.** $26\sqrt[3]{x^2}$ **21.** $\sqrt[6]{a^3b^2}$

23. $9 + 12\sqrt{x} + 4x$ **25.** $7\sqrt{2}$ **27.** $\dfrac{25\sqrt{2}}{12}$ **29.** $\dfrac{\sqrt{2b}}{2b}$ **31.** $72\sqrt{2}$ **33.** 10   **35.** 14   **37.** 15.2

39. $x^{2n-1} + (xy)^{n-1} + x^n y^{n-2} + y^{2n-3}$ **41.** $\dfrac{27x^3y^6}{8}$ **43.** $8x^9y^6$ **45.** $\dfrac{3}{w^2}$ **47.** $\dfrac{1}{3x}$ **49.** $\dfrac{1}{x} - \dfrac{2}{y^2}$ **51.** $\dfrac{1}{9x^2} + \dfrac{y^8}{4x^4}$

53. $p^{2a-1} + (pq)^{a-1} + p^a q^{a-2} + q^{2a-3}$ **55.** $\dfrac{p^2}{q}$ **57.** $\dfrac{r^2}{s^3}$ **59.** 1   **61.** $V = 36\pi r^3$ **63.** $B = \sqrt{A}\sqrt{C}$

◆◆◆ CHAPTER 14 ◆◆◆

Exercise 1

1. $0, \dfrac{2}{5}$ **3.** $0, \dfrac{9}{2}$ **5.** ± 3 **7.** ± 1 **9.** $0, 0.355$ **11.** $3, -5$ **13.** $5, -4$ **15.** $2, -1$ **17.** $\dfrac{5}{2}, -1$ **19.** $\dfrac{3}{2}, -4$

21. $\dfrac{1}{5}, -3$ **23.** $\dfrac{1}{2}, -\dfrac{2}{3}$ **25.** $9, -4$ **27.** $1, -\dfrac{1}{6}$ **29.** $-a, -3a$ **31.** $-2b, -\dfrac{4b}{3}$ **33.** $x^2 - 11x + 28 = 0$

35. $15x^2 - 19x + 6 = 0$ **37.** $6, \dfrac{15}{4}$ **39.** 4 (-21 doesn't check)

Exercise 2
1. $4 \pm \sqrt{14}$ **3.** $(-7 \pm \sqrt{61})/2$ **5.** $(3 \pm \sqrt{89})/8$ **7.** $(-3 \pm \sqrt{21})/3$ **9.** $(-7 \pm \sqrt{129})/8$ **11.** $0.485, -6.74$

Exercise 3
1. $3.17, 8.83$ **3.** $3.89, -4.87$ **5.** $1.96, -5.96$ **7.** $0.401, -0.485$ **9.** $0.170, -0.599$ **11.** $2.87, 0.465$ **13.** $0.907, -2.57$ **15.** $5.87, -1.87$ **17.** $12.0, -14.0$ **19.** $5.08, 10.9$ **21.** $2.81, -1.61$ **23.** $8.37, -2.10$ **25.** $0.132, -15.1$ **27.** $2.31, 0.623$ **29.** Real, unequal   **31.** Not real   **33.** Not real

Exercise 4
1. $\dfrac{2}{3}$ or $\dfrac{3}{2}$ **3.** 4 and 11   **5.** 5 and 15   **7.** 4×6 m   **9.** 15×30 cm   **11.** 162×200 m   **13.** 2.26 cm   **15.** 6.6 cm
17. 146 mi./h   **19.** 80 km/h   **21.** 9 h and 12 h   **23.** 7 m/day   **25.** At each end   **27.** 0.594 s and 8.59 s
29. 125 Ω and 655 Ω **31.** 0.381 A and 0.769 A   **33.** -0.5 A and 0.1 A   **35.** -1.1 V

Exercise 5

1.
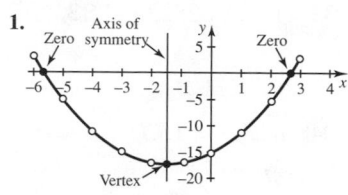
3.
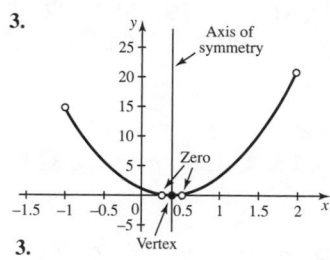
5. $10\overline{0}$ m; 80.0 m; 11.7 m; 40.0 m   **7.** 4.47 s

9. 41.7 m; 16.7 m; 51.4 m; 17.1 m; 32.0 m; 16.0 m   **11.** 20 cm   **13.** 1.95 m

Exercise 6

1. $\sqrt[3]{2}$, $\sqrt[3]{4}$ **3.** 1   **5.** 1 (0 doesn't check)   **7.** -1, $-\dfrac{1}{27}$ **9.** 1, -64 **11.** $\dfrac{\sqrt[3]{2}}{8}$ (8.69 doesn't check)

13. 625 (256 doesn't check)   **15.** 496 (243 doesn't check)

Exercise 7

1. 3, -5 **3.** 2, -1 **5.** $\dfrac{3}{2}$, -4 **7.** $\dfrac{4}{3}$, $-\dfrac{8}{3}$

Exercise 8

1. (6, 2)   **3.** (3, 2), (-5, 6)   **5.** (2, 3), (-46, 15)   **7.** (2.61, -1.20), (-2.61, -1.20)   **9.** (4.86, 0.845), (4.86, -0.845)
11. (4, -7), (7, -4)

Review Problems

1. 6, -1 **3.** ±2, ±3 **5.** 0, 5   **7.** 0, -2 **9.** 5, $-\dfrac{2}{3}$ **11.** 1, $\dfrac{1}{2}$ **13.** ±3 **15.** 1.79, -2.79 **17.** 4, 9

19. 0.692, -0.803 **21.** $\pm\sqrt{10}$ **23.** $\dfrac{1}{2}$, -2 **25.** 0, 9   **27.** 1, 512   **29.** ±5 **31.** 3.54, -2.64 **33.** 0.777, -2.49

35. 0.716, -2.32 **37.** 1.49, -1.26 **39.** (7.41, -0.41), (-0.41, 7.41)   **41.** 2, 4, -3 **43.** (1.68, 1.16), (-2.68, 3.34)
45. 202 bags   **47.** 15×30 m   **49.** 3 km/h   **51.** $1\frac{1}{2}$ s   **53.** 5.0 cm $\times$ 16 cm or 8.0 cm $\times$ 10 cm   **55.** 3.56 km/h
57. 12 m $\times$ 12 m

◆◆◆ **CHAPTER 15** ◆◆◆

Exercise 1

1. II   **3.** IV   **5.** I   **7.** II   **9.** II or III   **11.** IV   **13.** III   **15.** pos   **17.** pos   **19.** pos   **21.** neg

	sin	cos	tan		r	sin	cos	tan	cot	sec	csc
23.	+	−	−	**29.**	12.65	0.949	−0.316	−3.00	−0.333	−3.16	1.05
25.	−	+	−	**31.**	17.0	−0.471	−0.882	0.533	1.88	−1.13	−2.12
27.	+	−	−	**33.**	5.00	0	1.00	0	—	1.00	—

sin   cos   tan

57. 0.8192   **59.** 1.906   **61.** 1.711   **63.** 1.111   **65.** 135°, 315°   **67.** 33.2°, 326.8°   **69.** 81.1°, 261.1°   **71.** 90°, 270°   **73.** 219.5°, 320.5°   **75.** 60.1°, 240.1°   **77.** 227.4°, 312.6°   **79.** 34.2°, 325.8°   **81.** 102.6°, 282.6°   **83.** 1/2 **85.** 0   **87.** 1   **89.** 0   **91.** 0   **93.** −1   **95.** 0   **97.** 1/2   **99.** 1   **101.** 0

Exercise 2

1. A 32.8° C 101° c 413   **3.** A 108° B 29.2° a 29.3   **5.** B 71.1° b 8.20 c 6.34   **7.** A 104° a 21.7 c 15.7 **9.** A 23.0° a 42.8 b 89.7   **11.** A 106.4° a 90.74 c 29.55   **13.** C 150° b 34.1 c 82.0

Exercise 3

1. A 44.2° B 29.8° c 21.7   **3.** B 48.7° C 63.0° a 22.6   **5.** A 56.9° B 95.8° c 70.1   **7.** B 25.5° C 19.5° a 452 **9.** A 45.6° B 80.4° C 54.0°   **11.** B 30.8° C 34.2° a 82.8   **13.** A 69.0° B 44.0° c 8.95   **15.** B 10.3° C 11.7 a 3.69 **17.** A 66.21° B 72.02° c 1052   **19.** A 69.7° B 51.7° C 58.6°

Exercise 4

1. 30.8 m and 85.6 m   **3.** 32.3°; 60.3°; 87.4°   **5.** N 48.8° W   **7.** S 59.4° E   **9.** 598 km   **11.** 28.3 m   **13.** 77.3 m; 131 m   **15.** 107 m   **17.** 337 mm   **19.** 33.7 cm   **21.** 131°; 88.2°; 73.5°   **23.** 85.8° and 94.2°   **25.** 3.49 m and 4.37 m   **27.** 110   **29.** −1/4   **31.** 1.2 m

Exercise 5

1. 521; 10°   **3.** 87.1; 31.9°   **5.** 6708; 41.16°   **7.** 37.3 N at 20.5°   **9.** 121 N at N 59.8° W   **11.** 1720 N at 29.8° from larger force   **13.** 44.2° and 20.7°   **15.** Wind: S 37.4° E; plane: S 84.8° W   **17.** 413 km/h at N 41.2° E   **19.** 632 km/h; 3.09°   **21.** 26.9 A; 32.3°   **23.** 39.8 at 26.2°

Review Problems

1. $A = 24.1°$; $B = 20.9°$; $c = 77.6$ **3.** $A = 61.6°$; $C = 80.0°$; $b = 1.30$ **5.** $B = 20.2°$; $C = 27.8°$; $a = 82.4$ **7.** IV **9.** II   **11.** neg   **13.** neg   **15.** neg   **17.** sin = −0.800; cos = −0.600; tan = 1.33   **19.** 1200 at 36.3°   **21.** 22.1 at 121°   **23.** 1.11 km   **25.** sin = 0.0872; cos = −0.9962; tan = −0.0875   **27.** sin = 0.7948; cos = −0.6069; tan = −1.3095   **29.** 0.4698   **31.** 1.469   **33.** S 3.5° E   **35.** 130.8°; 310.8°   **37.** 47.5°; 132.5°   **39.** 80.0°; 280.0° **41.** 695 N; 17.0°   **43.** −224 cm/s; 472 cm/s   **45.** 22.42

◆◆◆ CHAPTER 16 ◆◆

Exercise 1

1. 0.834 rad   **3.** 0.6152 rad   **5.** 9.74 rad   **7.** 0.279 rev   **9.** 0.497 rev   **11.** 0.178 rev   **13.** 162°   **15.** 21.3° **17.** 65.3°   **19.** $\dfrac{\pi}{3}$ **21.** $\dfrac{11\pi}{30}$ **23.** $\dfrac{7\pi}{10}$ **25.** $\dfrac{13\pi}{30}$ **27.** $\dfrac{20\pi}{9}$ **29.** $\dfrac{9\pi}{20}$ **31.** $22\frac{1}{2}°$ **33.** 147°   **35.** 20° **37.** 158°   **39.** 24°   **41.** 15°   **43.** 0.8660   **45.** 0.4863   **47.** −0.3090   **49.** 1.067   **51.** −2.747   **53.** −0.8090 **55.** −0.2582   **57.** 0.5854   **59.** 0.8129   **61.** 1.337   **63.** 0.2681   **65.** 1.309   **67.** 1.116   **69.** 0.5000 **71.** 0.1585   **73.** 19.1 cm^2 **75.** 485 cm^2 **77.** 3.86 in.   **79.** 2280 cm^2

Exercise 2

1. deg/s = 607 mm   **3.** s = 230 ft   **5.** s = 43.5 in.   **7.** θ = 2.21 rad   **9.** θ = 1.24 rad   **11.** r = 0.824 cm **13.** r = 125 ft.   **15.** 6090 km   **17.** 589 m   **19.** 24.5 cm   **21.** 3.2 × 10 km   **23.** 0.251 m   **25.** r = 355 mm; R = 710 mm; θ = 60.8°   **27.** 9780 km   **29.** 14.1 cm   **31.** 87.6065 ft

Exercise 3

1. deg/s = 11 100;   **3.** rev/min = 12.9; rad/s = 1.35;   **5.** rev/min = 8.02; rad/s = 0.840;   **7.** 4350° **9.** 0.007 49 s   **11.** 353 rev/min   **13.** 1.62 m/s   **15.** 1660 km/h   **17.** 57.6 rad/s

Review Problems

1. 77.1°   **3.** 20°   **5.** 165°   **7.** 2.42 rad/s   **9.** $\dfrac{5\pi}{3}$ **11.** $\dfrac{23\pi}{18}$ **13.** 0.3420   **15.** 1.000   **17.** −0.018 27

19. 191 rev/min   **21.** 336 mi./h   **23.** 1.1089   **25.** −0.8812   **27.** −1.0291   **29.** 0.9777   **31.** 0.9097 **33.** 222 mm   **35.** 2830 cm^3

◆◆◆ CHAPTER 17 ◆◆

Exercise 1

1. $P = 2$ s; $f = 1/2$ cycle/s; amplitude $= 7$

3.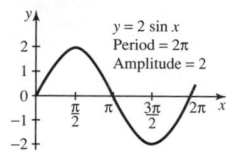

5.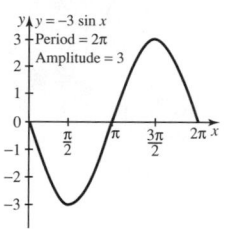

7.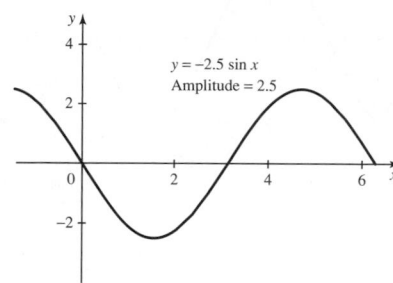

9.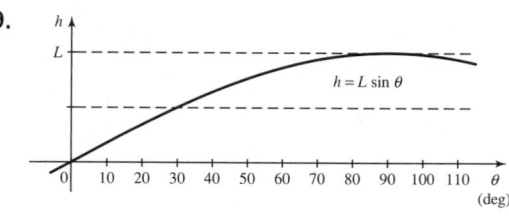

11.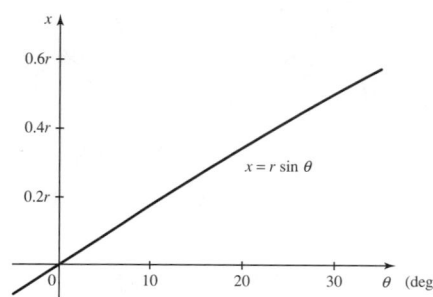

Exercise 2

1.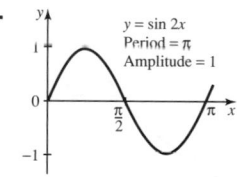

3.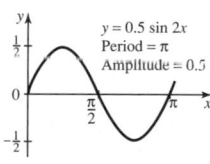

5.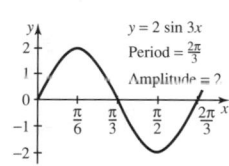

7.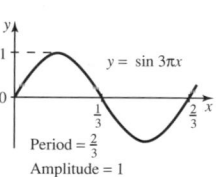

9.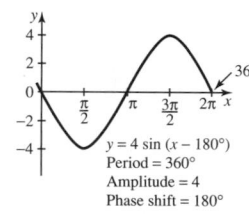

11.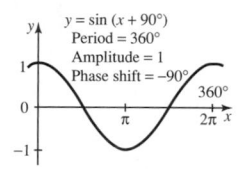

13.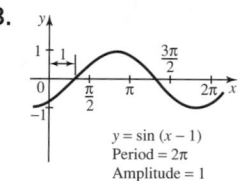

15.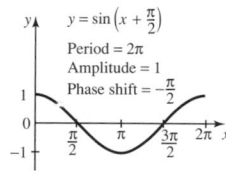

17. (a) $y = \sin 2x$ (b) $y = \sin(x - \pi/2)$ (c) $y = 2 \sin(x - \pi)$ **19.** $y = -2 \sin\left(\dfrac{x}{3} + \dfrac{\pi}{12}\right)$

21.   **23.**

Exercise 4

1.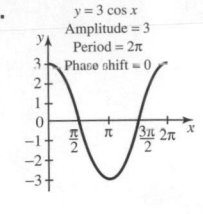

3.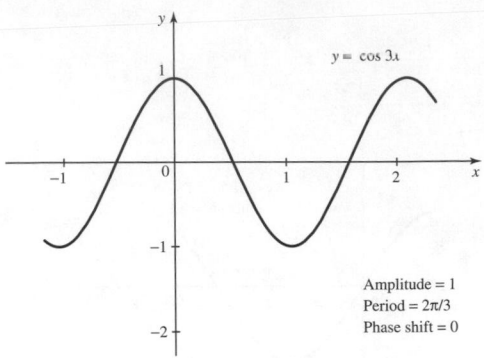

5.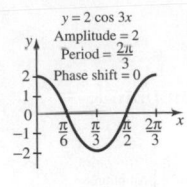

7.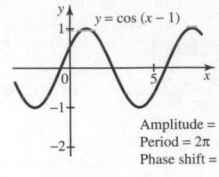

9.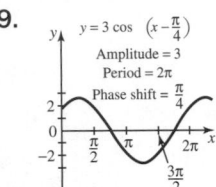

11.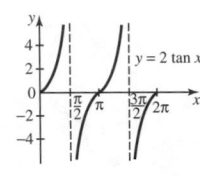

13.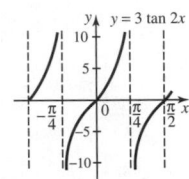

15.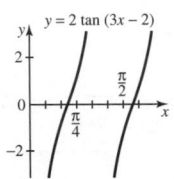

17.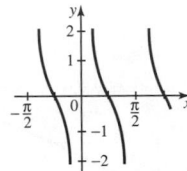

19.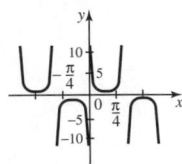

21.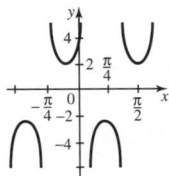

23.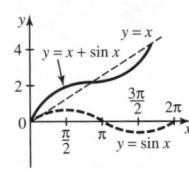

25.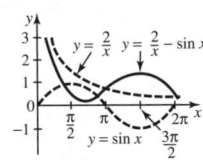

27.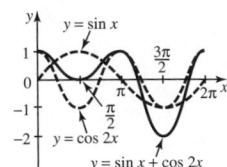

29.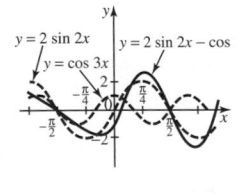

31.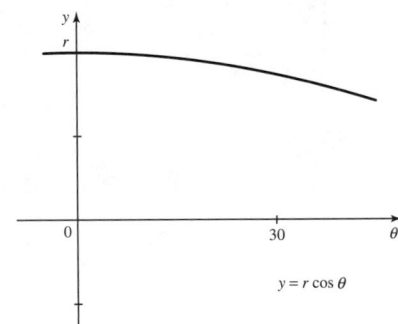

33.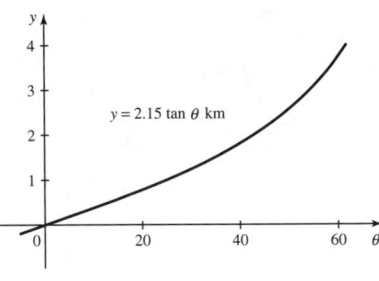

35.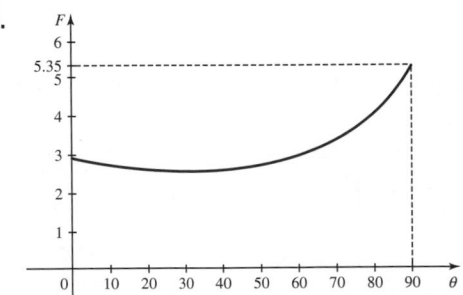

Exercise 5

1. $P = 0.0147$ s; $\omega = 427$ rad/s   **3.** $P = 0.0002$ s; $\omega = 31{,}400$ rad/s   **5.** $f = 8$ Hz; $\omega = 50.3$ rad/s   **7.** 3.33 s
9. $P = 0.0138$ s; $f = 72.4$ Hz   **11.** $P = 0.0126$ s; $f = 79.6$ Hz   **13.** $P = 400$ ms; amplitude $= 10$; $\phi = 1.1$ rad
15. $y = 5 \sin(750t + 15°)$

17. 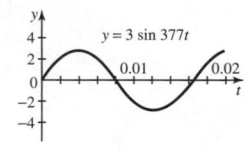 **19.**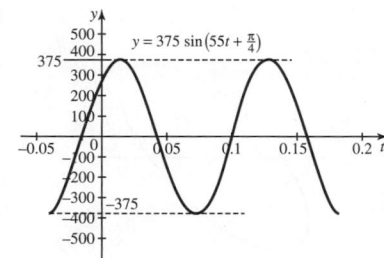

21. $y = 35.7 \sin(\omega t + 45.4°)$ **23.** $y = 843 \sin(\omega t - 28.2°)$ **25.** $y = 9.40 \sin(\omega t + 51.7°)$

27. $y = 660 \sin(\omega t + 56.5°)$

29. -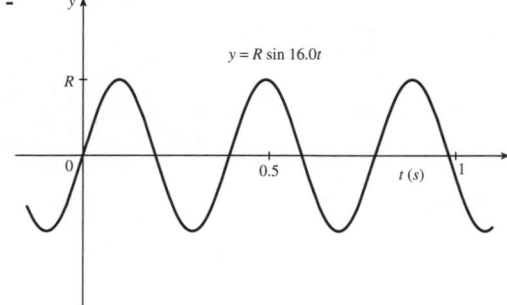

31. $v_{max} = 4.27$ V; $P = 13.6$ ms; $f = 73.7$ Hz; $\phi = 27° = 0.471$ rad; $v(0.12) = -2.11$ V **33.** $i = 49.2 \sin(220t + 63.2°)$ mA

Exercise 6

1.
3.
5.
7.
9.
11.
13.
15.
17.
19.
21.
23.
25.

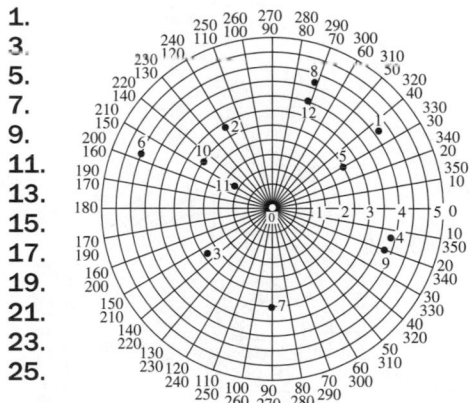

27. 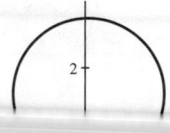 **29.** 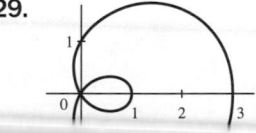 **31.** 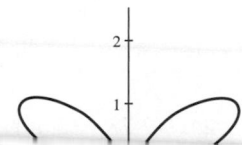 **33.**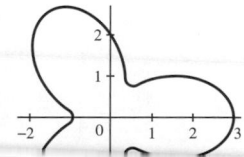

Exercise 7

7.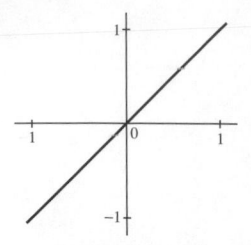

9.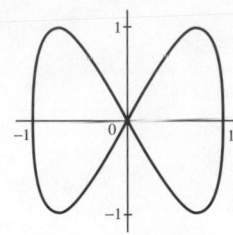

11.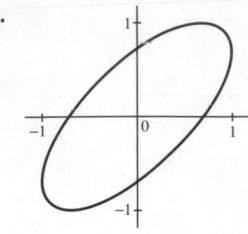

13.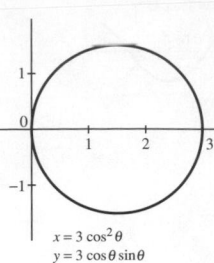

$x = 3\cos^2\theta$
$y = 3\cos\theta\sin\theta$

15.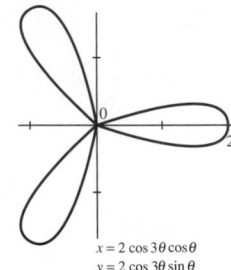

$x = 2\cos 3\theta\cos\theta$
$y = 2\cos 3\theta\sin\theta$

17. $x = (3\sin(3t)-2)\cos(t),\ y = (3\sin(3t)-2)\sin(t),$

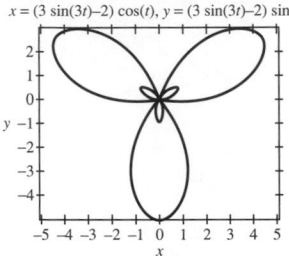

Review Problems

1.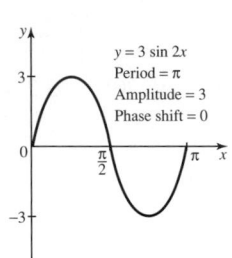

$y = 3\sin 2x$
Period $= \pi$
Amplitude $= 3$
Phase shift $= 0$

3.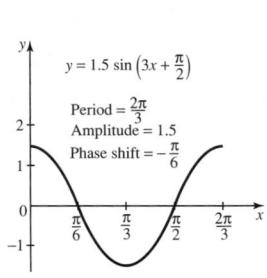

$y = 1.5\sin\left(3x + \dfrac{\pi}{2}\right)$
Period $= \dfrac{2\pi}{3}$
Amplitude $= 1.5$
Phase shift $= -\dfrac{\pi}{6}$

5.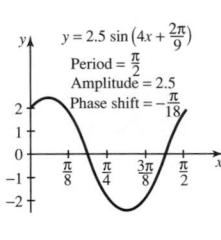

$y = 2.5\sin\left(4x + \dfrac{2\pi}{9}\right)$
Period $= \dfrac{\pi}{2}$
Amplitude $= 2.5$
Phase shift $= -\dfrac{\pi}{18}$

7. $y = 5\sin\left(\dfrac{2x}{3} + \dfrac{\pi}{9}\right)$

9.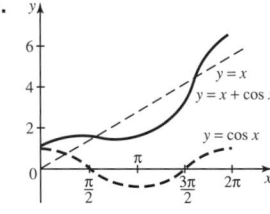

$y = x$
$y = x + \cos x$
$y = \cos x$

11. and 13.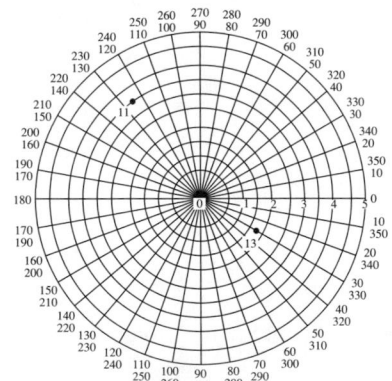

15.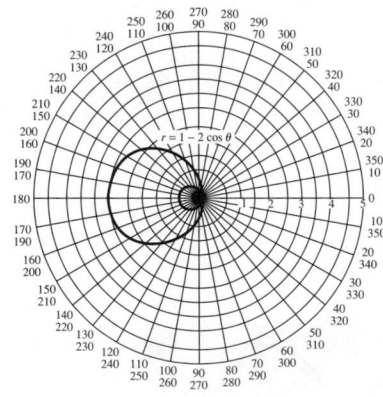

$r = 1 - 2\cos\theta$

17. $(6.52, 144°)$ **19.** $(2.54, 2.82)$ **21.** $(2.73, -2.64)$ **23.** $x^2 + y^2 = 2x$ **25.** $r = \dfrac{1}{(5\cos\theta + 2\sin\theta)}$

27. $P = 0.00833$ s; $\omega = 754$ rad/s   **29.** 3.33 s   **31.**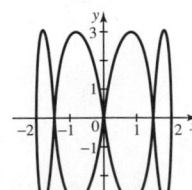

33. $(a)\ y = 101\sin(\omega t - 15.2°)$
$(b)\ y = 3.16\sin(\omega t + 56.8°)$

35.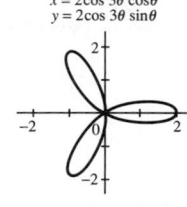
$x = 2\cos 3\theta \cos\theta$
$y = 2\cos 3\theta \sin\theta$

37.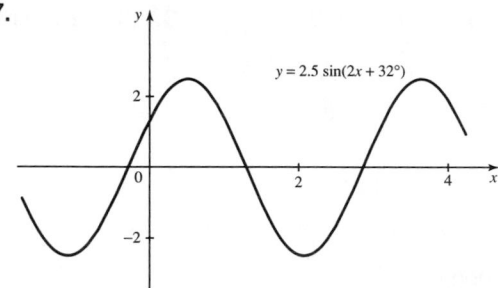
$y = 2.5 \sin(2x + 32°)$

39.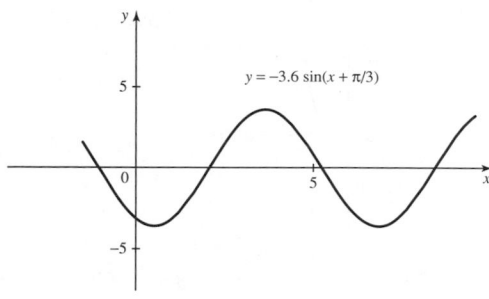
$y = -3.6 \sin(x + \pi/3)$

◆◆◆ **CHAPTER 18** ◆◆◆

Exercise 1

1. $\dfrac{\sin x - 1}{\cos x}$ **3.** $\dfrac{1}{\cos\theta}$ **5.** $\dfrac{1}{\cos\theta}$ **7.** $-\tan^2 x$ **9.** $\sin\theta$ **11.** $\sec\theta$ **13.** $\tan x$ **15.** $\cos\theta$ **17.** $\sin\theta$ **19.** 1

21. 1   **23.** $\sin x$ **25.** $\sec x$ **27.** $\tan^2 x$ **29.** $-\tan^2\theta$

Exercise 2

1. $\frac{1}{2}(\sqrt{3}\sin\theta + \cos\theta)$ **3.** $\frac{1}{2}(\sin x \sqrt{3}\cos x)$ **5.** $-\sin x$ **7.** $\sin\theta\cos 2\phi + \cos\theta\sin 2\phi$

9. $(\tan 2\theta - \tan 3\alpha)/(1 + \tan 2\theta\tan 3\alpha)$ **11.** $-\cos\theta - \sin\theta$ **13.** $-\sin x$ **33.** $y = 11{,}200\sin(\omega t + 41.0°)$

35. $y = 112\sin(\omega t + 41.4°)$

Exercise 3

1. 1   **3.** $\sin 2x$ **5.** 2

Exercise 5

1. $30°, 150°$ **3.** $45°, 225°$ **5.** $60°, 120°, 240°, 300°$ **7.** $90°$ **9.** $30°, 150°, 210°, 330°$ **11.** $45°, 135°, 225°, 315°$

13. $0°, 45°, 180°, 225°$ **15.** $60°, 120°, 240°, 300°$ **17.** $60°, 180°, 300°$ **19.** $120°, 240°$ **21.** $0°, 60°, 120°, 180°,$
$240°, 300°$ **23.** $135°, 315°$ **25.** $0°, 60°, 180°, 300°$

Exercise 6

1. $22.0°$ **3.** $-78.5°$ **5.** $81.8°$ **7.** $75.4°$ **9.** $33.2°$

Review Problems

13. 1   **15.** $2\csc^2\theta$ **17.** 1   **19.** 1   **21.** 1   **23.** $-\tan\theta$ **25.** $120°, 240°$ **27.** $45°, 225°$ **29.** $15°, 45°, 75°, 135°,$
$195°, 225°, 255°, 315°$ **31.** $90°, 306.9°$ **33.** $189.3°, 350.7°$ **35.** $60°, 300°$ **37.** $55.6°$ **39.** $74.3°$
41. $78.2\sin(\omega t + 52.5°)$

◆◆◆ **CHAPTER 19** ◆◆◆

Exercise 1

11. $-5a$ **13.** $(x + 2)/x$ **15.** ± 12 **17.** ± 15 **19.** $30, 42$ **21.** $24, 32$

13. 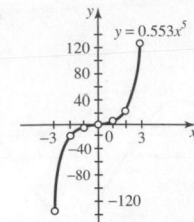   **15.** 396 m   **17.** 4.03 s   **19.** 765 W   **21.** 1.73   **23.** f8   **25.** f13   **27.** 0.234 m³

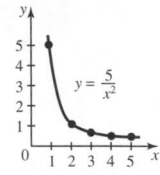

29. 236 t   **31.** 65 m²   **33.** 3.67 cm.   **35.** 28 900 m²

Exercise 4

1. 1420   **3.** y is halved.   **5.** $x = 91.5$; $y = 61.5$ **7.** $x = 9236$; $y = 1443$ **9.** 41.4% increase   **11.**

13. 707 kPa   **15.** 44.9 pN   **17.** 79 lb.   **19.** 200 lx   **21.** 5.31 m   **23.** 0.909   **25.** $33\frac{1}{3}$% increase

Exercise 5

1. 352   **3.** 4.2% increase   **5.** $w = 116$; $x = 43.5$; $y = 121$ **7.** 13.9   **9.** $y = 4.08x^{3/2}/w$ **11.** $13\frac{3}{4}$% decrease
13. 1/3   **15.** 9/4   **17.** 169 W   **19.** 3.58 m³   **21.** 3.0 weeks   **23.** 5940 kg   **25.** 394 times/s   **27.** 2.21

Review Problems

1. 1040   **3.** 10.4% increase   **5.** 121 L/min   **7.** 1.73   **9.** 29 y   **11.** Shortened by 7.813 mm   **13.** 8.5   **15.** 22 h
52 min   **17.** 347 000 km   **19.** 42 kg   **21.** 2.8   **23.** 45% increase   **25.** 469 U.S. gal.   **27.** 102 m²   **29.** 2.2 cm
31. 17.6 m²   **33.** 12.7 t   **35.** 144 oz.   **37.** $608   **39.** 1.5   **41.** 761 cm²

◆◆◆ CHAPTER 20 ◆◆

Exercise 1

1. 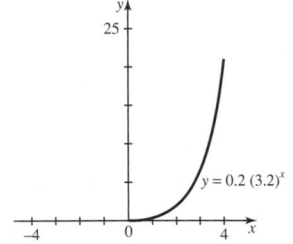   **3.** 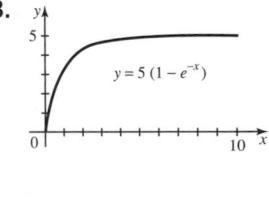   **5.** 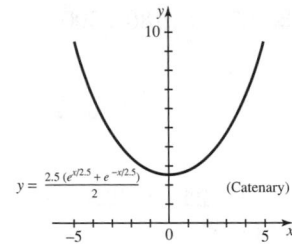   **7.** $4852   **9.** (*a*) $6.73;
(*b*) $7.33; (*c*) $7.39
11. $1823

Exercise 2

1. 284 units   **3.** 12.4 million   **5.** 18.2 million barrels   **7.** 506 °C   **9.** 0.101 A   **11.** 559 °C   **13.** 29.46 kPa
15. 72 mA   **17.** $3\overline{0}$%

Exercise 3

1. $\log_3 81 = 4$ **3.** $\log_4 4096 = 6$ **5.** $\log_x 995 = 5$ **7.** $10^2 = 100$ **9.** $5^3 = 125$ **11.** $3^{57} = x$ **13.** $5^y = x$
15. 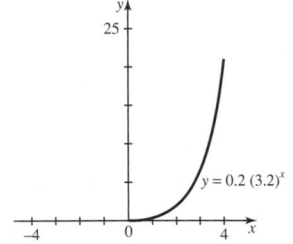   **17.** 1.4409   **19.** 0.7731   **21.** 1.6839   **23.** 2.9222   **25.** 1.4378   **27.** −2.1325
29. 4.5437   **31.** 38.4   **33.** 187 000   **35.** 1.39×10^{-3} **37.** 5.93×10^{-3} **39.** 9530
41. 1.07×10^{-4} **43.** 3.8774   **45.** 7.7685   **47.** −0.1684   **49.** −4.7843   **51.** −9.9825
53. 10.5558   **55.** 17.22   **57.** 2.408   **59.** 0.6942   **61.** 2.550   **63.** 7.572×10^6
65. 1.063×10^{-8}

Exercise 4

1. $\log 2 - \log 3$ **3.** $\log a + \log b$ **5.** $\log x + \log y + \log z$ **7.** $\log 3 + \log x - \log 4$ **9.** $-\log 2 - \log x$

11. $\log a + \log b + \log c - \log d$ **13.** $\log 12$ **15.** $\log (3/2)$ **17.** $\log 27$ **19.** $\log \dfrac{a^3 c^4}{b^2}$ **21.** $\log \dfrac{xy^2 z^3}{ab^2 c^3}$

23. $\log[\sqrt{x + 2}(x - 2)]$ **25.** $2^x = xy^2$ **27.** $a^2 - 2b^2 = 0$ **29.** $p - q = 100$ **31.** 1 **33.** 2 **35.** x **37.** $3y$

39. 549.0 **41.** 322 **43.** 2.735 **45.** 12.58 **47.** 3.634 **49.** 3.63 **51.** 1.639 **53.** 2.28 **55.** 195

57. -8.80 **59.** 5.46

Exercise 5

1. 2.81 **3.** 16.1 **5.** 2.46 **7.** 1.17 **9.** 5.10 **11.** 1.49 **13.** 0.239 **15.** -3.44 **17.** 1.39 **19.** 0

21. $3e^{4.159x}$ **23.** 0.0147 s **25.** 1.69 s **27.** 1.5 s **29.** 8790 m **31.** 4.62 rad **33.** 2.4 y **35.** 7.95 y **37.** 9.9 y

39. 3.4%

Exercise 6

1. 2 **3.** 1/3 **5.** 2/3 **7.** 2/3 **9.** 2 **11.** 3 **13.** 25 **15.** 6 **17.** 3/2 **19.** 47.5 **21.** $-11.0, 9.05$

23. 10/3 **25.** 22 **27.** 0.928 **29.** 0.916 **31.** 12 **33.** 101 **35.** 50 000 **37.** 7.0 y **39.** 3.5

41. 90.29 kPa **43.** $13\overline{0}0$ kW **45.** -3.01 dB **47.** 35.2y **49.** 10^{-7} **51.** $4\overline{0}$

Exercise 7

1. 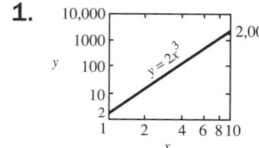 **3.** 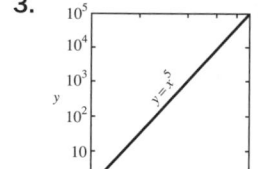 **5.** 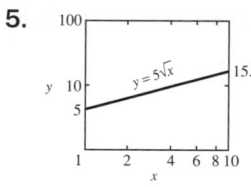 **7.**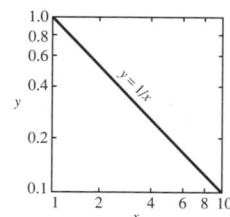

9. 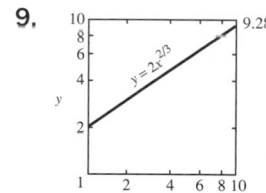 **11.** $y = 2000x^{-2.97}$ 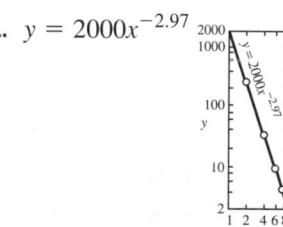 **13.** 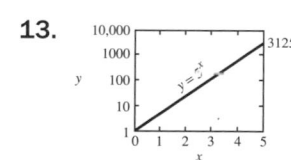 **15.**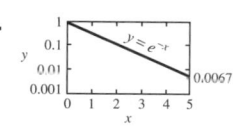

17. 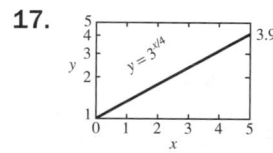 **19.** 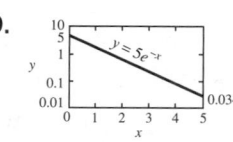 **21.** $y = 2.5^x$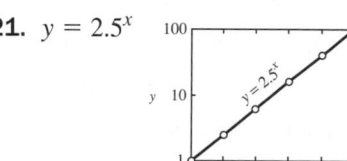

23. $i = 16.5v^{0.598}$ 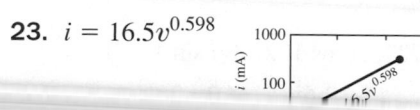 **25.** $p = 460v^{-1.06}$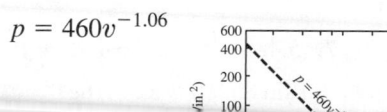

31. 2.101   **33.** 13.44   **35.** 3.17   **37.** 1.00×10^{-3} **39.** \$2071   **41.** 828 rev/min   **43.** 23 y

45.   **47.** $t = 100e^{-0.077t}$

✦✦✦ CHAPTER 21 ✦✦✦

Exercise 1

1. $j3$ **3.** $\dfrac{j}{2}$ **5.** $4 + j2$ **7.** $-5 + \dfrac{j3}{2}$ **9.** $j5$ **11.** $-1 + j$ **13.** $2a + j2$ **15.** $\dfrac{3}{4} + \dfrac{j}{6}$ **17.** $4.03 + j1.20$

19. j **21.** j **23.** $j14$ **25.** -15 **27.** $-j96$ **29.** -25 **31.** $6 - j8$ **33.** $8 + j20$ **35.** $36 + j8$ **37.** $42 - j39$

39. $21 - j20$ **41.** $2 + j3$ **43.** $p - jq$ **45.** $n + jm$ **47.** $j2$ **49.** 2 **51.** $2 + j$ **53.** $-\dfrac{15}{2} - \dfrac{j}{2}$ **55.** $\dfrac{11}{34} - \dfrac{j41}{34}$

57. $0.833 \pm j1.28$ **59.** $1.00 \pm j2.24$ **61.** $(x + j3)(x - j3)$ **63.** $(2y + jz)(2y - jz)$

Exercise 2

1.
3.
5.
7.
9.

Exercise 3

1. $6.40\underline{/38.7°}$; $6.40(\cos 38.7° + j \sin 38.7°)$ **3.** $5\underline{/323°}$; $5(\cos 323° + j \sin 323°)$ **5.** $5.39\underline{/202°}$;
$5.39(\cos 202° + j \sin 202°)$ **7.** $10.3\underline{/209°}$; $10.3(\cos 209° + j \sin 209°)$ **9.** $8.06\underline{/240°}$; $8.06(\cos 240° + j \sin 240°)$
11. $1.63 + j2.52$; $3\underline{/57°}$ **13.** $-7.79 + j4.50$; $9\underline{/150°}$ **15.** $3.70 + j4.01$; $5.46\underline{/47.3°}$ **17.** $4.64 + j7.71$;
$9(\cos 59° + j \sin 59°)$ **19.** $4.21 - j5.59$; $7(\cos 307° + j \sin 307°)$ **21.** $15(\cos 40° + j \sin 40°)$
23. $16.1(\cos 54.9° + j \sin 54.9°)$ **25.** $56\underline{/60°}$ **27.** $2(\cos 25° + j \sin 25°)$ **29.** $4.70(\cos 50.2° + j \sin 50.2°)$
31. $10\underline{/60°}$ **33.** $8(\cos 45° + j \sin 45°)$ **35.** $49\underline{/40°}$ **37.** $7.55\underline{/26°}$; $7.55\underline{/206°}$ **39.** $3.36\underline{/24.3°}$; $3.36\underline{/144.3°}$;
$3.36\underline{/264.3°}$ **41.** $13.8 + j7.41$; $-13.8 - j7.41$

Exercise 4

1. $3.61e^{j0.983}$ **3.** $3e^{j0.873}$ **5.** $2.5e^{j\pi/6}$ **7.** $5.4e^{j\pi/12}$ **9.** $-4.95 + j0.706$; $5\underline{/172°}$; $5(\cos 172° + j \sin 172°)$
11. $0.156 + j2.19$; $2.2\underline{/85.9°}$; $2.2(\cos 85.9° + j \sin 85.9°)$ **13.** $18e^{j6}$ **15.** $21e^{j4}$ **17.** $3.6e^{j7}$ **19.** $3e^{j3}$
21. $11e^{j3}$ **23.** $3e^{j}$ **25.** $9e^{j10}$ **27.** $8e^{j3}$

Exercise 5

1. $4.59 + j5.28$; $7.00\underline{/49.0°}$ **3.** $54.8 + j185$; $193\underline{/73.5°}$ **5.** $-31.2 + j23.3$; $39.0\underline{/143°}$ **7.** $8 - j2$ **9.** $38.3 + j60.2$
11. $8.67 + j6.30$ **13.** $29 - j29$ **15.** $12(\cos 32° - j \sin 32°)$ **17.** $3.08\underline{/32°}$

Exercise 6

1. $177\underline{/25°}$ **3.** $40\underline{/-90°}$ **5.** $102\underline{/0°}$ **7.** $212 \sin \omega t$ **9.** $424 \sin(\omega t - 90°)$ **11.** $11 \sin \omega t$ **13.** $155 + j0$; $155\underline{/0°}$
15. $0 - j18$; $18\underline{/270°}$ **17.** $72 - j42$; $83.4\underline{/-30.3°}$ **19.** $552 - j572$; $795\underline{/-46.0°}$ **21.** (a) $v = 603 \sin(\omega t + 85.3°)$;
(b) $v = 293 \sin(\omega t - 75.5°)$

Review Problems

1. $3 + j2$; $3.61\underline{/33.7°}$; $3.61(\cos 33.7° + j \sin 33.7°)$; $3.61e^{j0.588}$ **3.** $-2 + j7$; $7.28\underline{/106°}$; $7.28(\cos 106° + j \sin 106°)$; $7.28e^{j1.85}$ **5.** j **7.** $9 + j2$ **9.** $60.8 + j45.6$ **11.** $11 + j7$ **13.** $12(\cos 38° + j \sin 38°)$ **15.** $33/17 - j21/17$
17. $2(\cos 45° + j \sin 45°)$ **19.**
21.

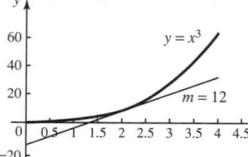

23. $7 - j24$ **25.** $125(\cos 30° + j \sin 30°)$ **27.** $32\underline{/32°}$ **29.** $0.884\underline{/-45°}$

◆◆◆ CHAPTER 22 ◆◆◆

Exercise 1

1. 3 **3.** 9 **5.** 6.60 **7.** 3.34 **9.** 6.22 **11.** -7.90 **13.** 26 **15.** 9.06 **17.** 4.66 **19.** 1.615 **21.** -1.080
23. $61.5°$ **25.** $110°$ **27.** $93.8°$ **29.** $-1/2$ **31.** 0.541 **33.** 0.1862 **35.** 0.276 **37.** $166°$ **39.** $164°$
41. $35.5°$ **43.** $125°$ **45.** **47.** **49.** $2.25x - y - 1.48 = 0$

51. $5.52x + 0.390y - 22.9 = 0$ **53.** $2x + y + 7 = 0$ **55.** $3x - y + 6 = 0$ **57.** $2x - 3y - 6.9 = 0$
59. $3x + 4y - 8 = 0$ **61.** $4x - y - 17 = 0$ **63.** $3x + y - 19 = 0$ **65.** $0.472x + y + 3.52 = 0$
67. $x - 0.400y + 4.30 = 0$ **69.** $x = -3$ **71.** $2x - y - 2 = 0$ **73.** $x + 2y - 2 = 0$ **75.** 78.6 mm **77.** 2008 ft.
79. 1260 m **81.** $-0.911°$ **83.** $33.7°$ **85.** $F = kL - kL_0$ **87.** (a) 10.5 m/s; (b) 64.3 m/s **89.** 150.1Ω
93. $P = 20.6 + 0.432x$; 21.8 ft. **95.** $t = -0.789x + 25.0$; $x = 31.7$ cm; $m = -0.789$ **97.** $y = P + t(S - P)/L$; \$5555

Exercise 2

1. $x^2 + y^2 = 49$ **3.** $(x - 2)^2 + (y - 3)^2 = 25$ **5.** $(x - 5)^2 + (y + 3)^2 = 16$ **7.** $C(0, 0)$; $r = 7$ **9.** $C(2, -4)$; $r = 4$
11. $C(3, -5)$; $r = 6$ **13.** $C(4, 0)$; $r = 4$ **15.** $C(5, -6)$; $r = 6$ **17.** $C(-3, 1)$; $r = 5$ **19.** $x^2 + y^2 - 2x - 2y - 23 = 0$
21. $4x + 3y - 25 = 0$ **23.** $4x - 3y + 10 = 0$ **25.** $(3, 0)$; $(2, 0)$; $(0, 1)$; $(0, 6)$ **27.** $(2, 3)$ and $(-3/2, -1/2)$ **31.** 2.49 m
33. $(x - 6)^2 + y^2 = 100$; 7.08 m

Exercise 3

1. $F(2, 0)$; $L = 8$ **3.** $F(0, -3/7)$; $L = 12/7$ **5.** $x^2 + 8y = 0$ **7.** $3y^2 = 4x$ **9.** $V(3, 5)$; $F(6, 5)$; $L = 12$; $y = 5$
11. $V(3, -1)$; $F(3, 5)$; $L = 24$; $x = 3$ **13.** $V(2, -1)$; $F(13/8, -1)$; $L = 3/2$; $y = -1$ **15.** $V(3/2, 5/4)$; $F(3/2, 1)$; $L = 1$;
$x = 3/2$ **17.** $y^2 - 4y - 8x + 12 = 0$ **19.** $y^2 - 2y - 4x + 9 = 0$ **21.** $2y^2 + 4y - x - 4 = 0$ **23.** $y^2 = 3(10^8)x$
25. 4.2 m **27.** $x^2 = -109y$; 22.9 m **29.** 0.562 m **31.** $x^2 = -1200(y - 75)$ **33.** 0.0716 ft.

Exercise 4

5. $V(0, \pm 4)$; $F(0, \pm 2)$ **7.** $\dfrac{x^2}{} + \dfrac{y^2}{} = 1$ **9.** $\dfrac{x^2}{26} + \dfrac{y^2}{4} = 1$

1. $V(\pm 4, 0)$; $F(\pm \sqrt{41}, 0)$;

5. $V(\pm 4, 0)$; $F(\pm 2\sqrt{5}, 0)$; $a = 4$; $b = 2$; slope $= \pm 1/2$ **7.** $\dfrac{x^2}{25} - \dfrac{y^2}{144} = 1$ **9.** $\dfrac{x^2}{9} - \dfrac{y^2}{7} = 1$ **11.** $\dfrac{y^2}{16} - \dfrac{x^2}{16} = 1$

13. $\dfrac{y^2}{25} - \dfrac{3x^2}{64} = 1$ **15.** $C(2, -1)$; $a = 5$, $b = 4$; $F(8.4, -1)$, $F'(-4.4, -1)$; $V(7, -1)$, $V'(-3, -1)$; slope $= \pm 4/5$

17. $C(2, -3)$; $a = 3$, $b = 4$; $F(7, -3)$, $F'(-3, -3)$; $V(5, -3)$, $V'(-1, -3)$; slope $= \pm 4/3$

19. $C(-1, -1)$; $a = 2$, $b = \sqrt{5}$; $F(-1, 2)$, $F'(-1, -4)$; $V(-1, 1)$, $V'(-1, -3)$; slope $= \pm 2/\sqrt{5}$

21. $\dfrac{(y-2)^2}{16} - \dfrac{(x-3)^2}{4} = 1$ **23.** $\dfrac{(x-1)^2}{4} - \dfrac{(y+2)^2}{12} = 1$ **25.** $xy = 36$ **27.** $\dfrac{x^2}{324} - \dfrac{y^2}{352} = 1$

29. $pv = 25\,000$

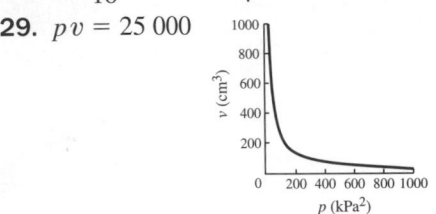

Review Problems

1. 4   **3.** -1.44 **5.** $147°$ **7.** $-2b/a$ **9.** $m = 3/2$; $b = -7/2$ **11.** $7x - 3y + 32 = 0$ **13.** $5x - y + 27 = 0$
15. $7x - 3y + 21 = 0$ **17.** $y = 5$ **19.** -3 **21.** $x + 3 = 0$ **23.** 23.0   **35.** Parabola; $V(-3, 3)$; $F(-3, 2)$
37. Ellipse; $C(4, 5)$; $a = 10$, $b = 6$; $F(4, 13)$, $(4, -3)$; $V(4, 15)$, $(4, -5)$ **39.** Circle; $C(0, 0)$, $r = 3$
41. $x^2 + y^2 - 8x + 6y = 0$ **43.** $(x-1)^2 - 9(y-1)^2 = 16$ **45.** $625x^2 - 84y^2 = 10\,000$ **47.** $(y+7)^2 = 4(x-3)$

49. $(-1, -3)$ and $(-3, -1)$ **51.** $\dfrac{x^2}{6.25} - \dfrac{y^2}{36} = 1$ **53.** 12.3 ft. and 83.7 ft.

❖❖❖ CHAPTER 23

Exercise 1

1. 2   **3.** 3   **5.** 6   **7.** 13   **9.** 12   **11.** 5   **13.** 103   **15.** 119   **17.** 101   **19.** 10   **21.** 0100 1000   **23.** 0101
1101   **25.** 1 0001 0010   **27.** 0111 0110   **29.** 10 0000 1011 0111   **31.** 1 0100 0011 0011 0100   **33.** 0.1
35. 0.11   **37.** 0.0100 1100   **39.** 0.1000 1100   **41.** 0.111   **43.** 0.0110 0011   **45.** 0.5   **47.** 0.25   **49.** 0.5625
51. 101.1   **53.** 100.011   **55.** 11 1011 0100.0111 1000   **57.** 1.5   **59.** 2.25   **61.** 25.406 25

Exercise 2

1. D   **3.** 9   **5.** 93   **7.** D8   **9.** 92A6   **11.** 9.3   **13.** 1.38   **15.** 0110 1111   **17.** 0100 1010   **19.** 0010 1111
0011 0101   **21.** 0100 0111 1010 0010   **23.** 0101.1111   **25.** 1001.1010 1010   **27.** 242   **29.** 51   **31.** 14 244
33. 62 068   **35.** 3.9375   **37.** 2748.867 1875   **39.** 27   **41.** 399   **43.** AB5   **45.** 6C8

Exercise 3

1. 6   **3.** 7   **5.** 63   **7.** 155   **9.** 111   **11.** 01 0110   **13.** 1 1001 0011   **15.** 1010 1010 0011
17. 0110 0110 0110 1000

Exercise 4

1. 0110 0010   **3.** 0010 0111 0100   **5.** 0100 0010.1001 0001   **7.** 9   **9.** 61   **11.** 36.8

Review Problems

1. 6.125   **3.** 0100 0101   **5.** 1264   **7.** 1 1010.111   **9.** 348   **11.** BE   **13.** 20AA   **15.** 134
17. 0100 1110 0101 1101   **19.** 0100 1010

❖❖❖ CHAPTER 24

Exercise 1

1. unconditional, nonlinear   **3.** conditional, nonlinear   **5.** conditional, linear   **7.** $x > 5$ and $x < 9$ **9.** $x > -11$
and $x \le 1$ **11.** $5 < x < 8$ **13.** $-1 \le x < 24$

15. 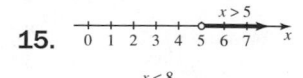   **17.** 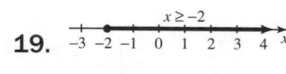   **19.** 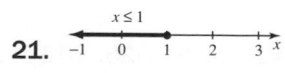   **21.**

15. $x > 5$ **17.** $x < 3$ **19.** $x \geq -2$ **21.** $x \leq 1$

23. 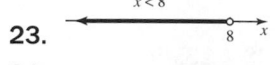   **25.** 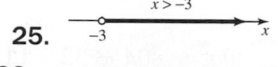   **27.** 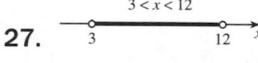   **29.**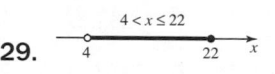

23. $x < 8$ **25.** $x > -3$ **27.** $3 < x < 12$ **29.** $4 < x \leq 22$

31. 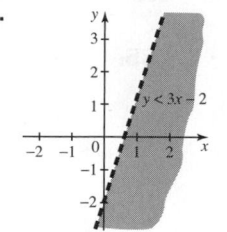   **33.** 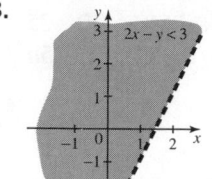   **35.** 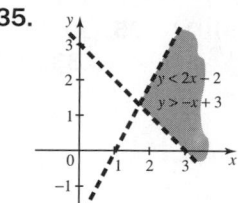   **37.**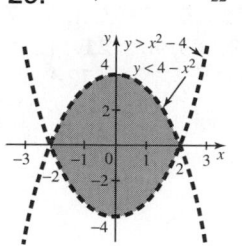

39.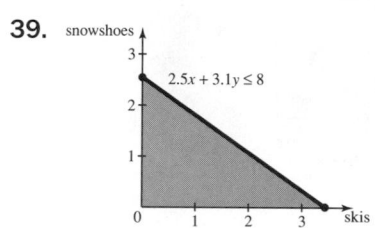

Exercise 2

1. $x > 9$ **3.** $-3 \leq x \leq 3$ **5.** $x < 1$ or $x > 3/2$ **7.** $x \geq 40/7$ **9.** $-2 < x < 5$ **11.** $-3 \leq x \leq 1$ **13.** $x > 3$ or $x < -3$ **15.** $x > 8/5$ or $x < -8/5$ **17.** $x \geq 10/3$ or $x \leq -9$ **19.** $x < 5$ **21.** $-1/2 \leq x \leq 7/8$ **23.** $x \leq -5$ or $x \geq 1$ **25.** $x \leq -2$ or $x \geq 3$ **27.** $25x + 155 \leq 350$; 7.8 h   **29.** 106 lenses per week

Exercise 3

1. $x = 1.5$; $y = 3.5$; $z = 8.5$ **3.** $x = 2.78$; $y = 3.89$; $z = 29.5$ **5.** 69 pulleys; 68 sprockets   **7.** 8 disk drives, 4 monitors

Review Problems

1. $x > -11$ and $x \leq 1$ **3.** $(-\infty, 8)$ **5.** $x > 8$ **7.** $x > 8$ **9.** $x < -3$ or $x > 11$

11. $x = 2.5$; $y = 0$; $z = 17.5$ **13.** $x \leq -5$ or $x \geq -3$ **15.** $x \leq -27$ or $x \geq 2$ **17.** $-1.73 < x \leq 8.82$
19. $0.253 < x < 1.18$

◆◆◆ **CHAPTER 25** ◆◆

Exercise 1

1. $3 + 6 + 9 + 12 + 15 + \cdots + 3n + \cdots$ **3.** $2 + \dfrac{3}{4} + \dfrac{4}{9} + \dfrac{5}{16} + \dfrac{6}{25} + \cdots + \dfrac{n+1}{n^2} + \cdots$ **5.** $u_n = 2n$; 8, 10

7. $u_n = \dfrac{2^n}{n+3}$; $\dfrac{32}{8}, \dfrac{64}{9}$ **9.** $u_n = u_{n-1} + 4$; 13, 17   **11.** $u_n = (u_{n-1})^2$; 6561, 43 046 721

Exercise 5

1. 720   **3.** 42   **5.** 35   **25.** $512\,512x^{10}$ **27.** $330x^4a^7$ **29.** $6435x^{14}$

Review Problems

1. 35   **3.** 6, 9, 12, 15   **5.** $2\frac{2}{5}, 3, 4, 6$ **7.** 85   **9.** $a^5 - 10a^4 + 40a^3 - 80a^2 + 80a - 32$ **11.** 48   **13.** $9 - x$
15. 8   **17.** 5/7   **19.** 49/90   **21.** 1215   **23.** -1024 **25.** 20, 50   **27.** 1440   **29.** 252
31. $128x^7/y^{14} - 448x^{13/2}/y^{11} + 672x^6/y^8 - 560x^{11/2}/y^5 + \cdots$ **33.** $1 - a/2 + 3a^2/8 - 5a^3/16 + \cdots$ **35.** $-26\,730a^4b^7$

••• CHAPTER 26 ••

Exercise 1

1. Discrete   **3.** Categorical   **5.** Categorical   **7.** 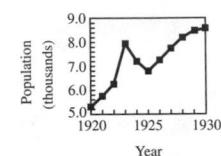   **9.**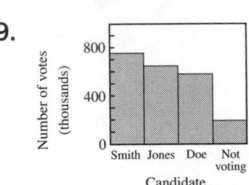

Exercise 2

1.

Range = 172 − 111 = 61

Class Midpt.	Class	Limits	1(b) Abs. Freq.	1(c) Rel. Freq. (%)	10(a) Cumulative Freq. Abs.	10(b) Cumulative Freq. Rel. (%)
113	110.5	115.5	3	7.5	3	7.5
118	115.5	120.5	4	10.0	7	17.5
123	120.5	125.5	1	2.5	8	20.0
128	125.5	130.5	3	7.5	11	27.5
133	130.5	135.5	0	0.0	11	27.5
138	135.5	140.5	2	5.0	13	32.5
143	140.5	145.5	2	5.0	15	37.5
148	145.5	150.5	6	15.0	21	52.5
153	150.5	155.5	7	17.5	28	70.0
158	155.5	160.6	2	5.0	30	75.0
163	160.5	165.5	5	12.5	35	87.5
168	165.5	170.5	3	7.5	38	95.0
173	170.5	175.5	2	5.0	40	100.0

3.

Range = 972 − 584 = 388

Class Midpt.	Class	Limits	3(b) Abs. Freq.	3(c) Rel. Freq. (%)	12(a) Cumulative Freq. Abs.	12(b) Cumulative Freq. Rel. (%)
525	500.1	550	0	0.0	0	0.0
575	550.1	600	1	3.3	1	3.3
625	600.1	650	2	6.7	3	10.0
675	650.1	700	4	13.3	7	23.3
725	700.1	750	3	10.0	10	33.3
775	750.1	800	7	23.3	17	56.7
825	800.1	850	1	3.3	18	60.0
875	850.1	900	7	23.3	25	83.3
925	900.1	950	4	13.3	29	96.7
975	950.1	1000	1	3.3	30	100.0

5.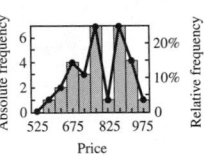

7.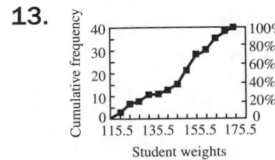

9.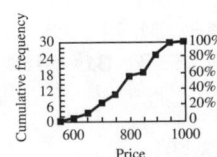

11.

Range = 99.2 − 48.4 = 50.8

Class Midpt.	Class	Limits	2(b) Abs. Freq.	2(c) Rel. Freq. (%)	11(a) Cumulative Freq. Abs.	11(b) Cumulative Freq. Rel. (%)
47.5	45.05	50.05	5	16.7	5	16.7
52.5	50.05	55.05	2	6.7	7	23.3
57.5	55.05	60.05	3	10.0	10	33.3
62.5	60.05	65.05	1	3.3	11	36.7
67.5	65.05	70.05	3	10.0	14	46.7
72.5	70.05	75.05	9	30.0	23	76.7
77.5	75.05	80.05	0	0.0	23	76.7
82.5	80.05	85.05	2	6.7	25	83.3
87.5	85.05	90.05	3	10.0	28	93.3
92.5	90.05	95.05	1	3.3	29	96.7
97.5	95.05	100.05	1	3.3	30	100.0

13.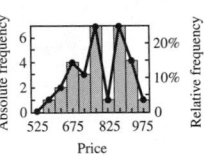

15.

17.

3									
4	8.5	8.9	9.6	8.4	9.4				
5	4.8	9.3	9.3	9.3	0.3				
6	1.4	9.3	6.3	9.3					
7	2.5	1.2	1.4	4.5	3.6	1.4	4.9	2.7	2.8
8	8.2	4.6	9.4	5.7	3.8				
9	9.2	2.4							
10									

Exercise 3

1. 77   **3.** 145 lb.   **5.** $796   **7.** 81.6   **9.** 157.5   **11.** None   **13.** 59.3 min   **15.** 76   **17.** 149 lb.   **19.** $792
21. 116, 142, 158, 164, 199   **23.** 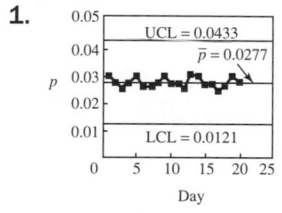   **25.** 83   **27.** 4.28, 7.40 and 10.02; 5.74   **29.** 697.4, 26.4
31. 201, 14.2

Exercise 4

1. 8/15   **3.** 0.375   **5.** 1/9   **7.** 1/36   **9.** 0.63   **11.** 5/18   **13.** 0.133   **15.** 0.230   **17.** 0.367
19. 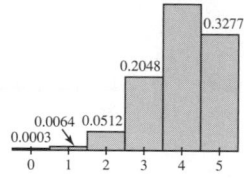   **21.** 0.117   **23.** 7.68×10^{-6}

Exercise 5

1. 0.4332   **3.** 0.2119   **5.** 620 students   **7.** 36 students   **9.** 1

Exercise 6

1. 69.47 ± 0.34 in.   **3.** 2.35 ± 0.24 in.   **5.** 164.0 ± 2.88 **7.** 16.31 ± 2.04 **9.** 0.250 ± 0.031

Exercise 7

1. 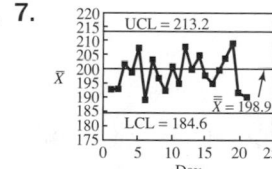   **3.**   **5., 7.**   **9., 11.**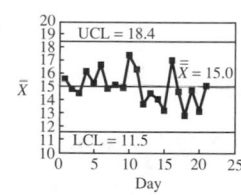

Exercise 8

1. 1.00   **3.** 1.00   **5.** $y = -4.79x - 14.5$

Review Problems

1. Continuous
3. Categorical
5.

19.

	Cumulative Absolute	Cumulative Relative		Cumulative Absolute	Cumulative Relative Frequency

21. $\bar{x} = 150$ **23.** 137 and 153 **25.** $s = 22.4$ **27.** 150 ± 7.0 **29.** 22.4 ± 5.00 **31.** 13

33. $Q_1 = 407, Q_2 = 718, Q_3 = 1055$; quartile range = 648 **35.**

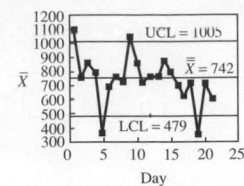

37. $-1.00; y = -2.31x + 18.1$

39. 0.0769 ± 0.0188

◆◆◆ CHAPTER 27 ◆◆◆

Exercise 1

1. 1 **3.** $1/5$ **5.** 10 **7.** 4 **9.** -8 **11.** $-4/5$ **13.** $-1/4$ **15.** 2 **17.** $1/5$ **19.** 0 **21.** 1 **23.** -4

25. $+\infty$ **27.** $+\infty$ **29.** $-\infty$ **31.** x^2 **33.** 0 **35.** 3 **37.** $3x - \dfrac{1}{x^2 - 4}$ **39.** $3x^2$ **41.** $2x - 2$

Exercise 2

1. 1.5 **3.** 3 **5.** 2 **7.** $2x$ **9.** $3x^2$ **11.** $-\dfrac{3}{x^2}$ **13.** $-\dfrac{1}{2\sqrt{3 - x}}$ **15.** -2 **17.** $\dfrac{3}{4}$ **19.** 2 **21.** 12 **23.** $4x$

25. -4 **27.** -16 **29.** 3 **31.** -5 **33.** 3

35.

x	Slope
5.00	
5.10	2.20
5.20	4.80
5.30	4.10
5.40	6.10
5.50	6.70
5.60	7.70
5.70	4.60

Exercise 3

1. 0 **3.** $7x^6$ **5.** $6x$ **7.** $-5x^{-6}$ **9.** $-1/x^2$ **11.** $-9x^{-4}$ **13.** $2.5x^{-2/3}$ **15.** $2x^{-1/2}$ **17.** $\dfrac{-51\sqrt{x}}{2}$ **19.** -2

21. $3 - 3x^2$ **23.** $9x^2 + 14x - 2$ **25.** a **27.** $x - x^6$ **29.** $(3/2)x^{-1/4} - x^{-5/4}$ **31.** $(8/3)x^{1/3} - 2x^{-1/3}$ **33.** $-4x^{-2}$

35. $6x^2$ **37.** 54 **39.** -8 **41.** $15x^4 + 2$ **43.** $10.4/x^3$ **45.** $6x + 2$ **47.** 4 **49.** 3 **51.** 13.5 **53.** $10t - 3$

55. $175t^2 - 63.8$ **57.** $\dfrac{3\sqrt{5w}}{2}$ **59.** $-341/t^5$

Exercise 4

1. $10(2x + 1)^4$ **3.** $24x(3x^2 + 2)^3 - 2$ **5.** $\dfrac{-3}{(2 - 5x)^{2/5}}$ **7.** $\dfrac{-4.30x}{(x^2 + a^2)^2}$ **9.** $-\dfrac{6x}{(x^2 + 2)^2}$ **11.** $\dfrac{2b}{x^2}\left(a - \dfrac{b}{x}\right)$

13. $\dfrac{-3x}{\sqrt{1 - 3x^2}}$ **15.** $\dfrac{-1}{\sqrt{1 - 2x}}$ **17.** $\dfrac{-3}{(4 - 9x)^{2/3}}$ **19.** $\dfrac{-1}{2\sqrt{(x + 1)^3}}$ **21.** $2(3x^5 + 2x)(15x^4 + 2)$

23. $\dfrac{28.8(4.8 - 7.2x^{-2})}{x^3}$ **25.** $2(5t^2 - 3t + 4)(10t - 3)$ **27.** $\dfrac{7.6 - 49.8t^2}{(8.3t^3 - 3.8t)^3}$ **29.** 118 000 **31.** 540

Exercise 5

1. $3x^2 - 3$ **3.** $42x + 44$ **5.** $5x^4 - 12x^2 + 4$ **7.** $x^2(x - 5)^2(2x - 1)(8x - 7)(32x - 105)$

9. $\dfrac{x}{\sqrt{1 + 2x}} + \sqrt{1 + 2x}$ **11.** $\dfrac{-2x^2}{\sqrt{3 - 4x}} + 2x\sqrt{3 - 4x}$ **13.** $\dfrac{bx}{2\sqrt{a + bx}} + \sqrt{a + bx}$

15. $\dfrac{2(3x+1)^3}{\sqrt{4x-2}} + 9(3x+1)^2\sqrt{4x-2}$ **17.** $(t+4)(20t^2+31t-12)$

19. $2(81.3t^3-73.8t)(t-47.2)+(t-47.2)^2(244t^2-73.8)$ **21.** $(2x^5+5x)^2+(2x-6)(2x^5+5x)(10x^4+5)$

23. $3x(x-1)^2(x-2)^2+(2x^2+2x)(x-2)^3+(x+1)^2(x-2)^3$ **25.** $(x-1)^{1/2}(x+1)^{3/2}+(3/2)(x^2-1)^{1/2}(x+2)+$

$\dfrac{(x+1)^{3/2}(x+2)}{2\sqrt{x+1}}$ **27.** $\dfrac{2}{(x+2)^2}$ **29.** $\dfrac{8x}{(4-x^2)^2}$ **31.** $\dfrac{-5}{(x-3)^2}$ **33.** $\dfrac{2-4x}{(x-1)^3}$ **35.** $\dfrac{1}{2\sqrt{x}(\sqrt{x}+1)^2}$

37. $\dfrac{-a^2}{(z^2-a^2)^{3/2}}$ **39.** -0.456 **41.** 0

Exercise 6

1. $6u^2\dfrac{du}{dw}$ **3.** $2y\dfrac{dy}{du}+3u^2$ **5.** $2x^3y\dfrac{dy}{dx}+3x^2y^2$ **7.** $\dfrac{3z}{\sqrt{3z^2+5}}\dfrac{dz}{dt}$ **9.** $2y-7$ **11.** $(6y-7)(y+3)^4$

13. $5/2$ **15.** $-y/x$ **17.** $2a/y$ **19.** $\dfrac{ay-x^2}{y^2-ax}$ **21.** $\dfrac{1+3x^2}{1+3y^2}$ **23.** $\dfrac{8xy}{3y-8x^2+4y^2}$ **25.** -0.436 **27.** $14/15$

29. $3x^2\,dx$ **31.** $\dfrac{2\,dx}{(x+1)^2}$ **33.** $2(x+1)(2x+3)^2(5x+6)\,dx$ **35.** $\dfrac{2x-13}{2\sqrt{x-4}(3-2x)^2}\,dx$ **37.** $\dfrac{-x^2}{2y^2}\,dx$

39. $-\dfrac{2\sqrt{y}}{3\sqrt{x}}\,dx$

Exercise 7

1. $36x^2-6x$ **3.** $\dfrac{8}{(x+2)^3}$ **5.** $\dfrac{-20}{(5-4x^2)^{3/2}}$ **7.** $4(x-3)^2+8(x-6)(x-3)$ **9.** 1.888

Review Problems

1. $\dfrac{2x}{(x^2+1)\sqrt{x^4-1}}$ **3.** $-\sqrt[3]{\dfrac{y}{x}}$ **5.** $8/5$ **7.** $\dfrac{-b^2x}{a^2y}$ **9.** -0.03516 **11.** $5-6x$ **13.** $3/64$ **15.** 178 **17.** 3

19. $-4(x^2-1)(x+3)^{-5}+2x(x+3)^{-4}$ **21.** $-8\dfrac{dw}{dx}+2w\dfrac{dw}{dx}$ **23.** $-24x$ **25.** $174.9t^2-63.8$ **27.** $12x-2$

29. $48x^2+8x-1$ **31.** $34x^{16}+48x^{11}-7-1/x^3$ **33.** $\dfrac{x^2-2x-3}{(x-1)^2}$ **35.** $72x^2+24x-14$ **37.** $12x^3\,dx$

39. $-\dfrac{x}{3y}\,dx$

◆◆◆ **CHAPTER 28** ◆◆

Exercise 3

1. 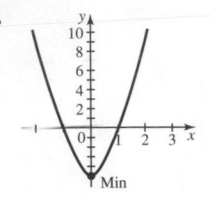 **3.** 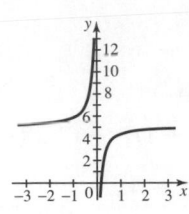 **5.** 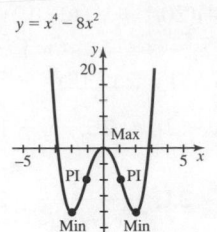 **7.** 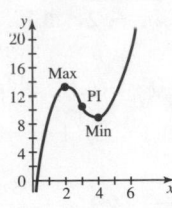 **9.**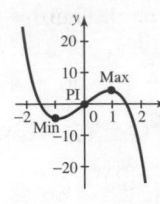

11. 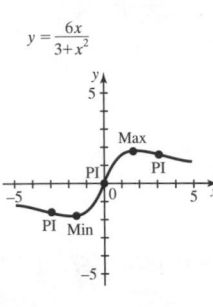 **13.** 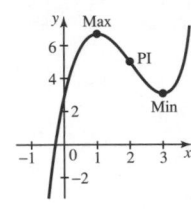 **15.** 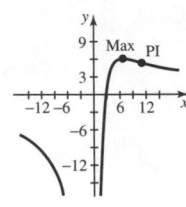 **17.** 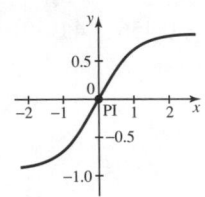 **19.**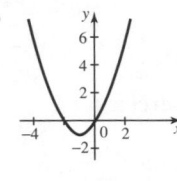

21. 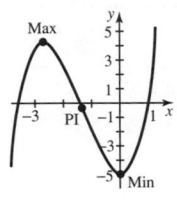 **23.** 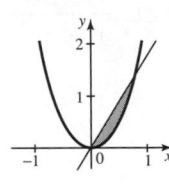 **25.**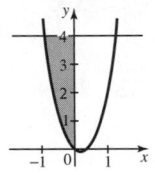

Exercise 4

1. 2.66   **3.** 3.06   **5.** 1.67   **7.** −1.72   **9.** $x(16 − 2x)^2 = 282$; 1.89 cm, 3.53 cm   **11.** 3.38 m

Review Problems

1. −2.21, 0.54, 1.68   **3.** max(1, 2); min(−1, −2)   **5.** $7x − y − 9 = 0$; $x + 7y − 37 = 0$ **7.** −7/3   **9.** Rising for $x > 0$; never falls   **11.** Downward   **13.** $34x − y = 44$; $x + 34y = 818$ **15.** 71.6°

◆◆◆ CHAPTER 29 ◆◆◆

Exercise 1

1. −0.175 kPa/cm³   **3.** 12.6 m³/m   **5.** 0.0234 s/cm   **7.** $\dfrac{dy}{dx} = \dfrac{wx}{6EI}(x^2 + 3L^2 − 3Lx)$ **9.** $i = dq/dt = 6.96t − 1.64$ A

11. 50.8 V   **13.** $i = 22.4 + 125t^2$ A; 804 A   **15.** 2.48 mA   **17.** 95.6 kV

Exercise 2

1. $v = 0$; $a = −16$ **3.** $v = −3$; $a = 18$ **5.** $v = −8$; $a = −32$ **7.** $v = 6\overline{0}$ m/s; $a = 9.8$ m/s²   **9.** 60 m/s
11. (a) 700 m/s; (b) 25 000 m; (c) 600 m/s   **13.** $s = 0$; $a = 32$ **15.** (a) 89.2° (b) $v_x = 0.0175$ m/s; $v_y = 1.25$ m/s
17. 42.3 cm/s at 320°   **19.** (a) $x = 32\,500$ ft.; $y = 29\,500$ ft. (b) $v_x = 4640$ ft./s; $v_y = 4110$ ft./s   **21.** 200 km/h
23. 6080 rad/s²   **25.** (a) 198 rad/s (b) 82.1 rad/s²

Exercise 3

1. 60 m/s   **3.** 5.37 km/h   **5.** 2.40 m/s   **7.** 2.24 m/s   **9.** 0.829 ft./s   **11.** 170 km/h   **13.** 8.33 m/s   **15.** 1.60 m/s
17. 0.101 cm²/s   **19.** 0.133 cm/min   **21.** −4.00 cm/min   **23.** 1.19 m/min   **25.** 0.0500 m/min   **27.** 21.0 kPa/s
29. 1.61 in./s

Exercise 4

1. 1   **3.** $6\frac{2}{3}$ and $13\frac{1}{3}$ **5.** 18.0 m × 24.0 m   **7.** 9.00 m × 18.0 m   **9.** $d = 6.51$ cm; $h = 3.26$ cm   **11.** 25.0 cm² **13.** 6
15. 8.66 cm.   **17.** (2, 2)   **19.** 5.00 cm × 4.35 cm   **21.** 9.90 units × 14.1 units   **23.** $r = 4.08$ cm; $h = 5.77$ cm
25. 12.0 m **27.** $r = 5.24$ m; $h = 5.26$ m   **29.** 26.0 cm × 36.7 cm.   **31.** 59.1 cm   **33.** 7.5 A   **35.** 5.0 A   **37.** 0.65
39. $r = 11.0$ ft.; $h = 26.3$ ft.; cost = $11,106

Review Problems

1. 190 m/s **3.** 4.0 km/h **5.** 3.33 cm^2/min **7.** $s/2$ **9.** 6.93 **11.** -2.26 lb./sq. in. per second **13.** $t = 4, s = 108$
15. 4 **17.** 4 and 6 **19.** 10 cm $\times$ 10 cm $\times$ 5 cm **21.** (a) 142 rad/s; (b) 37.0 rad/s^2 **23.** 36.4 A

◆◆◆ **CHAPTER 30** ◆◆◆

Exercise 1

1. $x + C$ **3.** $\frac{1}{2}x^2 + C$ **5.** $-\frac{1}{x} + C$ **7.** $\frac{3x^2}{2} + C$ **9.** $\frac{3}{4}x^4 + C$ **11.** $\frac{x^{n+1}}{n+1} + C$ **13.** $2\sqrt{x} + C$

15. $x^4 + C$ **17.** $x^{7/2} + C$ **19.** $x^2 + 2x + C$ **21.** $x^{4/3} + C$ **23.** $u^2 + C$ **25.** $\frac{8}{3}s^{3/2} + C$

27. $\frac{6x^{5/2}}{5} - \frac{4x^{3/2}}{3} + C$ **29.** $2x^2 - 4\sqrt{x} + C$ **31.** $s - \frac{3s^2}{2} + s^3 - \frac{s^4}{4} + C$ **33.** $y = \frac{4x^3}{3} + C$

35. $y = -\frac{1}{2x^2} + C$ **37.** $s = \frac{3t^{1/3}}{2} + C$

Exercise 2

1. $\frac{1}{4}(x^4 + 1)^4 + C$ **3.** $(x^2 + 2x)^2 + C$ **5.** $\frac{1}{1-x} + C$ **7.** $(x^3 + 1)^3 + C$ **9.** $-\frac{2}{3}(1 - y^3)^{1/2} + C$

11. $\frac{2}{9-x^2} + C$ **13.** $3\ln|x| + C$ **15.** $\frac{5}{3}\ln|z^3 + 3| + C$ **17.** $x + \ln|x| + C$ **19.** $y = -\frac{1}{8}(1 - x^2)^4 + C$

21. $y = \frac{1}{8}(2x^3 + x^2)^4 + C$ **23.** $v = \left(-\frac{5}{3}\right)(7 - t^2)^{3/2} + C$ **25.** $y = 7\ln|x| + C$ **27.** $y = \frac{1}{2}\ln|x^2 + 5| + C$

29. $-\frac{1}{12}\ln\left|\frac{3t-2}{3t+2}\right| + C$ **31.** $\frac{x}{2}\sqrt{25 - 9x^2} + \frac{25}{6}\text{Sin}^{-1}\left(\frac{3x}{5}\right) + C$ **33.** $\frac{1}{3}\text{Tan}^{-1}\frac{x}{3} + C$ **35.** $\frac{1}{12}\text{Tan}^{-1}\left(\frac{4x}{3}\right) + C$

37. $-\frac{1}{12}\ln\left|\frac{2x-3}{2x+3}\right| + C$ **39.** $\frac{1}{4}\ln\left|\frac{x-2}{x+2}\right| + C$ **41.** $\frac{x}{4}\sqrt{x^2 - 4} - \ln|x + \sqrt{x^2 - 4}| + C$

43. $\frac{5}{2}\text{Arcsin } x^2 + C$

Exercise 3

1. $y = 3x + C$ 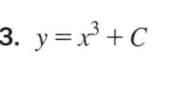 **3.** $y = x^3 + C$ 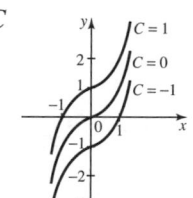 **5.** $x^3 - 3y + 2 = 0$ **7.** 17 **9.** $6y = x^3 - 6x - 9$
11. $6x + xy = 6$

Exercise 4

1. 1.50 **3.** 60.7 **5.** 37.3 **7.** 72.7 **9.** 2.30 **11.** 0.112 **13.** 0.732

Review Problems

1. $\dfrac{3x^{2/3}}{2} + C$ **3.** $1.03y^3 + C$ **5.** $\dfrac{4x\sqrt{x}}{3} + C$ **7.** $\dfrac{1}{4}(x^4 - 2x^3)^2 + C$ **9.** $\ln|x + 5| + C$ **11.** $81\frac{2}{3}$ **13.** $\dfrac{a^2}{6}$

15. 25.03 **17.** 28.75 **19.** $-\dfrac{1}{12}\ln\left|\dfrac{2x - 3}{2x + 3}\right| + C$ **21.** $\dfrac{1}{6}\operatorname{Arctan}\left(\dfrac{2x}{3}\right) + C$ **23.** 0.3583 **25.** $N = 100e^{t/2}$

27. 7.283 **29.** $16/3$

◆◆◆ CHAPTER 31 ◆◆◆

Exercise 1

1. $v = 4t - \dfrac{1}{3}t^3 - 1$ **3.** (a). $x = \dfrac{1}{3}t^3 - 4t, y = 2t^2$ (b). $9.6\,\text{cm}$ **5.** $s = 20t - 16t^2$ **7.** $v_x = 1670\,\text{cm/s}; v_y = 10\overline{0}\,\text{cm/s}$

9. $v_x = 344\,\text{cm/s}; v_y = 2250\,\text{cm/s}$ **11.** $18\,\overline{0}00\,\text{rev/}$ **13.** $15.7\,\text{rad/s}$

Exercise 2

1. $12.1\,\text{C}$ **3.** $78.4\,\text{C}$ **5.** $2.26\,\text{V}$ **7.** $4.26\,\text{A}$ **9.** $20.2\,\text{A}$

Exercise 3

1. 2.797 **3.** 16 **5.** $42\dfrac{2}{3}$ **7.** $1\dfrac{1}{6}$ **9.** $10\dfrac{2}{3}$ **11.** 72.4 **13.** 32.83 **15.** $-2\dfrac{2}{3}$ **17.** 18.75 **19.** 36 **21.** 2.70

23. $4/3$ **33.** $12\,800\,\text{cm}^3$ **35.** $3.75\,\text{m}^3$ **37.** $9.00\,\text{cm}^2$

Exercise 4

1. 57.4 **3.** π **5.** 0.666 **7.** 0.479 **9.** 60.3 **11.** 32π **13.** 40.2 **15.** 3.35 **17.** 102 **19.** $1.65\,\text{m}^3$ **21.** $0.683\,\text{m}^3$

Review Problems

1. 31.4 **3.** 19.8 **5.** 25.6 **7.** 2.13 **9.** $19\,300\,\text{rad/s}; 15\,400\,\text{rev}$ **11.** $V_x = \dfrac{t^2}{2} + 4; V_y = \dfrac{5}{2}t^2 + 15; x = \dfrac{t^3}{6} + 4t + 1;$

$y = \dfrac{5t^3}{6} + 15t + 1$ **13.** 151 **15.** 442 **17.** 12π **19.** $3.30\,\text{A}$

◆◆◆ CHAPTER 32 ◆◆◆

Exercise 1

1. 0.704 **3.** 37.7 **5.** 5.92 **7.** 9.29 **9.** 9.07 **11.** $329\,\text{m}$ **13.** $222\,\text{m}$ **15.** $10.1\,\text{m}$

Exercise 2

1. 32.1 **3.** 154 **5.** 36.2 **7.** 3.77 **9.** 1570 **11.** 36.2 **13.** $4\pi r^2$ **15.** $0.521\,\text{m}^2$

Exercise 3

1. $(5/4, -3/2)$ **3.** $\bar{y} = 3.22$ **5.** $(2.40, 0)$ **7.** $\bar{y} = 27/10$ **9.** $\bar{x} = 1.067$ **11.** $\bar{x} = 9/20; \bar{y} = 9/20$
13. $\bar{x} = 3/10$ **15.** $\bar{x} = 1.25$ **17.** 5 **19.** 5 **21.** $h/4$ **23.** $1.38\,\text{m}$ **25.** $36.1\,\text{mm}$

Exercise 4

1. $47.1\,\text{t}$ **3.** $82.5\,\text{t}$ **5.** $1000\,\text{kg, or } 1.00\,\text{t}$ **7.** $2160\,\text{lb.}$

Exercise 5

1. $30\overline{0}\,\text{N·cm}$ **3.** $32\,\text{N·cm}$ **5.** $1.08 \times 10^5\,\text{N·m}$ **7.** $5.71 \times 10^5\,\text{ft.·lb.}$ **9.** $1.69 \times 10^6\,\text{N·m}$ **11.** $k/100\,\text{N·m}$
13. $28\,900\,\text{N·m}$

Exercise 6

1. $1/12$ **3.** 2.62 **5.** 19.5 **7.** $2/9$ **9.** $10.1m$ **11.** $89.4m$ **13.** $\dfrac{3Mr^2}{10}$ **15.** $\dfrac{4Mp^2}{3}$

Review Problems

1. 4.24 **3.** (1.70, 1.27) **5.** 4.66×10^5 N·m **7.** 52.3 kN **9.** 21.8 N·cm **11.** 21.2 kN **13.** 51.5 **15.** 141
17. 410

••• CHAPTER 33 •••

Exercise 1

1. $\cos x$ **3.** $-3\sin x \cos^2 x$ **5.** $3\cos 3x$ **7.** $\cos^2 x - \sin^2 x$ **9.** $3.75(\cos x - x\sin x)$ **11.** $-2\sin(\pi - x)\cos(\pi - x)$
13. $2\cos 2\theta \cos \theta - \sin 2\theta \sin \theta$ **15.** $2\sin x \cos^2 x - \sin^3 x$ **17.** $1.23(2\cos 3x \sin x \cos x - 3\sin^2 x \sin 3x)$
19. $\dfrac{-\sin 2t}{\sqrt{\cos 2t}}$ **21.** $-\sin x$ **23.** $-2\sin x - x\cos x$ **25.** $-\dfrac{\pi^2}{4}$ **27.** $\dfrac{y\cos x + \cos y - y}{x\sin y - \sin x + x}$ **29.** $\sec(x + y) - 1$
31. -0.4161 **33.** 1.382 **35.** $\max(\pi/2, 1); \min(3\pi/2, -1); PI(\pi, 0), (\pi, 0), (2\pi, 0)$ **37.** $\max(2.50, 5); \min(5.64, -5);$
$PI(0.927, 0), (4.07, 0)$ **39.** 0.515 **41.** 1.404

Exercise 2

1. $2\sec^2 2x$ **3.** $-15\csc 3x \cot 3x$ **5.** $6.50x \sec^2 x^2$ **7.** $-21x^2 \csc x^3 \cot x^3$ **9.** $x\sec^2 x + \tan x$
11. $5\csc 6x - 30x\csc 6x \cot 6x$ **13.** $2\sin \theta \sec^2 2\theta + \tan 2\theta \cos \theta$ **15.** $5\csc 3t \sec^2 t - 15\tan t \csc 3t \cot 3t$
17. -294 **19.** 853 **21.** $6\sec^2 x \tan x$ **23.** 18.73 **25.** $-y\sec x \csc x$ **27.** -1 **29.** $3.43x - y - 1.87$
31. $\max(\pi/4, 0.571); \min(3\pi/4, 5.71); PI(0, 0), (\pi, 2\pi)$ **33.** $v = 23.5$ cm/s; $a = -19.1$ cm/s^2 **35.** $-3.44°/$min
37. 15.6 ft. **39.** 31° **41.** 6.25 m

Exercise 3

1. $\mathrm{Sin}^{-1} x + \dfrac{x}{\sqrt{1 - x^2}}$ **3.** $\dfrac{-1}{\sqrt{a^2 - x^2}}$ **5.** $\dfrac{\cos x + \sin x}{\sqrt{1 + \sin 2x}}$ **7.** $-\dfrac{t^2}{\sqrt{1 - t^2}} + 2t\,\mathrm{Cos}^{-1} t$ **9.** $\dfrac{1}{1 + 2x + 2x^2}$

11. $\dfrac{-a}{a^2 + x^2}$ **13.** $\dfrac{-1}{x\sqrt{4x^2 - 1}}$ **15.** $2t\,\mathrm{Arcsin}\dfrac{t}{2} + \dfrac{t^2}{\sqrt{4 - t^2}}$ **17.** $\dfrac{1}{\sqrt{a^2 - x^2}}$ **19.** -0.285 **21.** -0.0537

Exercise 4

1. $\dfrac{1}{x}\log e$ **3.** $\dfrac{3}{x \ln b}$ **5.** $\dfrac{(5 + 9x)\log e}{5x + 6x^2}$ **7.** $\log\left(\dfrac{2}{x}\right) - \log e$ **9.** $\dfrac{1}{x}$ **11.** $\dfrac{2x - 3}{x^2 - 3x}$ **13.** $8.25 + 2.75\ln 1.02x^3$

15. $\dfrac{1}{x^2(x + 5)} - \dfrac{2\ln(x + 5)}{x^3}$ **17.** $\dfrac{1}{2t - 10}$ **19.** $\cot x$ **21.** $\cos x(1 + \ln \sin x)$ **23.** $\dfrac{\sin x}{1 + \ln y}$ **25.** $\dfrac{x + y - 1}{x + y + 1}$

27. $-y$ **29.** $-\dfrac{a^2}{x^2\sqrt{a^2 - x^2}}$ **31.** $x^{\sin x}[(\sin x)/x + \cos x \ln x]$ **33.** $(\mathrm{Arccos}\, x)^x\left(\ln \mathrm{Arccos}\, x - \dfrac{x}{\sqrt{1 - x^2}\,\mathrm{Arccos}\, x}\right)$

35. 1 **37.** 0.3474 **39.** 128° **41.** $\min(1/e, -1/e)$ **43.** $\min(e, e), PI(e^2, \tfrac{1}{2}e^2)$ **45.** 1.37 **47.** 0.607
49. 1.2 per min **51.** -0.0054 dB/day

Exercise 5

1. $2(3^{2x})\ln 3$ **3.** $10^{2x+3}(1 + 2x\ln 10)$ **5.** $2x(2^{x^2})\ln 2$ **7.** $2e^{2x}$ **9.** $e^{x + e^x}$ **11.** $-\dfrac{xe^{\sqrt{1 - x^2}}}{\sqrt{1 - x^2}}$ **13.** $-\dfrac{2}{e^x}$

Review Problems

1. $\frac{1}{2}(e^{x/a} + e^{-x/a})$ **3.** $\frac{4}{\sqrt{x}} \sec^2 \sqrt{x}$ **5.** $\frac{4x}{16x^2 + 1} + \text{Arctan } 4x$ **7.** $\frac{\cos 2x}{y}$ **9.** $x \cos x + \sin x$

11. $3x^2 \cos x - x^3 \sin x$ **13.** $\frac{x \cos x - \sin x}{x^2}$ **15.** $\frac{1 + 3x^2}{x^3 + x} \log e$ **17.** $\frac{1}{\sqrt{x^2 + a^2}}$ **19.** $-3 \csc 3x \cot 3x$

21. $\sec^2 x$ **23.** $\frac{6x^2 + 1}{2x^3 + x}$ **25.** $2x \text{ Arccos } x - \frac{x^2}{\sqrt{1 - x^2}}$ **27.** $(8.96, 0.291)$ **29.** $x = \pi/4$ **31.** 0.0538

33. $y - 1/2 = \frac{\sqrt{3}}{2}(x - \frac{\pi}{6})$ **35.** $x = \pi/4$ **37.** 0.517 **39.** $630\,°C$ at 5.49 h   **41.** $t = 1.1$ s

⬥⬥⬥ CHAPTER 34 ⬥⬥⬥

Exercise 1

1. $\frac{a^{5x}}{5 \ln a} + C$ **3.** $\frac{5^{7x}}{7 \ln 5} + C$ **5.** $\frac{a^{3y}}{3 \ln a} + C$ **7.** $4e^x + C$ **9.** $\frac{e^{x^2}}{2} + C$ **11.** $27\,122$ **13.** $2e^{\sqrt{x}} + C$

15. 0.7580 **17.** $2e^{\sqrt{x-2}} + C$ **19.** $ae^{x/a} - ae^{-x/a} + C$ **21.** $x \ln 3x - x + C$ **23.** 1.273 **25.** 6.106

27. 198.0 **29.** 4.51 **31.** 7.21 **33.** 0.930 **35.** $2.12, 0.718$

Exercise 2

1. $-\frac{1}{3} \cos 3x + C$ **3.** $-\frac{1}{5} \ln|\cos 5\theta| + C$ **5.** $\frac{1}{4} \ln|\sec 4x + \tan 4x| + C$ **7.** $-\frac{1}{3} \ln|\cos 9\theta| + C$

9. $-\frac{1}{2} \cos x^2 + C$ **11.** $-\frac{1}{3} \ln|\cos \theta^3| + C$ **13.** $-\cos(x + 1) + C$ **15.** $\frac{1}{5} \ln|\cos(4 - 5\theta)| + C$

17. $\frac{1}{8} \ln|\sec(4x^2 - 3) + \tan(4x^2 - 3)| + C$ **19.** 0.859 **21.** 2 **23.** 2 **25.** 2 **27.** $8/\pi$ **29.** 3 **31.** 14.4

33. $(\pi/2, \pi/8)$

Exercise 3

1. 12 **3.** 4.08 **5.** $\frac{1}{2}$ **7.** 7.81 **9.** 19.1 **11.** 1.08

Exercise 4

1. $\sin x - x \cos x + C$ **3.** $x \tan x + \ln|\cos x| + C$ **5.** $x \sin x + \cos x + C$ **7.** $\frac{x^2}{4} - \frac{x}{12} \sin 6x - \frac{1}{72} \cos 6x + C$

9. 1.24 **11.** $-\frac{1}{x + 1}(\ln|x + 1| + 1) + C$ **13.** 5217 **15.** $-\frac{x^2}{3}(1 - x^2)^{3/2} - \frac{2}{15}(1 - x^2)^{3/2} + C$ or $-\frac{1}{15}(1 - x^2)^{3/2}$

$(3x^2 + 2) + C$ **17.** $\frac{x}{\sqrt{1 - x^2}} - \text{Sin}^{-1} x + C$ **19.** $\sin x(\ln|\sin x| - 1) + C$ **21.** $-x^2e^{-x} - 2xe^{-x} - 2e^{-x} + C$

23. $\frac{e^x}{2}(\sin x - \cos x) + C$ **25.** -0.0813 **27.** $\frac{e^{-x}}{\pi^2 + 1}(\pi \sin \pi x - \cos \pi x) + C$

Exercise 5

1. $\ln\left|\frac{x(x - 2)}{(x + 1)^2}\right| + C$ **3.** 0.2939 **5.** $\ln\left|\frac{(x - 1)^{1/2}}{x^{1/3}(x + 3)^{1/6}}\right| + C$ **7.** -0.739 **9.** $\frac{9 - 4x}{2(x - 2)^2} + C$ **11.** 0.0637

13. 0.1782 **15.** 1.946 **17.** $\frac{1}{4} \ln\left|\frac{x + 1}{x - 1}\right| - \frac{1}{2} \text{Arctan } x + C$ **19.** 0.6676 **21.** $\frac{x^2}{2} - 4 \ln|x^2 + 4| - \frac{8}{(x^2 + 4)} + C$

Exercise 6

1. $2\sqrt{x} - 2 \ln|1 + \sqrt{x}| + C$ **3.** $\frac{3}{2} \ln|x^{2/3} - 1| + C$ **5.** $\frac{3}{5}\sqrt[3]{(1 + x)^5} - \frac{3}{2}\sqrt[3]{(1 + x)^2} + C$ or

$\dfrac{3}{10}(1 + x)^{2/3}(2x - 3) + C$ **7.** $-\dfrac{2}{147}\sqrt{2 - 7x}\,(7x + 4) + C$ **9.** $\dfrac{6x^2 + 6x + 1}{12(4x + 1)^{3/2}} + C$

11. $2\sqrt{x + 2} + \sqrt{2}\,\text{Arctan}\sqrt{\dfrac{x + 2}{2}} + C$ **13.** $\ln\left|\dfrac{1 - \sqrt{1 - x^2}}{x}\right| + C$ **15.** $\dfrac{3}{28}(1 + x)^{4/3}(4x - 3) + C$

17. 0.2375 **19.** 0.1042

Exercise 7

1. $\dfrac{x}{\sqrt{5 - x^2}} + C$ **3.** $\dfrac{1}{2}\ln\left|\dfrac{x}{2 + \sqrt{x^2 + 4}}\right| + C$ or $\dfrac{1}{2}\ln\left|\dfrac{\sqrt{x^2 + 4} - 2}{x}\right| + C$ **5.** $\dfrac{-\sqrt{4 - x^2}}{4x} + C$ **7.** 0.3199

9. $-\dfrac{\sqrt{16 - x^2}}{x} - \text{Arcsin}\dfrac{x}{4} + C$ **11.** 16.49 **13.** $\dfrac{\sqrt{x^2 - 7}}{7x} + C$

Exercise 8

1. $1/2$ **3.** e **5.** π **7.** 2 **9.** Diverges   **11.** π **13.** π **15.** $1/s^2$

Exercise 9

1. 1.80 **3.** 66.0 **5.** 2.00 **7.** $1/3$ **9.** 5960 **11.** 260 **13.** 1450 m^2 **15.** $20\,000\text{ m}^3$ **17.** 86.6 kW

Review Problems

1. $-\dfrac{1}{5}\cot^5 x - \dfrac{1}{3}\cot^3 x + C$ **3.** $\dfrac{1}{12}\ln\left|\dfrac{3x - 2}{3x + 2}\right| + C$ **5.** $\dfrac{x^2}{2}(\ln x^2 - 1) + C$ **7.** -0.1808

9. $\dfrac{3}{2}\ln|x^2 + 9| - \dfrac{1}{3}\text{Arctan}\dfrac{x}{3} + C$ **11.** $\ln\left|\dfrac{\sqrt{1 + x^2} - 1}{x}\right| + C$ **13.** $\text{Tan}^{-1}(2x - 1) + C$

15. $x \tan x + \ln|\cos x| + C$ **17.** $\dfrac{1}{2}\ln|x^2 + x + 1| + \sqrt{3}\,\text{Arctan}\left(\dfrac{2x + 1}{\sqrt{3}}\right) + C$ **19.** -0.4139 **21.** 1.493

23. $\dfrac{1}{6}\ln\left|\dfrac{3x - 1}{3x + 1}\right| + C$ **25.** $\dfrac{4x - 1}{12(1 - x)^4} + C$ **27.** $-\csc x - \sin x + C$ **29.** $\dfrac{1}{6}(3x^2 - 2)\ln(3x^2 - 2) - \dfrac{x^2}{2} + C$

31. -0.0112 **33.** 0.2877 **35.** 2.172 **37.** $3\sqrt[3]{2}$ **39.** Diverges   **41.** $1/2$

◆◆◆ **CHAPTER 35** ◆◆

Exercise 1

1. First order, first degree, ordinary   **3.** Third order, first degree, ordinary   **5.** Second order, fourth degree, ordinary

7. $y = \dfrac{7x^2}{2} + C$ **9.** $y = \dfrac{7x^2}{3} - \dfrac{5x}{3} + C$ **11.** $y = \dfrac{x^3}{3} + C$

Exercise 4

1. $xy - 7x = C$ **3.** $xy - 3x = C$ **5.** $4xy - x^2 = C$ **7.** $\dfrac{y}{x} - 3x = C$ **9.** $x^3 + 2xy = C$ **11.** $x^2 + y^2 - 4xy = C$

13. $\dfrac{x}{y} = 2x + C$ **15.** $2y^3 = x + Cy$ **17.** $4xy^2 - 3x^2 = C$ **19.** $2x^2 - xy = 15$ **21.** $3y^3 - y = 2x$

23. $3x^2 + 3y^2 - 4xy = 8$

Exercise 5

1. $x(x - 3y)^2 = C$ **3.** $x \ln y - y = Cx$ **5.** $y^3 = x^3(3 \ln x + C)$ **7.** $x(x - 3y)^2 = 4$ **9.** $(y - 2x)^2(x + y) = 27$

Exercise 6

1. $y = 2x + \dfrac{C}{x}$ **3.** $y = x^3 + \dfrac{C}{x}$ **5.** $y = 1 + Ce^{-x^3/3}$ **7.** $y = \dfrac{C}{\sqrt{x}} - \dfrac{3}{x}$ **9.** $y = x + Cx^2$ **11.** $(1 + x^2)^2 y =$

$2 \ln x + x^2 + C$ **13.** $y = \dfrac{C}{xe^{x^2/2}}$ **15.** $y = \dfrac{e^x}{2} + \dfrac{C}{e^x}$ **17.** $y = (4x + C)e^{2x}$ **19.** $x^2 y = 2 \ln^2 x + C$

21. $y = \dfrac{1}{2}\left(\sin x - \cos x\right) + Ce^{-x}$ **23.** $y = \sin x + \cos x + Ce^{-x}$ **25.** $xy\left(C - \dfrac{3x^2}{2}\right) = 1$ **27.** $y = \dfrac{1}{x^2 + Ce^{-x}}$

29. $xy - 2x^2 = 3$ **31.** $y = \dfrac{5x}{2} - \dfrac{1}{2x}$ **33.** $y = \tan x + \sqrt{2} \sin x$

Exercise 7

1. $y = x^2 + 3x - 1$ **3.** $y^2 = x^2 + 7$ **5.** $y = 0.812e^x - x - 1$ **7.** $xy = 4$ **9.** $y = \dfrac{e^x + e^{-x}}{2}$ **11.** $y = Cx$

13. $3xy^2 + x^3 = k$

Exercise 8

3. 6.6 million bbl./d   **5.** 5.40 °C   **7.** 695 °C   **13.** 4.58 m   **15.** 2.03 s   **17.** 2.35 s

Exercise 9

7. 231 mA; 0.888 V   **9.** 46.4 mA   **11.** 237 mA   **13.** 139 mA

Exercise 10

1. $y = \dfrac{5}{2}x^2 + C_1 x + C_2$ **3.** $y = 3e^x + C_1 x + C_2$ **5.** $y = \dfrac{x^4}{12} + x$

Exercise 11

1. $y = C_1 e^x + C_2 e^{5x}$ **3.** $y = C_1 e^x + C_2 e^{2x}$ **5.** $y = C_1 e^{3x} + C_2 e^{-2x}$ **7.** $y = C_1 + C_2 e^{2x/5}$ **9.** $y = C_1 e^{2x/3} + C_2 e^{-3x/2}$
11. $y = (C_1 + C_2 x)e^{2x}$ **13.** $y = C_1 e^x + C_2 x e^x$ **15.** $y = (C_1 + C_2 x)e^{-2x}$ **17.** $y = C_1 e^{-x} + C_2 x e^{-x}$
19. $y = e^{-2x}(C_1 \cos 3x + C_2 \sin 3x)$ **21.** $y = e^{3x}(C_1 \cos 4x + C_2 \sin 4x)$ **23.** $y = C_1 \cos 2x + C_2 \sin 2x$
25. $y = e^{2x}(C_1 \cos x + C_2 \sin x)$ **27.** $y = 3xe^{-3x}$ **29.** $y = 5e^x - 14xe^x$ **31.** $y = 1 + e^{2x}$
33. $y = (1/4)e^{2x} + (3/4)e^{-2x}$ **35.** $y = e^{-x} \sin x$ **37.** $y = C_1 e^x + C_2 e^{-x} + C_3 e^{2x}$ **39.** $y = C_1 e^x + C_2 e^{2x} + C_3 e^{3x}$
41. $y = C_1 e^x + C_2 e^{-x} + C_3 e^{3x}$ **43.** $y = (C_1 + C_2 x)e^{x/2} + C_3 e^{-x}$

Exercise 12

1. $y = C_1 e^{2x} + C_2 e^{-2x} - 3$ **3.** $y = C_1 e^{2x} + C_2 e^{-x} - 2x + 1$ **5.** $y = C_1 e^{2x} + C_2 e^{-2x} - (1/4)x^3 - (5/8)x$
7. $y = C_1 e^{2x} + C_2 e^{-x} - 3e^x$ **9.** $y = C_1 e^{2x} + C_2 e^{-2x} + e^x - x$ **11.** $y = e^{-2x}(C_1 + C_2 x) + (2e^{2x} + x - 1)/4$
13. $y = C_1 \cos 2x + C_2 \sin 2x - \dfrac{x \cos 2x}{4}$ **15.** $y = (C_1 + C_2 x)e^{-x} + (1/2) \sin x$ **17.** $y = C_1 \cos x + C_2 \sin x +$
$\cos 2x + x \sin x$ **19.** $y = C_1 \cos x + C_2 \sin x + (e^x/5)(\sin x - 2 \cos x)$ **21.** $y = e^{2x}[C_1 \sin x + C_2 \cos x - (x/2)\cos x]$
23. $y = \dfrac{e^{4x}}{2} - 2x - \dfrac{1}{2}$ **25.** $y = \dfrac{1 - \cos 2x}{2}$ **27.** $y = e^x(x^2 - 1)$ **29.** $y = x \cos x$

Exercise 13

13. 9.82 Hz   **15.** 0.496 μF   **17.** 10 A

Review Problems

1. $y = \dfrac{e^x}{2x} + \dfrac{C}{xe^x}$ **3.** $y = \sin x + \cos x + Ce^{-x}$ **5.** $x^3 + 2xy = C$ **7.** $x \ln y - y = Cx$ **9.** $y \ln|1 - x| + Cy = 1$

11. $x^2 + \dfrac{x}{y} + \ln y = C$ **13.** $\sin x + \cos y = 1.707$ **15.** $x + 2y = 2xy^2$ **17.** 14.6 rev/s   **19.** $y = -2 \ln|x| + C$

21. $y = 2x \ln x + 2x$ **23.** 4740 g   **25.** $y = C_1 e^{2x} + C_2 e^{-x} - (5/2)x + 5/4$ **27.** $y = C_1 e^{5x} + C_2 e^{-x} - (1/5)e^{4x}$

29. $y = (11e^{5x} - 30x + 14)/25$ **31.** $y = C_1 e^{9x} + C_2 e^{-2x}$ **33.** $y = C_1 e^{2x} + C_2 e^{3x} + C_3 e^x$

35. $y = C_1 \cos \sqrt{2}x + C_2 \sin \sqrt{2}x - (3/2) \sin 2x$ **37.** $y = 4e^{-x} - 3e^{-2x}$ **39.** $y = C_1 e^x + C_2 e^{-4x}$

41. $y = C_1 e^{-5.19x} + C_2 e^{0.193x}$ **43.** $5y = 2(e^x - e^{-4x})$ **45.** $y = C_1 e^{-3x} + C_2 xe^{-3x}$ **47.** $y = C_1 e^{x/3} + C_2 xe^{x/3}$

49. $i = 4.66e^{-18.75t} \sin 258t$ A   **51.** $i = 98.1 \sin 510t$ mA

◆◆◆ **CHAPTER 36** ◆◆

Exercise 1

1. $6/s$ **3.** $2/s^3$ **5.** $\dfrac{s}{s^2 + 25}$ **7.** $2/s^3 + 4/s$ **9.** $\dfrac{3}{s - 1} + \dfrac{2}{s + 1}$ **11.** $\dfrac{2}{s^2 + 4} + \dfrac{s}{s^2 + 9}$ **13.** $\dfrac{5(s - 3)}{(s - 3)^2 + 25}$

15. $\dfrac{5}{s - 1} - \dfrac{6s}{(s^2 + 9)^2}$ **17.** $s\mathscr{L}[y] - 1 + 2\mathscr{L}[y]$ **19.** $s\mathscr{L}[y] - 4\mathscr{L}[y]$ **21.** $s^2\mathscr{L}[y] + 3s\mathscr{L}[y] - \mathscr{L}[y] - s - 6$

23. $2s^2\mathscr{L}[y] + 3s\mathscr{L}[y] + \mathscr{L}[y] - 4s - 12$

Exercise 2

1. 1   **3.** t^2 **5.** $2 \sin 2t$ **7.** $3 \cos \sqrt{2}t$ **9.** $e^{9t}(1 + 13t)$ **11.** $\dfrac{5}{3}e^{-2t} \sin 3t$ **13.** $\cos t + 1$ **15.** $e^t + e^{-2t} + t - 2$

17. $\dfrac{2}{3}e^t - \dfrac{2}{3}e^{-2t}$ **19.** $2(3e^{-3t} - 2e^{-2t})$

Exercise 3

1. $y = e^{3t}$ **3.** $y = -1 - t/2 + e^{t/2}$ **5.** $y = (21/4)e^{2t/3} - 9/4 - (3/2)t - (1/2)t^2$ **7.** $y = \frac{1}{4}\sin 2t + \frac{1}{4}\cos 2t - \frac{1}{4}e^{-2t}$

9. $y = (1/2)e^t - (1/2)e^{-3t}$ **11.** $y = 4/3 + t - (1/3)e^{-3t}$ **13.** $y = 4t + 3 \cos 0.707t - 5.66 \sin 0.707t$

15. $y = t/3 - 4/9 + (9/2)e^{-t} - (37/18)e^{-3t}$ **17.** $y = 12 - 9.9e^{-t/3} - 0.1 \cos t - 0.3 \sin t$ **19.** $y = (1/2)t^2 e^t$

21. $y = 0.111e^{-t} \cos \sqrt{2}t + 0.0393e^{-t} \sin \sqrt{2}t - 0.111e^t + 0.167te^t$ **23.** 5.68 m/s   **25.** 5.7×10^{29} bacteria

27. $x = 1.02e^{-6.25t} - 1.02 \cos 2.25t + 2.84 \sin 2.25t$

Exercise 5

1. 0.644   **3.** 1.899   **5.** 7.26   **7.** 3.12   **9.** 13.8

Exercise 6

1. 2.50   **3.** 12.1   **5.** 9.03   **7.** −1.27   **9.** 3.27

Review Problems

◆◆◆ CHAPTER 37 ◆◆◆

Exercise 1

1. Converges   **3.** Diverges   **5.** Cannot tell   **7.** Cannot tell   **9.** Converges   **11.** Diverges   **13.** Test fails

Exercise 2

1. $-1 < x < 1$ **3.** $x = 0$ **5.** $-1 < x < 1$ **21.** 1.625   **23.** 0.1827   **25.** 1.049

Exercise 3

11. 0.91   **13.** 0.688 353   **15.** 2.0887   **17.** -0.92889 **19.** 0.2240

Exercise 4

21. $1 - \dfrac{x^2}{2!} + \dfrac{x^4}{4!} - \dfrac{x^6}{6!} + \cdots$ **23.** $1 - x + x^2 - x^3 + \cdots$ **25.** $-x - \dfrac{x^2}{2} - \dfrac{x^3}{3} - \dfrac{x^4}{4} - \cdots$ **27.** 2.98   **29.** 0.0342

Exercise 5

1. $2 + 3\cos x + 2\cos 2x + \cos 3x + 4\sin x + 3\sin 2x + 2\sin 3x$

Exercise 6

1. Neither   **3.** Even   **5.** Neither   **7.** Yes   **9.** No   **11.** No

Exercise 8

1. $y = -4.92\cos x + 2.15\cos 3x + 0.95\cos 5x + 10.16\sin x + 3.10\sin 3x + 0.79\sin 5x$

3. $y = 3.92\cos x - 4.75\cos 3x + 1.20\cos 5x + 31.50\sin x - 3.31\sin 3x + 1.60\sin 5x$

Review Problems

1. 1.345 **3.** 0.476 25 **5.** $-1 < x < 1$ **7.** $2 + 8 + 18 + 32 + 50 + \cdots$ **11.** $\cdots + 34 + 47 + \cdots + (n^2 - 2) + \cdots$
13. 1.316   **15.** $\cdots + 2207 + 4\,870\,847 + \cdots ; u_1 = 3, u_{n+1} = (u_n)^2 - 2$ **19.** Odd   **21.** Odd   **23.** Yes   **25.** Yes
31. $y = -4.12\cos x + 2.06\cos 3x + 0.57\cos 5x + 11.25\sin x + 2.70\sin 3x + 0.64\sin 5x + \cdots$

Index

Index of Applications